Fundamental Equations of Dynamics

KINEMATICS

Particle Rectilinear Motion

Variable a

$$a = \frac{dv}{dt}$$

$$v = \frac{ds}{dt}$$

$$a\,ds = v\,dv$$

Constant $a = a_c$

$$v = v_0 + a_c t$$

$$s = s_0 + v_0 t + \tfrac{1}{2}a_c t^2$$

$$v^2 = v_0^2 + 2a_c(s - s_0)$$

Particle Curvilinear Motion

x, y, z Coordinates

$$v_x = \dot{x} \quad a_x = \ddot{x}$$
$$v_y = \dot{y} \quad a_y = \ddot{y}$$
$$v_z = \dot{z} \quad a_z = \ddot{z}$$

r, θ, z Coordinates

$$v_r = \dot{r} \quad a_r = \ddot{r} - r\dot{\theta}^2$$
$$v_\theta = r\dot{\theta} \quad a_\theta = r\ddot{\theta} + 2\dot{r}\dot{\theta}$$
$$v_z = \dot{z} \quad a_z = \ddot{z}$$

n, t, b Coordinates

$$v = \dot{s}$$

$$a_t = \dot{v} = v\frac{dv}{ds}$$

$$a_n = \frac{v^2}{\rho} \quad \rho = \frac{[1 + (dy/dx)^2]^{3/2}}{|d^2y/dx^2|}$$

Relative Motion

$$\mathbf{v}_B = \mathbf{v}_A + \mathbf{v}_{B/A} \qquad \mathbf{a}_B = \mathbf{a}_A + \mathbf{a}_{B/A}$$

Rigid Body Motion About a Fixed Axis

Variable α

$$\alpha = \frac{d\omega}{dt}$$

$$\omega = \frac{d\theta}{dt}$$

$$\omega\,d\omega = \alpha\,d\theta$$

Constant $\alpha = \alpha_c$

$$\omega = \omega_0 + \alpha_c t$$

$$\theta = \theta_0 + \omega_0 t + \tfrac{1}{2}\alpha_c t^2$$

$$\omega^2 = \omega_0^2 + 2\alpha_c(\theta - \theta_0)$$

For Point P

$$s = \theta r \quad v = \omega r \quad a_t = \alpha r \quad a_n = \omega^2 r$$

Relative General Plane Motion—Translating Axes

$$\mathbf{v}_B = \mathbf{v}_A + \mathbf{v}_{B/A\,(\text{pin})} \qquad \mathbf{a}_B = \mathbf{a}_A + \mathbf{a}_{B/A\,(\text{pin})}$$

Relative General Plane Motion—Trans. and Rot. Axis

$$\mathbf{v}_B = \mathbf{v}_A + \mathbf{\Omega} \times \mathbf{r}_{B/A} + (\mathbf{v}_{B/A})_{xyz}$$

$$\mathbf{a}_B = \mathbf{a}_A + \dot{\mathbf{\Omega}} \times \mathbf{r}_{B/A} + \mathbf{\Omega} \times (\mathbf{\Omega} \times \mathbf{r}_{B/A}) + 2\mathbf{\Omega} \times (\mathbf{v}_{B/A})_{xyz} + (\mathbf{a}_{B/A})_{xyz}$$

KINETICS

Mass Moment of Inertia $I = \int r^2\,dm$

Parallel-Axis Theorem $I = I_G + md^2$

Radius of Gyration $k = \sqrt{\dfrac{I}{m}}$

Equations of Motion

Particle	$\Sigma\mathbf{F} = m\mathbf{a}$
Rigid Body	$\Sigma F_x = m(a_G)_x$
(Plane Motion)	$\Sigma F_y = m(a_G)_y$
	$\Sigma M_G = I_G\alpha$ or $\Sigma M_P = \Sigma(\mathcal{M}_k)_P$

Principle of Work and Energy

$$T_1 + \Sigma U_{1-2} = T_2$$

Kinetic Energy

Particle	$T = \tfrac{1}{2}mv^2$
Rigid Body (Plane Motion)	$T = \tfrac{1}{2}mv_G^2 + \tfrac{1}{2}I_G\omega^2$

Work

Variable force	$U_F = \int F\cos\theta\,ds$
Constant force	$U_F = (F_c\cos\theta)\,\Delta s$
Weight	$U_W = -W\Delta y$
Spring	$U_s = -\left(\tfrac{1}{2}ks_2^2 - \tfrac{1}{2}ks_1^2\right)$
Couple moment	$U_M = M\Delta\theta$

Power and Efficiency

$$P = \frac{dU}{dt} = \mathbf{F}\cdot\mathbf{v} \qquad \varepsilon = \frac{P_{\text{out}}}{P_{\text{in}}} = \frac{U_{\text{out}}}{U_{\text{in}}}$$

Conservation of Energy Theorem

$$T_1 + V_1 = T_2 + V_2$$

Potential Energy

$$V = V_g + V_e, \text{ where } V_g = \pm Wy, \ V_e = +\tfrac{1}{2}ks^2$$

Principle of Linear Impulse and Momentum

Particle	$m\mathbf{v}_1 + \Sigma\int \mathbf{F}\,dt = m\mathbf{v}_2$
Rigid Body	$m(\mathbf{v}_G)_1 + \Sigma\int \mathbf{F}\,dt = m(\mathbf{v}_G)_2$

Conservation of Linear Momentum

$$\Sigma(\text{syst. } m\mathbf{v})_1 = \Sigma(\text{syst. } m\mathbf{v})_2$$

Coefficient of Restitution $e = \dfrac{(v_B)_2 - (v_A)_2}{(v_A)_1 - (v_B)_1}$

Principle of Angular Impulse and Momentum

Particle	$(\mathbf{H}_O)_1 + \Sigma\int \mathbf{M}_O\,dt = (\mathbf{H}_O)_2$
	where $H_O = (d)(mv)$
Rigid Body (Plane motion)	$(\mathbf{H}_G)_1 + \Sigma\int \mathbf{M}_G\,dt = (\mathbf{H}_G)_2$
	where $H_G = I_G\omega$
	$(\mathbf{H}_O)_1 + \Sigma\int \mathbf{M}_O\,dt = (\mathbf{H}_O)_2$
	where $H_O = I_O\omega$

Conservation of Angular Momentum

$$\Sigma(\text{syst. } \mathbf{H})_1 = \Sigma(\text{syst. } \mathbf{H})_2$$

SI Prefixes

Multiple	Exponential Form	Prefix	SI Symbol
1 000 000 000	10^9	giga	G
1 000 000	10^6	mega	M
1 000	10^3	kilo	k

Submultiple			
0.001	10^{-3}	milli	m
0.000 001	10^{-6}	micro	μ
0.000 000 001	10^{-9}	nano	n

Conversion Factors (FPS) to (SI)

Quantity	Unit of Measurement (FPS)	Equals	Unit of Measurement (SI)
Force	lb		4.448 N
Mass	slug		14.59 kg
Length	ft		0.3048 m

Conversion Factors (FPS)

1 ft = 12 in. (inches)
1 mi. (mile) = 5280 ft
1 kip (kilopound) = 1000 lb
1 ton = 2000 lb

STATICS AND DYNAMICS

FOURTEENTH EDITION

ENGINEERING MECHANICS

STATICS AND DYNAMICS

FOURTEENTH EDITION

R. C. HIBBELER

PEARSON

Hoboken Boston Columbus San Francisco New York Indianapolis
London Toronto Sydney Singapore Tokyo Montreal Dubai Madrid
Hong Kong Mexico City Munich Paris Amsterdam Cape Town

Library of Congress Cataloging-in-Publication Data on File

Vice President and Editorial Director, ECS: Marcia Horton
Senior Editor: Norrin Dias
Editorial Assistant: Michelle Bayman
Program and Project Management Team Lead: Scott Disanno
Program Manager: Sandra L. Rodriguez
Project Manager: Rose Kernan
Cover Designer: Black Horse Designs
Art Editor: Gregory Dulles
Senior Digital Producer: Felipe Gonzalez
Operations Specialist: Maura Zaldivar-Garcia
Product Marketing Manager: Bram Van Kempen
Field Marketing Manager: Demetrius Hall
Marketing Assistant: Jon Bryant

Cover Image: Fotog/Getty Images

Pearson Education Ltd., *London*
Pearson Education Australia Pty. Ltd., *Sydney*
Pearson Education Singapore, Pte. Ltd.
Pearson Education North Asia Ltd., *Hong Kong*
Pearson Education Canada, Inc., *Toronto*
Pearson Educación de Mexico, S.A. de C.V.
Pearson Education—Japan, *Tokyo*
Pearson Education Malaysia, Pte. Ltd.
Pearson Education, Inc., *Hoboken, New Jersey*

Printed in the United States of America

PEARSON

ISBN-10: 0-13-391542-5
ISBN-13: 978-0-13-391542-6

To the Student

With the hope that this work will stimulate
an interest in Engineering Mechanics
and provide an acceptable guide to its understanding.

PREFACE

The main purpose of this book is to provide the student with a clear and thorough presentation of the theory and application of engineering mechanics. To achieve this objective, this work has been shaped by the comments and suggestions of hundreds of reviewers in the teaching profession, as well as many of the author's students.

New to this Edition

Preliminary Problems. This new feature can be found throughout the text, and is given just before the Fundamental Problems. The intent here is to test the student's conceptual understanding of the theory. Normally the solutions require little or no calculation, and as such, these problems provide a basic understanding of the concepts before they are applied numerically. All the solutions are given in the back of the text.

Expanded Important Points Sections. Summaries have been added which reinforce the reading material and highlights the important definitions and concepts of the sections.

Re-writing of Text Material. Further clarification of concepts has been included in this edition, and important definitions are now in boldface throughout the text to highlight their importance.

End-of-Chapter Review Problems. All the review problems now have solutions given in the back, so that students can check their work when studying for exams, and reviewing their skills when the chapter is finished.

New Photos. The relevance of knowing the subject matter is reflected by the real-world applications depicted in the over 60 new or updated photos placed throughout the book. These photos generally are used to explain how the relevant principles apply to real-world situations and how materials behave under load.

New Problems. There are approximately 30% new problems that have been added to this edition, which involve applications to many different fields of engineering.

Hallmark Features

Besides the new features mentioned above, other outstanding features that define the contents of the text include the following.

Organization and Approach. Each chapter is organized into well-defined sections that contain an explanation of specific topics, illustrative example problems, and a set of homework problems. The topics within each section are placed into subgroups defined by boldface titles. The purpose of this is to present a structured method for introducing each new definition or concept and to make the book convenient for later reference and review.

Chapter Contents. Each chapter begins with an illustration demonstrating a broad-range application of the material within the chapter. A bulleted list of the chapter contents is provided to give a general overview of the material that will be covered.

Emphasis on Free-Body Diagrams. Drawing a free-body diagram is particularly important when solving problems, and for this reason this step is strongly emphasized throughout the book. In particular, special sections and examples are devoted to show how to draw free-body diagrams. Specific homework problems have also been added to develop this practice.

Procedures for Analysis. A general procedure for analyzing any mechanical problem is presented at the end of the first chapter. Then this procedure is customized to relate to specific types of problems that are covered throughout the book. This unique feature provides the student with a logical and orderly method to follow when applying the theory. The example problems are solved using this outlined method in order to clarify its numerical application. Realize, however, that once the relevant principles have been mastered and enough confidence and judgment have been obtained, the student can then develop his or her own procedures for solving problems.

Important Points. This feature provides a review or summary of the most important concepts in a section and highlights the most significant points that should be realized when applying the theory to solve problems.

Fundamental Problems. These problem sets are selectively located just after most of the example problems. They provide students with simple applications of the concepts, and therefore, the chance to develop their problem-solving skills before attempting to solve any of the standard problems that follow. In addition, they can be used for preparing for exams, and they can be used at a later time when preparing for the Fundamentals in Engineering Exam.

Conceptual Understanding. Through the use of photographs placed throughout the book, theory is applied in a simplified way in order to illustrate some of its more important conceptual features and instill the physical meaning of many of the terms used in the equations. These simplified applications increase interest in the subject matter and better prepare the student to understand the examples and solve problems.

Homework Problems.   Apart from the Fundamental and Conceptual type problems mentioned previously, other types of problems contained in the book include the following:

- **Free-Body Diagram Problems.** Some sections of the book contain introductory problems that only require drawing the free-body diagram for the specific problems within a problem set. These assignments will impress upon the student the importance of mastering this skill as a requirement for a complete solution of any equilibrium problem.

- **General Analysis and Design Problems.** The majority of problems in the book depict realistic situations encountered in engineering practice. Some of these problems come from actual products used in industry. It is hoped that this realism will both stimulate the student's interest in engineering mechanics and provide a means for developing the skill to reduce any such problem from its physical description to a model or symbolic representation to which the principles of mechanics may be applied.

Throughout the book, there is an approximate balance of problems using either SI or FPS units. Furthermore, in any set, an attempt has been made to arrange the problems in order of increasing difficulty except for the end of chapter review problems, which are presented in random order.

- **Computer Problems.** An effort has been made to include some problems that may be solved using a numerical procedure executed on either a desktop computer or a programmable pocket calculator. The intent here is to broaden the student's capacity for using other forms of mathematical analysis without sacrificing the time needed to focus on the application of the principles of mechanics. Problems of this type, which either can or must be solved using numerical procedures, are identified by a "square" symbol (■) preceding the problem number.

The many homework problems in this edition, have been placed into two different categories. Problems that are simply indicated by a problem number have an answer and in some cases an additional numerical result given in the back of the book. An asterisk (*) before every fourth problem number indicates a problem without an answer.

Accuracy.   As with the previous editions, apart from the author, the accuracy of the text and problem solutions has been thoroughly checked by four other parties: Scott Hendricks, Virginia Polytechnic Institute and State University; Karim Nohra, University of South Florida; Kurt Norlin, Bittner Development Group; and finally Kai Beng, a practicing engineer, who in addition to accuracy review provided suggestions for problem development.

Contents

Statics

The book is divided into 11 chapters, in which the principles are first applied to simple, then to more complicated situations. In a general sense, each principle is applied first to a particle, then a rigid body subjected to a coplanar system of forces, and finally to three-dimensional force systems acting on a rigid body.

Chapter 1 begins with an introduction to mechanics and a discussion of units. The vector properties of a concurrent force system are introduced in Chapter 2. This theory is then applied to the equilibrium of a particle in Chapter 3. Chapter 4 contains a general discussion of both concentrated and distributed force systems and the methods used to simplify them. The principles of rigid-body equilibrium are developed in Chapter 5 and then applied to specific problems involving the equilibrium of trusses, frames, and machines in Chapter 6, and to the analysis of internal forces in beams and cables in Chapter 7. Applications to problems involving frictional forces are discussed in Chapter 8, and topics related to the center of gravity and centroid are treated in Chapter 9. If time permits, sections involving more advanced topics, indicated by stars (★), may be covered. Most of these topics are included in Chapter 10 (area and mass moments of inertia) and Chapter 11 (virtual work and potential energy). Note that this material also provides a suitable reference for basic principles when it is discussed in more advanced courses. Finally, Appendix A provides a review and list of mathematical formulas needed to solve the problems in the book.

Alternative Coverage. At the discretion of the instructor, some of the material may be presented in a different sequence with no loss of continuity. For example, it is possible to introduce the concept of a force and all the necessary methods of vector analysis by first covering Chapter 2 and Section 4.2 (the cross product). Then after covering the rest of Chapter 4 (force and moment systems), the equilibrium methods of Chapters 3 and 5 can be discussed.

Dynamics

The book is divided into 11 chapters, in which the principles are first applied to simple, then to more complicated situations.

The kinematics of a particle is discussed in Chapter 12, followed by a discussion of particle kinetics in Chapter 13 (Equation of Motion), Chapter 14 (Work and Energy), and Chapter 15 (Impulse and Momentum). The concepts of particle dynamics contained in these four chapters are then summarized in a "review" section, and the student is given the chance to identify and solve a variety of problems. A similar sequence of presentation is given for the planar motion of a rigid body: Chapter 16 (Planar Kinematics), Chapter 17 (Equations of Motion), Chapter 18 (Work and Energy), and Chapter 19 (Impulse and Momentum), followed by a summary and review set of problems for these chapters.

If time permits, some of the material involving three-dimensional rigid-body motion may be included in the course. The kinematics and kinetics of this motion are discussed in Chapters 20 and 21, respectively. Chapter 22 (Vibrations) may be included if the student has the necessary mathematical background. Sections of the book that are considered to be beyond the scope of the basic dynamics course are indicated by a star (★) and may be omitted. Note that this material also provides a suitable reference for basic principles when it is discussed in more advanced courses. Finally, Appendix A provides a list of mathematical formulas needed to solve the problems in the book, Appendix B provides a brief review of vector analysis, and Appendix C reviews application of the chain rule.

Alternative Coverage. At the discretion of the instructor, it is possible to cover Chapters 12 through 19 in the following order with no loss in continuity: Chapters 12 and 16 (Kinematics), Chapters 13 and 17 (Equations of Motion), Chapter 14 and 18 (Work and Energy), and Chapters 15 and 19 (Impulse and Momentum).

Acknowledgments

The author has endeavored to write this book so that it will appeal to both the student and instructor. Through the years, many people have helped in its development, and I will always be grateful for their valued suggestions and comments. Specifically, I wish to thank all the individuals who have contributed their comments relative to preparing the fourteenth edition of this work, and in particular, R. Bankhead of Highline Community College, K. Cook-Chennault of Rutgers, the State University of New Jersey, E. Erisman, College of Lake County Illinois, M. Freeman of the University of Alabama, A. Itani of the University of Nevada, Y. Laio of Arizona State University, H. Lu of University of Texas at Dallas, T. Miller of Oregon State University, J. Morgan of Texas A & M University, R. Neptune of the University of Texas, I. Orabi of the University of New Haven, M. Reynolds of the University of Arkansas, N. Schulz of the University of Portland, C. Sulzbach of the Colorado School of Mines, T. Tan, University of Memphis, R. Viesca of Tufts University, G. Young, Oklahoma State University, and P. Ziehl of the University of South Carolina.

There are a few other people that I also feel deserve particular recognition. These include comments sent to me by J. Dix, H. Kuhlman, S. Larwood, D. Pollock, and H. Wenzel. A long-time friend and associate, Kai Beng Yap, was of great help to me in preparing and checking problem solutions. A special note of thanks also goes to Kurt Norlin of Bittner Development Group in this regard. During the production process I am thankful for the assistance of Martha McMaster, my copy editor, and Rose Kernan, my production editor as well as my wife, Conny, who have helped prepare the manuscript for publication.

Lastly, many thanks are extended to all my students and to members of the teaching profession who have freely taken the time to e-mail me their suggestions and comments. Since this list is too long to mention, it is hoped that those who have given help in this manner will accept this anonymous recognition.

I would greatly appreciate hearing from you if at any time you have any comments, suggestions, or problems related to any matters regarding this edition.

Russell Charles Hibbeler
hibbeler@bellsouth.net

your work...

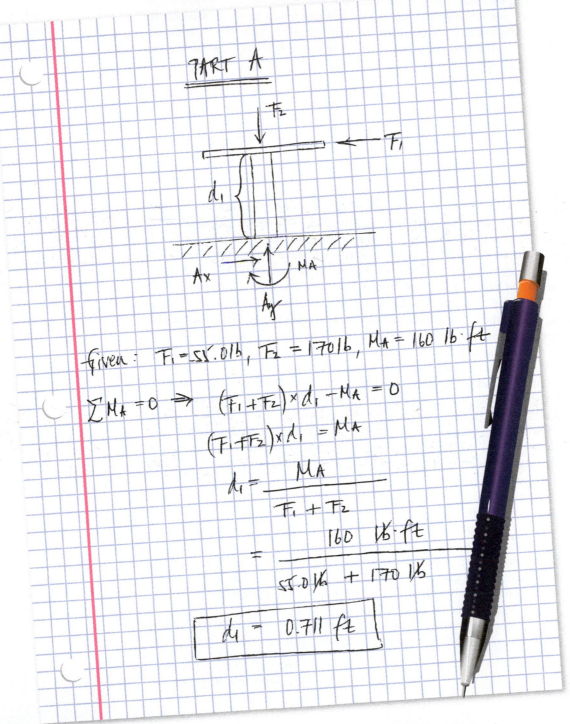

PART A

Given: $F_1 = 55.0\,lb$, $F_2 = 170\,lb$, $M_A = 160\,lb\cdot ft$

$$\sum M_A = 0 \implies (F_1 + F_2) \times d_1 - M_A = 0$$

$$(F_1 + F_2) \times d_1 = M_A$$

$$d_1 = \frac{M_A}{F_1 + F_2}$$

$$= \frac{160\ lb\cdot ft}{55.0\ lb + 170\ lb}$$

$$\boxed{d_1 = 0.711\ ft}$$

your answer specific feedback

Express your answer numerically in feet to three significant figures.

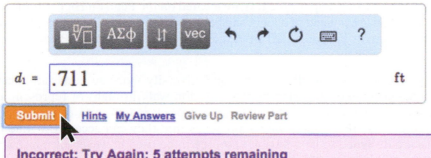

$d_1 = $ `.711`   ft

Submit   Hints   My Answers   Give Up   Review Part

Incorrect; Try Again; 5 attempts remaining

The sum of the two forces do not contribute to the moment about point A. The magnitude of the moment about A is equal to the force multiplied by the perpendicular distance between point A and the line of action of the force. What is the perpendicular distance between each force's line of action and point A?

www.MasteringEngineering®.com

Resources for Instructors

- **MasteringEngineering.** This online Tutorial Homework program allows you to integrate dynamic homework with automatic grading and adaptive tutoring. MasteringEngineering allows you to easily track the performance of your entire class on an assignment-by-assignment basis, or the detailed work of an individual student.

- **Instructor's Solutions Manual.** This supplement provides complete solutions supported by problem statements and problem figures. The fourteenth edition manual was revised to improve readability and was triple accuracy checked. The Instructor's Solutions Manual is available on Pearson Higher Education website: www.pearsonhighered.com.

- **Instructor's Resource.** Visual resources to accompany the text are located on the Pearson Higher Education website: www.pearsonhighered.com. If you are in need of a login and password for this site, please contact your local Pearson representative. Visual resources include all art from the text, available in PowerPoint slide and JPEG format.

- **Video Solutions.** Developed by Professor Edward Berger, Purdue University, video solutions are located in the study area of MasteringEngineering and offer step-by-step solution walkthroughs of representative homework problems from each section of the text. Make efficient use of class time and office hours by showing students the complete and concise problem-solving approaches that they can access any time and view at their own pace. The videos are designed to be a flexible resource to be used however each instructor and student prefers. A valuable tutorial resource, the videos are also helpful for student self-evaluation as students can pause the videos to check their understanding and work alongside the video. Access the videos at www.masteringengineering.com.

Resources for Students

- **MasteringEngineering.** Tutorial homework problems emulate the instructor's office-hour environment, guiding students through engineering concepts with self-paced individualized coaching. These in-depth tutorial homework problems are designed to coach students with feedback specific to their errors and optional hints that break problems down into simpler steps.

- **Statics Study Pack.** This supplement contains chapter-by-chapter study materials and a Free-Body Diagram Workbook.

- **Dynamics Study Pack.** This supplement contains chapter-by-chapter study materials and a Free-Body Diagram Workbook.

- **Video Solutions.** Complete, step-by-step solution walkthroughs of representative homework problems from each section. Videos offer fully worked solutions that show every step of representative homework problems—this helps students make vital connections between concepts.

- **Statics Practice Problems Workbook.** This workbook contains additional worked problems. The problems are partially solved and are designed to help guide students through difficult topics.

- **Dynamics Practice Problems Workbook.** This workbook contains additional worked problems. The problems are partially solved and are designed to help guide students through difficult topics.

Ordering Options

The *Statics and Dynamics Study Packs* and MasteringEngineering resources are available as stand-alone items for student purchase and are also available packaged with the texts. The ISBN for each valuepack is as follows:

- *Engineering Mechanics: Statics* with Study Pack: ISBN 0134136683
- *Engineering Mechanics: Statics* Plus MasteringEngineering with Pearson eText—Access Card Package: ISBN: 0134160681
- *Engineering Mechanics: Dynamics* with Study Pack: ISBN: 0134116658
- *Engineering Mechanics: Dynamics* Plus MasteringEngineering with Pearson eText — Access Card Package: ISBN: 0134116992

Custom Solutions

Please contact your local Pearson Sales Representative for more details about custom options or visit

www.pearsonlearningsolutions.com, keyword: Hibbeler.

CONTENTS

1
General Principles 3

2
Force Vectors 17

3
Equilibrium of a Particle 87

4
Force System Resultants 121

16
Planar Kinematics of a Rigid Body 319

17
Planar Kinetics of a Rigid Body: Force and Acceleration 409

18
Planar Kinetics of a Rigid Body: Work and Energy 473

19
Planar Kinetics of a Rigid Body: Impulse and Momentum 517

20
Three-Dimensional Kinematics of a Rigid Body 561

21
Three-Dimensional Kinetics of a Rigid Body 591

22
Vibrations 643

Appendix

CREDITS

Chapter opening images are credited as follows:

Chapter 1, Andrew Peacock/Lonely Planet Images/Getty Images
Chapter 2, Vasiliy Koval/Fotolia
Chapter 3, Igor Tumarkin/ITPS/Shutterstock
Chapter 4, Rolf Adlercreutz/Alamy
Chapter 5, YuryZap/Shutterstock
Chapter 6, Tim Scrivener/Alamy
Chapter 7, Tony Freeman/Science Source
Chapter 8, Pavel Polkovnikov/Shutterstock
Chapter 9, Heather Reeder/Shutterstock
Chapter 10, Michael N. Paras/AGE Fotostock/Alamy
Chapter 11, John Kershaw/Alamy
Chapter 12, Lars Johansson/Fotolia
Chapter 13, Migel/Shutterstock
Chapter 14, Oliver Furrer/Ocean/Corbis
Chapter 15, David J. Green/Alamy
Chapter 16, TFoxFoto/Shutterstock
Chapter 17, Surasaki/Fotolia
Chapter 18, Arinahabich/Fotolia
Chapter 19, Hellen Sergeyeva/Fotolia
Chapter 20, Philippe Psaila/Science Source
Chapter 21, Derek Watt/Alamy
Chapter 22, Daseaford/Fotolia

ENGINEERING MECHANICS

STATICS

FOURTEENTH EDITION

Chapter 1

(© Andrew Peacock/Lonely Planet Images/Getty Images)

Large cranes such as this one are required to lift extremely large loads. Their design is based on the basic principles of statics and dynamics, which form the subject matter of engineering mechanics.

General Principles

1.1   Mechanics

Mechanics is a branch of the physical sciences that is concerned with the state of rest or motion of bodies that are subjected to the action of forces. In general, this subject can be subdivided into three branches: *rigid-body mechanics, deformable-body mechanics*, and *fluid mechanics*. In this book we will study rigid-body mechanics since it is a basic requirement for the study of the mechanics of deformable bodies and the mechanics of fluids. Furthermore, rigid-body mechanics is essential for the design and analysis of many types of structural members, mechanical components, or electrical devices encountered in engineering.

Rigid-body mechanics is divided into two areas: statics and dynamics. *Statics* deals with the equilibrium of bodies, that is, those that are either at rest or move with a constant velocity; whereas *dynamics* is concerned with the accelerated motion of bodies. We can consider statics as a special case of dynamics, in which the acceleration is zero; however, statics deserves separate treatment in engineering education since many objects are designed with the intention that they remain in equilibrium.

Historical Development.   The subject of statics developed very early in history because its principles can be formulated simply from measurements of geometry and force. For example, the writings of Archimedes (287–212 B.C.) deal with the principle of the lever. Studies of the pulley, inclined plane, and wrench are also recorded in ancient writings—at times when the requirements for engineering were limited primarily to building construction.

Since the principles of dynamics depend on an accurate measurement of time, this subject developed much later. Galileo Galilei (1564–1642) was one of the first major contributors to this field. His work consisted of experiments using pendulums and falling bodies. The most significant contributions in dynamics, however, were made by Isaac Newton (1642–1727), who is noted for his formulation of the three fundamental laws of motion and the law of universal gravitational attraction. Shortly after these laws were postulated, important techniques for their application were developed by other scientists and engineers, some of whom will be mentioned throughout the text.

1.2   Fundamental Concepts

Before we begin our study of engineering mechanics, it is important to understand the meaning of certain fundamental concepts and principles.

Basic Quantities.   The following four quantities are used throughout mechanics.

Length.   *Length* is used to locate the position of a point in space and thereby describe the size of a physical system. Once a standard unit of length is defined, one can then use it to define distances and geometric properties of a body as multiples of this unit.

Time.   *Time* is conceived as a succession of events. Although the principles of statics are time independent, this quantity plays an important role in the study of dynamics.

Mass.   *Mass* is a measure of a quantity of matter that is used to compare the action of one body with that of another. This property manifests itself as a gravitational attraction between two bodies and provides a measure of the resistance of matter to a change in velocity.

Force.   In general, *force* is considered as a "push" or "pull" exerted by one body on another. This interaction can occur when there is direct contact between the bodies, such as a person pushing on a wall, or it can occur through a distance when the bodies are physically separated. Examples of the latter type include gravitational, electrical, and magnetic forces. In any case, a force is completely characterized by its magnitude, direction, and point of application.

Idealizations.   Models or idealizations are used in mechanics in order to simplify application of the theory. Here we will consider three important idealizations.

Particle.   A *particle* has a mass, but a size that can be neglected. For example, the size of the earth is insignificant compared to the size of its orbit, and therefore the earth can be modeled as a particle when studying its orbital motion. When a body is idealized as a particle, the principles of mechanics reduce to a rather simplified form since the geometry of the body *will not be involved* in the analysis of the problem.

Rigid Body.   A *rigid body* can be considered as a combination of a large number of particles in which all the particles remain at a fixed distance from one another, both before and after applying a load. This model is important because the body's shape does not change when a load is applied, and so we do not have to consider the type of material from which the body is made. In most cases the actual deformations occurring in structures, machines, mechanisms, and the like are relatively small, and the rigid-body assumption is suitable for analysis.

Concentrated Force.   A *concentrated force* represents the effect of a loading which is assumed to act at a point on a body. We can represent a load by a concentrated force, provided the area over which the load is applied is very small compared to the overall size of the body. An example would be the contact force between a wheel and the ground.

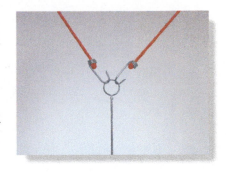

Three forces act on the ring. Since these forces all meet at a point, then for any force analysis, we can assume the ring to be represented as a particle. (© Russell C. Hibbeler)

Steel is a common engineering material that does not deform very much under load. Therefore, we can consider this railroad wheel to be a rigid body acted upon by the concentrated force of the rail. (© Russell C. Hibbeler)

1

Newton's Three Laws of Motion.

Engineering mechanics is formulated on the basis of Newton's three laws of motion, the validity of which is based on experimental observation. These laws apply to the motion of a particle as measured from a *nonaccelerating* reference frame. They may be briefly stated as follows.

First Law.   A particle originally at rest, or moving in a straight line with constant velocity, tends to remain in this state provided the particle is *not* subjected to an unbalanced force, Fig. 1–1*a*.

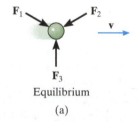

Equilibrium

(a)

Second Law.   A particle acted upon by an *unbalanced force* **F** experiences an acceleration **a** that has the same direction as the force and a magnitude that is directly proportional to the force, Fig. 1–1*b*.* If **F** is applied to a particle of mass *m*, this law may be expressed mathematically as

$$\mathbf{F} = m\mathbf{a} \tag{1–1}$$

Accelerated motion

(b)

Third Law.   The mutual forces of action and reaction between two particles are equal, opposite, and collinear, Fig. 1–1*c*.

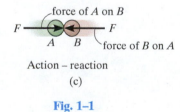

force of A on B

F F

A B force of B on A

Action – reaction

(c)

Fig. 1–1

---

*Stated another way, the unbalanced force acting on the particle is proportional to the time rate of change of the particle's linear momentum.

Newton's Law of Gravitational Attraction. Shortly after formulating his three laws of motion, Newton postulated a law governing the gravitational attraction between any two particles. Stated mathematically,

$$F = G\frac{m_1 m_2}{r^2} \qquad (1–2)$$

where

F = force of gravitation between the two particles

G = universal constant of gravitation; according to experimental evidence, $G = 66.73(10^{-12})$ m³/(kg·s²)

m_1, m_2 = mass of each of the two particles

r = distance between the two particles

Weight. According to Eq. 1–2, any two particles or bodies have a mutual attractive (gravitational) force acting between them. In the case of a particle located at or near the surface of the earth, however, the only gravitational force having any sizable magnitude is that between the earth and the particle. Consequently, this force, termed the **weight**, will be the only gravitational force considered in our study of mechanics.

From Eq. 1–2, we can develop an approximate expression for finding the weight W of a particle having a mass $m_1 = m$. If we assume the earth to be a nonrotating sphere of constant density and having a mass $m_2 = M_e$, then if r is the distance between the earth's center and the particle, we have

$$W = G\frac{m M_e}{r^2}$$

Letting $g = GM_e/r^2$ yields

$$\boxed{W = mg} \qquad (1–3)$$

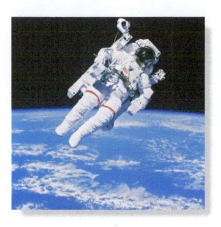

The astronaut's weight is diminished since she is far removed from the gravitational field of the earth. (© NikoNomad/ Shutterstock)

By comparison with $\mathbf{F} = m\mathbf{a}$, we can see that g is the acceleration due to gravity. Since it depends on r, then the weight of a body is *not* an absolute quantity. Instead, its magnitude is determined from where the measurement was made. For most engineering calculations, however, g is determined at sea level and at a latitude of 45°, which is considered the "standard location."

1.3 Units of Measurement

The four basic quantities—length, time, mass, and force—are not all independent from one another; in fact, they are *related* by Newton's second law of motion, $\mathbf{F} = m\mathbf{a}$. Because of this, the *units* used to measure these quantities cannot *all* be selected arbitrarily. The equality $\mathbf{F} = m\mathbf{a}$ is maintained only if three of the four units, called **base units**, are *defined* and the fourth unit is then *derived* from the equation.

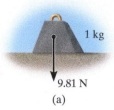

1 kg

9.81 N

(a)

SI Units.
The International System of units, abbreviated SI after the French "Système International d'Unités," is a modern version of the metric system which has received worldwide recognition. As shown in Table 1–1, the SI system defines length in meters (m), time in seconds (s), and mass in kilograms (kg). The unit of force, called a **newton** (N), is *derived* from $\mathbf{F} = m\mathbf{a}$. Thus, 1 newton is equal to a force required to give 1 kilogram of mass an acceleration of 1 m/s^2 (N = $kg \cdot m/s^2$).

If the weight of a body located at the "standard location" is to be determined in newtons, then Eq. 1–3 must be applied. Here measurements give $g = 9.806\,65$ m/s^2; however, for calculations, the value $g = 9.81$ m/s^2 will be used. Thus,

$$W = mg \qquad (g = 9.81 \text{ m/s}^2) \tag{1–4}$$

Therefore, a body of mass 1 kg has a weight of 9.81 N, a 2-kg body weighs 19.62 N, and so on, Fig. 1–2a.

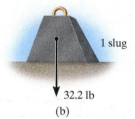

1 slug

32.2 lb

(b)

Fig. 1–2

U.S. Customary.
In the U.S. Customary system of units (FPS) length is measured in feet (ft), time in seconds (s), and force in pounds (lb), Table 1–1. The unit of mass, called a **slug**, is *derived* from $\mathbf{F} = m\mathbf{a}$. Hence, 1 slug is equal to the amount of matter accelerated at 1 ft/s^2 when acted upon by a force of 1 lb (slug = $lb \cdot s^2/ft$).

Therefore, if the measurements are made at the "standard location," where $g = 32.2$ ft/s^2, then from Eq. 1–3,

$$m = \frac{W}{g} \qquad (g = 32.2 \text{ ft/s}^2) \tag{1–5}$$

And so a body weighing 32.2 lb has a mass of 1 slug, a 64.4-lb body has a mass of 2 slugs, and so on, Fig. 1–2b.

TABLE 1–1 Systems of Units				
Name	Length	Time	Mass	Force
International System of Units SI	meter	second	kilogram	newton*
	m	s	kg	N $\left(\dfrac{kg \cdot m}{s^2}\right)$
U.S. Customary FPS	foot	second	slug*	pound
	ft	s	$\left(\dfrac{lb \cdot s^2}{ft}\right)$	lb

*Derived unit.

Conversion of Units. Table 1–2 provides a set of direct conversion factors between FPS and SI units for the basic quantities. Also, in the FPS system, recall that 1 ft = 12 in. (inches), 5280 ft = 1 mi (mile), 1000 lb = 1 kip (kilo-pound), and 2000 lb = 1 ton.

TABLE 1–2 Conversion Factors			
Quantity	Unit of Measurement (FPS)	Equals	Unit of Measurement (SI)
Force	lb		4.448 N
Mass	slug		14.59 kg
Length	ft		0.3048 m

1.4 The International System of Units

The SI system of units is used extensively in this book since it is intended to become the worldwide standard for measurement. Therefore, we will now present some of the rules for its use and some of its terminology relevant to engineering mechanics.

Prefixes. When a numerical quantity is either very large or very small, the units used to define its size may be modified by using a prefix. Some of the prefixes used in the SI system are shown in Table 1–3. Each represents a multiple or submultiple of a unit which, if applied successively, moves the decimal point of a numerical quantity to every third place.* For example, 4 000 000 N = 4 000 kN (kilo-newton) = 4 MN (mega-newton), or 0.005 m = 5 mm (milli-meter). Notice that the SI system does not include the multiple deca (10) or the submultiple centi (0.01), which form part of the metric system. Except for some volume and area measurements, the use of these prefixes is to be avoided in science and engineering.

TABLE 1–3 Prefixes			
	Exponential Form	Prefix	SI Symbol
Multiple			
1 000 000 000	10^9	giga	G
1 000 000	10^6	mega	M
1 000	10^3	kilo	k
Submultiple			
0.001	10^{-3}	milli	m
0.000 001	10^{-6}	micro	μ
0.000 000 001	10^{-9}	nano	n

*The kilogram is the only base unit that is defined with a prefix.

Rules for Use.   Here are a few of the important rules that describe the proper use of the various SI symbols:

- Quantities defined by several units which are multiples of one another are separated by a *dot* to avoid confusion with prefix notation, as indicated by $N = kg \cdot m/s^2 = kg \cdot m \cdot s^{-2}$. Also, $m \cdot s$ (meter-second), whereas ms (milli-second).

- The exponential power on a unit having a prefix refers to both the unit *and* its prefix. For example, $\mu N^2 = (\mu N)^2 = \mu N \cdot \mu N$. Likewise, mm^2 represents $(mm)^2 = mm \cdot mm$.

- With the exception of the base unit the kilogram, in general avoid the use of a prefix in the denominator of composite units. For example, do not write N/mm, but rather kN/m; also, m/mg should be written as Mm/kg.

- When performing calculations, represent the numbers in terms of their *base or derived units* by converting all prefixes to powers of 10. The final result should then be expressed using a *single prefix*. Also, after calculation, it is best to keep numerical values between 0.1 and 1000; otherwise, a suitable prefix should be chosen. For example,

$$(50 \text{ kN})(60 \text{ nm}) = \left[50(10^3) \text{ N} \right] \left[60(10^{-9}) \text{ m} \right]$$
$$= 3000(10^{-6}) \text{ N} \cdot \text{m} = 3(10^{-3}) \text{ N} \cdot \text{m} = 3 \text{ mN} \cdot \text{m}$$

1.5   Numerical Calculations

Numerical work in engineering practice is most often performed by using handheld calculators and computers. It is important, however, that the answers to any problem be reported with justifiable accuracy using appropriate significant figures. In this section we will discuss these topics together with some other important aspects involved in all engineering calculations.

Computers are often used in engineering for advanced design and analysis. (© Blaize Pascall/Alamy)

Dimensional Homogeneity.   The terms of any equation used to describe a physical process must be *dimensionally homogeneous*; that is, each term must be expressed in the same units. Provided this is the case, all the terms of an equation can then be combined if numerical values are substituted for the variables. Consider, for example, the equation $s = vt + \frac{1}{2}at^2$, where, in SI units, s is the position in meters, m, t is time in seconds, s, v is velocity in m/s and a is acceleration in m/s^2. Regardless of how this equation is evaluated, it maintains its dimensional homogeneity. In the form stated, each of the three terms is expressed in meters $\left[m, (m/s)s, (m/s^2)s^2 \right]$ or solving for a, $a = 2s/t^2 - 2v/t$, the terms are each expressed in units of $m/s^2 \left[m/s^2, m/s^2, (m/s)/s \right]$.

Keep in mind that problems in mechanics always involve the solution of dimensionally homogeneous equations, and so this fact can then be used as a partial check for algebraic manipulations of an equation.

Significant Figures.

The number of significant figures contained in any number determines the accuracy of the number. For instance, the number 4981 contains four significant figures. However, if zeros occur at the end of a whole number, it may be unclear as to how many significant figures the number represents. For example, 23 400 might have three (234), four (2340), or five (23 400) significant figures. To avoid these ambiguities, we will use *engineering notation* to report a result. This requires that numbers be rounded off to the appropriate number of significant digits and then expressed in multiples of (10^3), such as (10^3), (10^6), or (10^{-9}). For instance, if 23 400 has five significant figures, it is written as $23.400(10^3)$, but if it has only three significant figures, it is written as $23.4(10^3)$.

If zeros occur at the beginning of a number that is less than one, then the zeros are not significant. For example, 0.008 21 has three significant figures. Using engineering notation, this number is expressed as $8.21(10^{-3})$. Likewise, 0.000 582 can be expressed as $0.582(10^{-3})$ or $582(10^{-6})$.

Rounding Off Numbers.

Rounding off a number is necessary so that the accuracy of the result will be the same as that of the problem data. As a general rule, any numerical figure ending in a number greater than five is rounded up and a number less than five is not rounded up. The rules for rounding off numbers are best illustrated by examples. Suppose the number 3.5587 is to be rounded off to *three* significant figures. Because the fourth digit (8) is *greater than* 5, the third number is rounded up to 3.56. Likewise 0.5896 becomes 0.590 and 9.3866 becomes 9.39. If we round off 1.341 to three significant figures, because the fourth digit (1) is *less than* 5, then we get 1.34. Likewise 0.3762 becomes 0.376 and 9.871 becomes 9.87. There is a special case for any number that ends in a 5. As a general rule, if the digit preceding the 5 is an *even number*, then this digit is *not* rounded up. If the digit preceding the 5 is an *odd number*, then it is rounded up. For example, 75.25 rounded off to three significant digits becomes 75.2, 0.1275 becomes 0.128, and 0.2555 becomes 0.256.

Calculations.

When a sequence of calculations is performed, it is best to store the intermediate results in the calculator. In other words, do not round off calculations until expressing the final result. This procedure maintains precision throughout the series of steps to the final solution. In this text we will generally round off the answers to three significant figures since most of the data in engineering mechanics, such as geometry and loads, may be reliably measured to this accuracy.

When solving problems, do the work as neatly as possible. Being neat will stimulate clear and orderly thinking, and vice versa. (© Russell C. Hibbeler)

1.6   General Procedure for Analysis

Attending a lecture, reading this book, and studying the example problems helps, but **the most effective way of learning the principles of engineering mechanics is to *solve problems***. To be successful at this, it is important to always present the work in a *logical* and *orderly manner*, as suggested by the following sequence of steps:

- Read the problem carefully and try to correlate the actual physical situation with the theory studied.

- Tabulate the problem data and *draw to a large scale* any necessary diagrams.

- Apply the relevant principles, generally in mathematical form. When writing any equations, be sure they are dimensionally homogeneous.

- Solve the necessary equations, and report the answer with no more than three significant figures.

- Study the answer with technical judgment and common sense to determine whether or not it seems reasonable.

Important Points

- Statics is the study of bodies that are at rest or move with constant velocity.

- A particle has a mass but a size that can be neglected, and a rigid body does not deform under load.

- A force is considered as a "push" or "pull" of one body on another.

- Concentrated forces are assumed to act at a point on a body.

- Newton's three laws of motion should be memorized.

- Mass is measure of a quantity of matter that does not change from one location to another. Weight refers to the gravitational attraction of the earth on a body or quantity of mass. Its magnitude depends upon the elevation at which the mass is located.

- In the SI system the unit of force, the newton, is a derived unit. The meter, second, and kilogram are base units.

- Prefixes G, M, k, m, μ, and n are used to represent large and small numerical quantities. Their exponential size should be known, along with the rules for using the SI units.

- Perform numerical calculations with several significant figures, and then report the final answer to three significant figures.

- Algebraic manipulations of an equation can be checked in part by verifying that the equation remains dimensionally homogeneous.

- Know the rules for rounding off numbers.

EXAMPLE 1.1

Convert 2 km/h to m/s How many ft/s is this?

SOLUTION

Since 1 km = 1000 m and 1 h = 3600 s, the factors of conversion are arranged in the following order, so that a cancellation of the units can be applied:

$$2 \text{ km/h} = \frac{2 \text{ km}}{\text{h}} \left(\frac{1000 \text{ m}}{\text{km}} \right) \left(\frac{1 \text{ h}}{3600 \text{ s}} \right)$$

$$= \frac{2000 \text{ m}}{3600 \text{ s}} = 0.556 \text{ m/s} \qquad Ans.$$

From Table 1–2, 1 ft = 0.3048 m. Thus,

$$0.556 \text{ m/s} = \left(\frac{0.556 \text{ m}}{\text{s}} \right) \left(\frac{1 \text{ ft}}{0.3048 \text{ m}} \right)$$

$$= 1.82 \text{ ft/s} \qquad Ans.$$

NOTE: Remember to round off the final answer to three significant figures.

EXAMPLE 1.2

Convert the quantities 300 lb · s and 52 slug/ft³ to appropriate SI units.

SOLUTION

Using Table 1–2, 1 lb = 4.448 N.

$$300 \text{ lb} \cdot \text{s} = 300 \text{ lb} \cdot \text{s} \left(\frac{4.448 \text{ N}}{1 \text{ lb}} \right)$$

$$= 1334.5 \text{ N} \cdot \text{s} = 1.33 \text{ kN} \cdot \text{s} \qquad Ans.$$

Since 1 slug = 14.59 kg and 1 ft = 0.3048 m, then

$$52 \text{ slug/ft}^3 = \frac{52 \text{ slug}}{\text{ft}^3} \left(\frac{14.59 \text{ kg}}{1 \text{ slug}} \right) \left(\frac{1 \text{ ft}}{0.3048 \text{ m}} \right)^3$$

$$= 26.8(10^3) \text{ kg/m}^3$$

$$= 26.8 \text{ Mg/m}^3 \qquad Ans.$$

EXAMPLE | 1.3

Evaluate each of the following and express with SI units having an appropriate prefix: (a) (50 mN)(6 GN), (b) (400 mm)(0.6 MN)^2, (c) 45 MN^3/900 Gg.

SOLUTION

First convert each number to base units, perform the indicated operations, then choose an appropriate prefix.

Part (a)

$$(50 \text{ mN})(6 \text{ GN}) = \left[50(10^{-3}) \text{ N} \right] \left[6(10^9) \text{ N} \right]$$

$$= 300(10^6) \text{ N}^2$$

$$= 300(10^6) \text{ N}^2 \left(\frac{1 \text{ kN}}{10^3 \text{ N}} \right) \left(\frac{1 \text{ kN}}{10^3 \text{ N}} \right)$$

$$= 300 \text{ kN}^2 \qquad \text{Ans.}$$

NOTE: Keep in mind the convention kN^2 = (kN)^2 = 10^6 N^2.

Part (b)

$$(400 \text{ mm})(0.6 \text{ MN})^2 = \left[400(10^{-3}) \text{ m} \right] \left[0.6(10^6) \text{ N} \right]^2$$

$$= \left[400(10^{-3}) \text{ m} \right] \left[0.36(10^{12}) \text{ N}^2 \right]$$

$$= 144(10^9) \text{ m} \cdot \text{N}^2$$

$$= 144 \text{ Gm} \cdot \text{N}^2 \qquad \text{Ans.}$$

We can also write

$$144(10^9) \text{ m} \cdot \text{N}^2 = 144(10^9) \text{ m} \cdot \text{N}^2 \left(\frac{1 \text{ MN}}{10^6 \text{ N}} \right) \left(\frac{1 \text{ MN}}{10^6 \text{ N}} \right)$$

$$= 0.144 \text{ m} \cdot \text{MN}^2 \qquad \text{Ans.}$$

Part (c)

$$\frac{45 \text{ MN}^3}{900 \text{ Gg}} = \frac{45(10^6 \text{ N})^3}{900(10^6) \text{ kg}}$$

$$= 50(10^9) \text{ N}^3/\text{kg}$$

$$= 50(10^9) \text{ N}^3 \left(\frac{1 \text{ kN}}{10^3 \text{ N}} \right)^3 \frac{1}{\text{kg}}$$

$$= 50 \text{ kN}^3/\text{kg} \qquad \text{Ans.}$$

PROBLEMS

The answers to all but every fourth problem (asterisk) are given in the back of the book.

1–1. What is the weight in newtons of an object that has a mass of (a) 8 kg, (b) 0.04 kg, and (c) 760 Mg?

1–2. Represent each of the following combinations of units in the correct SI form: (a) kN/μs, (b) Mg/mN, and (c) MN/(kg · ms).

1–3. Represent each of the following combinations of units in the correct SI form: (a) Mg/ms, (b) N/mm, (c) mN/(kg · μs).

*1–4. Convert: (a) 200 lb · ft to N · m, (b) 350 lb/ft³ to kN/m³, (c) 8 ft/h to mm/s. Express the result to three significant figures. Use an appropriate prefix.

1–5. Represent each of the following as a number between 0.1 and 1000 using an appropriate prefix: (a) 45 320 kN, (b) 568(10^5) mm, and (c) 0.00563 mg.

1–6. Round off the following numbers to three significant figures: (a) 58 342 m, (b) 68.534 s, (c) 2553 N, and (d) 7555 kg.

1–7. Represent each of the following quantities in the correct SI form using an appropriate prefix: (a) 0.000 431 kg, (b) 35.3(10^3) N, (c) 0.005 32 km.

*1–8. Represent each of the following combinations of units in the correct SI form using an appropriate prefix: (a) Mg/mm, (b) mN/μs, (c) μm · Mg.

1–9. Represent each of the following combinations of units in the correct SI form using an appropriate prefix: (a) m/ms, (b) μkm, (c) ks/mg, and (d) km · μN.

1–10. Represent each of the following combinations of units in the correct SI form: (a) GN · μm, (b) kg/μm, (c) N/ks², and (d) kN/μs.

1–11. Represent each of the following with SI units having an appropriate prefix: (a) 8653 ms, (b) 8368 N, (c) 0.893 kg.

*1–12. Evaluate each of the following to three significant figures and express each answer in SI units using an appropriate prefix: (a) (684 μm)/(43 ms), (b) (28 ms)(0.0458 Mm)/(348 mg), (c) (2.68 mm)(426 Mg).

1–13. The density (mass/volume) of aluminum is 5.26 slug/ft³. Determine its density in SI units. Use an appropriate prefix.

1–14. Evaluate each of the following to three significant figures and express each answer in SI units using an appropriate prefix: (a) (212 mN)², (b) (52 800 ms)², and (c) [548(10^6)]$^{1/2}$ ms.

1–15. Using the SI system of units, show that Eq. 1–2 is a dimensionally homogeneous equation which gives F in newtons. Determine to three significant figures the gravitational force acting between two spheres that are touching each other. The mass of each sphere is 200 kg and the radius is 300 mm.

*1–16. The *pascal* (Pa) is actually a very small unit of pressure. To show this, convert 1 Pa = 1 N/m² to lb/ft². Atmosphere pressure at sea level is 14.7 lb/in². How many pascals is this?

1–17. Water has a density of 1.94 slug/ft³. What is the density expressed in SI units? Express the answer to three significant figures.

1–18. Evaluate each of the following to three significant figures and express each answer in SI units using an appropriate prefix: (a) 354 mg(45 km)/(0.0356 kN), (b) (0.004 53 Mg)(201 ms), (c) 435 MN/23.2 mm.

1–19. A concrete column has a diameter of 350 mm and a length of 2 m. If the density (mass/volume) of concrete is 2.45 Mg/m³, determine the weight of the column in pounds.

*1–20. If a man weighs 155 lb on earth, specify (a) his mass in slugs, (b) his mass in kilograms, and (c) his weight in newtons. If the man is on the moon, where the acceleration due to gravity is g_m = 5.30 ft/s², determine (d) his weight in pounds, and (e) his mass in kilograms.

1–21. Two particles have a mass of 8 kg and 12 kg, respectively. If they are 800 mm apart, determine the force of gravity acting between them. Compare this result with the weight of each particle.

Chapter 2

(© Vasiliy Koval/Fotolia)

This electric transmission tower is stabilized by cables that exert forces on the tower at their points of connection. In this chapter we will show how to express these forces as Cartesian vectors, and then determine their resultant.

Force Vectors

CHAPTER OBJECTIVES

- To show how to add forces and resolve them into components using the Parallelogram Law.

- To express force and position in Cartesian vector form and explain how to determine the vector's magnitude and direction.

- To introduce the dot product in order to use it to find the angle between two vectors or the projection of one vector onto another.

2.1 Scalars and Vectors

Many physical quantities in engineering mechanics are measured using either scalars or vectors.

Scalar. A *scalar* is any positive or negative physical quantity that can be completely specified by its *magnitude*. Examples of scalar quantities include length, mass, and time.

Vector. A *vector* is any physical quantity that requires both a *magnitude* and a *direction* for its complete description. Examples of vectors encountered in statics are force, position, and moment. A vector is shown graphically by an arrow. The length of the arrow represents the *magnitude* of the vector, and the angle θ between the vector and a fixed axis defines the *direction of its line of action*. The head or tip of the arrow indicates the *sense of direction* of the vector, Fig. 2–1.

In print, vector quantities are represented by boldface letters such as **A**, and the magnitude of a vector is italicized, A. For handwritten work, it is often convenient to denote a vector quantity by simply drawing an arrow above it, $\vec{A}$.

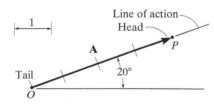

Fig. 2–1

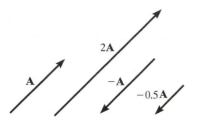

Scalar multiplication and division

Fig. 2–2

2.2   Vector Operations

Multiplication and Division of a Vector by a Scalar.   If a vector is multiplied by a positive scalar, its magnitude is increased by that amount. Multiplying by a negative scalar will also change the directional sense of the vector. Graphic examples of these operations are shown in Fig. 2–2.

Vector Addition.   When adding two vectors together it is important to account for both their magnitudes and their directions. To do this we must use the ***parallelogram law of addition***. To illustrate, the two ***component vectors*** **A** and **B** in Fig. 2–3a are added to form a ***resultant vector*** **R** = **A** + **B** using the following procedure:

- First join the tails of the components at a point to make them concurrent, Fig. 2–3b.

- From the head of **B**, draw a line parallel to **A**. Draw another line from the head of **A** that is parallel to **B**. These two lines intersect at point *P* to form the adjacent sides of a parallelogram.

- The diagonal of this parallelogram that extends to *P* forms **R**, which then represents the resultant vector **R** = **A** + **B**, Fig. 2–3c.

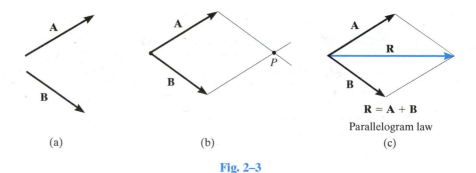

Fig. 2–3

We can also add **B** to **A**, Fig. 2–4a, using the ***triangle rule***, which is a special case of the parallelogram law, whereby vector **B** is added to vector **A** in a "head-to-tail" fashion, i.e., by connecting the head of **A** to the tail of **B**, Fig. 2–4b. The resultant **R** extends from the tail of **A** to the head of **B**. In a similar manner, **R** can also be obtained by adding **A** to **B**, Fig. 2–4c. By comparison, it is seen that vector addition is commutative; in other words, the vectors can be added in either order, i.e., **R** = **A** + **B** = **B** + **A**.

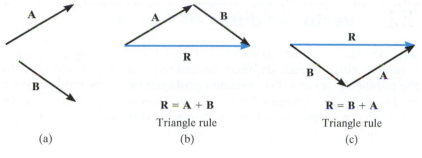

$R = A + B$	$R = B + A$
Triangle rule	Triangle rule
(a) (b)	(c)

Fig. 2–4

As a special case, if the two vectors **A** and **B** are ***collinear***, i.e., both have the same line of action, the parallelogram law reduces to an *algebraic* or *scalar addition* $R = A + B$, as shown in Fig. 2–5.

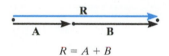

$$R = A + B$$

Addition of collinear vectors

Fig. 2–5

Vector Subtraction.
The resultant of the *difference* between two vectors **A** and **B** of the same type may be expressed as

$$R' = A - B = A + (-B)$$

This vector sum is shown graphically in Fig. 2–6. Subtraction is therefore defined as a special case of addition, so the rules of vector addition also apply to vector subtraction.

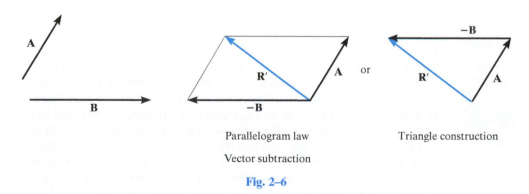

Parallelogram law

Vector subtraction

Fig. 2–6

2.3   Vector Addition of Forces

Experimental evidence has shown that a force is a vector quantity since it has a specified magnitude, direction, and sense and it adds according to the parallelogram law. Two common problems in statics involve either finding the resultant force, knowing its components, or resolving a known force into two components. We will now describe how each of these problems is solved using the parallelogram law.

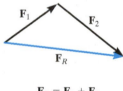

The parallelogram law must be used to determine the resultant of the two forces acting on the hook. (© Russell C. Hibbeler)

Finding a Resultant Force.   The two component forces F_1 and F_2 acting on the pin in Fig. 2–7a can be added together to form the resultant force $F_R = F_1 + F_2$, as shown in Fig. 2–7b. From this construction, or using the triangle rule, Fig. 2–7c, we can apply the law of cosines or the law of sines to the triangle in order to obtain the magnitude of the resultant force and its direction.

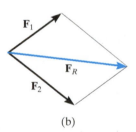

$$F_R = F_1 + F_2$$

(a)   (b)   (c)

Fig. 2–7

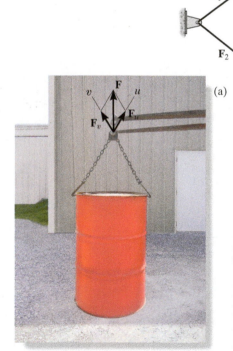

Using the parallelogram law the supporting force F can be resolved into components acting along the u and v axes. (© Russell C. Hibbeler)

Finding the Components of a Force.   Sometimes it is necessary to resolve a force into two *components* in order to study its pulling or pushing effect in two specific directions. For example, in Fig. 2–8a, F is to be resolved into two components along the two members, defined by the u and v axes. In order to determine the magnitude of each component, a parallelogram is constructed first, by drawing lines starting from the tip of F, one line parallel to u, and the other line parallel to v. These lines then intersect with the v and u axes, forming a parallelogram. The force components F_u and F_v are then established by simply joining the tail of F to the intersection points on the u and v axes, Fig. 2–8b. This parallelogram can then be reduced to a triangle, which represents the triangle rule, Fig. 2–8c. From this, the law of sines can then be applied to determine the unknown magnitudes of the components.

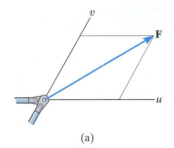

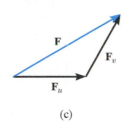

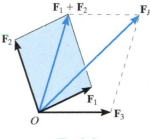

(a)   (b)   (c)

Fig. 2–8

Addition of Several Forces.   If more than two forces are to be added, successive applications of the parallelogram law can be carried out in order to obtain the resultant force. For example, if three forces $\mathbf{F}_1$, $\mathbf{F}_2$, $\mathbf{F}_3$ act at a point O, Fig. 2–9, the resultant of any two of the forces is found, say, $\mathbf{F}_1 + \mathbf{F}_2$—and then this resultant is added to the third force, yielding the resultant of all three forces; i.e., $\mathbf{F}_R = (\mathbf{F}_1 + \mathbf{F}_2) + \mathbf{F}_3$. Using the parallelogram law to add more than two forces, as shown here, often requires extensive geometric and trigonometric calculation to determine the numerical values for the magnitude and direction of the resultant. Instead, problems of this type are easily solved by using the "rectangular-component method," which is explained in Sec. 2.4.

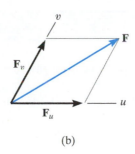

Fig. 2–9

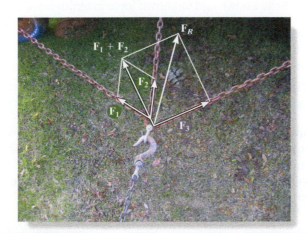

The resultant force $\mathbf{F}_R$ on the hook requires the addition of $\mathbf{F}_1 + \mathbf{F}_2$, then this resultant is added to $\mathbf{F}_3$. (© Russell C. Hibbeler)

2

Important Points

- A scalar is a positive or negative number.

- A vector is a quantity that has a magnitude, direction, and sense.

- Multiplication or division of a vector by a scalar will change the magnitude of the vector. The sense of the vector will change if the scalar is negative.

- As a special case, if the vectors are collinear, the resultant is formed by an algebraic or scalar addition.

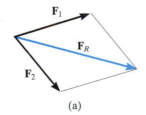

(a)

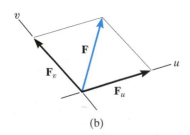

(b)

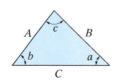

(c)

Fig. 2–10

Procedure for Analysis

Problems that involve the addition of two forces can be solved as follows:

Parallelogram Law.

- Two "component" forces F_1 and F_2 in Fig. 2–10a add according to the parallelogram law, yielding a *resultant* force F_R that forms the diagonal of the parallelogram.

- If a force F is to be resolved into *components* along two axes u and v, Fig. 2–10b, then start at the head of force F and construct lines parallel to the axes, thereby forming the parallelogram. The sides of the parallelogram represent the components, F_u and F_v.

- Label all the known and unknown force magnitudes and the angles on the sketch and identify the two unknowns as the magnitude and direction of F_R, or the magnitudes of its components.

Trigonometry.

- Redraw a half portion of the parallelogram to illustrate the triangular head-to-tail addition of the components.

- From this triangle, the magnitude of the resultant force can be determined using the law of cosines, and its direction is determined from the law of sines. The magnitudes of two force components are determined from the law of sines. The formulas are given in Fig. 2–10c.

EXAMPLE 2.1

The screw eye in Fig. 2–11a is subjected to two forces, $\mathbf{F}_1$ and $\mathbf{F}_2$. Determine the magnitude and direction of the resultant force.

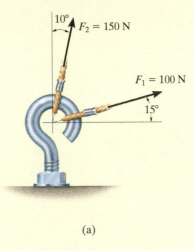

(a)

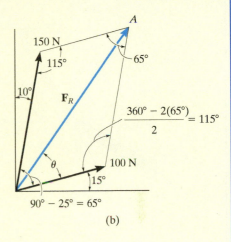

(b)

SOLUTION

Parallelogram Law.   The parallelogram is formed by drawing a line from the head of $\mathbf{F}_1$ that is parallel to $\mathbf{F}_2$, and another line from the head of $\mathbf{F}_2$ that is parallel to $\mathbf{F}_1$. The resultant force $\mathbf{F}_R$ extends to where these lines intersect at point A, Fig. 2–11b. The two unknowns are the magnitude of $\mathbf{F}_R$ and the angle θ (theta).

Trigonometry.   From the parallelogram, the vector triangle is constructed, Fig. 2–11c. Using the law of cosines

$$F_R = \sqrt{(100 \text{ N})^2 + (150 \text{ N})^2 - 2(100 \text{ N})(150 \text{ N}) \cos 115°}$$

$$= \sqrt{10\,000 + 22\,500 - 30\,000(-0.4226)} = 212.6 \text{ N}$$

$$= 213 \text{ N} \qquad \qquad \textit{Ans.}$$

Applying the law of sines to determine θ,

$$\frac{150 \text{ N}}{\sin \theta} = \frac{212.6 \text{ N}}{\sin 115°} \qquad \sin \theta = \frac{150 \text{ N}}{212.6 \text{ N}} (\sin 115°)$$

$$\theta = 39.8°$$

Thus, the direction ϕ (phi) of $\mathbf{F}_R$, measured from the horizontal, is

$$\phi = 39.8° + 15.0° = 54.8° \qquad \qquad \textit{Ans.}$$

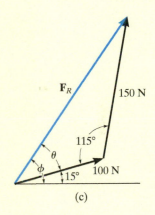

Fig. 2–11

NOTE: The results seem reasonable, since Fig. 2–11b shows $\mathbf{F}_R$ to have a magnitude larger than its components and a direction that is between them.

EXAMPLE 2.2

Resolve the horizontal 600-lb force in Fig. 2–12a into components acting along the u and v axes and determine the magnitudes of these components.

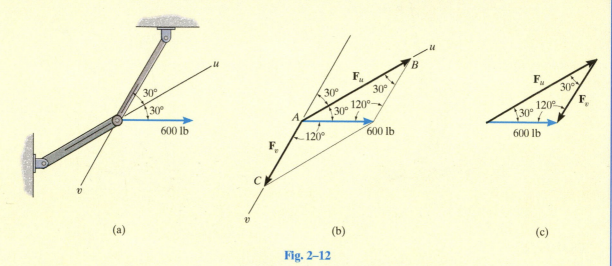

(a)   (b)   (c)

Fig. 2–12

SOLUTION

The parallelogram is constructed by extending a line from the *head* of the 600-lb force parallel to the v axis until it intersects the u axis at point B, Fig. 2–12b. The arrow from A to B represents $\mathbf{F}_u$. Similarly, the line extended from the head of the 600-lb force drawn parallel to the u axis intersects the v axis at point C, which gives $\mathbf{F}_v$.

The vector addition using the triangle rule is shown in Fig. 2–12c. The two unknowns are the magnitudes of $\mathbf{F}_u$ and $\mathbf{F}_v$. Applying the law of sines,

$$\frac{F_u}{\sin 120°} = \frac{600 \text{ lb}}{\sin 30°}$$

$$F_u = 1039 \text{ lb} \qquad \text{Ans.}$$

$$\frac{F_v}{\sin 30°} = \frac{600 \text{ lb}}{\sin 30°}$$

$$F_v = 600 \text{ lb} \qquad \text{Ans.}$$

NOTE: The result for F_u shows that sometimes a component can have a greater magnitude than the resultant.

EXAMPLE | 2.3

Determine the magnitude of the component force **F** in Fig. 2–13*a* and the magnitude of the resultant force **F**$_R$ if **F**$_R$ is directed along the positive *y* axis.

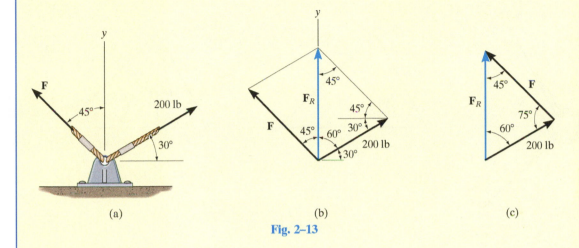

(a)   (b)   (c)

Fig. 2–13

SOLUTION

The parallelogram law of addition is shown in Fig. 2–13*b*, and the triangle rule is shown in Fig. 2–13*c*. The magnitudes of **F**$_R$ and **F** are the two unknowns. They can be determined by applying the law of sines.

$$\frac{F}{\sin 60°} = \frac{200 \text{ lb}}{\sin 45°}$$

$$F = 245 \text{ lb} \qquad\qquad Ans.$$

$$\frac{F_R}{\sin 75°} = \frac{200 \text{ lb}}{\sin 45°}$$

$$F_R = 273 \text{ lb} \qquad\qquad Ans.$$

EXAMPLE | 2.4

It is required that the resultant force acting on the eyebolt in Fig. 2–14a be directed along the positive x axis and that $\mathbf{F}_2$ have a *minimum* magnitude. Determine this magnitude, the angle θ, and the corresponding resultant force.

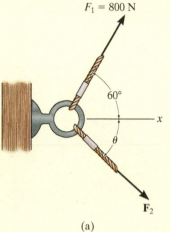

(a)

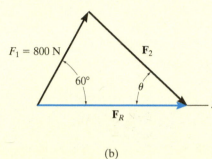

(b)

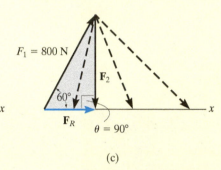

(c)

Fig. 2–14

SOLUTION

The triangle rule for $\mathbf{F}_R = \mathbf{F}_1 + \mathbf{F}_2$ is shown in Fig. 2–14b. Since the magnitudes (lengths) of $\mathbf{F}_R$ and $\mathbf{F}_2$ are not specified, then $\mathbf{F}_2$ can actually be any vector that has its head touching the line of action of $\mathbf{F}_R$, Fig. 2–14c. However, as shown, the magnitude of $\mathbf{F}_2$ is a *minimum* or the shortest length when its line of action is *perpendicular* to the line of action of $\mathbf{F}_R$, that is, when

$$\theta = 90° \qquad \qquad Ans.$$

Since the vector addition now forms the shaded right triangle, the two unknown magnitudes can be obtained by trigonometry.

$$F_R = (800 \text{ N})\cos 60° = 400 \text{ N} \qquad Ans.$$
$$F_2 = (800 \text{ N})\sin 60° = 693 \text{ N} \qquad Ans.$$

It is strongly suggested that you test yourself on the solutions to these examples, by covering them over and then trying to draw the parallelogram law, and thinking about how the sine and cosine laws are used to determine the unknowns. Then before solving any of the problems, try to solve the Preliminary Problems and some of the Fundamental Problems given on the next pages. The solutions and answers to these are given in the back of the book. Doing this throughout the book will help immensely in developing your problem-solving skills.

PRELIMINARY PROBLEMS

Partial solutions and answers to all Preliminary Problems are given in the back of the book.

P2–1. In each case, construct the parallelogram law to show $\mathbf{F}_R = \mathbf{F}_1 + \mathbf{F}_2$. Then establish the triangle rule, where $\mathbf{F}_R = \mathbf{F}_1 + \mathbf{F}_2$. Label all known and unknown sides and internal angles.

P2–2. In each case, show how to resolve the force $\mathbf{F}$ into components acting along the u and v axes using the parallelogram law. Then establish the triangle rule to show $\mathbf{F}_R = \mathbf{F}_u + \mathbf{F}_v$. Label all known and unknown sides and interior angles.

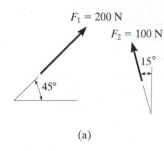

(a)

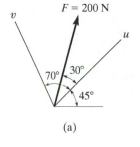

(a)

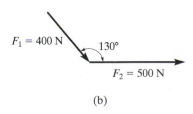

(b)

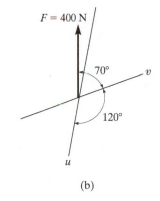

(b)

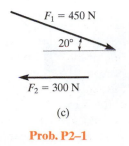

(c)

Prob. P2–1

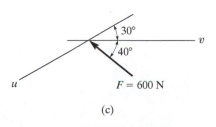

(c)

Prob. P2–2

FUNDAMENTAL PROBLEMS

Partial solutions and answers to all Fundamental Problems are given in the back of the book.

F2–1. Determine the magnitude of the resultant force acting on the screw eye and its direction measured clockwise from the x axis.

Prob. F2–1

F2–2. Two forces act on the hook. Determine the magnitude of the resultant force.

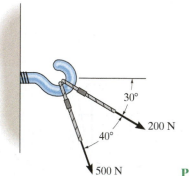

Prob. F2–2

F2–3. Determine the magnitude of the resultant force and its direction measured counterclockwise from the positive x axis.

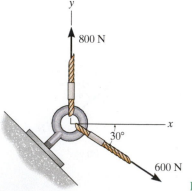

Prob. F2–3

F2–4. Resolve the 30-lb force into components along the u and v axes, and determine the magnitude of each of these components.

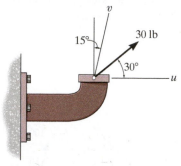

Prob. F2–4

F2–5. The force $F = 450$ lb acts on the frame. Resolve this force into components acting along members AB and AC, and determine the magnitude of each component.

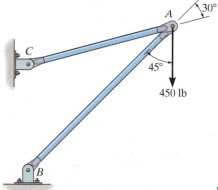

Prob. F2–5

F2–6. If force F is to have a component along the u axis of $F_u = 6$ kN, determine the magnitude of F and the magnitude of its component F_v along the v axis.

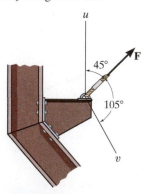

Prob. F2–6

PROBLEMS

2–1. If $\theta = 60°$ and $F = 450$ N, determine the magnitude of the resultant force and its direction, measured counterclockwise from the positive x axis.

2–2. If the magnitude of the resultant force is to be 500 N, directed along the positive y axis, determine the magnitude of force **F** and its direction θ.

*2–4. The vertical force **F** acts downward at A on the two-membered frame. Determine the magnitudes of the two components of **F** directed along the axes of AB and AC. Set $F = 500$ N.

2–5. Solve Prob. 2–4 with $F = 350$ lb.

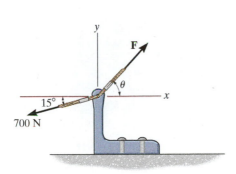

Probs. 2–1/2

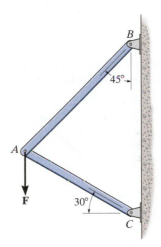

Probs. 2–4/5

2–3. Determine the magnitude of the resultant force $\mathbf{F}_R = \mathbf{F}_1 + \mathbf{F}_2$ and its direction, measured counterclockwise from the positive x axis.

2–6. Determine the magnitude of the resultant force $\mathbf{F}_R = \mathbf{F}_1 + \mathbf{F}_2$ and its direction, measured clockwise from the positive u axis.

2–7. Resolve the force $\mathbf{F}_1$ into components acting along the u and v axes and determine the magnitudes of the components.

*2–8. Resolve the force $\mathbf{F}_2$ into components acting along the u and v axes and determine the magnitudes of the components.

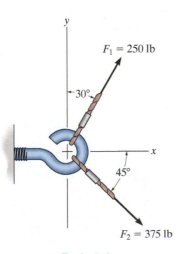

Prob. 2–3

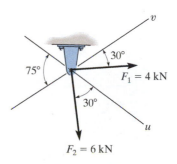

Probs. 2–6/7/8

2–9. If the resultant force acting on the support is to be 1200 lb, directed horizontally to the right, determine the force **F** in rope *A* and the corresponding angle θ.

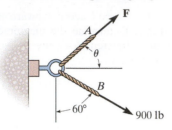

Prob. 2–9

2–10. Determine the magnitude of the resultant force and its direction, measured counterclockwise from the positive *x* axis.

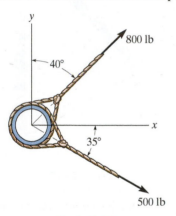

Prob. 2–10

2–11. The plate is subjected to the two forces at *A* and *B* as shown. If θ = 60°, determine the magnitude of the resultant of these two forces and its direction measured clockwise from the horizontal.

*2–12. Determine the angle θ for connecting member *A* to the plate so that the resultant force of **F**$_A$ and **F**$_B$ is directed horizontally to the right. Also, what is the magnitude of the resultant force?

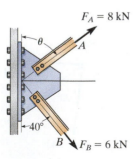

Probs. 2–11/12

2–13. The force acting on the gear tooth is *F* = 20 lb. Resolve this force into two components acting along the lines *aa* and *bb*.

2–14. The component of force **F** acting along line *aa* is required to be 30 lb. Determine the magnitude of **F** and its component along line *bb*.

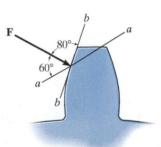

Probs. 2–13/14

2–15. Force **F** acts on the frame such that its component acting along member *AB* is 650 lb, directed from *B* towards *A*, and the component acting along member *BC* is 500 lb, directed from *B* towards *C*. Determine the magnitude of **F** and its direction θ. Set φ = 60°.

*2–16. Force **F** acts on the frame such that its component acting along member *AB* is 650 lb, directed from *B* towards *A*. Determine the required angle φ (0° ≤ φ ≤ 45°) and the component acting along member *BC*. Set *F* = 850 lb and θ = 30°.

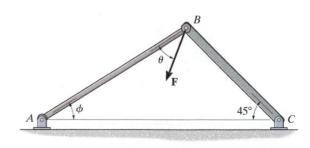

Probs. 2–15/16

2–17. Determine the magnitude and direction of the resultant $\mathbf{F}_R = \mathbf{F}_1 + \mathbf{F}_2 + \mathbf{F}_3$ of the three forces by first finding the resultant $\mathbf{F}' = \mathbf{F}_1 + \mathbf{F}_2$ and then forming $\mathbf{F}_R = \mathbf{F}' + \mathbf{F}_3$.

2–18. Determine the magnitude and direction of the resultant $\mathbf{F}_R = \mathbf{F}_1 + \mathbf{F}_2 + \mathbf{F}_3$ of the three forces by first finding the resultant $\mathbf{F}' = \mathbf{F}_2 + \mathbf{F}_3$ and then forming $\mathbf{F}_R = \mathbf{F}' + \mathbf{F}_1$.

2–21. Determine the magnitude and direction of the resultant force, $\mathbf{F}_R$ measured counterclockwise from the positive x axis. Solve the problem by first finding the resultant $\mathbf{F}' = \mathbf{F}_1 + \mathbf{F}_2$ and then forming $\mathbf{F}_R = \mathbf{F}' + \mathbf{F}_3$.

2–22. Determine the magnitude and direction of the resultant force, measured counterclockwise from the positive x axis. Solve l by first finding the resultant $\mathbf{F}' = \mathbf{F}_2 + \mathbf{F}_3$ and then forming $\mathbf{F}_R = \mathbf{F}' + \mathbf{F}_1$.

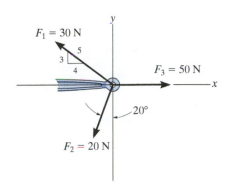

Probs. 2–17/18

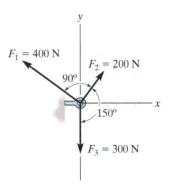

Probs. 2–21/22

2–19. Determine the design angle θ ($0° \le \theta \le 90°$) for strut AB so that the 400-lb horizontal force has a component of 500 lb directed from A towards C. What is the component of force acting along member AB? Take $\phi = 40°$.

***2–20.** Determine the design angle ϕ ($0° \le \phi \le 90°$) between struts AB and AC so that the 400-lb horizontal force has a component of 600 lb which acts up to the left, in the same direction as from B towards A. Take $\theta = 30°$.

2–23. Two forces act on the screw eye. If $F_1 = 400$ N and $F_2 = 600$ N, determine the angle θ ($0° \le \theta \le 180°$) between them, so that the resultant force has a magnitude of $F_R = 800$ N.

***2–24.** Two forces $\mathbf{F}_1$ and $\mathbf{F}_2$ act on the screw eye. If their lines of action are at an angle θ apart and the magnitude of each force is $F_1 = F_2 = F$, determine the magnitude of the resultant force $\mathbf{F}_R$ and the angle between $\mathbf{F}_R$ and $\mathbf{F}_1$.

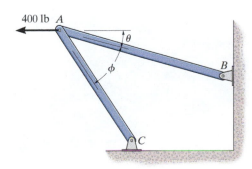

Probs. 2–19/20

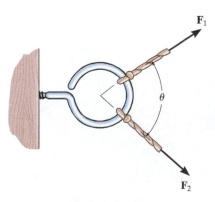

Probs. 2–23/24

2–25. If $F_1 = 30$ lb and $F_2 = 40$ lb, determine the angles θ and ϕ so that the resultant force is directed along the positive x axis and has a magnitude of $F_R = 60$ lb.

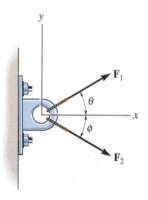

Prob. 2–25

2–26. Determine the magnitude and direction θ of $\mathbf{F}_A$ so that the resultant force is directed along the positive x axis and has a magnitude of 1250 N.

2–27. Determine the magnitude and direction, measured counterclockwise from the positive x axis, of the resultant force acting on the ring at O, if $F_A = 750$ N and $\theta = 45°$.

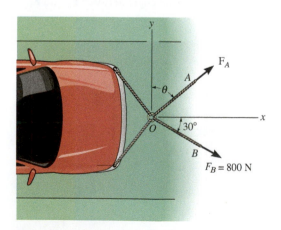

Probs. 2–26/27

*2–28. Determine the magnitude of force **F** so that the resultant $\mathbf{F}_R$ of the three forces is as small as possible. What is the minimum magnitude of $\mathbf{F}_R$?

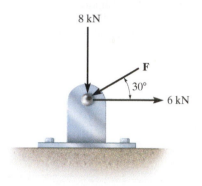

Prob. 2–28

2–29. If the resultant force of the two tugboats is 3 kN, directed along the positive x axis, determine the required magnitude of force $\mathbf{F}_B$ and its direction θ.

2–30. If $\mathbf{F}_B = 3$ kN and $\theta = 45°$, determine the magnitude of the resultant force of the two tugboats and its direction measured clockwise form the positive x axis.

2–31. If the resultant force of the two tugboats is required to be directed towards the positive x axis, and $\mathbf{F}_B$ is to be a minimum, determine the magnitude of $\mathbf{F}_R$ and $\mathbf{F}_B$ and the angle θ.

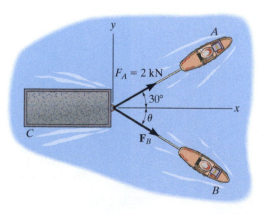

Probs. 2–29/30/31

2.4   Addition of a System of Coplanar Forces

When a force is resolved into two components along the x and y axes, the components are then called ***rectangular components***. For analytical work we can represent these components in one of two ways, using either scalar or Cartesian vector notation.

Scalar Notation.   The rectangular components of force **F** shown in Fig. 2–15*a* are found using the parallelogram law, so that $\mathbf{F} = \mathbf{F}_x + \mathbf{F}_y$. Because these components form a right triangle, they can be determined from

$$F_x = F \cos \theta \quad \text{and} \quad F_y = F \sin \theta$$

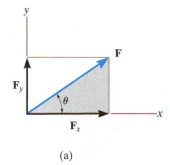

(a)

Instead of using the angle θ, however, the direction of **F** can also be defined using a small "slope" triangle, as in the example shown in Fig. 2–15*b*. Since this triangle and the larger shaded triangle are similar, the proportional length of the sides gives

$$\frac{F_x}{F} = \frac{a}{c}$$

or

$$F_x = F\left(\frac{a}{c}\right)$$

and

$$\frac{F_y}{F} = \frac{b}{c}$$

or

$$F_y = -F\left(\frac{b}{c}\right)$$

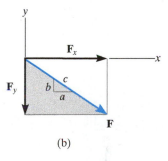

(b)

Fig. 2–15

Here the y component is a *negative scalar* since $\mathbf{F}_y$ is directed along the negative y axis.

It is important to keep in mind that this positive and negative scalar notation is to be used only for computational purposes, not for graphical representations in figures. Throughout the book, the *head of a vector arrow* in *any figure* indicates the sense of the vector *graphically*; algebraic signs are not used for this purpose. Thus, the vectors in Figs. 2–15*a* and 2–15*b* are designated by using boldface (vector) notation.* Whenever italic symbols are written near vector arrows in figures, they indicate the *magnitude* of the vector, which is *always* a *positive* quantity.

---

*Negative signs are used only in figures with boldface notation when showing equal but opposite pairs of vectors, as in Fig. 2–2.

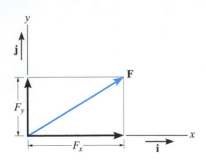

Fig. 2–16

Cartesian Vector Notation.

It is also possible to represent the x and y components of a force in terms of Cartesian unit vectors **i** and **j**. They are called unit vectors because they have a dimensionless magnitude of 1, and so they can be used to designate the *directions* of the x and y axes, respectively, Fig. 2–16.*

Since the *magnitude* of each component of **F** is *always a positive quantity*, which is represented by the (positive) scalars F_x and F_y, then we can express **F** as a **Cartesian vector**,

$$\mathbf{F} = F_x \mathbf{i} + F_y \mathbf{j}$$

Coplanar Force Resultants.

We can use either of the two methods just described to determine the resultant of several **coplanar forces**, i.e., forces that all lie in the same plane. To do this, each force is first resolved into its x and y components, and then the respective components are added using *scalar algebra* since they are collinear. The resultant force is then formed by adding the resultant components using the parallelogram law. For example, consider the three concurrent forces in Fig. 2–17a, which have x and y components shown in Fig. 2–17b. Using Cartesian vector notation, each force is first represented as a Cartesian vector, i.e.,

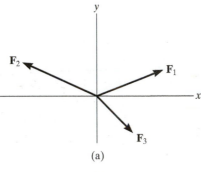

(a)

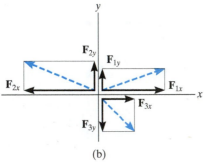

(b)

Fig. 2–17

$$\mathbf{F}_1 = F_{1x}\mathbf{i} + F_{1y}\mathbf{j}$$
$$\mathbf{F}_2 = -F_{2x}\mathbf{i} + F_{2y}\mathbf{j}$$
$$\mathbf{F}_3 = F_{3x}\mathbf{i} - F_{3y}\mathbf{j}$$

The vector resultant is therefore

$$\begin{aligned}\mathbf{F}_R &= \mathbf{F}_1 + \mathbf{F}_2 + \mathbf{F}_3 \\ &= F_{1x}\mathbf{i} + F_{1y}\mathbf{j} - F_{2x}\mathbf{i} + F_{2y}\mathbf{j} + F_{3x}\mathbf{i} - F_{3y}\mathbf{j} \\ &= (F_{1x} - F_{2x} + F_{3x})\mathbf{i} + (F_{1y} + F_{2y} - F_{3y})\mathbf{j} \\ &= (F_{Rx})\mathbf{i} + (F_{Ry})\mathbf{j}\end{aligned}$$

If *scalar notation* is used, then indicating the positive directions of components along the x and y axes with symbolic arrows, we have

$$\xrightarrow{+} \qquad (F_R)_x = F_{1x} - F_{2x} + F_{3x}$$
$$+\uparrow \qquad (F_R)_y = F_{1y} + F_{2y} - F_{3y}$$

These are the *same* results as the **i** and **j** components of $\mathbf{F}_R$ determined above.

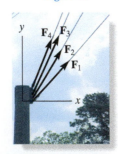

The resultant force of the four cable forces acting on the post can be determined by adding algebraically the separate x and y components of each cable force. This resultant $\mathbf{F}_R$ produces the *same pulling effect* on the post as all four cables. (© Russell C. Hibbeler)

*For handwritten work, unit vectors are usually indicated using a circumflex, e.g., $\hat{\imath}$ and $\hat{\jmath}$. Also, realize that F_x and F_y in Fig. 2–16 represent the *magnitudes* of the components, which are *always positive scalars*. The directions are defined by **i** and **j**. If instead we used scalar notation, then F_x and F_y could be positive or negative scalars, since they would account for *both* the magnitude and direction of the components.

We can represent the components of the resultant force of any number of coplanar forces symbolically by the algebraic sum of the x and y components of all the forces, i.e.,

$$\begin{aligned} (F_R)_x &= \Sigma F_x \\ (F_R)_y &= \Sigma F_y \end{aligned} \qquad (2\text{--}1)$$

Once these components are determined, they may be sketched along the x and y axes with their proper sense of direction, and the resultant force can be determined from vector addition, as shown in Fig. 2–17c. From this sketch, the magnitude of $\mathbf{F}_R$ is then found from the Pythagorean theorem; that is,

$$F_R = \sqrt{(F_R)_x^2 + (F_R)_y^2}$$

Also, the angle θ, which specifies the direction of the resultant force, is determined from trigonometry:

$$\theta = \tan^{-1} \left| \frac{(F_R)_y}{(F_R)_x} \right|$$

The above concepts are illustrated numerically in the examples which follow.

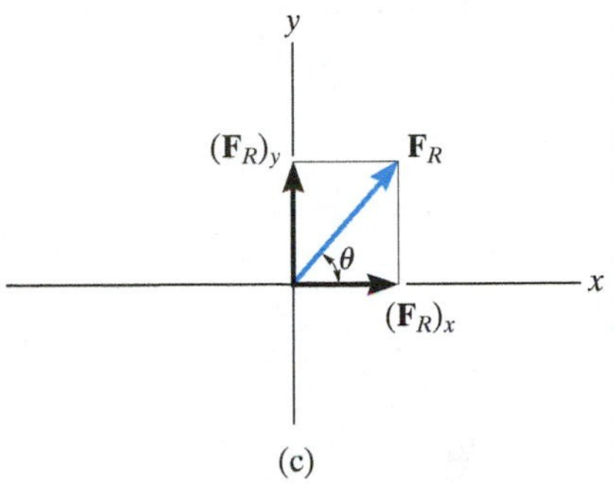

(c)

Fig. 2–17 (cont.)

Important Points

- The resultant of several coplanar forces can easily be determined if an x, y coordinate system is established and the forces are resolved along the axes.

- The direction of each force is specified by the angle its line of action makes with one of the axes, or by a slope triangle.

- The orientation of the x and y axes is arbitrary, and their positive direction can be specified by the Cartesian unit vectors $\mathbf{i}$ and $\mathbf{j}$.

- The x and y components of the *resultant force* are simply the algebraic addition of the components of all the coplanar forces.

- The magnitude of the resultant force is determined from the Pythagorean theorem, and when the resultant components are sketched on the x and y axes, Fig. 2–17c, the direction θ can be determined from trigonometry.

EXAMPLE 2.5

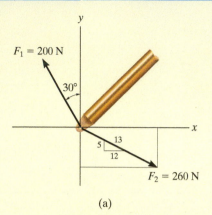

(a)

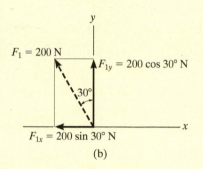

(b)

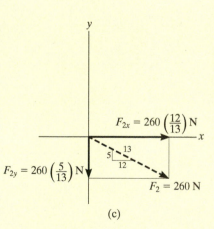

(c)

Fig. 2–18

Determine the x and y components of $\mathbf{F}_1$ and $\mathbf{F}_2$ acting on the boom shown in Fig. 2–18a. Express each force as a Cartesian vector.

SOLUTION

Scalar Notation. By the parallelogram law, $\mathbf{F}_1$ is resolved into x and y components, Fig. 2–18b. Since $\mathbf{F}_{1x}$ acts in the $-x$ direction, and $\mathbf{F}_{1y}$ acts in the $+y$ direction, we have

$$F_{1x} = -200 \sin 30° \text{ N} = -100 \text{ N} = 100 \text{ N} \leftarrow \qquad Ans.$$

$$F_{1y} = 200 \cos 30° \text{ N} = 173 \text{ N} = 173 \text{ N} \uparrow \qquad Ans.$$

The force $\mathbf{F}_2$ is resolved into its x and y components, as shown in Fig. 2–18c. Here the *slope* of the line of action for the force is indicated. From this "slope triangle" we could obtain the angle θ, e.g., $\theta = \tan^{-1}\left(\frac{5}{12}\right)$, and then proceed to determine the magnitudes of the components in the same manner as for $\mathbf{F}_1$. The easier method, however, consists of using proportional parts of similar triangles, i.e.,

$$\frac{F_{2x}}{260 \text{ N}} = \frac{12}{13} \qquad F_{2x} = 260 \text{ N} \left(\frac{12}{13}\right) = 240 \text{ N}$$

Similarly,

$$F_{2y} = 260 \text{ N} \left(\frac{5}{13}\right) = 100 \text{ N}$$

Notice how the magnitude of the *horizontal component*, $\mathbf{F}_{2x}$, was obtained by multiplying the force magnitude by the ratio of the *horizontal leg* of the slope triangle divided by the hypotenuse; whereas the magnitude of the *vertical component*, $\mathbf{F}_{2y}$, was obtained by multiplying the force magnitude by the ratio of the *vertical leg* divided by the hypotenuse. Hence, using scalar notation to represent these components, we have

$$F_{2x} = 240 \text{ N} = 240 \text{ N} \rightarrow \qquad Ans.$$

$$F_{2y} = -100 \text{ N} = 100 \text{ N} \downarrow \qquad Ans.$$

Cartesian Vector Notation. Having determined the magnitudes and directions of the components of each force, we can express each force as a Cartesian vector.

$$\mathbf{F}_1 = \{-100\mathbf{i} + 173\mathbf{j}\} \text{ N} \qquad Ans.$$

$$\mathbf{F}_2 = \{240\mathbf{i} - 100\mathbf{j}\} \text{ N} \qquad Ans.$$

EXAMPLE | 2.6

The link in Fig. 2–19a is subjected to two forces $\mathbf{F}_1$ and $\mathbf{F}_2$. Determine the magnitude and direction of the resultant force.

SOLUTION I

Scalar Notation. First we resolve each force into its x and y components, Fig. 2–19b, then we sum these components algebraically.

$\xrightarrow{+} (F_R)_x = \Sigma F_x;$ $(F_R)_x = 600 \cos 30° \text{ N} - 400 \sin 45° \text{ N}$
$= 236.8 \text{ N} \rightarrow$

$+\uparrow (F_R)_y = \Sigma F_y;$ $(F_R)_y = 600 \sin 30° \text{ N} + 400 \cos 45° \text{ N}$
$= 582.8 \text{ N} \uparrow$

The resultant force, shown in Fig. 2–19c, has a *magnitude* of

$$F_R = \sqrt{(236.8 \text{ N})^2 + (582.8 \text{ N})^2}$$
$$= 629 \text{ N} \qquad \qquad \textit{Ans.}$$

From the vector addition,

$$\theta = \tan^{-1}\left(\frac{582.8 \text{ N}}{236.8 \text{ N}}\right) = 67.9° \qquad \textit{Ans.}$$

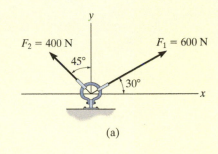

(a)

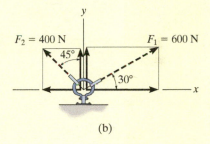

(b)

SOLUTION II

Cartesian Vector Notation. From Fig. 2–19b, each force is first expressed as a Cartesian vector.

$$\mathbf{F}_1 = \{600 \cos 30°\mathbf{i} + 600 \sin 30°\mathbf{j}\} \text{ N}$$
$$\mathbf{F}_2 = \{-400 \sin 45°\mathbf{i} + 400 \cos 45°\mathbf{j}\} \text{ N}$$

Then,

$$\mathbf{F}_R = \mathbf{F}_1 + \mathbf{F}_2 = (600 \cos 30° \text{ N} - 400 \sin 45° \text{ N})\mathbf{i}$$
$$+ (600 \sin 30° \text{ N} + 400 \cos 45° \text{ N})\mathbf{j}$$
$$= \{236.8\mathbf{i} + 582.8\mathbf{j}\} \text{ N}$$

The magnitude and direction of $\mathbf{F}_R$ are determined in the same manner as before.

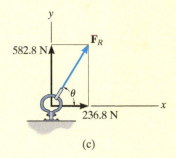

(c)

Fig. 2–19

NOTE: Comparing the two methods of solution, notice that the use of scalar notation is more efficient since the components can be found *directly*, without first having to express each force as a Cartesian vector before adding the components. Later, however, we will show that Cartesian vector analysis is very beneficial for solving three-dimensional problems.

EXAMPLE 2.7

The end of the boom O in Fig. 2–20a is subjected to three concurrent and coplanar forces. Determine the magnitude and direction of the resultant force.

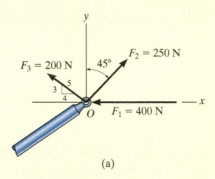

(a)

SOLUTION

Each force is resolved into its x and y components, Fig. 2–20b. Summing the x components, we have

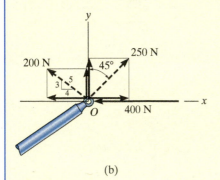

(b)

$$\xrightarrow{+}(F_R)_x = \Sigma F_x; \qquad (F_R)_x = -400 \text{ N} + 250 \sin 45° \text{ N} - 200\left(\tfrac{4}{5}\right) \text{ N}$$

$$= -383.2 \text{ N} = 383.2 \text{ N} \leftarrow$$

The negative sign indicates that F_{Rx} acts to the left, i.e., in the negative x direction, as noted by the small arrow. Obviously, this occurs because F_1 and F_3 in Fig. 2–20b contribute a greater pull to the left than F_2 which pulls to the right. Summing the y components yields

$$+\uparrow(F_R)_y = \Sigma F_y; \qquad (F_R)_y = 250 \cos 45° \text{ N} + 200\left(\tfrac{3}{5}\right) \text{ N}$$

$$= 296.8 \text{ N} \uparrow$$

The resultant force, shown in Fig. 2–20c, has a *magnitude* of

$$F_R = \sqrt{(-383.2 \text{ N})^2 + (296.8 \text{ N})^2}$$

$$= 485 \text{ N} \qquad\qquad\qquad Ans.$$

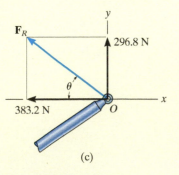

(c)

Fig. 2–20

From the vector addition in Fig. 2–20c, the direction angle θ is

$$\theta = \tan^{-1}\left(\frac{296.8}{383.2}\right) = 37.8° \qquad\qquad Ans.$$

NOTE: Application of this method is more convenient, compared to using two applications of the parallelogram law, first to add $\mathbf{F}_1$ and $\mathbf{F}_2$ then adding $\mathbf{F}_3$ to this resultant.

FUNDAMENTAL PROBLEMS

F2–7. Resolve each force acting on the post into its x and y components.

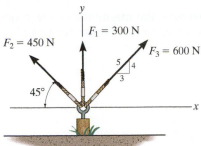

Prob. F2–7

F2–8. Determine the magnitude and direction of the resultant force.

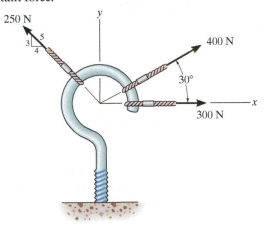

Prob. F2–8

F2–9. Determine the magnitude of the resultant force acting on the corbel and its direction θ measured counterclockwise from the x axis.

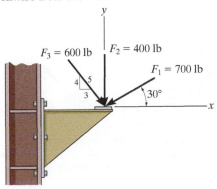

Prob. F2–9

F2–10. If the resultant force acting on the bracket is to be 750 N directed along the positive x axis, determine the magnitude of **F** and its direction θ.

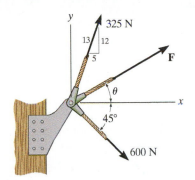

Prob. F2–10

F2–11. If the magnitude of the resultant force acting on the bracket is to be 80 lb directed along the u axis, determine the magnitude of **F** and its direction θ.

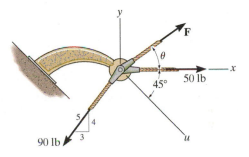

Prob. F2–11

F2–12. Determine the magnitude of the resultant force and its direction θ measured counterclockwise from the positive x axis.

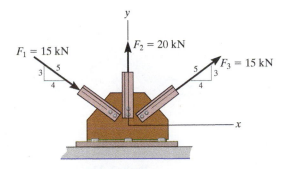

Prob. F2–12

PROBLEMS

*2–32.   Determine the magnitude of the resultant force and its direction, measured counterclockwise from the positive x axis.

2–34.   Resolve $\mathbf{F}_1$ and $\mathbf{F}_2$ into their x and y components.

2–35.   Determine the magnitude of the resultant force and its direction measured counterclockwise from the positive x axis.

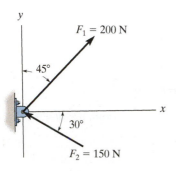

Prob. 2–32

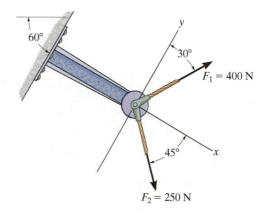

Probs. 2–34/35

2–33.   Determine the magnitude of the resultant force and its direction, measured clockwise from the positive x axis.

*2–36.   Resolve each force acting on the *gusset plate* into its x and y components, and express each force as a Cartesian vector.

2–37.   Determine the magnitude of the resultant force acting on the plate and its direction, measured counter-clockwise from the positive x axis.

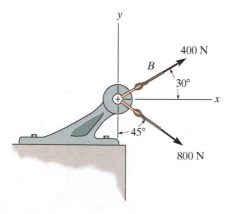

Prob. 2–33

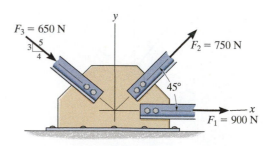

Probs. 2–36/37

2–38. Express each of the three forces acting on the support in Cartesian vector form and determine the magnitude of the resultant force and its direction, measured clockwise from positive *x* axis.

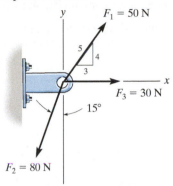

Prob. 2–38

2–39. Determine the *x* and *y* components of **F**₁ and **F**₂.

*2–40. Determine the magnitude of the resultant force and its direction, measured counterclockwise from the positive *x* axis.

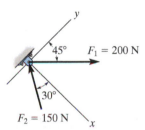

Probs. 2–39/40

2–41. Determine the magnitude of the resultant force and its direction, measured counterclockwise from the positive *x* axis.

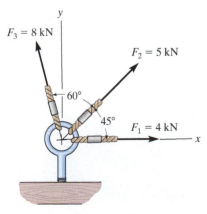

Prob. 2–41

2–42. Express **F**₁, **F**₂, and **F**₃ as Cartesian vectors.

2–43. Determine the magnitude of the resultant force and its direction, measured counterclockwise from the positive *x* axis.

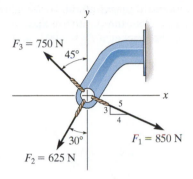

Probs. 2–42/43

*2–44. Determine the magnitude of the resultant force and its direction, measured clockwise from the positive *x* axis.

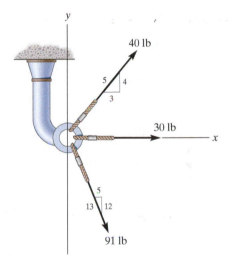

Prob. 2–44

2–45. Determine the magnitude and direction θ of the resultant force **F**$_R$. Express the result in terms of the magnitudes of the components **F**₁ and **F**₂ and the angle ϕ.

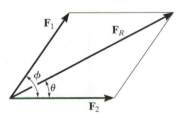

Prob. 2–45

2–46. Determine the magnitude and orientation θ of $\mathbf{F}_B$ so that the resultant force is directed along the positive y axis and has a magnitude of 1500 N.

2–47. Determine the magnitude and orientation, measured counterclockwise from the positive y axis, of the resultant force acting on the bracket, if $F_B = 600$ N and $\theta = 20°$.

2–50. Express $\mathbf{F}_1$, $\mathbf{F}_2$, and $\mathbf{F}_3$ as Cartesian vectors.

2–51. Determine the magnitude of the resultant force and its direction, measured counterclockwise from the positive x axis.

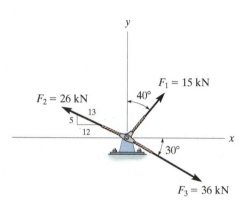

Probs. 2–50/51

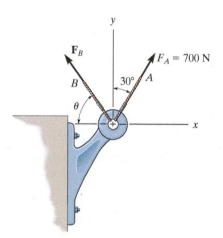

Probs. 2–46/47

*2–48. Three forces act on the bracket. Determine the magnitude and direction θ of $\mathbf{F}_1$ so that the resultant force is directed along the positive x' axis and has a magnitude of 800 N.

2–49. If $F_1 = 300$ N and $\theta = 10°$, determine the magnitude and direction, measured counterclockwise from the positive x' axis, of the resultant force acting on the bracket.

*2–52. Determine the x and y components of each force acting on the *gusset plate* of a bridge truss. Show that the resultant force is zero.

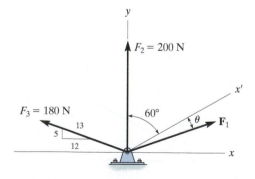

Probs. 2–48/49

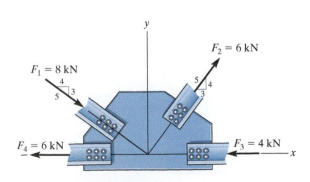

Prob. 2–52

2–53. Express $\mathbf{F}_1$ and $\mathbf{F}_2$ as Cartesian vectors.

2–54. Determine the magnitude of the resultant force and its direction measured counterclockwise from the positive x axis.

*2–56. If the magnitude of the resultant force acting on the bracket is to be 450 N directed along the positive u axis, determine the magnitude of $\mathbf{F}_1$ and its direction ϕ.

2–57. If the resultant force acting on the bracket is required to be a minimum, determine the magnitudes of $\mathbf{F}_1$ and the resultant force. Set $\phi = 30°$.

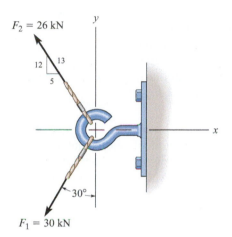

Probs. 2–53/54

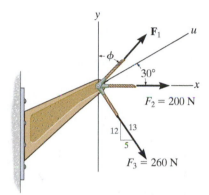

Probs. 2–56/57

2–58. Three forces act on the bracket. Determine the magnitude and direction θ of $\mathbf{F}$ so that the resultant force is directed along the positive x' axis and has a magnitude of 8 kN.

2–59. If $F = 5$ kN and $\theta = 30°$, determine the magnitude of the resultant force and its direction, measured counterclockwise from the positive x axis.

2–55. Determine the magnitude of force $\mathbf{F}$ so that the resultant force of the three forces is as small as possible. What is the magnitude of the resultant force?

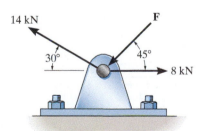

Prob. 2–55

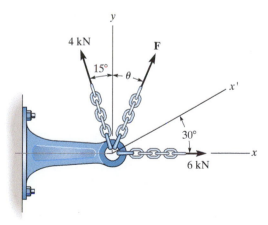

Probs. 2–58/59

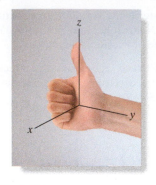

2.5   Cartesian Vectors

The operations of vector algebra, when applied to solving problems in *three dimensions*, are greatly simplified if the vectors are first represented in Cartesian vector form. In this section we will present a general method for doing this; then in the next section we will use this method for finding the resultant force of a system of concurrent forces.

Right-Handed Coordinate System. We will use a right-handed coordinate system to develop the theory of vector algebra that follows. A rectangular coordinate system is said to be **right-handed** if the thumb of the right hand points in the direction of the positive z axis when the right-hand fingers are curled about this axis and directed from the positive x towards the positive y axis, Fig. 2–21.

Rectangular Components of a Vector. A vector $\mathbf{A}$ may have one, two, or three rectangular components along the x, y, z coordinate axes, depending on how the vector is oriented relative to the axes. In general, though, when $\mathbf{A}$ is directed within an octant of the x, y, z frame, Fig. 2–22, then by two successive applications of the parallelogram law, we may resolve the vector into components as $\mathbf{A} = \mathbf{A}' + \mathbf{A}_z$ and then $\mathbf{A}' = \mathbf{A}_x + \mathbf{A}_y$. Combining these equations, to eliminate $\mathbf{A}'$, $\mathbf{A}$ is represented by the vector sum of its *three* rectangular components,

$$\mathbf{A} = \mathbf{A}_x + \mathbf{A}_y + \mathbf{A}_z \tag{2–2}$$

Cartesian Unit Vectors. In three dimensions, the set of Cartesian unit vectors, $\mathbf{i}, \mathbf{j}, \mathbf{k}$, is used to designate the directions of the x, y, z axes, respectively. As stated in Sec. 2–4, the *sense* (or arrowhead) of these vectors will be represented analytically by a plus or minus sign, depending on whether they are directed along the positive or negative $x, y,$ or z axes. The positive Cartesian unit vectors are shown in Fig. 2–23.

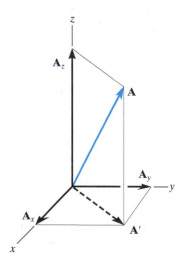

Fig. 2–22

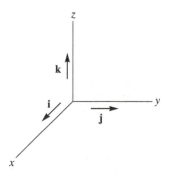

Fig. 2–23

Cartesian Vector Representation.   Since the three components of **A** in Eq. 2–2 act in the positive **i**, **j**, and **k** directions, Fig. 2–24, we can write **A** in Cartesian vector form as

$$A = A_x i + A_y j + A_z k \qquad (2\text{–}3)$$

There is a distinct advantage to writing vectors in this manner. Separating the *magnitude* and *direction* of each *component vector* will simplify the operations of vector algebra, particularly in three dimensions.

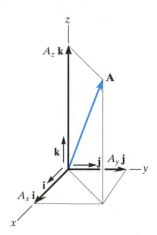

Fig. 2–24

Magnitude of a Cartesian Vector.   It is always possible to obtain the magnitude of **A** provided it is expressed in Cartesian vector form. As shown in Fig. 2–25, from the blue right triangle, $A = \sqrt{A'^2 + A_z^2}$, and from the gray right triangle, $A' = \sqrt{A_x^2 + A_y^2}$. Combining these equations to eliminate A' yields

$$A = \sqrt{A_x^2 + A_y^2 + A_z^2} \qquad (2\text{–}4)$$

*Hence, the magnitude of **A** is equal to the positive square root of the sum of the squares of its components.*

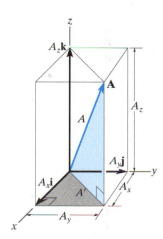

Fig. 2–25

Coordinate Direction Angles.   We will define the *direction* of **A** by the ***coordinate direction angles*** α (alpha), β (beta), and γ (gamma), measured between the *tail* of **A** and the *positive x, y, z* axes provided they are located at the tail of **A**, Fig. 2–26. Note that regardless of where **A** is directed, each of these angles will be between 0° and 180°.

To determine α, β, and γ, consider the projection of **A** onto the *x, y, z* axes, Fig. 2–27. Referring to the colored right triangles shown in the figure, we have

$$\cos\alpha = \frac{A_x}{A} \quad \cos\beta = \frac{A_y}{A} \quad \cos\gamma = \frac{A_z}{A} \qquad (2\text{–}5)$$

These numbers are known as the ***direction cosines*** of **A**. Once they have been obtained, the coordinate direction angles α, β, γ can then be determined from the inverse cosines.

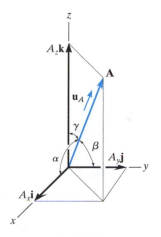

Fig. 2–26

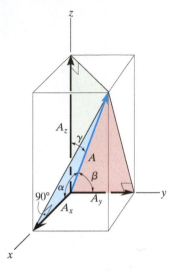

Fig. 2–27

An easy way of obtaining these direction cosines is to form a unit vector $\mathbf{u}_A$ in the direction of A, Fig. 2–26. If $\mathbf{A}$ is expressed in Cartesian vector form, $\mathbf{A} = A_x\mathbf{i} + A_y\mathbf{j} + A_z\mathbf{k}$, then $\mathbf{u}_A$ will have a magnitude of one and be dimensionless provided $\mathbf{A}$ is divided by its magnitude, i.e.,

$$\mathbf{u}_A = \frac{\mathbf{A}}{A} = \frac{A_x}{A}\mathbf{i} + \frac{A_y}{A}\mathbf{j} + \frac{A_z}{A}\mathbf{k} \tag{2–6}$$

where $A = \sqrt{A_x^2 + A_y^2 + A_z^2}$. By comparison with Eqs. 2–5, it is seen that *the* $\mathbf{i}, \mathbf{j}, \mathbf{k}$ *components of* $\mathbf{u}_A$ *represent the direction cosines of* $\mathbf{A}$, i.e.,

$$\mathbf{u}_A = \cos\alpha\,\mathbf{i} + \cos\beta\,\mathbf{j} + \cos\gamma\,\mathbf{k} \tag{2–7}$$

Since the magnitude of a vector is equal to the positive square root of the sum of the squares of the magnitudes of its components, and $\mathbf{u}_A$ has a magnitude of one, then from the above equation an important relation among the direction cosines can be formulated as

$$\cos^2\alpha + \cos^2\beta + \cos^2\gamma = 1 \tag{2–8}$$

Here we can see that if only *two* of the coordinate angles are known, the third angle can be found using this equation.

Finally, if the magnitude and coordinate direction angles of $\mathbf{A}$ are known, then $\mathbf{A}$ may be expressed in Cartesian vector form as

$$\begin{aligned}\mathbf{A} &= A\mathbf{u}_A \\ &= A\cos\alpha\,\mathbf{i} + A\cos\beta\,\mathbf{j} + A\cos\gamma\,\mathbf{k} \\ &= A_x\mathbf{i} + A_y\mathbf{j} + A_z\mathbf{k}\end{aligned} \tag{2–9}$$

Transverse and Azmuth Angles.   Sometimes, the direction of $\mathbf{A}$ can be specified using two angles, namely, a ***transverse angle*** θ and an ***azmuth angle*** ϕ (phi), such as shown in Fig. 2–28. The components of $\mathbf{A}$ can then be determined by applying trigonometry first to the light blue right triangle, which yields

$$A_z = A\cos\phi$$

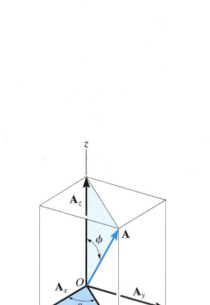

Fig. 2–28

and

$$A' = A\sin\phi$$

Now applying trigonometry to the dark blue right triangle,

$$A_x = A'\cos\theta = A\sin\phi\cos\theta$$

$$A_y = A'\sin\theta = A\sin\phi\sin\theta$$

Therefore **A** written in Cartesian vector form becomes

$$\mathbf{A} = A \sin \phi \cos \theta \, \mathbf{i} + A \sin \phi \sin \theta \, \mathbf{j} + A \cos \phi \, \mathbf{k}$$

You should not memorize this equation, rather it is important to understand how the components were determined using trigonometry.

2.6 Addition of Cartesian Vectors

The addition (or subtraction) of two or more vectors is greatly simplified if the vectors are expressed in terms of their Cartesian components. For example, if $\mathbf{A} = A_x\mathbf{i} + A_y\mathbf{j} + A_z\mathbf{k}$ and $\mathbf{B} = B_x\mathbf{i} + B_y\mathbf{j} + B_z\mathbf{k}$, Fig. 2–29, then the resultant vector, **R**, has components which are the scalar sums of the $\mathbf{i}, \mathbf{j}, \mathbf{k}$ components of **A** and **B**, i.e.,

$$\mathbf{R} = \mathbf{A} + \mathbf{B} = (A_x + B_x)\mathbf{i} + (A_y + B_y)\mathbf{j} + (A_z + B_z)\mathbf{k}$$

If this is generalized and applied to a system of several concurrent forces, then the force resultant is the vector sum of all the forces in the system and can be written as

$$\boxed{\mathbf{F}_R = \Sigma\mathbf{F} = \Sigma F_x\mathbf{i} + \Sigma F_y\mathbf{j} + \Sigma F_z\mathbf{k}} \qquad (2\text{–}10)$$

Here ΣF_x, ΣF_y, and ΣF_z represent the algebraic sums of the respective x, y, z or $\mathbf{i}, \mathbf{j}, \mathbf{k}$ components of each force in the system.

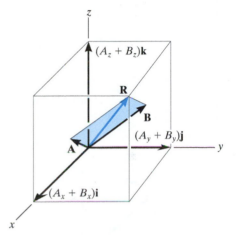

Fig. 2–29

Important Points

- A Cartesian vector **A** has $\mathbf{i}, \mathbf{j}, \mathbf{k}$ components along the x, y, z axes. If **A** is known, its magnitude is defined by $A = \sqrt{A_x^2 + A_y^2 + A_z^2}$.

- The direction of a Cartesian vector can be defined by the three angles α, β, γ, measured from the *positive x, y, z* axes to the *tail* of the vector. To find these angles formulate a unit vector in the direction of **A**, i.e., $\mathbf{u}_A = \mathbf{A}/A$, and determine the inverse cosines of its components. Only two of these angles are independent of one another; the third angle is found from $\cos^2\alpha + \cos^2\beta + \cos^2\gamma = 1$.

- The direction of a Cartesian vector can also be specified using a transverse angle θ and azimuth angle ϕ.

Cartesian vector analysis provides a convenient method for finding both the resultant force and its components in three dimensions. (© Russell C. Hibbeler)

EXAMPLE | **2.8**

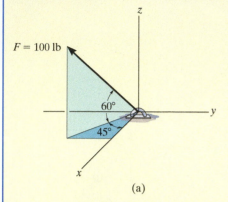

(a)

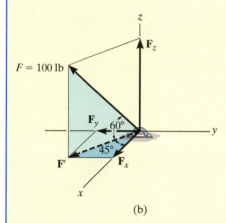

(b)

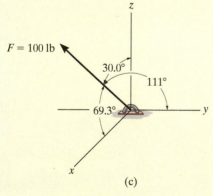

(c)

Fig. 2–30

Express the force **F** shown in Fig. 2–30*a* as a Cartesian vector.

SOLUTION
The angles of 60° and 45° defining the direction of **F** are *not* coordinate direction angles. Two successive applications of the parallelogram law are needed to resolve **F** into its *x, y, z* components. First $\mathbf{F} = \mathbf{F'} + \mathbf{F}_z$, then $\mathbf{F'} = \mathbf{F}_x + \mathbf{F}_y$, Fig. 2–30*b*. By trigonometry, the magnitudes of the components are

$$F_z = 100 \sin 60° \text{ lb} = 86.6 \text{ lb}$$

$$F' = 100 \cos 60° \text{ lb} = 50 \text{ lb}$$

$$F_x = F' \cos 45° = 50 \cos 45° \text{ lb} = 35.4 \text{ lb}$$

$$F_y = F' \sin 45° = 50 \sin 45° \text{ lb} = 35.4 \text{ lb}$$

Realizing that $\mathbf{F}_y$ has a direction defined by –**j**, we have

$$\mathbf{F} = \{35.4\mathbf{i} - 35.4\mathbf{j} + 86.6\mathbf{k}\} \text{ lb} \qquad \textit{Ans.}$$

To show that the magnitude of this vector is indeed 100 lb, apply Eq. 2–4,

$$F = \sqrt{F_x^2 + F_y^2 + F_z^2}$$

$$= \sqrt{(35.4)^2 + (35.4)^2 + (86.6)^2} = 100 \text{ lb}$$

If needed, the coordinate direction angles of **F** can be determined from the components of the unit vector acting in the direction of **F**. Hence,

$$\mathbf{u} = \frac{\mathbf{F}}{F} = \frac{F_x}{F}\mathbf{i} + \frac{F_y}{F}\mathbf{j} + \frac{F_z}{F}\mathbf{k}$$

$$= \frac{35.4}{100}\mathbf{i} - \frac{35.4}{100}\mathbf{j} + \frac{86.6}{100}\mathbf{k}$$

$$= 0.354\mathbf{i} - 0.354\mathbf{j} + 0.866\mathbf{k}$$

so that

$$\alpha = \cos^{-1}(0.354) = 69.3°$$

$$\beta = \cos^{-1}(-0.354) = 111°$$

$$\gamma = \cos^{-1}(0.866) = 30.0°$$

These results are shown in Fig. 2–30*c*.

EXAMPLE 2.9

Two forces act on the hook shown in Fig. 2–31a. Specify the magnitude of F_2 and its coordinate direction angles so that the resultant force $\mathbf{F}_R$ acts along the positive y axis and has a magnitude of 800 N.

SOLUTION

To solve this problem, the resultant force $\mathbf{F}_R$ and its two components, $\mathbf{F}_1$ and $\mathbf{F}_2$, will each be expressed in Cartesian vector form. Then, as shown in Fig. 2–31b, it is necessary that $\mathbf{F}_R = \mathbf{F}_1 + \mathbf{F}_2$.

Applying Eq. 2–9,

$$\mathbf{F}_1 = F_1 \cos \alpha_1 \mathbf{i} + F_1 \cos \beta_1 \mathbf{j} + F_1 \cos \gamma_1 \mathbf{k}$$

$$= 300 \cos 45° \, \mathbf{i} + 300 \cos 60° \, \mathbf{j} + 300 \cos 120° \, \mathbf{k}$$

$$= \{212.1\mathbf{i} + 150\mathbf{j} - 150\mathbf{k}\} \, N$$

$$\mathbf{F}_2 = F_{2x}\mathbf{i} + F_{2y}\mathbf{j} + F_{2z}\mathbf{k}$$

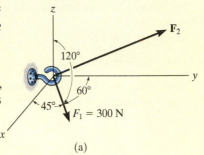

(a)

Since $\mathbf{F}_R$ has a magnitude of 800 N and acts in the $+\mathbf{j}$ direction,

$$\mathbf{F}_R = (800 \text{ N})(+\mathbf{j}) = \{800\mathbf{j}\} \, N$$

We require

$$\mathbf{F}_R = \mathbf{F}_1 + \mathbf{F}_2$$

$$800\mathbf{j} = 212.1\mathbf{i} + 150\mathbf{j} - 150\mathbf{k} + F_{2x}\mathbf{i} + F_{2y}\mathbf{j} + F_{2z}\mathbf{k}$$

$$800\mathbf{j} = (212.1 + F_{2x})\mathbf{i} + (150 + F_{2y})\mathbf{j} + (-150 + F_{2z})\mathbf{k}$$

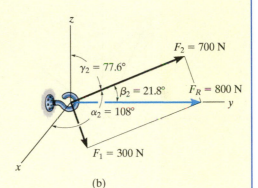

(b)

Fig. 2–31

To satisfy this equation the $\mathbf{i}, \mathbf{j}, \mathbf{k}$ components of $\mathbf{F}_R$ must be equal to the corresponding $\mathbf{i}, \mathbf{j}, \mathbf{k}$ components of $(\mathbf{F}_1 + \mathbf{F}_2)$. Hence,

$$0 = 212.1 + F_{2x} \qquad F_{2x} = -212.1 \text{ N}$$

$$800 = 150 + F_{2y} \qquad F_{2y} = 650 \text{ N}$$

$$0 = -150 + F_{2z} \qquad F_{2z} = 150 \text{ N}$$

The magnitude of $\mathbf{F}_2$ is thus

$$F_2 = \sqrt{(-212.1 \text{ N})^2 + (650 \text{ N})^2 + (150 \text{ N})^2}$$

$$= 700 \text{ N} \qquad\qquad\qquad\qquad \textit{Ans.}$$

We can use Eq. 2–9 to determine $\alpha_2, \beta_2, \gamma_2$.

$$\cos \alpha_2 = \frac{-212.1}{700}; \qquad \alpha_2 = 108° \qquad \textit{Ans.}$$

$$\cos \beta_2 = \frac{650}{700}; \qquad \beta_2 = 21.8° \qquad \textit{Ans.}$$

$$\cos \gamma_2 = \frac{150}{700}; \qquad \gamma_2 = 77.6° \qquad \textit{Ans.}$$

These results are shown in Fig. 2–31b.

PRELIMINARY PROBLEMS

P2–3.   Sketch the following forces on the x, y, z coordinate axes. Show α, β, γ.

a) $\mathbf{F} = \{50\mathbf{i} + 60\mathbf{j} - 10\mathbf{k}\}$ kN

b) $\mathbf{F} = \{-40\mathbf{i} - 80\mathbf{j} + 60\mathbf{k}\}$ kN

P2–4.   In each case, establish $\mathbf{F}$ as a Cartesian vector, and find the magnitude of $\mathbf{F}$ and the direction cosine of β.

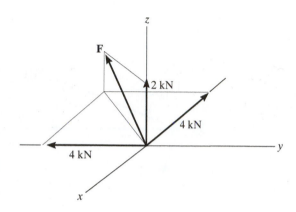

(a)

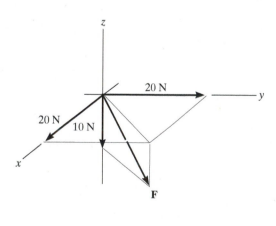

(b)

Prob. P2–4

P2–5.   Show how to resolve each force into its x, y, z components. Set up the calculation used to find the magnitude of each component.

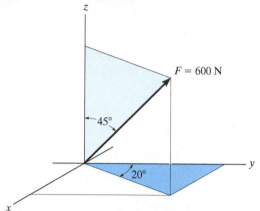

(a)

$F = 500$ N

(b)

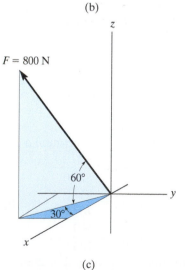

(c)

Prob. P2–5

FUNDAMENTAL PROBLEMS

F2–13. Determine the coordinate direction angles of the force.

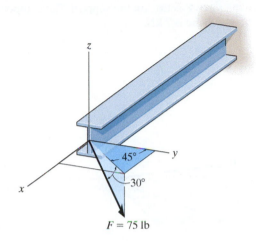

$F = 75$ lb

Prob. F2–13

F2–14. Express the force as a Cartesian vector.

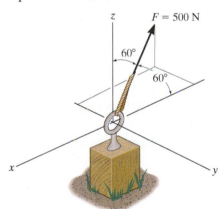

$F = 500$ N

Prob. F2–14

F2–15. Express the force as a Cartesian vector.

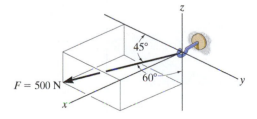

$F = 500$ N

Prob. F2–15

F2–16. Express the force as a Cartesian vector.

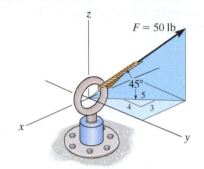

$F = 50$ lb

Prob. F2–16

F2–17. Express the force as a Cartesian vector.

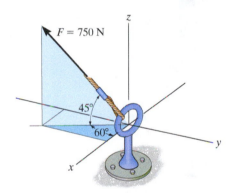

$F = 750$ N

Prob. F2–17

F2–18. Determine the resultant force acting on the hook.

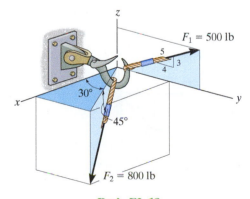

$F_1 = 500$ lb

$F_2 = 800$ lb

Prob. F2–18

PROBLEMS

*2–60. The force **F** has a magnitude of 80 lb and acts within the octant shown. Determine the magnitudes of the x, y, z components of **F**.

2–62. Determine the magnitude and coordinate direction angles of the force **F** acting on the support. The component of **F** in the x–y plane is 7 kN.

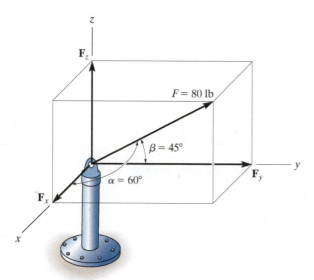

Prob. 2–60

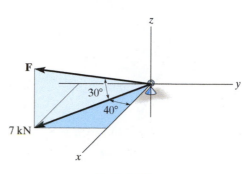

Prob. 2–62

2–61. The bolt is subjected to the force **F**, which has components acting along the x, y, z axes as shown. If the magnitude of **F** is 80 N, and $\alpha = 60°$ and $\gamma = 45°$, determine the magnitudes of its components.

2–63. Determine the magnitude and coordinate direction angles of the resultant force and sketch this vector on the coordinate system.

*2–64. Specify the coordinate direction angles of $\mathbf{F}_1$ and $\mathbf{F}_2$ and express each force as a Cartesian vector.

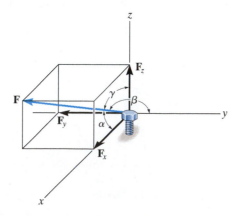

Prob. 2–61

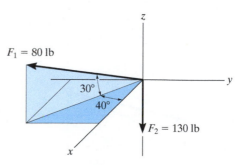

Probs. 2–63/64

2–65. The screw eye is subjected to the two forces shown. Express each force in Cartesian vector form and then determine the resultant force. Find the magnitude and coordinate direction angles of the resultant force.

2–66. Determine the coordinate direction angles of $\mathbf{F}_1$.

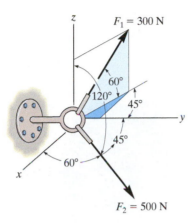

Probs. 2–65/66

2–67. Determine the magnitude and coordinate direction angles of $\mathbf{F}_3$ so that the resultant of the three forces acts along the positive y axis and has a magnitude of 600 lb.

***2–68.** Determine the magnitude and coordinate direction angles of $\mathbf{F}_3$ so that the resultant of the three forces is zero.

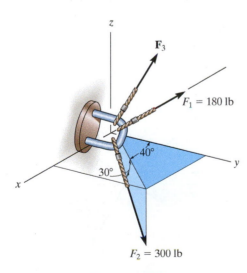

Probs. 2–67/68

2–69. Determine the magnitude and coordinate direction angles of the resultant force, and sketch this vector on the coordinate system.

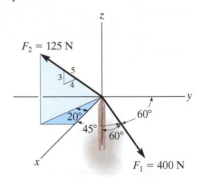

Prob. 2–69

2–70. Determine the magnitude and coordinate direction angles of the resultant force, and sketch this vector on the coordinate system.

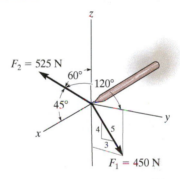

Prob. 2–70

2–71. Specify the magnitude and coordinate direction angles α_1, β_1, γ_1 of $\mathbf{F}_1$ so that the resultant of the three forces acting on the bracket is $\mathbf{F}_R = \{-350\mathbf{k}\}$ lb. Note that $\mathbf{F}_3$ lies in the x–y plane.

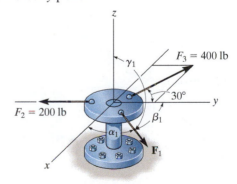

Prob. 2–71

*2–72. Two forces $\mathbf{F}_1$ and $\mathbf{F}_2$ act on the screw eye. If the resultant force $\mathbf{F}_R$ has a magnitude of 150 lb and the coordinate direction angles shown, determine the magnitude of $\mathbf{F}_2$ and its coordinate direction angles.

2–75. The spur gear is subjected to the two forces caused by contact with other gears. Express each force as a Cartesian vector.

*2–76. The spur gear is subjected to the two forces caused by contact with other gears. Determine the resultant of the two forces and express the result as a Cartesian vector.

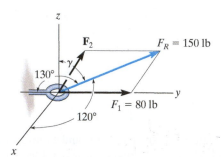

Prob. 2–72

2–73. Express each force in Cartesian vector form.

2–74. Determine the magnitude and coordinate direction angles of the resultant force, and sketch this vector on the coordinate system.

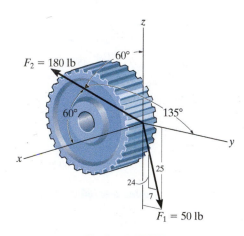

Probs. 2–75/76

2–77. Determine the magnitude and coordinate direction angles of the resultant force, and sketch this vector on the coordinate system.

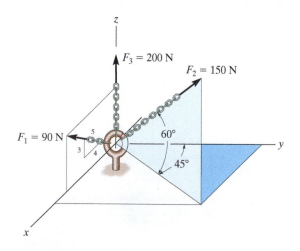

Probs. 2–73/74

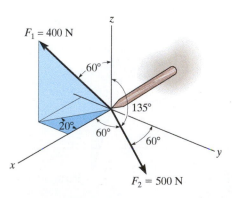

Prob. 2–77

2–78. The two forces $\mathbf{F}_1$ and $\mathbf{F}_2$ acting at A have a resultant force of $\mathbf{F}_R = \{-100\mathbf{k}\}$ lb. Determine the magnitude and coordinate direction angles of $\mathbf{F}_2$.

2–79. Determine the coordinate direction angles of the force $\mathbf{F}_1$ and indicate them on the figure.

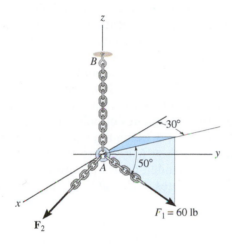

Probs. 2–78/79

*2–80. The bracket is subjected to the two forces shown. Express each force in Cartesian vector form and then determine the resultant force $\mathbf{F}_R$. Find the magnitude and coordinate direction angles of the resultant force.

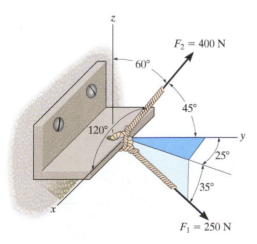

Prob. 2–80

2–81. If the coordinate direction angles for $\mathbf{F}_3$ are $\alpha_3 = 120°$, $\beta_3 = 60°$ and $\gamma_3 = 45°$, determine the magnitude and coordinate direction angles of the resultant force acting on the eyebolt.

2–82. If the coordinate direction angles for $\mathbf{F}_3$ are $\alpha_3 = 120°$, $\beta_3 = 45°$, and $\gamma_3 = 60°$, determine the magnitude and coordinate direction angles of the resultant force acting on the eyebolt.

2–83. If the direction of the resultant force acting on the eyebolt is defined by the unit vector $\mathbf{u}_{F_R} = \cos 30°\mathbf{j} + \sin 30°\mathbf{k}$, determine the coordinate direction angles of $\mathbf{F}_3$ and the magnitude of $\mathbf{F}_R$.

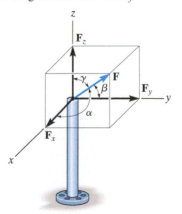

Probs. 2–81/82/83

*2–84. The pole is subjected to the force $\mathbf{F}$, which has components acting along the x, y, z axes as shown. If the magnitude of $\mathbf{F}$ is 3 kN, $\beta = 30°$, and $\gamma = 75°$, determine the magnitudes of its three components.

2–85. The pole is subjected to the force $\mathbf{F}$ which has components $F_x = 1.5$ kN and $F_z = 1.25$ kN. If $\beta = 75°$, determine the magnitudes of $\mathbf{F}$ and $\mathbf{F}_y$.

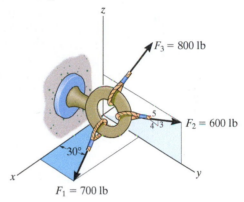

Probs. 2–84/85

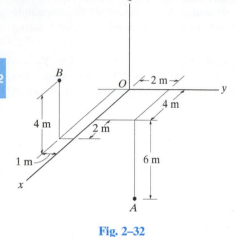

Fig. 2–32

2.7   Position Vectors

In this section we will introduce the concept of a position vector. It will be shown that this vector is of importance in formulating a Cartesian force vector directed between two points in space.

x, y, z Coordinates.

Throughout the book we will use a *right-handed* coordinate system to reference the location of points in space. We will also use the convention followed in many technical books, which requires the positive *z* axis to be directed *upward* (the zenith direction) so that it measures the height of an object or the altitude of a point. The *x, y* axes then lie in the horizontal plane, Fig. 2–32. Points in space are located relative to the origin of coordinates, *O*, by successive measurements along the *x, y, z* axes. For example, the coordinates of point *A* are obtained by starting at *O* and measuring $x_A = +4$ m along the *x* axis, then $y_A = +2$ m along the *y* axis, and finally $z_A = -6$ m along the *z* axis, so that $A(4 \text{ m}, 2 \text{ m}, -6 \text{ m})$. In a similar manner, measurements along the *x, y, z* axes from *O* to *B* yield the coordinates of *B*, that is, $B(6 \text{ m}, -1 \text{ m}, 4 \text{ m})$.

Position Vector.

A *position vector* **r** is defined as a fixed vector which locates a point in space relative to another point. For example, if **r** extends from the origin of coordinates, *O*, to point $P(x, y, z)$, Fig. 2–33*a*, then **r** can be expressed in Cartesian vector form as

$$\mathbf{r} = x\mathbf{i} + y\mathbf{j} + z\mathbf{k}$$

Note how the head-to-tail vector addition of the three components yields vector **r**, Fig. 2–33*b*. Starting at the origin *O*, one "travels" *x* in the $+\mathbf{i}$ direction, then *y* in the $+\mathbf{j}$ direction, and finally *z* in the $+\mathbf{k}$ direction to arrive at point $P(x, y, z)$.

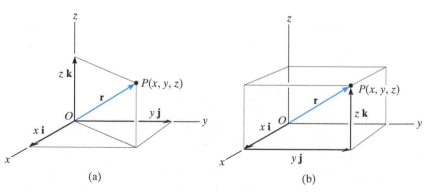

(a)   (b)

Fig. 2–33

In the more general case, the position vector may be directed from point A to point B in space, Fig. 2–34a. This vector is also designated by the symbol $\mathbf{r}$. As a matter of convention, we will *sometimes* refer to this vector with *two subscripts* to indicate from and to the point where it is directed. Thus, $\mathbf{r}$ can also be designated as $\mathbf{r}_{AB}$. Also, note that $\mathbf{r}_A$ and $\mathbf{r}_B$ in Fig. 2–34a are referenced with only one subscript since they extend from the origin of coordinates.

From Fig. 2–34a, by the head-to-tail vector addition, using the triangle rule, we require

$$\mathbf{r}_A + \mathbf{r} = \mathbf{r}_B$$

Solving for $\mathbf{r}$ and expressing $\mathbf{r}_A$ and $\mathbf{r}_B$ in Cartesian vector form yields

$$\mathbf{r} = \mathbf{r}_B - \mathbf{r}_A = (x_B\mathbf{i} + y_B\mathbf{j} + z_B\mathbf{k}) - (x_A\mathbf{i} + y_A\mathbf{j} + z_A\mathbf{k})$$

or

$$\mathbf{r} = (x_B - x_A)\mathbf{i} + (y_B - y_A)\mathbf{j} + (z_B - z_A)\mathbf{k} \qquad (2\text{–}11)$$

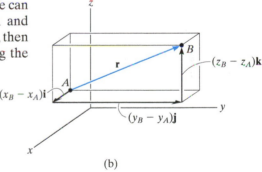

(a)

(b)

Fig. 2–34

Thus, the $\mathbf{i}$, $\mathbf{j}$, $\mathbf{k}$ *components of the position vector* $\mathbf{r}$ *may be formed by taking the coordinates of the tail of the vector* $A(x_A, y_A, z_A)$ *and subtracting them from the corresponding coordinates of the head* $B(x_B, y_B, z_B)$. We can also form these components *directly*, Fig. 2–34b, by starting at A and moving through a distance of $(x_B - x_A)$ along the positive x axis ($+\mathbf{i}$), then $(y_B - y_A)$ along the positive y axis ($+\mathbf{j}$), and finally $(z_B - z_A)$ along the positive z axis ($+\mathbf{k}$) to get to B.

If an x, y, z coordinate system is established, then the coordinates of two points A and B on the cable can be determined. From this the position vector $\mathbf{r}$ acting along the cable can be formulated. Its magnitude represents the distance from A to B, and its unit vector, $\mathbf{u} = \mathbf{r}/r$, gives the direction defined by α, β, γ. (© Russell C. Hibbeler)

EXAMPLE | 2.10

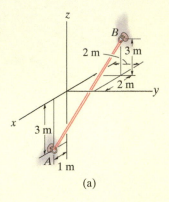

(a)

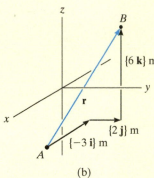

(b)

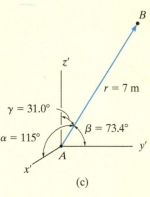

(c)

Fig. 2–35

An elastic rubber band is attached to points A and B as shown in Fig. 2–35a. Determine its length and its direction measured from A toward B.

SOLUTION

We first establish a position vector from A to B, Fig. 2–35b. In accordance with Eq. 2–11, the coordinates of the tail $A(1 \text{ m}, 0, -3 \text{ m})$ are subtracted from the coordinates of the head $B(-2 \text{ m}, 2 \text{ m}, 3 \text{ m})$, which yields

$$\mathbf{r} = [-2 \text{ m} - 1 \text{ m}]\mathbf{i} + [2 \text{ m} - 0]\mathbf{j} + [3 \text{ m} - (-3 \text{ m})]\mathbf{k}$$
$$= \{-3\mathbf{i} + 2\mathbf{j} + 6\mathbf{k}\} \text{ m}$$

These components of $\mathbf{r}$ can also be determined *directly* by realizing that they represent the direction and distance one must travel along each axis in order to move from A to B, i.e., along the x axis $\{-3\mathbf{i}\}$ m, along the y axis $\{2\mathbf{j}\}$ m, and finally along the z axis $\{6\mathbf{k}\}$ m.

The length of the rubber band is therefore

$$r = \sqrt{(-3 \text{ m})^2 + (2 \text{ m})^2 + (6 \text{ m})^2} = 7 \text{ m} \qquad \textit{Ans.}$$

Formulating a unit vector in the direction of $\mathbf{r}$, we have

$$\mathbf{u} = \frac{\mathbf{r}}{r} = -\frac{3}{7}\mathbf{i} + \frac{2}{7}\mathbf{j} + \frac{6}{7}\mathbf{k}$$

The components of this unit vector give the coordinate direction angles

$$\alpha = \cos^{-1}\left(-\frac{3}{7}\right) = 115° \qquad \textit{Ans.}$$

$$\beta = \cos^{-1}\left(\frac{2}{7}\right) = 73.4° \qquad \textit{Ans.}$$

$$\gamma = \cos^{-1}\left(\frac{6}{7}\right) = 31.0° \qquad \textit{Ans.}$$

NOTE: These angles are measured from the *positive axes* of a localized coordinate system placed at the tail of $\mathbf{r}$, as shown in Fig. 2–35c.

2.8 Force Vector Directed Along a Line

Quite often in three-dimensional statics problems, the direction of a force is specified by two points through which its line of action passes. Such a situation is shown in Fig. 2–36, where the force **F** is directed along the cord AB. We can formulate **F** as a Cartesian vector by realizing that it has the *same direction* and *sense* as the position vector **r** directed from point A to point B on the cord. This common direction is specified by the ***unit vector*** $\mathbf{u} = \mathbf{r}/r$. Hence,

$$\mathbf{F} = F\mathbf{u} = F\left(\frac{\mathbf{r}}{r}\right) = F\left(\frac{(x_B - x_A)\mathbf{i} + (y_B - y_A)\mathbf{j} + (z_B - z_A)\mathbf{k}}{\sqrt{(x_B - x_A)^2 + (y_B - y_A)^2 + (z_B - z_A)^2}}\right)$$

Although we have represented **F** symbolically in Fig. 2–36, note that it has *units of force*, unlike **r**, which has units of length.

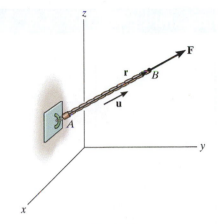

Fig. 2–36

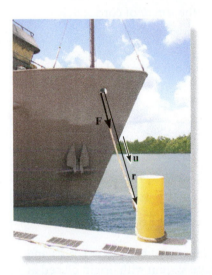

The force **F** acting along the rope can be represented as a Cartesian vector by establishing x, y, z axes and first forming a position vector **r** along the length of the rope. Then the corresponding unit vector $\mathbf{u} = \mathbf{r}/r$ that defines the direction of both the rope and the force can be determined. Finally, the magnitude of the force is combined with its direction, $\mathbf{F} = F\mathbf{u}$. (© Russell C. Hibbeler)

Important Points

- A position vector locates one point in space relative to another point.

- The easiest way to formulate the components of a position vector is to determine the distance and direction that must be traveled along the x, y, z directions—going from the tail to the head of the vector.

- A force **F** acting in the direction of a position vector **r** can be represented in Cartesian form if the unit vector **u** of the position vector is determined and it is multiplied by the magnitude of the force, i.e., $\mathbf{F} = F\mathbf{u} = F(\mathbf{r}/r)$.

EXAMPLE | 2.11

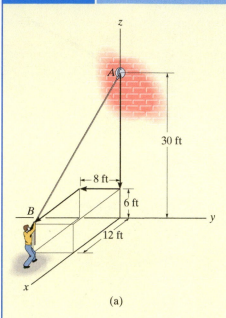

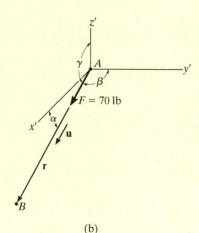

(a)

(b)

Fig. 2–37

The man shown in Fig. 2–37a pulls on the cord with a force of 70 lb. Represent this force acting on the support A as a Cartesian vector and determine its direction.

SOLUTION

Force $\mathbf{F}$ is shown in Fig. 2–37b. The *direction* of this vector, $\mathbf{u}$, is determined from the position vector $\mathbf{r}$, which extends from A to B. Rather than using the coordinates of the end points of the cord, $\mathbf{r}$ can be determined *directly* by noting in Fig. 2–37a that one must travel from A $\{-24\mathbf{k}\}$ ft, then $\{-8\mathbf{j}\}$ ft, and finally $\{12\mathbf{i}\}$ ft to get to B. Thus,

$$\mathbf{r} = \{12\mathbf{i} - 8\mathbf{j} - 24\mathbf{k}\}\ \text{ft}$$

The magnitude of $\mathbf{r}$, which represents the *length* of cord AB, is

$$r = \sqrt{(12\ \text{ft})^2 + (-8\ \text{ft})^2 + (-24\ \text{ft})^2} = 28\ \text{ft}$$

Forming the unit vector that defines the direction and sense of both $\mathbf{r}$ and $\mathbf{F}$, we have

$$\mathbf{u} = \frac{\mathbf{r}}{r} = \frac{12}{28}\mathbf{i} - \frac{8}{28}\mathbf{j} - \frac{24}{28}\mathbf{k}$$

Since $\mathbf{F}$ has a *magnitude* of 70 lb and a *direction* specified by $\mathbf{u}$, then

$$\mathbf{F} = F\mathbf{u} = 70\ \text{lb}\left(\frac{12}{28}\mathbf{i} - \frac{8}{28}\mathbf{j} - \frac{24}{28}\mathbf{k}\right)$$

$$= \{30\mathbf{i} - 20\mathbf{j} - 60\mathbf{k}\}\ \text{lb} \qquad \textit{Ans.}$$

The coordinate direction angles are measured between $\mathbf{r}$ (or $\mathbf{F}$) and the *positive axes* of a localized coordinate system with origin placed at A, Fig. 2–37b. From the components of the unit vector:

$$\alpha = \cos^{-1}\left(\frac{12}{28}\right) = 64.6° \qquad \textit{Ans.}$$

$$\beta = \cos^{-1}\left(\frac{-8}{28}\right) = 107° \qquad \textit{Ans.}$$

$$\gamma = \cos^{-1}\left(\frac{-24}{28}\right) = 149° \qquad \textit{Ans.}$$

NOTE: These results make sense when compared with the angles identified in Fig. 2–37b.

EXAMPLE 2.12

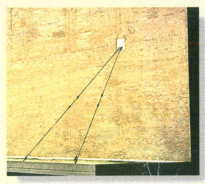

(© Russell C. Hibbeler)

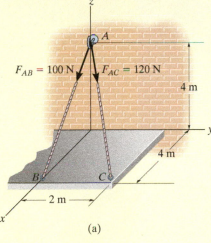

(a)

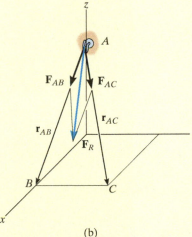

(b)

Fig. 2–38

The roof is supported by cables as shown in the photo. If the cables exert forces $F_{AB} = 100$ N and $F_{AC} = 120$ N on the wall hook at A as shown in Fig. 2–38a, determine the resultant force acting at A. Express the result as a Cartesian vector.

SOLUTION

The resultant force $\mathbf{F}_R$ is shown graphically in Fig. 2–38b. We can express this force as a Cartesian vector by first formulating $\mathbf{F}_{AB}$ and $\mathbf{F}_{AC}$ as Cartesian vectors and then adding their components. The directions of $\mathbf{F}_{AB}$ and $\mathbf{F}_{AC}$ are specified by forming unit vectors $\mathbf{u}_{AB}$ and $\mathbf{u}_{AC}$ along the cables. These unit vectors are obtained from the associated position vectors $\mathbf{r}_{AB}$ and $\mathbf{r}_{AC}$. With reference to Fig. 2–38a, to go from A to B, we must travel $\{-4\mathbf{k}\}$ m, and then $\{4\mathbf{i}\}$ m. Thus,

$$\mathbf{r}_{AB} = \{4\mathbf{i} - 4\mathbf{k}\} \text{ m}$$

$$r_{AB} = \sqrt{(4 \text{ m})^2 + (-4 \text{ m})^2} = 5.66 \text{ m}$$

$$\mathbf{F}_{AB} = F_{AB}\left(\frac{\mathbf{r}_{AB}}{r_{AB}}\right) = (100 \text{ N})\left(\frac{4}{5.66}\mathbf{i} - \frac{4}{5.66}\mathbf{k}\right)$$

$$\mathbf{F}_{AB} = \{70.7\mathbf{i} - 70.7\mathbf{k}\} \text{ N}$$

To go from A to C, we must travel $\{-4\mathbf{k}\}$ m, then $\{2\mathbf{j}\}$ m, and finally $\{4\mathbf{i}\}$. Thus,

$$\mathbf{r}_{AC} = \{4\mathbf{i} + 2\mathbf{j} - 4\mathbf{k}\} \text{ m}$$

$$r_{AC} = \sqrt{(4 \text{ m})^2 + (2 \text{ m})^2 + (-4 \text{ m})^2} = 6 \text{ m}$$

$$\mathbf{F}_{AC} = F_{AC}\left(\frac{\mathbf{r}_{AC}}{r_{AC}}\right) = (120 \text{ N})\left(\frac{4}{6}\mathbf{i} + \frac{2}{6}\mathbf{j} - \frac{4}{6}\mathbf{k}\right)$$

$$= \{80\mathbf{i} + 40\mathbf{j} - 80\mathbf{k}\} \text{ N}$$

The resultant force is therefore

$$\mathbf{F}_R = \mathbf{F}_{AB} + \mathbf{F}_{AC} = \{70.7\mathbf{i} - 70.7\mathbf{k}\} \text{ N} + \{80\mathbf{i} + 40\mathbf{j} - 80\mathbf{k}\} \text{ N}$$

$$= \{151\mathbf{i} + 40\mathbf{j} - 151\mathbf{k}\} \text{ N} \qquad \textit{Ans.}$$

EXAMPLE | 2.13

The force in Fig. 2–39a acts on the hook. Express it as a Cartesian vector.

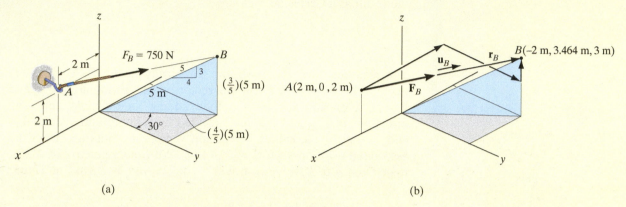

Fig. 2–39

SOLUTION

As shown in Fig. 2–39b, the coordinates for points A and B are

$$A(2 \text{ m}, 0, 2 \text{ m})$$

and

$$B\left[-\left(\frac{4}{5}\right)5 \sin 30° \text{ m}, \left(\frac{4}{5}\right)5 \cos 30° \text{ m}, \left(\frac{3}{5}\right)5 \text{ m}\right]$$

or

$$B(-2 \text{ m}, 3.464 \text{ m}, 3 \text{ m})$$

Therefore, to go from A to B, one must travel {−4i} m, then {3.464j} m, and finally {1k} m. Thus,

$$\mathbf{u}_B = \left(\frac{\mathbf{r}_B}{r_B}\right) = \frac{\{-4\mathbf{i} + 3.464\mathbf{j} + 1\mathbf{k}\} \text{ m}}{\sqrt{(-4 \text{ m})^2 + (3.464 \text{ m})^2 + (1 \text{ m})^2}}$$

$$= -0.7428\mathbf{i} + 0.6433\mathbf{j} + 0.1857\mathbf{k}$$

Force $\mathbf{F}_B$ expressed as a Cartesian vector becomes

$$\mathbf{F}_B = F_B \mathbf{u}_B = (750 \text{ N})(-0.74281\mathbf{i} + 0.6433\mathbf{j} + 0.1857\mathbf{k})$$

$$= \{-557\mathbf{i} + 482\mathbf{j} + 139\mathbf{k}\} \text{ N} \qquad \textit{Ans.}$$

PRELIMINARY PROBLEMS

P2–6. In each case, establish a position vector from point A to point B.

P2–7. In each case, express **F** as a Cartesian vector.

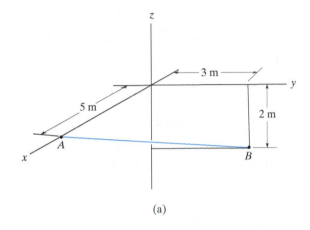

(a)

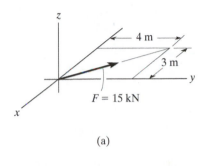

(a)

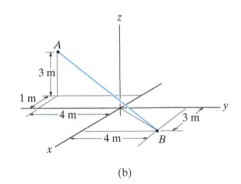

(b)

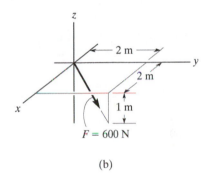

(b)

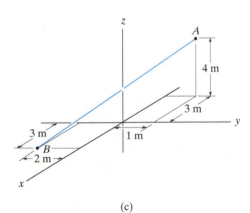

(c)

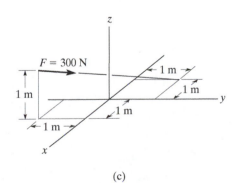

(c)

Prob. P2–6

Prob. P2–7

FUNDAMENTAL PROBLEMS

F2–19. Express the position vector $\mathbf{r}_{AB}$ in Cartesian vector form, then determine its magnitude and coordinate direction angles.

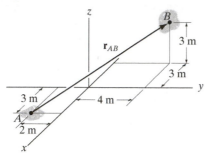

Prob. F2–19

F2–20. Determine the length of the rod and the position vector directed from A to B. What is the angle θ?

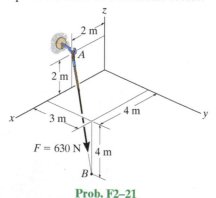

Prob. F2–20

F2–21. Express the force as a Cartesian vector.

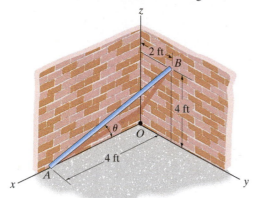

Prob. F2–21

F2–22. Express the force as a Cartesian vector.

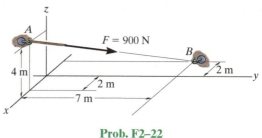

Prob. F2–22

F2–23. Determine the magnitude of the resultant force at A.

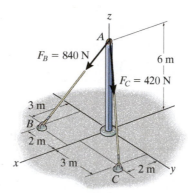

Prob. F2–23

F2–24. Determine the resultant force at A.

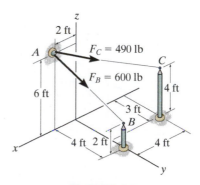

Prob. F2–24

PROBLEMS

2–86. Determine the length of the connecting rod AB by first formulating a Cartesian position vector from A to B and then determining its magnitude.

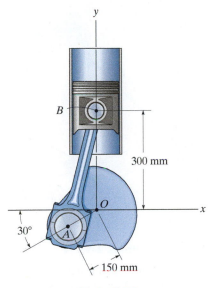

Prob. 2–86

2–87. Express force $\mathbf{F}$ as a Cartesian vector; then determine its coordinate direction angles.

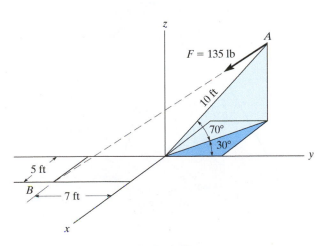

Prob. 2–87

*2–88. Express each of the forces in Cartesian vector form and determine the magnitude and coordinate direction angles of the resultant force.

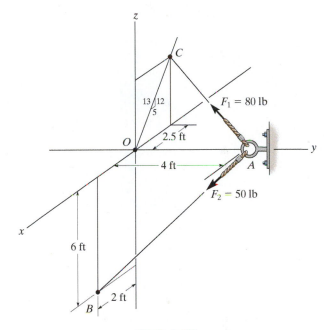

Prob. 2–88

2–89. If $\mathbf{F} = \{350\mathbf{i} - 250\mathbf{j} - 450\mathbf{k}\}$ N and cable AB is 9 m long, determine the x, y, z coordinates of point A.

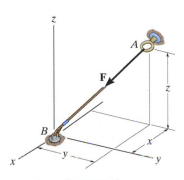

Prob. 2–89

2–90. The 8-m-long cable is anchored to the ground at A. If $x = 4$ m and $y = 2$ m, determine the coordinate z to the highest point of attachment along the column.

2–91. The 8-m-long cable is anchored to the ground at A. If $z = 5$ m, determine the location $+x, +y$ of point A. Choose a value such that $x = y$.

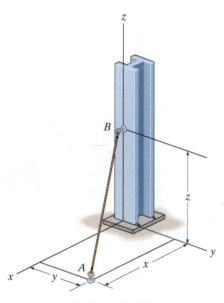

Probs. 2–90/91

*2–92. Express each of the forces in Cartesian vector form and determine the magnitude and coordinate direction angles of the resultant force.

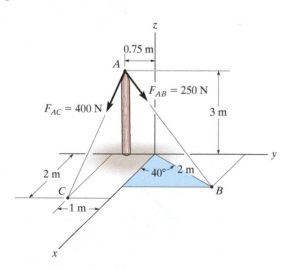

Prob. 2–92

2–93. If $F_B = 560$ N and $F_C = 700$ N, determine the magnitude and coordinate direction angles of the resultant force acting on the flag pole.

2–94. If $F_B = 700$ N, and $F_C = 560$ N, determine the magnitude and coordinate direction angles of the resultant force acting on the flag pole.

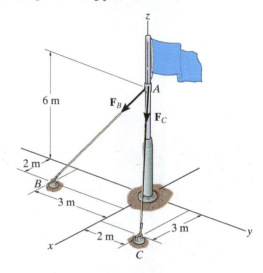

Probs. 2–93/94

2–95. The plate is suspended using the three cables which exert the forces shown. Express each force as a Cartesian vector.

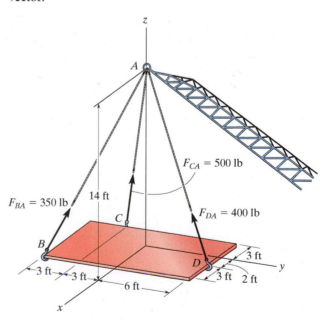

Prob. 2–95

*2–96. The three supporting cables exert the forces shown on the sign. Represent each force as a Cartesian vector.

2–97. Determine the magnitude and coordinate direction angles of the resultant force of the two forces acting on the sign at point *A*.

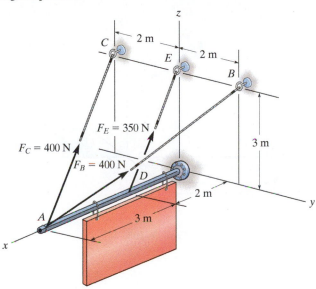

Probs. 2–96/97

2–98. The force **F** has a magnitude of 80 lb and acts at the midpoint *C* of the thin rod. Express the force as a Cartesian vector.

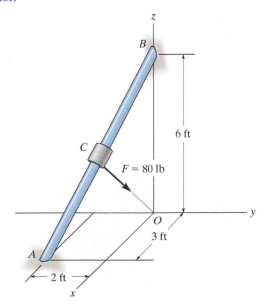

Prob. 2–98

2–99. The load at *A* creates a force of 60 lb in wire *AB*. Express this force as a Cartesian vector acting on *A* and directed toward *B* as shown.

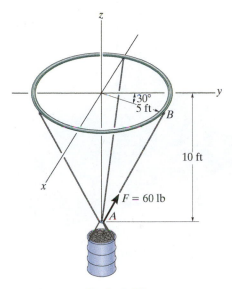

Prob. 2–99

*2–100. Determine the magnitude and coordinate direction angles of the resultant force acting at point *A* on the post.

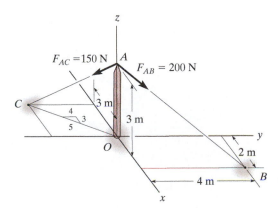

Prob. 2–100

2–101. The two mooring cables exert forces on the stern of a ship as shown. Represent each force as as Cartesian vector and determine the magnitude and coordinate direction angles of the resultant.

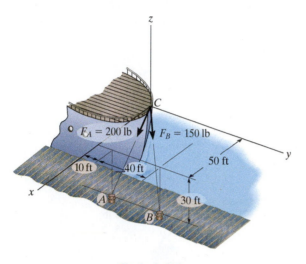

Prob. 2–101

2–102. The engine of the lightweight plane is supported by struts that are connected to the space truss that makes up the structure of the plane. The anticipated loading in two of the struts is shown. Express each of those forces as Cartesian vector.

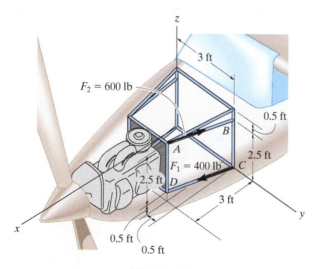

Prob. 2–102

2–103. Determine the magnitude and coordinate direction angles of the resultant force.

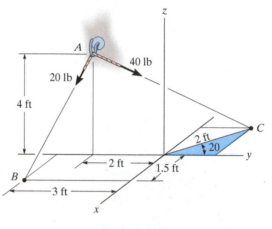

Prob. 2–103

*2–104. If the force in each cable tied to the bin is 70 lb, determine the magnitude and coordinate direction angles of the resultant force.

2–105. If the resultant of the four forces is $F_R = \{-360\mathbf{k}\}$ lb, determine the tension developed in each cable. Due to symmetry, the tension in the four cables is the same.

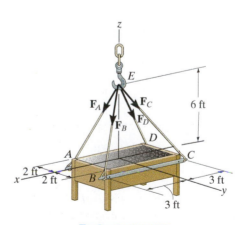

Probs. 2–104/105

2.9 Dot Product

Occasionally in statics one has to find the angle between two lines or the components of a force parallel and perpendicular to a line. In two dimensions, these problems can readily be solved by trigonometry since the geometry is easy to visualize. In three dimensions, however, this is often difficult, and consequently vector methods should be employed for the solution. The dot product, which defines a particular method for "multiplying" two vectors, can be used to solve the above-mentioned problems.

The *dot product* of vectors **A** and **B**, written **A** · **B** and read "**A** dot **B**," is defined as the product of the magnitudes of **A** and **B** and the cosine of the angle θ between their tails, Fig. 2–40. Expressed in equation form,

$$\boxed{\mathbf{A} \cdot \mathbf{B} = AB \cos \theta} \qquad (2\text{–}12)$$

where $0° \leq \theta \leq 180°$. The dot product is often referred to as the *scalar product* of vectors since the result is a *scalar* and not a vector.

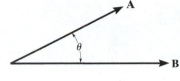

Fig. 2–40

Laws of Operation.

1. Commutative law: $\mathbf{A} \cdot \mathbf{B} = \mathbf{B} \cdot \mathbf{A}$
2. Multiplication by a scalar: $a\,(\mathbf{A} \cdot \mathbf{B}) = (a\mathbf{A}) \cdot \mathbf{B} = \mathbf{A} \cdot (a\mathbf{B})$
3. Distributive law: $\mathbf{A} \cdot (\mathbf{B} + \mathbf{D}) = (\mathbf{A} \cdot \mathbf{B}) + (\mathbf{A} \cdot \mathbf{D})$

It is easy to prove the first and second laws by using Eq. 2–12. The proof of the distributive law is left as an exercise (see Prob. 2–112).

Cartesian Vector Formulation.
Equation 2–12 must be used to find the dot product for any two Cartesian unit vectors. For example, $\mathbf{i} \cdot \mathbf{i} = (1)(1) \cos 0° = 1$ and $\mathbf{i} \cdot \mathbf{j} = (1)(1) \cos 90° = 0$. If we want to find the dot product of two general vectors **A** and **B** that are expressed in Cartesian vector form, then we have

$$\begin{aligned}
\mathbf{A} \cdot \mathbf{B} &= (A_x\mathbf{i} + A_y\mathbf{j} + A_z\mathbf{k}) \cdot (B_x\mathbf{i} + B_y\mathbf{j} + B_z\mathbf{k}) \\
&= A_xB_x(\mathbf{i} \cdot \mathbf{i}) + A_xB_y(\mathbf{i} \cdot \mathbf{j}) + A_xB_z(\mathbf{i} \cdot \mathbf{k}) \\
&\quad + A_yB_x(\mathbf{j} \cdot \mathbf{i}) + A_yB_y(\mathbf{j} \cdot \mathbf{j}) + A_yB_z(\mathbf{j} \cdot \mathbf{k}) \\
&\quad + A_zB_x(\mathbf{k} \cdot \mathbf{i}) + A_zB_y(\mathbf{k} \cdot \mathbf{j}) + A_zB_z(\mathbf{k} \cdot \mathbf{k})
\end{aligned}$$

Carrying out the dot-product operations, the final result becomes

$$\boxed{\mathbf{A} \cdot \mathbf{B} = A_xB_x + A_yB_y + A_zB_z} \qquad (2\text{–}13)$$

Thus, to determine the dot product of two Cartesian vectors, multiply their corresponding x, y, z components and sum these products algebraically. Note that the result will be either a positive or negative *scalar*, or it could be zero.

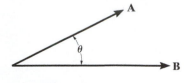

Fig. 2–40 (Repeated)

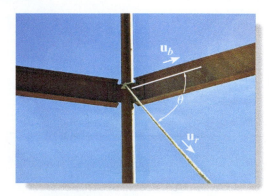

The angle θ between the rope and the beam can be determined by formulating unit vectors along the beam and rope and then using the dot product $\mathbf{u}_b \cdot \mathbf{u}_r = (1)(1)\cos\theta$. (© Russell C. Hibbeler)

The projection of the cable force **F** along the beam can be determined by first finding the unit vector $\mathbf{u}_b$ that defines this direction. Then apply the dot product, $F_b = \mathbf{F} \cdot \mathbf{u}_b$. (© Russell C. Hibbeler)

Applications.   The dot product has two important applications in mechanics.

- ***The angle formed between two vectors or intersecting lines.***   The angle θ between the tails of vectors **A** and **B** in Fig. 2–40 can be determined from Eq. 2–12 and written as

$$\theta = \cos^{-1}\left(\frac{\mathbf{A}\cdot\mathbf{B}}{AB}\right) \quad 0° \le \theta \le 180°$$

Here $\mathbf{A}\cdot\mathbf{B}$ is found from Eq. 2–13. In particular, notice that if $\mathbf{A}\cdot\mathbf{B} = 0, \theta = \cos^{-1} 0 = 90°$ so that **A** will be *perpendicular* to **B**.

- ***The components of a vector parallel and perpendicular to a line.***   The component of vector **A** parallel to or collinear with the line aa in Fig. 2–40 is defined by A_a where $A_a = A\cos\theta$. This component is sometimes referred to as the ***projection*** of **A** onto the line, since a *right angle* is formed in the construction. If the *direction* of the line is specified by the unit vector $\mathbf{u}_a$, then since $u_a = 1$, we can determine the magnitude of A_a directly from the dot product (Eq. 2–12); i.e.,

$$A_a = A\cos\theta = \mathbf{A}\cdot\mathbf{u}_a$$

*Hence, the scalar projection of **A** along a line is determined from the dot product of **A** and the unit vector $\mathbf{u}_a$ which defines the direction of the line.* Notice that if this result is positive, then $\mathbf{A}_a$ has a directional sense which is the same as $\mathbf{u}_a$, whereas if A_a is a negative scalar, then $\mathbf{A}_a$ has the opposite sense of direction to $\mathbf{u}_a$.

The component $\mathbf{A}_a$ represented as a *vector* is therefore

$$\mathbf{A}_a = A_a\,\mathbf{u}_a$$

The component of **A** that is *perpendicular* to line aa can also be obtained, Fig. 2–41. Since $\mathbf{A} = \mathbf{A}_a + \mathbf{A}_\perp$, then $\mathbf{A}_\perp = \mathbf{A} - \mathbf{A}_a$. There are two possible ways of obtaining $A_\perp$. One way would be to determine θ from the dot product, $\theta = \cos^{-1}(\mathbf{A}\cdot\mathbf{u}_A/A)$, then $A_\perp = A\sin\theta$. Alternatively, if A_a is known, then by Pythagorean's theorem we can also write $A_\perp = \sqrt{A^2 - A_a{}^2}$.

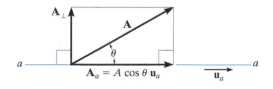

Fig. 2–41

Important Points

- The dot product is used to determine the angle between two vectors or the projection of a vector in a specified direction.

- If vectors $\mathbf{A}$ and $\mathbf{B}$ are expressed in Cartesian vector form, the dot product is determined by multiplying the respective x, y, z scalar components and algebraically adding the results, i.e., $\mathbf{A} \cdot \mathbf{B} = A_x B_x + A_y B_y + A_z B_z$.

- From the definition of the dot product, the angle formed between the tails of vectors $\mathbf{A}$ and $\mathbf{B}$ is $\theta = \cos^{-1}(\mathbf{A} \cdot \mathbf{B}/AB)$.

- The magnitude of the projection of vector $\mathbf{A}$ along a line aa whose direction is specified by $\mathbf{u}_a$ is determined from the dot product $A_a = \mathbf{A} \cdot \mathbf{u}_a$.

EXAMPLE   2.14

Determine the magnitudes of the projection of the force $\mathbf{F}$ in Fig. 2–42 onto the u and v axes.

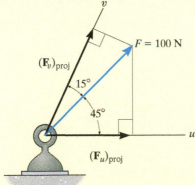

Fig. 2–42

SOLUTION

Projections of Force.   The graphical representation of the *projections* is shown in Fig. 2–42. From this figure, the magnitudes of the projections of $\mathbf{F}$ onto the u and v axes can be obtained by trigonometry:

$$(F_u)_{\text{proj}} = (100 \text{ N})\cos 45° = 70.7 \text{ N} \qquad Ans.$$

$$(F_v)_{\text{proj}} = (100 \text{ N})\cos 15° = 96.6 \text{ N} \qquad Ans.$$

NOTE: These projections are not equal to the magnitudes of the components of force $\mathbf{F}$ along the u and v axes found from the parallelogram law. They will only be equal if the u and v axes are *perpendicular* to one another.

EXAMPLE | **2.15**

The frame shown in Fig. 2–43a is subjected to a horizontal force $\mathbf{F} = \{300\mathbf{j}\}$ N. Determine the magnitudes of the components of this force parallel and perpendicular to member AB.

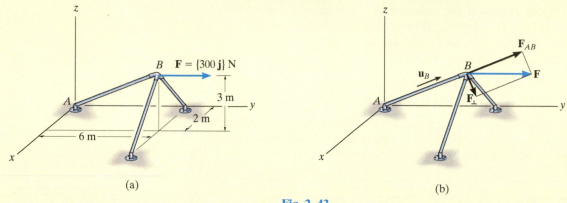

Fig. 2–43

SOLUTION

The magnitude of the component of $\mathbf{F}$ along AB is equal to the dot product of $\mathbf{F}$ and the unit vector $\mathbf{u}_B$, which defines the direction of AB, Fig. 2–43b. Since

$$\mathbf{u}_B = \frac{\mathbf{r}_B}{r_B} = \frac{2\mathbf{i} + 6\mathbf{j} + 3\mathbf{k}}{\sqrt{(2)^2 + (6)^2 + (3)^2}} = 0.286\mathbf{i} + 0.857\mathbf{j} + 0.429\mathbf{k}$$

then

$$\begin{aligned}
F_{AB} = F\cos\theta = \mathbf{F} \cdot \mathbf{u}_B &= (300\mathbf{j}) \cdot (0.286\mathbf{i} + 0.857\mathbf{j} + 0.429\mathbf{k}) \\
&= (0)(0.286) + (300)(0.857) + (0)(0.429) \\
&= 257.1 \text{ N} \hspace{3cm} \textit{Ans.}
\end{aligned}$$

Since the result is a positive scalar, $\mathbf{F}_{AB}$ has the same sense of direction as $\mathbf{u}_B$, Fig. 2–43b.

Expressing $\mathbf{F}_{AB}$ in Cartesian vector form, we have

$$\begin{aligned}
\mathbf{F}_{AB} = F_{AB}\mathbf{u}_B &= (257.1 \text{ N})(0.286\mathbf{i} + 0.857\mathbf{j} + 0.429\mathbf{k}) \\
&= \{73.5\mathbf{i} + 220\mathbf{j} + 110\mathbf{k}\} \text{ N} \hspace{2cm} \textit{Ans.}
\end{aligned}$$

The perpendicular component, Fig. 2–43b, is therefore

$$\begin{aligned}
\mathbf{F}_\perp = \mathbf{F} - \mathbf{F}_{AB} &= 300\mathbf{j} - (73.5\mathbf{i} + 220\mathbf{j} + 110\mathbf{k}) \\
&= \{-73.5\mathbf{i} + 79.6\mathbf{j} - 110\mathbf{k}\} \text{ N}
\end{aligned}$$

Its magnitude can be determined either from this vector or by using the Pythagorean theorem, Fig. 2–43b:

$$\begin{aligned}
F_\perp = \sqrt{F^2 - F_{AB}^2} &= \sqrt{(300 \text{ N})^2 - (257.1 \text{ N})^2} \\
&= 155 \text{ N} \hspace{4cm} \textit{Ans.}
\end{aligned}$$

EXAMPLE | 2.16

The pipe in Fig. 2–44a is subjected to the force of $F = 80$ lb. Determine the angle θ between $\mathbf{F}$ and the pipe segment BA and the projection of $\mathbf{F}$ along this segment.

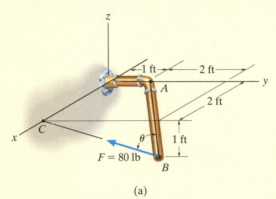

(a)

SOLUTION

Angle θ.   First we will establish position vectors from B to A and B to C; Fig. 2–44b. Then we will determine the angle θ between the tails of these two vectors.

$$\mathbf{r}_{BA} = \{-2\mathbf{i} - 2\mathbf{j} + 1\mathbf{k}\} \text{ ft}, \ r_{BA} = 3 \text{ ft}$$
$$\mathbf{r}_{BC} = \{-3\mathbf{j} + 1\mathbf{k}\} \text{ ft}, \ r_{BC} = \sqrt{10} \, ft$$

Thus,

$$\cos \theta = \frac{\mathbf{r}_{BA} \cdot \mathbf{r}_{BC}}{r_{BA}r_{BC}} = \frac{(-2)(0) + (-2)(-3) + (1)(1)}{3\sqrt{10}} = 0.7379$$

$$\theta = 42.5° \qquad\qquad\qquad \textit{Ans.}$$

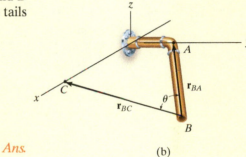

(b)

Components of F.   The component of $\mathbf{F}$ along BA is shown in Fig. 2–44c. We must first formulate the unit vector along BA and force $\mathbf{F}$ as Cartesian vectors.

$$\mathbf{u}_{BA} = \frac{\mathbf{r}_{BA}}{r_{BA}} = \frac{(-2\mathbf{i} - 2\mathbf{j} + 1\mathbf{k})}{3} = -\frac{2}{3}\mathbf{i} - \frac{2}{3}\mathbf{j} + \frac{1}{3}\mathbf{k}$$

$$\mathbf{F} = 80 \text{ lb}\left(\frac{\mathbf{r}_{BC}}{r_{BC}}\right) = 80\left(\frac{-3\mathbf{j} + 1\mathbf{k}}{\sqrt{10}}\right) = -75.89\mathbf{j} + 25.30\mathbf{k}$$

Thus,

$$F_{BA} = \mathbf{F} \cdot \mathbf{u}_{BA} = (-75.89\mathbf{j} + 25.30\mathbf{k}) \cdot \left(-\frac{2}{3}\mathbf{i} - \frac{2}{3}\mathbf{j} + \frac{1}{3}\mathbf{k}\right)$$

$$= 0\left(-\frac{2}{3}\right) + (-75.89)\left(-\frac{2}{3}\right) + (25.30)\left(\frac{1}{3}\right)$$

$$= 59.0 \text{ lb} \qquad\qquad\qquad \textit{Ans.}$$

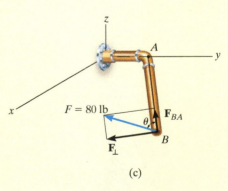

(c)

Fig. 2–44

NOTE: Since θ has been calculated, then also, $F_{BA} = F\cos\theta = 80$ lb $\cos 42.5° = 59.0$ lb.

PRELIMINARY PROBLEMS

P2–8. In each case, set up the dot product to find the angle θ. Do not calculate the result.

P2–9. In each case, set up the dot product to find the magnitude of the projection of the force $\mathbf{F}$ along a-a axes. Do not calculate the result.

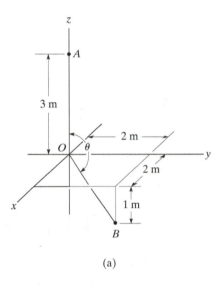

(a)

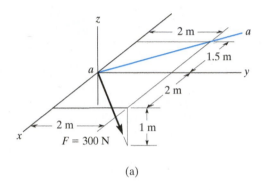

(a)

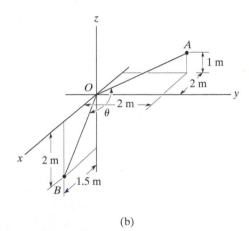

(b)

Prob. P2–8

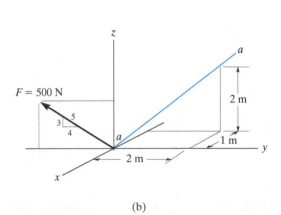

(b)

Prob. P2–9

FUNDAMENTAL PROBLEMS

F2–25. Determine the angle θ between the force and the line AO.

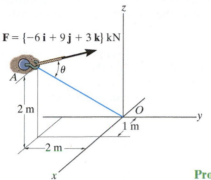

Prob. F2–25

F2–26. Determine the angle θ between the force and the line AB.

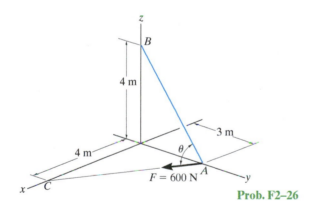

Prob. F2–26

F2–27. Determine the angle θ between the force and the line OA.

F2–28. Determine the projected component of the force along the line OA.

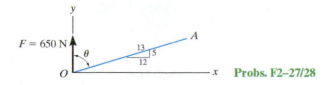

Probs. F2–27/28

F2–29. Find the magnitude of the projected component of the force along the pipe AO.

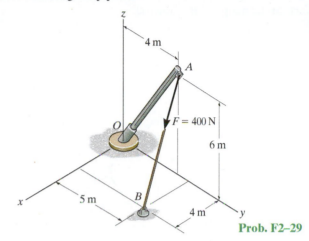

Prob. F2–29

F2–30. Determine the components of the force acting parallel and perpendicular to the axis of the pole.

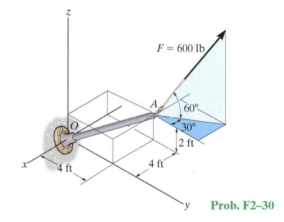

Prob. F2–30

F2–31. Determine the magnitudes of the components of the force $F = 56$ N acting along and perpendicular to line AO.

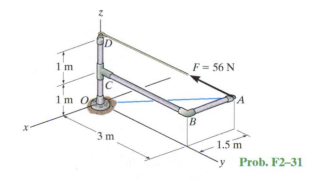

Prob. F2–31

PROBLEMS

2–106.   Express the force **F** in Cartesian vector form if it acts at the midpoint B of the rod.

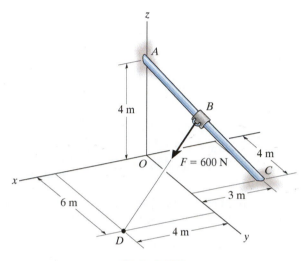

Prob. 2–106

2–107.   Express force **F** in Cartesian vector form if point B is located 3 m along the rod from end C.

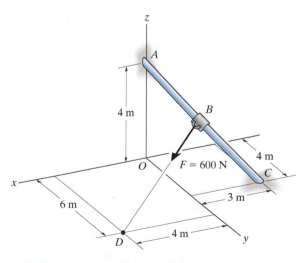

Prob. 2–107

*2–108.   The chandelier is supported by three chains which are concurrent at point O. If the force in each chain has a magnitude of 60 lb, express each force as a Cartesian vector and determine the magnitude and coordinate direction angles of the resultant force.

2–109.   The chandelier is supported by three chains which are concurrent at point O. If the resultant force at O has a magnitude of 130 lb and is directed along the negative z axis, determine the force in each chain.

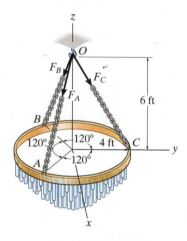

Probs. 2–108/109

2–110.   The window is held open by chain AB. Determine the length of the chain, and express the 50-lb force acting at A along the chain as a Cartesian vector and determine its coordinate direction angles.

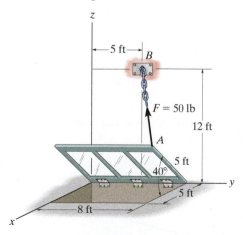

Prob. 2–110

2–111.   The window is held open by cable *AB*. Determine the length of the cable and express the 30-N force acting at *A* along the cable as a Cartesian vector.

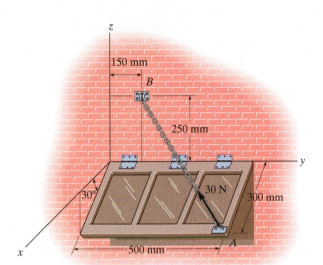

Prob. 2–111

2–114.   Determine the angle θ between the two cables.

2–115.   Determine the magnitude of the projection of the force $\mathbf{F}_1$ along cable *AC*.

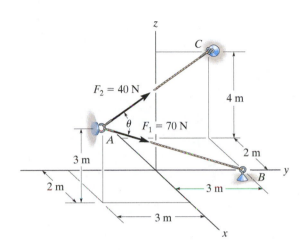

Probs. 2–114/115

*2–112.   Given the three vectors **A, B,** and **D,** show that $\mathbf{A} \cdot (\mathbf{B} + \mathbf{D}) = (\mathbf{A} \cdot \mathbf{B}) + (\mathbf{A} \cdot \mathbf{D})$.

2–113.   Determine the magnitudes of the components of $F = 600$ N acting along and perpendicular to segment *DE* of the pipe assembly.

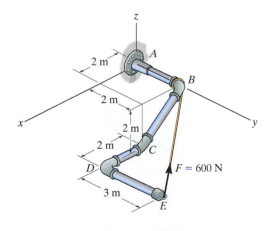

Probs. 2–112/113

*2–116.   Determine the angle θ between the *y* axis of the pole and the wire *AB*.

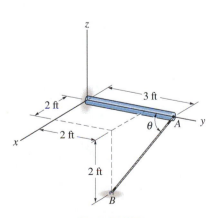

Prob. 2–116

2–117. Determine the magnitudes of the projected components of the force $\mathbf{F} = [60\mathbf{i} + 12\mathbf{j} - 40\mathbf{k}]$ N along the cables AB and AC.

2–118. Determine the angle θ between cables AB and AC.

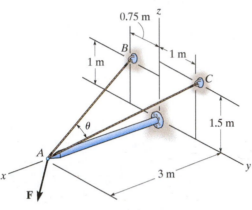

Probs. 2–117/118

2–119. A force of $\mathbf{F} = \{-40\mathbf{k}\}$ lb acts at the end of the pipe. Determine the magnitudes of the components $\mathbf{F}_1$ and $\mathbf{F}_2$ which are directed along the pipe's axis and perpendicular to it.

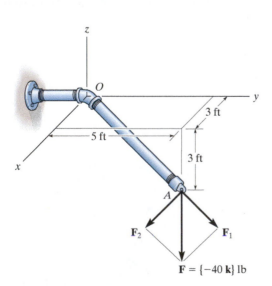

Prob. 2–119

*2–120. Two cables exert forces on the pipe. Determine the magnitude of the projected component of $\mathbf{F}_1$ along the line of action of $\mathbf{F}_2$.

2–121. Determine the angle θ between the two cables attached to the pipe.

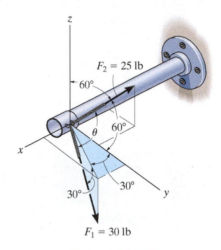

Probs. 2–120/121

2–122. Determine the angle θ between the cables AB and AC.

2–123. Determine the magnitude of the projected component of the force $\mathbf{F} = \{400\mathbf{i} - 200\mathbf{j} + 500\mathbf{k}\}$ N acting along the cable BA.

*2–124. Determine the magnitude of the projected component of the force $\mathbf{F} = \{400\mathbf{i} - 200\mathbf{j} + 500\mathbf{k}\}$ N acting along the cable CA.

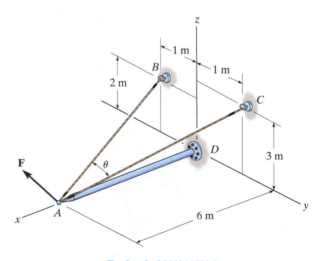

Probs. 2–122/123/124

2–125. Determine the magnitude of the projection of force $F = 600$ N along the u axis.

*2–128.** Determine the angle θ between BA and BC.

2–129. Determine the magnitude of the projected component of the 3 kN force acting along the axis BC of the pipe.

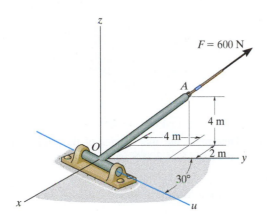

Prob. 2–125

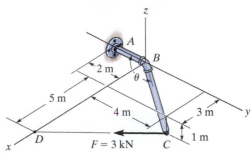

Probs. 2–128/129

2–126. Determine the magnitude of the projected component of the 100-lb force acting along the axis BC of the pipe.

2–127. Determine the angle θ between pipe segments BA and BC.

2–130. Determine the angles θ and ϕ made between the axes OA of the flag pole and AB and AC, respectively, of each cable.

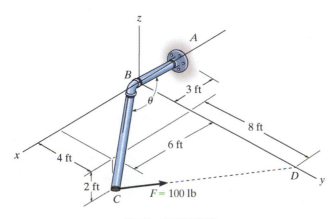

Probs. 2–126/127

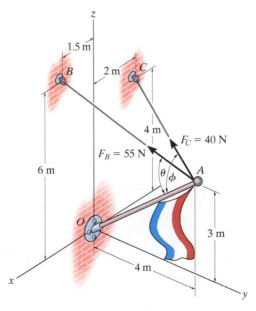

Prob. 2–130

2–131. Determine the magnitudes of the components of **F** acting along and perpendicular to segment *BC* of the pipe assembly.

*2–132. Determine the magnitude of the projected component of **F** along *AC*. Express this component as a Cartesian vector.

2–133. Determine the angle θ between the pipe segments *BA* and *BC*.

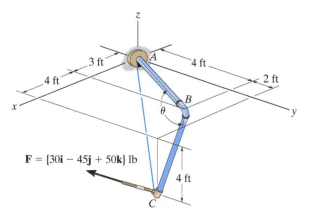

Probs. 2–131/132/133

2–134. If the force *F* = 100 N lies in the plane *DBEC*, which is parallel to the *x–z* plane, and makes an angle of 10° with the extended line *DB* as shown, determine the angle that **F** makes with the diagonal *AB* of the crate.

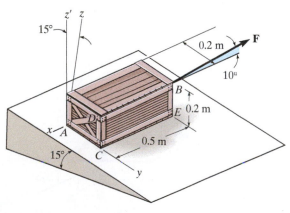

Prob. 2–134

2–135. Determine the magnitudes of the components of the force *F* = 90 lb acting parallel and perpendicular to diagonal *AB* of the crate.

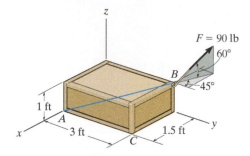

Prob. 2–135

*2–136. Determine the magnitudes of the projected components of the force *F* = 300 N acting along the *x* and *y* axes.

2–137. Determine the magnitude of the projected component of the force *F* = 300 N acting along line *OA*.

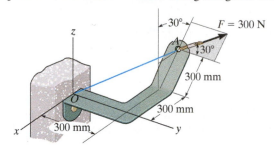

Probs. 2–136/137

2–138. Determine the angle θ between the two cables.

2–139. Determine the projected component of the force *F* = 12 lb acting in the direction of cable *AC*. Express the result as a Cartesian vector.

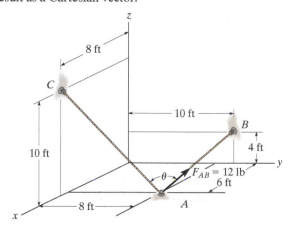

Probs. 2–138/139

CHAPTER REVIEW

<table>
<tr>
<td>

A scalar is a positive or negative number; e.g., mass and temperature.

A vector has a magnitude and direction, where the arrowhead represents the sense of the vector.

</td>
<td></td>
<td>

</td>
</tr>
<tr>
<td>

Multiplication or division of a vector by a scalar will change only the magnitude of the vector. If the scalar is negative, the sense of the vector will change so that it acts in the opposite sense.

</td>
<td></td>
<td>

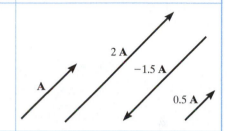

</td>
</tr>
<tr>
<td>

If vectors are collinear, the resultant is simply the algebraic or scalar addition.

</td>
<td>

$$R = A + B$$

</td>
<td>

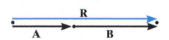

</td>
</tr>
<tr>
<td>

Parallelogram Law

Two forces add according to the parallelogram law. The *components* form the sides of the parallelogram and the *resultant* is the diagonal.

To find the components of a force along any two axes, extend lines from the head of the force, parallel to the axes, to form the components.

</td>
<td></td>
<td>

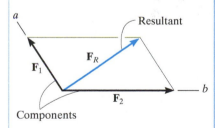

</td>
</tr>
<tr>
<td>

To obtain the components of the resultant, show how the forces add by tip-to-tail using the triangle rule, and then use the law of cosines and the law of sines to calculate their values.

</td>
<td>

$$F_R = \sqrt{F_1^2 + F_2^2 - 2\,F_1 F_2 \cos \theta_R}$$

$$\frac{F_1}{\sin \theta_1} = \frac{F_2}{\sin \theta_2} = \frac{F_R}{\sin \theta_R}$$

</td>
<td>

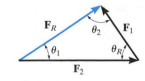

</td>
</tr>
</table>

2

Rectangular Components: Two Dimensions

Vectors $\mathbf{F}_x$ and $\mathbf{F}_y$ are rectangular components of $\mathbf{F}$.

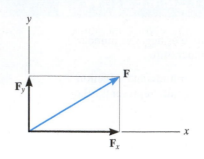

The resultant force is determined from the algebraic sum of its components.

$$(F_R)_x = \Sigma F_x$$

$$(F_R)_y = \Sigma F_y$$

$$F_R = \sqrt{(F_R)_x^2 + (F_R)_y^2}$$

$$\theta = \tan^{-1}\left|\frac{(F_R)_y}{(F_R)_x}\right|$$

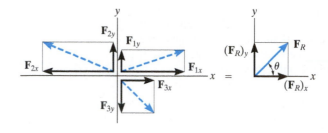

Cartesian Vectors

The unit vector $\mathbf{u}$ has a length of 1, no units, and it points in the direction of the vector $\mathbf{F}$.

$$\mathbf{u} = \frac{\mathbf{F}}{F}$$

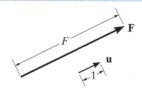

A force can be resolved into its Cartesian components along the x, y, z axes so that $\mathbf{F} = F_x\mathbf{i} + F_y\mathbf{j} + F_z\mathbf{k}$.

The magnitude of $\mathbf{F}$ is determined from the positive square root of the sum of the squares of its components.

$$F = \sqrt{F_x^2 + F_y^2 + F_z^2}$$

The coordinate direction angles α, β, γ are determined by formulating a unit vector in the direction of $\mathbf{F}$. The x, y, z components of $\mathbf{u}$ represent $\cos\alpha, \cos\beta, \cos\gamma$.

$$\mathbf{u} = \frac{\mathbf{F}}{F} = \frac{F_x}{F}\mathbf{i} + \frac{F_y}{F}\mathbf{j} + \frac{F_z}{F}\mathbf{k}$$

$$\mathbf{u} = \cos\alpha\,\mathbf{i} + \cos\beta\,\mathbf{j} + \cos\gamma\,\mathbf{k}$$

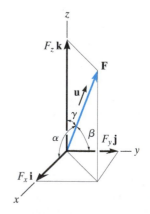

The coordinate direction angles are related so that only two of the three angles are independent of one another.

$$\cos^2 \alpha + \cos^2 \beta + \cos^2 \gamma = 1$$

To find the resultant of a concurrent force system, express each force as a Cartesian vector and add the $\mathbf{i}, \mathbf{j}, \mathbf{k}$ components of all the forces in the system.

$$\mathbf{F}_R = \Sigma \mathbf{F} = \Sigma F_x \mathbf{i} + \Sigma F_y \mathbf{j} + \Sigma F_z \mathbf{k}$$

Position and Force Vectors

A position vector locates one point in space relative to another. The easiest way to formulate the components of a position vector is to determine the distance and direction that one must travel along the $x, y,$ and z directions—going from the tail to the head of the vector.

$$\mathbf{r} = (x_B - x_A)\mathbf{i}$$
$$+ (y_B - y_A)\mathbf{j}$$
$$+ (z_B - z_A)\mathbf{k}$$

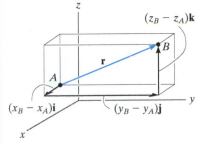

If the line of action of a force passes through points A and B, then the force acts in the same direction as the position vector $\mathbf{r}$, which is defined by the unit vector $\mathbf{u}$. The force can then be expressed as a Cartesian vector.

$$\mathbf{F} = F\mathbf{u} = F\left(\frac{\mathbf{r}}{r}\right)$$

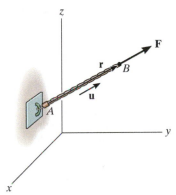

Dot Product

The dot product between two vectors $\mathbf{A}$ and $\mathbf{B}$ yields a scalar. If $\mathbf{A}$ and $\mathbf{B}$ are expressed in Cartesian vector form, then the dot product is the sum of the products of their $x, y,$ and z components.

$$\mathbf{A} \cdot \mathbf{B} = AB \cos \theta$$
$$= A_x B_x + A_y B_y + A_z B_z$$

The dot product can be used to determine the angle between $\mathbf{A}$ and $\mathbf{B}$.

$$\theta = \cos^{-1}\left(\frac{\mathbf{A} \cdot \mathbf{B}}{AB}\right)$$

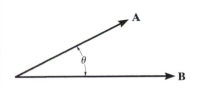

The dot product is also used to determine the projected component of a vector $\mathbf{A}$ onto an axis aa defined by its unit vector $\mathbf{u}_a$.

$$\mathbf{A}_a = A \cos \theta \, \mathbf{u}_a = (\mathbf{A} \cdot \mathbf{u}_a)\mathbf{u}_a$$

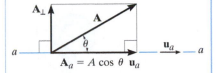

REVIEW PROBLEMS

Partial solutions and answers to all Review Problems are given in the back of the book.

R2–1. Determine the magnitude of the resultant force $\mathbf{F}_R$ and its direction, measured clockwise from the positive *u* axis.

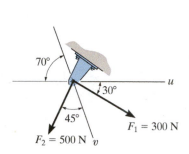

Prob. R2–1

R2–3. Determine the magnitude of the resultant force acting on the *gusset plate* of the bridge truss.

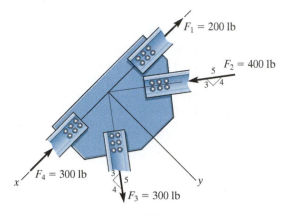

Prob. R2–3

R2–2. Resolve **F** into components along the *u* and *v* axes and determine the magnitudes of these components.

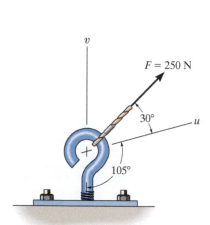

Prob. R2–2

R2–4. The cable at the end of the crane boom exerts a force of 250 lb on the boom as shown. Express **F** as a Cartesian vector.

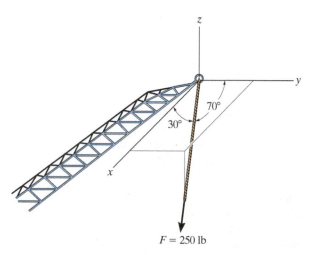

Prob. R2–4

R2–5. The cable attached to the tractor at B exerts a force of 350 lb on the framework. Express this force as a Cartesian vector.

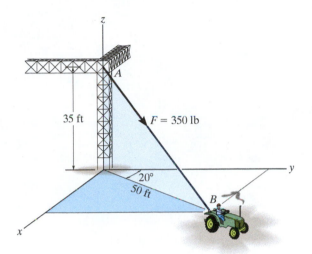

Prob. R2–5

R2–6. Express $\mathbf{F}_1$ and $\mathbf{F}_2$ as Cartesian vectors.

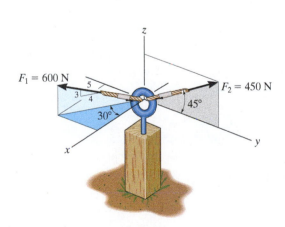

Prob. R2–6

R2–7. Determine the angle θ between the edges of the sheet-metal bracket.

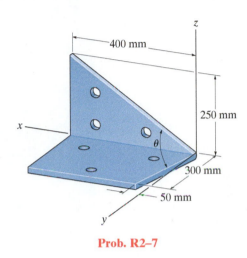

Prob. R2–7

R2–8. Determine the projection of the force $\mathbf{F}$ along the pole.

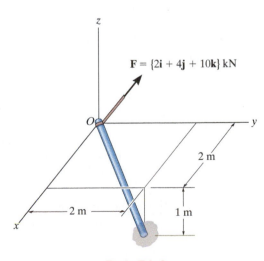

Prob. R2–8

Chapter 3

(© Igor Tumarkin/ITPS/Shutterstock)

When this load is lifted at constant velocity, or is just suspended, then it is in a state of equilibrium. In this chapter we will study equilibrium for a particle and show how these ideas can be used to calculate the forces in cables used to hold suspended loads.

Equilibrium of a Particle

CHAPTER OBJECTIVES

■ To introduce the concept of the free-body diagram for a particle.

■ To show how to solve particle equilibrium problems using the equations of equilibrium.

3.1 Condition for the Equilibrium of a Particle

A particle is said to be in *equilibrium* if it remains at rest if originally at rest, or has a constant velocity if originally in motion. Most often, however, the term "equilibrium" or, more specifically, "static equilibrium" is used to describe an object at rest. To maintain equilibrium, it is *necessary* to satisfy Newton's first law of motion, which requires the *resultant force* acting on a particle to be equal to *zero*. This condition is stated by the *equation of equilibrium*,

$$\Sigma \mathbf{F} = \mathbf{0} \qquad\qquad (3\text{–}1)$$

where $\Sigma \mathbf{F}$ is the vector *sum of all the forces* acting on the particle.

Not only is Eq. 3–1 a necessary condition for equilibrium, it is also a *sufficient* condition. This follows from Newton's second law of motion, which can be written as $\Sigma \mathbf{F} = m\mathbf{a}$. Since the force system satisfies Eq. 3–1, then $m\mathbf{a} = \mathbf{0}$, and therefore the particle's acceleration $\mathbf{a} = \mathbf{0}$. Consequently, the particle indeed moves with constant velocity or remains at rest.

3.2   The Free-Body Diagram

To apply the equation of equilibrium, we must account for *all* the known and unknown forces ($\Sigma\mathbf{F}$) which act *on* the particle. The best way to do this is to think of the particle as isolated and "free" from its surroundings. A drawing that shows the particle with *all* the forces that act on it is called a *free-body diagram (FBD)*.

Before presenting a formal procedure as to how to draw a free-body diagram, we will first consider three types of supports often encountered in particle equilibrium problems.

(a)

Springs.   If a *linearly elastic spring* (or cord) of undeformed length l_0 is used to support a particle, the length of the spring will change in direct proportion to the force $\mathbf{F}$ acting on it, Fig. 3–1*a*. A characteristic that defines the "elasticity" of a spring is the *spring constant* or *stiffness k*.

The magnitude of force exerted on a linearly elastic spring which has a stiffness k and is deformed (elongated or compressed) a distance $s = l - l_0$, measured from its *unloaded* position, is

$$F = ks \qquad\qquad (3\text{--}2)$$

If s is positive, causing an elongation, then $\mathbf{F}$ must pull on the spring; whereas if s is negative, causing a shortening, then $\mathbf{F}$ must push on it. For example, if the spring in Fig. 3–1*a* has an unstretched length of 0.8 m and a stiffness $k = 500 \text{ N/m}$ and it is stretched to a length of 1 m, so that $s = l - l_0 = 1 \text{ m} - 0.8 \text{ m} = 0.2 \text{ m}$, then a force $F = ks = 500 \text{ N/m}(0.2 \text{ m}) = 100 \text{ N}$ is needed.

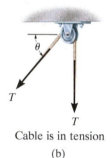

Cable is in tension

(b)

Fig. 3–1

Cables and Pulleys.   Unless otherwise stated throughout this book, except in Sec. 7.4, all cables (or cords) will be assumed to have negligible weight and they cannot stretch. Also, a cable can support *only* a tension or "pulling" force, and this force always acts in the direction of the cable. In Chapter 5, it will be shown that the tension force developed in a continuous cable which passes over a frictionless pulley must have a *constant* magnitude to keep the cable in equilibrium. Hence, for any angle θ, shown in Fig. 3–1*b*, the cable is subjected to a constant tension T throughout its length.

Smooth Contact.   If an object rests on a *smooth surface*, then the surface will exert a force on the object that is normal to the surface at the point of contact. An example of this is shown in Fig. 3–2*a*. In addition to this normal force $\mathbf{N}$, the cylinder is also subjected to its weight $\mathbf{W}$ and the force $\mathbf{T}$ of the cord. Since these three forces are concurrent at the center of the cylinder, Fig. 3–2*b*, we can apply the equation of equilibrium to this "particle," which is the same as applying it to the cylinder.

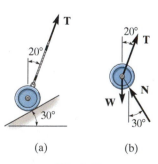

(a)   (b)

Fig. 3–2

Procedure for Drawing a Free-Body Diagram

Since we must account for *all the forces acting on the particle* when applying the equations of equilibrium, the importance of first drawing a free-body diagram cannot be overemphasized. To construct a free-body diagram, the following three steps are necessary.

Draw Outlined Shape.

Imagine the particle to be *isolated* or cut "free" from its surroundings. This requires *removing* all the supports and drawing the particle's outlined shape.

Show All Forces.

Indicate on this sketch *all* the forces that act *on the particle*. These forces can be *active forces*, which tend to set the particle in motion, or they can be *reactive forces* which are the result of the constraints or supports that tend to prevent motion. To account for all these forces, it may be helpful to trace around the particle's boundary, carefully noting each force acting on it.

Identify Each Force.

The forces that are *known* should be labeled with their proper magnitudes and directions. Letters are used to represent the magnitudes and directions of forces that are unknown.

The bucket is held in equilibrium by the cable, and instinctively we know that the force in the cable must equal the weight of the bucket. By drawing a free-body diagram of the bucket we can understand why this is so. This diagram shows that there are only two forces *acting on the bucket*, namely, its weight **W** and the force **T** of the cable. For equilibrium, the resultant of these forces must be equal to zero, and so $T = W$. (© Russell C. Hibbeler)

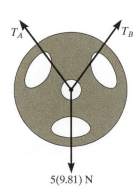

5(9.81) N

The 5-kg plate is suspended by two straps *A* and *B*. To find the force in each strap we should consider the free-body diagram of the plate. As noted, the three forces acting on it are concurrent at the center. (© Russell C. Hibbeler)

EXAMPLE | 3.1

The sphere in Fig. 3–3a has a mass of 6 kg and is supported as shown. Draw a free-body diagram of the sphere, the cord CE, and the knot at C.

$\mathbf{F}_{CE}$ (Force of cord CE acting on sphere)

(Forces of smooth planes acting on sphere)

$\mathbf{N}_A$ $\mathbf{N}_B$

30°   30°

58.9 N
(Weight or gravity acting on sphere)

(b)

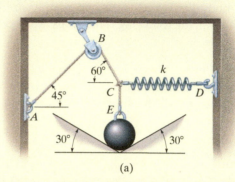

(a)

SOLUTION

Sphere.   Once the supports are *removed*, we can see that there are four forces acting on the sphere, namely, its weight, 6 kg (9.81 m/s²) = 58.9 N, the force of cord CE, and the two normal forces caused by the smooth inclined planes. The free-body diagram is shown in Fig. 3–3b.

Cord CE.   When the cord CE is isolated from its surroundings, its free-body diagram shows only two forces acting on it, namely, the force of the sphere and the force of the knot, Fig. 3–3c. Notice that $\mathbf{F}_{CE}$ shown here is equal but opposite to that shown in Fig. 3–3b, a consequence of Newton's third law of action–reaction. Also, $\mathbf{F}_{CE}$ and $\mathbf{F}_{EC}$ pull on the cord and keep it in tension so that it doesn't collapse. For equilibrium, $F_{CE} = F_{EC}$.

Knot.   The knot at C is subjected to three forces, Fig. 3–3d. They are caused by the cords CBA and CE and the spring CD. As required, the free-body diagram shows all these forces labeled with their magnitudes and directions. It is important to recognize that the weight of the sphere does not directly act on the knot. Instead, the cord CE subjects the knot to this force.

$\mathbf{F}_{EC}$ (Force of knot acting on cord CE)

$\mathbf{F}_{CE}$ (Force of sphere acting on cord CE)

(c)

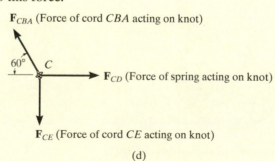

$\mathbf{F}_{CBA}$ (Force of cord CBA acting on knot)

60°   C

$\mathbf{F}_{CD}$ (Force of spring acting on knot)

$\mathbf{F}_{CE}$ (Force of cord CE acting on knot)

(d)

Fig. 3–3

3.3 Coplanar Force Systems

If a particle is subjected to a system of coplanar forces that lie in the x–y plane, as in Fig. 3–4, then each force can be resolved into its $\mathbf{i}$ and $\mathbf{j}$ components. For equilibrium, these forces must sum to produce a zero force resultant, i.e.,

$$\Sigma \mathbf{F} = \mathbf{0}$$
$$\Sigma F_x \mathbf{i} + \Sigma F_y \mathbf{j} = \mathbf{0}$$

For this vector equation to be satisfied, the resultant force's x and y components must both be equal to zero. Hence,

$$\boxed{\begin{aligned}\Sigma F_x &= 0 \\ \Sigma F_y &= 0\end{aligned}} \qquad (3\text{--}3)$$

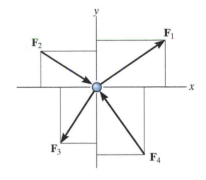

Fig. 3–4

These two equations can be solved for at most two unknowns, generally represented as angles and magnitudes of forces shown on the particle's free-body diagram.

When applying each of the two equations of equilibrium, we must account for the sense of direction of any component by using an *algebraic sign* which corresponds to the arrowhead direction of the component along the x or y axis. It is important to note that if a force has an *unknown magnitude*, then the arrowhead sense of the force on the free-body diagram can be *assumed*. Then if the *solution* yields a *negative scalar*, this indicates that the sense of the force is opposite to that which was assumed.

For example, consider the free-body diagram of the particle subjected to the two forces shown in Fig. 3–5. Here it is *assumed* that the *unknown force* $\mathbf{F}$ acts to the right, that is, in the positive x direction, to maintain equilibrium. Applying the equation of equilibrium along the x axis, we have

$$\xrightarrow{+} \Sigma F_x = 0; \qquad\qquad +F + 10\,\text{N} = 0$$

Both terms are "positive" since both forces act in the positive x direction. When this equation is solved, $F = -10\,\text{N}$. Here the *negative sign* indicates that $\mathbf{F}$ must act to the left to hold the particle in equilibrium, Fig. 3–5. Notice that if the $+x$ axis in Fig. 3–5 were directed to the left, both terms in the above equation would be negative, but again, after solving, $F = -10\,\text{N}$, indicating that $\mathbf{F}$ would have to be directed to the left.

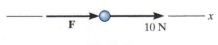

Fig. 3–5

Important Points

The first step in solving any equilibrium problem is to draw the particle's free-body diagram. This requires *removing all the supports* and isolating or freeing the particle from its surroundings and then showing all the forces that act on it.

Equilibrium means the particle is at rest or moving at constant velocity. In two dimensions, the necessary and sufficient conditions for equilibrium require $\Sigma F_x = 0$ and $\Sigma F_y = 0$.

Procedure for Analysis

Coplanar force equilibrium problems for a particle can be solved using the following procedure.

Free-Body Diagram.
- Establish the x, y axes in any suitable orientation.

- Label all the known and unknown force magnitudes and directions on the diagram.

- The sense of a force having an unknown magnitude can be assumed.

Equations of Equilibrium.
- Apply the equations of equilibrium, $\Sigma F_x = 0$ and $\Sigma F_y = 0$. For convenience, arrows can be written alongside each equation to define the positive directions.

- Components are positive if they are directed along a positive axis, and negative if they are directed along a negative axis.

- If more than two unknowns exist and the problem involves a spring, apply $F = ks$ to relate the spring force to the deformation s of the spring.

- Since the magnitude of a force is always a positive quantity, then if the solution for a force yields a negative result, this indicates that its sense is the reverse of that shown on the free-body diagram.

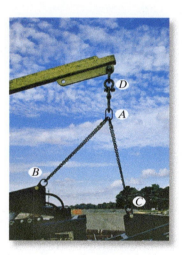

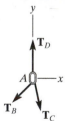

The chains exert three forces on the ring at A, as shown on its free-body diagram. The ring will not move, or will move with constant velocity, provided the summation of these forces along the x and along the y axis equals zero. If one of the three forces is known, the magnitudes of the other two forces can be obtained from the two equations of equilibrium. (© Russell C. Hibbeler)

EXAMPLE 3.2

Determine the tension in cables BA and BC necessary to support the 60-kg cylinder in Fig. 3–6a.

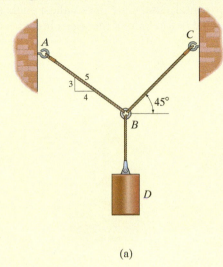

(a)

$T_{BD} = 60\ (9.81)\ \text{N}$

$60\ (9.81)\ \text{N}$

(b)

SOLUTION

Free-Body Diagram. Due to equilibrium, the weight of the cylinder causes the tension in cable BD to be $T_{BD} = 60(9.81)$ N, Fig. 3–6b. The forces in cables BA and BC can be determined by investigating the equilibrium of ring B. Its free-body diagram is shown in Fig. 3–6c. The magnitudes of $\mathbf{T}_A$ and $\mathbf{T}_C$ are unknown, but their directions are known.

Equations of Equilibrium. Applying the equations of equilibrium along the x and y axes, we have

$\xrightarrow{+} \Sigma F_x = 0; \qquad T_C \cos 45° - \left(\tfrac{4}{5}\right)T_A = 0 \qquad (1)$

$+\uparrow \Sigma F_y = 0; \qquad T_C \sin 45° + \left(\tfrac{3}{5}\right)T_A - 60(9.81)\ \text{N} = 0 \qquad (2)$

Equation (1) can be written as $T_A = 0.8839T_C$. Substituting this into Eq. (2) yields

$$T_C \sin 45° + \left(\tfrac{3}{5}\right)(0.8839T_C) - 60(9.81)\ \text{N} = 0$$

so that

$$T_C = 475.66\ \text{N} = 476\ \text{N} \qquad\qquad \textit{Ans.}$$

Substituting this result into either Eq. (1) or Eq. (2), we get

$$T_A = 420\ \text{N} \qquad\qquad \textit{Ans.}$$

NOTE: The accuracy of these results, of course, depends on the accuracy of the data, i.e., measurements of geometry and loads. For most engineering work involving a problem such as this, the data as measured to three significant figures would be sufficient.

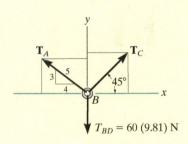

$T_{BD} = 60\ (9.81)\ \text{N}$

(c)

Fig. 3–6

EXAMPLE 3.3

The 200-kg crate in Fig. 3–7a is suspended using the ropes AB and AC. Each rope can withstand a maximum force of 10 kN before it breaks. If AB always remains horizontal, determine the smallest angle θ to which the crate can be suspended before one of the ropes breaks.

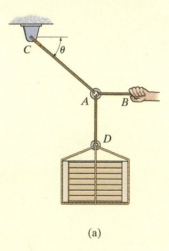

(a)

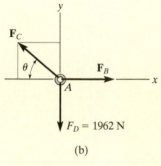

(b)

Fig. 3–7

SOLUTION

Free-Body Diagram.   We will study the equilibrium of ring A. There are three forces acting on it, Fig. 3–7b. The magnitude of $\mathbf{F}_D$ is equal to the weight of the crate, i.e., $F_D = 200\,(9.81)\ \text{N} = 1962\ \text{N} < 10\ \text{kN}$.

Equations of Equilibrium.   Applying the equations of equilibrium along the x and y axes,

$$\xrightarrow{+}\ \Sigma F_x = 0; \qquad -F_C \cos\theta + F_B = 0; \quad F_C = \frac{F_B}{\cos\theta} \tag{1}$$

$$+\uparrow\Sigma F_y = 0; \qquad F_C \sin\theta - 1962\ \text{N} = 0 \tag{2}$$

From Eq. (1), F_C is always greater than F_B since $\cos\theta \leq 1$. Therefore, rope AC will reach the maximum tensile force of 10 kN *before* rope AB. Substituting $F_C = 10$ kN into Eq. (2), we get

$$[10(10^3)\,\text{N}]\sin\theta - 1962\ \text{N} = 0$$

$$\theta = \sin^{-1}(0.1962) = 11.31° = 11.3° \qquad \textit{Ans.}$$

The force developed in rope AB can be obtained by substituting the values for θ and F_C into Eq. (1).

$$10(10^3)\ \text{N} = \frac{F_B}{\cos 11.31°}$$

$$F_B = 9.81\ \text{kN}$$

EXAMPLE | 3.4

Determine the required length of cord AC in Fig. 3–8a so that the 8-kg lamp can be suspended in the position shown. The *undeformed* length of spring AB is $l'_{AB} = 0.4$ m, and the spring has a stiffness of $k_{AB} = 300$ N/m.

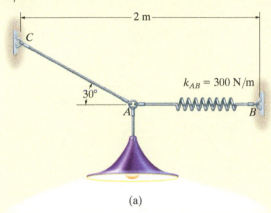

(a)

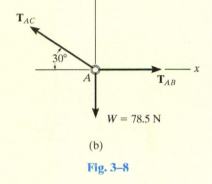

(b)

Fig. 3–8

SOLUTION

If the force in spring AB is known, the stretch of the spring can be found using $F = ks$. From the problem geometry, it is then possible to calculate the required length of AC.

Free-Body Diagram.   The lamp has a weight $W = 8(9.81) = 78.5$ N and so the free-body diagram of the ring at A is shown in Fig. 3–8b.

Equations of Equilibrium.   Using the x, y axes,

$$\xrightarrow{+} \Sigma F_x = 0; \qquad T_{AB} - T_{AC}\cos 30° = 0$$
$$+\uparrow \Sigma F_y = 0; \qquad T_{AC}\sin 30° - 78.5\ \text{N} = 0$$

Solving, we obtain

$$T_{AC} = 157.0\ \text{N}$$
$$T_{AB} = 135.9\ \text{N}$$

The stretch of spring AB is therefore

$$T_{AB} = k_{AB}s_{AB}; \qquad 135.9\ \text{N} = 300\ \text{N/m}(s_{AB})$$
$$s_{AB} = 0.453\ \text{m}$$

so the stretched length is

$$l_{AB} = l'_{AB} + s_{AB}$$
$$l_{AB} = 0.4\ \text{m} + 0.453\ \text{m} = 0.853\ \text{m}$$

The horizontal distance from C to B, Fig. 3–8a, requires

$$2\ \text{m} = l_{AC}\cos 30° + 0.853\ \text{m}$$
$$l_{AC} = 1.32\ \text{m} \qquad\qquad \textit{Ans.}$$

PRELIMINARY PROBLEMS

P3–1. In each case, draw a free-body diagram of the ring at *A* and identify each force.

P3–2. Write the two equations of equilibrium, $\Sigma F_x = 0$ and $\Sigma F_y = 0$. Do not solve.

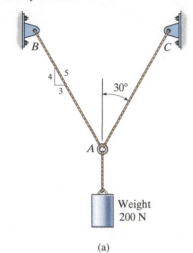

(a)

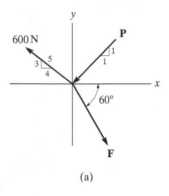

(a)

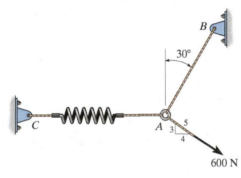

(b)

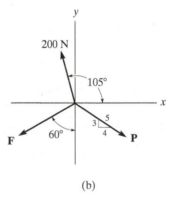

(b)

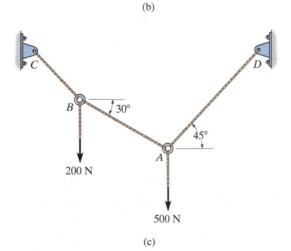

(c)

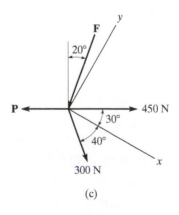

(c)

FUNDAMENTAL PROBLEMS

All problem solutions must include an FBD.

F3–1. The crate has a weight of 550 lb. Determine the force in each supporting cable.

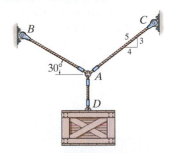

Prob. F3–1

F3–2. The beam has a weight of 700 lb. Determine the shortest cable *ABC* that can be used to lift it if the maximum force the cable can sustain is 1500 lb.

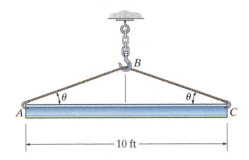

Prob. F3–2

F3–3. If the 5-kg block is suspended from the pulley *B* and the sag of the cord is *d* = 0.15 m, determine the force in cord *ABC*. Neglect the size of the pulley.

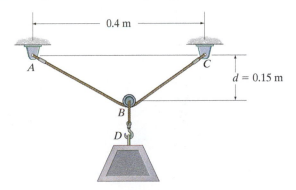

Prob. F3–3

F3–4. The block has a mass of 5 kg and rests on the smooth plane. Determine the unstretched length of the spring.

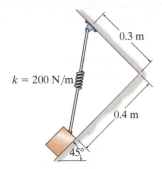

Prob. F3–4

F3–5. If the mass of cylinder *C* is 40 kg, determine the mass of cylinder *A* in order to hold the assembly in the position shown.

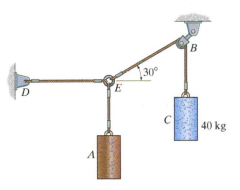

Prob. F3–5

F3–6. Determine the tension in cables *AB*, *BC*, and *CD*, necessary to support the 10-kg and 15-kg traffic lights at *B* and *C*, respectively. Also, find the angle *θ*.

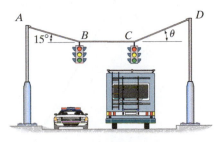

Prob. F3–6

PROBLEMS

All problem solutions must include an FBD.

3–1. The members of a truss are pin connected at joint O. Determine the magnitudes of $\mathbf{F}_1$ and $\mathbf{F}_2$ for equilibrium. Set $\theta = 60°$.

3–2. The members of a truss are pin connected at joint O. Determine the magnitude of $\mathbf{F}_1$ and its angle θ for equilibrium. Set $F_2 = 6$ kN.

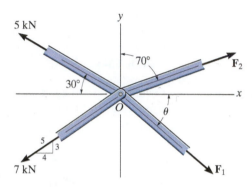

Probs. 3–1/2

3–3. Determine the magnitude and direction θ of $\mathbf{F}$ so that the particle is in equilibrium.

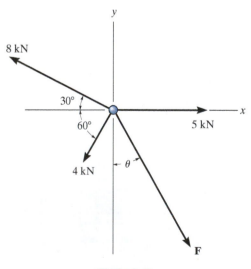

Prob. 3–3

*3–4. The bearing consists of rollers, symmetrically confined within the housing. The bottom one is subjected to a 125-N force at its contact A due to the load on the shaft. Determine the normal reactions N_B and N_C on the bearing at its contact points B and C for equilibrium.

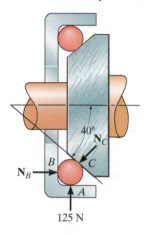

Prob. 3–4

3–5. The members of a truss are connected to the gusset plate. If the forces are concurrent at point O, determine the magnitudes of $\mathbf{F}$ and $\mathbf{T}$ for equilibrium. Take $\theta = 90°$.

3–6. The gusset plate is subjected to the forces of three members. Determine the tension force in member C and its angle θ for equilibrium. The forces are concurrent at point O. Take $F = 8$ kN.

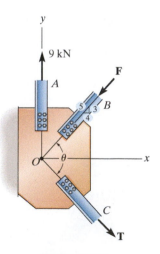

Probs. 3–5/6

3–7. The man attempts to pull down the tree using the cable and *small* pulley arrangement shown. If the tension in *AB* is 60 lb, determine the tension in cable *CAD* and the angle θ which the cable makes at the pulley.

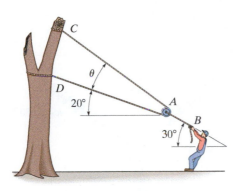

Prob. 3–7

3–9. Determine the maximum force **F** that can be supported in the position shown if each chain can support a maximum tension of 600 lb before it fails.

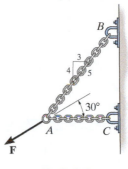

Prob. 3–9

***3–8.** The cords *ABC* and *BD* can each support a maximum load of 100 lb. Determine the maximum weight of the crate, and the angle θ for equilibrium.

3–10. The block has a weight of 20 lb and is being hoisted at uniform velocity. Determine the angle θ for equilibrium and the force in cord *AB*.

3–11. Determine the maximum weight W of the block that can be suspended in the position shown if cords *AB* and *CAD* can each support a maximum tension of 80 lb. Also, what is the angle θ for equilibrium?

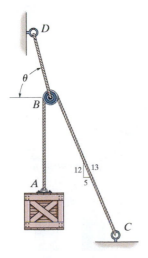

Prob. 3–8

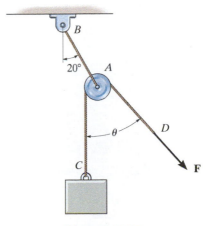

Probs. 3–10/11

3–12. The lift sling is used to hoist a container having a mass of 500 kg. Determine the force in each of the cables AB and AC as a function of θ. If the maximum tension allowed in each cable is 5 kN, determine the shortest length of cables AB and AC that can be used for the lift. The center of gravity of the container is located at G.

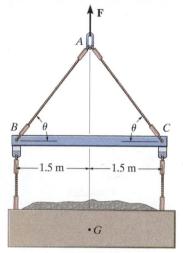

Prob. 3–12

3–13. A nuclear-reactor vessel has a weight of $500(10^3)$ lb. Determine the horizontal compressive force that the spreader bar AB exerts on point A and the force that each cable segment CA and AD exert on this point while the vessel is hoisted upward at constant velocity.

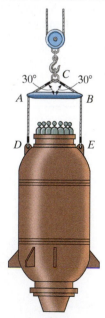

Prob. 3–13

3–14. Determine the stretch in each spring for equilibrium of the 2-kg block. The springs are shown in the equilibrium position.

3–15. The unstretched length of spring AB is 3 m. If the block is held in the equilibrium position shown, determine the mass of the block at D.

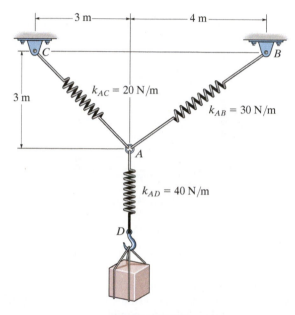

Probs. 3–14/15

*3–16. Determine the mass of each of the two cylinders if they cause a sag of $s = 0.5$ m when suspended from the rings at A and B. Note that $s = 0$ when the cylinders are removed.

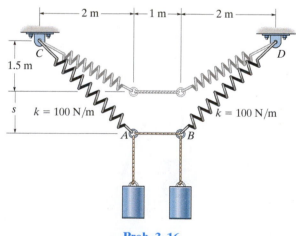

Prob. 3–16

3–17. Determine the stiffness k_T of the single spring such that the force **F** will stretch it by the same amount s as the force **F** stretches the two springs. Express k_T in terms of stiffness k_1 and k_2 of the two springs.

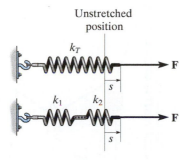

Unstretched position

Prob. 3–17

3–18. If the spring DB has an unstretched length of 2 m, determine the stiffness of the spring to hold the 40-kg crate in the position shown.

3–19. Determine the unstretched length of DB to hold the 40-kg crate in the position shown. Take $k = 180$ N/m.

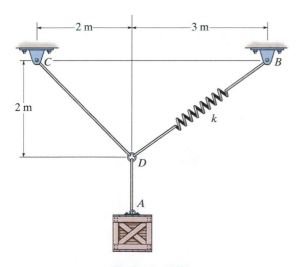

Probs. 3–18/19

*3–20. A vertical force $P = 10$ lb is applied to the ends of the 2-ft cord AB and spring AC. If the spring has an unstretched length of 2 ft, determine the angle θ for equilibrium. Take $k = 15$ lb/ft.

3–21. Determine the unstretched length of spring AC if a force $P = 80$ lb causes the angle $\theta = 60°$ for equilibrium. Cord AB is 2 ft long. Take $k = 50$ lb/ft.

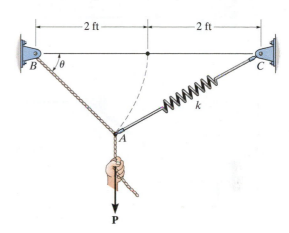

Probs. 3–20/21

3–22. The springs BA and BC each have a stiffness of 500 N/m and an unstretched length of 3 m. Determine the horizontal force **F** applied to the cord which is attached to the small ring B so that the displacement of AB from the wall is $d = 1.5$ m.

3–23. The springs BA and BC each have a stiffness of 500 N/m and an unstretched length of 3 m. Determine the displacement d of the cord from the wall when a force $F = 175$ N is applied to the cord.

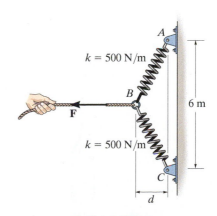

Probs. 3–22/23

*3–24. Determine the distances x and y for equilibrium if $F_1 = 800$ N and $F_2 = 1000$ N.

3–25. Determine the magnitude of F_1 and the distance y if $x = 1.5$ m and $F_2 = 1000$ N.

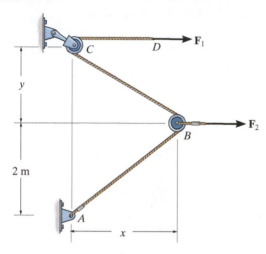

Probs. 3–24/25

3–26. The 30-kg pipe is supported at A by a system of five cords. Determine the force in each cord for equilibrium.

3–27. Each cord can sustain a maximum tension of 500 N. Determine the largest mass of pipe that can be supported.

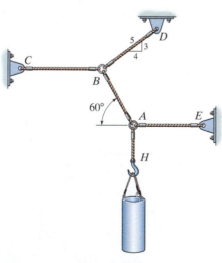

Probs. 3–26/27

*3–28. The street-lights at A and B are suspended from the two poles as shown. If each light has a weight of 50 lb, determine the tension in each of the three supporting cables and the required height h of the pole DE so that cable AB is horizontal.

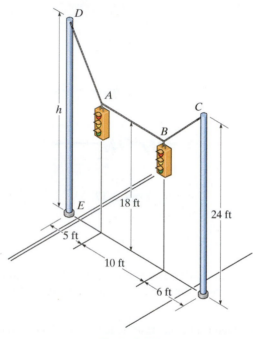

Prob. 3–28

3–29. Determine the tension developed in each cord required for equilibrium of the 20-kg lamp.

3–30. Determine the maximum mass of the lamp that the cord system can support so that no single cord develops a tension exceeding 400 N.

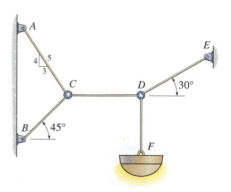

Probs. 3–29/30

3–31. Blocks D and E have a mass of 4 kg and 6 kg, respectively. If $x = 2$ m determine the force **F** and the sag s for equilibrium.

*3–32. Blocks D and E have a mass of 4 kg and 6 kg, respectively. If $F = 80$ N, determine the sag s and distance x for equilibrium.

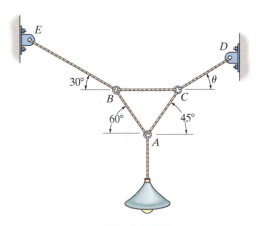

Probs. 3–31/32

3–33. The lamp has a weight of 15 lb and is supported by the six cords connected together as shown. Determine the tension in each cord and the angle θ for equilibrium. Cord BC is horizontal.

3–34. Each cord can sustain a maximum tension of 20 lb. Determine the largest weight of the lamp that can be supported. Also, determine θ of cord DC for equilibrium.

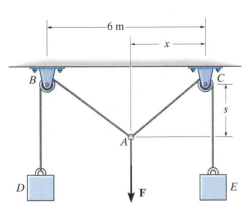

Probs. 3–33/34

3–35. The ring of negligible size is subjected to a vertical force of 200 lb. Determine the required length l of cord AC such that the tension acting in AC is 160 lb. Also, what is the force in cord AB? *Hint:* Use the equilibrium condition to determine the required angle θ for attachment, then determine l using trigonometry applied to triangle ABC.

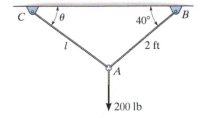

Prob. 3–35

*3–36. Cable ABC has a length of 5 m. Determine the position x and the tension developed in ABC required for equilibrium of the 100-kg sack. Neglect the size of the pulley at B.

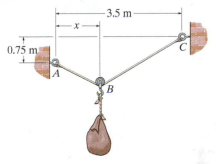

Prob. 3–36

3–37. A 4-kg sphere rests on the smooth parabolic surface. Determine the normal force it exerts on the surface and the mass m_B of block B needed to hold it in the equilibrium position shown.

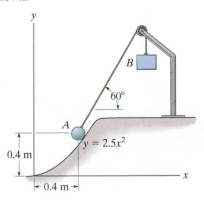

Prob. 3–37

3–38. Determine the forces in cables AC and AB needed to hold the 20-kg ball D in equilibrium. Take $F = 300$ N and $d = 1$ m.

3–39. The ball D has a mass of 20 kg. If a force of $F = 100$ N is applied horizontally to the ring at A, determine the dimension d so that the force in cable AC is zero.

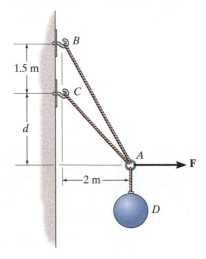

Probs. 3–38/39

*3–40. The 200-lb uniform container is suspended by means of a 6-ft-long cable, which is attached to the sides of the tank and passes over the small pulley located at O. If the cable can be attached at either points A and B, or C and D, determine which attachment produces the least amount of tension in the cable. What is this tension?

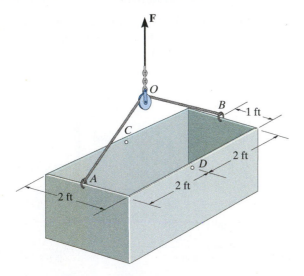

Prob. 3–40

3–41. The single elastic cord ABC is used to support the 40-lb load. Determine the position x and the tension in the cord that is required for equilibrium. The cord passes through the smooth ring at B and has an unstretched length of 6ft and stiffness of $k = 50$ lb/ft.

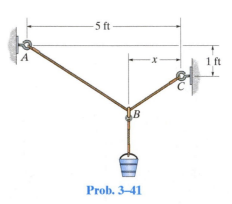

Prob. 3–41

3–42. A "scale" is constructed with a 4-ft-long cord and the 10-lb block D. The cord is fixed to a pin at A and passes over two *small* pulleys. Determine the weight of the suspended block B if the system is in equilibrium when $s = 1.5$ ft.

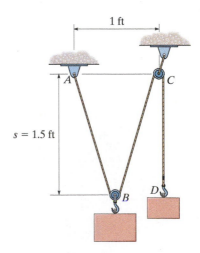

Prob. 3–42

CONCEPTUAL PROBLEMS

C3–1. The concrete wall panel is hoisted into position using the two cables *AB* and *AC* of equal length. Establish appropriate dimensions and use an equilibrium analysis to show that the longer the cables the less the force in each cable.

Prob. C3–1 (© Russell C. Hibbeler)

C3–2. The hoisting cables *BA* and *BC* each have a length of 20 ft. If the maximum tension that can be supported by each cable is 900 lb, determine the maximum distance *AC* between them in order to lift the uniform 1200-lb truss with constant velocity.

Prob. C3–2 (© Russell C. Hibbeler)

C3–3. The device *DB* is used to pull on the chain *ABC* to hold a door closed on the bin. If the angle between *AB* and *BC* is 30°, determine the angle between *DB* and *BC* for equilibrium.

Prob. C3–3 (© Russell C. Hibbeler)

C3–4. Chain *AB* is 1 m long and chain *AC* is 1.2 m long. If the distance *BC* is 1.5 m, and *AB* can support a maximum force of 2 kN, whereas *AC* can support a maximum force of 0.8 kN, determine the largest vertical force *F* that can be applied to the link at *A*.

Prob. C3–4 (© Russell C. Hibbeler)

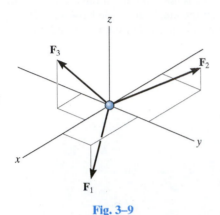

Fig. 3–9

3.4   Three-Dimensional Force Systems

In Section 3.1 we stated that the necessary and sufficient condition for particle equilibrium is

$$\Sigma \mathbf{F} = \mathbf{0} \qquad (3\text{–}4)$$

In the case of a three-dimensional force system, as in Fig. 3–9, we can resolve the forces into their respective $\mathbf{i}$, $\mathbf{j}$, $\mathbf{k}$ components, so that $\Sigma F_x \mathbf{i} + \Sigma F_y \mathbf{j} + \Sigma F_z \mathbf{k} = \mathbf{0}$. To satisfy this equation we require

$$\begin{aligned} \Sigma F_x &= 0 \\ \Sigma F_y &= 0 \\ \Sigma F_z &= 0 \end{aligned} \qquad (3\text{–}5)$$

These three equations state that the *algebraic sum* of the components of all the forces acting on the particle along each of the coordinate axes must be zero. Using them we can solve for at most three unknowns, generally represented as coordinate direction angles or magnitudes of forces shown on the particle's free-body diagram.

Procedure for Analysis

Three-dimensional force equilibrium problems for a particle can be solved using the following procedure.

Free-Body Diagram.

- Establish the x, y, z axes in any suitable orientation.

- Label all the known and unknown force magnitudes and directions on the diagram.

- The sense of a force having an unknown magnitude can be assumed.

Equations of Equilibrium.

- Use the scalar equations of equilibrium, $\Sigma F_x = 0$, $\Sigma F_y = 0$, $\Sigma F_z = 0$, in cases where it is easy to resolve each force into its x, y, z components.

- If the three-dimensional geometry appears difficult, then first express each force on the free-body diagram as a Cartesian vector, substitute these vectors into $\Sigma \mathbf{F} = \mathbf{0}$, and then set the $\mathbf{i}$, $\mathbf{j}$, $\mathbf{k}$ components equal to zero.

- If the solution for a force yields a negative result, this indicates that its sense is the reverse of that shown on the free-body diagram.

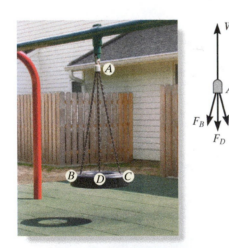

The joint at A is subjected to the force from the support as well as forces from each of the three chains. If the tire and any load on it have a weight W, then the force at the support will be $\mathbf{W}$, and the three scalar equations of equilibrium can be applied to the free-body diagram of the joint in order to determine the chain forces, $\mathbf{F}_B$, $\mathbf{F}_C$, and $\mathbf{F}_D$. (© Russell C. Hibbeler)

EXAMPLE 3.5

A 90-lb load is suspended from the hook shown in Fig. 3–10a. If the load is supported by two cables and a spring having a stiffness $k = 500$ lb/ft, determine the force in the cables and the stretch of the spring for equilibrium. Cable AD lies in the x–y plane and cable AC lies in the x–z plane.

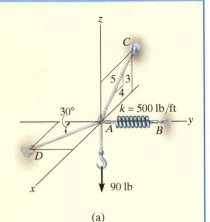

(a)

SOLUTION

The stretch of the spring can be determined once the force in the spring is determined.

Free-Body Diagram. The connection at A is chosen for the equilibrium analysis since the cable forces are concurrent at this point. The free-body diagram is shown in Fig. 3–10b.

Equations of Equilibrium. By inspection, each force can easily be resolved into its x, y, z components, and therefore the three scalar equations of equilibrium can be used. Considering components directed along each positive axis as "positive," we have

$$\Sigma F_x = 0; \qquad F_D \sin 30° - \left(\tfrac{4}{5}\right) F_C = 0 \qquad (1)$$

$$\Sigma F_y = 0; \qquad -F_D \cos 30° + F_B = 0 \qquad (2)$$

$$\Sigma F_z = 0; \qquad \left(\tfrac{3}{5}\right) F_C - 90 \text{ lb} = 0 \qquad (3)$$

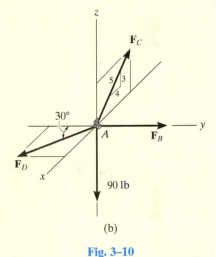

(b)

Fig. 3–10

Solving Eq. (3) for F_C, then Eq. (1) for F_D, and finally Eq. (2) for F_B, yields

$$F_C = 150 \text{ lb} \qquad \qquad Ans.$$

$$F_D = 240 \text{ lb} \qquad \qquad Ans.$$

$$F_B = 207.8 \text{ lb} = 208 \text{ lb} \qquad Ans.$$

The stretch of the spring is therefore

$$F_B = k s_{AB}$$

$$207.8 \text{ lb} = (500 \text{ lb/ft})(s_{AB})$$

$$s_{AB} = 0.416 \text{ ft} \qquad \qquad Ans.$$

NOTE: Since the results for all the cable forces are positive, each cable is in tension; that is, it pulls on point A as expected, Fig. 3–10b.

EXAMPLE | **3.6**

The 10-kg lamp in Fig. 3–11a is suspended from the three equal-length cords. Determine its smallest vertical distance s from the ceiling if the force developed in any cord is not allowed to exceed 50 N.

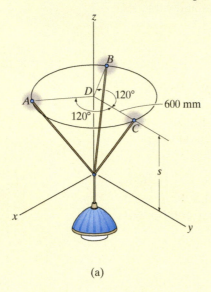

(a)

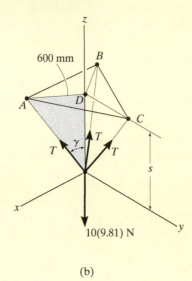

(b)

Fig. 3–11

SOLUTION

Free-Body Diagram. Due to symmetry, Fig. 3–11b, the distance $DA = DB = DC = 600$ mm. It follows that from $\Sigma F_x = 0$ and $\Sigma F_y = 0$, the tension T in each cord will be the same. Also, the angle between each cord and the z axis is γ.

Equation of Equilibrium. Applying the equilibrium equation along the z axis, with $T = 50$ N, we have

$$\Sigma F_z = 0; \qquad 3[(50\ \text{N})\cos\gamma] - 10(9.81)\ \text{N} = 0$$

$$\gamma = \cos^{-1}\frac{98.1}{150} = 49.16°$$

From the shaded triangle shown in Fig. 3–11b,

$$\tan 49.16° = \frac{600\ \text{mm}}{s}$$

$$s = 519\ \text{mm} \qquad\qquad Ans.$$

EXAMPLE 3.7

Determine the force in each cable used to support the 40-lb crate shown in Fig. 3–12a.

SOLUTION

Free-Body Diagram. As shown in Fig. 3–12b, the free-body diagram of point A is considered in order to "expose" the three unknown forces in the cables.

Equations of Equilibrium. First we will express each force in Cartesian vector form. Since the coordinates of points B and C are $B(-3 \text{ ft}, -4 \text{ ft}, 8 \text{ ft})$ and $C(-3 \text{ ft}, 4 \text{ ft}, 8 \text{ ft})$, we have

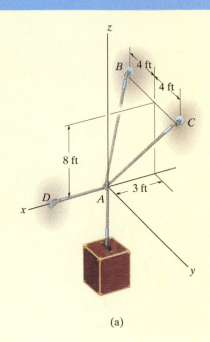

(a)

$$\mathbf{F}_B = F_B \left[\frac{-3\mathbf{i} - 4\mathbf{j} + 8\mathbf{k}}{\sqrt{(-3)^2 + (-4)^2 + (8)^2}} \right]$$

$$= -0.318F_B\mathbf{i} - 0.424F_B\mathbf{j} + 0.848F_B\mathbf{k}$$

$$\mathbf{F}_C = F_C \left[\frac{-3\mathbf{i} + 4\mathbf{j} + 8\mathbf{k}}{\sqrt{(-3)^2 + (4)^2 + (8)^2}} \right]$$

$$= -0.318F_C\mathbf{i} + 0.424F_C\mathbf{j} + 0.848F_C\mathbf{k}$$

$$\mathbf{F}_D = F_D\mathbf{i}$$

$$\mathbf{W} = \{-40\mathbf{k}\} \text{ lb}$$

Equilibrium requires

$$\Sigma\mathbf{F} = \mathbf{0}; \qquad \mathbf{F}_B + \mathbf{F}_C + \mathbf{F}_D + \mathbf{W} = \mathbf{0}$$

$$-0.318F_B\mathbf{i} - 0.424F_B\mathbf{j} + 0.848F_B\mathbf{k}$$

$$-0.318F_C\mathbf{i} + 0.424F_C\mathbf{j} + 0.848F_C\mathbf{k} + F_D\mathbf{i} - 40\mathbf{k} = \mathbf{0}$$

Equating the respective $\mathbf{i}, \mathbf{j}, \mathbf{k}$ components to zero yields

$$\Sigma F_x = 0; \qquad -0.318F_B - 0.318F_C + F_D = 0 \qquad (1)$$

$$\Sigma F_y = 0; \qquad -0.424F_B + 0.424F_C = 0 \qquad (2)$$

$$\Sigma F_z = 0; \qquad 0.848F_B + 0.848F_C - 40 = 0 \qquad (3)$$

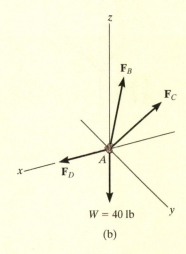

(b)

Fig. 3–12

Equation (2) states that $F_B = F_C$. Thus, solving Eq. (3) for F_B and F_C and substituting the result into Eq. (1) to obtain F_D, we have

$$F_B = F_C = 23.6 \text{ lb} \qquad \textit{Ans.}$$

$$F_D = 15.0 \text{ lb} \qquad \textit{Ans.}$$

EXAMPLE 3.8

Determine the tension in each cord used to support the 100-kg crate shown in Fig. 3–13a.

SOLUTION

Free-Body Diagram.   The force in each of the cords can be determined by investigating the equilibrium of point A. The free-body diagram is shown in Fig. 3–13b. The weight of the crate is $W = 100(9.81) = 981$ N.

Equations of Equilibrium.   Each force on the free-body diagram is first expressed in Cartesian vector form. Using Eq. 2–9 for $\mathbf{F}_C$ and noting point $D(-1$ m, 2 m, 2 m$)$ for $\mathbf{F}_D$, we have

$$\mathbf{F}_B = F_B \mathbf{i}$$

$$\mathbf{F}_C = F_C \cos 120°\mathbf{i} + F_C \cos 135°\mathbf{j} + F_C \cos 60°\mathbf{k}$$

$$= -0.5F_C\mathbf{i} - 0.707F_C\mathbf{j} + 0.5F_C\mathbf{k}$$

$$\mathbf{F}_D = F_D\left[\frac{-1\mathbf{i} + 2\mathbf{j} + 2\mathbf{k}}{\sqrt{(-1)^2 + (2)^2 + (2)^2}}\right]$$

$$= -0.333F_D\mathbf{i} + 0.667F_D\mathbf{j} + 0.667F_D\mathbf{k}$$

$$\mathbf{W} = \{-981\mathbf{k}\}\ \text{N}$$

Equilibrium requires

$$\Sigma\mathbf{F} = \mathbf{0};\qquad \mathbf{F}_B + \mathbf{F}_C + \mathbf{F}_D + \mathbf{W} = \mathbf{0}$$

$$F_B\mathbf{i} - 0.5F_C\mathbf{i} - 0.707F_C\mathbf{j} + 0.5F_C\mathbf{k}$$

$$-0.333F_D\mathbf{i} + 0.667F_D\mathbf{j} + 0.667F_D\mathbf{k} - 981\mathbf{k} = \mathbf{0}$$

Equating the respective $\mathbf{i}, \mathbf{j}, \mathbf{k}$ components to zero,

$$\Sigma F_x = 0;\qquad\qquad F_B - 0.5F_C - 0.333F_D = 0 \qquad (1)$$

$$\Sigma F_y = 0;\qquad\qquad -0.707F_C + 0.667F_D = 0 \qquad (2)$$

$$\Sigma F_z = 0;\qquad\qquad 0.5F_C + 0.667F_D - 981 = 0 \qquad (3)$$

Solving Eq. (2) for F_D in terms of F_C and substituting this into Eq. (3) yields F_C. F_D is then determined from Eq. (2). Finally, substituting the results into Eq. (1) gives F_B. Hence,

$$F_C = 813\ \text{N} \qquad\qquad\qquad Ans.$$

$$F_D = 862\ \text{N} \qquad\qquad\qquad Ans.$$

$$F_B = 694\ \text{N} \qquad\qquad\qquad Ans.$$

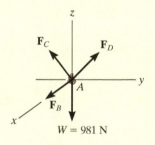

(a)

(b)

Fig. 3–13

FUNDAMENTAL PROBLEMS

All problem solutions must include an FBD.

F3–7. Determine the magnitude of forces $\mathbf{F}_1$, $\mathbf{F}_2$, $\mathbf{F}_3$, so that the particle is held in equilibrium.

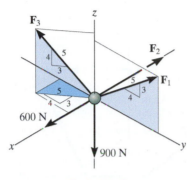

600 N

900 N

Prob. F3–7

F3–8. Determine the tension developed in cables AB, AC, and AD.

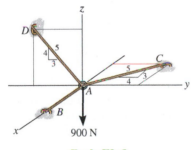

900 N

Prob. F3–8

F3–9. Determine the tension developed in cables AB, AC, and AD.

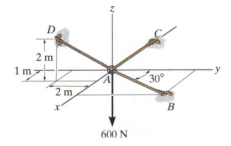

600 N

Prob. F3–9

F3–10. Determine the tension developed in cables AB, AC, and AD.

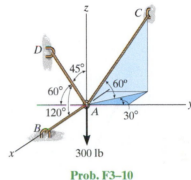

300 lb

Prob. F3–10

F3–11. The 150-lb crate is supported by cables AB, AC, and AD. Determine the tension in these wires.

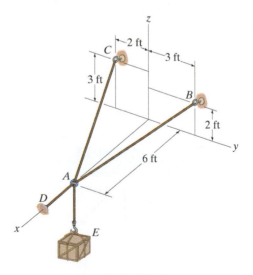

Prob. F3–11

PROBLEMS

All problem solutions must include an FBD.

3–43. The three cables are used to support the 40-kg flowerpot. Determine the force developed in each cable for equilibrium.

3–45. If the bucket and its contents have a total weight of 20 lb, determine the force in the supporting cables DA, DB, and DC.

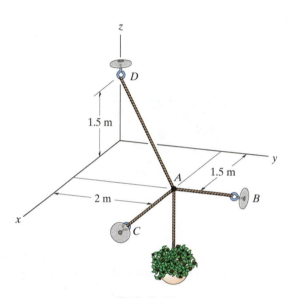

Prob. 3–43

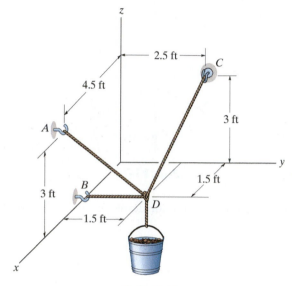

Prob. 3–45

*3–44. Determine the magnitudes of $\mathbf{F}_1$, $\mathbf{F}_2$, and $\mathbf{F}_3$ for equilibrium of the particle.

3–46. Determine the stretch in each of the two springs required to hold the 20-kg crate in the equilibrium position shown. Each spring has an unstretched length of 2 m and a stiffness of $k = 300$ N/m.

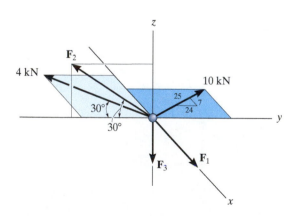

Prob. 3–44

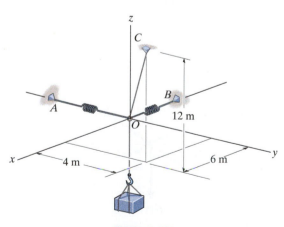

Prob. 3–46

3–47. Determine the force in each cable needed to support the 20-kg flowerpot.

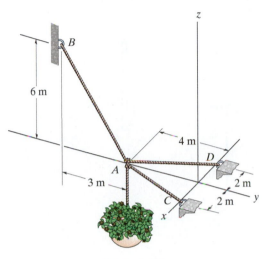

Prob. 3–47

3–50. Determine the force in each cable if $F = 500$ lb.

3–51. Determine the greatest force **F** that can be applied to the ring if each cable can support a maximum force of 800 lb.

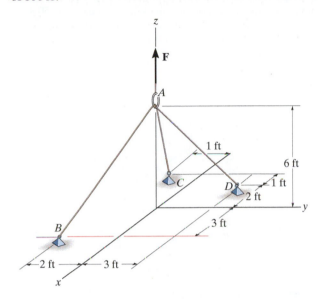

Probs. 3–50/51

*3–48. Determine the tension in the cables in order to support the 100-kg crate in the equilibrium position shown.

3–49. Determine the maximum mass of the crate so that the tension developed in any cable does not exceeded 3 kN.

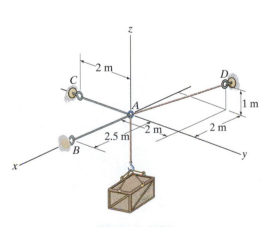

Probs. 3–48/49

*3–52. Determine the tension developed in cables AB and AC and the force developed along strut AD for equilibrium of the 400-lb crate.

3–53. If the tension developed in each cable cannot exceed 300 lb, determine the largest weight of the crate that can be supported. Also, what is the force developed along strut AD?

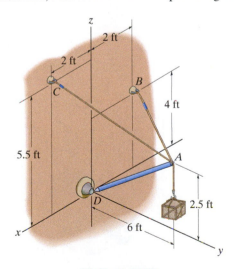

Probs. 3–52/53

3–54. Determine the tension developed in each cable for equilibrium of the 300-lb crate.

3–55. Determine the maximum weight of the crate that can be suspended from cables AB, AC, and AD so that the tension developed in any one of the cables does not exceed 250 lb.

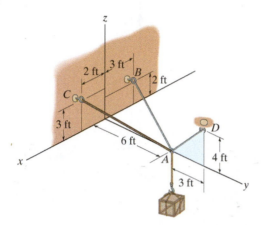

Probs. 3–54/55

***3–56.** The 25-kg flowerpot is supported at A by the three cords. Determine the force acting in each cord for equilibrium.

3–57. If each cord can sustain a maximum tension of 50 N before it fails, determine the greatest weight of the flowerpot the cords can support.

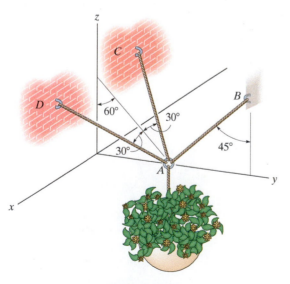

Probs. 3–56/57

3–58. Determine the tension developed in the three cables required to support the traffic light, which has a mass of 15 kg. Take $h = 4$ m.

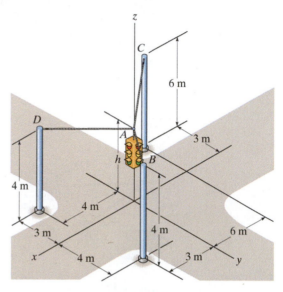

Prob. 3–58

3–59. Determine the tension developed in the three cables required to support the traffic light, which has a mass of 20 kg. Take $h = 3.5$ m.

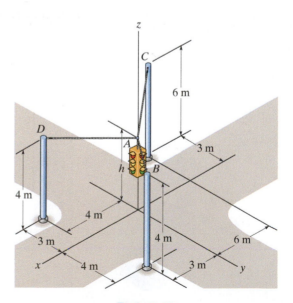

Prob. 3–59

*3–60. The 800-lb cylinder is supported by three chains as shown. Determine the force in each chain for equilibrium. Take $d = 1$ ft.

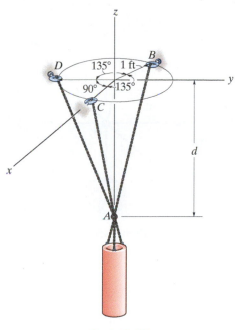

Prob. 3–60

3–61. Determine the tension in each cable for equilibrium.

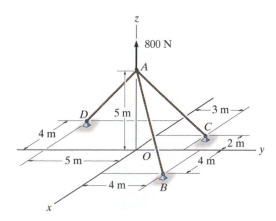

Prob. 3–61

3–62. If the maximum force in each rod can not exceed 1500 N, determine the greatest mass of the crate that can be supported.

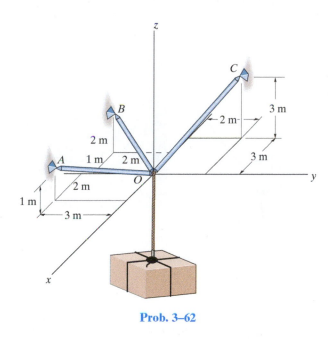

Prob. 3–62

3–63. The crate has a mass of 130 kg. Determine the tension developed in each cable for equilibrium.

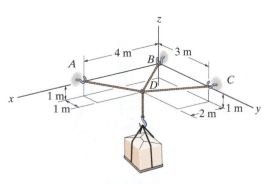

Prob. 3–63

*3–64. If cable AD is tightened by a turnbuckle and develops a tension of 1300 lb, determine the tension developed in cables AB and AC and the force developed along the antenna tower AE at point A.

3–66. Determine the tension developed in cables AB, AC, and AD required for equilibrium of the 300-lb crate.

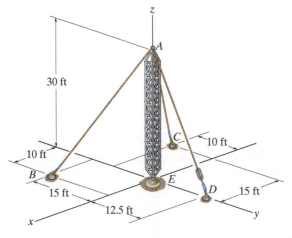

Prob. 3–64

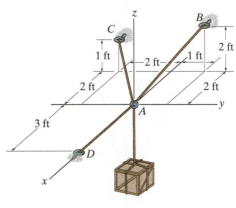

Prob. 3–66

3–65. If the tension developed in either cable AB or AC can not exceed 1000 lb, determine the maximum tension that can be developed in cable AD when it is tightened by the turnbuckle. Also, what is the force developed along the antenna tower at point A?

3–67. Determine the maximum weight of the crate so that the tension developed in any cable does not exceed 450 lb.

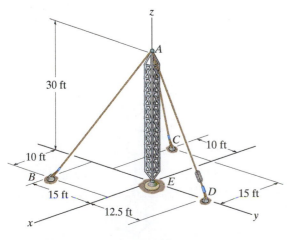

Prob. 3–65

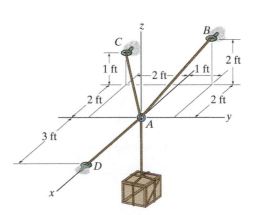

Prob. 3–67

CHAPTER REVIEW

Particle Equilibrium

When a particle is at rest or moves with constant velocity, it is in equilibrium. This requires that all the forces acting on the particle form a zero resultant force.

In order to account for all the forces that act on a particle, it is necessary to draw its free-body diagram. This diagram is an outlined shape of the particle that shows all the forces listed with their known or unknown magnitudes and directions.

$$F_R = \Sigma F = 0$$

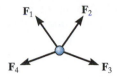

Two Dimensions

If the problem involves a linearly elastic spring, then the stretch or compression s of the spring can be related to the force applied to it.

$$F = ks$$

The tensile force developed in a *continuous cable* that passes over a frictionless pulley must have a *constant* magnitude throughout the cable to keep the cable in equilibrium.

The two scalar equations of force equilibrium can be applied with reference to an established x, y coordinate system.

$$\Sigma F_x = 0$$
$$\Sigma F_y = 0$$

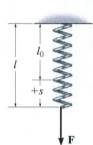

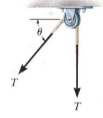

Three Dimensions

If the three-dimensional geometry is difficult to visualize, then the equilibrium equation should be applied using a Cartesian vector analysis. This requires first expressing each force on the free-body diagram as a Cartesian vector. When the forces are summed and set equal to zero, then the $\mathbf{i}$, $\mathbf{j}$, and $\mathbf{k}$ components are also zero.

$$\Sigma \mathbf{F} = \mathbf{0}$$

$$\Sigma F_x = 0$$
$$\Sigma F_y = 0$$
$$\Sigma F_z = 0$$

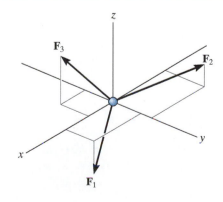

REVIEW PROBLEMS

All problem solutions must include an FBD.

R3–1.   The pipe is held in place by the vise. If the bolt exerts a force of 50 lb on the pipe in the direction shown, determine the forces F_A and F_B that the smooth contacts at A and B exert on the pipe.

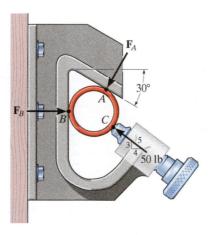

Prob. R3–1

R3–2.   Determine the maximum weight of the engine that can be supported without exceeding a tension of 450 lb in chain AB and 480 lb in chain AC.

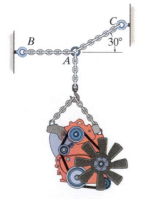

Prob. R3–2

R3–3.   Determine the maximum weight of the flowerpot that can be supported without exceeding a cable tension of 50 lb in either cable AB or AC.

Prob. R3–3

R3–4.   When y is zero, the springs sustain a force of 60 lb. Determine the magnitude of the applied vertical forces **F** and **−F** required to pull point A away from point B a distance of $y = 2$ ft. The ends of cords CAD and CBD are attached to rings at C and D.

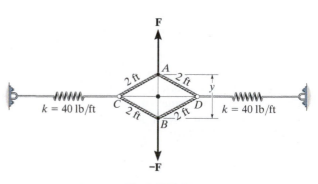

Prob. R3–4

R3–5. The joint of a space frame is subjected to four member forces. Member *OA* lies in the *x–y* plane and member *OB* lies in the *y–z* plane. Determine the force acting in each of the members required for equilibrium of the joint.

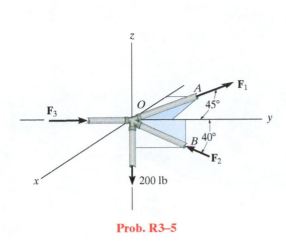

Prob. R3–5

R3–6. Determine the magnitudes of $\mathbf{F}_1$, $\mathbf{F}_2$, and $\mathbf{F}_3$ for equilibrium of the particle.

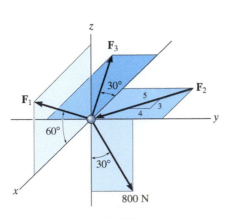

Prob. R3–6

R3–7. Determine the force in each cable needed to support the 500-lb load.

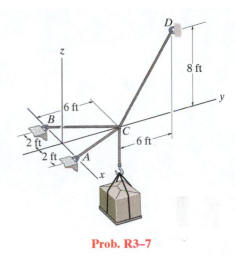

Prob. R3–7

R3–8. If cable *AB* is subjected to a tension of 700 N, determine the tension in cables *AC* and *AD* and the magnitude of the vertical force $\mathbf{F}$.

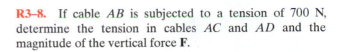

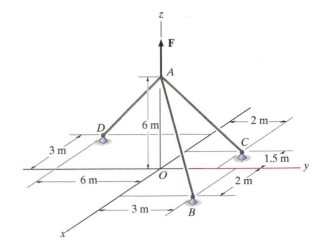

Prob. R3–8

Chapter 4

(© Rolf Adlercreutz/Alamy)

The force applied to this wrench will produce rotation or a tendency for rotation. This effect is called a moment, and in this chapter we will study how to determine the moment of a system of forces and calculate their resultants.

Force System Resultants

CHAPTER OBJECTIVES

- To discuss the concept of the moment of a force and show how to calculate it in two and three dimensions.
- To provide a method for finding the moment of a force about a specified axis.
- To define the moment of a couple.
- To show how to find the resultant effect of a nonconcurrent force system.
- To indicate how to reduce a simple distributed loading to a resultant force acting at a specified location.

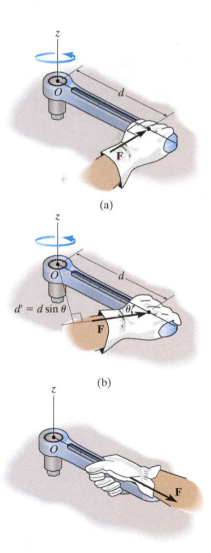

(a)

(b)

(c)

Fig. 4–1

4.1 Moment of a Force— Scalar Formulation

When a force is applied to a body it will produce a tendency for the body to rotate about a point that is not on the line of action of the force. This tendency to rotate is sometimes called a *torque*, but most often it is called the moment of a force or simply the *moment*. For example, consider a wrench used to unscrew the bolt in Fig. 4–1a. If a force is applied to the handle of the wrench it will tend to turn the bolt about point O (or the z axis). The magnitude of the moment is directly proportional to the magnitude of $\mathbf{F}$ and the perpendicular distance or *moment arm d*. The larger the force or the longer the moment arm, the greater the moment or turning effect. Note that if the force $\mathbf{F}$ is applied at an angle $\theta \neq 90°$, Fig. 4–1b, then it will be more difficult to turn the bolt since the moment arm $d' = d \sin \theta$ will be smaller than d. If $\mathbf{F}$ is applied along the wrench, Fig. 4–1c, its moment arm will be zero since the line of action of $\mathbf{F}$ will intersect point O (the z axis). As a result, the moment of $\mathbf{F}$ about O is also zero and no turning can occur.

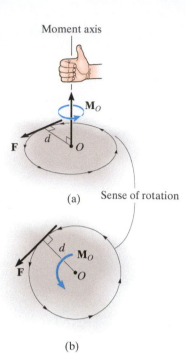

Moment axis

$\mathbf{M}_O$

F

d

O

(a)   Sense of rotation

d

$\mathbf{M}_O$

F

O

(b)

Fig. 4–2

We can generalize the above discussion and consider the force **F** and point O which lie in the shaded plane as shown in Fig. 4–2a. The moment $\mathbf{M}_O$ about point O, or about an axis passing through O and perpendicular to the plane, is a *vector quantity* since it has a specified magnitude and direction.

Magnitude.
The magnitude of $\mathbf{M}_O$ is

$$M_O = Fd \qquad (4\text{–}1)$$

where d is the **moment arm** or *perpendicular distance* from the axis at point O to the line of action of the force. Units of moment magnitude consist of force times distance, e.g., $N \cdot m$ or $lb \cdot ft$.

Direction.
The direction of $\mathbf{M}_O$ is defined by its **moment axis**, which is perpendicular to the plane that contains the force **F** and its moment arm d. The right-hand rule is used to establish the sense of direction of $\mathbf{M}_O$. According to this rule, the natural curl of the fingers of the right hand, as they are drawn towards the palm, represent the rotation, or if no movement is possible, there is a tendency for rotation caused by the moment. As this action is performed, the thumb of the right hand will give the directional sense of $\mathbf{M}_O$, Fig. 4–2a. Notice that the moment vector is represented three-dimensionally by a curl around an arrow. In two dimensions this vector is represented only by the curl as in Fig. 4–2b. Since in this case the moment will tend to cause a counterclockwise rotation, the moment vector is actually directed out of the page.

Resultant Moment.
For two-dimensional problems, where all the forces lie within the x–y plane, Fig. 4–3, the resultant moment $(\mathbf{M}_R)_O$ about point O (the z axis) can be determined by *finding the algebraic sum* of the moments caused by all the forces in the system. As a convention, we will generally consider *positive moments* as *counterclockwise* since they are directed along the positive z axis (out of the page). *Clockwise moments* will be *negative*. Doing this, the directional sense of each moment can be represented by a *plus* or *minus* sign. Using this sign convention, with a symbolic curl to define the positive direction, the resultant moment in Fig. 4–3 is therefore

$$\zeta + (M_R)_o = \Sigma Fd; \qquad (M_R)_o = F_1 d_1 - F_2 d_2 + F_3 d_3$$

If the numerical result of this sum is a positive scalar, $(\mathbf{M}_R)_o$ will be a counterclockwise moment (out of the page); and if the result is negative, $(\mathbf{M}_R)_o$ will be a clockwise moment (into the page).

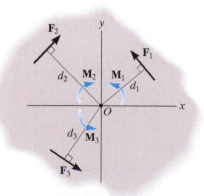

$\mathbf{F}_2$

y

$\mathbf{F}_1$

d_2 $\mathbf{M}_2$ $\mathbf{M}_1$

d_1

O x

d_3 $\mathbf{M}_3$

$\mathbf{F}_3$

Fig. 4–3

EXAMPLE | 4.1

For each case illustrated in Fig. 4–4, determine the moment of the force about point O.

SOLUTION *(SCALAR ANALYSIS)*

The line of action of each force is extended as a dashed line in order to establish the moment arm d. Also illustrated is the tendency of rotation of the member as caused by the force. Furthermore, the orbit of the force about O is shown as a colored curl. Thus,

Fig. 4–4a $M_O = (100 \text{ N})(2 \text{ m}) = 200 \text{ N} \cdot \text{m} \;\downarrow$ *Ans.*

Fig. 4–4b $M_O = (50 \text{ N})(0.75 \text{ m}) = 37.5 \text{ N} \cdot \text{m} \;\downarrow$ *Ans.*

Fig. 4–4c $M_O = (40 \text{ lb})(4 \text{ ft} + 2 \cos 30° \text{ ft}) = 229 \text{ lb} \cdot \text{ft} \;\downarrow$ *Ans.*

Fig. 4–4d $M_O = (60 \text{ lb})(1 \sin 45° \text{ ft}) = 42.4 \text{ lb} \cdot \text{ft} \;\circlearrowright$ *Ans.*

Fig. 4–4e $M_O = (7 \text{ kN})(4 \text{ m} - 1 \text{ m}) = 21.0 \text{ kN} \cdot \text{m} \;\circlearrowright$ *Ans.*

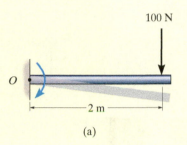

(a)

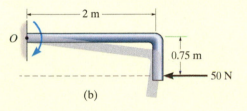

(b)

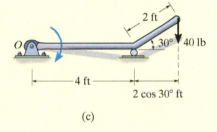

(c)

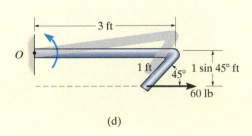

(d)

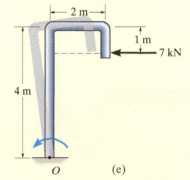

(e)

Fig. 4–4

EXAMPLE | **4.2**

Determine the resultant moment of the four forces acting on the rod shown in Fig. 4–5 about point O.

y

50 N

2 m | 2 m

60 N

x

30°

O

3 m

20 N

40 N

Fig. 4–5

SOLUTION

Assuming that positive moments act in the $+\mathbf{k}$ direction, i.e., counterclockwise, we have

$$\zeta + (M_R)_o = \Sigma Fd;$$

$$(M_R)_o = -50\text{ N}(2\text{ m}) + 60\text{ N}(0) + 20\text{ N}(3\sin 30°\text{ m})$$

$$-40\text{ N}(4\text{ m} + 3\cos 30°\text{ m})$$

$$(M_R)_o = -334\text{ N}\cdot\text{m} = 334\text{ N}\cdot\text{m}\,\zeta \qquad \textit{Ans.}$$

For this calculation, note how the moment-arm distances for the 20-N and 40-N forces are established from the extended (dashed) lines of action of each of these forces.

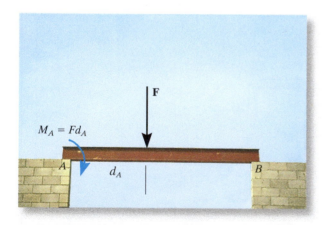

$M_A = Fd_A$

F

A

d_A

B

As illustrated by the example problems, the moment of a force does not always cause a rotation. For example, the force **F** tends to rotate the beam clockwise about its support at A with a moment $M_A = Fd_A$. The actual rotation would occur if the support at B were removed. (© Russell C. Hibbeler)

$\mathbf{F}_H$

O

$\mathbf{F}_N$

The ability to remove the nail will require the moment of $\mathbf{F}_H$ about point O to be larger than the moment of the force $\mathbf{F}_N$ about O that is needed to pull the nail out. (© Russell C. Hibbeler)

4.2   Cross Product

The moment of a force will be formulated using Cartesian vectors in the next section. Before doing this, however, it is first necessary to expand our knowledge of vector algebra and introduce the cross-product method of vector multiplication, first used by Willard Gibbs in lectures given in the late 19th century.

The ***cross product*** of two vectors **A** and **B** yields the vector **C**, which is written

$$\mathbf{C} = \mathbf{A} \times \mathbf{B} \qquad\qquad (4\text{--}2)$$

and is read "**C** equals **A** cross **B**."

Magnitude.   The *magnitude* of **C** is defined as the product of the magnitudes of **A** and **B** and the sine of the angle θ between their tails $(0° \leq \theta \leq 180°)$. Thus, $C = AB \sin \theta$.

Direction.   Vector **C** has a *direction* that is perpendicular to the plane containing **A** and **B** such that **C** is specified by the right-hand rule; i.e., curling the fingers of the right hand from vector **A** (cross) to vector **B**, the thumb points in the direction of **C**, as shown in Fig. 4–6.

Knowing both the magnitude and direction of **C**, we can write

$$\mathbf{C} = \mathbf{A} \times \mathbf{B} = (AB \sin \theta)\mathbf{u}_C \qquad\qquad (4\text{--}3)$$

where the scalar $AB \sin\theta$ defines the *magnitude* of **C** and the unit vector $\mathbf{u}_C$ defines the *direction* of **C**. The terms of Eq. 4–3 are illustrated graphically in Fig. 4–6.

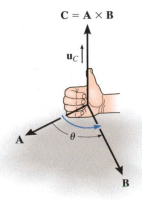

$$\mathbf{C} = \mathbf{A} \times \mathbf{B}$$

Fig. 4–6

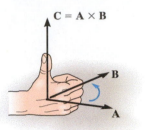

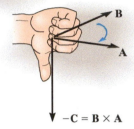

Fig. 4–7

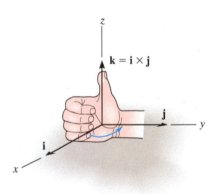

Fig. 4–8

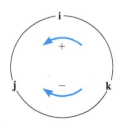

Fig. 4–9

Laws of Operation.

- The commutative law is *not* valid; i.e., $\mathbf{A} \times \mathbf{B} \neq \mathbf{B} \times \mathbf{A}$. Rather,

$$\mathbf{A} \times \mathbf{B} = -\mathbf{B} \times \mathbf{A}$$

This is shown in Fig. 4–7 by using the right-hand rule. The cross product $\mathbf{B} \times \mathbf{A}$ yields a vector that has the same magnitude but acts in the opposite direction to $\mathbf{C}$; i.e., $\mathbf{B} \times \mathbf{A} = -\mathbf{C}$.

- If the cross product is multiplied by a scalar a, it obeys the associative law;

$$a(\mathbf{A} \times \mathbf{B}) = (a\mathbf{A}) \times \mathbf{B} = \mathbf{A} \times (a\mathbf{B}) = (\mathbf{A} \times \mathbf{B})a$$

This property is easily shown since the magnitude of the resultant vector ($|a|AB \sin \theta$) and its direction are the same in each case.

- The vector cross product also obeys the distributive law of addition,

$$\mathbf{A} \times (\mathbf{B} + \mathbf{D}) = (\mathbf{A} \times \mathbf{B}) + (\mathbf{A} \times \mathbf{D})$$

- The proof of this identity is left as an exercise (see Prob. 4–1). It is important to note that *proper order* of the cross products must be maintained, since they are not commutative.

Cartesian Vector Formulation.

Equation 4–3 may be used to find the cross product of any pair of Cartesian unit vectors. For example, to find $\mathbf{i} \times \mathbf{j}$, the magnitude of the resultant vector is $(i)(j)(\sin 90°) = (1)(1)(1) = 1$, and its direction is determined using the right-hand rule. As shown in Fig. 4–8, the resultant vector points in the $+\mathbf{k}$ direction. Thus, $\mathbf{i} \times \mathbf{j} = (1)\mathbf{k}$. In a similar manner,

$$\mathbf{i} \times \mathbf{j} = \mathbf{k} \quad \mathbf{i} \times \mathbf{k} = -\mathbf{j} \quad \mathbf{i} \times \mathbf{i} = 0$$
$$\mathbf{j} \times \mathbf{k} = \mathbf{i} \quad \mathbf{j} \times \mathbf{i} = -\mathbf{k} \quad \mathbf{j} \times \mathbf{j} = 0$$
$$\mathbf{k} \times \mathbf{i} = \mathbf{j} \quad \mathbf{k} \times \mathbf{j} = -\mathbf{i} \quad \mathbf{k} \times \mathbf{k} = 0$$

These results should *not* be memorized; rather, it should be clearly understood how each is obtained by using the right-hand rule and the definition of the cross product. A simple scheme shown in Fig. 4–9 is helpful for obtaining the same results when the need arises. If the circle is constructed as shown, then "crossing" two unit vectors in a *counterclockwise* fashion around the circle yields the *positive* third unit vector; e.g., $\mathbf{k} \times \mathbf{i} = \mathbf{j}$. "Crossing" *clockwise*, a *negative* unit vector is obtained; e.g., $\mathbf{i} \times \mathbf{k} = -\mathbf{j}$.

Let us now consider the cross product of two general vectors **A** and **B** which are expressed in Cartesian vector form. We have

$$\mathbf{A} \times \mathbf{B} = (A_x\mathbf{i} + A_y\mathbf{j} + A_z\mathbf{k}) \times (B_x\mathbf{i} + B_y\mathbf{j} + B_z\mathbf{k})$$

$$= A_xB_x(\mathbf{i} \times \mathbf{i}) + A_xB_y(\mathbf{i} \times \mathbf{j}) + A_xB_z(\mathbf{i} \times \mathbf{k})$$

$$+ A_yB_x(\mathbf{j} \times \mathbf{i}) + A_yB_y(\mathbf{j} \times \mathbf{j}) + A_yB_z(\mathbf{j} \times \mathbf{k})$$

$$+ A_zB_x(\mathbf{k} \times \mathbf{i}) + A_zB_y(\mathbf{k} \times \mathbf{j}) + A_zB_z(\mathbf{k} \times \mathbf{k})$$

Carrying out the cross-product operations and combining terms yields

$$\mathbf{A} \times \mathbf{B} = (A_yB_z - A_zB_y)\mathbf{i} - (A_xB_z - A_zB_x)\mathbf{j} + (A_xB_y - A_yB_x)\mathbf{k} \qquad (4\text{–}4)$$

This equation may also be written in a more compact determinant form as

$$\mathbf{A} \times \mathbf{B} = \begin{vmatrix} \mathbf{i} & \mathbf{j} & \mathbf{k} \\ A_x & A_y & A_z \\ B_x & B_y & B_z \end{vmatrix} \qquad (4\text{–}5)$$

Thus, to find the cross product of any two Cartesian vectors **A** and **B**, it is necessary to expand a determinant whose first row of elements consists of the unit vectors **i**, **j**, and **k** and whose second and third rows represent the *x, y, z* components of the two vectors **A** and **B**, respectively.*

*A determinant having three rows and three columns can be expanded using three minors, each of which is multiplied by one of the three terms in the first row. There are four elements in each minor, for example,

$$\begin{vmatrix} A_{11} & A_{12} \\ A_{21} & A_{22} \end{vmatrix}$$

By *definition*, this determinant notation represents the terms $(A_{11}A_{22} - A_{12}A_{21})$, which is simply the product of the two elements intersected by the arrow slanting downward to the right $(A_{11}A_{22})$ *minus* the product of the two elements intersected by the arrow slanting downward to the left $(A_{12}A_{21})$. For a 3×3 determinant, such as Eq. 4–5, the three minors can be generated in accordance with the following scheme:

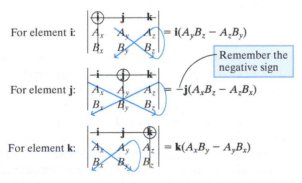

For element **i**: $\begin{vmatrix} \oplus & \mathbf{j} & \mathbf{k} \\ A_x & A_y & A_z \\ B_x & B_y & B_z \end{vmatrix} = \mathbf{i}(A_yB_z - A_zB_y)$

Remember the negative sign

For element **j**: $\begin{vmatrix} \mathbf{i} & \oplus & \mathbf{k} \\ A_x & A_y & A_z \\ B_x & B_y & B_z \end{vmatrix} = -\mathbf{j}(A_xB_z - A_zB_x)$

For element **k**: $\begin{vmatrix} \mathbf{i} & \mathbf{j} & \oplus \\ A_x & A_y & A_z \\ B_x & B_y & B_z \end{vmatrix} = \mathbf{k}(A_xB_y - A_yB_x)$

Adding the results and noting that the **j** element *must include the minus sign* yields the expanded form of **A** × **B** given by Eq. 4–4.

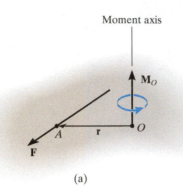

(a)

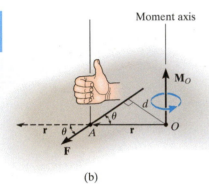

(b)

Fig. 4–10

4.3   Moment of a Force—Vector Formulation

The moment of a force $\mathbf{F}$ about point O, or actually about the moment axis passing through O and perpendicular to the plane containing O and $\mathbf{F}$, Fig. 4–10a, can be expressed using the vector cross product, namely,

$$\mathbf{M}_O = \mathbf{r} \times \mathbf{F} \qquad (4\text{–}6)$$

Here $\mathbf{r}$ represents a position vector directed *from* O to *any point* on the line of action of $\mathbf{F}$. We will now show that indeed the moment $\mathbf{M}_O$, when determined by this cross product, has the proper magnitude and direction.

Magnitude.   The magnitude of the cross product is defined from Eq. 4–3 as $M_O = rF \sin \theta$, where the angle θ is measured between the *tails* of $\mathbf{r}$ and $\mathbf{F}$. To establish this angle, $\mathbf{r}$ must be treated as a sliding vector so that θ can be constructed properly, Fig. 4–10b. Since the moment arm $d = r \sin \theta$, then

$$M_O = rF \sin \theta = F(r \sin \theta) = Fd$$

which agrees with Eq. 4–1.

Direction.   The direction and sense of $\mathbf{M}_O$ in Eq. 4–6 are determined by the right-hand rule as it applies to the cross product. Thus, sliding $\mathbf{r}$ to the dashed position and curling the right-hand fingers from $\mathbf{r}$ toward $\mathbf{F}$, "$\mathbf{r}$ cross $\mathbf{F}$," the thumb is directed upward or perpendicular to the plane containing $\mathbf{r}$ and $\mathbf{F}$ and this is in the *same direction* as $\mathbf{M}_O$, the moment of the force about point O, Fig. 4–10b. Note that the "curl" of the fingers, like the curl around the moment vector, indicates the sense of rotation caused by the force. Since the cross product does not obey the commutative law, the order of $\mathbf{r} \times \mathbf{F}$ must be maintained to produce the correct sense of direction for $\mathbf{M}_O$.

Principle of Transmissibility.   The cross product operation is often used in three dimensions since the perpendicular distance or moment arm from point O to the line of action of the force is not needed. In other words, we can use any position vector $\mathbf{r}$ measured from point O to any point on the line of action of the force $\mathbf{F}$, Fig. 4–11. Thus,

$$\mathbf{M}_O = \mathbf{r}_1 \times \mathbf{F} = \mathbf{r}_2 \times \mathbf{F} = \mathbf{r}_3 \times \mathbf{F}$$

Since $\mathbf{F}$ can be applied at any point along its line of action and still create this *same moment* about point O, then $\mathbf{F}$ can be considered a *sliding vector*. This property is called the *principle of transmissibility* of a force.

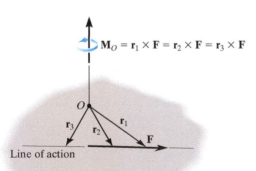

Fig. 4–11

Cartesian Vector Formulation.   If we establish x, y, z coordinate axes, then the position vector **r** and force **F** can be expressed as Cartesian vectors, Fig. 4–12a. Applying Eq. 4–5 we have

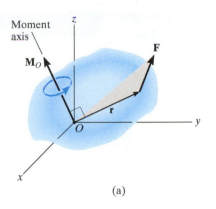

Moment axis

(a)

$$M_O = r \times F = \begin{vmatrix} i & j & k \\ r_x & r_y & r_z \\ F_x & F_y & F_z \end{vmatrix} \qquad (4\text{–}7)$$

where

r_x, r_y, r_z represent the x, y, z components of the position vector drawn from point O to *any point* on the line of action of the force

F_x, F_y, F_z represent the x, y, z components of the force vector

If the determinant is expanded, then like Eq. 4–4 we have

$$M_O = (r_yF_z - r_zF_y)i - (r_xF_z - r_zF_x)j + (r_xF_y - r_yF_x)k \qquad (4\text{–}8)$$

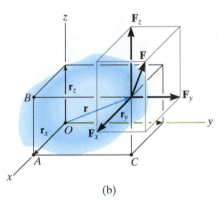

(b)

Fig. 4–12

The physical meaning of these three moment components becomes evident by studying Fig. 4–12b. For example, the **i** component of M_O can be determined from the moments of F_x, F_y, and F_z about the x axis. The component F_x does *not* create a moment or tendency to cause turning about the x axis since this force is *parallel* to the x axis. The line of action of F_y passes through point B, and so the magnitude of the moment of F_y about point A on the x axis is r_zF_y. By the right-hand rule this component acts in the *negative* **i** direction. Likewise, F_z passes through point C and so it contributes a moment component of r_yF_z**i** about the x axis. Thus, $(M_O)_x = (r_yF_z - r_zF_y)$ as shown in Eq. 4–8. As an exercise, establish the **j** and **k** components of M_O in this manner and show that indeed the expanded form of the determinant, Eq. 4–8, represents the moment of **F** about point O. Once M_O is determined, realize that it will always be *perpendicular* to the shaded plane containing vectors **r** and **F**, Fig. 4–12a.

Resultant Moment of a System of Forces.   If a body is acted upon by a system of forces, Fig. 4–13, the resultant moment of the forces about point O can be determined by vector addition of the moment of each force. This resultant can be written symbolically as

$$(M_R)_o = \Sigma(r \times F) \qquad (4\text{–}9)$$

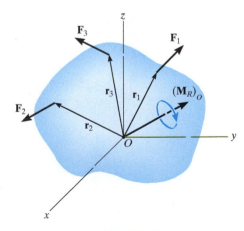

Fig. 4–13

EXAMPLE | **4.3**

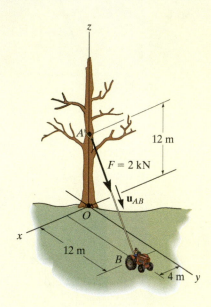

(a)

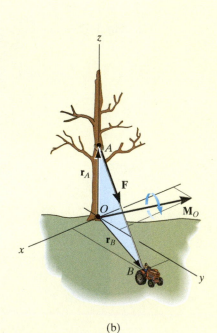

(b)

Fig. 4–14

Determine the moment produced by the force **F** in Fig. 4–14*a* about point *O*. Express the result as a Cartesian vector.

SOLUTION

As shown in Fig. 4–14*b*, either $\mathbf{r}_A$ or $\mathbf{r}_B$ can be used to determine the moment about point *O*. These position vectors are

$$\mathbf{r}_A = \{12\mathbf{k}\} \text{ m} \quad \text{and} \quad \mathbf{r}_B = \{4\mathbf{i} + 12\mathbf{j}\} \text{ m}$$

Force **F** expressed as a Cartesian vector is

$$\mathbf{F} = F\mathbf{u}_{AB} = 2 \text{ kN} \left[\frac{\{4\mathbf{i} + 12\mathbf{j} - 12\mathbf{k}\} \text{ m}}{\sqrt{(4 \text{ m})^2 + (12 \text{ m})^2 + (-12 \text{ m})^2}} \right]$$

$$= \{0.4588\mathbf{i} + 1.376\mathbf{j} - 1.376\mathbf{k}\} \text{ kN}$$

Thus

$$\mathbf{M}_O = \mathbf{r}_A \times \mathbf{F} = \begin{vmatrix} \mathbf{i} & \mathbf{j} & \mathbf{k} \\ 0 & 0 & 12 \\ 0.4588 & 1.376 & -1.376 \end{vmatrix}$$

$$= [0(-1.376) - 12(1.376)]\mathbf{i} - [0(-1.376) - 12(0.4588)]\mathbf{j}$$
$$+ [0(1.376) - 0(0.4588)]\mathbf{k}$$

$$= \{-16.5\mathbf{i} + 5.51\mathbf{j}\} \text{ kN} \cdot \text{m} \qquad \textit{Ans.}$$

or

$$\mathbf{M}_O = \mathbf{r}_B \times \mathbf{F} = \begin{vmatrix} \mathbf{i} & \mathbf{j} & \mathbf{k} \\ 4 & 12 & 0 \\ 0.4588 & 1.376 & -1.376 \end{vmatrix}$$

$$= [12(-1.376) - 0(1.376)]\mathbf{i} - [4(-1.376) - 0(0.4588)]\mathbf{j}$$
$$+ [4(1.376) - 12(0.4588)]\mathbf{k}$$

$$= \{-16.5\mathbf{i} + 5.51\mathbf{j}\} \text{ kN} \cdot \text{m} \qquad \textit{Ans.}$$

NOTE: As shown in Fig. 4–14*b*, $\mathbf{M}_O$ acts perpendicular to the plane that contains **F**, $\mathbf{r}_A$, and $\mathbf{r}_B$. Had this problem been worked using $M_O = Fd$, notice the difficulty that would arise in obtaining the moment arm *d*.

EXAMPLE 4.4

Two forces act on the rod shown in Fig. 4–15a. Determine the resultant moment they create about the flange at O. Express the result as a Cartesian vector.

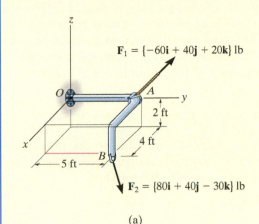

$\mathbf{F}_1 = \{-60\mathbf{i} + 40\mathbf{j} + 20\mathbf{k}\}$ lb

$\mathbf{F}_2 = \{80\mathbf{i} + 40\mathbf{j} - 30\mathbf{k}\}$ lb

(a)

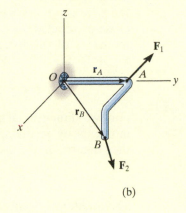

(b)

SOLUTION

Position vectors are directed from point O to each force as shown in Fig. 4–15b. These vectors are

$$\mathbf{r}_A = \{5\mathbf{j}\} \text{ ft}$$

$$\mathbf{r}_B = \{4\mathbf{i} + 5\mathbf{j} - 2\mathbf{k}\} \text{ ft}$$

The resultant moment about O is therefore

$$(\mathbf{M}_R)_O = \Sigma(\mathbf{r} \times \mathbf{F})$$

$$= \mathbf{r}_A \times \mathbf{F}_1 + \mathbf{r}_B \times \mathbf{F}_2$$

$$= \begin{vmatrix} \mathbf{i} & \mathbf{j} & \mathbf{k} \\ 0 & 5 & 0 \\ -60 & 40 & 20 \end{vmatrix} + \begin{vmatrix} \mathbf{i} & \mathbf{j} & \mathbf{k} \\ 4 & 5 & -2 \\ 80 & 40 & -30 \end{vmatrix}$$

$$= [5(20) - 0(40)]\mathbf{i} - [0]\mathbf{j} + [0(40) - (5)(-60)]\mathbf{k}$$

$$+ [5(-30) - (-2)(40)]\mathbf{i} - [4(-30) - (-2)(80)]\mathbf{j} + [4(40) - 5(80)]\mathbf{k}$$

$$= \{30\mathbf{i} - 40\mathbf{j} + 60\mathbf{k}\} \text{ lb} \cdot \text{ft} \qquad \textit{Ans.}$$

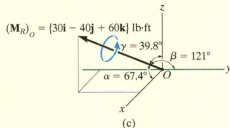

$(\mathbf{M}_R)_O = \{30\mathbf{i} - 40\mathbf{j} + 60\mathbf{k}\}$ lb·ft

$\gamma = 39.8°$

$\beta = 121°$

$\alpha = 67.4°$

(c)

Fig. 4–15

NOTE: This result is shown in Fig. 4–15c. The coordinate direction angles were determined from the unit vector for $(\mathbf{M}_R)_O$. Realize that the two forces tend to cause the rod to rotate about the moment axis in the manner shown by the curl indicated on the moment vector.

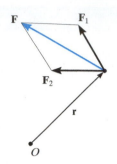

Fig. 4–16

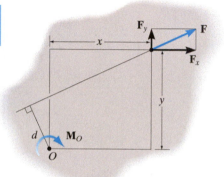

Fig. 4–17

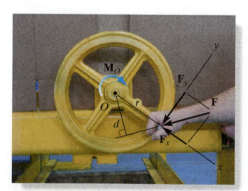

The moment of the force about point O is $M_O = Fd$. But it is easier to find this moment using $M_O = F_x(0) + F_yr = F_yr$. (© Russell C. Hibbeler)

4.4   Principle of Moments

A concept often used in mechanics is the *principle of moments*, which is sometimes referred to as **Varignon's theorem** since it was orginally developed by the French mathematician Pierre Varignon (1654–1722). It states that *the moment of a force about a point is equal to the sum of the moments of the components of the force about the point*. This theorem can be proven easily using the vector cross product since the cross product obeys the *distributive law*. For example, consider the moments of the force **F** and two of its components about point O, Fig. 4–16. Since $\mathbf{F} = \mathbf{F}_1 + \mathbf{F}_2$ we have

$$\mathbf{M}_O = \mathbf{r} \times \mathbf{F} = \mathbf{r} \times (\mathbf{F}_1 + \mathbf{F}_2) = \mathbf{r} \times \mathbf{F}_1 + \mathbf{r} \times \mathbf{F}_2$$

For two-dimensional problems, Fig. 4–17, we can use the principle of moments by resolving the force into its rectangular components and then determine the moment using a scalar analysis. Thus,

$$M_O = F_xy - F_yx$$

This method is generally easier than finding the same moment using $M_O = Fd$.

Important Points

- The moment of a force creates the tendency of a body to turn about an axis passing through a specific point O.

- Using the right-hand rule, the sense of rotation is indicated by the curl of the fingers, and the thumb is directed along the moment axis, or line of action of the moment.

- The magnitude of the moment is determined from $M_O = Fd$, where d is called the moment arm, which represents the perpendicular or shortest distance from point O to the line of action of the force.

- In three dimensions the vector cross product is used to determine the moment, i.e., $\mathbf{M}_O = \mathbf{r} \times \mathbf{F}$. Remember that $\mathbf{r}$ is directed *from* point O *to any point* on the line of action of **F**.

EXAMPLE 4.5

Determine the moment of the force in Fig. 4–18a about point O.

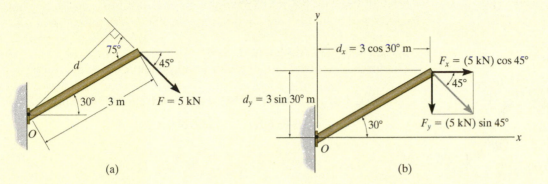

(a)   (b)

SOLUTION I

The moment arm d in Fig. 4–18a can be found from trigonometry.

$$d = (3 \text{ m}) \sin 75° = 2.898 \text{ m}$$

Thus,

$$M_O = Fd = (5 \text{ kN})(2.898 \text{ m}) = 14.5 \text{ kN} \cdot \text{m} \, \curvearrowright \qquad \textit{Ans.}$$

Since the force tends to rotate or orbit clockwise about point O, the moment is directed into the page.

SOLUTION II

The x and y components of the force are indicated in Fig. 4–18b. Considering counterclockwise moments as positive, and applying the principle of moments, we have

$$\curvearrowleft + M_O = -F_x d_y - F_y d_x$$

$$= -(5 \cos 45° \text{ kN})(3 \sin 30° \text{ m}) - (5 \sin 45° \text{ kN})(3 \cos 30° \text{ m})$$

$$= -14.5 \text{ kN} \cdot \text{m} = 14.5 \text{ kN} \cdot \text{m} \, \curvearrowright \qquad \textit{Ans.}$$

SOLUTION III

The x and y axes can be set parallel and perpendicular to the rod's axis as shown in Fig. 4–18c. Here $\mathbf{F}_x$ produces no moment about point O since its line of action passes through this point. Therefore,

$$\curvearrowleft + M_O = -F_y d_x$$

$$= -(5 \sin 75° \text{ kN})(3 \text{ m})$$

$$= -14.5 \text{ kN} \cdot \text{m} = 14.5 \text{ kN} \cdot \text{m} \, \curvearrowright \qquad \textit{Ans.}$$

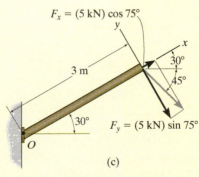

(c)

Fig. 4–18

EXAMPLE 4.6

Force **F** acts at the end of the angle bracket in Fig. 4–19a. Determine the moment of the force about point O.

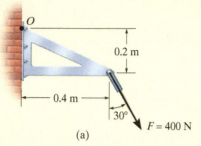

(a)

SOLUTION I *(SCALAR ANALYSIS)*
The force is resolved into its x and y components, Fig. 4–19b, then

$$\zeta + M_O = 400 \sin 30° \text{ N}(0.2 \text{ m}) - 400 \cos 30° \text{ N}(0.4 \text{ m})$$

$$= -98.6 \text{ N} \cdot \text{m} = 98.6 \text{ N} \cdot \text{m} \; \rotatebox{0}{$\curvearrowright$}$$

or

$$M_O = \{-98.6\mathbf{k}\} \text{ N} \cdot \text{m} \qquad\qquad Ans.$$

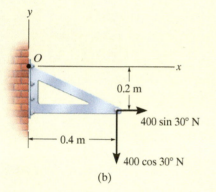

(b)

SOLUTION II *(VECTOR ANALYSIS)*
Using a Cartesian vector approach, the force and position vectors, Fig. 4–19c, are

$$\mathbf{r} = \{0.4\mathbf{i} - 0.2\mathbf{j}\} \text{ m}$$

$$\mathbf{F} = \{400 \sin 30°\mathbf{i} - 400 \cos 30°\mathbf{j}\} \text{ N}$$

$$= \{200.0\mathbf{i} - 346.4\mathbf{j}\} \text{ N}$$

The moment is therefore

$$M_O = \mathbf{r} \times \mathbf{F} = \begin{vmatrix} \mathbf{i} & \mathbf{j} & \mathbf{k} \\ 0.4 & -0.2 & 0 \\ 200.0 & -346.4 & 0 \end{vmatrix}$$

$$= 0\mathbf{i} - 0\mathbf{j} + [0.4(-346.4) - (-0.2)(200.0)]\mathbf{k}$$

$$= \{-98.6\mathbf{k}\} \text{ N} \cdot \text{m} \qquad\qquad Ans.$$

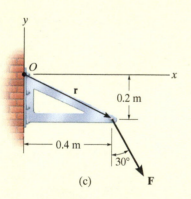

(c)

Fig. 4–19

NOTE: It is seen that the scalar analysis (Solution I) provides a more *convenient method* for analysis than Solution II since the direction of the moment and the moment arm for each component force are easy to establish. Hence, this method is generally recommended for solving problems displayed in two dimensions, whereas a Cartesian vector analysis is generally recommended only for solving three-dimensional problems.

PRELIMINARY PROBLEMS

P4–1. In each case, determine the moment of the force about point O.

P4–2. In each case, set up the determinant to find the moment of the force about point P.

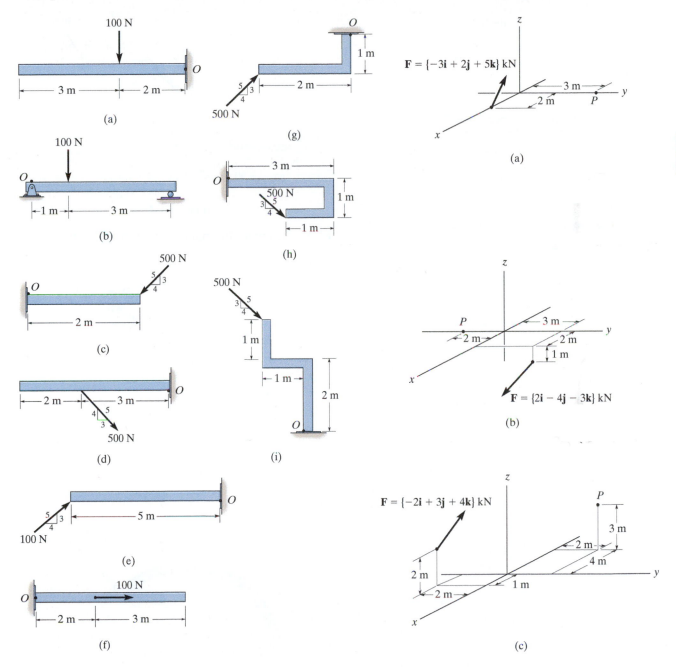

(a)

(b)

(c)

(d)

(e)

(f)

(g)

(h)

(i)

$\mathbf{F} = \{-3\mathbf{i} + 2\mathbf{j} + 5\mathbf{k}\} \text{ kN}$

(a)

$\mathbf{F} = \{2\mathbf{i} - 4\mathbf{j} - 3\mathbf{k}\} \text{ kN}$

(b)

$\mathbf{F} = \{-2\mathbf{i} + 3\mathbf{j} + 4\mathbf{k}\} \text{ kN}$

(c)

Prob. P4–1

Prob. P4–2

FUNDAMENTAL PROBLEMS

F4–1. Determine the moment of the force about point O.

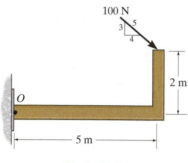

Prob. F4–1

F4–2. Determine the moment of the force about point O.

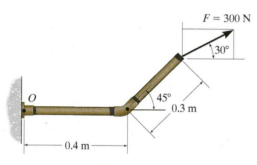

Prob. F4–2

F4–3. Determine the moment of the force about point O.

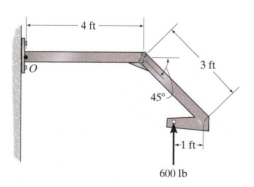

Prob. F4–3

F4–4. Determine the moment of the force about point O. Neglect the thickness of the member.

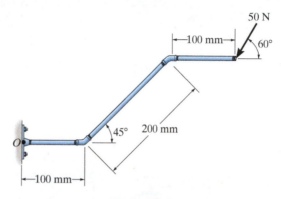

Prob. F4–4

F4–5. Determine the moment of the force about point O.

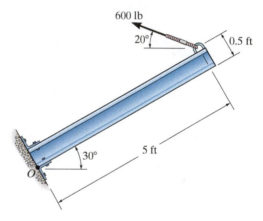

Prob. F4–5

F4–6. Determine the moment of the force about point O.

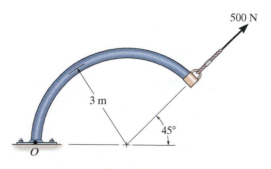

Prob. F4–6

F4–7. Determine the resultant moment produced by the forces about point O.

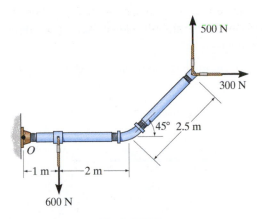

500 N

300 N

45° 2.5 m

O

1 m   2 m

600 N

Prob. F4–7

F4–8. Determine the resultant moment produced by the forces about point O.

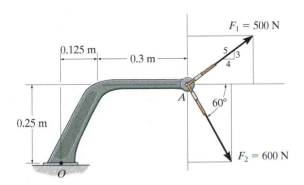

$F_1 = 500$ N

0.125 m

0.3 m

5 3 4

0.25 m

A 60°

O

$F_2 = 600$ N

Prob. F4–8

F4–9. Determine the resultant moment produced by the forces about point O.

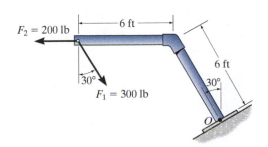

$F_2 = 200$ lb

6 ft

6 ft

30°

$F_1 = 300$ lb   30°

O

Prob. F4–9

F4–10. Determine the moment of force $\mathbf{F}$ about point O. Express the result as a Cartesian vector.

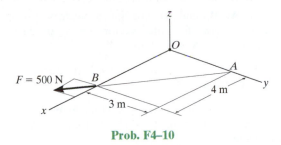

z

O

A

$F = 500$ N B

4 m

3 m

x

y

Prob. F4–10

F4–11. Determine the moment of force $\mathbf{F}$ about point O. Express the result as a Cartesian vector.

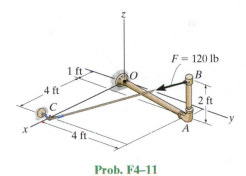

z

1 ft

4 ft O $F = 120$ lb

B

C

2 ft

x 4 ft A y

Prob. F4–11

F4–12. If the two forces $\mathbf{F}_1 = \{100\mathbf{i} - 120\mathbf{j} + 75\mathbf{k}\}$ lb and $\mathbf{F}_2 = \{-200\mathbf{i} + 250\mathbf{j} + 100\mathbf{k}\}$ lb act at A, determine the resultant moment produced by these forces about point O. Express the result as a Cartesian vector.

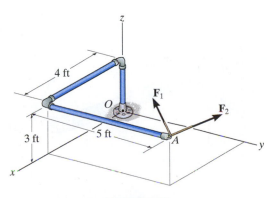

z

4 ft

F_1

O F_2

3 ft   5 ft

A

x y

Prob. F4–12

PROBLEMS

4–1. If **A**, **B**, and **D** are given vectors, prove the distributive law for the vector cross product, i.e., $\mathbf{A} \times (\mathbf{B} + \mathbf{D}) = (\mathbf{A} \times \mathbf{B}) + (\mathbf{A} \times \mathbf{D})$.

4–2. Prove the triple scalar product identity $\mathbf{A} \cdot (\mathbf{B} \times \mathbf{C}) = (\mathbf{A} \times \mathbf{B}) \cdot \mathbf{C}$.

4–3. Given the three nonzero vectors **A**, **B**, and **C**, show that if $\mathbf{A} \cdot (\mathbf{B} \times \mathbf{C}) = 0$, the three vectors *must* lie in the same plane.

*4–4. Determine the moment about point A of each of the three forces acting on the beam.

4–5. Determine the moment about point B of each of the three forces acting on the beam.

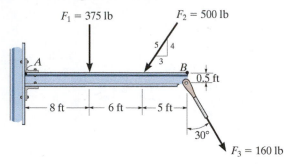

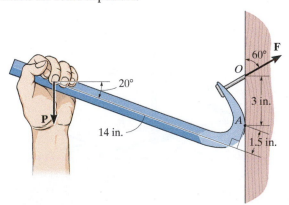

Probs. 4–4/5

4–6. The crowbar is subjected to a vertical force of $P = 25$ lb at the grip, whereas it takes a force of $F = 155$ lb at the claw to pull the nail out. Find the moment of each force about point A and determine if **P** is sufficient to pull out the nail. The crowbar contacts the board at point A.

Prob. 4–6

4–7. Determine the moment of each of the three forces about point A.

*4–8. Determine the moment of each of the three forces about point B.

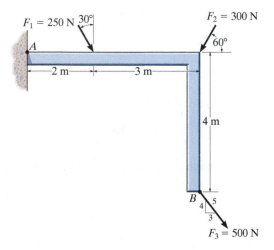

Probs. 4–7/8

4–9. Determine the moment of each force about the bolt located at A. Take $F_B = 40$ lb, $F_C = 50$ lb.

4–10. If $F_B = 30$ lb and $F_C = 45$ lb, determine the resultant moment about the bolt located at A.

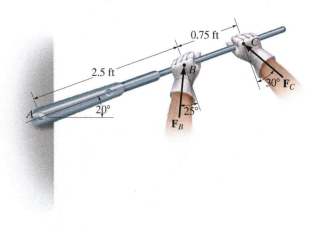

Probs. 4–9/10

4–11. The towline exerts a force of $P = 6$ kN at the end of the 8-m-long crane boom. If $\theta = 30°$, determine the placement x of the hook at B so that this force creates a maximum moment about point O. What is this moment?

*4–12. The towline exerts a force of $P = 6$ kN at the end of the 8-m-long crane boom. If $x = 10$ m, determine the position θ of the boom so that this force creates a maximum moment about point O. What is this moment?

4–15. Two men exert forces of $F = 80$ lb and $P = 50$ lb on the ropes. Determine the moment of each force about A. Which way will the pole rotate, clockwise or counterclockwise?

*4–16. If the man at B exerts a force of $P = 30$ lb on his rope, determine the magnitude of the force $\mathbf{F}$ the man at C must exert to prevent the pole from rotating, i.e., so the resultant moment about A of both forces is zero.

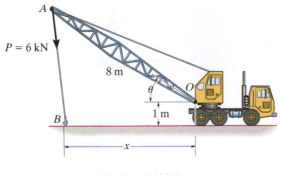

Probs. 4–11/12

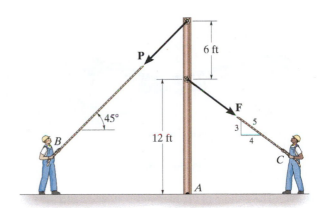

Probs. 4–15/16

4–13. The 20-N horizontal force acts on the handle of the socket wrench. What is the moment of this force about point B. Specify the coordinate direction angles α, β, γ of the moment axis.

4–14. The 20-N horizontal force acts on the handle of the socket wrench. Determine the moment of this force about point O. Specify the coordinate direction angles α, β, γ of the moment axis.

4–17. The torque wrench ABC is used to measure the moment or torque applied to a bolt when the bolt is located at A and a force is applied to the handle at C. The mechanic reads the torque on the scale at B. If an extension AO of length d is used on the wrench, determine the required scale reading if the desired torque on the bolt at O is to be M.

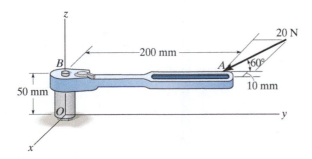

Probs. 4–13/14

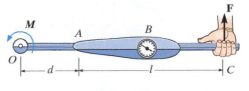

Prob. 4–17

4–18. The tongs are used to grip the ends of the drilling pipe P. Determine the torque (moment) M_P that the applied force $F = 150$ lb exerts on the pipe about point P as a function of θ. Plot this moment M_P versus θ for $0 \leq \theta \leq 90°$.

4–19. The tongs are used to grip the ends of the drilling pipe P. If a torque (moment) of $M_P = 800$ lb · ft is needed at P to turn the pipe, determine the cable force F that must be applied to the tongs. Set $\theta = 30°$.

4–22. Old clocks were constructed using a *fusee B* to drive the gears and watch hands. The purpose of the fusee is to increase the leverage developed by the mainspring A as it uncoils and thereby loses some of its tension. The mainspring can develop a torque (moment) $T_s = k\theta$, where $k = 0.015$ N · m/rad is the torsional stiffness and θ is the angle of twist of the spring in radians. If the torque T_f developed by the fusee is to remain constant as the mainspring winds down, and $x = 10$ mm when $\theta = 4$ rad, determine the required radius of the fusee when $\theta = 3$ rad.

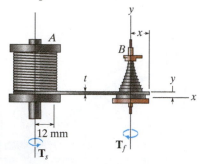

Prob. 4–22

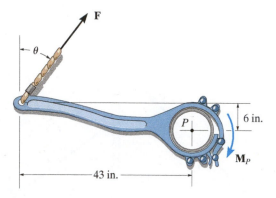

Probs. 4–18/19

*4–20. The handle of the hammer is subjected to the force of $F = 20$ lb. Determine the moment of this force about the point A.

4–21. In order to pull out the nail at B, the force $\mathbf{F}$ exerted on the handle of the hammer must produce a clockwise moment of 500 lb · in. about point A. Determine the required magnitude of force $\mathbf{F}$.

4–23. The tower crane is used to hoist the 2-Mg load upward at constant velocity. The 1.5-Mg jib BD, 0.5-Mg jib BC, and 6-Mg counterweight C have centers of mass at G_1, G_2, and G_3, respectively. Determine the resultant moment produced by the load and the weights of the tower crane jibs about point A and about point B.

*4–24. The tower crane is used to hoist a 2-Mg load upward at constant velocity. The 1.5-Mg jib BD and 0.5-Mg jib BC have centers of mass at G_1 and G_2, respectively. Determine the required mass of the counterweight C so that the resultant moment produced by the load and the weight of the tower crane jibs about point A is zero. The center of mass for the counterweight is located at G_3.

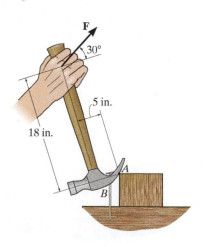

Probs. 4–20/21

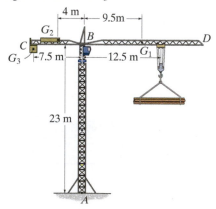

Probs. 4–23/24

4–25. If the 1500-lb boom AB, the 200-lb cage BCD, and the 175-lb man have centers of gravity located at points G_1, G_2, and G_3, respectively, determine the resultant moment produced by each weight about point A.

4–26. If the 1500-lb boom AB, the 200-lb cage BCD, and the 175-lb man have centers of gravity located at points G_1, G_2, and G_3, respectively, determine the resultant moment produced by all the weights about point A.

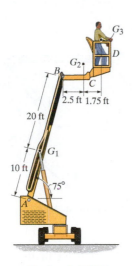

Probs. 4–25/26

4–27. Determine the moment of the force **F** about point O. Express the result as a Cartesian vector.

*4–28. Determine the moment of the force **F** about point P. Express the result as a Cartesian vector.

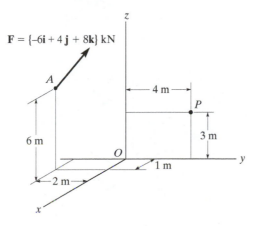

Probs. 4–27/28

4–29. The force $\mathbf{F} = \{400\mathbf{i} - 100\mathbf{j} - 700\mathbf{k}\}$ lb acts at the end of the beam. Determine the moment of this force about point O.

4–30. The force $\mathbf{F} = \{400\mathbf{i} - 100\mathbf{j} - 700\mathbf{k}\}$ lb acts at the end of the beam. Determine the moment of this force about point A.

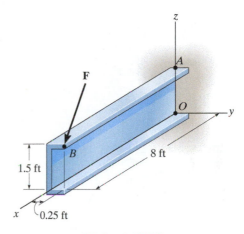

Probs. 4–29/30

4–31. Determine the moment of the force **F** about point P. Express the result as a Cartesian vector.

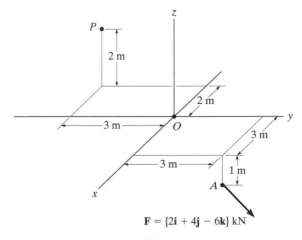

Prob. 4–31

*4–32. The pipe assembly is subjected to the force of $\mathbf{F} = \{600\mathbf{i} + 800\mathbf{j} - 500\mathbf{k}\}$ N. Determine the moment of this force about point A.

4–33. The pipe assembly is subjected to the force of $\mathbf{F} = \{600\mathbf{i} + 800\mathbf{j} - 500\mathbf{k}\}$ N. Determine the moment of this force about point B.

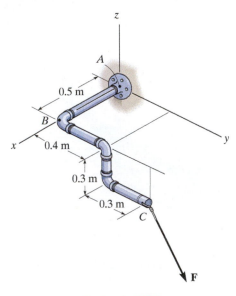

Probs. 4–32/33

4–34. Determine the moment of the force of $F = 600$ N about point A.

4–35. Determine the smallest force F that must be applied along the rope in order to cause the curved rod, which has a radius of 4 m, to fail at the support A. This requires a moment of $M = 1500$ N · m to be developed at A.

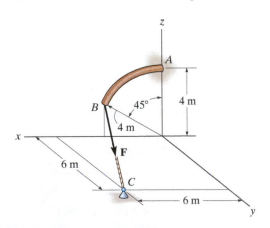

Probs. 4–34/35

*4–36. Determine the coordinate direction angles α, β, γ of force $\mathbf{F}$, so that the moment of $\mathbf{F}$ about O is zero.

4–37. Determine the moment of force $\mathbf{F}$ about point O. The force has a magnitude of 800 N and coordinate direction angles of $\alpha = 60°$, $\beta = 120°$, $\gamma = 45°$. Express the result as a Cartesian vector.

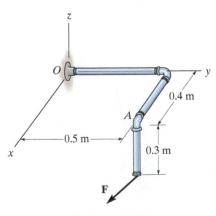

Probs. 4–36/37

4–38. Determine the moment of the force $\mathbf{F}$ about the door hinge at A. Express the result as a Cartesian vector.

4–39. Determine the moment of the force $\mathbf{F}$ about the door hinge at B. Express the result as a Cartesian vector.

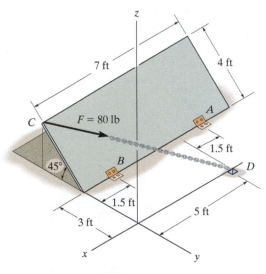

Probs. 4–38/39

*4–40. The curved rod has a radius of 5 ft. If a force of 60 lb acts at its end as shown, determine the moment of this force about point C.

4–41. Determine the smallest force F that must be applied along the rope in order to cause the curved rod, which has a radius of 5 ft, to fail at the support C. This requires a moment of $M = 80$ lb · ft to be developed at C.

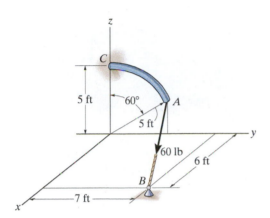

Probs. 4–40/41

4–42. A 20-N horizontal force is applied perpendicular to the handle of the socket wrench. Determine the magnitude and the coordinate direction angles of the moment created by this force about point O.

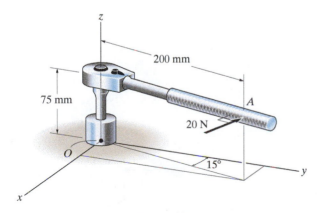

Prob. 4–42

4–43. The pipe assembly is subjected to the 80-N force. Determine the moment of this force about point A.

*4–44. The pipe assembly is subjected to the 80-N force. Determine the moment of this force about point B.

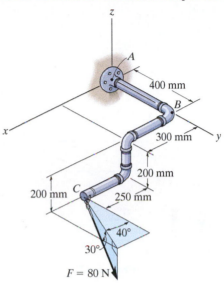

Probs. 4–43/44

4–45. A force of $F = \{6\mathbf{i} - 2\mathbf{j} + 1\mathbf{k}\}$ kN produces a moment of $M_O = \{4\mathbf{i} + 5\mathbf{j} - 14\mathbf{k}\}$ kN · m about the origin, point O. If the force acts at a point having an x coordinate of $x = 1$ m, determine the y and z coordinates. *Note:* The figure shows $\mathbf{F}$ and $\mathbf{M}_O$ in an arbitrary position.

4–46. The force $\mathbf{F} = \{6\mathbf{i} + 8\mathbf{j} + 10\mathbf{k}\}$ N creates a moment about point O of $\mathbf{M}_O = \{-14\mathbf{i} + 8\mathbf{j} + 2\mathbf{k}\}$ N · m. If the force passes through a point having an x coordinate of 1 m, determine the y and z coordinates of the point. Also, realizing that $M_O = Fd$, determine the perpendicular distance d from point O to the line of action of $\mathbf{F}$. *Note:* The figure shows $\mathbf{F}$ and $\mathbf{M}_O$ in an arbitrary position.

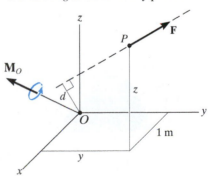

Probs. 4–45/46

4–47. A force **F** having a magnitude of $F = 100$ N acts along the diagonal of the parallelepiped. Determine the moment of **F** about the point A, using $\mathbf{M}_A = \mathbf{r}_B \times \mathbf{F}$ and $\mathbf{M}_A = \mathbf{r}_C \times \mathbf{F}$.

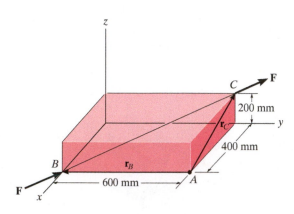

Prob. 4–47

4–50. Strut AB of the 1-m-diameter hatch door exerts a force of 450 N on point B. Determine the moment of this force about point O.

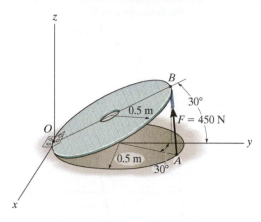

Prob. 4–50

*4–48. Force **F** acts perpendicular to the inclined plane. Determine the moment produced by **F** about point A. Express the result as a Cartesian vector.

4–49. Force **F** acts perpendicular to the inclined plane. Determine the moment produced by **F** about point B. Express the result as a Cartesian vector.

4–51. Using a ring collar, the 75-N force can act in the vertical plane at various angles θ. Determine the magnitude of the moment it produces about point A, plot the result of M (ordinate) versus θ (abscissa) for $0° \le \theta \le 180°$, and specify the angles that give the maximum and minimum moment.

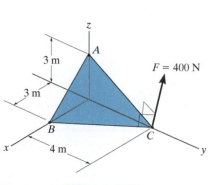

Probs. 4–48/49

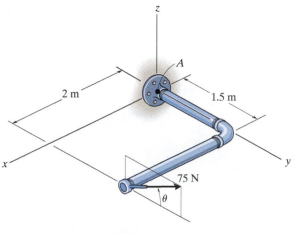

Prob. 4–51

4.5   Moment of a Force about a Specified Axis

Sometimes, the moment produced by a force about a *specified axis* must be determined. For example, suppose the lug nut at O on the car tire in Fig. 4–20a needs to be loosened. The force applied to the wrench will create a tendency for the wrench and the nut to rotate about the *moment axis* passing through O; however, the nut can only rotate about the y axis. Therefore, to determine the turning effect, only the y component of the moment is needed, and the total moment produced is not important. To determine this component, we can use either a scalar or vector analysis.

Scalar Analysis.   To use a scalar analysis in the case of the lug nut in Fig. 4–20a, the moment arm, or perpendicular distance from the axis to the line of action of the force, is $d_y = d \cos \theta$. Thus, the moment of **F** about the y axis is $M_y = F d_y = F(d \cos \theta)$. According to the right-hand rule, $\mathbf{M}_y$ is directed along the positive y axis as shown in the figure. In general, for any axis a, the moment is

$$M_a = F d_a \qquad (4\text{–}10)$$

(© Russell C. Hibbeler)

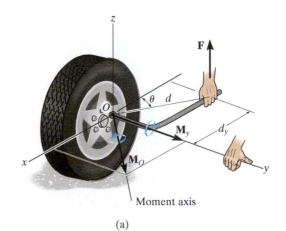

(a)

Fig. 4–20

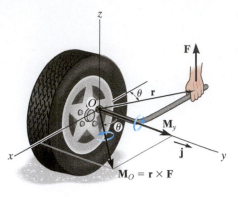

(b)

Fig. 4–20 (cont.)

4

Vector Analysis.   To find the moment of force **F** in Fig. 4–20b about the y axis using a vector analysis, we must first determine the moment of the force about *any point O* on the y axis by applying Eq. 4–7, $M_O = r \times F$. The component M_y along the y axis is the *projection* of M_O onto the y axis. It can be found using the *dot product* discussed in Chapter 2, so that $M_y = j \cdot M_O = j \cdot (r \times F)$, where **j** is the unit vector for the y axis.

We can generalize this approach by letting u_a be the unit vector that specifies the direction of the a axis shown in Fig. 4–21. Then the moment of **F** about a point O on the axis is $M_O = r \times F$, and the projection of this moment onto the a axis is $M_a = u_a \cdot (r \times F)$. This combination is referred to as the *scalar triple product*. If the vectors are written in Cartesian form, we have

$$M_a = [u_{a_x}i + u_{a_y}j + u_{a_z}k] \cdot \begin{vmatrix} i & j & k \\ r_x & r_y & r_z \\ F_x & F_y & F_z \end{vmatrix}$$

$$= u_{a_x}(r_yF_z - r_zF_y) - u_{a_y}(r_xF_z - r_zF_x) + u_{a_z}(r_xF_y - r_yF_x)$$

This result can also be written in the form of a determinant, making it easier to memorize.*

$$M_a = u_a \cdot (r \times F) = \begin{vmatrix} u_{a_x} & u_{a_y} & u_{a_z} \\ r_x & r_y & r_z \\ F_x & F_y & F_z \end{vmatrix} \qquad (4\text{–}11)$$

where

$u_{a_x}, u_{a_y}, u_{a_z}$ represent the x, y, z components of the unit vector defining the direction of the a axis

r_x, r_y, r_z represent the x, y, z components of the position vector extended from *any point O* on the a axis to *any point A* on the line of action of the force

F_x, F_y, F_z represent the x, y, z components of the force vector.

When M_a is evaluated from Eq. 4–11, it will yield a positive or negative scalar. The sign of this scalar indicates the sense of direction of M_a along the a axis. If it is positive, then M_a will have the same sense as u_a, whereas if it is negative, then M_a will act opposite to u_a. Once the a axis is established, point your right-hand thumb in the direction of M_a, and the curl of your fingers will indicate the sense of twist about the axis, Fig. 4–21.

Fig. 4–21

*Take a minute to expand this determinant, to show that it will yield the above result.

Provided M_a is determined, we can then express $\mathbf{M}_a$ as a Cartesian vector, namely,

$$\mathbf{M}_a = M_a \mathbf{u}_a \tag{4–12}$$

The examples which follow illustrate numerical applications of the above concepts.

Important Points

- The moment of a force about a specified axis can be determined provided the perpendicular distance d_a from the force line of action to the axis can be determined. $M_a = F d_a$.

- If vector analysis is used, $M_a = \mathbf{u}_a \cdot (\mathbf{r} \times \mathbf{F})$, where $\mathbf{u}_a$ defines the direction of the axis and $\mathbf{r}$ is extended from *any point* on the axis to *any point* on the line of action of the force.

- If M_a is calculated as a negative scalar, then the sense of direction of $\mathbf{M}_a$ is opposite to $\mathbf{u}_a$.

- The moment $\mathbf{M}_a$ expressed as a Cartesian vector is determined from $\mathbf{M}_a = M_a \mathbf{u}_a$.

EXAMPLE 4.7

Determine the resultant moment of the three forces in Fig. 4–22 about the x axis, the y axis, and the z axis.

SOLUTION

A force that is *parallel* to a coordinate axis or has a line of action that passes through the axis does *not* produce any moment or tendency for turning about that axis. Therefore, defining the positive direction of the moment of a force according to the right-hand rule, as shown in the figure, we have

$$M_x = (60 \text{ lb})(2 \text{ ft}) + (50 \text{ lb})(2 \text{ ft}) + 0 = 220 \text{ lb} \cdot \text{ft} \qquad Ans.$$

$$M_y = 0 - (50 \text{ lb})(3 \text{ ft}) - (40 \text{ lb})(2 \text{ ft}) = -230 \text{ lb} \cdot \text{ft} \qquad Ans.$$

$$M_z = 0 + 0 - (40 \text{ lb})(2 \text{ ft}) = -80 \text{ lb} \cdot \text{ft} \qquad Ans.$$

The negative signs indicate that $\mathbf{M}_y$ and $\mathbf{M}_z$ act in the $-y$ and $-z$ directions, respectively.

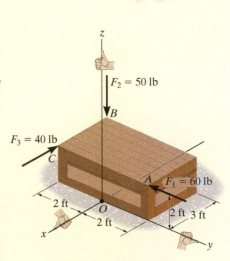

Fig. 4–22

EXAMPLE 4.8

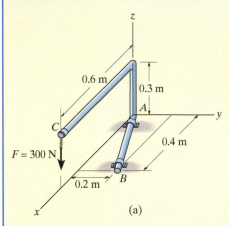

(a)

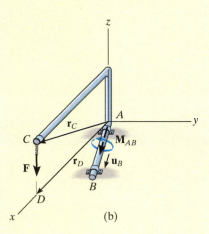

Fig. 4–23

Determine the moment $\mathbf{M}_{AB}$ produced by the force $\mathbf{F}$ in Fig. 4–23a, which tends to rotate the rod about the AB axis.

SOLUTION

A vector analysis using $M_{AB} = \mathbf{u}_B \cdot (\mathbf{r} \times \mathbf{F})$ will be considered for the solution rather than trying to find the moment arm or perpendicular distance from the line of action of $\mathbf{F}$ to the AB axis. Each of the terms in the equation will now be identified.

Unit vector $\mathbf{u}_B$ defines the direction of the AB axis of the rod, Fig. 4–23b, where

$$\mathbf{u}_B = \frac{\mathbf{r}_B}{r_B} = \frac{\{0.4\mathbf{i} + 0.2\mathbf{j}\}\ \text{m}}{\sqrt{(0.4\ \text{m})^2 + (0.2\ \text{m})^2}} = 0.8944\mathbf{i} + 0.4472\mathbf{j}$$

Vector $\mathbf{r}$ is directed from *any point* on the AB axis to *any point* on the line of action of the force. For example, position vectors $\mathbf{r}_C$ and $\mathbf{r}_D$ are suitable, Fig. 4–23b. (Although not shown, $\mathbf{r}_{BC}$ or $\mathbf{r}_{BD}$ can also be used.) For simplicity, we choose $\mathbf{r}_D$, where

$$\mathbf{r}_D = \{0.6\mathbf{i}\}\ \text{m}$$

The force is

$$\mathbf{F} = \{-300\mathbf{k}\}\ \text{N}$$

Substituting these vectors into the determinant form and expanding, we have

$$M_{AB} = \mathbf{u}_B \cdot (\mathbf{r}_D \times \mathbf{F}) = \begin{vmatrix} 0.8944 & 0.4472 & 0 \\ 0.6 & 0 & 0 \\ 0 & 0 & -300 \end{vmatrix}$$

$$= 0.8944[0(-300) - 0(0)] - 0.4472[0.6(-300) - 0(0)]$$
$$+ 0[0.6(0) - 0(0)]$$

$$= 80.50\ \text{N} \cdot \text{m}$$

This positive result indicates that the sense of $\mathbf{M}_{AB}$ is in the same direction as $\mathbf{u}_B$.

Expressing $\mathbf{M}_{AB}$ in Fig. 4–23b as a Cartesian vector yields

$$\mathbf{M}_{AB} = M_{AB}\mathbf{u}_B = (80.50\ \text{N} \cdot \text{m})(0.8944\mathbf{i} + 0.4472\mathbf{j})$$
$$= \{72.0\mathbf{i} + 36.0\mathbf{j}\}\ \text{N} \cdot \text{m} \qquad \textit{Ans.}$$

NOTE: If axis AB is defined using a unit vector directed from B toward A, then in the above formulation $-\mathbf{u}_B$ would have to be used. This would lead to $M_{AB} = -80.50\ \text{N} \cdot \text{m}$. Consequently, $\mathbf{M}_{AB} = M_{AB}(-\mathbf{u}_B)$, and the same result would be obtained.

EXAMPLE 4.9

Determine the magnitude of the moment of force **F** about segment *OA* of the pipe assembly in Fig. 4–24*a*.

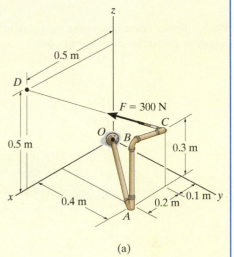

SOLUTION

The moment of **F** about the *OA* axis is determined from $M_{OA} = \mathbf{u}_{OA} \cdot (\mathbf{r} \times \mathbf{F})$, where **r** is a position vector extending from any point on the *OA* axis to any point on the line of action of **F**. As indicated in Fig. 4–24*b*, either $\mathbf{r}_{OD}$, $\mathbf{r}_{OC}$, $\mathbf{r}_{AD}$, or $\mathbf{r}_{AC}$ can be used; however, $\mathbf{r}_{OD}$ will be considered since it will simplify the calculation.

The unit vector $\mathbf{u}_{OA}$, which specifies the direction of the *OA* axis, is

$$\mathbf{u}_{OA} = \frac{\mathbf{r}_{OA}}{r_{OA}} = \frac{\{0.3\mathbf{i} + 0.4\mathbf{j}\}\ \text{m}}{\sqrt{(0.3\ \text{m})^2 + (0.4\ \text{m})^2}} = 0.6\mathbf{i} + 0.8\mathbf{j}$$

and the position vector $\mathbf{r}_{OD}$ is

$$\mathbf{r}_{OD} = \{0.5\mathbf{i} + 0.5\mathbf{k}\}\ \text{m}$$

The force **F** expressed as a Cartesian vector is

$$\mathbf{F} = F\left(\frac{\mathbf{r}_{CD}}{r_{CD}}\right)$$

$$= (300\ \text{N})\left[\frac{\{0.4\mathbf{i} - 0.4\mathbf{j} + 0.2\mathbf{k}\}\ \text{m}}{\sqrt{(0.4\ \text{m})^2 + (-0.4\ \text{m})^2 + (0.2\ \text{m})^2}}\right]$$

$$= \{200\mathbf{i} - 200\mathbf{j} + 100\mathbf{k}\}\ \text{N}$$

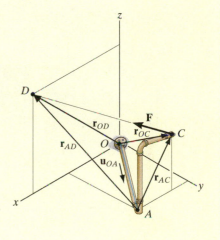

(b)

Fig. 4–24

Therefore,

$$M_{OA} = \mathbf{u}_{OA} \cdot (\mathbf{r}_{OD} \times \mathbf{F})$$

$$= \begin{vmatrix} 0.6 & 0.8 & 0 \\ 0.5 & 0 & 0.5 \\ 200 & -200 & 100 \end{vmatrix}$$

$$= 0.6[0(100) - (0.5)(-200)] - 0.8[0.5(100) - (0.5)(200)] + 0$$

$$= 100\ \text{N} \cdot \text{m} \qquad\qquad\qquad\qquad\qquad \textit{Ans.}$$

PRELIMINARY PROBLEMS

P4–3. In each case, determine the resultant moment of the forces acting about the x, y, and z axes.

P4–4. In each case, set up the determinant needed to find the moment of the force about the a–a axes.

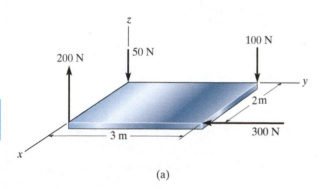

(a)

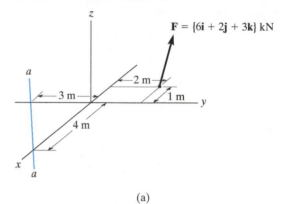

(a)

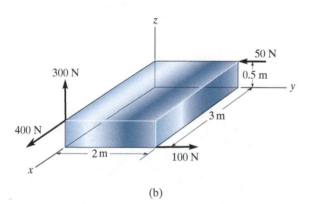

(b)

(b)

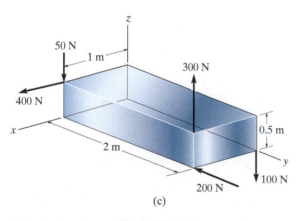

(c)

Prob. P4–3

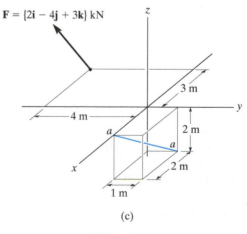

(c)

Prob. P4–4

FUNDAMENTAL PROBLEMS

F4–13. Determine the magnitude of the moment of the force $\mathbf{F} = \{300\mathbf{i} - 200\mathbf{j} + 150\mathbf{k}\}$ N about the x axis.

F4–14. Determine the magnitude of the moment of the force $\mathbf{F} = \{300\mathbf{i} - 200\mathbf{j} + 150\mathbf{k}\}$ N about the OA axis.

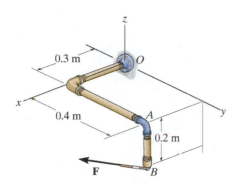

Probs. F4–13/14

F4–15. Determine the magnitude of the moment of the 200-N force about the x axis. Solve the problem using both a scalar and a vector analysis.

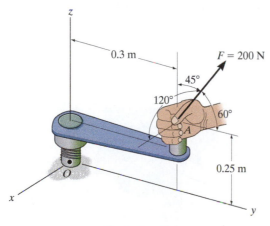

Prob. F4–15

F4–16. Determine the magnitude of the moment of the force about the y axis.

$$\mathbf{F} = \{30\mathbf{i} - 20\mathbf{j} + 50\mathbf{k}\} \text{ N}$$

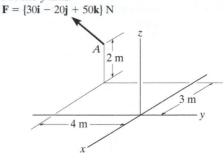

Prob. F4–16

F4–17. Determine the moment of the force $\mathbf{F} = \{50\mathbf{i} - 40\mathbf{j} + 20\mathbf{k}\}$ lb about the AB axis. Express the result as a Cartesian vector.

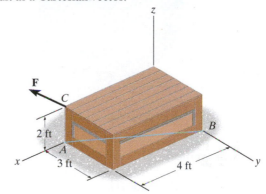

Prob. F4–17

F4–18. Determine the moment of force $\mathbf{F}$ about the x, the y, and the z axes. Solve the problem using both a scalar and a vector analysis.

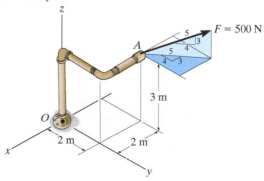

Prob. F4–18

PROBLEMS

*4–52. The lug nut on the wheel of the automobile is to be removed using the wrench and applying the vertical force of $F = 30$ N at A. Determine if this force is adequate, provided 14 N·m of torque about the x axis is initially required to turn the nut. If the 30-N force can be applied at A in any other direction, will it be possible to turn the nut?

4–53. Solve Prob. 4–52 if the cheater pipe AB is slipped over the handle of the wrench and the 30-N force can be applied at any point and in any direction on the assembly.

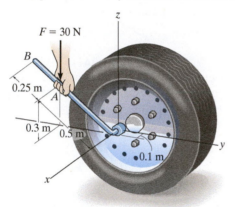

Probs. 4–52/53

4–54. The A-frame is being hoisted into an upright position by the vertical force of $F = 80$ lb. Determine the moment of this force about the y' axis passing through points A and B when the frame is in the position shown.

4–55. The A-frame is being hoisted into an upright position by the vertical force of $F = 80$ lb. Determine the moment of this force about the x axis when the frame is in the position shown.

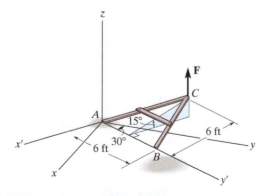

Probs. 4–54/55

*4–56. Determine the magnitude of the moments of the force $\mathbf{F}$ about the $x, y,$ and z axes. Solve the problem (a) using a Cartesian vector approach and (b) using a scalar approach.

4–57. Determine the moment of this force $\mathbf{F}$ about an axis extending between A and C. Express the result as a Cartesian vector.

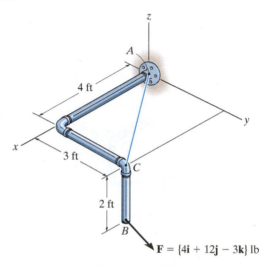

Probs. 4–56/57

4–58. The board is used to hold the end of a four-way lug wrench in the position shown when the man applies a force of $F = 100$ N. Determine the magnitude of the moment produced by this force about the x axis. Force $\mathbf{F}$ lies in a vertical plane.

4–59. The board is used to hold the end of a four-way lug wrench in position. If a torque of 30 N·m about the x axis is required to tighten the nut, determine the required magnitude of the force $\mathbf{F}$ that the man's foot must apply on the end of the wrench in order to turn it. Force $\mathbf{F}$ lies in a vertical plane.

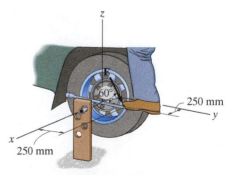

Probs. 4–58/59

*4–60. The A-frame is being hoisted into an upright position by the vertical force of $F = 80$ lb. Determine the moment of this force about the y axis when the frame is in the position shown.

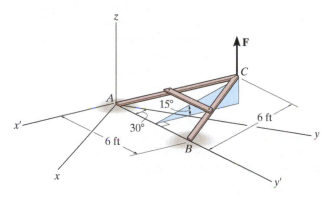

Prob. 4–60

4–61. Determine the magnitude of the moment of the force $\mathbf{F} = \{50\mathbf{i} - 20\mathbf{j} - 80\mathbf{k}\}$ N about the base line AB of the tripod.

4–62. Determine the magnitude of the moment of the force $\mathbf{F} = \{50\mathbf{i} - 20\mathbf{j} - 80\mathbf{k}\}$ N about the base line BC of the tripod.

4–63. Determine the magnitude of the moment of the force $\mathbf{F} = \{50\mathbf{i} - 20\mathbf{j} - 80\mathbf{k}\}$ N about the base line CA of the tripod.

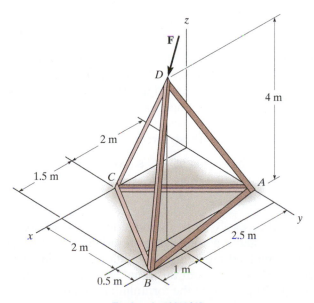

Probs. 4–61/62/63

*4–64. A horizontal force of $\mathbf{F} = \{-50\mathbf{i}\}$ N is applied perpendicular to the handle of the pipe wrench. Determine the moment that this force exerts along the axis OA (z axis) of the pipe assembly. Both the wrench and pipe assembly, $OABC$, lie in the y-z plane. *Suggestion:* Use a scalar analysis.

4–65. Determine the magnitude of the horizontal force $\mathbf{F} = -F\mathbf{i}$ acting on the handle of the wrench so that this force produces a component of the moment along the OA axis (z axis) of the pipe assembly of $\mathbf{M}_z = \{4\mathbf{k}\}$ N · m. Both the wrench and the pipe assembly, $OABC$, lie in the y-z plane. *Suggestion:* Use a scalar analysis.

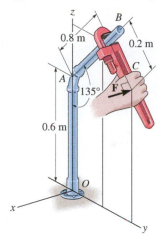

Probs. 4–64/65

4–66. The force of $F = 30$ N acts on the bracket as shown. Determine the moment of the force about the a–a axis of the pipe if $\alpha = 60°$, $\beta = 60°$, and $\gamma = 45°$. Also, determine the coordinate direction angles of F in order to produce the maximum moment about the a–a axis. What is this moment?

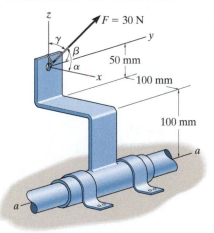

Prob. 4–66

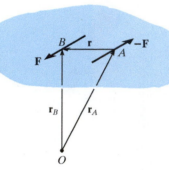

Fig. 4–25

Fig. 4–26

M

Fig. 4–27

4.6   Moment of a Couple

A *couple* is defined as two parallel forces that have the same magnitude, but opposite directions, and are separated by a perpendicular distance d, Fig. 4–25. Since the resultant force is zero, the only effect of a couple is to produce an actual rotation, or if no movement is possible, there is a tendency of rotation in a specified direction. For example, imagine that you are driving a car with both hands on the steering wheel and you are making a turn. One hand will push up on the wheel while the other hand pulls down, which causes the steering wheel to rotate.

The moment produced by a couple is called a *couple moment*. We can determine its value by finding the sum of the moments of both couple forces about *any* arbitrary point. For example, in Fig. 4–26, position vectors r_A and r_B are directed from point O to points A and B lying on the line of action of $-F$ and F. The couple moment determined about O is therefore

$$M = r_B \times F + r_A \times -F = (r_B - r_A) \times F$$

However $r_B = r_A + r$ or $r = r_B - r_A$, so that

$$M = r \times F \qquad (4\text{–}13)$$

This result indicates that a couple moment is a *free vector*, i.e., it can act at *any point* since M depends *only* upon the position vector r directed *between* the forces and *not* the position vectors r_A and r_B, directed from the arbitrary point O to the forces. This concept is unlike the moment of a force, which requires a definite point (or axis) about which moments are determined.

Scalar Formulation.   The moment of a couple, M, Fig. 4–27, is defined as having a *magnitude* of

$$\boxed{M = Fd} \qquad (4\text{–}14)$$

where F is the magnitude of one of the forces and d is the perpendicular distance or moment arm between the forces. The *direction* and sense of the couple moment are determined by the right-hand rule, where the thumb indicates this direction when the fingers are curled with the sense of rotation caused by the couple forces. In all cases, M will act perpendicular to the plane containing these forces.

Vector Formulation.   The moment of a couple can also be expressed by the vector cross product using Eq. 4–13, i.e.,

$$\boxed{M = r \times F} \qquad (4\text{–}15)$$

Application of this equation is easily remembered if one thinks of taking the moments of both forces about a point lying on the line of action of one of the forces. For example, if moments are taken about point A in Fig. 4–26, the moment of $-F$ is *zero* about this point, and the moment of F is defined from Eq. 4–15. Therefore, in the formulation r is crossed with the force F to which it is directed.

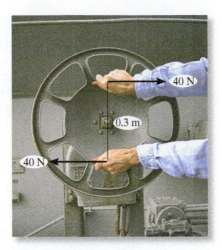

Fig. 4–28 (© Russell C. Hibbeler)

Equivalent Couples.   If two couples produce a moment with the *same magnitude and direction*, then these two couples are *equivalent*. For example, the two couples shown in Fig. 4–28 are *equivalent* because each couple moment has a magnitude of $M = 30\ \text{N}(0.4\ \text{m}) = 40\ \text{N}(0.3\ \text{m}) = 12\ \text{N} \cdot \text{m}$, and each is directed into the plane of the page. Notice that larger forces are required in the second case to create the same turning effect because the hands are placed closer together. Also, if the wheel was connected to the shaft at a point other than at its center, then the wheel would still turn when each couple is applied since the 12 N · m couple is a free vector.

Resultant Couple Moment.   Since couple moments are vectors, their resultant can be determined by vector addition. For example, consider the couple moments $\mathbf{M}_1$ and $\mathbf{M}_2$ acting on the pipe in Fig. 4–29*a*. Since each couple moment is a free vector, we can join their tails at any arbitrary point and find the resultant couple moment, $\mathbf{M}_R = \mathbf{M}_1 + \mathbf{M}_2$ as shown in Fig. 4–29*b*.

If more than two couple moments act on the body, we may generalize this concept and write the vector resultant as

$$\mathbf{M}_R = \Sigma(\mathbf{r} \times \mathbf{F}) \qquad (4\text{–}16)$$

These concepts are illustrated numerically in the examples that follow. In general, problems projected in two dimensions should be solved using a scalar analysis since the moment arms and force components are easy to determine.

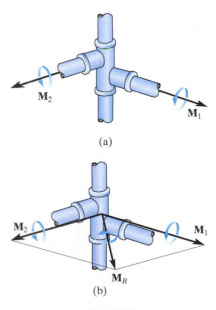

Fig. 4–29

Steering wheels on vehicles have been made smaller than on older vehicles because power steering does not require the driver to apply a large couple moment to the rim of the wheel. (© Russell C. Hibbeler)

Important Points

- A couple moment is produced by two noncollinear forces that are equal in magnitude but opposite in direction. Its effect is to produce pure rotation, or tendency for rotation in a specified direction.

- A couple moment is a free vector, and as a result it causes the same rotational effect on a body regardless of where the couple moment is applied to the body.

- The moment of the two couple forces can be determined about *any point*. For convenience, this point is often chosen on the line of action of one of the forces in order to eliminate the moment of this force about the point.

- In three dimensions the couple moment is often determined using the vector formulation, $\mathbf{M} = \mathbf{r} \times \mathbf{F}$, where $\mathbf{r}$ is directed from *any point* on the line of action of one of the forces to *any point* on the line of action of the other force $\mathbf{F}$.

- A resultant couple moment is simply the vector sum of all the couple moments of the system.

EXAMPLE 4.10

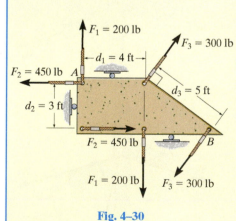

Fig. 4–30

Determine the resultant couple moment of the three couples acting on the plate in Fig. 4–30.

SOLUTION

As shown the perpendicular distances between each pair of couple forces are $d_1 = 4$ ft, $d_2 = 3$ ft, and $d_3 = 5$ ft. Considering counterclockwise couple moments as positive, we have

$$\zeta + M_R = \Sigma M; \quad M_R = -F_1 d_1 + F_2 d_2 - F_3 d_3$$

$$= -(200 \text{ lb})(4 \text{ ft}) + (450 \text{ lb})(3 \text{ ft}) - (300 \text{ lb})(5 \text{ ft})$$

$$= -950 \text{ lb} \cdot \text{ft} = 950 \text{ lb} \cdot \text{ft} \, \rangle \qquad \textit{Ans.}$$

The negative sign indicates that $\mathbf{M}_R$ has a clockwise rotational sense.

EXAMPLE 4.11

Determine the magnitude and direction of the couple moment acting on the gear in Fig. 4–31a.

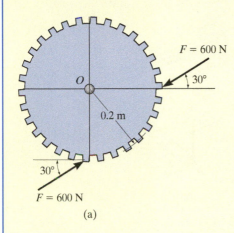

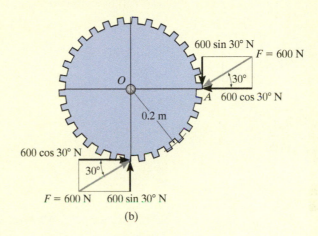

(a)

(b)

SOLUTION

The easiest solution requires resolving each force into its components as shown in Fig. 4–31b. The couple moment can be determined by summing the moments of these force components about any point, for example, the center O of the gear or point A. If we consider counterclockwise moments as positive, we have

$$\zeta + M = \Sigma M_O; \quad M = (600 \cos 30° \text{ N})(0.2 \text{ m}) - (600 \sin 30° \text{ N})(0.2 \text{ m})$$
$$= 43.9 \text{ N} \cdot \text{m} \, \zeta \qquad\qquad\qquad\qquad Ans.$$

or

$$\zeta + M = \Sigma M_A; \quad M = (600 \cos 30° \text{ N})(0.2 \text{ m}) - (600 \sin 30° \text{ N})(0.2 \text{ m})$$
$$= 43.9 \text{ N} \cdot \text{m} \, \zeta \qquad\qquad\qquad\qquad Ans.$$

This positive result indicates that $\mathbf{M}$ has a counterclockwise rotational sense, so it is directed outward, perpendicular to the page.

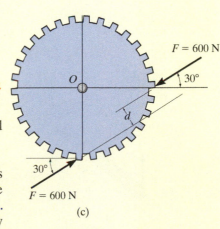

NOTE: The same result can also be obtained using $M = Fd$, where d is the perpendicular distance between the lines of action of the couple forces, Fig. 4–31c. However, the computation for d is more involved. Realize that the couple moment is a free vector and can act at any point on the gear and produce the same turning effect about point O.

(c)

Fig. 4–31

EXAMPLE 4.12

Determine the couple moment acting on the pipe shown in Fig. 4–32a.
Segment AB is directed 30° below the x–y plane.

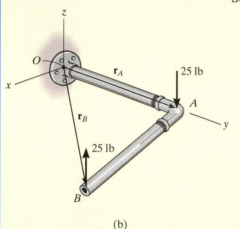

(b)

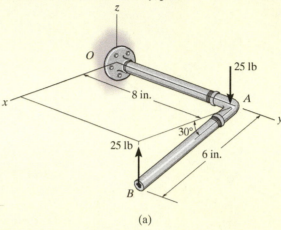

(a)

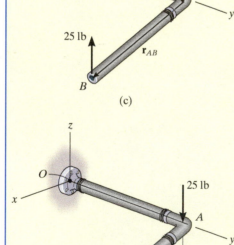

(c)

(d)

Fig. 4–32

SOLUTION I (VECTOR ANALYSIS)

The moment of the two couple forces can be found about *any point*. If point O is considered, Fig. 4–32b, we have

$$\mathbf{M} = \mathbf{r}_A \times (-25\mathbf{k}) + \mathbf{r}_B \times (25\mathbf{k})$$
$$= (8\mathbf{j}) \times (-25\mathbf{k}) + (6\cos 30°\mathbf{i} + 8\mathbf{j} - 6\sin 30°\mathbf{k}) \times (25\mathbf{k})$$
$$= -200\mathbf{i} - 129.9\mathbf{j} + 200\mathbf{i}$$
$$= \{-130\mathbf{j}\}\ \text{lb} \cdot \text{in.} \qquad \textit{Ans.}$$

It is *easier* to take moments of the couple forces about a point lying on the line of action of one of the forces, e.g., point A, Fig. 4–32c. In this case the moment of the force at A is zero, so that

$$\mathbf{M} = \mathbf{r}_{AB} \times (25\mathbf{k})$$
$$= (6\cos 30°\mathbf{i} - 6\sin 30°\mathbf{k}) \times (25\mathbf{k})$$
$$= \{-130\mathbf{j}\}\ \text{lb} \cdot \text{in.} \qquad \textit{Ans.}$$

SOLUTION II (SCALAR ANALYSIS)

Although this problem is shown in three dimensions, the geometry is simple enough to use the scalar equation $M = Fd$. The perpendicular distance between the lines of action of the couple forces is $d = 6\cos 30° = 5.196$ in., Fig. 4–32d. Hence, taking moments of the forces about either point A or point B yields

$$M = Fd = 25\ \text{lb}\ (5.196\ \text{in.}) = 129.9\ \text{lb} \cdot \text{in.}$$

Applying the right-hand rule, $\mathbf{M}$ acts in the $-\mathbf{j}$ direction. Thus,

$$\mathbf{M} = \{-130\mathbf{j}\}\ \text{lb} \cdot \text{in.} \qquad \textit{Ans.}$$

EXAMPLE | 4.13

Replace the two couples acting on the pipe column in Fig. 4–33*a* by a resultant couple moment.

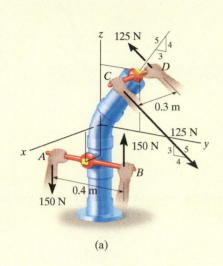

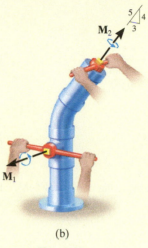

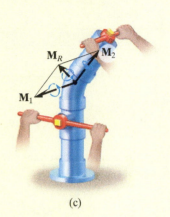

(a)   (b)   (c)

Fig. 4–33

SOLUTION *(VECTOR ANALYSIS)*

The couple moment $\mathbf{M}_1$, developed by the forces at A and B, can easily be determined from a scalar formulation.

$$M_1 = Fd = 150 \text{ N}(0.4 \text{ m}) = 60 \text{ N} \cdot \text{m}$$

By the right-hand rule, $\mathbf{M}_1$ acts in the $+\mathbf{i}$ direction, Fig. 4–33*b*. Hence,

$$\mathbf{M}_1 = \{60\mathbf{i}\} \text{ N} \cdot \text{m}$$

Vector analysis will be used to determine $\mathbf{M}_2$, caused by forces at C and D. If moments are calculated about point D, Fig. 4–33*a*, $\mathbf{M}_2 = \mathbf{r}_{DC} \times \mathbf{F}_C$, then

$$\mathbf{M}_2 = \mathbf{r}_{DC} \times \mathbf{F}_C = (0.3\mathbf{i}) \times \left[125\left(\tfrac{4}{5}\right)\mathbf{j} - 125\left(\tfrac{3}{5}\right)\mathbf{k} \right]$$

$$= (0.3\mathbf{i}) \times [100\mathbf{j} - 75\mathbf{k}] = 30(\mathbf{i} \times \mathbf{j}) - 22.5(\mathbf{i} \times \mathbf{k})$$

$$= \{22.5\mathbf{j} + 30\mathbf{k}\} \text{ N} \cdot \text{m}$$

Since $\mathbf{M}_1$ and $\mathbf{M}_2$ are free vectors, they may be moved to some arbitrary point and added vectorially, Fig. 4–33*c*. The resultant couple moment becomes

$$\mathbf{M}_R = \mathbf{M}_1 + \mathbf{M}_2 = \{60\mathbf{i} + 22.5\mathbf{j} + 30\mathbf{k}\} \text{ N} \cdot \text{m} \qquad \textit{Ans.}$$

FUNDAMENTAL PROBLEMS

F4–19.   Determine the resultant couple moment acting on the beam.

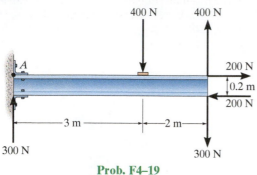

Prob. F4–19

F4–20.   Determine the resultant couple moment acting on the triangular plate.

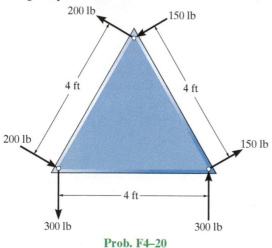

Prob. F4–20

F4–21.   Determine the magnitude of **F** so that the resultant couple moment acting on the beam is 1.5 kN · m clockwise.

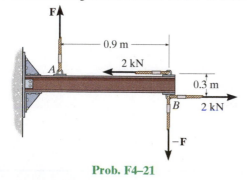

Prob. F4–21

F4–22.   Determine the couple moment acting on the beam.

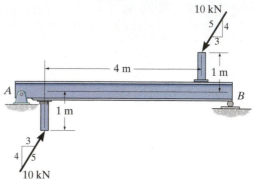

Prob. F4–22

F4–23.   Determine the resultant couple moment acting on the pipe assembly.

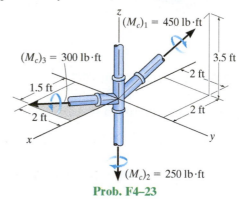

Prob. F4–23

F4–24.   Determine the couple moment acting on the pipe assembly and express the result as a Cartesian vector.

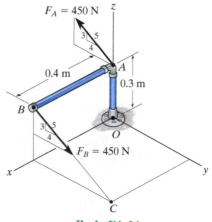

Prob. F4–24

PROBLEMS

4–67. A clockwise couple $M = 5\,\text{N}\cdot\text{m}$ is resisted by the shaft of the electric motor. Determine the magnitude of the reactive forces $-\mathbf{R}$ and $\mathbf{R}$ which act at supports A and B so that the resultant of the two couples is zero.

4–69. If the resultant couple of the three couples acting on the triangular block is to be zero, determine the magnitude of forces $\mathbf{F}$ and $\mathbf{P}$.

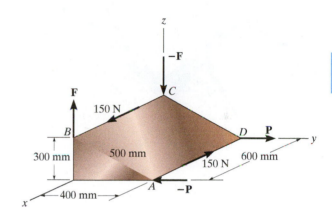

Prob. 4–69

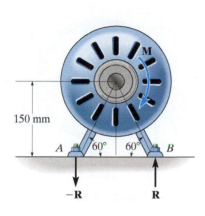

Prob. 4–67

4–70. Two couples act on the beam. If $F = 125$ lb, determine the resultant couple moment.

***4–68.** A twist of $4\,\text{N}\cdot\text{m}$ is applied to the handle of the screwdriver. Resolve this couple moment into a pair of couple forces $\mathbf{F}$ exerted on the handle and $\mathbf{P}$ exerted on the blade.

4–71. Two couples act on the beam. Determine the magnitude of $\mathbf{F}$ so that the resultant couple moment is $450\,\text{lb}\cdot\text{ft}$, counterclockwise. Where on the beam does the resultant couple moment act?

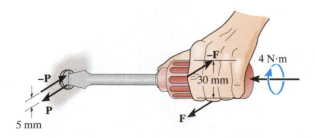

Prob. 4–68

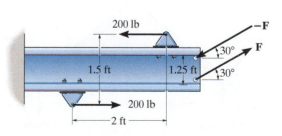

Probs. 4–70/71

*4–72. Determine the magnitude of the couple forces **F** so that the resultant couple moment on the crank is zero.

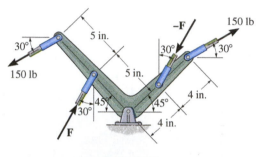

Prob. 4–72

4–74. The man tries to open the valve by applying the couple forces of $F = 75$ N to the wheel. Determine the couple moment produced.

4–75. If the valve can be opened with a couple moment of 25 N · m, determine the required magnitude of each couple force which must be applied to the wheel.

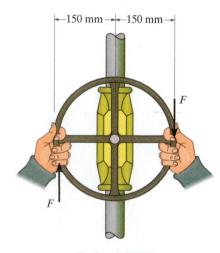

Probs. 4–74/75

4–73. The ends of the triangular plate are subjected to three couples. Determine the magnitude of the force **F** so that the resultant couple moment is 400 N · m clockwise.

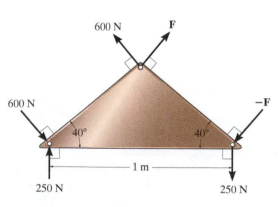

Prob. 4–73

*4–76. Determine the magnitude of **F** so that the resultant couple moment is 12 kN · m, counterclockwise. Where on the beam does the resultant couple moment act?

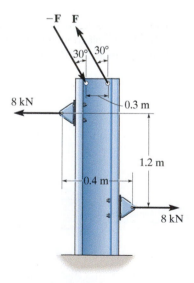

Prob. 4–76

4–77. Two couples act on the beam as shown. If $F = 150$ lb, determine the resultant couple moment.

4–78. Two couples act on the beam as shown. Determine the magnitude of **F** so that the resultant couple moment is 300 lb · ft counterclockwise. Where on the beam does the resultant couple act?

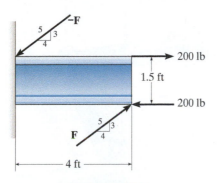

Probs. 4–77/78

4–79. Two couples act on the frame. If the resultant couple moment is to be zero, determine the distance d between the 80-lb couple forces.

*4–80. Two couples act on the frame. If $d = 4$ ft, determine the resultant couple moment. Compute the result by resolving each force into x and y components and (a) finding the moment of each couple (Eq. 4–13) and (b) summing the moments of all the force components about point A.

4–81. Two couples act on the frame. If $d = 4$ ft, determine the resultant couple moment. Compute the result by resolving each force into x and y components and (a) finding the moment of each couple (Eq. 4–13) and (b) summing the moments of all the force components about point B.

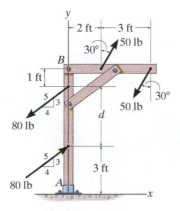

Probs. 4–79/80/81

4–82. Express the moment of the couple acting on the pipe assembly in Cartesian vector form. What is the magnitude of the couple moment?

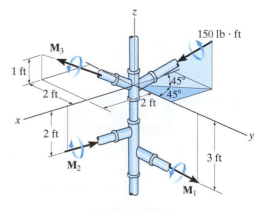

Prob. 4–82

4–83. If $M_1 = 180$ lb · ft, $M_2 = 90$ lb · ft, and $M_3 = 120$ lb · ft, determine the magnitude and coordinate direction angles of the resultant couple moment.

*4–84. Determine the magnitudes of couple moments $\mathbf{M}_1$, $\mathbf{M}_2$, and $\mathbf{M}_3$ so that the resultant couple moment is zero.

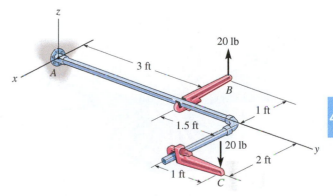

Probs. 4–83/84

4–85. The gears are subjected to the couple moments shown. Determine the magnitude and coordinate direction angles of the resultant couple moment.

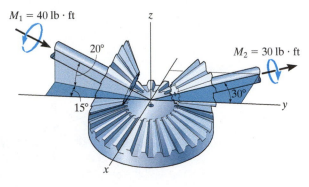

Prob. 4–85

4–86. Determine the required magnitude of the couple moments M_2 and M_3 so that the resultant couple moment is zero.

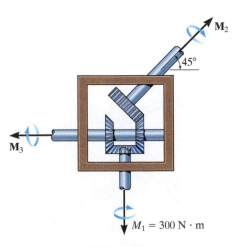

Prob. 4–86

4–87. Determine the resultant couple moment of the two couples that act on the assembly. Specify its magnitude and coordinate direction angles.

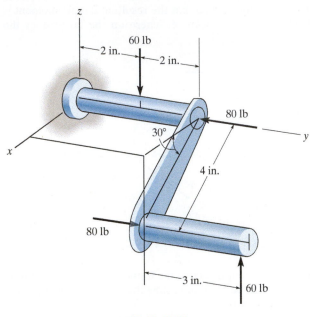

Prob. 4–87

*4–88. Express the moment of the couple acting on the frame in Cartesian vector form. The forces are applied perpendicular to the frame. What is the magnitude of the couple moment? Take $F = 50$ N.

4–89. In order to turn over the frame, a couple moment is applied as shown. If the component of this couple moment along the x axis is $M_x = \{-20\mathbf{i}\}$ N · m, determine the magnitude F of the couple forces.

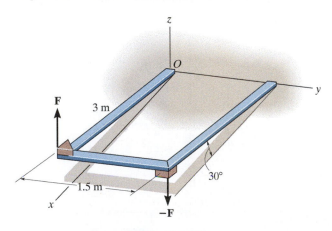

Probs. 4–88/89

4–90. Express the moment of the couple acting on the pipe in Cartesian vector form. What is the magnitude of the couple moment? Take $F = 125$ N.

4–91. If the couple moment acting on the pipe has a magnitude of 300 N · m, determine the magnitude F of the forces applied to the wrenches.

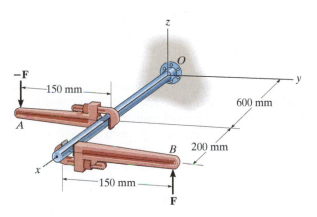

Probs. 4–90/91

4–92. If $F = 80$ N, determine the magnitude and coordinate direction angles of the couple moment. The pipe assembly lies in the x–y plane.

4–93. If the magnitude of the couple moment acting on the pipe assembly is 50 N · m, determine the magnitude of the couple forces applied to each wrench. The pipe assembly lies in the x–y plane.

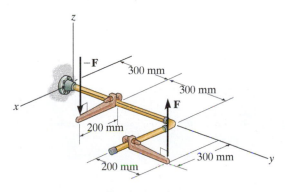

Probs. 4–92/93

4–94. Express the moment of the couple acting on the rod in Cartesian vector form. What is the magnitude of the couple moment?

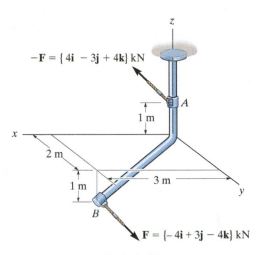

Prob. 4–94

4–95. If $F_1 = 100$ N, $F_2 = 120$ N, and $F_3 = 80$ N, determine the magnitude and coordinate direction angles of the resultant couple moment.

4–96. Determine the required magnitude of F_1, F_2, and F_3 so that the resultant couple moment is $(M_c)_R = [50\mathbf{i} - 45\mathbf{j} - 20\mathbf{k}]$ N · m.

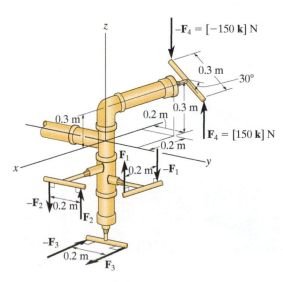

Probs. 4–95/96

4.7   Simplification of a Force and Couple System

Sometimes it is convenient to reduce a system of forces and couple moments acting on a body to a simpler form by replacing it with an ***equivalent system***, consisting of a single resultant force acting at a specific point and a resultant couple moment. A system is equivalent if the ***external effects*** it produces on a body are the same as those caused by the original force and couple moment system. In this context, the external effects of a system refer to the *translating and rotating motion* of the body if the body is free to move, or it refers to the *reactive forces* at the supports if the body is held fixed.

For example, consider holding the stick in Fig. 4–34*a*, which is subjected to the force **F** at point *A*. If we attach a pair of equal but opposite forces **F** and −**F** at point *B*, which is *on the line of action* of **F**, Fig. 4–34*b*, we observe that −**F** at *B* and **F** at *A* will cancel each other, leaving only **F** at *B*, Fig. 4–34*c*. Force **F** has now been moved from *A* to *B* without modifying its *external effects* on the stick; i.e., the reaction at the grip remains the same. This demonstrates the ***principle of transmissibility***, which states that a force acting on a body (stick) is a ***sliding vector*** since it can be applied at any point along its line of action.

We can also use the above procedure to move a force to a point that is *not* on the line of action of the force. If **F** is applied perpendicular to the stick, as in Fig. 4–35*a*, then we can attach a pair of equal but opposite forces **F** and −**F** to *B*, Fig. 4–35*b*. Force **F** is now applied at *B*, and the other two forces, **F** at *A* and −**F** at *B*, form a couple that produces the couple moment $M = Fd$, Fig. 4–35*c*. Therefore, the force **F** can be moved from *A* to *B* provided a couple moment **M** is added to maintain an equivalent system. This couple moment is determined by taking the moment of **F** about *B*. Since **M** is actually a *free vector*, it can act at any point on the stick. In both cases the systems are equivalent, which causes a downward force **F** and clockwise couple moment $M = Fd$ to be felt at the grip.

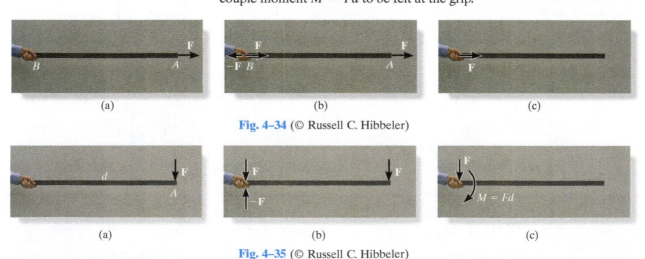

(a)   (b)   (c)

Fig. 4–34 (© Russell C. Hibbeler)

(a)   (b)   (c)

Fig. 4–35 (© Russell C. Hibbeler)

System of Forces and Couple Moments.   Using the above method, a system of several forces and couple moments acting on a body can be reduced to an equivalent single resultant force acting at a point O and a resultant couple moment. For example, in Fig. 4–36a, O is not on the line of action of $\mathbf{F}_1$, and so this force can be moved to point O provided a couple moment $(\mathbf{M}_O)_1 = \mathbf{r}_1 \times \mathbf{F}$ is added to the body. Similarly, the couple moment $(\mathbf{M}_O)_2 = \mathbf{r}_2 \times \mathbf{F}_2$ should be added to the body when we move $\mathbf{F}_2$ to point O. Finally, since the couple moment $\mathbf{M}$ is a free vector, it can just be moved to point O. By doing this, we obtain the equivalent system shown in Fig. 4–36b, which produces the same external effects (support reactions) on the body as that of the force and couple system shown in Fig. 4–36a. If we sum the forces and couple moments, we obtain the resultant force $\mathbf{F}_R = \mathbf{F}_1 + \mathbf{F}_2$ and the resultant couple moment $(\mathbf{M}_R)_O = \mathbf{M} + (\mathbf{M}_O)_1 + (\mathbf{M}_O)_2$, Fig. 4–36$c$.

Notice that $\mathbf{F}_R$ is independent of the location of point O since it is simply a summation of the forces. However, $(\mathbf{M}_R)_O$ depends upon this location since the moments $\mathbf{M}_1$ and $\mathbf{M}_2$ are determined using the position vectors $\mathbf{r}_1$ and $\mathbf{r}_2$, which extend from O to each force. Also note that $(\mathbf{M}_R)_O$ is a free vector and can act at *any point* on the body, although point O is generally chosen as its point of application.

We can generalize the above method of reducing a force and couple system to an equivalent resultant force $\mathbf{F}_R$ acting at point O and a resultant couple moment $(\mathbf{M}_R)_O$ by using the following two equations.

$$\mathbf{F}_R = \Sigma\mathbf{F}$$
$$(\mathbf{M}_R)_O = \Sigma\mathbf{M}_O + \Sigma\mathbf{M} \tag{4–17}$$

The first equation states that the resultant force of the system is equivalent to the sum of all the forces; and the second equation states that the resultant couple moment of the system is equivalent to the sum of all the couple moments $\Sigma\mathbf{M}$ plus the moments of all the forces $\Sigma\mathbf{M}_O$ about point O. If the force system lies in the x–y plane and any couple moments are perpendicular to this plane, then the above equations reduce to the following three scalar equations.

$$(F_R)_x = \Sigma F_x$$
$$(F_R)_y = \Sigma F_y$$
$$(M_R)_O = \Sigma M_O + \Sigma M \tag{4–18}$$

Here the resultant force is determined from the vector sum of its two components $(F_R)_x$ and $(F_R)_y$.

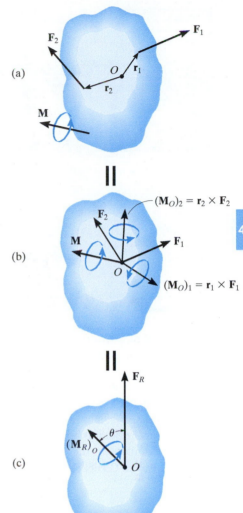

(a)

(b)

(c)

$(\mathbf{M}_O)_2 = \mathbf{r}_2 \times \mathbf{F}_2$

$(\mathbf{M}_O)_1 = \mathbf{r}_1 \times \mathbf{F}_1$

Fig. 4–36

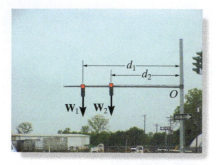

The weights of these traffic lights can be replaced by their equivalent resultant force $W_R = W_1 + W_2$ and a couple moment $(M_R)_O = W_1 d_1 + W_2 d_2$ at the support, O. In both cases the support must provide the same resistance to translation and rotation in order to keep the member in the horizontal position. (© Russell C. Hibbeler)

Important Points

- Force is a sliding vector, since it will create the same external effects on a body when it is applied at any point P along its line of action. This is called the principle of transmissibility.

- A couple moment is a free vector since it will create the same external effects on a body when it is applied at any point P on the body.

- When a force is moved to another point P that is not on its line of action, it will create the same external effects on the body if a couple moment is also applied to the body. The couple moment is determined by taking the moment of the force about point P.

Procedure for Analysis

The following points should be kept in mind when simplifying a force and couple moment system to an equivalent resultant force and couple system.

- Establish the coordinate axes with the origin located at point O and the axes having a selected orientation.

Force Summation.

- If the force system is *coplanar*, resolve each force into its x and y components. If a component is directed along the positive x or y axis, it represents a positive scalar; whereas if it is directed along the negative x or y axis, it is a negative scalar.

- In three dimensions, represent each force as a Cartesian vector before summing the forces.

Moment Summation.

- When determining the moments of a *coplanar* force system about point O, it is generally advantageous to use the principle of moments, i.e., determine the moments of the components of each force, rather than the moment of the force itself.

- In three dimensions use the vector cross product to determine the moment of each force about point O. Here the position vectors extend from O to any point on the line of action of each force.

EXAMPLE | 4.14

Replace the force and couple system shown in Fig. 4–37a by an equivalent resultant force and couple moment acting at point O.

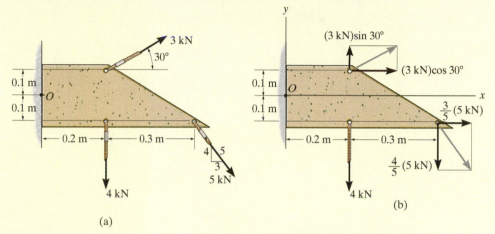

(a)

(b)

SOLUTION

Force Summation.   The 3 kN and 5 kN forces are resolved into their x and y components as shown in Fig. 4–37b. We have

$$\xrightarrow{+} (F_R)_x = \Sigma F_x; \qquad (F_R)_x = (3 \text{ kN}) \cos 30° + \left(\tfrac{3}{5}\right)(5 \text{ kN}) = 5.598 \text{ kN} \rightarrow$$

$$+\uparrow (F_R)_y = \Sigma F_y; \qquad (F_R)_y = (3 \text{ kN}) \sin 30° - \left(\tfrac{4}{5}\right)(5 \text{ kN}) - 4 \text{ kN} = -6.50 \text{ kN} = 6.50 \text{ kN} \downarrow$$

Using the Pythagorean theorem, Fig. 4–37c, the magnitude of $\mathbf{F}_R$ is

$$F_R = \sqrt{(F_R)_x{}^2 + (F_R)_y{}^2} = \sqrt{(5.598 \text{ kN})^2 + (6.50 \text{ kN})^2} = 8.58 \text{ kN} \quad \textit{Ans.}$$

Its direction θ is

$$\theta = \tan^{-1}\left(\frac{(F_R)_y}{(F_R)_x}\right) = \tan^{-1}\left(\frac{6.50 \text{ kN}}{5.598 \text{ kN}}\right) = 49.3° \qquad \textit{Ans.}$$

Moment Summation.   The moments of 3 kN and 5 kN about point O will be determined using their x and y components. Referring to Fig. 4–37b, we have

$$\zeta + (M_R)_O = \Sigma M_O;$$

$$(M_R)_O = (3 \text{ kN}) \sin 30°(0.2 \text{ m}) - (3 \text{ kN}) \cos 30°(0.1 \text{ m}) + \left(\tfrac{3}{5}\right)(5 \text{ kN})(0.1 \text{ m})$$

$$- \left(\tfrac{4}{5}\right)(5 \text{ kN})(0.5 \text{ m}) - (4 \text{ kN})(0.2 \text{ m})$$

$$= -2.46 \text{ kN} \cdot \text{m} = 2.46 \text{ kN} \cdot \text{m} \, \zeta \qquad \textit{Ans.}$$

This clockwise moment is shown in Fig. 4–37c.

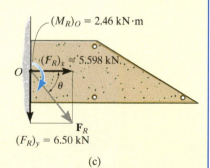

(c)

Fig. 4–37

NOTE: Realize that the resultant force and couple moment in Fig. 4–37c will produce the same external effects or reactions at the supports as those produced by the force system, Fig. 4–37a.

EXAMPLE   4.15

Replace the force and couple system acting on the member in Fig. 4–38*a* by an equivalent resultant force and couple moment acting at point *O*.

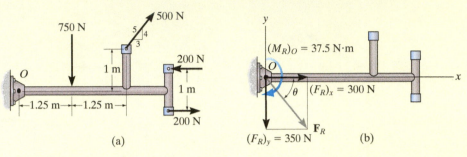

Fig. 4–38

SOLUTION

Force Summation.   Since the couple forces of 200 N are equal but opposite, they produce a zero resultant force, and so it is not necessary to consider them in the force summation. The 500-N force is resolved into its *x* and *y* components, thus,

$$\xrightarrow{+} (F_R)_x = \Sigma F_x; \quad (F_R)_x = \left(\tfrac{3}{5}\right)(500 \text{ N}) = 300 \text{ N} \rightarrow$$

$$+\uparrow (F_R)_y = \Sigma F_y; \quad (F_R)_y = (500 \text{ N})\left(\tfrac{4}{5}\right) - 750 \text{ N} = -350 \text{ N} = 350 \text{ N} \downarrow$$

From Fig. 4–15*b*, the magnitude of $\mathbf{F}_R$ is

$$F_R = \sqrt{(F_R)_x^2 + (F_R)_y^2}$$

$$= \sqrt{(300 \text{ N})^2 + (350 \text{ N})^2} = 461 \text{ N} \qquad \textit{Ans.}$$

And the angle θ is

$$\theta = \tan^{-1}\left(\frac{(F_R)_y}{(F_R)_x}\right) = \tan^{-1}\left(\frac{350 \text{ N}}{300 \text{ N}}\right) = 49.4° \qquad \textit{Ans.}$$

Moment Summation.   Since the couple moment is a free vector, it can act at any point on the member. Referring to Fig. 4–38*a*, we have

$$\zeta + (M_R)_O = \Sigma M_O + \Sigma M$$

$$(M_R)_O = (500 \text{ N}) \left(\tfrac{4}{5}\right)(2.5 \text{ m}) - (500 \text{ N}) \left(\tfrac{3}{5}\right)(1 \text{ m})$$

$$- (750 \text{ N})(1.25 \text{ m}) + 200 \text{ N} \cdot \text{m}$$

$$= -37.5 \text{ N} \cdot \text{m} = 37.5 \text{ N} \cdot \text{m} \, \rangle \qquad \textit{Ans.}$$

This clockwise moment is shown in Fig. 4–38*b*.

EXAMPLE 4.16

The structural member is subjected to a couple moment $\mathbf{M}$ and forces $\mathbf{F}_1$ and $\mathbf{F}_2$ in Fig. 4–39a. Replace this system by an equivalent resultant force and couple moment acting at its base, point O.

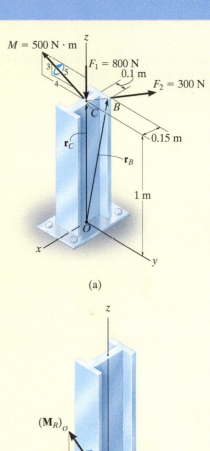

(a)

SOLUTION (*VECTOR ANALYSIS*)

The three-dimensional aspects of the problem can be simplified by using a Cartesian vector analysis. Expressing the forces and couple moment as Cartesian vectors, we have

$$\mathbf{F}_1 = \{-800\mathbf{k}\} \text{ N}$$

$$\mathbf{F}_2 = (300 \text{ N})\mathbf{u}_{CB}$$

$$= (300 \text{ N})\left(\frac{\mathbf{r}_{CB}}{r_{CB}}\right)$$

$$= 300 \text{ N}\left[\frac{\{-0.15\mathbf{i} + 0.1\mathbf{j}\} \text{ m}}{\sqrt{(-0.15 \text{ m})^2 + (0.1 \text{ m})^2}}\right] = \{-249.6\mathbf{i} + 166.4\mathbf{j}\} \text{ N}$$

$$\mathbf{M} = -500\left(\tfrac{4}{5}\right)\mathbf{j} + 500\left(\tfrac{3}{5}\right)\mathbf{k} = \{-400\mathbf{j} + 300\mathbf{k}\} \text{ N} \cdot \text{m}$$

Force Summation.

$$\mathbf{F}_R = \Sigma\mathbf{F}; \qquad \mathbf{F}_R = \mathbf{F}_1 + \mathbf{F}_2 = -800\mathbf{k} - 249.6\mathbf{i} + 166.4\mathbf{j}$$

$$= \{-250\mathbf{i} + 166\mathbf{j} - 800\mathbf{k}\} \text{ N} \qquad \qquad Ans.$$

Moment Summation.

$$(\mathbf{M}_R)_o = \Sigma\mathbf{M} + \Sigma\mathbf{M}_O$$

$$(\mathbf{M}_R)_o = \mathbf{M} + \mathbf{r}_C \times \mathbf{F}_1 + \mathbf{r}_B \times \mathbf{F}_2$$

$$(\mathbf{M}_R)_o = (-400\mathbf{j} + 300\mathbf{k}) + (1\mathbf{k}) \times (-800\mathbf{k}) + \begin{vmatrix} \mathbf{i} & \mathbf{j} & \mathbf{k} \\ -0.15 & 0.1 & 1 \\ -249.6 & 166.4 & 0 \end{vmatrix}$$

$$= (-400\mathbf{j} + 300\mathbf{k}) + (0) + (-166.4\mathbf{i} - 249.6\mathbf{j})$$

$$= \{-166\mathbf{i} - 650\mathbf{j} + 300\mathbf{k}\} \text{ N} \cdot \text{m} \qquad \qquad Ans.$$

The results are shown in Fig. 4–39b.

(b)

Fig. 4–39

4

PRELIMINARY PROBLEM

P4–5. In each case, determine the x and y components of the resultant force and the resultant couple moment at point O.

4

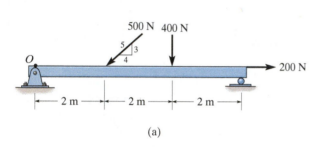

(a)

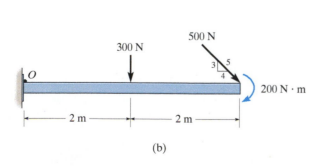

(b)

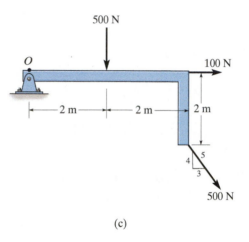

(c)

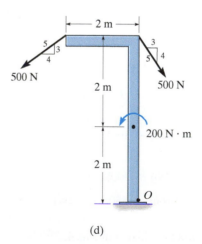

(d)

Prob. P4–5

FUNDAMENTAL PROBLEMS

F4–25. Replace the loading system by an equivalent resultant force and couple moment acting at point A.

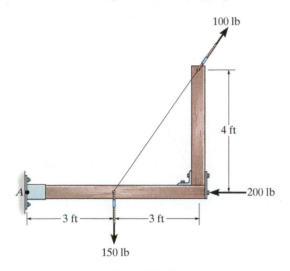

Prob. F4–25

F4–26. Replace the loading system by an equivalent resultant force and couple moment acting at point A.

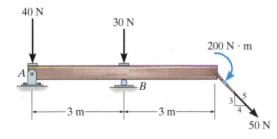

Prob. F4–26

F4–27. Replace the loading system by an equivalent resultant force and couple moment acting at point A.

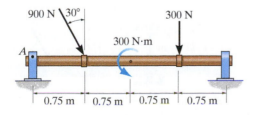

Prob. F4–27

F4–28. Replace the loading system by an equivalent resultant force and couple moment acting at point A.

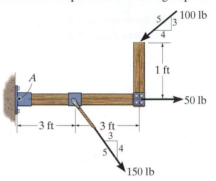

Prob. F4–28

F4–29. Replace the loading system by an equivalent resultant force and couple moment acting at point O.

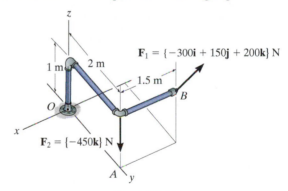

Prob. F4–29

F4–30. Replace the loading system by an equivalent resultant force and couple moment acting at point O.

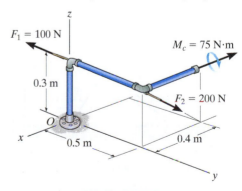

Prob. F4–30

PROBLEMS

4–97. Replace the force system by an equivalent resultant force and couple moment at point O.

4–98. Replace the force system by an equivalent resultant force and couple moment at point P.

4–101. Replace the loading system acting on the beam by an equivalent resultant force and couple moment at point O.

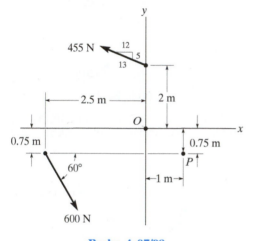

Probs. 4–97/98

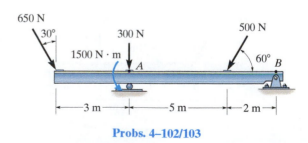

Prob. 4–101

4–99. Replace the force system acting on the beam by an equivalent force and couple moment at point A.

*4–100. Replace the force system acting on the beam by an equivalent force and couple moment at point B.

4–102. Replace the loading system acting on the post by an equivalent resultant force and couple moment at point A.

4–103. Replace the loading system acting on the post by an equivalent resultant force and couple moment at point B.

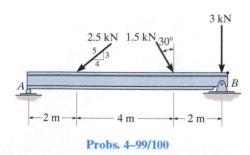

Probs. 4–99/100

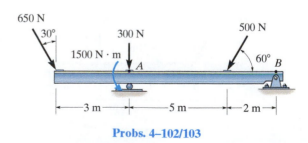

Probs. 4–102/103

*4–104. Replace the force system acting on the post by a resultant force and couple moment at point O.

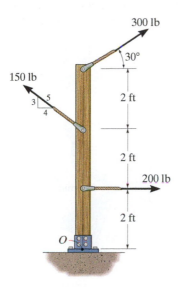

Prob. 4–104

4–105. Replace the force system acting on the frame by an equivalent resultant force and couple moment acting at point A.

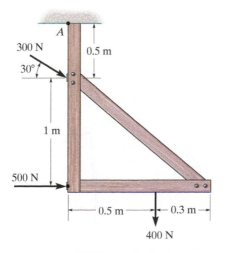

Prob. 4–105

4–106. The forces $\mathbf{F}_1 = \{-4\mathbf{i} + 2\mathbf{j} - 3\mathbf{k}\}$ kN and $\mathbf{F}_2 = \{3\mathbf{i} - 4\mathbf{j} - 2\mathbf{k}\}$ kN act on the end of the beam. Replace these forces by an equivalent force and couple moment acting at point O.

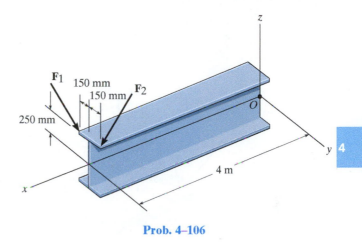

Prob. 4–106

4–107. A biomechanical model of the lumbar region of the human trunk is shown. The forces acting in the four muscle groups consist of $F_R = 35$ N for the rectus, $F_O = 45$ N for the oblique, $F_L = 23$ N for the lumbar latissimus dorsi, and $F_E = 32$ N for the erector spinae. These loadings are symmetric with respect to the y–z plane. Replace this system of parallel forces by an equivalent force and couple moment acting at the spine, point O. Express the results in Cartesian vector form.

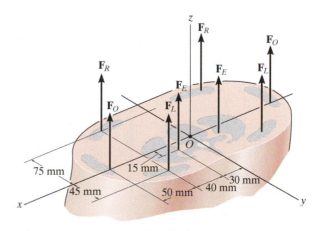

Prob. 4–107

*4–108. Replace the force system by an equivalent resultant force and couple moment at point O. Take $\mathbf{F}_3 = \{-200\mathbf{i} + 500\mathbf{j} - 300\mathbf{k}\}$ N.

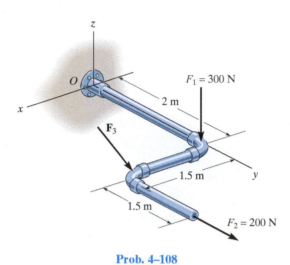

Prob. 4–108

4–110. Replace the force of $F = 80$ N acting on the pipe assembly by an equivalent resultant force and couple moment at point A.

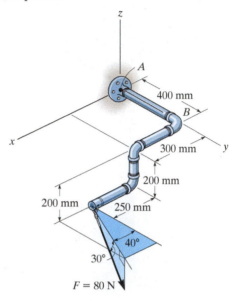

Prob. 4–110

4–109. Replace the loading by an equivalent resultant force and couple moment at point O.

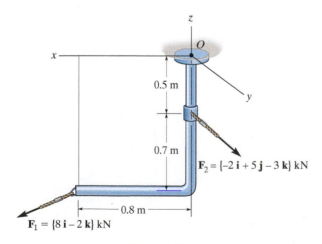

Prob. 4–109

4–111. The belt passing over the pulley is subjected to forces $\mathbf{F}_1$ and $\mathbf{F}_2$, each having a magnitude of 40 N. $\mathbf{F}_1$ acts in the $-\mathbf{k}$ direction. Replace these forces by an equivalent force and couple moment at point A. Express the result in Cartesian vector form. Set $\theta = 0°$ so that $\mathbf{F}_2$ acts in the $-\mathbf{j}$ direction.

*4–112. The belt passing over the pulley is subjected to two forces $\mathbf{F}_1$ and $\mathbf{F}_2$, each having a magnitude of 40 N. $\mathbf{F}_1$ acts in the $-\mathbf{k}$ direction. Replace these forces by an equivalent force and couple moment at point A. Express the result in Cartesian vector form. Take $\theta = 45°$.

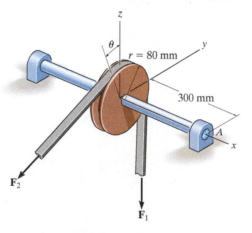

Probs. 4–111/112

4.8   Further Simplification of a Force and Couple System

In the preceding section, we developed a way to reduce a force and couple moment system acting on a rigid body into an equivalent resultant force $\mathbf{F}_R$ acting at a specific point O and a resultant couple moment $(\mathbf{M}_R)_O$. The force system can be further reduced to an equivalent single resultant force provided the lines of action of $\mathbf{F}_R$ and $(\mathbf{M}_R)_O$ are *perpendicular* to each other. Because of this condition, concurrent, coplanar, and parallel force systems can be further simplified.

Concurrent Force System.   Since a *concurrent force system* is one in which the lines of action of all the forces intersect at a common point O, Fig. 4–40a, then the force system produces no moment about this point. As a result, the equivalent system can be represented by a single resultant force $\mathbf{F}_R = \Sigma\mathbf{F}$ acting at O, Fig. 4–40b.

Coplanar Force System.   In the case of a *coplanar force system*, the lines of action of all the forces lie in the same plane, Fig. 4–41a, and so the resultant force $\mathbf{F}_R = \Sigma\mathbf{F}$ of this system also lies in this plane. Furthermore, the moment of each of the forces about any point O is directed perpendicular to this plane. Thus, the resultant moment $(\mathbf{M}_R)_O$ and resultant force $\mathbf{F}_R$ will be *mutually perpendicular*, Fig. 4–41b. The resultant moment can be replaced by moving the resultant force $\mathbf{F}_R$ a perpendicular or moment arm distance d away from point O such that $\mathbf{F}_R$ produces the *same moment* $(\mathbf{M}_R)_O$ about point O, Fig. 4–41c. This distance d can be determined from the scalar equation $(M_R)_O = F_R d = \Sigma M_O$ or $d = (M_R)_O/F_R$.

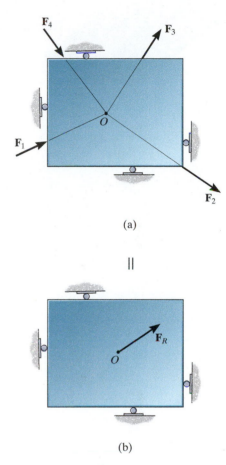

(a)

||

(b)

Fig. 4–40

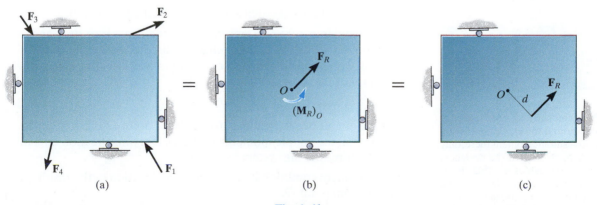

(a)   =   (b)   =   (c)

Fig. 4–41

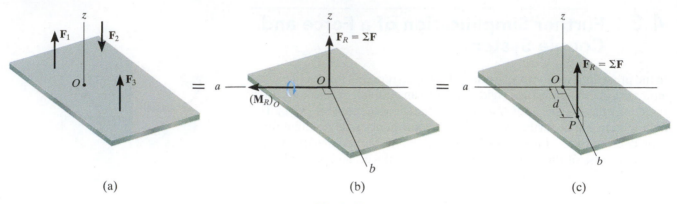

(a)   (b)   (c)

Fig. 4–42

Parallel Force System.

The *parallel force system* shown in Fig. 4–42a consists of forces that are all parallel to the z axis. Thus, the resultant force $F_R = \Sigma F$ at point O must also be parallel to this axis, Fig. 4–42b. The moment produced by each force lies in the plane of the plate, and so the resultant couple moment, $(M_R)_O$, will also lie in this plane, along the moment axis a since F_R and $(M_R)_O$ are mutually perpendicular. As a result, the force system can be further reduced to an equivalent single resultant force F_R, acting through point P located on the perpendicular b axis, Fig. 4–42c. The distance d along this axis from point O requires $(M_R)_O = F_R d = \Sigma M_O$ or $d = \Sigma M_O / F_R$.

Procedure for Analysis

The technique used to reduce a coplanar or parallel force system to a single resultant force follows a similar procedure outlined in the previous section.

- Establish the x, y, z, axes and locate the resultant force F_R an arbitrary distance away from the origin of the coordinates.

Force Summation.

- The resultant force is equal to the sum of all the forces in the system.
- For a coplanar force system, resolve each force into its x and y components. Positive components are directed along the positive x and y axes, and negative components are directed along the negative x and y axes.

Moment Summation.

- The moment of the resultant force about point O is equal to the sum of all the couple moments in the system plus the moments of all the forces in the system about O.
- This moment condition is used to find the location of the resultant force from point O.

The four cable forces are all concurrent at point O on this bridge tower. Consequently they produce no resultant moment there, only a resultant force F_R. Note that the designers have positioned the cables so that F_R is directed *along* the bridge tower directly to the support, so that it does not cause any bending of the tower. (© Russell C. Hibbeler)

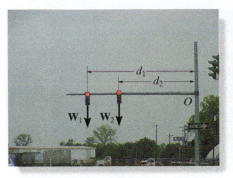

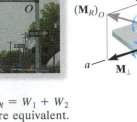
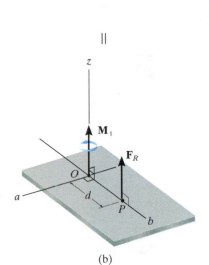

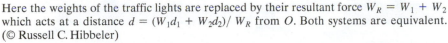

Here the weights of the traffic lights are replaced by their resultant force $W_R = W_1 + W_2$ which acts at a distance $d = (W_1 d_1 + W_2 d_2)/ W_R$ from O. Both systems are equivalent. (© Russell C. Hibbeler)

(a)

Reduction to a Wrench.

In general, a three-dimensional force and couple moment system will have an equivalent resultant force $\mathbf{F}_R$ acting at point O and a resultant couple moment $(\mathbf{M}_R)_O$ that are *not perpendicular* to one another, as shown in Fig. 4–43a. Although a force system such as this cannot be further reduced to an equivalent single resultant force, the resultant couple moment $(\mathbf{M}_R)_O$ can be resolved into components parallel and perpendicular to the line of action of $\mathbf{F}_R$, Fig. 4–43a. If this appears difficult to do in three dimensions, use the dot product to get $\mathbf{M}_{\parallel} = (\mathbf{M}_R) \cdot \mathbf{u}_{F_R}$ and then $\mathbf{M}_{\perp} = \mathbf{M}_R - \mathbf{M}_{\parallel}$. The perpendicular component $\mathbf{M}_{\perp}$ can be replaced if we move $\mathbf{F}_R$ to point P, a distance d from point O along the b axis, Fig. 4–43b. As shown, this axis is perpendicular to both the a axis and the line of action of $\mathbf{F}_R$. The location of P can be determined from $d = M_{\perp}/F_R$. Finally, because $\mathbf{M}_{\parallel}$ is a free vector, it can be moved to point P, Fig. 4–43c. This combination of a resultant force $\mathbf{F}_R$ and collinear couple moment $\mathbf{M}_{\parallel}$ will tend to translate and rotate the body about its axis and is referred to as a **wrench** or **screw**. A wrench is the simplest system that can represent any general force and couple moment system acting on a body.

(b)

Important Point

- A concurrent, coplanar, or parallel force system can always be reduced to a single resultant force acting at a specific point P. For any other type of force system, the simplest reduction is a wrench, which consists of resultant force and collinear couple moment acting at a specific point P.

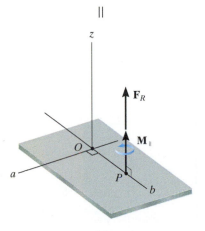

Fig. 4–43

EXAMPLE | 4.17

Replace the force and couple moment system acting on the beam in Fig. 4–44a by an equivalent resultant force, and find where its line of action intersects the beam, measured from point O.

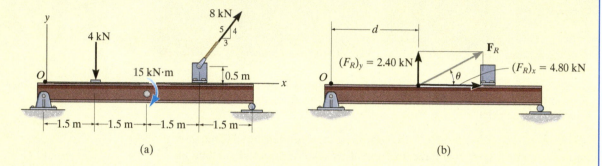

(a)   (b)

Fig. 4–44

SOLUTION

Force Summation. Summing the force components,

$$\xrightarrow{+} (F_R)_x = \Sigma F_x; \qquad (F_R)_x = 8 \text{ kN}\left(\tfrac{3}{5}\right) = 4.80 \text{ kN} \rightarrow$$

$$+\uparrow(F_R)_y = \Sigma F_y; \qquad (F_R)_y = -4 \text{ kN} + 8 \text{ kN}\left(\tfrac{4}{5}\right) = 2.40 \text{ kN} \uparrow$$

From Fig. 4–44b, the magnitude of $\mathbf{F}_R$ is

$$F_R = \sqrt{(4.80 \text{ kN})^2 + (2.40 \text{ kN})^2} = 5.37 \text{ kN} \qquad \textit{Ans.}$$

The angle θ is

$$\theta = \tan^{-1}\left(\frac{2.40 \text{ kN}}{4.80 \text{ kN}}\right) = 26.6° \qquad \textit{Ans.}$$

Moment Summation. We must equate the moment of $\mathbf{F}_R$ about point O in Fig. 4–44b to the sum of the moments of the force and couple moment system about point O in Fig. 4–44a. Since the line of action of $(\mathbf{F}_R)_x$ acts through point O, *only $(\mathbf{F}_R)_y$ produces a moment* about this point. Thus,

$$\zeta + (M_R)_O = \Sigma M_O; \qquad 2.40 \text{ kN}(d) = -(4 \text{ kN})(1.5 \text{ m}) - 15 \text{ kN} \cdot \text{m}$$

$$-\left[8 \text{ kN}\left(\tfrac{3}{5}\right)\right](0.5 \text{ m}) + \left[8 \text{ kN}\left(\tfrac{4}{5}\right)\right](4.5 \text{ m})$$

$$d = 2.25 \text{ m} \qquad \textit{Ans.}$$

EXAMPLE 4.18

The jib crane shown in Fig. 4–45a is subjected to three coplanar forces. Replace this loading by an equivalent resultant force and specify where the resultant's line of action intersects the column AB and boom BC.

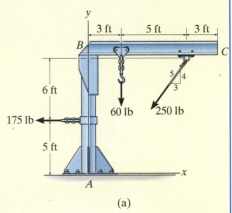

(a)

SOLUTION

Force Summation. Resolving the 250-lb force into x and y components and summing the force components yields

$$\xrightarrow{+} (F_R)_x = \Sigma F_x; \quad (F_R)_x = -250 \text{ lb}\left(\tfrac{3}{5}\right) - 175 \text{ lb} = -325 \text{ lb} = 325 \text{ lb} \leftarrow$$

$$+\uparrow (F_R)_y = \Sigma F_y; \quad (F_R)_y = -250 \text{ lb}\left(\tfrac{4}{5}\right) - 60 \text{ lb} = -260 \text{ lb} = 260 \text{ lb} \downarrow$$

As shown by the vector addition in Fig. 4–45b,

$$F_R = \sqrt{(325 \text{ lb})^2 + (260 \text{ lb})^2} = 416 \text{ lb} \qquad \text{Ans.}$$

$$\theta = \tan^{-1}\left(\frac{260 \text{ lb}}{325 \text{ lb}}\right) = 38.7° \nearrow \qquad \text{Ans.}$$

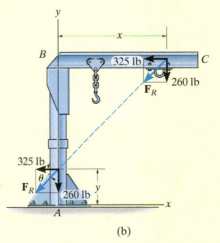

(b)

Fig. 4–45

Moment Summation. Moments will be summed about point A. Assuming the line of action of $\mathbf{F}_R$ intersects AB at a distance y from A, Fig. 4–45b, we have

$$\zeta + (M_R)_A = \Sigma M_A; \quad 325 \text{ lb } (y) + 260 \text{ lb } (0)$$

$$= 175 \text{ lb } (5 \text{ ft}) - 60 \text{ lb } (3 \text{ ft}) + 250 \text{ lb}\left(\tfrac{3}{5}\right)(11 \text{ ft}) - 250 \text{ lb}\left(\tfrac{4}{5}\right)(8 \text{ ft})$$

$$y = 2.29 \text{ ft} \qquad \text{Ans.}$$

By the principle of transmissibility, $\mathbf{F}_R$ can be placed at a distance x where it intersects BC, Fig. 4–45b. In this case we have

$$\zeta + (M_R)_A = \Sigma M_A; \quad 325 \text{ lb } (11 \text{ ft}) - 260 \text{ lb } (x)$$

$$= 175 \text{ lb } (5 \text{ ft}) - 60 \text{ lb } (3 \text{ ft}) + 250 \text{ lb}\left(\tfrac{3}{5}\right)(11 \text{ ft}) - 250 \text{ lb}\left(\tfrac{4}{5}\right)(8 \text{ ft})$$

$$x = 10.9 \text{ ft} \qquad \text{Ans.}$$

EXAMPLE   4.19

The slab in Fig. 4–46a is subjected to four parallel forces. Determine the magnitude and direction of a resultant force equivalent to the given force system, and locate its point of application on the slab.

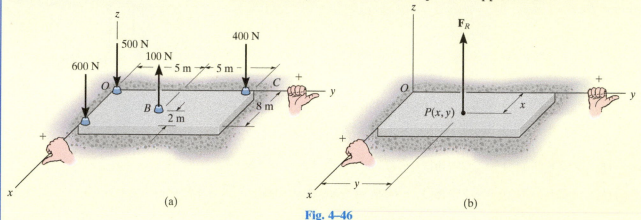

Fig. 4–46

SOLUTION (SCALAR ANALYSIS)

Force Summation.   From Fig. 4–46a, the resultant force is

$$+\uparrow F_R = \Sigma F; \qquad F_R = -600 \text{ N} + 100 \text{ N} - 400 \text{ N} - 500 \text{ N}$$

$$= -1400 \text{ N} = 1400 \text{ N}\downarrow \qquad Ans.$$

Moment Summation.   We require the moment about the x axis of the resultant force, Fig. 4–46b, to be equal to the sum of the moments about the x axis of all the forces in the system, Fig. 4–46a. The moment arms are determined from the y coordinates, since these coordinates represent the *perpendicular distances* from the x axis to the lines of action of the forces. Using the right-hand rule, we have

$$(M_R)_x = \Sigma M_x;$$

$$-(1400 \text{ N})y = 600 \text{ N}(0) + 100 \text{ N}(5 \text{ m}) - 400 \text{ N}(10 \text{ m}) + 500 \text{ N}(0)$$

$$-1400y = -3500 \qquad y = 2.50 \text{ m} \qquad Ans.$$

In a similar manner, a moment equation can be written about the y axis using moment arms defined by the x coordinates of each force.

$$(M_R)_y = \Sigma M_y;$$

$$(1400 \text{ N})x = 600 \text{ N}(8 \text{ m}) - 100 \text{ N}(6 \text{ m}) + 400 \text{ N}(0) + 500 \text{ N}(0)$$

$$1400x = 4200$$

$$x = 3 \text{ m} \qquad Ans.$$

NOTE: A force of $F_R = 1400$ N placed at point $P(3.00$ m, 2.50 m) on the slab, Fig. 4–46b, is therefore equivalent to the parallel force system acting on the slab in Fig. 4–46a.

EXAMPLE | 4.20

Replace the force system in Fig. 4–47a by an equivalent resultant force and specify its point of application on the pedestal.

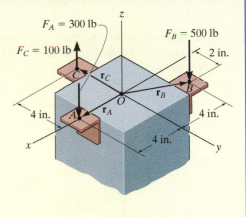

(a)

SOLUTION

Force Summation.   Here we will demonstrate a vector analysis. Summing forces,

$$\mathbf{F}_R = \Sigma\mathbf{F};\ \mathbf{F}_R = \mathbf{F}_A + \mathbf{F}_B + \mathbf{F}_C$$

$$= \{-300\mathbf{k}\}\ \text{lb} + \{-500\mathbf{k}\}\ \text{lb} + \{100\mathbf{k}\}\ \text{lb}$$

$$= \{-700\mathbf{k}\}\ \text{lb} \qquad\qquad Ans.$$

Location.   Moments will be summed about point O. The resultant force $\mathbf{F}_R$ is assumed to act through point $P\ (x, y, 0)$, Fig. 4–47b. Thus

$$(\mathbf{M}_R)_O = \Sigma\mathbf{M}_O;$$

$$\mathbf{r}_P \times \mathbf{F}_R = (\mathbf{r}_A \times \mathbf{F}_A) + (\mathbf{r}_B \times \mathbf{F}_B) + (\mathbf{r}_C \times \mathbf{F}_C)$$

$$(x\mathbf{i} + y\mathbf{j}) \times (-700\mathbf{k}) = [(4\mathbf{i}) \times (-300\mathbf{k})]$$

$$+ [(-4\mathbf{i} + 2\mathbf{j}) \times (-500\mathbf{k})] + [(-4\mathbf{j}) \times (100\mathbf{k})]$$

$$-700x(\mathbf{i} \times \mathbf{k}) - 700y(\mathbf{j} \times \mathbf{k}) = -1200(\mathbf{i} \times \mathbf{k}) + 2000(\mathbf{i} \times \mathbf{k})$$

$$- 1000(\mathbf{j} \times \mathbf{k}) - 400(\mathbf{j} \times \mathbf{k})$$

$$700x\mathbf{j} - 700y\mathbf{i} = 1200\mathbf{j} - 2000\mathbf{j} - 1000\mathbf{i} - 400\mathbf{i}$$

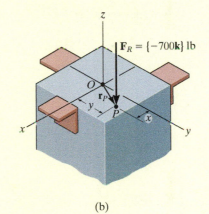

(b)

Fig. 4–47

Equating the $\mathbf{i}$ and $\mathbf{j}$ components,

$$-700y = -1400 \qquad\qquad (1)$$

$$y = 2\ \text{in.} \qquad\qquad Ans.$$

$$700x = -800 \qquad\qquad (2)$$

$$x = -1.14\ \text{in.} \qquad\qquad Ans.$$

The negative sign indicates that the x coordinate of point P is negative.

NOTE: It is also possible to establish Eq. 1 and 2 directly by summing moments about the x and y axes. Using the right-hand rule, we have

$$(M_R)_x = \Sigma M_x; \qquad -700y = -100\ \text{lb}(4\ \text{in.}) - 500\ \text{lb}(2\ \text{in.})$$

$$(M_R)_y = \Sigma M_y; \qquad 700x = 300\ \text{lb}(4\ \text{in.}) - 500\ \text{lb}(4\ \text{in.})$$

PRELIMINARY PROBLEMS

P4–6. In each case, determine the x and y components of the resultant force and specify the distance where this force acts from point O.

P4–7. In each case, determine the resultant force and specify its coordinates x and y where it acts on the x–y plane.

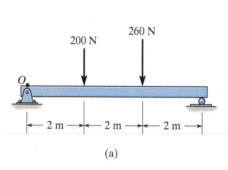

(a)

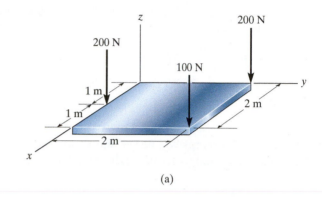

(a)

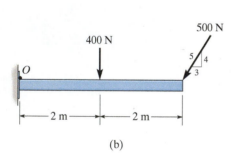

(b)

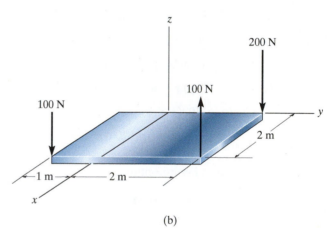

(b)

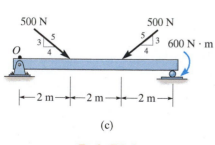

(c)

Prob. P4–6

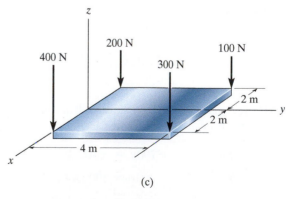

(c)

Prob. P4–7

FUNDAMENTAL PROBLEMS

F4–31. Replace the loading system by an equivalent resultant force and specify where the resultant's line of action intersects the beam measured from O.

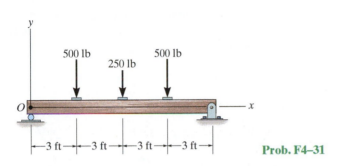

Prob. F4–31

F4–32. Replace the loading system by an equivalent resultant force and specify where the resultant's line of action intersects the member measured from A.

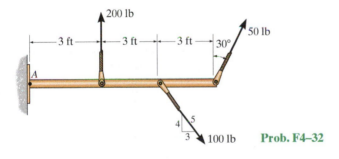

Prob. F4–32

F4–33. Replace the loading system by an equivalent resultant force and specify where the resultant's line of action intersects the horizontal segment of the member measured from A.

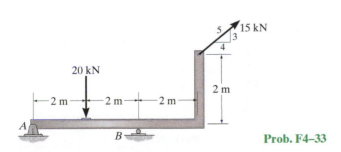

Prob. F4–33

F4–34. Replace the loading system by an equivalent resultant force and specify where the resultant's line of action intersects the member AB measured from A.

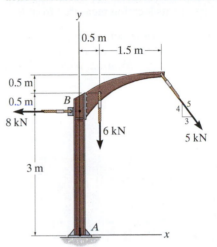

Prob. F4–34

F4–35. Replace the loading shown by an equivalent single resultant force and specify the x and y coordinates of its line of action.

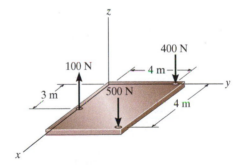

Prob. F4–35

F4–36. Replace the loading shown by an equivalent single resultant force and specify the x and y coordinates of its line of action.

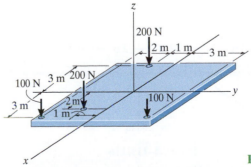

Prob. F4–36

PROBLEMS

4–113. The weights of the various components of the truck are shown. Replace this system of forces by an equivalent resultant force and specify its location measured from *B*.

4–114. The weights of the various components of the truck are shown. Replace this system of forces by an equivalent resultant force and specify its location measured from point *A*.

4–117. Replace the loading acting on the beam by a single resultant force. Specify where the force acts, measured from end *A*.

4–118. Replace the loading acting on the beam by a single resultant force. Specify where the force acts, measured from *B*.

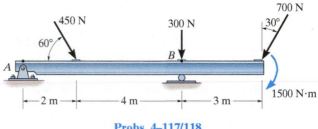

Probs. 4–117/118

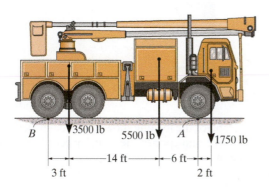

Probs. 4–113/114

4–119. Replace the loading on the frame by a single resultant force. Specify where its line of action intersects a vertical line along member *AB*, measured from *A*.

4–115. Replace the three forces acting on the shaft by a single resultant force. Specify where the force acts, measured from end *A*.

*4–116. Replace the three forces acting on the shaft by a single resultant force. Specify where the force acts, measured from end *B*.

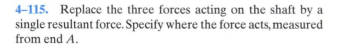

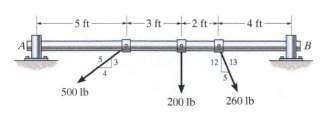

Probs. 4–115/116

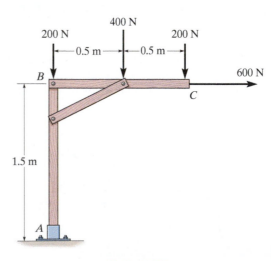

Prob. 4–119

*4–120. Replace the loading on the frame by a single resultant force. Specify where its line of action intersects a vertical line along member *AB*, measured from *A*.

4–121. Replace the loading on the frame by a single resultant force. Specify where its line of action intersects a horizontal line along member *CB*, measured from end *C*.

*4–124. Replace the parallel force system acting on the plate by a resultant force and specify its location on the *x-z* plane.

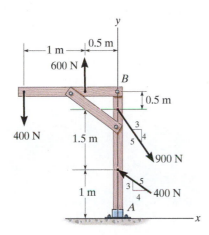

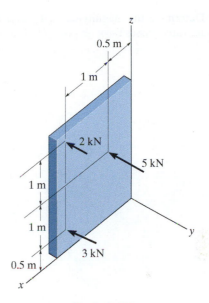

Prob. 4–124

Probs. 4–120/121

4–122. Replace the force system acting on the post by a resultant force, and specify where its line of action intersects the post *AB* measured from point *A*.

4–123. Replace the force system acting on the post by a resultant force, and specify where its line of action intersects the post *AB* measured from point *B*.

4–125. Replace the force and couple system acting on the frame by an equivalent resultant force and specify where the resultant's line of action intersects member *AB*, measured from *A*.

4–126. Replace the force and couple system acting on the frame by an equivalent resultant force and specify where the resultant's line of action intersects member *BC*, measured from *B*.

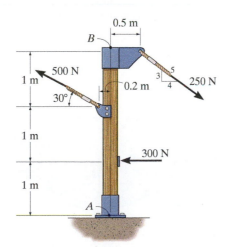

Probs. 4–122/123

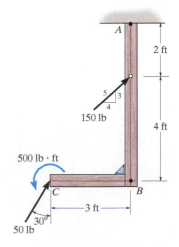

Probs. 4–125/126

4–127. If $F_A = 7$ kN and $F_B = 5$ kN, represent the force system acting on the corbels by a resultant force, and specify its location on the x–y plane.

*4–128. Determine the magnitudes of $\mathbf{F}_A$ and $\mathbf{F}_B$ so that the resultant force passes through point O of the column.

4–130. The building slab is subjected to four parallel column loadings. Determine the equivalent resultant force and specify its location (x, y) on the slab. Take $F_1 = 8$ kN and $F_2 = 9$ kN.

4–131. The building slab is subjected to four parallel column loadings. Determine $\mathbf{F}_1$ and $\mathbf{F}_2$ if the resultant force acts through point (12 m, 10 m).

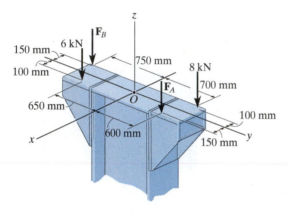

Probs. 4–127/128

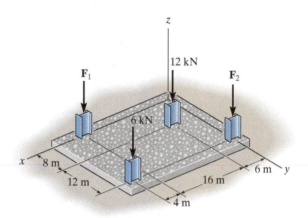

Probs. 4–130/131

4–129. The tube supports the four parallel forces. Determine the magnitudes of forces $\mathbf{F}_C$ and $\mathbf{F}_D$ acting at C and D so that the equivalent resultant force of the force system acts through the midpoint O of the tube.

*4–132. If $F_A = 40$ kN and $F_B = 35$ kN, determine the magnitude of the resultant force and specify the location of its point of application (x, y) on the slab.

4–133. If the resultant force is required to act at the center of the slab, determine the magnitude of the column loadings $\mathbf{F}_A$ and $\mathbf{F}_B$ and the magnitude of the resultant force.

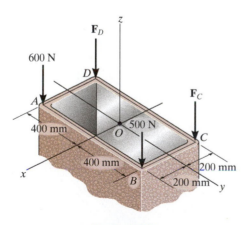

Prob. 4–129

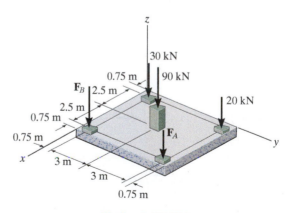

Probs. 4–132/133

4–134. Replace the two wrenches and the force, acting on the pipe assembly, by an equivalent resultant force and couple moment at point O.

*4–136. Replace the five forces acting on the plate by a wrench. Specify the magnitude of the force and couple moment for the wrench and the point $P(x, z)$ where the wrench intersects the x–z plane.

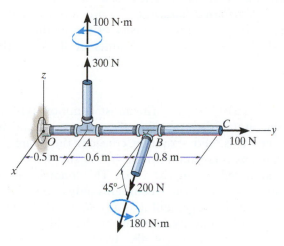

Prob. 4–134

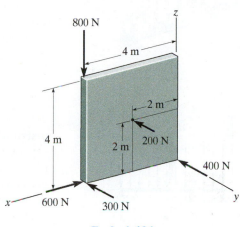

Prob. 4–136

4–135. Replace the force system by a wrench and specify the magnitude of the force and couple moment of the wrench and the point where the wrench intersects the x–z plane.

4–137. Replace the three forces acting on the plate by a wrench. Specify the magnitude of the force and couple moment for the wrench and the point $P(x, y)$ where the wrench intersects the plate.

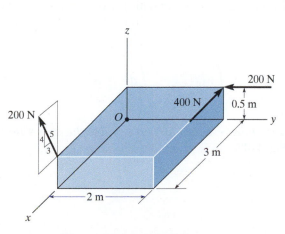

Prob. 4–135

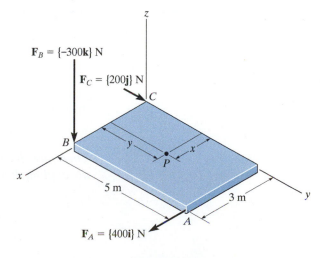

Prob. 4–137

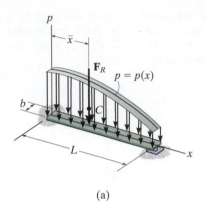

(a)

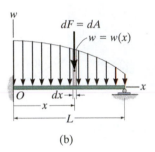

(b)

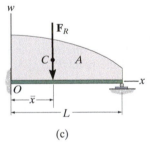

(c)

Fig. 4–48

4.9   Reduction of a Simple Distributed Loading

Sometimes, a body may be subjected to a loading that is distributed over its surface. For example, the pressure of the wind on the face of a sign, the pressure of water within a tank, or the weight of sand on the floor of a storage container, are all *distributed loadings*. The pressure exerted at each point on the surface indicates the intensity of the loading. It is measured using pascals Pa (or N/m^2) in SI units or lb/ft^2 in the U.S. Customary system.

Loading Along a Single Axis.   The most common type of distributed loading encountered in engineering practice can be represented along a single axis.* For example, consider the beam (or plate) in Fig. 4–48a that has a constant width and is subjected to a pressure loading that varies only along the x axis. This loading can be described by the function $p = p(x) \, N/m^2$. It contains only one variable x, and for this reason, we can also represent it as a *coplanar distributed load*. To do so, we multiply the loading function by the width b m of the beam, so that $w(x) = p(x)b \, N/m$, Fig. 4–48b. Using the methods of Sec. 4.8, we can replace this coplanar parallel force system with a single equivalent resultant force $\mathbf{F}_R$ acting at a specific location on the beam, Fig. 4–48c.

Magnitude of Resultant Force.   From Eq. 4–17 ($F_R = \Sigma F$), the magnitude of $\mathbf{F}_R$ is equivalent to the sum of all the forces in the system. In this case integration must be used since there is an infinite number of parallel forces $d\mathbf{F}$ acting on the beam, Fig. 4–48b. Since $d\mathbf{F}$ is acting on an element of length dx, and $w(x)$ is a force per unit length, then $dF = w(x) \, dx = dA$. In other words, the magnitude of $d\mathbf{F}$ is determined from the colored differential *area* dA under the loading curve. For the entire length L,

$$+\downarrow F_R = \Sigma F; \qquad \boxed{F_R = \int_L w(x) \, dx = \int_A dA = A} \qquad (4\text{–}19)$$

Therefore, the magnitude of the resultant force is equal to the area A under the loading diagram, Fig. 4–48c.

---

*The more general case of a surface loading acting on a body is considered in Sec. 9.5.

Location of Resultant Force. Applying Eq. 4–17 ($M_{R_O} = \Sigma M_O$), the location $\bar{x}$ of the line of action of $\mathbf{F}_R$ can be determined by equating the moments of the force resultant and the parallel force distribution about point O (the y axis). Since $d\mathbf{F}$ produces a moment of $x\, dF = xw(x)\, dx$ about O, Fig. 4–48b, then for the entire length, Fig. 4–48c,

$$\zeta + (M_R)_O = \Sigma M_O; \qquad\qquad -\bar{x}F_R = -\int_L xw(x)\, dx$$

Solving for $\bar{x}$, using Eq. 4–19, we have

$$\bar{x} = \frac{\displaystyle\int_L xw(x)\, dx}{\displaystyle\int_L w(x)\, dx} = \frac{\displaystyle\int_A x\, dA}{\displaystyle\int_A dA} \qquad (4\text{–}20)$$

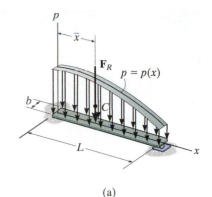

(a)

This coordinate $\bar{x}$, locates the geometric center or **centroid** of the *area under the distributed loading. In other words, the resultant force has a line of action which passes through the centroid C (geometric center) of the area under the loading diagram*, Fig. 4–48c. Detailed treatment of the integration techniques for finding the location of the centroid for areas is given in Chapter 9. In many cases, however, the distributed-loading diagram is in the shape of a rectangle, triangle, or some other simple geometric form. The centroid location for such common shapes does not have to be determined from the above equation but can be obtained directly from the tabulation given on the inside back cover.

Once $\bar{x}$ is determined, $\mathbf{F}_R$ by symmetry passes through point $(\bar{x}, 0)$ on the surface of the beam, Fig. 4–48a. *Therefore, in this case the resultant force has a magnitude equal to the volume under the loading curve* $p = p(x)$ *and a line of action which passes through the centroid (geometric center) of this volume.*

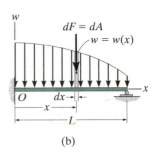

(b)

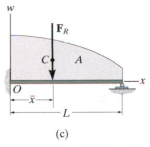

(c)

Fig. 4–48 (Repeated)

Important Points

- Coplanar distributed loadings are defined by using a loading function $w = w(x)$ that indicates the intensity of the loading along the length of a member. This intensity is measured in N/m or lb/ft.

- The external effects caused by a coplanar distributed load acting on a body can be represented by a single resultant force.

- This resultant force is equivalent to the *area* under the loading diagram, and has a line of action that passes through the *centroid* or geometric center of this area.

The pile of brick creates an approximate triangular distributed loading on the board. (© Russell C. Hibbeler)

EXAMPLE | **4.21**

Determine the magnitude and location of the equivalent resultant force acting on the shaft in Fig. 4–49a.

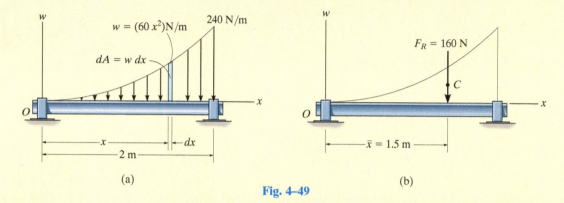

(a)   (b)

Fig. 4–49

SOLUTION

Since $w = w(x)$ is given, this problem will be solved by integration.

The differential element has an area $dA = w \, dx = 60x^2 \, dx$. Applying Eq. 4–19,

$+\downarrow F_R = \Sigma F;$

$$F_R = \int_A dA = \int_0^{2\,m} 60x^2 \, dx = 60\left(\frac{x^3}{3}\right)\Big|_0^{2\,m} = 60\left(\frac{2^3}{3} - \frac{0^3}{3}\right)$$

$$= 160 \text{ N} \hspace{5cm} \textit{Ans.}$$

The location $\bar{x}$ of $\mathbf{F}_R$ measured from O, Fig. 4–49b, is determined from Eq. 4–20.

$$\bar{x} = \frac{\displaystyle\int_A x \, dA}{\displaystyle\int_A dA} = \frac{\displaystyle\int_0^{2\,m} x(60x^2) \, dx}{160 \text{ N}} = \frac{60\left(\dfrac{x^4}{4}\right)\Big|_0^{2\,m}}{160 \text{ N}} = \frac{60\left(\dfrac{2^4}{4} - \dfrac{0^4}{4}\right)}{160 \text{ N}}$$

$$= 1.5 \text{ m} \hspace{5cm} \textit{Ans.}$$

NOTE: These results can be checked by using the table on the inside back cover, where it is shown that the formula for an exparabolic area of length a, height b, and shape shown in Fig. 4–49a, is

$$A = \frac{ab}{3} = \frac{2 \text{ m}(240 \text{ N/m})}{3} = 160 \text{ N and } \bar{x} = \frac{3}{4}a = \frac{3}{4}(2 \text{ m}) = 1.5 \text{ m}$$

EXAMPLE 4.22

A distributed loading of $p = (800x)$ Pa acts over the top surface of the beam shown in Fig. 4–50a. Determine the magnitude and location of the equivalent resultant force.

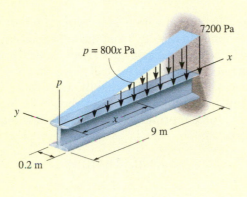

(a)

SOLUTION

Since the loading intensity is uniform along the width of the beam (the y axis), the loading can be viewed in two dimensions as shown in Fig. 4–50b. Here

$$w = (800x \text{ N/m}^2)(0.2 \text{ m})$$

$$= (160x) \text{ N/m}$$

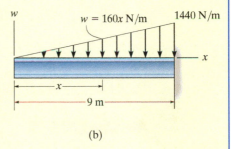

(b)

At $x = 9$ m, note that $w = 1440$ N/m. Although we may again apply Eqs. 4–19 and 4–20 as in the previous example, it is simpler to use the table on the inside back cover.

The magnitude of the resultant force is equivalent to the area of the triangle.

$$F_R = \tfrac{1}{2}(9 \text{ m})(1440 \text{ N/m}) = 6480 \text{ N} = 6.48 \text{ kN} \qquad Ans.$$

The line of action of $\mathbf{F}_R$ passes through the *centroid* C of this triangle. Hence,

$$\bar{x} = 9 \text{ m} - \tfrac{1}{3}(9 \text{ m}) = 6 \text{ m} \qquad Ans.$$

The results are shown in Fig. 4–50c.

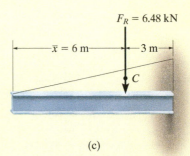

(c)

NOTE: We may also view the resultant $\mathbf{F}_R$ as *acting* through the *centroid* of the *volume* of the loading diagram $p = p(x)$ in Fig. 4–50a. Hence $\mathbf{F}_R$ intersects the x–y plane at the point (6 m, 0). Furthermore, the magnitude of $\mathbf{F}_R$ is equal to the volume under the loading diagram; i.e.,

$$F_R = V = \tfrac{1}{2}(7200 \text{ N/m}^2)(9 \text{ m})(0.2 \text{ m}) = 6.48 \text{ kN} \qquad Ans.$$

Fig. 4–50

EXAMPLE | 4.23

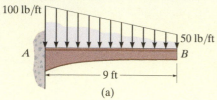

(a)

The granular material exerts the distributed loading on the beam as shown in Fig. 4–51a. Determine the magnitude and location of the equivalent resultant of this load.

SOLUTION

The area of the loading diagram is a *trapezoid*, and therefore the solution can be obtained directly from the area and centroid formulas for a trapezoid listed on the inside back cover. Since these formulas are not easily remembered, instead we will solve this problem by using "composite" areas. Here we will divide the trapezoidal loading into a rectangular and triangular loading as shown in Fig. 4–51b. The magnitude of the force represented by each of these loadings is equal to its associated *area*,

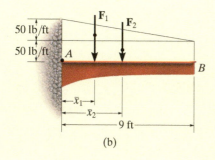

(b)

$$F_1 = \tfrac{1}{2}(9 \text{ ft})(50 \text{ lb/ft}) = 225 \text{ lb}$$

$$F_2 = (9 \text{ ft})(50 \text{ lb/ft}) = 450 \text{ lb}$$

The lines of action of these parallel forces act through the respective *centroids* of their associated areas and therefore intersect the beam at

$$\bar{x}_1 = \tfrac{1}{3}(9 \text{ ft}) = 3 \text{ ft}$$

$$\bar{x}_2 = \tfrac{1}{2}(9 \text{ ft}) = 4.5 \text{ ft}$$

The two parallel forces F_1 and F_2 can be reduced to a single resultant F_R. The magnitude of F_R is

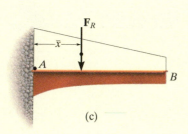

(c)

$$+\downarrow F_R = \Sigma F; \qquad F_R = 225 + 450 = 675 \text{ lb} \qquad \textit{Ans.}$$

We can find the location of F_R with reference to point A, Figs. 4–51b and 4–51c. We require

$$\zeta + (M_R)_A = \Sigma M_A; \qquad \bar{x}(675) = 3(225) + 4.5(450)$$

$$\bar{x} = 4 \text{ ft} \qquad \textit{Ans.}$$

NOTE: The trapezoidal area in Fig. 4–51a can also be divided into two triangular areas as shown in Fig. 4–51d. In this case

$$F_3 = \tfrac{1}{2}(9 \text{ ft})(100 \text{ lb/ft}) = 450 \text{ lb}$$

$$F_4 = \tfrac{1}{2}(9 \text{ ft})(50 \text{ lb/ft}) = 225 \text{ lb}$$

and

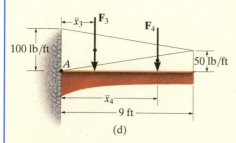

(d)

Fig. 4–51

$$\bar{x}_3 = \tfrac{1}{3}(9 \text{ ft}) = 3 \text{ ft}$$

$$\bar{x}_4 = 9 \text{ ft} - \tfrac{1}{3}(9 \text{ ft}) = 6 \text{ ft}$$

Using these results, show that again $F_R = 675 \text{ lb}$ and $\bar{x} = 4 \text{ ft}$.

FUNDAMENTAL PROBLEMS

F4–37. Determine the resultant force and specify where it acts on the beam measured from A.

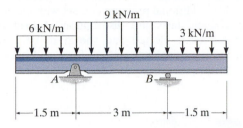

Prob. F4–37

F4–38. Determine the resultant force and specify where it acts on the beam measured from A.

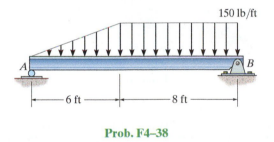

Prob. F4–38

F4–39. Determine the resultant force and specify where it acts on the beam measured from A.

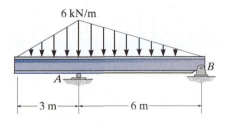

Prob. F4–39

F4–40. Determine the resultant force and specify where it acts on the beam measured from A.

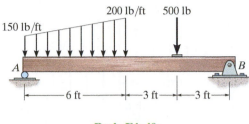

Prob. F4–40

F4–41. Determine the resultant force and specify where it acts on the beam measured from A.

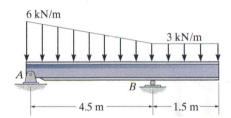

Prob. F4–41

F4–42. Determine the resultant force and specify where it acts on the beam measured from A.

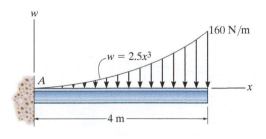

Prob. F4–42

PROBLEMS

4–138. Replace the loading by an equivalent resultant force and couple moment acting at point O.

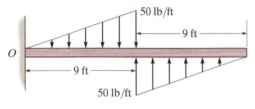

Prob. 4–138

4–139. Replace the distributed loading with an equivalent resultant force, and specify its location on the beam measured from point O.

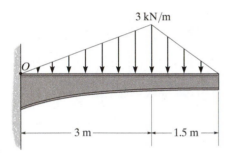

Prob. 4–139

*4–140. Replace the loading by an equivalent resultant force and specify its location on the beam, measured from point A.

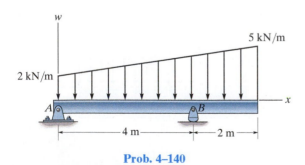

Prob. 4–140

4–141. Currently eighty-five percent of all neck injuries are caused by rear-end car collisions. To alleviate this problem, an automobile seat restraint has been developed that provides additional pressure contact with the cranium. During dynamic tests the distribution of load on the cranium has been plotted and shown to be parabolic. Determine the equivalent resultant force and its location, measured from point A.

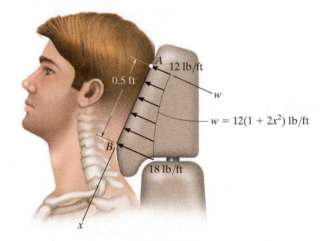

Prob. 4–141

4–142. Replace the distributed loading by an equivalent resultant force, and specify its location on the beam, measured from the pin at A.

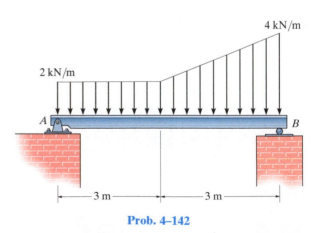

Prob. 4–142

4–143. Replace this loading by an equivalent resultant force and specify its location, measured from point O.

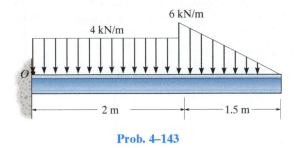

Prob. 4–143

*4–144. The distribution of soil loading on the bottom of a building slab is shown. Replace this loading by an equivalent resultant force and specify its location, measured from point O.

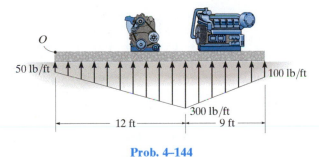

Prob. 4–144

4–145. Replace the loading by an equivalent resultant force and couple moment acting at point O.

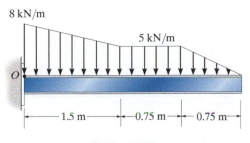

Prob. 4–145

4–146. Replace the distributed loading by an equivalent resultant force and couple moment acting at point A.

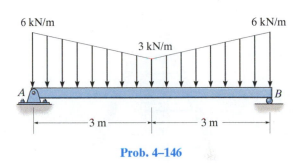

Prob. 4–146

4–147. Determine the length b of the triangular load and its position a on the beam such that the equivalent resultant force is zero and the resultant couple moment is 8 kN · m clockwise.

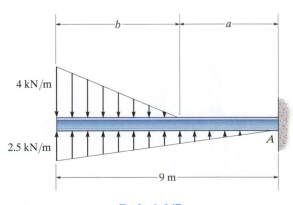

Prob. 4–147

4

*4–148. The form is used to cast a concrete wall having a width of 5 m. Determine the equivalent resultant force the wet concrete exerts on the form AB if the pressure distribution due to the concrete can be approximated as shown. Specify the location of the resultant force, measured from point B.

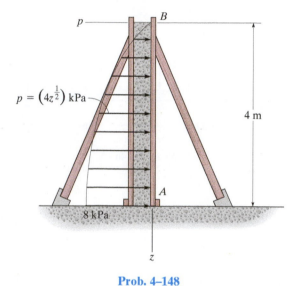

$p = \left(4z^{\frac{1}{2}}\right)$ kPa

Prob. 4–148

4–149. If the soil exerts a trapezoidal distribution of load on the bottom of the footing, determine the intensities w_1 and w_2 of this distribution needed to support the column loadings.

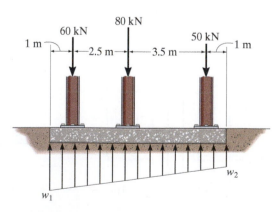

Prob. 4–149

4–150. Replace the loading by an equivalent force and couple moment acting at point O.

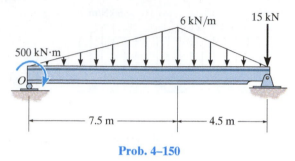

Prob. 4–150

4–151. Replace the loading by a single resultant force, and specify the location of the force measured from point O.

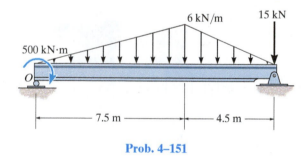

Prob. 4–151

*4–152. Replace the loading by an equivalent resultant force and couple moment acting at point A.

4–153. Replace the loading by a single resultant force, and specify its location on the beam measured from point A.

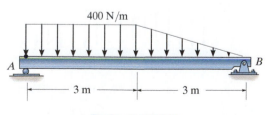

Probs. 4–152/153

4–154. Replace the distributed loading by an equivalent resultant force and specify where its line of action intersects a horizontal line along member AB, measured from A.

4–155. Replace the distributed loading by an equivalent resultant force and specify where its line of action intersects a vertical line along member BC, measured from C.

4–157. Determine the equivalent resultant force and couple moment at point O.

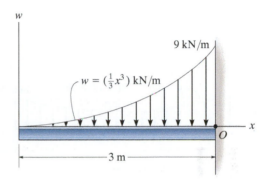

Prob. 4–157

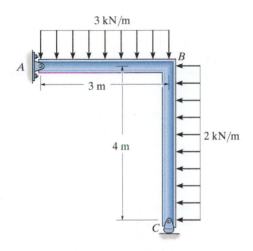

Probs. 4–154/155

4–158. Determine the magnitude of the equivalent resultant force and its location, measured from point O.

*4–156. Determine the length b of the triangular load and its position a on the beam such that the equivalent resultant force is zero and the resultant couple moment is 8 kN · m clockwise.

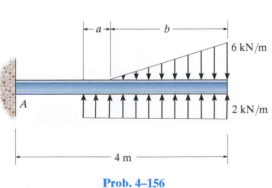

Prob. 4–156

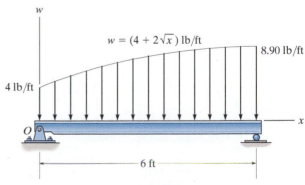

Prob. 4–158

4–159. The distributed load acts on the shaft as shown. Determine the magnitude of the equivalent resultant force and specify its location, measured from the support, A.

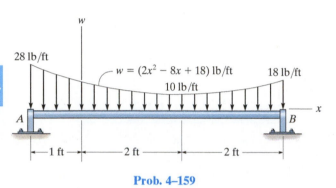

Prob. 4–159

4–161. Replace the loading by an equivalent resultant force and couple moment acting at point O.

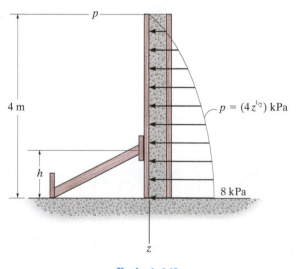

$w = w_0 \cos\left(\frac{\pi}{2L}x\right)$

Prob. 4–161

4–162. Wet concrete exerts a pressure distribution along the wall of the form. Determine the resultant force of this distribution and specify the height h where the bracing strut should be placed so that it lies through the line of action of the resultant force. The wall has a width of 5 m.

***4–160.** Replace the distributed loading with an equivalent resultant force, and specify its location on the beam measured from point A.

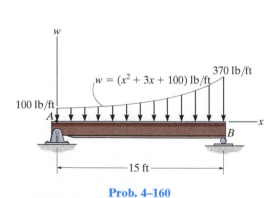

Prob. 4–160

$p = (4z^{1/2})$ kPa

Prob. 4–162

CHAPTER REVIEW

Moment of Force—Scalar Definition

A force produces a turning effect or moment about a point O that does not lie on its line of action. In scalar form, the moment *magnitude* is the product of the force and the moment arm or perpendicular distance from point O to the line of action of the force.

$$M_O = Fd$$

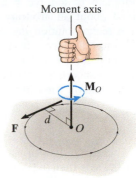

Moment axis

The *direction* of the moment is defined using the right-hand rule. $\mathbf{M}_O$ always acts along an axis perpendicular to the plane containing $\mathbf{F}$ and d, and passes through the point O.

Rather than finding d, it is normally easier to resolve the force into its x and y components, determine the moment of each component about the point, and then sum the results. This is called the principle of moments.

$$M_O = Fd = F_x y - F_y x$$

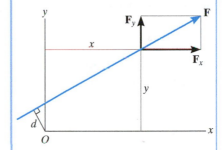

Moment of a Force—Vector Definition

Since three-dimensional geometry is generally more difficult to visualize, the vector cross product should be used to determine the moment. Here $\mathbf{M}_O = \mathbf{r} \times \mathbf{F}$, where $\mathbf{r}$ is a position vector that extends from point O to any point A, B, or C on the line of action of $\mathbf{F}$.

$$\mathbf{M}_O = \mathbf{r}_A \times \mathbf{F} = \mathbf{r}_B \times \mathbf{F} = \mathbf{r}_C \times \mathbf{F}$$

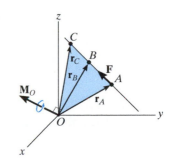

If the position vector $\mathbf{r}$ and force $\mathbf{F}$ are expressed as Cartesian vectors, then the cross product results from the expansion of a determinant.

$$\mathbf{M}_O = \mathbf{r} \times \mathbf{F} = \begin{vmatrix} \mathbf{i} & \mathbf{j} & \mathbf{k} \\ r_x & r_y & r_z \\ F_x & F_y & F_z \end{vmatrix}$$

Moment about an Axis

If the moment of a force $\mathbf{F}$ is to be determined about an arbitrary axis a, then for a scalar solution the moment arm, or shortest distance d_a from the line of action of the force to the axis must be used. This distance is perpendicular to both the axis and the force line of action.

$$M_a = F d_a$$

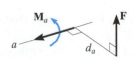

Note that when the line of action of $\mathbf{F}$ intersects the axis, then the moment of $\mathbf{F}$ about the axis is zero. Also, when the line of action of $\mathbf{F}$ is parallel to the axis, the moment of $\mathbf{F}$ about the axis is zero.

In three dimensions, the scalar triple product should be used. Here $\mathbf{u}_a$ is the unit vector that specifies the direction of the axis, and $\mathbf{r}$ is a position vector that is directed from any point on the axis to any point on the line of action of the force. If M_a is calculated as a negative scalar, then the sense of direction of $\mathbf{M}_a$ is opposite to $\mathbf{u}_a$.

$$M_a = \mathbf{u}_a \cdot (\mathbf{r} \times \mathbf{F}) = \begin{vmatrix} u_{a_x} & u_{a_y} & u_{a_z} \\ r_x & r_y & r_z \\ F_x & F_y & F_z \end{vmatrix}$$

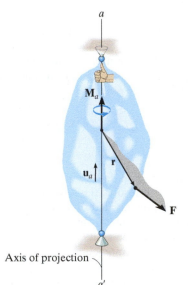

Axis of projection

Couple Moment

A couple consists of two equal but opposite forces that act a perpendicular distance d apart. Couples tend to produce a rotation without translation.

The magnitude of the couple moment is $M = Fd$, and its direction is established using the right-hand rule.

$$M = Fd$$

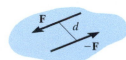

If the vector cross product is used to determine the moment of a couple, then $\mathbf{r}$ extends from any point on the line of action of one of the forces to any point on the line of action of the other force $\mathbf{F}$ that is used in the cross product.

$$\mathbf{M} = \mathbf{r} \times \mathbf{F}$$

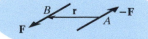

Simplification of a Force and Couple System

Any system of forces and couples can be reduced to a single resultant force and resultant couple moment acting at a point. The resultant force is the sum of all the forces in the system, $F_R = \Sigma F$, and the resultant couple moment is equal to the sum of all the moments of the forces about the point and couple moments. $M_{R_O} = \Sigma M_O + \Sigma M$.

Further simplification to a single resultant force is possible provided the force system is concurrent, coplanar, or parallel. To find the location of the resultant force from a point, it is necessary to equate the moment of the resultant force about the point to the moment of the forces and couples in the system about the same point.

If the resultant force and couple moment at a point are not perpendicular to one another, then this system can be reduced to a wrench, which consists of the resultant force and collinear couple moment.

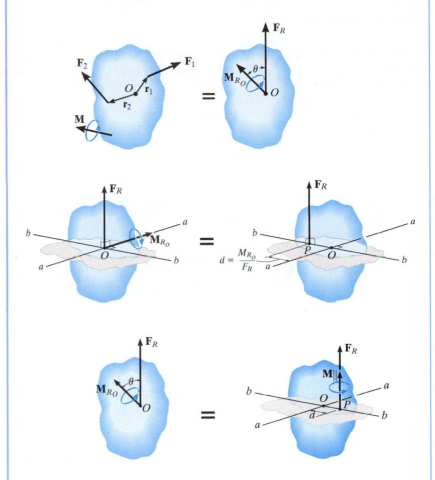

Coplanar Distributed Loading

A simple distributed loading can be represented by its resultant force, which is equivalent to the *area* under the loading curve. This resultant has a line of action that passes through the *centroid* or geometric center of the area or volume under the loading diagram.

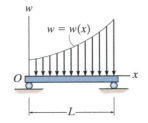
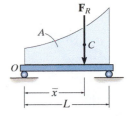

REVIEW PROBLEMS

R4–1. The boom has a length of 30 ft, a weight of 800 lb, and mass center at G. If the maximum moment that can be developed by a motor at A is $M = 20(10^3)$ lb·ft, determine the maximum load W, having a mass center at G', that can be lifted.

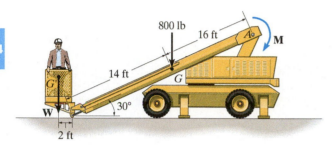

Prob. R4–1

R4–2. Replace the force $\mathbf{F}$ having a magnitude of $F = 50$ lb and acting at point A by an equivalent force and couple moment at point C.

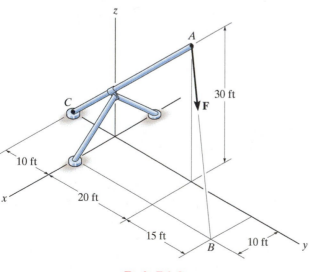

Prob. R4–2

R4–3. The hood of the automobile is supported by the strut AB, which exerts a force of $F = 24$ lb on the hood. Determine the moment of this force about the hinged axis y.

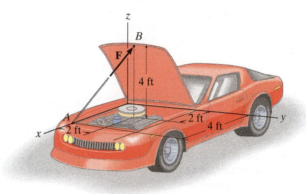

Prob. R4–3

R4–4. Friction on the concrete surface creates a couple moment of $M_O = 100$ N·m on the blades of the trowel. Determine the magnitude of the couple forces so that the resultant couple moment on the trowel is zero. The forces lie in the horizontal plane and act perpendicular to the handle of the trowel.

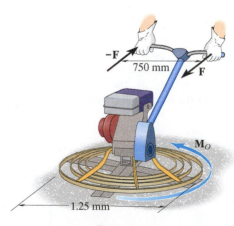

Prob. R4–4

R4–5. Replace the force and couple system by an equivalent force and couple moment at point P.

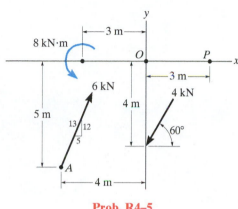

Prob. R4–5

R4–6. Replace the force system acting on the frame by a resultant force, and specify where its line of action intersects member AB, measured from point A.

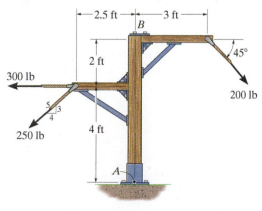

Prob. R4–6

R4–7. The building slab is subjected to four parallel column loadings. Determine the equivalent resultant force and specify its location (x, y) on the slab. Take $F_1 = 30$ kN, $F_2 = 40$ kN.

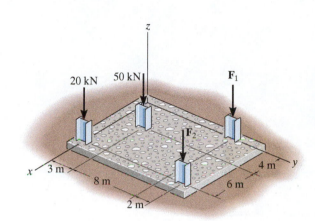

Prob. R4–7

R4–8. Replace the distributed loading by an equivalent resultant force, and specify its location on the beam, measured from the pin at C.

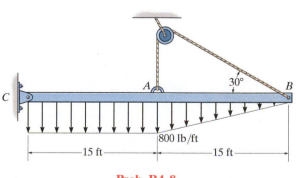

Prob. R4–8

4

Chapter 5

(© YuryZap/Shutterstock)

It is important to be able to determine the forces in the cables used to support this boom to ensure that it does not fail. In this chapter we will study how to apply equilibrium methods to determine the forces acting on the supports of a rigid body such as this.

Equilibrium of a Rigid Body

CHAPTER OBJECTIVES

- To develop the equations of equilibrium for a rigid body.
- To introduce the concept of the free-body diagram for a rigid body.
- To show how to solve rigid-body equilibrium problems using the equations of equilibrium.

5.1 Conditions for Rigid-Body Equilibrium

In this section, we will develop both the necessary and sufficient conditions for the equilibrium of the rigid body in Fig. 5–1a. As shown, this body is subjected to an external force and couple moment system that is the result of the effects of gravitational, electrical, magnetic, or contact forces caused by adjacent bodies. The internal forces caused by interactions between particles within the body are not shown in this figure because these forces occur in equal but opposite collinear pairs and hence will cancel out, a consequence of Newton's third law.

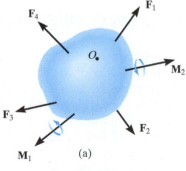

(a)

Fig. 5–1

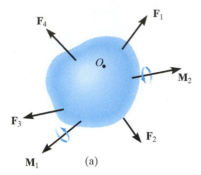

(a)

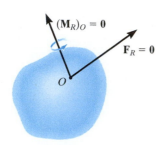

(b)

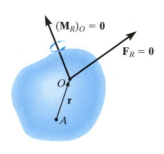

(c)

Fig. 5–1 (cont.)

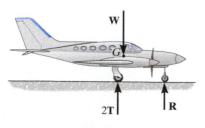

Fig. 5–2

Using the methods of the previous chapter, the force and couple moment system acting on a body can be reduced to an equivalent resultant force and resultant couple moment at any arbitrary point O on or off the body, Fig. 5–1b. If this resultant force and couple moment are both equal to zero, then the body is said to be in *equilibrium*. Mathematically, the equilibrium of a body is expressed as

$$\mathbf{F}_R = \Sigma\mathbf{F} = \mathbf{0}$$
$$(\mathbf{M}_R)_O = \Sigma\mathbf{M}_O = \mathbf{0}$$

(5–1)

The first of these equations states that the sum of the forces acting on the body is equal to *zero*. The second equation states that the sum of the moments of all the forces in the system about point O, added to all the couple moments, is equal to *zero*. These two equations are not only necessary for equilibrium, they are also sufficient. To show this, consider summing moments about some other point, such as point A in Fig. 5–1c. We require

$$\Sigma\mathbf{M}_A = \mathbf{r} \times \mathbf{F}_R + (\mathbf{M}_R)_O = \mathbf{0}$$

Since $\mathbf{r} \neq \mathbf{0}$, this equation is satisfied if Eqs. 5–1 are satisfied, namely $\mathbf{F}_R = \mathbf{0}$ and $(\mathbf{M}_R)_O = \mathbf{0}$.

When applying the equations of equilibrium, we will assume that the body remains rigid. In reality, however, all bodies deform when subjected to loads. Although this is the case, most engineering materials such as steel and concrete are very rigid and so their deformation is usually very small. Therefore, when applying the equations of equilibrium, we can generally assume that the body will remain *rigid* and *not deform* under the applied load without introducing any significant error. This way the direction of the applied forces and their moment arms with respect to a fixed reference remain the same both before and after the body is loaded.

EQUILIBRIUM IN TWO DIMENSIONS

In the first part of the chapter, we will consider the case where the force system acting on a rigid body lies in or may be projected onto a *single* plane and, furthermore, any couple moments acting on the body are directed perpendicular to this plane. This type of force and couple system is often referred to as a two-dimensional or *coplanar* force system. For example, the airplane in Fig. 5–2 has a plane of symmetry through its center axis, and so the loads acting on the airplane are symmetrical with respect to this plane. Thus, each of the two wing tires will support the same load $\mathbf{T}$, which is represented on the side (two-dimensional) view of the plane as $2\mathbf{T}$.

5.2 Free-Body Diagrams

Successful application of the equations of equilibrium requires a complete specification of *all* the known and unknown external forces that act *on* the body. The best way to account for these forces is to draw a ***free-body diagram***. This diagram is a sketch of the outlined shape of the body, which represents it as being *isolated* or "free" from its surroundings, i.e., a "free body." On this sketch it is necessary to show *all* the forces and couple moments that the surroundings exert *on the body* so that these effects can be accounted for when the equations of equilibrium are applied. *A thorough understanding of how to draw a free-body diagram is of primary importance for solving problems in mechanics.*

Support Reactions.   Before presenting a formal procedure as to how to draw a free-body diagram, we will first consider the various types of reactions that occur at supports and points of contact between bodies subjected to coplanar force systems. As a general rule,

- A support prevents the translation of a body in a given direction by exerting a force on the body in the opposite direction.

- A support prevents the rotation of a body in a given direction by exerting a couple moment on the body in the opposite direction.

For example, let us consider three ways in which a horizontal member, such as a beam, is supported at its end. One method consists of a *roller* or cylinder, Fig. 5–3a. Since this support only prevents the beam from *translating* in the vertical direction, the roller will only exert a *force* on the beam in this direction, Fig. 5–3b.

The beam can be supported in a more restrictive manner by using a *pin*, Fig. 5–3c. The pin passes through a hole in the beam and two leaves which are fixed to the ground. Here the pin can prevent *translation* of the beam in *any direction* ϕ, Fig. 5–3d, and so the pin must exert a *force* **F** on the beam in the opposite direction. For purposes of analysis, it is generally easier to represent this resultant force **F** by its two rectangular components $\mathbf{F}_x$ and $\mathbf{F}_y$, Fig. 5–3e. If F_x and F_y are known, then F and ϕ can be calculated.

The most restrictive way to support the beam would be to use a *fixed support* as shown in Fig. 5–3f. This support will prevent both *translation and rotation* of the beam. To do this a *force and couple moment* must be developed on the beam at its point of connection, Fig. 5–3g. As in the case of the pin, the force is usually represented by its rectangular components $\mathbf{F}_x$ and $\mathbf{F}_y$.

Table 5–1 lists other common types of supports for bodies subjected to coplanar force systems. (In all cases the angle θ is assumed to be known.) Carefully study each of the symbols used to represent these supports and the types of reactions they exert on their contacting members.

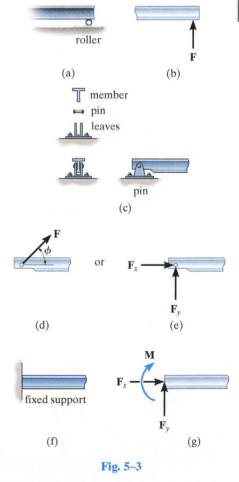

Fig. 5–3

TABLE 5–1 Supports for Rigid Bodies Subjected to Two-Dimensional Force Systems

Types of Connection	Reaction	Number of Unknowns
(1) cable		One unknown. The reaction is a tension force which acts away from the member in the direction of the cable.
(2) weightless link	or	One unknown. The reaction is a force which acts along the axis of the link.
(3) roller		One unknown. The reaction is a force which acts perpendicular to the surface at the point of contact.
(4) rocker		One unknown. The reaction is a force which acts perpendicular to the surface at the point of contact.
(5) smooth contacting surface		One unknown. The reaction is a force which acts perpendicular to the surface at the point of contact.
(6) roller or pin in confined smooth slot	or	One unknown. The reaction is a force which acts perpendicular to the slot.
(7) member pin connected to collar on smooth rod	or	One unknown. The reaction is a force which acts perpendicular to the rod.

continued

TABLE 5-1 Continued

Types of Connection	Reaction	Number of Unknowns
(8) smooth pin or hinge	F_y or F, ϕ F_x	Two unknowns. The reactions are two components of force, or the magnitude and direction ϕ of the resultant force. Note that ϕ and θ are not necessarily equal [usually not, unless the rod shown is a link as in (2)].
(9) member fixed connected to collar on smooth rod	F M	Two unknowns. The reactions are the couple moment and the force which acts perpendicular to the rod.
(10) fixed support	F_y, F_x, M or F, ϕ, M	Three unknowns. The reactions are the couple moment and the two force components, or the couple moment and the magnitude and direction ϕ of the resultant force.

Typical examples of actual supports are shown in the following sequence of photos. The numbers refer to the connection types in Table 5–1.

(© Russell C. Hibbeler)

The cable exerts a force on the bracket in the direction of the cable. (1)

The rocker support for this bridge girder allows horizontal movement so the bridge is free to expand and contract due to a change in temperature. (4) (© Russell C. Hibbeler)

This concrete girder rests on the ledge that is assumed to act as a smooth contacting surface. (5) (© Russell C. Hibbeler)

Typical pin support for a beam. (8) (© Russell C. Hibbeler)

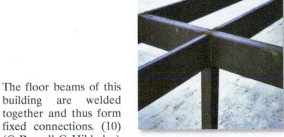

The floor beams of this building are welded together and thus form fixed connections. (10) (© Russell C. Hibbeler)

Internal Forces.   As stated in Sec. 5.1, the internal forces that act between adjacent particles in a body always occur in collinear pairs such that they have the same magnitude and act in opposite directions (Newton's third law). Since these forces cancel each other, they will not create an *external effect* on the body. It is for this reason that the internal forces should not be included on the free-body diagram if the entire body is to be considered. For example, the engine shown in Fig. 5–4*a* has a free-body diagram shown in Fig. 5–4*b*. The internal forces between all its connected parts, such as the screws and bolts, will cancel out because they form equal and opposite collinear pairs. Only the external forces T_1 and T_2, exerted by the chains and the engine weight W, are shown on the free-body diagram.

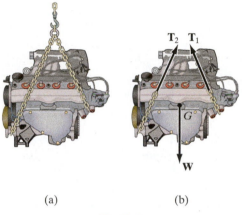

(a)   (b)

Fig. 5–4

Weight and the Center of Gravity.   When a body is within a gravitational field, then each of its particles has a specified weight. It was shown in Sec. 4.8 that such a system of forces can be reduced to a single resultant force acting through a specified point. We refer to this force resultant as the *weight* W of the body and to the location of its point of application as the **center of gravity**. The methods used for its determination will be developed in Chapter 9.

In the examples and problems that follow, if the weight of the body is important for the analysis, this force will be reported in the problem statement. Also, when the body is *uniform* or made from the same material, the center of gravity will be located at the body's *geometric center* or *centroid*; however, if the body consists of a nonuniform distribution of material, or has an unusual shape, then the location of its center of gravity G will be given.

Idealized Models.   When an engineer performs a force analysis of any object, he or she considers a corresponding analytical or idealized model that gives results that approximate as closely as possible the actual situation. To do this, careful choices have to be made so that selection of the type of supports, the material behavior, and the object's dimensions can be justified. This way one can feel confident that any design or analysis will yield results which can be trusted. In complex cases this process may require developing several different models of the object that must be analyzed. In any case, this selection process requires both skill and experience.

The following two cases illustrate what is required to develop a proper model. In Fig. 5–5a, the steel beam is to be used to support the three roof joists of a building. For a force analysis it is reasonable to assume the material (steel) is rigid since only very small deflections will occur when the beam is loaded. A bolted connection at *A* will allow for any slight rotation that occurs here when the load is applied, and so a *pin* can be considered for this support. At *B* a *roller* can be considered since this support offers no resistance to horizontal movement. Building code is used to specify the roof loading *A* so that the joist loads **F** can be calculated. These forces will be larger than any actual loading on the beam since they account for extreme loading cases and for dynamic or vibrational effects. Finally, the weight of the beam is generally neglected when it is small compared to the load the beam supports. The idealized model of the beam is therefore shown with average dimensions *a, b, c,* and *d* in Fig. 5–5b.

As a second case, consider the lift boom in Fig. 5–6a. By inspection, it is supported by a pin at *A* and by the hydraulic cylinder *BC*, which can be approximated as a weightless link. The material can be assumed rigid, and with its density known, the weight of the boom and the location of its center of gravity *G* are determined. When a design loading **P** is specified, the idealized model shown in Fig. 5–6b can be used for a force analysis. Average dimensions (not shown) are used to specify the location of the loads and the supports.

Idealized models of specific objects will be given in some of the examples throughout the text. It should be realized, however, that each case represents the reduction of a practical situation using simplifying assumptions like the ones illustrated here.

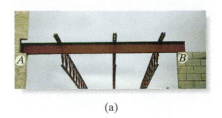

(a)

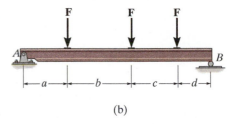

(b)

Fig. 5–5 (© Russell C. Hibbeler)

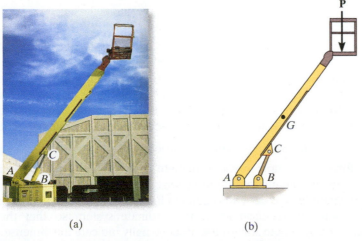

(a)   (b)

Fig. 5–6 (© Russell C. Hibbeler)

Important Points

- No equilibrium problem should be solved without *first drawing the free-body diagram*, so as to account for all the forces and couple moments that act on the body.

- If a support *prevents translation* of a body in a particular direction, then the support, when it is removed, exerts a *force* on the body in that direction.

- If *rotation is prevented*, then the support, when it is removed, exerts a *couple moment* on the body.

- Study Table 5–1.

- Internal forces are *never shown* on the free-body diagram since they occur in equal but opposite collinear pairs and therefore cancel out.

- The weight of a body is an external force, and its effect is represented by a single resultant force acting through the body's center of gravity *G*.

- *Couple moments* can be placed anywhere on the free-body diagram since they are *free vectors. Forces* can act at any point along their lines of action since they are *sliding vectors*.

Procedure for Analysis

To construct a free-body diagram for a rigid body or any group of bodies considered as a single system, the following steps should be performed:

Draw Outlined Shape.

Imagine the body to be *isolated* or cut "free" from its constraints and connections and draw (sketch) its outlined shape. Be sure to *remove all the supports* from the body.

Show All Forces and Couple Moments.

Identify all the known and unknown *external forces* and couple moments that *act on the body*. Those generally encountered are due to (1) applied loadings, (2) reactions occurring at the supports or at points of contact with other bodies (see Table 5–1), and (3) the weight of the body. To account for all these effects, it may help to trace over the boundary, carefully noting each force or couple moment acting on it.

Identify Each Loading and Give Dimensions.

The forces and couple moments that are known should be labeled with their proper magnitudes and directions. Letters are used to represent the magnitudes and direction angles of forces and couple moments that are unknown. Establish an *x, y* coordinate system so that these unknowns, A_x, A_y, etc., can be identified. Finally, indicate the dimensions of the body necessary for calculating the moments of forces.

EXAMPLE 5.1

Draw the free-body diagram of the uniform beam shown in Fig. 5–7a. The beam has a mass of 100 kg.

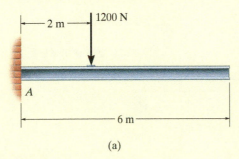

(a)

SOLUTION

The free-body diagram of the beam is shown in Fig. 5–7b. Since the support at *A* is fixed, the wall exerts three reactions *on the beam*, denoted as $\mathbf{A}_x$, $\mathbf{A}_y$, and $\mathbf{M}_A$. The magnitudes of these reactions are *unknown*, and their sense has been *assumed*. The weight of the beam, $W = 100(9.81) \text{ N} = 981 \text{ N}$, acts through the beam's center of gravity *G*, which is 3 m from *A* since the beam is uniform.

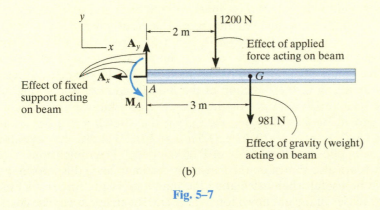

(b)

Fig. 5–7

EXAMPLE 5.2

Draw the free-body diagram of the foot lever shown in Fig. 5–8a. The operator applies a vertical force to the pedal so that the spring is stretched 1.5 in. and the force on the link at B is 20 lb.

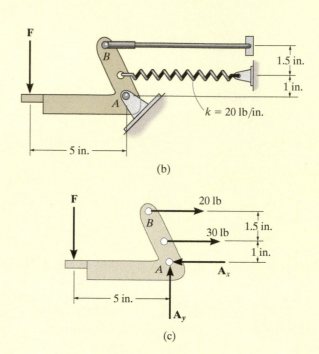

(b)

(a)

Fig. 5–8 (© Russell C. Hibbeler)

(c)

SOLUTION

By inspection of the photo the lever is loosely bolted to the frame at A and so this bolt acts as a pin. (See (8) in Table 5–1.) Although not shown here the link at B is pinned at both ends and so it is like (2) in Table 5–1. After making the proper measurements, the idealized model of the lever is shown in Fig. 5–8b. From this, the free-body diagram is shown in Fig. 5–8c. Since the pin at A is *removed*, it exerts force components A_x and A_y on the lever. The link exerts a force of 20 lb, acting in the direction of the link. In addition the spring also exerts a horizontal force on the lever. If the stiffness is measured and found to be $k = 20$ lb/in., then since the stretch $s = 1.5$ in., using Eq. 3–2, $F_s = ks = 20$ lb/in. (1.5 in.) = 30 lb. Finally, the operator's shoe applies a vertical force of **F** on the pedal. The dimensions of the lever are also shown on the free-body diagram, since this information will be useful when calculating the moments of the forces. As usual, the senses of the unknown forces at A have been assumed. The correct senses will become apparent after solving the equilibrium equations.

EXAMPLE 5.3

Two smooth pipes, each having a mass of 300 kg, are supported by the forked tines of the tractor in Fig. 5–9a. Draw the free-body diagrams for each pipe and both pipes together.

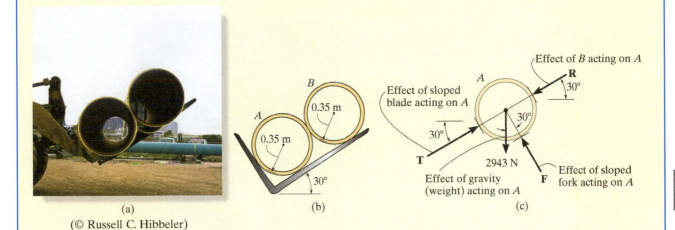

(a)

(© Russell C. Hibbeler)

(b)

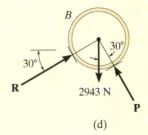

Effect of *B* acting on *A*

Effect of sloped blade acting on *A*

Effect of gravity (weight) acting on *A*

Effect of sloped fork acting on *A*

2943 N

(c)

SOLUTION

The idealized model from which we must draw the free-body diagrams is shown in Fig. 5–9b. Here the pipes are identified, the dimensions have been added, and the physical situation reduced to its simplest form.

Removing the surfaces of contact, the free-body diagram for pipe *A* is shown in Fig. 5–9c. Its weight is $W = 300(9.81) \text{ N} = 2943 \text{ N}$. Assuming all contacting surfaces are *smooth*, the reactive forces **T**, **F**, **R** act in a direction *normal* to the tangent at their surfaces of contact.

The free-body diagram of the isolated pipe *B* is shown in Fig. 5–9d. Can you identify each of the three forces acting *on this pipe*? In particular, note that **R**, representing the force of *A* on *B*, Fig. 5–9d, is equal and opposite to **R** representing the force of *B* on *A*, Fig. 5–9c. This is a consequence of Newton's third law of motion.

The free-body diagram of both pipes combined ("system") is shown in Fig. 5–9e. Here the contact force **R**, which acts between *A* and *B*, is considered as an *internal* force and hence is not shown on the free-body diagram. That is, it represents a pair of equal but opposite collinear forces which cancel each other.

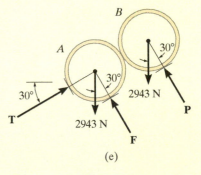

2943 N

(d)

2943 N

2943 N

(e)

Fig. 5–9

EXAMPLE | **5.4**

Draw the free-body diagram of the unloaded platform that is suspended off the edge of the oil rig shown in Fig. 5–10a. The platform has a mass of 200 kg.

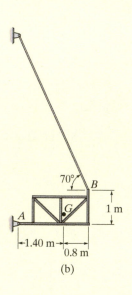

(b)

(a)

Fig. 5–10 (© Russell C. Hibbeler)

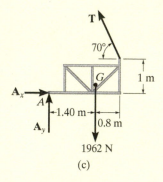

(c)

SOLUTION

The idealized model of the platform will be considered in two dimensions because by observation the loading and the dimensions are all symmetrical about a vertical plane passing through its center, Fig. 5–10b. The connection at A is considered to be a pin, and the cable supports the platform at B. The direction of the cable and average dimensions of the platform are listed, and the center of gravity G has been determined. It is from this model that we have drawn the free-body diagram shown in Fig. 5–10c. The platform's weight is $200(9.81) = 1962$ N. The supports have been *removed*, and the force components $\mathbf{A}_x$ and $\mathbf{A}_y$ along with the cable force $\mathbf{T}$ represent the reactions that *both* pins and *both* cables exert on the platform, Fig. 5–10a. As a result, half their magnitudes are developed on each side of the platform.

PROBLEMS

5–1. Draw the free-body diagram for the following problems.

a) The cantilevered beam in Prob. 5–10.

b) The beam in Prob. 5–11.

c) The beam in Prob. 5–12.

d) The beam in Prob. 5–14.

5–2. Draw the free-body diagram for the following problems.

a) The truss in Prob. 5–15.

b) The beam in Prob. 5–16.

c) The man and load in Prob. 5–17.

d) The beam in Prob. 5–18.

5–3. Draw the free-body diagram for the following problems.

a) The man and beam in Prob. 5–19.

b) The rod in Prob. 5–20.

c) The rod in Prob. 5–21.

d) The beam in Prob. 5–22.

*5–4. Draw the free-body diagram for the following problems.

a) The beam in Prob. 5–25.

b) The crane and boom in Prob. 5–26.

c) The bar in Prob. 5–27.

d) The rod in Prob. 5–28.

5–5. Draw the free-body diagram for the following problems.

a) The boom in Prob. 5–32.

b) The jib crane in Prob. 5–33.

c) The smooth pipe in Prob. 5–35.

d) The beam in Prob. 5–36.

5–6. Draw the free-body diagram for the following problems.

a) The jib crane in Prob. 5–37.

b) The bar in Prob. 5–39.

c) The bulkhead in Prob. 5–41.

d) The boom in Prob. 5–42.

5–7. Draw the free-body diagram for the following problems.

a) The rod in Prob. 5–44.

b) The hand truck and load when it is lifted in Prob. 5–45.

c) The beam in Prob. 5–47.

d) The cantilever footing in Prob. 5–51.

*5–8. Draw the free-body diagram for the following problems.

a) The beam in Prob. 5–52.

b) The boy and diving board in Prob. 5–53.

c) The rod in Prob. 5–54.

d) The rod in Prob. 5–56.

5–9. Draw the free-body diagram for the following problems.

a) The beam in Prob. 5–57.

b) The rod in Prob. 5–59.

c) The bar in Prob. 5–60.

5

5.3   Equations of Equilibrium

In Sec. 5.1 we developed the two equations which are both necessary and sufficient for the equilibrium of a rigid body, namely, $\Sigma \mathbf{F} = \mathbf{0}$ and $\Sigma \mathbf{M}_O = \mathbf{0}$. When the body is subjected to a system of forces, which all lie in the x–y plane, then the forces can be resolved into their x and y components. Consequently, the conditions for equilibrium in two dimensions are

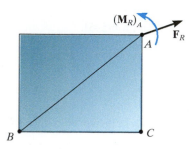

$$\begin{aligned} \Sigma F_x &= 0 \\ \Sigma F_y &= 0 \\ \Sigma M_O &= 0 \end{aligned} \tag{5–2}$$

Here ΣF_x and ΣF_y represent, respectively, the algebraic sums of the x and y components of all the forces acting on the body, and ΣM_O represents the algebraic sum of the couple moments and the moments of all the force components about the z axis, which is perpendicular to the x–y plane and passes through the arbitrary point O.

(a)

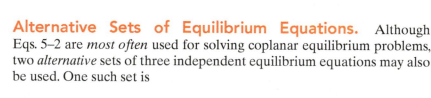

Alternative Sets of Equilibrium Equations. Although Eqs. 5–2 are *most often* used for solving coplanar equilibrium problems, two *alternative* sets of three independent equilibrium equations may also be used. One such set is

$$\begin{aligned} \Sigma F_x &= 0 \\ \Sigma M_A &= 0 \\ \Sigma M_B &= 0 \end{aligned} \tag{5–3}$$

(b)

When using these equations it is required that a line passing through points A and B is *not parallel* to the y axis. To prove that Eqs. 5–3 provide the *conditions* for equilibrium, consider the free-body diagram of the plate shown in Fig. 5–11a. Using the methods of Sec. 4.7, all the forces on the free-body diagram may be replaced by an equivalent resultant force $\mathbf{F}_R = \Sigma \mathbf{F}$, acting at point A, and a resultant couple moment $\left(\mathbf{M}_R \right)_A = \Sigma \mathbf{M}_A$, Fig. 5–11b. If $\Sigma M_A = 0$ is satisfied, it is necessary that $\left(\mathbf{M}_R \right)_A = \mathbf{0}$. Furthermore, in order that $\mathbf{F}_R$ satisfy $\Sigma F_x = 0$, it must have *no component* along the x axis, and therefore $\mathbf{F}_R$ must be parallel to the y axis, Fig. 5–11c. Finally, if it is required that $\Sigma M_B = 0$, where B does not lie on the line of action of $\mathbf{F}_R$, then $\mathbf{F}_R = \mathbf{0}$. Since Eqs. 5–3 show that both of these resultants are zero, indeed the body in Fig. 5–11a must be in equilibrium.

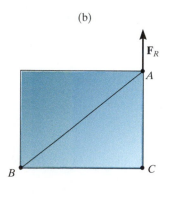

(c)

Fig. 5–11

A second alternative set of equilibrium equations is

$$\Sigma M_A = 0$$
$$\Sigma M_B = 0 \qquad\qquad (5\text{–}4)$$
$$\Sigma M_C = 0$$

Here it is necessary that points A, B, and C do not lie on the same line. To prove that these equations, when satisfied, ensure equilibrium, consider again the free-body diagram in Fig. 5–11b. If $\Sigma M_A = 0$ is to be satisfied, then $(\mathbf{M}_R)_A = \mathbf{0}$. $\Sigma M_C = 0$ is satisfied if the line of action of $\mathbf{F}_R$ passes through point C as shown in Fig. 5–11c. Finally, if we require $\Sigma M_B = 0$, it is necessary that $\mathbf{F}_R = \mathbf{0}$, and so the plate in Fig. 5–11a must then be in equilibrium.

Procedure for Analysis

Coplanar force equilibrium problems for a rigid body can be solved using the following procedure.

Free-Body Diagram.

- Establish the x, y coordinate axes in any suitable orientation.

- Remove all supports and draw an outlined shape of the body.

- Show all the forces and couple moments acting on the body.

- Label all the loadings and specify their directions relative to the x or y axis. The sense of a force or couple moment having an *unknown* magnitude but known line of action can be *assumed*.

- Indicate the dimensions of the body necessary for computing the moments of forces.

Equations of Equilibrium.

- Apply the moment equation of equilibrium, $\Sigma M_O = 0$, about a point (O) that lies at the intersection of the lines of action of two unknown forces. In this way, the moments of these unknowns are zero about O, and a *direct solution* for the third unknown can be determined.

- When applying the force equilibrium equations, $\Sigma F_x = 0$ and $\Sigma F_y = 0$, orient the x and y axes along lines that will provide the simplest resolution of the forces into their x and y components.

- If the solution of the equilibrium equations yields a negative scalar for a force or couple moment magnitude, this indicates that the sense is opposite to that which was assumed on the free-body diagram.

EXAMPLE | 5.5

Determine the horizontal and vertical components of reaction on the beam caused by the pin at B and the rocker at A as shown in Fig. 5–12a. Neglect the weight of the beam.

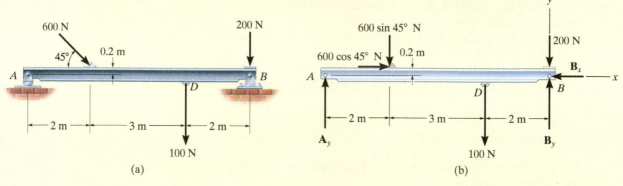

(a)

(b)

Fig. 5–12

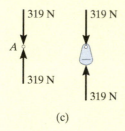

(c)

SOLUTION

Free-Body Diagram.   The supports are *removed*, and the free-body diagram of the beam is shown in Fig. 5–12b. (See Example 5.1.) For simplicity, the 600-N force is represented by its x and y components as shown in Fig. 5–12b.

Equations of Equilibrium.   Summing forces in the x direction yields

$$\xrightarrow{+} \Sigma F_x = 0; \qquad 600 \cos 45° \text{ N} - B_x = 0$$

$$B_x = 424 \text{ N} \qquad\qquad Ans.$$

A direct solution for $\mathbf{A}_y$ can be obtained by applying the moment equation $\Sigma M_B = 0$ about point B.

$$\zeta + \Sigma M_B = 0; \qquad 100 \text{ N}(2 \text{ m}) + (600 \sin 45° \text{ N})(5 \text{ m})$$

$$- (600 \cos 45° \text{ N})(0.2 \text{ m}) - A_y(7 \text{ m}) = 0$$

$$A_y = 319 \text{ N} \qquad\qquad Ans.$$

Summing forces in the y direction, using this result, gives

$$+\uparrow \Sigma F_y = 0; \qquad 319 \text{ N} - 600 \sin 45° \text{ N} - 100 \text{ N} - 200 \text{ N} + B_y = 0$$

$$B_y = 405 \text{ N} \qquad\qquad Ans.$$

NOTE: Remember, the support forces in Fig. 5–12b are the result of pins that *act on the beam*. The opposite forces act on the pins. For example, Fig. 5–12c shows the equilibrium of the pin at A and the rocker.

EXAMPLE 5.6

The cord shown in Fig. 5–13a supports a force of 100 lb and wraps over the frictionless pulley. Determine the tension in the cord at C and the horizontal and vertical components of reaction at pin A.

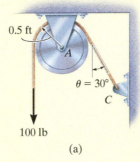

0.5 ft

A

$\theta = 30°$

C

100 lb

(a)

Fig. 5–13

SOLUTION

Free-Body Diagrams. The free-body diagrams of the cord and pulley are shown in Fig. 5–13b. Note that the principle of action, equal but opposite reaction must be carefully observed when drawing each of these diagrams: the cord exerts an unknown load distribution p on the pulley at the contact surface, whereas the pulley exerts an equal but opposite effect on the cord. For the solution, however, it is simpler to *combine* the free-body diagrams of the pulley and this portion of the cord, so that the distributed load becomes *internal* to this "system" and is therefore eliminated from the analysis, Fig. 5–13c.

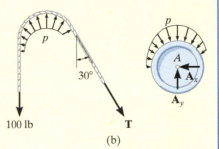

(b)

Equations of Equilibrium. Summing moments about point A to eliminate A_x and A_y, Fig. 5–13c, we have

$\zeta + \Sigma M_A = 0;$ 100 lb (0.5 ft) $-$ T(0.5 ft) = 0

$T = 100$ lb   *Ans.*

Using this result,

$\xrightarrow{+} \Sigma F_x = 0;$ $-A_x + 100 \sin 30°$ lb = 0

$A_x = 50.0$ lb   *Ans.*

$+\uparrow \Sigma F_y = 0;$ $A_y - 100$ lb $- 100 \cos 30°$ lb = 0

$A_y = 187$ lb   *Ans.*

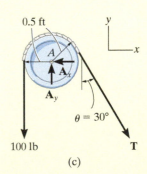

0.5 ft

A

A_x

A_y

$\theta = 30°$

100 lb   T

(c)

NOTE: From the moment equation, it is seen that the tension remains *constant* as the cord passes over the pulley. (This of course is true for *any* angle θ at which the cord is directed and for *any radius r* of the pulley.)

EXAMPLE 5.7

The member shown in Fig. 5–14a is pin connected at A and rests against a smooth support at B. Determine the horizontal and vertical components of reaction at the pin A.

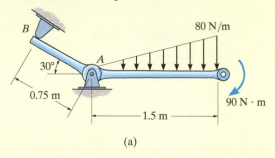

(a)

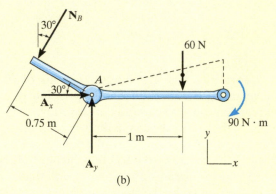

(b)

Fig. 5–14

SOLUTION

Free-Body Diagram. As shown in Fig. 5–14b, the supports are removed and the reaction N_B is perpendicular to the member at B. Also, horizontal and vertical components of reaction are represented at A. The resultant of the distributed loading is $\frac{1}{2}$(1.5 m)(80 N/m) = 60 N. It acts through the centroid of the triangle, 1 m from A as shown.

Equations of Equilibrium. Summing moments about A, we obtain a direct solution for N_B,

$\zeta + \Sigma M_A = 0;$ $-90 \text{ N} \cdot \text{m} - 60 \text{ N}(1 \text{ m}) + N_B(0.75 \text{ m}) = 0$

$$N_B = 200 \text{ N}$$

Using this result,

$\xrightarrow{+} \Sigma F_x = 0;$ $A_x - 200 \sin 30° \text{ N} = 0$

$$A_x = 100 \text{ N} \qquad \textit{Ans.}$$

$+\uparrow \Sigma F_y = 0;$ $A_y - 200 \cos 30° \text{ N} - 60 \text{ N} = 0$

$$A_y = 233 \text{ N} \qquad \textit{Ans.}$$

EXAMPLE 5.8

The box wrench in Fig. 5–15a is used to tighten the bolt at A. If the wrench does not turn when the load is applied to the handle, determine the torque or moment applied to the bolt and the force of the wrench on the bolt.

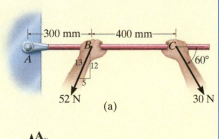

(a)

SOLUTION

Free-Body Diagram. The free-body diagram for the wrench is shown in Fig. 5–15b. Since the bolt acts as a "fixed support," when it is removed, it exerts force components A_x and A_y and a moment M_A on the wrench at A.

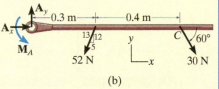

(b)

Fig. 5–15

Equations of Equilibrium.

$$\xrightarrow{+} \Sigma F_x = 0; \qquad A_x - 52\left(\tfrac{5}{13}\right) \text{N} + 30 \cos 60° \text{ N} = 0$$

$$A_x = 5.00 \text{ N} \qquad\qquad \text{Ans.}$$

$$+\uparrow \Sigma F_y = 0; \qquad A_y - 52\left(\tfrac{12}{13}\right) \text{N} - 30 \sin 60° \text{ N} = 0$$

$$A_y = 74.0 \text{ N} \qquad\qquad \text{Ans.}$$

$$\zeta + \Sigma M_A = 0; \quad M_A - \left[52\left(\tfrac{12}{13}\right)\text{N}\right](0.3 \text{ m}) - (30 \sin 60° \text{ N})(0.7 \text{ m}) = 0$$

$$M_A = 32.6 \text{ N} \cdot \text{m} \qquad\qquad \text{Ans.}$$

Note that M_A must be *included* in this moment summation. This couple moment is a free vector and represents the twisting resistance of the bolt on the wrench. By Newton's third law, the wrench exerts an equal but opposite moment or torque on the bolt. Furthermore, the resultant force on the wrench is

$$F_A = \sqrt{(5.00)^2 + (74.0)^2} = 74.1 \text{ N} \qquad\qquad \text{Ans.}$$

NOTE: Although only *three* independent equilibrium equations can be written for a rigid body, it is a good practice to *check* the calculations using a fourth equilibrium equation. For example, the above computations may be verified in part by summing moments about point C:

$$\zeta + \Sigma M_C = 0; \quad \left[52\left(\tfrac{12}{13}\right)\text{N}\right](0.4 \text{ m}) + 32.6 \text{ N} \cdot \text{m} - 74.0 \text{ N}(0.7 \text{ m}) = 0$$

$$19.2 \text{ N} \cdot \text{m} + 32.6 \text{ N} \cdot \text{m} - 51.8 \text{ N} \cdot \text{m} = 0$$

EXAMPLE 5.9

Determine the horizontal and vertical components of reaction on the member at the pin A, and the normal reaction at the roller B in Fig. 5–16a.

SOLUTION

Free-Body Diagram. All the supports are removed and so the free-body diagram is shown in Fig. 5–16b. The pin at A exerts two components of reaction on the member, $\mathbf{A}_x$ and $\mathbf{A}_y$.

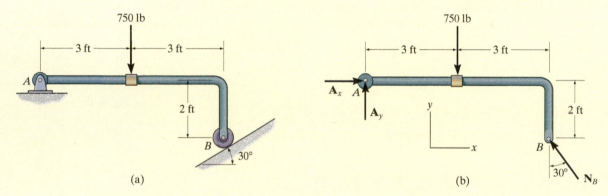

(a)

(b)

Fig. 5–16

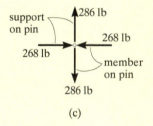

(c)

Equations of Equilibrium. The reaction N_B can be obtained *directly* by summing moments about point A, since $\mathbf{A}_x$ and $\mathbf{A}_y$ produce no moment about A.

$$\zeta +\Sigma M_A = 0;$$

$$[N_B \cos 30°](6 \text{ ft}) - [N_B \sin 30°](2 \text{ ft}) - 750 \text{ lb}(3 \text{ ft}) = 0$$

$$N_B = 536.2 \text{ lb} = 536 \text{ lb} \qquad Ans.$$

Using this result,

$$\xrightarrow{+} \Sigma F_x = 0; \qquad A_x - (536.2 \text{ lb}) \sin 30° = 0$$

$$A_x = 268 \text{ lb} \qquad Ans.$$

$$+\uparrow\Sigma F_y = 0; \qquad A_y + (536.2 \text{ lb}) \cos 30° - 750 \text{ lb} = 0$$

$$A_y = 286 \text{ lb} \qquad Ans.$$

Details of the equilibrium of the pin at A are shown in Fig. 5–16c.

EXAMPLE 5.10

The uniform smooth rod shown in Fig. 5–17a is subjected to a force and couple moment. If the rod is supported at A by a smooth wall and at B and C either at the top or bottom by rollers, determine the reactions at these supports. Neglect the weight of the rod.

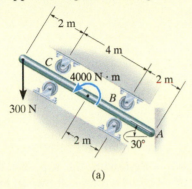

(a)

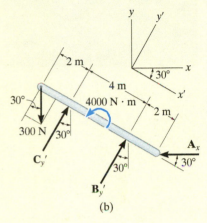

(b)

Fig. 5–17

SOLUTION

Free-Body Diagram. Removing the supports as shown in Fig. 5–17b, all the reactions act normal to the surfaces of contact since these surfaces are smooth. The reactions at B and C are shown acting in the positive y' direction. This assumes that only the rollers located on the bottom of the rod are used for support.

Equations of Equilibrium. Using the x, y coordinate system in Fig. 5–17b, we have

$$\xrightarrow{+} \Sigma F_x = 0; \qquad C_{y'} \sin 30° + B_{y'} \sin 30° - A_x = 0 \qquad (1)$$

$$+\uparrow \Sigma F_y = 0; \qquad -300 \text{ N} + C_{y'} \cos 30° + B_{y'} \cos 30° = 0 \qquad (2)$$

$$\zeta + \Sigma M_A = 0; \qquad -B_{y'}(2 \text{ m}) + 4000 \text{ N} \cdot \text{m} - C_{y'}(6 \text{ m})$$
$$+ (300 \cos 30° \text{ N})(8 \text{ m}) = 0 \qquad (3)$$

When writing the moment equation, it should be noted that the line of action of the force component 300 sin 30° N passes through point A, and therefore this force is not included in the moment equation.

Solving Eqs. 2 and 3 simultaneously, we obtain

$$B_{y'} = -1000.0 \text{ N} = -1 \text{ kN} \qquad \textit{Ans.}$$
$$C_{y'} = 1346.4 \text{ N} = 1.35 \text{ kN} \qquad \textit{Ans.}$$

Since $B_{y'}$ is a negative scalar, the sense of $\mathbf{B}_{y'}$ is opposite to that shown on the free-body diagram in Fig. 5–17b. Therefore, the top roller at B serves as the support rather than the bottom one. Retaining the negative sign for $B_{y'}$ (Why?) and substituting the results into Eq. 1, we obtain

$$1346.4 \sin 30° \text{ N} + (-1000.0 \sin 30° \text{ N}) - A_x = 0$$
$$A_x = 173 \text{ N} \qquad \textit{Ans.}$$

EXAMPLE 5.11

(a)

(© Russell C. Hibbeler)

The uniform truck ramp shown in Fig. 5–18a has a weight of 400 lb and is pinned to the body of the truck at each side and held in the position shown by the two side cables. Determine the tension in the cables.

SOLUTION

The idealized model of the ramp, which indicates all necessary dimensions and supports, is shown in Fig. 5–18b. Here the center of gravity is located at the midpoint since the ramp is considered to be uniform.

Free-Body Diagram.   Removing the supports from the idealized model, the ramp's free-body diagram is shown in Fig. 5–18c.

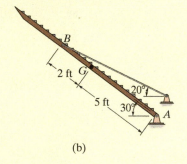

(b)

Equations of Equilibrium.   Summing moments about point A will yield a direct solution for the cable tension. Using the principle of moments, there are several ways of determining the moment of $\mathbf{T}$ about A. If we use x and y components, with $\mathbf{T}$ applied at B, we have

$$\zeta + \Sigma M_A = 0; \quad -T\cos 20°(7\sin 30° \text{ ft}) + T\sin 20°(7\cos 30° \text{ ft})$$

$$+ 400 \text{ lb }(5\cos 30° \text{ ft}) = 0$$

$$T = 1425 \text{ lb}$$

We can also determine the moment of $\mathbf{T}$ about A by resolving it into components along and perpendicular to the ramp at B. Then the moment of the component along the ramp will be zero about A, so that

$$\zeta + \Sigma M_A = 0; \quad -T\sin 10°(7 \text{ ft}) + 400 \text{ lb }(5\cos 30° \text{ ft}) = 0$$

$$T = 1425 \text{ lb}$$

Since there are two cables supporting the ramp,

$$T' = \frac{T}{2} = 712 \text{ lb} \qquad \textit{Ans.}$$

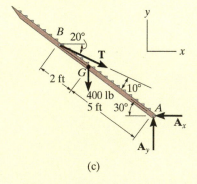

(c)

Fig. 5–18

NOTE: As an exercise, show that $A_x = 1339$ lb and $A_y = 887$ lb.

EXAMPLE 5.12

Determine the support reactions on the member in Fig. 5–19a. The collar at A is fixed to the member and can slide vertically along the vertical shaft.

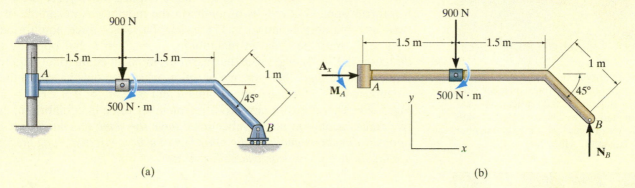

(a)   (b)

Fig. 5–19

SOLUTION

Free-Body Diagram.   Removing the supports, the free-body diagram of the member is shown in Fig. 5–19b. The collar exerts a horizontal force A_x and moment M_A on the member. The reaction N_B of the roller on the member is vertical.

Equations of Equilibrium.   The forces A_x and N_B can be determined directly from the force equations of equilibrium.

$\xrightarrow{+} \Sigma F_x = 0;$ $\qquad\qquad A_x = 0$ *Ans.*

$+\uparrow \Sigma F_y = 0;$ $\qquad\qquad N_B - 900\ \text{N} = 0$

$\qquad\qquad\qquad N_B = 900\ \text{N}$ *Ans.*

The moment M_A can be determined by summing moments either about point A or point B.

$\zeta + \Sigma M_A = 0;$

$M_A - 900\ \text{N}(1.5\ \text{m}) - 500\ \text{N} \cdot \text{m} + 900\ \text{N}\ [3\ \text{m} + (1\ \text{m})\cos 45°] = 0$

$\qquad\qquad M_A = -1486\ \text{N} \cdot \text{m} = 1.49\ \text{kN} \cdot \text{m} \zeta$ *Ans.*

or

$\zeta + \Sigma M_B = 0; \quad M_A + 900\ \text{N}\ [1.5\ \text{m} + (1\ \text{m})\cos 45°] - 500\ \text{N} \cdot \text{m} = 0$

$\qquad\qquad M_A = -1486\ \text{N} \cdot \text{m} = 1.49\ \text{kN} \cdot \text{m} \zeta$ *Ans.*

The negative sign indicates that M_A has the opposite sense of rotation to that shown on the free-body diagram.

5.4   Two- and Three-Force Members

The solutions to some equilibrium problems can be simplified by recognizing members that are subjected to only two or three forces.

Two-Force Members.   As the name implies, a *two-force member* has forces applied at only two points on the member. An example of a two-force member is shown in Fig. 5–20a. To satisfy force equilibrium, $\mathbf{F}_A$ and $\mathbf{F}_B$ must be equal in magnitude, $F_A = F_B = F$, but opposite in direction ($\Sigma\mathbf{F} = \mathbf{0}$), Fig. 5–20b. Furthermore, moment equilibrium requires that $\mathbf{F}_A$ and $\mathbf{F}_B$ share the same line of action, which can only happen if they are directed along the line joining points A and B ($\Sigma\mathbf{M}_A = \mathbf{0}$ or $\Sigma\mathbf{M}_B = \mathbf{0}$), Fig. 5–20c. Therefore, for any two-force member to be in equilibrium, the two forces acting on the member *must have the same magnitude, act in opposite directions, and have the same line of action, directed along the line joining the two points where these forces act.*

The hydraulic cylinder AB is a typical example of a two-force member since it is pin connected at its ends and, provided its weight is neglected, only the pin forces act on this member. (© Russell C. Hibbeler)

The link used for this railroad car brake is a three-force member. Since the force $\mathbf{F}_B$ in the tie rod at B and $\mathbf{F}_C$ from the link at C are parallel, then for equilibrium the resultant force $\mathbf{F}_A$ at the pin A must also be parallel with these two forces. (© Russell C. Hibbeler)

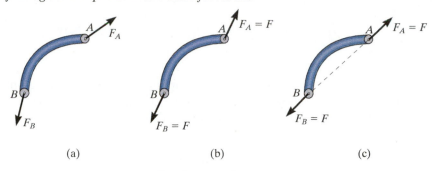

(a)   (b)   (c)

Two-force member

Fig. 5–20

Three-Force Members.   If a member is subjected to only *three forces*, it is called a *three-force member*. Moment equilibrium can be satisfied only if the three forces form a *concurrent* or *parallel* force system. To illustrate, consider the member subjected to the three forces $\mathbf{F}_1$, $\mathbf{F}_2$, and $\mathbf{F}_3$, shown in Fig. 5–21a. If the lines of action of $\mathbf{F}_1$ and $\mathbf{F}_2$ intersect at point O, then the line of action of $\mathbf{F}_3$ must *also* pass through point O so that the forces satisfy $\Sigma\mathbf{M}_O = \mathbf{0}$. As a special case, if the three forces are all parallel, Fig. 5–21b, the location of the point of intersection, O, will approach infinity.

The boom and bucket on this lift is a three-force member, provided its weight is neglected. Here the lines of action of the weight of the worker, $\mathbf{W}$, and the force of the two-force member (hydraulic cylinder) at B, $\mathbf{F}_B$, intersect at O. For moment equilibrium, the resultant force at the pin A, $\mathbf{F}_A$, must also be directed towards O. (© Russell C. Hibbeler)

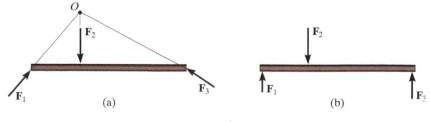

(a)   (b)

Three-force member

Fig. 5–21

EXAMPLE 5.13

The lever *ABC* is pin supported at *A* and connected to a short link *BD* as shown in Fig. 5–22a. If the weight of the members is negligible, determine the force of the pin on the lever at *A*.

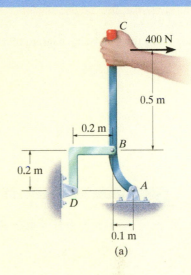

(a)

SOLUTION

Free-Body Diagrams. As shown in Fig. 5–22b, the short link *BD* is a *two-force member*, so the *resultant forces* from the pins *D* and *B* must be equal, opposite, and collinear. Although the magnitude of the force is unknown, the line of action is known since it passes through *B* and *D*.

Lever *ABC* is a *three-force member*, and therefore, in order to satisfy moment equilibrium, the three nonparallel forces acting on it must be concurrent at *O*, Fig. 5–22c. In particular, note that the force **F** on the lever at *B* is equal but opposite to the force **F** acting at *B* on the link. Why? The distance *CO* must be 0.5 m since the lines of action of **F** and the 400-N force are known.

Equations of Equilibrium. By requiring the force system to be concurrent at *O*, since $\Sigma M_O = 0$, the angle θ which defines the line of action of $\mathbf{F}_A$ can be determined from trigonometry,

$$\theta = \tan^{-1}\left(\frac{0.7}{0.4}\right) = 60.3°$$

Using the *x*, *y* axes and applying the force equilibrium equations,

$$\xrightarrow{+} \Sigma F_x = 0; \qquad F_A \cos 60.3° - F \cos 45° + 400 \text{ N} = 0$$

$$+\uparrow \Sigma F_y = 0; \qquad F_A \sin 60.3° - F \sin 45° = 0$$

Solving, we get

$$F_A = 1.07 \text{ kN} \qquad\qquad Ans.$$

$$F = 1.32 \text{ kN}$$

NOTE: We can also solve this problem by representing the force at *A* by its two components $\mathbf{A}_x$ and $\mathbf{A}_y$ and applying $\Sigma M_A = 0$, $\Sigma F_x = 0$, $\Sigma F_y = 0$ to the lever. Once A_x and A_y are determined, we can get F_A and θ.

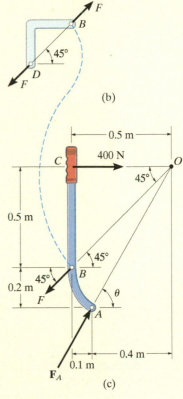

(b)

(c)

Fig. 5–22

PRELIMINARY PROBLEMS

P5–1. Draw the free-body diagram of each object.

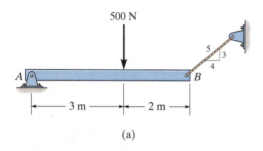

(a)

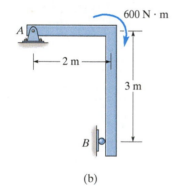

(b)

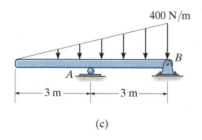

(c)

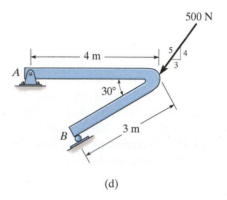

(d)

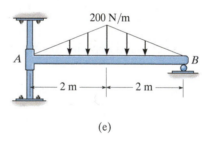

(e)

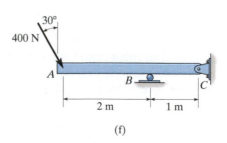

(f)

Prob. P5–1

FUNDAMENTAL PROBLEMS

All problem solutions must include an FBD.

F5–1. Determine the horizontal and vertical components of reaction at the supports. Neglect the thickness of the beam.

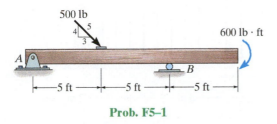

Prob. F5–1

F5–2. Determine the horizontal and vertical components of reaction at the pin *A* and the reaction on the beam at *C*.

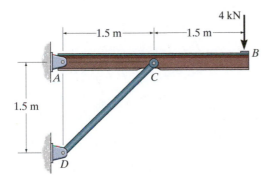

Prob. F5–2

F5–3. The truss is supported by a pin at *A* and a roller at *B*. Determine the support reactions.

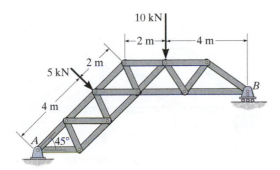

Prob. F5–3

F5–4. Determine the components of reaction at the fixed support *A*. Neglect the thickness of the beam.

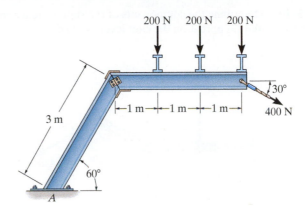

Prob. F5–4

F5–5. The 25-kg bar has a center of mass at *G*. If it is supported by a smooth peg at *C*, a roller at *A*, and cord *AB*, determine the reactions at these supports.

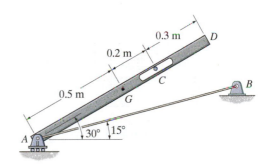

Prob. F5–5

F5–6. Determine the reactions at the smooth contact points *A*, *B*, and *C* on the bar.

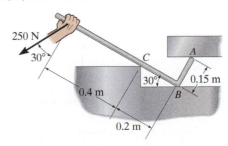

Prob. F5–6

PROBLEMS

All problem solutions must include an FBD.

5–10. Determine the components of the support reactions at the fixed support A on the cantilevered beam.

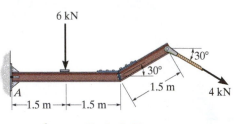

Prob. 5–10

5–11. Determine the reactions at the supports.

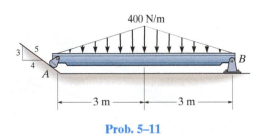

Prob. 5–11

*5–12. Determine the horizontal and vertical components of reaction at the pin A and the reaction of the rocker B on the beam.

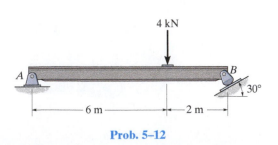

Prob. 5–12

5–13. Determine the reactions at the supports.

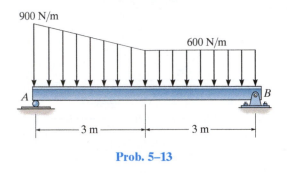

Prob. 5–13

5–14. Determine the reactions at the supports.

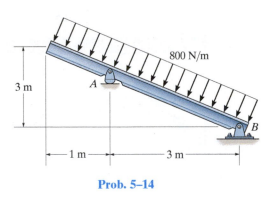

Prob. 5–14

5–15. Determine the reactions at the supports.

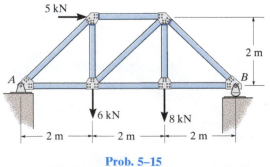

Prob. 5–15

*5–16. Determine the tension in the cable and the horizontal and vertical components of reaction of the pin A. The pulley at D is frictionless and the cylinder weighs 80 lb.

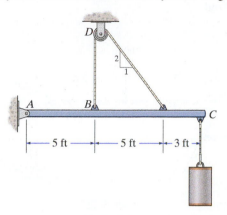

Prob. 5–16

5–17. The man attempts to support the load of boards having a weight W and a center of gravity at G. If he is standing on a smooth floor, determine the smallest angle θ at which he can hold them up in the position shown. Neglect his weight.

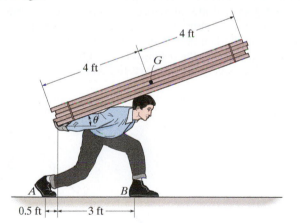

Prob. 5–17

5–18. Determine the components of reaction at the supports A and B on the rod.

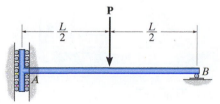

Prob. 5–18

5–19. The man has a weight W and stands at the center of the plank. If the planes at A and B are smooth, determine the tension in the cord in terms of W and θ.

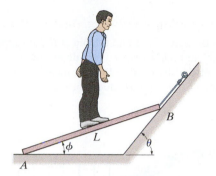

Prob. 5–19

*5–20. A uniform glass rod having a length L is placed in the smooth hemispherical bowl having a radius r. Determine the angle of inclination θ for equilibrium.

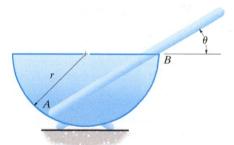

Prob. 5–20

5–21. The uniform rod AB has a mass of 40 kg. Determine the force in the cable when the rod is in the position shown. There is a smooth collar at A.

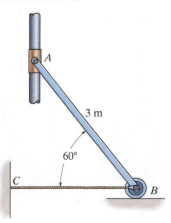

Prob. 5–21

5–22. If the intensity of the distributed load acting on the beam is $w = 3$ kN/m, determine the reactions at the roller A and pin B.

5–23. If the roller at A and the pin at B can support a load up to 4 kN and 8 kN, respectively, determine the maximum intensity of the distributed load w, measured in kN/m, so that failure of the supports does not occur.

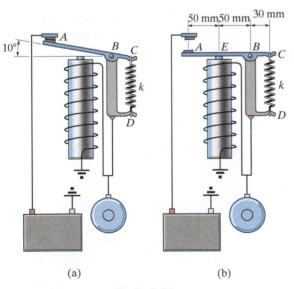

Probs. 5–22/23

*5–24. The relay regulates voltage and current. Determine the force in the spring CD, which has a stiffness of $k = 120$ N/m, so that it will allow the armature to make contact at A in figure (a) with a vertical force of 0.4 N. Also, determine the force in the spring when the coil is energized and attracts the armature to E, figure (b), thereby breaking contact at A.

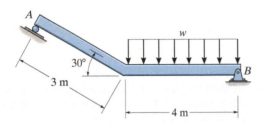

Prob. 5–24

5–25. Determine the reactions on the bent rod which is supported by a smooth surface at B and by a collar at A, which is fixed to the rod and is free to slide over the fixed inclined rod.

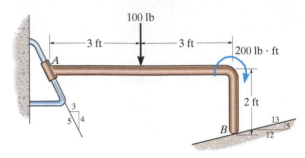

Prob. 5–25

5–26. The mobile crane is symmetrically supported by two outriggers at A and two at B in order to relieve the suspension of the truck upon which it rests and to provide greater stability. If the crane boom and truck have a mass of 18 Mg and center of mass at G_1, and the boom has a mass of 1.8 Mg and a center of mass at G_2, determine the vertical reactions at each of the four outriggers as a function of the boom angle θ when the boom is supporting a load having a mass of 1.2 Mg. Plot the results measured from $\theta = 0°$ to the critical angle where tipping starts to occur.

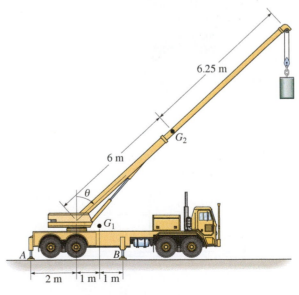

Prob. 5–26

5–27. Determine the reactions acting on the smooth uniform bar, which has a mass of 20 kg.

5–29. Determine the force P needed to pull the 50-kg roller over the smooth step. Take $\theta = 30°$.

5–30. Determine the magnitude and direction θ of the minimum force P needed to pull the 50-kg roller over the smooth step.

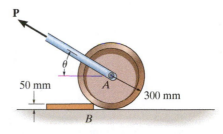

Probs. 5–29/30

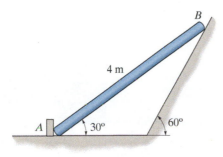

Prob. 5–27

5–31. The operation of the fuel pump for an automobile depends on the reciprocating action of the rocker arm ABC, which is pinned at B and is spring loaded at A and D. When the smooth cam C is in the position shown, determine the horizontal and vertical components of force at the pin and the force along the spring DF for equilibrium. The vertical force acting on the rocker arm at A is $F_A = 60$ N, and at C it is $F_C = 125$ N.

*5–28. A linear *torsional spring* deforms such that an applied couple moment M is related to the spring's rotation θ in radians by the equation $M = (20\,\theta)$ N · m. If such a spring is attached to the end of a pin-connected uniform 10-kg rod, determine the angle θ for equilibrium. The spring is undeformed when $\theta = 0°$.

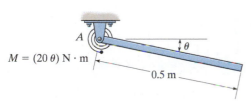

Prob. 5–28

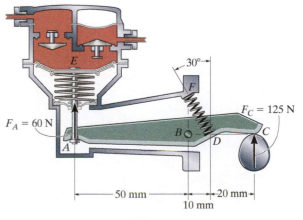

Prob. 5–31

*5–32. Determine the magnitude of force at the pin A and in the cable BC needed to support the 500-lb load. Neglect the weight of the boom AB.

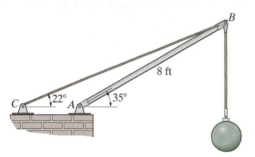

Prob. 5–32

5–33. The dimensions of a jib crane, which is manufactured by the Basick Co., are given in the figure. If the crane has a mass of 800 kg and a center of mass at G, and the maximum rated force at its end is $F = 15$ kN, determine the reactions at its bearings. The bearing at A is a journal bearing and supports only a horizontal force, whereas the bearing at B is a thrust bearing that supports both horizontal and vertical components.

5–34. The dimensions of a jib crane, which is manufactured by the Basick Co., are given in the figure. The crane has a mass of 800 kg and a center of mass at G. The bearing at A is a journal bearing and can support a horizontal force, whereas the bearing at B is a thrust bearing that supports both horizontal and vertical components. Determine the maximum load F that can be suspended from its end if the selected bearings at A and B can sustain a maximum resultant load of 24 kN and 34 kN, respectively.

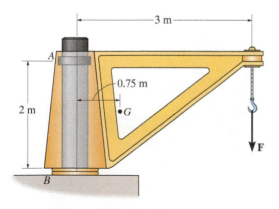

Probs. 5–33/34

5–35. The smooth pipe rests against the opening at the points of contact A, B, and C. Determine the reactions at these points needed to support the force of 300 N. Neglect the pipe's thickness in the calculation.

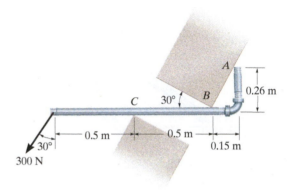

Prob. 5–35

*5–36. The beam of negligible weight is supported horizontally by two springs. If the beam is horizontal and the springs are unstretched when the load is removed, determine the angle of tilt of the beam when the load is applied.

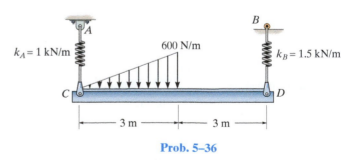

Prob. 5–36

5–37. The cantilevered jib crane is used to support the load of 780 lb. If $x = 5$ ft, determine the reactions at the supports. Note that the supports are collars that allow the crane to rotate freely about the vertical axis. The collar at B supports a force in the vertical direction, whereas the one at A does not.

5–38. The cantilevered jib crane is used to support the load of 780 lb. If the trolley T can be placed anywhere between $1.5 \text{ ft} \le x \le 7.5 \text{ ft}$, determine the maximum magnitude of reaction at the supports A and B. Note that the supports are collars that allow the crane to rotate freely about the vertical axis. The collar at B supports a force in the vertical direction, whereas the one at A does not.

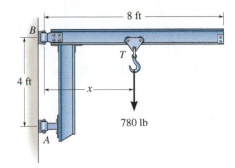

Probs. 5–37/38

5–39. The bar of negligible weight is supported by two springs, each having a stiffness $k = 100$ N/m. If the springs are originally unstretched, and the force is vertical as shown, determine the angle θ the bar makes with the horizontal, when the 30-N force is applied to the bar.

*5–40. Determine the stiffness k of each spring so that the 30-N force causes the bar to tip $\theta = 15°$ when the force is applied. Originally the bar is horizontal and the springs are unstretched. Neglect the weight of the bar.

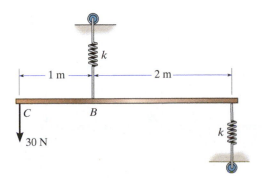

Probs. 5–39/40

5–41. The bulk head AD is subjected to both water and soil-backfill pressures. Assuming AD is "pinned" to the ground at A, determine the horizontal and vertical reactions there and also the required tension in the ground anchor BC necessary for equilibrium. The bulk head has a mass of 800 kg.

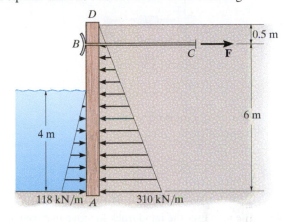

Prob. 5–41

5–42. The boom supports the two vertical loads. Neglect the size of the collars at D and B and the thickness of the boom, and compute the horizontal and vertical components of force at the pin A and the force in cable CB. Set $F_1 = 800$ N and $F_2 = 350$ N.

5–43. The boom is intended to support two vertical loads, $\mathbf{F}_1$ and $\mathbf{F}_2$. If the cable CB can sustain a maximum load of 1500 N before it fails, determine the critical loads if $F_1 = 2F_2$. Also, what is the magnitude of the maximum reaction at pin A?

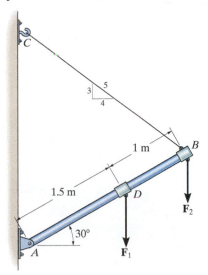

Probs. 5–42/43

*5–44. The 10-kg uniform rod is pinned at end A. If it is also subjected to a couple moment of 50 N · m, determine the smallest angle θ for equilibrium. The spring is unstretched when $\theta = 0$, and has a stiffness of $k = 60$ N/m.

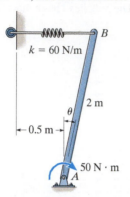

$k = 60$ N/m

B

2 m

θ

0.5 m

50 N · m

A

Prob. 5–44

5–45. The man uses the hand truck to move material up the step. If the truck and its contents have a mass of 50 kg with center of gravity at G, determine the normal reaction on both wheels and the magnitude and direction of the minimum force required at the grip B needed to lift the load.

0.4 m

B

0.5 m

0.2 m

G

0.4 m

60°

0.4 m

A 0.1 m

Prob. 5–45

5–46. Three uniform books, each having a weight W and length a, are stacked as shown. Determine the maximum distance d that the top book can extend out from the bottom one so the stack does not topple over.

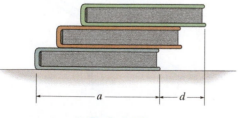

a d

Prob. 5–46

5–47. Determine the reactions at the pin A and the tension in cord BC. Set $F = 40$ kN. Neglect the thickness of the beam.

*5–48. If rope BC will fail when the tension becomes 50 kN, determine the greatest vertical load F that can be applied to the beam at B. What is the magnitude of the reaction at A for this loading? Neglect the thickness of the beam.

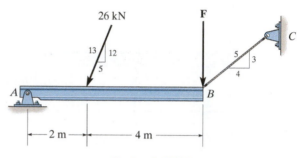

26 kN   **F**

13   12

5

A B

5   C

3

4

2 m   4 m

Probs. 5–47/48

5–49. The rigid metal strip of negligible weight is used as part of an electromagnetic switch. If the stiffness of the springs at A and B is $k = 5$ N/m and the strip is originally horizontal when the springs are unstretched, determine the smallest force F needed to close the contact gap at C.

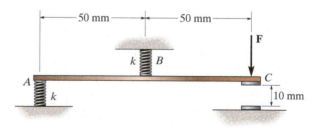

50 mm   50 mm

F

k B

A C

k 10 mm

Prob. 5–49

5–50. The rigid metal strip of negligible weight is used as part of an electromagnetic switch. Determine the maximum stiffness k of the springs at A and B so that the contact at C closes when the vertical force developed there is $F = 0.5$ N. Originally the strip is horizontal as shown.

*5–52. The uniform beam has a weight W and length l and is supported by a pin at A and a cable BC. Determine the horizontal and vertical components of reaction at A and the tension in the cable necessary to hold the beam in the position shown.

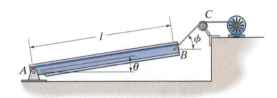

Prob. 5–52

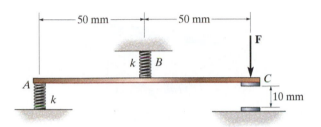

Prob. 5–50

5–53. A boy stands out at the end of the diving board, which is supported by two springs A and B, each having a stiffness of $k = 15$ kN/m. In the position shown the board is horizontal. If the boy has a mass of 40 kg, determine the angle of tilt which the board makes with the horizontal after he jumps off. Neglect the weight of the board and assume it is rigid.

5–51. The cantilever footing is used to support a wall near its edge A so that it causes a uniform soil pressure under the footing. Determine the uniform distribution loads, w_A and w_B, measured in lb/ft at pads A and B, necessary to support the wall forces of 8000 lb and 20 000 lb.

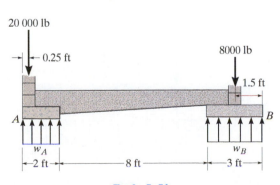

Prob. 5–51

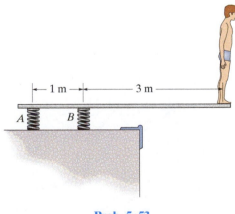

Prob. 5–53

5–54. The 30-N uniform rod has a length of $l = 1$ m. If $s = 1.5$ m, determine the distance h of placement at the end A along the smooth wall for equilibrium.

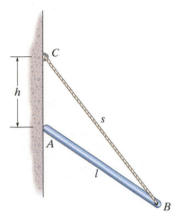

Prob. 5–54

5–55. The uniform rod has a length l and weight W. It is supported at one end A by a smooth wall and the other end by a cord of length s which is attached to the wall as shown. Determine the placement h for equilibrium.

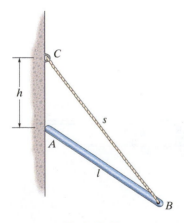

Prob. 5–55

*5–56. The uniform rod of length L and weight W is supported on the smooth planes. Determine its position θ for equilibrium. Neglect the thickness of the rod.

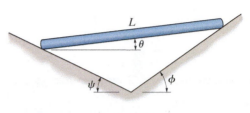

Prob. 5–56

5–57. The beam is subjected to the two concentrated loads. Assuming that the foundation exerts a linearly varying load distribution on its bottom, determine the load intensities w_1 and w_2 for equilibrium if $P = 500$ lb and $L = 12$ ft.

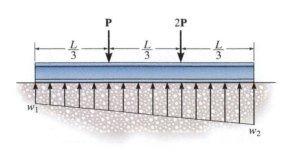

Prob. 5–57

5–58. The beam is subjected to the two concentrated loads. Assuming that the foundation exerts a linearly varying load distribution on its bottom, determine the load intensities w_1 and w_2 for equilibrium in terms of the parameters shown.

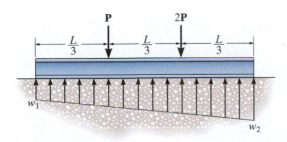

Prob. 5–58

5–59. The rod supports a weight of 200 lb and is pinned at its end A. If it is also subjected to a couple moment of 100 lb $\cdot$ ft, determine the angle θ for equilibrium. The spring has an unstretched length of 2 ft and a stiffness of $k = 50$ lb/ft.

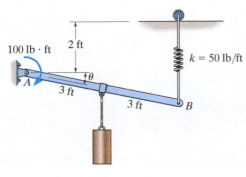

Prob. 5–59

*5–60. Determine the distance d for placement of the load **P** for equilibrium of the smooth bar in the position θ as shown. Neglect the weight of the bar.

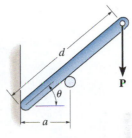

Prob. 5–60

5–61. If $d = 1$ m, and $\theta = 30°$, determine the normal reaction at the smooth supports and the required distance a for the placement of the roller if $P = 600$ N. Neglect the weight of the bar.

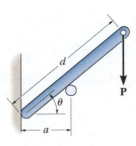

Prob. 5–61

CONCEPTUAL PROBLEMS

C5–1. The tie rod is used to support this overhang at the entrance of a building. If it is pin connected to the building wall at A and to the center of the overhang B, determine if the force in the rod will increase, decrease, or remain the same if (a) the support at A is moved to a lower position D, and (b) the support at B is moved to the outer position C. Explain your answer with an equilibrium analysis, using dimensions and loads. Assume the overhang is pin supported from the building wall.

Prob. C5–1 (© Russell C. Hibbeler)

C5–2. The man attempts to pull the four wheeler up the incline and onto the trailer. From the position shown, is it more effective to pull on the rope at A, or would it be better to pull on the rope at B? Draw a free-body diagram for each case, and do an equilibrium analysis to explain your answer. Use appropriate numerical values to do your calculations.

Prob. C5–2 (© Russell C. Hibbeler)

C5–3. Like all aircraft, this jet plane rests on three wheels. Why not use an additional wheel at the tail for better support? (Can you think of any other reason for not including this wheel?) If there was a fourth tail wheel, draw a free-body diagram of the plane from a side (2 D) view, and show why one would not be able to determine all the wheel reactions using the equations of equilibrium.

Prob. C5–3 (© Russell C. Hibbeler)

C5–4. Where is the best place to arrange most of the logs in the wheelbarrow so that it minimizes the amount of force on the backbone of the person transporting the load? Do an equilibrium analysis to explain your answer.

Prob. C5–4 (© Russell C. Hibbeler)

EQUILIBRIUM IN THREE DIMENSIONS

5.5 Free-Body Diagrams

The first step in solving three-dimensional equilibrium problems, as in the case of two dimensions, is to draw a free-body diagram. Before we can do this, however, it is first necessary to discuss the types of reactions that can occur at the supports.

Support Reactions. The reactive forces and couple moments acting at various types of supports and connections, when the members are viewed in three dimensions, are listed in Table 5–2. It is important to recognize the symbols used to represent each of these supports and to understand clearly how the forces and couple moments are developed. As in the two-dimensional case:

- A force is developed by a support that restricts the translation of its attached member.
- A couple moment is developed when rotation of the attached member is prevented.

For example, in Table 5–2, item (4), the ball-and-socket joint prevents any translation of the connecting member; therefore, a force must act on the member at the point of connection. This force has three components having unknown magnitudes, F_x, F_y, F_z. Provided these components are known, one can obtain the magnitude of force, $F = \sqrt{F_x^2 + F_y^2 + F_z^2}$, and the force's orientation defined by its coordinate direction angles α, β, γ, Eqs. 2–5.* Since the connecting member is allowed to rotate freely about *any* axis, no couple moment is resisted by a ball-and-socket joint.

It should be noted that the *single* bearing supports in items (5) and (7), the *single* pin (8), and the *single* hinge (9) are shown to resist both force and couple-moment components. If, however, these supports are used in conjunction with *other* bearings, pins, or hinges to hold a rigid body in equilibrium and the supports are *properly aligned* when connected to the body, then the *force reactions* at these supports *alone* are adequate for supporting the body. In other words, the couple moments become redundant and are not shown on the free-body diagram. The reason for this should become clear after studying the examples which follow.

---

* The three unknowns may also be represented as an unknown force magnitude F and two unknown coordinate direction angles. The third direction angle is obtained using the identity $\cos^2 \alpha + \cos^2 \beta + \cos^2 \gamma = 1$, Eq. 2–8

TABLE 5–2 Supports for Rigid Bodies Subjected to Three-Dimensional Force Systems

Types of Connection	Reaction	Number of Unknowns
(1) cable		One unknown. The reaction is a force which acts away from the member in the known direction of the cable.
(2) smooth surface support	$\mathbf{F}$	One unknown. The reaction is a force which acts perpendicular to the surface at the point of contact.
(3) roller	$\mathbf{F}$	One unknown. The reaction is a force which acts perpendicular to the surface at the point of contact.
(4) ball and socket	$\mathbf{F}_z$ $\mathbf{F}_x$ $\mathbf{F}_y$	Three unknowns. The reactions are three rectangular force components.
(5) single journal bearing	$\mathbf{M}_z$ $\mathbf{F}_z$ $\mathbf{M}_x$ $\mathbf{F}_x$	Four unknowns. The reactions are two force and two couple-moment components which act perpendicular to the shaft. Note: The couple moments are *generally not applied* if the body is supported elsewhere. See the examples.

continued

TABLE 5–2 Continued

Types of Connection	Reaction	Number of Unknowns
(6) single journal bearing with square shaft	M_z F_z M_y M_x F_x	Five unknowns. The reactions are two force and three couple-moment components. *Note*: The couple moments *are generally not applied* if the body is supported elsewhere. See the examples.
(7) single thrust bearing	M_z F_y F_z M_x F_x	Five unknowns. The reactions are three force and two couple-moment components. *Note*: The couple moments *are generally not applied* if the body is supported elsewhere. See the examples.
(8) single smooth pin	M_z F_z F_y M_y F_x	Five unknowns. The reactions are three force and two couple-moment components. *Note*: The couple moments *are generally not applied* if the body is supported elsewhere. See the examples.
(9) single hinge	M_z F_z F_y F_x M_x	Five unknowns. The reactions are three force and two couple-moment components. *Note*: The couple moments *are generally not applied* if the body is supported elsewhere. See the examples.
(10) fixed support	M_z F_z F_x F_y M_y M_x	Six unknowns. The reactions are three force and three couple-moment components.

Typical examples of actual supports that are referenced to Table 5–2 are shown in the following sequence of photos.

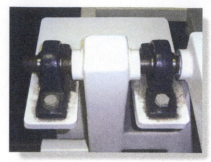

The journal bearings support the ends of the shaft. (5) (© Russell C. Hibbeler)

This ball-and-socket joint provides a connection for the housing of an earth grader to its frame. (4) (© Russell C. Hibbeler)

This thrust bearing is used to support the drive shaft on a machine. (7) (© Russell C. Hibbeler)

This pin is used to support the end of the strut used on a tractor. (8) (© Russell C. Hibbeler)

Free-Body Diagrams. The general procedure for establishing the free-body diagram of a rigid body has been outlined in Sec. 5.2. Essentially it requires first "isolating" the body by drawing its outlined shape. This is followed by a careful *labeling* of *all* the forces and couple moments with reference to an established *x, y, z* coordinate system. As a general rule, it is suggested to show the unknown components of reaction as acting on the free-body diagram in the *positive sense*. In this way, if any negative values are obtained, they will indicate that the components act in the negative coordinate directions.

EXAMPLE 5.14

Consider the two rods and plate, along with their associated free-body diagrams, shown in Fig. 5–23. The x, y, z axes are established on the diagram and the unknown reaction components are indicated in the *positive sense*. The weight is neglected.

SOLUTION

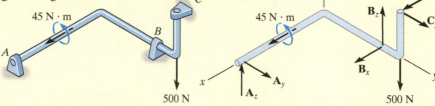

Properly aligned journal bearings at *A, B, C*.

The force reactions developed by the bearings are *sufficient* for equilibrium since they prevent the shaft from rotating about each of the coordinate axes. No couple moments at each bearing are developed.

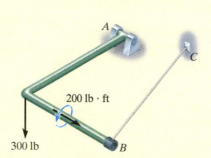

Pin at *A* and cable *BC*.

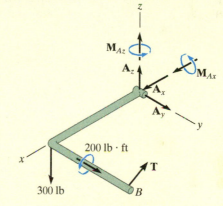

Moment components are developed by the pin on the rod to prevent rotation about the *x* and *z* axes.

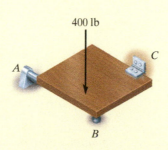

Properly aligned journal bearing at *A* and hinge at *C*. Roller at *B*.

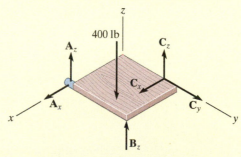

Only force reactions are developed by the bearing and hinge on the plate to prevent rotation about each coordinate axis. No moments are developed at the hinge.

Fig. 5–23

5.6   Equations of Equilibrium

As stated in Sec. 5.1, the conditions for equilibrium of a rigid body subjected to a three-dimensional force system require that both the *resultant* force and *resultant* couple moment acting on the body be equal to *zero*.

Vector Equations of Equilibrium.   The two conditions for equilibrium of a rigid body may be expressed mathematically in vector form as

$$\Sigma \mathbf{F} = 0$$
$$\Sigma \mathbf{M}_O = 0 \qquad (5-5)$$

where $\Sigma \mathbf{F}$ is the vector sum of all the external forces acting on the body and $\Sigma \mathbf{M}_O$ is the sum of the couple moments and the moments of all the forces about any point O located either on or off the body.

Scalar Equations of Equilibrium.   If all the external forces and couple moments are expressed in Cartesian vector form and substituted into Eqs. 5–5, we have

$$\Sigma \mathbf{F} = \Sigma F_x \mathbf{i} + \Sigma F_y \mathbf{j} + \Sigma F_z \mathbf{k} = 0$$

$$\Sigma \mathbf{M}_O = \Sigma M_x \mathbf{i} + \Sigma M_y \mathbf{j} + \Sigma M_z \mathbf{k} = 0$$

Since the $\mathbf{i}, \mathbf{j}$, and $\mathbf{k}$ components are independent from one another, the above equations are satisfied provided

$$\Sigma F_x = 0$$
$$\Sigma F_y = 0 \qquad (5-6a)$$
$$\Sigma F_z = 0$$

and

$$\Sigma M_x = 0$$
$$\Sigma M_y = 0 \qquad (5-6b)$$
$$\Sigma M_z = 0$$

These *six scalar equilibrium equations* may be used to solve for at most six unknowns shown on the free-body diagram. Equations 5–6a require the sum of the external force components acting in the x, y, and z directions to be zero, and Eqs. 5–6b require the sum of the moment components about the x, y, and z axes to be zero.

5.7   Constraints and Statical Determinacy

To ensure the equilibrium of a rigid body, it is not only necessary to satisfy the equations of equilibrium, but the body must also be properly held or constrained by its supports. Some bodies may have more supports than are necessary for equilibrium, whereas others may not have enough or the supports may be arranged in a particular manner that could cause the body to move. Each of these cases will now be discussed.

Redundant Constraints.   When a body has redundant supports, that is, more supports than are necessary to hold it in equilibrium, it becomes statically indeterminate. *Statically indeterminate* means that there will be more unknown loadings on the body than equations of equilibrium available for their solution. For example, the beam in Fig. 5–24a and the pipe assembly in Fig. 5–24b, shown together with their free-body diagrams, are both statically indeterminate because of additional (or redundant) support reactions. For the beam there are five unknowns, M_A, A_x, A_y, B_y, and C_y, for which only three equilibrium equations can be written ($\Sigma F_x = 0$, $\Sigma F_y = 0$, and $\Sigma M_O = 0$, Eq. 5–2). The pipe assembly has eight unknowns, for which only six equilibrium equations can be written, Eqs. 5–6.

The additional equations needed to solve statically indeterminate problems of the type shown in Fig. 5–24 are generally obtained from the deformation conditions at the points of support. These equations involve the physical properties of the body which are studied in subjects dealing with the mechanics of deformation, such as "mechanics of materials."*

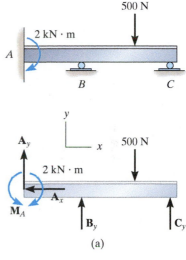

(a)

Fig. 5–24

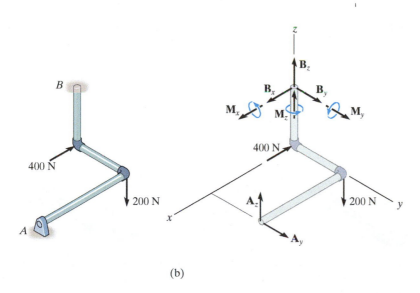

(b)

---

* See R. C. Hibbeler, *Mechanics of Materials*, 8th edition, Pearson Education/Prentice Hall, Inc.

Improper Constraints.   Having the same number of unknown reactive forces as available equations of equilibrium does not always guarantee that a body will be stable when subjected to a particular loading. For example, the pin support at A and the roller support at B for the beam in Fig. 5–25a are placed in such a way that the lines of action of the reactive forces are *concurrent* at point A. Consequently, the applied loading $\mathbf{P}$ will cause the beam to rotate slightly about A, and so the beam is improperly constrained, $\Sigma M_A \neq 0$.

In three dimensions, a body will be improperly constrained if the lines of action of all the reactive forces intersect a common axis. For example, the reactive forces at the ball-and-socket supports at A and B in Fig. 5–25b all intersect the axis passing through A and B. Since the moments of these forces about A and B are all zero, then the loading $\mathbf{P}$ will rotate the member about the AB axis, $\Sigma M_{AB} \neq 0$.

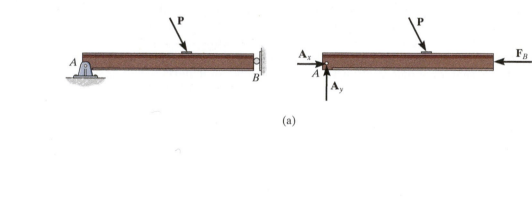

(a)

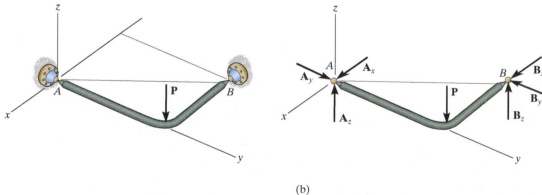

(b)

Fig. 5–25

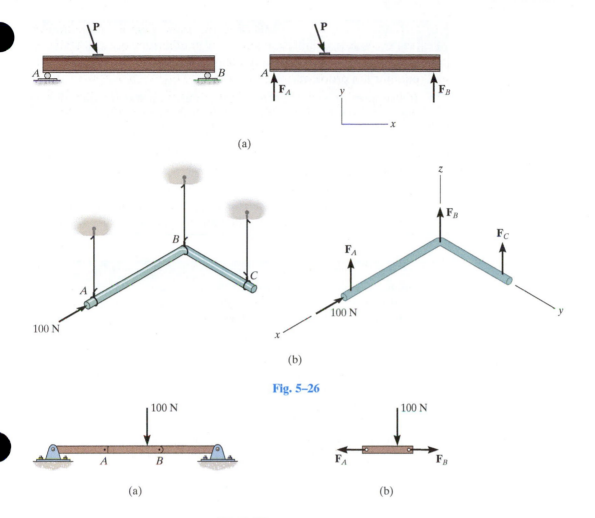

(a)

(b)

Fig. 5–26

(a)

(b)

Fig. 5–27

Another way in which improper constraining leads to instability occurs when the *reactive forces* are all *parallel*. Two- and three-dimensional examples of this are shown in Fig. 5–26. In both cases, the summation of forces along the x axis will not equal zero.

In some cases, a body may have *fewer* reactive forces than equations of equilibrium that must be satisfied. The body then becomes only *partially constrained*. For example, consider member *AB* in Fig. 5–27*a* with its corresponding free-body diagram in Fig. 5–27*b*. Here $\Sigma F_y = 0$ will not be satisfied for the loading conditions and therefore equilibrium will not be maintained.

To summarize these points, a body is considered *improperly constrained* if all the reactive forces intersect at a common point or pass through a common axis, or if all the reactive forces are parallel. In engineering practice, these situations should be avoided at all times since they will cause an unstable condition.

Stability is always an important concern when operating a crane, not only when lifting a load, but also when moving it about. (© Russell C. Hibbeler)

Important Points

- Always draw the free-body diagram first when solving any equilibrium problem.
- If a support *prevents translation* of a body in a specific direction, then the support exerts a *force* on the body in that direction.
- If a support *prevents rotation about an axis*, then the support exerts a *couple moment* on the body about the axis.
- If a body is subjected to more unknown reactions than available equations of equilibrium, then the problem is *statically indeterminate*.
- A stable body requires that the lines of action of the reactive forces do not intersect a common axis and are not parallel to one another.

Procedure for Analysis

Three-dimensional equilibrium problems for a rigid body can be solved using the following procedure.

Free-Body Diagram.

- Draw an outlined shape of the body.
- Show all the forces and couple moments acting on the body.
- Establish the origin of the *x, y, z* axes at a convenient point and orient the axes so that they are parallel to as many of the external forces and moments as possible.
- Label all the loadings and specify their directions. In general, show all the *unknown* components having a *positive sense* along the *x, y, z* axes.
- Indicate the dimensions of the body necessary for computing the moments of forces.

Equations of Equilibrium.

- If the *x, y, z* force and moment components seem easy to determine, then apply the six scalar equations of equilibrium; otherwise use the vector equations.
- It is not necessary that the set of axes chosen for force summation coincide with the set of axes chosen for moment summation. Actually, an axis in any arbitrary direction may be chosen for summing forces and moments.
- Choose the direction of an axis for moment summation such that it intersects the lines of action of as many unknown forces as possible. Realize that the moments of forces passing through points on this axis and the moments of forces which are parallel to the axis will then be zero.
- If the solution of the equilibrium equations yields a negative scalar for a force or couple moment magnitude, it indicates that the sense is opposite to that assumed on the free-body diagram.

EXAMPLE 5.15

The homogeneous plate shown in Fig. 5–28a has a mass of 100 kg and is subjected to a force and couple moment along its edges. If it is supported in the horizontal plane by a roller at A, a ball-and-socket joint at B, and a cord at C, determine the components of reaction at these supports.

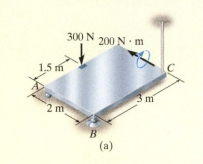

(a)

SOLUTION (*SCALAR ANALYSIS*)

Free-Body Diagram. There are five unknown reactions acting on the plate, as shown in Fig. 5–28b. Each of these reactions is assumed to act in a positive coordinate direction.

Equations of Equilibrium. Since the three-dimensional geometry is rather simple, a *scalar analysis* provides a *direct solution* to this problem. A force summation along each axis yields

$$\Sigma F_x = 0; \qquad B_x = 0 \qquad\qquad Ans.$$

$$\Sigma F_y = 0; \qquad B_y = 0 \qquad\qquad Ans.$$

$$\Sigma F_z = 0; \qquad A_z + B_z + T_C - 300\text{ N} - 981\text{ N} = 0 \qquad (1)$$

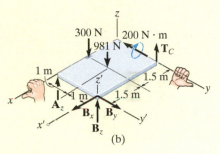

(b)

Fig. 5–28

Recall that the moment of a force about an axis is equal to the product of the force magnitude and the perpendicular distance (moment arm) from the line of action of the force to the axis. Also, forces that are parallel to an axis or pass through it create no moment about the axis. Hence, summing moments about the positive x and y axes, we have

$$\Sigma M_x = 0; \qquad T_C(2\text{ m}) - 981\text{ N}(1\text{ m}) + B_z(2\text{ m}) = 0 \qquad (2)$$

$$\Sigma M_y = 0; \qquad 300\text{ N}(1.5\text{ m}) + 981\text{ N}(1.5\text{ m}) - B_z(3\text{ m}) - A_z(3\text{ m})$$

$$- 200\text{ N}\cdot\text{m} = 0 \qquad (3)$$

The components of the force at B can be eliminated if moments are summed about the x' and y' axes. We obtain

$$\Sigma M_{x'} = 0; \qquad 981\text{ N}(1\text{ m}) + 300\text{ N}(2\text{ m}) - A_z(2\text{ m}) = 0 \qquad (4)$$

$$\Sigma M_{y'} = 0; \qquad -300\text{ N}(1.5\text{ m}) - 981\text{ N}(1.5\text{ m}) - 200\text{ N}\cdot\text{m}$$

$$+ T_C(3\text{ m}) = 0 \qquad (5)$$

Solving Eqs. 1 through 3 or the more convenient Eqs. 1, 4, and 5 yields

$$A_z = 790\text{ N} \quad B_z = -217\text{ N} \quad T_C = 707\text{ N} \qquad\qquad Ans.$$

The negative sign indicates that $\mathbf{B}_z$ acts downward.

NOTE: The solution of this problem does not require a summation of moments about the z axis. The plate is partially constrained since the supports cannot prevent it from turning about the z axis if a force is applied to it in the x–y plane.

EXAMPLE 5.16

Determine the components of reaction that the ball-and-socket joint at A, the smooth journal bearing at B, and the roller support at C exert on the rod assembly in Fig. 5–29a.

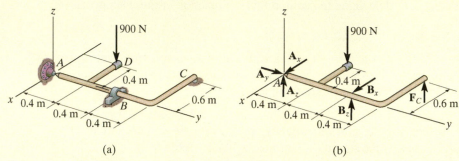

(a)   (b)

Fig. 5–29

SOLUTION (*SCALAR ANALYSIS*)

Free-Body Diagram.   As shown on the free-body diagram, Fig. 5–29b, the reactive forces of the supports will prevent the assembly from rotating about each coordinate axis, and so the journal bearing at B only exerts reactive forces on the member. No couple moments are required.

Equations of Equilibrium.   Because all the forces are either horizontal or vertical, it is convenient to use a scalar analysis. A direct solution for A_y can be obtained by summing forces along the y axis.

$$\Sigma F_y = 0; \qquad A_y = 0 \hspace{3cm} Ans.$$

The force F_C can be determined directly by summing moments about the y axis.

$$\Sigma M_y = 0; \qquad F_C(0.6\text{ m}) - 900\text{ N}(0.4\text{ m}) = 0$$
$$F_C = 600\text{ N} \hspace{3cm} Ans.$$

Using this result, B_z can be determined by summing moments about the x axis.

$$\Sigma M_x = 0; \qquad B_z(0.8\text{ m}) + 600\text{ N}(1.2\text{ m}) - 900\text{ N}(0.4\text{ m}) = 0$$
$$B_z = -450\text{ N} \hspace{3cm} Ans.$$

The negative sign indicates that $\mathbf{B}_z$ acts downward. The force B_x can be found by summing moments about the z axis.

$$\Sigma M_z = 0; \qquad -B_x(0.8\text{ m}) = 0 \quad B_x = 0 \hspace{2cm} Ans.$$

Thus,

$$\Sigma F_x = 0; \qquad A_x + 0 = 0 \qquad A_x = 0 \hspace{2.5cm} Ans.$$

Finally, using the results of B_z and F_C.

$$\Sigma F_z = 0; \qquad A_z + (-450\text{ N}) + 600\text{ N} - 900\text{ N} = 0$$
$$A_z = 750\text{ N} \hspace{3cm} Ans.$$

EXAMPLE 5.17

The boom is used to support the 75-lb flowerpot in Fig. 5–30a. Determine the tension developed in wires AB and AC.

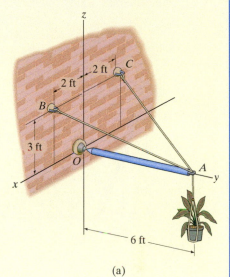

(a)

Fig. 5–30

SOLUTION (*VECTOR ANALYSIS*)

Free-Body Diagram. The free-body diagram of the boom is shown in Fig. 5–30b.

Equations of Equilibrium. Here the cable forces are directed at angles with the coordinate axes, so we will use a vector analysis.

$$\mathbf{F}_{AB} = F_{AB}\left(\frac{\mathbf{r}_{AB}}{r_{AB}}\right) = F_{AB}\left(\frac{\{2\mathbf{i} - 6\mathbf{j} + 3\mathbf{k}\}\text{ ft}}{\sqrt{(2\text{ ft})^2 + (-6\text{ ft})^2 + (3\text{ ft})^2}}\right)$$

$$= \tfrac{2}{7}F_{AB}\mathbf{i} - \tfrac{6}{7}F_{AB}\mathbf{j} + \tfrac{3}{7}F_{AB}\mathbf{k}$$

$$\mathbf{F}_{AC} = F_{AC}\left(\frac{\mathbf{r}_{AC}}{r_{AC}}\right) = F_{AC}\left(\frac{\{-2\mathbf{i} - 6\mathbf{j} + 3\mathbf{k}\}\text{ ft}}{\sqrt{(-2\text{ ft})^2 + (-6\text{ ft})^2 + (3\text{ ft})^2}}\right)$$

$$= -\tfrac{2}{7}F_{AC}\mathbf{i} - \tfrac{6}{7}F_{AC}\mathbf{j} + \tfrac{3}{7}F_{AC}\mathbf{k}$$

We can eliminate the force reaction at O by writing the moment equation of equilibrium about point O.

$$\Sigma \mathbf{M}_O = 0; \qquad \mathbf{r}_A \times (\mathbf{F}_{AB} + \mathbf{F}_{AC} + \mathbf{W}) = 0$$

$$(6\mathbf{j}) \times \left[\left(\tfrac{2}{7}F_{AB}\mathbf{i} - \tfrac{6}{7}F_{AB}\mathbf{j} + \tfrac{3}{7}F_{AB}\mathbf{k}\right) + \left(-\tfrac{2}{7}F_{AC}\mathbf{i} - \tfrac{6}{7}F_{AC}\mathbf{j} + \tfrac{3}{7}F_{AC}\mathbf{k}\right) + (-75\mathbf{k})\right] = 0$$

$$\left(\tfrac{18}{7}F_{AB} + \tfrac{18}{7}F_{AC} - 450\right)\mathbf{i} + \left(-\tfrac{12}{7}F_{AB} + \tfrac{12}{7}F_{AC}\right)\mathbf{k} = 0$$

$$\Sigma M_x = 0; \qquad \tfrac{18}{7}F_{AB} + \tfrac{18}{7}F_{AC} - 450 = 0 \qquad (1)$$

$$\Sigma M_y = 0; \qquad 0 = 0$$

$$\Sigma M_z = 0; \qquad -\tfrac{12}{7}F_{AB} + \tfrac{12}{7}F_{AC} = 0 \qquad (2)$$

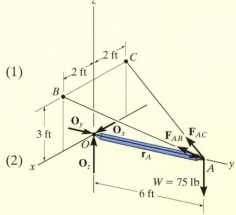

(b)

Solving Eqs. (1) and (2) simultaneously,

$$F_{AB} = F_{AC} = 87.5\text{ lb} \qquad \qquad \textit{Ans.}$$

EXAMPLE 5.18

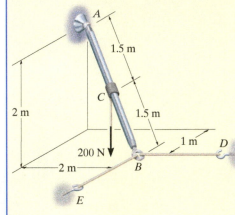

(a)

Rod AB shown in Fig. 5–31a is subjected to the 200-N force. Determine the reactions at the ball-and-socket joint A and the tension in the cables BD and BE. The collar at C is fixed to the rod.

SOLUTION (VECTOR ANALYSIS)

Free-Body Diagram.   Fig. 5–31b.

Equations of Equilibrium.   Representing each force on the free-body diagram in Cartesian vector form, we have

$$\mathbf{F}_A = A_x\mathbf{i} + A_y\mathbf{j} + A_z\mathbf{k}$$
$$\mathbf{T}_E = T_E\mathbf{i}$$
$$\mathbf{T}_D = T_D\mathbf{j}$$
$$\mathbf{F} = \{-200\mathbf{k}\}\ \text{N}$$

Applying the force equation of equilibrium.

$$\Sigma\mathbf{F} = \mathbf{0}; \qquad \mathbf{F}_A + \mathbf{T}_E + \mathbf{T}_D + \mathbf{F} = \mathbf{0}$$

$$(A_x + T_E)\mathbf{i} + (A_y + T_D)\mathbf{j} + (A_z - 200)\mathbf{k} = \mathbf{0}$$

$$\Sigma F_x = 0; \qquad A_x + T_E = 0 \qquad (1)$$
$$\Sigma F_y = 0; \qquad A_y + T_D = 0 \qquad (2)$$
$$\Sigma F_z = 0; \qquad A_z - 200 = 0 \qquad (3)$$

Summing moments about point A yields

$$\Sigma\mathbf{M}_A = \mathbf{0}; \qquad \mathbf{r}_C \times \mathbf{F} + \mathbf{r}_B \times (\mathbf{T}_E + \mathbf{T}_D) = \mathbf{0}$$

Since $\mathbf{r}_C = \frac{1}{2}\mathbf{r}_B$, then

$$(0.5\mathbf{i} + 1\mathbf{j} - 1\mathbf{k}) \times (-200\mathbf{k}) + (1\mathbf{i} + 2\mathbf{j} - 2\mathbf{k}) \times (T_E\mathbf{i} + T_D\mathbf{j}) = \mathbf{0}$$

Expanding and rearranging terms gives

$$(2T_D - 200)\mathbf{i} + (-2T_E + 100)\mathbf{j} + (T_D - 2T_E)\mathbf{k} = \mathbf{0}$$

$$\Sigma M_x = 0; \qquad 2T_D - 200 = 0 \qquad (4)$$
$$\Sigma M_y = 0; \qquad -2T_E + 100 = 0 \qquad (5)$$
$$\Sigma M_z = 0; \qquad T_D - 2T_E = 0 \qquad (6)$$

Solving Eqs. 1 through 5, we get

$$T_D = 100\ \text{N} \qquad \qquad \textit{Ans.}$$
$$T_E = 50\ \text{N} \qquad \qquad \textit{Ans.}$$
$$A_x = -50\ \text{N} \qquad \qquad \textit{Ans.}$$
$$A_y = -100\ \text{N} \qquad \qquad \textit{Ans.}$$
$$A_z = 200\ \text{N} \qquad \qquad \textit{Ans.}$$

(b)

Fig. 5–31

NOTE: The negative sign indicates that $\mathbf{A}_x$ and $\mathbf{A}_y$ have a sense which is opposite to that shown on the free-body diagram, Fig. 5–31b. Also, notice that Eqs. 1–6 can be set up *directly* using a scalar analysis.

EXAMPLE 5.19

The bent rod in Fig. 5–32a is supported at A by a journal bearing, at D by a ball-and-socket joint, and at B by means of cable BC. Using only *one equilibrium equation*, obtain a direct solution for the tension in cable BC. The bearing at A is capable of exerting force components only in the z and y directions since it is properly aligned on the shaft. In other words, no couple moments are required at this support.

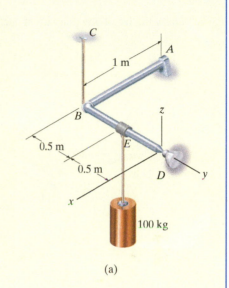

(a)

SOLUTION (*VECTOR ANALYSIS*)

Free-Body Diagram.   As shown in Fig. 5–32b, there are six unknowns.

Equations of Equilibrium.   The cable tension T_B may be obtained *directly* by summing moments about an axis that passes through points D and A. Why? The direction of this axis is defined by the unit vector **u**, where

$$\mathbf{u} = \frac{\mathbf{r}_{DA}}{r_{DA}} = -\frac{1}{\sqrt{2}}\mathbf{i} - \frac{1}{\sqrt{2}}\mathbf{j}$$
$$= -0.7071\mathbf{i} - 0.7071\mathbf{j}$$

Hence, the sum of the moments about this axis is zero provided

$$\Sigma M_{DA} = \mathbf{u} \cdot \Sigma (\mathbf{r} \times \mathbf{F}) = 0$$

Here **r** represents a position vector drawn from *any point* on the axis DA to any point on the line of action of force **F** (see Eq. 4–11). With reference to Fig. 5–32b, we can therefore write

$$\mathbf{u} \cdot (\mathbf{r}_B \times \mathbf{T}_B + \mathbf{r}_E \times \mathbf{W}) = 0$$
$$(-0.7071\mathbf{i} - 0.7071\mathbf{j}) \cdot \left[(-1\mathbf{j}) \times (T_B\mathbf{k})\right.$$
$$\left. + (-0.5\mathbf{j}) \times (-981\mathbf{k})\right] = 0$$
$$(-0.7071\mathbf{i} - 0.7071\mathbf{j}) \cdot [(-T_B + 490.5)\mathbf{i}] = 0$$
$$-0.7071(-T_B + 490.5) + 0 + 0 = 0$$
$$T_B = 490.5 \text{ N} \qquad\qquad Ans.$$

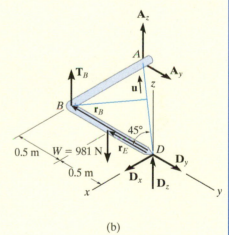

(b)

Fig. 5–32

NOTE: Since the moment arms from the axis to $\mathbf{T}_B$ and **W** are easy to obtain, we can also determine this result using a scalar analysis. As shown in Fig. 5–32b,

$$\Sigma M_{DA} = 0; \quad T_B(1 \text{ m } \sin 45°) - 981 \text{ N}(0.5 \text{ m } \sin 45°) = 0$$
$$T_B = 490.5 \text{ N} \qquad\qquad Ans.$$

PRELIMINARY PROBLEMS

P5–2. Draw the free-body diagram of each object.

P5–3. In each case, write the moment equations about the x, y, and z axes.

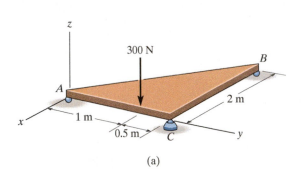

(a)

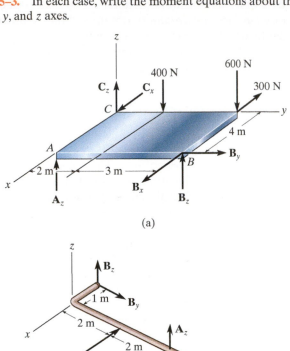

(a)

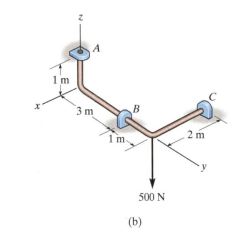

(b)

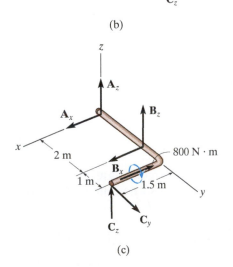

(b)

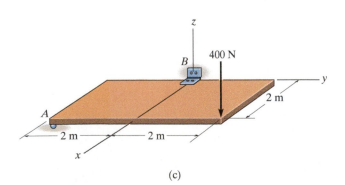

(c)

Prob. P5–2

(c)

Prob. P5–3

FUNDAMENTAL PROBLEMS

All problem solutions must include an FBD.

F5–7. The uniform plate has a weight of 500 lb. Determine the tension in each of the supporting cables.

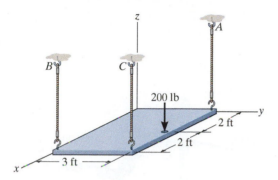

Prob. F5–7

F5–8. Determine the reactions at the roller support A, the ball-and-socket joint D, and the tension in cable BC for the plate.

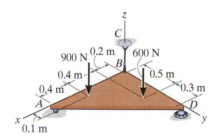

Prob. F5–8

F5–9. The rod is supported by smooth journal bearings at A, B, and C and is subjected to the two forces. Determine the reactions at these supports.

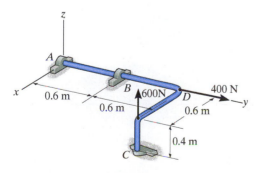

Prob. F5–9

F5–10. Determine the support reactions at the smooth journal bearings A, B, and C of the pipe assembly.

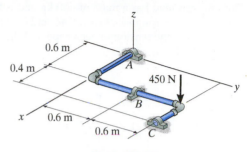

Prob. F5–10

F5–11. Determine the force developed in the short link BD, and the tension in the cords CE and CF, and the reactions of the ball-and-socket joint A on the block.

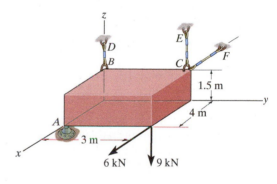

Prob. F5–11

F5–12. Determine the components of reaction that the thrust bearing A and cable BC exert on the bar.

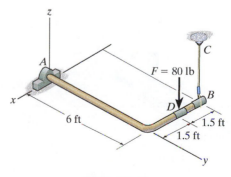

Prob. F5–12

5

PROBLEMS

All problem solutions must include an FBD.

5–62.   The uniform load has a mass of 600 kg and is lifted using a uniform 30-kg strongback beam *BAC* and four ropes as shown. Determine the tension in each rope and the force that must be applied at *A*.

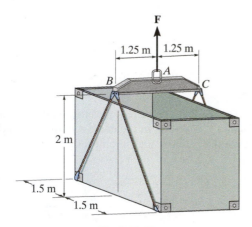

Prob. 5–62

5–63.   Due to an unequal distribution of fuel in the wing tanks, the centers of gravity for the airplane fuselage *A* and wings *B* and *C* are located as shown. If these components have weights $W_A = 45\,000$ lb, $W_B = 8000$ lb, and $W_C = 6000$ lb, determine the normal reactions of the wheels *D*, *E*, and *F* on the ground.

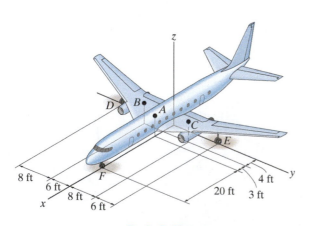

Prob. 5–63

***5–64.**   Determine the components of reaction at the fixed support *A*. The 400 N, 500 N, and 600 N forces are parallel to the *x*, *y*, and *z* axes, respectively.

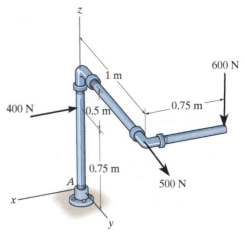

Prob. 5–64

5–65.   The 50-lb mulching machine has a center of gravity at *G*. Determine the vertical reactions at the wheels *C* and *B* and the smooth contact point *A*.

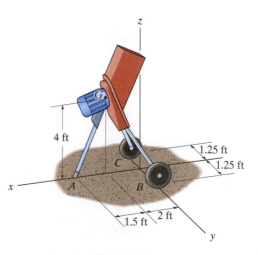

Prob. 5–65

5–66. The smooth uniform rod AB is supported by a ball-and-socket joint at A, the wall at B, and cable BC. Determine the components of reaction at A, the tension in the cable, and the normal reaction at B if the rod has a mass of 20 kg.

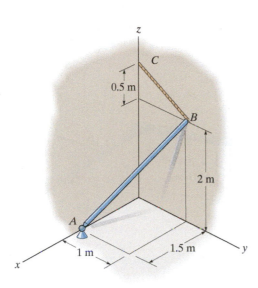

Prob. 5–66

5–67. The uniform concrete slab has a mass of 2400 kg. Determine the tension in each of the three parallel supporting cables when the slab is held in the horizontal plane as shown.

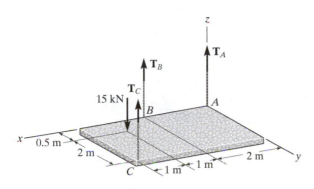

Prob. 5–67

*5–68. The 100-lb door has its center of gravity at G. Determine the components of reaction at hinges A and B if hinge B resists only forces in the x and y directions and A resists forces in the x, y, z directions.

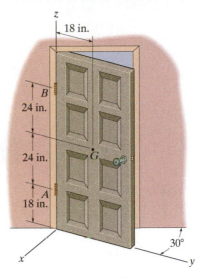

Prob. 5–68

5–69. Determine the tension in each cable and the components of reaction at D needed to support the load.

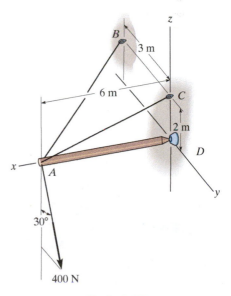

Prob. 5–69

5–70. The stiff-leg derrick used on ships is supported by a ball-and-socket joint at D and two cables BA and BC. The cables are attached to a smooth collar ring at B, which allows rotation of the derrick about z axis. If the derrick supports a crate having a mass of 200 kg, determine the tension in the cables and the x, y, z components of reaction at D.

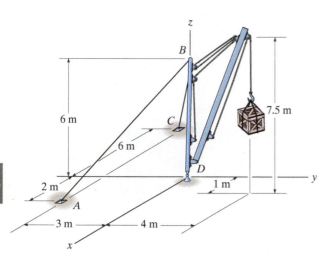

Prob. 5–70

5–71. Determine the components of reaction at the ball-and-socket joint A and the tension in each cable necessary for equilibrium of the rod.

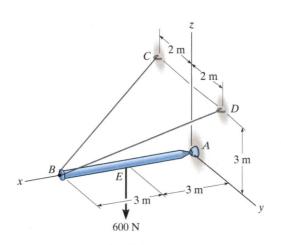

Prob. 5–71

*5–72. Determine the components of reaction at the ball-and-socket joint A and the tension in the supporting cables DB and DC.

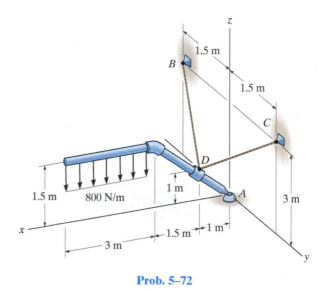

Prob. 5–72

5–73. The bent rod is supported at A, B, and C by smooth journal bearings. Determine the components of reaction at the bearings if the rod is subjected to the force $F = 800$ N. The bearings are in proper alignment and exert only force reactions on the rod.

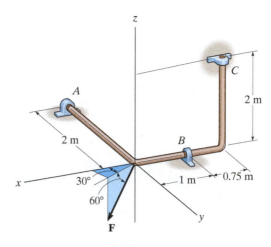

Prob. 5–73

5–74. The bent rod is supported at A, B, and C by smooth journal bearings. Determine the magnitude of **F** which will cause the positive x component of reaction at the bearing C to be $C_x = 50$ N. The bearings are in proper alignment and exert only force reactions on the rod.

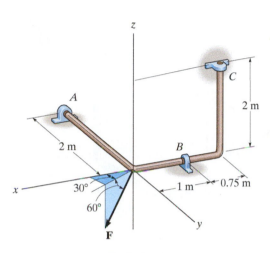

Prob. 5–74

5–75. Member AB is supported by a cable BC and at A by a *square* rod which fits loosely through the square hole in the collar fixed to the member as shown. Determine the components of reaction at A and the tension in the cable needed to hold the rod in equilibrium.

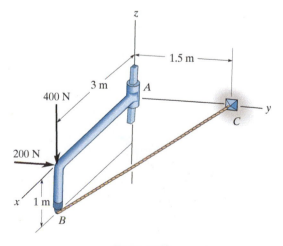

Prob. 5–75

***5–76.** The member is supported by a pin at A and cable BC. Determine the components of reaction at these supports if the cylinder has a mass of 40 kg.

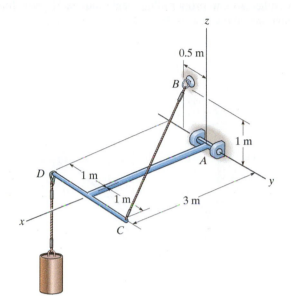

Prob. 5–76

5–77. The member is supported by a square rod which fits loosely through the smooth square hole of the attached collar at A and by a roller at B. Determine the components of reaction at these supports when the member is subjected to the loading shown.

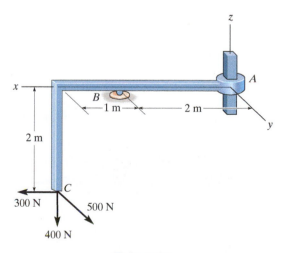

Prob. 5–77

5–78. The bent rod is supported at A, B, and C by smooth journal bearings. Compute the x, y, z components of reaction at the bearings if the rod is subjected to forces $F_1 = 300$ lb and $F_2 = 250$ lb. $\mathbf{F}_1$ lies in the y–z plane. The bearings are in proper alignment and exert only force reactions on the rod.

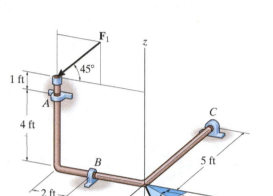

Prob. 5–78

5–79. The bent rod is supported at A, B, and C by smooth journal bearings. Determine the magnitude of $\mathbf{F}_2$ which will cause the reaction $\mathbf{C}_y$ at the bearing C to be equal to zero. The bearings are in proper alignment and exert only force reactions on the rod. Set $F_1 = 300$ lb.

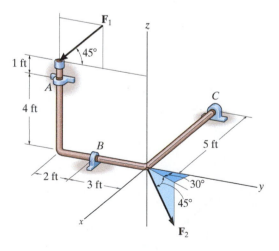

Prob. 5–79

***5–80.** The bar AB is supported by two smooth collars. At A the connection is with a ball-and-socket joint and at B it is a rigid attachment. If a 50-lb load is applied to the bar, determine the x, y, z components of reaction at A and B.

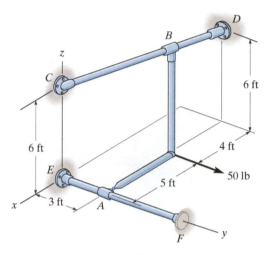

Prob. 5–80

5–81. The rod has a weight of 6 lb/ft. If it is supported by a ball-and-socket joint at C and a journal bearing at D, determine the x, y, z components of reaction at these supports and the moment M that must be applied along the axis of the rod to hold it in the position shown.

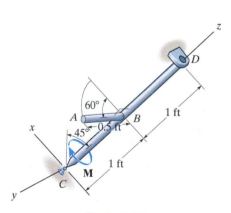

Prob. 5–81

5–82. The sign has a mass of 100 kg with center of mass at G. Determine the x, y, z components of reaction at the ball-and-socket joint A and the tension in wires BC and BD.

*5–84. Both pulleys are fixed to the shaft and as the shaft turns with constant angular velocity, the power of pulley A is transmitted to pulley B. Determine the horizontal tension **T** in the belt on pulley B and the x, y, z components of reaction at the journal bearing C and thrust bearing D if $\theta = 45°$. The bearings are in proper alignment and exert only force reactions on the shaft.

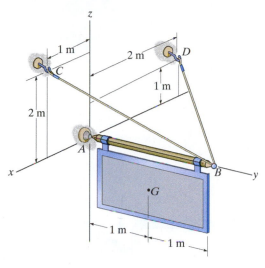

Prob. 5–82

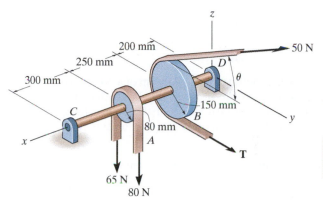

Prob. 5–84

5–83. Both pulleys are fixed to the shaft and as the shaft turns with constant angular velocity, the power of pulley A is transmitted to pulley B. Determine the horizontal tension **T** in the belt on pulley B and the x, y, z components of reaction at the journal bearing C and thrust bearing D if $\theta = 0°$. The bearings are in proper alignment and exert only force reactions on the shaft.

5–85. Member AB is supported by a cable BC and at A by a *square* rod which fits loosely through the square hole at the end joint of the member as shown. Determine the components of reaction at A and the tension in the cable needed to hold the 800-lb cylinder in equilibrium.

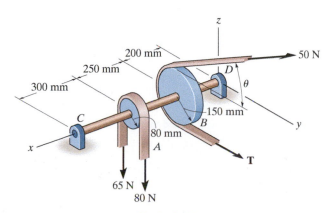

Prob. 5–83

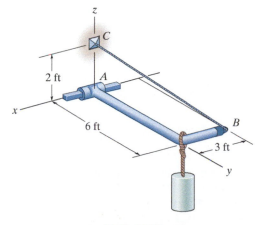

Prob. 5–85

5

CHAPTER REVIEW

Equilibrium

A body in equilibrium is at rest or can translate with constant velocity.

$$\Sigma F = 0$$

$$\Sigma M = 0$$

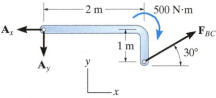

Two Dimensions

Before analyzing the equilibrium of a body, it is first necessary to draw its free-body diagram. This is an outlined shape of the body, which shows all the forces and couple moments that act on it.

Couple moments can be placed anywhere on a free-body diagram since they are free vectors. Forces can act at any point along their line of action since they are sliding vectors.

Angles used to resolve forces, and dimensions used to take moments of the forces, should also be shown on the free-body diagram.

Some common types of supports and their reactions are shown below in two dimensions.

Remember that a support will exert a force on the body in a particular direction if it prevents translation of the body in that direction, and it will exert a couple moment on the body if it prevents rotation.

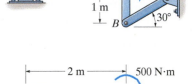

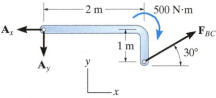

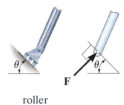

roller

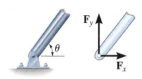

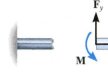

smooth pin or hinge

fixed support

The three scalar equations of equilibrium can be applied when solving problems in two dimensions, since the geometry is easy to visualize.

$$\Sigma F_x = 0$$
$$\Sigma F_y = 0$$
$$\Sigma M_O = 0$$

5

For the most direct solution, try to sum forces along an axis that will eliminate as many unknown forces as possible. Sum moments about a point A that passes through the line of action of as many unknown forces as possible.

$\Sigma F_x = 0$;

$A_x - P_2 = 0 \quad A_x = P_2$

$\Sigma M_A = 0$;

$P_2 d_2 + B_y d_B - P_1 d_1 = 0$

$B_y = \dfrac{P_1 d_1 - P_2 d_2}{d_B}$

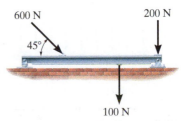

Three Dimensions

Some common types of supports and their reactions are shown here in three dimensions.

roller

ball and socket

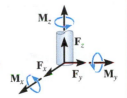

fixed support

In three dimensions, it is often advantageous to use a Cartesian vector analysis when applying the equations of equilibrium. To do this, first express each known and unknown force and couple moment shown on the free-body diagram as a Cartesian vector. Then set the force summation equal to zero. Take moments about a point O that lies on the line of action of as many unknown force components as possible. From point O direct position vectors to each force, and then use the cross product to determine the moment of each force.

$\Sigma \mathbf{F} = \mathbf{0}$

$\Sigma \mathbf{M}_O = \mathbf{0}$

The six scalar equations of equilibrium are established by setting the respective $\mathbf{i}$, $\mathbf{j}$, and $\mathbf{k}$ components of these force and moment summations equal to zero.

$\Sigma F_x = 0 \qquad \Sigma M_x = 0$

$\Sigma F_y = 0 \qquad \Sigma M_y = 0$

$\Sigma F_z = 0 \qquad \Sigma M_z = 0$

Determinacy and Stability

If a body is supported by a minimum number of constraints to ensure equilibrium, then it is statically determinate. If it has more constraints than required, then it is statically indeterminate.

To properly constrain the body, the reactions must not all be parallel to one another or concurrent.

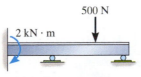

Statically indeterminate, five reactions, three equilibrium equations

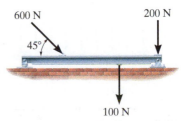

Proper constraint, statically determinate

REVIEW PROBLEMS

All problem solutions must include an FBD.

R5–1. If the roller at B can sustain a maximum load of 3 kN, determine the largest magnitude of each of the three forces **F** that can be supported by the truss.

R5–3. Determine the normal reaction at the roller A and horizontal and vertical components at pin B for equilibrium of the member.

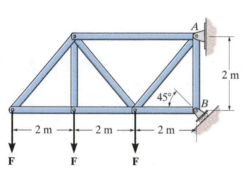

Prob. R5–1

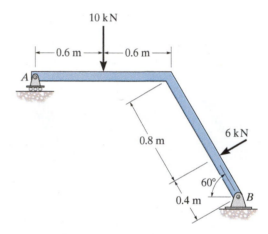

Prob. R5–3

R5–2. Determine the reactions at the supports A and B for equilibrium of the beam.

R5–4. Determine the horizontal and vertical components of reaction at the pin at A and the reaction of the roller at B on the lever.

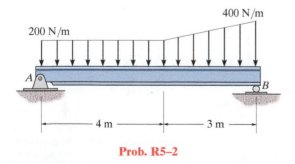

Prob. R5–2

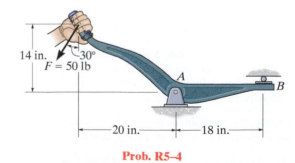

Prob. R5–4

R5–5. Determine the x, y, z components of reaction at the fixed wall A. The 150-N force is parallel to the z axis and the 200-N force is parallel to the y axis.

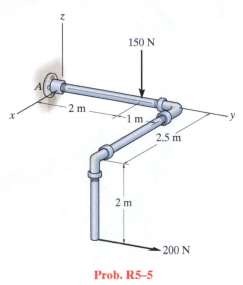

Prob. R5–5

R5–6. A vertical force of 80 lb acts on the crankshaft. Determine the horizontal equilibrium force **P** that must be applied to the handle and the x, y, z components of reaction at the journal bearing A and thrust bearing B. The bearings are properly aligned and exert only force reactions on the shaft.

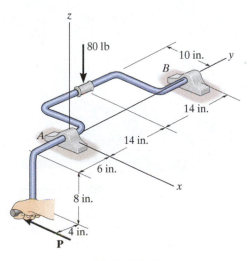

Prob. R5–6

R5–7. Determine the x, y, z components of reaction at the ball supports B and C and the ball-and-socket A (not shown) for the uniformly loaded plate.

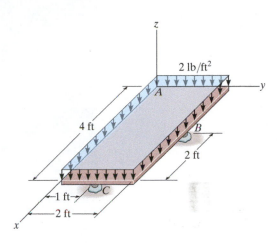

Prob. R5–7

R5–8. Determine the x and z components of reaction at the journal bearing A and the tension in cords BC and BD necessary for equilibrium of the rod.

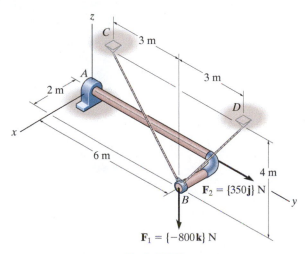

Prob. R5–8

Chapter 6

(© Tim Scrivener/Alamy)

In order to design the many parts of this boom assembly it is required that we know the forces that they must support. In this chapter we will show how to analyze such structures using the equations of equilibrium.

Structural Analysis

6.1 Simple Trusses

A *truss* is a structure composed of slender members joined together at their end points. The members commonly used in construction consist of wooden struts or metal bars. In particular, *planar* trusses lie in a single plane and are often used to support roofs and bridges. The truss shown in Fig. 6–1a is an example of a typical roof-supporting truss. In this figure, the roof load is transmitted to the truss *at the joints* by means of a series of *purlins*. Since this loading acts in the same plane as the truss, Fig. 6–1b, the analysis of the forces developed in the truss members will be two-dimensional.

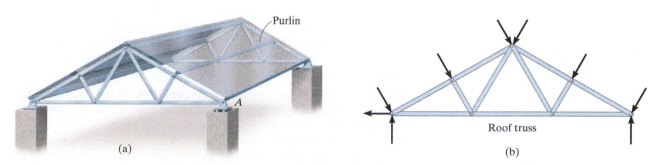

Purlin

Roof truss

(a)

(b)

Fig. 6–1

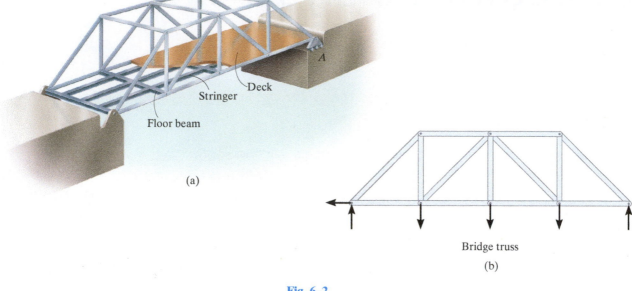

(a)

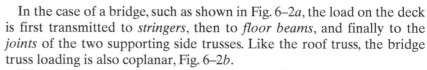

Bridge truss

(b)

Fig. 6–2

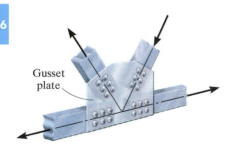

Gusset plate

(a)

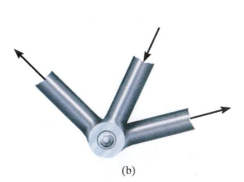

(b)

Fig. 6–3

In the case of a bridge, such as shown in Fig. 6–2*a*, the load on the deck is first transmitted to *stringers*, then to *floor beams*, and finally to the *joints* of the two supporting side trusses. Like the roof truss, the bridge truss loading is also coplanar, Fig. 6–2*b*.

When bridge or roof trusses extend over large distances, a rocker or roller is commonly used for supporting one end, for example, joint *A* in Figs. 6–1*a* and 6–2*a*. This type of support allows freedom for expansion or contraction of the members due to a change in temperature or application of loads.

Assumptions for Design.   To design both the members and the connections of a truss, it is necessary first to determine the *force* developed in each member when the truss is subjected to a given loading. To do this we will make two important assumptions:

- *All loadings are applied at the joints.* In most situations, such as for bridge and roof trusses, this assumption is true. Frequently the weight of the members is neglected because the force supported by each member is usually much larger than its weight. However, if the weight is to be included in the analysis, it is generally satisfactory to apply it as a vertical force, with half of its magnitude applied at each end of the member.

- *The members are joined together by smooth pins.* The joint connections are usually formed by bolting or welding the ends of the members to a common plate, called a *gusset plate*, as shown in Fig. 6–3*a*, or by simply passing a large bolt or pin through each of the members, Fig. 6–3*b*. We can assume these connections act as pins provided the center lines of the joining members are *concurrent*, as in Fig. 6–3.

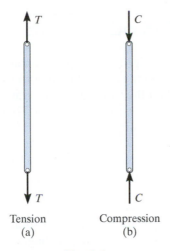

Tension
(a)

Compression
(b)

Fig. 6–4

Because of these two assumptions, *each truss member will act as a two-force member*, and therefore the force acting at each end of the member will be directed along the axis of the member. If the force tends to *elongate* the member, it is a ***tensile force*** (T), Fig. 6–4*a*; whereas if it tends to *shorten* the member, it is a ***compressive force*** (C), Fig. 6–4*b*. In the actual design of a truss it is important to state whether the nature of the force is tensile or compressive. Often, compression members must be made *thicker* than tension members because of the buckling or column effect that occurs when a member is in compression.

Simple Truss.   If three members are pin connected at their ends, they form a *triangular truss* that will be *rigid*, Fig. 6–5. Attaching two more members and connecting these members to a new joint *D* forms a larger truss, Fig. 6–6. This procedure can be repeated as many times as desired to form an even larger truss. If a truss can be constructed by expanding the basic triangular truss in this way, it is called a ***simple truss***.

The use of metal gusset plates in the construction of these Warren trusses is clearly evident. (© Russell C. Hibbeler)

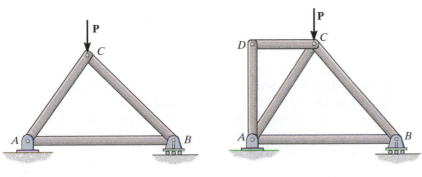

Fig. 6–5

Fig. 6–6

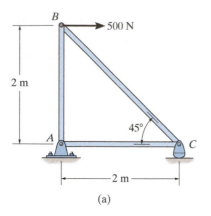

(a)

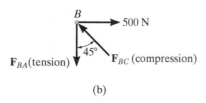

(b)

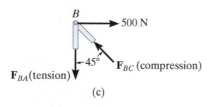

(c)

Fig. 6–7

The forces in the members of this simple roof truss can be determined using the method of joints. (© Russell C. Hibbeler)

6.2   The Method of Joints

In order to analyze or design a truss, it is necessary to determine the force in each of its members. One way to do this is to use the *method of joints*. This method is based on the fact that if the entire truss is in equilibrium, then each of its joints is also in equilibrium. Therefore, if the free-body diagram of each joint is drawn, the force equilibrium equations can then be used to obtain the member forces acting on each joint. Since the members of a *plane truss* are straight two-force members lying in a single plane, each joint is subjected to a force system that is *coplanar and concurrent*. As a result, only $\Sigma F_x = 0$ and $\Sigma F_y = 0$ need to be satisfied for equilibrium.

For example, consider the pin at joint B of the truss in Fig. 6–7a. Three forces act on the pin, namely, the 500-N force and the forces exerted by members BA and BC. The free-body diagram of the pin is shown in Fig. 6–7b. Here, $\mathbf{F}_{BA}$ is "pulling" on the pin, which means that member BA is in *tension;* whereas $\mathbf{F}_{BC}$ is "pushing" on the pin, and consequently member BC is in *compression*. These effects are clearly demonstrated by isolating the joint with small segments of the member connected to the pin, Fig. 6–7c. The pushing or pulling on these small segments indicates the effect of the member being either in compression or tension.

When using the method of joints, always start at a joint having at least one known force and at most two unknown forces, as in Fig. 6–7b. In this way, application of $\Sigma F_x = 0$ and $\Sigma F_y = 0$ yields two algebraic equations which can be solved for the two unknowns. When applying these equations, the correct sense of an unknown member force can be determined using one of two possible methods.

- The *correct* sense of direction of an unknown member force can, in many cases, be determined "by inspection." For example, $\mathbf{F}_{BC}$ in Fig. 6–7b must push on the pin (compression) since its horizontal component, $F_{BC} \sin 45°$, must balance the 500-N force ($\Sigma F_x = 0$). Likewise, $\mathbf{F}_{BA}$ is a tensile force since it balances the vertical component, $F_{BC} \cos 45°$ ($\Sigma F_y = 0$). In more complicated cases, the sense of an unknown member force can be *assumed*; then, after applying the equilibrium equations, the assumed sense can be verified from the numerical results. A *positive* answer indicates that the sense is *correct*, whereas a *negative* answer indicates that the sense shown on the free-body diagram must be *reversed*.

- *Always assume* the *unknown member forces* acting on the joint's free-body diagram to be in *tension*; i.e., the forces "pull" on the pin. If this is done, then numerical solution of the equilibrium equations will yield *positive scalars for members in tension and negative scalars for members in compression*. Once an unknown member force is found, use its *correct* magnitude and sense (T or C) on subsequent joint free-body diagrams.

Important Points

- Simple trusses are composed of triangular elements. The members are assumed to be pin connected at their ends and loads applied at the joints.

- If a truss is in equilibrium, then each of its joints is in equilibrium. The internal forces in the members become external forces when the free-body diagram of each joint of the truss is drawn. A force pulling on a joint is caused by tension in a member, and a force pushing on a joint is caused by compression.

Procedure for Analysis

The following procedure provides a means for analyzing a truss using the method of joints.

- Draw the free-body diagram of a joint having at least one known force and at most two unknown forces. (If this joint is at one of the supports, then it may be necessary first to calculate the external reactions at the support.)

- Use one of the two methods described above for establishing the sense of an unknown force.

- Orient the x and y axes such that the forces on the free-body diagram can be easily resolved into their x and y components and then apply the two force equilibrium equations $\Sigma F_x = 0$ and $\Sigma F_y = 0$. Solve for the two unknown member forces and verify their correct sense.

- Using the calculated results, continue to analyze each of the other joints. Remember that a member in *compression* "pushes" on the joint and a member in *tension* "pulls" on the joint. Also, be sure to choose a joint having at most two unknowns and at least one known force.

6

EXAMPLE   6.1

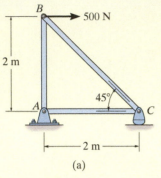

(a)

Determine the force in each member of the truss shown in Fig. 6–8a and indicate whether the members are in tension or compression.

SOLUTION

Since we should have no more than two unknown forces at the joint and at least one known force acting there, we will begin our analysis at joint B.

(b)

Joint B. The free-body diagram of the joint at B is shown in Fig. 6–8b. Applying the equations of equilibrium, we have

$$\xrightarrow{+} \Sigma F_x = 0; \qquad 500 \text{ N} - F_{BC} \sin 45° = 0 \quad F_{BC} = 707.1 \text{ N (C)} \quad \textit{Ans.}$$
$$+\uparrow \Sigma F_y = 0; \qquad F_{BC} \cos 45° - F_{BA} = 0 \quad F_{BA} = 500 \text{ N (T)} \quad \textit{Ans.}$$

Since the force in member BC has been calculated, we can proceed to analyze joint C to determine the force in member CA and the support reaction at the rocker.

(c)

Joint C. From the free-body diagram of joint C, Fig. 6–8c, we have

$$\xrightarrow{+} \Sigma F_x = 0; \quad -F_{CA} + 707.1 \cos 45° \text{ N} = 0 \quad F_{CA} = 500 \text{ N (T)} \quad \textit{Ans.}$$
$$+\uparrow \Sigma F_y = 0; \qquad C_y - 707.1 \sin 45° \text{ N} = 0 \quad C_y = 500 \text{ N} \quad \textit{Ans.}$$

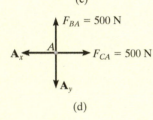

(d)

Joint A. Although it is not necessary, we can determine the components of the support reactions at joint A using the results of F_{CA} and F_{BA}. From the free-body diagram, Fig. 6–8d, we have

$$\xrightarrow{+} \Sigma F_x = 0; \qquad 500 \text{ N} - A_x = 0 \qquad A_x = 500 \text{ N}$$
$$+\uparrow \Sigma F_y = 0; \qquad 500 \text{ N} - A_y = 0 \qquad A_y = 500 \text{ N}$$

NOTE: The results of the analysis are summarized in Fig. 6–8e. Note that the free-body diagram of each joint (or pin) shows the effects of all the connected members and external forces applied to the joint, whereas the free-body diagram of each member shows only the effects of the end joints on the member.

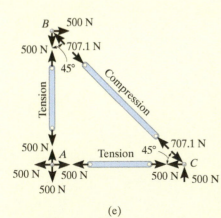

(e)

Fig. 6–8

EXAMPLE 6.2

Determine the forces acting in all the members of the truss shown in Fig. 6–9a.

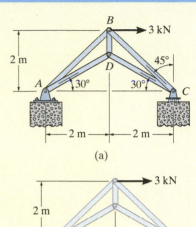

(a)

SOLUTION

By inspection, there are more than two unknowns at each joint. Consequently, the support reactions on the truss must first be determined. Show that they have been correctly calculated on the free-body diagram in Fig. 6–9b. We can now begin the analysis at joint C. Why?

Joint C. From the free-body diagram, Fig. 6–9c,

$$\xrightarrow{+}\ \Sigma F_x = 0; \qquad -F_{CD}\cos 30° + F_{CB}\sin 45° = 0$$
$$+\uparrow \Sigma F_y = 0; \qquad 1.5 \text{ kN} + F_{CD}\sin 30° - F_{CB}\cos 45° = 0$$

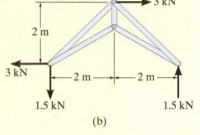

(b)

These two equations must be solved *simultaneously* for each of the two unknowns. Note, however, that a *direct solution* for one of the unknown forces may be obtained by applying a force summation along an axis that is *perpendicular* to the direction of the other unknown force. For example, summing forces along the y' axis, which is perpendicular to the direction of $\mathbf{F}_{CD}$, Fig. 6–9d, yields a *direct solution* for F_{CB}.

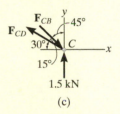

(c)

$$+\nearrow \Sigma F_{y'} = 0; \qquad 1.5\cos 30° \text{ kN} - F_{CB}\sin 15° = 0$$
$$F_{CB} = 5.019 \text{ kN} = 5.02 \text{ kN (C)} \qquad \textit{Ans.}$$

Then,

$$+\searrow \Sigma F_{x'} = 0;$$

$$-F_{CD} + 5.019\cos 15° - 1.5\sin 30° = 0; \quad F_{CD} = 4.10 \text{ kN (T)} \quad \textit{Ans.}$$

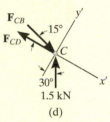

(d)

Joint D. We can now proceed to analyze joint D. The free-body diagram is shown in Fig. 6–9e.

$$\xrightarrow{+}\ \Sigma F_x = 0; \qquad -F_{DA}\cos 30° + 4.10\cos 30° \text{ kN} = 0$$
$$F_{DA} = 4.10 \text{ kN} \quad (T) \qquad \textit{Ans.}$$

$$+\uparrow \Sigma F_y = 0; \qquad F_{DB} - 2(4.10\sin 30° \text{ kN}) = 0$$
$$F_{DB} = 4.10 \text{ kN} \quad (T) \qquad \textit{Ans.}$$

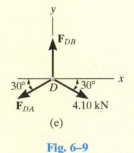

(e)

NOTE: The force in the last member, *BA*, can be obtained from joint B or joint A. As an exercise, draw the free-body diagram of joint B, sum the forces in the horizontal direction, and show that $F_{BA} = 0.776$ kN (C).

Fig. 6–9

EXAMPLE   6.3

Determine the force in each member of the truss shown in Fig. 6–10a. Indicate whether the members are in tension or compression.

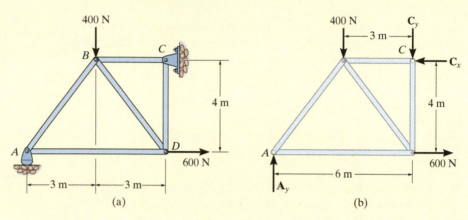

Fig. 6–10

SOLUTION

Support Reactions. No joint can be analyzed until the support reactions are determined, because each joint has at least three unknown forces acting on it. A free-body diagram of the entire truss is given in Fig. 6–10b. Applying the equations of equilibrium, we have

$$\xrightarrow{+} \Sigma F_x = 0; \qquad 600 \text{ N} - C_x = 0 \qquad\qquad C_x = 600 \text{ N}$$

$$\zeta + \Sigma M_C = 0; \qquad -A_y(6 \text{ m}) + 400 \text{ N}(3 \text{ m}) + 600 \text{ N}(4 \text{ m}) = 0$$

$$A_y = 600 \text{ N}$$

$$+\uparrow \Sigma F_y = 0; \qquad 600 \text{ N} - 400 \text{ N} - C_y = 0 \qquad C_y = 200 \text{ N}$$

The analysis can now start at either joint A or C. The choice is arbitrary since there are one known and two unknown member forces acting on the pin at each of these joints.

Joint A. (Fig. 6–10c). As shown on the free-body diagram, $\mathbf{F}_{AB}$ is assumed to be compressive and $\mathbf{F}_{AD}$ is tensile. Applying the equations of equilibrium, we have

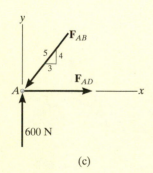

$$+\uparrow \Sigma F_y = 0; \qquad 600 \text{ N} - \tfrac{4}{5} F_{AB} = 0 \qquad F_{AB} = 750 \text{ N} \quad \text{(C)} \qquad \textit{Ans.}$$

$$\xrightarrow{+} \Sigma F_x = 0; \qquad F_{AD} - \tfrac{3}{5}(750 \text{ N}) = 0 \qquad F_{AD} = 450 \text{ N} \quad \text{(T)} \qquad \textit{Ans.}$$

Joint D. (Fig. 6–10d). Using the result for F_{AD} and summing forces in the horizontal direction, Fig. 6–10d, we have

$$\xrightarrow{+} \Sigma F_x = 0; \qquad -450\ \text{N} + \tfrac{3}{5}F_{DB} + 600\ \text{N} = 0 \qquad F_{DB} = -250\ \text{N}$$

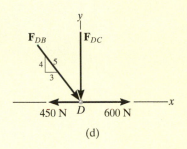

(d)

The negative sign indicates that $\mathbf{F}_{DB}$ acts in the *opposite sense* to that shown in Fig. 6–10d.* Hence,

$$F_{DB} = 250\ \text{N (T)} \qquad\qquad Ans.$$

To determine $\mathbf{F}_{DC}$, we can either correct the sense of $\mathbf{F}_{DB}$ on the free-body diagram, and then apply $\Sigma F_y = 0$, or apply this equation and retain the negative sign for F_{DB}, i.e.,

$$+\uparrow \Sigma F_y = 0; \qquad -F_{DC} - \tfrac{4}{5}(-250\ \text{N}) = 0 \qquad F_{DC} = 200\ \text{N} \quad (C) \quad Ans.$$

Joint C. (Fig. 6–10e).

$$\xrightarrow{+} \Sigma F_x = 0; \qquad F_{CB} - 600\ \text{N} = 0 \qquad F_{CB} = 600\ \text{N} \quad (C) \quad Ans.$$
$$+\uparrow \Sigma F_y = 0; \qquad\qquad 200\ \text{N} - 200\ \text{N} \equiv 0 \quad \text{(check)}$$

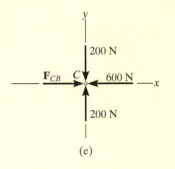

(e)

NOTE: The analysis is summarized in Fig. 6–10f, which shows the free-body diagram for each joint and member.

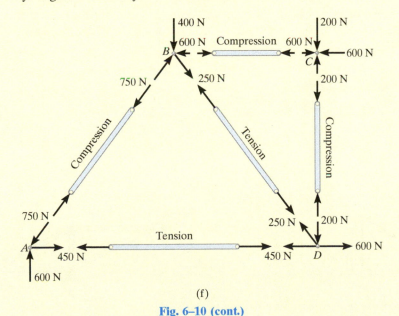

(f)

Fig. 6–10 (cont.)

*The proper sense could have been determined by inspection, prior to applying $\Sigma F_x = 0$.

6.3   Zero-Force Members

Truss analysis using the method of joints is greatly simplified if we can first identify those members which support *no loading*. These *zero-force members* are used to increase the stability of the truss during construction and to provide added support if the loading is changed.

The zero-force members of a truss can generally be found *by inspection* of each of the joints. For example, consider the truss shown in Fig. 6–11a. If a free-body diagram of the pin at joint A is drawn, Fig. 6–11b, it is seen that members AB and AF are zero-force members. (We could not have come to this conclusion if we had considered the free-body diagrams of joints F or B simply because there are five unknowns at each of these joints.) In a similar manner, consider the free-body diagram of joint D, Fig. 6–11c. Here again it is seen that DC and DE are zero-force members. From these observations, we can conclude that *if only two non-collinear members form a truss joint and no external load or support reaction is applied to the joint, the two members must be zero-force members.* The load on the truss in Fig. 6–11a is therefore supported by only five members as shown in Fig. 6–11d.

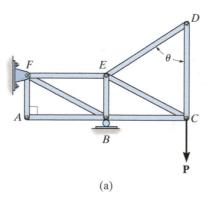

(a)

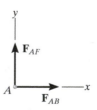

$$\xrightarrow{+} \Sigma F_x = 0; \quad F_{AB} = 0$$
$$+\uparrow \Sigma F_y = 0; \quad F_{AF} = 0$$

(b)

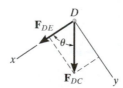

$+\searrow \Sigma F_y = 0; \ F_{DC} \sin \theta = 0; \quad F_{DC} = 0 \text{ since } \sin \theta \neq 0$
$+\swarrow \Sigma F_x = 0; F_{DE} + 0 = 0; \quad F_{DE} = 0$

(c)

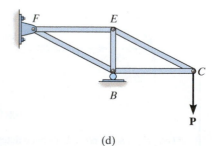

(d)

Fig. 6–11

Now consider the truss shown in Fig. 6–12*a*. The free-body diagram of the pin at joint *D* is shown in Fig. 6–12*b*. By orienting the *y* axis along members *DC* and *DE* and the *x* axis along member *DA*, it is seen that *DA* is a zero-force member. Note that this is also the case for member *CA*, Fig. 6–12*c*. In general then, *if three members form a truss joint for which two of the members are collinear, the third member is a zero-force member provided no external force or support reaction has a component that acts along this member.* The truss shown in Fig. 6–12*d* is therefore suitable for supporting the load **P**.

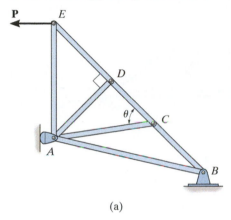

(a)

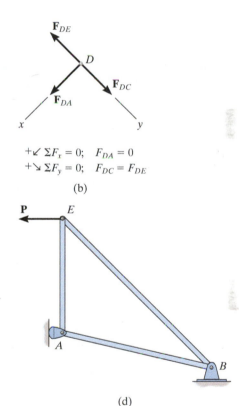

$$+\nwarrow \Sigma F_x = 0; \quad F_{DA} = 0$$
$$+\searrow \Sigma F_y = 0; \quad F_{DC} = F_{DE}$$

(b)

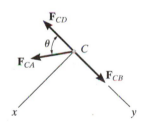

$$+\swarrow \Sigma F_x = 0; \quad F_{CA}\sin\theta = 0; \quad F_{CA} = 0 \text{ since } \sin\theta \neq 0;$$
$$+\searrow \Sigma F_y = 0; \quad F_{CB} = F_{CD}$$

(c)

(d)

Fig. 6–12

Important Point

- Zero-force members support no load; however, they are necessary for stability, and are available when additional loadings are applied to the joints of the truss. These members can usually be identified by inspection. They occur at joints where only two members are connected and no external load acts along either member. Also, at joints having two collinear members, a third member will be a zero-force member if no external force components act along this member.

EXAMPLE 6.4

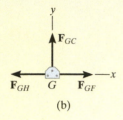

(b)

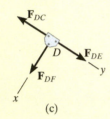

(c)

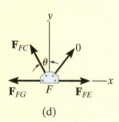

(d)

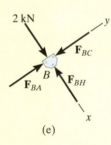

(e)

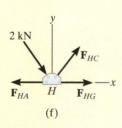

(f)

Using the method of joints, determine all the zero-force members of the *Fink roof truss* shown in Fig. 6–13a. Assume all joints are pin connected.

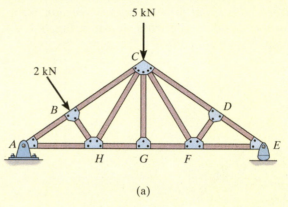

(a)

Fig. 6–13

SOLUTION

Look for joint geometries that have three members for which two are collinear. We have

Joint G.   (Fig. 6–13b).

$$+\uparrow \Sigma F_y = 0; \qquad\qquad F_{GC} = 0 \qquad\qquad Ans.$$

Realize that we could not conclude that *GC* is a zero-force member by considering joint *C*, where there are five unknowns. The fact that *GC* is a zero-force member means that the 5-kN load at *C* must be supported by members *CB*, *CH*, *CF*, and *CD*.

Joint D.   (Fig. 6–13c).

$$+\swarrow \Sigma F_x = 0; \qquad\qquad F_{DF} = 0 \qquad\qquad Ans.$$

Joint F.   (Fig. 6–13d).

$$+\uparrow \Sigma F_y = 0; \quad F_{FC} \cos\theta = 0 \quad \text{Since } \theta \neq 90°, \quad F_{FC} = 0 \qquad Ans.$$

NOTE: If joint *B* is analyzed, Fig. 6–13e,

$$+\searrow \Sigma F_x = 0; \qquad\qquad 2 \text{ kN} - F_{BH} = 0 \quad F_{BH} = 2 \text{ kN} \quad (C)$$

Also, F_{HC} must satisfy $\Sigma F_y = 0$, Fig. 6–13f, and therefore *HC* is *not* a zero-force member.

P6–1. In each case, calculate the support reactions and then draw the free-body diagrams of joints *A*, *B*, and *C* of the truss.

P6–2. Identify the zero-force members in each truss.

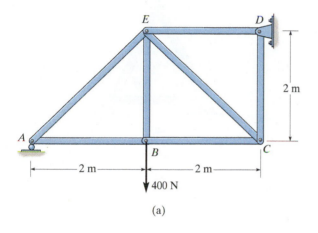

(a)

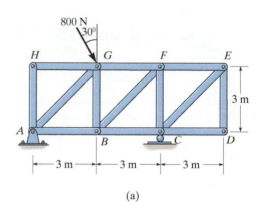

(a)

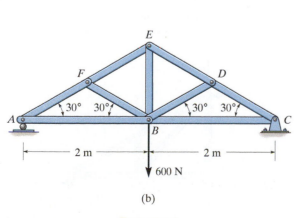

(b)

Prob. P6–1

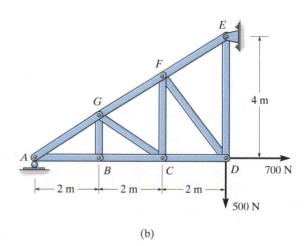

(b)

Prob. P6–2

FUNDAMENTAL PROBLEMS

All problem solutions must include FBDs.

F6–1. Determine the force in each member of the truss. State if the members are in tension or compression.

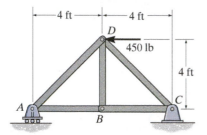

Prob. F6–1

F6–2. Determine the force in each member of the truss. State if the members are in tension or compression.

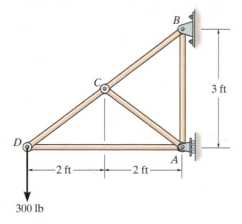

Prob. F6–2

F6–3. Determine the force in each member of the truss. State if the members are in tension or compression.

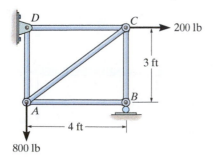

Prob. F6–3

F6–4. Determine the greatest load P that can be applied to the truss so that none of the members are subjected to a force exceeding either 2 kN in tension or 1.5 kN in compression.

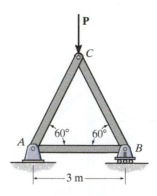

Prob. F6–4

F6–5. Identify the zero-force members in the truss.

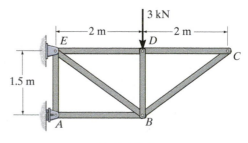

Prob. F6–5

F6–6. Determine the force in each member of the truss. State if the members are in tension or compression.

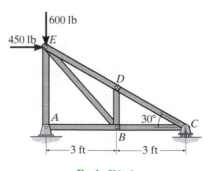

Prob. F6–6

All problem solutions must include FBDs.

6–1. Determine the force in each member of the truss and state if the members are in tension or compression. Set $P_1 = 20$ kN, $P_2 = 10$ kN.

6–2. Determine the force in each member of the truss and state if the members are in tension or compression. Set $P_1 = 45$ kN, $P_2 = 30$ kN.

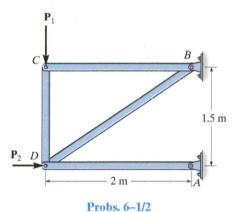

Probs. 6–1/2

6–3. Determine the force in each member of the truss. State if the members are in tension or compression.

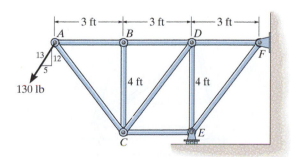

Prob. 6–3

***6–4.** Determine the force in each member of the truss and state if the members are in tension or compression.

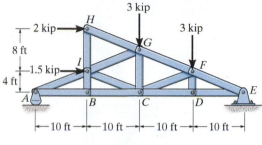

Prob. 6–4

6–5. Determine the force in each member of the truss, and state if the members are in tension or compression. Set $\theta = 0°$.

6–6. Determine the force in each member of the truss, and state if the members are in tension or compression. Set $\theta = 30°$.

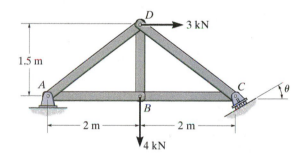

Probs. 6–5/6

6–7. Determine the force in each member of the truss and state if the members are in tension or compression.

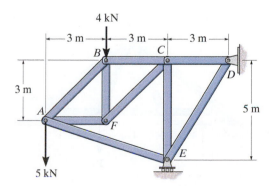

Prob. 6–7

*6–8. Determine the force in each member of the truss and state if the members are in tension or compression.

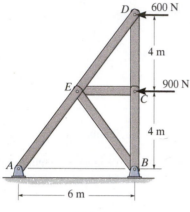

Prob. 6–8

6–9. Determine the force in each member of the truss and state if the members are in tension or compression. Set $P_1 = 3$ kN, $P_2 = 6$ kN.

6–10. Determine the force in each member of the truss and state if the members are in tension or compression. Set $P_1 = 6$ kN, $P_2 = 9$ kN.

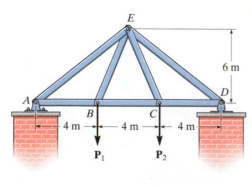

Probs. 6–9/10

6–11. Determine the force in each member of the *Pratt truss*, and state if the members are in tension or compression.

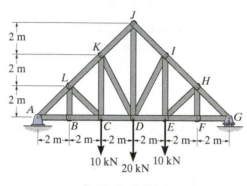

Prob. 6–11

*6–12. Determine the force in each member of the truss and state if the members are in tension or compression.

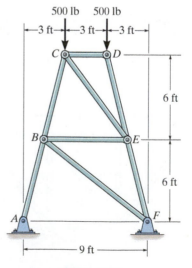

Prob. 6–12

6–13. Determine the force in each member of the truss in terms of the load P and state if the members are in tension or compression.

6–14. Members AB and BC can each support a maximum compressive force of 800 lb, and members AD, DC, and BD can support a maximum tensile force of 1500 lb. If $a = 10$ ft, determine the greatest load P the truss can support.

6–15. Members AB and BC can each support a maximum compressive force of 800 lb, and members AD, DC, and BD can support a maximum tensile force of 2000 lb. If $a = 6$ ft, determine the greatest load P the truss can support.

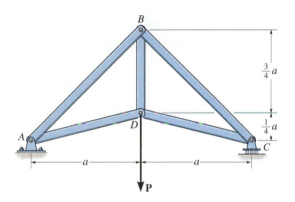

Probs. 6–13/14/15

*6–16. Determine the force in each member of the truss. State whether the members are in tension or compression. Set $P = 8$ kN.

6–17. If the maximum force that any member can support is 8 kN in tension and 6 kN in compression, determine the maximum force P that can be supported at joint D.

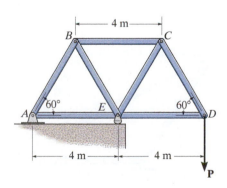

Probs. 6–16/17

6–18. Determine the force in each member of the truss and state if the members are in tension or compression. Set $P_1 = 10$ kN, $P_2 = 8$ kN.

6–19. Determine the force in each member of the truss and state if the members are in tension or compression. Set $P_1 = 8$ kN, $P_2 = 12$ kN.

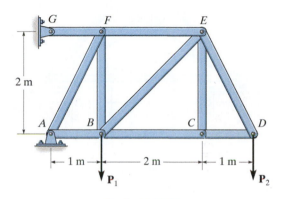

Probs. 6–18/19

*6–20. Determine the force in each member of the truss and state if the members are in tension or compression. Set $P_1 = 9$ kN, $P_2 = 15$ kN.

6–21. Determine the force in each member of the truss and state if the members are in tension or compression. Set $P_1 = 30$ kN, $P_2 = 15$ kN.

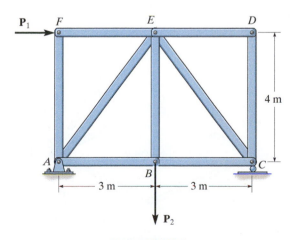

Probs. 6–20/21

6–22. Determine the force in each member of the double scissors truss in terms of the load P and state if the members are in tension or compression.

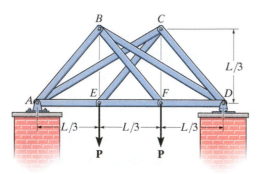

Prob. 6–22

6–23. Determine the force in each member of the truss in terms of the load P and state if the members are in tension or compression.

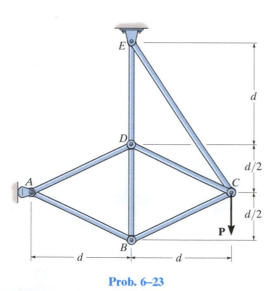

Prob. 6–23

*6–24. The maximum allowable tensile force in the members of the truss is $(F_t)_{max} = 5$ kN, and the maximum allowable compressive force is $(F_c)_{max} = 3$ kN. Determine the maximum magnitude of load **P** that can be applied to the truss. Take $d = 2$ m.

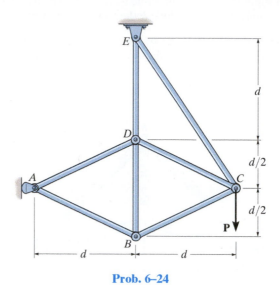

Prob. 6–24

6–25. Determine the force in each member of the truss in terms of the external loading and state if the members are in tension or compression. Take $P = 2$ kN.

6–26. The maximum allowable tensile force in the members of the truss is $(F_t)_{max} = 5$ kN, and the maximum allowable compressive force is $(F_c)_{max} = 3$ kN. Determine the maximum magnitude P of the two loads that can be applied to the truss.

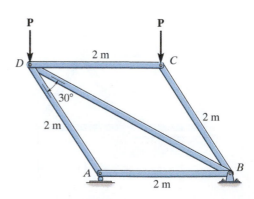

Probs. 6–25/26

6.4 The Method of Sections

When we need to find the force in only a few members of a truss, we can analyze the truss using the ***method of sections***. It is based on the principle that if the truss is in equilibrium then any segment of the truss is also in equilibrium. For example, consider the two truss members shown on the left in Fig. 6–14. If the forces within the members are to be determined, then an imaginary section, indicated by the blue line, can be used to cut each member into two parts and thereby "expose" each internal force as "external" to the free-body diagrams shown on the right. Clearly, it can be seen that equilibrium requires that the member in tension (T) be subjected to a "pull," whereas the member in compression (C) is subjected to a "push."

The method of sections can also be used to "cut" or section the members of an entire truss. If the section passes through the truss and the free-body diagram of either of its two parts is drawn, we can then apply the equations of equilibrium to that part to determine the member forces at the "cut section." Since only *three* independent equilibrium equations ($\Sigma F_x = 0$, $\Sigma F_y = 0$, $\Sigma M_O = 0$) can be applied to the free-body diagram of any segment, then we should try to select a section that, in general, passes through not more than *three* members in which the forces are unknown. For example, consider the truss in Fig. 6–15a. If the forces in members BC, GC, and GF are to be determined, then section aa would be appropriate. The free-body diagrams of the two segments are shown in Figs. 6–15b and 6–15c. Note that the line of action of each member force is specified from the *geometry* of the truss, since the force in a member is along its axis. Also, the member forces acting on one part of the truss are equal but opposite to those acting on the other part—Newton's third law. Members BC and GC are assumed to be in *tension* since they are subjected to a "pull," whereas GF in *compression* since it is subjected to a "push."

The three unknown member forces $\mathbf{F}_{BC}$, $\mathbf{F}_{GC}$, and $\mathbf{F}_{GF}$ can be obtained by applying the three equilibrium equations to the free-body diagram in Fig. 6–15b. If, however, the free-body diagram in Fig. 6–15c is considered, the three support reactions $\mathbf{D}_x$, $\mathbf{D}_y$ and $\mathbf{E}_x$ will have to be known, because only three equations of equilibrium are available. (This, of course, is done in the usual manner by considering a free-body diagram of the *entire truss*.)

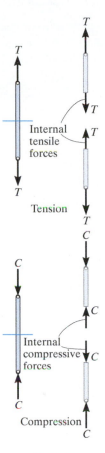

Tension

Compression

Fig. 6–14

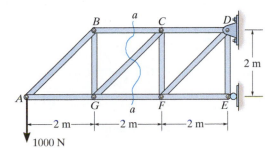

Fig. 6–15

The forces in selected members of this Pratt truss can readily be determined using the method of sections. (© Russell C. Hibbeler)

When applying the equilibrium equations, we should carefully consider ways of writing the equations so as to yield a *direct solution* for each of the unknowns, rather than having to solve simultaneous equations. For example, using the truss segment in Fig. 6–15b and summing moments about C would yield a direct solution for $\mathbf{F}_{GF}$ since $\mathbf{F}_{BC}$ and $\mathbf{F}_{GC}$ create zero moment about C. Likewise, $\mathbf{F}_{BC}$ can be directly obtained by summing moments about G. Finally, $\mathbf{F}_{GC}$ can be found directly from a force summation in the vertical direction since $\mathbf{F}_{GF}$ and $\mathbf{F}_{BC}$ have no vertical components. This ability to *determine directly* the force in a particular truss member is one of the main advantages of using the method of sections.*

As in the method of joints, there are two ways in which we can determine the correct sense of an unknown member force:

- The correct sense of an unknown member force can in many cases be determined "by inspection." For example, $\mathbf{F}_{BC}$ is a tensile force as represented in Fig. 6–15b since moment equilibrium about G requires that $\mathbf{F}_{BC}$ create a moment opposite to that of the 1000-N force. Also, $\mathbf{F}_{GC}$ is tensile since its vertical component must balance the 1000-N force which acts downward. In more complicated cases, the sense of an unknown member force may be *assumed*. If the solution yields a *negative* scalar, it indicates that the force's sense is *opposite* to that shown on the free-body diagram.

- *Always assume* that the unknown member forces at the cut section are *tensile* forces, i.e., "pulling" on the member. By doing this, the numerical solution of the equilibrium equations will yield *positive scalars for members in tension and negative scalars for members in compression.*

*Notice that if the method of joints were used to determine, say, the force in member GC, it would be necessary to analyze joints A, B, and G in sequence.

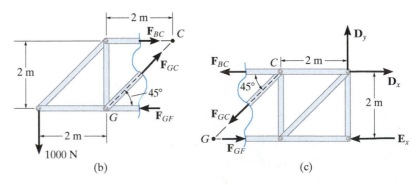

(b)   (c)

Fig. 6–15 (cont.)

Important Point

- If a truss is in equilibrium, then each of its segments is in equilibrium. The internal forces in the members become external forces when the free-body diagram of a segment of the truss is drawn. A force pulling on a member causes tension in the member, and a force pushing on a member causes compression.

Simple trusses are often used in the construction of large cranes in order to reduce the weight of the boom and tower. (© Russell C. Hibbeler)

Procedure for Analysis

The forces in the members of a truss may be determined by the method of sections using the following procedure.

Free-Body Diagram.

- Make a decision on how to "cut" or section the truss through the members where forces are to be determined.

- Before isolating the appropriate section, it may first be necessary to determine the truss's support reactions. If this is done then the three equilibrium equations will be available to solve for member forces at the section.

- Draw the free-body diagram of that segment of the sectioned truss which has the least number of forces acting on it.

- Use one of the two methods described above for establishing the sense of the unknown member forces.

Equations of Equilibrium.

- Moments should be summed about a point that lies at the intersection of the lines of action of two unknown forces, so that the third unknown force can be determined directly from the moment equation.

- If two of the unknown forces are *parallel*, forces may be summed *perpendicular* to the direction of these unknowns to determine *directly* the third unknown force.

EXAMPLE | 6.5

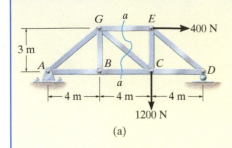

(a)

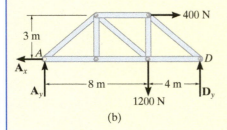

(b)

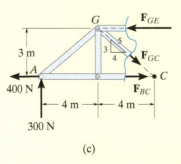

(c)

Fig. 6–16

Determine the force in members GE, GC, and BC of the truss shown in Fig. 6–16a. Indicate whether the members are in tension or compression.

SOLUTION

Section aa in Fig. 6–16a has been chosen since it cuts through the *three* members whose forces are to be determined. In order to use the method of sections, however, it is *first* necessary to determine the external reactions at A or D. Why? A free-body diagram of the entire truss is shown in Fig. 6–16b. Applying the equations of equilibrium, we have

$$\xrightarrow{+} \Sigma F_x = 0; \qquad 400\ \text{N} - A_x = 0 \qquad A_x = 400\ \text{N}$$

$$\zeta + \Sigma M_A = 0; \qquad -1200\ \text{N}(8\ \text{m}) - 400\ \text{N}(3\ \text{m}) + D_y(12\ \text{m}) = 0$$

$$D_y = 900\ \text{N}$$

$$+\uparrow \Sigma F_y = 0; \qquad A_y - 1200\ \text{N} + 900\ \text{N} = 0 \qquad A_y = 300\ \text{N}$$

Free-Body Diagram.   For the analysis the free-body diagram of the left portion of the sectioned truss will be used, since it involves the least number of forces, Fig. 6–16c.

Equations of Equilibrium.   Summing moments about point G eliminates $\mathbf{F}_{GE}$ and $\mathbf{F}_{GC}$ and yields a direct solution for F_{BC}.

$$\zeta + \Sigma M_G = 0; \quad -300\ \text{N}(4\ \text{m}) - 400\ \text{N}(3\ \text{m}) + F_{BC}(3\ \text{m}) = 0$$

$$F_{BC} = 800\ \text{N} \quad \text{(T)} \qquad\qquad Ans.$$

In the same manner, by summing moments about point C we obtain a direct solution for F_{GE}.

$$\zeta + \Sigma M_C = 0; \quad -300\ \text{N}(8\ \text{m}) + F_{GE}(3\ \text{m}) = 0$$

$$F_{GE} = 800\ \text{N} \quad \text{(C)} \qquad\qquad Ans.$$

Since $\mathbf{F}_{BC}$ and $\mathbf{F}_{GE}$ have no vertical components, summing forces in the y direction directly yields F_{GC}, i.e.,

$$+\uparrow \Sigma F_y = 0; \qquad 300\ \text{N} - \tfrac{3}{5}F_{GC} = 0$$

$$F_{GC} = 500\ \text{N} \quad \text{(T)} \qquad\qquad Ans.$$

NOTE: Here it is possible to tell, by inspection, the proper direction for each unknown member force. For example, $\Sigma M_C = 0$ requires $\mathbf{F}_{GE}$ to be *compressive* because it must balance the moment of the 300-N force about C.

EXAMPLE 6.6

Determine the force in member *CF* of the truss shown in Fig. 6–17a. Indicate whether the member is in tension or compression. Assume each member is pin connected.

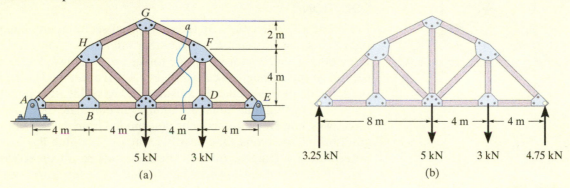

(a)

(b)

Fig. 6–17

SOLUTION

Free-Body Diagram. Section *aa* in Fig. 6–17a will be used since this section will "expose" the internal force in member *CF* as "external" on the free-body diagram of either the right or left portion of the truss. It is first necessary, however, to determine the support reactions on either the left or right side. Verify the results shown on the free-body diagram in Fig. 6–17b.

The free-body diagram of the right portion of the truss, which is the easiest to analyze, is shown in Fig. 6–17c. There are three unknowns, F_{FG}, F_{CF}, and F_{CD}.

Equations of Equilibrium. We will apply the moment equation about point *O* in order to eliminate the two unknowns F_{FG} and F_{CD}. The location of point *O* measured from *E* can be determined from proportional triangles, i.e., $4/(4 + x) = 6/(8 + x)$, $x = 4$ m. Or, stated in another manner, the slope of member *GF* has a drop of 2 m to a horizontal distance of 4 m. Since *FD* is 4 m, Fig. 6–17c, then from *D* to *O* the distance must be 8 m.

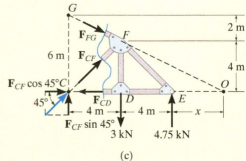

(c)

An easy way to determine the moment of $\mathbf{F}_{CF}$ about point *O* is to use the principle of transmissibility and slide $\mathbf{F}_{CF}$ to point *C*, and then resolve $\mathbf{F}_{CF}$ into its two rectangular components. We have

$$\zeta + \Sigma M_O = 0;$$

$$-F_{CF} \sin 45°(12 \text{ m}) + (3 \text{ kN})(8 \text{ m}) - (4.75 \text{ kN})(4 \text{ m}) = 0$$

$$F_{CF} = 0.589 \text{ kN} \quad (C) \qquad \qquad \textit{Ans.}$$

EXAMPLE | **6.7**

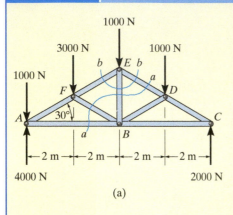

(a)

Determine the force in member *EB* of the roof truss shown in Fig. 6–18a. Indicate whether the member is in tension or compression.

SOLUTION

Free-Body Diagrams.   By the method of sections, any imaginary section that cuts through *EB*, Fig. 6–18a, will also have to cut through three other members for which the forces are unknown. For example, section *aa* cuts through *ED, EB, FB*, and *AB*. If a free-body diagram of the left side of this section is considered, Fig. 6–18b, it is possible to obtain F_{ED} by summing moments about *B* to eliminate the other three unknowns; however, F_{EB} cannot be determined from the remaining two equilibrium equations. One possible way of obtaining F_{EB} is first to determine F_{ED} from section *aa*, then use this result on section *bb*, Fig. 6–18a, which is shown in Fig. 6–18c. Here the force system is concurrent and our sectioned free-body diagram is the same as the free-body diagram for the joint at *E*.

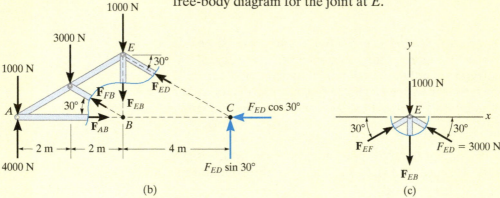

(b)   (c)

Fig. 6–18

Equations of Equilibrium.   In order to determine the moment of F_{ED} about point *B*, Fig. 6–18b, we will use the principle of transmissibility and slide the force to point *C* and then resolve it into its rectangular components as shown. Therefore,

$$\zeta + \Sigma M_B = 0; \qquad 1000 \text{ N}(4 \text{ m}) + 3000 \text{ N}(2 \text{ m}) - 4000 \text{ N}(4 \text{ m})$$

$$+ F_{ED} \sin 30°(4 \text{ m}) = 0$$

$$F_{ED} = 3000 \text{ N} \quad \text{(C)}$$

Considering now the free-body diagram of section *bb*, Fig. 6–18c, we have

$$\xrightarrow{+} \Sigma F_x = 0; \qquad F_{EF} \cos 30° - 3000 \cos 30° \text{ N} = 0$$

$$F_{EF} = 3000 \text{ N} \quad \text{(C)}$$

$$+\uparrow \Sigma F_y = 0; \qquad 2(3000 \sin 30° \text{ N}) - 1000 \text{ N} - F_{EB} = 0$$

$$F_{EB} = 2000 \text{ N} \quad \text{(T)} \qquad \qquad \textit{Ans.}$$

FUNDAMENTAL PROBLEMS

All problem solutions must include FBDs.

F6–7. Determine the force in members *BC*, *CF*, and *FE*. State if the members are in tension or compression.

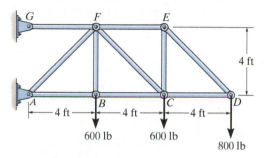

Prob. F6–7

F6–8. Determine the force in members *LK*, *KC*, and *CD* of the Pratt truss. State if the members are in tension or compression.

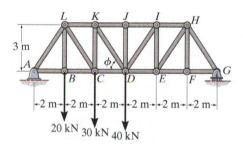

Prob. F6–8

F6–9. Determine the force in members *KJ*, *KD*, and *CD* of the Pratt truss. State if the members are in tension or compression.

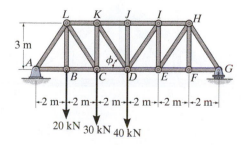

Prob. F6–9

F6–10. Determine the force in members *EF*, *CF*, and *BC* of the truss. State if the members are in tension or compression.

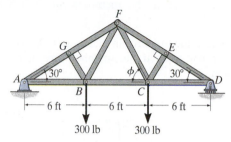

Prob. F6–10

F6–11. Determine the force in members *GF*, *GD*, and *CD* of the truss. State if the members are in tension or compression.

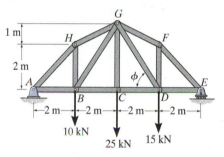

Prob. F6–11

F6–12. Determine the force in members *DC*, *HI*, and *JI* of the truss. State if the members are in tension or compression. *Suggestion:* Use the sections shown.

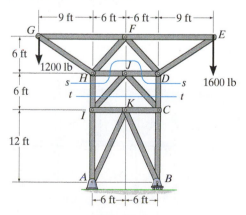

Prob. F6–12

PROBLEMS

All problem solutions must include FBDs.

6–27. Determine the force in members DC, HC, and HI of the truss, and state if the members are in tension or compression.

*6–28. Determine the force in members ED, EH, and GH of the truss, and state if the members are in tension or compression.

6–31. Determine the force in members CD, CJ, KJ, and DJ of the truss which serves to support the deck of a bridge. State if these members are in tension or compression.

*6–32. Determine the force in members EI and JI of the truss which serves to support the deck of a bridge. State if these members are in tension or compression.

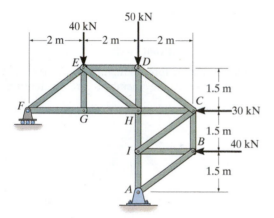

Probs. 6–27/28

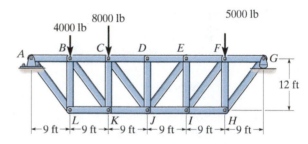

Probs. 6–31/32

6–29. Determine the force in members HG, HE and DE of the truss, and state if the members are in tension or compression.

6–30. Determine the force in members CD, HI, and CH of the truss, and state if the members are in tension or compression.

6–33. The *Howe truss* is subjected to the loading shown. Determine the force in members GF, CD, and GC, and state if the members are in tension or compression.

6–34. The *Howe truss* is subjected to the loading shown. Determine the force in members GH, BC, and BG of the truss and state if the members are in tension or compression.

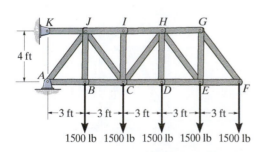

Probs. 6–29/30

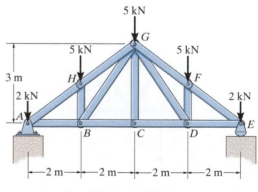

Probs. 6–33/34

6–35. Determine the force in members *EF*, *CF*, and *BC*, and state if the members are in tension or compression.

*6–36. Determine the force in members *AF*, *BF*, and *BC*, and state if the members are in tension or compression.

6–39. Determine the force in members *BC*, *HC*, and *HG*. After the truss is sectioned use a single equation of equilibrium for the calculation of each force. State if these members are in tension or compression.

*6–40. Determine the force in members *CD*, *CF*, and *CG* and state if these members are in tension or compression.

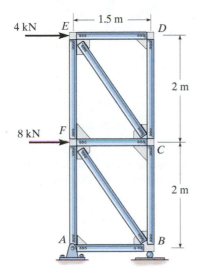

Probs. 6–35/36

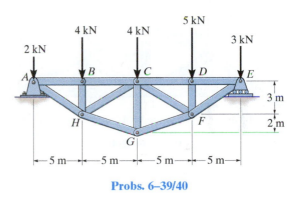

Probs. 6–39/40

6–37. Determine the force in members *EF*, *BE*, *BC* and *BF* of the truss and state if these members are in tension or compression. Set $P_1 = 9$ kN, $P_2 = 12$ kN, and $P_3 = 6$ kN.

6–38. Determine the force in members *BC*, *BE*, and *EF* of the truss and state if these members are in tension or compression. Set $P_1 = 6$ kN, $P_2 = 9$ kN, and $P_3 = 12$ kN.

6–41. Determine the force developed in members *FE*, *EB*, and *BC* of the truss and state if these members are in tension or compression.

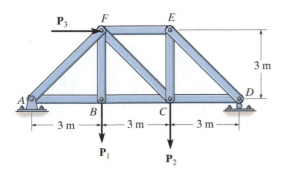

Probs. 6–37/38

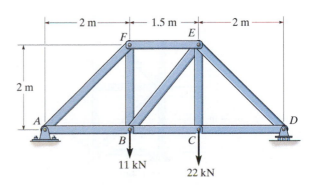

Prob. 6–41

6–42. Determine the force in members *BC*, *HC*, and *HG*. State if these members are in tension or compression.

6–43. Determine the force in members *CD*, *CJ*, *GJ*, and *CG* and state if these members are in tension or compression.

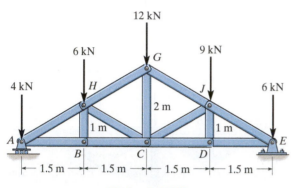

Probs. 6–42/43

6–46. Determine the force in members *BC*, *CH*, *GH*, and *CG* of the truss and state if the members are in tension or compression.

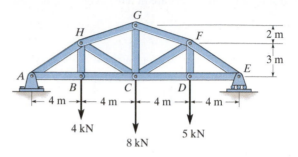

Prob. 6–46

6–47. Determine the force in members *CD*, *CJ*, and *KJ* and state if these members are in tension or compression.

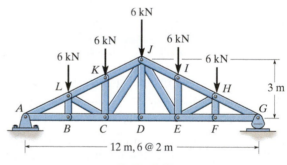

Prob. 6–47

***6–44.** Determine the force in members *BE*, *EF*, and *CB*, and state if the members are in tension or compression.

6–45. Determine the force in members *BF*, *BG*, and *AB*, and state if the members are in tension or compression.

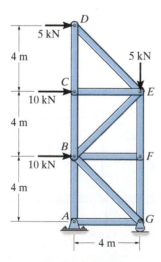

Probs. 6–44/45

***6–48.** Determine the force in members *JK*, *CJ*, and *CD* of the truss, and state if the members are in tension or compression.

6–49. Determine the force in members *HI*, *FI*, and *EF* of the truss, and state if the members are in tension or compression.

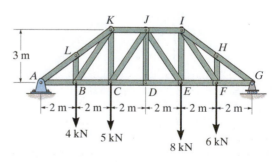

Probs. 6–48/49

*6.5   Space Trusses

A **space truss** consists of members joined together at their ends to form a stable three-dimensional structure. The simplest form of a space truss is a **tetrahedron**, formed by connecting six members together, as shown in Fig. 6–19. Any additional members added to this basic element would be redundant in supporting the force **P**. A *simple space truss* can be built from this basic tetrahedral element by adding three additional members and a joint, and continuing in this manner to form a system of multiconnected tetrahedrons.

Fig. 6–19

Assumptions for Design.   The members of a space truss may be treated as two-force members provided the external loading is applied at the joints and the joints consist of ball-and-socket connections. These assumptions are justified if the welded or bolted connections of the joined members intersect at a common point and the weight of the members can be neglected. In cases where the weight of a member is to be included in the analysis, it is generally satisfactory to apply it as a vertical force, half of its magnitude applied at each end of the member.

> ## Procedure for Analysis
>
> Either the method of joints or the method of sections can be used to determine the forces developed in the members of a simple space truss.
>
> ### Method of Joints.
>
> If the forces in *all* the members of the truss are to be determined, then the method of joints is most suitable for the analysis. Here it is necessary to apply the three equilibrium equations $\Sigma F_x = 0$, $\Sigma F_y = 0$, $\Sigma F_z = 0$ to the forces acting at each joint. Remember that the solution of many simultaneous equations can be avoided if the force analysis begins at a joint having at least one known force and at most three unknown forces. Also, if the three-dimensional geometry of the force system at the joint is hard to visualize, it is recommended that a Cartesian vector analysis be used for the solution.
>
> ### Method of Sections.
>
> If only a *few* member forces are to be determined, the method of sections can be used. When an imaginary section is passed through a truss and the truss is separated into two parts, the force system acting on one of the segments must satisfy the *six* equilibrium equations: $\Sigma F_x = 0$, $\Sigma F_y = 0$, $\Sigma F_z = 0$, $\Sigma M_x = 0$, $\Sigma M_y = 0$, $\Sigma M_z = 0$ (Eqs. 5–6). By proper choice of the section and axes for summing forces and moments, many of the unknown member forces in a space truss can be computed *directly*, using a single equilibrium equation.

Typical roof-supporting space truss. Notice the use of ball-and-socket joints for the connections. (© Russell C. Hibbeler)

For economic reasons, large electrical transmission towers are often constructed using space trusses. (© Russell C. Hibbeler)

EXAMPLE 6.8

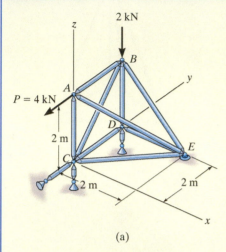

(a)

Determine the forces acting in the members of the space truss shown in Fig. 6–20a. Indicate whether the members are in tension or compression.

SOLUTION

Since there are one known force and three unknown forces acting at joint A, the force analysis of the truss will begin at this joint.

Joint A. (Fig. 6–20b). Expressing each force acting on the free-body diagram of joint A as a Cartesian vector, we have

$$\mathbf{P} = \{-4\mathbf{j}\}\ \text{kN}, \qquad \mathbf{F}_{AB} = F_{AB}\mathbf{j}, \quad \mathbf{F}_{AC} = -F_{AC}\mathbf{k},$$

$$\mathbf{F}_{AE} = F_{AE}\left(\frac{\mathbf{r}_{AE}}{r_{AE}}\right) = F_{AE}(0.577\mathbf{i} + 0.577\mathbf{j} - 0.577\mathbf{k})$$

For equilibrium,

$$\Sigma \mathbf{F} = \mathbf{0}; \qquad \mathbf{P} + \mathbf{F}_{AB} + \mathbf{F}_{AC} + \mathbf{F}_{AE} = \mathbf{0}$$

$$-4\mathbf{j} + F_{AB}\mathbf{j} - F_{AC}\mathbf{k} + 0.577F_{AE}\mathbf{i} + 0.577F_{AE}\mathbf{j} - 0.577F_{AE}\mathbf{k} = \mathbf{0}$$

$$\Sigma F_x = 0; \qquad\qquad\qquad 0.577F_{AE} = 0$$

$$\Sigma F_y = 0; \qquad -4 + F_{AB} + 0.577F_{AE} = 0$$

$$\Sigma F_z = 0; \qquad\qquad -F_{AC} - 0.577F_{AE} = 0$$

$$F_{AC} = F_{AE} = 0 \qquad\qquad Ans.$$

$$F_{AB} = 4\ \text{kN} \quad (\text{T}) \qquad\qquad Ans.$$

Since F_{AB} is known, joint B can be analyzed next.

Joint B. (Fig. 6–20c).

$$\Sigma F_x = 0; \qquad\qquad\qquad F_{BE}\frac{1}{\sqrt{2}} = 0$$

$$\Sigma F_y = 0; \qquad\qquad -4 + F_{CB}\frac{1}{\sqrt{2}} = 0$$

$$\Sigma F_z = 0; \qquad -2 + F_{BD} - F_{BE}\frac{1}{\sqrt{2}} + F_{CB}\frac{1}{\sqrt{2}} = 0$$

$$F_{BE} = 0, \qquad F_{CB} = 5.65\ \text{kN (C)} \qquad F_{BD} = 2\ \text{kN (T)} \qquad Ans.$$

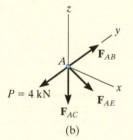

(b)

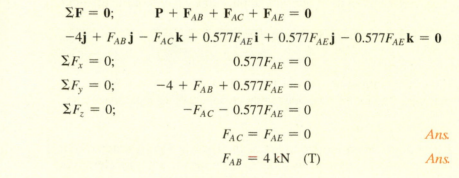

(c)

Fig. 6–20

The *scalar* equations of equilibrium can now be applied to the forces acting on the free-body diagrams of joints D and C. Show that

$$F_{DE} = F_{DC} = F_{CE} = 0 \qquad\qquad Ans.$$

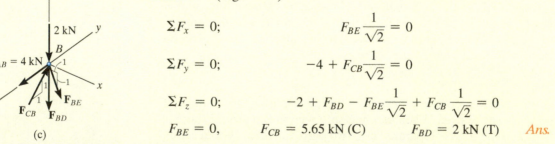

PROBLEMS

All problem solutions must include FBDs.

6–50. Determine the force developed in each member of the space truss and state if the members are in tension or compression. The crate has a weight of 150 lb.

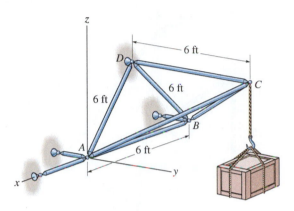

Prob. 6–50

6–51. Determine the force in each member of the space truss and state if the members are in tension or compression. *Hint:* The support reaction at E acts along member EB. Why?

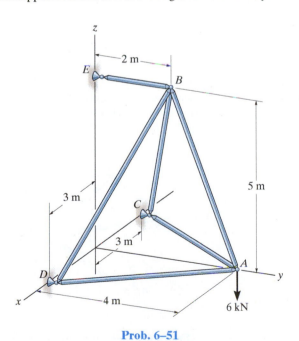

Prob. 6–51

*6–52. Determine the force in each member of the space truss and state if the members are in tension or compression. The truss is supported by ball-and-socket joints at $A, B, C,$ and D.

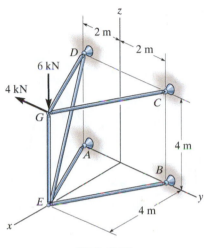

Prob. 6–52

6–53. The space truss supports a force $\mathbf{F} = \{-500\mathbf{i} + 600\mathbf{j} + 400\mathbf{k}\}$ lb. Determine the force in each member, and state if the members are in tension or compression.

6–54. The space truss supports a force $\mathbf{F} = \{600\mathbf{i} + 450\mathbf{j} - 750\mathbf{k}\}$ lb. Determine the force in each member, and state if the members are in tension or compression.

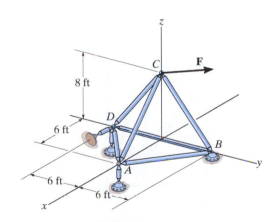

Probs. 6–53/54

6–55. Determine the force in members *EF, AF,* and *DF* of the space truss and state if the members are in tension or compression. The truss is supported by short links at *A, B, D,* and *E.*

6–57. The space truss is supported by a ball-and-socket joint at *D* and short links at *C* and *E.* Determine the force in each member and state if the members are in tension or compression. Take $F_1 = \{-500k\}$ lb and $F_2 = \{400j\}$ lb.

6–58. The space truss is supported by a ball-and-socket joint at *D* and short links at *C* and *E.* Determine the force in each member and state if the members are in tension or compression. Take $F_1 = \{200i + 300j - 500k\}$ lb and $F_2 = \{400j\}$ lb.

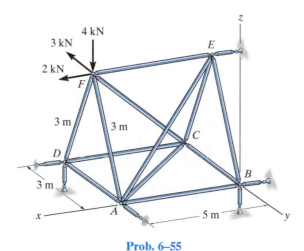

Prob. 6–55

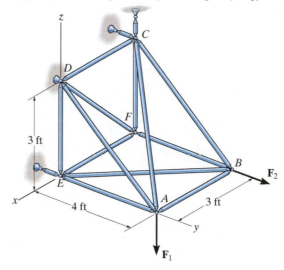

Probs. 6–57/58

6–59. Determine the force in each member of the space truss and state if the members are in tension or compression. The truss is supported by ball-and-socket joints at *A, B,* and *E.* Set $F = \{800j\}$ N. *Hint*: The support reaction at *E* acts along member *EC.* Why?

*6–56. The space truss is used to support the forces at joints *B* and *D.* Determine the force in each member and state if the members are in tension or compression.

*6–60. Determine the force in each member of the space truss and state if the members are in tension or compression. The truss is supported by ball-and-socket joints at *A, B,* and *E.* Set $F = \{-200i + 400j\}$ N. *Hint*: The support reaction at *E* acts along member *EC.* Why?

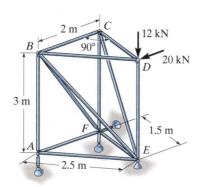

Prob. 6–56

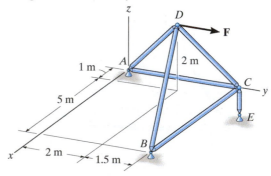

Probs. 6–59/60

6.6 Frames and Machines

Frames and machines are two types of structures which are often composed of pin-connected *multiforce members*, i.e., members that are subjected to more than two forces. *Frames* are used to support loads, whereas *machines* contain moving parts and are designed to transmit and alter the effect of forces. Provided a frame or machine contains no more supports or members than are necessary to prevent its collapse, the forces acting at the joints and supports can be determined by applying the equations of equilibrium to each of its members. Once these forces are obtained, it is then possible to *design* the size of the members, connections, and supports using the theory of mechanics of materials and an appropriate engineering design code.

This crane is a typical example of a framework. (© Russell C. Hibbeler)

Free-Body Diagrams. In order to determine the forces acting at the joints and supports of a frame or machine, the structure must be disassembled and the free-body diagrams of its parts must be drawn. The following important points *must* be observed:

- Isolate each part by drawing its *outlined shape*. Then show all the forces and/or couple moments that act on the part. Make sure to *label* or *identify* each known and unknown force and couple moment with reference to an established *x, y* coordinate system. Also, indicate any dimensions used for taking moments. Most often the equations of equilibrium are easier to apply if the forces are represented by their rectangular components. As usual, the sense of an unknown force or couple moment can be assumed.

- Identify all the two-force members in the structure and represent their free-body diagrams as having two equal but opposite collinear forces acting at their points of application. (See Sec. 5.4.) By recognizing the two-force members, we can avoid solving an unnecessary number of equilibrium equations.

- Forces common to *any* two *contacting* members act with equal magnitudes but opposite sense on the respective members. If the two members are treated as a *"system" of connected members*, then these forces are *"internal"* and are *not shown* on the *free-body diagram of the system*; however, if the free-body diagram of *each member* is drawn, the forces are *"external"* and *must* be shown as equal in magnitude and opposite in direction on each of the two free-body diagrams.

Common tools such as these pliers act as simple machines. Here the applied force on the handles creates a much larger force at the jaws. (© Russell C. Hibbeler)

The following examples graphically illustrate how to draw the free-body diagrams of a dismembered frame or machine. In all cases, the weight of the members is neglected.

EXAMPLE | 6.9

For the frame shown in Fig. 6–21a, draw the free-body diagram of (a) each member, (b) the pins at B and A, and (c) the two members connected together.

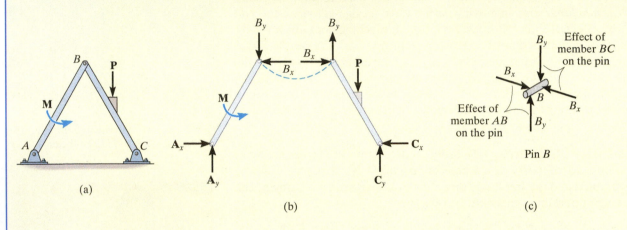

(a)

(b)

(c)

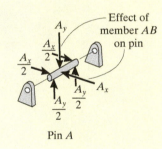

Pin A

(d)

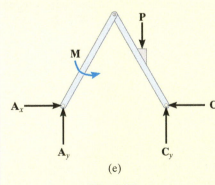

(e)

Fig. 6–21

SOLUTION

Part (a). By inspection, members BA and BC are *not* two-force members. Instead, as shown on the free-body diagrams, Fig. 6–21b, BC is subjected to a force from each of the pins at B and C and the external force **P**. Likewise, AB is subjected to a force from each of the pins at A and B and the external couple moment **M**. The pin forces are represented by their x and y components.

Part (b). The pin at B is subjected to only *two forces*, i.e., the force of member BC and the force of member AB. For *equilibrium* these forces (or their respective components) must be equal but opposite, Fig. 6–21c. Realize that Newton's third law is applied between the pin and its connected members, i.e., the effect of the pin on the two members, Fig. 6–21b, and the equal but opposite effect of the two members on the pin, Fig. 6–21c. In the same manner, there are three forces on pin A, Fig. 6–21d, caused by the force components of member AB and each of the two pin leafs.

Part (c). The free-body diagram of both members connected together, yet removed from the supporting pins at A and C, is shown in Fig. 6–21e. The force components $\mathbf{B}_x$ and $\mathbf{B}_y$ are *not shown* on this diagram since they are *internal* forces (Fig. 6–21b) and therefore cancel out. Also, to be consistent when later applying the equilibrium equations, the unknown force components at A and C must act in the *same sense* as those shown in Fig. 6–21b.

EXAMPLE | 6.10

A constant tension in the conveyor belt is maintained by using the device shown in Fig. 6–22a. Draw the free-body diagrams of the frame and the cylinder (or pulley) that the belt surrounds. The suspended block has a weight of W.

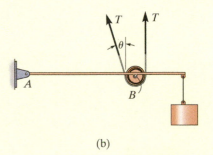

(b)

(a)

Fig. 6–22 (© Russell C. Hibbeler)

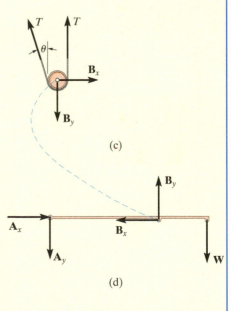

(c)

(d)

SOLUTION

The idealized model of the device is shown in Fig. 6–22b. Here the angle θ is assumed to be known. From this model, the free-body diagrams of the pulley and frame are shown in Figs. 6–22c and 6–22d, respectively. Note that the force components $\mathbf{B}_x$ and $\mathbf{B}_y$ that the pin at B exerts on the pulley must be equal but opposite to the ones acting on the frame. See Fig. 6–21c of Example 6.9.

EXAMPLE | 6.11

For the frame shown in Fig. 6–23a, draw the free-body diagrams of (a) the entire frame including the pulleys and cords, (b) the frame without the pulleys and cords, and (c) each of the pulleys.

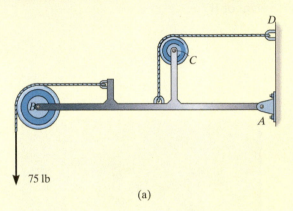

(a)

SOLUTION

Part (a). When the entire frame including the pulleys and cords is considered, the interactions at the points where the pulleys and cords are connected to the frame become pairs of *internal* forces which cancel each other and therefore are not shown on the free-body diagram, Fig. 6–23b.

Part (b). When the cords and pulleys are removed, their effect *on the frame* must be shown, Fig. 6–23c.

Part (c). The force components B_x, B_y, C_x, C_y of the pins on the pulleys, Fig. 6–23d, are equal but opposite to the force components exerted by the pins on the frame, Fig. 6–23c. See Example 6.9.

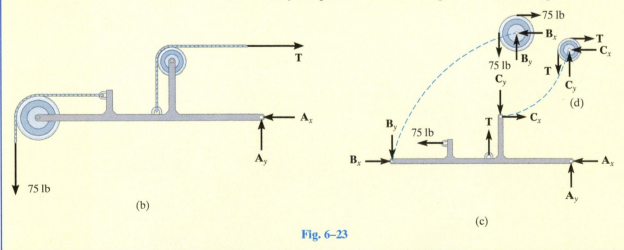

(b)

(c)

(d)

Fig. 6–23

EXAMPLE | 6.12

Draw the free-body diagrams of the members of the backhoe, shown in the photo, Fig. 6–24a. The bucket and its contents have a weight W.

SOLUTION

The idealized model of the assembly is shown in Fig. 6–24b. By inspection, members AB, BC, BE, and HI are all two-force members since they are pin connected at their end points and no other forces act on them. The free-body diagrams of the bucket and the stick are shown in Fig. 6–24c. Note that pin C is subjected to only two forces, whereas the pin at B is subjected to three forces, Fig. 6–24d. The free-body diagram of the entire assembly is shown in Fig. 6–24e.

Fig. 6–24 (© Russell C. Hibbeler)

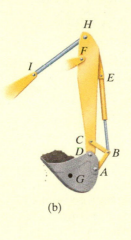

(b)

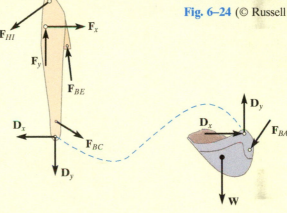

(c)

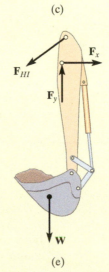

(d)

(e)

EXAMPLE | 6.13

Draw the free-body diagram of each part of the smooth piston and link mechanism used to crush recycled cans, Fig. 6–25a.

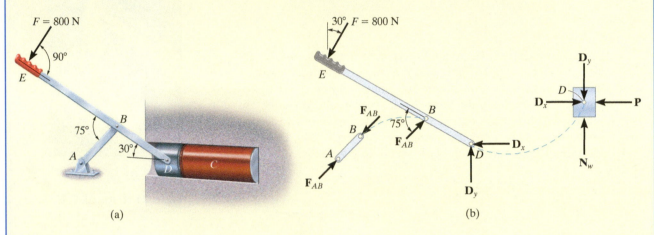

(a)

(b)

Fig. 6–25

SOLUTION

By inspection, member *AB* is a two-force member. The free-body diagrams of the three parts are shown in Fig. 6–25b. Since the pins at *B* and *D connect only two parts together*, the forces there are shown as equal but opposite on the separate free-body diagrams of their connected members. In particular, four components of force act on the piston: D_x and D_y represent the effect of the pin (or lever *EBD*), N_w is the *resultant force* of the wall support, and **P** is the resultant compressive force caused by the can *C*. The directional sense of each of the unknown forces is assumed, and the correct sense will be established after the equations of equilibrium are applied.

NOTE: A free-body diagram of the entire assembly is shown in Fig. 6–25c. Here the forces between the components are internal and are not shown on the free-body diagram.

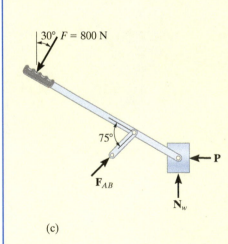

(c)

Before proceeding, it is highly recommended that you cover the solutions of these examples and attempt to draw the requested free-body diagrams. When doing so, make sure the work is neat and that all the forces and couple moments are properly labeled.

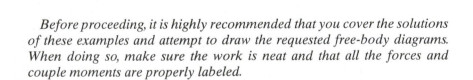

Procedure for Analysis

The joint reactions on frames or machines (structures) composed of multiforce members can be determined using the following procedure.

Free-Body Diagram.

- Draw the free-body diagram of the entire frame or machine, a portion of it, or each of its members. The choice should be made so that it leads to the most direct solution of the problem.

- Identify the two-force members. Remember that regardless of their shape, they have equal but opposite collinear forces acting at their ends.

- When the free-body diagram of a group of members of a frame or machine is drawn, the forces between the connected parts of this group are internal forces and are not shown on the free-body diagram of the group.

- Forces common to two members which are in contact act with equal magnitude but opposite sense on the respective free-body diagrams of the members.

- In many cases it is possible to tell by inspection the proper sense of the unknown forces acting on a member; however, if this seems difficult, the sense can be assumed.

- Remember that once the free-body diagram is drawn, a couple moment is a free vector and can act at any point on the diagram. Also, a force is a sliding vector and can act at any point along its line of action.

Equations of Equilibrium.

- Count the number of unknowns and compare it to the total number of equilibrium equations that are available. In two dimensions, there are three equilibrium equations that can be written for each member.

- Sum moments about a point that lies at the intersection of the lines of action of as many of the unknown forces as possible.

- If the solution of a force or couple moment magnitude is found to be negative, it means the sense of the force is the reverse of that shown on the free-body diagram.

EXAMPLE | 6.14

Determine the tension in the cables and also the force **P** required to support the 600-N force using the frictionless pulley system shown in Fig. 6–26*a*.

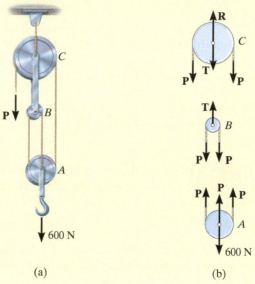

(a)   (b)

Fig. 6–26

SOLUTION

Free-Body Diagram.   A free-body diagram of each pulley *including* its pin and a portion of the contacting cable is shown in Fig. 6–26*b*. Since the cable is *continuous*, it has a *constant tension P* acting throughout its length. The link connection between pulleys *B* and *C* is a two-force member, and therefore it has an unknown tension *T* acting on it. Notice that the *principle of action, equal but opposite reaction* must be carefully observed for forces **P** and **T** when the *separate* free-body diagrams are drawn.

Equations of Equilibrium.   The three unknowns are obtained as follows:

Pulley A

$+\uparrow \Sigma F_y = 0;$ $3P - 600 \text{ N} = 0$ $P = 200 \text{ N}$ *Ans.*

Pulley B

$+\uparrow \Sigma F_y = 0;$ $T - 2P = 0$ $T = 400 \text{ N}$ *Ans.*

Pulley C

$+\uparrow \Sigma F_y = 0;$ $R - 2P - T = 0$ $R = 800 \text{ N}$ *Ans.*

EXAMPLE | 6.15

A 500-kg elevator car in Fig. 6–27a is being hoisted by motor A using the pulley system shown. If the car is traveling with a constant speed, determine the force developed in the two cables. Neglect the mass of the cable and pulleys.

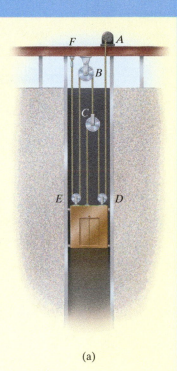

(a)

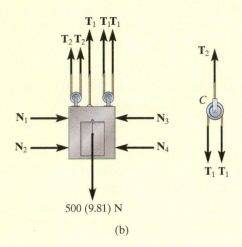

$\mathbf{T}_1 \ \mathbf{T}_1\mathbf{T}_1$

$\mathbf{T}_2 \ \mathbf{T}_2$

$\mathbf{N}_1 \rightarrow \qquad \leftarrow \mathbf{N}_3$

$\mathbf{N}_2 \rightarrow \qquad \leftarrow \mathbf{N}_4$

500 (9.81) N

(b)

$\mathbf{T}_2$

C

$\mathbf{T}_1 \ \mathbf{T}_1$

Fig. 6–27

SOLUTION

Free-Body Diagram.   We can solve this problem using the free-body diagrams of the elevator car and pulley C, Fig. 6–27b. The tensile forces developed in the cables are denoted as T_1 and T_2.

Equations of Equilibrium.   For pulley C,

$$+\uparrow \Sigma F_y = 0; \qquad T_2 - 2T_1 = 0 \qquad \text{or} \qquad T_2 = 2T_1 \qquad (1)$$

For the elevator car,

$$+\uparrow \Sigma F_y = 0; \qquad 3T_1 + 2T_2 - 500(9.81) \text{ N} = 0 \qquad (2)$$

Substituting Eq. (1) into Eq. (2) yields

$$3T_1 + 2(2T_1) - 500(9.81) \text{ N} = 0$$

$$T_1 = 700.71 \text{ N} = 701 \text{ N} \qquad \qquad Ans.$$

Substituting this result into Eq. (1),

$$T_2 = 2(700.71) \text{ N} = 1401 \text{ N} = 1.40 \text{ kN} \qquad Ans.$$

EXAMPLE | 6.16

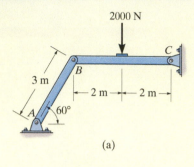

(a)

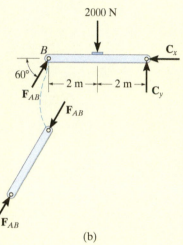

(b)

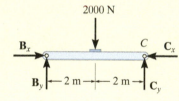

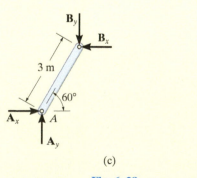

(c)

Fig. 6–28

Determine the horizontal and vertical components of force which the pin at C exerts on member BC of the frame in Fig. 6–28a.

SOLUTION I

Free-Body Diagrams.   By inspection it can be seen that AB is a two-force member. The free-body diagrams are shown in Fig. 6–28b.

Equations of Equilibrium.   The *three unknowns* can be determined by applying the three equations of equilibrium to member BC.

$\zeta + \Sigma M_C = 0$; $2000 \text{ N}(2 \text{ m}) - (F_{AB} \sin 60°)(4 \text{ m}) = 0$ $F_{AB} = 1154.7 \text{ N}$

$\xrightarrow{+} \Sigma F_x = 0$; $1154.7 \cos 60° \text{ N} - C_x = 0$ $C_x = 577 \text{ N}$ *Ans.*

$+\uparrow \Sigma F_y = 0$; $1154.7 \sin 60° \text{ N} - 2000 \text{ N} + C_y = 0$

$$C_y = 1000 \text{ N} \qquad \textit{Ans.}$$

SOLUTION II

Free-Body Diagrams.   If one does not recognize that AB is a two-force member, then more work is involved in solving this problem. The free-body diagrams are shown in Fig. 6–28c.

Equations of Equilibrium.   The *six unknowns* are determined by applying the three equations of equilibrium to each member.

Member AB

$\zeta + \Sigma M_A = 0$; $B_x(3 \sin 60° \text{ m}) - B_y(3 \cos 60° \text{ m}) = 0$ $\qquad$ (1)

$\xrightarrow{+} \Sigma F_x = 0$; $A_x - B_x = 0$ $\qquad$ (2)

$+\uparrow \Sigma F_y = 0$; $A_y - B_y = 0$ $\qquad$ (3)

Member BC

$\zeta + \Sigma M_C = 0$; $2000 \text{ N}(2 \text{ m}) - B_y(4 \text{ m}) = 0$ $\qquad$ (4)

$\xrightarrow{+} \Sigma F_x = 0$; $B_x - C_x = 0$ $\qquad$ (5)

$+\uparrow \Sigma F_y = 0$; $B_y - 2000 \text{ N} + C_y = 0$ $\qquad$ (6)

The results for C_x and C_y can be determined by solving these equations in the following sequence: 4, 1, 5, then 6. The results are

$$B_y = 1000 \text{ N}$$
$$B_x = 577 \text{ N}$$
$$C_x = 577 \text{ N} \qquad \textit{Ans.}$$
$$C_y = 1000 \text{ N} \qquad \textit{Ans.}$$

By comparison, Solution I is simpler since the requirement that F_{AB} in Fig. 6–28b be equal, opposite, and collinear at the ends of member AB automatically satisfies Eqs. 1, 2, and 3 above and therefore eliminates the need to write these equations. *As a result, save yourself some time and effort by always identifying the two-force members before starting the analysis!*

EXAMPLE 6.17

The compound beam shown in Fig. 6–29a is pin connected at *B*. Determine the components of reaction at its supports. Neglect its weight and thickness.

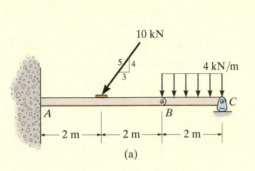

(a)

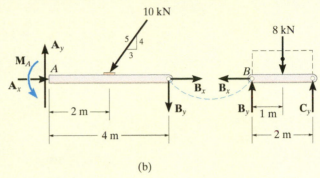

(b)

Fig. 6–29

SOLUTION

Free-Body Diagrams. By inspection, if we consider a free-body diagram of the *entire beam ABC*, there will be three unknown reactions at *A* and one at *C*. These four unknowns cannot all be obtained from the three available equations of equilibrium, and so for the solution it will become necessary to dismember the beam into its two segments, as shown in Fig. 6–29b.

Equations of Equilibrium. The six unknowns are determined as follows:

Segment BC

$$\xrightarrow{+} \Sigma F_x = 0; \qquad\qquad\qquad\qquad B_x = 0$$

$$\zeta + \Sigma M_B = 0; \qquad\qquad -8 \text{ kN}(1 \text{ m}) + C_y(2 \text{ m}) = 0$$

$$+\uparrow \Sigma F_y = 0; \qquad\qquad\qquad B_y - 8 \text{ kN} + C_y = 0$$

Segment AB

$$\xrightarrow{+} \Sigma F_x = 0; \qquad\qquad A_x - (10 \text{ kN})\left(\tfrac{3}{5}\right) + B_x = 0$$

$$\zeta + \Sigma M_A = 0; \qquad M_A - (10 \text{ kN})\left(\tfrac{4}{5}\right)(2 \text{ m}) - B_y(4 \text{ m}) = 0$$

$$+\uparrow \Sigma F_y = 0; \qquad\qquad A_y - (10 \text{ kN})\left(\tfrac{4}{5}\right) - B_y = 0$$

Solving each of these equations successively, using previously calculated results, we obtain

$A_x = 6$ kN	$A_y = 12$ kN	$M_A = 32$ kN·m	*Ans.*
$B_x = 0$	$B_y = 4$ kN		
$C_y = 4$ kN			*Ans.*

EXAMPLE | 6.18

The two planks in Fig. 6–30a are connected together by cable BC and a smooth spacer DE. Determine the reactions at the smooth supports A and F, and also find the force developed in the cable and spacer.

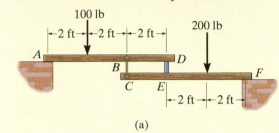

(a)

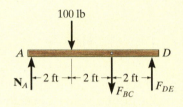 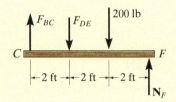

(b)

Fig. 6–30

SOLUTION

Free-Body Diagrams. The free-body diagram of each plank is shown in Fig. 6–30b. It is important to apply Newton's third law to the interaction forces F_{BC} and F_{DE} as shown.

Equations of Equilibrium. For plank AD,

$$\zeta + \Sigma M_A = 0; \qquad F_{DE}(6 \text{ ft}) - F_{BC}(4 \text{ ft}) - 100 \text{ lb} (2 \text{ ft}) = 0$$

For plank CF,

$$\zeta + \Sigma M_F = 0; \qquad F_{DE}(4 \text{ ft}) - F_{BC}(6 \text{ ft}) + 200 \text{ lb} (2 \text{ ft}) = 0$$

Solving simultaneously,

$$F_{DE} = 140 \text{ lb} \quad F_{BC} = 160 \text{ lb} \qquad\qquad Ans.$$

Using these results, for plank AD,

$$+\uparrow \Sigma F_y = 0; \qquad N_A + 140 \text{ lb} - 160 \text{ lb} - 100 \text{ lb} = 0$$

$$N_A = 120 \text{ lb} \qquad\qquad Ans.$$

And for plank CF,

$$+\uparrow \Sigma F_y = 0; \qquad N_F + 160 \text{ lb} - 140 \text{ lb} - 200 \text{ lb} = 0$$

$$N_F = 180 \text{ lb} \qquad\qquad Ans.$$

NOTE: Draw the free-body diagram of the system of *both* planks and apply $\Sigma M_A = 0$ to determine N_F. Then use the free-body diagram of CEF to determine F_{DE} and F_{BC}.

EXAMPLE | 6.19

The 75-kg man in Fig. 6–31a attempts to lift the 40-kg uniform beam off the roller support at B. Determine the tension developed in the cable attached to B and the normal reaction of the man on the beam when this is about to occur.

SOLUTION

Free-Body Diagrams.   The tensile force in the cable will be denoted as T_1. The free-body diagrams of the pulley E, the man, and the beam are shown in Fig. 6–31b. Since the man must lift the beam off the roller B then $N_B = 0$. When drawing each of these diagrams, it is very important to apply Newton's third law.

Equations of Equilibrium.   Using the free-body diagram of pulley E,

$$+\uparrow \Sigma F_y = 0; \qquad 2T_1 - T_2 = 0 \quad \text{or} \quad T_2 = 2T_1 \qquad (1)$$

Referring to the free-body diagram of the man using this result,

$$+\uparrow \Sigma F_y = 0 \qquad N_m + 2T_1 - 75(9.81) \text{ N} = 0 \qquad (2)$$

Summing moments about point A on the beam,

$$\zeta + \Sigma M_A = 0; \; T_1(3 \text{ m}) - N_m(0.8 \text{ m}) - [40(9.81) \text{ N}] (1.5 \text{ m}) = 0 \qquad (3)$$

Solving Eqs. 2 and 3 simultaneously for T_1 and N_m, then using Eq. (1) for T_2, we obtain

$$T_1 = 256 \text{ N} \qquad N_m = 224 \text{ N} \qquad T_2 = 512 \text{ N} \qquad \textit{Ans.}$$

SOLUTION II

A direct solution for T_1 can be obtained by considering the beam, the man, and pulley E as a *single system*. The free-body diagram is shown in Fig. 6–31c. Thus,

$$\zeta + \Sigma M_A = 0; \qquad 2T_1(0.8 \text{ m}) - [75(9.81) \text{ N}](0.8 \text{ m})$$

$$- [40(9.81) \text{ N}](1.5 \text{ m}) + T_1(3 \text{ m}) = 0$$

$$T_1 = 256 \text{ N} \qquad \textit{Ans.}$$

With this result Eqs. 1 and 2 can then be used to find N_m and T_2.

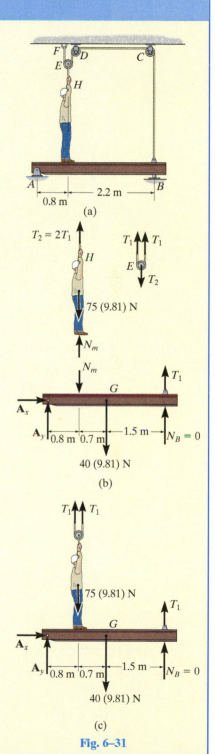

(a)

(b)

(c)

Fig. 6–31

EXAMPLE 6.20

The smooth disk shown in Fig. 6–32a is pinned at D and has a weight of 20 lb. Neglecting the weights of the other members, determine the horizontal and vertical components of reaction at pins B and D.

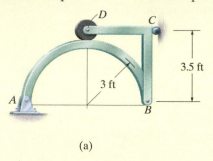

(a)

SOLUTION

Free-Body Diagrams.   The free-body diagrams of the entire frame and each of its members are shown in Fig. 6–32b.

Equations of Equilibrium.   The eight unknowns can of course be obtained by applying the eight equilibrium equations to each member—three to member AB, three to member BCD, and two to the disk. (Moment equilibrium is automatically satisfied for the disk.) If this is done, however, all the results can be obtained only from a simultaneous solution of some of the equations. (Try it and find out.) To avoid this situation, it is best first to determine the three support reactions on the *entire* frame; then, using these results, the remaining five equilibrium equations can be applied to two other parts in order to solve successively for the other unknowns.

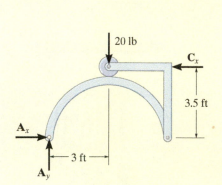

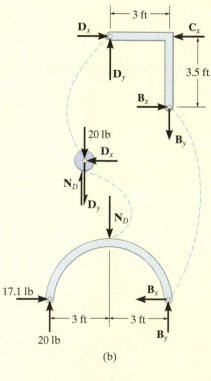

(b)

Fig. 6–32

Entire Frame

$\zeta + \Sigma M_A = 0;$ $-20 \text{ lb } (3 \text{ ft}) + C_x(3.5 \text{ ft}) = 0$ $C_x = 17.1 \text{ lb}$

$\xrightarrow{+} \Sigma F_x = 0;$ $A_x - 17.1 \text{ lb} = 0$ $A_x = 17.1 \text{ lb}$

$+\uparrow \Sigma F_y = 0;$ $A_y - 20 \text{ lb} = 0$ $A_y = 20 \text{ lb}$

Member AB

$\xrightarrow{+} \Sigma F_x = 0;$ $17.1 \text{ lb} - B_x = 0$ $B_x = 17.1 \text{ lb}$ *Ans.*

$\zeta + \Sigma M_B = 0;$ $-20 \text{ lb } (6 \text{ ft}) + N_D(3 \text{ ft}) = 0$ $N_D = 40 \text{ lb}$

$+\uparrow \Sigma F_y = 0;$ $20 \text{ lb} - 40 \text{ lb} + B_y = 0$ $B_y = 20 \text{ lb}$ *Ans.*

Disk

$\xrightarrow{+} \Sigma F_x = 0;$ $D_x = 0$ *Ans.*

$+\uparrow \Sigma F_y = 0;$ $40 \text{ lb} - 20 \text{ lb} - D_y = 0$ $D_y = 20 \text{ lb}$ *Ans.*

EXAMPLE 6.21

The frame in Fig. 6–33a supports the 50-kg cylinder. Determine the horizontal and vertical components of reaction at A and the force at C.

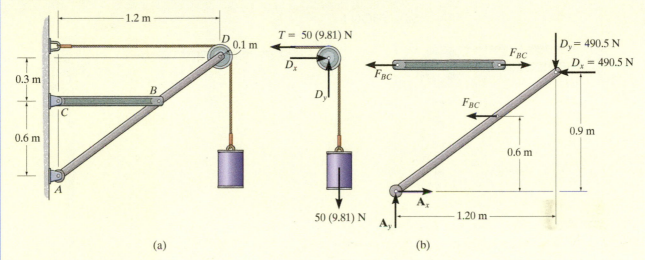

(a)   (b)

Fig. 6–33

SOLUTION

Free-Body Diagrams. The free-body diagram of pulley D, along with the cylinder and a portion of the cord (a system), is shown in Fig. 6–33b. Member BC is a two-force member as indicated by its free-body diagram. The free-body diagram of member ABD is also shown.

Equations of Equilibrium. We will begin by analyzing the equilibrium of the pulley. The moment equation of equilibrium is automatically satisfied with $T = 50(9.81)$ N, and so

$$\xrightarrow{+} \Sigma F_x = 0; \qquad D_x - 50(9.81)\text{ N} = 0 \quad D_x = 490.5\text{ N}$$

$$+\uparrow \Sigma F_y = 0; \qquad D_y - 50(9.81)\text{ N} = 0 \quad D_y = 490.5\text{ N} \qquad \textit{Ans.}$$

Using these results, F_{BC} can be determined by summing moments about point A on member ABD.

$$\zeta+\Sigma M_A = 0; \; F_{BC}(0.6\text{ m}) + 490.5\text{ N}(0.9\text{ m}) - 490.5\text{ N}(1.20\text{ m}) = 0$$

$$F_{BC} = 245.25\text{ N} \qquad \textit{Ans.}$$

Now A_x and A_y can be determined by summing forces.

$$\xrightarrow{+} \Sigma F_x = 0; \quad A_x - 245.25\text{ N} - 490.5\text{ N} = 0 \quad A_x = 736\text{ N} \qquad \textit{Ans.}$$

$$+\uparrow \Sigma F_y = 0; \qquad\qquad A_y - 490.5\text{ N} = 0 \quad A_y = 490.5\text{ N} \qquad \textit{Ans.}$$

EXAMPLE | **6.22**

Determine the force the pins at A and B exert on the two-member frame shown in Fig. 6–34a.

SOLUTION I

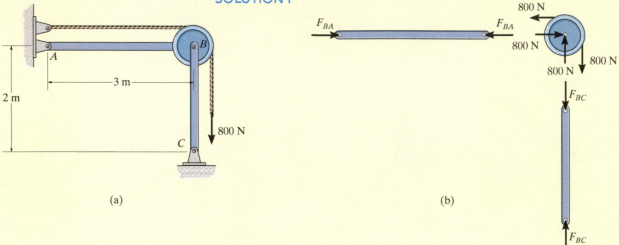

(a)

(b)

Free-Body Diagrams. By inspection AB and BC are two-force members. Their free-body diagrams, along with that of the pulley, are shown in Fig. 6–34b. In order to solve this problem we must also include the free-body diagram of the pin at B because this pin connects all *three members* together, Fig. 6–34c.

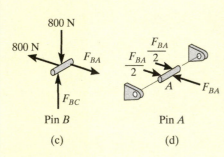

Pin B

(c)

Pin A

(d)

Equations of Equilibrium: Apply the equations of force equilibrium to pin B.

$$\xrightarrow{+}\Sigma F_x = 0; \quad F_{BA} - 800\ \text{N} = 0; \quad F_{BA} = 800\ \text{N} \qquad \textit{Ans.}$$

$$+\uparrow\Sigma F_y = 0; \quad F_{BC} - 800\ \text{N} = 0; \quad F_{BC} = 800\ \text{N} \qquad \textit{Ans.}$$

NOTE: The free-body diagram of the pin at A, Fig. 6–34d, indicates how the force F_{AB} is balanced by the force $(F_{AB}/2)$ exerted on the pin by each of the two pin leaves.

SOLUTION II

Free-Body Diagram. If we realize that AB and BC are two-force members, then the free-body diagram of the *entire frame* produces an easier solution, Fig. 6–34e. The force equations of equilibrium are the same as those above. Note that moment equilibrium will be satisfied, regardless of the radius of the pulley.

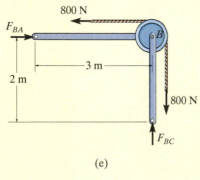

(e)

Fig. 6–34

PRELIMINARY PROBLEMS

P6–3. In each case, identify any two-force members, and then draw the free-body diagrams of each member of the frame.

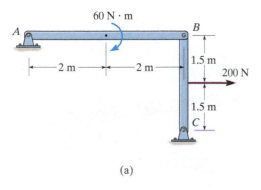

(a)

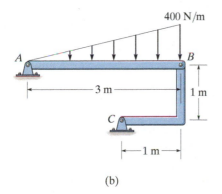

(b)

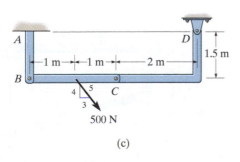

(c)

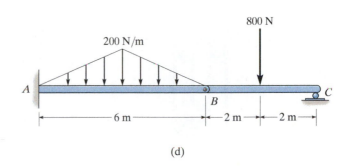

(d)

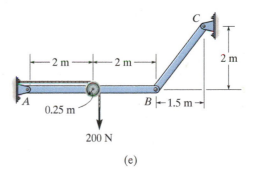

(e)

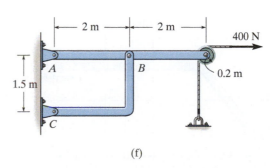

(f)

Prob. P6–3

FUNDAMENTAL PROBLEMS

All problem solutions must include FBDs.

F6–13.   Determine the force P needed to hold the 60-lb weight in equilibrium.

F6–15.   If a 100-N force is applied to the handles of the pliers, determine the clamping force exerted on the smooth pipe B and the magnitude of the resultant force that one of the members exerts on pin A.

Prob. F6–13

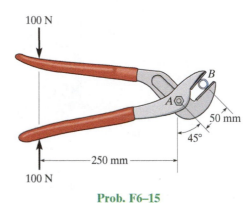

Prob. F6–15

F6–14.   Determine the horizontal and vertical components of reaction at pin C.

F6–16.   Determine the horizontal and vertical components of reaction at pin C.

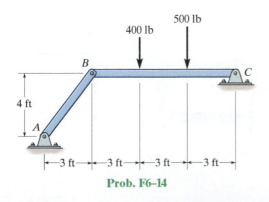

Prob. F6–14

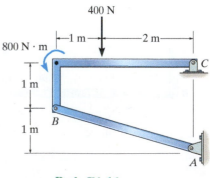

Prob. F6–16

F6–17. Determine the normal force that the 100-lb plate A exerts on the 30-lb plate B.

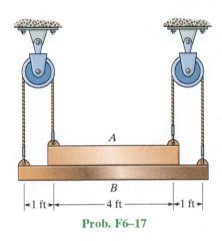

Prob. F6–17

F6–19. Determine the components of reaction at A and B.

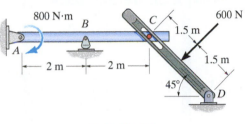

Prob. F6–19

F6–18. Determine the force P needed to lift the load. Also, determine the proper placement x of the hook for equilibrium. Neglect the weight of the beam.

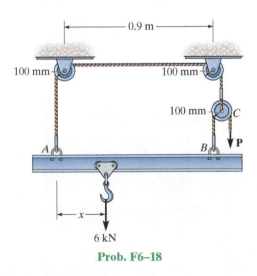

Prob. F6–18

F6–20. Determine the reactions at D.

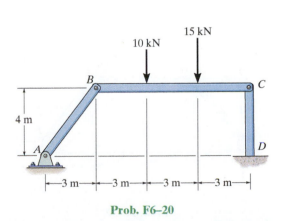

Prob. F6–20

F6–21. Determine the components of reaction at A and C.

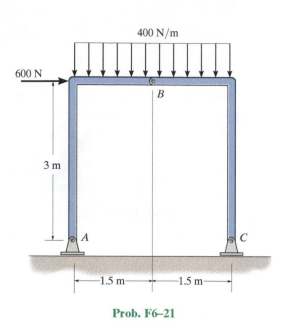

400 N/m

600 N

B

3 m

A

C

—1.5 m— —1.5 m—

Prob. F6–21

F6–22. Determine the components of reaction at C.

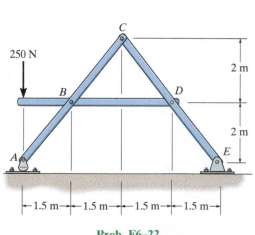

C

250 N

B

D

2 m

2 m

A

E

—1.5 m——1.5 m——1.5 m——1.5 m—

Prob. F6–22

F6–23. Determine the components of reaction at E.

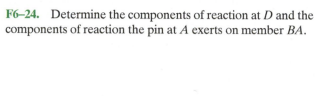

4 kN/m

A

B

2 m

E

D

C

—1.5 m— —1.5 m—

5 kN

Prob. F6–23

F6–24. Determine the components of reaction at D and the components of reaction the pin at A exerts on member BA.

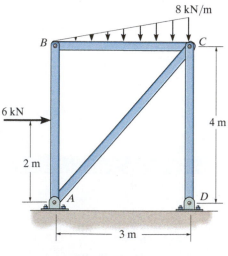

8 kN/m

B

C

6 kN

4 m

2 m

A

D

3 m

Prob. F6–24

PROBLEMS

All problem solutions must include FBDs.

6–61. Determine the force **P** required to hold the 100-lb weight in equilibrium.

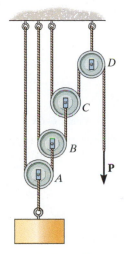

Prob. 6–61

6–62. In each case, determine the force **P** required to maintain equilibrium. The block weighs 100 lb.

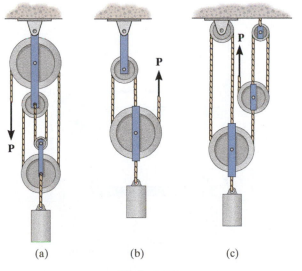

(a)   (b)   (c)

Prob. 6–62

6–63. Determine the force **P** required to hold the 50-kg mass in equilibrium.

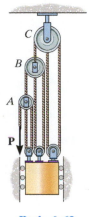

Prob. 6–63

*6–64.** Determine the force **P** required to hold the 150-kg crate in equilibrium.

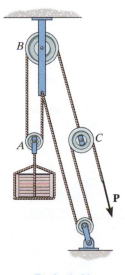

Prob. 6–64

6–65. Determine the horizontal and vertical components of force that pins A and B exert on the frame.

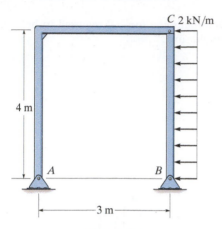

Prob. 6–65

6–66. Determine the horizontal and vertical components of force at pins A and D.

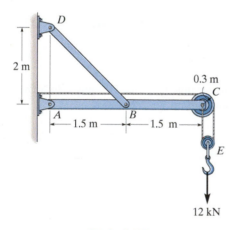

Prob. 6–66

6–67. Determine the force that the smooth roller C exerts on member AB. Also, what are the horizontal and vertical components of reaction at pin A? Neglect the weight of the frame and roller.

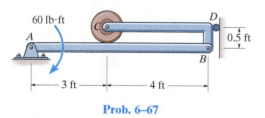

Prob. 6–67

*6–68. The bridge frame consists of three segments which can be considered pinned at A, D, and E, rocker supported at C and F, and roller supported at B. Determine the horizontal and vertical components of reaction at all these supports due to the loading shown.

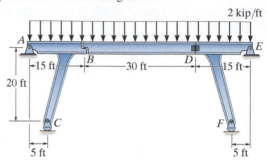

Prob. 6–68

6–69. Determine the reactions at supports A and B.

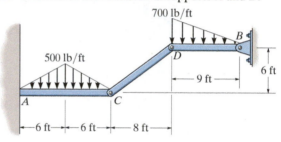

Prob. 6–69

6–70. Determine the horizontal and vertical components of force at pins B and C. The suspended cylinder has a mass of 75 kg.

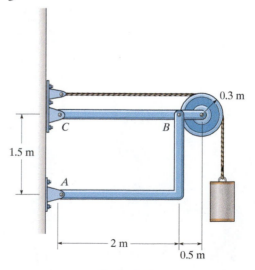

Prob. 6–70

6–71. Determine the reactions at the supports A, C, and E of the compound beam.

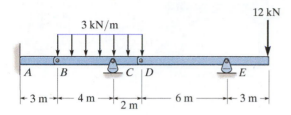

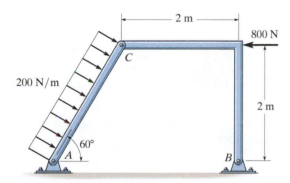

Prob. 6–71

*6–72. Determine the resultant force at pins A, B, and C on the three-member frame.

Prob. 6–72

6–73. Determine the reactions at the supports at A, E, and B of the compound beam.

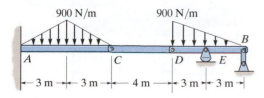

Prob. 6–73

6–74. The wall crane supports a load of 700 lb. Determine the horizontal and vertical components of reaction at the pins A and D. Also, what is the force in the cable at the winch W?

6–75. The wall crane supports a load of 700 lb. Determine the horizontal and vertical components of reaction at the pins A and D. Also, what is the force in the cable at the winch W? The jib ABC has a weight of 100 lb and member BD has a weight of 40 lb. Each member is uniform and has a center of gravity at its center.

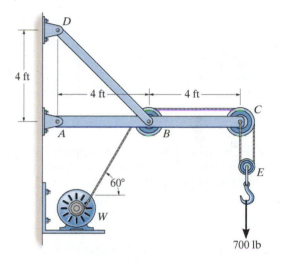

Probs. 6–74/75

*6–76. Determine the horizontal and vertical components of force which the pins at A and B exert on the frame.

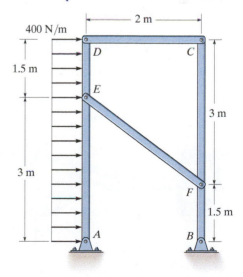

Prob. 6–76

6–77. The two-member structure is connected at C by a pin, which is fixed to BDE and passes through the smooth slot in member AC. Determine the horizontal and vertical components of reaction at the supports.

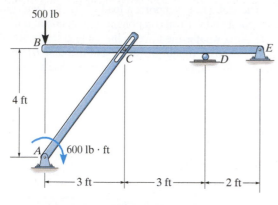

Prob. 6–77

6–78. The compound beam is pin supported at B and supported by rockers at A and C. There is a hinge (pin) at D. Determine the reactions at the supports.

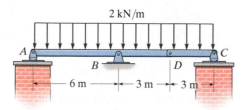

Prob. 6–78

6–79. When a force of 2 lb is applied to the handles of the brad squeezer, it pulls in the smooth rod AB. Determine the force $\mathbf{P}$ exerted on each of the smooth brads at C and D.

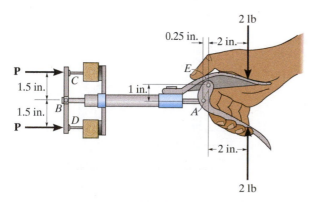

Prob. 6–79

*6–80.** The toggle clamp is subjected to a force $\mathbf{F}$ at the handle. Determine the vertical clamping force acting at E.

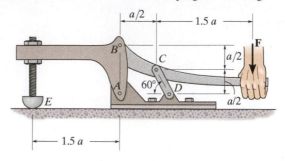

Prob. 6–80

6–81. The hoist supports the 125-kg engine. Determine the force the load creates in member DB and in member FB, which contains the hydraulic cylinder H.

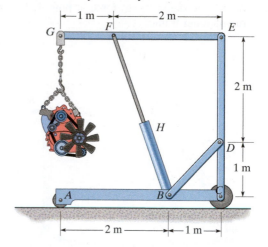

Prob. 6–81

6–82. A 5-lb force is applied to the handles of the vise grip. Determine the compressive force developed on the smooth bolt shank A at the jaws.

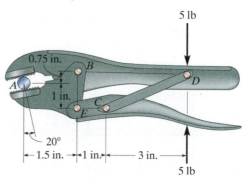

Prob. 6–82

6–83. Determine the force in members *FD* and *DB* of the frame. Also, find the horizontal and vertical components of reaction the pin at *C* exerts on member *ABC* and member *EDC*.

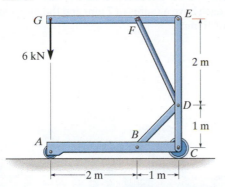

Prob. 6–83

*6–84. Determine the force that the smooth 20-kg cylinder exerts on members *AB* and *CDB*. Also, what are the horizontal and vertical components of reaction at pin *A*?

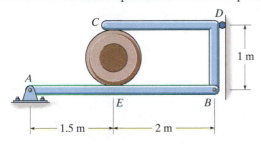

Prob. 6–84

6–85. The three power lines exert the forces shown on the pin-connected members at joints *B*, *C*, and *D*, which in turn are pin connected to the poles *AH* and *EG*. Determine the force in the guy cable *AI* and the pin reaction at the support *H*.

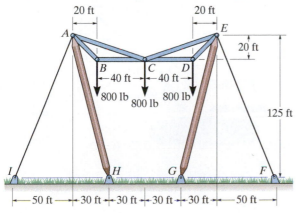

Prob. 6–85

6–86. The pumping unit is used to recover oil. When the walking beam *ABC* is horizontal, the force acting in the wireline at the well head is 250 lb. Determine the torque **M** which must be exerted by the motor in order to overcome this load. The horse-head *C* weighs 60 lb and has a center of gravity at G_C. The walking beam *ABC* has a weight of 130 lb and a center of gravity at G_B, and the counterweight has a weight of 200 lb and a center of gravity at G_W. The pitman, *AD*, is pin connected at its ends and has negligible weight.

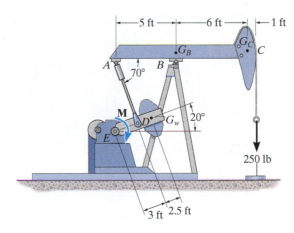

Prob. 6–86

6–87. Determine the force that the jaws *J* of the metal cutters exert on the smooth cable *C* if 100-N forces are applied to the handles. The jaws are pinned at *E* and *A*, and *D* and *B*. There is also a pin at *F*.

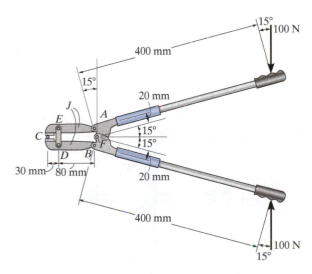

Prob. 6–87

*6–88. The machine shown is used for forming metal plates. It consists of two toggles ABC and DEF, which are operated by the hydraulic cylinder H. The toggles push the movable bar G forward, pressing the plate p into the cavity. If the force which the plate exerts on the head is $P = 12$ kN, determine the force F in the hydraulic cylinder when $\theta = 30°$.

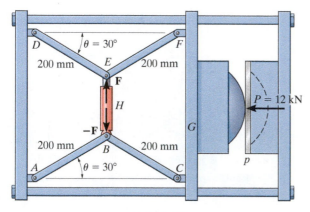

Prob. 6–88

6–89. Determine the horizontal and vertical components of force which pin C exerts on member ABC. The 600-N load is applied to the pin.

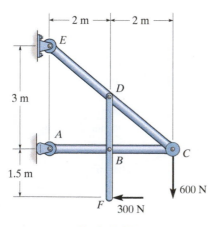

Prob. 6–89

6–90. The pipe cutter is clamped around the pipe P. If the wheel at A exerts a normal force of $F_A = 80$ N on the pipe, determine the normal forces of wheels B and C on the pipe. Also compute the pin reaction on the wheel at C. The three wheels each have a radius of 7 mm and the pipe has an outer radius of 10 mm.

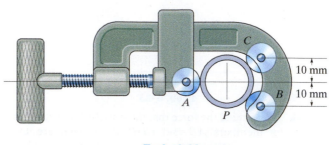

Prob. 6–90

6–91. Determine the force created in the hydraulic cylinders EF and AD in order to hold the shovel in equilibrium. The shovel load has a mass of 1.25 Mg and a center of gravity at G. All joints are pin connected.

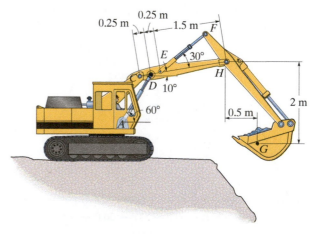

Prob. 6–91

*6–92. Determine the horizontal and vertical components of force at pin B and the normal force the pin at C exerts on the smooth slot. Also, determine the moment and horizontal and vertical reactions of force at A. There is a pulley at E.

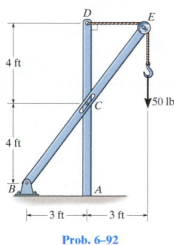

Prob. 6–92

6–93. The constant moment of 50 N · m is applied to the crank shaft. Determine the compressive force P that is exerted on the piston for equilibrium as a function of θ. Plot the results of P (vertical axis) versus θ (horizontal axis) for 0° ≤ θ ≤ 90°.

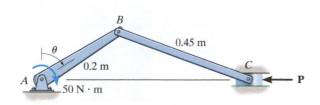

Prob. 6–93

6–94. Five coins are stacked in the smooth plastic container shown. If each coin weighs 0.0235 lb, determine the normal reactions of the bottom coin on the container at points A and B.

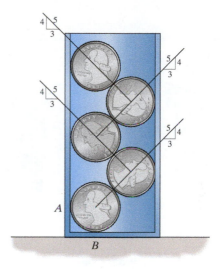

Prob. 6–94

6–95. The nail cutter consists of the handle and the two cutting blades. Assuming the blades are pin connected at B and the surface at D is smooth, determine the normal force on the fingernail when a force of 1 lb is applied to the handles as shown. The pin AC slides through a smooth hole at A and is attached to the bottom member at C.

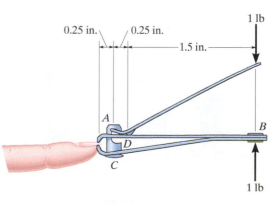

Prob. 6–95

*6–96. A man having a weight of 175 lb attempts to hold himself using one of the two methods shown. Determine the total force he must exert on bar AB in each case and the normal reaction he exerts on the platform at C. Neglect the weight of the platform.

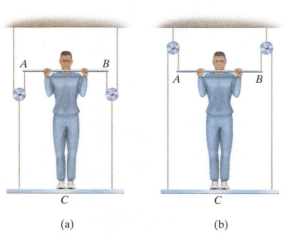

(a)   (b)

Prob. 6–96

6–97. A man having a weight of 175 lb attempts to hold himself using one of the two methods shown. Determine the total force he must exert on bar AB in each case and the normal reaction he exerts on the platform at C. The platform has a weight of 30 lb.

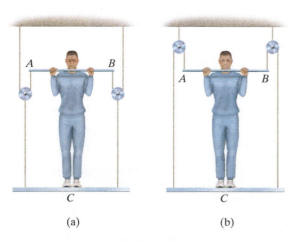

(a)   (b)

Prob. 6–97

6–98. The two-member frame is pin connected at E. The cable is attached to D, passes over the smooth peg at C, and supports the 500-N load. Determine the horizontal and vertical reactions at each pin.

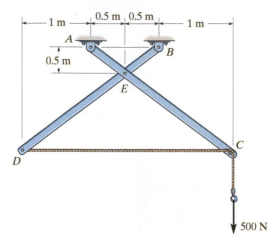

500 N

Prob. 6–98

6–99. If the 300-kg drum has a center of mass at point G, determine the horizontal and vertical components of force acting at pin A and the reactions on the smooth pads C and D. The grip at B on member DAB resists both horizontal and vertical components of force at the rim of the drum.

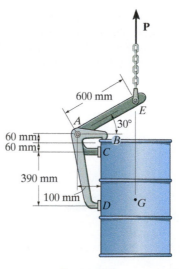

Prob. 6–99

*6–100. Operation of exhaust and intake valves in an automobile engine consists of the cam C, push rod DE, rocker arm EFG which is pinned at F, and a spring and valve, V. If the compression in the spring is 20 mm when the valve is open as shown, determine the normal force acting on the cam lobe at C. Assume the cam and bearings at H, I, and J are smooth. The spring has a stiffness of 300 N/m.

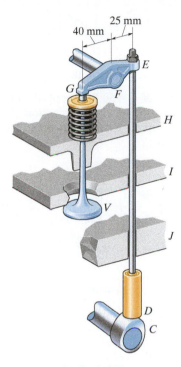

Prob. 6–100

6–101. If a clamping force of 300 N is required at A, determine the amount of force **F** that must be applied to the handle of the toggle clamp.

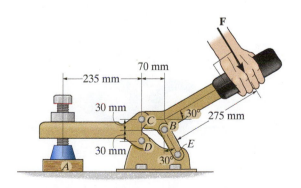

Prob. 6–101

6–102. If a force of $F = 350$ N is applied to the handle of the toggle clamp, determine the resulting clamping force at A.

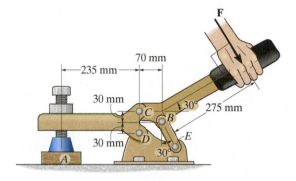

Prob. 6–102

6–103. Determine the horizontal and vertical components of force that the pins at A and B exert on the frame.

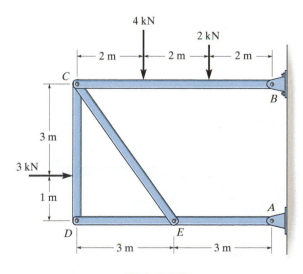

Prob. 6–103

*6–104. The hydraulic crane is used to lift the 1400-lb load. Determine the force in the hydraulic cylinder *AB* and the force in links *AC* and *AD* when the load is held in the position shown.

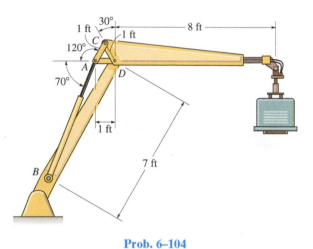

Prob. 6–104

6–105. Determine force **P** on the cable if the spring is compressed 0.025 m when the mechanism is in the position shown. The spring has a stiffness of *k* = 6 kN/m.

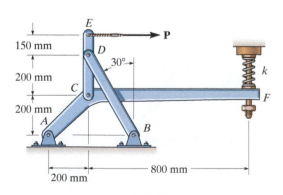

Prob. 6–105

6–106. If *d* = 0.75 ft and the spring has an unstretched length of 1 ft, determine the force *F* required for equilibrium.

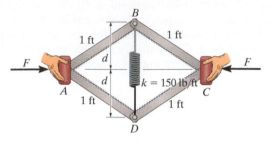

Prob. 6–106

6–107. If a force of *F* = 50 lb is applied to the pads at *A* and *C*, determine the smallest dimension *d* required for equilibrium if the spring has an unstretched length of 1 ft.

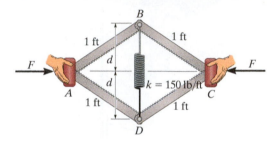

Prob. 6–107

*6–108. The skid-steer loader has a mass of 1.18 Mg, and in the position shown the center of mass is at G_1. If there is a 300-kg stone in the bucket, with center of mass at G_2, determine the reactions of each pair of wheels *A* and *B* on the ground and the force in the hydraulic cylinder *CD* and at the pin *E*. There is a similar linkage on each side of the loader.

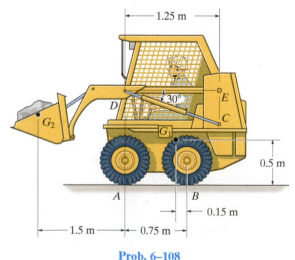

Prob. 6–108

6–109. Determine the force **P** on the cable if the spring is compressed 0.5 in. when the mechanism is in the position shown. The spring has a stiffness of $k = 800$ lb/ft.

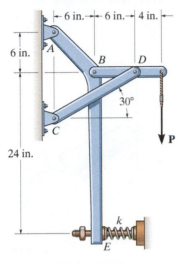

Prob. 6–109

6–110. The spring has an unstretched length of 0.3 m. Determine the angle θ for equilibrium if the uniform bars each have a mass of 20 kg.

6–111. The spring has an unstretched length of 0.3 m. Determine the mass m of each uniform bar if each angle $\theta = 30°$ for equilibrium.

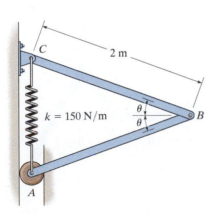

Probs. 6–110/111

*6–112. The piston C moves vertically between the two smooth walls. If the spring has a stiffness of $k = 15$ lb/in., and is unstretched when $\theta = 0°$, determine the couple **M** that must be applied to AB to hold the mechanism in equilibrium when $\theta = 30°$.

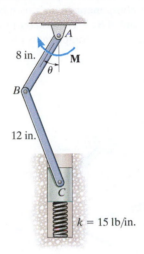

Prob. 6–112

6–113. The platform scale consists of a combination of third and first class levers so that the load on one lever becomes the effort that moves the next lever. Through this arrangement, a small weight can balance a massive object. If $x = 450$ mm, determine the required mass of the counterweight S required to balance a 90-kg load, L.

6–114. The platform scale consists of a combination of third and first class levers so that the load on one lever becomes the effort that moves the next lever. Through this arrangement, a small weight can balance a massive object. If $x = 450$ mm, and the mass of the counterweight S is 2 kg, determine the mass of the load L required to maintain the balance.

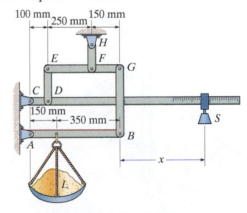

Probs. 6–113/114

6–115. The four-member "A" frame is supported at A and E by smooth collars and at G by a pin. All the other joints are ball-and-sockets. If the pin at G will fail when the resultant force there is 800 N, determine the largest vertical force P that can be supported by the frame. Also, what are the x, y, z force components which member BD exerts on members EDC and ABC? The collars at A and E and the pin at G only exert force components on the frame.

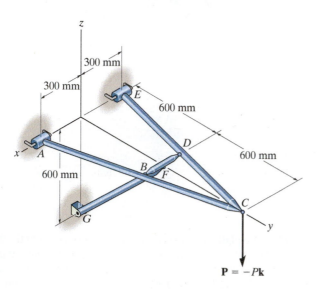

Prob. 6–115

*6–116. The structure is subjected to the loadings shown. Member AB is supported by a ball-and-socket at A and smooth collar at B. Member CD is supported by a pin at C. Determine the x, y, z components of reaction at A and C.

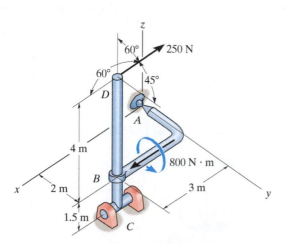

Prob. 6–116

6–117. The structure is subjected to the loading shown. Member AD is supported by a cable AB and roller at C and fits through a smooth circular hole at D. Member ED is supported by a roller at D and a pole that fits in a smooth snug circular hole at E. Determine the x, y, z components of reaction at E and the tension in cable AB.

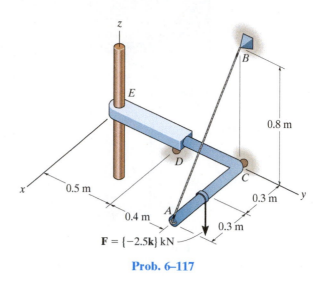

Prob. 6–117

6–118. The three pin-connected members shown in the *top view* support a downward force of 60 lb at G. If only vertical forces are supported at the connections B, C, E and pad supports A, D, F, determine the reactions at each pad.

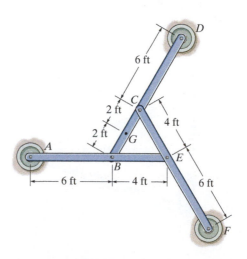

Prob. 6–118

CHAPTER REVIEW

Simple Truss

A simple truss consists of triangular elements connected together by pinned joints. The forces within its members can be determined by assuming the members are all two-force members, connected concurrently at each joint. The members are either in tension or compression, or carry no force.

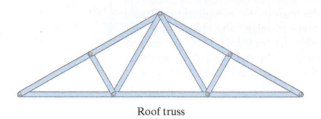

Roof truss

Method of Joints

The method of joints states that if a truss is in equilibrium, then each of its joints is also in equilibrium. For a plane truss, the concurrent force system at each joint must satisfy force equilibrium.

To obtain a numerical solution for the forces in the members, select a joint that has a free-body diagram with at most two unknown forces and one known force. (This may require first finding the reactions at the supports.)

Once a member force is determined, use its value and apply it to an adjacent joint.

Remember that forces that are found to *pull* on the joint are *tensile forces*, and those that *push* on the joint are *compressive forces*.

To avoid a simultaneous solution of two equations, set one of the coordinate axes along the line of action of one of the unknown forces and sum forces perpendicular to this axis. This will allow a direct solution for the other unknown.

The analysis can also be simplified by first identifying all the zero-force members.

$$\Sigma F_x = 0$$

$$\Sigma F_y = 0$$

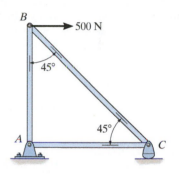

Method of Sections

The method of sections states that if a truss is in equilibrium, then each segment of the truss is also in equilibrium. Pass a section through the truss and the member whose force is to be determined. Then draw the free-body diagram of the sectioned part having the least number of forces on it.

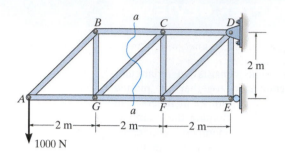

Sectioned members subjected to *pulling* are in *tension*, and those that are subjected to *pushing* are in *compression*.

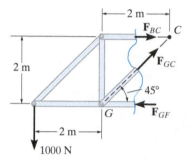

Three equations of equilibrium are available to determine the unknowns.

$$\Sigma F_x = 0$$
$$\Sigma F_y = 0$$
$$\Sigma M_O = 0$$

If possible, sum forces in a direction that is perpendicular to two of the three unknown forces. This will yield a direct solution for the third force.

$$+\uparrow \Sigma F_y = 0$$
$$-1000 \text{ N} + F_{GC} \sin 45° = 0$$
$$F_{GC} = 1.41 \text{ kN (T)}$$

Sum moments about the point where the lines of action of two of the three unknown forces intersect, so that the third unknown force can be determined directly.

$$\zeta + \Sigma M_C = 0$$
$$1000 \text{ N}(4 \text{ m}) - F_{GF} (2 \text{ m}) = 0$$
$$F_{GF} = 2 \text{ kN (C)}$$

Space Truss

A space truss is a three-dimensional truss built from tetrahedral elements, and is analyzed using the same methods as for plane trusses. The joints are assumed to be ball-and-socket connections.

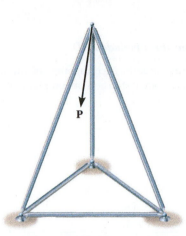

Frames and Machines

Frames and machines are structures that contain one or more multiforce members, that is, members with three or more forces or couples acting on them. Frames are designed to support loads, and machines transmit and alter the effect of forces.

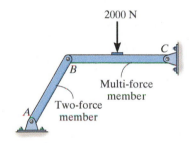

The forces acting at the joints of a frame or machine can be determined by drawing the free-body diagrams of each of its members or parts. The principle of action–reaction should be carefully observed when indicating these forces on the free-body diagram of each adjacent member or pin. For a coplanar force system, there are three equilibrium equations available for each member.

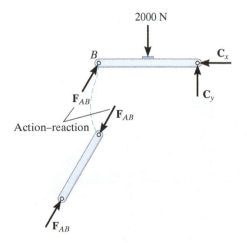

To simplify the analysis, be sure to recognize all two-force members. They have equal but opposite collinear forces at their ends.

REVIEW PROBLEMS

All problem solutions must include FBDs.

R6–1. Determine the force in each member of the truss and state if the members are in tension or compression.

R6–3. Determine the force in member *GJ* and *GC* of the truss and state if the members are in tension or compression.

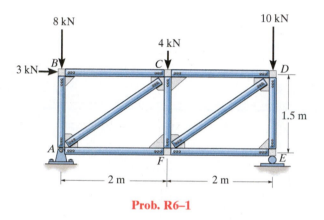

Prob. R6–1

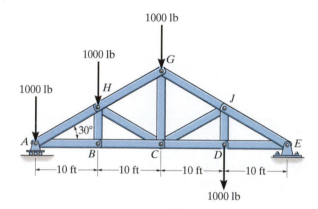

Prob. R6–3

R6–2. Determine the force in each member of the truss and state if the members are in tension or compression.

R6–4. Determine the force in members *GF*, *FB*, and *BC* of the *Fink truss* and state if the members are in tension or compression.

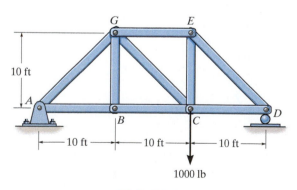

Prob. R6–2

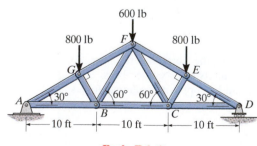

Prob. R6–4

6

R6–5. Determine the force in members AB, AD, and AC of the space truss and state if the members are in tension or compression.

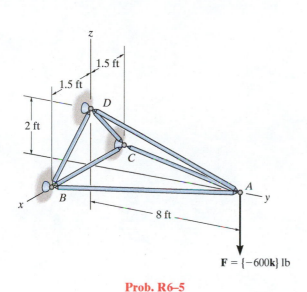

Prob. R6–5

R6–6. Determine the horizontal and vertical components of force that the pins A and B exert on the two-member frame.

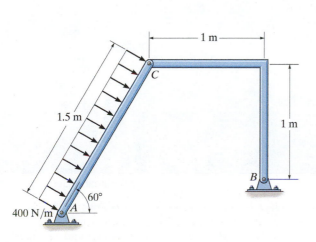

Prob. R6–6

R6–7. Determine the horizontal and vertical components of force at pins A and C of the two-member frame.

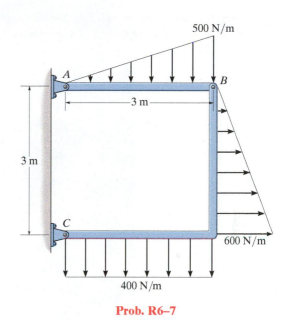

Prob. R6–7

R6–8. Determine the resultant forces at pins B and C on member ABC of the four-member frame.

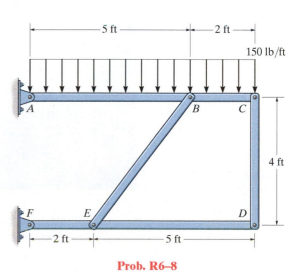

Prob. R6–8

6

Chapter 7

(© Tony Freeman/Science Source)

When external loads are placed upon these beams and columns, the loads within them must be determined if they are to be properly designed. In this chapter we will study how to determine these internal loadings.

Internal Forces

7.1 Internal Loadings Developed in Structural Members

To design a structural or mechanical member it is necessary to know the loading acting within the member in order to be sure the material can resist this loading. Internal loadings can be determined by using the *method of sections*. To illustrate this method, consider the cantilever beam in Fig. 7–1a. If the internal loadings acting on the cross section at point B are to be determined, we must pass an imaginary section a–a perpendicular to the axis of the beam through point B and then separate the beam into two segments. The internal loadings acting at B will then be exposed and become *external* on the free-body diagram of each segment, Fig. 7–1b.

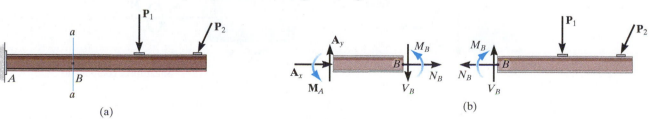

(a)

(b)

Fig. 7–1

In each case, the link on the backhoe is a two-force member. In the top photo it is subjected to both bending and an axial load at its center. It is more efficient to make the member straight, as in the bottom photo; then only an axial force acts within the member. (© Russell C. Hibbeler)

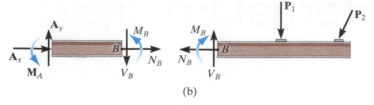

(b)

Fig. 7–1 (Repeated)

The force component $\mathbf{N}_B$ that acts *perpendicular* to the cross section is termed the ***normal force***. The force component $\mathbf{V}_B$ that is tangent to the cross section is called the ***shear force***, and the couple moment $\mathbf{M}_B$ is referred to as the ***bending moment***. The force components prevent the relative translation between the two segments, and the couple moment prevents the relative rotation. According to Newton's third law, these loadings must act in opposite directions on each segment, as shown in Fig. 7–1b. They can be determined by applying the equations of equilibrium to the free-body diagram of either segment. In this case, however, the right segment is the better choice since it does not involve the unknown support reactions at A. A direct solution for $\mathbf{N}_B$ is obtained by applying $\Sigma F_x = 0$, $\mathbf{V}_B$ is obtained from $\Sigma F_y = 0$, and $\mathbf{M}_B$ can be obtained by applying $\Sigma M_B = 0$, since the moments of $\mathbf{N}_B$ and $\mathbf{V}_B$ about B are zero.

In two dimensions, we have shown that three internal loading resultants exist, Fig. 7–2a; however in three dimensions, a general resultant internal force and couple moment resultant will act at the section. The x, y, z components of these loadings are shown in Fig. 7–2b. Here $\mathbf{N}_y$ is the *normal force*, and $\mathbf{V}_x$ and $\mathbf{V}_z$ are *shear force components*. $\mathbf{M}_y$ is a *torsional or twisting moment*, and $\mathbf{M}_x$ and $\mathbf{M}_z$ are *bending moment components*. For most applications, these *resultant loadings* will act at the geometric center or centroid (C) of the section's cross-sectional area. Although the magnitude for each loading generally will be different at various points along the axis of the member, the method of sections can always be used to determine their values.

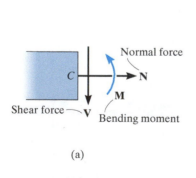

(a)

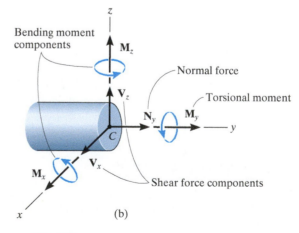

(b)

Fig. 7–2

Sign Convention. For problems in two dimensions engineers generally use a sign convention to report the three internal loadings **N, V,** and **M**. Although this sign convention can be arbitrarily assigned, the one that is widely accepted will be used here, Fig. 7–3. The normal force is said to be positive if it creates *tension*, a positive shear force will cause the beam segment on which it acts to rotate clockwise, and a positive bending moment will tend to bend the segment on which it acts in a concave upward manner. Loadings that are opposite to these are considered negative.

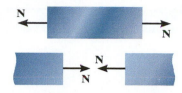

Positive normal force

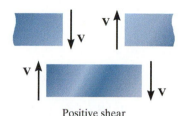

Positive shear

Important Point

- There can be four types of resultant internal loads in a member. They are the normal and shear forces and the bending and torsional moments. These loadings generally vary from point to point. They can be determined using the method of sections.

Positive moment

Fig. 7–3

Procedure for Analysis

The method of sections can be used to determine the internal loadings on the cross section of a member using the following procedure.

Support Reactions.

- Before the member is sectioned, it may first be necessary to determine its support reactions.

Free-Body Diagram.

- It is important to *keep* all distributed loadings, couple moments, and forces acting on the member in their *exact locations, then* pass an imaginary section through the member, perpendicular to its axis at the point where the internal loadings are to be determined.

- After the section is made, draw a free-body diagram of the segment that has the least number of loads on it, and indicate the components of the internal force and couple moment resultants at the cross section acting in their positive directions in accordance with the established sign convention.

Equations of Equilibrium.

- Moments should be summed at the section. This way the normal and shear forces at the section are eliminated, and we can obtain a direct solution for the moment.

- If the solution of the equilibrium equations yields a negative scalar, the sense of the quantity is opposite to that shown on the free-body diagram.

The designer of this shop crane realized the need for additional reinforcement around the joint at *A* in order to prevent severe internal bending of the joint when a large load is suspended from the chain hoist. (© Russell C. Hibbeler)

EXAMPLE 7.1

Determine the normal force, shear force, and bending moment acting just to the left, point B, and just to the right, point C, of the 6-kN force on the beam in Fig. 7–4a.

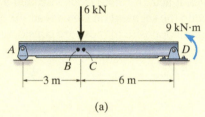

(a)

SOLUTION

Support Reactions.   The free-body diagram of the beam is shown in Fig. 7–4b. When determining the *external reactions*, realize that the 9-kN · m couple moment is a free vector and therefore it can be placed *anywhere* on the free-body diagram of the entire beam. Here we will only determine $\mathbf{A}_y$, since the left segments will be used for the analysis.

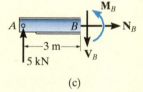

(b)

$$\zeta + \Sigma M_D = 0; \quad 9 \text{ kN} \cdot \text{m} + (6 \text{ kN})(6 \text{ m}) - A_y(9 \text{ m}) = 0$$
$$A_y = 5 \text{ kN}$$

Free-Body Diagrams.   The free-body diagrams of the left segments AB and AC of the beam are shown in Figs. 7–4c and 7–4d. In this case the 9-kN · m couple moment is *not included* on these diagrams since it must be kept in its *original position* until *after* the section is made and the appropriate segment is isolated.

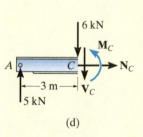

(c)

Equations of Equilibrium.

Segment AB

$$\xrightarrow{+} \Sigma F_x = 0; \qquad\qquad\qquad N_B = 0 \qquad\qquad Ans.$$
$$+\uparrow \Sigma F_y = 0; \qquad 5 \text{ kN} - V_B = 0 \quad V_B = 5 \text{ kN} \qquad Ans.$$
$$\zeta + \Sigma M_B = 0; \quad -(5 \text{ kN})(3 \text{ m}) + M_B = 0 \quad M_B = 15 \text{ kN} \cdot \text{m} \qquad Ans.$$

Segment AC

$$\xrightarrow{+} \Sigma F_x = 0; \qquad\qquad\qquad N_C = 0 \qquad\qquad Ans.$$
$$+\uparrow \Sigma F_y = 0; \quad 5 \text{ kN} - 6 \text{ kN} - V_C = 0 \quad V_C = -1 \text{ kN} \qquad Ans.$$
$$\zeta + \Sigma M_C = 0; \quad -(5 \text{ kN})(3 \text{ m}) + M_C = 0 \quad M_C = 15 \text{ kN} \cdot \text{m} \qquad Ans.$$

(d)

Fig. 7–4

NOTE: The negative sign indicates that $\mathbf{V}_C$ acts in the opposite sense to that shown on the free-body diagram. Also, the moment arm for the 5-kN force in both cases is approximately 3 m since B and C are "almost" coincident.

EXAMPLE 7.2

Determine the normal force, shear force, and bending moment at C of the beam in Fig. 7–5a.

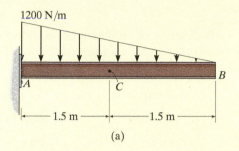

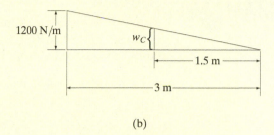

(a)   (b)

Fig. 7–5

SOLUTION

Free-Body Diagram. It is not necessary to find the support reactions at A since segment BC of the beam can be used to determine the internal loadings at C. The intensity of the triangular distributed load at C is determined using similar triangles from the geometry shown in Fig. 7–5b, i.e.,

$$w_C = (1200 \text{ N/m}) \left(\frac{1.5 \text{ m}}{3 \text{ m}} \right) = 600 \text{ N/m}$$

The distributed load acting on segment BC can now be replaced by its resultant force, and its location is indicated on the free-body diagram, Fig. 7–5c.

Equations of Equilibrium.

$$\xrightarrow{+} \Sigma F_x = 0; \qquad\qquad N_C = 0 \qquad\qquad\qquad \textit{Ans.}$$

$$+\uparrow \Sigma F_y = 0; \qquad V_C - \tfrac{1}{2}(600 \text{ N/m})(1.5 \text{ m}) = 0$$

$$V_C = 450 \text{ N} \qquad\qquad \textit{Ans.}$$

$$\zeta + \Sigma M_C = 0; \quad -M_C - \tfrac{1}{2}(600 \text{ N/m})(1.5 \text{ m})(0.5 \text{ m}) = 0$$

$$M_C = -225 \text{ N} \qquad\qquad \textit{Ans.}$$

The negative sign indicates that $\mathbf{M}_C$ acts in the opposite sense to that shown on the free-body diagram.

EXAMPLE | **7.3**

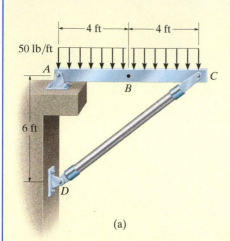

(a)

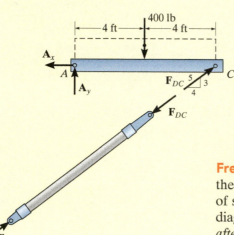

(b)

Fig. 7–6

Determine the normal force, shear force, and bending moment acting at point B of the two-member frame shown in Fig. 7–6a.

SOLUTION

Support Reactions.   A free-body diagram of each member is shown in Fig. 7–6b. Since CD is a two-force member, the equations of equilibrium need to be applied only to member AC.

$$\zeta +\Sigma M_A = 0; \quad -400 \text{ lb } (4 \text{ ft}) + \left(\tfrac{3}{5}\right) F_{DC}(8 \text{ ft}) = 0 \quad F_{DC} = 333.3 \text{ lb}$$

$$\overset{+}{\to}\Sigma F_x = 0; \quad\quad\quad -A_x + \left(\tfrac{4}{5}\right)(333.3 \text{ lb}) = 0 \quad A_x = 266.7 \text{ lb}$$

$$+\uparrow\Sigma F_y = 0; \quad A_y - 400 \text{ lb} + \left(\tfrac{3}{5}\right)(333.3 \text{ lb}) = 0 \quad A_y = 200 \text{ lb}$$

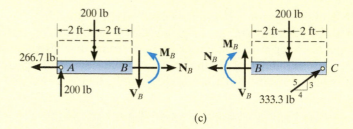

(c)

Free-Body Diagrams.   Passing an imaginary section perpendicular to the axis of member AC through point B yields the free-body diagrams of segments AB and BC shown in Fig. 7–6c. When constructing these diagrams it is important to keep the distributed loading where it is until *after the section is made*. Only then can it be replaced by a single resultant force.

Equations of Equilibrium.   Applying the equations of equilibrium to segment AB, we have

$$\overset{+}{\to}\Sigma F_x = 0; \quad\quad N_B - 266.7 \text{ lb} = 0 \quad N_B = 267 \text{ lb} \quad\quad Ans.$$

$$+\uparrow\Sigma F_y = 0; \quad 200 \text{ lb} - 200 \text{ lb} - V_B = 0 \quad V_B = 0 \quad\quad Ans.$$

$$\zeta +\Sigma M_B = 0; \quad M_B - 200 \text{ lb } (4 \text{ ft}) + 200 \text{ lb } (2 \text{ ft}) = 0$$

$$M_B = 400 \text{ lb} \cdot \text{ft} \quad\quad Ans.$$

NOTE: As an exercise, try to obtain these same results using segment BC.

EXAMPLE 7.4

Determine the normal force, shear force, and bending moment acting at point E of the frame loaded as shown in Fig. 7–7a.

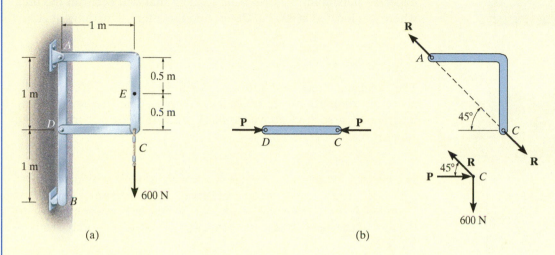

(a)

(b)

SOLUTION

Support Reactions. By inspection, members AC and CD are two-force members, Fig. 7–7b. In order to determine the internal loadings at E, we must first determine the force **R** acting at the end of member AC. To obtain it, we will analyze the equilibrium of the pin at C.

Summing forces in the vertical direction on the pin, Fig. 7–7b, we have

$$+\uparrow \Sigma F_y = 0; \qquad R \sin 45° - 600 \text{ N} = 0 \qquad R = 848.5 \text{ N}$$

Free-Body Diagram. The free-body diagram of segment CE is shown in Fig. 7–7c.

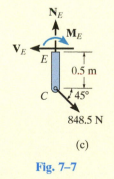

Equations of Equilibrium.

$$\xrightarrow{+} \Sigma F_x = 0; \qquad 848.5 \cos 45° \text{ N} - V_E = 0 \qquad V_E = 600 \text{ N} \qquad Ans.$$

$$+\uparrow \Sigma F_y = 0; \qquad -848.5 \sin 45° \text{ N} + N_E = 0 \qquad N_E = 600 \text{ N} \qquad Ans.$$

$$\zeta + \Sigma M_E = 0; \qquad 848.5 \cos 45° \text{ N}(0.5 \text{ m}) - M_E = 0$$

$$M_E = 300 \text{ N} \cdot \text{m} \qquad Ans.$$

(c)

Fig. 7–7

NOTE: These results indicate a poor design. Member AC should be *straight* (from A to C) so that bending within the member is *eliminated*. If AC were straight then the internal force would only create tension in the member.

EXAMPLE 7.5

(a)

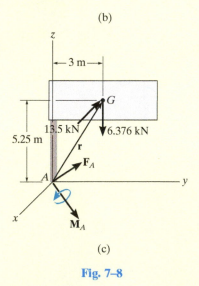

(b)

(c)

Fig. 7–8

The uniform sign shown in Fig. 7–8a has a mass of 650 kg and is supported on the fixed column. Design codes indicate that the expected maximum uniform wind loading that will occur in the area where it is located is 900 Pa. Determine the internal loadings at A.

SOLUTION

The idealized model for the sign is shown in Fig. 7–8b. Here the necessary dimensions are indicated. We can consider the free-body diagram of a section above point A since it does not involve the support reactions.

Free-Body Diagram.   The sign has a weight of $W = 650(9.81)\ N = 6.376\ kN$, and the wind creates a resultant force of $F_w = 900\ N/m^2(6\ m)(2.5\ m) = 13.5\ kN$, which acts perpendicular to the face of the sign. These loadings are shown on the free-body diagram, Fig. 7–8c.

Equations of Equilibrium.   Since the problem is three dimensional, a vector analysis will be used.

$$\Sigma F = 0; \qquad\qquad F_A - 13.5i - 6.376k = 0$$

$$F_A = \{13.5i + 6.38k\}\ kN \qquad\qquad \textit{Ans.}$$

$$\Sigma M_A = 0; \qquad\qquad M_A + r \times (F_w + W) = 0$$

$$M_A + \begin{vmatrix} i & j & k \\ 0 & 3 & 5.25 \\ -13.5 & 0 & -6.376 \end{vmatrix} = 0$$

$$M_A = \{19.1i + 70.9j - 40.5k\}\ kN \cdot m \qquad \textit{Ans.}$$

NOTE: Here $F_{A_z} = \{6.38k\}$ kN represents the normal force, whereas $F_{A_x} = \{13.5i\}$ kN is the shear force. Also, the torsional moment is $M_{A_z} = \{-40.5k\}$ kN · m, and the bending moment is determined from its components $M_{A_x} = \{19.1i\}$ kN · m and $M_{A_y} = \{70.9j\}$ kN · m; i.e., $(M_b)_A = \sqrt{(M_A)_x^2 + (M_A)_y^2} = 73.4$ kN · m.

(© Russell C. Hibbeler)

PRELIMINARY PROBLEMS

P7–1. In each case, calculate the reaction at A and then draw the free-body diagram of segment AB of the beam in order to determine the internal loading at B.

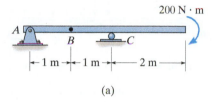

(a)

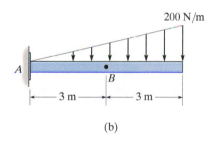

(b)

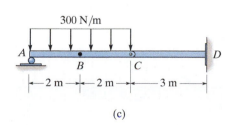

(c)

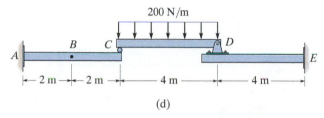

(d)

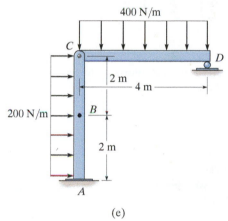

(e)

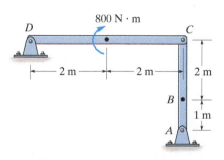

(f)

FUNDAMENTAL PROBLEMS

All problem solutions must include FBDs.

F7–1. Determine the normal force, shear force, and moment at point *C*.

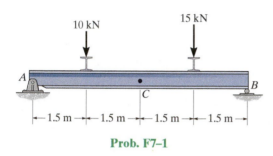

Prob. F7–1

F7–2. Determine the normal force, shear force, and moment at point *C*.

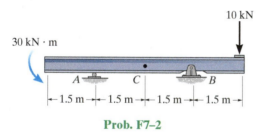

Prob. F7–2

F7–3. Determine the normal force, shear force, and moment at point *C*.

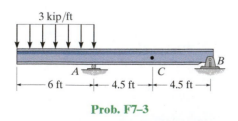

Prob. F7–3

F7–4. Determine the normal force, shear force, and moment at point *C*.

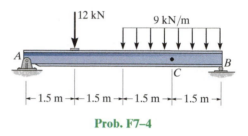

Prob. F7–4

F7–5. Determine the normal force, shear force, and moment at point *C*.

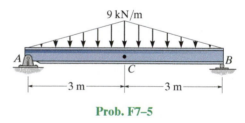

Prob. F7–5

F7–6. Determine the normal force, shear force, and moment at point *C*. Assume *A* is pinned and *B* is a roller.

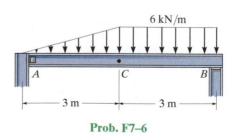

Prob. F7–6

PROBLEMS

All problem solutions must include FBDs.

7–1. Determine the shear force and moment at points C and D.

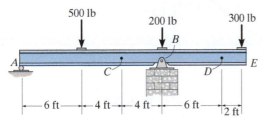

Prob. 7–1

7–2. Determine the internal normal force and shear force, and the bending moment in the beam at points C and D. Assume the support at B is a roller. Point C is located just to the right of the 8-kip load.

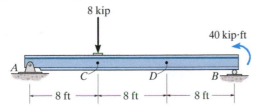

Prob. 7–2

7–3. Two beams are attached to the column such that structural connections transmit the loads shown. Determine the internal normal force, shear force, and moment acting in the column at a section passing horizontally through point A.

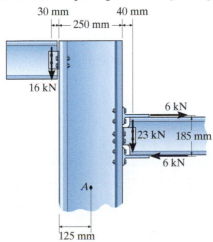

Prob. 7–3

***7–4.** The beam weighs 280 lb/ft. Determine the internal normal force, shear force, and moment at point C.

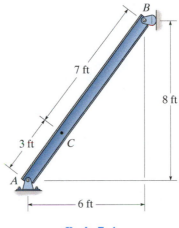

Prob. 7–4

7–5. The pliers are used to grip the tube at B. If a force of 20 lb is applied to the handles, determine the internal shear force and moment a point C. Assume the jaws of the pliers exert only normal forces on the tube.

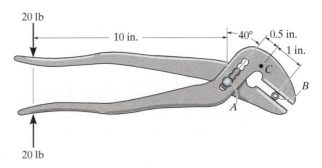

Prob. 7–5

7–6. Determine the distance a as a fraction of the beam's length L for locating the roller support so that the moment in the beam at B is zero.

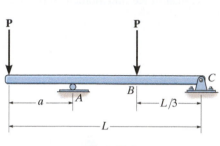

Prob. 7–6

7–7. Determine the internal shear force and moment acting at point C in the beam.

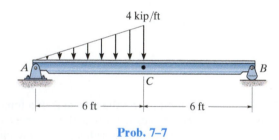

Prob. 7–7

*7–8. Determine the internal shear force and moment acting at point C in the beam.

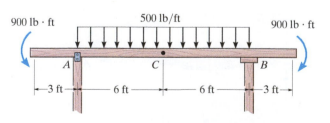

Prob. 7–8

7–9. Determine the normal force, shear force, and moment at a section passing through point C. Take $P = 8$ kN.

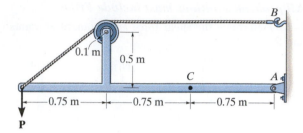

Prob. 7–9

7–10. The cable will fail when subjected to a tension of 2 kN. Determine the largest vertical load P the frame will support and calculate the internal normal force, shear force, and moment at a section passing through point C for this loading.

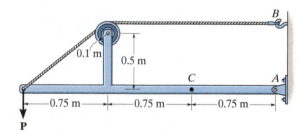

Prob. 7–10

7–11. Determine the internal normal force, shear force, and moment at points C and D of the beam.

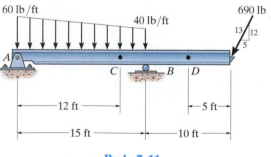

Prob. 7–11

*7–12. Determine the distance a between the bearings in terms of the shaft's length L so that the moment in the *symmetric* shaft is zero at its center.

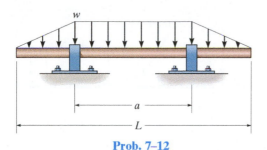

Prob. 7–12

7–13. Determine the internal normal force, shear force, and moment in the beam at sections passing through points D and E. Point D is located just to the left of the 5-kip load.

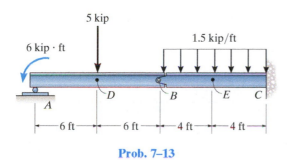

Prob. 7–13

7–14. The shaft is supported by a journal bearing at A and a thrust bearing at B. Determine the normal force, shear force, and moment at a section passing through (a) point C, which is just to the right of the bearing at A, and (b) point D, which is just to the left of the 3000-lb force.

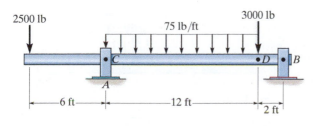

Prob. 7–14

7–15. Determine the internal normal force, shear force, and moment at point C.

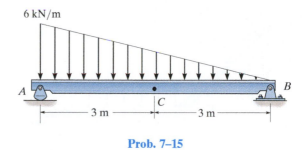

Prob. 7–15

*7–16. Determine the internal normal force, shear force, and moment at point C of the beam.

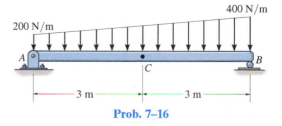

Prob. 7–16

7–17. The cantilevered rack is used to support each end of a smooth pipe that has a total weight of 300 lb. Determine the normal force, shear force, and moment that act in the arm at its fixed support A along a vertical section.

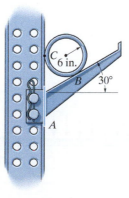

Prob. 7–17

7–18. Determine the internal normal force, shear force, and the moment at points C and D.

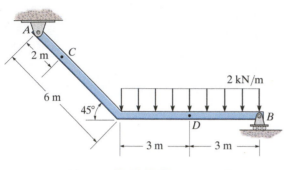

Prob. 7–18

7–19. Determine the internal normal force, shear force, and moment at point C.

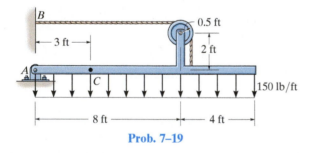

Prob. 7–19

*7–20. Rod AB is fixed to a smooth collar D, which slides freely along the vertical guide. Determine the internal normal force, shear force, and moment at point C, which is located just to the left of the 60-lb concentrated load.

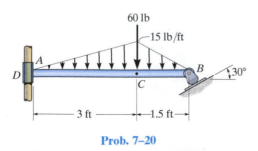

Prob. 7–20

7–21. Determine the internal normal force, shear force, and moment at points E and F of the compound beam. Point E is located just to the left of 800 N force.

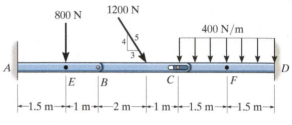

Prob. 7–21

7–22. Determine the internal normal force, shear force, and moment at points D and E in the overhang beam. Point D is located just to the left of the roller support at B, where the couple moment acts.

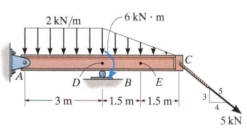

Prob. 7–22

7–23. Determine the internal normal force, shear force, and moment at point C.

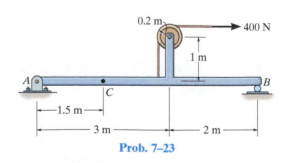

Prob. 7–23

*7–24. Determine the ratio of a/b for which the shear force will be zero at the midpoint C of the beam.

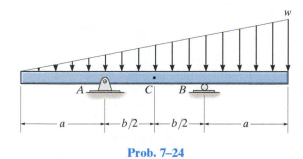

Prob. 7–24

7–25. Determine the normal force, shear force, and moment in the beam at sections passing through points D and E. Point E is just to the right of the 3-kip load.

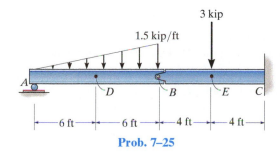

Prob. 7–25

7–26. Determine the internal normal force, shear force, and bending moment at point C.

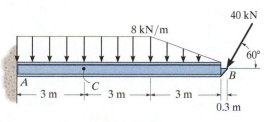

Prob. 7–26

7–27. Determine the internal normal force, shear force, and moment at point C.

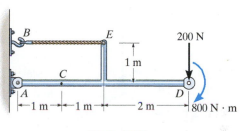

Prob. 7–27

*7–28. Determine the internal normal force, shear force, and moment at points C and D in the simply supported beam. Point D is located just to the left of the 10-kN concentrated load.

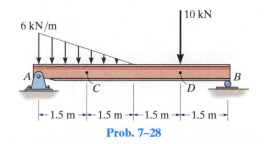

Prob. 7–28

7

7–29. Determine the normal force, shear force, and moment acting at a section passing through point *C*.

7–30. Determine the normal force, shear force, and moment acting at a section passing through point *D*.

*7–32. Determine the internal normal force, shear force, and moment at point *D*.

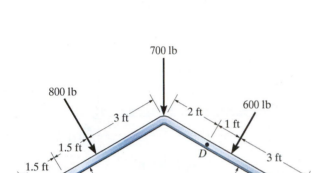

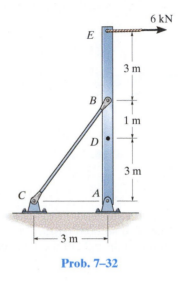

Prob. 7–32

Probs. 7–29/30

7–33. Determine the internal normal force, shear force, and moment at point *D* of the two-member frame.

7–34. Determine the internal normal force, shear force, and moment at point *E*.

7–31. Determine the internal normal force, shear force, and moment acting at points *D* and *E* of the frame.

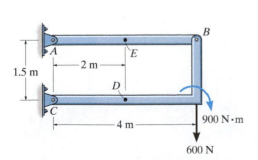

Prob. 7–31

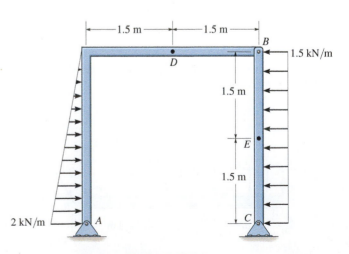

Probs. 7–33/34

7–35. The strongback or lifting beam is used for materials handling. If the suspended load has a weight of 2 kN and a center of gravity of G, determine the placement d of the padeyes on the top of the beam so that there is no moment developed within the length AB of the beam. The lifting bridle has two legs that are positioned at 45°, as shown.

7–37. Determine the internal normal force, shear force, and moment at point D of the two-member frame.

7–38. Determine the internal normal force, shear force, and moment at point E of the two-member frame.

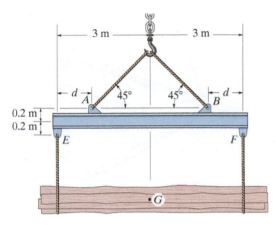

Prob. 7–35

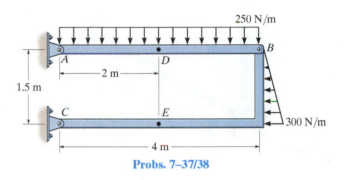
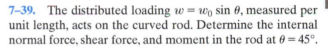

Probs. 7–37/38

*7–36. Determine the internal normal force, shear force, and moment acting at points B and C on the curved rod.

7–39. The distributed loading $w = w_0 \sin \theta$, measured per unit length, acts on the curved rod. Determine the internal normal force, shear force, and moment in the rod at $\theta = 45°$.

*7–40. Solve Prob. 7–39 for $\theta = 120°$.

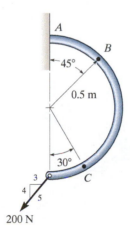

Prob. 7–36

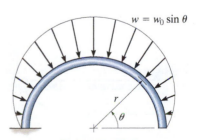

Probs. 7–39/40

7–41. Determine the x, y, z components of force and moment at point C in the pipe assembly. Neglect the weight of the pipe. Take $F_1 = \{350i - 400j\}$ lb and $F_2 = \{-300j + 150k\}$ lb.

7–43. Determine the x, y, z components of internal loading at a section passing through point B in the pipe assembly. Neglect the weight of the pipe. Take $F_1 = \{200i - 100j - 400k\}$ N and $F_2 = \{300i - 500k\}$ N.

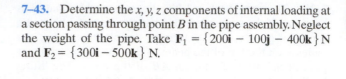

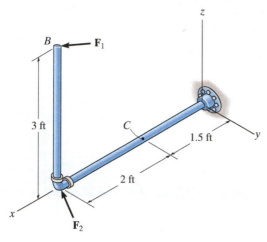

Prob. 7–41

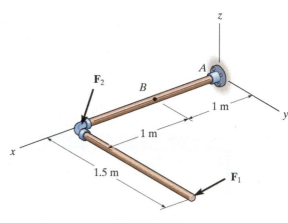

Prob. 7–43

7–42. Determine the x, y, z components of force and moment at point C in the pipe assembly. Neglect the weight of the pipe. The load acting at (0, 3.5 ft, 3 ft) is $F_1 = \{-24i - 10k\}$ lb and $M = \{-30k\}$ lb · ft and at point (0, 3.5 ft, 0) $F_2 = \{-80i\}$ lb.

*7–44. Determine the x, y, z components of internal loading at a section passing through point B in the pipe assembly. Neglect the weight of the pipe. Take $F_1 = \{100i - 200j - 300k\}$ N and $F_2 = \{100i + 500j\}$ N.

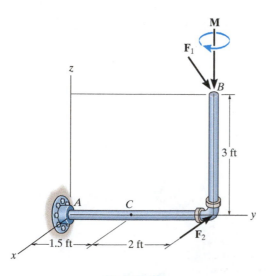

Prob. 7–42

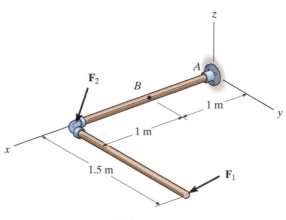

Prob. 7–44

*7.2   Shear and Moment Equations and Diagrams

Beams are structural members designed to support loadings applied perpendicular to their axes. In general, they are long and straight and have a constant cross-sectional area. They are often classified as to how they are supported. For example, a **simply supported beam** is pinned at one end and roller supported at the other, as in Fig. 7–9a, whereas a **cantilevered beam** is fixed at one end and free at the other. The actual design of a beam requires a detailed knowledge of the *variation* of the internal shear force V and bending moment M acting at *each point* along the axis of the beam.*

These *variations* of V and M along the beam's axis can be obtained by using the method of sections discussed in Sec. 7.1. In this case, however, it is necessary to section the beam at an arbitrary distance x from one end and then apply the equations of equilibrium to the segment having the length x. Doing this we can then obtain V and M as functions of x.

In general, the internal shear and bending-moment functions will be discontinuous, or their slopes will be discontinuous, at points where a distributed load changes or where concentrated forces or couple moments are applied. Because of this, these functions must be determined for *each segment* of the beam located between any two discontinuities of loading. For example, segments having lengths x_1, x_2, and x_3 will have to be used to describe the variation of V and M along the length of the beam in Fig. 7–9a. These functions will be valid *only* within regions from 0 to a for x_1, from a to b for x_2, and from b to L for x_3. If the resulting functions of x are plotted, the graphs are termed the **shear diagram** and **bending-moment diagram**, Fig. 7–9b and Fig. 7–9c, respectively.

To save on material and thereby produce an efficient design, these beams, also called girders, have been tapered, since the internal moment in the beam will be larger at the supports, or piers, than at the center of the span. (© Russell C. Hibbeler)

7

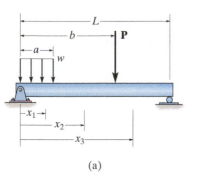

(a)

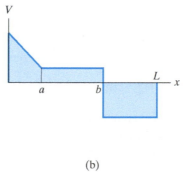

(b)

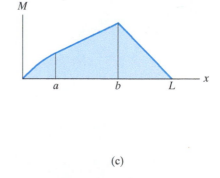

(c)

Fig. 7–9

---

*The internal normal force is not considered for two reasons. In most cases, the loads applied to a beam act perpendicular to the beam's axis and hence produce only an internal shear force and bending moment. And for design purposes, the beam's resistance to shear, and particularly to bending, is more important than its ability to resist a normal force.

Positive shear

Positive moment

Beam sign convention

Fig. 7–10

The shelving arms must be designed to resist the internal loading in the arms caused by the lumber. (© Russell C. Hibbeler)

Important Points

- Shear and moment diagrams for a beam provide graphical descriptions of how the internal shear and moment vary throughout the beam's length.

- To obtain these diagrams, the method of sections is used to determine V and M as functions of x. These results are then plotted. If the load on the beam suddenly changes, then regions between each load must be selected to obtain each function of x.

Procedure for Analysis

The shear and bending-moment diagrams for a beam can be constructed using the following procedure.

Support Reactions.

- Determine all the reactive forces and couple moments acting on the beam and resolve all the forces into components acting perpendicular and parallel to the beam's axis.

Shear and Moment Functions.

- Specify separate coordinates x having an origin at the beam's left end and extending to regions of the beam *between* concentrated forces and/or couple moments, or where the distributed loading is continuous.

- Section the beam at each distance x and draw the free-body diagram of one of the segments. Be sure **V** and **M** are shown acting in their *positive sense*, in accordance with the sign convention given in Fig. 7–10.

- The shear V is obtained by summing forces perpendicular to the beam's axis, and the moment M is obtained by summing moments about the sectioned end of the segment.

Shear and Moment Diagrams.

- Plot the shear diagram (V versus x) and the moment diagram (M versus x). If computed values of the functions describing V and M are *positive*, the values are plotted above the x axis, whereas *negative* values are plotted below the x axis.

EXAMPLE 7.6

Draw the shear and moment diagrams for the shaft shown in Fig. 7–11a. The support at A is a thrust bearing and the support at C is a journal bearing.

SOLUTION

Support Reactions.   The support reactions are shown on the shaft's free-body diagram, Fig. 7–11d.

Shear and Moment Functions.   The shaft is sectioned at an arbitrary distance x from point A, extending within the region AB, and the free-body diagram of the left segment is shown in Fig. 7–11b. The unknowns $\mathbf{V}$ and $\mathbf{M}$ are assumed to act in the *positive sense* on the right-hand face of the segment according to the established sign convention. Applying the equilibrium equations yields

$$+\uparrow \Sigma F_y = 0; \qquad\qquad V = 2.5 \text{ kN} \qquad\qquad (1)$$

$$\zeta + \Sigma M = 0; \qquad\qquad M = 2.5x \text{ kN} \cdot \text{m} \qquad\qquad (2)$$

A free-body diagram for a left segment of the shaft extending from A a distance x, within the region BC is shown in Fig. 7–11c. As always, $\mathbf{V}$ and $\mathbf{M}$ are shown acting in the positive sense. Hence,

$$+\uparrow \Sigma F_y = 0; \qquad 2.5 \text{ kN} - 5 \text{ kN} - V = 0$$

$$V = -2.5 \text{ kN} \qquad\qquad (3)$$

$$\zeta + \Sigma M = 0; \qquad M + 5 \text{ kN}(x - 2 \text{ m}) - 2.5 \text{ kN}(x) = 0$$

$$M = (10 - 2.5x) \text{ kN} \cdot \text{m} \qquad\qquad (4)$$

Shear and Moment Diagrams.   When Eqs. 1 through 4 are plotted within the regions in which they are valid, the shear and moment diagrams shown in Fig. 7–11d are obtained. The shear diagram indicates that the internal shear force is always 2.5 kN (positive) within segment AB. Just to the right of point B, the shear force changes sign and remains at a constant value of -2.5 kN for segment BC. The moment diagram starts at zero, increases linearly to point B at $x = 2$ m, where $M_{max} = 2.5$ kN(2 m) $= 5$ kN $\cdot$ m, and thereafter decreases back to zero.

NOTE: It is seen in Fig. 7–11d that the graphs of the shear and moment diagrams "jump" or changes abruptly where the concentrated force acts, i.e., at points A, B, and C. For this reason, as stated earlier, it is necessary to express both the shear and moment functions separately for regions between concentrated loads. It should be realized, however, that all loading discontinuities are mathematical, arising from the *idealization of a concentrated force and couple moment*. Physically, loads are always applied over a finite area, and if the actual load variation could be accounted for, the shear and moment diagrams would then be continuous over the shaft's entire length.

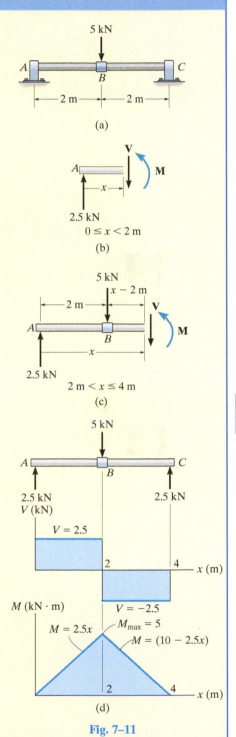

(a)

(b)

(c)

(d)

Fig. 7–11

EXAMPLE | 7.7

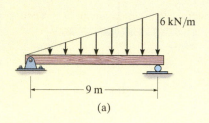

(a)

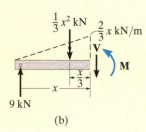

(b)

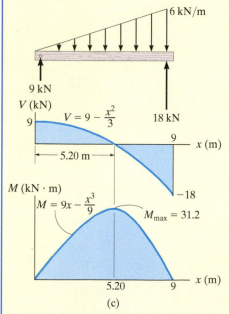

(c)

Fig. 7–12

Draw the shear and moment diagrams for the beam shown in Fig. 7–12a.

SOLUTION

Support Reactions.   The support reactions are shown on the beam's free-body diagram, Fig. 7–12c.

Shear and Moment Functions.   A free-body diagram for a left segment of the beam having a length x is shown in Fig. 7–12b. Due to proportional triangles, the distributed loading acting at the end of this segment has an intensity of $w/x = 6/9$ or $w = (2/3)x$. It is replaced by a resultant force *after* the segment is isolated as a free-body diagram. The *magnitude* of the resultant force is equal to $\frac{1}{2}(x)\left(\frac{2}{3}x\right) = \frac{1}{3}x^2$. This force *acts through the centroid* of the distributed loading area, a distance $\frac{1}{3}x$ from the right end. Applying the two equations of equilibrium yields

$$+\uparrow\Sigma F_y = 0; \qquad\qquad 9 - \frac{1}{3}x^2 - V = 0$$

$$V = \left(9 - \frac{x^2}{3}\right) \text{kN} \qquad (1)$$

$$\zeta + \Sigma M = 0; \qquad M + \frac{1}{3}x^2\left(\frac{x}{3}\right) - 9x = 0$$

$$M = \left(9x - \frac{x^3}{9}\right) \text{kN} \cdot \text{m} \qquad (2)$$

Shear and Moment Diagrams.   The shear and moment diagrams shown in Fig. 7–12c are obtained by plotting Eqs. 1 and 2.
     The point of *zero shear* can be found using Eq. 1:

$$V = 9 - \frac{x^2}{3} = 0$$

$$x = 5.20 \text{ m}$$

NOTE: It will be shown in Sec. 7.3 that this value of x happens to represent the point on the beam where the *maximum moment* occurs. Using Eq. 2, we have

$$M_{\text{max}} = \left(9(5.20) - \frac{(5.20)^3}{9}\right) \text{kN} \cdot \text{m}$$

$$= 31.2 \text{ kN} \cdot \text{m}$$

F7–7. Determine the shear and moment as a function of x, and then draw the shear and moment diagrams.

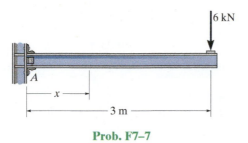

6 kN

A

x

3 m

Prob. F7–7

F7–10. Determine the shear and moment as a function of x, and then draw the shear and moment diagrams.

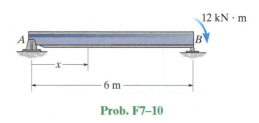

12 kN · m

A B

x

6 m

Prob. F7–10

F7–8. Determine the shear and moment as a function of x, and then draw the shear and moment diagrams.

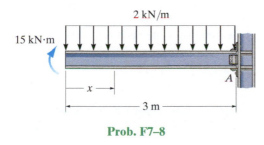

2 kN/m

15 kN·m

A

x

3 m

Prob. F7–8

F7–11. Determine the shear and moment as a function of x, where $0 \le x < 3$ m and 3 m $< x \le 6$ m, and then draw the shear and moment diagrams.

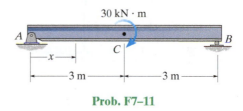

30 kN · m

A B

C

x

3 m   3 m

Prob. F7–11

F7–9. Determine the shear and moment as a function of x, and then draw the shear and moment diagrams.

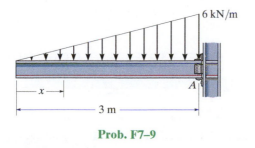

6 kN/m

A

x

3 m

Prob. F7–9

F7–12. Determine the shear and moment as a function of x, where $0 \le x < 3$ m and 3 m $< x \le 6$ m, and then draw the shear and moment diagrams.

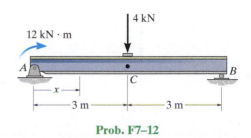

4 kN

12 kN · m

A B

C

x

3 m   3 m

Prob. F7–12

7

PROBLEMS

7–45. Draw the shear and moment diagrams for the shaft (a) in terms of the parameters shown; (b) set $P = 9$ kN, $a = 2$ m, $L = 6$ m. There is a thrust bearing at A and a journal bearing at B.

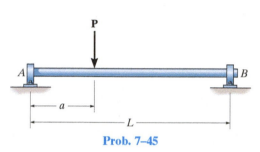

Prob. 7–45

7–46. Draw the shear and moment diagrams for the beam (a) in terms of the parameters shown; (b) set $P = 800$ lb, $a = 5$ ft, $L = 12$ ft.

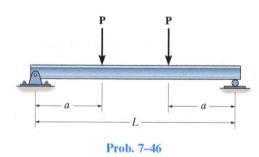

Prob. 7–46

7–47. Draw the shear and moment diagrams for the beam (a) in terms of the parameters shown; (b) set $P = 600$ lb, $a = 5$ ft, $b = 7$ ft.

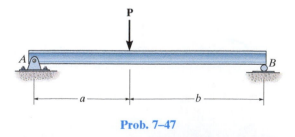

Prob. 7–47

*7–48. Draw the shear and moment diagrams for the cantilevered beam.

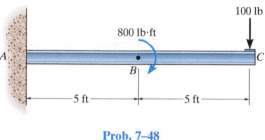

Prob. 7–48

7–49. Draw the shear and moment diagrams of the beam (a) in terms of the parameters shown; (b) set $M_0 = 500$ N · m, $L = 8$ m.

7–50. If $L = 9$ m, the beam will fail when the maximum shear force is $V_{max} = 5$ kN or the maximum bending moment is $M_{max} = 2$ kN · m. Determine the magnitude M_0 of the largest couple moments it will support.

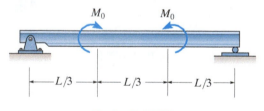

Probs. 7–49/50

7–51. Draw the shear and moment diagrams for the beam.

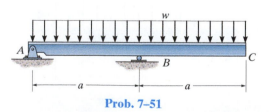

Prob. 7–51

*7–52.   Draw the shear and moment diagrams for the beam.

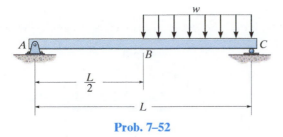

Prob. 7–52

7–53.   Draw the shear and bending-moment diagrams for the beam.

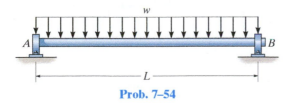

Prob. 7–53

7–54.   The shaft is supported by a smooth thrust bearing at *A* and a smooth journal bearing at *B*. Draw the shear and moment diagrams for the shaft (a) in terms of the parameters shown; (b) set w = 500 lb/ft, L = 10 ft.

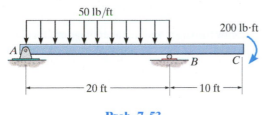

Prob. 7–54

7–55.   Draw the shear and moment diagrams for the beam.

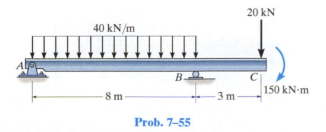

Prob. 7–55

*7–56.   Draw the shear and moment diagrams for the beam.

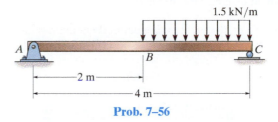

Prob. 7–56

7–57.   Draw the shear and moment diagrams for the compound beam. The beam is pin connected at *E* and *F*.

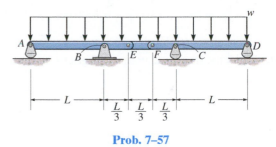

Prob. 7–57

7–58.   Draw the shear and bending-moment diagrams for each of the two segments of the compound beam.

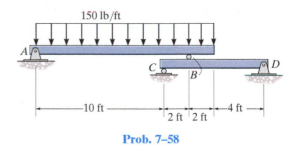

Prob. 7–58

7–59.   Draw the shear and moment diagrams for the beam.

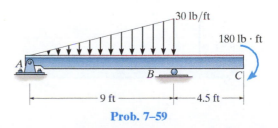

Prob. 7–59

*7–60. The shaft is supported by a smooth thrust bearing at *A* and a smooth journal bearing at *B*. Draw the shear and moment diagrams for the shaft.

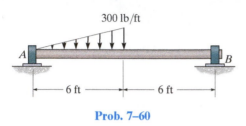

Prob. 7–60

7–61. Draw the shear and moment diagrams for the beam.

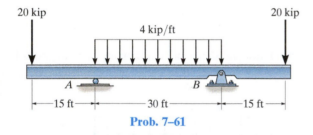

Prob. 7–61

7–62. The beam will fail when the maximum internal moment is M_{max}. Determine the position x of the concentrated force **P** and its smallest magnitude that will cause failure.

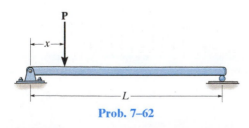

Prob. 7–62

7–63. Draw the shear and moment diagrams for the beam.

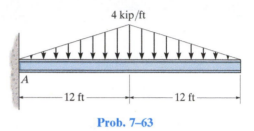

Prob. 7–63

*7–64. Draw the shear and moment diagrams for the beam.

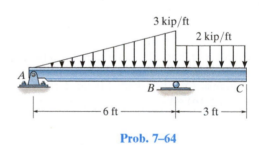

Prob. 7–64

7–65. Draw the shear and moment diagrams for the beam.

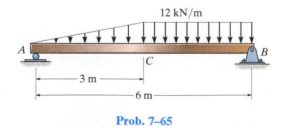

Prob. 7–65

7

7–66. Draw the shear and moment diagrams for the beam.

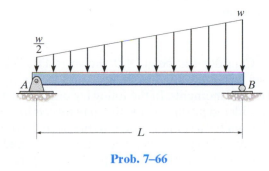

Prob. 7–66

7–67. Determine the internal normal force, shear force, and moment in the curved rod as a function of θ. The force **P** acts at the constant angle ϕ.

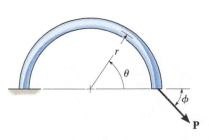

Prob. 7–67

*7–68. The quarter circular rod lies in the horizontal plane and supports a vertical force **P** at its end. Determine the magnitudes of the components of the internal shear force, moment, and torque acting in the rod as a function of the angle θ.

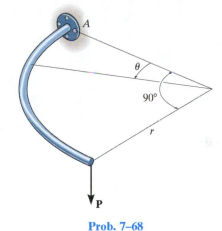

Prob. 7–68

7–69. Express the internal shear and moment components acting in the rod as a function of y, where $0 \le y \le 4$ ft.

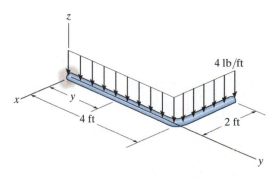

Prob. 7–69

In order to design the beam used to support these power lines, it is important to first draw the shear and moment diagrams for the beam. (© Russell C. Hibbeler)

*7.3   Relations between Distributed Load, Shear, and Moment

If a beam is subjected to several concentrated forces, couple moments, and distributed loads, the method of constructing the shear and bending-moment diagrams discussed in Sec. 7.2 may become quite tedious. In this section a simpler method for constructing these diagrams is discussed—a method based on differential relations that exist between the load, shear, and bending moment.

Distributed Load.

Consider the beam AD shown in Fig. 7–13a, which is subjected to an arbitrary load $w = w(x)$ and a series of concentrated forces and couple moments. In the following discussion, the *distributed load* will be considered *positive* when the *loading acts upward* as shown. A free-body diagram for a small segment of the beam having a length Δx is chosen at a point x along the beam which is *not* subjected to a concentrated force or couple moment, Fig. 7–13b. Hence any results obtained will not apply at these points of concentrated loading. The internal shear force and bending moment shown on the free-body diagram are assumed to act in the *positive sense* according to the established sign convention. Note that both the shear force and moment acting on the right-hand face must be increased by a small, finite amount in order to keep the segment in equilibrium. The distributed loading has been replaced by a resultant force $\Delta F = w(x)\,\Delta x$ that acts at a fractional distance $k(\Delta x)$ from the right end, where $0 < k < 1$ [for example, if $w(x)$ is *uniform*, $k = \frac{1}{2}$].

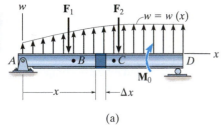

(a)

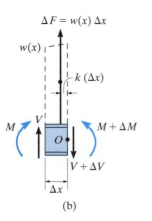

(b)

Fig. 7–13

Relation between the Distributed Load and Shear.

If we apply the force equation of equilibrium to the segment, then

$$+\uparrow \Sigma F_y = 0; \qquad V + w(x)\Delta x - (V + \Delta V) = 0$$
$$\Delta V = w(x)\Delta x$$

Dividing by Δx, and letting $\Delta x \rightarrow 0$, we get

$$\frac{dV}{dx} = w(x)$$

$$\frac{\text{Slope of}}{\text{shear diagram}} = \frac{\text{Distributed load}}{\text{intensity}}$$

(7–1)

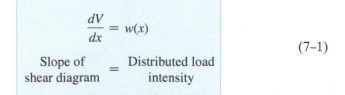

If we rewrite the above equation in the form $dV = w(x)dx$ and perform an integration between any two points B and C on the beam, we see that

$$\Delta V = \int w(x)\, dx$$

$$\frac{\text{Change in}}{\text{shear}} = \frac{\text{Area under}}{\text{loading curve}}$$

(7–2)

Relation between the Shear and Moment.

If we apply the moment equation of equilibrium about point O on the free-body diagram in Fig. 7–13b, we get

$$\zeta + \Sigma M_O = 0; \quad (M + \Delta M) - [w(x)\Delta x]\, k\Delta x - V\Delta x - M = 0$$
$$\Delta M = V\Delta x + k\, w(x)\Delta x^2$$

Dividing both sides of this equation by Δx, and letting $\Delta x \to 0$, yields

$$\frac{dM}{dx} = V$$

$$\frac{\text{Slope of}}{\text{moment diagram}} = \text{Shear}$$

(7–3)

In particular, notice that a maximum bending moment $|M|_{max}$ will occur at the point where the slope $dM/dx = 0$, since this is where the shear is equal to zero.

If Eq. 7–3 is rewritten in the form $dM = \int V\, dx$ and integrated between any two points B and C on the beam, we have

$$\Delta M = \int V\, dx$$

$$\frac{\text{Change in}}{\text{moment}} = \frac{\text{Area under}}{\text{shear diagram}}$$

(7–4)

As stated previously, the above equations do not apply at points where a *concentrated* force or couple moment acts. These two special cases create *discontinuities* in the shear and moment diagrams, and as a result, each deserves separate treatment.

Force.

A free-body diagram of a small segment of the beam in Fig. 7–13a, taken from under one of the forces, is shown in Fig. 7–14a. Here force equilibrium requires

$$+\uparrow \Sigma F_y = 0; \qquad \Delta V = F \qquad (7\text{–}5)$$

Since the *change in shear is positive*, the shear diagram will "jump" *upward when* **F** *acts upward* on the beam. Likewise, the jump in shear (ΔV) is downward when **F** acts downward.

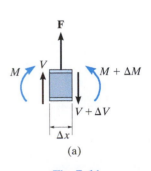

(a)

Fig. 7–14

$$M_0$$

$$M \quad V \quad M + \Delta M$$

$$V + \Delta V$$

$$\Delta x$$

(b)

Fig. 7–14 (cont.)

Couple Moment.

If we remove a segment of the beam in Fig. 7–13a that is located at the couple moment M_0, the free-body diagram shown in Fig. 7–14b results. In this case letting $\Delta x \rightarrow 0$, moment equilibrium requires

$$\zeta + \Sigma M = 0; \qquad\qquad \Delta M = M_0 \qquad\qquad (7\text{–}6)$$

Thus, the *change in moment is positive*, or the moment diagram will "jump" *upward if* M_0 *is clockwise.* Likewise, the jump ΔM is downward when M_0 is counterclockwise.

The examples which follow illustrate application of the above equations when used to construct the shear and moment diagrams. After working through these examples, it is recommended that you also go back and solve Examples 7.6 and 7.7 using this method.

This concrete beam is used to support the deck. Its size and the placement of steel reinforcement within it can be determined once the shear and moment diagrams have been established. (© Russell C. Hibbeler)

Important Points

- The slope of the shear diagram at a point is equal to the intensity of the distributed loading, where positive distributed loading is upward, i.e., $dV/dx = w(x)$.

- The change in the shear ΔV between two points is equal to *the area* under the distributed-loading curve between the points.

- If a concentrated force acts upward on the beam, the shear will jump upward by the same amount.

- The slope of the moment diagram at a point is equal to the shear, i.e., $dM/dx = V$.

- The change in the moment ΔM between two points is equal to the *area* under the shear diagram between the two points.

- If a *clock*wise couple moment acts on the beam, the shear will not be affected; however, the moment diagram will jump *upward* by the amount of the moment.

- Points of *zero shear* represent points of *maximum or minimum moment* since $dM/dx = 0$.

- Because two integrations of $w = w(x)$ are involved to first determine the change in shear, $\Delta V = \int w(x)\, dx$, then to determine the change in moment, $\Delta M = \int V\, dx$, then if the loading curve $w = w(x)$ is a polynomial of degree n, $V = V(x)$ will be a curve of degree $n + 1$, and $M = M(x)$ will be a curve of degree $n + 2$.

EXAMPLE 7.8

Draw the shear and moment diagrams for the cantilever beam in Fig. 7–15a.

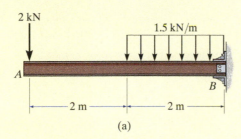

Fig. 7–15

SOLUTION

The support reactions at the fixed support B are shown in Fig. 7–15b.

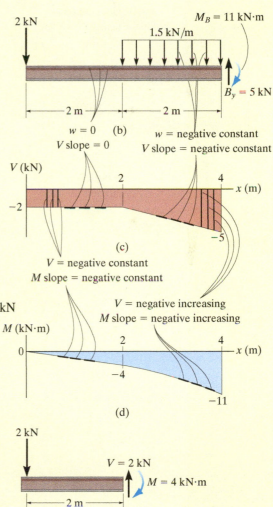

Shear Diagram.   The shear at end A is -2 kN. This value is plotted at $x = 0$, Fig. 7–15c. Notice how the shear diagram is constructed by following the slopes defined by the loading w. The shear at $x = 4$ m is -5 kN, the reaction on the beam. This value can be verified by finding the area under the distributed loading; i.e.,

$$V|_{x=4\,m} = V|_{x=2\,m} + \Delta V = -2\text{ kN} - (1.5\text{ kN/m})(2\text{ m}) = -5\text{ kN}$$

Moment Diagram.   The moment of zero at $x=0$ is plotted in Fig. 7–15d. Construction of the moment diagram is based on knowing that its slope is equal to the shear at each point. The change of moment from $x = 0$ to $x = 2$ m is determined from the area under the shear diagram. Hence, the moment at $x = 2$ m is

$$M|_{x=2\,m} = M|_{x=0} + \Delta M = 0 + [-2\text{ kN}(2\text{ m})] = -4\text{ kN} \cdot \text{m}$$

This same value can be determined from the method of sections, Fig. 7–15e.

EXAMPLE 7.9

Draw the shear and moment diagrams for the overhang beam in Fig. 7–16a.

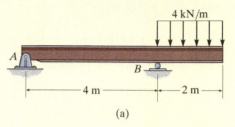

4 kN/m

A

B

4 m   2 m

(a)

Fig. 7–16

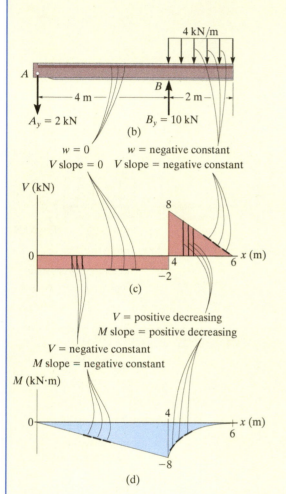

4 kN/m

A

B

4 m   2 m

$A_y = 2$ kN $B_y = 10$ kN

(b)

w = 0   w = negative constant
V slope = 0   V slope = negative constant

V (kN)

8

0   4   6   x (m)

−2

(c)

V = positive decreasing
M slope = positive decreasing

V = negative constant
M slope = negative constant

M (kN·m)

4

0   6   x (m)

−8

(d)

SOLUTION

The support reactions are shown in Fig. 7–16b.

Shear Diagram.   The shear of −2 kN at end A of the beam is plotted at x = 0, Fig. 7–16c. The slopes are determined from the loading and from this the shear diagram is constructed, as indicated in the figure. In particular, notice the positive jump of 10 kN at x = 4 m due to the force B_y, as indicated in the figure.

Moment Diagram.   The moment of zero at x = 0 is plotted, Fig. 7–16d, then following the behavior of the slope found from the shear diagram, the moment diagram is constructed. The moment at x = 4 m is found from the area under the shear diagram.

$$M|_{x=4\,m} = M|_{x=0} + \Delta M = 0 + [-2\text{ kN}(4\text{ m})] = -8\text{ kN}\cdot\text{m}$$

We can also obtain this value by using the method of sections, as shown in Fig. 7–16e.

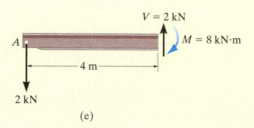

V = 2 kN

A   M = 8 kN·m

4 m

2 kN

(e)

EXAMPLE 7.10

The shaft in Fig. 7–17a is supported by a thrust bearing at A and a journal bearing at B. Draw the shear and moment diagrams.

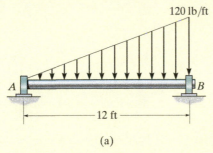

120 lb/ft

A   B

|— 12 ft —|

(a)

Fig. 7–17

SOLUTION

The support reactions are shown in Fig. 7–17b.

Shear Diagram. As shown in Fig. 7–17c, the shear at $x=0$ is +240. Following the slope defined by the loading, the shear diagram is constructed, where at B its value is −480 lb. Since the shear changes sign, the point where $V=0$ must be located. To do this we will use the method of sections. The free-body diagram of the left segment of the shaft, sectioned at an arbitrary position x within the region $0 \le x < 12$ ft, is shown in Fig. 7–17e. Notice that the intensity of the distributed load at x is $w=10x$, which has been found by proportional triangles, i.e., $120/12 = w/x$.

Thus, for $V=0$,

$$+\uparrow \Sigma F_y = 0; \qquad 240\text{ lb} - \tfrac{1}{2}(10x)x = 0$$

$$x = 6.93 \text{ ft}$$

Moment Diagram. The moment diagram starts at 0 since there is no moment at A, then it is constructed based on the slope as determined from the shear diagram. The maximum moment occurs at $x = 6.93$ ft, where the shear is equal to zero, since $dM/dx = V = 0$, Fig. 7–17e,

$$\zeta + \Sigma M = 0;$$
$$M_{\text{max}} + \tfrac{1}{2}[(10)(6.93)]\,6.93\left(\tfrac{1}{3}(6.93)\right) - 240(6.93) = 0$$

$$M_{\text{max}} = 1109 \text{ lb} \cdot \text{ft}$$

Finally, notice how integration, first of the loading w which is linear, produces a shear diagram which is parabolic, and then a moment diagram which is cubic.

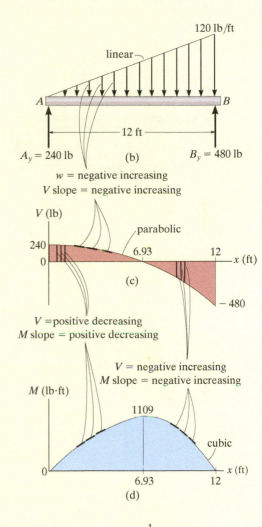

120 lb/ft

linear

A   B

|— 12 ft —|

$A_y = 240$ lb   (b) $B_y = 480$ lb

w = negative increasing
V slope = negative increasing

V (lb)

parabolic

240

0   6.93   12 x (ft)

(c)

− 480

V = positive decreasing
M slope = positive decreasing

V = negative increasing
M slope = negative increasing

M (lb·ft)

1109

cubic

0   6.93   12 x (ft)

(d)

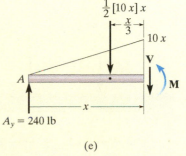

$\tfrac{1}{2}[10\,x]\,x$

$\tfrac{x}{3}$

10 x

A   V

x M

$A_y = 240$ lb

(e)

7

FUNDAMENTAL PROBLEMS

F7–13. Draw the shear and moment diagrams for the beam.

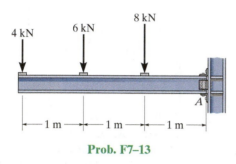

Prob. F7–13

F7–14. Draw the shear and moment diagrams for the beam.

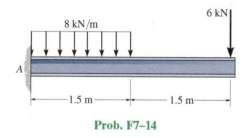

Prob. F7–14

F7–15. Draw the shear and moment diagrams for the beam.

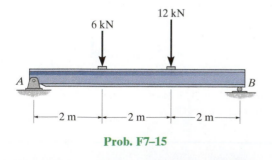

Prob. F7–15

F7–16. Draw the shear and moment diagrams for the beam.

Prob. F7–16

F7–17. Draw the shear and moment diagrams for the beam.

Prob. F7–17

F7–18. Draw the shear and moment diagrams for the beam.

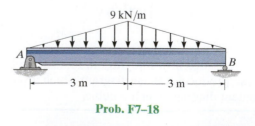

Prob. F7–18

PROBLEMS

7–70. Draw the shear and moment diagrams for the beam.

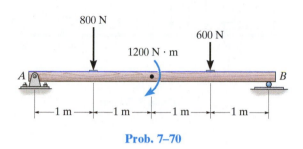

Prob. 7–70

7–71. Draw the shear and moment diagrams for the beam.

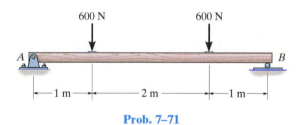

Prob. 7–71

*7–72. Draw the shear and moment diagrams for the beam. The support at *A* offers no resistance to vertical load.

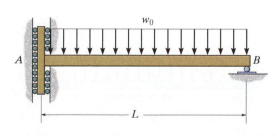

Prob. 7–72

7–73. Draw the shear and moment diagrams for the simply-supported beam.

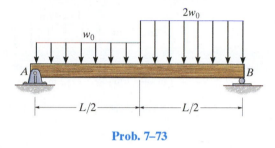

Prob. 7–73

7–74. Draw the shear and moment diagrams for the beam. The supports at *A* and *B* are a thrust bearing and journal bearing, respectively.

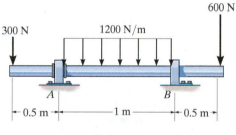

Prob. 7–74

7–75. Draw the shear and moment diagrams for the beam.

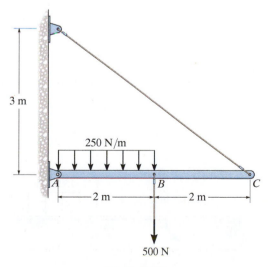

Prob. 7–75

*7–76. Draw the shear and moment diagrams for the beam.

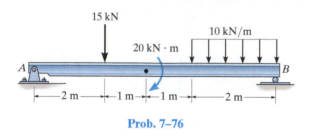

15 kN

10 kN/m

20 kN · m

A   B

2 m — 1 m — 1 m — 2 m

Prob. 7–76

7–79. Draw the shear and moment diagrams for the shaft. The support at *A* is a journal bearing and at *B* it is a thrust bearing.

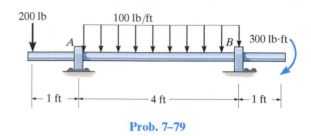

200 lb

100 lb/ft

A   B   300 lb·ft

1 ft — 4 ft — 1 ft

Prob. 7–79

7–77. Draw the shear and moment diagrams for the beam.

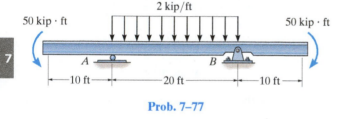

2 kip/ft

50 kip · ft   50 kip · ft

A   B

10 ft — 20 ft — 10 ft

Prob. 7–77

*7–80. Draw the shear and moment diagrams for the beam.

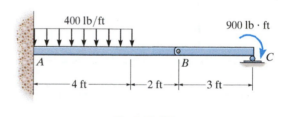

400 lb/ft

900 lb · ft

A   B   C

4 ft — 2 ft — 3 ft

Prob. 7–80

7–78. Draw the shear and moment diagrams for the beam.

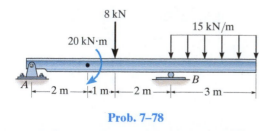

8 kN

15 kN/m

20 kN·m

A   B

2 m — 1 m — 2 m — 3 m

Prob. 7–78

7–81. The beam consists of three segments pin connected at *B* and *E*. Draw the shear and moment diagrams for the beam.

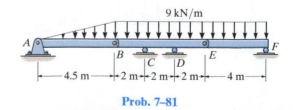

9 kN/m

A

B   C   D   E   F

4.5 m — 2 m — 2 m — 2 m — 4 m

Prob. 7–81

7–82.   Draw the shear and moment diagrams for the beam. The supports at *A* and *B* are a thrust and journal bearing, respectively.

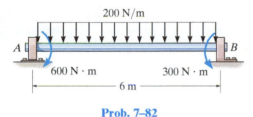

Prob. 7–82

7–83.   Draw the shear and moment diagrams for the beam.

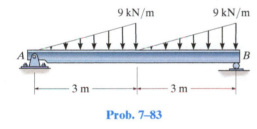

Prob. 7–83

*7–84.   Draw the shear and moment diagrams for the beam.

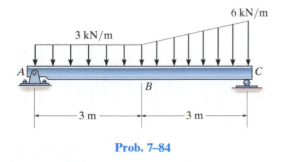

Prob. 7–84

7–85.   Draw the shear and moment diagrams for the beam.

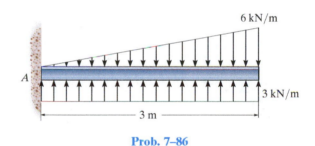

Prob. 7–85

7–86.   Draw the shear and moment diagrams for the beam.

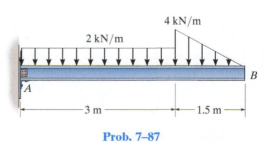

Prob. 7–86

7–87.   Draw the shear and moment diagrams for the beam.

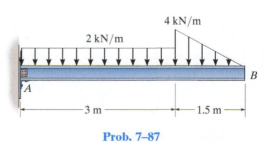

Prob. 7–87

7

*7–88.   Draw the shear and moment diagrams for the beam.

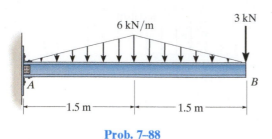

Prob. 7–88

7–89.   Draw the shear and moment diagrams for the beam.

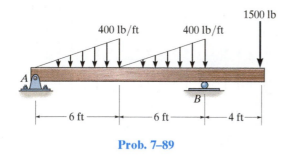

Prob. 7–89

7–90.   Draw the shear and moment diagrams for the beam.

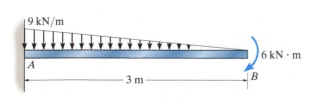

Prob. 7–90

7–91.   Draw the shear and moment diagrams for the beam.

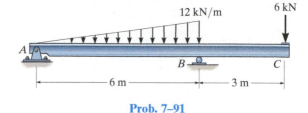

Prob. 7–91

*7–92.   Draw the shear and moment diagrams for the beam.

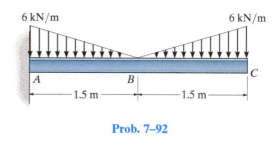

Prob. 7–92

7–93.   Draw the shear and moment diagrams for the beam.

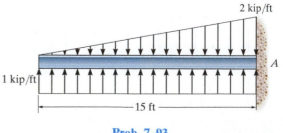

Prob. 7–93

*7.4   Cables

Flexible cables and chains combine strength with lightness and often are used in structures for support and to transmit loads from one member to another. When used to support suspension bridges and trolley wheels, cables form the main load-carrying element of the structure. In the force analysis of such systems, the weight of the cable itself may be neglected because it is often small compared to the load it carries. On the other hand, when cables are used as transmission lines and guys for radio antennas and derricks, the cable weight may become important and must be included in the structural analysis.

Three cases will be considered in the analysis that follows. In each case we will make the assumption that the cable is *perfectly flexible* and *inextensible*. Due to its flexibility, the cable offers no resistance to bending, and therefore, the tensile force acting in the cable is always tangent to the cable at points along its length. Being inextensible, the cable has a constant length both before and after the load is applied. As a result, once the load is applied, the geometry of the cable remains unchanged, and the cable or a segment of it can be treated as a rigid body.

Each of the cable segments remains approximately straight as they support the weight of these traffic lights. (© Russell C. Hibbeler)

Cable Subjected to Concentrated Loads.

When a cable of negligible weight supports several concentrated loads, the cable takes the form of several straight-line segments, each of which is subjected to a constant tensile force. Consider, for example, the cable shown in Fig. 7–18, where the distances h, L_1, L_2, and L_3 and the loads $\mathbf{P}_1$ and $\mathbf{P}_2$ are known. The problem here is to determine the *nine unknowns* consisting of the tension in each of the *three* segments, the *four* components of reaction at A and B, and the *two* sags y_C and y_D at points C and D. For the solution we can write *two* equations of force equilibrium at each of points A, B, C, and D. This results in a total of *eight equations.** To complete the solution, we need to know something about the geometry of the cable in order to obtain the necessary ninth equation. For example, if the cable's total *length L* is specified, then the Pythagorean theorem can be used to relate each of the three segmental lengths, written in terms of h, y_C, y_D, L_1, L_2, and L_3, to the total length L. Unfortunately, this type of problem cannot be solved easily by hand. Another possibility, however, is to specify one of the sags, either y_C or y_D, instead of the cable length. By doing this, the equilibrium equations are then sufficient for obtaining the unknown forces and the remaining sag. Once the sag at each point of loading is obtained, the length of the cable can then be determined by trigonometry. The following example illustrates a procedure for performing the equilibrium analysis for a problem of this type.

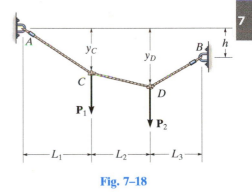

Fig. 7–18

*As will be shown in the following example, the eight equilibrium equations *also* can be written for the entire cable, or any part thereof. But *no more* than *eight* independent equations are available.

EXAMPLE 7.11

Determine the tension in each segment of the cable shown in Fig. 7–19a.

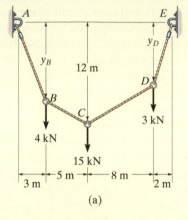

(a)

SOLUTION

By inspection, there are four unknown external reactions (A_x, A_y, E_x, and E_y) and four unknown cable tensions, one in each cable segment. These eight unknowns along with the two unknown sags y_B and y_D can be determined from *ten* available equilibrium equations. One method is to apply the force equations of equilibrium ($\Sigma F_x = 0$, $\Sigma F_y = 0$) to each of the five points A through E. Here, however, we will take a more direct approach.

Consider the free-body diagram for the entire cable, Fig. 7–19b. Thus,

$$\xrightarrow{+} \Sigma F_x = 0; \qquad\qquad -A_x + E_x = 0$$

$$\zeta + \Sigma M_E = 0;$$

$$-A_y(18\text{ m}) + 4\text{ kN }(15\text{ m}) + 15\text{ kN }(10\text{ m}) + 3\text{ kN }(2\text{ m}) = 0$$

$$A_y = 12\text{ kN}$$

$$+\uparrow \Sigma F_y = 0; \qquad 12\text{ kN} - 4\text{ kN} - 15\text{ kN} - 3\text{ kN} + E_y = 0$$

$$E_y = 10\text{ kN}$$

Since the sag $y_C = 12$ m is known, we will now consider the leftmost section, which cuts cable BC, Fig. 7–19c.

$$\zeta + \Sigma M_C = 0; \; A_x(12\text{ m}) - 12\text{ kN }(8\text{ m}) + 4\text{ kN }(5\text{ m}) = 0$$

$$A_x = E_x = 6.33\text{ kN}$$

$$\xrightarrow{+} \Sigma F_x = 0; \qquad T_{BC}\cos\theta_{BC} - 6.33\text{ kN} = 0$$

$$+\uparrow \Sigma F_y = 0; \quad 12\text{ kN} - 4\text{ kN} - T_{BC}\sin\theta_{BC} = 0$$

Thus,

$$\theta_{BC} = 51.6°$$

$$T_{BC} = 10.2\text{ kN} \qquad\qquad \textit{Ans.}$$

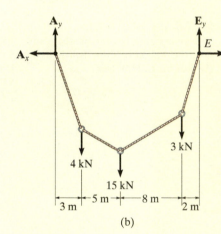

(b)

(c)

Fig. 7–19

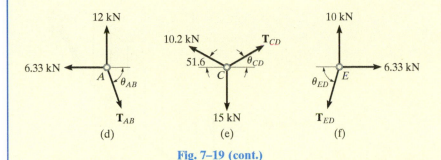

Fig. 7–19 (cont.)

   Proceeding now to analyze the equilibrium of points A, C, and E in sequence, we have

Point A.   (Fig. 7–19d).

$$\xrightarrow{+}\Sigma F_x = 0;\qquad T_{AB}\cos\theta_{AB} - 6.33\text{ kN} = 0$$

$$+\uparrow\Sigma F_y = 0;\qquad -T_{AB}\sin\theta_{AB} + 12\text{ kN} = 0$$

$$\theta_{AB} = 62.2°$$

$$T_{AB} = 13.6\text{ kN}\qquad\qquad Ans.$$

Point C.   (Fig. 7–19e).

$$\xrightarrow{+}\Sigma F_x = 0;\qquad T_{CD}\cos\theta_{CD} - 10.2\cos 51.6°\text{ kN} = 0$$

$$+\uparrow\Sigma F_y = 0;\qquad T_{CD}\sin\theta_{CD} + 10.2\sin 51.6°\text{ kN} - 15\text{ kN} = 0$$

$$\theta_{CD} = 47.9°$$

$$T_{CD} = 9.44\text{ kN}\qquad\qquad Ans.$$

Point E.   (Fig. 7–19f).

$$\xrightarrow{+}\Sigma F_x = 0;\qquad 6.33\text{ kN} - T_{ED}\cos\theta_{ED} = 0$$

$$+\uparrow\Sigma F_y = 0;\qquad 10\text{ kN} - T_{ED}\sin\theta_{ED} = 0$$

$$\theta_{ED} = 57.7°$$

$$T_{ED} = 11.8\text{ kN}\qquad\qquad Ans.$$

NOTE: By comparison, the maximum cable tension is in segment AB since this segment has the greatest slope (θ) and it is required that for any cable segment the horizontal component $T\cos\theta = A_x = E_x$ (a constant). Also, since the slope angles that the cable segments make with the horizontal have now been determined, it is possible to determine the sags y_B and y_D, Fig. 7–19a, using trigonometry.

The cable and suspenders are used to support the uniform load of a gas pipe which crosses the river. (© Russell C. Hibbeler)

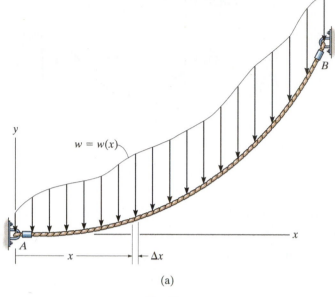

(a)

Fig. 7–20

Cable Subjected to a Distributed Load.
Let us now consider the weightless cable shown in Fig. 7–20a, which is subjected to a distributed loading $w = w(x)$ that is *measured in the x direction*. The free-body diagram of a small segment of the cable having a length Δs is shown in Fig. 7–20b. Since the tensile force changes in both magnitude and direction along the cable's length, we will denote this change on the free-body diagram by ΔT. Finally, the distributed load is represented by its resultant force $w(x)(\Delta x)$, which acts at a fractional distance $k(\Delta x)$ from point O, where $0 < k < 1$. Applying the equations of equilibrium, we have

$$\overset{+}{\rightarrow} \Sigma F_x = 0; \qquad -T\cos\theta + (T + \Delta T)\cos(\theta + \Delta\theta) = 0$$

$$+\uparrow \Sigma F_y = 0; \qquad -T\sin\theta - w(x)(\Delta x) + (T + \Delta T)\sin(\theta + \Delta\theta) = 0$$

$$\curvearrowleft + \Sigma M_O = 0; \qquad w(x)(\Delta x)k(\Delta x) - T\cos\theta\,\Delta y + T\sin\theta\,\Delta x = 0$$

Dividing each of these equations by Δx and taking the limit as $\Delta x \to 0$, and therefore $\Delta y \to 0$, $\Delta\theta \to 0$, and $\Delta T \to 0$, we obtain

$$\frac{d(T\cos\theta)}{dx} = 0 \qquad\qquad (7\text{–}7)$$

$$\frac{d(T\sin\theta)}{dx} - w(x) = 0 \qquad\qquad (7\text{–}8)$$

$$\frac{dy}{dx} = \tan\theta \qquad\qquad (7\text{–}9)$$

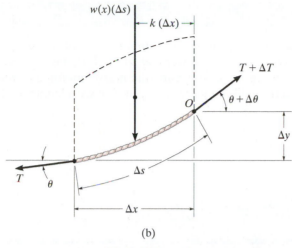

(b)

Fig. 7–20 (cont.)

Integrating Eq. 7–7, we have

$$T \cos \theta = \text{constant} = F_H \qquad (7\text{--}10)$$

where F_H represents the horizontal component of tensile force at *any point* along the cable.

Integrating Eq. 7–8 gives

$$T \sin \theta = \int w(x)\, dx \qquad (7\text{--}11)$$

Dividing Eq. 7–11 by Eq. 7–10 eliminates T. Then, using Eq. 7–9, we can obtain the slope of the cable.

$$\tan \theta = \frac{dy}{dx} = \frac{1}{F_H} \int w(x)\, dx$$

Performing a second integration yields

$$y = \frac{1}{F_H} \int \left(\int w(x)\, dx \right) dx \qquad (7\text{--}12)$$

This equation is used to determine the curve for the cable, $y = f(x)$. The horizontal force component F_H and the additional two constants, say C_1 and C_2, resulting from the integration are determined by applying the boundary conditions for the curve.

The cables of the suspension bridge exert very large forces on the tower and the foundation block which have to be accounted for in their design. (© Russell C. Hibbeler)

EXAMPLE   7.12

The cable of a suspension bridge supports half of the uniform road surface between the two towers at A and B, Fig. 7–21a. If this distributed loading is w_0, determine the maximum force developed in the cable and the cable's required length. The span length L and sag h are known.

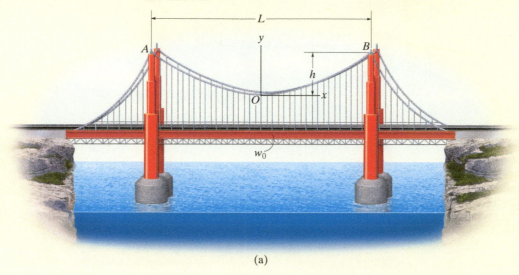

(a)

Fig. 7–21

SOLUTION

We can determine the unknowns in the problem by first finding the equation of the curve that defines the shape of the cable using Eq. 7–12. For reasons of symmetry, the origin of coordinates has been placed at the cable's center. Noting that $w(x) = w_0$, we have

$$y = \frac{1}{F_H} \int \left(\int w_0 \, dx \right) dx$$

Performing the two integrations gives

$$y = \frac{1}{F_H} \left(\frac{w_0 x^2}{2} + C_1 x + C_2 \right) \tag{1}$$

The constants of integration may be determined using the boundary conditions $y = 0$ at $x = 0$ and $dy/dx = 0$ at $x = 0$. Substituting into Eq. 1 and its derivative yields $C_1 = C_2 = 0$. The equation of the curve then becomes

$$y = \frac{w_0}{2F_H} x^2 \tag{2}$$

This is the equation of a *parabola*. The constant F_H may be obtained using the boundary condition $y = h$ at $x = L/2$. Thus,

$$F_H = \frac{w_0 L^2}{8h} \tag{3}$$

Therefore, Eq. 2 becomes

$$y = \frac{4h}{L^2} x^2 \tag{4}$$

Since F_H is known, the tension in the cable may now be determined using Eq. 7–10, written as $T = F_H/\cos\theta$. For $0 \le \theta < \pi/2$, the maximum tension will occur when θ is *maximum*, i.e., at point B, Fig. 7–21a. From Eq. 2, the slope at this point is

$$\left.\frac{dy}{dx}\right|_{x=L/2} = \tan\theta_{max} = \left.\frac{w_0}{F_H}x\right|_{x=L/2}$$

or

$$\theta_{max} = \tan^{-1}\left(\frac{w_0 L}{2F_H}\right) \tag{5}$$

Therefore,

$$T_{max} = \frac{F_H}{\cos(\theta_{max})} \tag{6}$$

Using the triangular relationship shown in Fig. 7–21b, which is based on Eq. 5, Eq. 6 may be written as

$$T_{max} = \frac{\sqrt{4F_H^2 + w_0^2 L^2}}{2}$$

Substituting Eq. 3 into the above equation yields

$$T_{max} = \frac{w_0 L}{2}\sqrt{1 + \left(\frac{L}{4h}\right)^2} \qquad Ans.$$

For a differential segment of cable length ds, we can write

$$ds = \sqrt{(dx)^2 + (dy)^2} = \sqrt{1 + \left(\frac{dy}{dx}\right)^2}\,dx$$

Hence, the total length of the cable can be determined by integration. Using Eq. 4, we have

$$\mathscr{L} = \int ds = 2\int_0^{L/2}\sqrt{1 + \left(\frac{8h}{L^2}x\right)^2}\,dx \tag{7}$$

Integrating yields

$$\mathscr{L} = \frac{L}{2}\left[\sqrt{1 + \left(\frac{4h}{L}\right)^2} + \frac{L}{4h}\sinh^{-1}\left(\frac{4h}{L}\right)\right] \qquad Ans.$$

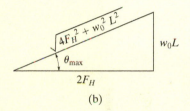

(b)

Fig. 7–21 (cont.)

7

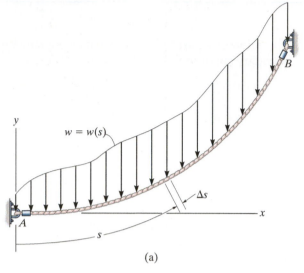

(a)

Fig. 7–22

Cable Subjected to Its Own Weight.

When the weight of a cable becomes important in the force analysis, the loading function along the cable will be a function of the arc length s rather than the projected length x. To analyze this problem, we will consider a generalized loading function $w = w(s)$ acting along the cable, as shown in Fig. 7–22a. The free-body diagram for a small segment Δs of the cable is shown in Fig. 7–22b. Applying the equilibrium equations to the force system on this diagram, one obtains relationships identical to those given by Eqs. 7–7 through 7–9, but with s replacing x in Eqs. 7–7 and 7–8. Therefore, we can show that

$$T \cos \theta = F_H$$

$$T \sin \theta = \int w(s)\, ds \tag{7–13}$$

$$\frac{dy}{dx} = \frac{1}{F_H} \int w(s)\, ds \tag{7–14}$$

To perform a direct integration of Eq. 7–14, it is necessary to replace dy/dx by ds/dx. Since

$$ds = \sqrt{dx^2 + dy^2}$$

then

$$\frac{dy}{dx} = \sqrt{\left(\frac{ds}{dx}\right)^2 - 1}$$

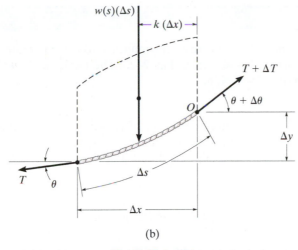

$w(s)(\Delta s)$

$k\,(\Delta x)$

$T + \Delta T$

$\theta + \Delta\theta$

O

Δy

Δs

T

θ

Δx

(b)

Fig. 7–22 (cont.)

Therefore,

$$\frac{ds}{dx} = \left[1 + \frac{1}{F_H^2}\left(\int w(s)\,ds\right)^2\right]^{1/2}$$

Separating the variables and integrating we obtain

$$x = \int \frac{ds}{\left[1 + \dfrac{1}{F_H^2}\left(\displaystyle\int w(s)\,ds\right)^2\right]^{1/2}} \qquad (7\text{–}15)$$

The two constants of integration, say C_1 and C_2, are found using the boundary conditions for the curve.

Electrical transmission towers must be designed to support the weights of the suspended power lines. The weight and length of the cables can be determined since they each form a catenary curve. (© Russell C. Hibbeler)

EXAMPLE | 7.13

Determine the deflection curve, the length, and the maximum tension in the uniform cable shown in Fig. 7–23. The cable has a weight per unit length of $w_0 = 5\ \text{N/m}$.

SOLUTION

For reasons of symmetry, the origin of coordinates is located at the center of the cable. The deflection curve is expressed as $y = f(x)$. We can determine it by first applying Eq. 7–15, where $w(s) = w_0$.

$$x = \int \frac{ds}{\left[1 + (1/F_H^2)\left(\int w_0\,ds\right)^2\right]^{1/2}}$$

Integrating the term under the integral sign in the denominator, we have

$$x = \int \frac{ds}{[1 + (1/F_H^2)(w_0 s + C_1)^2]^{1/2}}$$

Substituting $u = (1/F_H)(w_0 s + C_1)$ so that $du = (w_0/F_H)\,ds$, a second integration yields

$$x = \frac{F_H}{w_0}(\sinh^{-1} u + C_2)$$

or

$$x = \frac{F_H}{w_0}\left\{\sinh^{-1}\left[\frac{1}{F_H}(w_0 s + C_1)\right] + C_2\right\} \qquad (1)$$

To evaluate the constants note that, from Eq. 7–14,

$$\frac{dy}{dx} = \frac{1}{F_H}\int w_0\,ds \quad \text{or} \quad \frac{dy}{dx} = \frac{1}{F_H}(w_0 s + C_1)$$

Since $dy/dx = 0$ at $s = 0$, then $C_1 = 0$. Thus,

$$\frac{dy}{dx} = \frac{w_0 s}{F_H} \qquad (2)$$

The constant C_2 may be evaluated by using the condition $s = 0$ at $x = 0$ in Eq. 1, in which case $C_2 = 0$. To obtain the deflection curve, solve for s in Eq. 1, which yields

$$s = \frac{F_H}{w_0}\sinh\left(\frac{w_0}{F_H}x\right) \qquad (3)$$

Now substitute into Eq. 2, in which case

$$\frac{dy}{dx} = \sinh\left(\frac{w_0}{F_H}x\right)$$

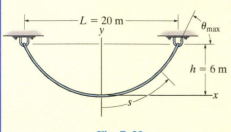

$L = 20\ \text{m}$

$h = 6\ \text{m}$

θ_{max}

Fig. 7–23

Hence,

$$y = \frac{F_H}{w_0}\cosh\left(\frac{w_0}{F_H}x\right) + C_3$$

If the boundary condition $y = 0$ at $x = 0$ is applied, the constant $C_3 = -F_H/w_0$, and therefore the deflection curve becomes

$$y = \frac{F_H}{w_0}\left[\cosh\left(\frac{w_0}{F_H}x\right) - 1\right] \qquad (4)$$

This equation defines the shape of a **catenary curve**. The constant F_H is obtained by using the boundary condition that $y = h$ at $x = L/2$, in which case

$$h = \frac{F_H}{w_0}\left[\cosh\left(\frac{w_0 L}{2F_H}\right) - 1\right] \qquad (5)$$

Since $w_0 = 5$ N/m, $h = 6$ m, and $L = 20$ m, Eqs. 4 and 5 become

$$y = \frac{F_H}{5\,\text{N/m}}\left[\cosh\left(\frac{5\,\text{N/m}}{F_H}x\right) - 1\right] \qquad (6)$$

$$6\,\text{m} = \frac{F_H}{5\,\text{N/m}}\left[\cosh\left(\frac{50\,\text{N}}{F_H}\right) - 1\right] \qquad (7)$$

Equation 7 can be solved for F_H by using a trial-and-error procedure. The result is

$$F_H = 45.9\,\text{N}$$

and therefore the deflection curve, Eq. 6, becomes

$$y = 9.19[\cosh(0.109x) - 1]\,\text{m} \qquad \textit{Ans.}$$

Using Eq. 3, with $x = 10$ m, the half-length of the cable is

$$\frac{\mathcal{L}}{2} = \frac{45.9\,\text{N}}{5\,\text{N/m}}\sinh\left[\frac{5\,\text{N/m}}{45.9\,\text{N}}(10\,\text{m})\right] = 12.1\,\text{m}$$

Hence,

$$\mathcal{L} = 24.2\,\text{m} \qquad \textit{Ans.}$$

Since $T = F_H/\cos\theta$, the maximum tension occurs when θ is maximum, i.e., at $s = \mathcal{L}/2 = 12.1$ m. Using Eq. 2 yields

$$\left.\frac{dy}{dx}\right|_{s=12.1\,\text{m}} = \tan\theta_{max} = \frac{5\,\text{N/m}(12.1\,\text{m})}{45.9\,\text{N}} = 1.32$$

$$\theta_{max} = 52.8°$$

And so,

$$T_{max} = \frac{F_H}{\cos\theta_{max}} = \frac{45.9\,\text{N}}{\cos 52.8°} = 75.9\,\text{N} \qquad \textit{Ans.}$$

PROBLEMS

7–94. The cable supports the three loads shown. Determine the sags y_B and y_D of B and D. Take $P_1 = 800$ N, $P_2 = 500$ N.

7–95. The cable supports the three loads shown. Determine the magnitude of $\mathbf{P}_1$ if $P_2 = 600$ N and $y_B = 3$ m. Also find sag y_D.

7–97. The cable supports the loading shown. Determine the distance x_B the force at B acts from A. Set $P = 800$ N.

7–98. The cable supports the loading shown. Determine the magnitude of the horizontal force $\mathbf{P}$ so that $x_B = 5$ m.

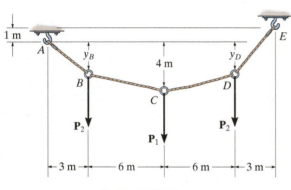

Probs. 7–94/95

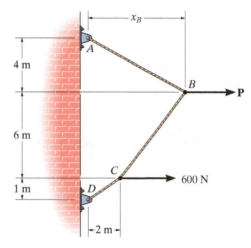

Probs. 7–97/98

***7–96.** Determine the tension in each segment of the cable and the cable's total length.

7–99. The cable supports the three loads shown. Determine the sags y_B and y_D of points B and D. Take $P_1 = 400$ lb, $P_2 = 250$ lb.

***7–100.** The cable supports the three loads shown. Determine the magnitude of $\mathbf{P}_1$ if $P_2 = 300$ lb and $y_B = 8$ ft. Also find the sag y_D.

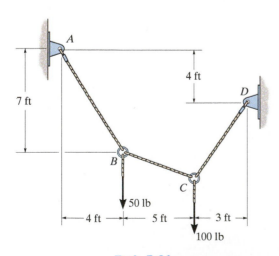

Prob. 7–96

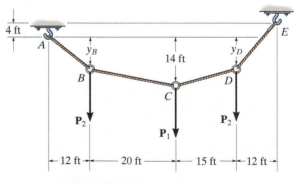

Probs. 7–99/100

7–101. Determine the force P needed to hold the cable in the position shown, i.e., so segment BC remains horizontal. Also, compute the sag y_B and the maximum tension in the cable.

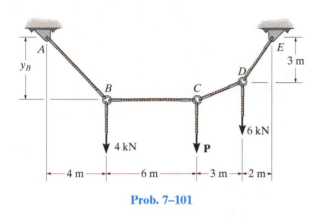

Prob. 7–101

7–102. Determine the maximum uniform loading w, measured in lb/ft, that the cable can support if it is capable of sustaining a maximum tension of 3000 lb before it will break.

7–103. The cable is subjected to a uniform loading of $w = 250$ lb/ft. Determine the maximum and minimum tension in the cable.

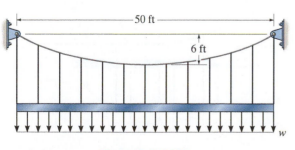

Probs. 7–102/103

*7–104. The cable AB is subjected to a uniform loading of 200 N/m. If the weight of the cable is neglected and the slope angles at points A and B are 30° and 60°, respectively, determine the curve that defines the cable shape and the maximum tension developed in the cable.

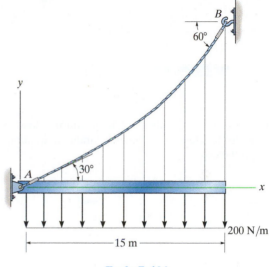

Prob. 7–104

7–105. If $x = 2$ ft and the crate weighs 300 lb, which cable segment AB, BC, or CD has the greatest tension? What is this force and what is the sag y_B?

7–106. If $y_B = 1.5$ ft, determine the largest weight of the crate and its placement x so that neither cable segment AB, BC, or CD is subjected to a tension that exceeds 200 lb.

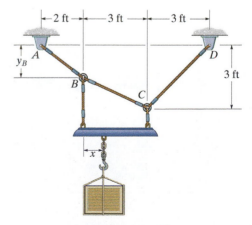

Probs. 7–105/106

7–107. The cable supports a girder which weighs 850 lb/ft. Determine the tension in the cable at points A, B, and C.

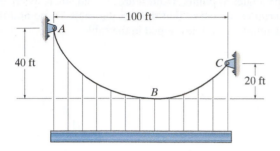

Prob. 7–107

*7–108. The cable is subjected to a uniform loading of $w = 200$ lb/ft. Determine the maximum and minimum tension in the cable.

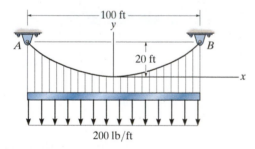

Prob. 7–108

7–109. If the pipe has a mass per unit length of 1500 kg/m, determine the maximum tension developed in the cable.

7–110. If the pipe has a mass per unit length of 1500 kg/m, determine the minimum tension developed in the cable.

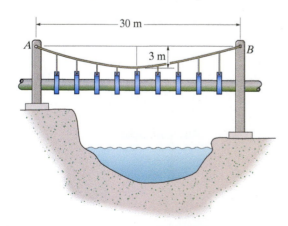

Probs. 7–109/110

7–111. Determine the maximum tension developed in the cable if it is subjected to the triangular distributed load.

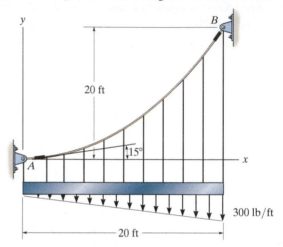

Prob. 7–111

*7–112. The cable will break when the maximum tension reaches $T_{max} = 10$ kN. Determine the minimum sag h if it supports the uniform distributed load of $w = 600$ N/m.

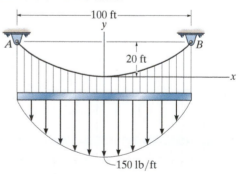

Prob. 7–112

7–113. The cable is subjected to the parabolic loading $w = 150(1 - (x/50)^2)$ lb/ft, where x is in ft. Determine the equation $y = f(x)$ which defines the cable shape AB and the maximum tension in the cable.

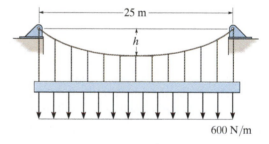

Prob. 7–113

7–114. The power transmission cable weighs 10 lb/ft. If the resultant horizontal force on tower BD is required to be zero, determine the sag h of cable BC.

7–115. The power transmission cable weighs 10 lb/ft. If $h = 10$ ft, determine the resultant horizontal and vertical forces the cables exert on tower BD.

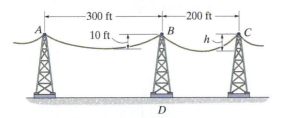

Probs. 7–114/115

*7–116. The man picks up the 52-ft chain and holds it just high enough so it is completely off the ground. The chain has points of attachment A and B that are 50 ft apart. If the chain has a weight of 3 lb/ft, and the man weighs 150 lb, determine the force he exerts on the ground. Also, how high h must he lift the chain? *Hint:* The slopes at A and B are zero.

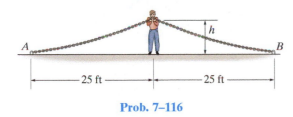

Prob. 7–116

7–117. The cable has a mass of 0.5 kg/m and is 25 m long. Determine the vertical and horizontal components of force it exerts on the top of the tower.

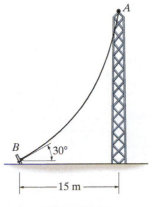

Prob. 7–117

7–118. A 50-ft cable is suspended between two points a distance of 15 ft apart and at the same elevation. If the minimum tension in the cable is 200 lb, determine the total weight of the cable and the maximum tension developed in the cable.

7–119. Show that the deflection curve of the cable discussed in Example 7.13 reduces to Eq. 4 in Example 7.12 when the *hyperbolic cosine function* is expanded in terms of a series and only the first two terms are retained. (The answer indicates that the *catenary* may be replaced by a *parabola* in the analysis of problems in which the sag is small. In this case, the cable weight is assumed to be uniformly distributed along the horizontal.)

*7–120. A telephone line (cable) stretches between two points which are 150 ft apart and at the same elevation. The line sags 5 ft and the cable has a weight of 0.3 lb/ft. Determine the length of the cable and the maximum tension in the cable.

7–121. A cable has a weight of 2 lb/ft. If it can span 100 ft and has a sag of 12 ft, determine the length of the cable. The ends of the cable are supported from the same elevation.

7–122. A cable has a weight of 3 lb/ft and is supported at points that are 500 ft apart and at the same elevation. If it has a length of 600 ft, determine the sag.

7–123. A cable has a weight of 5 lb/ft. If it can span 300 ft and has a sag of 15 ft, determine the length of the cable. The ends of the cable are supported at the same elevation.

*7–124. The 10 kg/m cable is suspended between the supports A and B. If the cable can sustain a maximum tension of 1.5 kN and the maximum sag is 3 m, determine the maximum distance L between the supports.

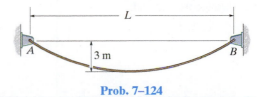

Prob. 7–124

CHAPTER REVIEW

Internal Loadings

If a coplanar force system acts on a member, then in general a resultant internal *normal force* **N**, *shear force* **V**, and *bending moment* **M** will act at any cross section along the member. For two-dimensional problems the positive directions of these loadings are shown in the figure.

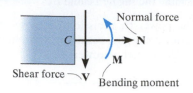

The resultant internal normal force, shear force, and bending moment are determined using the method of sections. To find them, the member is sectioned at the point C where the internal loadings are to be determined. A free-body diagram of one of the sectioned parts is then drawn and the internal loadings are shown in their positive directions.

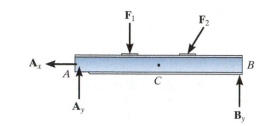

The resultant normal force is determined by summing forces normal to the cross section. The resultant shear force is found by summing forces tangent to the cross section, and the resultant bending moment is found by summing moments about the geometric center or centroid of the cross-sectional area.

$$\Sigma F_x = 0$$
$$\Sigma F_y = 0$$
$$\Sigma M_C = 0$$

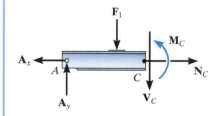

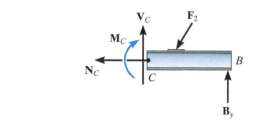

If the member is subjected to a three-dimensional loading, then, in general, a *torsional moment* will also act on the cross section. It can be determined by summing moments about an axis that is perpendicular to the cross section and passes through its centroid.

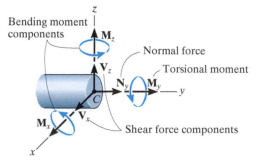

Shear and Moment Diagrams

To construct the shear and moment diagrams for a member, it is necessary to section the member at an arbitrary point, located a distance x from the left end.

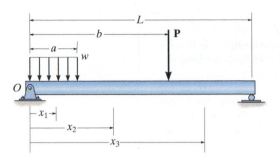

If the external loading consists of changes in the distributed load, or a series of concentrated forces and couple moments act on the member, then different expressions for V and M must be determined within regions between any load discontinuities.

The unknown shear and moment are indicated on the cross section in the positive direction according to the established sign convention, and then the internal shear and moment are determined as functions of x.

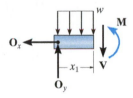

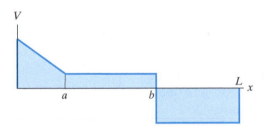

Each of the functions of the shear and moment is then plotted to create the shear and moment diagrams.

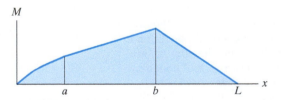

Relations between Shear and Moment

It is possible to plot the shear and moment diagrams quickly by using differential relationships that exist between the distributed loading w, V and M.

The slope of the shear diagram is equal to the distributed loading at any point. The slope is positive if the distributed load acts upward, and vice-versa.

$$\frac{dV}{dx} = w$$

The slope of the moment diagram is equal to the shear at any point. The slope is positive if the shear is positive, or vice-versa.

$$\frac{dM}{dx} = V$$

The change in shear between any two points is equal to the area under the distributed loading between the points.

$$\Delta V = \int w \, dx$$

The change in the moment is equal to the area under the shear diagram between the points.

$$\Delta M = \int V \, dx$$

Cables

When a flexible and inextensible cable is subjected to a series of concentrated forces, then the analysis of the cable can be performed by using the equations of equilibrium applied to free-body diagrams of either segments or points of application of the loading.

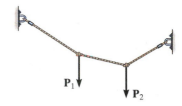

If external distributed loads or the weight of the cable are to be considered, then the shape of the cable must be determined by first analyzing the forces on a differential segment of the cable and then integrating this result. The two constants, say C_1 and C_2, resulting from the integration are determined by applying the boundary conditions for the cable.

$$y = \frac{1}{F_H} \int \left(\int w(x) \, dx \right) dx$$

Distributed load

$$x = \int \frac{ds}{\left[1 + \frac{1}{F_H^2} \left(\int w(s) \, ds \right)^2 \right]^{1/2}}$$

Cable weight

REVIEW PROBLEMS

All problem solutions must include FBDs.

R7–1. Determine the internal normal force, shear force, and moment at points D and E of the frame.

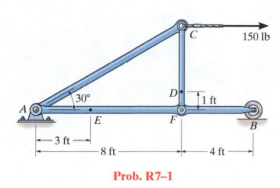

Prob. R7–1

R7–2. Determine the normal force, shear force, and moment at points B and C of the beam.

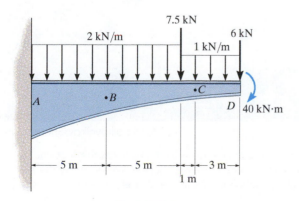

Prob. R7–2

R7–3. Draw the shear and moment diagrams for the beam.

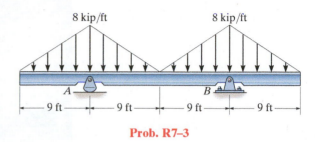

Prob. R7–3

R7–4. Draw the shear and moment diagrams for the beam.

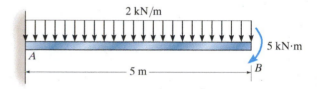

Prob. R7–4

R7–5. Draw the shear and moment diagrams for the beam.

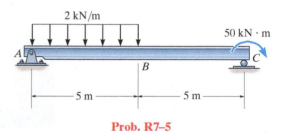

Prob. R7–5

R7–6. A chain is suspended between points at the same elevation and spaced a distance of 60 ft apart. If it has a weight per unit length of 0.5 lb/ft and the sag is 3 ft, determine the maximum tension in the chain.

7

Chapter 8

(© Pavel Polkovnikov/Shutterstock)

The effective design of this brake requires that it resist the frictional forces developed between it and the wheel. In this chapter we will study dry friction, and show how to analyze friction forces for various engineering applications.

Friction

CHAPTER OBJECTIVES

■ To introduce the concept of dry friction and show how to analyze the equilibrium of rigid bodies subjected to this force.

■ To present specific applications of frictional force analysis on wedges, screws, belts, and bearings.

■ To investigate the concept of rolling resistance.

8.1   Characteristics of Dry Friction

Friction is a force that resists the movement of two contacting surfaces that slide relative to one another. This force always acts *tangent* to the surface at the points of contact and is directed so as to oppose the possible or existing motion between the surfaces.

In this chapter, we will study the effects of **dry friction**, which is sometimes called *Coulomb friction* since its characteristics were studied extensively by the French physicist Charles-Augustin de Coulomb in 1781. Dry friction occurs between the contacting surfaces of bodies when there is no lubricating fluid.*

The heat generated by the abrasive action of friction can be noticed when using this grinder to sharpen a metal blade. (© Russell C. Hibbeler)

*Another type of friction, called fluid friction, is studied in fluid mechanics.

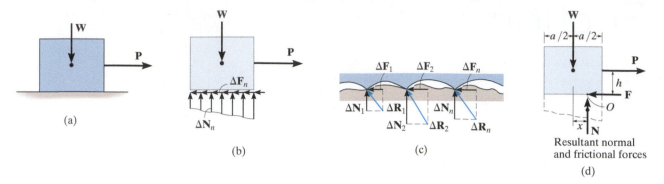

(a)   (b)   (c)   (d)

Resultant normal
and frictional forces

Fig. 8–1

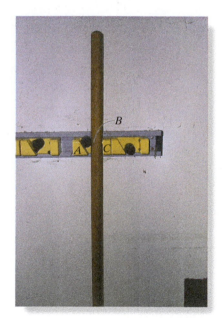

Regardless of the weight of the rake or shovel that is suspended, the device has been designed so that the small roller holds the handle in equilibrium due to frictional forces that develop at the points of contact, A, B, C. (© Russell C. Hibbeler)

Theory of Dry Friction. The theory of dry friction can be explained by considering the effects caused by pulling horizontally on a block of uniform weight $\mathbf{W}$ which is resting on a rough horizontal surface that is *nonrigid or deformable*, Fig. 8–1a. The upper portion of the block, however, can be considered rigid. As shown on the free-body diagram of the block, Fig. 8–1b, the floor exerts an uneven *distribution* of both *normal force* $\Delta\mathbf{N}_n$ and *frictional force* $\Delta\mathbf{F}_n$ along the contacting surface. For equilibrium, the normal forces must act *upward* to balance the block's weight $\mathbf{W}$, and the frictional forces act to the left to prevent the applied force $\mathbf{P}$ from moving the block to the right. Close examination of the contacting surfaces between the floor and block reveals how these frictional and normal forces develop, Fig. 8–1c. It can be seen that many microscopic irregularities exist between the two surfaces and, as a result, reactive forces $\Delta\mathbf{R}_n$ are developed at each point of contact.* As shown, each reactive force contributes both a frictional component $\Delta\mathbf{F}_n$ and a normal component $\Delta\mathbf{N}_n$.

Equilibrium. The effect of the *distributed* normal and frictional loadings is indicated by their *resultants* $\mathbf{N}$ and $\mathbf{F}$ on the free-body diagram, Fig. 8–1d. Notice that $\mathbf{N}$ acts a distance x to the right of the line of action of $\mathbf{W}$, Fig. 8–1d. This location, which coincides with the centroid or geometric center of the normal force distribution in Fig. 8–1b, is necessary in order to balance the "tipping effect" caused by $\mathbf{P}$. For example, if $\mathbf{P}$ is applied at a height h from the surface, Fig. 8–1d, then moment equilibrium about point O is satisfied if $Wx = Ph$ or $x = Ph/W$.

*Besides mechanical interactions as explained here, which is referred to as a classical approach, a detailed treatment of the nature of frictional forces must also include the effects of temperature, density, cleanliness, and atomic or molecular attraction between the contacting surfaces. See J. Krim, *Scientific American*, October, 1996.

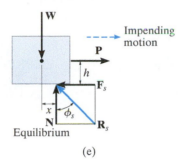

Fig. 8–1 (cont.)

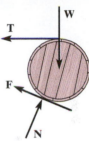

Impending Motion. In cases where the surfaces of contact are rather "slippery," the frictional force **F** may *not* be great enough to balance **P**, and consequently the block will tend to slip. In other words, as P is slowly increased, F correspondingly increases until it attains a certain *maximum value F_s*, called the *limiting static frictional force*, Fig. 8–1e. When this value is reached, the block is in *unstable equilibrium* since any further increase in P will cause the block to move. Experimentally, it has been determined that this limiting static frictional force F_s is *directly proportional* to the resultant normal force N. Expressed mathematically,

Some objects, such as this barrel, may not be on the verge of slipping, and therefore the friction force **F** must be determined strictly from the equations of equilibrium. (© Russell C. Hibbeler)

$$F_s = \mu_s N \tag{8–1}$$

where the constant of proportionality, μ_s (mu "sub" s), is called the *coefficient of static friction*.

Thus, when the block is on the *verge of sliding*, the normal force **N** and frictional force **F$_s$** combine to create a resultant **R$_s$**, Fig. 8–1e. The angle ϕ_s (phi "sub" s) that **R$_s$** makes with **N** is called the *angle of static friction*. From the figure,

$$\phi_s = \tan^{-1}\left(\frac{F_s}{N}\right) = \tan^{-1}\left(\frac{\mu_s N}{N}\right) = \tan^{-1}\mu_s$$

Typical values for μ_s are given in Table 8–1. Note that these values can vary since experimental testing was done under variable conditions of roughness and cleanliness of the contacting surfaces. For applications, therefore, it is important that both caution and judgment be exercised when selecting a coefficient of friction for a given set of conditions. When a more accurate calculation of F_s is required, the coefficient of friction should be determined directly by an experiment that involves the two materials to be used.

Table 8–1 Typical Values for μ_s	
Contact Materials	Coefficient of Static Friction (μ_s)
Metal on ice	0.03–0.05
Wood on wood	0.30–0.70
Leather on wood	0.20–0.50
Leather on metal	0.30–0.60
Copper on copper	0.74–1.21

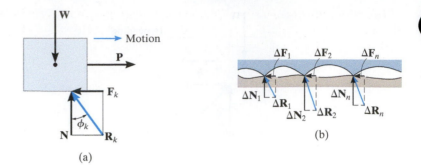

Fig. 8–2

Motion. If the magnitude of **P** acting on the block is increased so that it becomes slightly greater than F_s, the frictional force at the contacting surface will drop to a smaller value F_k, called the *kinetic frictional force*. The block will begin to slide with increasing speed, Fig. 8–2a. As this occurs, the block will "ride" on top of these peaks at the points of contact, as shown in Fig. 8–2b. The continued breakdown of the surface is the dominant mechanism creating kinetic friction.

Experiments with sliding blocks indicate that the magnitude of the kinetic friction force is directly proportional to the magnitude of the resultant normal force, expressed mathematically as

$$F_k = \mu_k N \qquad (8–2)$$

Here the constant of proportionality, μ_k, is called the **coefficient of kinetic friction**. Typical values for μ_k are approximately 25 percent *smaller* than those listed in Table 8–1 for μ_s.

As shown in Fig. 8–2a, in this case, the resultant force at the surface of contact, **R**$_k$, has a line of action defined by ϕ_k. This angle is referred to as the **angle of kinetic friction**, where

$$\phi_k = \tan^{-1}\left(\frac{F_k}{N}\right) = \tan^{-1}\left(\frac{\mu_k N}{N}\right) = \tan^{-1}\mu_k$$

By comparison, $\phi_s \geq \phi_k$.

The above effects regarding friction can be summarized by referring to the graph in Fig. 8–3, which shows the variation of the frictional force F versus the applied load P. Here the frictional force is categorized in three different ways:

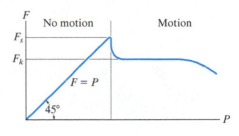

Fig. 8–3

- F is a *static frictional force* if equilibrium is maintained.

- F is a *limiting static frictional force* F_s when it reaches a maximum value needed to maintain equilibrium.

- F is a *kinetic frictional force* F_k when sliding occurs at the contacting surface.

Notice also from the graph that for very large values of P or for high speeds, aerodynamic effects will cause F_k and likewise μ_k to begin to decrease.

Characteristics of Dry Friction.

As a result of *experiments* that pertain to the foregoing discussion, we can state the following rules which apply to bodies subjected to dry friction.

- The frictional force acts *tangent* to the contacting surfaces in a direction *opposed* to the *motion* or tendency for motion of one surface relative to another.

- The maximum static frictional force F_s that can be developed is independent of the area of contact, provided the normal pressure is not very low nor great enough to severely deform or crush the contacting surfaces of the bodies.

- The maximum static frictional force is generally greater than the kinetic frictional force for any two surfaces of contact. However, if one of the bodies is moving with a *very low velocity* over the surface of another, F_k becomes approximately equal to F_s, i.e., $\mu_s \approx \mu_k$.

- When *slipping* at the surface of contact is *about to occur*, the maximum static frictional force is proportional to the normal force, such that $F_s = \mu_s N$.

- When *slipping* at the surface of contact is *occurring*, the kinetic frictional force is proportional to the normal force, such that $F_k = \mu_k N$.

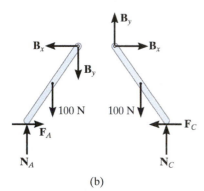

(a)

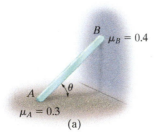

(b)

Fig. 8–4

8.2 Problems Involving Dry Friction

If a rigid body is in equilibrium when it is subjected to a system of forces that includes the effect of friction, the force system must satisfy not only the equations of equilibrium but *also* the laws that govern the frictional forces.

Types of Friction Problems.   In general, there are three types of static problems involving dry friction. They can easily be classified once free-body diagrams are drawn and the total number of unknowns are identified and compared with the total number of available equilibrium equations.

No Apparent Impending Motion.   Problems in this category are strictly equilibrium problems, which require the number of unknowns to be *equal* to the number of available equilibrium equations. Once the frictional forces are determined from the solution, however, their numerical values must be checked to be sure they satisfy the inequality $F \le \mu_s N$; otherwise, slipping will occur and the body will not remain in equilibrium. A problem of this type is shown in Fig. 8–4a. Here we must determine the frictional forces at A and C to check if the equilibrium position of the two-member frame can be maintained. If the bars are uniform and have known weights of 100 N each, then the free-body diagrams are as shown in Fig. 8–4b. There are six unknown force components which can be determined *strictly* from the six equilibrium equations (three for each member). Once F_A, N_A, F_C, and N_C are determined, then the bars will remain in equilibrium provided $F_A \le 0.3N_A$ and $F_C \le 0.5N_C$ are satisfied.

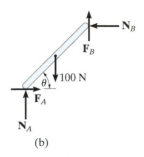

(a)

(b)

Fig. 8–5

Impending Motion at All Points of Contact.   In this case the total number of unknowns will *equal* the total number of available equilibrium equations *plus* the total number of available frictional equations, $F = \mu N$. When *motion is impending* at the points of contact, then $F_s = \mu_s N$; whereas if the body is *slipping*, then $F_k = \mu_k N$. For example, consider the problem of finding the smallest angle θ at which the 100-N bar in Fig. 8–5a can be placed against the wall without slipping. The free-body diagram is shown in Fig. 8–5b. Here the *five* unknowns are determined from the *three* equilibrium equations and *two* static frictional equations which apply at *both* points of contact, so that $F_A = 0.3N_A$ and $F_B = 0.4N_B$.

Impending Motion at Some Points of Contact. Here the number of unknowns will be *less* than the number of available equilibrium equations plus the number of available frictional equations or conditional equations for tipping. As a result, several possibilities for motion or impending motion will exist and the problem will involve a determination of the kind of motion which actually occurs. For example, consider the two-member frame in Fig. 8–6a. In this problem we wish to determine the horizontal force P needed to cause movement. If each member has a weight of 100 N, then the free-body diagrams are as shown in Fig. 8–6b. There are *seven* unknowns. For a unique solution we must satisfy the *six* equilibrium equations (three for each member) and only *one* of two possible static frictional equations. This means that as P increases it will either cause slipping at A and no slipping at C, so that $F_A = 0.3N_A$ and $F_C \le 0.5N_C$; or slipping occurs at C and no slipping at A, in which case $F_C = 0.5N_C$ and $F_A \le 0.3N_A$. The actual situation can be determined by calculating P for each case and then choosing the case for which P is *smaller*. If in both cases the *same value* for P is calculated, which would be highly improbable, then slipping at both points occurs simultaneously; i.e., the *seven unknowns* would satisfy *eight equations*.

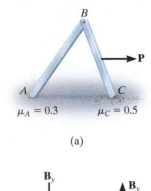

(a)

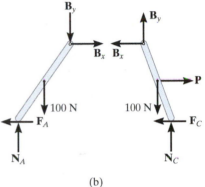

(b)

Fig. 8–6

Equilibrium Versus Frictional Equations. Whenever we solve a problem such as the one in Fig. 8–4, where the friction force F is to be an "equilibrium force" and satisfies the inequality $F < \mu_s N$, then we can assume the sense of direction of F on the free-body diagram. The correct sense is made known *after* solving the equations of equilibrium for F. If F is a negative scalar the sense of $\mathbf{F}$ is the reverse of that which was assumed. This convenience of *assuming* the sense of $\mathbf{F}$ is possible because the equilibrium equations equate to zero the *components of vectors* acting in the *same direction*. However, in cases where the frictional equation $F = \mu N$ is used in the solution of a problem, as in the case shown in Fig. 8–5, then the convenience of *assuming* the sense of $\mathbf{F}$ is *lost*, since the frictional equation relates only the *magnitudes* of two *perpendicular* vectors. Consequently, $\mathbf{F}$ *must always* be shown acting with its *correct sense* on the free-body diagram, *whenever* the frictional equation is used for the solution of a problem.

Depending upon where the man pushes on the crate, it will either tip or slip. (© Russell C. Hibbeler)

Important Points

- Friction is a tangential force that resists the movement of one surface relative to another.

- If no sliding occurs, the maximum value for the friction force is equal to the product of the coefficient of static friction and the normal force at the surface.

- If sliding occurs at a slow speed, then the friction force is the product of the coefficient of kinetic friction and the normal force at the surface.

- There are three types of static friction problems. Each of these problems is analyzed by first drawing the necessary free-body diagrams, and then applying the equations of equilibrium, while satisfying the conditions of friction or the possibility of tipping.

Procedure for Analysis

Equilibrium problems involving dry friction can be solved using the following procedure.

Free-Body Diagrams.

- Draw the necessary free-body diagrams, and unless it is stated in the problem that impending motion or slipping occurs, *always* show the frictional forces as unknowns (i.e., *do not assume* $F = \mu N$).

- Determine the number of unknowns and compare this with the number of available equilibrium equations.

- If there are more unknowns than equations of equilibrium, it will be necessary to apply the frictional equation at some, if not all, points of contact to obtain the extra equations needed for a complete solution.

- If the equation $F = \mu N$ is to be used, it will be necessary to show **F** acting in the correct sense of direction on the free-body diagram.

Equations of Equilibrium and Friction.

- Apply the equations of equilibrium and the necessary frictional equations (or conditional equations if tipping is possible) and solve for the unknowns.

- If the problem involves a three-dimensional force system such that it becomes difficult to obtain the force components or the necessary moment arms, apply the equations of equilibrium using Cartesian vectors.

8

EXAMPLE 8.1

The uniform crate shown in Fig. 8–7a has a mass of 20 kg. If a force $P = 80$ N is applied to the crate, determine if it remains in equilibrium. The coefficient of static friction is $\mu_s = 0.3$.

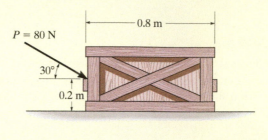

(a)

Fig. 8–7

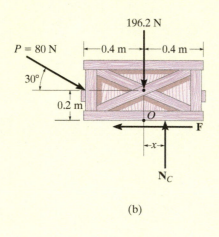

(b)

SOLUTION

Free-Body Diagram. As shown in Fig. 8–7b, the *resultant* normal force N_C must act a distance x from the crate's center line in order to counteract the tipping effect caused by **P**. There are *three unknowns*, F, N_C, and x, which can be determined strictly from the *three* equations of equilibrium.

Equations of Equilibrium.

$$\xrightarrow{+} \Sigma F_x = 0; \qquad\qquad\qquad 80 \cos 30° \text{ N} - F = 0$$

$$+\uparrow \Sigma F_y = 0; \qquad -80 \sin 30° \text{ N} + N_C - 196.2 \text{ N} = 0$$

$$\zeta + \Sigma M_O = 0; \quad 80 \sin 30° \text{ N}(0.4 \text{ m}) - 80 \cos 30° \text{ N}(0.2 \text{ m}) + N_C(x) = 0$$

Solving,

$$F = 69.3 \text{ N}$$

$$N_C = 236.2 \text{ N}$$

$$x = -0.00908 \text{ m} = -9.08 \text{ mm}$$

Since x is negative it indicates the *resultant* normal force acts (slightly) to the *left* of the crate's center line. No tipping will occur since $x < 0.4$ m. Also, the *maximum* frictional force which can be developed at the surface of contact is $F_{max} = \mu_s N_C = 0.3(236.2 \text{ N}) = 70.9$ N. Since $F = 69.3$ N < 70.9 N, the crate will *not slip*, although it is very close to doing so.

EXAMPLE | 8.2

It is observed that when the bed of the dump truck is raised to an angle of $\theta = 25°$ the vending machines will begin to slide off the bed, Fig. 8–8a. Determine the static coefficient of friction between a vending machine and the surface of the truckbed.

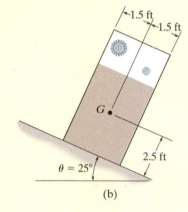

(a)

(© Russell C. Hibbeler)

SOLUTION

An idealized model of a vending machine resting on the truckbed is shown in Fig. 8–8b. The dimensions have been measured and the center of gravity has been located. We will assume that the vending machine weighs W.

Free-Body Diagram.   As shown in Fig. 8–8c, the dimension x is used to locate the position of the resultant normal force $\mathbf{N}$. There are four unknowns, N, F, μ_s, and x.

Equations of Equilibrium.

$$+\searrow\Sigma F_x = 0; \qquad\qquad W\sin 25° - F = 0 \qquad\qquad (1)$$

$$+\nearrow\Sigma F_y = 0; \qquad\qquad N - W\cos 25° = 0 \qquad\qquad (2)$$

$$\zeta +\Sigma M_O = 0; \quad -W\sin 25°(2.5 \text{ ft}) + W\cos 25°(x) = 0 \qquad (3)$$

Since slipping impends at $\theta = 25°$, using Eqs. 1 and 2, we have

$$F_s = \mu_s N; \qquad W\sin 25° = \mu_s(W\cos 25°)$$

$$\mu_s = \tan 25° = 0.466 \qquad\qquad\qquad\qquad Ans.$$

The angle of $\theta = 25°$ is referred to as the **angle of repose**, and by comparison, it is equal to the angle of static friction, $\theta = \phi_s$. Notice from the calculation that θ is independent of the weight of the vending machine, and so knowing θ provides a convenient method for determining the coefficient of static friction.

NOTE: From Eq. 3, we find $x = 1.17$ ft. Since 1.17 ft < 1.5 ft, indeed the vending machine will slip before it can tip as observed in Fig. 8–8a.

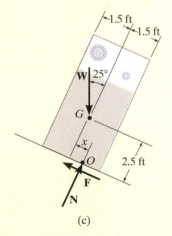

(b)

(c)

Fig. 8–8

EXAMPLE | 8.3

The uniform 10-kg ladder in Fig. 8–9a rests against the smooth wall at B, and the end A rests on the rough horizontal plane for which the coefficient of static friction is $\mu_s = 0.3$. Determine the angle of inclination θ of the ladder and the normal reaction at B if the ladder is on the verge of slipping.

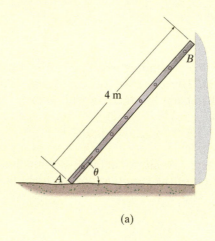

(a)

Fig. 8–9

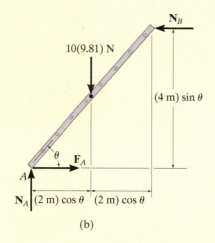

(b)

SOLUTION

Free-Body Diagram.   As shown on the free-body diagram, Fig. 8–9b, the frictional force $\mathbf{F}_A$ must act to the right since impending motion at A is to the left.

Equations of Equilibrium and Friction.   Since the ladder is on the verge of slipping, then $F_A = \mu_s N_A = 0.3 N_A$. By inspection, N_A can be obtained directly.

$$+\uparrow\Sigma F_y = 0; \qquad\qquad N_A - 10(9.81)\text{ N} = 0 \qquad\qquad N_A = 98.1\text{ N}$$

Using this result, $F_A = 0.3(98.1\text{ N}) = 29.43\text{ N}$. Now N_B can be found.

$$\xrightarrow{+}\Sigma F_x = 0; \qquad\qquad 29.43\text{ N} - N_B = 0$$

$$N_B = 29.43\text{ N} = 29.4\text{ N} \qquad\qquad Ans.$$

Finally, the angle θ can be determined by summing moments about point A.

$$\zeta+\Sigma M_A = 0; \qquad (29.43\text{ N})(4\text{ m})\sin\theta - [10(9.81)\text{ N}](2\text{ m})\cos\theta = 0$$

$$\frac{\sin\theta}{\cos\theta} = \tan\theta = 1.6667$$

$$\theta = 59.04° = 59.0° \qquad\qquad Ans.$$

EXAMPLE 8.4

Beam *AB* is subjected to a uniform load of 200 N/m and is supported at *B* by post *BC*, Fig. 8–10a. If the coefficients of static friction at *B* and *C* are $\mu_B = 0.2$ and $\mu_C = 0.5$, determine the force **P** needed to pull the post out from under the beam. Neglect the weight of the members and the thickness of the beam.

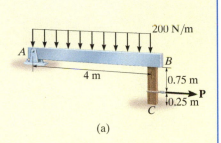
(a)

SOLUTION

Free-Body Diagrams. The free-body diagram of the beam is shown in Fig. 8–10b. Applying $\Sigma M_A = 0$, we obtain $N_B = 400$ N. This result is shown on the free-body diagram of the post, Fig. 8–10c. Referring to this member, the *four* unknowns F_B, P, F_C, and N_C are determined from the *three* equations of equilibrium and *one* frictional equation applied either at *B* or *C*.

Equations of Equilibrium and Friction.

$$\xrightarrow{+} \Sigma F_x = 0; \qquad P - F_B - F_C = 0 \qquad (1)$$
$$+\uparrow \Sigma F_y = 0; \qquad N_C - 400\text{ N} = 0 \qquad (2)$$
$$\zeta + \Sigma M_C = 0; \qquad -P(0.25\text{ m}) + F_B(1\text{ m}) = 0 \qquad (3)$$

(Post Slips at *B* and Rotates about *C*.) This requires $F_C \le \mu_C N_C$ and

$$F_B = \mu_B N_B; \qquad F_B = 0.2(400\text{ N}) = 80\text{ N}$$

Using this result and solving Eqs. 1 through 3, we obtain

$$P = 320\text{ N}$$
$$F_C = 240\text{ N}$$
$$N_C = 400\text{ N}$$

Since $F_C = 240$ N $> \mu_C N_C = 0.5(400$ N$) = 200$ N, slipping at *C* occurs. Thus the other case of movement must be investigated.

(Post Slips at *C* and Rotates about *B*.) Here $F_B \le \mu_B N_B$ and

$$F_C = \mu_C N_C; \qquad F_C = 0.5 N_C \qquad (4)$$

Solving Eqs. 1 through 4 yields
$$P = 267\text{ N} \qquad \qquad Ans.$$
$$N_C = 400\text{ N}$$
$$F_C = 200\text{ N}$$
$$F_B = 66.7\text{ N}$$

Obviously, this case occurs first since it requires a *smaller* value for *P*.

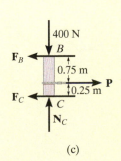
(c)

Fig. 8–10

(b)

EXAMPLE | 8.5

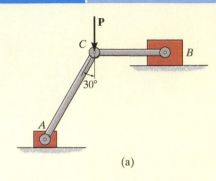

(a)

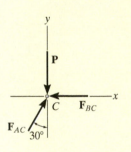

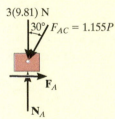

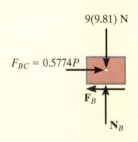

(b)

Fig. 8–11

Blocks A and B have a mass of 3 kg and 9 kg, respectively, and are connected to the weightless links shown in Fig. 8–11a. Determine the largest vertical force $\mathbf{P}$ that can be applied at the pin C without causing any movement. The coefficient of static friction between the blocks and the contacting surfaces is $\mu_s = 0.3$.

SOLUTION

Free-Body Diagram. The links are two-force members and so the free-body diagrams of pin C and blocks A and B are shown in Fig. 8–11b. Since the horizontal component of $\mathbf{F}_{AC}$ tends to move block A to the left, $\mathbf{F}_A$ must act to the right. Similarly, $\mathbf{F}_B$ must act to the left to oppose the tendency of motion of block B to the right, caused by $\mathbf{F}_{BC}$. There are seven unknowns and six available force equilibrium equations, two for the pin and two for each block, so that *only one* frictional equation is needed.

Equations of Equilibrium and Friction. The force in links AC and BC can be related to P by considering the equilibrium of pin C.

$+\uparrow\Sigma F_y = 0;$ $F_{AC}\cos 30° - P = 0;$ $F_{AC} = 1.155P$

$\xrightarrow{+}\Sigma F_x = 0;$ $1.155P\sin 30° - F_{BC} = 0;$ $F_{BC} = 0.5774P$

Using the result for F_{AC}, for block A,

$\xrightarrow{+}\Sigma F_x = 0;$ $F_A - 1.155P\sin 30° = 0;$ $F_A = 0.5774P$ (1)

$+\uparrow\Sigma F_y = 0;$ $N_A - 1.155P\cos 30° - 3(9.81\text{ N}) = 0;$

$$N_A = P + 29.43\text{ N} \tag{2}$$

Using the result for F_{BC}, for block B,

$\xrightarrow{+}\Sigma F_x = 0;$ $(0.5774P) - F_B = 0;$ $F_B = 0.5774P$ (3)

$+\uparrow\Sigma F_y = 0;$ $N_B - 9(9.81)\text{ N} = 0;$ $N_B = 88.29\text{ N}$

Movement of the system may be caused by the initial slipping of *either* block A or block B. If we assume that block A slips first, then

$$F_A = \mu_s N_A = 0.3 N_A \tag{4}$$

Substituting Eqs. 1 and 2 into Eq. 4,

$$0.5774P = 0.3(P + 29.43)$$

$$P = 31.8\text{ N} \qquad\qquad Ans.$$

Substituting this result into Eq. 3, we obtain $F_B = 18.4$ N. Since the maximum static frictional force at B is $(F_B)_{\max} = \mu_s N_B = 0.3(88.29\text{ N}) = 26.5\text{ N} > F_B$, block B will not slip. Thus, the above assumption is correct. Notice that if the inequality were not satisfied, we would have to assume slipping of block B and then solve for P.

PRELIMINARY PROBLEMS

P8–1. Determine the friction force at the surface of contact.

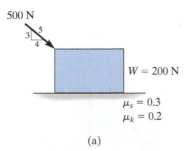

500 N

3 / 5 / 4

W = 200 N

$\mu_s = 0.3$
$\mu_k = 0.2$

(a)

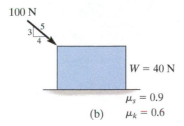

100 N

3 / 5 / 4

W = 40 N

$\mu_s = 0.9$
(b) $\mu_k = 0.6$

Prob. P8–1

P8–2. Determine **M** to cause impending motion of the cylinder.

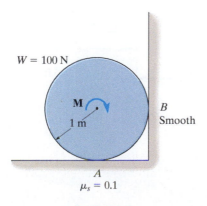

W = 100 N

M

1 m

B
Smooth

A
$\mu_s = 0.1$

Prob. P8–2

P8–3. Determine the force P to move block B.

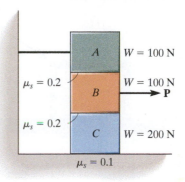

A   W = 100 N

$\mu_s = 0.2$ B   W = 100 N
→ **P**

$\mu_s = 0.2$ C   W = 200 N

$\mu_s = 0.1$

Prob. P8–3

P8–4. Determine the force P needed to cause impending motion of the block.

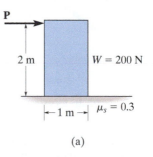

P →

2 m   W = 200 N

← 1 m → $\mu_s = 0.3$

(a)

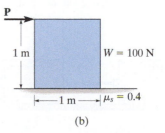

P →

1 m   W = 100 N

← 1 m → $\mu_s = 0.4$

(b)

Prob. P8–4

8

FUNDAMENTAL PROBLEMS

All problem solutions must include FBDs.

F8–1. Determine the friction developed between the 50-kg crate and the ground if a) $P = 200$ N, and b) $P = 400$ N. The coefficients of static and kinetic friction between the crate and the ground are $\mu_s = 0.3$ and $\mu_k = 0.2$.

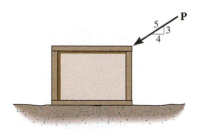

Prob. F8–1

F8–2. Determine the minimum force P to prevent the 30-kg rod AB from sliding. The contact surface at B is smooth, whereas the coefficient of static friction between the rod and the wall at A is $\mu_s = 0.2$.

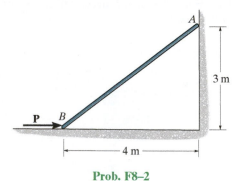

Prob. F8–2

F8–3. Determine the maximum force P that can be applied without causing the two 50-kg crates to move. The coefficient of static friction between each crate and the ground is $\mu_s = 0.25$.

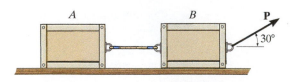

Prob. F8–3

F8–4. If the coefficient of static friction at contact points A and B is $\mu_s = 0.3$, determine the maximum force P that can be applied without causing the 100-kg spool to move.

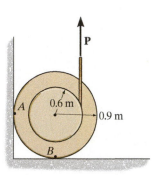

Prob. F8–4

F8–5. Determine the maximum force P that can be applied without causing movement of the 250-lb crate that has a center of gravity at G. The coefficient of static friction at the floor is $\mu_s = 0.4$.

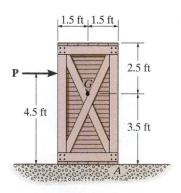

Prob. F8–5

F8–6. Determine the minimum coefficient of static friction between the uniform 50-kg spool and the wall so that the spool does not slip.

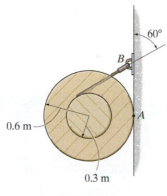

0.6 m

0.3 m

Prob. F8–6

F8–8. If the coefficient of static friction at all contacting surfaces is μ_s, determine the inclination θ at which the identical blocks, each of weight W, begin to slide.

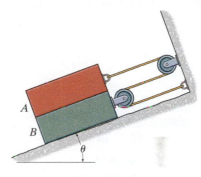

Prob. F8–8

F8–7. Blocks A, B, and C have weights of 50 N, 25 N, and 15 N, respectively. Determine the smallest horizontal force P that will cause impending motion. The coefficient of static friction between A and B is $\mu_s = 0.3$, between B and C, $\mu_s' = 0.4$, and between block C and the ground, $\mu_s'' = 0.35$.

F8–9. Blocks A and B have a mass of 7 kg and 10 kg, respectively. Using the coefficients of static friction indicated, determine the largest force P which can be applied to the cord without causing motion. There are pulleys at C and D.

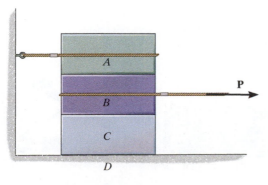

Prob. F8–7

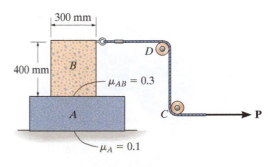

300 mm

400 mm

$\mu_{AB} = 0.3$

$\mu_A = 0.1$

Prob. F8–9

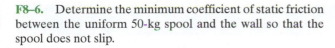

PROBLEMS

All problem solutions must include FBDs.

8–1. Determine the maximum force P the connection can support so that no slipping occurs between the plates. There are four bolts used for the connection and each is tightened so that it is subjected to a tension of 4 kN. The coefficient of static friction between the plates is $\mu_s = 0.4$.

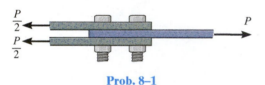

Prob. 8–1

8–2. The tractor exerts a towing force $T = 400$ lb. Determine the normal reactions at each of the two front and two rear tires and the tractive frictional force **F** on each rear tire needed to pull the load forward at constant velocity. The tractor has a weight of 7500 lb and a center of gravity located at G_T. An additinal weight of 600 lb is added to its front having a center of gravity at G_A. Take $\mu_s = 0.4$. The front wheels are free to roll.

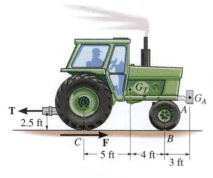

Prob. 8–2

8–3. The mine car and its contents have a total mass of 6 Mg and a center of gravity at G. If the coefficient of static friction between the wheels and the tracks is $\mu_s = 0.4$ when the wheels are locked, find the normal force acting on the front wheels at B and the rear wheels at A when the brakes at both A and B are locked. Does the car move?

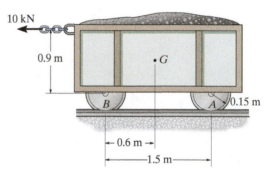

Prob. 8–3

*8–4. The winch on the truck is used to hoist the garbage bin onto the bed of the truck. If the loaded bin has a weight of 8500 lb and center of gravity at G, determine the force in the cable needed to begin the lift. The coefficients of static friction at A and B are $\mu_A = 0.3$ and $\mu_B = 0.2$, respectively. Neglect the height of the support at A.

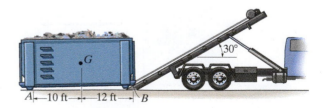

Prob. 8–4

8–5. The automobile has a mass of 2 Mg and center of mass at G. Determine the towing force $\mathbf{F}$ required to move the car if the back brakes are locked, and the front wheels are free to roll. Take $\mu_s = 0.3$.

8–6. The automobile has a mass of 2 Mg and center of mass at G. Determine the towing force $\mathbf{F}$ required to move the car. Both the front and rear brakes are locked. Take $\mu_s = 0.3$.

*8–8. The block brake consists of a pin-connected lever and friction block at B. The coefficient of static friction between the wheel and the lever is $\mu_s = 0.3$, and a torque of 5 N·m is applied to the wheel. Determine if the brake can hold the wheel stationary when the force applied to the lever is (a) $P = 30$ N, (b) $P = 70$ N.

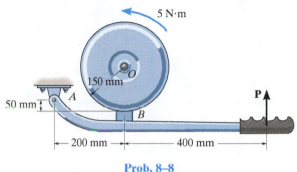

Prob. 8–8

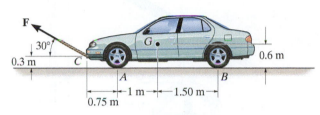

Probs. 8–5/6

8–7. The block brake consists of a pin-connected lever and friction block at B. The coefficient of static friction between the wheel and the lever is $\mu_s = 0.3$, and a torque of 5 N·m is applied to the wheel. Determine if the brake can hold the wheel stationary when the force applied to the lever is (a) $P = 30$ N, (b) $P = 70$ N.

8–9. The pipe of weight W is to be pulled up the inclined plane of slope α using a force $\mathbf{P}$. If $\mathbf{P}$ acts at an angle ϕ, show that for slipping $P = W \sin(\alpha + \theta)/\cos(\phi - \theta)$, where θ is the angle of static friction; $\theta = \tan^{-1} \mu_s$.

8–10. Determine the angle ϕ at which the applied force $\mathbf{P}$ should act on the pipe so that the magnitude of $\mathbf{P}$ is as small as possible for pulling the pipe up the incline. What is the corresponding value of P? The pipe weighs W and the slope α is known. Express the answer in terms of the angle of kinetic friction, $\theta = \tan^{-1} \mu_k$.

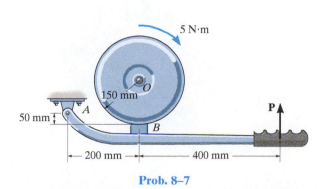

Prob. 8–7

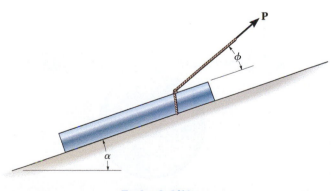

Probs. 8–9/10

8–11. Determine the maximum weight W the man can lift with constant velocity using the pulley system, without and then with the "leading block" or pulley at A. The man has a weight of 200 lb and the coefficient of static friction between his feet and the ground is $\mu_s = 0.6$.

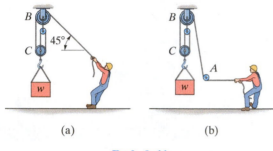

(a)   (b)

Prob. 8–11

8–13. If a torque of $M = 300$ N·m is applied to the flywheel, determine the force that must be developed in the hydraulic cylinder CD to prevent the flywheel from rotating. The coefficient of static friction between the friction pad at B and the flywheel is $\mu_s = 0.4$.

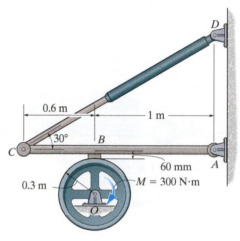

Prob. 8–13

*8–12. The block brake is used to stop the wheel from rotating when the wheel is subjected to a couple moment $\mathbf{M}_0$. If the coefficient of static friction between the wheel and the block is μ_s, determine the smallest force P that should be applied.

8–14. The car has a mass of 1.6 Mg and center of mass at G. If the coefficient of static friction between the shoulder of the road and the tires is $\mu_s = 0.4$, determine the greatest slope θ the shoulder can have without causing the car to slip or tip over if the car travels along the shoulder at constant velocity.

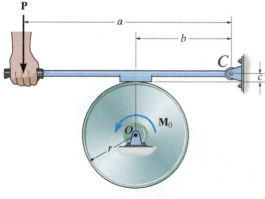

Prob. 8–12

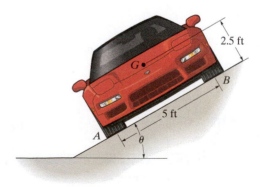

Prob. 8–14

8–15. The log has a coefficient of state friction of $\mu_s = 0.3$ with the ground and a weight of 40 lb/ft. If a man can pull on the rope with a maximum force of 80 lb, determine the greatest length l of log he can drag.

80 lb

Prob. 8–15

*8–16.** The 180-lb man climbs up the ladder and stops at the position shown after he senses that the ladder is on the verge of slipping. Determine the inclination θ of the ladder if the coefficient of static friction between the friction pad A and the ground is $\mu_s = 0.4$. Assume the wall at B is smooth. The center of gravity for the man is at G. Neglect the weight of the ladder.

8–17. The 180-lb man climbs up the ladder and stops at the position shown after he senses that the ladder is on the verge of slipping. Determine the coefficient of static friction between the friction pad at A and ground if the inclination of the ladder is $\theta = 60°$ and the wall at B is smooth. The center of gravity for the man is at G. Neglect the weight of the ladder.

8–18. The spool of wire having a weight of 300 lb rests on the ground at B and against the wall at A. Determine the force P required to begin pulling the wire horizontally off the spool. The coefficient of static friction between the spool and its points of contact is $\mu_s = 0.25$.

8–19. The spool of wire having a weight of 300 lb rests on the ground at B and against the wall at A. Determine the normal force acting on the spool at A if $P = 300$ lb. The coefficient of static friction between the spool and the ground at B is $\mu_s = 0.35$. The wall at A is smooth.

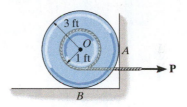

Probs. 8–18/19

*8–20.** The ring has a mass of 0.5 kg and is resting on the surface of the table. In an effort to move the ring a normal force **P** from the finger is exerted on it. If this force is directed towards the ring's center O as shown, determine its magnitude when the ring is on the verge of slipping at A. The coefficient of static friction at A is $\mu_A = 0.2$ and at B, $\mu_B = 0.3$.

8

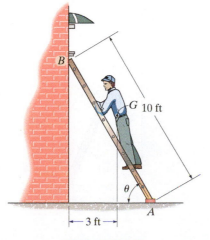

Probs. 8–16/17

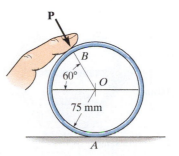

Prob. 8–20

8–21. A man attempts to support a stack of books horizontally by applying a compressive force of $F = 120$ N to the ends of the stack with his hands. If each book has a mass of 0.95 kg, determine the greatest number of books that can be supported in the stack. The coefficient of static friction between his hands and a book is $(\mu_s)_h = 0.6$ and between any two books $(\mu_s)_b = 0.4$.

$F = 120$ N $F = 120$ N

Prob. 8–21

8–22. The tongs are used to lift the 150-kg crate, whose center of mass is at G. Determine the least coefficient of static friction at the pivot blocks so that the crate can be lifted.

8–23. The beam is supported by a pin at A and a roller at B which has negligible weight and a radius of 15 mm. If the coefficient of static friction is $\mu_B = \mu_C = 0.3$, determine the largest angle θ of the incline so that the roller does not slip for any force **P** applied to the beam.

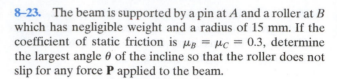

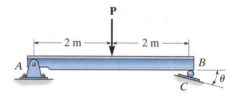

Prob. 8–23

*8–24. The uniform thin pole has a weight of 30 lb and a length of 26 ft. If it is placed against the smooth wall and on the rough floor in the position $d = 10$ ft, will it remain in this position when it is released? The coefficient of static friction is $\mu_s = 0.3$.

8–25. The uniform pole has a weight of 30 lb and a length of 26 ft. Determine the maximum distance d it can be placed from the smooth wall and not slip. The coefficient of static friction between the floor and the pole is $\mu_s = 0.3$.

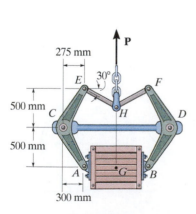

Prob. 8–22

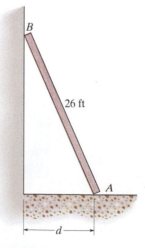

Probs. 8–24/25

8–26. The block brake is used to stop the wheel from rotating when the wheel is subjected to a couple moment $M_0 = 360$ N · m. If the coefficient of static friction between the wheel and the block is $\mu_s = 0.6$, determine the smallest force P that should be applied.

8–27. Solve Prob. 8–26 if the couple moment M_0 is applied counterclockwise.

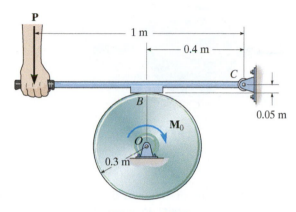

Probs. 8–26/27

8–29. The friction pawl is pinned at A and rests against the wheel at B. It allows freedom of movement when the wheel is rotating counterclockwise about C. Clockwise rotation is prevented due to friction of the pawl which tends to bind the wheel. If $(\mu_s)_B = 0.6$, determine the design angle θ which will prevent clockwise motion for any value of applied moment M. *Hint*: Neglect the weight of the pawl so that it becomes a two-force member.

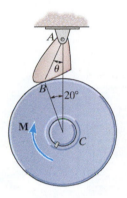

Prob. 8–29

8–30. Two blocks A and B have a weight of 10 lb and 6 lb, respectively. They are resting on the incline for which the coefficients of static friction are $\mu_A = 0.15$ and $\mu_B = 0.25$. Determine the incline angle θ for which both blocks begin to slide. Also find the required stretch or compression in the connecting spring for this to occur. The spring has a stiffness of $k = 2$ lb/ft.

8–31. Two blocks A and B have a weight of 10 lb and 6 lb, respectively. They are resting on the incline for which the coefficients of static friction are $\mu_A = 0.15$ and $\mu_B = 0.25$. Determine the angle θ which will cause motion of one of the blocks. What is the friction force under each of the blocks when this occurs? The spring has a stiffness of $k = 2$ lb/ft and is originally unstretched.

***8–28.** A worker walks up the sloped roof that is defined by the curve $y = (5e^{0.01x})$ ft, where x is in feet. Determine how high h he can go without slipping. The coefficient of static friction is $\mu_s = 0.6$.

Prob. 8–28

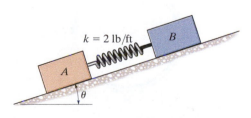

Probs. 8–30/31

8

*8–32. Determine the smallest force P that must be applied in order to cause the 150-lb uniform crate to move. The coefficent of static friction between the crate and the floor is $\mu_s = 0.5$.

8–33. The man having a weight of 200 lb pushes horizontally on the crate. If the coefficient of static friction between the 450-lb crate and the floor is $\mu_s = 0.3$ and between his shoes and the floor is $\mu'_s = 0.6$, determine if he can move the crate.

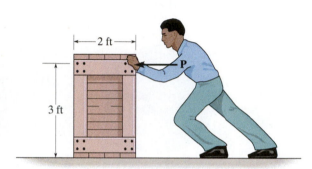

Probs. 8–32/33

8–34. The uniform hoop of weight W is subjected to the horizontal force P. Determine the coefficient of static friction between the hoop and the surface of A and B if the hoop is on the verge of rotating.

8–35. Determine the maximum horizontal force P that can be applied to the 30-lb hoop without causing it to rotate. The coefficient of static friction between the hoop and the surfaces A and B is $\mu_s = 0.2$. Take $r = 300$ mm.

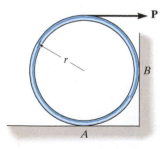

Probs. 8–34/35

*8–36. Determine the minimum force P needed to push the tube E up the incline. The force acts parallel to the plane, and the coefficients of static friction at the contacting surfaces are $\mu_A = 0.2$, $\mu_B = 0.3$, and $\mu_C = 0.4$. The 100-kg roller and 40-kg tube each have a radius of 150 mm.

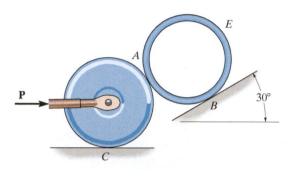

Prob. 8–36

8–37. The coefficients of static and kinetic friction between the drum and brake bar are $\mu_s = 0.4$ and $\mu_k = 0.3$, respectively. If $M = 50$ N·m and $P = 85$ N, determine the horizontal and vertical components of reaction at the pin O. Neglect the weight and thickness of the brake. The drum has a mass of 25 kg.

8–38. The coefficient of static friction between the drum and brake bar is $\mu_s = 0.4$. If the moment $M = 35$ N·m, determine the smallest force P that needs to be applied to the brake bar in order to prevent the drum from rotating. Also determine the corresponding horizontal and vertical components of reaction at pin O. Neglect the weight and thickness of the brake bar. The drum has a mass of 25 kg.

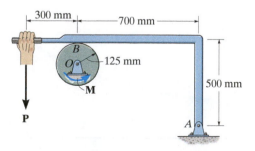

Probs. 8–37/38

8–39. Determine the smallest coefficient of static friction at both *A* and *B* needed to hold the uniform 100-lb bar in equilibrium. Neglect the thickness of the bar. Take $\mu_A = \mu_B = \mu$.

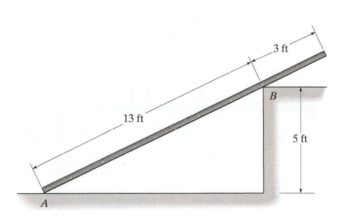

3 ft

13 ft

B

5 ft

A

Prob. 8–39

8–41. If the coefficient of static friction at *A* and *B* is $\mu_s = 0.6$, determine the maximum angle θ so that the frame remains in equilibrium, regardless of the mass of the cylinder. Neglect the mass of the rods.

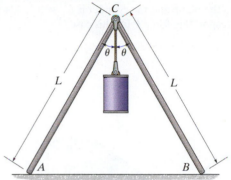

C

θ | θ

L | *L*

A | *B*

Prob. 8–41

8–42. The 100-kg disk rests on a surface for which $\mu_B = 0.2$. Determine the smallest vertical force **P** that can be applied tangentially to the disk which will cause motion to impend.

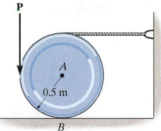

P

A

0.5 m

B

Prob. 8–42

***8–40.** If $\theta = 30°$, determine the minimum coefficient of static friction at *A* and *B* so that equilibrium of the supporting frame is maintained regardless of the mass of the cylinder. Neglect the mass of the rods.

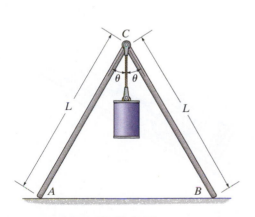

C

θ | θ

L | *L*

A | *B*

Prob. 8–40

8–43. Investigate whether the equilibrium can be maintained. The uniform block has a mass of 500 kg, and the coefficient of static friction is $\mu_s = 0.3$.

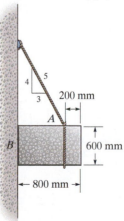

4 | 5

3 | 200 mm

A

B | 600 mm

800 mm

Prob. 8–43

8

*8–44. The homogenous semicylinder has a mass of 20 kg and mass center at G. If force **P** is applied at the edge, and $r = 300\,\text{mm}$, determine the angle θ at which the semicylinder is on the verge of slipping. The coefficient of static friction between the plane and the cylinder is $\mu_s = 0.3$. Also, what is the corresponding force P for this case?

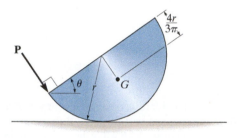

Prob. 8–44

8–46. The beam AB has a negligible mass and thickness and is subjected to a triangular distributed loading. It is supported at one end by a pin and at the other end by a post having a mass of 50 kg and negligible thickness. Determine the two coefficients of static friction at B and at C so that when the magnitude of the applied force is increased to $P = 150\,\text{N}$, the post slips at both B and C simultaneously.

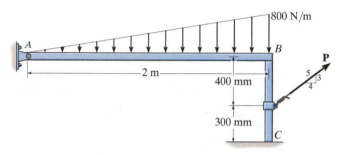

Prob. 8–46

8–45. The beam AB has a negligible mass and thickness and is subjected to a triangular distributed loading. It is supported at one end by a pin and at the other end by a post having a mass of 50 kg and negligible thickness. Determine the minimum force P needed to move the post. The coefficients of static friction at B and C are $\mu_B = 0.4$ and $\mu_C = 0.2$, respectively.

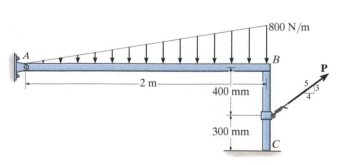

Prob. 8–45

8–47. Crates A and B weigh 200 lb and 150 lb, respectively. They are connected together with a cable and placed on the inclined plane. If the angle θ is gradually increased, determine θ when the crates begin to slide. The coefficients of static friction between the crates and the plane are $\mu_A = 0.25$ and $\mu_B = 0.35$.

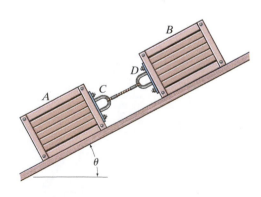

Prob. 8–47

*8–48. Two blocks A and B, each having a mass of 5 kg, are connected by the linkage shown. If the coefficient of static friction at the contacting surfaces is $\mu_s = 0.5$, determine the largest force P that can be applied to pin C of the linkage without causing the blocks to move. Neglect the weight of the links.

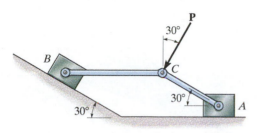

Prob. 8–48

8–49. The uniform crate has a mass of 150 kg. If the coefficient of static friction between the crate and the floor is $\mu_s = 0.2$, determine whether the 85-kg man can move the crate. The coefficient of static friction between his shoes and the floor is $\mu_s' = 0.4$. Assume the man only exerts a horizontal force on the crate.

8–50. The uniform crate has a mass of 150 kg. If the coefficient of static friction between the crate and the floor is $\mu_s = 0.2$, determine the smallest mass of the man so he can move the crate. The coefficient of static friction between his shoes and the floor is $\mu_s' = 0.45$. Assume the man exerts only a horizontal force on the crate.

8–51. Beam AB has a negligible mass and thickness, and supports the 200-kg uniform block. It is pinned at A and rests on the top of a post, having a mass of 20 kg and negligible thickness. Determine the minimum force P needed to move the post. The coefficients of static friction at B and C are $\mu_B = 0.4$ and $\mu_C = 0.2$, respectively.

*8–52. Beam AB has a negligible mass and thickness, and supports the 200-kg uniform block. It is pinned at A and rests on the top of a post, having a mass of 20 kg and negligible thickness. Determine the two coefficients of static friction at B and at C so that when the magnitude of the applied force is increased to $P = 300$ N, the post slips at both B and C simultaneously.

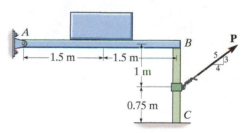

Probs. 8–51/52

8–53. Determine the smallest couple moment that can be applied to the 150-lb wheel that will cause impending motion. The uniform concrete block has a weight of 300 lb. The coefficients of static friction are $\mu_A = 0.2$, $\mu_B = 0.3$, and between the concrete block and the floor, $\mu = 0.4$.

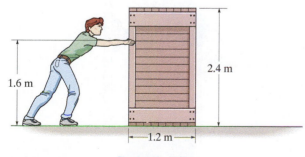

Probs. 8–49/50

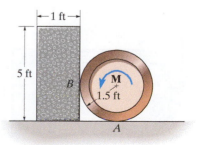

Prob. 8–53

8–54. Determine the greatest angle θ so that the ladder does not slip when it supports the 75-kg man in the position shown. The surface is rather slippery, where the coefficient of static friction at A and B is $\mu_s = 0.3$.

*8–56. The disk has a weight W and lies on a plane that has a coefficient of static friction μ. Determine the maximum height h to which the plane can be lifted without causing the disk to slip.

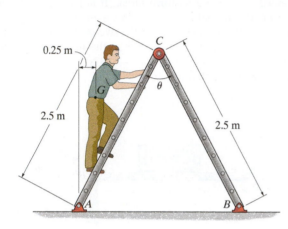

Prob. 8–54

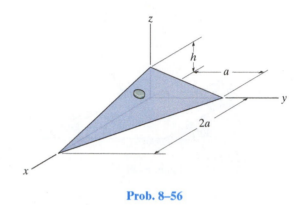

Prob. 8–56

8–57. The man has a weight of 200 lb, and the coefficient of static friction between his shoes and the floor is $\mu_s = 0.5$. Determine where he should position his center of gravity G at d in order to exert the maximum horizontal force on the door. What is this force?

8–55. The wheel weighs 20 lb and rests on a surface for which $\mu_B = 0.2$. A cord wrapped around it is attached to the top of the 30-lb homogeneous block. If the coefficient of static friction at D is $\mu_D = 0.3$, determine the smallest vertical force that can be applied tangentially to the wheel which will cause motion to impend.

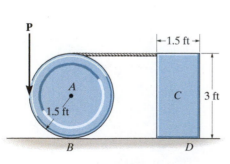

Prob. 8–55

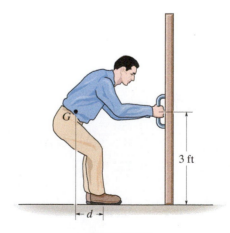

Prob. 8–57

CONCEPTUAL PROBLEMS

C8–1. Draw the free-body diagrams of each of the two members of this friction tong used to lift the 100-kg block.

C8–1 (© Russell C. Hibbeler)

C8–2. Show how to find the force needed to move the top block. Use reasonable data and use an equilibrium analysis to explain your answer.

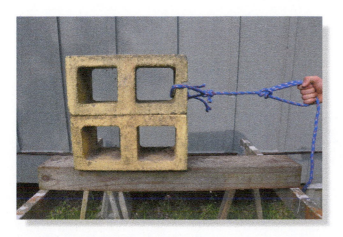

C8–2 (© Russell C. Hibbeler)

C8–3. The rope is used to tow the refrigerator. Is it best to pull slightly up on the rope as shown, pull horizontally, or pull somewhat downwards? Also, is it best to attach the rope at a high position as shown, or at a lower position? Do an equilibrium analysis to explain your answer.

C8–4. The rope is used to tow the refrigerator. In order to prevent yourself from slipping while towing, is it best to pull up as shown, pull horizontally, or pull downwards on the rope? Do an equilibrium analysis to explain your answer.

C8–3/4 (© Russell C. Hibbeler)

C8–5. Explain how to find the maximum force this man can exert on the vehicle. Use reasonable data and use an equilibrium analysis to explain your answer.

C8–5 (© Russell C. Hibbeler)

8

8.3   Wedges

A *wedge* is a simple machine that is often used to transform an applied force into much larger forces, directed at approximately right angles to the applied force. Wedges also can be used to slightly move or adjust heavy loads.

Consider, for example, the wedge shown in Fig. 8–12a, which is used to *lift* the block by applying a force to the wedge. Free-body diagrams of the block and wedge are shown in Fig. 8–12b. Here we have excluded the weight of the wedge since it is usually *small* compared to the weight W of the block. Also, note that the frictional forces F_1 and F_2 must oppose the motion of the wedge. Likewise, the frictional force F_3 of the wall on the block must act downward so as to oppose the block's upward motion. The locations of the resultant normal forces are not important in the force analysis since neither the block nor wedge will "tip." Hence the moment equilibrium equations will not be considered. There are seven unknowns, consisting of the applied force P, needed to cause motion of the wedge, and six normal and frictional forces. The seven available equations consist of four force equilibrium equations, $\Sigma F_x = 0$, $\Sigma F_y = 0$ applied to the wedge and block, and three frictional equations, $F = \mu N$, applied at each surface of contact.

If the block is to be *lowered*, then the frictional forces will all act in a sense opposite to that shown in Fig. 8–12b. Provided the coefficient of friction is very *small* or the wedge angle θ is *large,* then the applied force P must act to the right to hold the block. Otherwise, P may have a reverse sense of direction in order to *pull* on the wedge to remove it. If P is *not applied* and friction forces hold the block in place, then the wedge is referred to as *self-locking.*

Wedges are often used to adjust the elevation of structural or mechanical parts. Also, they provide stability for objects such as this pipe. (© Russell C. Hibbeler)

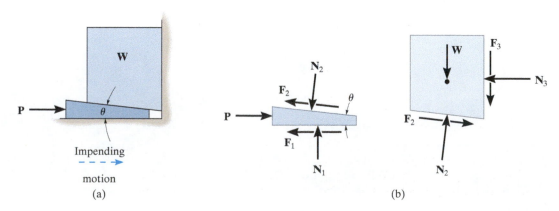

Impending
motion

(a)

(b)

Fig. 8–12

EXAMPLE | 8.6

The uniform stone in Fig. 8–13a has a mass of 500 kg and is held in the horizontal position using a wedge at B. If the coefficient of static friction is $\mu_s = 0.3$ at the surfaces of contact, determine the minimum force P needed to remove the wedge. Assume that the stone does not slip at A.

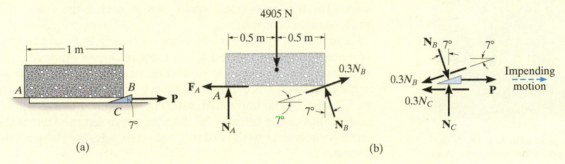

(a)   (b)

Fig. 8–13

SOLUTION

The minimum force P requires $F = \mu_s N$ at the surfaces of contact with the wedge. The free-body diagrams of the stone and wedge are shown in Fig. 8–13b. On the wedge the friction force opposes the impending motion, and on the stone at A, $F_A \leq \mu_s N_A$, since slipping does not occur there. There are five unknowns. Three equilibrium equations for the stone and two for the wedge are available for solution. From the free-body diagram of the stone,

$$\zeta + \Sigma M_A = 0; \quad -4905 \text{ N}(0.5 \text{ m}) + (N_B \cos 7° \text{ N})(1 \text{ m})$$
$$+ (0.3 N_B \sin 7° \text{ N})(1 \text{ m}) = 0$$
$$N_B = 2383.1 \text{ N}$$

Using this result for the wedge, we have

$$+\uparrow \Sigma F_y = 0; \quad N_C - 2383.1 \cos 7° \text{ N} - 0.3(2383.1 \sin 7° \text{ N}) = 0$$
$$N_C = 2452.5 \text{ N}$$

$$\overset{+}{\rightarrow} \Sigma F_x = 0; \quad 2383.1 \sin 7° \text{ N} - 0.3(2383.1 \cos 7° \text{ N}) +$$
$$P - 0.3(2452.5 \text{ N}) = 0$$
$$P = 1154.9 \text{ N} = 1.15 \text{ kN} \qquad\qquad \textit{Ans.}$$

NOTE: Since P is positive, indeed the wedge must be pulled out. If P were zero, the wedge would remain in place (self-locking) and the frictional forces developed at B and C would satisfy $F_B < \mu_s N_B$ and $F_C < \mu_s N_C$.

8.4   Frictional Forces on Screws

In most cases, screws are used as fasteners; however, in many types of machines they are incorporated to transmit power or motion from one part of the machine to another. A **square-threaded screw** is commonly used for the latter purpose, especially when large forces are applied along its axis. In this section, we will analyze the forces acting on square-threaded screws. The analysis of other types of screws, such as the V-thread, is based on these same principles.

For analysis, a square-threaded screw, as in Fig. 8–14, can be considered a cylinder having an inclined square ridge or **thread** wrapped around it. If we unwind the thread by one revolution, as shown in Fig. 8–14b, the slope or the **lead angle** θ is determined from $\theta = \tan^{-1}(l/2\pi r)$. Here l and $2\pi r$ are the vertical and horizontal distances between A and B, where r is the mean radius of the thread. The distance l is called the **lead** of the screw and it is equivalent to the distance the screw advances when it turns one revolution.

Upward Impending Motion.   Let us now consider the case of the square-threaded screw jack in Fig. 8–15 that is subjected to upward impending motion caused by the applied torsional moment *M. A free-body diagram of the *entire unraveled thread h* in contact with the jack can be represented as a block, as shown in Fig. 8–16a. The force **W** is the vertical force acting on the thread or the axial force applied to the shaft, Fig. 8–15, and M/r is the resultant horizontal force produced by the couple moment M about the axis of the shaft. The reaction **R** of the groove on the thread has both frictional and normal components, where $F = \mu_s N$. The angle of static friction is $\phi_s = \tan^{-1}(F/N) = \tan^{-1}\mu_s$. Applying the force equations of equilibrium along the horizontal and vertical axes, we have

$$\xrightarrow{+}\ \Sigma F_x = 0; \qquad M/r - R\sin(\theta + \phi_s) = 0$$

$$+\uparrow\Sigma F_y = 0; \qquad R\cos(\theta + \phi_s) - W = 0$$

Eliminating R from these equations, we obtain

$$\boxed{M = rW\tan(\theta + \phi_s)} \qquad (8\text{–}3)$$

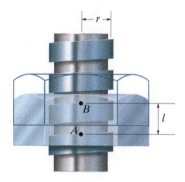

Square-threaded screws find applications on valves, jacks, and vises, where particularly large forces must be developed along the axis of the screw. (© Russell C. Hibbeler)

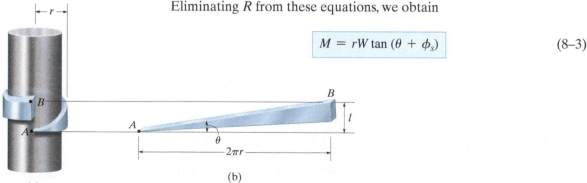

(a)

(b)

Fig. 8–14

*For applications, **M** is developed by applying a horizontal force **P** at a right angle to the end of a lever that would be fixed to the screw.

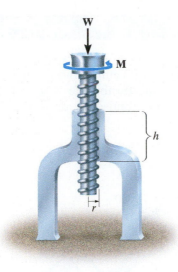

Fig. 8–15

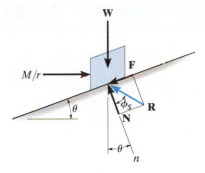

Upward screw motion
(a)

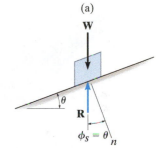

Self-locking screw ($\theta = \phi_S$)
(on the verge of rotating downward)

(b)

Self-Locking Screw.

A screw is said to be *self-locking* if it remains in place under any axial load **W** when the moment **M** is removed. For this to occur, the direction of the frictional force must be reversed so that **R** acts on the other side of **N**. Here the angle of static friction ϕ_s becomes greater than or equal to θ, Fig. 8–16d. If $\phi_s = \theta$, Fig. 8–16b, then **R** will act vertically to balance **W**, and the screw will be on the verge of winding downward.

Downward Impending Motion, $(\theta > \phi_s)$.

If the screw is not self-locking, it is necessary to apply a moment **M′** to prevent the screw from winding downward. Here, a horizontal force M'/r is required to push against the thread to prevent it from sliding down the plane, Fig. 8–16c. Using the same procedure as before, the magnitude of the moment **M′** required to prevent this unwinding is

$$M' = rW \tan(\theta - \phi_s) \qquad (8\text{--}4)$$

Downward Impending Motion, $(\phi_s > \theta)$.

If a screw is self-locking, a couple moment **M″** must be applied to the screw in the opposite direction to wind the screw downward ($\phi_s > \theta$). This causes a reverse horizontal force M''/r that pushes the thread down as indicated in Fig. 8–16d. In this case, we obtain

$$M'' = rW \tan(\phi_s - \theta) \qquad (8\text{--}5)$$

If *motion of the screw* occurs, Eqs. 8–3, 8–4, and 8–5 can be applied by simply replacing ϕ_s with ϕ_k.

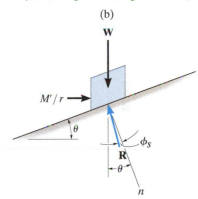

Downward screw motion ($\theta > \phi_S$)
(c)

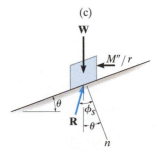

Downward screw motion ($\theta < \phi_S$)
(d)

Fig. 8–16

EXAMPLE | 8.7

The turnbuckle shown in Fig. 8–17 has a square thread with a mean radius of 5 mm and a lead of 2 mm. If the coefficient of static friction between the screw and the turnbuckle is $\mu_s = 0.25$, determine the moment **M** that must be applied to draw the end screws closer together.

(© Russell C. Hibbeler)

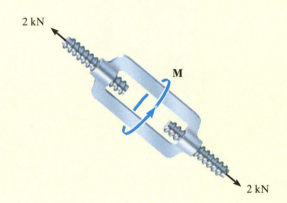

Fig. 8–17

SOLUTION

The moment can be obtained by applying Eq. 8–3. Since friction at *two screws* must be overcome, this requires

$$M = 2[rW \tan(\theta + \phi_s)] \qquad (1)$$

Here $W = 2000$ N, $\phi_s = \tan^{-1}\mu_s = \tan^{-1}(0.25) = 14.04°$, $r = 5$ mm, and $\theta = \tan^{-1}(l/2\pi r) = \tan^{-1}(2 \text{ mm}/[2\pi(5 \text{ mm})]) = 3.64°$. Substituting these values into Eq. 1 and solving gives

$$M = 2[(2000 \text{ N})(5 \text{ mm}) \tan(14.04° + 3.64°)]$$

$$= 6374.7 \text{ N} \cdot \text{mm} = 6.37 \text{ N} \cdot \text{m} \qquad \textit{Ans.}$$

NOTE: When the moment is *removed*, the turnbuckle will be self-locking; i.e., it will not unscrew since $\phi_s > \theta$.

PROBLEMS

8–58. Determine the largest angle θ that will cause the wedge to be self-locking regardless of the magnitude of horizontal force P applied to the blocks. The coefficient of static friction between the wedge and the blocks is $\mu_s = 0.3$. Neglect the weight of the wedge.

Prob. 8–58

8–59. If the beam AD is loaded as shown, determine the horizontal force P which must be applied to the wedge in order to remove it from under the beam. The coefficients of static friction at the wedge's top and bottom surfaces are $\mu_{CA} = 0.25$ and $\mu_{CB} = 0.35$, respectively. If $P = 0$, is the wedge self-locking? Neglect the weight and size of the wedge and the thickness of the beam.

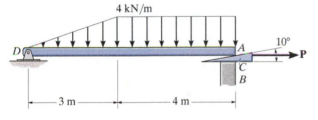

Prob. 8–59

*8–60. The wedge is used to level the member. Determine the horizontal force **P** that must be applied to begin to push the wedge forward. The coefficient of static friction between the wedge and the two surfaces of contact is $\mu_s = 0.2$. Neglect the weight of the wedge.

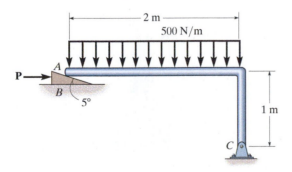

Prob. 8–60

8–61. The two blocks used in a measuring device have negligible weight. If the spring is compressed 5 in. when in the position shown, determine the smallest axial force P which the adjustment screw must exert on B in order to start the movement of B downward. The end of the screw is *smooth* and the coefficient of static friction at all other points of contact is $\mu_s = 0.3$.

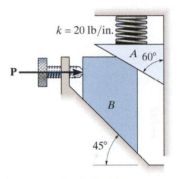

Prob. 8–61

8–62. If $P = 250$ N, determine the required minimum compression in the spring so that the wedge will not move to the right. Neglect the weight of A and B. The coefficient of static friction for all contacting surfaces is $\mu_s = 0.35$. Neglect friction at the rollers.

8–63. Determine the minimum applied force **P** required to move wedge A to the right. The spring is compressed a distance of 175 mm. Neglect the weight of A and B. The coefficient of static friction for all contacting surfaces is $\mu_s = 0.35$. Neglect friction at the rollers.

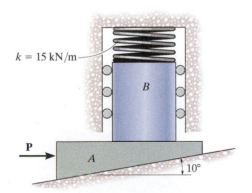

Probs. 8–62/63

*8–64. If the coefficient of static friction between all the surfaces of contact is μ_s, determine the force **P** that must be applied to the wedge in order to lift the block having a weight W.

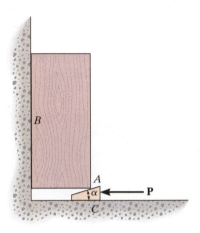

Prob. 8–64

8–65. Determine the smallest force P needed to lift the 3000-lb load. The coefficient of static friction between A and C and between B and D is $\mu_s = 0.3$, and between A and B $\mu'_s = 0.4$. Neglect the weight of each wedge.

8–66. Determine the reversed horizontal force $-$**P** needed to pull out wedge A. The coefficient of static friction between A and C and between B and D is $\mu_s = 0.2$, and between A and B $\mu'_s = 0.1$. Neglect the weight of each wedge.

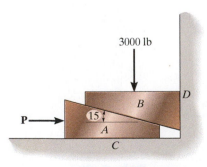

Probs. 8–65/66

8–67. If the clamping force at G is 900 N, determine the horizontal force **F** that must be applied perpendicular to the handle of the lever at E. The mean diameter and lead of both single square-threaded screws at C and D are 25 mm and 5 mm, respectively. The coefficient of static friction is $\mu_s = 0.3$.

*8–68. If a horizontal force of $F = 50$ N is applied perpendicular to the handle of the lever at E, determine the clamping force developed at G. The mean diameter and lead of the single square-threaded screw at C and D are 25 mm and 5 mm, respectively. The coefficient of static friction is $\mu_s = 0.3$.

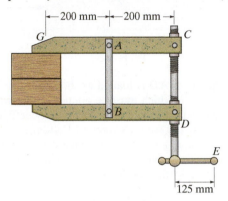

Probs. 8–67/68

8–69. The column is used to support the upper floor. If a force $F = 80$ N is applied perpendicular to the handle to tighten the screw, determine the compressive force in the column. The square-threaded screw on the jack has a coefficient of static friction of $\mu_s = 0.4$, mean diameter of 25 mm, and a lead of 3 mm.

8–70. If the force **F** is removed from the handle of the jack in Prob. 8–69, determine if the screw is self-locking.

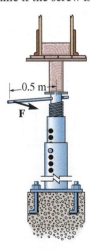

Probs. 8–69/70

8–71. If couple forces of $F = 10$ lb are applied perpendicular to the lever of the clamp at A and B, determine the clamping force on the boards. The single square-threaded screw of the clamp has a mean diameter of 1 in. and a lead of 0.25 in. The coefficient of static friction is $\mu_s = 0.3$.

***8–72.** If the clamping force on the boards is 600 lb, determine the required magnitude of the couple forces that must be applied perpendicular to the lever AB of the clamp at A and B in order to loosen the screw. The single square-threaded screw has a mean diameter of 1 in. and a lead of 0.25 in. The coefficient of static friction is $\mu_s = 0.3$.

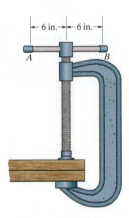

Probs. 8–71/72

8–73. Prove that the lead l must be less than $2\pi r\mu_s$ for the jack screw shown in Fig. 8–15 to be "self-locking."

8–74. The square-threaded bolt is used to join two plates together. If the bolt has a mean diameter of $d = 20$ mm and a lead of $l = 3$ mm, determine the smallest torque M required to loosen the bolt if the tension in the bolt is $T = 40$ kN. The coefficient of static friction between the threads and the bolt is $\mu_s = 0.15$.

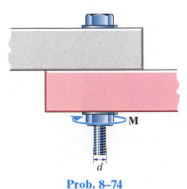

Prob. 8–74

8–75. The shaft has a square-threaded screw with a lead of 8 mm and a mean radius of 15 mm. If it is in contact with a plate gear having a mean radius of 30 mm, determine the resisting torque **M** on the plate gear which can be overcome if a torque of 7 N · m is applied to the shaft. The coefficient of static friction at the screw is $\mu_B = 0.2$. Neglect friction of the bearings located at A and B.

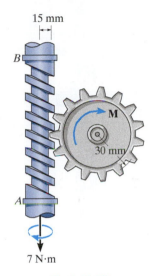

Prob. 8–75

***8–76.** If couple forces of $F = 35$ N are applied to the handle of the machinist's vise, determine the compressive force developed in the block. Neglect friction at the bearing A. The guide at B is smooth. The single square-threaded screw has a mean radius of 6 mm and a lead of 8 mm, and the coefficient of static friction is $\mu_s = 0.27$.

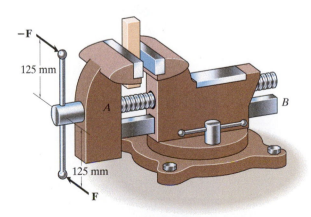

Prob. 8–76

8–77. The square-threaded screw has a mean diameter of 20 mm and a lead of 4 mm. If the weight of the plate A is 5 lb, determine the smallest coefficient of static friction between the screw and the plate so that the plate does not travel down the screw when the plate is suspended as shown.

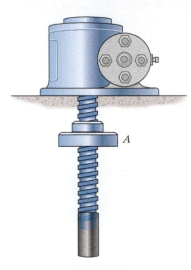

Prob. 8–77

8–78. The device is used to pull the battery cable terminal C from the post of a battery. If the required pulling force is 85 lb, determine the torque **M** that must be applied to the handle on the screw to tighten it. The screw has square threads, a mean diameter of 0.2 in., a lead of 0.08 in., and the coefficient of static friction is $\mu_s = 0.5$.

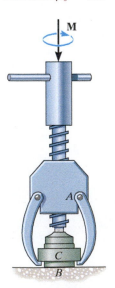

Prob. 8–78

8–79. Determine the clamping force on the board A if the screw is tightened with a torque of $M = 8$ N · m. The square-threaded screw has a mean radius of 10 mm and a lead of 3 mm, and the coefficient of static friction is $\mu_s = 0.35$.

*8–80. If the required clamping force at the board A is to be 2 kN, determine the torque M that must be applied to the screw to tighten it down. The square-threaded screw has a mean radius of 10 mm and a lead of 3 mm, and the coefficient of static friction is $\mu_s = 0.35$.

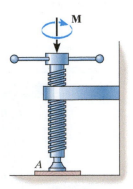

Probs. 8–79/80

8–81. If a horizontal force of $P = 100$ N is applied perpendicular to the handle of the lever at A, determine the compressive force **F** exerted on the material. Each single square-threaded screw has a mean diameter of 25 mm and a lead of 7.5 mm. The coefficient of static friction at all contacting surfaces of the wedges is $\mu_s = 0.2$, and the coefficient of static friction at the screw is $\mu_s' = 0.15$.

8–82. Determine the horizontal force **P** that must be applied perpendicular to the handle of the lever at A in order to develop a compressive force of 12 kN on the material. Each single square-threaded screw has a mean diameter of 25 mm and a lead of 7.5 mm. The coefficient of static friction at all contacting surfaces of the wedges is $\mu_s = 0.2$, and the coefficient of static friction at the screw is $\mu_s' = 0.15$.

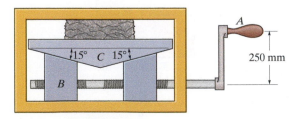

Probs. 8–81/82

8.5 Frictional Forces on Flat Belts

Whenever belt drives or band brakes are designed, it is necessary to determine the frictional forces developed between the belt and its contacting surface. In this section we will analyze the frictional forces acting on a flat belt, although the analysis of other types of belts, such as the V-belt, is based on similar principles.

Consider the flat belt shown in Fig. 8–18a, which passes over a fixed curved surface. The total angle of belt-to-surface contact in radians is β, and the coefficient of friction between the two surfaces is μ. We wish to determine the tension T_2 in the belt, which is needed to pull the belt counterclockwise over the surface, and thereby overcome both the frictional forces at the surface of contact and the tension T_1 in the other end of the belt. Obviously, $T_2 > T_1$.

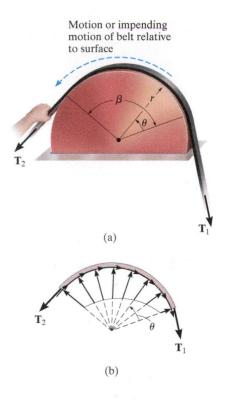

Motion or impending motion of belt relative to surface

(a)

Frictional Analysis. A free-body diagram of the belt segment in contact with the surface is shown in Fig. 8–18b. As shown, the normal and frictional forces, acting at different points along the belt, will vary both in magnitude and direction. Due to this *unknown* distribution, the analysis of the problem will first require a study of the forces acting on a differential element of the belt.

A free-body diagram of an element having a length ds is shown in Fig. 8–18c. Assuming either impending motion or motion of the belt, the magnitude of the frictional force $dF = \mu \, dN$. This force opposes the sliding motion of the belt, and so it will increase the magnitude of the tensile force acting in the belt by dT. Applying the two force equations of equilibrium, we have

$$\searrow +\Sigma F_x = 0; \qquad T\cos\left(\frac{d\theta}{2}\right) + \mu\, dN - (T + dT)\cos\left(\frac{d\theta}{2}\right) = 0$$

$$+\nearrow\Sigma F_y = 0; \qquad dN - (T + dT)\sin\left(\frac{d\theta}{2}\right) - T\sin\left(\frac{d\theta}{2}\right) = 0$$

Since $d\theta$ is of *infinitesimal size*, $\sin(d\theta/2) = d\theta/2$ and $\cos(d\theta/2) = 1$. Also, the *product* of the two infinitesimals dT and $d\theta/2$ may be neglected when compared to infinitesimals of the first order. As a result, these two equations become

$$\mu\, dN = dT$$

and

$$dN = T\, d\theta$$

Eliminating dN yields

$$\frac{dT}{T} = \mu\, d\theta$$

(b)

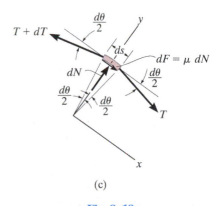

(c)

Fig. 8–18

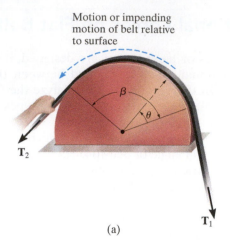

Motion or impending
motion of belt relative
to surface

T_2

T_1

(a)

Fig. 8–18 (Repeated)

Flat or V-belts are often used to transmit
the torque developed by a motor to a
wheel attached to a pump, fan, or blower.
(© Russell C. Hibbeler)

8

Integrating this equation between all the points of contact that the belt makes with the drum, and noting that $T = T_1$ at $\theta = 0$ and $T = T_2$ at $\theta = \beta$, yields

$$\int_{T_1}^{T_2} \frac{dT}{T} = \mu \int_0^\beta d\theta$$

$$\ln \frac{T_2}{T_1} = \mu\beta$$

Solving for T_2, we obtain

$$\boxed{T_2 = T_1 e^{\mu\beta}} \qquad (8\text{–}6)$$

where

T_2, T_1 = belt tensions; T_1 opposes the direction of motion (or impending motion) of the belt measured relative to the surface, while T_2 acts in the direction of the relative belt motion (or impending motion); because of friction, $T_2 > T_1$

μ = coefficient of static or kinetic friction between the belt and the surface of contact

β = angle of belt-to-surface contact, measured in radians

e = 2.718 . . . , base of the natural logarithm

Note that T_2 is *independent* of the *radius* of the drum, and instead it is a function of the angle of belt to surface contact, β. As a result, this equation is valid for flat belts passing over any curved contacting surface.

EXAMPLE 8.8

The maximum tension that can be developed in the cord shown in Fig. 8–19a is 500 N. If the pulley at A is free to rotate and the coefficient of static friction at the fixed drums B and C is $\mu_s = 0.25$, determine the largest mass of the cylinder that can be lifted by the cord.

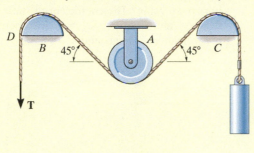

(a)

SOLUTION

Lifting the cylinder, which has a weight $W = mg$, causes the cord to move counterclockwise over the drums at B and C; hence, the maximum tension T_2 in the cord occurs at D. Thus, $F = T_2 = 500$ N. A section of the cord passing over the drum at B is shown in Fig. 8–19b. Since $180° = \pi$ rad the angle of contact between the drum and the cord is $\beta = (135°/180°)\pi = 3\pi/4$ rad. Using Eq. 8–6, we have

$$T_2 = T_1 e^{\mu_s \beta}; \qquad 500\ \text{N} = T_1 e^{0.25[(3/4)\pi]}$$

Hence,

$$T_1 = \frac{500\ \text{N}}{e^{0.25[(3/4)\pi]}} = \frac{500\ \text{N}}{1.80} = 277.4\ \text{N}$$

Since the pulley at A is free to rotate, equilibrium requires that the tension in the cord remains the *same* on both sides of the pulley.

The section of the cord passing over the drum at C is shown in Fig. 8–19c. The weight $W < 277.4$ N. Why? Applying Eq. 8–6, we obtain

$$T_2 = T_1 e^{\mu_s \beta}; \qquad 277.4\ \text{N} = We^{0.25[(3/4)\pi]}$$

$$W = 153.9\ \text{N}$$

so that

$$m = \frac{W}{g} = \frac{153.9\ \text{N}}{9.81\ \text{m/s}^2}$$

$$= 15.7\ \text{kg} \qquad \textit{Ans.}$$

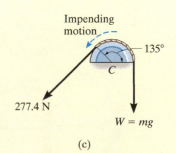

(b)

(c)

Fig. 8–19

8

PROBLEMS

8–83. A cylinder having a mass of 250 kg is to be supported by the cord that wraps over the pipe. Determine the smallest vertical force **F** needed to support the load if the cord passes (a) once over the pipe, $\beta = 180°$, and (b) two times over the pipe, $\beta = 540°$. Take $\mu_s = 0.2$.

*8–84. A cylinder having a mass of 250 kg is to be supported by the cord that wraps over the pipe. Determine the largest vertical force **F** that can be applied to the cord without moving the cylinder. The cord passes (a) once over the pipe, $\beta = 180°$, and (b) two times over the pipe, $\beta = 540°$. Take $\mu_s = 0.2$.

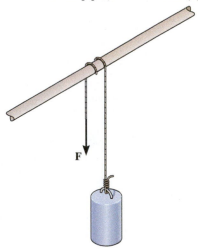

Probs. 8–83/84

8–85. A 180-lb farmer tries to restrain the cow from escaping by wrapping the rope two turns around the tree trunk as shown. If the cow exerts a force of 250 lb on the rope, determine if the farmer can successfully restrain the cow. The coefficient of static friction between the rope and the tree trunk is $\mu_s = 0.15$, and between the farmer's shoes and the ground $\mu_s' = 0.3$.

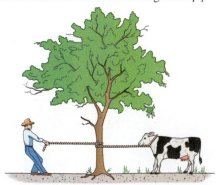

Prob. 8–85

8–86. The 100-lb boy at A is suspended from the cable that passes over the quarter circular cliff rock. Determine if it is possible for the 185-lb woman to hoist him up; and if this is possible, what smallest force must she exert on the horizontal cable? The coefficient of static friction between the cable and the rock is $\mu_s = 0.2$, and between the shoes of the woman and the ground $\mu_s' = 0.8$.

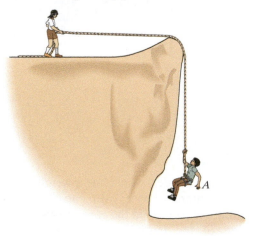

Prob. 8–86

8–87. The 100-lb boy at A is suspended from the cable that passes over the quarter circular cliff rock. What horizontal force must the woman at A exert on the cable in order to let the boy descend at constant velocity? The coefficients of static and kinetic friction between the cable and the rock are $\mu_s = 0.4$ and $\mu_k = 0.35$, respectively.

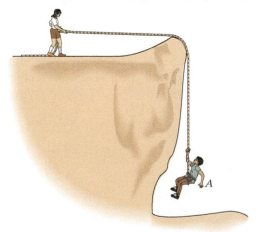

Prob. 8–87

*8–88. The uniform concrete pipe has a weight of 800 lb and is unloaded slowly from the truck bed using the rope and skids shown. If the coefficient of kinetic friction between the rope and pipe is $\mu_k = 0.3$, determine the force the worker must exert on the rope to lower the pipe at constant speed. There is a pulley at B, and the pipe does not slip on the skids. The lower portion of the rope is parallel to the skids.

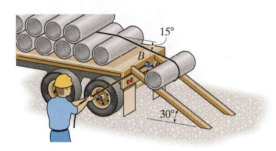

Prob. 8–88

8–89. A cable is attached to the 20-kg plate B, passes over a fixed peg at C, and is attached to the block at A. Using the coefficients of static friction shown, determine the smallest mass of block A so that it will prevent sliding motion of B down the plane.

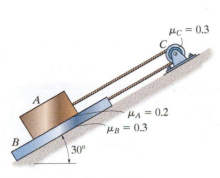

Prob. 8–89

8–90. The smooth beam is being hoisted using a rope that is wrapped around the beam and passes through a ring at A as shown. If the end of the rope is subjected to a tension **T** and the coefficient of static friction between the rope and ring is $\mu_s = 0.3$, determine the smallest angle of θ for equilibrium.

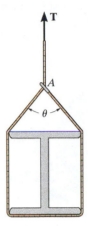

Prob. 8–90

8–91. The boat has a weight of 500 lb and is held in position off the side of a ship by the spars at A and B. A man having a weight of 130 lb gets in the boat, wraps a rope around an overhead boom at C, and ties it to the end of the boat as shown. If the boat is disconnected from the spars, determine the *minimum number of half turns* the rope must make around the boom so that the boat can be safely lowered into the water at constant velocity. Also, what is the normal force between the boat and the man? The coefficient of kinetic friction between the rope and the boom is $\mu_s = 0.15$. *Hint:* The problem requires that the normal force between the man's feet and the boat be as small as possible.

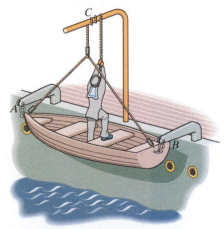

Prob. 8–91

8

*8–92.   Determine the force P that must be applied to the handle of the lever so that the wheel is on the verge of turning if $M = 300 \text{ N} \cdot \text{m}$. The coefficient of static friction between the belt and the wheel is $\mu_s = 0.3$.

8–93.   If a force of $P = 30 \text{ N}$ is applied to the handle of the lever, determine the largest couple moment $\mathbf{M}$ that can be resisted so that the wheel does not turn. The coefficient of static friction between the belt and the wheel is $\mu_s = 0.3$.

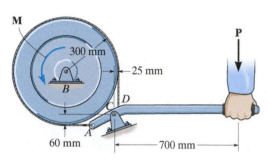

Probs. 8–92/93

8–94.   A minimum force of $P = 50 \text{ lb}$ is required to hold the cylinder from slipping against the belt and the wall. Determine the weight of the cylinder if the coefficient of friction between the belt and cylinder is $\mu_s = 0.3$ and slipping does not occur at the wall.

8–95.   The cylinder weighs 10 lb and is held in equilibrium by the belt and wall. If slipping does not occur at the wall, determine the minimum vertical force P which must be applied to the belt for equilibrium. The coefficient of static friction between the belt and the cylinder is $\mu_s = 0.25$.

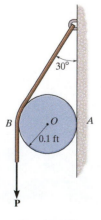

Probs. 8–94/95

*8–96.   Determine the maximum and the minimum values of weight W which may be applied without causing the 50-lb block to slip. The coefficient of static friction between the block and the plane is $\mu_s = 0.2$, and between the rope and the drum D is $\mu'_s = 0.3$.

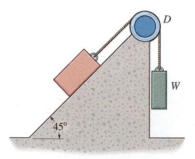

Prob. 8–96

8–97.   Granular material, having a density of 1.5 Mg/m^3, is transported on a conveyor belt that slides over the fixed surface, having a coefficient of kinetic friction of $\mu_k = 0.3$. Operation of the belt is provided by a motor that supplies a torque $\mathbf{M}$ to wheel A. The wheel at B is free to turn, and the coefficient of static friction between the wheel at A and the belt is $\mu_A = 0.4$. If the belt is subjected to a pretension of 300 N when no load is on the belt, determine the greatest volume V of material that is permitted on the belt at any time without allowing the belt to stop. What is the torque $\mathbf{M}$ required to drive the belt when it is subjected to this maximum load?

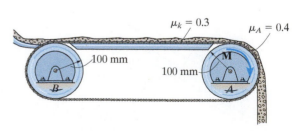

Prob. 8–97

8–98. Show that the frictional relationship between the belt tensions, the coefficient of friction μ, and the angular contacts α and β for the V-belt is $T_2 = T_1 e^{\mu\beta/\sin(\alpha/2)}$.

*8–100. Blocks A and B have a mass of 7 kg and 10 kg, respectively. Using the coefficients of static friction indicated, determine the largest vertical force P which can be applied to the cord without causing motion.

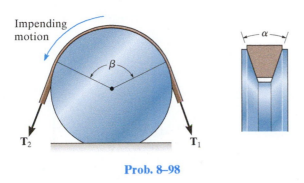

Prob. 8–98

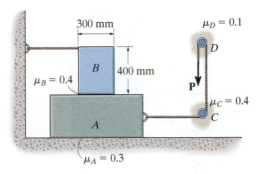

Prob. 8–100

8–99. The wheel is subjected to a torque of $M = 50\,\text{N} \cdot \text{m}$. If the coefficient of kinetic friction between the band brake and the rim of the wheel is $\mu_k = 0.3$, determine the smallest horizontal force P that must be applied to the lever to stop the wheel.

8–101. The uniform bar AB is supported by a rope that passes over a frictionless pulley at C and a fixed peg at D. If the coefficient of static friction between the rope and the peg is $\mu_D = 0.3$, determine the smallest distance x from the end of the bar at which a 20-N force may be placed and not cause the bar to move.

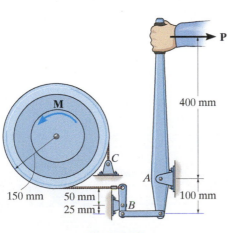

Prob. 8–99

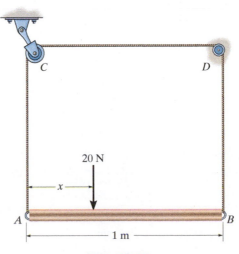

Prob. 8–101

8–102. The belt on the portable dryer wraps around the drum D, idler pulley A, and motor pulley B. If the motor can develop a maximum torque of $M = 0.80\,\text{N}\cdot\text{m}$, determine the smallest spring tension required to hold the belt from slipping. The coefficient of static friction between the belt and the drum and motor pulley is $\mu_s = 0.3$.

8–105. A 10-kg cylinder D, which is attached to a small pulley B, is placed on the cord as shown. Determine the largest angles θ so that the cord does not slip over the peg at C. The cylinder at E also has a mass of 10 kg, and the coefficient of static friction between the cord and the peg is $\mu_s = 0.1$.

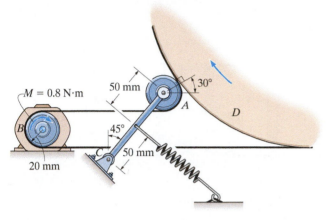

Prob. 8–102

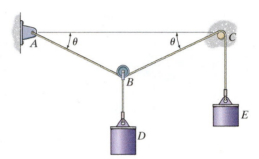

Prob. 8–105

8–103. Blocks A and B weigh 50 lb and 30 lb, respectively. Using the coefficients of static friction indicated, determine the greatest weight of block D without causing motion.

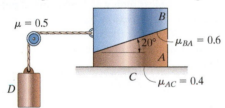

Prob. 8–103

***8–104.** The 20-kg motor has a center of gravity at G and is pin connected at C to maintain a tension in the drive belt. Determine the smallest counterclockwise twist or torque M that must be supplied by the motor to turn the disk B if wheel A locks and causes the belt to slip over the disk. No slipping occurs at A. The coefficient of static friction between the belt and the disk is $\mu_s = 0.3$.

8–106. A conveyer belt is used to transfer granular material and the frictional resistance on the top of the belt is $F = 500\,\text{N}$. Determine the smallest stretch of the spring attached to the moveable axle of the idle pulley B so that the belt does not slip at the drive pulley A when the torque M is applied. What minimum torque M is required to keep the belt moving? The coefficient of static friction between the belt and the wheel at A is $\mu_s = 0.2$.

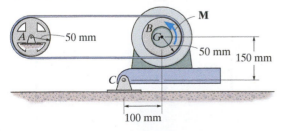

Prob. 8–104

Prob. 8–106

*8.6   Frictional Forces on Collar Bearings, Pivot Bearings, and Disks

Pivot and *collar bearings* are commonly used in machines to support an *axial load* on a rotating shaft. Typical examples are shown in Fig. 8–20. Provided these bearings are not lubricated, or are only partially lubricated, the laws of dry friction may be applied to determine the moment needed to turn the shaft when it supports an axial force.

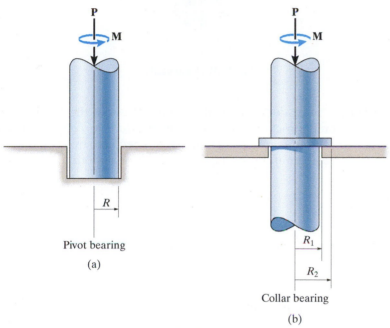

Pivot bearing

(a)

Collar bearing

(b)

Fig. 8–20

Frictional Analysis.   The collar bearing on the shaft shown in Fig. 8–21 is subjected to an axial force **P** and has a total bearing or contact area $\pi(R_2^2 - R_1^2)$. Provided the bearing is new and evenly supported, then the normal pressure p on the bearing will be *uniformly distributed* over this area. Since $\Sigma F_z = 0$, then p, measured as a force per unit area, is $p = P/\pi(R_2^2 - R_1^2)$.

The moment needed to cause impending rotation of the shaft can be determined from moment equilibrium about the z axis. A differential area element $dA = (r\,d\theta)(dr)$, shown in Fig. 8–21, is subjected to both a normal force $dN = p\,dA$ and an associated frictional force,

$$dF = \mu_s\,dN = \mu_s p\,dA = \frac{\mu_s P}{\pi(R_2^2 - R_1^2)}dA$$

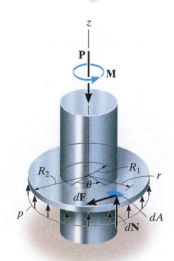

Fig. 8–21

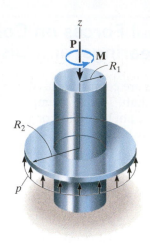

Fig. 8–21 (Repeated)

The normal force does not create a moment about the z axis of the shaft; however, the frictional force does; namely, $dM = r\,dF$. Integration is needed to compute the applied moment **M** needed to overcome all the frictional forces. Therefore, for impending rotational motion,

$$\Sigma M_z = 0; \qquad\qquad M - \int_A r\,dF = 0$$

Substituting for dF and dA and integrating over the entire bearing area yields

$$M = \int_{R_1}^{R_2}\int_0^{2\pi} r\left[\frac{\mu_s P}{\pi(R_2^2 - R_1^2)}\right](r\,d\theta\,dr) = \frac{\mu_s P}{\pi(R_2^2 - R_1^2)}\int_{R_1}^{R_2} r^2\,dr\int_0^{2\pi} d\theta$$

or

$$M = \frac{2}{3}\mu_s P\left(\frac{R_2^3 - R_1^3}{R_2^2 - R_1^2}\right) \qquad\qquad (8\text{--}7)$$

The moment developed at the end of the shaft, when it is *rotating* at constant speed, can be found by substituting μ_k for μ_s in Eq. 8–7.

In the case of a pivot bearing, Fig. 8–20a, then $R_2 = R$ and $R_1 = 0$, and Eq. 8–7 reduces to

$$M = \frac{2}{3}\mu_s PR \qquad\qquad (8\text{--}8)$$

Remember that Eqs. 8–7 and 8–8 apply only for bearing surfaces subjected to *constant pressure*. If the pressure is not uniform, a variation of the pressure as a function of the bearing area must be determined before integrating to obtain the moment. The following example illustrates this concept.

The motor that turns the disk of this sanding machine develops a torque that must overcome the frictional forces acting on the disk. (© Russell C. Hibbeler)

EXAMPLE | 8.9

The uniform bar shown in Fig. 8–22a has a weight of 4 lb. If it is assumed that the normal pressure acting at the contacting surface varies linearly along the length of the bar as shown, determine the couple moment **M** required to rotate the bar. Assume that the bar's width is negligible in comparison to its length. The coefficient of static friction is equal to $\mu_s = 0.3$.

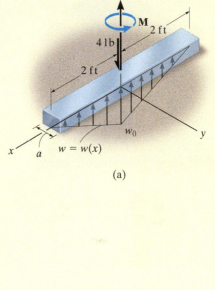

(a)

SOLUTION

A free-body diagram of the bar is shown in Fig. 8–22b. The intensity w_0 of the distributed load at the center ($x = 0$) is determined from vertical force equilibrium, Fig. 8–22a.

$$+\uparrow \Sigma F_z = 0; \qquad -4 \text{ lb} + 2\left[\frac{1}{2}\left(2 \text{ ft}\right)w_0\right] = 0 \qquad w_0 = 2 \text{ lb/ft}$$

Since $w = 0$ at $x = 2$ ft, the distributed load expressed as a function of x is

$$w = (2 \text{ lb/ft})\left(1 - \frac{x}{2 \text{ ft}}\right) = 2 - x$$

The magnitude of the normal force acting on a differential segment of area having a length dx is therefore

$$dN = w \, dx = (2 - x)dx$$

The magnitude of the frictional force acting on the same element of area is

$$dF = \mu_s \, dN = 0.3(2 - x)dx$$

Hence, the moment created by this force about the z axis is

$$dM = x \, dF = 0.3(2x - x^2)dx$$

The summation of moments about the z axis of the bar is determined by integration, which yields

$$\Sigma M_z = 0; \qquad M - 2\int_0^2 (0.3)(2x - x^2)\, dx = 0$$

$$M = 0.6\left(x^2 - \frac{x^3}{3}\right)\Big|_0^2$$

$$M = 0.8 \text{ lb} \cdot \text{ft} \qquad \textit{Ans.}$$

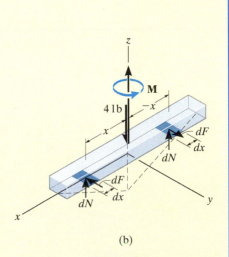

(b)

Fig. 8–22

8.7   Frictional Forces on Journal Bearings

When a shaft or axle is subjected to lateral loads, a *journal bearing* is commonly used for support. Provided the bearing is not lubricated, or is only partially lubricated, a reasonable analysis of the frictional resistance on the bearing can be based on the laws of dry friction.

Frictional Analysis.   A typical journal-bearing support is shown in Fig. 8–23*a*. As the shaft rotates, the contact point moves up the wall of the bearing to some point *A* where slipping occurs. If the vertical load acting at the end of the shaft is **P**, then the bearing reactive force **R** acting at *A* will be equal and opposite to **P**, Fig. 8–23*b*. The moment needed to maintain constant rotation of the shaft can be found by summing moments about the *z* axis of the shaft; i.e.,

$$\Sigma M_z = 0; \qquad M - (R \sin \phi_k)r = 0$$

or

$$M = Rr \sin \phi_k \qquad (8\text{–}9)$$

where ϕ_k is the angle of kinetic friction defined by $\tan \phi_k = F/N = \mu_k N/N = \mu_k$. In Fig. 8–23*c*, it is seen that $r \sin \phi_k = r_f$. The dashed circle with radius r_f is called the ***friction circle***, and as the shaft rotates, the reaction **R** will always be tangent to it. If the bearing is partially lubricated, μ_k is small, and therefore $\sin \phi_k \approx \tan \phi_k \approx \mu_k$. Under these conditions, a reasonable *approximation* to the moment needed to overcome the frictional resistance becomes

$$M \approx Rr\mu_k \qquad (8\text{–}10)$$

Notice that to minimize friction the bearing radius *r* should be as small as possible. In practice, however, this type of journal bearing is not suitable for long service since friction between the shaft and bearing will eventually wear down the surfaces. Instead, designers will incorporate "ball bearings" or "rollers" in journal bearings to minimize frictional losses.

Unwinding the cable from this spool requires overcoming friction from the supporting shaft. (© Russell C. Hibbeler)

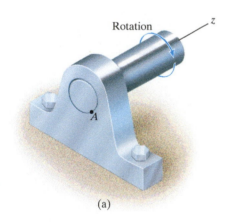

(a)

Fig. 8–23

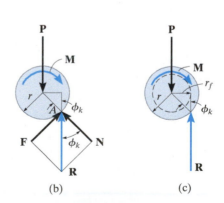

(b)   (c)

EXAMPLE | 8.10

The 100-mm-diameter pulley shown in Fig. 8–24a fits loosely on a 10-mm-diameter shaft for which the coefficient of static friction is $\mu_s = 0.4$. Determine the minimum tension T in the belt needed to (a) raise the 100-kg block and (b) lower the block. Assume that no slipping occurs between the belt and pulley and neglect the weight of the pulley.

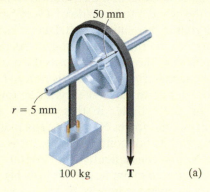

50 mm

$r = 5$ mm

100 kg   **T**   (a)

SOLUTION

Part (a).   A free-body diagram of the pulley is shown in Fig. 8–24b. When the pulley is subjected to belt tensions of 981 N each, it makes contact with the shaft at point P_1. As the tension T is *increased*, the contact point will move around the shaft to point P_2 before motion impends. From the figure, the friction circle has a radius $r_f = r \sin \phi_s$. Using the simplification that $\sin \phi_s \approx \tan \phi_s \approx \mu_s$ then $r_f \approx r\mu_s = (5 \text{ mm})(0.4) = 2$ mm, so that summing moments about P_2 gives

$\zeta + \Sigma M_{P_2} = 0;$ $981 \text{ N}(52 \text{ mm}) - T(48 \text{ mm}) = 0$
$\qquad\qquad T = 1063 \text{ N} = 1.06 \text{ k N}$ *Ans.*

If a more exact analysis is used, then $\phi_s = \tan^{-1} 0.4 = 21.8°$. Thus, the radius of the friction circle would be $r_f = r \sin \phi_s = 5 \sin 21.8° = 1.86$ mm. Therefore,

$\zeta + \Sigma M_{P_2} = 0;$

$\qquad 981 \text{ N}(50 \text{ mm} + 1.86 \text{ mm}) - T(50 \text{ mm} - 1.86 \text{ mm}) = 0$
$\qquad\qquad T = 1057 \text{ N} = 1.06 \text{ kN}$ *Ans.*

Part (b).   When the block is lowered, the resultant force **R** acting on the shaft passes through point as shown in Fig. 8–24c. Summing moments about this point yields

$\zeta + \Sigma M_{P_3} = 0;$ $981 \text{ N}(48 \text{ mm}) - T(52 \text{ mm}) = 0$
$\qquad\qquad T = 906 \text{ N}$ *Ans.*

NOTE: Using the approximate analysis, the difference between raising and lowering the block is thus 157 N.

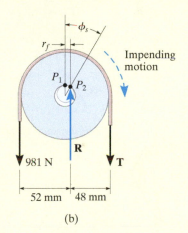

ϕ_s

r_f

Impending motion

P_1 P_2

R

981 N   **T**

52 mm   48 mm

(b)

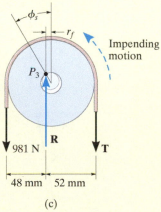

ϕ_s

r_f

Impending motion

P_3

R

981 N   **T**

48 mm   52 mm

(c)

Fig. 8–24

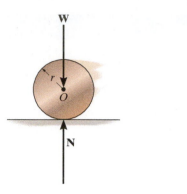

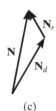

Rigid surface of contact

(a)

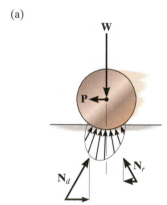

Soft surface of contact

(b)

(c)

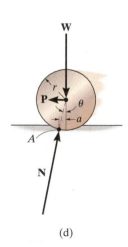

(d)

Fig. 8–25

*8.8   Rolling Resistance

When a *rigid* cylinder rolls at constant velocity along a *rigid* surface, the normal force exerted by the surface on the cylinder acts perpendicular to the tangent at the point of contact, as shown in Fig. 8–25a. Actually, however, no materials are perfectly rigid, and therefore the reaction of the surface on the cylinder consists of a distribution of normal pressure. For example, consider the cylinder to be made of a very hard material, and the surface on which it rolls to be relatively soft. Due to its weight, the cylinder compresses the surface underneath it, Fig. 8–25b. As the cylinder rolls, the surface material in front of the cylinder *retards* the motion since it is being *deformed*, whereas the material in the rear is *restored* from the deformed state and therefore tends to *push* the cylinder forward. The normal pressures acting on the cylinder in this manner are represented in Fig. 8–25b by their resultant forces $\mathbf{N}_d$ and $\mathbf{N}_r$. The magnitude of the force of *deformation*, $\mathbf{N}_d$, and its horizontal component is *always greater* than that of *restoration*, $\mathbf{N}_r$, and consequently a horizontal driving force $\mathbf{P}$ must be applied to the cylinder to maintain the motion. Fig. 8–25b.*

Rolling resistance is caused primarily by this effect, although it is also, to a lesser degree, the result of surface adhesion and relative microsliding between the surfaces of contact. Because the actual force $\mathbf{P}$ needed to overcome these effects is difficult to determine, a simplified method will be developed here to explain one way engineers have analyzed this phenomenon. To do this, we will consider the resultant of the *entire* normal pressure, $\mathbf{N} = \mathbf{N}_d + \mathbf{N}_r$, acting on the cylinder, Fig. 8–25c. As shown in Fig. 8–25d, this force acts at an angle θ with the vertical. To keep the cylinder in equilibrium, i.e., rolling at a constant rate, it is necessary that $\mathbf{N}$ be *concurrent* with the driving force $\mathbf{P}$ and the weight $\mathbf{W}$. Summing moments about point A gives $Wa = P(r\cos\theta)$. Since the deformations are generally very small in relation to the cylinder's radius, $\cos\theta \approx 1$; hence,

$$Wa \approx Pr$$

or

$$\boxed{P \approx \frac{Wa}{r}} \qquad (8\text{–}11)$$

The distance a is termed the **coefficient of rolling resistance**, which has the dimension of length. For instance, $a \approx 0.5$ mm for a wheel rolling on a rail, both of which are made of mild steel. For hardened steel ball

---

*Actually, the deformation force $\mathbf{N}_d$ causes *energy* to be stored in the material as its magnitude is increased, whereas the restoration force $\mathbf{N}_r$, as its magnitude is decreased, allows some of this energy to be released. The remaining energy is *lost* since it is used to heat up the surface, and if the cylinder's weight is very large, it accounts for permanent deformation of the surface. Work must be done by the horizontal force $\mathbf{P}$ to make up for this loss.

bearings on steel, $a \approx 0.1$ mm. Experimentally, though, this factor is difficult to measure, since it depends on such parameters as the rate of rotation of the cylinder, the elastic properties of the contacting surfaces, and the surface finish. For this reason, little reliance is placed on the data for determining a. The analysis presented here does, however, indicate why a heavy load (W) offers greater resistance to motion (P) than a light load under the same conditions. Furthermore, since Wa/r is generally very small compared to $\mu_k W$, the force needed to *roll* a cylinder over the surface will be much less than that needed to *slide* it across the surface. It is for this reason that a roller or ball bearings are often used to minimize the frictional resistance between moving parts.

Rolling resistance of railroad wheels on the rails is small since steel is very stiff. By comparison, the rolling resistance of the wheels of a tractor in a wet field is very large. (© Russell C. Hibbeler)

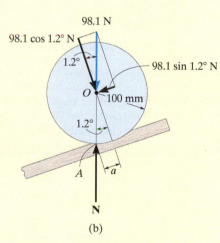

EXAMPLE 8.11

A 10-kg steel wheel shown in Fig. 8–26a has a radius of 100 mm and rests on an inclined plane made of soft wood. If θ is increased so that the wheel begins to roll down the incline with constant velocity when $\theta = 1.2°$, determine the coefficient of rolling resistance.

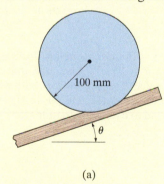

(a)

98.1 N

98.1 cos 1.2° N

1.2°

98.1 sin 1.2° N

O 100 mm

1.2°

A a

N

(b)

Fig. 8–26

SOLUTION

As shown on the free-body diagram, Fig. 8–26b, when the wheel has impending motion, the normal reaction **N** acts at point A defined by the dimension a. Resolving the weight into components parallel and perpendicular to the incline, and summing moments about point A, yields

$\zeta + \Sigma M_A = 0;$

$-(98.1 \cos 1.2° \text{ N})(a) + (98.1 \sin 1.2° \text{ N})(100 \cos 1.2° \text{ mm}) = 0$

Solving, we obtain

$$a = 2.09 \text{ mm} \qquad \textit{Ans.}$$

8

PROBLEMS

8–107. The collar bearing uniformly supports an axial force of $P = 5$ kN. If the coefficient of static friction is $\mu_s = 0.3$, determine the smallest torque M required to overcome friction.

***8–108.** The collar bearing uniformly supports an axial force of $P = 8$ kN. If a torque of $M = 200$ N · m is applied to the shaft and causes it to rotate at constant velocity, determine the coefficient of kinetic friction at the surface of contact.

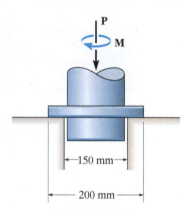

Probs. 8–107/108

8–109. The floor-polishing machine rotates at a constant angular velocity. If it has a weight of 80 lb, determine the couple forces F the operator must apply to the handles to hold the machine stationary. The coefficient of kinetic friction between the floor and brush is $\mu_k = 0.3$. Assume the brush exerts a uniform pressure on the floor.

Prob. 8–109

8–110. The *double-collar bearing* is subjected to an axial force $P = 4$ kN. Assuming that collar A supports $0.75P$ and collar B supports $0.25P$, both with a uniform distribution of pressure, determine the maximum frictional moment M that may be resisted by the bearing. Take $\mu_s = 0.2$ for both collars.

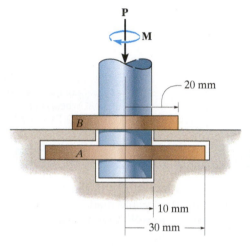

Prob. 8–110

8–111. The *double-collar bearing* is subjected to an axial force $P = 16$ kN. Assuming that collar A supports $0.75P$ and collar B supports $0.25P$, both with a uniform distribution of pressure, determine the smallest torque M that must be applied to overcome friction. Take $\mu_s = 0.2$ for both collars.

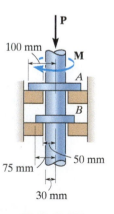

Prob. 8–111

*8–112. The pivot bearing is subjected to a pressure distribution at its surface of contact which varies as shown. If the coefficient of static friction is μ, determine the torque M required to overcome friction if the shaft supports an axial force **P**.

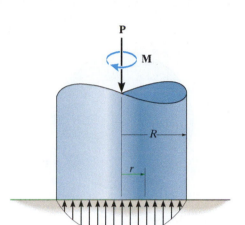

Prob. 8–112

8–113. The conical bearing is subjected to a constant pressure distribution at its surface of contact. If the coefficient of static friction is μ_s, determine the torque M required to overcome friction if the shaft supports an axial force **P**.

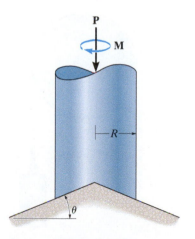

Prob. 8–113

8–114. The 4-in.-diameter shaft is held in the hole such that the normal pressure acting around the shaft varies linearly with its depth as shown. Determine the frictional torque that must be overcome to rotate the shaft. Take $\mu_s = 0.2$.

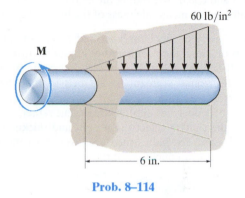

Prob. 8–114

8–115. The plate clutch consists of a flat plate A that slides over the rotating shaft S. The shaft is fixed to the driving plate gear B. If the gear C, which is in mesh with B, is subjected to a torque of $M = 0.8$ N·m, determine the smallest force P, that must be applied via the control arm, to stop the rotation. The coefficient of static friction between the plates A and D is $\mu_s = 0.4$. Assume the bearing pressure between A and D to be uniform.

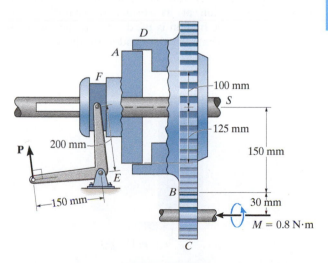

Prob. 8–115

8

*8–116. The collar fits *loosely* around a fixed shaft that has a radius of 2 in. If the coefficient of kinetic friction between the shaft and the collar is $\mu_k = 0.3$, determine the force P on the horizontal segment of the belt so that the collar rotates *counterclockwise* with a constant angular velocity. Assume that the belt does not slip on the collar; rather, the collar slips on the shaft. Neglect the weight and thickness of the belt and collar. The radius, measured from the center of the collar to the mean thickness of the belt, is 2.25 in.

8–117. The collar fits *loosely* around a fixed shaft that has a radius of 2 in. If the coefficient of kinetic friction between the shaft and the collar is $\mu_k = 0.3$, determine the force P on the horizontal segment of the belt so that the collar rotates *clockwise* with a constant angular velocity. Assume that the belt does not slip on the collar; rather, the collar slips on the shaft. Neglect the weight and thickness of the belt and collar. The radius, measured from the center of the collar to the mean thickness of the belt, is 2.25 in.

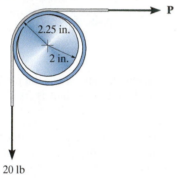

Probs. 8–116/117

8–118. The pivot bearing is subjected to a parabolic pressure distribution at its surface of contact. If the coefficient of static friction is μ_k, determine the torque M required to overcome friction and turn the shaft if it supports an axial force P.

Prob. 8–118

8–119. A disk having an outer diameter of 120 mm fits loosely over a fixed shaft having a diameter of 30 mm. If the coefficient of static friction between the disk and the shaft is $\mu_s = 0.15$ and the disk has a mass of 50 kg, determine the smallest vertical force F acting on the rim which must be applied to the disk to cause it to slip over the shaft.

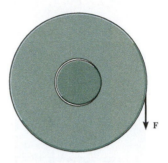

Prob. 8–119

*8–120. The 4-lb pulley has a diameter of 1 ft and the axle has a diameter of 1 in. If the coefficient of kinetic friction between the axle and the pulley is $\mu_k = 0.20$, determine the vertical force P on the rope required to lift the 20-lb block at constant velocity.

8–121. Solve Prob. 8–120 if the force P is applied horizontally to the left.

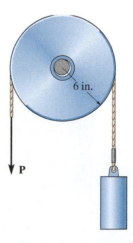

Probs. 8–120/121

8–122. Determine the tension **T** in the belt needed to overcome the tension of 200 lb created on the other side. Also, what are the normal and frictional components of force developed on the collar bushing? The coefficient of static friction is $\mu_s = 0.21$.

8–123. If a tension force $T = 215$ lb is required to pull the 200-lb force around the collar bushing, determine the coefficient of static friction at the contacting surface. The belt does not slip on the collar.

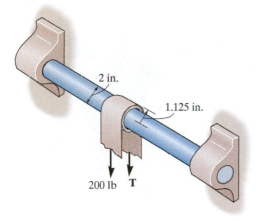

Probs. 8–122/123

*8–124. The uniform disk fits loosely over a fixed shaft having a diameter of 40 mm. If the coefficient of static friction between the disk and the shaft is $\mu_s = 0.15$, determine the smallest vertical force P, acting on the rim, which must be applied to the disk to cause it to slip on the shaft. The disk has a mass of 20 kg.

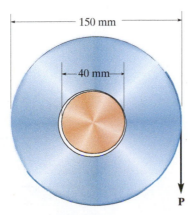

Prob. 8–124

8–125. The 5-kg skateboard rolls down the 5° slope at constant speed. If the coefficient of kinetic friction between the 12.5-mm-diameter axles and the wheels is $\mu_k = 0.3$, determine the radius of the wheels. Neglect rolling resistance of the wheels on the surface. The center of mass for the skateboard is at G.

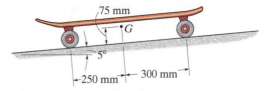

Prob. 8–125

8–126. The bell crank fits loosely into a 0.5-in-diameter pin. Determine the required force P which is just sufficient to rotate the bell crank clockwise. The coefficient of static friction between the pin and the bell crank is $\mu_s = 0.3$.

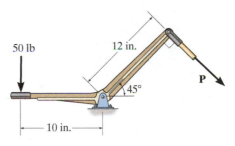

Prob. 8–126

8–127. The bell crank fits loosely into a 0.5-in-diameter pin. If $P = 41$ lb, the bell crank is then on the verge of rotating counterclockwise. Determine the coefficient of static friction between the pin and the bell crank.

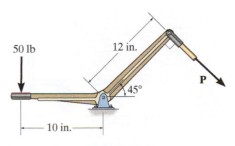

Prob. 8–127

8

*8–128. The vehicle has a weight of 2600 lb and center of gravity at G. Determine the horizontal force **P** that must be applied to overcome the rolling resistance of the wheels. The coefficient of rolling resistance is 0.5 in. The tires have a diameter of 2.75 ft.

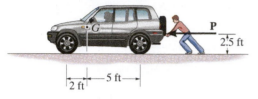

Prob. 8–128

8–129. The tractor has a weight of 16 000 lb and the coefficient of rolling resistance is $a = 2$ in. Determine the force **P** needed to overcome rolling resistance at all four wheels and push it forward.

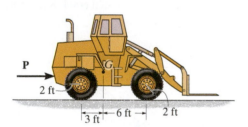

Prob. 8–129

8–130. The handcart has wheels with a diameter of 6 in. If a crate having a weight of 1500 lb is placed on the cart, determine the force P that must be applied to the handle to overcome the rolling resistance. The coefficient of rolling resistance is 0.04 in. Neglect the weight of the cart.

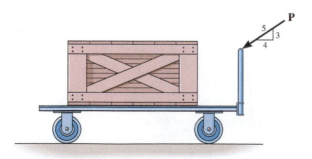

Prob. 8–130

8–131. The cylinder is subjected to a load that has a weight W. If the coefficients of rolling resistance for the cylinder's top and bottom surfaces are a_A and a_B, respectively, show that a horizontal force having a magnitude of $P = [W(a_A + a_B)]/2r$ is required to move the load and thereby roll the cylinder forward. Neglect the weight of the cylinder.

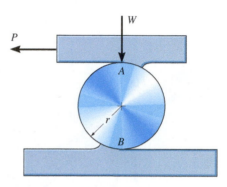

Prob. 8–131

*8–132. The 1.4-Mg machine is to be moved over a level surface using a series of rollers for which the coefficient of rolling resistance is 0.5 mm at the ground and 0.2 mm at the bottom surface of the machine. Determine the appropriate diameter of the rollers so that the machine can be pushed forward with a horizontal force of $P = 250$ N. *Hint:* Use the result of Prob. 8–131.

Prob. 8–132

CHAPTER REVIEW

Dry Friction

Frictional forces exist between two rough surfaces of contact. These forces act on a body so as to oppose its motion or tendency of motion.

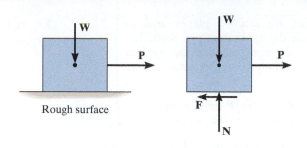

Rough surface

A static frictional force approaches a maximum value of $F_s = \mu_s N$, where μ_s is the *coefficient of static friction*. In this case, motion between the contacting surfaces is *impending*.

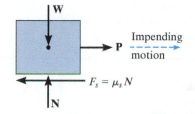

Impending motion

$$F_s = \mu_s N$$

If slipping occurs, then the friction force remains essentially constant and equal to $F_k = \mu_k N$. Here μ_k is the *coefficient of kinetic friction*.

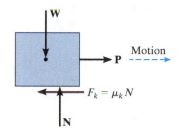

Motion

$$F_k = \mu_k N$$

The solution of a problem involving friction requires first drawing the free-body diagram of the body. If the unknowns cannot be determined strictly from the equations of equilibrium, and the possibility of slipping occurs, then the friction equation should be applied at the appropriate points of contact in order to complete the solution.

It may also be possible for slender objects, like crates, to tip over, and this situation should also be investigated.

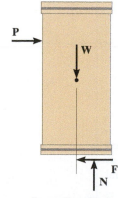

Impending slipping
$$F = \mu_s N$$

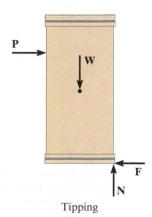

Tipping

Wedges

Wedges are inclined planes used to increase the application of a force. The two force equilibrium equations are used to relate the forces acting on the wedge.

An applied force **P** must push on the wedge to move it to the right.

If the coefficients of friction between the surfaces are large enough, then **P** can be removed, and the wedge will be self-locking and remain in place.

$$\Sigma F_x = 0$$
$$\Sigma F_y = 0$$

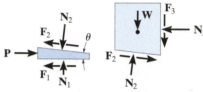

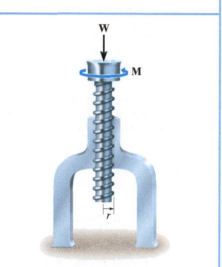

Screws

Square-threaded screws are used to move heavy loads. They represent an inclined plane, wrapped around a cylinder.

The moment needed to turn a screw depends upon the coefficient of friction and the screw's lead angle θ.

If the coefficient of friction between the surfaces is large enough, then the screw will support the load without tending to turn, i.e., it will be self-locking.

$$M = rW \tan(\theta + \phi_s)$$
Upward Impending Screw Motion

$$M' = rW \tan(\theta - \phi_s)$$
Downward Impending Screw Motion
$$\theta > \phi_s$$

$$M'' = rW \tan(\phi_s - \theta)$$
Downward Screw Motion
$$\phi_s > \theta$$

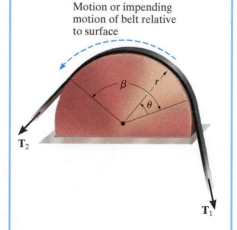

Flat Belts

The force needed to move a flat belt over a rough curved surface depends only on the angle of belt contact, β, and the coefficient of friction.

$$T_2 = T_1 e^{\mu\beta}$$
$$T_2 > T_1$$

8

Collar Bearings and Disks

The frictional analysis of a collar bearing or disk requires looking at a differential element of the contact area. The normal force acting on this element is determined from force equilibrium along the shaft, and the moment needed to turn the shaft at a constant rate is determined from moment equilibrium about the shaft's axis.

If the pressure on the surface of a collar bearing is uniform, then integration gives the result shown.

$$M = \frac{2}{3}\mu_s P\left(\frac{R_2^3 - R_1^3}{R_2^2 - R_1^2}\right)$$

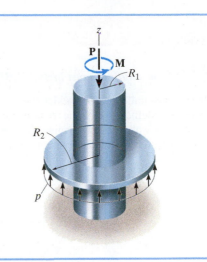

Journal Bearings

When a moment is applied to a shaft in a nonlubricated or partially lubricated journal bearing, the shaft will tend to roll up the side of the bearing until slipping occurs. This defines the radius of a friction circle, and from it the moment needed to turn the shaft can be determined.

$$M = Rr\sin\phi_k$$

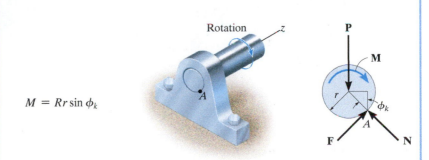

Rolling Resistance

The resistance of a wheel to rolling over a surface is caused by localized *deformation* of the two materials in contact. This causes the resultant normal force acting on the rolling body to be inclined so that it provides a component that acts in the opposite direction of the applied force **P** causing the motion. This effect is characterized using the *coefficient of rolling resistance, a*, which is determined from experiment.

$$P \approx \frac{Wa}{r}$$

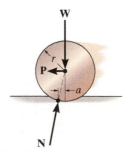

REVIEW PROBLEMS

All problem solutions must include FBDs.

R8–1. The uniform 20-lb ladder rests on the rough floor for which the coefficient of static friction is $\mu_s = 0.4$ and against the smooth wall at B. Determine the horizontal force P the man must exert on the ladder in order to cause it to move.

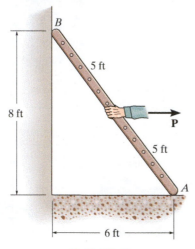

Prob. R8–1

R8–2. The uniform 60-kg crate C rests uniformly on a 10-kg dolly D. If the front casters of the dolly at A are locked to prevent rolling while the casters at B are free to roll, determine the maximum force P that may be applied without causing motion of the crate. The coefficient of static friction between the casters and the floor is $\mu_f = 0.35$ and between the dolly and the crate, $\mu_d = 0.5$.

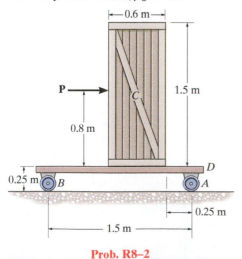

Prob. R8–2

R8–3. A 35-kg disk rests on an inclined surface for which $\mu_s = 0.2$. Determine the maximum vertical force P that may be applied to bar AB without causing the disk to slip at C. Neglect the mass of the bar.

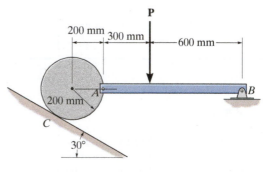

Prob. R8–3

R8–4. The cam is subjected to a couple moment of 5 N · m. Determine the minimum force P that should be applied to the follower in order to hold the cam in the position shown. The coefficient of static friction between the cam and the follower is $\mu = 0.4$. The guide at A is smooth.

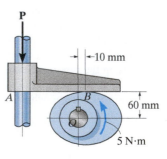

Prob. R8–4

R8–5. The three stone blocks have weights of $W_A = 600$ lb, $W_B = 150$ lb, and $W_C = 500$ lb. Determine the smallest horizontal force P that must be applied to block C in order to move this block. The coefficient of static friction between the blocks is $\mu_s = 0.3$, and between the floor and each block $\mu_s' = 0.5$.

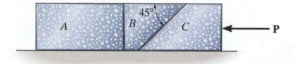

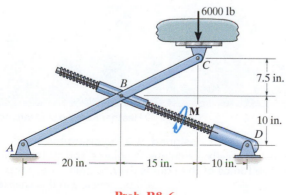

Prob. R8–5

R8–6. The jacking mechanism consists of a link that has a square-threaded screw with a mean diameter of 0.5 in. and a lead of 0.20 in., and the coefficient of static friction is $\mu_s = 0.4$. Determine the torque M that should be applied to the screw to start lifting the 6000-lb load acting at the end of member ABC.

Prob. R8–6

R8–7. The uniform 50-lb beam is supported by the rope that is attached to the end of the beam, wraps over the rough peg, and is then connected to the 100-lb block. If the coefficient of static friction between the beam and the block, and between the rope and the peg, is $\mu_s = 0.4$, determine the maximum distance that the block can be placed from A and still remain in equilibrium. Assume the block will not tip.

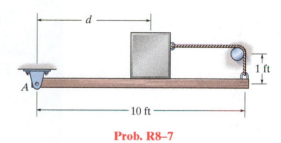

Prob. R8–7

R8–8. The hand cart has wheels with a diameter of 80 mm. If a crate having a mass of 500 kg is placed on the cart so that each wheel carries an equal load, determine the horizontal force P that must be applied to the handle to overcome the rolling resistance. The coefficient of rolling resistance is 2 mm. Neglect the mass of the cart.

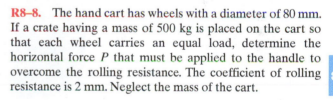

Prob. R8–8

Chapter 9

(© Heather Reeder/Shutterstock)

When a tank of any shape is designed, it is important to be able to determine its center of gravity, calculate its volume and surface area, and determine the forces of the liquids they contain. These topics will be covered in this chapter.

Center of Gravity and Centroid

9.1   Center of Gravity, Center of Mass, and the Centroid of a Body

Knowing the resultant or total weight of a body and its location is important when considering the effect this force produces on the body. The point of location is called the center of gravity, and in this section we will show how to find it for an irregularly shaped body. We will then extend this method to show how to find the body's center of mass, and its geometric center or centroid.

Center of Gravity.   A body is composed of an infinite number of particles of differential size, and so if the body is located within a gravitational field, then each of these particles will have a weight dW. These weights will form a parallel force system, and the resultant of this system is the total weight of the body, which passes through a single point called the *center of gravity*, G*.

---

*In a strict sense this is true as long as the gravity field is assumed to have the same magnitude and direction everywhere. Although the actual force of gravity is directed toward the center of the earth, and this force varies with its distance from the center, for most engineering applications we can assume the gravity field has the same magnitude and direction everywhere.

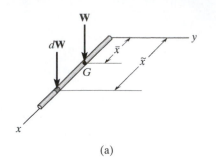

(a)

To show how to determine the location of the center of gravity, consider the rod in Fig. 9–1a, where the segment having the weight dW is located at the arbitrary position $\tilde{x}$. Using the methods outlined in Sec. 4.8, the total weight of the rod is the sum of the weights of all of its particles, that is

$$+\downarrow F_R = \Sigma F_z; \qquad\qquad W = \int dW$$

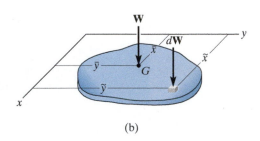

(b)

The location of the center of gravity, measured from the y axis, is determined by equating the moment of W about the y axis, Fig. 9–1b, to the sum of the moments of the weights of all its particles about this same axis. Therefore,

$$(M_R)_y = \Sigma M_y; \qquad\qquad \bar{x}W = \int \tilde{x}\,dW$$

$$\bar{x} = \frac{\int \tilde{x}\,dW}{\int dW}$$

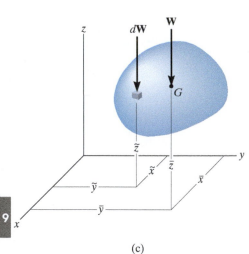

(c)

Fig. 9–1

In a similar manner, if the body represents a plate, Fig. 9–1b, then a moment balance about the x and y axes would be required to determine the location $(\bar{x}, \bar{y})$ of point G. Finally we can generalize this idea to a three-dimensional body, Fig. 9–1c, and perform a moment balance about all three axes to locate G for any rotated position of the axes. This results in the following equations.

$$\bar{x} = \frac{\int \tilde{x}\,dW}{\int dW} \qquad \bar{y} = \frac{\int \tilde{y}\,dW}{\int dW} \qquad \bar{z} = \frac{\int \tilde{z}\,dW}{\int dW} \qquad (9\text{–}1)$$

where

$\bar{x}, \bar{y}, \bar{z}$ are the coordinates of the center of gravity G.
$\tilde{x}, \tilde{y}, \tilde{z}$ are the coordinates of an arbitrary particle in the body.

Center of Mass of a Body.   In order to study the *dynamic response* or accelerated motion of a body, it becomes important to locate the body's *center of mass* C_m, Fig. 9–2. This location can be determined by substituting $dW = g\, dm$ into Eqs. 9–1. Provided g is constant, it cancels out, and so

$$\bar{x} = \frac{\int \tilde{x}\, dm}{\int dm} \qquad \bar{y} = \frac{\int \tilde{y}\, dm}{\int dm} \qquad \bar{z} = \frac{\int \tilde{z}\, dm}{\int dm} \qquad (9\text{–}2)$$

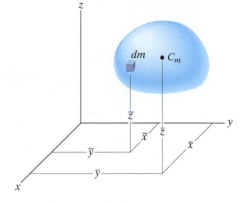

Fig. 9–2

Centroid of a Volume.   If the body in Fig. 9–3 is made from a *homogeneous material*, then its density ρ (rho) will be *constant*. Therefore, a differential element of volume dV has a mass $dm = \rho\, dV$. Substituting this into Eqs. 9–2 and canceling out ρ, we obtain formulas that locate the *centroid C* or geometric center of the body; namely

$$\bar{x} = \frac{\int_V \tilde{x}\, dV}{\int_V dV} \qquad \bar{y} = \frac{\int_V \tilde{y}\, dV}{\int_V dV} \qquad \bar{z} = \frac{\int_V \tilde{z}\, dV}{\int_V dV} \qquad (9\text{–}3)$$

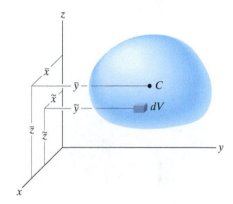

Fig. 9–3

These equations represent a balance of the moments of the volume of the body. Therefore, if the volume possesses two planes of symmetry, then its centroid must lie along the line of intersection of these two planes. For example, the cone in Fig. 9–4 has a centroid that lies on the y axis so that $\bar{x} = \bar{z} = 0$. The location $\bar{y}$ can be found using a single integration by choosing a differential element represented by a *thin disk* having a thickness dy and radius $r = z$. Its volume is $dV = \pi r^2\, dy = \pi z^2\, dy$ and its centroid is at $\tilde{x} = 0$, $\tilde{y} = y$, $\tilde{z} = 0$.

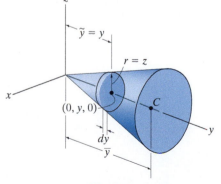

Fig. 9–4

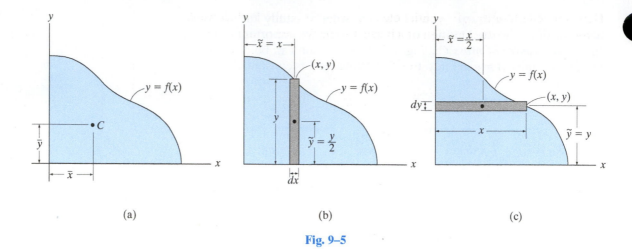

Fig. 9–5

Integration must be used to determine the location of the center of gravity of this lamp post due to the curvature of the member. (© Russell C. Hibbeler)

Centroid of an Area. If an area lies in the x–y plane and is bounded by the curve $y = f(x)$, as shown in Fig. 9–5a, then its centroid will be in this plane and can be determined from integrals similar to Eqs. 9–3, namely,

$$\bar{x} = \frac{\int_A \tilde{x}\, dA}{\int_A dA} \qquad \bar{y} = \frac{\int_A \tilde{y}\, dA}{\int_A dA} \qquad (9\text{–}4)$$

These integrals can be evaluated by performing a *single integration* if we use a *rectangular strip* for the differential area element. For example, if a vertical strip is used, Fig. 9–5b, the area of the element is $dA = y\, dx$, and its centroid is located at $\tilde{x} = x$ and $\tilde{y} = y/2$. If we consider a horizontal strip, Fig. 9–5c, then $dA = x\, dy$, and its centroid is located at $\tilde{x} = x/2$ and $\tilde{y} = y$.

Centroid of a Line. If a line segment (or rod) lies within the x–y plane and it can be described by a thin curve $y = f(x)$, Fig. 9–6a, then its centroid is determined from

$$\bar{x} = \frac{\int_L \tilde{x}\, dL}{\int_L dL} \qquad \bar{y} = \frac{\int_L \tilde{y}\, dL}{\int_L dL} \qquad (9\text{–}5)$$

Here, the length of the differential element is given by the Pythagorean theorem, $dL = \sqrt{(dx)^2 + (dy)^2}$, which can also be written in the form

$$dL = \sqrt{\left(\frac{dx}{dx}\right)^2 dx^2 + \left(\frac{dy}{dx}\right)^2 dx^2}$$

$$= \left(\sqrt{1 + \left(\frac{dy}{dx}\right)^2}\right) dx$$

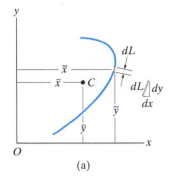

(a)

or

$$dL = \sqrt{\left(\frac{dx}{dy}\right)^2 dy^2 + \left(\frac{dy}{dy}\right)^2 dy^2}$$

$$= \left(\sqrt{\left(\frac{dx}{dy}\right)^2 + 1}\right) dy$$

Either one of these expressions can be used; however, for application, the one that will result in a simpler integration should be selected. For example, consider the rod in Fig. 9–6b, defined by $y = 2x^2$. The length of the element is $dL = \sqrt{1 + (dy/dx)^2}\, dx$, and since $dy/dx = 4x$, then $dL = \sqrt{1 + (4x)^2}\, dx$. The centroid for this element is located at $\tilde{x} = x$ and $\tilde{y} = y$.

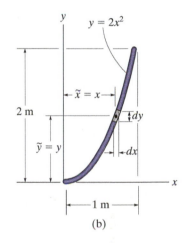

(b)

Fig. 9–6

Important Points

- The centroid represents the geometric center of a body. This point coincides with the center of mass or the center of gravity only if the material composing the body is uniform or homogeneous.

- Formulas used to locate the center of gravity or the centroid simply represent a balance between the sum of moments of all the parts of the system and the moment of the "resultant" for the system.

- In some cases the centroid is located at a point that is not on the object, as in the case of a ring, where the centroid is at its center. Also, this point will lie on any axis of symmetry for the body, Fig. 9–7.

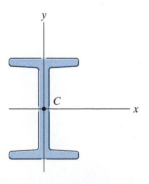

Fig. 9–7

Procedure for Analysis

The center of gravity or centroid of an object or shape can be determined by single integrations using the following procedure.

Differential Element.

- Select an appropriate coordinate system, specify the coordinate axes, and then choose a differential element for integration.

- For lines the element is represented by a differential line segment of length dL.

- For areas the element is generally a rectangle of area dA, having a finite length and differential width.

- For volumes the element can be a circular disk of volume dV, having a finite radius and differential thickness.

- Locate the element so that it touches the arbitrary point (x, y, z) on the curve that defines the boundary of the shape.

Size and Moment Arms.

- Express the length dL, area dA, or volume dV of the element in terms of the coordinates describing the curve.

- Express the moment arms $\tilde{x}, \tilde{y}, \tilde{z}$ for the centroid or center of gravity of the element in terms of the coordinates describing the curve.

Integrations.

- Substitute the formulations for $\tilde{x}, \tilde{y}, \tilde{z}$ and dL, dA, or dV into the appropriate equations (Eqs. 9–1 through 9–5).

- Express the function in the integrand in terms of the *same variable as the differential thickness of the element*.

- The limits of the integral are defined from the two extreme locations of the element's differential thickness, so that when the elements are "summed" or the integration performed, the entire region is covered.

9

EXAMPLE | 9.1

Locate the centroid of the rod bent into the shape of a parabolic arc as shown in Fig. 9–8.

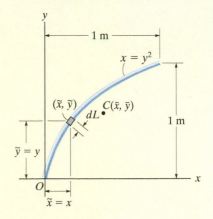

SOLUTION

Differential Element.   The differential element is shown in Fig. 9–8. It is located on the curve at the *arbitrary point* (x, y).

Area and Moment Arms.   The differential element of length dL can be expressed in terms of the differentials dx and dy using the Pythagorean theorem.

$$dL = \sqrt{(dx)^2 + (dy)^2} = \sqrt{\left(\frac{dx}{dy}\right)^2 + 1}\, dy$$

Since $x = y^2$, then $dx/dy = 2y$. Therefore, expressing dL in terms of y and dy, we have

$$dL = \sqrt{(2y)^2 + 1}\, dy$$

As shown in Fig. 9–8, the centroid of the element is located at $\tilde{x} = x$, $\tilde{y} = y$.

Integrations.   Applying Eq. 9–5 and using the integration formula to evaluate the integrals, we get

$$\bar{x} = \frac{\displaystyle\int_L \tilde{x}\, dL}{\displaystyle\int_L dL} = \frac{\displaystyle\int_0^{1\,m} x\sqrt{4y^2 + 1}\, dy}{\displaystyle\int_0^{1\,m} \sqrt{4y^2 + 1}\, dy} = \frac{\displaystyle\int_0^{1\,m} y^2\sqrt{4y^2 + 1}\, dy}{\displaystyle\int_0^{1\,m} \sqrt{4y^2 + 1}\, dy}$$

$$= \frac{0.6063}{1.479} = 0.410 \text{ m} \qquad\qquad\qquad\qquad\qquad \textit{Ans.}$$

$$\bar{y} = \frac{\displaystyle\int_L \tilde{y}\, dL}{\displaystyle\int_L dL} = \frac{\displaystyle\int_0^{1\,m} y\sqrt{4y^2 + 1}\, dy}{\displaystyle\int_0^{1\,m} \sqrt{4y^2 + 1}\, dy} = \frac{0.8484}{1.479} = 0.574 \text{ m} \qquad \textit{Ans.}$$

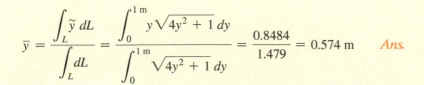

NOTE: These results for C seem reasonable when they are plotted on Fig. 9–8.

EXAMPLE   9.2

Locate the centroid of the circular wire segment shown in Fig. 9–9.

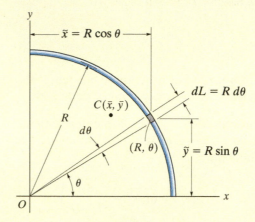

Fig. 9–9

SOLUTION

Polar coordinates will be used to solve this problem since the arc is circular.

Differential Element.   A differential circular arc is selected as shown in the figure. This element lies on the curve at (R, θ).

Length and Moment Arm.   The length of the differential element is $dL = R \, d\theta$, and its centroid is located at $\tilde{x} = R \cos \theta$ and $\tilde{y} = R \sin \theta$.

Integrations.   Applying Eqs. 9–5 and integrating with respect to θ, we obtain

$$\bar{x} = \frac{\displaystyle\int_L \tilde{x} \, dL}{\displaystyle\int_L dL} = \frac{\displaystyle\int_0^{\pi/2} (R \cos \theta) R \, d\theta}{\displaystyle\int_0^{\pi/2} R \, d\theta} = \frac{R^2 \displaystyle\int_0^{\pi/2} \cos \theta \, d\theta}{R \displaystyle\int_0^{\pi/2} d\theta} = \frac{2R}{\pi} \qquad Ans.$$

$$\bar{y} = \frac{\displaystyle\int_L \tilde{y} \, dL}{\displaystyle\int_L dL} = \frac{\displaystyle\int_0^{\pi/2} (R \sin \theta) R \, d\theta}{\displaystyle\int_0^{\pi/2} R \, d\theta} = \frac{R^2 \displaystyle\int_0^{\pi/2} \sin \theta \, d\theta}{R \displaystyle\int_0^{\pi/2} d\theta} = \frac{2R}{\pi} \qquad Ans.$$

NOTE: As expected, the two coordinates are numerically the same due to the symmetry of the wire.

EXAMPLE | **9.3**

Determine the distance $\bar{y}$ measured from the x axis to the centroid of the area of the triangle shown in Fig. 9–10.

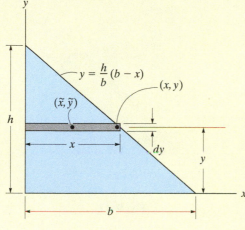

Fig. 9–10

SOLUTION

Differential Element.   Consider a rectangular element having a thickness dy, and located in an arbitrary position so that it intersects the boundary at (x, y), Fig. 9–10.

Area and Moment Arms.   The area of the element is $dA = x\,dy = \dfrac{b}{h}(h - y)\,dy$, and its centroid is located a distance $\tilde{y} = y$ from the x axis.

Integration.   Applying the second of Eqs. 9–4 and integrating with respect to y yields

$$\bar{y} = \frac{\displaystyle\int_A \tilde{y}\,dA}{\displaystyle\int_A dA} = \frac{\displaystyle\int_0^h y\left[\frac{b}{h}(h - y)\,dy\right]}{\displaystyle\int_0^h \frac{b}{h}(h - y)\,dy} = \frac{\frac{1}{6}bh^2}{\frac{1}{2}bh}$$

$$= \frac{h}{3} \qquad\qquad\qquad\qquad \textit{Ans.}$$

NOTE: This result is valid for any shape of triangle. It states that the centroid is located at one-third the height, measured from the base of the triangle.

EXAMPLE 9.4

Locate the centroid for the area of a quarter circle shown in Fig. 9–11.

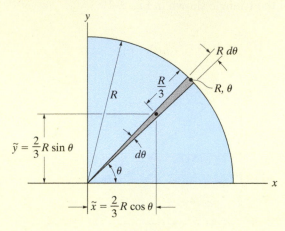

Fig. 9–11

SOLUTION

Differential Element. Polar coordinates will be used, since the boundary is circular. We choose the element in the shape of a *triangle*, Fig. 9–11. (Actually the shape is a circular sector; however, neglecting higher-order differentials, the element becomes triangular.) The element intersects the curve at point (R, θ).

Area and Moment Arms. The area of the element is

$$dA = \tfrac{1}{2}(R)(R\,d\theta) = \frac{R^2}{2}d\theta$$

and using the results of Example 9.3, the centroid of the (triangular) element is located at $\widetilde{x} = \tfrac{2}{3}R\cos\theta$, $\widetilde{y} = \tfrac{2}{3}R\sin\theta$.

Integrations. Applying Eqs. 9–4 and integrating with respect to θ, we obtain

$$\bar{x} = \frac{\displaystyle\int_A \widetilde{x}\,dA}{\displaystyle\int_A dA} = \frac{\displaystyle\int_0^{\pi/2}\left(\frac{2}{3}R\cos\theta\right)\frac{R^2}{2}d\theta}{\displaystyle\int_0^{\pi/2}\frac{R^2}{2}d\theta} = \frac{\left(\dfrac{2}{3}R\right)\displaystyle\int_0^{\pi/2}\cos\theta\,d\theta}{\displaystyle\int_0^{\pi/2}d\theta} = \frac{4R}{3\pi} \quad Ans.$$

$$\bar{y} = \frac{\displaystyle\int_A \widetilde{y}\,dA}{\displaystyle\int_A dA} = \frac{\displaystyle\int_0^{\pi/2}\left(\frac{2}{3}R\sin\theta\right)\frac{R^2}{2}d\theta}{\displaystyle\int_0^{\pi/2}\frac{R^2}{2}d\theta} = \frac{\left(\dfrac{2}{3}R\right)\displaystyle\int_0^{\pi/2}\sin\theta\,d\theta}{\displaystyle\int_0^{\pi/2}d\theta} = \frac{4R}{3\pi} \quad Ans.$$

EXAMPLE 9.5

Locate the centroid of the area shown in Fig. 9–12a.

SOLUTION I

Differential Element. A differential element of thickness dx is shown in Fig. 9–12a. The element intersects the curve at the *arbitrary point* (x, y), and so it has a height y.

Area and Moment Arms. The area of the element is $dA = y\, dx$, and its centroid is located at $\tilde{x} = x$, $\tilde{y} = y/2$.

Integrations. Applying Eqs. 9–4 and integrating with respect to x yields

$$\bar{x} = \frac{\int_A \tilde{x}\, dA}{\int_A dA} = \frac{\int_0^{1\,m} xy\, dx}{\int_0^{1\,m} y\, dx} = \frac{\int_0^{1\,m} x^3\, dx}{\int_0^{1\,m} x^2\, dx} = \frac{0.250}{0.333} = 0.75\ m \qquad Ans.$$

$$\bar{y} = \frac{\int_A \tilde{y}\, dA}{\int_A dA} = \frac{\int_0^{1\,m} (y/2)y\, dx}{\int_0^{1\,m} y\, dx} = \frac{\int_0^{1\,m} (x^2/2)x^2\, dx}{\int_0^{1\,m} x^2\, dx} = \frac{0.100}{0.333} = 0.3\ m \quad Ans.$$

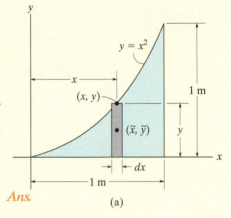

(a)

SOLUTION II

Differential Element. The differential element of thickness dy is shown in Fig. 9–12b. The element intersects the curve at the *arbitrary point* (x, y), and so it has a length $(1 - x)$.

Area and Moment Arms. The area of the element is $dA = (1 - x)\, dy$, and its centroid is located at

$$\tilde{x} = x + \left(\frac{1 - x}{2}\right) = \frac{1 + x}{2}, \quad \tilde{y} = y$$

Integrations. Applying Eqs. 9–4 and integrating with respect to y, we obtain

$$\bar{x} = \frac{\int_A \tilde{x}\, dA}{\int_A dA} = \frac{\int_0^{1\,m} [(1 + x)/2](1 - x)\, dy}{\int_0^{1\,m} (1 - x)\, dy} = \frac{\frac{1}{2}\int_0^{1\,m} (1 - y)\, dy}{\int_0^{1\,m} (1 - \sqrt{y})\, dy} = \frac{0.250}{0.333} = 0.75\ m \quad Ans.$$

$$\bar{y} = \frac{\int_A \tilde{y}\, dA}{\int_A dA} = \frac{\int_0^{1\,m} y(1 - x)\, dy}{\int_0^{1\,m} (1 - x)\, dy} = \frac{\int_0^{1\,m} (y - y^{3/2})\, dy}{\int_0^{1\,m} (1 - \sqrt{y})\, dy} = \frac{0.100}{0.333} = 0.3\ m \qquad Ans.$$

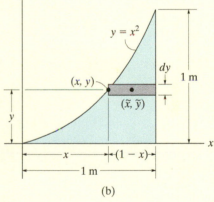

(b)

Fig. 9–12

NOTE: Plot these results and notice that they seem reasonable. Also, for this problem, elements of thickness dx offer a simpler solution.

EXAMPLE | **9.6**

Locate the centroid of the semi-elliptical area shown in Fig. 9–13a.

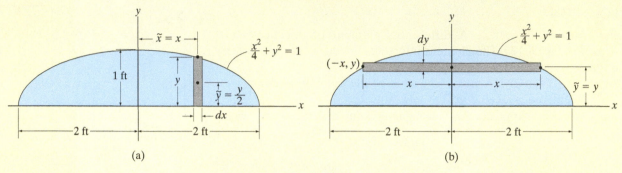

Fig. 9–13

SOLUTION I

Differential Element.   The rectangular differential element parallel to the y axis shown shaded in Fig. 9–13a will be considered. This element has a thickness of dx and a height of y.

Area and Moment Arms.   Thus, the area is $dA = y \, dx$, and its centroid is located at $\tilde{x} = x$ and $\tilde{y} = y/2$.

Integration.   Since the area is symmetrical about the y axis,

$$\bar{x} = 0 \qquad\qquad Ans.$$

Applying the second of Eqs. 9–4 with $y = \sqrt{1 - \dfrac{x^2}{4}}$, we have

$$\bar{y} = \frac{\displaystyle\int_A \tilde{y} \, dA}{\displaystyle\int_A dA} = \frac{\displaystyle\int_{-2\,\text{ft}}^{2\,\text{ft}} \frac{y}{2}(y \, dx)}{\displaystyle\int_{-2\,\text{ft}}^{2\,\text{ft}} y \, dx} = \frac{\dfrac{1}{2}\displaystyle\int_{-2\,\text{ft}}^{2\,\text{ft}} \left(1 - \frac{x^2}{4}\right)dx}{\displaystyle\int_{-2\,\text{ft}}^{2\,\text{ft}} \sqrt{1 - \frac{x^2}{4}} \, dx} = \frac{4/3}{\pi} = 0.424 \text{ ft} \quad Ans.$$

SOLUTION II

Differential Element.   The shaded rectangular differential element of thickness dy and width $2x$, parallel to the x axis, will be considered, Fig. 9–13b.

Area and Moment Arms.   The area is $dA = 2x \, dy$, and its centroid is at $\tilde{x} = 0$ and $\tilde{y} = y$.

Integration.   Applying the second of Eqs. 9–4, with $x = 2\sqrt{1 - y^2}$, we have

$$\bar{y} = \frac{\displaystyle\int_A \tilde{y} \, dA}{\displaystyle\int_A dA} = \frac{\displaystyle\int_0^{1\,\text{ft}} y(2x \, dy)}{\displaystyle\int_0^{1\,\text{ft}} 2x \, dy} = \frac{\displaystyle\int_0^{1\,\text{ft}} 4y\sqrt{1 - y^2} \, dy}{\displaystyle\int_0^{1\,\text{ft}} 4\sqrt{1 - y^2} \, dy} = \frac{4/3}{\pi} \text{ ft} = 0.424 \text{ ft} \quad Ans.$$

EXAMPLE   9.7

Locate the $\bar{y}$ centroid for the paraboloid of revolution, shown in Fig. 9–14.

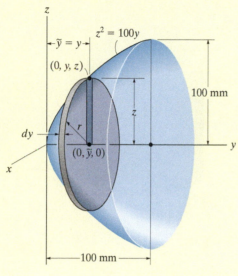

Fig. 9–14

SOLUTION

Differential Element.   An element having the shape of a *thin disk* is chosen. This element has a thickness dy, it intersects the generating curve at the *arbitrary point* $(0, y, z)$, and so its radius is $r = z$.

Volume and Moment Arm.   The volume of the element is $dV = (\pi z^2)\, dy$, and its centroid is located at $\tilde{y} = y$.

Integration.   Applying the second of Eqs. 9–3 and integrating with respect to y yields.

$$\bar{y} = \frac{\displaystyle\int_V \tilde{y}\, dV}{\displaystyle\int_V dV} = \frac{\displaystyle\int_0^{100\ mm} y(\pi z^2)\, dy}{\displaystyle\int_0^{100\ mm} (\pi z^2)\, dy} = \frac{100\pi \displaystyle\int_0^{100\ mm} y^2\, dy}{100\pi \displaystyle\int_0^{100\ mm} y\, dy} = 66.7 \text{ mm} \qquad Ans.$$

EXAMPLE | 9.8

Determine the location of the center of mass of the cylinder shown in Fig. 9–15 if its density varies directly with the distance from its base, i.e., $\rho = 200z$ kg/m^3.

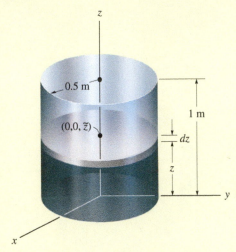

Fig. 9–15

SOLUTION
For reasons of material symmetry,

$$\bar{x} = \bar{y} = 0 \qquad\qquad Ans.$$

Differential Element.   A disk element of radius 0.5 m and thickness dz is chosen for integration, Fig. 9–15, since the *density of the entire element is constant* for a given value of z. The element is located along the z axis at the *arbitrary point* $(0, 0, z)$.

Volume and Moment Arm.   The volume of the element is $dV = \pi(0.5)^2 \, dz$, and its centroid is located at $\tilde{z} = z$.

Integrations.   Using the third of Eqs. 9–2 with $dm = \rho \, dV$ and integrating with respect to z, noting that $\rho = 200z$, we have

$$\bar{z} = \frac{\displaystyle\int_V \tilde{z}\rho \, dV}{\displaystyle\int_V \rho \, dV} = \frac{\displaystyle\int_0^{1\,m} z(200z)\left[\pi(0.5)^2 \, dz\right]}{\displaystyle\int_0^{1\,m} (200z)\pi(0.5)^2 \, dz}$$

$$= \frac{\displaystyle\int_0^{1\,m} z^2 \, dz}{\displaystyle\int_0^{1\,m} z \, dz} = 0.667 \text{ m} \qquad\qquad Ans.$$

PRELIMINARY PROBLEM

P9–1.   In each case, use the element shown and specify $\tilde{x}$, $\tilde{y}$, and dA.

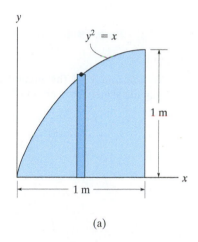

(a)

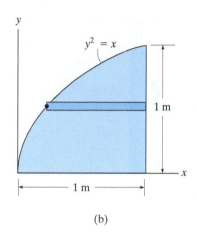

(b)

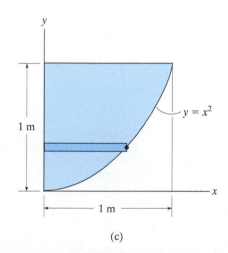

(c)

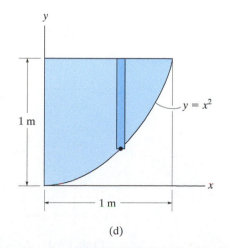

(d)

Prob. P9–1

FUNDAMENTAL PROBLEMS

F9–1. Determine the centroid $(\bar{x}, \bar{y})$ of the shaded area.

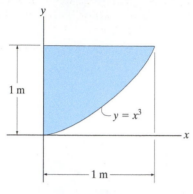

Prob. F9–1

F9–2. Determine the centroid $(\bar{x}, \bar{y})$ of the shaded area.

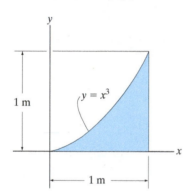

Prob. F9–2

F9–3. Determine the centroid $\bar{y}$ of the shaded area.

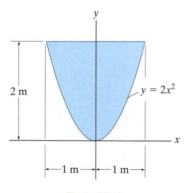

Prob. F9–3

F9–4. Locate the center of mass $\bar{x}$ of the straight rod if its mass per unit length is given by $m = m_0(1 + x^2/L^2)$.

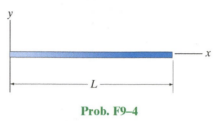

Prob. F9–4

F9–5. Locate the centroid $\bar{y}$ of the homogeneous solid formed by revolving the shaded area about the y axis.

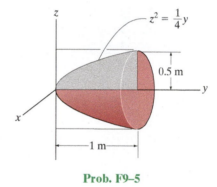

Prob. F9–5

F9–6. Locate the centroid $\bar{z}$ of the homogeneous solid formed by revolving the shaded area about the z axis.

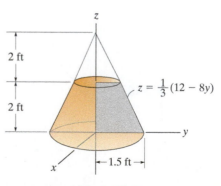

Prob. F9–6

PROBLEMS

9–1. Locate the center of mass of the homogeneous rod bent into the shape of a circular arc.

9–3. Locate the center of gravity $\bar{x}$ of the homogeneous rod. If the rod has a weight per unit length of 100 N/m, determine the vertical reaction at A and the x and y components of reaction at the pin B.

*9–4. Locate the center of gravity $\bar{y}$ of the homogeneous rod.

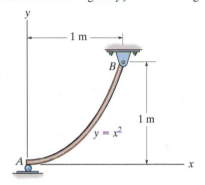

Probs. 9–3/4

9–5. Determine the distance $\bar{y}$ to the center of gravity of the homogeneous rod.

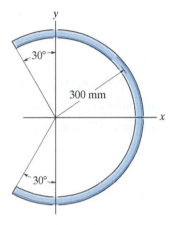

Prob. 9–1

9–2. Determine the location $(\bar{x}, \bar{y})$ of the centroid of the wire.

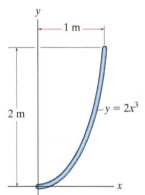

Prob. 9–5

9–6. Locate the centroid $\bar{y}$ of the area.

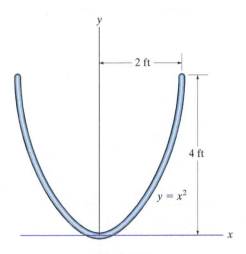

Prob. 9–2

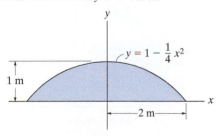

Prob. 9–6

9–7. Locate the centroid $\bar{x}$ of the parabolic area.

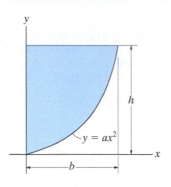

$y = ax^2$

Prob. 9–7

*9–8. Locate the centroid of the shaded area.

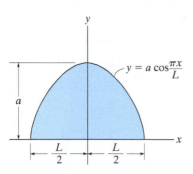

$y = a\cos\dfrac{\pi x}{L}$

Prob. 9–8

9–9. Locate the centroid $\bar{x}$ of the shaded area.

9–10. Locate the centroid $\bar{y}$ of the shaded area.

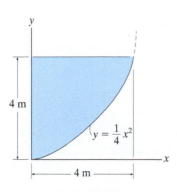

$y = \dfrac{1}{4}x^2$

Probs. 9–9/10

9–11. Locate the centroid $\bar{x}$ of the area.

*9–12. Locate the centroid $\bar{y}$ of the area.

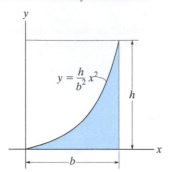

$y = \dfrac{h}{b^2}x^2$

Probs. 9–11/12

9–13. Locate the centroid $\bar{x}$ of the area.

9–14. Locate the centroid $\bar{y}$ of the area.

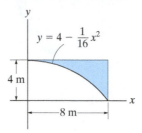

$y = 4 - \dfrac{1}{16}x^2$

Probs. 9–13/14

9–15. Locate the centroid $\bar{x}$ of the shaded area. Solve the problem by evaluating the integrals using Simpson's rule.

*9–16. Locate the centroid $\bar{y}$ of the shaded area. Solve the problem by evaluating the integrals using Simpson's rule.

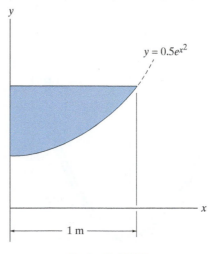

$y = 0.5e^{x^2}$

Probs. 9–15/16

9–17. Locate the centroid $\bar{y}$ of the area.

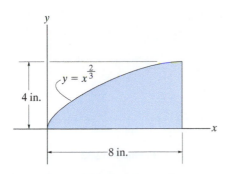

Prob. 9–17

9–18. Locate the centroid $\bar{x}$ of the area.

9–19. Locate the centroid $\bar{y}$ of the area.

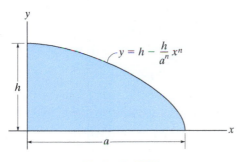

Probs. 9–18/19

*9–20. Locate the centroid $\bar{y}$ of the shaded area.

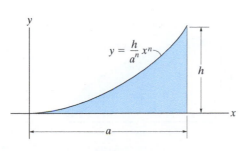

Prob. 9–20

9–21. Locate the centroid $\bar{x}$ of the shaded area.

9–22. Locate the centroid $\bar{y}$ of the shaded area.

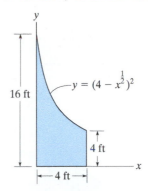

Probs. 9–21/22

9–23. Locate the centroid $\bar{x}$ of the shaded area.

*9–24. Locate the centroid $\bar{y}$ of the shaded area.

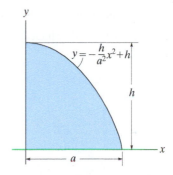

Probs. 9–23/24

9–25. The plate has a thickness of 0.25 ft and a specific weight of $\gamma = 180$ lb/ft^3. Determine the location of its center of gravity. Also, find the tension in each of the cords used to support it.

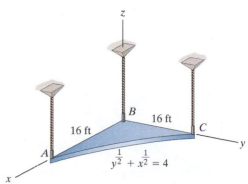

Prob. 9–25

9–26. Locate the centroid $\bar{x}$ of the shaded area.

9–27. Locate the centroid $\bar{y}$ of the shaded area.

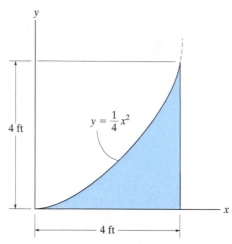

$$y = \frac{1}{4}x^2$$

4 ft

4 ft

Probs. 9–26/27

*9–28. Locate the centroid $\bar{x}$ of the shaded area.

9–29. Locate the centroid $\bar{y}$ of the shaded area.

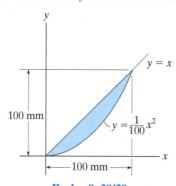

$y = x$

100 mm

$$y = \frac{1}{100}x^2$$

100 mm

Probs. 9–28/29

9–30. Locate the centroid $\bar{x}$ of the shaded area.

9–31. Locate the centroid $\bar{y}$ of the shaded area.

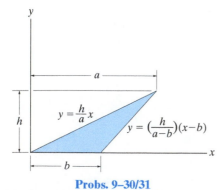

a

$$y = \frac{h}{a}x$$

$$y = \left(\frac{h}{a-b}\right)(x-b)$$

h

b

Probs. 9–30/31

*9–32. Locate the centroid $\bar{x}$ of the area.

9–33. Locate the centroid $\bar{y}$ of the area.

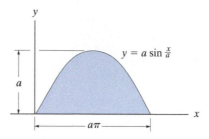

$$y = a\sin\frac{x}{a}$$

a

$a\pi$

Probs. 9–32/33

9–34. The steel plate is 0.3 m thick and has a density of 7850 kg/m³. Determine the location of its center of mass. Also find the reactions at the pin and roller support.

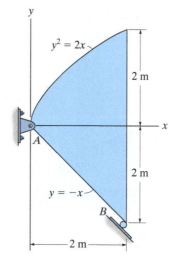

$y^2 = 2x$

2 m

A

2 m

$y = -x$

B

2 m

Prob. 9–34

9–35. Locate the centroid $\bar{x}$ of the shaded area.

*9–36. Locate the centroid $\bar{y}$ of the shaded area.

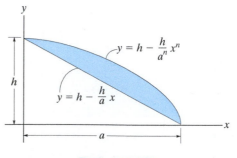

$$y = h - \frac{h}{a^n}x^n$$

h

$$y = h - \frac{h}{a}x$$

a

Probs. 9–35/36

9–37. Locate the centroid $\bar{x}$ of the circular sector.

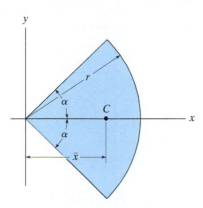

Prob. 9–37

9–38. Determine the location $\bar{r}$ of the centroid C for the loop of the lemniscate, $r^2 = 2a^2\cos 2\theta$, $(-45° \leq \theta \leq 45°)$.

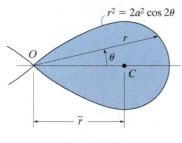

Prob. 9–38

9–39. Locate the center of gravity of the volume. The material is homogeneous.

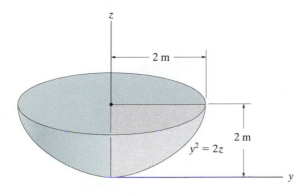

Prob. 9–39

*9–40. Locate the centroid $\bar{y}$ of the paraboloid.

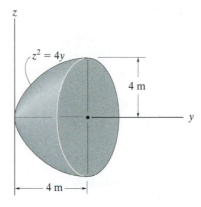

$z^2 = 4y$

4 m

4 m

Prob. 9–40

9–41. Locate the centroid $\bar{z}$ of the frustum of the right-circular cone.

Prob. 9–41

9–42. Determine the centroid $\bar{y}$ of the solid.

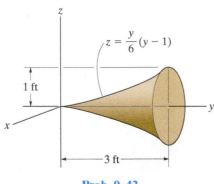

$$z = \frac{y}{6}(y-1)$$

1 ft

3 ft

Prob. 9–42

9–43. Locate the centroid of the quarter-cone.

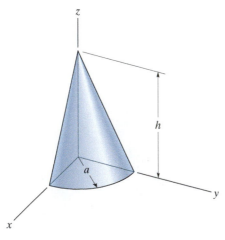

h

a

Prob. 9–43

*9–44. The hemisphere of radius r is made from a stack of very thin plates such that the density varies with height, $\rho = kz$, where k is a constant. Determine its mass and the distance $\bar{z}$ to the center of mass G.

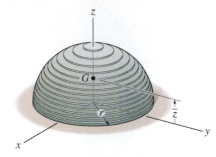

G

r

$\bar{z}$

Prob. 9–44

9–45. Locate the centroid $\bar{z}$ of the volume.

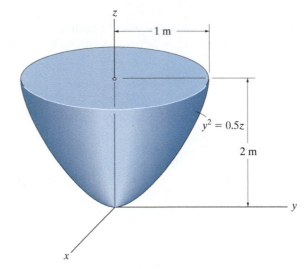

1 m

$y^2 = 0.5z$

2 m

Prob. 9–45

9–46. Locate the centroid of the ellipsoid of revolution.

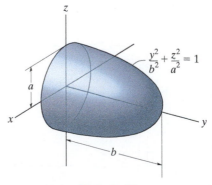

$$\frac{y^2}{b^2} + \frac{z^2}{a^2} = 1$$

a

b

Prob. 9–46

9–47.   Locate the center of gravity $\bar{z}$ of the solid.

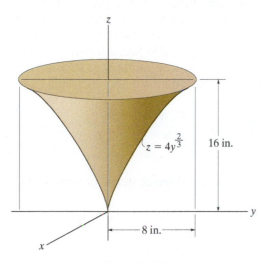

Prob. 9–47

9–49.   Locate the centroid $\bar{z}$ of the spherical segment.

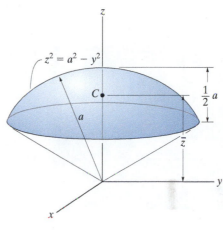

Prob. 9–49

9–50.   Determine the location $\bar{z}$ of the centroid for the tetrahedron. *Suggestion:* Use a triangular "plate" element parallel to the x–y plane and of thickness dz.

*9–48.   Locate the center of gravity $\bar{y}$ of the volume. The material is homogeneous.

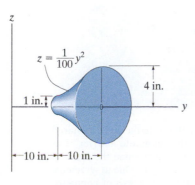

Prob. 9–48

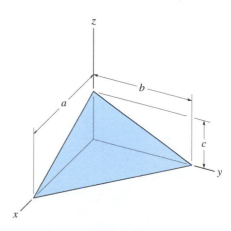

Prob. 9–50

9.2   Composite Bodies

A *composite body* consists of a series of connected "simpler" shaped bodies, which may be rectangular, triangular, semicircular, etc. Such a body can often be sectioned or divided into its composite parts and, provided the *weight* and location of the center of gravity of each of these parts are known, we can then eliminate the need for integration to determine the center of gravity for the entire body. The method for doing this follows the same procedure outlined in Sec. 9.1. Formulas analogous to Eqs. 9–1 result; however, rather than account for an infinite number of differential weights, we have instead a finite number of weights. Therefore,

$$\bar{x} = \frac{\Sigma \tilde{x}W}{\Sigma W} \quad \bar{y} = \frac{\Sigma \tilde{y}W}{\Sigma W} \quad \bar{z} = \frac{\Sigma \tilde{z}W}{\Sigma W} \qquad (9\text{–}6)$$

Here

$\bar{x}, \bar{y}, \bar{z}$ represent the coordinates of the center of gravity G of the composite body.

$\tilde{x}, \tilde{y}, \tilde{z}$ represent the coordinates of the center of gravity of each composite part of the body.

ΣW is the sum of the weights of all the composite parts of the body, or simply the total weight of the body.

When the body has a *constant density or specific weight*, the center of gravity *coincides* with the centroid of the body. The centroid for composite lines, areas, and volumes can be found using relations analogous to Eqs. 9–6; however, the W's are replaced by L's, A's, and V's, respectively. Centroids for common shapes of lines, areas, shells, and volumes that often make up a composite body are given in the table on the inside back cover.

A stress analysis of this angle requires that the centroid of its cross-sectional area be located. (© Russell C. Hibbeler)

In order to determine the force required to tip over this concrete barrier it is first necessary to determine the location of its center of gravity G. Due to symmetry, G will lie on the vertical axis of symmetry. (© Russell C. Hibbeler)

Procedure for Analysis

The location of the center of gravity of a body or the centroid of a composite geometrical object represented by a line, area, or volume can be determined using the following procedure.

Composite Parts.

- Using a sketch, divide the body or object into a finite number of composite parts that have simpler shapes.
- If a composite body has a *hole*, or a geometric region having no material, then consider the composite body without the hole and consider the hole as an *additional* composite part having *negative* weight or size.

Moment Arms.

- Establish the coordinate axes on the sketch and determine the coordinates $\tilde{x}$, $\tilde{y}$, $\tilde{z}$ of the center of gravity or centroid of each part.

Summations.

- Determine $\bar{x}$, $\bar{y}$, $\bar{z}$ by applying the center of gravity equations, Eqs. 9–6, or the analogous centroid equations.
- If an object is *symmetrical* about an axis, the centroid of the object lies on this axis.

If desired, the calculations can be arranged in tabular form, as indicated in the following three examples.

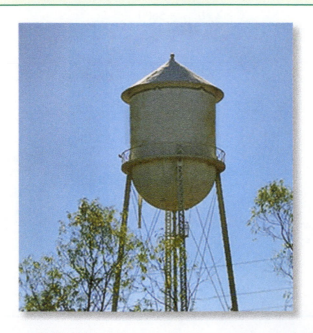

The center of gravity of this water tank can be determined by dividing it into composite parts and applying Eqs. 9–6. (© Russell C. Hibbeler)

EXAMPLE | 9.9

Locate the centroid of the wire shown in Fig. 9–16a.

SOLUTION

Composite Parts.   The wire is divided into three segments as shown in Fig. 9–16b.

Moment Arms.   The location of the centroid for each segment is determined and indicated in the figure. In particular, the centroid of segment ① is determined either by integration or by using the table on the inside back cover.

Summations.   For convenience, the calculations can be tabulated as follows:

Segment	L (mm)	$\tilde{x}$ (mm)	$\tilde{y}$ (mm)	$\tilde{z}$ (mm)	$\tilde{x}L$ (mm²)	$\tilde{y}L$ (mm²)	$\tilde{z}L$ (mm²)
1	$\pi(60) = 188.5$	60	-38.2	0	11 310	-7200	0
2	40	0	20	0	0	800	0
3	20	0	40	-10	0	800	-200
	$\Sigma L = 248.5$				$\Sigma\tilde{x}L = 11\ 310$	$\Sigma\tilde{y}L = -5600$	$\Sigma\tilde{z}L = -200$

Thus,

$$\bar{x} = \frac{\Sigma\tilde{x}L}{\Sigma L} = \frac{11\ 310}{248.5} = 45.5 \text{ mm} \qquad Ans.$$

$$\bar{y} = \frac{\Sigma\tilde{y}L}{\Sigma L} = \frac{-5600}{248.5} = -22.5 \text{ mm} \qquad Ans.$$

$$\bar{z} = \frac{\Sigma\tilde{z}L}{\Sigma L} = \frac{-200}{248.5} = -0.805 \text{ mm} \qquad Ans.$$

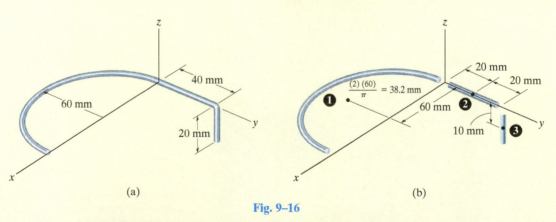

(a)

(b)

Fig. 9–16

EXAMPLE | 9.10

Locate the centroid of the plate area shown in Fig. 9–17a.

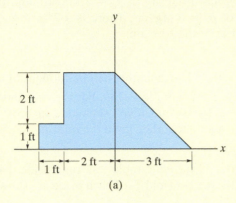

(a)

Fig. 9–17

SOLUTION

Composite Parts.   The plate is divided into three segments as shown in Fig. 9–17b. Here the area of the small rectangle ③ is considered "negative" since it must be subtracted from the larger one ②.

Moment Arms.   The centroid of each segment is located as indicated in the figure. Note that the $\tilde{x}$ coordinates of ② and ③ are *negative*.

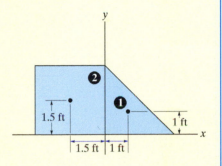

Summations.   Taking the data from Fig. 9–17b, the calculations are tabulated as follows:

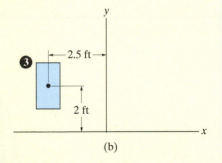

Segment	A (ft²)	$\tilde{x}$ (ft)	$\tilde{y}$ (ft)	$\tilde{x}A$ (ft³)	$\tilde{y}A$ (ft³)
1	$\frac{1}{2}(3)(3) = 4.5$	1	1	4.5	4.5
2	$(3)(3) = 9$	−1.5	1.5	−13.5	13.5
3	$-(2)(1) = -2$	−2.5	2	5	−4
	$\Sigma A = 11.5$			$\Sigma \tilde{x}A = -4$	$\Sigma \tilde{y}A = 14$

Thus,

$$\bar{x} = \frac{\Sigma \tilde{x}A}{\Sigma A} = \frac{-4}{11.5} = -0.348 \text{ ft} \qquad Ans.$$

$$\bar{y} = \frac{\Sigma \tilde{y}A}{\Sigma A} = \frac{14}{11.5} = 1.22 \text{ ft} \qquad Ans.$$

(b)

NOTE: If these results are plotted in Fig. 9–17a, the location of point C seems reasonable.

EXAMPLE 9.11

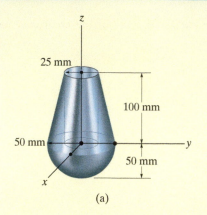

(a)

Fig. 9–18

Locate the center of mass of the assembly shown in Fig. 9–18a. The conical frustum has a density of $\rho_c = 8\ \text{Mg/m}^3$, and the hemisphere has a density of $\rho_h = 4\ \text{Mg/m}^3$. There is a 25-mm-radius cylindrical hole in the center of the frustum.

SOLUTION

Composite Parts. The assembly can be thought of as consisting of four segments as shown in Fig. 9–18b. For the calculations, ③ and ④ must be considered as "negative" segments in order that the four segments, when added together, yield the total composite shape shown in Fig. 9–18a.

Moment Arm. Using the table on the inside back cover, the computations for the centroid $\tilde{z}$ of each piece are shown in the figure.

Summations. Because of *symmetry*, note that

$$\bar{x} = \bar{y} = 0 \qquad\qquad Ans.$$

Since $W = mg$, and g is constant, the third of Eqs. 9–6 becomes $\bar{z} = \Sigma\tilde{z}m/\Sigma m$. The mass of each piece can be computed from $m = \rho V$ and used for the calculations. Also, $1\ \text{Mg/m}^3 = 10^{-6}\ \text{kg/mm}^3$, so that

Segment	m (kg)	$\tilde{z}$ (mm)	$\tilde{z}m$ (kg · mm)
1	$8(10^{-6})\left(\frac{1}{3}\right)\pi(50)^2(200) = 4.189$	50	209.440
2	$4(10^{-6})\left(\frac{2}{3}\right)\pi(50)^3 = 1.047$	-18.75	-19.635
3	$-8(10^{-6})\left(\frac{1}{3}\right)\pi(25)^2(100) = -0.524$	$100 + 25 = 125$	-65.450
4	$-8(10^{-6})\pi(25)^2(100) = -1.571$	50	-78.540
	$\Sigma m = 3.142$		$\Sigma\tilde{z}m = 45.815$

Thus, $\tilde{z} = \dfrac{\Sigma\tilde{z}m}{\Sigma m} = \dfrac{45.815}{3.142} = 14.6\ \text{mm}$ *Ans.*

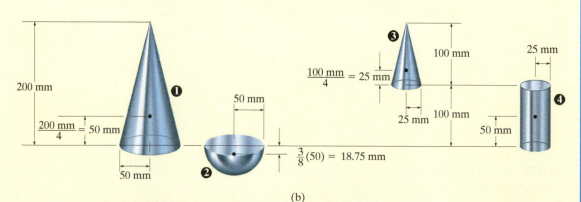

(b)

FUNDAMENTAL PROBLEMS

F9–7. Locate the centroid $(\bar{x}, \bar{y}, \bar{z})$ of the wire bent in the shape shown.

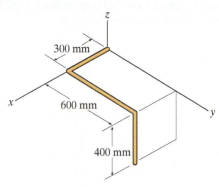

Prob. F9–7

F9–8. Locate the centroid $\bar{y}$ of the beam's cross-sectional area.

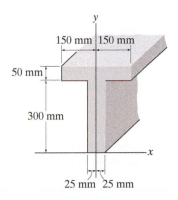

Prob. F9–8

F9–9. Locate the centroid $\bar{y}$ of the beam's cross-sectional area.

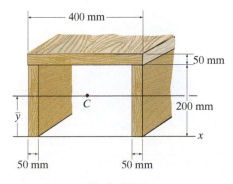

Prob. F9–9

F9–10. Locate the centroid $(\bar{x}, \bar{y})$ of the cross-sectional area.

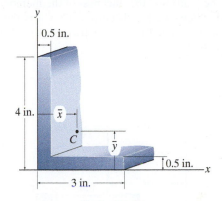

Prob. F9–10

F9–11. Locate the center of mass $(\bar{x}, \bar{y}, \bar{z})$ of the homogeneous solid block.

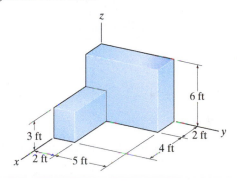

Prob. F9–11

F9–12. Determine the center of mass $(\bar{x}, \bar{y}, \bar{z})$ of the homogeneous solid block.

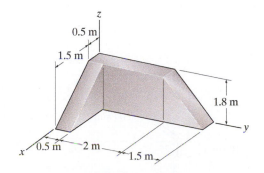

Prob. F9–12

9

PROBLEMS

9–51. The truss is made from five members, each having a length of 4 m and a mass of 7 kg/m. If the mass of the gusset plates at the joints and the thickness of the members can be neglected, determine the distance d to where the hoisting cable must be attached, so that the truss does not tip (rotate) when it is lifted.

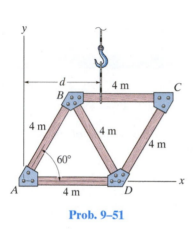

Prob. 9–51

*9–52. Determine the location $(\bar{x}, \bar{y}, \bar{z})$ of the centroid of the homogeneous rod.

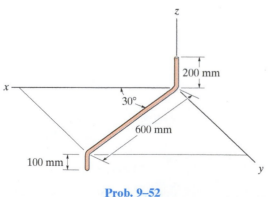

Prob. 9–52

9–53. A rack is made from roll-formed sheet steel and has the cross section shown. Determine the location $(\bar{x}, \bar{y})$ of the centroid of the cross section. The dimensions are indicated at the center thickness of each segment.

Prob. 9–53

9–54. Locate the centroid $(\bar{x}, \bar{y})$ of the metal cross section. Neglect the thickness of the material and slight bends at the corners.

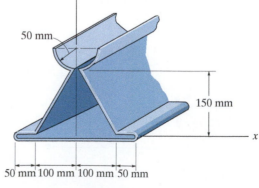

Prob. 9–54

9–55. Locate the center of gravity $(\bar{x}, \bar{y}, \bar{z})$ of the homogeneous wire.

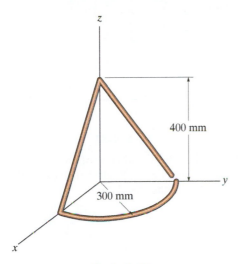

Prob. 9–55

9–57. Locate the center of gravity $G(\bar{x}, \bar{y})$ of the streetlight. Neglect the thickness of each segment. The mass per unit length of each segment is as follows: $\rho_{AB} = 12$ kg/m, $\rho_{BC} = 8$ kg/m, $\rho_{CD} = 5$ kg/m, and $\rho_{DE} = 2$ kg/m.

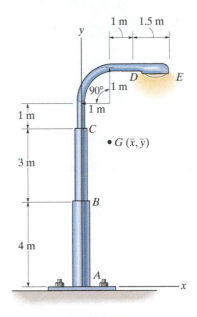

Prob. 9–57

*9–56. The steel and aluminum plate assembly is bolted together and fastened to the wall. Each plate has a constant width in the z direction of 200 mm and thickness of 20 mm. If the density of A and B is $\rho_s = 7.85$ Mg/m³, and for C, $\rho_{al} = 2.71$ Mg/m³, determine the location $\bar{x}$ of the center of mass. Neglect the size of the bolts.

9–58. Determine the location $\bar{y}$ of the centroidal axis $\bar{x}$–$\bar{x}$ of the beam's cross-sectional area. Neglect the size of the corner welds at A and B for the calculation.

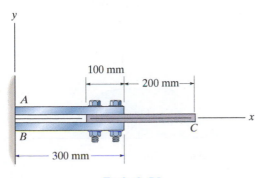

Prob. 9–56

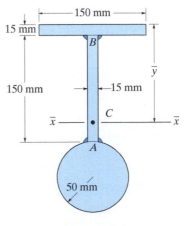

Prob. 9–58

9–59. Locate the centroid ($\bar{x}$, $\bar{y}$) of the shaded area.

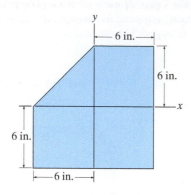

Prob. 9–59

*9–60. Locate the centroid $\bar{y}$ for the beam's cross-sectional area.

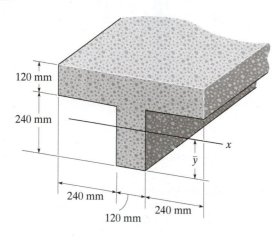

Prob. 9–60

9–61. Determine the location $\bar{y}$ of the centroid C of the beam having the cross-sectional area shown.

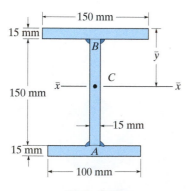

Prob. 9–61

9–62. Locate the centroid ($\bar{x}$, $\bar{y}$) of the shaded area.

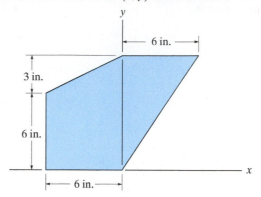

Prob. 9–62

9–63. Determine the location $\bar{y}$ of the centroid of the beam's cross-sectional area. Neglect the size of the corner welds at A and B for the calculation.

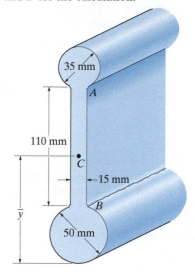

Prob. 9–63

*9–64. Locate the centroid ($\bar{x}$, $\bar{y}$) of the shaded area.

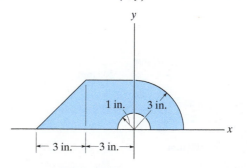

Prob. 9–64

9–65. Determine the location $(\bar{x}, \bar{y})$ of the centroid C of the area.

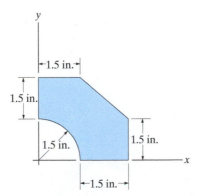

Prob. 9–65

9–66. Determine the location $\bar{y}$ of the centroid C for a beam having the cross-sectional area shown. The beam is symmetric with respect to the y axis.

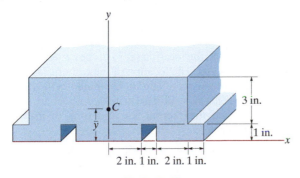

Prob. 9–66

9–67. Locate the centroid $\bar{y}$ of the cross-sectional area of the beam constructed from a channel and a plate. Assume all corners are square and neglect the size of the weld at A.

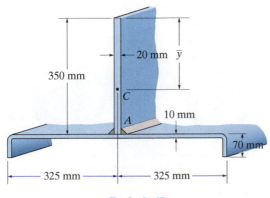

Prob. 9–67

*9–68. A triangular plate made of homogeneous material has a constant thickness that is very small. If it is folded over as shown, determine the location $\bar{y}$ of the plate's center of gravity G.

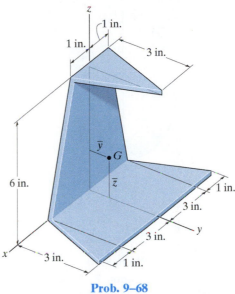

Prob. 9–68

9–69. A triangular plate made of homogeneous material has a constant thickness that is very small. If it is folded over as shown, determine the location $\bar{z}$ of the plate's center of gravity G.

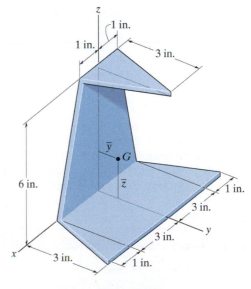

Prob. 9–69

9–70. Locate the center of mass $\bar{z}$ of the forked level which is made from a homogeneous material and has the dimensions shown.

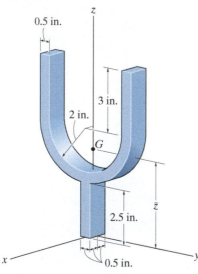

Prob. 9–70

9–71. Determine the location $\bar{x}$ of the centroid C of the shaded area that is part of a circle having a radius r.

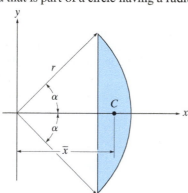

Prob. 9–71

*9–72. A toy skyrocket consists of a solid conical top, $\rho_i = 600 \text{ kg/m}^3$, a hollow cylinder, $\rho_c = 400 \text{ kg/m}^3$, and a stick having a circular cross section, $\rho_s = 300 \text{ kg/m}^3$. Determine the length of the stick, x, so that the center of gravity G of the skyrocket is located along line aa.

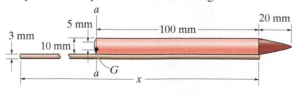

Prob. 9–72

9–73. Locate the centroid $\bar{y}$ for the cross-sectional area of the angle.

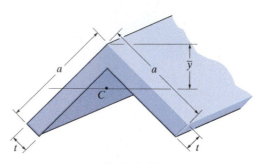

Prob. 9–73

9–74. Determine the location $(\bar{x}, \bar{y})$ of the center of gravity of the three-wheeler. The location of the center of gravity of each component and its weight are tabulated in the figure. If the three-wheeler is symmetrical with respect to the x–y plane, determine the normal reaction each of its wheels exerts on the ground.

1. Rear wheels	18 lb
2. Mechanical components	85 lb
3. Frame	120 lb
4. Front wheel	8 lb

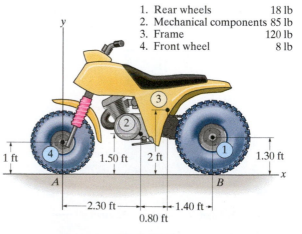

Prob. 9–74

9–75. Locate the center of mass ($\bar{x}$, $\bar{y}$, $\bar{z}$) of the homogeneous block assembly.

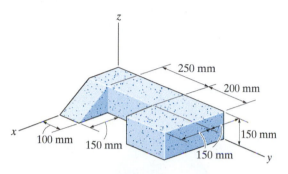

Prob. 9–75

9–78. The wooden table is made from a square board having a weight of 15 lb. Each of the legs weighs 2 lb and is 3 ft long. Determine how high its center of gravity is from the floor. Also, what is the angle, measured from the horizontal, through which its top surface can be tilted on two of its legs before it begins to overturn? Neglect the thickness of each leg.

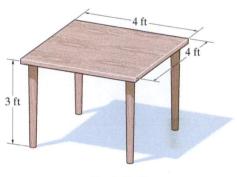

Prob. 9–78

*9–76. The sheet metal part has the dimensions shown. Determine the location ($\bar{x}$, $\bar{y}$, $\bar{z}$) of its centroid.

9–77. The sheet metal part has a weight per unit area of 2 lb/ft^2 and is supported by the smooth rod and the cord at C. If the cord is cut, the part will rotate about the y axis until it reaches equilibrium. Determine the equilibrium angle of tilt, measured downward from the negative x axis, that AD makes with the $-x$ axis.

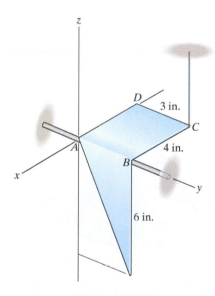

Probs. 9–76/77

9–79. The buoy is made from two homogeneous cones each having a radius of 1.5 ft. If $h = 1.2$ ft, find the distance $\bar{z}$ to the buoy's center of gravity G.

*9–80. The buoy is made from two homogeneous cones each having a radius of 1.5 ft. If it is required that the buoy's center of gravity G be located at $\bar{z} = 0.5$ ft, determine the height h of the top cone.

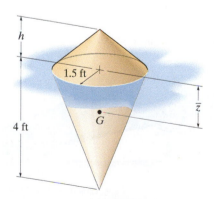

Probs. 9–79/80

9–81. The assembly is made from a steel hemisphere, $\rho_{st} = 7.80 \text{ Mg/m}^3$, and an aluminum cylinder, $\rho_{al} = 2.70 \text{ Mg/m}^3$. Determine the mass center of the assembly if the height of the cylinder is $h = 200$ mm.

9–82. The assembly is made from a steel hemisphere, $\rho_{st} = 7.80 \text{ Mg/m}^3$, and an aluminum cylinder, $\rho_{al} = 2.70 \text{ Mg/m}^3$. Determine the height h of the cylinder so that the mass center of the assembly is located at $\bar{z} = 160$ mm.

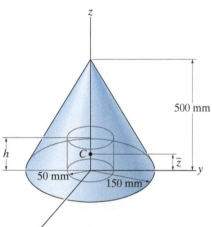

Probs. 9–81/82

9–83. The car rests on four scales and in this position the scale readings of both the front and rear tires are shown by F_A and F_B. When the rear wheels are elevated to a height of 3 ft above the front scales, the new readings of the front wheels are also recorded. Use this data to compute the location $\bar{x}$ and $\bar{y}$ to the center of gravity G of the car. The tires each have a diameter of 1.98 ft.

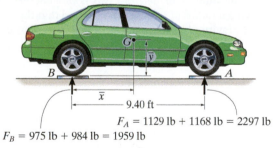

$F_A = 1129 \text{ lb} + 1168 \text{ lb} = 2297 \text{ lb}$

$F_B = 975 \text{ lb} + 984 \text{ lb} = 1959 \text{ lb}$

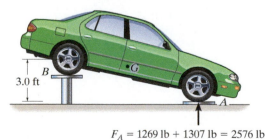

$F_A = 1269 \text{ lb} + 1307 \text{ lb} = 2576 \text{ lb}$

Prob. 9–83

*9–84. Determine the distance h to which a 100-mm-diameter hole must be bored into the base of the cone so that the center of mass of the resulting shape is located at $\bar{z} = 115$ mm. The material has a density of 8 Mg/m^3.

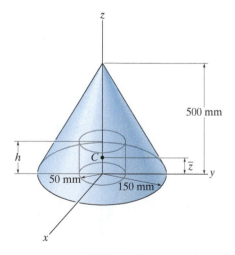

Prob. 9–84

9–85. Determine the distance $\bar{z}$ to the centroid of the shape that consists of a cone with a hole of height $h = 50$ mm bored into its base.

Prob. 9–85

9–86. Locate the center of mass $\bar{z}$ of the assembly. The cylinder and the cone are made from materials having densities of 5 Mg/m³ and 9 Mg/m³, respectively.

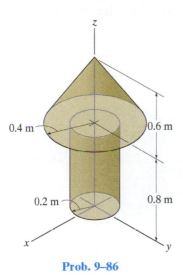

Prob. 9–86

9–87. Major floor loadings in a shop are caused by the weights of the objects shown. Each force acts through its respective center of gravity G. Locate the center of gravity $(\bar{x}, \bar{y})$ of all these components.

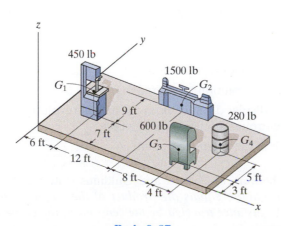

Prob. 9–87

*9–88. The assembly consists of a 20-in. wooden dowel rod and a tight-fitting steel collar. Determine the distance $\bar{x}$ to its center of gravity if the specific weights of the materials are $\gamma_w = 150$ lb/ft³ and $\gamma_{st} = 490$ lb/ft³. The radii of the dowel and collar are shown.

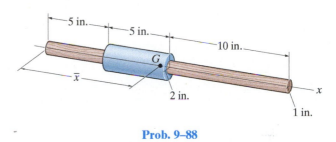

Prob. 9–88

9–89. The composite plate is made from both steel (A) and brass (B) segments. Determine the mass and location $(\bar{x}, \bar{y}, \bar{z})$ of its mass center G. Take $\rho_{st} = 7.85$ Mg/m³ and $\rho_{br} = 8.74$ Mg/m³.

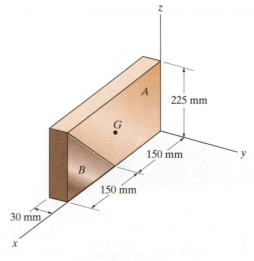

Prob. 9–89

*9.3   Theorems of Pappus and Guldinus

The two *theorems of Pappus and Guldinus* are used to find the surface area and volume of any body of revolution. They were first developed by Pappus of Alexandria during the fourth century A.D. and then restated at a later time by the Swiss mathematician Paul Guldin or Guldinus (1577–1643).

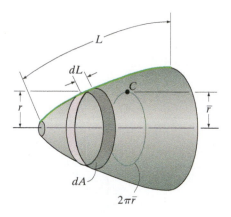

Fig. 9–19

Surface Area.   If we revolve a *plane curve* about an axis that does not intersect the curve we will generate a *surface area of revolution*. For example, the surface area in Fig. 9–19 is formed by revolving the curve of length L about the horizontal axis. To determine this surface area, we will first consider the differential line element of length dL. If this element is revolved 2π radians about the axis, a ring having a surface area of $dA = 2\pi r\, dL$ will be generated. Thus, the surface area of the entire body is $A = 2\pi \int r\, dL$. Since $\int r\, dL = \bar{r}L$ (Eq. 9–5), then $A = 2\pi \bar{r}L$. If the curve is revolved only through an angle θ (radians), then

$$\boxed{A = \theta \bar{r} L} \qquad\qquad (9–7)$$

where

$\quad A$ = surface area of revolution

$\quad \theta$ = angle of revolution measured in radians, $\theta \le 2\pi$

$\quad \bar{r}$ = perpendicular distance from the axis of revolution to the centroid of the generating curve

$\quad L$ = length of the generating curve

Therefore the first theorem of Pappus and Guldinus states that *the area of a surface of revolution equals the product of the length of the generating curve and the distance traveled by the centroid of the curve in generating the surface area.*

The amount of material used on this storage building can be estimated by using the first theorem of Pappus and Guldinus to determine its surface area. (© Russell C. Hibbeler)

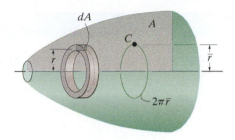

Fig. 9–20

Volume.

A *volume* can be generated by revolving a *plane area* about an axis that does not intersect the area. For example, if we revolve the shaded area A in Fig. 9–20 about the horizontal axis, it generates the volume shown. This volume can be determined by first revolving the differential element of area dA 2π radians about the axis, so that a ring having the volume $dV = 2\pi r\, dA$ is generated. The entire volume is then $V = 2\pi \int r\, dA$. However, $\int r\, dA = \bar{r}A$, Eq. 9–4, so that $V = 2\pi\bar{r}A$. If the area is only revolved through an angle θ (radians), then

$$V = \theta\bar{r}A \tag{9–8}$$

where

V = volume of revolution

θ = angle of revolution measured in radians, $\theta \leq 2\pi$

$\bar{r}$ = perpendicular distance from the axis of revolution to the centroid of the generating area

A = generating area

Therefore the second theorem of Pappus and Guldinus states that *the volume of a body of revolution equals the product of the generating area and the distance traveled by the centroid of the area in generating the volume.*

Composite Shapes.

We may also apply the above two theorems to lines or areas that are composed of a series of composite parts. In this case the total surface area or volume generated is the addition of the surface areas or volumes generated by each of the composite parts. If the perpendicular distance from the axis of revolution to the centroid of each composite part is $\tilde{r}$, then

$$A = \theta\Sigma(\tilde{r}L) \tag{9–9}$$

and

$$V = \theta\Sigma(\tilde{r}A) \tag{9–10}$$

Application of the above theorems is illustrated numerically in the following examples.

The volume of fertilizer contained within this silo can be determined using the second theorem of Pappus and Guldinus. (© Russell C. Hibbeler)

EXAMPLE | 9.12

Show that the surface area of a sphere is $A = 4\pi R^2$ and its volume is $V = \frac{4}{3}\pi R^3$.

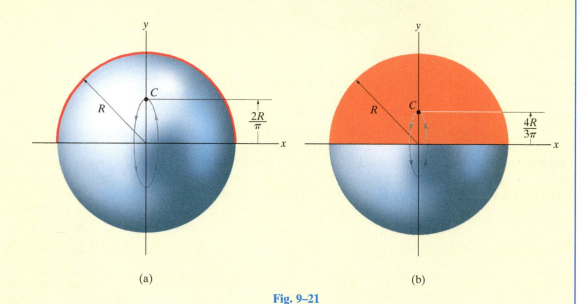

(a)   (b)

Fig. 9–21

SOLUTION

Surface Area.   The surface area of the sphere in Fig. 9–21a is generated by revolving a semicircular *arc* about the *x* axis. Using the table on the inside back cover, it is seen that the centroid of this arc is located at a distance $\bar{r} = 2R/\pi$ from the axis of revolution (*x* axis). Since the centroid moves through an angle of $\theta = 2\pi$ rad to generate the sphere, then applying Eq. 9–7 we have

$$A = \theta\bar{r}L; \qquad\qquad A = 2\pi\left(\frac{2R}{\pi}\right)\pi R = 4\pi R^2 \qquad\qquad Ans.$$

Volume.   The volume of the sphere is generated by revolving the semicircular *area* in Fig. 9–21b about the *x* axis. Using the table on the inside back cover to locate the centroid of the area, i.e., $\bar{r} = 4R/3\pi$, and applying Eq. 9–8, we have

$$V = \theta\bar{r}A; \qquad\qquad V = 2\pi\left(\frac{4R}{3\pi}\right)\left(\frac{1}{2}\pi R^2\right) = \frac{4}{3}\pi R^3 \qquad\qquad Ans.$$

EXAMPLE | 9.13

Determine the surface area and volume of the full solid in Fig. 9–22a.

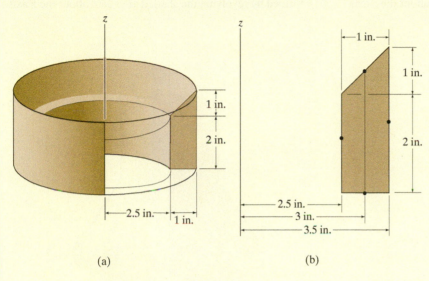

(a)   (b)

Fig. 9–22

SOLUTION

Surface Area. The surface area is generated by revolving the four line segments shown in Fig. 9–22b 2π radians about the z axis. The distances from the centroid of each segment to the z axis are also shown in the figure. Applying Eq. 9–7 yields

$$A = 2\pi\Sigma\bar{r}L = 2\pi[(2.5 \text{ in.})(2 \text{ in.}) + (3 \text{ in.})\left(\sqrt{(1 \text{ in.})^2 + (1 \text{ in.})^2}\right)$$
$$+ (3.5 \text{ in.})(3 \text{ in.}) + (3 \text{ in.})(1 \text{ in.})]$$
$$= 143 \text{ in}^2 \qquad\qquad\qquad Ans.$$

Volume. The volume of the solid is generated by revolving the two area segments shown in Fig. 9–22c 2π radians about the z axis. The distances from the centroid of each segment to the z axis are also shown in the figure. Applying Eq. 9–10, we have

$$V = 2\pi\Sigma\bar{r}A = 2\pi\left\{(3.1667 \text{ in.})\left[\frac{1}{2}(1 \text{ in.})(1 \text{ in.})\right] + (3 \text{ in.})[(2 \text{ in.})(1 \text{ in.})]\right\}$$
$$= 47.6 \text{ in}^3 \qquad\qquad\qquad Ans.$$

2.5 in. + $(\frac{2}{3})$(1 in.) = 3.1667 in.

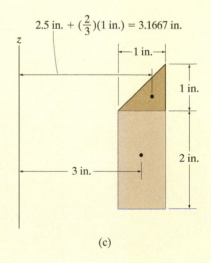

(c)

9

FUNDAMENTAL PROBLEMS

F9–13. Determine the surface area and volume of the solid formed by revolving the shaded area 360° about the z axis.

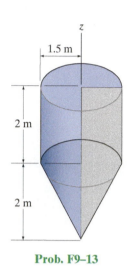

1.5 m

2 m

2 m

Prob. F9–13

F9–15. Determine the surface area and volume of the solid formed by revolving the shaded area 360° about the z axis.

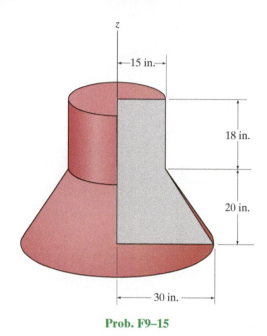

15 in.

18 in.

20 in.

30 in.

Prob. F9–15

F9–14. Determine the surface area and volume of the solid formed by revolving the shaded area 360° about the z axis.

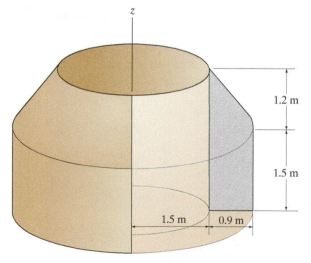

1.2 m

1.5 m

1.5 m   0.9 m

Prob. F9–14

F9–16. Determine the surface area and volume of the solid formed by revolving the shaded area 360° about the z axis.

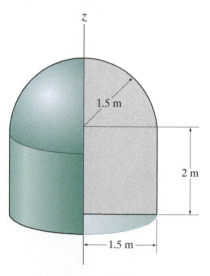

1.5 m

2 m

1.5 m

Prob. F9–16

PROBLEMS

9–90. Determine the volume of the silo which consists of a cylinder and hemispherical cap. Neglect the thickness of the plates.

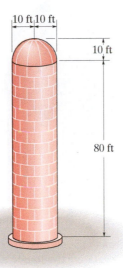

10 ft ┊ 10 ft

10 ft

80 ft

Prob. 9–90

9–91. Determine the outside surface area of the storage tank.

*9–92. Determine the volume of the storage tank.

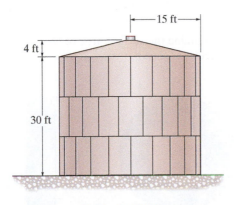

15 ft

4 ft

30 ft

Probs. 9–91/92

9–93. Determine the surface area of the concrete seawall, excluding its bottom.

9–94. A circular seawall is made of concrete. Determine the total weight of the wall if the concrete has a specific weight of $\gamma_c = 150 \text{ lb/ft}^3$.

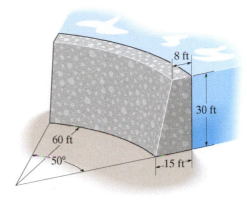

8 ft

30 ft

60 ft

50°

15 ft

Probs. 9–93/94

9–95. A ring is generated by rotating the quarter circular area about the x axis. Determine its volume.

*9–96. A ring is generated by rotating the quarter circular area about the x axis. Determine its surface area.

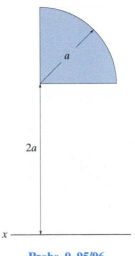

a

$2a$

x

Probs. 9–95/96

9–97. Determine the volume of concrete needed to construct the curb.

9–98. Determine the surface area of the curb. Do not include the area of the ends in the calculation.

9–101. The water-supply tank has a hemispherical bottom and cylindrical sides. Determine the weight of water in the tank when it is filled to the top at C. Take $\gamma_w = 62.4 \text{ lb/ft}^3$.

9–102. Determine the number of gallons of paint needed to paint the outside surface of the water-supply tank, which consists of a hemispherical bottom, cylindrical sides, and conical top. Each gallon of paint can cover 250 ft^2.

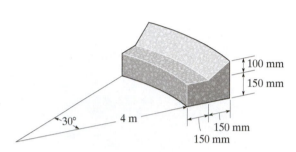

Probs. 9–97/98

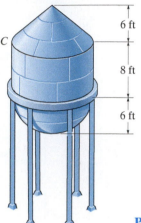

Probs. 9–101/102

9–103. Determine the surface area and the volume of the ring formed by rotating the square about the vertical axis.

9–99. A ring is formed by rotating the area 360° about the $\bar{x}-\bar{x}$ axes. Determine its surface area.

*9–100. A ring is formed by rotating the area 360° about the $\bar{x}-\bar{x}$ axes. Determine its volume.

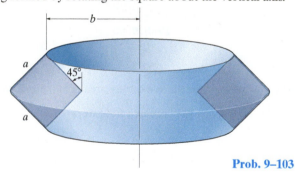

Prob. 9–103

*9–104. Determine the surface area of the ring. The cross section is circular as shown.

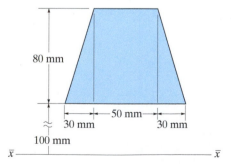

Probs. 9–99/100

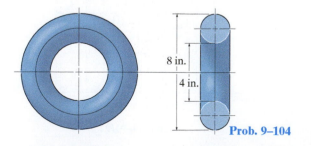

Prob. 9–104

9–105. The heat exchanger radiates thermal energy at the rate of 2500 kJ/h for each square meter of its surface area. Determine how many joules (J) are radiated within a 5-hour period.

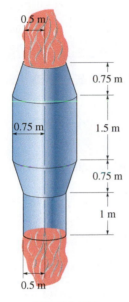

0.5 m
0.75 m
0.75 m
1.5 m
0.75 m
1 m
0.5 m

Prob. 9–105

9–106. Determine the interior surface area of the brake piston. It consists of a full circular part. Its cross section is shown in the figure.

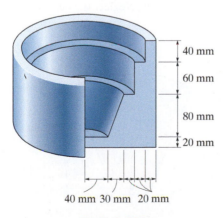

40 mm
60 mm
80 mm
20 mm
40 mm 30 mm 20 mm

Prob. 9–106

9–107. The suspension bunker is made from plates which are curved to the natural shape which a completely flexible membrane would take if subjected to a full load of coal. This curve may be approximated by a parabola, $y = 0.2x^2$. Determine the weight of coal which the bunker would contain when completely filled. Coal has a specific weight of $\gamma = 50$ lb/ft^3, and assume there is a 20% loss in volume due to air voids. Solve the problem by integration to determine the cross-sectional area of ABC; then use the second theorem of Pappus–Guldinus to find the volume.

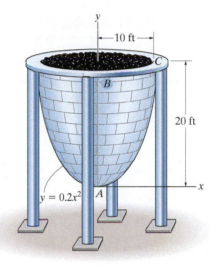

10 ft
20 ft
$y = 0.2x^2$

Prob. 9–107

*9–108. Determine the height h to which liquid should be poured into the cup so that it contacts three-fourths the surface area on the inside of the cup. Neglect the cup's thickness for the calculation.

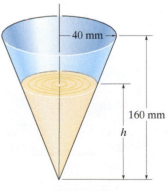

40 mm
160 mm
h

Prob. 9–108

9–109. Determine the surface area of the roof of the structure if it is formed by rotating the parabola about the y axis.

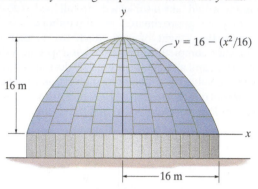

$y = 16 - (x^2/16)$

16 m

16 m

Prob. 9–109

9–110. A steel wheel has a diameter of 840 mm and a cross section as shown in the figure. Determine the total mass of the wheel if $\rho = 5$ Mg/m³.

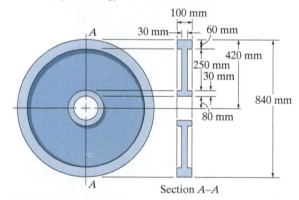

100 mm

A

30 mm — 60 mm

420 mm

250 mm

30 mm

840 mm

80 mm

A

Section $A–A$

Prob. 9–110

9–111. Half the cross section of the steel housing is shown in the figure. There are six 10-mm-diameter bolt holes around its rim. Determine its mass. The density of steel is 7.85 Mg/m³. The housing is a full circular part.

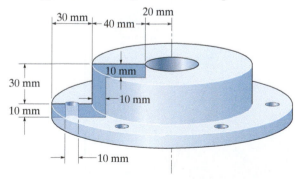

30 mm

20 mm

40 mm

10 mm

30 mm

10 mm

10 mm

10 mm

Prob. 9–111

*9–112. The water tank has a paraboloid-shaped roof. If one liter of paint can cover 3 m² of the tank, determine the number of liters required to coat the roof.

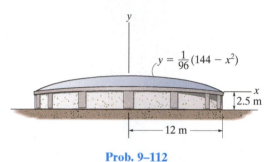

$y = \dfrac{1}{96}(144 - x^2)$

x

2.5 m

12 m

Prob. 9–112

9–113. Determine the volume of material needed to make the casting.

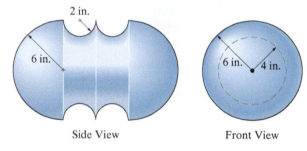

2 in.

6 in.

6 in.   4 in.

Side View   Front View

Prob. 9–113

9–114. Determine the height h to which liquid should be poured into the cup so that it contacts half the surface area on the inside of the cup. Neglect the cup's thickness for the calculation.

30 mm

50 mm

h

10 mm

Prob. 9–114

*9.4   Resultant of a General Distributed Loading

In Sec. 4.9, we discussed the method used to simplify a two-dimensional distributed loading to a single resultant force acting at a specific point. In this section we will generalize this method to include flat surfaces that have an arbitrary shape and are subjected to a variable load distribution. Consider, for example, the flat plate shown in Fig. 9–23a, which is subjected to the loading defined by $p = p(x, y)$ Pa, where 1 Pa (pascal) $= 1$ N/m². Knowing this function, we can determine the resultant force $\mathbf{F}_R$ acting on the plate and its location $(\bar{x}, \bar{y})$, Fig. 9–23b.

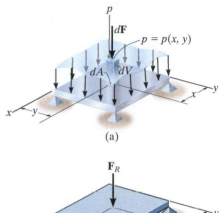

(a)

(b)

Fig. 9–23

Magnitude of Resultant Force.   The force $d\mathbf{F}$ acting on the differential area dA m² of the plate, located at the arbitrary point (x, y), has a magnitude of $dF = [p(x, y) \text{ N/m}^2](dA \text{ m}^2) = [p(x, y) \, dA]$ N. Notice that $p(x, y) \, dA = dV$, the colored differential *volume element* shown in Fig. 9–23a. The *magnitude* of $\mathbf{F}_R$ is the sum of the differential forces acting over the plate's *entire surface area A*. Thus:

$$F_R = \Sigma F; \qquad F_R = \int_A p(x, y) \, dA = \int_V dV = V \qquad (9\text{–}11)$$

This result indicates that the *magnitude of the resultant force is equal to the total volume under the distributed-loading diagram*.

Location of Resultant Force.   The location $(\bar{x}, \bar{y})$ of $\mathbf{F}_R$ is determined by setting the moments of $\mathbf{F}_R$ equal to the moments of all the differential forces $d\mathbf{F}$ about the respective y and x axes: From Figs. 9–23a and 9–23b, using Eq. 9–11, this results in

The resultant of a wind loading that is distributed on the front or side walls of this building must be calculated using integration in order to design the framework that holds the building together. (© Russell C. Hibbeler)

$$\bar{x} = \frac{\int_A x p(x, y) \, dA}{\int_A p(x, y) \, dA} = \frac{\int_V x \, dV}{\int_V dV} \qquad \bar{y} = \frac{\int_A y p(x, y) \, dA}{\int_A p(x, y) \, dA} = \frac{\int_V y \, dV}{\int_V dV} \qquad (9\text{–}12)$$

Hence, the *line of action of the resultant force passes through the geometric center or centroid of the volume under the distributed-loading diagram*.

*9.5   Fluid Pressure

According to Pascal's law, a fluid at rest creates a pressure p at a point that is the *same* in *all* directions. The magnitude of p, measured as a force per unit area, depends on the specific weight γ or mass density ρ of the fluid and the depth z of the point from the fluid surface.* The relationship can be expressed mathematically as

$$p = \gamma z = \rho g z \qquad (9\text{–}13)$$

where g is the acceleration due to gravity. This equation is valid only for fluids that are assumed *incompressible*, as in the case of most liquids. Gases are compressible fluids, and since their density changes significantly with both pressure and temperature, Eq. 9–13 cannot be used.

To illustrate how Eq. 9–13 is applied, consider the submerged plate shown in Fig. 9–24. Three points on the plate have been specified. Since point B is at depth z_1 from the liquid surface, the *pressure* at this point has a magnitude $p_1 = \gamma z_1$. Likewise, points C and D are both at depth z_2; hence, $p_2 = \gamma z_2$. In all cases, the pressure acts *normal* to the surface area dA located at the specified point.

Using Eq. 9–13 and the results of Sec. 9.4, it is possible to determine the resultant force caused by a liquid and specify its location on the surface of a submerged plate. Three different shapes of plates will now be considered.

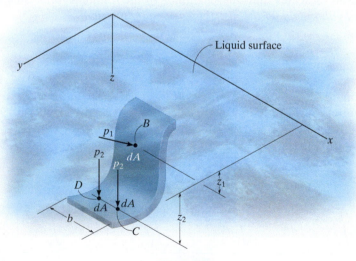

Fig. 9–24

*In particular, for water $\gamma = 62.4 \text{ lb/ft}^3$, or $\gamma = \rho g = 9810 \text{ N/m}^3$ since $\rho = 1000 \text{ kg/m}^3$ and $g = 9.81 \text{ m/s}^2$.

Flat Plate of Constant Width. A flat rectangular plate of constant width, which is submerged in a liquid having a specific weight γ, is shown in Fig. 9–25a. Since pressure varies linearly with depth, Eq. 9–13, the distribution of pressure over the plate's surface is represented by a trapezoidal volume having an intensity of $p_1 = \gamma z_1$ at depth z_1 and $p_2 = \gamma z_2$ at depth z_2. As noted in Sec. 9.4, the magnitude of the *resultant force* $\mathbf{F}_R$ is equal to the *volume* of this loading diagram and $\mathbf{F}_R$ has a *line of action* that passes through the volume's centroid C. Hence, $\mathbf{F}_R$ does *not* act at the centroid of the plate; rather, it acts at point P, called the *center of pressure*.

Since the plate has a *constant width*, the loading distribution may also be viewed in two dimensions, Fig. 9–25b. Here the loading intensity is measured as force/length and varies linearly from $w_1 = bp_1 = b\gamma z_1$ to $w_2 = bp_2 = b\gamma z_2$. The magnitude of $\mathbf{F}_R$ in this case equals the trapezoidal *area*, and $\mathbf{F}_R$ has a *line of action* that passes through the area's *centroid C*. For numerical applications, the area and location of the centroid for a trapezoid are tabulated on the inside back cover.

The walls of the tank must be designed to support the pressure loading of the liquid that is contained within it. (© Russell C. Hibbeler)

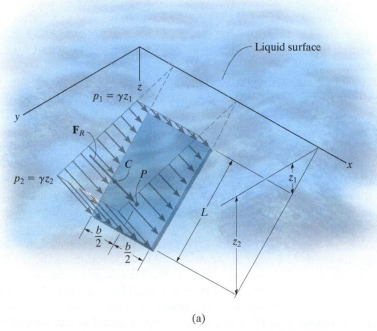

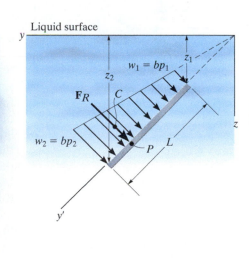

(a)   (b)

Fig. 9–25

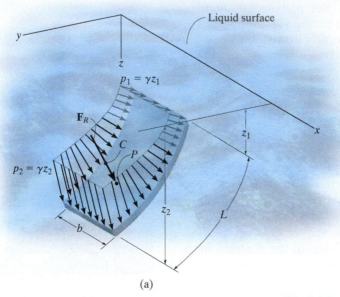

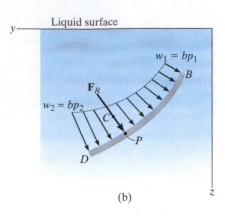

(a)

(b)

Fig. 9–26

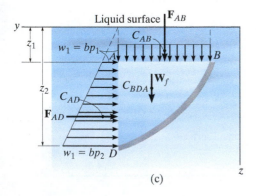

(c)

Curved Plate of Constant Width.

When a submerged plate of constant width is curved, the pressure acting normal to the plate continually changes both its magnitude and direction, and therefore calculation of the magnitude of $\mathbf{F}_R$ and its location P is more difficult than for a flat plate. Three- and two-dimensional views of the loading distribution are shown in Figs. 9–26a and 9–26b, respectively. Although integration can be used to solve this problem, a simpler method exists. This method requires separate calculations for the horizontal and vertical *components* of $\mathbf{F}_R$.

For example, the distributed loading acting on the plate can be represented by the *equivalent loading* shown in Fig. 9–26c. Here the plate supports the weight of liquid W_f contained within the block BDA. This force has a magnitude $W_f = (\gamma b)(\text{area}_{BDA})$ and acts through the centroid of BDA. In addition, there are the pressure distributions caused by the liquid acting along the vertical and horizontal sides of the block. Along the vertical side AD, the force $\mathbf{F}_{AD}$ has a magnitude equal to the area of the trapezoid. It acts through the centroid C_{AD} of this area. The distributed loading along the horizontal side AB is *constant* since all points lying in this plane are at the same depth from the surface of the liquid. The magnitude of $\mathbf{F}_{AB}$ is simply the area of the rectangle. This force acts through the centroid C_{AB} or at the midpoint of AB. Summing these three forces yields $\mathbf{F}_R = \Sigma\mathbf{F} = \mathbf{F}_{AD} + \mathbf{F}_{AB} + \mathbf{W}_f$. Finally, the location of the center of pressure P on the plate is determined by applying $M_R = \Sigma M$, which states that the moment of the resultant force about a convenient reference point such as D or B, in Fig. 9–26b, is equal to the sum of the moments of the three forces in Fig. 9–26c about this same point.

Flat Plate of Variable Width.   The pressure distribution acting on the surface of a submerged plate having a variable width is shown in Fig. 9–27. If we consider the force $d\mathbf{F}$ acting on the differential area strip dA, parallel to the x axis, then its magnitude is $dF = p\,dA$. Since the depth of dA is z, the pressure on the element is $p = \gamma z$. Therefore, $dF = (\gamma z)dA$ and so the resultant force becomes

$$F_R = \int dF = \gamma \int z\,dA$$

If the depth to the centroid C' of the area is $\bar{z}$, Fig. 9–27, then, $\int z\,dA = \bar{z}A$. Substituting, we have

$$F_R = \gamma\bar{z}A \qquad (9\text{–}14)$$

In other words, *the magnitude of the resultant force acting on any flat plate is equal to the product of the area A of the plate and the pressure $p = \gamma\bar{z}$ at the depth of the area's centroid C'*. As discussed in Sec. 9.4, this force is also equivalent to the volume under the pressure distribution. Realize that its line of action passes through the centroid C of this *volume* and intersects the plate at the center of pressure P, Fig. 9–27. Notice that the location of C' does not coincide with the location of P.

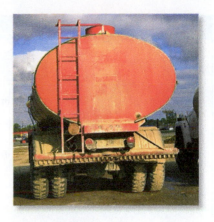

The resultant force of the water pressure and its location on the elliptical back plate of this tank truck must be determined by integration. (© Russell C. Hibbeler)

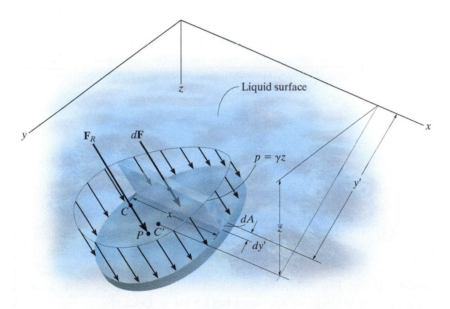

Fig. 9–27

EXAMPLE 9.14

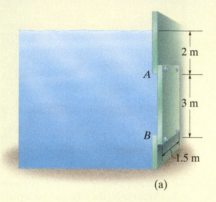

(a)

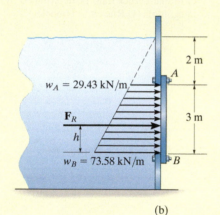

(b)

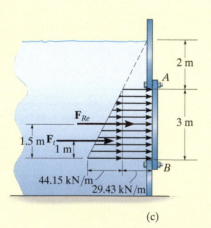

(c)

Fig. 9–28

Determine the magnitude and location of the resultant hydrostatic force acting on the submerged rectangular plate AB shown in Fig. 9–28a. The plate has a width of 1.5 m; $\rho_w = 1000$ kg/m^3.

SOLUTION I
The water pressures at depths A and B are

$$p_A = \rho_w g z_A = (1000 \text{ kg/m}^3)(9.81 \text{ m/s}^2)(2 \text{ m}) = 19.62 \text{ kPa}$$

$$p_B = \rho_w g z_B = (1000 \text{ kg/m}^3)(9.81 \text{ m/s}^2)(5 \text{ m}) = 49.05 \text{ kPa}$$

Since the plate has a constant width, the pressure loading can be viewed in two dimensions, as shown in Fig. 9–28b. The intensities of the load at A and B are

$$w_A = b p_A = (1.5 \text{ m})(19.62 \text{ kPa}) = 29.43 \text{ kN/m}$$

$$w_B = b p_B = (1.5 \text{ m})(49.05 \text{ kPa}) = 73.58 \text{ kN/m}$$

From the table on the inside back cover, the magnitude of the resultant force $\mathbf{F}_R$ created by this distributed load is

$$F_R = \text{area of a trapezoid} = \tfrac{1}{2}(3)(29.4 + 73.6) = 154.5 \text{ kN} \qquad \textit{Ans.}$$

This force acts through the centroid of this area,

$$h = \frac{1}{3}\left(\frac{2(29.43) + 73.58}{29.43 + 73.58}\right)(3) = 1.29 \text{ m} \qquad \textit{Ans.}$$

measured upward from B, Fig. 9–31b.

SOLUTION II
The same results can be obtained by considering two components of $\mathbf{F}_R$, defined by the triangle and rectangle shown in Fig. 9–28c. Each force acts through its associated centroid and has a magnitude of

$$F_{Re} = (29.43 \text{ kN/m})(3 \text{ m}) = 88.3 \text{ kN}$$

$$F_t = \tfrac{1}{2}(44.15 \text{ kN/m})(3 \text{ m}) = 66.2 \text{ kN}$$

Hence,

$$F_R = F_{Re} + F_t = 88.3 + 66.2 = 154.5 \text{ kN} \qquad \textit{Ans.}$$

The location of $\mathbf{F}_R$ is determined by summing moments about B, Figs. 9–28b and c, i.e.,

$$\zeta + (M_R)_B = \Sigma M_B; \quad (154.5)h = 88.3(1.5) + 66.2(1)$$

$$h = 1.29 \text{ m} \qquad \textit{Ans.}$$

NOTE: Using Eq. 9–14, the resultant force can be calculated as $F_R = \gamma \bar{z} A = (9810 \text{ N/m}^3)(3.5 \text{ m})(3 \text{ m})(1.5 \text{ m}) = 154.5 \text{ kN}$.

EXAMPLE 9.15

Determine the magnitude of the resultant hydrostatic force acting on the surface of a seawall shaped in the form of a parabola, as shown in Fig. 9–29a. The wall is 5 m long; $\rho_w = 1020 \text{ kg/m}^3$.

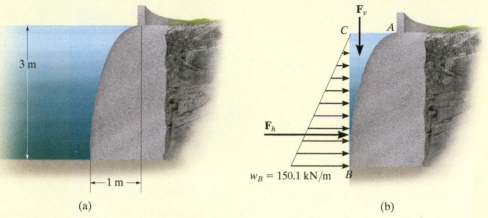

(a)   (b)

Fig. 9–29

SOLUTION

The horizontal and vertical components of the resultant force will be calculated, Fig. 9–29b. Since

$$p_B = \rho_w g z_B = (1020 \text{ kg/m}^3)(9.81 \text{ m/s}^2)(3 \text{ m}) = 30.02 \text{ kPa}$$

then

$$w_B = b p_B = 5 \text{ m}(30.02 \text{ kPa}) = 150.1 \text{ kN/m}$$

Thus,

$$F_h = \tfrac{1}{2}(3 \text{ m})(150.1 \text{ kN/m}) = 225.1 \text{ kN}$$

The area of the parabolic section ABC can be determined using the formula for a parabolic area $A = \tfrac{1}{3}ab$. Hence, the weight of water within this 5-m-long region is

$$F_v = (\rho_w g b)(\text{area}_{ABC})$$

$$= (1020 \text{ kg/m}^3)(9.81 \text{ m/s}^2)(5 \text{ m})\left[\tfrac{1}{3}(1 \text{ m})(3 \text{ m})\right] = 50.0 \text{ kN}$$

The resultant force is therefore

$$F_R = \sqrt{F_h^2 + F_v^2} = \sqrt{(225.1 \text{ kN})^2 + (50.0 \text{ kN})^2}$$

$$= 231 \text{ kN} \qquad\qquad Ans.$$

EXAMPLE | 9.16

Determine the magnitude and location of the resultant force acting on the triangular end plates of the water trough shown in Fig. 9–30a; $\rho_w = 1000 \text{ kg/m}^3$.

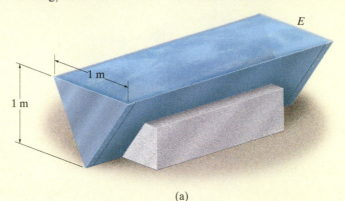

(a)

SOLUTION

The pressure distribution acting on the end plate E is shown in Fig. 9–30b. The magnitude of the resultant force is equal to the volume of this loading distribution. We will solve the problem by integration. Choosing the differential volume element shown in the figure, we have

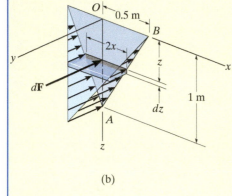

(b)

Fig. 9–30

$$dF = dV = p \, dA = \rho_w g z (2x \, dz) = 19 \, 620 z x \, dz$$

The equation of line AB is

$$x = 0.5(1 - z)$$

Hence, substituting and integrating with respect to z from $z = 0$ to $z = 1$ m yields

$$F = V = \int_V dV = \int_0^{1 \text{ m}} (19 \, 620) z [0.5(1 - z)] \, dz$$

$$= 9810 \int_0^{1 \text{ m}} (z - z^2) \, dz = 1635 \text{ N} = 1.64 \text{ kN} \qquad \textit{Ans.}$$

This resultant passes through the *centroid of the volume*. Because of symmetry,

$$\bar{x} = 0 \qquad \textit{Ans.}$$

Since $\tilde{z} = z$ for the volume element, then

$$\bar{z} = \frac{\int_V \tilde{z} \, dV}{\int_V dV} = \frac{\int_0^{1 \text{ m}} z(19 \, 620) z [0.5(1 - z)] \, dz}{1635} = \frac{9810 \int_0^{1 \text{ m}} (z^2 - z^3) \, dz}{1635}$$

$$= 0.5 \text{ m} \qquad \textit{Ans.}$$

NOTE: We can also determine the resultant force by applying Eq. 9–14, $F_R = \gamma \bar{z} A = \left(9810 \text{ N/m}^3 \right) \left(\frac{1}{3} \right) (1 \text{ m}) \left[\frac{1}{2} (1 \text{ m})(1 \text{ m}) \right] = 1.64 \text{ kN}.$

FUNDAMENTAL PROBLEMS

F9–17. Determine the magnitude of the hydrostatic force acting per meter length of the wall. Water has a density of $\rho = 1 \text{ Mg/m}^3$.

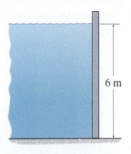

Prob. F9–17

F9–18. Determine the magnitude of the hydrostatic force acting on gate AB, which has a width of 4 ft. The specific weight of water is $\gamma = 62.4 \text{ lb/ft}^3$.

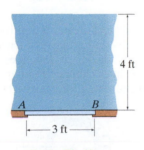

Prob. F9–18

F9–19. Determine the magnitude of the hydrostatic force acting on gate AB, which has a width of 1.5 m. Water has a density of $\rho = 1 \text{ Mg/m}^3$.

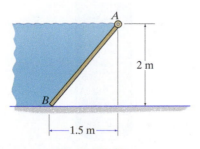

Prob. F9–19

F9–20. Determine the magnitude of the hydrostatic force acting on gate AB, which has a width of 2 m. Water has a density of $\rho = 1 \text{ Mg/m}^3$.

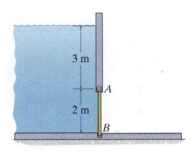

Prob. F9–20

F9–21. Determine the magnitude of the hydrostatic force acting on gate AB, which has a width of 2 ft. The specific weight of water is $\gamma = 62.4 \text{ lb/ft}^3$.

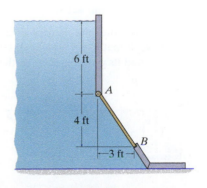

Prob. F9–21

PROBLEMS

9–115. The pressure loading on the plate varies uniformly along each of its edges. Determine the magnitude of the resultant force and the coordinates $(\bar{x}, \bar{y})$ of the point where the line of action of the force intersects the plate. *Hint:* The equation defining the boundary of the load has the form $p = ax + by + c$, where the constants a, b, and c have to be determined.

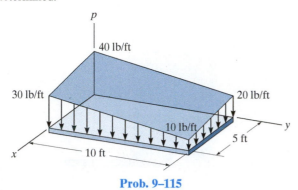

Prob. 9–115

*9–116. The load over the plate varies linearly along the sides of the plate such that $p = (12 - 6x + 4y)$ kPa. Determine the magnitude of the resultant force and the coordinates $(\bar{x}, \bar{y})$ of the point where the line of action of the force intersects the plate.

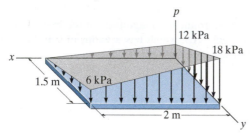

Prob. 9–116

9–117. The load over the plate varies linearly along the sides of the plate such that $p = \frac{2}{3}[x(4 - y)]$ kPa. Determine the resultant force and its position $(\bar{x}, \bar{y})$ on the plate.

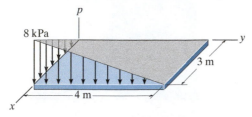

Prob. 9–117

9–118. The rectangular plate is subjected to a distributed load over its *entire surface*. The load is defined by the expression $p = p_0 \sin(\pi x/a) \sin(\pi y/b)$, where p_0 represents the pressure acting at the center of the plate. Determine the magnitude and location of the resultant force acting on the plate.

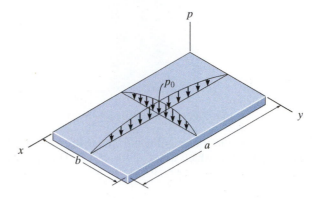

Prob. 9–118

9–119. A wind loading creates a positive pressure on one side of the chimney and a negative (suction) pressure on the other side, as shown. If this pressure loading acts uniformly along the chimney's length, determine the magnitude of the resultant force created by the wind.

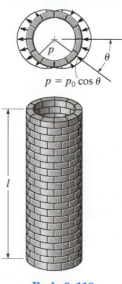

Prob. 9–119

*9–120. When the tide water A subsides, the tide gate automatically swings open to drain the marsh B. For the condition of high tide shown, determine the horizontal reactions developed at the hinge C and stop block D. The length of the gate is 6 m and its height is 4 m. $\rho_w = 1.0\ \mathrm{Mg/m^3}$.

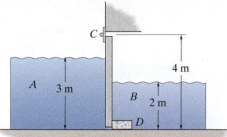

Prob. 9–120

9–121. The tank is filled with water to a depth of $d = 4$ m. Determine the resultant force the water exerts on side A and side B of the tank. If oil instead of water is placed in the tank, to what depth d should it reach so that it creates the same resultant forces? $\rho_o = 900\ \mathrm{kg/m^3}$ and $\rho_w = 1000\ \mathrm{kg/m^3}$.

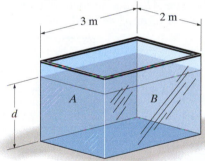

Prob. 9–121

9–122. The concrete "gravity" dam is held in place by its own weight. If the density of concrete is $\rho_c = 2.5\ \mathrm{Mg/m^3}$, and water has a density of $\rho_w = 1.0\ \mathrm{Mg/m^3}$, determine the smallest dimension d that will prevent the dam from overturning about its end A.

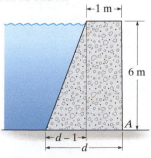

Prob. 9–122

9–123. The factor of safety for tipping of the concrete dam is defined as the ratio of the stabilizing moment due to the dam's weight divided by the overturning moment about O due to the water pressure. Determine this factor if the concrete has a density of $\rho_{conc} = 2.5\ \mathrm{Mg/m^3}$ and for water $\rho_w = 1\ \mathrm{Mg/m^3}$.

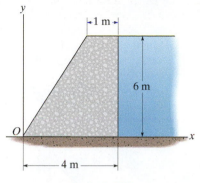

Prob. 9–123

*9–124. The concrete dam in the shape of a quarter circle. Determine the magnitude of the resultant hydrostatic force that acts on the dam per meter of length. The density of water is $\rho_w = 1\ \mathrm{Mg/m^3}$.

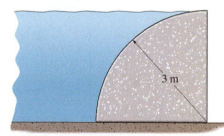

Prob. 9–124

9–125. The tank is used to store a liquid having a density of 80 lb/ft³. If it is filled to the top, determine the magnitude of force the liquid exerts on each of its two sides $ABDC$ and $BDFE$.

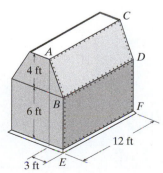

Prob. 9–125

9

9–126. The parabolic plate is subjected to a fluid pressure that varies linearly from 0 at its top to 100 lb/ft at its bottom B. Determine the magnitude of the resultant force and its location on the plate.

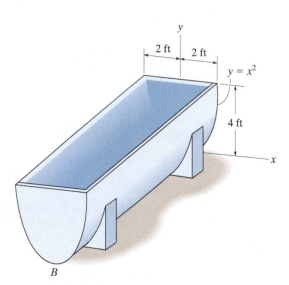

Prob. 9–126

9–127. The 2-m-wide rectangular gate is pinned at its center A and is prevented from rotating by the block at B. Determine the reactions at these supports due to hydrostatic pressure. $\rho_w = 1.0 \text{ Mg/m}^3$.

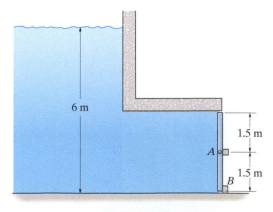

Prob. 9–127

*9–128. The tank is filled with a liquid that has a density of 900 kg/m^3. Determine the resultant force that it exerts on the elliptical end plate, and the location of the center of pressure, measured from the x axis.

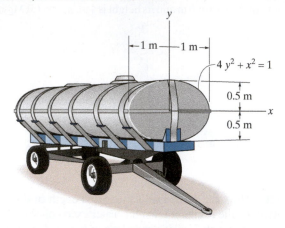

Prob. 9–128

9–129. Determine the magnitude of the resultant force acting on the gate ABC due to hydrostatic pressure. The gate has a width of 1.5 m. $\rho_w = 1.0 \text{ Mg/m}^3$.

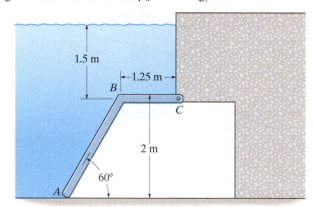

Prob. 9–129

9–130. The semicircular drainage pipe is filled with water. Determine the resultant horizontal and vertical force components that the water exerts on the side AB of the pipe per foot of pipe length; $\gamma_w = 62.4 \text{ lb/ft}^3$.

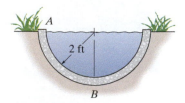

Prob. 9–130

CHAPTER REVIEW

Center of Gravity and Centroid

The *center of gravity G* represents a point where the weight of the body can be considered concentrated. The distance from an axis to this point can be determined from a balance of moments, which requires that the moment of the weight of all the particles of the body about this axis must equal the moment of the entire weight of the body about the axis.

$$\bar{x} = \frac{\int \tilde{x}\, dW}{\int dW}$$

$$\bar{y} = \frac{\int \tilde{y}\, dW}{\int dW}$$

$$\bar{z} = \frac{\int \tilde{z}\, dW}{\int dW}$$

The center of mass will coincide with the center of gravity provided the acceleration of gravity is constant.

$$\bar{x} = \frac{\int_L \tilde{x}\, dL}{\int_L dL} \qquad \bar{y} = \frac{\int_L \tilde{y}\, dL}{\int_L dL} \qquad \bar{z} = \frac{\int_L \tilde{z}\, dL}{\int_L dL}$$

The *centroid* is the location of the geometric center for the body. It is determined in a similar manner, using a moment balance of geometric elements such as line, area, or volume segments. For bodies having a continuous shape, moments are summed (integrated) using differential elements.

$$\bar{x} = \frac{\int_A \tilde{x}\, dA}{\int_A dA} \qquad \bar{y} = \frac{\int_A \tilde{y}\, dA}{\int_A dA} \qquad \bar{z} = \frac{\int_A \tilde{z}\, dA}{\int_A dA}$$

$$\bar{x} = \frac{\int_V \tilde{x}\, dV}{\int_V dV} \qquad \bar{y} = \frac{\int_V \tilde{y}\, dV}{\int_A dV} \qquad \bar{z} = \frac{\int_V \tilde{z}\, dV}{\int_V dV}$$

The center of mass will coincide with the centroid provided the material is homogeneous, i.e., the density of the material is the same throughout. The centroid will always lie on an axis of symmetry.

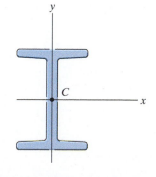

Composite Body

If the body is a composite of several shapes, each having a known location for its center of gravity or centroid, then the location of the center of gravity or centroid of the body can be determined from a discrete summation using its composite parts.

$$\bar{x} = \frac{\Sigma \tilde{x} W}{\Sigma W}$$

$$\bar{y} = \frac{\Sigma \tilde{y} W}{\Sigma W}$$

$$\bar{z} = \frac{\Sigma \tilde{z} W}{\Sigma W}$$

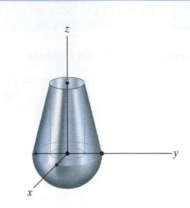

Theorems of Pappus and Guldinus

The theorems of Pappus and Guldinus can be used to determine the surface area and volume of a body of revolution.

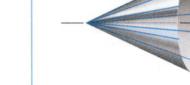

The *surface area* equals the product of the length of the generating curve and the distance traveled by the centroid of the curve needed to generate the area.

$$A = \theta \bar{r} L$$

The *volume* of the body equals the product of the generating area and the distance traveled by the centroid of this area needed to generate the volume.

$$V = \theta \bar{r} A$$

General Distributed Loading

The magnitude of the resultant force is equal to the total volume under the distributed-loading diagram. The line of action of the resultant force passes through the geometric center or centroid of this volume.

$$F_R = \int_A p(x, y)\, dA = \int_V dV$$

$$\bar{x} = \frac{\displaystyle\int_V x\, dV}{\displaystyle\int_V dV}$$

$$\bar{y} = \frac{\displaystyle\int_V y\, dV}{\displaystyle\int_V dV}$$

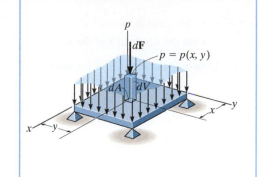

Fluid Pressure

The pressure developed by a liquid at a point on a submerged surface depends upon the depth of the point and the density of the liquid in accordance with Pascal's law, $p = \rho g h = \gamma h$. This pressure will create a *linear distribution* of loading on a flat vertical or inclined surface.

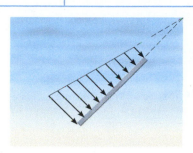

If the surface is horizontal, then the loading will be *uniform*.

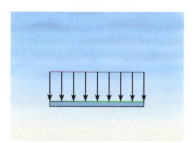

In any case, the resultants of these loadings can be determined by finding the volume under the loading curve or using $F_R = \gamma \bar{z} A$, where $\bar{z}$ is the depth to the centroid of the plate's area. The line of action of the resultant force passes through the centroid of the volume of the loading diagram and acts at a point P on the plate called the center of pressure.

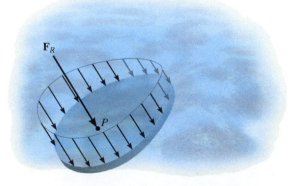

REVIEW PROBLEMS

R9–1. Locate the centroid $\bar{x}$ of the area.

R9–2. Locate the centroid $\bar{y}$ of the area.

R9–4. Locate the centroid of the rod.

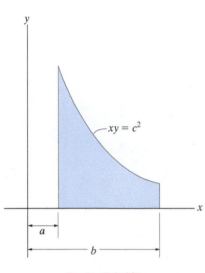

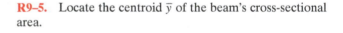

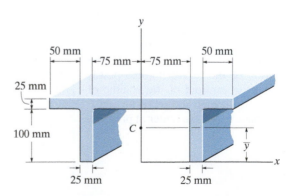

Prob. R9–4

Probs. R9–1/2

R9–5. Locate the centroid $\bar{y}$ of the beam's cross-sectional area.

R9–3. Locate the centroid $\bar{z}$ of the hemisphere.

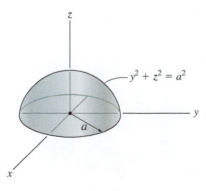

Prob. R9–3

Prob. R9–5

R9–6. A circular V-belt has an inner radius of 600 mm and a cross-sectional area as shown. Determine the surface area of the belt.

R9–7. A circular V-belt has an inner radius of 600 mm and a cross-sectional area as shown. Determine the volume of material required to make the belt.

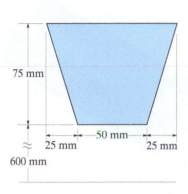

Probs. R9–6/7

R9–9. The gate AB is 8 m wide. Determine the horizontal and vertical components of force acting on the pin at B and the vertical reaction at the smooth support A; $\rho_w = 1.0 \text{ Mg/m}^3$.

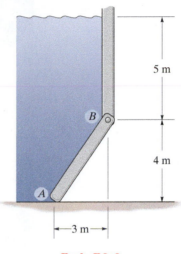

Prob. R9–9

R9–8. The rectangular bin is filled with coal, which creates a pressure distribution along wall A that varies as shown, i.e., $p = 4z^{1/3}$ lb/ft^2, where z is in feet. Determine the resultant force created by the coal, and its location, measured from the top surface of the coal.

R9–10. Determine the magnitude of the resultant hydrostatic force acting per foot of length on the seawall; $\gamma_w = 62.4$ lb/ft^3.

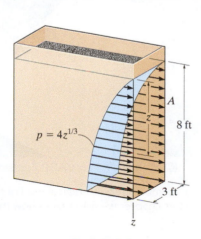

Prob. R9–8

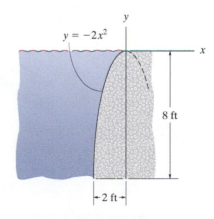

Prob. R9–10

Chapter 10

(© Michael N. Paras/AGE Fotostock/Alamy)

The design of these structural members requires calculation of their cross-sectional moment of inertia. In this chapter we will discuss how this is done.

Moments of Inertia

CHAPTER OBJECTIVES

■ To develop a method for determining the moment of inertia for an area.

■ To introduce the product of inertia and show how to determine the maximum and minimum moments of inertia for an area.

■ To discuss the mass moment of inertia.

10.1 Definition of Moments of Inertia for Areas

Whenever a distributed load acts perpendicular to an area and its intensity varies linearly, the calculation of the moment of the loading about an axis will involve an integral of the form $\int y^2 dA$. For example, consider the plate in Fig. 10–1, which is submerged in a fluid and subjected to the pressure p. As discussed in Sec. 9.5, this pressure varies linearly with depth, such that $p = \gamma y$, where γ is the specific weight of the fluid. Thus, the force acting on the differential area dA of the plate is $dF = p\, dA = (\gamma y)dA$. The *moment* of this force about the x axis is therefore $dM = y\, dF = \gamma y^2 dA$, and so integrating dM over the entire area of the plate yields $M = \gamma \int y^2 dA$. The integral $\int y^2 dA$ is sometimes referred to as the "second moment" of the area about an axis (the x axis), but more often it is called the ***moment of inertia of the area***. The word "inertia" is used here since the formulation is similar to the mass moment of inertia, $\int y^2 dm$, which is a dynamical property described in Sec. 10.8. Although for an area this integral has no physical meaning, it often arises in formulas used in fluid mechanics, mechanics of materials, structural mechanics, and mechanical design, and so the engineer needs to be familiar with the methods used to determine the moment of inertia.

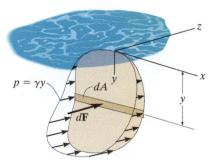

Fig. 10–1

Moment of Inertia.   By definition, the moments of inertia of a differential area dA about the x and y axes are $dI_x = y^2\, dA$ and $dI_y = x^2\, dA$, respectively, Fig. 10–2. For the entire area A the **moments of inertia** are determined by integration; i.e.,

$$
I_x = \int_A y^2\, dA
$$
$$
I_y = \int_A x^2\, dA
$$

(10–1)

We can also formulate this quantity for dA about the "pole" O or z axis, Fig. 10–2. This is referred to as the **polar moment of inertia**. It is defined as $dJ_O = r^2\, dA$, where r is the perpendicular distance from the pole (z axis) to the element dA. For the entire area the *polar moment of inertia* is

$$
J_O = \int_A r^2\, dA = I_x + I_y
$$

(10–2)

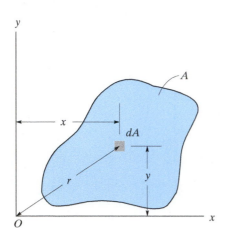

Fig. 10–2

This relation between J_O and I_x, I_y is possible since $r^2 = x^2 + y^2$, Fig. 10–2.

From the above formulations it is seen that I_x, I_y, and J_O will *always* be *positive* since they involve the product of distance squared and area. Furthermore, the units for moment of inertia involve length raised to the fourth power, e.g., m^4, mm^4, or ft^4, $in.^4$.

10.2   Parallel-Axis Theorem for an Area

The **parallel-axis theorem** can be used to find the moment of inertia of an area about *any axis* that is parallel to an axis passing through the centroid and about which the moment of inertia is known. To develop this theorem, we will consider finding the moment of inertia of the shaded area shown in Fig. 10–3 about the x axis. To start, we choose a differential element dA located at an arbitrary distance y' from the *centroidal x' axis*. If the distance between the parallel x and x' axis is d_y, then the moment of inertia of dA about the x axis is $dI_x = (y' + d_y)^2\, dA$. For the entire area,

$$
I_x = \int_A (y' + d_y)^2\, dA
$$

$$
= \int_A y'^2\, dA + 2d_y \int_A y'\, dA + d_y^2 \int_A dA
$$

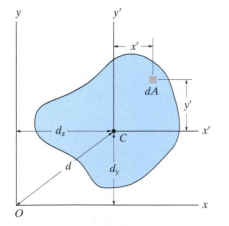

Fig. 10–3

The first integral represents the moment of inertia of the area about the centroidal axis, $\bar{I}_{x'}$. The second integral is zero since the x' axis passes through the area's centroid C; i.e., $\int y' \, dA = \bar{y}' \int dA = 0$ since $\bar{y}' = 0$. Since the third integral represents the total area A, the final result is therefore

$$I_x = \bar{I}_{x'} + Ad_y^2 \qquad (10\text{–}3)$$

A similar expression can be written for I_y; i.e.,

$$I_y = \bar{I}_{y'} + Ad_x^2 \qquad (10\text{–}4)$$

And finally, for the polar moment of inertia, since $\bar{J}_C = \bar{I}_{x'} + \bar{I}_{y'}$ and $d^2 = d_x^2 + d_y^2$, we have

$$J_O = \bar{J}_C + Ad^2 \qquad (10\text{–}5)$$

The form of each of these three equations states that *the moment of inertia for an area about an axis is equal to its moment of inertia about a parallel axis passing through the area's centroid plus the product of the area and the square of the perpendicular distance between the axes.*

In order to predict the strength and deflection of this beam, it is necessary to calculate the moment of inertia of the beam's cross-sectional area. (© Russell C. Hibbeler)

10.3   Radius of Gyration of an Area

The *radius of gyration* of an area about an axis has units of length and is a quantity that is often used for the design of columns in structural mechanics. Provided the areas and moments of inertia are *known*, the radii of gyration are determined from the formulas

$$k_x = \sqrt{\frac{I_x}{A}}$$

$$k_y = \sqrt{\frac{I_y}{A}} \qquad (10\text{–}6)$$

$$k_O = \sqrt{\frac{J_O}{A}}$$

The form of these equations is easily remembered since it is similar to that for finding the moment of inertia for a differential area about an axis. For example, $I_x = k_x^2 A$; whereas for a differential area, $dI_x = y^2 \, dA$.

10

Important Points

- The moment of inertia is a geometric property of an area that is used to determine the strength of a structural member or the location of a resultant pressure force acting on a plate submerged in a fluid. It is sometimes referred to as the second moment of the area about an axis, because the distance from the axis to each area element is squared.

- If the moment of inertia of an area is known about its centroidal axis, then the moment of inertia about a corresponding parallel axis can be determined using the parallel-axis theorem.

Procedure for Analysis

In most cases the moment of inertia can be determined using a single integration. The following procedure shows two ways in which this can be done.

- If the curve defining the boundary of the area is expressed as $y = f(x)$, then select a rectangular differential element such that it has a finite length and differential width.

- The element should be located so that it intersects the curve at the *arbitrary point* (x, y).

Case 1.

- Orient the element so that its length is *parallel* to the axis about which the moment of inertia is computed. This situation occurs when the rectangular element shown in Fig. 10–4a is used to determine I_x for the area. Here the entire element is at a distance y from the x axis since it has a thickness dy. Thus $I_x = \int y^2 dA$. To find I_y, the element is oriented as shown in Fig. 10–4b. This element lies at the *same* distance x from the y axis so that $I_y = \int x^2 dA$.

Case 2.

- The length of the element can be oriented *perpendicular* to the axis about which the moment of inertia is computed; however, Eq. 10–1 *does not apply* since all points on the element will *not* lie at the same moment-arm distance from the axis. For example, if the rectangular element in Fig. 10–4a is used to determine I_y, it will first be necessary to calculate the moment of inertia of the *element* about an axis parallel to the y axis that passes through the element's centroid, and then determine the moment of inertia of the *element* about the y axis using the parallel-axis theorem. Integration of this result will yield I_y. See Examples 10.2 and 10.3.

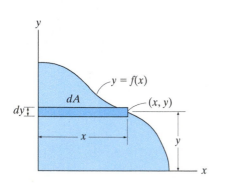

(a)

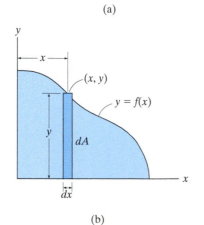

(b)

Fig. 10–4

EXAMPLE | 10.1

Determine the moment of inertia for the rectangular area shown in Fig. 10–5 with respect to (a) the centroidal x' axis, (b) the axis x_b passing through the base of the rectangle, and (c) the pole or z' axis perpendicular to the x'–y' plane and passing through the centroid C.

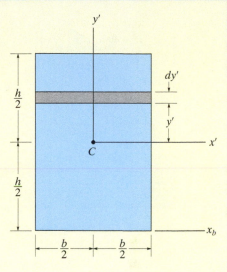

SOLUTION (CASE 1)

Part (a). The differential element shown in Fig. 10–5 is chosen for integration. Because of its location and orientation, the *entire element* is at a distance y' from the x' axis. Here it is necessary to integrate from $y' = -h/2$ to $y' = h/2$. Since $dA = b\,dy'$, then

$$\bar{I}_{x'} = \int_A y'^2\,dA = \int_{-h/2}^{h/2} y'^2(b\,dy') = b\int_{-h/2}^{h/2} y'^2\,dy'$$

$$\bar{I}_{x'} = \frac{1}{12}bh^3 \qquad \qquad Ans.$$

Fig. 10–5

Part (b). The moment of inertia about an axis passing through the base of the rectangle can be obtained by using the above result of part (a) and applying the parallel-axis theorem, Eq. 10–3.

$$I_{x_b} = \bar{I}_{x'} + A d_y^2$$

$$= \frac{1}{12}bh^3 + bh\left(\frac{h}{2}\right)^2 = \frac{1}{3}bh^3 \qquad \qquad Ans.$$

Part (c). To obtain the polar moment of inertia about point C, we must first obtain $\bar{I}_{y'}$, which may be found by interchanging the dimensions b and h in the result of part (a), i.e.,

$$\bar{I}_{y'} = \frac{1}{12}hb^3$$

Using Eq. 10–2, the polar moment of inertia about C is therefore

$$\bar{J}_C = \bar{I}_{x'} + \bar{I}_{y'} = \frac{1}{12}bh(h^2 + b^2) \qquad \qquad Ans.$$

10

EXAMPLE   10.2

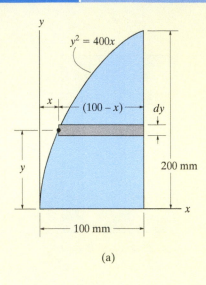

(a)

Determine the moment of inertia for the shaded area shown in Fig. 10–6a about the x axis.

SOLUTION I (CASE 1)

A differential element of area that is *parallel* to the x axis, as shown in Fig. 10–6a, is chosen for integration. Since this element has a thickness dy and intersects the curve at the *arbitrary point* (x, y), its area is $dA = (100 - x) dy$. Furthermore, the element lies at the same distance y from the x axis. Hence, integrating with respect to y, from $y = 0$ to $y = 200$ mm, yields

$$I_x = \int_A y^2 \, dA = \int_0^{200 \text{ mm}} y^2(100 - x) \, dy$$

$$= \int_0^{200 \text{ mm}} y^2 \left(100 - \frac{y^2}{400} \right) dy = \int_0^{200 \text{ mm}} \left(100y^2 - \frac{y^4}{400} \right) dy$$

$$= 107(10^6) \text{ mm}^4 \qquad\qquad\qquad\qquad\qquad\qquad\qquad \textit{Ans.}$$

SOLUTION II (CASE 2)

A differential element *parallel* to the y axis, as shown in Fig. 10–6b, is chosen for integration. It intersects the curve at the *arbitrary point* (x, y). In this case, all points of the element do *not* lie at the same distance from the x axis, and therefore the parallel-axis theorem must be used to determine the *moment of inertia of the element* with respect to this axis. For a rectangle having a base b and height h, the moment of inertia about its centroidal axis has been determined in part (a) of Example 10.1. There it was found that $\bar{I}_{x'} = \frac{1}{12}bh^3$. For the differential element shown in Fig. 10–6b, $b = dx$ and $h = y$, and thus $d\bar{I}_{x'} = \frac{1}{12}dx \, y^3$. Since the centroid of the element is $\tilde{y} = y/2$ from the x axis, the moment of inertia of the element about this axis is

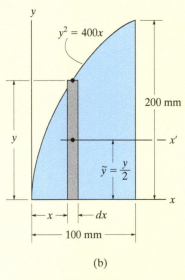

(b)

Fig. 10–6

$$dI_x = d\bar{I}_{x'} + dA \, \tilde{y}^2 = \frac{1}{12}dx \, y^3 + y \, dx \left(\frac{y}{2} \right)^2 = \frac{1}{3}y^3 \, dx$$

(This result can also be concluded from part (b) of Example 10.1.) Integrating with respect to x, from $x = 0$ to $x = 100$ mm, yields

$$I_x = \int dI_x = \int_0^{100 \text{ mm}} \frac{1}{3}y^3 \, dx = \int_0^{100 \text{ mm}} \frac{1}{3}(400x)^{3/2} \, dx$$

$$= 107(10^6) \text{ mm}^4 \qquad\qquad\qquad\qquad\qquad\qquad\qquad \textit{Ans.}$$

EXAMPLE | 10.3

Determine the moment of inertia with respect to the x axis for the circular area shown in Fig. 10–7a.

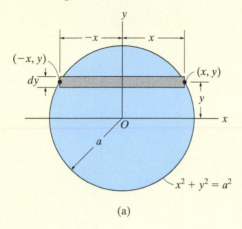

(a)

SOLUTION I (CASE 1)

Using the differential element shown in Fig. 10–7a, since $dA = 2x\,dy$, we have

$$I_x = \int_A y^2\, dA = \int_A y^2(2x)\, dy$$

$$= \int_{-a}^{a} y^2\left(2\sqrt{a^2 - y^2}\right) dy = \frac{\pi a^4}{4} \qquad \textit{Ans.}$$

SOLUTION II (CASE 2)

When the differential element shown in Fig. 10–7b is chosen, the centroid for the element happens to lie on the x axis, and since $\bar{I}_{x'} = \frac{1}{12}bh^3$ for a rectangle, we have

$$dI_x = \frac{1}{12}dx(2y)^3$$

$$= \frac{2}{3}y^3\, dx$$

Integrating with respect to x yields

$$I_x = \int_{-a}^{a}\frac{2}{3}(a^2 - x^2)^{3/2}\, dx = \frac{\pi a^4}{4} \qquad \textit{Ans.}$$

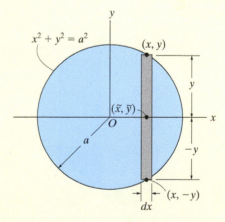

(b)

Fig. 10–7

NOTE: By comparison, Solution I requires much less computation. Therefore, if an integral using a particular element appears difficult to evaluate, try solving the problem using an element oriented in the other direction.

FUNDAMENTAL PROBLEMS

F10–1. Determine the moment of inertia of the shaded area about the *x* axis.

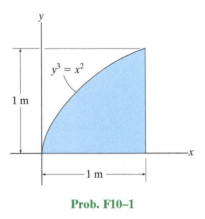

Prob. F10–1

F10–3. Determine the moment of inertia of the shaded area about the *y* axis.

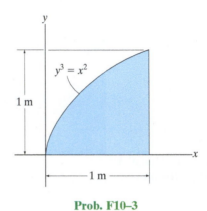

Prob. F10–3

F10–2. Determine the moment of inertia of the shaded area about the *x* axis.

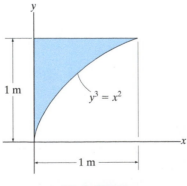

Prob. F10–2

F10–4. Determine the moment of inertia of the shaded area about the *y* axis.

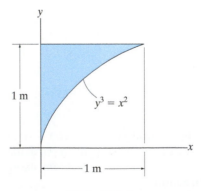

Prob. F10–4

10

PROBLEMS

10–1. Determine the moment of inertia about the x axis.

10–2. Determine the moment of inertia about the y axis.

10–5. Determine the moment of inertia for the shaded area about the x axis.

10–6. Determine the moment of inertia for the shaded area about the y axis.

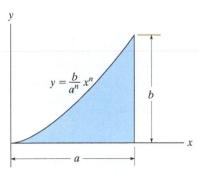

Probs. 10–1/2

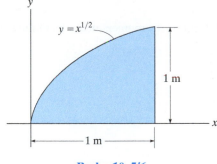

Probs. 10–5/6

10–3. Determine the moment of inertia for the shaded area about the x axis.

*10–4. Determine the moment of inertia for the shaded area about the y axis.

10–7. Determine the moment of inertia for the shaded area about the x axis.

*10–8. Determine the moment of inertia for the shaded area about the y axis.

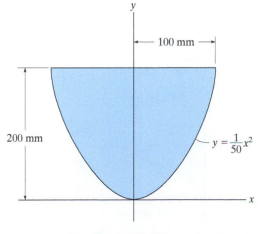

Probs. 10–3/4

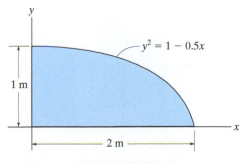

Probs. 10–7/8

10

10–9. Determine the moment of inertia of the area about the x axis. Solve the problem in two ways, using rectangular differential elements: (a) having a thickness dx and (b) having a thickness of dy.

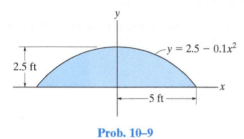

Prob. 10–9

10–10. Determine the moment of inertia of the area about the x axis.

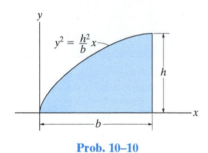

Prob. 10–10

10–11. Determine the moment of inertia for the shaded area about the x axis.

*10–12. Determine the moment of inertia for the shaded area about the y axis.

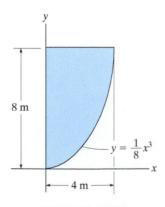

Probs. 10–11/12

10–13. Determine the moment of inertia about the x axis.

10–14. Determine the moment of inertia about the y axis.

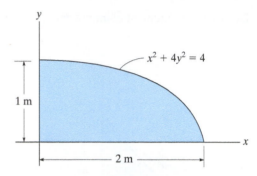

Probs. 10–13/14

10–15. Determine the moment of inertia for the shaded area about the x axis.

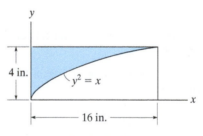

Prob. 10–15

*10–16. Determine the moment of inertia for the shaded area about the y axis.

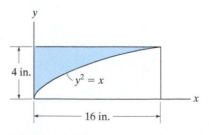

Prob. 10–16

10–17. Determine the moment of inertia for the shaded area about the x axis.

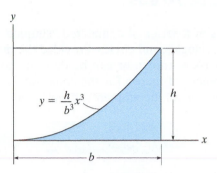

$$y = \frac{h}{b^3}x^3$$

h

b

Prob. 10–17

10–18. Determine the moment of inertia for the shaded area about the y axis.

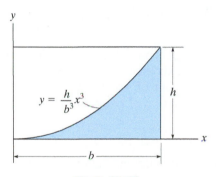

$$y = \frac{h}{b^3}x^3$$

h

b

Prob. 10–18

10–19. Determine the moment of inertia for the shaded area about the x axis.

10–20. Determine the moment of inertia for the shaded area about the y axis.

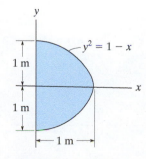

$y^2 = 1 - x$

1 m

1 m

1 m

Probs. 10–19/20

10–21. Determine the moment of inertia for the shaded area about the x axis.

10–22. Determine the moment of inertia for the shaded area about the y axis.

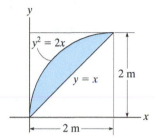

$y^2 = 2x$

$y = x$

2 m

2 m

Probs. 10–21/22

10–23. Determine the moment of inertia for the shaded area about the x axis.

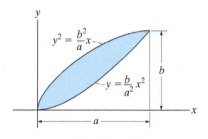

$$y^2 = \frac{b^2}{a}x$$

$$y = \frac{b}{a^2}x^2$$

b

a

Prob. 10–23

10–24. Determine the moment of inertia for the shaded area about the y axis.

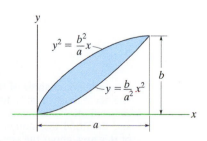

$$y^2 = \frac{b^2}{a}x$$

$$y = \frac{b}{a^2}x^2$$

b

a

Prob. 10–24

10

10.4   Moments of Inertia for Composite Areas

A composite area consists of a series of connected "simpler" parts or shapes, such as rectangles, triangles, and circles. Provided the moment of inertia of each of these parts is known or can be determined about a common axis, then the moment of inertia for the composite area about this axis equals the *algebraic sum* of the moments of inertia of all its parts.

Procedure for Analysis

The moment of inertia for a composite area about a reference axis can be determined using the following procedure.

Composite Parts.

- Using a sketch, divide the area into its composite parts and indicate the perpendicular distance from the centroid of each part to the reference axis.

Parallel-Axis Theorem.

- If the centroidal axis for each part does not coincide with the reference axis, the parallel-axis theorem, $I = \bar{I} + Ad^2$, should be used to determine the moment of inertia of the part about the reference axis. For the calculation of $\bar{I}$, use the table on the inside back cover.

Summation.

- The moment of inertia of the entire area about the reference axis is determined by summing the results of its composite parts about this axis.

- If a composite part has an empty region (hole), its moment of inertia is found by subtracting the moment of inertia of this region from the moment of inertia of the entire part including the region.

For design or analysis of this T-beam, engineers must be able to locate the centroid of its cross-sectional area, and then find the moment of inertia of this area about the centroidal axis. (© Russell C. Hibbeler)

EXAMPLE 10.4

Determine the moment of inertia of the area shown in Fig. 10–8a about the x axis.

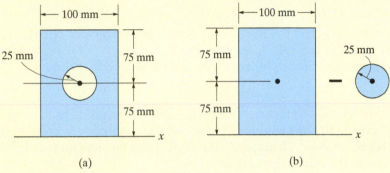

Fig. 10–8

SOLUTION

Composite Parts.   The area can be obtained by *subtracting* the circle from the rectangle shown in Fig. 10–8b. The centroid of each area is located in the figure.

Parallel-Axis Theorem.   The moments of inertia about the x axis are determined using the parallel-axis theorem and the geometric properties formulae for circular and rectangular areas $I_x = \frac{1}{4}\pi r^4$; $I_x = \frac{1}{12}bh^3$, found on the inside back cover.

Circle

$$I_x = \bar{I}_{x'} + A d_y^2$$

$$= \frac{1}{4}\pi(25)^4 + \pi(25)^2(75)^2 = 11.4(10^6) \text{ mm}^4$$

Rectangle

$$I_x = \bar{I}_{x'} + A d_y^2$$

$$= \frac{1}{12}(100)(150)^3 + (100)(150)(75)^2 = 112.5(10^6) \text{ mm}^4$$

Summation.   The moment of inertia for the area is therefore

$$I_x = -11.4(10^6) + 112.5(10^6)$$

$$= 101(10^6) \text{ mm}^4 \qquad\qquad Ans.$$

10

EXAMPLE | **10.5**

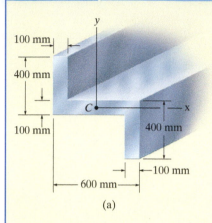

100 mm

400 mm

100 mm

C ──── x

400 mm

100 mm

600 mm

(a)

Determine the moments of inertia for the cross-sectional area of the member shown in Fig. 10–9a about the x and y centroidal axes.

SOLUTION

Composite Parts.   The cross section can be subdivided into the three rectangular areas A, B, and D shown in Fig. 10–9b. For the calculation, the centroid of each of these rectangles is located in the figure.

Parallel-Axis Theorem.   From the table on the inside back cover, or Example 10.1, the moment of inertia of a rectangle about its centroidal axis is $\bar{I} = \frac{1}{12}bh^3$. Hence, using the parallel-axis theorem for rectangles A and D, the calculations are as follows:

Rectangles A and D

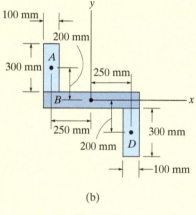

100 mm

200 mm

A

300 mm

250 mm

B ──── x

250 mm

200 mm D

300 mm

100 mm

(b)

Fig. 10–9

$$I_x = \bar{I}_{x'} + A d_y^2 = \frac{1}{12}(100)(300)^3 + (100)(300)(200)^2$$

$$= 1.425(10^9) \text{ mm}^4$$

$$I_y = \bar{I}_{y'} + A d_x^2 = \frac{1}{12}(300)(100)^3 + (100)(300)(250)^2$$

$$= 1.90(10^9) \text{ mm}^4$$

Rectangle B

$$I_x = \frac{1}{12}(600)(100)^3 = 0.05(10^9) \text{ mm}^4$$

$$I_y = \frac{1}{12}(100)(600)^3 = 1.80(10^9) \text{ mm}^4$$

Summation.   The moments of inertia for the entire cross section are thus

$$I_x = 2[1.425(10^9)] + 0.05(10^9)$$

$$= 2.90(10^9) \text{ mm}^4 \qquad \qquad Ans.$$

$$I_y = 2[1.90(10^9)] + 1.80(10^9)$$

$$= 5.60(10^9) \text{ mm}^4 \qquad \qquad Ans.$$

10

FUNDAMENTAL PROBLEMS

F10–5. Determine the moment of inertia of the beam's cross-sectional area about the centroidal x and y axes.

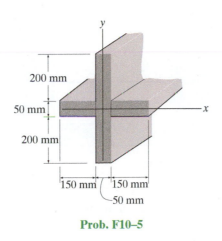

Prob. F10–5

F10–7. Determine the moment of inertia of the cross-sectional area of the channel with respect to the y axis.

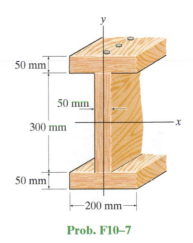

Prob. F10–7

F10–6. Determine the moment of inertia of the beam's cross-sectional area about the centroidal x and y axes.

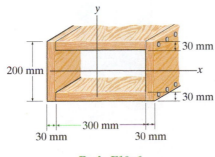

Prob. F10–6

F10–8. Determine the moment of inertia of the cross-sectional area of the T-beam with respect to the x' axis passing through the centroid of the cross section.

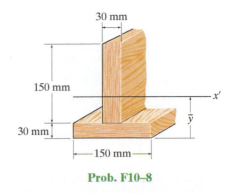

Prob. F10–8

10

PROBLEMS

10–25. Determine the moment of inertia of the composite area about the x axis.

10–26. Determine the moment of inertia of the composite area about the y axis.

*10–28. Determine the location $\bar{y}$ of the centroid of the channel's cross-sectional area and then calculate the moment of inertia of the area about this axis.

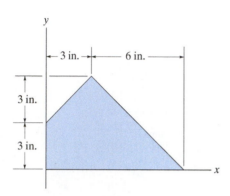

Probs. 10–25/26

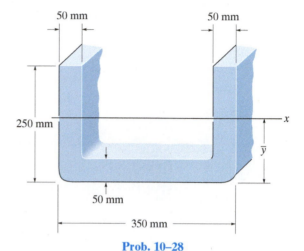

Prob. 10–28

10–27. The polar moment of inertia for the area is $J_C = 642 \,(10^6)$ mm^4, about the z' axis passing through the centroid C. The moment of inertia about the y' axis is $264 \,(10^6)$ mm^4, and the moment of inertia about the x axis is $938 \,(10^6)$ mm^4. Determine the area A.

10–29. Determine $\bar{y}$, which locates the centroidal axis x' for the cross-sectional area of the T-beam, and then find the moments of inertia $I_{x'}$ and $I_{y'}$.

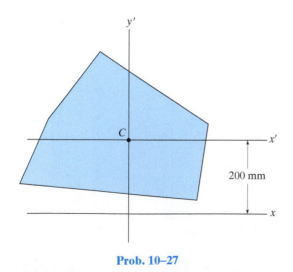

Prob. 10–27

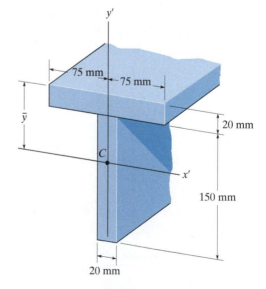

Prob. 10–29

10–30. Determine the moment of inertia for the beam's cross-sectional area about the x axis.

10–31. Determine the moment of inertia for the beam's cross-sectional area about the y axis.

10–34. Determine the moment of inertia of the beam's cross-sectional area about the y axis.

10–35. Determine $\overline{y}$, which locates the centroidal axis x' for the cross-sectional area of the T-beam, and then find the moment of inertia about the x' axis.

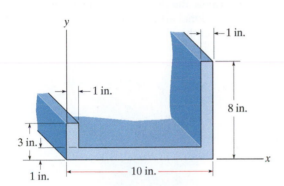

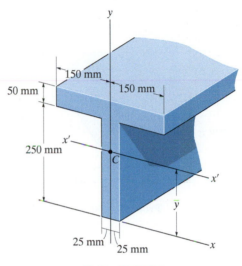

Probs. 10–30/31

Probs. 10–34/35

*10–32. Determine the moment of inertia I_x of the shaded area about the x axis.

10–33. Determine the moment of inertia I_x of the shaded area about the y axis.

*10–36. Determine the moment of inertia about the x axis.

10–37. Determine the moment of inertia about the y axis.

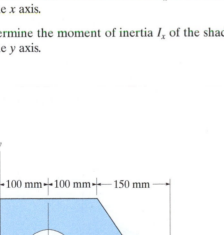

Probs. 10–32/33

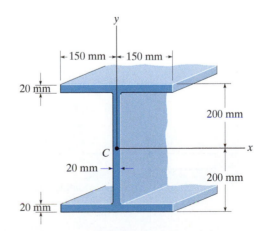

Probs. 10–36/37

10–38. Determine the moment of inertia of the shaded area about the x axis.

10–39. Determine the moment of inertia of the shaded area about the y axis.

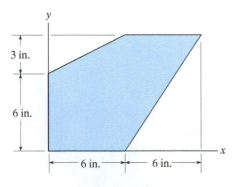

Probs. 10–38/39

*10–40. Determine the distance $\bar{y}$ to the centroid of the beam's cross-sectional area; then find the moment of inertia about the centroidal x' axis.

10–41. Determine the moment of inertia for the beam's cross-sectional area about the y axis.

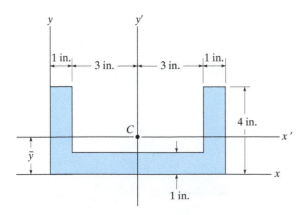

Probs. 10–40/41

10–42. Determine the moment of inertia of the beam's cross-sectional area about the x axis.

10–43. Determine the moment of inertia of the beam's cross-sectional area about the y axis.

*10–44. Determine the distance $\bar{y}$ to the centroid C of the beam's cross-sectional area and then compute the moment of inertia $\bar{I}_{x'}$ about the x' axis.

10–45. Determine the distance $\bar{x}$ to the centroid C of the beam's cross-sectional area and then compute the moment of inertia $\bar{I}_{y'}$ about the y' axis.

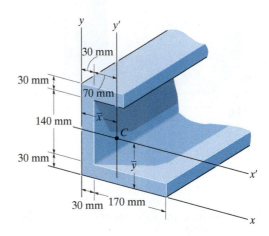

Probs. 10–42/43/44/45

10–46. Determine the moment of inertia for the shaded area about the x axis.

10–47. Determine the moment of inertia for the shaded area about the y axis.

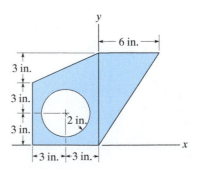

Probs. 10–46/47

*10–48. Determine the moment of inertia of the parallelogram about the x' axis, which passes through the centroid C of the area.

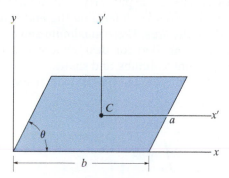

Prob. 10–48

10–49. Determine the moment of inertia of the parallelogram about the y' axis, which passes through the centroid C of the area.

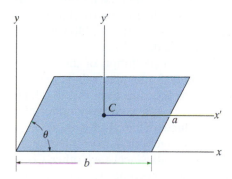

Prob. 10–49

10–50. Locate the centroid $\bar{y}$ of the cross section and determine the moment of inertia of the section about the x' axis.

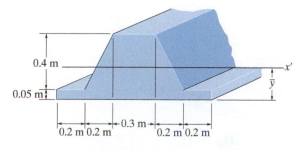

Prob. 10–50

10–51. Determine the moment of inertia for the beam's cross-sectional area about the x' axis passing through the centroid C of the cross section.

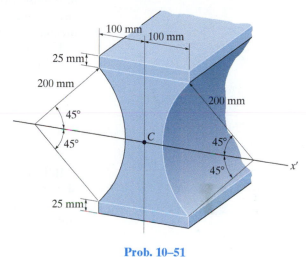

Prob. 10–51

*10–52. Determine the moment of inertia of the area about the x axis.

10–53. Determine the moment of inertia of the area about the y axis.

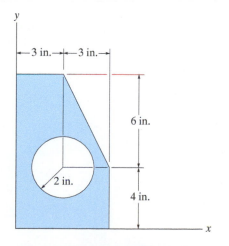

Probs. 10–52/53

*10.5   Product of Inertia for an Area

It will be shown in the next section that the property of an area, called the product of inertia, is required in order to determine the *maximum* and *minimum* moments of inertia for the area. These maximum and minimum values are important properties needed for designing structural and mechanical members such as beams, columns, and shafts.

The **product of inertia** of the area in Fig. 10–10 with respect to the *x* and *y* axes is defined as

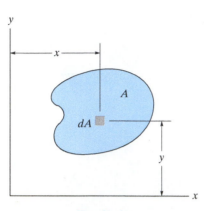

Fig. 10–10

$$I_{xy} = \int_A xy\, dA \qquad (10\text{–}7)$$

If the element of area chosen has a differential size in two directions, as shown in Fig. 10–10, a double integration must be performed to evaluate I_{xy}. Most often, however, it is easier to choose an element having a differential size or thickness in only one direction in which case the evaluation requires only a single integration (see Example 10.6).

Like the moment of inertia, the product of inertia has units of length raised to the fourth power, e.g., m^4, mm^4 or ft^4, in^4. However, since x or y may be negative, the product of inertia may either be positive, negative, or zero, depending on the location and orientation of the coordinate axes. For example, the product of inertia I_{xy} for an area will be *zero* if either the x or y axis is an axis of *symmetry* for the area, as in Fig. 10–11. Here every element dA located at point (x, y) has a corresponding element dA located at $(x, -y)$. Since the products of inertia for these elements are, respectively, $xy\, dA$ and $-xy\, dA$, the algebraic sum or integration of all the elements that are chosen in this way will cancel each other. Consequently, the product of inertia for the total area becomes zero. It also follows from the definition of I_{xy} that the "sign" of this quantity depends on the quadrant where the area is located. As shown in Fig. 10–12, if the area is rotated from one quadrant to another, the sign of I_{xy} will change.

The effectiveness of this beam to resist bending can be determined once its moments of inertia and its product of inertia are known. (© Russell C. Hibbeler)

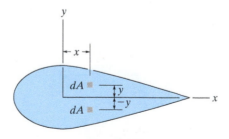

Fig. 10–11

10

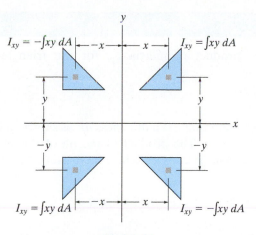

Fig. 10–12

Parallel-Axis Theorem.   Consider the shaded area shown in Fig. 10–13, where x' and y' represent a set of axes passing through the *centroid* of the area, and x and y represent a corresponding set of parallel axes. Since the product of inertia of dA with respect to the x and y axes is $dI_{xy} = (x' + d_x)(y' + d_y)\, dA$, then for the entire area,

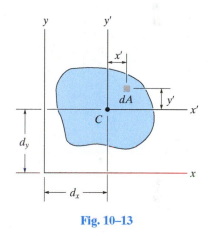

Fig. 10–13

$$I_{xy} = \int_A (x' + d_x)(y' + d_y)\, dA$$

$$= \int_A x'y'\, dA + d_x \int_A y'\, dA + d_y \int_A x'\, dA + d_x d_y \int_A dA$$

The first term on the right represents the product of inertia for the area with respect to the centroidal axes, $\bar{I}_{x'y'}$. The integrals in the second and third terms are zero since the moments of the area are taken about the centroidal axis. Realizing that the fourth integral represents the entire area A, the parallel-axis theorem for the product of inertia becomes

$$\boxed{I_{xy} = \bar{I}_{x'y'} + A d_x d_y} \qquad (10\text{–}8)$$

It is important that the *algebraic signs* for d_x and d_y be maintained when applying this equation.

EXAMPLE | 10.6

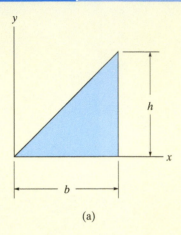

(a)

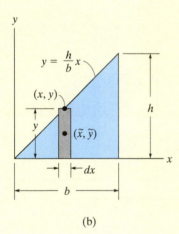

(b)

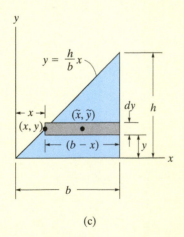

(c)

Fig. 10–14

Determine the product of inertia I_{xy} for the triangle shown in Fig. 10–14a.

SOLUTION I

A differential element that has a thickness dx, as shown in Fig. 10–14b, has an area $dA = y\,dx$. The product of inertia of this element with respect to the x *and* y axes is determined using the parallel-axis theorem.

$$dI_{xy} = d\bar{I}_{x'y'} + dA\,\tilde{x}\,\tilde{y}$$

where $\tilde{x}$ and $\tilde{y}$ locate the *centroid* of the element or the origin of the x', y' axes. (See Fig. 10–13.) Since $d\bar{I}_{x'y'} = 0$, due to symmetry, and $\tilde{x} = x$, $\tilde{y} = y/2$, then

$$dI_{xy} = 0 + (y\,dx)x\left(\frac{y}{2}\right) = \left(\frac{h}{b}x\,dx\right)x\left(\frac{h}{2b}x\right)$$

$$= \frac{h^2}{2b^2}x^3\,dx$$

Integrating with respect to x from $x = 0$ to $x = b$ yields

$$I_{xy} = \frac{h^2}{2b^2}\int_0^b x^3\,dx = \frac{b^2h^2}{8} \qquad \text{Ans.}$$

SOLUTION II

The differential element that has a thickness dy, as shown in Fig. 10–14c, can also be used. Its area is $dA = (b - x)\,dy$. The *centroid* is located at point $\tilde{x} = x + (b-x)/2 = (b+x)/2$, $\tilde{y} = y$, so the product of inertia of the element becomes

$$dI_{xy} = d\bar{I}_{x'y'} + dA\,\tilde{x}\,\tilde{y}$$

$$= 0 + (b - x)\,dy\left(\frac{b+x}{2}\right)y$$

$$= \left(b - \frac{b}{h}y\right)dy\left[\frac{b + (b/h)y}{2}\right]y = \frac{1}{2}y\left(b^2 - \frac{b^2}{h^2}y^2\right)dy$$

Integrating with respect to y from $y = 0$ to $y = h$ yields

$$I_{xy} = \frac{1}{2}\int_0^h y\left(b^2 - \frac{b^2}{h^2}y^2\right)dy = \frac{b^2h^2}{8} \qquad \text{Ans.}$$

EXAMPLE | 10.7

Determine the product of inertia for the cross-sectional area of the member shown in Fig. 10–15a, about the x and y centroidal axes.

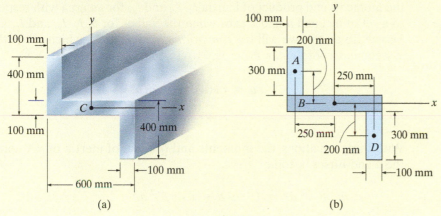

Fig. 10–15

SOLUTION
As in Example 10.5, the cross section can be subdivided into three composite rectangular areas A, B, and D, Fig. 10–15b. The coordinates for the centroid of each of these rectangles are shown in the figure. Due to symmetry, the product of inertia of *each rectangle* is *zero* about a set of x', y' axes that passes through the centroid of each rectangle. Using the parallel-axis theorem, we have

Rectangle A

$$I_{xy} = \bar{I}_{x'y'} + A d_x d_y$$
$$= 0 + (300)(100)(-250)(200) = -1.50(10^9) \text{ mm}^4$$

Rectangle B

$$I_{xy} = \bar{I}_{x'y'} + A d_x d_y$$
$$= 0 + 0 = 0$$

Rectangle D

$$I_{xy} = \bar{I}_{x'y'} + A d_x d_y$$
$$= 0 + (300)(100)(250)(-200) = -1.50(10^9) \text{ mm}^4$$

The product of inertia for the entire cross section is therefore

$$I_{xy} = -1.50(10^9) + 0 - 1.50(10^9) = -3.00(10^9) \text{ mm}^4 \quad \textit{Ans.}$$

NOTE: This negative result is due to the fact that rectangles A and D have centroids located with negative x and negative y coordinates, respectively.

10

*10.6   Moments of Inertia for an Area about Inclined Axes

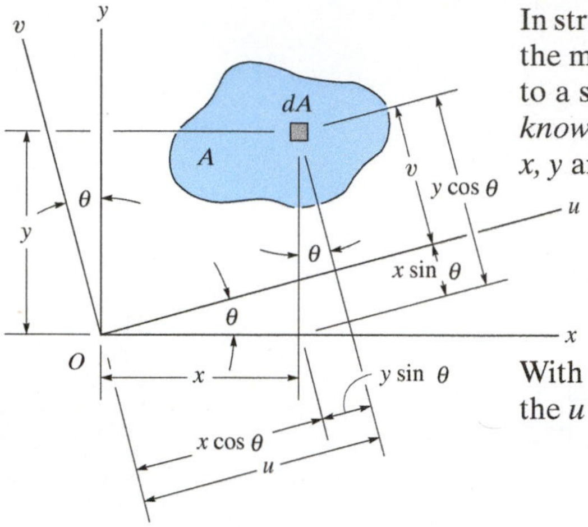

In structural and mechanical design, it is sometimes necessary to calculate the moments and product of inertia I_u, I_v, and I_{uv} for an area with respect to a set of inclined u and v axes when the values for θ, I_x, I_y, and I_{xy} are *known*. To do this we will use *transformation equations* which relate the x, y and u, v coordinates. From Fig. 10–16, these equations are

$$u = x \cos \theta + y \sin \theta$$

$$v = y \cos \theta - x \sin \theta$$

With these equations, the moments and product of inertia of dA about the u and v axes become

$$dI_u = v^2 \, dA = (y \cos \theta - x \sin \theta)^2 \, dA$$

$$dI_v = u^2 \, dA = (x \cos \theta + y \sin \theta)^2 \, dA$$

$$dI_{uv} = uv \, dA = (x \cos \theta + y \sin \theta)(y \cos \theta - x \sin \theta) \, dA$$

Expanding each expression and integrating, realizing that $I_x = \int y^2 \, dA$, $I_y = \int x^2 \, dA$, and $I_{xy} = \int xy \, dA$, we obtain

$$I_u = I_x \cos^2 \theta + I_y \sin^2 \theta - 2I_{xy} \sin \theta \cos \theta$$

$$I_v = I_x \sin^2 \theta + I_y \cos^2 \theta + 2I_{xy} \sin \theta \cos \theta$$

$$I_{uv} = I_x \sin \theta \cos \theta - I_y \sin \theta \cos \theta + I_{xy}(\cos^2 \theta - \sin^2 \theta)$$

Using the trigonometric identities $\sin 2\theta = 2 \sin \theta \cos \theta$ and $\cos 2\theta = \cos^2 \theta - \sin^2 \theta$ we can simplify the above expressions, in which case

$$I_u = \frac{I_x + I_y}{2} + \frac{I_x - I_y}{2} \cos 2\theta - I_{xy} \sin 2\theta$$

$$I_v = \frac{I_x + I_y}{2} - \frac{I_x - I_y}{2} \cos 2\theta + I_{xy} \sin 2\theta \qquad (10\text{–}9)$$

$$I_{uv} = \frac{I_x - I_y}{2} \sin 2\theta + I_{xy} \cos 2\theta$$

Notice that if the first and second equations are added together, we can show that the polar moment of inertia about the z axis passing through point O is, as expected, *independent* of the orientation of the u and v axes; i.e.,

$$J_O = I_u + I_v = I_x + I_y$$

Fig. 10–16

Principal Moments of Inertia.   Equations 10–9 show that I_u, I_v, and I_{uv} depend on the angle of inclination, θ, of the u, v axes. We will now determine the orientation of these axes about which the moments of inertia for the area are maximum and minimum. This particular set of axes is called the *principal axes* of the area, and the corresponding moments of inertia with respect to these axes are called the **principal moments of inertia**. In general, there is a set of principal axes for every chosen origin O. However, for structural and mechanical design, the origin O is located at the centroid of the area.

The angle which defines the orientation of the principal axes can be found by differentiating the first of Eqs. 10–9 with respect to θ and setting the result equal to zero. Thus,

$$\frac{dI_u}{d\theta} = -2\left(\frac{I_x - I_y}{2}\right)\sin 2\theta - 2I_{xy}\cos 2\theta = 0$$

Therefore, at $\theta = \theta_p$,

$$\tan 2\theta_p = \frac{-I_{xy}}{(I_x - I_y)/2} \qquad (10–10)$$

The two roots θ_{p_1} and θ_{p_2} of this equation are $90°$ apart, and so they each specify the inclination of one of the principal axes. In order to substitute them into Eq. 10–9, we must first find the sine and cosine of $2\theta_{p_1}$ and $2\theta_{p_2}$. This can be done using these ratios from the triangles shown in Fig. 10–17, which are based on Eq. 10–10.

Substituting each of the sine and cosine ratios into the first or second of Eqs. 10–9 and simplifying, we obtain

$$I_{\substack{max \\ min}} = \frac{I_x + I_y}{2} \pm \sqrt{\left(\frac{I_x - I_y}{2}\right)^2 + I_{xy}^2} \qquad (10–11)$$

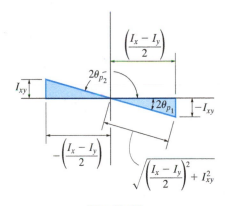

Fig. 10–17

Depending on the sign chosen, this result gives the maximum or minimum moment of inertia for the area. Furthermore, if the above trigonometric relations for θ_{p_1} and θ_{p_2} are substituted into the third of Eqs. 10–9, it can be shown that $I_{uv} = 0$; that is, the *product of inertia with respect to the principal axes is zero*. Since it was indicated in Sec. 10.6 that the product of inertia is zero with respect to any symmetrical axis, it therefore follows that *any symmetrical axis represents a principal axis of inertia for the area.*

EXAMPLE 10.8

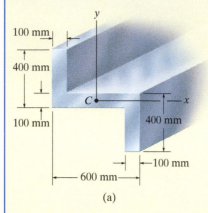

(a)

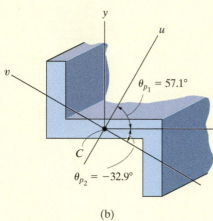

(b)

Fig. 10–18

Determine the principal moments of inertia and the orientation of the principal axes for the cross-sectional area of the member shown in Fig. 10–18a with respect to an axis passing through the centroid.

SOLUTION

The moments and product of inertia of the cross section with respect to the x, y axes have been determined in Examples 10.5 and 10.7. The results are

$$I_x = 2.90(10^9) \text{ mm}^4 \quad I_y = 5.60(10^9) \text{ mm}^4 \quad I_{xy} = -3.00(10^9) \text{ mm}^4$$

Using Eq. 10–10, the angles of inclination of the principal axes u and v are

$$\tan 2\theta_p = \frac{-I_{xy}}{(I_x - I_y)/2} = \frac{-[-3.00(10^9)]}{[2.90(10^9) - 5.60(10^9)]/2} = -2.22$$

$$2\theta_p = -65.8° \text{ and } 114.2°$$

Thus, by inspection of Fig. 10–18b,

$$\theta_{p_2} = -32.9° \quad \text{and} \quad \theta_{p_1} = 57.1° \qquad \text{Ans.}$$

The principal moments of inertia with respect to these axes are determined from Eq. 10–11. Hence,

$$I_{\substack{\max \\ \min}} = \frac{I_x + I_y}{2} \pm \sqrt{\left(\frac{I_x - I_y}{2}\right)^2 + I_{xy}^2}$$

$$= \frac{2.90(10^9) + 5.60(10^9)}{2}$$

$$\pm \sqrt{\left[\frac{2.90(10^9) - 5.60(10^9)}{2}\right]^2 + [-3.00(10^9)]^2}$$

$$I_{\substack{\max \\ \min}} = 4.25(10^9) \pm 3.29(10^9)$$

or

$$I_{\max} = 7.54(10^9) \text{ mm}^4 \quad I_{\min} = 0.960(10^9) \text{ mm}^4 \qquad \text{Ans.}$$

NOTE: The maximum moment of inertia, $I_{\max} = 7.54(10^9) \text{ mm}^4$, occurs with respect to the u axis since *by inspection* most of the cross-sectional area is farthest away from this axis. Or, stated in another manner, $I_{\max}$ occurs about the u axis since this axis is located within $\pm 45°$ of the y axis, which has the larger value of I ($I_y > I_x$). Also, this can be concluded by substituting the data with $\theta = 57.1°$ into the first of Eqs. 10–9 and solving for I_u.

10

*10.7   Mohr's Circle for Moments of Inertia

Equations 10–9 to 10–11 have a graphical solution that is convenient to use and generally easy to remember. Squaring the first and third of Eqs. 10–9 and adding, it is found that

$$\left(I_u - \frac{I_x + I_y}{2}\right)^2 + I_{uv}^2 = \left(\frac{I_x - I_y}{2}\right)^2 + I_{xy}^2$$

Here I_x, I_y, and I_{xy} are *known constants*. Thus, the above equation may be written in compact form as

$$(I_u - a)^2 + I_{uv}^2 = R^2$$

When this equation is plotted on a set of axes that represent the respective moment of inertia and the product of inertia, as shown in Fig. 10–19, the resulting graph represents a *circle* of radius

$$R = \sqrt{\left(\frac{I_x - I_y}{2}\right)^2 + I_{xy}^2}$$

and having its center located at point $(a, 0)$, where $a = (I_x + I_y)/2$. The circle so constructed is called **Mohr's circle**, named after the German engineer Otto Mohr (1835–1918).

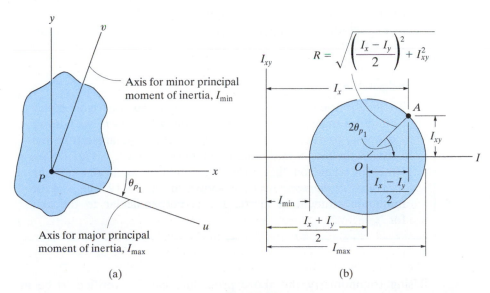

(a)   (b)

Fig. 10–19

10

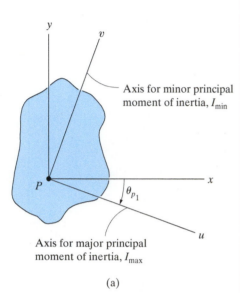

Axis for minor principal
moment of inertia, I_{min}

θ_{p_1}

Axis for major principal
moment of inertia, I_{max}

(a)

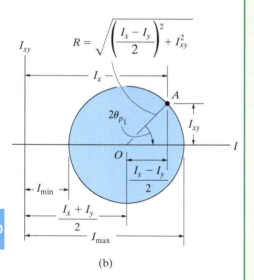

$$R = \sqrt{\left(\frac{I_x - I_y}{2}\right)^2 + I_{xy}^2}$$

(b)

Fig. 10–19 (Repeated)

Procedure for Analysis

The main purpose in using Mohr's circle here is to have a convenient means for finding the principal moments of inertia for an area. The following procedure provides a method for doing this.

Determine I_x, I_y, and I_{xy}.

• Establish the x, y axes and determine I_x, I_y, and I_{xy}, Fig. 10–19a.

Construct the Circle.

• Construct a rectangular coordinate system such that the horizontal axis represents the moment of inertia I, and the vertical axis represents the product of inertia I_{xy}, Fig. 10–19b.

• Determine the center of the circle, O, which is located at a distance $(I_x + I_y)/2$ from the origin, and plot the reference point A having coordinates (I_x, I_{xy}). Remember, I_x is always positive, whereas I_{xy} can be either positive or negative.

• Connect the reference point A with the center of the circle and determine the distance OA by trigonometry. This distance represents the radius of the circle, Fig. 10–19b. Finally, draw the circle.

Principal Moments of Inertia.

• The points where the circle intersects the I axis give the values of the principal moments of inertia I_{min} and I_{max}. Notice that, as expected, the *product of inertia will be zero at these points*, Fig. 10–19b.

Principal Axes.

• To find the orientation of the major principal axis, use trigonometry to find the angle $2\theta_{p_1}$, *measured from the radius OA to the positive I axis*, Fig. 10–19b. This angle represents *twice* the angle from the x axis to the axis of maximum moment of inertia I_{max}, Fig. 10–19a. Both the angle on the circle, $2\theta_{p_1}$, and the angle θ_{p_1} *must be measured in the same sense*, as shown in Fig. 10–19. The axis for minimum moment of inertia I_{min} is perpendicular to the axis for I_{max}.

Using trigonometry, the above procedure can be verified to be in accordance with the equations developed in Sec. 10.6.

EXAMPLE | 10.9

Using Mohr's circle, determine the principal moments of inertia and the orientation of the major principal axes for the cross-sectional area of the member shown in Fig. 10–20a, with respect to an axis passing through the centroid.

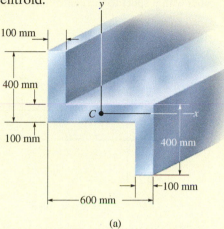

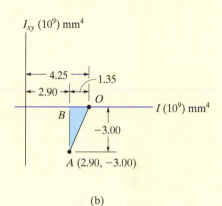

(a)

(b)

SOLUTION

Determine I_x, I_y, I_{xy}. The moments and product of inertia have been determined in Examples 10.5 and 10.7 with respect to the x, y axes shown in Fig. 10–20a. The results are $I_x = 2.90(10^9)$ mm⁴, $I_y = 5.60(10^9)$ mm⁴, and $I_{xy} = -3.00(10^9)$ mm⁴.

Construct the Circle. The I and I_{xy} axes are shown in Fig. 10–20b. The center of the circle, O, lies at a distance $(I_x + I_y)/2 = (2.90 + 5.60)/2 = 4.25$ from the origin. When the reference point $A(I_x, I_{xy})$ or $A(2.90, -3.00)$ is connected to point O, the radius OA is determined from the triangle OBA using the Pythagorean theorem.

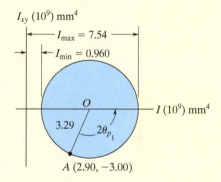

$$OA = \sqrt{(1.35)^2 + (-3.00)^2} = 3.29$$

The circle is constructed in Fig. 10–20c.

Principal Moments of Inertia. The circle intersects the I axis at points (7.54, 0) and (0.960, 0). Hence,

$$I_{max} = (4.25 + 3.29)10^9 = 7.54(10^9) \text{ mm}^4 \qquad \textit{Ans.}$$

$$I_{min} = (4.25 - 3.29)10^9 = 0.960(10^9) \text{ mm}^4 \qquad \textit{Ans.}$$

Principal Axes. As shown in Fig. 10–20c, the angle $2\theta_{p_1}$ is determined from the circle by measuring counterclockwise from OA to the direction of the *positive I* axis. Hence,

$$2\theta_{p_1} = 180° - \sin^{-1}\left(\frac{|BA|}{|OA|}\right) = 180° - \sin^{-1}\left(\frac{3.00}{3.29}\right) = 114.2°$$

The principal axis for $I_{max} = 7.54(10^9)$ mm⁴ is therefore oriented at an angle $\theta_{p_1} = 57.1°$, measured *counterclockwise*, from the *positive x* axis to the *positive u* axis. The v axis is perpendicular to this axis. The results are shown in Fig. 10–20d.

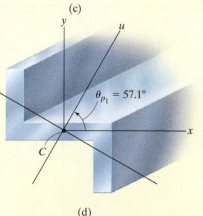

(d)

Fig. 10–20

10

PROBLEMS

10–54. Determine the product of inertia of the thin strip of area with respect to the x and y axes. The strip is oriented at an angle θ from the x axis. Assume that $t \ll l$.

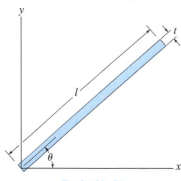

Prob. 10–54

10–55. Determine the product of inertia of the shaded area with respect to the x and y axes.

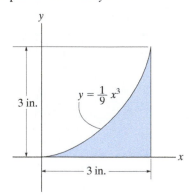

Prob. 10–55

*__10–56.__ Determine the product of inertia for the shaded portion of the parabola with respect to the x and y axes.

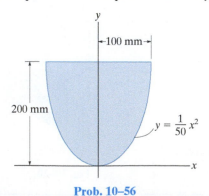

Prob. 10–56

10–57. Determine the product of inertia of the shaded area with respect to the x and y axes, and then use the parallel-axis theorem to find the product of inertia of the area with respect to the centroidal x' and y' axes.

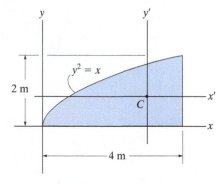

Prob. 10–57

10–58. Determine the product of inertia for the parabolic area with respect to the x and y axes.

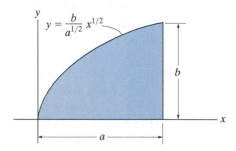

Prob. 10–58

10–59. Determine the product of inertia of the shaded area with respect to the x and y axes.

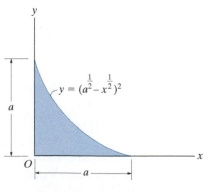

Prob. 10–59

*10–60. Determine the product of inertia of the shaded area with respect to the x and y axes.

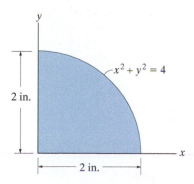

Prob. 10–60

10–61. Determine the product of inertia of the shaded area with respect to the x and y axes.

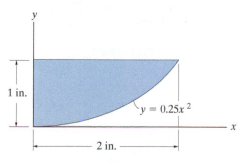

Prob. 10–61

10–62. Determine the product of inertia for the beam's cross-sectional area with respect to the x and y axes.

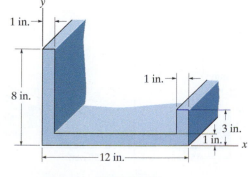

Prob. 10–62

10–63. Determine the moments of inertia of the shaded area with respect to the u and v axes.

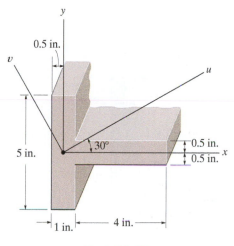

Prob. 10–63

*10–64. Determine the product of inertia for the beam's cross-sectional area with respect to the u and v axes.

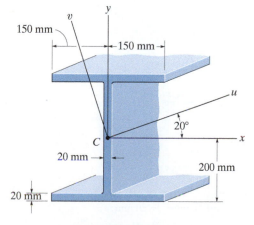

Prob. 10–64

10

10–65. Determine the product of inertia for the shaded area with respect to the x and y axes.

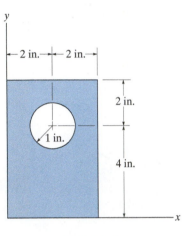

Prob. 10–65

10–66. Determine the product of inertia of the cross-sectional area with respect to the x and y axes.

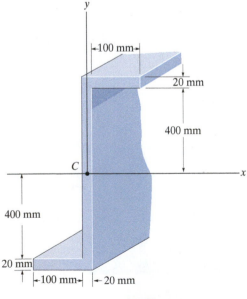

Prob. 10–66

10–67. Determine the location $(\bar{x}, \bar{y})$ to the centroid C of the angle's cross-sectional area, and then compute the product of inertia with respect to the x' and y' axes.

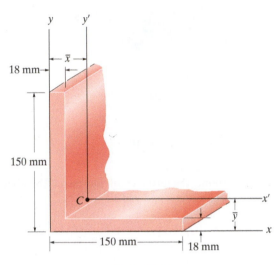

Prob. 10–67

*10–68. Determine the distance $\bar{y}$ to the centroid of the area and then calculate the moments of inertia I_u and I_v of the channel's cross-sectional area. The u and v axes have their origin at the centroid C. For the calculation, assume all corners to be square.

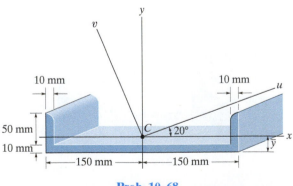

Prob. 10–68

10–69. Determine the moments of inertia I_u, I_v and the product of inertia I_{uv} for the beam's cross-sectional area. Take $\theta = 45°$.

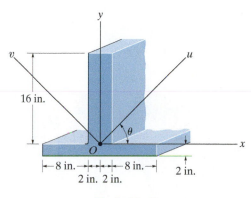

Prob. 10–69

10–70. Determine the moments of inertia I_u, I_v and the product of inertia I_{uv} for the rectangular area. The u and v axes pass through the centroid C.

10–71. Solve Prob. 10–70 using Mohr's circle. *Hint*: To solve, find the coordinates of the point $P(I_u, I_{uv})$ on the circle, measured counterclockwise from the radial line OA. (See Fig. 10–19.) The point $Q(I_v, -I_{uv})$ is on the opposite side of the circle.

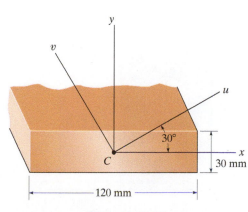

Probs. 10–70/71

*10–72. Determine the directions of the principal axes having an origin at point O, and the principal moments of inertia for the triangular area about the axes.

10–73. Solve Prob. 10–72 using Mohr's circle.

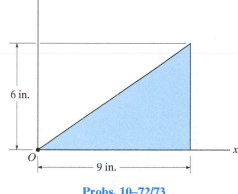

Probs. 10–72/73

10–74. Determine the orientation of the principal axes having an origin at point C, and the principal moments of inertia of the cross section about these axes.

10–75. Solve Prob. 10–74 using Mohr's circle.

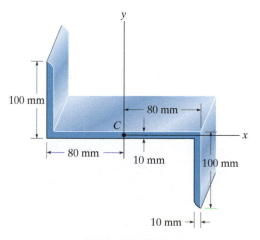

Probs. 10–74/75

10

*10–76.   Determine the orientation of the principal axes having an origin at point O, and the principal moments of inertia for the rectangular area about these axes.

10–77.   Solve Prob. 10–76 using Mohr's circle.

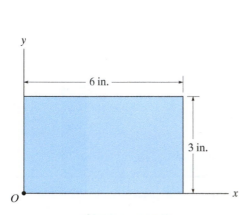

Probs. 10–76/77

10–78.   The area of the cross section of an airplane wing has the following properties about the x and y axes passing through the centroid C: $\bar{I}_x = 450 \text{ in}^4$, $\bar{I}_y = 1730 \text{ in}^4$, $\bar{I}_{xy} = 138 \text{ in}^4$. Determine the orientation of the principal axes and the principal moments of inertia.

10–79.   Solve Prob. 10–78 using Mohr's circle.

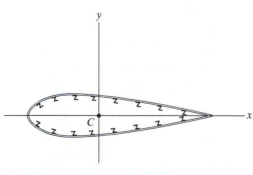

Probs. 10–78/79

*10–80.   Determine the moments and product of inertia for the shaded area with respect to the u and v axes.

10–81.   Solve Prob. 10–80 using Mohr's circle.

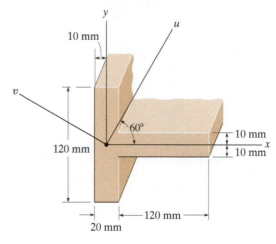

Probs. 10–80/81

10–82.   Determine the directions of the principal axes with origin located at point O, and the principal moments of inertia for the area about these axes.

10–83.   Solve Prob. 10–82 using Mohr's circle.

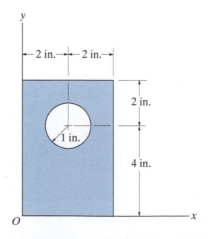

Probs. 10–82/83

10.8   Mass Moment of Inertia

The mass moment of inertia of a body is a measure of the body's resistance to angular acceleration. Since it is used in dynamics to study rotational motion, methods for its calculation will now be discussed.*

Consider the rigid body shown in Fig. 10–21. We define the *mass moment of inertia* of the body about the *z* axis as

$$I = \int_m r^2 \, dm \tag{10–12}$$

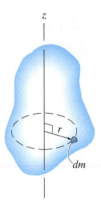

Here *r* is the perpendicular distance from the axis to the arbitrary element *dm*. Since the formulation involves *r*, the value of *I* is *unique* for each axis about which it is computed. The axis which is generally chosen, however, passes through the body's mass center *G*. Common units used for its measurement are $kg \cdot m^2$ or $slug \cdot ft^2$.

If the body consists of material having a density ρ, then $dm = \rho \, dV$, Fig. 10–22a. Substituting this into Eq. 10–12, the body's moment of inertia is then computed using *volume elements* for integration; i.e.,

$$I = \int_V r^2 \rho \, dV \tag{10–13}$$

Fig. 10–21

For most applications, ρ will be a *constant*, and so this term may be factored out of the integral, and the integration is then purely a function of geometry.

$$I = \rho \int_V r^2 \, dV \tag{10–14}$$

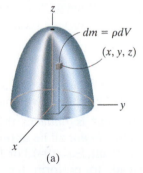

$dm = \rho dV$
(x, y, z)

(a)

Fig. 10–22

*Another property of the body, which measures the symmetry of the body's mass with respect to a coordinate system, is the mass product of inertia. This property most often applies to the three-dimensional motion of a body and is discussed in *Engineering Mechanics: Dynamics* (Chapter 21).

10

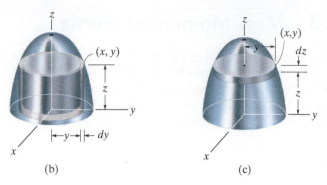

(b)   (c)

Fig. 10–22 (cont'd)

Procedure for Analysis

If a body is symmetrical with respect to an axis, as in Fig. 10–22, then its mass moment of inertia about the axis can be determined by using a single integration. Shell and disk elements are used for this purpose.

Shell Element.

- If a *shell element* having a height z, radius y, and thickness dy is chosen for integration, Fig. 10–22b, then its volume is $dV = (2\pi y)(z)\,dy$.

- This element can be used in Eq. 10–13 or 10–14 for determining the moment of inertia I_z of the body about the z axis since the *entire element*, due to its "thinness," lies at the *same* perpendicular distance $r = y$ from the z axis (see Example 10.10).

Disk Element.

- If a disk element having a radius y and a thickness dz is chosen for integration, Fig. 10–22c, then its volume is $dV = (\pi y^2)\,dz$.

- In this case the element is *finite* in the radial direction, and consequently its points *do not* all lie at the *same radial distance r* from the z axis. As a result, Eqs. 10–13 or 10–14 *cannot* be used to determine I_z. Instead, to perform the integration using this element, it is first necessary to determine the moment of inertia *of the element* about the z axis and then integrate this result (see Example 10.11).

EXAMPLE | 10.10

Determine the mass moment of inertia of the cylinder shown in Fig. 10–23a about the z axis. The density of the material, ρ, is constant.

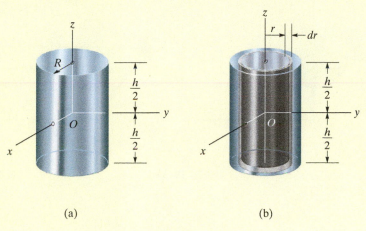

(a)   (b)

Fig. 10–23

SOLUTION

Shell Element.   This problem will be solved using the *shell element* in Fig. 10–23b and thus only a single integration is required. The volume of the element is $dV = (2\pi r)(h)\, dr$, and so its mass is $dm = \rho\, dV = \rho(2\pi hr\, dr)$. Since the *entire element* lies at the same distance r from the z axis, the moment of inertia *of the element* is

$$dI_z = r^2\, dm = \rho 2\pi hr^3\, dr$$

Integrating over the entire cylinder yields

$$I_z = \int_m r^2\, dm = \rho 2\pi h \int_0^R r^3\, dr = \frac{\rho\pi}{2} R^4 h$$

Since the mass of the cylinder is

$$m = \int_m dm = \rho 2\pi h \int_0^R r\, dr = \rho\pi hR^2$$

then

$$I_z = \frac{1}{2} mR^2 \qquad\qquad\qquad \textit{Ans.}$$

10

EXAMPLE | **10.11**

If the density of the solid in Fig. 10–24a is 5 slug/ft³, determine the mass moment of inertia about the *y* axis.

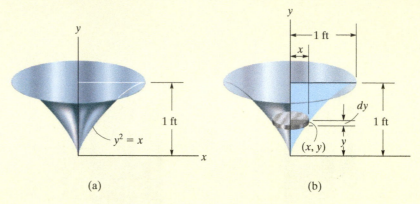

Fig. 10–24

SOLUTION

Disk Element. The moment of inertia will be determined using this *disk element*, as shown in Fig. 10–24b. Here the element intersects the curve at the arbitrary point (x, y) and has a mass

$$dm = \rho \, dV = \rho(\pi x^2) \, dy$$

Although all points on the element are *not* located at the same distance from the *y* axis, it is still possible to determine the moment of inertia dI_y *of the element* about the *y* axis. In the previous example it was shown that the moment of inertia of a homogeneous cylinder about its longitudinal axis is $I = \frac{1}{2}mR^2$, where m and R are the mass and radius of the cylinder. Since the height of the cylinder is not involved in this formula, we can also use this result for a disk. Thus, for the disk element in Fig. 10–24b, we have

$$dI_y = \frac{1}{2}(dm)x^2 = \frac{1}{2}[\rho(\pi x^2) \, dy]x^2$$

Substituting $x = y^2$, $\rho = 5$ slug/ft³, and integrating with respect to y, from $y = 0$ to $y = 1$ ft, yields the moment of inertia for the entire solid.

$$I_y = \frac{5\pi}{2} \int_0^{1 \, \text{ft}} x^4 \, dy = \frac{5\pi}{2} \int_0^{1 \, \text{ft}} y^8 \, dy = 0.873 \text{ slug} \cdot \text{ft}^2 \quad \textit{Ans.}$$

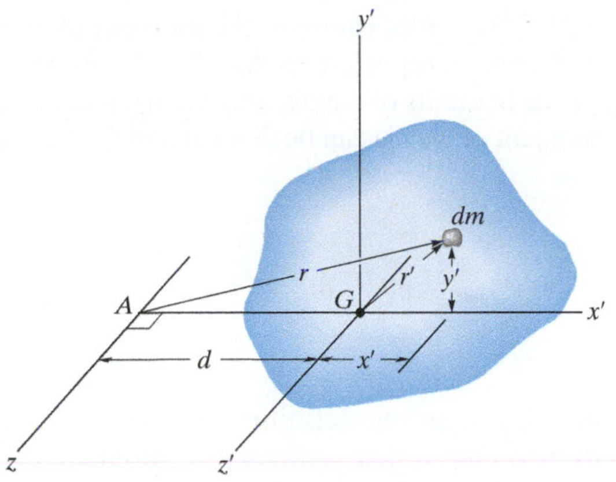

Fig. 10–25

Parallel-Axis Theorem.   If the moment of inertia of the body about an axis passing through the body's mass center is known, then the moment of inertia about any other *parallel axis* can be determined by using the *parallel-axis theorem*. To derive this theorem, consider the body shown in Fig. 10–25. The z' axis passes through the mass center G, whereas the corresponding *parallel z axis* lies at a constant distance d away. Selecting the differential element of mass dm, which is located at point (x', y'), and using the Pythagorean theorem, $r^2 = (d + x')^2 + y'^2$, the moment of inertia of the body about the z axis is

$$I = \int_m r^2\, dm = \int_m [(d + x')^2 + y'^2]\, dm$$

$$= \int_m (x'^2 + y'^2)\, dm + 2d \int_m x'\, dm + d^2 \int_m dm$$

Since $r'^2 = x'^2 + y'^2$, the first integral represents I_G. The second integral is equal to *zero*, since the z' axis passes through the body's mass center, i.e., $\int x'\, dm = \bar{x} \int dm = 0$ since $\bar{x} = 0$. Finally, the third integral is the total mass m of the body. Hence, the moment of inertia about the z axis becomes

$$\boxed{I = I_G + md^2} \qquad\qquad (10\text{–}15)$$

where

    I_G = moment of inertia about the z' axis passing through the mass center G

    m = mass of the body

    d = distance between the parallel axes

Radius of Gyration. Occasionally, the moment of inertia of a body about a specified axis is reported in handbooks using the *radius of gyration, k.* This value has units of length, and when it and the body's mass m are known, the moment of inertia can be determined from the equation

$$I = mk^2 \quad \text{or} \quad k = \sqrt{\frac{I}{m}} \tag{10–16}$$

Note the *similarity* between the definition of k in this formula and r in the equation $dI = r^2\,dm$, which defines the moment of inertia of a differential element of mass dm of the body about an axis.

Composite Bodies. If a body is constructed from a number of simple shapes such as disks, spheres, and rods, the moment of inertia of the body about any axis z can be determined by adding algebraically the moments of inertia of all the composite shapes calculated about the same axis. Algebraic addition is necessary since a composite part must be considered as a negative quantity if it has already been included within another part—as in the case of a "hole" subtracted from a solid plate. Also, the parallel-axis theorem is needed for the calculations if the center of mass of each composite part does not lie on the z axis. For calculations, a table of some simple shapes is given on the inside back cover.

This flywheel, which operates a metal cutter, has a large moment of inertia about its center. Once it begins rotating it is difficult to stop it and therefore a uniform motion can be effectively transferred to the cutting blade. (© Russell C. Hibbeler)

EXAMPLE | 10.12

If the plate shown in Fig. 10–26a has a density of 8000 kg/m³ and a thickness of 10 mm, determine its mass moment of inertia about an axis perpendicular to the page and passing through the pin at O.

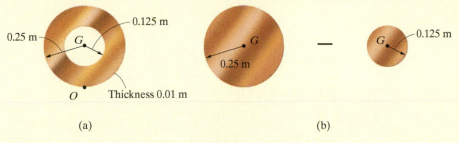

Fig. 10–26

SOLUTION

The plate consists of two composite parts, the 250-mm-radius disk *minus* a 125-mm-radius disk, Fig. 10–26b. The moment of inertia about O can be determined by finding the moment of inertia of each of these parts about O and then *algebraically* adding the results. The calculations are performed by using the parallel-axis theorem in conjunction with the mass moment of inertia formula for a circular disk, $I_G = \frac{1}{2}mr^2$, as found on the inside back cover.

Disk. The moment of inertia of a disk about an axis perpendicular to the plane of the disk and passing through G is $I_G = \frac{1}{2}mr^2$. The mass center of both disks is 0.25 m from point O. Thus,

$$m_d = \rho_d V_d = 8000 \text{ kg/m}^3 \, [\pi(0.25 \text{ m})^2(0.01 \text{ m})] = 15.71 \text{ kg}$$

$$(I_O)_d = \frac{1}{2}m_d r_d^2 + m_d d^2$$

$$= \frac{1}{2}(15.71 \text{ kg})(0.25 \text{ m})^2 + (15.71 \text{ kg})(0.25 \text{ m})^2$$

$$= 1.473 \text{ kg} \cdot \text{m}^2$$

Hole. For the smaller disk (hole), we have

$$m_h = \rho_h V_h = 8000 \text{ kg/m}^3 \, [\pi(0.125 \text{ m})^2(0.01 \text{ m})] = 3.93 \text{ kg}$$

$$(I_O)_h = \frac{1}{2}m_h r_h^2 + m_h d^2$$

$$= \frac{1}{2}(3.93 \text{ kg})(0.125 \text{ m})^2 + (3.93 \text{ kg})(0.25 \text{ m})^2$$

$$= 0.276 \text{ kg} \cdot \text{m}^2$$

The moment of inertia of the plate about the pin is therefore

$$I_O = (I_O)_d - (I_O)_h$$

$$= 1.473 \text{ kg} \cdot \text{m}^2 - 0.276 \text{ kg} \cdot \text{m}^2$$

$$= 1.20 \text{ kg} \cdot \text{m}^2 \qquad\qquad \textit{Ans.}$$

10

EXAMPLE | 10.13

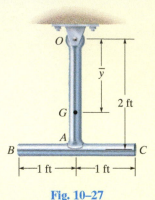

Fig. 10–27

The pendulum in Fig. 10–27 consists of two thin rods each having a weight of 10 lb. Determine the pendulum's mass moment of inertia about an axis passing through (a) the pin at O, and (b) the mass center G of the pendulum.

SOLUTION

Part (a). Using the table on the inside back cover, the moment of inertia of rod OA about an axis perpendicular to the page and passing through the end point O of the rod is $I_O = \frac{1}{3}ml^2$. Hence,

$$(I_{OA})_O = \frac{1}{3}ml^2 = \frac{1}{3}\left(\frac{10 \text{ lb}}{32.2 \text{ ft/s}^2}\right)(2 \text{ ft})^2 = 0.414 \text{ slug} \cdot \text{ft}^2$$

Realize that this same value may be determined using $I_G = \frac{1}{12}ml^2$ and the parallel-axis theorem; i.e.,

$$(I_{OA})_O = \frac{1}{12}ml^2 + md^2 = \frac{1}{12}\left(\frac{10 \text{ lb}}{32.2 \text{ ft/s}^2}\right)(2 \text{ ft})^2 + \frac{10 \text{ lb}}{32.2 \text{ ft/s}^2}(1 \text{ ft})^2$$

$$= 0.414 \text{ slug} \cdot \text{ft}^2$$

For rod BC we have

$$(I_{BC})_O = \frac{1}{12}ml^2 + md^2 = \frac{1}{12}\left(\frac{10 \text{ lb}}{32.2 \text{ ft/s}^2}\right)(2 \text{ ft})^2 + \frac{10 \text{ lb}}{32.2 \text{ ft/s}^2}(2 \text{ ft})^2$$

$$= 1.346 \text{ slug} \cdot \text{ft}^2$$

The moment of inertia of the pendulum about O is therefore

$$I_O = 0.414 + 1.346 = 1.76 \text{ slug} \cdot \text{ft}^2 \qquad \textit{Ans.}$$

Part (b). The mass center G will be located relative to the pin at O. Assuming this distance to be $\bar{y}$, Fig. 10–27, and using the formula for determining the mass center, we have

$$\bar{y} = \frac{\Sigma \tilde{y}m}{\Sigma m} = \frac{1(10/32.2) + 2(10/32.2)}{(10/32.2) + (10/32.2)} = 1.50 \text{ ft}$$

The moment of inertia I_G may be computed in the same manner as I_O, which requires successive applications of the parallel-axis theorem in order to transfer the moments of inertia of rods OA and BC to G. A more direct solution, however, involves applying the parallel-axis theorem using the result for I_O determined above; i.e.,

$$I_O = I_G + md^2; \qquad 1.76 \text{ slug} \cdot \text{ft}^2 = I_G + \left(\frac{20 \text{ lb}}{32.2 \text{ ft/s}^2}\right)(1.50 \text{ ft})^2$$

$$I_G = 0.362 \text{ slug} \cdot \text{ft}^2 \qquad \textit{Ans.}$$

10

PROBLEMS

*10–84. Determine the moment of inertia of the thin ring about the z axis. The ring has a mass m.

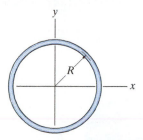

Prob. 10–84

10–85. Determine the moment of inertia of the ellipsoid with respect to the x axis and express the result in terms of the mass m of the ellipsoid. The material has a constant density ρ.

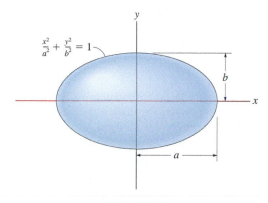

Prob. 10–85

10–86. Determine the radius of gyration k_x of the paraboloid. The density of the material is $\rho = 5 \text{ Mg/m}^3$.

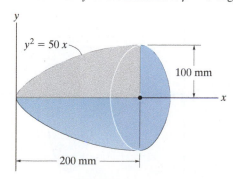

$y^2 = 50\,x$

100 mm

200 mm

Prob. 10–86

10–87. The paraboloid is formed by revolving the shaded area around the x axis. Determine the moment of inertia about the x axis and express the result in terms of the total mass m of the paraboloid. The material has a constant density ρ.

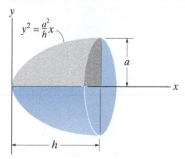

$y^2 = \dfrac{a^2}{h}x$

Prob. 10–87

*10–88. Determine the moment of inertia of the homogenous triangular prism with respect to the y axis. Express the result in terms of the mass m of the prism. *Hint*: For integration, use thin plate elements parallel to the x–y plane having a thickness of dz.

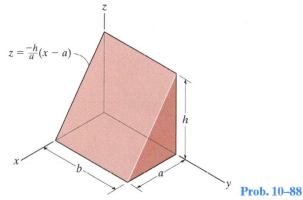

$z = \dfrac{-h}{a}(x - a)$

Prob. 10–88

10–89. Determine the moment of inertia of the semiellipsoid with respect to the x axis and express the result in terms of the mass m of the semiellipsoid. The material has a constant density ρ.

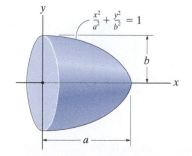

$\dfrac{x^2}{a^2} + \dfrac{y^2}{b^2} = 1$

Prob. 10–89

10

10–90. Determine the radius of gyration k_x of the solid formed by revolving the shaded area about x axis. The density of the material is ρ.

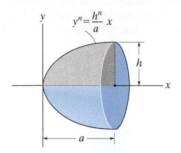

$$y^n = \frac{h^n}{a} x$$

Prob. 10–90

10–91. The concrete shape is formed by rotating the shaded area about the y axis. Determine the moment of inertia I_y. The specific weight of concrete is $\gamma = 150 \text{ lb/ft}^3$.

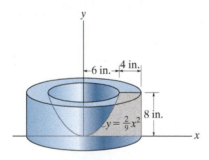

$$y = \frac{2}{9} x^2$$

Prob. 10–91

*10–92. Determine the moment of inertia I_x of the sphere and express the result in terms of the total mass m of the sphere. The sphere has a constant density ρ.

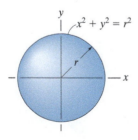

$$x^2 + y^2 = r^2$$

Prob. 10–92

10–93. The right circular cone is formed by revolving the shaded area around the x axis. Determine the moment of inertia I_x and express the result in terms of the total mass m of the cone. The cone has a constant density ρ.

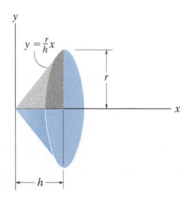

$$y = \frac{r}{h} x$$

Prob. 10–93

10–94. Determine the mass moment of inertia I_y of the solid formed by revolving the shaded area around the y axis. The total mass of the solid is 1500 kg.

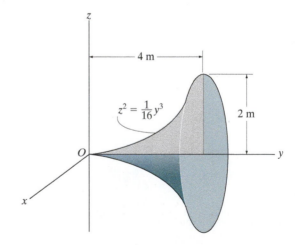

$$z^2 = \frac{1}{16} y^3$$

Prob. 10–94

10–95. The slender rods have a mass of 4 kg/m. Determine the moment of inertia of the assembly about an axis perpendicular to the page and passing through point A.

10–97. Determine the moment of inertia I_z of the frustum of the cone which has a conical depression. The material has a density of 200 kg/m³.

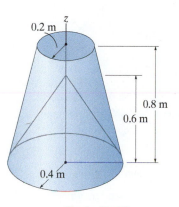

0.2 m

z

0.8 m

0.6 m

0.4 m

Prob. 10–97

A

200 mm

← 100 mm → ← 100 mm →

Prob. 10–95

10–98. The pendulum consists of the 3-kg slender rod and the 5-kg thin plate. Determine the location $\bar{y}$ of the center of mass G of the pendulum; then find the mass moment of inertia of the pendulum about an axis perpendicular to the page and passing through G.

*10–96. The pendulum consists of a 8-kg circular disk A, a 2-kg circular disk B, and a 4-kg slender rod. Determine the radius of gyration of the pendulum about an axis perpendicular to the page and passing through point O.

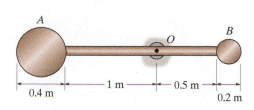

A

O

B

0.4 m

← 1 m → ← 0.5 m →

0.2 m

Prob. 10–96

O

$\bar{y}$

2 m

G

0.5 m

← 1 m →

Prob. 10–98

10

10–99. Determine the mass moment of inertia of the thin plate about an axis perpendicular to the page and passing through point O. The material has a mass per unit area of 20 kg/m².

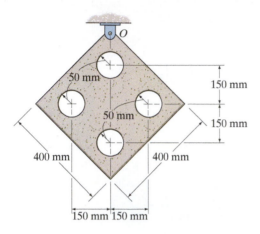

50 mm

50 mm

150 mm

150 mm

400 mm   400 mm

150 mm   150 mm

Prob. 10–99

*10–100. The pendulum consists of a plate having a weight of 12 lb and a slender rod having a weight of 4 lb. Determine the radius of gyration of the pendulum about an axis perpendicular to the page and passing through point O.

1 ft

1 ft

3 ft

2 ft

O

Prob. 10–100

10–101. If the large ring, small ring and each of the spokes weigh 100 lb, 15 lb, and 20 lb, respectively, determine the mass moment of inertia of the wheel about an axis perpendicular to the page and passing through point A.

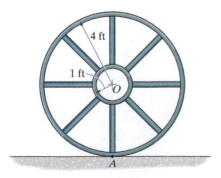

4 ft

1 ft

O

A

Prob. 10–101

10–102. Determine the mass moment of inertia of the assembly about the z axis. The density of the material is 7.85 Mg/m³.

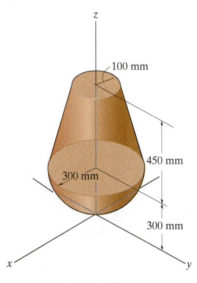

z

100 mm

450 mm

300 mm

300 mm

x

y

Prob. 10–102

10–103. Each of the three slender rods has a mass m. Determine the moment of inertia of the assembly about an axis that is perpendicular to the page and passes through the center point O.

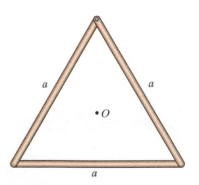

a a

$\bullet O$

a

Prob. 10–103

*10–104. The thin plate has a mass per unit area of 10 kg/m². Determine its mass moment of inertia about the *y* axis.

10–105. The thin plate has a mass per unit area of 10 kg/m². Determine its mass moment of inertia about the *z* axis.

*10–108. The pendulum consists of two slender rods *AB* and *OC* which have a mass of 3 kg/m. The thin plate has a mass of 12 kg/m². Determine the location $\bar{y}$ of the center of mass *G* of the pendulum, then calculate the moment of inertia of the pendulum about an axis perpendicular to the page and passing through *G*.

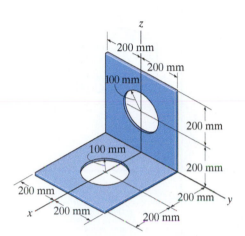

Probs. 10–104/105

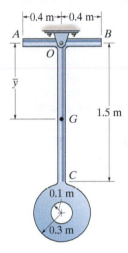

Prob. 10–108

10–106. Determine the moment of inertia of the assembly about an axis that is perpendicular to the page and passes through the center of mass *G*. The material has a specific weight of $\gamma = 90$ lb/ft³.

10–107. Determine the moment of inertia of the assembly about an axis that is perpendicular to the page and passes through point *O*. The material has a specific weight of $\gamma = 90$ lb/ft³.

10–109. Determine the moment of inertia I_z of the frustum of the cone which has a conical depression. The material has a density of 200 kg/m³.

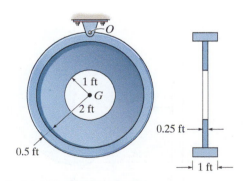

Probs. 10–106/107

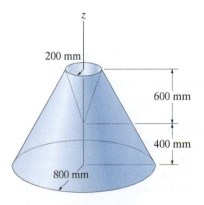

Prob. 10–109

10

CHAPTER REVIEW

Area Moment of Inertia

The *area moment of inertia* represents the second moment of the area about an axis. It is frequently used in formulas related to the strength and stability of structural members or mechanical elements.

If the area shape is irregular but can be described mathematically, then a differential element must be selected and integration over the entire area must be performed to determine the moment of inertia.

$$I_y = \int_A x^2\, dA$$

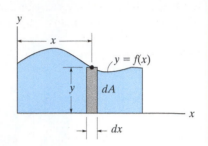

Parallel-Axis Theorem

If the moment of inertia for an area is known about a centroidal axis, then its moment of inertia about a parallel axis can be determined using the parallel-axis theorem.

$$I = \bar{I} + Ad^2$$

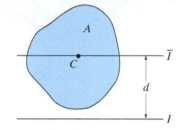

Composite Area

If an area is a composite of common shapes, as found on the inside back cover, then its moment of inertia is equal to the algebraic sum of the moments of inertia of each of its parts.

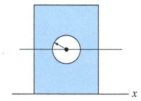

Product of Inertia

The *product of inertia* of an area is used in formulas to determine the orientation of an axis about which the moment of inertia for the area is a maximum or minimum.

$$I_{xy} = \int_A xy\, dA$$

If the product of inertia for an area is known with respect to its centroidal x', y' axes, then its value can be determined with respect to any x, y axes using the parallel-axis theorem for the product of inertia.

$$I_{xy} = \bar{I}_{x'y'} + Ad_xd_y$$

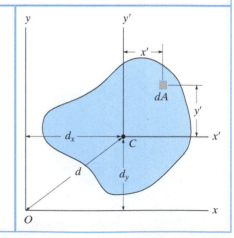

Principal Moments of Inertia

Provided the moments of inertia, I_x and I_y, and the product of inertia, I_{xy}, are known, then the transformation formulas, or Mohr's circle, can be used to determine the maximum and minimum or *principal moments of inertia* for the area, as well as finding the orientation of the principal axes of inertia.

$$I_{\substack{max \\ min}} = \frac{I_x + I_y}{2} \pm \sqrt{\left(\frac{I_x - I_y}{2}\right)^2 + I_{xy}^2}$$

$$\tan 2\theta_p = \frac{-I_{xy}}{(I_x - I_y)/2}$$

Mass Moment of Inertia

The *mass moment of inertia* is a property of a body that measures its resistance to a change in its rotation. It is defined as the "second moment" of the mass elements of the body about an axis.

$$I = \int_m r^2 \, dm$$

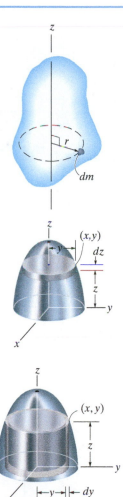

For homogeneous bodies having axial symmetry, the mass moment of inertia can be determined by a single integration, using a disk or shell element.

$$I = \rho \int_V r^2 \, dV$$

The mass moment of inertia of a composite body is determined by using tabular values of its composite shapes, found on the inside back cover, along with the parallel-axis theorem.

$$I = I_G + md^2$$

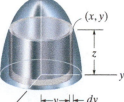

REVIEW PROBLEMS

R10–1. Determine the moment of inertia for the shaded area about the *x* axis.

R10–3. Determine the area moment of inertia of the shaded area about the *y* axis.

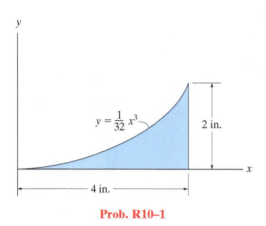

$$y = \frac{1}{32} x^3$$

2 in.

4 in.

Prob. R10–1

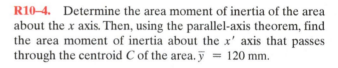

$$4y = 4 - x^2$$

1 ft

2 ft

Prob. R10–3

R10–4. Determine the area moment of inertia of the area about the *x* axis. Then, using the parallel-axis theorem, find the area moment of inertia about the *x'* axis that passes through the centroid *C* of the area. $\bar{y} = 120$ mm.

R10–2. Determine the moment of inertia for the shaded area about the *x* axis.

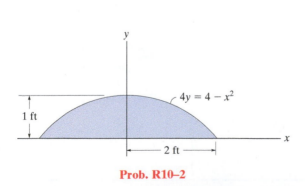

$$4y = 4 - x^2$$

1 ft

2 ft

Prob. R10–2

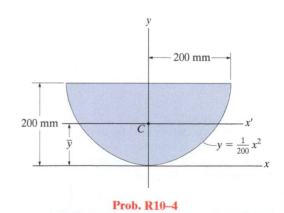

200 mm

200 mm

$$y = \frac{1}{200} x^2$$

Prob. R10–4

R10–5. Determine the area moment of inertia of the triangular area about (a) the x axis, and (b) the centroidal x' axis.

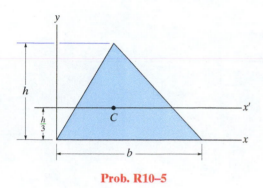

Prob. R10–5

R10–7. Determine the area moment of inertia of the beam's cross-sectional area about the x axis which passes through the centroid C.

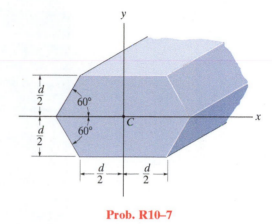

Prob. R10–7

R10–6. Determine the product of inertia of the shaded area with respect to the x and y axes.

R10–8. Determine the mass moment of inertia I_x of the body and express the result in terms of the total mass m of the body. The density is constant.

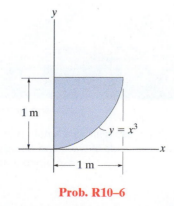

Prob. R10–6

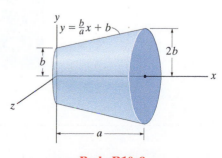

Prob. R10–8

10

Chapter 11

(© John Kershaw/Alamy)

Equilibrium and stability of this scissors lift as a function of its position can be determined using the methods of work and energy, which are explained in this chapter.

Virtual Work

- To introduce the principle of virtual work and show how it applies to finding the equilibrium configuration of a system of pin-connected members.

- To establish the potential-energy function and use the potential-energy method to investigate the type of equilibrium or stability of a rigid body or system of pin-connected members.

11.1   Definition of Work

The principle of virtual work was proposed by the Swiss mathematician Jean Bernoulli in the eighteenth century. It provides an alternative method for solving problems involving the equilibrium of a particle, a rigid body, or a system of connected rigid bodies. Before we discuss this principle, however, we must first define the work produced by a force and by a couple moment.

Work of a Force.   A force does work when it undergoes a displacement in the direction of its line of action. Consider, for example, the force $\mathbf{F}$ in Fig. 11–1a that undergoes a differential displacement $d\mathbf{r}$. If θ is the angle between the force and the displacement, then the component of $\mathbf{F}$ in

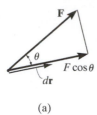

(a)

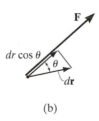

(b)

Fig. 11–1

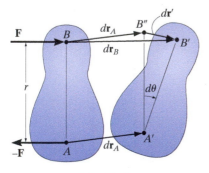

Fig. 11–2

the direction of the displacement is $F \cos \theta$. And so the work produced by **F** is

$$dU = F \, dr \cos \theta$$

Notice that this expression is also the product of the force F and the component of displacement in the direction of the force, $dr \cos \theta$, Fig. 11–1b. If we use the definition of the dot product (Eq. 2–11) the work can also be written as

$$dU = \mathbf{F} \cdot d\mathbf{r}$$

As the above equations indicate, work is a **scalar**, and like other scalar quantities, it has a magnitude that can either be *positive* or *negative*.

In the SI system, the unit of work is a **joule** (J), which is the work produced by a 1-N force that displaces through a distance of 1 m in the direction of the force (1 J = 1 N · m). The unit of work in the FPS system is the foot-pound (ft · lb), which is the work produced by a 1-lb force that displaces through a distance of 1 ft in the direction of the force.

The moment of a force has this same combination of units; however, the concepts of moment and work are in no way related. A moment is a vector quantity, whereas work is a scalar.

Work of a Couple Moment.   The rotation of a couple moment also produces work. Consider the rigid body in Fig. 11–2, which is acted upon by the couple forces **F** and –**F** that produce a couple moment **M** having a magnitude $M = Fr$. When the body undergoes the differential displacement shown, points A and B move $d\mathbf{r}_A$ and $d\mathbf{r}_B$ to their final positions A' and B', respectively. Since $d\mathbf{r}_B = d\mathbf{r}_A + d\mathbf{r}'$, this movement can be thought of as a *translation* $d\mathbf{r}_A$, where A and B move to A' and B'', and a *rotation* about A', where the body rotates through the angle $d\theta$ about A. The couple forces do no work during the translation $d\mathbf{r}_A$ because each force undergoes the same amount of displacement in opposite directions, thus canceling out the work. During rotation, however, **F** is displaced $dr' = r \, d\theta$, and so it does work $dU = F \, dr' = F \, r \, d\theta$. Since $M = Fr$, the work of the couple moment **M** is therefore

$$dU = M \, d\theta$$

If **M** and $d\theta$ have the same sense, the work is *positive*; however, if they have the opposite sense, the work will be *negative*.

Virtual Work. The definitions of the work of a force and a couple have been presented in terms of *actual movements* expressed by differential displacements having magnitudes of dr and $d\theta$. Consider now an *imaginary* or **virtual movement** of a body in static equilibrium, which indicates a displacement or rotation that is *assumed* and *does not actually exist*. These movements are first-order differential quantities and will be denoted by the symbols δr and $\delta\theta$ (delta r and delta θ), respectively. The *virtual work* done by a force having a virtual displacement δr is

$$\delta U = F \cos \theta \, \delta r \qquad (11\text{–}1)$$

Similarly, when a couple undergoes a virtual rotation $\delta\theta$ in the plane of the couple forces, the *virtual work* is

$$\delta U = M \, \delta\theta \qquad (11\text{–}2)$$

11.2 Principle of Virtual Work

The **principle of virtual** work states that if a body is in equilibrium, then the algebraic sum of the virtual work done by all the forces and couple moments acting on the body is zero for any virtual displacement of the body. Thus,

$$\delta U = 0 \qquad (11\text{–}3)$$

For example, consider the free-body diagram of the particle (ball) that rests on the floor, Fig. 11–3. If we "imagine" the ball to be displaced downwards a virtual amount δy, then the weight does positive virtual work, $W \, \delta y$, and the normal force does negative virtual work, $-N \, \delta y$. For equilibrium the total virtual work must be zero, so that $\delta U = W \, \delta y - N \, \delta y = (W - N) \, \delta y = 0$. Since $\delta y \neq 0$, then $N = W$ as required by applying $\Sigma F_y = 0$.

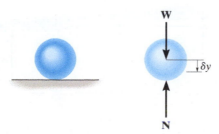

Fig. 11–3

11

In a similar manner, we can also apply the virtual-work equation $\delta U = 0$ to a rigid body subjected to a coplanar force system. Here, separate virtual translations in the x and y directions, and a virtual rotation about an axis perpendicular to the x–y plane that passes through an arbitrary point O, will correspond to the three equilibrium equations, $\Sigma F_x = 0$, $\Sigma F_y = 0$, and $\Sigma M_O = 0$. When writing these equations, it is *not necessary* to include the work done by the *internal forces* acting within the body since a rigid body *does not deform* when subjected to an external loading, and furthermore, when the body moves through a virtual displacement, the internal forces occur in equal but opposite collinear pairs, so that the corresponding work done by each pair of forces will cancel.

To demonstrate an application, consider the simply supported beam in Fig. 11–4a. When the beam is given a virtual rotation $\delta\theta$ about point B, Fig. 11–4b, the only forces that do work are $\mathbf{P}$ and $\mathbf{A}_y$. Since $\delta y = l\,\delta\theta$ and $\delta y' = (l/2)\,\delta\theta$, the virtual work equation for this case is $\delta U = A_y(l\,\delta\theta) - P(l/2)\,\delta\theta = (A_y l - Pl/2)\,\delta\theta = 0$. Since $\delta\theta \neq 0$, then $A_y = P/2$. Excluding $\delta\theta$, notice that the terms in parentheses actually represent the application of $\Sigma M_B = 0$.

As seen from the above two examples, no added advantage is gained by solving particle and rigid-body equilibrium problems using the principle of virtual work. This is because for each application of the virtual-work equation, the virtual displacement, common to every term, factors out, leaving an equation that could have been obtained in a more *direct manner* by simply applying an equation of equilibrium.

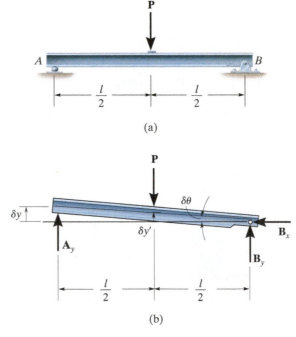

(a)

(b)

Fig. 11–4

11.3   Principle of Virtual Work for a System of Connected Rigid Bodies

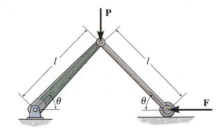

The method of virtual work is particularly effective for solving equilibrium problems that involve a system of several *connected* rigid bodies, such as the ones shown in Fig. 11–5.

Each of these systems is said to have only one degree of freedom since the arrangement of the links can be completely specified using only one coordinate θ. In other words, with this single coordinate and the length of the members, we can locate the position of the forces **F** and **P**.

In this text, we will only consider the application of the principle of virtual work to systems containing one degree of freedom.* Because they are less complicated, they will serve as a way to approach the solution of more complex problems involving systems with many degrees of freedom. The procedure for solving problems involving a system of frictionless connected rigid bodies follows.

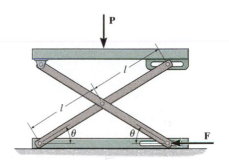

Fig. 11–5

Important Points

- A force does work when it moves through a displacement in the direction of the force. A couple moment does work when it moves through a collinear rotation. Specifically, positive work is done when the force or couple moment and its displacement have the same sense of direction.

- The principle of virtual work is generally used to determine the equilibrium configuration for a system of multiple connected members.

- A virtual displacement is imaginary; i.e., it does not really happen. It is a differential displacement that is given in the positive direction of a position coordinate.

- Forces or couple moments that do not virtually displace do no virtual work.

This scissors lift has one degree of freedom. Without the need for dismembering the mechanism, the force in the hydraulic cylinder *AB* required to provide the lift can be determined *directly* by using the principle of virtual work. (© Russell C. Hibbeler)

---

*This method of applying the principle of virtual work is sometimes called the *method of virtual displacements* because a virtual displacement is applied, resulting in the calculation of a real force. Although it is not used here, we can also apply the principle of virtual work as a *method of virtual forces*. This method is often used to apply a virtual force and then determine the displacements of points on deformable bodies. See R. C. Hibbeler, *Mechanics of Materials*, 8th edition, Pearson/Prentice Hall, 2011.

Procedure for Analysis

Free-Body Diagram.

- Draw the free-body diagram of the entire system of connected bodies and define the *coordinate q*.

- Sketch the "deflected position" of the system on the free-body diagram when the system undergoes a *positive virtual* displacement δq.

Virtual Displacements.

- Indicate *position coordinates s*, each measured from a *fixed point* on the free-body diagram. These coordinates are directed to the forces that do work.

- Each of these coordinate axes should be *parallel* to the line of action of the force to which it is directed, so that the virtual work along the coordinate axis can be calculated.

- Relate each of the position coordinates *s* to the coordinate *q*; then *differentiate* these expressions in order to express each virtual displacement δs in terms of δq.

Virtual-Work Equation.

- Write the *virtual-work equation* for the system assuming that, whether possible or not, each position coordinate *s* undergoes a *positive* virtual displacement δs. If a force or couple moment is in the same direction as the positive virtual displacement, the work is positive. Otherwise, it is negative.

- Express the work of *each* force and couple moment in the equation in terms of δq.

- Factor out this common displacement from all the terms, and solve for the unknown force, couple moment, or equilibrium position *q*.

11

EXAMPLE | 11.1

Determine the angle θ for equilibrium of the two-member linkage shown in Fig. 11–6a. Each member has a mass of 10 kg.

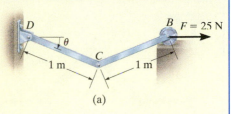

(a)

SOLUTION

Free-Body Diagram.   The system has only one degree of freedom since the location of both links can be specified by the single coordinate, $(q =)\,\theta$. As shown on the free-body diagram in Fig. 11–6b, when θ has a *positive* (clockwise) virtual rotation $\delta\theta$, only the force $\mathbf{F}$ and the two 98.1-N weights do work. (The reactive forces $\mathbf{D}_x$ and $\mathbf{D}_y$ are fixed, and $\mathbf{B}_y$ does not displace along its line of action.)

Virtual Displacements.   If the origin of coordinates is established at the *fixed* pin support D, then the position of $\mathbf{F}$ and $\mathbf{W}$ can be specified by the *position coordinates* x_B and y_w. In order to determine the work, note that, as required, these coordinates are parallel to the lines of action of their associated forces. Expressing these position coordinates in terms of θ and taking the derivatives yields

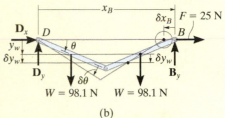

(b)

Fig. 11–6

$$x_B = 2(1\cos\theta)\text{ m} \qquad \delta x_B = -2\sin\theta\;\delta\theta\text{ m} \qquad (1)$$
$$y_w = \tfrac{1}{2}(1\sin\theta)\text{ m} \qquad \delta y_w = 0.5\cos\theta\;\delta\theta\text{ m} \qquad (2)$$

It is seen by the *signs* of these equations, and indicated in Fig. 11–6b, that an *increase* in θ (i.e., $\delta\theta$) causes a *decrease* in x_B and an *increase* in y_w.

Virtual-Work Equation.   If the virtual displacements δx_B and δy_w were *both positive*, then the forces $\mathbf{W}$ and $\mathbf{F}$ would do positive work since the forces and their corresponding displacements would have the same sense. Hence, the virtual-work equation for the displacement $\delta\theta$ is

$$\delta U = 0; \qquad W\,\delta y_w + W\,\delta y_w + F\,\delta x_B = 0 \qquad (3)$$

Substituting Eqs. 1 and 2 into Eq. 3 in order to relate the virtual displacements to the common virtual displacement $\delta\theta$ yields

$$98.1(0.5\cos\theta\;\delta\theta) + 98.1(0.5\cos\theta\;\delta\theta) + 25(-2\sin\theta\;\delta\theta) = 0$$

Notice that the "negative work" done by $\mathbf{F}$ (force in the opposite sense to displacement) has actually been *accounted* for in the above equation by the "negative sign" of Eq. 1. Factoring out the *common displacement* $\delta\theta$ and solving for θ, noting that $\delta\theta \neq 0$, yields

$$(98.1\cos\theta - 50\sin\theta)\,\delta\theta = 0$$

$$\theta = \tan^{-1}\frac{98.1}{50} = 63.0° \qquad\qquad Ans.$$

NOTE: If this problem had been solved using the equations of equilibrium, it would be necessary to dismember the links and apply three scalar equations to *each* link. The principle of virtual work, by means of calculus, has eliminated this task so that the answer is obtained directly.

EXAMPLE 11.2

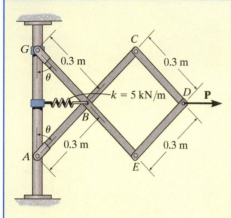

(a)

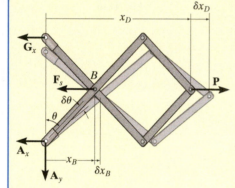

(b)

Fig. 11–7

Determine the required force P in Fig. 11–7a needed to maintain equilibrium of the scissors linkage when $\theta = 60°$. The spring is unstretched when $\theta = 30°$. Neglect the mass of the links.

SOLUTION

Free-Body Diagram.   Only $\mathbf{F}_s$ and $\mathbf{P}$ do work when θ undergoes a *positive* virtual displacement $\delta\theta$, Fig. 11–7b. For the arbitrary position θ, the spring is stretched $(0.3\text{ m})\sin\theta - (0.3\text{ m})\sin 30°$, so that

$$F_s = ks = 5000\text{ N/m }[(0.3\text{ m})\sin\theta - (0.3\text{ m})\sin 30°]$$

$$= (1500\sin\theta - 750)\text{ N}$$

Virtual Displacements.   The position coordinates, x_B and x_D, measured from the *fixed point A*, are used to locate $\mathbf{F}_s$ and $\mathbf{P}$. These coordinates are parallel to the line of action of their corresponding forces. Expressing x_B and x_D in terms of the angle θ using trigonometry,

$$x_B = (0.3\text{ m})\sin\theta$$

$$x_D = 3[(0.3\text{ m})\sin\theta] = (0.9\text{ m})\sin\theta$$

Differentiating, we obtain the virtual displacements of points B and D.

$$\delta x_B = 0.3\cos\theta\,\delta\theta \qquad (1)$$

$$\delta x_D = 0.9\cos\theta\,\delta\theta \qquad (2)$$

Virtual-Work Equation.   Force $\mathbf{P}$ does positive work since it acts in the positive sense of its virtual displacement. The spring force $\mathbf{F}_s$ does negative work since it acts opposite to its positive virtual displacement. Thus, the virtual-work equation becomes

$$\delta U = 0; \qquad -F_s\,\delta x_B + P\delta x_D = 0$$

$$-[1500\sin\theta - 750]\,(0.3\cos\theta\,\delta\theta) + P\,(0.9\cos\theta\,\delta\theta) = 0$$

$$[0.9P + 225 - 450\sin\theta]\cos\theta\,\delta\theta = 0$$

Since $\cos\theta\,\delta\theta \neq 0$, then this equation requires

$$P = 500\sin\theta - 250$$

When $\theta = 60°$,

$$P = 500\sin 60° - 250 = 183\text{ N} \qquad \textit{Ans.}$$

11

EXAMPLE 11.3

If the box in Fig. 11–8a has a mass of 10 kg, determine the couple moment M needed to maintain equilibrium when $\theta = 60°$. Neglect the mass of the members.

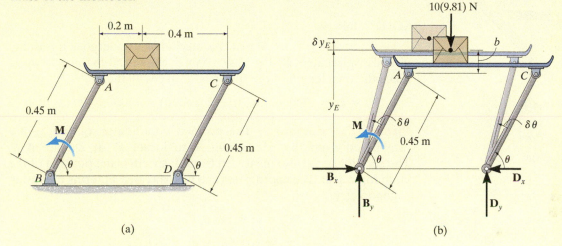

Fig. 11–8

SOLUTION

Free-Body Diagram. When θ undergoes a positive virtual displacement $\delta\theta$, only the couple moment **M** and the weight of the box do work, Fig. 11–8b.

Virtual Displacements. The position coordinate y_E, measured from the *fixed point B*, locates the weight, 10(9.81) N. Here,

$$y_E = (0.45 \text{ m}) \sin\theta + b$$

where b is a constant distance. Differentiating this equation, we obtain

$$\delta y_E = 0.45 \text{ m} \cos\theta \ \delta\theta \qquad (1)$$

Virtual-Work Equation. The virtual-work equation becomes

$$\delta U = 0; \qquad M \ \delta\theta - [10(9.81) \text{ N}]\delta y_E = 0$$

Substituting Eq. 1 into this equation

$$M \ \delta\theta - 10(9.81) \text{ N}(0.45 \text{ m} \cos\theta \ \delta\theta) = 0$$
$$\delta\theta(M - 44.145 \cos\theta) = 0$$

Since $\delta\theta \neq 0$, then

$$M - 44.145 \cos\theta = 0$$

Since it is required that $\theta = 60°$, then

$$M = 44.145 \cos 60° = 22.1 \text{ N} \cdot \text{m} \qquad \textit{Ans.}$$

11

EXAMPLE | 11.4

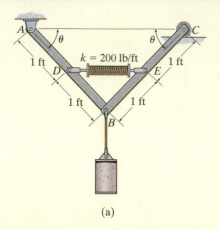

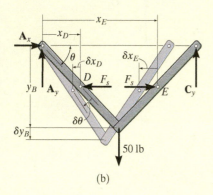

(a)

(b)

Fig. 11–9

The mechanism in Fig. 11–9a supports the 50-lb cylinder. Determine the angle θ for equilibrium if the spring has an unstretched length of 2 ft when $\theta = 0°$. Neglect the mass of the members.

SOLUTION

Free-Body Diagram. When the mechanism undergoes a positive virtual displacement $\delta\theta$, Fig. 11–9b, only $\mathbf{F}_s$ and the 50-lb force do work. Since the final length of the spring is 2(1 ft cos θ), then

$$F_s = ks = (200 \text{ lb/ft})(2 \text{ ft} - 2 \text{ ft} \cos\theta) = (400 - 400\cos\theta)\text{ lb}$$

Virtual Displacements. The position coordinates x_D and x_E are established from the *fixed point A* to locate $\mathbf{F}_s$ at D and at E. The coordinate y_B, also measured from A, specifies the position of the 50-lb force at B. The coordinates can be expressed in terms of θ using trigonometry.

$$x_D = (1 \text{ ft}) \cos\theta$$
$$x_E = 3[(1 \text{ ft}) \cos\theta] = (3 \text{ ft}) \cos\theta$$
$$y_B = (2 \text{ ft}) \sin\theta$$

Differentiating, we obtain the virtual displacements of points D, E, and B as

$$\delta x_D = -1 \sin\theta\ \delta\theta \tag{1}$$
$$\delta x_E = -3 \sin\theta\ \delta\theta \tag{2}$$
$$\delta y_B = 2 \cos\theta\ \delta\theta \tag{3}$$

Virtual-Work Equation. The virtual-work equation is written as if all virtual displacements are positive, thus

$$\delta U = 0; \qquad F_s\,\delta x_E + 50\,\delta y_B - F_s\,\delta x_D = 0$$

$$(400 - 400\cos\theta)(-3\sin\theta\ \delta\theta) + 50(2\cos\theta\ \delta\theta)$$

$$- (400 - 400\cos\theta)(-1\sin\theta\ \delta\theta) = 0$$

$$\delta\theta(800\sin\theta\cos\theta - 800\sin\theta + 100\cos\theta) = 0$$

Since $\delta\theta \neq 0$, then

$$800\sin\theta\cos\theta - 800\sin\theta + 100\cos\theta = 0$$

Solving by trial and error,

$$\theta = 34.9° \qquad\qquad Ans.$$

FUNDAMENTAL PROBLEMS

F11–1. Determine the required magnitude of force **P** to maintain equilibrium of the linkage at $\theta = 60°$. Each link has a mass of 20 kg.

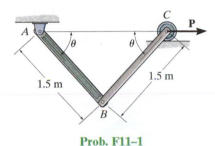

Prob. F11–1

F11–2. Determine the magnitude of force **P** required to hold the 50-kg smooth rod in equilibrium at $\theta = 60°$.

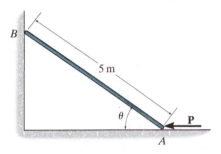

Prob. F11–2

F11–3. The linkage is subjected to a force of $P = 2$ kN. Determine the angle θ for equilibrium. The spring is unstretched when $\theta = 0°$. Neglect the mass of the links.

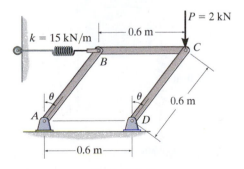

Prob. F11–3

F11–4. The linkage is subjected to a force of $P = 6$ kN. Determine the angle θ for equilibrium. The spring is unstretched at $\theta = 60°$. Neglect the mass of the links.

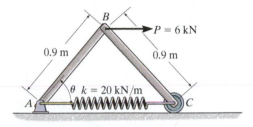

Prob. F11–4

F11–5. Determine the angle θ where the 50-kg bar is in equilibrium. The spring is unstretched at $\theta = 60°$.

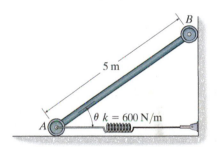

Prob. F11–5

F11–6. The scissors linkage is subjected to a force of $P = 150$ N. Determine the angle θ for equilibrium. The spring is unstretched at $\theta = 0°$. Neglect the mass of the links.

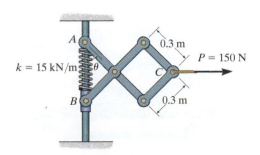

Prob. F11–6

11

PROBLEMS

11–1. Use the method of virtual work to determine the tension in cable AC. The lamp weighs 10 lb.

11–3. If a force of $P = 5$ lb is applied to the handle of the mechanism, determine the force the screw exerts on the cork of the bottle. The screw is attached to the pin at A and passes through the collar that is attached to the bottle neck at B.

Prob. 11–1

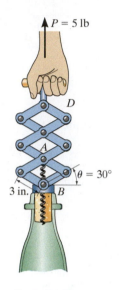

Prob. 11–3

11–2. The scissors jack supports a load **P**. Determine the axial force in the screw necessary for equilibrium when the jack is in the position θ. Each of the four links has a length L and is pin connected at its center. Points B and D can move horizontally.

***11–4.** The disk has a weight of 10 lb and is subjected to a vertical force $P = 8$ lb and a couple moment $M = 8$ lb · ft. Determine the disk's rotation θ if the end of the spring wraps around the periphery of the disk as the disk turns. The spring is originally unstretched.

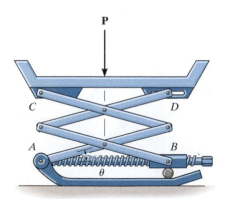

Prob. 11–2

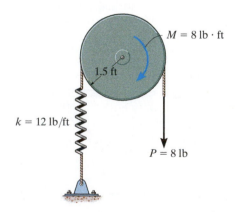

Prob. 11–4

11–5. The punch press consists of the ram R, connecting rod AB, and a flywheel. If a torque of $M = 75 \, \text{N} \cdot \text{m}$ is applied to the flywheel, determine the force **F** applied at the ram to hold the rod in the position $\theta = 60°$.

11–6. The flywheel is subjected to a torque of $M = 75 \, \text{N} \cdot \text{m}$. Determine the horizontal compressive force F and plot the result of F (ordinate) versus the equilibrium position θ (abscissa) for $0° \le \theta \le 180°$.

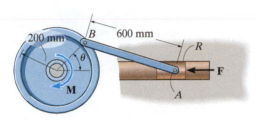

Probs. 11–5/6

11–7. When $\theta = 20°$, the 50-lb uniform block compresses the two vertical springs 4 in. If the uniform links AB and CD each weigh 10 lb, determine the magnitude of the applied couple moments **M** needed to maintain equilibrium when $\theta = 20°$.

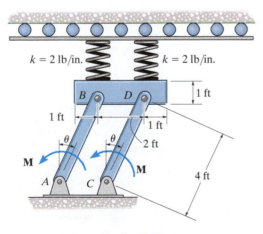

Prob. 11–7

*11–8. The bar is supported by the spring and smooth collar that allows the spring to be always perpendicular to the bar for any angle θ. If the unstretched length of the spring is l_0, determine the force P needed to hold the bar in the equilibrium position θ. Neglect the weight of the bar.

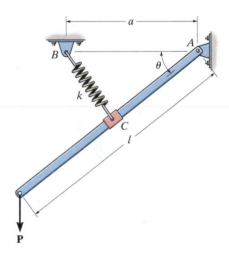

Prob. 11–8

11–9. The 4-ft members of the mechanism are pin connected at their centers. If vertical forces $P_1 = P_2 = 30 \, \text{lb}$ act at C and E as shown, determine the angle θ for equilibrium. The spring is unstretched when $\theta = 45°$. Neglect the weight of the members.

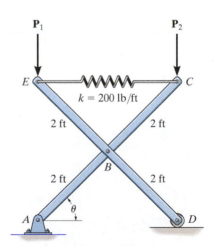

Prob. 11–9

11

11–10. The thin rod of weight W rests against the smooth wall and floor. Determine the magnitude of force **P** needed to hold it in equilibrium for a given angle θ.

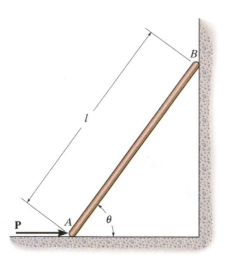

Prob. 11–10

11–11. If each of the three links of the mechanism have a mass of 4 kg, determine the angle θ for equilibrium. The spring, which always remains vertical, is unstretched when $\theta = 0°$.

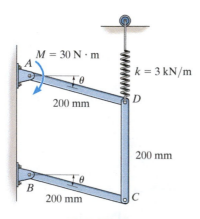

Prob. 11–11

*11–12. The disk is subjected to a couple moment M. Determine the disk's rotation θ required for equilibrium. The end of the spring wraps around the periphery of the disk as the disk turns. The spring is originally unstretched.

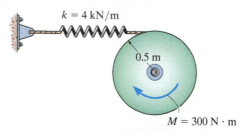

Prob. 11–12

11–13. A 5-kg uniform serving table is supported on each side by pairs of two identical links, AB and CD, and springs CE. If the bowl has a mass of 1 kg, determine the angle θ where the table is in equilibrium. The springs each have a stiffness of $k = 200\,\text{N/m}$ and are unstretched when $\theta = 90°$. Neglect the mass of the links.

11–14. A 5-kg uniform serving table is supported on each side by two pairs of identical links, AB and CD, and springs CE. If the bowl has a mass of 1 kg and is in equilibrium when $\theta = 45°$, determine the stiffness k of each spring. The springs are unstretched when $\theta = 90°$. Neglect the mass of the links.

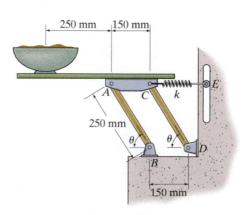

Probs. 11–13/14

11

11–15. The service window at a fast-food restaurant consists of glass doors that open and close automatically using a motor which supplies a torque **M** to each door. The far ends, *A* and *B*, move along the horizontal guides. If a food tray becomes stuck between the doors as shown, determine the horizontal force the doors exert on the tray at the position *θ*.

11–17. When *θ* = 30°, the 25-kg uniform block compresses the two horizontal springs 100 mm. Determine the magnitude of the applied couple moments **M** needed to maintain equilibrium. Take *k* = 3 kN/m and neglect the mass of the links.

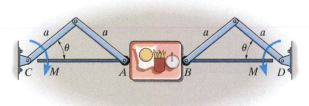

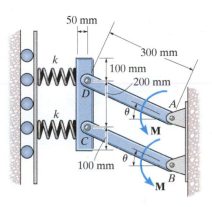

Prob. 11–15

Prob. 11–17

*****11–16.** The members of the mechanism are pin connected. If a vertical force of 800 N acts at *A*, determine the angle *θ* for equilibrium. The spring is unstretched when *θ* = 0°. Neglect the mass of the links.

11–18. The "Nuremberg scissors" is subjected to a horizontal force of *P* = 600 N. Determine the angle *θ* for equilibrium. The spring has a stiffness of *k* = 15 kN/m and is unstretched when *θ* = 15°.

11–19. The "Nuremberg scissors" is subjected to a horizontal force of *P* = 600 N. Determine the stiffness *k* of the spring for equilibrium when *θ* = 60°. The spring is unstretched when *θ* = 15°.

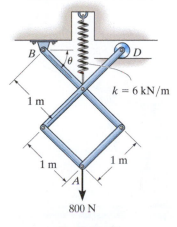

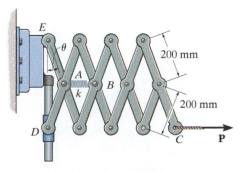

Prob. 11–16

Probs. 11–18/19

11

*11–20. The crankshaft is subjected to a torque of $M = 50$ N·m. Determine the horizontal compressive force F applied to the piston for equilibrium when $\theta = 60°$.

11–21. The crankshaft is subjected to a torque of $M = 50$ N·m. Determine the horizontal compressive force F and plot the result of F (ordinate) versus θ (abscissa) for $0° \le \theta \le 90°$.

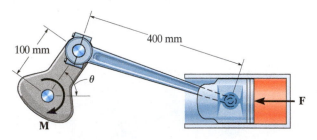

Probs. 11–20/21

11–22. The spring is unstretched when $\theta = 0°$. If $P = 8$ lb, determine the angle θ for equilibrium. Due to the roller guide, the spring always remains vertical. Neglect the weight of the links.

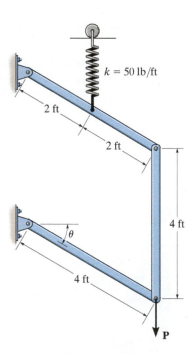

Prob. 11–22

11–23. Determine the weight of block G required to balance the differential lever when the 20-lb load F is placed on the pan. The lever is in balance when the load and block are not on the lever. Take $x = 12$ in.

*11–24. If the load F weighs 20 lb and the block G weighs 2 lb, determine its position x for equilibrium of the differential lever. The lever is in balance when the load and block are not on the lever.

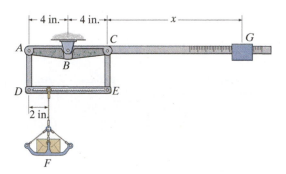

Probs. 11–23/24

11–25. The dumpster has a weight W and a center of gravity at G. Determine the force in the hydraulic cylinder needed to hold it in the general position θ.

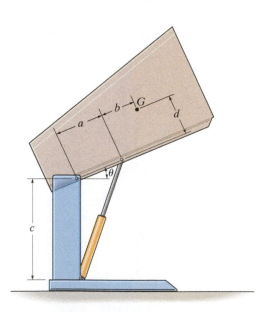

Prob. 11–25

*11.4   Conservative Forces

When a force does work that depends only upon the initial and final positions of the force, and it is *independent* of the path it travels, then the force is referred to as a ***conservative force***. The weight of a body and the force of a spring are two examples of conservative forces.

Weight.   Consider a block of weight **W** that travels along the path in Fig. 11–10a. When it is displaced up the path by an amount $d\mathbf{r}$, then the work is $dU = \mathbf{W} \cdot d\mathbf{r}$ or $dU = -W(dr\cos\theta) = -W\,dy$, as shown in Fig. 11–10b. In this case, the work is *negative* since **W** acts in the opposite sense of dy. Thus, if the block moves from A to B, through the vertical displacement h, the work is

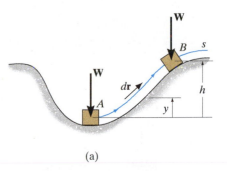

(a)

$$U = -\int_0^h W\,dy = -Wh$$

The weight of a body is therefore a conservative force, since the work done by the weight depends only on the *vertical displacement* of the body, and is independent of the path along which the body travels.

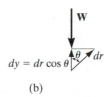

$dy = dr\cos\theta$

(b)

Fig. 11–10

Spring Force.   Now consider the linearly elastic spring in Fig. 11–11, which undergoes a displacement ds. The work done by the spring force on the block is $dU = -F_s\,ds = -ks\,ds$. The work is *negative* because $\mathbf{F}_s$ acts in the opposite sense to that of ds. Thus, the work of $\mathbf{F}_s$ when the block is displaced from $s = s_1$ to $s = s_2$ is

$$U = -\int_{s_1}^{s_2} ks\,ds = -\left(\tfrac{1}{2}ks_2^2 - \tfrac{1}{2}ks_1^2\right)$$

Here the work depends only on the spring's initial and final positions, s_1 and s_2, measured from the spring's unstretched position. Since this result is independent of the path taken by the block as it moves, then a spring force is also a *conservative force*.

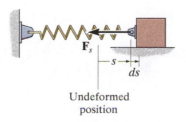

F_s

s

ds

Undeformed
position

Fig. 11–11

11

Friction.   In contrast to a conservative force, consider the force of *friction* exerted on a sliding body by a fixed surface. The work done by the frictional force depends on the path; the longer the path, the greater the work. Consequently, frictional forces are *nonconservative*, and most of the work done by them is dissipated from the body in the form of heat.

*11.5   Potential Energy

A conservative force can give the body the capacity to do work. This capacity, measured as **potential energy**, depends on the location or "position" of the body measured relative to a fixed reference position or datum.

Gravitational Potential Energy.   If a body is located a distance *y above* a fixed horizontal reference or datum as in Fig. 11–12, the weight of the body has *positive* gravitational potential energy V_g since **W** has the capacity of doing positive work when the body is moved back down to the datum. Likewise, if the body is located a distance *y below* the datum, V_g is *negative* since the weight does negative work when the body is moved back up to the datum. At the datum, $V_g = 0$.

Measuring *y* as *positive upward*, the gravitational potential energy of the body's weight **W** is therefore

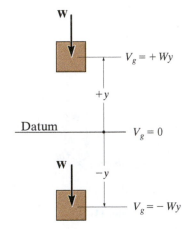

$$V_g = Wy \qquad (11\text{–}4)$$

Fig. 11–12

Elastic Potential Energy.   When a spring is either elongated or compressed by an amount *s* from its unstretched position (the datum), the energy stored in the spring is called *elastic potential energy*. It is determined from

$$V_e = \tfrac{1}{2} ks^2 \qquad (11\text{–}5)$$

This energy is always a positive quantity since the spring force acting on the attached body does *positive* work on the body as the force returns the body to the spring's unstretched position, Fig. 11–13.

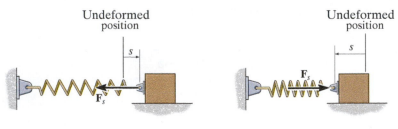

$$V_e = +\tfrac{1}{2} ks^2$$

Fig. 11–13

11

Potential Function.   In the general case, if a body is subjected to *both* gravitational and elastic forces, the *potential energy or potential function V* of the body can be expressed as the algebraic sum

$$V = V_g + V_e \qquad (11\text{--}6)$$

where measurement of V depends on the location of the body with respect to a selected datum in accordance with Eqs. 11–4 and 11–5.

In particular, if a *system* of frictionless connected rigid bodies has a *single degree of freedom,* such that its vertical distance from the datum is defined by the coordinate q, then the potential function for the system can be expressed as $V = V(q)$. The work done by all the weight and spring forces acting on the system in moving it from q_1 to q_2, is measured by the *difference* in V; i.e.,

$$U_{1-2} = V(q_1) - V(q_2) \qquad (11\text{--}7)$$

For example, the potential function for a system consisting of a block of weight **W** supported by a spring, as in Fig. 11–14, can be expressed in terms of the coordinate $(q =\ ) y$, measured from a fixed datum located at the unstretched length of the spring. Here

$$V = V_g + V_e$$
$$= -Wy + \tfrac{1}{2}ky^2 \qquad (11\text{--}8)$$

If the block moves from y_1 to y_2, then applying Eq. 11–7 the work of **W** and $\mathbf{F}_s$ is

$$U_{1-2} = V(y_1) - V(y_2) = -W(y_1 - y_2) + \tfrac{1}{2}ky_1^2 - \tfrac{1}{2}ky_2^2$$

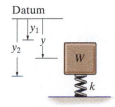

(a)

Fig. 11–14

*11.6   Potential-Energy Criterion for Equilibrium

If a frictionless connected system has one degree of freedom, and its position is defined by the coordinate q, then if it displaces from q to $q + dq$, Eq. 11–7 becomes

$$dU = V(q) - V(q + dq)$$

or

$$dU = -dV$$

If the system is in equilibrium and undergoes a *virtual displacement* δq, rather than an actual displacement dq, then the above equation becomes $\delta U = -\delta V$. However, the principle of virtual work requires that $\delta U = 0$, and therefore, $\delta V = 0$, and so we can write $\delta V = (dV/dq)\delta q = 0$. Since $\delta q \neq 0$, this expression becomes

$$\boxed{\frac{dV}{dq} = 0} \qquad (11-9)$$

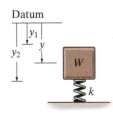

The counterweight at A balances the weight of the deck B of this simple lift bridge. By applying the method of potential energy we can analyze the equilibrium state of the bridge. (© Russell C. Hibbeler)

Hence, *when a frictionless connected system of rigid bodies is in equilibrium, the first derivative of its potential function is zero.* For example, using Eq. 11–8 we can determine the equilibrium position for the spring and block in Fig. 11–14a. We have

$$\frac{dV}{dy} = -W + ky = 0$$

(a)

Hence, the equilibrium position $y = y_{eq}$ is

(b)

Fig. 11–14 (cont'd)

$$y_{eq} = \frac{W}{k}$$

Of course, this *same result* can be obtained by applying $\Sigma F_y = 0$ to the forces acting on the free-body diagram of the block, Fig. 11–14b.

*11.7   Stability of Equilibrium Configuration

The potential function V of a system can also be used to investigate the stability of the equilibrium configuration, which is classified as *stable*, *neutral*, or *unstable*.

Stable Equilibrium.   A system is said to be in **stable equilibrium** if a system has a tendency to return to its original position when a small displacement is given to the system. The potential energy of the system in this case is at its *minimum*. A simple example is shown in Fig. 11–15a. When the disk is given a small displacement, its center of gravity G will always move (rotate) back to its equilibrium position, which is at the *lowest point* of its path. This is where the potential energy of the disk is at its *minimum*.

Neutral Equilibrium.   A system is said to be in **neutral equilibrium** if the system still remains in equilibrium when the system is given a small displacement away from its original position. In this case, the potential energy of the system is *constant*. Neutral equilibrium is shown in Fig. 11–15b, where a disk is pinned at G. Each time the disk is rotated, a new equilibrium position is established and the potential energy remains unchanged.

Unstable Equilibrium.   A system is said to be in **unstable equilibrium** if it has a tendency to be *displaced farther away* from its original equilibrium position when it is given a small displacement. The potential energy of the system in this case is a *maximum*. An unstable equilibrium position of the disk is shown in Fig. 11–15c. Here the disk will rotate away from its equilibrium position when its center of gravity is slightly displaced. At this *highest point*, its potential energy is at a *maximum*.

One-Degree-of-Freedom System.   If a system has only one degree of freedom, and its position is defined by the coordinate q, then the potential function V for the system in terms of q can be plotted, Fig. 11–16.

During high winds and when going around a curve, these sugar-cane trucks can become unstable and tip over since their center of gravity is high off the road when they are fully loaded. (© Russell C. Hibbeler)

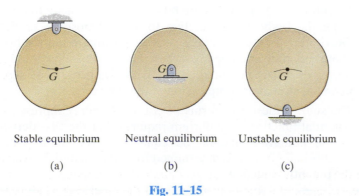

Stable equilibrium	Neutral equilibrium	Unstable equilibrium
(a)	(b)	(c)

Fig. 11–15

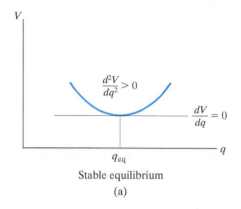

$\dfrac{d^2V}{dq^2} > 0$

$\dfrac{dV}{dq} = 0$

q_{eq}

Stable equilibrium

(a)

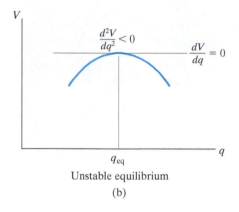

$\dfrac{d^2V}{dq^2} < 0$

$\dfrac{dV}{dq} = 0$

q_{eq}

Unstable equilibrium

(b)

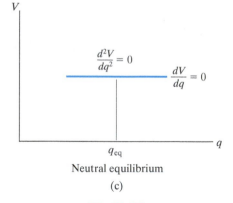

$\dfrac{d^2V}{dq^2} = 0$

$\dfrac{dV}{dq} = 0$

q_{eq}

Neutral equilibrium

(c)

Fig. 11–16

Provided the system is in *equilibrium*, then dV/dq, which represents the slope of this function, must be equal to zero. An investigation of stability at the equilibrium configuration therefore requires that the second derivative of the potential function be evaluated.

If d^2V/dq^2 is greater than zero, Fig. 11–16a, the potential energy of the system will be a *minimum*. This indicates that the equilibrium configuration is *stable*. Thus,

$$\frac{dV}{dq} = 0, \qquad \frac{d^2V}{dq^2} > 0 \qquad \text{stable equilibrium} \qquad (11\text{–}10)$$

If d^2V/dq^2 is less than zero, Fig. 11–16b, the potential energy of the system will be a *maximum*. This indicates an *unstable* equilibrium configuration. Thus,

$$\frac{dV}{dq} = 0, \qquad \frac{d^2V}{dq^2} < 0 \qquad \text{unstable equilibrium} \qquad (11\text{–}11)$$

Finally, if d^2V/dq^2 is equal to zero, it will be necessary to investigate the higher order derivatives to determine the stability. The equilibrium configuration will be *stable* if the first non-zero derivative is of an *even* order and it is *positive*. Likewise, the equilibrium will be *unstable* if this first non-zero derivative is odd or if it is even and negative. If all the higher order derivatives are *zero*, the system is said to be in *neutral equilibrium*, Fig. 11–16c. Thus,

$$\frac{dV}{dq} = \frac{d^2V}{dq^2} = \frac{d^3V}{dq^3} = \cdots = 0 \qquad \text{neutral equilibrium} \quad (11\text{–}12)$$

This condition occurs only if the potential-energy function for the system is constant at or around the neighborhood of q_{eq}.

Important Points

- A conservative force does work that is independent of the path through which the force moves. Examples include the weight and the spring force.

- Potential energy provides the body with the capacity to do work when the body moves relative to a fixed position or datum. Gravitational potential energy can be positive when the body is above a datum, and negative when the body is below the datum. Spring or elastic potential energy is always positive. It depends upon the stretch or compression of the spring.

- The sum of these two forms of potential energy represents the potential function. Equilibrium requires that the first derivative of the potential function be equal to zero. Stability at the equilibrium position is determined from the second derivative of the potential function.

Procedure for Analysis

Using potential-energy methods, the equilibrium positions and the stability of a body or a system of connected bodies having a single degree of freedom can be obtained by applying the following procedure.

Potential Function.

- Sketch the system so that it is in the *arbitrary position* specified by the coordinate q.

- Establish a horizontal *datum* through a *fixed point** and express the gravitational potential energy V_g in terms of the weight W of each member and its vertical distance y from the datum, $V_g = Wy$.

- Express the elastic potential energy V_e of the system in terms of the stretch or compression, s, of any connecting spring, $V_e = \frac{1}{2}ks^2$.

- Formulate the potential function $V = V_g + V_e$ and express the *position coordinates* y and s in terms of the single coordinate q.

Equilibrium Position.

- The equilibrium position of the system is determined by taking the first derivative of V and setting it equal to zero, $dV/dq = 0$.

Stability.

- Stability at the equilibrium position is determined by evaluating the second or higher-order derivatives of V.

- If the second derivative is greater than zero, the system is stable; if all derivatives are equal to zero, the system is in neutral equilibrium; and if the second derivative is less than zero, the system is unstable.

*The location of the datum is *arbitrary*, since only the *changes* or differentials of V are required for investigation of the equilibrium position and its stability.

EXAMPLE 11.5

The uniform link shown in Fig. 11–17a has a mass of 10 kg. If the spring is unstretched when $\theta = 0°$, determine the angle θ for equilibrium and investigate the stability at the equilibrium position.

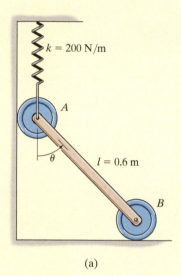

$k = 200$ N/m

A

θ

$l = 0.6$ m

B

(a)

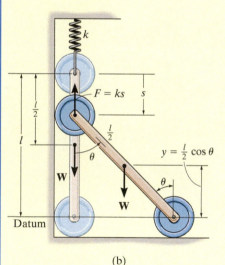

k

$F = ks$ s

$\frac{l}{2}$

l

$\frac{l}{2}$

θ

W

θ

W

$y = \frac{l}{2} \cos \theta$

θ

Datum

(b)

Fig. 11–17

SOLUTION

Potential Function. The datum is established at the bottom of the link, Fig. 11–17b. When the link is located in the arbitrary position θ, the spring increases its potential energy by stretching and the weight decreases its potential energy. Hence,

$$V = V_e + V_g = \frac{1}{2}ks^2 + Wy$$

Since $l = s + l \cos \theta$ or $s = l(1 - \cos \theta)$, and $y = (l/2) \cos \theta$, then

$$V = \frac{1}{2}kl^2(1 - \cos \theta)^2 + W\left(\frac{l}{2}\cos \theta\right)$$

Equilibrium Position. The first derivative of V is

$$\frac{dV}{d\theta} = kl^2(1 - \cos \theta)\sin \theta - \frac{Wl}{2}\sin \theta = 0$$

or

$$l\left[kl(1 - \cos \theta) - \frac{W}{2}\right]\sin \theta = 0$$

This equation is satisfied provided

$$\sin \theta = 0 \qquad \theta = 0° \qquad\qquad \textit{Ans.}$$

or

$$\theta = \cos^{-1}\left(1 - \frac{W}{2kl}\right) = \cos^{-1}\left[1 - \frac{10(9.81)}{2(200)(0.6)}\right] = 53.8° \qquad \textit{Ans.}$$

Stability. The second derivative of V is

$$\frac{d^2V}{d\theta^2} = kl^2(1 - \cos \theta)\cos \theta + kl^2 \sin \theta \sin \theta - \frac{Wl}{2}\cos \theta$$

$$= kl^2(\cos \theta - \cos 2\theta) - \frac{Wl}{2}\cos \theta$$

Substituting values for the constants, with $\theta = 0°$ and $\theta = 53.8°$, yields

$$\frac{d^2V}{d\theta^2}\bigg|_{\theta=0°} = 200(0.6)^2(\cos 0° - \cos 0°) - \frac{10(9.81)(0.6)}{2}\cos 0°$$

$$= -29.4 < 0 \qquad \text{(unstable equilibrium at } \theta = 0°) \qquad \textit{Ans.}$$

$$\frac{d^2V}{d\theta^2}\bigg|_{\theta=53.8°} = 200(0.6)^2(\cos 53.8° - \cos 107.6°) - \frac{10(9.81)(0.6)}{2}\cos 53.8°$$

$$= 46.9 > 0 \qquad \text{(stable equilibrium at } \theta = 53.8°) \qquad \textit{Ans.}$$

11

EXAMPLE   11.6

If the spring AD in Fig. 11–18a has a stiffness of 18 kN/m and is unstretched when $\theta = 60°$, determine the angle θ for equilibrium. The load has a mass of 1.5 Mg. Investigate the stability at the equilibrium position.

SOLUTION

Potential Energy.   The gravitational potential energy for the load with respect to the fixed datum, shown in Fig. 11–18b, is

$$V_g = mgy = 1500(9.81) \text{ N}[(4 \text{ m}) \sin \theta + h] = 58\,860 \sin \theta + 14\,715h$$

where h is a constant distance. From the geometry of the system, the elongation of the spring when the load is on the platform is $s = (4 \text{ m}) \cos \theta - (4 \text{ m}) \cos 60° = (4 \text{ m}) \cos \theta - 2 \text{ m}$.
   Thus, the elastic potential energy of the system is

$$V_e = \tfrac{1}{2}ks^2 = \tfrac{1}{2}(18\,000 \text{ N/m})(4 \text{ m} \cos \theta - 2 \text{ m})^2 = 9000(4 \cos \theta - 2)^2$$

The potential energy function for the system is therefore

$$V = V_g + V_e = 58\,860 \sin \theta + 14\,715h + 9000(4 \cos \theta - 2)^2 \quad (1)$$

Equilibrium.   When the system is in equilibrium,

$$\frac{dV}{d\theta} = 58\,860 \cos \theta + 18\,000(4 \cos \theta - 2)(-4 \sin \theta) = 0$$

$$58\,860 \cos \theta - 288\,000 \sin \theta \cos \theta + 144\,000 \sin \theta = 0$$

Since $\sin 2\theta = 2 \sin \theta \cos \theta$,

$$58\,860 \cos \theta - 144\,000 \sin 2\theta + 144\,000 \sin \theta = 0$$

Solving by trial and error,

$$\theta = 28.18° \text{ and } \theta = 45.51° \qquad \textit{Ans.}$$

Stability.   Taking the second derivative of Eq. 1,

$$\frac{d^2V}{d\theta^2} = -58\,860 \sin \theta - 288\,000 \cos 2\theta + 144\,000 \cos \theta$$

Substituting $\theta = 28.18°$ yields

$$\frac{d^2V}{d\theta^2} = -60\,402 < 0 \qquad \text{Unstable} \qquad \textit{Ans.}$$

And for $\theta = 45.51°$,

$$\frac{d^2V}{d\theta^2} = 64\,073 > 0 \qquad \text{Stable} \qquad \textit{Ans.}$$

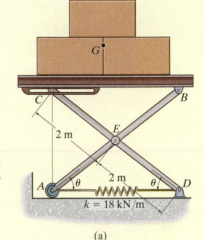

(a)

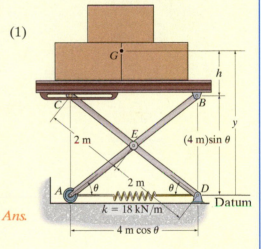

(b)

Fig. 11–18

11

EXAMPLE | 11.7

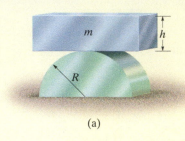

(a)

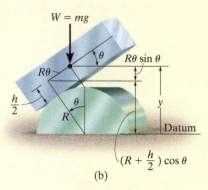

(b)

Fig. 11–19

The uniform block having a mass m rests on the top surface of the half cylinder, Fig. 11–19a. Show that this is a condition of unstable equilibrium if $h > 2R$.

SOLUTION

Potential Function.   The datum is established at the base of the cylinder, Fig. 11–19b. If the block is displaced by an amount θ from the equilibrium position, the potential function is

$$V = V_e + V_g$$

$$= 0 + mgy$$

From Fig. 11–19b,

$$y = \left(R + \frac{h}{2}\right)\cos\theta + R\theta\sin\theta$$

Thus,

$$V = mg\left[\left(R + \frac{h}{2}\right)\cos\theta + R\theta\sin\theta\right]$$

Equilibrium Position.

$$\frac{dV}{d\theta} = mg\left[-\left(R + \frac{h}{2}\right)\sin\theta + R\sin\theta + R\theta\cos\theta\right] = 0$$

$$= mg\left(-\frac{h}{2}\sin\theta + R\theta\cos\theta\right) = 0$$

Note that $\theta = 0°$ satisfies this equation.

Stability.   Taking the second derivative of V yields

$$\frac{d^2V}{d\theta^2} = mg\left(-\frac{h}{2}\cos\theta + R\cos\theta - R\theta\sin\theta\right)$$

At $\theta = 0°$,

$$\left.\frac{d^2V}{d\theta^2}\right|_{\theta=0°} = -mg\left(\frac{h}{2} - R\right)$$

Since all the constants are positive, the block is in unstable equilibrium provided $h > 2R$, because then $d^2V/d\theta^2 < 0$.

11

PROBLEMS

11–26. The potential energy of a one-degree-of-freedom system is defined by $V = (20x^3 - 10x^2 - 25x - 10)$ ft $\cdot$ lb, where x is in ft. Determine the equilibrium positions and investigate the stability for each position.

11–27. If the potential function for a conservative one-degree-of-freedom system is $V = (12 \sin 2\theta + 15 \cos \theta)$ J, where $0° < \theta < 180°$, determine the positions for equilibrium and investigate the stability at each of these positions.

*11–28. If the potential function for a conservative one-degree-of-freedom system is $V = (8x^3 - 2x^2 - 10)$ J, where x is given in meters, determine the positions for equilibrium and investigate the stability at each of these positions.

11–29. If the potential function for a conservative one-degree-of-freedom system is $V = (10 \cos 2\theta + 25 \sin \theta)$ J, where $0° < \theta < 180°$, determine the positions for equilibrium and investigate the stability at each of these positions.

11–30. If the potential energy for a conservative one-degree-of-freedom system is expressed by the relation $V = (4x^3 - x^2 - 3x + 10)$ ft $\cdot$ lb, where x is given in feet, determine the equilibrium positions and investigate the stability at each position.

11–31. The uniform link AB, has a mass of 3 kg and is pin connected at both of its ends. The rod BD, having negligible weight, passes through a swivel block at C. If the spring has a stiffness of $k = 100 \text{N/m}$ and is unstretched when $\theta = 0°$, determine the angle θ for equilibrium and investigate the stability at the equilibrium position. Neglect the size of the swivel block.

*11–32. The spring of the scale has an unstretched length of a. Determine the angle θ for equilibrium when a weight W is supported on the platform. Neglect the weight of the members. What value W would be required to keep the scale in neutral equilibrium when $\theta = 0°$?

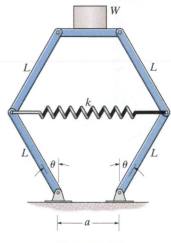

Prob. 11–32

11–33. The uniform bar has a mass of 80 kg. Determine the angle θ for equilibrium and investigate the stability of the bar when it is in this position. The spring has an unstretched length when $\theta = 90°$.

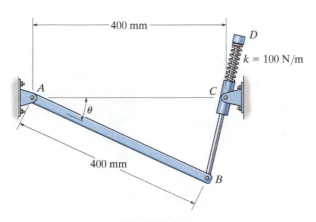

Prob. 11–31

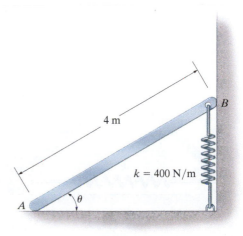

Prob. 11–33

11

11–34. The uniform bar AD has a mass of 20 kg. If the attached spring is unstretched when $\theta = 90°$, determine the angle θ for equilibrium. Note that the spring always remains in the vertical position due to the roller guide. Investigate the stability of the bar when it is in the equilibrium position.

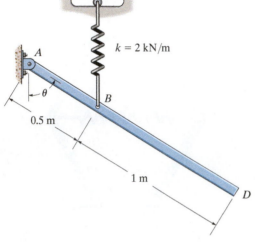

Prob. 11–34

11–35.   The two bars each have a weight of 8 lb. Determine the required stiffness k of the spring so that the two bars are in equilibrium when $\theta = 30°$. The spring has an unstretched length of 1 ft.

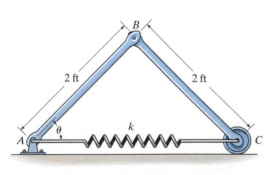

Prob. 11–35

*11–36. Determine the angle θ for equilibrium and investigate the stability at this position. The bars each have a mass of 3 kg and the suspended block D has a mass of 7 kg. Cord DC has a total length of 1 m.

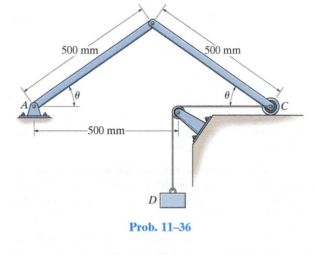

Prob. 11–36

11–37. Determine the angle θ for equilibrium and investigate the stability at this position. The bars each have a mass of 10 kg and the spring has an unstretched length of 100 mm.

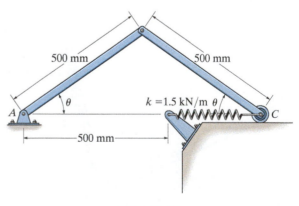

Prob. 11–37

11–38. The two bars each have a mass of 8 kg. Determine the required stiffness k of the spring so that the two bars are in equilibrium when $\theta = 60°$. The spring has an unstretched length of 1 m. Investigate the stability of the system at the equilibrium position.

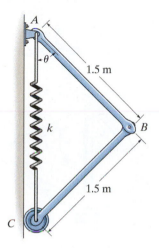

Prob. 11–38

11–39. A spring with a torsional stiffness k is attached to the hinge at B. It is unstretched when the rod assembly is in the vertical position. Determine the weight W of the block that results in neutral equilibrium. *Hint:* Establish the potential energy function for a small angle θ, i.e., approximate $\sin \theta \approx 0$, and $\cos \theta \approx 1 - \theta^2/2$.

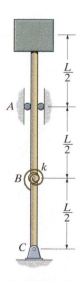

Prob. 11–39

***11–40.** A conical hole is drilled into the bottom of the cylinder, which is supported on the fulcrum at A. Determine the minimum distance d in order for it to remain in stable equilibrium.

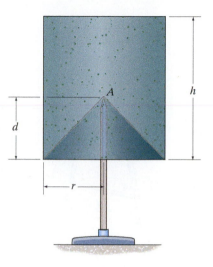

Prob. 11–40

11–41. The uniform rod has a mass of 100 kg. If the spring is unstretched when $\theta = 60°$, determine the angle θ for equilibrium and investigate the stability at the equilibrium position. The spring is always in the horizontal position due to the roller guide at B.

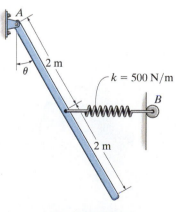

Prob. 11–41

11

11–42. Each bar has a mass per length of m_0. Determine the angles θ and ϕ at which they are suspended in equilibrium. The contact at A is smooth, and both are pin connected at B.

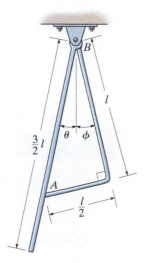

Prob. 11–42

11–43. The truck has a mass of 20 Mg and a mass center at G. Determine the steepest grade θ along which it can park without overturning and investigate the stability in this position.

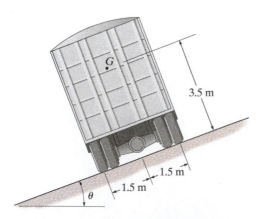

Prob. 11–43

*11–44. The small postal scale consists of a counterweight W_1, connected to the members having negligible weight. Determine the weight W_2 that is on the pan in terms of the angles θ and ϕ and the dimensions shown. All members are pin connected.

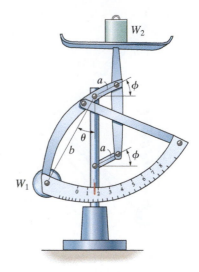

Prob. 11–44

11–45. A 3-lb weight is attached to the end of rod ABC. If the rod is supported by a smooth slider block at C and rod BD, determine the angle θ for equilibrium. Neglect the weight of the rods and the slider.

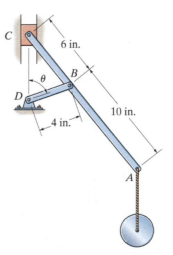

Prob. 11–45

11

11–46. If the uniform rod OA has a mass of 12 kg, determine the mass m that will hold the rod in equilibrium when $\theta = 30°$. Point C is coincident with B when OA is horizontal. Neglect the size of the pulley at B.

*11–48. The bent rod has a weight of 5 lb/ft. A pivot is attached at its center A and the rod is balanced as shown. Determine the length L of its vertical segments so that it remains in neutral equilibrium. Neglect the thickness of the rod.

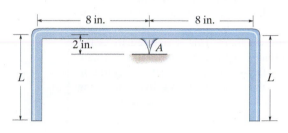

Prob. 11–48

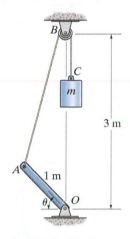

Prob. 11–46

11–47. The cylinder is made of two materials such that it has a mass of m and a center of gravity at point G. Show that when G lies above the centroid C of the cylinder, the equilibrium is unstable.

11–49. The triangular block of weight W rests on the smooth corners which are a distance a apart. If the block has three equal sides of length d, determine the angle θ for equilibrium.

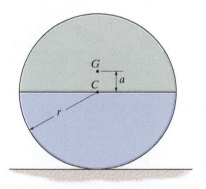

Prob. 11–47

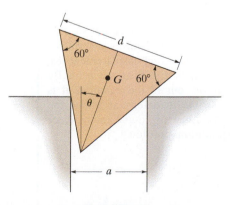

Prob. 11–49

11

CHAPTER REVIEW

Principle of Virtual Work

The forces on a body will do *virtual work* when the body undergoes an *imaginary* differential displacement or rotation.

δy, $\delta y'$ –virtual displacements

$\delta\theta$ –virtual rotation

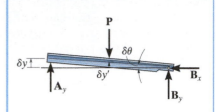

For equilibrium, the sum of the virtual work done by all the forces acting on the body must be equal to zero for any virtual displacement. This is referred to as the *principle of virtual work*, and it is useful for finding the equilibrium configuration for a mechanism or a reactive force acting on a series of connected members.

$$\delta U = 0$$

If the system of connected members has one degree of freedom, then its position can be specified by one independent coordinate, such as θ.

To apply the principle of virtual work, it is first necessary to use *position coordinates* to locate all the forces and moments on the mechanism that will do work when the mechanism undergoes a virtual movement $\delta\theta$.

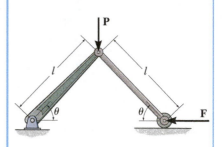

The coordinates are related to the independent coordinate θ and then these expressions are differentiated in order to relate the *virtual* coordinate displacements to the virtual displacement $\delta\theta$.

Finally, the equation of virtual work is written for the mechanism in terms of the common virtual displacement $\delta\theta$, and then it is set equal to zero. By factoring $\delta\theta$ out of the equation, it is then possible to determine either the unknown force or couple moment, or the equilibrium position θ.

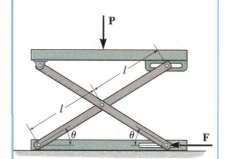

11

Potential-Energy Criterion for Equilibrium

When a system is subjected only to conservative forces, such as weight and spring forces, then the equilibrium configuration can be determined using the *potential-energy function V* for the system.

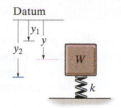

Datum

$$V = V_g + V_e = -Wy + \tfrac{1}{2} ky^2$$

The potential-energy function is established by expressing the weight and spring potential energy for the system in terms of the independent coordinate q.

Once the potential-energy function is formulated, its first derivative is set equal to zero. The solution yields the equilibrium position q_{eq} for the system.

$$\frac{dV}{dq} = 0$$

The stability of the system can be investigated by taking the second derivative of V.

$$\frac{dV}{dq} = 0, \quad \frac{d^2V}{dq^2} > 0 \quad \text{stable equilibrium}$$

$$\frac{dV}{dq} = 0, \quad \frac{d^2V}{dq^2} < 0 \quad \text{unstable equilibrium}$$

$$\frac{dV}{dq} = \frac{d^2V}{dq^2} = \frac{d^3V}{dq^3} = \cdots = 0 \quad \text{neutral equilibrium}$$

11

REVIEW PROBLEMS

R11–1. The toggle joint is subjected to the load P. Determine the compressive force F it creates on the cylinder at A as a function of θ.

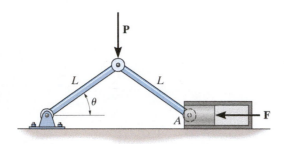

Prob. R11–1

R11–3. The punch press consists of the ram R, connecting rod AB, and a flywheel. If a torque of $M = 50\,\text{N}\cdot\text{m}$ is applied to the flywheel, determine the force F applied at the ram to hold the rod in the position $\theta = 60°$.

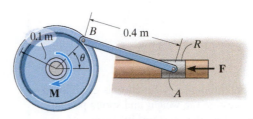

Prob. R11–3

R11–2. The uniform links AB and BC each weigh 2 lb and the cylinder weighs 20 lb. Determine the horizontal force P required to hold the mechanism in the position when $\theta = 45°$. The spring has an unstretched length of 6 in.

R11–4. The uniform bar AB weighs 10 lb. If the attached spring is unstretched when $\theta = 90°$, use the method of virtual work and determine the angle θ for equilibrium. Note that the spring always remains in the vertical position due to the roller guide.

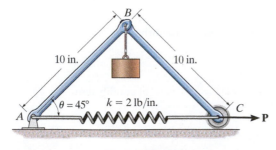

Prob. R11–2

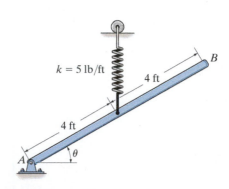

Prob. R11–4

R11–5. The spring has an unstretched length of 0.3 m. Determine the angle θ for equilibrium if the uniform links each have a mass of 5 kg.

R11–7. The uniform bar AB weighs 100 lb. If both springs DE and BC are unstretched when $\theta = 90°$, determine the angle θ for equilibrium using the principle of potential energy. Investigate the stability at the equilibrium position. Both springs always act in the horizontal position because of the roller guides at C and E.

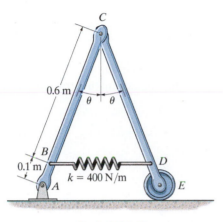

Prob. R11–5

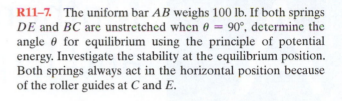
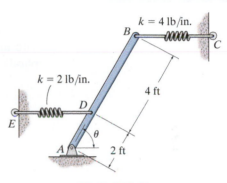

Prob. R11–7

R11–6. Determine the angle θ for equilibrium and investigate the stability of the mechanism in this position. The spring has a stiffness of $k = 1.5 \, \text{kN/m}$ and is unstretched when $\theta = 90°$. The block A has a mass of 40 kg. Neglect the mass of the links.

R11–8. The spring attached to the mechanism has an unstretched length when $\theta = 90°$. Determine the position θ for equilibrium and investigate the stability of the mechanism at this position. Disk A is pin connected to the frame at B and has a weight of 20 lb. Neglect the weight of the bars.

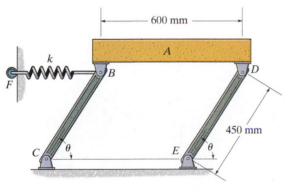

Prob. R11–6

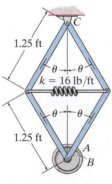

Prob. R11–8

11

Mathematical Review and Expressions

Geometry and Trigonometry Review

The angles θ in Fig. A–1 are equal between the transverse and two parallel lines.

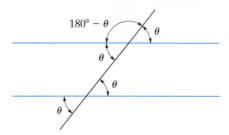

Fig. A–1

For a line and its normal, the angles θ in Fig. A–2 are equal.

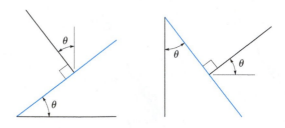

Fig. A–2

For the circle in Fig. A–3, $s = \theta r$, so that when $\theta = 360° = 2\pi$ rad then the circumference is $s = 2\pi r$. Also, since $180° = \pi$ rad, then θ (rad) $= (\pi/180°)\theta°$. The area of the circle is $A = \pi r^2$.

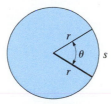

Fig. A–3

The sides of a similar triangle can be obtained by proportion as in Fig. A–4, where $\dfrac{a}{A} = \dfrac{b}{B} = \dfrac{c}{C}$.

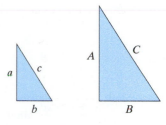

Fig. A–4

For the right triangle in Fig. A–5, the Pythagorean theorem is

$$h = \sqrt{(o)^2 + (a)^2}$$

The trigonometric functions are

$$\sin\theta = \frac{o}{h}$$

$$\cos\theta = \frac{a}{h}$$

$$\tan\theta = \frac{o}{a}$$

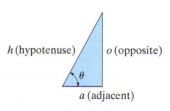

Fig. A–5

This is easily remembered as "soh, cah, toa", i.e., the sine is the opposite over the hypotenuse, etc. The other trigonometric functions follow from this.

$$\csc\theta = \frac{1}{\sin\theta} = \frac{h}{o}$$

$$\sec\theta = \frac{1}{\cos\theta} = \frac{h}{a}$$

$$\cot\theta = \frac{1}{\tan\theta} = \frac{a}{o}$$

A

Trigonometric Identities

$$\sin^2 \theta + \cos^2 \theta = 1$$

$$\sin(\theta \pm \phi) = \sin \theta \cos \phi \pm \cos \theta \sin \phi$$

$$\sin 2\theta = 2 \sin \theta \cos \theta$$

$$\cos(\theta \pm \phi) = \cos \theta \cos \phi \mp \sin \theta \sin \phi$$

$$\cos 2\theta = \cos^2 \theta - \sin^2 \theta$$

$$\cos \theta = \pm\sqrt{\frac{1 + \cos 2\theta}{2}}, \ \sin \theta = \pm\sqrt{\frac{1 - \cos 2\theta}{2}}$$

$$\tan \theta = \frac{\sin \theta}{\cos \theta}$$

$$1 + \tan^2 \theta = \sec^2 \theta \qquad 1 + \cot^2 \theta = \csc^2 \theta$$

Quadratic Formula

If $ax^2 + bx + c = 0$, then $x = \dfrac{-b \pm \sqrt{b^2 - 4ac}}{2a}$

Hyperbolic Functions

$$\sinh x = \frac{e^x - e^{-x}}{2},$$

$$\cosh x = \frac{e^x + e^{-x}}{2},$$

$$\tanh x = \frac{\sinh x}{\cosh x}$$

Power-Series Expansions

$$\sin x = x - \frac{x^3}{3!} + \cdots, \ \cos x = 1 - \frac{x^2}{2!} + \cdots$$

$$\sinh x = x + \frac{x^3}{3!} + \cdots, \ \cosh x = 1 + \frac{x^2}{2!} + \cdots$$

Derivatives

$$\frac{d}{dx}(u^n) = nu^{n-1}\frac{du}{dx} \qquad \frac{d}{dx}(\sin u) = \cos u \frac{du}{dx}$$

$$\frac{d}{dx}(uv) = u\frac{dv}{dx} + v\frac{du}{dx} \qquad \frac{d}{dx}(\cos u) = -\sin u \frac{du}{dx}$$

$$\frac{d}{dx}\left(\frac{u}{v}\right) = \frac{v\dfrac{du}{dx} - u\dfrac{dv}{dx}}{v^2} \qquad \frac{d}{dx}(\tan u) = \sec^2 u \frac{du}{dx}$$

$$\frac{d}{dx}(\cot u) = -\csc^2 u \frac{du}{dx} \qquad \frac{d}{dx}(\sinh u) = \cosh u \frac{du}{dx}$$

$$\frac{d}{dx}(\sec u) = \tan u \sec u \frac{du}{dx} \qquad \frac{d}{dx}(\cosh u) = \sinh u \frac{du}{dx}$$

$$\frac{d}{dx}(\csc u) = -\csc u \cot u \frac{du}{dx}$$

A

Integrals

$$\int x^n\, dx = \frac{x^{n+1}}{n+1} + C, n \neq -1$$

$$\int \frac{dx}{a+bx} = \frac{1}{b}\ln(a+bx) + C$$

$$\int \frac{dx}{a+bx^2} = \frac{1}{2\sqrt{-ab}}\ln\left[\frac{a+x\sqrt{-ab}}{a-x\sqrt{-ab}}\right] + C,$$
$$ab < 0$$

$$\int \frac{x\,dx}{a+bx^2} = \frac{1}{2b}\ln(bx^2+a) + C$$

$$\int \frac{x^2\,dx}{a+bx^2} = \frac{x}{b} - \frac{a}{b\sqrt{ab}}\tan^{-1}\frac{x\sqrt{ab}}{a} + C, ab > 0$$

$$\int \sqrt{a+bx}\, dx = \frac{2}{3b}\sqrt{(a+bx)^3} + C$$

$$\int x\sqrt{a+bx}\, dx = \frac{-2(2a-3bx)\sqrt{(a+bx)^3}}{15b^2} + C$$

$$\int x^2\sqrt{a+bx}\, dx =$$

$$\frac{2(8a^2 - 12abx + 15b^2x^2)\sqrt{(a+bx)^3}}{105b^3} + C$$

$$\int \sqrt{a^2-x^2}\, dx = \frac{1}{2}\left[x\sqrt{a^2-x^2} + a^2\sin^{-1}\frac{x}{a}\right] + C,$$
$$a > 0$$

$$\int x\sqrt{a^2-x^2}\, dx = -\frac{1}{3}\sqrt{(a^2-x^2)^3} + C$$

$$\int x^2\sqrt{a^2-x^2}\, dx = -\frac{x}{4}\sqrt{(a^2-x^2)^3}$$

$$+ \frac{a^2}{8}\left(x\sqrt{a^2-x^2} + a^2\sin^{-1}\frac{x}{a}\right) + C, a > 0$$

$$\int \sqrt{x^2 \pm a^2}\, dx =$$

$$\frac{1}{2}\left[x\sqrt{x^2 \pm a^2} \pm a^2\ln\left(x + \sqrt{x^2 \pm a^2}\right)\right] + C$$

$$\int x\sqrt{x^2 \pm a^2}\, dx = \frac{1}{3}\sqrt{(x^2 \pm a^2)^3} + C$$

$$\int x^2\sqrt{x^2 \pm a^2}\, dx = \frac{x}{4}\sqrt{(x^2 \pm a^2)^3}$$

$$\mp \frac{a^2}{8}x\sqrt{x^2 \pm a^2} - \frac{a^4}{8}\ln\left(x + \sqrt{x^2 \pm a^2}\right) + C$$

$$\int \frac{dx}{\sqrt{a+bx}} = \frac{2\sqrt{a+bx}}{b} + C$$

$$\int \frac{x\,dx}{\sqrt{x^2 \pm a^2}} = \sqrt{x^2 \pm a^2} + C$$

$$\int \frac{dx}{\sqrt{a+bx+cx^2}} = \frac{1}{\sqrt{c}}\ln\left[\sqrt{a+bx+cx^2} + \right.$$

$$\left. x\sqrt{c} + \frac{b}{2\sqrt{c}}\right] + C, c > 0$$

$$= \frac{1}{\sqrt{-c}}\sin^{-1}\left(\frac{-2cx-b}{\sqrt{b^2-4ac}}\right) + C, c < 0$$

$$\int \sin x\, dx = -\cos x + C$$

$$\int \cos x\, dx = \sin x + C$$

$$\int x\cos(ax)\, dx = \frac{1}{a^2}\cos(ax) + \frac{x}{a}\sin(ax) + C$$

$$\int x^2\cos(ax)\, dx = \frac{2x}{a^2}\cos(ax) + \frac{a^2x^2-2}{a^3}\sin(ax) + C$$

$$\int e^{ax}\, dx = \frac{1}{a}e^{ax} + C$$

$$\int xe^{ax}\, dx = \frac{e^{ax}}{a^2}(ax-1) + C$$

$$\int \sinh x\, dx = \cosh x + C$$

$$\int \cosh x\, dx = \sinh x + C$$

Fundamental Problems
Partial Solutions And Answers

Chapter 2

F2–1.

$$F_R = \sqrt{(2 \text{ kN})^2 + (6 \text{ kN})^2 - 2(2 \text{ kN})(6 \text{ kN}) \cos 105°}$$
$$= 6.798 \text{ kN} = 6.80 \text{ kN} \qquad \qquad Ans.$$
$$\frac{\sin \phi}{6 \text{ kN}} = \frac{\sin 105°}{6.798 \text{ kN}}, \quad \phi = 58.49°$$
$$\theta = 45° + \phi = 45° + 58.49° = 103° \qquad Ans.$$

F2–2. $\quad F_R = \sqrt{200^2 + 500^2 - 2(200)(500) \cos 140°}$
$$= 666 \text{ N} \qquad \qquad Ans.$$

F2–3. $\quad F_R = \sqrt{600^2 + 800^2 - 2(600)(800) \cos 60°}$
$$= 721.11 \text{ N} = 721 \text{ N} \qquad \qquad Ans.$$
$$\frac{\sin \alpha}{800} = \frac{\sin 60°}{721.11}; \quad \alpha = 73.90°$$
$$\phi = \alpha - 30° = 73.90° - 30° = 43.9° \qquad Ans.$$

F2–4.
$$\frac{F_u}{\sin 45°} = \frac{30}{\sin 105°}; \quad F_u = 22.0 \text{ lb} \qquad Ans.$$
$$\frac{F_v}{\sin 30°} = \frac{30}{\sin 105°}; \quad F_v = 15.5 \text{ lb} \qquad Ans.$$

F2–5.
$$\frac{F_{AB}}{\sin 105°} = \frac{450}{\sin 30°}$$
$$F_{AB} = 869 \text{ lb} \qquad \qquad Ans.$$
$$\frac{F_{AC}}{\sin 45°} = \frac{450}{\sin 30°}$$
$$F_{AC} = 636 \text{ lb} \qquad \qquad Ans.$$

F2–6.
$$\frac{F}{\sin 30°} = \frac{6}{\sin 105°} \quad F = 3.11 \text{ kN} \qquad Ans.$$
$$\frac{F_v}{\sin 45°} = \frac{6}{\sin 105°} \quad F_v = 4.39 \text{ kN} \qquad Ans.$$

F2–7. $\quad (F_1)_x = 0 \quad (F_1)_y = 300 \text{ N} \qquad \qquad Ans.$
$$(F_2)_x = -(450 \text{ N}) \cos 45° = -318 \text{ N} \qquad Ans.$$
$$(F_2)_y = (450 \text{ N}) \sin 45° = 318 \text{ N} \qquad Ans.$$
$$(F_3)_x = \left(\tfrac{3}{5}\right) 600 \text{ N} = 360 \text{ N} \qquad Ans.$$
$$(F_3)_y = \left(\tfrac{4}{5}\right) 600 \text{ N} = 480 \text{ N} \qquad Ans.$$

F2–8. $\quad F_{Rx} = 300 + 400 \cos 30° - 250\left(\tfrac{4}{5}\right) = 446.4 \text{ N}$
$$F_{Ry} = 400 \sin 30° + 250\left(\tfrac{3}{5}\right) = 350 \text{ N}$$
$$F_R = \sqrt{(446.4)^2 + 350^2} = 567 \text{ N} \qquad Ans.$$
$$\theta = \tan^{-1} \tfrac{350}{446.4} = 38.1° \angle \qquad Ans.$$

F2–9.

$\xrightarrow{+}(F_R)_x = \Sigma F_x;$
$$(F_R)_x = -(700 \text{ lb}) \cos 30° + 0 + \left(\tfrac{3}{5}\right)(600 \text{ lb})$$
$$= -246.22 \text{ lb}$$
$+\uparrow(F_R)_y = \Sigma F_y;$
$$(F_R)_y = -(700 \text{ lb}) \sin 30° - 400 \text{ lb} - \left(\tfrac{4}{5}\right)(600 \text{ lb})$$
$$= -1230 \text{ lb}$$
$$F_R = \sqrt{(246.22 \text{ lb})^2 + (1230 \text{ lb})^2} = 1254 \text{ lb} \qquad Ans.$$
$$\phi = \tan^{-1}\left(\tfrac{1230 \text{ lb}}{246.22 \text{ lb}}\right) = 78.68°$$
$$\theta = 180° + \phi = 180° + 78.68° = 259° \qquad Ans.$$

F2–10. $\xrightarrow{+}(F_R)_x = \Sigma F_x;$
$$750 \text{ N} = F \cos \theta + \left(\tfrac{5}{13}\right)(325 \text{ N}) + (600 \text{ N}) \cos 45°$$
$+\uparrow(F_R)_y = \Sigma F_y;$
$$0 = F \sin \theta + \left(\tfrac{12}{13}\right)(325 \text{ N}) - (600 \text{ N}) \sin 45°$$
$$\tan \theta = 0.6190 \quad \theta = 31.76° = 31.8° \angle \qquad Ans.$$
$$F = 236 \text{ N} \qquad \qquad Ans.$$

F2–11. $\xrightarrow{+}(F_R)_x = \Sigma F_x;$
$$(80 \text{ lb}) \cos 45° = F \cos \theta + 50 \text{ lb} - \left(\tfrac{3}{5}\right)90 \text{ lb}$$
$+\uparrow(F_R)_y = \Sigma F_y;$
$$-(80 \text{ lb}) \sin 45° = F \sin \theta - \left(\tfrac{4}{5}\right)(90 \text{ lb})$$
$$\tan \theta = 0.2547 \quad \theta = 14.29° = 14.3° \angle \qquad Ans.$$
$$F = 62.5 \text{ lb} \qquad \qquad Ans.$$

F2–12. $\quad (F_R)_x = 15\left(\tfrac{4}{5}\right) + 0 + 15\left(\tfrac{4}{5}\right) = 24 \text{ kN} \rightarrow$
$$(F_R)_y = 15\left(\tfrac{3}{5}\right) + 20 - 15\left(\tfrac{3}{5}\right) = 20 \text{ kN} \uparrow$$
$$F_R = 31.2 \text{ kN} \qquad \qquad Ans.$$
$$\theta = 39.8° \qquad \qquad Ans.$$

F2–13. $\quad F_x = 75 \cos 30° \sin 45° = 45.93 \text{ lb}$
$$F_y = 75 \cos 30° \cos 45° = 45.93 \text{ lb}$$
$$F_z = -75 \sin 30° = -37.5 \text{ lb}$$
$$\alpha = \cos^{-1}\left(\tfrac{45.93}{75}\right) = 52.2° \qquad Ans.$$
$$\beta = \cos^{-1}\left(\tfrac{45.93}{75}\right) = 52.2° \qquad Ans.$$
$$\gamma = \cos^{-1}\left(\tfrac{-37.5}{75}\right) = 120° \qquad Ans.$$

F2–14. $\cos \beta = \sqrt{1 - \cos^2 120° - \cos^2 60°} = \pm 0.7071$

Require $\beta = 135°$.

$\mathbf{F} = F\mathbf{u}_F = (500 \text{ N})(-0.5\mathbf{i} - 0.7071\mathbf{j} + 0.5\mathbf{k})$

$= \{-250\mathbf{i} - 354\mathbf{j} + 250\mathbf{k}\} \text{ N}$ *Ans.*

F2–15. $\cos^2\alpha + \cos^2 135° + \cos^2 120° = 1$

$\alpha = 60°$

$\mathbf{F} = F\mathbf{u}_F = (500 \text{ N})(0.5\mathbf{i} - 0.7071\mathbf{j} - 0.5\mathbf{k})$

$= \{250\mathbf{i} - 354\mathbf{j} - 250\mathbf{k}\} \text{ N}$ *Ans.*

F2–16. $F_z = (50 \text{ lb}) \sin 45° = 35.36 \text{ lb}$

$F' = (50 \text{ lb}) \cos 45° = 35.36 \text{ lb}$

$F_x = \left(\frac{3}{5}\right)(35.36 \text{ lb}) = 21.21 \text{ lb}$

$F_y = \left(\frac{4}{5}\right)(35.36 \text{ lb}) = 28.28 \text{ lb}$

$\mathbf{F} = \{-21.2\mathbf{i} + 28.3\mathbf{j} + 35.4\mathbf{k}\} \text{ lb}$ *Ans.*

F2–17. $F_z = (750 \text{ N}) \sin 45° = 530.33 \text{ N}$

$F' = (750 \text{ N}) \cos 45° = 530.33 \text{ N}$

$F_x = (530.33 \text{ N}) \cos 60° = 265.2 \text{ N}$

$F_y = (530.33 \text{ N}) \sin 60° = 459.3 \text{ N}$

$\mathbf{F}_2 = \{265\mathbf{i} - 459\mathbf{j} + 530\mathbf{k}\} \text{ N}$ *Ans.*

F2–18. $\mathbf{F}_1 = \left(\frac{4}{5}\right)(500 \text{ lb}) \mathbf{j} + \left(\frac{3}{5}\right)(500 \text{ lb})\mathbf{k}$

$= \{400\mathbf{j} + 300\mathbf{k}\} \text{ lb}$

$\mathbf{F}_2 = [(800 \text{ lb}) \cos 45°] \cos 30° \mathbf{i}$
$+ [(800 \text{ lb}) \cos 45°] \sin 30°\mathbf{j}$
$+ (800 \text{ lb}) \sin 45° (-\mathbf{k})$

$= \{489.90\mathbf{i} + 282.84\mathbf{j} - 565.69\mathbf{k}\} \text{ lb}$

$\mathbf{F}_R = \mathbf{F}_1 + \mathbf{F}_2 = \{490\mathbf{i} + 683\mathbf{j} - 266\mathbf{k}\} \text{ lb}$ *Ans.*

F2–19. $\mathbf{r}_{AB} = \{-6\mathbf{i} + 6\mathbf{j} + 3\mathbf{k}\} \text{ m}$ *Ans.*

$r_{AB} = \sqrt{(-6 \text{ m})^2 + (6 \text{ m})^2 + (3 \text{ m})^2} = 9 \text{ m}$ *Ans.*

$\alpha = 132°, \quad \beta = 48.2°, \quad \gamma = 70.5°$ *Ans.*

F2–20. $\mathbf{r}_{AB} = \{-4\mathbf{i} + 2\mathbf{j} + 4\mathbf{k}\} \text{ ft}$ *Ans.*

$r_{AB} = \sqrt{(-4 \text{ ft})^2 + (2 \text{ ft})^2 + (4 \text{ ft})^2} = 6 \text{ ft}$ *Ans.*

$\alpha = \cos^{-1}\left(\frac{-4 \text{ ft}}{6 \text{ ft}}\right) = 131.8°$

$\theta = 180° - 131.8° = 48.2°$ *Ans.*

F2–21. $\mathbf{r}_{AB} = \{2\mathbf{i} + 3\mathbf{j} - 6\mathbf{k}\} \text{ m}$

$\mathbf{F}_{AB} = F_{AB}\mathbf{u}_{AB}$

$= (630 \text{ N})\left(\frac{2}{7}\mathbf{i} + \frac{3}{7}\mathbf{j} - \frac{6}{7}\mathbf{k}\right)$

$= \{180\mathbf{i} + 270\mathbf{j} - 540\mathbf{k}\} \text{ N}$ *Ans.*

F2–22. $\mathbf{F} = F\mathbf{u}_{AB} = 900\text{N}\left(-\frac{4}{9}\mathbf{i} + \frac{7}{9}\mathbf{j} - \frac{4}{9}\mathbf{k}\right)$

$= \{-400\mathbf{i} + 700\mathbf{j} - 400\mathbf{k}\} \text{ N}$ *Ans.*

F2–23. $\mathbf{F}_B = F_B\mathbf{u}_B$

$= (840 \text{ N})\left(\frac{3}{7}\mathbf{i} - \frac{2}{7}\mathbf{j} - \frac{6}{7}\mathbf{k}\right)$

$= \{360\mathbf{i} - 240\mathbf{j} - 720\mathbf{k}\} \text{ N}$

$\mathbf{F}_C = F_C\mathbf{u}_C$

$= (420 \text{ N})\left(\frac{2}{7}\mathbf{i} + \frac{3}{7}\mathbf{j} - \frac{6}{7}\mathbf{k}\right)$

$= \{120\mathbf{i} + 180\mathbf{j} - 360\mathbf{k}\} \text{ N}$

$F_R = \sqrt{(480 \text{ N})^2 + (-60 \text{ N})^2 + (-1080 \text{ N})^2}$

$= 1.18 \text{ kN}$ *Ans.*

F2–24. $\mathbf{F}_B = F_B\mathbf{u}_B$

$= (600 \text{ lb})\left(-\frac{1}{3}\mathbf{i} + \frac{2}{3}\mathbf{j} - \frac{2}{3}\mathbf{k}\right)$

$= \{-200\mathbf{i} + 400\mathbf{j} - 400\mathbf{k}\} \text{ lb}$

$\mathbf{F}_C = F_C\mathbf{u}_C$

$= (490 \text{ lb})\left(-\frac{6}{7}\mathbf{i} + \frac{3}{7}\mathbf{j} - \frac{2}{7}\mathbf{k}\right)$

$= \{-420\mathbf{i} + 210\mathbf{j} - 140\mathbf{k}\} \text{ lb}$

$\mathbf{F}_R = \mathbf{F}_B + \mathbf{F}_C = \{-620\mathbf{i} + 610\mathbf{j} - 540\mathbf{k}\} \text{ lb}$ *Ans.*

F2–25. $\mathbf{u}_{AO} = -\frac{1}{3}\mathbf{i} + \frac{2}{3}\mathbf{j} - \frac{2}{3}\mathbf{k}$

$\mathbf{u}_F = -0.5345\mathbf{i} + 0.8018\mathbf{j} + 0.2673\mathbf{k}$

$\theta = \cos^{-1}(\mathbf{u}_{AO} \cdot \mathbf{u}_F) = 57.7°$ *Ans.*

F2–26. $\mathbf{u}_{AB} = -\frac{3}{5}\mathbf{j} + \frac{4}{5}\mathbf{k}$

$\mathbf{u}_F = \frac{4}{5}\mathbf{i} - \frac{3}{5}\mathbf{j}$

$\theta = \cos^{-1}(\mathbf{u}_{AB} \cdot \mathbf{u}_F) = 68.9°$ *Ans.*

F2–27. $\mathbf{u}_{OA} = \frac{12}{13}\mathbf{i} + \frac{5}{13}\mathbf{j}$

$\mathbf{u}_{OA} \cdot \mathbf{j} = u_{OA}(1) \cos \theta$

$\cos \theta = \frac{5}{13}; \quad \theta = 67.4°$ *Ans.*

F2–28. $\mathbf{u}_{OA} = \frac{12}{13}\mathbf{i} + \frac{5}{13}\mathbf{j}$

$\mathbf{F} = F\mathbf{u}_F = [650\mathbf{j}] \text{ N}$

$F_{OA} = \mathbf{F} \cdot \mathbf{u}_{OA} = 250 \text{ N}$

$\mathbf{F}_{OA} = F_{OA} \mathbf{u}_{OA} = \{231\mathbf{i} + 96.2\mathbf{j}\} \text{ N}$ *Ans.*

F2–29.
$$\mathbf{F} = (400\ \text{N})\frac{\{4\mathbf{i} + 1\mathbf{j} - 6\mathbf{k}\}\ \text{m}}{\sqrt{(4\ \text{m})^2 + (1\ \text{m})^2 + (-6\ \text{m})^2}}$$
$$= \{219.78\mathbf{i} + 54.94\mathbf{j} - 329.67\mathbf{k}\}\ \text{N}$$
$$\mathbf{u}_{AO} = \frac{\{-4\mathbf{j} - 6\mathbf{k}\}\ \text{m}}{\sqrt{(-4\ \text{m})^2 + (-6\ \text{m})^2}}$$
$$= -0.5547\mathbf{j} - 0.8321\mathbf{k}$$
$$(F_{AO})_{\text{proj}} = \mathbf{F} \cdot \mathbf{u}_{AO} = 244\ \text{N} \qquad Ans.$$

F2–30.
$$\mathbf{F} = [(-600\ \text{lb})\cos 60°]\sin 30°\ \mathbf{i}$$
$$+ [(600\ \text{lb})\cos 60°]\cos 30°\ \mathbf{j}$$
$$+ [(600\ \text{lb})\sin 60°]\ \mathbf{k}$$
$$= \{-150\mathbf{i} + 259.81\mathbf{j} + 519.62\mathbf{k}\}\ \text{lb}$$
$$\mathbf{u}_A = -\tfrac{2}{3}\mathbf{i} + \tfrac{2}{3}\mathbf{j} + \tfrac{1}{3}\mathbf{k}$$
$$(F_A)_{\text{par}} = \mathbf{F} \cdot \mathbf{u}_A = 446.41\ \text{lb} = 446\ \text{lb} \qquad Ans.$$
$$(F_A)_{\text{per}} = \sqrt{(600\ \text{lb})^2 - (446.41\ \text{lb})^2}$$
$$= 401\ \text{lb} \qquad Ans.$$

F2–31.
$$\mathbf{F} = 56\ \text{N}(\tfrac{3}{7}\mathbf{i} - \tfrac{6}{7}\mathbf{j} + \tfrac{2}{7}\mathbf{k})$$
$$= \{24\mathbf{i} - 48\mathbf{j} + 16\mathbf{k}\}\ \text{N}$$
$$(F_{AO})_\parallel = \mathbf{F} \cdot \mathbf{u}_{AO} = (24\mathbf{i} - 48\mathbf{j} + 16\mathbf{k}) \cdot (\tfrac{3}{7}\mathbf{i} - \tfrac{6}{7}\mathbf{j} - \tfrac{2}{7}\mathbf{k})$$
$$= 46.86\ \text{N} = 46.9\ \text{N} \qquad Ans.$$
$$(F_{AO})_\perp = \sqrt{F^2 - (F_{AO})_\parallel} = \sqrt{(56)^2 - (46.86)^2}$$
$$= 30.7\ \text{N} \qquad Ans.$$

Chapter 3

F3–1.
$$\xrightarrow{+}\Sigma F_x = 0;\ \tfrac{4}{5}F_{AC} - F_{AB}\cos 30° = 0$$
$$+\uparrow\Sigma F_y = 0;\ \tfrac{3}{5}F_{AC} + F_{AB}\sin 30° - 550 = 0$$
$$F_{AB} = 478\ \text{lb} \qquad Ans.$$
$$F_{AC} = 518\ \text{lb} \qquad Ans.$$

F3–2.
$$+\uparrow\Sigma F_y = 0;\ -2(1500)\sin\theta + 700 = 0$$
$$\theta = 13.5°$$
$$L_{ABC} = 2\left(\frac{5\ \text{ft}}{\cos 13.5°}\right) = 10.3\ \text{ft} \qquad Ans.$$

F3–3.
$$\xrightarrow{+}\Sigma F_x = 0;\ T\cos\theta - T\cos\phi = 0$$
$$\phi = \theta$$
$$+\uparrow\Sigma F_y = 0;\ 2T\sin\theta - 49.05\ \text{N} = 0$$
$$\theta = \tan^{-1}\left(\frac{0.15\ \text{m}}{0.2\ \text{m}}\right) = 36.87°$$
$$T = 40.9\ \text{N} \qquad Ans.$$

F3–4.
$$+\nearrow\Sigma F_x = 0;\ \tfrac{4}{5}(F_{sp}) - 5(9.81)\sin 45° = 0$$
$$F_{sp} = 43.35\ \text{N}$$
$$F_{sp} = k(l - l_0);\ 43.35 = 200(0.5 - l_0)$$
$$l_0 = 0.283\ \text{m} \qquad Ans.$$

F3–5.
$$+\uparrow\Sigma F_y = 0;\ (392.4\ \text{N})\sin 30° - m_A(9.81) = 0$$
$$m_A = 20\ \text{kg} \qquad Ans.$$

F3–6.
$$+\uparrow\Sigma F_y = 0;\ T_{AB}\sin 15° - 10(9.81)\ \text{N} = 0$$
$$T_{AB} = 379.03\ \text{N} = 379\ \text{N} \qquad Ans.$$
$$\xrightarrow{+}\Sigma F_x = 0;\ T_{BC} - 379.03\ \text{N}\cos 15° = 0$$
$$T_{BC} = 366.11\ \text{N} = 366\ \text{N} \qquad Ans.$$
$$\xrightarrow{+}\Sigma F_x = 0;\ T_{CD}\cos\theta - 366.11\ \text{N} = 0$$
$$+\uparrow\Sigma F_y = 0;\ T_{CD}\sin\theta - 15(9.81)\ \text{N} = 0$$
$$T_{CD} = 395\ \text{N} \qquad Ans.$$
$$\theta = 21.9° \qquad Ans.$$

F3–7.
$$\Sigma F_x = 0;\ [(\tfrac{3}{5})F_3](\tfrac{3}{5}) + 600\ \text{N} - F_2 = 0 \qquad (1)$$
$$\Sigma F_y = 0;\ (\tfrac{4}{5})F_1 - [(\tfrac{3}{5})F_3](\tfrac{4}{5}) = 0 \qquad (2)$$
$$\Sigma F_z = 0;\ (\tfrac{4}{5})F_3 + (\tfrac{3}{5})F_1 - 900\ \text{N} = 0 \qquad (3)$$
$$F_3 = 776\ \text{N} \qquad Ans.$$
$$F_1 = 466\ \text{N} \qquad Ans.$$
$$F_2 = 879\ \text{N} \qquad Ans.$$

F3–8.
$$\Sigma F_z = 0;\ F_{AD}(\tfrac{4}{5}) - 900 = 0$$
$$F_{AD} = 1125\ \text{N} = 1.125\ \text{kN} \qquad Ans.$$
$$\Sigma F_y = 0;\ F_{AC}(\tfrac{4}{5}) - 1125(\tfrac{3}{5}) = 0$$
$$F_{AC} = 843.75\ \text{N} = 844\ \text{N} \qquad Ans.$$
$$\Sigma F_x = 0;\ F_{AB} - 843.75(\tfrac{3}{5}) = 0$$
$$F_{AB} = 506.25\ \text{N} = 506\ \text{N} \qquad Ans.$$

F3–9.
$$\mathbf{F}_{AD} = F_{AD}\left(\frac{\mathbf{r}_{AD}}{r_{AD}}\right) = \tfrac{1}{3}F_{AD}\mathbf{i} - \tfrac{2}{3}F_{AD}\mathbf{j} + \tfrac{2}{3}F_{AD}\mathbf{k}$$
$$\Sigma F_z = 0;\qquad \tfrac{2}{3}F_{AD} - 600 = 0$$
$$F_{AD} = 900\ \text{N} \qquad Ans.$$
$$\Sigma F_y = 0;\qquad F_{AB}\cos 30° - \tfrac{2}{3}(900) = 0$$
$$F_{AB} = 692.82\ \text{N} = 693\ \text{N} \qquad Ans.$$
$$\Sigma F_x = 0;\qquad \tfrac{1}{3}(900) + 692.82\sin 30° - F_{AC} = 0$$
$$F_{AC} = 646.41\ \text{N} = 646\ \text{N} \qquad Ans.$$

F3–10.
$$\mathbf{F}_{AC} = F_{AC}\{-\cos 60°\sin 30°\ \mathbf{i}$$
$$+ \cos 60°\cos 30°\ \mathbf{j} + \sin 60°\ \mathbf{k}\}$$
$$= -0.25F_{AC}\mathbf{i} + 0.4330F_{AC}\mathbf{j} + 0.8660F_{AC}\mathbf{k}$$
$$\mathbf{F}_{AD} = F_{AD}\{\cos 120°\ \mathbf{i} + \cos 120°\ \mathbf{j} + \cos 45°\ \mathbf{k}\}$$
$$= -0.5F_{AD}\mathbf{i} - 0.5F_{AD}\mathbf{j} + 0.7071F_{AD}\mathbf{k}$$
$$\Sigma F_y = 0;\ 0.4330F_{AC} - 0.5F_{AD} = 0$$
$$\Sigma F_z = 0;\ 0.8660F_{AC} + 0.7071F_{AD} - 300 = 0$$
$$F_{AD} = 175.74\ \text{lb} = 176\ \text{lb} \qquad Ans.$$
$$F_{AC} = 202.92\ \text{lb} = 203\ \text{lb} \qquad Ans.$$
$$\Sigma F_x = 0;\ F_{AB} - 0.25(202.92) - 0.5(175.74) = 0$$
$$F_{AB} = 138.60\ \text{lb} = 139\ \text{lb} \qquad Ans.$$

F3–11. $\mathbf{F}_B = F_B\left(\dfrac{\mathbf{r}_{AB}}{r_{AB}}\right)$

$= F_B\left[\dfrac{\{-6\mathbf{i} + 3\mathbf{j} + 2\mathbf{k}\}\ \text{ft}}{\sqrt{(-6\ \text{ft})^2 + (3\ \text{ft})^2 + (2\ \text{ft})^2}}\right]$

$= -\tfrac{6}{7}F_B\mathbf{i} + \tfrac{3}{7}F_B\mathbf{j} + \tfrac{2}{7}F_B\mathbf{k}$

$\mathbf{F}_C = F_C\left(\dfrac{\mathbf{r}_{AC}}{r_{AC}}\right)$

$= F_C\left[\dfrac{\{-6\mathbf{i} - 2\mathbf{j} + 3\mathbf{k}\}\ \text{ft}}{\sqrt{(-6\ \text{ft})^2 + (-2\ \text{ft})^2 + (3\text{ft})^2}}\right]$

$= -\tfrac{6}{7}F_C\mathbf{i} - \tfrac{2}{7}F_C\mathbf{j} + \tfrac{3}{7}F_C\mathbf{k}$

$\mathbf{F}_D = F_D\mathbf{i}$

$\mathbf{W} = \{-150\mathbf{k}\}\ \text{lb}$

$\Sigma F_x = 0;\quad -\tfrac{6}{7}F_B - \tfrac{6}{7}F_C + F_D = 0$ (1)

$\Sigma F_y = 0;\quad \tfrac{3}{7}F_B - \tfrac{2}{7}F_C = 0$ (2)

$\Sigma F_z = 0;\quad \tfrac{2}{7}F_B + \tfrac{3}{7}F_C - 150 = 0$ (3)

$F_B = 162\ \text{lb}$ *Ans.*

$F_C = 1.5(162\ \text{lb}) = 242\ \text{lb}$ *Ans.*

$F_D = 346.15\ \text{lb} = 346\ \text{lb}$ *Ans.*

Chapter 4

F4–1. $\zeta + M_O = -\left(\tfrac{4}{5}\right)(100\ \text{N})(2\ \text{m}) - \left(\tfrac{3}{5}\right)(100\ \text{N})(5\ \text{m})$

$= -460\ \text{N}\cdot\text{m} = 460\ \text{N}\cdot\text{m}\ \zeta$ *Ans.*

F4–2. $\zeta + M_O = [(300\ \text{N})\sin 30°][0.4\ \text{m} + (0.3\ \text{m})\cos 45°]$

$- [(300\ \text{N})\cos 30°][(0.3\ \text{m})\sin 45°]$

$= 36.7\ \text{N}\cdot\text{m}$ *Ans.*

F4–3. $\zeta + M_O = (600\ \text{lb})(4\ \text{ft} + (3\ \text{ft})\cos 45° - 1\ \text{ft})$

$= 3.07\ \text{kip}\cdot\text{ft}$ *Ans.*

F4–4. $\zeta + M_O = 50\sin 60°\,(0.1 + 0.2\cos 45° + 0.1)$

$- 50\cos 60°(0.2\sin 45°)$

$= 11.2\ \text{N}\cdot\text{m}$ *Ans.*

F4–5. $\zeta + M_O = 600\sin 50°\,(5) + 600\cos 50°\,(0.5)$

$= 2.49\ \text{kip}\cdot\text{ft}$ *Ans.*

F4–6. $\zeta + M_O = 500\sin 45°\,(3 + 3\cos 45°)$

$- 500\cos 45°\,(3\sin 45°)$

$= 1.06\ \text{kN}\cdot\text{m}$ *Ans.*

F4–7. $\zeta + (M_R)_O = \Sigma Fd;$

$(M_R)_O = -(600\ \text{N})(1\ \text{m})$

$+ (500\ \text{N})[3\ \text{m} + (2.5\ \text{m})\cos 45°]$

$- (300\ \text{N})[(2.5\ \text{m})\sin 45°]$

$= 1254\ \text{N}\cdot\text{m} = 1.25\ \text{kN}\cdot\text{m}$ *Ans.*

F4–8. $\zeta + (M_R)_O = \Sigma Fd;$

$(M_R)_O = \left[\left(\tfrac{3}{5}\right)500\ \text{N}\right](0.425\ \text{m})$

$- \left[\left(\tfrac{4}{5}\right)500\ \text{N}\right](0.25\ \text{m})$

$- [(600\ \text{N})\cos 60°](0.25\ \text{m})$

$- [(600\ \text{N})\sin 60°](0.425\ \text{m})$

$= -268\ \text{N}\cdot\text{m} = 268\ \text{N}\cdot\text{m}\ \zeta$ *Ans.*

F4–9. $\zeta + (M_R)_O = \Sigma Fd;$

$(M_R)_O = (300\cos 30°\ \text{lb})(6\ \text{ft} + 6\sin 30°\ \text{ft})$

$- (300\sin 30°\ \text{lb})(6\cos 30°\ \text{ft})$

$+ (200\ \text{lb})(6\cos 30°\ \text{ft})$

$= 2.60\ \text{kip}\cdot\text{ft}$ *Ans.*

F4–10. $\mathbf{F} = F\mathbf{u}_{AB} = 500\ \text{N}\left(\tfrac{4}{5}\mathbf{i} - \tfrac{3}{5}\mathbf{j}\right) = \{400\mathbf{i} - 300\mathbf{j}\}\ \text{N}$

$\mathbf{M}_O = \mathbf{r}_{OA} \times \mathbf{F} = \{3\mathbf{j}\}\ \text{m} \times \{400\mathbf{i} - 300\mathbf{j}\}\ \text{N}$

$= \{-1200\mathbf{k}\}\ \text{N}\cdot\text{m}$ *Ans.*

or

$\mathbf{M}_O = \mathbf{r}_{OB} \times \mathbf{F} = \{4\mathbf{i}\}\ \text{m} \times \{400\mathbf{i} - 300\mathbf{j}\}\ \text{N}$

$= \{-1200\mathbf{k}\}\ \text{N}\cdot\text{m}$ *Ans.*

F4–11. $\mathbf{F} = F\mathbf{u}_{BC}$

$= 120\ \text{lb}\left[\dfrac{\{4\mathbf{i} - 4\mathbf{j} - 2\mathbf{k}\}\ \text{ft}}{\sqrt{(4\ \text{ft})^2 + (-4\ \text{ft})^2 + (-2\ \text{ft})^2}}\right]$

$= \{80\mathbf{i} - 80\mathbf{j} - 40\mathbf{k}\}\ \text{lb}$

$\mathbf{M}_O = \mathbf{r}_C \times \mathbf{F} = \begin{vmatrix} \mathbf{i} & \mathbf{j} & \mathbf{k} \\ 5 & 0 & 0 \\ 80 & -80 & -40 \end{vmatrix}$

$= \{200\mathbf{j} - 400\mathbf{k}\}\ \text{lb}\cdot\text{ft}$ *Ans.*

or

$\mathbf{M}_O = \mathbf{r}_B \times \mathbf{F} = \begin{vmatrix} \mathbf{i} & \mathbf{j} & \mathbf{k} \\ 1 & 4 & 2 \\ 80 & -80 & -40 \end{vmatrix}$

$= \{200\mathbf{j} - 400\mathbf{k}\}\ \text{lb}\cdot\text{ft}$ *Ans.*

F4–12. $\mathbf{F}_R = \mathbf{F}_1 + \mathbf{F}_2$

$= \{(100 - 200)\mathbf{i} + (-120 + 250)\mathbf{j}$

$+ (75 + 100)\mathbf{k}\}\ \text{lb}$

$= \{-100\mathbf{i} + 130\mathbf{j} + 175\mathbf{k}\}\ \text{lb}$

$(\mathbf{M}_R)_O = \mathbf{r}_A \times \mathbf{F}_R = \begin{vmatrix} \mathbf{i} & \mathbf{j} & \mathbf{k} \\ 4 & 5 & 3 \\ -100 & 130 & 175 \end{vmatrix}$

$= \{485\mathbf{i} - 1000\mathbf{j} + 1020\mathbf{k}\}\ \text{lb}\cdot\text{ft}$ *Ans.*

F4–13. $M_x = \mathbf{i} \cdot (\mathbf{r}_{OB} \times \mathbf{F}) = \begin{vmatrix} 1 & 0 & 0 \\ 0.3 & 0.4 & -0.2 \\ 300 & -200 & 150 \end{vmatrix}$

$= 20 \text{ N} \cdot \text{m}$ *Ans.*

F4–14. $\mathbf{u}_{OA} = \dfrac{\mathbf{r}_A}{r_A} = \dfrac{\{0.3\mathbf{i} + 0.4\mathbf{j}\} \text{ m}}{\sqrt{(0.3 \text{ m})^2 + (0.4 \text{ m})^2}} = 0.6\,\mathbf{i} + 0.8\,\mathbf{j}$

$M_{OA} = \mathbf{u}_{OA} \cdot (\mathbf{r}_{AB} \times \mathbf{F}) = \begin{vmatrix} 0.6 & 0.8 & 0 \\ 0 & 0 & -0.2 \\ 300 & -200 & 150 \end{vmatrix}$

$= -72 \text{ N} \cdot \text{m}$ *Ans.*

$\left| M_{OA} \right| = 72 \text{ N} \cdot \text{m}$

F4–15. Scalar Analysis

The magnitudes of the force components are

$F_x = \left| 200 \cos 120° \right| = 100 \text{ N}$

$F_y = 200 \cos 60° = 100 \text{ N}$

$F_z = 200 \cos 45° = 141.42 \text{ N}$

$M_x = -F_y(z) + F_z(y)$

$\quad = -(100 \text{ N})(0.25 \text{ m}) + (141.42 \text{ N})(0.3 \text{ m})$

$\quad = 17.4 \text{ N} \cdot \text{m}$ *Ans.*

Vector Analysis

$M_x = \begin{vmatrix} 1 & 0 & 0 \\ 0 & 0.3 & 0.25 \\ -100 & 100 & 141.42 \end{vmatrix} = 17.4 \text{ N} \cdot \text{m}$ *Ans.*

F4–16. $M_y = \mathbf{j} \cdot (\mathbf{r}_A \times \mathbf{F}) = \begin{vmatrix} 0 & 1 & 0 \\ -3 & -4 & 2 \\ 30 & -20 & 50 \end{vmatrix}$

$= 210 \text{ N} \cdot \text{m}$ *Ans.*

F4–17. $\mathbf{u}_{AB} = \dfrac{\mathbf{r}_{AB}}{r_{AB}} = \dfrac{\{-4\mathbf{i} + 3\mathbf{j}\} \text{ ft}}{\sqrt{(-4 \text{ ft})^2 + (3 \text{ ft})^2}} = -0.8\mathbf{i} + 0.6\mathbf{j}$

$M_{AB} = \mathbf{u}_{AB} \cdot (\mathbf{r}_{AC} \times \mathbf{F})$

$= \begin{vmatrix} -0.8 & 0.6 & 0 \\ 0 & 0 & 2 \\ 50 & -40 & 20 \end{vmatrix} = -4 \text{ lb} \cdot \text{ft}$

$\mathbf{M}_{AB} = M_{AB}\mathbf{u}_{AB} = \{3.2\mathbf{i} - 2.4\mathbf{j}\} \text{ lb} \cdot \text{ft}$ *Ans.*

F4–18. Scalar Analysis

The magnitudes of the force components are

$F_x = \left(\tfrac{3}{5}\right)\left[\tfrac{4}{5}(500)\right] = 240 \text{ N}$

$F_y = \tfrac{4}{5}\left[\tfrac{4}{5}(500)\right] = 320 \text{ N}$

$F_z = \tfrac{3}{5}(500) = 300 \text{ N}$

$M_x = -320(3) + 300(2) = -360 \text{ N} \cdot \text{m}$ *Ans.*

$M_y = -240(3) - 300(-2) = -120 \text{ N} \cdot \text{m}$ *Ans.*

$M_z = 240(2) - 320(2) = -160 \text{ N} \cdot \text{m}$ *Ans.*

Vector Analysis

$\mathbf{F} = \{-240\mathbf{i} + 320\mathbf{j} + 300\mathbf{k}\} \text{ N}$

$\mathbf{r}_{OA} = \{-2\mathbf{i} + 2\mathbf{j} + 3\mathbf{k}\} \text{ m}$

$M_x = \mathbf{i} \cdot (\mathbf{r}_{OA} \times \mathbf{F}) = -360 \text{ N} \cdot \text{m}$

$M_y = \mathbf{j} \cdot (\mathbf{r}_{OA} \times \mathbf{F}) = -120 \text{ N} \cdot \text{m}$

$M_z = \mathbf{k} \cdot (\mathbf{r}_{OA} \times \mathbf{F}) = -160 \text{ N} \cdot \text{m}$

F4–19. $\zeta + M_{C_R} = \Sigma M_A = 400(3) - 400(5) + 300(5)$

$\quad + 200(0.2) = 740 \text{ N} \cdot \text{m}$ *Ans.*

Also,

$\zeta + M_{C_R} = 300(5) - 400(2) + 200(0.2)$

$\quad = 740 \text{ N} \cdot \text{m}$ *Ans.*

F4–20. $\zeta + M_{C_R} = 300(4) + 200(4) + 150(4)$

$\quad = 2600 \text{ lb} \cdot \text{ft}$ *Ans.*

F4–21. $\zeta + (M_B)_R = \Sigma M_B$

$-1.5 \text{ kN} \cdot \text{m} = (2 \text{ kN})(0.3 \text{ m}) - F(0.9 \text{ m})$

$F = 2.33 \text{ kN}$ *Ans.*

F4–22. $\zeta + M_C = 10\left(\tfrac{3}{5}\right)(2) - 10\left(\tfrac{4}{5}\right)(4) = -20 \text{ kN} \cdot \text{m}$

$= 20 \text{ kN} \cdot \text{m} \curvearrowright$ *Ans.*

F4–23. $\mathbf{u}_1 = \dfrac{\mathbf{r}_1}{r_1} = \dfrac{\{-2\mathbf{i} + 2\mathbf{j} + 3.5\mathbf{k}\} \text{ ft}}{\sqrt{(-2 \text{ ft})^2 + (2 \text{ ft})^2 + (3.5 \text{ ft})^2}}$

$= -\tfrac{2}{4.5}\mathbf{i} + \tfrac{2}{4.5}\mathbf{j} + \tfrac{3.5}{4.5}\mathbf{k}$

$\mathbf{u}_2 = -\mathbf{k}$

$\mathbf{u}_3 = \tfrac{1.5}{2.5}\mathbf{i} - \tfrac{2}{2.5}\mathbf{j}$

$(\mathbf{M}_c)_1 = (M_c)_1 \mathbf{u}_1$

$\quad = (450 \text{ lb} \cdot \text{ft})\left(-\tfrac{2}{4.5}\mathbf{i} + \tfrac{2}{4.5}\mathbf{j} + \tfrac{3.5}{4.5}\mathbf{k}\right)$

$\quad = \{-200\mathbf{i} + 200\mathbf{j} + 350\mathbf{k}\} \text{ lb} \cdot \text{ft}$

$(\mathbf{M}_c)_2 = (M_c)_2 \mathbf{u}_2 = (250 \text{ lb} \cdot \text{ft})(-\mathbf{k})$

$\quad = \{-250\mathbf{k}\} \text{ lb} \cdot \text{ft}$

$(\mathbf{M}_c)_3 = (M_c)_3 \mathbf{u}_3 = (300 \text{ lb} \cdot \text{ft})\left(\tfrac{1.5}{2.5}\mathbf{i} - \tfrac{2}{2.5}\mathbf{j}\right)$

$\quad = \{180\mathbf{i} - 240\mathbf{j}\} \text{ lb} \cdot \text{ft}$

$(\mathbf{M}_c)_R = \Sigma \mathbf{M}_c;$

$(\mathbf{M}_c)_R = \{-20\mathbf{i} - 40\mathbf{j} + 100\mathbf{k}\} \text{ lb} \cdot \text{ft}$ *Ans.*

F4–24. $\mathbf{F}_B = \left(\frac{4}{5}\right)(450 \text{ N})\mathbf{j} - \left(\frac{3}{5}\right)(450 \text{ N})\mathbf{k}$

$= \{360\mathbf{j} - 270\mathbf{k}\} \text{ N}$

$$\mathbf{M}_c = \mathbf{r}_{AB} \times \mathbf{F}_B = \begin{vmatrix} \mathbf{i} & \mathbf{j} & \mathbf{k} \\ 0.4 & 0 & 0 \\ 0 & 360 & -270 \end{vmatrix}$$

$= \{108\mathbf{j} + 144\mathbf{k}\} \text{ N} \cdot \text{m}$ *Ans.*

Also,

$\mathbf{M}_c = (\mathbf{r}_A \times \mathbf{F}_A) + (\mathbf{r}_B \times \mathbf{F}_B)$

$$= \begin{vmatrix} \mathbf{i} & \mathbf{j} & \mathbf{k} \\ 0 & 0 & 0.3 \\ 0 & -360 & 270 \end{vmatrix} + \begin{vmatrix} \mathbf{i} & \mathbf{j} & \mathbf{k} \\ 0.4 & 0 & 0.3 \\ 0 & 360 & -270 \end{vmatrix}$$

$= \{108\mathbf{j} + 144\mathbf{k}\} \text{ N} \cdot \text{m}$ *Ans.*

F4–25. $\overset{+}{\leftarrow} F_{Rx} = \Sigma F_x; \; F_{Rx} = 200 - \frac{3}{5}(100) = 140 \text{ lb}$

$+\downarrow F_{Ry} = \Sigma F_y; \; F_{Ry} = 150 - \frac{4}{5}(100) = 70 \text{ lb}$

$F_R = \sqrt{140^2 + 70^2} = 157 \text{ lb}$ *Ans.*

$\theta = \tan^{-1}\left(\frac{70}{140}\right) = 26.6°$ *Ans.*

$\circlearrowleft + M_{A_R} = \Sigma M_A;$

$M_{A_R} = \frac{3}{5}(100)(4) - \frac{4}{5}(100)(6) + 150(3)$

$M_{R_A} = 210 \text{ lb} \cdot \text{ft}$ *Ans.*

F4–26. $\overset{+}{\rightarrow} F_{Rx} = \Sigma F_x; \quad F_{Rx} = \frac{4}{5}(50) = 40 \text{ N}$

$+\downarrow F_{Ry} = \Sigma F_y; \quad F_{Ry} = 40 + 30 + \frac{3}{5}(50)$

$= 100 \text{ N}$

$F_R = \sqrt{(40)^2 + (100)^2} = 108 \text{ N}$ *Ans.*

$\theta = \tan^{-1}\left(\frac{100}{40}\right) = 68.2°$ *Ans.*

$\circlearrowleft + M_{A_R} = \Sigma M_A;$

$M_{A_R} = 30(3) + \frac{3}{5}(50)(6) + 200$

$= 470 \text{ N} \cdot \text{m}$ *Ans.*

F4–27. $\overset{+}{\rightarrow} (F_R)_x = \Sigma F_x;$

$(F_R)_x = 900 \sin 30° = 450 \text{ N} \rightarrow$

$+\uparrow (F_R)_y = \Sigma F_y;$

$(F_R)_y = -900 \cos 30° - 300$

$= -1079.42 \text{ N} = 1079.42 \text{ N} \downarrow$

$F_R = \sqrt{450^2 + 1079.42^2}$

$= 1169.47 \text{ N} = 1.17 \text{ kN}$ *Ans.*

$\theta = \tan^{-1}\left(\frac{1079.42}{450}\right) = 67.4°$ *Ans.*

$\circlearrowleft + (M_R)_A = \Sigma M_A;$

$(M_R)_A = 300 - 900 \cos 30° (0.75) - 300(2.25)$

$= -959.57 \text{ N} \cdot \text{m}$

$= 960 \text{ N} \cdot \text{m}$ *Ans.*

F4–28. $\overset{+}{\rightarrow} (F_R)_x = \Sigma F_x;$

$(F_R)_x = 150\left(\frac{3}{5}\right) + 50 - 100\left(\frac{4}{5}\right) = 60 \text{ lb} \rightarrow$

$+\uparrow (F_R)_y = \Sigma F_y;$

$(F_R)_y = -150\left(\frac{4}{5}\right) - 100\left(\frac{3}{5}\right)$

$= -180 \text{ lb} = 180 \text{ lb} \downarrow$

$F_R = \sqrt{60^2 + 180^2} = 189.74 \text{ lb} = 190 \text{ lb}$ *Ans.*

$\theta = \tan^{-1}\left(\frac{180}{60}\right) = 71.6°$ *Ans.*

$\circlearrowleft + (M_R)_A = \Sigma M_A;$

$(M_R)_A = 100\left(\frac{4}{5}\right)(1) - 100\left(\frac{3}{5}\right)(6) - 150\left(\frac{4}{5}\right)(3)$

$= -640 = 640 \text{ lb} \cdot \text{ft}$ *Ans.*

F4–29. $\mathbf{F}_R = \Sigma \mathbf{F};$

$\mathbf{F}_R = \mathbf{F}_1 + \mathbf{F}_2$

$= (-300\mathbf{i} + 150\mathbf{j} + 200\mathbf{k}) + (-450\mathbf{k})$

$= \{-300\mathbf{i} + 150\mathbf{j} - 250\mathbf{k}\} \text{ N}$ *Ans.*

$\mathbf{r}_{OA} = (2 - 0)\mathbf{j} = \{2\mathbf{j}\} \text{ m}$

$\mathbf{r}_{OB} = (-1.5 - 0)\mathbf{i} + (2 - 0)\mathbf{j} + (1 - 0)\mathbf{k}$

$= \{-1.5\mathbf{i} + 2\mathbf{j} + 1\mathbf{k}\} \text{ m}$

$(\mathbf{M}_R)_O = \Sigma \mathbf{M};$

$(\mathbf{M}_R)_O = \mathbf{r}_{OB} \times \mathbf{F}_1 + \mathbf{r}_{OA} \times \mathbf{F}_2$

$$= \begin{vmatrix} \mathbf{i} & \mathbf{j} & \mathbf{k} \\ -1.5 & 2 & 1 \\ -300 & 150 & 200 \end{vmatrix} + \begin{vmatrix} \mathbf{i} & \mathbf{j} & \mathbf{k} \\ 0 & 2 & 0 \\ 0 & 0 & -450 \end{vmatrix}$$

$= \{-650\mathbf{i} + 375\mathbf{k}\} \text{ N} \cdot \text{m}$ *Ans.*

F4–30. $\mathbf{F}_1 = \{-100\mathbf{j}\} \text{ N}$

$\mathbf{F}_2 = (200 \text{ N})\left[\dfrac{\{-0.4\mathbf{i} - 0.3\mathbf{k}\} \text{ m}}{\sqrt{(-0.4 \text{ m})^2 + (-0.3 \text{ m})^2}}\right]$

$= \{-160\mathbf{i} - 120\mathbf{k}\} \text{ N}$

$\mathbf{M}_c = \{-75\mathbf{i}\} \text{ N} \cdot \text{m}$

$\mathbf{F}_R = \{-160\mathbf{i} - 100\mathbf{j} - 120\mathbf{k}\} \text{ N}$ *Ans.*

$(\mathbf{M}_R)_O = (0.3\mathbf{k}) \times (-100\mathbf{j})$

$$+ \begin{vmatrix} \mathbf{i} & \mathbf{j} & \mathbf{k} \\ 0 & 0.5 & 0.3 \\ -160 & 0 & -120 \end{vmatrix} + (-75\mathbf{i})$$

$= \{-105\mathbf{i} - 48\mathbf{j} + 80\mathbf{k}\} \text{ N} \cdot \text{m}$ *Ans.*

F4–31. $+\downarrow F_R = \Sigma F_y; \quad F_R = 500 + 250 + 500$

$= 1250 \text{ lb}$ *Ans.*

$\circlearrowleft + F_R x = \Sigma M_O;$

$1250(x) = 500(3) + 250(6) + 500(9)$

$x = 6 \text{ ft}$ *Ans.*

F4–32. $\xrightarrow{+} (F_R)_x = \Sigma F_x;$

$(F_R)_x = 100\left(\frac{3}{5}\right) + 50 \sin 30° = 85 \text{ lb} \rightarrow$

$+\uparrow(F_R)_y = \Sigma F_y;$

$(F_R)_y = 200 + 50 \cos 30° - 100\left(\frac{4}{5}\right)$

$= 163.30 \text{ lb}\uparrow$

$F_R = \sqrt{85^2 + 163.30^2} = 184 \text{ lb}$

$\theta = \tan^{-1}\left(\frac{163.30}{85}\right) = 62.5° \; \measuredangle$ *Ans.*

$\zeta +(M_R)_A = \Sigma M_A;$

$163.30(d) = 200(3) - 100\left(\frac{4}{5}\right)(6) + 50 \cos 30°(9)$

$d = 3.12 \text{ ft}$ *Ans.*

F4–33. $\xrightarrow{+} (F_R)_x = \Sigma F_x;$

$(F_R)_x = 15\left(\frac{4}{5}\right) = 12 \text{ kN} \rightarrow$

$+\uparrow(F_R)_y = \Sigma F_y;$

$(F_R)_y = -20 + 15\left(\frac{3}{5}\right) = -11 \text{ kN} = 11 \text{ kN}\downarrow$

$F_R = \sqrt{12^2 + 11^2} = 16.3 \text{ kN}$ *Ans.*

$\theta = \tan^{-1}\left(\frac{11}{12}\right) = 42.5° \; \measuredangle$ *Ans.*

$\zeta +(M_R)_A = \Sigma M_A;$

$-11(d) = -20(2) - 15\left(\frac{4}{5}\right)(2) + 15\left(\frac{3}{5}\right)(6)$

$d = 0.909 \text{ m}$ *Ans.*

F4–34. $\xrightarrow{+}(F_R)_x = \Sigma F_x;$

$(F_R)_x = \left(\frac{3}{5}\right) 5 \text{ kN} - 8 \text{ kN}$

$= -5 \text{ kN} = 5 \text{ kN} \leftarrow$

$+\uparrow(F_R)_y = \Sigma F_y;$

$(F_R)_y = -6 \text{ kN} - \left(\frac{4}{5}\right) 5 \text{ kN}$

$= -10 \text{ kN} = 10 \text{ kN}\downarrow$

$F_R = \sqrt{5^2 + 10^2} = 11.2 \text{ kN}$ *Ans.*

$\theta = \tan^{-1}\left(\frac{10 \text{ kN}}{5 \text{ kN}}\right) = 63.4° \; \measuredangle$ *Ans.*

$\zeta +(M_R)_A = \Sigma M_A;$

$5 \text{ kN}(d) = 8 \text{ kN}(3 \text{ m}) - 6 \text{ kN}(0.5 \text{ m})$

$- \left[\left(\frac{4}{5}\right)5 \text{ kN}\right](2 \text{ m})$

$- \left[\left(\frac{3}{5}\right)5 \text{ kN}\right](4 \text{ m})$

$d = 0.2 \text{ m}$ *Ans.*

F4–35. $+\downarrow F_R = \Sigma F_z;$ $F_R = 400 + 500 - 100$

$= 800 \text{ N}$ *Ans.*

$M_{Rx} = \Sigma M_x;\; -800y = -400(4) - 500(4)$

$y = 4.50 \text{ m}$ *Ans.*

$M_{Ry} = \Sigma M_y;\;\; 800x = 500(4) - 100(3)$

$x = 2.125 \text{ m}$ *Ans.*

F4–36. $+\downarrow F_R = \Sigma F_z;$

$F_R = 200 + 200 + 100 + 100$

$= 600 \text{ N}$ *Ans.*

$\zeta +M_{Rx} = \Sigma M_x;$

$-600y = 200(1) + 200(1) + 100(3) - 100(3)$

$y = -0.667 \text{ m}$ *Ans.*

$\zeta +M_{Ry} = \Sigma M_y;$

$600x = 100(3) + 100(3) + 200(2) - 200(3)$

$x = 0.667 \text{ m}$ *Ans.*

F4–37. $+\uparrow F_R = \Sigma F_y;$

$-F_R = -6(1.5) - 9(3) - 3(1.5)$

$F_R = 40.5 \text{ kN}\downarrow$ *Ans.*

$\zeta +(M_R)_A = \Sigma M_A;$

$-40.5(d) = 6(1.5)(0.75)$

$- 9(3)(1.5) - 3(1.5)(3.75)$

$d = 1.25 \text{ m}$ *Ans.*

F4–38. $F_R = \frac{1}{2}(6)(150) + 8(150) = 1650 \text{ lb}$ *Ans.*

$\zeta +M_{A_R} = \Sigma M_A;$

$1650d = \left[\frac{1}{2}(6)(150)\right](4) + [8(150)](10)$

$d = 8.36 \text{ ft}$ *Ans.*

F4–39. $+\uparrow F_R = \Sigma F_y;$

$-F_R = -\frac{1}{2}(6)(3) - \frac{1}{2}(6)(6)$

$F_R = 27 \text{ kN}\downarrow$ *Ans.*

$\zeta +(M_R)_A = \Sigma M_A;$

$-27(d) = \frac{1}{2}(6)(3)(1) - \frac{1}{2}(6)(6)(2)$

$d = 1 \text{ m}$ *Ans.*

F4–40. $+\downarrow F_R = \Sigma F_y;$

$F_R = \frac{1}{2}(50)(6) + 150(6) + 500$

$= 1550 \text{ lb}$ *Ans.*

$\zeta +M_{A_R} = \Sigma M_A;$

$1550d = \left[\frac{1}{2}(50)(6)\right](4) + [150(6)](3) + 500(9)$

$d = 5.03 \text{ ft}$ *Ans.*

F4–41. $+\uparrow F_R = \Sigma F_y;$

$-F_R = -\frac{1}{2}(3)(4.5) - 3(6)$

$F_R = 24.75 \text{ kN}\downarrow$ *Ans.*

$\zeta +(M_R)_A = \Sigma M_A;$

$-24.75(d) = -\frac{1}{2}(3)(4.5)(1.5) - 3(6)(3)$

$d = 2.59 \text{ m}$ *Ans.*

F4–42. $F_R = \int w(x)\,dx = \int_0^4 2.5x^3\,dx = 160 \text{ N}$

$\zeta + M_{A_R} = \Sigma M_A;$

$$\bar{x} = \frac{\int x w(x)\,dx}{\int w(x)\,dx} = \frac{\int_0^4 2.5x^4\,dx}{160} = 3.20 \text{ m } \textit{Ans.}$$

Chapter 5

F5–1. $\xrightarrow{+}\Sigma F_x = 0; \quad -A_x + 500\left(\frac{3}{5}\right) = 0$

$A_x = 300 \text{ lb}$ *Ans.*

$\zeta + \Sigma M_A = 0; \quad B_y(10) - 500\left(\frac{4}{5}\right)(5) - 600 = 0$

$B_y = 260 \text{ lb}$ *Ans.*

$+\uparrow\Sigma F_y = 0; \quad A_y + 260 - 500\left(\frac{4}{5}\right) = 0$

$A_y = 140 \text{ lb}$ *Ans.*

F5–2. $\zeta + \Sigma M_A = 0;$

$F_{CD}\sin 45°(1.5 \text{ m}) - 4 \text{ kN}(3 \text{ m}) = 0$

$F_{CD} = 11.31 \text{ kN} = 11.3 \text{ kN}$ *Ans.*

$\xrightarrow{+}\Sigma F_x = 0; \quad A_x + (11.31 \text{ kN})\cos 45° = 0$

$A_x = -8 \text{ kN} = 8 \text{ kN} \leftarrow$ *Ans.*

$+\uparrow\Sigma F_y = 0;$

$A_y + (11.31 \text{ kN})\sin 45° - 4 \text{ kN} = 0$

$A_y = -4 \text{ kN} = 4 \text{ kN} \downarrow$ *Ans.*

F5–3. $\zeta + \Sigma M_A = 0;$

$N_B[6 \text{ m} + (6 \text{ m})\cos 45°]$

$\quad - 10 \text{ kN}[2 \text{ m} + (6 \text{ m})\cos 45°]$

$\quad - 5 \text{ kN}(4 \text{ m}) = 0$

$N_B = 8.047 \text{ kN} = 8.05 \text{ kN}$ *Ans.*

$\xrightarrow{+}\Sigma F_x = 0;$

$(5 \text{ kN})\cos 45° - A_x = 0$

$A_x = 3.54 \text{ kN}$ *Ans.*

$+\uparrow\Sigma F_y = 0;$

$A_y + 8.047 \text{ kN} - (5 \text{ kN})\sin 45° - 10 \text{ kN} = 0$

$A_y = 5.49 \text{ kN}$ *Ans.*

F5–4. $\xrightarrow{+}\Sigma F_x = 0; \quad -A_x + 400\cos 30° = 0$

$A_x = 346 \text{ N}$ *Ans.*

$+\uparrow\Sigma F_y = 0;$

$A_y - 200 - 200 - 200 - 400\sin 30° = 0$

$A_y = 800 \text{ N}$ *Ans.*

$\zeta + \Sigma M_A = 0;$

$M_A - 200(2.5) - 200(3.5) - 200(4.5)$

$\quad - 400\sin 30°(4.5) - 400\cos 30°(3\sin 60°) = 0$

$M_A = 3.90 \text{ kN} \cdot \text{m}$ *Ans.*

F5–5. $\zeta + \Sigma M_A = 0;$

$N_C(0.7 \text{ m}) - [25(9.81) \text{ N}](0.5 \text{ m})\cos 30° = 0$

$N_C = 151.71 \text{ N} = 152 \text{ N}$ *Ans.*

$\xrightarrow{+}\Sigma F_x = 0;$

$T_{AB}\cos 15° - (151.71 \text{ N})\cos 60° = 0$

$T_{AB} = 78.53 \text{ N} = 78.5 \text{ N}$ *Ans.*

$+\uparrow\Sigma F_y = 0;$

$F_A + (78.53 \text{ N})\sin 15°$

$\quad + (151.71 \text{ N})\sin 60° - 25(9.81) \text{ N} = 0$

$F_A = 93.5 \text{ N}$ *Ans.*

F5–6. $\xrightarrow{+}\Sigma F_x = 0;$

$N_C\sin 30° - (250 \text{ N})\sin 60° = 0$

$N_C = 433.0 \text{ N} = 433 \text{ N}$ *Ans.*

$\zeta + \Sigma M_B = 0;$

$-N_A\sin 30°(0.15 \text{ m}) - 433.0 \text{ N}(0.2 \text{ m})$

$\quad + [(250 \text{ N})\cos 30°](0.6 \text{ m}) = 0$

$N_A = 577.4 \text{ N} = 577 \text{ N}$ *Ans.*

$+\uparrow\Sigma F_y = 0;$

$N_B - 577.4 \text{ N} + (433.0 \text{ N})\cos 30°$

$\quad - (250 \text{ N})\cos 60° = 0$

$N_B = 327 \text{ N}$ *Ans.*

F5–7. $\Sigma F_z = 0;$

$T_A + T_B + T_C - 200 - 500 = 0$

$\Sigma M_x = 0;$

$T_A(3) + T_C(3) - 500(1.5) - 200(3) = 0$

$\Sigma M_y = 0;$

$-T_B(4) - T_C(4) + 500(2) + 200(2) = 0$

$T_A = 350 \text{ lb}, T_B = 250 \text{ lb}, T_C = 100 \text{ lb}$ *Ans.*

F5–8. $\Sigma M_y = 0;$

$600 \text{ N}(0.2 \text{ m}) + 900 \text{ N}(0.6 \text{ m}) - F_A(1 \text{ m}) = 0$

$F_A = 660 \text{ N}$ *Ans.*

$\Sigma M_x = 0;$

$D_z(0.8 \text{ m}) - 600 \text{ N}(0.5 \text{ m}) - 900 \text{ N}(0.1 \text{ m}) = 0$

$D_z = 487.5 \text{ N}$ *Ans.*

$\Sigma F_x = 0; \quad D_x = 0$ *Ans.*

$\Sigma F_y = 0; \quad D_y = 0$ *Ans.*

$\Sigma F_z = 0;$

$T_{BC} + 660 \text{ N} + 487.5 \text{ N} - 900 \text{ N} - 600 \text{ N} = 0$

$T_{BC} = 352.5 \text{ N}$ *Ans.*

F5–9. $\Sigma F_y = 0;$ $400\text{ N} + C_y = 0;$

$$C_y = -400\text{ N} \qquad Ans.$$

$\Sigma M_y = 0;$ $-C_x(0.4\text{ m}) - 600\text{ N}(0.6\text{ m}) = 0$

$$C_x = -900\text{ N} \qquad Ans.$$

$\Sigma M_x = 0;$ $B_z(0.6\text{ m}) + 600\text{ N}(1.2\text{ m})$

$$+ (-400\text{ N})(0.4\text{ m}) = 0$$

$$B_z = -933.3\text{ N} \qquad Ans.$$

$\Sigma M_z = 0;$

$$-B_x(0.6\text{ m}) - (-900\text{ N})(1.2\text{ m})$$

$$+ (-400\text{ N})(0.6\text{ m}) = 0$$

$$B_x = 1400\text{ N} \qquad Ans.$$

$\Sigma F_x = 0;$ $1400\text{ N} + (-900\text{ N}) + A_x = 0$

$$A_x = -500\text{ N} \qquad Ans.$$

$\Sigma F_z = 0;$ $A_z - 933.3\text{ N} + 600\text{ N} = 0$

$$A_z = 333.3\text{ N} \qquad Ans.$$

F5–10. $\Sigma F_x = 0;$ $B_x = 0$ $Ans.$

$\Sigma M_z = 0;$

$C_y(0.4\text{ m} + 0.6\text{ m}) = 0$ $C_y = 0$ $Ans.$

$\Sigma F_y = 0;$ $A_y + 0 = 0$ $A_y = 0$ $Ans.$

$\Sigma M_x = 0;$ $C_z(0.6\text{ m} + 0.6\text{ m}) + B_z(0.6\text{ m})$

$$- 450\text{ N}(0.6\text{ m} + 0.6\text{ m}) = 0$$

$1.2C_z + 0.6B_z - 540 = 0$

$\Sigma M_y = 0;$ $-C_z(0.6\text{ m} + 0.4\text{ m})$

$$- B_z(0.6\text{ m}) + 450\text{ N}(0.6\text{ m}) = 0$$

$$-C_z - 0.6B_z + 270 = 0$$

$C_z = 1350\text{ N}$ $B_z = -1800\text{ N}$ $Ans.$

$\Sigma F_z = 0;$

$A_z + 1350\text{ N} + (-1800\text{ N}) - 450\text{ N} = 0$

$$A_z = 900\text{ N} \qquad Ans.$$

F5–11. $\Sigma F_y = 0;$ $A_y = 0$ $Ans.$

$\Sigma M_x = 0;$ $-9(3) + F_{CE}(3) = 0$

$$F_{CE} = 9\text{ kN} \qquad Ans.$$

$\Sigma M_z = 0;$ $F_{CF}(3) - 6(3) = 0$

$$F_{CF} = 6\text{ kN} \qquad Ans.$$

$\Sigma M_y = 0;$ $9(4) - A_z(4) - 6(1.5) = 0$

$$A_z = 6.75\text{ kN} \qquad Ans.$$

$\Sigma F_x = 0;$ $A_x + 6 - 6 = 0$ $A_x = 0$ $Ans.$

$\Sigma F_z = 0;$ $F_{DB} + 9 - 9 + 6.75 = 0$

$$F_{DB} = -6.75\text{ kN} \qquad Ans.$$

F5–12. $\Sigma F_x = 0;$ $A_x = 0$ $Ans.$

$\Sigma F_y = 0;$ $A_y = 0$ $Ans.$

$\Sigma F_z = 0;$ $A_z + F_{BC} - 80 = 0$

$\Sigma M_x = 0;$ $(M_A)_x + 6F_{BC} - 80(6) = 0$

$\Sigma M_y = 0;$ $3F_{BC} - 80(1.5) = 0$ $F_{BC} = 40\text{ lb}$ $Ans.$

$\Sigma M_z = 0;$ $(M_A)_z = 0$ $Ans.$

$A_z = 40\text{ lb}$ $(M_A)_x = 240\text{ lb}\cdot\text{ft}$ $Ans.$

Chapter 6

F6–1. *Joint A.*

$+\uparrow\Sigma F_y = 0;$ $225\text{ lb} - F_{AD}\sin 45° = 0$

$F_{AD} = 318.20\text{ lb} = 318\text{ lb (C)}$ $Ans.$

$\xrightarrow{+}\Sigma F_x = 0;$ $F_{AB} - (318.20\text{ lb})\cos 45° = 0$

$F_{AB} = 225\text{ lb (T)}$ $Ans.$

Joint B.

$\xrightarrow{+}\Sigma F_x = 0;$ $F_{BC} - 225\text{ lb} = 0$

$F_{BC} = 225\text{ lb (T)}$ $Ans.$

$+\uparrow\Sigma F_y = 0;$ $F_{BD} = 0$ $Ans.$

Joint D.

$\xrightarrow{+}\Sigma F_x = 0;$

$F_{CD}\cos 45° + (318.20\text{ lb})\cos 45° - 450\text{ lb} = 0$

$F_{CD} = 318.20\text{ lb} = 318\text{ lb (T)}$ $Ans.$

F6–2. *Joint D.*

$+\uparrow\Sigma F_y = 0;$ $\frac{3}{5}F_{CD} - 300 = 0;$

$F_{CD} = 500\text{ lb (T)}$ $Ans.$

$\xrightarrow{+}\Sigma F_x = 0;$ $-F_{AD} + \frac{4}{5}(500) = 0$

$F_{AD} = 400\text{ lb (C)}$ $Ans.$

$F_{BC} = 500\text{ lb (T)},$ $F_{AC} = F_{AB} = 0$ $Ans.$

F6–3. $D_x = 200\text{ lb},$ $D_y = 650\text{ lb},$ $B_y = 150\text{ lb}$

Joint B.

$\xrightarrow{+}\Sigma F_x = 0;$ $F_{BA} = 0$ $Ans.$

$+\uparrow\Sigma F_y = 0;$ $150 - F_{BC} = 0;$ $F_{BC} = 150\text{ lb (C)}$ $Ans.$

Joint A.

$\xrightarrow{+}\Sigma F_x = 0;$ $F_{AC}\left(\frac{4}{5}\right) = 0;$ $F_{AC} = 0$ $Ans.$

$+\uparrow\Sigma F_y = 0;$ $F_{AD} - 800 = 0;$ $F_{AD} = 800\text{ lb (T)}$ $Ans.$

Joint C.

$\xrightarrow{+}\Sigma F_x = 0;$ $-F_{CD} + 200 = 0;$ $F_{CD} = 200\text{ lb (T)}$ $Ans.$

F6–4. *Joint C.*

$+\uparrow\Sigma F_y = 0;$ $2F\cos 30° - P = 0$

$F_{AC} = F_{BC} = F = \frac{P}{2\cos 30°} = 0.5774P$ (C)

Joint B.

$\xrightarrow{+}\Sigma F_x = 0;\ 0.5774P\cos 60° - F_{AB} = 0$

$F_{AB} = 0.2887P$ (T)

$F_{AB} = 0.2887P = 2$ kN

$P = 6.928$ kN

$F_{AC} = F_{BC} = 0.5774P = 1.5$ kN

$P = 2.598$ kN

The *smaller value* of P is chosen,

$P = 2.598$ kN $= 2.60$ kN   *Ans.*

F6–5. $F_{CB} = 0$ *Ans.*

$F_{CD} = 0$ *Ans.*

$F_{AE} = 0$ *Ans.*

$F_{DE} = 0$ *Ans.*

F6–6. *Joint C.*

$+\uparrow\Sigma F_y = 0;$ $259.81\text{ lb} - F_{CD}\sin 30° = 0$

$F_{CD} = 519.62$ lb $= 520$ lb (C)   *Ans.*

$\xrightarrow{+}\Sigma F_x = 0;$ $(519.62\text{ lb})\cos 30° - F_{BC} = 0$

$F_{BC} = 450$ lb (T)   *Ans.*

Joint D.

$+\nearrow\Sigma F_{y'} = 0;$ $F_{BD}\cos 30° = 0$ $F_{BD} = 0$ *Ans.*

$+\searrow\Sigma F_{x'} = 0;$ $F_{DE} - 519.62\text{ lb} = 0$

$F_{DE} = 519.62$ lb $= 520$ lb (C)   *Ans.*

Joint B.

$\uparrow\Sigma F_y = 0;$ $F_{BE}\sin\phi = 0$ $F_{BE} = 0$ *Ans.*

$\xrightarrow{+}\Sigma F_x = 0;$ $450\text{ lb} - F_{AB} = 0$

$F_{AB} = 450$ lb (T)   *Ans.*

Joint A.

$+\uparrow\Sigma F_y = 0;$ $340.19\text{ lb} - F_{AE} = 0$

$F_{AE} = 340$ lb (C)   *Ans.*

F6–7. $+\uparrow\Sigma F_y = 0;\ F_{CF}\sin 45° - 600 - 800 = 0$

$F_{CF} = 1980$ lb (T)   *Ans.*

$\zeta+\Sigma M_C = 0;\ F_{FE}(4) - 800(4) = 0$

$F_{FE} = 800$ lb (T)   *Ans.*

$\zeta+\Sigma M_F = 0;\ F_{BC}(4) - 600(4) - 800(8) = 0$

$F_{BC} = 2200$ lb (C)   *Ans.*

F6–8. $\zeta+\Sigma M_A = 0;$ $G_y(12\text{ m}) - 20\text{ kN}(2\text{ m})$

$- 30\text{ kN}(4\text{ m}) - 40\text{ kN}(6\text{ m}) = 0$

$G_y = 33.33$ kN

$+\uparrow\Sigma F_y = 0;\ F_{KC} + 33.33\text{ kN} - 40\text{ kN} = 0$

$F_{KC} = 6.67$ kN (C)   *Ans.*

$\zeta+\Sigma M_K = 0;$

$33.33\text{ kN}(8\text{ m}) - 40\text{ kN}(2\text{ m}) - F_{CD}(3\text{ m}) = 0$

$F_{CD} = 62.22$ kN $= 62.2$ kN (T)   *Ans.*

$\xrightarrow{+}\Sigma F_x = 0;$ $F_{LK} - 62.22\text{ kN} = 0$

$F_{LK} = 62.2$ kN (C)   *Ans.*

F6–9. From the geometry of the truss,

$\phi = \tan^{-1}(3\text{ m}/2\text{ m}) = 56.31°.$

$\zeta+\Sigma M_K = 0;$

$33.33\text{ kN}(8\text{ m}) - 40\text{ kN}(2\text{ m}) - F_{CD}(3\text{ m}) = 0$

$F_{CD} = 62.2$ kN (T)   *Ans.*

$\zeta+\Sigma M_D = 0;$ $33.33\text{ kN}(6\text{ m}) - F_{KJ}(3\text{ m}) = 0$

$F_{KJ} = 66.7$ kN (C)   *Ans.*

$+\uparrow\Sigma F_y = 0;$

$33.33\text{ kN} - 40\text{ kN} + F_{KD}\sin 56.31° = 0$

$F_{KD} = 8.01$ kN (T)   *Ans.*

F6–10. From the geometry of the truss,

$\tan\phi = \frac{(9\text{ ft})\tan 30°}{3\text{ ft}} = 1.732\quad \phi = 60°$

$\zeta+\Sigma M_C = 0;$

$F_{EF}\sin 30°(6\text{ ft}) + 300\text{ lb}(6\text{ ft}) = 0$

$F_{EF} = -600$ lb $= 600$ lb (C)   *Ans.*

$\zeta+\Sigma M_D = 0;$

$300\text{ lb}(6\text{ ft}) - F_{CF}\sin 60°(6\text{ ft}) = 0$

$F_{CF} = 346.41$ lb $= 346$ lb (T)   *Ans.*

$\zeta+\Sigma M_F = 0;$

$300\text{ lb}(9\text{ ft}) - 300\text{ lb}(3\text{ ft}) - F_{BC}(9\text{ ft})\tan 30° = 0$

$F_{BC} = 346.41$ lb $= 346$ lb (T)   *Ans.*

F6–11. From the geometry of the truss,

$\theta = \tan^{-1}(1\text{ m}/2\text{ m}) = 26.57°$

$\phi = \tan^{-1}(3\text{ m}/2\text{ m}) = 56.31°.$

The location of O can be found using similar triangles.

$$\frac{1\text{ m}}{2\text{ m}} = \frac{2\text{ m}}{2\text{ m} + x}$$

$$4\text{ m} = 2\text{ m} + x$$

$$x = 2\text{ m}$$

$\zeta + \Sigma M_G = 0;$

$26.25 \text{ kN}(4 \text{ m}) - 15 \text{ kN}(2 \text{ m}) - F_{CD}(3 \text{ m}) = 0$

$\qquad F_{CD} = 25 \text{ kN (T)} \qquad\qquad$ *Ans.*

$\zeta + \Sigma M_D = 0;$

$26.25 \text{ kN}(2 \text{ m}) - F_{GF} \cos 26.57°(2 \text{ m}) = 0$

$\qquad F_{GF} = 29.3 \text{ kN (C)} \qquad\qquad$ *Ans.*

$\zeta + \Sigma M_O = 0; \quad 15 \text{ kN}(4 \text{ m}) - 26.25 \text{ kN}(2 \text{ m})$

$\qquad\qquad\qquad\qquad - F_{GD} \sin 56.31°(4 \text{ m}) = 0$

$\qquad F_{GD} = 2.253 \text{ kN} = 2.25 \text{ kN (T)} \qquad$ *Ans.*

F6–12. $\zeta + \Sigma M_H = 0;$

$F_{DC}(12 \text{ ft}) + 1200 \text{ lb}(9 \text{ ft}) - 1600 \text{ lb}(21 \text{ ft}) = 0$

$\qquad F_{DC} = 1900 \text{ lb (C)} \qquad\qquad$ *Ans.*

$\zeta + \Sigma M_D = 0;$

$1200 \text{ lb}(21 \text{ ft}) - 1600 \text{ lb}(9 \text{ ft}) - F_{HI}(12 \text{ ft}) = 0$

$\qquad F_{HI} = 900 \text{ lb (C)} \qquad\qquad$ *Ans.*

$\zeta + \Sigma M_C = 0; \quad F_{JI} \cos 45°(12 \text{ ft}) + 1200 \text{ lb}(21 \text{ ft})$

$\qquad\qquad\qquad - 900 \text{ lb}(12 \text{ ft}) - 1600 \text{ lb}(9 \text{ ft}) = 0$

$\qquad F_{JI} = 0 \qquad\qquad$ *Ans.*

F6–13. $+\uparrow\Sigma F_y = 0; \quad 3P - 60 = 0$

$\qquad P = 20 \text{ lb} \qquad\qquad$ *Ans.*

F6–14. $\zeta + \Sigma M_C = 0;$

$-\left(\frac{4}{5}\right)(F_{AB})(9) + 400(6) + 500(3) = 0$

$\qquad F_{AB} = 541.67 \text{ lb}$

$\xrightarrow{+}\Sigma F_x = 0; -C_x + \frac{3}{5}(541.67) = 0$

$\qquad C_x = 325 \text{ lb} \qquad\qquad$ *Ans.*

$+\uparrow\Sigma F_y = 0; C_y + \frac{4}{5}(541.67) - 400 - 500 = 0$

$\qquad C_y = 467 \text{ lb} \qquad\qquad$ *Ans.*

F6–15. $\zeta + \Sigma M_A = 0; 100 \text{ N}(250 \text{ mm}) - N_B(50 \text{ mm}) = 0$

$\qquad N_B = 500 \text{ N} \qquad\qquad$ *Ans.*

$\xrightarrow{+}\Sigma F_x = 0; \quad (500 \text{ N}) \sin 45° - A_x = 0$

$\qquad A_x = 353.55 \text{ N}$

$+\uparrow\Sigma F_y = 0; A_y - 100 \text{ N} - (500 \text{ N}) \cos 45° = 0$

$\qquad A_y = 453.55 \text{ N}$

$F_A = \sqrt{(353.55 \text{ N})^2 + (453.55 \text{ N})^2}$

$\qquad = 575 \text{ N} \qquad\qquad$ *Ans.*

F6–16. $\zeta + \Sigma M_C = 0;$

$400(2) + 800 - F_{BA}\left(\frac{3}{\sqrt{10}}\right)(1)$

$\qquad\qquad\qquad\qquad - F_{BA}\left(\frac{1}{\sqrt{10}}\right)(3) = 0$

$\qquad F_{BA} = 843.27 \text{ N}$

$\xrightarrow{+}\Sigma F_x = 0; C_x - 843.27\left(\frac{3}{\sqrt{10}}\right) = 0$

$\qquad C_x = 800 \text{ N} \qquad\qquad$ *Ans.*

$+\uparrow\Sigma F_y = 0; C_y + 843.27\left(\frac{1}{\sqrt{10}}\right) - 400 = 0$

$\qquad C_y = 133 \text{ N} \qquad\qquad$ *Ans.*

F6–17.   Plate A:

$+\uparrow\Sigma F_y = 0; \quad 2T + N_{AB} - 100 = 0$

Plate B:

$+\uparrow\Sigma F_y = 0; \quad 2T - N_{AB} - 30 = 0$

$\qquad T = 32.5 \text{ lb}, N_{AB} = 35 \text{ lb} \qquad$ *Ans.*

F6–18.   Pulley C:

$+\uparrow\Sigma F_y = 0; \quad T - 2P = 0; T = 2P$

Beam:

$+\uparrow\Sigma F_y = 0; \quad 2P + P - 6 = 0$

$\qquad P = 2 \text{ kN} \qquad\qquad$ *Ans.*

$\zeta + \Sigma M_A = 0; \quad 2(1) - 6(x) = 0$

$\qquad x = 0.333 \text{ m} \qquad\qquad$ *Ans.*

F6–19.   Member CD

$\zeta + \Sigma M_D = 0; \quad 600(1.5) - N_C(3) = 0$

$\qquad N_C = 300 \text{ N}$

Member ABC

$\zeta + \Sigma M_A = 0; \quad -800 + B_y(2) - (300 \sin 45°)\, 4 = 0$

$\qquad B_y = 824.26 = 824 \text{ N} \qquad$ *Ans.*

$\xrightarrow{+}\Sigma F_x = 0; \quad A_x - 300 \cos 45° = 0;$

$\qquad A_x = 212 \text{ N} \qquad\qquad$ *Ans.*

$+\uparrow\Sigma F_y = 0; \quad -A_y + 824.26 - 300 \sin 45° = 0;$

$\qquad A_y = 612 \text{ N} \qquad\qquad$ *Ans.*

F6–20. AB is a two-force member.

Member BC

$\zeta + \Sigma M_c = 0; \quad 15(3) + 10(6) - F_{BC}\left(\frac{4}{5}\right)(9) = 0$

$\qquad F_{BC} = 14.58 \text{ kN}$

$\xrightarrow{+}\Sigma F_x = 0; \quad (14.58)\left(\frac{3}{5}\right) - C_x = 0;$

$\qquad C_x = 8.75 \text{ kN}$

$+\uparrow\Sigma F_y = 0; \quad (14.58)\left(\frac{4}{5}\right) - 10 - 15 + C_y = 0;$

$\qquad C_y = 13.3 \text{ kN}$

Member CD

$\xrightarrow{+}\Sigma F_x = 0; \qquad 8.75 - D_x = 0; \quad D_x = 8.75 \text{ kN} \quad$ *Ans.*

$+\uparrow\Sigma F_y = 0; \qquad -13.3 + D_y = 0; \quad D_y = 13.3 \text{ kN} \quad$ *Ans.*

$\zeta + \Sigma M_D = 0; \; -8.75(4) + M_D = 0; \; M_D = 35 \text{ kN} \cdot \text{m} \;$ *Ans.*

F6–21. Entire frame

$\zeta + \Sigma M_A = 0;\quad -600(3) - [400(3)](1.5) + C_y(3) = 0$

$\qquad\qquad C_y = 1200\ \text{N} \qquad\qquad Ans.$

$+\uparrow\Sigma F_y = 0;\quad A_y - 400(3) + 1200 = 0$

$\qquad\qquad A_y = 0 \qquad\qquad Ans.$

$\xrightarrow{+}\Sigma F_x = 0;\quad 600 - A_x - C_x = 0$

Member AB

$\zeta + \Sigma M_B = 0;\quad 400(1.5)(0.75) - A_x(3) = 0$

$\qquad\qquad A_x = 150\ \text{N} \qquad\qquad Ans.$

$\qquad\qquad C_x = 450\ \text{N} \qquad\qquad Ans.$

These same results can be obtained by considering members AB and BC.

F6–22. Entire frame

$\zeta + \Sigma M_E = 0;\quad 250(6) - A_y(6) = 0$

$\qquad\qquad A_y = 250\ \text{N}$

$\xrightarrow{+}\Sigma F_x = 0;\quad E_x = 0$

$+\uparrow\Sigma F_y = 0;\quad 250 - 250 + E_y = 0;\quad E_y = 0$

Member BD

$\zeta + \Sigma M_D = 0;\quad 250(4.5) - B_y(3) = 0;$

$\qquad\qquad B_y = 375\ \text{N}$

Member ABC

$\zeta + \Sigma M_C = 0;\; -250(3) + 375(1.5) + B_x(2) = 0$

$\qquad\qquad B_x = 93.75\ \text{N}$

$\xrightarrow{+}\Sigma F_x = 0;\quad C_x - B_x = 0;\quad C_x = 93.75\ \text{N} \quad Ans.$

$+\uparrow\Sigma F_y = 0;\quad 250 - 375 + C_y = 0;\quad C_y = 125\ \text{N} \quad Ans.$

F6–23. AD, CB are two-force members.

Member AB

$\zeta + \Sigma M_A = 0;\quad -[\tfrac{1}{2}(3)(4)](1.5) + B_y(3) = 0$

$\qquad\qquad B_y = 3\ \text{kN}$

Since BC is a two-force member $C_y = B_y = 3\ \text{kN}$ and $C_x = 0\ (\Sigma M_B = 0)$.

Member EDC

$\zeta + \Sigma M_E = 0;\quad F_{DA}(\tfrac{4}{5})(1.5) - 5(3) - 3(3) = 0$

$\qquad\qquad F_{DA} = 20\ \text{kN}$

$\xrightarrow{+}\Sigma F_x = 0;\quad E_x - 20(\tfrac{3}{5}) = 0;\quad E_x = 12\ \text{kN} \quad Ans.$

$+\uparrow\Sigma F_y = 0;\quad -E_y + 20(\tfrac{4}{5}) - 5 - 3 = 0;$

$\qquad\qquad E_y = 8\ \text{kN} \qquad\qquad Ans.$

F6–24. AC and DC are two-force members.

Member BC

$\zeta + \Sigma M_C = 0;\quad [\tfrac{1}{2}(3)(8)](1) - B_y(3) = 0$

$\qquad\qquad B_y = 4\ \text{kN}$

Member BA

$\zeta + \Sigma M_B = 0;\quad 6(2) - A_x(4) = 0$

$\qquad\qquad A_x = 3\ \text{kN} \qquad\qquad Ans.$

$+\uparrow\Sigma F_y = 0;\quad -4\ \text{kN} + A_y = 0;\quad A_y = 4\ \text{kN}\ Ans.$

Entire Frame

$\zeta + \Sigma M_A = 0;\quad -6(2) - [\tfrac{1}{2}(3)(8)](2) + D_y(3) = 0$

$\qquad\qquad D_y = 12\ \text{kN} \qquad\qquad Ans.$

Since DC is a two-force member $(\Sigma M_C = 0)$ then

$\qquad\qquad D_x = 0 \qquad\qquad Ans.$

Chapter 7

F7–1. $\zeta + \Sigma M_A = 0;\quad B_y(6) - 10(1.5) - 15(4.5) = 0$

$\qquad\qquad B_y = 13.75\ \text{kN}$

$\xrightarrow{+}\Sigma F_x = 0;\quad N_C = 0 \qquad\qquad Ans.$

$+\uparrow\Sigma F_y = 0;\quad V_C + 13.75 - 15 = 0$

$\qquad\qquad V_C = 1.25\ \text{kN} \qquad\qquad Ans.$

$\zeta + \Sigma M_C = 0;\quad 13.75(3) - 15(1.5) - M_C = 0$

$\qquad\qquad M_C = 18.75\ \text{kN} \cdot \text{m} \qquad\qquad Ans.$

F7–2. $\zeta + \Sigma M_B = 0;\quad 30 - 10(1.5) - A_y(3) = 0$

$\qquad\qquad A_y = 5\ \text{kN}$

$\xrightarrow{+}\Sigma F_x = 0;\quad N_C = 0 \qquad\qquad Ans.$

$+\uparrow\Sigma F_y = 0;\quad 5 - V_C = 0$

$\qquad\qquad V_C = 5\ \text{kN} \qquad\qquad Ans.$

$\zeta + \Sigma M_C = 0;\quad M_C + 30 - 5(1.5) = 0$

$\qquad\qquad M_C = -22.5\ \text{kN} \cdot \text{m} \qquad\qquad Ans.$

F7–3. $\xrightarrow{+}\Sigma F_x = 0;\quad B_x = 0$

$\zeta + \Sigma M_A = 0;\quad 3(6)(3) - B_y(9) = 0$

$\qquad\qquad B_y = 6\ \text{kip}$

$\xrightarrow{+}\Sigma F_x = 0;\quad N_C = 0 \qquad\qquad Ans.$

$\zeta + \uparrow\Sigma F_y = 0;\quad V_C - 6 = 0$

$\qquad\qquad V_C = 6\ \text{kip} \qquad\qquad Ans.$

$\zeta + \Sigma M_C = 0;\quad -M_C - 6(4.5) = 0$

$\qquad\qquad M_C = -27\ \text{kip} \cdot \text{ft} \qquad\qquad Ans.$

F7–4. $\zeta + \Sigma M_A = 0;\quad B_y(6) - 12(1.5) - 9(3)(4.5) = 0$

$\qquad\qquad B_y = 23.25\ \text{kN}$

$\xrightarrow{+}\Sigma F_x = 0;\quad N_C = 0 \qquad\qquad Ans.$

$+\uparrow\Sigma F_y = 0;\quad V_C + 23.25 - 9(1.5) = 0$

$\qquad\qquad V_C = -9.75\ \text{kN} \qquad\qquad Ans.$

$\zeta + \Sigma M_C = 0;$

$23.25(1.5) - 9(1.5)(0.75) - M_C = 0$

$\qquad\qquad M_C = 24.75\ \text{kN} \cdot \text{m} \qquad\qquad Ans.$

F7–5. $\zeta + \Sigma M_A = 0$; $B_y(6) - \frac{1}{2}(9)(6)(3) = 0$
$$B_y = 13.5 \text{ kN}$$
$\overset{+}{\rightarrow} \Sigma F_x = 0$; $N_C = 0$ *Ans.*
$+\uparrow \Sigma F_y = 0$; $V_C + 13.5 - \frac{1}{2}(9)(3) = 0$
$$V_C = 0 \qquad \text{Ans.}$$
$\zeta + \Sigma M_C = 0$; $13.5(3) - \frac{1}{2}(9)(3)(1) - M_C = 0$
$$M_C = 27 \text{ kN} \cdot \text{m} \qquad \text{Ans.}$$

F7–6. $\zeta + \Sigma M_A = 0$;
$$B_y(6) - \frac{1}{2}(6)(3)(2) - 6(3)(4.5) = 0$$
$$B_y = 16.5 \text{ kN}$$
$\overset{+}{\rightarrow} \Sigma F_x = 0$; $N_C = 0$ *Ans.*
$+\uparrow \Sigma F_y = 0$; $V_C + 16.5 - 6(3) = 0$
$$V_C = 1.50 \text{ kN} \qquad \text{Ans.}$$
$\zeta + \Sigma M_C = 0$; $16.5(3) - 6(3)(1.5) - M_C = 0$
$$M_C = 22.5 \text{ kN} \cdot \text{m} \qquad \text{Ans.}$$

F7–7. $+\uparrow \Sigma F_y = 0$; $6 - V = 0$ $V = 6 \text{kN}$
$\zeta + \Sigma M_O = 0$; $M + 18 - 6x = 0$
$$M = (6x - 18) \text{ kN} \cdot \text{m}$$

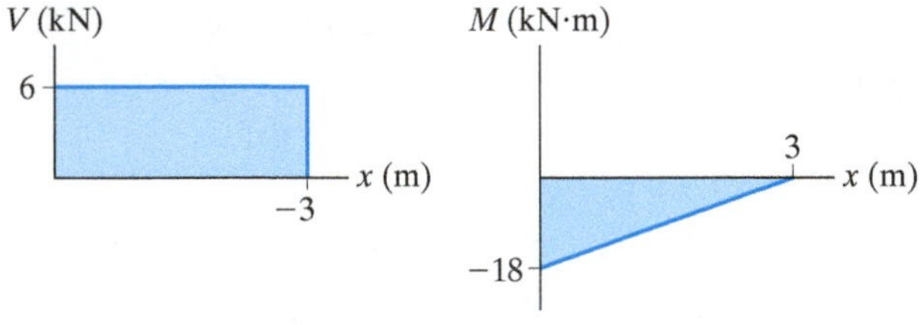

Fig. F7–7

F7–8. $+\uparrow \Sigma F_y = 0$; $-V - 2x = 0$
$$V = (-2x) \text{ kN}$$
$\zeta + \Sigma M_O = 0$; $M + 2x\left(\frac{x}{2}\right) - 15 = 0$
$$M = (15 - x^2) \text{ kN} \cdot \text{m}$$

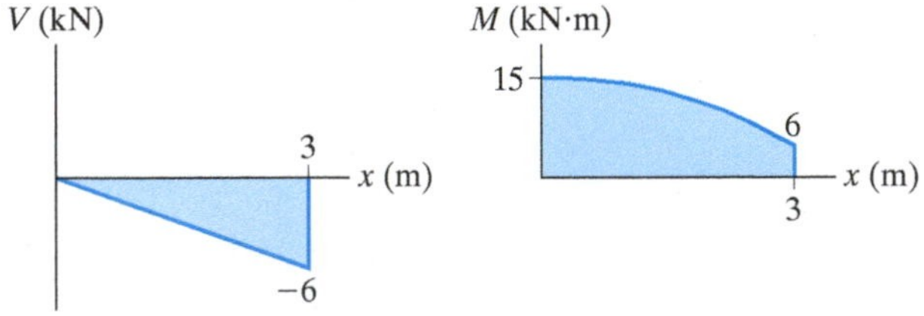

Fig. F7–8

F7–9. $+\uparrow \Sigma F_y = 0$; $-V - \frac{1}{2}(2x)(x) = 0$
$$V = -(x^2) \text{ kN}$$
$\zeta + \Sigma M_O = 0$; $M + \frac{1}{2}(2x)(x)(\frac{x}{3}) = 0$
$$M = -(\frac{1}{3}x^3) \text{ kN} \cdot \text{m}$$

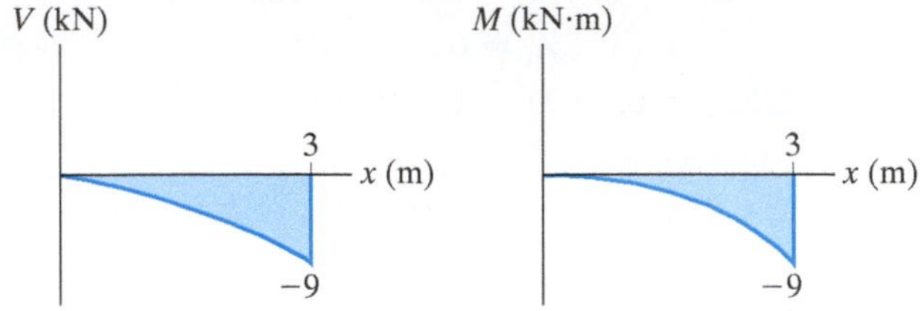

Fig. F7–9

F7–10. $+\uparrow \Sigma F_y = 0$; $-V - 2x = 0$
$$V = -2 \text{ kN}$$
$\zeta + \Sigma M_O = 0$; $M + 2x = 0$
$$M = (-2x) \text{ kN} \cdot \text{m}$$

Fig. F7–10

F7–11. Region $3 \text{ m} \leq x < 3 \text{ m}$
$+\uparrow \Sigma F_y = 0$; $-V - 5 = 0$ $V = -5 \text{ kN}$
$\zeta + \Sigma M_O = 0$; $M + 5x = 0$
$$M = (-5x) \text{ kN} \cdot \text{m}$$
Region $0 < x \leq 6 \text{ m}$
$+\uparrow \Sigma F_y = 0$; $V + 5 = 0$ $V = -5 \text{ kN}$
$\zeta + \Sigma M_O = 0$; $5(6 - x) - M = 0$
$$M = (5(6 - x)) \text{ kN} \cdot \text{m}$$

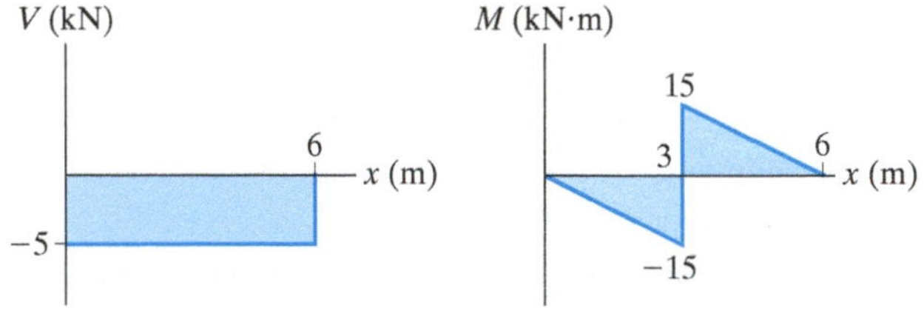

Fig. F7–11

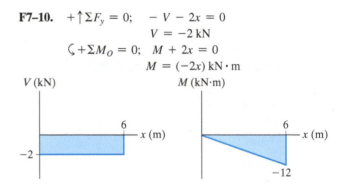

F7–12. Region $0 \leq x < 3$ m

$+\uparrow \Sigma F_y = 0;$ $V = 0$

$\zeta + \Sigma M_O = 0;$ $M - 12 = 0$

$M = 12$ kN $\cdot$ m

Region 3 m $< x \leq 6$ m

$+\uparrow \Sigma F_y = 0;$ $V + 4 = 0$ $V = -4$ kN

$\zeta + \Sigma M_O = 0;$ $4(6 - x) - M = 0$

$M = \big(4(6 - x)\big)$ kN $\cdot$ m

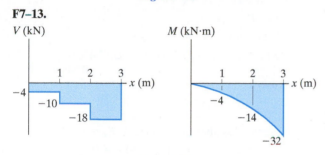

Fig. F7–12

F7–13.

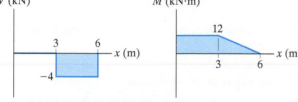

Fig. F7–13

F7–14.

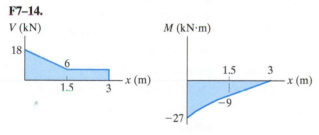

Fig. F7–14

F7–15.

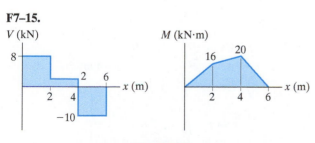

Fig. F7–15

F7–16.

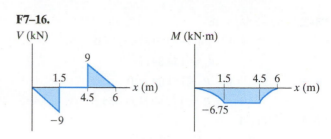

Fig. F7–16

F7–17.

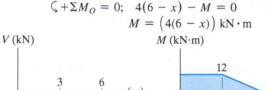

Fig. F7–17

F7–18.

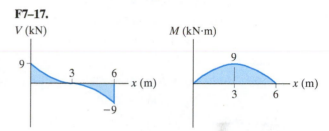

Fig. F7–18

Chapter 8

F8–1. a) $+\uparrow \Sigma F_y = 0;$ $N - 50(9.81) - 200\big(\tfrac{3}{5}\big) = 0$

$N = 610.5$ N

$\xrightarrow{+} \Sigma F_x = 0;$ $F - 200\big(\tfrac{4}{5}\big) = 0$

$F = 160$ N

$F < F_{max} = \mu_s N = 0.3(610.5) = 183.15$ N,

therefore $F = 160$ N   *Ans.*

b) $+\uparrow \Sigma F_y = 0;$ $N - 50(9.81) - 400\big(\tfrac{3}{5}\big) = 0$

$N = 730.5$ N

$\xrightarrow{+} \Sigma F_x = 0;$ $F - 400\big(\tfrac{4}{5}\big) = 0$

$F = 320$ N

$F > F_{max} = \mu_s N = 0.3(730.5) = 219.15$ N

Block slips

$F = \mu_s N = 0.2(730.5) = 146$ N   *Ans.*

F8–2. $\zeta + \Sigma M_B = 0;$

$$N_A(3) + 0.2N_A(4) - 30(9.81)(2) = 0$$

$$N_A = 154.89 \text{ N}$$

$\xrightarrow{+} \Sigma F_x = 0; \quad P - 154.89 = 0$

$$P = 154.89 \text{ N} = 155 \text{ N} \qquad Ans.$$

F8–3. Crate A

$+\uparrow \Sigma F_y = 0; \quad N_A - 50(9.81) = 0$

$$N_A = 490.5 \text{ N}$$

$\xrightarrow{+} \Sigma F_x = 0; \quad T - 0.25(490.5) = 0$

$$T = 122.62 \text{ N}$$

Crate B

$+\uparrow \Sigma F_y = 0; \quad N_B + P \sin 30° - 50(9.81) = 0$

$$N_B = 490.5 - 0.5P$$

$\xrightarrow{+} \Sigma F_x = 0;$

$P \cos 30° - 0.25(490.5 - 0.5\,P) - 122.62 = 0$

$$P = 247 \text{ N} \qquad Ans.$$

F8–4. $\xrightarrow{+} \Sigma F_x = 0; \quad N_A - 0.3N_B = 0$

$+\uparrow \Sigma F_y = 0;$

$N_B + 0.3N_A + P - 100(9.81) = 0$

$\zeta + \Sigma M_O = 0;$

$P(0.6) - 0.3N_B(0.9) - 0.3\,N_A(0.9) = 0$

$N_A = 175.70 \text{ N} \qquad N_B = 585.67 \text{ N}$

$$P = 343 \text{ N} \qquad Ans.$$

F8–5. If slipping occurs:

$+\uparrow \Sigma F_y = 0; \quad N_c - 250 \text{ lb} = 0; \ N_c = 250 \text{ lb}$

$\xrightarrow{+} \Sigma F_x = 0; \quad P - 0.4(250) = 0; \ P = 100 \text{ lb}$

If tipping occurs:

$\zeta + \Sigma M_A = 0; \quad -P(4.5) + 250(1.5) = 0$

$$P = 83.3 \text{ lb} \qquad Ans.$$

F8–6.
$\zeta + \Sigma M_A = 0; \quad 490.5(0.6) - T\cos 60°(0.3 \cos 60° + 0.6)$

$$- \ T \sin 60° \ (0.3 \sin 60°) = 0$$

$$T = 490.5 \text{ N}$$

$\xrightarrow{+} \Sigma F_x = 0; \quad 490.5 \sin 60° - N_A = 0; \ N_A = 424.8 \text{ N}$

$+\uparrow \Sigma F_y = 0; \quad \mu_s(424.8) + 490.5 \cos 60° - 490.5 = 0$

$$\mu_s = 0.577 \qquad Ans.$$

F8–7. A will not move. Assume B is about to slip on C and A, and C is stationary.

$\xrightarrow{+} \Sigma F_x = 0; \quad P - 0.3(50) - 0.4(75); \quad P = 45 \text{ N}$

Assume C is about to slip and B does not slip on C, but is about to slip at A.

$\xrightarrow{+} \Sigma F_x = 0; \quad P - 0.3(50) - 0.35(90) = 0$

$$P = 46.5 \text{ N} > 45 \text{ N}$$

$$P = 45 \text{ N} \qquad Ans.$$

F8–8. A is about to move down the plane and B moves upward.

Block A

$+\nwarrow \Sigma F_y = 0; \quad N = W \cos \theta$

$+\nearrow \Sigma F_x = 0; \quad T + \mu_s(W \cos \theta) - W \sin \theta = 0$

$T = W \sin \theta - \mu_s W \cos \theta \qquad (1)$

Block B

$+\nwarrow \Sigma F_y = 0; \quad N' = 2W \cos \theta$

$+\nearrow \Sigma F_x = 0; \quad 2T - \mu_s W \cos \theta - \mu_s(2W \cos \theta)$
$$- \ W \sin \theta = 0$$

Using Eq.(1);

$$\theta = \tan^{-1} 5\mu_s \qquad Ans.$$

F8–9. Assume B is about to slip on A, $F_B = 0.3\,N_B$.

$\xrightarrow{+} \Sigma F_x = 0; \quad P - 0.3(10)(9.81) = 0$

$$P = 29.4 \text{ N}$$

Assume B is about to tip on A, $x = 0$.

$\zeta + \Sigma M_O = 0; \quad 10(9.81)(0.15) - P(0.4) = 0$

$$P = 36.8 \text{ N}$$

Assume A is about to slip, $F_A = 0.1\,N_A$.

$\xrightarrow{+} \Sigma F_x = 0 \quad P - 0.1[7(9.81) + 10(9.81)] = 0$

$$P = 16.7 \text{ N}$$

Choose the smallest result. $P = 16.7 \text{ N} \qquad Ans.$

Chapter 9

F9–1. $\bar{x} = \dfrac{\displaystyle\int_A \tilde{x}\, dA}{\displaystyle\int_A dA} = \dfrac{\dfrac{1}{2}\displaystyle\int_0^{1\,m} y^{2/3}\, dy}{\displaystyle\int_0^{1\,m} y^{1/3} dy} = 0.4 \text{ m} \qquad Ans.$

$\bar{y} = \dfrac{\displaystyle\int_A \tilde{y}\, dA}{\displaystyle\int_A dA} = \dfrac{\displaystyle\int_0^{1\,m} y^{4/3}\, dy}{\displaystyle\int_0^{1\,m} y^{1/3} dy} = 0.571 \text{ m} \qquad Ans.$

F9–2. $\bar{x} = \dfrac{\displaystyle\int_A \tilde{x}\, dA}{\displaystyle\int_A dA} = \dfrac{\displaystyle\int_0^{1\,\text{m}} x(x^3\, dx)}{\displaystyle\int_0^{1\,\text{m}} x^3\, dx}$

$= 0.8\ \text{m}$ *Ans.*

$\bar{y} = \dfrac{\displaystyle\int_A \tilde{y}\, dA}{\displaystyle\int_A dA} = \dfrac{\displaystyle\int_0^{1\,\text{m}} \frac{1}{2}x^3\left(x^3\, dx\right)}{\displaystyle\int_0^{1\,\text{m}} x^3\, dx}$

$= 0.286\ \text{m}$ *Ans.*

F9–3. $\bar{y} = \dfrac{\displaystyle\int_A \tilde{y}\, dA}{\displaystyle\int_A dA} = \dfrac{\displaystyle\int_0^{2\,\text{m}} y\left(2\left(\dfrac{y^{1/2}}{\sqrt{2}}\right)\right)dy}{\displaystyle\int_0^{2\,\text{m}} 2\left(\dfrac{y^{1/2}}{\sqrt{2}}\right)dy}$

$= 1.2\ \text{m}$ *Ans.*

F9–4. $\bar{x} = \dfrac{\displaystyle\int_m \tilde{x}\, dm}{\displaystyle\int_m dm} = \dfrac{\displaystyle\int_0^L x\left[m_0\left(1 + \dfrac{x^2}{L^2}\right)dx\right]}{\displaystyle\int_0^L m_0\left(1 + \dfrac{x^2}{L^2}\right)dx}$

$= \dfrac{9}{16}L$ *Ans.*

F9–5. $\bar{y} = \dfrac{\displaystyle\int_V \tilde{y}\, dV}{\displaystyle\int_V dV} = \dfrac{\displaystyle\int_0^{1\,\text{m}} y\left(\dfrac{\pi}{4}y\, dy\right)}{\displaystyle\int_0^{1\,\text{m}} \dfrac{\pi}{4}y\, dy}$

$= 0.667\ \text{m}$ *Ans.*

F9–6. $\bar{z} = \dfrac{\displaystyle\int_V \tilde{z}\, dV}{\displaystyle\int_V dV} = \dfrac{\displaystyle\int_0^{2\,\text{ft}} z\left[\dfrac{9\pi}{64}(4 - z)^2\, dz\right]}{\displaystyle\int_0^{2\,\text{ft}} \dfrac{9\pi}{64}(4 - z)^2\, dz}$

$= 0.786\ \text{ft}$ *Ans.*

F9–7. $\bar{x} = \dfrac{\Sigma \tilde{x} L}{\Sigma L} = \dfrac{150(300) + 300(600) + 300(400)}{300 + 600 + 400}$

$= 265\ \text{mm}$ *Ans.*

$\bar{y} = \dfrac{\Sigma \tilde{y} L}{\Sigma L} = \dfrac{0(300) + 300(600) + 600(400)}{300 + 600 + 400}$

$= 323\ \text{mm}$ *Ans.*

$\bar{z} = \dfrac{\Sigma \tilde{z} L}{\Sigma L} = \dfrac{0(300) + 0(600) + (-200)(400)}{300 + 600 + 400}$

$= -61.5\ \text{mm}$ *Ans.*

F9–8. $\bar{y} = \dfrac{\Sigma \tilde{y} A}{\Sigma A} = \dfrac{150[300(50)] + 325[50(300)]}{300(50) + 50(300)}$

$= 237.5\ \text{mm}$ *Ans.*

F9–9. $\bar{y} = \dfrac{\Sigma \tilde{y} A}{\Sigma A} = \dfrac{100[2(200)(50)] + 225[50(400)]}{2(200)(50) + 50(400)}$

$= 162.5\ \text{mm}$ *Ans.*

F9–10. $\bar{x} = \dfrac{\Sigma \tilde{x} A}{\Sigma A} = \dfrac{0.25[4(0.5)] + 1.75[0.5(2.5)]}{4(0.5) + 0.5(2.5)}$

$= 0.827\ \text{in.}$ *Ans.*

$\bar{y} = \dfrac{\Sigma \tilde{y} A}{\Sigma A} = \dfrac{2[4(0.5)] + 0.25[(0.5)(2.5)]}{4(0.5) + (0.5)(2.5)}$

$= 1.33\ \text{in.}$ *Ans.*

F9–11. $\bar{x} = \dfrac{\Sigma \tilde{x} V}{\Sigma V} = \dfrac{1[2(7)(6)] + 4[4(2)(3)]}{2(7)(6) + 4(2)(3)}$

$= 1.67\ \text{ft}$ *Ans.*

$\bar{y} = \dfrac{\Sigma \tilde{y} V}{\Sigma V} = \dfrac{3.5[2(7)(6)] + 1[4(2)(3)]}{2(7)(6) + 4(2)(3)}$

$= 2.94\ \text{ft}$ *Ans.*

$\bar{z} = \dfrac{\Sigma \tilde{z} V}{\Sigma V} = \dfrac{3[2(7)(6)] + 1.5[4(2)(3)]}{2(7)(6) + 4(2)(3)}$

$= 2.67\ \text{ft}$ *Ans.*

F9–12. $\bar{x} = \dfrac{\Sigma \tilde{x} V}{\Sigma V}$

$= \dfrac{0.25[0.5(2.5)(1.8)] + 0.25\left[\frac{1}{2}(1.5)(1.8)(0.5)\right] + (1.0)\left[\frac{1}{2}(1.5)(1.8)(0.5)\right]}{0.5(2.5)(1.8) + \frac{1}{2}(1.5)(1.8)(0.5) + \frac{1}{2}(1.5)(1.8)(0.5)}$

$= 0.391\ \text{m}$ *Ans.*

$\bar{y} = \dfrac{\Sigma \tilde{y} V}{\Sigma V} = \dfrac{5.00625}{3.6} = 1.39\ \text{m}$ *Ans.*

$\bar{z} = \dfrac{\Sigma \tilde{z} V}{\Sigma V} = \dfrac{2.835}{3.6} = 0.7875\ \text{m}$ *Ans.*

F9–13. $A = 2\pi \Sigma \tilde{r} L$

$= 2\pi\left[0.75(1.5) + 1.5(2) + 0.75\sqrt{(1.5)^2 + (2)^2}\right]$

$= 37.7\ \text{m}^2$ *Ans.*

$V = 2\pi \Sigma \tilde{r} A$

$= 2\pi\left[0.75(1.5)(2) + 0.5\left(\frac{1}{2}\right)(1.5)(2)\right]$

$= 18.8\ \text{m}^3$ *Ans.*

F9–14. $A = 2\pi \Sigma \tilde{r} L$

$= 2\pi \left[1.95\sqrt{(0.9)^2 + (1.2)^2} + 2.4(1.5) + 1.95(0.9) + 1.5(2.7) \right]$

$= 77.5 \text{ m}^2$ *Ans.*

$V = 2\pi \Sigma \tilde{r} A$

$= 2\pi \left[1.8 \left(\tfrac{1}{2} \right)(0.9)(1.2) + 1.95(0.9)(1.5) \right]$

$= 22.6 \text{ m}^3$ *Ans.*

F9–15. $A = 2\pi \Sigma \tilde{r} L$

$= 2\pi \left[7.5(15) + 15(18) + 22.5\sqrt{15^2 + 20^2} + 15(30) \right]$

$= 8765 \text{ in.}^2$ *Ans.*

$V = 2\pi \Sigma \tilde{r} A$

$= 2\pi \left[7.5(15)(38) + 20 \left(\tfrac{1}{2} \right)(15)(20) \right]$

$= 45\,710 \text{ in.}^3$ *Ans.*

F9–16. $A = 2\pi \Sigma \tilde{r} L$

$= 2\pi \left[\tfrac{2(1.5)}{\pi} \left(\tfrac{\pi(1.5)}{2} \right) + 1.5(2) + 0.75(1.5) \right]$

$= 40.1 \text{ m}^2$ *Ans.*

$V = 2\pi \Sigma \tilde{r} A$

$= 2\pi \left[\tfrac{4(1.5)}{3\pi} \left(\tfrac{\pi(1.5^2)}{4} \right) + 0.75(1.5)(2) \right]$

$= 21.2 \text{ m}^3$ *Ans.*

F9–17. $w_b = \rho_w g h b = 1000(9.81)(6)(1)$

$= 58.86 \text{ kN/m}$

$F_R = \tfrac{1}{2}(58.76)(6) = 176.58 \text{ kN} = 177 \text{ kN}$ *Ans.*

F9–18. $w_b = \gamma_w h b = 62.4\,(4)(4) = 998.4 \text{ lb/ft}$

$F_R = 998.4(3) = 3.00 \text{ kip}$ *Ans.*

F9–19. $w_b = \rho_w g h_B b = 1000(9.81)(2)(1.5)$

$= 29.43 \text{ kN/m}$

$F_R = \tfrac{1}{2}(29.43) \left(\sqrt{(1.5)^2 + (2)^2} \right)$

$= 36.8 \text{ kN}$ *Ans.*

F9–20. $w_A = \rho_w g h_A b = 1000(9.81)(3)(2)$

$= 58.86 \text{ kN/m}$

$w_B = \rho_w g h_B b = 1000(9.81)(5)(2)$

$= 98.1 \text{ kN/m}$

$F_R = \tfrac{1}{2}(58.86 + 98.1)(2) = 157 \text{ kN}$ *Ans.*

F9–21. $w_A = \gamma_w h_A b = 62.4(6)(2) = 748.8 \text{ lb/ft}$

$w_B = \gamma_w h_B b = 62.4(10)(2) = 1248 \text{ lb/ft}$

$F_R = \tfrac{1}{2}(748.8 + 1248) \left(\sqrt{(3)^2 + (4)^2} \right)$

$= 4.99 \text{ kip}$ *Ans.*

Chapter 10

F10–1.

$I_x = \int_A y^2 \, dA = \int_0^{1 \text{ m}} y^2 \left[\left(1 - y^{3/2} \right) dy \right] = 0.111 \text{ m}^4$ *Ans.*

F10–2.

$I_x = \int_A y^2 \, dA = \int_0^{1 \text{ m}} y^2 \left(y^{3/2} \, dy \right) = 0.222 \text{ m}^4$ *Ans.*

F10–3.

$I_y = \int_A x^2 \, dA = \int_0^{1 \text{ m}} x^2 \left(x^{2/3} \right) dx = 0.273 \text{ m}^4$ *Ans.*

F10–4.

$I_y = \int_A x^2 \, dA = \int_0^{1 \text{ m}} x^2 \left[(1 - x^{2/3}) \, dx \right] = 0.0606 \text{ m}^4$ *Ans.*

F10–5. $I_x = \left[\tfrac{1}{12}(50) \left(450^3 \right) + 0 \right] + \left[\tfrac{1}{12}(300) \left(50^3 \right) + 0 \right]$

$= 383 \left(10^6 \right) \text{ mm}^4$ *Ans.*

$I_y = \left[\tfrac{1}{12}(450) \left(50^3 \right) + 0 \right]$

$+ 2 \left[\tfrac{1}{12}(50) \left(150^3 \right) + (150)(50)(100)^2 \right]$

$= 183 \left(10^6 \right) \text{ mm}^4$ *Ans.*

F10–6. $I_x = \tfrac{1}{12}(360) \left(200^3 \right) - \tfrac{1}{12}(300) \left(140^3 \right)$

$= 171 \left(10^6 \right) \text{ mm}^4$ *Ans.*

$I_y = \tfrac{1}{12}(200) \left(360^3 \right) - \tfrac{1}{12}(140) \left(300^3 \right)$

$= 463 \left(10^6 \right) \text{ mm}^4$ *Ans.*

F10–7. $I_y = 2 \left[\tfrac{1}{12}(50) \left(200^3 \right) + 0 \right]$

$+ \left[\tfrac{1}{12}(300) \left(50^3 \right) + 0 \right]$

$= 69.8 \, (10^6) \text{ mm}^4$ *Ans.*

F10–8.

$\bar{y} = \dfrac{\Sigma \tilde{y} A}{\Sigma A} = \dfrac{15(150)(30) + 105(30)(150)}{150(30) + 30(150)} = 60 \text{ mm}$

$\bar{I}_{x'} = \Sigma (\bar{I} + A d^2)$

$= \left[\tfrac{1}{12}(150)(30)^3 + (150)(30)(60 - 15)^2 \right]$

$+ \left[\tfrac{1}{12}(30)(150)^3 + 30(150)(105 - 60)^2 \right]$

$= 27.0 \, (10^6) \text{ mm}^4$ *Ans.*

Chapter 11

F11–1. $y_G = 0.75 \sin \theta$ $\delta y_G = 0.75 \cos \theta \, \delta\theta$

$x_C = 2(1.5) \cos \theta$ $\delta x_C = -3 \sin \theta \, \delta\theta$

$\delta U = 0; \quad 2W\delta y_G + P\delta x_C = 0$

     $(294.3 \cos \theta - 3P \sin \theta)\delta\theta = 0$

$P = 98.1 \cot \theta \big|_{\theta=60°} = 56.6 \text{ N}$ *Ans.*

F11–2. $x_A = 5 \cos \theta$ $\delta x_A = -5 \sin \theta \, \delta\theta$

$y_G = 2.5 \sin \theta$ $\delta y_G = 2.5 \cos \theta \, \delta\theta$

$\delta U = 0; \quad\quad -P\delta x_A + (-W\delta y_G) = 0$

     $(5P \sin \theta - 1226.25 \cos \theta)\delta\theta = 0$

$P = 245.25 \cot \theta \big|_{\theta=60°} = 142 \text{ N}$ *Ans.*

F11–3. $x_B = 0.6 \sin \theta$ $\delta x_B = 0.6 \cos \theta \, \delta\theta$

$y_C = 0.6 \cos \theta$ $\delta y_C = -0.6 \sin \theta \, \delta\theta$

$\delta U = 0; \quad\quad -F_{sp}\delta x_B + (-P\delta y_C) = 0$

     $-9(10^3) \sin \theta \, (0.6 \cos \theta \, \delta\theta)$

     $- 2000(-0.6 \sin \theta \, \delta\theta) = 0$

$\sin \theta = 0 \quad\quad \theta = 0°$ *Ans.*

     $-5400 \cos \theta + 1200 = 0$

$\theta = 77.16° = 77.2°$ *Ans.*

F11–4. $x_B = 0.9 \cos \theta$ $\delta x_B = -0.9 \sin \theta \, \delta\theta$

$x_C = 2(0.9 \cos \theta)$ $\delta x_C = -1.8 \sin \theta \, \delta\theta$

$\delta U = 0; \quad P\delta x_B + \left(-F_{sp} \, \delta x_C\right) = 0$

$6(10^3)(-0.9 \sin \theta \, \delta\theta)$

$- 36(10^3)(\cos \theta - 0.5)(-1.8 \sin \theta \, \delta\theta) = 0$

$\sin \theta \, (64\,800 \cos \theta - 37\,800)\delta\theta = 0$

$\sin \theta = 0 \quad\quad \theta = 0°$ *Ans.*

     $64\,800 \cos \theta - 37\,800 = 0$

     $\theta = 54.31° = 54.3°$ *Ans.*

F11–5. $y_G = 2.5 \sin \theta$ $\delta y_G = 2.5 \cos \theta \, \delta\theta$

$x_A = 5 \cos \theta$ $\delta x_C = -5 \sin \theta \, \delta\theta$

$\delta U = 0; \quad\quad \left(-F_{sp}\delta x_A\right) - W\delta y_G = 0$

$(15\,000 \sin \theta \cos \theta - 7500 \sin \theta$

     $- 1226.25 \cos \theta)\delta\theta = 0$

$\theta = 56.33° = 56.3°$ *Ans.*

or $\theta = 9.545° = 9.55°$ *Ans.*

F11–6. $F_{sp} = 15\,000(0.6 - 0.6 \cos \theta)$

$x_C = 3[0.3 \sin \theta]$ $\delta x_C = 0.9 \cos \theta \, \delta\theta$

$y_B = 2[0.3 \cos \theta]$ $\delta y_B = -0.6 \sin \theta \, \delta\theta$

$\delta U = 0; \quad P\delta x_C + F_{sp}\delta y_B = 0$

     $(135 \cos \theta - 5400 \sin \theta + 5400 \sin \theta \cos \theta)\delta\theta = 0$

$\theta = 20.9°$ *Ans.*

Preliminary Problems
Statics Solutions

Chapter 2

2–1.

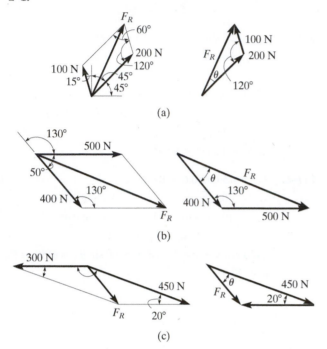

(a)

(b)

(c)

2–2.

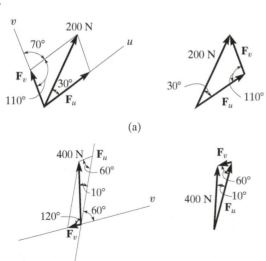

(a)

(b)

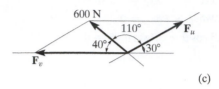

(c)

2–3.

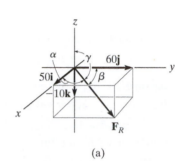

(a)

(b)

2–4.   a) $\mathbf{F} = \{-4\mathbf{i} - 4\mathbf{j} + 2\mathbf{k}\}$ kN

$F = \sqrt{(4)^2 + (-4)^2 + (2)^2} = 6$ kN

$\cos \beta = \dfrac{-2}{3}$

b) $\mathbf{F} = \{20\mathbf{i} + 20\mathbf{j} - 10\mathbf{k}\}$ N

$F = \sqrt{(20)^2 + (20)^2 + (-10)^2} = 30$ N

$\cos \beta = \dfrac{2}{3}$

2–5.

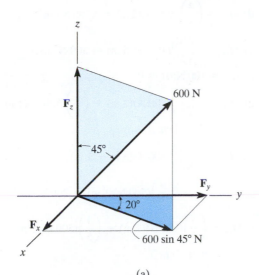

(a)

$$F_x = (600 \sin 45°) \sin 20° \text{ N}$$
$$F_y = (600 \sin 45°) \cos 20° \text{ N}$$
$$F_z = 600 \cos 45° \text{ N}$$

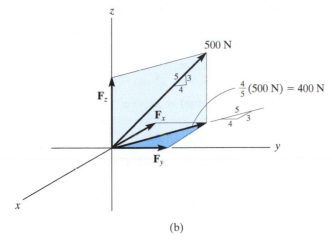

(b)

$$F_x = -\frac{3}{5}(400) \text{ N}$$
$$F_y = \frac{4}{5}(400) \text{ N}$$
$$F_z = \frac{3}{5}(500) \text{ N}$$

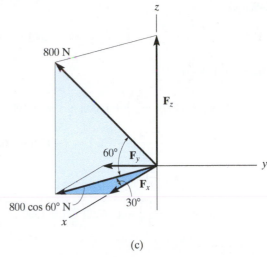

(c)

$$F_x = 800 \cos 60° \cos 30° \text{ N}$$
$$F_y = -800 \cos 60° \sin 30° \text{ N}$$
$$F_z = 800 \sin 60° \text{ N}$$

2–6. a) $\mathbf{r}_{AB} = \{-5\mathbf{i} + 3\mathbf{j} - 2\mathbf{k}\}$ m

b) $\mathbf{r}_{AB} = \{4\mathbf{i} + 8\mathbf{j} - 3\mathbf{k}\}$ m

c) $\mathbf{r}_{AB} = \{6\mathbf{i} - 3\mathbf{j} - 4\mathbf{k}\}$ m

2–7. a) $\mathbf{F} = 15 \text{ kN}\left(\dfrac{-3}{5}\mathbf{i} + \dfrac{4}{5}\mathbf{j}\right) = \{-9\mathbf{i} + 12\mathbf{j}\}$ kN

b) $\mathbf{F} = 600 \text{ N}\left(\dfrac{2}{3}\mathbf{i} + \dfrac{2}{3}\mathbf{j} - \dfrac{1}{3}\mathbf{k}\right)$

$\qquad = \{400\mathbf{i} + 400\mathbf{j} - 200\mathbf{k}\}$ N

c) $\mathbf{F} = 300 \text{ N}\left(-\dfrac{2}{3}\mathbf{i} + \dfrac{2}{3}\mathbf{j} - \dfrac{1}{3}\mathbf{k}\right)$

$\qquad = \{-200\mathbf{i} + 200\mathbf{j} - 100\mathbf{k}\}$ N

2–8. a) $\mathbf{r}_A = \{3\mathbf{k}\}$ m, $r_A = 3$ m

$\qquad \mathbf{r}_B = \{2\mathbf{i} + 2\mathbf{j} - 1\mathbf{k}\}$ m, $r_B = 3$ m

$\qquad \mathbf{r}_A \cdot \mathbf{r}_B = 0(2) + 0(2) + (3)(-1) = -3 \text{ m}^2$

$\qquad \mathbf{r}_A \cdot \mathbf{r}_B = r_A r_B \cos \theta$

$\qquad\quad -3 = 3(3) \cos \theta$

b) $\mathbf{r}_A = \{-2\mathbf{i} + 2\mathbf{j} + 1\mathbf{k}\}$ m, $r_A = 3$ m

$\qquad \mathbf{r}_B = \{1.5\mathbf{i} - 2\mathbf{k}\}$ m, $r_B = 2.5$ m

$\qquad \mathbf{r}_A \cdot \mathbf{r}_B = (-2)(1.5) + 2(0) + (1)(-2) = -5 \text{ m}^2$

$\qquad \mathbf{r}_A \cdot \mathbf{r}_B = r_A r_B \cos \theta$

$\qquad\quad -5 = 3(2.5) \cos \theta$

2–9. a)

$$\mathbf{F} = 300 \text{ N}\left(\frac{2}{3}\mathbf{i} + \frac{2}{3}\mathbf{j} - \frac{1}{3}\mathbf{k}\right) = \{200\mathbf{i} + 200\mathbf{j} - 100\mathbf{k}\} \text{ N}$$

$$\mathbf{u}_a = -\frac{3}{5}\mathbf{i} + \frac{4}{5}\mathbf{j}$$

$$F_a = \mathbf{F} \cdot \mathbf{u}_a = (200)\left(-\frac{3}{5}\right) + (200)\left(\frac{4}{5}\right) + (-100)\left(0\right)$$

b) $\mathbf{F} = 500 \text{ N}\left(-\frac{4}{5}\mathbf{j} + \frac{3}{5}\mathbf{k}\right) = \{-400\mathbf{j} + 300\mathbf{k}\} \text{ N}$

$$\mathbf{u}_a = -\frac{1}{3}\mathbf{i} + \frac{2}{3}\mathbf{j} + \frac{2}{3}\mathbf{k}$$

$$F_a = \mathbf{F} \cdot \mathbf{u}_a = (0)\left(-\frac{1}{3}\right) + (-400)\left(\frac{2}{3}\right) + (300)\left(\frac{2}{3}\right)$$

Chapter 3

3–1.

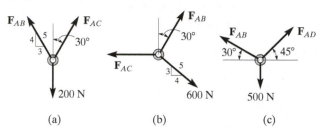

(a)   (b)   (c)

3–2. a) $\Sigma F_x = 0;$ $F \cos 60° - P\left(\dfrac{1}{\sqrt{2}}\right) - 600\left(\dfrac{4}{5}\right) = 0$

$\Sigma F_y = 0;$ $-F \sin 60° - P\left(\dfrac{1}{\sqrt{2}}\right) + 600\left(\dfrac{3}{5}\right) = 0$

b) $\Sigma F_x = 0;$ $P\left(\dfrac{4}{5}\right) - F \sin 60° - 200 \sin 15° = 0$

$\Sigma F_y = 0;$ $-P\left(\dfrac{3}{5}\right) - F \cos 60° + 200 \cos 15° = 0$

c) $\Sigma F_x = 0;$

$300 \cos 40° + 450 \cos 30° - P \cos 30° + F \sin 10° = 0$

$\Sigma F_y = 0;$

$-300 \sin 40° + 450 \sin 30° - P \sin 30° - F \cos 10° = 0$

Chapter 4

4–1. a) $M_O = 100 \text{ N}(2 \text{ m}) = 200 \text{ N} \cdot \text{m} \circlearrowright$

b) $M_O = -100 \text{ N}(1 \text{ m}) = 100 \text{ N} \cdot \text{m} \circlearrowleft$

c) $M_O = -\left(\dfrac{3}{5}\right)(500 \text{ N})(2 \text{ m}) = 600 \text{ N} \cdot \text{m} \circlearrowleft$

d) $M_O = \left(\dfrac{4}{5}\right)(500 \text{ N})(3 \text{ m}) = 1200 \text{ N} \cdot \text{m} \circlearrowright$

e) $M_O = -\left(\dfrac{3}{5}\right)(100 \text{ N})(5 \text{ m}) = 300 \text{ N} \cdot \text{m} \circlearrowleft$

f) $M_O = 100 \text{ N}(0) = 0$

g) $M_O = -\left(\dfrac{3}{5}\right)(500 \text{ N})(2 \text{ m}) + \left(\dfrac{4}{5}\right)(500 \text{ N})(1 \text{ m})$

   $= 200 \text{ N} \cdot \text{m} \circlearrowleft$

h) $M_O = -\left(\dfrac{3}{5}\right)(500 \text{ N})(3 \text{ m} - 1 \text{ m})$

   $+ \left(\dfrac{4}{5}\right)(500 \text{ N})(1 \text{ m}) = 200 \text{ N} \cdot \text{m} \circlearrowleft$

i) $M_O = \left(\dfrac{3}{5}\right)(500 \text{ N})(1 \text{ m}) - \left(\dfrac{4}{5}\right)(500 \text{ N})(3 \text{ m})$

   $= 900 \text{ N} \cdot \text{m} \circlearrowleft$

4–2. $\mathbf{M}_P = \begin{vmatrix} \mathbf{i} & \mathbf{j} & \mathbf{k} \\ 2 & -3 & 0 \\ -3 & 2 & 5 \end{vmatrix}$ $\mathbf{M}_P = \begin{vmatrix} \mathbf{i} & \mathbf{j} & \mathbf{k} \\ 2 & 5 & -1 \\ 2 & -4 & -3 \end{vmatrix}$

$\mathbf{M}_P = \begin{vmatrix} \mathbf{i} & \mathbf{j} & \mathbf{k} \\ 5 & -4 & -1 \\ -2 & 3 & 4 \end{vmatrix}$

4–3. a) $M_x = -(100 \text{ N})(3 \text{ m}) = -300 \text{ N} \cdot \text{m}$

$M_y = -(200 \text{ N})(2 \text{ m}) = -400 \text{ N} \cdot \text{m}$

$M_z = -(300 \text{ N})(2 \text{ m}) = -600 \text{ N} \cdot \text{m}$

b) $M_x = (50 \text{ N})(0.5 \text{ m}) = 25 \text{ N} \cdot \text{m}$

$M_y = (400 \text{ N})(0.5 \text{ m}) - (300 \text{ N})(3 \text{ m}) = -700 \text{ N} \cdot \text{m}$

$M_z = (100 \text{ N})(3 \text{ m}) = 300 \text{ N} \cdot \text{m}$

c) $M_x = (300 \text{ N})(2 \text{ m}) - (100 \text{ N})(2 \text{ m}) = 400 \text{ N} \cdot \text{m}$

$M_y = -(300 \text{ N})(1 \text{ m}) + (50 \text{ N})(1 \text{ m})$

   $+ (400 \text{ N})(0.5 \text{ m}) = 250 \text{ N} \cdot \text{m}$

$M_z = -(200 \text{ N})(1 \text{ m}) = -200 \text{ N} \cdot \text{m}$

4–4. a)

$$M_a = \begin{vmatrix} -\dfrac{4}{5} & -\dfrac{3}{5} & 0 \\ -5 & 2 & 0 \\ 6 & 2 & 3 \end{vmatrix} = \begin{vmatrix} -\dfrac{4}{5} & -\dfrac{3}{5} & 0 \\ -1 & 5 & 0 \\ 6 & 2 & 3 \end{vmatrix}$$

b)

$$M_a = \begin{vmatrix} -\dfrac{1}{\sqrt{2}} & \dfrac{1}{\sqrt{2}} & 0 \\ 3 & 4 & -2 \\ 2 & -4 & 3 \end{vmatrix} = \begin{vmatrix} -\dfrac{1}{\sqrt{2}} & \dfrac{1}{\sqrt{2}} & 0 \\ 5 & 2 & -2 \\ 2 & -4 & 3 \end{vmatrix}$$

c) $M_a = \begin{vmatrix} \frac{2}{3} & -\frac{1}{3} & \frac{2}{3} \\ -5 & -4 & 0 \\ 2 & -4 & 3 \end{vmatrix} = \begin{vmatrix} \frac{2}{3} & -\frac{1}{3} & \frac{2}{3} \\ -3 & -5 & 2 \\ 2 & -4 & 3 \end{vmatrix}$

4–5. a) $\xrightarrow{+} (F_R)_x = \Sigma F_x;$

$(F_R)_x = -\left(\frac{4}{5}\right)500\text{ N} + 200\text{ N} = -200\text{ N}$

$+\uparrow (F_R)_y = \Sigma F_y;$

$(F_R)_y = -\frac{3}{5}(500\text{ N}) - 400\text{ N} = -700\text{ N}$

$\zeta + (M_R)_O = \Sigma M_O;$

$(M_R)_O = -\left(\frac{3}{5}\right)(500\text{ N})(2\text{ m}) - 400\text{ N}(4\text{ m})$

$= -2200\text{ N} \cdot \text{m}$

b) $\xrightarrow{+} (F_R)_x = \Sigma F_x;$

$(F_R)_x = \left(\frac{4}{5}\right)(500\text{ N}) = 400\text{ N}$

$+\uparrow (F_R)_y = \Sigma F_y;$

$(F_R)_y = -(300\text{ N}) - \left(\frac{3}{5}\right)(500\text{ N}) = -600\text{ N}$

$\zeta + (M_R)_O = \Sigma M_O;$

$(M_R)_O = -(300\text{ N})(2\text{ m}) - \left(\frac{3}{5}\right)(500\text{ N})(4\text{ m})$

$- 200\text{ N} \cdot \text{m} = -2000\text{ N} \cdot \text{m}$

c) $\xrightarrow{+} (F_R)_x = \Sigma F_x;$

$(F_R)_x = \left(\frac{3}{5}\right)(500\text{ N}) + 100\text{ N} = 400$

$+\uparrow (F_R)_y = \Sigma F_y;$

$(F_R)_y = -(500\text{ N}) - \left(\frac{4}{5}\right)(500\text{ N}) = -900\text{ N}$

$\zeta + (M_R)_O = \Sigma M_O;$

$(M_R)_O = -(500\text{ N})(2\text{ m}) - \left(\frac{4}{5}\right)(500\text{ N})(4\text{ m})$

$+ \left(\frac{3}{5}\right)(500\text{ N})(2\text{ m}) = -2000\text{ N} \cdot \text{m}$

d) $\xrightarrow{+} (F_R)_x = \Sigma F_x;$

$(F_R)_x = -\left(\frac{4}{5}\right)(500\text{ N}) + \left(\frac{3}{5}\right)(500\text{ N}) = -100\text{ N}$

$+\uparrow (F_R)_y = \Sigma F_y;$

$(F_R)_y = -\left(\frac{3}{5}\right)(500\text{ N}) - \left(\frac{4}{5}\right)(500\text{ N}) = -700\text{ N}$

$\zeta + (M_R)_O = \Sigma M_O;$

$(M_R)_O = \left(\frac{4}{5}\right)(500\text{ N})(4\text{ m}) + \left(\frac{3}{5}\right)(500\text{ N})(2\text{ m})$

$- \left(\frac{3}{5}\right)(500\text{ N})(4\text{ m}) + 200\text{ N} \cdot \text{m} = 1200\text{ N} \cdot \text{m}$

4–6. a) $\xrightarrow{+} (F_R)_x = \Sigma F_x; \quad (F_R)_x = 0$

$+\uparrow (F_R)_y = \Sigma F_y;$

$(F_R)_y = -200\text{ N} - 260\text{ N} = -460\text{ N}$

$\zeta + (F_R)_y d = \Sigma M_O;$

$-(460\text{ N})d = -(200\text{ N})(2\text{ m}) - (260\text{ N})(4\text{ m})$

$d = 3.13\text{ m}$

Note: Although 460 N acts downward, this is *not* why $-(460\text{ N})d$ is negative. It is because the *moment* of 460 N about O is negative.

b) $\xrightarrow{+} (F_R)_x = \Sigma F_x;$

$(F_R)_x = -\left(\frac{3}{5}\right)(500\text{ N}) = -300\text{ N}$

$+\uparrow (F_R)_y = \Sigma F_y;$

$(F_R)_y = -400\text{ N} - \left(\frac{4}{5}\right)(500\text{ N}) = -800\text{ N}$

$\zeta + (F_R)_y d = \Sigma M_O;$

$-(800\text{ N})d = -(400\text{ N})(2\text{ m}) - \left(\frac{4}{5}\right)(500\text{ N})(4\text{ m})$

$d = 3\text{ m}$

c) $\xrightarrow{+} (F_R)_x = \Sigma F_x;$

$(F_R)_x = \left(\frac{4}{5}\right)(500\text{ N}) - \left(\frac{4}{5}\right)(500\text{ N}) = 0$

$+\uparrow (F_R)_y = \Sigma F_y;$

$(F_R)_y = -\left(\frac{3}{5}\right)(500\text{ N}) - \left(\frac{3}{5}\right)(500\text{ N}) = -600\text{ N}$

$\zeta + (F_R)_y d = \Sigma M_O;$

$-(600\text{ N})d = -\left(\frac{3}{5}\right)(500\text{ N})(2\text{ m}) - \left(\frac{3}{5}\right)(500\text{ N})(4\text{ m})$

$- 600\text{ N} \cdot \text{m}$

$d = 4\text{ m}$

4–7. a) $+\downarrow F_R = \Sigma F_z;$

$F_R = 200\text{ N} + 100\text{ N} + 200\text{ N} = 500\text{ N}$

$(M_R)_x = \Sigma M_x;$

$-(500\text{ N})y = -(100\text{ N})(2\text{ m}) - (200\text{ N})(2\text{ m})$

$y = 1.20\text{ m}$

$(M_R)_y = \Sigma M_y;$

$(500\text{ N})x = (100\text{ N})(2\text{ m}) + (200\text{ N})(1\text{ m})$

$x = 0.80\text{ m}$

b) $+\downarrow F_R = \Sigma F_z$;

$F_R = 100\text{ N} - 100\text{ N} + 200\text{ N} = 200\text{ N}$

$(M_R)_x = \Sigma M_x$;

$-(200\text{ N})y = (100\text{ N})(1\text{ m}) + (100\text{ N})(2\text{ m})$
$\quad - (200\text{ N})(2\text{ m})$

$\quad\quad\quad y = 0.5\text{ m}$

$(M_R)_y = \Sigma M_y$;

$(200\text{ N})x = -(100\text{ N})(2\text{ m}) + (100\text{ N})(2\text{ m})$
$\quad\quad\quad x = 0$

c) $+\downarrow F_R = \Sigma F_z$;

$F_R = 400\text{ N} + 300\text{ N} + 200\text{ N} + 100\text{ N} = 1000\text{ N}$

$(M_R)_x = \Sigma M_x$;

$-(1000\text{ N})y = -(300\text{ N})(4\text{ m}) - (100\text{ N})(4\text{ m})$
$\quad\quad\quad\quad y = 1.6\text{ m}$

$(M_R)_y = \Sigma M_y$;

$(1000\text{ N})x = (400\text{ N})(2\text{ m}) + (300\text{ N})(2\text{ m})$
$\quad\quad\quad\quad - (200\text{ N})(2\text{ m}) - (100\text{ N})(2\text{ m})$
$\quad\quad\quad x = 0.8\text{ m}$

Chapter 5

5–1.

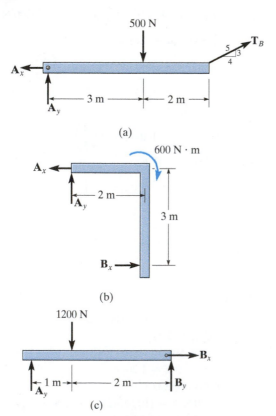

(a)

(b)

(c)

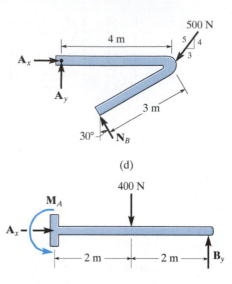

(d)

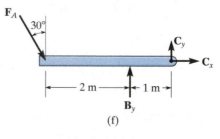

(e)

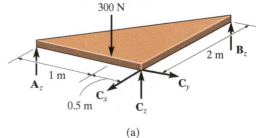

(f)

5–2.

(a)

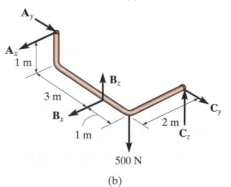

(b)

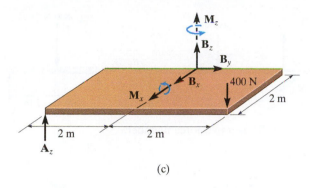

(c)

5–3. a) $\Sigma M_x = 0$;

$-(400\text{ N})(2\text{ m}) - (600\text{ N})(5\text{ m}) + B_z(5\text{ m}) = 0$

$\Sigma M_y = 0$; $-A_z(4\text{ m}) - B_z(4\text{ m}) = 0$

$\Sigma M_z = 0$; $B_y(4\text{ m}) - B_x(5\text{ m})$

$+ (300\text{ N})(5\text{ m}) = 0$

b) $\Sigma M_x = 0$; $A_z(4\text{ m}) + C_z(6\text{ m}) = 0$

$\Sigma M_y = 0$; $B_z(1\text{ m}) - C_z(1\text{ m}) = 0$

$\Sigma M_z = 0$; $-B_y(1\text{ m}) + (300\text{ N})(2\text{ m})$

$- A_x(4\text{ m}) + C_y(1\text{ m}) = 0$

c) $\Sigma M_x = 0$; $B_z(2\text{ m}) + C_z(3\text{ m}) - 800\text{ N}\cdot\text{m} = 0$

$\Sigma M_y = 0$; $-C_z(1.5\text{ m}) = 0$

$\Sigma M_z = 0$; $-B_x(2\text{ m}) + C_y(1.5\text{ m}) = 0$

Chapter 6

6–1. a) $A_y = 200\text{ N}, D_x = 0, D_y = 200\text{ N}$

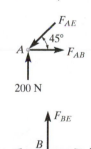

b) $A_y = 300\text{ N}, C_x = 0, C_y = 300\text{ N}$

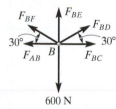

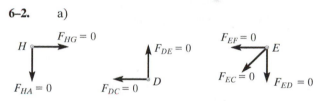

6–2.   a)

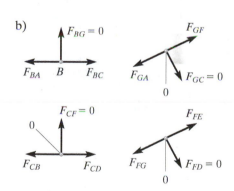

6–3.   a)

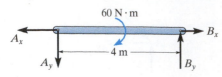

b) *CB* is a two-force member.

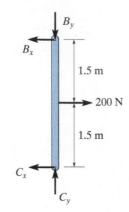

c) *CD* is a two-force member.

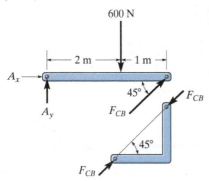

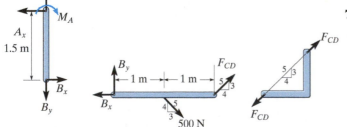

d)

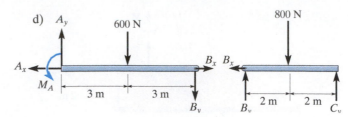

BC is a two-force member.

e)

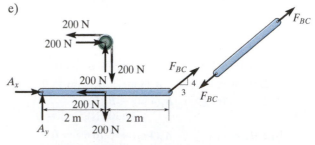

BC is a two-force member.

f)

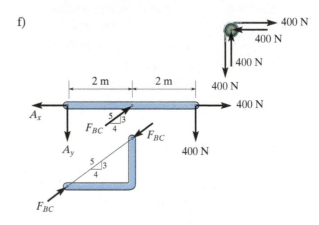

Chapter 7

7–1.

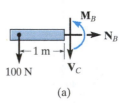

(a)

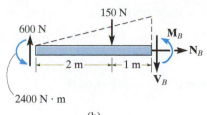

(b)

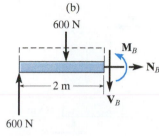

(c)

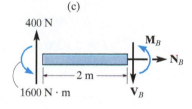

(d)

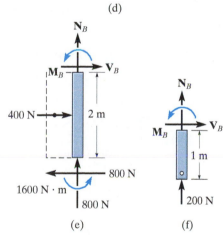

(e)   (f)

Chapter 8

8–1.   a)

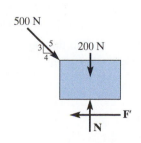

$\xrightarrow{+} \Sigma F_x = 0;$

$\left(\dfrac{4}{5}\right)(500\ \text{N}) - F' = 0,\ F' = 400\ \text{N}$

$+\uparrow \Sigma F_y = 0;$

$N - 200\ \text{N} - \left(\dfrac{3}{5}\right)(500\ \text{N}) = 0,\ N = 500\ \text{N}$

$F_{\max} = 0.3(500\ \text{N}) = 150\ \text{N} < 400\ \text{N}$

Slipping $F = \mu_k N = 0.2(500\ \text{N}) = 100\ \text{N}$ *Ans.*

b)

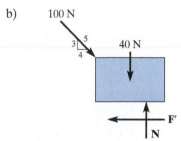

$\xrightarrow{+} \Sigma F_x = 0;$

$\dfrac{4}{5}(100\ \text{N}) - F' = 0;\ F' = 80\ \text{N}$

$+\uparrow \Sigma F_y = 0;$

$N - 40\ \text{N} - \left(\dfrac{3}{5}\right)(100\ \text{N}) = 0;\ N = 100\ \text{N}$

$F_{\max} = 0.9(100\ \text{N}) = 90\ \text{N} > 80\ \text{N}$

$F = F' = 80\ \text{N}$ *Ans.*

8–2.

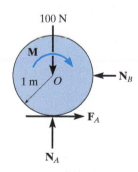

Require $F_A = 0.1 N_A$

$+\uparrow \Sigma F_y = 0;$ $N_A - 100\ \text{N} = 0$

$N_A = 100\ \text{N}$

$F_A = 0.1(100\ \text{N}) = 10\ \text{N}$

$\zeta + \Sigma M_O = 0;$ $-M + (10\ \text{N})(1\ \text{m}) = 0$

$M = 10\ \text{N} \cdot \text{m}$

8–3.   a) Slipping must occur between A and B.

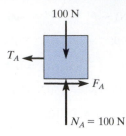

$$F_A = 0.2(100 \text{ N}) = 20 \text{ N}$$

b) Assume B slips on C and C does not slip.

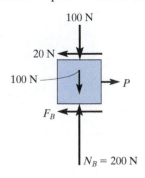

$$F_B = 0.2(200 \text{ N}) = 40 \text{ N}$$

$$\xrightarrow{+} \Sigma F_x = 0; \quad P - 20 \text{ N} - 40 \text{ N} = 0$$

$$P = 60 \text{ N}$$

c) Assume C slips and B does not slip on C.

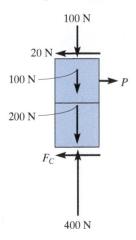

$$F_C = 0.1(400 \text{ N}) = 40 \text{ N}$$

$$\xrightarrow{+} \Sigma F_x = 0; \quad P - 20 \text{ N} - 40 \text{ N} = 0$$

$$P = 60 \text{ N}$$

Therefore, $P = 60 \text{ N}$ *Ans.*

8–4.

a)

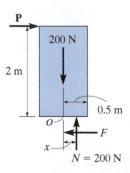

Assume slipping, $F = 0.3(200 \text{ N}) = 60 \text{ N}$

$$\xrightarrow{+} \Sigma F_x = 0; \quad P - 60 \text{ N} = 0; \quad P = 60 \text{ N}$$

$$\zeta + \Sigma M_O = 0; \quad 200 \text{ N}(x) - (60 \text{ N})(2 \text{ m}) = 0$$

$$x = 0.6 \text{ m} > 0.5 \text{ m}$$

Block tips, $x = 0.5 \text{ m}$

$$\zeta + \Sigma M_O = 0 \quad (200 \text{ N})(0.5 \text{ m}) - P(2 \text{ m}) = 0$$

$$P = 50 \text{ N}$$ *Ans.*

b)

100 N

P

1 m

0.5 m

F

x

$N = 100 \text{ N}$

Assume slipping, $F = 0.4(100 \text{ N}) = 40 \text{ N}$

$$\xrightarrow{+} \Sigma F_x = 0; \quad P - 40 \text{ N} = 0; \quad P = 40 \text{ N}$$

$$\zeta + \Sigma M_O = 0; \quad (100 \text{ N})(x) - (40 \text{ N})(1 \text{ m}) = 0$$

$$x = 0.4 \text{ m} < 0.5 \text{ m}$$

No tipping

$$P = 40 \text{ N}$$ *Ans.*

Chapter 9

9–1. a)

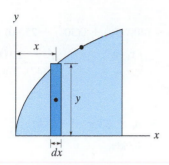

$$\tilde{x} = x$$

$$\tilde{y} = \frac{y}{2} = \frac{\sqrt{x}}{2}$$

$$dA = y\,dx = \sqrt{x}\,dx$$

c)

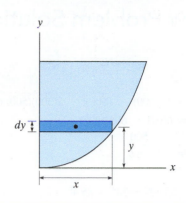

$$\tilde{x} = \frac{x}{2} = \frac{\sqrt{y}}{2}$$

$$\tilde{y} = y$$

$$dA = x\,dy = \sqrt{y}\,dy$$

b)

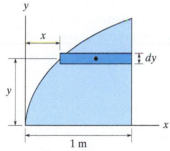

$$\tilde{x} = x + \left(\frac{1-x}{2}\right) = \frac{1+x}{2} = \frac{1+y^2}{2}$$

$$\tilde{y} = y$$

$$dA = (1-x)\,dy = (1-y^2)\,dy$$

d)

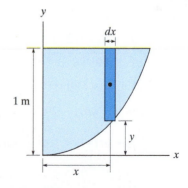

$$\tilde{x} = x$$

$$\tilde{y} = y + \left(\frac{1-y}{2}\right) = \frac{1+y}{2} = \frac{1+x^2}{2}$$

$$dA = (1-y)\,dx = (1-x^2)\,dx$$

Review Problem Solutions

Chapter 2

R2–1. $F_R = \sqrt{(300)^2 + (500)^2 - 2(300)(500)\cos 95°}$

$\qquad = 605.1 = 605 \text{ N}$ *Ans.*

$\qquad \dfrac{605.1}{\sin 95°} = \dfrac{500}{\sin \theta}$

$\qquad \theta = 55.40°$

$\qquad \phi = 55.40° + 30° = 85.4°$ *Ans.*

R2–2. $\dfrac{F_{1v}}{\sin 30°} = \dfrac{250}{\sin 105°}$ $F_{1v} = 129 \text{ N}$ *Ans.*

$\qquad \dfrac{F_{1u}}{\sin 45°} = \dfrac{250}{\sin 105°}$ $F_{1u} = 183 \text{ N}$ *Ans.*

R2–3. $F_{Rx} = F_{1x} + F_{2x} + F_{3x} + F_{4x}$

$\qquad F_{Rx} = -200 + 320 + 180 - 300 = 0$

$\qquad F_{Ry} = F_{1y} + F_{2y} + F_{3y} + F_{4y}$

$\qquad F_{Ry} = 0 - 240 + 240 + 0 = 0$

$\qquad$ Thus, $F_R = 0$ *Ans.*

R2–4. $\cos^2 30° + \cos^2 70° + \cos^2 \gamma = 1$

$\qquad \cos \gamma = \pm 0.3647$

$\qquad \gamma = 68.61°$ or $111.39°$

$\qquad$ By inspection, $\gamma = 111.39°$.

$\qquad \mathbf{F} = 250\{\cos 30°\mathbf{i} + \cos 70°\mathbf{j} + \cos 111.39°\}$ lb

$\qquad = \{217\mathbf{i} + 85.5\mathbf{j} - 91.2\mathbf{k}\}$ lb   *Ans.*

R2–5. $\mathbf{r} = \{50 \sin 20°\mathbf{i} + 50 \cos 20°\mathbf{j} - 35\mathbf{k}\}$ ft

$\qquad r = \sqrt{(17.10)^2 + (46.98)^2 + (-35)^2} = 61.03$ ft

$\qquad \mathbf{u} = \dfrac{\mathbf{r}}{r} = (0.280\mathbf{i} + 0.770\mathbf{j} - 0.573\mathbf{k})$

$\qquad \mathbf{F} = F\mathbf{u} = \{98.1\mathbf{i} + 269\mathbf{j} - 201\mathbf{k}\}$ lb   *Ans.*

R2–6. $\mathbf{F}_1 = 600\left(\dfrac{4}{5}\right)\cos 30°(+\mathbf{i}) + 600\left(\dfrac{4}{5}\right)\sin 30°(-\mathbf{j})$

$\qquad\qquad\qquad + 600\left(\dfrac{3}{5}\right)(+\mathbf{k})$

$\qquad = \{415.69\mathbf{i} - 240\mathbf{j} + 360\mathbf{k}\}$ N   *Ans.*

$\qquad \mathbf{F}_2 = 0\mathbf{i} + 450 \cos 45°(+\mathbf{j}) + 450 \sin 45°(+\mathbf{k})$

$\qquad = \{318.20\mathbf{j} + 318.20\mathbf{k}\}$ N   *Ans.*

R2–7. $\mathbf{r}_1 = \{400\mathbf{i} + 250\mathbf{k}\}$ mm; $r_1 = 471.70$ mm

$\qquad \mathbf{r}_2 = \{50\mathbf{i} + 300\mathbf{j}\}$ mm; $r_2 = 304.14$ mm

648

$\mathbf{r}_1 \cdot \mathbf{r}_2 = (400)(50) + 0(300) + 250(0) = 20\,000$

$\theta = \cos^{-1}\left(\dfrac{\mathbf{r}_1 \cdot \mathbf{r}_2}{r_1 r_2}\right) = \cos^{-1}\left(\dfrac{20\,000}{(471.70)(304.14)}\right)$

$\qquad = 82.0°$ *Ans.*

R2–8. $F_{\text{Proj}} = \mathbf{F} \cdot \mathbf{u}_v = (2\mathbf{i} + 4\mathbf{j} + 10\mathbf{k}) \cdot \left(\dfrac{2}{3}\mathbf{i} + \dfrac{2}{3}\mathbf{j} - \dfrac{1}{3}\mathbf{k}\right)$

$\qquad F_{\text{Proj}} = 0.667$ kN

Chapter 3

R3–1. $\xrightarrow{+} \Sigma F_x = 0;$ $F_B - F_A \cos 60° - 50\left(\dfrac{4}{5}\right) = 0$

$\qquad +\uparrow \Sigma F_y = 0;$ $-F_A \sin 60° + 50\left(\dfrac{3}{5}\right) = 0$

$\qquad F_A = 34.6$ lb $F_B = 57.3$ lb   *Ans.*

R3–2. $\xrightarrow{+} \Sigma F_x = 0;$ $F_{AC} \cos 30° - F_{AB} = 0$ (1)

$\qquad +\uparrow \Sigma F_y = 0;$ $F_{AC} \sin 30° - W = 0$ (2)

$\qquad$ Assuming cable AB reaches the maximum tension $F_{AB} = 450$ lb.

$\qquad$ From Eq. (1) $F_{AC} \cos 30° - 450 = 0$

$\qquad\qquad F_{AC} = 519.6$ lb > 480 lb   (No Good)

$\qquad$ Assuming cable AC reaches the maximum tension $F_{AC} = 480$ lb.

$\qquad$ From Eq. (1) $480 \cos 30° - F_{AB} = 0$

$\qquad\qquad F_{AB} = 415.7$ lb < 450 lb   (OK)

$\qquad$ From Eq. (2) $480 \sin 30° - W = 0$ $W = 240$ lb

$\qquad\qquad\qquad\qquad\qquad\qquad\qquad\qquad\qquad$ *Ans.*

R3–3. $\xrightarrow{+} \Sigma F_x = 0;$ $F_{AC} \sin 30° - F_{AB}\left(\dfrac{3}{5}\right) = 0$

$\qquad\qquad F_{AC} = 1.20 F_{AB}$ (1)

$\qquad +\uparrow \Sigma F_y = 0;$ $F_{AC} \cos 30° + F_{AB}\left(\dfrac{4}{5}\right) - W = 0$

$\qquad\qquad 0.8660 F_{AC} + 0.8 F_{AB} = W$ (2)

$\qquad$ Since $F_{AC} > F_{AB}$, failure will occur first at cable AC with $F_{AC} = 50$ lb. Then solving Eqs. (1) and (2) yields

$\qquad\qquad F_{AB} = 41.67$ lb

$\qquad\qquad W = 76.6$ lb   *Ans.*

R3–4. $s_1 = \dfrac{60}{40} = 1.5 \text{ ft}$

$+\uparrow \Sigma F_y = 0; \qquad F - 2\left(\dfrac{1}{2}T\right) = 0; \qquad F = T$

$\xrightarrow{+} \Sigma F_x = 0; \qquad -F_s + 2\left(\dfrac{\sqrt{3}}{2}\right)F = 0$

$\qquad\qquad\qquad F_s = 1.732F$

Final stretch is $1.5 + 0.268 = 1.768 \text{ ft}$

$\qquad\qquad 40(1.768) = 1.732F$

$\qquad\qquad\qquad F = 40.8 \text{ lb} \qquad\qquad\qquad Ans.$

R3–5. $\Sigma F_x = 0; \qquad -F_1 \sin 45° = 0 \qquad F_1 = 0 \qquad Ans.$

$\qquad \Sigma F_z = 0; \qquad F_2 \sin 40° - 200 = 0$

$\qquad\qquad\qquad F_2 = 311.14 \text{ lb} = 311 \text{ lb} \qquad Ans.$

Using the results $F_1 = 0$ and $F_2 = 311.14 \text{ lb}$ and then summing forces along the y axis, we have

$\qquad \Sigma F_y = 0; \qquad F_3 - 311.14 \cos 40° = 0$

$\qquad\qquad\qquad F_3 = 238 \text{ lb} \qquad\qquad\qquad Ans.$

R3–6. $\mathbf{F}_1 = F_1\{\cos 60°\mathbf{i} + \sin 60°\mathbf{k}\}$

$\qquad = \{0.5F_1\mathbf{i} + 0.8660F_1\mathbf{k}\} \text{ N}$

$\qquad \mathbf{F}_2 = F_2\left\{\dfrac{3}{5}\mathbf{i} - \dfrac{4}{5}\mathbf{j}\right\}$

$\qquad = \{0.6 F_2\mathbf{i} - 0.8 F_2\mathbf{j}\} \text{ N}$

$\qquad \mathbf{F}_3 = F_3\{-\cos 30°\mathbf{i} - \sin 30°\mathbf{j}\}$

$\qquad = \{-0.8660F_3\mathbf{i} - 0.5F_3\mathbf{j}\} \text{ N}$

$\qquad \Sigma F_x = 0; \qquad 0.5F_1 + 0.6F_2 - 0.8660F_3 = 0$

$\qquad \Sigma F_y = 0; \qquad -0.8F_2 - 0.5F_3 + 800 \sin 30° = 0$

$\qquad \Sigma F_z = 0; \qquad 0.8660F_1 - 800 \cos 30° = 0$

$\qquad F_1 = 800 \text{ N} \qquad F_2 = 147 \text{ N} \qquad F_3 = 564 \text{ N} \qquad Ans.$

R3–7. $\Sigma F_x = 0; \quad F_{CA}\left(\dfrac{1}{\sqrt{10}}\right) - F_{CB}\left(\dfrac{1}{\sqrt{10}}\right) = 0$

$\qquad \Sigma F_y = 0; \quad -F_{CA}\left(\dfrac{3}{\sqrt{10}}\right) - F_{CB}\left(\dfrac{3}{\sqrt{10}}\right) + F_{CD}\left(\dfrac{3}{5}\right) = 0$

$\qquad \Sigma F_z = 0; \qquad -500 + F_{CD}\left(\dfrac{4}{5}\right) = 0$

Solving:

$\qquad F_{CD} = 625 \text{ lb} \qquad F_{CA} = F_{CB} = 198 \text{ lb}$

R3–8. $\mathbf{F}_{AB} = 700\left(\dfrac{2\mathbf{i} + 3\mathbf{j} - 6\mathbf{k}}{\sqrt{2^2 + 3^2 + (-6)^2}}\right)$

$\qquad = \{200\mathbf{i} + 300\mathbf{j} - 600\mathbf{k}\} \text{ N}$

$\mathbf{F}_{AC} = F_{AC}\left(\dfrac{-1.5\mathbf{i} + 2\mathbf{j} - 6\mathbf{k}}{\sqrt{(-1.5)^2 + 2^2 + (-6)^2}}\right)$

$\qquad = -0.2308F_{AC}\mathbf{i} + 0.3077F_{AC}\mathbf{j} - 0.9231F_{AC}\mathbf{k}$

$\mathbf{F}_{AD} = F_{AD}\left(\dfrac{-3\mathbf{i} - 6\mathbf{j} - 6\mathbf{k}}{\sqrt{(-3)^2 + (-6)^2 + (-6)^2}}\right)$

$\qquad = -0.3333F_{AD}\mathbf{i} - 0.6667F_{AD}\mathbf{j} - 0.6667F_{AD}\mathbf{k}$

$\mathbf{F} = F\mathbf{k}$

$\qquad \Sigma\mathbf{F} = 0; \qquad \mathbf{F}_{AB} + \mathbf{F}_{AC} + \mathbf{F}_{AD} + \mathbf{F} = 0$

$(200 - 0.2308F_{AC} - 0.3333F_{AD})\mathbf{i}$

$\qquad + (300 + 0.3077F_{AC} - 0.6667F_{AD})\mathbf{j}$

$\qquad + (-600 - 0.9231F_{AC} - 0.6667F_{AD} + F)\mathbf{k} = \mathbf{0}$

$200 - 0.2308F_{AC} - 0.3333F_{AD} = 0$

$300 + 0.3077F_{AC} - 0.6667F_{AD} = 0$

$-600 - 0.9231F_{AC} - 0.6667F_{AD} + F = 0$

$F_{AC} - 130 \text{ N} \qquad F_{AD} = 510 \text{ N}$

$\qquad\qquad F = 1060 \text{ N} = 1.06 \text{ kN} \qquad Ans.$

Chapter 4

R4–1. $20(10^3) = 800(16 \cos 30°) + W(30 \cos 30° + 2)$

$\qquad\qquad\qquad W = 319 \text{ lb} \qquad\qquad Ans.$

R4–2. $\mathbf{F}_R = 50 \text{ lb}\left[\dfrac{(10\mathbf{i} + 15\mathbf{j} - 30\mathbf{k})}{\sqrt{(10)^2 + (15)^2 + (-30)^2}}\right]$

$\qquad \mathbf{F}_R = \{14.3\mathbf{i} + 21.4\mathbf{j} - 42.9\mathbf{k}\} \text{ lb} \qquad Ans.$

$(\mathbf{M}_R)_C = \mathbf{r}_{CB} \times \mathbf{F} = \begin{vmatrix} \mathbf{i} & \mathbf{j} & \mathbf{k} \\ 10 & 45 & 0 \\ 14.29 & 21.43 & -42.86 \end{vmatrix}$

$\qquad = \{-1929\mathbf{i} + 428.6\mathbf{j} - 428.6\mathbf{k}\} \text{ lb} \cdot \text{ft} \qquad Ans.$

R4–3. $\mathbf{r} = \{4\mathbf{i}\} \text{ ft}$

$\qquad \mathbf{F} = 24 \text{ lb}\left(\dfrac{-2\mathbf{i} + 2\mathbf{j} + 4\mathbf{k}}{\sqrt{(-2)^2 + (2)^2 + (4)^2}}\right)$

$\qquad = \{-9.80\mathbf{i} + 9.80\mathbf{j} + 19.60\mathbf{k}\} \text{ lb}$

$M_y = \begin{vmatrix} 0 & 1 & 0 \\ 4 & 0 & 0 \\ -9.80 & 9.80 & 19.60 \end{vmatrix} = -78.4 \text{ lb} \cdot \text{ft}$

$\qquad \mathbf{M}_y = \{-78.4\mathbf{j}\} \text{ lb} \cdot \text{ft} \qquad\qquad Ans.$

R4–4. $(M_c)_R = \Sigma M_z; \qquad 0 = 100 - 0.75F$

$\qquad\qquad\qquad F = 133 \text{ N} \qquad\qquad\qquad Ans.$

R4–5. $\xrightarrow{+} \Sigma F_{Rx} = \Sigma F_x;$ $F_{Rx} = 6\left(\dfrac{5}{13}\right) - 4\cos 60°$

$$= 0.30769 \text{ kN}$$

$+\uparrow \Sigma F_{Ry} = \Sigma F_y;$ $F_{Ry} = 6\left(\dfrac{12}{13}\right) - 4\sin 60°$

$$= 2.0744 \text{ kN}$$

$F_R = \sqrt{(0.30769)^2 + (2.0744)^2} = 2.10 \text{ kN}$ *Ans.*

$\theta = \tan^{-1}\left[\dfrac{2.0744}{0.30769}\right] = 81.6° \measuredangle$ *Ans.*

$\zeta + M_P = \Sigma M_P;$ $M_P = 8 - 6\left(\dfrac{12}{13}\right)(7) + 6\left(\dfrac{5}{13}\right)(5)$

$$- 4\cos 60°(4) + 4\sin 60°(3)$$

$$M_P = -16.8 \text{ kN} \cdot \text{m}$$

$$= 16.8 \text{ kN} \cdot \text{m} \curvearrowright$$ *Ans.*

R4–6. $\xrightarrow{+} \Sigma (F_R)_x = \Sigma F_x;$ $(F_R)_x = 200\cos 45° - 250\left(\dfrac{4}{5}\right)$

$$- 300 = -358.58 \text{ lb} = 358.58 \text{ lb} \leftarrow$$

$+\uparrow (F_R)_y = \Sigma F_y;$ $(F_R)_y = -200\sin 45° - 250\left(\dfrac{3}{5}\right)$

$$= -291.42 \text{ lb} = 291.42 \text{ lb} \downarrow$$

$F_R = \sqrt{(F_R)_x^2 + (F_R)_y^2} = \sqrt{358.58^2 + 291.42^2}$

$$= 462.07 \text{ lb} = 462 \text{ lb}$$ *Ans.*

$\theta = \tan^{-1}\left[\dfrac{(F_R)_y}{(F_R)_x}\right] = \tan^{-1}\left[\dfrac{291.42}{358.58}\right] = 39.1° \nearrow$ *Ans.*

$\zeta + (M_R)_A = \Sigma M_A;$ $358.58(d) = 250\left(\dfrac{3}{5}\right)(2.5) + 250\left(\dfrac{4}{5}\right)(4)$

$$+ 300(4) - 200\cos 45°(6) - 200\sin 45°(3)$$

$$d = 3.07 \text{ ft}$$ *Ans.*

R4–7. $+\uparrow F_R = \Sigma F_z;$ $F_R = -20 - 50 - 30 - 40$

$$= -140 \text{ kN} = 140 \text{ kN} \downarrow$$ *Ans.*

$(M_R)_x = \Sigma M_x;$ $-140y = -50(3) - 30(11) - 40(13)$

$$y = 7.14 \text{ m}$$ *Ans.*

$(M_R)_y = \Sigma M_y;$ $140x = 50(4) + 20(10) + 40(10)$

$$x = 5.71 \text{ m}$$ *Ans.*

R4–8. $+\downarrow F_R = \Sigma F;$ $F_R = 12\,000 + 6000 = 18\,000 \text{ lb}$

$$F_R = 18.0 \text{ kip}$$ *Ans.*

$\zeta + M_{RC} = \Sigma M_C;$ $18\,000x = 12\,000(7.5) + 6000(20)$

$$x = 11.7 \text{ ft}$$ *Ans.*

Chapter 5

R5–1. $\zeta + \Sigma M_A = 0: F(6) + F(4) + F(2) - 3\cos 45°(2) = 0$

$$F = 0.3536 \text{ kN} = 354 \text{ N}$$ *Ans.*

R5–2. $\zeta + \Sigma M_A = 0;$ $N_B(7) - 1400(3.5) - 300(6) = 0$

$$N_B = 957.14 \text{ N} = 957 \text{ N}$$ *Ans.*

$+\uparrow \Sigma F_y = 0;$ $A_y - 1400 - 300 + 957 = 0$ $A_y = 743 \text{ N}$

$\xrightarrow{+} \Sigma F_x = 0;$ $A_x = 0$ *Ans.*

R5–3. $\zeta + \Sigma M_A = 0;$ $10(0.6 + 1.2\cos 60°) + 6(0.4)$

$$- N_A(1.2 + 1.2\cos 60°) = 0$$

$$N_A = 8.00 \text{ kN}$$ *Ans.*

$\xrightarrow{+} \Sigma F_x = 0;$ $B_x - 6\cos 30° = 0;$ $B_x = 5.20 \text{ kN}$ *Ans.*

$+\uparrow \Sigma F_y = 0;$ $B_y + 8.00 - 6\sin 30° - 10 = 0$

$$B_y = 5.00 \text{ kN}$$ *Ans.*

R5–4. $\zeta + \Sigma M_A = 0;$ $50\cos 30°(20) + 50\sin 30°(14)$

$$- F_B(18) = 0$$

$$F_B = 67.56 \text{ lb} = 67.6 \text{ lb}$$ *Ans.*

$\xrightarrow{+} \Sigma F_x = 0;$ $A_x - 50\sin 30° = 0$

$$A_x = 25 \text{ lb}$$ *Ans.*

$+\uparrow \Sigma F_y = 0;$ $A_y - 50\cos 30° - 67.56 = 0$

$$A_y = 110.86 \text{ lb} = 111 \text{ lb}$$ *Ans.*

R5–5. $\Sigma F_x = 0;$ $A_x = 0$ *Ans.*

$\Sigma F_y = 0;$ $A_y + 200 = 0$

$$A_y = -200 \text{ N}$$ *Ans.*

$\Sigma F_z = 0;$ $A_z - 150 = 0$

$$A_z = 150 \text{ N}$$ *Ans.*

$\Sigma M_x = 0;$ $-150(2) + 200(2) - (M_A)_x = 0$

$$(M_A)_x = 100 \text{ N} \cdot \text{m}$$ *Ans.*

$\Sigma M_y = 0;$ $(M_A)_y = 0$ *Ans.*

$\Sigma M_z = 0;$ $200(2.5) - (M_A)_z = 0$

$$(M_A)_z = 500 \text{ N} \cdot \text{m}$$ *Ans.*

R5–6.

$\Sigma M_y = 0;$ $P(8) - 80(10) = 0$ $P = 100 \text{ lb}$ *Ans.*

$\Sigma M_x = 0;$ $B_z(28) - 80(14) = 0$ $B_z = 40 \text{ lb}$ *Ans.*

$\Sigma M_z = 0;$ $-B_x(28) - 100(10) = 0$ $B_x = -35.7 \text{ lb}$ *Ans.*

$\Sigma F_x = 0;$ $A_x + (-35.7) - 100 = 0$ $A_x = 136 \text{ lb}$ *Ans.*

$\Sigma F_y = 0;$ $B_y = 0$ *Ans.*

$\Sigma F_z = 0;$ $A_z + 40 - 80 = 0$ $A_z = 40 \text{ lb}$ *Ans.*

R5–7. $W = (4 \text{ ft})(2 \text{ ft})(2 \text{ lb/ft}^2) = 16 \text{ lb}$

$\Sigma F_x = 0; \qquad A_x = 0$ *Ans.*

$\Sigma F_y = 0; \qquad A_y = 0$ *Ans.*

$\Sigma F_z = 0; \qquad A_z + B_z + C_z - 16 = 0$

$\Sigma M_x = 0; \qquad 2B_z - 16(1) + C_z(1) = 0$

$\Sigma M_y = 0; \qquad -B_z(2) + 16(2) - C_z(4) = 0$

$\qquad\qquad\qquad A_z + B_z + C_z = 5.33 \text{ lb}$ *Ans.*

R5–8.

$\Sigma F_x = 0; \qquad A_x = 0$ *Ans.*

$\Sigma F_y = 0; \qquad 350 - 0.6 F_{BC} + 0.6 F_{BD} = 0$

$\Sigma F_z = 0; \qquad A_z - 800 + 0.8 F_{BC} + 0.8 F_{BD} = 0$

$\Sigma M_x = 0; \qquad (M_A)_x + 0.8 F_{BD}(6) + 0.8 F_{BC}(6) - 800(6) = 0$

$\Sigma M_y = 0; \qquad 800(2) - 0.8 F_{BC}(2) - 0.8 F_{BD}(2) = 0$

$\Sigma M_z = 0; \qquad (M_A)_z - 0.6 F_{BC}(2) + 0.6 F_{BD}(2) = 0$

$\qquad\qquad\qquad F_{BD} = 208 \text{ N}$ *Ans.*

$\qquad\qquad\qquad F_{BC} = 792 \text{ N}$ *Ans.*

$\qquad\qquad\qquad A_z = 0$ *Ans.*

$\qquad\qquad\qquad (M_A)_x = 0$ *Ans.*

$\qquad\qquad\qquad (M_A)_z = 700 \text{ N} \cdot \text{m}$ *Ans.*

Chapter 6

R6–1.   Joint B:

$\xrightarrow{+} \Sigma F_x = 0; \qquad F_{BC} = 3 \text{ kN (C)}$ *Ans.*

$+\uparrow \Sigma F_y = 0; \qquad F_{BA} = 8 \text{ kN (C)}$ *Ans.*

Joint A:

$+\uparrow \Sigma F_y = 0; \qquad 8.875 - 8 - \dfrac{3}{5} F_{AC} = 0$

$\qquad\qquad\qquad F_{AC} = 1.458 = 1.46 \text{ kN (C)}$ *Ans.*

$\xrightarrow{+} \Sigma F_x = 0; \qquad F_{AF} - 3 - \dfrac{4}{5}(1.458) = 0$

$\qquad\qquad\qquad F_{AF} = 4.17 \text{ kN (T)}$ *Ans.*

Joint C:

$\xrightarrow{+} \Sigma F_x = 0; \qquad 3 + \dfrac{4}{5}(1.458) - F_{CD} = 0$

$\qquad\qquad\qquad F_{CD} = 4.167 = 4.17 \text{ kN (C)}$ *Ans.*

$+\uparrow \Sigma F_y = 0; \qquad F_{CF} - 4 + \dfrac{3}{5}(1.458) = 0$

$\qquad\qquad\qquad F_{CF} = 3.125 = 3.12 \text{ kN (C)}$ *Ans.*

Joint E:

$\xrightarrow{+} \Sigma F_x = 0; \qquad F_{EF} = 0$ *Ans.*

$+\uparrow \Sigma F_y = 0; \qquad F_{ED} = 13.125 = 13.1 \text{ kN (C)}$ *Ans.*

Joint D:

$+\uparrow \Sigma F_y = 0; \qquad 13.125 - 10 - \dfrac{3}{5} F_{DF} = 0$

$\qquad\qquad\qquad F_{DF} = 5.21 \text{ kN (T)}$ *Ans.*

R6–2.   Joint A:

$\xrightarrow{+} \Sigma F_x = 0; \qquad F_{AB} - F_{AG} \cos 45° = 0$

$+\uparrow \Sigma F_y = 0; \qquad 333.3 - F_{AG} \sin 45° = 0$

$\qquad\qquad\qquad F_{AG} = 471 \text{ lb (C)}$ *Ans.*

$\qquad\qquad\qquad F_{AB} = 333.3 = 333 \text{ lb (T)}$ *Ans.*

Joint B:

$\xrightarrow{+} \Sigma F_x = 0; \qquad F_{BC} = 333.3 = 333 \text{ lb (T)}$ *Ans.*

$+\uparrow \Sigma F_y = 0; \qquad F_{GB} = 0$ *Ans.*

Joint D:

$\xrightarrow{+} \Sigma F_x = 0; \qquad -F_{DC} + F_{DE} \cos 45° = 0$ *Ans.*

$+\uparrow \Sigma F_y = 0; \qquad 666.7 - F_{DE} \sin 45° = 0$

$\qquad\qquad\qquad F_{DE} = 942.9 \text{ lb} = 943 \text{ lb (C)}$ *Ans.*

$\qquad\qquad\qquad F_{DC} = 666.7 \text{ lb} = 667 \text{ lb (T)}$ *Ans.*

Joint E:

$\xrightarrow{+} \Sigma F_x = 0; \qquad -942.9 \sin 45° + F_{EG} = 0$

$+\uparrow \Sigma F_y = 0; \qquad -F_{EC} + 942.9 \cos 45° = 0$

$\qquad\qquad\qquad F_{EC} = 666.7 \text{ lb} = 667 \text{ lb (T)}$ *Ans.*

$\qquad\qquad\qquad F_{EG} = 666.7 \text{ lb} = 667 \text{ lb (C)}$ *Ans.*

Joint C:

$+\uparrow \Sigma F_y = 0; \qquad F_{GC} \cos 45° + 666.7 - 1000 = 0$

$\qquad\qquad\qquad F_{GC} = 471 \text{ lb (T)}$ *Ans.*

R6–3. $\quad \zeta + \Sigma M_C = 0; \qquad -1000(10) + 1500(20)$

$\qquad\qquad\qquad\qquad - F_{GJ} \cos 30°(20 \tan 30°) = 0$

$\qquad\qquad\qquad F_{GJ} = 2.00 \text{ kip (C)}$ *Ans.*

$+\uparrow \Sigma F_y = 0; \qquad -1000 + 2(2000 \cos 60°) - F_{GC} = 0$

$\qquad\qquad\qquad F_{GC} = 1.00 \text{ kip (T)}$ *Ans.*

R6–4.

$+\uparrow \Sigma F_y = 0; \qquad 2A_y - 800 - 600 - 800 = 0 \quad A_y = 1100 \text{ lb}$

$\xrightarrow{+} \Sigma F_x = 0; \qquad A_x = 0$

$\zeta + \Sigma M_B = 0; \qquad F_{GF} \sin 30°(10) + 800(10 - 10 \cos^2 30°)$

$\qquad\qquad\qquad - 1100(10) = 0$

$\qquad\qquad\qquad F_{GF} = 1800 \text{ lb (C)} = 1.80 \text{ kip (C)}$ *Ans.*

$\zeta + \Sigma M_A = 0; \qquad F_{FB} \sin 60°(10) - 800(10 \cos^2 30°) = 0$

$\qquad\qquad\qquad F_{FB} = 692.82 \text{ lb (T)} = 693 \text{ lb (T)}$ *Ans.*

$\zeta + \Sigma M_F = 0; \qquad F_{BC}(15 \tan 30°) + 800(15 - 10 \cos^2 30°)$

$\qquad\qquad\qquad - 1100(15) = 0$

$\qquad\qquad\qquad F_{BC} = 1212.43 \text{ lb (T)} = 1.21 \text{ kip (T)}$ *Ans.*

R6–5. Joint A:

$$\Sigma F_z = 0; \quad F_{AD}\left(\frac{2}{\sqrt{68}}\right) - 600 = 0$$

$$F_{AD} = 2473.86 \text{ lb (T)} = 2.47 \text{ kip (T)} \qquad Ans.$$

$$\Sigma F_x = 0; \quad F_{AC}\left(\frac{1.5}{\sqrt{66.25}}\right) - F_{AB}\left(\frac{1.5}{\sqrt{66.25}}\right) = 0$$

$$F_{AC} = F_{AB}$$

$$\Sigma F_y = 0; \quad F_{AC}\left(\frac{8}{\sqrt{66.25}}\right) + F_{AB}\left(\frac{8}{\sqrt{66.25}}\right)$$

$$- 2473.86\left(\frac{8}{\sqrt{68}}\right) = 0$$

$$0.9829 F_{AC} + 0.9829 F_{AB} = 2400$$

$$F_{AC} = F_{AB} = 1220.91 \text{ lb (C)} = 1.22 \text{ kip (C)} \qquad Ans.$$

R6–6. CB is a two force member.

Member AC:

$$\zeta + \Sigma M_A = 0; \quad -600(0.75) + 1.5(F_{CB} \sin 75°) = 0$$

$$F_{CB} = 310.6$$

$$B_x = B_y = 310.6\left(\frac{1}{\sqrt{2}}\right) = 220 \text{ N} \qquad Ans.$$

$$\xrightarrow{+} \Sigma F_x = 0; \quad -A_x + 600 \sin 60° - 310.6 \cos 45° = 0$$

$$A_x = 300 \text{ N} \qquad Ans.$$

$$+\uparrow \Sigma F_y = 0; \quad A_y - 600 \cos 60° + 310.6 \sin 45° = 0$$

$$A_y = 80.4 \text{ N} \qquad Ans.$$

R6–7. Member AB:

$$\zeta + \Sigma M_A = 0; \quad -750(2) + B_y(3) = 0$$

$$B_y = 500 \text{ N}$$

Member BC:

$$\zeta + \Sigma M_C = 0; \quad -1200(1.5) - 900(1) + B_x(3) - 500(3) = 0$$

$$B_x = 1400 \text{ N}$$

$$+\uparrow \Sigma F_y = 0; \quad A_y - 750 + 500 = 0$$

$$A_y = 250 \text{ N} \qquad Ans.$$

Member AB:

$$\xrightarrow{+} \Sigma F_x = 0; \quad -A_x + 1400 = 0$$

$$A_x = 1400 \text{ N} = 1.40 \text{ kN} \qquad Ans.$$

Member BC:

$$\xrightarrow{+} \Sigma F_x = 0; \quad C_x + 900 - 1400 = 0$$

$$C_x = 500 \text{ N} \qquad Ans.$$

$$+\uparrow \Sigma F_y = 0; \quad -500 - 1200 + C_y = 0$$

$$C_y = 1700 \text{ N} = 1.70 \text{ kN} \qquad Ans.$$

R6–8. $\zeta + \Sigma M_B = 0; \quad F_{CD}(7) - \frac{4}{5}F_{BE}(2) = 0$

$$\zeta + \Sigma M_A = 0; \quad -150(7)(3.5) + \frac{4}{5}F_{BE}(5) - F_{CD}(7) = 0$$

$$F_{BE} = 1531 \text{ lb} = 1.53 \text{ kip} \qquad Ans.$$

$$F_{CD} = 350 \text{ lb} \qquad Ans.$$

Chapter 7

R7–1. $\zeta + \Sigma M_A = 0; \quad F_{CD}(8) - 150(8 \tan 30°) = 0$

$$F_{CD} = 86.60 \text{ lb}$$

Since member CF is a two-force member,

$$V_D = M_D = 0 \qquad Ans.$$

$$N_D = F_{CD} = 86.6 \text{ lb} \qquad Ans.$$

$$\zeta + \Sigma M_A = 0; \quad B_y(12) - 150(8 \tan 30°) = 0$$

$$B_y = 57.735 \text{ lb}$$

$$\xrightarrow{+} \Sigma F_x = 0; \quad N_E = 0 \qquad Ans.$$

$$+\uparrow \Sigma F_y = 0; \quad V_E + 57.735 - 86.60 = 0$$

$$V_E = 28.9 \text{ lb} \qquad Ans.$$

$$\zeta + \Sigma M_E = 0; \quad 57.735(9) - 86.60(5) - M_E = 0$$

$$M_E = 86.6 \text{ lb} \cdot \text{ft} \qquad Ans.$$

R7–2. Segment DC

$$\xrightarrow{+} \Sigma F_x = 0; \quad N_C = 0 \qquad Ans.$$

$$+\uparrow \Sigma F_y = 0; \quad V_C - 3.00 - 6 = 0 \quad V_C = 9.00 \text{ kN} \qquad Ans.$$

$$\zeta + \Sigma M_C = 0; \quad -M_C - 3.00(1.5) - 6(3) - 40 = 0$$

$$M_C = -62.5 \text{ kN} \cdot \text{m} \qquad Ans.$$

Segment DB

$$\xrightarrow{+} \Sigma F_x = 0; \quad N_B = 0 \qquad Ans.$$

$$+\uparrow \Sigma F_y = 0; \quad V_B - 10.0 - 7.5 - 4.00 - 6 = 0$$

$$V_B = 27.5 \text{ kN} \qquad Ans.$$

$$\zeta + \Sigma M_B = 0; \quad -M_B - 10.0(2.5) - 7.5(5)$$

$$-4.00(7) - 6(9) - 40 = 0$$

$$M_B = -184.5 \text{ kN} \cdot \text{m} \qquad Ans.$$

R7–3.

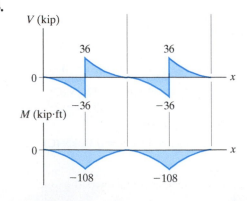

R7–4.

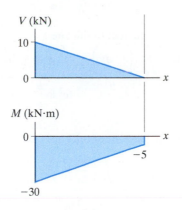

R7–5.

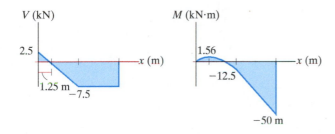

R7–6.

At $x = 30$ ft; $y = 3$ ft; $3 = \dfrac{F_H}{0.5}\left[\cosh\left(\dfrac{0.5}{F_H}(30)\right) - 1\right]$

$$F_H = 75.25 \text{ lb}$$

$\tan\theta_{max} = \dfrac{dy}{dx}\Big|_{x=30 \text{ ft}} = \sinh\left(\dfrac{0.5(30)}{75.25}\right)$ $\theta_{max} = 11.346°$

$T_{max} = \dfrac{F_H}{\cos\theta_{max}} = \dfrac{75.25}{\cos 11.346°} = 76.7 \text{ lb}$ *Ans.*

Chapter 8

R8–1. Assume that the ladder slips at A:

$F_A = 0.4 N_A$

$+\uparrow\Sigma F_y = 0;$ $N_A - 20 = 0$

$\qquad N_A = 20 \text{ lb}$

$\qquad F_A = 0.4(20) = 8 \text{ lb}$

$\zeta +\Sigma M_B = 0;$ $P(4) - 20(3) + 20(6) - 8(8) = 0$

$\qquad P = 1 \text{ lb}$ *Ans.*

$\xrightarrow{+}\Sigma F_x = 0;$ $N_B + 1 - 8 = 0$

$\qquad N_B = 7 \text{ lb} > 0$ **OK**

The ladder will remain in contact with the wall.

R8–2. Crate

$+\uparrow\Sigma F_y = 0;$ $N_d - 588.6 = 0$ $N_d = 588.6 \text{ N}$

$\xrightarrow{+}\Sigma F_x = 0;$ $P - F_d = 0$ (1)

$\zeta +\Sigma M_A = 0;$ $588.6(x) - P(0.8) = 0$ (2)

Crate and dolly

$+\uparrow\Sigma F_y = 0;$ $N_B + N_A - 588.6 - 98.1 = 0$ (3)

$\xrightarrow{+}\Sigma F_x = 0;$ $P - F_A = 0$ (4)

$\zeta +\Sigma M_B = 0;$ $N_A(1.5) - P(1.05)$

$\qquad\qquad - 588.6(0.95) - 98.1(0.75) = 0$ (5)

Friction: Assuming the crate slips on dolly, then $F_d = \mu_{sd}N_d = 0.5(588.6) = 294.3 \text{ N}$. Solving Eqs. (1) and (2)

$$P = 294.3 \text{ N}\qquad x = 0.400 \text{ m}$$

Since $x > 0.3$ m, the crate tips on the dolly. If this is the case $x = 0.3$ m. Solving Eqs. (1) and (2) with $x = 0.3$ m yields

$$P = 220.725 \text{ N}$$

$$F_d = 220.725 \text{ N}$$

Assuming the dolly slips at A, then $F_A = \mu_{sf}N_A = 0.35N_A$. Substituting this value into Eqs. (3), (4), and (5) and solving, we have

$$N_A = 559 \text{ N}\qquad N_B = 128 \text{ N}$$

$$P = 195.6 \text{ N} = 196 \text{ N } (Controls)\qquad Ans.$$

R8–3. Bar

$\zeta +\Sigma M_B = 0;$ $P(600) - A_y(900) = 0$ $A_y = 0.6667P$

Disk

$+\uparrow\Sigma F_y = 0;$ $N_C \sin 60° - F_C \sin 30°$

$\qquad\qquad - 0.6667P - 343.35 = 0$ (1)

$\zeta +\Sigma M_O = 0;$ $F_C(200) - 0.6667P(200) = 0$ (2)

Friction: If the disk is on the verge of moving, slipping would have to occur at point C. Hence, $F_C = \mu_s N_C = 0.2N_C$. Substituting this into Eqs. (1) and (2) and solving, we have

$$P = 182 \text{ N}\qquad Ans.$$

$$N_C = 606.60 \text{ N}$$

R8–4. Cam:

$\zeta +\Sigma M_O = 0;$ $5 - 0.4 N_B(0.06) - 0.01(N_B) = 0$

$\qquad N_B = 147.06 \text{ N}$

Follower:

$+\uparrow\Sigma F_y = 0;$ $147.06 - P = 0$

$\qquad P = 147 \text{ N}$ *Ans.*

R8–5. $\xrightarrow{+} \Sigma F_x = 0;$　$-P + 0.5(1250) = 0$

$$P = 625 \text{ lb}$$

Assume block B slips up and block A does not move.

Block A:

$\xrightarrow{+} \Sigma F_x = 0;$　$F_A - N'' = 0$

$+\uparrow \Sigma F_y = 0;$　$N_A - 600 - 0.3N'' = 0$

Block B:

$\xrightarrow{+} \Sigma F_x = 0;$　$N'' - N' \cos 45° - 0.3\,N' \sin 45° = 0$

$+\uparrow \Sigma F_y = 0;$　$N' \sin 45 - 0.3\,N' \cos 45° - 150 - 0.3\,N''$
　　　　　　　$= 0$

Block C:

$\xrightarrow{+} \Sigma F_x = 0;$　$0.3\,N'\cos 45 - N'\cos 45 - 0.5\,N_C - P = 0$

$+\uparrow \Sigma F_y = 0;$　$N_C - N'\sin 45 - 0.3\,N' \sin 45 - 500 = 0$

Solving

$N'' = 629.0 \text{ lb}, \; N' = 684.3 \text{ lb}, \; N_C = 838.7 \text{ lb}, \; P = 1048 \text{ lb},$

$N_A = 411.3 \text{ lb}$

$F_A = 629.0 \text{ lb} > 0.5 \,(411.3) = 205.6 \text{ lb}$　　No good

All blocks slip at the same time:　　$P = 625 \text{ lb}$　*Ans.*

R8–6.　$\alpha = \tan^{-1}\left(\dfrac{10}{25}\right) = 21.80°$

$\zeta + \Sigma M_A = 0;$　$-6000\,(35) + F_{BD} \cos 21.80°(10)$
　　　　　　　　　　$+ F_{BD} \sin 21.80°(20) = 0$

$$F_{BD} = 12\,565 \text{ lb}$$

$$\phi_s = \tan^{-1}(0.4) = 21.80°$$

$$\theta = \tan^{-1}\left(\frac{0.2}{2\pi(0.25)}\right) = 7.256°$$

$$M = Wr \tan(\theta + \phi)$$

$$M = 12\,565\,(0.25) \tan(7.256° + 21.80°)$$

$$M = 1745 \text{ lb} \cdot \text{in} = 145 \text{ lb} \cdot \text{ft}$$　　**Ans.**

R8–7.　Block:

$+\uparrow \Sigma F_y = 0;$　$N - 100 = 0$

　　　　　　$N - 100 \text{ lb}$

$\xrightarrow{+} \Sigma F_x = 0;$　$T_1 = 0.4(100) = 0$

　　　　　　$T_1 = 40 \text{ lb}$

$T_2 = T_1 e^{\mu\beta};$　$T_2 = 40e^{0.4\left(\frac{\pi}{2}\right)} = 74.978 \text{ lb}$

System:

$\zeta + \Sigma M_A = 0;$　$-100(d) - 40(1) - 50(5) + 74.978(10) = 0$

　　　　　　　　$d = 4.60 \text{ ft}$　*Ans.*

R8–8.　$P \approx \dfrac{Wa}{r}$

　　　$= 500(9.81)\left(\dfrac{2}{40}\right)$

　　　$P = 245 \text{ N}$　　*Ans.*

Chapter 9

R9–1.　Using an element of thickness dx,

$$\bar{x} = \frac{\displaystyle\int_A \tilde{x}\,dA}{\displaystyle\int_A dA} = \frac{\displaystyle\int_a^b x\left(\frac{c^2}{x}\,dx\right)}{c^2 \ln \dfrac{b}{a}} = \frac{\displaystyle\int_a^b c^2\,dx}{c^2 \ln \dfrac{b}{a}} = \frac{c^2 x \Big|_a^b}{c^2 \ln \dfrac{b}{a}} = \frac{b - a}{\ln \dfrac{b}{a}}$$

Ans.

R9–2.　Using an element of thickness dx,

$$\bar{y} = \frac{\displaystyle\int_A y\,dA}{\displaystyle\int_A dA} = \frac{\displaystyle\int_a^b \left(\frac{c^2}{2x}\right)\left(\frac{c^2}{x}\,dx\right)}{c^2 \ln \dfrac{b}{a}} = \frac{\displaystyle\int_a^b \frac{c^4}{2x^2}\,dx}{c^2 \ln \dfrac{b}{a}}$$

$$= \frac{-\dfrac{c^4}{2x}\Big|_a^b}{c^2 \ln \dfrac{b}{a}} = \frac{c^2(b - a)}{2ab \ln \dfrac{b}{a}}$$

Ans.

R9–3.　$\bar{z} = \dfrac{\displaystyle\int_v \tilde{z}\,dV}{\displaystyle\int_v dV} = \dfrac{\displaystyle\int_0^a z\,[\pi(a^2 - z^2)\,dz]}{\displaystyle\int_0^a \pi(a^2 - z^2)\,dz}$

$$= \frac{\pi\left(\dfrac{a^2 z^2}{2} - \dfrac{z^4}{4}\right)\Big|_0^a}{\pi\left(a^2 z - \dfrac{z^3}{3}\right)\Big|_0^a} = \frac{3}{8}a$$

Ans.

R9–4.　$\Sigma \tilde{x}L = 0(4) + 2(\pi)(2) = 12.5664 \text{ ft}^2$

$\Sigma \tilde{y}L = 0(4) + \dfrac{2(2)}{\pi}(\pi)(2) = 8 \text{ ft}^2$

$\Sigma \tilde{z}L = 2(4) + 0(\pi)(2) = 8 \text{ ft}^2$

$\Sigma L = 4 + \pi(2) = 10.2832 \text{ ft}$

$\bar{x} = \dfrac{\Sigma \tilde{x}L}{\Sigma L} = \dfrac{12.5664}{10.2832} = 1.22 \text{ ft}$　*Ans.*

$\bar{y} = \dfrac{\Sigma \tilde{y}L}{\Sigma L} = \dfrac{8}{10.2832} = 0.778 \text{ ft}$　*Ans.*

$\bar{z} = \dfrac{\Sigma \tilde{z}L}{\Sigma L} = \dfrac{8}{10.2832} = 0.778 \text{ ft}$　*Ans.*

R9–5.

Segment	$A\,(\text{mm}^2)$	$\tilde{y}\,(\text{mm})$	$\tilde{y}A\,(\text{mm}^3)$
1	300(25)	112.5	843 750
2	100(50)	50	250 000
Σ	12 500		1 093 750

Thus,

$$\bar{y} = \frac{\Sigma \tilde{y} A}{\Sigma A} = \frac{1\,093\,750}{12\,500} = 87.5 \text{ mm} \qquad Ans.$$

R9–6.

$$A = \Sigma \theta \tilde{r} L$$

$$= 2\pi \left[0.6\,(0.05) + 2(0.6375)\sqrt{(0.025)^2 + (0.075)^2}\right.$$
$$\left. + 0.675\,(0.1)\right]$$

$$= 1.25 \text{ m}^2 \qquad Ans.$$

R9–7.

$$V = \Sigma \theta \tilde{r} A$$

$$= 2\pi \left[2\,(0.65)\left(\frac{1}{2}\,(0.025)(0.075)\right) + 0.6375(0.05)(0.075)\right]$$

$$= 0.0227 \text{ m}^3 \qquad Ans.$$

R9–8. $\quad dF = \displaystyle\int dA = 4z^{\frac{1}{3}}(3)dz$

$$F = 12 \int_0^x z^{\frac{1}{3}}\,dz = 12\left[\frac{3}{4}z^{\frac{4}{3}}\right]_0^8 = 144 \text{ lb} \qquad Ans.$$

$$\int_A z\,dF = 12 \int_0^8 z^{\frac{4}{3}}\,dz = 12\left[\frac{3}{7}z^{\frac{7}{3}}\right]_0^8 = 658.29 \text{ lb} \cdot \text{ft}$$

$$\bar{z} = \frac{658.29}{144} = 4.57 \text{ ft} \qquad Ans.$$

R9–9.

$$p_a = 1.0(10^3)(9.81)(9) = 88\,290 \text{ N/m}^2 = 88.29 \text{ kN/m}^2$$

$$p_b = 1.0(10^3)(9.81)(5) = 49\,050 \text{ N/m}^2 = 49.05 \text{ kN/m}^2$$

Thus,

$$w_A = 88.29(8) = 706.32 \text{ kN/m}$$

$$w_B = 49.05(8) = 392.40 \text{ kN/m}$$

$$F_{R_1} = 392.4(5) = 1962.0 \text{ kN}$$

$$F_{R_2} = \frac{1}{2}\,(706.32 - 392.4)\,(5) = 784.8 \text{ kN}$$

$$\zeta + \Sigma M_B = 0; \quad 1962.0(2.5) + 784.8(3.333) - A_y(3) = 0$$

$$A_y = 2507 \text{ kN} = 2.51 \text{ MN} \qquad Ans.$$

$$\xrightarrow{+}\Sigma F_x = 0; \quad 784.8\left(\frac{4}{5}\right) + 1962\left(\frac{4}{5}\right) - B_x = 0$$

$$B_x = 2197 \text{ kN} = 2.20 \text{ MN} \qquad Ans.$$

$$+\uparrow\Sigma F_y = 0; \quad 2507 - 784.8\left(\frac{3}{5}\right) - 1962\left(\frac{3}{5}\right) - B_y = 0$$

$$B_y = 859 \text{ kN} \qquad Ans.$$

R9–10.

$$A = \int_A dA = \int_{-2}^a -y\,dx = \int_{-2}^a 2x^2\,dx = \frac{2}{3}x^3\Big|_{-2}^0 = 5.333 \text{ ft}^2$$

$$w = b\,\gamma\,h = 1(62.4)(8) = 499.2 \text{ lb} \cdot \text{ft}$$

$$F_y = 5.333(1)(62.4) = 332.8 \text{ lb}$$

$$F_x = \frac{1}{2}\,(499.2)(8) = 1997 \text{ lb}$$

$$F_N = \sqrt{(332.8)^2 + (1997)^2} = 2024 \text{ lb} = 2.02 \text{ kip} \qquad Ans.$$

Chapter 10

R10–1.

$$I_x = \int_A y^2\,dA = \int_0^2 y^2(4 - x)dy = \int_0^2 y^2\big(4 - (32)^{\frac{1}{3}}y^{\frac{1}{3}}\big)dy$$

$$= 1.07 \text{ in}^4 \qquad Ans.$$

R10–2.

$$I_x = \int_A y^2\,dA = \int_0^1 y^2(2x\,dy) = \int_0^1 y^2\big(4(1 - y)^{\frac{1}{2}}\big)dy$$

$$= 0.610 \text{ ft}^4 \qquad Ans.$$

R10–3.

$$I_y = \int_A x^2\,dA = 2\int_0^2 x^2(y\,dx) = 2\int_0^2 x^2(1 - 0.25\,x^2)dx$$

$$= 2.13 \text{ ft}^4 \qquad Ans.$$

R10–4. $\quad dI_{xy} = d\bar{I}_{x'y'} + dA\bar{x}\,\bar{y} = 0 + \big(y^{\frac{1}{3}}\,dy\big)\left(\frac{1}{2}y^{\frac{1}{3}}\right)(y)$

$$= \frac{1}{2}\,y^{\frac{5}{3}}\,d\,y$$

$$I_{xy} = \int dI_{xy} = \int_0^{1\text{ m}} \frac{1}{2}\,y^{\frac{5}{3}}\,dy = \frac{3}{16}\,y^{\frac{8}{3}}\Big|_0^{1\text{ m}} = 0.1875 \text{ m}^4 \quad Ans.$$

R10–5. $\quad \dfrac{s}{h - y} = \dfrac{b}{h}, \quad s = \dfrac{b}{h}(h - y)$

(a) $dA = s\,dy = \left[\dfrac{b}{h}(h - y)\right]dy$

$$I_x = \int y^2\,dA = \int_0^h y^2\left[\frac{b}{h}(h - y)\right]dy = \frac{bh^3}{12} \qquad Ans.$$

(b) $I_x = \bar{I}_{x'} + A\,d^2 \quad \dfrac{bh^3}{12} = \bar{I}_{x'} + \dfrac{1}{2}bh\left(\dfrac{h}{3}\right)^2 \quad I_x = \dfrac{bh^3}{36}$ $Ans.$

R10–6. $\quad dI_{xy} = dI_{x'y'} + dA\,\bar{x}\,\bar{y}$

$$= 0 + \big(y^{\frac{1}{3}}dy\big)\left(\frac{1}{2}y^{\frac{1}{3}}\right)(y)$$

$$= \frac{1}{2}\,y^{\frac{5}{3}}\,dy$$

$$I_{xy} = \int dI_{xy} = \int_0^{1\,m} \frac{1}{2} y^{\frac{5}{3}} dy = \frac{3}{16} y^{\frac{8}{3}} \Big|_0^{1\,m}$$

$$= 0.1875 \text{ m}^4 \qquad \text{Ans.}$$

R10–7. $I_y = \left[\frac{1}{12}(d)(d^3) + 0 \right] + 4 \left[\frac{1}{36}(0.2887d)\left(\frac{d}{2}\right)^3 \right.$

$$\left. + \frac{1}{2}(0.2887d)\left(\frac{d}{2}\right)\left(\frac{d}{6}\right)^2 \right]$$

$$= 0.0954d^4 \qquad \text{Ans.}$$

R10–8. $d I_x = \frac{1}{2} \rho \pi y^4 dx = \frac{1}{2} \rho \pi \left(\frac{b^4}{a^4} x^4 + \frac{4b^4}{a^3} x^3 + \frac{6b^4}{a^2} x^2 \right.$

$$\left. + \frac{4b^4}{a} x + b^4 \right) dx$$

$$I_x = \int dI_x = \frac{1}{2} \rho \pi \int_0^a \left(\frac{b^4}{a^4} x^4 + \frac{4b^4}{a^3} x^3 + \frac{6b^4}{a^2} x^2 \right.$$

$$\left. + \frac{4b^4}{a} x + b^4 \right) dx$$

$$= \frac{31}{10} \rho \pi ab^4$$

$$m = \int_m dm = \int_0^a \rho \pi y^2 dx$$

$$= \rho \pi \int_0^a \left(\frac{b^2}{a^2} x^2 + \frac{2b^2}{a} x + b^2 \right) dx$$

$$= \frac{7}{3} \rho \pi ab^2$$

$$I_x = \frac{93}{70} mb^2 \qquad \text{Ans.}$$

Chapter 11

R11–1. $x = 2L \cos \theta$

$\delta x = -2L \sin \theta \, \delta \theta$

$y = L \sin \theta$

$\delta y = L \cos \theta \, \delta \theta$

$\delta U = 0; \quad -P \delta y - F \delta x = 0$

$-PL \cos \theta \delta \theta - F(-2L \sin \theta) \delta \theta = 0$

$-P \cos \theta + 2F \sin \theta = 0$

$$F = \frac{P}{2 \tan \theta} \qquad \text{Ans.}$$

R11–2. $y_B = 10 \sin \theta \qquad \delta y_B = 10 \cos \theta \delta \theta$

$y_D = 5 \sin \theta \qquad \delta y_D = 5 \cos \theta \delta \theta$

$x_C = 2(10 \cos \theta) \qquad \delta x_C = -20 \sin \theta \delta \theta$

$\delta U = 0; \quad -F_{sp} \delta x_C - 2(2 \delta y_D - 20 \delta y_B + P \delta x_C = 0$

$(20 F_{sp} \sin \theta - 20P \sin \theta - 220 \cos \theta) \delta \theta = 0$

However, from the spring formula,

$F_{sp} = kx = 2[2(10 \cos \theta) - 6] = 40 \cos \theta - 12.$

Substituting

$(800 \sin \theta \cos \theta - 240 \sin \theta - 220 \cos \theta - 20P \sin \theta) \, \delta \theta = 0$

Since $\delta \theta \neq 0$, then

$800 \sin \theta \cos \theta - 240 \sin \theta - 220 \cos \theta - 20P \sin \theta = 0$

$P = 40 \cos \theta - 11 \cot \theta - 12$

At the equilibrium position, $\theta = 45°$. Then

$P = 40 \cos 45° - 11 \cot 45° - 12 = 5.28 \text{ lb} \qquad \text{Ans.}$

R11–3. Using the law of cosines,

$0.4^2 = x_A^2 + 0.1^2 - 2(x_A)(0.1) \cos \theta$

Differentiating,

$0 = 2x_A \delta x_A - 0.2 \delta x_A \cos \theta + 0.2 x_A \sin \theta \delta \theta$

$$\delta x_A = \frac{0.2 x_A \sin \theta}{0.2 \cos \theta - 2x_A} \delta \theta$$

$\delta U = 0; \qquad -F \delta x_A - 50 \delta \theta = 0$

$$\left(\frac{0.2 x_A \sin \theta}{0.2 \cos \theta - 2x_A} F - 50 \right) \delta \theta = 0$$

Since $\delta \theta \neq 0$, then

$$\frac{0.2 x_A \sin \theta}{0.2 \cos \theta - 2x_A} F - 50 = 0$$

$$F = \frac{50(0.2 \cos \theta - 2x_A)}{0.2 x_A \sin \theta}$$

At the equilibrium position, $\theta = 60°$,

$0.4^2 = x_A^2 + 0.1^2 - 2(x_A)(0.1) \cos 60°$

$x_A = 0.4405 \text{ m}$

$$F = -\frac{50 \left[0.2 \cos 60° - 2(0.4405) \right]}{0.2(0.4405) \sin 60°} = 512 \text{ N} \qquad \text{Ans.}$$

R11–4. $y = 4 \sin \theta$

$\delta y = 4 \cos \theta \, \delta \theta$

$F_s = 5(4 - 4 \sin \theta)$

$\delta U = 0; \qquad -10 \delta y + F_s \delta y = 0$

$[-10 + 20(1 - \sin \theta)](4 \cos \theta \, \delta \theta) = 0$

$\cos \theta = 0 \quad \text{and} \quad 10 - 20 \sin \theta = 0$

$\theta = 90° \qquad\qquad \theta = 30° \qquad \text{Ans.}$

R11–5. $x_B = 0.1 \sin \theta \qquad \delta x_B = 0.1 \cos \theta \delta \theta$

$x_D = 2(0.7 \sin \theta) - 0.1 \sin \theta = 1.3 \sin \theta \quad \delta x_D = 1.3 \cos \theta \delta \theta$

$y_G = 0.35 \cos \theta \quad \delta y_G = -0.35 \sin \theta \delta \theta$

$\delta U = 0;$ $2(-49.05\delta y_G) + F_{sp}(\delta x_B - \delta x_D) = 0$

$(34.335 \sin \theta - 1.2F_{sp} \cos \theta)\delta\theta = 0$

However, from the spring formula,

$F_{sp} = kx = 400\left[2(0.6 \sin \theta) - 0.3\right] = 480 \sin \theta - 120.$

Substituting,

$(34.335 \sin \theta - 576 \sin \theta \cos \theta + 144 \cos \theta)\delta\theta = 0$

Since $\delta\theta \neq 0$, then

$34.335 \sin \theta - 576 \sin \theta \cos \theta + 144 \cos \theta = 0$

$\theta = 15.5°$ *Ans.*

and $\theta = 85.4°$ *Ans.*

R11–6.

$V_g = mgy = 40(9.81)(0.45 \sin \theta + b) = 176.58 \sin \theta + 392.4\,b$

$V_e = \frac{1}{2}(1500)(0.45 \cos \theta)^2 = 151.875 \cos^2 \theta$

$V = V_g + V_e = 176.58 \sin \theta + 151.875 \cos^2 \theta + 392.4\,b$

$\frac{dV}{d\theta} = 176.58 \cos \theta - 303.75 \cos \theta \sin \theta = 0$

$\cos \theta(176.58 - 303.75 \sin \theta) = 0$

$\cos \theta = 0$ $\theta = 90°$ *Ans.*

$\theta = 35.54° = 35.5°$ *Ans.*

$\frac{d^2V}{d^2\theta} = -176.58 \sin \theta - 303.75 \cos 2\theta$

At $\theta = 90°$, $\left.\frac{d^2V}{d^2\theta}\right|_{\theta-d°} = -176.58 \sin 90° - 303.75 \cos 180°$

$= 127.17 > 0$

$= 127.17 > 0$ Stable   *Ans.*

At $\theta = 35.54°$, $\left.\frac{d^2V}{d^2\theta}\right|_{\theta=35.54°} = -176.58 \sin 35.54°$

$- 303.75 \cos 71.09°$

$= -201.10 < 0$ Unstable   *Ans.*

R11–7. $V = V_e + V_g$

$= \frac{1}{2}(24)(2 \cos \theta)^2 + \frac{1}{2}(48)(6 \cos \theta)^2$

$+ 100(3 \sin \theta)$

$= 912 \cos^2 \theta + 300 \sin \theta$

$\frac{dV}{d\theta} = -1824 \sin \theta \cos \theta + 300 \cos \theta = 0$

$\frac{dV}{d\theta} = -912 \sin 2\theta + 300 \cos \theta = 0$

$\theta = 90°$ or $\theta = 9.467°$

$\frac{d^2V}{d\theta^2} = -1824 \cos 2\theta - 300 \sin \theta$

$\left.\frac{d^2V}{d\theta^2}\right|_{\theta=90°} = -1824 \cos 180° - 300 \sin 90°$

$= 1524 > 0$ *Ans.*

$\left.\frac{d^2V}{d\theta^2}\right|_{\theta=9.467°} = -1824 \cos 18.933° - 300 \sin 9.467°$

$= 1774.7 < 0$

Thus, the system is in unstable equilibrium at $\theta = 9.47°$.   *Ans.*

R11–8. $V = V_e + V_g$

$= \frac{1}{2}kx^2 - Wy$

$= \frac{1}{2}(16)(2.5 - 2.5 \sin \theta)^2 - 20(2.5 \cos \theta)$

$= 50 \sin^2 \theta - 100 \sin \theta - 50 \cos \theta + 50$

$\frac{dV}{d\theta} = 100 \sin \theta \cos \theta - 100 \cos \theta + 50 \sin \theta = 0$

$\frac{dV}{d\theta} = 50 \sin 2\theta - 100 \cos \theta + 50 \sin \theta = 0$

$\theta = 37.77° = 37.8°$

$\frac{d^2V}{d\theta^2} = 100 \cos 2\theta + 100 \sin \theta + 50 \cos \theta$

$\left.\frac{d^2V}{d\theta^2}\right|_{\theta=37.77°} = 100 \cos 75.55° + 100 \sin 37.77° + 50 \cos 37.77°$

$= 125.7 > 0$

Thus, the system is in stable equilibrium at $\theta = 37.8°$ *Ans.*

Answers to Selected Problems

Chapter 1

1–1.
  a.   78.5 N
  b.   0.392 mN
  c.   7.46 MN

1–2.
  a.   GN/s
  b.   Gg/N
  c.   GN/(kg·s)

1–3.
  a.   Gg/s
  b.   kN/m
  c.   kN/(kg·s)

1–5.
  a.   45.3 MN
  b.   56.8 km
  c.   5.63 μg

1–6.
  a.   58.3 km
  b.   68.5 s
  c.   2.55 kN
  d.   7.56 mg

1–7.
  a.   0.431 g
  b.   35.3 kN
  c.   5.32 m

1–9.
  a.   km/s
  b.   mm
  c.   Gs/kg
  d.   mm·N

1–10.
  a.   kN·m
  b.   Gg/m
  c. μN/s^2
  d.   GN/s

1–11.
  a.   8.653 s
  b.   8.368 kN
  c.   893 g

1–13.   2.71 Mg/m^3

1–14.
  a.   44.9(10)$^{-3}$ N^2
  b.   2.79(10^3) s^2
  c.   23.4 s

1–15.   7.41 μN

1–17.   1.00 Mg/m^3

1–18.
  a.   0.447 kg·m/N
  b.   0.911 kg·s
  c.   18.8 GN/m

1–19.   1.04 kip

1–21. $F = 10.0$ nN, $W_1 = 78.5$ N, $W_2 = 118$ N

Chapter 2

2–1. $F_R = 497$ N, $\phi = 155°$

2–2. $F = 960$ N, $\theta = 45.2°$

2–3. $F_R = 393$ lb, $\phi = 353°$

2–5. $F_{AB} = 314$ lb, $F_{AC} = 256$ lb

2–6. $\phi = 1.22°$

2–7. $(F_1)_v = 2.93$ kN, $(F_1)_u = 2.07$ kN

2–9. $F = 616$ lb, $\theta = 46.9°$

2–10. $F_R = 980$ lb, $\phi = 19.4°$

2–11. $F_R = 10.8$ kN, $\phi = 3.16°$

2–13. $F_a = 30.6$ lb, $F_b = 26.9$ lb

2–14. $F = 19.6$ lb, $F_b = 26.4$ lb

2–15. $F = 917$ lb, $\theta = 31.8°$

2–17. $F_R = 19.2$ N, $\theta = 2.37°$ ◿

2–18. $F_R = 19.2$ N, $\theta = 2.37°$ ◿

2–19. $\theta = 53.5°$, $F_{AB} = 621$ lb

2–21. $F_R = 257$ N, $\phi = 163°$

2–22. $F_R = 257$ N, $\phi = 163°$

2–23. $\theta = 75.5°$

2–25. $\theta = 36.3°$, $\phi = 26.4°$

2–26. $\theta = 54.3°$, $F_A = 686$ N

2–27. $F_R = 1.23$ kN, $\theta = 6.08°$

2–29. $F_B = 1.61$ kN, $\theta = 38.3°$

2–30. $F_R = 4.01$ kN, $\phi = 16.2°$

2–31. $\theta = 90°$, $F_B = 1$ kN, $F_R = 1.73$ kN

2–33. $F_R = 983$ N, $\theta = 21.8°$

2–34. $\mathbf{F}_1 = \{200\mathbf{i} + 346\mathbf{j}\}$ N, $\mathbf{F}_2 = \{177\mathbf{i} - 177\mathbf{j}\}$ N

2–35. $F_R = 413$ N, $\theta = 24.2°$

2–37. $F_R = 1.96$ kN, $\theta = 4.12°$

2–38. $\mathbf{F}_1 = \{30\mathbf{i} + 40\mathbf{j}\}$ N, $\mathbf{F}_2 = \{-20.7\mathbf{i} - 77.3\mathbf{j}\}$ N,
     $\mathbf{F}_3 = \{30\mathbf{i}\}$, $F_R = 54.2$ N, $\theta = 43.5°$

2–39. $F_{1x} = 141$ N, $F_{1y} = 141$ N, $F_{2x} = -130$ N,
     $F_{2y} = 75$ N

2–41. $F_R = 12.5$ kN, $\theta = 64.1°$

2–42. $\mathbf{F}_1 = \{680\mathbf{i} - 510\mathbf{j}\}$ N, $\mathbf{F}_2 = \{-312\mathbf{i} - 541\mathbf{j}\}$ N,
     $\mathbf{F}_3 = \{-530\mathbf{i} + 530\mathbf{j}\}$ N

2–43. $F_R = 546$ N, $\theta = 253°$

2–45. $F_R = \sqrt{F_1^2 + F_2^2 + 2F_1F_2 \cos\phi}$,
$$\theta = \tan^{-1}\left(\frac{F_1 \sin\phi}{F_2 + F_1 \cos\phi}\right)$$

2–46. $\theta = 68.6°$, $F_B = 960$ N

2–47. $F_R = 839$ N, $\theta = 14.8°$

2–49. $F_R = 389$ N, $\phi' = 42.7°$

2–50. $\mathbf{F}_1 = \{9.64\mathbf{i} + 11.5\mathbf{j}\}$ kN, $\mathbf{F}_2 = \{-24\mathbf{i} + 10\mathbf{j}\}$ kN,
     $\mathbf{F}_3 = \{31.2\mathbf{i} - 18\mathbf{j}\}$ kN

2–51. $F_R = 17.2$ kN, $\theta = 11.7°$

2–53. $\mathbf{F}_1 = \{-15.0\mathbf{i} - 26.0\mathbf{j}\}$ kN,
     $\mathbf{F}_2 = \{-10.0\mathbf{i} + 24.0\mathbf{j}\}$ kN

2–54. $F_R = 25.1$ kN, $\theta = 185°$

2–55. $F = 2.03$ kN, $F_R = 7.87$ kN

2–57. $F_R = 380$ N, $F_1 = 57.8$ N

2–58. $\theta = 86.0°$, $F = 1.97$ kN

2–59. $F_R = 11.1$ kN, $\theta = 47.7°$

2–61. $F_x = 40$ N, $F_y = 40$ N, $F_z = 56.6$ N

2–62. $\alpha = 48.4°$, $\beta = 124°$, $\gamma = 60°$, $F = 8.08$ kN

2–63. $F_R = 114$ lb, $\alpha = 62.1°$, $\beta = 113°$, $\gamma = 142°$

2–65. $F_1 = \{-106i + 106j + 260k\}$ N,
$F_2 = \{250i + 354j - 250k\}$ N,
$F_R = \{144i + 460j + 9.81k\}$ N, $F_R = 482$ N,
$\alpha = 72.6°$, $\beta = 17.4°$, $\gamma = 88.8°$

2–66. $\alpha_1 = 111°$, $\beta_1 = 69.3°$, $\gamma_1 = 30.0°$

2–67. $F_3 = 428$ lb, $\alpha = 88.3°$, $\beta = 20.6°$, $\gamma = 69.5°$

2–69. $F_R = 430$ N, $\alpha = 28.9°$, $\beta = 67.3°$, $\gamma = 107°$

2–70. $F_R = 384$ N, $\cos \alpha = 14.8°$, $\cos \beta = 88.9°$,
$\cos \gamma = 105°$

2–71. $F_1 = 429$ lb, $\alpha_1 = 62.2°$, $\beta_1 = 110°$, $\gamma_1 = 145°$

2–73. $F_1 = \{72.0i + 54.0k\}$ N,
$F_2 = \{53.0i + 53.0j + 130k\}$ N, $F_3 = \{200k\}$

2–74. $F_R = 407$ N, $\alpha = 72.1°$, $\beta = 82.5°$, $\gamma = 19.5°$

2–75. $F_1 = \{14.0j - 48.0k\}$ lb,
$F_2 = \{90i - 127j + 90k\}$ lb

2–77. $F_R = 610$ N, $\alpha = 19.4°$, $\beta = 77.5°$, $\gamma = 105°$

2–78. $F_2 = 66.4$ lb, $\alpha = 59.8°$, $\beta = 107°$, $\gamma = 144°$

2–79. $\alpha = 124°$, $\beta = 71.3°$, $\gamma = 140°$

2–81. $F_R = 1.55$ kip, $\alpha = 82.4°$, $\beta = 37.6°$, $\gamma = 53.4°$

2–82. $F_R = 1.60$ kN, $\alpha = 82.6°$, $\beta = 29.4°$, $\gamma = 61.7°$

2–83. $\alpha_3 = 139°$,
$\beta_3 = 128°$, $\gamma_3 = 102°$, $F_{R1} = 387$ N,
$\beta_3 = 60.7°$, $\gamma_3 = 64.4°$, $F_{R2} = 1.41$ kN

2–85. $F = 2.02$ kN, $F_y = 0.523$ kN

2–86. $r_{AB} = 397$ mm

2–87. $F = \{59.4i - 88.2j - 83.2k\}$ lb, $\alpha = 63.9°$,
$\beta = 131°$, $\gamma = 128°$

2–89. $x = 5.06$ m, $y = 3.61$ m, $z = 6.51$ m

2–90. $z = 6.63$ m

2–91. $x = y = 4.42$ m

2–93. $F_R = 1.17$ kN, $\alpha = 66.9°$, $\beta = 92.0°$, $\gamma = 157°$

2–94. $F_R = 1.17$ kN, $\alpha = 68.0°$, $\beta = 96.8°$, $\gamma = 157°$

2–95. $F_{BA} = \{-109i + 131j + 306k\}$ lb,
$F_{CA} = \{103i + 103j + 479k\}$ lb,
$F_{DA} = \{-52.1i - 156j + 365k\}$ lb

2–97. $F_R = 757$ N, $\alpha = 149°$, $\beta = 90.0°$, $\gamma = 59.0°$

2–98. $F = \{-34.3i + 22.9j - 68.6k\}$ lb

2–99. $F = \{13.4i + 23.2j + 53.7k\}$ lb

2–101. $F_A = \{169i + 33.8j - 101k\}$ lb,
$F_B = \{97.6i + 97.6j - 58.6k\}$ lb,
$F_R = 338$ lb, $\alpha = 37.8°$,
$\beta = 67.1°$, $\gamma = 118°$

2–102. $F_1 = \{389i - 64.9j + 64.9k\}$ lb,
$F_2 = \{-584i + 97.3j - 97.3k\}$ lb

2–103. $F_R = 52.2$ lb, $\alpha = 87.8°$, $\beta = 63.7°$, $\gamma = 154°$

2–105. $F = 105$ lb

2–106. $F = \{466i + 339j - 169k\}$ N

2–107. $F = \{476i + 329j - 159k\}$ N

2–109. $F = 52.1$ lb

2–110. $r_{AB} = 10.0$ ft,
$F = \{-19.1i - 14.9j + 43.7k\}$ lb, $\alpha = 112°$,
$\beta = 107°$, $\gamma = 29.0°$

2–111. $r_{AB} = 592$ mm, $F = \{-13.2i - 17.7j + 20.3k\}$ N

2–113. $(F_{ED})_\| = 334$ N, $(F_{ED})_\perp = 498$ N

2–114. $\theta = 36.4°$

2–115. $(F_1)_{AC} = 56.3$ N

2–117. $|\text{Proj } F_{AB}| = 70.5$ N, $|\text{Proj } F_{AC}| = 65.1$ N

2–118. $\theta = 31.0°$

2–119. $F_1 = 18.3$ lb, $F_2 = 35.6$ lb

2–121. $\theta = 100°$

2–122. $\theta = 19.2°$

2–123. $F_{BA} = 187$ N

2–125. $F_u = 246$ N

2–126. $F_\| = 10.5$ lb

2–127. $\theta = 142°$

2–129. $F_\| = 0.182$ kN

2–130. $\theta = 74.4°$, $\phi = 55.4°$

2–131. $(F_{BC})_\| = 28.3$ lb, $(F_{BC})_\perp = 68.0$ lb

2–133. $\theta = 132°$

2–134. $\theta = 23.4°$

2–135. $[(F)_{AB}]_\| = 63.2$ lb, $[(F)_{AB}]_\perp = 64.1$ lb

2–137. $F_{OA} = 242$ N

2–138. $\theta = 82.9°$

2–139. $\text{Proj } F_{AB} = \{0.229i - 0.916j + 1.15k\}$ lb

Chapter 3

3–1. $F_2 = 9.60$ kN, $F_1 = 1.83$ kN

3–2. $\theta = 4.69°$, $F_1 = 4.31$ kN

3–3. $\theta = 82.2°$, $F = 3.96$ kN

3–5. $T = 7.20$ kN, $F = 5.40$ kN

3–6. $T = 7.66$ kN, $\theta = 70.1°$

3–7. $\theta = 20°$, $T = 30.5$ lb

3–9. $F = 960$ lb

3–10. $\theta = 40°$, $T_{AB} = 37.6$ lb

3–11. $\theta = 40°$, $W = 42.6$ lb

3–13. $F_{CA} = 500(10^3)$ lb, $F_{AB} = 433(10^3)$ lb,
$F_{AD} = 250(10^3)$ lb

3–14. $x_{AD} = 0.4905$ m, $x_{AC} = 0.793$ m, $x_{AB} = 0.467$ m

3–15. $m = 8.56$ kg

3–17. $\dfrac{1}{k_T} = \dfrac{1}{k_1} + \dfrac{1}{k_2}$

3–18. $k = 176$ N/m

3–19. $l_0 = 2.03$ m

3–21. $l = 2.66$ ft

3–22. $F = 158$ N

3–23. $d = 1.56$ m

3–25. $y = 2$ m, $F_1 = 833$ N

3–26. $T_{HA} = 294\,\text{N}$, $T_{AB} = 340\,\text{N}$, $T_{AE} = 170\,\text{N}$, $T_{BD} = 490\,\text{N}$, $T_{BC} = 562\,\text{N}$

3–27. $m = 26.7\,\text{kg}$

3–29. $F_{DE} = 392\,\text{N}$, $F_{CD} = 340\,\text{N}$, $F_{CB} = 275\,\text{N}$, $F_{CA} = 243\,\text{N}$

3–30. $m = 20.4\,\text{kg}$

3–31. $s = 3.38\,\text{m}$, $F = 76.0\,\text{N}$

3–33. $T_{AB} = 11.0\,\text{lb}$, $T_{AC} = 7.76\,\text{lb}$, $T_{BC} = 11.0\,\text{lb}$, $T_{BE} = 19.0\,\text{lb}$, $T_{CD} = 17.4\,\text{lb}$, $\theta = 18.4°$

3–34. $\theta = 18.4°$, $W = 15.8\,\text{lb}$

3–35. $F_{AB} = 175\,\text{lb}$, $l = 2.34\,\text{ft}$, or $F_{AB} = 82.4\,\text{lb}$, $l = 1.40\,\text{ft}$

3–37. $m_B = 3.58\,\text{kg}$, $N = 19.7\,\text{N}$

3–38. $F_{AB} = 98.6\,\text{N}$, $F_{AC} = 267\,\text{N}$

3–39. $d = 2.42\,\text{m}$

3–41. $T = 30.6\,\text{lb}$, $x = 1.92\,\text{ft}$

3–42. $W_B = 18.3\,\text{lb}$

3–43. $F_{AD} = 763\,\text{N}$, $F_{AC} = 392\,\text{N}$, $F_{AB} = 523\,\text{N}$

3–45. $F_{DA} = 10.0\,\text{lb}$, $F_{DB} = 1.11\,\text{lb}$, $F_{DC} = 15.6\,\text{lb}$

3–46. $s_{OB} = 327\,\text{mm}$, $s_{OA} = 218\,\text{mm}$

3–47. $F_{AB} = 219\,\text{N}$, $F_{AC} = F_{AD} = 54.8\,\text{N}$

3–49. $m = 102\,\text{kg}$

3–50. $F_{AC} = 113\,\text{lb}$, $F_{AB} = 257\,\text{lb}$, $F_{AD} = 210\,\text{lb}$

3–51. $F = 1558\,\text{lb}$

3–53. $F_{AD} = 557\,\text{lb}$, $W = 407\,\text{lb}$

3–54. $F_{AB} = 79.2\,\text{lb}$, $F_{AC} = 119\,\text{lb}$, $F_{AD} = 283\,\text{lb}$

3–55. $W_C = 265\,\text{lb}$

3–57. $W = 55.8\,\text{N}$

3–58. $F_{AB} = 441\,\text{N}$, $F_{AC} = 515\,\text{N}$, $F_{AD} = 221\,\text{N}$

3–59. $F_{AB} = 348\,\text{N}$, $F_{AC} = 413\,\text{N}$, $F_{AD} = 174\,\text{N}$

3–61. $F_{AC} = 85.8\,\text{N}$, $F_{AB} = 578\,\text{N}$, $F_{AD} = 565\,\text{N}$

3–62. $m = 88.5\,\text{kg}$

3–63. $F_{AD} = 1.56\,\text{kN}$, $F_{BD} = 521\,\text{N}$, $F_{CD} = 1.28\,\text{kN}$

3–65. $F_{AE} = 2.91\,\text{kip}$, $F = 1.61\,\text{kip}$

3–66. $F_{AB} = 360\,\text{lb}$, $F_{AC} = 180\,\text{lb}$, $F_{AD} = 360\,\text{lb}$

3–67. $W = 375\,\text{lb}$

Chapter 4

4–5. $(M_{F_1})_B = 4.125\,\text{kip}\cdot\text{ft} \,\circlearrowright$, $(M_{F_2})_B = 2.00\,\text{kip}\cdot\text{ft} \,\circlearrowright$, $(M_{F_3})_B = 40.0\,\text{lb}\cdot\text{ft} \,\circlearrowright$

4–6. $M_P = 341\,\text{in.}\cdot\text{lb} \,\circlearrowright$, $M_F = 403\,\text{in.}\cdot\text{lb} \,\circlearrowleft$, Not sufficient

4–7. $(M_{F_1})_A = 433\,\text{N}\cdot\text{m} \,\circlearrowleft$, $(M_{F_2})_A = 1.30\,\text{kN}\cdot\text{m} \,\circlearrowleft$, $(M_{F_3})_A = 800\,\text{N}\cdot\text{m} \,\circlearrowleft$

4–9. $M_B = 90.6\,\text{lb}\cdot\text{ft} \,\circlearrowleft$, $M_C = 141\,\text{lb}\cdot\text{ft} \,\circlearrowright$

4–10. $M_A = 195\,\text{lb}\cdot\text{ft} \,\circlearrowright$

4–11. $(M_O)_{max} = 48.0\,\text{kN}\cdot\text{m} \,\circlearrowright$, $x = 9.81\,\text{m}$

4–13. $\mathbf{M}_B = \{-3.36\mathbf{k}\}\,\text{N}\cdot\text{m}$, $\alpha = 90°$, $\beta = 90°$, $\gamma = 180°$

4–14. $\mathbf{M}_O = \{0.5\mathbf{i} + 0.866\mathbf{j} - 3.36\mathbf{k}\}\,\text{N}\cdot\text{m}$, $\alpha = 81.8°$, $\beta = 75.7°$, $\gamma = 163°$

4–15. $(M_A)_C = 768\,\text{lb}\cdot\text{ft} \,\circlearrowleft$ $(M_A)_B = 636\,\text{lb}\cdot\text{ft} \,\circlearrowright$ Clockwise

4–17. $m = \left(\dfrac{l}{d+l}\right)M$

4–18. $M_P = (537.5\cos\theta + 75\sin\theta)\,\text{lb}\cdot\text{ft}$

4–19. $F = 239\,\text{lb}$

4–21. $F = 27.6\,\text{lb}$

4–22. $r = 13.3\,\text{mm}$

4–23. $(M_R)_A = (M_R)_B = 76.0\,\text{kN}\cdot\text{m} \,\circlearrowright$

4–25. $(M_{AB})_A = 3.88\,\text{kip}\cdot\text{ft} \,\circlearrowleft$, $(M_{BCD})_A = 2.05\,\text{kip}\cdot\text{ft} \,\circlearrowleft$, $(M_{man})_A = 2.10\,\text{kip}\cdot\text{ft} \,\circlearrowleft$

4–26. $(M_R)_A = 8.04\,\text{kip}\cdot\text{ft} \,\circlearrowleft$

4–27. $\mathbf{M}_O = \{-40\mathbf{i} - 44\mathbf{j} - 8\mathbf{k}\}\,\text{kN}\cdot\text{m}$

4–29. $\mathbf{M}_O = \{-25\mathbf{i} + 6200\mathbf{j} - 900\mathbf{k}\}\,\text{lb}\cdot\text{ft}$

4–30. $\mathbf{M}_A = \{-175\mathbf{i} + 5600\mathbf{j} - 900\mathbf{k}\}\,\text{lb}\cdot\text{ft}$

4–31. $\mathbf{M}_P = \{-24\mathbf{i} + 24\mathbf{j} + 8\mathbf{k}\}\,\text{kN}\cdot\text{m}$

4–33. $\mathbf{M}_B = \{-110\mathbf{i} - 180\mathbf{j} - 420\mathbf{k}\}\,\text{N}\cdot\text{m}$

4–34. $\mathbf{M}_A = \{574\mathbf{i} + 350\mathbf{j} + 1385\mathbf{k}\}\,\text{N}\cdot\text{m}$

4–35. $F = 585\,\text{N}$

4–37. $\mathbf{M}_O = \{163\mathbf{i} - 346\mathbf{j} - 360\mathbf{k}\}\,\text{N}\cdot\text{m}$

4–38. $\mathbf{M}_A = \{-82.9\mathbf{i} + 41.5\mathbf{j} + 232\mathbf{k}\}\,\text{lb}\cdot\text{ft}$

4–39. $\mathbf{M}_B = \{-82.9\mathbf{i} - 96.8\mathbf{j} - 52.8\mathbf{k}\}\,\text{lb}\cdot\text{ft}$

4–41. $F = 18.6\,\text{lb}$

4–42. $M_O = 4.27\,\text{N}\cdot\text{m}$, $\alpha = 95.2°$, $\beta = 110°$, $\gamma = 20.6°$

4–43. $\mathbf{M}_A = \{-5.39\mathbf{i} + 13.1\mathbf{j} + 11.4\mathbf{k}\}\,\text{N}\cdot\text{m}$

4–45. $y = 2\,\text{m}$, $z = 1\,\text{m}$

4–46. $y = 1\,\text{m}$, $z = 3\,\text{m}$, $d = 1.15\,\text{m}$

4–47. $\mathbf{M}_A = \{-16.0\mathbf{i} - 32.1\mathbf{k}\}\,\text{N}\cdot\text{m}$

4–49. $\mathbf{M}_B = \{1.00\mathbf{i} + 0.750\mathbf{j} - 1.56\mathbf{k}\}\,\text{kN}\cdot\text{m}$

4–50. $\mathbf{M}_O = \{373\mathbf{i} - 99.9\mathbf{j} + 173\mathbf{k}\}\,\text{N}\cdot\text{m}$

4–51. $\theta_{max} = 90°$, $\theta_{min} = 0, 180°$

4–53. Yes, yes

4–54. $M_{y'} = 464\,\text{lb}\cdot\text{ft}$

4–55. $M_x = 440\,\text{lb}\cdot\text{ft}$

4–57. $\mathbf{M}_{AC} = \{11.5\mathbf{i} + 8.64\mathbf{j}\}\,\text{lb}\cdot\text{ft}$

4–58. $M_x = 21.7\,\text{N}\cdot\text{m}$

4–59. $F = 139\,\text{N}$

4–61. $M_{AB} = 136\,\text{N}\cdot\text{m}$

4–62. $M_{BC} = 165\,\text{N}\cdot\text{m}$

4–63. $M_{CA} = 226\,\text{N}\cdot\text{m}$

4–65. $F = 5.66\,\text{N}$

4–66. $M_a = 4.37\,\text{N}\cdot\text{m}$, $\alpha = 33.7°$, $\beta = 90°$, $\gamma = 56.3°$, $M = 5.41\,\text{N}\cdot\text{m}$

4–67. $R = 28.9\,\text{N}$

4–69. $F = 75\,\text{N}$, $P = 100\,\text{N}$

4–70. $(M_R)_C = 435\,\text{lb}\cdot\text{ft} \,\circlearrowright$

4–71. $F = 139\,\text{lb}$

4–73. $F = 830\,\text{N}$

4–74. $M_C = 22.5 \text{ N} \cdot \text{m} \, \circlearrowright$

4–75. $F = 83.3 \text{ N}$

4–77. $(M_R)_C = 240 \text{ lb} \cdot \text{ft} \, \circlearrowright$

4–78. $F = 167 \text{ lb}$. Resultant couple can act anywhere.

4–79. $d = 2.03 \text{ ft}$

4–81. $M_C = 126 \text{ lb} \cdot \text{ft} \, \circlearrowright$

4–82. $\mathbf{M}_C = \{-50\mathbf{i} + 60\mathbf{j}\} \text{ lb} \cdot \text{ft}$

4–83. $M_R = 96.0 \text{ lb} \cdot \text{ft}, \alpha = 47.4°, \beta = 74.9°, \gamma = 133°$

4–85. $M_R = 64.0 \text{ lb} \cdot \text{ft}, \alpha = 94.7°, \beta = 13.2°, \gamma = 102°$

4–86. $M_2 = 424 \text{ N} \cdot \text{m}, M_3 = 300 \text{ N} \cdot \text{m}$

4–87. $M_R = 576 \text{ lb} \cdot \text{in.}, \alpha = 37.0°, \beta = 111°, \gamma = 61.2°$

4–89. $F = 15.4 \text{ N}$

4–90. $M_C = 45.1 \text{ N} \cdot \text{m}$

4–91. $F = 832 \text{ N}$

4–93. $F = 98.1 \text{ N}$

4–94. $\mathbf{M}_C = \{-2\mathbf{i} + 20\mathbf{j} + 17\mathbf{k}\} \text{ kN} \cdot \text{m},$
$M_C = 26.3 \text{ kN} \cdot \text{m}$

4–95. $(M_C)_R = 71.9 \text{ N} \cdot \text{m}, \alpha = 44.2°, \beta = 131°, \gamma = 103°$

4–97. $F_R = 365 \text{ N}, \theta = 70.8° \, \triangleright, (M_R)_O = 2364 \text{ N} \cdot \text{m} \, \circlearrowright$

4–98. $F_R = 365 \text{ N}, \theta = 70.8° \, \triangleright, (M_R)_P = 2799 \text{ N} \cdot \text{m} \, \circlearrowright$

4–99. $F_R = 5.93 \text{ kN}, \theta = 77.8° \, \triangleright, M_{R_A} = 34.8 \text{ kN} \cdot \text{m} \, \circlearrowright$

4–101. $F_R = 294 \text{ N}, \theta = 40.1° \, \triangleright,$
$M_{RO} = 39.6 \text{ N} \cdot \text{m} \, \circlearrowright$

4–102. $F_R = 1.30 \text{ kN}, \theta = 86.7° \, \triangledown,$
$(M_R)_A = 1.02 \text{ kN} \cdot \text{m} \, \circlearrowright$

4–103. $F_R = 1.30 \text{ kN}, \theta = 86.7° \, \triangledown,$
$(M_R)_B = 10.1 \text{ kN} \cdot \text{m} \, \circlearrowright$

4–105. $F_R = 938 \text{ N}, \theta = 35.9° \, \triangledown, (M_R)_A = 680 \text{ N} \cdot \text{m} \, \circlearrowright$

4–106. $\mathbf{M}_{RO} = \{0.650\mathbf{i} + 19.75\mathbf{j} - 9.05\mathbf{k}\} \text{ kN} \cdot \text{m}$

4–107. $\mathbf{F}_R = \{270\mathbf{k}\} \text{ N}, \mathbf{M}_{RO} = \{-2.22\mathbf{i}\} \text{ N} \cdot \text{m}$

4–109. $\mathbf{F}_R = \{6\mathbf{i} + 5\mathbf{j} - 5\mathbf{k}\} \text{ kN},$
$(\mathbf{M}_R)_O = \{2.5\mathbf{i} - 7\mathbf{j}\} \text{ kN} \cdot \text{m}$

4–110. $\mathbf{F}_R = \{44.5\mathbf{i} + 53.1\mathbf{j} - 40.0\mathbf{k}\} \text{ N},$
$\mathbf{M}_{RA} = \{-5.39\mathbf{i} + 13.1\mathbf{j} + 11.4\mathbf{k}\} \text{ N} \cdot \text{m}$

4–111. $\mathbf{F}_R = \{-40\mathbf{j} - 40\mathbf{k}\} \text{ N},$
$\mathbf{M}_{RA} = \{-12\mathbf{j} + 12\mathbf{k}\} \text{ N} \cdot \text{m}$

4–113. $F_R = 10.75 \text{ kip} \downarrow, d = 13.7 \text{ ft}$

4–114. $F_R = 10.75 \text{ kip} \downarrow, d = 9.26 \text{ ft}$

4–115. $F = 798 \text{ lb}, 67.9° \, \triangleright, x = 7.43 \text{ ft}$

4–117. $F = 1302 \text{ N}, \theta = 84.5° \, \triangleright, x = 7.36 \text{ m}$

4–118. $F = 1302 \text{ N}, \theta = 84.5° \, \triangleright,$
$x = 1.36 \text{ m (to the right)}$

4–119. $F_R = 1000 \text{ N}, \theta = 53.1° \, \triangledown, d = 2.17 \text{ m}$

4–121. $F_R = 356 \text{ N}, \theta = 51.8°, d = 1.75 \text{ m}$

4–122. $F_R = 542 \text{ N}, \theta = 10.6° \, \triangleleft, d = 0.827 \text{ m}$

4–123. $F_R = 542 \text{ N}, \theta = 10.6° \, \triangleleft, d = 2.17 \text{ m}$

4–125. $F_R = 197 \text{ lb}, \theta = 42.6° \, \triangleleft, d = 5.24 \text{ ft}$

4–126. $F_R = 197 \text{ lb}, \theta = 42.6° \, \triangleleft, d = 0.824 \text{ ft}$

4–127. $F_R = 26 \text{ kN}, y = 82.7 \text{ mm}, x = 3.85 \text{ mm}$

4–129. $F_C = 600 \text{ N}, F_D = 500 \text{ N}$

4–130. $F_R = 35 \text{ kN}, y = 11.3 \text{ m}, x = 11.5 \text{ m}$

4–131. $F_1 = 27.6 \text{ kN}, F_2 = 24.0 \text{ kN}$

4–133. $F_A = 30 \text{ kN}, F_B = 20 \text{ kN}, F_R = 190 \text{ kN}$

4–134. $\mathbf{F}_R = \{141\mathbf{i} + 100\mathbf{j} + 159\mathbf{k}\} \text{ N},$
$\mathbf{M}_{R_O} = \{122\mathbf{i} - 183\mathbf{k}\} \text{ N} \cdot \text{m}$

4–135. $F_R = 379 \text{ N}, M_R = 590 \text{ N} \cdot \text{m}, z = 2.68 \text{ m},$
$x = -2.76 \text{ m}$

4–137. $F_R = 539 \text{ N}, M_R = 1.45 \text{ kN} \cdot \text{m}, x = 1.21 \text{ m},$
$y = 3.59 \text{ m}$

4–138. $F_R = 0, M_{RO} = 1.35 \text{ kip} \cdot \text{ft}$

4–139. $F_R = 6.75 \text{ kN}, \bar{x} = 2.5 \text{ m}$

4–141. $F_R = 7 \text{ lb}, \bar{x} = 0.268 \text{ ft}$

4–142. $F_R = 15.0 \text{ kN}, d = 3.40 \text{ m}$

4–143. $F_R = 12.5 \text{ kN}, d = 1.54 \text{ m}$

4–145. $F_R = 15.4 \text{ kN}, (M_R)_O = 18.5 \text{ kN} \cdot \text{m} \, \circlearrowright$

4–146. $F_R = 27.0 \text{ kN}, (M_R)_A = 81.0 \text{ kN} \cdot \text{m} \, \circlearrowright$

4–147. $a = 1.54 \text{ m}$

4–149. $w_2 = 17.2 \text{ kN/m}, w_1 = 30.3 \text{ kN/m}$

4–150. $F_R = 51.0 \text{ kN} \downarrow, M_{R_O} = 914 \text{ kN} \cdot \text{m} \, \circlearrowright$

4–151. $F_R = 51.0 \text{ kN} \downarrow, d = 17.9 \text{ m}$

4–153. $F_R = 1.80 \text{ kN}, d = 2.33 \text{ m}$

4–154. $F_R = 12.0 \text{ kN}, \theta = 48.4° \, \triangleright, d = 3.28 \text{ m}$

4–155. $F_R = 12.0 \text{ kN}, \theta = 48.4° \, \triangleright, d = 3.69 \text{ m}$

4–157. $F_R = 6.75 \text{ kN}, (M_R)_O = 4.05 \text{ kN} \cdot \text{m} \, \circlearrowright$

4–158. $F_R = 43.6 \text{ lb}, x = 3.27 \text{ ft}$

4–159. $d = 2.22 \text{ ft}$

4–161. $F_R = \dfrac{2Lw_0}{\pi}, (M_R)_O = \left(\dfrac{2\pi - 4}{\pi^2}\right)w_0L^2 \, \circlearrowright$

4–162. $F_R = 107 \text{ kN}, h = 1.60 \text{ m}$

Chapter 5

5–10. $A_x = 3.46 \text{ kN}, A_y = 8 \text{ kN}, M_A = 20.2 \text{ kN} \cdot \text{m}$

5–11. $N_A = 750 \text{ N}, B_y = 600 \text{ N}, B_x = 450 \text{ N}$

5–13. $N_A = 2.175 \text{ kN}, B_y = 1.875 \text{ kN}, B_x = 0$

5–14. $N_A = 3.33 \text{ kN}, B_x = 2.40 \text{ kN}, B_y = 133 \text{ N}$

5–15. $A_y = 5.00 \text{ kN}, N_B = 9.00 \text{ kN}, A_x = 5.00 \text{ kN}$

5–17. $\theta = 41.4°$

5–18. $A_x = 0, B_y = P, M_A = \dfrac{PL}{2}$

5–19. $T = \dfrac{W}{2} \sin \theta$

5–21. $T_{BC} = 113 \text{ N}$

5–22. $N_A = 3.71 \text{ kN}, B_x = 1.86 \text{ kN}, B_y = 8.78 \text{ kN}$

5–23. $w = 2.67 \text{ kN/m}$

5–25. $N_A = 39.7 \text{ lb}, N_B = 82.5 \text{ lb}, M_A = 106 \text{ lb} \cdot \text{ft}$

5–26. $\theta = 70.3°, N'_A = (29.4 - 31.3 \sin \theta) \text{ kN},$
$N'_B = (73.6 + 31.3 \sin \theta) \text{ kN}$

5–27. $N_B = 98.1 \text{ N}, A_x = 85.0 \text{ N}, A_y = 147 \text{ N}$

5–29. $P = 272 \text{ N}$

5–30. $P_{min} = 271 \text{ N}$

5–31. $F_B = 86.6 \text{ N}, B_x = 43.3 \text{ N}, B_y = 110 \text{ N}$

5–33. $A_x = 25.4 \text{ kN}, B_y = 22.8 \text{ kN}, B_x = 25.4 \text{ kN}$

5–34. $F = 14.0 \text{ kN}$

5–35. $N_A = 173$ N, $N_C = 416$ N, $N_B = 69.2$ N

5–37. $N_A = 975$ lb, $B_x = 975$ lb, $B_y = 780$ lb

5–38. $A_x = 1.46$ kip, $F_B = 1.66$ kip

5–39. $\theta = 17.5°$

5–41. $F = 311$ kN, $A_x = 460$ kN, $A_y = 7.85$ kN

5–42. $F_{CB} = 782$ N, $A_x = 625$ N, $A_y = 681$ N

5–43. $F_2 = 724$ N, $F_1 = 1.45$ kN, $F_A = 1.75$ kN

5–45. $P = 660$ N, $N_A = 442$ N, $\theta = 48.0°$ ◺

5–46. $d = \dfrac{3a}{4}$

5–47. $F_{BC} = 80$ kN, $A_x = 54$ kN, $A_y = 16$ kN

5–49. $F_C = 10$ mN

5–50. $k = 250$ N/m

5–51. $w_B = 2.19$ kip/ft, $w_A = 10.7$ kip/ft

5–53. $\alpha = 10.4°$

5–54. $h = 0.645$ m

5–55. $h = \sqrt{\dfrac{s^2 - l^2}{3}}$

5–57. $w_1 = 83.3$ lb/ft, $w_2 = 167$ lb/ft

5–58. $w_1 = \dfrac{2P}{L}, w_2 = \dfrac{4P}{L}$

5–59. $\theta = 23.2°, 85.2°$

5–61. $N_A = 346$ N, $N_B = 693$ N, $a = 0.650$ m

5–62. $T = 1.84$ kN, $F = 6.18$ kN

5–63. $R_D = 22.6$ kip, $R_E = 22.6$ kip, $R_F = 13.7$ kip

5–65. $N_A = 28.6$ lb, $N_B = 10.7$ lb, $N_C = 10.7$ lb

5–66. $T_{BC} = 43.9$ N, $N_B = 58.9$ N, $A_x = 58.9$ N, $A_y = 39.2$ N, $A_z = 177$ N

5–67. $T_C = 14.8$ kN, $T_B = 16.5$ kN, $T_A = 7.27$ kN

5–69. $F_{AB} = 467$ N, $F_{AC} = 674$ N, $D_x = 1.04$ kN, $D_y = 0, D_z = 0$

5–70. $T_{BA} = 2.00$ kN, $T_{BC} = 1.35$ kN, $D_x = 0.327$ kN, $D_y = 1.31$ kN, $D_z = 4.58$ kN

5–71. $F_{BD} = F_{BC} = 350$ N, $A_x = 600$ N, $A_y = 0, A_z = 300$ N

5–73. $C_y = 800$ N, $B_z = 107$ N, $B_y = 600$ N, $C_x = 53.6$ N, $A_x = 400$ N, $A_z = 800$ N

5–74. $F = 746$ N

5–75. $T_{BC} = 1.40$ kN, $A_y = 800$ N, $A_x = 1.20$ kN, $(M_A)_x = 600$ N·m, $(M_A)_y = 1.20$ kN·m, $(M_A)_z = 2.40$ kN·m

5–77. $A_x = 300$ N, $A_y = 500$ N, $N_B = 400$ N, $(M_A)_x = 1.00$ kN·m, $(M_A)_y = 200$ N·m, $(M_A)_z = 1.50$ kN·m

5–78. $A_x = 633$ lb, $A_y = -141$ lb, $B_x = -721$ lb, $B_z = 895$ lb, $C_y = 200$ lb, $C_z = -506$ lb

5–79. $F_2 = 674$ lb

5–81. $C_z = 10.6$ lb, $D_y = -0.230$ lb, $C_y = 0.230$ lb, $D_x = 5.17$ lb, $C_x = 5.44$ lb, $M = 0.459$ lb·ft

5–82. $F_{BD} = 294$ N, $F_{BC} = 589$ N, $A_x = 0$, $A_y = 589$N, $A_z = 490.5$ N

5–83. $T = 58.0$ N, $C_z = 87.0$ N, $C_y = 28.8$ N, $D_x = 0$, $D_y = 79.2$ N, $D_z = 58.0$ N

5–85. $F_{BC} = 0, A_y = 0, A_z = 800$ lb, $(M_A)_x = 4.80$ kip·ft, $(M_A)_y = 0, (M_A)_z = 0$

Chapter 6

6–1. $F_{CB} = 0$, $F_{CD} = 20.0$ kN (C), $F_{DB} = 33.3$ kN (T), $F_{DA} = 36.7$ kN (C)

6–2. $F_{CB} = 0$, $F_{CD} = 45.0$ kN (C), $F_{DB} = 75.0$ kN (T), $F_{DA} = 90.0$ kN (C)

6–3. $F_{AC} = 150$ lb (C), $F_{AB} = 140$ lb (T), $F_{BD} = 140$ lb (T), $F_{BC} = 0$, $F_{CD} = 150$ lb (T), $F_{CE} = 180$ lb (C), $F_{DE} = 120$ lb (C), $F_{DF} = 230$ lb (T), $F_{EF} = 300$ lb (C)

6–5. $F_{CD} = 5.21$ kN (C), $F_{CB} = 4.17$ kN (T), $F_{AD} = 1.46$ kN (C), $F_{AB} = 4.17$ kN (T), $F_{BD} = 4$ kN (T)

6–6. $F_{CD} = 5.21$ kN (C), $F_{CB} = 2.36$ kN (T), $F_{AD} = 1.46$ kN (C), $F_{AB} = 2.36$ kN (T), $F_{BD} = 4$ kN (T)

6–7. $F_{DE} = 16.3$ kN (C), $F_{DC} = 8.40$ kN (T), $F_{EA} = 8.85$ kN (C), $F_{EC} = 6.20$ kN (C), $F_{CF} = 8.77$ kN (T), $F_{CB} = 2.20$ kN (T), $F_{BA} = 3.11$ kN (T), $F_{BF} = 6.20$ kN (C), $F_{FA} = 6.20$ kN (T)

6–9. $F_{AE} = 5.66$ kN (C), $F_{AB} = 4.00$ kN (T), $F_{DE} = 7.07$ kN (C), $F_{DC} = 5.00$ kN (T), $F_{BE} = 3.16$ kN (T), $F_{BC} = 3.00$ kN (T), $F_{CE} = 6.32$ kN (T)

6–10. $F_{AE} = 9.90$ kN (C), $F_{AB} = 7.00$ kN (T), $F_{DE} = 11.3$ kN (C), $F_{DC} = 8.00$ kN (T), $F_{BE} = 6.32$ kN (T), $F_{BC} = 5.00$ kN (T), $F_{CE} = 9.49$ kN (T)

6–11. $F_{JD} = 33.3$ kN (T), $F_{AL} = F_{GH} = F_{LK} = F_{HI} = 28.3$ kN (C), $F_{AB} = F_{GF} = F_{BC} = F_{FE} = F_{CD} = F_{ED} = $
    20 kN (T), $F_{BL} = F_{FH} = F_{LC} = F_{HE} = 0$, $F_{CK} = F_{EI} = 10$ kN (T), $F_{KJ} = F_{IJ} = 23.6$ kN (C), $F_{KD} = F_{ID} = 7.45$ kN (C)

6–13. $F_{CD} = F_{AD} = 0.687P$ (T), $F_{CB} = F_{AB} = 0.943P$ (C), $F_{DB} = 1.33P$ (T)

6–14. $P_{max} = 849$ lb

6–15. $P_{max} = 849$ lb

6–17. $P = 5.20$ kN

6–18. $F_{DE} = 8.94$ kN (T), $F_{DC} = 4.00$ kN (C), $F_{CB} = 4.00$ kN (C), $F_{CE} = 0$, $F_{EB} = 11.3$ kN (C), $F_{EF} = 12.0$ kN (T), $F_{BA} = 12.0$ kN (C), $F_{BF} = 18.0$ kN (T), $F_{FA} = 20.1$ kN (C), $F_{FG} = 21.0$ kN (T)

6–19. $F_{DE} = 13.4$ kN (T), $F_{DC} = 6.00$ kN (C),
$F_{CB} = 6.00$ kN (C), $F_{CE} = 0$, $F_{EB} = 17.0$ kN (C),
$F_{EF} = 18.0$ kN (T), $F_{BA} = 18.0$ kN (C),
$F_{BF} = 20.0$ kN (T), $F_{FA} = 22.4$ kN (C),
$F_{FG} = 28.0$ kN (T)

6–21. $F_{DE} = F_{DC} = F_{FA} = 0$, $F_{CE} = 34.4$ kN (C),
$F_{CB} = 20.6$ kN (T), $F_{BA} = 20.6$ kN (T),
$F_{BE} = 15.0$ kN (T), $F_{FE} = 30.0$ kN (C),
$F_{EA} = 15.6$ kN (T)

6–22. $F_{FE} = 0.667P$ (T), $F_{FD} = 1.67P$ (T),
$F_{AB} = 0.471P$ (C), $F_{AE} = 1.67P$ (T),
$F_{AC} = 1.49P$ (C), $F_{BF} = 1.41P$ (T),
$F_{BD} = 1.49P$ (C), $F_{EC} = 1.41P$ (T),
$F_{CD} = 0.471P$ (C)

6–23. $F_{EC} = 1.20P$ (T), $F_{ED} = 0$,
$F_{AB} = F_{AD} = 0.373P$ (C), $F_{DC} = 0.373P$ (C),
$F_{DB} = 0.333P$ (T), $F_{BC} = 0.373P$ (C)

6–25. $F_{CB} = 2.31$ kN (C), $F_{CD} = 1.15$ kN (C),
$F_{DB} = 4.00$ kN (T), $F_{DA} = 4.62$ kN (C),
$F_{AB} = 2.31$ kN (C)

6–26. $P_{max} = 1.30$ kN

6–27. $F_{HI} = 42.5$ kN (T), $F_{HC} = 100$ kN (T),
$F_{DC} = 125$ kN (C)

6–29. $F_{HG} = 1125$ lb (T), $F_{DE} = 3375$ lb (C),
$F_{EH} = 3750$ lb (T)

6–30. $F_{CD} = 3375$ lb (C), $F_{HI} = 6750$ lb (T),
$F_{CH} = 5625$ lb (C)

6–31. $F_{KJ} = 11.25$ kip (T), $F_{CD} = 9.375$ kip (C),
$F_{CJ} = 3.125$ kip (C), $F_{DJ} = 0$

6–33. $F_{GF} = 12.5$ kN (C), $F_{CD} = 6.67$ kN (T), $F_{GC} = 0$

6–34. $F_{GH} = 12.5$ kN (C), $F_{BG} = 6.01$ kN (T),
$F_{BC} = 6.67$ kN (T)

6–35. $F_{BC} = 5.33$ kN (C), $F_{EF} = 5.33$ kN (T),
$F_{CF} = 4.00$ kN (T)

6–37. $F_{EF} = 14.0$ kN (C), $F_{BC} = 13.0$ kN (T),
$F_{BE} = 1.41$ kN (T), $F_{BF} = 8.00$ kN (T)

6–38. $F_{EF} = 15.0$ kN (C), $F_{BC} = 12.0$ kN (T),
$F_{BE} = 4.24$ kN (T)

6–39. $F_{BC} = 10.4$ kN (C), $F_{HG} = 9.16$ kN (T),
$F_{HC} = 2.24$ kN (T)

6–41. $F_{BC} = 18.0$ kN (T), $F_{FE} = 15.0$ kN (C),
$F_{EB} = 5.00$ kN (C)

6–42. $F_{HG} = 17.6$ kN (C), $F_{HC} = 5.41$ kN (C),
$F_{BC} = 19.1$ kN (T)

6–43. $F_{GJ} = 17.6$ kN (C), $F_{CJ} = 8.11$ kN (C),
$F_{CD} = 21.4$ kN (T), $F_{CG} = 7.50$ kN (T)

6–45. $F_{BF} = 0$, $F_{BG} = 35.4$ kN (C), $F_{AB} = 45$ kN (T)

6–46. $F_{BC} = 11.0$ kN (C), $F_{GH} = 11.2$ kN (C),
$F_{CH} = 1.25$ kN (C), $F_{CG} = 10.0$ kN (T)

6–47. $F_{CD} = 18.0$ kN (T), $F_{CJ} = 10.8$ kN (T),
$F_{KJ} = 26.8$ kN (T)

6–49. $F_{EF} = 12.9$ kN (T), $F_{FI} = 7.21$ kN (T),
$F_{HI} = 21.1$ kN (C)

6–50. $F_{CA} = F_{CB} = 122$ lb (C), $F_{CD} = 173$ lb (T),
$F_{BD} = 86.6$ lb (T), $F_{BA} = 0$, $F_{DA} = 86.6$ lb (T)

6–51. $F_{AB} = 6.46$ kN (T), $F_{AC} = F_{AD} = 1.50$ kN (C),
$F_{BC} = F_{BD} = 3.70$ kN (C), $F_{BE} = 4.80$ kN (T)

6–53. $F_{CA} = 833$ lb (T), $F_{CB} = 667$ lb (C),
$F_{CD} = 333$ lb (T), $F_{AD} = F_{AB} = 354$ lb (C),
$F_{DB} = 50$ lb (T)

6–54. $F_{CA} = 1000$ lb (C), $F_{CD} = 406$ lb (T),
$F_{CB} = 344$ lb (C), $F_{AB} = F_{AD} = 424$ lb (T),
$F_{DB} = 544$ lb (C)

6–55. $F_{DF} = 5.31$ kN (C), $F_{EF} = 2.00$ kN (T),
$F_{AF} = 0.691$ kN (T)

6–57. $F_{BF} = 0$, $F_{BC} = 0$, $F_{BE} = 500$ lb (T),
$F_{AB} = 300$ lb (C), $F_{AC} = 583$ lb (T),
$F_{AD} = 333$ lb (T), $F_{AE} = 667$ lb (C), $F_{DE} = 0$,
$F_{EF} = 300$ lb (C), $F_{CD} = 300$ lb (C),
$F_{CF} = 300$ lb (C), $F_{DF} = 424$ lb (T)

6–58. $F_{BF} = 0$, $F_{BC} = 0$, $F_{BE} = 500$ lb (T),
$F_{AB} = 300$ lb (C), $F_{AC} = 972$ lb (T), $F_{AD} = 0$,
$F_{AE} = 367$ lb (C), $F_{DE} = 0$, $F_{EF} = 300$ lb (C),
$F_{CD} = 500$ lb (C), $F_{CF} = 300$ lb (C),
$F_{DF} = 424$ lb (T)

6–59. $F_{AD} = 686$ N (T), $F_{BD} = 0$, $F_{CD} = 615$ N (C),
$F_{BC} = 229$ N (T), $F_{AC} = 343$ N (T),
$F_{EC} = 457$ N (C)

6–61. $P = 12.5$ lb

6–62. **a.** $P = 25.0$ lb, **b.** $P = 33.3$ lb, **c.** $P = 11.1$ lb

6–63. $P = 18.9$ N

6–65. $B_x = 4.00$ kN, $B_y = 5.33$ kN, $A_x = 4.00$ kN,
$A_y = 5.33$ kN

6–66. $A_x = 24.0$ kN, $A_y = 12.0$ kN, $D_x = 18.0$ kN,
$D_y = 24.0$ kN

6–67. $A_x = 120$ lb, $A_y = 0$, $N_C = 15.0$ lb

6–69. $B_x = 2.80$ kip, $B_y = 1.05$ kip, $A_x = 2.80$ kip,
$A_y = 5.10$ kip, $M_A = 43.2$ kip·ft

6–70. $C_y = 184$ N, $C_x = 490.5$ N, $B_x = 1.23$ kN,
$B_y = 920$ kN

6–71. $N_E = 18.0$ kN, $N_C = 4.50$ kN, $A_x = 0$,
$A_y = 7.50$ kN, $M_A = 22.5$ kN·m

6–73. $N_E = 3.60$ kN, $N_B = 900$ N, $A_x = 0$,
$A_y = 2.70$ kN, $M_A = 8.10$ kN·m

6–74. $T = 350$ lb, $A_y = 700$ lb, $A_x = 1.88$ kip,
$D_x = 1.70$ kip, $D_y = 1.70$ kip

6–75. $T = 350$ lb, $A_y = 700$ lb, $D_x = 1.82$ kip,
$D_y = 1.84$ kip, $A_x = 2.00$ kip

6–77. $A_x = 96$ lb, $A_y = 72$ lb, $D_y = 2.18$ kip,
$E_x = 96.0$ lb, $E_y = 1.61$ kip

6–78. $N_C = 3.00$ kN, $N_A = 3.00$ kN,
$B_y = 18.0$ kN, $B_x = 0$

6–79. $N_C = N_D = 2$ lb

6–81. $F_{FB} = 1.94$ kN, $F_{BD} = 2.60$ kN

6–82. $N_A = 36.0$ lb

6–83. $F_{FD} = 20.1$ kN, $F_{BD} = 25.5$ kN,
Member EDC: $C_x' = 18.0$ kN, $C_y' = 12.0$ kN,
Member ABC: $C_y'' = 12.0$ kN, $C_x'' = 18.0$ kN

6–85. $T_{AI} = 2.88$ kip, $F_H = 3.99$ kip

6–86. $M = 314$ lb·ft

6–87. $F_C = 19.6$ kN

6–89. $C_x = 650$ N, $C_y = 0$

6–90. $N_B = N_C = 49.5$ N

6–91. $F_{EF} = 8.18$ kN (T), $F_{AD} = 158$ kN (C)

6–93. $P(\theta) = \dfrac{250\sqrt{2.25^2 - \cos^2\theta}}{\sin\theta\cos\theta + \sqrt{2.25^2 - \cos^2\theta}\cdot\cos\theta}$

6–94. $N_B = 0.1175$ lb, $N_A = 0.0705$ lb

6–95. $F_N = 5.25$ lb

6–97. **a.** $F = 205$ lb, $N_C = 380$ lb,
b. $F = 102$ lb, $N_C = 72.5$ lb

6–98. $E_y = 1.00$ kN, $E_x = 3.00$ kN, $B_x = 2.50$ kN,
$B_y = 1.00$ kN, $A_x = 2.50$ kN, $A_y = 500$ N

6–99. $N_C = 12.7$ kN, $A_x = 12.7$ kN, $A_y = 2.94$ kN,
$N_D = 1.05$ kN

6–101. $F = 370$ N

6–102. $N_A = 284$ N

6–103. $B_y = 2.67$ kN, $B_x = 4.25$ kN,
$A_y = 3.33$ kN, $A_x = 7.25$ kN

6–105. $P = 198$ N

6–106. $F = 66.1$ lb

6–107. $d = 0.638$ ft

6–109. $P = 46.9$ lb

6–110. $\theta = 23.7°$

6–111. $m = 26.0$ kg

6–113. $m_S = 1.71$ kg

6–114. $m_L = 106$ kg

6–115. $P = 283$ N, $B_x = D_x = 42.5$ N,
$B_y = D_y = 283$ N, $B_z = D_z = 283$ N

6–117. $M_{Ex} = 0.5$ kN·m, $M_{Ey} = 0$, $E_y = 0$, $E_x = 0$

6–118. $F_D = 20.8$ lb, $F_F = 14.7$ lb, $F_A = 24.5$ lb

Chapter 7

7–1. $N_C = 0$, $V_C = -386$ lb, $M_C = -857$ lb·ft,
$N_D = 0$, $V_D = 300$ lb, $M_D = -600$ lb·ft

7–2. $N_C = 0$, $V_C = -1.00$ kip, $M_C = 56.0$ kip·ft,
$N_D = 0$, $V_D = -1.00$ kip, $M_D = 48.0$ kip·ft

7–3. $V_A = 0$, $N_A = -39$ kN, $M_A = -2.425$ kN·m

7–5. $V_C = -133$ lb, $M_C = 133$ lb·in.

7–6. $a = \dfrac{L}{3}$

7–7. $V_C = -4.00$ kip, $M_C = 24.0$ kip·ft

7–9. $N_C = -30$ kN, $V_C = -8$ kN, $M_C = 6$ kN·m

7–10. $P = 0.533$ kN, $N_C = -2$ kN, $V_C = -0.533$ kN,
$M_C = 0.400$ kN·m

7–11. $N_C = 265$ lb, $V_C = -649$ lb, $M_C = -4.23$ kip·ft,
$N_D = -265$ lb, $V_D = 637$ lb, $M_D = -3.18$ kip·ft

7–13. $N_D = 0$, $V_D = 3.00$ kip, $M_D = 12.0$ kip·ft,
$N_E = 0$, $V_E = -8.00$ kip, $M_E = -20.0$ kip·ft

7–14. $M_C = -15.0$ kip·ft, $N_C = 0$, $V_C = 2.01$ kip,
$M_D = 3.77$ kip·ft, $N_D = 0$, $V_D = 1.11$ kip

7–15. $N_C = 0$, $V_C = -1.50$ kN, $M_C = 13.5$ kN·m

7–17. $N_A = 86.6$ lb, $V_A = 150$ lb, $M_A = 1.80$ kip·in.

7–18. $V_C = 2.49$ kN, $N_C = 2.49$ kN, $M_C = 4.97$ kN·m,
$N_D = 0$, $V_D = -2.49$ kN, $M_D = 16.5$ kN·m

7–19. $N_C = -4.32$ kip, $V_C = 1.35$ kip, $M_C = 4.72$ kip·ft

7–21. $N_E = 720$ N, $V_E = 1.12$ kN, $M_E = -320$ N·m,
$N_F = 0$, $V_F = -1.24$ kN, $M_F = -1.41$ kN·m

7–22. $N_D = 4$ kN, $V_D = -9$ kN, $M_D = -18$ kN·m,
$N_E = 4$ kN, $V_E = 3.75$ kN, $M_E = -4.875$ kN·m

7–23. $N_C = 400$ N, $V_C = -96$ N, $M_C = -144$ N·m

7–25. $N_D = 0$, $V_D = 0.75$ kip, $M_D = 13.5$ kip·ft,
$N_E = 0$, $V_E = -9$ kip, $M_E = -24.0$ kip·ft

7–26. $N_C = -20.0$ kN, $V_C = 70.6$ kN,
$M_C = -302$ kN·m

7–27. $N_C = -1.60$ kN, $V_C = 200$ N, $M_C = 200$ N·m

7–29. $N_C = -406$ lb, $V_C = 903$ lb, $M_C = 1.35$ kip·ft

7–30. $N_D = -464$ lb, $V_D = -203$ lb, $M_D = 2.61$ kip·ft

7–31. $N_E = 2.20$ kN, $V_E = 0$, $M_E = 0$,
$N_D = -2.20$ kN, $V_D = 600$ N, $M_D = 1.20$ kN·m

7–33. $N_D = -2.25$ kN, $V_D = 1.25$ kN, -1.88 kN·m

7–34. $N_E = 1.25$ kN, $V_E = 0$, $M_B = 1.69$ kN·m

7–35. $d = 0.200$ m

7–37. $N_D = 1.26$ kN, $V_D = 0$, $M_D = 500$ N·m

7–38. $N_E = -1.48$ kN, $V_E = 500$ N, $M_E = 1000$ N·m

7–39. $V = 0.278\,w_0 r$, $N = 0.0759\,w_0 r$,
$M = 0.0759\,w_0 r^2$

7–41. $N_C = -350$ lb, $(V_C)_y = 700$ lb, $(V_C)_z = -150$ lb,
$(M_C)_x = -1.20$ kip·ft, $(M_C)_y = -750$ lb·ft,
$(M_C)_z = 1.40$ kip·ft

7–42. $(V_C)_x = 104$ lb, $N_C = 0$, $(V_C)_z = 10$ lb,
$(M_C)_x = 20$ lb·ft, $(M_C)_y = 72$ lb·ft,
$(M_C)_z = -178$ lb·ft

7–43. $N_x = -500$ N, $V_y = 100$ N, $V_z = 900$ N,
$M_x = 600$ N·m, $M_y = -900$ N·m,
$M_z = 400$ N·m

7–45. **a.** $0 \le x < a$: $V = \left(1 - \dfrac{a}{L}\right)P$,

$M = \left(1 - \dfrac{a}{L}\right)Px$,

$a < x \le L$: $V = -\left(\dfrac{a}{L}\right)P$,

$M = P\left(a - \dfrac{a}{L}x\right)$

b. $0 \le x < 2$ m: $V = 6$ kN, $M = \{6x\}$ kN $\cdot$ m
2 m $< x \le 6$ m: $V = -3$ kN,
$M = \{18 - 3x\}$ kN $\cdot$ m

7–46. a. For $0 \le x < a$, $V = P$, $M = Px$,
For $a < x < L - a$, $V = 0$, $M = Pa$,
For $L - a < x \le L$, $V = -P$,
$M = P(L - x)$

b. For $0 \le x < 5$ ft, $V = 800$ lb,
$M = 800x$ lb $\cdot$ ft,
For 5 ft $< x < 7$ ft, $V = 0$,
$M = 4000$ lb $\cdot$ ft,
For 7 ft $< x \le 12$ ft, $V = -800$ lb,
$M = (9600 - 800x)$ lb $\cdot$ ft

7–47. a. For $0 \le x < a$, $V = \dfrac{Pb}{a+b}$, $M = \dfrac{Pb}{a+b}x$,
For $a < x \le a + b$, $V = -\dfrac{Pa}{a+b}$,
$M = Pa - \dfrac{Pa}{a+b}x$,

b. For $0 \le x < 5$ ft, $V = 350$ lb,
$M = 350x$ lb $\cdot$ ft,
For 5 ft $< x \le 12$ ft, $V = -250$ lb,
$M = 3000 - 250x$ lb $\cdot$ ft

7–49. $0 \le x < \dfrac{L}{3}$: $V = 0$, $M = 0$,
$\dfrac{L}{3} < x < \dfrac{2L}{3}$: $V = 0$, $M = M_0$,
$\dfrac{2L}{3} < x \le L$: $V = 0$, $M = 0$,
$0 \le x < \dfrac{8}{3}$ m: $V = 0$, $M = 0$,
$\dfrac{8}{3}$ m $< x < \dfrac{16}{3}$ m: $V = 0$, $M = 500$ N $\cdot$ m,
$\dfrac{16}{3}$ m $< x \le 8$ m: $V = 0$, $M = 0$

7–50. $M_{max} = 2$ kN $\cdot$ m

7–51. $0 \le x < a$: $V = -wx$, $M = -\dfrac{w}{2}x^2$
$a < x \le 2a$: $V = w(2a - x)$,
$M = 2wax - 2wa^2 - \dfrac{w}{2}x^2$

7–53. For $0 \le x < 20$ ft, $V = \{490 - 50.0x\}$ lb,
$M = \{490x - 25.0x^2\}$ lb $\cdot$ ft,
For 20 ft $< x \le 30$ ft, $V = 0$, $M = -200$ lb $\cdot$ ft

7–54. a. $V = \dfrac{w}{2}(L - 2x)$, $M = \dfrac{w}{2}(Lx - x^2)$

b. $V = (2500 - 500x)$ lb,
$M = (2500x - 250x^2)$ lb $\cdot$ ft

7–55. For $0 \le x < 8$ m, $V = (133.75 - 40x)$ kN,
$M = (133.75x - 20x^2)$ kN $\cdot$ m,
For 8 m $< x \le 11$ m, $V = 20$ kN,
$M = (20x - 370)$ kN $\cdot$ m

7–57. For $0 \le x < L$, $V = \dfrac{w}{18}(7L - 18x)$,
$M = \dfrac{w}{18}(7Lx - 9x^2)$,
For $L < x < 2L$,
$V = \dfrac{w}{2}(3L - 2x)$, $M = \dfrac{w}{18}(27Lx - 20L^2 - 9x^2)$,
For $2L < x \le 3L$, $V = \dfrac{w}{18}(47L - 18x)$,
$M = \dfrac{w}{18}(47Lx - 9x^2 - 60L^2)$

7–58. Member AB: For $0 \le x < 12$ ft,
$V = \{875 - 150x\}$ lb,
$M = \{875x - 75.0x^2\}$ lb $\cdot$ ft,
For 12 ft $< x \le 14$ ft, $V = \{2100 - 150x\}$ lb,
$M = \{-75.0x^2 + 2100x - 14700\}$ lb $\cdot$ ft,
Member CBD: For $0 \le x < 2$ ft, $V = 919$ lb,
$M = \{919x\}$ lb $\cdot$ ft, For 2 ft $< x \le 8$ ft,
$V = -306$ lb, $M = \{2450 - 306x\}$ lb $\cdot$ ft

7–59. For $0 \le x < 9$ ft, $V = 25 - 1.67x^2$,
$M = 25x - 0.556x^3$
For 9 ft $< x \le 13.5$ ft, $V = 0$, $M = -180$

7–61. $x = 15^-$, $V = -20$, $M = -300$,
$x = 30^+$, $V = 0$, $M = 150$,
$x = 45^-$, $V = -60$, $M = -300$

7–62. $x = \dfrac{L}{2}$, $P = \dfrac{4M_{max}}{L}$

7–63. $0 \le x \le 12$ ft: $V = \left\{48.0 - \dfrac{x^2}{6}\right\}$ kip,
$M = \left\{48.0x - \dfrac{x^3}{18} - 576\right\}$ kip $\cdot$ ft,
$12 < x \le 24$ ft: $V = \left\{\dfrac{1}{6}(24 - x)^2\right\}$ kip,
$M = \left\{-\dfrac{1}{18}(24 - x)^3\right\}$ kip $\cdot$ ft

7–65. For $0 \le x < 3$ m, $V = \left\{21.0 - 2x^2\right\}$ kN,
$M = \left\{21.0x - \dfrac{2}{3}x^3\right\}$ kN $\cdot$ m,
For 3 m $< x \le 6$ m, $V = \{39.0 - 12x\}$ kN,
$M = \{-6x^2 + 39x - 18\}$ kN $\cdot$ m

7–66. $V = \dfrac{w}{12L}(4L^2 - 6Lx - 3x^2)$,
$M = \dfrac{w}{12L}(4L^2x - 3Lx^2 - x^3)$, $M_{max} = 0.0940\, wL^2$

7–67. $N = P\sin(\theta + \phi)$, $V = -P\cos(\theta + \phi)$,
$M = Pr[\sin(\theta + \phi) - \sin\phi]$

7–69. $V_x = 0$, $V_z = \{24.0 - 4y\}$ lb,
$M_x = \{2y^2 - 24y + 64.0\}$ lb $\cdot$ ft,
$M_y = 8.00$ lb $\cdot$ ft, $M_z = 0$

7–70. $x = 1^-$, $V = 450$ N, $M = 450$ N $\cdot$ m,
$x = 3^+$, $V = -950$ N, $M = 950$ N $\cdot$ m

7–71. $x = 1^-$, $V = 600$ N, $M = 600$ N $\cdot$ m

7–74. $x = 0.5^+$, $V = 450$ N, $M = -150$ N $\cdot$ m, $x = 1.5^-$, $V = -750$ N, $M = -300$ N $\cdot$ m

7–75. $x = 2^+$, $V = -375$ N, $M = 750$ N $\cdot$ m

7–77. $x = 10^+$, $V = 20.0$ kip, $M = -50.0$ kip $\cdot$ ft

7–78. $x = 2^+$, $V = -14.3$, $M = -8.6$

7–79. $x = 1^+$, $V = 175$, $M = -200$, $x = 5^-$, $V = -225$, $M = -300$

7–81. $x = 4.5^-$, $V = -31.5$ kN, $M = -45.0$ kN $\cdot$ m, $x = 8.5^+$, $V = 36.0$ kN, $M = -54.0$ kN $\cdot$ m

7–82. $x = 2.75$, $V = 0$, $M = 1356$ N $\cdot$ m

7–83. $x = 3$, $V = -2.25$ kN, $M = 20.25$ kN $\cdot$ m

7–85. $x = 3^+$, $V = 1800$ lb, $M = -900$ lb $\cdot$ ft $x = 6$, $V = 0$, $M = 1800$ lb $\cdot$ ft

7–86. $x = 1.5$, $V = 2.25$ kN, $M = -2.25$ kN $\cdot$ m

7–87. $x = 3$, $V = 3.00$ kN, $M = -1.50$ kN $\cdot$ m

7–89. $x = \sqrt{15}$, $V = 0$, $M = 1291$ lb $\cdot$ ft $x = 12^-$, $V = -1900$ lb, $M = -6000$ lb $\cdot$ ft

7–90. $x = 0$, $V = 13.5$ kN, $M = -9.5$ kN $\cdot$ m

7–91. $x = 3$, $V = 0$, $M = 18.0$ kN $\cdot$ m $x = 6^-$, $V = -27.0$ kN, $M = -18.0$ kN $\cdot$ m

7–93. $x = 15$, $V = 0$, $M = 37.5$ kip $\cdot$ ft

7–94. $y_B = 2.22$ m, $y_D = 1.55$ m

7–95. $P_1 = 320$ N, $y_D = 2.33$ m

7–97. $x_B = 5.39$ m

7–98. $P = 700$ N

7–99. $y_B = 8.67$ ft, $y_D = 7.04$ ft

7–101. $y_B = 3.53$ m, $P = 0.8$ kN, $T_{max} = T_{DE} = 8.17$ kN

7–102. $w = 51.9$ lb/ft

7–103. $T_{max} = 14.4$ kip, $T_{min} = 13.0$ kip

7–105. $T_{AB} = T_{CD} = 212$ lb (max), $y_B = 2$ ft

7–106. $x = 2.57$ ft, $W = 247$ lb

7–107. $T_A = 61.7$ kip, $T_B = 36.5$ kip, $T_C = 50.7$ kip

7–109. $T_{max} = 594$ kN

7–110. $T_{min} = 552$ kN

7–111. $T_{max} = 3.63$ kip

7–113. $y = \dfrac{x^2}{7813}\left(75 - \dfrac{x^2}{200}\right)$, $T_{max} = 9.28$ kip

7–114. $h = 4.44$ ft

7–115. $(F_h)_R = 6.25$ kip, $(F_v)_R = 2.51$ kip

7–117. $(F_v)_A = 165$ N, $(F_h)_A = 73.9$ N

7–118. $W = 4.00$ kip, $T_{max} = 2.01$ kip

7–121. $l = 104$ ft

7–122. $h = 146$ ft

7–123. $L = 302$ ft

Chapter 8

8–1. $P = 12.8$ kN

8–2. $N_B = 2.43$ kip, $N_C = 1.62$ kip, $F = 200$ lb

8–3. $N_A = 16.5$ kN, $N_B = 42.3$ kN, It does not move.

8–5. $F = 2.76$ kN

8–6. $F = 5.79$ kN

8–7. **a.** No
b. Yes

8–10. $\phi = \theta$, $P = W \sin(\alpha + \theta)$

8–11. **a.** $W = 318$ lb
b. $W = 360$ lb

8–13. $F_{CD} = 3.05$ kN

8–14. $\theta = 21.8°$

8–15. $l = 26.7$ ft

8–17. $\mu_s = 0.231$

8–18. $P = 1350$ lb

8–19. $N_A = 200$ lb

8–21. $n = 12$

8–22. $\mu_s = 0.595$

8–23. $\theta = 33.4°$

8–25. $d = 13.4$ ft

8–26. $P = 740$ N

8–27. $P = 860$ N

8–29. $\theta = 11.0°$

8–30. $\theta = 10.6°$, $x = 0.184$ ft

8–31. $\theta = 8.53°$, $F_A = 1.48$ lb, $F_B = 0.890$ lb

8–33. No

8–34. If $P = \dfrac{1}{2} W$, $\mu_s = \dfrac{1}{3}$

If $P \neq \dfrac{1}{2} W$,

$$\mu_s = \frac{(P + W) - \sqrt{(W + 7P)(W - P)}}{2(2P - W)}$$

for $0 < P < W$

8–35. $P = 8.18$ lb

8–37. $O_y = 400$ N, $O_x = 46.4$ N

8–38. $P = 350$ N, $O_y = 945$ N, $O_x = 280$ N

8–39. $\mu_s = 0.230$

8–41. $\theta = 31.0°$

8–42. $P = 654$ N

8–43. The block fails to be in equilibrium.

8–45. $P = 355$ N

8–46. $\mu_C = 0.0734$, $\mu_B = 0.0964$

8–47. $\theta = 16.3°$

8–49. Yes

8–50. $m = 66.7$ kg

8–51. $P = 408$ N

8–53. $M = 55.2$ lb $\cdot$ ft

8–54. $\theta = 33.4°$

8–55. $P = 13.3$ lb

8–57. $P = 100$ N, $d = 1.50$ ft

8–58. $\theta = 33.4°$

8–59. $P = 5.53$ kN, yes

8–61. $P = 39.6$ lb

8–62. $x = 18.3$ mm

8–63. $P = 2.39$ kN

8–65. $P = 4.05$ kip

8–66. $P = 106$ lb

8–67. $F = 66.7$ N

8–69. $W = 7.19$ kN

8–70. The screw is self-locking.

8–71. $P = 617$ lb

8–74. $M = 40.6$ N·m

8–75. $M = 48.3$ N·m

8–77. $\mu_s = 0.0637$

8–78. $M = 5.69$ lb·in.

8–79. $F = 1.98$ kN

8–81. $F = 11.6$ kN

8–82. $P = 104$ N

8–83. **a.** $F = 1.31$ kN

      b. $F = 372$ N

8–85. He will successfully restrain the cow.

8–86. Yes, it is possible.

      $F = 137$ lb

8–87. $T_1 = 57.7$ lb

8–89. $m_A = 2.22$ kg

8–90. $\theta = 99.2°$

8–91. $n = 3$ half turns, $N_m = 6.74$ lb

8–93. $M = 458$ N·m

8–94. $W = 9.17$ lb

8–95. $P = 78.7$ lb

8–97. $M = 75.4$ N·m, $V = 0.171$ m^3

8–99. $P = 53.6$ N

8–101. $x = 0.384$ m

8–102. $F_s = 85.4$ N

8–103. $W_D = 12.7$ lb

8–105. $\theta_{max} = 38.8°$

8–106. $M = 50.0$ N·m, $x = 286$ mm

8–107. $M = 132$ N·m

8–109. $F = 10.7$ lb

8–110. $M = 16.1$ N·m

8–111. $M = 237$ N·m

8–113. $M = \dfrac{2\mu_s PR}{3 \cos \theta}$

8–114. $T = 905$ lb·in.

8–115. $P = 118$ N

8–117. $P = 29.0$ lb

8–118. $M = \dfrac{8}{15}\mu_s PR$

8–119. $F = 18.9$ N

8–121. $P = 20.5$ lb

8–122. $T = 289$ lb, $N = 479$ lb, $F = 101$ lb

8–123. $\mu_s = 0.0407$

8–125. $r = 20.6$ mm

8–126. $P = 42.2$ lb

8–127. $\mu_s = 0.411$

8–129. $P = 1333$ lb

8–130. $P = 25.3$ lb

Chapter 9

9–1. $\bar{x} = 124$ mm, $\bar{y} = 0$

9–2. $\bar{x} = 0$, $\bar{y} = 1.82$ ft

9–3. $\bar{x} = 0.574$ m, $B_x = 0$, $A_y = 63.1$ N, $B_y = 84.8$ N

9–5. $\bar{y} = 0.857$ m

9–6. $\bar{y} = \dfrac{2}{5}$ m

9–7. $\bar{x} = \dfrac{3}{8}a$

9–9. $\bar{x} = \dfrac{3}{2}$ m

9–10. $\bar{y} = \dfrac{12}{5}$ m

9–11. $\bar{x} = \dfrac{3}{4}b$

9–13. $\bar{x} = 6$ m

9–14. $\bar{y} = 2.8$ m

9–15. $\bar{x} = 0.398$ m

9–17. $\bar{y} = 1.43$ in.

9–18. $\bar{x} = \dfrac{a(1 + n)}{2(2 + n)}$

9–19. $\bar{y} = \dfrac{hn}{2n + 1}$

9–21. $\bar{x} = 1\dfrac{3}{5}$ ft

9–22. $\bar{y} = 4\dfrac{8}{55}$ ft

9–23. $\bar{x} = \dfrac{3}{8}a$

9–25. $\bar{x} = 3.20$ ft, $\bar{y} = 3.20$ ft, $T_A = 384$ lb, $T_C = 384$ lb, $T_B = 1.15$ kip

9–26. $\bar{x} = 3$ ft

9–27. $\bar{y} = \dfrac{6}{5}$ ft

9–29. $\bar{y} = 40.0$ mm

9–30. $\bar{x} = \dfrac{1}{3}(a + b)$

9–31. $\bar{y} = \dfrac{h}{3}$

9–33. $\bar{y} = \dfrac{\pi a}{8}$

9–34. $\bar{x} = 1.26$ m, $\bar{y} = 0.143$ m, $N_B = 47.9$ kN, $A_x = 33.9$ kN, $A_y = 73.9$ kN

9–35. $\bar{x} = \left[\dfrac{2(n + 1)}{3(n + 2)}\right]a$

9–37. $\bar{x} = \dfrac{2}{3}\left(\dfrac{r \sin \alpha}{\alpha}\right)$

9–38. $\bar{x} = 0.785\,a$

9–39. $\bar{x} = \bar{y} = 0$, $\bar{z} = \dfrac{4}{3}$ m

9–41. $\bar{z} = \dfrac{R^2 + 3r^2 + 2rR}{4(R^2 + r^2 + rR)} h$

9–42. $\bar{y} = 2.61 \text{ ft}$

9–43. $\bar{z} = \dfrac{h}{4}, \bar{x} = \bar{y} = \dfrac{a}{\pi}$

9–45. $\bar{z} = \dfrac{4}{3} \text{ m}$

9–46. $\bar{y} = \dfrac{3}{8} b, \bar{x} = \bar{z} = 0$

9–47. $\bar{z} = 12.8 \text{ in.}$

9–49. $\bar{z} = 0.675a$

9–50. $\bar{z} = \dfrac{c}{4}$

9–51. $d = 3 \text{ m}$

9–53. $\bar{x} = 24.4 \text{ mm}, \bar{y} = 40.6 \text{ mm}$

9–54. $\bar{x} = 0, \bar{y} = 58.3 \text{ mm}$

9–55. $\bar{x} = 112 \text{ mm}, \bar{y} = 112 \text{ mm}, \bar{z} = 136 \text{ mm}$

9–57. $\bar{x} = 0.200 \text{ m}, \bar{y} = 4.37 \text{ m}$

9–58. $\bar{y} = 154 \text{ mm}$

9–59. $\bar{x} = 0.571 \text{ in.}, \bar{y} = -0.571 \text{ in.}$

9–61. $\bar{y} = 79.7 \text{ mm}$

9–62. $\bar{x} = -1.00 \text{ in.}, \bar{y} = 4.625 \text{ in.}$

9–63. $\bar{y} = 85.9 \text{ mm}$

9–65. $\bar{x} = 1.57 \text{ in.}, \bar{y} = 1.57 \text{ in.}$

9–66. $\bar{y} = 2 \text{ in.}$

9–67. $\bar{y} = 272 \text{ mm}$

9–69. $\bar{z} = 1.625 \text{ in.}$

9–70. $\bar{z} = 4.32 \text{ in.}$

9–71. $\bar{x} = \dfrac{\frac{2}{3} r \sin^3 \alpha}{\alpha - \frac{\sin 2\alpha}{2}}$

9–73. $\bar{y} = \dfrac{\sqrt{2}(a^2 + at - t^2)}{2(2a - t)}$

9–74. $\bar{x} = 2.81 \text{ ft}, \bar{y} = 1.73 \text{ ft}, N_B = 72.1 \text{ lb},$
$N_A = 86.9 \text{ lb}$

9–75. $\bar{x} = 120 \text{ mm}, \bar{y} = 305 \text{ mm}, \bar{z} = 73.4 \text{ mm}$

9–77. $\theta = 53.1°$

9–78. $\bar{z} = 2.48 \text{ ft}, \theta = 38.9°$

9–79. $\bar{z} = 0.70 \text{ ft}$

9–81. $\bar{z} = 122 \text{ mm}$

9–82. $h = 385 \text{ mm}$

9–83. $\bar{x} = 5.07 \text{ ft}, \bar{y} = 3.80 \text{ ft}$

9–85. $\bar{z} = 128 \text{ mm}$

9–86. $\bar{z} = 754 \text{ mm}$

9–87. $\bar{x} = 19.0 \text{ ft}, \bar{y} = 11.0 \text{ ft}$

9–89. $\Sigma m = 16.4 \text{ kg}, \bar{x} = 153 \text{ mm},$
$\bar{y} = -15 \text{ mm}, \bar{z} = 111 \text{ mm}$

9–90. $V = 27.2(10^3) \text{ ft}^3$

9–91. $A = 3.56 (10^3) \text{ ft}^2$

9–93. $A = 4856 \text{ ft}^2$

9–94. $W = 3.12(10^6) \text{ lb}$

9–95. $V = \dfrac{\pi(6\pi + 4)}{6} a^3$

9–97. $V = 0.114 \text{ m}^3$

9–98. $A = 2.25 \text{ m}^2$

9–99. $A = 276(10^3) \text{ mm}^2$

9–101. $W = 84.7 \text{ kip}$

9–102. Number of gal. $= 2.75 \text{ gal}$

9–103. $A = 8\pi ba, V = 2\pi ba^2$

9–105. $Q = 205 \text{ MJ}$

9–106. $A = 119(10^3) \text{ mm}^2$

9–107. $W = 126 \text{ kip}$

9–109. $A = 1365 \text{ m}^2$

9–110. $m = 138 \text{ kg}$

9–111. $m = 2.68 \text{ kg}$

9–113. $V = 1.40(10^3) \text{ in}^3$

9–114. $h = 29.9 \text{ mm}$

9–115. $F_R = 1250 \text{ lb}, \bar{x} = 2.33 \text{ ft}, \bar{y} = 4.33 \text{ ft}$

9–117. $F_R = 24.0 \text{ kN},$
$\bar{x} = 2.00 \text{ m}, \bar{y} = 1.33 \text{ m}$

9–118. $F_R = \dfrac{4ab}{\pi^2} p_0, \bar{x} = \dfrac{a}{2}, \bar{y} = \dfrac{b}{2}$

9–119. $F_{Rx} = 2rl p_0 \left(\dfrac{\pi}{2}\right), F_R = \pi l r p_0$

9–121. For water: $F_{R_A} = 157 \text{ kN}, F_{R_B} = 235 \text{ kN}$
For oil: $d = 4.22 \text{ m}$

9–122. $d = 2.61 \text{ m}$

9–123. F.S. $= 2.71$

9–125. $F_1 = 9.60 \text{ kip}, F_2 = 40.3 \text{ kip}$

9–126. $F_R = 427 \text{ lb}, \bar{y} = 1.71 \text{ ft}, \bar{x} = 0$

9–127. $F_B = 29.4 \text{ kN}, F_A = 235 \text{ kN}$

9–129. $F = 102 \text{ kN}$

9–130. $F_{R_v} = 196 \text{ lb}, F_{R_h} = 125 \text{ lb}$

Chapter 10

10–1. $I_x = \dfrac{ab^3}{3(3n + 1)}$

10–2. $I_y = \dfrac{a^3 b}{n + 3}$

10–3. $I_x = 457(10^6) \text{ mm}^4$

10–5. $I_x = 0.133 \text{ m}^4$

10–6. $I_y = 0.286 \text{ m}^4$

10–7. $I_x = 0.267 \text{ m}^4$

10–9. $I_x = 23.8 \text{ ft}^4$

10–10. $I_x = \dfrac{2}{15} bh^3$

10–11. $I_x = 614 \text{ m}^4$

10–13. $I_x = \dfrac{\pi}{8} \text{ m}^4$

10–14. $I_y = \dfrac{\pi}{2} \text{ m}^4$

10–15. $I_x = 205$ in^4

10–17. $I_x = \dfrac{1}{30}bh^3$

10–18. $I_y = \dfrac{b^3h}{6}$

10–19. $I_x = 0.267$ m^4

10–21. $I_x = 0.8$ m^4

10–22. $I_y = 0.571$ m^4

10–23. $I_x = \dfrac{3ab^3}{35}$

10–25. $I_x = 209$ in^4

10–26. $I_y = 533$ in^4

10–27. $A = 14.0(10^3)$ mm^2

10–29. $\bar{y} = 52.5$ mm, $I_{x'} = 16.6(10^6)$ mm^4, $I_{y'} = 5.725(10^6)$ mm^4

10–30. $I_x = 182$ in^4

10–31. $I_y = 966$ in^4

10–33. $I_y = 2.03(10^9)$ mm^4

10–34. $I_y = 115(10^6)$ mm^4

10–35. $\bar{y} = 207$ mm, $\bar{I}_{x'} = 222(10^6)$ mm^4

10–37. $I_y = 90.2(10^6)$ mm^4

10–38. $I_x = 1971$ in^4

10–39. $I_y = 2376$ in^4

10–41. $I_y = 341$ in^4

10–42. $I_x = 154(10^6)$ mm^4

10–43. $I_y = 91.3(10^6)$ mm^4

10–45. $\bar{x} = 61.6$ mm, $\bar{I}_{y'} = 41.2(10^6)$ mm^4

10–46. $I_x = 1845$ in^4

10–47. $I_y = 522$ in^4

10–49. $I_{y'} = \dfrac{ab\sin\theta}{12}(b^2 + a^2\cos^2\theta)$

10–50. $\bar{y} = 0.181$ m, $\bar{I}_{x'} = 4.23(10^{-3})$ m^4

10–51. $\bar{I}_{x'} = 520(10^6)$ mm^4

10–53. $I_y = 365$ in^4

10–54. $I_{xy} = \dfrac{1}{3}tl^3\sin 2\theta$

10–55. $I_{xy} = 5.06$ in^4

10–57. $I_{xy} = 10.7$ m^4, $\bar{I}_{x'y'} = 1.07$ m^4

10–58. $I_{xy} = \dfrac{1}{6}a^2b^2$

10–59. $I_{xy} = \dfrac{a^4}{280}$

10–61. $I_{xy} = 0.667$ in^4

10–62. $I_{xy} = 97.8$ in^4

10–63. $I_u = 15.75$ in^4, $I_u = 25.75$ in^4

10–65. $I_{xy} = 119$ in^4

10–66. $I_{xy} = 98.4(10^6)$ mm^4

10–67. $\bar{x} = \bar{y} = 44.1$ mm, $I_{x'y'} = -6.26(10^6)$ mm^4

10–69. $I_u = 3.47(10^3)$ in^4, $I_v = 3.47(10^3)$ in^4, $I_{uv} = 2.05(10^3)$ in^4

10–70. $I_u = 1.28(10^6)$ mm^4, $I_v = 3.31(10^6)$ mm^4, $I_{uv} = -1.75(10^6)$ mm^4

10–71. $I_u = 1.28(10^6)$ mm^4, $I_{uv} = -1.75(10^6)$ mm^4, $I_v = 3.31(10^6)$ mm^4

10–73. $I_{max} = 1219$ in^4, $I_{min} = 36.3$ in^4, $(\theta_p)_2 = 19.0°$ ↷, $(\theta_p)_1 = 71.0°$ ↶

10–74. $I_{max} = 17.4(10^6)$ mm^4, $I_{min} = 1.84(10^6)$ mm^4 $(\theta_p)_1 = 60.0°$, $(\theta_p)_2 = -30.0°$

10–75. $I_{max} = 17.4(10^6)$ mm^4, $I_{min} = 1.84(10^6)$ mm^4, $(\theta_p)_2 = 30.0°$ ↶, $(\theta_p)_1 = 60.0°$ ↷

10–77. $I_{max} = 250$ in^4, $I_{min} = 20.4$ in^4, $(\theta_p)_2 = 22.5°$ ↷ $(\theta_p)_1 = 67.5°$ ↶

10–78. $\theta = 6.08°$, $I_{max} = 1.74(10^3)$ in^4, $I_{min} = 435$ in^4

10–79. $\theta = 6.08°$, $I_{max} = 1.74(10^3)$ in^4, $I_{min} = 435$ in^4

10–81. $I_u = 11.8(10^6)$ mm^4, $I_{uv} = -5.09(10^6)$ mm^4, $I_v = 5.90(10^6)$ mm^4

10–82. $\theta_{p1} = -31.4°$, $\theta_{p2} = 58.6°$, $I_{max} = 309$ in^4, $I_{min} = 42.1$ in^4

10–83. $I_{max} = 309$ in^4, $I_{min} = 42.1$ in^4, $\theta_{p1} = -31.4°$, $\theta_{p2} = 58.6°$

10–85. $I_x = \dfrac{2}{5}mb^2$

10–86. $k_x = 57.7$ mm

10–87. $I_x = \dfrac{1}{3}ma^2$

10–89. $I_x = \dfrac{2}{5}mb^2$

10–90. $k_x = \sqrt{\dfrac{n+2}{2(n+4)}}\,h$

10–91. $I_y = 2.25$ slug · ft^2

10–93. $I_x = \dfrac{3}{10}mr^2$

10–94. $I_y = 1.71(10^3)$ kg · m^2

10–95. $I_A = 0.0453$ kg · m^2

10–97. $I_z = 1.53$ kg · m^2

10–98. $\bar{y} = 1.78$ m, $I_G = 4.45$ kg · m^2

10–99. $I_O = 0.276$ kg · m^2

10–101. $I_A = 222$ slug · ft^2

10–102. $I_z = 29.4$ kg · m^2

10–103. $I_O = \dfrac{1}{2}ma^2$

10–105. $I_z = 0.113$ kg · m^2

10–106. $I_G = 118$ slug · ft^2

10–107. $I_O = 282$ slug · ft^2

10–109. $I_z = 34.2$ kg · m^2

Chapter 11

11–1. $F_{AC} = 7.32$ lb

11–2. $F = 2P\cot\theta$

11–3. $F_S = 15$ lb

11–5. $F = 369$ N

11–7. $M = 52.0\ \text{lb} \cdot \text{ft}$

11–9. $\theta = 16.6°,\ \theta = 35.8°$

11–10. $P = \dfrac{W}{2} \cot \theta$

11–11. $\theta = 23.8°,\ \theta = 72.3°$

11–13. $\theta = 90°,\ \theta = 36.1°$

11–14. $k = 166\ \text{N/m}$

11–15. $F = \dfrac{M}{2a \sin \theta}$

11–17. $M = 13.1\ \text{N} \cdot \text{m}$

11–18. $\theta = 41.2°$

11–19. $k = 9.88\ \text{kN/m}$

11–21. $F = \dfrac{500 \sqrt{0.04 \cos^2 \theta + 0.6}}{(0.2 \cos \theta + \sqrt{0.04 \cos^2 \theta + 0.6}) \sin \theta}$

11–22. $\theta = 9.21°$

11–23. $W_G = 2.5\ \text{lb}$

11–25. $F = \dfrac{W(a + b - d \tan \theta)}{ac} \sqrt{a^2 + c^2 + 2ac \sin \theta}$

11–26. $x = -0.5\ \text{ft unstable},\ x = 0.833\ \text{ft stable}$

11–27. Unstable at $\theta = 34.6°$, stable at $\theta = 145°$

11–29. $\theta = 38.7°$ unstable, $\theta = 90°$ stable, $\theta = 141°$ unstable

11–30. $x = -0.424\ \text{ft unstable},\ x = 0.590\ \text{ft stable}$

11–31. $\theta = 20.2°$, stable

11–33. Unstable equilibrium at $\theta = 90°$
Stable equilibrium at $\theta = 49.0°$

11–34. Unstable equilibrium at $\theta = 0°$
Stable equilibrium at $\theta = 72.9°$

11–35. $k = 2.81\ \text{lb/ft}$

11–37. Stable equilibrium at $\theta = 51.2°$
Unstable equilibrium at $\theta = 4.71°$

11–38. $k = 157\ \text{N/m}$
Stable equilibrium at $\theta = 60°$

11–39. $W = \dfrac{8k}{3L}$

11–41. Stable equilibrium at $\theta = 24.6°$

11–42. $\phi = 17.4°,\ \theta = 9.18°$

11–43. Unstable equilibrium at $\theta = 23.2°$

11–45. $\theta = 0°,\ \theta = 33.0°$

11–46. $m = 5.29\ \text{kg}$

11–49. $\theta = 0°,\ \theta = \cos^{-1}\left(\dfrac{d}{4a}\right)$

Index

ENGINEERING MECHANICS

DYNAMICS

FOURTEENTH EDITION

Chapter 12

(© Lars Johansson/Fotolia)

Although each of these boats is rather large, from a distance their motion can be analyzed as if each were a particle.

Kinematics of a Particle

CHAPTER OBJECTIVES

■ To introduce the concepts of position, displacement, velocity, and acceleration.

■ To study particle motion along a straight line and represent this motion graphically.

■ To investigate particle motion along a curved path using different coordinate systems.

■ To present an analysis of dependent motion of two particles.

■ To examine the principles of relative motion of two particles using translating axes.

12.1 Introduction

Mechanics is a branch of the physical sciences that is concerned with the state of rest or motion of bodies subjected to the action of forces. Engineering mechanics is divided into two areas of study, namely, statics and dynamics. *Statics* is concerned with the equilibrium of a body that is either at rest or moves with constant velocity. Here we will consider *dynamics*, which deals with the accelerated motion of a body. The subject of dynamics will be presented in two parts: *kinematics*, which treats only the geometric aspects of the motion, and *kinetics*, which is the analysis of the forces causing the motion. To develop these principles, the dynamics of a particle will be discussed first, followed by topics in rigid-body dynamics in two and then three dimensions.

Historically, the principles of dynamics developed when it was possible to make an accurate measurement of time. Galileo Galilei (1564–1642) was one of the first major contributors to this field. His work consisted of experiments using pendulums and falling bodies. The most significant contributions in dynamics, however, were made by Isaac Newton (1642–1727), who is noted for his formulation of the three fundamental laws of motion and the law of universal gravitational attraction. Shortly after these laws were postulated, important techniques for their application were developed by Euler, D'Alembert, Lagrange, and others.

There are many problems in engineering whose solutions require application of the principles of dynamics. Typically the structural design of any vehicle, such as an automobile or airplane, requires consideration of the motion to which it is subjected. This is also true for many mechanical devices, such as motors, pumps, movable tools, industrial manipulators, and machinery. Furthermore, predictions of the motions of artificial satellites, projectiles, and spacecraft are based on the theory of dynamics. With further advances in technology, there will be an even greater need for knowing how to apply the principles of this subject.

Problem Solving.   Dynamics is considered to be more involved than statics since both the forces applied to a body and its motion must be taken into account. Also, many applications require using calculus, rather than just algebra and trigonometry. In any case, the most effective way of learning the principles of dynamics is *to solve problems*. To be successful at this, it is necessary to present the work in a logical and orderly manner as suggested by the following sequence of steps:

1. Read the problem carefully and try to correlate the actual physical situation with the theory you have studied.
2. Draw any necessary diagrams and tabulate the problem data.
3. Establish a coordinate system and apply the relevant principles, generally in mathematical form.
4. Solve the necessary equations algebraically as far as practical; then, use a consistent set of units and complete the solution numerically. Report the answer with no more significant figures than the accuracy of the given data.
5. Study the answer using technical judgment and common sense to determine whether or not it seems reasonable.
6. Once the solution has been completed, review the problem. Try to think of other ways of obtaining the same solution.

In applying this general procedure, do the work as neatly as possible. Being neat generally stimulates clear and orderly thinking, and vice versa.

12.2   Rectilinear Kinematics: Continuous Motion

We will begin our study of dynamics by discussing the kinematics of a particle that moves along a rectilinear or straight-line path. Recall that a *particle* has a mass but negligible size and shape. Therefore we must limit application to those objects that have dimensions that are of no consequence in the analysis of the motion. In most problems, we will be interested in bodies of finite size, such as rockets, projectiles, or vehicles. Each of these objects can be considered as a particle, as long as the motion is characterized by the motion of its mass center and any rotation of the body is neglected.

Rectilinear Kinematics.   The kinematics of a particle is characterized by specifying, at any given instant, the particle's position, velocity, and acceleration.

Position.   The straight-line path of a particle will be defined using a single coordinate axis s, Fig. 12–1a. The origin O on the path is a fixed point, and from this point the *position coordinate s* is used to specify the location of the particle at any given instant. The magnitude of s is the distance from O to the particle, usually measured in meters (m) or feet (ft), and the sense of direction is defined by the algebraic sign on s. Although the choice is arbitrary, in this case s is positive since the coordinate axis is positive to the right of the origin. Likewise, it is negative if the particle is located to the left of O. Realize that *position is a vector quantity* since it has both magnitude and direction. Here, however, it is being represented by the algebraic scalar s, rather than in boldface **s**, since the direction always remains along the coordinate axis.

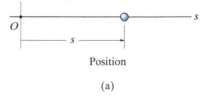

Position

(a)

Displacement.   The *displacement* of the particle is defined as the *change in its position*. For example, if the particle moves from one point to another, Fig. 12–1b, the displacement is

$$\Delta s = s' - s$$

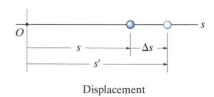

Displacement

(b)

Fig. 12–1

In this case Δs is *positive* since the particle's final position is to the *right* of its initial position, i.e., $s' > s$. Likewise, if the final position were to the *left* of its initial position, Δs would be *negative*.

   The displacement of a particle is also a *vector quantity*, and it should be distinguished from the distance the particle travels. Specifically, the *distance traveled* is a *positive scalar* that represents the total length of path over which the particle travels.

12

Velocity.   If the particle moves through a displacement Δs during the time interval Δt, the ***average velocity*** of the particle during this time interval is

$$v_{\text{avg}} = \frac{\Delta s}{\Delta t}$$

If we take smaller and smaller values of Δt, the magnitude of Δs becomes smaller and smaller. Consequently, the ***instantaneous velocity*** is a vector defined as $v = \lim_{\Delta t \to 0}(\Delta s/\Delta t)$, or

$(\overset{+}{\rightarrow})$
$$v = \frac{ds}{dt} \qquad (12\text{--}1)$$

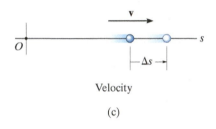

Velocity

(c)

Since Δt or dt is always positive, the sign used to define the *sense* of the velocity is the same as that of Δs or ds. For example, if the particle is moving to the *right*, Fig. 12–1c, the velocity is *positive;* whereas if it is moving to the *left*, the velocity is *negative.* (This is emphasized here by the arrow written at the left of Eq. 12–1.) The *magnitude* of the velocity is known as the ***speed***, and it is generally expressed in units of m/s or ft/s.

Occasionally, the term "average speed" is used. The ***average speed*** is always a positive scalar and is defined as the total distance traveled by a particle, s_T, divided by the elapsed time Δt; i.e.,

$$(v_{\text{sp}})_{\text{avg}} = \frac{s_T}{\Delta t}$$

For example, the particle in Fig. 12–1d travels along the path of length s_T in time Δt, so its average speed is $(v_{\text{sp}})_{\text{avg}} = s_T/\Delta t$, but its average velocity is $v_{\text{avg}} = -\Delta s/\Delta t$.

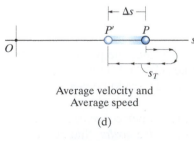

Average velocity and
Average speed

(d)

Fig. 12–1 (cont.)

Acceleration.   Provided the velocity of the particle is known at two points, the ***average acceleration*** of the particle during the time interval Δt is defined as

$$a_{\text{avg}} = \frac{\Delta v}{\Delta t}$$

Here Δv represents the difference in the velocity during the time interval Δt, i.e., $\Delta v = v' - v$, Fig. 12–1e.

The ***instantaneous acceleration*** at time t is a *vector* that is found by taking smaller and smaller values of Δt and corresponding smaller and smaller values of Δv, so that $a = \lim_{\Delta t \to 0} (\Delta v / \Delta t)$, or

(↦)
$$a = \frac{dv}{dt} \qquad (12\text{–}2)$$

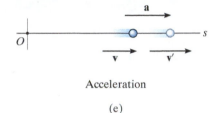

Acceleration

(e)

Substituting Eq. 12–1 into this result, we can also write

(↦)
$$a = \frac{d^2 s}{dt^2}$$

Both the average and instantaneous acceleration can be either positive or negative. In particular, when the particle is *slowing down*, or its speed is decreasing, the particle is said to be ***decelerating***. In this case, v' in Fig. 12–1f is *less* than v, and so $\Delta v = v' - v$ will be negative. Consequently, a will also be negative, and therefore it will act to the *left*, in the *opposite sense* to v. Also, notice that if the particle is originally at rest, then it can have an acceleration if a moment later it has a velocity v'; and, if the *velocity is constant*, then the *acceleration is zero* since $\Delta v = v - v = 0$. Units commonly used to express the magnitude of acceleration are m/s^2 or ft/s^2.

Finally, an important differential relation involving the displacement, velocity, and acceleration along the path may be obtained by eliminating the time differential dt between Eqs. 12–1 and 12–2. We have

$$dt = \frac{ds}{v} = \frac{dv}{a}$$

or

(↦)
$$\boxed{a\,ds = v\,dv} \qquad (12\text{–}3)$$

Although we have now produced three important kinematic equations, realize that the above equation is not independent of Eqs. 12–1 and 12–2.

Deceleration

(f)

Fig. 12–1 (cont.)

12

When the ball is released, it has zero velocity but an acceleration of 9.81 m/s². (© R.C. Hibbeler)

Constant Acceleration, $a = a_c$. When the acceleration is constant, each of the three kinematic equations $a_c = dv/dt$, $v = ds/dt$, and $a_c\,ds = v\,dv$ can be integrated to obtain formulas that relate a_c, v, s, and t.

Velocity as a Function of Time.   Integrate $a_c = dv/dt$, assuming that initially $v = v_0$ when $t = 0$.

$$\int_{v_0}^{v} dv = \int_{0}^{t} a_c\,dt$$

($\xrightarrow{+}$)
$$\boxed{v = v_0 + a_c t}$$
Constant Acceleration
(12–4)

Position as a Function of Time.   Integrate $v = ds/dt = v_0 + a_c t$, assuming that initially $s = s_0$ when $t = 0$.

$$\int_{s_0}^{s} ds = \int_{0}^{t} (v_0 + a_c t)\,dt$$

($\xrightarrow{+}$)
$$\boxed{s = s_0 + v_0 t + \tfrac{1}{2} a_c t^2}$$
Constant Acceleration
(12–5)

Velocity as a Function of Position.   Either solve for t in Eq. 12–4 and substitute into Eq. 12–5, or integrate $v\,dv = a_c\,ds$, assuming that initially $v = v_0$ at $s = s_0$.

$$\int_{v_0}^{v} v\,dv = \int_{s_0}^{s} a_c\,ds$$

($\xrightarrow{+}$)
$$\boxed{v^2 = v_0^2 + 2a_c(s - s_0)}$$
Constant Acceleration
(12–6)

The algebraic signs of s_0, v_0, and a_c, used in the above three equations, are determined from the positive direction of the s axis as indicated by the arrow written at the left of each equation. Remember that these equations are useful *only when the acceleration is constant and when $t = 0$, $s = s_0$, $v = v_0$.* A typical example of constant accelerated motion occurs when a body falls freely toward the earth. If air resistance is neglected and the distance of fall is short, then the *downward* acceleration of the body when it is close to the earth is constant and approximately 9.81 m/s² or 32.2 ft/s². The proof of this is given in Example 13.2.

Important Points

- Dynamics is concerned with bodies that have accelerated motion.
- Kinematics is a study of the geometry of the motion.
- Kinetics is a study of the forces that cause the motion.
- Rectilinear kinematics refers to straight-line motion.
- Speed refers to the magnitude of velocity.
- Average speed is the total distance traveled divided by the total time. This is different from the average velocity, which is the displacement divided by the time.
- A particle that is slowing down is decelerating.
- A particle can have an acceleration and yet have zero velocity.
- The relationship $a\,ds = v\,dv$ is derived from $a = dv/dt$ and $v = ds/dt$, by eliminating dt.

During the time this vvvvket undergoes rectilinear motion, its altitude as a function of time can be measured and expressed as $s = s(t)$. Its velocity can then be found using $v = ds/dt$, and its acceleration can be determined from $a = dv/dt$. (© NASA)

Procedure for Analysis

Coordinate System.

- Establish a position coordinate s along the path and specify its *fixed origin* and positive direction.
- Since motion is along a straight line, the vector quantities position, velocity, and acceleration can be represented as algebraic scalars. For analytical work the sense of s, v, and a is then defined by their *algebraic signs*.
- The positive sense for each of these scalars can be indicated by an arrow shown alongside each kinematic equation as it is applied.

Kinematic Equations.

- If a relation is known between any *two* of the four variables a, v, s, and t, then a third variable can be obtained by using one of the kinematic equations, $a = dv/dt$, $v = ds/dt$ or $a\,ds = v\,dv$, since each equation relates all three variables.*
- Whenever integration is performed, it is important that the position and velocity be known at a given instant in order to evaluate either the constant of integration if an indefinite integral is used, or the limits of integration if a definite integral is used.
- Remember that Eqs. 12–4 through 12–6 have only limited use. These equations apply *only* when the *acceleration is constant* and the initial conditions are $s = s_0$ and $v = v_0$ when $t = 0$.

*Some standard differentiation and integration formulas are given in Appendix A.

EXAMPLE 12.1

The car on the left in the photo and in Fig. 12–2 moves in a straight line such that for a short time its velocity is defined by $v = (3t^2 + 2t)$ ft/s, where t is in seconds. Determine its position and acceleration when $t = 3$ s. When $t = 0$, $s = 0$.

Fig. 12–2

(© R.C. Hibbeler)

SOLUTION

Coordinate System. The position coordinate extends from the fixed origin O to the car, positive to the right.

Position. Since $v = f(t)$, the car's position can be determined from $v = ds/dt$, since this equation relates v, s, and t. Noting that $s = 0$ when $t = 0$, we have*

$$(\xrightarrow{+}) \qquad v = \frac{ds}{dt} = (3t^2 + 2t)$$

$$\int_0^s ds = \int_0^t (3t^2 + 2t)dt$$

$$s \Big|_0^s = t^3 + t^2 \Big|_0^t$$

$$s = t^3 + t^2$$

When $t = 3$ s,

$$s = (3)^3 + (3)^2 = 36 \text{ ft} \qquad \qquad Ans.$$

Acceleration. Since $v = f(t)$, the acceleration is determined from $a = dv/dt$, since this equation relates a, v, and t.

$$(\xrightarrow{+}) \qquad a = \frac{dv}{dt} = \frac{d}{dt}(3t^2 + 2t)$$

$$= 6t + 2$$

When $t = 3$ s,

$$a = 6(3) + 2 = 20 \text{ ft/s}^2 \rightarrow \qquad \qquad Ans.$$

NOTE: The formulas for constant acceleration *cannot* be used to solve this problem, because the acceleration is a function of time.

*The *same result* can be obtained by evaluating a constant of integration C rather than using definite limits on the integral. For example, integrating $ds = (3t^2 + 2t)dt$ yields $s = t^3 + t^2 + C$. Using the condition that at $t = 0$, $s = 0$, then $C = 0$.

EXAMPLE 12.2

A small projectile is fired vertically *downward* into a fluid medium with an initial velocity of 60 m/s. Due to the drag resistance of the fluid the projectile experiences a deceleration of $a = (-0.4v^3)$ m/s^2, where v is in m/s. Determine the projectile's velocity and position 4 s after it is fired.

SOLUTION

Coordinate System.   Since the motion is downward, the position coordinate is positive downward, with origin located at O, Fig. 12–3.

Velocity.   Here $a = f(v)$ and so we must determine the velocity as a function of time using $a = dv/dt$, since this equation relates v, a, and t. (Why not use $v = v_0 + a_c t$?) Separating the variables and integrating, with $v_0 = 60$ m/s when $t = 0$, yields

$(+\downarrow)$

Fig. 12–3

$$a = \frac{dv}{dt} = -0.4v^3$$

$$\int_{60 \text{ m/s}}^{v} \frac{dv}{-0.4v^3} = \int_0^t dt$$

$$\frac{1}{-0.4}\left(\frac{1}{-2}\right)\frac{1}{v^2}\Big|_{60}^{v} = t - 0$$

$$\frac{1}{0.8}\left[\frac{1}{v^2} - \frac{1}{(60)^2}\right] = t$$

$$v = \left\{\left[\frac{1}{(60)^2} + 0.8t\right]^{-1/2}\right\} \text{ m/s}$$

Here the positive root is taken, since the projectile will continue to move downward. When $t = 4$ s,

$$v = 0.559 \text{ m/s} \downarrow \qquad \textit{Ans.}$$

Position.   Knowing $v = f(t)$, we can obtain the projectile's position from $v = ds/dt$, since this equation relates s, v, and t. Using the initial condition $s = 0$, when $t = 0$, we have

$(+\downarrow)$

$$v = \frac{ds}{dt} = \left[\frac{1}{(60)^2} + 0.8t\right]^{-1/2}$$

$$\int_0^s ds = \int_0^t \left[\frac{1}{(60)^2} + 0.8t\right]^{-1/2} dt$$

$$s = \frac{2}{0.8}\left[\frac{1}{(60)^2} + 0.8t\right]^{1/2}\Big|_0^t$$

$$s = \frac{1}{0.4}\left\{\left[\frac{1}{(60)^2} + 0.8t\right]^{1/2} - \frac{1}{60}\right\} \text{ m}$$

When $t = 4$ s,

$$s = 4.43 \text{ m} \qquad \textit{Ans.}$$

12

EXAMPLE | 12.3

During a test a rocket travels upward at 75 m/s, and when it is 40 m from the ground its engine fails. Determine the maximum height s_B reached by the rocket and its speed just before it hits the ground. While in motion the rocket is subjected to a constant downward acceleration of 9.81 m/s² due to gravity. Neglect the effect of air resistance.

SOLUTION

Coordinate System.   The origin O for the position coordinate s is taken at ground level with positive upward, Fig. 12–4.

Maximum Height.   Since the rocket is traveling *upward*, $v_A = +75$ m/s when $t = 0$. At the maximum height $s = s_B$ the velocity $v_B = 0$. For the entire motion, the acceleration is $a_c = -9.81$ m/s² (negative since it acts in the *opposite* sense to positive velocity or positive displacement). Since a_c is *constant* the rocket's position may be related to its velocity at the two points A and B on the path by using Eq. 12–6, namely,

$(+\uparrow)$ $\qquad v_B^2 = v_A^2 + 2a_c(s_B - s_A)$

$\qquad 0 = (75 \text{ m/s})^2 + 2(-9.81 \text{ m/s}^2)(s_B - 40 \text{ m})$

$\qquad s_B = 327 \text{ m}$ 　　　　　　　　　　　　　*Ans.*

Velocity.   To obtain the velocity of the rocket just before it hits the ground, we can apply Eq. 12–6 between points B and C, Fig. 12–4.

$(+\uparrow)$ $\qquad v_C^2 = v_B^2 + 2a_c(s_C - s_B)$

$\qquad\qquad = 0 + 2(-9.81 \text{ m/s}^2)(0 - 327 \text{ m})$

$\qquad v_C = -80.1 \text{ m/s} = 80.1 \text{ m/s} \downarrow$ 　　　*Ans.*

The negative root was chosen since the rocket is moving downward. Similarly, Eq. 12–6 may also be applied between points A and C, i.e.,

$(+\uparrow)$ $\qquad v_C^2 = v_A^2 + 2a_c(s_C - s_A)$

$\qquad\qquad = (75 \text{ m/s})^2 + 2(-9.81 \text{ m/s}^2)(0 - 40 \text{ m})$

$\qquad v_C = -80.1 \text{ m/s} = 80.1 \text{ m/s} \downarrow$ 　　　*Ans.*

NOTE: It should be realized that the rocket is subjected to a *deceleration* from A to B of 9.81 m/s², and then from B to C it is *accelerated* at this rate. Furthermore, even though the rocket momentarily comes to *rest* at B ($v_B = 0$) the acceleration at B is still 9.81 m/s² downward!

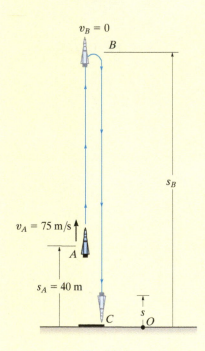

$v_B = 0$

B

s_B

$v_A = 75$ m/s

A

$s_A = 40$ m

C　s　O

Fig. 12–4

EXAMPLE 12.4

A metallic particle is subjected to the influence of a magnetic field as it travels downward through a fluid that extends from plate A to plate B, Fig. 12–5. If the particle is released from rest at the midpoint C, $s = 100$ mm, and the acceleration is $a = (4s)$ m/s^2, where s is in meters, determine the velocity of the particle when it reaches plate B, $s = 200$ mm, and the time it takes to travel from C to B.

SOLUTION

Coordinate System.   As shown in Fig. 12–5, s is positive downward, measured from plate A.

Velocity.   Since $a = f(s)$, the velocity as a function of position can be obtained by using $v\,dv = a\,ds$. Realizing that $v = 0$ at $s = 0.1$ m, we have

$(+\downarrow)$
$$v\,dv = a\,ds$$
$$\int_0^v v\,dv = \int_{0.1\,\text{m}}^s 4s\,ds$$
$$\frac{1}{2}v^2\Big|_0^v = \frac{4}{2}s^2\Big|_{0.1\,\text{m}}^s$$
$$v = 2(s^2 - 0.01)^{1/2}\ \text{m/s} \qquad (1)$$

At $s = 200$ mm $= 0.2$ m,

$$v_B = 0.346\ \text{m/s} = 346\ \text{mm/s} \downarrow \qquad \textit{Ans.}$$

The positive root is chosen since the particle is traveling downward, i.e., in the $+s$ direction.

Time.   The time for the particle to travel from C to B can be obtained using $v = ds/dt$ and Eq. 1, where $s = 0.1$ m when $t = 0$. From Appendix A,

$(+\downarrow)$
$$ds = v\,dt$$
$$= 2(s^2 - 0.01)^{1/2}dt$$
$$\int_{0.1}^s \frac{ds}{(s^2 - 0.01)^{1/2}} = \int_0^t 2\,dt$$
$$\ln\left(\sqrt{s^2 - 0.01} + s\right)\Big|_{0.1}^s = 2t\Big|_0^t$$
$$\ln\left(\sqrt{s^2 - 0.01} + s\right) + 2.303 = 2t$$

At $s = 0.2$ m,

$$t = \frac{\ln\left(\sqrt{(0.2)^2 - 0.01} + 0.2\right) + 2.303}{2} = 0.658\ \text{s} \qquad \textit{Ans.}$$

NOTE: The formulas for constant acceleration cannot be used here because the acceleration changes with position, i.e., $a = 4s$.

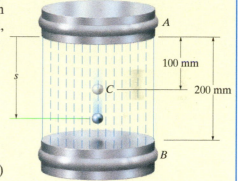

Fig. 12–5

EXAMPLE | 12.5

A particle moves along a horizontal path with a velocity of $v = (3t^2 - 6t)$ m/s, where t is the time in seconds. If it is initially located at the origin O, determine the distance traveled in 3.5 s, and the particle's average velocity and average speed during the time interval.

SOLUTION

Coordinate System.   Here positive motion is to the right, measured from the origin O, Fig. 12–6a.

Distance Traveled.   Since $v = f(t)$, the position as a function of time may be found by integrating $v = ds/dt$ with $t = 0$, $s = 0$.

$$(\xrightarrow{+})$$
$$ds = v\, dt$$
$$= (3t^2 - 6t)\, dt$$
$$\int_0^s ds = \int_0^t (3t^2 - 6t)\, dt$$
$$s = (t^3 - 3t^2) \text{ m} \qquad (1)$$

In order to determine the distance traveled in 3.5 s, it is necessary to investigate the path of motion. If we consider a graph of the velocity function, Fig. 12–6b, then it reveals that for $0 < t < 2$ s the velocity is *negative*, which means the particle is traveling to the *left*, and for $t > 2$ s the velocity is *positive*, and hence the particle is traveling to the *right*. Also, note that $v = 0$ at $t = 2$ s. The particle's position when $t = 0$, $t = 2$ s, and $t = 3.5$ s can be determined from Eq. 1. This yields

$$s|_{t=0} = 0 \quad s|_{t=2\,s} = -4.0 \text{ m} \quad s|_{t=3.5\,s} = 6.125 \text{ m}$$

The path is shown in Fig. 12–6a. Hence, the distance traveled in 3.5 s is

$$s_T = 4.0 + 4.0 + 6.125 = 14.125 \text{ m} = 14.1 \text{ m} \qquad \textit{Ans.}$$

Velocity.   The *displacement* from $t = 0$ to $t = 3.5$ s is

$$\Delta s = s|_{t=3.5\,s} - s|_{t=0} = 6.125 \text{ m} - 0 = 6.125 \text{ m}$$

and so the average velocity is

$$v_{\text{avg}} = \frac{\Delta s}{\Delta t} = \frac{6.125 \text{ m}}{3.5 \text{ s} - 0} = 1.75 \text{ m/s} \rightarrow \qquad \textit{Ans.}$$

The average speed is defined in terms of the *distance traveled* s_T. This positive scalar is

$$(v_{\text{sp}})_{\text{avg}} = \frac{s_T}{\Delta t} = \frac{14.125 \text{ m}}{3.5 \text{ s} - 0} = 4.04 \text{ m/s} \qquad \textit{Ans.}$$

NOTE: In this problem, the acceleration is $a = dv/dt = (6t - 6)$ m/s^2, which is not constant.

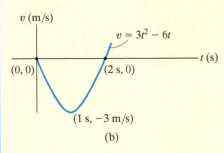

$s = -4.0$ m | $s = 6.125$ m

O

$t = 2$ s $t = 0$ s $t = 3.5$ s

(a)

v (m/s)

$v = 3t^2 - 6t$

$(0,0)$ $(2 \text{ s}, 0)$ t (s)

$(1 \text{ s}, -3 \text{ m/s})$

(b)

Fig. 12–6

It is highly suggested that you test yourself on the solutions to these examples, by covering them over and then trying to think about which equations of kinematics must be used and how they are applied in order to determine the unknowns. Then before solving any of the problems, try and solve some of the Preliminary and Fundamental Problems which follow. The solutions and answers to all these problems are given in the back of the book. **Doing this throughout the book will help immensely in understanding how to apply the theory, and thereby develop your problem-solving skills.**

PRELIMINARY PROBLEM

P12–1.

a) If $s = (2t^3)$ m, where t is in seconds, determine v when $t = 2$ s.

b) If $v = (5s)$ m/s, where s is in meters, determine a at $s = 1$ m.

c) If $v = (4t + 5)$ m/s, where t is in seconds, determine a when $t = 2$ s.

d) If $a = 2$ m/s^2, determine v when $t = 2$ s if $v = 0$ when $t = 0$.

e) If $a = 2$ m/s^2, determine v at $s = 4$ m if $v = 3$ m/s at $s = 0$.

f) If $a = (s)$ m/s^2, where s is in meters, determine v when $s = 5$ m if $v = 0$ at $s = 4$ m.

g) If $a = 4$ m/s^2, determine s when $t = 3$ s if $v = 2$ m/s and $s = 2$ m when $t = 0$.

h) If $a = (8t^2)$ m/s^2, determine v when $t = 1$ s if $v = 0$ at $t = 0$.

i) If $s = (3t^2 + 2)$ m, determine v when $t = 2$ s.

j) When $t = 0$ the particle is at A. In four seconds it travels to B, then in another six seconds it travels to C. Determine the average velocity and the average speed. The origin of the coordinate is at O.

Prob. P12–1

FUNDAMENTAL PROBLEMS

F12–1. Initially, the car travels along a straight road with a speed of 35 m/s. If the brakes are applied and the speed of the car is reduced to 10 m/s in 15 s, determine the constant deceleration of the car.

Prob. F12–1

F12–2. A ball is thrown vertically upward with a speed of 15 m/s. Determine the time of flight when it returns to its original position.

Prob. F12–2

F12–3. A particle travels along a straight line with a velocity of $v = (4t - 3t^2)$ m/s, where t is in seconds. Determine the position of the particle when $t = 4$ s. $s = 0$ when $t = 0$.

F12–4. A particle travels along a straight line with a speed $v = (0.5t^3 - 8t)$ m/s, where t is in seconds. Determine the acceleration of the particle when $t = 2$ s.

F12–5. The position of the particle is given by $s = (2t^2 - 8t + 6)$ m, where t is in seconds. Determine the time when the velocity of the particle is zero, and the total distance traveled by the particle when $t = 3$ s.

Prob. F12–5

F12–6. A particle travels along a straight line with an acceleration of $a = (10 - 0.2s)$ m/s^2, where s is measured in meters. Determine the velocity of the particle when $s = 10$ m if $v = 5$ m/s at $s = 0$.

Prob. F12–6

F12–7. A particle moves along a straight line such that its acceleration is $a = (4t^2 - 2)$ m/s^2, where t is in seconds. When $t = 0$, the particle is located 2 m to the left of the origin, and when $t = 2$ s, it is 20 m to the left of the origin. Determine the position of the particle when $t = 4$ s.

F12–8. A particle travels along a straight line with a velocity of $v = (20 - 0.05s^2)$ m/s, where s is in meters. Determine the acceleration of the particle at $s = 15$ m.

PROBLEMS

12–1.   Starting from rest, a particle moving in a straight line has an acceleration of $a = (2t - 6)$ m/s^2, where t is in seconds. What is the particle's velocity when $t = 6$ s, and what is its position when $t = 11$ s?

12–2.   If a particle has an initial velocity of $v_0 = 12$ ft/s to the right, at $s_0 = 0$, determine its position when $t = 10$ s, if $a = 2$ ft/s^2 to the left.

12–3.   A particle travels along a straight line with a velocity $v = (12 - 3t^2)$ m/s, where t is in seconds. When $t = 1$ s, the particle is located 10 m to the left of the origin. Determine the acceleration when $t = 4$ s, the displacement from $t = 0$ to $t = 10$ s, and the distance the particle travels during this time period.

*12–4.   A particle travels along a straight line with a constant acceleration. When $s = 4$ ft, $v = 3$ ft/s and when $s = 10$ ft, $v = 8$ ft/s. Determine the velocity as a function of position.

12–5.   The velocity of a particle traveling in a straight line is given by $v = (6t - 3t^2)$ m/s, where t is in seconds. If $s = 0$ when $t = 0$, determine the particle's deceleration and position when $t = 3$ s. How far has the particle traveled during the 3-s time interval, and what is its average speed?

12–6.   The position of a particle along a straight line is given by $s = (1.5t^3 - 13.5t^2 + 22.5t)$ ft, where t is in seconds. Determine the position of the particle when $t = 6$ s and the total distance it travels during the 6-s time interval. *Hint*: Plot the path to determine the total distance traveled.

12–7.   A particle moves along a straight line such that its position is defined by $s = (t^2 - 6t + 5)$ m. Determine the average velocity, the average speed, and the acceleration of the particle when $t = 6$ s.

*12–8.   A particle is moving along a straight line such that its position is defined by $s = (10t^2 + 20)$ mm, where t is in seconds. Determine (a) the displacement of the particle during the time interval from $t = 1$ s to $t = 5$ s, (b) the average velocity of the particle during this time interval, and (c) the acceleration when $t = 1$ s.

12–9.   The acceleration of a particle as it moves along a straight line is given by $a = (2t - 1)$ m/s^2, where t is in seconds. If $s = 1$ m and $v = 2$ m/s when $t = 0$, determine the particle's velocity and position when $t = 6$ s. Also, determine the total distance the particle travels during this time period.

12–10.   A particle moves along a straight line with an acceleration of $a = 5/(3s^{1/3} + s^{5/2})$ m/s^2, where s is in meters. Determine the particle's velocity when $s = 2$ m, if it starts from rest when $s = 1$ m. Use a numerical method to evaluate the integral.

12–11.   A particle travels along a straight-line path such that in 4 s it moves from an initial position $s_A = -8$ m to a position $s_B = +3$ m. Then in another 5 s it moves from s_B to $s_C = -6$ m. Determine the particle's average velocity and average speed during the 9-s time interval.

*12–12.   Traveling with an initial speed of 70 km/h, a car accelerates at 6000 km/h^2 along a straight road. How long will it take to reach a speed of 120 km/h? Also, through what distance does the car travel during this time?

12–13.   Tests reveal that a normal driver takes about 0.75 s before he or she can *react* to a situation to avoid a collision. It takes about 3 s for a driver having 0.1% alcohol in his system to do the same. If such drivers are traveling on a straight road at 30 mph (44 ft/s) and their cars can decelerate at 2 ft/s^2, determine the shortest stopping distance d for each from the moment they see the pedestrians. *Moral*: If you must drink, please don't drive!

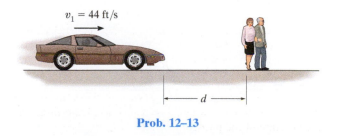

$v_1 = 44$ ft/s

Prob. 12–13

12–14.   The position of a particle along a straight-line path is defined by $s = (t^3 - 6t^2 - 15t + 7)$ ft, where t is in seconds. Determine the total distance traveled when $t = 10$ s. What are the particle's average velocity, average speed, and the instantaneous velocity and acceleration at this time?

12–15.   A particle is moving with a velocity of v_0 when $s = 0$ and $t = 0$. If it is subjected to a deceleration of $a = -kv^3$, where k is a constant, determine its velocity and position as functions of time.

*12–16.   A particle is moving along a straight line with an initial velocity of 6 m/s when it is subjected to a deceleration of $a = (-1.5v^{1/2})$ m/s^2, where v is in m/s. Determine how far it travels before it stops. How much time does this take?

12–17.   Car B is traveling a distance d ahead of car A. Both cars are traveling at 60 ft/s when the driver of B suddenly applies the brakes, causing his car to decelerate at 12 ft/s^2. It takes the driver of car A 0.75 s to react (this is the normal reaction time for drivers). When he applies his brakes, he decelerates at 15 ft/s^2. Determine the minimum distance d be tween the cars so as to avoid a collision.

Prob. 12–17

12–18.   The acceleration of a rocket traveling upward is given by $a = (6 + 0.02s)$ m/s^2, where s is in meters. Determine the time needed for the rocket to reach an altitude of $s = 100$ m. Initially, $v = 0$ and $s = 0$ when $t = 0$.

Prob. 12–18

12–19.   A train starts from rest at station A and accelerates at 0.5 m/s^2 for 60 s. Afterwards it travels with a constant velocity for 15 min. It then decelerates at 1 m/s^2 until it is brought to rest at station B. Determine the distance between the stations.

*12–20.   The velocity of a particle traveling along a straight line is $v = (3t^2 - 6t)$ ft/s, where t is in seconds. If $s = 4$ ft when $t = 0$, determine the position of the particle when $t = 4$ s. What is the total distance traveled during the time interval $t = 0$ to $t = 4$ s? Also, what is the acceleration when $t = 2$ s?

12–21.   A freight train travels at $v = 60(1 - e^{-t})$ ft/s, where t is the elapsed time in seconds. Determine the distance traveled in three seconds, and the acceleration at this time.

Prob. 12–21

12–22. A sandbag is dropped from a balloon which is ascending vertically at a constant speed of 6 m/s. If the bag is released with the same upward velocity of 6 m/s when $t = 0$ and hits the ground when $t = 8$ s, determine the speed of the bag as it hits the ground and the altitude of the balloon at this instant.

12–23. A particle is moving along a straight line such that its acceleration is defined as $a = (-2v)$ m/s^2, where v is in meters per second. If $v = 20$ m/s when $s = 0$ and $t = 0$, determine the particle's position, velocity, and acceleration as functions of time.

*12–24. The acceleration of a particle traveling along a straight line is $a = \dfrac{1}{4} s^{1/2}$ m/s^2, where s is in meters. If $v = 0$, $s = 1$ m when $t = 0$, determine the particle's velocity at $s = 2$ m.

12–25. If the effects of atmospheric resistance are accounted for, a freely falling body has an acceleration defined by the equation $a = 9.81[1 - v^2 (10^{-4})]$ m/s^2, where v is in m/s and the positive direction is downward. If the body is released from rest at a *very high altitude*, determine (a) the velocity when $t = 5$ s, and (b) the body's terminal or maximum attainable velocity (as $t \to \infty$).

12–26. The acceleration of a particle along a straight line is defined by $a = (2t - 9)$ m/s^2, where t is in seconds. At $t = 0$, $s = 1$ m and $v = 10$ m/s. When $t = 9$ s, determine (a) the particle's position, (b) the total distance traveled, and (c) the velocity.

12–27. When a particle falls through the air, its initial acceleration $a = g$ diminishes until it is zero, and there-after it falls at a constant or terminal velocity v_f. If this variation of the acceleration can be expressed as $a = (g/v_f^2) (v_f^2 - v^2)$, determine the time needed for the velocity to become $v = v_f/2$. Initially the particle falls from rest.

*12–28. Two particles A and B start from rest at the origin $s = 0$ and move along a straight line such that $a_A = (6t - 3)$ ft/s^2 and $a_B = (12t^2 - 8)$ ft/s^2, where t is in seconds. Determine the distance between them when $t = 4$ s and the total distance each has traveled in $t = 4$ s.

12–29. A ball A is thrown vertically upward from the top of a 30-m-high building with an initial velocity of 5 m/s. At the same instant another ball B is thrown upward from the ground with an initial velocity of 20 m/s. Determine the height from the ground and the time at which they pass.

12–30. A sphere is fired downwards into a medium with an initial speed of 27 m/s. If it experiences a deceleration of $a = (-6t)$ m/s^2, where t is in seconds, determine the distance traveled before it stops.

12–31. The velocity of a particle traveling along a straight line is $v = v_0 - ks$, where k is constant. If $s = 0$ when $t = 0$, determine the position and acceleration of the particle as a function of time.

*12–32. Ball A is thrown vertically upwards with a velocity of v_0. Ball B is thrown upwards from the same point with the same velocity t seconds later. Determine the elapsed time $t < 2v_0/g$ from the instant ball A is thrown to when the balls pass each other, and find the velocity of each ball at this instant.

12–33. As a body is projected to a high altitude above the earth's *surface*, the variation of the acceleration of gravity with respect to altitude y must be taken into account. Neglecting air resistance, this acceleration is determined from the formula $a = -g_0[R^2/(R + y)^2]$, where g_0 is the constant gravitational acceleration at sea level, R is the radius of the earth, and the positive direction is measured upward. If $g_0 = 9.81$ m/s^2 and $R = 6356$ km, determine the minimum initial velocity (escape velocity) at which a projectile should be shot vertically from the earth's surface so that it does not fall back to the earth. *Hint:* This requires that $v = 0$ as $y \to \infty$.

12–34. Accounting for the variation of gravitational acceleration a with respect to altitude y (see Prob. 12–36), derive an equation that relates the velocity of a freely falling particle to its altitude. Assume that the particle is released from rest at an altitude y_0 from the earth's surface. With what velocity does the particle strike the earth if it is released from rest at an altitude $y_0 = 500$ km? Use the numerical data in Prob. 12–33.

12

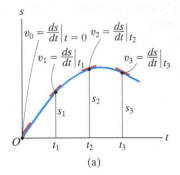

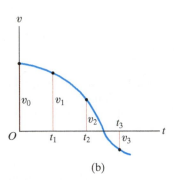

Fig. 12–7

12.3   Rectilinear Kinematics: Erratic Motion

When a particle has erratic or changing motion then its position, velocity, and acceleration *cannot* be described by a single continuous mathematical function along the entire path. Instead, a series of functions will be required to specify the motion at different intervals. For this reason, it is convenient to represent the motion as a graph. If a graph of the motion that relates any two of the variables s, v, a, t can be drawn, then this graph can be used to construct subsequent graphs relating two other variables since the variables are related by the differential relationships $v = ds/dt$, $a = dv/dt$, or $a\,ds = v\,dv$. Several situations occur frequently.

The s–t, v–t, and a–t Graphs.   To construct the v–t graph given the s–t graph, Fig. 12–7a, the equation $v = ds/dt$ should be used, since it relates the variables s and t to v. This equation states that

$$\frac{ds}{dt} = v$$

$$\text{slope of} \atop s\text{–}t \text{ graph} = \text{velocity}$$

For example, by measuring the slope on the s–t graph when $t = t_1$, the velocity is v_1, which is plotted in Fig. 12–7b. The v–t graph can be constructed by plotting this and other values at each instant.

The a–t graph can be constructed from the v–t graph in a similar manner, Fig. 12–8, since

$$\frac{dv}{dt} = a$$

$$\text{slope of} \atop v\text{–}t \text{ graph} = \text{acceleration}$$

Examples of various measurements are shown in Fig. 12–8a and plotted in Fig. 12–8b.

If the s–t curve for each interval of motion can be expressed by a mathematical function $s = s(t)$, then the equation of the v–t graph for the same interval can be obtained by differentiating this function with respect to time since $v = ds/dt$. Likewise, the equation of the a–t graph for the same interval can be determined by differentiating $v = v(t)$ since $a = dv/dt$. Since differentiation reduces a polynomial of degree n to that of degree $n - 1$, then if the s–t graph is parabolic (a second-degree curve), the v–t graph will be a sloping line (a first-degree curve), and the a–t graph will be a constant or a horizontal line (a zero-degree curve).

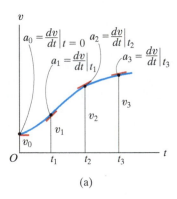

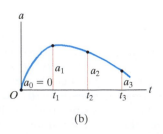

Fig. 12–8

If the a–t graph is given, Fig. 12–9a, the v–t graph may be constructed using $a = dv/dt$, written as

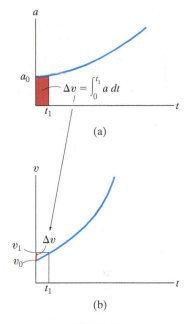

$$\Delta v = \int a \, dt$$

$$\text{change in} \atop \text{velocity} \; = \; {\text{area under} \atop a\text{–}t \text{ graph}}$$

Hence, to construct the v–t graph, we begin with the particle's initial velocity v_0 and then add to this small increments of area (Δv) determined from the a–t graph. In this manner successive points, $v_1 = v_0 + \Delta v$, etc., for the v–t graph are determined, Fig. 12–9b. Notice that an algebraic addition of the area increments of the a–t graph is necessary, since areas lying above the t axis correspond to an increase in v ("positive" area), whereas those lying below the axis indicate a decrease in v ("negative" area).

Similarly, if the v–t graph is given, Fig. 12–10a, it is possible to determine the s–t graph using $v = ds/dt$, written as

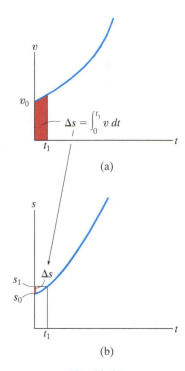

$$\Delta s = \int v \, dt$$

$$\text{displacement} \; = \; {\text{area under} \atop v\text{–}t \text{ graph}}$$

In the same manner as stated above, we begin with the particle's initial position s_0 and add (algebraically) to this small area increments Δs determined from the v–t graph, Fig. 12–10b.

If segments of the a–t graph can be described by a series of equations, then each of these equations can be *integrated* to yield equations describing the corresponding segments of the v–t graph. In a similar manner, the s–t graph can be obtained by integrating the equations which describe the segments of the v–t graph. As a result, if the a–t graph is linear (a first-degree curve), integration will yield a v–t graph that is parabolic (a second-degree curve) and an s–t graph that is cubic (third-degree curve).

12

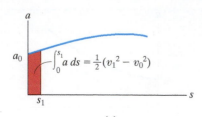

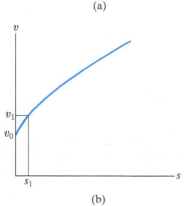

(a)

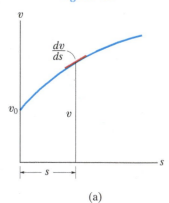

(b)

Fig. 12–11

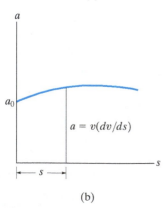

(a)

(b)

Fig. 12–12

The v–s and a–s Graphs.   If the a–s graph can be constructed, then points on the v–s graph can be determined by using $v\,dv = a\,ds$. Integrating this equation between the limits $v = v_0$ at $s = s_0$ and $v = v_1$ at $s = s_1$, we have,

$$\tfrac{1}{2}(v_1^2 - v_0^2) = \int_{s_0}^{s_1} a\,ds$$

area under
a–s graph

Therefore, if the red area in Fig. 12–11a is determined, and the initial velocity v_0 at $s_0 = 0$ is known, then $v_1 = \left(2\int_0^{s_1} a\,ds + v_0^2\right)^{1/2}$, Fig. 12–11$b$. Successive points on the v–s graph can be constructed in this manner.

If the v–s graph is known, the acceleration a at any position s can be determined using $a\,ds = v\,dv$, written as

$$a = v\left(\frac{dv}{ds}\right)$$

velocity times
acceleration = slope of
v–s graph

Thus, at any point (s, v) in Fig. 12–12a, the slope dv/ds of the v–s graph is measured. Then with v and dv/ds known, the value of a can be calculated, Fig. 12–12b.

The v–s graph can also be constructed from the a–s graph, or vice versa, by approximating the known graph in various intervals with mathematical functions, $v = f(s)$ or $a = g(s)$, and then using $a\,ds = v\,dv$ to obtain the other graph.

EXAMPLE | 12.6

A bicycle moves along a straight road such that its position is described by the graph shown in Fig. 12–13a. Construct the v–t and a–t graphs for $0 \leq t \leq 30$ s.

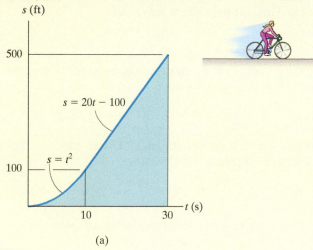

(a)

SOLUTION

v–t Graph. Since $v = ds/dt$, the v–t graph can be determined by differentiating the equations defining the s–t graph, Fig. 12–13a. We have

$$0 \leq t < 10 \text{ s}; \qquad s = (t^2) \text{ ft} \qquad v = \frac{ds}{dt} = (2t) \text{ ft/s}$$

$$10 \text{ s} < t \leq 30 \text{ s}; \qquad s = (20t - 100) \text{ ft} \qquad v = \frac{ds}{dt} = 20 \text{ ft/s}$$

The results are plotted in Fig. 12–13b. We can also obtain specific values of v by measuring the *slope* of the s–t graph at a given instant. For example, at $t = 20$ s, the slope of the s–t graph is determined from the straight line from 10 s to 30 s, i.e.,

$$t = 20 \text{ s}; \qquad v = \frac{\Delta s}{\Delta t} = \frac{500 \text{ ft} - 100 \text{ ft}}{30 \text{ s} - 10 \text{ s}} = 20 \text{ ft/s}$$

a–t Graph. Since $a = dv/dt$, the a–t graph can be determined by differentiating the equations defining the lines of the v–t graph. This yields

$$0 \leq t < 10 \text{ s}; \qquad v = (2t) \text{ ft/s} \qquad a = \frac{dv}{dt} = 2 \text{ ft/s}^2$$

$$10 < t \leq 30 \text{ s}; \qquad v = 20 \text{ ft/s} \qquad a = \frac{dv}{dt} = 0$$

The results are plotted in Fig. 12–13c.

NOTE: Show that $a = 2 \text{ ft/s}^2$ when $t = 5$ s by measuring the slope of the v–t graph.

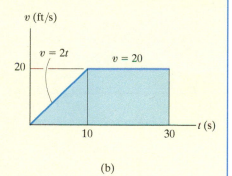

(b)

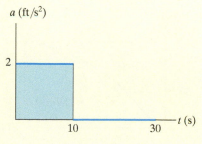

(c)

Fig. 12–13

EXAMPLE | 12.7

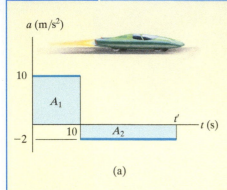

(a)

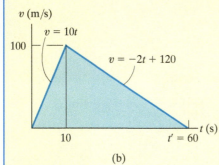

(b)

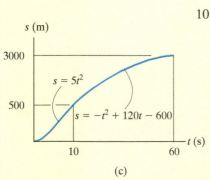

(c)

Fig. 12–14

The car in Fig. 12–14a starts from rest and travels along a straight track such that it accelerates at 10 m/s^2 for 10 s, and then decelerates at 2 m/s^2. Draw the v–t and s–t graphs and determine the time t' needed to stop the car. How far has the car traveled?

SOLUTION

v–t Graph. Since $dv = a\,dt$, the v–t graph is determined by integrating the straight-line segments of the a–t graph. Using the *initial condition* $v = 0$ when $t = 0$, we have

$$0 \le t < 10 \text{ s}; \quad a = (10) \text{ m/s}^2; \quad \int_0^v dv = \int_0^t 10\,dt, \quad v = 10t$$

When $t = 10 \text{ s}$, $v = 10(10) = 100 \text{ m/s}$. Using this as the *initial condition* for the next time period, we have

$$10 \text{ s} < t \le t'; a = (-2) \text{ m/s}^2; \int_{100 \text{ m/s}}^v dv = \int_{10 \text{ s}}^t -2\,dt, v = (-2t + 120) \text{ m/s}$$

When $t = t'$ we require $v = 0$. This yields, Fig. 12–14b,

$$t' = 60 \text{ s} \qquad\qquad Ans.$$

A more direct solution for t' is possible by realizing that the area under the a–t graph is equal to the change in the car's velocity. We require $\Delta v = 0 = A_1 + A_2$, Fig. 12–14a. Thus

$$0 = 10 \text{ m/s}^2(10 \text{ s}) + (-2 \text{ m/s}^2)(t' - 10 \text{ s})$$
$$t' = 60 \text{ s} \qquad\qquad Ans.$$

s–t Graph. Since $ds = v\,dt$, integrating the equations of the v–t graph yields the corresponding equations of the s–t graph. Using the *initial condition* $s = 0$ when $t = 0$, we have

$$0 \le t \le 10 \text{ s}; \quad v = (10t) \text{ m/s}; \quad \int_0^s ds = \int_0^t 10t\,dt, \quad s = (5t^2) \text{ m}$$

When $t = 10 \text{ s}$, $s = 5(10)^2 = 500 \text{ m}$. Using this *initial condition*,

$$10 \text{ s} \le t \le 60 \text{ s}; v = (-2t + 120) \text{ m/s}; \int_{500 \text{ m}}^s ds = \int_{10 \text{ s}}^t (-2t + 120)\,dt$$
$$s - 500 = -t^2 + 120t - [-(10)^2 + 120(10)]$$
$$s = (-t^2 + 120t - 600) \text{ m}$$

When $t' = 60 \text{ s}$, the position is

$$s = -(60)^2 + 120(60) - 600 = 3000 \text{ m} \qquad Ans.$$

The s–t graph is shown in Fig. 12–14c.

NOTE: A direct solution for s is possible when $t' = 60 \text{ s}$, since the *triangular* area under the v–t graph would yield the displacement $\Delta s = s - 0$ from $t = 0$ to $t' = 60 \text{ s}$. Hence,

$$\Delta s = \tfrac{1}{2}(60 \text{ s})(100 \text{ m/s}) = 3000 \text{ m} \qquad Ans.$$

EXAMPLE 12.8

The v–s graph describing the motion of a motorcycle is shown in Fig. 12–15a. Construct the a–s graph of the motion and determine the time needed for the motorcycle to reach the position $s = 400$ ft.

SOLUTION

a–s Graph. Since the equations for segments of the v–s graph are given, the a–s graph can be determined using $a\, ds = v\, dv$.

$0 \le s < 200$ ft; $\qquad v = (0.2s + 10)$ ft/s

$$a = v\frac{dv}{ds} = (0.2s + 10)\frac{d}{ds}(0.2s + 10) = 0.04s + 2$$

200 ft $< s \le 400$ ft; $\qquad v = 50$ ft/s

$$a = v\frac{dv}{ds} = (50)\frac{d}{ds}(50) = 0$$

The results are plotted in Fig. 12–15b.

Time. The time can be obtained using the v–s graph and $v = ds/dt$, because this equation relates v, s, and t. For the first segment of motion, $s = 0$ when $t = 0$, so

$0 \le s < 200$ ft; $\qquad v = (0.2s + 10)$ ft/s; $\qquad dt = \dfrac{ds}{v} = \dfrac{ds}{0.2s + 10}$

$$\int_0^t dt = \int_0^s \frac{ds}{0.2s + 10}$$

$$t = (5\ln(0.2s + 10) - 5\ln 10)\ \text{s}$$

At $s = 200$ ft, $t = 5\ln[0.2(200) + 10] - 5\ln 10 = 8.05$ s. Therefore, using these initial conditions for the second segment of motion,

200 ft $< s \le 400$ ft; $\qquad v = 50$ ft/s; $\qquad dt = \dfrac{ds}{v} = \dfrac{ds}{50}$

$$\int_{8.05\,\text{s}}^t dt = \int_{200\,\text{m}}^s \frac{ds}{50};$$

$$t - 8.05 = \frac{s}{50} - 4; \quad t = \left(\frac{s}{50} + 4.05\right)\text{s}$$

Therefore, at $s = 400$ ft,

$$t = \frac{400}{50} + 4.05 = 12.0\ \text{s} \qquad\qquad \textit{Ans.}$$

NOTE: The graphical results can be checked in part by calculating slopes. For example, at $s = 0$, $a = v(dv/ds) = 10(50 - 10)/200 = 2$ m/s^2. Also, the results can be checked in part by inspection. The v–s graph indicates the initial increase in velocity (acceleration) followed by constant velocity ($a = 0$).

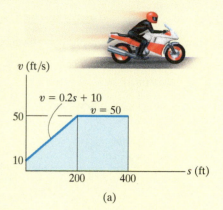

v (ft/s)

$v = 0.2s + 10$

$v = 50$

50

10

$200 \qquad 400$ $\quad s$ (ft)

(a)

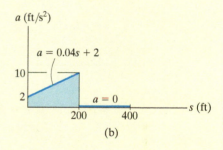

a (ft/s^2)

$a = 0.04s + 2$

10

2

$a = 0$

$200 \qquad 400$ $\quad s$ (ft)

(b)

Fig. 12–15

12

PRELIMINARY PROBLEM

P12–2.

a) Draw the *s*–*t* and *a*–*t* graphs if *s* = 0 when *t* = 0.

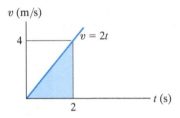

b) Draw the *a*–*t* and *v*–*t* graphs.

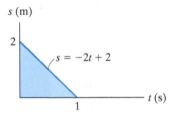

c) Draw the *v*–*t* and *s*–*t* graphs if *v* = 0, *s* = 0 when *t* = 0.

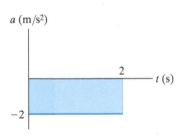

d) Determine *s* and *a* when *t* = 3 s if *s* = 0 when *t* = 0.

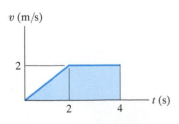

e) Draw the *v*–*t* graph if *v* = 0 when *t* = 0. Find the equation *v* = *f*(*t*) for each segment.

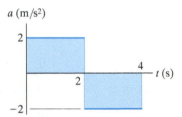

f) Determine *v* at *s* = 2 m if *v* = 1 m/s at *s* = 0.

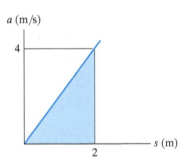

g) Determine *a* at *s* = 1 m.

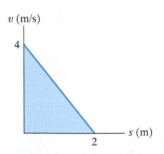

Prob. P12–2

12

FUNDAMENTAL PROBLEMS

F12–9.   The particle travels along a straight track such that its position is described by the s–t graph. Construct the v–t graph for the same time interval.

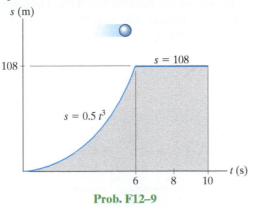

Prob. F12–9

F12–10.   A van travels along a straight road with a velocity described by the graph. Construct the s–t and a–t graphs during the same period. Take $s = 0$ when $t = 0$.

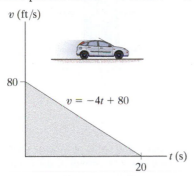

Prob. F12–10

F12–11.   A bicycle travels along a straight road where its velocity is described by the v–s graph. Construct the a–s graph for the same interval.

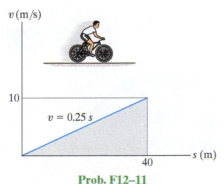

Prob. F12–11

F12–12.   The sports car travels along a straight road such that its acceleration is described by the graph. Construct the v–s graph for the same interval and specify the velocity of the car when $s = 10$ m and $s = 15$ m.

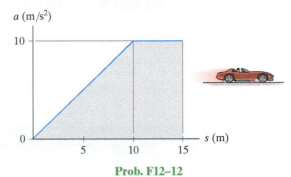

Prob. F12–12

F12–13.   The dragster starts from rest and has an acceleration described by the graph. Construct the v–t graph for the time interval $0 \le t \le t'$, where t' is the time for the car to come to rest.

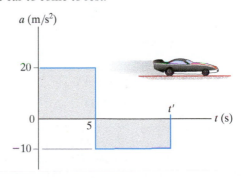

Prob. F12–13

F12–14.   The dragster starts from rest and has a velocity described by the graph. Construct the s–t graph during the time interval $0 \le t \le 15$ s. Also, determine the total distance traveled during this time interval.

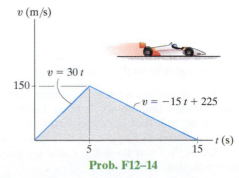

Prob. F12–14

PROBLEMS

12–35. A freight train starts from rest and travels with a constant acceleration of 0.5 ft/s². After a time t' it maintains a constant speed so that when $t = 160$ s it has traveled 2000 ft. Determine the time t' and draw the v–t graph for the motion.

***12–36.** The s–t graph for a train has been experimentally determined. From the data, construct the v–t and a–t graphs for the motion; $0 \leq t \leq 40$ s. For $0 \leq t \leq 30$ s, the curve is $s = (0.4t^2)$ m, and then it becomes straight for $t \geq 30$ s.

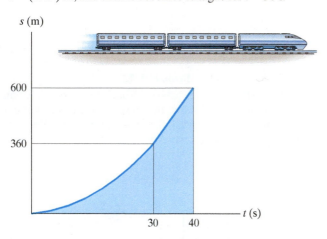

Prob. 12–36

12–37. Two rockets start from rest at the same elevation. Rocket A accelerates vertically at 20 m/s² for 12 s and then maintains a constant speed. Rocket B accelerates at 15 m/s² until reaching a constant speed of 150 m/s. Construct the a–t, v–t, and s–t graphs for each rocket until $t = 20$ s. What is the distance between the rockets when $t = 20$ s?

12–38. A particle starts from $s = 0$ and travels along a straight line with a velocity $v = (t^2 - 4t + 3)$ m/s, where t is in seconds. Construct the v–t and a–t graphs for the time interval $0 \leq t \leq 4$ s.

12–39. If the position of a particle is defined by $s = [2 \sin (\pi/5)t + 4]$ m, where t is in seconds, construct the s–t, v–t, and a–t graphs for $0 \leq t \leq 10$ s.

***12–40.** An airplane starts from rest, travels 5000 ft down a runway, and after uniform acceleration, takes off with a speed of 162 mi/h. It then climbs in a straight line with a uniform acceleration of 3 ft/s² until it reaches a constant speed of 220 mi/h. Draw the s–t, v–t, and a–t graphs that describe the motion.

12–41. The elevator starts from rest at the first floor of the building. It can accelerate at 5 ft/s² and then decelerate at 2 ft/s². Determine the shortest time it takes to reach a floor 40 ft above the ground. The elevator starts from rest and then stops. Draw the a–t, v–t, and s–t graphs for the motion.

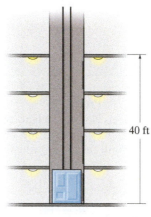

Prob. 12–41

12–42. The velocity of a car is plotted as shown. Determine the total distance the car moves until it stops ($t = 80$ s). Construct the a–t graph.

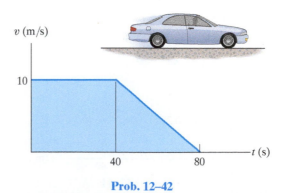

Prob. 12–42

12–43. The motion of a jet plane just after landing on a runway is described by the a–t graph. Determine the time t' when the jet plane stops. Construct the v–t and s–t graphs for the motion. Here $s = 0$, and $v = 300$ ft/s when $t = 0$.

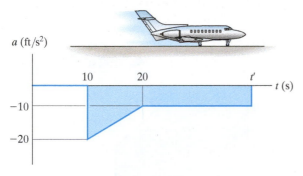

Prob. 12–43

12–46. The a–s graph for a rocket moving along a straight track has been experimentally determined. If the rocket starts at $s = 0$ when $v = 0$, determine its speed when it is at $s = 75$ ft, and 125 ft, respectively. Use Simpson's rule with $n = 100$ to evaluate v at $s = 125$ ft.

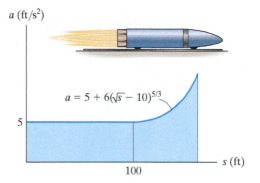

$$a = 5 + 6(\sqrt{s} - 10)^{5/3}$$

Prob. 12–46

*12–44. The v–t graph for a particle moving through an electric field from one plate to another has the shape shown in the figure. The acceleration and deceleration that occur are constant and both have a magnitude of 4 m/s². If the plates are spaced 200 mm apart, determine the maximum velocity v_{max} and the time t' for the particle to travel from one plate to the other. Also draw the s–t graph. When $t = t'/2$ the particle is at $s = 100$ mm.

12–45. The v–t graph for a particle moving through an electric field from one plate to another has the shape shown in the figure, where $t' = 0.2$ s and $v_{max} = 10$ m/s. Draw the s–t and a–t graphs for the particle. When $t = t'/2$ the particle is at $s = 0.5$ m.

12–47. A two-stage rocket is fired vertically from rest at $s = 0$ with the acceleration as shown. After 30 s the first stage, A, burns out and the second stage, B, ignites. Plot the v–t and s–t graphs which describe the motion of the second stage for $0 \le t \le 60$ s.

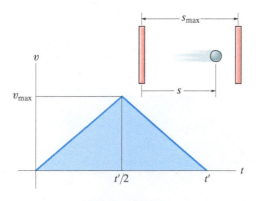

Probs. 12–44/45

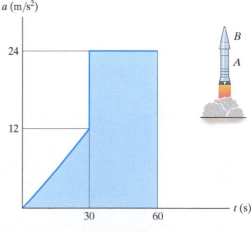

Prob. 12–47

*12–48. The race car starts from rest and travels along a straight road until it reaches a speed of 26 m/s in 8 s as shown on the v–t graph. The flat part of the graph is caused by shifting gears. Draw the a–t graph and determine the maximum acceleration of the car.

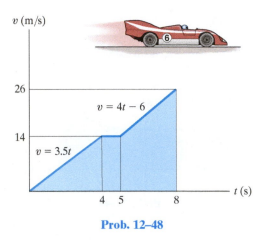

v (m/s)

26

$v = 4t - 6$

14

$v = 3.5t$

4 5   8 t (s)

Prob. 12–48

12–49. The jet car is originally traveling at a velocity of 10 m/s when it is subjected to the acceleration shown. Determine the car's maximum velocity and the time t' when it stops. When $t = 0, s = 0$.

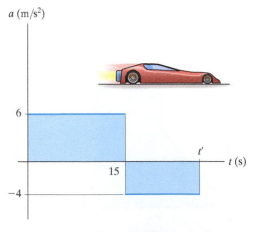

a (m/s^2)

6

t'

t (s)

15

−4

Prob. 12–49

12–50. The car starts from rest at $s = 0$ and is subjected to an acceleration shown by the a–s graph. Draw the v–s graph and determine the time needed to travel 200 ft.

a (ft/s^2)

12

$a = -0.04s + 24$

6

300   450 s (ft)

Prob. 12–50

12–51. The v–t graph for a train has been experimentally determined. From the data, construct the s–t and a–t graphs for the motion for $0 \le t \le 180$ s. When $t = 0, s = 0$.

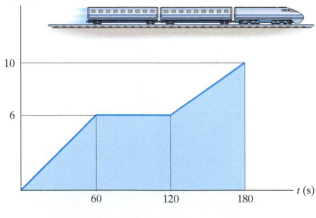

v (m/s)

10

6

60   120   180 t (s)

Prob. 12–51

*12–52. A motorcycle starts from rest at $s = 0$ and travels along a straight road with the speed shown by the v–t graph. Determine the total distance the motorcycle travels until it stops when $t = 15$ s. Also plot the a–t and s–t graphs.

12–53. A motorcycle starts from rest at $s = 0$ and travels along a straight road with the speed shown by the v–t graph. Determine the motorcycle's acceleration and position when $t = 8$ s and $t = 12$ s.

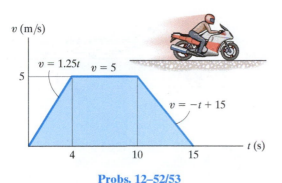

Probs. 12–52/53

12–54. The v–t graph for the motion of a car as it moves along a straight road is shown. Draw the s–t and a–t graphs. Also determine the average speed and the distance traveled for the 15-s time interval. When $t = 0$, $s = 0$.

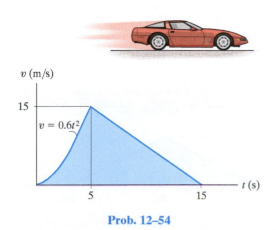

Prob. 12–54

12–55. An airplane lands on the straight runway, originally traveling at 110 ft/s when $s = 0$. If it is subjected to the decelerations shown, determine the time t' needed to stop the plane and construct the s–t graph for the motion.

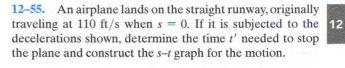

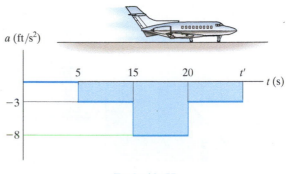

Prob. 12–55

*12–56. Starting from rest at $s = 0$, a boat travels in a straight line with the acceleration shown by the a–s graph. Determine the boat's speed when $s = 50$ ft, 100 ft, and 150 ft.

12–57. Starting from rest at $s = 0$, a boat travels in a straight line with the acceleration shown by the a–s graph. Construct the v–s graph.

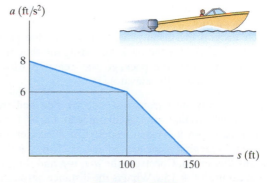

Probs. 12–56/57

12–58. A two-stage rocket is fired vertically from rest with the acceleration shown. After 15 s the first stage A burns out and the second stage B ignites. Plot the v–t and s–t graphs which describe the motion of the second stage for $0 \le t \le 40$ s.

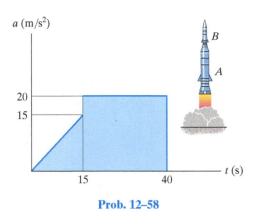

Prob. 12–58

12–59. The speed of a train during the first minute has been recorded as follows:

t (s)	0	20	40	60
v (m/s)	0	16	21	24

Plot the v–t graph, approximating the curve as straight-line segments between the given points. Determine the total distance traveled.

*12–60. A man riding upward in a freight elevator accidentally drops a package off the elevator when it is 100 ft from the ground. If the elevator maintains a constant upward speed of 4 ft/s, determine how high the elevator is from the ground the instant the package hits the ground. Draw the v–t curve for the package during the time it is in motion. Assume that the package was released with the same upward speed as the elevator.

12–61. Two cars start from rest side by side and travel along a straight road. Car A accelerates at 4 m/s² for 10 s and then maintains a constant speed. Car B accelerates at 5 m/s² until reaching a constant speed of 25 m/s and then maintains this speed. Construct the a–t, v–t, and s–t graphs for each car until $t = 15$ s. What is the distance between the two cars when $t = 15$ s?

12–62. If the position of a particle is defined as $s = (5t - 3t^2)$ ft, where t is in seconds, construct the s–t, v–t, and a–t graphs for $0 \le t \le 10$ s.

12–63. From experimental data, the motion of a jet plane while traveling along a runway is defined by the v–t graph. Construct the s–t and a–t graphs for the motion. When $t = 0, s = 0$.

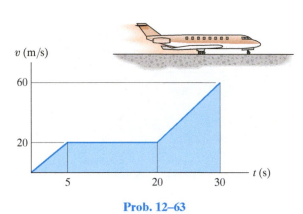

Prob. 12–63

*12–64. The motion of a train is described by the a–s graph shown. Draw the v–s graph if $v = 0$ at $s = 0$.

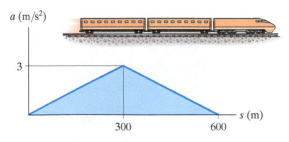

Prob. 12–64

12–65. The jet plane starts from rest at $s = 0$ and is subjected to the acceleration shown. Determine the speed of the plane when it has traveled 1000 ft. Also, how much time is required for it to travel 1000 ft?

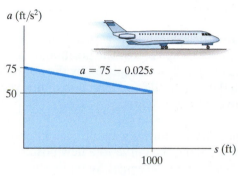

a (ft/s^2)

75

$a = 75 - 0.025s$

50

1000

s (ft)

Prob. 12–65

12–66. The boat travels along a straight line with the speed described by the graph. Construct the s–t and a–s graphs. Also, determine the time required for the boat to travel a distance $s = 400$ m if $s = 0$ when $t = 0$.

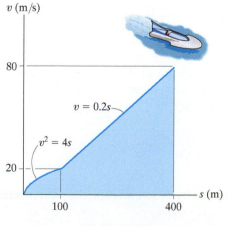

v (m/s)

80

$v = 0.2s$

$v^2 = 4s$

20

100   400

s (m)

Prob. 12–66

12–67. The v–s graph of a cyclist traveling along a straight road is shown. Construct the a–s graph.

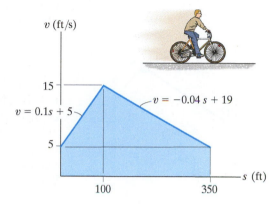

v (ft/s)

15

$v = 0.1s + 5$

$v = -0.04 s + 19$

5

100   350

s (ft)

Prob. 12–67

*12–68.** The v–s graph for a test vehicle is shown. Determine its acceleration when $s = 100$ m and when $s = 175$ m.

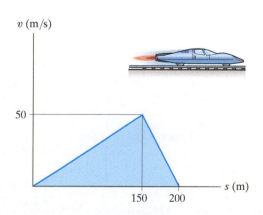

v (m/s)

50

150   200

s (m)

Prob. 12–68

12

12.4   General Curvilinear Motion

Curvilinear motion occurs when a particle moves along a curved path. Since this path is often described in three dimensions, vector analysis will be used to formulate the particle's position, velocity, and acceleration.* In this section the general aspects of curvilinear motion are discussed, and in subsequent sections we will consider three types of coordinate systems often used to analyze this motion.

Position.   Consider a particle located at a point on a space curve defined by the path function $s(t)$, Fig. 12–16a. The position of the particle, measured from a fixed point O, will be designated by the *position vector* $\mathbf{r} = \mathbf{r}(t)$. Notice that both the magnitude and direction of this vector will change as the particle moves along the curve.

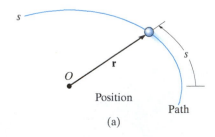

Position
(a)

Displacement.   Suppose that during a small time interval Δt the particle moves a distance Δs along the curve to a new position, defined by $\mathbf{r}' = \mathbf{r} + \Delta\mathbf{r}$, Fig. 12–16b. The *displacement* $\Delta\mathbf{r}$ represents the change in the particle's position and is determined by vector subtraction; i.e., $\Delta\mathbf{r} = \mathbf{r}' - \mathbf{r}$.

Velocity.   During the time Δt, the *average velocity* of the particle is

$$\mathbf{v}_{avg} = \frac{\Delta\mathbf{r}}{\Delta t}$$

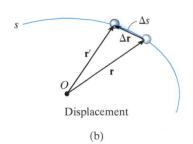

Displacement
(b)

The *instantaneous velocity* is determined from this equation by letting $\Delta t \rightarrow 0$, and consequently the direction of $\Delta\mathbf{r}$ approaches the *tangent* to the curve. Hence, $\mathbf{v} = \lim_{\Delta t \to 0}(\Delta\mathbf{r}/\Delta t)$ or

$$\mathbf{v} = \frac{d\mathbf{r}}{dt} \qquad (12\text{–}7)$$

Since $d\mathbf{r}$ will be tangent to the curve, the *direction* of $\mathbf{v}$ is also *tangent to the curve*, Fig. 12–16c. The *magnitude* of $\mathbf{v}$, which is called the *speed*, is obtained by realizing that the length of the straight line segment $\Delta\mathbf{r}$ in Fig. 12–16b approaches the arc length Δs as $\Delta t \rightarrow 0$, we have $v = \lim_{\Delta t \to 0}(\Delta r/\Delta t) = \lim_{\Delta t \to 0}(\Delta s/\Delta t)$, or

$$v = \frac{ds}{dt} \qquad (12\text{–}8)$$

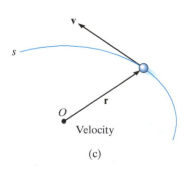

Velocity
(c)

Fig. 12–16

Thus, the *speed* can be obtained by differentiating the path function s with respect to time.

*A summary of some of the important concepts of vector analysis is given in Appendix B.

Acceleration.   If the particle has a velocity **v** at time *t* and a velocity **v**′ = **v** + Δ**v** at *t* + Δ*t*, Fig. 12–16*d*, then the *average acceleration* of the particle during the time interval Δ*t* is

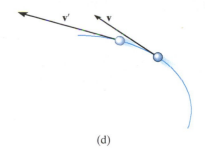

(d)

$$\mathbf{a}_{\text{avg}} = \frac{\Delta \mathbf{v}}{\Delta t}$$

where Δ**v** = **v**′ − **v**. To study this time rate of change, the two velocity vectors in Fig. 12–16*d* are plotted in Fig. 12–16*e* such that their tails are located at the fixed point *O*′ and their arrowheads touch points on a curve. This curve is called a *hodograph*, and when constructed, it describes the locus of points for the arrowhead of the velocity vector in the same manner as the *path s* describes the locus of points for the arrowhead of the position vector, Fig. 12–16*a*.

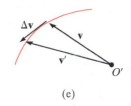

(e)

To obtain the *instantaneous acceleration*, let Δ*t* → 0 in the above equation. In the limit Δ**v** will approach the *tangent to the hodograph*, and so $\mathbf{a} = \lim\limits_{\Delta t \to 0} (\Delta \mathbf{v}/\Delta t)$, or

$$\mathbf{a} = \frac{d\mathbf{v}}{dt} \qquad (12\text{–}9)$$

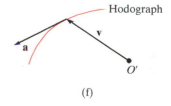

(f)

Substituting Eq. 12–7 into this result, we can also write

$$\mathbf{a} = \frac{d^2\mathbf{r}}{dt^2}$$

By definition of the derivative, **a** acts *tangent to the hodograph*, Fig. 12–16*f*, and, *in general it is not tangent to the path of motion*, Fig. 12–16*g*. To clarify this point, realize that Δ**v** and consequently **a** must account for the change made in *both* the magnitude *and* direction of the velocity **v** as the particle moves from one point to the next along the path, Fig. 12–16*d*. However, in order for the particle to follow any curved path, the directional change always "swings" the velocity vector toward the "inside" or "concave side" of the path, and therefore **a** *cannot* remain tangent to the path. In summary, **v** is always tangent to the *path* and **a** is always tangent to the *hodograph*.

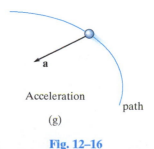

Acceleration

(g)

Fig. 12–16

12

12.5   Curvilinear Motion: Rectangular Components

Occasionally the motion of a particle can best be described along a path that can be expressed in terms of its x, y, z coordinates.

Position.   If the particle is at point (x, y, z) on the curved path s shown in Fig. 12–17a, then its location is defined by the *position vector*

$$\mathbf{r} = x\mathbf{i} + y\mathbf{j} + z\mathbf{k} \tag{12–10}$$

When the particle moves, the x, y, z components of $\mathbf{r}$ will be functions of time; i.e., $x = x(t)$, $y = y(t)$, $z = z(t)$, so that $\mathbf{r} = \mathbf{r}(t)$.

At any instant the *magnitude* of $\mathbf{r}$ is defined from Eq. B–3 in Appendix B as

$$r = \sqrt{x^2 + y^2 + z^2}$$

And the *direction* of $\mathbf{r}$ is specified by the unit vector $\mathbf{u}_r = \mathbf{r}/r$.

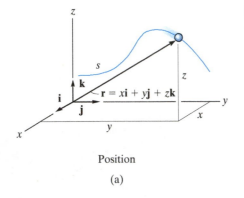

Position

(a)

Velocity.   The first time derivative of $\mathbf{r}$ yields the velocity of the particle. Hence,

$$\mathbf{v} = \frac{d\mathbf{r}}{dt} = \frac{d}{dt}(x\mathbf{i}) + \frac{d}{dt}(y\mathbf{j}) + \frac{d}{dt}(z\mathbf{k})$$

When taking this derivative, it is necessary to account for changes in *both* the magnitude and direction of each of the vector's components. For example, the derivative of the $\mathbf{i}$ component of $\mathbf{r}$ is

$$\frac{d}{dt}(x\mathbf{i}) = \frac{dx}{dt}\mathbf{i} + x\frac{d\mathbf{i}}{dt}$$

The second term on the right side is zero, provided the x, y, z reference frame is *fixed*, and therefore the *direction* (and the *magnitude*) of $\mathbf{i}$ does not change with time. Differentiation of the $\mathbf{j}$ and $\mathbf{k}$ components may be carried out in a similar manner, which yields the final result,

$$\mathbf{v} = \frac{d\mathbf{r}}{dt} = v_x\mathbf{i} + v_y\mathbf{j} + v_z\mathbf{k} \tag{12–11}$$

where

$$v_x = \dot{x} \quad v_y = \dot{y} \quad v_z = \dot{z} \tag{12–12}$$

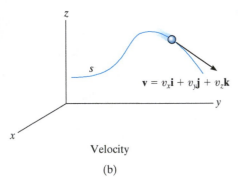

Velocity

(b)

Fig. 12–17

The "dot" notation $\dot{x}$, $\dot{y}$, $\dot{z}$ represents the first time derivatives of $x = x(t)$, $y = y(t)$, $z = z(t)$, respectively.
  The velocity has a *magnitude* that is found from

$$v = \sqrt{v_x^2 + v_y^2 + v_z^2}$$

and a *direction* that is specified by the unit vector $\mathbf{u}_v = \mathbf{v}/v$. As discussed in Sec. 12.4, this direction is *always tangent to the path*, as shown in Fig. 12–17b.

Acceleration.   The acceleration of the particle is obtained by taking the first time derivative of Eq. 12–11 (or the second time derivative of Eq. 12–10). We have

$$\boxed{\mathbf{a} = \frac{d\mathbf{v}}{dt} = a_x\mathbf{i} + a_y\mathbf{j} + a_z\mathbf{k}} \qquad (12\text{–}13)$$

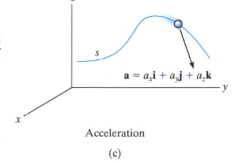

Acceleration

(c)

where

$$\boxed{\begin{aligned} a_x &= \dot{v}_x = \ddot{x} \\ a_y &= \dot{v}_y = \ddot{y} \\ a_z &= \dot{v}_z = \ddot{z} \end{aligned}} \qquad (12\text{–}14)$$

Here a_x, a_y, a_z represent, respectively, the first time derivatives of $v_x = v_x(t)$, $v_y = v_y(t)$, $v_z = v_z(t)$, or the second time derivatives of the functions $x = x(t)$, $y = y(t)$, $z = z(t)$.
  The acceleration has a *magnitude*

$$a = \sqrt{a_x^2 + a_y^2 + a_z^2}$$

and a *direction* specified by the unit vector $\mathbf{u}_a = \mathbf{a}/a$. Since $\mathbf{a}$ represents the time rate of *change* in both the magnitude and direction of the velocity, in general $\mathbf{a}$ will *not* be tangent to the path, Fig. 12–17c.

Important Points

- Curvilinear motion can cause changes in *both* the magnitude and direction of the position, velocity, and acceleration vectors.

- The velocity vector is always directed *tangent* to the path.

- In general, the acceleration vector is *not* tangent to the path, but rather, it is tangent to the hodograph.

- If the motion is described using rectangular coordinates, then the components along each of the axes do not change direction, only their magnitude and sense (algebraic sign) will change.

- By considering the component motions, the change in magnitude and direction of the particle's position and velocity are automatically taken into account.

Procedure for Analysis

Coordinate System.

- A rectangular coordinate system can be used to solve problems for which the motion can conveniently be expressed in terms of its *x, y, z* components.

Kinematic Quantities.

- Since *rectilinear motion* occurs along *each coordinate axis*, the motion along each axis is found using $v = ds/dt$ and $a = dv/dt$; or in cases where the motion is not expressed as a function of time, the equation $a \, ds = v \, dv$ can be used.

- In two dimensions, the equation of the path $y = f(x)$ can be used to relate the *x* and *y* components of velocity and acceleration by applying the chain rule of calculus. A review of this concept is given in Appendix C.

- Once the *x, y, z* components of **v** and **a** have been determined, the magnitudes of these vectors are found from the Pythagorean theorem, Eq. B-3, and their coordinate direction angles from the components of their unit vectors, Eqs. B-4 and B-5.

EXAMPLE 12.9

At any instant the horizontal position of the weather balloon in Fig. 12–18a is defined by $x = (8t)$ ft, where t is in seconds. If the equation of the path is $y = x^2/10$, determine the magnitude and direction of the velocity and the acceleration when $t = 2$ s.

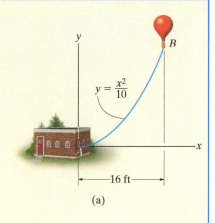

(a)

SOLUTION

Velocity.   The velocity component in the x direction is

$$v_x = \dot{x} = \frac{d}{dt}(8t) = 8 \text{ ft/s} \rightarrow$$

To find the relationship between the velocity components we will use the chain rule of calculus. When $t = 2$ s, $x = 8(2) = 16$ ft, Fig. 12–18a, and so

$$v_y = \dot{y} = \frac{d}{dt}(x^2/10) = 2x\dot{x}/10 = 2(16)(8)/10 = 25.6 \text{ ft/s} \uparrow$$

When $t = 2$ s, the magnitude of velocity is therefore

$$v = \sqrt{(8 \text{ ft/s})^2 + (25.6 \text{ ft/s})^2} = 26.8 \text{ ft/s} \qquad Ans.$$

The direction is tangent to the path, Fig. 12–18b, where

$$\theta_v = \tan^{-1}\frac{v_y}{v_x} = \tan^{-1}\frac{25.6}{8} = 72.6° \qquad Ans.$$

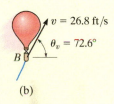

(b)

Acceleration.   The relationship between the acceleration components is determined using the chain rule. (See Appendix C.) We have

$$a_x = \dot{v}_x = \frac{d}{dt}(8) = 0$$

$$a_y = \dot{v}_y = \frac{d}{dt}(2x\dot{x}/10) = 2(\dot{x})\dot{x}/10 + 2x(\ddot{x})/10$$

$$= 2(8)^2/10 + 2(16)(0)/10 = 12.8 \text{ ft/s}^2 \uparrow$$

Thus,

$$a = \sqrt{(0)^2 + (12.8)^2} = 12.8 \text{ ft/s}^2 \qquad Ans.$$

The direction of **a**, as shown in Fig. 12–18c, is

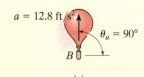

(c)

Fig. 12–18

$$\theta_a = \tan^{-1}\frac{12.8}{0} = 90° \qquad Ans.$$

NOTE: It is also possible to obtain v_y and a_y by first expressing $y = f(t) = (8t)^2/10 = 6.4t^2$ and then taking successive time derivatives.

12

EXAMPLE   12.10

(© R.C. Hibbeler)

For a short time, the path of the plane in Fig. 12–19a is described by $y = (0.001x^2)$ m. If the plane is rising with a constant upward velocity of 10 m/s, determine the magnitudes of the velocity and acceleration of the plane when it reaches an altitude of $y = 100$ m.

SOLUTION

When $y = 100$ m, then $100 = 0.001x^2$ or $x = 316.2$ m. Also, due to constant velocity $v_y = 10$ m/s, so

$$y = v_y t; \qquad 100 \text{ m} = (10 \text{ m/s}) t \qquad t = 10 \text{ s}$$

Velocity. Using the chain rule (see Appendix C) to find the relationship between the velocity components, we have

$$y = 0.001x^2$$

$$v_y = \dot{y} = \frac{d}{dt}(0.001x^2) = (0.002x)\dot{x} = 0.002xv_x \qquad (1)$$

Thus

$$10 \text{ m/s} = 0.002(316.2 \text{ m})(v_x)$$
$$v_x = 15.81 \text{ m/s}$$

y

$y = 0.001x^2$

100 m

x

(a)

The magnitude of the velocity is therefore

$$v = \sqrt{v_x^2 + v_y^2} = \sqrt{(15.81 \text{ m/s})^2 + (10 \text{ m/s})^2} = 18.7 \text{ m/s} \qquad Ans.$$

Acceleration. Using the chain rule, the time derivative of Eq. (1) gives the relation between the acceleration components.

$$a_y = \dot{v}_y = (0.002\dot{x})\dot{x} + 0.002x(\ddot{x}) = 0.002(v_x^2 + xa_x)$$

y

v_y v
a
v_x

100 m

x

(b)

Fig. 12–19

When $x = 316.2$ m, $v_x = 15.81$ m/s, $\dot{v}_y = a_y = 0$,

$$0 = 0.002\big[(15.81 \text{ m/s})^2 + 316.2 \text{ m}(a_x)\big]$$
$$a_x = -0.791 \text{ m/s}^2$$

The magnitude of the plane's acceleration is therefore

$$a = \sqrt{a_x^2 + a_y^2} = \sqrt{(-0.791 \text{ m/s}^2)^2 + (0 \text{ m/s}^2)^2}$$
$$= 0.791 \text{ m/s}^2 \qquad Ans.$$

These results are shown in Fig. 12–19b.

12.6   Motion of a Projectile

The free-flight motion of a projectile is often studied in terms of its rectangular components. To illustrate the kinematic analysis, consider a projectile launched at point (x_0, y_0), with an initial velocity of $\mathbf{v}_0$, having components $(\mathbf{v}_0)_x$ and $(\mathbf{v}_0)_y$, Fig. 12–20. When air resistance is neglected, the only force acting on the projectile is its weight, which causes the projectile to have a *constant downward acceleration* of approximately $a_c = g = 9.81 \text{ m/s}^2$ or $g = 32.2 \text{ ft/s}^2$.*

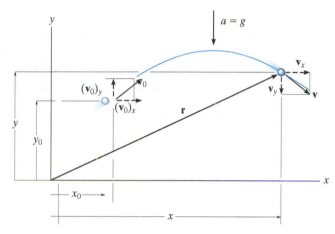

Fig. 12–20

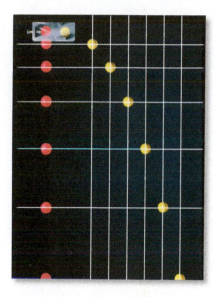

Each picture in this sequence is taken after the same time interval. The red ball falls from rest, whereas the yellow ball is given a horizontal velocity when released. Both balls accelerate downward at the same rate, and so they remain at the same elevation at any instant. This acceleration causes the difference in elevation between the balls to increase between successive photos. Also, note the horizontal distance between successive photos of the yellow ball is constant since the velocity in the horizontal direction remains constant. (© R.C. Hibbeler)

Horizontal Motion. Since $a_x = 0$, application of the constant acceleration equations, 12–4 to 12–6, yields

$(\xrightarrow{+})$ $\qquad v = v_0 + a_c t$ $\qquad\qquad v_x = (v_0)_x$

$(\xrightarrow{+})$ $\qquad x = x_0 + v_0 t + \frac{1}{2}a_c t^2;$ $\qquad x = x_0 + (v_0)_x t$

$(\xrightarrow{+})$ $\qquad v^2 = v_0^2 + 2a_c(x - x_0);$ $\qquad v_x = (v_0)_x$

The first and last equations indicate that *the horizontal component of velocity always remains constant during the motion.*

Vertical Motion. Since the positive y axis is directed upward, then $a_y = -g$. Applying Eqs. 12–4 to 12–6, we get

$(+\uparrow)$ $\qquad v = v_0 + a_c t;$ $\qquad\qquad v_y = (v_0)_y - gt$

$(+\uparrow)$ $\qquad y = y_0 + v_0 t + \frac{1}{2}a_c t^2;$ $\qquad y = y_0 + (v_0)_y t - \frac{1}{2}gt^2$

$(+\uparrow)$ $\qquad v^2 = v_0^2 + 2a_c(y - y_0);$ $\qquad v_y^2 = (v_0)_y^2 - 2g(y - y_0)$

Recall that the last equation can be formulated on the basis of eliminating the time t from the first two equations, and therefore *only two of the above three equations are independent of one another.*

*This assumes that the earth's gravitational field does not vary with altitude.

12

Once thrown, the basketball follows a parabolic trajectory. (© R.C. Hibbeler)

To summarize, problems involving the motion of a projectile can have at most three unknowns since only three independent equations can be written; that is, *one* equation in the *horizontal direction* and *two* in the *vertical direction*. Once $\mathbf{v}_x$ and $\mathbf{v}_y$ are obtained, the resultant velocity $\mathbf{v}$, which is *always tangent* to the path, can be determined by the *vector sum* as shown in Fig. 12–20.

Procedure for Analysis

Coordinate System.

- Establish the fixed x, y coordinate axes and sketch the trajectory of the particle. Between any *two points* on the path specify the given problem data and identify the *three unknowns*. In all cases the acceleration of gravity acts downward and equals 9.81 m/s^2 or 32.2 ft/s^2. The particle's initial and final velocities should be represented in terms of their x and y components.

- Remember that positive and negative position, velocity, and acceleration components always act in accordance with their associated coordinate directions.

Kinematic Equations.

- Depending upon the known data and what is to be determined, a choice should be made as to which three of the following four equations should be applied between the two points on the path to obtain the most direct solution to the problem.

Horizontal Motion.

- The *velocity* in the horizontal or x direction is *constant*, i.e., $v_x = (v_0)_x$, and

$$x = x_0 + (v_0)_x t$$

Vertical Motion.

- In the vertical or y direction *only two* of the following three equations can be used for solution.

$$v_y = (v_0)_y + a_c t$$

$$y = y_0 + (v_0)_y t + \tfrac{1}{2} a_c t^2$$

$$v_y^2 = (v_0)_y^2 + 2a_c(y - y_0)$$

For example, if the particle's final velocity v_y is not needed, then the first and third of these equations will not be useful.

Gravel falling off the end of this conveyor belt follows a path that can be predicted using the equations of constant acceleration. In this way the location of the accumulated pile can be determined. Rectangular coordinates are used for the analysis since the acceleration is only in the vertical direction. (© R.C. Hibbeler)

EXAMPLE 12.11

A sack slides off the ramp, shown in Fig. 12–21, with a horizontal velocity of 12 m/s. If the height of the ramp is 6 m from the floor, determine the time needed for the sack to strike the floor and the range R where sacks begin to pile up.

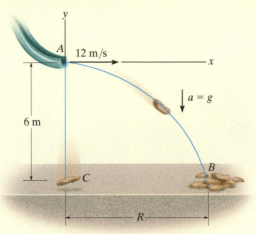

Fig. 12–21

SOLUTION

Coordinate System. The origin of coordinates is established at the beginning of the path, point A, Fig. 12–21. The initial velocity of a sack has components $(v_A)_x = 12$ m/s and $(v_A)_y = 0$. Also, between points A and B the acceleration is $a_y = -9.81$ m/s². Since $(v_B)_x = (v_A)_x = 12$ m/s, the three unknowns are $(v_B)_y$, R, and the time of flight t_{AB}. Here we do not need to determine $(v_B)_y$.

Vertical Motion. The vertical distance from A to B is known, and therefore we can obtain a direct solution for t_{AB} by using the equation

$(+\uparrow)$ $$y_B = y_A + (v_A)_y t_{AB} + \tfrac{1}{2} a_c t_{AB}^2$$

$$-6 \text{ m} = 0 + 0 + \tfrac{1}{2}(-9.81 \text{ m/s}^2) t_{AB}^2$$

$$t_{AB} = 1.11 \text{ s} \qquad \qquad \textit{Ans.}$$

Horizontal Motion. Since t_{AB} has been calculated, R is determined as follows:

$(\xrightarrow{+})$ $$x_B = x_A + (v_A)_x t_{AB}$$

$$R = 0 + 12 \text{ m/s} (1.11 \text{ s})$$

$$R = 13.3 \text{ m} \qquad \qquad \textit{Ans.}$$

NOTE: The calculation for t_{AB} also indicates that if a sack were released *from rest* at A, it would take the same amount of time to strike the floor at C, Fig. 12–21.

12

EXAMPLE | 12.12

The chipping machine is designed to eject wood chips at $v_O = 25$ ft/s as shown in Fig. 12–22. If the tube is oriented at 30° from the horizontal, determine how high, h, the chips strike the pile if at this instant they land on the pile 20 ft from the tube.

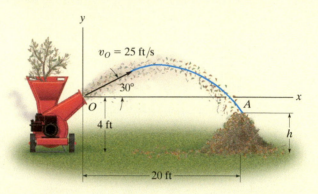

Fig. 12–22

SOLUTION

Coordinate System. When the motion is analyzed between points O and A, the three unknowns are the height h, time of flight t_{OA}, and vertical component of velocity $(v_A)_y$. [Note that $(v_A)_x = (v_O)_x$.] With the origin of coordinates at O, Fig. 12–22, the initial velocity of a chip has components of

$$(v_O)_x = (25 \cos 30°) \text{ ft/s} = 21.65 \text{ ft/s} \rightarrow$$

$$(v_O)_y = (25 \sin 30°) \text{ ft/s} = 12.5 \text{ ft/s} \uparrow$$

Also, $(v_A)_x = (v_O)_x = 21.65$ ft/s and $a_y = -32.2$ ft/s². Since we do not need to determine $(v_A)_y$, we have

Horizontal Motion.

$(\xrightarrow{+})$
$$x_A = x_O + (v_O)_x t_{OA}$$
$$20 \text{ ft} = 0 + (21.65 \text{ ft/s}) t_{OA}$$
$$t_{OA} = 0.9238 \text{ s}$$

Vertical Motion. Relating t_{OA} to the initial and final elevations of a chip, we have

$(+\uparrow)$ $\quad y_A = y_O + (v_O)_y t_{OA} + \frac{1}{2} a_c t_{OA}^2$

$(h - 4 \text{ ft}) = 0 + (12.5 \text{ ft/s})(0.9238 \text{ s}) + \frac{1}{2}(-32.2 \text{ ft/s}^2)(0.9238 \text{ s})^2$

$$h = 1.81 \text{ ft} \qquad\qquad\qquad\qquad\qquad\qquad\qquad \textit{Ans.}$$

NOTE: We can determine $(v_A)_y$ by using $(v_A)_y = (v_O)_y + a_c t_{OA}$.

EXAMPLE | 12.13

The track for this racing event was designed so that riders jump off the slope at 30°, from a height of 1 m. During a race it was observed that the rider shown in Fig. 12–23a remained in mid air for 1.5 s. Determine the speed at which he was traveling off the ramp, the horizontal distance he travels before striking the ground, and the maximum height he attains. Neglect the size of the bike and rider.

(© R.C. Hibbeler)

(a)

SOLUTION

Coordinate System. As shown in Fig. 12–23b, the origin of the coordinates is established at A. Between the end points of the path AB the three unknowns are the initial speed v_A, range R, and the vertical component of velocity $(v_B)_y$.

Vertical Motion. Since the time of flight and the vertical distance between the ends of the path are known, we can determine v_A.

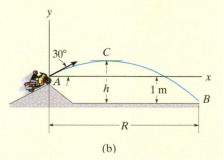

(b)

Fig. 12–23

$$(+\uparrow) \qquad y_B = y_A + (v_A)_y t_{AB} + \tfrac{1}{2} a_c t_{AB}^2$$
$$-1\ \text{m} = 0 + v_A \sin 30°(1.5\ \text{s}) + \tfrac{1}{2}(-9.81\ \text{m/s}^2)(1.5\ \text{s})^2$$
$$v_A = 13.38\ \text{m/s} = 13.4\ \text{m/s} \qquad\qquad Ans.$$

Horizontal Motion. The range R can now be determined.

$$(\xrightarrow{+}) \qquad x_B = x_A + (v_A)_x t_{AB}$$
$$R = 0 + 13.38 \cos 30°\ \text{m/s}(1.5\ \text{s})$$
$$= 17.4\ \text{m} \qquad\qquad Ans.$$

In order to find the maximum height h we will consider the path AC, Fig. 12–23b. Here the three unknowns are the time of flight t_{AC}, the horizontal distance from A to C, and the height h. At the maximum height $(v_C)_y = 0$, and since v_A is known, we can determine h *directly* without considering t_{AC} using the following equation.

$$(v_C)_y^2 = (v_A)_y^2 + 2a_c[y_C - y_A]$$
$$0^2 = (13.38 \sin 30°\ \text{m/s})^2 + 2(-9.81\ \text{m/s}^2)[(h - 1\ \text{m}) - 0]$$
$$h = 3.28\ \text{m} \qquad\qquad Ans.$$

NOTE: Show that the bike will strike the ground at B with a velocity having components of

$$(v_B)_x = 11.6\ \text{m/s} \rightarrow, \quad (v_B)_y = 8.02\ \text{m/s} \downarrow$$

12

PRELIMINARY PROBLEMS

P12–3. Use the chain-rule and find $\dot{y}$ and $\ddot{y}$ in terms of $x, \dot{x}$ and $\ddot{x}$ if

a) $y = 4x^2$

b) $y = 3e^x$

c) $y = 6 \sin x$

P12–4. The particle travels from A to B. Identify the three unknowns, and write the three equations needed to solve for them.

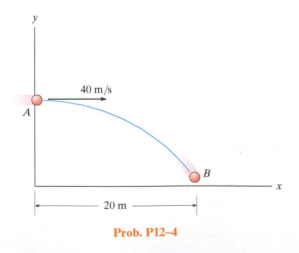

Prob. P12–4

P12–5. The particle travels from A to B. Identify the three unknowns, and write the three equations needed to solve for them.

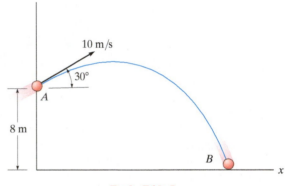

Prob. P12–5

P12–6. The particle travels from A to B. Identify the three unknowns, and write the three equations needed to solve for them.

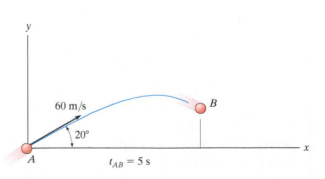

Prob. P12–6

FUNDAMENTAL PROBLEMS

F12–15. If the x and y components of a particle's velocity are $v_x = (32t)$ m/s and $v_y = 8$ m/s, determine the equation of the path $y = f(x)$, if $x = 0$ and $y = 0$ when $t = 0$.

F12–16. A particle is traveling along the straight path. If its position along the x axis is $x = (8t)$ m, where t is in seconds, determine its speed when $t = 2$ s.

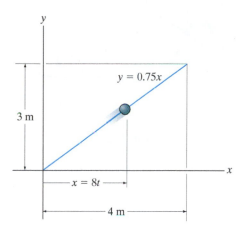

$y = 0.75x$

3 m

$x = 8t$

4 m

Prob. F12–16

F12–17. A particle is constrained to travel along the path. If $x = (4t^4)$ m, where t is in seconds, determine the magnitude of the particle's velocity and acceleration when $t = 0.5$ s.

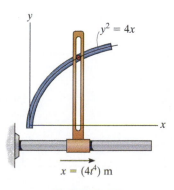

$y^2 = 4x$

$x = (4t^4)$ m

Prob. F12–17

F12–18. A particle travels along a straight-line path $y = 0.5x$. If the x component of the particle's velocity is $v_x = (2t^2)$ m/s, where t is in seconds, determine the magnitude of the particle's velocity and acceleration when $t = 4$ s.

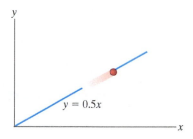

$y = 0.5x$

Prob. F12–18

F12–19. A particle is traveling along the parabolic path $y = 0.25x^2$. If $x = 8$ m, $v_x = 8$ m/s, and $a_x = 4$ m/s^2 when $t = 2$ s, determine the magnitude of the particle's velocity and acceleration at this instant.

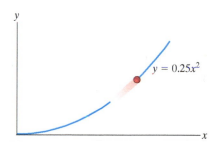

$y = 0.25x^2$

Prob. F12–19

F12–20. The box slides down the slope described by the equation $y = (0.05x^2)$ m, where x is in meters. If the box has x components of velocity and acceleration of $v_x = -3$ m/s and $a_x = -1.5$ m/s^2 at $x = 5$ m, determine the y components of the velocity and the acceleration of the box at this instant.

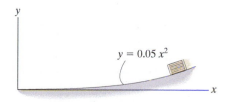

$y = 0.05 x^2$

Prob. F12–20

12

F12–21. The ball is kicked from point A with the initial velocity $v_A = 10$ m/s. Determine the maximum height h it reaches.

F12–22. The ball is kicked from point A with the initial velocity $v_A = 10$ m/s. Determine the range R, and the speed when the ball strikes the ground.

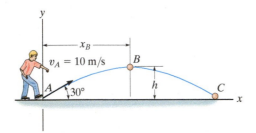

Prob. F12–21/22

F12–23. Determine the speed at which the basketball at A must be thrown at the angle of 30° so that it makes it to the basket at B.

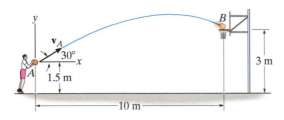

Prob. F12–23

F12–24. Water is sprayed at an angle of 90° from the slope at 20 m/s. Determine the range R.

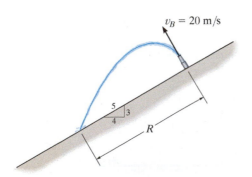

Prob. F12–24

F12–25. A ball is thrown from A. If it is required to clear the wall at B, determine the minimum magnitude of its initial velocity $\mathbf{v}_A$.

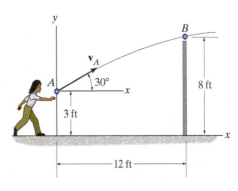

Prob. F12–25

F12–26. A projectile is fired with an initial velocity of $v_A = 150$ m/s off the roof of the building. Determine the range R where it strikes the ground at B.

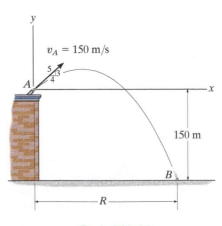

Prob. F12–26

PROBLEMS

12–69. If the velocity of a particle is defined as $\mathbf{v}(t) = \{0.8t^2\mathbf{i} + 12t^{1/2}\mathbf{j} + 5\mathbf{k}\}$ m/s, determine the magnitude and coordinate direction angles α, β, γ of the particle's acceleration when $t = 2$ s.

12–70. The velocity of a particle is $\mathbf{v} = \{3\mathbf{i} + (6 - 2t)\mathbf{j}\}$ m/s, where t is in seconds. If $\mathbf{r} = \mathbf{0}$ when $t = 0$, determine the displacement of the particle during the time interval $t = 1$ s to $t = 3$ s.

12–71. A particle, originally at rest and located at point (3 ft, 2 ft, 5 ft), is subjected to an acceleration of $\mathbf{a} = \{6t\mathbf{i} + 12t^2\mathbf{k}\}$ ft/s². Determine the particle's position (x, y, z) at $t = 1$ s.

*12–72. The velocity of a particle is given by $v = \{16t^2\mathbf{i} + 4t^3\mathbf{j} + (5t + 2)\mathbf{k}\}$ m/s, where t is in seconds. If the particle is at the origin when $t = 0$, determine the magnitude of the particle's acceleration when $t = 2$ s. Also, what is the x, y, z coordinate position of the particle at this instant?

12–73. The water sprinkler, positioned at the base of a hill, releases a stream of water with a velocity of 15 ft/s as shown. Determine the point $B(x, y)$ where the water strikes the ground on the hill. Assume that the hill is defined by the equation $y = (0.05x^2)$ ft and neglect the size of the sprinkler.

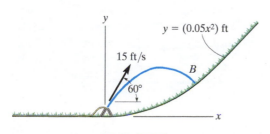

Prob. 12–73

12–74. A particle, originally at rest and located at point (3 ft, 2 ft, 5 ft), is subjected to an acceleration $\mathbf{a} = \{6t\mathbf{i} + 12t^2\mathbf{k}\}$ ft/s². Determine the particle's position (x, y, z) when $t = 2$ s.

12–75. A particle travels along the curve from A to B in 2 s. It takes 4 s for it to go from B to C and then 3 s to go from C to D. Determine its average speed when it goes from A to D.

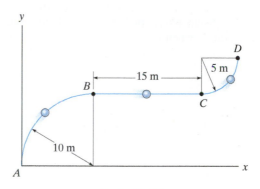

Prob. 12–75

*12–76. A particle travels along the curve from A to B in 5 s. It takes 8 s for it to go from B to C and then 10 s to go from C to A. Determine its average speed when it goes around the closed path.

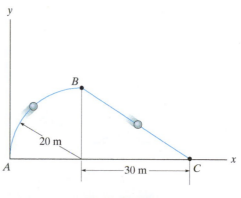

Prob. 12–76

12

12–77. The position of a crate sliding down a ramp is given by $x = (0.25t^3)$ m, $y = (1.5t^2)$ m, $z = (6 - 0.75t^{5/2})$ m, where t is in seconds. Determine the magnitude of the crate's velocity and acceleration when $t = 2$ s.

12–78. A rocket is fired from rest at $x = 0$ and travels along a parabolic trajectory described by $y^2 = [120(10^3)x]$ m. If the x component of acceleration is $a_x = \left(\frac{1}{4}t^2\right)$ m/s^2, where t is in seconds, determine the magnitude of the rocket's velocity and acceleration when $t = 10$ s.

12–79. The particle travels along the path defined by the parabola $y = 0.5x^2$. If the component of velocity along the x axis is $v_x = (5t)$ ft/s, where t is in seconds, determine the particle's distance from the origin O and the magnitude of its acceleration when $t = 1$ s. When $t = 0$, $x = 0$, $y = 0$.

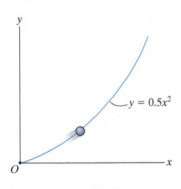

$y = 0.5x^2$

Prob. 12–79

*12–80.** The motorcycle travels with constant speed v_0 along the path that, for a short distance, takes the form of a sine curve. Determine the x and y components of its velocity at any instant on the curve.

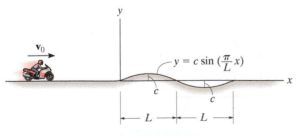

$y = c \sin\left(\frac{\pi}{L}x\right)$

Prob. 12–80

12–81. A particle travels along the curve from A to B in 1 s. If it takes 3 s for it to go from A to C, determine its *average velocity* when it goes from B to C.

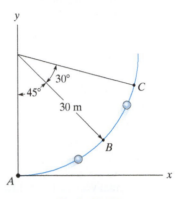

$30°$
$45°$
30 m

Prob. 12–81

12–82. The roller coaster car travels down the helical path at constant speed such that the parametric equations that define its position are $x = c \sin kt$, $y = c \cos kt$, $z = h - bt$, where c, h, and b are constants. Determine the magnitudes of its velocity and acceleration.

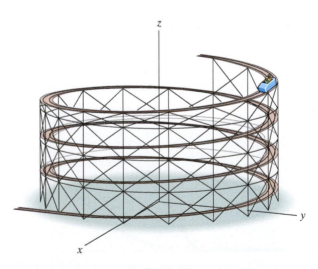

Prob. 12–82

12–83. Pegs A and B are restricted to move in the elliptical slots due to the motion of the slotted link. If the link moves with a constant speed of 10 m/s, determine the magnitude of the velocity and acceleration of peg A when $x = 1$ m.

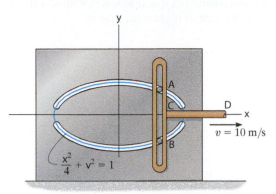

Prob. 12–83

12–84. The van travels over the hill described by $y = (-1.5(10^{-3})x^2 + 15)$ ft. If it has a constant speed of 75 ft/s, determine the x and y components of the van's velocity and acceleration when $x = 50$ ft.

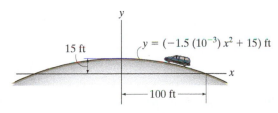

Prob. 12–84

12–85. The flight path of the helicopter as it takes off from A is defined by the parametric equations $x = (2t^2)$ m and $y = (0.04t^3)$ m, where t is the time in seconds. Determine the distance the helicopter is from point A and the magnitudes of its velocity and acceleration when $t = 10$ s.

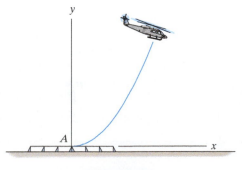

Prob. 12–85

12–86. Determine the minimum initial velocity v_0 and the corresponding angle θ_0 at which the ball must be kicked in order for it to just cross over the 3-m high fence.

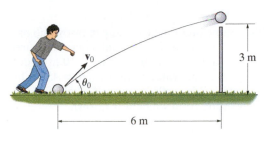

Prob. 12–86

12–87. The catapult is used to launch a ball such that it strikes the wall of the building at the maximum height of its trajectory. If it takes 1.5 s to travel from A to B, determine the velocity $\mathbf{v}_A$ at which it was launched, the angle of release θ, and the height h.

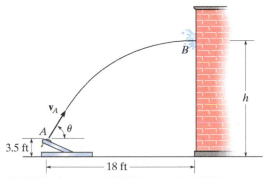

Prob. 12–87

12–88. Neglecting the size of the ball, determine the magnitude v_A of the basketball's initial velocity and its velocity when it passes through the basket.

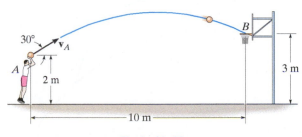

Prob. 12–88

12–89. The girl at A can throw a ball at $v_A = 10$ m/s. Calculate the maximum possible range $R = R_{max}$ and the associated angle θ at which it should be thrown. Assume the ball is caught at B at the same elevation from which it is thrown.

12–90. Show that the girl at A can throw the ball to the boy at B by launching it at equal angles measured up or down from a 45° inclination. If $v_A = 10$ m/s, determine the range R if this value is 15°, i.e., $\theta_1 = 45° - 15° = 30°$ and $\theta_2 = 45° + 15° = 60°$. Assume the ball is caught at the same elevation from which it is thrown.

12–93. A golf ball is struck with a velocity of 80 ft/s as shown. Determine the distance d to where it will land.

12–94. A golf ball is struck with a velocity of 80 ft/s as shown. Determine the speed at which it strikes the ground at B and the time of flight from A to B.

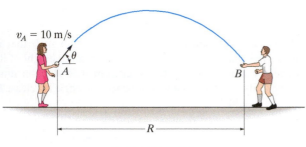

Probs. 12–89/90

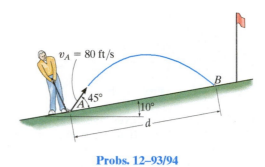

Probs. 12–93/94

12–91. The ball at A is kicked with a speed $v_A = 80$ ft/s and at an angle $\theta_A = 30°$. Determine the point $(x, -y)$ where it strikes the ground. Assume the ground has the shape of a parabola as shown.

*12–92. The ball at A is kicked such that $\theta_A = 30°$. If it strikes the ground at B having coordinates $x = 15$ ft, $y = -9$ ft, determine the speed at which it is kicked and the speed at which it strikes the ground.

12–95. The basketball passed through the hoop even though it barely cleared the hands of the player B who attempted to block it. Neglecting the size of the ball, determine the magnitude v_A of its initial velocity and the height h of the ball when it passes over player B.

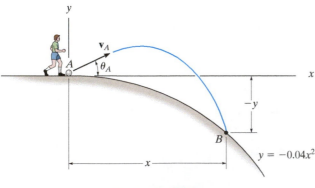

Probs. 12–91/92

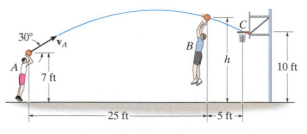

Prob. 12–95

*12–96.   It is observed that the skier leaves the ramp A at an angle $\theta_A = 25°$ with the horizontal. If he strikes the ground at B, determine his initial speed v_A and the time of flight t_{AB}.

12–97.   It is observed that the skier leaves the ramp A at an angle $\theta_A = 25°$ with the horizontal. If he strikes the ground at B, determine his initial speed v_A and the speed at which he strikes the ground.

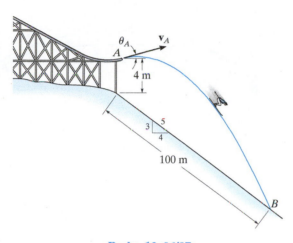

Probs. 12–96/97

12–98.   Determine the horizontal velocity v_A of a tennis ball at A so that it just clears the net at B. Also, find the distance s where the ball strikes the ground.

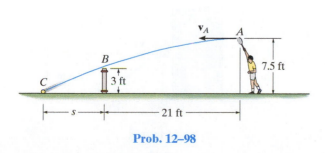

Prob. 12–98

12–99.   The missile at A takes off from rest and rises vertically to B, where its fuel runs out in 8 s. If the acceleration varies with time as shown, determine the missile's height h_B and speed v_B. If by internal controls the missile is then suddenly pointed 45° as shown, and allowed to travel in free flight, determine the maximum height attained, h_C, and the range R to where it crashes at D.

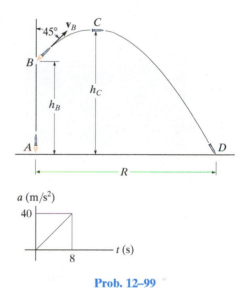

Prob. 12–99

*12–100.   The projectile is launched with a velocity $\mathbf{v}_0$. Determine the range R, the maximum height h attained, and the time of flight. Express the results in terms of the angle θ and v_0. The acceleration due to gravity is g.

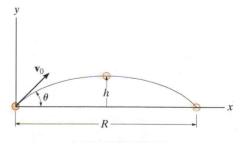

Prob. 12–100

12–101. The drinking fountain is designed such that the nozzle is located from the edge of the basin as shown. Determine the maximum and minimum speed at which water can be ejected from the nozzle so that it does not splash over the sides of the basin at *B* and *C*.

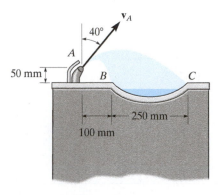

Prob. 12–101

12–102. If the dart is thrown with a speed of 10 m/s, determine the shortest possible time before it strikes the target. Also, what is the corresponding angle θ_A at which it should be thrown, and what is the velocity of the dart when it strikes the target?

12–103. If the dart is thrown with a speed of 10 m/s, determine the longest possible time when it strikes the target. Also, what is the corresponding angle θ_A at which it should be thrown, and what is the velocity of the dart when it strikes the target?

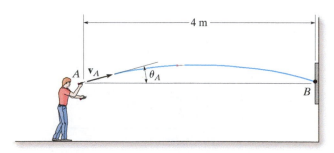

Probs. 12–102/103

*12–104. The man at *A* wishes to throw two darts at the target at *B* so that they arrive at the *same time*. If each dart is thrown with a speed of 10 m/s, determine the angles θ_C and θ_D at which they should be thrown and the time between each throw. Note that the first dart must be thrown at θ_C ($> \theta_D$), then the second dart is thrown at θ_D.

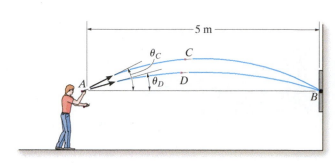

Prob. 12–104

12–105. The velocity of the water jet discharging from the orifice can be obtained from $v = \sqrt{2gh}$, where $h = 2$ m is the depth of the orifice from the free water surface. Determine the time for a particle of water leaving the orifice to reach point *B* and the horizontal distance *x* where it hits the surface.

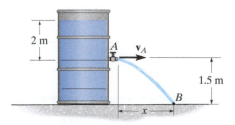

Prob. 12–105

12–106. The snowmobile is traveling at 10 m/s when it leaves the embankment at A. Determine the time of flight from A to B and the range R of the trajectory.

*12–108. The baseball player A hits the baseball at $v_A = 40$ ft/s and $\theta_A = 60°$ from the horizontal. When the ball is directly overhead of player B he begins to run under it. Determine the constant speed at which B must run and the distance d in order to make the catch at the same elevation at which the ball was hit.

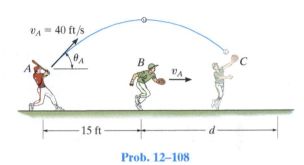

Prob. 12–108

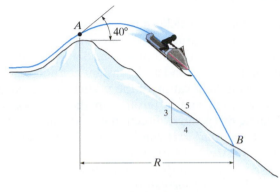

Prob. 12–106

12–107. The fireman wishes to direct the flow of water from his hose to the fire at B. Determine two possible angles θ_1 and θ_2 at which this can be done. Water flows from the hose at $v_A = 80$ ft/s.

12–109. The catapult is used to launch a ball such that it strikes the wall of the building at the maximum height of its trajectory. If it takes 1.5 s to travel from A to B, determine the velocity $\mathbf{v}_A$ at which it was launched, the angle of release θ, and the height h.

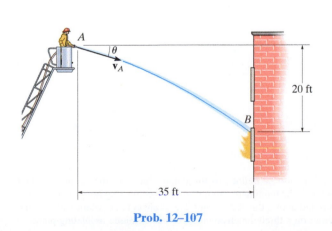

Prob. 12–107

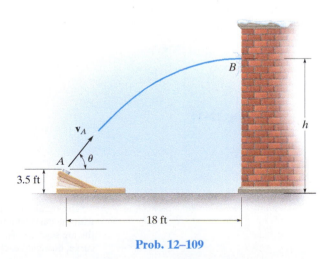

Prob. 12–109

12

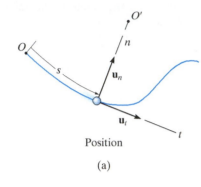

Position

(a)

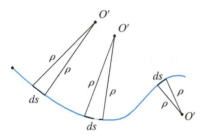

Radius of curvature

(b)

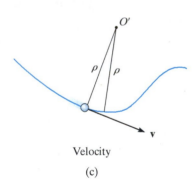

Velocity

(c)

Fig. 12–24

12.7   Curvilinear Motion: Normal and Tangential Components

When the path along which a particle travels is *known*, then it is often convenient to describe the motion using *n* and *t* coordinate axes which act normal and tangent to the path, respectively, and at the instant considered have their *origin located at the particle*.

Planar Motion.   Consider the particle shown in Fig. 12–24a, which moves in a plane along a fixed curve, such that at a given instant it is at position *s*, measured from point *O*. We will now consider a coordinate system that has its origin on the curve, and at the instant considered this origin happens to *coincide* with the location of the particle. The *t* axis is *tangent* to the curve at the point and is positive in the direction of *increasing s*. We will designate this positive direction with the unit vector $\mathbf{u}_t$. A unique choice for the *normal axis* can be made by noting that geometrically the curve is constructed from a series of differential arc segments *ds*, Fig. 12–24b. Each segment *ds* is formed from the arc of an associated circle having a *radius of curvature ρ* (rho) and *center of curvature O′*. The normal axis *n* is perpendicular to the *t* axis with its positive sense directed *toward* the center of curvature *O′*, Fig. 12–24a. This positive direction, which is *always* on the concave side of the curve, will be designated by the unit vector $\mathbf{u}_n$. The plane which contains the *n* and *t* axes is referred to as the embracing or *osculating plane*, and in this case it is fixed in the plane of motion.*

Velocity.   Since the particle moves, *s* is a function of time. As indicated in Sec. 12.4, the particle's velocity **v** has a *direction* that is *always tangent to the path*, Fig. 12–24c, and a *magnitude* that is determined by taking the time derivative of the path function $s = s(t)$, i.e., $v = ds/dt$ (Eq. 12–8). Hence

$$\mathbf{v} = v\mathbf{u}_t \tag{12–15}$$

where

$$v = \dot{s} \tag{12–16}$$

---

*The osculating plane may also be defined as the plane which has the greatest contact with the curve at a point. It is the limiting position of a plane contacting both the point and the arc segment *ds*. As noted above, the osculating plane is always coincident with a plane curve; however, each point on a three-dimensional curve has a unique osculating plane.

Acceleration. The acceleration of the particle is the time rate of change of the velocity. Thus,

$$\mathbf{a} = \dot{\mathbf{v}} = \dot{v}\mathbf{u}_t + v\dot{\mathbf{u}}_t \tag{12-17}$$

In order to determine the time derivative $\dot{\mathbf{u}}_t$, note that as the particle moves along the arc ds in time dt, $\mathbf{u}_t$ preserves its magnitude of unity; however, its *direction* changes, and becomes $\mathbf{u}'_t$, Fig. 12–24d. As shown in Fig. 12–24e, we require $\mathbf{u}'_t = \mathbf{u}_t + d\mathbf{u}_t$. Here $d\mathbf{u}_t$ stretches between the arrowheads of $\mathbf{u}_t$ and $\mathbf{u}'_t$, which lie on an infinitesimal arc of radius $u_t = 1$. Hence, $d\mathbf{u}_t$ has a *magnitude* of $du_t = (1)\, d\theta$, and its *direction* is defined by $\mathbf{u}_n$. Consequently, $d\mathbf{u}_t = d\theta \mathbf{u}_n$, and therefore the time derivative becomes $\dot{\mathbf{u}}_t = \dot{\theta}\mathbf{u}_n$. Since $ds = \rho\, d\theta$, Fig. 12–24d, then $\dot{\theta} = \dot{s}/\rho$, and therefore

$$\dot{\mathbf{u}}_t = \dot{\theta}\mathbf{u}_n = \frac{\dot{s}}{\rho}\mathbf{u}_n = \frac{v}{\rho}\mathbf{u}_n$$

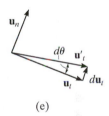

(d)

Substituting into Eq. 12–17, $\mathbf{a}$ can be written as the sum of its two components,

$$\boxed{\mathbf{a} = a_t\mathbf{u}_t + a_n\mathbf{u}_n} \tag{12-18}$$

where

$$\boxed{a_t = \dot{v}} \quad \text{or} \quad \boxed{a_t\, ds = v\, dv} \tag{12-19}$$

(e)

and

$$\boxed{a_n = \frac{v^2}{\rho}} \tag{12-20}$$

These two mutually perpendicular components are shown in Fig. 12–24f. Therefore, the *magnitude* of acceleration is the positive value of

$$a = \sqrt{a_t^2 + a_n^2} \tag{12-21}$$

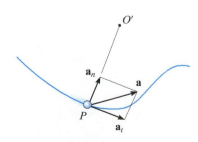

Acceleration

(f)

Fig. 12–24 (cont.)

As the boy swings upward with a velocity **v**, his motion can be analyzed using *n–t* coordinates. As he rises, the magnitude of his velocity (speed) is decreasing, and so a_t will be negative. The rate at which the direction of his velocity changes is a_n, which is always positive, that is, towards the center of rotation. (© R.C. Hibbeler)

To better understand these results, consider the following two special cases of motion.

1. If the particle moves along a straight line, then $\rho \rightarrow \infty$ and from Eq. 12–20, $a_n = 0$. Thus $a = a_t = \dot{v}$, and we can conclude that the *tangential component of acceleration represents the time rate of change in the magnitude of the velocity.*

2. If the particle moves along a curve with a constant speed, then $a_t = \dot{v} = 0$ and $a = a_n = v^2/\rho$. Therefore, the *normal component of acceleration represents the time rate of change in the direction of the velocity.* Since $\mathbf{a}_n$ *always* acts towards the center of curvature, this component is sometimes referred to as the *centripetal* (or center seeking) *acceleration.*

As a result of these interpretations, a particle moving along the curved path in Fig. 12–25 will have accelerations directed as shown.

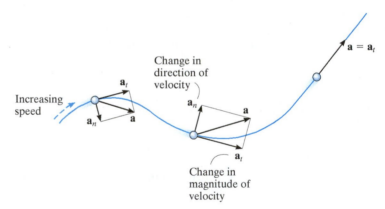

Fig. 12–25

Three-Dimensional Motion.

If the particle moves along a space curve, Fig. 12–26, then at a given instant the *t* axis is uniquely specified; however, an infinite number of straight lines can be constructed normal to the tangent axis. As in the case of planar motion, we will choose the positive *n* axis directed toward the path's center of curvature O'. This axis is referred to as the *principal normal* to the curve. With the *n* and *t* axes so defined, Eqs. 12–15 through 12–21 can be used to determine **v** and **a**. Since $\mathbf{u}_t$ and $\mathbf{u}_n$ are always perpendicular to one another and lie in the osculating plane, for spatial motion a third unit vector, $\mathbf{u}_b$, defines the *binormal axis b* which is perpendicular to $\mathbf{u}_t$ and $\mathbf{u}_n$, Fig. 12–26.

Since the three unit vectors are related to one another by the vector cross product, e.g., $\mathbf{u}_b = \mathbf{u}_t \times \mathbf{u}_n$, Fig. 12–26, it may be possible to use this relation to establish the direction of one of the axes, if the directions of the other two are known. For example, no motion occurs in the $\mathbf{u}_b$ direction, and if this direction and $\mathbf{u}_t$ are known, then $\mathbf{u}_n$ can be determined, where in this case $\mathbf{u}_n = \mathbf{u}_b \times \mathbf{u}_t$, Fig. 12–26. Remember, though, that $\mathbf{u}_n$ is always on the concave side of the curve.

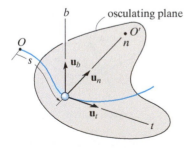

Fig. 12–26

Procedure for Analysis

Coordinate System.

- Provided the *path* of the particle is *known*, we can establish a set of *n* and *t* coordinates having a *fixed origin,* which is coincident with the particle at the instant considered.
- The positive tangent axis acts in the direction of motion and the positive normal axis is directed toward the path's center of curvature.

Velocity.

- The particle's *velocity* is always tangent to the path.
- The magnitude of velocity is found from the time derivative of the path function.

$$v = \dot{s}$$

Tangential Acceleration.

- The tangential component of acceleration is the result of the time rate of change in the *magnitude* of velocity. This component acts in the positive *s* direction if the particle's speed is increasing or in the opposite direction if the speed is decreasing.
- The relations between a_t, v, t, and s are the same as for rectilinear motion, namely,

$$a_t = \dot{v} \quad a_t\,ds = v\,dv$$

- If a_t is constant, $a_t = (a_t)_c$, the above equations, when integrated, yield

$$s = s_0 + v_0 t + \tfrac{1}{2}(a_t)_c t^2$$
$$v = v_0 + (a_t)_c t$$
$$v^2 = v_0^2 + 2(a_t)_c(s - s_0)$$

Normal Acceleration.

- The normal component of acceleration is the result of the time rate of change in the *direction* of the velocity. This component is *always* directed toward the center of curvature of the path, i.e., along the positive *n* axis.
- The magnitude of this component is determined from

$$a_n = \frac{v^2}{\rho}$$

- If the path is expressed as $y = f(x)$, the radius of curvature ρ at any point on the path is determined from the equation

$$\rho = \frac{[1 + (dy/dx)^2]^{3/2}}{|d^2y/dx^2|}$$

The derivation of this result is given in any standard calculus text.

Once the rotation is constant, the riders will then have only a normal component of acceleration. (© R.C. Hibbeler)

Motorists traveling along this cloverleaf interchange experience a normal acceleration due to the change in direction of their velocity. A tangential component of acceleration occurs when the cars' speed is increased or decreased. (© R.C. Hibbeler)

12

EXAMPLE   12.14

When the skier reaches point A along the parabolic path in Fig. 12–27a, he has a speed of 6 m/s which is increasing at 2 m/s^2. Determine the direction of his velocity and the direction and magnitude of his acceleration at this instant. Neglect the size of the skier in the calculation.

SOLUTION

Coordinate System.   Although the path has been expressed in terms of its x and y coordinates, we can still establish the origin of the n, t axes at the fixed point A on the path and determine the components of $\mathbf{v}$ and $\mathbf{a}$ along these axes, Fig. 12–27a.

Velocity.   By definition, the velocity is always directed tangent to the path. Since $y = \frac{1}{20}x^2$, $dy/dx = \frac{1}{10}x$, then at $x = 10$ m, $dy/dx = 1$. Hence, at A, $\mathbf{v}$ makes an angle of $\theta = \tan^{-1}1 = 45°$ with the x axis, Fig. 12–27b. Therefore,

$$v_A = 6 \text{ m/s} \quad 45° \nearrow \qquad\qquad Ans.$$

The acceleration is determined from $\mathbf{a} = \dot{v}\mathbf{u}_t + (v^2/\rho)\mathbf{u}_n$. However, it is first necessary to determine the radius of curvature of the path at A (10 m, 5 m). Since $d^2y/dx^2 = \frac{1}{10}$, then

$$\rho = \frac{[1 + (dy/dx)^2]^{3/2}}{|d^2y/dx^2|} = \frac{\left[1 + \left(\frac{1}{10}x\right)^2\right]^{3/2}}{\left|\frac{1}{10}\right|}\Bigg|_{x=10 \text{ m}} = 28.28 \text{ m}$$

The acceleration becomes

$$\mathbf{a}_A = \dot{v}\mathbf{u}_t + \frac{v^2}{\rho}\mathbf{u}_n$$

$$= 2\mathbf{u}_t + \frac{(6 \text{ m/s})^2}{28.28 \text{ m}}\mathbf{u}_n$$

$$= \{2\mathbf{u}_t + 1.273\mathbf{u}_n\} \text{ m/s}^2$$

As shown in Fig. 12–27b,

$$a = \sqrt{(2 \text{ m/s}^2)^2 + (1.273 \text{ m/s}^2)^2} = 2.37 \text{ m/s}^2$$

$$\phi = \tan^{-1}\frac{2}{1.273} = 57.5°$$

Thus, $45° + 90° + 57.5° - 180° = 12.5°$ so that,

$$a = 2.37 \text{ m/s}^2 \quad 12.5° \nearrow \qquad\qquad Ans.$$

NOTE: By using n, t coordinates, we were able to readily solve this problem through the use of Eq. 12–18, since it accounts for the separate changes in the magnitude and direction of $\mathbf{v}$.

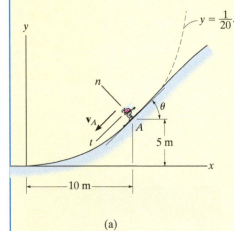

(a)

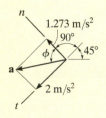

(b)

Fig. 12–27

EXAMPLE | 12.15

A race car C travels around the horizontal circular track that has a radius of 300 ft, Fig. 12–28. If the car increases its speed at a constant rate of 7 ft/s^2, starting from rest, determine the time needed for it to reach an acceleration of 8 ft/s^2. What is its speed at this instant?

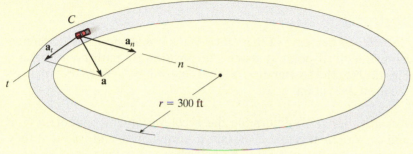

Fig. 12–28

SOLUTION

Coordinate System.   The origin of the n and t axes is coincident with the car at the instant considered. The t axis is in the direction of motion, and the positive n axis is directed toward the center of the circle. This coordinate system is selected since the path is known.

Acceleration.   The magnitude of acceleration can be related to its components using $a = \sqrt{a_t^2 + a_n^2}$. Here $a_t = 7$ ft/s^2. Since $a_n = v^2/\rho$, the velocity as a function of time must be determined first.

$$v = v_0 + (a_t)_c t$$
$$v = 0 + 7t$$

Thus

$$a_n = \frac{v^2}{\rho} = \frac{(7t)^2}{300} = 0.163t^2 \text{ ft/s}^2$$

The time needed for the acceleration to reach 8 ft/s^2 is therefore

$$a = \sqrt{a_t^2 + a_n^2}$$
$$8 \text{ ft/s}^2 = \sqrt{(7 \text{ ft/s}^2)^2 + (0.163t^2)^2}$$

Solving for the positive value of t yields

$$0.163t^2 = \sqrt{(8 \text{ ft/s}^2)^2 - (7 \text{ ft/s}^2)^2}$$
$$t = 4.87 \text{ s} \qquad\qquad Ans.$$

Velocity.   The speed at time $t = 4.87$ s is

$$v = 7t = 7(4.87) = 34.1 \text{ ft/s} \qquad\qquad Ans.$$

NOTE: Remember the velocity will always be tangent to the path, whereas the acceleration will be directed within the curvature of the path.

12

EXAMPLE | 12.16

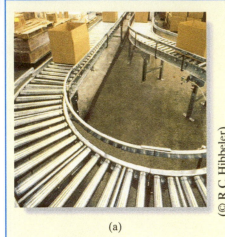

(a)

@ R.C. Hibbeler

The boxes in Fig. 12–29a travel along the industrial conveyor. If a box as in Fig. 12–29b starts from rest at A and increases its speed such that $a_t = (0.2t)$ m/s², where t is in seconds, determine the magnitude of its acceleration when it arrives at point B.

SOLUTION

Coordinate System. The position of the box at any instant is defined from the fixed point A using the position or path coordinate s, Fig. 12–29b. The acceleration is to be determined at B, so the origin of the n, t axes is at this point.

Acceleration. To determine the acceleration components $a_t = \dot{v}$ and $a_n = v^2/\rho$, it is first necessary to formulate v and $\dot{v}$ so that they may be evaluated at B. Since $v_A = 0$ when $t = 0$, then

$$a_t = \dot{v} = 0.2t \tag{1}$$

$$\int_0^v dv = \int_0^t 0.2t \, dt$$

$$v = 0.1t^2 \tag{2}$$

The time needed for the box to reach point B can be determined by realizing that the position of B is $s_B = 3 + 2\pi(2)/4 = 6.142$ m, Fig. 12–29b, and since $s_A = 0$ when $t = 0$ we have

$$v = \frac{ds}{dt} = 0.1t^2$$

$$\int_0^{6.142 \text{ m}} ds = \int_0^{t_B} 0.1t^2 dt$$

$$6.142 \text{ m} = 0.0333t_B^3$$

$$t_B = 5.690 \text{s}$$

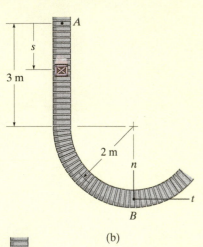

(b)

Substituting into Eqs. 1 and 2 yields

$$(a_B)_t = \dot{v}_B = 0.2(5.690) = 1.138 \text{ m/s}^2$$

$$v_B = 0.1(5.69)^2 = 3.238 \text{ m/s}$$

At B, $\rho_B = 2$ m, so that

$$(a_B)_n = \frac{v_B^2}{\rho_B} = \frac{(3.238 \text{ m/s})^2}{2 \text{ m}} = 5.242 \text{ m/s}^2$$

The magnitude of $\mathbf{a}_B$, Fig. 12–29c, is therefore

$$a_B = \sqrt{(1.138 \text{ m/s}^2)^2 + (5.242 \text{ m/s}^2)^2} = 5.36 \text{ m/s}^2 \quad \textit{Ans.}$$

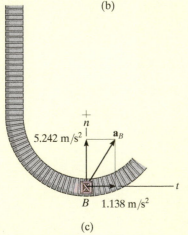

(c)

Fig. 12–29

PRELIMINARY PROBLEM

P12–7.

a) Determine the acceleration at the instant shown.

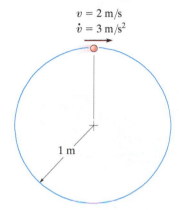

b) Determine the increase in speed and the normal component of acceleration at $s = 2$ m. At $s = 0$, $v = 0$.

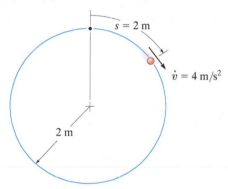

c) Determine the acceleration at the instant shown. The particle has a constant speed of 2 m/s.

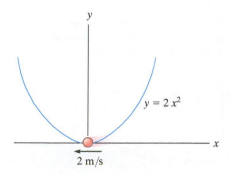

d) Determine the normal and tangential components of acceleration at $s = 0$ if $v = (4s + 1)$ m/s, where s is in meters.

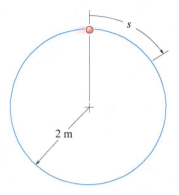

e) Determine the acceleration at $s = 2$ m if $\dot{v} = (2\,s)$ m/s^2, where s is in meters. At $s = 0$, $v = 1$ m/s.

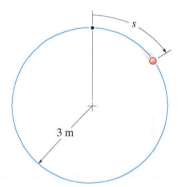

f. Determine the acceleration when $t = 1$ s if $v = (4t^2 + 2)$ m/s, where t is in seconds.

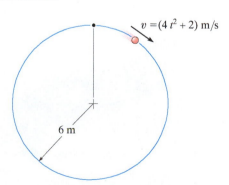

FUNDAMENTAL PROBLEMS

F12–27. The boat is traveling along the circular path with a speed of $v = (0.0625t^2)$ m/s, where t is in seconds. Determine the magnitude of its acceleration when $t = 10$ s.

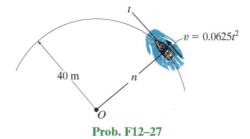

Prob. F12–27

F12–28. The car is traveling along the road with a speed of $v = (2\,s)$ m/s, where s is in meters. Determine the magnitude of its acceleration when $s = 10$ m.

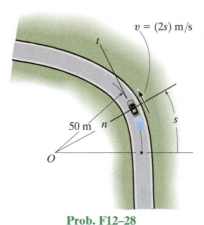

Prob. F12–28

F12–29. If the car decelerates uniformly along the curved road from 25 m/s at A to 15 m/s at C, determine the acceleration of the car at B.

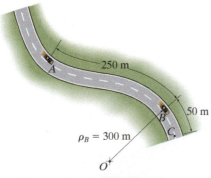

Prob. F12–29

F12–30. When $x = 10$ ft, the crate has a speed of 20 ft/s which is increasing at 6 ft/s². Determine the direction of the crate's velocity and the magnitude of the crate's acceleration at this instant.

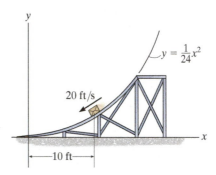

Prob. F12–30

F12–31. If the motorcycle has a deceleration of $a_t = -(0.001s)$ m/s² and its speed at position A is 25 m/s, determine the magnitude of its acceleration when it passes point B.

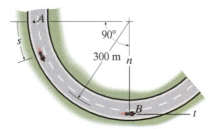

Prob. F12–31

F12–32. The car travels up the hill with a speed of $v = (0.2s)$ m/s, where s is in meters, measured from A. Determine the magnitude of its acceleration when it is at point $s = 50$ m, where $\rho = 500$ m.

Prob. F12–32

PROBLEMS

12–110.   An automobile is traveling on a curve having a radius of 800 ft. If the acceleration of the automobile is 5 ft/s², determine the constant speed at which the automobile is traveling.

12–111.   Determine the maximum constant speed a race car can have if the acceleration of the car cannot exceed 7.5 m/s² while rounding a track having a radius of curvature of 200 m.

*12–112.   A boat has an initial speed of 16 ft/s. If it then increases its speed along a circular path of radius $\rho = 80$ ft at the rate of $\dot{v} = (1.5s)$ ft/s, where s is in feet, determine the time needed for the boat to travel $s = 50$ ft.

12–113.   The position of a particle is defined by $\mathbf{r} = \{4(t - \sin t)\mathbf{i} + (2t^2 - 3)\mathbf{j}\}$ m, where t is in seconds and the argument for the sine is in radians. Determine the speed of the particle and its normal and tangential components of acceleration when $t = 1$ s.

12–114.   The automobile has a speed of 80 ft/s at point A and an acceleration having a magnitude of 10 ft/s², acting in the direction shown. Determine the radius of curvature of the path at point A and the tangential component of acceleration.

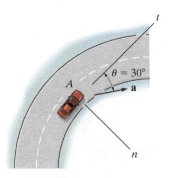

Prob. 12–114

12–115.   The automobile is originally at rest at $s = 0$. If its speed is increased by $\dot{v} = (0.05t^2)$ ft/s², where t is in seconds, determine the magnitudes of its velocity and acceleration when $t = 18$ s.

*12–116.   The automobile is originally at rest $s = 0$. If it then starts to increase its speed at $\dot{v} = (0.05t^2)$ ft/s², where t is in seconds, determine the magnitudes of its velocity and acceleration at $s = 550$ ft.

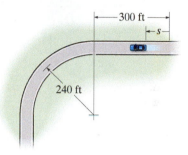

Probs. 12–115/116

12–117.   The two cars A and B travel along the circular path at constant speeds $v_A = 80$ ft/s and $v_B = 100$ ft/s, respectively. If they are at the positions shown when $t = 0$, determine the time when the cars are side by side, and the time when they are 90° apart.

12–118.   Cars A and B are traveling around the circular race track. At the instant shown, A has a speed of 60 ft/s and is increasing its speed at the rate of 15 ft/s² until it travels for a distance of 100π ft, after which it maintains a constant speed. Car B has a speed of 120 ft/s and is decreasing its speed at 15 ft/s² until it travels a distance of 65π ft, after which it maintains a constant speed. Determine the time when they come side by side.

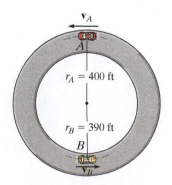

Probs. 12–117/118

12–119. The satellite S travels around the earth in a circular path with a constant speed of 20 Mm/h. If the acceleration is 2.5 m/s^2, determine the altitude h. Assume the earth's diameter to be 12 713 km.

12–121. The car passes point A with a speed of 25 m/s after which its speed is defined by $v = (25 - 0.15s)$ m/s. Determine the magnitude of the car's acceleration when it reaches point B, where $s = 51.5$ m and $x = 50$ m.

12–122. If the car passes point A with a speed of 20 m/s and begins to increase its speed at a constant rate of $a_t = 0.5$ m/s^2, determine the magnitude of the car's acceleration when $s = 101.68$ m and $x = 0$.

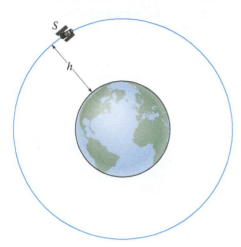

Prob. 12–119

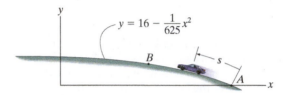

Probs. 12–121/122

*12–120. The car travels along the circular path such that its speed is increased by $a_t = (0.5e^t)$ m/s^2, where t is in seconds. Determine the magnitudes of its velocity and acceleration after the car has traveled $s = 18$ m starting from rest. Neglect the size of the car.

12–123. The motorcycle is traveling at 1 m/s when it is at A. If the speed is then increased at $\dot{v} = 0.1$ m/s^2, determine its speed and acceleration at the instant $t = 5$ s.

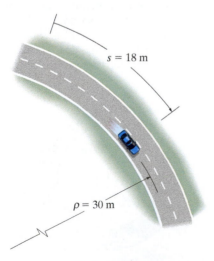

Prob. 12–120

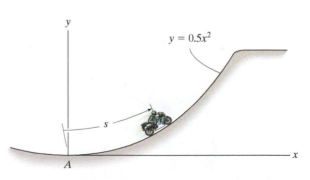

Prob. 12–123

*12–124. The box of negligible size is sliding down along a curved path defined by the parabola $y = 0.4x^2$. When it is at $A(x_A = 2\text{ m}, y_A = 1.6\text{ m})$, the speed is $v = 8$ m/s and the increase in speed is $dv/dt = 4$ m/s^2. Determine the magnitude of the acceleration of the box at this instant.

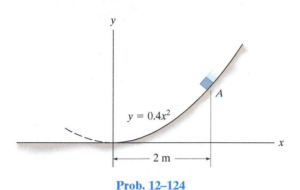

$y = 0.4x^2$

2 m

Prob. 12–124

12–125. The car travels around the circular track having a radius of $r = 300$ m such that when it is at point A it has a velocity of 5 m/s, which is increasing at the rate of $\dot{v} = (0.06t)$ m/s^2, where t is in seconds. Determine the magnitudes of its velocity and acceleration when it has traveled one-third the way around the track.

12–126. The car travels around the portion of a circular track having a radius of $r = 500$ ft such that when it is at point A it has a velocity of 2 ft/s, which is increasing at the rate of $\dot{v} = (0.002t)$ ft/s^2, where t is in seconds. Determine the magnitudes of its velocity and acceleration when it has traveled three-fourths the way around the track.

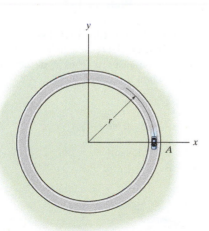

Probs. 12–125/126

12–127. At a given instant the train engine at E has a speed of 20 m/s and an acceleration of 14 m/s^2 acting in the direction shown. Determine the rate of increase in the train's speed and the radius of curvature ρ of the path.

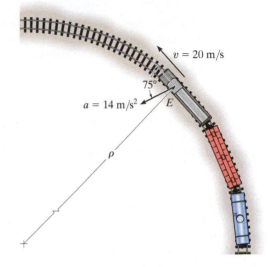

$v = 20$ m/s

$75°$

$a = 14$ m/s^2 E

ρ

Prob. 12–127

*12–128. The car has an initial speed $v_0 = 20$ m/s. If it increases its speed along the circular track at $s = 0$, $a_t = (0.8s)$ m/s^2, where s is in meters, determine the time needed for the car to travel $s = 25$ m.

12–129. The car starts from rest at $s = 0$ and increases its speed at $a_t = 4$ m/s^2. Determine the time when the magnitude of acceleration becomes 20 m/s^2. At what position s does this occur?

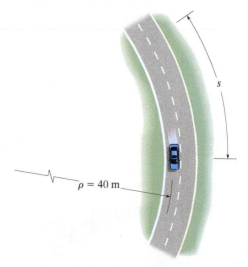

s

$\rho = 40$ m

Probs. 12–128/129

12–130. A boat is traveling along a circular curve having a radius of 100 ft. If its speed at $t = 0$ is 15 ft/s and is increasing at $\dot{v} = (0.8t)$ ft/s², determine the magnitude of its acceleration at the instant $t = 5$ s.

12–131. A boat is traveling along a circular path having a radius of 20 m. Determine the magnitude of the boat's acceleration when the speed is $v = 5$ m/s and the rate of increase in the speed is $\dot{v} = 2$ m/s².

*12–132. Starting from rest, a bicyclist travels around a horizontal circular path, $\rho = 10$ m, at a speed of $v = (0.09t^2 + 0.1t)$ m/s, where t is in seconds. Determine the magnitudes of his velocity and acceleration when he has traveled $s = 3$ m.

12–133. A particle travels around a circular path having a radius of 50 m. If it is initially traveling with a speed of 10 m/s and its speed then increases at a rate of $\dot{v} = (0.05\, v)$ m/s², determine the magnitude of the particle's acceleration four seconds later.

12–134. The motorcycle is traveling at a constant speed of 60 km/h. Determine the magnitude of its acceleration when it is at point A.

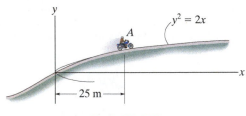

Prob. 12–134

12–135. When $t = 0$, the train has a speed of 8 m/s, which is increasing at 0.5 m/s². Determine the magnitude of the acceleration of the engine when it reaches point A, at $t = 20$ s. Here the radius of curvature of the tracks is $\rho_A = 400$ m.

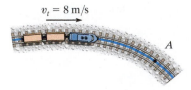

Prob. 12–135

*12–136. At a given instant the jet plane has a speed of 550 m/s and an acceleration of 50 m/s² acting in the direction shown. Determine the rate of increase in the plane's speed, and also the radius of curvature ρ of the path.

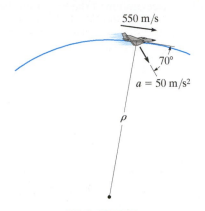

Prob. 12–136

12–137. The ball is ejected horizontally from the tube with a speed of 8 m/s. Find the equation of the path, $y = f(x)$, and then find the ball's velocity and the normal and tangential components of acceleration when $t = 0.25$ s.

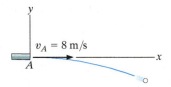

Prob. 12–137

12–138. The motorcycle is traveling at 40 m/s when it is at A. If the speed is then decreased at $\dot{v} = -(0.05\, s)$ m/s², where s is in meters measured from A, determine its speed and acceleration when it reaches B.

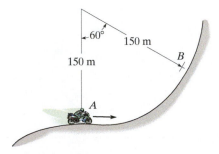

Prob. 12–138

12–139. Cars move around the "traffic circle" which is in the shape of an ellipse. If the speed limit is posted at 60 km/h, determine the minimum acceleration experienced by the passengers.

*12–140. Cars move around the "traffic circle" which is in the shape of an ellipse. If the speed limit is posted at 60 km/h, determine the maximum acceleration experienced by the passengers.

12–142. The race car has an initial speed $v_A = 15$ m/s at A. If it increases its speed along the circular track at the rate $a_t = (0.4s)$ m/s^2, where s is in meters, determine the time needed for the car to travel 20 m. Take $\rho = 150$ m.

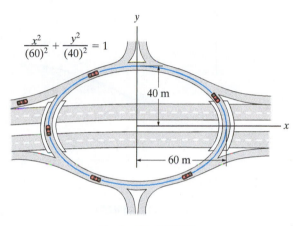

$$\frac{x^2}{(60)^2} + \frac{y^2}{(40)^2} = 1$$

40 m

60 m

Probs. 12–139/140

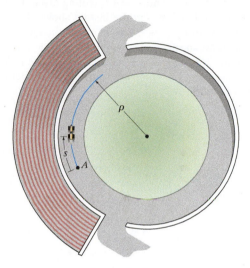

Prob. 12–142

12–141. A package is dropped from the plane which is flying with a constant horizontal velocity of $v_A = 150$ ft/s. Determine the normal and tangential components of acceleration and the radius of curvature of the path of motion (a) at the moment the package is released at A, where it has a horizontal velocity of $v_A = 150$ ft/s, and (b) *just before* it strikes the ground at B.

12–143. The motorcycle travels along the elliptical track at a constant speed v. Determine its greatest acceleration if $a > b$.

*12–144. The motorcycle travels along the elliptical track at a constant speed v. Determine its smallest acceleration if $a > b$.

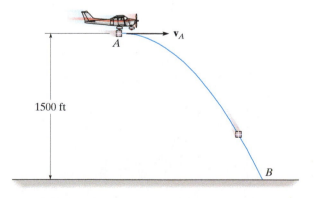

v_A

A

1500 ft

B

Prob. 12–141

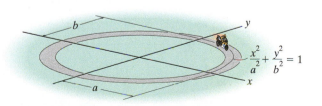

b

y

$$\frac{x^2}{a^2} + \frac{y^2}{b^2} = 1$$

a

x

Probs. 12–143/144

12–145. Particles A and B are traveling counter-clockwise around a circular track at a constant speed of 8 m/s. If at the instant shown the speed of A begins to increase by $(a_t)_A = (0.4s_A)$ m/s^2, where s_A is in meters, determine the distance measured counterclockwise along the track from B to A when $t = 1$ s. What is the magnitude of the acceleration of each particle at this instant?

12–146. Particles A and B are traveling around a circular track at a speed of 8 m/s at the instant shown. If the speed of B is increasing by $(a_t)_B = 4$ m/s^2, and at the same instant A has an increase in speed of $(a_t)_A = 0.8t$ m/s^2, determine how long it takes for a collision to occur. What is the magnitude of the acceleration of each particle just before the collision occurs?

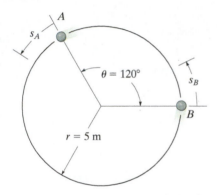

Probs. 12–145/146

12–147. The jet plane is traveling with a speed of 120 m/s which is decreasing at 40 m/s^2 when it reaches point A. Determine the magnitude of its acceleration when it is at this point. Also, specify the direction of flight, measured from the x axis.

*12–148. The jet plane is traveling with a constant speed of 110 m/s along the curved path. Determine the magnitude of the acceleration of the plane at the instant it reaches point $A(y = 0)$.

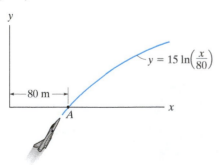

Probs. 12–147/148

12–149. The train passes point B with a speed of 20 m/s which is decreasing at $a_t = -0.5$ m/s^2. Determine the magnitude of acceleration of the train at this point.

12–150. The train passes point A with a speed of 30 m/s and begins to decrease its speed at a constant rate of $a_t = -0.25$ m/s^2. Determine the magnitude of the acceleration of the train when it reaches point B, where $s_{AB} = 412$ m.

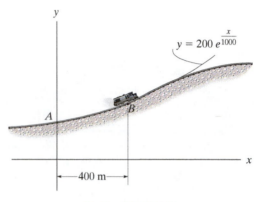

Probs. 12–149/150

12–151. The particle travels with a constant speed of 300 mm/s along the curve. Determine the particle's acceleration when it is located at point (200 mm, 100 mm) and sketch this vector on the curve.

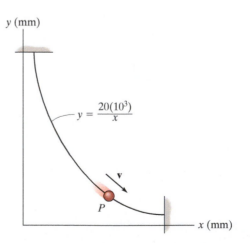

Prob. 12–151

*12–152. A particle P travels along an elliptical spiral path such that its position vector $\mathbf{r}$ is defined by $\mathbf{r} = \{2\cos(0.1t)\mathbf{i} + 1.5\sin(0.1t)\mathbf{j} + (2t)\mathbf{k}\}$ m, where t is in seconds and the arguments for the sine and cosine are given in radians. When $t = 8$ s, determine the coordinate direction angles α, β, and γ, which the binormal axis to the osculating plane makes with the x, y, and z axes. *Hint:* Solve for the velocity $\mathbf{v}_P$ and acceleration $\mathbf{a}_P$ of the particle in terms of their $\mathbf{i}, \mathbf{j}, \mathbf{k}$ components. The binormal is parallel to $\mathbf{v}_P \times \mathbf{a}_P$. Why?

12–153. The motion of a particle is defined by the equations $x = (2t + t^2)$ m and $y = (t^2)$ m, where t is in seconds. Determine the normal and tangential components of the particle's velocity and acceleration when $t = 2$ s.

12–154. If the speed of the crate at A is 15 ft/s, which is increasing at a rate $\dot{v} = 3$ ft/s^2, determine the magnitude of the acceleration of the crate at this instant.

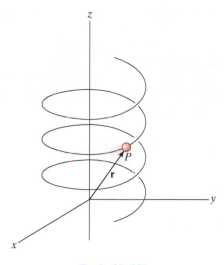

Prob. 12–152

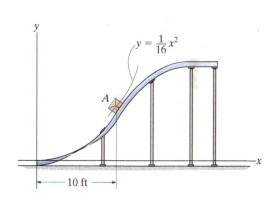

Prob. 12–154

12.8   Curvilinear Motion: Cylindrical Components

Sometimes the motion of the particle is constrained on a path that is best described using cylindrical coordinates. If motion is restricted to the plane, then polar coordinates are used.

Polar Coordinates.   We can specify the location of the particle shown in Fig. 12–30a using a *radial coordinate r*, which extends outward from the fixed origin O to the particle, and a *transverse coordinate θ*, which is the counterclockwise angle between a fixed reference line and the r axis. The angle is generally measured in degrees or radians, where 1 rad = $180°/\pi$. The positive directions of the r and θ coordinates are defined by the unit vectors $\mathbf{u}_r$ and $\mathbf{u}_\theta$, respectively. Here $\mathbf{u}_r$ is in the direction of increasing r when θ is held fixed, and $\mathbf{u}_\theta$ is in a direction of increasing θ when r is held fixed. Note that these directions are perpendicular to one another.

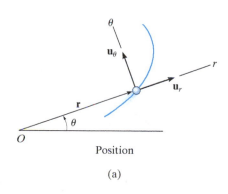

Position

(a)

Fig. 12–30

12

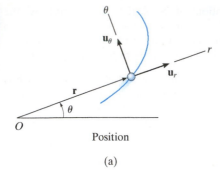

Position

(a)

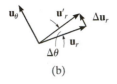

(b)

Position. At any instant the position of the particle, Fig. 12–30a, is defined by the position vector

$$\mathbf{r} = r\mathbf{u}_r \qquad (12\text{--}22)$$

Velocity. The instantaneous velocity $\mathbf{v}$ is obtained by taking the time derivative of $\mathbf{r}$. Using a dot to represent the time derivative, we have

$$\mathbf{v} = \dot{\mathbf{r}} = \dot{r}\mathbf{u}_r + r\dot{\mathbf{u}}_r$$

To evaluate $\dot{\mathbf{u}}_r$, notice that $\mathbf{u}_r$ only changes its direction with respect to time, since by definition the magnitude of this vector is always one unit. Hence, during the time Δt, a change Δr will not cause a change in the direction of $\mathbf{u}_r$; however, a change $\Delta \theta$ will cause $\mathbf{u}_r$ to become $\mathbf{u}'_r$, where $\mathbf{u}'_r = \mathbf{u}_r + \Delta \mathbf{u}_r$, Fig. 12–30b. The time change in $\mathbf{u}_r$ is then $\Delta \mathbf{u}_r$. For small angles $\Delta \theta$ this vector has a magnitude $\Delta u_r \approx 1(\Delta \theta)$ and acts in the $\mathbf{u}_\theta$ direction. Therefore, $\Delta \mathbf{u}_r = \Delta \theta \mathbf{u}_\theta$, and so

$$\dot{\mathbf{u}}_r = \lim_{\Delta t \to 0} \frac{\Delta \mathbf{u}_r}{\Delta t} = \left(\lim_{\Delta t \to 0} \frac{\Delta \theta}{\Delta t} \right) \mathbf{u}_\theta$$

$$\dot{\mathbf{u}}_r = \dot{\theta}\mathbf{u}_\theta \qquad (12\text{--}23)$$

Substituting into the above equation, the velocity can be written in component form as

$$\boxed{\mathbf{v} = v_r\mathbf{u}_r + v_\theta\mathbf{u}_\theta} \qquad (12\text{--}24)$$

where

$$\boxed{\begin{aligned} v_r &= \dot{r} \\ v_\theta &= r\dot{\theta} \end{aligned}} \qquad (12\text{--}25)$$

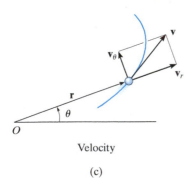

Velocity

(c)

Fig. 12–30 (cont.)

These components are shown graphically in Fig. 12–30c. The *radial component* v_r is a measure of the rate of increase or decrease in the length of the radial coordinate, i.e., $\dot{r}$; whereas the *transverse component* v_θ can be interpreted as the rate of motion along the circumference of a circle having a radius r. In particular, the term $\dot{\theta} = d\theta/dt$ is called the *angular velocity*, since it indicates the time rate of change of the angle θ. Common units used for this measurement are rad/s.

Since v_r and v_θ are mutually perpendicular, the *magnitude* of velocity or speed is simply the positive value of

$$v = \sqrt{(\dot{r})^2 + (r\dot{\theta})^2} \qquad (12\text{--}26)$$

and the *direction* of $\mathbf{v}$ is, of course, tangent to the path, Fig. 12–30c.

Acceleration.   Taking the time derivatives of Eq. 12–24, using Eqs. 12–25, we obtain the particle's instantaneous acceleration,

$$\mathbf{a} = \dot{\mathbf{v}} = \ddot{r}\mathbf{u}_r + \dot{r}\dot{\mathbf{u}}_r + \dot{r}\dot{\theta}\mathbf{u}_\theta + r\ddot{\theta}\mathbf{u}_\theta + r\dot{\theta}\dot{\mathbf{u}}_\theta$$

To evaluate $\dot{\mathbf{u}}_\theta$, it is necessary only to find the change in the direction of $\mathbf{u}_\theta$ since its magnitude is always unity. During the time Δt, a change Δr will not change the direction of $\mathbf{u}_\theta$, however, a change $\Delta\theta$ will cause $\mathbf{u}_\theta$ to become $\mathbf{u}'_\theta$, where $\mathbf{u}'_\theta = \mathbf{u}_\theta + \Delta\mathbf{u}_\theta$, Fig. 12–30d. The time change in $\mathbf{u}_\theta$ is thus $\Delta\mathbf{u}_\theta$. For small angles this vector has a magnitude $\Delta u_\theta \approx 1(\Delta\theta)$ and acts in the $-\mathbf{u}_r$ direction; i.e., $\Delta\mathbf{u}_\theta = -\Delta\theta\mathbf{u}_r$. Thus,

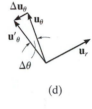

(d)

$$\dot{\mathbf{u}}_\theta = \lim_{\Delta t \to 0}\frac{\Delta\mathbf{u}_\theta}{\Delta t} = -\left(\lim_{\Delta t \to 0}\frac{\Delta\theta}{\Delta t}\right)\mathbf{u}_r$$

$$\dot{\mathbf{u}}_\theta = -\dot{\theta}\mathbf{u_r} \qquad (12\text{–}27)$$

Substituting this result and Eq. 12–23 into the above equation for **a**, we can write the acceleration in component form as

$$\boxed{\mathbf{a} = a_r\mathbf{u}_r + a_\theta\mathbf{u}_\theta} \qquad (12\text{–}28)$$

where

$$\boxed{\begin{aligned} a_r &= \ddot{r} - r\dot{\theta}^2 \\ a_\theta &= r\ddot{\theta} + 2\dot{r}\dot{\theta} \end{aligned}} \qquad (12\text{–}29)$$

The term $\ddot{\theta} = d^2\theta/dt^2 = d/dt(d\theta/dt)$ is called the *angular acceleration* since it measures the change made in the angular velocity during an instant of time. Units for this measurement are rad/s^2.

Since $\mathbf{a}_r$ and $\mathbf{a}_\theta$ are always perpendicular, the *magnitude* of acceleration is simply the positive value of

$$a = \sqrt{(\ddot{r} - r\dot{\theta}^2)^2 + (r\ddot{\theta} + 2\dot{r}\dot{\theta})^2} \qquad (12\text{–}30)$$

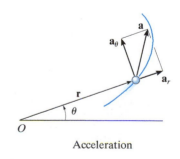

Acceleration

(e)

The *direction* is determined from the vector addition of its two components. In general, **a** will *not* be tangent to the path, Fig. 12–30e.

Fig. 12–31

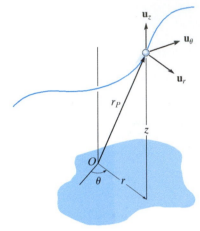

The spiral motion of this girl can be followed by using cylindrical components. Here the radial coordinate r is constant, the transverse coordinate θ will increase with time as the girl rotates about the vertical, and her altitude z will decrease with time. (© R.C. Hibbeler)

Cylindrical Coordinates.

If the particle moves along a space curve as shown in Fig. 12–31, then its location may be specified by the three *cylindrical coordinates, r, θ, z*. The z coordinate is identical to that used for rectangular coordinates. Since the unit vector defining its direction, $\mathbf{u}_z$, is constant, the time derivatives of this vector are zero, and therefore the position, velocity, and acceleration of the particle can be written in terms of its cylindrical coordinates as follows:

$$\mathbf{r}_P = r\mathbf{u}_r + z\mathbf{u}_z$$

$$\mathbf{v} = \dot{r}\mathbf{u}_r + r\dot{\theta}\mathbf{u}_\theta + \dot{z}\mathbf{u}_z \tag{12–31}$$

$$\mathbf{a} = (\ddot{r} - r\dot{\theta}^2)\mathbf{u}_r + (r\ddot{\theta} + 2\dot{r}\dot{\theta})\mathbf{u}_\theta + \ddot{z}\mathbf{u}_z \tag{12–32}$$

Time Derivatives.

The above equations require that we obtain the time derivatives $\dot{r}, \ddot{r}, \dot{\theta}$, and $\ddot{\theta}$ in order to evaluate the r and θ components of $\mathbf{v}$ and $\mathbf{a}$. Two types of problems generally occur:

1. If the polar coordinates are specified as time parametric equations, $r = r(t)$ and $\theta = \theta(t)$, then the time derivatives can be found directly.

2. If the time-parametric equations are not given, then the path $r = f(\theta)$ must be known. Using the chain rule of calculus we can then find the relation between $\dot{r}$ and $\dot{\theta}$, and between $\ddot{r}$ and $\ddot{\theta}$. Application of the chain rule, along with some examples, is explained in Appendix C.

Procedure for Analysis

Coordinate System.

- Polar coordinates are a suitable choice for solving problems when data regarding the angular motion of the radial coordinate r is given to describe the particle's motion. Also, some paths of motion can conveniently be described in terms of these coordinates.

- To use polar coordinates, the origin is established at a fixed point, and the radial line r is directed to the particle.

- The transverse coordinate θ is measured from a fixed reference line to the radial line.

Velocity and Acceleration.

- Once r and the four time derivatives $\dot{r}, \ddot{r}, \dot{\theta}$, and $\ddot{\theta}$ have been evaluated at the instant considered, their values can be substituted into Eqs. 12–25 and 12–29 to obtain the radial and transverse components of $\mathbf{v}$ and $\mathbf{a}$.

- If it is necessary to take the time derivatives of $r = f(\theta)$, then the chain rule of calculus must be used. See Appendix C.

- Motion in three dimensions requires a simple extension of the above procedure to include $\dot{z}$ and $\ddot{z}$.

EXAMPLE | 12.17

The amusement park ride shown in Fig. 12–32a consists of a chair that is rotating in a horizontal circular path of radius r such that the arm OB has an angular velocity $\dot\theta$ and angular acceleration $\ddot\theta$. Determine the radial and transverse components of velocity and acceleration of the passenger. Neglect his size in the calculation.

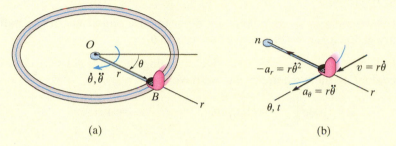

(a) (b)

Fig. 12–32

SOLUTION

Coordinate System.   Since the angular motion of the arm is reported, polar coordinates are chosen for the solution, Fig. 12–32a. Here θ is not related to r, since the radius is constant for all θ.

Velocity and Acceleration.   It is first necessary to specify the first and second time derivatives of r and θ. Since r is *constant*, we have

$$r = r \qquad \dot r = 0 \qquad \ddot r = 0$$

Thus,

$$v_r = \dot r = 0 \qquad\qquad Ans.$$
$$v_\theta = r\dot\theta \qquad\qquad Ans.$$
$$a_r = \ddot r - r\dot\theta^2 = -r\dot\theta^2 \qquad\qquad Ans.$$
$$a_\theta = r\ddot\theta + 2\dot r\dot\theta = r\ddot\theta \qquad\qquad Ans.$$

These results are shown in Fig. 12–32b.

NOTE: The n, t axes are also shown in Fig. 12–32b, which in this special case of circular motion happen to be *collinear* with the r and θ axes, respectively. Since $v = v_\theta = v_t = r\dot\theta$, then by comparison,

$$-a_r = a_n = \frac{v^2}{\rho} = \frac{(r\dot\theta)^2}{r} = r\dot\theta^2$$

$$a_\theta = a_t = \frac{dv}{dt} = \frac{d}{dt}(r\dot\theta) = \frac{dr}{dt}\dot\theta + r\frac{d\dot\theta}{dt} = 0 + r\ddot\theta$$

12

EXAMPLE | **12.18**

The rod OA in Fig. 12–33a rotates in the horizontal plane such that $\theta = (t^3)$ rad. At the same time, the collar B is sliding outward along OA so that $r = (100t^2)$ mm. If in both cases t is in seconds, determine the velocity and acceleration of the collar when $t = 1$ s.

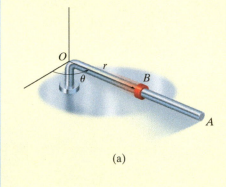

(a)

SOLUTION

Coordinate System. Since time-parametric equations of the path are given, it is not necessary to relate r to θ.

Velocity and Acceleration. Determining the time derivatives and evaluating them when $t = 1$ s, we have

$$r = 100t^2 \Big|_{t=1\,s} = 100 \text{ mm} \quad \theta = t^3 \Big|_{t=1\,s} = 1 \text{ rad} = 57.3°$$

$$\dot{r} = 200t \Big|_{t=1\,s} = 200 \text{ mm/s} \quad \dot{\theta} = 3t^2 \Big|_{t=1\,s} = 3 \text{ rad/s}$$

$$\ddot{r} = 200 \Big|_{t=1\,s} = 200 \text{ mm/s}^2 \quad \ddot{\theta} = 6t \Big|_{t=1\,s} = 6 \text{ rad/s}^2.$$

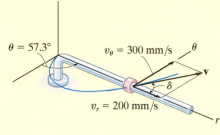

(b)

As shown in Fig. 12–33b,

$$\mathbf{v} = \dot{r}\mathbf{u}_r + r\dot{\theta}\mathbf{u}_\theta$$
$$= 200\mathbf{u}_r + 100(3)\mathbf{u}_\theta = \{200\mathbf{u}_r + 300\mathbf{u}_\theta\} \text{ mm/s}$$

The magnitude of $\mathbf{v}$ is

$$v = \sqrt{(200)^2 + (300)^2} = 361 \text{ mm/s} \qquad \textit{Ans.}$$

$$\delta = \tan^{-1}\left(\frac{300}{200}\right) = 56.3° \quad \delta + 57.3° = 114° \qquad \textit{Ans.}$$

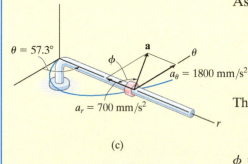

(c)

Fig. 12–33

As shown in Fig. 12–33c,

$$\mathbf{a} = (\ddot{r} - r\dot{\theta}^2)\mathbf{u}_r + (r\ddot{\theta} + 2\dot{r}\dot{\theta})\mathbf{u}_\theta$$
$$= [200 - 100(3)^2]\mathbf{u}_r + [100(6) + 2(200)3]\mathbf{u}_\theta$$
$$= \{-700\mathbf{u}_r + 1800\mathbf{u}_\theta\} \text{ mm/s}^2$$

The magnitude of $\mathbf{a}$ is

$$a = \sqrt{(-700)^2 + (1800)^2} = 1930 \text{ mm/s}^2 \qquad \textit{Ans.}$$

$$\phi = \tan^{-1}\left(\frac{1800}{700}\right) = 68.7° \quad (180° - \phi) + 57.3° = 169° \qquad \textit{Ans.}$$

NOTE: The velocity is tangent to the path; however, the acceleration is directed within the curvature of the path, as expected.

EXAMPLE   12.19

The searchlight in Fig. 12–34a casts a spot of light along the face of a wall that is located 100 m from the searchlight. Determine the magnitudes of the velocity and acceleration at which the spot appears to travel across the wall at the instant $\theta = 45°$. The searchlight rotates at a constant rate of $\dot{\theta} = 4$ rad/s.

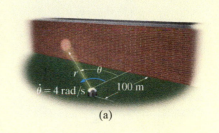

(a)

SOLUTION

Coordinate System.   Polar coordinates will be used to solve this problem since the angular rate of the searchlight is given. To find the necessary time derivatives it is first necessary to relate r to θ. From Fig. 12–34a,

$$r = 100/\cos\theta = 100 \sec\theta$$

Velocity and Acceleration.   Using the chain rule of calculus, noting that $d(\sec\theta) = \sec\theta\tan\theta\, d\theta$, and $d(\tan\theta) = \sec^2\theta\, d\theta$, we have

$$\dot{r} = 100(\sec\theta\tan\theta)\dot{\theta}$$
$$\ddot{r} = 100(\sec\theta\tan\theta)\dot{\theta}(\tan\theta)\dot{\theta} + 100\sec\theta(\sec^2\theta)\dot{\theta}(\dot{\theta})$$
$$+ 100\sec\theta\tan\theta(\ddot{\theta})$$
$$= 100\sec\theta\tan^2\theta\,(\dot{\theta})^2 + 100\sec^3\theta\,(\dot{\theta})^2 + 100(\sec\theta\tan\theta)\ddot{\theta}$$

Since $\dot{\theta} = 4$ rad/s = constant, then $\ddot{\theta} = 0$, and the above equations, when $\theta = 45°$, become

$$r = 100\sec 45° = 141.4$$
$$\dot{r} = 400\sec 45°\tan 45° = 565.7$$
$$\ddot{r} = 1600\,(\sec 45°\tan^2 45° + \sec^3 45°) = 6788.2$$

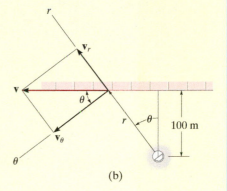

(b)

As shown in Fig. 12–34b,

$$\mathbf{v} = \dot{r}\mathbf{u}_r + r\dot{\theta}\mathbf{u}_\theta$$
$$= 565.7\mathbf{u}_r + 141.4(4)\mathbf{u}_\theta$$
$$= \{565.7\mathbf{u}_r + 565.7\mathbf{u}_\theta\}\ \text{m/s}$$
$$v = \sqrt{v_r^2 + v_\theta^2} = \sqrt{(565.7)^2 + (565.7)^2}$$
$$= 800\ \text{m/s} \qquad\qquad Ans.$$

As shown in Fig. 12–34c,

$$\mathbf{a} = (\ddot{r} - r\dot{\theta}^2)\mathbf{u}_r + (r\ddot{\theta} + 2\dot{r}\dot{\theta})\mathbf{u}_\theta$$
$$= [6788.2 - 141.4(4)^2]\mathbf{u}_r + [141.4(0) + 2(565.7)4]\mathbf{u}_\theta$$
$$= \{4525.5\mathbf{u}_r + 4525.5\mathbf{u}_\theta\}\ \text{m/s}^2$$
$$a = \sqrt{a_r^2 + a_\theta^2} = \sqrt{(4525.5)^2 + (4525.5)^2}$$
$$= 6400\ \text{m/s}^2 \qquad\qquad Ans.$$

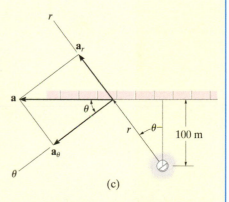

(c)

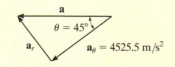

(d)

NOTE: It is also possible to find a without having to calculate $\ddot{r}$ (or a_r). As shown in Fig. 12–34d, since $a_\theta = 4525.5$ m/s^2, then by vector resolution, $a = 4525.5/\cos 45° = 6400$ m/s^2.

Fig. 12–34

12

EXAMPLE | 12.20

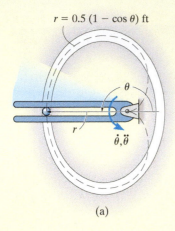

$r = 0.5 (1 - \cos \theta)$ ft

(a)

Due to the rotation of the forked rod, the ball in Fig. 12–35a travels around the slotted path, a portion of which is in the shape of a cardioid, $r = 0.5(1 - \cos \theta)$ ft, where θ is in radians. If the ball's velocity is $v = 4$ ft/s and its acceleration is $a = 30$ ft/s² at the instant $\theta = 180°$, determine the angular velocity $\dot{\theta}$ and angular acceleration $\ddot{\theta}$ of the fork.

SOLUTION

Coordinate System.   This path is most unusual, and mathematically it is best expressed using polar coordinates, as done here, rather than rectangular coordinates. Also, since $\dot{\theta}$ and $\ddot{\theta}$ must be determined, then r, θ coordinates are an obvious choice.

Velocity and Acceleration.   The time derivatives of r and θ can be determined using the chain rule.

$$r = 0.5(1 - \cos \theta)$$

$$\dot{r} = 0.5(\sin \theta)\dot{\theta}$$

$$\ddot{r} = 0.5(\cos \theta)\dot{\theta}(\dot{\theta}) + 0.5(\sin \theta)\ddot{\theta}$$

Evaluating these results at $\theta = 180°$, we have

$$r = 1 \text{ ft} \qquad \dot{r} = 0 \qquad \ddot{r} = -0.5\dot{\theta}^2$$

Since $v = 4$ ft/s, using Eq. 12–26 to determine $\dot{\theta}$ yields

$$v = \sqrt{(\dot{r})^2 + (r\dot{\theta})^2}$$

$$4 = \sqrt{(0)^2 + (1\dot{\theta})^2}$$

$$\dot{\theta} = 4 \text{ rad/s} \qquad\qquad\qquad Ans.$$

In a similar manner, $\ddot{\theta}$ can be found using Eq. 12–30.

$$a = \sqrt{(\ddot{r} - r\dot{\theta}^2)^2 + (r\ddot{\theta} + 2\dot{r}\dot{\theta})^2}$$

$$30 = \sqrt{[-0.5(4)^2 - 1(4)^2]^2 + [1\ddot{\theta} + 2(0)(4)]^2}$$

$$(30)^2 = (-24)^2 + \ddot{\theta}^2$$

$$\ddot{\theta} = 18 \text{ rad/s}^2 \qquad\qquad\qquad Ans.$$

Vectors **a** and **v** are shown in Fig. 12–35b.

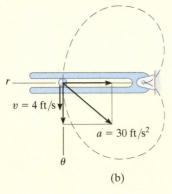

r

$v = 4$ ft/s

$a = 30$ ft/s²

θ

(b)

Fig. 12–35

NOTE: At this location, the θ and t (tangential) axes will coincide. The $+n$ (normal) axis is directed to the right, opposite to $+r$.

FUNDAMENTAL PROBLEMS

F12–33. The car has a speed of 55 ft/s. Determine the angular velocity $\dot{\theta}$ of the radial line OA at this instant.

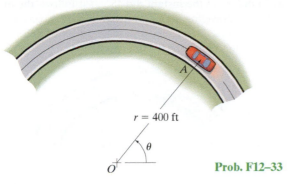

$r = 400$ ft

Prob. F12–33

F12–34. The platform is rotating about the vertical axis such that at any instant its angular position is $\theta = (4t^{3/2})$ rad, where t is in seconds. A ball rolls outward along the radial groove so that its position is $r = (0.1t^3)$ m, where t is in seconds. Determine the magnitudes of the velocity and acceleration of the ball when $t = 1.5$ s.

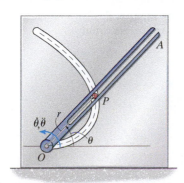

Prob. F12–34

F12–35. Peg P is driven by the fork link OA along the curved path described by $r = (2\theta)$ ft. At the instant $\theta = \pi/4$ rad, the angular velocity and angular acceleration of the link are $\dot{\theta} = 3$ rad/s and $\ddot{\theta} = 1$ rad/s². Determine the magnitude of the peg's acceleration at this instant.

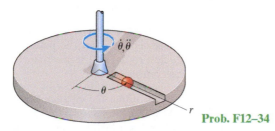

Prob. F12–35

F12–36. Peg P is driven by the forked link OA along the path described by $r = e^{\theta}$, where r is in meters. When $\theta = \frac{\pi}{4}$ rad, the link has an angular velocity and angular acceleration of $\dot{\theta} = 2$ rad/s and $\ddot{\theta} = 4$ rad/s². Determine the radial and transverse components of the peg's acceleration at this instant.

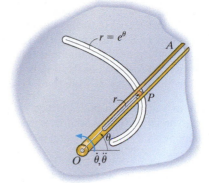

Prob. F12–36

F12–37. The collars are pin connected at B and are free to move along rod OA and the curved guide OC having the shape of a cardioid, $r = [0.2(1 + \cos \theta)]$ m. At $\theta = 30°$, the angular velocity of OA is $\dot{\theta} = 3$ rad/s. Determine the magnitude of the velocity of the collars at this point.

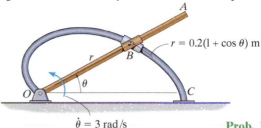

$\dot{\theta} = 3$ rad/s   **Prob. F12–37**

F12–38. At the instant $\theta = 45°$, the athlete is running with a constant speed of 2 m/s. Determine the angular velocity at which the camera must turn in order to follow the motion.

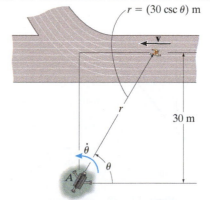

Prob. F12–38

PROBLEMS

12–155. A particle is moving along a circular path having a radius of 4 in. such that its position as a function of time is given by $\theta = \cos 2t$, where θ is in radians and t is in seconds. Determine the magnitude of the acceleration of the particle when $\theta = 30°$.

*12–156. For a short time a rocket travels up and to the right at a constant speed of 800 m/s along the parabolic path $y = 600 - 35x^2$. Determine the radial and transverse components of velocity of the rocket at the instant $\theta = 60°$, where θ is measured counterclockwise from the x axis.

12–157. A particle moves along a path defined by polar coordinates $r = (2e^t)$ ft and $\theta = (8t^2)$ rad, where t is in seconds. Determine the components of its velocity and acceleration when $t = 1$ s.

12–158. An airplane is flying in a straight line with a velocity of 200 mi/h and an acceleration of 3 mi/h². If the propeller has a diameter of 6 ft and is rotating at a constant angular rate of 120 rad/s, determine the magnitudes of velocity and acceleration of a particle located on the tip of the propeller.

12–159. The small washer is sliding down the cord OA. When it is at the midpoint, its speed is 28 m/s and its acceleration is 7 m/s². Express the velocity and acceleration of the washer at this point in terms of its cylindrical components.

*12–160. A radar gun at O rotates with the angular velocity of $\dot\theta = 0.1$ rad/s and angular acceleration of $\ddot\theta = 0.025$ rad/s², at the instant $\theta = 45°$, as it follows the motion of the car traveling along the circular road having a radius of $r = 200$ m. Determine the magnitudes of velocity and acceleration of the car at this instant.

$r = 200$ m

θ

O

Prob. 12–160

12–161. If a particle moves along a path such that $r = (2 \cos t)$ ft and $\theta = (t/2)$ rad, where t is in seconds, plot the path $r = f(\theta)$ and determine the particle's radial and transverse components of velocity and acceleration.

12–162. If a particle moves along a path such that $r = (e^{at})$ m and $\theta = t$, where t is in seconds, plot the path $r = f(\theta)$, and determine the particle's radial and transverse components of velocity and acceleration.

12–163. The car travels along the circular curve having a radius $r = 400$ ft. At the instant shown, its angular rate of rotation is $\dot\theta = 0.025$ rad/s, which is decreasing at the rate $\ddot\theta = -0.008$ rad/s². Determine the radial and transverse components of the car's velocity and acceleration at this instant and sketch these components on the curve.

*12–164. The car travels along the circular curve of radius $r = 400$ ft with a constant speed of $v = 30$ ft/s. Determine the angular rate of rotation $\dot\theta$ of the radial line r and the magnitude of the car's acceleration.

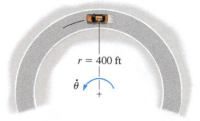

$r = 400$ ft

$\dot\theta$

Probs. 12–163/164

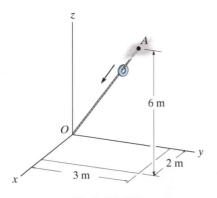

z

A

6 m

O

y

2 m

3 m

x

Prob. 12–159

12–165. The time rate of change of acceleration is referred to as the *jerk*, which is often used as a means of measuring passenger discomfort. Calculate this vector, $\dot{\mathbf{a}}$, in terms of its cylindrical components, using Eq. 12–32.

12–166. A particle is moving along a circular path having a radius of 6 in. such that its position as a function of time is given by $\theta = \sin 3t$, where θ and the argument for the sine are in radians, and t is in seconds. Determine the magnitude of the acceleration of the particle at $\theta = 30°$. The particle starts from rest at $\theta = 0°$.

12–167. The slotted link is pinned at O, and as a result of the constant angular velocity $\dot{\theta} = 3$ rad/s it drives the peg P for a short distance along the spiral guide $r = (0.4\,\theta)$ m, where θ is in radians. Determine the radial and transverse components of the velocity and acceleration of P at the instant $\theta = \pi/3$ rad.

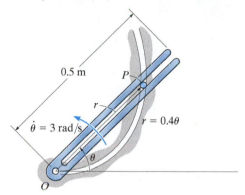

Prob. 12–167

*12–168. For a short time the bucket of the backhoe traces the path of the cardioid $r = 25(1 - \cos\theta)$ ft. Determine the magnitudes of the velocity and acceleration of the bucket when $\theta = 120°$ if the boom is rotating with an angular velocity of $\dot{\theta} = 2$ rad/s and an angular acceleration of $\ddot{\theta} = 0.2$ rad/s² at the instant shown.

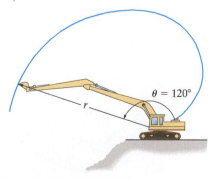

Prob. 12–168

12–169. The slotted link is pinned at O, and as a result of the constant angular velocity $\dot{\theta} = 3$ rad/s it drives the peg P for a short distance along the spiral guide $r = (0.4\,\theta)$ m, where θ is in radians. Determine the velocity and acceleration of the particle at the instant it leaves the slot in the link, i.e., when $r = 0.5$ m.

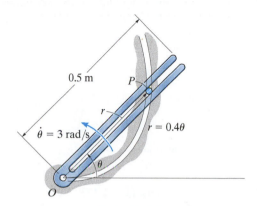

Prob. 12–169

12–170. A particle moves in the $x-y$ plane such that its position is defined by $r = \{2t\mathbf{i} + 4t^2\mathbf{j}\}$ ft, where t is in seconds. Determine the radial and transverse components of the particle's velocity and acceleration when $t = 2$ s.

12–171. At the instant shown, the man is twirling a hose over his head with an angular velocity $\dot{\theta} = 2$ rad/s and an angular acceleration $\ddot{\theta} = 3$ rad/s². If it is assumed that the hose lies in a horizontal plane, and water is flowing through it at a constant rate of 3 m/s, determine the magnitudes of the velocity and acceleration of a water particle as it exits the open end, $r = 1.5$ m.

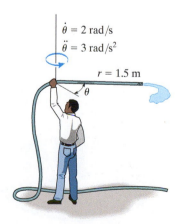

Prob. 12–171

*12–172. The rod OA rotates clockwise with a constant angular velocity of 6 rad/s. Two pin-connected slider blocks, located at B, move freely on OA and the curved rod whose shape is a limaçon described by the equation $r = 200(2 - \cos \theta)$ mm. Determine the speed of the slider blocks at the instant $\theta = 150°$.

12–173. Determine the magnitude of the acceleration of the slider blocks in Prob. 12–172 when $\theta = 150°$.

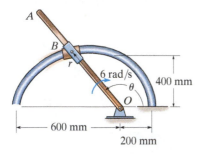

Probs. 12–172/173

12–174. A double collar C is pin connected together such that one collar slides over a fixed rod and the other slides over a rotating rod. If the geometry of the fixed rod for a short distance can be defined by a lemniscate, $r^2 = (4 \cos 2\theta)$ ft^2, determine the collar's radial and transverse components of velocity and acceleration at the instant $\theta = 0°$ as shown. Rod OA is rotating at a constant rate of $\dot\theta = 6$ rad/s.

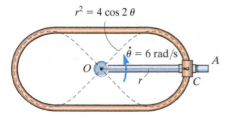

Prob. 12–174

12–175. A block moves outward along the slot in the platform with a speed of $\dot r = (4t)$ m/s, where t is in seconds. The platform rotates at a constant rate of 6 rad/s. If the block starts from rest at the center, determine the magnitudes of its velocity and acceleration when $t = 1$ s.

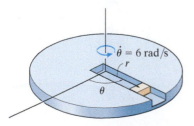

Prob. 12–175

*12–176. The car travels around the circular track with a constant speed of 20 m/s. Determine the car's radial and transverse components of velocity and acceleration at the instant $\theta = \pi/4$ rad.

12–177. The car travels around the circular track such that its transverse component is $\theta = (0.006t^2)$ rad, where t is in seconds. Determine the car's radial and transverse components of velocity and acceleration at the instant $t = 4$ s.

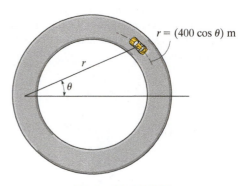

Probs. 12–176/177

12–178. The car travels along a road which for a short distance is defined by $r = (200/\theta)$ ft, where θ is in radians. If it maintains a constant speed of $v = 35$ ft/s, determine the radial and transverse components of its velocity when $\theta = \pi/3$ rad.

Prob. 12–178

12–179. A horse on the merry-go-round moves according to the equations $r = 8$ ft, $\theta = (0.6t)$ rad, and $z = (1.5 \sin \theta)$ ft, where t is in seconds. Determine the cylindrical components of the velocity and acceleration of the horse when $t = 4$ s.

***12–180.** A horse on the merry-go-round moves according to the equations $r = 8$ ft, $\dot\theta = 2$ rad/s and $z = (1.5 \sin \theta)$ ft, where t is in seconds. Determine the maximum and minimum magnitudes of the velocity and acceleration of the horse during the motion.

12–183. A truck is traveling along the horizontal circular curve of radius $r = 60$ m with a constant speed $v = 20$ m/s. Determine the angular rate of rotation $\dot\theta$ of the radial line r and the magnitude of the truck's acceleration.

***12–184.** A truck is traveling along the horizontal circular curve of radius $r = 60$ m with a speed of 20 m/s which is increasing at 3 m/s^2, Determine the truck's radial and transverse components of acceleration.

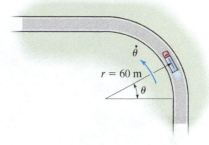

Probs. 12–183/184

Probs. 12–179/180

12–181. If the slotted arm AB rotates counterclockwise with a constant angular velocity of $\dot\theta = 2$ rad/s, determine the magnitudes of the velocity and acceleration of peg P at $\theta = 30°$. The peg is constrained to move in the slots of the fixed bar CD and rotating bar AB.

12–182. The peg is constrained to move in the slots of the fixed bar CD and rotating bar AB. When $\theta = 30°$, the angular velocity and angular acceleration of arm AB are $\dot\theta = 2$ rad/s and $\ddot\theta = 3$ rad/s^2, respectively. Determine the magnitudes of the velocity and acceleration of the peg P at this instant.

12–185. The rod OA rotates counterclockwise with a constant angular velocity of $\dot\theta = 5$ rad/s. Two pin-connected slider blocks, located at B, move freely on OA and the curved rod whose shape is a limaçon described by the equation $r = 100(2 - \cos \theta)$ mm. Determine the speed of the slider blocks at the instant $\theta = 120°$.

12–186. Determine the magnitude of the acceleration of the slider blocks in Prob. 12–185 when $\theta = 120°$.

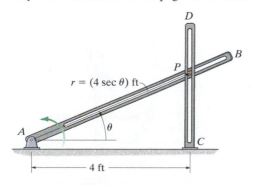

Probs. 12–181/182

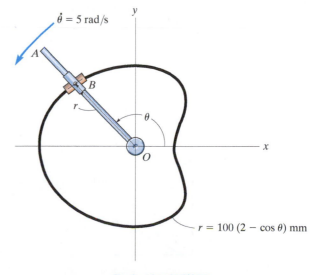

Probs. 12–185/186

12–187. The searchlight on the boat anchored 2000 ft from shore is turned on the automobile, which is traveling along the straight road at a constant speed 80 ft/s, Determine the angular rate of rotation of the light when the automobile is $r = 3000$ ft from the boat.

*12–188. If the car in Prob. 12–187 is accelerating at 15 ft/s² at the instant $r = 3000$ ft determine the required angular acceleration $\ddot{\theta}$ of the light at this instant.

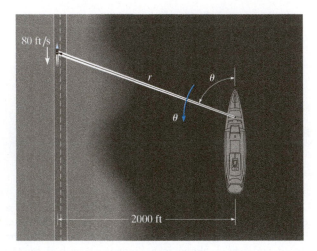

80 ft/s

r

θ

θ

2000 ft

Probs. 12–187/188

12–189. A particle moves along an Archimedean spiral $r = (8\theta)$ ft, where θ is given in radians. If $\dot{\theta} = 4$ rad/s (constant), determine the radial and transverse components of the particle's velocity and acceleration at the instant $\theta = \pi/2$ rad. Sketch the curve and show the components on the curve.

12–190. Solve Prob. 12–189 if the particle has an angular acceleration $\ddot{\theta} = 5$ rad/s² when $\dot{\theta} = 4$ rad/s at $\theta = \pi/2$ rad.

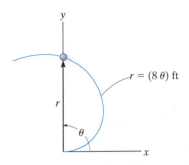

y

$r = (8\,\theta)$ ft

r

θ

x

Probs. 12–189/190

12–191. The arm of the robot moves so that $r = 3$ ft is constant, and its grip A moves along the path $z = (3 \sin 4\theta)$ ft, where θ is in radians. If $\theta = (0.5t)$ rad, where t is in seconds, determine the magnitudes of the grip's velocity and acceleration when $t = 3$ s.

*12–192. For a short time the arm of the robot is extending such that $\dot{r} = 1.5$ ft/s when $r = 3$ ft, $z = (4t^2)$ ft, and $\theta = 0.5t$ rad, where t is in seconds. Determine the magnitudes of the velocity and acceleration of the grip A when $t = 3$ s.

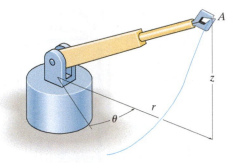

A

z

r

θ

Probs. 12–191/192

12–193. The double collar C is pin connected together such that one collar slides over the fixed rod and the other slides over the rotating rod AB. If the angular velocity of AB is given as $\dot{\theta} = (e^{0.5\,t^2})$ rad/s, where t is in seconds, and the path defined by the fixed rod is $r = |(0.4 \sin \theta + 0.2)|$ m, determine the radial and transverse components of the collar's velocity and acceleration when $t = 1$ s. When $t = 0, \theta = 0$. Use Simpson's rule with $n = 50$ to determine θ at $t = 1$ s.

12–194. The double collar C is pin connected together such that one collar slides over the fixed rod and the other slides over the rotating rod AB. If the mechanism is to be designed so that the largest speed given to the collar is 6 m/s, determine the required constant angular velocity $\dot{\theta}$ of rod AB. The path defined by the fixed rod is $r = (0.4 \sin \theta + 0.2)$ m.

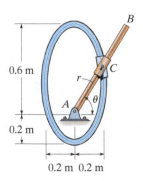

B

0.6 m

r

C

A

θ

0.2 m

0.2 m 0.2 m

Probs. 12–193/194

12.9   Absolute Dependent Motion Analysis of Two Particles

In some types of problems the motion of one particle will *depend* on the corresponding motion of another particle. This dependency commonly occurs if the particles, here represented by blocks, are interconnected by inextensible cords which are wrapped around pulleys. For example, the movement of block A downward along the inclined plane in Fig. 12–36 will cause a corresponding movement of block B up the other incline. We can show this mathematically by first specifying the location of the blocks using *position coordinates* s_A and s_B. Note that each of the coordinate axes is (1) measured from a *fixed* point (O) or *fixed* datum line, (2) measured along each inclined plane *in the direction of motion* of each block, and (3) has a positive sense from the fixed datums to A and to B. If the total cord length is l_T, the two position coordinates are related by the equation

$$s_A + l_{CD} + s_B = l_T$$

Here l_{CD} is the length of the cord passing over arc CD. Taking the time derivative of this expression, realizing that l_{CD} and l_T remain *constant*, while s_A and s_B measure the segments of the cord that change in length, we have

$$\frac{ds_A}{dt} + \frac{ds_B}{dt} = 0 \qquad \text{or} \qquad v_B = -v_A$$

The negative sign indicates that when block A has a velocity downward, i.e., in the direction of positive s_A, it causes a corresponding upward velocity of block B; i.e., B moves in the negative s_B direction.

In a similar manner, time differentiation of the velocities yields the relation between the accelerations, i.e.,

$$a_B = -a_A$$

A more complicated example is shown in Fig. 12–37a. In this case, the position of block A is specified by s_A, and the position of the *end* of the cord from which block B is suspended is defined by s_B. As above, we have chosen position coordinates which (1) have their origin at fixed points or datums, (2) are measured in the direction of motion of each block, and (3) from the fixed datums are positive to the right for s_A and positive downward for s_B. During the motion, the length of the red colored segments of the cord in Fig. 12–37a *remains constant*. If l represents the total length of cord minus these segments, then the position coordinates can be related by the equation

$$2s_B + h + s_A = l$$

Since l and h are constant during the motion, the two time derivatives yield

$$2v_B = -v_A \qquad 2a_B = -a_A$$

Hence, when B moves downward ($+s_B$), A moves to the left ($-s_A$) with twice the motion.

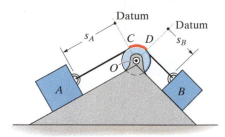

Fig. 12–36

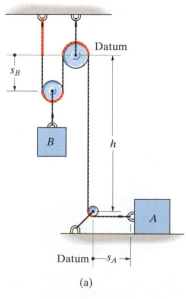

(a)

Fig. 12–37

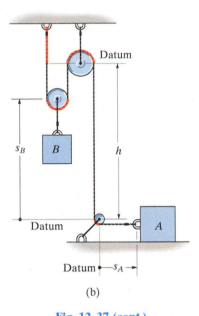

Fig. 12–37 (cont.)

This example can also be worked by defining the position of block B from the center of the bottom pulley (a fixed point), Fig. 12–37b. In this case

$$2(h - s_B) + h + s_A = l$$

Time differentiation yields

$$2v_B = v_A \qquad 2a_B = a_A$$

Here the signs are the same. Why?

Procedure for Analysis

The above method of relating the dependent motion of one particle to that of another can be performed using algebraic scalars or position coordinates provided each particle moves along a rectilinear path. When this is the case, only the magnitudes of the velocity and acceleration of the particles will change, not their line of direction.

Position-Coordinate Equation.

- Establish each position coordinate with an origin located at a *fixed* point or datum.

- It is *not necessary* that the *origin* be the *same* for each of the coordinates; however, it is *important* that each coordinate axis selected be directed along the *path of motion* of the particle.

- Using geometry or trigonometry, relate the position coordinates to the total length of the cord, l_T, or to that portion of cord, l, which *excludes* the segments that do not change length as the particles move—such as arc segments wrapped over pulleys.

- If a problem involves a *system* of two or more cords wrapped around pulleys, then the position of a point on one cord must be related to the position of a point on another cord using the above procedure. Separate equations are written for a fixed length of each cord of the system and the positions of the two particles are then related by these equations (see Examples 12.22 and 12.23).

Time Derivatives.

- Two successive time derivatives of the position-coordinate equations yield the required velocity and acceleration equations which relate the motions of the particles.

- The signs of the terms in these equations will be consistent with those that specify the positive and negative sense of the position coordinates.

The cable is wrapped around the pulleys on this crane in order to reduce the required force needed to hoist a load. (© R.C. Hibbeler)

EXAMPLE | 12.21

Determine the speed of block A in Fig. 12–38 if block B has an upward speed of 6 ft/s.

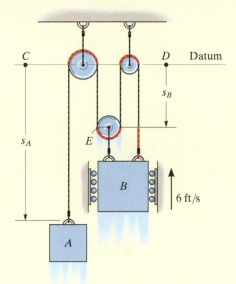

Fig. 12–38

SOLUTION

Position-Coordinate Equation.   There is *one cord* in this system having segments which change length. Position coordinates s_A and s_B will be used since each is measured from a fixed point (C or D) and extends along each block's *path of motion*. In particular, s_B is directed to point E since motion of B and E is the *same*.

The red colored segments of the cord in Fig. 12–38 remain at a constant length and do not have to be considered as the blocks move. The remaining length of cord, l, is also constant and is related to the changing position coordinates s_A and s_B by the equation

$$s_A + 3s_B = l$$

Time Derivative.   Taking the time derivative yields

$$v_A + 3v_B = 0$$

so that when $v_B = -6$ ft/s (upward),

$$v_A = 18 \text{ ft/s} \downarrow \qquad \qquad \textit{Ans.}$$

EXAMPLE   12.22

Determine the speed of A in Fig. 12–39 if B has an upward speed of 6 ft/s.

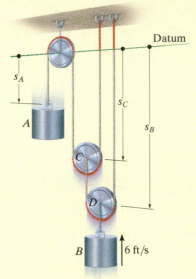

Fig. 12–39

SOLUTION

Position-Coordinate Equation.   As shown, the positions of blocks A and B are defined using coordinates s_A and s_B. Since the system has *two cords* with segments that change length, it will be necessary to use a third coordinate, s_C, in order to relate s_A to s_B. In other words, the length of one of the cords can be expressed in terms of s_A and s_C, and the length of the other cord can be expressed in terms of s_B and s_C.

The red colored segments of the cords in Fig. 12–39 do not have to be considered in the analysis. Why? For the remaining cord lengths, say l_1 and l_2, we have

$$s_A + 2s_C = l_1 \qquad s_B + (s_B - s_C) = l_2$$

Time Derivative.   Taking the time derivative of these equations yields

$$v_A + 2v_C = 0 \qquad 2v_B - v_C = 0$$

Eliminating v_C produces the relationship between the motions of each cylinder.

$$v_A + 4v_B = 0$$

so that when $v_B = -6$ ft/s (upward),

$$v_A = +24 \text{ ft/s} = 24 \text{ ft/s} \downarrow \qquad \qquad Ans.$$

EXAMPLE | **12.23**

Determine the speed of block B in Fig. 12–40 if the end of the cord at A is pulled down with a speed of 2 m/s.

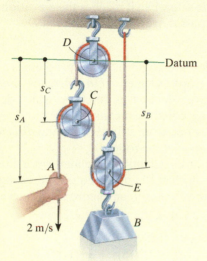

Fig. 12–40

SOLUTION

Position-Coordinate Equation.   The position of point A is defined by s_A, and the position of block B is specified by s_B since point E on the pulley will have the *same motion* as the block. Both coordinates are measured from a horizontal datum passing through the *fixed* pin at pulley D. Since the system consists of *two* cords, the coordinates s_A and s_B cannot be related directly. Instead, by establishing a third position coordinate, s_C, we can now express the length of one of the cords in terms of s_B and s_C, and the length of the other cord in terms of s_A, s_B, and s_C.

Excluding the red colored segments of the cords in Fig. 12–40, the remaining constant cord lengths l_1 and l_2 (along with the hook and link dimensions) can be expressed as

$$s_C + s_B = l_1$$
$$(s_A - s_C) + (s_B - s_C) + s_B = l_2$$

Time Derivative.   The time derivative of each equation gives

$$v_C + v_B = 0$$
$$v_A - 2v_C + 2v_B = 0$$

Eliminating v_C, we obtain

$$v_A + 4v_B = 0$$

so that when $v_A = 2$ m/s (downward),

$$v_B = -0.5 \text{ m/s} = 0.5 \text{ m/s} \uparrow \qquad \qquad Ans.$$

EXAMPLE | **12.24**

A man at A is hoisting a safe S as shown in Fig. 12–41 by walking to the right with a constant velocity $v_A = 0.5$ m/s. Determine the velocity and acceleration of the safe when it reaches the elevation of 10 m. The rope is 30 m long and passes over a small pulley at D.

SOLUTION

Position-Coordinate Equation. This problem is unlike the previous examples since rope segment DA changes *both direction and magnitude*. However, the ends of the rope, which define the positions of C and A, are specified by means of the x and y coordinates since they must be measured from a fixed point and *directed along the paths of motion* of the ends of the rope.

The x and y coordinates may be related since the rope has a fixed length $l = 30$ m, which at all times is equal to the length of segment DA plus CD. Using the Pythagorean theorem to determine l_{DA}, we have $l_{DA} = \sqrt{(15)^2 + x^2}$; also, $l_{CD} = 15 - y$. Hence,

$$l = l_{DA} + l_{CD}$$
$$30 = \sqrt{(15)^2 + x^2} + (15 - y)$$
$$y = \sqrt{225 + x^2} - 15 \tag{1}$$

Time Derivatives. Taking the time derivative, using the chain rule (see Appendix C), where $v_S = dy/dt$ and $v_A = dx/dt$, yields

$$v_S = \frac{dy}{dt} = \left[\frac{1}{2} \frac{2x}{\sqrt{225 + x^2}}\right] \frac{dx}{dt}$$

$$= \frac{x}{\sqrt{225 + x^2}} v_A \tag{2}$$

At $y = 10$ m, x is determined from Eq. 1, i.e., $x = 20$ m. Hence, from Eq. 2 with $v_A = 0.5$ m/s,

$$v_S = \frac{20}{\sqrt{225 + (20)^2}}(0.5) = 0.4 \text{ m/s} = 400 \text{ mm/s} \uparrow \qquad Ans.$$

The acceleration is determined by taking the time derivative of Eq. 2. Since v_A is constant, then $a_A = dv_A/dt = 0$, and we have

$$a_S = \frac{d^2y}{dt^2} = \left[\frac{-x(dx/dt)}{(225 + x^2)^{3/2}}\right]xv_A + \left[\frac{1}{\sqrt{225 + x^2}}\right]\left(\frac{dx}{dt}\right)v_A + \left[\frac{1}{\sqrt{225 + x^2}}\right]x\frac{dv_A}{dt} = \frac{225v_A^2}{(225 + x^2)^{3/2}}$$

At $x = 20$ m, with $v_A = 0.5$ m/s, the acceleration becomes

$$a_S = \frac{225(0.5 \text{ m/s})^2}{[225 + (20 \text{ m})^2]^{3/2}} = 0.00360 \text{ m/s}^2 = 3.60 \text{ mm/s}^2 \uparrow \qquad Ans.$$

NOTE: The constant velocity at A causes the other end C of the rope to have an acceleration since $\mathbf{v}_A$ causes segment DA to change its direction as well as its length.

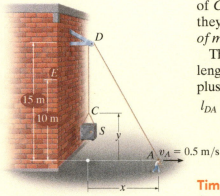

Fig. 12–41

12.10   Relative-Motion of Two Particles Using Translating Axes

Throughout this chapter the absolute motion of a particle has been determined using a single fixed reference frame. There are many cases, however, where the path of motion for a particle is complicated, so that it may be easier to analyze the motion in parts by using two or more frames of reference. For example, the motion of a particle located at the tip of an airplane propeller, while the plane is in flight, is more easily described if one observes first the motion of the airplane from a fixed reference and then superimposes (vectorially) the circular motion of the particle measured from a reference attached to the airplane.

In this section *translating frames of reference* will be considered for the analysis.

Position.   Consider particles A and B, which move along the arbitrary paths shown in Fig. 12–42. The *absolute position* of each particle, $\mathbf{r}_A$ and $\mathbf{r}_B$, is measured from the common origin O of the *fixed* x, y, z reference frame. The origin of a second frame of reference x', y', z' is attached to and moves with particle A. The axes of this frame are *only permitted to translate* relative to the fixed frame. The position of B measured relative to A is denoted by the *relative-position vector* $\mathbf{r}_{B/A}$. Using vector addition, the three vectors shown in Fig. 12–42 can be related by the equation

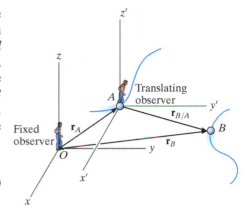

Fig. 12–42

$$\mathbf{r}_B = \mathbf{r}_A + \mathbf{r}_{B/A} \qquad (12\text{–}33)$$

Velocity.   An equation that relates the velocities of the particles is determined by taking the time derivative of the above equation; i.e.,

$$\mathbf{v}_B = \mathbf{v}_A + \mathbf{v}_{B/A} \qquad (12\text{–}34)$$

Here $\mathbf{v}_B = d\mathbf{r}_B/dt$ and $\mathbf{v}_A = d\mathbf{r}_A/dt$ refer to *absolute velocities*, since they are observed from the fixed frame; whereas the *relative velocity* $\mathbf{v}_{B/A} = d\mathbf{r}_{B/A}/dt$ is observed from the translating frame. It is important to note that since the x', y', z' axes translate, the *components* of $\mathbf{r}_{B/A}$ will *not* change direction and therefore the time derivative of these components will only have to account for the change in their magnitudes. Equation 12–34 therefore states that the velocity of B is equal to the velocity of A plus (vectorially) the velocity of "B with respect to A," as measured by the *translating observer* fixed in the x', y', z' reference frame.

Acceleration.   The time derivative of Eq. 12–34 yields a similar vector relation between the *absolute* and *relative accelerations* of particles *A* and *B*.

$$\mathbf{a}_B = \mathbf{a}_A + \mathbf{a}_{B/A} \qquad (12\text{–}35)$$

Here $\mathbf{a}_{B/A}$ is the acceleration of *B* as seen by the observer located at *A* and translating with the x', y', z' reference frame.*

Procedure for Analysis

- When applying the relative velocity and acceleration equations, it is first necessary to specify the particle *A* that is the origin for the translating x', y', z' axes. Usually this point has a *known* velocity or acceleration.

- Since vector addition forms a triangle, there can be at most *two unknowns*, represented by the magnitudes and/or directions of the vector quantities.

- These unknowns can be solved for either graphically, using trigonometry (law of sines, law of cosines), or by resolving each of the three vectors into rectangular or Cartesian components, thereby generating a set of scalar equations.

The pilots of these close-flying planes must be aware of their relative positions and velocities at all times in order to avoid a collision. (© R.C. Hibbeler)

---

* An easy way to remember the setup of these equations is to note the "cancellation" of the subscript *A* between the two terms, e.g., $\mathbf{a}_B = \mathbf{a}_{\cancel{A}} + \mathbf{a}_{B/\cancel{A}}$.

EXAMPLE 12.25

A train travels at a constant speed of 60 mi/h and crosses over a road as shown in Fig. 12–43a. If the automobile A is traveling at 45 mi/h along the road, determine the magnitude and direction of the velocity of the train relative to the automobile.

SOLUTION I

Vector Analysis.   The relative velocity $\mathbf{v}_{T/A}$ is measured from the translating x', y' axes attached to the automobile, Fig. 12–43a. It is determined from $\mathbf{v}_T = \mathbf{v}_A + \mathbf{v}_{T/A}$. Since $\mathbf{v}_T$ and $\mathbf{v}_A$ are known in *both* magnitude and direction, the unknowns become the x and y components of $\mathbf{v}_{T/A}$. Using the x, y axes in Fig. 12–43a, we have

$$\mathbf{v}_T = \mathbf{v}_A + \mathbf{v}_{T/A}$$

$$60\mathbf{i} = (45 \cos 45°\mathbf{i} + 45 \sin 45°\mathbf{j}) + \mathbf{v}_{T/A}$$

$$\mathbf{v}_{T/A} = \{28.2\mathbf{i} - 31.8\mathbf{j}\} \text{ mi/h}$$

The magnitude of $\mathbf{v}_{T/A}$ is thus

$$v_{T/A} = \sqrt{(28.2)^2 + (-31.8)^2} = 42.5 \text{ mi/h} \qquad \textit{Ans.}$$

From the direction of each component, Fig. 12–43b, the direction of $\mathbf{v}_{T/A}$ is

$$\tan \theta = \frac{(v_{T/A})_y}{(v_{T/A})_x} = \frac{31.8}{28.2}$$

$$\theta = 48.5° \searrow \qquad \textit{Ans.}$$

Note that the vector addition shown in Fig. 12–43b indicates the correct sense for $\mathbf{v}_{T/A}$. This figure anticipates the answer and can be used to check it.

SOLUTION II

Scalar Analysis.   The unknown components of $\mathbf{v}_{T/A}$ can also be determined by applying a scalar analysis. We will assume these components act in the *positive* x and y directions. Thus,

$$\mathbf{v}_T = \mathbf{v}_A + \mathbf{v}_{T/A}$$

$$\begin{bmatrix} 60 \text{ mi/h} \\ \rightarrow \end{bmatrix} = \begin{bmatrix} 45 \text{ mi/h} \\ \nearrow 45° \end{bmatrix} + \begin{bmatrix} (v_{T/A})_x \\ \rightarrow \end{bmatrix} + \begin{bmatrix} (v_{T/A})_y \\ \uparrow \end{bmatrix}$$

Resolving each vector into its x and y components yields

$$(\xrightarrow{+}) \qquad 60 = 45 \cos 45° + (v_{T/A})_x + 0$$

$$(+\uparrow) \qquad 0 = 45 \sin 45° + 0 + (v_{T/A})_y$$

Solving, we obtain the previous results,

$$(v_{T/A})_x = 28.2 \text{ mi/h} = 28.2 \text{ mi/h} \rightarrow$$

$$(v_{T/A})_y = -31.8 \text{ mi/h} = 31.8 \text{ mi/h} \downarrow$$

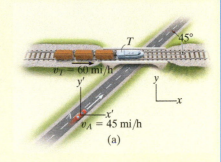

(a)

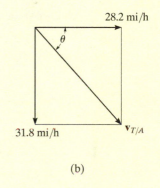

(b)

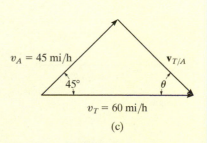

(c)

Fig. 12–43

12

EXAMPLE   12.26

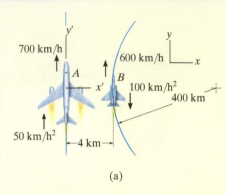

(a)

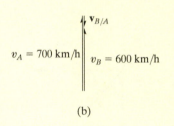

(b)

Plane A in Fig. 12–44a is flying along a straight-line path, whereas plane B is flying along a circular path having a radius of curvature of $\rho_B = 400$ km. Determine the velocity and acceleration of B as measured by the pilot of A.

SOLUTION

Velocity.   The origin of the x and y axes are located at an arbitrary fixed point. Since the motion relative to plane A is to be determined, the *translating frame of reference x', y'* is attached to it, Fig. 12–44a. Applying the relative-velocity equation in scalar form since the velocity vectors of both planes are parallel at the instant shown, we have

$$(+\uparrow) \qquad\qquad v_B = v_A + v_{B/A}$$

$$600 \text{ km/h} = 700 \text{ km/h} + v_{B/A}$$

$$v_{B/A} = -100 \text{ km/h} = 100 \text{ km/h} \downarrow \qquad Ans.$$

The vector addition is shown in Fig. 12–44b.

Acceleration.   Plane B has both tangential and normal components of acceleration since it is flying along a *curved path*. From Eq. 12–20, the magnitude of the normal component is

$$(a_B)_n = \frac{v_B^2}{\rho} = \frac{(600 \text{ km/h})^2}{400 \text{ km}} = 900 \text{ km/h}^2$$

Applying the relative-acceleration equation gives

$$\mathbf{a}_B = \mathbf{a}_A + \mathbf{a}_{B/A}$$

$$900\mathbf{i} - 100\mathbf{j} = 50\mathbf{j} + \mathbf{a}_{B/A}$$

Thus,

$$\mathbf{a}_{B/A} = \{900\mathbf{i} - 150\mathbf{j}\} \text{ km/h}^2$$

From Fig. 12–44c, the magnitude and direction of $\mathbf{a}_{B/A}$ are therefore

$$a_{B/A} = 912 \text{ km/h}^2 \quad \theta = \tan^{-1}\frac{150}{900} = 9.46° \quad \text{↘} \qquad Ans.$$

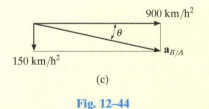

(c)

Fig. 12–44

NOTE: The solution to this problem was possible using a translating frame of reference, since the pilot in plane A is "translating." Observation of the motion of plane A with respect to the pilot of plane B, however, must be obtained using a *rotating* set of axes attached to plane B. (This assumes, of course, that the pilot of B is fixed in the rotating frame, so he does not turn his eyes to follow the motion of A.) The analysis for this case is given in Example 16.21.

EXAMPLE 12.27

At the instant shown in Fig. 12–45a, cars A and B are traveling with speeds of 18 m/s and 12 m/s, respectively. Also at this instant, A has a decrease in speed of 2 m/s², and B has an increase in speed of 3 m/s². Determine the velocity and acceleration of B with respect to A.

SOLUTION

Velocity. The fixed x, y axes are established at an arbitrary point on the ground and the translating x', y' axes are attached to car A, Fig. 12–45a. Why? The relative velocity is determined from $\mathbf{v}_B = \mathbf{v}_A + \mathbf{v}_{B/A}$. What are the two unknowns? Using a Cartesian vector analysis, we have

$$\mathbf{v}_B = \mathbf{v}_A + \mathbf{v}_{B/A}$$

$$-12\mathbf{j} = (-18\cos 60°\mathbf{i} - 18\sin 60°\mathbf{j}) + \mathbf{v}_{B/A}$$

$$\mathbf{v}_{B/A} = \{9\mathbf{i} + 3.588\mathbf{j}\}\ \text{m/s}$$

Thus,

$$v_{B/A} = \sqrt{(9)^2 + (3.588)^2} = 9.69\ \text{m/s} \qquad \textit{Ans.}$$

Noting that $\mathbf{v}_{B/A}$ has $+\mathbf{i}$ and $+\mathbf{j}$ components, Fig. 12–45b, its direction is

$$\tan\theta = \frac{(v_{B/A})_y}{(v_{B/A})_x} = \frac{3.588}{9}$$

$$\theta = 21.7° \qquad \textit{Ans.}$$

Acceleration. Car B has both tangential and normal components of acceleration. Why? The magnitude of the normal component is

$$(a_B)_n = \frac{v_B^2}{\rho} = \frac{(12\ \text{m/s})^2}{100\ \text{m}} = 1.440\ \text{m/s}^2$$

Applying the equation for relative acceleration yields

$$\mathbf{a}_B = \mathbf{a}_A + \mathbf{a}_{B/A}$$

$$(-1.440\mathbf{i} - 3\mathbf{j}) = (2\cos 60°\mathbf{i} + 2\sin 60°\mathbf{j}) + \mathbf{a}_{B/A}$$

$$\mathbf{a}_{B/A} = \{-2.440\mathbf{i} - 4.732\mathbf{j}\}\ \text{m/s}^2$$

Here $\mathbf{a}_{B/A}$ has $-\mathbf{i}$ and $-\mathbf{j}$ components. Thus, from Fig. 12–45c,

$$a_{B/A} = \sqrt{(2.440)^2 + (4.732)^2} = 5.32\ \text{m/s}^2 \qquad \textit{Ans.}$$

$$\tan\phi = \frac{(a_{B/A})_y}{(a_{B/A})_x} = \frac{4.732}{2.440}$$

$$\phi = 62.7° \qquad \textit{Ans.}$$

NOTE: Is it possible to obtain the relative acceleration of $\mathbf{a}_{A/B}$ using this method? Refer to the comment made at the end of Example 12.26.

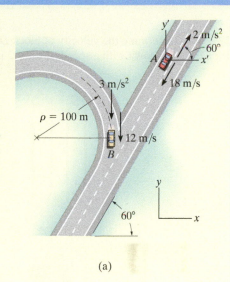

(a)

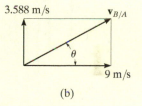

(b)

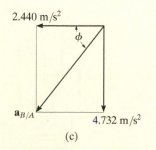

(c)

Fig. 12–45

FUNDAMENTAL PROBLEMS

F12–39. Determine the velocity of block D if end A of the rope is pulled down with a speed of $v_A = 3$ m/s.

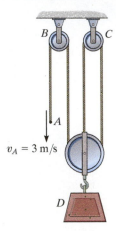

$v_A = 3$ m/s

Prob. F12–39

F12–40. Determine the velocity of block A if end B of the rope is pulled down with a speed of 6 m/s.

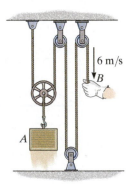

6 m/s

Prob. F12–40

F12–41. Determine the velocity of block A if end B of the rope is pulled down with a speed of 1.5 m/s.

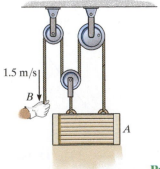

1.5 m/s

Prob. F12–41

F12–42. Determine the velocity of block A if end F of the rope is pulled down with a speed of $v_F = 3$ m/s.

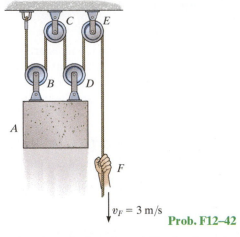

$v_F = 3$ m/s

Prob. F12–42

F12–43. Determine the velocity of car A if point P on the cable has a speed of 4 m/s when the motor M winds the cable in.

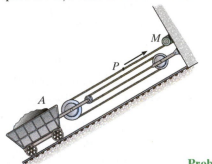

Prob. F12–43

F12–44. Determine the velocity of cylinder B if cylinder A moves downward with a speed of $v_A = 4$ ft/s.

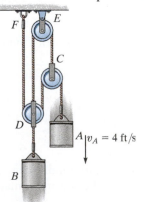

A $v_A = 4$ ft/s

Prob. F12–44

F12–45. Car *A* is traveling with a constant speed of 80 km/h due north, while car *B* is traveling with a constant speed of 100 km/h due east. Determine the velocity of car *B* relative to car *A*.

F12–47. The boats *A* and *B* travel with constant speeds of $v_A = 15$ m/s and $v_B = 10$ m/s when they leave the pier at *O* at the same time. Determine the distance between them when $t = 4$ s.

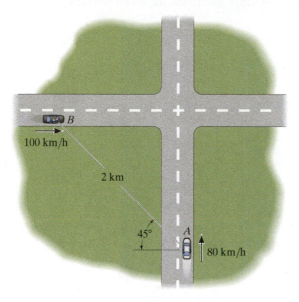

Prob. F12–45

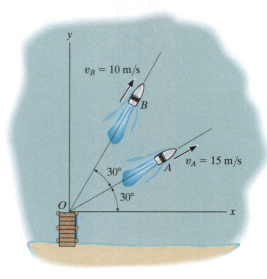

Prob. F12–47

F12–46. Two planes *A* and *B* are traveling with the constant velocities shown. Determine the magnitude and direction of the velocity of plane *B* relative to plane *A*.

F12–48. At the instant shown, cars *A* and *B* are traveling at the speeds shown. If *B* is accelerating at 1200 km/h² while *A* maintains a constant speed, determine the velocity and acceleration of *A* with respect to *B*.

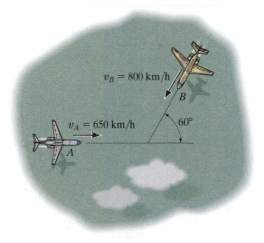

Prob. F12–46

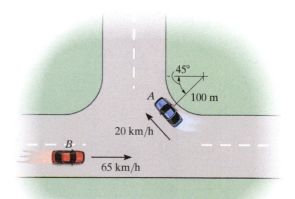

Prob. F12–48

PROBLEMS

12–195. If the end of the cable at A is pulled down with a speed of 2 m/s, determine the speed at which block B rises.

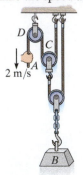

Prob. 12–195

*12–196. The motor at C pulls in the cable with an acceleration $a_C = (3t^2)$ m/s^2, where t is in seconds. The motor at D draws in its cable at $a_D = 5$ m/s^2. If both motors start at the same instant from rest when $d = 3$ m, determine (a) the time needed for $d = 0$, and (b) the velocities of blocks A and B when this occurs.

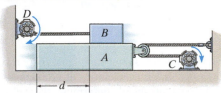

Prob. 12–196

12–197. The pulley arrangement shown is designed for hoisting materials. If BC *remains fixed* while the plunger P is pushed downward with a speed of 4 ft/s, determine the speed of the load at A.

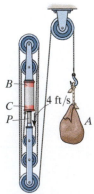

Prob. 12–197

12–198. If the end of the cable at A is pulled down with a speed of 5 m/s, determine the speed at which block B rises.

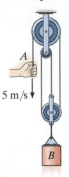

Prob. 12–198

12–199. Determine the displacement of the log if the truck at C pulls the cable 4 ft to the right.

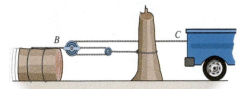

Prob. 12–199

*12–200. Determine the constant speed at which the cable at A must be drawn in by the motor in order to hoist the load 6 m in 1.5 s.

12–201. Starting from rest, the cable can be wound onto the drum of the motor at a rate of $v_A = (3t^2)$ m/s, where t is in seconds. Determine the time needed to lift the load 7 m.

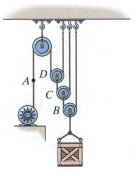

Probs. 12–200/201

12–202.   If the end A of the cable is moving at $v_A = 3$ m/s, determine the speed of block B.

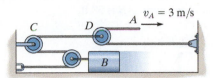

Prob. 12–202

12–203.   Determine the time needed for the load at B to attain a speed of 10 m/s, starting from rest, if the cable is drawn into the motor with an acceleration of 3 m/s².

*12–204.   The cable at A is being drawn toward the motor at $v_A = 8$ m/s. Determine the velocity of the block.

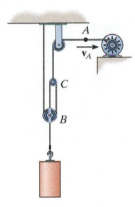

Probs. 12–203/204

12–205.   If block A of the pulley system is moving downward at 6 ft/s while block C is moving down at 18 ft/s, determine the relative velocity of block B with respect to C.

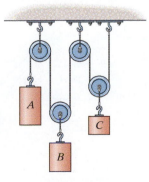

Prob. 12–205

12–206.   Determine the speed of the block at B.

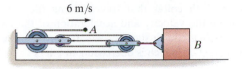

Prob. 12–206

12–207.   Determine the speed of block A if the end of the rope is pulled down with a speed of 4 m/s.

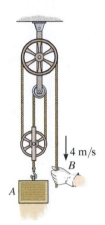

Prob. 12–207

*12–208.   The motor draws in the cable at C with a constant velocity of $v_C = 4$ m/s. The motor draws in the cable at D with a constant acceleration of $a_D = 8$ m/s². If $v_D = 0$ when $t = 0$, determine (a) the time needed for block A to rise 3 m, and (b) the relative velocity of block A with respect to block B when this occurs.

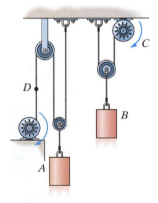

Prob. 12–208

12–209. The cord is attached to the pin at C and passes over the two pulleys at A and D. The pulley at A is attached to the smooth collar that travels along the vertical rod. Determine the velocity and acceleration of the end of the cord at B if at the instant $s_A = 4$ ft the collar is moving upward at 5 ft/s, which is decreasing at 2 ft/s².

12–210. The 16-ft-long cord is attached to the pin at C and passes over the two pulleys at A and D. The pulley at A is attached to the smooth collar that travels along the vertical rod. When $s_B = 6$ ft, the end of the cord at B is pulled downward with a velocity of 4 ft/s and is given an acceleration of 3 ft/s². Determine the velocity and acceleration of the collar at this instant.

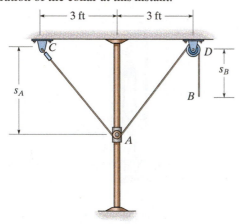

Probs. 12–209/210

12–211. The roller at A is moving with a velocity of $v_A = 4$ m/s and has an acceleration of $a_A = 2$ m/s² when $x_A = 3$ m. Determine the velocity and acceleration of block B at this instant.

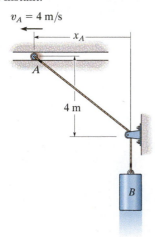

Prob. 12–211

*12–212. The girl at C stands near the edge of the pier and pulls in the rope *horizontally* at a constant speed of 6 ft/s. Determine how fast the boat approaches the pier at the instant the rope length AB is 50 ft.

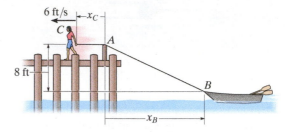

Prob. 12–212

12–213. If the hydraulic cylinder H draws in rod BC at 2 ft/s, determine the speed of slider A.

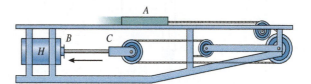

Prob. 12–213

12–214. At the instant shown, the car at A is traveling at 10 m/s around the curve while increasing its speed at 5 m/s². The car at B is traveling at 18.5 m/s along the straightaway and increasing its speed at 2 m/s². Determine the relative velocity and relative acceleration of A with respect to B at this instant.

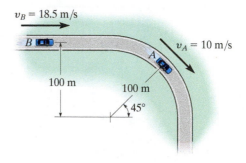

Prob. 12–214

12–215. The motor draws in the cord at B with an acceleration of $a_B = 2$ m/s^2. When $s_A = 1.5$ m, $v_B = 6$ m/s. Determine the velocity and acceleration of the collar at this instant.

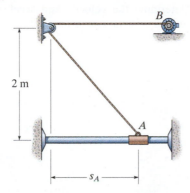

2 m

s_A

Prob. 12–215

*12–216. If block B is moving down with a velocity v_B and has an acceleration a_B, determine the velocity and acceleration of block A in terms of the parameters shown.

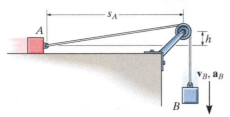

s_A

A

h

$\mathbf{v}_B, \mathbf{a}_B$

B

Prob. 12–216

12–217. The crate C is being lifted by moving the roller at A downward with a constant speed of $v_A = 2$ m/s along the guide. Determine the velocity and acceleration of the crate at the instant $s = 1$ m. When the roller is at B, the crate rests on the ground. Neglect the size of the pulley in the calculation. *Hint:* Relate the coordinates x_C and x_A using the problem geometry, then take the first and second time derivatives.

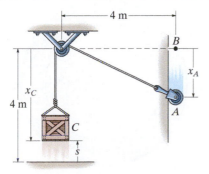

4 m

B

x_A

x_C

4 m

A

C

s

Prob. 12–217

12–218. Two planes, A and B, are flying at the same altitude. If their velocities are $v_A = 500$ km/h and $v_B = 700$ km/h such that the angle between their straight-line courses is $\theta = 60°$, determine the velocity of plane B with respect to plane A.

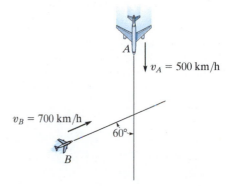

A

$v_A = 500$ km/h

$v_B = 700$ km/h

$60°$

B

Prob. 12–218

12–219. At the instant shown, cars A and B are traveling at speeds of 55 mi/h and 40 mi/h, respectively. If B is increasing its speed by 1200 mi/h^2, while A maintains a constant speed, determine the velocity and acceleration of B with respect to A. Car B moves along a curve having a radius of curvature of 0.5 mi.

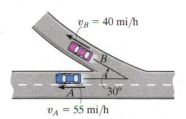

$v_B = 40$ mi/h

B

$30°$

$v_A = 55$ mi/h

Prob. 12–219

*12–220. The boat can travel with a speed of 16 km/h in still water. The point of destination is located along the dashed line. If the water is moving at 4 km/h, determine the bearing angle θ at which the boat must travel to stay on course.

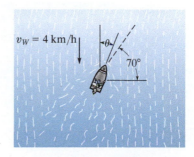

$v_W = 4$ km/h

θ

$70°$

Prob. 12–220

12–221. Two boats leave the pier P at the same time and travel in the directions shown. If $v_A = 40$ ft/s and $v_B = 30$ ft/s, determine the velocity of boat A relative to boat B. How long after leaving the pier will the boats be 1500 ft apart?

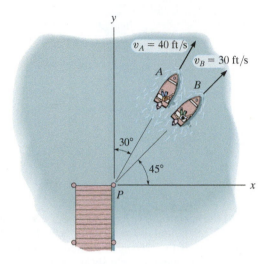

Prob. 12–221

12–222. A car is traveling north along a straight road at 50 km/h. An instrument in the car indicates that the wind is coming from the east. If the car's speed is 80 km/h, the instrument indicates that the wind is coming from the northeast. Determine the speed and direction of the wind.

12–223. Two boats leave the shore at the same time and travel in the directions shown. If $v_A = 10$ m/s and $v_B = 15$ m/s, determine the velocity of boat A with respect to boat B. How long after leaving the shore will the boats be 600 m apart?

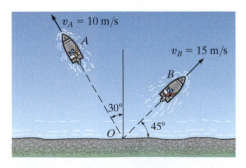

Prob. 12–223

*12–224. At the instant shown, car A has a speed of 20 km/h, which is being increased at the rate of 300 km/h² as the car enters the expressway. At the same instant, car B is decelerating at 250 km/h² while traveling forward at 100 km/h. Determine the velocity and acceleration of A with respect to B.

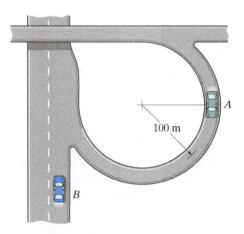

Prob. 12–224

12–225. Cars A and B are traveling around the circular race track. At the instant shown, A has a speed of 90 ft/s and is increasing its speed at the rate of 15 ft/s², whereas B has a speed of 105 ft/s and is decreasing its speed at 25 ft/s². Determine the relative velocity and relative acceleration of car A with respect to car B at this instant.

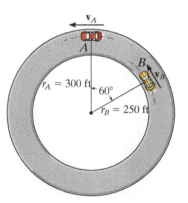

Prob. 12–225

12–226. A man walks at 5 km/h in the direction of a 20 km/h wind. If raindrops fall vertically at 7 km/h in *still air*, determine direction in which the drops appear to fall with respect to the man.

Prob. 12–226

12–227. At the instant shown, cars A and B are traveling at velocities of 40 m/s and 30 m/s, respectively. If B is increasing its velocity by 2 m/s², while A maintains a constant velocity, determine the velocity and acceleration of B with respect to A. The radius of curvature at B is $\rho_B = 200$ m.

*12–228. At the instant shown, cars A and B are traveling at velocities of 40 m/s and 30 m/s, respectively. If A is increasing its speed at 4 m/s², whereas the speed of B is decreasing at 3 m/s², determine the velocity and acceleration of B with respect to A. The radius of curvature at B is $\rho_B = 200$ m.

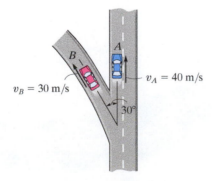

Probs. 12–227/228

12–229. A passenger in an automobile observes that raindrops make an angle of 30° with the horizontal as the auto travels forward with a speed of 60 km/h. Compute the terminal (constant) velocity $\mathbf{v}_r$ of the rain if it is assumed to fall vertically.

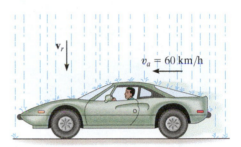

Prob. 12–229

12–230. A man can swim at 4 ft/s in still water. He wishes to cross the 40-ft-wide river to point B, 30 ft downstream. If the river flows with a velocity of 2 ft/s, determine the speed of the man and the time needed to make the crossing. *Note:* While in the water he must not direct himself toward point B to reach this point. Why?

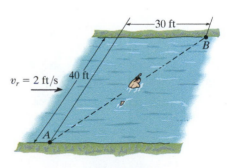

Prob. 12–230

12

12–231. The ship travels at a constant speed of $v_s = 20$ m/s and the wind is blowing at a speed of $v_w = 10$ m/s, as shown. Determine the magnitude and direction of the horizontal component of velocity of the smoke coming from the smoke stack as it appears to a passenger on the ship.

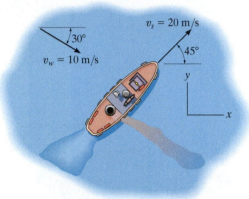

Prob. 12–231

*12–232. The football player at A throws the ball in the y–z plane at a speed $v_A = 50$ ft/s and an angle $\theta_A = 60°$ with the horizontal. At the instant the ball is thrown, the player is at B and is running with constant speed along the line BC in order to catch it. Determine this speed, v_B, so that he makes the catch at the same elevation from which the ball was thrown.

12–233. The football player at A throws the ball in the y–z plane with a speed $v_A = 50$ ft/s and an angle $\theta_A = 60°$ with the horizontal. At the instant the ball is thrown, the player is at B and is running at a constant speed of $v_B = 23$ ft/s along the line BC. Determine if he can reach point C, which has the same elevation as A, before the ball gets there.

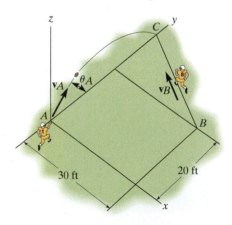

Probs. 12–232/233

12–234. At a given instant the football player at A throws a football C with a velocity of 20 m/s in the direction shown. Determine the constant speed at which the player at B must run so that he can catch the football at the same elevation at which it was thrown. Also calculate the relative velocity and relative acceleration of the football with respect to B at the instant the catch is made. Player B is 15 m away from A when A starts to throw the football.

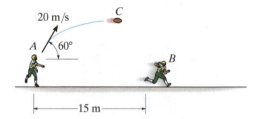

Prob. 12–234

12–235. At the instant shown, car A travels along the straight portion of the road with a speed of 25 m/s. At this same instant car B travels along the circular portion of the road with a speed of 15 m/s. Determine the velocity of car B relative to car A.

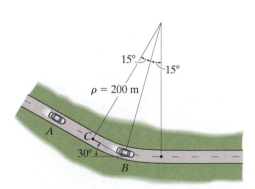

Prob. 12–235

CONCEPTUAL PROBLEMS

C12–1. If you measured the time it takes for the construction elevator to go from A to B, then B to C, and then C to D, and you also know the distance between each of the points, how could you determine the average velocity and average acceleration of the elevator as it ascends from A to D? Use numerical values to explain how this can be done.

Prob. C12–1 (© R.C. Hibbeler)

C12–2. If the sprinkler at A is 1 m from the ground, then scale the necessary measurements from the photo to determine the approximate velocity of the water jet as it flows from the nozzle of the sprinkler.

Prob. C12–2 (© R.C. Hibbeler)

C12–3. The basketball was thrown at an angle measured from the horizontal to the man's outstretched arm. If the basket is 3 m from the ground, make appropriate measurements in the photo and determine if the ball located as shown will pass through the basket.

Prob. C12–3 (© R.C. Hibbeler)

C12–4. The pilot tells you the wingspan of her plane and her constant airspeed. How would you determine the acceleration of the plane at the moment shown? Use numerical values and take any necessary measurements from the photo.

Prob. C12–4 (© R.C. Hibbeler)

12

CHAPTER REVIEW

Rectilinear Kinematics

Rectilinear kinematics refers to motion along a straight line. A position coordinate s specifies the location of the particle on the line, and the displacement Δs is the change in this position.

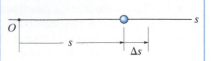

The average velocity is a vector quantity, defined as the displacement divided by the time interval.

$$v_{avg} = \frac{\Delta s}{\Delta t}$$

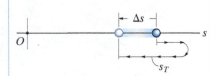

The average speed is a scalar, and is the total distance traveled divided by the time of travel.

$$(v_{sp})_{avg} = \frac{s_T}{\Delta t}$$

The time, position, velocity, and acceleration are related by three differential equations.

$$a = \frac{dv}{dt}, \quad v = \frac{ds}{dt}, \quad a\,ds = v\,dv$$

If the acceleration is known to be constant, then the differential equations relating time, position, velocity, and acceleration can be integrated.

$$v = v_0 + a_c t$$
$$s = s_0 + v_0 t + \tfrac{1}{2} a_c t^2$$
$$v^2 = v_0^2 + 2a_c(s - s_0)$$

Graphical Solutions

If the motion is erratic, then it can be described by a graph. If one of these graphs is given, then the others can be established using the differential relations between a, v, s, and t.

$$a = \frac{dv}{dt},$$
$$v = \frac{ds}{dt},$$
$$a\,ds = v\,dv$$

Curvilinear Motion, x, y, z

Curvilinear motion along the path can be resolved into rectilinear motion along the x, y, z axes. The equation of the path is used to relate the motion along each axis.

$$v_x = \dot{x} \quad a_x = \dot{v}_x$$

$$v_y = \dot{y} \quad a_y = \dot{v}_y$$

$$v_z = \dot{z} \quad a_z = \dot{v}_z$$

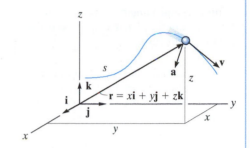

Projectile Motion

Free-flight motion of a projectile follows a parabolic path. It has a constant velocity in the horizontal direction, and a constant downward acceleration of $g = 9.81$ m/s^2 or 32.2 ft/s^2 in the vertical direction. Any two of the three equations for constant acceleration apply in the vertical direction, and in the horizontal direction only one equation applies.

$$(+\uparrow) \quad v_y = (v_0)_y + a_c t$$

$$(+\uparrow) \quad y = y_0 + (v_0)_y t + \tfrac{1}{2} a_c t^2$$

$$(+\uparrow) \quad v_y^2 = (v_0)_y^2 + 2a_c(y - y_0)$$

$$(\overset{+}{\rightarrow}) \quad x = x_0 + (v_0)_x t$$

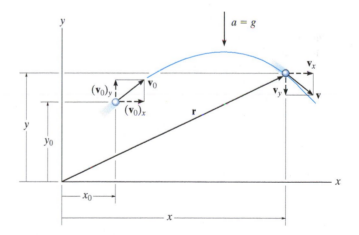

12

Curvilinear Motion n, t

If normal and tangential axes are used for the analysis, then **v** is always in the positive t direction.

The acceleration has two components. The tangential component, $\mathbf{a}_t$, accounts for the change in the magnitude of the velocity; a slowing down is in the negative t direction, and a speeding up is in the positive t direction. The normal component $\mathbf{a}_n$ accounts for the change in the direction of the velocity. This component is always in the positive n direction.

$$a_t = \dot{v} \quad \text{or} \quad a_t\,ds = v\,dv$$

$$a_n = \frac{v^2}{\rho}$$

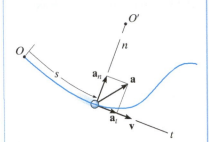

Curvilinear Motion r, θ

If the path of motion is expressed in polar coordinates, then the velocity and acceleration components can be related to the time derivatives of r and θ.

To apply the time-derivative equations, it is necessary to determine $r, \dot{r}, \ddot{r}, \dot{\theta}, \ddot{\theta}$ at the instant considered. If the path $r = f(\theta)$ is given, then the chain rule of calculus must be used to obtain time derivatives. (See Appendix C.)

Once the data are substituted into the equations, then the algebraic sign of the results will indicate the direction of the components of **v** or **a** along each axis.

$$v_r = \dot{r}$$
$$v_\theta = r\dot{\theta}$$

$$a_r = \ddot{r} - r\dot{\theta}^2$$
$$a_\theta = r\ddot{\theta} + 2\dot{r}\dot{\theta}$$

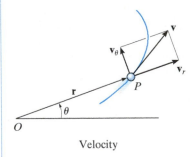

Velocity

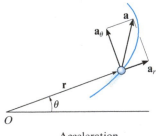

Acceleration

Absolute Dependent Motion of Two Particles

The dependent motion of blocks that are suspended from pulleys and cables can be related by the geometry of the system. This is done by first establishing position coordinates, measured from a fixed origin to each block. Each coordinate must be directed along the line of motion of a block.

Using geometry and/or trigonometry, the coordinates are then related to the cable length in order to formulate a position coordinate equation.

$$2s_B + h + s_A = l$$

The first time derivative of this equation gives a relationship between the velocities of the blocks, and a second time derivative gives the relation between their accelerations.

$$2v_B = -v_A$$

$$2a_B = -a_A$$

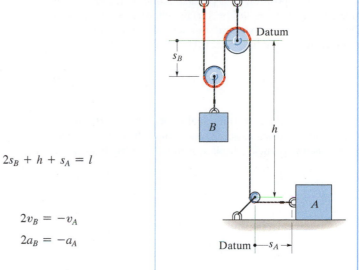

Relative-Motion Analysis Using Translating Axes

If two particles A and B undergo independent motions, then these motions can be related to their relative motion using a *translating set of axes* attached to one of the particles (A).

$$\mathbf{r}_B = \mathbf{r}_A + \mathbf{r}_{B/A}$$

For planar motion, each vector equation produces two scalar equations, one in the x, and the other in the y direction. For solution, the vectors can be expressed in Cartesian form, or the x and y scalar components can be written directly.

$$\mathbf{v}_B = \mathbf{v}_A + \mathbf{v}_{B/A}$$

$$\mathbf{a}_B = \mathbf{a}_A + \mathbf{a}_{B/A}$$

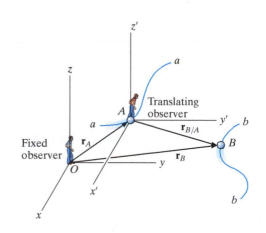

REVIEW PROBLEMS

R12–1. The position of a particle along a straight line is given by $s = (t^3 - 9t^2 + 15t)$ ft, where t is in seconds. Determine its maximum acceleration and maximum velocity during the time interval $0 \le t \le 10$ s.

R12–2. If a particle has an initial velocity $v_0 = 12$ ft/s to the right, and a constant acceleration of 2 ft/s^2 to the left, determine the particle's displacement in 10 s. Originally $s_0 = 0$.

R12–3. A projectile, initially at the origin, moves along a straight-line path through a fluid medium such that its velocity is $v = 1800(1 - e^{-0.3t})$ mm/s where t is in seconds. Determine the displacement of the projectile during the first 3 s.

R12–4. The v–t graph of a car while traveling along a road is shown. Determine the acceleration when $t = 2.5$ s, 10 s, and 25 s. Also if $s = 0$ when $t = 0$, find the position when $t = 5$ s, 20 s, and 30 s.

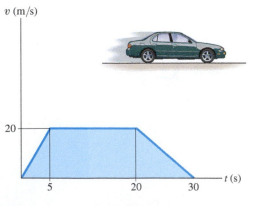

Prob. R12–4

R12–5. A car traveling along the straight portions of the road has the velocities indicated in the figure when it arrives at points A, B, and C. If it takes 3 s to go from A to B, and then 5 s to go from B to C, determine the average acceleration between points A and B and between points A and C.

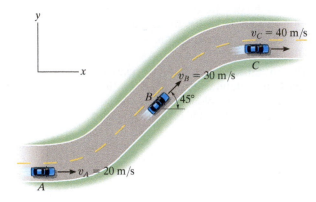

Prob. R12–5

R12–6. From a videotape, it was observed that a player kicked a football 126 ft during a measured time of 3.6 seconds. Determine the initial speed of the ball and the angle θ at which it was kicked.

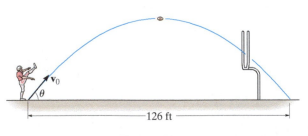

Prob. R12–6

R12–7. The truck travels in a circular path having a radius of 50 m at a speed of $v = 4$ m/s. For a short distance from $s = 0$, its speed is increased by $\dot{v} = (0.05s)$ m/s², where s is in meters. Determine its speed and the magnitude of its acceleration when it has moved $s = 10$ m.

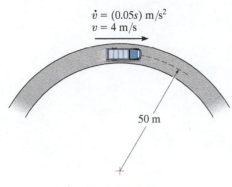

$$\dot{v} = (0.05s) \text{ m/s}^2$$
$$v = 4 \text{ m/s}$$

50 m

Prob. R12–7

R12–8. Car B turns such that its speed is increased by $(a_t)_B = (0.5e^t)$ m/s², where t is in seconds. If the car starts from rest when $\theta = 0°$, determine the magnitudes of its velocity and acceleration when $t = 2$ s. Neglect the size of the car.

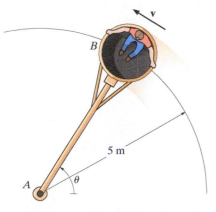

5 m

Prob. R12–8

R12–9. A particle is moving along a circular path of 2-m radius such that its position as a function of time is given by $\theta = (5t^2)$ rad, where t is in seconds. Determine the magnitude of the particle's acceleration when $\theta = 30°$. The particle starts from rest when $\theta = 0°$.

R12–10. Determine the time needed for the load at B to attain a speed of 8 m/s, starting from rest, if the cable is drawn into the motor with an acceleration of 0.2 m/s².

Prob. R12–10

R12–11. Two planes, A and B, are flying at the same altitude. If their velocities are $v_A = 600$ km/h and $v_B = 500$ km/h such that the angle between their straight-line courses is $\theta = 75°$, determine the velocity of plane B with respect to plane A.

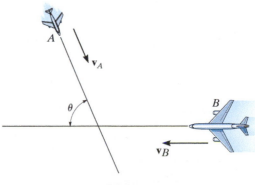

Prob. R12–11

Chapter 13

(© Migel/Shutterstock)

A car driving along this road will be subjected to forces that create both normal and tangential accelerations. In this chapter we will study how these forces are related to the accelerations they create.

Kinetics of a Particle: Force and Acceleration

CHAPTER OBJECTIVES

■ To state Newton's Second Law of Motion and to define mass and weight.

■ To analyze the accelerated motion of a particle using the equation of motion with different coordinate systems.

■ To investigate central-force motion and apply it to problems in space mechanics.

13.1   Newton's Second Law of Motion

Kinetics is a branch of dynamics that deals with the relationship between the change in motion of a body and the forces that cause this change. The basis for kinetics is Newton's second law, which states that when an *unbalanced force* acts on a particle, the particle will *accelerate* in the direction of the force with a magnitude that is proportional to the force.

This law can be verified experimentally by applying a known unbalanced force **F** to a particle, and then measuring the acceleration **a**. Since the force and acceleration are directly proportional, the constant of proportionality, m, may be determined from the ratio $m = F/a$. This positive scalar m is called the *mass* of the particle. Being constant during any acceleration, m provides a quantitative measure of the resistance of the particle to a change in its velocity, that is its inertia.

The jeep leans backward due to its inertia, which resists its forward acceleration. (© R.C. Hibbeler)

If the mass of the particle is m, Newton's second law of motion may be written in mathematical form as

$$\mathbf{F} = m\mathbf{a}$$

The above equation, which is referred to as the *equation of motion*, is one of the most important formulations in mechanics.* As previously stated, its validity is based solely on *experimental evidence*. In 1905, however, Albert Einstein developed the theory of relativity and placed limitations on the use of Newton's second law for describing general particle motion. Through experiments it was proven that *time* is not an absolute quantity as assumed by Newton; and as a result, the equation of motion fails to predict the exact behavior of a particle, especially when the particle's speed approaches the speed of light (0.3 Gm/s). Developments of the theory of quantum mechanics by Erwin Schrödinger and others indicate further that conclusions drawn from using this equation are also invalid when particles are the size of an atom and move close to one another. For the most part, however, these requirements regarding particle speed and size are not encountered in engineering problems, so their effects will not be considered in this book.

Newton's Law of Gravitational Attraction.

Shortly after formulating his three laws of motion, Newton postulated a law governing the mutual attraction between any two particles. In mathematical form this law can be expressed as

$$F = G\frac{m_1 m_2}{r^2} \tag{13–1}$$

where

$$
\begin{aligned}
F &= \text{force of attraction between the two particles} \\
G &= \text{universal constant of gravitation; according to experimental} \\
&\quad \text{evidence } G = 66.73(10^{-12}) \text{ m}^3/(\text{kg} \cdot \text{s}^2) \\
m_1, m_2 &= \text{mass of each of the two particles} \\
r &= \text{distance between the centers of the two particles}
\end{aligned}
$$

---

*Since m is constant, we can also write $\mathbf{F} = d(m\mathbf{v})/dt$, where $m\mathbf{v}$ is the particle's linear momentum. Here the unbalanced force acting on the particle is proportional to the time rate of change of the particle's linear momentum.

In the case of a particle located at or near the surface of the earth, the only gravitational force having any sizable magnitude is that between the earth and the particle. This force is termed the "weight" and, for our purpose, it will be the only gravitational force considered.

From Eq. 13–1, we can develop a general expression for finding the weight W of a particle having a mass $m_1 = m$. Let $m_2 = M_e$ be the mass of the earth and r the distance between the earth's center and the particle. Then, if $g = GM_e/r^2$, we have

$$W = mg$$

By comparison with $F = ma$, we term g the acceleration due to gravity. For most engineering calculations g is measured at a point on the surface of the earth at sea level, and at a latitude of 45°, which is considered the "standard location." Here the values $g = 9.81 \text{ m/s}^2 = 32.2 \text{ ft/s}^2$ will be used for calculations.

In the SI system the mass of the body is specified in kilograms, and the weight must be calculated using the above equation, Fig. 13–1a. Thus,

$$W = mg \text{ (N)} \quad (g = 9.81 \text{ m/s}^2) \qquad (13\text{–}2)$$

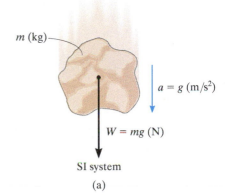

SI system

(a)

As a result, a body of mass 1 kg has a weight of 9.81 N; a 2-kg body weighs 19.62 N; and so on.

In the FPS system the weight of the body is specified in pounds. The mass is measured in slugs, a term derived from "sluggish" which refers to the body's inertia. It must be calculated, Fig. 13–1b, using

$$m = \frac{W}{g} \text{ (slug)} \quad (g = 32.2 \text{ ft/s}^2) \qquad (13\text{–}3)$$

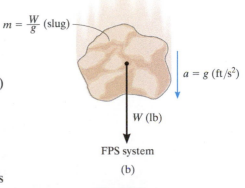

FPS system

(b)

Therefore, a body weighing 32.2 lb has a mass of 1 slug; a 64.4-lb body has a mass of 2 slugs; and so on.

Fig. 13–1

13

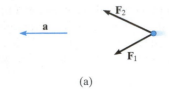

(a)

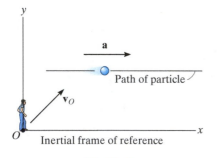

Free-body diagram		Kinetic diagram

(b)   **Fig. 13–2**   (c)

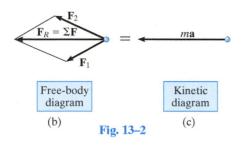

Fig. 13–3

13.2   The Equation of Motion

When more than one force acts on a particle, the resultant force is determined by a vector summation of all the forces; i.e., $\mathbf{F}_R = \Sigma \mathbf{F}$. For this more general case, the equation of motion may be written as

$$\boxed{\Sigma \mathbf{F} = m\mathbf{a}} \qquad (13\text{–}4)$$

To illustrate application of this equation, consider the particle shown in Fig. 13–2a, which has a mass m and is subjected to the action of two forces, $\mathbf{F}_1$ and $\mathbf{F}_2$. We can graphically account for the magnitude and direction of each force acting on the particle by drawing the particle's *free-body diagram*, Fig. 13–2b. Since the *resultant* of these forces *produces* the vector $m\mathbf{a}$, its magnitude and direction can be represented graphically on the *kinetic diagram*, shown in Fig. 13–2c.* The equal sign written between the diagrams symbolizes the *graphical* equivalency between the free-body diagram and the kinetic diagram; i.e., $\Sigma \mathbf{F} = m\mathbf{a}$.† In particular, note that if $\mathbf{F}_R = \Sigma \mathbf{F} = \mathbf{0}$, then the acceleration is also zero, so that the particle will either remain at *rest* or move along a straight-line path with *constant velocity*. Such are the conditions of *static equilibrium*, Newton's first law of motion.

Inertial Reference Frame.   When applying the equation of motion, it is important that the acceleration of the particle be measured with respect to a reference frame that is *either fixed or translates with a constant velocity*. In this way, the observer will not accelerate and measurements of the particle's acceleration will be the *same* from *any* reference of this type. Such a frame of reference is commonly known as a *Newtonian* or *inertial reference frame*, Fig. 13–3.

When studying the motions of rockets and satellites, it is justifiable to consider the inertial reference frame as fixed to the stars, whereas dynamics problems concerned with motions on or near the surface of the earth may be solved by using an inertial frame which is assumed fixed to the earth. Even though the earth both rotates about its own axis and revolves about the sun, the accelerations created by these rotations are relatively small and so they can be neglected for most applications.

---

*Recall the free-body diagram considers the particle to be free of its surrounding supports and shows all the forces acting on the particle. The kinetic diagram pertains to the particle's motion as caused by the forces.

†The equation of motion can also be rewritten in the form $\Sigma \mathbf{F} - m\mathbf{a} = \mathbf{0}$. The vector $-m\mathbf{a}$ is referred to as the *inertia force vector*. If it is treated in the same way as a "force vector," then the state of "equilibrium" created is referred to as *dynamic equilibrium*. This method of application, which will not be used in this text, is often referred to as the *D'Alembert principle*, named after the French mathematician Jean le Rond d'Alembert.

We are all familiar with the sensation one feels when sitting in a car that is subjected to a forward acceleration. Often people think this is caused by a "force" which acts on them and tends to push them back in their seats; however, this is not the case. Instead, this sensation occurs due to their inertia or the resistance of their mass to a change in velocity.

Consider the passenger who is strapped to the seat of a rocket sled. Provided the sled is at rest or is moving with constant velocity, then no force is exerted on his back as shown on his free-body diagram.

Keystone/Hulton Archive/Getty Images

When the thrust of the rocket engine causes the sled to accelerate, then the seat upon which he is sitting exerts a force **F** on him which pushes him forward with the sled. In the photo, notice that the inertia of his head resists this change in motion (acceleration), and so his head moves back against the seat and his face, which is nonrigid, tends to distort backward.

Keystone/Hulton Archive/Getty Images

Upon deceleration the force of the seatbelt **F′** tends to pull his body to a stop, but his head leaves contact with the back of the seat and his face distorts forward, again due to his inertia or tendency to continue to move forward. No force is pulling him forward, although this is the sensation he receives.

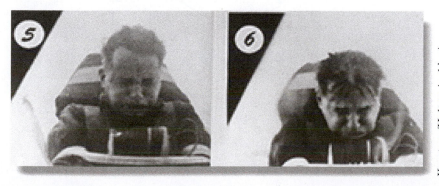

Keystone/Hulton Archive/Getty Images

13.3   Equation of Motion for a System of Particles

The equation of motion will now be extended to include a system of particles isolated within an enclosed region in space, as shown in Fig. 13–4a. In particular, there is no restriction in the way the particles are connected, so the following analysis applies equally well to the motion of a solid, liquid, or gas system.

At the instant considered, the arbitrary i-th particle, having a mass m_i, is subjected to a system of internal forces and a resultant external force. The *internal force*, represented symbolically as $\mathbf{f}_i$, is the resultant of all the forces the other particles exert on the ith particle. The *resultant external force* $\mathbf{F}_i$ represents, for example, the effect of gravitational, electrical, magnetic, or contact forces between the ith particle and adjacent bodies or particles *not* included within the system.

The free-body and kinetic diagrams for the ith particle are shown in Fig. 13–4b. Applying the equation of motion,

$$\Sigma \mathbf{F} = m\mathbf{a}; \qquad\qquad \mathbf{F}_i + \mathbf{f}_i = m_i\mathbf{a}_i$$

When the equation of motion is applied to each of the other particles of the system, similar equations will result. And, if all these equations are added together *vectorially*, we obtain

$$\Sigma \mathbf{F}_i + \Sigma \mathbf{f}_i = \Sigma m_i\mathbf{a}_i$$

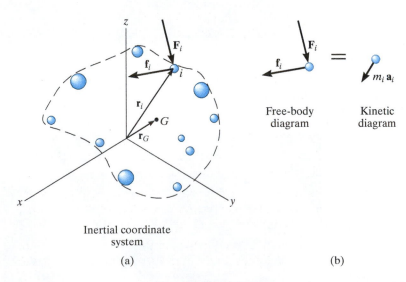

Inertial coordinate system

(a)

Free-body diagram

Kinetic diagram

(b)

Fig. 13–4

The summation of the internal forces, if carried out, will equal zero, since internal forces between any two particles occur in equal but opposite collinear pairs. Consequently, only the sum of the external forces will remain, and therefore the equation of motion, written for the system of particles, becomes

$$\Sigma \mathbf{F}_i = \Sigma m_i \mathbf{a}_i \qquad (13\text{–}5)$$

If $\mathbf{r}_G$ is a position vector which locates the *center of mass G* of the particles, Fig. 13–4a, then by definition of the center of mass, $m\mathbf{r}_G = \Sigma m_i \mathbf{r}_i$, where $m = \Sigma m_i$ is the total mass of all the particles. Differentiating this equation twice with respect to time, assuming that no mass is entering or leaving the system, yields

$$m\mathbf{a}_G = \Sigma m_i \mathbf{a}_i$$

Substituting this result into Eq. 13–5, we obtain

$$\boxed{\Sigma \mathbf{F} = m\mathbf{a}_G} \qquad (13\text{–}6)$$

Hence, the sum of the external forces acting on the system of particles is equal to the total mass of the particles times the acceleration of its center of mass G. Since in reality all particles must have a finite size to possess mass, Eq. 13–6 justifies application of the equation of motion to a *body* that is represented as a single particle.

Important Points

- The equation of motion is based on experimental evidence and is valid only when applied within an inertial frame of reference.

- The equation of motion states that the *unbalanced force* on a particle causes it to *accelerate*.

- An inertial frame of reference does not rotate, rather its axes either translate with constant velocity or are at rest.

- Mass is a property of matter that provides a quantitative measure of its resistance to a change in velocity. It is an absolute quantity and so it does not change from one location to another.

- Weight is a force that is caused by the earth's gravitation. It is not absolute; rather it depends on the altitude of the mass from the earth's surface.

13.4   Equations of Motion: Rectangular Coordinates

When a particle moves relative to an inertial x, y, z frame of reference, the forces acting on the particle, as well as its acceleration, can be expressed in terms of their $\mathbf{i}, \mathbf{j}, \mathbf{k}$ components, Fig. 13–5. Applying the equation of motion, we have

$$\Sigma \mathbf{F} = m\mathbf{a}; \qquad \Sigma F_x \mathbf{i} + \Sigma F_y \mathbf{j} + \Sigma F_z \mathbf{k} = m(a_x \mathbf{i} + a_y \mathbf{j} + a_z \mathbf{k})$$

For this equation to be satisfied, the respective $\mathbf{i}, \mathbf{j}, \mathbf{k}$ components on the left side must equal the corresponding components on the right side. Consequently, we may write the following three scalar equations:

$$\begin{aligned} \Sigma F_x &= ma_x \\ \Sigma F_y &= ma_y \\ \Sigma F_z &= ma_z \end{aligned} \qquad (13\text{–}7)$$

In particular, if the particle is constrained to move only in the x–y plane, then the first two of these equations are used to specify the motion.

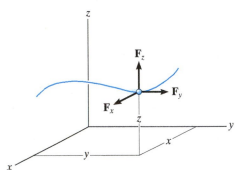

Fig. 13–5

*It is a convention in this text always to use the kinetic diagram as a graphical aid when developing the proofs and theory. The particle's acceleration or its components will be shown as blue colored vectors near the free-body diagram in the examples.

Equations of Motion.

- If the forces can be resolved directly from the free-body diagram, apply the equations of motion in their scalar component form.

- If the geometry of the problem appears complicated, which often occurs in three dimensions, Cartesian vector analysis can be used for the solution.

- *Friction.* If a moving particle contacts a rough surface, it may be necessary to use the *frictional equation*, which relates the frictional and normal forces $\mathbf{F}_f$ and $\mathbf{N}$ acting at the surface of contact by using the coefficient of kinetic friction, i.e., $F_f = \mu_k N$. Remember that $\mathbf{F}_f$ always acts on the free-body diagram such that it opposes the motion of the particle relative to the surface it contacts. If the particle is *on the verge* of relative motion, then the coefficient of static friction should be used.

- *Spring.* If the particle is connected to an *elastic spring* having negligible mass, the spring force F_s can be related to the deformation of the spring by the equation $F_s = ks$. Here k is the spring's stiffness measured as a force per unit length, and s is the stretch or compression defined as the difference between the deformed length l and the undeformed length l_0, i.e., $s = l - l_0$.

Kinematics.

- If the velocity or position of the particle is to be found, it will be necessary to apply the necessary kinematic equations once the particle's acceleration is determined from $\Sigma\mathbf{F} = m\mathbf{a}$.

- If *acceleration* is a function of time, use $a = dv/dt$ and $v = ds/dt$ which, when integrated, yield the particle's velocity and position, respectively.

- If *acceleration* is a function of displacement, integrate $a\,ds = v\,dv$ to obtain the velocity as a function of position.

- If *acceleration is constant*, use $v = v_0 + a_c t$, $s = s_0 + v_0 t + \frac{1}{2}a_c t^2$, $v^2 = v_0^2 + 2a_c(s - s_0)$ to determine the velocity or position of the particle.

- If the problem involves the dependent motion of several particles, use the method outlined in Sec. 12.9 to relate their accelerations. In all cases, make sure the positive inertial coordinate directions used for writing the kinematic equations are the same as those used for writing the equations of motion; otherwise, simultaneous solution of the equations will result in errors.

- If the solution for an unknown vector component yields a negative scalar, it indicates that the component acts in the direction opposite to that which was assumed.

EXAMPLE | **13.1**

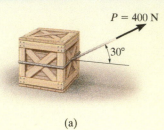

P = 400 N

30°

(a)

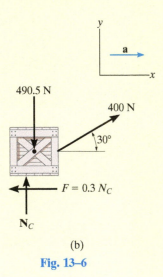

490.5 N

400 N

30°

F = 0.3 N_C

N_C

(b)

Fig. 13–6

The 50-kg crate shown in Fig. 13–6a rests on a horizontal surface for which the coefficient of kinetic friction is $\mu_k = 0.3$. If the crate is subjected to a 400-N towing force as shown, determine the velocity of the crate in 3 s starting from rest.

SOLUTION
Using the equations of motion, we can relate the crate's acceleration to the force causing the motion. The crate's velocity can then be determined using kinematics.

Free-Body Diagram.   The weight of the crate is $W = mg = 50 \text{ kg} (9.81 \text{ m/s}^2) = 490.5$ N. As shown in Fig. 13–6b, the frictional force has a magnitude $F = \mu_k N_C$ and acts to the left, since it opposes the motion of the crate. The acceleration **a** is assumed to act horizontally, in the positive x direction. There are two unknowns, namely N_C and a.

Equations of Motion.   Using the data shown on the free-body diagram, we have

$$\xrightarrow{+} \Sigma F_x = ma_x; \quad 400 \cos 30° - 0.3N_C = 50a \quad (1)$$
$$+\uparrow \Sigma F_y = ma_y; \quad N_C - 490.5 + 400 \sin 30° = 0 \quad (2)$$

Solving Eq. 2 for N_C, substituting the result into Eq. 1, and solving for a yields

$$N_C = 290.5 \text{ N}$$
$$a = 5.185 \text{ m/s}^2$$

Kinematics.   Notice that the acceleration is *constant*, since the applied force **P** is constant. Since the initial velocity is zero, the velocity of the crate in 3 s is

$$(\xrightarrow{+}) \quad v = v_0 + a_c t = 0 + 5.185(3)$$
$$= 15.6 \text{ m/s} \rightarrow \quad \text{*Ans.*}$$

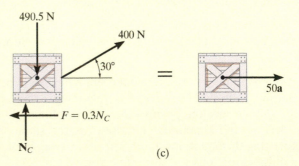

(c)

NOTE: We can also use the alternative procedure of drawing the crate's free-body *and* kinetic diagrams, Fig. 13–6c, prior to applying the equations of motion.

EXAMPLE | **13.2**

A 10-kg projectile is fired vertically upward from the ground, with an initial velocity of 50 m/s, Fig. 13–7a. Determine the maximum height to which it will travel if (a) atmospheric resistance is neglected; and (b) atmospheric resistance is measured as $F_D = (0.01v^2)$ N, where v is the speed of the projectile at any instant, measured in m/s.

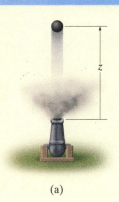

(a)

SOLUTION

In both cases the known force on the projectile can be related to its acceleration using the equation of motion. Kinematics can then be used to relate the projectile's acceleration to its position.

Part (a) Free-Body Diagram.   As shown in Fig. 13–7b, the projectile's weight is $W = mg = 10(9.81) = 98.1$ N. We will assume the unknown acceleration **a** acts upward in the *positive z* direction.

Equation of Motion.

$$+\uparrow \Sigma F_z = ma_z; \qquad -98.1 = 10a, \qquad a = -9.81 \text{ m/s}^2$$

The result indicates that the projectile, like every object having free-flight motion near the earth's surface, is subjected to a *constant* downward acceleration of 9.81 m/s².

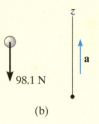

(b)

Kinematics.   Initially, $z_0 = 0$ and $v_0 = 50$ m/s, and at the maximum height $z = h$, $v = 0$. Since the acceleration is *constant*, then

$$(+\uparrow) \qquad v^2 = v_0^2 + 2a_c(z - z_0)$$
$$0 = (50)^2 + 2(-9.81)(h - 0)$$
$$h = 127 \text{ m} \qquad \qquad \textit{Ans.}$$

Part (b) Free-Body Diagram.   Since the force $F_D = (0.01v^2)$ N tends to retard the upward motion of the projectile, it acts downward as shown on the free-body diagram, Fig. 13–7c.

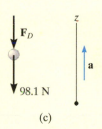

(c)

Fig. 13–7

Equation of Motion.

$$+\uparrow \Sigma F_z = ma_z; \quad -0.01v^2 - 98.1 = 10a, \quad a = -(0.001v^2 + 9.81)$$

Kinematics.   Here the acceleration is *not constant* since F_D depends on the velocity. Since $a = f(v)$, we can relate a to position using

$$(+\uparrow) \, a \, dz = v \, dv; \qquad -(0.001v^2 + 9.81) \, dz = v \, dv$$

Separating the variables and integrating, realizing that initially $z_0 = 0$, $v_0 = 50$ m/s (positive upward), and at $z = h$, $v = 0$, we have

$$\int_0^h dz = -\int_{50 \text{ m/s}}^0 \frac{v \, dv}{0.001v^2 + 9.81} = -500 \ln(v^2 + 9810) \Big|_{50 \text{ m/s}}^0$$

$$h = 114 \text{ m} \qquad \qquad \textit{Ans.}$$

NOTE: The answer indicates a lower elevation than that obtained in part (a) due to atmospheric resistance or drag.

EXAMPLE | 13.3

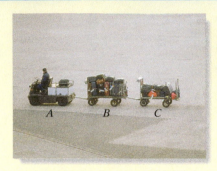

(© R.C. Hibbeler)

The baggage truck *A* shown in the photo has a weight of 900 lb and tows a 550-lb cart *B* and a 325-lb cart *C*. For a short time the driving frictional force developed at the wheels of the truck is $F_A = (40t)$ lb, where *t* is in seconds. If the truck starts from rest, determine its speed in 2 seconds. Also, what is the horizontal force acting on the coupling between the truck and cart *B* at this instant? Neglect the size of the truck and carts.

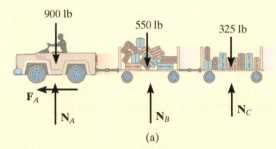

(a)

SOLUTION

Free-Body Diagram.   As shown in Fig. 13–8*a*, it is the frictional driving force that gives both the truck and carts an acceleration. Here we have considered all three vehicles as a single system.

Equation of Motion.   Only motion in the horizontal direction has to be considered.

$$\xleftarrow{+} \Sigma F_x = ma_x; \qquad 40t = \left(\frac{900 + 550 + 325}{32.2}\right)a$$

$$a = 0.7256t$$

Kinematics.   Since the acceleration is a function of time, the velocity of the truck is obtained using $a = dv/dt$ with the initial condition that $v_0 = 0$ at $t = 0$. We have

$$\int_0^v dv = \int_0^{2\,s} 0.7256t\, dt; \qquad v = 0.3628t^2 \Big|_0^{2\,s} = 1.45 \text{ ft/s} \qquad \textit{Ans.}$$

Free-Body Diagram.   In order to determine the force between the truck and cart *B*, we will consider a free-body diagram of the truck so that we can "expose" the coupling force **T** as external to the free-body diagram, Fig. 13–8*b*.

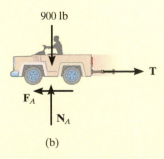

(b)

Fig. 13–8

Equation of Motion.   When $t = 2$ s, then

$$\xleftarrow{+} \Sigma F_x = ma_x: \qquad 40(2) - T = \left(\frac{900}{32.2}\right)[0.7256(2)]$$

$$T = 39.4 \text{ lb} \qquad \textit{Ans.}$$

NOTE: Try and obtain this same result by considering a free-body diagram of carts *B* and *C* as a single system.

EXAMPLE 13.4

A smooth 2-kg collar, shown in Fig. 13–9a, is attached to a spring having a stiffness $k = 3$ N/m and an unstretched length of 0.75 m. If the collar is released from rest at A, determine its acceleration and the normal force of the rod on the collar at the instant $y = 1$ m.

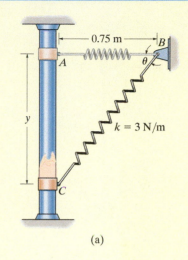

(a)

SOLUTION

Free-Body Diagram.   The free-body diagram of the collar when it is located at the arbitrary position y is shown in Fig. 13–9b. Furthermore, the collar is *assumed* to be accelerating so that "**a**" acts downward in the *positive y* direction. There are four unknowns, namely, N_C, F_s, a, and θ.

Equations of Motion.

$$\xrightarrow{+} \Sigma F_x = ma_x; \qquad -N_C + F_s \cos \theta = 0 \qquad (1)$$

$$+\downarrow \Sigma F_y = ma_y; \qquad 19.62 - F_s \sin \theta = 2a \qquad (2)$$

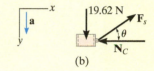

(b)

Fig. 13–9

From Eq. 2 it is seen that the acceleration depends on the magnitude and direction of the spring force. Solution for N_C and a is possible once F_s and θ are known.

The magnitude of the spring force is a function of the stretch s of the spring; i.e., $F_s = ks$. Here the unstretched length is $AB = 0.75$ m, Fig. 13–9a; therefore, $s = CB - AB = \sqrt{y^2 + (0.75)^2} - 0.75$. Since $k = 3$ N/m, then

$$F_s = ks = 3 \left(\sqrt{y^2 + (0.75)^2} - 0.75 \right) \qquad (3)$$

From Fig. 13–9a, the angle θ is related to y by trigonometry.

$$\tan \theta = \frac{y}{0.75}$$

Substituting $y = 1$ m into Eqs. 3 and 4 yields $F_s = 1.50$ N and $\theta = 53.1°$. Substituting these results into Eqs. 1 and 2, we obtain

$$N_C = 0.900 \text{ N} \qquad\qquad\qquad Ans.$$

$$a = 9.21 \text{ m/s}^2 \downarrow \qquad\qquad\qquad Ans.$$

NOTE: This is not a case of constant acceleration, since the spring force changes both its magnitude and direction as the collar moves downward.

EXAMPLE | 13.5

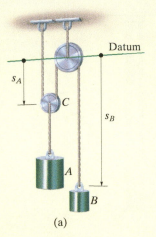

(a)

T T

$2T$

(b)

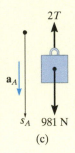

$2T$

$\mathbf{a}_A$

s_A 981 N

(c)

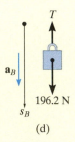

T

$\mathbf{a}_B$

196.2 N

s_B

(d)

Fig. 13–10

The 100-kg block A shown in Fig. 13–10a is released from rest. If the masses of the pulleys and the cord are neglected, determine the velocity of the 20-kg block B in 2 s.

SOLUTION

Free-Body Diagrams.   Since the mass of the pulleys is *neglected*, then for pulley C, $ma = 0$ and we can apply $\Sigma F_y = 0$, as shown in Fig. 13–10b. The free-body diagrams for blocks A and B are shown in Fig. 13–10c and d, respectively. Notice that for A to remain stationary $T = 490.5$ N, whereas for B to remain static $T = 196.2$ N. Hence A will move down while B moves up. Although this is the case, we will assume both blocks accelerate downward, in the direction of $+s_A$ and $+s_B$. The three unknowns are T, a_A, and a_B.

Equations of Motion.   Block A,

$$+\downarrow \Sigma F_y = ma_y; \qquad\qquad 981 - 2T = 100a_A \qquad\qquad (1)$$

Block B,

$$+\downarrow \Sigma F_y = ma_y; \qquad\qquad 196.2 - T = 20a_B \qquad\qquad (2)$$

Kinematics.   The necessary third equation is obtained by relating a_A to a_B using a dependent motion analysis, discussed in Sec. 12.9. The coordinates s_A and s_B in Fig. 13–10a measure the positions of A and B from the fixed datum. It is seen that

$$2s_A + s_B = l$$

where l is constant and represents the total vertical length of cord. Differentiating this expression twice with respect to time yields

$$2a_A = -a_B \qquad\qquad (3)$$

Notice that when writing Eqs. 1 to 3, the *positive direction was always assumed downward*. It is very important to be *consistent* in this assumption since we are seeking a simultaneous solution of equations. The results are

$$T = 327.0 \text{ N}$$
$$a_A = 3.27 \text{ m/s}^2$$
$$a_B = -6.54 \text{ m/s}^2$$

Hence when block A accelerates *downward*, block B accelerates *upward* as expected. Since a_B is constant, the velocity of block B in 2 s is thus

$$(+\downarrow) \qquad\qquad\qquad v = v_0 + a_B t$$
$$= 0 + (-6.54)(2)$$
$$= -13.1 \text{ m/s} \qquad\qquad Ans.$$

The negative sign indicates that block B is moving upward.

PRELIMINARY PROBLEMS

P13–1. The 10-kg block is subjected to the forces shown. In each case, determine its velocity when $t = 2$ s if $v = 0$ when $t = 0$.

P13–3. Determine the initial acceleration of the 10-kg smooth collar. The spring has an unstretched length of 1 m.

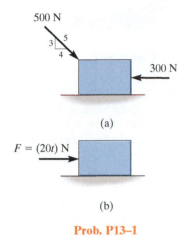

(a)

(b)

Prob. P13–1

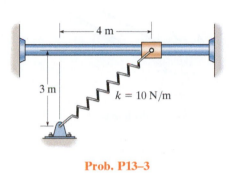

Prob. P13–3

P13–2. The 10-kg block is subjected to the forces shown. In each case, determine its velocity at $s = 8$ m if $v = 3$ m/s at $s = 0$. Motion occurs to the right.

P13–4. Write the equations of motion in the x and y directions for the 10-kg block.

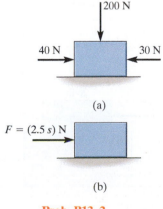

(a)

(b)

Prob. P13–2

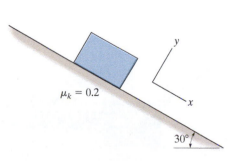

Prob. P13–4

FUNDAMENTAL PROBLEMS

F13–1. The motor winds in the cable with a constant acceleration, such that the 20-kg crate moves a distance $s = 6$ m in 3 s, starting from rest. Determine the tension developed in the cable. The coefficient of kinetic friction between the crate and the plane is $\mu_k = 0.3$.

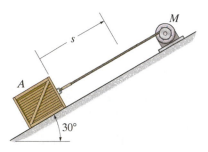

Prob. F13–1

F13–2. If motor M exerts a force of $F = (10t^2 + 100)$ N on the cable, where t is in seconds, determine the velocity of the 25-kg crate when $t = 4$ s. The coefficients of static and kinetic friction between the crate and the plane are $\mu_s = 0.3$ and $\mu_k = 0.25$, respectively. The crate is initially at rest.

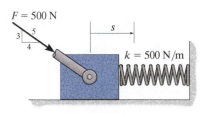

Prob. F13–2

F13–3. A spring of stiffness $k = 500$ N/m is mounted against the 10-kg block. If the block is subjected to the force of $F = 500$ N, determine its velocity at $s = 0.5$ m. When $s = 0$, the block is at rest and the spring is uncompressed. The contact surface is smooth.

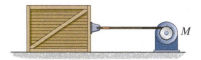

Prob. F13–3

F13–4. The 2-Mg car is being towed by a winch. If the winch exerts a force of $T = 100(s + 1)$ N on the cable, where s is the displacement of the car in meters, determine the speed of the car when $s = 10$ m, starting from rest. Neglect rolling resistance of the car.

Prob. F13–4

F13–5. The spring has a stiffness $k = 200$ N/m and is unstretched when the 25-kg block is at A. Determine the acceleration of the block when $s = 0.4$ m. The contact surface between the block and the plane is smooth.

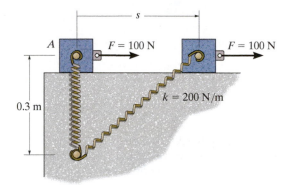

Prob. F13–5

F13–6. Block B rests upon a smooth surface. If the coefficients of static and kinetic friction between A and B are $\mu_s = 0.4$ and $\mu_k = 0.3$, respectively, determine the acceleration of each block if $P = 6$ lb.

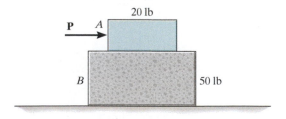

Prob. F13–6

PROBLEMS

13–1. The 6-lb particle is subjected to the action of its weight and forces $\mathbf{F}_1 = \{2\mathbf{i} + 6\mathbf{j} - 2t\mathbf{k}\}$ lb, $\mathbf{F}_2 = \{t^2\mathbf{i} - 4t\mathbf{j} - 1\mathbf{k}\}$ lb, and $\mathbf{F}_3 = \{-2t\mathbf{i}\}$ lb, where t is in seconds. Determine the distance the ball is from the origin 2 s after being released from rest.

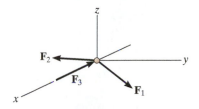

Prob. 13–1

13–2. The two boxcars A and B have a weight of 20 000 lb and 30 000 lb, respectively. If they are freely coasting down the incline when the brakes are applied to all the wheels of car A, determine the force in the coupling C between the two cars. The coefficient of kinetic friction between the wheels of A and the tracks is $\mu_k = 0.5$. The wheels of car B are free to roll. Neglect their mass in the calculation. *Suggestion:* Solve the problem by representing single resultant normal forces acting on A and B, respectively.

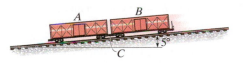

Prob. 13–2

13–3. If the coefficient of kinetic friction between the 50-kg crate and the ground is $\mu_k = 0.3$, determine the distance the crate travels and its velocity when $t = 3$ s. The crate starts from rest, and $P = 200$ N.

*13–4. If the 50-kg crate starts from rest and achieves a velocity of $v = 4$ m/s when it travels a distance of 5 m to the right, determine the magnitude of force $\mathbf{P}$ acting on the crate. The coefficient of kinetic friction between the crate and the ground is $\mu_k = 0.3$.

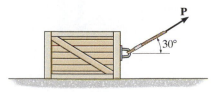

Probs. 13–3/4

13–5. If blocks A and B of mass 10 kg and 6 kg respectively, are placed on the inclined plane and released, determine the force developed in the link. The coefficients of kinetic friction between the blocks and the inclined plane are $\mu_A = 0.1$ and $\mu_B = 0.3$. Neglect the mass of the link.

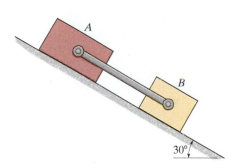

Prob. 13–5

13–6. The 10-lb block has a speed of 4 ft/s when the force of $F = (8t^2)$ lb is applied. Determine the velocity of the block when $t = 2$ s. The coefficient of kinetic friction at the surface is $\mu_k = 0.2$.

13–7. The 10-lb block has a speed of 4 ft/s when the force of $F = (8t^2)$ lb is applied. Determine the velocity of the block when it moves $s = 30$ ft. The coefficient of kinetic friction at the surface is $\mu_s = 0.2$.

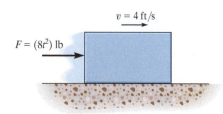

Probs. 13–6/7

*13–8. The speed of the 3500-lb sports car is plotted over the 30-s time period. Plot the variation of the traction force **F** needed to cause the motion.

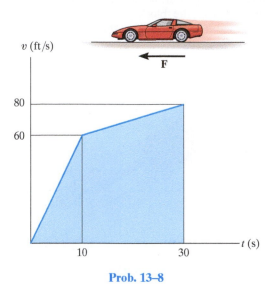

Prob. 13–8

13–9. The conveyor belt is moving at 4 m/s. If the coefficient of static friction between the conveyor and the 10-kg package B is $\mu_s = 0.2$, determine the shortest time the belt can stop so that the package does not slide on the belt.

13–10. The conveyor belt is designed to transport packages of various weights. Each 10-kg package has a coefficient of kinetic friction $\mu_k = 0.15$. If the speed of the conveyor is 5 m/s, and then it suddenly stops, determine the distance the package will slide on the belt before coming to rest.

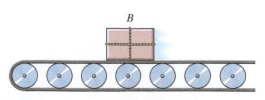

Probs. 13–9/10

13–11. Determine the time needed to pull the cord at B down 4 ft starting from rest when a force of 10 lb is applied to the cord. Block A weighs 20 lb. Neglect the mass of the pulleys and cords.

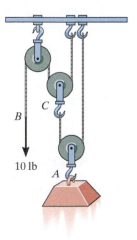

Prob. 13–11

*13–12. Cylinder B has a mass m and is hoisted using the cord and pulley system shown. Determine the magnitude of force **F** as a function of the block's vertical position y so that when **F** is applied the block rises with a constant acceleration $\mathbf{a}_B$. Neglect the mass of the cord and pulleys.

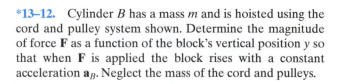

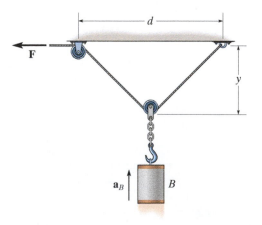

Prob. 13–12

13–13. Block A has a weight of 8 lb and block B has a weight of 6 lb. They rest on a surface for which the coefficient of kinetic friction is $\mu_k = 0.2$. If the spring has a stiffness of $k = 20$ lb/ft, and it is compressed 0.2 ft, determine the acceleration of each block just after they are released.

Prob. 13–13

13–14. The 2-Mg truck is traveling at 15 m/s when the brakes on all its wheels are applied, causing it to skid for a distance of 10 m before coming to rest. Determine the constant horizontal force developed in the coupling C, and the frictional force developed between the tires of the truck and the road during this time. The total mass of the boat and trailer is 1 Mg.

Prob. 13–14

13–15. The motor lifts the 50-kg crate with an acceleration of 6 m/s². Determine the components of force reaction and the couple moment at the fixed support A.

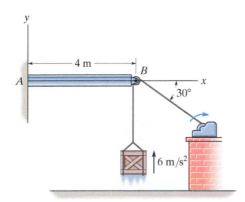

Prob. 13–15

***13–16.** The 75-kg man pushes on the 150-kg crate with a horizontal force $\mathbf{F}$. If the coefficients of static and kinetic friction between the crate and the surface are $\mu_s = 0.3$ and $\mu_k = 0.2$, and the coefficient of static friction between the man's shoes and the surface is $\mu_s = 0.8$, show that the man is able to move the crate. What is the greatest acceleration the man can give the crate?

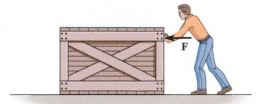

Prob. 13–16

13–17. Determine the acceleration of the blocks when the system is released. The coefficient of kinetic friction is μ_k, and the mass of each block is m. Neglect the mass of the pulleys and cord.

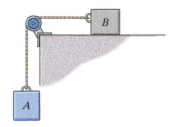

Prob. 13–17

13–18. A 40-lb suitcase slides from rest 20 ft down the smooth ramp. Determine the point where it strikes the ground at C. How long does it take to go from A to C?

13–19. Solve Prob. 13–18 if the suitcase has an initial velocity down the ramp of $v_A = 10$ ft/s and the coefficient of kinetic friction along AB is $\mu_k = 0.2$.

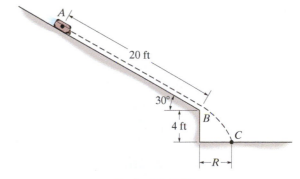

Probs. 13–18/19

13

*13–20. The conveyor belt delivers each 12-kg crate to the ramp at A such that the crate's speed is $v_A = 2.5$ m/s, directed down *along* the ramp. If the coefficient of kinetic friction between each crate and the ramp is $\mu_k = 0.3$, determine the speed at which each crate slides off the ramp at B. Assume that no tipping occurs. Take $\theta = 30°$.

13–21. The conveyor belt delivers each 12-kg crate to the ramp at A such that the crate's speed is $v_A = 2.5$ m/s, directed down *along* the ramp. If the coefficient of kinetic friction between each crate and the ramp is $\mu_k = 0.3$, determine the smallest incline θ of the ramp so that the crates will slide off and fall into the cart.

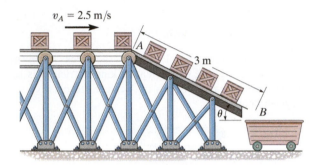

Probs. 13–20/21

13–22. The 50-kg block A is released from rest. Determine the velocity of the 15-kg block B in 2 s.

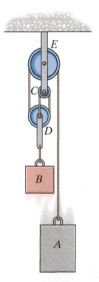

Prob. 13–22

13–23. If the supplied force $F = 150$ N, determine the velocity of the 50-kg block A when it has risen 3 m, starting from rest.

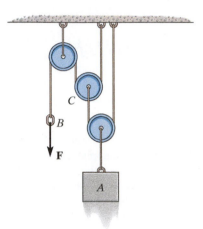

Prob. 13–23

*13–24. A 60-kg suitcase slides from rest 5 m down the smooth ramp. Determine the distance R where it strikes the ground at B. How long does it take to go from A to B?

13–25. Solve Prob. 13–24 if the suitcase has an initial velocity down the ramp of $v_A = 2$ m/s, and the coefficient of kinetic friction along AC is $\mu_k = 0.2$.

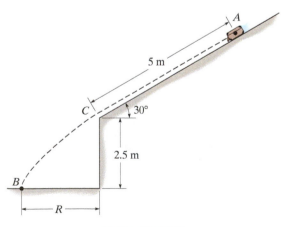

Probs. 13–24/25

13–26. The 1.5 Mg sports car has a tractive force of $F = 4.5$ kN. If it produces the velocity described by v-t graph shown, plot the air resistance R versus t for this time period.

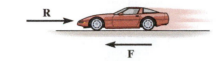

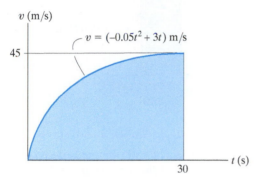

Prob. 13–26

13–27. The conveyor belt is moving downward at 4 m/s. If the coefficient of static friction between the conveyor and the 15-kg package B is $\mu_s = 0.8$, determine the shortest time the belt can stop so that the package does not slide on the belt.

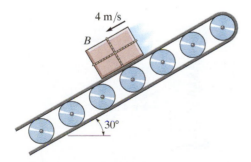

Prob. 13–27

*13–28. At the instant shown the 100-lb block A is moving down the plane at 5 ft/s while being attached to the 50-lb block B. If the coefficient of kinetic friction between the block and the incline is $\mu_k = 0.2$, determine the acceleration of A and the distance A slides before it stops. Neglect the mass of the pulleys and cables.

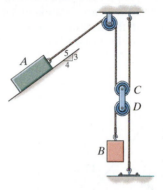

Prob. 13–28

13–29. The force exerted by the motor on the cable is shown in the graph. Determine the velocity of the 200-lb crate when $t = 2.5$ s.

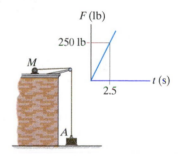

Prob. 13–29

13–30. The force of the motor M on the cable is shown in the graph. Determine the velocity of the 400-kg crate A when $t = 2$ s.

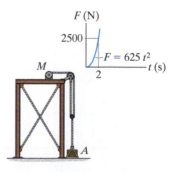

Prob. 13–30

13–31.   The tractor is used to lift the 150-kg load B with the 24-m-long rope, boom, and pulley system. If the tractor travels to the right at a constant speed of 4 m/s, determine the tension in the rope when $s_A = 5$ m. When $s_A = 0$, $s_B = 0$.

*13–32.   The tractor is used to lift the 150-kg load B with the 24-m-long rope, boom, and pulley system. If the tractor travels to the right with an acceleration of 3 m/s² and has a velocity of 4 m/s at the instant $s_A = 5$ m, determine the tension in the rope at this instant. When $s_A = 0$, $s_B = 0$.

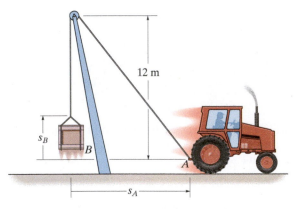

Probs. 13–31/32

13–33.   Block A and B each have a mass m. Determine the largest horizontal force $\mathbf{P}$ which can be applied to B so that it will not slide on A. Also, what is the corresponding acceleration? The coefficient of static friction between A and B is μ_s. Neglect any friction between A and the horizontal surface.

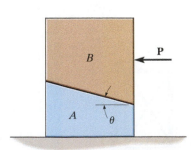

Prob. 13–33

13–34.   The 4-kg smooth cylinder is supported by the spring having a stiffness of $k_{AB} = 120$ N/m. Determine the velocity of the cylinder when it moves downward $s = 0.2$ m from its equilibrium position, which is caused by the application of the force $F = 60$ N.

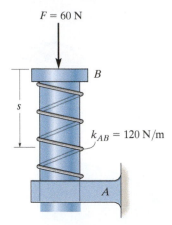

Prob. 13–34

13–35.   The coefficient of static friction between the 200-kg crate and the flat bed of the truck is $\mu_s = 0.3$. Determine the shortest time for the truck to reach a speed of 60 km/h, starting from rest with constant acceleration, so that the crate does not slip.

Prob. 13–35

*13–36.   The 2-lb collar C fits loosely on the smooth shaft. If the spring is unstretched when $s = 0$ and the collar is given a velocity of 15 ft/s, determine the velocity of the collar when $s = 1$ ft.

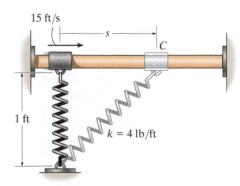

Prob. 13–36

13–37. The 10-kg block A rests on the 50-kg plate B in the position shown. Neglecting the mass of the rope and pulley, and using the coefficients of kinetic friction indicated, determine the time needed for block A to slide 0.5 m *on the plate* when the system is released from rest.

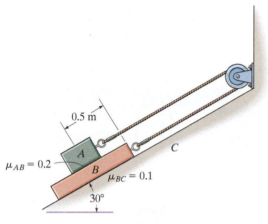

Prob. 13–37

13–38. The 300-kg bar B, originally at rest, is being towed over a series of small rollers. Determine the force in the cable when $t = 5$ s, if the motor M is drawing in the cable for a short time at a rate of $v = (0.4t^2)$ m/s, where t is in seconds ($0 \le t \le 6$ s). How far does the bar move in 5 s? Neglect the mass of the cable, pulley, and the rollers.

13–39. An electron of mass m is discharged with an initial horizontal velocity of $\mathbf{v}_0$. If it is subjected to two fields of force for which $F_x = F_0$ and $F_y = 0.3F_0$, where F_0 is constant, determine the equation of the path, and the speed of the electron at any time t.

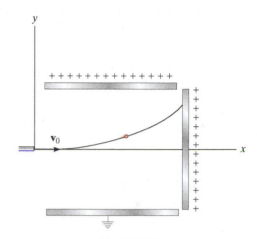

Prob. 13–39

*13–40. The 400-lb cylinder at A is hoisted using the motor and the pulley system shown. If the speed of point B on the cable is increased at a constant rate from zero to $v_B = 10$ ft/s in $t = 5$ s, determine the tension in the cable at B to cause the motion.

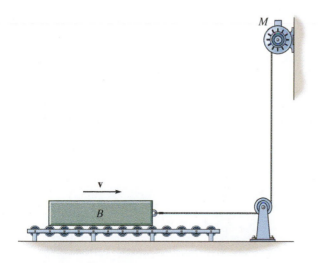

Prob. 13–38

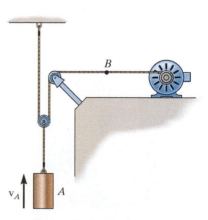

Prob. 13–40

13–41. Block A has a mass m_A and is attached to a spring having a stiffness k and unstretched length l_0. If another block B, having a mass m_B, is pressed against A so that the spring deforms a distance d, determine the distance both blocks slide on the smooth surface before they begin to separate. What is their velocity at this instant?

13–42. Block A has a mass m_A and is attached to a spring having a stiffness k and unstretched length l_0. If another block B, having a mass m_B, is pressed against A so that the spring deforms a distance d, show that for separation to occur it is necessary that $d > 2\mu_k g(m_A + m_B)/k$, where μ_k is the coefficient of kinetic friction between the blocks and the ground. Also, what is the distance the blocks slide on the surface before they separate?

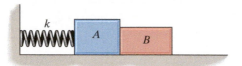

Probs. 13–41/42

13–43. A parachutist having a mass m opens his parachute from an at-rest position at a very high altitude. If the atmospheric drag resistance is $F_D = kv^2$, where k is a constant, determine his velocity when he has fallen for a time t. What is his velocity when he lands on the ground? This velocity is referred to as the *terminal velocity*, which is found by letting the time of fall $t \to \infty$.

Prob. 13–43

*13–44. If the motor draws in the cable with an acceleration of 3 m/s², determine the reactions at the supports A and B. The beam has a uniform mass of 30 kg/m, and the crate has a mass of 200 kg. Neglect the mass of the motor and pulleys.

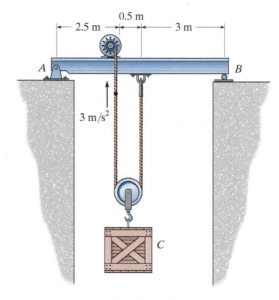

Prob. 13–44

13–45. If the force exerted on cable AB by the motor is $F = (100t^{3/2})$ N, where t is in seconds, determine the 50-kg crate's velocity when $t = 5$ s. The coefficients of static and kinetic friction between the crate and the ground are $\mu_s = 0.4$ and $\mu_k = 0.3$, respectively. Initially the crate is at rest.

Prob. 13–45

13–46. Blocks A and B each have a mass m. Determine the largest horizontal force $\mathbf{P}$ which can be applied to B so that A will not move relative to B. All surfaces are smooth.

13–47. Blocks A and B each have a mass m. Determine the largest horizontal force $\mathbf{P}$ which can be applied to B so that A will not slip on B. The coefficient of static friction between A and B is μ_s. Neglect any friction between B and C.

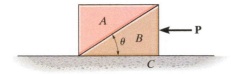

Probs. 13–46/47

*13–48. The smooth block B of negligible size has a mass m and rests on the horizontal plane. If the board AC pushes on the block at an angle θ with a constant acceleration $\mathbf{a}_0$, determine the velocity of the block along the board and the distance s the block moves along the board as a function of time t. The block starts from rest when $s = 0$, $t = 0$.

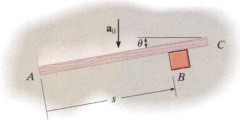

Prob. 13–48

13–49. If a horizontal force $P = 12$ lb is applied to block A determine the acceleration of the block B. Neglect friction.

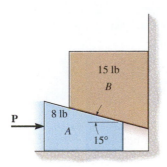

Prob. 13–49

13–50. A freight elevator, including its load, has a mass of 1 Mg. It is prevented from rotating due to the track and wheels mounted along its sides. If the motor M develops a constant tension $T = 4$ kN in its attached cable, determine the velocity of the elevator when it has moved upward 6 m starting from rest. Neglect the mass of the pulleys and cables.

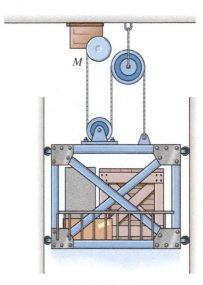

Prob. 13–50

13–51. The block A has a mass m_A and rests on the pan B, which has a mass m_B. Both are supported by a spring having a stiffness k that is attached to the bottom of the pan and to the ground. Determine the distance d the pan should be pushed down from the equilibrium position and then released from rest so that separation of the block will take place from the surface of the pan at the instant the spring becomes unstretched.

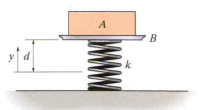

Prob. 13–51

13

13.5   Equations of Motion: Normal and Tangential Coordinates

When a particle moves along a curved path which is known, the equation of motion for the particle may be written in the tangential, normal, and binormal directions, Fig. 13–11. Note that there is no motion of the particle in the binormal direction, since the particle is constrained to move along the path. We have

$$\Sigma \mathbf{F} = m\mathbf{a}$$

$$\Sigma F_t \mathbf{u}_t + \Sigma F_n \mathbf{u}_n + \Sigma F_b \mathbf{u}_b = m\mathbf{a}_t + m\mathbf{a}_n$$

This equation is satisfied provided

$$\Sigma F_t = ma_t$$
$$\Sigma F_n = ma_n \qquad\qquad (13\text{–}8)$$
$$\Sigma F_b = 0$$

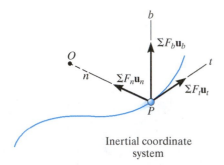

Inertial coordinate
system

Fig. 13–11

Recall that $a_t\,(= dv/dt)$ represents the time rate of change in the magnitude of velocity. So if $\Sigma \mathbf{F}_t$ acts in the direction of motion, the particle's speed will increase, whereas if it acts in the opposite direction, the particle will slow down. Likewise, $a_n\,(= v^2/\rho)$ represents the time rate of change in the velocity's direction. It is caused by $\Sigma \mathbf{F}_n$, which *always* acts in the positive n direction, i.e., toward the path's center of curvature. For this reason it is often referred to as the *centripetal force*.

As a roller coaster falls downward along the track, the cars have both a normal and a tangential component of acceleration. (© R.C. Hibbeler)

Procedure for Analysis

When a problem involves the motion of a particle along a *known curved path*, normal and tangential coordinates should be considered for the analysis since the acceleration components can be readily formulated. The method for applying the equations of motion, which relate the forces to the acceleration, has been outlined in the procedure given in Sec. 13.4. Specifically, for t, n, b coordinates it may be stated as follows:

Free-Body Diagram.

* Establish the inertial t, n, b coordinate system at the particle and draw the particle's free-body diagram.

* The particle's normal acceleration $\mathbf{a}_n$ *always* acts in the positive n direction.

* If the tangential acceleration $\mathbf{a}_t$ is unknown, assume it acts in the positive t direction.

* There is no acceleration in the b direction.

* Identify the unknowns in the problem.

Equations of Motion.

* Apply the equations of motion, Eq. 13–8.

Kinematics.

* Formulate the tangential and normal components of acceleration; i.e., $a_t = dv/dt$ or $a_t = v\,dv/ds$ and $a_n = v^2/\rho$.

* If the path is defined as $y = f(x)$, the radius of curvature at the point where the particle is located can be obtained from $\rho = [1 + (dy/dx)^2]^{3/2}/|d^2y/dx^2|$.

The unbalanced force of the rope on the skier gives him a normal component of acceleration. (© R.C. Hibbeler)

13

EXAMPLE 13.6

Determine the banking angle θ for the race track so that the wheels of the racing cars shown in Fig. 13–12a will not have to depend upon friction to prevent any car from sliding up or down the track. Assume the cars have negligible size, a mass m, and travel around the curve of radius ρ with a constant speed v.

(© R.C. Hibbeler)

(a)

SOLUTION

Before looking at the following solution, give some thought as to why it should be solved using t, n, b coordinates.

Free-Body Diagram. As shown in Fig. 13–12b, and as stated in the problem, no frictional force acts on the car. Here N_C represents the *resultant* of the ground on all four wheels. Since a_n can be calculated, the unknowns are N_C and θ.

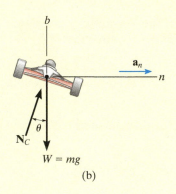

(b)

Fig. 13–12

Equations of Motion.   Using the n, b axes shown,

$$\xrightarrow{+} \Sigma F_n = ma_n; \qquad N_C \sin \theta = m\frac{v^2}{\rho} \qquad (1)$$

$$+\uparrow \Sigma F_b = 0; \qquad N_C \cos \theta - mg = 0 \qquad (2)$$

Eliminating N_C and m from these equations by dividing Eq. 1 by Eq. 2, we obtain

$$\tan \theta = \frac{v^2}{g\rho}$$

$$\theta = \tan^{-1}\left(\frac{v^2}{g\rho}\right) \qquad \qquad Ans.$$

NOTE: The result is independent of the mass of the car. Also, a force summation in the tangential direction is of no consequence to the solution. If it were considered, then $a_t = dv/dt = 0$, since the car moves with *constant speed*. A further analysis of this problem is discussed in Prob. 21–53.

EXAMPLE | 13.7

The 3-kg disk D is attached to the end of a cord as shown in Fig. 13–13a. The other end of the cord is attached to a ball-and-socket joint located at the center of a platform. If the platform rotates rapidly, and the disk is placed on it and released from rest as shown, determine the time it takes for the disk to reach a speed great enough to break the cord. The maximum tension the cord can sustain is 100 N, and the coefficient of kinetic friction between the disk and the platform is $\mu_k = 0.1$.

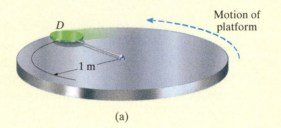

(a)

SOLUTION

Free-Body Diagram. The frictional force has a magnitude $F = \mu_k N_D = 0.1 N_D$ and a sense of direction that opposes the *relative motion* of the disk with respect to the platform. It is this force that gives the disk a tangential component of acceleration causing v to increase, thereby causing T to increase until it reaches 100 N. The weight of the disk is $W = 3(9.81) = 29.43$ N. Since a_n can be related to v, the unknowns are N_D, a_t, and v.

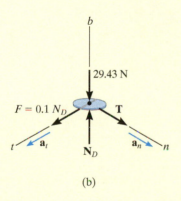

(b)

Fig. 13–13

Equations of Motion.

$$\Sigma F_n = m a_n; \qquad\qquad T = 3\left(\frac{v^2}{1}\right) \qquad (1)$$

$$\Sigma F_t = m a_t; \qquad\qquad 0.1 N_D = 3 a_t \qquad (2)$$

$$\Sigma F_b = 0; \qquad\quad N_D - 29.43 = 0 \qquad (3)$$

Setting $T = 100$ N, Eq. 1 can be solved for the critical speed v_{cr} of the disk needed to break the cord. Solving all the equations, we obtain

$$N_D = 29.43 \text{ N}$$

$$a_t = 0.981 \text{ m/s}^2$$

$$v_{cr} = 5.77 \text{ m/s}$$

Kinematics.   Since a_t is *constant*, the time needed to break the cord is

$$v_{cr} = v_0 + a_t t$$

$$5.77 = 0 + (0.981)t$$

$$t = 5.89 \text{ s} \qquad\qquad\qquad Ans.$$

EXAMPLE 13.8

Design of the ski jump shown in the photo requires knowing the type of forces that will be exerted on the skier and her approximate trajectory. If in this case the jump can be approximated by the parabola shown in Fig. 13–14a, determine the normal force on the 150-lb skier the instant she arrives at the end of the jump, point A, where her velocity is 65 ft/s. Also, what is her acceleration at this point?

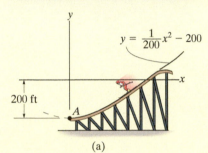

(© R.C. Hibbeler)

SOLUTION

Why consider using n, t coordinates to solve this problem?

Free-Body Diagram. Since $dy/dx = x/100|_{x=0} = 0$, the slope at A is horizontal. The free-body diagram of the skier when she is at A is shown in Fig. 13–14b. Since the path is *curved*, there are two components of acceleration, $\mathbf{a}_n$ and $\mathbf{a}_t$. Since a_n can be calculated, the unknowns are a_t and N_A.

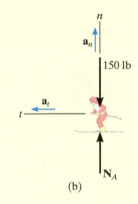

(a)

Equations of Motion.

$$+\uparrow \Sigma F_n = ma_n; \qquad N_A - 150 = \frac{150}{32.2}\left(\frac{(65)^2}{\rho}\right) \qquad (1)$$

$$\xleftarrow{+} \Sigma F_t = ma_t; \qquad 0 = \frac{150}{32.2}a_t \qquad (2)$$

The radius of curvature ρ for the path must be determined at point $A(0, -200 \text{ ft})$. Here $y = \frac{1}{200}x^2 - 200$, $dy/dx = \frac{1}{100}x$, $d^2y/dx^2 = \frac{1}{100}$, so that at $x = 0$,

$$\rho = \frac{[1 + (dy/dx)^2]^{3/2}}{|d^2y/dx^2|}\bigg|_{x=0} = \frac{[1 + (0)^2]^{3/2}}{|\frac{1}{100}|} = 100 \text{ ft}$$

Substituting this into Eq. 1 and solving for N_A, we obtain

$$N_A = 347 \text{ lb} \qquad \textit{Ans.}$$

Kinematics. From Eq. 2,

$$a_t = 0$$

Thus,

$$a_n = \frac{v^2}{\rho} = \frac{(65)^2}{100} = 42.2 \text{ ft/s}^2$$

$$a_A = a_n = 42.2 \text{ ft/s}^2 \uparrow \qquad \textit{Ans.}$$

(b)

Fig. 13–14

NOTE: Apply the equation of motion in the y direction and show that when the skier is in midair her downward acceleration is 32.2 ft/s².

EXAMPLE 13.9

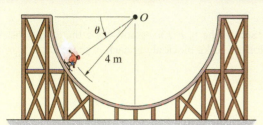

(a)

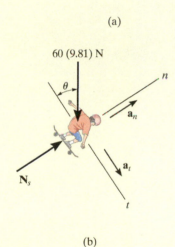

(b)

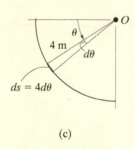

(c)

Fig. 13–15

The 60-kg skateboarder in Fig. 13–15a coasts down the circular track. If he starts from rest when $\theta = 0°$, determine the magnitude of the normal reaction the track exerts on him when $\theta = 60°$. Neglect his size for the calculation.

SOLUTION

Free-Body Diagram.   The free-body diagram of the skateboarder when he is at an *arbitrary position* θ is shown in Fig. 13–15b. At $\theta = 60°$ there are three unknowns, N_s, a_t, and a_n (or v).

Equations of Motion.

$$+\nearrow \Sigma F_n = ma_n; \quad N_s - [60(9.81)\text{N}]\sin\theta = (60\text{ kg})\left(\frac{v^2}{4\text{ m}}\right) \quad (1)$$

$$+\searrow \Sigma F_t = ma_t; \qquad [60(9.81)\text{N}]\cos\theta = (60\text{ kg})\,a_t$$

$$a_t = 9.81\cos\theta$$

Kinematics.   Since a_t is expressed in terms of θ, the equation $v\,dv = a_t\,ds$ must be used to determine the speed of the skateboarder when $\theta = 60°$. Using the geometric relation $s = \theta r$, where $ds = r\,d\theta = (4\text{ m})\,d\theta$, Fig. 13–15c, and the initial condition $v = 0$ at $\theta = 0°$, we have,

$$v\,dv = a_t\,ds$$

$$\int_0^v v\,dv = \int_0^{60°} 9.81\cos\theta(4\,d\theta)$$

$$\left.\frac{v^2}{2}\right|_0^v = 39.24\sin\theta\Big|_0^{60°}$$

$$\frac{v^2}{2} - 0 = 39.24(\sin 60° - 0)$$

$$v^2 = 67.97\text{ m}^2/\text{s}^2$$

Substituting this result and $\theta = 60°$ into Eq. (1), yields

$$N_s = 1529.23\text{ N} = 1.53\text{ kN} \qquad \textit{Ans.}$$

PRELIMINARY PROBLEMS

13

P13–5. Set up the *n, t* axes and write the equations of motion for the 10-kg block along each of these axes.

P13–6. Set up the *n, b, t* axes and write the equations of motion for the 10-kg block along each of these axes.

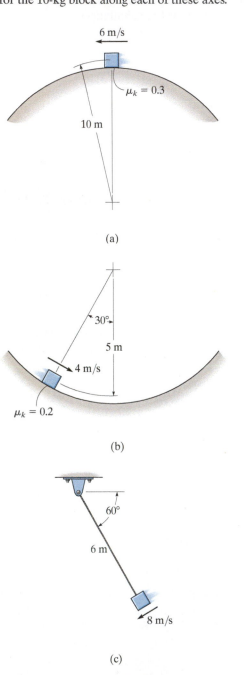

6 m/s

$\mu_k = 0.3$

10 m

(a)

30°

5 m

4 m/s

$\mu_k = 0.2$

(b)

60°

6 m

8 m/s

(c)

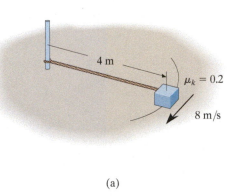

4 m

$\mu_k = 0.2$

8 m/s

(a)

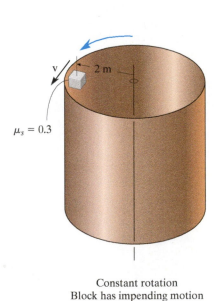

v   2 m

$\mu_s = 0.3$

Constant rotation
Block has impending motion

(b)

Prob. P13–5

Prob. P13–6

FUNDAMENTAL PROBLEMS

F13–7. The block rests at a distance of 2 m from the center of the platform. If the coefficient of static friction between the block and the platform is $\mu_s = 0.3$, determine the maximum speed which the block can attain before it begins to slip. Assume the angular motion of the disk is slowly increasing.

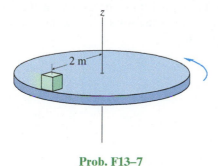

Prob. F13–7

F13–8. Determine the maximum speed that the jeep can travel over the crest of the hill and not lose contact with the road.

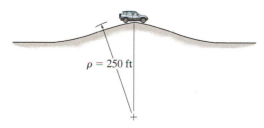

Prob. F13–8

F13–9. A pilot weighs 150 lb and is traveling at a constant speed of 120 ft/s. Determine the normal force he exerts on the seat of the plane when he is upside down at A. The loop has a radius of curvature of 400 ft.

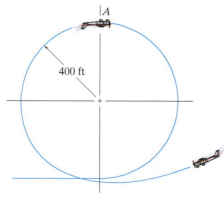

Prob. F13–9

F13–10. The sports car is traveling along a 30° banked road having a radius of curvature of $\rho = 500$ ft. If the coefficient of static friction between the tires and the road is $\mu_s = 0.2$, determine the maximum safe speed so no slipping occurs. Neglect the size of the car.

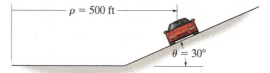

Prob. F13–10

F13–11. If the 10-kg ball has a velocity of 3 m/s when it is at the position A, along the vertical path, determine the tension in the cord and the increase in the speed of the ball at this position.

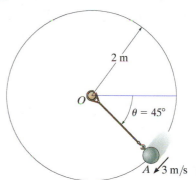

Prob. F13–11

F13–12. The motorcycle has a mass of 0.5 Mg and a negligible size. It passes point A traveling with a speed of 15 m/s, which is increasing at a constant rate of 1.5 m/s². Determine the resultant frictional force exerted by the road on the tires at this instant.

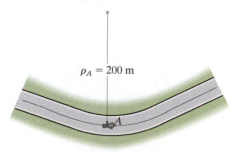

Prob. F13–12

PROBLEMS

*13–52.** A girl, having a mass of 15 kg, sits motionless relative to the surface of a horizontal platform at a distance of $r = 5$ m from the platform's center. If the angular motion of the platform is *slowly* increased so that the girl's tangential component of acceleration can be neglected, determine the maximum speed which the girl will have before she begins to slip off the platform. The coefficient of static friction between the girl and the platform is $\mu = 0.2$.

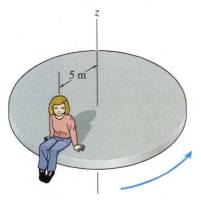

Prob. 13–52

13–53. The 2-kg block *B* and 15-kg cylinder *A* are connected to a light cord that passes through a hole in the center of the smooth table. If the block is given a speed of $v = 10$ m/s, determine the radius r of the circular path along which it travels.

13–54. The 2-kg block *B* and 15-kg cylinder *A* are connected to a light cord that passes through a hole in the center of the smooth table. If the block travels along a circular path of radius $r = 1.5$ m, determine the speed of the block.

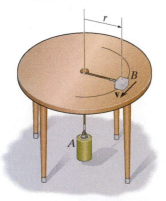

Probs. 13–53/54

13–55. Determine the maximum constant speed at which the pilot can travel around the vertical curve having a radius of curvature $\rho = 800$ m, so that he experiences a maximum acceleration $a_n = 8g = 78.5$ m/s². If he has a mass of 70 kg, determine the normal force he exerts on the seat of the airplane when the plane is traveling at this speed and is at its lowest point.

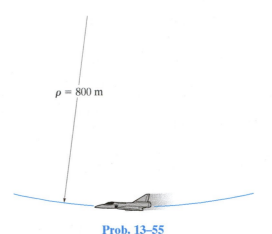

$\rho = 800$ m

Prob. 13–55

*13–56.** Cartons having a mass of 5 kg are required to move along the assembly line at a constant speed of 8 m/s. Determine the smallest radius of curvature, ρ, for the conveyor so the cartons do not slip. The coefficients of static and kinetic friction between a carton and the conveyor are $\mu_s = 0.7$ and $\mu_k = 0.5$, respectively.

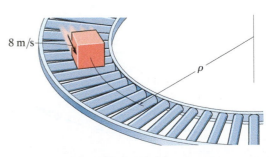

8 m/s

Prob. 13–56

13–57. The collar A, having a mass of 0.75 kg, is attached to a spring having a stiffness of $k = 200$ N/m. When rod BC rotates about the vertical axis, the collar slides outward along the smooth rod DE. If the spring is unstretched when $s = 0$, determine the constant speed of the collar in order that $s = 100$ mm. Also, what is the normal force of the rod on the collar? Neglect the size of the collar.

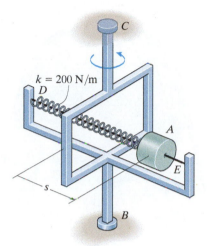

Prob. 13–57

13–58. The 2-kg spool S fits loosely on the inclined rod for which the coefficient of static friction is $\mu_s = 0.2$. If the spool is located 0.25 m from A, determine the minimum constant speed the spool can have so that it does not slip down the rod.

13–59. The 2-kg spool S fits loosely on the inclined rod for which the coefficient of static friction is $\mu_s = 0.2$. If the spool is located 0.25 m from A, determine the maximum constant speed the spool can have so that it does not slip up the rod.

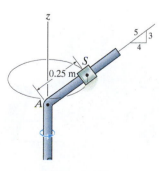

Probs. 13–58/59

*13–60.** At the instant $\theta = 60°$, the boy's center of mass G has a downward speed $v_G = 15$ ft/s. Determine the rate of increase in his speed and the tension in each of the two supporting cords of the swing at this instant. The boy has a weight of 60 lb. Neglect his size and the mass of the seat and cords.

13–61. At the instant $\theta = 60°$, the boy's center of mass G is momentarily at rest. Determine his speed and the tension in each of the two supporting cords of the swing when $\theta = 90°$. The boy has a weight of 60 lb. Neglect his size and the mass of the seat and cords.

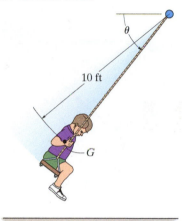

Probs. 13–60/61

13–62. A girl having a mass of 25 kg sits at the edge of the merry-go-round so her center of mass G is at a distance of 1.5 m from the axis of rotation. If the angular motion of the platform is *slowly* increased so that the girl's tangential component of acceleration can be neglected, determine the maximum speed which she can have before she begins to slip off the merry-go-round. The coefficient of static friction between the girl and the merry-go-round is $\mu_s = 0.3$.

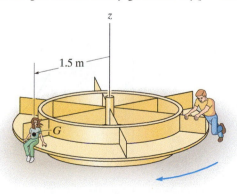

Prob. 13–62

13–63. The pendulum bob B has a weight of 5 lb and is released from rest in the position shown, $\theta = 0°$. Determine the tension in string BC just after the bob is released, $\theta = 0°$, and also at the instant the bob reaches $\theta = 45°$. Take $r = 3$ ft.

*13–64. The pendulum bob B has a mass m and is released from rest when $\theta = 0°$. Determine the tension in string BC immediately afterwards, and also at the instant the bob reaches the arbitrary position θ.

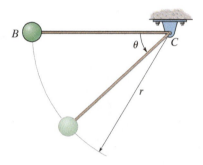

Probs. 13–63/64

13–65. Determine the constant speed of the passengers on the amusement-park ride if it is observed that the supporting cables are directed at $\theta = 30°$ from the vertical. Each chair including its passenger has a mass of 80 kg. Also, what are the components of force in the n, t, and b directions which the chair exerts on a 50-kg passenger during the motion?

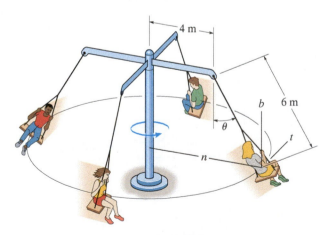

Prob. 13–65

13–66. A motorcyclist in a circus rides his motorcycle within the confines of the hollow sphere. If the coefficient of static friction between the wheels of the motorcycle and the sphere is $\mu_s = 0.4$, determine the minimum speed at which he must travel if he is to ride along the wall when $\theta = 90°$. The mass of the motorcycle and rider is 250 kg, and the radius of curvature to the center of gravity is $\rho = 20$ ft. Neglect the size of the motorcycle for the calculation.

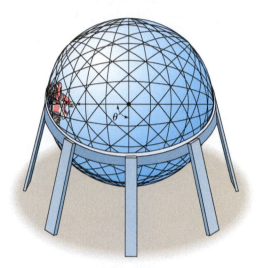

Prob. 13–66

13–67. The vehicle is designed to combine the feel of a motorcycle with the comfort and safety of an automobile. If the vehicle is traveling at a constant speed of 80 km/h along a circular curved road of radius 100 m, determine the tilt angle θ of the vehicle so that only a normal force from the seat acts on the driver. Neglect the size of the driver.

Prob. 13–67

*13–68.   The 0.8-Mg car travels over the hill having the shape of a parabola. If the driver maintains a constant speed of 9 m/s, determine both the resultant normal force and the resultant frictional force that all the wheels of the car exert on the road at the instant it reaches point A. Neglect the size of the car.

13–69.   The 0.8-Mg car travels over the hill having the shape of a parabola. When the car is at point A, it is traveling at 9 m/s and increasing its speed at 3 m/s². Determine both the resultant normal force and the resultant frictional force that all the wheels of the car exert on the road at this instant. Neglect the size of the car.

13–71.   The 150-lb man lies against the cushion for which the coefficient of static friction is $\mu_s = 0.5$. Determine the resultant normal and frictional forces the cushion exerts on him if, it due to rotation about the z axis, he has a constant speed $v = 20$ ft/s. Neglect the size of the man. Take $\theta = 60°$.

*13–72.   The 150-lb man lies against the cushion for which the coefficient of static friction is $\mu_s = 0.5$. If he rotates about the z axis with a constant speed $v = 30$ ft/s, determine the smallest angle θ of the cushion at which he will begin to slip off.

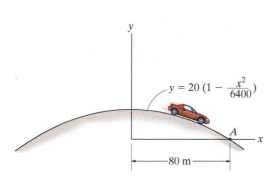

$$y = 20 \left(1 - \frac{x^2}{6400}\right)$$

— 80 m —

Probs. 13–68/69

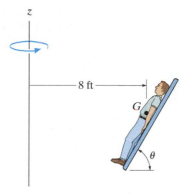

8 ft

Probs. 13–71/72

13–70.   The package has a weight of 5 lb and slides down the chute. When it reaches the curved portion AB, it is traveling at 8 ft/s ($\theta = 0°$). If the chute is smooth, determine the speed of the package when it reaches the intermediate point C ($\theta = 30°$) and when it reaches the horizontal plane ($\theta = 45°$). Also, find the normal force on the package at C.

13–73.   Determine the maximum speed at which the car with mass m can pass over the top point A of the vertical curved road and still maintain contact with the road. If the car maintains this speed, what is the normal reaction the road exerts on the car when it passes the lowest point B on the road?

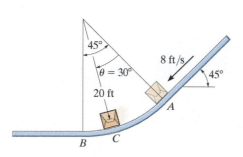

45°
$\theta = 30°$
8 ft/s
45°
20 ft
A
B C

Prob. 13–70

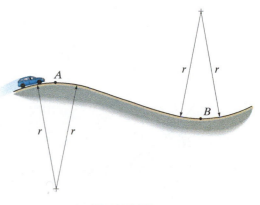

A
r r
B
r r

Prob. 13–73

13–74. Determine the maximum constant speed at which the 2-Mg car can travel over the crest of the hill at A without leaving the surface of the road. Neglect the size of the car in the calculation.

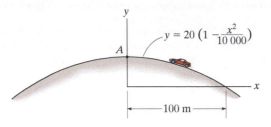

$$y = 20\left(1 - \frac{x^2}{10\,000}\right)$$

100 m

Prob. 13–74

13–75. The box has a mass m and slides down the smooth chute having the shape of a parabola. If it has an initial velocity of v_0 at the origin determine its velocity as a function of x. Also, what is the normal force on the box, and the tangential acceleration as a function of x?

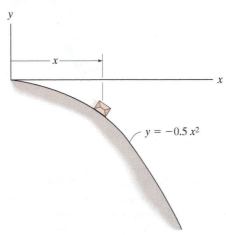

$$y = -0.5\,x^2$$

Prob. 13–75

*13–76. Prove that if the block is released from rest at point B of a smooth path of *arbitrary shape*, the speed it attains when it reaches point A is equal to the speed it attains when it falls freely through a distance h; i.e., $v = \sqrt{2gh}$.

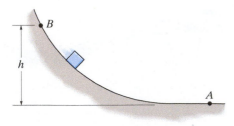

Prob. 13–76

13–77. The cylindrical plug has a weight of 2 lb and it is free to move within the confines of the smooth pipe. The spring has a stiffness $k = 14$ lb/ft and when no motion occurs the distance $d = 0.5$ ft. Determine the force of the spring on the plug when the plug is at rest with respect to the pipe. The plug is traveling with a constant speed of 15 ft/s, which is caused by the rotation of the pipe about the vertical axis.

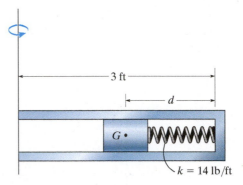

3 ft

d

G

$k = 14$ lb/ft

Prob. 13–77

13–78. When crossing an intersection, a motorcyclist encounters the slight bump or crown caused by the intersecting road. If the crest of the bump has a radius of curvature $\rho = 50$ ft, determine the maximum constant speed at which he can travel without leaving the surface of the road. Neglect the size of the motorcycle and rider in the calculation. The rider and his motorcycle have a total weight of 450 lb.

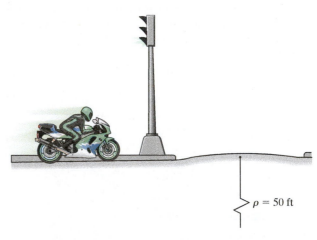

$\rho = 50$ ft

Prob. 13–78

13–79. The airplane, traveling at a constant speed of 50 m/s, is executing a horizontal turn. If the plane is banked at $\theta = 15°$, when the pilot experiences only a normal force on the seat of the plane, determine the radius of curvature ρ of the turn. Also, what is the normal force of the seat on the pilot if he has a mass of 70 kg.

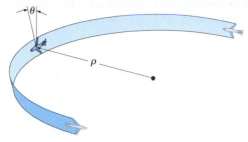

Prob. 13–79

*13–80. The 2-kg pendulum bob moves in the vertical plane with a velocity of 8 m/s when $\theta = 0°$. Determine the initial tension in the cord and also at the instant the bob reaches $\theta = 30°$. Neglect the size of the bob.

13–81. The 2-kg pendulum bob moves in the vertical plane with a velocity of 6 m/s when $\theta = 0°$. Determine the angle θ where the tension in the cord becomes zero.

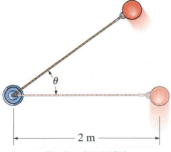

Probs. 13–80/81

13–82. The 8-kg sack slides down the smooth ramp. If it has a speed of 1.5 m/s when $y = 0.2$ m, determine the normal reaction the ramp exerts on the sack and the rate of increase in the speed of sack at this instant.

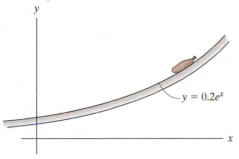

Prob. 13–82

13–83. The ball has a mass m and is attached to the cord of length l. The cord is tied at the top to a swivel and the ball is given a velocity $\mathbf{v}_0$. Show that the angle θ which the cord makes with the vertical as the ball travels around the circular path must satisfy the equation $\tan \theta \sin \theta = v_0^2/gl$. Neglect air resistance and the size of the ball.

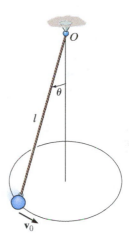

Prob. 13–83

*13–84. The 2-lb block is released from rest at A and slides down along the smooth cylindrical surface. If the attached spring has a stiffness $k = 2$ lb/ft, determine its unstretched length so that it does not allow the block to leave the surface until $\theta = 60°$.

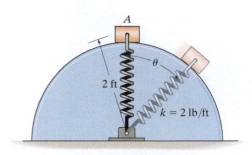

Prob. 13–84

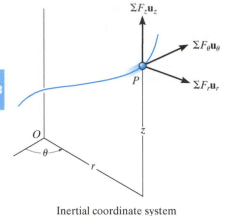

Inertial coordinate system

Fig. 13–16

13.6   Equations of Motion: Cylindrical Coordinates

When all the forces acting on a particle are resolved into cylindrical components, i.e., along the unit-vector directions $\mathbf{u}_r$, $\mathbf{u}_\theta$, $\mathbf{u}_z$, Fig. 13–16, the equation of motion can be expressed as

$$\Sigma\mathbf{F} = m\mathbf{a}$$

$$\Sigma F_r\mathbf{u}_r + \Sigma F_\theta\mathbf{u}_\theta + \Sigma F_z\mathbf{u}_z = ma_r\mathbf{u}_r + ma_\theta\mathbf{u}_\theta + ma_z\mathbf{u}_z$$

To satisfy this equation, we require

$$\begin{array}{c} \Sigma F_r = ma_r \\[4pt] \Sigma F_\theta = ma_\theta \\[4pt] \Sigma F_z = ma_z \end{array} \qquad (13\text{–}9)$$

If the particle is constrained to move only in the r–θ plane, then only the first two of Eq. 13–9 are used to specify the motion.

Tangential and Normal Forces.   The most straightforward type of problem involving cylindrical coordinates requires the determination of the resultant force components ΣF_r, ΣF_θ, ΣF_z which cause a particle to move with a *known* acceleration. If, however, the particle's accelerated motion is not completely specified at the given instant, then some information regarding the directions or magnitudes of the forces acting on the particle must be known or calculated in order to solve Eqs. 13–9. For example, the force **P** causes the particle in Fig. 13–17*a* to move along a path $r = f(\theta)$. The *normal force* **N** which the path exerts on the particle is always *perpendicular to the tangent of the path*, whereas the frictional force **F** always acts along the tangent in the opposite direction of motion. The *directions* of **N** and **F** can be specified relative to the radial coordinate by using the angle ψ (psi), Fig. 13–17*b*, which is defined between the *extended* radial line and the tangent to the curve.

Motion of the roller coaster along this spiral can be studied using cylindrical coordinates. (© R.C. Hibbeler)

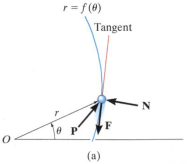

(a)

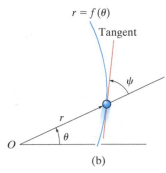

(b)

Fig. 13–17

This angle can be obtained by noting that when the particle is displaced a distance ds along the path, Fig. 13–17c, the component of displacement in the radial direction is dr and the component of displacement in the transverse direction is $r\,d\theta$. Since these two components are mutually perpendicular, the angle ψ can be determined from $\tan\psi = r\,d\theta/dr$, or

$$\tan\psi = \frac{r}{dr/d\theta} \tag{13–10}$$

If ψ is calculated as a positive quantity, it is measured from the *extended radial line* to the tangent in a counterclockwise sense or in the positive direction of θ. If it is negative, it is measured in the opposite direction to positive θ. For example, consider the cardioid $r = a(1 + \cos\theta)$, shown in Fig. 13–18. Because $dr/d\theta = -a\sin\theta$, then when $\theta = 30°$, $\tan\psi = a(1 + \cos 30°)/(-a\sin 30°) = -3.732$, or $\psi = -75°$, measured clockwise, opposite to $+\theta$ as shown in the figure.

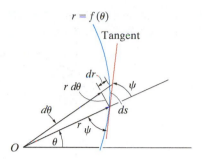

$r = f(\theta)$

(c)

Fig. 13–17

Procedure for Analysis

Cylindrical or polar coordinates are a suitable choice for the analysis of a problem for which data regarding the angular motion of the radial line r are given, or in cases where the path can be conveniently expressed in terms of these coordinates. Once these coordinates have been established, the equations of motion can then be applied in order to relate the forces acting on the particle to its acceleration components. The method for doing this has been outlined in the procedure for analysis given in Sec. 13.4. The following is a summary of this procedure.

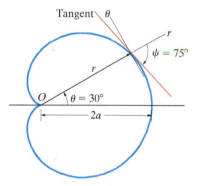

Fig. 13–18

Free-Body Diagram.

- Establish the r, θ, z inertial coordinate system and draw the particle's free-body diagram.
- Assume that $\mathbf{a}_r$, $\mathbf{a}_\theta$, $\mathbf{a}_z$ act in the positive directions of r, θ, z if they are unknown.
- Identify all the unknowns in the problem.

Equations of Motion.

- Apply the equations of motion, Eq. 13–9.

Kinematics.

- Use the methods of Sec. 12.8 to determine r and the time derivatives $\dot{r}$, $\ddot{r}$, $\dot{\theta}$, $\ddot{\theta}$, $\ddot{z}$, and then evaluate the acceleration components $a_r = \ddot{r} - r\dot{\theta}^2$, $a_\theta = r\ddot{\theta} + 2\dot{r}\dot{\theta}$, $a_z = \ddot{z}$.
- If any of the acceleration components is computed as a negative quantity, it indicates that it acts in its negative coordinate direction.
- When taking the time derivatives of $r = f(\theta)$, it is very important to use the chain rule of calculus, which is discussed in Appendix C.

EXAMPLE | **13.10**

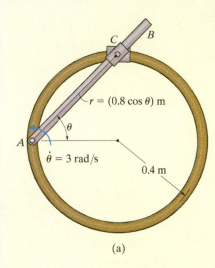

$r = (0.8 \cos \theta)$ m

θ

$\dot{\theta} = 3$ rad/s

0.4 m

(a)

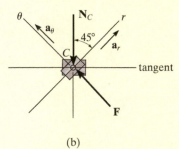

(b)

Fig. 13–19

The smooth 0.5-kg double-collar in Fig. 13–19a can freely slide on arm *AB* and the circular guide rod. If the arm rotates with a constant angular velocity of $\dot{\theta} = 3$ rad/s, determine the force the arm exerts on the collar at the instant $\theta = 45°$. Motion is in the horizontal plane.

SOLUTION

Free-Body Diagram.   The normal reaction $\mathbf{N}_C$ of the circular guide rod and the force $\mathbf{F}$ of arm *AB* act on the collar in the plane of motion, Fig. 13–19b. Note that $\mathbf{F}$ acts perpendicular to the axis of arm *AB*, that is, in the direction of the θ axis, while $\mathbf{N}_C$ acts perpendicular to the tangent of the circular path at $\theta = 45°$. The four unknowns are N_C, F, a_r, a_θ.

Equations of Motion.

$$+\nearrow\Sigma F_r = ma_r: \qquad\qquad -N_C \cos 45° = (0.5 \text{ kg}) \, a_r \qquad\qquad (1)$$

$$+\nwarrow\Sigma F_\theta = ma_\theta: \qquad\qquad F - N_C \sin 45° = (0.5 \text{ kg}) \, a_\theta \qquad\qquad (2)$$

Kinematics.   Using the chain rule (see Appendix C), the first and second time derivatives of r when $\theta = 45°$, $\dot{\theta} = 3$ rad/s, $\ddot{\theta} = 0$, are

$$r = 0.8 \cos \theta = 0.8 \cos 45° = 0.5657 \text{ m}$$

$$\dot{r} = -0.8 \sin \theta \, \dot{\theta} = -0.8 \sin 45°(3) = -1.6971 \text{ m/s}$$

$$\ddot{r} = -0.8 \left[\sin \theta \, \ddot{\theta} + \cos \theta \, \dot{\theta}^2 \right]$$

$$= -0.8[\sin 45°(0) + \cos 45°(3^2)] = -5.091 \text{ m/s}^2$$

We have

$$a_r = \ddot{r} - r\dot{\theta}^2 = -5.091 \text{ m/s}^2 - (0.5657 \text{ m})(3 \text{ rad/s})^2 = -10.18 \text{ m/s}^2$$

$$a_\theta = r\ddot{\theta} + 2\dot{r}\dot{\theta} = (0.5657 \text{ m})(0) + 2(-1.6971 \text{ m/s})(3 \text{ rad/s})$$

$$= -10.18 \text{ m/s}^2$$

Substituting these results into Eqs. (1) and (2) and solving, we get

$$N_C = 7.20 \text{ N}$$

$$F = 0 \qquad\qquad\qquad\qquad \textit{Ans.}$$

EXAMPLE | 13.11

The smooth 2-kg cylinder C in Fig. 13–20a has a pin P through its center which passes through the slot in arm OA. If the arm is forced to rotate in the *vertical plane* at a constant rate $\dot{\theta} = 0.5$ rad/s, determine the force that the arm exerts on the peg at the instant $\theta = 60°$.

SOLUTION

Why is it a good idea to use polar coordinates to solve this problem?

Free-Body Diagram. The free-body diagram for the cylinder is shown in Fig. 13–20b. The force on the peg, $\mathbf{F}_P$, acts perpendicular to the slot in the arm. As usual, $\mathbf{a}_r$ and $\mathbf{a}_\theta$ are assumed to act in the directions of *positive r* and θ, respectively. Identify the four unknowns.

Equations of Motion. Using the data in Fig. 13–20b, we have

$$+\swarrow\Sigma F_r = ma_r; \quad 19.62 \sin\theta - N_C \sin\theta = 2a_r \quad (1)$$
$$+\searrow\Sigma F_\theta = ma_\theta; \quad 19.62 \cos\theta + F_P - N_C \cos\theta = 2a_\theta \quad (2)$$

Kinematics. From Fig. 13–20a, r can be related to θ by the equation

$$r = \frac{0.4}{\sin\theta} = 0.4 \csc\theta$$

Since $d(\csc\theta) = -(\csc\theta \cot\theta)\, d\theta$ and $d(\cot\theta) = -(\csc^2\theta)\, d\theta$, then r and the necessary time derivatives become

$$\dot{\theta} = 0.5 \qquad r = 0.4 \csc\theta$$
$$\ddot{\theta} = 0 \qquad \dot{r} = -0.4(\csc\theta\cot\theta)\dot{\theta}$$
$$= -0.2 \csc\theta\cot\theta$$
$$\ddot{r} = -0.2(-\csc\theta\cot\theta)(\dot{\theta})\cot\theta - 0.2\csc\theta(-\csc^2\theta)\dot{\theta}$$
$$= 0.1\csc\theta(\cot^2\theta + \csc^2\theta)$$

Evaluating these formulas at $\theta = 60°$, we get

$$\dot{\theta} = 0.5 \qquad r = 0.462$$
$$\ddot{\theta} = 0 \qquad \dot{r} = -0.133$$
$$\ddot{r} = 0.192$$
$$a_r = \ddot{r} - r\dot{\theta}^2 = 0.192 - 0.462(0.5)^2 = 0.0770$$
$$a_\theta = r\ddot{\theta} + 2\dot{r}\dot{\theta} = 0 + 2(-0.133)(0.5) = -0.133$$

Substituting these results into Eqs. 1 and 2 with $\theta = 60°$ and solving yields

$$N_C = 19.4 \text{ N} \qquad F_P = -0.356 \text{ N} \qquad \textit{Ans.}$$

The negative sign indicates that $\mathbf{F}_P$ acts opposite to the direction shown in Fig. 13–20b.

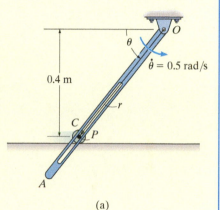

(a)

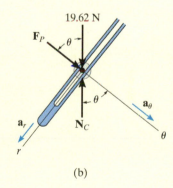

(b)

Fig. 13–20

EXAMPLE | 13.12

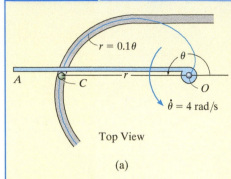

Top View

(a)

A can C, having a mass of 0.5 kg, moves along a grooved horizontal slot shown in Fig. 13–21a. The slot is in the form of a spiral, which is defined by the equation $r = (0.1\theta)$ m, where θ is in radians. If the arm OA rotates with a constant rate $\dot{\theta} = 4$ rad/s in the horizontal plane, determine the force it exerts on the can at the instant $\theta = \pi$ rad. Neglect friction and the size of the can.

SOLUTION

Free-Body Diagram.   The driving force $\mathbf{F}_C$ acts perpendicular to the arm OA, whereas the normal force of the wall of the slot on the can, $\mathbf{N}_C$, acts perpendicular to the tangent to the curve at $\theta = \pi$ rad, Fig. 13–21b. As usual, $\mathbf{a}_r$ and $\mathbf{a}_\theta$ are assumed to act in the *positive directions* of r and θ, respectively. Since the path is specified, the angle ψ which the extended radial line r makes with the tangent, Fig. 13–21c, can be determined from Eq. 13–10. We have $r = 0.1\theta$, so that $dr/d\theta = 0.1$, and therefore

$$\tan\psi = \frac{r}{dr/d\theta} = \frac{0.1\theta}{0.1} = \theta$$

When $\theta = \pi$, $\psi = \tan^{-1}\pi = 72.3°$, so that $\phi = 90° - \psi = 17.7°$, as shown in Fig. 13–21c. Identify the four unknowns in Fig. 13–21b.

Equations of Motion.   Using $\phi = 17.7°$ and the data shown in Fig. 13–21b, we have

$$\xrightarrow{+}\ \Sigma F_r = ma_r; \qquad\qquad N_C\cos 17.7° = 0.5a_r \qquad (1)$$

$$+\!\downarrow \Sigma F_\theta = ma_\theta; \qquad\qquad F_C - N_C\sin 17.7° = 0.5a_\theta \qquad (2)$$

Kinematics.   The time derivatives of r and θ are

$$\dot{\theta} = 4\text{ rad/s} \qquad\qquad r = 0.1\theta$$

$$\ddot{\theta} = 0 \qquad\qquad \dot{r} = 0.1\dot{\theta} = 0.1(4) = 0.4\text{ m/s}$$

$$\ddot{r} = 0.1\ddot{\theta} = 0$$

At the instant $\theta = \pi$ rad,

$$a_r = \ddot{r} - r\dot{\theta}^2 = 0 - 0.1(\pi)(4)^2 = -5.03\text{ m/s}^2$$

$$a_\theta = r\ddot{\theta} + 2\dot{r}\dot{\theta} = 0 + 2(0.4)(4) = 3.20\text{ m/s}^2$$

Substituting these results into Eqs. 1 and 2 and solving yields

$$N_C = -2.64\text{ N}$$

$$F_C = 0.800\text{ N} \qquad\qquad Ans.$$

What does the negative sign for N_C indicate?

(b)

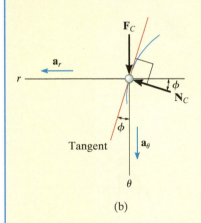

(c)

Fig. 13–21

FUNDAMENTAL PROBLEMS

F13–13. Determine the constant angular velocity $\dot{\theta}$ of the vertical shaft of the amusement ride if $\phi = 45°$. Neglect the mass of the cables and the size of the passengers.

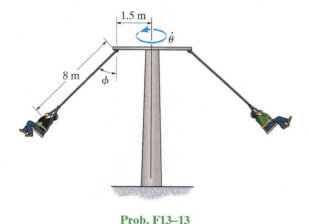

Prob. F13–13

F13–14. The 0.2-kg ball is blown through the smooth vertical circular tube whose shape is defined by $r = (0.6 \sin \theta)$ m, where θ is in radians. If $\theta = (\pi t^2)$ rad, where t is in seconds, determine the magnitude of force **F** exerted by the blower on the ball when $t = 0.5$ s.

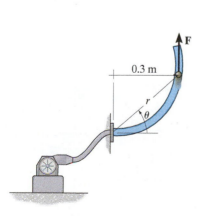

Prob. F13–14

F13–15. The 2-Mg car is traveling along the curved road described by $r = (50e^{2\theta})$ m, where θ is in radians. If a camera is located at A and it rotates with an angular velocity of $\dot{\theta} = 0.05$ rad/s and an angular acceleration of $\ddot{\theta} = 0.01$ rad/s² at the instant $\theta = \frac{\pi}{6}$ rad, determine the resultant friction force developed between the tires and the road at this instant.

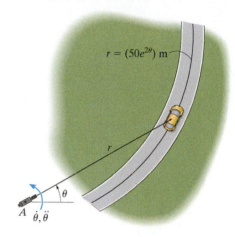

Prob. F13–15

F13–16. The 0.2-kg pin P is constrained to move in the smooth curved slot, which is defined by the lemniscate $r = (0.6 \cos 2\theta)$ m. Its motion is controlled by the rotation of the slotted arm OA, which has a constant clockwise angular velocity of $\dot{\theta} = -3$ rad/s. Determine the force arm OA exerts on the pin P when $\theta = 0°$. Motion is in the vertical plane.

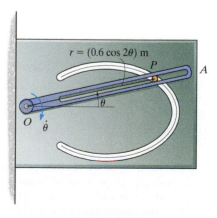

Prob. F13–16

PROBLEMS

13–85. The spring-held follower AB has a weight of 0.75 lb and moves back and forth as its end rolls on the contoured surface of the cam, where $r = 0.2$ ft and $z = (0.1 \sin 2\theta)$ ft. If the cam is rotating at a constant rate of 6 rad/s, determine the force at the end A of the follower when $\theta = 45°$. In this position the spring is compressed 0.4 ft. Neglect friction at the bearing C.

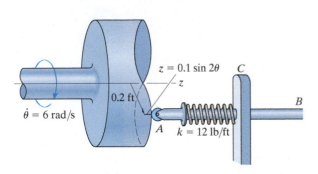

Prob. 13–85

13–86. Determine the magnitude of the resultant force acting on a 5-kg particle at the instant $t = 2$ s, if the particle is moving along a horizontal path defined by the equations $r = (2t + 10)$ m and $\theta = (1.5t^2 - 6t)$ rad, where t is in seconds.

13–87. The path of motion of a 5-lb particle in the horizontal plane is described in terms of polar coordinates as $r = (2t + 1)$ ft and $\theta = (0.5t^2 - t)$ rad, where t is in seconds. Determine the magnitude of the unbalanced force acting on the particle when $t = 2$ s.

*13–88. Rod OA rotates counterclockwise with a constant angular velocity of $\dot{\theta} = 5$ rad/s. The double collar B is pin-connected together such that one collar slides over the rotating rod and the other slides over the *horizontal* curved rod, of which the shape is described by the equation $r = 1.5(2 - \cos \theta)$ ft. If both collars weigh 0.75 lb, determine the normal force which the curved rod exerts on one collar at the instant $\theta = 120°$. Neglect friction.

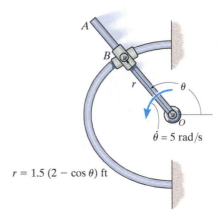

Prob. 13–88

13–89. The boy of mass 40 kg is sliding down the spiral slide at a constant speed such that his position, measured from the top of the chute, has components $r = 1.5$ m, $\theta = (0.7t)$ rad, and $z = (-0.5t)$ m, where t is in seconds. Determine the components of force $\mathbf{F}_r$, $\mathbf{F}_\theta$, and $\mathbf{F}_z$ which the slide exerts on him at the instant $t = 2$ s. Neglect the size of the boy.

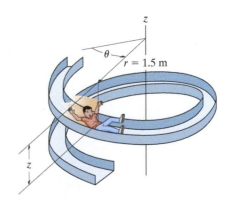

Prob. 13–89

13–90. The 40-kg boy is sliding down the smooth spiral slide such that $z = -2$ m/s and his speed is 2 m/s. Determine the r, θ, z components of force the slide exerts on him at this instant. Neglect the size of the boy.

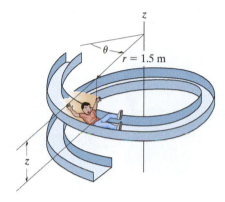

Prob. 13–90

13–91. Using a forked rod, a 0.5-kg smooth peg P is forced to move along the *vertical slotted* path $r = (0.5\theta)$ m, where θ is in radians. If the angular position of the arm is $\theta = (\frac{\pi}{8}t^2)$ rad, where t is in seconds, determine the force of the rod on the peg and the normal force of the slot on the peg at the instant $t = 2$ s. The peg is in contact with only *one edge* of the rod and slot at any instant.

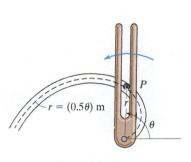

Prob. 13–91

*13–92. The arm is rotating at a rate of $\dot{\theta} = 4$ rad/s when $\ddot{\theta} = 3$ rad/s^2 and $\theta = 180°$. Determine the force it must exert on the 0.5-kg smooth cylinder if it is confined to move along the slotted path. Motion occurs in the horizontal plane.

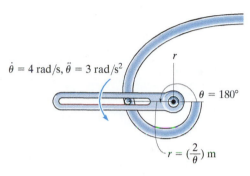

$\dot{\theta} = 4$ rad/s, $\ddot{\theta} = 3$ rad/s^2

$\theta = 180°$

$r = \left(\frac{2}{\theta}\right)$ m

Prob. 13–92

13–93. If arm OA rotates with a constant clockwise angular velocity of $\dot{\theta} = 1.5$ rad/s, determine the force arm OA exerts on the smooth 4-lb cylinder B when $\theta = 45°$.

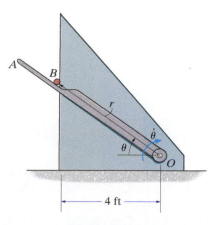

4 ft

Prob. 13–93

13–94. Determine the normal and frictional driving forces that the partial spiral track exerts on the 200-kg motorcycle at the instant $\theta = \frac{5}{3}\pi$ rad, $\dot{\theta} = 0.4$ rad/s, $\ddot{\theta} = 0.8$ rad/s². Neglect the size of the motorcycle.

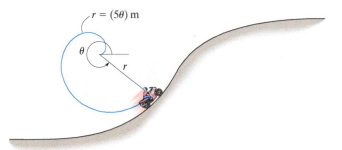

$r = (5\theta)$ m

Prob. 13–94

13–95. A smooth can C, having a mass of 3 kg, is lifted from a feed at A to a ramp at B by a rotating rod. If the rod maintains a constant angular velocity of $\dot{\theta} = 0.5$ rad/s, determine the force which the rod exerts on the can at the instant $\theta = 30°$. Neglect the effects of friction in the calculation and the size of the can so that $r = (1.2 \cos \theta)$ m. The ramp from A to B is circular, having a radius of 600 mm.

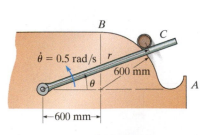

B
C
$\dot{\theta} = 0.5$ rad/s r
600 mm
θ
A
←600 mm→

Prob. 13–95

*13–96. The spring-held follower AB has a mass of 0.5 kg and moves back and forth as its end rolls on the contoured surface of the cam, where $r = 0.15$ m and $z = (0.02 \cos 2\theta)$ m. If the cam is rotating at a constant rate of 30 rad/s, determine the force component F_z at the end A of the follower when $\theta = 30°$. The spring is uncompressed when $\theta = 90°$. Neglect friction at the bearing C.

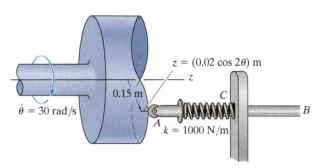

$z = (0.02 \cos 2\theta)$ m
z
0.15 m
C
$\dot{\theta} = 30$ rad/s
B
A
$k = 1000$ N/m

Prob. 13–96

13–97. The spring-held follower AB has a mass of 0.5 kg and moves back and forth as its end rolls on the contoured surface of the cam, where $r = 0.15$ m and $z = (0.02 \cos 2\theta)$ m. If the cam is rotating at a constant rate of 30 rad/s, determine the maximum and minimum force components F_z the follower exerts on the cam if the spring is uncompressed when $\theta = 90°$.

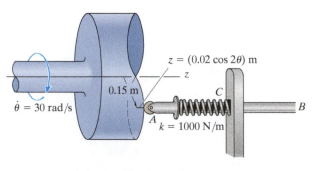

$z = (0.02 \cos 2\theta)$ m
z
0.15 m
C
$\dot{\theta} = 30$ rad/s
B
A
$k = 1000$ N/m

Prob. 13–97

13–98. The particle has a mass of 0.5 kg and is confined to move along the smooth vertical slot due to the rotation of the arm OA. Determine the force of the rod on the particle and the normal force of the slot on the particle when $\theta = 30°$. The rod is rotating with a constant angular velocity $\dot{\theta} = 2$ rad/s. Assume the particle contacts only one side of the slot at any instant.

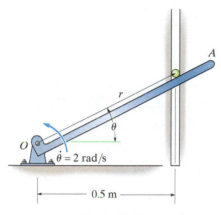

Prob. 13–98

13–99. A car of a roller coaster travels along a track which for a short distance is defined by a conical spiral, $r = \frac{3}{4}z$, $\theta = -1.5z$, where r and z are in meters and θ in radians. If the angular motion $\dot{\theta} = 1$ rad/s is always maintained, determine the r, θ, z components of reaction exerted on the car by the track at the instant $z = 6$ m. The car and passengers have a total mass of 200 kg.

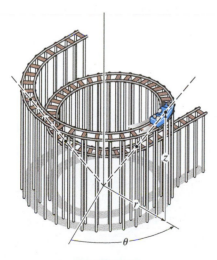

Prob. 13–99

***13–100.** The 0.5-lb ball is guided along the vertical circular path $r = 2r_c \cos \theta$ using the arm OA. If the arm has an angular velocity $\dot{\theta} = 0.4$ rad/s and an angular acceleration $\ddot{\theta} = 0.8$ rad/s² at the instant $\theta = 30°$, determine the force of the arm on the ball. Neglect friction and the size of the ball. Set $r_c = 0.4$ ft.

13–101. The ball of mass m is guided along the vertical circular path $r = 2r_c \cos \theta$ using the arm OA. If the arm has a constant angular velocity $\dot{\theta}_0$, determine the angle $\theta \leq 45°$ at which the ball starts to leave the surface of the semicylinder. Neglect friction and the size of the ball.

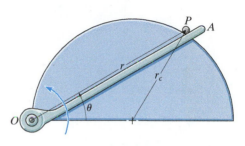

Probs. 13–100/101

13–102. Using a forked rod, a smooth cylinder P, having a mass of 0.4 kg, is forced to move along the *vertical slotted path* $r = (0.6\theta)$ m, where θ is in radians. If the cylinder has a constant speed of $v_C = 2$ m/s, determine the force of the rod and the normal force of the slot on the cylinder at the instant $\theta = \pi$ rad. Assume the cylinder is in contact with only *one edge* of the rod and slot at any instant. *Hint:* To obtain the time derivatives necessary to compute the cylinder's acceleration components a_r and a_θ, take the first and second time derivatives of $r = 0.6\theta$. Then, for further information, use Eq. 12–26 to determine $\dot{\theta}$. Also, take the time derivative of Eq. 12–26, noting that $\dot{v} = 0$ to determine $\ddot{\theta}$.

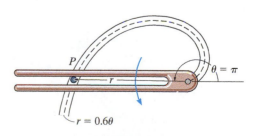

Prob. 13–102

13–103. The pilot of the airplane executes a vertical loop which in part follows the path of a cardioid, $r = 200(1 + \cos\theta)$ m, where θ is in radians. If his speed at A is a constant $v_p = 85$ m/s, determine the vertical reaction the seat of the plane exerts on the pilot when the plane is at A. He has a mass of 80 kg. *Hint:* To determine the time derivatives necessary to calculate the acceleration components a_r and a_θ, take the first and second time derivatives of $r = 200(1 + \cos\theta)$. Then, for further information, use Eq. 12–26 to determine $\dot\theta$.

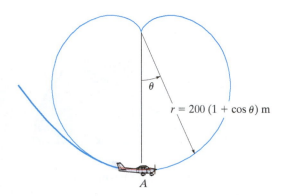

Prob. 13–103

*13–104.** The collar has a mass of 2 kg and travels along the smooth horizontal rod defined by the equiangular spiral $r = (e^\theta)$ m, where θ is in radians. Determine the tangential force F and the normal force N acting on the collar when $\theta = 45°$, if the force F maintains a constant angular motion $\dot\theta = 2$ rad/s.

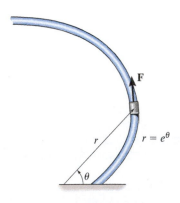

Prob. 13–104

13–105. The particle has a mass of 0.5 kg and is confined to move along the smooth horizontal slot due to the rotation of the arm OA. Determine the force of the rod on the particle and the normal force of the slot on the particle when $\theta = 30°$. The rod is rotating with a constant angular velocity $\dot\theta = 2$ rad/s. Assume the particle contacts only one side of the slot at any instant.

13–106. Solve Prob. 13–105 if the arm has an angular acceleration of $\ddot\theta = 3$ rad/s² when $\dot\theta = 2$ rad/s at $\theta = 30°$.

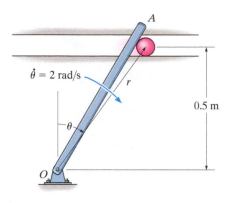

Probs. 13–105/106

13–107. The forked rod is used to move the smooth 2-lb particle around the horizontal path in the shape of a limaçon, $r = (2 + \cos\theta)$ ft. If $\theta = (0.5t^2)$ rad, where t is in seconds, determine the force which the rod exerts on the particle at the instant $t = 1$ s. The fork and path contact the particle on only one side.

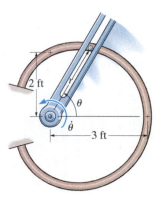

Prob. 13–107

*13–108. The collar, which has a weight of 3 lb, slides along the smooth rod lying in the *horizontal plane* and having the shape of a parabola $r = 4/(1 - \cos \theta)$, where θ is in radians and r is in feet. If the collar's angular rate is constant and equals $\dot{\theta} = 4$ rad/s, determine the tangential retarding force P needed to cause the motion and the normal force that the collar exerts on the rod at the instant $\theta = 90°$.

Prob. 13–108

13–109. Rod OA rotates counterclockwise at a constant angular rate $\dot{\theta} = 4$ rad/s. The double collar B is pin-connected together such that one collar slides over the rotating rod and the other collar slides over the circular rod described by the equation $r = (1.6 \cos \theta)$ m. If *both* collars have a mass of 0.5 kg, determine the force which the circular rod exerts on one of the collars and the force that OA exerts on the other collar at the instant $\theta = 45°$. Motion is in the horizontal plane.

13–110. Solve Prob. 13–109 if motion is in the vertical plane.

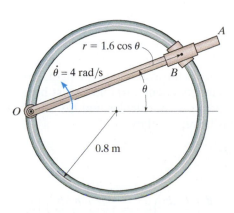

Probs. 13–109/110

13–111. A 0.2-kg spool slides down along a smooth rod. If the rod has a constant angular rate of rotation $\dot{\theta} = 2$ rad/s in the vertical plane, show that the equations of motion for the spool are $\ddot{r} - 4r - 9.81 \sin \theta = 0$ and $0.8\ddot{r} + N_s - 1.962 \cos \theta = 0$, where N_s is the magnitude of the normal force of the rod on the spool. Using the methods of differential equations, it can be shown that the solution of the first of these equations is $r = C_1 e^{-2t} + C_2 e^{2t} - (9.81/8) \sin 2t$. If r, $\dot{r}$, and θ are zero when $t = 0$, evaluate the constants C_1 and C_2 determine r at the instant $\theta = \pi/4$ rad.

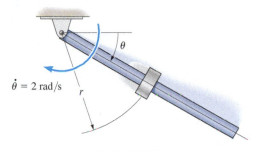

Prob. 13–111

*13–112. The pilot of an airplane executes a vertical loop which in part follows the path of a "four-leaved rose," $r = (-600\cos 2\theta)$ ft, where θ is in radians. If his speed is a constant $v_P = 80$ ft/s, determine the vertical reaction the seat of the plane exerts on the pilot when the plane is at A. He weights 130 lb. *Hint*: To determine the time derivatives necessary to compute the acceleration components a_r, and a_θ, take the first and second time derivatives of $r = 400(1 + \cos\theta)$. Then, for further information, use Eq. 12–26 to determine $\dot{\theta}$. Also, take the time derivative of Eq. 12–26, noting that $\dot{v}_P = 0$ to determine $\ddot{\theta}$.

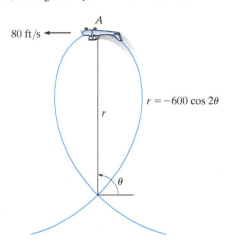

Prob. 13–112

*13.7   Central-Force Motion and Space Mechanics

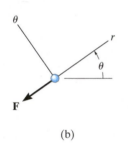

$dA = \frac{1}{2}r^2 d\theta$

(a)

(b)

Fig. 13–22

If a particle is moving only under the influence of a force having a line of action which is always directed toward a fixed point, the motion is called *central-force motion*. This type of motion is commonly caused by electrostatic and gravitational forces.

In order to analyze the motion, we will consider the particle P shown in Fig. 13–22a, which has a mass m and is acted upon only by the central force **F**. The free-body diagram for the particle is shown in Fig. 13–22b. Using polar coordinates (r, θ), the equations of motion, Eq. 13–9, become

$$\Sigma F_r = ma_r;$$
$$-F = m\left[\frac{d^2r}{dt^2} - r\left(\frac{d\theta}{dt}\right)^2\right] \tag{13–11}$$

$$\Sigma F_\theta = ma_\theta; \qquad 0 = m\left(r\frac{d^2\theta}{dt^2} + 2\frac{dr}{dt}\frac{d\theta}{dt}\right)$$

The second of these equations may be written in the form

$$\frac{1}{r}\left[\frac{d}{dt}\left(r^2\frac{d\theta}{dt}\right)\right] = 0$$

so that integrating yields

$$r^2\frac{d\theta}{dt} = h \tag{13–12}$$

Here h is the constant of integration.

From Fig. 13–22a notice that the shaded area described by the radius r, as r moves through an angle $d\theta$, is $dA = \frac{1}{2}r^2\,d\theta$. If the *areal velocity* is defined as

$$\frac{dA}{dt} = \frac{1}{2}r^2\frac{d\theta}{dt} = \frac{h}{2} \tag{13–13}$$

then it is seen that the areal velocity for a particle subjected to central-force motion is *constant*. In other words, the particle will sweep out equal segments of area per unit of time as it travels along the path. To obtain the *path of motion*, $r = f(\theta)$, the independent variable t must be eliminated from Eqs. 13–11. Using the chain rule of calculus and Eq. 13–12, the time derivatives of Eqs. 13–11 may be replaced by

$$\frac{dr}{dt} = \frac{dr}{d\theta}\frac{d\theta}{dt} = \frac{h}{r^2}\frac{dr}{d\theta}$$

$$\frac{d^2r}{dt^2} = \frac{d}{dt}\left(\frac{h}{r^2}\frac{dr}{d\theta}\right) = \frac{d}{d\theta}\left(\frac{h}{r^2}\frac{dr}{d\theta}\right)\frac{d\theta}{dt} = \left[\frac{d}{d\theta}\left(\frac{h}{r^2}\frac{dr}{d\theta}\right)\right]\frac{h}{r^2}$$

Substituting a new dependent variable (xi) $\xi = 1/r$ into the second equation, we have

$$\frac{d^2r}{dt^2} = -h^2\xi^2\frac{d^2\xi}{d\theta^2}$$

Also, the square of Eq. 13–12 becomes

$$\left(\frac{d\theta}{dt}\right)^2 = h^2\xi^4$$

Substituting these two equations into the first of Eqs. 13–11 yields

$$-h^2\xi^2\frac{d^2\xi}{d\theta^2} - h^2\xi^3 = -\frac{F}{m}$$

or

$$\frac{d^2\xi}{d\theta^2} + \xi = \frac{F}{mh^2\xi^2} \tag{13–14}$$

This satellite is subjected to a central force and its orbital motion can be closely predicted using the equations developed in this section. (UniversalImagesGroup/ Getty Images)

This differential equation defines the path over which the particle travels when it is subjected to the central force **F**.*

For application, the force of gravitational attraction will be considered. Some common examples of central-force systems which depend on gravitation include the motion of the moon and artificial satellites about the earth, and the motion of the planets about the sun. As a typical problem in space mechanics, consider the trajectory of a space satellite or space vehicle launched into free-flight orbit with an initial velocity $\mathbf{v}_0$, Fig. 13–23. It will be assumed that this velocity is initially *parallel* to the tangent at the surface of the earth, as shown in the figure.† Just after the satellite is released into free flight, the only force acting on it is the gravitational force of the earth. (Gravitational attractions involving other bodies such as the moon or sun will be neglected, since for orbits close to the earth their effect is small in comparison with the earth's gravitation.) According to Newton's law of gravitation, force **F** will always act between the mass centers of the earth and the satellite, Fig. 13–23. From Eq. 13–1, this force of attraction has a magnitude of

$$F = G\frac{M_e m}{r^2}$$

where M_e and m represent the mass of the earth and the satellite, respectively, G is the gravitational constant, and r is the distance between

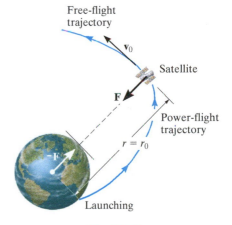

Fig. 13–23

---

*In the derivation, **F** is considered positive when it is directed toward point O. If **F** is oppositely directed, the right side of Eq. 13–14 should be negative.

†The case where $\mathbf{v}_0$ acts at some initial angle θ to the tangent is best described using the conservation of angular momentum.

the mass centers. To obtain the orbital path, we set $\xi = 1/r$ in the foregoing equation and substitute the result into Eq. 13–14. We obtain

$$\frac{d^2\xi}{d\theta^2} + \xi = \frac{GM_e}{h^2} \tag{13–15}$$

This second-order differential equation has constant coefficients and is nonhomogeneous. The solution is the sum of the complementary and particular solutions given by

$$\xi = \frac{1}{r} = C\cos(\theta - \phi) + \frac{GM_e}{h^2} \tag{13–16}$$

This equation represents the *free-flight trajectory* of the satellite. It is the equation of a conic section expressed in terms of polar coordinates.

A geometric interpretation of Eq. 13–16 requires knowledge of the equation for a conic section. As shown in Fig. 13–24, a conic section is defined as the locus of a point P that moves in such a way that the ratio of its distance to a *focus*, or fixed point F, to its perpendicular distance to a fixed line DD called the *directrix*, is constant. This constant ratio will be denoted as e and is called the *eccentricity*. By definition

$$e = \frac{FP}{PA}$$

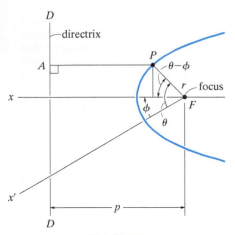

Fig. 13–24

From Fig. 13–24,

$$FP = r = e(PA) = e[p - r\cos(\theta - \phi)]$$

or

$$\frac{1}{r} = \frac{1}{p}\cos(\theta - \phi) + \frac{1}{ep}$$

Comparing this equation with Eq. 13–16, it is seen that the fixed distance from the focus to the directrix is

$$p = \frac{1}{C} \tag{13–17}$$

And the eccentricity of the conic section for the trajectory is

$$e = \frac{Ch^2}{GM_e} \tag{13–18}$$

Provided the polar angle θ is measured from the x axis (an axis of symmetry since it is perpendicular to the directrix), the angle ϕ is zero, Fig. 13–24, and therefore Eq. 13–16 reduces to

$$\frac{1}{r} = C \cos \theta + \frac{GM_e}{h^2} \tag{13–19}$$

The constants h and C are determined from the data obtained for the position and velocity of the satellite at the end of the *power-flight trajectory*. For example, if the initial height or distance to the space vehicle is r_0, measured from the center of the earth, and its initial speed is v_0 at the beginning of its free flight, Fig. 13–25, then the constant h may be obtained from Eq. 13–12. When $\theta = \phi = 0°$, the velocity $\mathbf{v}_0$ has no radial component; therefore, from Eq. 12–25, $v_0 = r_0(d\theta/dt)$, so that

$$h = r_0^2 \frac{d\theta}{dt}$$

or

$$h = r_0 v_0 \tag{13–20}$$

To determine C, use Eq. 13–19 with $\theta = 0°$, $r = r_0$, and substitute Eq. 13–20 for h:

$$C = \frac{1}{r_0}\left(1 - \frac{GM_e}{r_0 v_0^2}\right) \tag{13–21}$$

The equation for the free-flight trajectory therefore becomes

$$\frac{1}{r} = \frac{1}{r_0}\left(1 - \frac{GM_e}{r_0 v_0^2}\right) \cos \theta + \frac{GM_e}{r_0^2 v_0^2} \tag{13–22}$$

The type of path traveled by the satellite is determined from the value of the eccentricity of the conic section as given by Eq. 13–18. If

$$\begin{array}{ll}
e = 0 & \text{free-flight trajectory is a circle} \\
e = 1 & \text{free-flight trajectory is a parabola} \\
e < 1 & \text{free-flight trajectory is an ellipse} \\
e > 1 & \text{free-flight trajectory is a hyperbola}
\end{array} \tag{13–23}$$

Parabolic Path.   Each of these trajectories is shown in Fig. 13–25. From the curves it is seen that when the satellite follows a parabolic path, it is "on the border" of never returning to its initial starting point. The initial launch velocity, $\mathbf{v}_0$, required for the satellite to follow a parabolic path is called the *escape velocity*. The speed, v_e, can be determined by using the second of Eqs. 13–23, $e = 1$, with Eqs. 13–18, 13–20, and 13–21. It is left as an exercise to show that

$$v_e = \sqrt{\frac{2GM_e}{r_0}} \tag{13–24}$$

Circular Orbit.   The speed v_c required to launch a satellite into a *circular orbit* can be found using the first of Eqs. 13–23, $e = 0$. Since e is related to h and C, Eq. 13–18, C must be zero to satisfy this equation (from Eq. 13–20, h cannot be zero); and therefore, using Eq. 13–21, we have

$$v_c = \sqrt{\frac{GM_e}{r_0}} \tag{13–25}$$

Provided r_0 represents a minimum height for launching, in which frictional resistance from the atmosphere is neglected, speeds at launch which are less than v_c will cause the satellite to reenter the earth's atmosphere and either burn up or crash, Fig. 13–25.

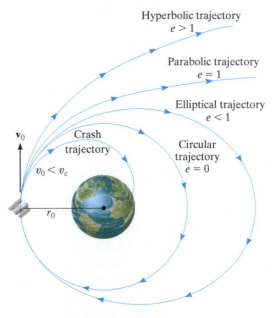

Fig. 13–25

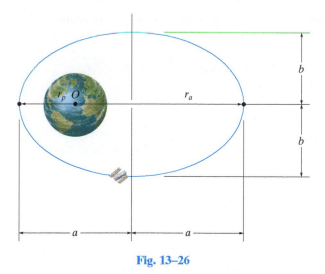

Fig. 13–26

Elliptical Orbit. All the trajectories attained by planets and most satellites are elliptical, Fig. 13–26. For a satellite's orbit about the earth, the *minimum distance* from the orbit to the center of the earth O (which is located at one of the foci of the ellipse) is r_p and can be found using Eq. 13–22 with $\theta = 0°$. Therefore;

$$r_p = r_0 \qquad (13\text{–}26)$$

This minimum distance is called the *perigee* of the orbit. The *apogee* or maximum distance r_a can be found using Eq. 13–22 with $\theta = 180°$.* Thus,

$$r_a = \frac{r_0}{(2GM_e/r_0v_0^2) - 1} \qquad (13\text{–}27)$$

With reference to Fig. 13–26, the half-length of the major axis of the ellipse is

$$a = \frac{r_p + r_a}{2} \qquad (13\text{–}28)$$

Using analytical geometry, it can be shown that the half-length of the minor axis is determined from the equation

$$b = \sqrt{r_p r_a} \qquad (13\text{–}29)$$

---

*Actually, the terminology perigee and apogee pertains only to orbits about the *earth.* If any other heavenly body is located at the focus of an elliptical orbit, the minimum and maximum distances are referred to respectively as the *periapsis* and *apoapsis* of the orbit.

13

Furthermore, by direct integration, the area of an ellipse is

$$A = \pi ab = \frac{\pi}{2}(r_p + r_a)\sqrt{r_p r_a} \qquad (13\text{–}30)$$

The areal velocity has been defined by Eq. 13–13, $dA/dt = h/2$. Integrating yields $A = hT/2$, where T is the *period* of time required to make one orbital revolution. From Eq. 13–30, the period is

$$T = \frac{\pi}{h}(r_p + r_a)\sqrt{r_p r_a} \qquad (13\text{–}31)$$

In addition to predicting the orbital trajectory of earth satellites, the theory developed in this section is valid, to a surprisingly close approximation, at predicting the actual motion of the planets traveling around the sun. In this case the mass of the sun, M_s, should be substituted for M_e when the appropriate formulas are used.

The fact that the planets do indeed follow elliptic orbits about the sun was discovered by the German astronomer Johannes Kepler in the early seventeenth century. His discovery was made *before* Newton had developed the laws of motion and the law of gravitation, and so at the time it provided important proof as to the validity of these laws. Kepler's laws, developed after 20 years of planetary observation, are summarized as follows:

1. Every planet travels in its orbit such that the line joining it to the center of the sun sweeps over equal areas in equal intervals of time, whatever the line's length.

2. The orbit of every planet is an ellipse with the sun placed at one of its foci.

3. The square of the period of any planet is directly proportional to the cube of the major axis of its orbit.

A mathematical statement of the first and second laws is given by Eqs. 13–13 and 13–22, respectively. The third law can be shown from Eq. 13–31 using Eqs. 13–19, 13–28, and 13–29. (See Prob. 13–117.)

PROBLEMS

In the following problems, except where otherwise indicated, assume that the radius of the earth is 6378 km, the earth's mass is $5.976(10^{24})$ kg, the mass of the sun is $1.99(10^{30})$ kg, and the gravitational constant is $G = 66.73(10^{-12})$ m^3/(kg·s^2).

13–113. The earth has an orbit with eccentricity 0.0167 around the sun. Knowing that the earth's minimum distance from the sun is $146(10^6)$ km, find the speed at which the earth travels when it is at this distance. Determine the equation in polar coordinates which describes the earth's orbit about the sun.

13–114. A communications satellite is in a circular orbit above the earth such that it always remains directly over a point on the earth's surface. As a result, the period of the satellite must equal the rotation of the earth, which is approximately 24 hours. Determine the satellite's altitude h above the earth's surface and its orbital speed.

13–115. The speed of a satellite launched into a circular orbit about the earth is given by Eq. 13–25. Determine the speed of a satellite launched parallel to the surface of the earth so that it travels in a circular orbit 800 km from the earth's surface.

*13–116. The rocket is in circular orbit about the earth at an altitude of 20 Mm. Determine the minimum increment in speed it must have in order to escape the earth's gravitational field.

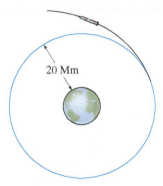

Prob. 13–116

13–117. Prove Kepler's third law of motion. *Hint:* Use Eqs. 13–19, 13–28, 13–29, and 13–31.

13–118. The satellite is moving in an elliptical orbit with an eccentricity $e = 0.25$. Determine its speed when it is at its maximum distance A and minimum distance B from the earth.

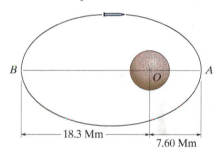

Prob. 13–118

13–119. The rocket is traveling in free flight along the elliptical orbit. The planet has no atmosphere, and its mass is 0.60 times that of the earth. If the rocket has the orbit shown, determine the rocket's speed when it is at A and at B.

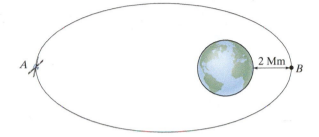

Prob. 13–119

*13–120. Determine the constant speed of satellite S so that it circles the earth with an orbit of radius $r = 15$ Mm. *Hint:* Use Eq. 13–1.

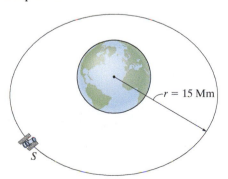

Prob. 13–120

13–121. The rocket is in free flight along an elliptical trajectory $A'A$. The planet has no atmosphere, and its mass is 0.70 times that of the earth. If the rocket has an apoapsis and periapsis as shown in the figure, determine the speed of the rocket when it is at point A.

13–123. The rocket is initially in free-flight circular orbit around the earth. Determine the speed of the rocket at A. What change in the speed at A is required so that it can move in an elliptical orbit to reach point A'?

*13–124. The rocket is in free-flight circular orbit around the earth. Determine the time needed for the rocket to travel from the innner orbit at A to the outer orbit at A'.

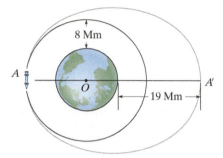

Probs. 13–123/124

Prob. 13–121

13–122. The Viking Explorer approaches the planet Mars on a parabolic trajectory as shown. When it reaches point A its velocity is 10 Mm/h. Determine r_0 and the required change in velocity at A so that it can then maintain a circular orbit as shown. The mass of Mars is 0.1074 times the mass of the earth.

13–125. A satellite is launched at its apogee with an initial velocity $v_0 = 2500$ mi/h parallel to the surface of the earth. Determine the required altitude (or range of altitudes) above the earth's surface for launching if the free-flight trajectory is to be (a) circular, (b) parabolic, (c) elliptical, with launch at apogee, and (d) hyperbolic. Take $G = 34.4(10^{-9})(\text{lb} \cdot \text{ft}^2)/\text{slug}^2$, $M_e = 409(10^{21})$ slug, the earth's radius $r_e = 3960$ mi, and 1 mi = 5280 ft.

13–126. The rocket is traveling around the earth in free flight along the elliptical orbit. If the rocket has the orbit shown, determine the speed of the rocket when it is at A and at B.

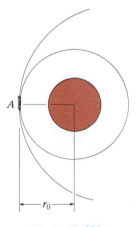

Prob. 13–122

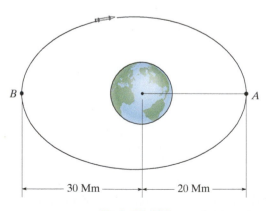

Prob. 13–126

13–127. An elliptical path of a satellite has an eccentricity $e = 0.130$. If it has a speed of 15 Mm/h when it is at perigee, P, determine its speed when it arrives at apogee, A. Also, how far is it from the earth's surface when it is at A?

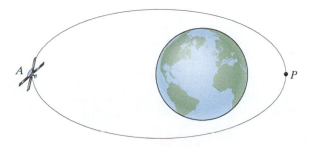

Prob. 13–127

*13–128. A rocket is in free-flight elliptical orbit around the planet Venus. Knowing that the periapsis and apoapsis of the orbit are 8 Mm and 26 Mm, respectively, determine (a) the speed of the rocket at point A', (b) the required speed it must attain at A just after braking so that it undergoes an 8-Mm free-flight circular orbit around Venus, and (c) the periods of both the circular and elliptical orbits. The mass of Venus is 0.816 times the mass of the earth.

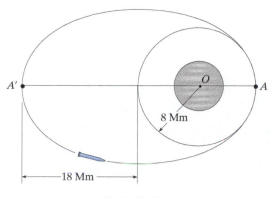

Prob. 13–128

13–129. The rocket is traveling in a free flight along an elliptical trajectory $A'A$. The planet has no atmosphere, and its mass is 0.60 times that of the earth. If the rocket has the orbit shown, determine the rocket's velocity when it is at point A.

13–130. If the rocket is to land on the surface of the planet, determine the required free-flight speed it must have at A' so that the landing occurs at B. How long does it take for the rocket to land, going from A' to B? The planet has no atmosphere, and its mass is 0.6 times that of the earth.

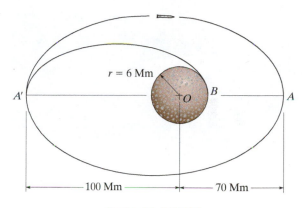

Probs. 13–129/130

13–131. The rocket is traveling around the earth in free flight along an elliptical orbit AC. If the rocket has the orbit shown, determine the rocket's velocity when it is at point A.

*13–132. The rocket is traveling around the earth in free flight along the elliptical orbit AC. Determine its change in speed when it reaches A so that it travels along the elliptical orbit AB.

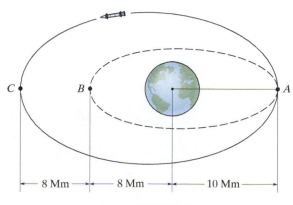

Probs. 13–131/132

CONCEPTUAL PROBLEMS

13

C13–1. If the box is released from rest at *A*, use numerical values to show how you would estimate the time for it to arrive at *B*. Also, list the assumptions for your analysis.

Prob. C13–1 (© R.C. Hibbeler)

C13–2. The tugboat has a known mass and its propeller provides a known maximum thrust. When the tug is fully powered you observe the time it takes for the tug to reach a speed of known value starting from rest. Show how you could determine the mass of the barge. Neglect the drag force of the water on the tug. Use numerical values to explain your answer.

Prob. C13–2 (© R.C. Hibbeler)

C13–3. Determine the smallest speed of each car *A* and *B* so that the passengers do not lose contact with the seat while the arms turn at a constant rate. What is the largest normal force of the seat on each passenger? Use numerical values to explain your answer.

Prob. C13–3 (© R.C. Hibbeler)

C13–4. Each car is pin connected at its ends to the rim of the wheel which turns at a constant speed. Using numerical values, show how to determine the resultant force the seat exerts on the passenger located in the top car *A*. The passengers are seated toward the center of the wheel. Also, list the assumptions for your analysis.

Prob. C13–4 (© R.C. Hibbeler)

CHAPTER REVIEW

Kinetics

Kinetics is the study of the relation between forces and the acceleration they cause. This relation is based on Newton's second law of motion, expressed mathematically as $\Sigma \mathbf{F} = m\mathbf{a}$.

Before applying the equation of motion, it is important to first draw the particle's *free-body diagram* in order to account for all of the forces that act on the particle. Graphically, this diagram is equal to the *kinetic diagram*, which shows the result of the forces, that is, the $m\mathbf{a}$ vector.

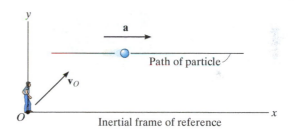

Free-body diagram

Kinetic diagram

Inertial Coordinate Systems

When applying the equation of motion, it is important to measure the acceleration from an inertial coordinate system. This system has axes that do not rotate but are either fixed or translate with a constant velocity. Various types of inertial coordinate systems can be used to apply $\Sigma \mathbf{F} = m\mathbf{a}$ in component form.

Path of particle

Inertial frame of reference

Rectangular *x, y, z* axes are used to describe the motion along each of the axes.

$$\Sigma F_x = ma_x, \ \Sigma F_y = ma_y, \ \Sigma F_z = ma_z$$

Normal, tangential, and binormal axes *n, t, b*, are often used when the path is known. Recall that $\mathbf{a}_n$ is always directed in the $+n$ direction. It indicates the change in the velocity direction. Also recall that $\mathbf{a}_t$ is tangent to the path. It indicates the change in the velocity magnitude.

$$\Sigma F_t = ma_t, \ \Sigma F_n = ma_n, \ \Sigma F_b = 0$$
$$a_t = dv/dt \quad \text{or} \quad a_t = v \, dv/ds$$

$$a_n = v^2/\rho \quad \text{where} \quad \rho = \frac{[1 + (dy/dx)^2]^{3/2}}{|d^2y/dx^2|}$$

Cylindrical coordinates are useful when angular motion of the radial line *r* is specified or when the path can conveniently be described with these coordinates.

$$\Sigma F_r = m(\ddot{r} - r\dot{\theta}^2)$$
$$\Sigma F_\theta = m(r\ddot{\theta} + 2\dot{r}\dot{\theta})$$
$$\Sigma F_z = m\ddot{z}$$

Central-Force Motion

When a single force acts upon a particle, such as during the free-flight trajectory of a satellite in a gravitational field, then the motion is referred to as central-force motion. The orbit depends upon the eccentricity *e*; and as a result, the trajectory can either be circular, parabolic, elliptical, or hyperbolic.

REVIEW PROBLEMS

13

R13–1. The van is traveling at 20 km/h when the coupling of the trailer at A fails. If the trailer has a mass of 250 kg and coasts 45 m before coming to rest, determine the constant horizontal force F created by rolling friction which causes the trailer to stop.

R13–3. Block B rests on a smooth surface. If the coefficients of friction between A and B are $\mu_s = 0.4$ and $\mu_k = 0.3$, determine the acceleration of each block if $F = 50$ lb.

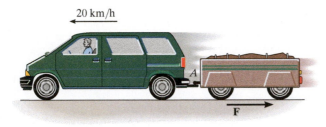

Prob. R13–1

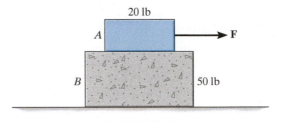

Prob. R13–3

R13–2. The motor M pulls in its attached rope with an acceleration $a_p = 6 \text{ m/s}^2$. Determine the towing force exerted by M on the rope in order to move the 50-kg crate up the inclined plane. The coefficient of kinetic friction between the crate and the plane is $\mu_k = 0.3$. Neglect the mass of the pulleys and rope.

R13–4. If the motor draws in the cable at a rate of $v = (0.05 s^{3/2})$ m/s, where s is in meters, determine the tension developed in the cable when $s = 10$ m. The crate has a mass of 20 kg, and the coefficient of kinetic friction between the crate and the ground is $\mu_k = 0.2$.

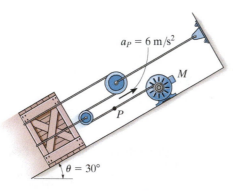

Prob. R13–2

Prob. R13–4

R13–5. The ball has a mass of 30 kg and a speed $v = 4$ m/s at the instant it is at its lowest point, $\theta = 0°$. Determine the tension in the cord and the rate at which the ball's speed is decreasing at the instant $\theta = 20°$. Neglect the size of the ball.

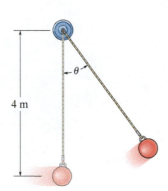

Prob. R13–5

R13–6. The bottle rests at a distance of 3 ft from the center of the horizontal platform. If the coefficient of static friction between the bottle and the platform is $\mu_s = 0.3$, determine the maximum speed that the bottle can attain before slipping. Assume the angular motion of the platform is slowly increasing.

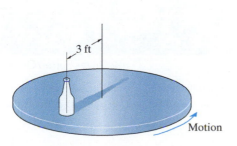

Prob. R13–6

R13–7. The 10-lb suitcase slides down the curved ramp for which the coefficient of kinetic friction is $\mu_k = 0.2$. If at the instant it reaches point A it has a speed of 5 ft/s, determine the normal force on the suitcase and the rate of increase of its speed.

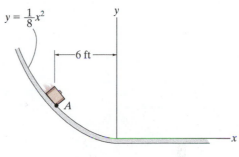

Prob. R13–7

R13–8. The spool, which has a mass of 4 kg, slides along the rotating rod. At the instant shown, the angular rate of rotation of the rod is $\dot{\theta} = 6$ rad/s and this rotation is increasing at $\ddot{\theta} = 2$ rad/s². At this same instant, the spool has a velocity of 3 m/s and an acceleration of 1 m/s², both measured relative to the rod and directed away from the center O when $r = 0.5$ m. Determine the radial frictional force and the normal force, both exerted by the rod on the spool at this instant.

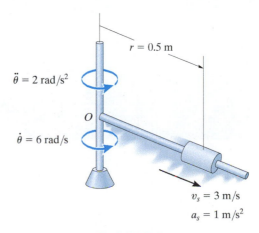

Prob. R13–8

(© Oliver Furrer/Ocean/Corbis)

As the woman falls, her energy will have to be absorbed by the bungee cord.
The principles of work and energy can be used to predict the motion.

Kinetics of a Particle: Work and Energy

CHAPTER OBJECTIVES

■ To develop the principle of work and energy and apply it to solve problems that involve force, velocity, and displacement.

■ To study problems that involve power and efficiency.

■ To introduce the concept of a conservative force and apply the theorem of conservation of energy to solve kinetic problems.

14.1   The Work of a Force

In this chapter, we will analyze motion of a particle using the concepts of work and energy. The resulting equation will be useful for solving problems that involve force, velocity, and displacement. Before we do this, however, we must first define the work of a force. Specifically, a force $\mathbf{F}$ will do *work* on a particle only when the particle undergoes a *displacement in the direction of the force*. For example, if the force $\mathbf{F}$ in Fig. 14–1 causes the particle to move along the path s from position $\mathbf{r}$ to a new position $\mathbf{r}'$, the displacement is then $d\mathbf{r} = \mathbf{r}' - \mathbf{r}$. The magnitude of $d\mathbf{r}$ is ds, the length of the differential segment along the path. If the angle between the tails of $d\mathbf{r}$ and $\mathbf{F}$ is θ, Fig. 14–1, then the work done by $\mathbf{F}$ is a *scalar quantity*, defined by

$$dU = F\, ds \cos \theta$$

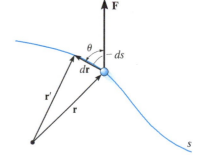

Fig. 14–1

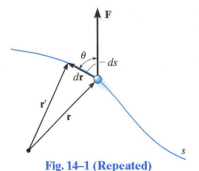

Fig. 14–1 (Repeated)

By definition of the dot product (see Eq. B–14) this equation can also be written as

$$dU = \mathbf{F} \cdot d\mathbf{r}$$

This result may be interpreted in one of two ways: either as the product of F and the component of displacement $ds \cos \theta$ in the direction of the force, or as the product of ds and the component of force, $F \cos \theta$, in the direction of displacement. Note that if $0° \le \theta < 90°$, then the force component and the displacement have the *same sense* so that the work is *positive;* whereas if $90° < \theta \le 180°$, these vectors will have *opposite sense*, and therefore the work is *negative*. Also, $dU = 0$ if the force is *perpendicular* to displacement, since $\cos 90° = 0$, or if the force is applied at a *fixed point*, in which case the displacement is zero.

The unit of work in SI units is the joule (J), which is the amount of work done by a one-newton force when it moves through a distance of one meter in the direction of the force (1 J = 1 N·m). In the FPS system, work is measured in units of foot-pounds (ft·lb), which is the work done by a one-pound force acting through a distance of one foot in the direction of the force.*

Work of a Variable Force.

If the particle acted upon by the force **F** undergoes a finite displacement along its path from $\mathbf{r}_1$ to $\mathbf{r}_2$ or s_1 to s_2, Fig. 14–2a, the work of force **F** is determined by integration. Provided F and θ can be expressed as a function of position, then

$$U_{1-2} = \int_{\mathbf{r}_1}^{\mathbf{r}_2} \mathbf{F} \cdot d\mathbf{r} = \int_{s_1}^{s_2} F \cos \theta \, ds \qquad (14\text{–}1)$$

Sometimes, this relation may be obtained by using experimental data to plot a graph of $F \cos \theta$ vs. s. Then the area under this graph bounded by s_1 and s_2 represents the total work, Fig. 14–2b.

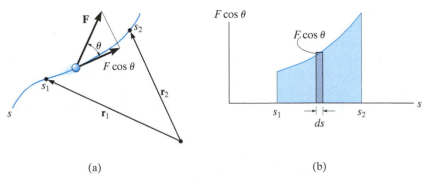

(a)

(b)

Fig. 14–2

---

*By convention, the units for the moment of a force or torque are written as lb·ft, to distinguish them from those used to signify work, ft·lb.

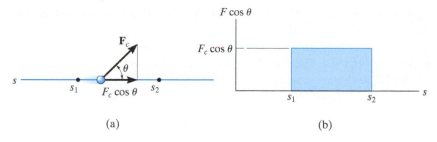

Fig. 14–3

Work of a Constant Force Moving Along a Straight Line.

If the force $\mathbf{F}_c$ has a constant magnitude and acts at a constant angle θ from its straight-line path, Fig. 14–3a, then the component of $\mathbf{F}_c$ in the direction of displacement is always $F_c \cos \theta$. The work done by $\mathbf{F}_c$ when the particle is displaced from s_1 to s_2 is determined from Eq. 14–1, in which case

$$U_{1-2} = F_c \cos \theta \int_{s_1}^{s_2} ds$$

or

$$U_{1-2} = F_c \cos \theta (s_2 - s_1) \qquad (14\text{–}2)$$

Here the work of $\mathbf{F}_c$ represents the *area of the rectangle* in Fig. 14–3b.

The crane must do work in order to hoist the weight of the pipe. (© R.C. Hibbeler)

Work of a Weight.

Consider a particle of weight $\mathbf{W}$, which moves up along the path s shown in Fig. 14–4 from position s_1 to position s_2. At an intermediate point, the displacement $d\mathbf{r} = dx\mathbf{i} + dy\mathbf{j} + dz\mathbf{k}$. Since $\mathbf{W} = -W\mathbf{j}$, applying Eq. 14–1 we have

$$U_{1-2} = \int \mathbf{F} \cdot d\mathbf{r} = \int_{\mathbf{r}_1}^{\mathbf{r}_2} (-W\mathbf{j}) \cdot (dx\mathbf{i} + dy\mathbf{j} + dz\mathbf{k})$$

$$= \int_{y_1}^{y_2} -W \, dy = -W(y_2 - y_1)$$

or

$$U_{1-2} = -W \, \Delta y \qquad (14\text{–}3)$$

Thus, the work is independent of the path and is equal to the magnitude of the particle's weight times its vertical displacement. In the case shown in Fig. 14–4 the work is *negative*, since W is downward and Δy is upward. Note, however, that if the particle is displaced *downward* $(-\Delta y)$, the work of the weight is *positive*. Why?

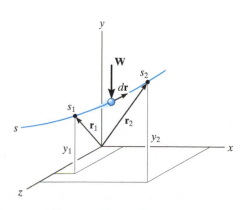

Fig. 14–4

Work of a Spring Force. If an elastic spring is elongated a distance ds, Fig. 14–5a, then the work done by the force that acts on the attached particle is $dU = -F_s ds = -ks\,ds$. The work is *negative* since $\mathbf{F}_s$ acts in the opposite sense to ds. If the particle displaces from s_1 to s_2, the work of $\mathbf{F}_s$ is then

$$U_{1-2} = \int_{s_1}^{s_2} F_s\,ds = \int_{s_1}^{s_2} -ks\,ds$$

$$\boxed{U_{1-2} = -\left(\tfrac{1}{2}ks_2^2 - \tfrac{1}{2}ks_1^2\right)} \qquad (14\text{–}4)$$

This work represents the trapezoidal area under the line $F_s = ks$, Fig. 14–5b.

A mistake in sign can be avoided when applying this equation if one simply notes the direction of the spring force acting on the particle and compares it with the sense of direction of displacement of the particle — if both are in the *same sense, positive work* results; if they are *opposite* to one another, the *work is negative*.

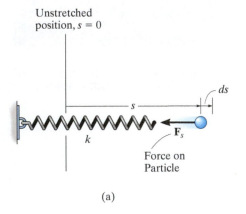

(a)

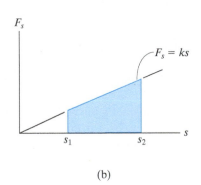

(b)

Fig. 14–5

The forces acting on the cart, as it is pulled a distance s up the incline, are shown on its free-body diagram. The constant towing force $\mathbf{T}$ does positive work of $U_T = (T\cos\phi)s$, the weight does negative work of $U_W = -(W\sin\theta)s$, and the normal force $\mathbf{N}$ does no work since there is no displacement of this force along its line of action. (© R.C. Hibbeler)

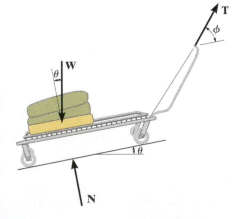

EXAMPLE 14.1

The 10-kg block shown in Fig. 14–6a rests on the smooth incline. If the spring is originally stretched 0.5 m, determine the total work done by all the forces acting on the block when a horizontal force $P = 400$ N pushes the block up the plane $s = 2$ m.

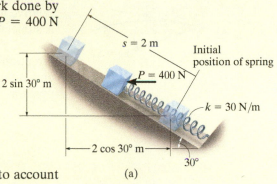

(a)

Fig. 14–6

SOLUTION

First the free-body diagram of the block is drawn in order to account for all the forces that act on the block, Fig. 14–6b.

Horizontal Force P. Since this force is *constant*, the work is determined using Eq. 14–2. The result can be calculated as the force times the component of displacement in the direction of the force; i.e.,

$$U_P = 400 \text{ N } (2 \text{ m } \cos 30°) = 692.8 \text{ J}$$

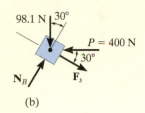

(b)

or the displacement times the component of force in the direction of displacement, i.e.,

$$U_P = 400 \text{ N } \cos 30°(2 \text{ m}) = 692.8 \text{ J}$$

Spring Force F_s. In the initial position the spring is stretched $s_1 = 0.5$ m and in the final position it is stretched $s_2 = 0.5$ m $+ 2$ m $= 2.5$ m. We require the work to be negative since the force and displacement are opposite to each other. The work of $\mathbf{F}_s$ is thus

$$U_s = -\left[\tfrac{1}{2}(30 \text{ N/m})(2.5 \text{ m})^2 - \tfrac{1}{2}(30 \text{ N/m})(0.5 \text{ m})^2\right] = -90 \text{ J}$$

Weight W. Since the weight acts in the opposite sense to its vertical displacement, the work is negative; i.e.,

$$U_W = -(98.1 \text{ N }) (2 \text{ m } \sin 30°) = -98.1 \text{ J}$$

Note that it is also possible to consider the component of weight in the direction of displacement; i.e.,

$$U_W = -(98.1 \sin 30° \text{ N}) (2 \text{ m}) = -98.1 \text{ J}$$

Normal Force N_B. This force does *no work* since it is *always* perpendicular to the displacement.

Total Work. The work of all the forces when the block is displaced 2 m is therefore

$$U_T = 692.8 \text{ J} - 90 \text{ J} - 98.1 \text{ J} = 505 \text{ J} \qquad \textit{Ans.}$$

14.2   Principle of Work and Energy

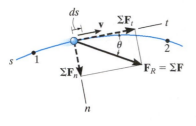

Fig. 14–7

Consider the particle in Fig. 14–7, which is located on the path defined relative to an inertial coordinate system. If the particle has a mass m and is subjected to a system of external forces represented by the resultant $\mathbf{F}_R = \Sigma\mathbf{F}$, then the equation of motion for the particle in the tangential direction is $\Sigma F_t = ma_t$. Applying the kinematic equation $a_t = v\,dv/ds$ and integrating both sides, assuming initially that the particle has a position $s = s_1$ and a speed $v = v_1$, and later at $s = s_2$, $v = v_2$, we have

$$\Sigma \int_{s_1}^{s_2} F_t\,ds = \int_{v_1}^{v_2} mv\,dv$$

$$\Sigma \int_{s_1}^{s_2} F_t\,ds = \tfrac{1}{2}mv_2^2 - \tfrac{1}{2}mv_1^2 \tag{14–5}$$

From Fig. 14–7, note that $\Sigma F_t = \Sigma F\cos\theta$, and since work is defined from Eq. 14–1, the final result can be written as

$$\Sigma U_{1-2} = \tfrac{1}{2}mv_2^2 - \tfrac{1}{2}mv_1^2 \tag{14–6}$$

This equation represents the *principle of work and energy* for the particle. The term on the left is the sum of the work done by *all* the forces acting on the particle as the particle moves from point 1 to point 2. The two terms on the right side, which are of the form $T = \tfrac{1}{2}mv^2$, define the particle's final and initial *kinetic energy*, respectively. Like work, kinetic energy is a *scalar* and has units of joules (J) and ft · lb. However, unlike work, which can be either positive or negative, the kinetic energy is *always positive*, regardless of the direction of motion of the particle.

When Eq. 14–6 is applied, it is often expressed in the form

$$\boxed{T_1 + \Sigma U_{1-2} = T_2} \tag{14–7}$$

which states that the particle's initial kinetic energy plus the work done by all the forces acting on the particle as it moves from its initial to its final position is equal to the particle's final kinetic energy.

As noted from the derivation, the principle of work and energy represents an integrated form of $\Sigma F_t = ma_t$, obtained by using the kinematic equation $a_t = v\,dv/ds$. As a result, this principle will provide a convenient *substitution* for $\Sigma F_t = ma_t$ when solving those types of kinetic problems which involve *force*, *velocity*, and *displacement* since these quantities are involved in Eq. 14–7. For application, it is suggested that the following procedure be used.

Procedure for Analysis

Work (Free-Body Diagram).

- Establish the inertial coordinate system and draw a free-body diagram of the particle in order to account for all the forces that do work on the particle as it moves along its path.

Principle of Work and Energy.

- Apply the principle of work and energy, $T_1 + \Sigma U_{1-2} = T_2$.

- The kinetic energy at the initial and final points is *always positive*, since it involves the speed squared $\left(T = \frac{1}{2}mv^2\right)$.

- A force does work when it moves through a displacement in the direction of the force.

- Work is *positive* when the force component is in the *same sense of direction* as its displacement, otherwise it is negative.

- Forces that are functions of displacement must be integrated to obtain the work. Graphically, the work is equal to the area under the force-displacement curve.

- The work of a weight is the product of the weight magnitude and the vertical displacement, $U_W = \pm Wy$. It is positive when the weight moves downwards.

- The work of a spring is of the form $U_s = \frac{1}{2}ks^2$, where k is the spring stiffness and s is the stretch or compression of the spring.

Numerical application of this procedure is illustrated in the examples following Sec. 14.3.

If an oncoming car strikes these crash barrels, the car's kinetic energy will be transformed into work, which causes the barrels, and to some extent the car, to be deformed. By knowing the amount of energy that can be absorbed by each barrel it is possible to design a crash cushion such as this. (© R.C. Hibbeler)

14.3   Principle of Work and Energy for a System of Particles

The principle of work and energy can be extended to include a system of particles isolated within an enclosed region of space as shown in Fig. 14–8. Here the arbitrary *i*th particle, having a mass m_i, is subjected to a resultant external force $\mathbf{F}_i$ and a resultant internal force $\mathbf{f}_i$ which all the other particles exert on the *i*th particle. If we apply the principle of work and energy to this and each of the other particles in the system, then since work and energy are scalar quantities, the equations can be summed algebraically, which gives

$$\Sigma T_1 + \Sigma U_{1-2} = \Sigma T_2 \qquad (14\text{–}8)$$

In this case, the initial kinetic energy of the system plus the work done by all the external and internal forces acting on the system is equal to the final kinetic energy of the system.

If the system represents a *translating rigid body*, or a series of connected translating bodies, then all the particles in each body will undergo the *same displacement*. Therefore, the work of all the internal forces will occur in equal but opposite collinear pairs and so it will cancel out. On the other hand, if the body is assumed to be *nonrigid*, the particles of the body may be displaced along *different paths*, and some of the energy due to force interactions would be given off and lost as heat or stored in the body if permanent deformations occur. We will discuss these effects briefly at the end of this section and in Sec. 15.4. Throughout this text, however, the principle of work and energy will be applied to problems where direct accountability of such energy losses does not have to be considered.

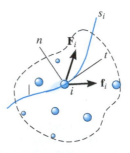

Inertial coordinate system

Fig. 14–8

Work of Friction Caused by Sliding.

A special class of problems will now be investigated which requires a careful application of Eq. 14–8. These problems involve cases where a body slides over the surface of another body in the presence of friction. Consider, for example, a block which is translating a distance *s* over a rough surface as shown in Fig. 14–9*a*. If the applied force **P** just balances the *resultant* frictional force $\mu_k N$, Fig. 14–9*b*, then due to equilibrium a constant velocity **v** is maintained, and one would expect Eq. 14–8 to be applied as follows:

$$\tfrac{1}{2}mv^2 + Ps - \mu_k Ns = \tfrac{1}{2}mv^2$$

Indeed this equation is satisfied if $P = \mu_k N$; however, as one realizes from experience, the sliding motion will *generate heat*, a form of energy which seems not to be accounted for in the work-energy equation. In order to explain this paradox and thereby more closely represent the nature of friction, we should actually model the block so that the surfaces of contact are *deformable* (nonrigid).* Recall that the rough portions at the bottom of the block act as "teeth," and when the block slides these teeth *deform slightly* and either break off or vibrate as they pull away from "teeth" at the contacting surface, Fig. 14–9*c*. As a result, frictional forces that act on the block at these points are displaced slightly, due to the localized deformations, and later they are replaced by other frictional forces as other points of contact are made. At any instant, the *resultant* **F** of all these frictional forces remains essentially constant, i.e., $\mu_k N$; however, due to the many *localized deformations*, the actual displacement s' of $\mu_k N$ is *not* the same as the displacement *s* of the applied force **P**. Instead, s' will be *less* than *s* ($s' < s$), and therefore the *external work* done by the resultant frictional force will be $\mu_k N s'$ and not $\mu_k N s$. The remaining amount of work, $\mu_k N(s - s')$, manifests itself as an increase in *internal energy*, which in fact causes the block's temperature to rise.

In summary then, Eq. 14–8 can be applied to problems involving sliding friction; however, it should be fully realized that the work of the resultant frictional force is not represented by $\mu_k N s$; instead, this term represents *both* the external work of friction ($\mu_k N s'$) *and* internal work $[\mu_k N(s - s')]$ which is converted into various forms of internal energy, such as heat.†

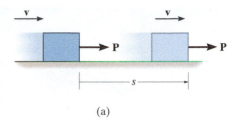

(a)

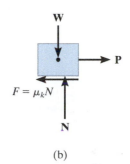

(b)

(c)

Fig. 14–9

14

---

*See Chapter 8 of *Engineering Mechanics: Statics.*

†See B. A. Sherwood and W. H. Bernard, "Work and Heat Transfer in the Presence of Sliding Friction," *Am. J. Phys.* 52, 1001 (1984).

EXAMPLE | 14.2

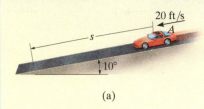

(a)

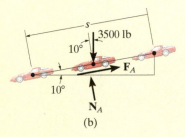

(b)

Fig. 14–10

The 3500-lb automobile shown in Fig. 14–10a travels down the 10° inclined road at a speed of 20 ft/s. If the driver jams on the brakes, causing his wheels to lock, determine how far *s* the tires skid on the road. The coefficient of kinetic friction between the wheels and the road is $\mu_k = 0.5$.

SOLUTION

This problem can be solved using the principle of work and energy, since it involves force, velocity, and displacement.

Work (Free-Body Diagram). As shown in Fig. 14–10b, the normal force $\mathbf{N}_A$ does no work since it never undergoes displacement along its line of action. The weight, 3500 lb, is displaced $s \sin 10°$ and does positive work. Why? The frictional force $\mathbf{F}_A$ does both external and internal work when it undergoes a displacement s. This work is negative since it is in the opposite sense of direction to the displacement. Applying the equation of equilibrium normal to the road, we have

$$+\nwarrow\Sigma F_n = 0; \qquad N_A - 3500 \cos 10° \text{ lb} = 0 \qquad N_A = 3446.8 \text{ lb}$$

Thus,

$$F_A = \mu_k N_A = 0.5\,(3446.8 \text{ lb}) = 1723.4 \text{ lb}$$

Principle of Work and Energy.

$$T_1 + \Sigma U_{1-2} = T_2$$

$$\frac{1}{2}\left(\frac{3500 \text{ lb}}{32.2 \text{ ft/s}^2}\right)(20 \text{ ft/s})^2 + 3500 \text{ lb}(s \sin 10°) - (1723.4 \text{ lb})s = 0$$

Solving for *s* yields

$$s = 19.5 \text{ ft} \qquad\qquad\qquad \textit{Ans.}$$

NOTE: If this problem is solved by using the equation of motion, *two steps* are involved. First, from the free-body diagram, Fig. 14–10b, the equation of motion is applied along the incline. This yields

$$+\swarrow\Sigma F_s = ma_s; \qquad 3500 \sin 10° \text{ lb} - 1723.4 \text{ lb} = \frac{3500 \text{ lb}}{32.2 \text{ ft/s}^2}a$$

$$a = -10.3 \text{ ft/s}^2$$

Then, since *a* is constant, we have

$$(+\swarrow) \quad v^2 = v_0^2 + 2a_c(s - s_0);$$

$$(0)^2 = (20 \text{ ft/s})^2 + 2(-10.3 \text{ ft/s}^2)(s - 0)$$

$$s = 19.5 \text{ ft} \qquad\qquad\qquad \textit{Ans.}$$

EXAMPLE 14.3

For a short time the crane in Fig. 14–11a lifts the 2.50-Mg beam with a force of $F = (28 + 3s^2)$ kN. Determine the speed of the beam when it has risen $s = 3$ m. Also, how much time does it take to attain this height starting from rest?

SOLUTION

We can solve part of this problem using the principle of work and energy since it involves force, velocity, and displacement. Kinematics must be used to determine the time. Note that at $s = 0$, $F = 28(10^3)\text{N} > W = 2.50(10^3)(9.81)\text{N}$, so motion will occur.

Work (Free-Body Diagram).   As shown on the free-body diagram, Fig. 14–11b, the lifting force **F** does positive work, which must be determined by integration since this force is a variable. Also, the weight is constant and will do negative work since the displacement is upward.

(a)

Principles of Work and Energy.

$$T_1 + \Sigma U_{1-2} = T_2$$

$$0 + \int_0^s (28 + 3s^2)(10^3)\, ds - (2.50)(10^3)(9.81)s = \tfrac{1}{2}(2.50)(10^3)v^2$$

$$28(10^3)s + (10^3)s^3 - 24.525(10^3)s = 1.25(10^3)v^2$$

$$v = (2.78s + 0.8s^3)^{\frac{1}{2}} \qquad (1)$$

When $s = 3$ m,

$$v = 5.47 \text{ m/s} \qquad\qquad Ans.$$

Kinematics.   Since we were able to express the velocity as a function of displacement, the time can be determined using $v = ds/dt$. In this case,

$$(2.78s + 0.8s^3)^{\frac{1}{2}} = \frac{ds}{dt}$$

$$t = \int_0^3 \frac{ds}{(2.78s + 0.8s^3)^{\frac{1}{2}}}$$

The integration can be performed numerically using a pocket calculator. The result is

$$t = 1.79 \text{ s} \qquad\qquad Ans.$$

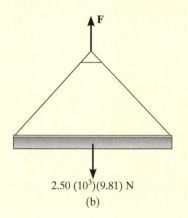

2.50 $(10^3)(9.81)$ N

(b)

Fig. 14–11

NOTE: The acceleration of the beam can be determined by integrating Eq. (1) using $v\, dv = a\, ds$, or more directly, by applying the equation of motion, $\Sigma F = ma$.

14

EXAMPLE | **14.4**

The platform P, shown in Fig. 14–12a, has negligible mass and is tied down so that the 0.4-m-long cords keep a 1-m-long spring compressed 0.6 m when *nothing* is on the platform. If a 2-kg block is placed on the platform and released from rest after the platform is pushed down 0.1 m, Fig. 14–12b, determine the maximum height h the block rises in the air, measured from the ground.

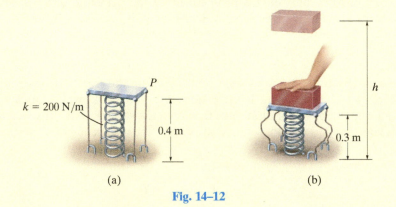

(a)   (b)

Fig. 14–12

SOLUTION

Work (Free-Body Diagram).   Since the block is released from rest and later reaches its maximum height, the initial and final velocities are zero. The free-body diagram of the block when it is still in contact with the platform is shown in Fig. 14–12c. Note that the weight does negative work and the spring force does positive work. Why? In particular, the *initial compression* in the spring is $s_1 = 0.6 \text{ m} + 0.1 \text{ m} = 0.7 \text{ m}$. Due to the cords, the spring's *final compression* is $s_2 = 0.6 \text{ m}$ (after the block leaves the platform). The bottom of the block rises from a height of $(0.4 \text{ m} - 0.1 \text{ m}) = 0.3 \text{ m}$ to a final height h.

Principle of Work and Energy.

$$T_1 + \Sigma U_{1-2} = T_2$$

$$\tfrac{1}{2}mv_1^2 + \left\{ -\left(\tfrac{1}{2}ks_2^2 - \tfrac{1}{2}ks_1^2\right) - W\,\Delta y \right\} = \tfrac{1}{2}mv_2^2$$

Note that here $s_1 = 0.7 \text{ m} > s_2 = 0.6 \text{ m}$ and so the work of the spring as determined from Eq. 14–4 will indeed be positive once the calculation is made. Thus,

$$0 + \left\{ -\left[\tfrac{1}{2}(200 \text{ N/m})(0.6 \text{ m})^2 - \tfrac{1}{2}(200 \text{ N/m})(0.7 \text{ m})^2 \right] \right.$$
$$\left. - (19.62 \text{ N})[h - (0.3 \text{ m})] \right\} = 0$$

Solving yields

$$h = 0.963 \text{ m} \qquad \qquad Ans.$$

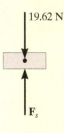

19.62 N

$\mathbf{F}_s$

(c)

EXAMPLE   14.5

The 40-kg boy in Fig. 14–13a slides down the smooth water slide. If he starts from rest at A, determine his speed when he reaches B and the normal reaction the slide exerts on the boy at this position.

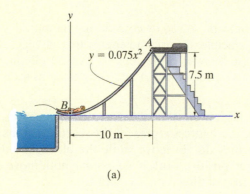

(a)

SOLUTION

Work (Free-Body Diagram).   As shown on the free-body diagram, Fig. 14–13b, there are two forces acting on the boy as he goes down the slide. Note that the normal force does no work.

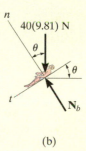

Principle of Work and Energy.

$$T_A + \Sigma U_{A-B} = T_B$$

$$0 + (40(9.81)\text{N})\,(7.5\text{ m}) = \tfrac{1}{2}(40\text{ kg})v_B^2$$

(b)

$$v_B = 12.13\text{ m/s} = 12.1\text{ m/s} \qquad Ans.$$

Equation of Motion.   Referring to the free-body diagram of the boy when he is at B, Fig. 14–13c, the normal reaction $\mathbf{N}_B$ can now be obtained by applying the equation of motion along the n axis. Here the radius of curvature of the path is

$$\rho_B = \frac{\left[1 + \left(\dfrac{dy}{dx}\right)^2\right]^{3/2}}{|d^2y/dx^2|} = \frac{\left[1 + (0.15x)^2\right]^{3/2}}{|0.15|}\Bigg|_{x=0} = 6.667\text{ m}$$

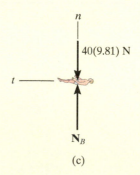

Thus,

$$+\uparrow\Sigma F_n = ma_n; \qquad N_B - 40(9.81)\text{ N} = 40\text{ kg}\left(\frac{(12.13\text{ m/s})^2}{6.667\text{ m}}\right)$$

$$N_B = 1275.3\text{ N} = 1.28\text{ kN} \qquad Ans.$$

(c)

Fig. 14–13

EXAMPLE | 14.6

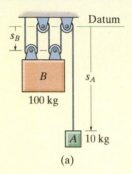

s_B

B
100 kg

s_A

A 10 kg

(a)

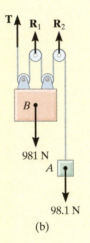

T R_1 R_2

B

981 N

A

98.1 N

(b)

Fig. 14–14

Blocks A and B shown in Fig. 14–14a have a mass of 10 kg and 100 kg, respectively. Determine the distance B travels when it is released from rest to the point where its speed becomes 2 m/s.

SOLUTION
This problem may be solved by considering the blocks separately and applying the principle of work and energy to each block. However, the work of the (unknown) cable tension can be eliminated from the analysis by considering blocks A and B together as a *single system*.

Work (Free-Body Diagram).   As shown on the free-body diagram of the system, Fig. 14–14b, the cable force **T** and reactions $\mathbf{R}_1$ and $\mathbf{R}_2$ do *no work*, since these forces represent the reactions at the supports and consequently they do not move while the blocks are displaced. The weights both do positive work if we *assume* both move downward, in the positive sense of direction of s_A and s_B.

Principle of Work and Energy.   Realizing the blocks are released from rest, we have

$$\Sigma T_1 + \Sigma U_{1-2} = \Sigma T_2$$

$$\left\{ \tfrac{1}{2}m_A(v_A)_1^2 + \tfrac{1}{2}m_B(v_B)_1^2 \right\} + \left\{ W_A\,\Delta s_A + W_B\,\Delta s_B \right\} =$$
$$\left\{ \tfrac{1}{2}m_A(v_A)_2^2 + \tfrac{1}{2}m_B(v_B)_2^2 \right\}$$

$$\{0 + 0\} + \left\{ 98.1\text{ N}\,(\Delta s_A) + 981\text{ N}\,(\Delta s_B) \right\} =$$
$$\left\{ \tfrac{1}{2}(10\text{ kg})(v_A)_2^2 + \tfrac{1}{2}(100\text{ kg})(2\text{ m/s})^2 \right\} \qquad (1)$$

Kinematics.   Using methods of kinematics, as discussed in Sec. 12.9, it may be seen from Fig. 14–14a that the total length l of all the vertical segments of cable may be expressed in terms of the position coordinates s_A and s_B as

$$s_A + 4s_B = l$$

Hence, a change in position yields the displacement equation

$$\Delta s_A + 4\,\Delta s_B = 0$$
$$\Delta s_A = -4\,\Delta s_B$$

Here we see that a downward displacement of one block produces an upward displacement of the other block. Note that Δs_A and Δs_B must have the *same* sign convention in both Eqs. 1 and 2. Taking the time derivative yields

$$v_A = -4v_B = -4(2\text{ m/s}) = -8\text{ m/s} \qquad (2)$$

Retaining the negative sign in Eq. 2 and substituting into Eq. 1 yields

$$\Delta s_B = 0.883\text{ m} \downarrow \qquad \qquad \textit{Ans.}$$

PRELIMINARY PROBLEMS

P14–1. Determine the work of the force when it displaces 2 m.

500 N

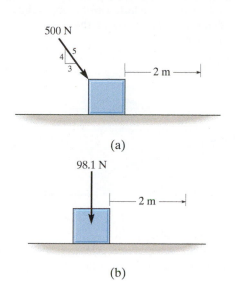

(a)

98.1 N

(b)

$F = (6 s^2)$ N

2 m

(c)

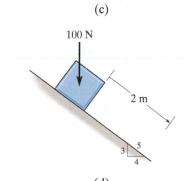

(d)

F

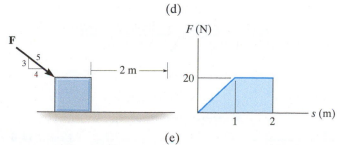

(e)

$k = 10$ N/m

2 m

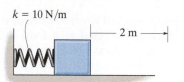

Spring is originally compressed 3 m.

(f)

100 N

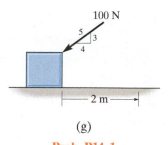

2 m

(g)

Prob. P14–1

P14–2. Determine the kinetic energy of the 10-kg block.

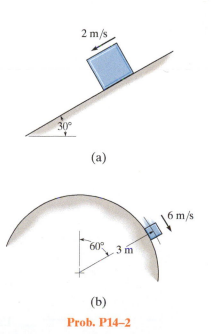

2 m/s

30°

(a)

6 m/s

60° 3 m

(b)

Prob. P14–2

FUNDAMENTAL PROBLEMS

F14–1. The spring is placed between the wall and the 10-kg block. If the block is subjected to a force of $F = 500$ N, determine its velocity when $s = 0.5$ m. When $s = 0$, the block is at rest and the spring is uncompressed. The contact surface is smooth.

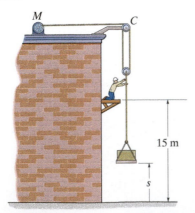

Prob. F14–1

F14–2. If the motor exerts a constant force of 300 N on the cable, determine the speed of the 20-kg crate when it travels $s = 10$ m up the plane, starting from rest. The coefficient of kinetic friction between the crate and the plane is $\mu_k = 0.3$.

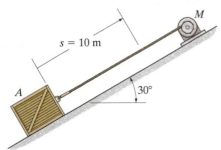

Prob. F14–2

F14–3. If the motor exerts a force of $F = (600 + 2s^2)$ N on the cable, determine the speed of the 100-kg crate when it rises to $s = 15$ m. The crate is initially at rest on the ground.

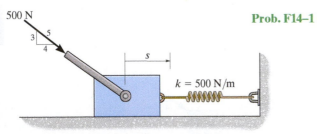

Prob. F14–3

F14–4. The 1.8-Mg dragster is traveling at 125 m/s when the engine is shut off and the parachute is released. If the drag force of the parachute can be approximated by the graph, determine the speed of the dragster when it has traveled 400 m.

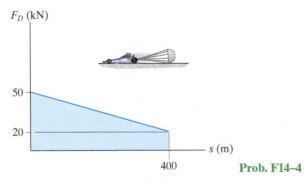

Prob. F14–4

F14–5. When $s = 0.6$ m, the spring is unstretched and the 10-kg block has a speed of 5 m/s down the smooth plane. Determine the distance s when the block stops.

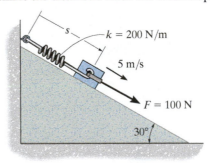

Prob. F14–5

F14–6. The 5-lb collar is pulled by a cord that passes around a small peg at C. If the cord is subjected to a constant force of $F = 10$ lb, and the collar is at rest when it is at A, determine its speed when it reaches B. Neglect friction.

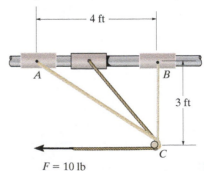

Prob. F14–6

PROBLEMS

14–1. The 20-kg crate is subjected to a force having a constant direction and a magnitude $F = 100$ N. When $s = 15$ m, the crate is moving to the right with a speed of 8 m/s. Determine its speed when $s = 25$ m. The coefficient of kinetic friction between the crate and the ground is $\mu_k = 0.25$.

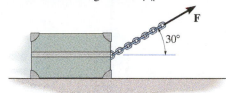

F

30°

Prob. 14–1

14–2. For protection, the barrel barrier is placed in front of the bridge pier. If the relation between the force and deflection of the barrier is $F = (90(10^3)x^{1/2})$ lb, where x is in ft, determine the car's maximum penetration in the barrier. The car has a weight of 4000 lb and it is traveling with a speed of 75 ft/s just before it hits the barrier.

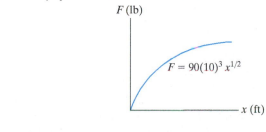

F (lb)

$F = 90(10)^3 \, x^{1/2}$

x (ft)

Prob. 14–2

14–3. The crate, which has a mass of 100 kg, is subjected to the action of the two forces. If it is originally at rest, determine the distance it slides in order to attain a speed of 6 m/s. The coefficient of kinetic friction between the crate and the surface is $\mu_k = 0.2$.

800 N

30°

1000 N

5 3 4

Prob. 14–3

*14–4. The 100-kg crate is subjected to the forces shown. If it is originally at rest, determine the distance it slides in order to attain a speed of $v = 8$ m/s. The coefficient of kinetic friction between the crate and the surface is $\mu_k = 0.2$.

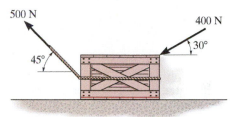

500 N

400 N

45°

30°

Prob. 14–4

14–5. Determine the required height h of the roller coaster so that when it is essentially at rest at the crest of the hill A it will reach a speed of 100 km/h when it comes to the bottom B. Also, what should be the minimum radius of curvature ρ for the track at B so that the passengers do not experience a normal force greater than $4mg = (39.24m)$ N? Neglect the size of the car and passenger.

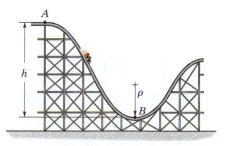

A

h

ρ

B

Prob. 14–5

14–6.   When the driver applies the brakes of a light truck traveling 40 km/h, it skids 3 m before stopping. How far will the truck skid if it is traveling 80 km/h when the brakes are applied?

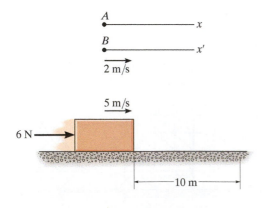

Prob. 14–6

14–7.   As indicated by the derivation, the principle of work and energy is valid for observers in *any* inertial reference frame. Show that this is so, by considering the 10-kg block which rests on the smooth surface and is subjected to a horizontal force of 6 N. If observer A is in a *fixed* frame x, determine the final speed of the block if it has an initial speed of 5 m/s and travels 10 m, both directed to the right and measured from the fixed frame. Compare the result with that obtained by an observer B, attached to the x' axis and moving at a constant velocity of 2 m/s relative to A. *Hint*: The distance the block travels will first have to be computed for observer B before applying the principle of work and energy.

*14–8.   A force of $F = 250$ N is applied to the end at B. Determine the speed of the 10-kg block when it has moved 1.5 m, starting from rest.

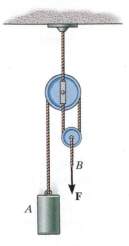

Prob. 14–8

14–9.   The "air spring" A is used to protect the support B and prevent damage to the conveyor-belt tensioning weight C in the event of a belt failure D. The force developed by the air spring as a function of its deflection is shown by the graph. If the block has a mass of 20 kg and is suspended a height $d = 0.4$ m above the top of the spring, determine the maximum deformation of the spring in the event the conveyor belt fails. Neglect the mass of the pulley and belt.

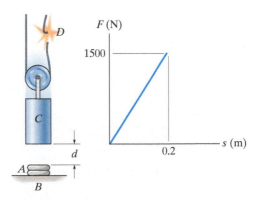

Prob. 14–7

Prob. 14–9

14–10. The force **F**, acting in a constant direction on the 20-kg block, has a magnitude which varies with the position s of the block. Determine how far the block must slide before its velocity becomes 15 m/s. When $s = 0$ the block is moving to the right at $v = 6$ m/s. The coefficient of kinetic friction between the block and surface is $\mu_k = 0.3$.

***14–12.** Design considerations for the bumper B on the 5-Mg train car require use of a nonlinear spring having the load-deflection characteristics shown in the graph. Select the proper value of k so that the maximum deflection of the spring is limited to 0.2 m when the car, traveling at 4 m/s, strikes the rigid stop. Neglect the mass of the car wheels.

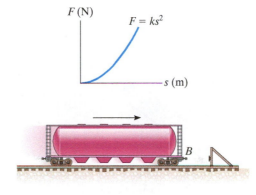

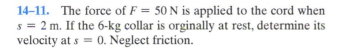

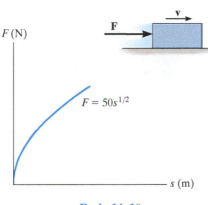

Prob. 14–10

Prob. 14–12

14–13. The 2-lb brick slides down a smooth roof, such that when it is at A it has a velocity of 5 ft/s. Determine the speed of the brick just before it leaves the surface at B, the distance d from the wall to where it strikes the ground, and the speed at which it hits the ground.

14–11. The force of $F = 50$ N is applied to the cord when $s = 2$ m. If the 6-kg collar is orginally at rest, determine its velocity at $s = 0$. Neglect friction.

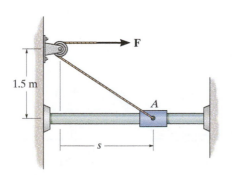

Prob. 14–11

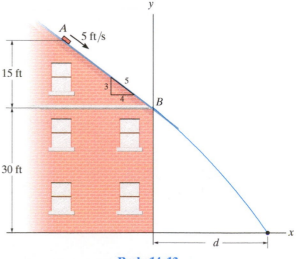

Prob. 14–13

14–14. Block *A* has a weight of 60 lb and block *B* has a weight of 10 lb. Determine the speed of block *A* after it moves 5 ft down the plane, starting from rest. Neglect friction and the mass of the cord and pulleys.

***14–16.** A small box of mass *m* is given a speed of $v = \sqrt{\frac{1}{4}gr}$ at the top of the smooth half cylinder. Determine the angle θ at which the box leaves the cylinder.

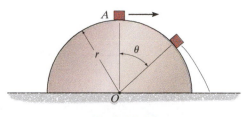

Prob. 14–16

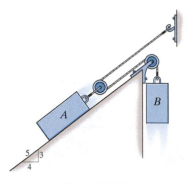

Prob. 14–14

14–17. If the cord is subjected to a constant force of *F* = 30 lb and the smooth 10-lb collar starts from rest at *A*, determine its speed when it passes point *B*. Neglect the size of pulley *C*.

14–15. The two blocks *A* and *B* have weights $W_A = 60$ lb and $W_B = 10$ lb. If the kinetic coefficient of friction between the incline and block *A* is $\mu_k = 0.2$, determine the speed of *A* after it moves 3 ft down the plane starting from rest. Neglect the mass of the cord and pulleys.

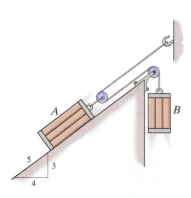

Prob. 14–15

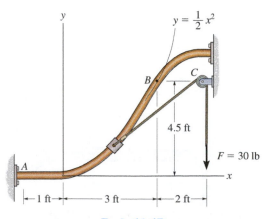

Prob. 14–17

14–18. When the 12-lb block A is released from rest it lifts the two 15-lb weights B and C. Determine the maximum distance A will fall before its motion is momentarily stopped. Neglect the weight of the cord and the size of the pulleys.

*14–20. The crash cushion for a highway barrier consists of a nest of barrels filled with an impact-absorbing material. The barrier stopping force is measured versus the vehicle penetration into the barrier. Determine the distance a car having a weight of 4000 lb will penetrate the barrier if it is originally traveling at 55 ft/s when it strikes the first barrel.

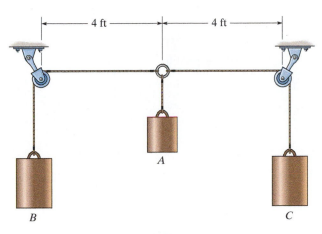

Prob. 14–18

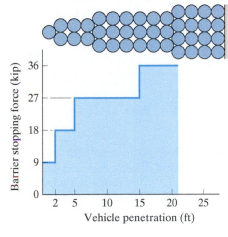

Prob. 14–20

14–19. If the cord is subjected to a constant force of $F = 300$ N and the 15-kg smooth collar starts from rest at A, determine the velocity of the collar when it reaches point B. Neglect the size of the pulley.

14–21. Determine the velocity of the 60-lb block A if the two blocks are released from rest and the 40-lb block B moves 2 ft up the incline. The coefficient of kinetic friction between both blocks and the inclined planes is $\mu_k = 0.10$.

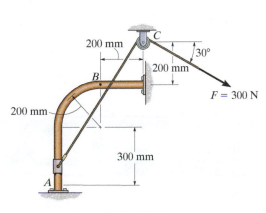

Prob. 14–19

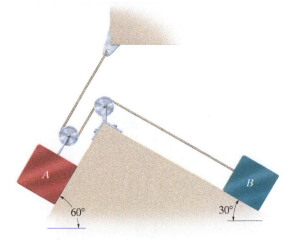

Prob. 14–21

14–22. The 25-lb block has an initial speed of $v_0 = 10$ ft/s when it is midway between springs A and B. After striking spring B, it rebounds and slides across the horizontal plane toward spring A, etc. If the coefficient of kinetic friction between the plane and the block is $\mu_k = 0.4$, determine the total distance traveled by the block before it comes to rest.

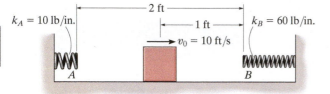

Prob. 14–22

14–23. The 8-kg block is moving with an initial speed of 5 m/s. If the coefficient of kinetic friction between the block and plane is $\mu_k = 0.25$, determine the compression in the spring when the block momentarily stops.

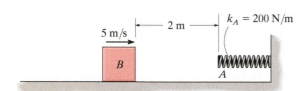

Prob. 14–23

***14–24.** At a given instant the 10-lb block A is moving downward with a speed of 6 ft/s. Determine its speed 2 s later. Block B has a weight of 4 lb, and the coefficient of kinetic friction between it and the horizontal plane is $\mu_k = 0.2$. Neglect the mass of the cord and pulleys.

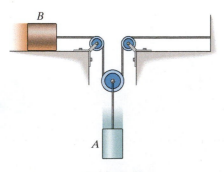

Prob. 14–24

14–25. The 5-lb cylinder is falling from A with a speed $v_A = 10$ ft/s onto the platform. Determine the maximum displacement of the platform, caused by the collision. The spring has an unstretched length of 1.75 ft and is originally kept in compression by the 1-ft long cables attached to the platform. Neglect the mass of the platform and spring and any energy lost during the collision.

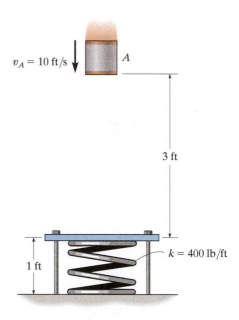

Prob. 14–25

14–26. The catapulting mechanism is used to propel the 10-kg slider A to the right along the smooth track. The propelling action is obtained by drawing the pulley attached to rod BC rapidly to the left by means of a piston P. If the piston applies a constant force $F = 20$ kN to rod BC such that it moves it 0.2 m, determine the speed attained by the slider if it was originally at rest. Neglect the mass of the pulleys, cable, piston, and rod BC.

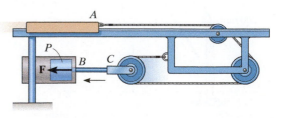

Prob. 14–26

14–27. The "flying car" is a ride at an amusement park which consists of a car having wheels that roll along a track mounted inside a rotating drum. By design the car cannot fall off the track, however motion of the car is developed by applying the car's brake, thereby gripping the car to the track and allowing it to move with a constant speed of the track, $v_t = 3$ m/s. If the rider applies the brake when going from B to A and then releases it at the top of the drum, A, so that the car coasts freely down along the track to B ($\theta = \pi$ rad), determine the speed of the car at B and the normal reaction which the drum exerts on the car at B. Neglect friction during the motion from A to B. The rider and car have a total mass of 250 kg and the center of mass of the car and rider moves along a circular path having a radius of 8 m.

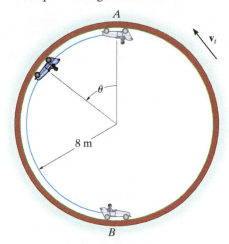

Prob. 14–27

***14–28.** The 10-lb box falls off the conveyor belt at 5-ft/s. If the coefficient of kinetic friction along AB is $\mu_k = 0.2$, determine the distance x when the box falls into the cart.

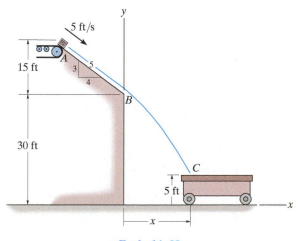

Prob. 14–28

14–29. The collar has a mass of 20 kg and slides along the smooth rod. Two springs are attached to it and the ends of the rod as shown. If each spring has an uncompressed length of 1 m and the collar has a speed of 2 m/s when $s = 0$, determine the maximum compression of each spring due to the back-and-forth (oscillating) motion of the collar.

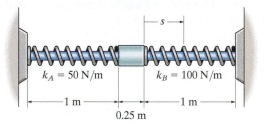

Prob. 14–29

14–30. The 30-lb box A is released from rest and slides down along the smooth ramp and onto the surface of a cart. If the cart is *prevented from moving*, determine the distance s from the end of the cart to where the box stops. The coefficient of kinetic friction between the cart and the box is $\mu_k = 0.6$.

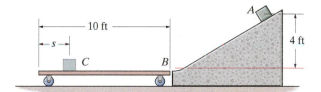

Prob. 14–30

14–31. Marbles having a mass of 5 g are dropped from rest at A through the smooth glass tube and accumulate in the can at C. Determine the placement R of the can from the end of the tube and the speed at which the marbles fall into the can. Neglect the size of the can.

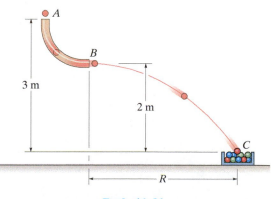

Prob. 14–31

*14–32. The block has a mass of 0.8 kg and moves within the smooth vertical slot. If it starts from rest when the *attached* spring is in the unstretched position at A, determine the *constant* vertical force F which must be applied to the cord so that the block attains a speed $v_B = 2.5$ m/s when it reaches B; $s_B = 0.15$ m. Neglect the size and mass of the pulley. *Hint:* The work of $\mathbf{F}$ can be determined by finding the difference Δl in cord lengths AC and BC and using $U_F = F \, \Delta l$.

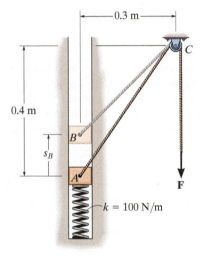

Prob. 14–32

14–33. The 10-lb block is pressed against the spring so as to compress it 2 ft when it is at A. If the plane is smooth, determine the distance d, measured from the wall, to where the block strikes the ground. Neglect the size of the block.

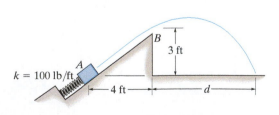

Prob. 14–33

14–34. The spring bumper is used to arrest the motion of the 4-lb block, which is sliding toward it at $v = 9$ ft/s. As shown, the spring is confined by the plate P and wall using cables so that its length is 1.5 ft. If the stiffness of the spring is $k = 50$ lb/ft, determine the required unstretched length of the spring so that the plate is not displaced more than 0.2 ft after the block collides into it. Neglect friction, the mass of the plate and spring, and the energy loss between the plate and block during the collision.

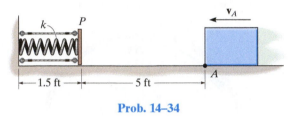

Prob. 14–34

14–35. When the 150-lb skier is at point A he has a speed of 5 ft/s. Determine his speed when he reaches point B on the smooth slope. For this distance the slope follows the cosine curve shown. Also, what is the normal force on his skis at B and his rate of increase in speed? Neglect friction and air resistance.

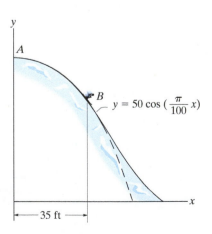

$$y = 50 \cos\left(\frac{\pi}{100} x\right)$$

Prob. 14–35

*14–36. The spring has a stiffness $k = 50$ lb/ft and an *unstretched length* of 2 ft. As shown, it is confined by the plate and wall using cables so that its length is 1.5 ft. A 4-lb block is given a speed v_A when it is at A, and it slides down the incline having a coefficient of kinetic friction $\mu_k = 0.2$. If it strikes the plate and pushes it forward 0.25 ft before stopping, determine its speed at A. Neglect the mass of the plate and spring.

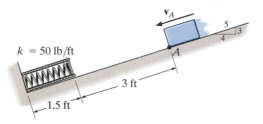

Prob. 14–36

14–37. If the track is to be designed so that the passengers of the roller coaster do not experience a normal force equal to zero or more than 4 times their weight, determine the limiting heights h_A and h_C so that this does not occur. The roller coaster starts from rest at position A. Neglect friction.

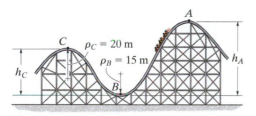

Prob. 14–37

14–38. If the 60-kg skier passes point A with a speed of 5 m/s, determine his speed when he reaches point B. Also find the normal force exerted on him by the slope at this point. Neglect friction.

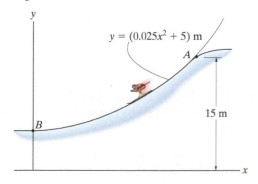

Prob. 14–38

14–39. If the 75-kg crate starts from rest at A, determine its speed when it reaches point B. The cable is subjected to a constant force of $F = 300$ N. Neglect friction and the size of the pulley.

*14–40. If the 75-kg crate starts from rest at A, and its speed is 6 m/s when it passes point B, determine the constant force **F** exerted on the cable. Neglect friction and the size of the pulley.

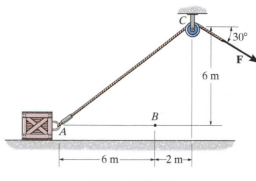

Probs. 14–39/40

14–41. A 2-lb block rests on the smooth semicylindrical surface. An elastic cord having a stiffness $k = 2$ lb/ft is attached to the block at B and to the base of the semicylinder at point C. If the block is released from rest at A ($\theta = 0°$), determine the unstretched length of the cord so the block begins to leave the semicylinder at the instant $\theta = 45°$. Neglect the size of the block.

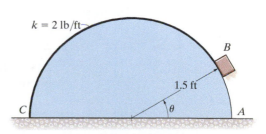

Prob. 14–41

14.4   Power and Efficiency

Power.   The term "power" provides a useful basis for choosing the type of motor or machine which is required to do a certain amount of work in a given time. For example, two pumps may each be able to empty a reservoir if given enough time; however, the pump having the larger power will complete the job sooner.

The *power* generated by a machine or engine that performs an amount of work dU within the time interval dt is therefore

$$P = \frac{dU}{dt} \qquad (14\text{–}9)$$

If the work dU is expressed as $dU = \mathbf{F} \cdot d\mathbf{r}$, then

$$P = \frac{dU}{dt} = \frac{\mathbf{F} \cdot d\mathbf{r}}{dt} = \mathbf{F} \cdot \frac{d\mathbf{r}}{dt}$$

or

$$P = \mathbf{F} \cdot \mathbf{v} \qquad (14\text{–}10)$$

Hence, power is a *scalar*, where in this formulation $\mathbf{v}$ represents the velocity of the particle which is acted upon by the force $\mathbf{F}$.

The basic units of power used in the SI and FPS systems are the watt (W) and horsepower (hp), respectively. These units are defined as

$$1\,\mathrm{W} = 1\,\mathrm{J/s} = 1\,\mathrm{N} \cdot \mathrm{m/s}$$

$$1\,\mathrm{hp} = 550\,\mathrm{ft} \cdot \mathrm{lb/s}$$

For conversion between the two systems of units, $1\,\mathrm{hp} = 746\,\mathrm{W}$.

The power output of this locomotive comes from the driving frictional force developed at its wheels. It is this force that overcomes the frictional resistance of the cars in tow and is able to lift the weight of the train up the grade. (© R.C. Hibbeler)

Efficiency.   The *mechanical efficiency* of a machine is defined as the ratio of the output of useful power produced by the machine to the input of power supplied to the machine. Hence,

$$\varepsilon = \frac{\text{power output}}{\text{power input}} \qquad (14\text{–}11)$$

If energy supplied to the machine occurs during the *same time interval* at which it is drawn, then the efficiency may also be expressed in terms of the ratio

$$\varepsilon = \frac{\text{energy output}}{\text{energy input}} \qquad (14\text{–}12)$$

Since machines consist of a series of moving parts, frictional forces will always be developed within the machine, and as a result, extra energy or power is needed to overcome these forces. Consequently, power output will be less than power input and so *the efficiency of a machine is always less than 1*.

The power supplied to a body can be determined using the following procedure.

Procedure for Analysis

• First determine the external force **F** acting on the body which causes the motion. This force is usually developed by a machine or engine placed either within or external to the body.

• If the body is accelerating, it may be necessary to draw its free-body diagram and apply the equation of motion ($\Sigma \mathbf{F} = m\mathbf{a}$) to determine **F**.

• Once **F** and the velocity **v** of the particle where **F** is applied have been found, the power is determined by multiplying the force magnitude with the component of velocity acting in the direction of **F**, (i.e., $P = \mathbf{F} \cdot \mathbf{v} = Fv \cos \theta$).

• In some problems the power may be found by calculating the work done by **F** per unit of time ($P_{avg} = \Delta U / \Delta t,$).

The power requirement of this hoist depends upon the vertical force **F** that acts on the elevator and causes it to move upward. If the velocity of the elevator is **v**, then the power output is $P = \mathbf{F} \cdot \mathbf{v}$. (© R.C. Hibbeler)

EXAMPLE | 14.7

The man in Fig. 14–15a pushes on the 50-kg crate with a force of $F = 150$ N. Determine the power supplied by the man when $t = 4$ s. The coefficient of kinetic friction between the floor and the crate is $\mu_k = 0.2$. Initially the create is at rest.

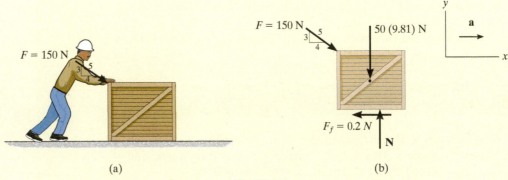

(a)

(b)

Fig. 14–15

SOLUTION

To determine the power developed by the man, the velocity of the 150-N force must be obtained first. The free-body diagram of the crate is shown in Fig. 14–15b. Applying the equation of motion,

$$+\uparrow \Sigma F_y = ma_y; \qquad N - \left(\tfrac{3}{5}\right)150\ \text{N} - 50(9.81)\ \text{N} = 0$$

$$N = 580.5\ \text{N}$$

$$\xrightarrow{+} \Sigma F_x = ma_x; \qquad \left(\tfrac{4}{5}\right)150\ \text{N} - 0.2(580.5\ \text{N}) = (50\ \text{kg})a$$

$$a = 0.078\ \text{m/s}^2$$

The velocity of the crate when $t = 4$ s is therefore

$$(\xrightarrow{+}) \qquad\qquad v = v_0 + a_c t$$

$$v = 0 + (0.078\ \text{m/s}^2)(4\ \text{s}) = 0.312\ \text{m/s}$$

The power supplied to the crate by the man when $t = 4$ s is therefore

$$P = \mathbf{F} \cdot \mathbf{v} = F_x v = \left(\tfrac{4}{5}\right)(150\ \text{N})(0.312\ \text{m/s})$$

$$= 37.4\ \text{W} \qquad\qquad\qquad Ans.$$

EXAMPLE 14.8

The motor M of the hoist shown in Fig. 14–16a lifts the 75-lb crate C so that the acceleration of point P is 4 ft/s^2. Determine the power that must be supplied to the motor at the instant P has a velocity of 2 ft/s. Neglect the mass of the pulley and cable and take $\varepsilon = 0.85$.

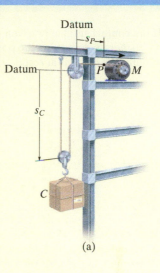

Datum

(a)

SOLUTION

In order to find the power output of the motor, it is first necessary to determine the tension in the cable since this force is developed by the motor.

From the free-body diagram, Fig. 14–16b, we have

$$+\downarrow \ \Sigma F_y = ma_y; \quad -2T + 75 \text{ lb} = \frac{75 \text{ lb}}{32.2 \text{ ft/s}^2}a_c \quad (1)$$

The acceleration of the crate can be obtained by using kinematics to relate it to the known acceleration of point P, Fig. 14–16a. Using the methods of absolute dependent motion, the coordinates s_C and s_P can be related to a constant portion of cable length l which is changing in the vertical and horizontal directions. We have $2s_C + s_P = l$. Taking the second time derivative of this equation yields

$$2a_C = -a_P \quad (2)$$

Since $a_P = +4$ ft/s^2, then $a_C = -(4 \text{ ft/s}^2)/2 = -2$ ft/s^2. What does the negative sign indicate? Substituting this result into Eq. 1 and *retaining* the negative sign since the acceleration in *both* Eq. 1 and Eq. 2 was considered positive downward, we have

(b)

Fig. 14–16

$$-2T + 75 \text{ lb} = \left(\frac{75 \text{ lb}}{32.2 \text{ ft/s}^2}\right)(-2 \text{ ft/s}^2)$$

$$T = 39.83 \text{ lb}$$

The power output, measured in units of horsepower, required to draw the cable in at a rate of 2 ft/s is therefore

$$P = \mathbf{T} \cdot \mathbf{v} = (39.83 \text{ lb})(2 \text{ ft/s})[1 \text{ hp}/(550 \text{ ft} \cdot \text{lb/s})]$$

$$= 0.1448 \text{ hp}$$

This *power output* requires that the motor provide a *power input* of

$$\text{power input} = \frac{1}{\varepsilon}(\text{power output})$$

$$= \frac{1}{0.85}(0.1448 \text{ hp}) = 0.170 \text{ hp} \qquad \textit{Ans.}$$

NOTE: Since the velocity of the crate is constantly changing, the power requirement is *instantaneous*.

FUNDAMENTAL PROBLEMS

F14–7. If the contact surface between the 20-kg block and the ground is smooth, determine the power of force **F** when $t = 4$ s. Initially, the block is at rest.

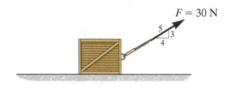

$F = 30$ N

5 3 4

Prob. F14–7

F14–8. If $F = (10\,s)$ N, where s is in meters, and the contact surface between the block and the ground is smooth, determine the power of force **F** when $s = 5$ m. When $s = 0$, the 20-kg block is moving at $v = 1$ m/s.

s $F = (10\,s)$ N

Prob. F14–8

F14–9. If the motor winds in the cable with a constant speed of $v = 3$ ft/s, determine the power supplied to the motor. The load weighs 100 lb and the efficiency of the motor is $\varepsilon = 0.8$. Neglect the mass of the pulleys.

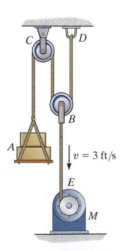

$v = 3$ ft/s

Prob. F14–9

F14–10. The coefficient of kinetic friction between the 20-kg block and the inclined plane is $\mu_k = 0.2$. If the block is traveling up the inclined plane with a constant velocity $v = 5$ m/s, determine the power of force **F**.

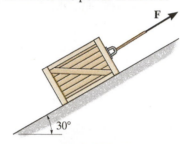

F

30°

Prob. F14–10

F14–11. If the 50-kg load A is hoisted by motor M so that the load has a constant velocity of 1.5 m/s, determine the power input to the motor, which operates at an efficiency $\varepsilon = 0.8$.

M

A 1.5 m/s

Prob. F14–11

F14–12. At the instant shown, point P on the cable has a velocity $v_P = 12$ m/s, which is increasing at a rate of $a_P = 6$ m/s^2. Determine the power input to motor M at this instant if it operates with an efficiency $\varepsilon = 0.8$. The mass of block A is 50 kg.

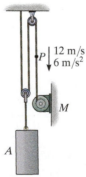

P 12 m/s
6 m/s^2

M

A

Prob. F14–12

PROBLEMS

14–42. The jeep has a weight of 2500 lb and an engine which transmits a power of 100 hp to *all* the wheels. Assuming the wheels do not slip on the ground, determine the angle θ of the largest incline the jeep can climb at a constant speed $v = 30$ ft/s.

Prob. 14–42

14–43. Determine the power input for a motor necessary to lift 300 lb at a constant rate of 5 ft/s. The efficiency of the motor is $\varepsilon = 0.65$.

*14–44. An automobile having a mass of 2 Mg travels up a 7° slope at a constant speed of $v = 100$ km/h. If mechanical friction and wind resistance are neglected, determine the power developed by the engine if the automobile has an efficiency $\varepsilon = 0.65$.

Prob. 14–44

14–45. The Milkin Aircraft Co. manufactures a turbojet engine that is placed in a plane having a weight of 13000 lb. If the engine develops a constant thrust of 5200 lb, determine the power output of the plane when it is just ready to take off with a speed of 600 mi/h.

14–46. To dramatize the loss of energy in an automobile, consider a car having a weight of 5000 lb that is traveling at 35 mi/h. If the car is brought to a stop, determine how long a 100-W light bulb must burn to expend the same amount of energy. (1 mi = 5280 ft.)

14–47. Escalator steps move with a constant speed of 0.6 m/s. If the steps are 125 mm high and 250 mm in length, determine the power of a motor needed to lift an average mass of 150 kg per step. There are 32 steps.

*14–48. The man having the weight of 150 lb is able to run up a 15-ft-high flight of stairs in 4 s. Determine the power generated. How long would a 100-W light bulb have to burn to expend the same amount of energy? *Conclusion*: Please turn off the lights when they are not in use!

15 ft

Prob. 14–48

14–49. The 2-Mg car increases its speed uniformly from rest to 25 m/s in 30 s up the inclined road. Determine the maximum power that must be supplied by the engine, which operates with an efficiency of $\varepsilon = 0.8$. Also, find the average power supplied by the engine.

1
10

Prob. 14–49

14–50. Determine the power output of the draw-works motor M necessary to lift the 600-lb drill pipe upward with a constant speed of 4 ft/s. The cable is tied to the top of the oil rig, wraps around the lower pulley, then around the top pulley, and then to the motor.

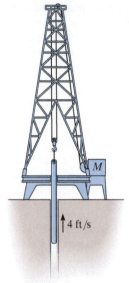

Prob. 14–50

14–51. The 1000-lb elevator is hoisted by the pulley system and motor M. If the motor exerts a constant force of 500 lb on the cable, determine the power that must be supplied to the motor at the instant the load has been hoisted $s = 15$ ft starting from rest. The motor has an efficiency of $\varepsilon = 0.65$.

Prob. 14–51

*14–52. The 50-lb crate is given a speed of 10 ft/s in $t = 4$ s starting from rest. If the acceleration is constant, determine the power that must be supplied to the motor when $t = 2$ s. The motor has an efficiency $\varepsilon = 0.65$. Neglect the mass of the pulley and cable.

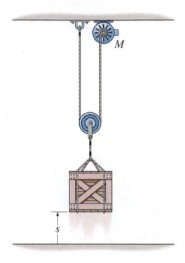

Prob. 14–52

14–53. The sports car has a mass of 2.3 Mg, and while it is traveling at 28 m/s the driver causes it to accelerate at 5 m/s². If the drag resistance on the car due to the wind is $F_D = (0.3v^2)$ N, where v is the velocity in m/s, determine the power supplied to the engine at this instant. The engine has a running efficiency of $\varepsilon = 0.68$.

14–54. The sports car has a mass of 2.3 Mg and accelerates at 6 m/s², starting from rest. If the drag resistance on the car due to the wind is $F_D = (10v)$ N, where v is the velocity in m/s, determine the power supplied to the engine when $t = 5$ s. The engine has a running efficiency of $\varepsilon = 0.68$.

Probs. 14–53/54

14–55. The elevator E and its freight have a total mass of 400 kg. Hoisting is provided by the motor M and the 60-kg block C. If the motor has an efficiency of $\varepsilon = 0.6$, determine the power that must be supplied to the motor when the elevator is hoisted upward at a constant speed of $v_E = 4$ m/s.

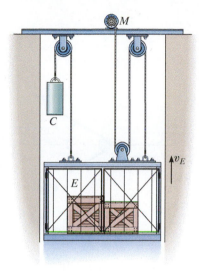

Prob. 14–55

*14–56. The 10-lb collar starts from rest at A and is lifted by applying a constant vertical force of $F = 25$ lb to the cord. If the rod is smooth, determine the power developed by the force at the instant $\theta = 60°$.

14–57. The 10-lb collar starts from rest at A and is lifted with a constant speed of 2 ft/s along the smooth rod. Determine the power developed by the force $\mathbf{F}$ at the instant shown.

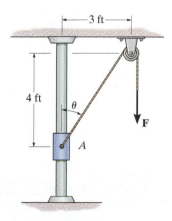

Prob. 14–57

14–58. The 50-lb block rests on the rough surface for which the coefficient of kinetic friction is $\mu_k = 0.2$. A force $F = (40 + s^2)$ lb, where s is in ft, acts on the block in the direction shown. If the spring is originally unstretched ($s = 0$) and the block is at rest, determine the power developed by the force the instant the block has moved $s = 1.5$ ft.

Prob. 14–58

14–59. The escalator steps move with a constant speed of 0.6 m/s. If the steps are 125 mm high and 250 mm in length, determine the power of a motor needed to lift an average mass of 150 kg per step. There are 32 steps.

*14–60. If the escalator in Prob. 14–46 is not moving, determine the constant speed at which a man having a mass of 80 kg must walk up the steps to generate 100 W of power—the same amount that is needed to power a standard light bulb.

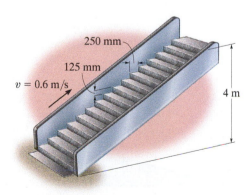

Probs. 14–59/60

14–61. If the jet on the dragster supplies a constant thrust of $T = 20$ kN, determine the power generated by the jet as a function of time. Neglect drag and rolling resistance, and the loss of fuel. The dragster has a mass of 1 Mg and starts from rest.

*14–64. The block has a weight of 80 lb and rests on the floor for which $\mu_k = 0.4$. If the motor draws in the cable at a constant rate of 6 ft/s, determine the output of the motor at the instant $\theta = 30°$. Neglect the mass of the cable and pulleys.

Prob. 14–61

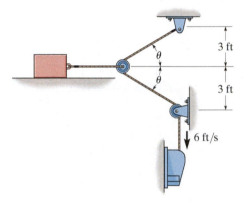

14–62. An athlete pushes against an exercise machine with a force that varies with time as shown in the first graph. Also, the velocity of the athlete's arm acting in the same direction as the force varies with time as shown in the second graph. Determine the power applied as a function of time and the work done in $t = 0.3$ s.

14–63. An athlete pushes against an exercise machine with a force that varies with time as shown in the first graph. Also, the velocity of the athlete's arm acting in the same direction as the force varies with time as shown in the second graph. Determine the maximum power developed during the 0.3-second time period.

Prob. 14–64

14–65. The block has a mass of 150 kg and rests on a surface for which the coefficients of static and kinetic friction are $\mu_s = 0.5$ and $\mu_k = 0.4$, respectively. If a force $F = (60t^2)$ N, where t is in seconds, is applied to the cable, determine the power developed by the force when $t = 5$ s. *Hint*: First determine the time needed for the force to cause motion.

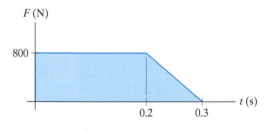

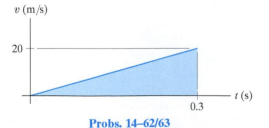

Probs. 14–62/63

Prob. 14–65

14.5   Conservative Forces and Potential Energy

Conservative Force.   If the work of a force is *independent of the path* and depends only on the force's initial and final positions on the path, then we can classify this force as a *conservative force*. Examples of conservative forces are the weight of a particle and the force developed by a spring. The work done by the weight depends *only* on the *vertical displacement* of the weight, and the work done by a spring force depends *only* on the spring's *elongation* or *compression*.

In contrast to a conservative force, consider the force of friction exerted *on a sliding object* by a fixed surface. The work done by the frictional force *depends on the path*—the longer the path, the greater the work. Consequently, *frictional forces are nonconservative*. The work is dissipated from the body in the form of heat.

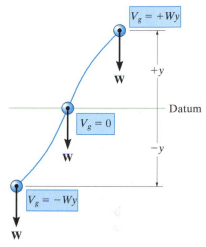

Gravitational potential energy

Fig. 14–17

Energy.   Energy is defined as the capacity for doing work. For example, if a particle is originally at rest, then the principle of work and energy states that $\Sigma U_{1\rightarrow 2} = T_2$. In other words, the kinetic energy is equal to the work that must be done on the particle to bring it from a state of rest to a speed v. Thus, the *kinetic energy* is a measure of the particle's *capacity to do work*, which is associated with the *motion* of the particle. When energy comes from the *position* of the particle, measured from a fixed datum or reference plane, it is called potential energy. Thus, *potential energy* is a measure of the amount of work a conservative force will do when it moves from a given position to the datum. In mechanics, the potential energy created by gravity (weight) and an elastic spring is important.

Gravitational Potential Energy.   If a particle is located a distance y *above* an arbitrarily selected datum, as shown in Fig. 14–17, the particle's weight $\mathbf{W}$ has positive *gravitational potential energy*, V_g, since $\mathbf{W}$ has the capacity of doing positive work when the particle is moved back down to the datum. Likewise, if the particle is located a distance y *below* the datum, V_g is negative since the weight does negative work when the particle is moved back up to the datum. At the datum $V_g = 0$.

In general, if y is *positive upward*, the gravitational potential energy of the particle of weight W is*

$$V_g = Wy \qquad\qquad (14\text{–}13)$$

Gravitational potential energy of this weight is increased as it is hoisted upward. (© R.C. Hibbeler)

---

*Here the weight is assumed to be *constant*. This assumption is suitable for small differences in elevation Δy. If the elevation change is significant, however, a variation of weight with elevation must be taken into account (see Prob. 14–82).

Elastic Potential Energy.   When an elastic spring is elongated or compressed a distance s from its unstretched position, elastic potential energy V_e can be stored in the spring. This energy is

$$V_e = +\tfrac{1}{2}ks^2 \tag{14–14}$$

Here V_e is *always positive* since, in the deformed position, the force of the spring has the *capacity* or "potential" for always doing positive work on the particle when the spring is returned to its unstretched position, Fig. 14–18.

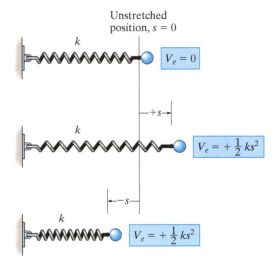

Elastic potential energy

Fig. 14–18

The weight of the sacks resting on this platform causes potential energy to be stored in the supporting springs. As each sack is removed, the platform will *rise* slightly since some of the potential energy within the springs will be transformed into an increase in gravitational potential energy of the remaining sacks. Such a device is useful for removing the sacks without having to bend over to pick them up as they are unloaded. (© R.C. Hibbeler)

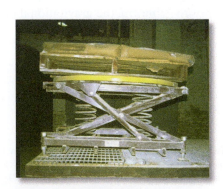

Potential Function.   In the general case, if a particle is subjected to both gravitational and elastic forces, the particle's potential energy can be expressed as a *potential function*, which is the algebraic sum

$$V = V_g + V_e \qquad (14\text{–}15)$$

Measurement of V depends on the location of the particle with respect to a selected datum in accordance with Eqs. 14–13 and 14–14.

The work done by a conservative force in moving the particle from one point to another point is measured by the *difference* of this function, i.e.,

$$U_{1-2} = V_1 - V_2 \qquad (14\text{–}16)$$

For example, the potential function for a particle of weight W suspended from a spring can be expressed in terms of its position, s, measured from a datum located at the unstretched length of the spring, Fig. 14–19. We have

$$V = V_g + V_e$$

$$= -Ws + \tfrac{1}{2}ks^2$$

If the particle moves from s_1 to a lower position s_2, then applying Eq. 14–16 it can be seen that the work of $\mathbf{W}$ and $\mathbf{F}_s$ is

$$U_{1-2} = V_1 - V_2 = \left(-Ws_1 + \tfrac{1}{2}ks_1^2\right) - \left(-Ws_2 + \tfrac{1}{2}ks_2^2\right)$$

$$= W(s_2 - s_1) - \left(\tfrac{1}{2}ks_2^2 - \tfrac{1}{2}ks_1^2\right)$$

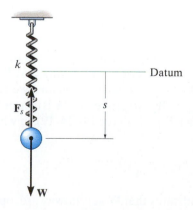

Datum

k

$\mathbf{F}_s$

s

$\mathbf{W}$

Fig. 14–19

When the displacement along the path is infinitesimal, i.e., from point (x, y, z) to $(x + dx, y + dy, z + dz)$, Eq. 14–16 becomes

$$dU = V(x, y, z) - V(x + dx, y + dy, z + dz)$$
$$= -dV(x, y, z) \qquad (14\text{–}17)$$

If we represent both the force and its displacement as Cartesian vectors, then the work can also be expressed as

$$dU = \mathbf{F} \cdot d\mathbf{r} = (F_x\mathbf{i} + F_y\mathbf{j} + F_z\mathbf{k}) \cdot (dx\mathbf{i} + dy\mathbf{j} + dz\mathbf{k})$$
$$= F_x\, dx + F_y\, dy + F_z\, dz$$

Substituting this result into Eq. 14–17 and expressing the differential $dV(x, y, z)$ in terms of its partial derivatives yields

$$F_x\, dx + F_y\, dy + F_z\, dz = -\left(\frac{\partial V}{\partial x}dx + \frac{\partial V}{\partial y}dy + \frac{\partial V}{\partial z}dz\right)$$

Since changes in x, y, and z are all independent of one another, this equation is satisfied provided

$$F_x = -\frac{\partial V}{\partial x}, \qquad F_y = -\frac{\partial V}{\partial y}, \qquad F_z = -\frac{\partial V}{\partial z} \qquad (14\text{–}18)$$

Thus,

$$\mathbf{F} = -\frac{\partial V}{\partial x}\mathbf{i} - \frac{\partial V}{\partial y}\mathbf{j} - \frac{\partial V}{\partial z}\mathbf{k}$$

$$= -\left(\frac{\partial}{\partial x}\mathbf{i} + \frac{\partial}{\partial y}\mathbf{j} + \frac{\partial}{\partial z}\mathbf{k}\right)V$$

or

$$\mathbf{F} = -\nabla V \qquad (14\text{–}19)$$

where ∇ (del) represents the vector operator $\nabla = (\partial/\partial x)\mathbf{i} + (\partial/\partial y)\mathbf{j} + (\partial/\partial z)\mathbf{k}$.

Equation 14–19 relates a force $\mathbf{F}$ to its potential function V and thereby provides a mathematical criterion for proving that $\mathbf{F}$ is conservative. For example, the gravitational potential function for a weight located a distance y above a datum is $V_g = Wy$. To prove that $\mathbf{W}$ is conservative, it is necessary to show that it satisfies Eq. 14–18 (or Eq. 14–19), in which case

$$F_y = -\frac{\partial V}{\partial y}; \qquad F_y = -\frac{\partial}{\partial y}(Wy) = -W$$

The negative sign indicates that $\mathbf{W}$ acts downward, opposite to positive y, which is upward.

14.6   Conservation of Energy

When a particle is acted upon by a system of *both* conservative and nonconservative forces, the portion of the work done by the *conservative forces* can be written in terms of the difference in their potential energies using Eq. 14–16, i.e., $(\Sigma U_{1-2})_{cons.} = V_1 - V_2$. As a result, the principle of work and energy can be written as

$$T_1 + V_1 + (\Sigma U_{1-2})_{noncons.} = T_2 + V_2 \qquad (14\text{–}20)$$

Here $(\Sigma U_{1-2})_{noncons.}$ represents the work of the nonconservative forces acting on the particle. If *only conservative forces* do work then we have

$$\boxed{T_1 + V_1 = T_2 + V_2} \qquad (14\text{–}21)$$

This equation is referred to as the *conservation of mechanical energy* or simply the *conservation of energy*. It states that during the motion the sum of the particle's kinetic and potential energies remains *constant*. For this to occur, kinetic energy must be transformed into potential energy, and vice versa. For example, if a ball of weight **W** is dropped from a height h above the ground (datum), Fig. 14–20, the potential energy of the ball is maximum before it is dropped, at which time its kinetic energy is zero. The total mechanical energy of the ball in its initial position is thus

$$E = T_1 + V_1 = 0 + Wh = Wh$$

When the ball has fallen a distance $h/2$, its speed can be determined by using $v^2 = v_0^2 + 2a_c(y - y_0)$, which yields $v = \sqrt{2g(h/2)} = \sqrt{gh}$. The energy of the ball at the mid-height position is therefore

$$E = T_2 + V_2 = \frac{1}{2}\frac{W}{g}\left(\sqrt{gh}\right)^2 + W\left(\frac{h}{2}\right) = Wh$$

Just before the ball strikes the ground, its potential energy is zero and its speed is $v = \sqrt{2gh}$. Here, again, the total energy of the ball is

$$E = T_3 + V_3 = \frac{1}{2}\frac{W}{g}\left(\sqrt{2gh}\right)^2 + 0 = Wh$$

Note that when the ball comes in contact with the ground, it deforms somewhat, and provided the ground is hard enough, the ball will rebound off the surface, reaching a new height h', which will be *less* than the height h from which it was first released. Neglecting air friction, the difference in height accounts for an energy loss, $E_l = W(h - h')$, which occurs during the collision. Portions of this loss produce noise, localized deformation of the ball and ground, and heat.

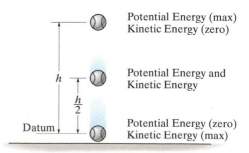

Fig. 14–20

System of Particles.   If a system of particles is *subjected only to conservative forces*, then an equation similar to Eq. 14–21 can be written for the particles. Applying the ideas of the preceding discussion, Eq. 14–8 ($\Sigma T_1 + \Sigma U_{1-2} = \Sigma T_2$) becomes

$$\Sigma T_1 + \Sigma V_1 = \Sigma T_2 + \Sigma V_2 \tag{14–22}$$

Here, the sum of the system's initial kinetic and potential energies is equal to the sum of the system's final kinetic and potential energies. In other words, $\Sigma T + \Sigma V = \text{const.}$

Procedure for Analysis

The conservation of energy equation can be used to solve problems involving *velocity, displacement*, and *conservative force systems*. It is generally *easier to apply* than the principle of work and energy because this equation requires specifying the particle's kinetic and potential energies at only *two points* along the path, rather than determining the work when the particle moves through a *displacement*. For application it is suggested that the following procedure be used.

Potential Energy.

- Draw two diagrams showing the particle located at its initial and final points along the path.

- If the particle is subjected to a vertical displacement, establish the fixed horizontal datum from which to measure the particle's gravitational potential energy V_g.

- Data pertaining to the elevation y of the particle from the datum and the stretch or compression s of any connecting springs can be determined from the geometry associated with the two diagrams.

- Recall $V_g = Wy$, where y is positive upward from the datum and negative downward from the datum; also for a spring, $V_e = \frac{1}{2}ks^2$, which is *always positive*.

Conservation of Energy.

- Apply the equation $T_1 + V_1 = T_2 + V_2$.

- When determining the kinetic energy, $T = \frac{1}{2}mv^2$, remember that the particle's speed v must be measured from an inertial reference frame.

EXAMPLE 14.9

The gantry structure in the photo is used to test the response of an airplane during a crash. As shown in Fig. 14–21a, the plane, having a mass of 8 Mg, is hoisted back until $\theta = 60°$, and then the pull-back cable AC is released when the plane is at rest. Determine the speed of the plane just before it crashes into the ground, $\theta = 15°$. Also, what is the maximum tension developed in the supporting cable during the motion? Neglect the size of the airplane and the effect of lift caused by the wings during the motion.

(© R.C. Hibbeler)

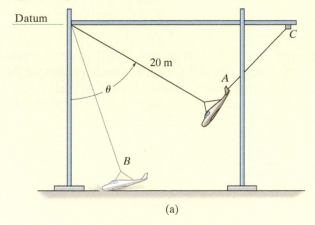

(a)

SOLUTION

Since the force of the cable does *no work* on the plane, it must be obtained using the equation of motion. First, however, we must determine the plane's speed at B.

Potential Energy.   For convenience, the datum has been established at the top of the gantry, Fig. 14–21a.

Conservation of Energy.

$$T_A + V_A = T_B + V_B$$

$$0 - 8000 \text{ kg } (9.81 \text{ m/s}^2)(20 \cos 60° \text{ m}) =$$

$$\tfrac{1}{2}(8000 \text{ kg})v_B^2 - 8000 \text{ kg } (9.81 \text{ m/s}^2)(20 \cos 15° \text{ m})$$

$$v_B = 13.52 \text{ m/s} = 13.5 \text{ m/s} \qquad \textit{Ans.}$$

Equation of Motion.   From the free-body diagram when the plane is at B, Fig. 14–21b, we have

$$+\nwarrow \quad \Sigma F_n = ma_n;$$

$$T - (8000(9.81) \text{ N}) \cos 15° = (8000 \text{ kg})\frac{(13.52 \text{ m/s})^2}{20 \text{ m}}$$

$$T = 149 \text{ kN} \qquad \textit{Ans.}$$

(b)

Fig. 14–21

EXAMPLE 14.10

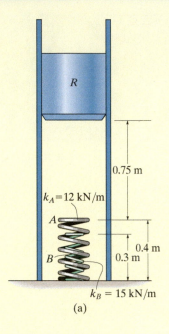

$k_A = 12$ kN/m

$k_B = 15$ kN/m

(a)

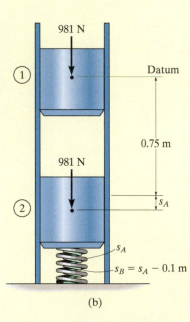

Datum

$s_B = s_A - 0.1$ m

(b)

Fig. 14–22

The ram R shown in Fig. 14–22a has a mass of 100 kg and is released from rest 0.75 m from the top of a spring, A, that has a stiffness $k_A = 12$ kN/m. If a second spring B, having a stiffness $k_B = 15$ kN/m, is "nested" in A, determine the maximum displacement of A needed to stop the downward motion of the ram. The unstretched length of each spring is indicated in the figure. Neglect the mass of the springs.

SOLUTION

Potential Energy.   We will *assume* that the ram compresses *both* springs at the instant it comes to rest. The datum is located through the center of gravity of the ram at its initial position, Fig. 14–22b. When the kinetic energy is reduced to zero ($v_2 = 0$), A is compressed a distance s_A and B compresses $s_B = s_A - 0.1$ m.

Conservation of Energy.

$$T_1 + V_1 = T_2 + V_2$$

$$0 + 0 = 0 + \left\{ \tfrac{1}{2}k_A s_A^2 + \tfrac{1}{2}k_B(s_A - 0.1)^2 - Wh \right\}$$

$$0 + 0 = 0 + \left\{ \tfrac{1}{2}(12\ 000\ \text{N/m})s_A^2 + \tfrac{1}{2}(15\ 000\ \text{N/m})(s_A - 0.1\ \text{m})^2 \right.$$
$$\left. - 981\ \text{N}\ (0.75\ \text{m} + s_A) \right\}$$

Rearranging the terms,

$$13\ 500 s_A^2 - 2481 s_A - 660.75 = 0$$

Using the quadratic formula and solving for the positive root, we have

$$s_A = 0.331\ \text{m} \qquad\qquad Ans.$$

Since $s_B = 0.331\ \text{m} - 0.1\ \text{m} = 0.231\ \text{m}$,   which   is   positive,   the assumption that *both* springs are compressed by the ram is correct.

NOTE: The second root, $s_A = -0.148$ m, does not represent the physical situation. Since positive s is measured downward, the negative sign indicates that spring A would have to be "extended" by an amount of 0.148 m to stop the ram.

EXAMPLE 14.11

A smooth 2-kg collar, shown in Fig. 14–23a, fits loosely on the vertical shaft. If the spring is unstretched when the collar is in the position A, determine the speed at which the collar is moving when $y = 1$ m, if (a) it is released from rest at A, and (b) it is released at A with an *upward* velocity $v_A = 2$ m/s.

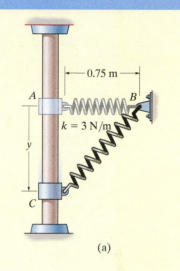

(a)

SOLUTION

Part (a) Potential Energy. For convenience, the datum is established through AB, Fig. 14–23b. When the collar is at C, the gravitational potential energy is $-(mg)y$, since the collar is *below* the datum, and the elastic potential energy is $\frac{1}{2}ks_{CB}^2$. Here $s_{CB} = 0.5$ m, which represents the *stretch* in the spring as shown in the figure.

Conservation of Energy.

$$T_A + V_A = T_C + V_C$$

$$0 + 0 = \tfrac{1}{2}mv_C^2 + \left\{\tfrac{1}{2}ks_{CB}^2 - mgy\right\}$$

$$0 + 0 = \left\{\tfrac{1}{2}(2 \text{ kg})v_C^2\right\} + \left\{\tfrac{1}{2}(3 \text{ N/m})(0.5 \text{ m})^2 - 2(9.81) \text{ N} (1 \text{ m})\right\}$$

$$v_C = 4.39 \text{ m/s} \downarrow \qquad\qquad\qquad Ans.$$

This problem can also be solved by using the equation of motion or the principle of work and energy. Note that for *both* of these methods the variation of the magnitude and direction of the spring force must be taken into account (see Example 13.4). Here, however, the above solution is clearly advantageous since the calculations depend *only* on data calculated at the initial and final points of the path.

Part (b) Conservation of Energy. If $v_A = 2$ m/s, using the data in Fig. 14–23b, we have

$$T_A + V_A = T_C + V_C$$

$$\tfrac{1}{2}mv_A^2 + 0 = \tfrac{1}{2}mv_C^2 + \left\{\tfrac{1}{2}ks_{CB}^2 - mgy\right\}$$

$$\tfrac{1}{2}(2 \text{ kg})(2 \text{ m/s})^2 + 0 = \tfrac{1}{2}(2 \text{ kg})v_C^2 + \left\{\tfrac{1}{2}(3 \text{ N/m})(0.5 \text{ m})^2\right.$$

$$\left. - 2(9.81) \text{ N} (1 \text{ m})\right\}$$

$$v_C = 4.82 \text{ m/s} \downarrow \qquad\qquad\qquad Ans.$$

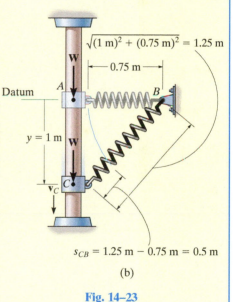

$s_{CB} = 1.25 \text{ m} - 0.75 \text{ m} = 0.5 \text{ m}$

(b)

Fig. 14–23

NOTE: The kinetic energy of the collar depends only on the *magnitude* of velocity, and therefore it is immaterial if the collar is moving up or down at 2 m/s when released at A.

PRELIMINARY PROBLEMS

P14–3. Determine the potential energy of the block that has a weight of 100 N.

P14–4. Determine the potential energy in the spring that has an unstretched length of 4 m.

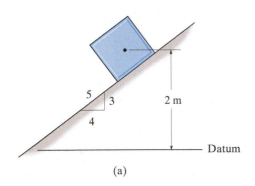

(a)

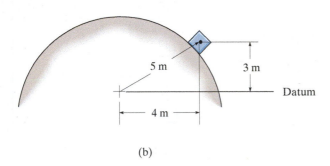

(b)

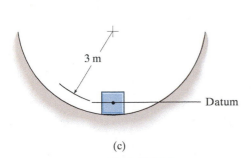

(c)

Prob. P14–3

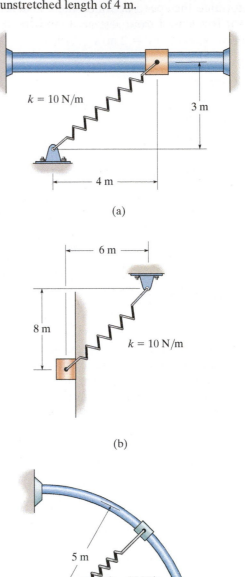

(a)

(b)

(c)

Prob. P14–4

FUNDAMENTAL PROBLEMS

F14–13. The 2-kg pendulum bob is released from rest when it is at A. Determine the speed of the bob and the tension in the cord when the bob passes through its lowest position, B.

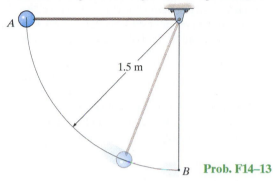

1.5 m

Prob. F14–13

F14–14. The 2-kg package leaves the conveyor belt at A with a speed of $v_A = 1$ m/s and slides down the smooth ramp. Determine the required speed of the conveyor belt at B so that the package can be delivered without slipping on the belt. Also, find the normal reaction the curved portion of the ramp exerts on the package at B if $\rho_B = 2$ m.

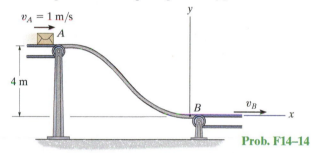

$v_A = 1$ m/s

4 m

Prob. F14–14

F14–15. The 2-kg collar is given a downward velocity of 4 m/s when it is at A. If the spring has an unstretched length of 1 m and a stiffness of $k = 30$ N/m, determine the velocity of the collar at $s = 1$ m.

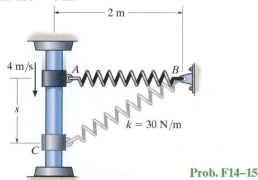

2 m

4 m/s

$k = 30$ N/m

Prob. F14–15

F14–16. The 5-lb collar is released from rest at A and travels along the frictionless guide. Determine the speed of the collar when it strikes the stop B. The spring has an unstretched length of 0.5 ft.

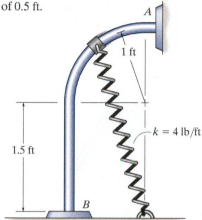

A

1 ft

$k = 4$ lb/ft

1.5 ft

B

Prob. F14–16

F14–17. The 75-lb block is released from rest 5 ft above the plate. Determine the compression of each spring when the block momentarily comes to rest after striking the plate. Neglect the mass of the plate. The springs are initially unstretched.

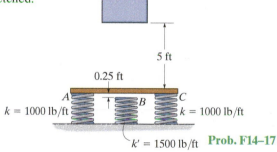

5 ft

0.25 ft

$k = 1000$ lb/ft A B C $k = 1000$ lb/ft

$k' = 1500$ lb/ft   **Prob. F14–17**

F14–18. The 4-kg collar C has a velocity of $v_A = 2$ m/s when it is at A. If the guide rod is smooth, determine the speed of the collar when it is at B. The spring has an unstretched length of $l_0 = 0.2$ m.

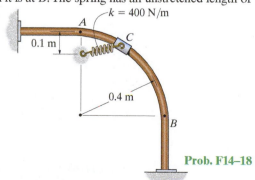

$k = 400$ N/m

A

0.1 m C

0.4 m

B

Prob. F14–18

PROBLEMS

14–66. The girl has a mass of 40 kg and center of mass at *G*. If she is swinging to a maximum height defined by $\theta = 60°$, determine the force developed along each of the four supporting posts such as *AB* at the instant $\theta = 0°$. The swing is centrally located between the posts.

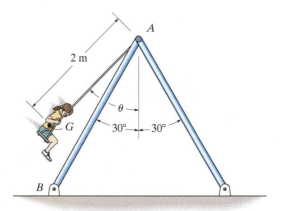

Prob. 14–66

14–67. The 30-lb block *A* is placed on top of two nested springs *B* and *C* and then pushed down to the position shown. If it is then released, determine the maximum height *h* to which it will rise.

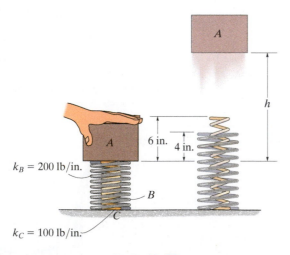

Prob. 14–67

14–68. The 5-kg collar has a velocity of 5 m/s to the right when it is at *A*. It then travels down along the smooth guide. Determine the speed of the collar when it reaches point *B*, which is located just before the end of the curved portion of the rod. The spring has an unstretched length of 100 mm and *B* is located just before the end of the curved portion of the rod.

14–69. The 5-kg collar has a velocity of 5 m/s to the right when it is at *A*. It then travels along the smooth guide. Determine its speed when its center reaches point *B* and the normal force it exerts on the rod at this point. The spring has an unstretched length of 100 mm and *B* is located just before the end of the curved portion of the rod.

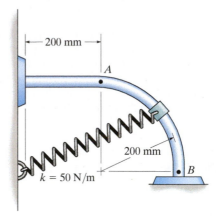

Probs. 14–68/69

14–70. The ball has a weight of 15 lb and is fixed to a rod having a negligible mass. If it is released from rest when $\theta = 0°$, determine the angle θ at which the compressive force in the rod becomes zero.

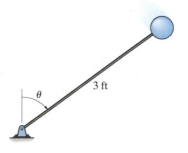

Prob. 14–70

14–71. The car C and its contents have a weight of 600 lb, whereas block B has a weight of 200 lb. If the car is released from rest, determine its speed when it travels 30 ft down the 20° incline. *Suggestion*: To measure the gravitational potential energy, establish separate datums at the initial elevations of B and C.

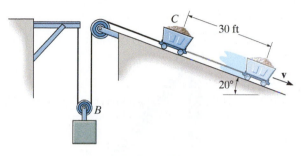

Prob. 14–71

*14–72. The roller coaster car has a mass of 700 kg, including its passenger. If it starts from the top of the hill A with a speed $v_A = 3$ m/s, determine the minimum height h of the hill crest so that the car travels around the inside loops without leaving the track. Neglect friction, the mass of the wheels, and the size of the car. What is the normal reaction on the car when the car is at B and when it is at C? Take $\rho_B = 7.5$ m and $\rho_C = 5$ m.

14–73. The roller coaster car has a mass of 700 kg, including its passenger. If it is released from rest at the top of the hill A, determine the minimum height h of the hill crest so that the car travels around both inside the loops without leaving the track. Neglect friction, the mass of the wheels, and the size of the car. What is the normal reaction on the car when the car is at B and when it is at C? Take $\rho_B = 7.5$ m and $\rho_C = 5$ m.

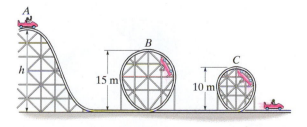

Probs. 14–72/73

14–74. The assembly consists of two blocks A and B which have a mass of 20 kg and 30 kg, respectively. Determine the speed of each block when B descends 1.5 m. The blocks are released from rest. Neglect the mass of the pulleys and cords.

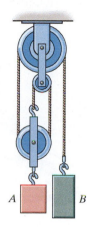

Prob. 14–74

14–75. The assembly consists of two blocks A and B, which have a mass of 20 kg and 30 kg, respectively. Determine the distance B must descend in order for A to achieve a speed of 3 m/s starting from rest.

Prob. 14–75

14

*14–76. The spring has a stiffness $k = 50 \text{ N/m}$ and an unstretched length of 0.3 m. If it is attached to the 2-kg smooth collar and the collar is released from rest at A ($\theta = 0°$), determine the speed of the collar when $\theta = 60°$. The motion occurs in the horizontal plane. Neglect the size of the collar.

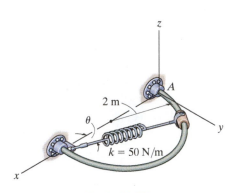

Prob. 14–76

14–77. The roller coaster car having a mass m is released from rest at point A. If the track is to be designed so that the car does not leave it at B, determine the required height h. Also, find the speed of the car when it reaches point C. Neglect friction.

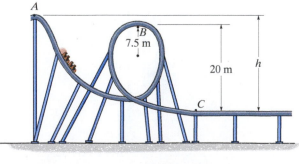

Prob. 14–77

14–78. The spring has a stiffness $k = 200 \text{ N/m}$ and an unstretched length of 0.5 m. If it is attached to the 3-kg smooth collar and the collar is released from rest at A, determine the speed of the collar when it reaches B. Neglect the size of the collar.

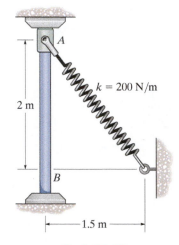

Prob. 14–78

14–79. A 2-lb block rests on the smooth semicylindrical surface at A. An elastic cord having a stiffness of $k = 2 \text{ lb/ft}$ is attached to the block at B and to the base of the semicylinder at C. If the block is released from rest at $\theta = 0$, A, determine the longest unstretched length of the cord so the block begins to leave the semicylinder at the instant $\theta = 45°$. Neglect the size of the block.

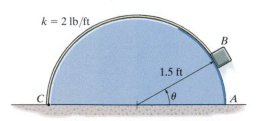

Prob. 14–79

*14–80. When $s = 0$, the spring on the firing mechanism is unstretched. If the arm is pulled back such that $s = 100$ mm and released, determine the speed of the 0.3-kg ball and the normal reaction of the circular track on the ball when $\theta = 60°$. Assume all surfaces of contact to be smooth. Neglect the mass of the spring and the size of the ball.

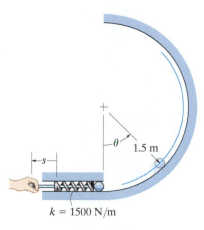

Prob. 14–80

14–81. When $s = 0$, the spring on the firing mechanism is unstretched. If the arm is pulled back such that $s = 100$ mm and released, determine the maximum angle θ the ball will travel without leaving the circular track. Assume all surfaces of contact to be smooth. Neglect the mass of the spring and the size of the ball.

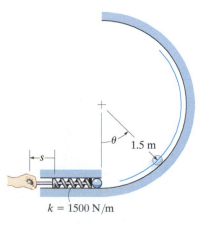

Prob. 14–81

14–82. If the mass of the earth is M_e, show that the gravitational potential energy of a body of mass m located a distance r from the center of the earth is $V_g = -GM_em/r$. Recall that the gravitational force acting between the earth and the body is $F = G(M_em/r^2)$, Eq. 13–1. For the calculation, locate the datum at $r \to \infty$. Also, prove that F is a conservative force.

14–83. A rocket of mass m is fired vertically from the surface of the earth, i.e., at $r = r_1$. Assuming that no mass is lost as it travels upward, determine the work it must do against gravity to reach a distance r_2. The force of gravity is $F = GM_em/r^2$ (Eq. 13–1), where M_e is the mass of the earth and r the distance between the rocket and the center of the earth.

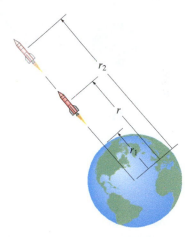

Probs. 14–82/83

*14–84. The 4-kg smooth collar has a speed of 3 m/s when it is at $s = 0$. Determine the maximum distance s it travels before it stops momentarily. The spring has an unstretched length of 1 m.

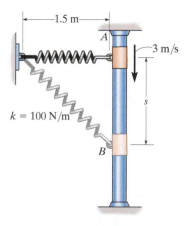

Prob. 14–84

14–85. A 60-kg satellite travels in free flight along an elliptical orbit such that at A, where $r_A = 20$ Mm, it has a speed $v_A = 40$ Mm/h. What is the speed of the satellite when it reaches point B, where $r_B = 80$ Mm? *Hint:* See Prob. 14–82, where $M_e = 5.976(10^{24})$ kg and $G = 66.73(10^{-12})$ m³/(kg·s²).

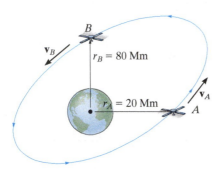

Prob. 14–85

14–86. The skier starts from rest at A and travels down the ramp. If friction and air resistance can be neglected, determine his speed v_B when he reaches B. Also, compute the distance s to where he strikes the ground at C, if he makes the jump traveling horizontally at B. Neglect the skier's size. He has a mass of 70 kg.

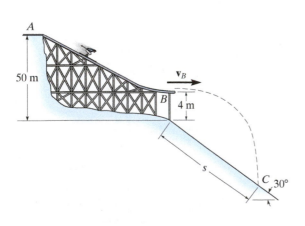

Prob. 14–86

14–87. The block has a mass of 20 kg and is released from rest when $s = 0.5$ m. If the mass of the bumpers A and B can be neglected, determine the maximum deformation of each spring due to the collision.

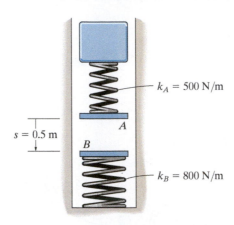

Prob. 14–87

*14–88. The 2-lb collar has a speed of 5 ft/s at A. The attached spring has an unstretched length of 2 ft and a stiffness of $k = 10$ lb/ft. If the collar moves over the smooth rod, determine its speed when it reaches point B, the normal force of the rod on the collar, and the rate of decrease in its speed.

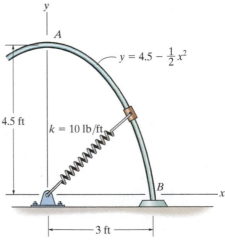

Prob. 14–88

14–89. When the 6-kg box reaches point A it has a speed of $v_A = 2$ m/s. Determine the angle θ at which it leaves the smooth circular ramp and the distance s to where it falls into the cart. Neglect friction.

14–91. When the 5-kg box reaches point A it has a speed $v_A = 10$ m/s. Determine how high the box reaches up the surface before it comes to a stop. Also, what is the resultant normal force on the surface at this point and the acceleration? Neglect friction and the size of the box.

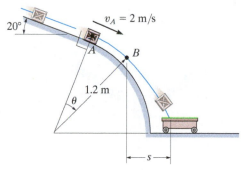

Prob. 14–89

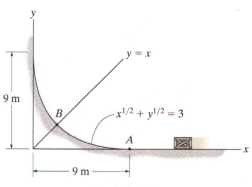

Prob. 14–91

14–90. When the 5-kg box reaches point A it has a speed $v_A = 10$ m/s. Determine the normal force the box exerts on the surface when it reaches point B. Neglect friction and the size of the box.

*14–92. The roller coaster car has a speed of 15 ft/s when it is at the crest of a vertical parabolic track. Determine the car's velocity and the normal force it exerts on the track when it reaches point B. Neglect friction and the mass of the wheels. The total weight of the car and the passengers is 350 lb.

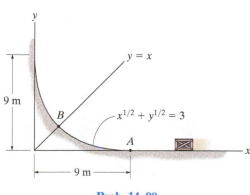

Prob. 14–90

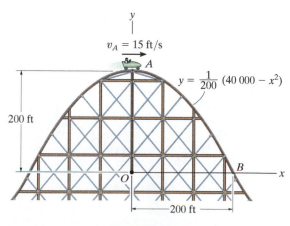

Prob. 14–92

14–93. The 10-kg sphere C is released from rest when $\theta = 0°$ and the tension in the spring is 100 N. Determine the speed of the sphere at the instant $\theta = 90°$. Neglect the mass of rod AB and the size of the sphere.

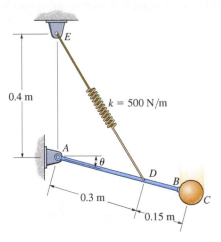

0.4 m

$k = 500$ N/m

A

θ

D

B

0.3 m

0.15 m

C

Prob. 14–93

14–94. A quarter-circular tube AB of mean radius r contains a smooth chain that has a mass per unit length of m_0. If the chain is released from rest from the position shown, determine its speed when it emerges completely from the tube.

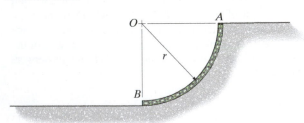

O

A

r

B

Prob. 14–94

14–95. The cylinder has a mass of 20 kg and is released from rest when $h = 0$. Determine its speed when $h = 3$ m. Each spring has a stiffness $k = 40$ N/m and an unstretched length of 2 m.

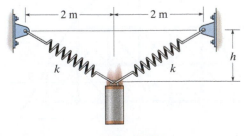

2 m   2 m

k k

h

Prob. 14–95

*14–96. If the 20-kg cylinder is released from rest at $h = 0$, determine the required stiffness k of each spring so that its motion is arrested or stops when $h = 0.5$ m. Each spring has an unstretched length of 1 m.

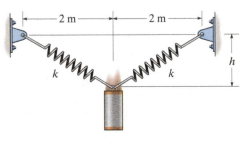

2 m   2 m

k k

h

Prob. 14–96

14–97. A pan of negligible mass is attached to two identical springs of stiffness $k = 250$ N/m. If a 10-kg box is dropped from a height of 0.5 m above the pan, determine the maximum vertical displacement d. Initially each spring has a tension of 50 N.

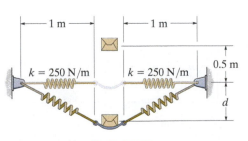

1 m   1 m

$k = 250$ N/m $k = 250$ N/m

0.5 m

d

Prob. 14–97

CONCEPTUAL PROBLEMS

C14–1. The roller coaster is momentarily at rest at *A*. Determine the approximate normal force it exerts on the track at *B*. Also determine its approximate acceleration at this point. Use numerical data, and take scaled measurements from the photo with a known height at *A*.

Prob. C14–1 (© R.C. Hibbeler)

C14–2. As the large ring rotates, the operator can apply a breaking mechanism that binds the cars to the ring, which then allows the cars to rotate with the ring. Assuming the passengers are not belted into the cars, determine the smallest speed of the ring (cars) so that no passenger will fall out. When should the operator release the brake so that the cars can achieve their greatest speed as they slide freely on the ring? Estimate the greatest normal force of the seat on a passenger when this speed is reached. Use numerical values to explain your answer.

Prob. C14–2 (© R.C. Hibbeler)

C14–3. The woman pulls the water balloon launcher back, stretching each of the four elastic cords. Estimate the maximum height and the maximum range of a ball placed within the container if it is released from the position shown. Use numerical values and any necessary measurements from the photo. Assume the unstretched length and stiffness of each cord is known.

Prob. C14–3 (© R.C. Hibbeler)

C14–4. The girl is momentarily at rest in the position shown. If the unstretched length and stiffness of each of the two elastic cords is known, determine approximately how far the girl descends before she again becomes momentarily at rest. Use numerical values and take any necessary measurements from the photo.

Prob. C14–4 (© R.C. Hibbeler)

14

CHAPTER REVIEW

Work of a Force

A force does work when it undergoes a displacement along its line of action. If the force varies with the displacement, then the work is $U = \int F \cos \theta \, ds$.

Graphically, this represents the area under the $F{-}s$ diagram.

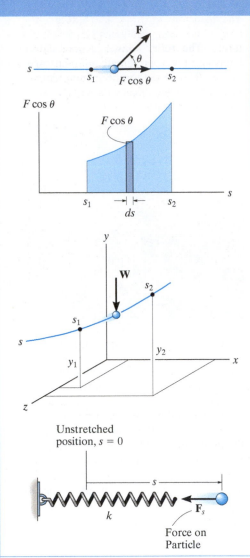

If the force is constant, then for a displacement Δs in the direction of the force, $U = F_c \, \Delta s$. A typical example of this case is the work of a weight, $U = -W \, \Delta y$. Here, Δy is the vertical displacement.

The work done by a spring force, $F = ks$, depends upon the stretch or compression s of the spring.

$$U = -\left(\tfrac{1}{2}ks_2^2 - \tfrac{1}{2}ks_1^2 \right)$$

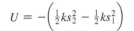

Force on Particle

The Principle of Work and Energy

If the equation of motion in the tangential direction, $\Sigma F_t = ma_t$, is combined with the kinematic equation, $a_t \, ds = v \, dv$, we obtain the principle of work and energy. This equation states that the initial kinetic energy T, plus the work done ΣU_{1-2} is equal to the final kinetic energy.

$$T_1 + \Sigma U_{1-2} = T_2$$

The principle of work and energy is useful for solving problems that involve force, velocity, and displacement. For application, the free-body diagram of the particle should be drawn in order to identify the forces that do work.

Power and Efficiency

Power is the time rate of doing work. For application, the force **F** creating the power and its velocity **v** must be specified.

$$P = \frac{dU}{dt}$$

$$P = \mathbf{F} \cdot \mathbf{v}$$

Efficiency represents the ratio of power output to power input. Due to frictional losses, it is always less than one.

$$\varepsilon = \frac{\text{power output}}{\text{power input}}$$

Conservation of Energy

A conservative force does work that is independent of its path. Two examples are the weight of a particle and the spring force.

Friction is a nonconservative force since the work depends upon the length of the path. The longer the path, the more work done.

The work done by a conservative force depends upon its position relative to a datum. When this work is referenced from a datum, it is called potential energy. For a weight, it is $V_g = \pm Wy$, and for a spring it is $V_e = +\frac{1}{2}ks^2$.

Mechanical energy consists of kinetic energy T and gravitational and elastic potential energies V. According to the conservation of energy, this sum is constant and has the same value at any position on the path. If only gravitational and spring forces cause motion of the particle, then the conservation-of-energy equation can be used to solve problems involving these conservative forces, displacement, and velocity.

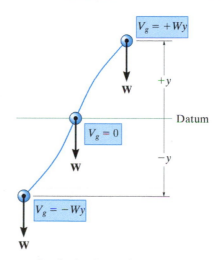

Gravitational potential energy

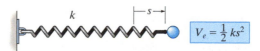

Elastic potential energy

$$T_1 + V_1 = T_2 + V_2$$

REVIEW PROBLEMS

R14–1. If a 150-lb crate is released from rest at A, determine its speed after it slides 30 ft down the plane. The coefficient of kinetic friction between the crate and plane is $\mu_k = 0.3$.

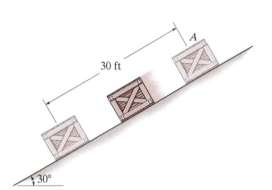

Prob. R14–1

R14–2. The small 2-lb collar starting from rest at A slides down along the smooth rod. During the motion, the collar is acted upon by a force $\mathbf{F} = \{10\mathbf{i} + 6y\mathbf{j} + 2z\mathbf{k}\}$ lb, where x, y, z are in feet. Determine the collar's speed when it strikes the wall at B.

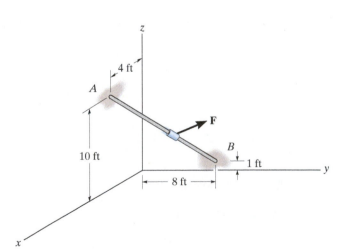

Prob. R14–2

R14–3. The block has a weight of 1.5 lb and slides along the smooth chute AB. It is released from rest at A, which has coordinates of $A(5 \text{ ft}, 0, 10 \text{ ft})$. Determine the speed at which it slides off at B, which has coordinates of $B(0, 8 \text{ ft}, 0)$.

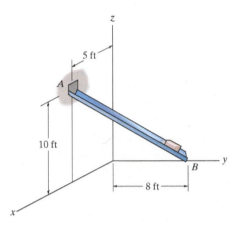

Prob. R14–3

R14–4. The block has a mass of 0.5 kg and moves within the smooth vertical slot. If the block starts from rest when the *attached* spring is in the unstretched position at A, determine the *constant* vertical force F which must be applied to the cord so that the block attains a speed $v_B = 2.5$ m/s when it reaches B; $s_B = 0.15$ m. Neglect the mass of the cord and pulley.

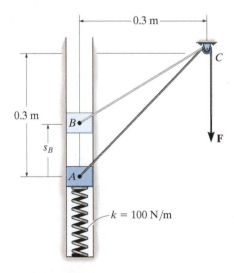

Prob. R14–4

R14–5. The crate, having a weight of 50 lb, is hoisted by the pulley system and motor M. If the crate starts from rest and, by constant acceleration, attains a speed of 12 ft/s after rising 10 ft, determine the power that must be supplied to the motor at the instant $s = 10$ ft. The motor has an efficiency $\varepsilon = 0.74$.

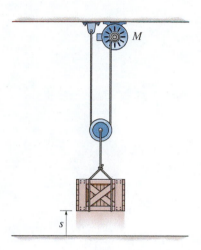

Prob. R14–5

R14–6. The 50-lb load is hoisted by the pulley system and motor M. If the motor exerts a constant force of 30 lb on the cable, determine the power that must be supplied to the motor if the load has been hoisted $s = 10$ ft starting from rest. The motor has an efficiency of $\varepsilon = 0.76$.

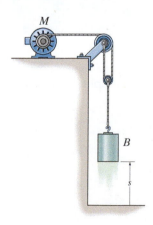

Prob. R14–6

R14–7. The collar of negligible size has a mass of 0.25 kg and is attached to a spring having an unstretched length of 100 mm. If the collar is released from rest at A and travels along the smooth guide, determine its speed just before it strikes B.

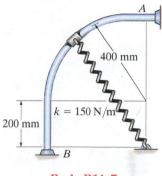

Prob. R14–7

R14–8. The blocks A and B weigh 10 and 30 lb, respectively. They are connected together by a light cord and ride in the frictionless grooves. Determine the speed of each block after block A moves 6 ft up along the plane. The blocks are released from rest.

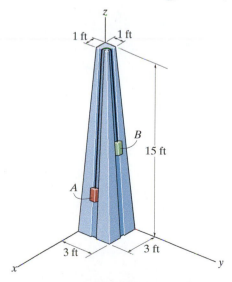

Prob. R14–8

Chapter 15

(© David J. Green/Alamy)

The design of the bumper cars used for this amusement park ride requires knowledge of the principles of impulse and momentum.

Kinetics of a Particle: Impulse and Momentum

CHAPTER OBJECTIVES

■ To develop the principle of linear impulse and momentum for a particle and apply it to solve problems that involve force, velocity, and time.

■ To study the conservation of linear momentum for particles.

■ To analyze the mechanics of impact.

■ To introduce the concept of angular impulse and momentum.

■ To solve problems involving steady fluid streams and propulsion with variable mass.

15.1 Principle of Linear Impulse and Momentum

In this section we will integrate the equation of motion with respect to time and thereby obtain the principle of impulse and momentum. The resulting equation will be useful for solving problems involving force, velocity, and time.

Using kinematics, the equation of motion for a particle of mass m can be written as

$$\Sigma \mathbf{F} = m\mathbf{a} = m\frac{d\mathbf{v}}{dt} \tag{15-1}$$

where $\mathbf{a}$ and $\mathbf{v}$ are both measured from an inertial frame of reference. Rearranging the terms and integrating between the limits $\mathbf{v} = \mathbf{v}_1$ at $t = t_1$ and $\mathbf{v} = \mathbf{v}_2$ at $t = t_2$, we have

$$\Sigma \int_{t_1}^{t_2} \mathbf{F}dt = m\int_{\mathbf{v}_1}^{\mathbf{v}_2} d\mathbf{v}$$

or

$$\Sigma \int_{t_1}^{t_2} \mathbf{F} dt = m\mathbf{v}_2 - m\mathbf{v}_1 \qquad (15\text{–}2)$$

This equation is referred to as the *principle of linear impulse and momentum*. From the derivation it can be seen that it is simply a time integration of the equation of motion. It provides a *direct means* of obtaining the particle's final velocity $\mathbf{v}_2$ after a specified time period when the particle's initial velocity is known and the forces acting on the particle are either constant or can be expressed as functions of time. By comparison, if $\mathbf{v}_2$ was determined using the equation of motion, a two-step process would be necessary; i.e., apply $\Sigma \mathbf{F} = m\mathbf{a}$ to obtain $\mathbf{a}$, then integrate $\mathbf{a} = d\mathbf{v}/dt$ to obtain $\mathbf{v}_2$.

Linear Momentum. Each of the two vectors of the form $\mathbf{L} = m\mathbf{v}$ in Eq. 15–2 is referred to as the particle's linear momentum. Since m is a positive scalar, the linear-momentum vector has the same direction as $\mathbf{v}$, and its magnitude mv has units of mass times velocity, e.g., $\text{kg} \cdot \text{m/s}$, or $\text{slug} \cdot \text{ft/s}$.

Linear Impulse. The integral $\mathbf{I} = \int \mathbf{F} \, dt$ in Eq. 15–2 is referred to as the *linear impulse*. This term is a vector quantity which measures the effect of a force during the time the force acts. Since time is a positive scalar, the impulse acts in the same direction as the force, and its magnitude has units of force times time, e.g., $\text{N} \cdot \text{s}$ or $\text{lb} \cdot \text{s}$.*

If the force is expressed as a function of time, the impulse can be determined by direct evaluation of the integral. In particular, if the force is constant in both magnitude and direction, the resulting impulse becomes

$$\mathbf{I} = \int_{t_1}^{t_2} \mathbf{F}_c \, dt = \mathbf{F}_c(t_2 - t_1).$$

Graphically the magnitude of the impulse can be represented by the shaded area under the curve of force versus time, Fig. 15–1. A constant force creates the shaded rectangular area shown in Fig. 15–2.

The impulse tool is used to remove the dent in the trailer fender. To do so its end is first screwed into a hole drilled in the fender, then the weight is gripped and jerked upwards, striking the stop ring. The impulse developed is transferred along the shaft of the tool and pulls suddenly on the dent. (© R.C. Hibbeler)

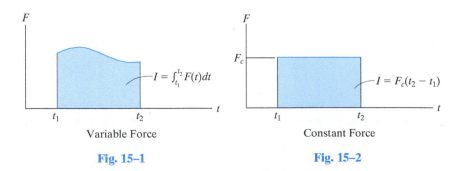

Variable Force

Fig. 15–1

Constant Force

Fig. 15–2

*Although the units for impulse and momentum are defined differently, it can be shown that Eq. 15–2 is dimensionally homogeneous.

Principle of Linear Impulse and Momentum. For problem solving, Eq. 15–2 will be rewritten in the form

$$m\mathbf{v}_1 + \Sigma \int_{t_1}^{t_2} \mathbf{F} \, dt = m\mathbf{v}_2 \qquad (15\text{–}3)$$

which states that the initial momentum of the particle at time t_1 plus the sum of all the impulses applied to the particle from t_1 to t_2 is equivalent to the final momentum of the particle at time t_2. These three terms are illustrated graphically on the *impulse and momentum diagrams* shown in Fig. 15–3. The two *momentum diagrams* are simply outlined shapes of the particle which indicate the direction and magnitude of the particle's initial and final momenta, $m\mathbf{v}_1$ and $m\mathbf{v}_2$. Similar to the free-body diagram, the *impulse diagram* is an outlined shape of the particle showing all the impulses that act on the particle when it is located at some intermediate point along its path.

If each of the vectors in Eq. 15–3 is resolved into its x, y, z components, we can write the following three scalar equations of linear impulse and momentum.

The study of many types of sports, such as golf, requires application of the principle of linear impulse and momentum. (© R.C. Hibbeler)

$$m(v_x)_1 + \Sigma \int_{t_1}^{t_2} F_x \, dt = m(v_x)_2$$

$$m(v_y)_1 + \Sigma \int_{t_1}^{t_2} F_y \, dt = m(v_y)_2 \qquad (15\text{–}4)$$

$$m(v_z)_1 + \Sigma \int_{t_1}^{t_2} F_z \, dt = m(v_z)_2$$

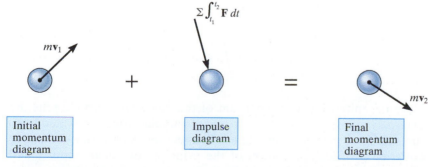

Initial momentum diagram · Impulse diagram · Final momentum diagram

Fig. 15–3

15.2   Principle of Linear Impulse and Momentum for a System of Particles

The principle of linear impulse and momentum for a system of particles moving relative to an inertial reference, Fig. 15–4, is obtained from the equation of motion applied to all the particles in the system, i.e.,

$$\Sigma \mathbf{F}_i = \Sigma m_i \frac{d\mathbf{v}_i}{dt} \tag{15–5}$$

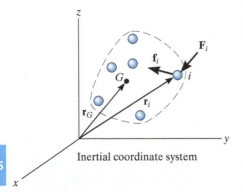

Inertial coordinate system

Fig. 15–4

The term on the left side represents only the sum of the *external forces* acting on the particles. Recall that the internal forces $\mathbf{f}_i$ acting between particles do not appear with this summation, since by Newton's third law they occur in equal but opposite collinear pairs and therefore cancel out. Multiplying both sides of Eq. 15–5 by dt and integrating between the limits $t = t_1, \mathbf{v}_i = (\mathbf{v}_i)_1$ and $t = t_2, \mathbf{v}_i = (\mathbf{v}_i)_2$ yields

$$\Sigma m_i(\mathbf{v}_i)_1 + \Sigma \int_{t_1}^{t_2} \mathbf{F}_i\, dt = \Sigma m_i(\mathbf{v}_i)_2 \tag{15–6}$$

This equation states that the initial linear momentum of the system plus the impulses of all the *external forces* acting on the system from t_1 to t_2 is equal to the system's final linear momentum.

Since the location of the mass center G of the system is determined from $m\mathbf{r}_G = \Sigma m_i \mathbf{r}_i$, where $m = \Sigma m_i$ is the total mass of all the particles, Fig. 15–4, then taking the time derivative, we have

$$m\mathbf{v}_G = \Sigma m_i \mathbf{v}_i$$

which states that the total linear momentum of the system of particles is equivalent to the linear momentum of a "fictitious" aggregate particle of mass $m = \Sigma m_i$ moving with the velocity of the mass center of the system. Substituting into Eq. 15–6 yields

$$m(\mathbf{v}_G)_1 + \Sigma \int_{t_1}^{t_2} \mathbf{F}_i\, dt = m(\mathbf{v}_G)_2 \tag{15–7}$$

Here the initial linear momentum of the aggregate particle plus the external impulses acting on the system of particles from t_1 to t_2 is equal to the aggregate particle's final linear momentum. As a result, the above equation justifies application of the principle of linear impulse and momentum to a system of particles that compose a rigid body.

Procedure for Analysis

The principle of linear impulse and momentum is used to solve problems involving *force, time*, and *velocity*, since these terms are involved in the formulation. For application it is suggested that the following procedure be used.*

Free-Body Diagram.

- Establish the *x, y, z* inertial frame of reference and draw the particle's free-body diagram in order to account for all the forces that produce impulses on the particle.

- The direction and sense of the particle's initial and final velocities should be established.

- If a vector is unknown, assume that the sense of its components is in the direction of the positive inertial coordinate(s).

- As an alternative procedure, draw the impulse and momentum diagrams for the particle as discussed in reference to Fig. 15–3.

Principle of Impulse and Momentum.

- In accordance with the established coordinate system, apply the principle of linear impulse and momentum, $m\mathbf{v}_1 + \Sigma \int_{t_1}^{t_2} \mathbf{F}\,dt = m\mathbf{v}_2$. If motion occurs in the *x–y* plane, the two scalar component equations can be formulated by either resolving the vector components of **F** from the free-body diagram, or by using the data on the impulse and momentum diagrams.

- Realize that every force acting on the particle's free-body diagram will create an impulse, even though some of these forces will do no work.

- Forces that are functions of time must be integrated to obtain the impulse. Graphically, the impulse is equal to the area under the force–time curve.

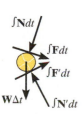

As the wheels of the pitching machine rotate, they apply frictional impulses to the ball, thereby giving it a linear momentum. These impulses are shown on the impulse diagram. Here both the frictional and normal impulses vary with time. By comparison, the weight impulse is constant and is very small since the time Δt the ball is in contact with the wheels is very small. (© R.C. Hibbeler)

*This procedure will be followed when developing the proofs and theory in the text.

EXAMPLE | **15.1**

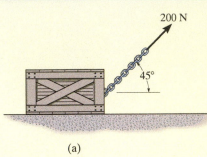

(a)

The 100-kg crate shown in Fig. 15–5a is originally at rest on the smooth horizontal surface. If a towing force of 200 N, acting at an angle of 45°, is applied for 10 s, determine the final velocity and the normal force which the surface exerts on the crate during this time interval.

SOLUTION

This problem can be solved using the principle of impulse and momentum since it involves force, velocity, and time.

Free-Body Diagram. See Fig. 15–5b. Since all the forces acting are *constant*, the impulses are simply the product of the force magnitude and 10 s $[\mathbf{I} = \mathbf{F}_c(t_2 - t_1)]$. Note the alternative procedure of drawing the crate's impulse and momentum diagrams, Fig. 15–5c.

Principle of Impulse and Momentum. Applying Eqs. 15–4 yields

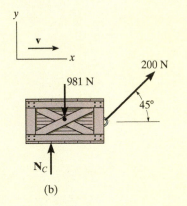

(b)

$(\xrightarrow{+})$
$$m(v_x)_1 + \Sigma \int_{t_1}^{t_2} F_x \, dt = m(v_x)_2$$

$$0 + 200 \text{ N} \cos 45°(10 \text{ s}) = (100 \text{ kg})v_2$$

$$v_2 = 14.1 \text{ m/s} \qquad \textit{Ans.}$$

$(+\uparrow)$
$$m(v_y)_1 + \Sigma \int_{t_1}^{t_2} F_y \, dt = m(v_y)_2$$

$$0 + N_C(10 \text{ s}) - 981 \text{ N}(10 \text{ s}) + 200 \text{ N} \sin 45°(10 \text{ s}) = 0$$

$$N_C = 840 \text{ N} \qquad \textit{Ans.}$$

NOTE: Since no motion occurs in the y direction, direct application of the equilibrium equation $\Sigma F_y = 0$ gives the same result for N_C. Try to solve the problem by first applying $\Sigma F_x = ma_x$, then $v = v_0 + a_c t$.

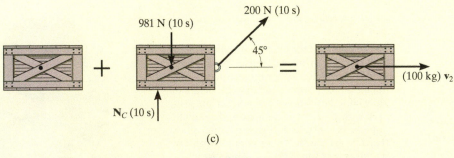

(c)

Fig. 15–5

EXAMPLE 15.2

The 50-lb crate shown in Fig. 15–6a is acted upon by a force having a variable magnitude $P = (20t)$ lb, where t is in seconds. Determine the crate's velocity 2 s after **P** has been applied. The initial velocity is $v_1 = 3$ ft/s down the plane, and the coefficient of kinetic friction between the crate and the plane is $\mu_k = 0.3$.

SOLUTION

Free-Body Diagram.   See Fig. 15–6b. Since the magnitude of force $P = 20t$ varies with time, the impulse it creates must be determined by integrating over the 2-s time interval.

Principle of Impulse and Momentum.   Applying Eqs. 15–4 in the x direction, we have

$$(+\swarrow) \qquad m(v_x)_1 + \Sigma \int_{t_1}^{t_2} F_x \, dt = m(v_x)_2$$

$$\frac{50 \text{ lb}}{32.2 \text{ ft/s}^2}(3 \text{ ft/s}) + \int_0^{2\text{ s}} 20t \, dt - 0.3N_C(2 \text{ s}) + (50 \text{ lb}) \sin 30°(2 \text{ s}) = \frac{50 \text{ lb}}{32.2 \text{ ft/s}^2} v_2$$

$$4.658 + 40 - 0.6N_C + 50 = 1.553v_2$$

The equation of equilibrium can be applied in the y direction. Why?

$$+\nwarrow \Sigma F_y = 0; \qquad N_C - 50 \cos 30° \text{ lb} = 0$$

Solving,

$$N_C = 43.30 \text{ lb}$$

$$v_2 = 44.2 \text{ ft/s} \swarrow \qquad\qquad \textit{Ans.}$$

NOTE: We can also solve this problem using the equation of motion. From Fig. 15–6b,

$$+\swarrow \Sigma F_x = ma_x; \quad 20t - 0.3(43.30) + 50 \sin 30° = \frac{50}{32.2}a$$

$$a = 12.88t + 7.734$$

Using kinematics

$$+\swarrow dv = a \, dt; \qquad \int_{3 \text{ ft/s}}^{v} dv = \int_0^{2\text{ s}} (12.88t + 7.734)dt$$

$$v = 44.2 \text{ ft/s} \qquad\qquad \textit{Ans.}$$

By comparison, application of the principle of impulse and momentum eliminates the need for using kinematics ($a = dv/dt$) and thereby yields an easier method for solution.

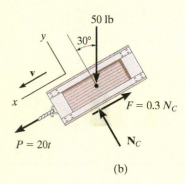

(a)

(b)

Fig. 15–6

15

EXAMPLE | 15.3

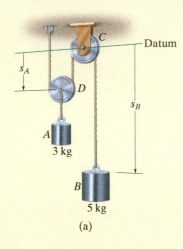

(a)

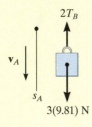

$T_A = 2T_B$

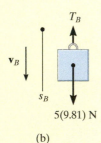

(b)

Fig. 15–7

Blocks A and B shown in Fig. 15–7a have a mass of 3 kg and 5 kg, respectively. If the system is released from rest, determine the velocity of block B in 6 s. Neglect the mass of the pulleys and cord.

SOLUTION

Free-Body Diagram.   See Fig. 15–7b. Since the weight of each block is constant, the cord tensions will also be constant. Furthermore, since the mass of pulley D is neglected, the cord tension $T_A = 2T_B$. Note that the blocks are both assumed to be moving downward in the positive coordinate directions, s_A and s_B.

Principle of Impulse and Momentum.
Block A:

$$(+\downarrow) \qquad\qquad m(v_A)_1 + \Sigma \int_{t_1}^{t_2} F_y\, dt = m(v_A)_2$$

$$0 - 2T_B(6\text{ s}) + 3(9.81)\text{ N}(6\text{ s}) = (3\text{ kg})(v_A)_2 \qquad (1)$$

Block B:

$$(+\downarrow) \qquad\qquad m(v_B)_1 + \Sigma \int_{t_1}^{t_2} F_y\, dt = m(v_B)_2$$

$$0 + 5(9.81)\text{ N}(6\text{ s}) - T_B(6\text{ s}) = (5\text{ kg})(v_B)_2 \qquad (2)$$

Kinematics.   Since the blocks are subjected to dependent motion, the velocity of A can be related to that of B by using the kinematic analysis discussed in Sec. 12.9. A horizontal datum is established through the fixed point at C, Fig. 15–7a, and the position coordinates, s_A and s_B, are related to the constant total length l of the vertical segments of the cord by the equation

$$2s_A + s_B = l$$

Taking the time derivative yields

$$2v_A = -v_B \qquad (3)$$

As indicated by the negative sign, when B moves downward A moves upward. Substituting this result into Eq. 1 and solving Eqs. 1 and 2 yields

$$(v_B)_2 = 35.8 \text{ m/s} \downarrow \qquad\qquad Ans.$$

$$T_B = 19.2 \text{ N}$$

NOTE: Realize that the *positive* (downward) direction for $\mathbf{v}_A$ and $\mathbf{v}_B$ is *consistent* in Figs. 15–7a and 15–7b and in Eqs. 1 to 3. This is important since we are seeking a simultaneous solution of equations.

PRELIMINARY PROBLEMS

15–1. Determine the impulse of the force for $t = 2$ s.

e)

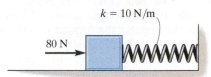

a)

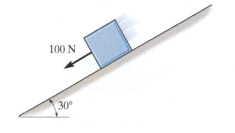

f)

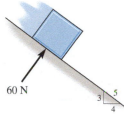

Prob. P15–1

b)

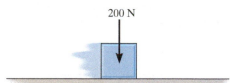

15–2. Determine the linear momentum of the 10-kg block.

a)

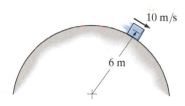

c)

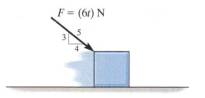

b)

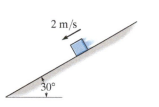

d)

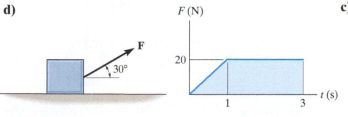

c)

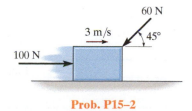

Prob. P15–2

15

FUNDAMENTAL PROBLEMS

F15–1. The 0.5-kg ball strikes the rough ground and rebounds with the velocities shown. Determine the magnitude of the impulse the ground exerts on the ball. Assume that the ball does not slip when it strikes the ground, and neglect the size of the ball and the impulse produced by the weight of the ball.

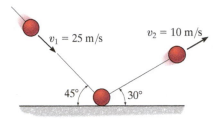

Prob. F15–1

F15–2. If the coefficient of kinetic friction between the 150-lb crate and the ground is $\mu_k = 0.2$, determine the speed of the crate when $t = 4$ s. The crate starts from rest and is towed by the 100-lb force.

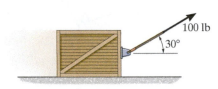

Prob. F15–2

F15–3. The motor exerts a force of $F = (20t^2)$ N on the cable, where t is in seconds. Determine the speed of the 25-kg crate when $t = 4$ s. The coefficients of static and kinetic friction between the crate and the plane are $\mu_s = 0.3$ and $\mu_k = 0.25$, respectively.

Prob. F15–3

F15–4. The wheels of the 1.5-Mg car generate the traction force **F** described by the graph. If the car starts from rest, determine its speed when $t = 6$ s.

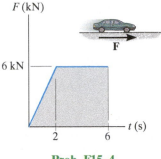

Prob. F15–4

F15–5. The 2.5-Mg four-wheel-drive SUV tows the 1.5-Mg trailer. The traction force developed at the wheels is $F_D = 9$ kN. Determine the speed of the truck in 20 s, starting from rest. Also, determine the tension developed in the coupling, A, between the SUV and the trailer. Neglect the mass of the wheels.

Prob. F15–5

F15–6. The 10-lb block A attains a velocity of 1 ft/s in 5 seconds, starting from rest. Determine the tension in the cord and the coefficient of kinetic friction between block A and the horizontal plane. Neglect the weight of the pulley. Block B has a weight of 8 lb.

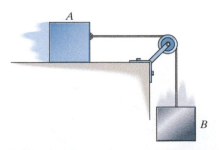

Prob. F15–6

PROBLEMS

15–1. A man kicks the 150-g ball such that it leaves the ground at an angle of 60° and strikes the ground at the same elevation a distance of 12 m away. Determine the impulse of his foot on the ball at A. Neglect the impulse caused by the ball's weight while it's being kicked.

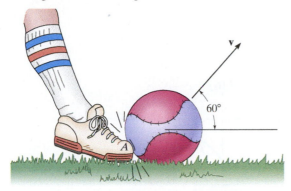

Prob. 15–1

15–2. A 20-lb block slides down a 30° inclined plane with an initial velocity of 2 ft/s. Determine the velocity of the block in 3 s if the coefficient of kinetic friction between the block and the plane is $\mu_k = 0.25$.

15–3. The uniform beam has a weight of 5000 lb. Determine the average tension in each of the two cables AB and AC if the beam is given an upward speed of 8 ft/s in 1.5 s starting from rest. Neglect the mass of the cables.

*15–4. Each of the cables can sustain a maximum tension of 5000 lb. If the uniform beam has a weight of 5000 lb, determine the shortest time possible to lift the beam with a speed of 10 ft/s starting from rest.

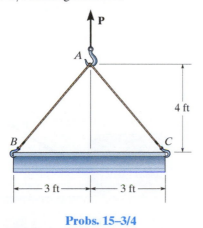

Probs. 15–3/4

15–5. A hockey puck is traveling to the left with a velocity of $v_1 = 10$ m/s when it is struck by a hockey stick and given a velocity of $v_2 = 20$ m/s as shown. Determine the magnitude of the net impulse exerted by the hockey stick on the puck. The puck has a mass of 0.2 kg.

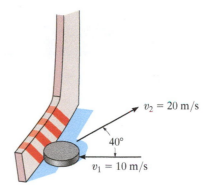

Prob. 15–5

15–6. A train consists of a 50-Mg engine and three cars, each having a mass of 30 Mg. If it takes 80 s for the train to increase its speed uniformly to 40 km/h, starting from rest, determine the force T developed at the coupling between the engine E and the first car A. The wheels of the engine provide a resultant frictional tractive force $\mathbf{F}$ which gives the train forward motion, whereas the car wheels roll freely. Also, determine F acting on the engine wheels.

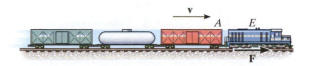

Prob. 15–6

15–7. Crates A and B weigh 100 lb and 50 lb, respectively. If they start from rest, determine their speed when $t = 5$ s. Also, find the force exerted by crate A on crate B during the motion. The coefficient of kinetic friction between the crates and the ground is $\mu_k = 0.25$.

15–9. The 200-kg crate rests on the ground for which the coefficients of static and kinetic friction are $\mu_s = 0.5$ and $\mu_k = 0.4$, respectively. The winch delivers a horizontal towing force T to its cable at A which varies as shown in the graph. Determine the speed of the crate when $t = 4$ s. Originally the tension in the cable is zero. *Hint:* First determine the force needed to begin moving the crate.

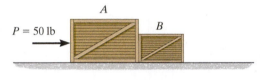

$P = 50$ lb

Prob. 15–7

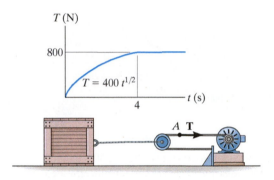

T (N)

800

$T = 400\ t^{1/2}$

4

t (s)

A T

Prob. 15–9

***15–8.** The automobile has a weight of 2700 lb and is traveling forward at 4 ft/s when it crashes into the wall. If the impact occurs in 0.06 s, determine the average impulsive force acting on the car. Assume the brakes are *not applied*. If the coefficient of kinetic friction between the wheels and the pavement is $\mu_k = 0.3$, calculate the impulsive force on the wall if the brakes *were applied* during the crash. The brakes are applied to all four wheels so that all the wheels slip.

15–10. The 50-kg crate is pulled by the constant force **P**. If the crate starts from rest and achieves a speed of 10 m/s in 5 s, determine the magnitude of **P**. The coefficient of kinetic friction between the crate and the ground is $\mu_k = 0.2$.

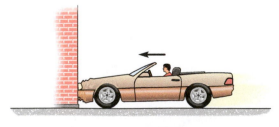

Prob. 15–8

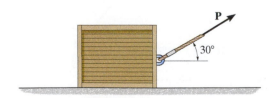

P

30°

Prob. 15–10

15–11.   During operation the jack hammer strikes the concrete surface with a force which is indicated in the graph. To achieve this the 2-kg spike S is fired into the surface at 90 m/s. Determine the speed of the spike just after rebounding.

15–14.   A tankcar has a mass of 20 Mg and is freely rolling to the right with a speed of 0.75 m/s. If it strikes the barrier, determine the horizontal impulse needed to stop the car if the spring in the bumper B has a stiffness (a) $k \rightarrow \infty$ (bumper is rigid), and (b) $k = 15$ kN/m.

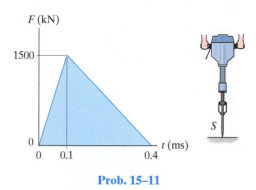

Prob. 15–11

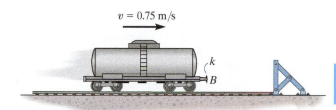

Prob. 15–14

***15–12.**   For a short period of time, the frictional driving force acting on the wheels of the 2.5-Mg van is $F_D = (600t^2)$ N, where t is in seconds. If the van has a speed of 20 km/h when $t = 0$, determine its speed when $t = 5$ s.

15–15.   The motor, M, pulls on the cable with a force $F = (10t^2 + 300)$ N, where t is in seconds. If the 100 kg crate is originally at rest at $t = 0$, determine its speed when $t = 4$ s. Neglect the mass of the cable and pulleys. *Hint:* First find the time needed to begin lifting the crate.

Prob. 15–12

15–13.   The 2.5-Mg van is traveling with a speed of 100 km/h when the brakes are applied and all four wheels lock. If the speed decreases to 40 km/h in 5 s, determine the coefficient of kinetic friction between the tires and the road.

Prob. 15–13

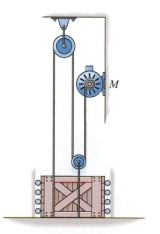

Prob. 15–15

*15–16. The choice of a seating material for moving vehicles depends upon its ability to resist shock and vibration. From the data shown in the graphs, determine the impulses created by a falling weight onto a sample of urethane foam and CONFOR foam.

15–18. The motor exerts a force F on the 40-kg crate as shown in the graph. Determine the speed of the crate when $t = 3$ s and when $t = 6$ s. When $t = 0$, the crate is moving downward at 10 m/s.

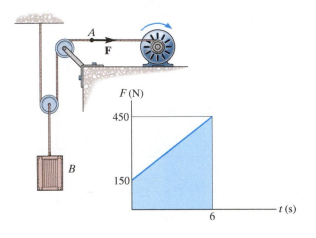

Prob. 15–16

Prob. 15–18

15–17. The towing force acting on the 400-kg safe varies as shown on the graph. Determine its speed, starting from rest, when $t = 8$ s. How far has it traveled during this time?

15–19. The 30-kg slider block is moving to the left with a speed of 5 m/s when it is acted upon by the forces $\mathbf{F}_1$ and $\mathbf{F}_2$. If these loadings vary in the manner shown on the graph, determine the speed of the block at $t = 6$ s. Neglect friction and the mass of the pulleys and cords.

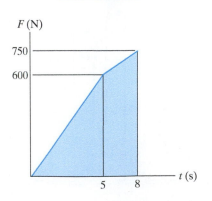

Prob. 15–17

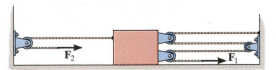

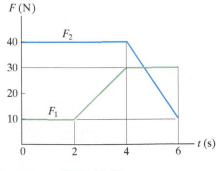

Prob. 15–19

*15–20. The 200-lb cabinet is subjected to the force $F = 20(t+1)$ lb where t is in seconds. If the cabinet is initially moving to the left with a velocity of 20 ft/s, determine its speed when $t = 5$ s. Neglect the size of the rollers.

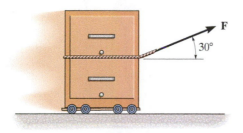

Prob. 15–20

15–21. If it takes 35 s for the 50-Mg tugboat to increase its speed uniformly to 25 km/h, starting from rest, determine the force of the rope on the tugboat. The propeller provides the propulsion force **F** which gives the tugboat forward motion, whereas the barge moves freely. Also, determine F acting on the tugboat. The barge has a mass of 75 Mg.

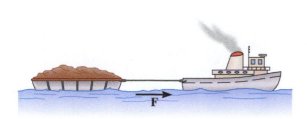

Prob. 15–21

15–22. The thrust on the 4-Mg rocket sled is shown in the graph. Determine the sleds maximum velocity and the distance the sled travels when $t = 35$ s. Neglect friction.

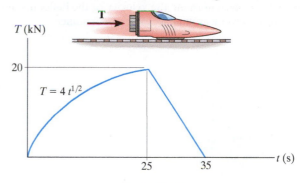

Prob. 15–22

15–23. The motor pulls on the cable at A with a force $F = (30 + t^2)$ lb, where t is in seconds. If the 34-lb crate is originally on the ground at $t = 0$, determine its speed in $t = 4$ s. Neglect the mass of the cable and pulleys. *Hint:* First find the time needed to begin lifting the crate.

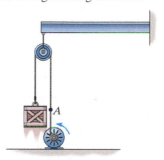

Prob. 15–23

*15–24. The motor pulls on the cable at A with a force $F = (e^{2t})$ lb, where t is in seconds. If the 34-lb crate is originally at rest on the ground at $t = 0$, determine the crate's velocity when $t = 2$ s. Neglect the mass of the cable and pulleys. *Hint:* First find the time needed to begin lifting the crate.

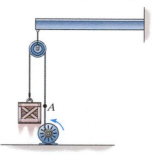

Prob. 15–24

15

15–25. The balloon has a total mass of 400 kg including the passengers and ballast. The balloon is rising at a constant velocity of 18 km/h when $h = 10$ m. If the man drops the 40-kg sand bag, determine the velocity of the balloon when the bag strikes the ground. Neglect air resistance.

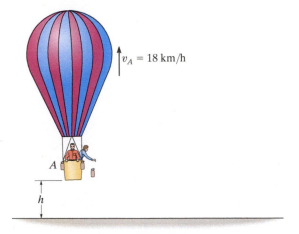

Prob. 15–25

15–26. As indicated by the derivation, the principle of impulse and momentum is valid for observers in *any* inertial reference frame. Show that this is so, by considering the 10-kg block which slides along the smooth surface and is subjected to a horizontal force of 6 N. If observer A is in a *fixed* frame x, determine the final speed of the block in 4 s if it has an initial speed of 5 m/s measured from the fixed frame. Compare the result with that obtained by an observer B, attached to the x' axis that moves at a constant velocity of 2 m/s relative to A.

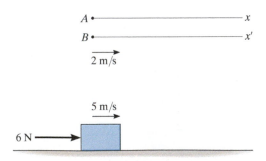

Prob. 15–26

15–27. The 20-kg crate is lifted by a force of $F = (100 + 5t^2)$ N, where t is in seconds. Determine the speed of the crate when $t = 3$ s, starting from rest.

*15–28. The 20-kg crate is lifted by a force of $F = (100 + 5t^2)$ N, where t is in seconds. Determine how high the crate has moved upward when $t = 3$ s, starting from rest.

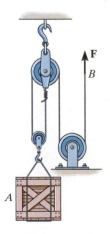

Probs. 15–27/28

15–29. In case of emergency, the gas actuator is used to move a 75-kg block B by exploding a charge C near a pressurized cylinder of negligible mass. As a result of the explosion, the cylinder fractures and the released gas forces the front part of the cylinder, A, to move B forward, giving it a speed of 200 mm/s in 0.4 s. If the coefficient of kinetic friction between B and the floor is $\mu_k = 0.5$, determine the impulse that the actuator imparts to B.

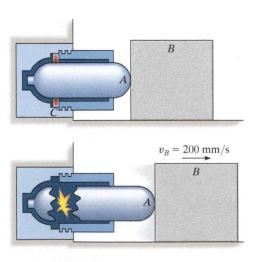

Prob. 15–29

15–30. A jet plane having a mass of 7 Mg takes off from an aircraft carrier such that the engine thrust varies as shown by the graph. If the carrier is traveling forward with a speed of 40 km/h, determine the plane's airspeed after 5 s.

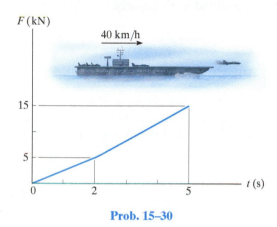

Prob. 15–30

15–31. Block A weighs 10 lb and block B weighs 3 lb. If B is moving downward with a velocity $(v_B)_1 = 3$ ft/s at $t = 0$, determine the velocity of A when $t = 1$ s. Assume that the horizontal plane is smooth. Neglect the mass of the pulleys and cords.

*****15–32.** Block A weighs 10 lb and block B weighs 3 lb. If B is moving downward with a velocity $(v_B)_1 = 3$ ft/s at $t = 0$, determine the velocity of A when $t = 1$ s. The coefficient of kinetic friction between the horizontal plane and block A is $\mu_A = 0.15$.

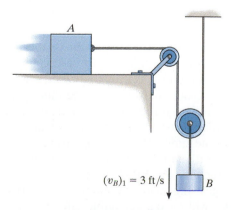

Probs. 15–31/32

15–33. The log has a mass of 500 kg and rests on the ground for which the coefficients of static and kinetic friction are $\mu_s = 0.5$ and $\mu_k = 0.4$, respectively. The winch delivers a horizontal towing force T to its cable at A which varies as shown in the graph. Determine the speed of the log when $t = 5$ s. Originally the tension in the cable is zero. *Hint:* First determine the force needed to begin moving the log.

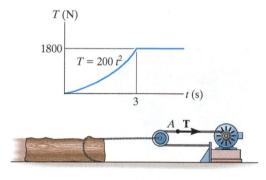

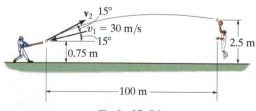

Prob. 15–33

15–34. The 0.15-kg baseball has a speed of $v = 30$ m/s just before it is struck by the bat. It then travels along the trajectory shown before the outfielder catches it. Determine the magnitude of the average impulsive force imparted to the ball if it is in contact with the bat for 0.75 ms.

Prob. 15–34

(© R.C. Hibbeler)

The hammer in the top photo applies an impulsive force to the stake. During this extremely short time of contact the weight of the stake can be considered nonimpulsive, and provided the stake is driven into soft ground, the impulse of the ground acting on the stake can also be considered nonimpulsive. By contrast, if the stake is used in a concrete chipper to break concrete, then two impulsive forces act on the stake: one at its top due to the chipper and the other on its bottom due to the rigidity of the concrete. (© R.C. Hibbeler)

15.3 Conservation of Linear Momentum for a System of Particles

When the sum of the *external impulses* acting on a system of particles is *zero*, Eq. 15–6 reduces to a simplified form, namely,

$$\Sigma m_i(\mathbf{v}_i)_1 = \Sigma m_i(\mathbf{v}_i)_2 \qquad (15\text{–}8)$$

This equation is referred to as the *conservation of linear momentum*. It states that the total linear momentum for a system of particles remains constant during the time period t_1 to t_2. Substituting $m\mathbf{v}_G = \Sigma m_i \mathbf{v}_i$ into Eq. 15–8, we can also write

$$(\mathbf{v}_G)_1 = (\mathbf{v}_G)_2 \qquad (15\text{–}9)$$

which indicates that the velocity $\mathbf{v}_G$ of the mass center for the system of particles does not change if no external impulses are applied to the system.

The conservation of linear momentum is often applied when particles collide or interact. For application, a careful study of the free-body diagram for the *entire* system of particles should be made in order to identify the forces which create either external or internal impulses and thereby determine in what direction(s) linear momentum is conserved. As stated earlier, the *internal impulses* for the system will always cancel out, since they occur in equal but opposite collinear pairs. If the time period over which the motion is studied is *very short*, some of the external impulses may also be neglected or considered approximately equal to zero. The forces causing these negligible impulses are called *nonimpulsive forces*. By comparison, forces which are very large and act for a very short period of time produce a significant change in momentum and are called *impulsive forces*. They, of course, cannot be neglected in the impulse–momentum analysis.

Impulsive forces normally occur due to an explosion or the striking of one body against another, whereas nonimpulsive forces may include the weight of a body, the force imparted by a slightly deformed spring having a relatively small stiffness, or for that matter, any force that is very small compared to other larger (impulsive) forces. When making this distinction between impulsive and nonimpulsive forces, it is important to realize that this only applies during the time t_1 to t_2. To illustrate, consider the effect of striking a tennis ball with a racket as shown in the photo. During the *very short* time of interaction, the force of the racket on the ball is impulsive since it changes the ball's momentum drastically. By comparison, the ball's weight will have a negligible effect on the change in momentum, and therefore it is nonimpulsive. Consequently, it can be neglected from an impulse–momentum analysis during this time. If an impulse–momentum analysis is considered during the much longer time of flight after the racket–ball interaction, then the impulse of the ball's weight is important since it, along with air resistance, causes the change in the momentum of the ball.

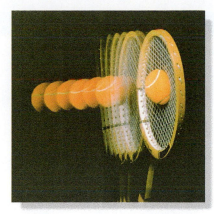

(© R.C. Hibbeler)

Procedure for Analysis

Generally, the principle of linear impulse and momentum or the conservation of linear momentum is applied to a *system of particles* in order to determine the final velocities of the particles *just after* the time period considered. By applying this principle to the entire system, the internal impulses acting within the system, which may be unknown, are *eliminated* from the analysis. For application it is suggested that the following procedure be used.

Free-Body Diagram.

- Establish the *x, y, z* inertial frame of reference and draw the free-body diagram for each particle of the system in order to identify the internal and external forces.

- The conservation of linear momentum applies to the system in a direction which either has no external forces or the forces can be considered nonimpulsive.

- Establish the direction and sense of the particles' initial and final velocities. If the sense is unknown, assume it is along a positive inertial coordinate axis.

- As an alternative procedure, draw the impulse and momentum diagrams for each particle of the system.

Momentum Equations.

- Apply the principle of linear impulse and momentum or the conservation of linear momentum in the appropriate directions.

- If it is necessary to determine the *internal impulse* $\int F \, dt$ acting on only one particle of a system, then the particle must be *isolated* (free-body diagram), and the principle of linear impulse and momentum must be applied *to this particle*.

- After the impulse is calculated, and provided the time Δt for which the impulse acts is known, then the *average impulsive force* F_{avg} can be determined from $F_{avg} = \int F \, dt / \Delta t$.

EXAMPLE | 15.4

15

The 15-Mg boxcar A is coasting at 1.5 m/s on the horizontal track when it encounters a 12-Mg tank car B coasting at 0.75 m/s toward it as shown in Fig. 15–8a. If the cars collide and couple together, determine (a) the speed of both cars just after the coupling, and (b) the average force between them if the coupling takes place in 0.8 s.

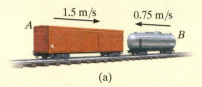

(a)

SOLUTION

Part (a) Free-Body Diagram.*   Here we have considered *both* cars as a single system, Fig. 15–8b. By inspection, momentum is conserved in the x direction since the coupling force $\mathbf{F}$ is *internal* to the system and will therefore cancel out. It is assumed both cars, when coupled, move at $\mathbf{v}_2$ in the positive x direction.

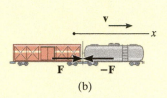

(b)

Conservation of Linear Momentum.

$$(\xrightarrow{+}) \qquad m_A(v_A)_1 + m_B(v_B)_1 = (m_A + m_B)v_2$$

$$(15\ 000\ \text{kg})(1.5\ \text{m/s}) - 12\ 000\ \text{kg}(0.75\ \text{m/s}) = (27\ 000\ \text{kg})v_2$$

$$v_2 = 0.5\ \text{m/s} \rightarrow \qquad\qquad Ans.$$

Part (b).   The average (impulsive) coupling force, $\mathbf{F}_{\text{avg}}$, can be determined by applying the principle of linear momentum to *either one* of the cars.

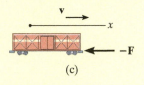

(c)

Fig. 15–8

Free-Body Diagram.   As shown in Fig. 15–8c, by isolating the boxcar the coupling force is *external* to the car.

Principle of Impulse and Momentum.   Since $\int F\, dt = F_{\text{avg}}\, \Delta t = F_{\text{avg}}(0.8\ \text{s})$, we have

$$(\xrightarrow{+}) \qquad m_A(v_A)_1 + \Sigma \int F\, dt = m_A v_2$$

$$(15\ 000\ \text{kg})(1.5\ \text{m/s}) - F_{\text{avg}}(0.8\ \text{s}) = (15\ 000\ \text{kg})(0.5\ \text{m/s})$$

$$F_{\text{avg}} = 18.8\ \text{kN} \qquad\qquad Ans.$$

NOTE: Solution was possible here since the boxcar's final velocity was obtained in Part (a). Try solving for F_{avg} by applying the principle of impulse and momentum to the tank car.

*Only horizontal forces are shown on the free-body diagram.

EXAMPLE | 15.5

The bumper cars A and B in Fig. 15–9a each have a mass of 150 kg and are coasting with the velocities shown before they freely collide head on. If no energy is lost during the collision, determine their velocities after collision.

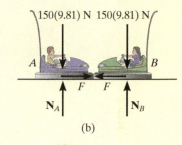

$(v_A)_1 = 3$ m/s $(v_B)_1 = 2$ m/s

(a)

SOLUTION

Free-Body Diagram.   The cars will be considered as a single system. The free-body diagram is shown in Fig. 15–9b.

Conservation of Momentum.

$(\xrightarrow{+})$ $m_A(v_A)_1 + m_B(v_B)_1 = m_A(v_A)_2 + m_B(v_B)_2$

$(150 \text{ kg})(3 \text{ m/s}) + (150 \text{ kg})(-2 \text{ m/s}) = (150 \text{ kg})(v_A)_2 + (150 \text{ kg})(v_B)_2$

$$(v_A)_2 = 1 - (v_B)_2 \qquad (1)$$

Conservation of Energy.   Since no energy is lost, the conservation of energy theorem gives

$$T_1 + V_1 = T_2 + V_2$$

$$\frac{1}{2}m_A(v_A)_1^2 + \frac{1}{2}m_B(v_B)_1^2 + 0 = \frac{1}{2}m_A(v_A)_2^2 + \frac{1}{2}m_B(v_B)_2^2 + 0$$

$$\frac{1}{2}(150 \text{ kg})(3 \text{ m/s})^2 + \frac{1}{2}(150 \text{ kg})(2 \text{ m/s})^2 + 0 = \frac{1}{2}(150 \text{ kg})(v_A)_2^2$$

$$+ \frac{1}{2}(150 \text{ kg})(v_B)_2^2 + 0$$

$$(v_A)_2^2 + (v_B)_2^2 = 13 \qquad (2)$$

Substituting Eq. (1) into (2) and simplifying, we get

$$(v_B)_2^2 - (v_B)_2 - 6 = 0$$

Solving for the two roots,

$$(v_B)_2 = 3 \text{ m/s} \qquad \text{and} \qquad (v_B)_2 = -2 \text{ m/s}$$

Since $(v_B)_2 = -2$ m/s refers to the velocity of B just *before* collision, then the velocity of B just after the collision must be

$$(v_B)_2 = 3 \text{ m/s} \rightarrow \qquad\qquad Ans.$$

Substituting this result into Eq. (1), we obtain

$$(v_A)_2 = 1 - 3 \text{ m/s} = -2 \text{ m/s} = 2 \text{ m/s} \leftarrow \qquad Ans.$$

150(9.81) N 150(9.81) N

N_A F F N_B

(b)

Fig. 15–9

15

EXAMPLE 15.6

(© R.C. Hibbeler)

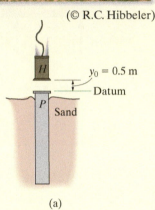

$y_0 = 0.5$ m

Datum

Sand

(a)

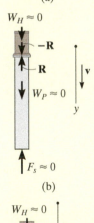

$W_H \approx 0$

$-R$

R

$W_P \approx 0$

$\mathbf{v}$

y

$F_s \approx 0$

(b)

$W_H \approx 0$

$\mathbf{v}$

R y

(c)

Fig. 15–10

An 800-kg rigid pile shown in Fig. 15–10a is driven into the ground using a 300-kg hammer. The hammer falls from rest at a height $y_0 = 0.5$ m and strikes the top of the pile. Determine the impulse which the pile exerts on the hammer if the pile is surrounded entirely by loose sand so that after striking, the hammer does *not* rebound off the pile.

SOLUTION

Conservation of Energy. The velocity at which the hammer strikes the pile can be determined using the conservation of energy equation applied to the hammer. With the datum at the top of the pile, Fig. 15–10a, we have

$$T_0 + V_0 = T_1 + V_1$$

$$\frac{1}{2}m_H(v_H)_0^2 + W_H y_0 = \frac{1}{2}m_H(v_H)_1^2 + W_H y_1$$

$$0 + 300(9.81)\,\text{N}(0.5\,\text{m}) = \frac{1}{2}(300\,\text{kg})(v_H)_1^2 + 0$$

$$(v_H)_1 = 3.132\,\text{m/s}$$

Free-Body Diagram. From the physical aspects of the problem, the free-body diagram of the hammer and pile, Fig. 15–10b, indicates that during the *short time* from *just before* to *just after* the *collision*, the weights of the hammer and pile and the resistance force $\mathbf{F}_s$ of the sand are all *nonimpulsive*. The impulsive force $\mathbf{R}$ is internal to the system and therefore cancels. Consequently, momentum is conserved in the vertical direction during this short time.

Conservation of Momentum. Since the hammer does not rebound off the pile just after collision, then $(v_H)_2 = (v_P)_2 = v_2$.

$$(+\downarrow)\qquad m_H(v_H)_1 + m_P(v_P)_1 = m_H v_2 + m_P v_2$$

$$(300\,\text{kg})(3.132\,\text{m/s}) + 0 = (300\,\text{kg})v_2 + (800\,\text{kg})v_2$$

$$v_2 = 0.8542\,\text{m/s}$$

Principle of Impulse and Momentum. The impulse which the pile imparts to the hammer can now be determined since v_2 is known. From the free-body diagram for the hammer, Fig. 15–10c, we have

$$(+\downarrow)\qquad m_H(v_H)_1 + \Sigma\int_{t_1}^{t_2} F_y\,dt = m_H v_2$$

$$(300\,\text{kg})(3.132\,\text{m/s}) - \int R\,dt = (300\,\text{kg})(0.8542\,\text{m/s})$$

$$\int R\,dt = 683\,\text{N}\cdot\text{s} \qquad\qquad Ans.$$

NOTE: The equal but opposite impulse acts on the pile. Try finding this impulse by applying the principle of impulse and momentum to the pile.

EXAMPLE 15.7

The 80-kg man can throw the 20-kg box horizontally at 4 m/s when standing on the ground. If instead he firmly stands in the 120-kg boat and throws the box, as shown in the photo, determine how far the boat will move in three seconds. Neglect water resistance.

(© R.C. Hibbeler)

SOLUTION

Free-Body Diagram. If the man, boat, and box are considered as a single system, the horizontal forces between the man and the boat and the man and the box become internal to the system, Fig. 15–11a, and so linear momentum will be conserved along the x axis.

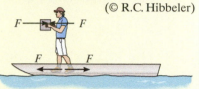

(a)

Conservation of Momentum. When writing the conservation of momentum equation, it is *important* that the velocities be measured from the same inertial coordinate system, assumed here to be fixed. From this coordinate system, we will assume that the boat and man go to the right while the box goes to the left, as shown in Fig. 15–11b.

Applying the conservation of linear momentum to the man, boat, box system,

(b)

$$(\xrightarrow{+}) \qquad 0 + 0 + 0 = (m_m + m_b)\,v_b - m_{\text{box}}\,v_{\text{box}}$$

$$0 = (80 \text{ kg} + 120 \text{ kg})\,v_b - (20 \text{ kg})\,v_{\text{box}}$$

$$v_{\text{box}} = 10\,v_b \qquad (1)$$

Fig. 15–11

Kinematics. Since the velocity of the box *relative* to the man (and boat), $v_{\text{box/b}}$, is known, then v_b can also be related to v_{box} using the relative velocity equation.

$$(\xrightarrow{+}) \qquad v_{\text{box}} = v_b + v_{\text{box/b}}$$

$$-v_{\text{box}} = v_b - 4 \text{ m/s} \qquad (2)$$

Solving Eqs. (1) and (2),

$$v_{\text{box}} = 3.64 \text{ m/s} \leftarrow$$

$$v_b = 0.3636 \text{ m/s} \rightarrow$$

The displacement of the boat in three seconds is therefore

$$s_b = v_b t = (0.3636 \text{ m/s})(3 \text{ s}) = 1.09 \text{ m} \qquad \textit{Ans.}$$

15

EXAMPLE | 15.8

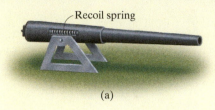

Recoil spring

(a)

The 1200-lb cannon shown in Fig. 15–12a fires an 8-lb projectile with a muzzle velocity of 1500 ft/s measured relative to the cannon. If firing takes place in 0.03 s, determine the recoil velocity of the cannon just after firing. The cannon support is fixed to the ground, and the horizontal recoil of the cannon is absorbed by two springs.

SOLUTION

Part (a) Free-Body Diagram.* As shown in Fig. 15–12b, we will consider the projectile and cannon as a single system, since the impulsive forces, **F** and −**F**, between the cannon and projectile are *internal* to the system and will therefore cancel from the analysis. Furthermore, during the time $\Delta t = 0.03$ s, the two recoil springs which are attached to the support each exert a *nonimpulsive force* **F**$_s$ on the cannon. This is because Δt is very short, so that during this time the cannon only moves through a *very small* distance s. Consequently, $F_s = ks \approx 0$, where k is the spring's stiffness, which is also considered to be relatively small. Hence it can be concluded that momentum for the system is conserved in the *horizontal direction*.

Conservation of Linear Momentum.

$$(\xrightarrow{+}) \qquad m_c(v_c)_1 + m_p(v_p)_1 = -m_c(v_c)_2 + m_p(v_p)_2$$

$$0 + 0 = -\left(\frac{1200 \text{ lb}}{32.2 \text{ ft/s}^2}\right)(v_c)_2 + \left(\frac{8 \text{ lb}}{32.2 \text{ ft/s}^2}\right)(v_p)_2$$

$$(v_p)_2 = 150 (v_c)_2 \qquad (1)$$

These unknown velocities are measured by a *fixed* observer. As in Example 15–7, they can also be related using the relative velocity equation.

$$\xrightarrow{+} \qquad (v_p)_2 = (v_c)_2 + v_{p/c}$$
$$(v_p)_2 = -(v_c)_2 + 1500 \text{ ft/s} \qquad (2)$$

Solving Eqs. (1) and (2) yields

$$(v_c)_2 = 9.93 \text{ ft/s} \qquad \qquad Ans.$$
$$(v_p)_2 = 1490 \text{ ft/s}$$

Apply the principle of impulse and momentum to the projectile (or the cannon) and show that the average impulsive force on the projectile is 12.3 kip.

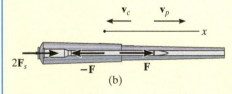

$\mathbf{v}_c$ $\mathbf{v}_p$

x

$2\mathbf{F}_s$ $-\mathbf{F}$ $\mathbf{F}$

(b)

Fig. 15–12

NOTE: If the cannon is firmly fixed to its support (no springs), the reactive force of the support on the cannon must be considered as an external impulse to the system, since the support would allow no movement of the cannon. In this case momentum is *not* conserved.

*Only horizontal forces are shown on the free-body diagram.

FUNDAMENTAL PROBLEMS

F15–7. The freight cars A and B have a mass of 20 Mg and 15 Mg, respectively. Determine the velocity of A after collision if the cars collide and rebound, such that B moves to the right with a speed of 2 m/s. If A and B are in contact for 0.5 s, find the average impulsive force which acts between them.

Prob. F15–7

F15–8. The cart and package have a mass of 20 kg and 5 kg, respectively. If the cart has a smooth surface and it is initially at rest, while the velocity of the package is as shown, determine the final common velocity of the cart and package after the impact.

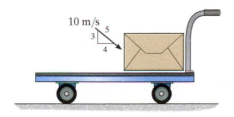

Prob. F15–8

F15–9. The 5-kg block A has an initial speed of 5 m/s as it slides down the smooth ramp, after which it collides with the stationary block B of mass 8 kg. If the two blocks couple together after collision, determine their common velocity immediately after collision.

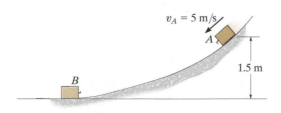

Prob. F15–9

F15–10. The spring is fixed to block A and block B is pressed against the spring. If the spring is compressed $s = 200$ mm and then the blocks are released, determine their velocity at the instant block B loses contact with the spring. The masses of blocks A and B are 10 kg and 15 kg, respectively.

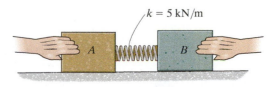

Prob. F15–10

F15–11. Blocks A and B have a mass of 15 kg and 10 kg, respectively. If A is stationary and B has a velocity of 15 m/s just before collision, and the blocks couple together after impact, determine the maximum compression of the spring.

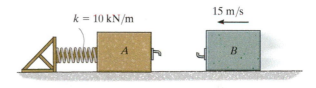

Prob. F15–11

F15–12. The cannon and support without a projectile have a mass of 250 kg. If a 20-kg projectile is fired from the cannon with a velocity of 400 m/s, measured *relative* to the cannon, determine the speed of the projectile as it leaves the barrel of the cannon. Neglect rolling resistance.

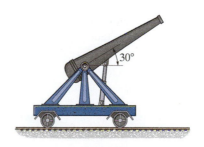

Prob. F15–12

PROBLEMS

15–35. The 5-Mg bus B is traveling to the right at 20 m/s. Meanwhile a 2-Mg car A is traveling at 15 m/s to the right. If the vehicles crash and become entangled, determine their common velocity just after the collision. Assume that the vehicles are free to roll during collision.

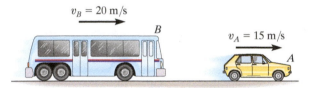

$v_B = 20$ m/s

B

$v_A = 15$ m/s

A

Prob. 15–35

*15–36. The 50-kg boy jumps on the 5-kg skateboard with a horizontal velocity of 5 m/s. Determine the distance s the boy reaches up the inclined plane before momentarily coming to rest. Neglect the skateboard's rolling resistance.

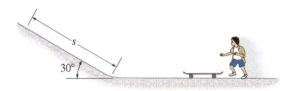

s

$30°$

Prob. 15–36

15–37. The 2.5-Mg pickup truck is towing the 1.5-Mg car using a cable as shown. If the car is initially at rest and the truck is coasting with a velocity of 30 km/h when the cable is slack, determine the common velocity of the truck and the car just after the cable becomes taut. Also, find the loss of energy.

30 km/h

Prob. 15–37

15–38. A railroad car having a mass of 15 Mg is coasting at 1.5 m/s on a horizontal track. At the same time another car having a mass of 12 Mg is coasting at 0.75 m/s in the opposite direction. If the cars meet and couple together, determine the speed of both cars just after the coupling. Find the difference between the total kinetic energy before and after coupling has occurred, and explain qualitatively what happened to this energy.

15–39. A ballistic pendulum consists of a 4-kg wooden block originally at rest, $\theta = 0°$. When a 2-g bullet strikes and becomes embedded in it, it is observed that the block swings upward to a maximum angle of $\theta = 6°$. Estimate the initial speed of the bullet.

θ 1.25 m

θ 1.25 m

Prob. 15–39

*15–40. The boy jumps off the flat car at A with a velocity of $v = 4$ ft/s relative to the car as shown. If he lands on the second flat car B, determine the final speed of both cars after the motion. Each car has a weight of 80 lb. The boy's weight is 60 lb. Both cars are originally at rest. Neglect the mass of the car's wheels.

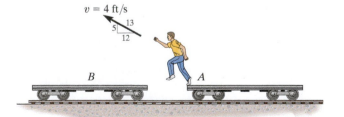

$v = 4$ ft/s

13
5
12

B A

Prob. 15–40

15–41. A 0.03-lb bullet traveling at 1300 ft/s strikes the 10-lb wooden block and exits the other side at 50 ft/s as shown. Determine the speed of the block just after the bullet exits the block, and also determine how far the block slides before it stops. The coefficient of kinetic friction between the block and the surface is $\mu_k = 0.5$.

15–42. A 0.03-lb bullet traveling at 1300 ft/s strikes the 10-lb wooden block and exits the other side at 50 ft/s as shown. Determine the speed of the block just after the bullet exits the block. Also, determine the average normal force on the block if the bullet passes through it in 1 ms, and the time the block slides before it stops. The coefficient of kinetic friction between the block and the surface is $\mu_k = 0.5$.

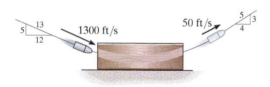

Probs. 15–41/42

15–43. The 20-g bullet is traveling at 400 m/s when it becomes embedded in the 2-kg stationary block. Determine the distance the block will slide before it stops. The coefficient of kinetic friction between the block and the plane is $\mu_k = 0.2$.

Prob. 15–43

*15–44. A toboggan having a mass of 10 kg starts from rest at A and carries a girl and boy having a mass of 40 kg and 45 kg, respectively. When the toboggan reaches the bottom of the slope at B, the boy is pushed off from the back with a horizontal velocity of $v_{b/t} = 2$ m/s, measured relative to the toboggan. Determine the velocity of the toboggan afterwards. Neglect friction in the calculation.

Prob. 15–44

15–45. The block of mass m travels at v_1 in the direction θ_1 shown at the top of the smooth slope. Determine its speed v_2 and its direction θ_2 when it reaches the bottom.

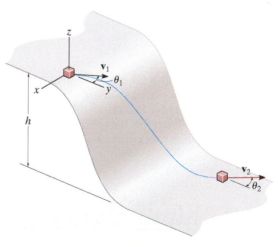

Prob. 15–45

15

15–46. The two blocks A and B each have a mass of 5 kg and are suspended from parallel cords. A spring, having a stiffness of $k = 60$ N/m, is attached to B and is compressed 0.3 m against A and B as shown. Determine the maximum angles θ and ϕ of the cords when the blocks are released from rest and the spring becomes unstretched.

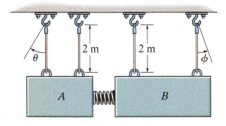

Prob. 15–46

15–47. The 30-Mg freight car A and 15-Mg freight car B are moving towards each other with the velocities shown. Determine the maximum compression of the spring mounted on car A. Neglect rolling resistance.

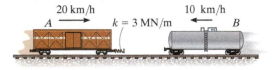

Prob. 15–47

*15–48. Blocks A and B have masses of 40 kg and 60 kg, respectively. They are placed on a smooth surface and the spring connected between them is stretched 2 m. If they are released from rest, determine the speeds of both blocks the instant the spring becomes unstretched.

Prob. 15–48

15–49. A boy A having a weight of 80 lb and a girl B having a weight of 65 lb stand motionless at the ends of the toboggan, which has a weight of 20 lb. If they exchange positions, A going to B and then B going to A's original position, determine the final position of the toboggan just after the motion. Neglect friction between the toboggan and the snow.

15–50. A boy A having a weight of 80 lb and a girl B having a weight of 65 lb stand motionless at the ends of the toboggan, which has a weight of 20 lb. If A walks to B and stops, and both walk back together to the original position of A, determine the final position of the toboggan just after the motion stops. Neglect friction between the toboggan and the snow.

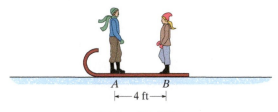

Probs. 15–49/50

15–51. The 10-Mg barge B supports a 2-Mg automobile A. If someone drives the automobile to the other side of the barge, determine how far the barge moves. Neglect the resistance of the water.

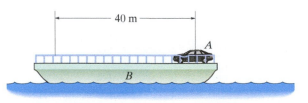

Prob. 15–51

*15–52. The free-rolling ramp has a mass of 40 kg. A 10-kg crate is released from rest at A and slides down 3.5 m to point B. If the surface of the ramp is smooth, determine the ramp's speed when the crate reaches B. Also, what is the velocity of the crate?

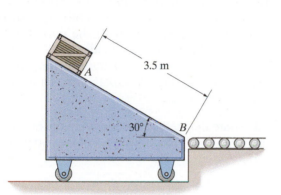

Prob. 15–52

15–53. Block A has a mass of 5 kg and is placed on the smooth triangular block B having a mass of 30 kg. If the system is released from rest, determine the distance B moves from point O when A reaches the bottom. Neglect the size of block A.

15–54. Solve Prob. 15–53 if the coefficient of kinetic friction between A and B is $\mu_k = 0.3$. Neglect friction between block B and the horizontal plane.

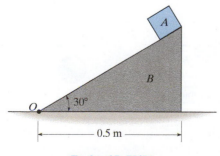

Probs. 15–53/54

15–55. The cart has a mass of 3 kg and rolls freely from A down the slope. When it reaches the bottom, a spring loaded gun fires a 0.5-kg ball out the back with a horizontal velocity of $v_{b/c} = 0.6$ m/s, measured relative to the cart. Determine the final velocity of the cart.

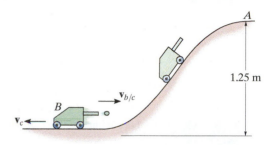

Prob. 15–55

*15–56. Two boxes A and B, each having a weight of 160 lb, sit on the 500-lb conveyor which is free to roll on the ground. If the belt starts from rest and begins to run with a speed of 3 ft/s, determine the final speed of the conveyor if (a) the boxes are not stacked and A falls off then B falls off, and (b) A is stacked on top of B and both fall off together.

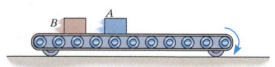

Prob. 15–56

15–57. The 10-kg block is held at rest on the smooth inclined plane by the stop block at A. If the 10-g bullet is traveling at 300 m/s when it becomes embedded in the 10-kg block, determine the distance the block will slide up along the plane before momentarily stopping.

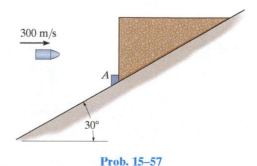

Prob. 15–57

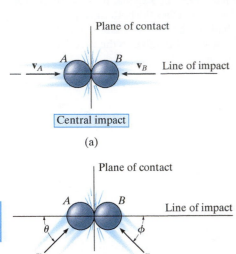

Plane of contact

$\mathbf{v}_A$ A B $\mathbf{v}_B$ Line of impact

Central impact

(a)

Plane of contact

A B Line of impact

θ ϕ

$\mathbf{v}_A$ $\mathbf{v}_B$

Oblique impact

(b)

Fig. 15–13

15.4 Impact

Impact occurs when two bodies collide with each other during a very *short* period of time, causing relatively large (impulsive) forces to be exerted between the bodies. The striking of a hammer on a nail, or a golf club on a ball, are common examples of impact loadings.

In general, there are two types of impact. *Central impact* occurs when the direction of motion of the mass centers of the two colliding particles is along a line passing through the mass centers of the particles. This line is called the *line of impact*, which is perpendicular to the plane of contact, Fig. 15–13a. When the motion of one or both of the particles make an angle with the line of impact, Fig. 15–13b, the impact is said to be *oblique impact*.

Central Impact.
To illustrate the method for analyzing the mechanics of impact, consider the case involving the central impact of the two particles A and B shown in Fig. 15–14.

- The particles have the initial momenta shown in Fig. 15–14a. Provided $(v_A)_1 > (v_B)_1$, collision will eventually occur.

- During the collision the particles must be thought of as *deformable* or nonrigid. The particles will undergo a *period of deformation* such that they exert an equal but opposite deformation impulse $\int \mathbf{P} \, dt$ on each other, Fig. 15–14b.

- Only at the instant of *maximum deformation* will both particles move with a common velocity $\mathbf{v}$, since their relative motion is zero, Fig. 15–14c.

- Afterward a *period of restitution* occurs, in which case the particles will either return to their original shape or remain permanently deformed. The equal but opposite *restitution impulse* $\int \mathbf{R} \, dt$ pushes the particles apart from one another, Fig. 15–14d. In reality, the physical properties of any two bodies are such that the deformation impulse will *always be greater* than that of restitution, i.e., $\int P \, dt > \int R \, dt$.

- Just after separation the particles will have the final momenta shown in Fig. 15–14e, where $(v_B)_2 > (v_A)_2$.

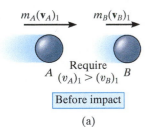

$m_A(\mathbf{v}_A)_1$ $m_B(\mathbf{v}_B)_1$

Require
A $(v_A)_1 > (v_B)_1$ B

Before impact

(a)

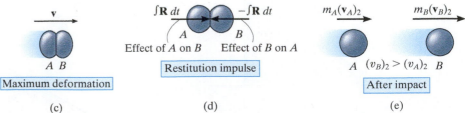

$\int \mathbf{P} \, dt$ $-\int \mathbf{P} \, dt$

A B

Effect of A on B Effect of B on A

Deformation impulse

(b)

$\mathbf{v}$

A B

Maximum deformation

(c)

$\int \mathbf{R} \, dt$ $-\int \mathbf{R} \, dt$

A B

Effect of A on B Effect of B on A

Restitution impulse

(d)

$m_A(\mathbf{v}_A)_2$ $m_B(\mathbf{v}_B)_2$

A $(v_B)_2 > (v_A)_2$ B

After impact

(e)

Fig. 15–14

In most problems the initial velocities of the particles will be *known*, and it will be necessary to determine their final velocities $(v_A)_2$ and $(v_B)_2$. In this regard, *momentum* for the *system of particles* is *conserved* since during collision the internal impulses of deformation and restitution *cancel*. Hence, referring to Fig. 15–14a and Fig. 15–14e we require

$$(\stackrel{+}{\rightarrow}) \qquad m_A(v_A)_1 + m_B(v_B)_1 = m_A(v_A)_2 + m_B(v_B)_2 \qquad (15\text{--}10)$$

In order to obtain a second equation necessary to solve for $(v_A)_2$ and $(v_B)_2$, we must apply the principle of impulse and momentum to *each particle*. For example, during the deformation phase for particle A, Figs. 15–14a, 15–14b, and 15–14c, we have

$$(\stackrel{+}{\rightarrow}) \qquad m_A(v_A)_1 - \int P\, dt = m_A v$$

For the restitution phase, Figs. 15–14c, 15–14d, and 15–14e,

$$(\stackrel{+}{\rightarrow}) \qquad m_A v - \int R\, dt = m_A(v_A)_2$$

The ratio of the restitution impulse to the deformation impulse is called the *coefficient of restitution*, e. From the above equations, this value for particle A is

$$e = \frac{\int R\, dt}{\int P\, dt} = \frac{v - (v_A)_2}{(v_A)_1 - v}$$

In a similar manner, we can establish e by considering particle B, Fig. 15–14. This yields

$$e = \frac{\int R\, dt}{\int P\, dt} = \frac{(v_B)_2 - v}{v - (v_B)_1}$$

If the unknown v is eliminated from the above two equations, the coefficient of restitution can be expressed in terms of the particles' initial and final velocities as

$$(\stackrel{+}{\rightarrow}) \qquad \boxed{e = \frac{(v_B)_2 - (v_A)_2}{(v_A)_1 - (v_B)_1}} \qquad (15\text{--}11)$$

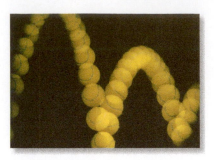

The quality of a manufactured tennis ball is measured by the height of its bounce, which can be related to its coefficient of restitution. Using the mechanics of oblique impact, engineers can design a separation device to remove substandard tennis balls from a production line. (© Gary S. Settles/Science Source)

Provided a value for e is specified, Eqs. 15–10 and 15–11 can be solved simultaneously to obtain $(v_A)_2$ and $(v_B)_2$. In doing so, however, it is important to carefully establish a sign convention for defining the positive direction for both $\mathbf{v}_A$ and $\mathbf{v}_B$ and then use it *consistently* when writing *both* equations. As noted from the application shown, and indicated symbolically by the arrow in parentheses, we have defined the positive direction to the right when referring to the motions of both A and B. Consequently, if a negative value results from the solution of either $(v_A)_2$ or $(v_B)_2$, it indicates motion is to the left.

Coefficient of Restitution.

From Figs. 15–14a and 15–14e, it is seen that Eq. 15–11 states that e is equal to the ratio of the relative velocity of the particles' separation *just after impact*, $(v_B)_2 - (v_A)_2$, to the relative velocity of the particles' approach *just before impact*, $(v_A)_1 - (v_B)_1$. By measuring these relative velocities experimentally, it has been found that e varies appreciably with impact velocity as well as with the size and shape of the colliding bodies. For these reasons the coefficient of restitution is reliable only when used with data which closely approximate the conditions which were known to exist when measurements of it were made. In general e has a value between zero and one, and one should be aware of the physical meaning of these two limits.

Elastic Impact ($e = 1$).

If the collision between the two particles is *perfectly elastic*, the deformation impulse $\left(\int \mathbf{P}\, dt \right)$ is equal and opposite to the restitution impulse $\left(\int \mathbf{R}\, dt \right)$. Although in reality this can never be achieved, $e = 1$ for an elastic collision.

The mechanics of pool depends upon application of the conservation of momentum and the coefficient of restitution. (© R.C. Hibbeler)

Plastic Impact ($e = 0$).

The impact is said to be *inelastic or plastic* when $e = 0$. In this case there is no restitution impulse $\left(\int \mathbf{R}\, dt = \mathbf{0} \right)$, so that after collision both particles couple or stick *together* and move with a common velocity.

From the above derivation it should be evident that the principle of work and energy cannot be used for the analysis of impact problems since it is not possible to know how the *internal forces* of deformation and restitution vary or displace during the collision. By knowing the particle's velocities before and after collision, however, the energy loss during collision can be calculated on the basis of the difference in the particle's kinetic energy. This energy loss, $\Sigma U_{1-2} = \Sigma T_2 - \Sigma T_1$, occurs because some of the initial kinetic energy of the particle is transformed into thermal energy as well as creating sound and localized deformation of the material when the collision occurs. In particular, if the impact is *perfectly elastic*, no energy is lost in the collision; whereas if the collision is *plastic*, the energy lost during collision is a maximum.

Procedure for Analysis (Central Impact)

In most cases the *final velocities* of two smooth particles are to be determined *just after* they are subjected to direct central impact. Provided the coefficient of restitution, the mass of each particle, and each particle's initial velocity *just before* impact are known, the solution to this problem can be obtained using the following two equations:

- The conservation of momentum applies to the system of particles, $\Sigma mv_1 = \Sigma mv_2$.
- The coefficient of restitution, $e = [(v_B)_2 - (v_A)_2]/[(v_A)_1 - (v_B)_1]$, relates the relative velocities of the particles along the line of impact, just before and just after collision.

When applying these two equations, the sense of an unknown velocity can be assumed. If the solution yields a negative magnitude, the velocity acts in the opposite sense.

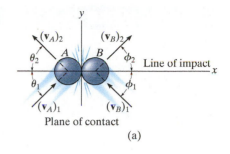

(a)

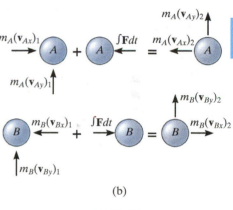

(b)

Fig. 15–15

Oblique Impact. When oblique impact occurs between two smooth particles, the particles move away from each other with velocities having unknown directions as well as unknown magnitudes. Provided the initial velocities are known, then four unknowns are present in the problem. As shown in Fig. 15–15a, these unknowns may be represented either as $(v_A)_2, (v_B)_2, \theta_2$, and ϕ_2, or as the x and y components of the final velocities.

Procedure for Analysis (Oblique Impact)

If the y axis is established within the plane of contact and the x axis along the line of impact, the impulsive forces of deformation and restitution act *only in the x direction*, Fig. 15–15b. By resolving the velocity or momentum vectors into components along the x and y axes, Fig. 15–15b, it is then possible to write four independent scalar equations in order to determine $(v_{Ax})_2, (v_{Ay})_2, (v_{Bx})_2$, and $(v_{By})_2$.

- Momentum of the system is conserved *along the line of impact, x axis*, so that $\Sigma m(v_x)_1 = \Sigma m(v_x)_2$.
- The coefficient of restitution, $e = [(v_{Bx})_2 - (v_{Ax})_2]/[(v_{Ax})_1 - (v_{Bx})_1]$, relates the relative-velocity *components* of the particles *along the line of impact (x axis)*.
- If these two equations are solved simultaneously, we obtain $(v_{Ax})_2$ and $(v_{Bx})_2$.
- Momentum of particle A is conserved along the y axis, perpendicular to the line of impact, since no impulse acts on particle A in this direction. As a result $m_A(v_{Ay})_1 = m_A(v_{Ay})_2$ or $(v_{Ay})_1 = (v_{Ay})_2$
- Momentum of particle B is conserved along the y axis, perpendicular to the line of impact, since no impulse acts on particle B in this direction. Consequently $(v_{By})_1 = (v_{By})_2$.

Application of these four equations is illustrated in Example 15.11.

EXAMPLE | 15.9

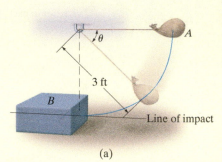

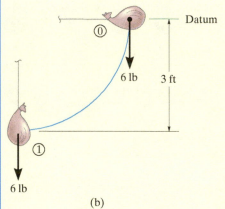

6 lb

(b)

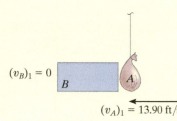

$(v_B)_1 = 0$

$(v_A)_1 = 13.90$ ft/s

Just before impact

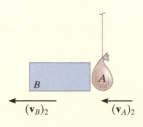

$(v_B)_2$ $(v_A)_2$

Just after impact

(c)

Fig. 15–16

The bag A, having a weight of 6 lb, is released from rest at the position $\theta = 0°$, as shown in Fig. 15–16a. After falling to $\theta = 90°$, it strikes an 18-lb box B. If the coefficient of restitution between the bag and box is $e = 0.5$, determine the velocities of the bag and box just after impact. What is the loss of energy during collision?

SOLUTION

This problem involves central impact. Why? Before analyzing the mechanics of the impact, however, it is first necessary to obtain the velocity of the bag *just before* it strikes the box.

Conservation of Energy. With the datum at $\theta = 0°$, Fig. 15–16b, we have

$$T_0 + V_0 = T_1 + V_1$$

$$0 + 0 = \frac{1}{2}\left(\frac{6 \text{ lb}}{32.2 \text{ ft/s}^2}\right)(v_A)_1^2 - 6 \text{ lb}(3 \text{ ft}); (v_A)_1 = 13.90 \text{ ft/s}$$

Conservation of Momentum. After impact we will assume A and B travel to the left. Applying the conservation of momentum to the system, Fig. 15–16c, we have

$(\pm)$ $m_B(v_B)_1 + m_A(v_A)_1 = m_B(v_B)_2 + m_A(v_A)_2$

$$0 + \left(\frac{6 \text{ lb}}{32.2 \text{ ft/s}^2}\right)(13.90 \text{ ft/s}) = \left(\frac{18 \text{ lb}}{32.2 \text{ ft/s}^2}\right)(v_B)_2 + \left(\frac{6 \text{ lb}}{32.2 \text{ ft/s}^2}\right)(v_A)_2$$

$$(v_A)_2 = 13.90 - 3(v_B)_2 \qquad (1)$$

Coefficient of Restitution. Realizing that for separation to occur after collision $(v_B)_2 > (v_A)_2$, Fig. 15–16c, we have

$(\pm)$ $e = \frac{(v_B)_2 - (v_A)_2}{(v_A)_1 - (v_B)_1};$ $0.5 = \frac{(v_B)_2 - (v_A)_2}{13.90 \text{ ft/s} - 0}$

$$(v_A)_2 = (v_B)_2 - 6.950 \qquad (2)$$

Solving Eqs. 1 and 2 simultaneously yields

$(v_A)_2 = -1.74 \text{ ft/s} = 1.74 \text{ ft/s} \rightarrow$ and $(v_B)_2 = 5.21 \text{ ft/s} \leftarrow$ *Ans.*

Loss of Energy. Applying the principle of work and energy to the bag and box just before and just after collision, we have

$$\Sigma U_{1-2} = T_2 - T_1;$$

$$\Sigma U_{1-2} = \left[\frac{1}{2}\left(\frac{18 \text{ lb}}{32.2 \text{ ft/s}^2}\right)(5.21 \text{ ft/s})^2 + \frac{1}{2}\left(\frac{6 \text{ lb}}{32.2 \text{ ft/s}^2}\right)(1.74 \text{ ft/s})^2\right]$$

$$- \left[\frac{1}{2}\left(\frac{6 \text{ lb}}{32.2 \text{ ft/s}^2}\right)(13.9 \text{ ft/s})^2\right]$$

$$\Sigma U_{1-2} = -10.1 \text{ ft} \cdot \text{lb} \qquad Ans.$$

NOTE: The energy loss occurs due to inelastic deformation during the collision.

EXAMPLE 15.10

Ball B shown in Fig. 15–17a has a mass of 1.5 kg and is suspended from the ceiling by a 1-m-long elastic cord. If the cord is *stretched* downward 0.25 m and the ball is released from rest, determine how far the cord stretches after the ball rebounds from the ceiling. The stiffness of the cord is $k = 800$ N/m, and the coefficient of restitution between the ball and ceiling is $e = 0.8$. The ball makes a central impact with the ceiling.

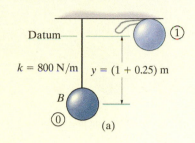

(a)

SOLUTION

First we must obtain the velocity of the ball *just before* it strikes the ceiling using energy methods, then consider the impulse and momentum between the ball and ceiling, and finally again use energy methods to determine the stretch in the cord.

Conservation of Energy. With the datum located as shown in Fig. 15–17a, realizing that initially $y = y_0 = (1 + 0.25)$ m $= 1.25$ m, we have

$$T_0 + V_0 = T_1 + V_1$$

$$\tfrac{1}{2}m(v_B)_0^2 - W_B y_0 + \tfrac{1}{2}ks^2 = \tfrac{1}{2}m(v_B)_1^2 + 0$$

$$0 - 1.5(9.81)\text{N}(1.25 \text{ m}) + \tfrac{1}{2}(800 \text{ N/m})(0.25 \text{ m})^2 = \tfrac{1}{2}(1.5 \text{ kg})(v_B)_1^2$$

$$(v_B)_1 = 2.968 \text{ m/s} \uparrow$$

The interaction of the ball with the ceiling will now be considered using the principles of impact.* Since an unknown portion of the mass of the ceiling is involved in the impact, the conservation of momentum for the ball–ceiling system will not be written. The "velocity" of this portion of ceiling is zero since it (or the earth) are assumed to remain at rest *both* before and after impact.

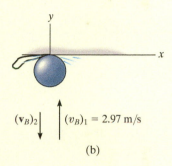

(b)

Coefficient of Restitution. Fig. 15–17b.

$$(+\uparrow) \quad e = \frac{(v_B)_2 - (v_A)_2}{(v_A)_1 - (v_B)_1}; \qquad 0.8 = \frac{(v_B)_2 - 0}{0 - 2.968 \text{ m/s}}$$

$$(v_B)_2 = -2.374 \text{ m/s} = 2.374 \text{ m/s} \downarrow$$

Conservation of Energy. The maximum stretch s_3 in the cord can be determined by again applying the conservation of energy equation to the ball just after collision. Assuming that $y = y_3 = (1 + s_3)$ m, Fig. 15–17c, then

$$T_2 + V_2 = T_3 + V_3$$

$$\tfrac{1}{2}m(v_B)_2^2 + 0 = \tfrac{1}{2}m(v_B)_3^2 - W_B y_3 + \tfrac{1}{2}ks_3^2$$

$$\tfrac{1}{2}(1.5 \text{ kg})(2.37 \text{ m/s})^2 = 0 - 9.81(1.5) \text{ N}(1 \text{ m} + s_3) + \tfrac{1}{2}(800 \text{ N/m})s_3^2$$

$$400s_3^2 - 14.715s_3 - 18.94 = 0$$

Solving this quadratic equation for the positive root yields

$$s_3 = 0.237 \text{ m} = 237 \text{ mm} \qquad Ans.$$

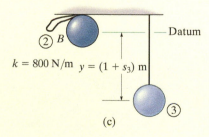

(c)

Fig. 15–17

*The weight of the ball is considered a nonimpulsive force.

EXAMPLE 15.11

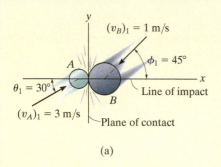

(a)

Two smooth disks A and B, having a mass of 1 kg and 2 kg, respectively, collide with the velocities shown in Fig. 15–18a. If the coefficient of restitution for the disks is $e = 0.75$, determine the x and y components of the final velocity of each disk just after collision.

SOLUTION

This problem involves *oblique impact*. Why? In order to solve it, we have established the x and y axes along the line of impact and the plane of contact, respectively, Fig. 15–18a.

Resolving each of the initial velocities into x and y components, we have

$$(v_{Ax})_1 = 3\cos 30° = 2.598 \text{ m/s} \qquad (v_{Ay})_1 = 3\sin 30° = 1.50 \text{ m/s}$$
$$(v_{Bx})_1 = -1\cos 45° = -0.7071 \text{ m/s} \quad (v_{By})_1 = -1\sin 45° = -0.7071 \text{ m/s}$$

The four unknown velocity components after collision are *assumed to act in the positive directions*, Fig. 15–18b. Since the impact occurs in the x direction (line of impact), the conservation of momentum for *both* disks can be applied in this direction. Why?

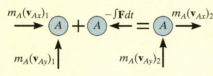

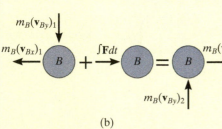

(b)

Conservation of "x" Momentum. In reference to the momentum diagrams, we have

$$(\xrightarrow{+}) \qquad m_A(v_{Ax})_1 + m_B(v_{Bx})_1 = m_A(v_{Ax})_2 + m_B(v_{Bx})_2$$
$$1 \text{ kg}(2.598 \text{ m/s}) + 2 \text{ kg}(-0.707 \text{ m/s}) = 1 \text{ kg}(v_{Ax})_2 + 2 \text{ kg}(v_{Bx})_2$$
$$(v_{Ax})_2 + 2(v_{Bx})_2 = 1.184 \qquad (1)$$

Coefficient of Restitution (x).

$$(\xrightarrow{+}) \qquad e = \frac{(v_{Bx})_2 - (v_{Ax})_2}{(v_{Ax})_1 - (v_{Bx})_1}; \quad 0.75 = \frac{(v_{Bx})_2 - (v_{Ax})_2}{2.598 \text{ m/s} - (-0.7071 \text{ m/s})}$$
$$(v_{Bx})_2 - (v_{Ax})_2 = 2.482 \qquad (2)$$

Solving Eqs. 1 and 2 for $(v_{Ax})_2$ and $(v_{Bx})_2$ yields

$$(v_{Ax})_2 = -1.26 \text{ m/s} = 1.26 \text{ m/s} \leftarrow \quad (v_{Bx})_2 = 1.22 \text{ m/s} \rightarrow \qquad Ans.$$

Conservation of "y" Momentum. The momentum of *each disk* is *conserved* in the y direction (plane of contact), since the disks are smooth and therefore *no* external impulse acts in this direction. From Fig. 15–18b,

$$(+\uparrow) \; m_A(v_{Ay})_1 = m_A(v_{Ay})_2; \quad (v_{Ay})_2 = 1.50 \text{ m/s} \uparrow \qquad Ans.$$
$$(+\uparrow) \; m_B(v_{By})_1 = m_B(v_{By})_2; \quad (v_{By})_2 = -0.707 \text{ m/s} = 0.707 \text{ m/s} \downarrow \; Ans.$$

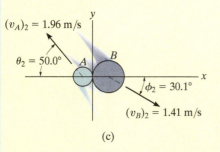

(c)

Fig. 15–18

NOTE: Show that when the velocity components are summed vectorially, one obtains the results shown in Fig. 15–18c.

FUNDAMENTAL PROBLEMS

F15–13. Determine the coefficient of restitution e between ball A and ball B. The velocities of A and B before and after the collision are shown.

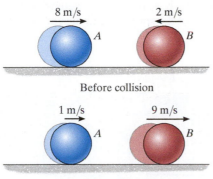

8 m/s   2 m/s

A B

Before collision

1 m/s   9 m/s

A B

After collision

Prob. F15–13

F15–14. The 15-Mg tank car A and 25-Mg freight car B travel toward each other with the velocities shown. If the coefficient of restitution between the bumpers is $e = 0.6$, determine the velocity of each car just after the collision.

5 m/s

A B 7 m/s

Prob. F15–14

F15–15. The 30-lb package A has a speed of 5 ft/s when it enters the smooth ramp. As it slides down the ramp, it strikes the 80-lb package B which is initially at rest. If the coefficient of restitution between A and B is $e = 0.6$, determine the velocity of B just after the impact.

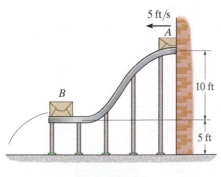

5 ft/s

A

B

10 ft

5 ft

Prob. F15–15

F15–16. The ball strikes the smooth wall with a velocity of $(v_b)_1 = 20$ m/s. If the coefficient of restitution between the ball and the wall is $e = 0.75$, determine the velocity of the ball just after the impact.

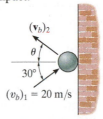

$(v_b)_2$

θ

$30°$

$(v_b)_1 = 20$ m/s

Prob. F15–16

F15–17. Disk A has a mass of 2 kg and slides on the smooth horizontal plane with a velocity of 3 m/s. Disk B has a mass of 11 kg and is initially at rest. If after impact A has a velocity of 1 m/s, parallel to the positive x axis, determine the speed of disk B after impact.

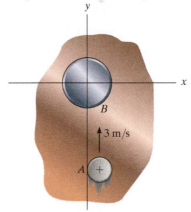

y

B

x

3 m/s

A

Prob. F15–17

F15–18. Two disks A and B each have a mass of 1 kg and the initial velocities shown just before they collide. If the coefficient of restitution is $e = 0.5$, determine their speeds just after impact.

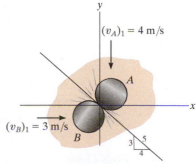

y

$(v_A)_1 = 4$ m/s

A

x

$(v_B)_1 = 3$ m/s

B

3
5
4

Prob. F15–18

PROBLEMS

15–58.   Disk A has a mass of 250 g and is sliding on a *smooth* horizontal surface with an initial velocity $(v_A)_1 = 2$ m/s. It makes a direct collision with disk B, which has a mass of 175 g and is originally at rest. If both disks are of the same size and the collision is perfectly elastic ($e = 1$), determine the velocity of each disk just after collision. Show that the kinetic energy of the disks before and after collision is the same.

15–59.   The 5-Mg truck and 2-Mg car are traveling with the free-rolling velocities shown just before they collide. After the collision, the car moves with a velocity of 15 km/h to the right *relative* to the truck. Determine the coefficient of restitution between the truck and car and the loss of energy due to the collision.

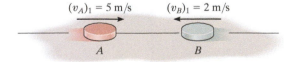

Prob. 15–59

*15–60.   Disk A has a mass of 2 kg and is sliding forward on the *smooth* surface with a velocity $(v_A)_1 = 5$ m/s when it strikes the 4-kg disk B, which is sliding towards A at $(v_B)_1 = 2$ m/s, with direct central impact. If the coefficient of restitution between the disks is $e = 0.4$, compute the velocities of A and B just after collision.

Prob. 15–60

15–61.   The 15-kg block A slides on the surface for which $\mu_k = 0.3$. The block has a velocity $v = 10$ m/s when it is $s = 4$ m from the 10-kg block B. If the unstretched spring has a stiffness $k = 1000$ N/m, determine the maximum compression of the spring due to the collision. Take $e = 0.6$.

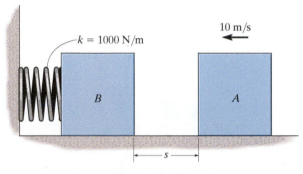

Prob. 15–61

15–62.   The four smooth balls each have the same mass m. If A and B are rolling forward with velocity $\mathbf{v}$ and strike C, explain why after collision C and D each move off with velocity $\mathbf{v}$. Why doesn't D move off with velocity $2\mathbf{v}$? The collision is elastic, $e = 1$. Neglect the size of each ball.

15–63.   The four balls each have the same mass m. If A and B are rolling forward with velocity $\mathbf{v}$ and strike C, determine the velocity of each ball after the first three collisions. Take $e = 0.5$ between each ball.

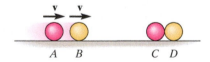

Probs. 15–62/63

*15–64.   Ball A has a mass of 3 kg and is moving with a velocity of 8 m/s when it makes a direct collision with ball B, which has a mass of 2 kg and is moving with a velocity of 4 m/s. If $e = 0.7$, determine the velocity of each ball just after the collision. Neglect the size of the balls.

Prob. 15–64

15–65. A 1-lb ball A is traveling horizontally at 20 ft/s when it strikes a 10-lb block B that is at rest. If the coefficient of restitution between A and B is $e = 0.6$, and the coefficient of kinetic friction between the plane and the block is $\mu_k = 0.4$, determine the time for the block B to stop sliding.

15–66. Block A, having a mass m, is released from rest, falls a distance h and strikes the plate B having a mass $2m$. If the coefficient of restitution between A and B is e, determine the velocity of the plate just after collision. The spring has a stiffness k.

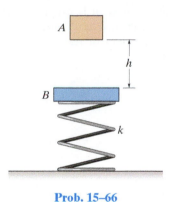

Prob. 15–66

15–67. The three balls each weigh 0.5 lb and have a coefficient of restitution of $e = 0.85$. If ball A is released from rest and strikes ball B and then ball B strikes ball C, determine the velocity of each ball after the second collision has occurred. The balls slide without friction.

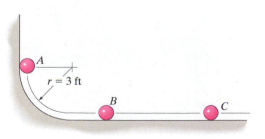

Prob. 15–67

*15–68. A pitching machine throws the 0.5-kg ball toward the wall with an initial velocity $v_A = 10$ m/s as shown. Determine (a) the velocity at which it strikes the wall at B, (b) the velocity at which it rebounds from the wall if $e = 0.5$, and (c) the distance s from the wall to where it strikes the ground at C.

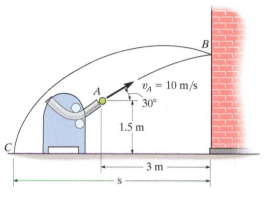

Prob. 15–68

15–69. A 300-g ball is kicked with a velocity of $v_A = 25$ m/s at point A as shown. If the coefficient of restitution between the ball and the field is $e = 0.4$, determine the magnitude and direction θ of the velocity of the rebounding ball at B.

Prob. 15–69

15–70. Two smooth spheres A and B each have a mass m. If A is given a velocity of v_0, while sphere B is at rest, determine the velocity of B just after it strikes the wall. The coefficient of restitution for any collision is e.

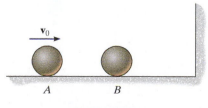

Prob. 15–70

15–71. It was observed that a tennis ball when served horizontally 7.5 ft above the ground strikes the smooth ground at B 20 ft away. Determine the initial velocity $\mathbf{v}_A$ of the ball and the velocity $\mathbf{v}_B$ (and θ) of the ball just after it strikes the court at B. Take $e = 0.7$.

*15–72. The tennis ball is struck with a horizontal velocity $\mathbf{v}_A$, strikes the smooth ground at B, and bounces upward at $\theta = 30°$. Determine the initial velocity $\mathbf{v}_A$, the final velocity $\mathbf{v}_B$, and the coefficient of restitution between the ball and the ground.

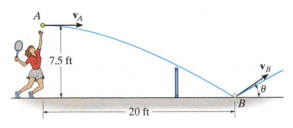

Probs. 15–71/72

15–73. Two smooth disks A and B each have a mass of 0.5 kg. If both disks are moving with the velocities shown when they collide, determine their final velocities just after collision. The coefficient of restitution is $e = 0.75$.

15–74. Two smooth disks A and B each have a mass of 0.5 kg. If both disks are moving with the velocities shown when they collide, determine the coefficient of restitution between the disks if after collision B travels along a line, 30° counterclockwise from the y axis.

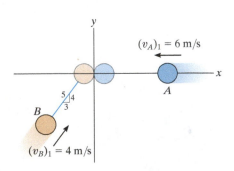

Probs. 15–73/74

15–75. The 0.5-kg ball is fired from the tube at A with a velocity of $v = 6$ m/s. If the coefficient of restitution between the ball and the surface is $e = 0.8$, determine the height h after it bounces off the surface.

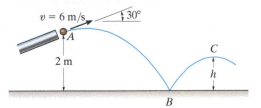

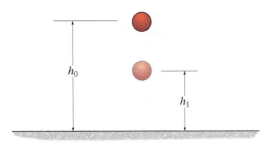

Prob. 15–75

*15–76. A ball of mass m is dropped vertically from a height h_0 above the ground. If it rebounds to a height of h_1, determine the coefficient of restitution between the ball and the ground.

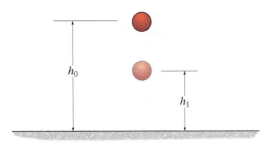

Prob. 15–76

15–77. The cue ball A is given an initial velocity $(v_A)_1 = 5$ m/s. If it makes a direct collision with ball B ($e = 0.8$), determine the velocity of B and the angle θ just after it rebounds from the cushion at C ($e' = 0.6$). Each ball has a mass of 0.4 kg. Neglect their size.

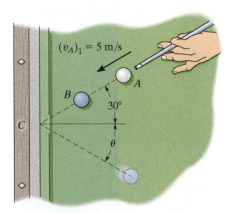

Prob. 15–77

15–78. Using a slingshot, the boy fires the 0.2-lb marble at the concrete wall, striking it at B. If the coefficient of restitution between the marble and the wall is $e = 0.5$, determine the speed of the marble after it rebounds from the wall.

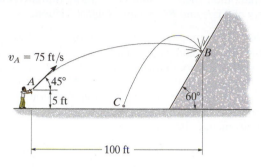

Prob. 15–78

15–79. The two disks A and B have a mass of 3 kg and 5 kg, respectively. If they collide with the initial velocities shown, determine their velocities just after impact. The coefficient of restitution is $e = 0.65$.

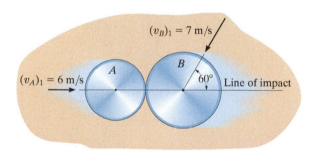

Prob. 15–79

*15–80. A ball of negligible size and mass m is given a velocity of $\mathbf{v}_0$ on the center of the cart which has a mass M and is originally at rest. If the coefficient of restitution between the ball and walls A and B is e, determine the velocity of the ball and the cart just after the ball strikes A. Also, determine the total time needed for the ball to strike A, rebound, then strike B, and rebound and then return to the center of the cart. Neglect friction.

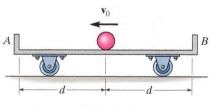

Prob. 15–80

15–81. The girl throws the 0.5-kg ball toward the wall with an initial velocity $v_A = 10$ m/s. Determine (a) the velocity at which it strikes the wall at B, (b) the velocity at which it rebounds from the wall if the coefficient of restitution $e = 0.5$, and (c) the distance s from the wall to where it strikes the ground at C.

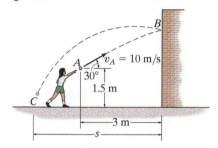

Prob. 15–81

15–82. The 20-lb box slides on the surface for which $\mu_k = 0.3$. The box has a velocity $v = 15$ ft/s when it is 2 ft from the plate. If it strikes the smooth plate, which has a weight of 10 lb and is held in position by an unstretched spring of stiffness $k = 400$ lb/ft, determine the maximum compression imparted to the spring. Take $e = 0.8$ between the box and the plate. Assume that the plate slides smoothly.

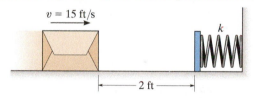

Prob. 15–82

15–83. The 10-lb collar B is at rest, and when it is in the position shown the spring is unstretched. If another 1-lb collar A strikes it so that B slides 4 ft on the smooth rod before momentarily stopping, determine the velocity of A just after impact, and the average force exerted between A and B during the impact if the impact occurs in 0.002 s. The coefficient of restitution between A and B is $e = 0.5$.

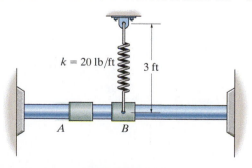

Prob. 15–83

*15–84. A ball is thrown onto a rough floor at an angle θ. If it rebounds at an angle ϕ and the coefficient of kinetic friction is μ, determine the coefficient of restitution e. Neglect the size of the ball. *Hint:* Show that during impact, the average impulses in the x and y directions are related by $I_x = \mu I_y$. Since the time of impact is the same, $F_x \Delta t = \mu F_y \Delta t$ or $F_x = \mu F_y$.

15–85. A ball is thrown onto a rough floor at an angle of $\theta = 45°$. If it rebounds at the same angle $\phi = 45°$, determine the coefficient of kinetic friction between the floor and the ball. The coefficient of restitution is $e = 0.6$. *Hint:* Show that during impact, the average impulses in the x and y directions are related by $I_x = \mu I_y$. Since the time of impact is the same, $F_x \Delta t = \mu F_y \Delta t$ or $F_x = \mu F_y$.

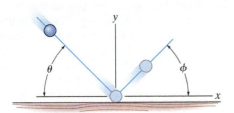

Probs. 15–84/85

15–86. Two smooth billiard balls A and B each have a mass of 200 g. If A strikes B with a velocity $(v_A)_1 = 1.5$ m/s as shown, determine their final velocities just after collision. Ball B is originally at rest and the coefficient of restitution is $e = 0.85$. Neglect the size of each ball.

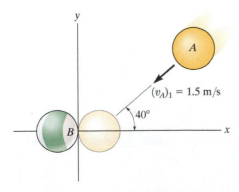

Prob. 15–86

15–87. The "stone" A used in the sport of curling slides over the ice track and strikes another "stone" B as shown. If each "stone" is smooth and has a weight of 47 lb, and the coefficient of restitution between the "stones" is $e = 0.8$, determine their speeds just after collision. Initially A has a velocity of 8 ft/s and B is at rest. Neglect friction.

*15–88. The "stone" A used in the sport of curling slides over the ice track and strikes another "stone" B as shown. If each "stone" is smooth and has a weight of 47 lb, and the coefficient of restitution between the "stone" is $e = 0.8$, determine the time required just after collision for B to slide off the runway. This requires the horizontal component of displacement to be 3 ft.

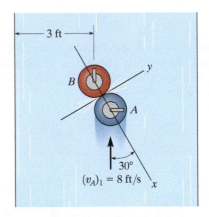

Probs. 15–87/88

15–89. Two smooth disks A and B have the initial velocities shown just before they collide. If they have masses $m_A = 4$ kg and $m_B = 2$ kg, determine their speeds just after impact. The coefficient of restitution is $e = 0.8$.

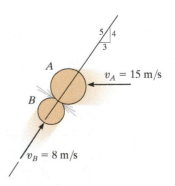

Prob. 15–89

15–90. Before a cranberry can make it to your dinner plate, it must pass a bouncing test which rates its quality. If cranberries having an $e \geq 0.8$ are to be accepted, determine the dimensions d and h for the barrier so that when a cranberry falls from rest at A it strikes the incline at B and bounces over the barrier at C.

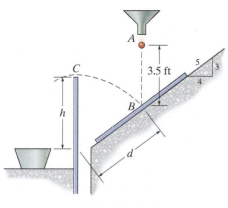

Prob. 15–90

*15–92. The two billiard balls A and B are originally in contact with one another when a third ball C strikes each of them at the same time as shown. If ball C remains at rest after the collision, determine the coefficient of restitution. All the balls have the same mass. Neglect the size of each ball.

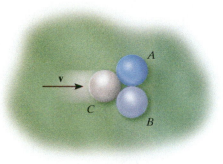

Prob. 15–92

15–91. The 200-g billiard ball is moving with a speed of 2.5 m/s when it strikes the side of the pool table at A. If the coefficient of restitution between the ball and the side of the table is $e = 0.6$, determine the speed of the ball just after striking the table twice, i.e., at A, then at B. Neglect the size of the ball.

15–93. Disks A and B have a mass of 15 kg and 10 kg, respectively. If they are sliding on a smooth horizontal plane with the velocities shown, determine their speeds just after impact. The coefficient of restitution between them is $e = 0.8$.

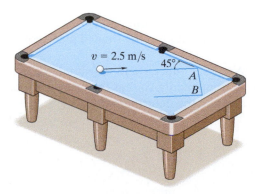

Prob. 15–91

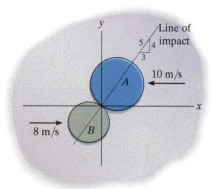

Prob. 15–93

15.5   Angular Momentum

The *angular momentum* of a particle about point O is defined as the "moment" of the particle's linear momentum about O. Since this concept is analogous to finding the moment of a force about a point, the angular momentum, $\mathbf{H}_O$, is sometimes referred to as the *moment of momentum*.

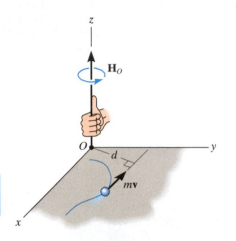

Fig. 15–19

Scalar Formulation.   If a particle moves along a curve lying in the x–y plane, Fig. 15–19, the angular momentum at any instant can be determined about point O (actually the z axis) by using a scalar formulation. The *magnitude* of $\mathbf{H}_O$ is

$$(H_O)_z = (d)(mv) \tag{15–12}$$

Here d is the moment arm or perpendicular distance from O to the line of action of $m\mathbf{v}$. Common units for $(H_O)_z$ are $\text{kg} \cdot \text{m}^2/\text{s}$ or $\text{slug} \cdot \text{ft}^2/\text{s}$. The *direction* of $\mathbf{H}_O$ is defined by the right-hand rule. As shown, the curl of the fingers of the right hand indicates the sense of rotation of $m\mathbf{v}$ about O, so that in this case the thumb (or $\mathbf{H}_O$) is directed perpendicular to the x–y plane along the $+z$ axis.

Vector Formulation.   If the particle moves along a space curve, Fig. 15–20, the vector cross product can be used to determine the *angular momentum* about O. In this case

$$\mathbf{H}_O = \mathbf{r} \times m\mathbf{v} \tag{15–13}$$

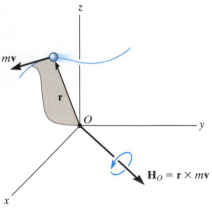

Fig. 15–20

Here $\mathbf{r}$ denotes a position vector drawn from point O to the particle. As shown in the figure, $\mathbf{H}_O$ is *perpendicular* to the shaded plane containing $\mathbf{r}$ and $m\mathbf{v}$.

In order to evaluate the cross product, $\mathbf{r}$ and $m\mathbf{v}$ should be expressed in terms of their Cartesian components, so that the angular momentum can be determined by evaluating the determinant:

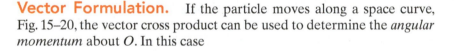

$$\mathbf{H}_O = \begin{vmatrix} \mathbf{i} & \mathbf{j} & \mathbf{k} \\ r_x & r_y & r_z \\ mv_x & mv_y & mv_z \end{vmatrix} \tag{15–14}$$

15.6   Relation Between Moment of a Force and Angular Momentum

The moments about point O of all the forces acting on the particle in Fig. 15–21a can be related to the particle's angular momentum by applying the equation of motion. If the mass of the particle is constant, we may write

$$\Sigma \mathbf{F} = m\dot{\mathbf{v}}$$

The moments of the forces about point O can be obtained by performing a cross-product multiplication of each side of this equation by the position vector $\mathbf{r}$, which is measured from the x, y, z inertial frame of reference. We have

$$\Sigma \mathbf{M}_O = \mathbf{r} \times \Sigma \mathbf{F} = \mathbf{r} \times m\dot{\mathbf{v}}$$

From Appendix B, the derivative of $\mathbf{r} \times m\mathbf{v}$ can be written as

$$\dot{\mathbf{H}}_O = \frac{d}{dt}(\mathbf{r} \times m\mathbf{v}) = \dot{\mathbf{r}} \times m\mathbf{v} + \mathbf{r} \times m\dot{\mathbf{v}}$$

The first term on the right side, $\dot{\mathbf{r}} \times m\mathbf{v} = m(\dot{\mathbf{r}} \times \dot{\mathbf{r}}) = \mathbf{0}$, since the cross product of a vector with itself is zero. Hence, the above equation becomes

$$\Sigma \mathbf{M}_O = \dot{\mathbf{H}}_O \qquad (15\text{–}15)$$

which states that *the resultant moment about point O of all the forces acting on the particle is equal to the time rate of change of the particle's angular momentum about point O.* This result is similar to Eq. 15–1, i.e.,

$$\Sigma \mathbf{F} = \dot{\mathbf{L}} \qquad (15\text{–}16)$$

Here $\mathbf{L} = m\mathbf{v}$, so that *the resultant force acting on the particle is equal to the time rate of change of the particle's linear momentum.*

From the derivations, it is seen that Eqs. 15–15 and 15–16 are actually another way of stating Newton's second law of motion. In other sections of this book it will be shown that these equations have many practical applications when extended and applied to problems involving either a system of particles or a rigid body.

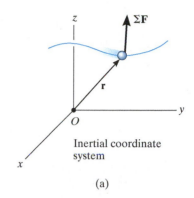

Inertial coordinate system

(a)

Fig. 15–21

15

System of Particles.   An equation having the same form as Eq. 15–15 may be derived for the system of particles shown in Fig. 15–21b. The forces acting on the arbitrary *i*th particle of the system consist of a resultant *external force* $\mathbf{F}_i$ and a resultant *internal force* $\mathbf{f}_i$. Expressing the moments of these forces about point O, using the form of Eq. 15–15, we have

$$(\mathbf{r}_i \times \mathbf{F}_i) + (\mathbf{r}_i \times \mathbf{f}_i) = (\dot{\mathbf{H}}_i)_O$$

Here $(\dot{\mathbf{H}}_i)_O$ is the time rate of change in the angular momentum of the *i*th particle about O. Similar equations can be written for each of the other particles of the system. When the results are summed vectorially, the result is

$$\Sigma(\mathbf{r}_i \times \mathbf{F}_i) + \Sigma(\mathbf{r}_i \times \mathbf{f}_i) = \Sigma(\dot{\mathbf{H}}_i)_O$$

The second term is zero since the internal forces occur in equal but opposite collinear pairs, and hence the moment of each pair about point O is zero. Dropping the index notation, the above equation can be written in a simplified form as

$$\Sigma\mathbf{M}_O = \dot{\mathbf{H}}_O \qquad (15–17)$$

which states that *the sum of the moments about point O of all the external forces acting on a system of particles is equal to the time rate of change of the total angular momentum of the system about point O.* Although O has been chosen here as the origin of coordinates, it actually can represent any *fixed point* in the inertial frame of reference.

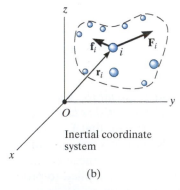

Inertial coordinate system

(b)

Fig. 15–21 (cont.)

EXAMPLE | 15.12

The box shown in Fig. 15–22a has a mass m and travels down the smooth circular ramp such that when it is at the angle θ it has a speed v. Determine its angular momentum about point O at this instant and the rate of increase in its speed, i.e., a_t.

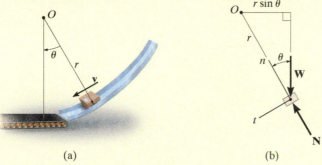

(a)   (b)

Fig. 15–22

SOLUTION

Since **v** is tangent to the path, applying Eq. 15–12 the angular momentum is

$$H_O = r m v \quveright \qquad \textit{Ans.}$$

The rate of increase in its speed (dv/dt) can be found by applying Eq. 15–15. From the free-body diagram of the box, Fig. 15–22b, it can be seen that only the weight $W = mg$ contributes a moment about point O. We have

$$\curvearrowleft + \Sigma M_O = \dot{H}_O; \qquad mg(r \sin \theta) = \frac{d}{dt}(r m v)$$

Since r and m are constant,

$$mgr \sin \theta = r m \frac{dv}{dt}$$

$$\frac{dv}{dt} = g \sin \theta \qquad \textit{Ans.}$$

NOTE: This same result can, of course, be obtained from the equation of motion applied in the tangential direction, Fig. 15–22b, i.e.,

$$+\swarrow \Sigma F_t = m a_t; \qquad mg \sin \theta = m\left(\frac{dv}{dt}\right)$$

$$\frac{dv}{dt} = g \sin \theta \qquad \textit{Ans.}$$

15.7   Principle of Angular Impulse and Momentum

Principle of Angular Impulse and Momentum.   If Eq. 15–15 is rewritten in the form $\Sigma \mathbf{M}_O \, dt = d\mathbf{H}_O$ and integrated, assuming that at time $t = t_1$, $\mathbf{H}_O = (\mathbf{H}_O)_1$ and at time $t = t_2$, $\mathbf{H}_O = (\mathbf{H}_O)_2$, we have

$$\Sigma \int_{t_1}^{t_2} \mathbf{M}_O \, dt = (\mathbf{H}_O)_2 - (\mathbf{H}_O)_1$$

or

$$(\mathbf{H}_O)_1 + \Sigma \int_{t_1}^{t_2} \mathbf{M}_O \, dt = (\mathbf{H}_O)_2 \qquad (15\text{–}18)$$

This equation is referred to as the *principle of angular impulse and momentum*. The initial and final angular momenta $(\mathbf{H}_O)_1$ and $(\mathbf{H}_O)_2$ are defined as the moment of the linear momentum of the particle $(\mathbf{H}_O = \mathbf{r} \times m\mathbf{v})$ at the instants t_1 and t_2, respectively. The second term on the left side, $\Sigma \int \mathbf{M}_O \, dt$, is called the *angular impulse*. It is determined by integrating, with respect to time, the moments of all the forces acting on the particle over the time period t_1 to t_2. Since the moment of a force about point O is $\mathbf{M}_O = \mathbf{r} \times \mathbf{F}$, the angular impulse may be expressed in vector form as

$$\text{angular impulse} = \int_{t_1}^{t_2} \mathbf{M}_O \, dt = \int_{t_1}^{t_2} (\mathbf{r} \times \mathbf{F}) \, dt \qquad (15\text{–}19)$$

Here $\mathbf{r}$ is a position vector which extends from point O to any point on the line of action of $\mathbf{F}$.

In a similar manner, using Eq. 15–18, the principle of angular impulse and momentum for a system of particles may be written as

$$\Sigma(\mathbf{H}_O)_1 + \Sigma \int_{t_1}^{t_2} \mathbf{M}_O \, dt = \Sigma(\mathbf{H}_O)_2 \qquad (15\text{–}20)$$

Here the first and third terms represent the angular momenta of all the particles $[\Sigma \mathbf{H}_O = \Sigma(\mathbf{r}_i \times m\mathbf{v}_i)]$ at the instants t_1 and t_2. The second term is the sum of the angular impulses given to all the particles from t_1 to t_2. Recall that these impulses are created only by the moments of the external forces acting on the system where, for the ith particle, $\mathbf{M}_O = \mathbf{r}_i \times \mathbf{F}_i$.

Vector Formulation.   Using impulse and momentum principles, it is therefore possible to write two equations which define the particle's motion, namely, Eqs. 15–3 and Eqs. 15–18, restated as

$$
\begin{aligned}
m\mathbf{v}_1 + \Sigma \int_{t_1}^{t_2} \mathbf{F}\, dt &= m\mathbf{v}_2 \\[2mm]
(\mathbf{H}_O)_1 + \Sigma \int_{t_1}^{t_2} \mathbf{M}_O\, dt &= (\mathbf{H}_O)_2
\end{aligned}
\tag{15–21}
$$

Scalar Formulation.   In general, the above equations can be expressed in x, y, z component form. If the particle is confined to move in the x–y plane, then three scalar equations can be written to express the motion, namely,

$$
\begin{aligned}
m(v_x)_1 + \Sigma \int_{t_1}^{t_2} F_x\, dt &= m(v_x)_2 \\[2mm]
m(v_y)_1 + \Sigma \int_{t_1}^{t_2} F_y\, dt &= m(v_y)_2 \\[2mm]
(H_O)_1 + \Sigma \int_{t_1}^{t_2} M_O\, dt &= (H_O)_2
\end{aligned}
\tag{15–22}
$$

The first two of these equations represent the principle of linear impulse and momentum in the x and y directions, which has been discussed in Sec. 15–1, and the third equation represents the principle of angular impulse and momentum about the z axis.

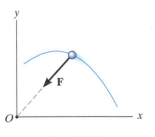

Fig. 15–23

Conservation of Angular Momentum.

When the angular impulses acting on a particle are all zero during the time t_1 to t_2, Eq. 15–18 reduces to the following simplified form:

$$(\mathbf{H}_O)_1 = (\mathbf{H}_O)_2 \qquad (15\text{–}23)$$

This equation is known as the *conservation of angular momentum*. It states that from t_1 to t_2 the particle's angular momentum remains constant. Obviously, if no external impulse is applied to the particle, both linear and angular momentum will be conserved. In some cases, however, the particle's angular momentum will be conserved and linear momentum may not. An example of this occurs when the particle is subjected *only* to a *central force* (see Sec. 13.7). As shown in Fig. 15–23, the impulsive central force $\mathbf{F}$ is always directed toward point O as the particle moves along the path. Hence, the angular impulse (moment) created by $\mathbf{F}$ about the z axis is always zero, and therefore angular momentum of the particle is conserved about this axis.

From Eq. 15–20, we can also write the conservation of angular momentum for a system of particles as

$$\Sigma(\mathbf{H}_O)_1 = \Sigma(\mathbf{H}_O)_2 \qquad (15\text{–}24)$$

In this case the summation must include the angular momenta of all particles in the system.

(© Petra Hilke/Fotolia)

Provided air resistance is neglected, the passengers on this amusement-park ride are subjected to a conservation of angular momentum about the z axis of rotation. As shown on the free-body diagram, the line of action of the normal force $\mathbf{N}$ of the seat on the passenger passes through this axis, and the passenger's weight $\mathbf{W}$ is parallel to it. Thus, no angular impulse acts around the z axis.

Procedure for Analysis

When applying the principles of angular impulse and momentum, or the conservation of angular momentum, it is suggested that the following procedure be used.

Free-Body Diagram.

- Draw the particle's free-body diagram in order to determine any axis about which angular momentum may be conserved. For this to occur, the moments of all the forces (or impulses) must either be parallel or pass through the axis so as to create zero moment throughout the time period t_1 to t_2.

- The direction and sense of the particle's initial and final velocities should also be established.

- An alternative procedure would be to draw the impulse and momentum diagrams for the particle.

Momentum Equations.

- Apply the principle of angular impulse and momentum, $(\mathbf{H}_O)_1 + \Sigma \int_{t_1}^{t_2} \mathbf{M}_O dt = (\mathbf{H}_O)_2$, or if appropriate, the conservation of angular momentum, $(\mathbf{H}_O)_1 = (\mathbf{H}_O)_2$.

EXAMPLE | 15.13

The 1.5-Mg car travels along the circular road as shown in Fig. 15–24a. If the traction force of the wheels on the road is $F = (150t^2)$ N, where t is in seconds, determine the speed of the car when $t = 5$ s. The car initially travels with a speed of 5 m/s. Neglect the size of the car.

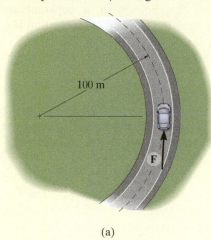

100 m

(a)

Free-Body Diagram.   The free-body diagram of the car is shown in Fig. 15–24b. If we apply the principle of angular impulse and momentum about the z axis, then the angular impulse created by the weight, normal force, and radial frictional force will be eliminated since they act parallel to the axis or pass through it.

Principle of Angular Impulse and Momentum.

$$(H_z)_1 + \Sigma \int_{t_1}^{t_2} M_z \, dt = (H_z)_2$$

$$r\, m_c(v_c)_1 + \int_{t_1}^{t_2} r\, F \, dt = r\, m_c(v_c)_2$$

$$(100 \text{ m})(1500 \text{ kg})(5 \text{ m/s}) + \int_0^{5\text{ s}} (100 \text{ m})[(150t^2) \text{ N}] \, dt$$

$$= (100 \text{ m})(1500 \text{ kg})(v_c)_2$$

$$750(10^3) + 5000t^3 \Big|_0^{5\text{ s}} = 150(10^3)(v_c)_2$$

$$(v_c)_2 = 9.17 \text{ m/s} \qquad \textit{Ans.}$$

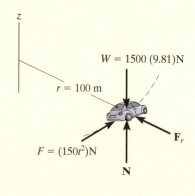

z

$W = 1500\,(9.81)$N

$r = 100$ m

$F = (150t^2)$N

$\mathbf{F}_r$

N

(b)

Fig. 15–24

EXAMPLE 15.14

(a)

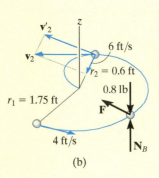

(b)

Fig. 15–25

The 0.8-lb ball B, shown in Fig. 15–25a, is attached to a cord which passes through a hole at A in a smooth table. When the ball is $r_1 = 1.75$ ft from the hole, it is rotating around in a circle such that its speed is $v_1 = 4$ ft/s. By applying the force F the cord is pulled downward through the hole with a constant speed $v_c = 6$ ft/s. Determine (a) the speed of the ball at the instant it is $r_2 = 0.6$ ft from the hole, and (b) the amount of work done by F in shortening the radial distance from r_1 to r_2. Neglect the size of the ball.

SOLUTION

Part (a) Free-Body Diagram. As the ball moves from r_1 to r_2, Fig. 15–25b, the cord force F on the ball always passes through the z axis, and the weight and N_B are parallel to it. Hence the moments, or angular impulses created by these forces, are all *zero* about this axis. Therefore, angular momentum is conserved about the z axis.

Conservation of Angular Momentum. The ball's velocity v_2 is resolved into two components. The radial component, 6 ft/s, is known; however, it produces zero angular momentum about the z axis. Thus,

$$H_1 = H_2$$

$$r_1 m_B v_1 = r_2 m_B v_2'$$

$$1.75 \text{ ft}\left(\frac{0.8 \text{ lb}}{32.2 \text{ ft/s}^2}\right) 4 \text{ ft/s} = 0.6 \text{ ft}\left(\frac{0.8 \text{ lb}}{32.2 \text{ ft/s}^2}\right) v_2'$$

$$v_2' = 11.67 \text{ ft/s}$$

The speed of the ball is thus

$$v_2 = \sqrt{(11.67 \text{ ft/s})^2 + (6 \text{ ft/s})^2}$$

$$= 13.1 \text{ ft/s}$$

Part (b). The only force that does work on the ball is F. (The normal force and weight do not move vertically.) The initial and final kinetic energies of the ball can be determined so that from the principle of work and energy we have

$$T_1 + \Sigma U_{1-2} = T_2$$

$$\frac{1}{2}\left(\frac{0.8 \text{ lb}}{32.2 \text{ ft/s}^2}\right)(4 \text{ ft/s})^2 + U_F = \frac{1}{2}\left(\frac{0.8 \text{ lb}}{32.2 \text{ ft/s}^2}\right)(13.1 \text{ ft/s})^2$$

$$U_F = 1.94 \text{ ft} \cdot \text{lb} \qquad Ans.$$

NOTE: The force F is not constant because the normal component of acceleration, $a_n = v^2/r$, changes as r changes.

EXAMPLE | 15.15

The 2-kg disk shown in Fig. 15–26a rests on a smooth horizontal surface and is attached to an elastic cord that has a stiffness $k_c = 20$ N/m and is initially unstretched. If the disk is given a velocity $(v_D)_1 = 1.5$ m/s, perpendicular to the cord, determine the rate at which the cord is being stretched and the speed of the disk at the instant the cord is stretched 0.2 m.

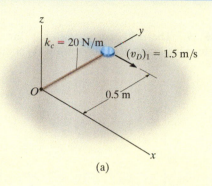

(a)

SOLUTION

Free-Body Diagram. After the disk has been launched, it slides along the path shown in Fig. 15–26b. By inspection, angular momentum about point O (or the z axis) is *conserved*, since none of the forces produce an angular impulse about this axis. Also, when the distance is 0.7 m, only the transverse component $(v_D')_2$ produces angular momentum of the disk about O.

Conservation of Angular Momentum. The component $(v_D')_2$ can be obtained by applying the conservation of angular momentum about O (the z axis).

$$(H_O)_1 = (H_O)_2$$

$$r_1 m_D (v_D)_1 = r_2 m_D (v_D')_2$$

$$0.5 \text{ m} (2 \text{ kg})(1.5 \text{ m/s}) = 0.7 \text{ m}(2 \text{ kg})(v_D')_2$$

$$(v_D')_2 = 1.071 \text{ m/s}$$

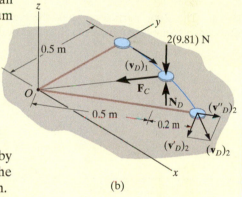

(b)

Fig. 15–26

Conservation of Energy. The speed of the disk can be obtained by applying the conservation of energy equation at the point where the disk was launched and at the point where the cord is stretched 0.2 m.

$$T_1 + V_1 = T_2 + V_2$$

$$\tfrac{1}{2}m_D(v_D)_1^2 + \tfrac{1}{2}kx_1^2 = \tfrac{1}{2}m_D(v_D)_2^2 + \tfrac{1}{2}kx_2^2$$

$$\tfrac{1}{2}(2 \text{ kg})(1.5 \text{ m/s})^2 + 0 = \tfrac{1}{2}(2 \text{ kg})(v_D)_2^2 + \tfrac{1}{2}(20 \text{ N/m})(0.2 \text{ m})^2$$

$$(v_D)_2 = 1.360 \text{ m/s} = 1.36 \text{ m/s} \qquad \textit{Ans.}$$

Having determined $(v_D)_2$ and its component $(v_D')_2$, the rate of stretch of the cord, or radial component, $(v_D'')_2$ is determined from the Pythagorean theorem,

$$(v_D'')_2 = \sqrt{(v_D)_2^2 - (v_D')_2^2}$$

$$= \sqrt{(1.360 \text{ m/s})^2 - (1.071 \text{ m/s})^2}$$

$$= 0.838 \text{ m/s} \qquad \textit{Ans.}$$

15

FUNDAMENTAL PROBLEMS

F15–19. The 2-kg particle *A* has the velocity shown. Determine its angular momentum $\mathbf{H}_O$ about point *O*.

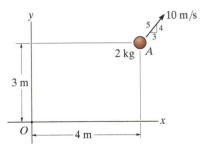

Prob. F15–19

F15–20. The 2-kg particle *A* has the velocity shown. Determine its angular momentum $\mathbf{H}_P$ about point *P*.

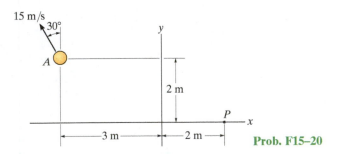

Prob. F15–20

F15–21. Initially the 5-kg block is moving with a constant speed of 2 m/s around the circular path centered at *O* on the smooth horizontal plane. If a constant tangential force $F = 5$ N is applied to the block, determine its speed when $t = 3$ s. Neglect the size of the block.

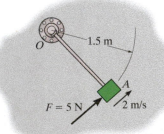

Prob. F15–21

F15–22. The 5-kg block is moving around the circular path centered at *O* on the smooth horizontal plane when it is subjected to the force $F = (10t)$ N, where *t* is in seconds. If the block starts from rest, determine its speed when $t = 4$ s. Neglect the size of the block. The force maintains the same constant angle tangent to the path.

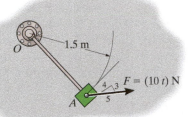

Prob. F15–22

F15–23. The 2-kg sphere is attached to the light rigid rod, which rotates in the *horizontal plane* centered at *O*. If the system is subjected to a couple moment $M = (0.9t^2)$ N·m, where *t* is in seconds, determine the speed of the sphere at the instant $t = 5$ s starting from rest.

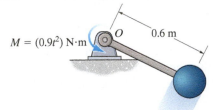

Prob. F15–23

F15–24. Two identical 10-kg spheres are attached to the light rigid rod, which rotates in the horizontal plane centered at pin *O*. If the spheres are subjected to tangential forces of $P = 10$ N, and the rod is subjected to a couple moment $M = (8t)$ N·m, where *t* is in seconds, determine the speed of the spheres at the instant $t = 4$ s. The system starts from rest. Neglect the size of the spheres.

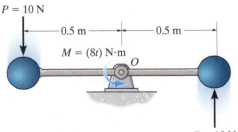

Prob. F15–24

PROBLEMS

15–94. Determine the angular momentum $\mathbf{H}_O$ of the 6-lb particle about point O.

15–95. Determine the angular momentum $\mathbf{H}_p$ of the 6-lb particle about point P.

15–98. Determine the angular momentum $\mathbf{H}_O$ of the 3-kg particle about point O.

15–99. Determine the angular momentum $\mathbf{H}_P$ of the 3-kg particle about point P.

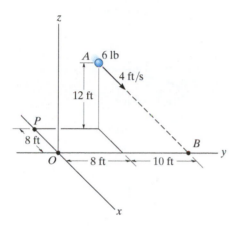

Probs. 15–94/95

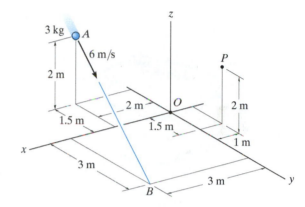

Probs. 15–98/99

*15–96. Determine the angular momentum $\mathbf{H}_o$ of each of the two particles about point O.

15–97. Determine the angular momentum $\mathbf{H}_p$ of each of the two particles about point P.

*15–100. Each ball has a negligible size and a mass of 10 kg and is attached to the end of a rod whose mass may be neglected. If the rod is subjected to a torque $M = (t^2 + 2) \,\text{N} \cdot \text{m}$, where t is in seconds, determine the speed of each ball when $t = 3$ s. Each ball has a speed $v = 2$ m/s when $t = 0$.

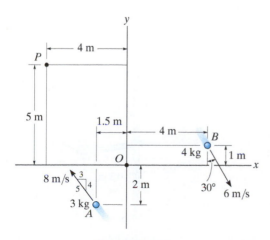

Probs. 15–96/97

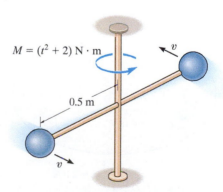

Prob. 15–100

15–101. The 800-lb roller-coaster car starts from rest on the track having the shape of a cylindrical helix. If the helix descends 8 ft for every one revolution, determine the speed of the car when $t = 4$ s. Also, how far has the car descended in this time? Neglect friction and the size of the car.

15–102. The 800-lb roller-coaster car starts from rest on the track having the shape of a cylindrical helix. If the helix descends 8 ft for every one revolution, determine the time required for the car to attain a speed of 60 ft/s. Neglect friction and the size of the car.

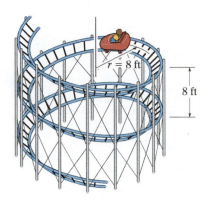

Probs. 15–101/102

15–103. A 4-lb ball B is traveling around in a circle of radius $r_1 = 3$ ft with a speed $(v_B)_1 = 6$ ft/s. If the attached cord is pulled down through the hole with a constant speed $v_r = 2$ ft/s, determine the ball's speed at the instant $r_2 = 2$ ft. How much work has to be done to pull down the cord? Neglect friction and the size of the ball.

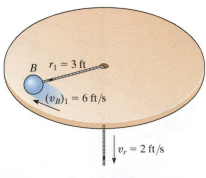

Prob. 15–103

*15–104. A 4-lb ball B is traveling around in a circle of radius $r_1 = 3$ ft with a speed $(v_B)_1 = 6$ ft/s. If the attached cord is pulled down through the hole with a constant speed $v_r = 2$ ft/s, determine how much time is required for the ball to reach a speed of 12 ft/s. How far r_2 is the ball from the hole when this occurs? Neglect friction and the size of the ball.

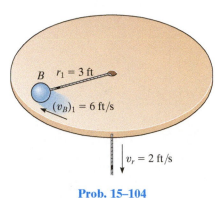

Prob. 15–104

15–105. The two blocks A and B each have a mass of 400 g. The blocks are fixed to the horizontal rods, and their initial velocity along the circular path is 2 m/s. If a couple moment of $M = (0.6)$ N · m is applied about CD of the frame, determine the speed of the blocks when $t = 3$ s. The mass of the frame is negligible, and it is free to rotate about CD. Neglect the size of the blocks.

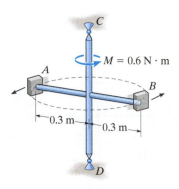

Prob. 15–105

15–106. A small particle having a mass m is placed inside the semicircular tube. The particle is placed at the position shown and released. Apply the principle of angular momentum about point O ($\Sigma M_O = \dot{H}_O$), and show that the motion of the particle is governed by the differential equation $\ddot{\theta} + (g/R) \sin \theta = 0$.

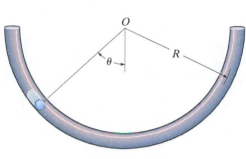

Prob. 15–106

*15–108. When the 2-kg bob is given a horizontal speed of 1.5 m/s, it begins to move around the horizontal circular path A. If the force $\mathbf{F}$ on the cord is increased, the bob rises and then moves around the horizontal circular path B. Determine the speed of the bob around path B. Also, find the work done by force $\mathbf{F}$.

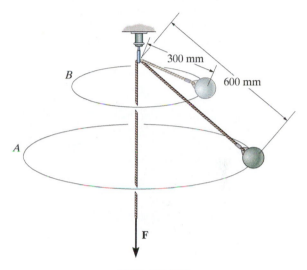

Prob. 15–108

15–107. If the rod of negligible mass is subjected to a couple moment of $M = (30t^2)$ N·m, and the engine of the car supplies a traction force of $F = (15t)$ N to the wheels, where t is in seconds, determine the speed of the car at the instant $t = 5$ s. The car starts from rest. The total mass of the car and rider is 150 kg. Neglect the size of the car.

15–109. The elastic cord has an unstretched length $l_0 = 1.5$ ft and a stiffness $k = 12$ lb/ft. It is attached to a fixed point at A and a block at B, which has a weight of 2 lb. If the block is released from rest from the position shown, determine its speed when it reaches point C after it slides along the smooth guide. After leaving the guide, it is launched onto the smooth *horizontal* plane. Determine if the cord becomes unstretched. Also, calculate the angular momentum of the block about point A, at any instant after it passes point C.

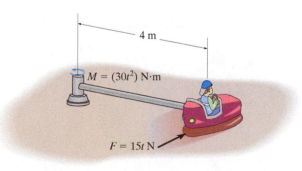

Prob. 15–107

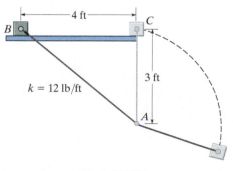

Prob. 15–109

15–110. The amusement park ride consists of a 200-kg car and passenger that are traveling at 3 m/s along a circular path having a radius of 8 m. If at $t = 0$, the cable OA is pulled in toward O at 0.5 m/s, determine the speed of the car when $t = 4$ s. Also, determine the work done to pull in the cable.

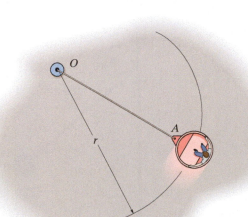

Prob. 15–110

15–111. A box having a weight of 8 lb is moving around in a circle of radius $r_A = 2$ ft with a speed of $(v_A)_1 = 5$ ft/s while connected to the end of a rope. If the rope is pulled inward with a constant speed of $v_r = 4$ ft/s, determine the speed of the box at the instant $r_B = 1$ ft. How much work is done after pulling in the rope from A to B? Neglect friction and the size of the box.

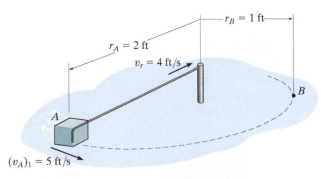

Prob. 15–111

*15–112. A toboggan and rider, having a total mass of 150 kg, enter horizontally tangent to a 90° circular curve with a velocity of $v_A = 70$ km/h. If the track is flat and banked at an angle of 60°, determine the speed v_B and the angle θ of "descent," measured from the horizontal in a vertical x–z plane, at which the toboggan exists at B. Neglect friction in the calculation.

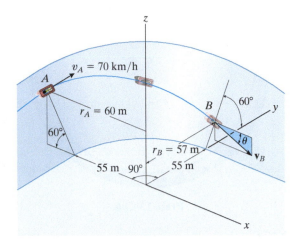

Prob. 15–112

15–113. An earth satellite of mass 700 kg is launched into a free-flight trajectory about the earth with an initial speed of $v_A = 10$ km/s when the distance from the center of the earth is $r_A = 15$ Mm. If the launch angle at this position is $\phi_A = 70°$, determine the speed v_B of the satellite and its closest distance r_B from the center of the earth. The earth has a mass $M_e = 5.976(10^{24})$ kg. *Hint:* Under these conditions, the satellite is subjected only to the earth's gravitational force, $F = GM_e m_s/r^2$, Eq. 13–1. For part of the solution, use the conservation of energy.

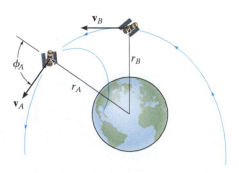

Prob. 15–113

15.8   Steady Flow of a Fluid Stream

Up to this point we have restricted our study of impulse and momentum principles to a system of particles contained within a *closed volume*. In this section, however, we will apply the principle of impulse and momentum to the steady mass flow of fluid particles entering into and then out of a *control volume*. This volume is defined as a region in space where fluid particles can flow into or out of the region. The size and shape of the control volume is frequently made to coincide with the solid boundaries and openings of a pipe, turbine, or pump. Provided the flow of the fluid into the control volume is equal to the flow out, then the flow can be classified as *steady flow*.

Principle of Impulse and Momentum.
Consider the steady flow of a fluid stream in Fig. 15–27a that passes through a pipe. The region within the pipe and its openings will be taken as the control volume. As shown, the fluid flows into and out of the control volume with velocities $\mathbf{v}_A$ and $\mathbf{v}_B$, respectively. The change in the direction of the fluid flow within the control volume is caused by an impulse produced by the resultant external force exerted on the control surface by the wall of the pipe. This resultant force can be determined by applying the principle of impulse and momentum to the control volume.

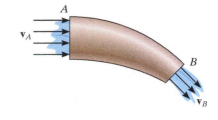

(a)

Fig. 15–27

The conveyor belt must supply frictional forces to the gravel that falls upon it in order to change the momentum of the gravel stream, so that it begins to travel along the belt. (© R.C. Hibbeler)

The air on one side of this fan is essentially at rest, and as it passes through the blades its momentum is increased. To change the momentum of the air flow in this manner, the blades must exert a horizontal thrust on the air stream. As the blades turn faster, the equal but opposite thrust of the air on the blades could overcome the rolling resistance of the wheels on the ground and begin to move the frame of the fan. (© R.C. Hibbeler)

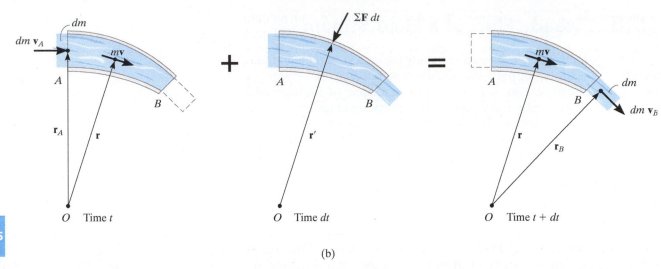

(b)

As indicated in Fig. 15–27b, a small amount of fluid having a mass dm is about to enter the control volume through opening A with a velocity of $\mathbf{v}_A$ at time t. Since the flow is considered steady, at time $t + dt$, the same amount of fluid will leave the control volume through opening B with a velocity $\mathbf{v}_B$. The momenta of the fluid entering and leaving the control volume are therefore $dm\ \mathbf{v}_A$ and $dm\ \mathbf{v}_B$, respectively. Also, during the time dt, the momentum of the fluid mass within the control volume remains constant and is denoted as $m\mathbf{v}$. As shown on the center diagram, the resultant external force exerted on the control volume produces the impulse $\Sigma \mathbf{F}\ dt$. If we apply the principle of linear impulse and momentum, we have

$$dm\ \mathbf{v}_A + m\mathbf{v} + \Sigma \mathbf{F}\ dt = dm\ \mathbf{v}_B + m\mathbf{v}$$

If $\mathbf{r}$, $\mathbf{r}_A$, $\mathbf{r}_B$ are position vectors measured from point O to the geometric centers of the control volume and the openings at A and B, Fig. 15–27b, then the principle of angular impulse and momentum about O becomes

$$\mathbf{r}_A \times dm\ \mathbf{v}_A + \mathbf{r} \times m\mathbf{v} + \mathbf{r}' \times \Sigma \mathbf{F}\ dt = \mathbf{r} \times m\mathbf{v} + \mathbf{r}_B \times dm\ \mathbf{v}_B$$

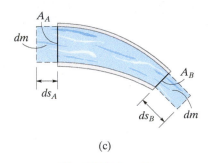

(c)

Fig. 15–27 (cont.)

Dividing both sides of the above two equations by dt and simplifying, we get

$$\Sigma \mathbf{F} = \frac{dm}{dt}(\mathbf{v}_B - \mathbf{v}_A) \qquad (15\text{–}25)$$

$$\Sigma \mathbf{M}_O = \frac{dm}{dt}(\mathbf{r}_B \times \mathbf{v}_B - \mathbf{r}_A \times \mathbf{v}_A) \qquad (15\text{–}26)$$

The term dm/dt is called the *mass flow*. It indicates the constant amount of fluid which flows either into or out of the control volume per unit of time. If the cross-sectional areas and densities of the fluid at the entrance A are A_A, ρ_A and at exit B, A_B, ρ_B, Fig. 15–27c, then for an incompressible fluid, the continuity of mass requires $dm = \rho dV = \rho_A(ds_A A_A) = \rho_B(ds_B A_B)$. Hence, during the time dt, since $v_A = ds_A/dt$ and $v_B = ds_B/dt$, we have $dm/dt = \rho_A v_A A_A = \rho_B v_B A_B$ or in general,

$$\frac{dm}{dt} = \rho v A = \rho Q \qquad (15\text{--}27)$$

The term $Q = vA$ measures the volume of fluid flow per unit of time and is referred to as the *discharge* or the *volumetric flow*.

Procedure for Analysis

Problems involving steady flow can be solved using the following procedure.

Kinematic Diagram.

- Identify the control volume. If it is *moving*, a *kinematic diagram* may be helpful for determining the entrance and exit velocities of the fluid flowing into and out of its openings since a *relative-motion analysis* of velocity will be involved.

- The measurement of velocities v_A and v_B must be made by an observer fixed in an inertial frame of reference.

- Once the velocity of the fluid flowing into the control volume is determined, the mass flow is calculated using Eq. 15–27.

Free-Body Diagram.

- Draw the free-body diagram of the control volume in order to establish the forces $\Sigma \mathbf{F}$ that act on it. These forces will include the support reactions, the weight of all solid parts and the fluid contained within the control volume, and the static gauge pressure forces of the fluid on the entrance and exit sections.* The gauge pressure is the pressure measured above atmospheric pressure, and so if an opening is exposed to the atmosphere, the gauge pressure there will be zero.

Equations of Steady Flow.

- Apply the equations of steady flow, Eq. 15–25 and 15–26, using the appropriate components of velocity and force shown on the kinematic and free-body diagrams.

* In the SI system, pressure is measured using the pascal (Pa), where $1\text{Pa} = 1 \text{ N}/\text{m}^2$.

EXAMPLE 15.16

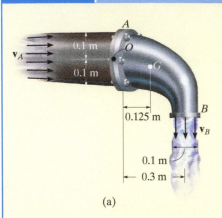

(a)

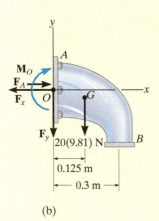

(b)

Fig. 15–28

Determine the components of reaction which the fixed pipe joint at A exerts on the elbow in Fig. 15–28a, if water flowing through the pipe is subjected to a static gauge pressure of 100 kPa at A. The discharge at B is $Q_B = 0.2 \text{ m}^3/\text{s}$. Water has a density $\rho_w = 1000 \text{ kg/m}^3$, and the water-filled elbow has a mass of 20 kg and center of mass at G.

SOLUTION

We will consider the control volume to be the outer surface of the elbow. Using a fixed inertial coordinate system, the velocity of flow at A and B and the mass flow rate can be obtained from Eq. 15–27. Since the density of water is constant, $Q_B = Q_A = Q$. Hence,

$$\frac{dm}{dt} = \rho_w Q = (1000 \text{ kg/m}^3)(0.2 \text{ m}^3/\text{s}) = 200 \text{ kg/s}$$

$$v_B = \frac{Q}{A_B} = \frac{0.2 \text{ m}^3/\text{s}}{\pi(0.05 \text{ m})^2} = 25.46 \text{ m/s} \downarrow$$

$$v_A = \frac{Q}{A_A} = \frac{0.2 \text{ m}^3/\text{s}}{\pi(0.1 \text{ m})^2} = 6.37 \text{ m/s} \rightarrow$$

Free-Body Diagram.   As shown on the free-body diagram of the control volume (elbow) Fig. 15–28b, the *fixed* connection at A exerts a resultant couple moment $\mathbf{M}_O$ and force components $\mathbf{F}_x$ and $\mathbf{F}_y$ on the elbow. Due to the static pressure of water in the pipe, the pressure force acting on the open control surface at A is $F_A = p_A A_A$. Since 1 kPa $= 1000 \text{ N/m}^2$,

$$F_A = p_A A_A = [100(10^3) \text{ N/m}^2][\pi(0.1 \text{ m})^2] = 3141.6 \text{ N}$$

There is no static pressure acting at B, since the water is discharged at atmospheric pressure; i.e., the pressure measured by a gauge at B is equal to zero, $p_B = 0$.

Equations of Steady Flow.

$$\xrightarrow{+} \Sigma F_x = \frac{dm}{dt}(v_{Bx} - v_{Ax}); \quad -F_x + 3141.6 \text{ N} = 200 \text{ kg/s}(0 - 6.37 \text{ m/s})$$

$$F_x = 4.41 \text{ kN} \qquad \qquad Ans.$$

$$+\uparrow \Sigma F_y = \frac{dm}{dt}(v_{By} - v_{Ay}); -F_y - 20(9.81) \text{ N} = 200 \text{ kg/s}(-25.46 \text{ m/s} - 0)$$

$$F_y = 4.90 \text{ kN} \qquad \qquad Ans.$$

If moments are summed about point O, Fig. 15–28b, then $\mathbf{F}_x$, $\mathbf{F}_y$, and the static pressure $\mathbf{F}_A$ are eliminated, as well as the moment of momentum of the water entering at A, Fig. 15–28a. Hence,

$$\zeta + \Sigma M_O = \frac{dm}{dt}(d_{OB}v_B - d_{OA}v_A)$$

$$M_O + 20(9.81) \text{ N} (0.125 \text{ m}) = 200 \text{ kg/s}[(0.3 \text{ m})(25.46 \text{ m/s}) - 0]$$

$$M_O = 1.50 \text{ kN} \cdot \text{m} \qquad \qquad Ans.$$

EXAMPLE 15.17

A 2-in.-diameter water jet having a velocity of 25 ft/s impinges upon a single moving blade, Fig. 15–29a. If the blade moves with a constant velocity of 5 ft/s away from the jet, determine the horizontal and vertical components of force which the blade is exerting on the water. What power does the water generate on the blade? Water has a specific weight of $\gamma_w = 62.4$ lb/ft³.

SOLUTION

Kinematic Diagram.   Here the control volume will be the stream of water on the blade. From a fixed inertial coordinate system, Fig. 15–29b, the rate at which water enters the control volume at A is

$$\mathbf{v}_A = \{25\mathbf{i}\}\ \text{ft/s}$$

The *relative-flow velocity* within the control volume is $\mathbf{v}_{w/cv} = \mathbf{v}_w - \mathbf{v}_{cv} = 25\mathbf{i} - 5\mathbf{i} = \{20\mathbf{i}\}$ ft/s. Since the control volume is moving with a velocity of $\mathbf{v}_{cv} = \{5\mathbf{i}\}$ ft/s, the velocity of flow at B measured from the fixed x, y axes is the vector sum, shown in Fig. 15–29b. Here,

$$\mathbf{v}_B = \mathbf{v}_{cv} + \mathbf{v}_{w/cv}$$

$$= \{5\mathbf{i} + 20\mathbf{j}\}\ \text{ft/s}$$

Thus, the mass flow of water *onto* the control volume that undergoes a momentum change is

$$\frac{dm}{dt} = \rho_w(v_{w/cv})A_A = \left(\frac{62.4}{32.2}\right)(20)\left[\pi\left(\frac{1}{12}\right)^2\right] = 0.8456\ \text{slug/s}$$

Free-Body Diagram.   The free-body diagram of the control volume is shown in Fig. 15–29c. The weight of the water will be neglected in the calculation, since this force will be small compared to the reactive components $\mathbf{F}_x$ and $\mathbf{F}_y$.

Equations of Steady Flow.

$$\Sigma \mathbf{F} = \frac{dm}{dt}(\mathbf{v}_B - \mathbf{v}_A)$$

$$-F_x\mathbf{i} + F_y\mathbf{j} = 0.8456(5\mathbf{i} + 20\mathbf{j} - 25\mathbf{i})$$

Equating the respective **i** and **j** components gives

$$F_x = 0.8456(20) = 16.9\ \text{lb} \leftarrow \qquad Ans.$$

$$F_y = 0.8456(20) = 16.9\ \text{lb} \uparrow \qquad Ans.$$

The water exerts equal but opposite forces on the blade.

Since the water force which causes the blade to move forward horizontally with a velocity of 5 ft/s is $F_x = 16.9$ lb, then from Eq. 14–10 the power is

$$P = \mathbf{F}\cdot\mathbf{v}; \qquad P = \frac{16.9\ \text{lb}(5\ \text{ft/s})}{550\ \text{hp}/(\text{ft}\cdot\text{lb/s})} = 0.154\ \text{hp}$$

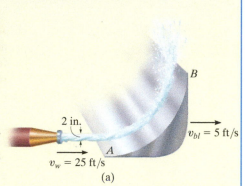

(a)

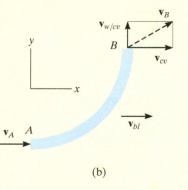

(b)

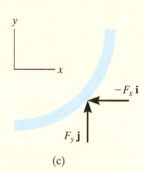

(c)

Fig. 15–29

*15.9   Propulsion with Variable Mass

A Control Volume That Loses Mass.   Consider a device such as a rocket which at an instant of time has a mass m and is moving forward with a velocity $\mathbf{v}$, Fig. 15–30a. At this same instant the amount of mass m_e is expelled from the device with a mass flow velocity $\mathbf{v}_e$. For the analysis, the control volume will include *both the mass m of the device and the expelled mass m_e*. The impulse and momentum diagrams for the control volume are shown in Fig. 15–30b. During the time dt, its velocity is increased from $\mathbf{v}$ to $\mathbf{v} + d\mathbf{v}$ since an amount of mass dm_e has been ejected and thereby gained in the exhaust. This increase in forward velocity, however, does not change the velocity $\mathbf{v}_e$ of the expelled mass, as seen by a fixed observer, since this mass moves with a constant velocity once it has been ejected. The impulses are created by $\Sigma \mathbf{F}_{cv}$, which represents the resultant of all the external forces, such as drag or weight, that *act on the control volume* in the direction of motion. This force resultant *does not include* the force which causes the control volume to move forward, since this force (called a *thrust*) is *internal to the control volume*; that is, the thrust acts with equal magnitude but opposite direction on the mass m of the device and the expelled exhaust mass m_e.* Applying the principle of impulse and momentum to the control volume, Fig. 15–30b, we have

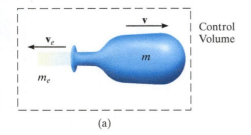

(a)

$$(\overset{+}{\rightarrow})\quad mv - m_e v_e + \Sigma F_{cv}\, dt = (m - dm_e)(v + dv) - (m_e + dm_e)v_e$$

or

$$\Sigma F_{cv}\, dt = -v\, dm_e + m\, dv - dm_e\, dv - v_e\, dm_e$$

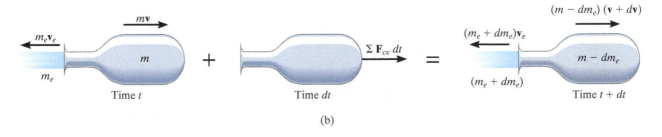

Time t Time dt Time $t + dt$

(b)

Fig. 15–30

---

*$\Sigma\mathbf{F}$ represents the external resultant force *acting on the control volume*, which is different from $\mathbf{F}$, the resultant force acting only on the device.

Without loss of accuracy, the third term on the right side may be neglected since it is a "second-order" differential. Dividing by dt gives

$$\Sigma F_{cv} = m\frac{dv}{dt} - (v + v_e)\frac{dm_e}{dt}$$

The velocity of the device as seen by an observer moving with the particles of the ejected mass is $v_{D/e} = (v + v_e)$, and so the final result can be written as

$$\boxed{\Sigma F_{cv} = m\frac{dv}{dt} - v_{D/e}\frac{dm_e}{dt}} \qquad (15\text{–}28)$$

Here the term dm_e/dt represents the rate at which mass is being ejected.

To illustrate an application of Eq. 15–28, consider the rocket shown in Fig. 15–31, which has a weight $\mathbf{W}$ and is moving upward against an atmospheric drag force $\mathbf{F}_D$. The control volume to be considered consists of the mass of the rocket and the mass of ejected gas m_e. Applying Eq. 15–28 gives

$$(+\uparrow) \qquad -F_D - W = \frac{W}{g}\frac{dv}{dt} - v_{D/e}\frac{dm_e}{dt}$$

The last term of this equation represents the *thrust* $\mathbf{T}$ which the engine exhaust exerts on the rocket, Fig. 15–31. Recognizing that $dv/dt = a$, we can therefore write

$$(+\uparrow) \qquad T - F_D - W = \frac{W}{g}a$$

If a free-body diagram of the rocket is drawn, it becomes obvious that this equation represents an application of $\Sigma\mathbf{F} = m\mathbf{a}$ for the rocket.

A Control Volume That Gains Mass.

A device such as a scoop or a shovel may gain mass as it moves forward. For example, the device shown in Fig. 15–32a has a mass m and moves forward with a velocity $\mathbf{v}$. At this instant, the device is collecting a particle stream of mass m_i. The flow velocity $\mathbf{v}_i$ of this injected mass is constant and independent of the velocity $\mathbf{v}$ such that $v > v_i$. The control volume to be considered here includes both the mass of the device and the mass of the injected particles.

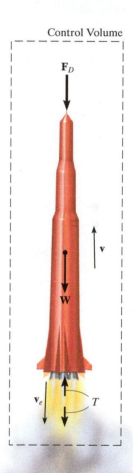

Fig. 15–31

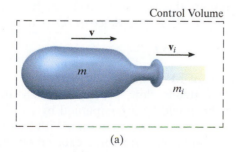

(a)

Fig. 15–32

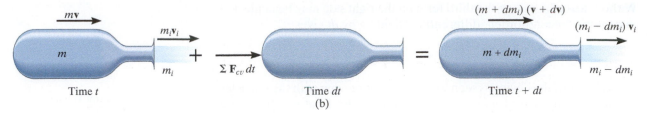

Time t Time dt Time $t + dt$

(b)

Fig. 15–32 (cont.)

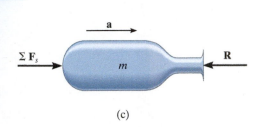

(c)

The impulse and momentum diagrams are shown in Fig. 15–32b. Along with an increase in mass dm_i gained by the device, there is an assumed increase in velocity $d\mathbf{v}$ during the time interval dt. This increase is caused by the impulse created by $\Sigma\mathbf{F}_{cv}$, the resultant of all the external forces *acting on the control volume* in the direction of motion. The force summation does not include the retarding force of the injected mass acting on the device. Why? Applying the principle of impulse and momentum to the control volume, we have

$$(\xrightarrow{+}) \qquad mv + m_i v_i + \Sigma F_{cv}\,dt = (m + dm_i)(v + dv) + (m_i - dm_i)v_i$$

Using the same procedure as in the previous case, we may write this equation as

$$\Sigma F_{cv} = m\frac{dv}{dt} + (v - v_i)\frac{dm_i}{dt}$$

Since the velocity of the device as seen by an observer moving with the particles of the injected mass is $v_{D/i} = (v - v_i)$, the final result can be written as

$$\boxed{\Sigma F_{cv} = m\frac{dv}{dt} + v_{D/i}\frac{dm_i}{dt}} \qquad (15\text{–}29)$$

where dm_i/dt is the rate of mass injected into the device. The last term in this equation represents the magnitude of force $\mathbf{R}$, which the injected mass *exerts on the device*, Fig. 15–32c. Since $dv/dt = a$, Eq. 15–29 becomes

$$\Sigma F_{cv} - R = ma$$

This is the application of $\Sigma\mathbf{F} = m\mathbf{a}$.

As in the case of steady flow, problems which are solved using Eqs. 15–28 and 15–29 should be accompanied by an identified control volume and the necessary free-body diagram. With this diagram one can then determine ΣF_{cv} and isolate the force exerted on the device by the particle stream.

The scraper box behind this tractor represents a device that gains mass. If the tractor maintains a constant velocity v, then $dv/dt = 0$ and, because the soil is originally at rest, $v_{D/i} = v$. Applying Eq. 15–29, the horizontal towing force on the scraper box is then $T = 0 + v(dm/dt)$, where dm/dt is the rate of soil accumulated in the box. (© R.C. Hibbeler)

EXAMPLE 15.18

The initial combined mass of a rocket and its fuel is m_0. A total mass m_f of fuel is consumed at a constant rate of $dm_e/dt = c$ and expelled at a constant speed of u relative to the rocket. Determine the maximum velocity of the rocket, i.e., at the instant the fuel runs out. Neglect the change in the rocket's weight with altitude and the drag resistance of the air. The rocket is fired vertically from rest.

(© NASA)

SOLUTION
Since the rocket loses mass as it moves upward, Eq. 15–28 can be used for the solution. The only *external force* acting on the *control volume* consisting of the rocket and a portion of the expelled mass is the weight **W**, Fig. 15–33. Hence,

$$+\uparrow \Sigma F_{cv} = m\frac{dv}{dt} - v_{D/e}\frac{dm_e}{dt}; \qquad -W = m\frac{dv}{dt} - uc \qquad (1)$$

The rocket's velocity is obtained by integrating this equation.

At any given instant t during the flight, the mass of the rocket can be expressed as $m = m_0 - (dm_e/dt)t = m_0 - ct$. Since $W = mg$, Eq. 1 becomes

$$-(m_0 - ct)g = (m_0 - ct)\frac{dv}{dt} - uc$$

Separating the variables and integrating, realizing that $v = 0$ at $t = 0$, we have

$$\int_0^v dv = \int_0^t \left(\frac{uc}{m_0 - ct} - g\right) dt$$

$$v = -u \ln(m_0 - ct) - gt \Big|_0^t = u \ln\left(\frac{m_0}{m_0 - ct}\right) - gt \qquad (2)$$

Note that liftoff requires the first term on the right to be greater than the second during the initial phase of motion. The time t' needed to consume all the fuel is

$$m_f = \left(\frac{dm_e}{dt}\right)t' = ct'$$

Hence,

$$t' = m_f/c$$

Substituting into Eq. 2 yields

$$v_{\max} = u \ln\left(\frac{m_0}{m_0 - m_f}\right) - \frac{gm_f}{c} \qquad \textit{Ans.}$$

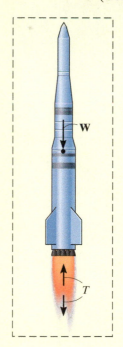

Fig. 15–33

EXAMPLE | 15.19

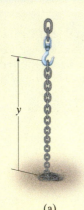

(a)

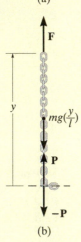

(b)

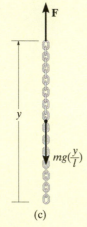

(c)

Fig. 15–34

A chain of length l, Fig. 15–34a, has a mass m. Determine the magnitude of force $\mathbf{F}$ required to (a) raise the chain with a constant speed v_c, starting from rest when $y = 0$; and (b) lower the chain with a constant speed v_c, starting from rest when $y = l$.

SOLUTION

Part (a). As the chain is raised, all the suspended links are given a sudden downward impulse by each added link which is lifted off the ground. Thus, the *suspended portion* of the chain may be considered as a device which is *gaining mass*. The control volume to be considered is the length of chain y which is suspended by $\mathbf{F}$ at any instant, including the next link which is about to be added but is still at rest, Fig. 15–34b. The forces acting on the control volume *exclude* the internal forces $\mathbf{P}$ and $-\mathbf{P}$, which act between the added link and the suspended portion of the chain. Hence, $\Sigma F_{cv} = F - mg(y/l)$.

To apply Eq. 15–29, it is also necessary to find the rate at which mass is being added to the system. The velocity $\mathbf{v}_c$ of the chain is equivalent to $\mathbf{v}_{D/i}$. Why? Since v_c is constant, $dv_c/dt = 0$ and $dy/dt = v_c$. Integrating, using the initial condition that $y = 0$ when $t = 0$, gives $y = v_c t$. Thus, the mass of the control volume at any instant is $m_{cv} = m(y/l) = m(v_c t/l)$, and therefore the *rate* at which mass is *added* to the suspended chain is

$$\frac{dm_i}{dt} = m\left(\frac{v_c}{l}\right)$$

Applying Eq. 15–29 using this data, we have

$$+\uparrow \Sigma F_{cv} = m\frac{dv_c}{dt} + v_{D/i}\frac{dm_i}{dt}$$

$$F - mg\left(\frac{y}{l}\right) = 0 + v_c m\left(\frac{v_c}{l}\right)$$

Hence,

$$F = (m/l)(gy + v_c^2)$$ *Ans.*

Part (b). When the chain is being lowered, the links which are expelled (given zero velocity) *do not* impart an impulse to the *remaining* suspended links. Why? Thus, the control volume in Part (a) will not be considered. Instead, the equation of motion will be used to obtain the solution. At time t the portion of chain still off the floor is y. The free-body diagram for a suspended portion of the chain is shown in Fig. 15–34c. Thus,

$$+\uparrow \Sigma F = ma; \qquad F - mg\left(\frac{y}{l}\right) = 0$$

$$F = mg\left(\frac{y}{l}\right)$$ *Ans.*

PROBLEMS

15–114. The fire boat discharges two streams of seawater, each at a flow of 0.25 m³/s and with a nozzle velocity of 50 m/s. Determine the tension developed in the anchor chain, needed to secure the boat. The density of seawater is $\rho_{sw} = 1020$ kg/m³.

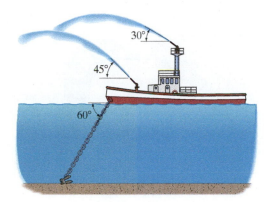

Prob. 15–114

15–115. The chute is used to divert the flow of water, $Q = 0.6$ m³/s. If the water has a cross-sectional area of 0.05 m², determine the force components at the pin D and roller C necessary for equilibrium. Neglect the weight of the chute and weight of the water on the chute. $\rho_w = 1$ Mg/m³.

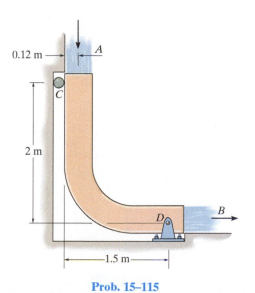

Prob. 15–115

*****15–116.** The 200-kg boat is powered by the fan which develops a slipstream having a diameter of 0.75 m. If the fan ejects air with a speed of 14 m/s, measured relative to the boat, determine the initial acceleration of the boat if it is initially at rest. Assume that air has a constant density of $\rho_w = 1.22$ kg/m³ and that the entering air is essentially at rest. Neglect the drag resistance of the water.

Prob. 15–116

15–117. The nozzle discharges water at a constant rate of 2 ft³/s. The cross-sectional area of the nozzle at A is 4 in², and at B the cross-sectional area is 12 in². If the static gauge pressure due to the water at B is 2 lb/in², determine the magnitude of force which must be applied by the coupling at B to hold the nozzle in place. Neglect the weight of the nozzle and the water within it. $\gamma_w = 62.4$ lb/ft³.

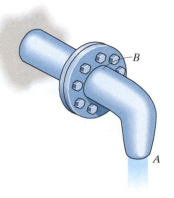

Prob. 15–117

15–118. The blade divides the jet of water having a diameter of 4 in. If one-half of the water flows to the right while the other half flows to the left, and the total flow is $Q = 1.5$ ft³/s, determine the vertical force exerted on the blade by the jet, $\gamma_\omega = 62.4$ lb/ft³.

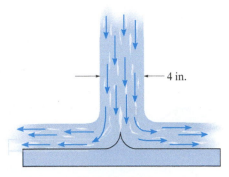

4 in.

Prob. 15–118

15–119. The blade divides the jet of water having a diameter of 3 in. If one-fourth of the water flows downward while the other three-fourths flows upward, and the total flow is $Q = 0.5$ ft³/s, determine the horizontal and vertical components of force exerted on the blade by the jet, $\gamma_w = 62.4$ lb/ft³.

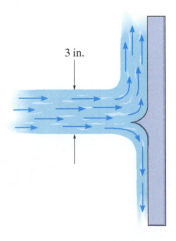

3 in.

Prob. 15–119

*15–120. The gauge pressure of water at A is 150.5 kPa. Water flows through the pipe at A with a velocity of 18 m/s, and out the pipe at B and C with the same velocity v. Determine the horizontal and vertical components of force exerted on the elbow necessary to hold the pipe assembly in equilibrium. Neglect the weight of water within the pipe and the weight of the pipe. The pipe has a diameter of 50 mm at A, and at B and C the diameter is 30 mm. $\rho_w = 1000$ kg/m³.

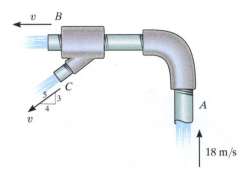

v B

C

5 3 4

A

v

18 m/s

Prob. 15–120

15–121. The gauge pressure of water at C is 40 lb/in². If water flows out of the pipe at A and B with velocities $v_A = 12$ ft/s and $v_B = 25$ ft/s, determine the horizontal and vertical components of force exerted on the elbow necessary to hold the pipe assembly in equilibrium. Neglect the weight of water within the pipe and the weight of the pipe. The pipe has a diameter of 0.75 in. at C, and at A and B the diameter is 0.5 in. $\gamma_w = 62.4$ lb/ft³.

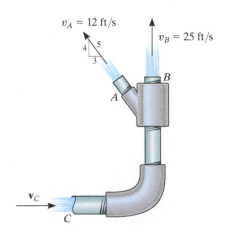

$v_A = 12$ ft/s

4 5 3

$v_B = 25$ ft/s

B

A

$\mathbf{v}_C$

C

Prob. 15–121

15–122. The fountain shoots water in the direction shown. If the water is discharged at 30° from the horizontal, and the cross-sectional area of the water stream is approximately 2 in², determine the force it exerts on the concrete wall at B. $\gamma_w = 62.4$ lb/ft³.

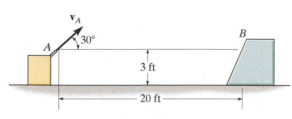

Prob. 15–122

15–123. A plow located on the front of a locomotive scoops up snow at the rate of 10 ft³/s and stores it in the train. If the locomotive is traveling at a constant speed of 12 ft/s, determine the resistance to motion caused by the shoveling. The specific weight of snow is $\gamma_s = 6$ lb/ft³.

*15–124. The boat has a mass of 180 kg and is traveling forward on a river with a constant velocity of 70 km/h, measured *relative* to the river. The river is flowing in the opposite direction at 5 km/h. If a tube is placed in the water, as shown, and it collects 40 kg of water in the boat in 80 s, determine the horizontal thrust T on the tube that is required to overcome the resistance due to the water collection and yet maintain the constant speed of the boat. $\rho_w = 1$ Mg/m³.

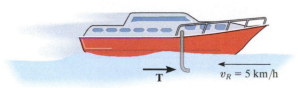

Prob. 15–124

15–125. Water is discharged from a nozzle with a velocity of 12 m/s and strikes the blade mounted on the 20-kg cart. Determine the tension developed in the cord, needed to hold the cart stationary, and the normal reaction of the wheels on the cart. The nozzle has a diameter of 50 mm and the density of water is $\rho_w = 1000$ kg/m³.

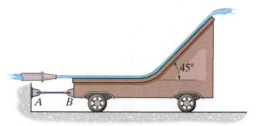

Prob. 15–125

15–126. A snowblower having a scoop S with a cross-sectional area of $A_s = 0.12$ m³ is pushed into snow with a speed of $v_s = 0.5$ m/s. The machine discharges the snow through a tube T that has a cross-sectional area of $A_T = 0.03$ m² and is directed 60° from the horizontal. If the density of snow is $\rho_s = 104$ kg/m³, determine the horizontal force P required to push the blower forward, and the resultant frictional force F of the wheels on the ground, necessary to prevent the blower from moving sideways. The wheels roll freely.

Prob. 15–126

15

15–127. The fan blows air at 6000 ft³/min. If the fan has a weight of 30 lb and a center of gravity at G, determine the smallest diameter d of its base so that it will not tip over. The specific weight of air is $\gamma = 0.076$ lb/ft³.

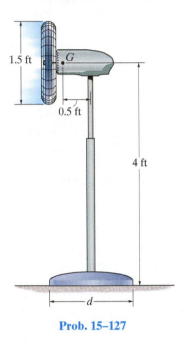

Prob. 15–127

15–129. The water flow enters below the hydrant at C at the rate of 0.75 m³/s. It is then divided equally between the two outlets at A and B. If the gauge pressure at C is 300 kPa, determine the horizontal and vertical force reactions and the moment reaction on the fixed support at C. The diameter of the two outlets at A and B is 75 mm, and the diameter of the inlet pipe at C is 150 mm. The density of water is $\rho_w = 1000$ kg/m³. Neglect the mass of the contained water and the hydrant.

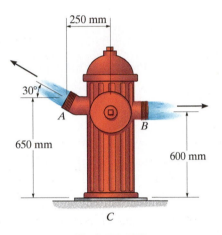

Prob. 15–129

*15–128. The nozzle has a diameter of 40 mm. If it discharges water uniformly with a downward velocity of 20 m/s against the fixed blade, determine the vertical force exerted by the water on the blade. $\rho_w = 1$ Mg/m³.

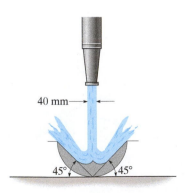

Prob. 15–128

15–130. Sand drops onto the 2-Mg empty rail car at 50 kg/s from a conveyor belt. If the car is initially coasting at 4 m/s, determine the speed of the car as a function of time.

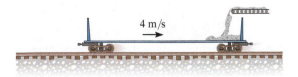

Prob. 15–130

15–131. Sand is discharged from the silo at A at a rate of 50 kg/s with a vertical velocity of 10 m/s onto the conveyor belt, which is moving with a constant velocity of 1.5 m/s. If the conveyor system and the sand on it have a total mass of 750 kg and center of mass at point G, determine the horizontal and vertical components of reaction at the pin support B and roller support A. Neglect the thickness of the conveyor.

15–133. The tractor together with the empty tank has a total mass of 4 Mg. The tank is filled with 2 Mg of water. The water is discharged at a constant rate of 50 kg/s with a constant velocity of 5 m/s, measured relative to the tractor. If the tractor starts from rest, and the rear wheels provide a resultant traction force of 250 N, determine the velocity and acceleration of the tractor at the instant the tank becomes empty.

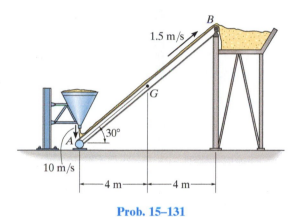

Prob. 15–131

Prob. 15–133

15–134. A rocket has an empty weight of 500 lb and carries 300 lb of fuel. If the fuel is burned at the rate of 15 lb/s and ejected with a relative velocity of 4400 ft/s, determine the maximum speed attained by the rocket starting from rest. Neglect the effect of gravitation on the rocket.

*15–132. Sand is deposited from a chute onto a conveyor belt which is moving at 0.5 m/s. If the sand is assumed to fall vertically onto the belt at A at the rate of 4 kg/s, determine the belt tension F_B to the right of A. The belt is free to move over the conveyor rollers and its tension to the left of A is $F_C = 400$ N.

Prob. 15–132

Prob. 15–134

15–135. A power lawn mower hovers very close over the ground. This is done by drawing air in at a speed of 6 m/s through an intake unit A, which has a cross-sectional area of $A_A = 0.25 \text{ m}^2$, and then discharging it at the ground, B, where the cross-sectional area is $A_B = 0.35 \text{ m}^2$. If air at A is subjected only to atmospheric pressure, determine the air pressure which the lawn mower exerts on the ground when the weight of the mower is freely supported and no load is placed on the handle. The mower has a mass of 15 kg with center of mass at G. Assume that air has a constant density of $\rho_a = 1.22 \text{ kg/m}^3$.

Prob. 15–135

*15–136.** The rocket car has a mass of 2 Mg (empty) and carries 120 kg of fuel. If the fuel is consumed at a constant rate of 6 kg/s and ejected from the car with a relative velocity of 800 m/s, determine the maximum speed attained by the car starting from rest. The drag resistance due to the atmosphere is $F_D = (6.8v^2)$ N, where v is the speed in m/s.

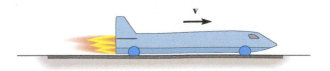

Prob. 15–136

15–137. If the chain is lowered at a constant speed $v = 4 \text{ ft/s}$, determine the normal reaction exerted on the floor as a function of time. The chain has a weight of 5 lb/ft and a total length of 20 ft.

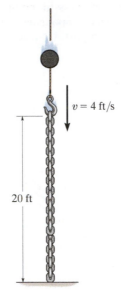

$v = 4 \text{ ft/s}$

20 ft

Prob. 15–137

15–138. The second stage of a two-stage rocket weighs 2000 lb (empty) and is launched from the first stage with a velocity of 3000 mi/h. The fuel in the second stage weighs 1000 lb. If it is consumed at the rate of 50 lb/s and ejected with a relative velocity of 8000 ft/s, determine the acceleration of the second stage just after the engine is fired. What is the rocket's acceleration just before all the fuel is consumed? Neglect the effect of gravitation.

15–139. The missile weighs 40 000 lb. The constant thrust provided by the turbojet engine is $T = 15\,000$ lb. Additional thrust is provided by *two* rocket boosters B. The propellant in each booster is burned at a constant rate of 150 lb/s, with a relative exhaust velocity of 3000 ft/s. If the mass of the propellant lost by the turbojet engine can be neglected, determine the velocity of the missile after the 4-s burn time of the boosters. The initial velocity of the missile is 300 mi/h.

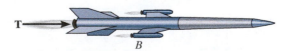

T ◀

B

Prob. 15–139

*15–140. The jet is traveling at a speed of 720 km/h. If the fuel is being spent at 0.8 kg/s, and the engine takes in air at 200 kg/s, whereas the exhaust gas (air and fuel) has a relative speed of 12 000 m/s, determine the acceleration of the plane at this instant. The drag resistance of the air is $F_D = (55\,v^2)$, where the speed is measured in m/s. The jet has a mass of 7 Mg.

720 km/h

Prob. 15–140

15–141. The rope has a mass m' per unit length. If the end length $y = h$ is draped off the edge of the table, and released, determine the velocity of its end A for any position y, as the rope uncoils and begins to fall.

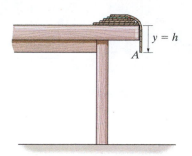

$y = h$

A

Prob. 15–141

15–142. The 12-Mg jet airplane has a constant speed of 950 km/h when it is flying along a horizontal straight line. Air enters the intake scoops S at the rate of 50 m³/s. If the engine burns fuel at the rate of 0.4 kg/s and the gas (air and fuel) is exhausted relative to the plane with a speed of 450 m/s, determine the resultant drag force exerted on the plane by air resistance. Assume that air has a constant density of 1.22 kg/m³. *Hint:* Since mass both enters and exits the plane, Eqs. 15–28 and 15–29 must be combined to yield

$$\Sigma F_s = m\frac{dv}{dt} - v_{D/e}\frac{dm_e}{dt} + v_{D/i}\frac{dm_i}{dt}.$$

$v = 950$ km/h

S

Prob. 15–142

15–143. The jet is traveling at a speed of 500 mi/h, 30° with the horizontal. If the fuel is being spent at 3 lb/s, and the engine takes in air at 400 lb/s, whereas the exhaust gas (air and fuel) has a relative speed of 32 800 ft/s, determine the acceleration of the plane at this instant. The drag resistance of the air is $F_D = (0.7v^2)$ lb, where the speed is measured in ft/s. The jet has a weight of 15 000 lb. *Hint:* See Prob. 15–142.

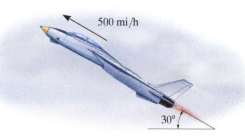

500 mi/h

30°

Prob. 15–143

*15–144. A four-engine commercial jumbo jet is cruising at a constant speed of 800 km/h in level flight when all four engines are in operation. Each of the engines is capable of discharging combustion gases with a velocity of 775 m/s relative to the plane. If during a test two of the engines, one on each side of the plane, are shut off, determine the new cruising speed of the jet. Assume that air resistance (drag) is proportional to the square of the speed, that is, $F_D = cv^2$, where c is a constant to be determined. Neglect the loss of mass due to fuel consumption.

Prob. 15–144

15

15–145. The 10-Mg helicopter carries a bucket containing 500 kg of water, which is used to fight fires. If it hovers over the land in a fixed position and then releases 50 kg/s of water at 10 m/s, measured relative to the helicopter, determine the initial upward acceleration the helicopter experiences as the water is being released.

Prob. 15–145

15–146. A rocket has an empty weight of 500 lb and carries 300 lb of fuel. If the fuel is burned at the rate of 1.5 lb/s and ejected with a velocity of 4400 ft/s relative to the rocket, determine the maximum speed attained by the rocket starting from rest. Neglect the effect of gravitation on the rocket.

15–147. Determine the magnitude of force **F** as a function of time, which must be applied to the end of the cord at A to raise the hook H with a constant speed $v = 0.4$ m/s. Initially the chain is at rest on the ground. Neglect the mass of the cord and the hook. The chain has a mass of 2 kg/m.

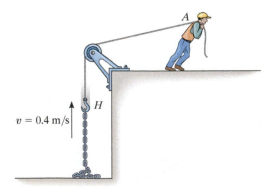

Prob. 15–147

*15–148. The truck has a mass of 50 Mg when empty. When it is unloading 5 m³ of sand at a constant rate of 0.8 m³/s, the sand flows out the back at a speed of 7 m/s, measured relative to the truck, in the direction shown. If the truck is free to roll, determine its initial acceleration just as the load begins to empty. Neglect the mass of the wheels and any frictional resistance to motion. The density of sand is $\rho_s = 1520$ kg/m³.

Prob. 15–148

15–149. The car has a mass m_0 and is used to tow the smooth chain having a total length l and a mass per unit of length m'. If the chain is originally piled up, determine the tractive force F that must be supplied by the rear wheels of the car, necessary to maintain a constant speed v while the chain is being drawn out.

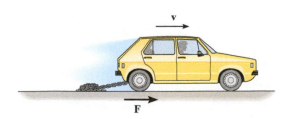

Prob. 15–149

CONCEPTUAL PROBLEMS

C15–1. The ball travels to the left when it is struck by the bat. If the ball then moves horizontally to the right, determine which measurements you could make in order to determine the net impulse given to the ball. Use numerical values to give an example of how this can be done.

Prob. C15–1 (© R.C. Hibbeler)

C15–2. The steel wrecking "ball" is suspended from the boom using an old rubber tire A. The crane operator lifts the ball then allows it to drop freely to break up the concrete. Explain, using appropriate numerical data, why it is a good idea to use the rubber tire for this work.

Prob. C15–2 (© R.C. Hibbeler)

C15–3. The train engine on the left, A, is at rest, and the one on the right, B, is coasting to the left. If the engines are identical, use numerical values to show how to determine the maximum compression in each of the spring bumpers that are mounted in the front of the engines. Each engine is free to roll.

Prob. C15–3 (© R.C. Hibbeler)

C15–4. Three train cars each have the same mass and are rolling freely when they strike the fixed bumper. Legs AB and BC on the bumper are pin connected at their ends and the angle BAC is 30° and BCA is 60°. Compare the average impulse in each leg needed to stop the motion if the cars have no bumper and if the cars have a spring bumper. Use appropriate numerical values to explain your answer.

Prob. C15–4 (© R.C. Hibbeler)

15

CHAPTER REVIEW

15

Impulse

An impulse is defined as the product of force and time. Graphically it represents the area under the F–t diagram. If the force is constant, then the impulse becomes $I = F_c(t_2 - t_1)$.

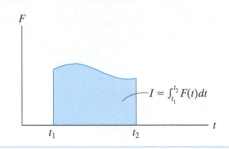

$$I = \int_{t_1}^{t_2} F(t)\,dt$$

Principle of Impulse and Momentum

When the equation of motion, $\Sigma \mathbf{F} = m\mathbf{a}$, and the kinematic equation, $a = dv/dt$, are combined, we obtain the principle of impulse and momentum. This is a vector equation that can be resolved into rectangular components and used to solve problems that involve force, velocity, and time. For application, the free-body diagram should be drawn in order to account for all the impulses that act on the particle.

$$m\mathbf{v}_1 + \Sigma \int_{t_1}^{t_2} \mathbf{F}\, dt = m\mathbf{v}_2$$

Conservation of Linear Momentum

If the principle of impulse and momentum is applied to a *system of particles*, then the collisions between the particles produce internal impulses that are equal, opposite, and collinear, and therefore cancel from the equation. Furthermore, if an external impulse is small, that is, the force is small and the time is short, then the impulse can be classified as nonimpulsive and can be neglected. Consequently, momentum for the system of particles is conserved.

The conservation-of-momentum equation is useful for finding the final velocity of a particle when internal impulses are exerted between two particles and the initial velocities of the particles is known. If the internal impulse is to be determined, then one of the particles is isolated and the principle of impulse and momentum is applied to this particle.

$$\Sigma m_i(\mathbf{v}_i)_1 = \Sigma m_i(\mathbf{v}_i)_2$$

Impact

When two particles A and B have a direct impact, the internal impulse between them is equal, opposite, and collinear. Consequently the conservation of momentum for this system applies along the line of impact.

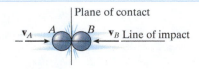

$$m_A(v_A)_1 + m_B(v_B)_1 = m_A(v_A)_2 + m_B(v_B)_2$$

If the final velocities are unknown, a second equation is needed for solution. We must use the coefficient of restitution, e. This experimentally determined coefficient depends upon the physical properties of the colliding particles. It can be expressed as the ratio of their relative velocity after collision to their relative velocity before collision. If the collision is elastic, no energy is lost and $e = 1$. For a plastic collision $e = 0$.

$$e = \frac{(v_B)_2 - (v_A)_2}{(v_A)_1 - (v_B)_1}$$

If the impact is oblique, then the conservation of momentum for the system and the coefficient-of-restitution equation apply along the line of impact. Also, conservation of momentum for each particle applies perpendicular to this line (plane of contact) because no impulse acts on the particles in this direction.

Plane of contact
Line of impact

Principle of Angular Impulse and Momentum

The moment of the linear momentum about an axis (z) is called the angular momentum.

$$(H_O)_z = (d)(mv)$$

The principle of angular impulse and momentum is often used to eliminate unknown impulses by summing the moments about an axis through which the lines of action of these impulses produce no moment. For this reason, a free-body diagram should accompany the solution.

$$(\mathbf{H}_O)_1 + \Sigma \int_{t_1}^{t_2} \mathbf{M}_O \, dt = (\mathbf{H}_O)_2$$

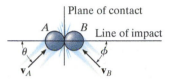

Steady Fluid Streams

Impulse-and-momentum methods are often used to determine the forces that a device exerts on the mass flow of a fluid—liquid or gas. To do so, a free-body diagram of the fluid mass in contact with the device is drawn in order to identify these forces. Also, the velocity of the fluid as it flows into and out of a control volume for the device is calculated. The equations of steady flow involve summing the forces and the moments to determine these reactions.

$$\Sigma \mathbf{F} = \frac{dm}{dt}(\mathbf{v}_B - \mathbf{v}_A)$$

$$\Sigma \mathbf{M}_O = \frac{dm}{dt}(\mathbf{r}_B \times \mathbf{v}_B - \mathbf{r}_A \times \mathbf{v}_A)$$

Propulsion with Variable Mass

Some devices, such as a rocket, lose mass as they are propelled forward. Others gain mass, such as a shovel. We can account for this mass loss or gain by applying the principle of impulse and momentum to a control volume for the device. From this equation, the force exerted on the device by the mass flow can then be determined.

$$\Sigma F_{cv} = m\frac{dv}{dt} - v_{D/e}\frac{dm_e}{dt}$$

Loses Mass

$$\Sigma F_{cv} = m\frac{dv}{dt} + v_{D/i}\frac{dm_i}{dt}$$

Gains Mass

FUNDAMENTAL REVIEW PROBLEMS

R15–1. Packages having a mass of 6 kg slide down a smooth chute and land horizontally with a speed of 3 m/s on the surface of a conveyor belt. If the coefficient of kinetic friction between the belt and a package is $\mu_k = 0.2$, determine the time needed to bring the package to rest on the belt if the belt is moving in the same direction as the package with a speed $v = 1$ m/s.

R15–3. A 20-kg block is originally at rest on a horizontal surface for which the coefficient of static friction is $\mu_s = 0.6$ and the coefficient of kinetic friction is $\mu_k = 0.5$. If a horizontal force F is applied such that it varies with time as shown, determine the speed of the block in 10 s. *Hint:* First determine the time needed to overcome friction and start the block moving.

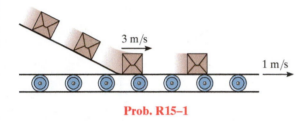

Prob. R15–1

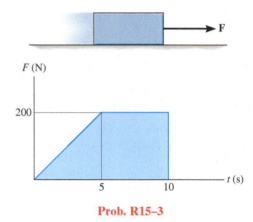

Prob. R15–3

R15–2. The 50-kg block is hoisted up the incline using the cable and motor arrangement shown. The coefficient of kinetic friction between the block and the surface is $\mu_k = 0.4$. If the block is initially moving up the plane at $v_0 = 2$ m/s, and at this instant ($t = 0$) the motor develops a tension in the cord of $T = (300 + 120\sqrt{t})$ N, where t is in seconds, determine the velocity of the block when $t = 2$ s.

R15–4. The three freight cars A, B, and C have masses of 10 Mg, 5 Mg, and 20 Mg, respectively. They are traveling along the track with the velocities shown. Car A collides with car B first, followed by car C. If the three cars couple together after collision, determine the common velocity of the cars after the two collisions have taken place.

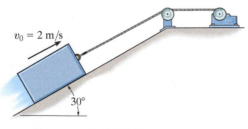

Prob. R15–2

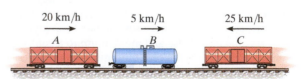

Prob. R15–4

R15–5. The 200-g projectile is fired with a velocity of 900 m/s towards the center of the 15-kg wooden block, which rests on a rough surface. If the projectile penetrates and emerges from the block with a velocity of 300 m/s, determine the velocity of the block just after the projectile emerges. How long does the block slide on the rough surface, after the projectile emerges, before it comes to rest again? The coefficient of kinetic friction between the surface and the block is $\mu_k = 0.2$.

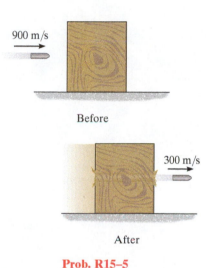

Before

After

Prob. R15–5

R15–6. Block A has a mass of 3 kg and is sliding on a rough horizontal surface with a velocity $(v_A)_1 = 2$ m/s when it makes a direct collision with block B, which has a mass of 2 kg and is originally at rest. If the collision is perfectly elastic ($e = 1$), determine the velocity of each block just after collision and the distance between the blocks when they stop sliding. The coefficient of kinetic friction between the blocks and the plane is $\mu_k = 0.3$.

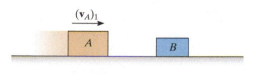

Prob. R15–6

R15–7. Two smooth billiard balls A and B have an equal mass of $m = 200$ g. If A strikes B with a velocity of $(v_A)_1 = 2$ m/s as shown, determine their final velocities just after collision. Ball B is originally at rest and the coefficient of restitution is $e = 0.75$.

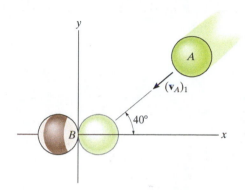

Prob. R15–7

R15–8. The small cylinder C has a mass of 10 kg and is attached to the end of a rod whose mass may be neglected. If the frame is subjected to a couple $M = (8t^2 + 5)$ N·m, where t is in seconds, and the cylinder is subjected to a force of 60 N, which is always directed as shown, determine the speed of the cylinder when $t = 2$ s. The cylinder has a speed $v_0 = 2$ m/s when $t = 0$.

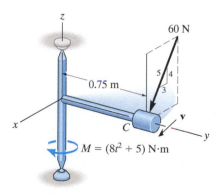

Prob. R15–8

Chapter 16

(© TFoxFoto/Shutterstock)

Kinematics is important for the design of the mechanism used on this dump truck.

Planar Kinematics of a Rigid Body

CHAPTER OBJECTIVES

- To classify the various types of rigid-body planar motion.
- To investigate rigid-body translation and angular motion about a fixed axis.
- To study planar motion using an absolute motion analysis.
- To provide a relative motion analysis of velocity and acceleration using a translating frame of reference.
- To show how to find the instantaneous center of zero velocity and determine the velocity of a point on a body using this method.
- To provide a relative-motion analysis of velocity and acceleration using a rotating frame of reference.

16.1   Planar Rigid-Body Motion

In this chapter, the planar kinematics of a rigid body will be discussed. This study is important for the design of gears, cams, and mechanisms used for many mechanical operations. Once the kinematics is thoroughly understood, then we can apply the equations of motion, which relate the forces on the body to the body's motion.

The *planar motion* of a body occurs when all the particles of a rigid body move along paths which are equidistant from a fixed plane. There are three types of rigid-body planar motion. In order of increasing complexity, they are

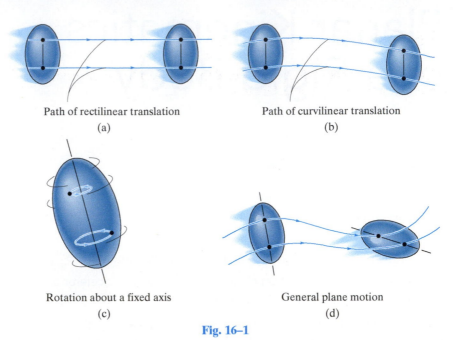

Path of rectilinear translation
(a)

Path of curvilinear translation
(b)

Rotation about a fixed axis
(c)

General plane motion
(d)

Fig. 16–1

- *Translation.* This type of motion occurs when a line in the body remains parallel to its original orientation throughout the motion. When the paths of motion for any two points on the body are parallel lines, the motion is called *rectilinear translation*, Fig. 16–1a. If the paths of motion are along curved lines, the motion is called *curvilinear translation*, Fig. 16–1b.

- *Rotation about a fixed axis.* When a rigid body rotates about a fixed axis, all the particles of the body, except those which lie on the axis of rotation, move along circular paths, Fig. 16–1c.

- *General plane motion.* When a body is subjected to general plane motion, it undergoes a combination of translation *and* rotation, Fig. 16–1d. The translation occurs within a reference plane, and the rotation occurs about an axis perpendicular to the reference plane.

In the following sections we will consider each of these motions in detail. Examples of bodies undergoing these motions are shown in Fig. 16–2.

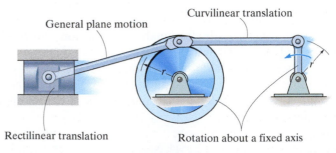

General plane motion

Curvilinear translation

Rectilinear translation

Rotation about a fixed axis

Fig. 16–2

16.2   Translation

Consider a rigid body which is subjected to either rectilinear or curvilinear translation in the *x–y* plane, Fig. 16–3.

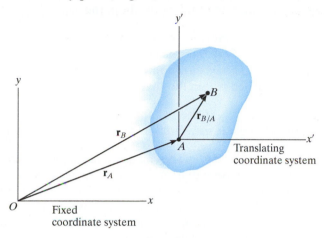

Fig. 16–3

Position.   The locations of points *A* and *B* on the body are defined with respect to fixed *x*, *y* reference frame using *position vectors* $\mathbf{r}_A$ and $\mathbf{r}_B$. The translating *x′*, *y′* coordinate system is *fixed in the body* and has its origin at *A*, hereafter referred to as the *base point*. The position of *B* with respect to *A* is denoted by the *relative-position vector* $\mathbf{r}_{B/A}$ ("$\mathbf{r}$ of *B* with respect to *A*"). By vector addition,

$$\mathbf{r}_B = \mathbf{r}_A + \mathbf{r}_{B/A}$$

Velocity.   A relation between the instantaneous velocities of *A* and *B* is obtained by taking the time derivative of this equation, which yields $\mathbf{v}_B = \mathbf{v}_A + d\mathbf{r}_{B/A}/dt$. Here $\mathbf{v}_A$ and $\mathbf{v}_B$ denote *absolute velocities* since these vectors are measured with respect to the *x, y* axes. The term $d\mathbf{r}_{B/A}/dt = \mathbf{0}$, since the *magnitude* of $\mathbf{r}_{B/A}$ is *constant* by definition of a rigid body, and because the body is translating the *direction* of $\mathbf{r}_{B/A}$ is also *constant*. Therefore,

$$\mathbf{v}_B = \mathbf{v}_A$$

Acceleration.   Taking the time derivative of the velocity equation yields a similar relationship between the instantaneous accelerations of *A* and *B*:

$$\mathbf{a}_B = \mathbf{a}_A$$

The above two equations indicate that *all points in a rigid body subjected to either rectilinear or curvilinear translation move with the same velocity and acceleration.* As a result, the kinematics of particle motion, discussed in Chapter 12, can also be used to specify the kinematics of points located in a translating rigid body.

Riders on this Ferris wheel are subjected to curvilinear translation, since the gondolas move in a circular path, yet it always remains in the upright position. (© R.C. Hibbeler)

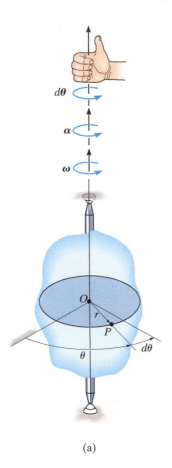

(a)

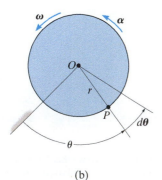

(b)

Fig. 16–4

16.3   Rotation about a Fixed Axis

When a body rotates about a fixed axis, any point P located in the body travels along a *circular path*. To study this motion it is first necessary to discuss the angular motion of the body about the axis.

Angular Motion.   Since a point is without dimension, it cannot have angular motion. *Only lines or bodies undergo angular motion.* For example, consider the body shown in Fig. 16–4a and the angular motion of a radial line r located within the shaded plane.

Angular Position.   At the instant shown, the *angular position* of r is defined by the angle θ, measured from a *fixed* reference line to r.

Angular Displacement.   The change in the angular position, which can be measured as a differential $d\theta$, is called the *angular displacement*.* This vector has a *magnitude* of $d\theta$, measured in degrees, radians, or revolutions, where 1 rev $= 2\pi$ rad. Since motion is about a *fixed axis*, the direction of $d\boldsymbol{\theta}$ is *always* along this axis. Specifically, the *direction* is determined by the right-hand rule; that is, the fingers of the right hand are curled with the sense of rotation, so that in this case the thumb, or $d\boldsymbol{\theta}$, points upward, Fig. 16–4a. In two dimensions, as shown by the top view of the shaded plane, Fig. 16–4b, both θ and $d\theta$ are counterclockwise, and so the thumb points outward from the page.

Angular Velocity.   The time rate of change in the angular position is called the *angular velocity* $\boldsymbol{\omega}$ (omega). Since $d\boldsymbol{\theta}$ occurs during an instant of time dt, then,

$(\circlearrowleft +)$

$$\omega = \frac{d\theta}{dt} \qquad (16–1)$$

This vector has a *magnitude* which is often measured in rad/s. It is expressed here in scalar form since its *direction* is also along the axis of rotation, Fig. 16–4a. When indicating the angular motion in the shaded plane, Fig. 16–4b, we can refer to the sense of rotation as clockwise or counterclockwise. Here we have *arbitrarily* chosen counterclockwise rotations as *positive* and indicated this by the curl shown in parentheses next to Eq. 16–1. Realize, however, that the directional sense of $\boldsymbol{\omega}$ is actually outward from the page.

*It is shown in Sec. 20.1 that finite rotations or finite angular displacements are *not* vector quantities, although differential rotations $d\boldsymbol{\theta}$ are vectors.

Angular Acceleration.   The *angular acceleration* $\boldsymbol{\alpha}$ (alpha) measures the time rate of change of the angular velocity. The *magnitude* of this vector is

$(\zeta+)$
$$\alpha = \frac{d\omega}{dt}$$
(16–2)

Using Eq. 16–1, it is also possible to express α as

$(\zeta+)$
$$\alpha = \frac{d^2\theta}{dt^2}$$
(16–3)

The gears used in the operation of a crane all rotate about fixed axes. Engineers must be able to relate their angular motions in order to properly design this gear system. (© R.C. Hibbeler)

The line of action of $\boldsymbol{\alpha}$ is the same as that for $\boldsymbol{\omega}$, Fig. 16–4a; however, its sense of *direction* depends on whether $\boldsymbol{\omega}$ is increasing or decreasing. If $\boldsymbol{\omega}$ is decreasing, then $\boldsymbol{\alpha}$ is called an *angular deceleration* and therefore has a sense of direction which is opposite to $\boldsymbol{\omega}$.

By eliminating dt from Eqs. 16–1 and 16–2, we obtain a differential relation between the angular acceleration, angular velocity, and angular displacement, namely,

$(\zeta+)$
$$\alpha \, d\theta = \omega \, d\omega$$
(16–4)

The similarity between the differential relations for angular motion and those developed for rectilinear motion of a particle ($v = ds/dt$, $a = dv/dt$, and $a\,ds = v\,dv$) should be apparent.

Constant Angular Acceleration.   If the angular acceleration of the body is *constant*, $\boldsymbol{\alpha} = \boldsymbol{\alpha}_c$, then Eqs. 16–1, 16–2, and 16–4, when integrated, yield a set of formulas which relate the body's angular velocity, angular position, and time. These equations are similar to Eqs. 12–4 to 12–6 used for rectilinear motion. The results are

$(\zeta+)$
$$\omega = \omega_0 + \alpha_c t$$
(16–5)

$(\zeta+)$
$$\theta = \theta_0 + \omega_0 t + \tfrac{1}{2}\alpha_c t^2$$
(16–6)

$(\zeta+)$
$$\omega^2 = \omega_0^2 + 2\alpha_c(\theta - \theta_0)$$
(16–7)

Constant Angular Acceleration

Here θ_0 and ω_0 are the initial values of the body's angular position and angular velocity, respectively.

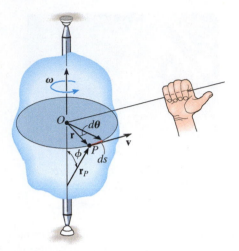

(c)

Fig. 16–4 (cont.)

16

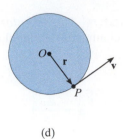

(d)

Motion of Point P.

As the rigid body in Fig. 16–4c rotates, point P travels along a *circular path* of radius r with center at point O. This path is contained within the shaded plane shown in top view, Fig. 16–4d.

Position and Displacement.

The position of P is defined by the position vector $\mathbf{r}$, which extends from O to P. If the body rotates $d\theta$ then P will displace $ds = r\,d\theta$.

Velocity.

The velocity of P has a magnitude which can be found by dividing $ds = r\,d\theta$ by dt so that

$$v = \omega r \qquad (16\text{–}8)$$

As shown in Figs. 16–4c and 16–4d, the *direction* of $\mathbf{v}$ is *tangent* to the circular path.

Both the magnitude and direction of $\mathbf{v}$ can also be accounted for by using the cross product of $\boldsymbol{\omega}$ and $\mathbf{r}_P$ (see Appendix B). Here, $\mathbf{r}_P$ is directed from *any point* on the axis of rotation to point P, Fig. 16–4c. We have

$$\mathbf{v} = \boldsymbol{\omega} \times \mathbf{r}_P \qquad (16\text{–}9)$$

The order of the vectors in this formulation is important, since the cross product is not commutative, i.e., $\boldsymbol{\omega} \times \mathbf{r}_P \neq \mathbf{r}_P \times \boldsymbol{\omega}$. Notice in Fig. 16–4c how the correct direction of $\mathbf{v}$ is established by the right-hand rule. The fingers of the right hand are curled from $\boldsymbol{\omega}$ toward $\mathbf{r}_P$ ($\boldsymbol{\omega}$ "cross" $\mathbf{r}_P$). The thumb indicates the correct direction of $\mathbf{v}$, which is tangent to the path in the direction of motion. From Eq. B–8, the magnitude of $\mathbf{v}$ in Eq. 16–9 is $v = \omega r_P \sin\phi$, and since $r = r_P \sin\phi$, Fig. 16–4c, then $v = \omega r$, which agrees with Eq. 16–8. As a special case, the position vector $\mathbf{r}$ can be chosen for $\mathbf{r}_P$. Here $\mathbf{r}$ lies in the plane of motion and again the velocity of point P is

$$\mathbf{v} = \boldsymbol{\omega} \times \mathbf{r} \qquad (16\text{–}10)$$

Acceleration.   The acceleration of P can be expressed in terms of its normal and tangential components. Applying Eq. 12–19 and Eq. 12–20, $a_t = dv/dt$ and $a_n = v^2/\rho$, where $\rho = r$, $v = \omega r$, and $\alpha = d\omega/dt$, we get

$$a_t = \alpha r \tag{16–11}$$

$$a_n = \omega^2 r \tag{16–12}$$

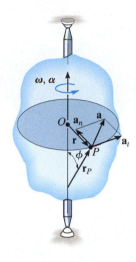

The *tangential component of acceleration*, Figs. 16–4e and 16–4f, represents the time rate of change in the velocity's magnitude. If the speed of P is increasing, then $\mathbf{a}_t$ acts in the same direction as $\mathbf{v}$; if the speed is decreasing, $\mathbf{a}_t$ acts in the opposite direction of $\mathbf{v}$; and finally, if the speed is constant, $\mathbf{a}_t$ is zero.

The *normal component of acceleration* represents the time rate of change in the velocity's direction. The *direction* of $\mathbf{a}_n$ is always toward O, the center of the circular path, Figs. 16–4e and 16–4f.

Like the velocity, the acceleration of point P can be expressed in terms of the vector cross product. Taking the time derivative of Eq. 16–9 we have

$$\mathbf{a} = \frac{d\mathbf{v}}{dt} = \frac{d\boldsymbol{\omega}}{dt} \times \mathbf{r}_P + \boldsymbol{\omega} \times \frac{d\mathbf{r}_P}{dt}$$

(e)

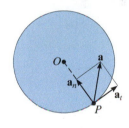

Recalling that $\boldsymbol{\alpha} = d\boldsymbol{\omega}/dt$, and using Eq. 16–9 ($d\mathbf{r}_P/dt = \mathbf{v} = \boldsymbol{\omega} \times \mathbf{r}_P$), yields

(f)

$$\mathbf{a} = \boldsymbol{\alpha} \times \mathbf{r}_P + \boldsymbol{\omega} \times (\boldsymbol{\omega} \times \mathbf{r}_P) \tag{16–13}$$

Fig. 16–4 (cont.)

From the definition of the cross product, the first term on the right has a magnitude $a_t = \alpha r_P \sin \phi = \alpha r$, and by the right-hand rule, $\boldsymbol{\alpha} \times \mathbf{r}_P$ is in the direction of $\mathbf{a}_t$, Fig. 16–4e. Likewise, the second term has a magnitude $a_n = \omega^2 r_P \sin \phi = \omega^2 r$, and applying the right-hand rule twice, first to determine the result $\mathbf{v}_P = \boldsymbol{\omega} \times \mathbf{r}_P$ then $\boldsymbol{\omega} \times \mathbf{v}_P$, it can be seen that this result is in the same direction as $\mathbf{a}_n$, shown in Fig. 16–4e. Noting that this is also the *same* direction as $-\mathbf{r}$, which lies in the plane of motion, we can express $\mathbf{a}_n$ in a much simpler form as $\mathbf{a}_n = -\omega^2 \mathbf{r}$. Hence, Eq. 16–13 can be identified by its two components as

$$\begin{aligned} \mathbf{a} &= \mathbf{a}_t + \mathbf{a}_n \\ &= \boldsymbol{\alpha} \times \mathbf{r} - \omega^2 \mathbf{r} \end{aligned} \tag{16–14}$$

Since $\mathbf{a}_t$ and $\mathbf{a}_n$ are perpendicular to one another, if needed the magnitude of acceleration can be determined from the Pythagorean theorem; namely, $a = \sqrt{a_n^2 + a_t^2}$, Fig. 16–4f.

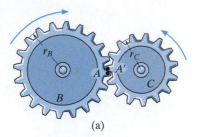

(a)

If two rotating bodies contact one another, then the *points in contact* move along *different circular paths*, and the velocity and *tangential components* of acceleration of the points will be the *same*: however, the *normal components* of acceleration will *not* be the same. For example, consider the two meshed gears in Fig. 16–5a. Point A is located on gear B and a coincident point A' is located on gear C. Due to the rotational motion, $\mathbf{v}_A = \mathbf{v}_{A'}$, Fig. 16–5b, and as a result, $\omega_B r_B = \omega_C r_C$ or $\omega_B = \omega_C(r_C/r_B)$. Also, from Fig. 16–5c, $(\mathbf{a}_A)_t = (\mathbf{a}_{A'})_t$, so that $\alpha_B = \alpha_C(r_C/r_B)$; however, since both points follow different circular paths, $(\mathbf{a}_A)_n \neq (\mathbf{a}_{A'})_n$ and therefore, as shown, $\mathbf{a}_A \neq \mathbf{a}_{A'}$.

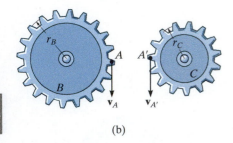

(b)

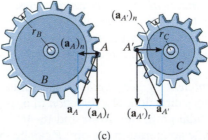

(c)

Fig. 16–5

Important Points

- A body can undergo two types of translation. During rectilinear translation all points follow parallel straight-line paths, and during curvilinear translation the points follow curved paths that are the same shape.

- All the points on a translating body move with the same velocity and acceleration.

- Points located on a body that rotates about a fixed axis follow circular paths.

- The relation $\alpha\, d\theta = \omega\, d\omega$ is derived from $\alpha = d\omega/dt$ and $\omega = d\theta/dt$ by eliminating dt.

- Once angular motions ω and α are known, the velocity and acceleration of any point on the body can be determined.

- The velocity always acts tangent to the path of motion.

- The acceleration has two components. The tangential acceleration measures the rate of change in the magnitude of the velocity and can be determined from $a_t = \alpha r$. The normal acceleration measures the rate of change in the direction of the velocity and can be determined from $a_n = \omega^2 r$.

Procedure for Analysis

The velocity and acceleration of a point located on a rigid body that is rotating about a fixed axis can be determined using the following procedure.

Angular Motion.

- Establish the positive sense of rotation about the axis of rotation and show it alongside each kinematic equation as it is applied.

- If a relation is known between any *two* of the four variables α, ω, θ, and t, then a third variable can be obtained by using one of the following kinematic equations which relates all three variables.

$$\omega = \frac{d\theta}{dt} \qquad \alpha = \frac{d\omega}{dt} \qquad \alpha\, d\theta = \omega\, d\omega$$

- If the body's angular acceleration is *constant*, then the following equations can be used:

$$\omega = \omega_0 + \alpha_c t$$

$$\theta = \theta_0 + \omega_0 t + \tfrac{1}{2}\alpha_c t^2$$

$$\omega^2 = \omega_0^2 + 2\alpha_c(\theta - \theta_0)$$

- Once the solution is obtained, the sense of θ, ω, and α is determined from the algebraic signs of their numerical quantities.

Motion of Point P.

- In most cases the velocity of P and its two components of acceleration can be determined from the scalar equations

$$v = \omega r$$

$$a_t = \alpha r$$

$$a_n = \omega^2 r$$

- If the geometry of the problem is difficult to visualize, the following vector equations should be used:

$$\mathbf{v} = \boldsymbol{\omega} \times \mathbf{r}_P = \boldsymbol{\omega} \times \mathbf{r}$$

$$\mathbf{a}_t = \boldsymbol{\alpha} \times \mathbf{r}_P = \boldsymbol{\alpha} \times \mathbf{r}$$

$$\mathbf{a}_n = \boldsymbol{\omega} \times (\boldsymbol{\omega} \times \mathbf{r}_P) = -\omega^2 \mathbf{r}$$

- Here $\mathbf{r}_P$ is directed from any point on the axis of rotation to point P, whereas $\mathbf{r}$ lies in the plane of motion of P. Either of these vectors, along with $\boldsymbol{\omega}$ and $\boldsymbol{\alpha}$, should be expressed in terms of its $\mathbf{i}, \mathbf{j}, \mathbf{k}$ components, and, if necessary, the cross products determined using a determinant expansion (see Eq. B–12).

16

EXAMPLE | 16.1

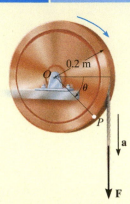

Fig. 16–6

A cord is wrapped around a wheel in Fig. 16–6, which is initially at rest when $\theta = 0$. If a force is applied to the cord and gives it an acceleration $a = (4t) \text{ m/s}^2$, where t is in seconds, determine, as a function of time, (a) the angular velocity of the wheel, and (b) the angular position of line OP in radians.

SOLUTION

Part (a). The wheel is subjected to rotation about a fixed axis passing through point O. Thus, point P on the wheel has motion about a circular path, and the acceleration of this point has *both* tangential and normal components. The tangential component is $(a_P)_t = (4t) \text{ m/s}^2$, since the cord is wrapped around the wheel and moves *tangent* to it. Hence the angular acceleration of the wheel is

$(\curvearrowleft +)$
$$(a_P)_t = \alpha r$$
$$(4t) \text{ m/s}^2 = \alpha(0.2 \text{ m})$$
$$\alpha = (20t) \text{ rad/s}^2 \,\curvearrowright$$

Using this result, the wheel's angular velocity ω can now be determined from $\alpha = d\omega/dt$, since this equation relates α, t, and ω. Integrating, with the initial condition that $\omega = 0$ when $t = 0$, yields

$(\curvearrowleft +)$
$$\alpha = \frac{d\omega}{dt} = (20t) \text{ rad/s}^2$$
$$\int_0^\omega d\omega = \int_0^t 20t \, dt$$
$$\omega = 10t^2 \text{ rad/s} \,\curvearrowright \qquad \text{Ans.}$$

Part (b). Using this result, the angular position θ of OP can be found from $\omega = d\theta/dt$, since this equation relates θ, ω, and t. Integrating, with the initial condition $\theta = 0$ when $t = 0$, we have

$(\curvearrowleft +)$
$$\frac{d\theta}{dt} = \omega = (10t^2) \text{ rad/s}$$
$$\int_0^\theta d\theta = \int_0^t 10t^2 \, dt$$
$$\theta = 3.33t^3 \text{ rad} \qquad \text{Ans.}$$

NOTE: We cannot use the equation of constant angular acceleration, since α is a function of time.

EXAMPLE 16.2

The motor shown in the photo is used to turn a wheel and attached blower contained within the housing. The details are shown in Fig. 16–7a. If the pulley A connected to the motor begins to rotate from rest with a constant angular acceleration of $\alpha_A = 2$ rad/s², determine the magnitudes of the velocity and acceleration of point P on the wheel, after the pulley has turned two revolutions. Assume the transmission belt does not slip on the pulley and wheel.

(© R.C. Hibbeler)

SOLUTION

Angular Motion.   First we will convert the two revolutions to radians. Since there are 2π rad in one revolution, then

$$\theta_A = 2 \text{ rev} \left(\frac{2\pi \text{ rad}}{1 \text{ rev}} \right) = 12.57 \text{ rad}$$

Since α_A is constant, the angular velocity of pulley A is therefore

$(\zeta +)$
$$\omega^2 = \omega_0^2 + 2\alpha_c(\theta - \theta_0)$$
$$\omega_A^2 = 0 + 2(2 \text{ rad/s}^2)(12.57 \text{ rad} - 0)$$
$$\omega_A = 7.090 \text{ rad/s}$$

The belt has the same speed and tangential component of acceleration as it passes over the pulley and wheel. Thus,

$$v = \omega_A r_A = \omega_B r_B; \quad 7.090 \text{ rad/s}(0.15 \text{ m}) = \omega_B(0.4 \text{ m})$$
$$\omega_B = 2.659 \text{ rad/s}$$
$$a_t = \alpha_A r_A = \alpha_B r_B; \quad 2 \text{ rad/s}^2(0.15 \text{ m}) = \alpha_B(0.4 \text{ m})$$
$$\alpha_B = 0.750 \text{ rad/s}^2$$

Motion of P.   As shown on the kinematic diagram in Fig. 16–7b, we have

$$v_P = \omega_B r_B = 2.659 \text{ rad/s}(0.4 \text{ m}) = 1.06 \text{ m/s} \qquad Ans.$$
$$(a_P)_t = \alpha_B r_B = 0.750 \text{ rad/s}^2(0.4 \text{ m}) = 0.3 \text{ m/s}^2$$
$$(a_P)_n = \omega_B^2 r_B = (2.659 \text{ rad/s})^2(0.4 \text{ m}) = 2.827 \text{ m/s}^2$$

Thus

$$a_P = \sqrt{(0.3 \text{ m/s}^2)^2 + (2.827 \text{ m/s}^2)^2} = 2.84 \text{ m/s}^2 \qquad Ans.$$

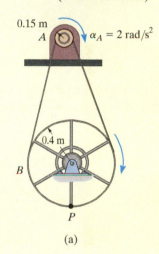

(a)

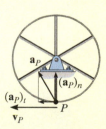

(b)

Fig. 16–7

FUNDAMENTAL PROBLEMS

F16–1. When the gear rotates 20 revolutions, it achieves an angular velocity of $\omega = 30$ rad/s, starting from rest. Determine its constant angular acceleration and the time required.

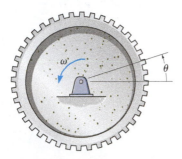

Prob. F16–1

F16–2. The flywheel rotates with an angular velocity of $\omega = (0.005\theta^2)$ rad/s, where θ is in radians. Determine the angular acceleration when it has rotated 20 revolutions.

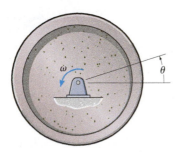

Prob. F16–2

F16–3. The flywheel rotates with an angular velocity of $\omega = (4\,\theta^{1/2})$ rad/s, where θ is in radians. Determine the time it takes to achieve an angular velocity of $\omega = 150$ rad/s. When $t = 0$, $\theta = 1$ rad.

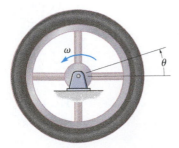

Prob. F16–3

F16–4. The bucket is hoisted by the rope that wraps around a drum wheel. If the angular displacement of the wheel is $\theta = (0.5t^3 + 15t)$ rad, where t is in seconds, determine the velocity and acceleration of the bucket when $t = 3$ s.

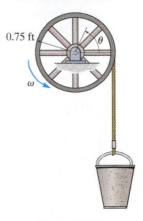

0.75 ft

Prob. F16–4

F16–5. A wheel has an angular acceleration of $\alpha = (0.5\,\theta)$ rad/s^2, where θ is in radians. Determine the magnitude of the velocity and acceleration of a point P located on its rim after the wheel has rotated 2 revolutions. The wheel has a radius of 0.2 m and starts at $\omega_0 = 2$ rad/s.

F16–6. For a short period of time, the motor turns gear A with a constant angular acceleration of $\alpha_A = 4.5$ rad/s^2, starting from rest. Determine the velocity of the cylinder and the distance it travels in three seconds. The cord is wrapped around pulley D which is rigidly attached to gear B.

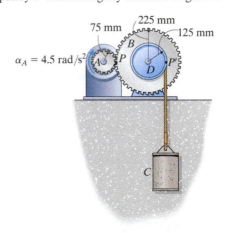

75 mm   225 mm   125 mm

$\alpha_A = 4.5$ rad/s^2

Prob. F16–6

PROBLEMS

16–1. The angular velocity of the disk is defined by $\omega = (5t^2 + 2)$ rad/s, where t is in seconds. Determine the magnitudes of the velocity and acceleration of point A on the disk when $t = 0.5$ s.

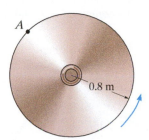

Prob. 16–1

16–2. The angular acceleration of the disk is defined by $\alpha = 3t^2 + 12$ rad/s, where t is in seconds. If the disk is originally rotating at $\omega_0 = 12$ rad/s, determine the magnitude of the velocity and the n and t components of acceleration of point A on the disk when $t = 2$ s.

16–3. The disk is originally rotating at $\omega_0 = 12$ rad/s. If it is subjected to a constant angular acceleration of $\alpha = 20$ rad/s², determine the magnitudes of the velocity and the n and t components of acceleration of point A at the instant $t = 2$ s.

*16–4. The disk is originally rotating at $\omega_0 = 12$ rad/s. If it is subjected to a constant angular acceleration of $\alpha = 20$ rad/s², determine the magnitudes of the velocity and the n and t components of acceleration of point B when the disk undergoes 2 revolutions.

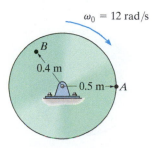

Probs. 16–2/3/4

16–5. The disk is driven by a motor such that the angular position of the disk is defined by $\theta = (20t + 4t^2)$ rad, where t is in seconds. Determine the number of revolutions, the angular velocity, and angular acceleration of the disk when $t = 90$ s.

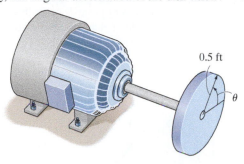

Prob. 16–5

16–6. A wheel has an initial clockwise angular velocity of 10 rad/s and a constant angular acceleration of 3 rad/s². Determine the number of revolutions it must undergo to acquire a clockwise angular velocity of 15 rad/s. What time is required?

16–7. If gear A rotates with a constant angular acceleration of $\alpha_A = 90$ rad/s², starting from rest, determine the time required for gear D to attain an angular velocity of 600 rpm. Also, find the number of revolutions of gear D to attain this angular velocity. Gears A, B, C, and D have radii of 15 mm, 50 mm, 25 mm, and 75 mm, respectively.

*16–8. If gear A rotates with an angular velocity of $\omega_A = (\theta_A + 1)$ rad/s, where θ_A is the angular displacement of gear A, measured in radians, determine the angular acceleration of gear D when $\theta_A = 3$ rad, starting from rest. Gears A, B, C, and D have radii of 15 mm, 50 mm, 25 mm, and 75 mm, respectively.

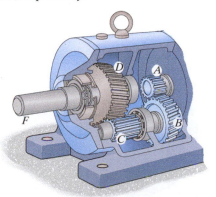

Probs. 16–7/8

16–9. At the instant $\omega_A = 5$ rad/s, pulley A is given an angular acceleration $\alpha = (0.8\theta)$ rad/s², where θ is in radians. Determine the magnitude of acceleration of point B on pulley C when A rotates 3 revolutions. Pulley C has an inner hub which is fixed to its outer one and turns with it.

16–10. At the instant $\omega_A = 5$ rad/s, pulley A is given a constant angular acceleration $\alpha_A = 6$ rad/s². Determine the magnitude of acceleration of point B on pulley C when A rotates 2 revolutions. Pulley C has an inner hub which is fixed to its outer one and turns with it.

*16–12. The power of a bus engine is transmitted using the belt-and-pulley arrangement shown. If the engine turns pulley A at $\omega_A = (20t + 40)$ rad/s, where t is in seconds, determine the angular velocities of the generator pulley B and the air-conditioning pulley C when $t = 3$ s.

16–13. The power of a bus engine is transmitted using the belt-and-pulley arrangement shown. If the engine turns pulley A at $\omega_A = 60$ rad/s, determine the angular velocities of the generator pulley B and the air-conditioning pulley C. The hub at D is rigidly *connected* to B and turns with it.

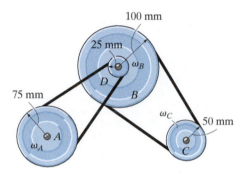

Probs. 16–12/13

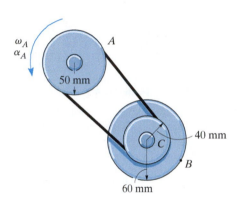

Probs. 16–9/10

16–11. The cord, which is wrapped around the disk, is given an acceleration of $a = (10t)$ m/s², where t is in seconds. Starting from rest, determine the angular displacement, angular velocity, and angular acceleration of the disk when $t = 3$ s.

16–14. The disk starts from rest and is given an angular acceleration $\alpha = (2t^2)$ rad/s², where t is in seconds. Determine the angular velocity of the disk and its angular displacement when $t = 4$ s.

16–15. The disk starts from rest and is given an angular acceleration $\alpha = (5t^{1/2})$ rad/s², where t is in seconds. Determine the magnitudes of the normal and tangential components of acceleration of a point P on the rim of the disk when $t = 2$ s.

*16–16. The disk starts at $\omega_0 = 1$ rad/s when $\theta = 0$, and is given an angular acceleration $\alpha = (0.3\theta)$ rad/s², where θ is in radians. Determine the magnitudes of the normal and tangential components of acceleration of a point P on the rim of the disk when $\theta = 1$ rev.

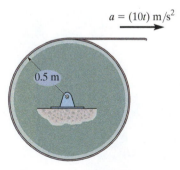

Prob. 16–11

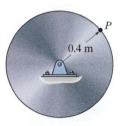

Probs. 16–14/15/16

16–17.   A motor gives gear A an angular acceleration of $\alpha_A = (2 + 0.006\ \theta^2)$ rad/s^2, where θ is in radians. If this gear is initially turning at $\omega_A = 15$ rad/s, determine the angular velocity of gear B after A undergoes an angular displacement of 10 rev.

16–18.   A motor gives gear A an angular acceleration of $\alpha_A = (2t^3)$ rad/s^2, where t is in seconds. If this gear is initially turning at $\omega_A = 15$ rad/s, determine the angular velocity of gear B when $t = 3$ s.

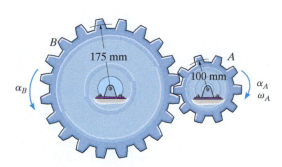

Probs. 16–17/18

*16–20.   A motor gives gear A an angular acceleration of $\alpha_A = (4t^3)$ rad/s^2, where t is in seconds. If this gear is initially turning at $(\omega_A)_0 = 20$ rad/s, determine the angular velocity of gear B when $t = 2$ s.

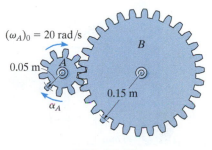

Prob. 16–20

16–19.   The vacuum cleaner's armature shaft S rotates with an angular acceleration of $\alpha = 4\omega^{3/4}$ rad/s^2, where ω is in rad/s. Determine the brush's angular velocity when $t = 4$ s, starting from $\omega_0 = 1$ rad/s, at $\theta = 0$. The radii of the shaft and the brush are 0.25 in. and 1 in., respectively. Neglect the thickness of the drive belt.

16–21.   The motor turns the disk with an angular velocity of $\omega = (5t^2 + 3t)$ rad/s, where t is in seconds. Determine the magnitudes of the velocity and the n and t components of acceleration of the point A on the disk when $t = 3$ s.

Prob. 16–19

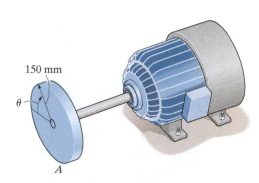

Prob. 16–21

16–22. If the motor turns gear A with an angular acceleration of $\alpha_A = 2 \text{ rad/s}^2$ when the angular velocity is $\omega_A = 20 \text{ rad/s}$, determine the angular acceleration and angular velocity of gear D.

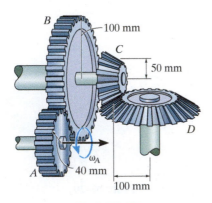

100 mm

C

50 mm

D

ω_A

40 mm

100 mm

Prob. 16–22

16–23. If the motor turns gear A with an angular acceleration of $\alpha_A = 3 \text{ rad/s}^2$ when the angular velocity is $\omega_A = 60 \text{ rad/s}$, determine the angular acceleration and angular velocity of gear D.

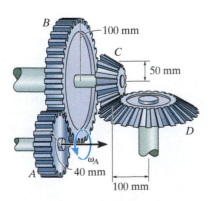

B

100 mm

C

50 mm

D

ω_A

40 mm

100 mm

Prob. 16–23

*16–24. The gear A on the drive shaft of the outboard motor has a radius $r_A = 0.5$ in. and the meshed pinion gear B on the propeller shaft has a radius $r_B = 1.2$ in. Determine the angular velocity of the propeller in $t = 1.5$ s, if the drive shaft rotates with an angular acceleration $\alpha = (400t^3) \text{ rad/s}^2$, where t is in seconds. The propeller is originally at rest and the motor frame does not move.

16–25. For the outboard motor in Prob. 16–24, determine the magnitude of the velocity and acceleration of point P located on the tip of the propeller at the instant $t = 0.75$ s.

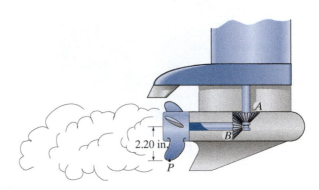

2.20 in.

A

B

P

Probs. 16–24/25

16–26. The pinion gear A on the motor shaft is given a constant angular acceleration $\alpha = 3 \text{ rad/s}^2$. If the gears A and B have the dimensions shown, determine the angular velocity and angular displacement of the output shaft C, when $t = 2$ s starting from rest. The shaft is fixed to B and turns with it.

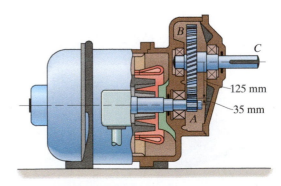

B

C

125 mm

35 mm

A

Prob. 16–26

16–27.   The gear A on the drive shaft of the outboard motor has a radius $r_A = 0.7$ in. and the meshed pinion gear B on the propeller shaft has a radius $r_B = 1.4$ in. Determine the angular velocity of the propeller in $t = 1.3$ s if the drive shaft rotates with an angular acceleration $\alpha = (300\sqrt{t})$ rad/s^2, where t is in seconds. The propeller is originally at rest and the motor frame does not move.

16–29.   A stamp S, located on the revolving drum, is used to label canisters. If the canisters are centered 200 mm apart on the conveyor, determine the radius r_A of the driving wheel A and the radius r_B of the conveyor belt drum so that for each revolution of the stamp it marks the top of a canister. How many canisters are marked per minute if the drum at B is rotating at $\omega_B = 0.2$ rad/s? Note that the driving belt is twisted as it passes between the wheels.

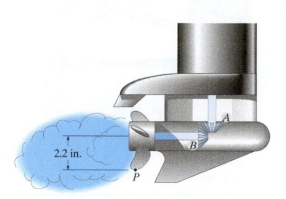

Prob. 16–27

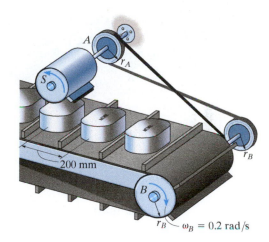

Prob. 16–29

*16–28.   The gear A on the drive shaft of the outboard motor has a radius $r_A = 0.7$ in. and the meshed pinion gear B on the propeller shaft has a radius $r_B = 1.4$ in. Determine the magnitudes of the velocity and acceleration of a point P located on the tip of the propeller at the instant $t = 0.75$ s. The drive shaft rotates with an angular acceleration $\alpha = (300\sqrt{t})$ rad/s^2, where t is in seconds. The propeller is originally at rest and the motor frame does not move.

16–30.   At the instant shown, gear A is rotating with a constant angular velocity of $\omega_A = 6$ rad/s. Determine the largest angular velocity of gear B and the maximum speed of point C.

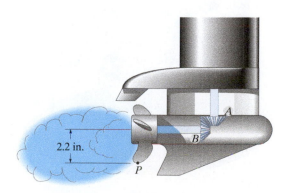

Prob. 16–28

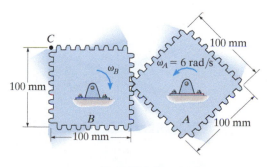

Prob. 16–30

16

16–31.   Determine the distance the load W is lifted in $t = 5$ s using the hoist. The shaft of the motor M turns with an angular velocity $\omega = 100(4 + t)$ rad/s, where t is in seconds.

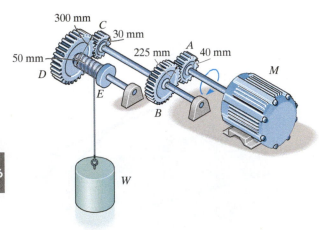

Prob. 16–31

*16–32.   The driving belt is twisted so that pulley B rotates in the opposite direction to that of drive wheel A. If A has a constant angular acceleration of $\alpha_A = 30$ rad/s^2, determine the tangential and normal components of acceleration of a point located at the rim of B when $t = 3$ s, starting from rest.

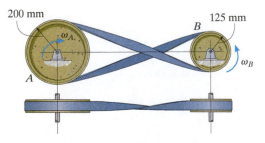

Prob. 16–32

16–33.   The driving belt is twisted so that pulley B rotates in the opposite direction to that of drive wheel A. If the angular displacement of A is $\theta_A = (5t^3 + 10t^2)$ rad, where t is in seconds, determine the angular velocity and angular acceleration of B when $t = 3$ s.

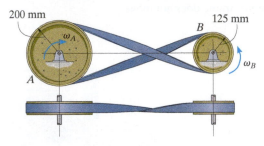

Prob. 16–33

16–34.   For a short time a motor of the random-orbit sander drives the gear A with an angular velocity of $\omega_A = 40(t^3 + 6t)$ rad/s, where t is in seconds. This gear is connected to gear B, which is fixed connected to the shaft CD. The end of this shaft is connected to the eccentric spindle EF and pad P, which causes the pad to orbit around shaft CD at a radius of 15 mm. Determine the magnitudes of the velocity and the tangential and normal components of acceleration of the spindle EF when $t = 2$ s after starting from rest.

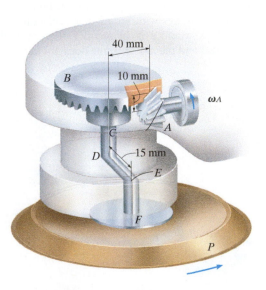

Prob. 16–34

16–35.   If the shaft and plate rotates with a constant angular velocity of $\omega = 14$ rad/s, determine the velocity and acceleration of point C located on the corner of the plate at the instant shown. Express the result in Cartesian vector form.

16–37.   The rod assembly is supported by ball-and-socket joints at A and B. At the instant shown it is rotating about the y axis with an angular velocity $\omega = 5$ rad/s and has an angular acceleration $\alpha = 8$ rad/s^2. Determine the magnitudes of the velocity and acceleration of point C at this instant. Solve the problem using Cartesian vectors and Eqs. 16–9 and 16–13.

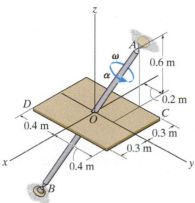

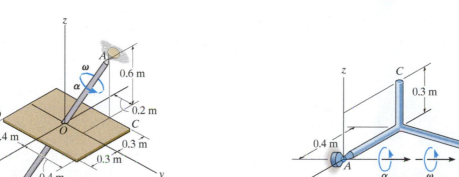

Prob. 16–35

Prob. 16–37

*16–36.   At the instant shown, the shaft and plate rotates with an angular velocity of $\omega = 14$ rad/s and angular acceleration of $\alpha = 7$ rad/s^2. Determine the velocity and acceleration of point D located on the corner of the plate at this instant. Express the result in Cartesian vector form.

16–38.   The sphere starts from rest at $\theta = 0°$ and rotates with an angular acceleration of $\alpha = (4\theta + 1)$ rad/s^2, where θ is in radians. Determine the magnitudes of the velocity and acceleration of point P on the sphere at the instant $\theta = 6$ rad.

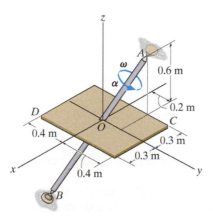

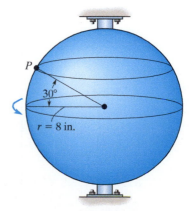

Prob. 16–36

Prob. 16–38

16.4   Absolute Motion Analysis

A body subjected to *general plane motion* undergoes a *simultaneous translation and rotation*. If the body is represented by a thin slab, the slab translates in the plane of the slab and rotates about an axis perpendicular to this plane. The motion can be completely specified by knowing *both* the angular rotation of a line fixed in the body and the motion of a point on the body. One way to relate these motions is to use a rectilinear position coordinate s to locate the point along its path and an angular position coordinate θ to specify the orientation of the line. The two coordinates are then related using the geometry of the problem. By *direct application* of the time-differential equations $v = ds/dt$, $a = dv/dt$, $\omega = d\theta/dt$, and $\alpha = d\omega/dt$, the *motion* of the point and the *angular motion* of the line can then be related. This procedure is similar to that used to solve dependent motion problems involving pulleys, Sec. 12.9. In some cases, this same procedure may be used to relate the motion of one body, undergoing either rotation about a fixed axis or translation, to that of a connected body undergoing general plane motion.

The dumping bin on the truck rotates about a fixed axis passing through the pin at A. It is operated by the extension of the hydraulic cylinder BC. The angular position of the bin can be specified using the angular position coordinate θ, and the position of point C on the bin is specified using the rectilinear position coordinate s. Since a and b are fixed lengths, then the two coordinates can be related by the cosine law, $s = \sqrt{a^2 + b^2 - 2ab \cos \theta}$. The time derivative of this equation relates the speed at which the hydraulic cylinder extends to the angular velocity of the bin. (© R.C. Hibbeler)

Procedure for Analysis

The velocity and acceleration of a point P undergoing rectilinear motion can be related to the angular velocity and angular acceleration of a line contained within a body using the following procedure.

Position Coordinate Equation.

- Locate point P on the body using a position coordinate s, which is measured from a *fixed origin* and is *directed along the straight-line path of motion* of point P.

- Measure from a fixed reference line the angular position θ of a line lying in the body.

- From the dimensions of the body, relate s to θ, $s = f(\theta)$, using geometry and/or trigonometry.

Time Derivatives.

- Take the first derivative of $s = f(\theta)$ with respect to time to get a relation between v and ω.

- Take the second time derivative to get a relation between a and α.

- In each case the chain rule of calculus must be used when taking the time derivatives of the position coordinate equation. See Appendix C.

EXAMPLE 16.3

The end of rod R shown in Fig. 16–8 maintains contact with the cam by means of a spring. If the cam rotates about an axis passing through point O with an angular acceleration α and angular velocity ω, determine the velocity and acceleration of the rod when the cam is in the arbitrary position θ.

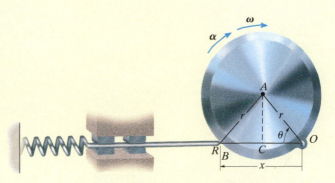

Fig. 16–8

16

SOLUTION

Position Coordinate Equation.   Coordinates θ and x are chosen in order to relate the *rotational motion* of the line segment OA on the cam to the *rectilinear translation* of the rod. These coordinates are measured from the *fixed point O* and can be related to each other using trigonometry. Since $OC = CB = r \cos \theta$, Fig. 16–8, then

$$x = 2r \cos \theta$$

Time Derivatives.   Using the chain rule of calculus, we have

$$\frac{dx}{dt} = -2r(\sin \theta)\frac{d\theta}{dt}$$
$$v = -2r\omega \sin \theta \qquad\qquad\qquad Ans.$$
$$\frac{dv}{dt} = -2r\left(\frac{d\omega}{dt}\right)\sin \theta - 2r\omega(\cos \theta)\frac{d\theta}{dt}$$
$$a = -2r(\alpha \sin \theta + \omega^2 \cos \theta) \qquad Ans.$$

NOTE: The negative signs indicate that v and a are opposite to the direction of positive x. This seems reasonable when you visualize the motion.

EXAMPLE 16.4

At a given instant, the cylinder of radius r, shown in Fig. 16–9, has an angular velocity $\boldsymbol{\omega}$ and angular acceleration $\boldsymbol{\alpha}$. Determine the velocity and acceleration of its center G if the cylinder rolls without slipping.

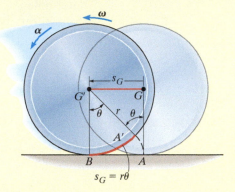

Fig. 16–9

SOLUTION

Position Coordinate Equation. The cylinder undergoes general plane motion since it simultaneously translates and rotates. By inspection, point G moves in a *straight line* to the left, from G to G', as the cylinder rolls, Fig. 16–9. Consequently its new position G' will be specified by the *horizontal* position coordinate s_G, which is measured from G to G'. Also, as the cylinder rolls (without slipping), the arc length $A'B$ on the rim which was in contact with the ground from A to B, is equivalent to s_G. Consequently, the motion requires the radial line GA to rotate θ to the position $G'A'$. Since the arc $A'B = r\theta$, then G travels a distance

$$s_G = r\theta$$

Time Derivatives. Taking successive time derivatives of this equation, realizing that r is constant, $\omega = d\theta/dt$, and $\alpha = d\omega/dt$, gives the necessary relationships:

$$s_G = r\theta$$
$$v_G = r\omega \qquad\qquad \text{Ans.}$$
$$a_G = r\alpha \qquad\qquad \text{Ans.}$$

NOTE: Remember that these relationships are valid only if the cylinder (disk, wheel, ball, etc.) rolls *without* slipping.

EXAMPLE 16.5

The large window in Fig. 16–10 is opened using a hydraulic cylinder AB. If the cylinder extends at a constant rate of 0.5 m/s, determine the angular velocity and angular acceleration of the window at the instant $\theta = 30°$.

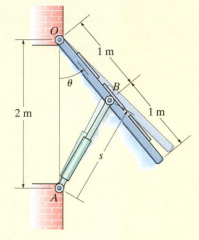

SOLUTION

Position Coordinate Equation. The angular motion of the window can be obtained using the coordinate θ, whereas the extension or motion *along the hydraulic cylinder* is defined using a coordinate s, which measures its length from the fixed point A to the moving point B. These coordinates can be related using the law of cosines, namely,

$$s^2 = (2 \text{ m})^2 + (1 \text{ m})^2 - 2(2 \text{ m})(1 \text{ m}) \cos \theta$$

$$s^2 = 5 - 4 \cos \theta \qquad (1)$$

When $\theta = 30°$,

$$s = 1.239 \text{ m}$$

Fig. 16–10

Time Derivatives. Taking the time derivatives of Eq. 1, we have

$$2s \frac{ds}{dt} = 0 - 4(-\sin \theta) \frac{d\theta}{dt}$$

$$s(v_s) = 2(\sin \theta)\omega \qquad (2)$$

Since $v_s = 0.5$ m/s, then at $\theta = 30°$,

$$(1.239 \text{ m})(0.5 \text{ m/s}) = 2 \sin 30°\omega$$

$$\omega = 0.6197 \text{ rad/s} = 0.620 \text{ rad/s} \qquad \textit{Ans.}$$

Taking the time derivative of Eq. 2 yields

$$\frac{ds}{dt} v_s + s \frac{dv_s}{dt} = 2(\cos \theta) \frac{d\theta}{dt} \omega + 2(\sin \theta) \frac{d\omega}{dt}$$

$$v_s^2 + sa_s = 2(\cos \theta)\omega^2 + 2(\sin \theta)\alpha$$

Since $a_s = dv_s/dt = 0$, then

$$(0.5 \text{ m/s})^2 + 0 = 2 \cos 30°(0.6197 \text{ rad/s})^2 + 2 \sin 30°\alpha$$

$$\alpha = -0.415 \text{ rad/s}^2 \qquad \textit{Ans.}$$

Because the result is negative, it indicates the window has an angular deceleration.

PROBLEMS

16–39. The end A of the bar is moving downward along the slotted guide with a constant velocity $\mathbf{v}_A$. Determine the angular velocity ω and angular acceleration α of the bar as a function of its position y.

16–41. At the instant $\theta = 50°$, the slotted guide is moving upward with an acceleration of 3 m/s² and a velocity of 2 m/s. Determine the angular acceleration and angular velocity of link AB at this instant. *Note:* The upward motion of the guide is in the negative y direction.

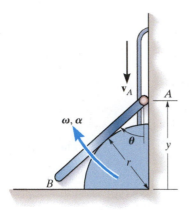

Prob. 16–39

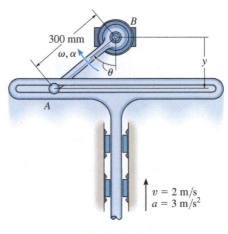

Prob. 16–41

***16–40.** At the instant $\theta = 60°$, the slotted guide rod is moving to the left with an acceleration of 2 m/s² and a velocity of 5 m/s. Determine the angular acceleration and angular velocity of link AB at this instant.

16–42. At the instant shown, $\theta = 60°$, and rod AB is subjected to a deceleration of 16 m/s² when the velocity is 10 m/s. Determine the angular velocity and angular acceleration of link CD at this instant.

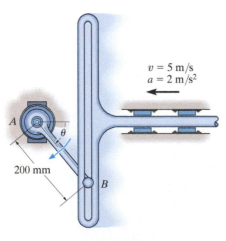

Prob. 16–40

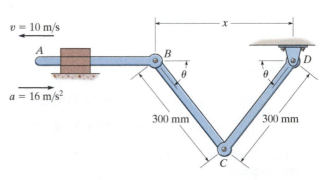

Prob. 16–42

16–43. The crank *AB* is rotating with a constant angular velocity of 4 rad/s. Determine the angular velocity of the connecting rod *CD* at the instant $\theta = 30°$.

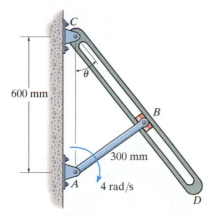

Prob. 16–43

*16–44. Determine the velocity and acceleration of the follower rod *CD* as a function of θ when the contact between the cam and follower is along the straight region *AB* on the face of the cam. The cam rotates with a constant counterclockwise angular velocity $\boldsymbol{\omega}$.

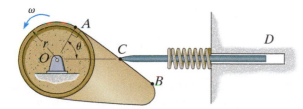

Prob. 16–44

16–45. Determine the velocity of rod *R* for any angle θ of the cam *C* if the cam rotates with a constant angular velocity $\boldsymbol{\omega}$. The pin connection at *O* does not cause an interference with the motion of *A* on *C*.

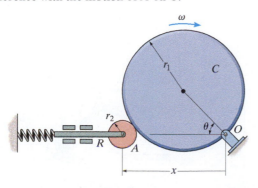

Prob. 16–45

16–46. The circular cam rotates about the fixed point *O* with a constant angular velocity $\boldsymbol{\omega}$. Determine the velocity v of the follower rod *AB* as a function of θ.

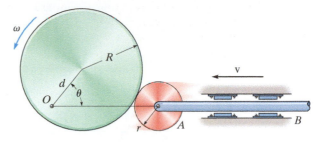

Prob. 16–46

16–47. Determine the velocity of the rod *R* for any angle θ of cam *C* as the cam rotates with a constant angular velocity $\boldsymbol{\omega}$. The pin connection at *O* does not cause an interference with the motion of plate *A* on *C*.

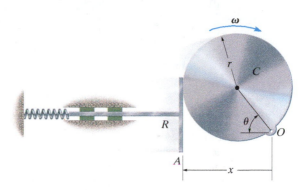

Prob. 16–47

*16–48. Determine the velocity and acceleration of the peg *A* which is confined between the vertical guide and the rotating slotted rod.

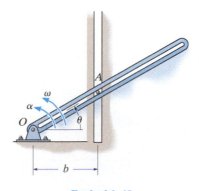

Prob. 16–48

16

16–49. Bar AB rotates uniformly about the fixed pin A with a constant angular velocity ω. Determine the velocity and acceleration of block C, at the instant $\theta = 60°$.

16–51. The pins at A and B are confined to move in the vertical and horizontal tracks. If the slotted arm is causing A to move downward at v_A, determine the velocity of B at the instant shown.

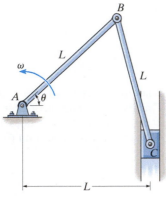

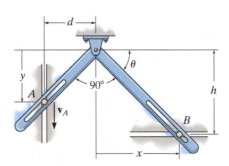

Prob. 16–51

Prob. 16–49

16–50. The center of the cylinder is moving to the left with a constant velocity v_0. Determine the angular velocity ω and angular acceleration α of the bar. Neglect the thickness of the bar.

*16–52. The crank AB has a constant angular velocity ω. Determine the velocity and acceleration of the slider at C as a function of θ. *Suggestion:* Use the x coordinate to express the motion of C and the ϕ coordinate for CB. $x = 0$ when $\phi = 0°$.

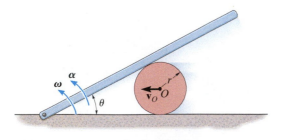

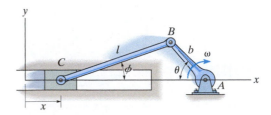

Prob. 16–50

Prob. 16–52

16–53. If the wedge moves to the left with a constant velocity **v**, determine the angular velocity of the rod as a function of θ.

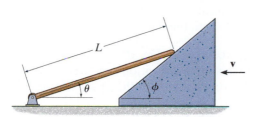

Prob. 16–53

16–55. Arm AB has an angular velocity of $\boldsymbol{\omega}$ and an angular acceleration of $\boldsymbol{\alpha}$. If no slipping occurs between the disk D and the fixed curved surface, determine the angular velocity and angular acceleration of the disk.

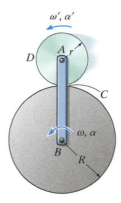

Prob. 16–55

16–54. The crate is transported on a platform which rests on rollers, each having a radius r. If the rollers do not slip, determine their angular velocity if the platform moves forward with a velocity **v**.

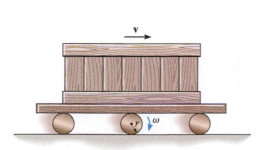

Prob. 16–54

*16–56. At the instant shown, the disk is rotating with an angular velocity of $\boldsymbol{\omega}$ and has an angular acceleration of $\boldsymbol{\alpha}$. Determine the velocity and acceleration of cylinder B at this instant. Neglect the size of the pulley at C.

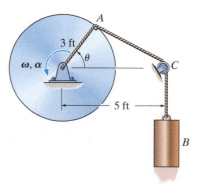

Prob. 16–56

16

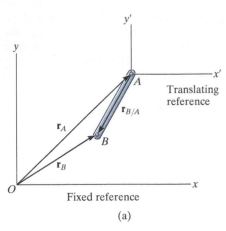

Translating reference

Fixed reference

(a)

Fig. 16–11

16.5   Relative-Motion Analysis: Velocity

The general plane motion of a rigid body can be described as a *combination* of translation and rotation. To view these "component" motions *separately* we will use a *relative-motion analysis* involving two sets of coordinate axes. The x, y coordinate system is fixed and measures the *absolute* position of two points A and B on the body, here represented as a bar, Fig. 16–11a. The origin of the x', y' coordinate system will be attached to the selected "base point" A, which generally has a *known* motion. The axes of this coordinate system *translate* with respect to the fixed frame but do not rotate with the bar.

Position.   The position vector $\mathbf{r}_A$ in Fig. 16–11a specifies the location of the "base point" A, and the relative-position vector $\mathbf{r}_{B/A}$ locates point B with respect to point A. By vector addition, the *position* of B is then

$$\mathbf{r}_B = \mathbf{r}_A + \mathbf{r}_{B/A}$$

Displacement.   During an instant of time dt, points A and B undergo displacements $d\mathbf{r}_A$ and $d\mathbf{r}_B$ as shown in Fig. 16–11b. If we consider the general plane motion by its component parts then the *entire bar* first *translates* by an amount $d\mathbf{r}_A$ so that A, the base point, moves to its *final position* and point B moves to B', Fig. 16–11c. The bar is then *rotated* about A by an amount $d\theta$ so that B' undergoes a *relative displacement* $d\mathbf{r}_{B/A}$ and thus moves to its final position B. Due to the rotation about A, $dr_{B/A} = r_{B/A}\, d\theta$, and the displacement of B is

$$d\mathbf{r}_B = d\mathbf{r}_A + d\mathbf{r}_{B/A}$$

due to rotation about A
due to translation of A
due to translation and rotation

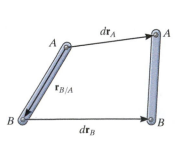

Time t Time $t + dt$

General plane motion

(b)

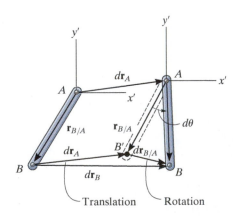

Translation   Rotation

(c)

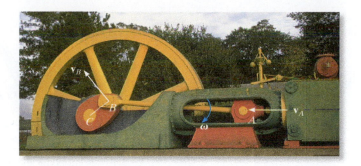

As slider block A moves horizontally to the left with a velocity $\mathbf{v}_A$, it causes crank CB to rotate counterclockwise, such that $\mathbf{v}_B$ is directed tangent to its circular path, i.e., upward to the left. The connecting rod AB is subjected to general plane motion, and at the instant shown it has an angular velocity $\boldsymbol{\omega}$. (© R.C. Hibbeler)

Velocity. To determine the relation between the velocities of points A and B, it is necessary to take the time derivative of the position equation, or simply divide the displacement equation by dt. This yields

$$\frac{d\mathbf{r}_B}{dt} = \frac{d\mathbf{r}_A}{dt} + \frac{d\mathbf{r}_{B/A}}{dt}$$

The terms $d\mathbf{r}_B/dt = \mathbf{v}_B$ and $d\mathbf{r}_A/dt = \mathbf{v}_A$ are measured with respect to the fixed x, y axes and represent the *absolute velocities* of points A and B, respectively. Since the relative displacement is caused by a rotation, the magnitude of the third term is $dr_{B/A}/dt = r_{B/A}\, d\theta/dt = r_{B/A}\dot{\theta} = r_{B/A}\omega$, where ω is the angular velocity of the body at the instant considered. We will denote this term as the *relative velocity* $\mathbf{v}_{B/A}$, since it represents the velocity of B with respect to A as measured by an observer fixed to the translating x', y' axes. In other words, *the bar appears to move as if it were rotating with an angular velocity $\boldsymbol{\omega}$ about the z' axis passing through A.* Consequently, $\mathbf{v}_{B/A}$ has a magnitude of $v_{B/A} = \omega r_{B/A}$ and a *direction* which is perpendicular to $\mathbf{r}_{B/A}$. We therefore have

$$\mathbf{v}_B = \mathbf{v}_A + \mathbf{v}_{B/A} \qquad (16\text{--}15)$$

where

$$\mathbf{v}_B = \text{velocity of point } B$$
$$\mathbf{v}_A = \text{velocity of the base point } A$$
$$\mathbf{v}_{B/A} = \text{velocity of } B \text{ with respect to } A$$

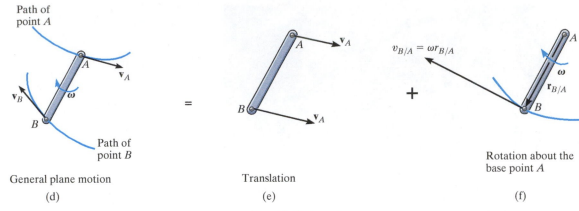

Path of
point A

Path of
point B

General plane motion

(d)

Translation

(e)

Rotation about the
base point A

(f)

Fig. 16–11 (cont.)

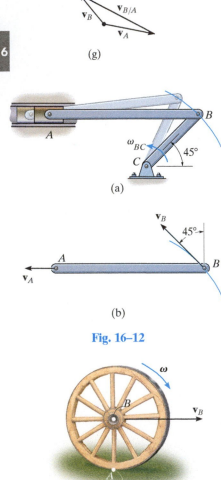

(g)

(a)

(b)

Fig. 16–12

Fig. 16–13

What the equation $\mathbf{v}_B = \mathbf{v}_A + \mathbf{v}_{B/A}$ states is that the velocity of B, Fig. 16–11d, is determined by considering the entire bar to translate with a velocity of $\mathbf{v}_A$, Fig. 16–11e, and rotate about A with an angular velocity $\boldsymbol{\omega}$, Fig. 16–11f. Vector addition of these two effects, applied to B, yields $\mathbf{v}_B$, as shown in Fig. 16–11g.

Since the relative velocity $\mathbf{v}_{B/A}$ represents the effect of *circular motion*, about A, this term can be expressed by the cross product $\mathbf{v}_{B/A} = \boldsymbol{\omega} \times \mathbf{r}_{B/A}$, Eq. 16–9. Hence, for application using Cartesian vector analysis, we can also write Eq. 16–15 as

$$\mathbf{v}_B = \mathbf{v}_A + \boldsymbol{\omega} \times \mathbf{r}_{B/A} \qquad (16\text{–}16)$$

where

$\mathbf{v}_B$ = velocity of B
$\mathbf{v}_A$ = velocity of the base point A
$\boldsymbol{\omega}$ = angular velocity of the body
$\mathbf{r}_{B/A}$ = position vector directed from A to B

The velocity equation 16–15 or 16–16 may be used in a practical manner to study the general plane motion of a rigid body which is either pin connected to or in contact with other moving bodies. When applying this equation, points A and B should generally be selected as points on the body which are pin-connected to other bodies, or as points in contact with adjacent bodies which have a *known motion*. For example, point A on link AB in Fig. 16–12a must move along a horizontal path, whereas point B moves on a circular path. The *directions* of $\mathbf{v}_A$ and $\mathbf{v}_B$ can therefore be established since they are always *tangent* to their paths of motion, Fig. 16–12b. In the case of the wheel in Fig. 16–13, which rolls *without slipping*, point A on the wheel can be selected at the ground. Here A (momentarily) has zero velocity since the ground does not move. Furthermore, the center of the wheel, B, moves along a horizontal path so that $\mathbf{v}_B$ is horizontal.

Procedure for Analysis

The relative velocity equation can be applied either by using Cartesian vector analysis, or by writing the x and y scalar component equations directly. For application, it is suggested that the following procedure be used.

Vector Analysis

Kinematic Diagram.

- Establish the directions of the fixed x, y coordinates and draw a kinematic diagram of the body. Indicate on it the velocities $\mathbf{v}_A$, $\mathbf{v}_B$ of points A and B, the angular velocity $\boldsymbol{\omega}$, and the relative-position vector $\mathbf{r}_{B/A}$.

- If the magnitudes of $\mathbf{v}_A$, $\mathbf{v}_B$, or $\boldsymbol{\omega}$ are unknown, the sense of direction of these vectors can be assumed.

Velocity Equation.

- To apply $\mathbf{v}_B = \mathbf{v}_A + \boldsymbol{\omega} \times \mathbf{r}_{B/A}$, express the vectors in Cartesian vector form and substitute them into the equation. Evaluate the cross product and then equate the respective $\mathbf{i}$ and $\mathbf{j}$ components to obtain two scalar equations.

- If the solution yields a *negative* answer for an *unknown* magnitude, it indicates the sense of direction of the vector is opposite to that shown on the kinematic diagram.

Scalar Analysis

Kinematic Diagram.

- If the velocity equation is to be applied in scalar form, then the magnitude and direction of the relative velocity $\mathbf{v}_{B/A}$ must be established. Draw a kinematic diagram such as shown in Fig. 16–11g, which shows the relative motion. Since the body is considered to be "pinned" momentarily at the base point A, the magnitude of $\mathbf{v}_{B/A}$ is $v_{B/A} = \omega r_{B/A}$. The sense of direction of $\mathbf{v}_{B/A}$ is always perpendicular to $\mathbf{r}_{B/A}$ in accordance with the rotational motion $\boldsymbol{\omega}$ of the body.*

Velocity Equation.

- Write Eq. 16–15 in symbolic form, $\mathbf{v}_B = \mathbf{v}_A + \mathbf{v}_{B/A}$, and underneath each of the terms represent the vectors graphically by showing their magnitudes and directions. The scalar equations are determined from the x and y components of these vectors.

*The notation $\mathbf{v}_B = \mathbf{v}_A + \mathbf{v}_{B/A(\text{pin})}$ may be helpful in recalling that A is "pinned."

EXAMPLE | **16.6**

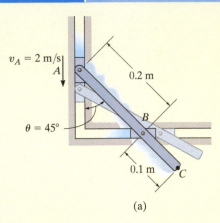

$v_A = 2$ m/s

A

0.2 m

B

$\theta = 45°$

0.1 m C

(a)

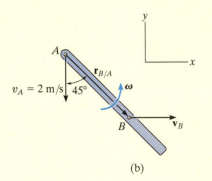

y

A

$\mathbf{r}_{B/A}$

$v_A = 2$ m/s 45°

ω

B

$\mathbf{v}_B$

x

(b)

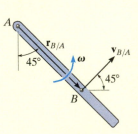

A

$\mathbf{r}_{B/A}$

45°

ω

$\mathbf{v}_{B/A}$

45°

B

Relative motion

(c)

Fig. 16–14

The link shown in Fig. 16–14a is guided by two blocks at A and B, which move in the fixed slots. If the velocity of A is 2 m/s downward, determine the velocity of B at the instant $\theta = 45°$.

SOLUTION I (VECTOR ANALYSIS)

Kinematic Diagram. Since points A and B are restricted to move along the fixed slots and $\mathbf{v}_A$ is directed downward, then velocity $\mathbf{v}_B$ must be directed horizontally to the right, Fig. 16–14b. This motion causes the link to rotate counterclockwise; that is, by the right-hand rule the angular velocity $\boldsymbol{\omega}$ is directed outward, perpendicular to the plane of motion.

Velocity Equation. Expressing each of the vectors in Fig. 16–14b in terms of their $\mathbf{i}, \mathbf{j}, \mathbf{k}$ components and applying Eq. 16–16 to A, the base point, and B, we have

$$\mathbf{v}_B = \mathbf{v}_A + \boldsymbol{\omega} \times \mathbf{r}_{B/A}$$
$$v_B\mathbf{i} = -2\mathbf{j} + [\omega\mathbf{k} \times (0.2 \sin 45°\mathbf{i} - 0.2 \cos 45°\mathbf{j})]$$
$$v_B\mathbf{i} = -2\mathbf{j} + 0.2\omega \sin 45°\mathbf{j} + 0.2\omega \cos 45°\mathbf{i}$$

Equating the $\mathbf{i}$ and $\mathbf{j}$ components gives

$$v_B = 0.2\omega \cos 45° \quad 0 = -2 + 0.2\omega \sin 45°$$

Thus,

$$\omega = 14.1 \text{ rad/s} \circlearrowright \qquad v_B = 2 \text{ m/s} \rightarrow \qquad Ans.$$

SOLUTION II (SCALAR ANALYSIS)

The kinematic diagram of the relative "circular motion" which produces $\mathbf{v}_{B/A}$ is shown in Fig. 16–14c. Here $v_{B/A} = \omega(0.2 \text{ m})$.
    Thus,

$$v_B = v_A + v_{B/A}$$

$$\begin{bmatrix} v_B \\ \rightarrow \end{bmatrix} = \begin{bmatrix} 2 \text{ m/s} \\ \downarrow \end{bmatrix} + \begin{bmatrix} \omega(0.2 \text{ m}) \\ \measuredangle 45° \end{bmatrix}$$

$(\xrightarrow{+})$ $\qquad\qquad v_B = 0 + \omega(0.2) \cos 45°$

$(+\uparrow)$ $\qquad\qquad 0 = -2 + \omega(0.2) \sin 45°$

The solution produces the above results.
    It should be emphasized that these results are *valid only* at the instant $\theta = 45°$. A recalculation for $\theta = 44°$ yields $v_B = 2.07$ m/s and $\omega = 14.4$ rad/s; whereas when $\theta = 46°$, $v_B = 1.93$ m/s and $\omega = 13.9$ rad/s, etc.

NOTE: Since v_A and ω are *known*, the velocity of any other point on the link can be determined. As an exercise, see if you can apply Eq. 16–16 to points A and C or to points B and C and show that when $\theta = 45°$, $v_C = 3.16$ m/s, directed at an angle of 18.4° up from the horizontal.

EXAMPLE 16.7

The cylinder shown in Fig. 16–15a rolls without slipping on the surface of a conveyor belt which is moving at 2 ft/s. Determine the velocity of point A. The cylinder has a clockwise angular velocity $\omega = 15$ rad/s at the instant shown.

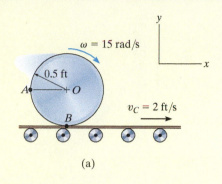

SOLUTION I (VECTOR ANALYSIS)

Kinematic Diagram. Since no slipping occurs, point B on the cylinder has the same velocity as the conveyor, Fig. 16–15b. Also, the angular velocity of the cylinder is known, so we can apply the velocity equation to B, the base point, and A to determine $\mathbf{v}_A$.

Velocity Equation

$$\mathbf{v}_A = \mathbf{v}_B + \boldsymbol{\omega} \times \mathbf{r}_{A/B}$$

$$(v_A)_x\mathbf{i} + (v_A)_y\mathbf{j} = 2\mathbf{i} + (-15\mathbf{k}) \times (-0.5\mathbf{i} + 0.5\mathbf{j})$$

$$(v_A)_x\mathbf{i} + (v_A)_y\mathbf{j} = 2\mathbf{i} + 7.50\mathbf{j} + 7.50\mathbf{i}$$

so that

$$(v_A)_x = 2 + 7.50 = 9.50 \text{ ft/s} \qquad (1)$$

$$(v_A)_y = 7.50 \text{ ft/s} \qquad (2)$$

Thus,

$$v_A = \sqrt{(9.50)^2 + (7.50)^2} = 12.1 \text{ ft/s} \qquad \textit{Ans.}$$

$$\theta = \tan^{-1}\frac{7.50}{9.50} = 38.3° \quad \measuredangle \qquad \textit{Ans.}$$

SOLUTION II (SCALAR ANALYSIS)

As an alternative procedure, the scalar components of $\mathbf{v}_A = \mathbf{v}_B + \mathbf{v}_{A/B}$ can be obtained directly. From the kinematic diagram showing the relative "circular" motion which produces $\mathbf{v}_{A/B}$, Fig. 16–15c, we have

$$v_{A/B} = \omega r_{A/B} = (15 \text{ rad/s})\left(\frac{0.5 \text{ ft}}{\cos 45°}\right) = 10.6 \text{ ft/s}$$

Thus,

$$\mathbf{v}_A = \mathbf{v}_B + \mathbf{v}_{A/B}$$

$$\begin{bmatrix} (v_A)_x \\ \rightarrow \end{bmatrix} + \begin{bmatrix} (v_A)_y \\ \uparrow \end{bmatrix} = \begin{bmatrix} 2 \text{ ft/s} \\ \rightarrow \end{bmatrix} + \begin{bmatrix} 10.6 \text{ ft/s} \\ \measuredangle 45° \end{bmatrix}$$

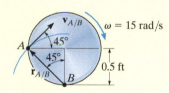

Relative motion
(c)

Fig. 16–15

Equating the x and y components gives the same results as before, namely,

$(\overset{+}{\rightarrow})$ $(v_A)_x = 2 + 10.6 \cos 45° = 9.50 \text{ ft/s}$

$(+\uparrow)$ $(v_A)_y = 0 + 10.6 \sin 45° = 7.50 \text{ ft/s}$

16

EXAMPLE 16.8

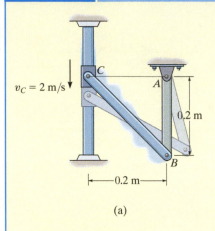

$v_C = 2$ m/s

0.2 m

0.2 m

(a)

The collar C in Fig. 16–16a is moving downward with a velocity of 2 m/s. Determine the angular velocity of CB at this instant.

SOLUTION I (VECTOR ANALYSIS)

Kinematic Diagram.   The downward motion of C causes B to move to the right along a curved path. Also, CB and AB rotate counterclockwise.

Velocity Equation.   Link CB (general plane motion): See Fig. 16–16b.

$$\mathbf{v}_B = \mathbf{v}_C + \boldsymbol{\omega}_{CB} \times \mathbf{r}_{B/C}$$

$$v_B\mathbf{i} = -2\mathbf{j} + \omega_{CB}\mathbf{k} \times (0.2\mathbf{i} - 0.2\mathbf{j})$$

$$v_B\mathbf{i} = -2\mathbf{j} + 0.2\omega_{CB}\mathbf{j} + 0.2\omega_{CB}\mathbf{i}$$

$$v_B = 0.2\omega_{CB} \tag{1}$$

$$0 = -2 + 0.2\omega_{CB} \tag{2}$$

$$\omega_{CB} = 10 \text{ rad/s} \circlearrowright \qquad \textit{Ans.}$$

$$v_B = 2 \text{ m/s} \rightarrow$$

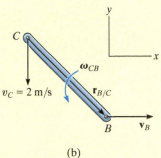

(b)

SOLUTION II (SCALAR ANALYSIS)

The scalar component equations of $\mathbf{v}_B = \mathbf{v}_C + \mathbf{v}_{B/C}$ can be obtained directly. The kinematic diagram in Fig. 16–16c shows the relative "circular" motion which produces $\mathbf{v}_{B/C}$. We have

$$\mathbf{v}_B = \mathbf{v}_C + \mathbf{v}_{B/C}$$

$$\begin{bmatrix} v_B \\ \rightarrow \end{bmatrix} = \begin{bmatrix} 2 \text{ m/s} \\ \downarrow \end{bmatrix} + \begin{bmatrix} \omega_{CB}(0.2\sqrt{2} \text{ m}) \\ \angle 45° \end{bmatrix}$$

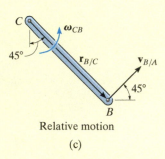

Relative motion

(c)

Resolving these vectors in the x and y directions yields

$$(\xrightarrow{+}) \qquad v_B = 0 + \omega_{CB}(0.2\sqrt{2} \cos 45°)$$

$$(+\uparrow) \qquad 0 = -2 + \omega_{CB}(0.2\sqrt{2} \sin 45°)$$

which is the same as Eqs. 1 and 2.

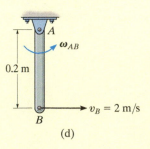

0.2 m

$v_B = 2$ m/s

(d)

Fig. 16–16

NOTE: Since link AB rotates about a fixed axis and v_B is known, Fig. 16–16d, its angular velocity is found from $v_B = \omega_{AB}r_{AB}$ or 2 m/s $= \omega_{AB}$ (0.2 m), $\omega_{AB} = 10$ rad/s.

PRELIMINARY PROBLEM

P16–1. Set up the relative velocity equation between points A and B.

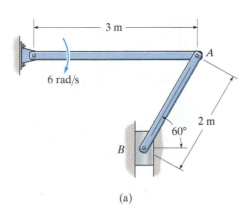

(a)

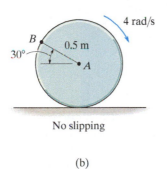

No slipping

(b)

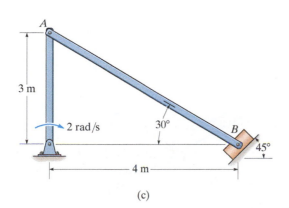

(c)

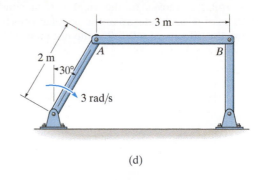

(d)

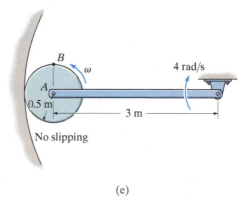

No slipping

(e)

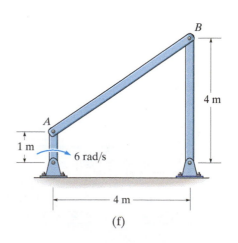

(f)

Prob. P16–1

16

FUNDAMENTAL PROBLEMS

F16–7. If roller *A* moves to the right with a constant velocity of $v_A = 3$ m/s, determine the angular velocity of the link and the velocity of roller *B* at the instant $\theta = 30°$.

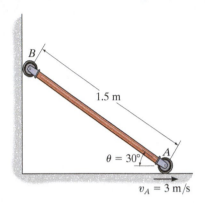

Prob. F16–7

F16–8. The wheel rolls without slipping with an angular velocity of $\omega = 10$ rad/s. Determine the magnitude of the velocity of point *B* at the instant shown.

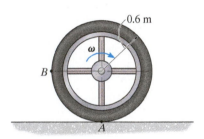

Prob. F16–8

F16–9. Determine the angular velocity of the spool. The cable wraps around the inner core, and the spool does not slip on the platform *P*.

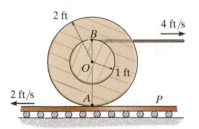

Prob. F16–9

F16–10. If crank *OA* rotates with an angular velocity of $\omega = 12$ rad/s, determine the velocity of piston *B* and the angular velocity of rod *AB* at the instant shown.

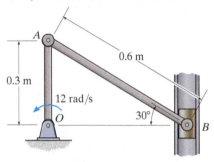

Prob. F16–10

F16–11. If rod *AB* slides along the horizontal slot with a velocity of 60 ft/s, determine the angular velocity of link *BC* at the instant shown.

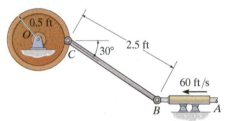

Prob. F16–11

F16–12. End *A* of the link has a velocity of $v_A = 3$ m/s. Determine the velocity of the peg at *B* at this instant. The peg is constrained to move along the slot.

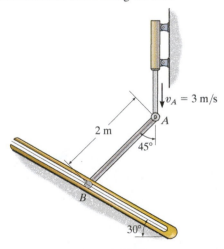

Prob. F16–12

PROBLEMS

16–57. At the instant shown the boomerang has an angular velocity $\omega = 4$ rad/s, and its mass center G has a velocity $v_G = 6$ in./s. Determine the velocity of point B at this instant.

16–59. The link AB has an angular velocity of 3 rad/s. Determine the velocity of block C and the angular velocity of link BC at the instant $\theta = 45°$. Also, sketch the position of link BC when $\theta = 60°, 45°$, and $30°$ to show its general plane motion.

Prob. 16–59

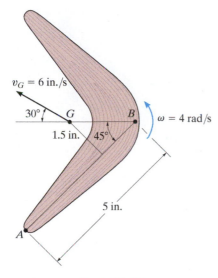

Prob. 16–57

*16–60.** The slider block C moves at 8 m/s down the inclined groove. Determine the angular velocities of links AB and BC, at the instant shown.

16–58. If the block at C is moving downward at 4 ft/s, determine the angular velocity of bar AB at the instant shown.

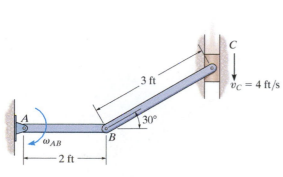

Prob. 16–58

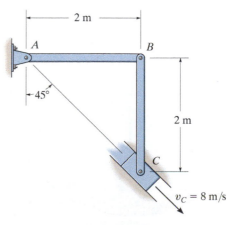

Prob. 16–60

16–61. Determine the angular velocity of links AB and BC at the instant $\theta = 30°$. Also, sketch the position of link BC when $\theta = 55°$, $45°$, and $30°$ to show its general plane motion.

16–63. If the angular velocity of link AB is $\omega_{AB} = 3$ rad/s, determine the velocity of the block at C and the angular velocity of the connecting link CB at the instant $\theta = 45°$ and $\phi = 30°$.

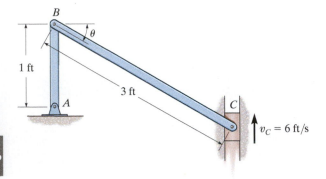

Prob. 16–61

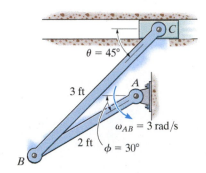

Prob. 16–63

16–62. The planetary gear A is pinned at B. Link BC rotates clockwise with an angular velocity of 8 rad/s, while the outer gear rack rotates counterclockwise with an angular velocity of 2 rad/s. Determine the angular velocity of gear A.

*16–64.** The pinion gear A rolls on the fixed gear rack B with an angular velocity $\omega = 4$ rad/s. Determine the velocity of the gear rack C.

16–65. The pinion gear rolls on the gear racks. If B is moving to the right at 8 ft/s and C is moving to the left at 4 ft/s, determine the angular velocity of the pinion gear and the velocity of its center A.

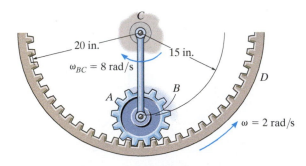

Prob. 16–62

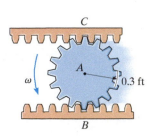

Probs. 16–64/65

16–66. Determine the angular velocity of the gear and the velocity of its center O at the instant shown.

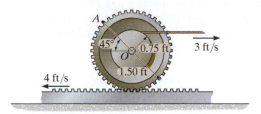

Prob. 16–66

16–67. Determine the velocity of point A on the rim of the gear at the instant shown.

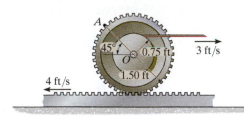

Prob. 16–67

*16–68. Knowing that angular velocity of link AB is $\omega_{AB} = 4$ rad/s, determine the velocity of the collar at C and the angular velocity of link CB at the instant shown. Link CB is horizontal at this instant.

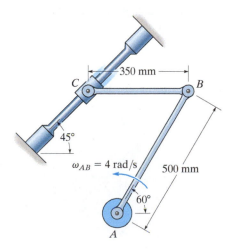

Prob. 16–68

16–69. Rod AB is rotating with an angular velocity of $\omega_{AB} = 60$ rad/s. Determine the velocity of the slider C at the instant $\theta = 60°$ and $\phi = 45°$. Also, sketch the position of bar BC when $\theta = 30°$, $60°$ and $90°$ to show its general plane motion.

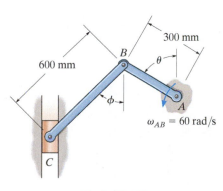

Prob. 16–69

16–70. The angular velocity of link AB is $\omega_{AB} = 5$ rad/s. Determine the velocity of block C and the angular velocity of link BC at the instant $\theta = 45°$ and $\phi = 30°$. Also, sketch the position of link CB when $\theta = 45°, 60°$, and $75°$ to show its general plane motion.

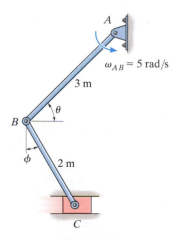

Prob. 16–70

16

16–71. The similar links AB and CD rotate about the fixed pins at A and C. If AB has an angular velocity $\omega_{AB} = 8$ rad/s, determine the angular velocity of BDP and the velocity of point P.

16–74. The epicyclic gear train consists of the sun gear A which is in mesh with the planet gear B. This gear has an inner hub C which is fixed to B and in mesh with the fixed ring gear R. If the connecting link DE pinned to B and C is rotating at $\omega_{DE} = 18$ rad/s about the pin at E, determine the angular velocities of the planet and sun gears.

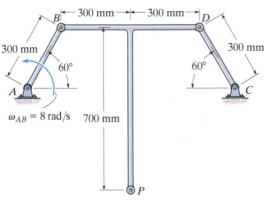

Prob. 16–71

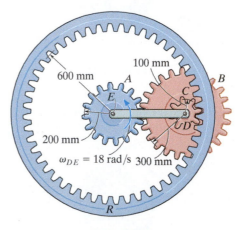

Prob. 16–74

16–75. If link AB is rotating at $\omega_{AB} = 3$ rad/s, determine the angular velocity of link CD at the instant shown.

*16–76. If link CD is rotating at $\omega_{CD} = 5$ rad/s, determine the angular velocity of link AB at the instant shown.

*16–72. If the slider block A is moving downward at $v_A = 4$ m/s, determine the velocities of blocks B and C at the instant shown.

16–73. If the slider block A is moving downward at $v_A = 4$ m/s, determine the velocity of point E at the instant shown.

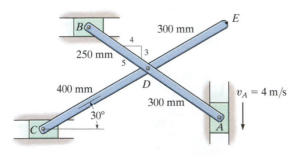

Probs. 16–72/73

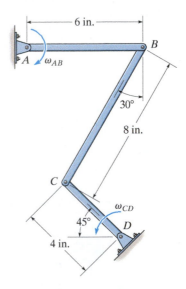

Probs. 16–75/76

16–77. The planetary gear system is used in an automatic transmission for an automobile. By locking or releasing certain gears, it has the advantage of operating the car at different speeds. Consider the case where the ring gear R is held fixed, $\omega_R = 0$, and the sun gear S is rotating at $\omega_S = 5$ rad/s. Determine the angular velocity of each of the planet gears P and shaft A.

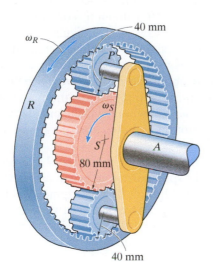

Prob. 16–77

16–78. If the ring gear A rotates clockwise with an angular velocity of $\omega_A = 30$ rad/s, while link BC rotates clockwise with an angular velocity of $\omega_{BC} = 15$ rad/s, determine the angular velocity of gear D.

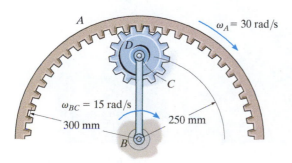

Prob. 16–78

16–79. The mechanism shown is used in a riveting machine. It consists of a driving piston A, three links, and a riveter which is attached to the slider block D. Determine the velocity of D at the instant shown, when the piston at A is traveling at $v_A = 20$ m/s.

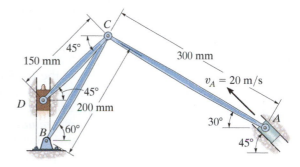

Prob. 16–79

*16–80. The mechanism is used on a machine for the manufacturing of a wire product. Because of the rotational motion of link AB and the, sliding of block F, the segmental gear lever DE undergoes general plane motion. If AB is rotating at $\omega_{AB} = 5$ rad/s, determine the velocity of point E at the instant shown.

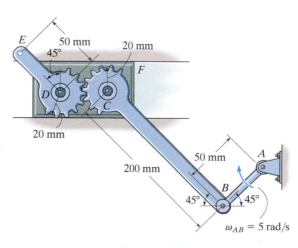

Prob. 16–80

16.6   Instantaneous Center of Zero Velocity

The velocity of any point B located on a rigid body can be obtained in a very direct way by choosing the base point A to be a point that has *zero velocity* at the instant considered. In this case, $\mathbf{v}_A = \mathbf{0}$, and therefore the velocity equation, $\mathbf{v}_B = \mathbf{v}_A + \boldsymbol{\omega} \times \mathbf{r}_{B/A}$, becomes $\mathbf{v}_B = \boldsymbol{\omega} \times \mathbf{r}_{B/A}$. For a body having general plane motion, point A so chosen is called the *instantaneous center of zero velocity (IC)*, and it lies on the *instantaneous axis of zero velocity*. This axis is always perpendicular to the plane of motion, and the intersection of the axis with this plane defines the location of the *IC*. Since point A coincides with the *IC*, then $\mathbf{v}_B = \boldsymbol{\omega} \times \mathbf{r}_{B/IC}$ and so point B moves momentarily about the *IC* in a *circular path;* in other words, the body appears to rotate about the instantaneous axis. The *magnitude* of $\mathbf{v}_B$ is simply $v_B = \omega r_{B/IC}$, where ω is the angular velocity of the body. Due to the circular motion, the *direction* of $\mathbf{v}_B$ must always be *perpendicular* to $\mathbf{r}_{B/IC}$.

For example, the *IC* for the bicycle wheel in Fig. 16–17 is at the contact point with the ground. There the spokes are somewhat visible, whereas at the top of the wheel they become blurred. If one imagines that the wheel is momentarily pinned at this point, the velocities of various points can be found using $v = \omega r$. Here the radial distances shown in the photo, Fig. 16–17, must be determined from the geometry of the wheel.

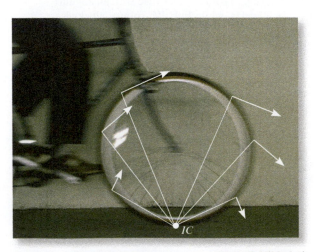

(© R.C. Hibbeler)

Fig. 16–17

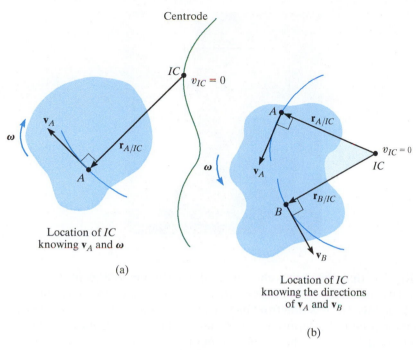

Location of *IC*
knowing $\mathbf{v}_A$ and $\boldsymbol{\omega}$

(a)

Location of *IC*
knowing the directions
of $\mathbf{v}_A$ and $\mathbf{v}_B$

(b)

Fig. 16–18

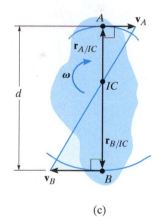

(c)

Location of the *IC*.

To locate the *IC* we can use the fact that the *velocity* of a point on the body is *always perpendicular* to the *relative-position vector* directed from the *IC* to the point. Several possibilities exist:

- *The velocity* $\mathbf{v}_A$ *of a point A on the body and the angular velocity* $\boldsymbol{\omega}$ *of the body are known*, Fig. 16–18a. In this case, the *IC* is located along the line drawn perpendicular to $\mathbf{v}_A$ at A, such that the distance from A to the *IC* is $r_{A/IC} = v_A/\omega$. Note that the *IC* lies up and to the right of A since $\mathbf{v}_A$ must cause a clockwise angular velocity $\boldsymbol{\omega}$ about the *IC*.

- *The lines of action of two nonparallel velocities* $\mathbf{v}_A$ *and* $\mathbf{v}_B$ *are known*, Fig. 16–18b. Construct at points A and B line segments that are perpendicular to $\mathbf{v}_A$ and $\mathbf{v}_B$. Extending these perpendiculars to their *point of intersection* as shown locates the *IC* at the instant considered.

- *The magnitude and direction of two parallel velocities* $\mathbf{v}_A$ *and* $\mathbf{v}_B$ *are known*. Here the location of the *IC* is determined by proportional triangles. Examples are shown in Fig. 16–18c and d. In both cases $r_{A/IC} = v_A/\omega$ and $r_{B/IC} = v_B/\omega$. If d is a known distance between points A and B, then in Fig. 16–18c, $r_{A/IC} + r_{B/IC} = d$ and in Fig. 16–18d, $r_{B/IC} - r_{A/IC} = d$.

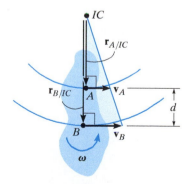

Location of *IC*
knowing $\mathbf{v}_A$ and $\mathbf{v}_B$

(d)

As the board slides downward to the left it is subjected to general plane motion. Since the directions of the velocities of its ends A and B are known, the IC is located as shown. At this instant the board will momentarily rotate about this point. Draw the board in several other positions and establish the IC for each case. (© R.C. Hibbeler)

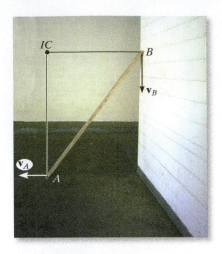

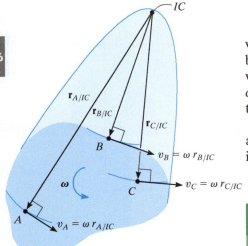

Fig. 16–19

Realize that the point chosen as the instantaneous center of zero velocity for the body *can only be used at the instant considered* since the body changes its position from one instant to the next. The locus of points which define the location of the IC during the body's motion is called a *centrode*, Fig. 16–18a, and so each point on the centrode acts as the IC for the body only for an instant.

Although the IC may be conveniently used to determine the velocity of any point in a body, it generally *does not have zero acceleration* and therefore it *should not* be used for finding the accelerations of points in a body.

Procedure for Analysis

The velocity of a point on a body which is subjected to general plane motion can be determined with reference to its instantaneous center of zero velocity provided the location of the IC is first established using one of the three methods described above.

- As shown on the kinematic diagram in Fig. 16–19, the body is imagined as "extended and pinned" at the IC so that, at the instant considered, it rotates about this pin with its angular velocity ω.

- The *magnitude* of velocity for each of the arbitrary points A, B, and C on the body can be determined by using the equation $v = \omega r$, where r is the radial distance from the IC to each point.

- The line of action of each velocity vector $\mathbf{v}$ is *perpendicular* to its associated radial line $\mathbf{r}$, and the velocity has a *sense of direction* which tends to move the point in a manner consistent with the angular rotation ω of the radial line, Fig. 16–19.

EXAMPLE | **16.9**

Show how to determine the location of the instantaneous center of zero velocity for (a) member BC shown in Fig. 16–20a; and (b) the link CB shown in Fig. 16–20c.

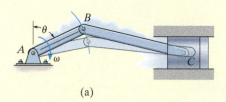

(a)

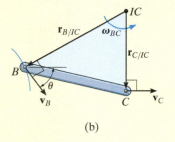

(b)

SOLUTION

Part (a).   As shown in Fig. 16–20a, point B moves in a circular path such that $\mathbf{v}_B$ is perpendicular to AB. Therefore, it acts at an angle θ from the horizontal as shown in Fig. 16–20b. The motion of point B causes the piston to move forward *horizontally* with a velocity $\mathbf{v}_C$. When lines are drawn perpendicular to $\mathbf{v}_B$ and $\mathbf{v}_C$, Fig. 16–20b, they intersect at the *IC*.

Part (b).   Points B and C follow circular paths of motion since links AB and DC are each subjected to rotation about a fixed axis, Fig. 16–20c. Since the velocity is always tangent to the path, at the instant considered, $\mathbf{v}_C$ on rod DC and $\mathbf{v}_B$ on rod AB are both directed vertically downward, along the axis of link CB, Fig. 16–20d. Radial lines drawn perpendicular to these two velocities form parallel lines which intersect at "infinity;" i.e., $r_{C/IC} \rightarrow \infty$ and $r_{B/IC} \rightarrow \infty$. Thus, $\omega_{CB} = (v_C/r_{C/IC}) \rightarrow 0$. As a result, link CB momentarily *translates*. An instant later, however, CB will move to a tilted position, causing the *IC* to move to some finite location.

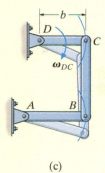

(c)

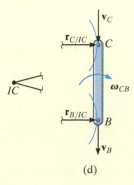

(d)

Fig. 16–20

EXAMPLE | **16.10**

Block D shown in Fig. 16–21a moves with a speed of 3 m/s. Determine the angular velocities of links BD and AB, at the instant shown.

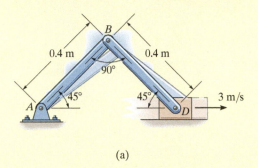

(a)

SOLUTION

As D moves to the right, it causes AB to rotate clockwise about point A. Hence, $\mathbf{v}_B$ is directed perpendicular to AB. The instantaneous center of zero velocity for BD is located at the intersection of the line segments drawn perpendicular to $\mathbf{v}_B$ and $\mathbf{v}_D$, Fig. 16–21b. From the geometry,

$$r_{B/IC} = 0.4 \tan 45° \text{ m} = 0.4 \text{ m}$$

$$r_{D/IC} = \frac{0.4 \text{ m}}{\cos 45°} = 0.5657 \text{ m}$$

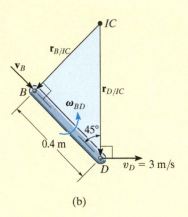

(b)

Since the magnitude of $\mathbf{v}_D$ is known, the angular velocity of link BD is

$$\omega_{BD} = \frac{v_D}{r_{D/IC}} = \frac{3 \text{ m/s}}{0.5657 \text{ m}} = 5.30 \text{ rad/s} \circlearrowright \qquad \textit{Ans.}$$

The velocity of B is therefore

$$v_B = \omega_{BD}(r_{B/IC}) = 5.30 \text{ rad/s} (0.4 \text{ m}) = 2.12 \text{ m/s} \quad \searrow 45°$$

From Fig. 16–21c, the angular velocity of AB is

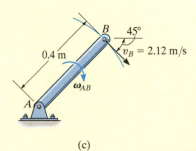

(c)

Fig. 16–21

$$\omega_{AB} = \frac{v_B}{r_{B/A}} = \frac{2.12 \text{ m/s}}{0.4 \text{ m}} = 5.30 \text{ rad/s} \circlearrowleft \qquad \textit{Ans.}$$

NOTE: Try to solve this problem by applying $\mathbf{v}_D = \mathbf{v}_B + \mathbf{v}_{D/B}$ to member BD.

EXAMPLE 16.11

The cylinder shown in Fig. 16–22a rolls without slipping between the two moving plates E and D. Determine the angular velocity of the cylinder and the velocity of its center C.

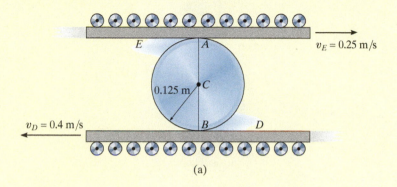

(a)

SOLUTION

Since no slipping occurs, the contact points A and B on the cylinder have the same velocities as the plates E and D, respectively. Furthermore, the velocities $\mathbf{v}_A$ and $\mathbf{v}_B$ are *parallel*, so that by the proportionality of right triangles the IC is located at a point on line AB, Fig. 16–22b. Assuming this point to be a distance x from B, we have

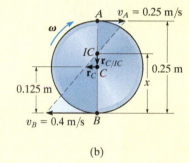

(b)

Fig. 16–22

$$v_B = \omega x; \qquad\qquad 0.4 \text{ m/s} = \omega x$$

$$v_A = \omega(0.25 \text{ m} - x); \qquad 0.25 \text{ m/s} = \omega(0.25 \text{ m} - x)$$

Dividing one equation into the other eliminates ω and yields

$$0.4(0.25 - x) = 0.25x$$

$$x = \frac{0.1}{0.65} = 0.1538 \text{ m}$$

Hence, the angular velocity of the cylinder is

$$\omega = \frac{v_B}{x} = \frac{0.4 \text{ m/s}}{0.1538 \text{ m}} = 2.60 \text{ rad/s} \,\circlearrowleft \qquad\qquad Ans.$$

The velocity of point C is therefore

$$v_C = \omega r_{C/IC} = 2.60 \text{ rad/s} \,(0.1538 \text{ m} - 0.125 \text{ m})$$

$$= 0.0750 \text{ m/s} \leftarrow \qquad\qquad Ans.$$

EXAMPLE   16.12

The crankshaft AB turns with a clockwise angular velocity of 10 rad/s, Fig. 16–23a. Determine the velocity of the piston at the instant shown.

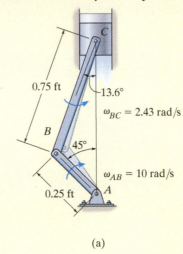

0.75 ft

13.6°

$\omega_{BC} = 2.43$ rad/s

B

45°

$\omega_{AB} = 10$ rad/s

0.25 ft A

(a)

SOLUTION

The crankshaft rotates about a fixed axis, and so the velocity of point B is

$$v_B = 10 \text{ rad/s } (0.25 \text{ ft}) = 2.50 \text{ ft/s } \measuredangle 45°$$

Since the directions of the velocities of B and C are known, then the location of the IC for the connecting rod BC is at the intersection of the lines extended from these points, perpendicular to $\mathbf{v}_B$ and $\mathbf{v}_C$, Fig. 16–23b. The magnitudes of $\mathbf{r}_{B/IC}$ and $\mathbf{r}_{C/IC}$ can be obtained from the geometry of the triangle and the law of sines, i.e.,

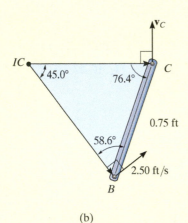

$\mathbf{v}_C$

IC 45.0°   76.4° C

58.6°   0.75 ft

2.50 ft/s

B

(b)

Fig. 16–23

$$\frac{0.75 \text{ ft}}{\sin 45°} = \frac{r_{B/IC}}{\sin 76.4°}$$

$$r_{B/IC} = 1.031 \text{ ft}$$

$$\frac{0.75 \text{ ft}}{\sin 45°} = \frac{r_{C/IC}}{\sin 58.6°}$$

$$r_{C/IC} = 0.9056 \text{ ft}$$

The rotational sense of $\boldsymbol{\omega}_{BC}$ must be the same as the rotation caused by $\mathbf{v}_B$ about the IC, which is counterclockwise. Therefore,

$$\omega_{BC} = \frac{v_B}{r_{B/IC}} = \frac{2.5 \text{ ft/s}}{1.031 \text{ ft}} = 2.425 \text{ rad/s}$$

Using this result, the velocity of the piston is

$$v_C = \omega_{BC} r_{C/IC} = (2.425 \text{ rad/s})(0.9056 \text{ ft}) = 2.20 \text{ ft/s} \quad \textit{Ans.}$$

PRELIMINARY PROBLEM

P16–2. Establish the location of the instantaneous center of zero velocity for finding the velocity of point B.

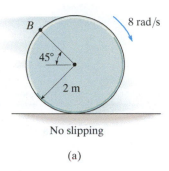

(a)

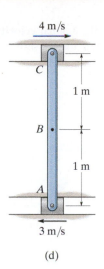

(d)

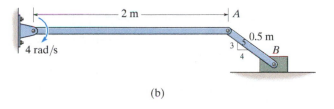

(b)

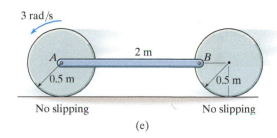

(e)

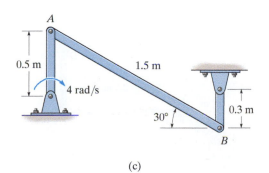

(c)

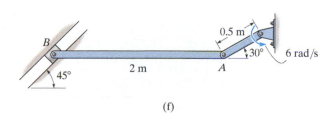

(f)

Prob. P16–2

FUNDAMENTAL PROBLEMS

F16–13. Determine the angular velocity of the rod and the velocity of point C at the instant shown.

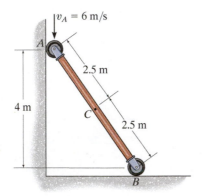

Prob. F16–13

F16–14. Determine the angular velocity of link BC and velocity of the piston C at the instant shown.

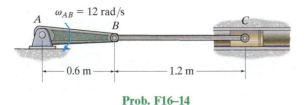

Prob. F16–14

F16–15. If the center O of the wheel is moving with a speed of $v_O = 6$ m/s, determine the velocity of point A on the wheel. The gear rack B is fixed.

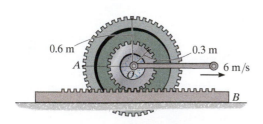

Prob. F16–15

F16–16. If cable AB is unwound with a speed of 3 m/s, and the gear rack C has a speed of 1.5 m/s, determine the angular velocity of the gear and the velocity of its center O.

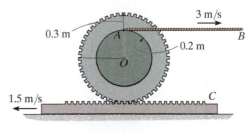

Prob. F16–16

F16–17. Determine the angular velocity of link BC and the velocity of the piston C at the instant shown.

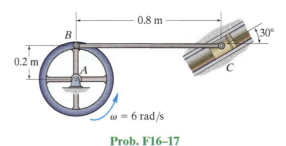

Prob. F16–17

F16–18. Determine the angular velocity of links BC and CD at the instant shown.

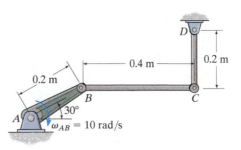

Prob. F16–18

PROBLEMS

16–81. In each case show graphically how to locate the instantaneous center of zero velocity of link *AB*. Assume the geometry is known.

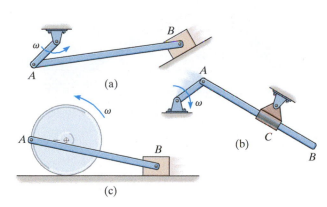

(a)

(b)

(c)

Prob. 16–81

16–82. Determine the angular velocity of link *AB* at the instant shown if block *C* is moving upward at 12 in/s.

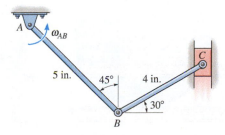

Prob. 16–82

16–83. The shaper mechanism is designed to give a slow cutting stroke and a quick return to a blade attached to the slider at *C*. Determine the angular velocity of the link *CB* at the instant shown, if the link *AB* is rotating at 4 rad/s.

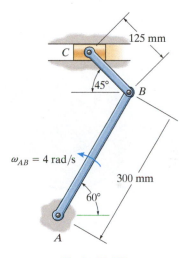

Prob. 16–83

*16–84. The conveyor belt is moving to the right at $v = 8$ ft/s, and at the same instant the cylinder is rolling counterclockwise at $\omega = 2$ rad/s without slipping. Determine the velocities of the cylinder's center *C* and point *B* at this instant.

16–85. The conveyor belt is moving to the right at $v = 12$ ft/s, and at the same instant the cylinder is rolling counterclockwise at $\omega = 6$ rad/s while its center has a velocity of 4 ft/s to the left. Determine the velocities of points *A* and *B* on the disk at this instant. Does the cylinder slip on the conveyor?

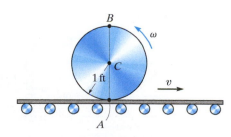

Probs. 16–84/85

16–86. As the cord unravels from the wheel's inner hub, the wheel is rotating at $\omega = 2$ rad/s at the instant shown. Determine the velocities of points A and B.

*16–88. If bar AB has an angular velocity $\omega_{AB} = 6$ rad/s, determine the velocity of the slider block C at the instant shown.

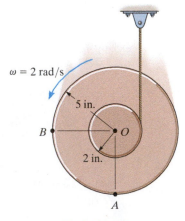

Prob. 16–86

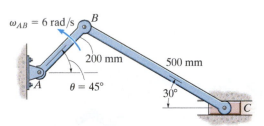

Prob. 16–88

16–87. If rod CD is rotating with an angular velocity $\omega_{CD} = 4$ rad/s, determine the angular velocities of rods AB and CB at the instant shown.

16–89. Show that if the rim of the wheel and its hub maintain contact with the three tracks as the wheel rolls, it is necessary that slipping occurs at the hub A if no slipping occurs at B. Under these conditions, what is the speed at A if the wheel has angular velocity ω?

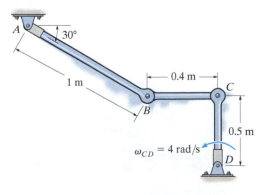

Prob. 16–87

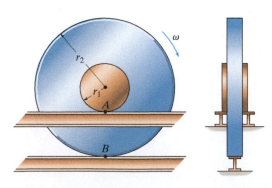

Prob. 16–89

16–90. Due to slipping, points A and B on the rim of the disk have the velocities shown. Determine the velocities of the center point C and point D at this instant.

16–91. Due to slipping, points A and B on the rim of the disk have the velocities shown. Determine the velocities of the center point C and point E at this instant.

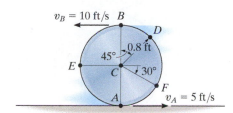

Probs. 16–90/91

*16–92. Member AB is rotating at $\omega_{AB} = 6$ rad/s. Determine the velocity of point D and the angular velocity of members BPD and CD.

16–93. Member AB is rotating at $\omega_{AB} = 6$ rad/s. Determine the velocity of point P, and the angular velocity of member BPD.

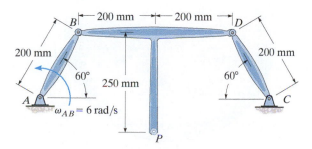

Probs. 16–92/93

16–94. The cylinder B rolls on the fixed cylinder A without slipping. If connected bar CD is rotating with an angular velocity $\omega_{CD} = 5$ rad/s, determine the angular velocity of cylinder B. Point C is a fixed point.

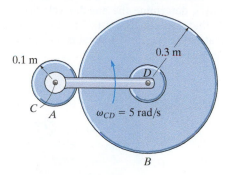

Prob. 16–94

16–95. As the car travels forward at 80 ft/s on a wet road, due to slipping, the rear wheels have an angular velocity $\omega = 100$ rad/s. Determine the speeds of points A, B, and C caused by the motion.

Prob. 16–95

*16–96. The pinion gear A rolls on the fixed gear rack B with an angular velocity $\omega = 8$ rad/s. Determine the velocity of the gear rack C.

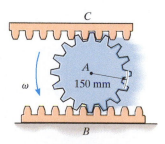

Prob. 16–96

16

16–97. If the hub gear H and ring gear R have angular velocities $\omega_H = 5$ rad/s and $\omega_R = 20$ rad/s, respectively, determine the angular velocity ω_S of the spur gear S and the angular velocity of its attached arm OA.

16–98. If the hub gear H has an angular velocity $\omega_H = 5$ rad/s, determine the angular velocity of the ring gear R so that the arm OA attached to the spur gear S remains stationary ($\omega_{OA} = 0$). What is the angular velocity of the spur gear?

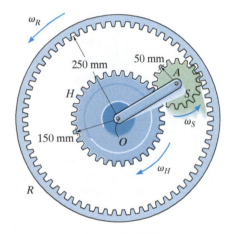

Probs. 16–97/98

16–99. The crankshaft AB rotates at $\omega_{AB} = 50$ rad/s about the fixed axis through point A, and the disk at C is held fixed in its support at E. Determine the angular velocity of rod CD at the instant shown.

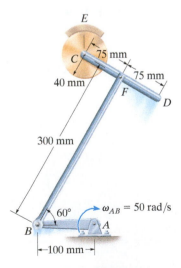

Prob. 16–99

*16–100. Cylinder A rolls on the *fixed cylinder* B without slipping. If bar CD is rotating with an angular velocity of $\omega_{CD} = 3$ rad/s, determine the angular velocity of A.

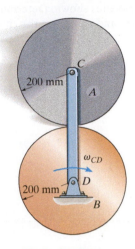

Prob. 16–100

16–101. The planet gear A is pin connected to the end of the link BC. If the link rotates about the fixed point B at 4 rad/s, determine the angular velocity of the ring gear R. The sun gear D is fixed from rotating.

16–102. Solve Prob. 16–101 if the sun gear D is rotating clockwise at $\omega_D = 5$ rad/s while link BC rotates counterclockwise at $\omega_{BC} = 4$ rad/s.

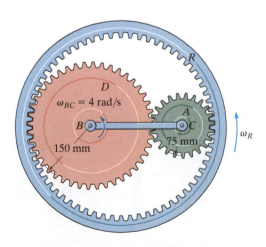

Probs. 16–101/102

16.7 Relative-Motion Analysis: Acceleration

An equation that relates the accelerations of two points on a bar (rigid body) subjected to general plane motion may be determined by differentiating $\mathbf{v}_B = \mathbf{v}_A + \mathbf{v}_{B/A}$ with respect to time. This yields

$$\frac{d\mathbf{v}_B}{dt} = \frac{d\mathbf{v}_A}{dt} + \frac{d\mathbf{v}_{B/A}}{dt}$$

The terms $d\mathbf{v}_B/dt = \mathbf{a}_B$ and $d\mathbf{v}_A/dt = \mathbf{a}_A$ are measured with respect to a set of *fixed x, y axes* and represent the *absolute accelerations* of points B and A. The last term represents the acceleration of B with respect to A as measured by an observer fixed to translating x', y' axes which have their origin at the base point A. In Sec. 16.5 it was shown that to this observer point B appears to move along a *circular arc* that has a radius of curvature $r_{B/A}$. Consequently, $\mathbf{a}_{B/A}$ can be expressed in terms of its tangential and normal components; i.e., $\mathbf{a}_{B/A} = (\mathbf{a}_{B/A})_t + (\mathbf{a}_{B/A})_n$, where $(a_{B/A})_t = \alpha r_{B/A}$ and $(a_{B/A})_n = \omega^2 r_{B/A}$. Hence, the relative-acceleration equation can be written in the form

$$\boxed{\mathbf{a}_B = \mathbf{a}_A + (\mathbf{a}_{B/A})_t + (\mathbf{a}_{B/A})_n} \qquad (16\text{–}17)$$

where

$\mathbf{a}_B$ = acceleration of point B

$\mathbf{a}_A$ = acceleration of point A

$(\mathbf{a}_{B/A})_t$ = tangential acceleration component of B with respect to A. The *magnitude* is $(a_{B/A})_t = \alpha r_{B/A}$, and the *direction* is perpendicular to $\mathbf{r}_{B/A}$.

$(\mathbf{a}_{B/A})_n$ = normal acceleration component of B with respect to A. The *magnitude* is $(a_{B/A})_n = \omega^2 r_{B/A}$, and the *direction* is always from B toward A.

The terms in Eq. 16–17 are represented graphically in Fig. 16–24. Here it is seen that at a given instant the acceleration of B, Fig. 16–24a, is determined by considering the bar to translate with an acceleration $\mathbf{a}_A$, Fig. 16–24b, and simultaneously rotate about the base point A with an instantaneous angular velocity $\boldsymbol{\omega}$ and angular acceleration $\boldsymbol{\alpha}$, Fig. 16–24c. Vector addition of these two effects, applied to B, yields $\mathbf{a}_B$, as shown in Fig. 16–24d. It should be noted from Fig. 16–24a that since points A and B move along *curved paths*, the accelerations of these points will have *both tangential and normal components*. (Recall that the acceleration of a point is *tangent to the path only* when the path is *rectilinear* or when it is an inflection point on a curve.)

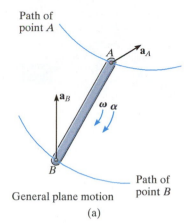

Path of point A

General plane motion

(a)

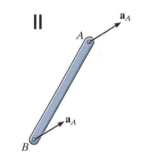

Translation

(b)

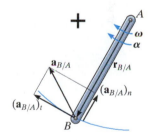

Rotation about the base point A

(c)

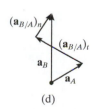

(d)

Fig. 16–24

Path of B

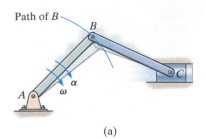

(a)

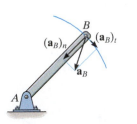

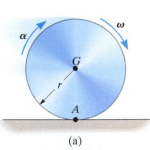

(b)

Fig. 16–25

Since the relative-acceleration components represent the effect of *circular motion* observed from translating axes having their origin at the base point A, these terms can be expressed as $(\mathbf{a}_{B/A})_t = \boldsymbol{\alpha} \times \mathbf{r}_{B/A}$ and $(\mathbf{a}_{B/A})_n = -\omega^2 \mathbf{r}_{B/A}$, Eq. 16–14. Hence, Eq. 16–17 becomes

$$\mathbf{a}_B = \mathbf{a}_A + \boldsymbol{\alpha} \times \mathbf{r}_{B/A} - \omega^2 \mathbf{r}_{B/A} \qquad (16\text{–}18)$$

where

$$\begin{aligned}
\mathbf{a}_B &= \text{acceleration of point } B\\
\mathbf{a}_A &= \text{acceleration of the base point } A\\
\boldsymbol{\alpha} &= \text{angular acceleration of the body}\\
\omega &= \text{angular velocity of the body}\\
\mathbf{r}_{B/A} &= \text{position vector directed from } A \text{ to } B
\end{aligned}$$

If Eq. 16–17 or 16–18 is applied in a practical manner to study the accelerated motion of a rigid body which is *pin connected* to two other bodies, it should be realized that points which are *coincident at the pin* move with the *same acceleration*, since the path of motion over which they travel is the *same*. For example, point B lying on either rod BA or BC of the crank mechanism shown in Fig. 16–25a has the same acceleration, since the rods are pin connected at B. Here the motion of B is along a *circular path*, so that $\mathbf{a}_B$ can be expressed in terms of its tangential and normal components. At the other end of rod BC point C moves along a *straight-lined path*, which is defined by the piston. Hence, $\mathbf{a}_C$ is horizontal, Fig. 16–25b.

Finally, consider a disk that rolls without slipping as shown in Fig. 16–26a. As a result, $v_A = 0$ and so from the kinematic diagram in Fig. 16–26b, the velocity of the mass center G is

$$\mathbf{v}_G = \mathbf{v}_A + \boldsymbol{\omega} \times \mathbf{r}_{G/A} = \mathbf{0} + (-\omega\mathbf{k}) \times (r\mathbf{j})$$

So that

$$v_G = \omega r \qquad (16\text{–}19)$$

This same result can also be determined using the IC method where point A is the *IC*.

Since G moves along a *straight line*, its acceleration in this case can be determined from the time derivative of its velocity.

$$\frac{dv_G}{dt} = \frac{d\omega}{dt} r$$

$$a_G = \alpha r \qquad (16\text{–}20)$$

These two important results were also obtained in Example 16–4. They apply as well to any circular object, such as a ball, gear, wheel, etc., that *rolls without slipping*.

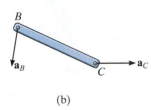

(a)

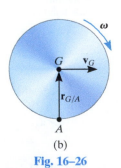

(b)

Fig. 16–26

Procedure for Analysis

The relative acceleration equation can be applied between any two points A and B on a body either by using a Cartesian vector analysis, or by writing the x and y scalar component equations directly.

Velocity Analysis.

- Determine the angular velocity $\boldsymbol{\omega}$ of the body by using a velocity analysis as discussed in Sec. 16.5 or 16.6. Also, determine the velocities $\mathbf{v}_A$ and $\mathbf{v}_B$ of points A and B if these points move along curved paths.

Vector Analysis
Kinematic Diagram.

- Establish the directions of the fixed x, y coordinates and draw the kinematic diagram of the body. Indicate on it $\mathbf{a}_A$, $\mathbf{a}_B$, $\boldsymbol{\omega}$, $\boldsymbol{\alpha}$, and $\mathbf{r}_{B/A}$.

- If points A and B move along curved paths, then their accelerations should be indicated in terms of their tangential and normal components, i.e., $\mathbf{a}_A = (\mathbf{a}_A)_t + (\mathbf{a}_A)_n$ and $\mathbf{a}_B = (\mathbf{a}_B)_t + (\mathbf{a}_B)_n$.

Acceleration Equation.

- To apply $\mathbf{a}_B = \mathbf{a}_A + \boldsymbol{\alpha} \times \mathbf{r}_{B/A} - \omega^2 \mathbf{r}_{B/A}$, express the vectors in Cartesian vector form and substitute them into the equation. Evaluate the cross product and then equate the respective $\mathbf{i}$ and $\mathbf{j}$ components to obtain two scalar equations.

- If the solution yields a negative answer for an unknown magnitude, it indicates that the sense of direction of the vector is opposite to that shown on the kinematic diagram.

Scalar Analysis
Kinematic Diagram.

- If the acceleration equation is applied in scalar form, then the magnitudes and directions of the relative-acceleration components $(\mathbf{a}_{B/A})_t$ and $(\mathbf{a}_{B/A})_n$ must be established. To do this draw a kinematic diagram such as shown in Fig. 16–24c. Since the body is considered to be momentarily "pinned" at the base point A, the magnitudes of these components are $(a_{B/A})_t = \alpha r_{B/A}$ and $(a_{B/A})_n = \omega^2 r_{B/A}$. Their sense of direction is established from the diagram such that $(\mathbf{a}_{B/A})_t$ acts perpendicular to $\mathbf{r}_{B/A}$, in accordance with the rotational motion $\boldsymbol{\alpha}$ of the body, and $(\mathbf{a}_{B/A})_n$ is directed from B toward A.*

Acceleration Equation.

- Represent the vectors in $\mathbf{a}_B = \mathbf{a}_A + (\mathbf{a}_{B/A})_t + (\mathbf{a}_{B/A})_n$ graphically by showing their magnitudes and directions underneath each term. The scalar equations are determined from the x and y components of these vectors.

*The notation $\mathbf{a}_B = \mathbf{a}_A + (\mathbf{a}_{B/A(\text{pin})})_t + (\mathbf{a}_{B/A(\text{pin})})_n$ may be helpful in recalling that A is assumed to be pinned.

The mechanism for a window is shown. Here CA rotates about a fixed axis through C, and AB undergoes general plane motion. Since point A moves along a curved path it has two components of acceleration, whereas point B moves along a straight track and the direction of its acceleration is specified. (© R.C. Hibbeler)

16

EXAMPLE | 16.13

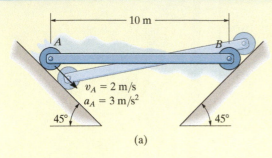

(a)

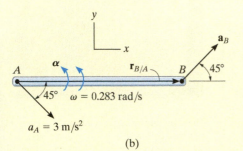

(b)

The rod AB shown in Fig. 16–27a is confined to move along the inclined planes at A and B. If point A has an acceleration of 3 m/s² and a velocity of 2 m/s, both directed down the plane at the instant the rod is horizontal, determine the angular acceleration of the rod at this instant.

SOLUTION I (VECTOR ANALYSIS)

We will apply the acceleration equation to points A and B on the rod. To do so it is first necessary to determine the angular velocity of the rod. Show that it is $\omega = 0.283$ rad/s $\circlearrowright$ using either the velocity equation or the method of instantaneous centers.

Kinematic Diagram.   Since points A and B both move along straight-line paths, they have *no* components of acceleration normal to the paths. There are two unknowns in Fig. 16–27b, namely, a_B and α.

Acceleration Equation.

$$\mathbf{a}_B = \mathbf{a}_A + \boldsymbol{\alpha} \times \mathbf{r}_{B/A} - \omega^2 \mathbf{r}_{B/A}$$

$$a_B \cos 45°\mathbf{i} + a_B \sin 45°\mathbf{j} = 3 \cos 45°\mathbf{i} - 3 \sin 45°\mathbf{j} + (\alpha\mathbf{k}) \times (10\mathbf{i}) - (0.283)^2(10\mathbf{i})$$

Carrying out the cross product and equating the **i** and **j** components yields

$$a_B \cos 45° = 3 \cos 45° - (0.283)^2(10) \tag{1}$$

$$a_B \sin 45° = -3 \sin 45° + \alpha(10) \tag{2}$$

Solving, we have

$$a_B = 1.87 \text{ m/s}^2 \measuredangle 45°$$

$$\alpha = 0.344 \text{ rad/s}^2 \circlearrowright \qquad \textit{Ans.}$$

SOLUTION II (SCALAR ANALYSIS)

From the kinematic diagram, showing the relative-acceleration components $(\mathbf{a}_{B/A})_t$ and $(\mathbf{a}_{B/A})_n$, Fig. 16–27c, we have

$$\mathbf{a}_B = \mathbf{a}_A + (\mathbf{a}_{B/A})_t + (\mathbf{a}_{B/A})_n$$

$$\begin{bmatrix} a_B \\ \measuredangle 45° \end{bmatrix} = \begin{bmatrix} 3 \text{ m/s}^2 \\ \measuredangle 45° \end{bmatrix} + \begin{bmatrix} \alpha(10 \text{ m}) \\ \uparrow \end{bmatrix} + \begin{bmatrix} (0.283 \text{ rad/s})^2(10 \text{ m}) \\ \leftarrow \end{bmatrix}$$

Equating the x and y components yields Eqs. 1 and 2, and the solution proceeds as before.

(c)

Fig. 16–27

EXAMPLE | 16.14

The disk rolls without slipping and has the angular motion shown in Fig. 16–28a. Determine the acceleration of point A at this instant.

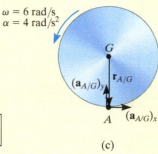

SOLUTION I (VECTOR ANALYSIS)

Kinematic Diagram.   Since no slipping occurs, applying Eq. 16–20,

$$a_G = \alpha r = (4 \text{ rad/s}^2)(0.5 \text{ ft}) = 2 \text{ ft/s}^2$$

Acceleration Equation.

We will apply the acceleration equation to points G and A, Fig. 16–28b,

$$\mathbf{a}_A = \mathbf{a}_G + \boldsymbol{\alpha} \times \mathbf{r}_{A/G} - \omega^2 \mathbf{r}_{A/G}$$
$$\mathbf{a}_A = -2\mathbf{i} + (4\mathbf{k}) \times (-0.5\mathbf{j}) - (6)^2(-0.5\mathbf{j})$$
$$= \{18\mathbf{j}\} \text{ ft/s}^2$$

SOLUTION II (SCALAR ANALYSIS)

Using the result for $a_G = 2 \text{ ft/s}^2$ determined above, and from the kinematic diagram, showing the relative motion $\mathbf{a}_{A/G}$, Fig. 16–28c, we have

$$\mathbf{a}_A = \mathbf{a}_G + (\mathbf{a}_{A/G})_x + (\mathbf{a}_{A/G})_y$$

$$\begin{bmatrix} (a_A)_x \\ \rightarrow \end{bmatrix} + \begin{bmatrix} (a_A)_y \\ \uparrow \end{bmatrix} = \begin{bmatrix} 2 \text{ ft/s}^2 \\ \leftarrow \end{bmatrix} + \begin{bmatrix} (4 \text{ rad/s}^2)(0.5 \text{ ft}) \\ \rightarrow \end{bmatrix} + \begin{bmatrix} (6 \text{ rad/s})^2(0.5 \text{ ft}) \\ \uparrow \end{bmatrix}$$

$\overset{+}{\rightarrow}$ $(a_A)_x = -2 + 2 = 0$

$+\uparrow$ $(a_A)_y = 18 \text{ ft/s}^2$

Therefore,

$$a_A = \sqrt{(0)^2 + (18 \text{ ft/s}^2)^2} = 18 \text{ ft/s}^2 \qquad \textit{Ans.}$$

Fig. 16–28

NOTE: The fact that $a_A = 18 \text{ ft/s}^2$ indicates that the instantaneous center of zero velocity, point A, is *not* a point of zero acceleration.

EXAMPLE   16.15

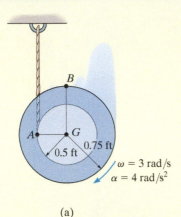

(a)

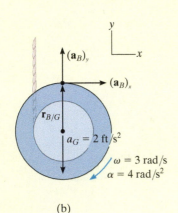

(b)

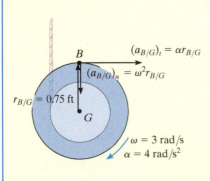

(c)

Fig. 16–29

The spool shown in Fig. 16–29a unravels from the cord, such that at the instant shown it has an angular velocity of 3 rad/s and an angular acceleration of 4 rad/s². Determine the acceleration of point B.

SOLUTION I (VECTOR ANALYSIS)

The spool "appears" to be rolling downward without slipping at point A. Therefore, we can use the results of Eq. 16–20 to determine the acceleration of point G, i.e.,

$$a_G = \alpha r = (4 \text{ rad/s}^2)(0.5 \text{ ft}) = 2 \text{ ft/s}^2$$

We will apply the acceleration equation to points G and B.

Kinematic Diagram.   Point B moves along a *curved path* having an *unknown* radius of curvature.* Its acceleration will be represented by its unknown x and y components as shown in Fig. 16–29b.

Acceleration Equation.

$$\mathbf{a}_B = \mathbf{a}_G + \boldsymbol{\alpha} \times \mathbf{r}_{B/G} - \omega^2 \mathbf{r}_{B/G}$$

$$(a_B)_x\mathbf{i} + (a_B)_y\mathbf{j} = -2\mathbf{j} + (-4\mathbf{k}) \times (0.75\mathbf{j}) - (3)^2(0.75\mathbf{j})$$

Equating the **i** and **j** terms, the component equations are

$$(a_B)_x = 4(0.75) = 3 \text{ ft/s}^2 \rightarrow \qquad\qquad (1)$$

$$(a_B)_y = -2 - 6.75 = -8.75 \text{ ft/s}^2 = 8.75 \text{ ft/s}^2 \downarrow \qquad (2)$$

The magnitude and direction of **a**$_B$ are therefore

$$a_B = \sqrt{(3)^2 + (8.75)^2} = 9.25 \text{ ft/s}^2 \qquad\qquad Ans.$$

$$\theta = \tan^{-1}\frac{8.75}{3} = 71.1° \searrow \qquad\qquad Ans.$$

SOLUTION II (SCALAR ANALYSIS)

This problem may be solved by writing the scalar component equations directly. The kinematic diagram in Fig. 16–29c shows the relative-acceleration components $(\mathbf{a}_{B/G})_t$ and $(\mathbf{a}_{B/G})_n$. Thus,

$$\mathbf{a}_B = \mathbf{a}_G + (\mathbf{a}_{B/G})_t + (\mathbf{a}_{B/G})_n$$

$$\begin{bmatrix} (a_B)_x \\ \rightarrow \end{bmatrix} + \begin{bmatrix} (a_B)_y \\ \uparrow \end{bmatrix}$$

$$= \begin{bmatrix} 2 \text{ ft/s}^2 \\ \downarrow \end{bmatrix} + \begin{bmatrix} 4 \text{ rad/s}^2 (0.75 \text{ ft}) \\ \rightarrow \end{bmatrix} + \begin{bmatrix} (3 \text{ rad/s})^2(0.75 \text{ ft}) \\ \downarrow \end{bmatrix}$$

The x and y components yield Eqs. 1 and 2 above.

*Realize that the path's radius of curvature ρ is not equal to the radius of the spool since the spool is not rotating about point G. Furthermore, ρ is not defined as the distance from A (IC) to B, since the location of the IC depends only on the velocity of a point and not the geometry of its path.

EXAMPLE | 16.16

The collar *C* in Fig. 16–30*a* moves downward with an acceleration of 1 m/s^2. At the instant shown, it has a speed of 2 m/s which gives links *CB* and *AB* an angular velocity $\omega_{AB} = \omega_{CB} = 10$ rad/s. (See Example 16.8.) Determine the angular accelerations of *CB* and *AB* at this instant.

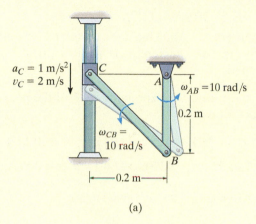

(a)

SOLUTION (VECTOR ANALYSIS)

Kinematic Diagram.   The kinematic diagrams of *both* links *AB* and *CB* are shown in Fig. 16–30*b*. To solve, we will apply the appropriate kinematic equation to each link.

Acceleration Equation.

Link *AB* (rotation about a fixed axis):

$$\mathbf{a}_B = \boldsymbol{\alpha}_{AB} \times \mathbf{r}_B - \omega_{AB}^2 \mathbf{r}_B$$
$$\mathbf{a}_B = (\alpha_{AB}\mathbf{k}) \times (-0.2\mathbf{j}) - (10)^2(-0.2\mathbf{j})$$
$$\mathbf{a}_B = 0.2\alpha_{AB}\mathbf{i} + 20\mathbf{j}$$

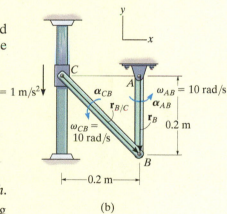

(b)

Fig. 16–30

Note that $\mathbf{a}_B$ has *n* and *t* components since it moves along a *circular path*.

Link *BC* (general plane motion): Using the result for $\mathbf{a}_B$ and applying Eq. 16–18, we have

$$\mathbf{a}_B = \mathbf{a}_C + \boldsymbol{\alpha}_{CB} \times \mathbf{r}_{B/C} - \omega_{CB}^2 \mathbf{r}_{B/C}$$
$$0.2\alpha_{AB}\mathbf{i} + 20\mathbf{j} = -1\mathbf{j} + (\alpha_{CB}\mathbf{k}) \times (0.2\mathbf{i} - 0.2\mathbf{j}) - (10)^2(0.2\mathbf{i} - 0.2\mathbf{j})$$
$$0.2\alpha_{AB}\mathbf{i} + 20\mathbf{j} = -1\mathbf{j} + 0.2\alpha_{CB}\mathbf{j} + 0.2\alpha_{CB}\mathbf{i} - 20\mathbf{i} + 20\mathbf{j}$$

Thus,

$$0.2\alpha_{AB} = 0.2\alpha_{CB} - 20$$
$$20 = -1 + 0.2\alpha_{CB} + 20$$

Solving,

$$\alpha_{CB} = 5 \text{ rad/s}^2 \; \circlearrowright \qquad \textit{Ans.}$$
$$\alpha_{AB} = -95 \text{ rad/s}^2 = 95 \text{ rad/s}^2 \; \circlearrowleft \qquad \textit{Ans.}$$

EXAMPLE 16.17

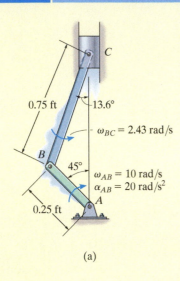

(a)

(b)

Fig. 16–31

The crankshaft AB turns with a clockwise angular acceleration of 20 rad/s^2, Fig. 16–31a. Determine the acceleration of the piston at the instant AB is in the position shown. At this instant $\omega_{AB} = 10$ rad/s and $\omega_{BC} = 2.43$ rad/s. (See Example 16.12.)

SOLUTION (VECTOR ANALYSIS)

Kinematic Diagram.   The kinematic diagrams for both AB and BC are shown in Fig. 16–31b. Here $\mathbf{a}_C$ is vertical since C moves along a straight-line path.

Acceleration Equation.   Expressing each of the position vectors in Cartesian vector form

$$\mathbf{r}_B = \{-0.25 \sin 45°\mathbf{i} + 0.25 \cos 45°\mathbf{j}\} \text{ ft} = \{-0.177\mathbf{i} + 0.177\mathbf{j}\} \text{ ft}$$

$$\mathbf{r}_{C/B} = \{0.75 \sin 13.6°\mathbf{i} + 0.75 \cos 13.6°\mathbf{j}\} \text{ ft} = \{0.177\mathbf{i} + 0.729\mathbf{j}\} \text{ ft}$$

Crankshaft AB (rotation about a fixed axis):

$$\mathbf{a}_B = \boldsymbol{\alpha}_{AB} \times \mathbf{r}_B - \omega_{AB}^2\mathbf{r}_B$$

$$= (-20\mathbf{k}) \times (-0.177\mathbf{i} + 0.177\mathbf{j}) - (10)^2(-0.177\mathbf{i} + 0.177\mathbf{j})$$

$$= \{21.21\mathbf{i} - 14.14\mathbf{j}\} \text{ ft/s}^2$$

Connecting Rod BC (general plane motion): Using the result for $\mathbf{a}_B$ and noting that $\mathbf{a}_C$ is in the vertical direction, we have

$$\mathbf{a}_C = \mathbf{a}_B + \boldsymbol{\alpha}_{BC} \times \mathbf{r}_{C/B} - \omega_{BC}^2\mathbf{r}_{C/B}$$

$$a_C\mathbf{j} = 21.21\mathbf{i} - 14.14\mathbf{j} + (\alpha_{BC}\mathbf{k}) \times (0.177\mathbf{i} + 0.729\mathbf{j}) - (2.43)^2(0.177\mathbf{i} + 0.729\mathbf{j})$$

$$a_C\mathbf{j} = 21.21\mathbf{i} - 14.14\mathbf{j} + 0.177\alpha_{BC}\mathbf{j} - 0.729\alpha_{BC}\mathbf{i} - 1.04\mathbf{i} - 4.30\mathbf{j}$$

$$0 = 20.17 - 0.729\alpha_{BC}$$

$$a_C = 0.177\alpha_{BC} - 18.45$$

Solving yields

$$\alpha_{BC} = 27.7 \text{ rad/s}^2 \circlearrowright$$

$$a_C = -13.5 \text{ ft/s}^2 \qquad\qquad Ans.$$

NOTE: Since the piston is moving upward, the negative sign for a_C indicates that the piston is decelerating, i.e., $\mathbf{a}_C = \{-13.5\mathbf{j}\}$ ft/s^2. This causes the speed of the piston to decrease until AB becomes vertical, at which time the piston is momentarily at rest.

PRELIMINARY PROBLEM

P16–3.   Set up the relative acceleration equation between points A and B. The angular velocity is given.

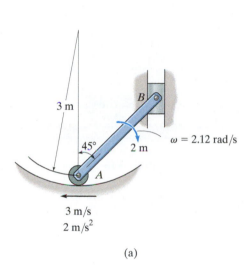

(a)

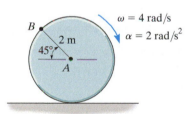

No slipping

(b)

3 rad/s
2 rad/s²

(c)

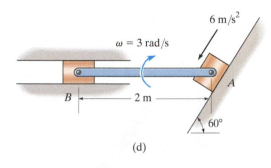

(d)

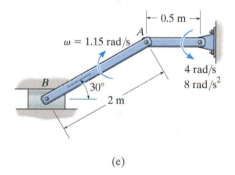

(e)

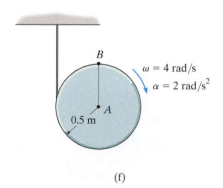

(f)

Prob. P16–3

FUNDAMENTAL PROBLEMS

F16–19. At the instant shown, end A of the rod has the velocity and acceleration shown. Determine the angular acceleration of the rod and acceleration of end B of the rod.

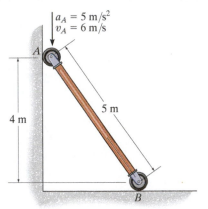

Prob. F16–19

F16–20. The gear rolls on the fixed rack with an angular velocity of $\omega = 12$ rad/s and angular acceleration of $\alpha = 6$ rad/s². Determine the acceleration of point A.

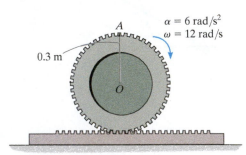

Prob. F16–20

F16–21. The gear rolls on the fixed rack B. At the instant shown, the center O of the gear moves with a velocity of $v_O = 6$ m/s and acceleration of $a_O = 3$ m/s². Determine the angular acceleration of the gear and acceleration of point A at this instant.

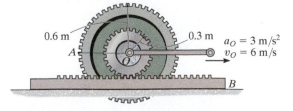

Prob. F16–21

F16–22. At the instant shown, cable AB has a velocity of 3 m/s and acceleration of 1.5 m/s², while the gear rack has a velocity of 1.5 m/s and acceleration of 0.75 m/s². Determine the angular acceleration of the gear at this instant.

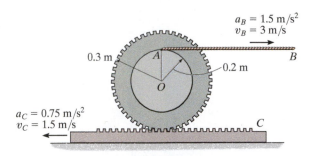

Prob. F16–22

F16–23. At the instant shown, the wheel rotates with an angular velocity of $\omega = 12$ rad/s and an angular acceleration of $\alpha = 6$ rad/s². Determine the angular acceleration of link BC at the instant shown.

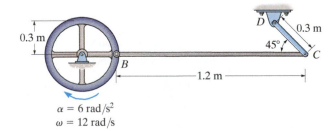

Prob. F16–23

F16–24. At the instant shown, wheel A rotates with an angular velocity of $\omega = 6$ rad/s and an angular acceleration of $\alpha = 3$ rad/s². Determine the angular acceleration of link BC and the acceleration of piston C.

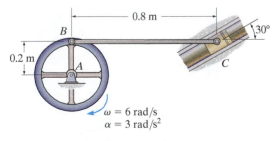

Prob. F16–24

PROBLEMS

16–103.   Bar AB has the angular motions shown. Determine the velocity and acceleration of the slider block C at this instant.

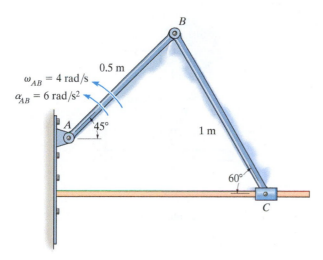

Prob. 16–103

16–106.   Member AB has the angular motions shown. Determine the velocity and acceleration of the slider block C at this instant.

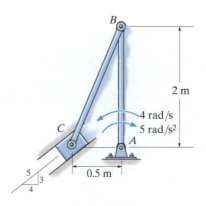

Prob. 16–106

16

*16–104.   At a given instant the bottom A of the ladder has an acceleration $a_A = 4$ ft/s^2 and velocity $v_A = 6$ ft/s, both acting to the left. Determine the acceleration of the top of the ladder, B, and the ladder's angular acceleration at this same instant.

16–105.   At a given instant the top B of the ladder has an acceleration $a_B = 2$ ft/s^2 and a velocity of $v_B = 4$ ft/s, both acting downward. Determine the acceleration of the bottom A of the ladder, and the ladder's angular acceleration at this instant.

16–107.   At a given instant the roller A on the bar has the velocity and acceleration shown. Determine the velocity and acceleration of the roller B, and the bar's angular velocity and angular acceleration at this instant.

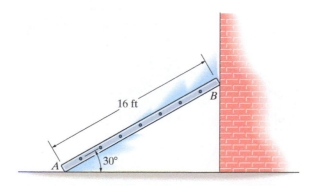

Probs. 16–104/105

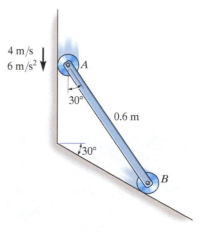

Prob. 16–107

*16–108. The rod is confined to move along the path due to the pins at its ends. At the instant shown, point A has the motion shown. Determine the velocity and acceleration of point B at this instant.

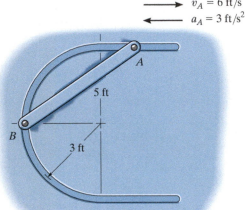

$v_A = 6 \text{ ft/s}$

$a_A = 3 \text{ ft/s}^2$

A

5 ft

B

3 ft

Prob. 16–108

16–110. The slider block has the motion shown. Determine the angular velocity and angular acceleration of the wheel at this instant.

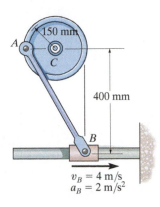

150 mm

A

C

400 mm

B

$v_B = 4 \text{ m/s}$

$a_B = 2 \text{ m/s}^2$

Prob. 16–110

16–111. At a given instant the slider block A is moving to the right with the motion shown. Determine the angular acceleration of link AB and the acceleration of point B at this instant.

16–109. Member AB has the angular motions shown. Determine the angular velocity and angular acceleration of members CB and DC.

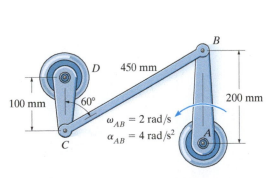

B

D　450 mm

100 mm　$60°$

200 mm

$\omega_{AB} = 2 \text{ rad/s}$

$\alpha_{AB} = 4 \text{ rad/s}^2$

A

C

Prob. 16–109

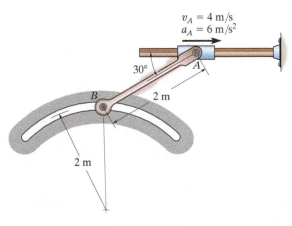

$v_A = 4 \text{ m/s}$

$a_A = 6 \text{ m/s}^2$

$30°$

A

B　2 m

2 m

Prob. 16–111

*16–112. Determine the angular acceleration of link CD if link AB has the angular velocity and angular acceleration shown.

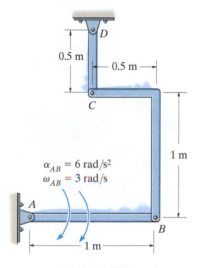

$\alpha_{AB} = 6 \text{ rad/s}^2$
$\omega_{AB} = 3 \text{ rad/s}$

Prob. 16–112

16–113. The reel of rope has the angular motion shown. Determine the velocity and acceleration of point A at the instant shown.

16–114. The reel of rope has the angular motion shown. Determine the velocity and acceleration of point B at the instant shown.

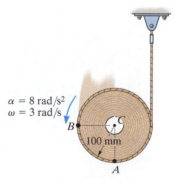

$\alpha = 8 \text{ rad/s}^2$
$\omega = 3 \text{ rad/s}$

Probs. 16–113/114

16–115. A cord is wrapped around the inner spool of the gear. If it is pulled with a constant velocity $\mathbf{v}$, determine the velocities and accelerations of points A and B. The gear rolls on the fixed gear rack.

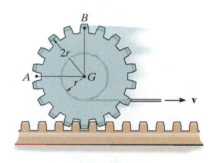

Prob. 16–115

*16–116. The disk has an angular acceleration $\alpha = 8 \text{ rad/s}^2$ and angular velocity $\omega = 3 \text{ rad/s}$ at the instant shown. If it does not slip at A, determine the acceleration of point B.

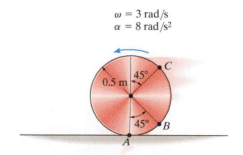

$\omega = 3 \text{ rad/s}$
$\alpha = 8 \text{ rad/s}^2$

Prob. 16–116

16

16–117. The disk has an angular acceleration $\alpha = 8$ rad/s² and angular velocity $\omega = 3$ rad/s at the instant shown. If it does not slip at A, determine the acceleration of point C.

16–119. The wheel rolls without slipping such that at the instant shown it has an angular velocity ω and angular acceleration α. Determine the velocity and acceleration of point B on the rod at this instant.

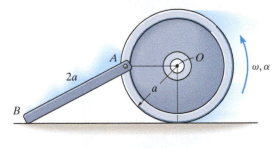

Prob. 16–119

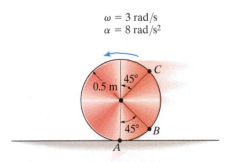

Prob. 16–117

***16–120.** The collar is moving downward with the motion shown. Determine the angular velocity and angular acceleration of the gear at the instant shown as it rolls along the fixed gear rack.

16–118. A single pulley having both an inner and outer rim is pin connected to the block at A. As cord CF unwinds from the inner rim of the pulley with the motion shown, cord DE unwinds from the outer rim. Determine the angular acceleration of the pulley and the acceleration of the block at the instant shown.

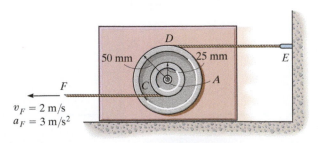

Prob. 16–118

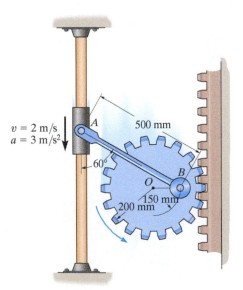

Prob. 16–120

16–121. The tied crank and gear mechanism gives rocking motion to crank *AC*, necessary for the operation of a printing press. If link *DE* has the angular motion shown, determine the respective angular velocities of gear *F* and crank *AC* at this instant, and the angular acceleration of crank *AC*.

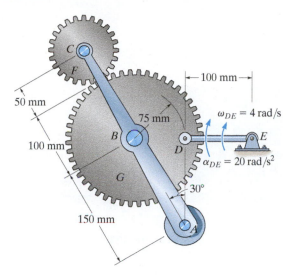

Prob. 16–121

16–122. If member *AB* has the angular motion shown, determine the angular velocity and angular acceleration of member *CD* at the instant shown.

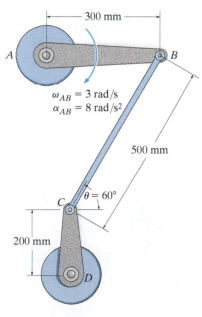

Prob. 16–122

16–123. If member *AB* has the angular motion shown, determine the velocity and acceleration of point *C* at the instant shown.

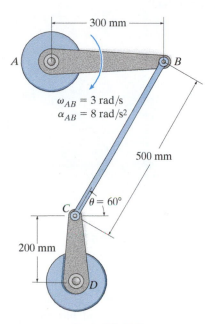

Prob. 16–123

*16–124. The disk rolls without slipping such that it has an angular acceleration of $\alpha = 4 \text{ rad/s}^2$ and angular velocity of $\omega = 2 \text{ rad/s}$ at the instant shown. Determine the acceleration of points *A* and *B* on the link and the link's angular acceleration at this instant. Assume point *A* lies on the periphery of the disk, 150 mm from *C*.

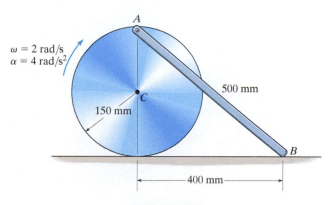

Prob. 16–124

16

16–125. The ends of the bar AB are confined to move along the paths shown. At a given instant, A has a velocity of $v_A = 4$ ft/s and an acceleration of $a_A = 7$ ft/s^2. Determine the angular velocity and angular acceleration of AB at this instant.

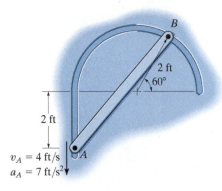

Prob. 16–125

16–126. The mechanism produces intermittent motion of link AB. If the sprocket S is turning with an angular acceleration $\alpha_S = 2$ rad/s^2 and has an angular velocity $\omega_S = 6$ rad/s at the instant shown, determine the angular velocity and angular acceleration of link AB at this instant. The sprocket S is mounted on a shaft which is *separate* from a collinear shaft attached to AB at A. The pin at C is attached to one of the chain links such that it moves vertically downward.

16–127. The slider block moves with a velocity of $v_B = 5$ ft/s and an acceleration of $a_B = 3$ ft/s^2. Determine the angular acceleration of rod AB at the instant shown.

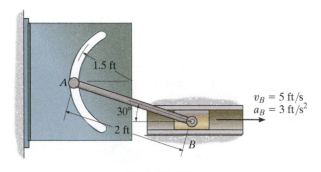

Prob. 16–127

*****16–128.** The slider block moves with a velocity of $v_B = 5$ ft/s and an acceleration of $a_B = 3$ ft/s^2. Determine the acceleration of A at the instant shown.

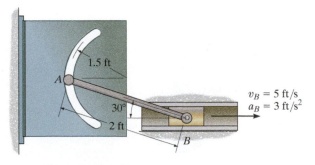

Prob. 16–128

Prob. 16–126

16.8 Relative-Motion Analysis using Rotating Axes

In the previous sections the relative-motion analysis for velocity and acceleration was described using a translating coordinate system. This type of analysis is useful for determining the motion of points on the *same* rigid body, or the motion of points located on several pin-connected bodies. In some problems, however, rigid bodies (mechanisms) are constructed such that *sliding* will occur at their connections. The kinematic analysis for such cases is best performed if the motion is analyzed using a coordinate system which both *translates* and *rotates*. Furthermore, this frame of reference is useful for analyzing the motions of two points on a mechanism which are *not* located in the *same* body and for specifying the kinematics of particle motion when the particle moves along a rotating path.

In the following analysis two equations will be developed which relate the velocity and acceleration of two points, one of which is the origin of a moving frame of reference subjected to both a translation and a rotation in the plane.*

Position.   Consider the two points A and B shown in Fig. 16–32a. Their location is specified by the position vectors $\mathbf{r}_A$ and $\mathbf{r}_B$, which are measured with respect to the fixed X, Y, Z coordinate system. As shown in the figure, the "base point" A represents the origin of the x, y, z coordinate system, which is assumed to be both translating and rotating with respect to the X, Y, Z system. The position of B with respect to A is specified by the relative-position vector $\mathbf{r}_{B/A}$. The components of this vector may be expressed either in terms of unit vectors along the X, Y axes, i.e., $\mathbf{I}$ and $\mathbf{J}$, or by unit vectors along the x, y axes, i.e., $\mathbf{i}$ and $\mathbf{j}$. For the development which follows, $\mathbf{r}_{B/A}$ will be measured with respect to the moving x, y frame of reference. Thus, if B has coordinates (x_B, y_B), Fig. 16–32a, then

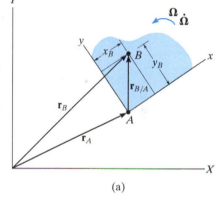

$$\mathbf{r}_{B/A} = x_B\mathbf{i} + y_B\mathbf{j}$$

Using vector addition, the three position vectors in Fig. 16–32a are related by the equation

$$\boxed{\mathbf{r}_B = \mathbf{r}_A + \mathbf{r}_{B/A}} \qquad (16\text{–}21)$$

At the instant considered, point A has a velocity $\mathbf{v}_A$ and an acceleration $\mathbf{a}_A$, while the angular velocity and angular acceleration of the x, y axes are $\boldsymbol{\Omega}$ (omega) and $\dot{\boldsymbol{\Omega}} = d\boldsymbol{\Omega}/dt$, respectively.

Fig. 16–32

*The more general, three-dimensional motion of the points is developed in Sec. 20.4.

Velocity. The velocity of point B is determined by taking the time derivative of Eq. 16–21, which yields

$$\mathbf{v}_B = \mathbf{v}_A + \frac{d\mathbf{r}_{B/A}}{dt} \qquad (16\text{–}22)$$

The last term in this equation is evaluated as follows:

$$\frac{d\mathbf{r}_{B/A}}{dt} = \frac{d}{dt}(x_B\mathbf{i} + y_B\mathbf{j})$$

$$= \frac{dx_B}{dt}\mathbf{i} + x_B\frac{d\mathbf{i}}{dt} + \frac{dy_B}{dt}\mathbf{j} + y_B\frac{d\mathbf{j}}{dt}$$

$$= \left(\frac{dx_B}{dt}\mathbf{i} + \frac{dy_B}{dt}\mathbf{j}\right) + \left(x_B\frac{d\mathbf{i}}{dt} + y_B\frac{d\mathbf{j}}{dt}\right) \qquad (16\text{–}23)$$

The two terms in the first set of parentheses represent the components of velocity of point B as measured by an observer attached to the moving x, y, z coordinate system. These terms will be denoted by vector $(\mathbf{v}_{B/A})_{xyz}$. In the second set of parentheses the instantaneous time rate of change of the unit vectors $\mathbf{i}$ and $\mathbf{j}$ is measured by an observer located in the fixed X, Y, Z coordinate system. These changes, $d\mathbf{i}$ and $d\mathbf{j}$, are due *only* to the *rotation $d\theta$* of the x, y, z axes, causing $\mathbf{i}$ to become $\mathbf{i}' = \mathbf{i} + d\mathbf{i}$ and $\mathbf{j}$ to become $\mathbf{j}' = \mathbf{j} + d\mathbf{j}$, Fig. 16–32b. As shown, the *magnitudes* of both $d\mathbf{i}$ and $d\mathbf{j}$ equal $1\ d\theta$, since $i = i' = j = j' = 1$. The *direction* of $d\mathbf{i}$ is defined by $+\mathbf{j}$, since $d\mathbf{i}$ is tangent to the path described by the arrowhead of $\mathbf{i}$ in the limit as $\Delta t \rightarrow dt$. Likewise, $d\mathbf{j}$ acts in the $-\mathbf{i}$ direction, Fig. 16–32b. Hence,

$$\frac{d\mathbf{i}}{dt} = \frac{d\theta}{dt}(\mathbf{j}) = \Omega\mathbf{j} \qquad \frac{d\mathbf{j}}{dt} = \frac{d\theta}{dt}(-\mathbf{i}) = -\Omega\mathbf{i}$$

Viewing the axes in three dimensions, Fig. 16–32c, and noting that $\mathbf{\Omega} = \Omega\mathbf{k}$, we can express the above derivatives in terms of the cross product as

$$\frac{d\mathbf{i}}{dt} = \mathbf{\Omega} \times \mathbf{i} \qquad \frac{d\mathbf{j}}{dt} = \mathbf{\Omega} \times \mathbf{j} \qquad (16\text{–}24)$$

Substituting these results into Eq. 16–23 and using the distributive property of the vector cross product, we obtain

$$\frac{d\mathbf{r}_{B/A}}{dt} = (\mathbf{v}_{B/A})_{xyz} + \mathbf{\Omega} \times (x_B\mathbf{i} + y_B\mathbf{j}) = (\mathbf{v}_{B/A})_{xyz} + \mathbf{\Omega} \times \mathbf{r}_{B/A} \qquad (16\text{–}25)$$

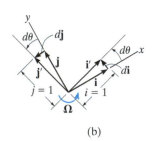

(b)

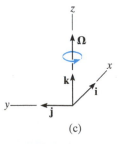

(c)

Fig. 16–32 (cont.)

Hence, Eq. 16–22 becomes

$$\mathbf{v}_B = \mathbf{v}_A + \mathbf{\Omega} \times \mathbf{r}_{B/A} + (\mathbf{v}_{B/A})_{xyz} \qquad (16\text{–}26)$$

where

$\mathbf{v}_B$ = velocity of B, measured from the X, Y, Z reference

$\mathbf{v}_A$ = velocity of the origin A of the x, y, z reference, measured from the X, Y, Z reference

$(\mathbf{v}_{B/A})_{xyz}$ = velocity of "B with respect to A," as measured by an observer attached to the rotating x, y, z reference

$\mathbf{\Omega}$ = angular velocity of the x, y, z reference, measured from the X, Y, Z reference

$\mathbf{r}_{B/A}$ = position of B with respect to A

Comparing Eq. 16–26 with Eq. 16–16 ($\mathbf{v}_B = \mathbf{v}_A + \mathbf{\Omega} \times \mathbf{r}_{B/A}$), which is valid for a translating frame of reference, it can be seen that the only difference between these two equations is represented by the term $(\mathbf{v}_{B/A})_{xyz}$.

When applying Eq. 16–26 it is often useful to understand what each of the terms represents. In order of appearance, they are as follows:

$\mathbf{v}_B \quad \begin{cases} \text{absolute velocity of } B \end{cases} \qquad \begin{cases} \text{motion of } B \text{ observed} \\ \text{from the } X, Y, Z \text{ frame} \end{cases}$

(equals)

$\mathbf{v}_A \quad \begin{cases} \text{absolute velocity of the} \\ \text{origin of } x, y, z \text{ frame} \end{cases}$

(plus)

$\mathbf{\Omega} \times \mathbf{r}_{B/A} \quad \begin{cases} \text{angular velocity effect caused} \\ \text{by rotation of } x, y, z \text{ frame} \end{cases}$

$\left. \begin{array}{c} \\ \\ \\ \end{array} \right\} \begin{array}{l} \text{motion of } x, y, z \text{ frame} \\ \text{observed from the} \\ X, Y, Z \text{ frame} \end{array}$

(plus)

$(\mathbf{v}_{B/A})_{xyz} \quad \begin{cases} \text{velocity of } B \\ \text{with respect to } A \end{cases} \qquad \begin{cases} \text{motion of } B \text{ observed} \\ \text{from the } x, y, z \text{ frame} \end{cases}$

Acceleration.   The acceleration of B, observed from the X, Y, Z coordinate system, may be expressed in terms of its motion measured with respect to the rotating system of coordinates by taking the time derivative of Eq. 16–26.

$$\frac{d\mathbf{v}_B}{dt} = \frac{d\mathbf{v}_A}{dt} + \frac{d\mathbf{\Omega}}{dt} \times \mathbf{r}_{B/A} + \mathbf{\Omega} \times \frac{d\mathbf{r}_{B/A}}{dt} + \frac{d(\mathbf{v}_{B/A})_{xyz}}{dt}$$

$$\mathbf{a}_B = \mathbf{a}_A + \dot{\mathbf{\Omega}} \times \mathbf{r}_{B/A} + \mathbf{\Omega} \times \frac{d\mathbf{r}_{B/A}}{dt} + \frac{d(\mathbf{v}_{B/A})_{xyz}}{dt} \qquad (16\text{–}27)$$

Here $\dot{\mathbf{\Omega}} = d\mathbf{\Omega}/dt$ is the angular acceleration of the x, y, z coordinate system. Since $\mathbf{\Omega}$ is always perpendicular to the plane of motion, then $\dot{\mathbf{\Omega}}$ measures *only the change in magnitude* of $\mathbf{\Omega}$. The derivative $d\mathbf{r}_{B/A}/dt$ is defined by Eq. 16–25, so that

$$\mathbf{\Omega} \times \frac{d\mathbf{r}_{B/A}}{dt} = \mathbf{\Omega} \times (\mathbf{v}_{B/A})_{xyz} + \mathbf{\Omega} \times (\mathbf{\Omega} \times \mathbf{r}_{B/A}) \qquad (16\text{–}28)$$

Finding the time derivative of $(\mathbf{v}_{B/A})_{xyz} = (v_{B/A})_x \mathbf{i} + (v_{B/A})_y \mathbf{j}$,

$$\frac{d(\mathbf{v}_{B/A})_{xyz}}{dt} = \left[\frac{d(v_{B/A})_x}{dt} \mathbf{i} + \frac{d(v_{B/A})_y}{dt} \mathbf{j} \right] + \left[(v_{B/A})_x \frac{d\mathbf{i}}{dt} + (v_{B/A})_y \frac{d\mathbf{j}}{dt} \right]$$

The two terms in the first set of brackets represent the components of acceleration of point B as measured by an observer attached to the rotating coordinate system. These terms will be denoted by $(\mathbf{a}_{B/A})_{xyz}$. The terms in the second set of brackets can be simplified using Eqs. 16–24.

$$\frac{d(\mathbf{v}_{B/A})_{xyz}}{dt} = (\mathbf{a}_{B/A})_{xyz} + \mathbf{\Omega} \times (\mathbf{v}_{B/A})_{xyz}$$

Substituting this and Eq. 16–28 into Eq. 16–27 and rearranging terms,

$$\boxed{\mathbf{a}_B = \mathbf{a}_A + \dot{\mathbf{\Omega}} \times \mathbf{r}_{B/A} + \mathbf{\Omega} \times (\mathbf{\Omega} \times \mathbf{r}_{B/A}) + 2\mathbf{\Omega} \times (\mathbf{v}_{B/A})_{xyz} + (\mathbf{a}_{B/A})_{xyz}}$$

$$(16\text{–}29)$$

where

$\mathbf{a}_B =$ acceleration of B, measured from the X, Y, Z reference

$\mathbf{a}_A =$ acceleration of the origin A of the x, y, z reference, measured from the X, Y, Z reference

$(\mathbf{a}_{B/A})_{xyz}, (\mathbf{v}_{B/A})_{xyz} =$ acceleration and velocity of B with respect to A, as measured by an observer attached to the rotating x, y, z reference

$\dot{\mathbf{\Omega}}, \mathbf{\Omega} =$ angular acceleration and angular velocity of the x, y, z reference, measured from the X, Y, Z reference

$\mathbf{r}_{B/A} =$ position of B with respect to A

If Eq. 16–29 is compared with Eq. 16–18, written in the form $\mathbf{a}_B = \mathbf{a}_A + \dot{\mathbf{\Omega}} \times \mathbf{r}_{B/A} + \mathbf{\Omega} \times (\mathbf{\Omega} \times \mathbf{r}_{B/A})$, which is valid for a translating frame of reference, it can be seen that the difference between these two equations is represented by the terms $2\mathbf{\Omega} \times (\mathbf{v}_{B/A})_{xyz}$ and $(\mathbf{a}_{B/A})_{xyz}$. In particular, $2\mathbf{\Omega} \times (\mathbf{v}_{B/A})_{xyz}$ is called the *Coriolis acceleration*, named after the French engineer G. C. Coriolis, who was the first to determine it. This term represents the difference in the acceleration of B as measured from nonrotating and rotating x, y, z axes. As indicated by the vector cross product, the Coriolis acceleration will *always* be perpendicular to both $\mathbf{\Omega}$ and $(\mathbf{v}_{B/A})_{xyz}$. It is an important component of the acceleration which must be considered whenever rotating reference frames are used. This often occurs, for example, when studying the accelerations and forces which act on rockets, long-range projectiles, or other bodies having motions whose measurements are significantly affected by the rotation of the earth.

The following interpretation of the terms in Eq. 16–29 may be useful when applying this equation to the solution of problems.

16

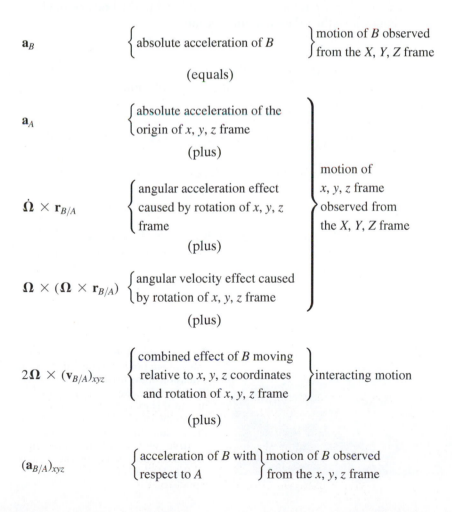

$\mathbf{a}_B$ $\left\{\begin{array}{l}\text{absolute acceleration of } B\end{array}\right.$ $\left.\begin{array}{l}\text{motion of } B \text{ observed} \\ \text{from the } X, Y, Z \text{ frame}\end{array}\right\}$

(equals)

$\mathbf{a}_A$ $\left\{\begin{array}{l}\text{absolute acceleration of the} \\ \text{origin of } x, y, z \text{ frame}\end{array}\right.$

(plus)

$\dot{\mathbf{\Omega}} \times \mathbf{r}_{B/A}$ $\left\{\begin{array}{l}\text{angular acceleration effect} \\ \text{caused by rotation of } x, y, z \\ \text{frame}\end{array}\right.$ $\left.\begin{array}{l}\text{motion of} \\ x, y, z \text{ frame} \\ \text{observed from} \\ \text{the } X, Y, Z \text{ frame}\end{array}\right\}$

(plus)

$\mathbf{\Omega} \times (\mathbf{\Omega} \times \mathbf{r}_{B/A})$ $\left\{\begin{array}{l}\text{angular velocity effect caused} \\ \text{by rotation of } x, y, z \text{ frame}\end{array}\right.$

(plus)

$2\mathbf{\Omega} \times (\mathbf{v}_{B/A})_{xyz}$ $\left\{\begin{array}{l}\text{combined effect of } B \text{ moving} \\ \text{relative to } x, y, z \text{ coordinates} \\ \text{and rotation of } x, y, z \text{ frame}\end{array}\right\}$ interacting motion

(plus)

$(\mathbf{a}_{B/A})_{xyz}$ $\left\{\begin{array}{l}\text{acceleration of } B \text{ with} \\ \text{respect to } A\end{array}\right.$ $\left.\begin{array}{l}\text{motion of } B \text{ observed} \\ \text{from the } x, y, z \text{ frame}\end{array}\right\}$

Procedure for Analysis

Equations 16–26 and 16–29 can be applied to the solution of problems involving the planar motion of particles or rigid bodies using the following procedure.

Coordinate Axes.

- Choose an appropriate location for the origin and proper orientation of the axes for both fixed X, Y, Z and moving x, y, z reference frames.

- Most often solutions are easily obtained if at the instant considered:
  1. the origins are coincident
  2. the corresponding axes are collinear
  3. the corresponding axes are parallel

- The moving frame should be selected fixed to the body or device along which the relative motion occurs.

Kinematic Equations.

- After defining the origin A of the moving reference and specifying the moving point B, Eqs. 16–26 and 16–29 should be written in symbolic form

$$\mathbf{v}_B = \mathbf{v}_A + \boldsymbol{\Omega} \times \mathbf{r}_{B/A} + (\mathbf{v}_{B/A})_{xyz}$$

$$\mathbf{a}_B = \mathbf{a}_A + \dot{\boldsymbol{\Omega}} \times \mathbf{r}_{B/A} + \boldsymbol{\Omega} \times (\boldsymbol{\Omega} \times \mathbf{r}_{B/A}) + 2\boldsymbol{\Omega} \times (\mathbf{v}_{B/A})_{xyz} + (\mathbf{a}_{B/A})_{xyz}$$

- The Cartesian components of all these vectors may be expressed along either the X, Y, Z axes or the x, y, z axes. The choice is arbitrary provided a consistent set of unit vectors is used.

- Motion of the moving reference is expressed by $\mathbf{v}_A$, $\mathbf{a}_A$, $\boldsymbol{\Omega}$, and $\dot{\boldsymbol{\Omega}}$; and motion of B with respect to the moving reference is expressed by $\mathbf{r}_{B/A}$, $(\mathbf{v}_{B/A})_{xyz}$, and $(\mathbf{a}_{B/A})_{xyz}$.

The rotation of the dumping bin of the truck about point C is operated by the extension of the hydraulic cylinder AB. To determine the rotation of the bin due to this extension, we can use the equations of relative motion and fix the x, y axes to the cylinder so that the relative motion of the cylinder's extension occurs along the y axis. (© R.C. Hibbeler)

EXAMPLE | 16.18

At the instant $\theta = 60°$, the rod in Fig. 16–33 has an angular velocity of 3 rad/s and an angular acceleration of 2 rad/s². At this same instant, collar C travels outward along the rod such that when $x = 0.2$ m the velocity is 2 m/s and the acceleration is 3 m/s², both measured relative to the rod. Determine the Coriolis acceleration and the velocity and acceleration of the collar at this instant.

SOLUTION

Coordinate Axes.   The origin of both coordinate systems is located at point O, Fig. 16–33. Since motion of the collar is reported relative to the rod, the moving x, y, z frame of reference is *attached* to the rod.

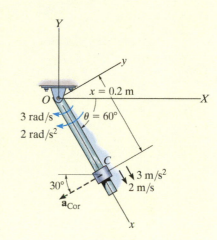

Fig. 16–33

Kinematic Equations.

$$\mathbf{v}_C = \mathbf{v}_O + \mathbf{\Omega} \times \mathbf{r}_{C/O} + (\mathbf{v}_{C/O})_{xyz} \qquad (1)$$

$$\mathbf{a}_C = \mathbf{a}_O + \dot{\mathbf{\Omega}} \times \mathbf{r}_{C/O} + \mathbf{\Omega} \times (\mathbf{\Omega} \times \mathbf{r}_{C/O}) + 2\mathbf{\Omega} \times (\mathbf{v}_{C/O})_{xyz} + (\mathbf{a}_{C/O})_{xyz} \qquad (2)$$

It will be simpler to express the data in terms of $\mathbf{i}, \mathbf{j}, \mathbf{k}$ component vectors rather than $\mathbf{I}, \mathbf{J}, \mathbf{K}$ components. Hence,

Motion of moving reference	Motion of C with respect to moving reference
$\mathbf{v}_O = 0$	$\mathbf{r}_{C/O} = \{0.2\mathbf{i}\}$ m
$\mathbf{a}_O = 0$	$(\mathbf{v}_{C/O})_{xyz} = \{2\mathbf{i}\}$ m/s
$\mathbf{\Omega} = \{-3\mathbf{k}\}$ rad/s	$(\mathbf{a}_{C/O})_{xyz} = \{3\mathbf{i}\}$ m/s²
$\dot{\mathbf{\Omega}} = \{-2\mathbf{k}\}$ rad/s²	

The Coriolis acceleration is defined as

$$\mathbf{a}_{\text{Cor}} = 2\mathbf{\Omega} \times (\mathbf{v}_{C/O})_{xyz} = 2(-3\mathbf{k}) \times (2\mathbf{i}) = \{-12\mathbf{j}\} \text{ m/s}^2 \qquad \textit{Ans.}$$

This vector is shown dashed in Fig. 16–33. If desired, it may be resolved into $\mathbf{I}, \mathbf{J}$ components acting along the X and Y axes, respectively.

The velocity and acceleration of the collar are determined by substituting the data into Eqs. 1 and 2 and evaluating the cross products, which yields

$$\mathbf{v}_C = \mathbf{v}_O + \mathbf{\Omega} \times \mathbf{r}_{C/O} + (\mathbf{v}_{C/O})_{xyz}$$

$$= 0 + (-3\mathbf{k}) \times (0.2\mathbf{i}) + 2\mathbf{i}$$

$$= \{2\mathbf{i} - 0.6\mathbf{j}\} \text{ m/s} \qquad \textit{Ans.}$$

$$\mathbf{a}_C = \mathbf{a}_O + \dot{\mathbf{\Omega}} \times \mathbf{r}_{C/O} + \mathbf{\Omega} \times (\mathbf{\Omega} \times \mathbf{r}_{C/O}) + 2\mathbf{\Omega} \times (\mathbf{v}_{C/O})_{xyz} + (\mathbf{a}_{C/O})_{xyz}$$

$$= 0 + (-2\mathbf{k}) \times (0.2\mathbf{i}) + (-3\mathbf{k}) \times [(-3\mathbf{k}) \times (0.2\mathbf{i})] + 2(-3\mathbf{k}) \times (2\mathbf{i}) + 3\mathbf{i}$$

$$= 0 - 0.4\mathbf{j} - 1.80\mathbf{i} - 12\mathbf{j} + 3\mathbf{i}$$

$$= \{1.20\mathbf{i} - 12.4\mathbf{j}\} \text{ m/s}^2 \qquad \textit{Ans.}$$

16

EXAMPLE | 16.19

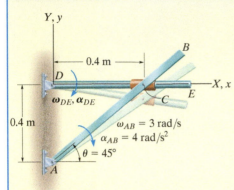

Fig. 16–34

Rod AB, shown in Fig. 16–34, rotates clockwise such that it has an angular velocity $\omega_{AB} = 3$ rad/s and angular acceleration $\alpha_{AB} = 4$ rad/s^2 when $\theta = 45°$. Determine the angular motion of rod DE at this instant. The collar at C is pin connected to AB and slides over rod DE.

SOLUTION

Coordinate Axes. The origin of both the fixed and moving frames of reference is located at D, Fig. 16–34. Furthermore, the x, y, z reference is attached to and rotates with rod DE so that the relative motion of the collar is easy to follow.

Kinematic Equations.

$$\mathbf{v}_C = \mathbf{v}_D + \boldsymbol{\Omega} \times \mathbf{r}_{C/D} + (\mathbf{v}_{C/D})_{xyz} \tag{1}$$

$$\mathbf{a}_C = \mathbf{a}_D + \dot{\boldsymbol{\Omega}} \times \mathbf{r}_{C/D} + \boldsymbol{\Omega} \times (\boldsymbol{\Omega} \times \mathbf{r}_{C/D}) + 2\boldsymbol{\Omega} \times (\mathbf{v}_{C/D})_{xyz} + (\mathbf{a}_{C/D})_{xyz} \tag{2}$$

All vectors will be expressed in terms of $\mathbf{i}, \mathbf{j}, \mathbf{k}$ components.

Motion of moving reference	Motion of C with respect to moving reference
$\mathbf{v}_D = \mathbf{0}$	$\mathbf{r}_{C/D} = \{0.4\mathbf{i}\}$ m
$\mathbf{a}_D = \mathbf{0}$	$(\mathbf{v}_{C/D})_{xyz} = (v_{C/D})_{xyz}\mathbf{i}$
$\boldsymbol{\Omega} = -\omega_{DE}\mathbf{k}$	$(\mathbf{a}_{C/D})_{xyz} = (a_{C/D})_{xyz}\mathbf{i}$
$\dot{\boldsymbol{\Omega}} = -\alpha_{DE}\mathbf{k}$	

Motion of C: Since the collar moves along a *circular path* of radius AC, its velocity and acceleration can be determined using Eqs. 16–9 and 16–14.

$$\mathbf{v}_C = \boldsymbol{\omega}_{AB} \times \mathbf{r}_{C/A} = (-3\mathbf{k}) \times (0.4\mathbf{i} + 0.4\mathbf{j}) = \{1.2\mathbf{i} - 1.2\mathbf{j}\} \text{ m/s}$$
$$\mathbf{a}_C = \boldsymbol{\alpha}_{AB} \times \mathbf{r}_{C/A} - \omega_{AB}^2 \mathbf{r}_{C/A}$$
$$= (-4\mathbf{k}) \times (0.4\mathbf{i} + 0.4\mathbf{j}) - (3)^2(0.4\mathbf{i} + 0.4\mathbf{j}) = \{-2\mathbf{i} - 5.2\mathbf{j}\} \text{ m/s}^2$$

Substituting the data into Eqs. 1 and 2, we have

$$\mathbf{v}_C = \mathbf{v}_D + \boldsymbol{\Omega} \times \mathbf{r}_{C/D} + (\mathbf{v}_{C/D})_{xyz}$$
$$1.2\mathbf{i} - 1.2\mathbf{j} = \mathbf{0} + (-\omega_{DE}\mathbf{k}) \times (0.4\mathbf{i}) + (v_{C/D})_{xyz}\mathbf{i}$$
$$1.2\mathbf{i} - 1.2\mathbf{j} = \mathbf{0} - 0.4\omega_{DE}\mathbf{j} + (v_{C/D})_{xyz}\mathbf{i}$$
$$(v_{C/D})_{xyz} = 1.2 \text{ m/s}$$
$$\omega_{DE} = 3 \text{ rad/s} \,\rotatebox{0}{$\circlearrowright$} \qquad\qquad Ans.$$

$$\mathbf{a}_C = \mathbf{a}_D + \dot{\boldsymbol{\Omega}} \times \mathbf{r}_{C/D} + \boldsymbol{\Omega} \times (\boldsymbol{\Omega} \times \mathbf{r}_{C/D}) + 2\boldsymbol{\Omega} \times (\mathbf{v}_{C/D})_{xyz} + (\mathbf{a}_{C/D})_{xyz}$$
$$-2\mathbf{i} - 5.2\mathbf{j} = \mathbf{0} + (-\alpha_{DE}\mathbf{k}) \times (0.4\mathbf{i}) + (-3\mathbf{k}) \times [(-3\mathbf{k}) \times (0.4\mathbf{i})]$$
$$+ 2(-3\mathbf{k}) \times (1.2\mathbf{i}) + (a_{C/D})_{xyz}\mathbf{i}$$
$$-2\mathbf{i} - 5.2\mathbf{j} = -0.4\alpha_{DE}\mathbf{j} - 3.6\mathbf{i} - 7.2\mathbf{j} + (a_{C/D})_{xyz}\mathbf{i}$$
$$(a_{C/D})_{xyz} = 1.6 \text{ m/s}^2$$
$$\alpha_{DE} = -5 \text{ rad/s}^2 = 5 \text{ rad/s}^2 \,\rotatebox{0}{$\circlearrowright$} \qquad\qquad Ans.$$

16

EXAMPLE   16.20

Planes A and B fly at the same elevation and have the motions shown in Fig. 16–35. Determine the velocity and acceleration of A as measured by the pilot of B.

SOLUTION

Coordinate Axes. Since the relative motion of A with respect to the pilot in B is being sought, the x, y, z axes are attached to plane B, Fig. 16–35. At the *instant* considered, the origin B coincides with the origin of the fixed X, Y, Z frame.

Kinematic Equations.

$$\mathbf{v}_A = \mathbf{v}_B + \mathbf{\Omega} \times \mathbf{r}_{A/B} + (\mathbf{v}_{A/B})_{xyz} \tag{1}$$

$$\mathbf{a}_A = \mathbf{a}_B + \dot{\mathbf{\Omega}} \times \mathbf{r}_{A/B} + \mathbf{\Omega} \times (\mathbf{\Omega} \times \mathbf{r}_{A/B}) + 2\mathbf{\Omega} \times (\mathbf{v}_{A/B})_{xyz} + (\mathbf{a}_{A/B})_{xyz} \tag{2}$$

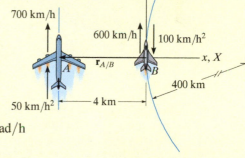

Motion of Moving Reference:

$$\mathbf{v}_B = \{600\mathbf{j}\} \text{ km/h}$$

$$(a_B)_n = \frac{v_B^2}{\rho} = \frac{(600)^2}{400} = 900 \text{ km/h}^2$$

$$\mathbf{a}_B = (\mathbf{a}_B)_n + (\mathbf{a}_B)_t = \{900\mathbf{i} - 100\mathbf{j}\} \text{ km/h}^2$$

$$\Omega = \frac{v_B}{\rho} = \frac{600 \text{ km/h}}{400 \text{ km}} = 1.5 \text{ rad/h} \;\curvearrowright \qquad \mathbf{\Omega} = \{-1.5\mathbf{k}\} \text{ rad/h}$$

$$\dot{\Omega} = \frac{(a_B)_t}{\rho} = \frac{100 \text{ km/h}^2}{400 \text{ km}} = 0.25 \text{ rad/h}^2 \;\curvearrowright \qquad \dot{\mathbf{\Omega}} = \{0.25\mathbf{k}\} \text{ rad/h}^2$$

Fig. 16–35

Motion of A with Respect to Moving Reference:

$$\mathbf{r}_{A/B} = \{-4\mathbf{i}\} \text{ km} \quad (\mathbf{v}_{A/B})_{xyz} = ? \quad (\mathbf{a}_{A/B})_{xyz} = ?$$

Substituting the data into Eqs. 1 and 2, realizing that $\mathbf{v}_A = \{700\mathbf{j}\}$ km/h and $\mathbf{a}_A = \{50\mathbf{j}\}$ km/h^2, we have

$$\mathbf{v}_A = \mathbf{v}_B + \mathbf{\Omega} \times \mathbf{r}_{A/B} + (\mathbf{v}_{A/B})_{xyz}$$

$$700\mathbf{j} = 600\mathbf{j} + (-1.5\mathbf{k}) \times (-4\mathbf{i}) + (\mathbf{v}_{A/B})_{xyz}$$

$$(\mathbf{v}_{A/B})_{xyz} = \{94\mathbf{j}\} \text{ km/h} \qquad \textit{Ans.}$$

$$\mathbf{a}_A = \mathbf{a}_B + \dot{\mathbf{\Omega}} \times \mathbf{r}_{A/B} + \mathbf{\Omega} \times (\mathbf{\Omega} \times \mathbf{r}_{A/B}) + 2\mathbf{\Omega} \times (\mathbf{v}_{A/B})_{xyz} + (\mathbf{a}_{A/B})_{xyz}$$

$$50\mathbf{j} = (900\mathbf{i} - 100\mathbf{j}) + (0.25\mathbf{k}) \times (-4\mathbf{i})$$

$$+ (-1.5\mathbf{k}) \times [(-1.5\mathbf{k}) \times (-4\mathbf{i})] + 2(-1.5\mathbf{k}) \times (94\mathbf{j}) + (\mathbf{a}_{A/B})_{xyz}$$

$$(\mathbf{a}_{A/B})_{xyz} = \{-1191\mathbf{i} + 151\mathbf{j}\} \text{ km/h}^2 \qquad \textit{Ans.}$$

NOTE: The solution of this problem should be compared with that of Example 12.26, where it is seen that $(v_{B/A})_{xyz} \neq (v_{A/B})_{xyz}$ and $(a_{B/A})_{xyz} \neq (a_{A/B})_{xyz}$.

PROBLEMS

16–129. At the instant shown, ball B is rolling along the slot in the disk with a velocity of 600 mm/s and an acceleration of 150 mm/s², both measured relative to the disk and directed away from O. If at the same instant the disk has the angular velocity and angular acceleration shown, determine the velocity and acceleration of the ball at this instant.

16–131. While the swing bridge is closing with a constant rotation of 0.5 rad/s, a man runs along the roadway at a constant speed of 5 ft/s relative to the roadway. Determine his velocity and acceleration at the instant $d = 15$ ft.

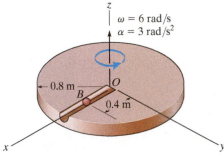

Prob. 16–129

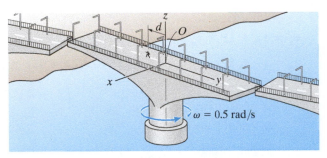

Prob. 16–131

16–130. The crane's telescopic boom rotates with the angular velocity and angular acceleration shown. At the same instant, the boom is extending with a constant speed of 0.5 ft/s, measured relative to the boom. Determine the magnitudes of the velocity and acceleration of point B at this instant.

*16–132. While the swing bridge is closing with a constant rotation of 0.5 rad/s, a man runs along the roadway such that when $d = 10$ ft he is running outward from the center at 5 ft/s with an acceleration of 2 ft/s², both measured relative to the roadway. Determine his velocity and acceleration at this instant.

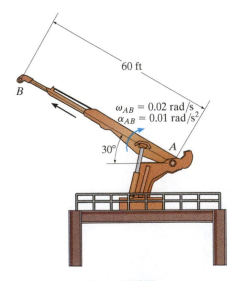

Prob. 16–130

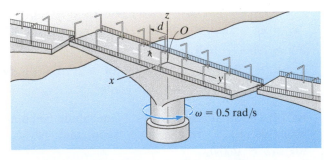

Prob. 16–132

16–133. Water leaves the impeller of the centrifugal pump with a velocity of 25 m/s and acceleration of 30 m/s², both measured relative to the impeller along the blade line AB. Determine the velocity and acceleration of a water particle at A as it leaves the impeller at the instant shown. The impeller rotates with a constant angular velocity of $\omega = 15$ rad/s.

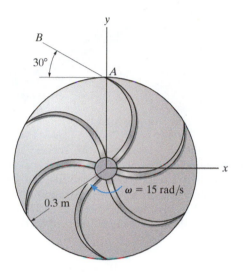

Prob. 16–133

16–134. Block A, which is attached to a cord, moves along the slot of a horizontal forked rod. At the instant shown, the cord is pulled down through the hole at O with an acceleration of 4 m/s² and its velocity is 2 m/s. Determine the acceleration of the block at this instant. The rod rotates about O with a constant angular velocity $\omega = 4$ rad/s.

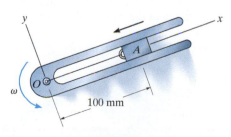

Prob. 16–134

16–135. Rod AB rotates counterclockwise with a constant angular velocity $\omega = 3$ rad/s. Determine the velocity of point C located on the double collar when $\theta = 30°$. The collar consists of two pin-connected slider blocks which are constrained to move along the circular path and the rod AB.

*16–136. Rod AB rotates counterclockwise with a constant angular velocity $\omega = 3$ rad/s. Determine the velocity and acceleration of point C located on the double collar when $\theta = 45°$. The collar consists of two pin-connected slider blocks which are constrained to move along the circular path and the rod AB.

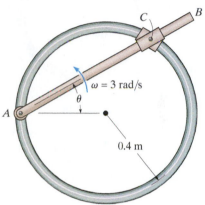

Probs. 16–135/136

16–137. Particles B and A move along the parabolic and circular paths, respectively. If B has a velocity of 7 m/s in the direction shown and its speed is increasing at 4 m/s², while A has a velocity of 8 m/s in the direction shown and its speed is decreasing at 6 m/s², determine the relative velocity and relative acceleration of B with respect to A.

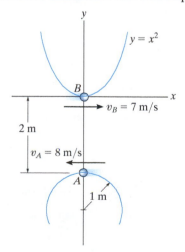

Prob. 16–137

16–138. Collar *B* moves to the left with a speed of 5 m/s, which is increasing at a constant rate of 1.5 m/s², relative to the hoop, while the hoop rotates with the angular velocity and angular acceleration shown. Determine the magnitudes of the velocity and acceleration of the collar at this instant.

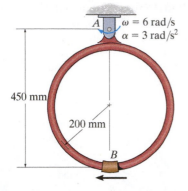

$\omega = 6$ rad/s
$\alpha = 3$ rad/s²

450 mm

200 mm

Prob. 16–138

16–139. Block *D* of the mechanism is confined to move within the slot of member *CB*. If link *AD* is rotating at a constant rate of $\omega_{AD} = 4$ rad/s, determine the angular velocity and angular acceleration of member *CB* at the instant shown.

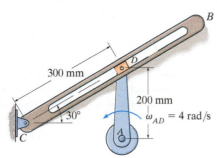

300 mm

200 mm

30°

$\omega_{AD} = 4$ rad/s

Prob. 16–139

*16–140. At the instant shown rod *AB* has an angular velocity $\omega_{AB} = 4$ rad/s and an angular acceleration $\alpha_{AB} = 2$ rad/s². Determine the angular velocity and angular acceleration of rod *CD* at this instant. The collar at *C* is pin connected to *CD* and slides freely along *AB*.

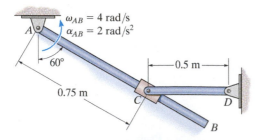

$\omega_{AB} = 4$ rad/s
$\alpha_{AB} = 2$ rad/s²

60°

—0.5 m—

0.75 m

Prob. 16–140

16–141. The collar *C* is pinned to rod *CD* while it slides on rod *AB*. If rod *AB* has an angular velocity of 2 rad/s and an angular acceleration of 8 rad/s², both acting counterclockwise, determine the angular velocity and the angular acceleration of rod *CD* at the instant shown.

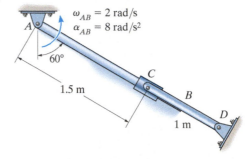

$\omega_{AB} = 2$ rad/s
$\alpha_{AB} = 8$ rad/s²

60°

1.5 m

1 m

Prob. 16–141

16–142. At the instant shown, the robotic arm *AB* is rotating counterclockwise at $\omega = 5$ rad/s and has an angular acceleration $\alpha = 2$ rad/s². Simultaneously, the grip *BC* is rotating counterclockwise at $\omega' = 6$ rad/s and $\alpha' = 2$ rad/s², both measured relative to a *fixed* reference. Determine the velocity and acceleration of the object held at the grip *C*.

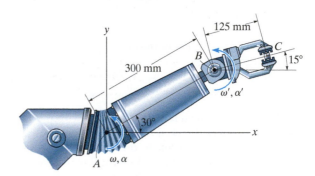

125 mm

300 mm

15°

ω', α'

30°

ω, α

Prob. 16–142

16–143. Peg *B* on the gear slides freely along the slot in link *AB*. If the gear's center *O* moves with the velocity and acceleration shown, determine the angular velocity and angular acceleration of the link at this instant.

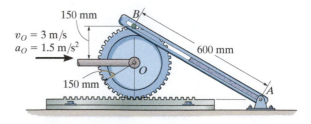

150 mm

$v_O = 3$ m/s
$a_O = 1.5$ m/s²

600 mm

150 mm

Prob. 16–143

*16–144.   The cars on the amusement-park ride rotate around the axle at A with a constant angular velocity $\omega_{A/f} = 2$ rad/s, measured relative to the frame AB. At the same time the frame rotates around the main axle support at B with a constant angular velocity $\omega_f = 1$ rad/s. Determine the velocity and acceleration of the passenger at C at the instant shown.

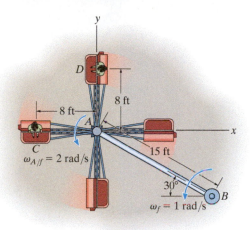

Prob. 16–144

16–145.   A ride in an amusement park consists of a rotating arm AB having a constant angular velocity $\omega_{AB} = 2$ rad/s point A and a car mounted at the end of the arm which has a constant angular velocity $\omega' = \{-0.5\mathbf{k}\}$ rad/s, measured relative to the arm. At the instant shown, determine the velocity and acceleration of the passenger at C.

16–146.   A ride in an amusement park consists of a rotating arm AB that has an angular acceleration of $\alpha_{AB} = 1$ rad/s^2 when $\omega_{AB} = 2$ rad/s at the instant shown. Also at this instant the car mounted at the end of the arm has an angular acceleration of $\alpha = \{-0.6\mathbf{k}\}$ rad/s^2 and angular velocity of $\omega' = \{-0.5\mathbf{k}\}$ rad/s, measured relative to the arm. Determine the velocity and acceleration of the passenger C at this instant.

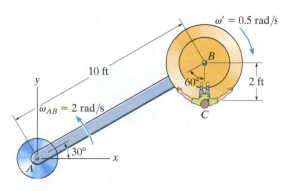

Probs. 16–145/146

16–147.   If the slider block C is fixed to the disk that has a constant counterclockwise angular velocity of 4 rad/s, determine the angular velocity and angular acceleration of the slotted arm AB at the instant shown.

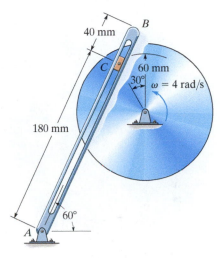

Prob. 16–147

*16–148.   At the instant shown, car A travels with a speed of 25 m/s, which is decreasing at a constant rate of 2 m/s^2, while car C travels with a speed of 15 m/s, which is increasing at a constant rate of 3 m/s. Determine the velocity and acceleration of car A with respect to car C.

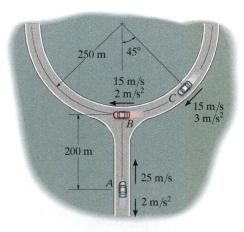

Prob. 16–148

16–149. At the instant shown, car B travels with a speed of 15 m/s, which is increasing at a constant rate of 2 m/s², while car C travels with a speed of 15 m/s, which is increasing at a constant rate of 3 m/s². Determine the velocity and acceleration of car B with respect to car C.

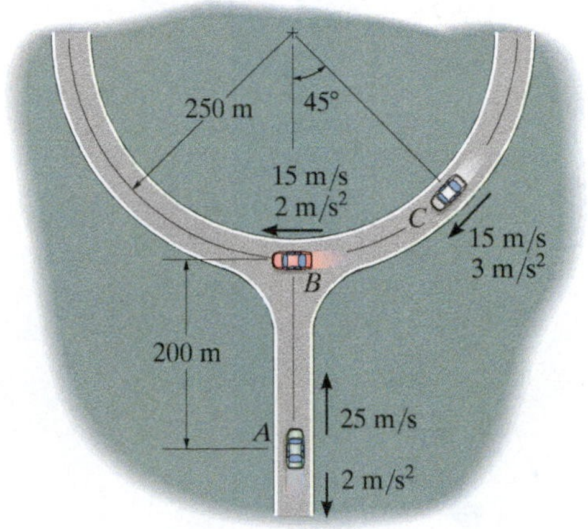

Prob. 16–149

16–150. The two-link mechanism serves to amplify angular motion. Link AB has a pin at B which is confined to move within the slot of link CD. If at the instant shown, AB (input) has an angular velocity of $\omega_{AB} = 2.5$ rad/s, determine the angular velocity of CD (output) at this instant.

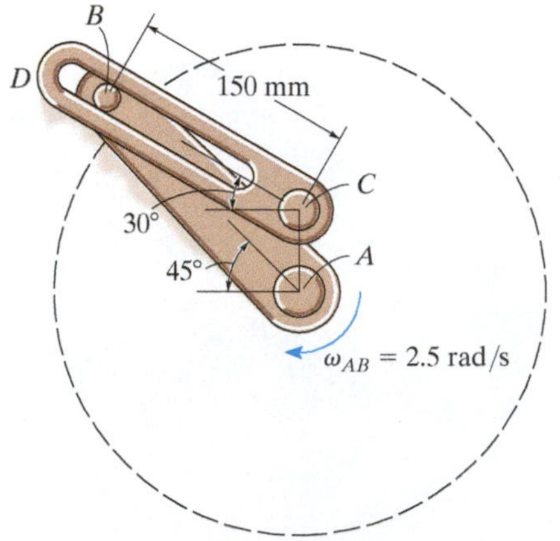

Prob. 16–150

16–151. The disk rotates with the angular motion shown. Determine the angular velocity and angular acceleration of the slotted link AC at this instant. The peg at B is fixed to the disk.

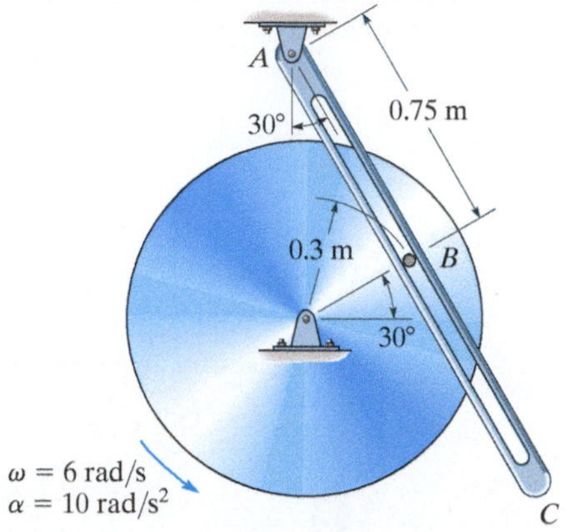

Prob. 16–151

***16–152.** The Geneva mechanism is used in a packaging system to convert constant angular motion into intermittent angular motion. The star wheel A makes one sixth of a revolution for each full revolution of the driving wheel B and the attached guide C. To do this, pin P, which is attached to B, slides into one of the radial slots of A, thereby turning wheel A, and then exits the slot. If B has a constant angular velocity of $\omega_B = 4$ rad/s, determine $\boldsymbol{\omega}_A$ and $\boldsymbol{\alpha}_A$ of wheel A at the instant shown.

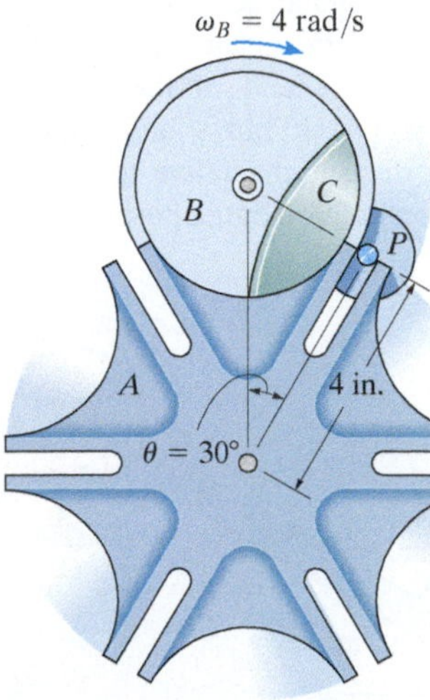

Prob. 16–152

CONCEPTUAL PROBLEMS

C16–1. An electric motor turns the tire at A at a constant angular velocity, and friction then causes the tire to roll without slipping on the inside rim of the Ferris wheel. Using appropriate numerical values, determine the magnitude of the velocity and acceleration of passengers in one of the baskets. Do passengers in the other baskets experience this same motion? Explain.

Prob. C16–1 (© R.C. Hibbeler)

C16–2. The crank AB turns counterclockwise at a constant rate ω causing the connecting arm CD and rocking beam DE to move. Draw a sketch showing the location of the IC for the connecting arm when $\theta = 0°, 90°, 180°,$ and $270°$. Also, how was the curvature of the head at E determined, and why is it curved in this way?

Prob. C16–2 (© R.C. Hibbeler)

C16–3. The bi-fold hangar door is opened by cables that move upward at a constant speed of 0.5 m/s. Determine the angular velocity of BC and the angular velocity of AB when $\theta = 45°$. Panel BC is pinned at C and has a height which is the same as the height of BA. Use appropriate numerical values to explain your result.

Prob. C16–3 (© R.C. Hibbeler)

C16–4. If the tires do not slip on the pavement, determine the points on the tire that have a maximum and minimum speed and the points that have a maximum and minimum acceleration. Use appropriate numerical values for the car's speed and tire size to explain your result.

Prob. C16–4 (© R.C. Hibbeler)

CHAPTER REVIEW

Rigid-Body Planar Motion

A rigid body undergoes three types of planar motion: translation, rotation about a fixed axis, and general plane motion.

Path of rectilinear translation

Translation

When a body has rectilinear translation, all the particles of the body travel along parallel straight-line paths. If the paths have the same radius of curvature, then curvilinear translation occurs. Provided we know the motion of one of the particles, then the motion of all of the others is also known.

Path of curvilinear translation

Rotation about a Fixed Axis

For this type of motion, all of the particles move along circular paths. Here, all line segments in the body undergo the same angular displacement, angular velocity, and angular acceleration.

Once the angular motion of the body is known, then the velocity of any particle a distance r from the axis can be obtained.

The acceleration of any particle has two components. The tangential component accounts for the change in the magnitude of the velocity, and the normal component accounts for the change in the velocity's direction.

Rotation about a fixed axis

$$\omega = d\theta/dt \qquad\qquad \omega = \omega_0 + \alpha_c t$$

$$\alpha = d\omega/dt \qquad \text{or} \qquad \theta = \theta_0 + \omega_0 t + \tfrac{1}{2}\alpha_c t^2$$

$$\alpha \, d\theta = \omega \, d\omega \qquad\qquad \omega^2 = \omega_0^2 + 2\alpha_c(\theta - \theta_0)$$

$$\text{Constant } \alpha_c$$

$$v = \omega r \qquad\qquad a_t = \alpha r, \quad a_n = \omega^2 r$$

General Plane Motion

When a body undergoes general plane motion, it simultaneously translates and rotates. There are several methods for analyzing this motion.

Absolute Motion Analysis
If the motion of a point on a body or the angular motion of a line is known, then it may be possible to relate this motion to that of another point or line using an absolute motion analysis. To do so, linear position coordinates s or angular position coordinates θ are established (measured from a fixed point or line). These position coordinates are then related using the geometry of the body. The time derivative of this equation gives the relationship between the velocities and/or the angular velocities. A second time derivative relates the accelerations and/or the angular accelerations.

General plane motion

Relative-Motion using Translating Axes

General plane motion can also be analyzed using a relative-motion analysis between two points A and B located on the body. This method considers the motion in parts: first a translation of the selected base point A, then a relative "rotation" of the body about point A, which is measured from a translating axis. Since the relative motion is viewed as circular motion about the base point, point B will have a velocity $\mathbf{v}_{B/A}$ that is tangent to the circle. It also has two components of acceleration, $(\mathbf{a}_{B/A})_t$ and $(\mathbf{a}_{B/A})_n$. It is also important to realize that $\mathbf{a}_A$ and $\mathbf{a}_B$ will have tangential and normal components if these points move along curved paths.

$$\mathbf{v}_B = \mathbf{v}_A + \boldsymbol{\omega} \times \mathbf{r}_{B/A}$$

$$\mathbf{a}_B = \mathbf{a}_A + \boldsymbol{\alpha} \times \mathbf{r}_{B/A} - \omega^2 \mathbf{r}_{B/A}$$

Instantaneous Center of Zero Velocity

If the base point A is selected as having zero velocity, then the relative velocity equation becomes $\mathbf{v}_B = \boldsymbol{\omega} \times \mathbf{r}_{B/A}$. In this case, motion appears as if the body rotates about an instantaneous axis passing through A.

The instantaneous center of rotation (*IC*) can be established provided the directions of the velocities of any two points on the body are known, or the velocity of a point and the angular velocity are known. Since a radial line r will always be perpendicular to each velocity, then the *IC* is at the point of intersection of these two radial lines. Its measured location is determined from the geometry of the body. Once it is established, then the velocity of any point P on the body can be determined from $v = \omega r$, where r extends from the *IC* to point P.

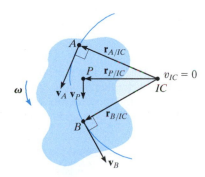

Relative Motion using Rotating Axes

Problems that involve connected members that slide relative to one another or points not located on the same body can be analyzed using a relative-motion analysis referenced from a rotating frame. This gives rise to the term $2\boldsymbol{\Omega} \times (\mathbf{v}_{B/A})_{xyz}$ that is called the Coriolis acceleration.

$$\mathbf{v}_B = \mathbf{v}_A + \boldsymbol{\Omega} \times \mathbf{r}_{B/A} + (\mathbf{v}_{B/A})_{xyz}$$

$$\mathbf{a}_B = \mathbf{a}_A + \dot{\boldsymbol{\Omega}} \times \mathbf{r}_{B/A} + \boldsymbol{\Omega} \times (\boldsymbol{\Omega} \times \mathbf{r}_{B/A}) + 2\boldsymbol{\Omega} \times (\mathbf{v}_{B/A})_{xyz} + (\mathbf{a}_{B/A})_{xyz}$$

16

REVIEW PROBLEMS

R16–1. The hoisting gear A has an initial angular velocity of 60 rad/s and a constant deceleration of 1 rad/s². Determine the velocity and deceleration of the block which is being hoisted by the hub on gear B when $t = 3$ s.

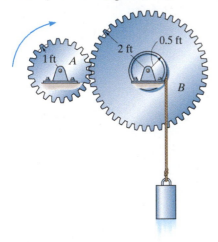

Prob. R16–1

R16–2. Starting at $(\omega_A)_0 = 3$ rad/s, when $\theta = 0$, $s = 0$, pulley A is given an angular acceleration $\alpha = (0.6\theta)$ rad/s², where θ is in radians. Determine the speed of block B when it has risen $s = 0.5$ m. The pulley has an inner hub D which is fixed to C and turns with it.

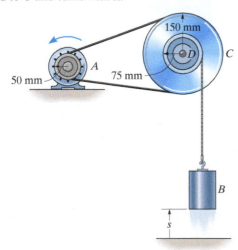

Prob. R16–2

R16–3. The board rests on the surface of two drums. At the instant shown, it has an acceleration of 0.5 m/s² to the right, while at the same instant points on the outer rim of each drum have an acceleration with a magnitude of 3 m/s². If the board does not slip on the drums, determine its speed due to the motion.

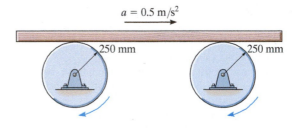

Prob. R16–3

R16–4. If bar AB has an angular velocity $\omega_{AB} = 6$ rad/s, determine the velocity of the slider block C at the instant shown.

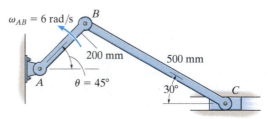

Prob. R16–4

R16–5. The center of the pulley is being lifted vertically with an acceleration of 4 m/s² at the instant it has a velocity of 2 m/s. If the cable does not slip on the pulley's surface, determine the accelerations of the cylinder B and point C on the pulley.

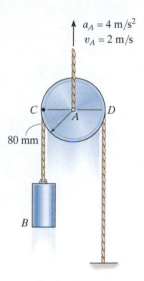

Prob. R16–5

R16–6. At the instant shown, link AB has an angular velocity $\omega_{AB} = 2$ rad/s and an angular acceleration $\alpha_{AB} = 6$ rad/s². Determine the acceleration of the pin at C and the angular acceleration of link CB at this instant, when $\theta = 60°$.

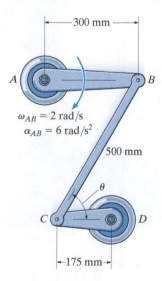

Prob. R16–6

R16–7. The disk is moving to the left such that it has an angular acceleration $\alpha = 8$ rad/s² and angular velocity $\omega = 3$ rad/s at the instant shown. If it does not slip at A, determine the acceleration of point B.

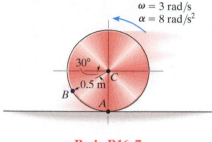

Prob. R16–7

R16–8. At the given instant member AB has the angular motions shown. Determine the velocity and acceleration of the slider block C at this instant.

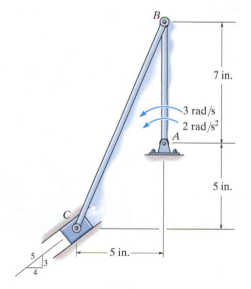

Prob. R16–8

Chapter 17

(© Surasaki/Fotolia)

Tractors and other heavy equipment can be subjected to severe loadings due to dynamic loadings as they accelerate. In this chapter we will show how to determine these loadings for planar motion.

Planar Kinetics of a Rigid Body: Force and Acceleration

CHAPTER OBJECTIVES

■ To introduce the methods used to determine the mass moment of inertia of a body.

■ To develop the planar kinetic equations of motion for a symmetric rigid body.

■ To discuss applications of these equations to bodies undergoing translation, rotation about a fixed axis, and general plane motion.

17.1   Mass Moment of Inertia

Since a body has a definite size and shape, an applied nonconcurrent force system can cause the body to both translate and rotate. The translational aspects of the motion were studied in Chapter 13 and are governed by the equation $\mathbf{F} = m\mathbf{a}$. It will be shown in the next section that the rotational aspects, caused by a moment $\mathbf{M}$, are governed by an equation of the form $\mathbf{M} = I\boldsymbol{\alpha}$. The symbol I in this equation is termed the mass moment of inertia. By comparison, the *moment of inertia* is a measure of the resistance of a body to *angular acceleration* ($\mathbf{M} = I\boldsymbol{\alpha}$) in the same way that *mass* is a measure of the body's resistance to *acceleration* ($\mathbf{F} = m\mathbf{a}$).

The flywheel on the engine of this tractor has a large moment of inertia about its axis of rotation. Once it is set into motion, it will be difficult to stop, and this in turn will prevent the engine from stalling and instead will allow it to maintain a constant power.

(© R.C. Hibbeler)

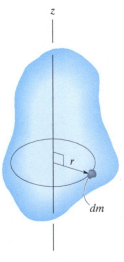

Fig. 17–1

We define the *moment of inertia* as the integral of the "second moment" about an axis of all the elements of mass dm which compose the body.* For example, the body's moment of inertia about the z axis in Fig. 17–1 is

$$I = \int_m r^2 \, dm \qquad (17\text{–}1)$$

Here the "moment arm" r is the perpendicular distance from the z axis to the arbitrary element dm. Since the formulation involves r, the value of I is different for each axis about which it is computed. In the study of planar kinetics, the axis chosen for analysis generally passes through the body's mass center G and is always perpendicular to the plane of motion. The moment of inertia about this axis will be denoted as I_G. Since r is squared in Eq. 17–1, the mass moment of inertia is always a *positive* quantity. Common units used for its measurement are kg·m^2 or slug·ft^2.

If the body consists of material having a variable density, $\rho = \rho(x,y,z)$, the elemental mass dm of the body can be expressed in terms of its density and volume as $dm = \rho \, dV$. Substituting dm into Eq. 17–1, the body's moment of inertia is then computed using *volume elements* for integration; i.e.,

$$I = \int_V r^2 \rho \, dV \qquad (17\text{–}2)$$

*Another property of the body, which measures the symmetry of the body's mass with respect to a coordinate system, is the product of inertia. This property applies to the three-dimensional motion of a body and will be discussed in Chapter 21.

In the special case of ρ being a *constant*, this term may be factored out of the integral, and the integration is then purely a function of geometry,

$$I = \rho \int_V r^2 \, dV \qquad (17\text{–}3)$$

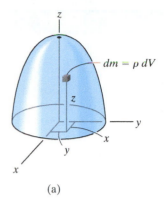

(a)

When the volume element chosen for integration has infinitesimal dimensions in all three directions, Fig. 17–2a, the moment of inertia of the body must be determined using "triple integration." The integration process can, however, be simplified to a *single integration* provided the chosen volume element has a differential size or thickness in only *one direction*. Shell or disk elements are often used for this purpose.

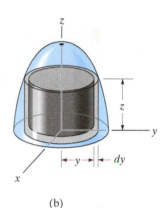

(b)

Procedure for Analysis

To obtain the moment of inertia by integration, we will consider only symmetric bodies having volumes which are generated by revolving a curve about an axis. An example of such a body is shown in Fig. 17–2a. Two types of differential elements can be chosen.

Shell Element.

- If a *shell element* having a height z, radius $r = y$, and thickness dy is chosen for integration, Fig. 17–2b, then the volume is $dV = (2\pi y)(z)dy$.

- This element may be used in Eq. 17–2 or 17–3 for determining the moment of inertia I_z of the body about the z axis, since the *entire element*, due to its "thinness," lies at the *same* perpendicular distance $r = y$ from the z axis (see Example 17.1).

Disk Element.

- If a disk element having a radius y and a thickness dz is chosen for integration, Fig. 17–2c, then the volume is $dV = (\pi y^2)dz$.

- This element is *finite* in the radial direction, and consequently its parts *do not* all lie at the *same radial distance* r from the z axis. As a result, Eq. 17–2 or 17–3 *cannot* be used to determine I_z directly. Instead, to perform the integration it is first necessary to determine the moment of inertia *of the element* about the z axis and then integrate this result (see Example 17.2).

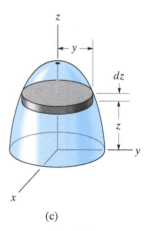

(c)

Fig. 17–2

17

EXAMPLE   17.1

Determine the moment of inertia of the cylinder shown in Fig. 17–3a about the z axis. The density of the material, ρ, is constant.

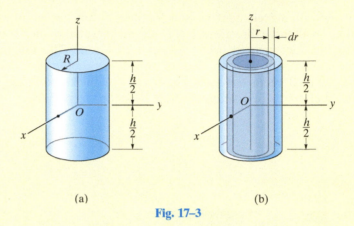

(a)   (b)

Fig. 17–3

SOLUTION

Shell Element.   This problem can be solved using the *shell element* in Fig. 17–3b and a single integration. The volume of the element is $dV = (2\pi r)(h)\, dr$, so that its mass is $dm = \rho dV = \rho(2\pi h r\, dr)$. Since the *entire element* lies at the same distance r from the z axis, the moment of inertia *of the element* is

$$dI_z = r^2 dm = \rho 2\pi h r^3\, dr$$

Integrating over the entire region of the cylinder yields

$$I_z = \int_m r^2\, dm = \rho 2\pi h \int_0^R r^3\, dr = \frac{\rho\pi}{2} R^4 h$$

The mass of the cylinder is

$$m = \int_m dm = \rho 2\pi h \int_0^R r\, dr = \rho\pi h R^2$$

so that

$$I_z = \frac{1}{2} m R^2 \qquad\qquad Ans.$$

EXAMPLE | 17.2

If the density of the material is 5 slug/ft^3, determine the moment of inertia of the solid in Fig. 17–4a about the y axis.

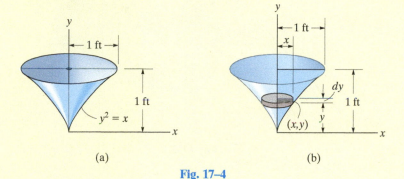

(a)   (b)

Fig. 17–4

SOLUTION

Disk Element. The moment of inertia will be found using a *disk element*, as shown in Fig. 17–4b. Here the element intersects the curve at the arbitrary point (x,y) and has a mass

$$dm = \rho \, dV = \rho(\pi x^2) \, dy$$

Although all portions of the element are *not* located at the same distance from the y axis, it is still possible to determine the moment of inertia dI_y *of the element* about the y axis. In the preceding example it was shown that the moment of inertia of a cylinder about its longitudinal axis is $I = \frac{1}{2}mR^2$, where m and R are the mass and radius of the cylinder. Since the height is not involved in this formula, the disk itself can be thought of as a cylinder. Thus, for the disk element in Fig. 17–4b, we have

$$dI_y = \tfrac{1}{2}(dm)x^2 = \tfrac{1}{2}[\rho(\pi x^2)\, dy]x^2$$

Substituting $x = y^2$, $\rho = 5$ slug/ft^3, and integrating with respect to y, from $y = 0$ to $y = 1$ ft, yields the moment of inertia for the entire solid.

$$I_y = \frac{\pi(5 \text{ slug/ft}^3)}{2} \int_0^{1\text{ ft}} x^4 \, dy = \frac{\pi(5)}{2} \int_0^{1\text{ ft}} y^8 \, dy = 0.873 \text{ slug} \cdot \text{ft}^2 \quad Ans.$$

17

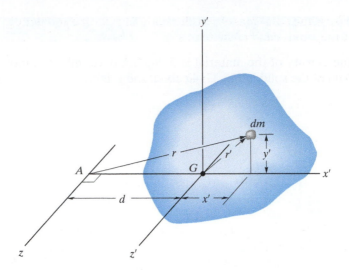

Fig. 17–5

Parallel-Axis Theorem.   If the moment of inertia of the body about an axis passing through the body's mass center is known, then the moment of inertia about any other *parallel axis* can be determined by using the *parallel-axis theorem*. This theorem can be derived by considering the body shown in Fig. 17–5. Here the z' axis passes through the mass center G, whereas the corresponding *parallel z axis* lies at a constant distance d away. Selecting the differential element of mass dm, which is located at point (x', y'), and using the Pythagorean theorem, $r^2 = (d + x')^2 + y'^2$, we can express the moment of inertia of the body about the z axis as

$$I = \int_m r^2 \, dm = \int_m [(d + x')^2 + y'^2] \, dm$$

$$= \int_m (x'^2 + y'^2) \, dm + 2d \int_m x' \, dm + d^2 \int_m dm$$

Since $r'^2 = x'^2 + y'^2$, the first integral represents I_G. The second integral equals *zero*, since the z' axis passes through the body's mass center, i.e., $\int x' \, dm = \bar{x}' m = 0$ since $\bar{x}' = 0$. Finally, the third integral represents the total mass m of the body. Hence, the moment of inertia about the z axis can be written as

$$I = I_G + md^2 \qquad (17\text{--}4)$$

where

I_G = moment of inertia about the z' axis passing through the mass center G

m = mass of the body

d = perpendicular distance between the parallel z and z' axes

Radius of Gyration.   Occasionally, the moment of inertia of a body about a specified axis is reported in handbooks using the *radius of gyration, k*. This is a geometrical property which has units of length. When it and the body's mass m are known, the body's moment of inertia is determined from the equation

$$I = mk^2 \quad \text{or} \quad k = \sqrt{\frac{I}{m}} \qquad (17\text{--}5)$$

Note the *similarity* between the definition of k in this formula and r in the equation $dI = r^2\, dm$, which defines the moment of inertia of an elemental mass dm of the body about an axis.

Composite Bodies.   If a body consists of a number of simple shapes such as disks, spheres, and rods, the moment of inertia of the body about any axis can be determined by adding algebraically the moments of inertia of all the composite shapes computed about the axis. Algebraic addition is necessary since a composite part must be considered as a negative quantity if it has already been counted as a piece of another part—for example, a "hole" subtracted from a solid plate. The parallel-axis theorem is needed for the calculations if the center of mass of each composite part does not lie on the axis. For the calculation, then, $I = \Sigma(I_G + md^2)$. Here I_G for each of the composite parts is determined by integration, or for simple shapes, such as rods and disks, it can be found from a table, such as the one given on the inside back cover of this book.

17

EXAMPLE   17.3

If the plate shown in Fig. 17–6a has a density of 8000 kg/m³ and a thickness of 10 mm, determine its moment of inertia about an axis directed perpendicular to the page and passing through point O.

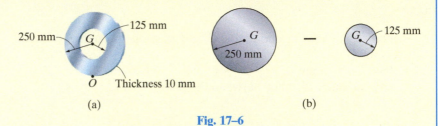

(a)   (b)

Fig. 17–6

SOLUTION

The plate consists of two composite parts, the 250-mm-radius disk *minus* a 125-mm-radius disk, Fig. 17–6b. The moment of inertia about O can be determined by computing the moment of inertia of each of these parts about O and then adding the results *algebraically*. The calculations are performed by using the parallel-axis theorem in conjunction with the data listed in the table on the inside back cover.

Disk. The moment of inertia of a disk about the centroidal axis perpendicular to the plane of the disk is $I_G = \frac{1}{2}mr^2$. The mass center of the disk is located at a distance of 0.25 m from point O. Thus,

$$m_d = \rho_d V_d = 8000 \text{ kg/m}^3 \, [\pi(0.25 \text{ m})^2(0.01 \text{ m})] = 15.71 \text{ kg}$$

$$(I_d)_O = \frac{1}{2}m_d r_d^2 + m_d d^2$$

$$= \frac{1}{2}(15.71 \text{ kg})(0.25 \text{ m})^2 + (15.71 \text{ kg})(0.25 \text{ m})^2$$

$$= 1.473 \text{ kg} \cdot \text{m}^2$$

Hole. For the 125-mm-radius disk (hole), we have

$$m_h = \rho_h V_h = 8000 \text{ kg/m}^3 \, [\pi(0.125 \text{ m})^2(0.01 \text{ m})] = 3.927 \text{ kg}$$

$$(I_h)_O = \frac{1}{2}m_h r_h^2 + m_h d^2$$

$$= \frac{1}{2}(3.927 \text{ kg})(0.125 \text{ m})^2 + (3.927 \text{ kg})(0.25 \text{ m})^2$$

$$= 0.276 \text{ kg} \cdot \text{m}^2$$

The moment of inertia of the plate about point O is therefore

$$I_O = (I_d)_O - (I_h)_O$$

$$= 1.473 \text{ kg} \cdot \text{m}^2 - 0.276 \text{ kg} \cdot \text{m}^2$$

$$= 1.20 \text{ kg} \cdot \text{m}^2 \qquad\qquad Ans.$$

EXAMPLE 17.4

The pendulum in Fig. 17–7 is suspended from the pin at O and consists of two thin rods. Rod OA weighs 10 lb, and BC weighs 8 lb. Determine the moment of inertia of the pendulum about an axis passing through (a) point O, and (b) the mass center G of the pendulum.

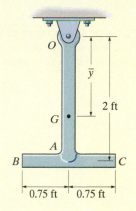

Fig. 17–7

SOLUTION

Part (a). Using the table on the inside back cover, the moment of inertia of rod OA about an axis perpendicular to the page and passing through point O of the rod is $I_O = \frac{1}{3}ml^2$. Hence,

$$(I_{OA})_O = \frac{1}{3}ml^2 = \frac{1}{3}\left(\frac{10\text{ lb}}{32.2\text{ ft/s}^2}\right)(2\text{ ft})^2 = 0.414\text{ slug}\cdot\text{ft}^2$$

This same value can be obtained using $I_G = \frac{1}{12}ml^2$ and the parallel-axis theorem.

$$(I_{OA})_O = \frac{1}{12}ml^2 + md^2 = \frac{1}{12}\left(\frac{10\text{ lb}}{32.2\text{ ft/s}^2}\right)(2\text{ ft})^2 + \left(\frac{10\text{ lb}}{32.2\text{ ft/s}^2}\right)(1\text{ ft})^2$$

$$= 0.414\text{ slug}\cdot\text{ft}^2$$

For rod BC we have

$$(I_{BC})_O = \frac{1}{12}ml^2 + md^2 = \frac{1}{12}\left(\frac{8\text{ lb}}{32.2\text{ ft/s}^2}\right)(1.5\text{ ft})^2 + \left(\frac{8\text{ lb}}{32.2\text{ ft/s}^2}\right)(2\text{ ft})^2$$

$$= 1.040\text{ slug}\cdot\text{ft}^2$$

The moment of inertia of the pendulum about O is therefore

$$I_O = 0.414 + 1.040 = 1.454 = 1.45\text{ slug}\cdot\text{ft}^2 \qquad \textit{Ans.}$$

Part (b). The mass center G will be located relative to point O. Assuming this distance to be $\bar{y}$, Fig. 17–7, and using the formula for determining the mass center, we have

$$\bar{y} = \frac{\Sigma \tilde{y}m}{\Sigma m} = \frac{1(10/32.2) + 2(8/32.2)}{(10/32.2) + (8/32.2)} = 1.444\text{ ft}$$

The moment of inertia I_G may be found in the same manner as I_O, which requires successive applications of the parallel-axis theorem to transfer the moments of inertia of rods OA and BC to G. A more direct solution, however, involves using the result for I_O, i.e.,

$$I_O = I_G + md^2; \quad 1.454\text{ slug}\cdot\text{ft}^2 = I_G + \left(\frac{18\text{ lb}}{32.2\text{ ft/s}^2}\right)(1.444\text{ ft})^2$$

$$I_G = 0.288\text{ slug}\cdot\text{ft}^2 \qquad \textit{Ans.}$$

17

PROBLEMS

17–1. Determine the moment of inertia I_y for the slender rod. The rod's density ρ and cross-sectional area A are constant. Express the result in terms of the rod's total mass m.

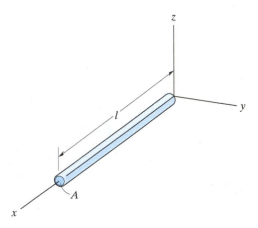

Prob. 17–1

17–2. The solid cylinder has an outer radius R, height h, and is made from a material having a density that varies from its center as $\rho = k + ar^2$, where k and a are constants. Determine the mass of the cylinder and its moment of inertia about the z axis.

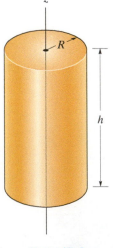

Prob. 17–2

17–3. Determine the moment of inertia of the thin ring about the z axis. The ring has a mass m.

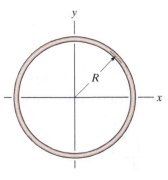

Prob. 17–3

*17–4.** The paraboloid is formed by revolving the shaded area around the x axis. Determine the radius of gyration k_x. The density of the material is $\rho = 5 \text{ Mg/m}^3$.

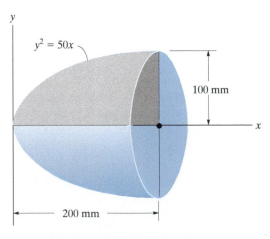

$y^2 = 50x$

100 mm

200 mm

Prob. 17–4

17–5. Determine the radius of gyration k_x of the body. The specific weight of the material is $\gamma = 380 \text{ lb/ft}^3$.

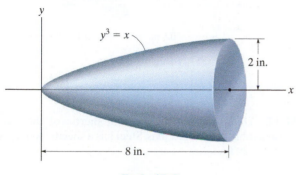

$y^3 = x$

2 in.

8 in.

Prob. 17–5

17–7. The frustum is formed by rotating the shaded area around the x axis. Determine the moment of inertia I_x and express the result in terms of the total mass m of the frustum. The frustum has a constant density ρ.

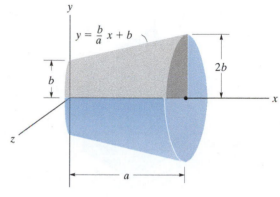

$y = \dfrac{b}{a}x + b$

b

$2b$

a

Prob. 17–7

17

17–6. The sphere is formed by revolving the shaded area around the x axis. Determine the moment of inertia I_x and express the result in terms of the total mass m of the sphere. The material has a constant density ρ.

*17–8.** The hemisphere is formed by rotating the shaded area around the y axis. Determine the moment of inertia I_y and express the result in terms of the total mass m of the hemisphere. The material has a constant density ρ.

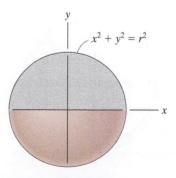

$x^2 + y^2 = r^2$

Prob. 17–6

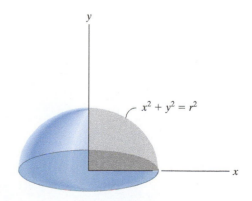

$x^2 + y^2 = r^2$

Prob. 17–8

17–9. Determine the moment of inertia of the homogeneous triangular prism with respect to the y axis. Express the result in terms of the mass m of the prism. *Hint:* For integration, use thin plate elements parallel to the x–y plane and having a thickness dz.

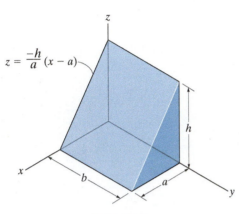

Prob. 17–9

17–10. The pendulum consists of a 4-kg circular plate and a 2-kg slender rod. Determine the radius of gyration of the pendulum about an axis perpendicular to the page and passing through point O.

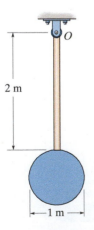

Prob. 17–10

17–11. The assembly is made of the slender rods that have a mass per unit length of 3 kg/m. Determine the mass moment of inertia of the assembly about an axis perpendicular to the page and passing through point O.

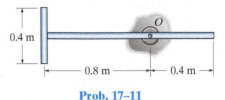

Prob. 17–11

*17–12. Determine the moment of inertia of the solid steel assembly about the x axis. Steel has a specific weight of $\gamma_{st} = 490$ lb/ft^3.

Prob. 17–12

17–13. The wheel consists of a thin ring having a mass of 10 kg and four spokes made from slender rods and each having a mass of 2 kg. Determine the wheel's moment of inertia about an axis perpendicular to the page and passing through point A.

Prob. 17–13

17–14. If the large ring, small ring and each of the spokes weigh 100 lb, 15 lb, and 20 lb, respectively, determine the mass moment of inertia of the wheel about an axis perpendicular to the page and passing through point A.

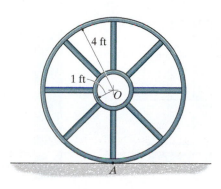

Prob. 17–14

17–15. Determine the moment of inertia about an axis perpendicular to the page and passing through the pin at O. The thin plate has a hole in its center. Its thickness is 50 mm, and the material has a density $\rho = 50 \text{ kg/m}^3$.

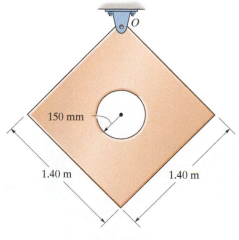

Prob. 17–15

*17–16. Determine the mass moment of inertia of the thin plate about an axis perpendicular to the page and passing through point O. The material has a mass per unit area of 20 kg/m^2.

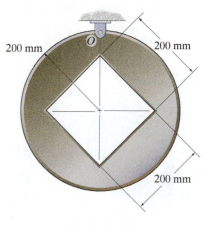

Prob. 17–16

17–17. Determine the location $\overline{y}$ of the center of mass G of the assembly and then calculate the moment of inertia about an axis perpendicular to the page and passing through G. The block has a mass of 3 kg and the semicylinder has a mass of 5 kg.

17–18. Determine the moment of inertia of the assembly about an axis perpendicular to the page and passing through point O. The block has a mass of 3 kg, and the semicylinder has a mass of 5 kg.

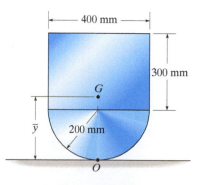

Probs. 17–17/18

17–19. Determine the moment of inertia of the wheel about an axis which is perpendicular to the page and passes through the center of mass G. The material has a specific weight $\gamma = 90$ lb/ft^3.

*17–20. Determine the moment of inertia of the wheel about an axis which is perpendicular to the page and passes through point O. The material has a specific weight $\gamma = 90$ lb/ft^3.

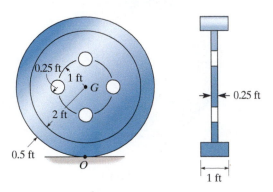

Probs. 17–19/20

17–21. The pendulum consists of the 3-kg slender rod and the 5-kg thin plate. Determine the location $\bar{y}$ of the center of mass G of the pendulum; then calculate the moment of inertia of the pendulum about an axis perpendicular to the page and passing through G.

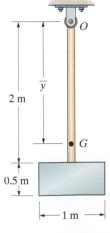

Prob. 17–21

17–22. Determine the moment of inertia of the overhung crank about the x axis. The material is steel having a density of $\rho = 7.85$ Mg/m^3.

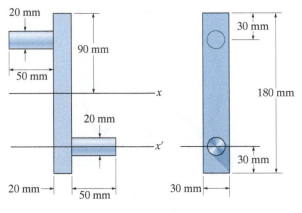

Prob. 17–22

17–23. Determine the moment of inertia of the overhung crank about the x' axis. The material is steel having a density of $\rho = 7.85$ Mg/m^3.

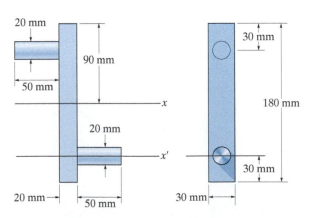

Prob. 17–23

17.2   Planar Kinetic Equations of Motion

In the following analysis we will limit our study of planar kinetics to rigid bodies which, along with their loadings, are considered to be *symmetrical* with respect to a fixed reference plane.* Since the motion of the body can be viewed within the reference plane, all the forces (and couple moments) acting on the body can then be projected onto the plane. An example of an arbitrary body of this type is shown in Fig. 17–8a. Here the *inertial frame of reference x, y, z* has its origin *coincident* with the arbitrary point P in the body. By definition, *these axes do not rotate and are either fixed or translate with constant velocity.*

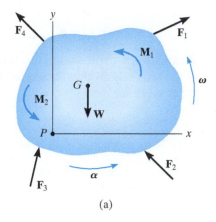

(a)

Fig. 17–8

Equation of Translational Motion.

The external forces acting on the body in Fig. 17–8a represent the effect of gravitational, electrical, magnetic, or contact forces between adjacent bodies. Since this force system has been considered previously in Sec. 13.3 for the analysis of a system of particles, the resulting Eq. 13–6 can be used here, in which case

$$\Sigma \mathbf{F} = m\mathbf{a}_G$$

This equation is referred to as the *translational equation of motion* for the mass center of a rigid body. It states that *the sum of all the external forces acting on the body is equal to the body's mass times the acceleration of its mass center G.*

For motion of the body in the x–y plane, the translational equation of motion may be written in the form of two independent scalar equations, namely,

$$\Sigma F_x = m(a_G)_x$$

$$\Sigma F_y = m(a_G)_y$$

*By doing this, the rotational equation of motion reduces to a rather simplified form. The more general case of body shape and loading is considered in Chapter 21.

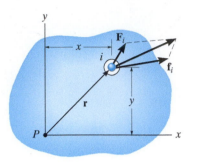

Particle free-body diagram

(b)

||

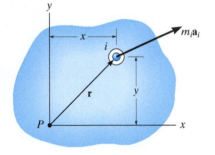

Particle kinetic diagram

(c)

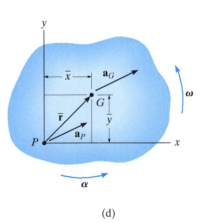

(d)

Fig. 17–8 (cont.)

Equation of Rotational Motion.

We will now determine the effects caused by the moments of the external force system computed about an axis perpendicular to the plane of motion (the z axis) and passing through point P. As shown on the free-body diagram of the ith particle, Fig. 17–8b, $\mathbf{F}_i$ represents the *resultant external force* acting on the particle, and $\mathbf{f}_i$ is the *resultant of the internal forces* caused by interactions with adjacent particles. If the particle has a mass m_i and its acceleration is $\mathbf{a}_i$, then its kinetic diagram is shown in Fig. 17–8c. Summing moments about point P, we require

$$\mathbf{r} \times \mathbf{F}_i + \mathbf{r} \times \mathbf{f}_i = \mathbf{r} \times m_i \mathbf{a}_i$$

or

$$(\mathbf{M}_P)_i = \mathbf{r} \times m_i \mathbf{a}_i$$

The moments about P can also be expressed in terms of the acceleration of point P, Fig. 17–8d. If the body has an angular acceleration $\boldsymbol{\alpha}$ and angular velocity $\boldsymbol{\omega}$, then using Eq. 16–18 we have

$$(\mathbf{M}_P)_i = m_i \mathbf{r} \times (\mathbf{a}_P + \boldsymbol{\alpha} \times \mathbf{r} - \omega^2 \mathbf{r})$$
$$= m_i[\mathbf{r} \times \mathbf{a}_P + \mathbf{r} \times (\boldsymbol{\alpha} \times \mathbf{r}) - \omega^2(\mathbf{r} \times \mathbf{r})]$$

The last term is zero, since $\mathbf{r} \times \mathbf{r} = \mathbf{0}$. Expressing the vectors with Cartesian components and carrying out the cross-product operations yields

$$(M_P)_i \mathbf{k} = m_i\{(x\mathbf{i} + y\mathbf{j}) \times [(a_P)_x \mathbf{i} + (a_P)_y \mathbf{j}]$$
$$+ (x\mathbf{i} + y\mathbf{j}) \times [\alpha \mathbf{k} \times (x\mathbf{i} + y\mathbf{j})]\}$$
$$(M_P)_i \mathbf{k} = m_i[-y(a_P)_x + x(a_P)_y + \alpha x^2 + \alpha y^2]\mathbf{k}$$
$$\zeta\, (M_P)_i = m_i[-y(a_P)_x + x(a_P)_y + \alpha r^2]$$

Letting $m_i \rightarrow dm$ and integrating with respect to the entire mass m of the body, we obtain the resultant moment equation

$$\zeta\, \Sigma M_P = -\left(\int_m y\, dm\right)(a_P)_x + \left(\int_m x\, dm\right)(a_P)_y + \left(\int_m r^2 dm\right)\alpha$$

Here ΣM_P represents only the moment of the *external forces* acting on the body about point P. The resultant moment of the internal forces is zero, since for the entire body these forces occur in equal and opposite collinear pairs and thus the moment of each pair of forces about P cancels. The integrals in the first and second terms on the right are used to locate the body's center of mass G with respect to P, since $\bar{y}m = \int y\, dm$ and $\bar{x}m = \int x\, dm$, Fig. 17–8d. Also, the last integral represents the body's moment of inertia about the z axis, i.e., $I_P = \int r^2 dm$. Thus,

$$\zeta\, \Sigma M_P = -\bar{y}m(a_P)_x + \bar{x}m(a_P)_y + I_P\alpha \qquad (17\text{–}6)$$

It is possible to reduce this equation to a simpler form if point P coincides with the mass center G for the body. If this is the case, then $\bar{x} = \bar{y} = 0$, and therefore*

$$\Sigma M_G = I_G \alpha \qquad (17\text{–}7)$$

This rotational equation of motion states that the sum of the moments of all the external forces about the body's mass center G is equal to the product of the moment of inertia of the body about an axis passing through G and the body's angular acceleration.

Equation 17–6 can also be rewritten in terms of the x and y components of $\mathbf{a}_G$ and the body's moment of inertia I_G. If point G is located at $(\bar{x}, \bar{y})$, Fig. 17–8d, then by the parallel-axis theorem, $I_P = I_G + m(\bar{x}^2 + \bar{y}^2)$. Substituting into Eq. 17–6 and rearranging terms, we get

$$\zeta\,\Sigma M_P = \bar{y}m[-(a_P)_x + \bar{y}\alpha] + \bar{x}m[(a_P)_y + \bar{x}\alpha] + I_G\alpha \qquad (17\text{–}8)$$

From the kinematic diagram of Fig. 17–8d, $\mathbf{a}_P$ can be expressed in terms of $\mathbf{a}_G$ as

$$\mathbf{a}_G = \mathbf{a}_P + \boldsymbol{\alpha} \times \bar{\mathbf{r}} - \omega^2 \bar{\mathbf{r}}$$

$$(a_G)_x \mathbf{i} + (a_G)_y \mathbf{j} = (a_P)_x \mathbf{i} + (a_P)_y \mathbf{j} + \alpha\mathbf{k} \times (\bar{x}\mathbf{i} + \bar{y}\mathbf{j}) - \omega^2(\bar{x}\mathbf{i} + \bar{y}\mathbf{j})$$

Carrying out the cross product and equating the respective $\mathbf{i}$ and $\mathbf{j}$ components yields the two scalar equations

$$(a_G)_x = (a_P)_x - \bar{y}\alpha - \bar{x}\omega^2$$
$$(a_G)_y = (a_P)_y + \bar{x}\alpha - \bar{y}\omega^2$$

From these equations, $[-(a_P)_x + \bar{y}\alpha] = [-(a_G)_x - \bar{x}\omega^2]$ and $[(a_P)_y + \bar{x}\alpha] = [(a_G)_y + \bar{y}\omega^2]$. Substituting these results into Eq. 17–8 and simplifying gives

$$\zeta\,\Sigma M_P = -\bar{y}m(a_G)_x + \bar{x}m(a_G)_y + I_G\alpha \qquad (17\text{–}9)$$

This important result indicates that when moments of the external forces shown on the free-body diagram are summed about point P, Fig. 17–8e, they are equivalent to the sum of the "kinetic moments" of the components of $m\mathbf{a}_G$ about P plus the "kinetic moment" of $I_G\boldsymbol{\alpha}$, Fig. 17–8f. In other words, when the "kinetic moments," $\Sigma(\mathcal{M}_k)_P$, are computed, Fig. 17–8f, the vectors $m(\mathbf{a}_G)_x$ and $m(\mathbf{a}_G)_y$ are treated as sliding vectors; that is, they can act at *any point along their line of action*. In a similar manner, $I_G\boldsymbol{\alpha}$ can be treated as a free vector and can therefore act at *any point*. It is important to keep in mind, however, that $m\mathbf{a}_G$ and $I_G\boldsymbol{\alpha}$ are not the same as a force or a couple moment. Instead, they are caused by the external effects of forces and couple moments acting on the body. With this in mind we can therefore write Eq. 17–9 in a more general form as

$$\Sigma M_P = \Sigma(\mathcal{M}_k)_P \qquad (17\text{–}10)$$

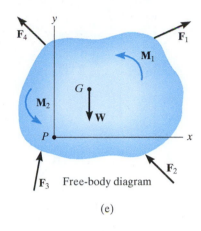

Free-body diagram

(e)

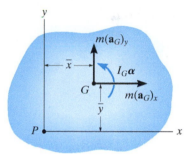

Kinetic diagram

(f)

Fig. 17–8 (cont.)

*It also reduces to this same simple form $\Sigma M_P = I_P\alpha$ if point P is a *fixed point* (see Eq. 17–16) or the acceleration of point P is directed along the line PG.

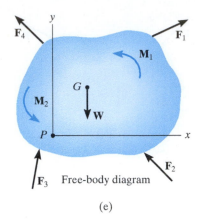

Free-body diagram

(e)

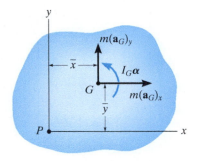

Kinetic diagram

(f)

Fig. 17–8 (cont.)

General Application of the Equations of Motion.

To summarize this analysis, *three* independent scalar equations can be written to describe the general plane motion of a symmetrical rigid body.

$$\Sigma F_x = m(a_G)_x$$
$$\Sigma F_y = m(a_G)_y$$
$$\Sigma M_G = I_G\alpha$$

or

$$\Sigma M_P = \Sigma(\mathcal{M}_k)_P \qquad (17\text{–}11)$$

When applying these equations, one should *always* draw a free-body diagram, Fig. 17–8e, in order to account for the terms involved in ΣF_x, ΣF_y, ΣM_G, or ΣM_P. In some problems it may also be helpful to draw the *kinetic diagram* for the body, Fig. 17–8f. This diagram graphically accounts for the terms $m(\mathbf{a}_G)_x$, $m(\mathbf{a}_G)_y$, and $I_G\boldsymbol{\alpha}$. It is especially convenient when used to determine the components of $m\mathbf{a}_G$ and the moment of these components in $\Sigma(\mathcal{M}_k)_P$.*

17.3   Equations of Motion: Translation

When the rigid body in Fig. 17–9a undergoes a *translation*, all the particles of the body have the *same acceleration*. Furthermore, $\boldsymbol{\alpha} = \mathbf{0}$, in which case the rotational equation of motion applied at point G reduces to a simplified form, namely, $\Sigma M_G = 0$. Application of this and the force equations of motion will now be discussed for each of the two types of translation.

Rectilinear Translation.

When a body is subjected to *rectilinear translation*, all the particles of the body (slab) travel along parallel straight-line paths. The free-body and kinetic diagrams are shown in Fig. 17–9b. Since $I_G\boldsymbol{\alpha} = \mathbf{0}$, only $m\mathbf{a}_G$ is shown on the kinetic diagram. Hence, the equations of motion which apply in this case become

$$\boxed{\begin{array}{l} \Sigma F_x = m(a_G)_x \\ \Sigma F_y = m(a_G)_y \\ \Sigma M_G = 0 \end{array}} \qquad (17\text{–}12)$$

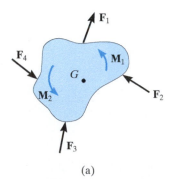

(a)

Fig. 17–9

*For this reason, the kinetic diagram will be used in the solution of an example problem whenever $\Sigma M_P = \Sigma(\mathcal{M}_k)_P$ is applied.

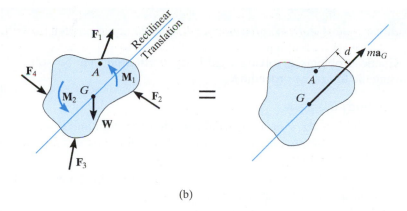

(b)

It is also possible to sum moments about other points on or off the body, in which case the moment of $m\mathbf{a}_G$ must be taken into account. For example, if point A is chosen, which lies at a perpendicular distance d from the line of action of $m\mathbf{a}_G$, the following moment equation applies:

$$\zeta + \Sigma M_A = \Sigma(\mathcal{M}_k)_A; \qquad \Sigma M_A = (ma_G)d$$

Here the sum of moments of the external forces and couple moments about A (ΣM_A, free-body diagram) equals the moment of $m\mathbf{a}_G$ about A ($\Sigma(\mathcal{M}_k)_A$, kinetic diagram).

Curvilinear Translation.

When a rigid body is subjected to *curvilinear translation*, all the particles of the body have the same accelerations as they travel along *curved paths* as noted in Sec.16.1. For analysis, it is often convenient to use an inertial coordinate system having an origin which coincides with the body's mass center at the instant considered, and axes which are oriented in the normal and tangential directions to the path of motion, Fig. 17–9c. The three scalar equations of motion are then

$$\Sigma F_n = m(a_G)_n$$
$$\Sigma F_t = m(a_G)_t \qquad (17\text{-}13)$$
$$\Sigma M_G = 0$$

If moments are summed about the arbitrary point B, Fig. 17–9c, then it is necessary to account for the moments, $\Sigma(\mathcal{M}_k)_B$, of the two components $m(\mathbf{a}_G)_n$ and $m(\mathbf{a}_G)_t$ about this point. From the kinetic diagram, h and e represent the perpendicular distances (or "moment arms") from B to the lines of action of the components. The required moment equation therefore becomes

$$\zeta + \Sigma M_B = \Sigma(\mathcal{M}_k)_B; \qquad \Sigma M_B = e[m(a_G)_t] - h[m(a_G)_n]$$

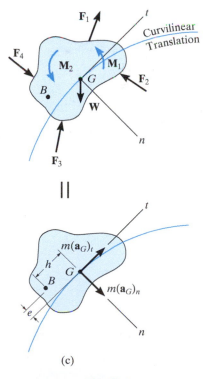

(c)

Fig. 17–9

The free-body and kinetic diagrams for this boat and trailer are drawn first in order to apply the equations of motion. Here the forces on the free-body diagram cause the effect shown on the kinetic diagram. If moments are summed about the mass center, G, then $\Sigma M_G = 0$. However, if moments are summed about point B then $\zeta + \Sigma M_B = ma_G(d)$. (© R.C. Hibbeler)

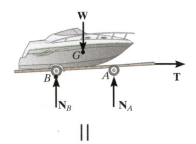

Procedure for Analysis

Kinetic problems involving rigid-body *translation* can be solved using the following procedure.

Free-Body Diagram.

- Establish the x, y or n, t inertial coordinate system and draw the free-body diagram in order to account for all the external forces and couple moments that act on the body.

- The direction and sense of the acceleration of the body's mass center $\mathbf{a}_G$ should be established.

- Identify the unknowns in the problem.

- If it is decided that the rotational equation of motion $\Sigma M_P = \Sigma(\mathcal{M}_k)_P$ is to be used in the solution, then consider drawing the kinetic diagram, since it graphically accounts for the components $m(\mathbf{a}_G)_x$, $m(\mathbf{a}_G)_y$ or $m(\mathbf{a}_G)_t$, $m(\mathbf{a}_G)_n$ and is therefore convenient for "visualizing" the terms needed in the moment sum $\Sigma(\mathcal{M}_k)_P$.

Equations of Motion.

- Apply the three equations of motion in accordance with the established sign convention.

- To simplify the analysis, the moment equation $\Sigma M_G = 0$ can be replaced by the more general equation $\Sigma M_P = \Sigma(\mathcal{M}_k)_P$, where point P is usually located at the intersection of the lines of action of as many unknown forces as possible.

- If the body is in contact with a *rough surface* and slipping occurs, use the friction equation $F = \mu_k N$. Remember, $\mathbf{F}$ always acts on the body so as to oppose the motion of the body relative to the surface it contacts.

Kinematics.

- Use kinematics to determine the velocity and position of the body.

- For rectilinear translation with *variable acceleration*
  $$a_G = dv_G/dt \quad a_G ds_G = v_G dv_G$$

- For rectilinear translation with *constant acceleration*
  $$v_G = (v_G)_0 + a_G t \quad v_G^2 = (v_G)_0^2 + 2a_G[s_G - (s_G)_0]$$
  $$s_G = (s_G)_0 + (v_G)_0 t + \tfrac{1}{2}a_G t^2$$

- For curvilinear translation
  $$(a_G)_n = v_G^2/\rho$$
  $$(a_G)_t = dv_G/dt \quad (a_G)_t ds_G = v_G dv_G$$

EXAMPLE 17.5

The car shown in Fig. 17–10*a* has a mass of 2 Mg and a center of mass at *G*. Determine the acceleration if the rear "driving" wheels are always slipping, whereas the front wheels are free to rotate. Neglect the mass of the wheels. The coefficient of kinetic friction between the wheels and the road is $\mu_k = 0.25$.

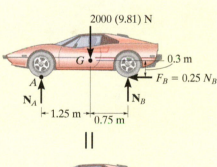

(a)

SOLUTION I

Free-Body Diagram. As shown in Fig. 17–10*b*, the rear-wheel frictional force $\mathbf{F}_B$ pushes the car forward, and since *slipping occurs*, $F_B = 0.25N_B$. The frictional forces acting on the *front wheels* are *zero*, since these wheels have negligible mass.* There are three unknowns in the problem, N_A, N_B, and a_G. Here we will sum moments about the mass center. The car (point *G*) accelerates to the left, i.e., in the negative *x* direction, Fig. 17–10*b*.

Equations of Motion.

$$\pm \Sigma F_x = m(a_G)_x; \qquad -0.25N_B = -(2000 \text{ kg})a_G \qquad (1)$$

$$+\uparrow \Sigma F_y = m(a_G)_y; \qquad N_A + N_B - 2000(9.81) \text{ N} = 0 \qquad (2)$$

$$\zeta +\Sigma M_G = 0; \quad -N_A(1.25 \text{ m}) - 0.25N_B(0.3 \text{ m}) + N_B(0.75 \text{ m}) = 0 \quad (3)$$

Solving,

$$a_G = 1.59 \text{ m/s}^2 \leftarrow \qquad \qquad Ans.$$

$$N_A = 6.88 \text{ kN}$$

$$N_B = 12.7 \text{ kN}$$

SOLUTION II

Free-Body and Kinetic Diagrams. If the "moment" equation is applied about point *A*, then the unknown N_A will be eliminated from the equation. To "visualize" the moment of $m\mathbf{a}_G$ about *A*, we will include the kinetic diagram as part of the analysis, Fig. 17–10*c*.

Equation of Motion.

$$\zeta +\Sigma M_A = \Sigma(\mathcal{M}_k)_A; \qquad N_B(2 \text{ m}) - [2000(9.81) \text{ N}](1.25 \text{ m}) =$$

$$(2000 \text{ kg})a_G(0.3 \text{ m})$$

Solving this and Eq. 1 for a_G leads to a simpler solution than that obtained from Eqs. 1 to 3.

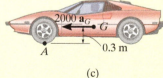

(b)

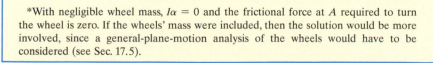

(c)

Fig. 17–10

*With negligible wheel mass, $I\alpha = 0$ and the frictional force at *A* required to turn the wheel is zero. If the wheels' mass were included, then the solution would be more involved, since a general-plane-motion analysis of the wheels would have to be considered (see Sec. 17.5).

EXAMPLE 17.6

(© R.C. Hibbeler)

The motorcycle shown in Fig. 17–11a has a mass of 125 kg and a center of mass at G_1, while the rider has a mass of 75 kg and a center of mass at G_2. Determine the minimum coefficient of static friction between the wheels and the pavement in order for the rider to do a "wheely," i.e., lift the front wheel off the ground as shown in the photo. What acceleration is necessary to do this? Neglect the mass of the wheels and assume that the front wheel is free to roll.

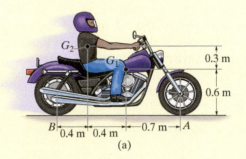

(a)

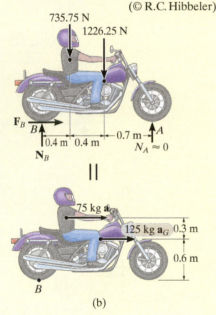

Fig. 17–11

SOLUTION

Free-Body and Kinetic Diagrams. In this problem we will consider both the motorcycle and the rider as a single *system*. It is possible first to determine the location of the center of mass for this "system" by using the equations $\bar{x} = \Sigma \tilde{x} m / \Sigma m$ and $\bar{y} = \Sigma \tilde{y} m / \Sigma m$. Here, however, we will consider the weight and mass of the motorcycle and rider separately as shown on the free-body and kinetic diagrams, Fig. 17–11b. Both of these parts move with the *same* acceleration. We have assumed that the front wheel is *about* to leave the ground, so that the normal reaction $N_A \approx 0$. The three unknowns in the problem are N_B, F_B, and a_G.

Equations of Motion.

$$\xrightarrow{+} \Sigma F_x = m(a_G)_x; \qquad\qquad F_B = (75 \text{ kg} + 125 \text{ kg})a_G \qquad (1)$$

$$+\uparrow \Sigma F_y = m(a_G)_y; \qquad\qquad N_B - 735.75 \text{ N} - 1226.25 \text{ N} = 0$$

$$\zeta + \Sigma M_B = \Sigma(\mathcal{M}_k)_B; \quad -(735.75 \text{ N})(0.4 \text{ m}) - (1226.25 \text{ N})(0.8 \text{ m}) =$$

$$-(75 \text{ kg } a_G)(0.9 \text{ m}) - (125 \text{ kg } a_G)(0.6 \text{ m}) \qquad (2)$$

Solving,

$$a_G = 8.95 \text{ m/s}^2 \rightarrow \qquad\qquad Ans.$$

$$N_B = 1962 \text{ N}$$

$$F_B = 1790 \text{ N}$$

Thus the minimum coefficient of static friction is

$$(\mu_s)_{\min} = \frac{F_B}{N_B} = \frac{1790 \text{ N}}{1962 \text{ N}} = 0.912 \qquad\qquad Ans.$$

EXAMPLE | **17.7**

The 100-kg beam *BD* shown in Fig. 17–12*a* is supported by two rods having negligible mass. Determine the force developed in each rod if at the instant $\theta = 30°$, $\omega = 6$ rad/s.

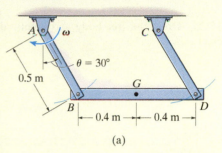

(a)

SOLUTION

Free-Body and Kinetic Diagrams.   The beam moves with *curvilinear translation* since all points on the beam move along circular paths, each path having the same radius of 0.5 m, but different centers of curvature. Using normal and tangential coordinates, the free-body and kinetic diagrams for the beam are shown in Fig. 17–12*b*. Because of the *translation*, *G* has the *same* motion as the pin at *B*, which is connected to both the rod and the beam. Note that the tangential component of acceleration acts downward to the left due to the clockwise direction of α, Fig. 17–12*c*. Furthermore, the normal component of acceleration is *always* directed toward the center of curvature (toward point *A* for rod *AB*). Since the angular velocity of *AB* is 6 rad/s when $\theta = 30°$, then

$$(a_G)_n = \omega^2 r = (6 \text{ rad/s})^2(0.5 \text{ m}) = 18 \text{ m/s}^2$$

The three unknowns are T_B, T_D, and $(a_G)_t$.

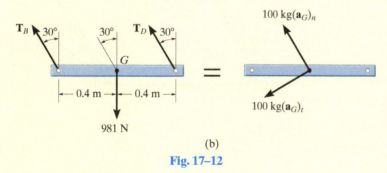

(b)

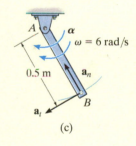

(c)

Fig. 17–12

Equations of Motion.

$$+\nwarrow\Sigma F_n = m(a_G)_n; \quad T_B + T_D - 981 \cos 30° \text{ N} = 100 \text{ kg}(18 \text{ m/s}^2) \quad (1)$$

$$+\swarrow\Sigma F_t = m(a_G)_t; \quad 981 \sin 30° = 100 \text{ kg}(a_G)_t \quad (2)$$

$$\zeta+\Sigma M_G = 0; \quad -(T_B \cos 30°)(0.4 \text{ m}) + (T_D \cos 30°)(0.4 \text{ m}) = 0 \quad (3)$$

Simultaneous solution of these three equations gives

$$T_B = T_D = 1.32 \text{ kN} \qquad\qquad Ans.$$

$$(a_G)_t = 4.905 \text{ m/s}^2$$

NOTE: It is also possible to apply the equations of motion along horizontal and vertical *x, y* axes, but the solution becomes more involved.

PRELIMINARY PROBLEMS

P17–1. Draw the free-body and kinetic diagrams of the object *AB*.

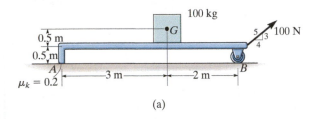

(a)

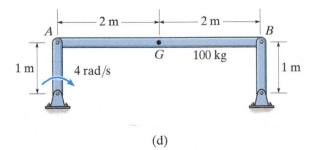

(d)

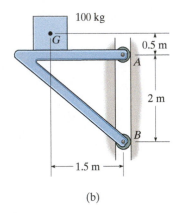

(b)

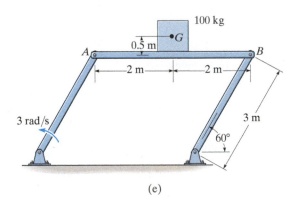

(e)

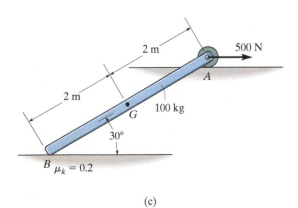

(c)

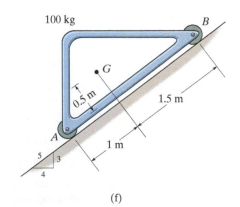

(f)

Prob. P17–1

P17–2. Draw the free-body and kinetic diagrams of the 100-kg object.

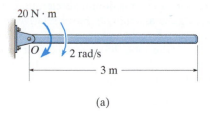

20 N · m

O   2 rad/s

3 m

(a)

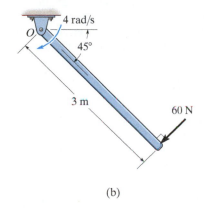

4 rad/s

O   45°

3 m

60 N

(b)

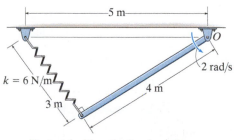

5 m

O

k = 6 N/m   2 rad/s

4 m

3 m

Unstretched length of spring is 1 m.

(c)

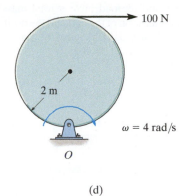

100 N

2 m

ω = 4 rad/s

O

(d)

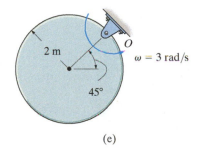

2 m

O   ω = 3 rad/s

45°

(e)

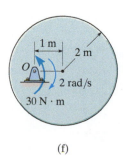

1 m   2 m

O   2 rad/s

30 N · m

(f)

FUNDAMENTAL PROBLEMS

F17–1. The cart and its load have a total mass of 100 kg. Determine the acceleration of the cart and the normal reactions on the pair of wheels at A and B. Neglect the mass of the wheels.

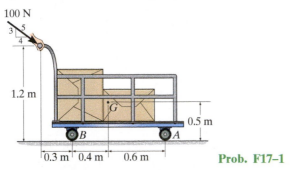

Prob. F17–1

F17–2. If the 80-kg cabinet is allowed to roll down the inclined plane, determine the acceleration of the cabinet and the normal reactions on the pair of rollers at A and B that have negligible mass.

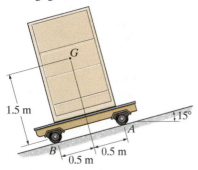

Prob. F17–2

F17–3. The 20-lb link AB is pinned to a moving frame at A and held in a vertical position by means of a string BC which can support a maximum tension of 10 lb. Determine the maximum acceleration of the frame without breaking the string. What are the corresponding components of reaction at the pin A?

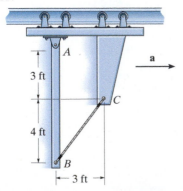

Prob. F17–3

F17–4. Determine the maximum acceleration of the truck without causing the assembly to move relative to the truck. Also what is the corresponding normal reaction on legs A and B? The 100-kg table has a mass center at G and the coefficient of static friction between the legs of the table and the bed of the truck is $\mu_s = 0.2$.

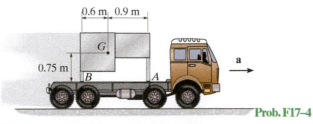

Prob. F17–4

F17–5. At the instant shown both rods of negligible mass swing with a counterclockwise angular velocity of $\omega = 5$ rad/s, while the 50-kg bar is subjected to the 100-N horizontal force. Determine the tension developed in the rods and the angular acceleration of the rods at this instant.

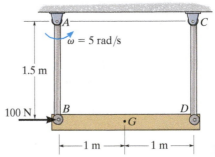

Prob. F17–5

F17–6. At the instant shown, link CD rotates with an angular velocity of $\omega = 6$ rad/s. If it is subjected to a couple moment $M = 450$ N · m, determine the force developed in link AB, the horizontal and vertical component of reaction on pin D, and the angular acceleration of link CD at this instant. The block has a mass of 50 kg and center of mass at G. Neglect the mass of links AB and CD.

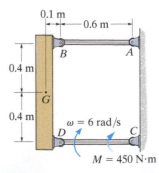

Prob. F17–6

PROBLEMS

*17–24. The door has a weight of 200 lb and a center of gravity at G. Determine how far the door moves in 2 s, starting from rest, if a man pushes on it at C with a horizontal force $F = 30$ lb. Also, find the vertical reactions at the rollers A and B.

17–25. The door has a weight of 200 lb and a center of gravity at G. Determine the constant force F that must be applied to the door to push it open 12 ft to the right in 5 s, starting from rest. Also, find the vertical reactions at the rollers A and B.

17–27. The sports car has a weight of 4500 lb and center of gravity at G. If it starts from rest it causes the rear wheels to slip as it accelerates. Determine how long it takes for it to reach a speed of 10 ft/s. Also, what are the normal reactions at each of the four wheels on the road? The coefficients of static and kinetic friction at the road are $\mu_s = 0.5$ and $\mu_k = 0.3$, respectively. Neglect the mass of the wheels.

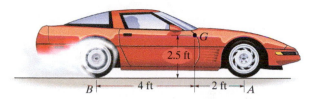

Prob. 17–27

*17–28. The assembly has a mass of 8 Mg and is hoisted using the boom and pulley system. If the winch at B draws in the cable with an acceleration of 2 m/s², determine the compressive force in the hydraulic cylinder needed to support the boom. The boom has a mass of 2 Mg and mass center at G.

17–29. The assembly has a mass of 4 Mg and is hoisted using the winch at B. Determine the greatest acceleration of the assembly so that the compressive force in the hydraulic cylinder supporting the boom does not exceed 180 kN. What is the tension in the supporting cable? The boom has a mass of 2 Mg and mass center at G.

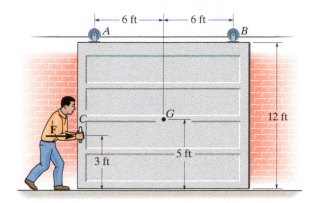

Probs. 17–24/25

17–26. The jet aircraft has a total mass of 22 Mg and a center of mass at G. Initially at take-off the engines provide a thrust $2T = 4$ kN and $T' = 1.5$ kN. Determine the acceleration of the plane and the normal reactions on the nose wheel at A and each of the *two* wing wheels located at B. Neglect the mass of the wheels and, due to low velocity, neglect any lift caused by the wings.

Prob. 17–26

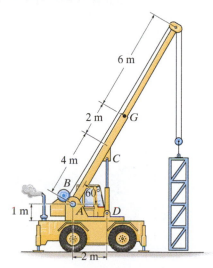

Probs. 17–28/29

17–30. The uniform girder AB has a mass of 8 Mg. Determine the internal axial, shear, and bending-moment loadings at the center of the girder if a crane gives it an upward acceleration of 3 m/s².

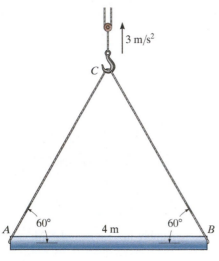

Prob. 17–30

17–31. A car having a weight of 4000 lb begins to skid and turn with the brakes applied to all four wheels. If the coefficient of kinetic friction between the wheels and the road is $\mu_k = 0.8$, determine the maximum critical height h of the center of gravity G such that the car does not overturn. Tipping will begin to occur after the car rotates 90° from its original direction of motion and, as shown in the figure, undergoes *translation* while skidding. *Hint*: Draw a free-body diagram of the car viewed from the front. When tipping occurs, the normal reactions of the wheels on the right side (or passenger side) are zero.

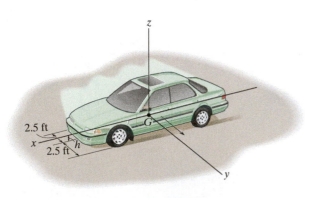

Prob. 17–31

*17–32. A force of $P = 300$ N is applied to the 60-kg cart. Determine the reactions at both the wheels at A and both the wheels at B. Also, what is the acceleration of the cart? The mass center of the cart is at G.

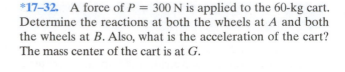

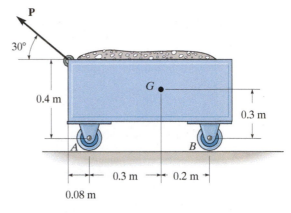

Prob. 17–32

17–33. Determine the largest force P that can be applied to the 60-kg cart, without causing one of the wheel reactions, either at A or at B, to be zero. Also, what is the acceleration of the cart? The mass center of the cart is at G.

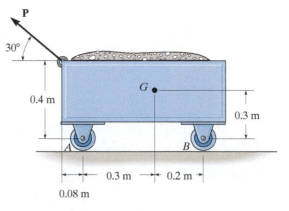

Prob. 17–33

17–34. The trailer with its load has a mass of 150-kg and a center of mass at G. If it is subjected to a horizontal force of $P = 600$ N, determine the trailer's acceleration and the normal force on the pair of wheels at A and at B. The wheels are free to roll and have negligible mass.

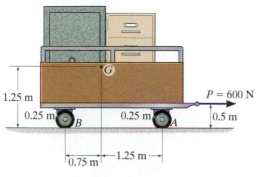

Prob. 17–34

17–35. The desk has a weight of 75 lb and a center of gravity at G. Determine its initial acceleration if a man pushes on it with a force $F = 60$ lb. The coefficient of kinetic friction at A and B is $\mu_k = 0.2$.

*17–36. The desk has a weight of 75 lb and a center of gravity at G. Determine the initial acceleration of a desk when the man applies enough force F to overcome the static friction at A and B. Also, find the vertical reactions on each of the two legs at A and at B. The coefficients of static and kinetic friction at A and B are $\mu_s = 0.5$ and $\mu_k = 0.2$, respectively.

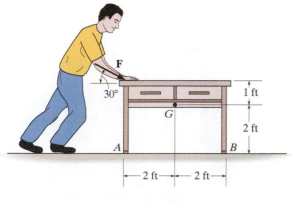

Probs. 17–35/36

17–37. The 150-kg uniform crate rests on the 10-kg cart. Determine the maximum force P that can be applied to the handle without causing the crate to tip on the cart. Slipping does not occur.

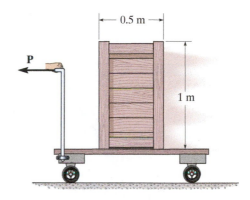

Prob. 17–37

17–38. The 150-kg uniform crate rests on the 10-kg cart. Determine the maximum force P that can be applied to the handle without causing the crate to slip or tip on the cart. The coefficient of static friction between the crate and cart is $\mu_s = 0.2$.

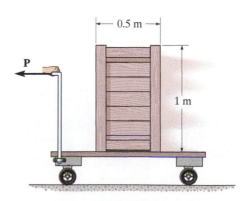

Prob. 17–38

17

17–39. The bar has a weight per length w and is supported by the smooth collar. If it is released from rest, determine the internal normal force, shear force, and bending moment in the bar as a function of x.

17–42. The uniform crate has a mass of 50 kg and rests on the cart having an inclined surface. Determine the smallest acceleration that will cause the crate either to tip or slip relative to the cart. What is the magnitude of this acceleration? The coefficient of static friction between the crate and cart is $\mu_s = 0.5$.

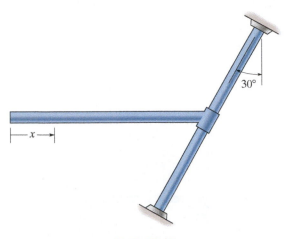

Prob. 17–39

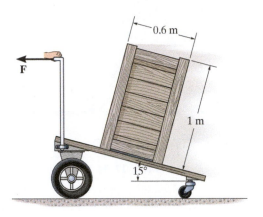

Prob. 17–42

*17–40. The smooth 180-lb pipe has a length of 20 ft and a negligible diameter. It is carried on a truck as shown. Determine the maximum acceleration which the truck can have without causing the normal reaction at A to be zero. Also determine the horizontal and vertical components of force which the truck exerts on the pipe at B.

17–41. The smooth 180-lb pipe has a length of 20 ft and a negligible diameter. It is carried on a truck as shown. If the truck accelerates at $a = 5 \text{ ft/s}^2$, determine the normal reaction at A and the horizontal and vertical components of force which the truck exerts on the pipe at B.

17–43. Determine the acceleration of the 150-lb cabinet and the normal reaction under the legs A and B if $P = 35$ lb. The coefficients of static and kinetic friction between the cabinet and the plane are $\mu_s = 0.2$ and $\mu_k = 0.15$, respectively. The cabinet's center of gravity is located at G.

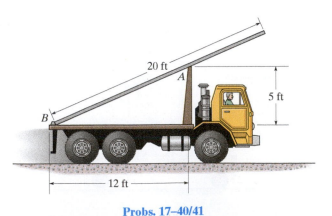

Probs. 17–40/41

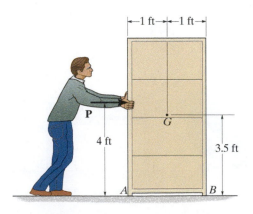

Prob. 17–43

*17–44. The uniform bar of mass m is pin connected to the collar, which slides along the smooth horizontal rod. If the collar is given a constant acceleration of $\mathbf{a}$, determine the bar's inclination angle θ. Neglect the collar's mass.

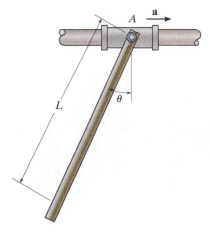

Prob. 17–44

17–45. The drop gate at the end of the trailer has a mass of 1.25 Mg and mass center at G. If it is supported by the cable AB and hinge at C, determine the tension in the cable when the truck begins to accelerate at 5 m/s². Also, what are the horizontal and vertical components of reaction at the hinge C?

17–46. The drop gate at the end of the trailer has a mass of 1.25 Mg and mass center at G. If it is supported by the cable AB and hinge at C, determine the maximum deceleration of the truck so that the gate does not begin to rotate forward. What are the horizontal and vertical components of reaction at the hinge C?

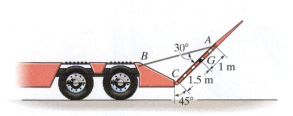

Probs. 17–45/46

17–47. The snowmobile has a weight of 250 lb, centered at G_1, while the rider has a weight of 150 lb, centered at G_2. If the acceleration is $a = 20$ ft/s², determine the maximum height h of G_2 of the rider so that the snowmobile's front skid does not lift off the ground. Also, what are the traction (horizontal) force and normal reaction under the rear tracks at A?

*17–48. The snowmobile has a weight of 250 lb, centered at G_1, while the rider has a weight of 150 lb, centered at G_2. If $h = 3$ ft, determine the snowmobile's maximum permissible acceleration $\mathbf{a}$ so that its front skid does not lift off the ground. Also, find the traction (horizontal) force and the normal reaction under the rear tracks at A.

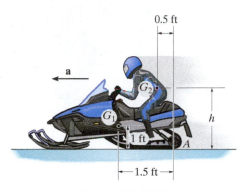

Probs. 17–47/48

17–49. If the cart's mass is 30 kg and it is subjected to a horizontal force of $P = 90$ N, determine the tension in cord AB and the horizontal and vertical components of reaction on end C of the uniform 15-kg rod BC.

17–50. If the cart's mass is 30 kg, determine the horizontal force P that should be applied to the cart so that the cord AB just becomes slack. The uniform rod BC has a mass of 15 kg.

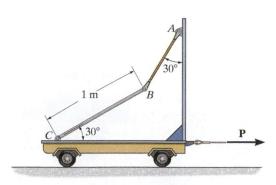

Probs. 17–49/50

17

17–51. The pipe has a mass of 800 kg and is being towed behind the truck. If the acceleration of the truck is $a_t = 0.5 \text{ m/s}^2$, determine the angle θ and the tension in the cable. The coefficient of kinetic friction between the pipe and the ground is $\mu_k = 0.1$.

17–53. The crate C has a weight of 150 lb and rests on the truck elevator for which the coefficient of static friction is $\mu_s = 0.4$. Determine the largest initial angular acceleration α, starting from rest, which the parallel links AB and DE can have without causing the crate to slip. No tipping occurs.

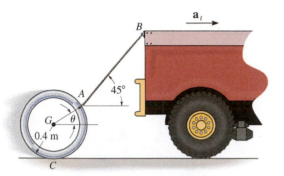

Prob. 17–51

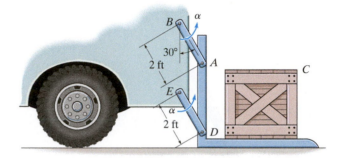

Prob. 17–53

*17–52. The pipe has a mass of 800 kg and is being towed behind a truck. If the angle $\theta = 30°$, determine the acceleration of the truck and the tension in the cable. The coefficient of kinetic friction between the pipe and the ground is $\mu_k = 0.1$.

17–54. The crate C has a weight of 150 lb and rests on the truck elevator. Determine the initial friction and normal force of the elevator on the crate if the parallel links are given an angular acceleration $\alpha = 2 \text{ rad/s}^2$ starting from rest.

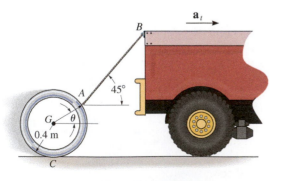

Prob. 17–52

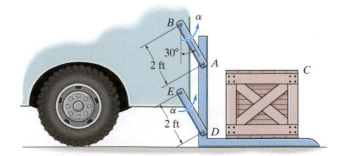

Prob. 17–54

17–55. The 100-kg uniform crate C rests on the elevator floor where the coefficient of static friction is $\mu_s = 0.4$. Determine the largest initial angular acceleration α, starting from rest at $\theta = 90°$, without causing the crate to slip. No tipping occurs.

*17–56. The two uniform 4-kg bars DC and EF are fixed (welded) together at E. Determine the normal force N_E, shear force V_E, and moment M_E, which DC exerts on EF at E if at the instant $\theta = 60°$ BC has an angular velocity $\omega = 2$ rad/s and an angular acceleration $\alpha = 4$ rad/s^2 as shown.

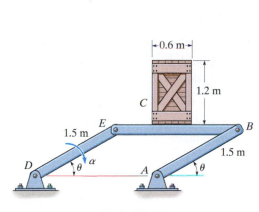

Prob. 17–55

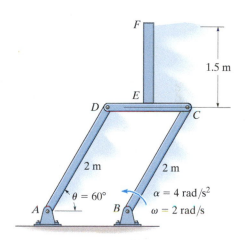

Prob. 17–56

17.4 Equations of Motion: Rotation about a Fixed Axis

Consider the rigid body (or slab) shown in Fig. 17–13a, which is constrained to rotate in the vertical plane about a fixed axis perpendicular to the page and passing through the pin at O. The angular velocity and angular acceleration are caused by the external force and couple moment system acting on the body. Because the body's center of mass G moves around a *circular path*, the acceleration of this point is best represented by its tangential and normal components. The *tangential component of acceleration* has a *magnitude* of $(a_G)_t = \alpha r_G$ and must act in a *direction* which is *consistent* with the body's angular acceleration α. The *magnitude* of the *normal component of acceleration* is $(a_G)_n = \omega^2 r_G$. This component is *always directed* from point G to O, regardless of the rotational sense of ω.

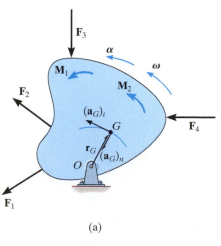

(a)

Fig. 17–13

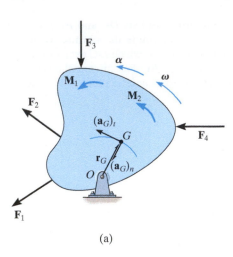

(a)

The free-body and kinetic diagrams for the body are shown in Fig. 17–13b. The two components $m(\mathbf{a}_G)_t$ and $m(\mathbf{a}_G)_n$, shown on the kinetic diagram, are associated with the tangential and normal components of acceleration of the body's mass center. The $I_G\boldsymbol{\alpha}$ vector acts in the same *direction* as $\boldsymbol{\alpha}$ and has a *magnitude* of $I_G\alpha$, where I_G is the body's moment of inertia calculated about an axis which is perpendicular to the page and passes through G. From the derivation given in Sec. 17.2, the equations of motion which apply to the body can be written in the form

$$\begin{aligned} \Sigma F_n &= m(a_G)_n = m\omega^2 r_G \\ \Sigma F_t &= m(a_G)_t = m\alpha r_G \\ \Sigma M_G &= I_G\alpha \end{aligned} \qquad (17\text{–}14)$$

The moment equation can be replaced by a moment summation about any arbitrary point P on or off the body provided one accounts for the moments $\Sigma(\mathcal{M}_k)_P$ produced by $I_G\boldsymbol{\alpha}$, $m(\mathbf{a}_G)_t$, and $m(\mathbf{a}_G)_n$ about the point.

Moment Equation About Point O.

Often it is convenient to sum moments about the pin at O in order to eliminate the *unknown* force $\mathbf{F}_O$. From the kinetic diagram, Fig. 17–13b, this requires

$$\zeta + \Sigma M_O = \Sigma(\mathcal{M}_k)_O; \qquad \Sigma M_O = r_G m(a_G)_t + I_G\alpha \qquad (17\text{–}15)$$

Note that the moment of $m(\mathbf{a}_G)_n$ is not included here since the line of action of this vector passes through O. Substituting $(a_G)_t = r_G\alpha$, we may rewrite the above equation as $\zeta + \Sigma M_O = (I_G + mr_G^2)\alpha$. From the parallel-axis theorem, $I_O = I_G + md^2$, and therefore the term in parentheses represents the *moment of inertia of the body about the fixed axis of rotation passing through O.* Consequently, we can write the three equations of motion for the body as

$$\begin{aligned} \Sigma F_n &= m(a_G)_n = m\omega^2 r_G \\ \Sigma F_t &= m(a_G)_t = m\alpha r_G \\ \Sigma M_O &= I_O\alpha \end{aligned} \qquad (17\text{–}16)$$

When using these equations, remember that "$I_O\alpha$" accounts for the "moment" of both $m(\mathbf{a}_G)_t$ and $I_G\boldsymbol{\alpha}$ about point O, Fig. 17–13b. In other words, $\Sigma M_O = \Sigma(\mathcal{M}_k)_O = I_O\alpha$, as indicated by Eqs. 17–15 and 17–16.

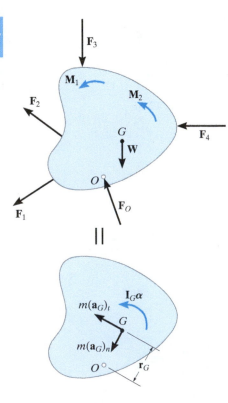

(b)

Fig. 17–13 (cont.)

*The result $\Sigma M_O = I_O\alpha$ can also be obtained *directly* from Eq. 17–6 by selecting point P to coincide with O, realizing that $(a_P)_x = (a_P)_y = 0$.

Procedure for Analysis

Kinetic problems which involve the rotation of a body about a fixed axis can be solved using the following procedure.

Free-Body Diagram.

- Establish the inertial n, t coordinate system and specify the direction and sense of the accelerations $(\mathbf{a}_G)_n$ and $(\mathbf{a}_G)_t$ and the angular acceleration $\boldsymbol{\alpha}$ of the body. Recall that $(\mathbf{a}_G)_t$ must act in a direction which is in accordance with the rotational sense of $\boldsymbol{\alpha}$, whereas $(\mathbf{a}_G)_n$ always acts toward the axis of rotation, point O.

- Draw the free-body diagram to account for all the external forces and couple moments that act on the body.

- Determine the moment of inertia I_G or I_O.

- Identify the unknowns in the problem.

- If it is decided that the rotational equation of motion $\Sigma M_P = \Sigma(\mathcal{M}_k)_P$ is to be used, i.e., P is a point other than G or O, then consider drawing the kinetic diagram in order to help "visualize" the "moments" developed by the components $m(\mathbf{a}_G)_n$, $m(\mathbf{a}_G)_t$, and $I_G\boldsymbol{\alpha}$ when writing the terms for the moment sum $\Sigma(\mathcal{M}_k)_P$.

Equations of Motion.

- Apply the three equations of motion in accordance with the established sign convention.

- If moments are summed about the body's mass center, G, then $\Sigma M_G = I_G\alpha$, since $(m\mathbf{a}_G)_t$ and $(m\mathbf{a}_G)_n$ create no moment about G.

- If moments are summed about the pin support O on the axis of rotation, then $(m\mathbf{a}_G)_n$ creates no moment about O, and it can be shown that $\Sigma M_O = I_O\alpha$.

Kinematics.

- Use kinematics if a complete solution cannot be obtained strictly from the equations of motion.

- If the angular acceleration is variable, use

$$\alpha = \frac{d\omega}{dt} \qquad \alpha\,d\theta = \omega\,d\omega \qquad \omega = \frac{d\theta}{dt}$$

- If the angular acceleration is constant, use

$$\omega = \omega_0 + \alpha_c t$$

$$\theta = \theta_0 + \omega_0 t + \tfrac{1}{2}\alpha_c t^2$$

$$\omega^2 = \omega_0^2 + 2\alpha_c(\theta - \theta_0)$$

17

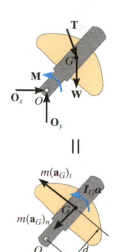

The crank on the oil-pumping rig undergoes rotation about a fixed axis which is caused by a driving torque $\mathbf{M}$ of the motor. The loadings shown on the free-body diagram cause the effects shown on the kinetic diagram. If moments are summed about the mass center, G, then $\Sigma M_G = I_G\alpha$. However, if moments are summed about point O, noting that $(a_G)_t = \alpha d$, then $\zeta + \Sigma M_O = I_G\alpha + m(a_G)_t d + m(a_G)_n(0) = (I_G + md^2)\alpha = I_O\alpha$. (© R.C. Hibbeler)

EXAMPLE 17.8

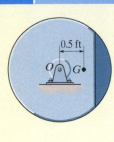

(a)

The unbalanced 50-lb flywheel shown in Fig. 17–14a has a radius of gyration of $k_G = 0.6$ ft about an axis passing through its mass center G. If it is released from rest, determine the horizontal and vertical components of reaction at the pin O.

SOLUTION

Free-Body and Kinetic Diagrams.   Since G moves in a circular path, it will have both normal and tangential components of acceleration. Also, since α, which is caused by the flywheel's weight, acts clockwise, the tangential component of acceleration must act downward. Why? Since $\omega = 0$, only $m(a_G)_t = m\alpha r_G$ and $I_G\alpha$ are shown on the kinetic diagram in Fig. 17–14b. Here, the moment of inertia about G is

$$I_G = mk_G^2 = (50\ \text{lb}/32.2\ \text{ft/s}^2)(0.6\ \text{ft})^2 = 0.559\ \text{slug}\cdot\text{ft}^2$$

The three unknowns are O_n, O_t, and α.

Equations of Motion.

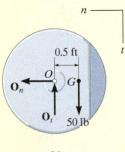

$$\xleftrightarrow{} \Sigma F_n = m\omega^2 r_G; \qquad\qquad O_n = 0 \qquad\qquad\qquad Ans.$$

$$+\downarrow\Sigma F_t = m\alpha r_G; \qquad -O_t + 50\ \text{lb} = \left(\frac{50\ \text{lb}}{32.2\ \text{ft/s}^2}\right)(\alpha)(0.5\ \text{ft}) \qquad (1)$$

$$\zeta +\Sigma M_G = I_G\alpha; \qquad O_t(0.5\ \text{ft}) = (0.5590\ \text{slug}\cdot\text{ft}^2)\alpha$$

Solving,

$$\alpha = 26.4\ \text{rad/s}^2 \quad O_t = 29.5\ \text{lb} \qquad\qquad Ans.$$

Moments can also be summed about point O in order to eliminate $\mathbf{O}_n$ and $\mathbf{O}_t$ and thereby obtain a *direct solution* for $\boldsymbol{\alpha}$, Fig. 17–14b. This can be done in one of *two* ways.

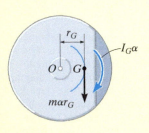

(b)

Fig. 17–14

$$\zeta +\Sigma M_O = \Sigma(\mathcal{M}_k)_O;$$

$$(50\ \text{lb})(0.5\ \text{ft}) = (0.5590\ \text{slug}\cdot\text{ft}^2)\alpha + \left[\left(\frac{50\ \text{lb}}{32.2\ \text{ft/s}^2}\right)\alpha(0.5\ \text{ft})\right](0.5\ \text{ft})$$

$$50\ \text{lb}(0.5\ \text{ft}) = 0.9472\alpha \qquad (2)$$

If $\Sigma M_O = I_O\alpha$ is applied, then by the parallel-axis theorem the moment of inertia of the flywheel about O is

$$I_O = I_G + mr_G^2 = 0.559 + \left(\frac{50}{32.2}\right)(0.5)^2 = 0.9472\ \text{slug}\cdot\text{ft}^2$$

Hence,

$$\zeta +\Sigma M_O = I_O\alpha; \quad (50\ \text{lb})(0.5\ \text{ft}) = (0.9472\ \text{slug}\cdot\text{ft}^2)\alpha$$

which is the same as Eq. 2. Solving for α and substituting into Eq. 1 yields the answer for O_t obtained previously.

EXAMPLE 17.9

At the instant shown in Fig. 17–15a, the 20-kg slender rod has an angular velocity of $\omega = 5$ rad/s. Determine the angular acceleration and the horizontal and vertical components of reaction of the pin on the rod at this instant.

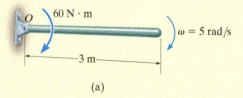

(a)

SOLUTION

Free-Body and Kinetic Diagrams. Fig. 17–15b. As shown on the kinetic diagram, point G moves around a circular path and so it has two components of acceleration. It is important that the tangential component $a_t = \alpha r_G$ act downward since it must be in accordance with the rotational sense of α. The three unknowns are O_n, O_t, and α.

Equation of Motion.

$$\overset{+}{\leftarrow} \Sigma F_n = m\omega^2 r_G; \qquad O_n = (20 \text{ kg})(5 \text{ rad/s})^2(1.5 \text{ m})$$

$$+\downarrow \Sigma F_t = m\alpha r_G; \qquad -O_t + 20(9.81)\text{N} = (20 \text{ kg})(\alpha)(1.5 \text{ m})$$

$$\zeta + \Sigma M_G = I_G\alpha; \qquad O_t(1.5 \text{ m}) + 60 \text{ N} \cdot \text{m} = \left[\tfrac{1}{12}(20 \text{ kg})(3 \text{ m})^2\right]\alpha$$

Solving

$$O_n = 750 \text{ N} \quad O_t = 19.05 \text{ N} \quad \alpha = 5.90 \text{ rad/s}^2 \qquad \textit{Ans.}$$

A more direct solution to this problem would be to sum moments about point O to eliminate $\mathbf{O}_n$ and $\mathbf{O}_t$ and obtain a *direct solution* for α. Here,

$$\zeta + \Sigma M_O = \Sigma(\mathcal{M}_k)_O; \quad 60 \text{ N} \cdot \text{m} + 20(9.81) \text{ N}(1.5 \text{ m}) =$$

$$\left[\tfrac{1}{12}(20 \text{ kg})(3 \text{ m})^2\right]\alpha + [20 \text{ kg}(\alpha)(1.5 \text{ m})](1.5 \text{ m})$$

$$\alpha = 5.90 \text{ rad/s}^2 \qquad \textit{Ans.}$$

Also, since $I_O = \tfrac{1}{3}ml^2$ for a slender rod, we can apply

$$\zeta + \Sigma M_O = I_O\alpha; \quad 60 \text{ N} \cdot \text{m} + 20(9.81) \text{ N}(1.5 \text{ m}) = \left[\tfrac{1}{3}(20 \text{ kg})(3 \text{ m})^2\right]\alpha$$

$$\alpha = 5.90 \text{ rad/s}^2 \qquad \textit{Ans.}$$

NOTE: By comparison, the last equation provides the simplest solution for α and *does not* require use of the kinetic diagram.

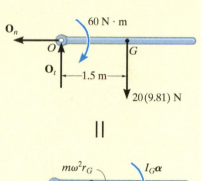

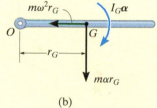

(b)

Fig. 17–15

17

EXAMPLE | 17.10

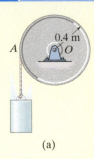

(a)

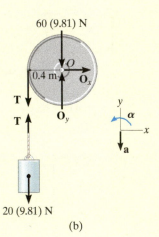

(b)

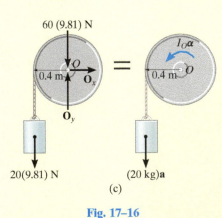

(c)

Fig. 17–16

The drum shown in Fig. 17–16a has a mass of 60 kg and a radius of gyration $k_O = 0.25$ m. A cord of negligible mass is wrapped around the periphery of the drum and attached to a block having a mass of 20 kg. If the block is released, determine the drum's angular acceleration.

SOLUTION I

Free-Body Diagram. Here we will consider the drum and block separately, Fig. 17–16b. Assuming the block accelerates *downward* at **a**, it creates a *counterclockwise* angular acceleration **α** of the drum. The moment of inertia of the drum is

$$I_O = mk_O^2 = (60 \text{ kg})(0.25 \text{ m})^2 = 3.75 \text{ kg} \cdot \text{m}^2$$

There are five unknowns, namely O_x, O_y, T, a, and α.

Equations of Motion. Applying the translational equations of motion $\Sigma F_x = m(a_G)_x$ and $\Sigma F_y = m(a_G)_y$ to the drum is of no consequence to the solution, since these equations involve the unknowns O_x and O_y. Thus, for the drum and block, respectively,

$$\zeta + \Sigma M_O = I_O \alpha; \qquad T(0.4 \text{ m}) = (3.75 \text{ kg} \cdot \text{m}^2)\alpha \qquad (1)$$

$$+\uparrow \Sigma F_y = m(a_G)_y; \qquad -20(9.81)\text{N} + T = -(20 \text{ kg})a \qquad (2)$$

Kinematics. Since the point of contact A between the cord and drum has a tangential component of acceleration **a**, Fig. 17–16a, then

$$\zeta + a = \alpha r; \qquad a = \alpha(0.4 \text{ m}) \qquad (3)$$

Solving the above equations,

$$T = 106 \text{ N} \quad a = 4.52 \text{ m/s}^2$$
$$\alpha = 11.3 \text{ rad/s}^2 \, \zeta \qquad \qquad Ans.$$

SOLUTION II

Free-Body and Kinetic Diagrams. The cable tension T can be eliminated from the analysis by considering the drum and block as a *single system*, Fig. 17–16c. The kinetic diagram is shown since moments will be summed about point O.

Equations of Motion. Using Eq. 3 and applying the moment equation about O to eliminate the unknowns O_x and O_y, we have

$$\zeta + \Sigma M_O = \Sigma(\mathcal{M}_k)_O; \qquad [20(9.81) \text{ N}] (0.4 \text{ m}) =$$
$$(3.75 \text{ kg} \cdot \text{m}^2)\alpha + [20 \text{ kg}(\alpha \, 0.4 \text{ m})](0.4 \text{ m})$$
$$\alpha = 11.3 \text{ rad/s}^2 \qquad \qquad Ans.$$

NOTE: If the block were *removed* and a force of 20(9.81) N were applied to the cord, show that $\alpha = 20.9$ rad/s². This value is larger since the block has an inertia, or resistance to acceleration.

EXAMPLE | 17.11

The slender rod shown in Fig. 17–17a has a mass m and length l and is released from rest when $\theta = 0°$. Determine the horizontal and vertical components of force which the pin at A exerts on the rod at the instant $\theta = 90°$.

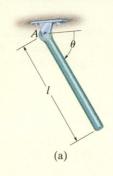

(a)

SOLUTION

Free-Body and Kinetic Diagrams. The free-body diagram for the rod in the general position θ is shown in Fig. 17–17b. For convenience, the force components at A are shown acting in the n and t directions. Note that $\boldsymbol{\alpha}$ acts clockwise and so $(\mathbf{a}_G)_t$ acts in the $+t$ direction.

The moment of inertia of the rod about point A is $I_A = \frac{1}{3}ml^2$.

Equations of Motion. Moments will be summed about A in order to eliminate A_n and A_t.

$$+\nwarrow\Sigma F_n = m\omega^2 r_G; \qquad A_n - mg\sin\theta = m\omega^2(l/2) \qquad (1)$$

$$+\swarrow\Sigma F_t = m\alpha r_G; \qquad A_t + mg\cos\theta = m\alpha(l/2) \qquad (2)$$

$$\zeta + \Sigma M_A = I_A\alpha; \qquad mg\cos\theta(l/2) = \left(\tfrac{1}{3}ml^2\right)\alpha \qquad (3)$$

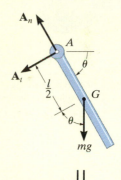

Kinematics. For a given angle θ there are four unknowns in the above three equations: A_n, A_t, ω, and α. As shown by Eq. 3, α is *not constant;* rather, it depends on the position θ of the rod. The necessary fourth equation is obtained using kinematics, where α and ω can be related to θ by the equation

$$(\zeta+) \qquad\qquad \omega\, d\omega = \alpha\, d\theta \qquad (4)$$

Note that the positive clockwise direction for this equation *agrees* with that of Eq. 3. This is important since we are seeking a simultaneous solution.

In order to solve for ω at $\theta = 90°$, eliminate α from Eqs. 3 and 4, which yields

$$\omega\, d\omega = (1.5g/l)\cos\theta\, d\theta$$

Since $\omega = 0$ at $\theta = 0°$, we have

$$\int_0^\omega \omega\, d\omega = (1.5g/l)\int_{0°}^{90°} \cos\theta\, d\theta$$

$$\omega^2 = 3g/l$$

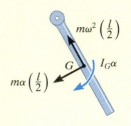

(b)

Fig. 17–17

Substituting this value into Eq. 1 with $\theta = 90°$ and solving Eqs. 1 to 3 yields

$$\alpha = 0$$

$$A_t = 0 \quad A_n = 2.5\, mg \qquad\qquad Ans.$$

NOTE: If $\Sigma M_A = \Sigma(\mathcal{M}_k)_A$ is used, one must account for the moments of $I_G\boldsymbol{\alpha}$ and $m(\mathbf{a}_G)_t$ about A.

FUNDAMENTAL PROBLEMS

F17–7. The 100-kg wheel has a radius of gyration about its center O of $k_O = 500$ mm. If the wheel starts from rest, determine its angular velocity in $t = 3$ s.

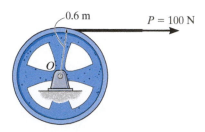

Prob. F17–7

F17–8. The 50-kg disk is subjected to the couple moment of $M = (9t)$ N $\cdot$ m, where t is in seconds. Determine the angular velocity of the disk when $t = 4$ s starting from rest.

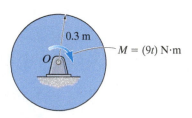

Prob. F17–8

F17–9. At the instant shown, the uniform 30-kg slender rod has a counterclockwise angular velocity of $\omega = 6$ rad/s. Determine the tangential and normal components of reaction of pin O on the rod and the angular acceleration of the rod at this instant.

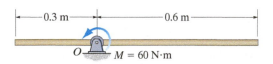

Prob. F17–9

F17–10. At the instant shown, the 30-kg disk has a counterclockwise angular velocity of $\omega = 10$ rad/s. Determine the tangential and normal components of reaction of the pin O on the disk and the angular acceleration of the disk at this instant.

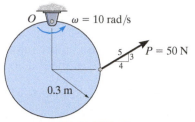

Prob. F17–10

F17–11. The uniform slender rod has a mass of 15 kg. Determine the horizontal and vertical components of reaction at the pin O, and the angular acceleration of the rod just after the cord is cut.

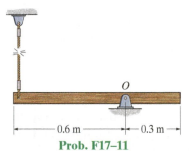

Prob. F17–11

F17–12. The uniform 30-kg slender rod is being pulled by the cord that passes over the small smooth peg at A. If the rod has a counterclockwise angular velocity of $\omega = 6$ rad/s at the instant shown, determine the tangential and normal components of reaction at the pin O and the angular acceleration of the rod.

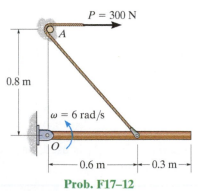

Prob. F17–12

PROBLEMS

17–57. The 10-kg wheel has a radius of gyration $k_A = 200$ mm. If the wheel is subjected to a moment $M = (5t)$ N · m, where t is in seconds, determine its angular velocity when $t = 3$ s starting from rest. Also, compute the reactions which the fixed pin A exerts on the wheel during the motion.

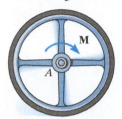

Prob. 17–57

17–58. The uniform 24-kg plate is released from rest at the position shown. Determine its initial angular acceleration and the horizontal and vertical reactions at the pin A.

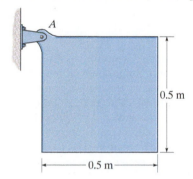

Prob. 17–58

17–59. The uniform slender rod has a mass m. If it is released from rest when $\theta = 0°$, determine the magnitude of the reactive force exerted on it by pin B when $\theta = 90°$.

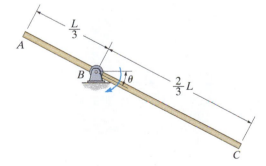

Prob. 17–59

*17–60. The bent rod has a mass of 2 kg/m. If it is released from rest in the position shown, determine its initial angular acceleration and the horizontal and vertical components of reaction at A.

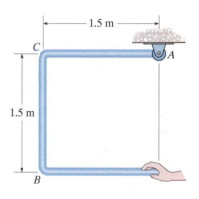

Prob. 17–60

17–61. If a horizontal force of $P = 100$ N is applied to the 300-kg reel of cable, determine its initial angular acceleration. The reel rests on rollers at A and B and has a radius of gyration of $k_O = 0.6$ m.

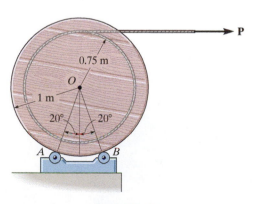

Prob. 17–61

17–62. The 10-lb bar is pinned at its center O and connected to a torsional spring. The spring has a stiffness $k = 5$ lb·ft/rad, so that the torque developed is $M = (5\theta)$ lb·ft, where θ is in radians. If the bar is released from rest when it is vertical at $\theta = 90°$, determine its angular velocity at the instant $\theta = 0°$.

17–63. The 10-lb bar is pinned at its center O and connected to a torsional spring. The spring has a stiffness $k = 5$ lb·ft/rad, so that the torque developed is $M = (5\theta)$ lb·ft, where θ is in radians. If the bar is released from rest when it is vertical at $\theta = 90°$, determine its angular velocity at the instant $\theta = 45°$.

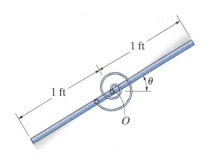

Probs. 17–62/63

*17–64.** A cord is wrapped around the outer surface of the 8-kg disk. If a force of $F = (\frac{1}{4}\theta^2)$ N, where θ is in radians, is applied to the cord, determine the disk's angular acceleration when it has turned 5 revolutions. The disk has an initial angular velocity of $\omega_0 = 1$ rad/s.

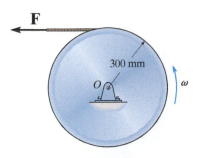

Prob. 17–64

17–65. Disk A has a weight of 5 lb and disk B has a weight of 10 lb. If no slipping occurs between them, determine the couple moment **M** which must be applied to disk A to give it an angular acceleration of 4 rad/s².

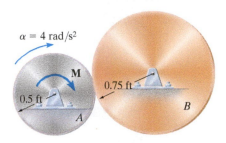

Prob. 17–65

17–66. The kinetic diagram representing the general rotational motion of a rigid body about a fixed axis passing through O is shown in the figure. Show that $I_G\alpha$ may be eliminated by moving the vectors $m(\mathbf{a}_G)_t$ and $m(\mathbf{a}_G)_n$ to point P, located a distance $r_{GP} = k_G^2/r_{OG}$ from the center of mass G of the body. Here k_G represents the radius of gyration of the body about an axis passing through G. The point P is called the *center of percussion* of the body.

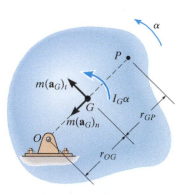

Prob. 17–66

17–67. If the cord at B suddenly fails, determine the horizontal and vertical components of the initial reaction at the pin A, and the angular acceleration of the 120-kg beam. Treat the beam as a uniform slender rod.

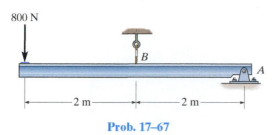

800 N

B

A

2 m — 2 m

Prob. 17–67

*17–68. The device acts as a pop-up barrier to prevent the passage of a vehicle. It consists of a 100-kg steel plate AC and a 200-kg counterweight solid concrete block located as shown. Determine the moment of inertia of the plate and block about the hinged axis through A. Neglect the mass of the supporting arms AB. Also, determine the initial angular acceleration of the assembly when it is released from rest at $\theta = 45°$.

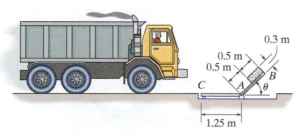

0.3 m
0.5 m
0.5 m
C A B
θ
1.25 m

Prob. 17–68

17–69. The 20-kg roll of paper has a radius of gyration $k_A = 90$ mm about an axis passing through point A. It is pin supported at both ends by two brackets AB. If the roll rests against a wall for which the coefficient of kinetic friction is $\mu_k = 0.2$ and a vertical force $F = 30$ N is applied to the end of the paper, determine the angular acceleration of the roll as the paper unrolls.

17–70. The 20-kg roll of paper has a radius of gyration $k_A = 90$ mm about an axis passing through point A. It is pin supported at both ends by two brackets AB. If the roll rests against a wall for which the coefficient of kinetic friction is $\mu_k = 0.2$, determine the constant vertical force F that must be applied to the roll to pull off 1 m of paper in $t = 3$ s starting from rest. Neglect the mass of paper that is removed.

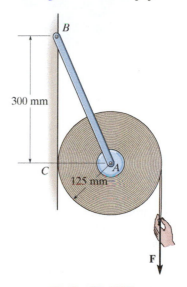

B

300 mm

C A
125 mm

F

Probs. 17–69/70

17–71. The reel of cable has a mass of 400 kg and a radius of gyration of $k_A = 0.75$ m. Determine its angular velocity when $t = 2$ s, starting from rest, if the force $P = (20t^2 + 80)$ N, when t is in seconds. Neglect the mass of the unwound cable, and assume it is always at a radius of 0.5 m.

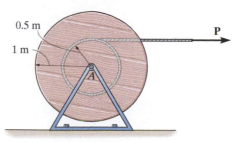

0.5 m
P
1 m
A

Prob. 17–71

17

*17–72. The 30-kg disk is originally spinning at $\omega = 125$ rad/s. If it is placed on the ground, for which the coefficient of kinetic friction is $\mu_C = 0.5$, determine the time required for the motion to stop. What are the horizontal and vertical components of force which the member AB exerts on the pin at A during this time? Neglect the mass of AB.

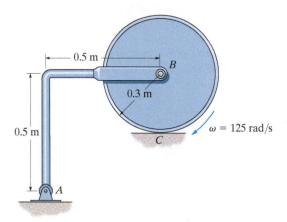

Prob. 17–72

17–74. The 5-kg cylinder is initially at rest when it is placed in contact with the wall B and the rotor at A. If the rotor always maintains a constant clockwise angular velocity $\omega = 6$ rad/s, determine the initial angular acceleration of the cylinder. The coefficient of kinetic friction at the contacting surfaces B and C is $\mu_k = 0.2$.

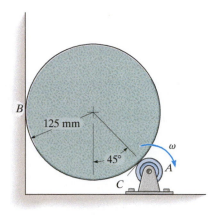

Prob. 17–74

17–73. Cable is unwound from a spool supported on small rollers at A and B by exerting a force $T = 300$ N on the cable. Compute the time needed to unravel 5 m of cable from the spool if the spool and cable have a total mass of 600 kg and a radius of gyration of $k_O = 1.2$ m. For the calculation, neglect the mass of the cable being unwound and the mass of the rollers at A and B. The rollers turn with no friction.

17–75. The wheel has a mass of 25 kg and a radius of gyration $k_B = 0.15$ m. It is originally spinning at $\omega = 40$ rad/s. If it is placed on the ground, for which the coefficient of kinetic friction is $\mu_C = 0.5$, determine the time required for the motion to stop. What are the horizontal and vertical components of reaction which the pin at A exerts on AB during this time? Neglect the mass of AB.

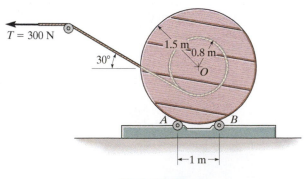

Prob. 17–73

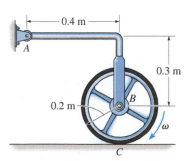

Prob. 17–75

*17–76. The 20-kg roll of paper has a radius of gyration $k_A = 120$ mm about an axis passing through point A. It is pin supported at both ends by two brackets AB. The roll rests on the floor, for which the coefficient of kinetic friction is $\mu_k = 0.2$. If a horizontal force $F = 60$ N is applied to the end of the paper, determine the initial angular acceleration of the roll as the paper unrolls.

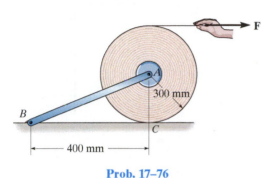

Prob. 17–76

17–77. Disk D turns with a constant clockwise angular velocity of 30 rad/s. Disk E has a weight of 60 lb and is initially at rest when it is brought into contact with D. Determine the time required for disk E to attain the same angular velocity as disk D. The coefficient of kinetic friction between the two disks is $\mu_k = 0.3$. Neglect the weight of bar BC.

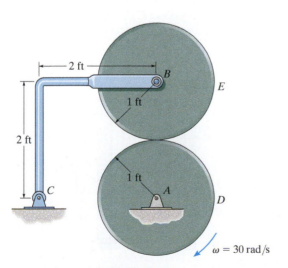

Prob. 17–77

17–78. Two cylinders A and B, having a weight of 10 lb and 5 lb, respectively, are attached to the ends of a cord which passes over a 3-lb pulley (disk). If the cylinders are released from rest, determine their speed in $t = 0.5$ s. The cord does not slip on the pulley. Neglect the mass of the cord. *Suggestion:* Analyze the "system" consisting of both the cylinders and the pulley.

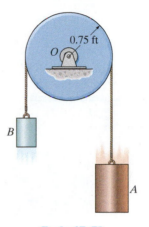

Prob. 17–78

17–79. The two blocks A and B have a mass of 5 kg and 10 kg, respectively. If the pulley can be treated as a disk of mass 3 kg and radius 0.15 m, determine the acceleration of block A. Neglect the mass of the cord and any slipping on the pulley.

*17–80. The two blocks A and B have a mass m_A and m_B, respectively, where $m_B > m_A$. If the pulley can be treated as a disk of mass M, determine the acceleration of block A. Neglect the mass of the cord and any slipping on the pulley.

Probs. 17–79/80

17–81. Determine the angular acceleration of the 25-kg diving board and the horizontal and vertical components of reaction at the pin A the instant the man jumps off. Assume that the board is uniform and rigid, and that at the instant he jumps off the spring is compressed a maximum amount of 200 mm, $\omega = 0$, and the board is horizontal. Take $k = 7$ kN/m.

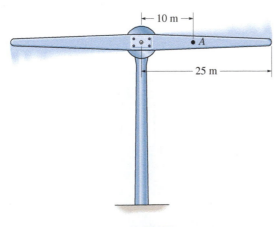

Prob. 17–81

17–82. The lightweight turbine consists of a rotor which is powered from a torque applied at its center. At the instant the rotor is horizontal it has an angular velocity of 15 rad/s and a clockwise angular acceleration of 8 rad/s². Determine the internal normal force, shear force, and moment at a section through A. Assume the rotor is a 50-m-long slender rod, having a mass of 3 kg/m.

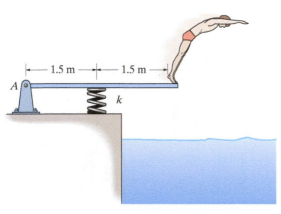

Prob. 17–82

17–83. The two-bar assembly is released from rest in the position shown. Determine the initial bending moment at the fixed joint B. Each bar has a mass m and length l.

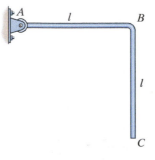

Prob. 17–83

*17–84. The armature (slender rod) AB has a mass of 0.2 kg and can pivot about the pin at A. Movement is controlled by the electromagnet E, which exerts a horizontal attractive force on the armature at B of $F_B = (0.2(10^{-3})l^{-2})$ N, where l in meters is the gap between the armature and the magnet at any instant. If the armature lies in the horizontal plane, and is originally at rest, determine the speed of the contact at B the instant $l = 0.01$ m. Originally $l = 0.02$ m.

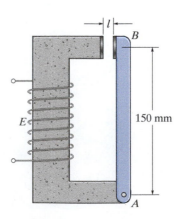

Prob. 17–84

17–85. The bar has a weight per length of w. If it is rotating in the vertical plane at a constant rate ω about point O, determine the internal normal force, shear force, and moment as a function of x and θ.

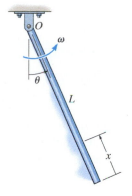

Prob. 17–85

17–87. The 100-kg pendulum has a center of mass at G and a radius of gyration about G of $k_G = 250$ mm. Determine the horizontal and vertical components of reaction on the beam by the pin A and the normal reaction of the roller B at the instant $\theta = 90°$ when the pendulum is rotating at $\omega = 8$ rad/s. Neglect the weight of the beam and the support.

*17–88. The 100-kg pendulum has a center of mass at G and a radius of gyration about G of $k_G = 250$ mm. Determine the horizontal and vertical components of reaction on the beam by the pin A and the normal reaction of the roller B at the instant $\theta = 0°$ when the pendulum is rotating at $\omega = 4$ rad/s. Neglect the weight of the beam and the support.

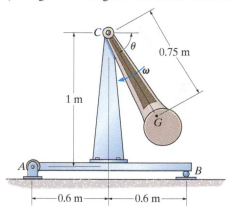

Probs. 17–87/88

17–89. The "Catherine wheel" is a firework that consists of a coiled tube of powder which is pinned at its center. If the powder burns at a constant rate of 20 g/s such as that the exhaust gases always exert a force having a constant magnitude of 0.3 N, directed tangent to the wheel, determine the angular velocity of the wheel when 75% of the mass is burned off. Initially, the wheel is at rest and has a mass of 100 g and a radius of $r = 75$ mm. For the calculation, consider the wheel to always be a thin disk.

17–86. The 4-kg slender rod is initially supported horizontally by a spring at B and pin at A. Determine the angular acceleration of the rod and the acceleration of the rod's mass center at the instant the 100-N force is applied.

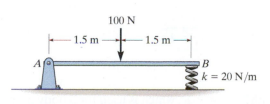

Prob. 17–86

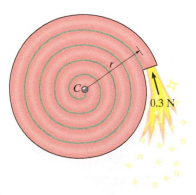

Prob. 17–89

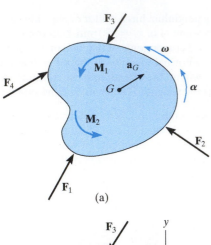

(a)

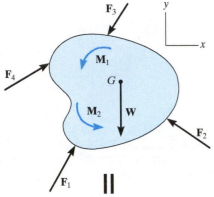

(b)

Fig. 17–18

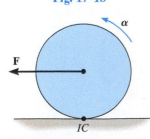

Fig. 17–19

17.5   Equations of Motion: General Plane Motion

The rigid body (or slab) shown in Fig. 17–18a is subjected to general plane motion caused by the externally applied force and couple-moment system. The free-body and kinetic diagrams for the body are shown in Fig. 17–18b. If an x and y inertial coordinate system is established as shown, the three equations of motion are

$$\begin{aligned} \Sigma F_x &= m(a_G)_x \\ \Sigma F_y &= m(a_G)_y \\ \Sigma M_G &= I_G \alpha \end{aligned} \qquad (17\text{–}17)$$

In some problems it may be convenient to sum moments about a point P other than G in order to eliminate as many unknown forces as possible from the moment summation. When used in this more general case, the three equations of motion are

$$\begin{aligned} \Sigma F_x &= m(a_G)_x \\ \Sigma F_y &= m(a_G)_y \\ \Sigma M_P &= \Sigma (\mathcal{M}_k)_P \end{aligned} \qquad (17\text{–}18)$$

Here $\Sigma (\mathcal{M}_k)_P$ represents the moment sum of $I_G \boldsymbol{\alpha}$ and $m\mathbf{a}_G$ (or its components) about P as determined by the data on the kinetic diagram.

Moment Equation About the IC. There is a particular type of problem that involves a uniform disk, or body of circular shape, that rolls on a rough surface *without slipping*, Fig. 17–19. If we sum the moments about the instantaneous center of zero velocity, then $\Sigma (\mathcal{M}_k)_{IC}$ becomes $I_{IC}\alpha$, so that

$$\Sigma M_{IC} = I_{IC}\alpha \qquad (17\text{–}19)$$

This result compares with $\Sigma M_O = I_O \alpha$, which is used for a body pinned at point O, Eq. 17–16. See Prob. 17–90.

(© R.C. Hibbeler)

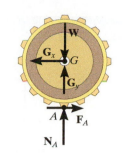

$=$

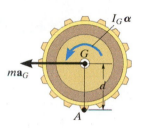

As the soil compactor, or "sheep's foot roller" moves forward, the roller has general plane motion. The forces shown on its free-body diagram cause the effects shown on the kinetic diagram. If moments are summed about the mass center, G, then $\Sigma M_G = I_G \alpha$. However, if moments are summed about point A (the IC) then $\zeta + \Sigma M_A = I_G \alpha + (m a_G) d = I_A \alpha$.

Procedure for Analysis

Kinetic problems involving general plane motion of a rigid body can be solved using the following procedure.

Free-Body Diagram.

- Establish the x, y inertial coordinate system and draw the free-body diagram for the body.
- Specify the direction and sense of the acceleration of the mass center, $\mathbf{a}_G$, and the angular acceleration $\boldsymbol{\alpha}$ of the body.
- Determine the moment of inertia I_G.
- Identify the unknowns in the problem.
- If it is decided that the rotational equation of motion $\Sigma M_P = \Sigma (\mathcal{M}_k)_P$ is to be used, then consider drawing the kinetic diagram in order to help "visualize" the "moments" developed by the components $m(\mathbf{a}_G)_x$, $m(\mathbf{a}_G)_y$, and $I_G \boldsymbol{\alpha}$ when writing the terms in the moment sum $\Sigma (\mathcal{M}_k)_P$.

Equations of Motion.

- Apply the three equations of motion in accordance with the established sign convention.
- When friction is present, there is the possibility for motion with no slipping or tipping. Each possibility for motion should be considered.

Kinematics.

- Use kinematics if a complete solution cannot be obtained strictly from the equations of motion.
- If the body's motion is *constrained* due to its supports, additional equations may be obtained by using $\mathbf{a}_B = \mathbf{a}_A + \mathbf{a}_{B/A}$, which relates the accelerations of any two points A and B on the body.
- When a wheel, disk, cylinder, or ball *rolls without slipping*, then $a_G = \alpha r$.

EXAMPLE | 17.12

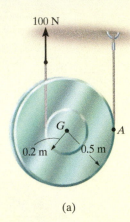

(a)

Determine the angular acceleration of the spool in Fig. 17–20a. The spool has a mass of 8 kg and a radius of gyration of $k_G = 0.35$ m. The cords of negligible mass are wrapped around its inner hub and outer rim.

SOLUTION I

Free-Body and Kinetic Diagrams. Fig. 17–20b. The 100-N force causes $\mathbf{a}_G$ to act upward. Also, $\boldsymbol{\alpha}$ acts clockwise, since the spool winds around the cord at A.

There are three unknowns T, a_G, and α. The moment of inertia of the spool about its mass center is

$$I_G = mk_G^2 = 8 \text{ kg}(0.35 \text{ m})^2 = 0.980 \text{ kg} \cdot \text{m}^2$$

Equations of Motion.

$$+\uparrow \Sigma F_y = m(a_G)_y; \qquad T + 100 \text{ N} - 78.48 \text{ N} = (8 \text{ kg})a_G \qquad (1)$$

$$\zeta + \Sigma M_G = I_G \alpha; \quad 100 \text{ N}(0.2 \text{ m}) - T(0.5 \text{ m}) = (0.980 \text{ kg} \cdot \text{m}^2)\alpha \qquad (2)$$

Kinematics. A complete solution is obtained if kinematics is used to relate a_G to α. In this case the spool "rolls without slipping" on the cord at A. Hence, we can use the results of Example 16.4 or 16.15 so that,

$$(\zeta +) \ a_G = \alpha r; \qquad\qquad a_G = \alpha (0.5 \text{ m}) \qquad (3)$$

Solving Eqs. 1 to 3, we have

$$\alpha = 10.3 \text{ rad/s}^2 \qquad\qquad Ans.$$
$$a_G = 5.16 \text{ m/s}^2$$
$$T = 19.8 \text{ N}$$

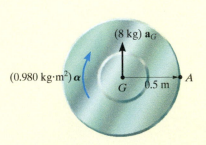

(b)

Fig. 17–20

SOLUTION II

Equations of Motion. We can eliminate the unknown T by summing moments about point A. From the free-body and kinetic diagrams Figs. 17–20b and 17–20c, we have

$$\zeta + \Sigma M_A = \Sigma(\mathcal{M}_k)_A; \qquad 100 \text{ N}(0.7 \text{ m}) - 78.48 \text{ N}(0.5 \text{ m})$$

$$= (0.980 \text{ kg} \cdot \text{m}^2)\alpha + [(8 \text{ kg})a_G](0.5 \text{ m})$$

Using Eq. (3),

$$\alpha = 10.3 \text{ rad/s}^2 \qquad\qquad Ans.$$

SOLUTION III

Equations of Motion. The simplest way to solve this problem is to realize that point A is the IC for the spool. Then Eq. 17–19 applies.

$$\zeta + \Sigma M_A = I_A \alpha; \quad (100 \text{ N})(0.7 \text{ m}) - (78.48 \text{ N})(0.5 \text{ m})$$

$$= [0.980 \text{ kg} \cdot \text{m}^2 + (8 \text{ kg})(0.5 \text{ m})^2]\alpha$$

$$\alpha = 10.3 \text{ rad/s}^2$$

EXAMPLE 17.13

The 50-lb wheel shown in Fig. 17–21 has a radius of gyration $k_G = 0.70$ ft. If a 35-lb·ft couple moment is applied to the wheel, determine the acceleration of its mass center G. The coefficients of static and kinetic friction between the wheel and the plane at A are $\mu_s = 0.3$ and $\mu_k = 0.25$, respectively.

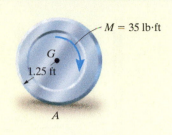

(a)

SOLUTION

Free-Body and Kinetic Diagrams. By inspection of Fig. 17–21b, it is seen that the couple moment causes the wheel to have a clockwise angular acceleration of $\boldsymbol{\alpha}$. As a result, the acceleration of the mass center, $\mathbf{a}_G$, is directed to the right. The moment of inertia is

$$I_G = mk_G^2 = \frac{50\ \text{lb}}{32.2\ \text{ft/s}^2}(0.70\ \text{ft})^2 = 0.7609\ \text{slug} \cdot \text{ft}^2$$

The unknowns are N_A, F_A, a_G, and α.

Equations of Motion.

$$\xrightarrow{+} \Sigma F_x = m(a_G)_x; \qquad F_A = \left(\frac{50\ \text{lb}}{32.2\ \text{ft/s}^2}\right)a_G \qquad (1)$$

$$+\uparrow \Sigma F_y = m(a_G)_y; \qquad N_A - 50\ \text{lb} = 0 \qquad (2)$$

$$\zeta + \Sigma M_G = I_G\alpha; \qquad 35\ \text{lb} \cdot \text{ft} - 1.25\ \text{ft}(F_A) = (0.7609\ \text{slug} \cdot \text{ft}^2)\alpha \qquad (3)$$

A fourth equation is needed for a complete solution.

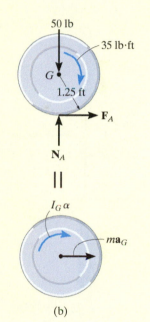

(b)

Fig. 17–21

Kinematics (No Slipping). If this assumption is made, then

$$(\zeta +) \qquad a_G = (1.25\ \text{ft})\alpha \qquad (4)$$

Solving Eqs. 1 to 4,

$$N_A = 50.0\ \text{lb} \qquad F_A = 21.3\ \text{lb}$$
$$\alpha = 11.0\ \text{rad/s}^2 \qquad a_G = 13.7\ \text{ft/s}^2$$

This solution requires that no slipping occurs, i.e., $F_A \leq \mu_s N_A$. However, since 21.3 lb > 0.3(50 lb) = 15 lb, the wheel slips as it rolls.

(Slipping). Equation 4 is not valid, and so $F_A = \mu_k N_A$, or

$$F_A = 0.25N_A \qquad (5)$$

Solving Eqs. 1 to 3 and 5 yields

$$N_A = 50.0\ \text{lb} \qquad F_A = 12.5\ \text{lb}$$
$$\alpha = 25.5\ \text{rad/s}^2$$
$$a_G = 8.05\ \text{ft/s}^2 \rightarrow \qquad \textit{Ans.}$$

EXAMPLE 17.14

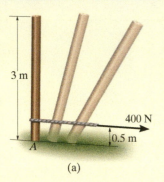

3 m

400 N

0.5 m

A

(a)

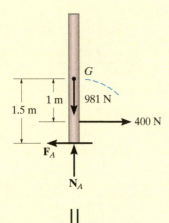

G

1 m   981 N

1.5 m

400 N

F_A

N_A

||

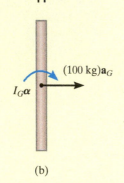

$(100 \text{ kg})\mathbf{a}_G$

$I_G\alpha$

(b)

Fig. 17–22

The uniform slender pole shown in Fig. 17–22a has a mass of 100 kg. If the coefficients of static and kinetic friction between the end of the pole and the surface are $\mu_s = 0.3$, and $\mu_k = 0.25$, respectively, determine the pole's angular acceleration at the instant the 400-N horizontal force is applied. The pole is originally at rest.

SOLUTION

Free-Body and Kinetic Diagrams. Figure 17–22b. The path of motion of the mass center G will be along an unknown curved path having a radius of curvature ρ, which is initially on a vertical line. However, there is no normal or y component of acceleration since the pole is originally at rest, i.e., $\mathbf{v}_G = \mathbf{0}$, so that $(a_G)_y = v_G^2/\rho = 0$. We will assume the mass center accelerates to the right and that the pole has a clockwise angular acceleration of $\boldsymbol{\alpha}$. The unknowns are N_A, F_A, a_G, and α.

Equation of Motion.

$$\xrightarrow{+} \Sigma F_x = m(a_G)_x; \qquad 400 \text{ N} - F_A = (100 \text{ kg})a_G \qquad (1)$$

$$+\uparrow \Sigma F_y = m(a_G)_y; \qquad N_A - 981 \text{ N} = 0 \qquad (2)$$

$$\zeta + \Sigma M_G = I_G\alpha; \; F_A(1.5 \text{ m}) - (400 \text{ N})(1 \text{ m}) = [\tfrac{1}{12}(100 \text{ kg})(3 \text{ m})^2]\alpha \quad (3)$$

A fourth equation is needed for a complete solution.

Kinematics (No Slipping). With this assumption, point A acts as a "pivot" so that α is clockwise, then a_G is directed to the right.

$$a_G = \alpha r_{AG}; \qquad a_G = (1.5 \text{ m})\,\alpha \qquad (4)$$

Solving Eqs. 1 to 4 yields

$$N_A = 981 \text{ N} \quad F_A = 300 \text{ N}$$

$$a_G = 1 \text{ m/s}^2 \quad \alpha = 0.667 \text{ rad/s}^2$$

The assumption of no slipping requires $F_A \leq \mu_s N_A$. However, $300 \text{ N} > 0.3(981 \text{ N}) = 294 \text{ N}$ and so the pole slips at A.

(Slipping). For this case Eq. 4 does *not* apply. Instead the frictional equation $F_A = \mu_k N_A$ must be used. Hence,

$$F_A = 0.25 N_A \qquad (5)$$

Solving Eqs. 1 to 3 and 5 simultaneously yields

$$N_A = 981 \text{ N} \quad F_A = 245 \text{ N} \quad a_G = 1.55 \text{ m/s}^2$$

$$\alpha = -0.428 \text{ rad/s}^2 = 0.428 \text{ rad/s}^2 \zeta \qquad \textit{Ans.}$$

EXAMPLE 17.15

The uniform 50-kg bar in Fig. 17–23a is held in the equilibrium position by cords AC and BD. Determine the tension in BD and the angular acceleration of the bar immediately after AC is cut.

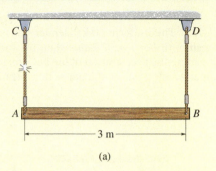

(a)

SOLUTION

Free-Body and Kinetic Diagrams. Fig. 17–23b. There are four unknowns, T_B, $(a_G)_x$, $(a_G)_y$, and α.

Equations of Motion.

$$\xrightarrow{+}\Sigma F_x = m(a_G)_x; \qquad 0 = 50 \text{ kg } (a_G)_x$$

$$(a_G)_x = 0$$

$$+\uparrow\Sigma F_y = m(a_G)_y; \quad T_B - 50(9.81)\text{N} = -50 \text{ kg } (a_G)_y \qquad (1)$$

$$\zeta+\Sigma M_G = I_G\alpha; \quad T_B(1.5 \text{ m}) = \left[\frac{1}{12}(50 \text{ kg})(3 \text{ m})^2\right]\alpha \qquad (2)$$

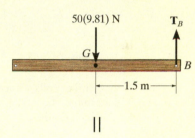

(b)

Kinematics. Since the bar is at rest just after the cable is cut, then its angular velocity and the velocity of point B at this instant are equal to zero. Thus $(a_B)_n = v_B^2/\rho_{BD} = 0$. Therefore, $\mathbf{a}_B$ only has a tangential component, which is directed along the x axis, Fig. 17–23c. Applying the relative acceleration equation to points G and B,

$$\mathbf{a}_G = \mathbf{a}_B + \boldsymbol{\alpha} \times \mathbf{r}_{G/B} - \omega^2\mathbf{r}_{G/B}$$

$$-(a_G)_y\mathbf{j} = a_B\mathbf{i} + (\alpha\mathbf{k}) \times (-1.5\mathbf{i}) - \mathbf{0}$$

$$-(a_G)_y\mathbf{j} = a_B\mathbf{i} - 1.5\alpha\mathbf{j}$$

Equating the $\mathbf{i}$ and $\mathbf{j}$ components of both sides of this equation,

$$0 = a_B$$

$$(a_G)_y = 1.5\alpha \qquad (3)$$

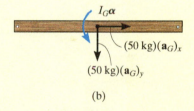

(c)

Fig. 17–23

Solving Eqs. (1) through (3) yields

$$\alpha = 4.905 \text{ rad/s}^2 \qquad \text{\textit{Ans.}}$$

$$T_B = 123 \text{ N} \qquad \text{\textit{Ans.}}$$

$$(a_G)_y = 7.36 \text{ m/s}^2$$

FUNDAMENTAL PROBLEMS

F17–13. The uniform 60-kg slender bar is initially at rest on a smooth *horizontal plane* when the forces are applied. Determine the acceleration of the bar's mass center and the angular acceleration of the bar at this instant.

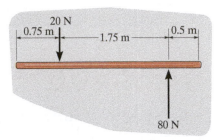

Prob. F17–13

F17–14. The 100-kg cylinder rolls without slipping on the horizontal plane. Determine the acceleration of its mass center and its angular acceleration.

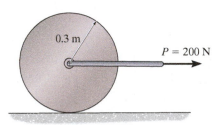

Prob. F17–14

F17–15. The 20-kg wheel has a radius of gyration about its center O of $k_O = 300$ mm. When the wheel is subjected to the couple moment, it slips as it rolls. Determine the angular acceleration of the wheel and the acceleration of the wheel's center O. The coefficient of kinetic friction between the wheel and the plane is $\mu_k = 0.5$.

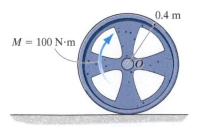

Prob. F17–15

F17–16. The 20-kg sphere rolls down the inclined plane without slipping. Determine the angular acceleration of the sphere and the acceleration of its mass center.

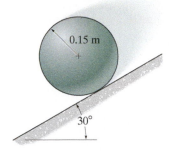

Prob. F17–16

F17–17. The 200-kg spool has a radius of gyration about its mass center of $k_G = 300$ mm. If the couple moment is applied to the spool and the coefficient of kinetic friction between the spool and the ground is $\mu_k = 0.2$, determine the angular acceleration of the spool, the acceleration of G and the tension in the cable.

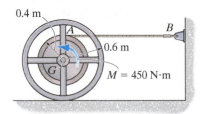

Prob. F17–17

F17–18. The 12-kg slender rod is pinned to a small roller A that slides freely along the slot. If the rod is released from rest at $\theta = 0°$, determine the angular acceleration of the rod and the acceleration of the roller immediately after the release.

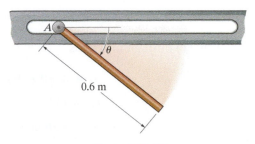

Prob. F17–18

PROBLEMS

17–90. If the disk in Fig. 17–19 *rolls without slipping*, show that when moments are summed about the instantaneous center of zero velocity, *IC*, it is possible to use the moment equation $\Sigma M_{IC} = I_{IC}\alpha$, where I_{IC} represents the moment of inertia of the disk calculated about the instantaneous axis of zero velocity.

17–91. The 20-kg punching bag has a radius of gyration about its center of mass G of $k_G = 0.4$ m. If it is initially at rest and is subjected to a horizontal force $F = 30$ N, determine the initial angular acceleration of the bag and the tension in the supporting cable AB.

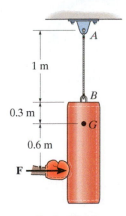

Prob. 17–91

*17–92. The uniform 150-lb beam is initially at rest when the forces are applied to the cables. Determine the magnitude of the acceleration of the mass center and the angular acceleration of the beam at this instant.

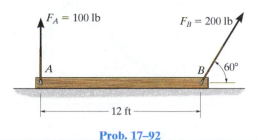

Prob. 17–92

17–93. The slender 12-kg bar has a clockwise angular velocity of $\omega = 2$ rad/s when it is in the position shown. Determine its angular acceleration and the normal reactions of the smooth surface A and B at this instant.

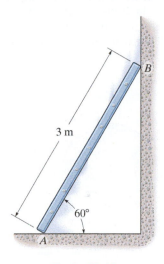

Prob. 17–93

17–94. The tire has a weight of 30 lb and a radius of gyration of $k_G = 0.6$ ft. If the coefficients of static and kinetic friction between the tire and the plane are $\mu_s = 0.2$ and $\mu_k = 0.15$, determine the tire's angular acceleration as it rolls down the incline. Set $\theta = 12°$.

17–95. The tire has a weight of 30 lb and a radius of gyration of $k_G = 0.6$ ft. If the coefficients of static and kinetic friction between the tire and the plane are $\mu_s = 0.2$ and $\mu_k = 0.15$, determine the maximum angle θ of the inclined plane so that the tire rolls without slipping.

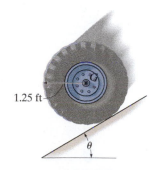

Probs. 17–94/95

17

*17–96. The spool has a mass of 100 kg and a radius of gyration of $k_G = 0.3$ m. If the coefficients of static and kinetic friction at A are $\mu_s = 0.2$ and $\mu_k = 0.15$, respectively, determine the angular acceleration of the spool if $P = 50$ N.

17–97. Solve Prob. 17–96 if the cord and force $P = 50$ N are directed vertically upwards.

17–98. The spool has a mass of 100 kg and a radius of gyration $k_G = 0.3$ m. If the coefficients of static and kinetic friction at A are $\mu_s = 0.2$ and $\mu_k = 0.15$, respectively, determine the angular acceleration of the spool if $P = 600$ N.

*17–100. A force of $F = 10$ N is applied to the 10-kg ring as shown. If slipping does not occur, determine the ring's initial angular acceleration, and the acceleration of its mass center, G. Neglect the thickness of the ring.

17–101. If the coefficient of static friction at C is $\mu_s = 0.3$, determine the largest force $\mathbf{F}$ that can be applied to the 5-kg ring, without causing it to slip. Neglect the thickness of the ring.

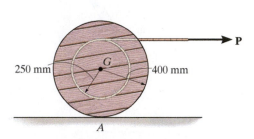

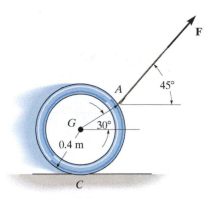

Probs. 17–96/97/98

Probs. 17–100/101

17–99. The 12-kg uniform bar is supported by a roller at A. If a horizontal force of $F = 80$ N is applied to the roller, determine the acceleration of the center of the roller at the instant the force is applied. Neglect the weight and the size of the roller.

17–102. The 25-lb slender rod has a length of 6 ft. Using a collar of negligible mass, its end A is confined to move along the smooth circular bar of radius $3\sqrt{2}$ ft. End B rests on the floor, for which the coefficient of kinetic friction is $\mu_B = 0.4$. If the bar is released from rest when $\theta = 30°$, determine the angular acceleration of the bar at this instant.

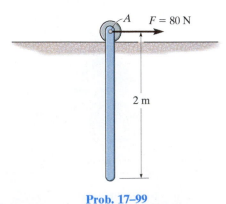

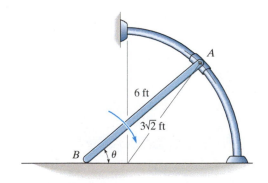

Prob. 17–99

Prob. 17–102

17–103. The 15-lb circular plate is suspended from a pin at A. If the pin is connected to a track which is given an acceleration $a_A = 5$ ft/s^2, determine the horizontal and vertical components of reaction at A and the angular acceleration of the plate. The plate is originally at rest.

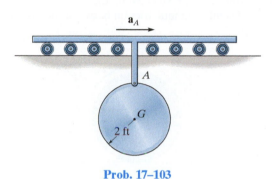

Prob. 17–103

*17–104. If $P = 30$ lb, determine the angular acceleration of the 50-lb roller. Assume the roller to be a uniform cylinder and that no slipping occurs.

17–105. If the coefficient of static friction between the 50-lb roller and the ground is $\mu_s = 0.25$, determine the maximum force P that can be applied to the handle, so that roller rolls on the ground without slipping. Also, find the angular acceleration of the roller. Assume the roller to be a uniform cylinder.

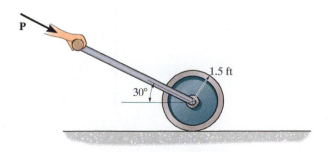

Probs. 17–104/105

17–106. The uniform bar of mass m and length L is balanced in the vertical position when the horizontal force **P** is applied to the roller at A. Determine the bar's initial angular acceleration and the acceleration of its top point B.

17–107. Solve Prob. 17–106 if the roller is removed and the coefficient of kinetic friction at the ground is μ_k.

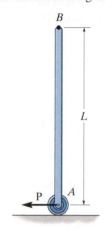

Probs. 17–106/107

*17–108. The semicircular disk having a mass of 10 kg is rotating at $\omega = 4$ rad/s at the instant $\theta = 60°$. If the coefficient of static friction at A is $\mu_s = 0.5$, determine if the disk slips at this instant.

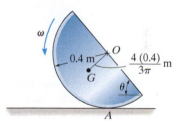

Prob. 17–108

17–109. The 500-kg concrete culvert has a mean radius of 0.5 m. If the truck has an acceleration of 3 m/s^2, determine the culvert's angular acceleration. Assume that the culvert does not slip on the truck bed, and neglect its thickness.

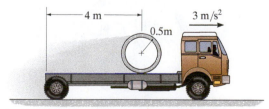

Prob. 17–109

17–110. The 15-lb disk rests on the 5-lb plate. A cord is wrapped around the periphery of the disk and attached to the wall at B. If a torque $M = 40$ lb · ft is applied to the disk, determine the angular acceleration of the disk and the time needed for the end C of the plate to travel 3 ft and strike the wall. Assume the disk does not slip on the plate and the plate rests on the surface at D having a coefficient of kinetic friction of $\mu_k = 0.2$. Neglect the mass of the cord.

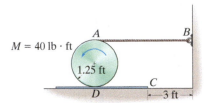

Prob. 17–110

17–111. The semicircular disk having a mass of 10 kg is rotating at $\omega = 4$ rad/s at the instant $\theta = 60°$. If the coefficient of static friction at A is $\mu_s = 0.5$, determine if the disk slips at this instant.

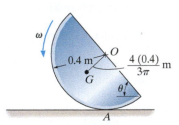

Prob. 17–111

*17–112. The circular concrete culvert rolls with an angular velocity of $\omega = 0.5$ rad/s when the man is at the position shown. At this instant the center of gravity of the culvert and the man is located at point G, and the radius of gyration about G is $k_G = 3.5$ ft. Determine the angular acceleration of the culvert. The combined weight of the culvert and the man is 500 lb. Assume that the culvert rolls without slipping, and the man does not move within the culvert.

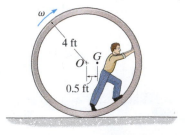

Prob. 17–112

17–113. The uniform disk of mass m is rotating with an angular velocity of ω_0 when it is placed on the floor. Determine the initial angular acceleration of the disk and the acceleration of its mass center. The coefficient of kinetic friction between the disk and the floor is μ_k.

17–114. The uniform disk of mass m is rotating with an angular velocity of ω_0 when it is placed on the floor. Determine the time before it starts to roll without slipping. What is the angular velocity of the disk at this instant? The coefficient of kinetic friction between the disk and the floor is μ_k.

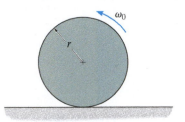

Probs. 17–113/114

17–115. A cord is wrapped around each of the two 10-kg disks. If they are released from rest determine the angular acceleration of each disk and the tension in the cord C. Neglect the mass of the cord.

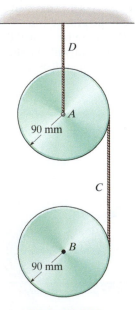

Prob. 17–115

*17–116. The disk of mass m and radius r rolls without slipping on the circular path. Determine the normal force which the path exerts on the disk and the disk's angular acceleration if at the instant shown the disk has an angular velocity of ω.

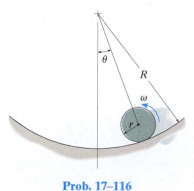

Prob. 17–116

17–117. The uniform beam has a weight W. If it is originally at rest while being supported at A and B by cables, determine the tension in cable A if cable B suddenly fails. Assume the beam is a slender rod.

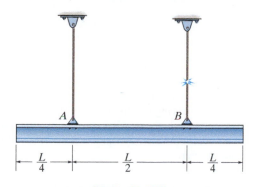

Prob. 17–117

17–118. The 500-lb beam is supported at A and B when it is subjected to a force of 1000 lb as shown. If the pin support at A suddenly fails, determine the beam's initial angular acceleration and the force of the roller support on the beam. For the calculation, assume that the beam is a slender rod so that its thickness can be neglected.

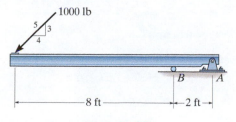

Prob. 17–118

17–119. The solid ball of radius r and mass m rolls without slipping down the 60° trough. Determine its angular acceleration.

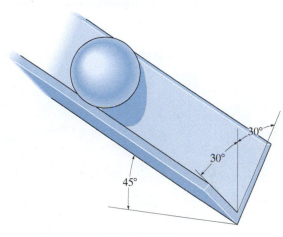

Prob. 17–119

*17–120. By pressing down with the finger at B, a thin ring having a mass m is given an initial velocity v_0 and a backspin ω_0 when the finger is released. If the coefficient of kinetic friction between the table and the ring is μ_k, determine the distance the ring travels forward before backspinning stops.

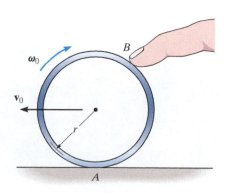

Prob. 17–120

CONCEPTUAL PROBLEMS

C17–1. The truck is used to pull the heavy container. To be most effective at providing traction to the rear wheels at *A*, is it best to keep the container where it is or place it at the front of the trailer? Use appropriate numerical values to explain your answer.

Prob. C17–1 (© R.C. Hibbeler)

C17–2. The tractor is about to tow the plane to the right. Is it possible for the driver to cause the front wheel of the plane to lift off the ground as he accelerates the tractor? Draw the free-body and kinetic diagrams and explain algebraically (letters) if and how this might be possible.

Prob. C17–2 (© R.C. Hibbeler)

C17–3. How can you tell the driver is accelerating this SUV? To explain your answer, draw the free-body and kinetic diagrams. Here power is supplied to the rear wheels. Would the photo look the same if power were supplied to the front wheels? Will the accelerations be the same? Use appropriate numerical values to explain your answers.

Prob. C17–3 (© R.C. Hibbeler)

C17–4. Here is something you should not try at home, at least not without wearing a helmet! Draw the free-body and kinetic diagrams and show what the rider must do to maintain this position. Use appropriate numerical values to explain your answer.

Prob. C17–4 (© R.C. Hibbeler)

CHAPTER REVIEW

Moment of Inertia

The moment of inertia is a measure of the resistance of a body to a change in its angular velocity. It is defined by $I = \int r^2 dm$ and will be different for each axis about which it is computed.

Many bodies are composed of simple shapes. If this is the case, then tabular values of I can be used, such as the ones given on the inside back cover of this book. To obtain the moment of inertia of a composite body about any specified axis, the moment of inertia of each part is determined about the axis and the results are added together. Doing this often requires use of the parallel-axis theorem.

$$I = I_G + md^2$$

Planar Equations of Motion

The equations of motion define the translational, and rotational motion of a rigid body. In order to account for all of the terms in these equations, a free-body diagram should always accompany their application, and for some problems, it may also be convenient to draw the kinetic diagram which shows $m\mathbf{a}_G$ and $I_G\boldsymbol{\alpha}$.

$$\Sigma F_x = m(a_G)_x$$
$$\Sigma F_y = m(a_G)_y$$
$$\Sigma M_G = 0$$

Rectilinear translation

$$\Sigma F_n = m(a_G)_n$$
$$\Sigma F_t = m(a_G)_t$$
$$\Sigma M_G = 0$$

Curvilinear translation

$$\Sigma F_n = m(a_G)_n = m\omega^2 r_G$$
$$\Sigma F_t = m(a_G)_t = m\alpha r_G$$
$$\Sigma M_G = I_G\alpha \text{ or } \Sigma M_O = I_O\alpha$$

Rotation About a Fixed Axis

$$\Sigma F_x = m(a_G)_x$$
$$\Sigma F_y = m(a_G)_y$$
$$\Sigma M_G = I_G\alpha \text{ or } \Sigma M_P = \Sigma(\mathcal{M}_k)_P$$

General Plane Motion

17

REVIEW PROBLEMS

R17–1. The handcart has a mass of 200 kg and center of mass at G. Determine the normal reactions at *each* of the wheels at A and B if a force $P = 50$ N is applied to the handle. Neglect the mass and rolling resistance of the wheels.

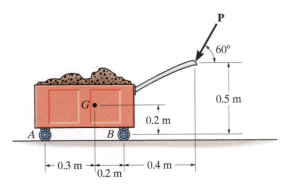

Prob. R17–1

R17–2. The two 3-lb rods EF and HI are fixed (welded) to the link AC at E. Determine the internal axial force E_x, shear force E_y, and moment M_E, which the bar AC exerts on FE at E if at the instant $\theta = 30°$ link AB has an angular velocity $\omega = 5$ rad/s and an angular acceleration $\alpha = 8$ rad/s^2 as shown.

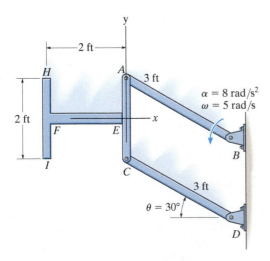

Prob. R17–2

R17–3. The car has a mass of 1.50 Mg and a mass center at G. Determine the maximum acceleration it can have if power is supplied only to the rear wheels. Neglect the mass of the wheels in the calculation, and assume that the wheels that do not receive power are free to roll. Also, assume that slipping of the powered wheels occurs, where the coefficient of kinetic friction is $\mu_k = 0.3$.

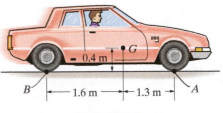

Prob. R17–3

R17–4. A 20-kg roll of paper, originally at rest, is pin-supported at its ends to bracket AB. The roll rest against a wall for which the coefficient of kinetic friction at C is $\mu_C = 0.3$. If a force of 40 N is applied uniformly to the end of the sheet, determine the initial angular acceleration of the roll and the tension in the bracket as the paper unwraps. For the calculation, treat the roll as a cylinder.

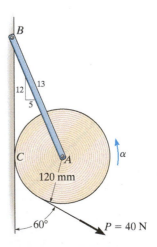

Prob. R17–4

R17–5. At the instant shown, two forces act on the 30-lb slender rod which is pinned at O. Determine the magnitude of force **F** and the initial angular acceleration of the rod so that the horizontal reaction which the *pin exerts on the rod* is 5 lb directed to the right.

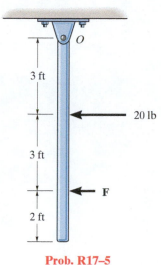

Prob. R17–5

R17–7. The spool and wire wrapped around its core have a mass of 20 kg and a centroidal radius of gyration $k_G = 250$ mm. if the coefficient of kinetic friction at the ground is $\mu_B = 0.1$, determine the angular acceleration of the spool when the 30-N · m couple moment is applied.

Prob. R17–7

17

R17–6. The pendulum consists of a 30-lb sphere and a 10-lb slender rod. Compute the reaction at the pin O just after the cord AB is cut.

R17–8. Determine the backspin ω which should be given to the 20-lb ball so that when its center is given an initial horizontal velocity $v_G = 20$ ft/s it stops spinning and translating at the same instant. The coefficient of kinetic friction is $\mu_A = 0.3$.

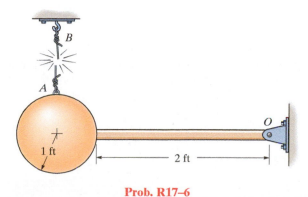

Prob. R17–6

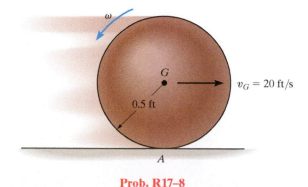

Prob. R17–8

Chapter 18 ●

(© Arinahabich/Fotolia)

Roller coasters must be able to coast over loops and through turns, and have enough energy to do so safely. Accurate calculation of this energy must account for the size of the car as it moves along the track.

Planar Kinetics of a Rigid Body: Work and Energy

CHAPTER OBJECTIVES

■ To develop formulations for the kinetic energy of a body, and define the various ways a force and couple do work.

■ To apply the principle of work and energy to solve rigid–body planar kinetic problems that involve force, velocity, and displacement.

■ To show how the conservation of energy can be used to solve rigid–body planar kinetic problems.

18.1 Kinetic Energy

In this chapter we will apply work and energy methods to solve planar motion problems involving force, velocity, and displacement. But first it will be necessary to develop a means of obtaining the body's kinetic energy when the body is subjected to translation, rotation about a fixed axis, or general plane motion.

To do this we will consider the rigid body shown in Fig. 18–1, which is represented here by a *slab* moving in the inertial x–y reference plane. An arbitrary ith particle of the body, having a mass dm, is located a distance r from the arbitrary point P. If at the *instant* shown the particle has a velocity $\mathbf{v}_i$, then the particle's kinetic energy is $T_i = \frac{1}{2}\, dm\, v_i^2$.

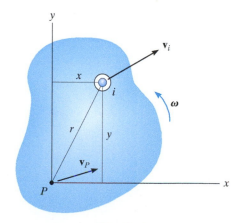

Fig. 18–1

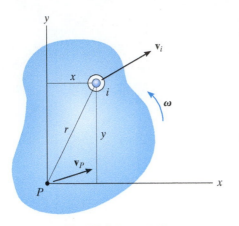

Fig. 18–1 (repeated)

The kinetic energy of the entire body is determined by writing similar expressions for each particle of the body and integrating the results, i.e.,

$$T = \frac{1}{2} \int_m dm \, v_i^2$$

This equation may also be expressed in terms of the velocity of point P. If the body has an angular velocity $\boldsymbol{\omega}$, then from Fig. 18–1 we have

$$\mathbf{v}_i = \mathbf{v}_P + \mathbf{v}_{i/P}$$
$$= (v_P)_x \mathbf{i} + (v_P)_y \mathbf{j} + \omega \mathbf{k} \times (x\mathbf{i} + y\mathbf{j})$$
$$= [(v_P)_x - \omega y]\mathbf{i} + [(v_P)_y + \omega x]\mathbf{j}$$

The square of the magnitude of $\mathbf{v}_i$ is thus

$$\mathbf{v}_i \cdot \mathbf{v}_i = v_i^2 = [(v_P)_x - \omega y]^2 + [(v_P)_y + \omega x]^2$$
$$= (v_P)_x^2 - 2(v_P)_x \omega y + \omega^2 y^2 + (v_P)_y^2 + 2(v_P)_y \omega x + \omega^2 x^2$$
$$= v_P^2 - 2(v_P)_x \omega y + 2(v_P)_y \omega x + \omega^2 r^2$$

Substituting this into the equation of kinetic energy yields

$$T = \frac{1}{2}\left(\int_m dm \right) v_P^2 - (v_P)_x \omega \left(\int_m y \, dm \right) + (v_P)_y \omega \left(\int_m x \, dm \right) + \frac{1}{2}\omega^2 \left(\int_m r^2 \, dm \right)$$

The first integral on the right represents the entire mass m of the body. Since $\bar{y}m = \int y \, dm$ and $\bar{x}m = \int x \, dm$, the second and third integrals locate the body's center of mass G with respect to P. The last integral represents the body's moment of inertia I_P, computed about the z axis passing through point P. Thus,

$$T = \tfrac{1}{2}mv_P^2 - (v_P)_x \omega \bar{y}m + (v_P)_y \omega \bar{x}m + \tfrac{1}{2}I_P\omega^2 \qquad (18\text{–}1)$$

As a special case, if point P coincides with the mass center G of the body, then $\bar{y} = \bar{x} = 0$, and therefore

$$T = \tfrac{1}{2}mv_G^2 + \tfrac{1}{2}I_G\omega^2 \qquad (18\text{–}2)$$

Both terms on the right side are *always positive*, since v_G and ω are squared. The first term represents the translational kinetic energy, referenced from the mass center, and the second term represents the body's rotational kinetic energy about the mass center.

Translation.   When a rigid body of mass m is subjected to either rectilinear or curvilinear *translation*, Fig. 18–2, the kinetic energy due to rotation is zero, since $\boldsymbol{\omega} = \mathbf{0}$. The kinetic energy of the body is therefore

$$T = \tfrac{1}{2}mv_G^2 \qquad (18\text{--}3)$$

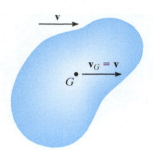

Translation

Fig. 18–2

Rotation about a Fixed Axis.   When a rigid body *rotates about a fixed axis* passing through point O, Fig. 18–3, the body has both *translational* and *rotational* kinetic energy so that

$$T = \tfrac{1}{2}mv_G^2 + \tfrac{1}{2}I_G\omega^2 \qquad (18\text{--}4)$$

The body's kinetic energy may also be formulated for this case by noting that $v_G = r_G\omega$, so that $T = \tfrac{1}{2}(I_G + mr_G^2)\omega^2$. By the parallel–axis theorem, the terms inside the parentheses represent the moment of inertia I_O of the body about an axis perpendicular to the plane of motion and passing through point O. Hence,*

$$T = \tfrac{1}{2}I_O\omega^2 \qquad (18\text{--}5)$$

From the derivation, this equation will give the same result as Eq. 18–4, since it accounts for *both* the translational and rotational kinetic energies of the body.

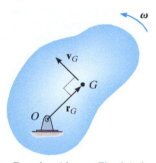

Rotation About a Fixed Axis

Fig. 18–3

General Plane Motion.   When a rigid body is subjected to general plane motion, Fig. 18–4, it has an angular velocity $\boldsymbol{\omega}$ and its mass center has a velocity $\mathbf{v}_G$. Therefore, the kinetic energy is

$$T = \tfrac{1}{2}mv_G^2 + \tfrac{1}{2}I_G\omega^2 \qquad (18\text{--}6)$$

This equation can also be expressed in terms of the body's motion about its instantaneous center of zero velocity i.e.,

$$T = \tfrac{1}{2}I_{IC}\omega^2 \qquad (18\text{--}7)$$

where I_{IC} is the moment of inertia of the body about its instantaneous center. The proof is similar to that of Eq. 18–5. (See Prob. 18–1.)

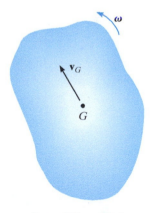

General Plane Motion

Fig. 18–4

*The similarity between this derivation and that of $\Sigma M_O = I_O\alpha$, should be noted. Also the same result can be obtained directly from Eq. 18–1 by selecting point P at O, realizing that $v_O = 0$.

The total kinetic energy of this soil compactor consists of the kinetic energy of the body or frame of the machine due to its translation, and the translational and rotational kinetic energies of the roller and the wheels due to their general plane motion. Here we exclude the additional kinetic energy developed by the moving parts of the engine and drive train. (© R.C. Hibbeler)

System of Bodies. Because energy is a scalar quantity, the total kinetic energy for a system of *connected* rigid bodies is the sum of the kinetic energies of all its moving parts. Depending on the type of motion, the kinetic energy of *each body* is found by applying Eq. 18–2 or the alternative forms mentioned above.

18.2   The Work of a Force

Several types of forces are often encountered in planar kinetics problems involving a rigid body. The work of each of these forces has been presented in Sec. 14.1 and is listed below as a summary.

Work of a Variable Force.
If an external force **F** acts on a body, the work done by the force when the body moves along the path *s*, Fig. 18–5, is

$$U_F = \int \mathbf{F} \cdot d\mathbf{r} = \int_s F \cos \theta \, ds \qquad (18\text{–}8)$$

Here θ is the angle between the "tails" of the force and the differential displacement. The integration must account for the variation of the force's direction and magnitude.

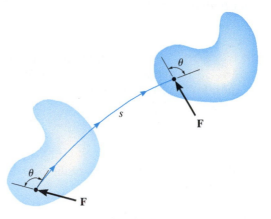

Fig. 18–5

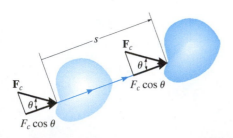

Fig. 18–6

Work of a Constant Force.
If an external force $\mathbf{F}_c$ acts on a body, Fig. 18–6, and maintains a constant magnitude F_c and constant direction θ, while the body undergoes a translation *s*, then the above equation can be integrated, so that the work becomes

$$U_{F_c} = (F_c \cos \theta)s \qquad (18\text{–}9)$$

Work of a Weight.   The weight of a body does work only when the body's center of mass G undergoes a *vertical displacement* Δy. If this displacement is *upward*, Fig. 18–7, the work is negative, since the weight is opposite to the displacement.

$$U_W = -W\,\Delta y \tag{18–10}$$

Likewise, if the displacement is *downward* $(-\Delta y)$ the work becomes *positive*. In both cases the elevation change is considered to be small so that $\mathbf{W}$, which is caused by gravitation, is constant.

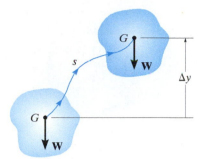

Fig. 18–7

Work of a Spring Force.   If a linear elastic spring is attached to a body, the spring force $F_s = ks$ *acting on the body* does work when the spring either stretches or compresses from s_1 to a *farther* position s_2. In both cases the work will be *negative* since the *displacement of the body* is in the opposite direction to the force, Fig. 18–8. The work is

$$U_s = -\left(\tfrac{1}{2}ks_2^2 - \tfrac{1}{2}ks_1^2\right) \tag{18–11}$$

where $|s_2| > |s_1|$.

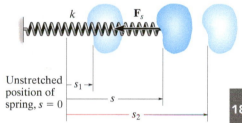

Fig. 18–8

Forces That Do No Work.   There are some external forces that do no work when the body is displaced. These forces act either at *fixed points* on the body, or they have a direction *perpendicular to their displacement*. Examples include the reactions at a pin support about which a body rotates, the normal reaction acting on a body that moves along a fixed surface, and the weight of a body when the center of gravity of the body moves in a *horizontal plane*, Fig. 18–9. A frictional force $\mathbf{F}_f$ acting on a round body as it *rolls without slipping* over a rough surface also does no work.* This is because, during any *instant of time* dt, $\mathbf{F}_f$ acts at a point on the body which has *zero velocity* (instantaneous center, IC) and so the work done by the force on the point is zero. In other words, the point is not displaced in the direction of the force during this instant. Since $\mathbf{F}_f$ contacts successive points for only an instant, the work of $\mathbf{F}_f$ will be zero.

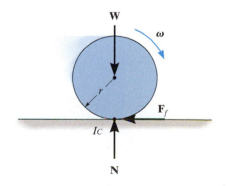

Fig. 18–9

*The work done by a frictional force *when the body slips* is discussed in Sec. 14.3.

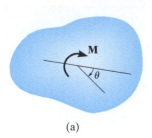

(a)

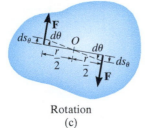

Translation
(b)

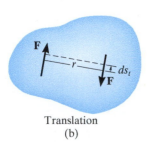

Rotation
(c)

Fig. 18–10

18.3   The Work of a Couple Moment

Consider the body in Fig. 18–10a, which is subjected to a couple moment $M = Fr$. If the body undergoes a differential displacement, then the work done by the couple forces can be found by considering the displacement as the sum of a separate translation plus rotation. When the body *translates*, the work of each force is produced only by the *component of displacement* along the line of action of the forces ds_t, Fig. 18–10b. Clearly the "positive" work of one force *cancels* the "negative" work of the other. When the body undergoes a differential rotation $d\theta$ about the arbitrary point O, Fig. 18–10c, then each force undergoes a displacement $ds_\theta = (r/2)\, d\theta$ in the direction of the force. Hence, the total work done is

$$dU_M = F\left(\frac{r}{2}d\theta\right) + F\left(\frac{r}{2}d\theta\right) = (Fr)\, d\theta$$
$$= M\, d\theta$$

The work is *positive* when **M** and $d\boldsymbol{\theta}$ have the *same sense of direction* and *negative* if these vectors are in the *opposite sense*.

When the body rotates in the plane through a finite angle θ measured in radians, from θ_1 to θ_2, the work of a couple moment is therefore

$$U_M = \int_{\theta_1}^{\theta_2} M\, d\theta \qquad (18\text{–}12)$$

If the couple moment **M** has a *constant magnitude*, then

$$U_M = M(\theta_2 - \theta_1) \qquad (18\text{–}13)$$

EXAMPLE 18.1

The bar shown in Fig. 18–11a has a mass of 10 kg and is subjected to a couple moment of $M = 50\,\text{N}\cdot\text{m}$ and a force of $P = 80\,\text{N}$, which is always applied perpendicular to the end of the bar. Also, the spring has an unstretched length of 0.5 m and remains in the vertical position due to the roller guide at B. Determine the total work done by all the forces acting on the bar when it has rotated downward from $\theta = 0°$ to $\theta = 90°$.

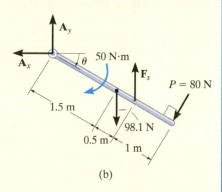

(a)

SOLUTION

First the free-body diagram of the bar is drawn in order to account for all the forces that act on it, Fig. 18–11b.

Weight W. Since the weight $10(9.81)\,\text{N} = 98.1\,\text{N}$ is displaced downward 1.5 m, the work is

$$U_W = 98.1\,\text{N}(1.5\,\text{m}) = 147.2\,\text{J}$$

Why is the work positive?

Couple Moment M. The couple moment rotates through an angle of $\theta = \pi/2$ rad. Hence,

$$U_M = 50\,\text{N}\cdot\text{m}(\pi/2) = 78.5\,\text{J}$$

Spring Force F_s. When $\theta = 0°$ the spring is stretched $(0.75\,\text{m} - 0.5\,\text{m}) = 0.25\,\text{m}$, and when $\theta = 90°$, the stretch is $(2\,\text{m} + 0.75\,\text{m}) - 0.5\,\text{m} = 2.25\,\text{m}$. Thus,

$$U_s = -\left[\tfrac{1}{2}(30\,\text{N/m})(2.25\,\text{m})^2 - \tfrac{1}{2}(30\,\text{N/m})(0.25\,\text{m})^2\right] = -75.0\,\text{J}$$

By inspection the spring does negative work on the bar since $\mathbf{F}_s$ acts in the opposite direction to displacement. This checks with the result.

Force P. As the bar moves downward, the force is displaced through a distance of $(\pi/2)(3\,\text{m}) = 4.712\,\text{m}$. The work is positive. Why?

$$U_P = 80\,\text{N}(4.712\,\text{m}) = 377.0\,\text{J}$$

Pin Reactions. Forces $\mathbf{A}_x$ and $\mathbf{A}_y$ do no work since they are not displaced.

Total Work. The work of all the forces when the bar is displaced is thus

$$U = 147.2\,\text{J} + 78.5\,\text{J} - 75.0\,\text{J} + 377.0\,\text{J} = 528\,\text{J} \qquad \textit{Ans.}$$

(b)

Fig. 18–11

18

The counterweight on this bascule bridge does positive work as the bridge is lifted and thereby cancels the negative work done by the weight of the bridge. (© R.C. Hibbeler)

18.4   Principle of Work and Energy

By applying the principle of work and energy developed in Sec. 14.2 to each of the particles of a rigid body and adding the results algebraically, since energy is a scalar, the principle of work and energy for a rigid body becomes

$$T_1 + \Sigma U_{1-2} = T_2 \qquad (18\text{–}14)$$

This equation states that the body's initial translational *and* rotational kinetic energy, plus the work done by all the external forces and couple moments acting on the body as the body moves from its initial to its final position, is equal to the body's final translational *and* rotational kinetic energy. Note that the work of the body's *internal forces* does not have to be considered. These forces occur in equal but opposite collinear pairs, so that when the body moves, the work of one force cancels that of its counterpart. Furthermore, since the body is rigid, *no relative movement* between these forces occurs, so that no internal work is done.

When several rigid bodies are pin connected, connected by inextensible cables, or in mesh with one another, Eq. 18–14 can be applied to the *entire system* of connected bodies. In all these cases the internal forces, which hold the various members together, do no work and hence are eliminated from the analysis.

The work of the torque or moment developed by the driving gears on the motors is transformed into kinetic energy of rotation of the drum. (© R.C. Hibbeler)

Procedure for Analysis

The principle of work and energy is used to solve kinetic problems that involve *velocity, force*, and *displacement*, since these terms are involved in the formulation. For application, it is suggested that the following procedure be used.

Kinetic Energy (Kinematic Diagrams).

- The kinetic energy of a body is made up of two parts. Kinetic energy of translation is referenced to the velocity of the mass center, $T = \frac{1}{2}mv_G^2$, and kinetic energy of rotation is determined using the moment of inertia of the body about the mass center, $T = \frac{1}{2}I_G\omega^2$. In the special case of rotation about a fixed axis (or rotation about the *IC*), these two kinetic energies are combined and can be expressed as $T = \frac{1}{2}I_O\omega^2$, where I_O is the moment of inertia about the axis of rotation.

- *Kinematic diagrams* for velocity may be useful for determining v_G and ω or for establishing a *relationship* between v_G and ω.*

Work (Free–Body Diagram).

- Draw a free–body diagram of the body when it is located at an intermediate point along the path in order to account for all the forces and couple moments which do work on the body as it moves along the path.

- A force does work when it moves through a displacement in the direction of the force.

- Forces that are functions of displacement must be integrated to obtain the work. Graphically, the work is equal to the area under the force–displacement curve.

- The work of a weight is the product of its magnitude and the vertical displacement, $U_W = Wy$. It is positive when the weight moves downwards.

- The work of a spring is of the form $U_s = \frac{1}{2}ks^2$, where k is the spring stiffness and s is the stretch or compression of the spring.

- The work of a couple is the product of the couple moment and the angle in radians through which it rotates, $U_M = M\theta$.

- Since *algebraic addition* of the work terms is required, it is important that the proper sign of each term be specified. Specifically, work is *positive* when the force (couple moment) is in the *same direction* as its displacement (rotation); otherwise, it is negative.

Principle of Work and Energy.

- Apply the principle of work and energy, $T_1 + \Sigma U_{1-2} = T_2$. Since this is a scalar equation, it can be used to solve for only one unknown when it is applied to a single rigid body.

*A brief review of Secs. 16.5 to 16.7 may prove helpful when solving problems, since computations for kinetic energy require a kinematic analysis of velocity.

18

EXAMPLE 18.2

The 30-kg disk shown in Fig. 18–12a is pin supported at its center. Determine the angle through which it must rotate to attain an angular velocity of 2 rad/s starting from rest. It is acted upon by a constant couple moment $M = 5$ N·m. The spring is orginally unstretched and its cord wraps around the rim of the disk.

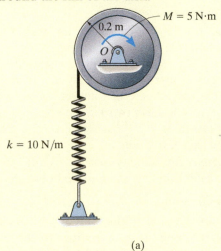

(a)

SOLUTION

Kinetic Energy. Since the disk rotates about a fixed axis, and it is initially at rest, then

$$T_1 = 0$$

$$T_2 = \tfrac{1}{2}I_O\omega_2^2 = \tfrac{1}{2}\left[\tfrac{1}{2}(30 \text{ kg})(0.2 \text{ m})^2\right](2 \text{ rad/s})^2 = 1.2 \text{ J}$$

Work (Free–Body Diagram). As shown in Fig. 18–12b, the pin reactions $\mathbf{O}_x$ and $\mathbf{O}_y$ and the weight (294.3 N) do no work, since they are not displaced. The *couple moment*, having a constant magnitude, does positive work $U_M = M\theta$ as the disk *rotates* through a clockwise angle of θ rad, and the spring does negative work $U_s = -\tfrac{1}{2}ks^2$.

Principle of Work and Energy.

$$\{T_1\} + \{\Sigma U_{1-2}\} = \{T_2\}$$

$$\{T_1\} + \left\{M\theta - \tfrac{1}{2}ks^2\right\} = \{T_2\}$$

$$\{0\} + \left\{(5 \text{ N·m})\theta - \frac{1}{2}(10 \text{ N/m})[\theta(0.2 \text{ m})]^2\right\} = \{1.2 \text{ J}\}$$

$$-0.2\theta^2 + 5\theta - 1.2 = 0$$

Solving this quadratic equation for the smallest positive root,

$$\theta = 0.2423 \text{ rad} = 0.2423 \text{ rad}\left(\frac{180°}{\pi \text{ rad}}\right) = 13.9° \qquad Ans.$$

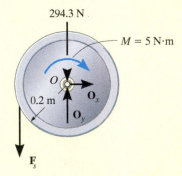

(b)

Fig. 18–12

EXAMPLE 18.3

The wheel shown in Fig. 18–13a weighs 40 lb and has a radius of gyration $k_G = 0.6$ ft about its mass center G. If it is subjected to a clockwise couple moment of 15 lb·ft and rolls from rest without slipping, determine its angular velocity after its center G moves 0.5 ft. The spring has a stiffness $k = 10$ lb/ft and is initially unstretched when the couple moment is applied.

$k = 10$ lb/ft

(a)

SOLUTION

Kinetic Energy (Kinematic Diagram).   Since the wheel is initially at rest,

$$T_1 = 0$$

The kinematic diagram of the wheel when it is in the final position is shown in Fig. 18–13b. The final kinetic energy is determined from

$$T_2 = \tfrac{1}{2} I_{IC} \omega_2^2$$

$$= \frac{1}{2}\left[\frac{40 \text{ lb}}{32.2 \text{ ft/s}^2}(0.6 \text{ ft})^2 + \left(\frac{40 \text{ lb}}{32.2 \text{ ft/s}^2} \right)(0.8 \text{ ft})^2 \right] \omega_2^2$$

$$T_2 = 0.6211\,\omega_2^2$$

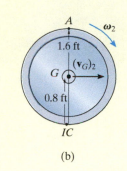

(b)

Work (Free–Body Diagram).   As shown in Fig. 18–13c, only the spring force $\mathbf{F}_s$ and the couple moment do work. The normal force does not move along its line of action and the frictional force does *no work*, since the wheel does not slip as it rolls.

The work of $\mathbf{F}_s$ is found using $U_s = -\tfrac{1}{2}ks^2$. Here the work is negative since $\mathbf{F}_s$ is in the opposite direction to displacement. Since the wheel does not slip when the center G moves 0.5 ft, then the wheel rotates $\theta = s_G/r_{G/IC} = 0.5 \text{ ft}/0.8 \text{ ft} = 0.625$ rad,   Fig. 18–13b. Hence, the spring stretches $s = \theta r_{A/IC} = (0.625 \text{ rad})(1.6 \text{ ft}) = 1$ ft.

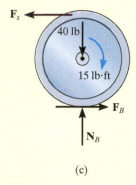

(c)

Fig. 18–13

Principle of Work and Energy.

$$\{T_1\} + \{\Sigma U_{1-2}\} = \{T_2\}$$

$$\{T_1\} + \{M\theta - \tfrac{1}{2}ks^2\} = \{T_2\}$$

$$\{0\} + \left\{ 15 \text{ lb·ft}(0.625 \text{ rad}) - \frac{1}{2}(10 \text{ lb/ft})(1 \text{ ft})^2 \right\} = \{0.6211\,\omega_2^2 \text{ ft·lb}\}$$

$$\omega_2 = 2.65 \text{ rad/s} \, \circlearrowright \qquad\qquad Ans.$$

EXAMPLE | **18.4**

(© R.C. Hibbeler)

The 700-kg pipe is equally suspended from the two tines of the fork lift shown in the photo. It is undergoing a swinging motion such that when $\theta = 30°$ it is momentarily at rest. Determine the normal and frictional forces acting on each tine which are needed to support the pipe at the instant $\theta = 0°$. Measurements of the pipe and the suspender are shown in Fig. 18–14a. Neglect the mass of the suspender and the thickness of the pipe.

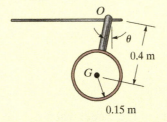

(a)

Fig. 18–14

SOLUTION

We must use the equations of motion to find the forces on the tines since these forces do no work. Before doing this, however, we will apply the principle of work and energy to determine the angular velocity of the pipe when $\theta = 0°$.

Kinetic Energy (Kinematic Diagram).   Since the pipe is originally at rest, then

$$T_1 = 0$$

The final kinetic energy may be computed with reference to either the fixed point O or the center of mass G. For the calculation we will consider the pipe to be a thin ring so that $I_G = mr^2$. If point G is considered, we have

$$T_2 = \tfrac{1}{2}m(v_G)_2^2 + \tfrac{1}{2}I_G\omega_2^2$$
$$= \tfrac{1}{2}(700 \text{ kg})[(0.4 \text{ m})\omega_2]^2 + \tfrac{1}{2}[700 \text{ kg}(0.15 \text{ m})^2]\omega_2^2$$
$$= 63.875\omega_2^2$$

If point O is considered then the parallel-axis theorem must be used to determine I_O. Hence,

$$T_2 = \tfrac{1}{2}I_O\omega_2^2 = \tfrac{1}{2}[700 \text{ kg}(0.15 \text{ m})^2 + 700 \text{ kg}(0.4 \text{ m})^2]\omega_2^2$$
$$= 63.875\omega_2^2$$

Work (Free-Body Diagram). Fig. 18–14b. The normal and frictional forces on the tines do no work since they do not move as the pipe swings. The weight does positive work since the weight moves downward through a vertical distance $\Delta y = 0.4 \text{ m} - 0.4 \cos 30° \text{ m} = 0.05359 \text{ m}$.

Principle of Work and Energy.

$$\{T_1\} + \{\Sigma U_{1-2}\} = \{T_2\}$$

$$\{0\} + \{700(9.81) \text{ N}(0.05359 \text{ m})\} = \{63.875\omega_2^2\}$$

$$\omega_2 = 2.400 \text{ rad/s}$$

Equations of Motion. Referring to the free-body and kinetic diagrams shown in Fig. 18–14c, and using the result for ω_2, we have

$$\overset{+}{\leftarrow}\Sigma F_t = m(a_G)_t; \quad F_T = (700 \text{ kg})(a_G)_t$$

$$+\uparrow\Sigma F_n = m(a_G)_n; \quad N_T - 700(9.81) \text{ N} = (700 \text{ kg})(2.400 \text{ rad/s})^2(0.4 \text{ m})$$

$$\zeta+\Sigma M_O = I_O\alpha; \quad 0 = [(700 \text{ kg})(0.15 \text{ m})^2 + (700 \text{ kg})(0.4 \text{ m})^2]\alpha$$

Since $(a_G)_t = (0.4 \text{ m})\alpha$, then

$$\alpha = 0, \quad (a_G)_t = 0$$

$$F_T = 0$$

$$N_T = 8.480 \text{ kN}$$

There are two tines used to support the load, therefore

$$F'_T = 0 \qquad\qquad\qquad\qquad \textit{Ans.}$$

$$N'_T = \frac{8.480 \text{ kN}}{2} = 4.24 \text{ kN} \qquad \textit{Ans.}$$

NOTE: Due to the swinging motion the tines are subjected to a *greater* normal force than would be the case if the load were static, in which case $N'_T = 700(9.81) \text{ N}/2 = 3.43 \text{ kN}$.

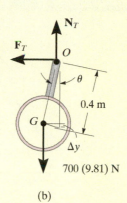

(b)

18

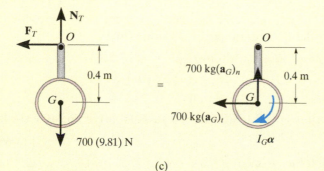

(c)

Fig. 18–14

EXAMPLE 18.5

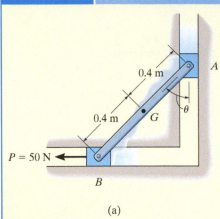

(a)

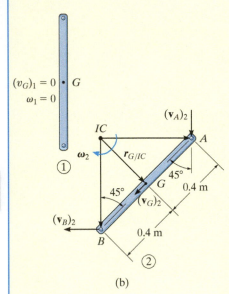

(b)

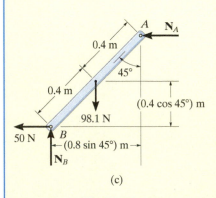

(c)

Fig. 18–15

The 10–kg rod shown in Fig. 18–15a is constrained so that its ends move along the grooved slots. The rod is initially at rest when $\theta = 0°$. If the slider block at B is acted upon by a horizontal force $P = 50$ N, determine the angular velocity of the rod at the instant $\theta = 45°$. Neglect friction and the mass of blocks A and B.

SOLUTION

Why can the principle of work and energy be used to solve this problem?

Kinetic Energy (Kinematic Diagrams).   Two kinematic diagrams of the rod, when it is in the initial position 1 and final position 2, are shown in Fig. 18–15b. When the rod is in position 1, $T_1 = 0$ since $(\mathbf{v}_G)_1 = \boldsymbol{\omega}_1 = \mathbf{0}$. In position 2 the angular velocity is $\boldsymbol{\omega}_2$ and the velocity of the mass center is $(\mathbf{v}_G)_2$. Hence, the kinetic energy is

$$T_2 = \tfrac{1}{2}m(v_G)_2^2 + \tfrac{1}{2}I_G\omega_2^2$$
$$= \tfrac{1}{2}(10 \text{ kg})(v_G)_2^2 + \tfrac{1}{2}\left[\tfrac{1}{12}(10 \text{ kg})(0.8 \text{ m})^2\right]\omega_2^2$$
$$= 5(v_G)_2^2 + 0.2667(\omega_2)^2$$

The two unknowns $(v_G)_2$ and ω_2 can be related from the instantaneous center of zero velocity for the rod. Fig. 18–15b. It is seen that as A moves downward with a velocity $(\mathbf{v}_A)_2$, B moves horizontally to the left with a velocity $(\mathbf{v}_B)_2$, Knowing these directions, the IC is located as shown in the figure. Hence,

$$(v_G)_2 = r_{G/IC}\omega_2 = (0.4 \tan 45° \text{ m})\omega_2$$
$$= 0.4\omega_2$$

Therefore,

$$T_2 = 0.8\omega_2^2 + 0.2667\omega_2^2 = 1.0667\omega_2^2$$

Of course, we can also determine this result using $T_2 = \tfrac{1}{2}I_{IC}\omega_2^2$.

Work (Free–Body Diagram).   Fig. 18–15c. The normal forces $\mathbf{N}_A$ and $\mathbf{N}_B$ do no work as the rod is displaced. Why? The 98.1-N weight is displaced a vertical distance of $\Delta y = (0.4 - 0.4 \cos 45°)$ m; whereas the 50-N force moves a horizontal distance of $s = (0.8 \sin 45°)$ m. Both of these forces do positive work. Why?

Principle of Work and Energy.

$$\{T_1\} + \{\Sigma U_{1-2}\} = \{T_2\}$$
$$\{T_1\} + \{W \Delta y + Ps\} = \{T_2\}$$
$$\{0\} + \{98.1 \text{ N}(0.4 \text{ m} - 0.4 \cos 45° \text{ m}) + 50 \text{ N}(0.8 \sin 45° \text{ m})\}$$
$$= \{1.0667\omega_2^2 \text{ J}\}$$

Solving for ω_2 gives

$$\omega_2 = 6.11 \text{ rad/s} \curvearrowright \qquad \textit{Ans.}$$

PRELIMINARY PROBLEM

P18–1.   Determine the kinetic energy of the 100-kg object.

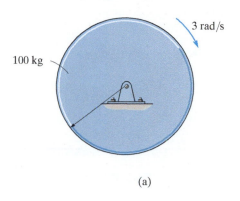

(a)

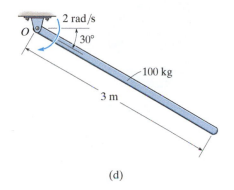

(d)

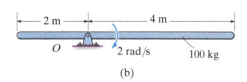

(b)

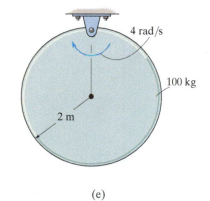

(e)

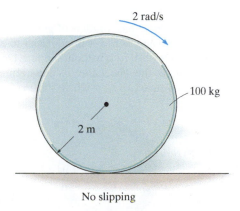

No slipping

(c)

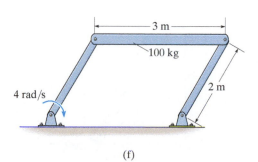

(f)

Prob. P18–1

FUNDAMENTAL PROBLEMS

F18–1. The 80-kg wheel has a radius of gyration about its mass center O of $k_O = 400$ mm. Determine its angular velocity after it has rotated 20 revolutions starting from rest.

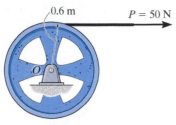

Prob. F18–1

F18–2. The uniform 50-lb slender rod is subjected to a couple moment of $M = 100$ lb·ft. If the rod is at rest when $\theta = 0°$, determine its angular velocity when $\theta = 90°$.

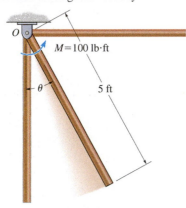

Prob. F18–2

F18–3. The uniform 50-kg slender rod is at rest in the position shown when $P = 600$ N is applied. Determine the angular velocity of the rod when the rod reaches the vertical position.

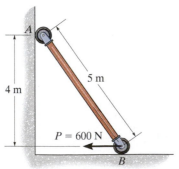

Prob. F18–3

F18–4. The 50-kg wheel is subjected to a force of 50 N. If the wheel starts from rest and rolls without slipping, determine its angular velocity after it has rotated 10 revolutions. The radius of gyration of the wheel about its mass center G is $k_G = 0.3$ m.

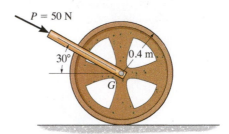

Prob. F18–4

F18–5. If the uniform 30-kg slender rod starts from rest at the position shown, determine its angular velocity after it has rotated 4 revolutions. The forces remain perpendicular to the rod.

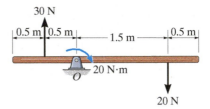

Prob. F18–5

F18–6. The 20-kg wheel has a radius of gyration about its center G of $k_G = 300$ mm. When it is subjected to a couple moment of $M = 50$ N·m, it rolls without slipping. Determine the angular velocity of the wheel after its mass center G has traveled through a distance of $s_G = 20$ m, starting from rest.

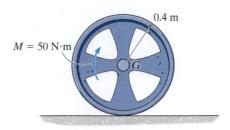

Prob. F18–6

18

PROBLEMS

18–1. At a given instant the body of mass m has an angular velocity ω and its mass center has a velocity v_G. Show that its kinetic energy can be represented as $T = \frac{1}{2}I_{IC}\omega^2$, where I_{IC} is the moment of inertia of the body determined about the instantaneous axis of zero velocity, located a distance $r_{G/IC}$ from the mass center as shown.

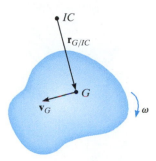

Prob. 18–1

18–2. The wheel is made from a 5-kg thin ring and two 2-kg slender rods. If the torsional spring attached to the wheel's center has a stiffness $k = 2$ N·m/rad, and the wheel is rotated until the torque $M = 25$ N·m is developed, determine the maximum angular velocity of the wheel if it is released from rest.

18–3. The wheel is made from a 5-kg thin ring and two 2-kg slender rods. If the torsional spring attached to the wheel's center has a stiffness $k = 2$ N·m/rad, so that the torque on the center of the wheel is $M = (2\theta)$ N·m, where θ is in radians, determine the maximum angular velocity of the wheel if it is rotated two revolutions and then released from rest.

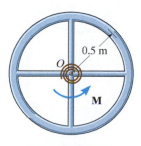

Probs. 18–2/3

*18–4. A force of $P = 60$ N is applied to the cable, which causes the 200-kg reel to turn since it is resting on the two rollers A and B of the dispenser. Determine the angular velocity of the reel after it has made two revolutions starting from rest. Neglect the mass of the rollers and the mass of the cable. Assume the radius of gyration of the reel about its center axis remains constant at $k_O = 0.6$ m.

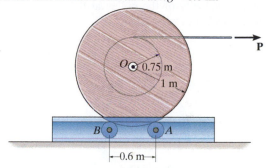

Prob. 18–4

18–5. A force of $P = 20$ N is applied to the cable, which causes the 175-kg reel to turn since it is resting on the two rollers A and B of the dispenser. Determine the angular velocity of the reel after it has made two revolutions starting from rest. Neglect the mass of the rollers and the mass of the cable. The radius of gyration of the reel about its center axis is $k_G = 0.42$ m.

18–6. A force of $P = 20$ N is applied to the cable, which causes the 175-kg reel to turn without slipping on the two rollers A and B of the dispenser. Determine the angular velocity of the reel after it has made two revolutions starting from rest. Neglect the mass of the cable. Each roller can be considered as an 18-kg cylinder, having a radius of 0.1 m. The radius of gyration of the reel about its center axis is $k_G = 0.42$ m.

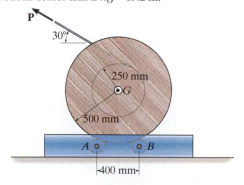

Probs. 18–5/6

18–7. The double pulley consists of two parts that are attached to one another. It has a weight of 50 lb and a centroidal radius of gyration of $k_O = 0.6$ ft and is turning with an angular velocity of 20 rad/s clockwise. Determine the kinetic energy of the system. Assume that neither cable slips on the pulley.

*18–8. The double pulley consists of two parts that are attached to one another. It has a weight of 50 lb and a centroidal radius of gyration of $k_O = 0.6$ ft and is turning with an angular velocity of 20 rad/s clockwise. Determine the angular velocity of the pulley at the instant the 20-lb weight moves 2 ft downward.

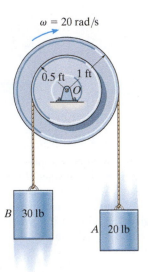

Probs. 18–7/8

18–9. The disk, which has a mass of 20 kg, is subjected to the couple moment of $M = (2\theta + 4)$ N·m, where θ is in radians. If it starts from rest, determine its angular velocity when it has made two revolutions.

Prob. 18–9

18–10. The spool has a mass of 40 kg and a radius of gyration of $k_O = 0.3$ m. If the 10-kg block is released from rest, determine the distance the block must fall in order for the spool to have an angular velocity $\omega = 15$ rad/s. Also, what is the tension in the cord while the block is in motion? Neglect the mass of the cord.

Prob. 18–10

18–11. The force of $T = 20$ N is applied to the cord of negligible mass. Determine the angular velocity of the 20-kg wheel when it has rotated 4 revolutions starting from rest. The wheel has a radius of gyration of $k_O = 0.3$ m.

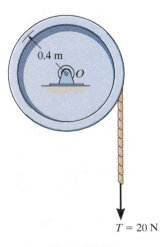

Prob. 18–11

*18–12.   Determine the velocity of the 50-kg cylinder after it has descended a distance of 2 m. Initially, the system is at rest. The reel has a mass of 25 kg and a radius of gyration about its center of mass A of $k_A = 125$ mm.

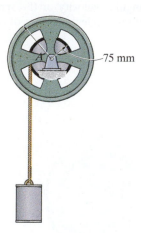

75 mm

Prob. 18–12

18–13.   The 10-kg uniform slender rod is suspended at rest when the force of $F = 150$ N is applied to its end. Determine the angular velocity of the rod when it has rotated 90° clockwise from the position shown. The force is always perpendicular to the rod.

18–14.   The 10-kg uniform slender rod is suspended at rest when the force of $F = 150$ N is applied to its end. Determine the angular velocity of the rod when it has rotated 180° clockwise from the position shown. The force is always perpendicular to the rod.

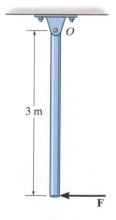

O

3 m

F

Probs. 18–13/14

18–15.   The pendulum consists of a 10-kg uniform disk and a 3-kg uniform slender rod. If it is released from rest in the position shown, determine its angular velocity when it rotates clockwise 90°.

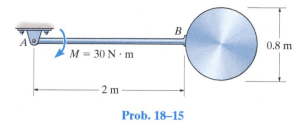

$M = 30$ N · m

A

B

0.8 m

2 m

Prob. 18–15

*18–16.   A motor supplies a constant torque $M = 6$ kN · m to the winding drum that operates the elevator. If the elevator has a mass of 900 kg, the counterweight C has a mass of 200 kg, and the winding drum has a mass of 600 kg and radius of gyration about its axis of $k = 0.6$ m, determine the speed of the elevator after it rises 5 m starting from rest. Neglect the mass of the pulleys.

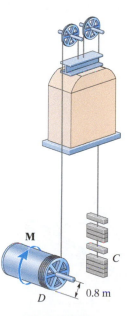

M

C

D   0.8 m

Prob. 18–16

18

18–17. The center O of the thin ring of mass m is given an angular velocity of ω_0. If the ring rolls without slipping, determine its angular velocity after it has traveled a distance of s down the plane. Neglect its thickness.

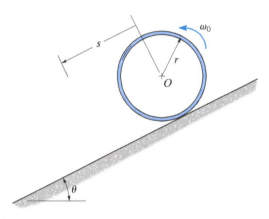

Prob. 18–17

18–18. The wheel has a mass of 100 kg and a radius of gyration of $k_O = 0.2$ m. A motor supplies a torque $M = (40\theta + 900)$ N·m, where θ is in radians, about the drive shaft at O. Determine the speed of the loading car, which has a mass of 300 kg, after it travels $s = 4$ m. Initially the car is at rest when $s = 0$ and $\theta = 0°$. Neglect the mass of the attached cable and the mass of the car's wheels.

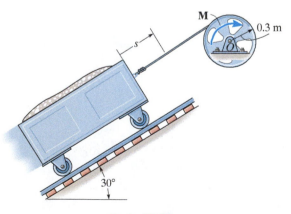

Prob. 18–18

18–19. The rotary screen S is used to wash limestone. When empty it has a mass of 800 kg and a radius of gyration of $k_G = 1.75$ m. Rotation is achieved by applying a torque of $M = 280$ N·m about the drive wheel at A. If no slipping occurs at A and the supporting wheel at B is free to roll, determine the angular velocity of the screen after it has rotated 5 revolutions. Neglect the mass of A and B.

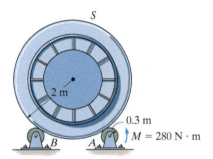

Prob. 18–19

*18–20. If $P = 200$ N and the 15-kg uniform slender rod starts from rest at $\theta = 0°$, determine the rod's angular velocity at the instant just before $\theta = 45°$.

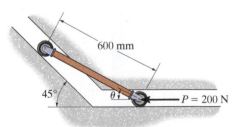

Prob. 18–20

18–21. A yo-yo has a weight of 0.3 lb and a radius of gyration of $k_O = 0.06$ ft. If it is released from rest, determine how far it must descend in order to attain an angular velocity $\omega = 70$ rad/s. Neglect the mass of the string and assume that the string is wound around the central peg such that the mean radius at which it unravels is $r = 0.02$ ft.

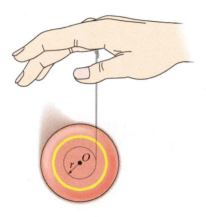

Prob. 18–21

18–22. If the 50-lb bucket, C, is released from rest, determine its velocity after it has fallen a distance of 10 ft. The windlass A can be considered as a 30-lb cylinder, while the spokes are slender rods, each having a weight of 2 lb. Neglect the pulley's weight.

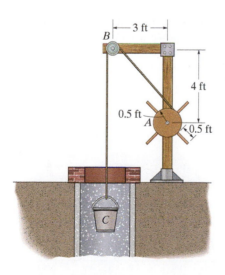

Prob. 18–22

18–23. The coefficient of kinetic friction between the 100-lb disk and the surface of the conveyor belt is $\mu_A = 0.2$. If the conveyor belt is moving with a speed of $v_C = 6$ ft/s when the disk is placed in contact with it, determine the number of revolutions the disk makes before it reaches a constant angular velocity.

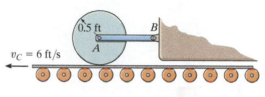

Prob. 18–23

*18–24.** The 30-kg disk is originally at rest, and the spring is unstretched. A couple moment of $M = 80$ N · m is then applied to the disk as shown. Determine its angular velocity when its mass center G has moved 0.5 m along the plane. The disk rolls without slipping.

18–25. The 30-kg disk is originally at rest, and the spring is unstretched. A couple moment $M = 80$ N · m is then applied to the disk as shown. Determine how far the center of mass of the disk travels along the plane before it momentarily stops. The disk rolls without slipping.

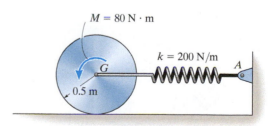

Probs. 18–24/25

18–26. Two wheels of negligible weight are mounted at corners A and B of the rectangular 75-lb plate. If the plate is released from rest at $\theta = 90°$, determine its angular velocity at the instant just before $\theta = 0°$.

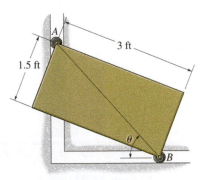

Prob. 18–26

18–27. The link AB is subjected to a couple moment of $M = 40$ N·m. If the ring gear C is fixed, determine the angular velocity of the 15-kg inner gear when the link has made two revolutions starting from rest. Neglect the mass of the link and assume the inner gear is a disk. Motion occurs in the vertical plane.

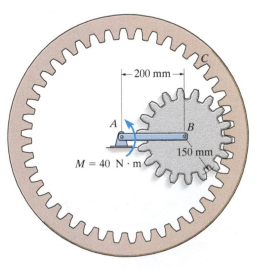

Prob. 18–27

***18–28.** The 10-kg rod AB is pin connected at A and subjected to a couple moment of $M = 15$ N·m. If the rod is released from rest when the spring is unstretched at $\theta = 30°$, determine the rod's angular velocity at the instant $\theta = 60°$. As the rod rotates, the spring always remains horizontal, because of the roller support at C.

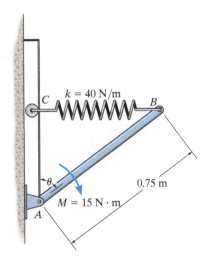

Prob. 18–28

18–29. The 10-lb sphere starts from rest at $\theta = 0°$ and rolls without slipping down the cylindrical surface which has a radius of 10 ft. Determine the speed of the sphere's center of mass at the instant $\theta = 45°$.

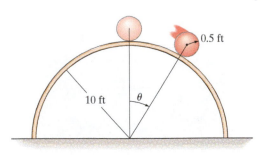

Prob. 18–29

18–30. Motor M exerts a constant force of $P = 750$ N on the rope. If the 100-kg post is at rest when $\theta = 0°$, determine the angular velocity of the post at the instant $\theta = 60°$. Neglect the mass of the pulley and its size, and consider the post as a slender rod.

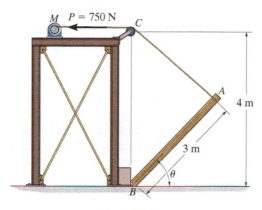

Prob. 18–30

18–31. The linkage consists of two 6-kg rods AB and CD and a 20-kg bar BD. When $\theta = 0°$, rod AB is rotating with an angular velocity $\omega = 2$ rad/s. If rod CD is subjected to a couple moment of $M = 30$ N $\cdot$ m, determine ω_{AB} at the instant $\theta = 90°$.

***18–32.** The linkage consists of two 6-kg rods AB and CD and a 20-kg bar BD. When $\theta = 0°$, rod AB is rotating with an angular velocity $\omega = 2$ rad/s. If rod CD is subjected to a couple moment $M = 30$ N $\cdot$ m, determine ω at the instant $\theta = 45°$.

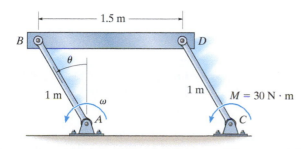

Probs. 18–31/32

18–33. The two 2-kg gears A and B are attached to the ends of a 3-kg slender bar. The gears roll within the fixed ring gear C, which lies in the horizontal plane. If a 10-N $\cdot$ m torque is applied to the center of the bar as shown, determine the number of revolutions the bar must rotate starting from rest in order for it to have an angular velocity of $\omega_{AB} = 20$ rad/s. For the calculation, assume the gears can be approximated by thin disks. What is the result if the gears lie in the vertical plane?

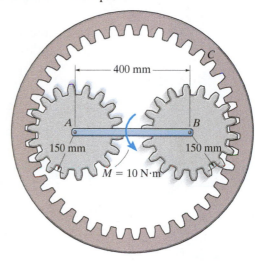

Prob. 18–33

18–34. The linkage consists of two 8-lb rods AB and CD and a 10-lb bar AD. When $\theta = 0°$, rod AB is rotating with an angular velocity $\omega_{AB} = 2$ rad/s. If rod CD is subjected to a couple moment $M = 15$ lb $\cdot$ ft and bar AD is subjected to a horizontal force $P = 20$ lb as shown, determine ω_{AB} at the instant $\theta = 90°$.

18–35. The linkage consists of two 8-lb rods AB and CD and a 10-lb bar AD. When $\theta = 0°$, rod AB is rotating with an angular velocity $\omega_{AB} = 2$ rad/s. If rod CD is subjected to a couple moment $M = 15$ lb $\cdot$ ft and bar AD is subjected to a horizontal force $P = 20$ lb as shown, determine ω_{AB} at the instant $\theta = 45°$.

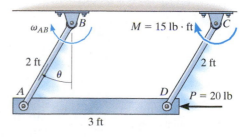

Probs. 18–34/35

18.5   Conservation of Energy

When a force system acting on a rigid body consists only of *conservative forces*, the conservation of energy theorem can be used to solve a problem that otherwise would be solved using the principle of work and energy. This theorem is often easier to apply since the work of a conservative force is *independent of the path* and depends only on the initial and final positions of the body. It was shown in Sec. 14.5 that the work of a conservative force can be expressed as the difference in the body's potential energy measured from an arbitrarily selected reference or datum.

Gravitational Potential Energy.   Since the total weight of a body can be considered concentrated at its center of gravity, the *gravitational potential energy* of the body is determined by knowing the height of the body's center of gravity above or below a horizontal datum.

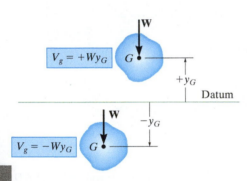

Gravitational potential energy

Fig. 18–16

$$V_g = Wy_G \qquad (18\text{--}15)$$

Here the potential energy is *positive* when y_G is positive upward, since the weight has the ability to do *positive work* when the body moves back to the datum, Fig. 18–16. Likewise, if G is located *below* the datum $(-y_G)$, the gravitational potential energy is *negative*, since the weight does *negative work* when the body returns to the datum.

Elastic Potential Energy.   The force developed by an elastic spring is also a conservative force. The *elastic potential energy* which a spring imparts to an attached body when the spring is stretched or compressed from an initial undeformed position $(s = 0)$ to a final position s, Fig. 18–17, is

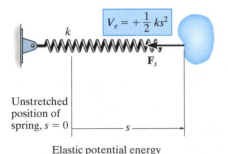

Elastic potential energy

Fig. 18–17

$$V_e = +\tfrac{1}{2}ks^2 \qquad (18\text{--}16)$$

In the deformed position, the spring force acting *on the body* always has the ability for doing positive work when the spring returns back to its original undeformed position (see Sec. 14.5).

Conservation of Energy.   In general, if a body is subjected to both gravitational and elastic forces, the total *potential energy* can be expressed as a potential function represented as the algebraic sum

$$V = V_g + V_e \qquad\qquad (18\text{–}17)$$

Here measurement of V depends upon the location of the body with respect to the selected datum.

Realizing that the work of conservative forces can be written as a difference in their potential energies, i.e., $(\Sigma U_{1-2})_{cons} = V_1 - V_2$, Eq. 14–16, we can rewrite the principle of work and energy for a rigid body as

$$T_1 + V_1 + (\Sigma U_{1-2})_{noncons} = T_2 + V_2 \qquad\qquad (18\text{–}18)$$

Here $(\Sigma U_{1-2})_{noncons}$ represents the work of the nonconservative forces such as friction. If this term is zero, then

$$T_1 + V_1 = T_2 + V_2 \qquad\qquad (18\text{–}19)$$

This equation is referred to as the conservation of mechanical energy. It states that the *sum* of the potential and kinetic energies of the body remains *constant* when the body moves from one position to another. It also applies to a system of smooth, pin-connected rigid bodies, bodies connected by inextensible cords, and bodies in mesh with other bodies. In all these cases the forces acting at the points of contact are *eliminated* from the analysis, since they occur in equal but opposite collinear pairs and each pair of forces moves through an equal distance when the system undergoes a displacement.

It is important to remember that only problems involving conservative force systems can be solved by using Eq. 18–19. As stated in Sec. 14.5, friction or other drag-resistant forces, which depend on velocity or acceleration, are nonconservative. The work of such forces is transformed into thermal energy used to heat up the surfaces of contact, and consequently this energy is dissipated into the surroundings and may not be recovered. Therefore, problems involving frictional forces can be solved by using either the principle of work and energy written in the form of Eq. 18–18, if it applies, or the equations of motion.

The torsional springs located at the top of the garage door wind up as the door is lowered. When the door is raised, the potential energy stored in the springs is then transferred into gravitational potential energy of the door's weight, thereby making it easy to open. (© R.C. Hibbeler)

18

Procedure for Analysis

The conservation of energy equation is used to solve problems involving *velocity, displacement*, and *conservative force systems*. For application it is suggested that the following procedure be used.

Potential Energy.

- Draw two diagrams showing the body located at its initial and final positions along the path.

- If the center of gravity, G, is subjected to a *vertical displacement*, establish a fixed horizontal datum from which to measure the body's gravitational potential energy V_g.

- Data pertaining to the elevation y_G of the body's center of gravity from the datum and the extension or compression of any connecting springs can be determined from the problem geometry and listed on the two diagrams.

- The potential energy is determined from $V = V_g + V_e$. Here $V_g = W y_G$, which can be positive or negative, and $V_e = \frac{1}{2} k s^2$, which is always positive.

Kinetic Energy.

- The kinetic energy of the body consists of two parts, namely translational kinetic energy, $T = \frac{1}{2} m v_G^2$, and rotational kinetic energy, $T = \frac{1}{2} I_G \omega^2$.

- Kinematic diagrams for velocity may be useful for establishing a *relationship* between v_G and ω.

Conservation of Energy.

- Apply the conservation of energy equation $T_1 + V_1 = T_2 + V_2$.

EXAMPLE 18.6

The 10-kg rod AB shown in Fig. 18–18a is confined so that its ends move in the horizontal and vertical slots. The spring has a stiffness of $k = 800$ N/m and is unstretched when $\theta = 0°$. Determine the angular velocity of AB when $\theta = 0°$, if the rod is released from rest when $\theta = 30°$. Neglect the mass of the slider blocks.

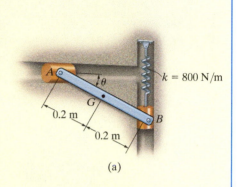

(a)

SOLUTION

Potential Energy. The two diagrams of the rod, when it is located at its initial and final positions, are shown in Fig. 18–18b. The datum, used to measure the gravitational potential energy, is placed in line with the rod when $\theta = 0°$.

When the rod is in position 1, the center of gravity G is located *below the datum* so its gravitational potential energy is *negative.* Furthermore, (positive) elastic potential energy is stored in the spring, since it is stretched a distance of $s_1 = (0.4 \sin 30°)$ m. Thus,

$$V_1 = -Wy_1 + \tfrac{1}{2}ks_1^2$$

$$= -(98.1 \text{ N})(0.2 \sin 30° \text{ m}) + \tfrac{1}{2}(800 \text{ N/m})(0.4 \sin 30° \text{ m})^2 = 6.19 \text{ J}$$

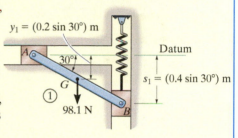

When the rod is in position 2, the potential energy of the rod is zero, since the center of gravity G is located at the datum, and the spring is unstretched, $s_2 = 0$. Thus,

$$V_2 = 0$$

Kinetic Energy. The rod is released from rest from position 1, thus $(\mathbf{v}_G)_1 = \boldsymbol{\omega}_1 = \mathbf{0}$, and so

$$T_1 = 0$$

In position 2, the angular velocity is $\boldsymbol{\omega}_2$ and the rod's mass center has a velocity of $(\mathbf{v}_G)_2$. Thus,

$$T_2 = \tfrac{1}{2}m(v_G)_2^2 + \tfrac{1}{2}I_G\omega_2^2$$

$$= \tfrac{1}{2}(10 \text{ kg})(v_G)_2^2 + \tfrac{1}{2}\left[\tfrac{1}{12}(10 \text{ kg})(0.4 \text{ m})^2\right]\omega_2^2$$

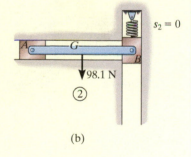

(b)

Using *kinematics*, $(\mathbf{v}_G)_2$ can be related to $\boldsymbol{\omega}_2$ as shown in Fig. 18–18c. At the instant considered, the instantaneous center of zero velocity (IC) for the rod is at point A; hence, $(v_G)_2 = (r_{G/IC})\omega_2 = (0.2 \text{ m})\omega_2$. Substituting into the above expression and simplifying (or using $\tfrac{1}{2}I_{IC}\omega_2^2$), we get

$$T_2 = 0.2667\omega_2^2$$

Conservation of Energy.

$$\{T_1\} + \{V_1\} = \{T_2\} + \{V_2\}$$

$$\{0\} + \{6.19 \text{ J}\} = \{0.2667\omega_2^2\} + \{0\}$$

$$\omega_2 = 4.82 \text{ rad/s} \circlearrowright \qquad \textit{Ans.}$$

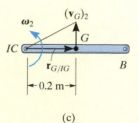

(c)

Fig. 18–18

18

EXAMPLE 18.7

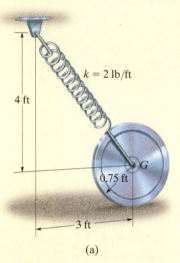

$k = 2$ lb/ft

4 ft

0.75 ft

G

3 ft

(a)

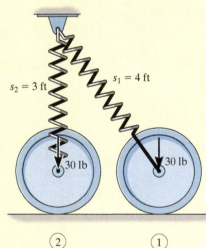

$s_2 = 3$ ft

$s_1 = 4$ ft

30 lb

30 lb

② ①

(b)

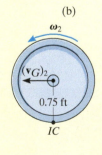

ω_2

$(\mathbf{v}_G)_2$

0.75 ft

IC

(c)

Fig. 18–19

The wheel shown in Fig. 18–19a has a weight of 30 lb and a radius of gyration of $k_G = 0.6$ ft. It is attached to a spring which has a stiffness $k = 2$ lb/ft and an unstretched length of 1 ft. If the disk is released from rest in the position shown and rolls without slipping, determine its angular velocity at the instant G moves 3 ft to the left.

SOLUTION

Potential Energy. Two diagrams of the wheel, when it at the initial and final positions, are shown in Fig. 18–19b. A gravitational datum is not needed here since the weight is not displaced vertically. From the problem geometry the spring is stretched $s_1 = \left(\sqrt{3^2 + 4^2} - 1\right) = 4$ ft in the initial position, and spring $s_2 = (4 - 1) = 3$ ft in the final position. Hence, the positive spring potential energy is

$$V_1 = \tfrac{1}{2}ks_1^2 = \tfrac{1}{2}(2 \text{ lb/ft})(4 \text{ ft})^2 = 16 \text{ ft} \cdot \text{lb}$$

$$V_2 = \tfrac{1}{2}ks_2^2 = \tfrac{1}{2}(2 \text{ lb/ft})(3 \text{ ft})^2 = 9 \text{ ft} \cdot \text{lb}$$

Kinetic Energy. The disk is released from rest and so $(\mathbf{v}_G)_1 = \mathbf{0}$, $\boldsymbol{\omega}_1 = \mathbf{0}$. Therefore,

$$T_1 = 0$$

Since the instantaneous center of zero velocity is at the ground, Fig. 18–19c, we have

$$T_2 = \frac{1}{2}I_{IC}\omega_2^2$$

$$= \frac{1}{2}\left[\left(\frac{30 \text{ lb}}{32.2 \text{ ft/s}^2}\right)(0.6 \text{ ft})^2 + \left(\frac{30 \text{ lb}}{32.2 \text{ ft/s}^2}\right)(0.75 \text{ ft})^2\right]\omega_2^2$$

$$= 0.4297\omega_2^2$$

Conservation of Energy.

$$\{T_1\} + \{V_1\} = \{T_2\} + \{V_2\}$$

$$\{0\} + \{16 \text{ ft} \cdot \text{lb}\} = \{0.4297\omega_2^2\} + \{9 \text{ ft} \cdot \text{lb}\}$$

$$\omega_2 = 4.04 \text{ rad/s} \,\circlearrowright \qquad \qquad Ans.$$

NOTE: If the principle of work and energy were used to solve this problem, then the work of the spring would have to be determined by considering both the change in magnitude and direction of the spring force.

EXAMPLE 18.8

The 10-kg homogeneous disk shown in Fig. 18–20a is attached to a uniform 5-kg rod AB. If the assembly is released from rest when $\theta = 60°$, determine the angular velocity of the rod when $\theta = 0°$. Assume that the disk rolls without slipping. Neglect friction along the guide and the mass of the collar at B.

SOLUTION

Potential Energy. Two diagrams for the rod and disk, when they are located at their initial and final positions, are shown in Fig. 18–20b. For convenience the datum passes through point A.

When the system is in position 1, only the rod's weight has positive potential energy. Thus,

$$V_1 = W_r y_1 = (49.05 \text{ N})(0.3 \sin 60° \text{ m}) = 12.74 \text{ J}$$

When the system is in position 2, both the weight of the rod and the weight of the disk have zero potential energy. Why? Thus,

$$V_2 = 0$$

Kinetic Energy. Since the entire system is at rest at the initial position,

$$T_1 = 0$$

In the final position the rod has an angular velocity $(\omega_r)_2$ and its mass center has a velocity $(v_G)_2$, Fig. 18–20c. Since the rod is *fully extended* in this position, the disk is momentarily at rest, so $(\omega_d)_2 = 0$ and $(v_A)_2 = 0$. For the rod $(v_G)_2$ can be related to $(\omega_r)_2$ from the instantaneous center of zero velocity, which is located at point A, Fig. 18–20c. Hence, $(v_G)_2 = r_{G/IC}(\omega_r)_2$ or $(v_G)_2 = 0.3(\omega_r)_2$. Thus,

$$T_2 = \frac{1}{2}m_r(v_G)_2^2 + \frac{1}{2}I_G(\omega_r)_2^2 + \frac{1}{2}m_d(v_A)_2^2 + \frac{1}{2}I_A(\omega_d)_2^2$$

$$= \frac{1}{2}(5 \text{ kg})[(0.3 \text{ m})(\omega_r)_2]^2 + \frac{1}{2}\left[\frac{1}{12}(5 \text{ kg})(0.6 \text{ m})^2\right](\omega_r)_2^2 + 0 + 0$$

$$= 0.3(\omega_r)_2^2$$

Conservation of Energy.

$$\{T_1\} + \{V_1\} = \{T_2\} + \{V_2\}$$

$$\{0\} + \{12.74 \text{ J}\} = \{0.3(\omega_r)_2^2\} + \{0\}$$

$$(\omega_r)_2 = 6.52 \text{ rad/s} \circlearrowleft \qquad \qquad Ans.$$

NOTE: We can also determine the final kinetic energy of the rod using $T_2 = \frac{1}{2}I_{IC}\omega_2^2$.

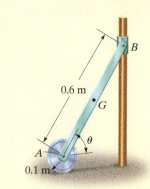

(a)

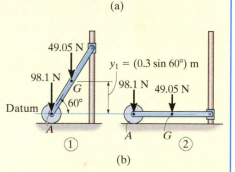

(b)

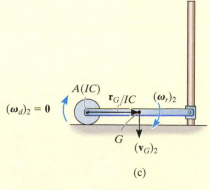

(c)

Fig. 18–20

FUNDAMENTAL PROBLEMS

F18–7. If the 30-kg disk is released from rest when $\theta = 0°$, determine its angular velocity when $\theta = 90°$.

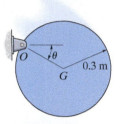

Prob. F18–7

F18–8. The 50-kg reel has a radius of gyration about its center O of $k_O = 300$ mm. If it is released from rest, determine its angular velocity when its center O has traveled 6 m down the smooth inclined plane.

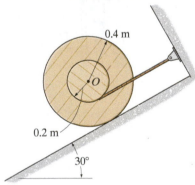

Prob. F18–8

F18–9. The 60-kg rod OA is released from rest when $\theta = 0°$. Determine its angular velocity when $\theta = 45°$. The spring remains vertical during the motion and is unstretched when $\theta = 0°$.

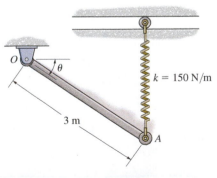

Prob. F18–9

F18–10. The 30-kg rod is released from rest when $\theta = 0°$. Determine the angular velocity of the rod when $\theta = 90°$. The spring is unstretched when $\theta = 0°$.

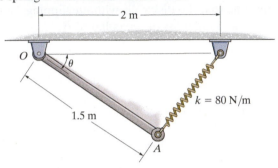

Prob. F18–10

F18–11. The 30-kg rod is released from rest when $\theta = 45°$. Determine the angular velocity of the rod when $\theta = 0°$. The spring is unstretched when $\theta = 45°$.

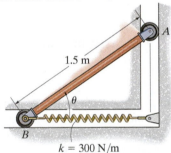

Prob. F18–11

F18–12. The 20-kg rod is released from rest when $\theta = 0°$. Determine its angular velocity when $\theta = 90°$. The spring has an unstretched length of 0.5 m.

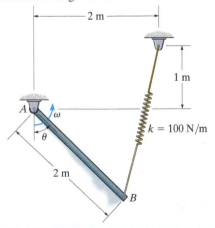

Prob. F18–12

PROBLEMS

*18–36. The assembly consists of a 3-kg pulley A and 10-kg pulley B. If a 2-kg block is suspended from the cord, determine the block's speed after it descends 0.5 m starting from rest. Neglect the mass of the cord and treat the pulleys as thin disks. No slipping occurs.

18–37. The assembly consists of a 3-kg pulley A and 10-kg pulley B. If a 2-kg block is suspended from the cord, determine the distance the block must descend, starting from rest, in order to cause B to have an angular velocity of 6 rad/s. Neglect the mass of the cord and treat the pulleys as thin disks. No slipping occurs.

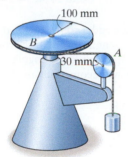

Probs. 18–36/37

18–38. The spool has a mass of 50 kg and a radius of gyration of $k_O = 0.280$ m. If the 20-kg block A is released from rest, determine the distance the block must fall in order for the spool to have an angular velocity $\omega = 5$ rad/s. Also, what is the tension in the cord while the block is in motion? Neglect the mass of the cord.

18–39. The spool has a mass of 50 kg and a radius of gyration of $k_O = 0.280$ m. If the 20-kg block A is released from rest, determine the velocity of the block when it descends 0.5 m.

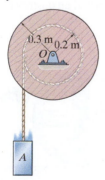

Probs. 18–38/39

*18–40. An automobile tire has a mass of 7 kg and radius of gyration of $k_G = 0.3$ m. If it is released from rest at A on the incline, determine its angular velocity when it reaches the horizontal plane. The tire rolls without slipping.

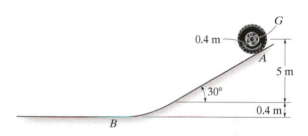

Prob. 18–40

18–41. The spool has a mass of 20 kg and a radius of gyration of $k_O = 160$ mm. If the 15-kg block A is released from rest, determine the distance the block must fall in order for the spool to have an angular velocity $\omega = 8$ rad/s. Also, what is the tension in the cord while the block is in motion? Neglect the mass of the cord.

18–42. The spool has a mass of 20 kg and a radius of gyration of $k_O = 160$ mm. If the 15-kg block A is released from rest, determine the velocity of the block when it descends 600 mm.

Probs. 18–41/42

18–43. A uniform ladder having a weight of 30 lb is released from rest when it is in the vertical position. If it is allowed to fall freely, determine the angle θ at which the bottom end A starts to slide to the right of A. For the calculation, assume the ladder to be a slender rod and neglect friction at A.

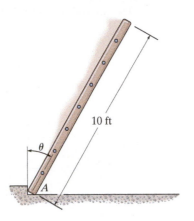

10 ft

θ

A

Prob. 18–43

***18–44.** Determine the speed of the 50-kg cylinder after it has descended a distance of 2 m, starting from rest. Gear A has a mass of 10 kg and a radius of gyration of 125 mm about its center of mass. Gear B and drum C have a combined mass of 30 kg and a radius of gyration about their center of mass of 150 mm.

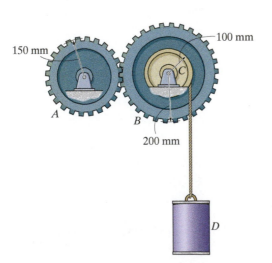

150 mm

100 mm

C

A

B

200 mm

D

Prob. 18–44

18–45. The 12-kg slender rod is attached to a spring, which has an unstretched length of 2 m. If the rod is released from rest when $\theta = 30°$, determine its angular velocity at the instant $\theta = 90°$.

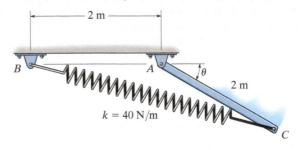

2 m

B A θ

2 m

$k = 40$ N/m

C

Prob. 18–45

18–46. The 12-kg slender rod is attached to a spring, which has an unstretched length of 2 m. If the rod is released from rest when $\theta = 30°$, determine the angular velocity of the rod the instant the spring becomes unstretched.

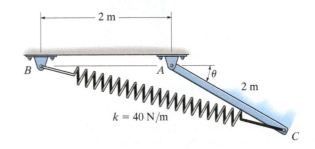

2 m

B A θ

2 m

$k = 40$ N/m

C

Prob. 18–46

18–47. The 40-kg wheel has a radius of gyration about its center of gravity G of $k_G = 250$ mm. If it rolls without slipping, determine its angular velocity when it has rotated clockwise 90° from the position shown. The spring AB has a stiffness $k = 100$ N/m and an unstretched length of 500 mm. The wheel is released from rest.

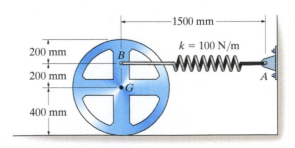

1500 mm

200 mm

200 mm

B $k = 100$ N/m

A

G

400 mm

Prob. 18–47

*18–48.   The assembly consists of two 10-kg bars which are pin connected. If the bars are released from rest when $\theta = 60°$, determine their angular velocities at the instant $\theta = 0°$. The 5-kg disk at C has a radius of 0.5 m and rolls without slipping.

18–49.   The assembly consists of two 10-kg bars which are pin connected. If the bars are released from rest when $\theta = 60°$, determine their angular velocities at the instant $\theta = 30°$. The 5-kg disk at C has a radius of 0.5 m and rolls without slipping.

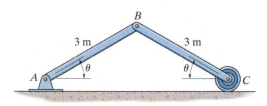

Prob. 18–48/49

18–50.   The compound disk pulley consists of a hub and attached outer rim. If it has a mass of 3 kg and a radius of gyration of $k_G = 45$ mm, determine the speed of block A after A descends 0.2 m from rest. Blocks A and B each have a mass of 2 kg. Neglect the mass of the cords.

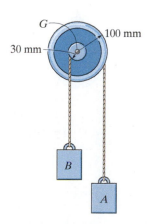

Prob. 18–50

18–51.   The uniform garage door has a mass of 150 kg and is guided along smooth tracks at its ends. Lifting is done using the two springs, each of which is attached to the anchor bracket at A and to the counterbalance shaft at B and C. As the door is raised, the springs begin to unwind from the shaft, thereby assisting the lift. If each spring provides a torsional moment of $M = (0.7\theta)$ N $\cdot$ m, where θ is in radians, determine the angle θ_0 at which both the left-wound and right-wound spring should be attached so that the door is completely balanced by the springs, i.e., when the door is in the vertical position and is given a slight force upward, the springs will lift the door along the side tracks to the horizontal plane with no final angular velocity. *Note*: The elastic potential energy of a torsional spring is $V_e = \frac{1}{2}k\theta^2$, where $M = k\theta$ and in this case $k = 0.7$ N $\cdot$ m/rad.

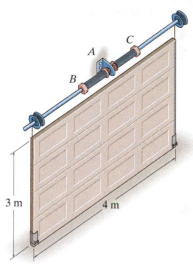

Prob. 18–51

*18–52.   The two 12-kg slender rods are pin connected and released from rest at the position $\theta = 60°$. If the spring has an unstretched length of 1.5 m, determine the angular velocity of rod BC, when the system is at the position $\theta = 0°$. Neglect the mass of the roller at C.

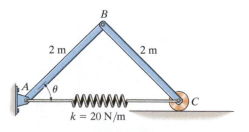

Prob. 18–52

18–53. The two 12-kg slender rods are pin connected and released from rest at the position $\theta = 60°$. If the spring has an unstretched length of 1.5 m, determine the angular velocity of rod BC, when the system is at the position $\theta = 30°$.

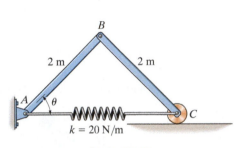

Prob. 18–53

18–54. If the 250-lb block is released from rest when the spring is unstretched, determine the velocity of the block after it has descended 5 ft. The drum has a weight of 50 lb and a radius of gyration of $k_O = 0.5$ ft about its center of mass O.

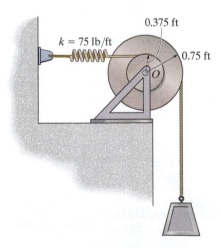

Prob. 18–54

18–55. The slender 15-kg bar is initially at rest and standing in the vertical position when the bottom end A is displaced slightly to the right. If the track in which it moves is smooth, determine the speed at which end A strikes the corner D. The bar is constrained to move in the vertical plane. Neglect the mass of the cord BC.

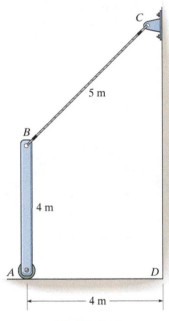

Prob. 18–55

*18–56. If the chain is released from rest from the position shown, determine the angular velocity of the pulley after the end B has risen 2 ft. The pulley has a weight of 50 lb and a radius of gyration of 0.375 ft about its axis. The chain weighs 6 lb/ft.

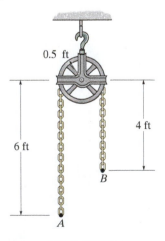

Prob. 18–56

18–57. If the gear is released from rest, determine its angular velocity after its center of gravity O has descended a distance of 4 ft. The gear has a weight of 100 lb and a radius of gyration about its center of gravity of $k = 0.75$ ft.

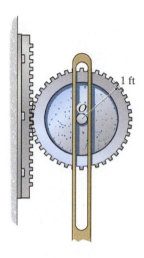

Prob. 18–57

18–58. The slender 6-kg bar AB is horizontal and at rest and the spring is unstretched. Determine the stiffness k of the spring so that the motion of the bar is momentarily stopped when it has rotated clockwise 90° after being released.

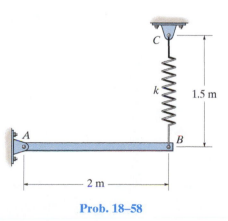

Prob. 18–58

18–59. The slender 6-kg bar AB is horizontal and at rest and the spring is unstretched. Determine the angular velocity of the bar when it has rotated clockwise 45° after being released. The spring has a stiffness of $k = 12$ N/m.

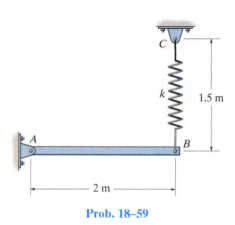

Prob. 18–59

*18–60. The pendulum consists of a 6-kg slender rod fixed to a 15-kg disk. If the spring has an unstretched length of 0.2 m, determine the angular velocity of the pendulum when it is released from rest and rotates clockwise 90° from the position shown. The roller at C allows the spring to always remain vertical.

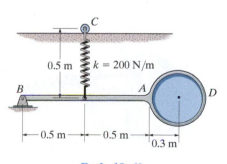

Prob. 18–60

18

18–61. The 500-g rod *AB* rests along the smooth inner surface of a hemispherical bowl. If the rod is released from rest from the position shown, determine its angular velocity at the instant it swings downward and becomes horizontal.

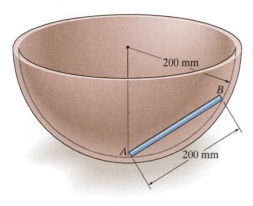

Prob. 18–61

18–62. The 50-lb wheel has a radius of gyration about its center of gravity *G* of $k_G = 0.7$ ft. If it rolls without slipping, determine its angular velocity when it has rotated clockwise 90° from the position shown. The spring *AB* has a stiffness $k = 1.20$ lb/ft and an unstretched length of 0.5 ft. The wheel is released from rest.

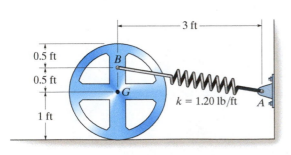

Prob. 18–62

18–63. The system consists of 60-lb and 20-lb blocks *A* and *B*, respectively, and 5-lb pulleys *C* and *D* that can be treated as thin disks. Determine the speed of block *A* after block *B* has risen 5 ft, starting from rest. Assume that the cord does not slip on the pulleys, and neglect the mass of the cord.

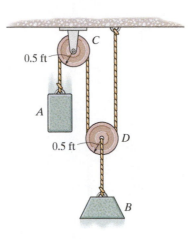

Prob. 18–63

*18–64. The door is made from one piece, whose ends move along the horizontal and vertical tracks. If the door is in the open position, $\theta = 0°$, and then released, determine the speed at which its end *A* strikes the stop at *C*. Assume the door is a 180-lb thin plate having a width of 10 ft.

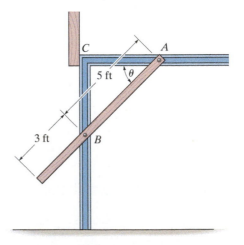

Prob. 18–64

18–65. The door is made from one piece, whose ends move along the horizontal and vertical tracks. If the door is in the open position, $\theta = 0°$, and then released, determine its angular velocity at the instant $\theta = 30°$. Assume the door is a 180-lb thin plate having a width of 10 ft.

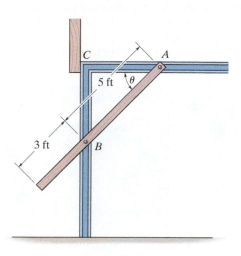

Prob. 18–65

18–66. The end A of the garage door AB travels along the horizontal track, and the end of member BC is attached to a spring at C. If the spring is originally unstretched, determine the stiffness k so that when the door falls downward from rest in the position shown, it will have zero angular velocity the moment it closes, i.e., when it and BC become vertical. Neglect the mass of member BC and assume the door is a thin plate having a weight of 200 lb and a width and height of 12 ft. There is a similar connection and spring on the other side of the door.

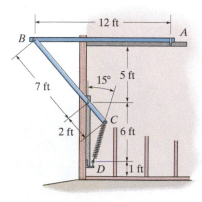

Prob. 18–66

18–67. The system consists of a 30-kg disk, 12-kg slender rod BA, and a 5-kg smooth collar A. If the disk rolls without slipping, determine the velocity of the collar at the instant $\theta = 0°$. The system is released from rest when $\theta = 45°$.

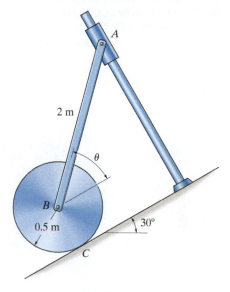

Prob. 18–67

***18–68.** The system consists of a 30-kg disk A, 12-kg slender rod BA, and a 5-kg smooth collar A. If the disk rolls without slipping, determine the velocity of the collar at the instant $\theta = 30°$. The system is released from rest when $\theta = 45°$.

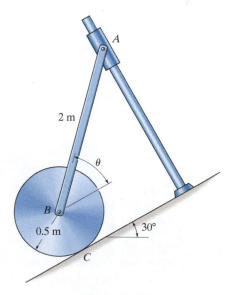

Prob. 18–68

18

CONCEPTUAL PROBLEMS

C18–1. The bicycle and rider start from rest at the top of the hill. Show how to determine the speed of the rider when he freely coasts down the hill. Use appropriate dimensions of the wheels, and the mass of the rider, frame and wheels of the bicycle to explain your results.

Prob. C18–1 (© R.C. Hibbeler)

C18–2. Two torsional springs, $M = k\theta$, are used to assist in opening and closing the hood of this truck. Assuming the springs are uncoiled ($\theta = 0°$) when the hood is opened, determine the stiffness k (N·m/rad) of each spring so that the hood can easily be lifted, i.e., practically no force applied to it, when it is closed in the unlocked position. Use appropriate numerical values to explain your result.

Prob. C18–2 (© R.C. Hibbeler)

C18–3. The operation of this garage door is assisted using two springs AB and side members BCD, which are pinned at C. Assuming the springs are unstretched when the door is in the horizontal (open) position and $ABCD$ is vertical, determine each spring stiffness k so that when the door falls to the vertical (closed) position, it will slowly come to a stop. Use appropriate numerical values to explain your result.

Prob. C18–3 (© R.C. Hibbeler)

C18–4. Determine the counterweight of A needed to balance the weight of the bridge deck when $\theta = 0°$. Show that this weight will maintain equilibrium of the deck by considering the potential energy of the system when the deck is in the arbitrary position θ. Both the deck and AB are horizontal when $\theta = 0°$. Neglect the weights of the other members. Use appropriate numerical values to explain this result.

Prob. C18–4 (© R.C. Hibbeler)

CHAPTER REVIEW

Kinetic Energy

The kinetic energy of a rigid body that undergoes planar motion can be referenced to its mass center. It includes a scalar sum of its translational and rotational kinetic energies.

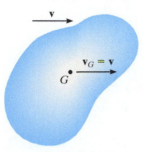

Translation

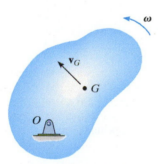

Rotation About a Fixed Axis

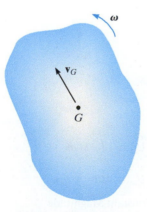

General Plane Motion

Translation

$$T = \tfrac{1}{2}mv_G^2$$

Rotation About a Fixed Axis

$$T = \tfrac{1}{2}mv_G^2 + \tfrac{1}{2}I_G\omega^2$$

or

$$T = \tfrac{1}{2}I_O\omega^2$$

General Plane Motion

$$T = \tfrac{1}{2}mv_G^2 + \tfrac{1}{2}I_G\omega^2$$

or

$$T = \tfrac{1}{2}I_{IC}\omega^2$$

18

Work of a Force and a Couple Moment

A force does work when it undergoes a displacement ds in the direction of the force. In particular, the frictional and normal forces that act on a cylinder or any circular body that rolls *without slipping* will do no work, since the normal force does not undergo a displacement and the frictional force acts on successive points on the surface of the body.

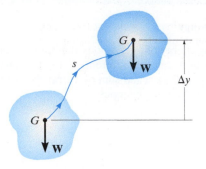

$$U_W = -W\Delta y$$

Weight

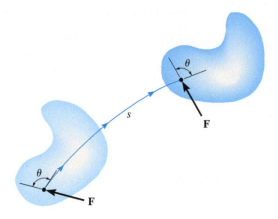

$$U_F = \int F \cos \theta \, ds$$

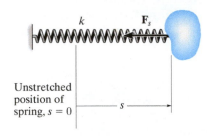

$$U = -\frac{1}{2} k s^2$$

Spring

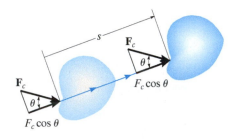

$$U_{F_c} = (F_c \cos \theta)s$$

Constant Force

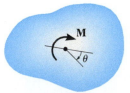

$$U_M = \int_{\theta_1}^{\theta_2} M \, d\theta$$

$$U_M = M(\theta_2 - \theta_1)$$

Constant Magnitude

Principle of Work and Energy

Problems that involve velocity, force, and displacement can be solved using the principle of work and energy. The kinetic energy is the sum of both its rotational and translational parts. For application, a free-body diagram should be drawn in order to account for the work of all of the forces and couple moments that act on the body as it moves along the path.

$$T_1 + \Sigma U_{1-2} = T_2$$

Conservation of Energy

If a rigid body is subjected only to conservative forces, then the conservation-of-energy equation can be used to solve the problem. This equation requires that the sum of the potential and kinetic energies of the body remain the same at any two points along the path.

$$T_1 + V_1 = T_2 + V_2$$

$$\text{where } V = V_g + V_e$$

The potential energy is the sum of the body's gravitational and elastic potential energies. The gravitational potential energy will be positive if the body's center of gravity is located above a datum. If it is below the datum, then it will be negative. The elastic potential energy is always positive, regardless if the spring is stretched or compressed.

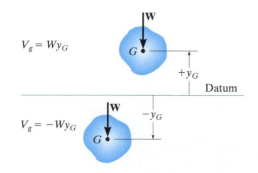

Gravitational potential energy

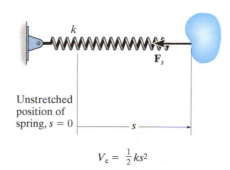

$$V_e = \tfrac{1}{2}ks^2$$

Elastic potential energy

REVIEW PROBLEMS

R18–1. The pendulum of the Charpy impact machine has a mass of 50 kg and a radius of gyration of $k_A = 1.75$ m. If it is released from rest when $\theta = 0°$, determine its angular velocity just before it strikes the specimen S, $\theta = 90°$

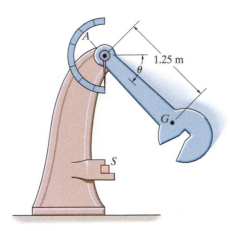

Prob. R18–1

R18–2. The 50-kg flywheel has a radius of gyration of $k_O = 200$ mm about its center of mass. If it is subjected to a torque of $M = (9\theta^{1/2} + 1)$ N·m, where θ is in radians, determine its angular velocity when it has rotated 5 revolutions, starting from rest.

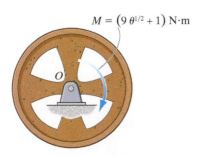

$$M = \left(9\,\theta^{1/2} + 1\right) \text{N·m}$$

Prob. R18–2

R18–3. The drum has a mass of 50 kg and a radius of gyration about the pin at O of $k_O = 0.23$ m. Starting from rest, the suspended 15-kg block B is allowed to fall 3 m without applying the brake ACD. Determine the speed of the block at this instant. If the coefficient of kinetic friction at the brake pad C is $\mu_k = 0.5$, determine the force **P** that must be applied at the brake handle which will then stop the block after it descends *another* 3 m. Neglect the thickness of the handle.

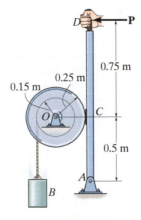

Prob. R18–3

R18–4. The spool has a mass of 60 kg and a radius of gyration of $k_G = 0.3$ m. If it is released from rest, determine how far its center descends down the smooth plane before it attains an angular velocity of $\omega = 6$ rad/s. Neglect the mass of the cord which is wound around the central core. The coefficient of kinetic friction between the spool and plane at A is $\mu_k = 0.2$.

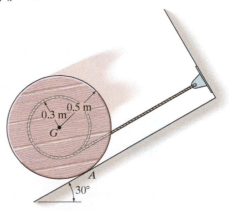

Prob. R18–4

18

R18–5. The gear rack has a mass of 6 kg, and the gears each have a mass of 4 kg and a radius of gyration of $k = 30$ mm at their centers. If the rack is originally moving downward at 2 m/s, when $s = 0$, determine the speed of the rack when $s = 600$ mm. The gears are free to turn about their centers A and B.

R18–7. The system consists of a 20-lb disk A, 4-lb slender rod BC, and a 1-lb smooth collar C. If the disk rolls without slipping, determine the velocity of the collar at the instant the rod becomes horizontal, i.e., $\theta = 0°$. The system is released from rest when $\theta = 45°$.

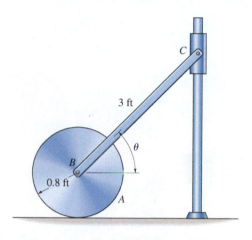

Prob. R18–7

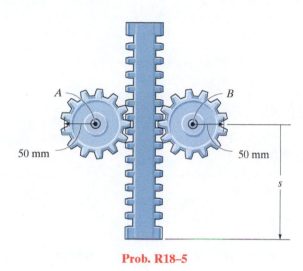

Prob. R18–5

R18-6. At the instant shown, the 50-lb bar rotates clockwise at 2 rad/s. The spring attached to its end always remains vertical due to the roller guide at C. If the spring has an unstretched length of 2 ft and a stiffness of $k = 6$ lb/ft, determine the angular velocity of the bar the instant it has rotated 30° clockwise.

R18–8. At the instant the spring becomes undeformed, the center of the 40-kg disk has a speed of 4 m/s. From this point determine the distance d the disk moves down the plane before momentarily stopping. The disk rolls without slipping.

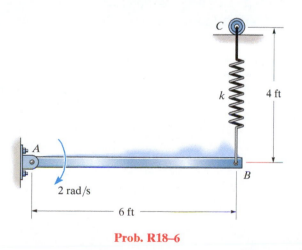

Prob. R18–6

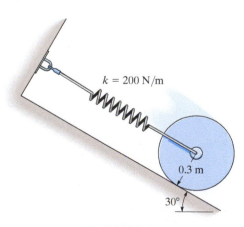

Prob. R18–8

Chapter 19

(© Hellen Sergeyeva/Fotolia)

The impulse that this tugboat imparts to this ship will cause it to turn in a manner that can be predicted by applying the principles of impulse and momentum.

Planar Kinetics of a Rigid Body: Impulse and Momentum

CHAPTER OBJECTIVES

- To develop formulations for the linear and angular momentum of a body.

- To apply the principles of linear and angular impulse and momentum to solve rigid-body planar kinetic problems that involve force, velocity, and time.

- To discuss application of the conservation of momentum.

- To analyze the mechanics of eccentric impact.

19.1 Linear and Angular Momentum

In this chapter we will use the principles of linear and angular impulse and momentum to solve problems involving force, velocity, and time as related to the planar motion of a rigid body. Before doing this, we will first formalize the methods for obtaining a body's linear and angular momentum, assuming the body is symmetric with respect to an inertial x–y reference plane.

Linear Momentum. The linear momentum of a rigid body is determined by summing vectorially the linear momenta of all the particles of the body, i.e., $\mathbf{L} = \Sigma m_i \mathbf{v}_i$. Since $\Sigma m_i \mathbf{v}_i = m\mathbf{v}_G$ (see Sec. 15.2) we can also write

$$\mathbf{L} = m\mathbf{v}_G \tag{19–1}$$

This equation states that the body's linear momentum is a vector quantity having a *magnitude* mv_G, which is commonly measured in units of $kg \cdot m/s$ or $slug \cdot ft/s$ and a *direction* defined by $\mathbf{v}_G$ the velocity of the body's mass center.

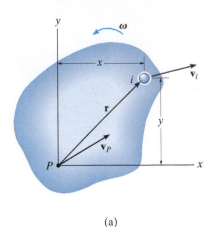

(a)

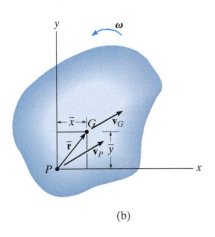

(b)

Fig. 19–1

Angular Momentum.

Consider the body in Fig. 19–1a, which is subjected to general plane motion. At the instant shown, the arbitrary point P has a known velocity $\mathbf{v}_P$, and the body has an angular velocity $\boldsymbol{\omega}$. Therefore the velocity of the ith particle of the body is

$$\mathbf{v}_i = \mathbf{v}_P + \mathbf{v}_{i/P} = \mathbf{v}_P + \boldsymbol{\omega} \times \mathbf{r}$$

The angular momentum of this particle about point P is equal to the "moment" of the particle's linear momentum about P, Fig. 19–1a. Thus,

$$(\mathbf{H}_P)_i = \mathbf{r} \times m_i \mathbf{v}_i$$

Expressing $\mathbf{v}_i$ in terms of $\mathbf{v}_P$ and using Cartesian vectors, we have

$$(H_P)_i \mathbf{k} = m_i(x\mathbf{i} + y\mathbf{j}) \times [(v_P)_x\mathbf{i} + (v_P)_y\mathbf{j} + \omega\mathbf{k} \times (x\mathbf{i} + y\mathbf{j})]$$
$$(H_P)_i = -m_i y(v_P)_x + m_i x(v_P)_y + m_i \omega r^2$$

Letting $m_i \rightarrow dm$ and integrating over the entire mass m of the body, we obtain

$$H_P = -\left(\int_m y\,dm\right)(v_P)_x + \left(\int_m x\,dm\right)(v_P)_y + \left(\int_m r^2\,dm\right)\omega$$

Here H_P represents the angular momentum of the body about an axis (the z axis) perpendicular to the plane of motion that passes through point P. Since $\bar{y}m = \int y\,dm$ and $\bar{x}m = \int x\,dm$, the integrals for the first and second terms on the right are used to locate the body's center of mass G with respect to P, Fig. 19–1b. Also, the last integral represents the body's moment of inertia about point P. Thus,

$$H_P = -\bar{y}m(v_P)_x + \bar{x}m(v_P)_y + I_P\omega \qquad (19\text{–}2)$$

This equation reduces to a simpler form if P coincides with the mass center G for the body,* in which case $\bar{x} = \bar{y} = 0$. Hence,

$$\boxed{H_G = I_G\omega} \qquad (19\text{–}3)$$

---

*It also reduces to the same simple form, $H_P = I_P\omega$, if point P is a *fixed point* (see Eq. 19–9) or the velocity of P is directed along the line PG.

Here the angular momentum of the body about G is equal to the product of the moment of inertia of the body about an axis passing through G and the body's angular velocity. Realize that $\mathbf{H}_G$ is a vector quantity having a *magnitude* $I_G\omega$, which is commonly measured in units of $kg \cdot m^2/s$ or $slug \cdot ft^2/s$, and a *direction* defined by $\boldsymbol{\omega}$, which is always perpendicular to the plane of motion.

Equation 19–2 can also be rewritten in terms of the x and y components of the velocity of the body's mass center, $(\mathbf{v}_G)_x$ and $(\mathbf{v}_G)_y$, and the body's moment of inertia I_G. Since G is located at coordinates $(\bar{x}, \bar{y})$, then by the parallel-axis theorem, $I_P = I_G + m(\bar{x}^2 + \bar{y}^2)$. Substituting into Eq. 19–2 and rearranging terms, we have

$$H_P = \bar{y}m[-(v_P)_x + \bar{y}\omega] + \bar{x}m[(v_P)_y + \bar{x}\omega] + I_G\omega \qquad (19\text{–}4)$$

From the kinematic diagram of Fig. 19–1b, $\mathbf{v}_G$ can be expressed in terms of $\mathbf{v}_P$ as

$$\mathbf{v}_G = \mathbf{v}_P + \boldsymbol{\omega} \times \bar{\mathbf{r}}$$

$$(v_G)_x\mathbf{i} + (v_G)_y\mathbf{j} = (v_P)_x\mathbf{i} + (v_P)_y\mathbf{j} + \omega\mathbf{k} \times (\bar{x}\mathbf{i} + \bar{y}\mathbf{j})$$

Carrying out the cross product and equating the respective $\mathbf{i}$ and $\mathbf{j}$ components yields the two scalar equations

$$(v_G)_x = (v_P)_x - \bar{y}\omega$$

$$(v_G)_y = (v_P)_y + \bar{x}\omega$$

Substituting these results into Eq. 19–4 yields

$$(\zeta +)H_P = -\bar{y}m(v_G)_x + \bar{x}m(v_G)_y + I_G\omega \qquad (19\text{–}5)$$

As shown in Fig. 19–1c, *this result indicates that when the angular momentum of the body is computed about point P, it is equivalent to the moment of the linear momentum $m\mathbf{v}_G$, or its components $m(\mathbf{v}_G)_x$ and $m(\mathbf{v}_G)_y$, about P plus the angular momentum $I_G\boldsymbol{\omega}$.* Using these results, we will now consider three types of motion.

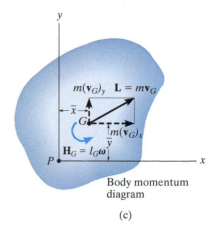

Body momentum diagram

(c)

Fig. 19–1

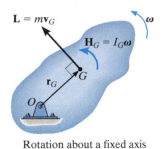

Translation

(a)

Translation.   When a rigid body is subjected to either rectilinear or curvilinear *translation*, Fig. 19–2a, then $\boldsymbol{\omega} = \mathbf{0}$ and its mass center has a velocity of $\mathbf{v}_G = \mathbf{v}$. Hence, the linear momentum, and the angular momentum about G, become

$$\boxed{\begin{aligned} L &= mv_G \\ H_G &= 0 \end{aligned}} \tag{19–6}$$

If the angular momentum is computed about some other point A, the "moment" of the linear momentum $\mathbf{L}$ must be found about the point. Since d is the "moment arm" as shown in Fig. 19–2a, then in accordance with Eq. 19–5, $H_A = (d)(mv_G) \circlearrowright$.

Rotation About a Fixed Axis.   When a rigid body is *rotating about a fixed axis*, Fig. 19–2b, the linear momentum, and the angular momentum about G, are

$$\boxed{\begin{aligned} L &= mv_G \\ H_G &= I_G\omega \end{aligned}} \tag{19–7}$$

Rotation about a fixed axis

(b)

Fig. 19–2

It is sometimes convenient to compute the angular momentum about point O. Noting that $\mathbf{L}$ (or $\mathbf{v}_G$) is always *perpendicular to* $\mathbf{r}_G$, we have

$$(\circlearrowleft +) \quad H_O = I_G\omega + r_G(mv_G) \tag{19–8}$$

Since $v_G = r_G\omega$, this equation can be written as $H_O = (I_G + mr_G^2)\omega$. Using the parallel-axis theorem,*

$$\boxed{H_O = I_O\omega} \tag{19–9}$$

For the calculation, then, either Eq. 19–8 or 19–9 can be used.

---

*The similarity between this derivation and that of Eq. 17–16 $(\Sigma M_O = I_O\alpha)$ and Eq. 18–5 $\left(T = \frac{1}{2}I_O\omega^2\right)$ should be noted. Also note that the same result can be obtained from Eq. 19–2 by selecting point P at O, realizing that $(v_O)_x = (v_O)_y = 0$.

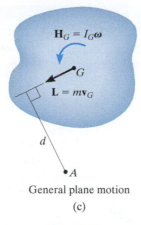

General plane motion

(c)

Fig. 19–2

General Plane Motion. When a rigid body is subjected to general plane motion, Fig. 19–2c, the linear momentum, and the angular momentum about G, become

$$L = mv_G$$
$$H_G = I_G\omega \qquad \text{(19–10)}$$

If the angular momentum is computed about point A, Fig. 19–2c, it is necessary to include the moment of $\mathbf{L}$ and $\mathbf{H}_G$ about this point. In this case,

$$(\curvearrowleft+)\quad H_A = I_G\omega + (d)(mv_G)$$

Here d is the moment arm, as shown in the figure.

   As a special case, if point A is the instantaneous center of zero velocity then, like Eq. 19–9, we can write the above equation in simplified form as

$$H_{IC} = I_{IC}\omega \qquad \text{(19–11)}$$

where I_{IC} is the moment of inertia of the body about the IC. (See Prob. 19–2.)

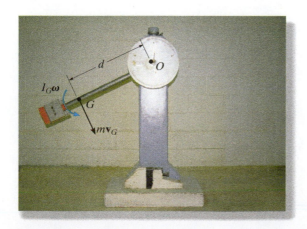

As the pendulum swings downward, its angular momentum about point O can be determined by computing the moment of $I_G\boldsymbol{\omega}$ and $m\mathbf{v}_G$ about O. This is $H_O = I_G\omega + (mv_G)d$. Since $v_G = \omega d$, then $H_O = I_G\omega + m(\omega d)d = (I_G + md^2)\omega = I_O\omega$.
(© R.C. Hibbeler)

EXAMPLE | 19.1

At a given instant the 5-kg slender bar has the motion shown in Fig. 19–3a. Determine its angular momentum about point G and about the IC at this instant.

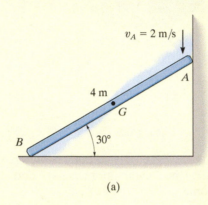

(a)

SOLUTION

Bar.   The bar undergoes *general plane motion*. The IC is established in Fig. 19–3b, so that

$$\omega = \frac{2 \text{ m/s}}{4 \text{ m cos } 30°} = 0.5774 \text{ rad/s}$$

$$v_G = (0.5774 \text{ rad/s})(2 \text{ m}) = 1.155 \text{ m/s}$$

Thus,

$$(\zeta +) H_G = I_G \omega = \left[\tfrac{1}{12}(5 \text{ kg})(4 \text{ m})^2 \right](0.5774 \text{ rad/s}) = 3.85 \text{ kg} \cdot \text{m}^2/\text{s} \, \zeta \quad Ans.$$

Adding $I_G\omega$ and the moment of mv_G about the IC yields

$$(\zeta +) H_{IC} = I_G \omega + d(mv_G)$$

$$= \left[\tfrac{1}{12}(5 \text{ kg})(4 \text{ m})^2 \right](0.5774 \text{ rad/s}) + (2 \text{ m})(5 \text{ kg})(1.155 \text{ m/s})$$

$$= 15.4 \text{ kg} \cdot \text{m}^2/\text{s} \, \zeta \quad\quad\quad Ans.$$

We can also use

$$(\zeta +) H_{IC} = I_{IC}\omega$$

$$= \left[\tfrac{1}{12}(5 \text{ kg})(4 \text{ m})^2 + (5 \text{ kg})(2 \text{ m})^2 \right](0.5774 \text{ rad/s})$$

$$= 15.4 \text{ kg} \cdot \text{m}^2/\text{s} \, \zeta \quad\quad\quad Ans.$$

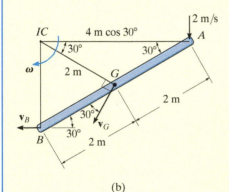

(b)

Fig. 19–3

19

19.2   Principle of Impulse and Momentum

Like the case for particle motion, the principle of impulse and momentum for a rigid body can be developed by *combining* the equation of motion with kinematics. The resulting equation will yield a *direct solution to problems involving force, velocity, and time.*

Principle of Linear Impulse and Momentum.

The equation of translational motion for a rigid body can be written as $\Sigma \mathbf{F} = m\mathbf{a}_G = m(d\mathbf{v}_G/dt)$. Since the mass of the body is constant,

$$\Sigma \mathbf{F} = \frac{d}{dt}(m\mathbf{v}_G)$$

Multiplying both sides by dt and integrating from $t = t_1$, $\mathbf{v}_G = (\mathbf{v}_G)_1$ to $t = t_2$, $\mathbf{v}_G = (\mathbf{v}_G)_2$ yields

$$\Sigma \int_{t_1}^{t_2} \mathbf{F}\, dt = m(\mathbf{v}_G)_2 - m(\mathbf{v}_G)_1$$

This equation is referred to as the *principle of linear impulse and momentum*. It states that the sum of all the impulses created by the *external force system* which acts on the body during the time interval t_1 to t_2 is equal to the change in the linear momentum of the body during this time interval, Fig. 19–4.

Principle of Angular Impulse and Momentum.

If the body has *general plane motion* then $\Sigma M_G = I_G \alpha = I_G(d\omega/dt)$. Since the moment of inertia is constant,

$$\Sigma M_G = \frac{d}{dt}(I_G \omega)$$

Multiplying both sides by dt and integrating from $t = t_1$, $\omega = \omega_1$ to $t = t_2$, $\omega = \omega_2$ gives

$$\Sigma \int_{t_1}^{t_2} M_G\, dt = I_G \omega_2 - I_G \omega_1 \qquad (19\text{–}12)$$

In a similar manner, for *rotation about a fixed axis* passing through point O, Eq. 17–16 ($\Sigma M_O = I_O \alpha$) when integrated becomes

$$\Sigma \int_{t_1}^{t_2} M_O\, dt = I_O \omega_2 - I_O \omega_1 \qquad (19\text{–}13)$$

Equations 19–12 and 19–13 are referred to as the *principle of angular impulse and momentum.* Both equations state that the sum of the angular impulses acting on the body during the time interval t_1 to t_2 is equal to the change in the body's angular momentum during this time interval.

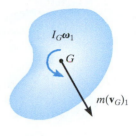

Initial momentum diagram

(a)

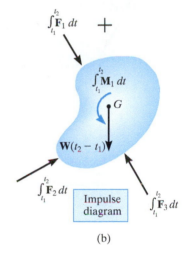

Impulse diagram

(b)

$\|$

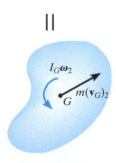

Final momentum diagram

(c)

Fig. 19–4

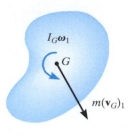

Initial momentum diagram

(a)

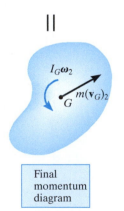

(b)

To summarize these concepts, if motion occurs in the *x*–*y* plane, the following *three scalar equations* can be written to describe the *planar motion* of the body.

$$m(v_{Gx})_1 + \Sigma \int_{t_1}^{t_2} F_x \, dt = m(v_{Gx})_2$$

$$m(v_{Gy})_1 + \Sigma \int_{t_1}^{t_2} F_y \, dt = m(v_{Gy})_2 \qquad (19\text{--}14)$$

$$I_G\omega_1 + \Sigma \int_{t_1}^{t_2} M_G \, dt = I_G\omega_2$$

The terms in these equations can be shown graphically by drawing a set of impulse and momentum diagrams for the body, Fig. 19–4. Note that the linear momentum $m\mathbf{v}_G$ is applied at the body's mass center, Figs. 19–4*a* and 19–4*c*; whereas the angular momentum $I_G\boldsymbol{\omega}$ is a free vector, and therefore, like a couple moment, it can be applied at any point on the body. When the impulse diagram is constructed, Fig. 19–4*b*, the forces **F** and moment **M** vary with time, and are indicated by the integrals. However, if **F** and **M** are *constant* integration of the impulses yields $\mathbf{F}(t_2 - t_1)$ and $\mathbf{M}(t_2 - t_1)$, respectively. Such is the case for the body's weight **W**, Fig. 19–4*b*.

Equations 19–14 can also be applied to an entire system of connected bodies rather than to each body separately. This eliminates the need to include interaction impulses which occur at the connections since they are *internal* to the system. The resultant equations may be written in symbolic form as

$$\left(\Sigma \frac{\text{syst. linear}}{\text{momentum}}\right)_{x1} + \left(\Sigma \frac{\text{syst. linear}}{\text{impulse}}\right)_{x(1-2)} = \left(\Sigma \frac{\text{syst. linear}}{\text{momentum}}\right)_{x2}$$

$$\left(\Sigma \frac{\text{syst. linear}}{\text{momentum}}\right)_{y1} + \left(\Sigma \frac{\text{syst. linear}}{\text{impulse}}\right)_{y(1-2)} = \left(\Sigma \frac{\text{syst. linear}}{\text{momentum}}\right)_{y2}$$

$$\left(\Sigma \frac{\text{syst. angular}}{\text{momentum}}\right)_{O1} + \left(\Sigma \frac{\text{syst. angular}}{\text{impulse}}\right)_{O(1-2)} = \left(\Sigma \frac{\text{syst. angular}}{\text{momentum}}\right)_{O2}$$

(19–15)

As indicated by the third equation, the system's angular momentum and angular impulse must be computed with respect to the *same reference point O* for all the bodies of the system.

||

Final momentum diagram

(c)

Fig. 19–4 (repeated)

19

Procedure For Analysis

Impulse and momentum principles are used to solve kinetic problems that involve *velocity, force*, and *time* since these terms are involved in the formulation.

Free-Body Diagram.

- Establish the *x, y, z* inertial frame of reference and draw the free-body diagram in order to account for all the forces and couple moments that produce impulses on the body.

- The direction and sense of the initial and final velocity of the body's mass center, $\mathbf{v}_G$, and the body's angular velocity $\boldsymbol{\omega}$ should be established. If any of these motions is unknown, assume that the sense of its components is in the direction of the positive inertial coordinates.

- Compute the moment of inertia I_G or I_O.

- As an alternative procedure, draw the impulse and momentum diagrams for the body or system of bodies. Each of these diagrams represents an outlined shape of the body which graphically accounts for the data required for each of the three terms in Eqs. 19–14 or 19–15, Fig. 19–4. These diagrams are particularly helpful in order to visualize the "moment" terms used in the principle of angular impulse and momentum, if application is about the *IC* or another point other than the body's mass center *G* or a fixed point *O*.

Principle of Impulse and Momentum.

- Apply the three scalar equations of impulse and momentum.

- The angular momentum of a rigid body rotating about a fixed axis is the moment of mv_G plus $I_G \boldsymbol{\omega}$ about the axis. This is equal to $H_O = I_O \omega$, where I_O is the moment of inertia of the body about the axis.

- All the forces acting on the body's free-body diagram will create an impulse; however, some of these forces will do no work.

- Forces that are functions of time must be integrated to obtain the impulse.

- The principle of angular impulse and momentum is often used to eliminate unknown impulsive forces that are parallel or pass through a common axis, since the moment of these forces is zero about this axis.

Kinematics.

- If more than three equations are needed for a complete solution, it may be possible to relate the velocity of the body's mass center to the body's angular velocity using *kinematics*. If the motion appears to be complicated, kinematic (velocity) diagrams may be helpful in obtaining the necessary relation.

19

EXAMPLE 19.2

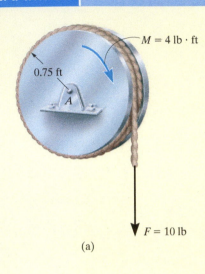

$M = 4$ lb·ft

0.75 ft

A

$F = 10$ lb

(a)

The 20-lb disk shown in Fig. 19–5a is acted upon by a constant couple moment of 4 lb·ft and a force of 10 lb which is applied to a cord wrapped around its periphery. Determine the angular velocity of the disk two seconds after starting from rest. Also, what are the force components of reaction at the pin?

SOLUTION

Since angular velocity, force, and time are involved in the problems, we will apply the principles of impulse and momentum to the solution.

Free-Body Diagram.   Fig. 19–5b. The disk's mass center does not move; however, the loading causes the disk to rotate clockwise. The moment of inertia of the disk about its fixed axis of rotation is

$$I_A = \frac{1}{2}mr^2 = \frac{1}{2}\left(\frac{20\ \text{lb}}{32.2\ \text{ft/s}^2}\right)(0.75\ \text{ft})^2 = 0.1747\ \text{slug}\cdot\text{ft}^2$$

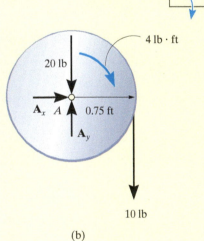

(b)

Fig. 19–5

Principle of Impulse and Momentum.

$(\xrightarrow{+})$

$$m(v_{Ax})_1 + \Sigma \int_{t_1}^{t_2} F_x\, dt = m(v_{Ax})_2$$

$$0 + A_x(2\ \text{s}) = 0$$

$(+\uparrow)$

$$m(v_{Ay})_1 + \Sigma \int_{t_1}^{t_2} F_y\, dt = m(v_{Ay})_2$$

$$0 + A_y(2\ \text{s}) - 20\ \text{lb}(2\ \text{s}) - 10\ \text{lb}(2\ \text{s}) = 0$$

$(\zeta+)$

$$I_A\omega_1 + \Sigma \int_{t_1}^{t_2} M_A\, dt = I_A\omega_2$$

$$0 + 4\ \text{lb}\cdot\text{ft}(2\ \text{s}) + [10\ \text{lb}(2\ \text{s})](0.75\ \text{ft}) = 0.1747\omega_2$$

Solving these equations yields

$$A_x = 0 \hspace{3cm} Ans.$$

$$A_y = 30\ \text{lb} \hspace{2.5cm} Ans.$$

$$\omega_2 = 132\ \text{rad/s} \,\zeta \hspace{2cm} Ans.$$

EXAMPLE 19.3

The 100-kg spool shown in Fig. 19–6a has a radius of gyration $k_G = 0.35$ m. A cable is wrapped around the central hub of the spool, and a horizontal force having a variable magnitude of $P = (t + 10)$ N is applied, where t is in seconds. If the spool is initially at rest, determine its angular velocity in 5 s. Assume that the spool rolls without slipping at A.

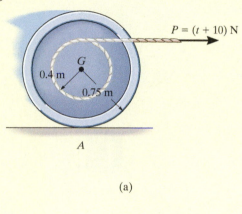

(a)

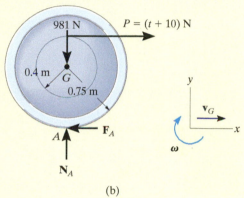

(b)

Fig. 19–6

SOLUTION

Free-Body Diagram.   From the free-body diagram, Fig. 19–6b, the *variable* force **P** will cause the friction force **F**$_A$ to be variable, and thus the impulses created by both **P** and **F**$_A$ must be determined by integration. Force **P** causes the mass center to have a velocity **v**$_G$ to the right, and so the spool has a clockwise angular velocity $\boldsymbol{\omega}$.

Principle of Impulse and Momentum.   A direct solution for $\boldsymbol{\omega}$ can be obtained by applying the principle of angular impulse and momentum about point A, the IC, in order to eliminate the unknown friction impulse.

$$(\circlearrowleft +) \qquad\qquad I_A\omega_1 + \Sigma \int M_A \, dt = I_A\omega_2$$

$$0 + \left[\int_0^{5\,\text{s}} (t + 10) \,\text{N}\, dt\right](0.75 \text{ m} + 0.4 \text{ m}) = [100 \text{ kg } (0.35 \text{ m})^2 + (100 \text{ kg})(0.75 \text{ m})^2]\omega_2$$

$$62.5(1.15) = 68.5\omega_2$$

$$\omega_2 = 1.05 \text{ rad/s} \circlearrowright \qquad\qquad\qquad \textit{Ans.}$$

NOTE: Try solving this problem by applying the principle of impulse and momentum about G and using the principle of linear impulse and momentum in the x direction.

EXAMPLE 19.4

The cylinder B, shown in Fig. 19–7a has a mass of 6 kg. It is attached to a cord which is wrapped around the periphery of a 20-kg disk that has a moment of inertia $I_A = 0.40 \text{ kg} \cdot \text{m}^2$. If the cylinder is initially moving downward with a speed of 2 m/s, determine its speed in 3 s. Neglect the mass of the cord in the calculation.

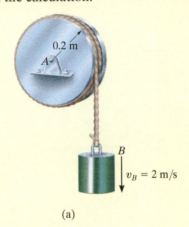

(a)

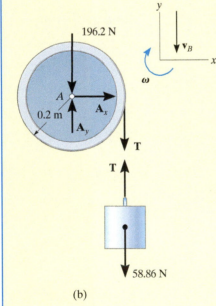

(b)

Fig. 19–7

SOLUTION I

Free-Body Diagram. The free-body diagrams of the cylinder and disk are shown in Fig. 19–7b. All the forces are *constant* since the weight of the cylinder causes the motion. The downward motion of the cylinder, $\mathbf{v}_B$, causes $\boldsymbol{\omega}$ of the disk to be clockwise.

Principle of Impulse and Momentum. We can eliminate $\mathbf{A}_x$ and $\mathbf{A}_y$ from the analysis by applying the principle of angular impulse and momentum about point A. Hence

Disk

$$(\zeta +) \qquad\qquad I_A\omega_1 + \Sigma \int M_A\, dt = I_A\omega_2$$

$$0.40 \text{ kg} \cdot \text{m}^2(\omega_1) + T(3 \text{ s})(0.2 \text{ m}) = (0.40 \text{ kg} \cdot \text{m}^2)\omega_2$$

Cylinder

$$(+\uparrow) \qquad\qquad m_B(v_B)_1 + \Sigma \int F_y\, dt = m_B(v_B)_2$$

$$-6 \text{ kg}(2 \text{ m/s}) + T(3 \text{ s}) - 58.86 \text{ N}(3 \text{ s}) = -6 \text{ kg}(v_B)_2$$

Kinematics. Since $\omega = v_B/r$, then $\omega_1 = (2 \text{ m/s})/(0.2 \text{ m}) = 10 \text{ rad/s}$ and $\omega_2 = (v_B)_2/0.2 \text{ m} = 5(v_B)_2$. Substituting and solving the equations simultaneously for $(v_B)_2$ yields

$$(v_B)_2 = 13.0 \text{ m/s} \downarrow \qquad\qquad Ans.$$

SOLUTION II

Impulse and Momentum Diagrams.   We can obtain $(v_B)_2$ *directly* by considering the *system* consisting of the cylinder, the cord, and the disk. The impulse and momentum diagrams have been drawn to clarify application of the principle of angular impulse and momentum about point A, Fig. 19–7c.

Principle of Angular Impulse and Momentum.   Realizing that $\omega_1 = 10$ rad/s and $\omega_2 = 5(v_B)_2$, we have

$$(\zeta +)\left(\sum{\text{syst. angular} \atop \text{momentum}}\right)_{A1} + \left(\sum{\text{syst. angular} \atop \text{impulse}}\right)_{A(1-2)} = \left(\sum{\text{syst. angular} \atop \text{momentum}}\right)_{A2}$$

$$(6\text{ kg})(2\text{ m/s})(0.2\text{ m}) + (0.40\text{ kg}\cdot\text{m}^2)(10\text{ rad/s}) + (58.86\text{ N})(3\text{ s})(0.2\text{ m})$$

$$= (6\text{ kg})(v_B)_2(0.2\text{ m}) + (0.40\text{ kg}\cdot\text{m}^2)[5(v_B)_2]$$

$$(v_B)_2 = 13.0\text{ m/s} \downarrow \qquad\qquad \textit{Ans.}$$

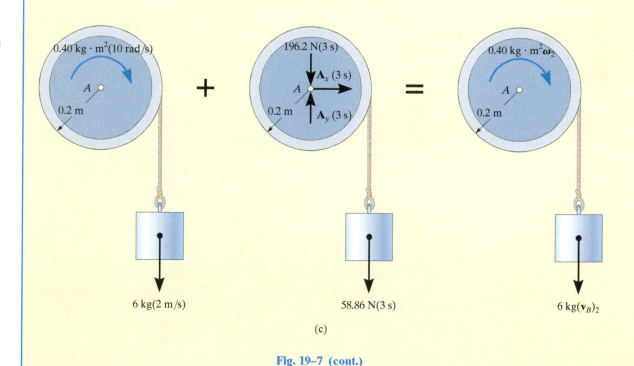

(c)

Fig. 19–7 (cont.)

EXAMPLE 19.5

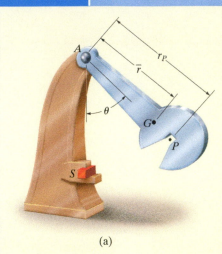

(a)

The Charpy impact test is used in materials testing to determine the energy absorption characteristics of a material during impact. The test is performed using the pendulum shown in Fig. 19–8a, which has a mass m, mass center at G, and a radius of gyration k_G about G. Determine the distance r_P from the pin at A to the point P where the impact with the specimen S should occur so that the horizontal force at the pin A is essentially zero during the impact. For the calculation, assume the specimen absorbs all the pendulum's kinetic energy gained during the time it falls and thereby stops the pendulum from swinging when $\theta = 0°$.

SOLUTION

Free-Body Diagram. As shown on the free-body diagram, Fig. 19–8b, the conditions of the problem require the horizontal force at A to be zero. Just before impact, the pendulum has a clockwise angular velocity $\boldsymbol{\omega}_1$, and the mass center of the pendulum is moving to the left at $(v_G)_1 = \bar{r}\omega_1$.

Principle of Impulse and Momentum. We will apply the principle of angular impulse and momentum about point A. Thus,

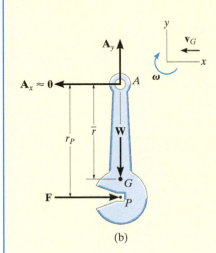

(b)

Fig. 19–8

$$I_A\omega_1 + \Sigma \int M_A \, dt = I_A\omega_2$$

$(\zeta +)$
$$I_A\omega_1 - \left(\int F \, dt\right)r_P = 0$$

$$m(v_G)_1 + \Sigma \int F \, dt = m(v_G)_2$$

$(\xrightarrow{+})$
$$-m(\bar{r}\omega_1) + \int F \, dt = 0$$

Eliminating the impulse $\int F \, dt$ and substituting $I_A = mk_G^2 + m\bar{r}^2$ yields

$$[mk_G^2 + m\bar{r}^2]\omega_1 - m(\bar{r}\omega_1)r_P = 0$$

Factoring out $m\omega_1$ and solving for r_P, we obtain

$$r_P = \bar{r} + \frac{k_G^2}{\bar{r}} \qquad \textit{Ans.}$$

NOTE: Point P, so defined, is called the *center of percussion*. By placing the striking point at P, the force developed at the pin will be minimized. Many sports rackets, clubs, etc. are designed so that collision with the object being struck occurs at the center of percussion. As a consequence, no "sting" or little sensation occurs in the hand of the player. (Also see Probs. 17–66 and 19–1.)

PRELIMINARY PROBLEMS

P19–1. Determine the angular momentum of the 100-kg disk or rod about point G and about point O.

P19–2. Determine the angular impulse about point O for $t = 3$ s.

a)

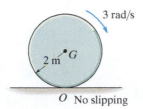

3 rad/s

2 m G

O No slipping

b)

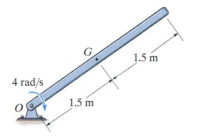

G

1.5 m

4 rad/s

1.5 m

O

c)

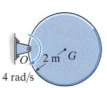

O 2 m G

4 rad/s

d)

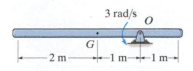

3 rad/s O

G

2 m—1 m—1 m

Prob. P19–1

a)

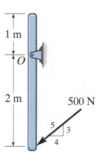

1 m

O

2 m

500 N

5 3 4

b)

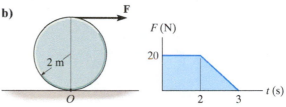

F

F (N)

2 m

O

20

2 3 t (s)

c)

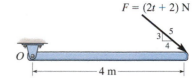

$F = (2t + 2)$ N

3 5 4

O

4 m

d)

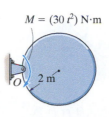

$M = (30\ t^2)$ N·m

O 2 m

Prob. P19–2

FUNDAMENTAL PROBLEMS

F19–1. The 60-kg wheel has a radius of gyration about its center O of $k_O = 300$ mm. If it is subjected to a couple moment of $M = (3t^2)$ N·m, where t is in seconds, determine the angular velocity of the wheel when $t = 4$ s, starting from rest.

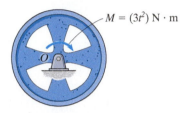

$M = (3t^2)$ N·m

Prob. F19–1

F19–2. The 300-kg wheel has a radius of gyration about its mass center O of $k_O = 400$ mm. If the wheel is subjected to a couple moment of $M = 300$ N·m, determine its angular velocity 6 s after it starts from rest and no slipping occurs. Also, determine the friction force that the ground applies to the wheel.

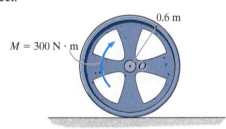

0.6 m

$M = 300$ N·m

Prob. F19–2

F19–3. If rod OA of negligible mass is subjected to the couple moment $M = 9$ N·m, determine the angular velocity of the 10-kg inner gear $t = 5$ s after it starts from rest. The gear has a radius of gyration about its mass center of $k_A = 100$ mm, and it rolls on the fixed outer gear, B. Motion occurs in the horizontal plane.

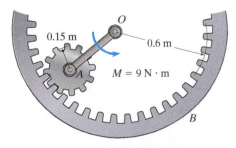

0.15 m O 0.6 m A $M = 9$ N·m B

Prob. F19–3

F19–4. Gears A and B of mass 10 kg and 50 kg have radii of gyration about their respective mass centers of $k_A = 80$ mm and $k_B = 150$ mm. If gear A is subjected to the couple moment $M = 10$ N·m when it is at rest, determine the angular velocity of gear B when $t = 5$ s.

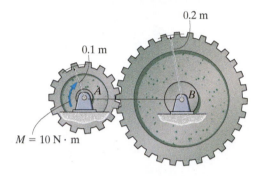

0.2 m   0.1 m A B $M = 10$ N·m

Prob. F19–4

F19–5. The 50-kg spool is subjected to a horizontal force of $P = 150$ N. If the spool rolls without slipping, determine its angular velocity 3 s after it starts from rest. The radius of gyration of the spool about its center of mass is $k_G = 175$ mm.

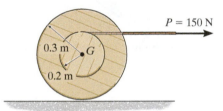

0.3 m G $P = 150$ N   0.2 m

Prob. F19–5

F19–6. The reel has a weight of 150 lb and a radius of gyration about its center of gravity of $k_G = 1.25$ ft. If it is subjected to a torque of $M = 25$ lb·ft, and starts from rest when the torque is applied, determine its angular velocity in 3 seconds. The coefficient of kinetic friction between the reel and the horizontal plane is $\mu_k = 0.15$.

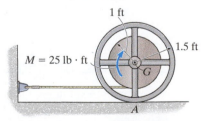

1 ft   1.5 ft $M = 25$ lb·ft G A

Prob. F19–6

19

PROBLEMS

19–1. The rigid body (slab) has a mass m and rotates with an angular velocity ω about an axis passing through the fixed point O. Show that the momenta of all the particles composing the body can be represented by a single vector having a magnitude mv_G and acting through point P, called the *center of percussion*, which lies at a distance $r_{P/G} = k_G^2/r_{G/O}$ from the mass center G. Here k_G is the radius of gyration of the body, computed about an axis perpendicular to the plane of motion and passing through G.

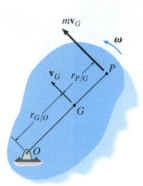

Prob. 19–1

19–2. At a given instant, the body has a linear momentum $\mathbf{L} = m\mathbf{v}_G$ and an angular momentum $\mathbf{H}_G = I_G\omega$ computed about its mass center. Show that the angular momentum of the body computed about the instantaneous center of zero velocity IC can be expressed as $\mathbf{H}_{IC} = I_{IC}\omega$, where I_{IC} represents the body's moment of inertia computed about the instantaneous axis of zero velocity. As shown, the IC is located at a distance $r_{G/IC}$ away from the mass center G.

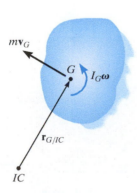

Prob. 19–2

19–3. Show that if a slab is rotating about a fixed axis perpendicular to the slab and passing through its mass center G, the angular momentum is the same when computed about any other point P.

Prob. 19–3

*19–4. The 40-kg disk is rotating at $\omega = 100$ rad/s. When the force $\mathbf{P}$ is applied to the brake as indicated by the graph. If the coefficient of kinetic friction at B is $\mu_k = 0.3$, determine the time t needed to stay the disk from rotating. Neglect the thickness of the brake.

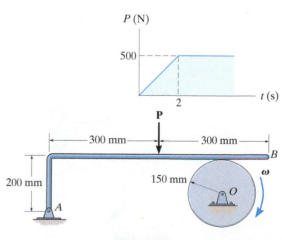

Prob. 19–4

19

19–5. The impact wrench consists of a slender 1-kg rod AB which is 580 mm long, and cylindrical end weights at A and B that each have a diameter of 20 mm and a mass of 1 kg. This assembly is free to turn about the handle and socket, which are attached to the lug nut on the wheel of a car. If the rod AB is given an angular velocity of 4 rad/s and it strikes the bracket C on the handle without rebounding, determine the angular impulse imparted to the lug nut.

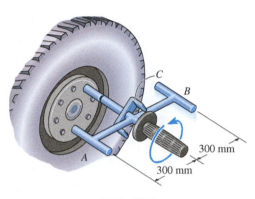

Prob. 19–5

19–6. The airplane is traveling in a straight line with a speed of 300 km/h, when the engines A and B produce a thrust of $T_A = 40$ kN and $T_B = 20$ kN, respectively. Determine the angular velocity of the airplane in $t = 5$ s. The plane has a mass of 200 Mg, its center of mass is located at G, and its radius of gyration about G is $k_G = 15$ m.

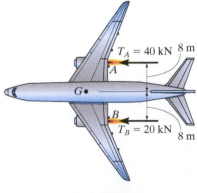

Prob. 19–6

19–7. The double pulley consists of two wheels which are attached to one another and turn at the same rate. The pulley has a mass of 15 kg and a radius of gyration of $k_O = 110$ mm. If the block at A has a mass of 40 kg, determine the speed of the block in 3 s after a constant force of 2 kN is applied to the rope wrapped around the inner hub of the pulley. The block is originally at rest.

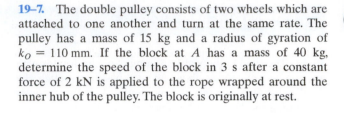
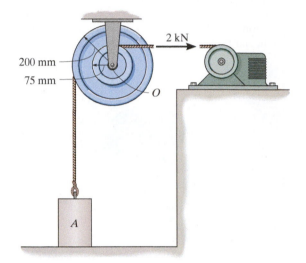

Prob. 19–7

*19–8. The assembly weighs 10 lb and has a radius of gyration $k_G = 0.6$ ft about its center of mass G. The kinetic energy of the assembly is 31 ft · lb when it is in the position shown. If it rolls counterclockwise on the surface without slipping, determine its linear momentum at this instant.

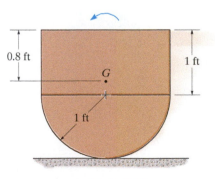

Prob. 19–8

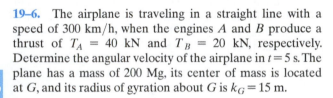

19–9. The disk has a weight of 10 lb and is pinned at its center O. If a vertical force of $P = 2$ lb is applied to the cord wrapped around its outer rim, determine the angular velocity of the disk in four seconds starting from rest. Neglect the mass of the cord.

19–11. The pulley has a weight of 8 lb and may be treated as a thin disk. A cord wrapped over its surface is subjected to forces $T_A = 4$ lb and $T_B = 5$ lb. Determine the angular velocity of the pulley when $t = 4$ s if it starts from rest when $t = 0$. Neglect the mass of the cord.

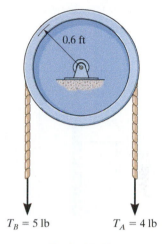

$T_B = 5$ lb $T_A = 4$ lb

Prob. 19–11

0.5 ft

O

P

Prob. 19–9

*19–12.** The 40-kg roll of paper rests along the wall where the coefficient of kinetic friction is $\mu_k = 0.2$. If a vertical force of $P = 40$ N is applied to the paper, determine the angular velocity of the roll when $t = 6$ s starting from rest. Neglect the mass of the unraveled paper and take the radius of gyration of the spool about the axle O to be $k_O = 80$ mm.

19–10. The 30-kg gear A has a radius of gyration about its center of mass O of $k_O = 125$ mm. If the 20-kg gear rack B is subjected to a force of $P = 200$ N, determine the time required for the gear to obtain an angular velocity of 20 rad/s, starting from rest. The contact surface between the gear rack and the horizontal plane is smooth.

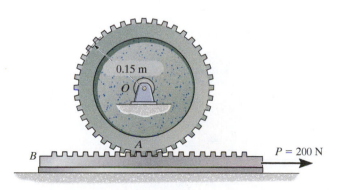

0.15 m

O

A

B $P = 200$ N

Prob. 19–10

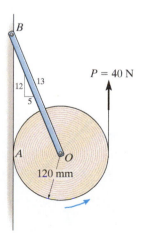

B

12 13

5

$P = 40$ N

A O

120 mm

Prob. 19–12

19–13. The slender rod has a mass m and is suspended at its end A by a cord. If the rod receives a horizontal blow giving it an impulse $\mathbf{I}$ at its bottom B, determine the location y of the point P about which the rod appears to rotate during the impact.

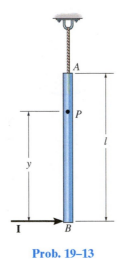

Prob. 19–13

19–14. The rod of length L and mass m lies on a smooth horizontal surface and is subjected to a force $\mathbf{P}$ at its end A as shown. Determine the location d of the point about which the rod begins to turn, i.e, the point that has zero velocity.

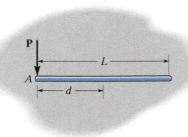

Prob. 19–14

19–15. A 4-kg disk A is mounted on arm BC, which has a negligible mass. If a torque of $M = (5e^{0.5t})\,\text{N} \cdot \text{m}$, where t is in seconds, is applied to the arm at C, determine the angular velocity of BC in 2 s starting from rest. Solve the problem assuming that (a) the disk is set in a smooth bearing at B so that it moves with curvilinear translation, (b) the disk is fixed to the shaft BC, and (c) the disk is given an initial freely spinning angular velocity of $\boldsymbol{\omega}_D = \{-80\mathbf{k}\}\,\text{rad/s}$ prior to application of the torque.

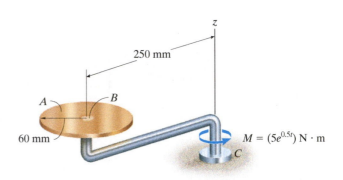

Prob. 19–15

*19–16. The frame of a tandem drum roller has a weight of 4000 lb excluding the two rollers. Each roller has a weight of 1500 lb and a radius of gyration about its axle of 1.25 ft. If a torque of $M = 300\,\text{lb} \cdot \text{ft}$ is supplied to the rear roller A, determine the speed of the drum roller 10 s later, starting from rest.

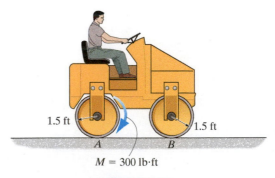

Prob. 19–16

19–17. The 100-lb wheel has a radius of gyration of $k_G = 0.75$ ft. If the upper wire is subjected to a tension of $T = 50$ lb, determine the velocity of the center of the wheel in 3 s, starting from rest. The coefficient of kinetic friction between the wheel and the surface is $\mu_k = 0.1$.

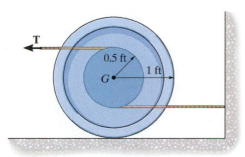

Prob. 19–17

19–19. The double pulley consists of two wheels which are attached to one another and turn at the same rate. The pulley has a mass of 15 kg and a radius of gyration $k_O = 110$ mm. If the block at A has a mass of 40 kg, determine the speed of the block in 3 s after a constant force $F = 2$ kN is applied to the rope wrapped around the inner hub of the pulley. The block is originally at rest. Neglect the mass of the rope.

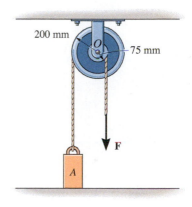

Prob. 19–19

19–18. The 4-kg slender rod rests on a smooth floor. If it is kicked so as to receive a horizontal impulse $I = 8$ N·s at point A as shown, determine its angular velocity and the speed of its mass center.

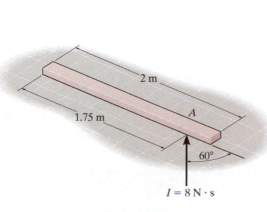

Prob. 19–18

*19–20. The 100-kg spool is resting on the inclined surface for which the coefficient of kinetic friction is $\mu_k = 0.1$. Determine the angular velocity of the spool when $t = 4$ s after it is released from rest. The radius of gyration about the mass center is $k_G = 0.25$ m.

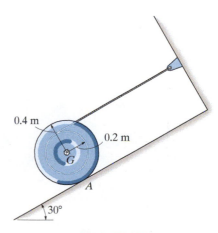

Prob. 19–20

19

19–21. The spool has a weight of 30 lb and a radius of gyration $k_O = 0.45$ ft. A cord is wrapped around its inner hub and the end subjected to a horizontal force $P = 5$ lb. Determine the spool's angular velocity in 4 s starting from rest. Assume the spool rolls without slipping.

19–23. The hoop (thin ring) has a mass of 5 kg and is released down the inclined plane such that it has a backspin $\omega = 8$ rad/s and its center has a velocity $v_G = 3$ m/s as shown. If the coefficient of kinetic friction between the hoop and the plane is $\mu_k = 0.6$, determine how long the hoop rolls before it stops slipping.

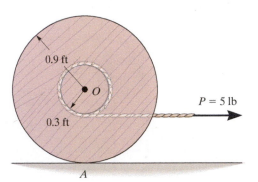

Prob. 19–21

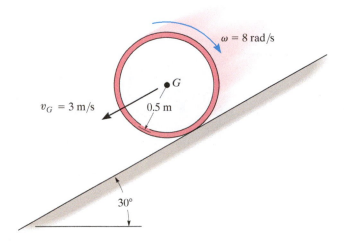

Prob. 19–23

19–22. The two gears A and B have weights and radii of gyration of $W_A = 15$ lb, $k_A = 0.5$ ft and $W_B = 10$ lb, $k_B = 0.35$ ft, respectively. If a motor transmits a couple moment to gear B of $M = 2(1 - e^{-0.5t})$ lb · ft, where t is in seconds, determine the angular velocity of gear A in $t = 5$ s, starting from rest.

*19–24. The 30-kg gear is subjected to a force of $P = (20t)$ N, where t is in seconds. Determine the angular velocity of the gear at $t = 4$ s, starting from rest. Gear rack B is fixed to the horizontal plane, and the gear's radius of gyration about its mass center O is $k_O = 125$ mm.

Prob. 19–22

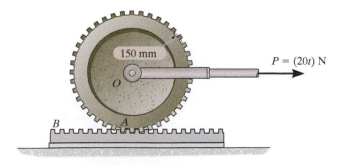

Prob. 19–24

19–25. The 30-lb flywheel A has a radius of gyration about its center of 4 in. Disk B weighs 50 lb and is coupled to the flywheel by means of a belt which does not slip at its contacting surfaces. If a motor supplies a counterclockwise torque to the flywheel of $M = (50t)$ lb·ft, where t is in seconds, determine the time required for the disk to attain an angular velocity of 60 rad/s starting from rest.

19–27. The double pulley consists of two wheels which are attached to one another and turn at the same rate. The pulley has a mass of 15 kg and a radius of gyration of $k_O = 110$ mm. If the block at A has a mass of 40 kg and the container at B has a mass of 85 kg, including its contents, determine the speed of the container when $t = 3$ s after it is released from rest.

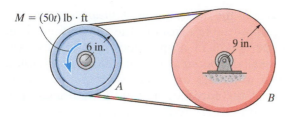

$M = (50t)$ lb · ft

6 in.

9 in.

A

B

Prob. 19–25

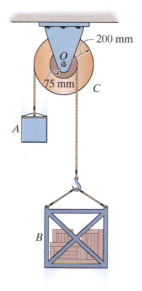

200 mm

O

75 mm C

A

B

Prob. 19–27

19–26. If the shaft is subjected to a torque of $M = (15t^2)$ N·m, where t is in seconds, determine the angular velocity of the assembly when $t = 3$ s, starting from rest. Rods AB and BC each have a mass of 9 kg.

***19–28.** The crate has a mass m_c. Determine the constant speed v_0 it acquires as it moves down the conveyor. The rollers each have a radius of r, mass m, and are spaced d apart. Note that friction causes each roller to rotate when the crate comes in contact with it.

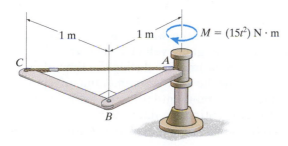

1 m   1 m $M = (15t^2)$ N · m

C

A

B

Prob. 19–26

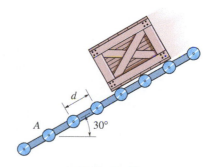

d

A 30°

Prob. 19–28

19

19.3   Conservation of Momentum

Conservation of Linear Momentum. If the sum of all the *linear impulses* acting on a system of connected rigid bodies is *zero* in a specific direction, then the linear momentum of the system is constant, or conserved in this direction, that is,

$$\left(\sum \begin{array}{c} \text{syst. linear} \\ \text{momentum} \end{array} \right)_1 = \left(\sum \begin{array}{c} \text{syst. linear} \\ \text{momentum} \end{array} \right)_2 \qquad (19\text{--}16)$$

This equation is referred to as the *conservation of linear momentum*.

Without introducing appreciable errors in the calculations, it may be possible to apply Eq. 19–16 in a specified direction for which the linear impulses are small or *nonimpulsive*. Specifically, nonimpulsive forces occur when small forces act over very short periods of time. Typical examples include the force of a slightly deformed spring, the initial contact force with soft ground, and in some cases the weight of the body.

Conservation of Angular Momentum. The angular momentum of a system of connected rigid bodies is conserved about the system's center of mass G, or a fixed point O, when the sum of all the angular impulses about these points is zero or appreciably small (nonimpulsive). The third of Eqs. 19–15 then becomes

$$\left(\sum \begin{array}{c} \text{syst. angular} \\ \text{momentum} \end{array} \right)_{O1} = \left(\sum \begin{array}{c} \text{syst. angular} \\ \text{momentum} \end{array} \right)_{O2} \qquad (19\text{--}17)$$

This equation is referred to as the *conservation of angular momentum*. In the case of a single rigid body, Eq. 19–17 applied to point G becomes $(I_G\omega)_1 = (I_G\omega)_2$. For example, consider a swimmer who executes a somersault after jumping off a diving board. By tucking his arms and legs in close to his chest, he *decreases* his body's moment of inertia and thus *increases* his angular velocity ($I_G\omega$ must be constant). If he straightens out just before entering the water, his body's moment of inertia is *increased*, and so his angular velocity *decreases*. Since the weight of his body creates a linear impulse during the time of motion, this example also illustrates how the angular momentum of a body can be conserved and yet the linear momentum is *not*. Such cases occur whenever the external forces creating the linear impulse pass through either the center of mass of the body or a fixed axis of rotation.

Procedure for Analysis

The conservation of linear or angular momentum should be applied using the following procedure.

Free-Body Diagram.

- Establish the x, y inertial frame of reference and draw the free-body diagram for the body or system of bodies during the time of impact. From this diagram classify each of the applied forces as being either "impulsive" or "nonimpulsive."

- By inspection of the free-body diagram, the *conservation of linear momentum* applies in a given direction when *no* external impulsive forces act on the body or system in that direction; whereas the *conservation of angular momentum* applies about a fixed point O or at the mass center G of a body or system of bodies when all the external impulsive forces acting on the body or system create zero moment (or zero angular impulse) about O or G.

- As an alternative procedure, draw the impulse and momentum diagrams for the body or system of bodies. These diagrams are particularly helpful in order to visualize the "moment" terms used in the conservation of angular momentum equation, when it has been decided that angular momenta are to be computed about a point other than the body's mass center G.

Conservation of Momentum.

- Apply the conservation of linear or angular momentum in the appropriate directions.

Kinematics.

- If the motion appears to be complicated, kinematic (velocity) diagrams may be helpful in obtaining the necessary kinematic relations.

19

EXAMPLE 19.6

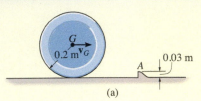

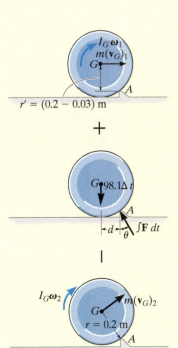

(a)

(b)

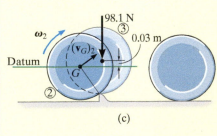

(c)

Fig. 19–9

The 10-kg wheel shown in Fig. 19–9a has a moment of inertia $I_G = 0.156 \, \text{kg} \cdot \text{m}^2$. Assuming that the wheel does not slip or rebound, determine the minimum velocity $\mathbf{v}_G$ it must have to just roll over the obstruction at A.

SOLUTION

Impulse and Momentum Diagrams. Since no slipping or rebounding occurs, the wheel essentially *pivots* about point A during contact. This condition is shown in Fig. 19–9b, which indicates, respectively, the momentum of the wheel *just before impact*, the impulses given to the wheel *during impact*, and the momentum of the wheel *just after impact*. Only two impulses (forces) act on the wheel. By comparison, the force at A is much greater than that of the weight, and since the time of impact is very short, the weight can be considered nonimpulsive. The impulsive force $\mathbf{F}$ at A has both an unknown magnitude and an unknown direction θ. To eliminate this force from the analysis, note that angular momentum about A is essentially *conserved* since $(98.1\Delta t)d \approx 0$.

Conservation of Angular Momentum. With reference to Fig. 19–9b,

$$(\zeta +) \qquad\qquad (H_A)_1 = (H_A)_2$$
$$r'm(v_G)_1 + I_G\omega_1 = rm(v_G)_2 + I_G\omega_2$$
$$(0.2 \, \text{m} - 0.03 \, \text{m})(10 \, \text{kg})(v_G)_1 + (0.156 \, \text{kg} \cdot \text{m}^2)(\omega_1) =$$
$$(0.2 \, \text{m})(10 \, \text{kg})(v_G)_2 + (0.156 \, \text{kg} \cdot \text{m}^2)(\omega_2)$$

Kinematics. Since no slipping occurs, in general $\omega = v_G/r = v_G/0.2 \, \text{m} = 5v_G$. Substituting this into the above equation and simplifying yields

$$(v_G)_2 = 0.8921(v_G)_1 \qquad\qquad (1)$$

Conservation of Energy.* In order to roll over the obstruction, the wheel must pass position 3 shown in Fig. 19–9c. Hence, if $(v_G)_2$ [or $(v_G)_1$] is to be a minimum, it is necessary that the kinetic energy of the wheel at position 2 be equal to the potential energy at position 3. Placing the datum through the center of gravity, as shown in the figure, and applying the conservation of energy equation, we have

$$\{T_2\} + \{V_2\} = \{T_3\} + \{V_3\}$$
$$\left\{ \tfrac{1}{2}(10 \, \text{kg})(v_G)_2^2 + \tfrac{1}{2}(0.156 \, \text{kg} \cdot \text{m}^2)\omega_2^2 \right\} + \{0\} =$$
$$\{0\} + \{(98.1 \, \text{N})(0.03 \, \text{m})\}$$

Substituting $\omega_2 = 5(v_G)_2$ and Eq. 1 into this equation, and solving,

$$(v_G)_1 = 0.729 \, \text{m/s} \rightarrow \qquad\qquad Ans.$$

---

*This principle *does not apply during impact*, since energy is *lost* during the collision. However, just after impact, as in Fig. 19–9c, it can be used.

EXAMPLE 19.7

The 5-kg slender rod shown in Fig. 19–10a is pinned at O and is initially at rest. If a 4-g bullet is fired into the rod with a velocity of 400 m/s, as shown in the figure, determine the angular velocity of the rod just after the bullet becomes embedded in it.

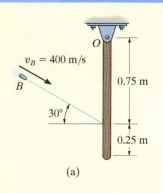

(a)

SOLUTION

Impulse and Momentum Diagrams. The impulse which the bullet exerts on the rod can be eliminated from the analysis, and the angular velocity of the rod just after impact can be determined by considering the bullet and rod as a single system. To clarify the principles involved, the impulse and momentum diagrams are shown in Fig. 19–10b. The momentum diagrams are drawn *just before and just after impact*. During impact, the bullet and rod exert equal but *opposite internal impulses* at A. As shown on the impulse diagram, the impulses that are external to the system are due to the reactions at O and the weights of the bullet and rod. Since the time of impact, Δt, is very short, the rod moves only a slight amount, and so the "moments" of the weight impulses about point O are essentially zero. Therefore angular momentum is conserved about this point.

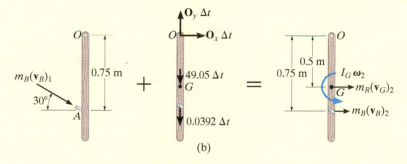

(b)

Conservation of Angular Momentum. From Fig. 19–10b, we have

$$(\zeta+) \qquad \Sigma(H_O)_1 = \Sigma(H_O)_2$$

$$m_B(v_B)_1 \cos 30°(0.75 \text{ m}) = m_B(v_B)_2(0.75 \text{ m}) + m_R(v_G)_2(0.5 \text{ m}) + I_G\omega_2$$

$$(0.004 \text{ kg})(400 \cos 30° \text{ m/s})(0.75 \text{ m}) =$$

$$(0.004 \text{ kg})(v_B)_2(0.75 \text{ m}) + (5 \text{ kg})(v_G)_2(0.5 \text{ m}) + \left[\tfrac{1}{12}(5 \text{ kg})(1 \text{ m})^2\right]\omega_2 \quad (1)$$

or

$$1.039 = 0.003(v_B)_2 + 2.50(v_G)_2 + 0.4167\omega_2$$

Kinematics. Since the rod is pinned at O, from Fig. 19–9c we have

$$(v_G)_2 = (0.5 \text{ m})\omega_2 \quad (v_B)_2 = (0.75 \text{ m})\omega_2$$

Substituting into Eq. 1 and solving yields

$$\omega_2 = 0.623 \text{ rad/s} \, \zeta \qquad\qquad Ans.$$

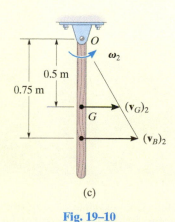

(c)

Fig. 19–10

*19.4   Eccentric Impact

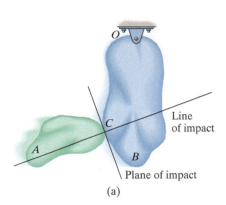

Line
of impact

Plane of impact

(a)

Fig. 19–11

The concepts involving central and oblique impact of particles were presented in Sec. 15.4. We will now expand this treatment and discuss the eccentric impact of two bodies. *Eccentric impact* occurs when the line connecting the *mass centers* of the two bodies *does not* coincide with the line of impact.* This type of impact often occurs when one or both of the bodies are constrained to rotate about a fixed axis. Consider, for example, the collision at C between the two bodies A and B, shown in Fig. 19–11a. It is assumed that just before collision B is rotating counterclockwise with an angular velocity $(\boldsymbol{\omega}_B)_1$, and the velocity of the contact point C located on A is $(\mathbf{u}_A)_1$. Kinematic diagrams for both bodies just before collision are shown in Fig. 19–11b. Provided the bodies are smooth, the *impulsive forces* they exert on each other *are directed along the line of impact*. Hence, the component of velocity of point C on body B, which is directed along the line of impact, is $(v_B)_1 = (\omega_B)_1 r$, Fig. 19–11b. Likewise, on body A the component of velocity $(\mathbf{u}_A)_1$ along the line of impact is $(\mathbf{v}_A)_1$. In order for a collision to occur, $(v_A)_1 > (v_B)_1$.

During the impact an equal but opposite impulsive force $\mathbf{P}$ is exerted between the bodies which *deforms* their shapes at the point of contact. The resulting impulse is shown on the impulse diagrams for both bodies, Fig. 19–11c. Note that the impulsive force at point C on the rotating body creates impulsive pin reactions at O. On these diagrams it is assumed that the impact creates forces which are much larger than the nonimpulsive weights of the bodies, which are not shown. When the deformation at point C is a maximum, C on both the bodies moves with a common velocity $\mathbf{v}$ along the line of impact, Fig. 19–11d. A period of *restitution* then occurs in which the bodies tend to regain their original shapes. The restitution phase creates an equal but opposite impulsive force $\mathbf{R}$ acting between the bodies as shown on the impulse diagram, Fig. 19–11e. After restitution the bodies move apart such that point C on body B has a velocity $(\mathbf{v}_B)_2$ and point C on body A has a velocity $(\mathbf{u}_A)_2$, Fig. 19–11f, where $(v_B)_2 > (v_A)_2$.

In general, a problem involving the impact of two bodies requires determining the *two unknowns* $(v_A)_2$ and $(v_B)_2$, assuming $(v_A)_1$ and $(v_B)_1$ are known (or can be determined using kinematics, energy methods, the equations of motion, etc.). To solve such problems, two equations must be written. The *first equation* generally involves application of *the conservation of angular momentum to the two bodies*. In the case of both bodies A and B, we can state that angular momentum is conserved about point O since the impulses at C are internal to the system and the impulses at O create zero moment (or zero angular impulse) about O. The *second equation* can be obtained using the definition of the *coefficient of restitution, e*, which is a ratio of the restitution impulse to the deformation impulse.

Here is an example of eccentric impact occurring between this bowling ball and pin. (© R.C. Hibbeler)

*When these lines coincide, central impact occurs and the problem can be analyzed as discussed in Sec. 15.4.

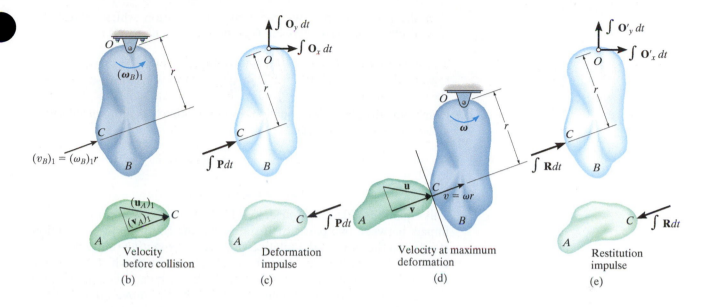

Velocity before collision
(b)

Deformation impulse
(c)

Velocity at maximum deformation
(d)

Restitution impulse
(e)

Is is important to realize, however, that *this analysis has only a very limited application in engineering, because values of e for this case have been found to be highly sensitive to the material, geometry, and the velocity of each of the colliding bodies.* To establish a useful form of the coefficient of restitution equation we must first apply the principle of angular impulse and momentum about point O to bodies B and A separately. Combining the results, we then obtain the necessary equation. Proceeding in this manner, the principle of impulse and momentum applied to body B from the time just before the collision to the instant of maximum deformation, Figs. 19–11*b*, 19–11*c*, and 19–11*d*, becomes

$$(\zeta +) \qquad I_O(\omega_B)_1 + r\int P\,dt = I_O\omega \qquad (19\text{–}18)$$

Here I_O is the moment of inertia of body B about point O. Similarly, applying the principle of angular impulse and momentum from the instant of maximum deformation to the time just after the impact, Figs. 19–11*d*, 19–11*e*, and 19–11*f*, yields

$$(\zeta +) \qquad I_O\omega + r\int R\,dt = I_O(\omega_B)_2 \qquad (19\text{–}19)$$

Solving Eqs. 19–18 and 19–19 for $\int P\,dt$ and $\int R\,dt$, respectively, and formulating e, we have

$$e = \frac{\displaystyle\int R\,dt}{\displaystyle\int P\,dt} = \frac{r(\omega_B)_2 - r\omega}{r\omega - r(\omega_B)_1} = \frac{(v_B)_2 - v}{v - (v_B)_1}$$

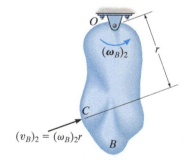

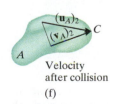

Velocity after collision
(f)

Fig. 19–11 (cont.)

In the same manner, we can write an equation which relates the magnitudes of velocity $(v_A)_1$ and $(v_A)_2$ of body A. The result is

$$e = \frac{v - (v_A)_2}{(v_A)_1 - v}$$

Combining the above two equations by eliminating the common velocity v yields the desired result, i.e.,

$(+\nearrow)$

$$\boxed{e = \frac{(v_B)_2 - (v_A)_2}{(v_A)_1 - (v_B)_1}}$$

(19–20)

This equation is identical to Eq. 15–11, which was derived for the central impact between two particles. It states that the coefficient of restitution is equal to the ratio of the relative velocity of *separation* of the points of contact (C) *just after impact* to the relative velocity at which the points *approach* one another *just* before impact. In deriving this equation, we assumed that the points of contact for both bodies move up and to the right *both* before and after impact. If motion of any one of the contacting points occurs down and to the left, the velocity of this point should be considered a negative quantity in Eq. 19–20.

During impact the columns of many highway signs are intended to break out of their supports and easily collapse at their joints. This is shown by the slotted connections at their base and the breaks at the column's midsection. (© R.C. Hibbeler)

EXAMPLE 19.8

The 10-lb slender rod is suspended from the pin at A, Fig. 19–12a. If a 2-lb ball B is thrown at the rod and strikes its center with a velocity of 30 ft/s, determine the angular velocity of the rod just after impact. The coefficient of restitution is $e = 0.4$.

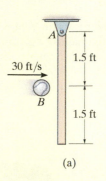

(a)

SOLUTION

Conservation of Angular Momentum. Consider the ball and rod as a system, Fig. 19–12b. Angular momentum is conserved about point A since the impulsive force between the rod and ball is *internal*. Also, the *weights* of the ball and rod are *nonimpulsive*. Noting the directions of the velocities of the ball and rod just after impact as shown on the kinematic diagram, Fig. 19–12c, we require

$$(\zeta+) \qquad\qquad (H_A)_1 = (H_A)_2$$

$$m_B(v_B)_1(1.5\text{ ft}) = m_B(v_B)_2(1.5\text{ ft}) + m_R(v_G)_2(1.5\text{ ft}) + I_G\omega_2$$

$$\left(\frac{2\text{ lb}}{32.2\text{ ft/s}^2}\right)(30\text{ ft/s})(1.5\text{ ft}) = \left(\frac{2\text{ lb}}{32.2\text{ ft/s}^2}\right)(v_B)_2(1.5\text{ ft}) +$$

$$\left(\frac{10\text{ lb}}{32.2\text{ ft/s}^2}\right)(v_G)_2(1.5\text{ ft}) + \left[\frac{1}{12}\left(\frac{10\text{ lb}}{32.2\text{ ft/s}^2}\right)(3\text{ ft})^2\right]\omega_2$$

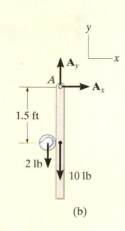

(b)

Since $(v_G)_2 = 1.5\omega_2$ then

$$2.795 = 0.09317(v_B)_2 + 0.9317\omega_2 \qquad (1)$$

Coefficient of Restitution. With reference to Fig. 19–12c, we have

$$(\overset{+}{\rightarrow}) \qquad e = \frac{(v_G)_2 - (v_B)_2}{(v_B)_1 - (v_G)_1} \qquad 0.4 = \frac{(1.5\text{ ft})\omega_2 - (v_B)_2}{30\text{ ft/s} - 0}$$

$$12.0 = 1.5\omega_2 - (v_B)_2 \qquad (2)$$

Solving Eqs. 1 and 2, yields

$$(v_B)_2 = -6.52\text{ ft/s} = 6.52\text{ ft/s} \leftarrow$$

$$\omega_2 = 3.65\text{ rad/s} \,\circlearrowright \qquad\qquad\qquad Ans.$$

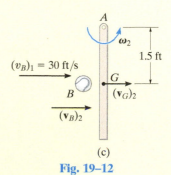

(c)

Fig. 19–12

PROBLEMS

19–29. The turntable T of a record player has a mass of 0.75 kg and a radius of gyration $k_z = 125$ mm. It is *turning freely* at $\omega_T = 2$ rad/s when a 50-g record (thin disk) falls on it. Determine the final angular velocity of the turntable just after the record stops slipping on the turntable.

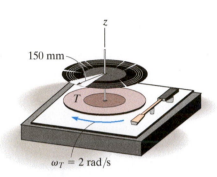

150 mm

$\omega_T = 2$ rad/s

Prob. 19–29

19–30. The 10-g bullet having a velocity of 800 m/s is fired into the edge of the 5-kg disk as shown. Determine the angular velocity of the disk just after the bullet becomes embedded into its edge. Also, calculate the angle θ the disk will swing when it stops. The disk is originally at rest. Neglect the mass of the rod AB.

19–31. The 10-g bullet having a velocity of 800 m/s is fired into the edge of the 5-kg disk as shown. Determine the angular velocity of the disk just after the bullet becomes embedded into its edge. Also, calculate the angle θ the disk will swing when it stops. The disk is originally at rest. The rod AB has a mass of 3 kg.

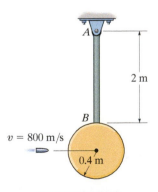

A

2 m

B

$v = 800$ m/s

0.4 m

Probs. 19–30/31

*19–32. The circular disk has a mass m and is suspended at A by the wire. If it receives a horizontal impulse $\mathbf{I}$ at its edge B, determine the location y of the point P about which the disk appears to rotate during the impact.

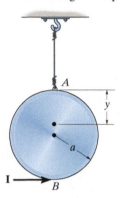

A

y

a

$\mathbf{I}$ B

Prob. 19–32

19–33. The 80-kg man is holding two dumbbells while standing on a turntable of negligible mass, which turns freely about a vertical axis. When his arms are fully extended, the turntable is rotating with an angular velocity of 0.5 rev/s. Determine the angular velocity of the man when he retracts his arms to the position shown. When his arms are fully extended, approximate each arm as a uniform 6-kg rod having a length of 650 mm, and his body as a 68-kg solid cylinder of 400-mm diameter. With his arms in the retracted position, assume the man is an 80-kg solid cylinder of 450-mm diameter. Each dumbbell consists of two 5-kg spheres of negligible size.

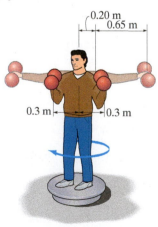

0.20 m
0.65 m

0.3 m   0.3 m

Prob. 19–33

19–34. The platform swing consists of a 200-lb flat plate suspended by four rods of negligible weight. When the swing is at rest, the 150-lb man jumps off the platform when his center of gravity G is 10 ft from the pin at A. This is done with a horizontal velocity of 5 ft/s, measured relative to the swing at the level of G. Determine the angular velocity he imparts to the swing just after jumping off.

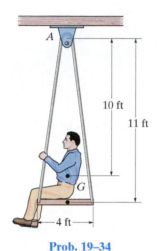

10 ft

11 ft

4 ft

Prob. 19–34

19–35. The 2-kg rod ACB supports the two 4-kg disks at its ends. If both disks are given a clockwise angular velocity $(\omega_A)_1 = (\omega_B)_1 = 5$ rad/s while the rod is held stationary and then released, determine the angular velocity of the rod after both disks have stopped spinning relative to the rod due to frictional resistance at the pins A and B. Motion is in the *horizontal plane*. Neglect friction at pin C.

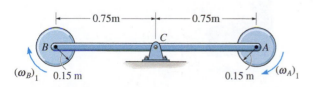

0.75m   0.75m

C

B A

$(\omega_B)_1$ 0.15 m   0.15 m $(\omega_A)_1$

Prob. 19–35

*19–36.** The satellite has a mass of 200 kg and a radius of gyration about z axis of $k_z = 0.1$ m, excluding the two solar panels A and B. Each solar panel has a mass of 15 kg and can be approximated as a thin plate. If the satellite is originally spinning about the z axis at a constant rate $\omega_z = 0.5$ rad/s when $\theta = 90°$, determine the rate of spin if both panels are raised and reach the upward position, $\theta = 0°$, at the same instant.

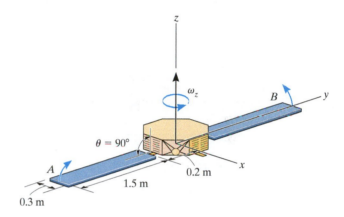

z

ω_z

B y

$\theta = 90°$

A

0.2 m

x

1.5 m

0.3 m

Prob. 19–36

19–37. Disk A has a weight of 20 lb. An inextensible cable is attached to the 10-lb weight and wrapped around the disk. The weight is dropped 2 ft before the slack is taken up. If the impact is perfectly elastic, i.e., $e = 1$, determine the angular velocity of the disk just after impact.

0.5 ft

A

10 lb

Prob. 19–37

19

19–38. The plank has a weight of 30 lb, center of gravity at G, and it rests on the two sawhorses at A and B. If the end D is raised 2 ft above the top of the sawhorses and is released from rest, determine how high end C will rise from the top of the sawhorses after the plank falls so that it rotates clockwise about A, strikes and pivots on the sawhorse at B, and rotates clockwise off the sawhorse at A.

*19–40. A thin rod of mass m has an angular velocity ω_0 while rotating on a smooth surface. Determine its new angular velocity just after its end strikes and hooks onto the peg and the rod starts to rotate about P without rebounding. Solve the problem (a) using the parameters given, (b) setting $m = 2$ kg, $\omega_0 = 4$ rad/s, $l = 1.5$ m.

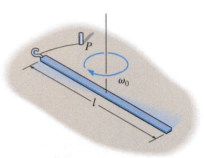

Prob. 19–40

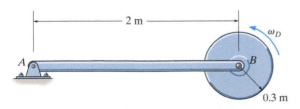

Prob. 19–38

19–41. Tests of impact on the fixed crash dummy are conducted using the 300-lb ram that is released from rest at $\theta = 30°$, and allowed to fall and strike the dummy at $\theta = 90°$. If the coefficient of restitution between the dummy and the ram is $e = 0.4$, determine the angle θ to which the ram will rebound before momentarily coming to rest.

19–39. The 12-kg rod AB is pinned to the 40-kg disk. If the disk is given an angular velocity $\omega_D = 100$ rad/s while the rod is held stationary, and the assembly is then released, determine the angular velocity of the rod after the disk has stopped spinning relative to the rod due to frictional resistance at the bearing B. Motion is in the *horizontal plane*. Neglect friction at the pin A.

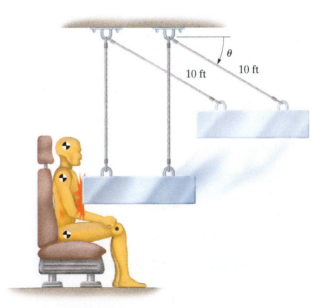

Prob. 19–41

Prob. 19–39

19–42. The vertical shaft is rotating with an angular velocity of 3 rad/s when $\theta = 0°$. If a force **F** is applied to the collar so that $\theta = 90°$, determine the angular velocity of the shaft. Also, find the work done by force **F**. Neglect the mass of rods *GH* and *EF* and the collars *I* and *J*. The rods *AB* and *CD* each have a mass of 10 kg.

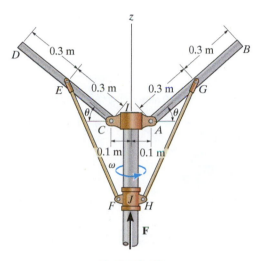

Prob. 19–42

19–43. The mass center of the 3-lb ball has a velocity of $(v_G)_1 = 6$ ft/s when it strikes the end of the smooth 5-lb slender bar which is at rest. Determine the angular velocity of the bar about the *z* axis just after impact if $e = 0.8$.

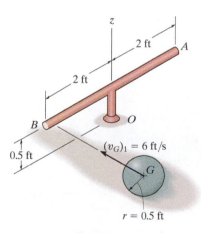

Prob. 19–43

*19–44. The pendulum consists of a slender 2-kg rod *AB* and 5-kg disk. It is released from rest without rotating. When it falls 0.3 m, the end *A* strikes the hook *S*, which provides a permanent connection. Determine the angular velocity of the pendulum after it has rotated 90°. Treat the pendulum's weight during impact as a nonimpulsive force.

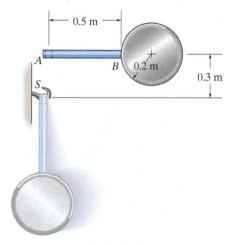

Prob. 19–44

19–45. The 10-lb block is sliding on the smooth surface when the corner *D* hits a stop block *S*. Determine the minimum velocity **v** the block should have which would allow it to tip over on its side and land in the position shown. Neglect the size of *S*. *Hint:* During impact consider the weight of the block to be nonimpulsive.

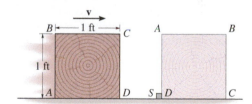

Prob. 19–45

19–46. Determine the height *h* at which a billiard ball of mass *m* must be struck so that no frictional force develops between it and the table at *A*. Assume that the cue *C* only exerts a horizontal force **P** on the ball.

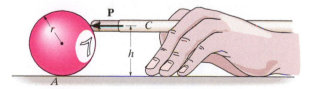

Prob. 19–46

19–47. The pendulum consists of a 15-kg solid ball and 6-kg rod. If it is released from rest when $\theta_1 = 90°$, determine the angle θ_2 after the ball strikes the wall, rebounds, and the pendulum swings up to the point of momentary rest. Take $e = 0.6$.

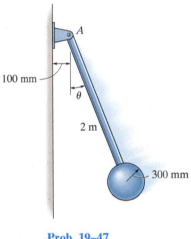

Prob. 19–47

19–48. The 4-lb rod AB is hanging in the vertical position. A 2-lb block, sliding on a smooth horizontal surface with a velocity of 12 ft/s, strikes the rod at its end B. Determine the velocity of the block immediately after the collision. The coefficient of restitution between the block and the rod at B is $e = 0.8$.

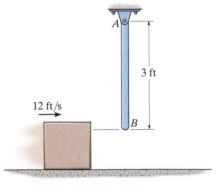

Prob. 19–48

19–49. The hammer consists of a 10-kg solid cylinder C and 6-kg uniform slender rod AB. If the hammer is released from rest when $\theta = 90°$ and strikes the 30-kg block D when $\theta = 0°$, determine the velocity of block D and the angular velocity of the hammer immediately after the impact. The coefficient of restitution between the hammer and the block is $e = 0.6$.

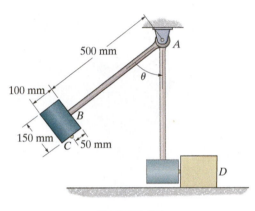

Prob. 19–49

19–50. The 20-kg disk strikes the step without rebounding. Determine the largest angular velocity ω_1 the disk can have and not lose contact with the step, A.

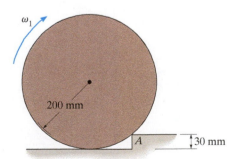

Prob. 19–50

19–51. The solid ball of mass m is dropped with a velocity $\mathbf{v}_1$ onto the edge of the rough step. If it rebounds horizontally off the step with a velocity $\mathbf{v}_2$, determine the angle θ at which contact occurs. Assume no slipping when the ball strikes the step. The coefficient of restitution is e.

19–53. The wheel has a mass of 50 kg and a radius of gyration of 125 mm about its center of mass G. If it rolls without slipping with an angular velocity of $\omega_1 = 5$ rad/s before it strikes the step at A, determine its angular velocity after it rolls over the step. The wheel does not lose contact with the step when it strikes it.

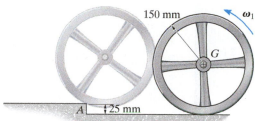

Prob. 19–53

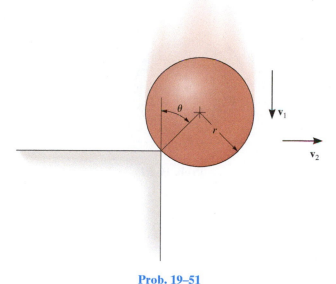

Prob. 19–51

*19–52.** The wheel has a mass of 50 kg and a radius of gyration of 125 mm about its center of mass G. Determine the minimum value of the angular velocity ω_1 of the wheel, so that it strikes the step at A without rebounding and then rolls over it without slipping.

19–54. The rod of mass m and length L is released from rest without rotating. When it falls a distance L, the end A strikes the hook S, which provides a permanent connection. Determine the angular velocity ω of the rod after it has rotated 90°. Treat the rod's weight during impact as a nonimpulsive force.

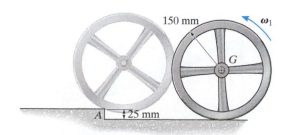

Prob. 19–52

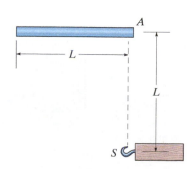

Prob. 19–54

19–55. The 15-lb rod *AB* is released from rest in the vertical position. If the coefficient or restitution between the floor and the cushion at *B* is $e = 0.7$, determine how high the end of the rod rebounds after impact with the floor.

19–57. A solid ball with a mass *m* is thrown on the ground such that at the instant of contact it has an angular velocity ω_1 and velocity components $(\mathbf{v}_G)_{x1}$ and $(\mathbf{v}_G)_{y1}$ as shown. If the ground is rough so no slipping occurs, determine the components of the velocity of its mass center just after impact. The coefficient of restitution is *e*.

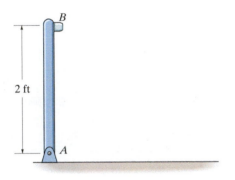

Prob. 19–55

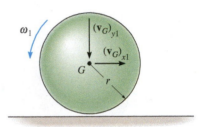

Prob. 19–57

*19–56.** A ball having a mass of 8 kg and initial speed of $v_1 = 0.2$ m/s rolls over a 30-mm-long depression. Assuming that the ball rolls off the edges of contact first *A*, then *B*, without slipping, determine its final velocity $\mathbf{v}_2$ when it reaches the other side.

19–58. The pendulum consists of a 10-lb solid ball and 4-lb rod. If it is released from rest when $\theta_0 = 0°$, determine the angle θ_1 of rebound after the ball strikes the wall and the pendulum swings up to the point of momentary rest. Take $e = 0.6$.

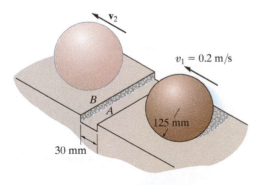

Prob. 19–56

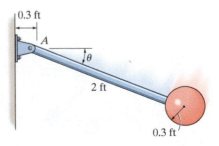

Prob. 19–58

CONCEPTUAL PROBLEMS

C19–1. The soil compactor moves forward at constant velocity by supplying power to the rear wheels. Use appropriate numerical data for the wheel, roller, and body and calculate the angular momentum of this system about point *A* at the ground, point *B* on the rear axle, and point *G*, the center of gravity for the system.

Prob. C19–1 (© R.C. Hibbeler)

C19–2. The swing bridge opens and closes by turning 90° using a motor located under the center of the deck at *A* that applies a torque **M** to the bridge. If the bridge was supported at its end *B*, would the same torque open the bridge at the same time, or would it open slower or faster? Explain your answer using numerical values and an impulse and momentum analysis. Also, what are the benefits of making the bridge have the variable depth as shown?

Prob. C19–2 (© R.C. Hibbeler)

C19–3. Why is it necessary to have the tail blade *B* on the helicopter that spins perpendicular to the spin of the main blade *A*? Explain your answer using numerical values and an impulse and momentum analysis.

Prob. C19–3 (© R.C. Hibbeler)

C19–4. The amusement park ride consists of two gondolas *A* and *B*, and counterweights *C* and *D* that swing in opposite directions. Using realistic dimensions and mass, calculate the angular momentum of this system for any angular position of the gondolas. Explain through analysis why it is a good idea to design this system to have counterweights with each gondola.

Prob. C19–4 (© R.C. Hibbeler)

19

CHAPTER REVIEW

Linear and Angular Momentum

The linear and angular momentum of a rigid body can be referenced to its mass center G.

If the angular momentum is to be determined about an axis other than the one passing through the mass center, then the angular momentum is determined by summing vector $\mathbf{H}_G$ and the moment of vector $\mathbf{L}$ about this axis.

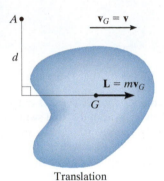

Translation

$$L = mv_G$$
$$H_G = 0$$
$$H_A = (mv_G)d$$

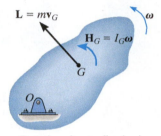

Rotation about a fixed axis

$$L = mv_G$$
$$H_G = I_G\omega$$
$$H_O = I_O\omega$$

General plane motion

$$L = mv_G$$
$$H_G = I_G\omega$$
$$H_A = I_G\omega + (mv_G)d$$

Principle of Impulse and Momentum

The principles of linear and angular impulse and momentum are used to solve problems that involve force, velocity, and time. Before applying these equations, it is important to establish the x, y, z inertial coordinate system. The free-body diagram for the body should also be drawn in order to account for all of the forces and couple moments that produce impulses on the body.

$$m(v_{Gx})_1 + \Sigma \int_{t_1}^{t_2} F_x \, dt = m(v_{Gx})_2$$

$$m(v_{Gy})_1 + \Sigma \int_{t_1}^{t_2} F_y \, dt = m(v_{Gy})_2$$

$$I_G\omega_1 + \Sigma \int_{t_1}^{t_2} M_G \, dt = I_G\omega_2$$

Conservation of Momentum

Provided the sum of the linear impulses acting on a system of connected rigid bodies is zero in a particular direction, then the linear momentum for the system is conserved in this direction. Conservation of angular momentum occurs if the impulses pass through an axis or are parallel to it. Momentum is also conserved if the external forces are small and thereby create nonimpulsive forces on the system. A free-body diagram should accompany any application in order to classify the forces as impulsive or nonimpulsive and to determine an axis about which the angular momentum may be conserved.

$$\left(\sum {}^{\text{syst. linear}}_{\text{momentum}} \right)_1 = \left(\sum {}^{\text{syst. linear}}_{\text{momentum}} \right)_2$$

$$\left(\sum {}^{\text{syst. angular}}_{\text{momentum}} \right)_{O1} = \left(\sum {}^{\text{syst. angular}}_{\text{momentum}} \right)_{O2}$$

Eccentric Impact

If the line of impact does not coincide with the line connecting the mass centers of two colliding bodies, then eccentric impact will occur. If the motion of the bodies just after the impact is to be determined, then it is necessary to consider a conservation of momentum equation for the system and use the coefficient of restitution equation.

$$e = \frac{(v_B)_2 - (v_A)_2}{(v_A)_1 - (v_B)_1}$$

19

REVIEW PROBLEMS

R19–1.   The cable is subjected to a force of $P = (10t^2)$ lb. where t is in seconds. Determine the angular velocity of the spool 3 s after **P** is applied, starting from rest. The spool has a weight of 150 lb and a radius of gyration of 1.25 ft about its center, O.

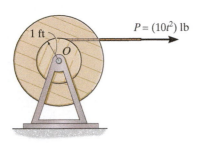

Prob. R19–1

R19–2.   The space capsule has a mass of 1200 kg and a moment of inertia $I_G = 900$ kg $\cdot$ m^2 about an axis passing through G and directed perpendicular to the page. If it is traveling forward with a speed $v_G = 800$ m/s and executes a turn by means of two jets, which provide a constant thrust of 400 N for 0.3 s, determine the capsule's angular velocity just after the jets are turned off.

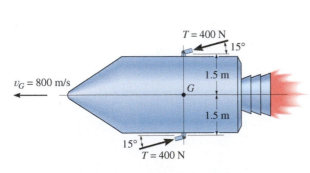

Prob. R19–2

R19–3.   The tire has a mass of 9 kg and a radius of gyration $k_O = 225$ mm. If it is released from rest and rolls down the plane without slipping, determine the speed of its center O when $t = 3$ s.

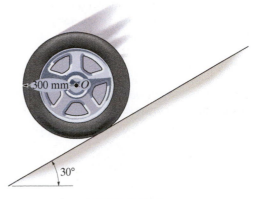

Prob. R19–3

R19–4.   The wheel having a mass of 100 kg and a radius of gyration about the z axis of $k_z = 300$ mm, rests on the smooth horizontal plane. If the belt is subjected to a force of $P = 200$ N, determine the angular velocity of the wheel and the speed of its center of mass O, three seconds after the force is applied.

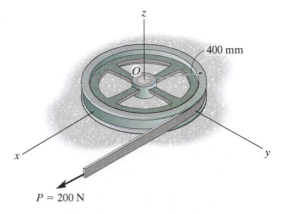

Prob. R19–4

R19–5. The spool has a weight of 30 lb and a radius of gyration $k_O = 0.65$ ft. If a force of 40 lb is applied to the cord at A, determine the angular velocity of the spool in $t = 3$ s starting from rest. Neglect the mass of the pulley and cord.

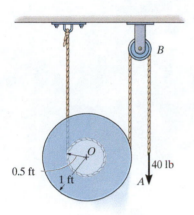

0.5 ft

1 ft

O

B

40 lb

A

Prob. R19–5

R19–6. Spool B is at rest and spool A is rotating at 6 rad/s when the slack in the cord connecting them is taken up. If the cord does not stretch, determine the angular velocity of each spool immediately after the cord is jerked tight. The spools A and B have weights and radii of gyration $W_A = 30$ lb, $k_A = 0.8$ ft, $W_B = 15$ lb, $k_B = 0.6$ ft, respectively.

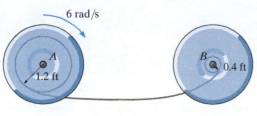

6 rad/s

A

1.2 ft

B

0.4 ft

Prob. R19–6

R19–7. A thin disk of mass m has an angular velocity ω_1 while rotating on a smooth surface. Determine its new angular velocity just after the hook at its edge strikes the peg P and the disk starts to rotate about P without rebounding.

P

r

ω_1

Prob. R19–7

R19–8. The space satellite has a mass of 125 kg and a moment of inertia $I_z = 0.940$ kg·m^2, excluding the four solar panels $A, B, C,$ and D. Each solar panel has a mass of 20 kg and can be approximated as a thin plate. If the satellite is originally spinning about the z axis at a constant rate $\omega_z = 0.5$ rad/s when $\theta = 90°$, determine the rate of spin if all the panels are raised and reach the upward position, $\theta = 0°$, at the same instant.

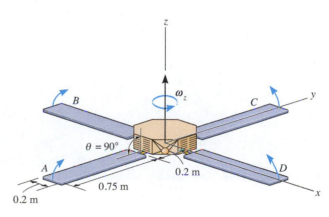

z

B

ω_z

C

y

$\theta = 90°$

A

0.75 m

0.2 m

0.2 m

D

x

Prob. R19–8

19

Chapter 20

(© Philippe Psaila/Science Source)

Design of industrial robots requires knowing the kinematics of their three-dimensional motions.

Three-Dimensional Kinematics of a Rigid Body

20.1   Rotation About a Fixed Point

When a rigid body rotates about a fixed point, the distance r from the point to a particle located on the body is the *same* for *any position* of the body. Thus, the path of motion for the particle lies on the *surface of a sphere* having a radius r and centered at the fixed point. Since motion along this path occurs only from a series of rotations made during a finite time interval, we will first develop a familiarity with some of the properties of rotational displacements.

The boom can rotate up and down, and because it is hinged at a point on the vertical axis about which it turns, it is subjected to rotation about a fixed point. (© R.C. Hibbeler)

Euler's Theorem.

Euler's theorem states that two "component" rotations about different axes passing through a point are equivalent to a single resultant rotation about an axis passing through the point. If more than two rotations are applied, they can be combined into pairs, and each pair can be further reduced and combined into one rotation.

Finite Rotations.

If component rotations used in Euler's theorem are *finite*, it is important that the *order* in which they are applied be maintained. To show this, consider the two finite rotations $\boldsymbol{\theta}_1 + \boldsymbol{\theta}_2$ applied to the block in Fig. 20–1a. Each rotation has a magnitude of 90° and a direction defined by the right-hand rule, as indicated by the arrow. The final position of the block is shown at the right. When these two rotations are applied in the order $\boldsymbol{\theta}_2 + \boldsymbol{\theta}_1$, as shown in Fig. 20–1$b$, the final position of the block is *not* the same as it is in Fig. 20–1a. Because *finite rotations* do not obey the commutative law of addition ($\boldsymbol{\theta}_1 + \boldsymbol{\theta}_2 \neq \boldsymbol{\theta}_2 + \boldsymbol{\theta}_1$), *they cannot be classified as vectors.* If smaller, yet finite, rotations had been used to illustrate this point, e.g., 10° instead of 90°, the *final position* of the block after each combination of rotations would also be different; however, in this case, the difference is only a small amount.

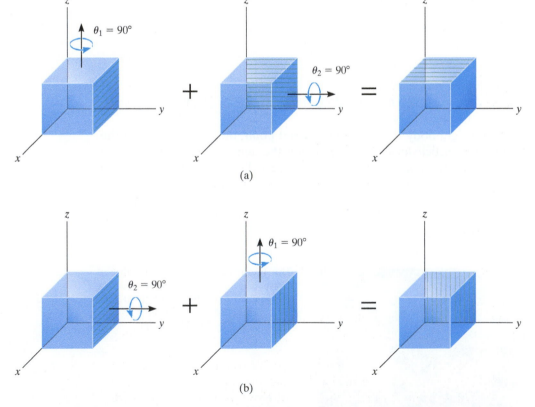

Fig. 20–1

Infinitesimal Rotations.   When defining the angular motions of a body subjected to three-dimensional motion, only rotations which are *infinitesimally small* will be considered. *Such rotations can be classified as vectors, since they can be added vectorially in any manner.* To show this, for purposes of simplicity let us consider the rigid body itself to be a sphere which is allowed to rotate about its central fixed point O, Fig. 20–2a. If we impose two infinitesimal rotations $d\boldsymbol{\theta}_1 + d\boldsymbol{\theta}_2$ on the body, it is seen that point P moves along the path $d\boldsymbol{\theta}_1 \times \mathbf{r} + d\boldsymbol{\theta}_2 \times \mathbf{r}$ and ends up at P'. Had the two successive rotations occurred in the order $d\boldsymbol{\theta}_2 + d\boldsymbol{\theta}_1$, then the resultant displacements of P would have been $d\boldsymbol{\theta}_2 \times \mathbf{r} + d\boldsymbol{\theta}_1 \times \mathbf{r}$. Since the vector cross product obeys the distributive law, by comparison $(d\boldsymbol{\theta}_1 + d\boldsymbol{\theta}_2) \times \mathbf{r} = (d\boldsymbol{\theta}_2 + d\boldsymbol{\theta}_1) \times \mathbf{r}$. Here infinitesimal rotations $d\boldsymbol{\theta}$ are vectors, since these quantities have both a magnitude and direction for which the order of (vector) addition is not important, i.e., $d\boldsymbol{\theta}_1 + d\boldsymbol{\theta}_2 = d\boldsymbol{\theta}_2 + d\boldsymbol{\theta}_1$. As a result, as shown in Fig. 20–2a, the two "component" rotations $d\boldsymbol{\theta}_1$ and $d\boldsymbol{\theta}_2$ are equivalent to a single resultant rotation $d\boldsymbol{\theta} = d\boldsymbol{\theta}_1 + d\boldsymbol{\theta}_2$, a consequence of Euler's theorem.

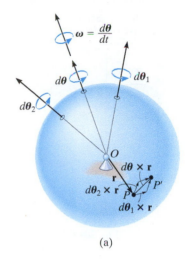

(a)

Angular Velocity.   If the body is subjected to an angular rotation $d\boldsymbol{\theta}$ about a fixed point, the angular velocity of the body is defined by the time derivative,

$$\boldsymbol{\omega} = \dot{\boldsymbol{\theta}} \qquad (20\text{–}1)$$

The line specifying the direction of $\boldsymbol{\omega}$, which is collinear with $d\boldsymbol{\theta}$, is referred to as the *instantaneous axis of rotation*, Fig. 20–2b. In general, this axis changes direction during each instant of time. Since $d\boldsymbol{\theta}$ is a vector quantity, so too is $\boldsymbol{\omega}$, and it follows from vector addition that if the body is subjected to two component angular motions, $\boldsymbol{\omega}_1 = \dot{\boldsymbol{\theta}}_1$ and $\boldsymbol{\omega}_2 = \dot{\boldsymbol{\theta}}_2$, the resultant angular velocity is $\boldsymbol{\omega} = \boldsymbol{\omega}_1 + \boldsymbol{\omega}_2$.

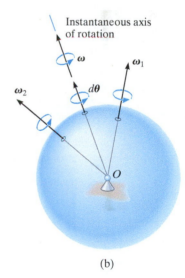

(b)

Fig. 20–2

Angular Acceleration.   The body's angular acceleration is determined from the time derivative of its angular velocity, i.e.,

$$\boldsymbol{\alpha} = \dot{\boldsymbol{\omega}} \qquad (20\text{–}2)$$

For motion about a fixed point, $\boldsymbol{\alpha}$ must account for a change in *both* the magnitude and direction of $\boldsymbol{\omega}$, so that, in general, $\boldsymbol{\alpha}$ is not directed along the instantaneous axis of rotation, Fig. 20–3.

As the direction of the instantaneous axis of rotation (or the line of action of $\boldsymbol{\omega}$) changes in space, the locus of the axis generates a fixed *space cone*, Fig. 20–4. If the change in the direction of this axis is viewed with respect to the rotating body, the locus of the axis generates a *body cone*.

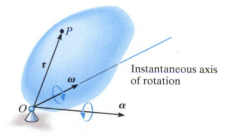

Fig. 20–3

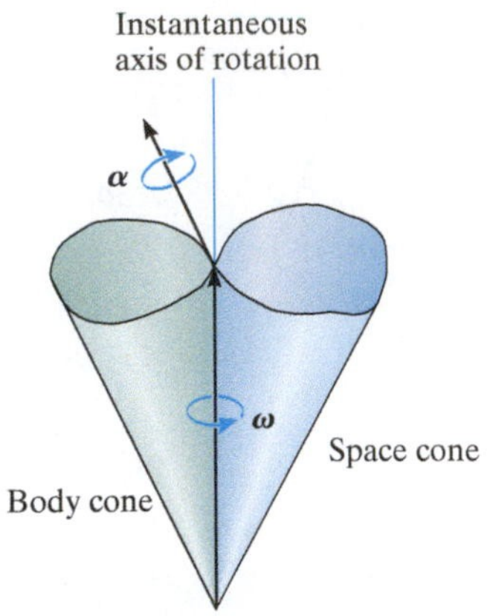

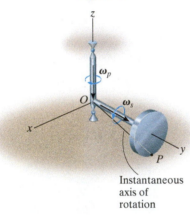

Fig. 20–4

(a)

(b)

Fig. 20–5

At any given instant, these cones meet along the instantaneous axis of rotation, and when the body is in motion, the body cone appears to roll either on the inside or the outside surface of the fixed space cone. Provided the paths defined by the open ends of the cones are described by the head of the $\boldsymbol{\omega}$ vector, then $\boldsymbol{\alpha}$ must act tangent to these paths at any given instant, since the time rate of change of $\boldsymbol{\omega}$ is equal to $\boldsymbol{\alpha}$. Fig. 20–4.

To illustrate this concept, consider the disk in Fig. 20–5a that spins about the rod at $\boldsymbol{\omega}_s$, while the rod and disk precess about the vertical axis at $\boldsymbol{\omega}_p$. The resultant angular velocity of the disk is therefore $\boldsymbol{\omega} = \boldsymbol{\omega}_s + \boldsymbol{\omega}_p$. Since both point O and the contact point P have zero velocity, then all points on a line between these points must have zero velocity. Thus, both $\boldsymbol{\omega}$ and the instantaneous axis of rotation are along OP. Therefore, as the disk rotates, this axis appears to move along the surface of the fixed space cone shown in Fig. 20–5b. If the axis is observed from the rotating disk, the axis then appears to move on the surface of the body cone. At any instant, though, these two cones meet each other along the axis OP. If $\boldsymbol{\omega}$ has a constant magnitude, then $\boldsymbol{\alpha}$ indicates only the change in the direction of $\boldsymbol{\omega}$, which is tangent to the cones at the tip of $\boldsymbol{\omega}$ as shown in Fig. 20–5b.

Velocity.   Once $\boldsymbol{\omega}$ is specified, the velocity of any point on a body rotating about a fixed point can be determined using the same methods as for a body rotating about a fixed axis. Hence, by the cross product,

$$\boxed{\mathbf{v} = \boldsymbol{\omega} \times \mathbf{r}} \tag{20–3}$$

Here $\mathbf{r}$ defines the position of the point measured from the fixed point O, Fig. 20–3.

Acceleration.   If $\boldsymbol{\omega}$ and $\boldsymbol{\alpha}$ are known at a given instant, the acceleration of a point can be obtained from the time derivative of Eq. 20–3, which yields

$$\boxed{\mathbf{a} = \boldsymbol{\alpha} \times \mathbf{r} + \boldsymbol{\omega} \times (\boldsymbol{\omega} \times \mathbf{r})} \tag{20–4}$$

*20.2   The Time Derivative of a Vector Measured from Either a Fixed or Translating-Rotating System

In many types of problems involving the motion of a body about a fixed point, the angular velocity $\boldsymbol{\omega}$ is specified in terms of its components. Then, if the angular acceleration $\boldsymbol{\alpha}$ of such a body is to be determined, it is often easier to compute the time derivative of $\boldsymbol{\omega}$ using a coordinate system that has a *rotation* defined by one or more of the components of $\boldsymbol{\omega}$. For example, in the case of the disk in Fig. 20–5a, where $\boldsymbol{\omega} = \boldsymbol{\omega}_s + \boldsymbol{\omega}_p$, the x, y, z axes can be given an angular velocity of $\boldsymbol{\omega}_p$. For this reason, and for other uses later, an equation will now be derived, which relates the time derivative of any vector $\mathbf{A}$ defined from a translating-rotating reference to its time derivative defined from a fixed reference.

Consider the x, y, z axes of the moving frame of reference to be rotating with an angular velocity $\boldsymbol{\Omega}$, which is measured from the fixed X, Y, Z axes, Fig. 20–6a. In the following discussion, it will be convenient to express vector $\mathbf{A}$ in terms of its $\mathbf{i}, \mathbf{j}, \mathbf{k}$ components, which define the directions of the moving axes. Hence,

$$\mathbf{A} = A_x\mathbf{i} + A_y\mathbf{j} + A_z\mathbf{k}$$

In general, the time derivative of $\mathbf{A}$ must account for the change in both its magnitude and direction. However, if this derivative is taken *with respect to the moving frame of reference*, only the change in the magnitudes of the components of $\mathbf{A}$ must be accounted for, since the directions of the components do not change with respect to the moving reference. Hence,

$$(\dot{\mathbf{A}})_{xyz} = \dot{A}_x\mathbf{i} + \dot{A}_y\mathbf{j} + \dot{A}_z\mathbf{k} \qquad (20\text{–}5)$$

When the time derivative of $\mathbf{A}$ is taken *with respect to the fixed frame of reference*, the *directions* of $\mathbf{i}, \mathbf{j}$, and $\mathbf{k}$ change only on account of the *rotation* $\boldsymbol{\Omega}$ of the axes and not their translation. Hence, in general,

$$\dot{\mathbf{A}} = \dot{A}_x\mathbf{i} + \dot{A}_y\mathbf{j} + \dot{A}_z\mathbf{k} + A_x\dot{\mathbf{i}} + A_y\dot{\mathbf{j}} + A_z\dot{\mathbf{k}}$$

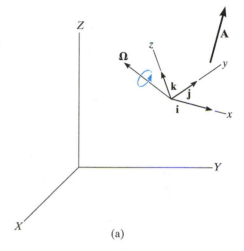

(a)

The time derivatives of the unit vectors will now be considered. For example, $\dot{\mathbf{i}} = d\mathbf{i}/dt$ represents only the change in the *direction* of $\mathbf{i}$ with respect to time, since $\mathbf{i}$ always has a magnitude of 1 unit. As shown in Fig. 20–6b, the change, $d\mathbf{i}$, is *tangent to the path* described by the arrowhead of $\mathbf{i}$ as $\mathbf{i}$ swings due to the rotation $\boldsymbol{\Omega}$. Accounting for both the magnitude and direction of $d\mathbf{i}$, we can therefore define $\dot{\mathbf{i}}$ using the cross product, $\dot{\mathbf{i}} = \boldsymbol{\Omega} \times \mathbf{i}$. In general, then

$$\dot{\mathbf{i}} = \boldsymbol{\Omega} \times \mathbf{i} \qquad \dot{\mathbf{j}} = \boldsymbol{\Omega} \times \mathbf{j} \qquad \dot{\mathbf{k}} = \boldsymbol{\Omega} \times \mathbf{k}$$

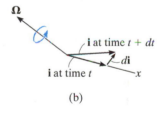

(b)

Fig. 20–6

These formulations were also developed in Sec. 16.8, regarding planar motion of the axes. Substituting these results into the above equation and using Eq. 20–5 yields

$$\boxed{\dot{\mathbf{A}} = (\dot{\mathbf{A}})_{xyz} + \boldsymbol{\Omega} \times \mathbf{A}} \qquad (20\text{–}6)$$

This result is important, and will be used throughout Sec. 20.4 and Chapter 21. It states that the time derivative of *any vector* $\mathbf{A}$ as observed from the fixed X, Y, Z frame of reference is equal to the time rate of change of $\mathbf{A}$ as observed from the x, y, z translating-rotating frame of reference, Eq. 20–5, plus $\boldsymbol{\Omega} \times \mathbf{A}$, the change of $\mathbf{A}$ caused by the rotation of the x, y, z frame. As a result, Eq. 20–6 should always be used whenever $\boldsymbol{\Omega}$ produces a change in the direction of $\mathbf{A}$ as seen from the X, Y, Z reference. If this change does not occur, i.e., $\boldsymbol{\Omega} = \mathbf{0}$, then $\dot{\mathbf{A}} = (\dot{\mathbf{A}})_{xyz}$, and so the time rate of change of $\mathbf{A}$ as observed from both coordinate systems will be the *same*.

20

EXAMPLE | 20.1

The disk shown in Fig. 20–7 spins about its axle with a constant angular velocity $\omega_s = 3$ rad/s, while the horizontal platform on which the disk is mounted rotates about the vertical axis at a constant rate $\omega_p = 1$ rad/s. Determine the angular acceleration of the disk and the velocity and acceleration of point A on the disk when it is in the position shown.

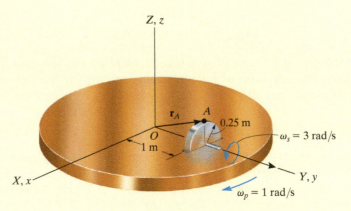

Fig. 20–7

SOLUTION

Point O represents a fixed point of rotation for the disk if one considers a hypothetical extension of the disk to this point. To determine the velocity and acceleration of point A, it is first necessary to determine the angular velocity $\boldsymbol{\omega}$ and angular acceleration $\boldsymbol{\alpha}$ of the disk, since these vectors are used in Eqs. 20–3 and 20–4.

Angular Velocity. The angular velocity, which is measured from X, Y, Z, is simply the vector addition of its two component motions. Thus,

$$\boldsymbol{\omega} = \boldsymbol{\omega}_s + \boldsymbol{\omega}_p = \{3\mathbf{j} - 1\mathbf{k}\} \text{ rad/s}$$

Angular Acceleration.   Since the magnitude of $\boldsymbol{\omega}$ is constant, only a change in its direction, as seen from the fixed reference, creates the angular acceleration $\boldsymbol{\alpha}$ of the disk. One way to obtain $\boldsymbol{\alpha}$ is to compute the time derivative of *each of the two components* of $\boldsymbol{\omega}$ using Eq. 20–6. At the instant shown in Fig. 20–7, imagine the fixed X, Y, Z and a rotating x, y, z frame to be coincident. If the rotating x, y, z frame is chosen to have an angular velocity of $\boldsymbol{\Omega} = \boldsymbol{\omega}_p = \{-1\mathbf{k}\}$ rad/s, then $\boldsymbol{\omega}_s$ will *always* be directed along the y (not Y) axis, and the time rate of change of $\boldsymbol{\omega}_s$ as seen from x, y, z is *zero*; i.e., $(\dot{\boldsymbol{\omega}}_s)_{xyz} = \mathbf{0}$ (the magnitude and direction of $\boldsymbol{\omega}_s$ is constant). Thus,

$$\dot{\boldsymbol{\omega}}_s = (\dot{\boldsymbol{\omega}}_s)_{xyz} + \boldsymbol{\omega}_p \times \boldsymbol{\omega}_s = \mathbf{0} + (-1\mathbf{k}) \times (3\mathbf{j}) = \{3\mathbf{i}\} \text{ rad/s}^2$$

By the same choice of axes rotation, $\boldsymbol{\Omega} = \boldsymbol{\omega}_p$, or even with $\boldsymbol{\Omega} = \mathbf{0}$, the time derivative $(\dot{\boldsymbol{\omega}}_p)_{xyz} = \mathbf{0}$, since $\boldsymbol{\omega}_p$ has a constant magnitude and direction with respect to x, y, z. Hence,

$$\dot{\boldsymbol{\omega}}_p = (\dot{\boldsymbol{\omega}}_p)_{xyz} + \boldsymbol{\omega}_p \times \boldsymbol{\omega}_p = \mathbf{0} + \mathbf{0} = \mathbf{0}$$

The angular acceleration of the disk is therefore

$$\boldsymbol{\alpha} = \dot{\boldsymbol{\omega}} = \dot{\boldsymbol{\omega}}_s + \dot{\boldsymbol{\omega}}_p = \{3\mathbf{i}\} \text{ rad/s}^2 \qquad \textit{Ans.}$$

Velocity and Acceleration.   Since $\boldsymbol{\omega}$ and $\boldsymbol{\alpha}$ have now been determined, the velocity and acceleration of point A can be found using Eqs. 20–3 and 20–4. Realizing that $\mathbf{r}_A = \{1\mathbf{j} + 0.25\mathbf{k}\}$ m, Fig. 20–7, we have

$$\mathbf{v}_A = \boldsymbol{\omega} \times \mathbf{r}_A = (3\mathbf{j} - 1\mathbf{k}) \times (1\mathbf{j} + 0.25\mathbf{k}) = \{1.75\mathbf{i}\} \text{ m/s} \qquad \textit{Ans.}$$

$$\mathbf{a}_A = \boldsymbol{\alpha} \times \mathbf{r}_A + \boldsymbol{\omega} \times (\boldsymbol{\omega} \times \mathbf{r}_A)$$

$$= (3\mathbf{i}) \times (1\mathbf{j} + 0.25\mathbf{k}) + (3\mathbf{j} - 1\mathbf{k}) \times [(3\mathbf{j} - 1\mathbf{k}) \times (1\mathbf{j} + 0.25\mathbf{k})]$$

$$= \{-2.50\mathbf{j} - 2.25\mathbf{k}\} \text{ m/s}^2 \qquad \textit{Ans.}$$

20

EXAMPLE | **20.2**

At the instant $\theta = 60°$, the gyrotop in Fig. 20–8 has three components of angular motion directed as shown and having magnitudes defined as:

$$\textit{Spin:}\ \ \omega_s = 10\ \text{rad/s, increasing at the rate of } 6\ \text{rad/s}^2$$
$$\textit{Nutation:}\ \ \omega_n = 3\ \text{rad/s, increasing at the rate of } 2\ \text{rad/s}^2$$
$$\textit{Precession:}\ \ \omega_p = 5\ \text{rad/s, increasing at the rate of } 4\ \text{rad/s}^2$$

Determine the angular velocity and angular acceleration of the top.

SOLUTION

Angular Velocity. The top rotates about the fixed point O. If the fixed and rotating frames are coincident at the instant shown, then the angular velocity can be expressed in terms of $\mathbf{i}, \mathbf{j}, \mathbf{k}$ components, with reference to the x, y, z frame; i.e.,

$$\boldsymbol{\omega} = -\omega_n\mathbf{i} + \omega_s \sin\theta\mathbf{j} + (\omega_p + \omega_s \cos\theta)\mathbf{k}$$
$$= -3\mathbf{i} + 10\sin 60°\mathbf{j} + (5 + 10\cos 60°)\mathbf{k}$$
$$= \{-3\mathbf{i} + 8.66\mathbf{j} + 10\mathbf{k}\}\ \text{rad/s} \qquad \textit{Ans.}$$

Angular Acceleration. As in the solution of Example 20.1, the angular acceleration $\boldsymbol{\alpha}$ will be determined by investigating separately the time rate of change of *each of the angular velocity components* as observed from the fixed X, Y, Z reference. We will choose an $\boldsymbol{\Omega}$ for the x, y, z reference so that the component of $\boldsymbol{\omega}$ being considered is viewed as having a *constant direction* when observed from x, y, z.

Careful examination of the motion of the top reveals that $\boldsymbol{\omega}_s$ has a *constant direction* relative to x, y, z if these axes rotate at $\boldsymbol{\Omega} = \boldsymbol{\omega}_n + \boldsymbol{\omega}_p$. Thus,

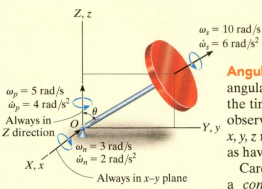

Z, z

$\omega_s = 10\ \text{rad/s}$
$\dot{\omega}_s = 6\ \text{rad/s}^2$

$\omega_p = 5\ \text{rad/s}$
$\dot{\omega}_p = 4\ \text{rad/s}^2$
Always in
Z direction

θ

O

Y, y

$\omega_n = 3\ \text{rad/s}$
$\dot{\omega}_n = 2\ \text{rad/s}^2$

X, x

Always in x–y plane

Fig. 20–8

$$\dot{\boldsymbol{\omega}}_s = (\dot{\boldsymbol{\omega}}_s)_{xyz} + (\boldsymbol{\omega}_n + \boldsymbol{\omega}_p) \times \boldsymbol{\omega}_s$$
$$= (6\sin 60°\mathbf{j} + 6\cos 60°\mathbf{k}) + (-3\mathbf{i} + 5\mathbf{k}) \times (10\sin 60°\mathbf{j} + 10\cos 60°\mathbf{k})$$
$$= \{-43.30\mathbf{i} + 20.20\mathbf{j} - 22.98\mathbf{k}\}\ \text{rad/s}^2$$

Since $\boldsymbol{\omega}_n$ *always* lies in the fixed X–Y plane, this vector has a *constant direction* if the motion is viewed from axes x, y, z having a rotation of $\boldsymbol{\Omega} = \boldsymbol{\omega}_p$ (not $\boldsymbol{\Omega} = \boldsymbol{\omega}_s + \boldsymbol{\omega}_p$). Thus,

$$\dot{\boldsymbol{\omega}}_n = (\dot{\boldsymbol{\omega}}_n)_{xyz} + \boldsymbol{\omega}_p \times \boldsymbol{\omega}_n = -2\mathbf{i} + (5\mathbf{k}) \times (-3\mathbf{i}) = \{-2\mathbf{i} - 15\mathbf{j}\}\,\text{rad/s}^2$$

Finally, the component $\boldsymbol{\omega}_p$ is *always directed* along the Z axis so that here it is not necessary to think of x, y, z as rotating, i.e., $\boldsymbol{\Omega} = \mathbf{0}$. Expressing the data in terms of the $\mathbf{i}, \mathbf{j}, \mathbf{k}$ components, we therefore have

$$\dot{\boldsymbol{\omega}}_p = (\dot{\boldsymbol{\omega}}_p)_{xyz} + \mathbf{0} \times \boldsymbol{\omega}_p = \{4\mathbf{k}\}\ \text{rad/s}^2$$

Thus, the angular acceleration of the top is

$$\boldsymbol{\alpha} = \dot{\boldsymbol{\omega}}_s + \dot{\boldsymbol{\omega}}_n + \dot{\boldsymbol{\omega}}_p = \{-45.3\mathbf{i} + 5.20\mathbf{j} - 19.0\mathbf{k}\}\ \text{rad/s}^2 \qquad \textit{Ans.}$$

20

20.3   General Motion

Shown in Fig. 20–9 is a rigid body subjected to general motion in three dimensions for which the angular velocity is $\boldsymbol{\omega}$ and the angular acceleration is $\boldsymbol{\alpha}$. If point A has a known motion of $\mathbf{v}_A$ and $\mathbf{a}_A$, the motion of any other point B can be determined by using a relative-motion analysis. In this section a *translating coordinate system* will be used to define the relative motion, and in the next section a reference that is both rotating and translating will be considered.

If the origin of the translating coordinate system x, y, z ($\boldsymbol{\Omega} = \mathbf{0}$) is located at the "base point" A, then, at the instant shown, the motion of the body can be regarded as the sum of an instantaneous translation of the body having a motion of $\mathbf{v}_A$, and $\mathbf{a}_A$, and a rotation of the body about an instantaneous axis passing through point A. Since the body is rigid, the motion of point B measured by an observer located at A is therefore the same as *the rotation of the body about a fixed point*. This relative motion occurs about the instantaneous axis of rotation and is defined by $\mathbf{v}_{B/A} = \boldsymbol{\omega} \times \mathbf{r}_{B/A}$, Eq. 20–3, and $\mathbf{a}_{B/A} = \boldsymbol{\alpha} \times \mathbf{r}_{B/A} + \boldsymbol{\omega} \times (\boldsymbol{\omega} \times \mathbf{r}_{B/A})$, Eq. 20–4. For translating axes, the relative motions are related to absolute motions by $\mathbf{v}_B = \mathbf{v}_A + \mathbf{v}_{B/A}$ and $\mathbf{a}_B = \mathbf{a}_A + \mathbf{a}_{B/A}$, Eqs. 16–15 and 16–17, so that the absolute velocity and acceleration of point B can be determined from the equations

$$\mathbf{v}_B = \mathbf{v}_A + \boldsymbol{\omega} \times \mathbf{r}_{B/A} \qquad (20\text{–}7)$$

and

$$\mathbf{a}_B = \mathbf{a}_A + \boldsymbol{\alpha} \times \mathbf{r}_{B/A} + \boldsymbol{\omega} \times (\boldsymbol{\omega} \times \mathbf{r}_{B/A}) \qquad (20\text{–}8)$$

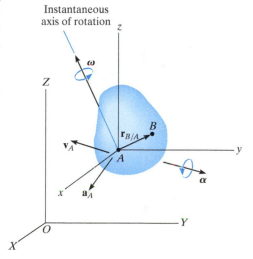

Fig. 20–9

These two equations are essentially the same as those describing the general plane motion of a rigid body, Eqs. 16–16 and 16–18. However, difficulty in application arises for three-dimensional motion, because $\boldsymbol{\alpha}$ now measures the change in *both* the magnitude and direction of $\boldsymbol{\omega}$.

Although this may be the case, a direct solution for $\mathbf{v}_B$ and $\mathbf{a}_B$ can be obtained by noting that $\mathbf{v}_{B/A} = \mathbf{v}_B - \mathbf{v}_A$, and so Eq. 20–7 becomes $\mathbf{v}_{B/A} = \boldsymbol{\omega} \times \mathbf{r}_{B/A}$. The cross product indicates that $\mathbf{v}_{B/A}$ is *perpendicular* to $\mathbf{r}_{B/A}$, and so, as noted by Eq. C–14 of Appendix C, we require

$$\mathbf{r}_{B/A} \cdot \mathbf{v}_{B/A} = 0 \qquad (20\text{–}9)$$

Taking the time derivative, we have

$$\mathbf{v}_{B/A} \cdot \mathbf{v}_{B/A} + \mathbf{r}_{B/A} \cdot \mathbf{a}_{B/A} = 0 \qquad (20\text{–}10)$$

Solution II of the following example illustrates application of this idea.

EXAMPLE 20.3

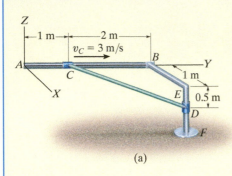

(a)

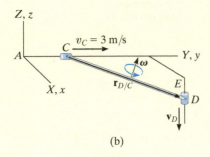

(b)

Fig. 20–10

If the collar at C in Fig. 20–10a moves toward B with a speed of 3 m/s, determine the velocity of the collar at D and the angular velocity of the bar at the instant shown. The bar is connected to the collars at its end points by ball-and-socket joints.

SOLUTION I

Bar CD is subjected to general motion. Why? The velocity of point D on the bar can be related to the velocity of point C by the equation

$$\mathbf{v}_D = \mathbf{v}_C + \boldsymbol{\omega} \times \mathbf{r}_{D/C}$$

The fixed and translating frames of reference are assumed to coincide at the instant considered, Fig. 20–10b. We have

$$\mathbf{v}_D = -v_D\mathbf{k} \qquad \mathbf{v}_C = \{3\mathbf{j}\} \text{ m/s}$$

$$\mathbf{r}_{D/C} = \{1\mathbf{i} + 2\mathbf{j} - 0.5\mathbf{k}\} \text{ m} \qquad \boldsymbol{\omega} = \omega_x\mathbf{i} + \omega_y\mathbf{j} + \omega_z\mathbf{k}$$

Substituting into the above equation we get

$$-v_D\mathbf{k} = 3\mathbf{j} + \begin{vmatrix} \mathbf{i} & \mathbf{j} & \mathbf{k} \\ \omega_x & \omega_y & \omega_z \\ 1 & 2 & -0.5 \end{vmatrix}$$

Expanding and equating the respective $\mathbf{i}, \mathbf{j}, \mathbf{k}$ components yields

$$-0.5\omega_y - 2\omega_z = 0 \tag{1}$$

$$0.5\omega_x + 1\omega_z + 3 = 0 \tag{2}$$

$$2\omega_x - 1\omega_y + v_D = 0 \tag{3}$$

These equations contain four unknowns.* A fourth equation can be written if the direction of $\boldsymbol{\omega}$ is specified. In particular, any component of $\boldsymbol{\omega}$ acting along the bar's axis has no effect on moving the collars. This is because the bar is *free to rotate* about its axis. Therefore, if $\boldsymbol{\omega}$ is specified as acting *perpendicular* to the axis of the bar, then $\boldsymbol{\omega}$ must have a *unique magnitude* to satisfy the above equations. Hence,

$$\boldsymbol{\omega} \cdot \mathbf{r}_{D/C} = (\omega_x\mathbf{i} + \omega_y\mathbf{j} + \omega_z\mathbf{k}) \cdot (1\mathbf{i} + 2\mathbf{j} - 0.5\mathbf{k}) = 0$$

$$1\omega_x + 2\omega_y - 0.5\omega_z = 0 \tag{4}$$

---

*Although this is the case, the magnitude of $\mathbf{v}_D$ can be obtained. For example, solve Eqs. 1 and 2 for ω_y and ω_x in terms of ω_z and substitute this into Eq. 3. Then ω_z will cancel out, which will allow a solution for v_D.

20

Solving Eqs. 1 through 4 simultaneously yields

$$\omega_x = -4.86 \text{ rad/s} \quad \omega_y = 2.29 \text{ rad/s} \quad \omega_z = -0.571 \text{ rad/s},$$
$$v_D = 12.0 \text{ m/s, so that} \qquad \omega = 5.40 \text{ rad/s} \qquad \textit{Ans.}$$

SOLUTION II

Applying Eq. 20–9, $\mathbf{v}_{D/C} = \mathbf{v}_D - \mathbf{v}_C = -v_D\mathbf{k} - 3\mathbf{j}$, so that

$$\mathbf{r}_{D/C} \cdot \mathbf{v}_{D/C} = (1\mathbf{i} + 2\mathbf{j} - 0.5\mathbf{k}) \cdot (-v_D\mathbf{k} - 3\mathbf{j}) = 0$$

$$(1)(0) + (2)(-3) + (-0.5)(-v_D) = 0$$

$$v_D = 12 \text{ m/s} \qquad \textit{Ans.}$$

Since $\boldsymbol{\omega}$ is *perpendicular* to $\mathbf{r}_{D/C}$ then $\mathbf{v}_{D/C} = \boldsymbol{\omega} \times \mathbf{r}_{D/C}$ or

$$v_{D/C} = \omega\, r_{D/C}$$

$$\sqrt{(-12)^2 + (-3)^2} = \omega\sqrt{(1)^2 + (2)^2 + (-0.5)^2}$$

$$\omega = 5.40 \text{ rad/s} \qquad \textit{Ans.}$$

PROBLEMS

20–1. The propeller of an airplane is rotating at a constant speed $\omega_x \mathbf{i}$, while the plane is undergoing a turn at a constant rate ω_t. Determine the angular acceleration of the propeller if (a) the turn is horizontal, i.e., $\omega_t \mathbf{k}$, and (b) the turn is vertical, downward, i.e., $\omega_t \mathbf{j}$.

20–2. The disk rotates about the z axis at a constant rate $\omega_z = 0.5$ rad/s without slipping on the horizontal plane. Determine the velocity and the acceleration of point A on the disk.

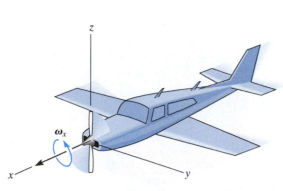

Prob. 20–1

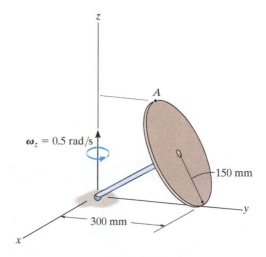

Prob. 20–2

20–3. The ladder of the fire truck rotates around the z axis with an angular velocity $\omega_1 = 0.15$ rad/s, which is increasing at 0.8 rad/s². At the same instant it is rotating upward at a constant rate $\omega_2 = 0.6$ rad/s. Determine the velocity and acceleration of point A located at the top of the ladder at this instant.

*20–4. The ladder of the fire truck rotates around the z axis with an angular velocity of $\omega_1 = 0.15$ rad/s, which is increasing at 0.2 rad/s². At the same instant it is rotating upward at $\omega_2 = 0.6$ rad/s while increasing at 0.4 rad/s². Determine the velocity and acceleration of point A located at the top of the ladder at this instant.

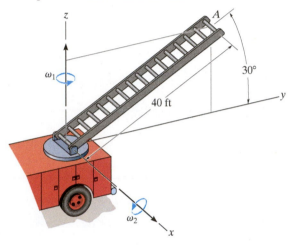

Probs. 20–3/4

20–5. If the plate gears A and B are rotating with the angular velocities shown, determine the angular velocity of gear C about the shaft DE. What is the angular velocity of DE about the y axis?

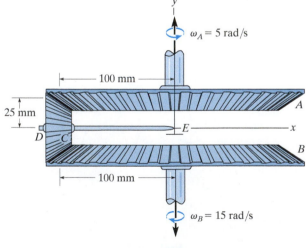

Prob. 20–5

20–6. The conical spool rolls on the plane without slipping. If the axle has an angular velocity of $\omega_1 = 3$ rad/s and an angular acceleration of $\alpha_1 = 2$ rad/s² at the instant shown, determine the angular velocity and angular acceleration of the spool at this instant.

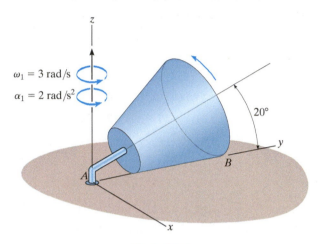

Prob. 20–6

20–7. At a given instant, the antenna has an angular motion $\omega_1 = 3$ rad/s and $\dot{\omega}_1 = 2$ rad/s² about the z axis. At this same instant $\theta = 30°$, the angular motion about the x axis is $\omega_2 = 1.5$ rad/s, and $\dot{\omega}_2 = 4$ rad/s². Determine the velocity and acceleration of the signal horn A at this instant. The distance from O to A is $d = 3$ ft.

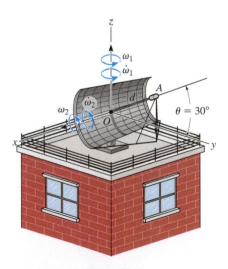

Prob. 20–7

*20–8. The disk rotates about the shaft S, while the shaft is turning about the z axis at a rate of $\omega_z = 4$ rad/s, which is increasing at 2 rad/s². Determine the velocity and acceleration of point A on the disk at the instant shown. No slipping occurs.

20–9. The disk rotates about the shaft S, while the shaft is turning about the z axis at a rate of $\omega_z = 4$ rad/s, which is increasing at 2 rad/s². Determine the velocity and acceleration of point B on the disk at the instant shown. No slipping occurs.

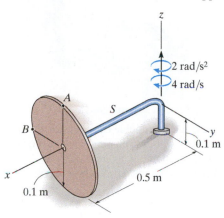

Probs. 20–8/9

20–10. The electric fan is mounted on a swivel support such that the fan rotates about the z axis at a constant rate of $\omega_z = 1$ rad/s and the fan blade is spinning at a constant rate $\omega_s = 60$ rad/s. If $\phi = 45°$ for the motion, determine the angular velocity and the angular acceleration of the blade.

20–11. The electric fan is mounted on a swivel support such that the fan rotates about the z axis at a constant rate of $\omega_z = 1$ rad/s and the fan blade is spinning at a constant rate $\omega_s = 60$ rad/s. If at the instant $\phi = 45°$, $\dot{\phi} = 2$ rad/s for the motion, determine the angular velocity and the angular acceleration of the blade.

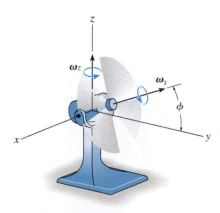

Probs. 20–10/11

*20–12. The drill pipe P turns at a constant angular rate $\omega_P = 4$ rad/s. Determine the angular velocity and angular acceleration of the conical rock bit, which rolls without slipping. Also, what are the velocity and acceleration of point A?

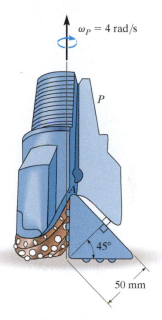

Prob. 20–12

20–13. The right circular cone rotates about the z axis at a constant rate of $\omega_1 = 4$ rad/s without slipping on the horizontal plane. Determine the magnitudes of the velocity and acceleration of points B and C.

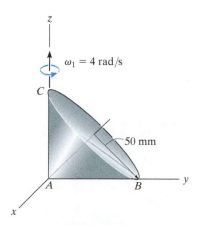

Prob. 20–13

20

20–14. The wheel is spinning about shaft AB with an angular velocity of $\omega_s = 10$ rad/s, which is increasing at a constant rate of $\dot{\omega}_s = 6$ rad/s^2, while the frame precesses about the z axis with an angular velocity of $\omega_p = 12$ rad/s, which is increasing at a constant rate of $\dot{\omega}_p = 3$ rad/s^2. Determine the velocity and acceleration of point C located on the rim of the wheel at this instant.

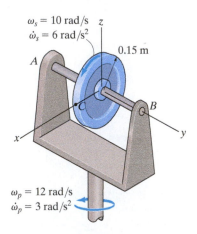

Prob. 20–14

20–15. At the instant shown, the tower crane rotates about the z axis with an angular velocity $\omega_1 = 0.25$ rad/s, which is increasing at 0.6 rad/s^2. The boom OA rotates downward with an angular velocity $\omega_2 = 0.4$ rad/s, which is increasing at 0.8 rad/s^2. Determine the velocity and acceleration of point A located at the end of the boom at this instant.

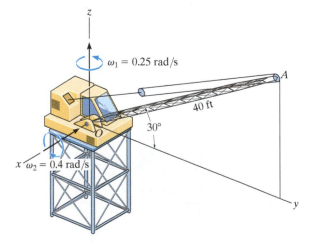

Prob. 20–15

*20–16. Gear A is fixed while gear B is free to rotate on the shaft S. If the shaft is turning about the z axis at $\omega_z = 5$ rad/s, while increasing at 2 rad/s^2, determine the velocity and acceleration of point P at the instant shown. The face of gear B lies in a vertical plane.

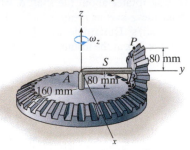

Prob. 20–16

20–17. The truncated double cone rotates about the z axis at $\omega_z = 0.4$ rad/s without slipping on the horizontal plane. If at this same instant ω_z is increasing at $\dot{\omega}_z = 0.5$ rad/s^2, determine the velocity and acceleration of point A on the cone.

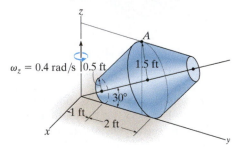

Prob. 20–17

20–18. Gear A is fixed to the crankshaft S, while gear C is fixed. Gear B and the propeller are free to rotate. The crankshaft is turning at 80 rad/s about its axis. Determine the magnitudes of the angular velocity of the propeller and the angular acceleration of gear B.

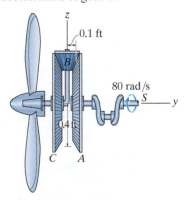

Prob. 20–18

20–19. Shaft BD is connected to a ball-and-socket joint at B, and a beveled gear A is attached to its other end. The gear is in mesh with a fixed gear C. If the shaft and gear A are *spinning* with a constant angular velocity $\omega_1 = 8$ rad/s, determine the angular velocity and angular acceleration of gear A.

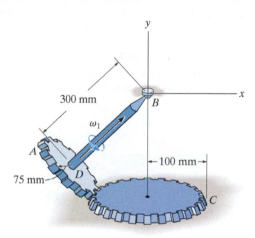

Prob. 20–19

*20–20. Gear B is driven by a motor mounted on turntable C. If gear A is held fixed, and the motor shaft rotates with a constant angular velocity of $\omega_y = 30$ rad/s, determine the angular velocity and angular acceleration of gear B.

20–21. Gear B is driven by a motor mounted on turntable C. If gear A and the motor shaft rotate with constant angular speeds of $\omega_A = \{10\mathbf{k}\}$ rad/s and $\omega_y = \{30\mathbf{j}\}$ rad/s, respectively, determine the angular velocity and angular acceleration of gear B.

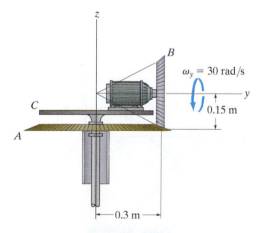

Probs. 20–20/21

20–22. The crane boom OA rotates about the z axis with a constant angular velocity of $\omega_1 = 0.15$ rad/s, while it is rotating downward with a constant angular velocity of $\omega_2 = 0.2$ rad/s. Determine the velocity and acceleration of point A located at the end of the boom at the instant shown.

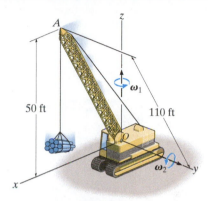

Prob. 20–22

20–23. The differential of an automobile allows the two rear wheels to rotate at different speeds when the automobile travels along a curve. For operation, the rear axles are attached to the wheels at one end and have beveled gears A and B on their other ends. The differential case D is placed over the left axle but can rotate about C independent of the axle. The case supports a pinion gear E on a shaft, which meshes with gears A and B. Finally, a ring gear G is *fixed* to the differential case so that the case rotates with the ring gear when the latter is driven by the drive pinion H. This gear, like the differential case, is free to rotate about the left wheel axle. If the drive pinion is turning at $\omega_H = 100$ rad/s and the pinion gear E is spinning about its shaft at $\omega_E = 30$ rad/s, determine the angular velocity, ω_A and ω_B, of each axle.

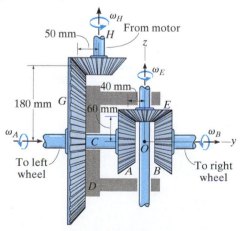

Prob. 20–23

*20–24. The end C of the plate rests on the horizontal plane, while end points A and B are restricted to move along the grooved slots. If at the instant shown A is moving downward with a constant velocity of $v_A = 4$ ft/s, determine the angular velocity of the plate and the velocities of points B and C.

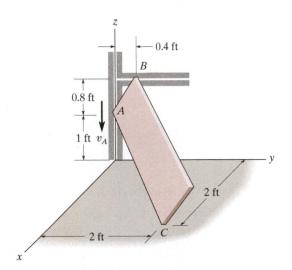

Prob. 20–24

20–25. Disk A rotates at a constant angular velocity of 10 rad/s. If rod BC is joined to the disk and a collar by ball-and-socket joints, determine the velocity of collar B at the instant shown. Also, what is the rod's angular velocity ω_{BC} if it is directed perpendicular to the axis of the rod?

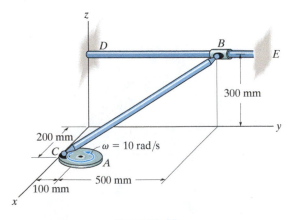

Prob. 20–25

20–26. Rod AB is attached to collars at its ends by using ball-and-socket joints. If collar A moves along the fixed rod at $v_A = 5$ m/s, determine the angular velocity of the rod and the velocity of collar B at the instant shown. Assume that the rod's angular velocity is directed perpendicular to the axis of the rod.

20–27. Rod AB is attached to collars at its ends by using ball-and-socket joints. If collar A moves along the fixed rod with a velocity of $v_A = 5$ m/s and has an acceleration $a_A = 2$ m/s² at the instant shown, determine the angular acceleration of the rod and the acceleration of collar B at this instant. Assume that the rod's angular velocity and angular acceleration are directed perpendicular to the axis of the rod.

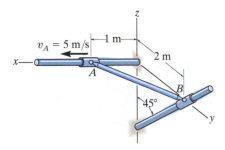

Probs. 20–26/27

*20–28. If the rod is attached with ball-and-socket joints to smooth collars A and B at its end points, determine the velocity of B at the instant shown if A is moving upward at a constant speed of $v_A = 5$ ft/s. Also, determine the angular velocity of the rod if it is directed perpendicular to the axis of the rod.

20–29. If the collar at A in Prob. 20–28 is moving upward with an acceleration of $\mathbf{a}_A = \{-2\mathbf{k}\}$ ft/s², at the instant its speed is $v_A = 5$ ft/s, determine the acceleration of the collar at B at this instant.

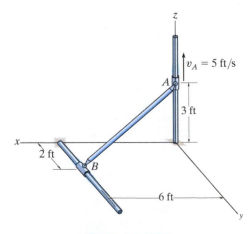

Probs. 20–28/29

20–30. Rod AB is attached to collars at its ends by ball-and-socket joints. If collar A has a speed $v_A = 4$ m/s, determine the speed of collar B at the instant $z = 2$ m. Assume the angular velocity of the rod is directed perpendicular to the rod.

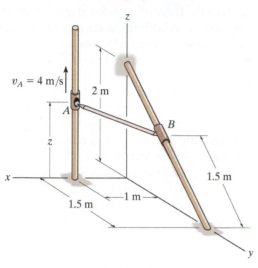

Prob. 20–30

20–31. The rod is attached to smooth collars A and B at its ends using ball-and-socket joints. Determine the speed of B at the instant shown if A is moving at $v_A = 8$ m/s. Also, determine the angular velocity of the rod if it is directed perpendicular to the axis of the rod.

***20–32.** If the collar A in Prob. 20–31 has a deceleration of $\mathbf{a}_A = \{-5\mathbf{k}\}$ m/s^2, at the instant shown, determine the acceleration of collar B at this instant.

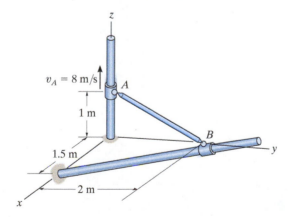

Probs. 20–31/32

20–33. Rod CD is attached to the rotating arms using ball-and-socket joints. If AC has the motion shown, determine the angular velocity of link BD at the instant shown.

20–34. Rod CD is attached to the rotating arms using ball-and-socket joints. If AC has the motion shown, determine the angular acceleration of link BD at this instant.

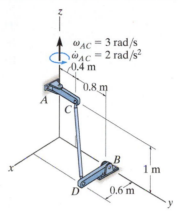

Probs. 20–33/34

20–35. Solve Prob. 20–28 if the connection at B consists of a pin as shown in the figure below, rather than a ball-and-socket joint. *Hint:* The constraint allows rotation of the rod both along the bar ($\mathbf{j}$ direction) and along the axis of the pin ($\mathbf{n}$ direction). Since there is no rotational component in the $\mathbf{u}$ direction, i.e., perpendicular to $\mathbf{n}$ and $\mathbf{j}$ where $\mathbf{u} = \mathbf{j} \times \mathbf{n}$, an additional equation for solution can be obtained from $\boldsymbol{\omega} \cdot \mathbf{u} = 0$. The vector $\mathbf{n}$ is in the same direction as $\mathbf{r}_{D/B} \times \mathbf{r}_{C/B}$.

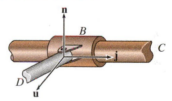

Prob. 20–35

***20–36.** Member ABC is pin connected at A and has a ball-and-socket joint at B. If the collar at B is moving along the inclined rod at $v_B = 8$ m/s, determine the velocity of point C at the instant shown. *Hint:* See Prob. 20–35.

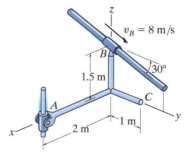

Prob. 20–36

20

*20.4   Relative-Motion Analysis Using Translating and Rotating Axes

The most general way to analyze the three-dimensional motion of a rigid body requires the use of x, y, z axes that both translate and rotate relative to a second frame X, Y, Z. This analysis also provides a means to determine the motions of two points A and B located on separate members of a mechanism, and the relative motion of one particle with respect to another when one or both particles are moving along *curved paths*.

As shown in Fig. 20–11, the locations of points A and B are specified relative to the X, Y, Z frame of reference by position vectors $\mathbf{r}_A$ and $\mathbf{r}_B$. The base point A represents the origin of the x, y, z coordinate system, which is translating and rotating with respect to X, Y, Z. At the instant considered, the velocity and acceleration of point A are $\mathbf{v}_A$ and $\mathbf{a}_A$, and the angular velocity and angular acceleration of the x, y, z axes are $\boldsymbol{\Omega}$ and $\dot{\boldsymbol{\Omega}} = d\boldsymbol{\Omega}/dt$. All these vectors are *measured* with respect to the X, Y, Z frame of reference, although they can be expressed in Cartesian component form along either set of axes.

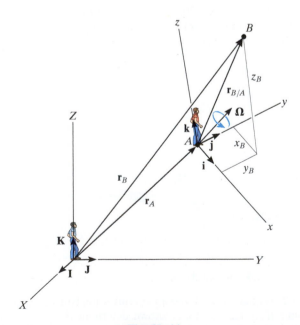

Fig. 20–11

Position.   If the position of "B with respect to A" is specified by the *relative-position vector* $\mathbf{r}_{B/A}$, Fig. 20–11, then, by vector addition,

$$\mathbf{r}_B = \mathbf{r}_A + \mathbf{r}_{B/A} \qquad (20\text{–}11)$$

where

$\mathbf{r}_B$ = position of B
$\mathbf{r}_A$ = position of the origin A
$\mathbf{r}_{B/A}$ = position of "B with respect to A"

Velocity.   The velocity of point B measured from X, Y, Z can be determined by taking the time derivative of Eq. 20–11,

$$\dot{\mathbf{r}}_B = \dot{\mathbf{r}}_A + \dot{\mathbf{r}}_{B/A}$$

The first two terms represent $\mathbf{v}_B$ and $\mathbf{v}_A$. The last term must be evaluated by applying Eq. 20–6, since $\mathbf{r}_{B/A}$ is measured with respect to a rotating reference. Hence,

$$\dot{\mathbf{r}}_{B/A} = (\dot{\mathbf{r}}_{B/A})_{xyz} + \boldsymbol{\Omega} \times \mathbf{r}_{B/A} = (\mathbf{v}_{B/A})_{xyz} + \boldsymbol{\Omega} \times \mathbf{r}_{B/A} \quad (20\text{–}12)$$

Therefore,

$$\mathbf{v}_B = \mathbf{v}_A + \boldsymbol{\Omega} \times \mathbf{r}_{B/A} + (\mathbf{v}_{B/A})_{xyz} \qquad (20\text{–}13)$$

where

$\mathbf{v}_B$ = velocity of B
$\mathbf{v}_A$ = velocity of the origin A of the x, y, z frame of reference
$(\mathbf{v}_{B/A})_{xyz}$ = velocity of "B with respect to A" as measured by an observer attached to the rotating x, y, z frame of reference
$\boldsymbol{\Omega}$ = angular velocity of the x, y, z frame of reference
$\mathbf{r}_{B/A}$ = position of "B with respect to A"

20

Acceleration.   The acceleration of point B measured from X, Y, Z is determined by taking the time derivative of Eq. 20–13.

$$\dot{\mathbf{v}}_B = \dot{\mathbf{v}}_A + \dot{\boldsymbol{\Omega}} \times \mathbf{r}_{B/A} + \boldsymbol{\Omega} \times \dot{\mathbf{r}}_{B/A} + \frac{d}{dt}(\mathbf{v}_{B/A})_{xyz}$$

The time derivatives defined in the first and second terms represent $\mathbf{a}_B$ and $\mathbf{a}_A$, respectively. The fourth term can be evaluated using Eq. 20–12, and the last term is evaluated by applying Eq. 20–6, which yields

$$\frac{d}{dt}(\mathbf{v}_{B/A})_{xyz} = (\dot{\mathbf{v}}_{B/A})_{xyz} + \boldsymbol{\Omega} \times (\mathbf{v}_{B/A})_{xyz} = (\mathbf{a}_{B/A})_{xyz} + \boldsymbol{\Omega} \times (\mathbf{v}_{B/A})_{xyz}$$

Here $(\mathbf{a}_{B/A})_{xyz}$ is the acceleration of B with respect to A measured from x, y, z. Substituting this result and Eq. 20–12 into the above equation and simplifying, we have

$$\mathbf{a}_B = \mathbf{a}_A + \dot{\boldsymbol{\Omega}} \times \mathbf{r}_{B/A} + \boldsymbol{\Omega} \times (\boldsymbol{\Omega} \times \mathbf{r}_{B/A}) + 2\boldsymbol{\Omega} \times (\mathbf{v}_{B/A})_{xyz} + (\mathbf{a}_{B/A})_{xyz}$$

$$(20\text{–}14)$$

where

$\mathbf{a}_B$ = acceleration of B

$\mathbf{a}_A$ = acceleration of the origin A of the x, y, z frame of reference

$(\mathbf{a}_{B/A})_{xyz}, (\mathbf{v}_{B/A})_{xyz}$ = relative acceleration and relative velocity of "B with respect to A" as measured by an observer attached to the rotating x, y, z frame of reference

$\dot{\boldsymbol{\Omega}}, \boldsymbol{\Omega}$ = angular acceleration and angular velocity of the x, y, z frame of reference

$\mathbf{r}_{B/A}$ = position of "B with respect to A"

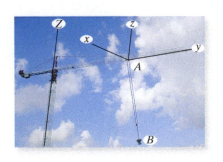

Complicated spatial motion of the concrete bucket B occurs due to the rotation of the boom about the Z axis, motion of the carriage A along the boom, and extension and swinging of the cable AB. A translating-rotating x, y, z coordinate system can be established on the carriage, and a relative-motion analysis can then be applied to study this motion. (© R.C. Hibbeler)

Equations 20–13 and 20–14 are identical to those used in Sec. 16.8 for analyzing relative plane motion.* In that case, however, application is simplified since $\boldsymbol{\Omega}$ and $\dot{\boldsymbol{\Omega}}$ have a *constant direction* which is always perpendicular to the plane of motion. For three-dimensional motion, $\dot{\boldsymbol{\Omega}}$ must be computed by using Eq. 20–6, since $\dot{\boldsymbol{\Omega}}$ depends on the change in *both* the magnitude and direction of $\boldsymbol{\Omega}$.

*Refer to Sec. 16.8 for an interpretation of the terms.

Procedure for Analysis

Three-dimensional motion of particles or rigid bodies can be analyzed with Eqs. 20–13 and 20–14 by using the following procedure.

Coordinate Axes.

- Select the location and orientation of the X, Y, Z and x, y, z coordinate axes. Most often solutions can be easily obtained if at the instant considered:

  (1) the origins are *coincident*

  (2) the axes are collinear

  (3) the axes are parallel

- If several components of angular velocity are involved in a problem, the calculations will be reduced if the x, y, z axes are selected such that only one component of angular velocity is observed with respect to this frame $(\boldsymbol{\Omega}_{xyz})$ and the frame rotates with $\boldsymbol{\Omega}$ defined by the other components of angular velocity.

Kinematic Equations.

- After the origin of the moving reference, A, is defined and the moving point B is specified, Eqs. 20–13 and 20–14 should then be written in symbolic form as

$$\mathbf{v}_B = \mathbf{v}_A + \boldsymbol{\Omega} \times \mathbf{r}_{B/A} + (\mathbf{v}_{B/A})_{xyz}$$

$$\mathbf{a}_B = \mathbf{a}_A + \dot{\boldsymbol{\Omega}} \times \mathbf{r}_{B/A} + \boldsymbol{\Omega} \times (\boldsymbol{\Omega} \times \mathbf{r}_{B/A}) + 2\boldsymbol{\Omega} \times (\mathbf{v}_{B/A})_{xyz} + (\mathbf{a}_{B/A})_{xyz}$$

- If $\mathbf{r}_A$ and $\boldsymbol{\Omega}$ appear to *change direction* when observed from the fixed X, Y, Z reference then use a set of primed reference axes, x', y', z' having a rotation $\boldsymbol{\Omega}' = \boldsymbol{\Omega}$. Equation 20–6 is then used to determine $\dot{\boldsymbol{\Omega}}$ and the motion $\mathbf{v}_A$ and $\mathbf{a}_A$ of the origin of the moving x, y, z axes.

- If $\mathbf{r}_{B/A}$ and $\boldsymbol{\Omega}_{xyz}$ appear to change direction as observed from x, y, z, then use a set of double-primed reference axes x'', y'', z'' having $\boldsymbol{\Omega}'' = \boldsymbol{\Omega}_{xyz}$ and apply Eq. 20–6 to determine $\dot{\boldsymbol{\Omega}}_{xyz}$ and the relative motion $(\mathbf{v}_{B/A})_{xyz}$ and $(\mathbf{a}_{B/A})_{xyz}$.

- After the final forms of $\dot{\boldsymbol{\Omega}}, \mathbf{v}_A, \mathbf{a}_A, \dot{\boldsymbol{\Omega}}_{xyz}, (\mathbf{v}_{B/A})_{xyz}$, and $(\mathbf{a}_{B/A})_{xyz}$ are obtained, numerical problem data can be substituted and the kinematic terms evaluated. The components of all these vectors can be selected either along the X, Y, Z or along the x, y, z axes. The choice is arbitrary, provided a consistent set of unit vectors is used.

20

EXAMPLE | 20.4

A motor and attached rod AB have the angular motions shown in Fig. 20–12. A collar C on the rod is located 0.25 m from A and is moving downward along the rod with a velocity of 3 m/s and an acceleration of 2 m/s². Determine the velocity and acceleration of C at this instant.

SOLUTION

Coordinate Axes.

The origin of the fixed X, Y, Z reference is chosen at the center of the platform, and the origin of the moving x, y, z frame at point A, Fig. 20–12. Since the collar is subjected to two components of angular motion, $\boldsymbol{\omega}_p$ and $\boldsymbol{\omega}_M$, it will be viewed as having an angular velocity of $\boldsymbol{\Omega}_{xyz} = \boldsymbol{\omega}_M$ in x, y, z. Therefore, the x, y, z axes will be attached to the platform so that $\boldsymbol{\Omega} = \boldsymbol{\omega}_p$.

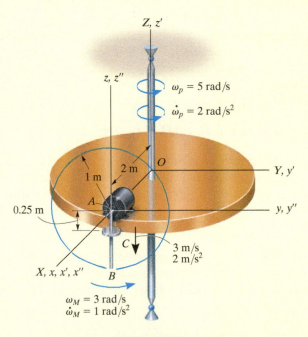

Fig. 20–12

Kinematic Equations. Equations 20–13 and 20–14, applied to points C and A, become

$$\mathbf{v}_C = \mathbf{v}_A + \boldsymbol{\Omega} \times \mathbf{r}_{C/A} + (\mathbf{v}_{C/A})_{xyz}$$

$$\mathbf{a}_C = \mathbf{a}_A + \dot{\boldsymbol{\Omega}} \times \mathbf{r}_{C/A} + \boldsymbol{\Omega} \times (\boldsymbol{\Omega} \times \mathbf{r}_{C/A}) + 2\boldsymbol{\Omega} \times (\mathbf{v}_{C/A})_{xyz} + (\mathbf{a}_{C/A})_{xyz}$$

Motion of A. Here $\mathbf{r}_A$ changes direction relative to X, Y, Z. To find the time derivatives of $\mathbf{r}_A$ we will use a set of x', y', z' axes coincident with the X, Y, Z axes that rotate at $\boldsymbol{\Omega}' = \boldsymbol{\omega}_p$. Thus,

$$\boldsymbol{\Omega} = \boldsymbol{\omega}_p = \{5\mathbf{k}\} \text{ rad/s } (\boldsymbol{\Omega} \text{ does not change direction relative to } X, Y, Z.)$$

$$\dot{\boldsymbol{\Omega}} = \dot{\boldsymbol{\omega}}_p = \{2\mathbf{k}\} \text{ rad/s}^2$$

$$\mathbf{r}_A = \{2\mathbf{i}\} \text{ m}$$

$$\mathbf{v}_A = \dot{\mathbf{r}}_A = (\dot{\mathbf{r}}_A)_{x'y'z'} + \boldsymbol{\omega}_p \times \mathbf{r}_A = 0 + 5\mathbf{k} \times 2\mathbf{i} = \{10\mathbf{j}\} \text{ m/s}$$

$$\mathbf{a}_A = \ddot{\mathbf{r}}_A = [(\ddot{\mathbf{r}}_A)_{x'y'z'} + \boldsymbol{\omega}_p \times (\dot{\mathbf{r}}_A)_{x'y'z'}] + \dot{\boldsymbol{\omega}}_p \times \mathbf{r}_A + \boldsymbol{\omega}_p \times \dot{\mathbf{r}}_A$$

$$= [0 + 0] + 2\mathbf{k} \times 2\mathbf{i} + 5\mathbf{k} \times 10\mathbf{j} = \{-50\mathbf{i} + 4\mathbf{j}\} \text{ m/s}^2$$

Motion of C with Respect to A. Here $\mathbf{r}_{C/A}$ changes direction relative to x, y, z, and so to find its time derivatives use a set of x'', y'', z'' axes that rotate at $\boldsymbol{\Omega}'' = \boldsymbol{\Omega}_{xyz} = \boldsymbol{\omega}_M$. Thus,

$$\boldsymbol{\Omega}_{xyz} = \boldsymbol{\omega}_M = \{3\mathbf{i}\} \text{ rad/s } (\boldsymbol{\Omega}_{xyz} \text{ does not change direction relative to } x, y, z.)$$

$$\dot{\boldsymbol{\Omega}}_{xyz} = \dot{\boldsymbol{\omega}}_M = \{1\mathbf{i}\} \text{ rad/s}^2$$

$$\mathbf{r}_{C/A} = \{-0.25\mathbf{k}\} \text{ m}$$

$$(\mathbf{v}_{C/A})_{xyz} = (\dot{\mathbf{r}}_{C/A})_{xyz} = (\dot{\mathbf{r}}_{C/A})_{x''y''z''} + \boldsymbol{\omega}_M \times \mathbf{r}_{C/A}$$

$$= -3\mathbf{k} + [3\mathbf{i} \times (-0.25\mathbf{k})] = \{0.75\mathbf{j} - 3\mathbf{k}\} \text{ m/s}$$

$$(\mathbf{a}_{C/A})_{xyz} = (\ddot{\mathbf{r}}_{C/A})_{xyz} = [(\ddot{\mathbf{r}}_{C/A})_{x''y''z''} + \boldsymbol{\omega}_M \times (\dot{\mathbf{r}}_{C/A})_{x''y''z''}] + \dot{\boldsymbol{\omega}}_M \times \mathbf{r}_{C/A} + \boldsymbol{\omega}_M \times (\dot{\mathbf{r}}_{C/A})_{xyz}$$

$$= [-2\mathbf{k} + 3\mathbf{i} \times (-3\mathbf{k})] + (1\mathbf{i}) \times (-0.25\mathbf{k}) + (3\mathbf{i}) \times (0.75\mathbf{j} - 3\mathbf{k})$$

$$= \{18.25\mathbf{j} + 0.25\mathbf{k}\} \text{ m/s}^2$$

Motion of C.

$$\mathbf{v}_C = \mathbf{v}_A + \boldsymbol{\Omega} \times \mathbf{r}_{C/A} + (\mathbf{v}_{C/A})_{xyz}$$

$$= 10\mathbf{j} + [5\mathbf{k} \times (-0.25\mathbf{k})] + (0.75\mathbf{j} - 3\mathbf{k})$$

$$= \{10.75\mathbf{j} - 3\mathbf{k}\} \text{ m/s} \qquad\qquad Ans.$$

$$\mathbf{a}_C = \mathbf{a}_A + \dot{\boldsymbol{\Omega}} \times \mathbf{r}_{C/A} + \boldsymbol{\Omega} \times (\boldsymbol{\Omega} \times \mathbf{r}_{C/A}) + 2\boldsymbol{\Omega} \times (\mathbf{v}_{C/A})_{xyz} + (\mathbf{a}_{C/A})_{xyz}$$

$$= (-50\mathbf{i} + 4\mathbf{j}) + [2\mathbf{k} \times (-0.25\mathbf{k})] + 5\mathbf{k} \times [5\mathbf{k} \times (-0.25\mathbf{k})]$$

$$+ 2[5\mathbf{k} \times (0.75\mathbf{j} - 3\mathbf{k})] + (18.25\mathbf{j} + 0.25\mathbf{k})$$

$$= \{-57.5\mathbf{i} + 22.25\mathbf{j} + 0.25\mathbf{k}\} \text{ m/s}^2 \qquad\qquad Ans.$$

20

EXAMPLE 20.5

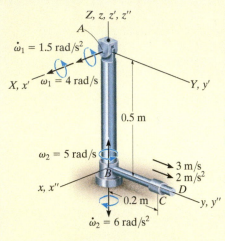

$\dot{\omega}_1 = 1.5 \text{ rad/s}^2$

$\omega_1 = 4 \text{ rad/s}$

0.5 m

$\omega_2 = 5 \text{ rad/s}$

3 m/s
2 m/s^2

0.2 m C

$\dot{\omega}_2 = 6 \text{ rad/s}^2$

Fig. 20–13

The pendulum shown in Fig. 20–13 consists of two rods; AB is pin supported at A and swings only in the Y–Z plane, whereas a bearing at B allows the attached rod BD to spin about rod AB. At a given instant, the rods have the angular motions shown. Also, a collar C, located 0.2 m from B, has a velocity of 3 m/s and an acceleration of 2 m/s^2 along the rod. Determine the velocity and acceleration of the collar at this instant.

SOLUTION I

Coordinate Axes. The origin of the fixed X, Y, Z frame will be placed at A. Motion of the collar is conveniently observed from B, so the origin of the x, y, z frame is located at this point. We will choose $\boldsymbol{\Omega} = \boldsymbol{\omega}_1$ and $\boldsymbol{\Omega}_{xyz} = \boldsymbol{\omega}_2$.

Kinematic Equations.

$$\mathbf{v}_C = \mathbf{v}_B + \boldsymbol{\Omega} \times \mathbf{r}_{C/B} + (\mathbf{v}_{C/B})_{xyz}$$

$$\mathbf{a}_C = \mathbf{a}_B + \dot{\boldsymbol{\Omega}} \times \mathbf{r}_{C/B} + \boldsymbol{\Omega} \times (\boldsymbol{\Omega} \times \mathbf{r}_{C/B}) + 2\boldsymbol{\Omega} \times (\mathbf{v}_{C/B})_{xyz} + (\mathbf{a}_{C/B})_{xyz}$$

Motion of B. To find the time derivatives of $\mathbf{r}_B$ let the x', y', z' axes rotate with $\boldsymbol{\Omega}' = \boldsymbol{\omega}_1$. Then

$$\boldsymbol{\Omega}' = \boldsymbol{\omega}_1 = \{4\mathbf{i}\} \text{ rad/s} \quad \dot{\boldsymbol{\Omega}}' = \dot{\boldsymbol{\omega}}_1 = \{1.5\mathbf{i}\} \text{ rad/s}^2$$

$$\mathbf{r}_B = \{-0.5\mathbf{k}\} \text{ m}$$

$$\mathbf{v}_B = \dot{\mathbf{r}}_B = (\dot{\mathbf{r}}_B)_{x'y'z'} + \boldsymbol{\omega}_1 \times \mathbf{r}_B = 0 + 4\mathbf{i} \times (-0.5\mathbf{k}) = \{2\mathbf{j}\} \text{ m/s}$$

$$\mathbf{a}_B = \ddot{\mathbf{r}}_B = [(\ddot{\mathbf{r}}_B)_{x'y'z'} + \boldsymbol{\omega}_1 \times (\dot{\mathbf{r}}_B)_{x'y'z'}] + \dot{\boldsymbol{\omega}}_1 \times \mathbf{r}_B + \boldsymbol{\omega}_1 \times \dot{\mathbf{r}}_B$$

$$= [0 + 0] + 1.5\mathbf{i} \times (-0.5\mathbf{k}) + 4\mathbf{i} \times 2\mathbf{j} = \{0.75\mathbf{j} + 8\mathbf{k}\} \text{ m/s}^2$$

Motion of C with Respect to B. To find the time derivatives of $\mathbf{r}_{C/B}$ relative to x, y, z, let the x'', y'', z'' axes rotate with $\boldsymbol{\Omega}_{xyz} = \boldsymbol{\omega}_2$. Then

$$\boldsymbol{\Omega}_{xyz} = \boldsymbol{\omega}_2 = \{5\mathbf{k}\} \text{ rad/s} \quad \dot{\boldsymbol{\Omega}}_{xyz} = \dot{\boldsymbol{\omega}}_2 = \{-6\mathbf{k}\} \text{ rad/s}^2$$

$$\mathbf{r}_{C/B} = \{0.2\mathbf{j}\} \text{ m}$$

$$(\mathbf{v}_{C/B})_{xyz} = (\dot{\mathbf{r}}_{C/B})_{xyz} = (\dot{\mathbf{r}}_{C/B})_{x''y''z''} + \boldsymbol{\omega}_2 \times \mathbf{r}_{C/B} = 3\mathbf{j} + 5\mathbf{k} \times 0.2\mathbf{j} = \{-1\mathbf{i} + 3\mathbf{j}\} \text{ m/s}$$

$$(\mathbf{a}_{C/B})_{xyz} = (\ddot{\mathbf{r}}_{C/B})_{xyz} = [(\ddot{\mathbf{r}}_{C/B})_{x''y''z''} + \boldsymbol{\omega}_2 \times (\dot{\mathbf{r}}_{C/B})_{x''y''z''}] + \dot{\boldsymbol{\omega}}_2 \times \mathbf{r}_{C/B} + \boldsymbol{\omega}_2 \times (\dot{\mathbf{r}}_{C/B})_{xyz}$$

$$= (2\mathbf{j} + 5\mathbf{k} \times 3\mathbf{j}) + (-6\mathbf{k} \times 0.2\mathbf{j}) + [5\mathbf{k} \times (-1\mathbf{i} + 3\mathbf{j})]$$

$$= \{-28.8\mathbf{i} - 3\mathbf{j}\} \text{ m/s}^2$$

Motion of C.

$$\mathbf{v}_C = \mathbf{v}_B + \boldsymbol{\Omega} \times \mathbf{r}_{C/B} + (\mathbf{v}_{C/B})_{xyz} = 2\mathbf{j} + 4\mathbf{i} \times 0.2\mathbf{j} + (-1\mathbf{i} + 3\mathbf{j})$$

$$= \{-1\mathbf{i} + 5\mathbf{j} + 0.8\mathbf{k}\} \text{ m/s} \qquad \textit{Ans.}$$

$$\mathbf{a}_C = \mathbf{a}_B + \dot{\boldsymbol{\Omega}} \times \mathbf{r}_{C/B} + \boldsymbol{\Omega} \times (\boldsymbol{\Omega} \times \mathbf{r}_{C/B}) + 2\boldsymbol{\Omega} \times (\mathbf{v}_{C/B})_{xyz} + (\mathbf{a}_{C/B})_{xyz}$$

$$= (0.75\mathbf{j} + 8\mathbf{k}) + (1.5\mathbf{i} \times 0.2\mathbf{j}) + [4\mathbf{i} \times (4\mathbf{i} \times 0.2\mathbf{j})]$$

$$+ 2[4\mathbf{i} \times (-1\mathbf{i} + 3\mathbf{j})] + (-28.8\mathbf{i} - 3\mathbf{j})$$

$$= \{-28.8\mathbf{i} - 5.45\mathbf{j} + 32.3\mathbf{k}\} \text{ m/s}^2 \qquad \textit{Ans.}$$

SOLUTION II

Coordinate Axes.   Here we will let the x, y, z axes rotate at

$$\mathbf{\Omega} = \boldsymbol{\omega}_1 + \boldsymbol{\omega}_2 = \{4\mathbf{i} + 5\mathbf{k}\} \text{ rad/s}$$

Then $\mathbf{\Omega}_{xyz} = \mathbf{0}$.

Motion of B.   From the constraints of the problem $\boldsymbol{\omega}_1$ does not change direction relative to X, Y, Z; however, the direction of $\boldsymbol{\omega}_2$ is changed by $\boldsymbol{\omega}_1$. Thus, to obtain $\dot{\mathbf{\Omega}}$ consider x', y', z' axes coincident with the X, Y, Z axes at A, so that $\mathbf{\Omega}' = \boldsymbol{\omega}_1$. Then taking the derivative of the components of $\mathbf{\Omega}$,

$$\dot{\mathbf{\Omega}} = \dot{\boldsymbol{\omega}}_1 + \dot{\boldsymbol{\omega}}_2 = [(\dot{\boldsymbol{\omega}}_1)_{x'y'z'} + \boldsymbol{\omega}_1 \times \boldsymbol{\omega}_1] + [(\dot{\boldsymbol{\omega}}_2)_{x'y'z'} + \boldsymbol{\omega}_1 \times \boldsymbol{\omega}_2]$$

$$= [1.5\mathbf{i} + \mathbf{0}] + [-6\mathbf{k} + 4\mathbf{i} \times 5\mathbf{k}] = \{1.5\mathbf{i} - 20\mathbf{j} - 6\mathbf{k}\} \text{ rad/s}^2$$

Also, $\boldsymbol{\omega}_1$ changes the direction of $\mathbf{r}_B$ so that the time derivatives of $\mathbf{r}_B$ can be found using the primed axes defined above. Hence,

$$\mathbf{v}_B = \dot{\mathbf{r}}_B = (\dot{\mathbf{r}}_B)_{x'y'z'} + \boldsymbol{\omega}_1 \times \mathbf{r}_B$$

$$= 0 + 4\mathbf{i} \times (-0.5\mathbf{k}) = \{2\mathbf{j}\} \text{ m/s}$$

$$\mathbf{a}_B = \ddot{\mathbf{r}}_B = [(\ddot{\mathbf{r}}_B)_{x'y'z'} + \boldsymbol{\omega}_1 \times (\dot{\mathbf{r}}_B)_{x'y'z'}] + \dot{\boldsymbol{\omega}}_1 \times \mathbf{r}_B + \boldsymbol{\omega}_1 \times \dot{\mathbf{r}}_B$$

$$= [0 + 0] + 1.5\mathbf{i} \times (-0.5\mathbf{k}) + 4\mathbf{i} \times 2\mathbf{j} = \{0.75\mathbf{j} + 8\mathbf{k}\} \text{ m/s}^2$$

Motion of C with Respect to B.

$$\mathbf{\Omega}_{xyz} = 0$$

$$\dot{\mathbf{\Omega}}_{xyz} = 0$$

$$\mathbf{r}_{C/B} = \{0.2\mathbf{j}\} \text{ m}$$

$$(\mathbf{v}_{C/B})_{xyz} = \{3\mathbf{j}\} \text{ m/s}$$

$$(\mathbf{a}_{C/B})_{xyz} = \{2\mathbf{j}\} \text{ m/s}^2$$

Motion of C.

$$\mathbf{v}_C = \mathbf{v}_B + \mathbf{\Omega} \times \mathbf{r}_{C/B} + (\mathbf{v}_{C/B})_{xyz}$$

$$= 2\mathbf{j} + [(4\mathbf{i} + 5\mathbf{k}) \times (0.2\mathbf{j})] + 3\mathbf{j}$$

$$= \{-1\mathbf{i} + 5\mathbf{j} + 0.8\mathbf{k}\} \text{ m/s} \qquad\qquad Ans.$$

$$\mathbf{a}_C = \mathbf{a}_B + \dot{\mathbf{\Omega}} \times \mathbf{r}_{C/B} + \mathbf{\Omega} \times (\mathbf{\Omega} \times \mathbf{r}_{C/B}) + 2\mathbf{\Omega} \times (\mathbf{v}_{C/B})_{xyz} + (\mathbf{a}_{C/B})_{xyz}$$

$$= (0.75\mathbf{j} + 8\mathbf{k}) + [(1.5\mathbf{i} - 20\mathbf{j} - 6\mathbf{k}) \times (0.2\mathbf{j})]$$

$$\quad + (4\mathbf{i} + 5\mathbf{k}) \times [(4\mathbf{i} + 5\mathbf{k}) \times 0.2\mathbf{j}] + 2[(4\mathbf{i} + 5\mathbf{k}) \times 3\mathbf{j}] + 2\mathbf{j}$$

$$= \{-28.8\mathbf{i} - 5.45\mathbf{j} + 32.3\mathbf{k}\} \text{ m/s}^2 \qquad\qquad Ans.$$

PROBLEMS

20–37. Solve Example 20.5 such that the x, y, z axes move with curvilinear translation, $\Omega = 0$ in which case the collar appears to have both an angular velocity $\Omega_{xyz} = \omega_1 + \omega_2$ and radial motion.

20–38. Solve Example 20.5 by fixing x, y, z axes to rod BD so that $\Omega = \omega_1 + \omega_2$. In this case the collar appears only to move radially outward along BD; hence $\Omega_{xyz} = 0$.

20–39. At the instant $\theta = 60°$, the telescopic boom AB of the construction lift is rotating with a constant angular velocity about the z axis of $\omega_1 = 0.5$ rad/s and about the pin at A with a constant angular speed of $\omega_2 = 0.25$ rad/s. Simultaneously, the boom is extending with a velocity of 1.5 ft/s, and it has an acceleration of 0.5 ft/s², both measured relative to the construction lift. Determine the velocity and acceleration of point B located at the end of the boom at this instant.

*20–40. At the instant $\theta = 60°$, the construction lift is rotating about the z axis with an angular velocity of $\omega_1 = 0.5$ rad/s and an angular acceleration of $\dot{\omega}_1 = 0.25$ rad/s² while the telescopic boom AB rotates about the pin at A with an angular velocity of $\omega_2 = 0.25$ rad/s and angular acceleration of $\dot{\omega}_2 = 0.1$ rad/s². Simultaneously, the boom is extending with a velocity of 1.5 ft/s, and it has an acceleration of 0.5 ft/s², both measured relative to the frame. Determine the velocity and acceleration of point B located at the end of the boom at this instant.

20–41. At the instant shown, the arm AB is rotating about the fixed pin A with an angular velocity $\omega_1 = 4$ rad/s and angular acceleration $\dot{\omega}_1 = 3$ rad/s². At this same instant, rod BD is rotating relative to rod AB with an angular velocity $\omega_2 = 5$ rad/s, which is increasing at $\dot{\omega}_2 = 7$ rad/s². Also, the collar C is moving along rod BD with a velocity of 3 m/s and an acceleration of 2 m/s², both measured relative to the rod. Determine the velocity and acceleration of the collar at this instant.

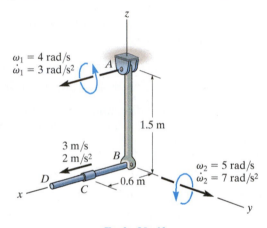

Prob. 20–41

20–42. At the instant $\theta = 30°$, the frame of the crane and the boom AB rotate with a constant angular velocity of $\omega_1 = 1.5$ rad/s and $\omega_2 = 0.5$ rad/s, respectively. Determine the velocity and acceleration of point B at this instant.

20–43. At the instant $\theta = 30°$, the frame of the crane is rotating with an angular velocity of $\omega_1 = 1.5$ rad/s and angular acceleration of $\dot{\omega}_1 = 0.5$ rad/s², while the boom AB rotates with an angular velocity of $\omega_2 = 0.5$ rad/s and angular acceleration of $\dot{\omega}_2 = 0.25$ rad/s². Determine the velocity and acceleration of point B at this instant.

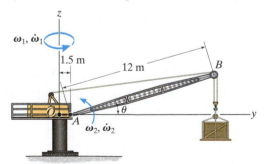

Probs. 20–42/43

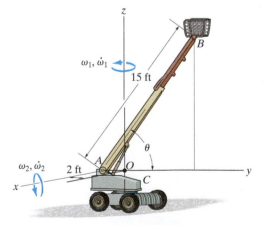

Probs. 20–39/40

*20–44. At the instant shown, the rod AB is rotating about the z axis with an angular velocity $\omega_1 = 4$ rad/s and an angular acceleration $\dot{\omega}_1 = 3$ rad/s². At this same instant, the circular rod has an angular motion relative to the rod as shown. If the collar C is moving down around the circular rod with a speed of 3 in./s, which is increasing at 8 in./s², both measured relative to the rod, determine the collar's velocity and acceleration at this instant.

20–46. At the instant shown, the industrial manipulator is rotating about the z axis at $\omega_1 = 5$ rad/s, and about joint B at $\omega_2 = 2$ rad/s. Determine the velocity and acceleration of the grip A at this instant, when $\phi = 30°$, $\theta = 45°$, and $r = 1.6$ m.

20–47. At the instant shown, the industrial manipulator is rotating about the z axis at $\omega_1 = 5$ rad/s, and $\dot{\omega}_1 = 2$ rad/s²; and about joint B at $\omega_2 = 2$ rad/s and $\dot{\omega}_2 = 3$ rad/s². Determine the velocity and acceleration of the grip A at this instant, when $\phi = 30°$, $\theta = 45°$, and $r = 1.6$ m.

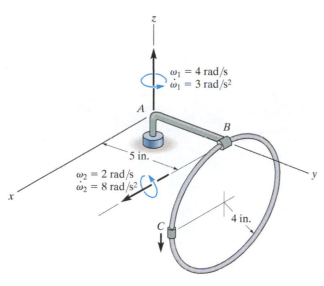

Prob. 20–44

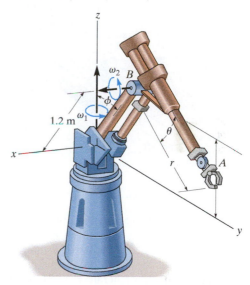

Probs. 20–46/47

20–45. The particle P slides around the circular hoop with a constant angular velocity of $\dot{\theta} = 6$ rad/s, while the hoop rotates about the x axis at a constant rate of $\omega = 4$ rad/s. If at the instant shown the hoop is in the x–y plane and the angle $\theta = 45°$, determine the velocity and acceleration of the particle at this instant.

*20–48. At the given instant, the rod is turning about the z axis with a constant angular velocity $\omega_1 = 3$ rad/s. At this same instant, the disk is spinning at $\omega_2 = 6$ rad/s when $\dot{\omega}_2 = 4$ rad/s², both measured *relative* to the rod. Determine the velocity and acceleration of point P on the disk at this instant.

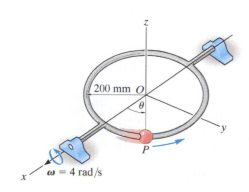

Prob. 20–45

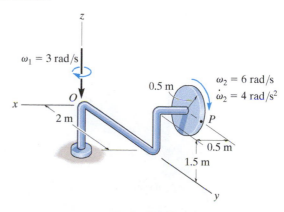

Prob. 20–48

20–49. At the instant shown, the backhoe is traveling forward at a constant speed $v_O = 2$ ft/s, and the boom ABC is rotating about the z axis with an angular velocity $\omega_1 = 0.8$ rad/s and an angular acceleration $\dot{\omega}_1 = 1.30$ rad/s². At this same instant the boom is rotating with $\omega_2 = 3$ rad/s when $\dot{\omega}_2 = 2$ rad/s², both measured relative to the frame. Determine the velocity and acceleration of point P on the bucket at this instant.

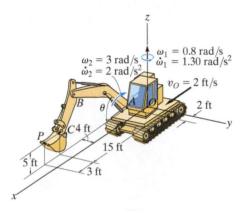

Prob. 20–49

20–50. At the instant shown, the arm OA of the conveyor belt is rotating about the z axis with a constant angular velocity $\omega_1 = 6$ rad/s, while at the same instant the arm is rotating upward at a constant rate $\omega_2 = 4$ rad/s. If the conveyor is running at a constant rate $\dot{r} = 5$ ft/s, determine the velocity and acceleration of the package P at the instant shown. Neglect the size of the package.

20–51. At the instant shown, the arm OA of the conveyor belt is rotating about the z axis with a constant angular velocity $\omega_1 = 6$ rad/s, while at the same instant the arm is rotating upward at a constant rate $\omega_2 = 4$ rad/s. If the conveyor is running at a rate $\dot{r} = 5$ ft/s, which is increasing at $\ddot{r} = 8$ ft/s², determine the velocity and acceleration of the package P at the instant shown. Neglect the size of the package.

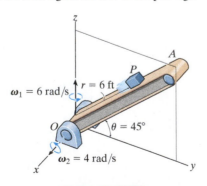

Probs. 20–50/51

*20–52.** The crane is rotating about the z axis with a constant rate $\omega_1 = 0.25$ rad/s, while the boom OA is rotating downward with a constant rate $\omega_2 = 0.4$ rad/s. Compute the velocity and acceleration of point A located at the top of the boom at the instant shown.

20–53. Solve Prob. 20–52 if the angular motions are increasing at $\dot{\omega}_1 = 0.4$ rad/s² and $\dot{\omega}_2 = 0.8$ rad/s² at the instant shown.

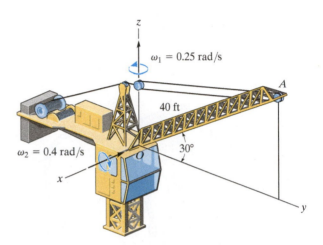

Probs. 20–52/53

20–54. At the instant shown, the arm AB is rotating about the fixed bearing with an angular velocity $\omega_1 = 2$ rad/s and angular acceleration $\dot{\omega}_1 = 6$ rad/s². At the same instant, rod BD is rotating relative to rod AB at $\omega_2 = 7$ rad/s, which is increasing at $\dot{\omega}_2 = 1$ rad/s². Also, the collar C is moving along rod BD with a velocity $\dot{r} = 2$ ft/s and a *deceleration* $\ddot{r} = -0.5$ ft/s², both measured relative to the rod. Determine the velocity and acceleration of the collar at this instant.

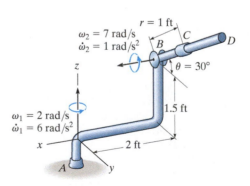

Prob. 20–54

CHAPTER REVIEW

Rotation About a Fixed Point

When a body rotates about a fixed point O, then points on the body follow a path that lies on the surface of a sphere centered at O.

Since the angular acceleration is a time rate of change in the angular velocity, then it is necessary to account for both the magnitude and directional changes of $\boldsymbol{\omega}$ when finding its time derivative. To do this, the angular velocity is often specified in terms of its component motions, such that the direction of some of these components will remain constant relative to rotating x, y, z axes. If this is the case, then the time derivative relative to the fixed axis can be determined using $\dot{\mathbf{A}} = (\dot{\mathbf{A}})_{xyz} + \boldsymbol{\Omega} \times \mathbf{A}$.

Once $\boldsymbol{\omega}$ and $\boldsymbol{\alpha}$ are known, the velocity and acceleration of any point P in the body can then be determined.

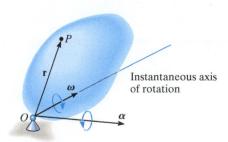

Instantaneous axis of rotation

$$v_P = \boldsymbol{\omega} \times \mathbf{r}$$

$$\mathbf{a}_P = \boldsymbol{\alpha} \times \mathbf{r} + \boldsymbol{\omega} \times (\boldsymbol{\omega} \times \mathbf{r})$$

General Motion

If the body undergoes general motion, then the motion of a point B on the body can be related to the motion of another point A using a relative motion analysis, with translating axes attached to A.

$$\mathbf{v}_B = \mathbf{v}_A + \boldsymbol{\omega} \times \mathbf{r}_{B/A}$$

$$\mathbf{a}_B = \mathbf{a}_A + \boldsymbol{\alpha} \times \mathbf{r}_{B/A} + \boldsymbol{\omega} \times (\boldsymbol{\omega} \times \mathbf{r}_{B/A})$$

Relative Motion Analysis Using Translating and Rotating Axes

The motion of two points A and B on a body, a series of connected bodies, or each point located on two different paths, can be related using a relative motion analysis with rotating and translating axes at A.

When applying the equations, to find $\mathbf{v}_B$ and $\mathbf{a}_B$, it is important to account for both the magnitude and directional changes of $\mathbf{r}_A$, $\mathbf{r}_{B/A}$, $\boldsymbol{\Omega}$, and $\boldsymbol{\Omega}_{xyz}$ when taking their time derivatives to find $\mathbf{v}_A$, $\mathbf{a}_A$, $(\mathbf{v}_{B/A})_{xyz}$, $(\mathbf{a}_{B/A})_{xyz}$, $\dot{\boldsymbol{\Omega}}$, and $\dot{\boldsymbol{\Omega}}_{xyz}$. To do this properly, one must use Eq. 20–6.

$$\mathbf{v}_B = \mathbf{v}_A + \boldsymbol{\Omega} \times \mathbf{r}_{B/A} + (\mathbf{v}_{B/A})_{xyz}$$

$$\mathbf{a}_B = \mathbf{a}_A + \dot{\boldsymbol{\Omega}} \times \mathbf{r}_{B/A} + \boldsymbol{\Omega} \times (\boldsymbol{\Omega} \times \mathbf{r}_{B/A}) + 2\boldsymbol{\Omega} \times (\mathbf{v}_{B/A})_{xyz} + (\mathbf{a}_{B/A})_{xyz}$$

20

Chapter 21

(© Derek Watt/Alamy)

The forces acting on each of these motorcycles can be determined using the equations of motion as discussed in this chapter.

Three-Dimensional Kinetics of a Rigid Body

CHAPTER OBJECTIVES

■ To introduce the methods for finding the moments of inertia and products of inertia of a body about various axes.

■ To show how to apply the principles of work and energy and linear and angular impulse and momentum to a rigid body having three-dimensional motion.

■ To develop and apply the equations of motion in three dimensions.

■ To study gyroscopic and torque-free motion.

*21.1   Moments and Products of Inertia

When studying the planar kinetics of a body, it was necessary to introduce the moment of inertia I_G, which was computed about an axis perpendicular to the plane of motion and passing through the body's mass center G. For the kinetic analysis of three-dimensional motion it will sometimes be necessary to calculate six inertial quantities. These terms, called the moments and products of inertia, describe in a particular way the distribution of mass for a body relative to a given coordinate system that has a specified orientation and point of origin.

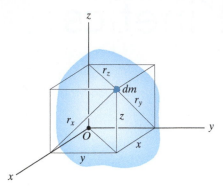

Fig. 21–1

Moment of Inertia.

Consider the rigid body shown in Fig. 21–1. The *moment of inertia* for a differential element dm of the body about any one of the three coordinate axes is defined as the product of the mass of the element and the square of the shortest distance from the axis to the element. For example, as noted in the figure, $r_x = \sqrt{y^2 + z^2}$, so that the mass moment of inertia of the element about the x axis is

$$dI_{xx} = r_x^2 \, dm = (y^2 + z^2) \, dm$$

The moment of inertia I_{xx} for the body can be determined by integrating this expression over the entire mass of the body. Hence, for each of the axes, we can write

$$
\begin{aligned}
I_{xx} &= \int_m r_x^2 \, dm = \int_m (y^2 + z^2) \, dm \\
I_{yy} &= \int_m r_y^2 \, dm = \int_m (x^2 + z^2) \, dm \\
I_{zz} &= \int_m r_z^2 \, dm = \int_m (x^2 + y^2) \, dm
\end{aligned}
\tag{21–1}
$$

Here it is seen that the moment of inertia is *always a positive quantity*, since it is the summation of the product of the mass dm, which is always positive, and the distances squared.

Product of Inertia.

The *product of inertia* for a differential element dm with respect to a set of *two orthogonal planes* is defined as the product of the mass of the element and the perpendicular (or shortest) distances from the planes to the element. For example, this distance is x to the y–z plane and it is y to the x–z plane, Fig. 21–1. The product of inertia dI_{xy} for the element is therefore

$$dI_{xy} = xy \, dm$$

Note also that $dI_{yx} = dI_{xy}$. By integrating over the entire mass, the products of inertia of the body with respect to each combination of planes can be expressed as

$$
\begin{aligned}
I_{xy} &= I_{yx} = \int_m xy \, dm \\
I_{yz} &= I_{zy} = \int_m yz \, dm \\
I_{xz} &= I_{zx} = \int_m xz \, dm
\end{aligned}
\tag{21–2}
$$

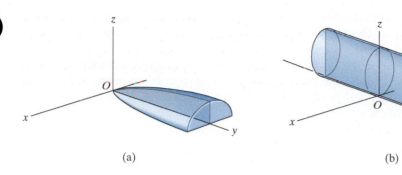

(a)   (b)

Fig. 21–2

Unlike the moment of inertia, which is always positive, the product of inertia may be positive, negative, or zero. The result depends on the algebraic signs of the two defining coordinates, which vary independently from one another. In particular, if either one or both of the orthogonal planes are *planes of symmetry* for the mass, the *product of inertia* with respect to these planes will be *zero*. In such cases, elements of mass will occur in *pairs* located on each side of the plane of symmetry. On one side of the plane the product of inertia for the element will be positive, while on the other side the product of inertia of the corresponding element will be negative, the sum therefore yielding zero. Examples of this are shown in Fig. 21–2. In the first case, Fig. 21–2a, the y–z plane is a plane of symmetry, and hence $I_{xy} = I_{xz} = 0$. Calculation of I_{yz} will yield a *positive* result, since all elements of mass are located using only positive y and z coordinates. For the cylinder, with the coordinate axes located as shown in Fig. 21–2b, the x–z and y–z planes are both planes of symmetry. Thus, $I_{xy} = I_{yz} = I_{zx} = 0$.

Parallel-Axis and Parallel-Plane Theorems.   The techniques of integration used to determine the moment of inertia of a body were described in Sec. 17.1. Also discussed were methods to determine the moment of inertia of a composite body, i.e., a body that is composed of simpler segments, as tabulated on the inside back cover. In both of these cases the *parallel-axis theorem* is often used for the calculations. This theorem, which was developed in Sec. 17.1, allows us to transfer the moment of inertia of a body from an axis passing through its mass center G to a parallel axis passing through some other point. If G has coordinates x_G, y_G, z_G defined with respect to the x, y, z axes, Fig. 21–3, then the parallel-axis equations used to calculate the moments of inertia about the x, y, z axes are

$$
\begin{aligned}
I_{xx} &= (I_{x'x'})_G + m(y_G^2 + z_G^2) \\
I_{yy} &= (I_{y'y'})_G + m(x_G^2 + z_G^2) \\
I_{zz} &= (I_{z'z'})_G + m(x_G^2 + y_G^2)
\end{aligned}
\tag{21–3}
$$

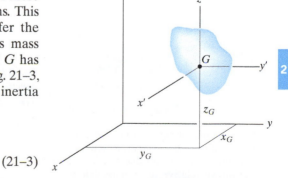

Fig. 21–3

21

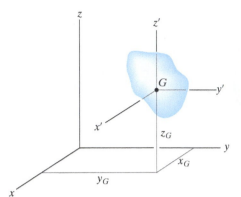

Fig. 21–3 (repeated)

The products of inertia of a composite body are computed in the same manner as the body's moments of inertia. Here, however, the *parallel-plane theorem* is important. This theorem is used to transfer the products of inertia of the body with respect to a set of three orthogonal planes passing through the body's mass center to a corresponding set of three parallel planes passing through some other point O. Defining the perpendicular distances between the planes as x_G, y_G, and z_G, Fig. 21–3, the parallel-plane equations can be written as

$$\begin{aligned} I_{xy} &= (I_{x'y'})_G + m x_G y_G \\ I_{yz} &= (I_{y'z'})_G + m y_G z_G \\ I_{zx} &= (I_{z'x'})_G + m z_G x_G \end{aligned} \qquad (21\text{–}4)$$

The derivation of these formulas is similar to that given for the parallel-axis equation, Sec. 17.1.

Inertia Tensor.   The inertial properties of a body are therefore completely characterized by nine terms, six of which are independent of one another. This set of terms is defined using Eqs. 21–1 and 21–2 and can be written as

$$\begin{pmatrix} I_{xx} & -I_{xy} & -I_{xz} \\ -I_{yx} & I_{yy} & -I_{yz} \\ -I_{zx} & -I_{zy} & I_{zz} \end{pmatrix}$$

This array is called an *inertia tensor*.* It has a unique set of values for a body when it is determined for each location of the origin O and orientation of the coordinate axes.

In general, for point O we can specify a unique axes inclination for which the products of inertia for the body are zero when computed with respect to these axes. When this is done, the inertia tensor is said to be "diagonalized" and may be written in the simplified form

$$\begin{pmatrix} I_x & 0 & 0 \\ 0 & I_y & 0 \\ 0 & 0 & I_z \end{pmatrix}$$

Here $I_x = I_{xx}$, $I_y = I_{yy}$, and $I_z = I_{zz}$ are termed the *principal moments of inertia* for the body, which are computed with respect to the *principal axes of inertia*. Of these three principal moments of inertia, one will be a maximum and another a minimum of the body's moment of inertia.

The dynamics of the space shuttle while it orbits the earth can be predicted only if its moments and products of inertia are known relative to its mass center. (©Ablestock/Getty Images)

*The negative signs are here as a consequence of the development of angular momentum, Eqs. 21–10.

The mathematical determination of the directions of principal axes of inertia will not be discussed here (see Prob. 21–22). However, there are many cases in which the principal axes can be determined by inspection. From the previous discussion it was noted that if the coordinate axes are oriented such that *two* of the three orthogonal planes containing the axes are planes of *symmetry* for the body, then all the products of inertia for the body are zero with respect to these coordinate planes, and hence these coordinate axes are principal axes of inertia. For example, the *x, y, z* axes shown in Fig. 21–2*b* represent the principal axes of inertia for the cylinder at point *O*.

Moment of Inertia About an Arbitrary Axis.

Consider the body shown in Fig. 21–4, where the nine elements of the inertia tensor have been determined with respect to the *x, y, z* axes having an origin at *O*. Here we wish to determine the moment of inertia of the body about the *Oa* axis, which has a direction defined by the unit vector $\mathbf{u}_a$. By definition $I_{Oa} = \int b^2 \, dm$, where *b* is the *perpendicular distance* from *dm* to *Oa*. If the position of *dm* is located using $\mathbf{r}$, then $b = r \sin\theta$, which represents the *magnitude* of the cross product $\mathbf{u}_a \times \mathbf{r}$. Hence, the moment of inertia can be expressed as

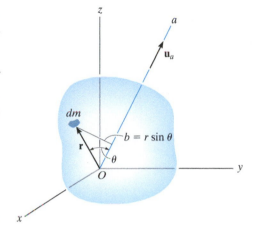

Fig. 21–4

$$I_{Oa} = \int_m |(\mathbf{u}_a \times \mathbf{r})|^2 dm = \int_m (\mathbf{u}_a \times \mathbf{r}) \cdot (\mathbf{u}_a \times \mathbf{r}) dm$$

Provided $\mathbf{u}_a = u_x \mathbf{i} + u_y \mathbf{j} + u_z \mathbf{k}$ and $\mathbf{r} = x\mathbf{i} + y\mathbf{j} + z\mathbf{k}$, then $\mathbf{u}_a \times \mathbf{r} = (u_y z - u_z y)\mathbf{i} + (u_z x - u_x z)\mathbf{j} + (u_x y - u_y x)\mathbf{k}$. After substituting and performing the dot-product operation, the moment of inertia is

$$I_{Oa} = \int_m [(u_y z - u_z y)^2 + (u_z x - u_x z)^2 + (u_x y - u_y x)^2] dm$$

$$= u_x^2 \int_m (y^2 + z^2) dm + u_y^2 \int_m (z^2 + x^2) dm + u_z^2 \int_m (x^2 + y^2) \, dm$$

$$- 2u_x u_y \int_m xy \, dm - 2u_y u_z \int_m yz \, dm - 2u_z u_x \int_m zx \, dm$$

Recognizing the integrals to be the moments and products of inertia of the body, Eqs. 21–1 and 21–2, we have

$$I_{Oa} = I_{xx} u_x^2 + I_{yy} u_y^2 + I_{zz} u_z^2 - 2I_{xy} u_x u_y - 2I_{yz} u_y u_z - 2I_{zx} u_z u_x \qquad (21\text{–}5)$$

Thus, if the inertia tensor is specified for the *x, y, z* axes, the moment of inertia of the body about the inclined *Oa* axis can be found. For the calculation, the direction cosines u_x, u_y, u_z of the axes must be determined. These terms specify the cosines of the coordinate direction angles α, β, γ made between the positive *Oa* axis and the positive *x, y, z* axes, respectively (see Appendix B).

21

EXAMPLE 21.1

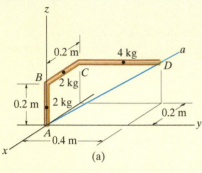

(a)

Fig. 21–5

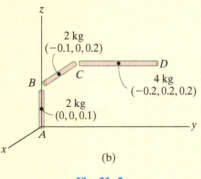

(b)

Determine the moment of inertia of the bent rod shown in Fig. 21–5a about the Aa axis. The mass of each of the three segments is given in the figure.

SOLUTION

Before applying Eq. 21–5, it is first necessary to determine the moments and products of inertia of the rod with respect to the x, y, z axes. This is done using the formula for the moment of inertia of a slender rod, $I = \frac{1}{12}ml^2$, and the parallel-axis and parallel-plane theorems, Eqs. 21–3 and 21–4. Dividing the rod into three parts and locating the mass center of each segment, Fig. 21–5b, we have

$$I_{xx} = \left[\tfrac{1}{12}(2)(0.2)^2 + 2(0.1)^2\right] + [0 + 2(0.2)^2]$$
$$+ \left[\tfrac{1}{12}(4)(0.4)^2 + 4((0.2)^2 + (0.2)^2)\right] = 0.480 \text{ kg} \cdot \text{m}^2$$

$$I_{yy} = \left[\tfrac{1}{12}(2)(0.2)^2 + 2(0.1)^2\right] + \left[\tfrac{1}{12}(2)(0.2)^2 + 2((-0.1)^2 + (0.2)^2)\right]$$
$$+ [0 + 4((-0.2)^2 + (0.2)^2)] = 0.453 \text{ kg} \cdot \text{m}^2$$

$$I_{zz} = \left[0 + 0\right] + \left[\tfrac{1}{12}(2)(0.2)^2 + 2(-0.1)^2\right] + \left[\tfrac{1}{12}(4)(0.4)^2 +\right.$$
$$\left.4((-0.2)^2 + (0.2)^2)\right] = 0.400 \text{ kg} \cdot \text{m}^2$$

$$I_{xy} = [0 + 0] + [0 + 0] + [0 + 4(-0.2)(0.2)] = -0.160 \text{ kg} \cdot \text{m}^2$$

$$I_{yz} = [0 + 0] + [0 + 0] + [0 + 4(0.2)(0.2)] = 0.160 \text{ kg} \cdot \text{m}^2$$

$$I_{zx} = [0 + 0] + [0 + 2(0.2)(-0.1)] +$$
$$[0 + 4(0.2)(-0.2)] = -0.200 \text{ kg} \cdot \text{m}^2$$

The Aa axis is defined by the unit vector

$$\mathbf{u}_{Aa} = \frac{\mathbf{r}_D}{r_D} = \frac{-0.2\mathbf{i} + 0.4\mathbf{j} + 0.2\mathbf{k}}{\sqrt{(-0.2)^2 + (0.4)^2 + (0.2)^2}} = -0.408\mathbf{i} + 0.816\mathbf{j} + 0.408\mathbf{k}$$

Thus,

$$u_x = -0.408 \quad u_y = 0.816 \quad u_z = 0.408$$

Substituting these results into Eq. 21–5 yields

$$I_{Aa} = I_{xx}u_x^2 + I_{yy}u_y^2 + I_{zz}u_z^2 - 2I_{xy}u_xu_y - 2I_{yz}u_yu_z - 2I_{zx}u_zu_x$$

$$= 0.480(-0.408)^2 + (0.453)(0.816)^2 + 0.400(0.408)^2$$

$$- 2(-0.160)(-0.408)(0.816) - 2(0.160)(0.816)(0.408)$$

$$- 2(-0.200)(0.408)(-0.408)$$

$$= 0.169 \text{ kg} \cdot \text{m}^2 \qquad \qquad \textit{Ans.}$$

PROBLEMS

21–1. Show that the sum of the moments of inertia of a body, $I_{xx} + I_{yy} + I_{zz}$, is independent of the orientation of the x, y, z axes and thus depends only on the location of the origin.

21–2. Determine the moment of inertia of the cone with respect to a vertical $\bar{y}$ axis passing through the cone's center of mass. What is the moment of inertia about a parallel axis y' that passes through the diameter of the base of the cone? The cone has a mass m.

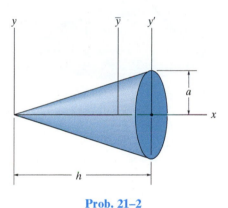

Prob. 21–2

21–3. Determine moment of inertia I_y of the solid formed by revolving the shaded area around the x axis. The density of the material is $\rho = 12 \text{ slug/ft}^3$.

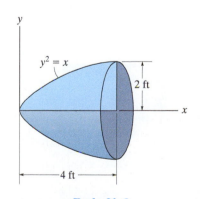

Prob. 21–3

*21–4. Determine the moments of inertia I_x and I_y of the paraboloid of revolution. The mass of the paraboloid is 20 slug.

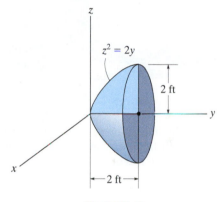

Prob. 21–4

21–5. Determine by direct integration the product of inertia I_{yz} for the homogeneous prism. The density of the material is ρ. Express the result in terms of the total mass m of the prism.

21–6. Determine by direct integration the product of inertia I_{xy} for the homogeneous prism. The density of the material is ρ. Express the result in terms of the total mass m of the prism.

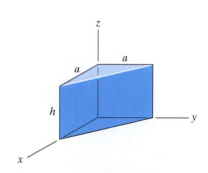

Probs. 21–5/6

21–7. Determine the product of inertia I_{xy} of the object formed by revolving the shaded area about the line $x = 5$ ft. Express the result in terms of the density of the material, ρ.

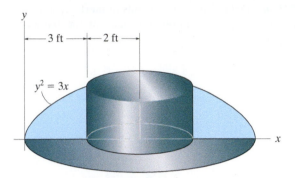

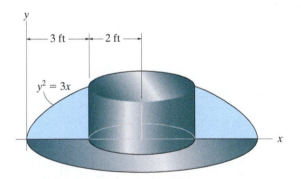

Prob. 21–7

*21–8. Determine the moment of inertia I_y of the object formed by revolving the shaded area about the line $x = 5$ ft. Express the result in terms of the density of the material, ρ.

Prob. 21–8

21–9. Determine the moment of inertia of the cone about the z' axis. The weight of the cone is 15 lb, the *height* is $h = 1.5$ ft and the radius is $r = 0.5$ ft.

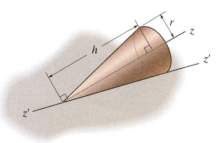

Prob. 21–9

21–10. Determine the radii of gyration k_x and k_y for the solid formed by revolving the shaded area about the y axis. The density of the material is ρ.

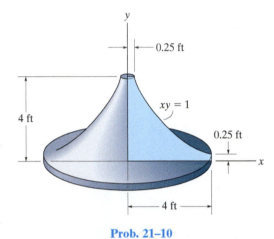

Prob. 21–10

21–11. Determine the moment of inertia of the cylinder with respect to the a–a axis of the cylinder. The cylinder has a mass m.

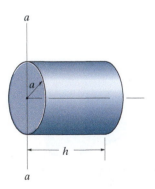

Prob. 21–11

*21–12. Determine the moment of inertia I_{xx} of the composite plate assembly. The plates have a specific weight of 6 lb/ft².

21–13. Determine the product of inertia I_{yz} of the composite plate assembly. The plates have a weight of 6 lb/ft².

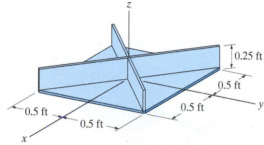

Probs. 21–12/13

21–14. Determine the products of inertia I_{xy}, I_{yz}, and I_{xz}, of the thin plate. The material has a density per unit area of 50 kg/m².

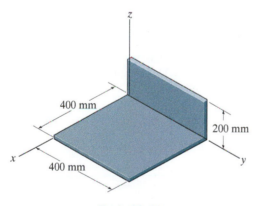

Prob. 21–14

21–15. Determine the moment of inertia of both the 1.5-kg rod and 4-kg disk about the z' axis.

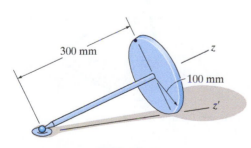

Prob. 21–15

*21–16. The bent rod has a mass of 3 kg/m. Determine the moment of inertia of the rod about the O–a axis.

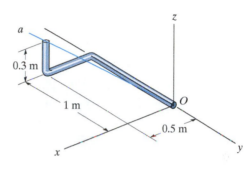

Prob. 21–16

21–17. The bent rod has a weight of 1.5 lb/ft. Locate the center of gravity $G(\bar{x}, \bar{y})$ and determine the principal moments of inertia $I_{x'}$, $I_{y'}$, and $I_{z'}$ of the rod with respect to the x', y', z' axes.

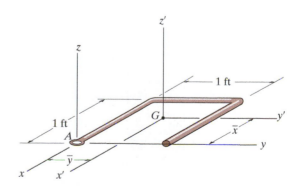

Prob. 21–17

21

21–18. Determine the moment of inertia of the rod-and-disk assembly about the x axis. The disks each have a weight of 12 lb. The two rods each have a weight of 4 lb, and their ends extend to the rims of the disks.

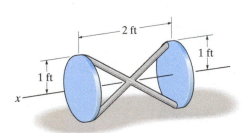

Prob. 21–18

*21–20. Determine the moment of inertia of the disk about the axis of shaft AB. The disk has a mass of 15 kg.

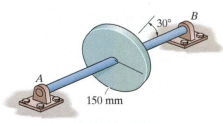

Prob. 21–20

21–21. The thin plate has a weight of 5 lb and each of the four rods weighs 3 lb. Determine the moment of inertia of the assembly about the z axis.

21–19. Determine the moment of inertia of the composite body about the aa axis. The cylinder weighs 20 lb, and each hemisphere weighs 10 lb.

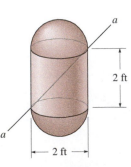

Prob. 21–19

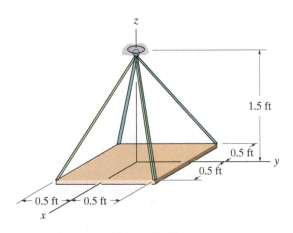

Prob. 21–21

21.2 Angular Momentum

In this section we will develop the necessary equations used to determine the angular momentum of a rigid body about an arbitrary point. These equations will provide a means for developing both the principle of impulse and momentum and the equations of rotational motion for a rigid body.

Consider the rigid body in Fig. 21–6, which has a mass m and center of mass at G. The X, Y, Z coordinate system represents an inertial frame of reference, and hence, its axes are fixed or translate with a constant velocity. The angular momentum as measured from this reference will be determined relative to the arbitrary point A. The position vectors $\mathbf{r}_A$ and $\boldsymbol{\rho}_A$ are drawn from the origin of coordinates to point A and from A to the ith particle of the body. If the particle's mass is m_i, the angular momentum about point A is

$$(\mathbf{H}_A)_i = \boldsymbol{\rho}_A \times m_i \mathbf{v}_i$$

where $\mathbf{v}_i$ represents the particle's velocity measured from the X, Y, Z coordinate system. If the body has an angular velocity $\boldsymbol{\omega}$ at the instant considered, $\mathbf{v}_i$ may be related to the velocity of A by applying Eq. 20–7, i.e.,

$$\mathbf{v}_i = \mathbf{v}_A + \boldsymbol{\omega} \times \boldsymbol{\rho}_A$$

Thus,

$$(\mathbf{H}_A)_i = \boldsymbol{\rho}_A \times m_i(\mathbf{v}_A + \boldsymbol{\omega} \times \boldsymbol{\rho}_A)$$

$$= (\boldsymbol{\rho}_A m_i) \times \mathbf{v}_A + \boldsymbol{\rho}_A \times (\boldsymbol{\omega} \times \boldsymbol{\rho}_A)m_i$$

Summing the moments of all the particles of the body requires an integration. Since $m_i \rightarrow dm$, we have

$$\mathbf{H}_A = \left(\int_m \boldsymbol{\rho}_A \, dm \right) \times \mathbf{v}_A + \int_m \boldsymbol{\rho}_A \times (\boldsymbol{\omega} \times \boldsymbol{\rho}_A) \, dm \qquad (21\text{–}6)$$

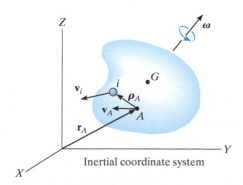

Inertial coordinate system

Fig. 21–6

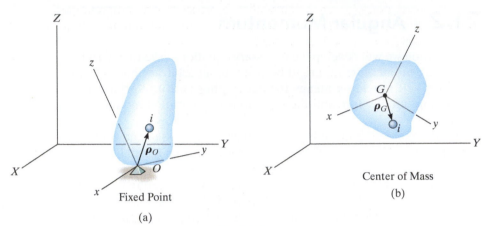

Fixed Point
(a)

Center of Mass
(b)

Fig. 21–7

Fixed Point O.

If A becomes a *fixed point* O in the body, Fig. 21–7a, then $\mathbf{v}_A = \mathbf{0}$ and Eq. 21–6 reduces to

$$\mathbf{H}_O = \int_m \boldsymbol{\rho}_O \times (\boldsymbol{\omega} \times \boldsymbol{\rho}_O)\, dm \qquad (21\text{–}7)$$

Center of Mass G.

If A is located at the *center of mass* G of the body, Fig. 21–7b, then $\int_m \boldsymbol{\rho}_A\, dm = \mathbf{0}$ and

$$\mathbf{H}_G = \int_m \boldsymbol{\rho}_G \times (\boldsymbol{\omega} \times \boldsymbol{\rho}_G)\, dm \qquad (21\text{–}8)$$

Arbitrary Point A.

In general, A can be a point other than O or G, Fig. 21–7c, in which case Eq. 21–6 may nevertheless be simplified to the following form (see Prob. 21–23).

$$\mathbf{H}_A = \boldsymbol{\rho}_{G/A} \times m\mathbf{v}_G + \mathbf{H}_G \qquad (21\text{–}9)$$

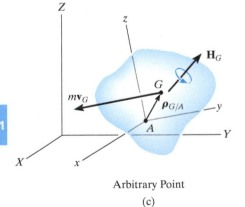

Arbitrary Point
(c)

Here the angular momentum consists of two parts—the moment of the linear momentum $m\mathbf{v}_G$ of the body about point A added (vectorially) to the angular momentum $\mathbf{H}_G$. Equation 21–9 can also be used to determine the angular momentum of the body about a fixed point O. The results, of course, will be the same as those found using the more convenient Eq. 21–7.

Rectangular Components of H.

To make practical use of Eqs. 21–7 through 21–9, the angular momentum must be expressed in terms of its scalar components. For this purpose, it is convenient to choose a second set of x, y, z axes having an arbitrary orientation relative to the X, Y, Z axes, Fig. 21–7, and for a general formulation, note that Eqs. 21–7 and 21–8 are both of the form

$$\mathbf{H} = \int_m \boldsymbol{\rho} \times (\boldsymbol{\omega} \times \boldsymbol{\rho})dm$$

Expressing $\mathbf{H}, \boldsymbol{\rho}$, and $\boldsymbol{\omega}$ in terms of x, y, z components, we have

$$H_x\mathbf{i} + H_y\mathbf{j} + H_z\mathbf{k} = \int_m (x\mathbf{i} + y\mathbf{j} + z\mathbf{k}) \times [(\omega_x\mathbf{i} + \omega_y\mathbf{j} + \omega_z\mathbf{k})$$
$$\times (x\mathbf{i} + y\mathbf{j} + z\mathbf{k})]dm$$

Expanding the cross products and combining terms yields

$$H_x\mathbf{i} + H_y\mathbf{j} + H_z\mathbf{k} = \left[\omega_x\int_m (y^2 + z^2)dm - \omega_y\int_m xy\,dm - \omega_z\int_m xz\,dm\right]\mathbf{i}$$
$$+ \left[-\omega_x\int_m xy\,dm + \omega_y\int_m (x^2 + z^2)dm - \omega_z\int_m yz\,dm\right]\mathbf{j}$$
$$+ \left[-\omega_x\int_m zx\,dm - \omega_y\int_m yz\,dm + \omega_z\int_m (x^2 + y^2)dm\right]\mathbf{k}$$

Equating the respective $\mathbf{i}, \mathbf{j}, \mathbf{k}$ components and recognizing that the integrals represent the moments and products of inertia, we obtain

$$\boxed{\begin{aligned} H_x &= I_{xx}\omega_x - I_{xy}\omega_y - I_{xz}\omega_z \\ H_y &= -I_{yx}\omega_x + I_{yy}\omega_y - I_{yz}\omega_z \\ H_z &= -I_{zx}\omega_x - I_{zy}\omega_y + I_{zz}\omega_z \end{aligned}} \qquad (21\text{–}10)$$

These equations can be simplified further if the x, y, z coordinate axes are oriented such that they become *principal axes of inertia* for the body at the point. When these axes are used, the products of inertia $I_{xy} = I_{yz} = I_{zx} = 0$, and if the principal moments of inertia about the x, y, z axes are represented as $I_x = I_{xx}, I_y = I_{yy}$, and $I_z = I_{zz}$, the three components of angular momentum become

$$\boxed{H_x = I_x\omega_x \quad H_y = I_y\omega_y \quad H_z = I_z\omega_z} \qquad (21\text{–}11)$$

21

The motion of the astronaut is controlled by use of small directional jets attached to his or her space suit. The impulses these jets provide must be carefully specified in order to prevent tumbling and loss of orientation. (© NASA)

Principle of Impulse and Momentum. Now that the formulation of the angular momentum for a body has been developed, the *principle of impulse and momentum*, as discussed in Sec. 19.2, can be used to solve kinetic problems which involve *force, velocity, and time*. For this case, the following two vector equations are available:

$$m(\mathbf{v}_G)_1 + \Sigma \int_{t_1}^{t_2} \mathbf{F}\, dt = m(\mathbf{v}_G)_2 \qquad (21\text{–}12)$$

$$(\mathbf{H}_O)_1 + \Sigma \int_{t_1}^{t_2} \mathbf{M}_O\, dt = (\mathbf{H}_O)_2 \qquad (21\text{–}13)$$

In three dimensions each vector term can be represented by three scalar components, and therefore a total of *six scalar equations* can be written. Three equations relate the linear impulse and momentum in the *x, y, z* directions, and the other three equations relate the body's angular impulse and momentum about the *x, y, z* axes. Before applying Eqs. 21–12 and 21–13 to the solution of problems, the material in Secs. 19.2 and 19.3 should be reviewed.

21.3   Kinetic Energy

In order to apply the principle of work and energy to solve problems involving general rigid body motion, it is first necessary to formulate expressions for the kinetic energy of the body. To do this, consider the rigid body shown in Fig. 21–8, which has a mass m and center of mass at G. The kinetic energy of the ith particle of the body having a mass m_i and velocity $\mathbf{v}_i$, measured relative to the inertial X, Y, Z frame of reference, is

$$T_i = \tfrac{1}{2} m_i v_i^2 = \tfrac{1}{2} m_i (\mathbf{v}_i \cdot \mathbf{v}_i)$$

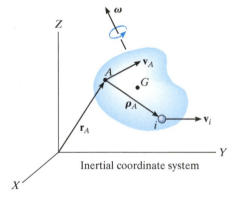

Inertial coordinate system

Fig. 21–8

Provided the velocity of an arbitrary point A in the body is known, $\mathbf{v}_i$ can be related to $\mathbf{v}_A$ by the equation $\mathbf{v}_i = \mathbf{v}_A + \boldsymbol{\omega} \times \boldsymbol{\rho}_A$, where $\boldsymbol{\omega}$ is the angular velocity of the body, measured from the X, Y, Z coordinate system, and $\boldsymbol{\rho}_A$ is a position vector extending from A to i. Using this expression, the kinetic energy for the particle can be written as

$$T_i = \tfrac{1}{2} m_i (\mathbf{v}_A + \boldsymbol{\omega} \times \boldsymbol{\rho}_A) \cdot (\mathbf{v}_A + \boldsymbol{\omega} \times \boldsymbol{\rho}_A)$$

$$= \tfrac{1}{2}(\mathbf{v}_A \cdot \mathbf{v}_A) m_i + \mathbf{v}_A \cdot (\boldsymbol{\omega} \times \boldsymbol{\rho}_A) m_i + \tfrac{1}{2}(\boldsymbol{\omega} \times \boldsymbol{\rho}_A) \cdot (\boldsymbol{\omega} \times \boldsymbol{\rho}_A) m_i$$

The kinetic energy for the entire body is obtained by summing the kinetic energies of all the particles of the body. This requires an integration. Since $m_i \rightarrow dm$, we get

$$T = \tfrac{1}{2} m(\mathbf{v}_A \cdot \mathbf{v}_A) + \mathbf{v}_A \cdot \left(\boldsymbol{\omega} \times \int_m \boldsymbol{\rho}_A\, dm \right) + \tfrac{1}{2} \int_m (\boldsymbol{\omega} \times \boldsymbol{\rho}_A) \cdot (\boldsymbol{\omega} \times \boldsymbol{\rho}_A)\, dm$$

The last term on the right can be rewritten using the vector identity $\mathbf{a} \times \mathbf{b} \cdot \mathbf{c} = \mathbf{a} \cdot \mathbf{b} \times \mathbf{c}$, where $\mathbf{a} = \boldsymbol{\omega}$, $\mathbf{b} = \boldsymbol{\rho}_A$, and $\mathbf{c} = \boldsymbol{\omega} \times \boldsymbol{\rho}_A$. The final result is

$$T = \tfrac{1}{2}m(\mathbf{v}_A \cdot \mathbf{v}_A) + \mathbf{v}_A \cdot \left(\boldsymbol{\omega} \times \int_m \boldsymbol{\rho}_A dm\right)$$
$$+ \tfrac{1}{2}\boldsymbol{\omega} \cdot \int_m \boldsymbol{\rho}_A \times (\boldsymbol{\omega} \times \boldsymbol{\rho}_A)dm \qquad (21\text{–}14)$$

This equation is rarely used because of the computations involving the integrals. Simplification occurs, however, if the reference point A is either a fixed point or the center of mass.

Fixed Point O.
If A is a *fixed point* O in the body, Fig. 21–7a, then $\mathbf{v}_A = \mathbf{0}$, and using Eq. 21–7, we can express Eq. 21–14 as

$$T = \tfrac{1}{2}\boldsymbol{\omega} \cdot \mathbf{H}_O$$

If the x, y, z axes represent the principal axes of inertia for the body, then $\boldsymbol{\omega} = \omega_x \mathbf{i} + \omega_y \mathbf{j} + \omega_z \mathbf{k}$ and $\mathbf{H}_O = I_x \omega_x \mathbf{i} + I_y \omega_y \mathbf{j} + I_z \omega_z \mathbf{k}$. Substituting into the above equation and performing the dot-product operations yields

$$T = \tfrac{1}{2}I_x \omega_x^2 + \tfrac{1}{2}I_y \omega_y^2 + \tfrac{1}{2}I_z \omega_z^2 \qquad (21\text{–}15)$$

Center of Mass G.
If A is located at the *center of mass* G of the body, Fig. 21–7b, then $\int \boldsymbol{\rho}_A\, dm = \mathbf{0}$ and, using Eq. 21–8, we can write Eq. 21–14 as

$$T = \tfrac{1}{2}mv_G^2 + \tfrac{1}{2}\boldsymbol{\omega} \cdot \mathbf{H}_G$$

In a manner similar to that for a fixed point, the last term on the right side may be represented in scalar form, in which case

$$T = \tfrac{1}{2}mv_G^2 + \tfrac{1}{2}I_x \omega_x^2 + \tfrac{1}{2}I_y \omega_y^2 + \tfrac{1}{2}I_z \omega_z^2 \qquad (21\text{–}16)$$

Here it is seen that the kinetic energy consists of two parts; namely, the translational kinetic energy of the mass center, $\tfrac{1}{2}mv_G^2$, and the body's rotational kinetic energy.

Principle of Work and Energy.
Having formulated the kinetic energy for a body, the *principle of work and energy* can be applied to solve kinetics problems which involve *force, velocity, and displacement*. For this case only one scalar equation can be written for each body, namely,

$$T_1 + \Sigma U_{1\text{-}2} = T_2 \qquad (21\text{–}17)$$

Before applying this equation, the material in Chapter 18 should be reviewed.

21

EXAMPLE | 21.2

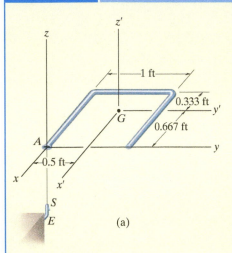

(a)

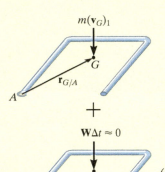

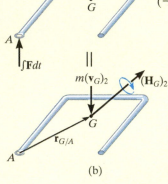

(b)

Fig. 21–9

The rod in Fig. 21–9a has a weight per unit length of 1.5 lb/ft. Determine its angular velocity just after the end A falls onto the hook at E. The hook provides a permanent connection for the rod due to the spring-lock mechanism S. Just before striking the hook the rod is falling downward with a speed $(v_G)_1 = 10$ ft/s.

SOLUTION

The principle of impulse and momentum will be used since impact occurs.

Impulse and Momentum Diagrams. Fig. 21–9b. During the short time Δt, the impulsive force $\mathbf{F}$ acting at A changes the momentum of the rod. (The impulse created by the rod's weight $\mathbf{W}$ during this time is small compared to $\int \mathbf{F} \, dt$, so that it can be neglected, i.e., the weight is a nonimpulsive force.) Hence, the angular momentum of the rod is *conserved* about point A since the moment of $\int \mathbf{F} \, dt$ about A is zero.

Conservation of Angular Momentum. Equation 21–9 must be used to find the angular momentum of the rod, since A does not become a *fixed point* until *after* the impulsive interaction with the hook. Thus, with reference to Fig. 21–9b, $(\mathbf{H}_A)_1 = (\mathbf{H}_A)_2$, or

$$\mathbf{r}_{G/A} \times m(\mathbf{v}_G)_1 = \mathbf{r}_{G/A} \times m(\mathbf{v}_G)_2 + (\mathbf{H}_G)_2 \qquad (1)$$

From Fig. 21–9a, $\mathbf{r}_{G/A} = \{-0.667\mathbf{i} + 0.5\mathbf{j}\}$ ft. Furthermore, the primed axes are principal axes of inertia for the rod because $I_{x'y'} = I_{x'z'} = I_{z'y'} = 0$. Hence, from Eqs. 21–11, $(\mathbf{H}_G)_2 = I_{x'}\omega_x\mathbf{i} + I_{y'}\omega_y\mathbf{j} + I_{z'}\omega_z\mathbf{k}$. The principal moments of inertia are $I_{x'} = 0.0272$ slug·ft^2, $I_{y'} = 0.0155$ slug·ft^2, $I_{z'} = 0.0427$ slug·ft^2 (see Prob. 21–17). Substituting into Eq. 1, we have

$$(-0.667\mathbf{i} + 0.5\mathbf{j}) \times \left[\left(\frac{4.5}{32.2}\right)(-10\mathbf{k})\right] = (-0.667\mathbf{i} + 0.5\mathbf{j}) \times \left[\left(\frac{4.5}{32.2}\right)(-v_G)_2\mathbf{k}\right]$$
$$+ \, 0.0272\omega_x\mathbf{i} + 0.0155\omega_y\mathbf{j} + 0.0427\omega_z\mathbf{k}$$

Expanding and equating the respective $\mathbf{i}, \mathbf{j}, \mathbf{k}$ components yields

$$-0.699 = -0.0699(v_G)_2 + 0.0272\omega_x \qquad (2)$$
$$-0.932 = -0.0932(v_G)_2 + 0.0155\omega_y \qquad (3)$$
$$0 = 0.0427\omega_z \qquad (4)$$

Kinematics. There are four unknowns in the above equations; however, another equation may be obtained by relating $\boldsymbol{\omega}$ to $(\mathbf{v}_G)_2$ using *kinematics*. Since $\omega_z = 0$ (Eq. 4) and after impact the rod rotates about the fixed point A, Eq. 20–3 can be applied, in which case $(\mathbf{v}_G)_2 = \boldsymbol{\omega} \times \mathbf{r}_{G/A}$, or

$$-(v_G)_2\mathbf{k} = (\omega_x\mathbf{i} + \omega_y\mathbf{j}) \times (-0.667\mathbf{i} + 0.5\mathbf{j})$$
$$-(v_G)_2 = 0.5\omega_x + 0.667\omega_y \qquad (5)$$

Solving Eqs. 2, 3 and 5 simultaneously yields

$$(\mathbf{v}_G)_2 = \{-8.41\mathbf{k}\} \text{ ft/s} \quad \boldsymbol{\omega} = \{-4.09\mathbf{i} - 9.55\mathbf{j}\} \text{ rad/s} \quad \textit{Ans.}$$

EXAMPLE | 21.3

A 5-N·m torque is applied to the vertical shaft CD shown in Fig. 21–10a, which allows the 10-kg gear A to turn freely about CE. Assuming that gear A starts from rest, determine the angular velocity of CD after it has turned two revolutions. Neglect the mass of shaft CD and axle CE and assume that gear A can be approximated by a thin disk. Gear B is fixed.

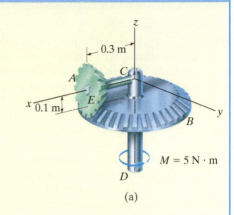

(a)

SOLUTION
The principle of work and energy may be used for the solution. Why?

Work.   If shaft CD, the axle CE, and gear A are considered as a system of connected bodies, only the applied torque $\mathbf{M}$ does work. For two revolutions of CD, this work is $\Sigma U_{1-2} = (5\text{ N·m})(4\pi\text{ rad}) = 62.83\text{ J}$.

Kinetic Energy.   Since the gear is initially at rest, its initial kinetic energy is zero. A kinematic diagram for the gear is shown in Fig. 21–10b. If the angular velocity of CD is taken as $\boldsymbol{\omega}_{CD}$, then the angular velocity of gear A is $\boldsymbol{\omega}_A = \boldsymbol{\omega}_{CD} + \boldsymbol{\omega}_{CE}$. The gear may be imagined as a portion of a massless extended body which is rotating about the *fixed point* C. The instantaneous axis of rotation for this body is along line CH, because both points C and H on the body (gear) have zero velocity and must therefore lie on this axis. This requires that the components $\boldsymbol{\omega}_{CD}$ and $\boldsymbol{\omega}_{CE}$ be related by the equation $\omega_{CD}/0.1\text{ m} = \omega_{CE}/0.3\text{ m}$ or $\omega_{CE} = 3\omega_{CD}$. Thus,

$$\boldsymbol{\omega}_A = -\omega_{CE}\mathbf{i} + \omega_{CD}\mathbf{k} = -3\omega_{CD}\mathbf{i} + \omega_{CD}\mathbf{k} \qquad (1)$$

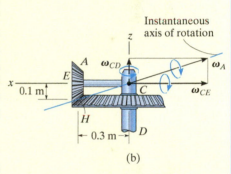

(b)

Fig. 21–10

The x, y, z axes in Fig. 21–10a represent *principal axes of inertia* at C for the gear. Since point C is a fixed point of rotation, Eq. 21–15 may be applied to determine the kinetic energy, i.e.,

$$T = \tfrac{1}{2}I_x\omega_x^2 + \tfrac{1}{2}I_y\omega_y^2 + \tfrac{1}{2}I_z\omega_z^2 \qquad (2)$$

Using the parallel-axis theorem, the moments of inertia of the gear about point C are as follows:

$$I_x = \tfrac{1}{2}(10\text{ kg})(0.1\text{ m})^2 = 0.05\text{ kg·m}^2$$

$$I_y = I_z = \tfrac{1}{4}(10\text{ kg})(0.1\text{ m})^2 + 10\text{ kg}(0.3\text{ m})^2 = 0.925\text{ kg·m}^2$$

Since $\omega_x = -3\omega_{CD}$, $\omega_y = 0$, $\omega_z = \omega_{CD}$, Eq. 2 becomes

$$T_A = \tfrac{1}{2}(0.05)(-3\omega_{CD})^2 + 0 + \tfrac{1}{2}(0.925)(\omega_{CD})^2 = 0.6875\omega_{CD}^2$$

Principle of Work and Energy.   Applying the principle of work and energy, we obtain

$$T_1 + \Sigma U_{1-2} = T_2$$

$$0 + 62.83 = 0.6875\omega_{CD}^2$$

$$\omega_{CD} = 9.56\text{ rad/s} \qquad \qquad Ans.$$

21

PROBLEMS

21–22. If a body contains *no planes of symmetry*, the principal moments of inertia can be determined mathematically. To show how this is done, consider the rigid body which is spinning with an angular velocity $\boldsymbol{\omega}$, directed along one of its principal axes of inertia. If the principal moment of inertia about this axis is I, the angular momentum can be expressed as $\mathbf{H} = I\boldsymbol{\omega} = I\omega_x\mathbf{i} + I\omega_y\mathbf{j} + I\omega_z\mathbf{k}$. The components of $\mathbf{H}$ may also be expressed by Eqs. 21–10, where the inertia tensor is assumed to be known. Equate the $\mathbf{i}, \mathbf{j}$, and $\mathbf{k}$ components of both expressions for $\mathbf{H}$ and consider ω_x, ω_y, and ω_z to be unknown. The solution of these three equations is obtained provided the determinant of the coefficients is zero. Show that this determinant, when expanded, yields the cubic equation

$$I^3 - (I_{xx} + I_{yy} + I_{zz})I^2$$

$$+ (I_{xx}I_{yy} + I_{yy}I_{zz} + I_{zz}I_{xx} - I_{xy}^2 - I_{yz}^2 - I_{zx}^2)I$$

$$- (I_{xx}I_{yy}I_{zz} - 2I_{xy}I_{yz}I_{zx} - I_{xx}I_{yz}^2 - I_{yy}I_{zx}^2 - I_{zz}I_{xy}^2) = 0$$

The three positive roots of I, obtained from the solution of this equation, represent the principal moments of inertia I_x, I_y, and I_z.

21–23. Show that if the angular momentum of a body is determined with respect to an arbitrary point A, then $\mathbf{H}_A$ can be expressed by Eq. 21–9. This requires substituting $\boldsymbol{\rho}_A = \boldsymbol{\rho}_G + \boldsymbol{\rho}_{G/A}$ into Eq. 21–6 and expanding, noting that $\int \boldsymbol{\rho}_G \, dm = \mathbf{0}$ by definition of the mass center and $\mathbf{v}_G = \mathbf{v}_A + \boldsymbol{\omega} \times \boldsymbol{\rho}_{G/A}$.

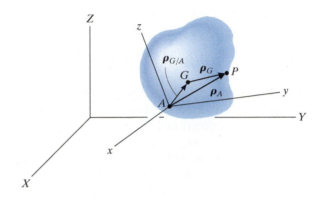

Prob. 21–23

*21–24. The 15-kg circular disk spins about its axle with a constant angular velocity of $\omega_1 = 10$ rad/s. Simultaneously, the yoke is rotating with a constant angular velocity of $\omega_2 = 5$ rad/s. Determine the angular momentum of the disk about its center of mass O, and its kinetic energy.

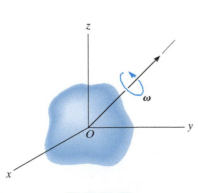

Prob. 21–22

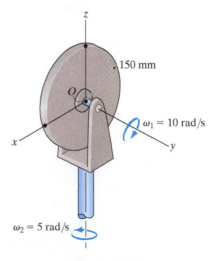

Prob. 21–24

21–25. The large gear has a mass of 5 kg and a radius of gyration of $k_z = 75$ mm. Gears B and C each have a mass of 200 g and a radius of gyration about the axis of their connecting shaft of 15 mm. If the gears are in mesh and C has an angular velocity of $\omega_c = \{15\mathbf{j}\}$ rad/s, determine the total angular momentum for the system of three gears about point A.

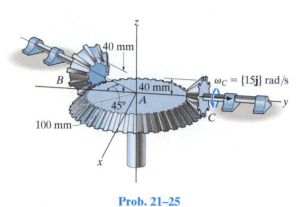

$\omega_C = \{15\mathbf{j}\}$ rad/s

40 mm

40 mm

45°

100 mm

Prob. 21–25

21–26. The circular disk has a weight of 15 lb and is mounted on the shaft AB at an angle of 45° with the horizontal. Determine the angular velocity of the shaft when $t = 3$ s if a constant torque $M = 2$ lb·ft is applied to the shaft. The shaft is originally spinning at $\omega_1 = 8$ rad/s when the torque is applied.

21–27. The circular disk has a weight of 15 lb and is mounted on the shaft AB at an angle of 45° with the horizontal. Determine the angular velocity of the shaft when $t = 2$ s if a torque $M = (4e^{0.1t})$ lb·ft, where t is in seconds, is applied to the shaft. The shaft is originally spinning at $\omega_1 = 8$ rad/s when the torque is applied.

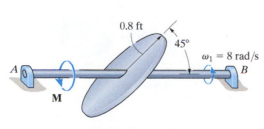

0.8 ft

45°

$\omega_1 = 8$ rad/s

A B

M

Probs. 21–26/27

*21–28. The rod assembly is supported at G by a ball-and-socket joint. Each segment has a mass of 0.5 kg/m. If the assembly is originally at rest and an impulse of $\mathbf{I} = \{-8\mathbf{k}\}$ N·s is applied at D, determine the angular velocity of the assembly just after the impact.

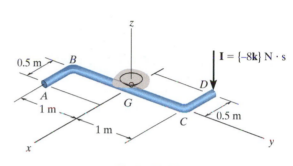

$\mathbf{I} = \{-8\mathbf{k}\}$ N·s

0.5 m B

A G D

1 m C 0.5 m

1 m

Prob. 21–28

21–29. The 4-lb rod AB is attached to the 1-lb collar at A and a 2-lb link BC using ball-and-socket joints. If the rod is released from rest in the position shown, determine the angular velocity of the link after it has rotated 180°.

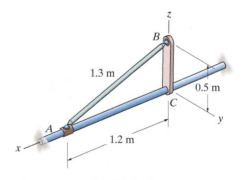

B

1.3 m

0.5 m

C

A

1.2 m

Prob. 21–29

21

21–30. The rod weighs 3 lb/ft and is suspended from parallel cords at A and B. If the rod has an angular velocity of 2 rad/s about the z axis at the instant shown, determine how high the center of the rod rises at the instant the rod momentarily stops swinging.

*21–32. The 2-kg thin disk is connected to the slender rod which is fixed to the ball-and-socket joint at A. If it is released from rest in the position shown, determine the spin of the disk about the rod when the disk reaches its lowest position. Neglect the mass of the rod. The disk rolls without slipping.

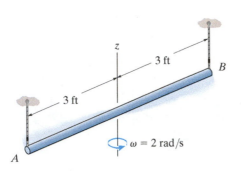

Prob. 21–30

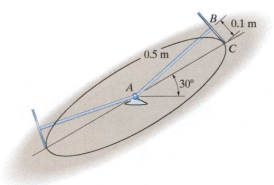

Prob. 21–32

21–31. The 4-lb rod AB is attached to the rod BC and collar A using ball-and-socket joints. If BC has a constant angular velocity of 2 rad/s, determine the kinetic energy of AB when it is in the position shown. Assume the angular velocity of AB is directed perpendicular to the axis of AB.

21–33. The 20-kg sphere rotates about the axle with a constant angular velocity of $\omega_s = 60$ rad/s. If shaft AB is subjected to a torque of $M = 50$ N·m, causing it to rotate, determine the value of ω_p after the shaft has turned 90° from the position shown. Initially, $\omega_p = 0$. Neglect the mass of arm CDE.

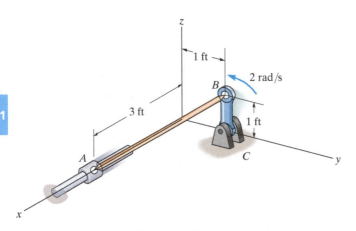

Prob. 21–31

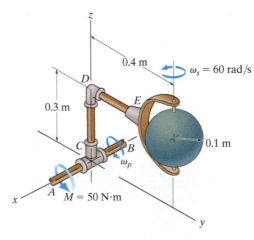

Prob. 21–33

21–34. The 200-kg satellite has its center of mass at point G. Its radii of gyration about the z', x', y' axes are $k_{z'} = 300$ mm, $k_{x'} = k_{y'} = 500$ mm, respectively. At the instant shown, the satellite rotates about the x', y', and z' axes with the angular velocity shown, and its center of mass G has a velocity of $v_G = \{-250i + 200j + 120k\}$ m/s. Determine the angular momentum of the satellite about point A at this instant.

21–35. The 200-kg satellite has its center of mass at point G. Its radii of gyration about the z', x', y' axes are $k_{z'} = 300$ mm, $k_{x'} = k_{y'} = 500$ mm, respectively. At the instant shown, the satellite rotates about the x', y', and z' axes with the angular velocity shown, and its center of mass G has a velocity of $v_G = \{-250i + 200j + 120k\}$ m/s. Determine the kinetic energy of the satellite at this instant.

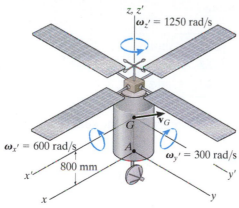

Probs. 21–34/35

*21–36. The 15-kg rectangular plate is free to rotate about the y axis because of the bearing supports at A and B. When the plate is balanced in the vertical plane, a 3-g bullet is fired into it, perpendicular to its surface, with a velocity $v = \{-2000i\}$ m/s. Compute the angular velocity of the plate at the instant it has rotated $180°$. If the bullet strikes corner D with the same velocity v, instead of at C, does the angular velocity remain the same? Why or why not?

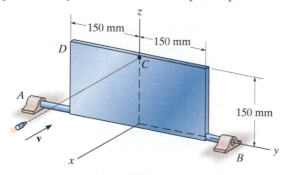

Prob. 21–36

21–37. The 5-kg thin plate is suspended at O using a ball-and-socket joint. It is rotating with a constant angular velocity $\omega = \{2k\}$ rad/s when the corner A strikes the hook at S, which provides a permanent connection. Determine the angular velocity of the plate immediately after impact.

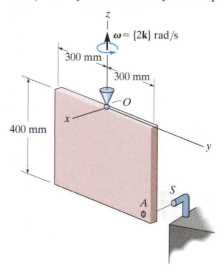

Prob. 21–37

21–38. Determine the kinetic energy of the 7-kg disk and 1.5-kg rod when the assembly is rotating about the z axis at $\omega = 5$ rad/s.

21–39. Determine the angular momentum H_z of the 7-kg disk and 1.5-kg rod when the assembly is rotating about the z axis at $\omega = 5$ rad/s.

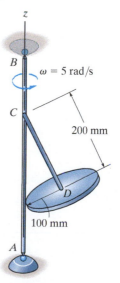

Probs. 21–38/39

*21.4 Equations of Motion

Having become familiar with the techniques used to describe both the inertial properties and the angular momentum of a body, we can now write the equations which describe the motion of the body in their most useful forms.

Equations of Translational Motion.   The *translational motion* of a body is defined in terms of the acceleration of the body's mass center, which is measured from an inertial X, Y, Z reference. The equation of translational motion for the body can be written in vector form as

$$\Sigma \mathbf{F} = m\mathbf{a}_G \qquad (21\text{–}18)$$

or by the three scalar equations

$$\begin{array}{c} \Sigma F_x = m(a_G)_x \\ \Sigma F_y = m(a_G)_y \\ \Sigma F_z = m(a_G)_z \end{array} \qquad (21\text{–}19)$$

Here, $\Sigma \mathbf{F} = \Sigma F_x \mathbf{i} + \Sigma F_y \mathbf{j} + \Sigma F_z \mathbf{k}$ represents the sum of all the external forces acting on the body.

Equations of Rotational Motion.   In Sec. 15.6, we developed Eq. 15–17, namely,

$$\Sigma \mathbf{M}_O = \dot{\mathbf{H}}_O \qquad (21\text{–}20)$$

which states that the sum of the moments of all the external forces acting on a system of particles (contained in a rigid body) about a fixed point O is equal to the time rate of change of the total angular momentum of the body about point O. When moments of the external forces acting on the particles are summed about the system's *mass center* G, one again obtains the same simple form of Eq. 21–20, relating the moment summation $\Sigma \mathbf{M}_G$ to the angular momentum $\mathbf{H}_G$. To show this, consider the system of particles in Fig. 21–11, where X, Y, Z represents an inertial frame of reference and the x, y, z axes, with origin at G, *translate* with respect to this frame. In general, G is *accelerating*, so by definition the translating frame is *not* an inertial reference. The angular momentum of the ith particle with respect to this frame is, however,

$$(\mathbf{H}_i)_G = \mathbf{r}_{i/G} \times m_i \mathbf{v}_{i/G}$$

where $\mathbf{r}_{i/G}$ and $\mathbf{v}_{i/G}$ represent the position and velocity of the ith particle with respect to G. Taking the time derivative we have

$$(\dot{\mathbf{H}}_i)_G = \dot{\mathbf{r}}_{i/G} \times m_i \mathbf{v}_{i/G} + \mathbf{r}_{i/G} \times m_i \dot{\mathbf{v}}_{i/G}$$

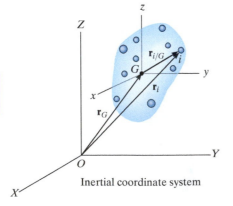

Inertial coordinate system

Fig. 21–11

By definition, $\mathbf{v}_{i/G} = \dot{\mathbf{r}}_{i/G}$. Thus, the first term on the right side is zero since the cross product of the same vectors is zero. Also, $\mathbf{a}_{i/G} = \dot{\mathbf{v}}_{i/G}$, so that

$$(\dot{\mathbf{H}}_i)_G = (\mathbf{r}_{i/G} \times m_i \mathbf{a}_{i/G})$$

Similar expressions can be written for the other particles of the body. When the results are summed, we get

$$\dot{\mathbf{H}}_G = \Sigma(\mathbf{r}_{i/G} \times m_i \mathbf{a}_{i/G})$$

Here $\dot{\mathbf{H}}_G$ is the time rate of change of the total angular momentum of the body computed about point G.

The relative acceleration for the ith particle is defined by the equation $\mathbf{a}_{i/G} = \mathbf{a}_i - \mathbf{a}_G$, where $\mathbf{a}_i$ and $\mathbf{a}_G$ represent, respectively, the accelerations of the ith particle and point G measured with respect to the *inertial frame of reference*. Substituting and expanding, using the distributive property of the vector cross product, yields

$$\dot{\mathbf{H}}_G = \Sigma(\mathbf{r}_{i/G} \times m_i \mathbf{a}_i) - (\Sigma m_i \mathbf{r}_{i/G}) \times \mathbf{a}_G$$

By definition of the mass center, the sum $(\Sigma m_i \mathbf{r}_{i/G}) = (\Sigma m_i)\bar{\mathbf{r}}$ is equal to zero, since the position vector $\bar{\mathbf{r}}$ relative to G is zero. Hence, the last term in the above equation is zero. Using the equation of motion, the product $m_i \mathbf{a}_i$ can be replaced by the resultant *external force* $\mathbf{F}_i$ acting on the ith particle. Denoting $\Sigma \mathbf{M}_G = \Sigma(\mathbf{r}_{i/G} \times \mathbf{F}_i)$, the final result can be written as

$$\Sigma \mathbf{M}_G = \dot{\mathbf{H}}_G \qquad (21\text{–}21)$$

The rotational equation of motion for the body will now be developed from either Eq. 21–20 or 21–21. In this regard, the scalar components of the angular momentum $\mathbf{H}_O$ or $\mathbf{H}_G$ are defined by Eqs. 21–10 or, if principal axes of inertia are used either at point O or G, by Eqs. 21–11. If these components are computed about x, y, z axes that are *rotating* with an angular velocity $\boldsymbol{\Omega}$ that is *different* from the body's angular velocity $\boldsymbol{\omega}$, then the time derivative $\dot{\mathbf{H}} = d\mathbf{H}/dt$, as used in Eqs. 21–20 and 21–21, must account for the rotation of the x, y, z axes as measured from the inertial X, Y, Z axes. This requires application of Eq. 20–6, in which case Eqs. 21–20 and 21–21 become

$$\Sigma \mathbf{M}_O = (\dot{\mathbf{H}}_O)_{xyz} + \boldsymbol{\Omega} \times \mathbf{H}_O$$
$$\Sigma \mathbf{M}_G = (\dot{\mathbf{H}}_G)_{xyz} + \boldsymbol{\Omega} \times \mathbf{H}_G \qquad (21\text{–}22)$$

Here $(\dot{\mathbf{H}})_{xyz}$ is the time rate of change of $\mathbf{H}$ measured from the x, y, z reference.

There are three ways in which one can define the motion of the x, y, z axes. Obviously, motion of this reference should be chosen so that it will yield the simplest set of moment equations for the solution of a particular problem.

21

x, y, z Axes Having Motion $\Omega = 0$. If the body has general motion, the *x, y, z* axes can be chosen with origin at *G*, such that the axes only *translate* relative to the inertial *X, Y, Z* frame of reference. Doing this simplifies Eq. 21–22, since $\Omega = 0$. However, the body may have a rotation ω about these axes, and therefore the moments and products of inertia of the body would have to be expressed as *functions of time*. In most cases this would be a difficult task, so that such a choice of axes has restricted application.

x, y, z Axes Having Motion $\Omega = \omega$. The *x, y, z* axes can be chosen such that they are *fixed in and move with the body*. The moments and products of inertia of the body relative to these axes will then be *constant* during the motion. Since $\Omega = \omega$, Eqs. 21–22 become

$$\Sigma \mathbf{M}_O = (\dot{\mathbf{H}}_O)_{xyz} + \boldsymbol{\omega} \times \mathbf{H}_O$$
$$\Sigma \mathbf{M}_G = (\dot{\mathbf{H}}_G)_{xyz} + \boldsymbol{\omega} \times \mathbf{H}_G$$

$$(21\text{–}23)$$

We can express each of these vector equations as three scalar equations using Eqs. 21–10. Neglecting the subscripts *O* and *G* yields

$$\Sigma M_x = I_{xx}\dot{\omega}_x - (I_{yy} - I_{zz})\omega_y\omega_z - I_{xy}(\dot{\omega}_y - \omega_z\omega_x)$$
$$- I_{yz}(\omega_y^2 - \omega_z^2) - I_{zx}(\dot{\omega}_z + \omega_x\omega_y)$$
$$\Sigma M_y = I_{yy}\dot{\omega}_y - (I_{zz} - I_{xx})\omega_z\omega_x - I_{yz}(\dot{\omega}_z - \omega_x\omega_y) \qquad (21\text{–}24)$$
$$- I_{zx}(\omega_z^2 - \omega_x^2) - I_{xy}(\dot{\omega}_x + \omega_y\omega_z)$$
$$\Sigma M_z = I_{zz}\dot{\omega}_z - (I_{xx} - I_{yy})\omega_x\omega_y - I_{zx}(\dot{\omega}_x - \omega_y\omega_z)$$
$$- I_{xy}(\omega_x^2 - \omega_y^2) - I_{yz}(\dot{\omega}_y + \omega_z\omega_x)$$

If the *x, y, z* axes are chosen as *principal axes of inertia*, the products of inertia are zero, $I_{xx} = I_x$, etc., and the above equations become

$$\boxed{\begin{aligned} \Sigma M_x &= I_x\dot{\omega}_x - (I_y - I_z)\omega_y\omega_z \\ \Sigma M_y &= I_y\dot{\omega}_y - (I_z - I_x)\omega_z\omega_x \\ \Sigma M_z &= I_z\dot{\omega}_z - (I_x - I_y)\omega_x\omega_y \end{aligned}} \qquad (21\text{–}25)$$

This set of equations is known historically as the *Euler equations of motion*, named after the Swiss mathematician Leonhard Euler, who first developed them. They apply *only* for moments summed about either point *O* or *G*.

21

When applying these equations it should be realized that $\dot{\omega}_x$, $\dot{\omega}_y$, $\dot{\omega}_z$ represent the time derivatives of the magnitudes of the x, y, z components of $\boldsymbol{\omega}$ as observed from x, y, z. To determine these components, it is first necessary to find ω_x, ω_y, ω_z when the x, y, z axes are oriented in a *general position* and *then* take the time derivative of the magnitude of these components, i.e., $(\dot{\boldsymbol{\omega}})_{xyz}$. However, since the x, y, z axes are rotating at $\boldsymbol{\Omega} = \boldsymbol{\omega}$, then from Eq. 20–6, it should be noted that $\dot{\boldsymbol{\omega}} = (\dot{\boldsymbol{\omega}})_{xyz} + \boldsymbol{\omega} \times \boldsymbol{\omega}$. Since $\boldsymbol{\omega} \times \boldsymbol{\omega} = \boldsymbol{0}$, then $\dot{\boldsymbol{\omega}} = (\dot{\boldsymbol{\omega}})_{xyz}$. This important result indicates that the time derivative of $\boldsymbol{\omega}$ with respect to the fixed X, Y, Z axes, that is $\dot{\boldsymbol{\omega}}$, can also be used to obtain $(\dot{\boldsymbol{\omega}})_{xyz}$. Generally this is the easiest way to determine the result. See Example 21.5.

x, *y*, *z* Axes Having Motion $\boldsymbol{\Omega} \neq \boldsymbol{\omega}$.

To simplify the calculations for the time derivative of $\boldsymbol{\omega}$, it is often convenient to choose the x, y, z axes having an angular velocity $\boldsymbol{\Omega}$ which is different from the angular velocity $\boldsymbol{\omega}$ of the body. This is particularly suitable for the analysis of spinning tops and gyroscopes which are *symmetrical* about their spinning axes.* When this is the case, the moments and products of inertia remain constant about the axis of spin.

Equations 21–22 are applicable for such a set of axes. Each of these two vector equations can be reduced to a set of three scalar equations which are derived in a manner similar to Eqs. 21–25,[†] i.e.,

$$\begin{aligned}
\Sigma M_x &= I_x\dot{\omega}_x - I_y\Omega_z\omega_y + I_z\Omega_y\omega_z \\
\Sigma M_y &= I_y\dot{\omega}_y - I_z\Omega_x\omega_z + I_x\Omega_z\omega_x \\
\Sigma M_z &= I_z\dot{\omega}_z - I_x\Omega_y\omega_x + I_y\Omega_x\omega_y
\end{aligned} \qquad (21\text{–}26)$$

Here Ω_x, Ω_y, Ω_z represent the x, y, z components of $\boldsymbol{\Omega}$, measured from the inertial frame of reference, and $\dot{\omega}_x$, $\dot{\omega}_y$, $\dot{\omega}_z$ must be determined relative to the x, y, z axes that have the rotation $\boldsymbol{\Omega}$. See Example 21.6.

Any one of these sets of moment equations, Eqs. 21–24, 21–25, or 21–26, represents a series of three first-order nonlinear differential equations. These equations are "coupled," since the angular-velocity components are present in all the terms. Success in determining the solution for a particular problem therefore depends upon what is unknown in these equations. Difficulty certainly arises when one attempts to solve for the unknown components of $\boldsymbol{\omega}$ when the external moments are functions of time. Further complications can arise if the moment equations are coupled to the three scalar equations of translational motion, Eqs. 21–19. This can happen because of the existence of kinematic constraints which relate the rotation of the body to the translation of its mass center, as in the case of a hoop which rolls

21

---

*A detailed discussion of such devices is given in Sec. 21.5.
[†]See Prob. 21–42.

without slipping. Problems that require the simultaneous solution of differential equations are generally solved using numerical methods with the aid of a computer. In many engineering problems, however, we are given information about the motion of the body and are required to determine the applied moments acting on the body. Most of these problems have direct solutions, so that there is no need to resort to computer techniques.

Procedure for Analysis

Problems involving the three-dimensional motion of a rigid body can be solved using the following procedure.

Free-Body Diagram.

- Draw a *free-body diagram* of the body at the instant considered and specify the x, y, z coordinate system. The origin of this reference must be located either at the body's mass center G, or at point O, considered fixed in an inertial reference frame and located either in the body or on a massless extension of the body.

- Unknown reactive force components can be shown having a positive sense of direction.

- Depending on the nature of the problem, decide what type of rotational motion $\boldsymbol{\Omega}$ the x, y, z coordinate system should have, i.e., $\boldsymbol{\Omega} = \mathbf{0}$, $\boldsymbol{\Omega} = \boldsymbol{\omega}$, or $\boldsymbol{\Omega} \neq \boldsymbol{\omega}$. When choosing, keep in mind that the moment equations are simplified when the axes move in such a manner that they represent principal axes of inertia for the body at all times.

- Compute the necessary moments and products of inertia for the body relative to the x, y, z axes.

Kinematics.

- Determine the x, y, z components of the body's angular velocity and find the time derivatives of $\boldsymbol{\omega}$.

- Note that if $\boldsymbol{\Omega} = \boldsymbol{\omega}$, then $\dot{\boldsymbol{\omega}} = (\dot{\boldsymbol{\omega}})_{xyz}$. Therefore we can either find the time derivative of $\boldsymbol{\omega}$ with respect to the X, Y, Z axes, $\dot{\boldsymbol{\omega}}$, and then determine its components $\dot{\omega}_x$, $\dot{\omega}_y$, $\dot{\omega}_z$, or we can find the components of $\boldsymbol{\omega}$ along the x, y, z axes, when the axes are oriented in a general position, and then take the time derivative of the magnitudes of these components, $(\dot{\boldsymbol{\omega}})_{xyz}$.

Equations of Motion.

- Apply either the two vector equations 21–18 and 21–22 or the six scalar component equations appropriate for the x, y, z coordinate axes chosen for the problem.

21

EXAMPLE 21.4

The gear shown in Fig. 21–12a has a mass of 10 kg and is mounted at an angle of 10° with the rotating shaft having negligible mass. If $I_z = 0.1 \text{ kg} \cdot \text{m}^2$, $I_x = I_y = 0.05 \text{ kg} \cdot \text{m}^2$, and the shaft is rotating with a constant angular velocity of $\omega = 30 \text{ rad/s}$, determine the components of reaction that the thrust bearing A and journal bearing B exert on the shaft at the instant shown.

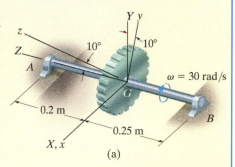

(a)

SOLUTION

Free-Body Diagram. Fig. 21–12b. The origin of the x, y, z coordinate system is located at the gear's center of mass G, which is also a fixed point. The axes are fixed in and rotate with the gear so that these axes will then always represent the principal axes of inertia for the gear. Hence $\mathbf{\Omega} = \mathbf{\omega}$.

Kinematics. As shown in Fig. 21–12c, the angular velocity $\mathbf{\omega}$ of the gear is constant in magnitude and is always directed along the axis of the shaft AB. Since this vector is measured from the X, Y, Z inertial frame of reference, for any position of the x, y, z axes,

$$\omega_x = 0 \quad \omega_y = -30 \sin 10° \quad \omega_z = 30 \cos 10°$$

These components remain constant for any general orientation of the x, y, z axes, and so $\dot{\omega}_x = \dot{\omega}_y = \dot{\omega}_z = 0$. Also note that since $\mathbf{\Omega} = \mathbf{\omega}$, then $\dot{\mathbf{\omega}} = (\dot{\mathbf{\omega}})_{xyz}$. Therefore, we can find these time derivatives relative to the X, Y, Z axes. In this regard $\mathbf{\omega}$ has a constant magnitude and direction $(+Z)$ since $\dot{\mathbf{\omega}} = \mathbf{0}$, and so $\dot{\omega}_x = \dot{\omega}_y = \dot{\omega}_z = 0$. Furthermore, since G is a fixed point, $(a_G)_x = (a_G)_y = (a_G)_z = 0$.

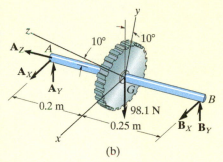

(b)

Equations of Motion. Applying Eqs. 21–25 $(\mathbf{\Omega} = \mathbf{\omega})$ yields

$$\Sigma M_x = I_x \dot{\omega}_x - (I_y - I_z)\omega_y\omega_z$$

$$-(A_Y)(0.2) + (B_Y)(0.25) = 0 - (0.05 - 0.1)(-30 \sin 10°)(30 \cos 10°)$$

$$-0.2A_Y + 0.25B_Y = -7.70 \quad (1)$$

$$\Sigma M_y = I_y \dot{\omega}_y - (I_z - I_x)\omega_z\omega_x$$

$$A_X(0.2) \cos 10° - B_X(0.25) \cos 10° = 0 - 0$$

$$A_X = 1.25B_X \quad (2)$$

$$\Sigma M_z = I_z \dot{\omega}_z - (I_x - I_y)\omega_x\omega_y$$

$$A_X(0.2) \sin 10° - B_X(0.25) \sin 10° = 0 - 0$$

$$A_X = 1.25B_X \text{ (check)}$$

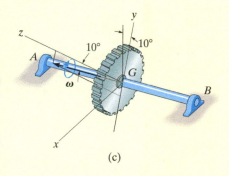

(c)

Fig. 21–12

Applying Eqs. 21–19, we have

$$\Sigma F_X = m(a_G)_X; \qquad A_X + B_X = 0 \quad (3)$$

$$\Sigma F_Y = m(a_G)_Y; \qquad A_Y + B_Y - 98.1 = 0 \quad (4)$$

$$\Sigma F_Z = m(a_G)_Z; \qquad A_Z = 0 \qquad \textit{Ans.}$$

Solving Eqs. 1 through 4 simultaneously gives

$$A_X = B_X = 0 \quad A_Y = 71.6 \text{ N} \quad B_Y = 26.5 \text{ N} \qquad \textit{Ans.}$$

EXAMPLE | 21.5

The airplane shown in Fig. 21–13*a* is in the process of making a steady *horizontal* turn at the rate of ω_p. During this motion, the propeller is spinning at the rate of ω_s. If the propeller has two blades, determine the moments which the propeller shaft exerts on the propeller at the instant the blades are in the vertical position. For simplicity, assume the blades to be a uniform slender bar having a moment of inertia I about an axis perpendicular to the blades passing through the center of the bar, and having zero moment of inertia about a longitudinal axis.

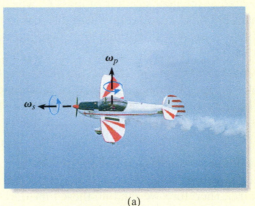

(© R.C. Hibbeler)

(a)

SOLUTION

Free-Body Diagram.   Fig. 21–13*b*. The reactions of the connecting shaft on the propeller are indicated by the resultants $\mathbf{F}_R$ and $\mathbf{M}_R$. (The propeller's weight is assumed to be negligible.) The x, y, z axes will be taken fixed to the propeller, since these axes always represent the principal axes of inertia for the propeller. Thus, $\boldsymbol{\Omega} = \boldsymbol{\omega}$. The moments of inertia I_x and I_y are equal ($I_x = I_y = I$) and $I_z = 0$.

Kinematics.   The angular velocity of the propeller observed from the X, Y, Z axes, coincident with the x, y, z axes, Fig. 21–13*c*, is $\boldsymbol{\omega} = \boldsymbol{\omega}_s + \boldsymbol{\omega}_p = \omega_s\mathbf{i} + \omega_p\mathbf{k}$, so that the x, y, z components of $\boldsymbol{\omega}$ are

$$\omega_x = \omega_s \qquad \omega_y = 0 \qquad \omega_z = \omega_p$$

Since $\boldsymbol{\Omega} = \boldsymbol{\omega}$, then $\dot{\boldsymbol{\omega}} = (\dot{\boldsymbol{\omega}})_{xyz}$. To find $\dot{\boldsymbol{\omega}}$, which is the time derivative with respect to the fixed X, Y, Z axes, we can use Eq. 20–6 since $\boldsymbol{\omega}$ changes direction relative to X, Y, Z. The time rate of change of each of these components $\dot{\boldsymbol{\omega}} = \dot{\boldsymbol{\omega}}_s + \dot{\boldsymbol{\omega}}_p$ relative to the X, Y, Z axes can be obtained by introducing a third coordinate system x', y', z', which has an angular velocity $\boldsymbol{\Omega}' = \boldsymbol{\omega}_p$ and is coincident with the X, Y, Z axes at the instant shown. Thus

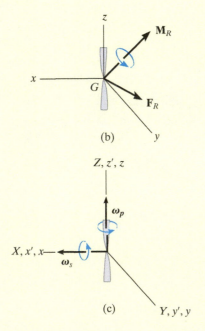

(b)

(c)

Fig. 21–13

$$\dot{\boldsymbol{\omega}} = (\dot{\boldsymbol{\omega}})_{x'\,y'\,z'} + \boldsymbol{\omega}_p \times \boldsymbol{\omega}$$

$$= (\dot{\boldsymbol{\omega}}_s)_{x'\,y'\,z'} + (\dot{\boldsymbol{\omega}}_p)_{x'\,y'\,z'} + \boldsymbol{\omega}_p \times (\boldsymbol{\omega}_s + \boldsymbol{\omega}_p)$$

$$= \mathbf{0} + \mathbf{0} + \boldsymbol{\omega}_p \times \boldsymbol{\omega}_s + \boldsymbol{\omega}_p \times \boldsymbol{\omega}_p$$

$$= \mathbf{0} + \mathbf{0} + \omega_p \mathbf{k} \times \omega_s \mathbf{i} + \mathbf{0} = \omega_p \omega_s \mathbf{j}$$

Since the X, Y, Z axes are coincident with the x, y, z axes at the instant shown, the components of $\dot{\boldsymbol{\omega}}$ along x, y, z are therefore

$$\dot{\omega}_x = 0 \quad \dot{\omega}_y = \omega_p \omega_s \quad \dot{\omega}_z = 0$$

These same results can also be determined by direct calculation of $(\dot{\boldsymbol{\omega}})_{xyz}$; however, this will involve a bit more work. To do this, it will be necessary to view the propeller (or the x, y, z axes) in some *general position* such as shown in Fig. 21–13d. Here the plane has turned through an angle ϕ (phi) and the propeller has turned through an angle ψ (psi) relative to the plane. Notice that $\boldsymbol{\omega}_p$ is always directed along the fixed Z axis and $\boldsymbol{\omega}_s$ follows the x axis. Thus the general components of $\boldsymbol{\omega}$ are

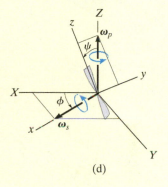

(d)

Fig. 21–13

$$\omega_x = \omega_s \quad \omega_y = \omega_p \sin \psi \quad \omega_z = \omega_p \cos \psi$$

Since ω_s and ω_p are constant, the time derivatives of these components become

$$\dot{\omega}_x = 0 \quad \dot{\omega}_y = \omega_p \cos \psi \, \dot{\psi} \quad \dot{\omega}_z = -\omega_p \sin \psi \, \dot{\psi}$$

But $\phi = \psi = 0°$ and $\dot{\psi} = \omega_s$ at the instant considered. Thus,

$$\omega_x = \omega_s \quad \omega_y = 0 \quad \omega_z = \omega_p$$

$$\dot{\omega}_x = 0 \quad \dot{\omega}_y = \omega_p \omega_s \quad \dot{\omega}_z = 0$$

which are the same results as those obtained previously.

Equations of Motion.   Using Eqs. 21–25, we have

$$\Sigma M_x = I_x \dot{\omega}_x - (I_y - I_z)\omega_y \omega_z = I(0) - (I - 0)(0)\omega_p$$

$$M_x = 0 \qquad\qquad\qquad\qquad \textit{Ans.}$$

$$\Sigma M_y = I_y \dot{\omega}_y - (I_z - I_x)\omega_z \omega_x = I(\omega_p \omega_s) - (0 - I)\omega_p \omega_s$$

$$M_y = 2I\omega_p \omega_s \qquad\qquad \textit{Ans.}$$

$$\Sigma M_z = I_z \dot{\omega}_z - (I_x - I_y)\omega_x \omega_y = 0(0) - (I - I)\omega_s(0)$$

$$M_z = 0 \qquad\qquad\qquad\qquad \textit{Ans.}$$

21

EXAMPLE 21.6

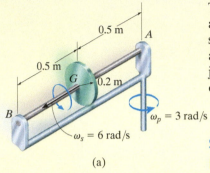

0.5 m

A

0.5 m

G   0.2 m

B

$\omega_p = 3$ rad/s

$\omega_s = 6$ rad/s

(a)

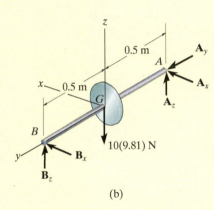

z

0.5 m A_y

A

x   0.5 m A_x

G

A_z

B

10(9.81) N

y B_x

B_z

(b)

Fig. 21–14

The 10-kg flywheel (or thin disk) shown in Fig. 21–14a rotates (spins) about the shaft at a constant angular velocity of $\omega_s = 6$ rad/s. At the same time, the shaft rotates (precessing) about the bearing at A with an angular velocity of $\omega_p = 3$ rad/s. If A is a thrust bearing and B is a journal bearing, determine the components of force reaction at each of these supports due to the motion.

SOLUTION I

Free-Body Diagram.   Fig. 21–14b. The origin of the x, y, z coordinate system is located at the center of mass G of the flywheel. Here we will let these coordinates have an angular velocity of $\Omega = \omega_p = \{3\mathbf{k}\}$ rad/s. Although the wheel spins relative to these axes, the moments of inertia *remain constant,** i.e.,

$$I_x = I_z = \tfrac{1}{4}(10 \text{ kg})(0.2 \text{ m})^2 = 0.1 \text{ kg} \cdot \text{m}^2$$

$$I_y = \tfrac{1}{2}(10 \text{ kg})(0.2 \text{ m})^2 = 0.2 \text{ kg} \cdot \text{m}^2$$

Kinematics.   From the coincident inertial X, Y, Z frame of reference, Fig. 21–14c, the flywheel has an angular velocity of $\boldsymbol{\omega} = \{6\mathbf{j} + 3\mathbf{k}\}$ rad/s, so that

$$\omega_x = 0 \quad \omega_y = 6 \text{ rad/s} \quad \omega_z = 3 \text{ rad/s}$$

The time derivative of $\boldsymbol{\omega}$ must be determined relative to the x, y, z axes. In this case both $\boldsymbol{\omega}_p$ and $\boldsymbol{\omega}_s$ do not change their magnitude or direction, and so

$$\dot{\omega}_x = 0 \quad \dot{\omega}_y = 0 \quad \dot{\omega}_z = 0$$

Equations of Motion.   Applying Eqs. 21–26 ($\Omega \neq \omega$) yields

$$\Sigma M_x = I_x \dot{\omega}_x - I_y \Omega_z \omega_y + I_z \Omega_y \omega_z$$

$$-A_z(0.5) + B_z(0.5) = 0 - (0.2)(3)(6) + 0 = -3.6$$

$$\Sigma M_y = I_y \dot{\omega}_y - I_z \Omega_x \omega_z + I_x \Omega_z \omega_x$$

$$0 = 0 - 0 + 0$$

$$\Sigma M_z = I_z \dot{\omega}_z - I_x \Omega_y \omega_x + I_y \Omega_x \omega_y$$

$$A_x(0.5) - B_x(0.5) = 0 - 0 + 0$$

*This would not be true for the propeller in Example 21.5.

Applying Eqs. 21–19, we have

$$\Sigma F_X = m(a_G)_X; \qquad A_x + B_x = 0$$

$$\Sigma F_Y = m(a_G)_Y; \qquad A_y = -10(0.5)(3)^2$$

$$\Sigma F_Z = m(a_G)_Z; \qquad A_z + B_z - 10(9.81) = 0$$

Solving these equations, we obtain

$$A_x = 0 \quad A_y = -45.0 \text{ N} \quad A_z = 52.6 \text{ N} \qquad \textit{Ans.}$$
$$B_x = 0 \qquad\qquad\qquad\quad B_z = 45.4 \text{ N} \qquad \textit{Ans.}$$

NOTE: If the precession $\boldsymbol{\omega}_p$ had not occurred, the z component of force at A and B would be equal to 49.05 N. In this case, however, the difference in these components is caused by the "gyroscopic moment" created whenever a spinning body precesses about another axis. We will study this effect in detail in the next section.

SOLUTION II

This example can also be solved using Euler's equations of motion, Eqs. 21–25. In this case $\boldsymbol{\Omega} = \boldsymbol{\omega} = \{6\mathbf{j} + 3\mathbf{k}\}$ rad/s, and the time derivative $(\dot{\boldsymbol{\omega}})_{xyz}$ can be conveniently obtained with reference to the fixed X, Y, Z axes since $\dot{\boldsymbol{\omega}} = (\dot{\boldsymbol{\omega}})_{xyz}$. This calculation can be performed by choosing x', y', z' axes to have an angular velocity of $\boldsymbol{\Omega}' = \boldsymbol{\omega}_p$, Fig. 21–14c, so that

$$\dot{\boldsymbol{\omega}} = (\dot{\boldsymbol{\omega}})_{x'y'z'} + \boldsymbol{\omega}_p \times \boldsymbol{\omega} = 0 + 3\mathbf{k} \times (6\mathbf{j} + 3\mathbf{k}) = \{-18\mathbf{i}\} \text{ rad/s}^2$$

$$\dot{\omega}_x = -18 \text{ rad/s} \quad \dot{\omega}_y = 0 \quad \dot{\omega}_z = 0$$

The moment equations then become

$$\Sigma M_x = I_x \dot{\omega}_x - (I_y - I_z)\omega_y\omega_z$$

$$-A_z(0.5) + B_z(0.5) = 0.1(-18) - (0.2 - 0.1)(6)(3) = -3.6$$

$$\Sigma M_y = I_y \dot{\omega}_y - (I_z - I_x)\omega_z\omega_x$$

$$0 = 0 - 0$$

$$\Sigma M_z = I_z \dot{\omega}_z - (I_x - I_y)\omega_x\omega_y$$

$$A_x(0.5) - B_x(0.5) = 0 - 0$$

The solution then proceeds as before.

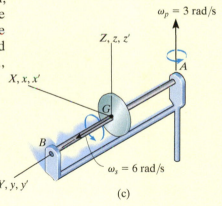

Fig. 21–14

PROBLEMS

*21–40. Derive the scalar form of the rotational equation of motion about the x axis if $\Omega \neq \omega$ and the moments and products of inertia of the body are *not constant* with respect to time.

21–41. Derive the scalar form of the rotational equation of motion about the x axis if $\Omega \neq \omega$ and the moments and products of inertia of the body are *constant* with respect to time.

21–42. Derive the Euler equations of motion for $\Omega \neq \omega$, i.e., Eqs. 21–26.

21–43. The 4-lb bar rests along the smooth corners of an open box. At the instant shown, the box has a velocity $v = \{3j\}$ ft/s and an acceleration $a = \{-6j\}$ ft/s². Determine the x, y, z components of force which the corners exert on the bar.

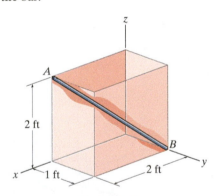

Prob. 21–43

*21–44. The uniform plate has a mass of $m = 2$ kg and is given a rotation of $\omega = 4$ rad/s about its bearings at A and B. If $a = 0.2$ m and $c = 0.3$ m, determine the vertical reactions at the instant shown. Use the x, y, z axes shown and note that $I_{zx} = -\left(\dfrac{mac}{12}\right)\left(\dfrac{c^2 - a^2}{c^2 + a^2}\right)$.

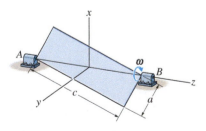

Prob. 21–44

21–45. If the shaft AB is rotating with a constant angular velocity of $\omega = 30$ rad/s, determine the X, Y, Z components of reaction at the thrust bearing A and journal bearing B at the instant shown. The disk has a weight of 15 lb. Neglect the weight of the shaft AB.

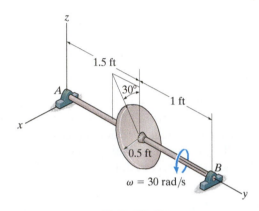

Prob. 21–45

21–46. The assembly is supported by journal bearings at A and B, which develop only y and z force reactions on the shaft. If the shaft is rotating in the direction shown at $\omega = \{2i\}$ rad/s, determine the reactions at the bearings when the assembly is in the position shown. Also, what is the shaft's angular acceleration? The mass per unit length of each rod is 5 kg/m.

21–47. The assembly is supported by journal bearings at A and B, which develop only y and z force reactions on the shaft. If the shaft A is subjected to a couple moment $M = \{40i\}$ N · m, and at the instant shown the shaft has an angular velocity of $\omega = \{2i\}$ rad/s, determine the reactions at the bearings of the assembly at this instant. Also, what is the shaft's angular acceleration? The mass per unit length of each rod is 5 kg/m.

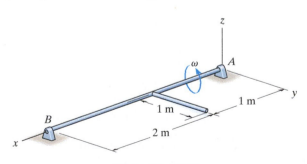

Probs. 21–46/47

*21–48. The man sits on a swivel chair which is rotating with a constant angular velocity of 3 rad/s. He holds the uniform 5-lb rod AB horizontal. He suddenly gives it an angular acceleration of 2 rad/s², measured relative to him, as shown. Determine the required force and moment components at the grip, A, necessary to do this. Establish axes at the rod's center of mass G, with $+z$ upward, and $+y$ directed along the axis of the rod toward A.

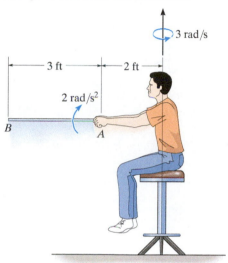

Prob. 21–48

21–49. The rod assembly is supported by a ball-and-socket joint at C and a journal bearing at D, which develops only x and y force reactions. The rods have a mass of 0.75 kg/m. Determine the angular acceleration of the rods and the components of reaction at the supports at the instant $\omega = 8$ rad/s as shown.

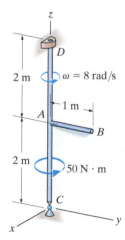

Prob. 21–49

21–50. The bent uniform rod ACD has a weight of 5 lb/ft and is supported at A by a pin and at B by a cord. If the vertical shaft rotates with a constant angular velocity $\omega = 20$ rad/s, determine the x, y, z components of force and moment developed at A and the tension in the cord.

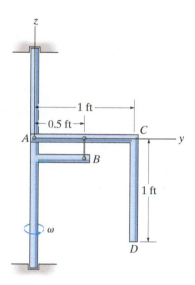

Prob. 21–50

21–51. The uniform hatch door, having a mass of 15 kg and a mass center at G, is supported in the horizontal plane by bearings at A and B. If a vertical force $F = 300$ N is applied to the door as shown, determine the components of reaction at the bearings and the angular acceleration of the door. The bearing at A will resist a component of force in the y direction, whereas the bearing at B will not. For the calculation, assume the door to be a thin plate and neglect the size of each bearing. The door is originally at rest.

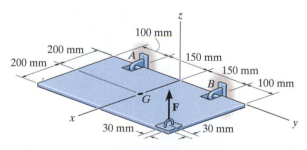

Prob. 21–51

21

*21–52. The 5-kg circular disk is mounted off center on a shaft which is supported by bearings at A and B. If the shaft is rotating at a constant rate of $\omega = 10$ rad/s, determine the vertical reactions at the bearings when the disk is in the position shown.

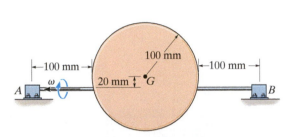

Prob. 21–52

21–53. Two uniform rods, each having a weight of 10 lb, are pin connected to the edge of a rotating disk. If the disk has a constant angular velocity $\omega_D = 4$ rad/s, determine the angle θ made by each rod during the motion, and the components of the force and moment developed at the pin A. *Suggestion*: Use the x, y, z axes oriented as shown.

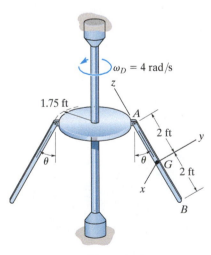

Prob. 21–53

21–54. The 10-kg disk turns around the shaft AB, while the shaft rotates about BC at a constant rate of $\omega_x = 5$ rad/s. If the disk does not slip, determine the normal and frictional force it exerts on the ground. Neglect the mass of shaft AB.

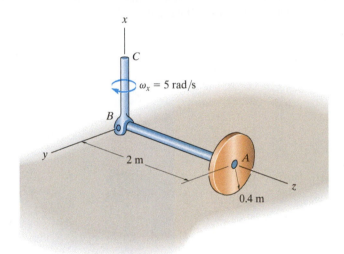

Prob. 21–54

21–55. The 20-kg disk is spinning on its axle at $\omega_s = 30$ rad/s, while the forked rod is turning at $\omega_1 = 6$ rad/s. Determine the x and z moment components the axle exerts on the disk during the motion.

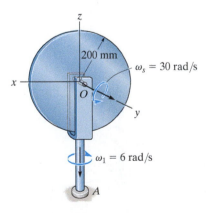

Prob. 21–55

21

*21–56. The 4-kg slender rod AB is pinned at A and held at B by a cord. The axle CD is supported at its ends by ball-and-socket joints and is rotating with a constant angular velocity of 2 rad/s. Determine the tension developed in the cord and the magnitude of force developed at the pin A.

21–58. The 15-lb cylinder is rotating about shaft AB with a constant angular speed $\omega = 4$ rad/s. If the supporting shaft at C, initially at rest, is given an angular acceleration $\alpha_C = 12$ rad/s^2, determine the components of reaction at the bearings A and B. The bearing at A cannot support a force component along the x axis, whereas the bearing at B does.

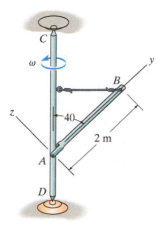

Prob. 21–56

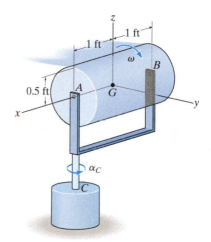

Prob. 21–58

21–57. The blades of a wind turbine spin about the shaft S with a constant angular speed of ω_s, while the frame precesses about the vertical axis with a constant angular speed of ω_p. Determine the x, y, and z components of moment that the shaft exerts on the blades as a function of θ. Consider each blade as a slender rod of mass m and length l.

21–59. The *thin rod* has a mass of 0.8 kg and a total length of 150 mm. It is rotating about its midpoint at a constant rate $\dot{\theta} = 6$ rad/s, while the table to which its axle A is fastened is rotating at 2 rad/s. Determine the x, y, z moment components which the axle exerts on the rod when the rod is in any position θ.

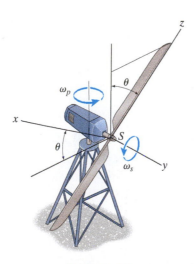

Prob. 21–57

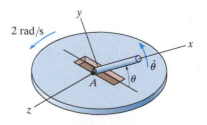

Prob. 21–59

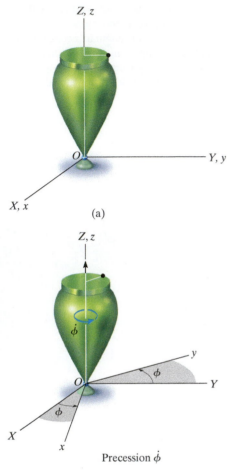

(a)

Precession $\dot{\phi}$

(b)

*21.5   Gyroscopic Motion

In this section we will develop the equations defining the motion of a body (top) which is symmetrical with respect to an axis and rotating about a fixed point. These equations also apply to the motion of a particularly interesting device, the gyroscope.

The body's motion will be analyzed using *Euler angles* ϕ, θ, ψ (phi, theta, psi). To illustrate how they define the position of a body, consider the top shown in Fig. 21–15a. To define its final position, Fig. 21–15d, a second set of x, y, z axes is fixed in the top. Starting with the X, Y, Z and x, y, z axes in coincidence, Fig. 21–15a, the final position of the top can be determined using the following three steps:

1. Rotate the top about the Z (or z) axis through an angle ϕ ($0 \le \phi < 2\pi$), Fig. 21–15b.

2. Rotate the top about the x axis through an angle θ ($0 \le \theta \le \pi$), Fig. 21–15c.

3. Rotate the top about the z axis through an angle ψ ($0 \le \psi < 2\pi$) to obtain the final position, Fig. 21–15d.

The sequence of these three angles, ϕ, θ, then ψ, must be maintained, since finite rotations are *not vectors* (see Fig. 20–1). Although this is the case, the differential rotations $d\phi$, $d\theta$, and $d\psi$ are vectors, and thus the angular velocity $\boldsymbol{\omega}$ of the top can be expressed in terms of the time derivatives of the Euler angles. The angular-velocity components $\dot{\phi}$, $\dot{\theta}$, and $\dot{\psi}$ are known as the *precession*, *nutation*, and *spin*, respectively.

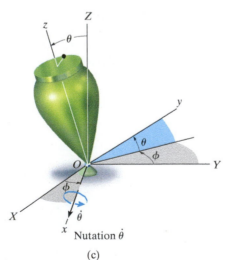

Nutation $\dot{\theta}$

(c)

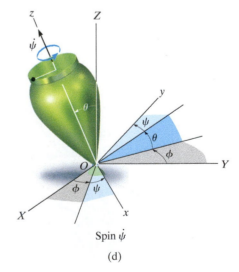

Spin $\dot{\psi}$

(d)

Fig. 21–15

Their positive directions are shown in Fig. 21–16. It is seen that these vectors are not all perpendicular to one another; however, $\boldsymbol{\omega}$ of the top can still be expressed in terms of these three components.

Since the body (top) is symmetric with respect to the z or spin axis, there is no need to attach the x, y, z axes to the top since the inertial properties of the top will remain constant with respect to this frame during the motion. Therefore $\boldsymbol{\Omega} = \boldsymbol{\omega}_p + \boldsymbol{\omega}_n$, Fig. 21–16. Hence, the angular velocity of the body is

$$\boldsymbol{\omega} = \omega_x\mathbf{i} + \omega_y\mathbf{j} + \omega_z\mathbf{k}$$

$$= \dot{\theta}\mathbf{i} + (\dot{\phi}\sin\theta)\mathbf{j} + (\dot{\phi}\cos\theta + \dot{\psi})\mathbf{k} \qquad (21\text{–}27)$$

And the angular velocity of the axes is

$$\boldsymbol{\Omega} = \Omega_x\mathbf{i} + \Omega_y\mathbf{j} + \Omega_z\mathbf{k}$$

$$= \dot{\theta}\mathbf{i} + (\dot{\phi}\sin\theta)\mathbf{j} + (\dot{\phi}\cos\theta)\mathbf{k} \qquad (21\text{–}28)$$

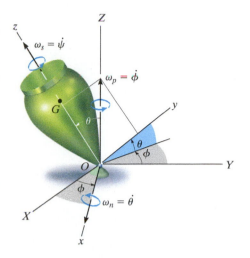

Fig. 21–16

Have the x, y, z axes represent principal axes of inertia for the top, and so the moments of inertia will be represented as $I_{xx} = I_{yy} = I$ and $I_{zz} = I_z$. Since $\boldsymbol{\Omega} \neq \boldsymbol{\omega}$, Eqs. 21–26 are used to establish the rotational equations of motion. Substituting into these equations the respective angular-velocity components defined by Eqs. 21–27 and 21–28, their corresponding time derivatives, and the moment of inertia components, yields

$$\Sigma M_x = I(\ddot{\theta} - \dot{\phi}^2\sin\theta\cos\theta) + I_z\dot{\phi}\sin\theta(\dot{\phi}\cos\theta + \dot{\psi})$$

$$\Sigma M_y = I(\ddot{\phi}\sin\theta + 2\dot{\phi}\dot{\theta}\cos\theta) - I_z\dot{\theta}(\dot{\phi}\cos\theta + \dot{\psi}) \qquad (21\text{–}29)$$

$$\Sigma M_z = I_z(\ddot{\psi} + \ddot{\phi}\cos\theta - \dot{\phi}\dot{\theta}\sin\theta)$$

Each moment summation applies only at the fixed point O or the center of mass G of the body. Since the equations represent a coupled set of nonlinear second-order differential equations, in general a closed-form solution may not be obtained. Instead, the Euler angles ϕ, θ, and ψ may be obtained graphically as functions of time using numerical analysis and computer techniques.

A special case, however, does exist for which simplification of Eqs. 21–29 is possible. Commonly referred to as *steady precession*, it occurs when the nutation angle θ, precession $\dot{\phi}$, and spin $\dot{\psi}$ all remain *constant*. Equations 21–29 then reduce to the form

$$\boxed{\Sigma M_x = -I\dot{\phi}^2\sin\theta\cos\theta + I_z\dot{\phi}\sin\theta(\dot{\phi}\cos\theta + \dot{\psi})} \qquad (21\text{–}30)$$

$$\Sigma M_y = 0$$

$$\Sigma M_z = 0$$

21

Equation 21–30 can be further simplified by noting that, from Eq. 21–27, $\omega_z = \dot{\phi} \cos \theta + \dot{\psi}$, so that

$$\Sigma M_x = -I\dot{\phi}^2 \sin \theta \cos \theta + I_z \dot{\phi} (\sin \theta) \omega_z$$

or

$$\Sigma M_x = \dot{\phi} \sin \theta (I_z \omega_z - I\dot{\phi} \cos \theta) \tag{21–31}$$

It is interesting to note what effects the spin $\dot{\psi}$ has on the moment about the x axis. To show this, consider the spinning rotor in Fig. 21–17. Here $\theta = 90°$, in which case Eq. 21–30 reduces to the form

$$\Sigma M_x = I_z \dot{\phi} \dot{\psi}$$

or

$$\Sigma M_x = I_z \Omega_y \omega_z \tag{21–32}$$

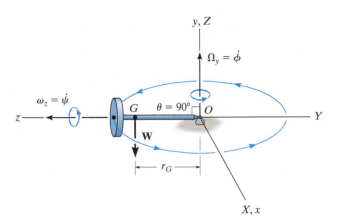

Fig. 21–17

From the figure it can be seen that $\boldsymbol{\Omega}_y$ and $\boldsymbol{\omega}_z$ act along their respective *positive axes* and therefore are mutually perpendicular. Instinctively, one would expect the rotor to fall down under the influence of gravity! However, this is not the case at all, provided the product $I_z \Omega_y \omega_z$ is correctly chosen to counterbalance the moment $\Sigma M_x = W r_G$ of the rotor's weight about O. This unusual phenomenon of rigid-body motion is often referred to as the *gyroscopic effect*.

Perhaps a more intriguing demonstration of the gyroscopic effect comes from studying the action of a *gyroscope*, frequently referred to as a *gyro*. A gyro is a rotor which spins at a very high rate about its axis of symmetry. This rate of spin is considerably greater than its precessional rate of rotation about the vertical axis. Hence, for all practical purposes, the angular momentum of the gyro can be assumed directed along its axis of spin. Thus, for the gyro rotor shown in Fig. 21–18, $\omega_z \gg \Omega_y$, and the magnitude of the angular momentum about point O, as determined from Eqs. 21–11, reduces to the form $H_O = I_z\omega_z$. Since both the magnitude and direction of $\mathbf{H}_O$ are constant as observed from x, y, z, direct application of Eq. 21–22 yields

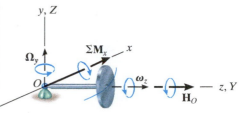

$$\boxed{\Sigma\mathbf{M}_x = \mathbf{\Omega}_y \times \mathbf{H}_O}$$

(21–33)

Fig. 21–18

Using the right-hand rule applied to the cross product, it can be seen that $\mathbf{\Omega}_y$ always swings $\mathbf{H}_O$ (or $\boldsymbol{\omega}_z$) toward the sense of $\Sigma\mathbf{M}_x$. In effect, the *change in direction* of the gyro's angular momentum, $d\mathbf{H}_O$, is equivalent to the angular impulse caused by the gyro's weight about O, i.e., $d\mathbf{H}_O = \Sigma\mathbf{M}_x \, dt$, Eq. 21–20. Also, since $H_O = I_z\omega_z$ and $\Sigma\mathbf{M}_x$, $\mathbf{\Omega}_y$, and $\mathbf{H}_O$ are mutually perpendicular, Eq. 21–33 reduces to Eq. 21–32.

When a gyro is mounted in gimbal rings, Fig. 21–19, it becomes *free* of external moments applied to its base. Thus, in theory, its angular momentum $\mathbf{H}$ will never precess but, instead, maintain its same fixed orientation along the axis of spin when the base is rotated. This type of gyroscope is called a *free gyro* and is useful as a gyrocompass when the spin axis of the gyro is directed north. In reality, the gimbal mechanism is never completely free of friction, so such a device is useful only for the local navigation of ships and aircraft. The gyroscopic effect is also useful as a means of stabilizing both the rolling motion of ships at sea and the trajectories of missiles and projectiles. Furthermore, this effect is of significant importance in the design of shafts and bearings for rotors which are subjected to forced precessions.

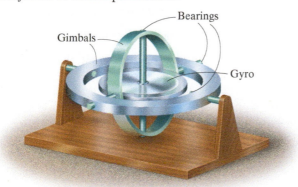

Fig. 21–19

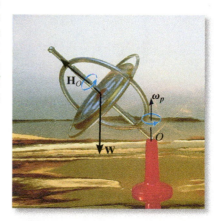

The spinning of the gyro within the frame of this toy gyroscope produces angular momentum $\mathbf{H}_O$, which is changing direction as the frame precesses $\boldsymbol{\omega}_p$ about the vertical axis. The gyroscope will not fall down since the moment of its weight $\mathbf{W}$ about the support is balanced by the change in the direction of $\mathbf{H}_O$. (© R.C. Hibbeler)

EXAMPLE 21.7

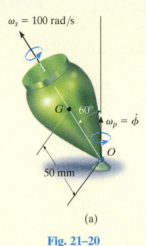

(a)

Fig. 21–20

The top shown in Fig. 21–20a has a mass of 0.5 kg and is precessing about the vertical axis at a constant angle of $\theta = 60°$. If it spins with an angular velocity $\omega_s = 100$ rad/s, determine the precession $\boldsymbol{\omega}_p$. Assume that the axial and transverse moments of inertia of the top are $0.45(10^{-3})$ kg·m² and $1.20(10^{-3})$ kg·m², respectively, measured with respect to the fixed point O.

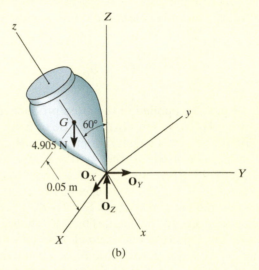

(b)

SOLUTION

Equation 21–30 will be used for the solution since the motion is *steady precession*. As shown on the free-body diagram, Fig. 21–20b, the coordinate axes are established in the usual manner, that is, with the positive z axis in the direction of spin, the positive Z axis in the direction of precession, and the positive x axis in the direction of the moment ΣM_x (refer to Fig. 21–16). Thus,

$$\Sigma M_x = -I\dot{\phi}^2 \sin\theta\cos\theta + I_z\dot{\phi}\sin\theta(\dot{\phi}\cos\theta + \dot{\psi})$$

$$4.905\text{ N}(0.05\text{ m})\sin 60° = -[1.20(10^{-3})\text{ kg·m}^2\,\dot{\phi}^2]\sin 60°\cos 60°$$
$$+ [0.45(10^{-3})\text{ kg·m}^2]\dot{\phi}\sin 60°(\dot{\phi}\cos 60° + 100\text{ rad/s})$$

or

$$\dot{\phi}^2 - 120.0\dot{\phi} + 654.0 = 0 \qquad (1)$$

Solving this quadratic equation for the precession gives

$$\dot{\phi} = 114\text{ rad/s} \quad \text{(high precession)} \qquad \textit{Ans.}$$

and

$$\dot{\phi} = 5.72\text{ rad/s} \quad \text{(low precession)} \qquad \textit{Ans.}$$

NOTE: In reality, low precession of the top would generally be observed, since high precession would require a larger kinetic energy.

EXAMPLE | 21.8

The 1-kg disk shown in Fig. 21–21a spins about its axis with a constant angular velocity $\omega_D = 70$ rad/s. The block at B has a mass of 2 kg, and by adjusting its position s one can change the precession of the disk about its supporting pivot at O while the shaft remains horizontal. Determine the position s that will enable the disk to have a constant precession $\omega_p = 0.5$ rad/s about the pivot. Neglect the weight of the shaft.

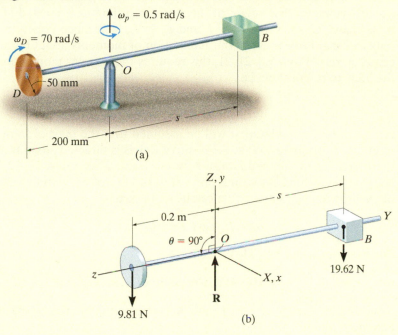

Fig. 21–21

SOLUTION
The free-body diagram of the assembly is shown in Fig. 21–21b. The origin for both the x, y, z and X, Y, Z coordinate systems is located at the fixed point O. In the conventional sense, the Z axis is chosen along the axis of precession, and the z axis is along the axis of spin, so that $\theta = 90°$. Since the precession is *steady*, Eq. 21–32 can be used for the solution.

$$\Sigma M_x = I_z \Omega_y \omega_z$$

Substituting the required data gives

$$(9.81 \text{ N}) (0.2 \text{ m}) - (19.62 \text{ N})s = \left[\tfrac{1}{2}(1 \text{ kg})(0.05 \text{ m})^2\right]0.5 \text{ rad/s}(-70 \text{ rad/s})$$

$$s = 0.102 \text{ m} = 102 \text{ mm} \qquad\qquad Ans.$$

21.6   Torque-Free Motion

When the only external force acting on a body is caused by gravity, the general motion of the body is referred to as *torque-free motion*. This type of motion is characteristic of planets, artificial satellites, and projectiles—provided air friction is neglected.

In order to describe the characteristics of this motion, the distribution of the body's mass will be assumed *axisymmetric*. The satellite shown in Fig. 21–22 is an example of such a body, where the z axis represents an axis of symmetry. The origin of the x, y, z coordinates is located at the mass center G, such that $I_{zz} = I_z$ and $I_{xx} = I_{yy} = I$. Since gravity is the only external force present, the summation of moments about the mass center is zero. From Eq. 21–21, this requires the angular momentum of the body to be constant, i.e.,

$$\mathbf{H}_G = \text{constant}$$

At the instant considered, it will be assumed that the inertial frame of reference is oriented so that the positive Z axis is directed along $\mathbf{H}_G$ and the y axis lies in the plane formed by the z and Z axes, Fig. 21–22. The Euler angle formed between Z and z is θ, and therefore, with this choice of axes the angular momentum can be expressed as

$$\mathbf{H}_G = H_G \sin \theta\, \mathbf{j} + H_G \cos \theta\, \mathbf{k}$$

Furthermore, using Eqs. 21–11, we have

$$\mathbf{H}_G = I\omega_x \mathbf{i} + I\omega_y \mathbf{j} + I_z \omega_z \mathbf{k}$$

Equating the respective $\mathbf{i}$, $\mathbf{j}$, and $\mathbf{k}$ components of the above two equations yields

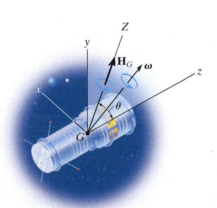

Fig. 21–22

$$\omega_x = 0 \quad \omega_y = \frac{H_G \sin\theta}{I} \quad \omega_z = \frac{H_G \cos\theta}{I_z} \qquad (21\text{–}34)$$

or

$$\boxed{\boldsymbol{\omega} = \frac{H_G \sin\theta}{I}\mathbf{j} + \frac{H_G \cos\theta}{I_z}\mathbf{k}} \qquad (21\text{–}35)$$

In a similar manner, equating the respective **i**, **j**, **k** components of Eq. 21–27 to those of Eq. 21–34, we obtain

$$\dot{\theta} = 0$$

$$\dot{\phi}\sin\theta = \frac{H_G \sin\theta}{I}$$

$$\dot{\phi}\cos\theta + \dot{\psi} = \frac{H_G \cos\theta}{I_z}$$

Solving, we get

$$\boxed{\begin{aligned} \theta &= \text{constant} \\ \dot{\phi} &= \frac{H_G}{I} \\ \dot{\psi} &= \frac{I - I_z}{I\,I_z}H_G \cos\theta \end{aligned}} \qquad (21\text{–}36)$$

Thus, for torque-free motion of an axisymmetrical body, the angle θ formed between the angular-momentum vector and the spin of the body remains constant. Furthermore, the angular momentum $\mathbf{H}_G$, precession $\dot{\phi}$, and spin $\dot{\psi}$ for the body remain constant at all times during the motion.

Eliminating H_G from the second and third of Eqs. 21–36 yields the following relation between the spin and precession:

$$\boxed{\dot{\psi} = \frac{I - I_z}{I_z}\dot{\phi}\cos\theta} \qquad (21\text{–}37)$$

21

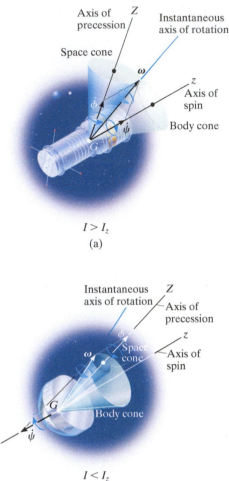

$I > I_z$

(a)

$I < I_z$

(b)

Fig. 21–23

These two components of angular motion can be studied by using the body and space cone models introduced in Sec. 20.1. The *space cone* defining the precession is fixed from rotating, since the precession has a fixed direction, while the outer surface of the *body cone* rolls on the space cone's outer surface. Try to imagine this motion in Fig. 21–23a. The interior angle of each cone is chosen such that the resultant angular velocity of the body is directed along the line of contact of the two cones. This line of contact represents the instantaneous axis of rotation for the body cone, and hence the angular velocity of both the body cone and the body must be directed along this line. Since the spin is a function of the moments of inertia I and I_z of the body, Eq. 21–36, the cone model in Fig. 21–23a is satisfactory for describing the motion, provided $I > I_z$. Torque-free motion which meets these requirements is called *regular precession*. If $I < I_z$, the spin is negative and the precession positive. This motion is represented by the satellite motion shown in Fig. 21–23b ($I < I_z$). The cone model can again be used to represent the motion; however, to preserve the correct vector addition of spin and precession to obtain the angular velocity $\boldsymbol{\omega}$, the inside surface of the body cone must roll on the outside surface of the (fixed) space cone. This motion is referred to as *retrograde precession*.

Satellites are often given a spin before they are launched. If their angular momentum is not collinear with the axis of spin, they will exhibit precession. In the photo on the left, regular precession will occur since $I > I_z$, and in the photo on the right, retrograde precession will occur since $I < I_z$.

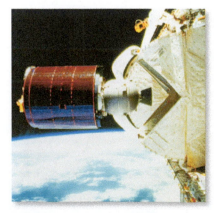

(© R.C. Hibbeler)

(© R.C. Hibbeler)

EXAMPLE | 21.9

The motion of a football is observed using a slow-motion projector. From the film, the spin of the football is seen to be directed 30° from the horizontal, as shown in Fig. 21–24a. Also, the football is precessing about the vertical axis at a rate $\dot{\phi} = 3$ rad/s. If the ratio of the axial to transverse moments of inertia of the football is $\frac{1}{3}$, measured with respect to the center of mass, determine the magnitude of the football's spin and its angular velocity. Neglect the effect of air resistance.

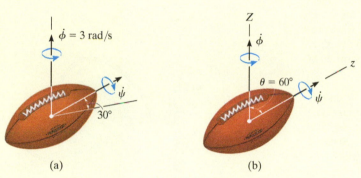

Fig. 21–24

SOLUTION

Since the weight of the football is the only force acting, the motion is torque-free. In the conventional sense, if the z axis is established along the axis of spin and the Z axis along the precession axis, as shown in Fig. 21–24b, then the angle $\theta = 60°$. Applying Eq. 21–37, the spin is

$$\dot{\psi} = \frac{I - I_z}{I_z}\dot{\phi}\cos\theta = \frac{I - \frac{1}{3}I}{\frac{1}{3}I}(3)\cos 60°$$

$$= 3 \text{ rad/s} \qquad \text{Ans.}$$

Using Eqs. 21–34, where $H_G = \dot{\phi}I$ (Eq. 21–36), we have

$$\omega_x = 0$$

$$\omega_y = \frac{H_G \sin\theta}{I} = \frac{3I \sin 60°}{I} = 2.60 \text{ rad/s}$$

$$\omega_z = \frac{H_G \cos\theta}{I_z} = \frac{3I \cos 60°}{\frac{1}{3}I} = 4.50 \text{ rad/s}$$

Thus,

$$\omega = \sqrt{(\omega_x)^2 + (\omega_y)^2 + (\omega_z)^2}$$

$$= \sqrt{(0)^2 + (2.60)^2 + (4.50)^2}$$

$$= 5.20 \text{ rad/s} \qquad \text{Ans.}$$

21

PROBLEMS

*21–60. Show that the angular velocity of a body, in terms of Euler angles ϕ, θ, and ψ, can be expressed as $\omega = (\dot{\phi} \sin \theta \sin \psi + \dot{\theta} \cos \psi)\mathbf{i} + (\dot{\phi} \sin \theta \cos \psi - \dot{\theta} \sin \psi)\mathbf{j} + (\dot{\phi} \cos \theta + \dot{\psi})\mathbf{k}$, where $\mathbf{i}, \mathbf{j}$, and $\mathbf{k}$ are directed along the x, y, z axes as shown in Fig. 21–15d.

21–61. A thin rod is initially coincident with the Z axis when it is given three rotations defined by the Euler angles $\phi = 30°$, $\theta = 45°$, and $\psi = 60°$. If these rotations are given in the order stated, determine the coordinate direction angles α, β, γ of the axis of the rod with respect to the X, Y, and Z axes. Are these directions the same for any order of the rotations? Why?

21–62. The gyroscope consists of a uniform 450-g disk D which is attached to the axle AB of negligible mass. The supporting frame has a mass of 180 g and a center of mass at G. If the disk is rotating about the axle at $\omega_D = 90$ rad/s, determine the constant angular velocity ω_p at which the frame precesses about the pivot point O. The frame moves in the horizontal plane.

21–63. The toy gyroscope consists of a rotor R which is attached to the frame of negligible mass. If it is observed that the frame is precessing about the pivot point O at $\omega_p = 2$ rad/s, determine the angular velocity ω_R of the rotor. The stem OA moves in the horizontal plane. The rotor has a mass of 200 g and a radius of gyration $k_{OA} = 20$ mm about OA.

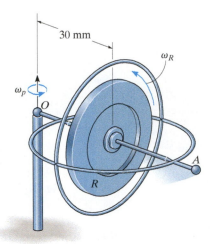

Prob. 21–63

*21–64. The top consists of a thin disk that has a weight of 8 lb and a radius of 0.3 ft. The rod has a negligible mass and a length of 0.5 ft. If the top is spinning with an angular velocity $\omega_s = 300$ rad/s, determine the steady-state precessional angular velocity ω_p of the rod when $\theta = 40°$.

21–65. Solve Prob. 21–64 when $\theta = 90°$.

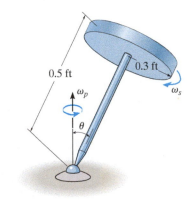

Probs. 21–64/65

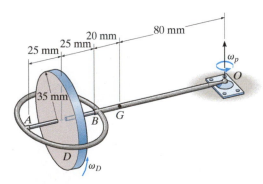

Prob. 21–62

21–66. The propeller on a single-engine airplane has a mass of 15 kg and a centroidal radius of gyration of 0.3 m computed about the axis of spin. When viewed from the front of the airplane, the propeller is turning clockwise at 350 rad/s about the spin axis. If the airplane enters a vertical curve having a radius of 80 m and is traveling at 200 km/h, determine the gyroscopic bending moment which the propeller exerts on the bearings of the engine when the airplane is in its lowest position.

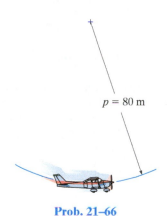

$p = 80$ m

Prob. 21–66

21–67. A wheel of mass m and radius r rolls with constant spin $\boldsymbol{\omega}$ about a circular path having a radius a. If the angle of inclination is θ, determine the rate of precession. Treat the wheel as a thin ring. No slipping occurs.

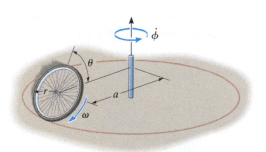

Prob. 21–67

*21–68. The conical top has a mass of 0.8 kg, and the moments of inertia are $I_x = I_y = 3.5(10^{-3})$ kg · m² and $I_z = 0.8(10^{-3})$ kg · m². If it spins freely in the ball-and socket joint at A with an angular velocity $\omega_s = 750$ rad/s, compute the precession of the top about the axis of the shaft AB.

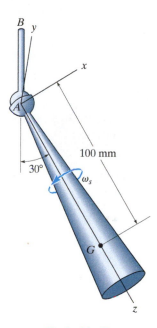

100 mm

30°

ω_s

Prob. 21–68

21–69. The top has a mass of 90 g, a center of mass at G, and a radius of gyration $k = 18$ mm about its axis of symmetry. About any transverse axis acting through point O the radius of gyration is $k_t = 35$ mm. If the top is connected to a ball-and-socket joint at O and the precession is $\omega_p = 0.5$ rad/s, determine the spin $\boldsymbol{\omega}_s$.

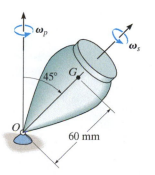

ω_p

ω_s

45° G

O 60 mm

Prob. 21–69

21

21–70. The 1-lb top has a center of gravity at point G. If it spins about its axis of symmetry and precesses about the vertical axis at constant rates of $\omega_s = 60 \text{ rad/s}$ and $\omega_p = 10 \text{ rad/s}$, respectively, determine the steady state angle θ. The radius of gyration of the top about the z axis is $k_z = 1$ in., and about the x and y axes it is $k_x = k_y = 4$ in.

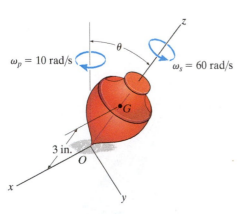

Prob. 21–70

21–71. The space capsule has a mass of 2 Mg, center of mass at G, and radii of gyration about its axis of symmetry (z axis) and its transverse axes (x or y axis) of $k_z = 2.75$ m and $k_x = k_y = 5.5$ m, respectively. If the capsule has the angular velocity shown, determine its precession $\dot{\phi}$ and spin $\dot{\psi}$. Indicate whether the precession is regular or retrograde. Also, draw the space cone and body cone for the motion.

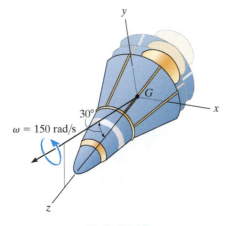

Prob. 21–71

*21–72. The 0.25 kg football is spinning at $\omega_z = 15 \text{ rad/s}$ as shown. If $\theta = 40°$, determine the precession about the z axis. The radius of gyration about the spin axis is $k_z = 0.042$ m, and about a transverse axis is $k_y = 0.13$ m.

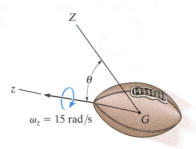

Prob. 21–72

21–73. The projectile shown is subjected to torque-free motion. The transverse and axial moments of inertia are I and I_z, respectively. If θ represents the angle between the precessional axis Z and the axis of symmetry z, and β is the angle between the angular velocity $\boldsymbol{\omega}$ and the z axis, show that β and θ are related by the equation $\tan \theta = (I/I_z) \tan \beta$.

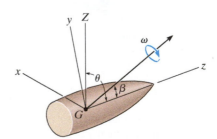

Prob. 21–73

21–74. The radius of gyration about an axis passing through the axis of symmetry of the 1.6-Mg space capsule is $k_z = 1.2$ m and about any transverse axis passing through the center of mass G, $k_t = 1.8$ m. If the capsule has a known steady-state precession of two revolutions per hour about the Z axis, determine the rate of spin about the z axis.

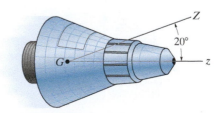

Prob. 21–74

21–75. The rocket has a mass of 4 Mg and radii of gyration $k_z = 0.85$ m and $k_x = k_y = 2.3$ m. It is initially spinning about the z axis at $\omega_z = 0.05$ rad/s when a meteoroid M strikes it at A and creates an impulse $\mathbf{I} = \{300\mathbf{i}\}$ N·s. Determine the axis of precession after the impact.

21–77. The satellite has a mass of 1.8 Mg, and about axes passing through the mass center G the axial and transverse radii of gyration are $k_z = 0.8$ m and $k_t = 1.2$ m, respectively. If it is spinning at $\omega_s = 6$ rad/s when it is launched, determine its angular momentum. Precession occurs about the Z axis.

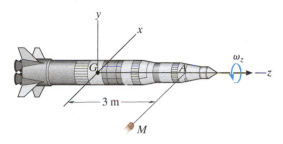

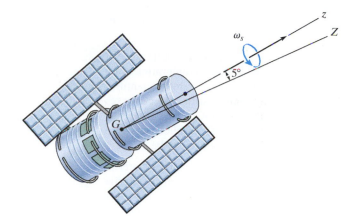

Prob. 21–75

Prob. 21–77

***21–76.** The football has a mass of 450 g and radii of gyration about its axis of symmetry (z axis) and its transverse axes (x or y axis) of $k_z = 30$ mm and $k_x = k_y = 50$ mm, respectively. If the football has an angular momentum of $H_G = 0.02$ kg·m²/s, determine its precession $\dot{\phi}$ and spin $\dot{\psi}$. Also, find the angle β that the angular velocity vector makes with the z axis.

21–78. The radius of gyration about an axis passing through the axis of symmetry of the 1.2-Mg satellite is $k_z = 1.4$ m, and about any transverse axis passing through the center of mass G, $k_t = 2.20$ m. If the satellite has a known spin of 2700 rev/h about the z axis, determine the steady-state precession about the z axis.

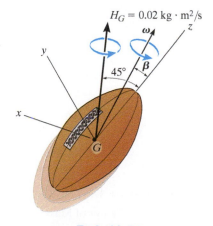

Prob. 21–76

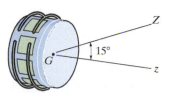

Prob. 21–78

CHAPTER REVIEW

Moments and Products of Inertia

A body has six components of inertia for any specified x, y, z axes. Three of these are moments of inertia about each of the axes, I_{xx}, I_{yy}, I_{zz}, and three are products of inertia, each defined from two orthogonal planes, I_{xy}, I_{yz}, I_{xz}. If either one or both of these planes are planes of symmetry, then the product of inertia with respect to these planes will be zero.

The moments and products of inertia can be determined by direct integration or by using tabulated values. If these quantities are to be determined with respect to axes or planes that do not pass through the mass center, then parallel-axis and parallel-plane theorems must be used.

Provided the six components of inertia are known, then the moment of inertia about any axis can be determined using the inertia transformation equation.

$$I_{xx} = \int_m r_x^2 \, dm = \int_m (y^2 + z^2) \, dm \qquad I_{xy} = I_{yx} = \int_m xy \, dm$$

$$I_{yy} = \int_m r_y^2 \, dm = \int_m (x^2 + z^2) \, dm \qquad I_{yz} = I_{zy} = \int_m yz \, dm$$

$$I_{zz} = \int_m r_z^2 \, dm = \int_m (x^2 + y^2) \, dm \qquad I_{xz} = I_{zx} = \int_m xz \, dm$$

$$I_{Oa} = I_{xx}u_x^2 + I_{yy}u_y^2 + I_{zz}u_z^2 - 2I_{xy}u_xu_y - 2I_{yz}u_yu_z - 2I_{zx}u_zu_x$$

Principal Moments of Inertia

At any point on or off the body, the x, y, z axes can be oriented so that the products of inertia will be zero. The resulting moments of inertia are called the principal moments of inertia. In general, one will be a maximum and the other a minimum.

$$\begin{pmatrix} I_x & 0 & 0 \\ 0 & I_y & 0 \\ 0 & 0 & I_z \end{pmatrix}$$

Principle of Impulse and Momentum

The angular momentum for a body can be determined about any arbitrary point A.

Once the linear and angular momentum for the body have been formulated, then the principle of impulse and momentum can be used to solve problems that involve force, velocity, and time.

$$m(\mathbf{v}_G)_1 + \Sigma \int_{t_1}^{t_2} \mathbf{F} \, dt = m(\mathbf{v}_G)_2$$

$$\mathbf{H}_O = \int_m \boldsymbol{\rho}_O \times (\boldsymbol{\omega} \times \boldsymbol{\rho}_O) \, dm$$

Fixed Point O

$$\mathbf{H}_G = \int_m \boldsymbol{\rho}_G \times (\boldsymbol{\omega} \times \boldsymbol{\rho}_G) \, dm$$

Center of Mass

$$\mathbf{H}_A = \boldsymbol{\rho}_{G/A} \times m\mathbf{v}_G + \mathbf{H}_G$$

Arbitrary Point

$$(\mathbf{H}_O)_1 + \Sigma \int_{t_1}^{t_2} \mathbf{M}_O \, dt = (\mathbf{H}_O)_2$$

where

$$H_x = I_{xx}\omega_x - I_{xy}\omega_y - I_{xz}\omega_z$$
$$H_y = -I_{yx}\omega_x + I_{yy}\omega_y - I_{yz}\omega_z$$
$$H_z = -I_{zx}\omega_x - I_{zy}\omega_y + I_{zz}\omega_z$$

Principle of Work and Energy

The kinetic energy for a body is usually determined relative to a fixed point or the body's mass center.

$$T = \tfrac{1}{2}I_x\omega_x^2 + \tfrac{1}{2}I_y\omega_y^2 + \tfrac{1}{2}I_z\omega_z^2 \qquad T = \tfrac{1}{2}mv_G^2 + \tfrac{1}{2}I_x\omega_x^2 + \tfrac{1}{2}I_y\omega_y^2 + \tfrac{1}{2}I_z\omega_z^2$$

Fixed Point   Center of Mass

These formulations can be used with the principle of work and energy to solve problems that involve force, velocity, and displacement.

$$T_1 + \Sigma U_{1-2} = T_2$$

Equations of Motion

There are three scalar equations of translational motion for a rigid body that moves in three dimensions.

$$\Sigma F_x = m(a_G)_x$$
$$\Sigma F_y = m(a_G)_y$$
$$\Sigma F_z = m(a_G)_z$$

The three scalar equations of rotational motion depend upon the motion of the x, y, z reference. Most often, these axes are oriented so that they are principal axes of inertia. If the axes are fixed in and move with the body so that $\Omega = \omega$, then the equations are referred to as the Euler equations of motion.

$$\Sigma M_x = I_x \dot{\omega}_x - (I_y - I_z)\omega_y\omega_z$$
$$\Sigma M_y = I_y \dot{\omega}_y - (I_z - I_x)\omega_z\omega_x$$
$$\Sigma M_z = I_z \dot{\omega}_z - (I_x - I_y)\omega_x\omega_y$$

$$\Omega = \omega$$

A free-body diagram should always accompany the application of the equations of motion.

$$\Sigma M_x = I_x \dot{\omega}_x - I_y \Omega_z\omega_y + I_z \Omega_y\omega_z$$
$$\Sigma M_y = I_y \dot{\omega}_y - I_z \Omega_x\omega_z + I_x \Omega_z\omega_x$$
$$\Sigma M_z = I_z \dot{\omega}_z - I_x \Omega_y\omega_x + I_y \Omega_x\omega_y$$

$$\Omega \neq \omega$$

Gyroscopic Motion

The angular motion of a gyroscope is best described using the three Euler angles ϕ, θ, and ψ. The angular velocity components are called the precession $\dot{\phi}$, the nutation $\dot{\theta}$, and the spin $\dot{\psi}$.

If $\dot{\theta} = 0$ and $\dot{\phi}$ and $\dot{\psi}$ are constant, then the motion is referred to as steady precession.

It is the spin of a gyro rotor that is responsible for holding a rotor from falling downward, and instead causing it to precess about a vertical axis. This phenomenon is called the gyroscopic effect.

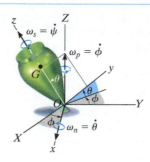

$$\Sigma M_x = -I\dot{\phi}^2 \sin\theta \cos\theta + I_z\dot{\phi} \sin\theta(\dot{\phi} \cos\theta + \dot{\psi})$$

$$\Sigma M_y = 0, \Sigma M_z = 0$$

Torque-Free Motion

A body that is only subjected to a gravitational force will have no moments on it about its mass center, and so the motion is described as torque-free motion. The angular momentum for the body about its mass center will remain constant. This causes the body to have both a spin and a precession. The motion depends upon the magnitude of the moment of inertia of a symmetric body about the spin axis, I_z, versus that about a perpendicular axis, I.

$$\theta = \text{constant}$$

$$\dot{\phi} = \frac{H_G}{I}$$

$$\dot{\psi} = \frac{I - I_z}{I I_z} H_G \cos\theta$$

21

Chapter 22

(© Daseaford/Fotolia)

The analysis of vibrations plays an important role in the study of the behavior of structures subjected to earthquakes.

Vibrations

*22.1 Undamped Free Vibration

A *vibration* is the oscillating motion of a body or system of connected bodies displaced from a position of equilibrium. In general, there are two types of vibration, free and forced. *Free vibration* occurs when the motion is maintained by gravitational or elastic restoring forces, such as the swinging motion of a pendulum or the vibration of an elastic rod. *Forced vibration* is caused by an external periodic or intermittent force applied to the system. Both of these types of vibration can either be damped or undamped. *Undamped* vibrations exclude frictional effects in the analysis. Since in reality both internal and external frictional forces are present, the motion of all vibrating bodies is actually *damped*.

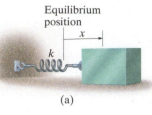

Equilibrium
position

x

k

(a)

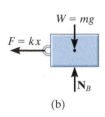

$W = mg$

$F = kx$

$\mathbf{N}_B$

(b)

Fig. 22–1

The simplest type of vibrating motion is undamped free vibration, represented by the block and spring model shown in Fig. 22–1a. Vibrating motion occurs when the block is released from a displaced position x so that the spring pulls on the block. The block will attain a velocity such that it will proceed to move out of equilibrium when $x = 0$, and provided the supporting surface is smooth, the block will oscillate back and forth.

The time-dependent path of motion of the block can be determined by applying the equation of motion to the block when it is in the displaced position x. The free-body diagram is shown in Fig. 22–1b. The elastic restoring force $F = kx$ is always directed toward the equilibrium position, whereas the acceleration **a** is assumed to act in the direction of *positive displacement*. Since $a = d^2x/dt^2 = \ddot{x}$, we have

$$\xrightarrow{+} \Sigma F_x = ma_x; \qquad -kx = m\ddot{x}$$

Note that the acceleration is proportional to the block's displacement. Motion described in this manner is called *simple harmonic motion*. Rearranging the terms into a "standard form" gives

$$\ddot{x} + \omega_n^2 x = 0 \qquad (22\text{–}1)$$

The constant ω_n, generally reported in rad/s, is called the *natural frequency*, and in this case

$$\boxed{\omega_n = \sqrt{\frac{k}{m}}} \qquad (22\text{–}2)$$

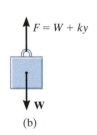

k

Equilibrium
position

y

(a)

$F = W + ky$

W

(b)

Fig. 22–2

Equation 22–1 can also be obtained by considering the block to be suspended so that the displacement y is measured from the block's *equilibrium position*, Fig. 22–2a. When the block is in equilibrium, the spring exerts an upward force of $F = W = mg$ on the block. Hence, when the block is displaced a distance y downward from this position, the magnitude of the spring force is $F = W + ky$, Fig. 22–2b. Applying the equation of motion gives

$$+\downarrow \Sigma F_y = ma_y; \qquad -W - ky + W = m\ddot{y}$$

or

$$\ddot{y} + \omega_n^2 y = 0$$

which is the same form as Eq. 22–1 and ω_n is defined by Eq. 22–2.

Equation 22–1 is a homogeneous, second-order, linear, differential equation with constant coefficients. It can be shown, using the methods of differential equations, that the general solution is

$$x = A \sin \omega_n t + B \cos \omega_n t \qquad (22\text{–}3)$$

Here A and B represent two constants of integration. The block's velocity and acceleration are determined by taking successive time derivatives, which yields

$$v = \dot{x} = A\omega_n \cos \omega_n t - B\omega_n \sin \omega_n t \qquad (22\text{–}4)$$

$$a = \ddot{x} = -A\omega_n^2 \sin \omega_n t - B\omega_n^2 \cos \omega_n t \qquad (22\text{–}5)$$

When Eqs. 22–3 and 22–5 are substituted into Eq. 22–1, the differential equation will be satisfied, showing that Eq. 22–3 is indeed the solution to Eq. 22–1.

The integration constants in Eq. 22–3 are generally determined from the initial conditions of the problem. For example, suppose that the block in Fig. 22–1a has been displaced a distance x_1 to the right from its equilibrium position and given an initial (positive) velocity $\mathbf{v}_1$ directed to the right. Substituting $x = x_1$ when $t = 0$ into Eq. 22–3 yields $B = x_1$. And since $v = v_1$ when $t = 0$, using Eq. 22–4 we obtain $A = v_1/\omega_n$. If these values are substituted into Eq. 22–3, the equation describing the motion becomes

$$x = \frac{v_1}{\omega_n} \sin \omega_n t + x_1 \cos \omega_n t \qquad (22\text{–}6)$$

Equation 22–3 may also be expressed in terms of simple sinusoidal motion. To show this, let

$$A = C \cos \phi \qquad (22\text{–}7)$$

and

$$B = C \sin \phi \qquad (22\text{–}8)$$

where C and ϕ are new constants to be determined in place of A and B. Substituting into Eq. 22–3 yields

$$x = C \cos \phi \sin \omega_n t + C \sin \phi \cos \omega_n t$$

And since $\sin(\theta + \phi) = \sin \theta \cos \phi + \cos \theta \sin \phi$, then

$$x = C \sin(\omega_n t + \phi) \qquad (22\text{–}9)$$

If this equation is plotted on an x versus $\omega_n t$ axis, the graph shown in Fig. 22–3 is obtained. The maximum displacement of the block from its equilibrium position is defined as the *amplitude* of vibration. From either the figure or Eq. 22–9 the amplitude is C. The angle ϕ is called the *phase angle* since it represents the amount by which the curve is displaced from the origin when $t = 0$. We can relate these two constants to A and B using Eqs. 22–7 and 22–8. Squaring and adding these two equations, the amplitude becomes

$$C = \sqrt{A^2 + B^2} \qquad (22\text{–}10)$$

If Eq. 22–8 is divided by Eq. 22–7, the phase angle is then

$$\phi = \tan^{-1}\frac{B}{A} \qquad (22\text{–}11)$$

Note that the sine curve, Eq. 22–9, completes one *cycle* in time $t = \tau$ (tau) when $\omega_n\tau = 2\pi$, or

$$\tau = \frac{2\pi}{\omega_n} \qquad (22\text{–}12)$$

This time interval is called a *period*, Fig. 22–3. Using Eq. 22–2, the period can also be represented as

$$\tau = 2\pi\sqrt{\frac{m}{k}} \qquad (22\text{–}13)$$

Finally, the *frequency* f is defined as the number of cycles completed per unit of time, which is the reciprocal of the period; that is,

$$f = \frac{1}{\tau} = \frac{\omega_n}{2\pi} \qquad (22\text{–}14)$$

or

$$f = \frac{1}{2\pi}\sqrt{\frac{k}{m}} \qquad (22\text{–}15)$$

The frequency is expressed in cycles/s. This ratio of units is called a *hertz* (Hz), where 1 Hz = 1 cycle/s = 2π rad/s.

When a body or system of connected bodies is given an initial displacement from its equilibrium position and released, it will vibrate with the *natural frequency*, ω_n. Provided the system has a single degree of freedom, that is, it requires only one coordinate to specify completely the position of the system at any time, then the vibrating motion will have the same characteristics as the simple harmonic motion of the block and spring just presented. Consequently, the motion is described by a differential equation of the same "standard form" as Eq. 22–1, i.e.,

$$\ddot{x} + \omega_n^2 x = 0 \qquad (22\text{–}16)$$

Hence, if the natural frequency ω_n is known, the period of vibration τ, frequency f, and other vibrating characteristics can be established using Eqs. 22–3 through 22–15.

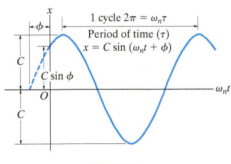

Fig. 22–3

Important Points

- Free vibration occurs when the motion is maintained by gravitational or elastic restoring forces.

- The amplitude is the maximum displacement of the body.

- The period is the time required to complete one cycle.

- The frequency is the number of cycles completed per unit of time, where 1 Hz = 1 cycle/s.

- Only one position coordinate is needed to describe the location of a one-degree-of-freedom system.

Procedure for Analysis

As in the case of the block and spring, the natural frequency ω_n of a body or system of connected bodies having a single degree of freedom can be determined using the following procedure:

Free-Body Diagram.

- Draw the free-body diagram of the body when the body is displaced a *small amount* from its equilibrium position.

- Locate the body with respect to its equilibrium position by using an appropriate *inertial coordinate q*. The acceleration of the body's mass center $\mathbf{a}_G$ or the body's angular acceleration $\boldsymbol{\alpha}$ should have an assumed sense of direction which is in the *positive direction* of the position coordinate.

- If the rotational equation of motion $\Sigma M_P = \Sigma(\mathcal{M}_k)_P$ is to be used, then it may be beneficial to also draw the kinetic diagram since it graphically accounts for the components $m(\mathbf{a}_G)_x$, $m(\mathbf{a}_G)_y$, and $I_G\boldsymbol{\alpha}$, and thereby makes it convenient for visualizing the terms needed in the moment sum $\Sigma(\mathcal{M}_k)_P$.

Equation of Motion.

- Apply the equation of motion to relate the elastic or gravitational *restoring* forces and couple moments acting on the body to the body's accelerated motion.

Kinematics.

- Using kinematics, express the body's accelerated motion in terms of the second time derivative of the position coordinate, $\ddot{q}$.

- Substitute the result into the equation of motion and determine ω_n by rearranging the terms so that the resulting equation is in the "standard form," $\ddot{q} + \omega_n^2 q = 0$.

22

EXAMPLE 22.1

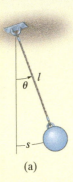

(a)

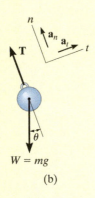

(b)

Fig. 22–4

Determine the period of oscillation for the simple pendulum shown in Fig. 22–4a. The bob has a mass m and is attached to a cord of length l. Neglect the size of the bob.

SOLUTION

Free-Body Diagram. Motion of the system will be related to the position coordinate ($q =$) θ, Fig. 22–4b. When the bob is displaced by a small angle θ, the *restoring force* acting on the bob is created by the tangential component of its weight, $mg \sin \theta$. Furthermore, $\mathbf{a}_t$ acts in the direction of *increasing s* (or θ).

Equation of Motion. Applying the equation of motion in the *tangential direction*, since it involves the restoring force, yields

$$+\nearrow \Sigma F_t = ma_t; \qquad -mg \sin \theta = ma_t \qquad (1)$$

Kinematics. $a_t = d^2s/dt^2 = \ddot{s}$. Furthermore, s can be related to θ by the equation $s = l\theta$, so that $a_t = l\ddot{\theta}$. Hence, Eq. 1 reduces to

$$\ddot{\theta} + \frac{g}{l} \sin \theta = 0 \qquad (2)$$

The solution of this equation involves the use of an elliptic integral. For *small displacements*, however, $\sin \theta \approx \theta$, in which case

$$\ddot{\theta} + \frac{g}{l} \theta = 0 \qquad (3)$$

Comparing this equation with Eq. 22–16 ($\ddot{x} + \omega_n^2 x = 0$), it is seen that $\omega_n = \sqrt{g/l}$. From Eq. 22–12, the period of time required for the bob to make one complete swing is therefore

$$\tau = \frac{2\pi}{\omega_n} = 2\pi \sqrt{\frac{l}{g}} \qquad \qquad Ans.$$

This interesting result, originally discovered by Galileo Galilei through experiment, indicates that the period depends only on the length of the cord and not on the mass of the pendulum bob or the angle θ.

NOTE: The solution of Eq. 3 is given by Eq. 22–3, where $\omega_n = \sqrt{g/l}$ and θ is substituted for x. Like the block and spring, the constants A and B in this problem can be determined if, for example, one knows the displacement and velocity of the bob at a given instant.

22

EXAMPLE 22.2

The 10-kg rectangular plate shown in Fig. 22–5a is suspended at its center from a rod having a torsional stiffness $k = 1.5 \text{ N} \cdot \text{m/rad}$. Determine the natural period of vibration of the plate when it is given a small angular displacement θ in the plane of the plate.

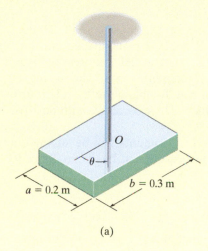

(a)

SOLUTION

Free-Body Diagram. Fig. 22–5b. Since the plate is displaced in its own plane, the torsional *restoring* moment created by the rod is $M = k\theta$. This moment acts in the direction opposite to the angular displacement θ. The angular acceleration $\ddot{\theta}$ acts in the direction of *positive* θ.

Equation of Motion.

$$\Sigma M_O = I_O \alpha; \qquad -k\theta = I_O \ddot{\theta}$$

or

$$\ddot{\theta} + \frac{k}{I_O}\theta = 0$$

Since this equation is in the "standard form," the natural frequency is $\omega_n = \sqrt{k/I_O}$.

From the table on the inside back cover, the moment of inertia of the plate about an axis coincident with the rod is $I_O = \frac{1}{12}m(a^2 + b^2)$. Hence,

$$I_O = \frac{1}{12}(10 \text{ kg})\left[(0.2 \text{ m})^2 + (0.3 \text{ m})^2\right] = 0.1083 \text{ kg} \cdot \text{m}^2$$

The natural period of vibration is therefore,

$$\tau = \frac{2\pi}{\omega_n} = 2\pi\sqrt{\frac{I_O}{k}} = 2\pi\sqrt{\frac{0.1083}{1.5}} = 1.69 \text{ s} \qquad Ans.$$

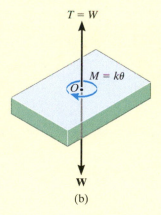

(b)

Fig. 22–5

22

EXAMPLE 22.3

The bent rod shown in Fig. 22–6a has a negligible mass and supports a 5-kg collar at its end. If the rod is in the equilibrium position shown, determine the natural period of vibration for the system.

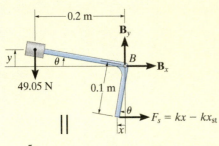

(a)

SOLUTION

Free-Body and Kinetic Diagrams. Fig. 22–6b. Here the rod is displaced by a small angle θ from the equilibrium position. Since the spring is subjected to an initial compression of x_{st} for equilibrium, then when the displacement $x > x_{st}$ the spring exerts a force of $F_s = kx - kx_{st}$ on the rod. To obtain the "standard form," Eq. 22–16, $5\mathbf{a}_y$ must act *upward*, which is in accordance with positive θ displacement.

Equation of Motion. Moments will be summed about point B to eliminate the unknown reaction at this point. Since θ is small,

$$\zeta + \Sigma M_B = \Sigma(\mathcal{M}_k)_B;$$

$$kx(0.1\text{ m}) - kx_{st}(0.1\text{ m}) + 49.05\text{ N}(0.2\text{ m}) = -(5\text{ kg})a_y(0.2\text{ m})$$

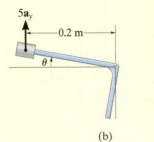

(b)

The second term on the left side, $-kx_{st}(0.1\text{ m})$, represents the moment created by the spring force which is necessary to hold the collar in *equilibrium*, i.e., at $x = 0$. Since this moment is equal and opposite to the moment $49.05\text{ N}(0.2\text{ m})$ created by the weight of the collar, these two terms cancel in the above equation, so that

$$kx(0.1) = -5a_y(0.2) \tag{1}$$

Kinematics. The deformation of the spring and the position of the collar can be related to the angle θ, Fig. 22–6c. Since θ is small, $x = (0.1\text{ m})\theta$ and $y = (0.2\text{ m})\theta$. Therefore, $a_y = \ddot{y} = 0.2\ddot{\theta}$. Substituting into Eq. 1 yields

$$400(0.1\theta)\, 0.1 = -5(0.2\ddot{\theta})0.2$$

Rewriting this equation in the "standard form" gives

$$\ddot{\theta} + 20\theta = 0$$

Compared with $\ddot{x} + \omega_n^2 x = 0$ (Eq. 22–16), we have

$$\omega_n^2 = 20 \quad \omega_n = 4.47\text{ rad/s}$$

The natural period of vibration is therefore

$$\tau = \frac{2\pi}{\omega_n} = \frac{2\pi}{4.47} = 1.40\text{ s} \qquad \textit{Ans.}$$

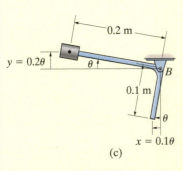

(c)

Fig. 22–6

EXAMPLE 22.4

A 10-lb block is suspended from a cord that passes over a 15-lb disk, as shown in Fig. 22–7a. The spring has a stiffness $k = 200$ lb/ft. Determine the natural period of vibration for the system.

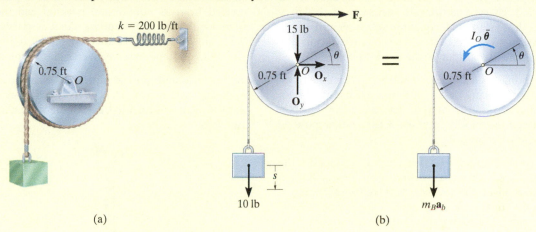

(a)   (b)

SOLUTION

Free-Body and Kinetic Diagrams. Fig. 22–7b. The *system* consists of the disk, which undergoes a rotation defined by the angle θ, and the block, which translates by an amount s. The vector $I_O\ddot{\theta}$ acts in the direction of *positive θ*, and consequently $m_B\mathbf{a}_b$ acts downward in the direction of *positive s*.

Equation of Motion. Summing moments about point O to eliminate the reactions $\mathbf{O}_x$ and $\mathbf{O}_y$, realizing that $I_O = \frac{1}{2}mr^2$, yields

$$\zeta + \Sigma M_O = \Sigma(\mathcal{M}_k)_O;$$

$$10 \text{ lb}(0.75 \text{ ft}) - F_s(0.75 \text{ ft})$$

$$= \frac{1}{2}\left(\frac{15 \text{ lb}}{32.2 \text{ ft/s}^2}\right)(0.75 \text{ ft})^2\,\ddot{\theta} + \left(\frac{10 \text{ lb}}{32.2 \text{ ft/s}^2}\right)a_b(0.75 \text{ ft}) \quad (1)$$

Kinematics. As shown on the kinematic diagram in Fig. 22–7c, a small positive displacement θ of the disk causes the block to lower by an amount $s = 0.75\theta$; hence, $a_b = \ddot{s} = 0.75\ddot{\theta}$. When $\theta = 0°$, the spring force required for *equilibrium* of the disk is 10 lb, acting to the right. For position θ, the spring force is $F_s = (200 \text{ lb/ft})(0.75\theta \text{ ft}) + 10 \text{ lb}$. Substituting these results into Eq. 1 and simplifying yields

$$\ddot{\theta} + 368\theta = 0$$

Hence,

$$\omega_n^2 = 368 \qquad \omega_n = 19.18 \text{ rad/s}$$

Therefore, the natural period of vibration is

$$\tau = \frac{2\pi}{\omega_n} = \frac{2\pi}{19.18} = 0.328 \text{ s} \qquad \textit{Ans.}$$

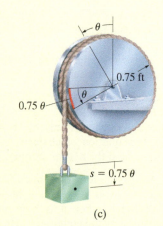

(c)

Fig. 22–7

22

PROBLEMS

22–1. A spring is stretched 175 mm by an 8-kg block. If the block is displaced 100 mm downward from its equilibrium position and given a downward velocity of 1.50 m/s, determine the differential equation which describes the motion. Assume that positive displacement is downward. Also, determine the position of the block when $t = 0.22$ s.

22–2. A spring has a stiffness of 800 N/m. If a 2-kg block is attached to the spring, pushed 50 mm above its equilibrium position, and released from rest, determine the equation that describes the block's motion. Assume that positive displacement is downward.

22–3. A spring is stretched 200 mm by a 15-kg block. If the block is displaced 100 mm downward from its equilibrium position and given a downward velocity of 0.75 m/s, determine the equation which describes the motion. What is the phase angle? Assume that positive displacement is downward.

*22–4. When a 20-lb weight is suspended from a spring, the spring is stretched a distance of 4 in. Determine the natural frequency and the period of vibration for a 10-lb weight attached to the same spring.

22–5. When a 3-kg block is suspended from a spring, the spring is stretched a distance of 60 mm. Determine the natural frequency and the period of vibration for a 0.2-kg block attached to the same spring.

22–6. An 8-kg block is suspended from a spring having a stiffness $k = 80$ N/m. If the block is given an upward velocity of 0.4 m/s when it is 90 mm above its equilibrium position, determine the equation which describes the motion and the maximum upward displacement of the block measured from the equilibrium position. Assume that positive displacement is measured downward.

22–7. A 2-lb weight is suspended from a spring having a stiffness $k = 2$ lb/in. If the weight is pushed 1 in. upward from its equilibrium position and then released from rest, determine the equation which describes the motion. What is the amplitude and the natural frequency of the vibration?

*22–8. A 6-lb weight is suspended from a spring having a stiffness $k = 3$ lb/in. If the weight is given an upward velocity of 20 ft/s when it is 2 in. above its equilibrium position, determine the equation which describes the motion and the maximum upward displacement of the weight, measured from the equilibrium position. Assume positive displacement is downward.

22–9. A 3-kg block is suspended from a spring having a stiffness of $k = 200$ N/m. If the block is pushed 50 mm upward from its equilibrium position and then released from rest, determine the equation that describes the motion. What are the amplitude and the natural frequency of the vibration? Assume that positive displacement is downward.

22–10. The uniform rod of mass m is supported by a pin at A and a spring at B. If B is given a small sideward displacement and released, determine the natural period of vibration.

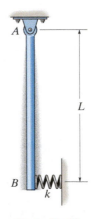

Prob. 22–10

22–11. While standing in an elevator, the man holds a pendulum which consists of an 18-in. cord and a 0.5-lb bob. If the elevator is descending with an acceleration $a = 4$ ft/s^2, determine the natural period of vibration for small amplitudes of swing.

Prob. 22–11

*22–12. Determine the natural period of vibration of the uniform bar of mass m when it is displaced downward slightly and released.

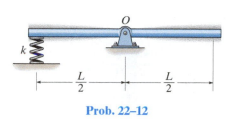

Prob. 22–12

22–14. The 20-lb rectangular plate has a natural period of vibration $\tau = 0.3$ s, as it oscillates around the axis of rod AB. Determine the torsional stiffness k, measured in lb · ft/rad, of the rod. Neglect the mass of the rod.

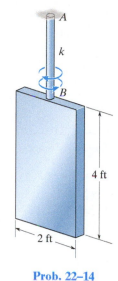

4 ft

2 ft

Prob. 22–14

22–13. The body of arbitrary shape has a mass m, mass center at G, and a radius of gyration about G of k_G. If it is displaced a slight amount θ from its equilibrium position and released, determine the natural period of vibration.

22–15. A platform, having an unknown mass, is supported by *four* springs, each having the same stiffness k. When nothing is on the platform, the period of vertical vibration is measured as 2.35 s; whereas if a 3-kg block is supported on the platform, the period of vertical vibration is 5.23 s. Determine the mass of a block placed on the (empty) platform which causes the platform to vibrate vertically with a period of 5.62 s. What is the stiffness k of each of the springs?

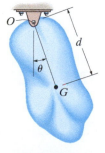

Prob. 22–13

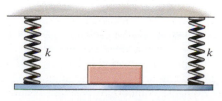

Prob. 22–15

22

*22–16. A block of mass m is suspended from two springs having a stiffness of k_1 and k_2, arranged a) parallel to each other, and b) as a series. Determine the equivalent stiffness of a single spring with the same oscillation characteristics and the period of oscillation for each case.

22–17. The 15-kg block is suspended from two springs having a different stiffness and arranged a) parallel to each other, and b) as a series. If the natural periods of oscillation of the parallel system and series system are observed to be 0.5 s and 1.5 s, respectively, determine the spring stiffnesses k_1 and k_2.

22–19. The slender rod has a mass of 0.2 kg and is supported at O by a pin and at its end A by two springs, each having a stiffness $k = 4$ N/m. The period of vibration of the rod can be set by fixing the 0.5-kg collar C to the rod at an appropriate location along its length. If the springs are originally unstretched when the rod is vertical, determine the position y of the collar so that the natural period of vibration becomes $\tau = 1$ s. Neglect the size of the collar.

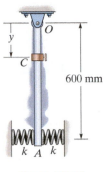

Prob. 22–19

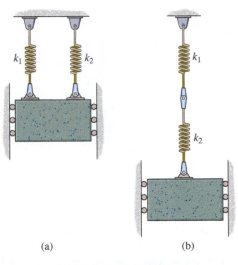

(a)   (b)

Probs. 22–16/17

22–18. The uniform beam is supported at its ends by two springs A and B, each having the same stiffness k. When nothing is supported on the beam, it has a period of vertical vibration of 0.83 s. If a 50-kg mass is placed at its center, the period of vertical vibration is 1.52 s. Compute the stiffness of each spring and the mass of the beam.

*22–20. A uniform board is supported on two wheels which rotate in opposite directions at a constant angular speed. If the coefficient of kinetic friction between the wheels and board is μ, determine the frequency of vibration of the board if it is displaced slightly, a distance x from the midpoint between the wheels, and released.

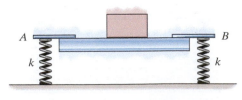

Prob. 22–18

Prob. 22–20

22–21. If the wire AB is subjected to a tension of 20 lb, determine the equation which describes the motion when the 5-lb weight is displaced 2 in. horizontally and released from rest.

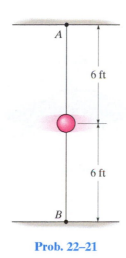

Prob. 22–21

22–22. The bar has a length l and mass m. It is supported at its ends by rollers of negligible mass. If it is given a small displacement and released, determine the natural frequency of vibration.

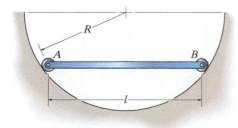

Prob. 22–22

22–23. The 20-kg disk, is pinned at its mass center O and supports the 4-kg block A. If the belt which passes over the disk is not allowed to slip at its contacting surface, determine the natural period of vibration of the system.

Prob. 22–23

*22–24. The 10-kg disk is pin connected at its mass center. Determine the natural period of vibration of the disk if the springs have sufficient tension in them to prevent the cord from slipping on the disk as it oscillates. *Hint:* Assume that the initial stretch in each spring is δ_O.

22–25. If the disk in Prob. 22–24 has a mass of 10 kg, determine the natural frequency of vibration. *Hint:* Assume that the initial stretch in each spring is δ_O.

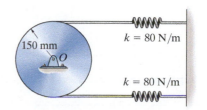

Probs. 22–24/25

22–26. A flywheel of mass m, which has a radius of gyration about its center of mass of k_O, is suspended from a circular shaft that has a torsional resistance of $M = C\theta$. If the flywheel is given a small angular displacement of θ and released, determine the natural period of oscillation.

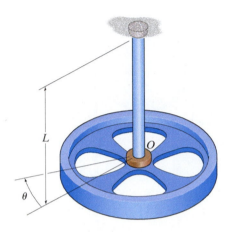

Prob. 22–26

22–27. The 6-lb weight is attached to the rods of negligible mass. Determine the natural frequency of vibration of the weight when it is displaced slightly from the equilibrium position and released.

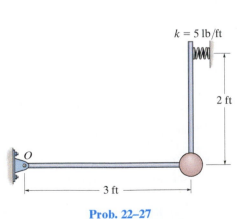

Prob. 22–27

***22–28.** The platform AB when empty has a mass of 400 kg, center of mass at G_1, and natural period of oscillation $\tau_1 = 2.38$ s. If a car, having a mass of 1.2 Mg and center of mass at G_2, is placed on the platform, the natural period of oscillation becomes $\tau_2 = 3.16$ s. Determine the moment of inertia of the car about an axis passing through G_2.

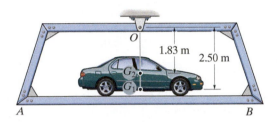

Prob. 22–28

22–29. The plate of mass m is supported by three symmetrically placed cords of length l as shown. If the plate is given a slight rotation about a vertical axis through its center and released, determine the natural period of oscillation.

Prob. 22–29

*22.2 Energy Methods

The simple harmonic motion of a body, discussed in the previous section, is due only to gravitational and elastic restoring forces acting on the body. Since these forces are *conservative*, it is also possible to use the conservation of energy equation to obtain the body's natural frequency or period of vibration. To show how to do this, consider again the block and spring model in Fig. 22–8. When the block is displaced x from the equilibrium position, the kinetic energy is $T = \frac{1}{2}mv^2 = \frac{1}{2}m\dot{x}^2$ and the potential energy is $V = \frac{1}{2}kx^2$. Since energy is conserved, it is necessary that

$$T + V = \text{constant}$$

$$\tfrac{1}{2}m\dot{x}^2 + \tfrac{1}{2}kx^2 = \text{constant} \qquad (22\text{–}17)$$

The differential equation describing the *accelerated motion* of the block can be obtained by *differentiating* this equation with respect to time; i.e.,

$$m\dot{x}\ddot{x} + kx\dot{x} = 0$$

$$\dot{x}(m\ddot{x} + kx) = 0$$

Since the velocity $\dot{x}$ is not *always* zero in a vibrating system,

$$\ddot{x} + \omega_n^2 x = 0 \qquad \omega_n = \sqrt{k/m}$$

which is the same as Eq. 22–1.

If the conservation of energy equation is written for a *system of connected bodies*, the natural frequency or the equation of motion can also be determined by time differentiation. It is *not necessary* to dismember the system to account for the internal forces because they do no work.

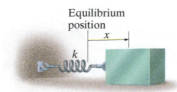

Equilibrium
position

x

k

Fig. 22–8

22

The suspension of a railroad car consists of a set of springs which are mounted between the frame of the car and the wheel truck. This will give the car a natural frequency of vibration which can be determined. (© R.C. Hibbeler)

Procedure for Analysis

The natural frequency ω_n of a body or system of connected bodies can be determined by applying the conservation of energy equation using the following procedure.

Energy Equation.

- Draw the body when it is displaced by a *small amount* from its equilibrium position and define the location of the body from its equilibrium position by an appropriate position coordinate q.

- Formulate the conservation of energy for the body, $T + V =$ constant, in terms of the position coordinate.

- In general, the kinetic energy must account for both the body's translational and rotational motion, $T = \frac{1}{2}mv_G^2 + \frac{1}{2}I_G\omega^2$, Eq. 18–2.

- The potential energy is the sum of the gravitational and elastic potential energies of the body, $V = V_g + V_e$, Eq. 18–17. In particular, V_g should be measured from a datum for which $q = 0$ (equilibrium position).

Time Derivative.

- Take the time derivative of the energy equation using the chain rule of calculus and factor out the common terms. The resulting differential equation represents the equation of motion for the system. The natural frequency of ω_n is obtained after rearranging the terms in the "standard form," $\ddot{q} + \omega_n^2 q = 0$.

22

EXAMPLE 22.5

The thin hoop shown in Fig. 22–9a is supported by the peg at O. Determine the natural period of oscillation for small amplitudes of swing. The hoop has a mass m.

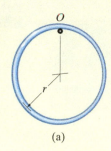

(a)

SOLUTION

Energy Equation. A diagram of the hoop when it is displaced a small amount $(q =) \theta$ from the equilibrium position is shown in Fig. 22–9b. Using the table on the inside back cover and the parallel-axis theorem to determine I_O, the kinetic energy is

$$T = \tfrac{1}{2}I_O\omega_n^2 = \tfrac{1}{2}[mr^2 + mr^2]\dot{\theta}^2 = mr^2\dot{\theta}^2$$

If a horizontal datum is placed through point O, then in the displaced position, the potential energy is

$$V = -mg(r \cos \theta)$$

The total energy in the system is

$$T + V = mr^2\dot{\theta}^2 - mgr \cos \theta$$

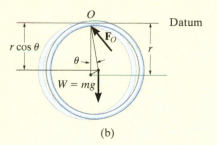

(b)

Fig. 22–9

Time Derivative.

$$mr^2(2\dot{\theta})\ddot{\theta} + mgr(\sin \theta)\dot{\theta} = 0$$

$$mr\dot{\theta}(2r\ddot{\theta} + g \sin \theta) = 0$$

Since $\dot{\theta}$ is not always equal to zero, from the terms in parentheses,

$$\ddot{\theta} + \frac{g}{2r}\sin \theta = 0$$

For small angle θ, $\sin \theta \approx \theta$.

$$\ddot{\theta} + \frac{g}{2r} \theta = 0$$

$$\omega_n = \sqrt{\frac{g}{2r}}$$

so that

$$\tau = \frac{2\pi}{\omega_n} = 2\pi\sqrt{\frac{2r}{g}} \qquad \qquad \textit{Ans.}$$

22

EXAMPLE 22.6

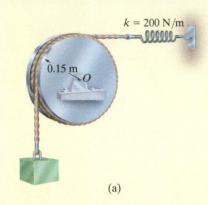

A 10-kg block is suspended from a cord wrapped around a 5-kg disk, as shown in Fig. 22–10a. If the spring has a stiffness $k = 200 \text{ N/m}$, determine the natural period of vibration for the system.

SOLUTION

Energy Equation.   A diagram of the block and disk when they are displaced by respective amounts s and θ from the equilibrium position is shown in Fig. 22–10b. Since $s = (0.15 \text{ m})\theta$, then $v_b \approx \dot{s} = (0.15 \text{ m})\dot{\theta}$. Thus, the kinetic energy of the system is

$$
\begin{aligned}
T &= \tfrac{1}{2}m_b v_b^2 + \tfrac{1}{2}I_O\omega_d^2 \\
&= \tfrac{1}{2}(10 \text{ kg})[(0.15 \text{ m})\dot{\theta}]^2 + \tfrac{1}{2}\left[\tfrac{1}{2}(5 \text{ kg})(0.15 \text{ m})^2\right](\dot{\theta})^2 \\
&= 0.1406(\dot{\theta})^2
\end{aligned}
$$

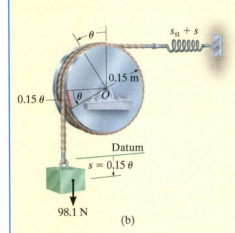

(a)

(b)

Fig. 22–10

Establishing the datum at the equilibrium position of the block and realizing that the spring stretches s_{st} for equilibrium, the potential energy is

$$
\begin{aligned}
V &= \tfrac{1}{2}k(s_{st} + s)^2 - Ws \\
&= \tfrac{1}{2}(200 \text{ N/m})[s_{st} + (0.15 \text{ m})\theta]^2 - 98.1 \text{ N}[(0.15 \text{ m})\theta]
\end{aligned}
$$

The total energy for the system is therefore,

$$
T + V = 0.1406(\dot{\theta})^2 + 100(s_{st} + 0.15\theta)^2 - 14.715\theta
$$

Time Derivative.

$$
0.28125(\dot{\theta})\ddot{\theta} + 200(s_{st} + 0.15\theta)0.15\dot{\theta} - 14.72\dot{\theta} = 0
$$

Since $s_{st} = 98.1/200 = 0.4905 \text{ m}$, the above equation reduces to the "standard form"

$$
\ddot{\theta} + 16\theta = 0
$$

so that

$$
\omega_n = \sqrt{16} = 4 \text{ rad/s}
$$

Thus,

$$
\tau = \frac{2\pi}{\omega_n} = \frac{2\pi}{4} = 1.57 \text{ s} \qquad \textit{Ans.}
$$

PROBLEMS

22–30. Determine the differential equation of motion of the 3-kg block when it is displaced slightly and released. The surface is smooth and the springs are originally unstretched.

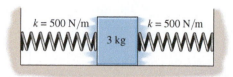

Prob. 22–30

22–31. Determine the natural period of vibration of the pendulum. Consider the two rods to be slender, each having a weight of 8 lb/ft.

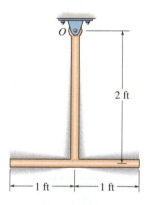

Prob. 22–31

*22–32.** Determine the natural period of vibration of the 10-lb semicircular disk.

Prob. 22–32

22–33. If the 20-kg wheel is displaced a small amount and released, determine the natural period of vibration. The radius of gyration of the wheel is $k_G = 0.36$ m. The wheel rolls without slipping.

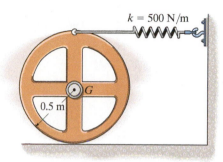

Prob. 22–33

22–34. Determine the differential equation of motion of the 3-kg spool. Assume that it does not slip at the surface of contact as it oscillates. The radius of gyration of the spool about its center of mass is $k_G = 125$ mm.

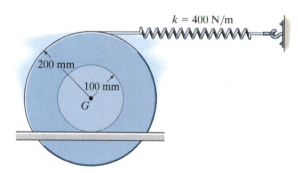

Prob. 22–34

22–35. Determine the natural period of vibration of the 3-kg sphere. Neglect the mass of the rod and the size of the sphere.

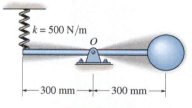

Prob. 22–35

22

*22–36. If the lower end of the 6-kg slender rod is displaced a small amount and released from rest, determine the natural frequency of vibration. Each spring has' a stiffness of $k = 200$ N/m and is unstretched when the rod is hanging vertically.

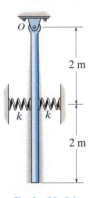

2 m

k k

2 m

Prob. 22–36

22–37. The disk has a weight of 30 lb and rolls without slipping on the horizontal surface as it oscillates about its equilibrium position. If the disk is displaced, by rolling it counterclockwise 0.2 rad, determine the equation which describes its oscillatory motion and the natural period when it is released.

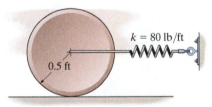

$k = 80$ lb/ft

0.5 ft

Prob. 22–37

22–38. The machine has a mass m and is uniformly supported by *four* springs, each having a stiffness k. Determine the natural period of vertical vibration.

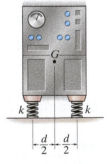

G

k k

$\dfrac{d}{2}$ $\dfrac{d}{2}$

Prob. 22–38

22–39. The slender rod has a weight of 4 lb/ft. If it is supported in the horizontal plane by a ball-and-socket joint at A and a cable at B, determine the natural frequency of vibration when the end B is given a small horizontal displacement and then released.

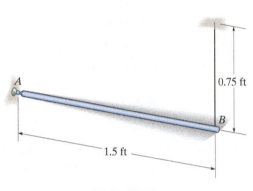

A

0.75 ft

B

1.5 ft

Prob. 22–39

*22–40. If the slender rod has a weight of 5 lb, determine the natural frequency of vibration. The springs are originally unstretched.

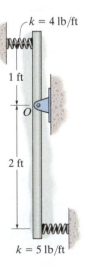

$k = 4$ lb/ft

1 ft

O

2 ft

$k = 5$ lb/ft

Prob. 22–40

*22.3   Undamped Forced Vibration

Undamped forced vibration is considered to be one of the most important types of vibrating motion in engineering. Its principles can be used to describe the motion of many types of machines and structures.

Periodic Force.   The block and spring shown in Fig. 22–11a provide a convenient model which represents the vibrational characteristics of a system subjected to a periodic force $F = F_0 \sin \omega_0 t$. This force has an amplitude of F_0 and a *forcing frequency* ω_0. The free-body diagram for the block when it is displaced a distance x is shown in Fig. 22–11b. Applying the equation of motion, we have

$$\xrightarrow{+} \Sigma F_x = ma_x; \qquad F_0 \sin \omega_0 t - kx = m\ddot{x}$$

or

$$\ddot{x} + \frac{k}{m}x = \frac{F_0}{m}\sin \omega_0 t \qquad (22\text{--}18)$$

This equation is a nonhomogeneous second-order differential equation. The general solution consists of a complementary solution, x_c, *plus* a particular solution, x_p.

The *complementary solution* is determined by setting the term on the right side of Eq. 22–18 equal to zero and solving the resulting homogeneous equation. The solution is defined by Eq. 22–9, i.e.,

$$x_c = C \sin(\omega_n t + \phi) \qquad (22\text{--}19)$$

where ω_n is the natural frequency, $\omega_n = \sqrt{k/m}$, Eq. 22–2.

Since the motion is periodic, the *particular solution* of Eq. 22–18 can be determined by assuming a solution of the form

$$x_p = X \sin \omega_0 t \qquad (22\text{--}20)$$

where X is a constant. Taking the second time derivative and substituting into Eq. 22–18 yields

$$-X\omega_0^2 \sin \omega_0 t + \frac{k}{m}(X \sin \omega_0 t) = \frac{F_0}{m}\sin \omega_0 t$$

Factoring out $\sin \omega_0 t$ and solving for X gives

$$X = \frac{F_0/m}{(k/m) - \omega_0^2} = \frac{F_0/k}{1 - (\omega_0/\omega_n)^2} \qquad (22\text{--}21)$$

Substituting into Eq. 22–20, we obtain the particular solution

$$x_p = \frac{F_0/k}{1 - (\omega_0/\omega_n)^2}\sin \omega_0 t \qquad (22\text{--}22)$$

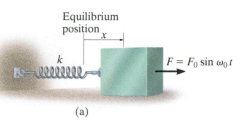

Equilibrium position

(a)

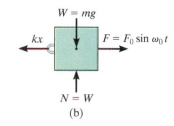

(b)

Fig. 22–11

Shaker tables provide forced vibration and are used to separate out granular materials. (© R.C. Hibbeler)

The *general solution* is therefore the sum of two sine functions having different frequencies.

$$x = x_c + x_p = C \sin(\omega_n t + \phi) + \frac{F_0/k}{1 - (\omega_0/\omega_n)^2} \sin \omega_0 t \qquad (22\text{–}23)$$

The *complementary solution* x_c defines the *free vibration*, which depends on the natural frequency $\omega_n = \sqrt{k/m}$ and the constants C and ϕ. The *particular solution* x_p describes the *forced vibration* of the block caused by the applied force $F = F_0 \sin \omega_0 t$. Since all vibrating systems are subject to *friction*, the free vibration, x_c, will in time dampen out. For this reason the free vibration is referred to as *transient*, and the forced vibration is called *steady-state*, since it is the only vibration that remains.

From Eq. 22–21 it is seen that the *amplitude* of forced or steady-state vibration depends on the *frequency ratio* ω_0/ω_n. If the *magnification factor* MF is defined as the ratio of the amplitude of steady-state vibration, X, to the static deflection, F_0/k, which would be produced by the amplitude of the periodic force F_0, then, from Eq. 22–21,

The soil compactor operates by forced vibration developed by an internal motor. It is important that the forcing frequency not be close to the natural frequency of vibration of the compactor, which can be determined when the motor is turned off; otherwise resonance will occur and the machine will become uncontrollable. (© R.C. Hibbeler)

$$MF = \frac{X}{F_0/k} = \frac{1}{1 - (\omega_0/\omega_n)^2} \qquad (22\text{-}24)$$

This equation is graphed in Fig. 22–12. Note that if the force or displacement is applied with a frequency close to the natural frequency of the system, i.e., $\omega_0/\omega_n \approx 1$, the amplitude of vibration of the block becomes extremely large. This occurs because the force **F** is applied to the block so that it always follows the motion of the block. This condition is called *resonance*, and in practice, resonating vibrations can cause tremendous stress and rapid failure of parts.*

Periodic Support Displacement.

Forced vibrations can also arise from the periodic excitation of the support of a system. The model shown in Fig. 22–13a represents the periodic vibration of a block which is caused by harmonic movement $\delta = \delta_0 \sin \omega_0 t$ of the support. The free-body diagram for the block in this case is shown in Fig. 22–13b. The displacement δ of the support is measured from the point of zero displacement, i.e., when the radial line OA coincides with OB. Therefore, general deformation of the spring is $(x - \delta_0 \sin \omega_0 t)$. Applying the equation of motion yields

$$\xrightarrow{+} F_x = ma_x; \qquad -k(x - \delta_0 \sin \omega_0 t) = m\ddot{x}$$

or

$$\ddot{x} + \frac{k}{m}x = \frac{k\delta_0}{m}\sin \omega_0 t \qquad (22\text{-}25)$$

By comparison, this equation is identical to the form of Eq. 22–18, *provided* F_0 is *replaced* by $k\delta_0$. If this substitution is made into the solutions defined by Eqs. 22–21 to 22–23, the results are appropriate for describing the motion of the block when subjected to the support displacement $\delta = \delta_0 \sin \omega_0 t$.

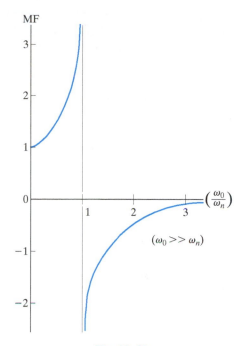

Fig. 22–12

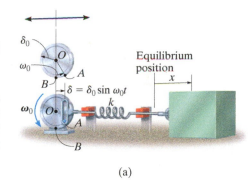

(a)

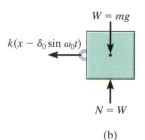

(b)

Fig. 22–13

---

*A swing has a natural period of vibration, as determined in Example 22.1. If someone pushes on the swing only when it reaches its highest point, neglecting drag or wind resistance, resonance will occur since the natural and forcing frequencies are the same.

EXAMPLE | 22.7

The instrument shown in Fig. 22–14 is rigidly attached to a platform P, which in turn is supported by *four* springs, each having a stiffness $k = 800 \text{ N/m}$. If the floor is subjected to a vertical displacement $\delta = 10 \sin(8t)$ mm, where t is in seconds, determine the amplitude of steady-state vibration. What is the frequency of the floor vibration required to cause resonance? The instrument and platform have a total mass of 20 kg.

Fig. 22–14

SOLUTION
The natural frequency is

$$\omega_n = \sqrt{\frac{k}{m}} = \sqrt{\frac{4(800 \text{ N/m})}{20 \text{ kg}}} = 12.65 \text{ rad/s}$$

The amplitude of steady-state vibration is found using Eq. 22–21, with $k\delta_0$ replacing F_0.

$$X = \frac{\delta_0}{1 - (\omega_0/\omega_n)^2} = \frac{10}{1 - [(8 \text{ rad/s})/(12.65 \text{ rad/s})]^2} = 16.7 \text{ mm} \quad Ans.$$

Resonance will occur when the amplitude of vibration X caused by the floor displacement approaches infinity. This requires

$$\omega_0 = \omega_n = 12.6 \text{ rad/s} \qquad Ans.$$

22

*22.4   Viscous Damped Free Vibration

The vibration analysis considered thus far has not included the effects of friction or damping in the system, and as a result, the solutions obtained are only in close agreement with the actual motion. Since all vibrations die out in time, the presence of damping forces should be included in the analysis.

In many cases damping is attributed to the resistance created by the substance, such as water, oil, or air, in which the system vibrates. Provided the body moves slowly through this substance, the resistance to motion is directly proportional to the body's speed. The type of force developed under these conditions is called a *viscous damping force*. The magnitude of this force is expressed by an equation of the form

$$F = c\dot{x} \tag{22–26}$$

where the constant c is called the *coefficient of viscous damping* and has units of $N \cdot s/m$ or $lb \cdot s/ft$.

The vibrating motion of a body or system having viscous damping can be characterized by the block and spring shown in Fig. 22–15a. The effect of damping is provided by the *dashpot* connected to the block on the right side. Damping occurs when the piston P moves to the right or left within the enclosed cylinder. The cylinder contains a fluid, and the motion of the piston is retarded since the fluid must flow around or through a small hole in the piston. The dashpot is assumed to have a coefficient of viscous damping c.

If the block is displaced a distance x from its equilibrium position, the resulting free-body diagram is shown in Fig. 22–15b. Both the spring and damping force oppose the forward motion of the block, so that applying the equation of motion yields

$$\xrightarrow{+} \Sigma F_x = ma_x; \qquad -kx - c\dot{x} = m\ddot{x}$$

or

$$m\ddot{x} + c\dot{x} + kx = 0 \tag{22–27}$$

This linear, second-order, homogeneous, differential equation has a solution of the form

$$x = e^{\lambda t}$$

where e is the base of the natural logarithm and λ (lambda) is a constant. The value of λ can be obtained by substituting this solution and its time derivatives into Eq. 22–27, which yields

$$m\lambda^2 e^{\lambda t} + c\lambda e^{\lambda t} + ke^{\lambda t} = 0$$

or

$$e^{\lambda t}(m\lambda^2 + c\lambda + k) = 0$$

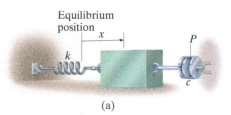

Equilibrium position

(a)

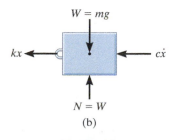

$W = mg$

$N = W$

(b)

Fig. 22–15

22

Since $e^{\lambda t}$ can never be zero, a solution is possible provided

$$m\lambda^2 + c\lambda + k = 0$$

Hence, by the quadratic formula, the two values of λ are

$$\lambda_1 = -\frac{c}{2m} + \sqrt{\left(\frac{c}{2m}\right)^2 - \frac{k}{m}}$$

$$\lambda_2 = -\frac{c}{2m} - \sqrt{\left(\frac{c}{2m}\right)^2 - \frac{k}{m}} \qquad (22\text{–}28)$$

The general solution of Eq. 22–27 is therefore a combination of exponentials which involves both of these roots. There are three possible combinations of λ_1 and λ_2 which must be considered. Before discussing these combinations, however, we will first define the *critical damping coefficient* c_c as the value of c which makes the radical in Eqs. 22–28 equal to zero; i.e.,

$$\left(\frac{c_c}{2m}\right)^2 - \frac{k}{m} = 0$$

or

$$c_c = 2m\sqrt{\frac{k}{m}} = 2m\omega_n \qquad (22\text{–}29)$$

Overdamped System.   When $c > c_c$, the roots λ_1 and λ_2 are both real. The general solution of Eq. 22–27 can then be written as

$$x = Ae^{\lambda_1 t} + Be^{\lambda_2 t} \qquad (22\text{–}30)$$

Motion corresponding to this solution is *nonvibrating*. The effect of damping is so strong that when the block is displaced and released, it simply creeps back to its original position without oscillating. The system is said to be *overdamped*.

Critically Damped System.   If $c = c_c$, then $\lambda_1 = \lambda_2 = -c_c/2m = -\omega_n$. This situation is known as *critical damping*, since it represents a condition where c has the smallest value necessary to cause the system to be nonvibrating. Using the methods of differential equations, it can be shown that the solution to Eq. 22–27 for critical damping is

$$x = (A + Bt)e^{-\omega_n t} \qquad (22\text{–}31)$$

Underdamped System.   Most often $c < c_c$, in which case the system is referred to as *underdamped*. In this case the roots λ_1 and λ_2 are complex numbers, and it can be shown that the general solution of Eq. 22–27 can be written as

$$x = D[e^{-(c/2m)t}\sin(\omega_d t + \phi)] \tag{22–32}$$

where D and ϕ are constants generally determined from the initial conditions of the problem. The constant ω_d is called the *damped natural frequency* of the system. It has a value of

$$\omega_d = \sqrt{\frac{k}{m} - \left(\frac{c}{2m}\right)^2} = \omega_n\sqrt{1 - \left(\frac{c}{c_c}\right)^2} \tag{22–33}$$

where the ratio c/c_c is called the *damping factor*.

The graph of Eq. 22–32 is shown in Fig. 22–16. The initial limit of motion, D, diminishes with each cycle of vibration, since motion is confined within the bounds of the exponential curve. Using the damped natural frequency ω_d, the period of damped vibration can be written as

$$\tau_d = \frac{2\pi}{\omega_d} \tag{22–34}$$

Since $\omega_d < \omega_n$, Eq. 22–33, the period of damped vibration, τ_d, will be greater than that of free vibration, $\tau = 2\pi/\omega_n$.

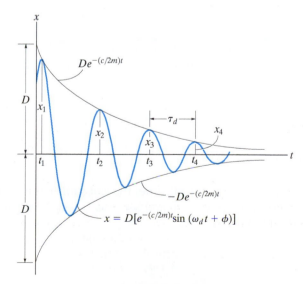

Fig. 22–16

*22.5   Viscous Damped Forced Vibration

The most general case of single-degree-of-freedom vibrating motion occurs when the system includes the effects of forced motion and induced damping. The analysis of this particular type of vibration is of practical value when applied to systems having significant damping characteristics.

If a dashpot is attached to the block and spring shown in Fig. 22–11a, the differential equation which describes the motion becomes

$$m\ddot{x} + c\dot{x} + kx = F_0 \sin \omega_0 t \qquad (22\text{–}35)$$

A similar equation can be written for a block and spring having a periodic support displacement, Fig. 22–13a, which includes the effects of damping. In that case, however, F_0 is replaced by $k\delta_0$. Since Eq. 22–35 is nonhomogeneous, the general solution is the sum of a complementary solution, x_c, and a particular solution, x_p. The complementary solution is determined by setting the right side of Eq. 22–35 equal to zero and solving the homogeneous equation, which is equivalent to Eq. 22–27. The solution is therefore given by Eq. 22–30, 22–31, or 22–32, depending on the values of λ_1 and λ_2. Because all systems are subjected to friction, then this solution will dampen out with time. Only the particular solution, which describes the *steady-state vibration* of the system, will remain. Since the applied forcing function is harmonic, the steady-state motion will also be harmonic. Consequently, the particular solution will be of the form

$$X_P = X' \sin(\omega_0 t - \phi') \qquad (22\text{–}36)$$

The constants X' and ϕ' are determined by taking the first and second time derivatives and substituting them into Eq. 22–35, which after simplification yields

$$-X'm\omega_0^2 \sin(\omega_0 t - \phi') +$$
$$X'c\omega_0 \cos(\omega_0 t - \phi') + X'k \sin(\omega_0 t - \phi') = F_0 \sin \omega_0 t$$

Since this equation holds for all time, the constant coefficients can be obtained by setting $\omega_0 t - \phi' = 0$ and $\omega_0 t - \phi' = \pi/2$, which causes the above equation to become

$$X'c\omega_0 = F_0 \sin \phi'$$
$$-X'm\omega_0^2 + X'k = F_0 \cos \phi'$$

The amplitude is obtained by squaring these equations, adding the results, and using the identity $\sin^2\phi' + \cos^2\phi' = 1$, which gives

$$X' = \frac{F_0}{\sqrt{(k - m\omega_0^2)^2 + c^2\omega_0^2}} \qquad (22\text{-}37)$$

Dividing the first equation by the second gives

$$\phi' = \tan^{-1}\left[\frac{c\omega_0}{k - m\omega_0^2}\right] \qquad (22\text{-}38)$$

Since $\omega_n = \sqrt{k/m}$ and $c_c = 2m\omega_n$, then the above equations can also be written as

$$X' = \frac{F_0/k}{\sqrt{[1 - (\omega_0/\omega_n)^2]^2 + [2(c/c_c)(\omega_0/\omega_n)]^2}}$$

$$\phi' = \tan^{-1}\left[\frac{2(c/c_c)(\omega_0/\omega_n)}{1 - (\omega_0/\omega_n)^2}\right] \qquad (22\text{-}39)$$

The angle ϕ' represents the phase difference between the applied force and the resulting steady-state vibration of the damped system.

The *magnification factor* MF has been defined in Sec. 22.3 as the ratio of the amplitude of deflection caused by the forced vibration to the deflection caused by a static force F_0. Thus,

$$\text{MF} = \frac{X'}{F_0/k} = \frac{1}{\sqrt{[1 - (\omega_0/\omega_n)^2]^2 + [2(c/c_c)(\omega_0/\omega_n)]^2}} \qquad (22\text{-}40)$$

The MF is plotted in Fig. 22–17 versus the frequency ratio ω_0/ω_n for various values of the damping factor c/c_c. It can be seen from this graph that the magnification of the amplitude increases as the damping factor decreases. Resonance obviously occurs only when the damping factor is zero and the frequency ratio equals 1.

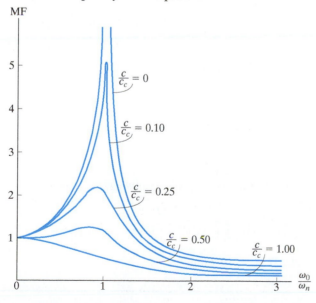

Fig. 22–17

EXAMPLE | 22.8

The 30-kg electric motor shown in Fig. 22–18 is confined to move vertically, and is supported by *four* springs, each spring having a stiffness of 200 N/m. If the rotor is unbalanced such that its effect is equivalent to a 4-kg mass located 60 mm from the axis of rotation, determine the amplitude of vibration when the rotor is turning at $\omega_0 = 10$ rad/s. The damping factor is $c/c_c = 0.15$.

Fig. 22–18

SOLUTION

The periodic force which causes the motor to vibrate is the centrifugal force due to the unbalanced rotor. This force has a constant magnitude of

$$F_0 = ma_n = mr\omega_0^2 = 4 \text{ kg}(0.06 \text{ m})(10 \text{ rad/s})^2 = 24 \text{ N}$$

The stiffness of the entire system of four springs is $k = 4(200 \text{ N/m}) = 800 \text{ N/m}$. Therefore, the natural frequency of vibration is

$$\omega_n = \sqrt{\frac{k}{m}} = \sqrt{\frac{800 \text{ N/m}}{30 \text{ kg}}} = 5.164 \text{ rad/s}$$

Since the damping factor is known, the steady-state amplitude can be determined from the first of Eqs. 22–39, i.e.,

$$X' = \frac{F_0/k}{\sqrt{[1 - (\omega_0/\omega_n)^2]^2 + [2(c/c_c)(\omega_0/\omega_n)]^2}}$$

$$= \frac{24/800}{\sqrt{[1 - (10/5.164)^2]^2 + [2(0.15)(10/5.164)]^2}}$$

$$= 0.0107 \text{ m} = 10.7 \text{ mm} \qquad\qquad Ans.$$

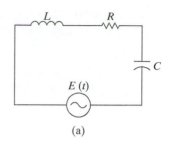

(a)

*22.6   Electrical Circuit Analogs

The characteristics of a vibrating mechanical system can be represented by an electric circuit. Consider the circuit shown in Fig. 22–19a, which consists of an inductor L, a resistor R, and a capacitor C. When a voltage $E(t)$ is applied, it causes a current of magnitude i to flow through the circuit. As the current flows past the inductor the voltage drop is $L(di/dt)$, when it flows across the resistor the drop is Ri, and when it arrives at the capacitor the drop is $(1/C)\int i\, dt$. Since current cannot flow past a capacitor, it is only possible to measure the charge q acting on the capacitor. The charge can, however, be related to the current by the equation $i = dq/dt$. Thus, the voltage drops which occur across the inductor, resistor, and capacitor become $L\, d^2q/dt^2$, $R\, dq/dt$, and q/C, respectively. According to Kirchhoff's voltage law, the applied voltage balances the sum of the voltage drops around the circuit. Therefore,

$$L\frac{d^2q}{dt^2} + R\frac{dq}{dt} + \frac{1}{C}q = E(t) \qquad (22\text{–}41)$$

Consider now the model of a single-degree-of-freedom mechanical system, Fig. 22–19b, which is subjected to both a general forcing function $F(t)$ and damping. The equation of motion for this system was established in the previous section and can be written as

$$m\frac{d^2x}{dt^2} + c\frac{dx}{dt} + kx = F(t) \qquad (22\text{–}42)$$

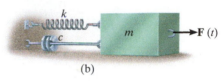

(b)

Fig. 22–19

By comparison, it is seen that Eqs. 22–41 and 22–42 have the same form, and hence mathematically the procedure of analyzing an electric circuit is the same as that of analyzing a vibrating mechanical system. The analogs between the two equations are given in Table 22–1.

This analogy has important application to experimental work, for it is much easier to simulate the vibration of a complex mechanical system using an electric circuit, which can be constructed on an analog computer, than to make an equivalent mechanical spring-and-dashpot model.

TABLE 22–1
Electrical–Mechanical Analogs

Electrical		Mechanical	
Electric charge	q	Displacement	x
Electric current	i	Velocity	dx/dt
Voltage	$E(t)$	Applied force	$F(t)$
Inductance	L	Mass	m
Resistance	R	Viscous damping coefficient	c
Reciprocal of capacitance	$1/C$	Spring stiffness	k

22

PROBLEMS

22–41. If the block-and-spring model is subjected to the periodic force $F = F_0 \cos \omega t$, show that the differential equation of motion is $\ddot{x} + (k/m)x = (F_0/m) \cos \omega t$, where x is measured from the equilibrium position of the block. What is the general solution of this equation?

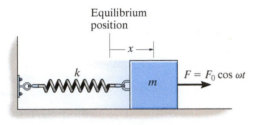

Equilibrium
position

$F = F_0 \cos \omega t$

Prob. 22–41

22–42. A block which has a mass m is suspended from a spring having a stiffness k. If an impressed downward vertical force $F = F_O$ acts on the weight, determine the equation which describes the position of the block as a function of time.

22–43. A 4-lb weight is attached to a spring having a stiffness $k = 10$ lb/ft. The weight is drawn downward a distance of 4 in. and released from rest. If the support moves with a vertical displacement $\delta = (0.5 \sin 4t)$ in., where t is in seconds, determine the equation which describes the position of the weight as a function of time.

*22–44. A 4-kg block is suspended from a spring that has a stiffness of $k = 600$ N/m. The block is drawn downward 50 mm from the equilibrium position and released from rest when $t = 0$. If the support moves with an impressed displacement of $\delta = (10 \sin 4t)$ mm, where t is in seconds, determine the equation that describes the vertical motion of the block. Assume positive displacement is downward.

22–45. Use a block-and-spring model like that shown in Fig. 22–13a, but suspended from a vertical position and subjected to a periodic support displacement $\delta = \delta_0 \sin \omega_0 t$, determine the equation of motion for the system, and obtain its general solution. Define the displacement y measured from the static equilibrium position of the block when $t = 0$.

22–46. A 5-kg block is suspended from a spring having a stiffness of 300 N/m. If the block is acted upon by a vertical force $F = (7 \sin 8t)$ N, where t is in seconds, determine the equation which describes the motion of the block when it is pulled down 100 mm from the equilibrium position and released from rest at $t = 0$. Assume that positive displacement is downward.

$k = 300$ N/m

$F = 7 \sin 8t$

Prob. 22–46

22–47. The uniform rod has a mass of m. If it is acted upon by a periodic force of $F = F_0 \sin \omega t$, determine the amplitude of the steady-state vibration.

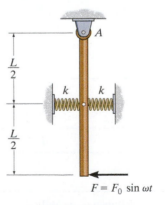

$\frac{L}{2}$

k k

$\frac{L}{2}$

$F = F_0 \sin \omega t$

Prob. 22–47

*22–48. The 30-lb block is attached to two springs having a stiffness of 10 lb/ft. A periodic force $F = (8 \cos 3t)$ lb, where t is in seconds, is applied to the block. Determine the maximum speed of the block after frictional forces cause the free vibrations to dampen out.

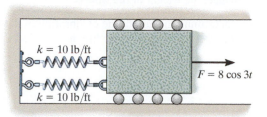

Prob. 22–48

22–49. The light elastic rod supports a 4-kg sphere. When an 18-N vertical force is applied to the sphere, the rod deflects 14 mm. If the wall oscillates with harmonic frequency of 2 Hz and has an amplitude of 15 mm, determine the amplitude of vibration for the sphere.

Prob. 22–49

22–50. Find the differential equation for small oscillations in terms of θ for the uniform rod of mass m. Also show that if $c < \sqrt{mk}/2$, then the system remains underdamped. The rod is in a horizontal position when it is in equilibrium.

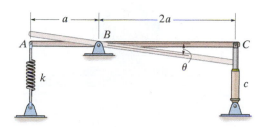

Prob. 22–50

22–51. The 40-kg block is attached to a spring having a stiffness of 800 N/m. A force $F = (100 \cos 2t)$ N, where t is in seconds is applied to the block. Determine the maximum speed of the block for the steady-state vibration.

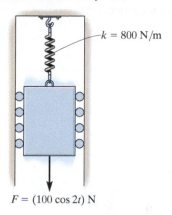

Prob. 22–51

*22–52. Using a block-and-spring model, like that shown in Fig. 22–13a, but suspended from a vertical position and subjected to a periodic support displacement of $\delta = \delta_0 \cos \omega_0 t$, determine the equation of motion for the system, and obtain its general solution. Define the displacement y measured from the static equilibrium position of the block when $t = 0$.

22–53. The fan has a mass of 25 kg and is fixed to the end of a horizontal beam that has a negligible mass. The fan blade is mounted eccentrically on the shaft such that it is equivalent to an unbalanced 3.5-kg mass located 100 mm from the axis of rotation. If the static deflection of the beam is 50 mm as a result of the weight of the fan, determine the angular velocity of the fan blade at which resonance will occur. *Hint:* See the first part of Example 22.8.

22–54. In Prob. 22–53, determine the amplitude of steady-state vibration of the fan if its angular velocity is 10 rad/s.

22–55. What will be the amplitude of steady-state vibration of the fan in Prob. 22–53 if the angular velocity of the fan blade is 18 rad/s? *Hint:* See the first part of Example 22.8.

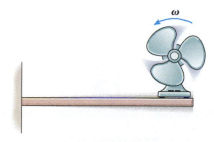

Probs. 22–53/54/55

22

*22–56. The small block at A has a mass of 4 kg and is mounted on the bent rod having negligible mass. If the rotor at B causes a harmonic movement $\delta_B = (0.1 \cos 15t)$ m, where t is in seconds, determine the steady-state amplitude of vibration of the block.

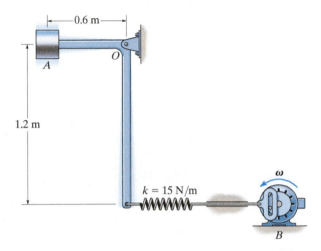

Prob. 22–56

22–57. The electric motor turns an eccentric flywheel which is equivalent to an unbalanced 0.25-lb weight located 10 in. from the axis of rotation. If the static deflection of the beam is 1 in. because of the weight of the motor, determine the angular velocity of the flywheel at which resonance will occur. The motor weighs 150 lb. Neglect the mass of the beam.

22–58. What will be the amplitude of steady-state vibration of the motor in Prob. 22–57 if the angular velocity of the flywheel is 20 rad/s?

22–59. Determine the angular velocity of the flywheel in Prob. 22–57 which will produce an amplitude of vibration of 0.25 in.

Probs. 22–57/58/59

*22–60. The 450-kg trailer is pulled with a constant speed over the surface of a bumpy road, which may be approximated by a cosine curve having an amplitude of 50 mm and wave length of 4 m. If the two springs s which support the trailer each have a stiffness of 800 N/m, determine the speed v which will cause the greatest vibration (resonance) of the trailer. Neglect the weight of the wheels.

22–61. Determine the amplitude of vibration of the trailer in Prob. 22–60 if the speed $v = 15$ km/h.

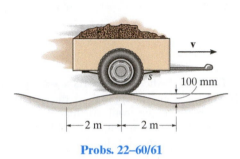

Probs. 22–60/61

22–62. The motor of mass M is supported by a simply supported beam of negligible mass. If block A of mass m is clipped onto the rotor, which is turning at constant angular velocity of ω, determine the amplitude of the steady-state vibration. *Hint:* When the beam is subjected to a concentrated force of P at its mid-span, it deflects $\delta = PL^3/48EI$ at this point. Here E is Young's modulus of elasticity, a property of the material, and I is the moment of inertia of the beam's cross-sectional area.

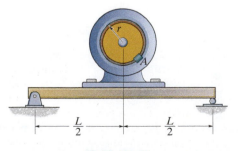

Prob. 22–62

22–63. The spring system is connected to a crosshead that oscillates vertically when the wheel rotates with a constant angular velocity of ω. If the amplitude of the steady-state vibration is observed to be 400 mm, and the springs each have a stiffness of $k = 2500$ N/m, determine the two possible values of ω at which the wheel must rotate. The block has a mass of 50 kg.

*22–64. The spring system is connected to a crosshead that oscillates vertically when the wheel rotates with a constant angular velocity of $\omega = 5$ rad/s. If the amplitude of the steady-state vibration is observed to be 400 mm, determine the two possible values of the stiffness k of the springs. The block has a mass of 50 kg.

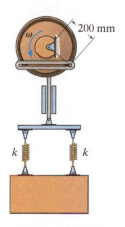

Probs. 22–63/64

22–65. A 7-lb block is suspended from a spring having a stiffness of $k = 75$ lb/ft. The support to which the spring is attached is given simple harmonic motion which may be expressed as $\delta = (0.15 \sin 2t)$ ft, where t is in seconds. If the damping factor is $c/c_c = 0.8$, determine the phase angle ϕ of forced vibration.

22–66. Determine the magnification factor of the block, spring, and dashpot combination in Prob. 22–65.

22–67. A block having a mass of 7 kg is suspended from a spring that has a stiffness $k = 600$ N/m. If the block is given an upward velocity of 0.6 m/s from its equilibrium position at $t = 0$, determine its position as a function of time. Assume that positive displacement of the block is downward and that motion takes place in a medium which furnishes a damping force $F = (50|v|)$ N, where v is in m/s.

*22–68. The 200-lb electric motor is fastened to the midpoint of the simply supported beam. It is found that the beam deflects 2 in. when the motor is not running. The motor turns an eccentric flywheel which is equivalent to an unbalanced weight of 1 lb located 5 in. from the axis of rotation. If the motor is turning at 100 rpm, determine the amplitude of steady-state vibration. The damping factor is $c/c_c = 0.20$. Neglect the mass of the beam.

Prob. 22–68

22–69. Two identical dashpots are arranged parallel to each other, as shown. Show that if the damping coefficient $c < \sqrt{mk}$, then the block of mass m will vibrate as an underdamped system.

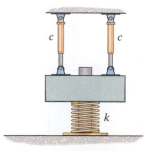

Prob. 22–69

22

22–70. The damping factor, c/c_c, may be determined experimentally by measuring the successive amplitudes of vibrating motion of a system. If two of these maximum displacements can be approximated by x_1 and x_2, as shown in Fig. 22–16, show that $\ln (x_1/x_2) = 2\pi(c/c_c)/\sqrt{1-(c/c_c)^2}$. The quantity $\ln (x_1/x_2)$ is called the *logarithmic decrement*.

22–71. If the amplitude of the 50-lb cylinder's steady-state vibration is 6 in., determine the wheel's angular velocity ω.

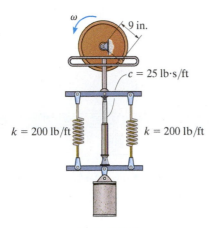

Prob. 22–71

*22–72. The block, having a weight of 12 lb, is immersed in a liquid such that the damping force acting on the block has a magnitude of $F = (0.7|v|)$ lb, where v is in ft/s. If the block is pulled down 0.62 ft and released from rest, determine the position of the block as a function of time. The spring has a stiffness of $k = 53$ lb/ft. Assume that positive displacement is downward.

22–73. The bar has a weight of 6 lb. If the stiffness of the spring is $k = 8$ lb/ft and the dashpot has a damping coefficient $c = 60$ lb·s/ft, determine the differential equation which describes the motion in terms of the angle θ of the bar's rotation. Also, what should be the damping coefficient of the dashpot if the bar is to be critically damped?

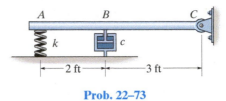

Prob. 22–73

22–74. A bullet of mass m has a velocity of $\mathbf{v}_0$ just before it strikes the target of mass M. If the bullet embeds in the target, and the vibration is to be critically damped, determine the dashpot's critical damping coefficient, and the springs' maximum compression. The target is free to move along the two horizontal guides that are "nested" in the springs.

22–75. A bullet of mass m has a velocity v_0 just before it strikes the target of mass M. If the bullet embeds in the target, and the dashpot's damping coefficient is $0 < c \ll c_c$, determine the springs' maximum compression. The target is free to move along the two horizontal guides that are "nested" in the springs.

Prob. 22–72

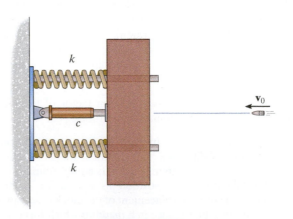

Probs. 22–74/75

*22–76. Determine the differential equation of motion for the damped vibratory system shown. What type of motion occurs? Take $k = 100$ N/m, $c = 200$ N·s/m, $m = 25$ kg.

22–78. Draw the electrical circuit that is equivalent to the mechanical system shown. What is the differential equation which describes the charge q in the circuit?

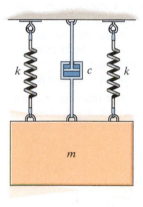

Prob. 22–78

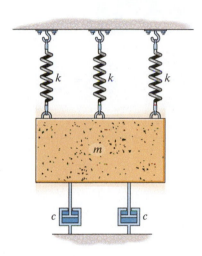

Prob. 22–76

22–79. Draw the electrical circuit that is equivalent to the mechanical system shown. Determine the differential equation which describes the charge q in the circuit.

22–77. Draw the electrical circuit that is equivalent to the mechanical system shown. Determine the differential equation which describes the charge q in the circuit.

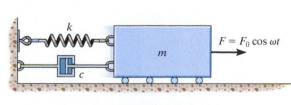

Prob. 22–77

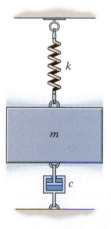

Prob. 22–79

22

CHAPTER REVIEW

Undamped Free Vibration

A body has free vibration when gravitational or elastic restoring forces cause the motion. This motion is undamped when friction forces are neglected. The periodic motion of an undamped, freely vibrating body can be studied by displacing the body from the equilibrium position and then applying the equation of motion along the path.

For a one-degree-of-freedom system, the resulting differential equation can be written in terms of its natural frequency ω_n.

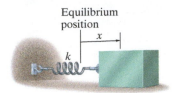

Equilibrium position

$$\ddot{x} + \omega_n^2 x = 0 \qquad \tau = \frac{2\pi}{\omega_n} \qquad f = \frac{1}{\tau} = \frac{\omega_n}{2\pi}$$

Energy Methods

Provided the restoring forces acting on the body are gravitational and elastic, then conservation of energy can also be used to determine its simple harmonic motion. To do this, the body is displaced a small amount from its equilibrium position, and an expression for its kinetic and potential energy is written. The time derivative of this equation can then be rearranged in the standard form $\ddot{x} + \omega_n^2 x = 0$.

Undamped Forced Vibration

When the equation of motion is applied to a body, which is subjected to a periodic force, or the support has a displacement with a frequency ω_0, then the solution of the differential equation consists of a complementary solution and a particular solution. The complementary solution is caused by the free vibration and can be neglected. The particular solution is caused by the forced vibration.

Resonance will occur if the natural frequency of vibration ω_n is equal to the forcing frequency ω_0. This should be avoided, since the motion will tend to become unbounded.

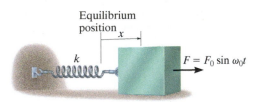

Equilibrium position

$F = F_0 \sin \omega_0 t$

$$x_p = \frac{F_0/k}{1 - (\omega_0/\omega_n)^2} \sin \omega_0 t$$

Viscous Damped Free Vibration

A viscous damping force is caused by fluid drag on the system as it vibrates. If the motion is slow, this drag force will be proportional to the velocity, that is, $F = c\dot{x}$. Here c is the coefficient of viscous damping. By comparing its value to the critical damping coefficient $c_c = 2m\omega_n$, we can specify the type of vibration that occurs. If $c > c_c$, it is an overdamped system; if $c = c_c$, it is a critically damped system; if $c < c_c$, it is an underdamped system.

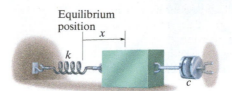

Viscous Damped Forced Vibration

The most general type of vibration for a one-degree-of-freedom system occurs when the system is damped and subjected to periodic forced motion. The solution provides insight as to how the damping factor, c/c_c, and the frequency ratio, ω_0/ω_n, influence the vibration.

Resonance is avoided provided $c/c_c \neq 0$ and $\omega_0/\omega_n \neq 1$.

Electrical Circuit Analogs

The vibrating motion of a complex mechanical system can be studied by modeling it as an electrical circuit. This is possible since the differential equations that govern the behavior of each system are the same.

22

Mathematical Expressions

Quadratic Formula

If $ax^2 + bx + c = 0$, then $x = \dfrac{-b \pm \sqrt{b^2 - 4ac}}{2a}$

Hyperbolic Functions

$\sinh x = \dfrac{e^x - e^{-x}}{2}$, $\cosh x = \dfrac{e^x + e^{-x}}{2}$, $\tanh x = \dfrac{\sinh x}{\cosh x}$

Trigonometric Identities

$\sin\theta = \dfrac{A}{C}$, $\csc\theta = \dfrac{C}{A}$

$\cos\theta = \dfrac{B}{C}$, $\sec\theta = \dfrac{C}{B}$

$\tan\theta = \dfrac{A}{B}$, $\cot\theta = \dfrac{B}{A}$

$\sin^2\theta + \cos^2\theta = 1$

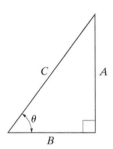

$\sin(\theta \pm \phi) = \sin\theta\cos\phi \pm \cos\theta\sin\phi$

$\sin 2\theta = 2\sin\theta\cos\theta$

$\cos(\theta \pm \phi) = \cos\theta\cos\phi \mp \sin\theta\sin\phi$

$\cos 2\theta = \cos^2\theta - \sin^2\theta$

$\cos\theta = \pm\sqrt{\dfrac{1 + \cos 2\theta}{2}}$, $\sin\theta = \pm\sqrt{\dfrac{1 - \cos 2\theta}{2}}$

$\tan\theta = \dfrac{\sin\theta}{\cos\theta}$

$1 + \tan^2\theta = \sec^2\theta \quad 1 + \cot^2\theta = \csc^2\theta$

Power-Series Expansions

$\sin x = x - \dfrac{x^3}{3!} + \cdots \qquad \sinh x = x + \dfrac{x^3}{3!} + \cdots$

$\cos x = 1 - \dfrac{x^2}{2!} + \cdots \qquad \cosh x = 1 + \dfrac{x^2}{2!} + \cdots$

Derivatives

$\dfrac{d}{dx}(u^n) = nu^{n-1}\dfrac{du}{dx}$

$\dfrac{d}{dx}(uv) = u\dfrac{dv}{dx} + v\dfrac{du}{dx}$

$\dfrac{d}{dx}\left(\dfrac{u}{v}\right) = \dfrac{v\dfrac{du}{dx} - u\dfrac{dv}{dx}}{v^2}$

$\dfrac{d}{dx}(\cot u) = -\csc^2 u\dfrac{du}{dx}$

$\dfrac{d}{dx}(\sec u) = \tan u\sec u\dfrac{du}{dx}$

$\dfrac{d}{dx}(\csc u) = -\csc u\cot u\dfrac{du}{dx}$

$\dfrac{d}{dx}(\sin u) = \cos u\dfrac{du}{dx}$

$\dfrac{d}{dx}(\cos u) = -\sin u\dfrac{du}{dx}$

$\dfrac{d}{dx}(\tan u) = \sec^2 u\dfrac{du}{dx}$

$\dfrac{d}{dx}(\sinh u) = \cosh u\dfrac{du}{dx}$

$\dfrac{d}{dx}(\cosh u) = \sinh u\dfrac{du}{dx}$

Integrals

$$\int x^n \, dx = \frac{x^{n+1}}{n+1} + C, \, n \neq -1$$

$$\int \frac{dx}{a+bx} = \frac{1}{b}\ln(a+bx) + C$$

$$\int \frac{dx}{a+bx^2} = \frac{1}{2\sqrt{-ba}}\ln\left[\frac{a+x\sqrt{-ab}}{a-x\sqrt{-ab}}\right] + C, \, ab < 0$$

$$\int \frac{x \, dx}{a+bx^2} = \frac{1}{2b}\ln(bx^2+a) + C$$

$$\int \frac{x^2 \, dx}{a+bx^2} = \frac{x}{b} - \frac{a}{b\sqrt{ab}}\tan^{-1}\frac{x\sqrt{ab}}{a} + C, \, ab > 0$$

$$\int \frac{dx}{a^2-x^2} = \frac{1}{2a}\ln\left[\frac{a+x}{a-x}\right] + C, \, a^2 > x^2$$

$$\int \sqrt{a+bx} \, dx = \frac{2}{3b}\sqrt{(a+bx)^3} + C$$

$$\int x\sqrt{a+bx} \, dx = \frac{-2(2a-3bx)\sqrt{(a+bx)^3}}{15b^2} + C$$

$$\int x^2\sqrt{a+bx} \, dx = \frac{2(8a^2-12abx+15b^2x^2)\sqrt{(a+bx)^3}}{105b^3} + C$$

$$\int \sqrt{a^2-x^2} \, dx = \frac{1}{2}\left[x\sqrt{a^2-x^2} + a^2\sin^{-1}\frac{x}{a}\right] + C, \, a > 0$$

$$\int x\sqrt{x^2 \pm a^2} \, dx = \frac{1}{3}\sqrt{(x^2 \pm a^2)^3} + C$$

$$\int x^2\sqrt{a^2-x^2} \, dx = -\frac{x}{4}\sqrt{(a^2-x^2)^3}$$

$$+ \frac{a^2}{8}\left(x\sqrt{a^2-x^2} + a^2\sin^{-1}\frac{x}{a}\right) + C, \, a > 0$$

$$\int \sqrt{x^2 \pm a^2} \, dx = \frac{1}{2}\left[x\sqrt{x^2 \pm a^2} \pm a^2\ln\left(x + \sqrt{x^2 \pm a^2}\right)\right] + C$$

$$\int x\sqrt{a^2-x^2} \, dx = -\frac{1}{3}\sqrt{(a^2-x^2)^3} + C$$

$$\int x^2\sqrt{x^2 \pm a^2} \, dx = \frac{x}{4}\sqrt{(x^2 \pm a^2)^3} \mp \frac{a^2}{8}x\sqrt{x^2 \pm a^2}$$

$$- \frac{a^4}{8}\ln\left(x + \sqrt{x^2 \pm a^2}\right) + C$$

$$\int \frac{dx}{\sqrt{a+bx}} = \frac{2\sqrt{a+bx}}{b} + C$$

$$\int \frac{x \, dx}{\sqrt{x^2 \pm a^2}} = \sqrt{x^2 \pm a^2} + C$$

$$\int \frac{dx}{\sqrt{a+bx+cx^2}} = \frac{1}{\sqrt{c}}\ln\left[\sqrt{a+bx+cx^2}\right.$$

$$\left. + x\sqrt{c} + \frac{b}{2\sqrt{c}}\right] + C, \, c > 0$$

$$= \frac{1}{\sqrt{-c}}\sin^{-1}\left(\frac{-2cx-b}{\sqrt{b^2-4ac}}\right) + C, \, c < 0$$

$$\int \sin x \, dx = -\cos x + C$$

$$\int \cos x \, dx = \sin x + C$$

$$\int x\cos(ax) \, dx = \frac{1}{a^2}\cos(ax) + \frac{x}{a}\sin(ax) + C$$

$$\int x^2\cos(ax) \, dx = \frac{2x}{a^2}\cos(ax)$$

$$+ \frac{a^2x^2-2}{a^3}\sin(ax) + C$$

$$\int e^{ax} \, dx = \frac{1}{a}e^{ax} + C$$

$$\int xe^{ax} \, dx = \frac{e^{ax}}{a^2}(ax-1) + C$$

$$\int \sinh x \, dx = \cosh x + C$$

$$\int \cosh x \, dx = \sinh x + C$$

A

Vector Analysis

The following discussion provides a brief review of vector analysis. A more detailed treatment of these topics is given in *Engineering Mechanics: Statics*.

Vector.   A vector, **A**, is a quantity which has magnitude and direction, and adds according to the parallelogram law. As shown in Fig. B–1, **A** = **B** + **C**, where **A** is the *resultant vector* and **B** and **C** are *component vectors*.

Unit Vector.   A unit vector, $\mathbf{u}_A$, has a magnitude of one "dimensionless" unit and acts in the same direction as **A**. It is determined by dividing **A** by its magnitude A, i.e,

$$\mathbf{u}_A = \frac{\mathbf{A}}{A} \tag{B–1}$$

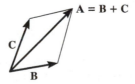

Fig. B–1

Cartesian Vector Notation.   The directions of the positive x, y, z axes are defined by the Cartesian unit vectors **i**, **j**, **k**, respectively.

As shown in Fig. B–2, vector **A** is formulated by the addition of its x, y, z components as

$$\mathbf{A} = A_x\mathbf{i} + A_y\mathbf{j} + A_z\mathbf{k} \qquad (B-2)$$

The *magnitude* of **A** is determined from

$$A = \sqrt{A_x^2 + A_y^2 + A_z^2} \qquad (B-3)$$

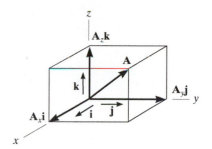

Fig. B–2

The *direction* of **A** is defined in terms of its *coordinate direction angles*, α, β, γ, measured from the *tail* of **A** to the *positive x, y, z* axes, Fig. B–3. These angles are determined from the *direction cosines* which represent the **i**, **j**, **k** components of the unit vector $\mathbf{u}_A$; i.e., from Eqs. B–1 and B–2

$$\mathbf{u}_A = \frac{A_x}{A}\mathbf{i} + \frac{A_y}{A}\mathbf{j} + \frac{A_z}{A}\mathbf{k} \qquad (B-4)$$

so that the direction cosines are

$$\cos \alpha = \frac{A_x}{A} \qquad \cos \beta = \frac{A_y}{A} \qquad \cos \gamma = \frac{A_z}{A} \qquad (B-5)$$

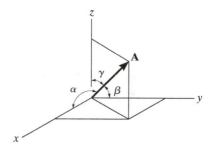

Fig. B–3

Hence, $\mathbf{u}_A = \cos \alpha \mathbf{i} + \cos \beta \mathbf{j} + \cos \gamma \mathbf{k}$, and using Eq. B–3, it is seen that

$$\cos^2 \alpha + \cos^2 \beta + \cos^2 \gamma = 1 \qquad (B-6)$$

The Cross Product.   The cross product of two vectors **A** and **B**, which yields the resultant vector **C**, is written as

$$\mathbf{C} = \mathbf{A} \times \mathbf{B} \qquad (B-7)$$

and reads **C** equals **A** "cross" **B**. The *magnitude* of **C** is

$$C = AB \sin \theta \qquad (B-8)$$

where θ is the angle made between the *tails* of **A** and **B** ($0° \leq \theta \leq 180°$). The *direction* of **C** is determined by the right-hand rule, whereby the fingers of the right hand are curled *from* **A** *to* **B** and the thumb points in the direction of **C**, Fig. B–4. This vector is perpendicular to the plane containing vectors **A** and **B**.

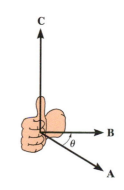

Fig. B–4

B

The vector cross product is *not* commutative, i.e., $\mathbf{A} \times \mathbf{B} \neq \mathbf{B} \times \mathbf{A}$. Rather,

$$\mathbf{A} \times \mathbf{B} = -\mathbf{B} \times \mathbf{A} \qquad (\text{B–9})$$

The distributive law is valid; i.e.,

$$\mathbf{A} \times (\mathbf{B} + \mathbf{D}) = \mathbf{A} \times \mathbf{B} + \mathbf{A} \times \mathbf{D} \qquad (\text{B–10})$$

And the cross product may be multiplied by a scalar m in any manner; i.e.,

$$m(\mathbf{A} \times \mathbf{B}) = (m\mathbf{A}) \times \mathbf{B} = \mathbf{A} \times (m\mathbf{B}) = (\mathbf{A} \times \mathbf{B})m \qquad (\text{B–11})$$

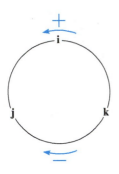

Fig. B–5

Equation B–7 can be used to find the cross product of any pair of Cartesian unit vectors. For example, to find $\mathbf{i} \times \mathbf{j}$, the magnitude is $(i)(j) \sin 90° = (1)(1)(1) = 1$, and its direction $+\mathbf{k}$ is determined from the right-hand rule, applied to $\mathbf{i} \times \mathbf{j}$, Fig. B–2. A simple scheme shown in Fig. B–5 may be helpful in obtaining this and other results when the need arises. If the circle is constructed as shown, then "crossing" two of the unit vectors in a *counterclockwise* fashion around the circle yields a *positive* third unit vector, e.g., $\mathbf{k} \times \mathbf{i} = \mathbf{j}$. Moving *clockwise*, a *negative* unit vector is obtained, e.g., $\mathbf{i} \times \mathbf{k} = -\mathbf{j}$.

If $\mathbf{A}$ and $\mathbf{B}$ are expressed in Cartesian component form, then the cross product, Eq. B–7, may be evaluated by expanding the determinant

$$\mathbf{C} = \mathbf{A} \times \mathbf{B} = \begin{vmatrix} \mathbf{i} & \mathbf{j} & \mathbf{k} \\ A_x & A_y & A_z \\ B_x & B_y & B_z \end{vmatrix} \qquad (\text{B–12})$$

which yields

$$\mathbf{C} = (A_y B_z - A_z B_y)\mathbf{i} - (A_x B_z - A_z B_x)\mathbf{j} + (A_x B_y - A_y B_x)\mathbf{k}$$

Recall that the cross product is used in statics to define the moment of a force $\mathbf{F}$ about point O, in which case

$$\mathbf{M}_O = \mathbf{r} \times \mathbf{F} \qquad (\text{B–13})$$

where $\mathbf{r}$ is a position vector directed from point O to *any point* on the line of action of $\mathbf{F}$.

B

The Dot Product.   The dot product of two vectors **A** and **B**, which yields a scalar, is defined as

$$\mathbf{A} \cdot \mathbf{B} = AB \cos \theta \qquad \text{(B–14)}$$

and reads **A** "dot" **B**. The angle θ is formed between the *tails* of **A** and **B** ($0° \leq \theta \leq 180°$).

The dot product is commutative; i.e.,

$$\mathbf{A} \cdot \mathbf{B} = \mathbf{B} \cdot \mathbf{A} \qquad \text{(B–15)}$$

The distributive law is valid; i.e.,

$$\mathbf{A} \cdot (\mathbf{B} + \mathbf{D}) = \mathbf{A} \cdot \mathbf{B} + \mathbf{A} \cdot \mathbf{D} \qquad \text{(B–16)}$$

And scalar multiplication can be performed in any manner, i.e.,

$$m(\mathbf{A} \cdot \mathbf{B}) = (m\mathbf{A}) \cdot \mathbf{B} = \mathbf{A} \cdot (m\mathbf{B}) = (\mathbf{A} \cdot \mathbf{B})m \qquad \text{(B–17)}$$

Using Eq. B–14, the dot product between any two Cartesian vectors can be determined. For example, $\mathbf{i} \cdot \mathbf{i} = (1)(1) \cos 0° = 1$ and $\mathbf{i} \cdot \mathbf{j} = (1)(1) \cos 90° = 0$.

If **A** and **B** are expressed in Cartesian component form, then the dot product, Eq. C–14, can be determined from

$$\mathbf{A} \cdot \mathbf{B} = A_x B_x + A_y B_y + A_z B_z \qquad \text{(B–18)}$$

The dot product may be used to determine the *angle θ formed between two vectors*. From Eq. B–14,

$$\theta = \cos^{-1}\left(\frac{\mathbf{A} \cdot \mathbf{B}}{AB}\right) \qquad \text{(B–19)}$$

B

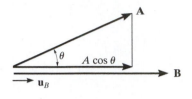

Fig. B–6

It is also possible to find the *component of a vector in a given direction* using the dot product. For example, the magnitude of the component (or projection) of vector **A** in the direction of **B**, Fig. B–6, is defined by $A \cos \theta$. From Eq. B–14, this magnitude is

$$A \cos \theta = \mathbf{A} \cdot \frac{\mathbf{B}}{B} = \mathbf{A} \cdot \mathbf{u}_B \tag{B–20}$$

where $\mathbf{u}_B$ represents a unit vector acting in the direction of **B**, Fig. B–6.

Differentiation and Integration of Vector Functions.   The rules for differentiation and integration of the sums and products of scalar functions also apply to vector functions. Consider, for example, the two vector functions $\mathbf{A}(s)$ and $\mathbf{B}(s)$. Provided these functions are smooth and continuous for all s, then

$$\frac{d}{ds}(\mathbf{A} + \mathbf{B}) = \frac{d\mathbf{A}}{ds} + \frac{d\mathbf{B}}{ds} \tag{B–21}$$

$$\int (\mathbf{A} + \mathbf{B})\, ds = \int \mathbf{A}\, ds + \int \mathbf{B}\, ds \tag{B–22}$$

For the cross product,

$$\frac{d}{ds}(\mathbf{A} \times \mathbf{B}) = \left(\frac{d\mathbf{A}}{ds} \times \mathbf{B} \right) + \left(\mathbf{A} \times \frac{d\mathbf{B}}{ds} \right) \tag{B–23}$$

Similarly, for the dot product,

$$\frac{d}{ds}(\mathbf{A} \cdot \mathbf{B}) = \frac{d\mathbf{A}}{ds} \cdot \mathbf{B} + \mathbf{A} \cdot \frac{d\mathbf{B}}{ds} \tag{B–24}$$

B

The Chain Rule

The chain rule of calculus can be used to determine the time derivative of a composite function. For example, if y is a function of x and x is a function of t, then we can find the derivative of y with respect to t as follows

$$\dot{y} = \frac{dy}{dt} = \frac{dy}{dx}\frac{dx}{dt} \qquad \text{(C–1)}$$

In other words, to find $\dot{y}$ we take the ordinary derivative (dy/dx) and multiply it by the time derivative (dx/dt).

If several variables are functions of time and they are multiplied together, then the product rule $d(uv) = du\, v + u\, dv$ must be used along with the chain rule when taking the time derivatives. Here are some examples.

EXAMPLE │ C–1

If $y = x^3$ and $x = t^4$, find $\ddot{y}$, the second derivative of y with respect to time.

SOLUTION
Using the chain rule, Eq. C–1,

$$\dot{y} = 3x^2\dot{x}$$

To obtain the second time derivative we must use the product rule since x and $\dot{x}$ are both functions of time, and also, for $3x^2$ the chain rule must be applied. Thus, with $u = 3x^2$ and $v = \dot{x}$, we have

$$\ddot{y} = [6x\dot{x}]\dot{x} + 3x^2[\ddot{x}]$$

$$= 3x[2\dot{x}^2 + x\ddot{x}]$$

Since $x = t^4$, then $\dot{x} = 4t^3$ and $\ddot{x} = 12t^2$ so that

$$\ddot{y} = 3(t^4)[2(4t^3)^2 + t^4(12t^2)]$$

$$= 132t^{10}$$

Note that this result can also be obtained by combining the functions, then taking the time derivatives, that is,

$$y = x^3 = (t^4)^3 = t^{12}$$

$$\dot{y} = 12t^{11}$$

$$\ddot{y} = 132t^{10}$$

EXAMPLE │ C–2

If $y = xe^x$, find $\ddot{y}$.

SOLUTION
Since x and e^x are both functions of time the product and chain rules must be applied. Have $u = x$ and $v = e^x$.

$$\dot{y} = [\dot{x}]e^x + x[e^x\dot{x}]$$

The second time derivative also requires application of the product and chain rules. Note that the product rule applies to the three time variables in the last term, i.e., x, e^x, and $\dot{x}$.

$$\ddot{y} = \{[\ddot{x}]e^x + \dot{x}[e^x\dot{x}]\} + \{[\dot{x}]e^x\dot{x} + x[e^x\dot{x}]\dot{x} + xe^x[\ddot{x}]\}$$

$$= e^x[\ddot{x}(1 + x) + \dot{x}^2(2 + x)]$$

If $x = t^2$ then $\dot{x} = 2t$, $\ddot{x} = 2$ so that in terms in t, we have

$$\ddot{y} = e^{t^2}[2(1 + t^2) + 4t^2(2 + t^2)]$$

EXAMPLE | C–3

If the path in radial coordinates is given as $r = 5\theta^2$, where θ is a known function of time, find $\ddot{r}$.

SOLUTION

First, using the chain rule then the chain and product rules where $u = 10\theta$ and $v = \dot{\theta}$, we have

$$r = 5\theta^2$$

$$\dot{r} = 10\theta\dot{\theta}$$

$$\ddot{r} = 10[(\dot{\theta})\dot{\theta} + \theta(\ddot{\theta})]$$

$$= 10\dot{\theta}^2 + 10\theta\ddot{\theta}$$

EXAMPLE | C–4

If $r^2 = 6\theta^3$, find $\ddot{r}$.

SOLUTION

Here the chain and product rules are applied as follows.

$$r^2 = 6\theta^3$$

$$2r\dot{r} = 18\theta^2\dot{\theta}$$

$$2[(\dot{r})\dot{r} + r(\ddot{r})] = 18[(2\theta\dot{\theta})\dot{\theta} + \theta^2(\ddot{\theta})]$$

$$\dot{r}^2 + r\ddot{r} = 9(2\theta\dot{\theta}^2 + \theta^2\ddot{\theta})$$

To find $\ddot{r}$ at a specified value of θ which is a known function of time, we can first find $\dot{\theta}$ and $\ddot{\theta}$. Then using these values, evaluate r from the first equation, $\dot{r}$ from the second equation and $\ddot{r}$ using the last equation.

C

Fundamental Problems
Partial Solutions And Answers

Chapter 12

F12–1. $v = v_0 + a_c t$
$10 = 35 + a_c(15)$
$a_c = -1.67 \text{ m/s}^2 = 1.67 \text{ m/s}^2 \leftarrow$ *Ans.*

F12–2. $s = s_0 + v_0 t + \frac{1}{2} a_c t^2$
$0 = 0 + 15t + \frac{1}{2}(-9.81)t^2$
$t = 3.06 \text{ s}$ *Ans.*

F12–3. $\quad ds = v \, dt$
$$\int_0^s ds = \int_0^t \left(4t - 3t^2\right) dt$$
$$s = \left(2t^2 - t^3\right) \text{m}$$
$$s = 2\left(4^2\right) - 4^3$$
$$= -32 \text{ m} = 32 \text{ m} \leftarrow \qquad Ans.$$

F12–4. $a = \frac{dv}{dt} = \frac{d}{dt}\left(0.5t^3 - 8t\right)$
$a = \left(1.5t^2 - 8\right) \text{ m/s}^2$
When $t = 2$ s,
$a = 1.5\left(2^2\right) - 8 = -2 \text{ m/s}^2 = 2 \text{ m/s}^2 \leftarrow$ *Ans.*

F12–5. $v = \frac{ds}{dt} = \frac{d}{dt}(2t^2 - 8t + 6) = (4t - 8) \text{ m/s}$
$v = 0 = (4t - 8)$
$t = 2 \text{ s}$ *Ans.*
$s|_{t=0} = 2\left(0^2\right) - 8(0) + 6 = 6 \text{ m}$
$s|_{t=2} = 2\left(2^2\right) - 8(2) + 6 = -2 \text{ m}$
$s|_{t=3} = 2\left(3^2\right) - 8(3) + 6 = 0 \text{ m}$
$(\Delta s)_{\text{Tot}} = 8 \text{ m} + 2 \text{ m} = 10 \text{ m}$ *Ans.*

F12–6. $\displaystyle\int v \, dv = \int a \, ds$
$$\int_{5 \text{ m/s}}^v v \, dv = \int_0^s (10 - 0.2s) ds$$
$$v = \left(\sqrt{20s - 0.2s^2 + 25}\right) \text{ m/s}$$
At $s = 10$ m,
$$v = \sqrt{20(10) - 0.2(10^2) + 25}$$
$$= 14.3 \text{ m/s} \rightarrow \qquad Ans.$$

F12–7. $v = \displaystyle\int (4t^2 - 2) \, dt$
$v = \frac{4}{3}t^3 - 2t + C_1$
$s = \displaystyle\int \left(\frac{4}{3}t^3 - 2t + C_1\right) dt$
$s = \frac{1}{3}t^4 - t^2 + C_1 t + C_2$
$t = 0, s = -2, C_2 = -2$
$t = 2, s = -20, C_1 = -9.67$
$t = 4, s = 28.7 \text{ m}$ *Ans.*

F12–8. $a = v\frac{dv}{ds}$
$= \left(20 - 0.05s^2\right)(-0.1s)$
At $s = 15$ m,
$a = -13.1 \text{ m/s}^2 = 13.1 \text{ m/s}^2 \leftarrow$ *Ans.*

F12–9. $v = \frac{ds}{dt} = \frac{d}{dt}\left(0.5t^3\right) = 1.5t^2$
$v = \frac{ds}{dt} = \frac{d}{dt}(108) = 0$ *Ans.*

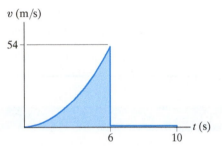

F12–10. $ds = v \, dt$
$$\int_0^s ds = \int_0^t (-4t + 80) \, dt$$
$$s = -2t^2 + 80t$$
$$a = \frac{dv}{dt} = \frac{d}{dt}(-4t + 80) = -4 \text{ ft/s}^2 = 4 \text{ ft/s}^2 \leftarrow$$
Also,
$$a = \frac{\Delta v}{\Delta t} = \frac{0 - 80 \text{ ft/s}}{20 \text{ s} - 0} = -4 \text{ ft/s}^2$$

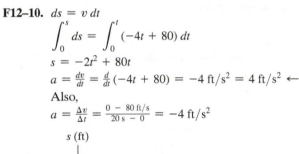

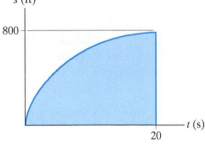

F12–11. $a \, ds = v \, dv$

$$a = v\frac{dv}{ds} = 0.25s\frac{d}{ds}(0.25s) = 0.0625s$$

$$a|_{s=40\,m} = 0.0625(40\text{ m}) = 2.5 \text{ m/s}^2 \rightarrow$$

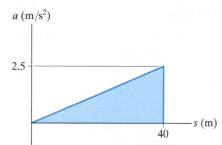

F12–12. For $0 \leq s \leq 10$ m

$$a = s$$

$$\int_0^v v \, dv = \int_0^s s \, ds$$

$$v = s$$

at $s = 10$ m, $v = 10$ m

For 10 m $\leq s \leq 15$

$$a = 10$$

$$\int_{10}^v v \, dv = \int_{10}^s 10 \, ds$$

$$\frac{1}{2}v^2 - 50 = 10s - 100$$

$$v = \sqrt{20s - 100}$$

at $s = 15$ m

$$v = 14.1 \text{ m/s} \qquad\qquad\qquad \textit{Ans.}$$

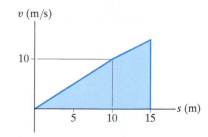

F12–13. $0 \leq t < 5$ s,

$$dv = a \, dt \qquad \int_0^v dv = \int_0^t 20 \, dt$$

$$v = (20t) \text{ m/s}$$

5 s $< t \leq t'$,

$$(\overset{+}{\rightarrow}) \quad dv = a \, dt \qquad \int_{100\,m/s}^v dv = \int_{5\,s}^t -10 \, dt$$

$$v - 100 = (50 - 10t) \text{ m/s},$$

$$0 = 150 - 10t'$$

$$t' = 15 \text{ s}$$

Also,

$$\Delta v = 0 = \text{Area under the } a{-}t \text{ graph}$$

$$0 = (20 \text{ m/s}^2)(5 \text{ s}) + [-(10 \text{ m/s})(t' - 5) \text{ s}]$$

$$t' = 15 \text{ s}$$

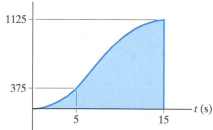

F12–14. $0 \leq t \leq 5$ s,

$$ds = v \, dt \qquad \int_0^s ds = \int_0^t 30t \, dt$$

$$s|_0^s = 15t^2|_0^t$$

$$s = (15t^2) \text{ m}$$

5 s $< t \leq 15$ s,

$$(\overset{+}{\rightarrow}) \, ds = v \, dt; \quad \int_{375\,m}^s ds = \int_{5\,s}^t (-15t + 225)dt$$

$$s = (-7.5t^2 + 225t - 562.5) \text{ m}$$

$$s = (-7.5)(15)^2 + 225(15) - 562.5 \text{ m}$$

$$= 1125 \text{ m} \qquad\qquad\qquad \textit{Ans.}$$

Also,

$$\Delta s = \text{Area under the } v{-}t \text{ graph}$$

$$= \tfrac{1}{2}(150 \text{ m/s})(15 \text{ s})$$

$$= 1125 \text{ m} \qquad\qquad\qquad \textit{Ans.}$$

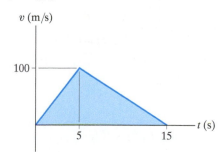

F12–15. $\int_0^x dx = \int_0^t 32t \, dt$

$$x = (16t^2) \text{ m} \qquad\qquad (1)$$

$$\int_0^y dy = \int_0^t 8 \, dt$$

$$t = \frac{y}{8} \qquad\qquad (2)$$

Substituting Eq. (2) into Eq. (1), get

$y = 2\sqrt{x}$ *Ans.*

F12–16. $y = 0.75(8t) = 6t$

$v_x = \dot{x} = \frac{dx}{dt} = \frac{d}{dt}(8t) = 8 \text{ m/s} \rightarrow$

$v_y = \dot{y} = \frac{dy}{dt} = \frac{d}{dt}(6t) = 6 \text{ m/s} \uparrow$

The magnitude of the particle's velocity is

$v = \sqrt{v_x^2 + v_y^2} = \sqrt{(8 \text{ m/s})^2 + (6 \text{ m/s})^2}$

$= 10 \text{ m/s}$ *Ans.*

F12–17. $y = (4t^2) \text{ m}$

$v_x = \dot{x} = \frac{d}{dt}(4t^4) = (16t^3) \text{ m/s} \rightarrow$

$v_y = \dot{y} = \frac{d}{dt}(4t^2) = (8t) \text{ m/s} \uparrow$

When $t = 0.5 \text{ s}$,

$v = \sqrt{v_x^2 + v_y^2} = \sqrt{(2 \text{ m/s})^2 + (4 \text{ m/s})^2}$

$= 4.47 \text{ m/s}$ *Ans.*

$a_x = \dot{v}_x = \frac{d}{dt}(16t^3) = (48t^2) \text{ m/s}^2$

$a_y = \dot{v}_y = \frac{d}{dt}(8t) = 8 \text{ m/s}^2$

When $t = 0.5 \text{ s}$,

$a = \sqrt{a_x^2 + a_y^2} = \sqrt{(12 \text{ m/s}^2)^2 + (8 \text{ m/s}^2)^2}$

$= 14.4 \text{ m/s}^2$ *Ans.*

F12–18. $y = 0.5x$

$\dot{y} = 0.5\dot{x}$

$v_y = t^2$

When $t = 4 \text{ s}$,

$v_x = 32 \text{ m/s} \qquad v_y = 16 \text{ m/s}$

$v = \sqrt{v_x^2 + v_y^2} = 35.8 \text{ m/s}$ *Ans.*

$a_x = \dot{v}_x = 4t$

$a_y = \dot{v}_y = 2t$

When $t = 4 \text{ s}$,

$a_x = 16 \text{ m/s}^2 \qquad a_y = 8 \text{ m/s}^2$

$a = \sqrt{a_x^2 + a_y^2} = \sqrt{16^2 + 8^2} = 17.9 \text{ m/s}^2$ *Ans.*

F12–19. $v_y = \dot{y} = 0.5 \, x \, \dot{x} = 0.5(8)(8) = 32 \text{ m/s}$

Thus,

$v = \sqrt{v_x^2 + v_y^2} = 33.0 \text{ m/s}$ *Ans.*

$a_y = \dot{v}_y = 0.5 \, \dot{x}^2 + 0.5 \, x\ddot{x}$

$= 0.5(8)^2 + 0.5(8)(4)$

$= 48 \text{ m/s}^2$

$a_x = 4 \text{ m/s}^2$

Thus,

$a = \sqrt{a_x^2 + a_y^2} = \sqrt{4^2 + 48^2} = 48.2 \text{ m/s}^2$ *Ans.*

F12–20. $\dot{y} = 0.1x\dot{x}$

$v_y = 0.1(5)(-3) = -1.5 \text{ m/s} = 1.5 \text{ m/s} \downarrow$ *Ans.*

$\ddot{y} = 0.1[\dot{x}\dot{x} + x\ddot{x}]$

$a_y = 0.1[(-3)^2 + 5(-1.5)] = 0.15 \text{ m/s}^2 \uparrow$ *Ans.*

F12–21. $(v_B)_y^2 = (v_A)_y^2 + 2a_y(y_B - y_A)$

$0^2 = (5 \text{ m/s})^2 + 2(-9.81 \text{ m/s}^2)(h - 0)$

$h = 1.27 \text{ m}$ *Ans.*

F12–22. $y_C = y_A + (v_A)_y t_{AC} + \frac{1}{2} a_y t_{AC}^2$

$0 = 0 + (5 \text{ m/s})t_{AC} + \frac{1}{2}(-9.81 \text{ m/s}^2)t_{AC}^2$

$t_{AC} = 1.0194 \text{ s}$

$(v_C)_y = (v_A)_y + a_y t_{AC}$

$(v_C)_y = 5 \text{ m/s} + (-9.81 \text{ m/s}^2)(1.0194 \text{ s})$

$= -5 \text{ m/s} = 5 \text{ m/s} \downarrow$

$v_C = \sqrt{(v_C)_x^2 + (v_C)_y^2}$

$= \sqrt{(8.660 \text{ m/s})^2 + (5 \text{ m/s})^2} = 10 \text{ m/s}$ *Ans.*

$R = x_A + (v_A)_x t_{AC} = 0 + (8.660 \text{ m/s})(1.0194 \text{ s})$

$= 8.83 \text{ m}$ *Ans.*

F12–23. $s = s_0 + v_0 t$

$10 = 0 + v_A \cos 30°t$

$s = s_0 + v_0 t + \frac{1}{2} a_c t^2$

$3 = 1.5 + v_A \sin 30°t + \frac{1}{2}(-9.81)t^2$

$t = 0.9334 \text{ s}, \quad v_A = 12.4 \text{ m/s}$ *Ans.*

F12–24. $s = s_0 + v_0 t$

$R\left(\frac{4}{5}\right) = 0 + 20\left(\frac{3}{5}\right)t$

$s = s_0 + v_0 t + \frac{1}{2} a_c t^2$

$-R\left(\frac{3}{5}\right) = 0 + 20\left(\frac{4}{5}\right)t + \frac{1}{2}(-9.81)t^2$

$t = 5.10 \text{ s}$

$R = 76.5 \text{ m}$ *Ans.*

F12–25. $x_B = x_A + (v_A)_x t_{AB}$

$12 \text{ ft} = 0 + (0.8660 \, v_A)t_{AB}$

$v_A t_{AB} = 13.856$ (1)

$y_B = y_A + (v_A)_y t_{AB} + \frac{1}{2} a_y t_{AB}^2$

$(8 - 3) \text{ ft} = 0 + 0.5v_A t_{AB} + \frac{1}{2}(-32.2 \text{ ft/s}^2)t_{AB}^2$

Using Eq. (1),

$5 = 0.5(13.856) - 16.1\, t_{AB}^2$

$t_{AB} = 0.3461 \text{ s}$

$v_A = 40.0 \text{ ft/s}$ *Ans.*

F12–26. $y_B = y_A + (v_A)_y t_{AB} + \frac{1}{2} a_y t_{AB}^2$

$-150 \text{ m} = 0 + (90 \text{ m/s})t_{AB} + \frac{1}{2}(-9.81 \text{ m/s}^2)t_{AB}^2$

$t_{AB} = 19.89 \text{ s}$

$x_B = x_A + (v_A)_x t_{AB}$

$R = 0 + 120 \text{ m/s}(19.89 \text{ s}) = 2386.37 \text{ m}$

$= 2.39 \text{ km}$ *Ans.*

F12–27. $a_t = \dot{v} = \frac{dv}{dt} = \frac{d}{dt}(0.0625t^2) = (0.125t) \text{ m/s}^2 \big|_{t=10\,\text{s}}$

$= 1.25 \text{ m/s}^2$

$a_n = \dfrac{v^2}{\rho} = \dfrac{(0.0625t^2)^2}{40 \text{ m}} = \left[97.656(10^{-6})t^4\right] \text{ m/s}^2 \big|_{t=10\,\text{s}}$

$= 0.9766 \text{ m/s}^2$

$a = \sqrt{a_t^2 + a_n^2} = \sqrt{(1.25 \text{ m/s}^2)^2 + (0.9766 \text{ m/s}^2)^2}$

$= 1.59 \text{ m/s}^2$ *Ans.*

F12–28. $v = 2s \big|_{s=10} = 20 \text{ m/s}$

$a_n = \dfrac{v^2}{\rho} = \dfrac{(20 \text{ m/s})^2}{50 \text{ m}} = 8 \text{ m/s}^2$

$a_t = v\dfrac{dv}{ds} = 4s \big|_{s=10} = 40 \text{ m/s}^2$

$a = \sqrt{a_t^2 + a_n^2} = \sqrt{(40 \text{ m/s}^2)^2 + (8 \text{ m/s}^2)^2}$

$= 40.8 \text{ m/s}^2$ *Ans.*

F12–29. $v_C^2 = v_A^2 + 2a_t(s_C - s_A)$

$(15 \text{ m/s})^2 = (25 \text{ m/s})^2 + 2a_t(300 \text{ m} - 0)$

$a_t = -0.6667 \text{ m/s}^2$

$v_B^2 = v_A^2 + 2a_t(s_B - s_A)$

$v_B^2 = (25 \text{ m/s})^2 + 2(-0.6667 \text{ m/s}^2)(250 \text{ m} - 0)$

$v_B = 17.08 \text{ m/s}$

$(a_B)_n = \dfrac{v_B^2}{\rho} = \dfrac{(17.08 \text{ m/s})^2}{300 \text{ m}} = 0.9722 \text{ m/s}^2$

$a_B = \sqrt{(a_B)_t^2 + (a_B)_n^2}$

$= \sqrt{(-0.6667 \text{ m/s}^2)^2 + (0.9722 \text{ m/s}^2)^2}$

$= 1.18 \text{ m/s}^2$ *Ans.*

F12–30. $\tan\theta = \dfrac{dy}{dx} = \dfrac{d}{dx}\left(\dfrac{1}{24}x^2\right) = \dfrac{1}{12}x$

$\theta = \tan^{-1}\left(\dfrac{1}{12}x\right)\Big|_{x=10\,\text{ft}}$

$= \tan^{-1}\left(\dfrac{10}{12}\right) = 39.81° = 39.8°\ \nearrow$ *Ans.*

$\rho = \dfrac{\left[1 + (dy/dx)^2\right]^{3/2}}{|d^2y/dx^2|} = \dfrac{\left[1 + \left(\frac{1}{12}x\right)^2\right]^{3/2}}{\left|\frac{1}{12}\right|}\Bigg|_{x=10\,\text{ft}}$

$= 26.468 \text{ ft}$

$a_n = \dfrac{v^2}{\rho} = \dfrac{(20 \text{ ft/s})^2}{26.468 \text{ ft}} = 15.11 \text{ ft/s}^2$

$a = \sqrt{(a_t)^2 + (a_n)^2} = \sqrt{(6 \text{ ft/s}^2)^2 + (15.11 \text{ ft/s}^2)^2}$

$= 16.3 \text{ ft/s}^2$ *Ans.*

F12–31. $(a_B)_t = -0.001s = (-0.001)(300 \text{ m})\left(\frac{\pi}{2}\text{ rad}\right) \text{ m/s}^2$

$= -0.4712 \text{ m/s}^2$

$v\, dv = a_t\, ds$

$\displaystyle\int_{25\,\text{m/s}}^{v_B} v\, dv = \int_0^{150\pi\,\text{m}} -0.001s\, ds$

$v_B = 20.07 \text{ m/s}$

$(a_B)_n = \dfrac{v_B^2}{\rho} = \dfrac{(20.07 \text{ m/s})^2}{300 \text{ m}} = 1.343 \text{ m/s}^2$

$a_B = \sqrt{(a_B)_t^2 + (a_B)_n^2}$

$= \sqrt{(-0.4712 \text{ m/s}^2)^2 + (1.343 \text{ m/s}^2)^2}$

$= 1.42 \text{ m/s}^2$ *Ans.*

F12–32. $a_t\, ds = v\, dv$

$a_t = v\dfrac{dv}{ds} = (0.2s)(0.2) = (0.04s) \text{ m/s}^2$

$a_t = 0.04(50 \text{ m}) = 2 \text{ m/s}^2$

$v = 0.2(50 \text{ m}) = 10 \text{ m/s}$

$a_n = \dfrac{v^2}{\rho} = \dfrac{(10 \text{ m/s})^2}{500 \text{ m}} = 0.2 \text{ m/s}^2$

$a = \sqrt{a_t^2 + a_n^2} = \sqrt{(2 \text{ m/s}^2)^2 + (0.2 \text{ m/s}^2)^2}$

$\quad = 2.01 \text{ m/s}^2$ *Ans.*

F12–33. $v_r = \dot{r} = 0$

$v_\theta = r\dot\theta = (400\dot\theta) \text{ ft/s}$

$v = \sqrt{v_r^2 + v_\theta^2}$

$55 \text{ ft/s} = \sqrt{0^2 + [(400\dot\theta) \text{ ft/s}]^2}$

$\dot\theta = 0.1375 \text{ rad/s}$ *Ans.*

F12–34. $r = 0.1t^3|_{t=1.5 \text{ s}} = 0.3375 \text{ m}$

$\dot{r} = 0.3t^2|_{t=1.5 \text{ s}} = 0.675 \text{ m/s}$

$\ddot{r} = 0.6t|_{t=1.5 \text{ s}} = 0.900 \text{ m/s}^2$

$\theta = 4t^{3/2}|_{t=1.5 \text{ s}} = 7.348 \text{ rad}$

$\dot\theta = 6t^{1/2}|_{t=1.5 \text{ s}} = 7.348 \text{ rad/s}$

$\ddot\theta = 3t^{-1/2}|_{t=1.5 \text{ s}} = 2.449 \text{ rad/s}^2$

$v_r = \dot{r} = 0.675 \text{ m/s}$

$v_\theta = r\dot\theta = (0.3375 \text{ m})(7.348 \text{ rad/s}) = 2.480 \text{ m/s}$

$a_r = \ddot{r} - r\dot\theta^2$

$\quad = (0.900 \text{ m/s}^2) - (0.3375 \text{ m})(7.348 \text{ rad/s})^2$

$\quad = -17.325 \text{ m/s}^2$

$a_\theta = r\ddot\theta + 2\dot{r}\dot\theta = (0.3375 \text{ m})(2.449 \text{ rad/s}^2)$

$\qquad + 2(0.675 \text{ m/s})(7.348 \text{ rad/s}) = 10.747 \text{ m/s}^2$

$v = \sqrt{v_r^2 + v_\theta^2}$

$\quad = \sqrt{(0.675 \text{ m/s})^2 + (2.480 \text{ m/s})^2}$

$\quad = 2.57 \text{ m/s}$ *Ans.*

$a = \sqrt{a_r^2 + a_\theta^2}$

$\quad = \sqrt{(-17.325 \text{ m/s}^2)^2 + (10.747 \text{ m/s}^2)^2}$

$\quad = 20.4 \text{ m/s}^2$ *Ans.*

F12–35. $r = 2\theta$

$\dot{r} = 2\dot\theta$

$\ddot{r} = 2\ddot\theta$

At $\theta = \pi/4 \text{ rad}$,

$r = 2\left(\dfrac{\pi}{4}\right) = \dfrac{\pi}{2} \text{ ft}$

$\dot{r} = 2(3 \text{ rad/s}) = 6 \text{ ft/s}$

$\ddot{r} = 2(1 \text{ rad/s}) = 2 \text{ ft/s}^2$

$a_r = \ddot{r} - r\dot\theta^2 = 2 \text{ ft/s}^2 - \left(\dfrac{\pi}{2} \text{ ft}\right)(3 \text{ rad/s})^2$

$\quad = -12.14 \text{ ft/s}^2$

$a_\theta = r\ddot\theta + 2\dot{r}\dot\theta$

$\quad = \left(\dfrac{\pi}{2} \text{ ft}\right)(1 \text{ rad/s}^2) + 2(6 \text{ ft/s})(3 \text{ rad/s})$

$\quad = 37.57 \text{ ft/s}^2$

$a = \sqrt{a_r^2 + a_\theta^2}$

$\quad = \sqrt{(-12.14 \text{ ft/s}^2)^2 + (37.57 \text{ ft/s}^2)^2}$

$\quad = 39.5 \text{ ft/s}^2$ *Ans.*

F12–36. $r = e^\theta$

$\dot{r} = e^\theta\dot\theta$

$\ddot{r} = e^\theta\ddot\theta + e^\theta\dot\theta^2$

$a_r = \ddot{r} - r\dot\theta^2 = (e^\theta\ddot\theta + e^\theta\dot\theta^2) - e^\theta\dot\theta^2 = e^{\pi/4}(4)$

$\quad = 8.77 \text{ m/s}^2$ *Ans.*

$a_\theta = r\ddot\theta + 2\dot{r}\dot\theta = (e^\theta\ddot\theta) + (2(e^\theta\dot\theta)\dot\theta) = e^\theta(\ddot\theta + 2\dot\theta^2)$

$\quad = e^{\pi/4}(4 + 2(2)^2)$

$\quad = 26.3 \text{ m/s}^2$ *Ans.*

F12–37. $r = [0.2(1 + \cos\theta)] \text{ m}|_{\theta=30°} = 0.3732 \text{ m}$

$\dot{r} = \left[-0.2(\sin\theta)\dot\theta\right] \text{ m/s}|_{\theta=30°}$

$\quad = -0.2\sin 30°(3 \text{ rad/s})$

$\quad = -0.3 \text{ m/s}$

$v_r = \dot{r} = -0.3 \text{ m/s}$

$v_\theta = r\dot\theta = (0.3732 \text{ m})(3 \text{ rad/s}) = 1.120 \text{ m/s}$

$v = \sqrt{v_r^2 + v_\theta^2} = \sqrt{(-0.3 \text{ m/s})^2 + (1.120 \text{ m/s})^2}$

$\quad = 1.16 \text{ m/s}$ *Ans.*

F12–38. $30 \text{ m} = r\sin\theta$

$r = \left(\dfrac{30 \text{ m}}{\sin\theta}\right) = (30\csc\theta) \text{ m}$

$r = (30\csc\theta)|_{\theta=45°} = 42.426 \text{ m}$

$\dot{r} = -30\csc\theta\cot\theta\,\dot\theta|_{\theta=45°} = -\left(42.426\dot\theta\right) \text{ m/s}$

$v_r = \dot{r} = -(42.426\dot\theta) \text{ m/s}$

$v_\theta = r\dot\theta = (42.426\dot\theta) \text{ m/s}$

$v = \sqrt{v_r^2 + v_\theta^2}$

$2 = \sqrt{(-42.426\dot\theta)^2 + (42.426\dot\theta)^2}$

$\dot\theta = 0.0333 \text{ rad/s}$ *Ans.*

F12–39. $l_T = 3s_D + s_A$

$0 = 3v_D + v_A$

$0 = 3v_D + 3 \text{ m/s}$

$v_D = -1 \text{ m/s} = 1 \text{ m/s} \uparrow$ *Ans.*

F12–40. $s_B + 2s_A + 2h = l$

$v_B + 2v_A = 0$

$6 + 2v_A = 0$ $v_A = -3$ m/s $= 3$ m/s ↑   *Ans.*

F12–41. $3s_A + s_B = l$

$3v_A + v_B = 0$

$3v_A + 1.5 = 0$ $v_A = -0.5$ m/s $= 0.5$ m/s ↑ *Ans.*

F12–42. $l_T = 4 s_A + s_F$

$0 = 4 v_A + v_F$

$0 = 4 v_A + 3$ m/s

$v_A = -0.75$ m/s $= 0.75$ m/s ↑   *Ans.*

F12–43. $s_A + 2(s_A - a) + (s_A - s_P) = l$

$4s_A - s_P = l + 2a$

$4v_A - v_P = 0$

$4v_A - (-4) = 0$

$4v_A + 4 = 0$ $v_A = -1$ m/s $= 1$ m/s ↗   *Ans.*

F12–44. $s_C + s_B = l_{CED}$ (1)

$(s_A - s_C) + (s_B - s_C) + s_B = l_{ACDF}$

$s_A + 2s_B - 2s_C = l_{ACDF}$ (2)

Thus

$v_C + v_B = 0$

$v_A + 2v_B - 2v_C = 0$

Eliminating v_C,

$v_A + 4v_B = 0$

Thus,

4 ft/s $+ 4v_B = 0$

$v_B = -1$ ft/s $= 1$ ft/s ↑   *Ans.*

F12–45. $\mathbf{v}_B = \mathbf{v}_A + \mathbf{v}_{B/A}$

$100\mathbf{i} = 80\mathbf{j} + \mathbf{v}_{B/A}$

$\mathbf{v}_{B/A} = 100\mathbf{i} - 80\mathbf{j}$

$v_{B/A} = \sqrt{(v_{B/A})_x^2 + (v_{B/A})_y^2}$

$= \sqrt{(100 \text{ km/h})^2 + (-80 \text{ km/h})^2}$

$= 128$ km/h   *Ans.*

$\theta = \tan^{-1}\left[\dfrac{(v_{B/A})_y}{(v_{B/A})_x}\right] = \tan^{-1}\left(\dfrac{80 \text{ km/h}}{100 \text{ km/h}}\right) = 38.7°$ ◺ *Ans.*

F12–46. $\mathbf{v}_B = \mathbf{v}_A + \mathbf{v}_{B/A}$

$(-400\mathbf{i} - 692.82\mathbf{j}) = (650\mathbf{i}) + \mathbf{v}_{B/A}$

$\mathbf{v}_{B/A} = [-1050\mathbf{i} - 692.82\mathbf{j}]$ km/h

$v_{B/A} = \sqrt{(v_{B/A})_x^2 + (v_{B/A})_y^2}$

$= \sqrt{(1050 \text{ km/h})^2 + (692.82 \text{ km/h})^2}$

$= 1258$ km/h   *Ans.*

$\theta = \tan^{-1}\left[\dfrac{(v_{B/A})_y}{(v_{B/A})_x}\right] = \tan^{-1}\left(\dfrac{692.82 \text{ km/h}}{1050 \text{ km/h}}\right) = 33.4°$ ◿ *Ans.*

F12–47. $\mathbf{v}_B = \mathbf{v}_A + \mathbf{v}_{B/A}$

$(5\mathbf{i} + 8.660\mathbf{j}) = (12.99\mathbf{i} + 7.5\mathbf{j}) + \mathbf{v}_{B/A}$

$\mathbf{v}_{B/A} = [-7.990\mathbf{i} + 1.160\mathbf{j}]$ m/s

$v_{B/A} = \sqrt{(-7.990 \text{ m/s})^2 + (1.160 \text{ m/s})^2}$

$= 8.074$ m/s

$d_{AB} = v_{B/A}t = (8.074 \text{ m/s})(4 \text{ s}) = 32.3$ m   *Ans.*

F12–48. $\mathbf{v}_A = \mathbf{v}_B + \mathbf{v}_{A/B}$

$-20 \cos 45°\mathbf{i} + 20 \sin 45°\mathbf{j} = 65\mathbf{i} + \mathbf{v}_{A/B}$

$\mathbf{v}_{A/B} = -79.14\mathbf{i} + 14.14\mathbf{j}$

$v_{A/B} = \sqrt{(-79.14)^2 + (14.14)^2}$

$= 80.4$ km/h   *Ans.*

$\mathbf{a}_A = \mathbf{a}_B + \mathbf{a}_{A/B}$

$\dfrac{(20)^2}{0.1} \cos 45°\mathbf{i} + \dfrac{(20)^2}{0.1} \sin 45°\mathbf{j} = 1200\mathbf{i} + \mathbf{a}_{A/B}$

$\mathbf{a}_{A/B} = 1628\mathbf{i} + 2828\mathbf{j}$

$a_{A/B} = \sqrt{(1628)^2 + (2828)^2}$

$= 3.26(10^3)$ km/h^2 *Ans.*

Chapter 13

F13–1. $s = s_0 + v_0 t + \frac{1}{2} a_c t^2$

6 m $= 0 + 0 + \frac{1}{2} a(3 \text{ s})^2$

$a = 1.333$ m/s^2

$\Sigma F_y = ma_y$; $N_A - 20(9.81)$ N $\cos 30° = 0$

$N_A = 169.91$ N

$\Sigma F_x = ma_x$; $T - 20(9.81)$ N $\sin 30°$

$- 0.3(169.91 \text{ N}) = (20 \text{ kg})(1.333 \text{ m/s}^2)$

$T = 176$ N   *Ans.*

F13–2. $(F_f)_{max} = \mu_s N_A = 0.3(245.25 \text{ N}) = 73.575$ N.

Since $F = 100$ N $> (F_f)_{max}$ when $t = 0$, the crate will start to move immediately after $\mathbf{F}$ is applied.

$+\uparrow \Sigma F_y = ma_y$; $N_A - 25(9.81)$ N $= 0$

$N_A = 245.25$ N

$\xrightarrow{+} \Sigma F_x = ma_x$;

$10t^2 + 100 - 0.25(245.25 \text{ N}) = (25 \text{ kg})a$

$a = (0.4t^2 + 1.5475)$ m/s^2

$dv = a \, dt$

$\int_0^v dv = \int_0^{4 \text{ s}} (0.4t^2 + 1.5475) dt$

$v = 14.7$ m/s →   *Ans.*

F13–3. $\xrightarrow{+} \Sigma F_x = ma_x;$

$\left(\frac{4}{5}\right)500\text{ N} - (500s)\text{N} = (10\text{ kg})a$

$a = (40 - 50s)\text{ m/s}^2$

$\qquad v\,dv = a\,ds$

$\displaystyle\int_0^v v\,dv = \int_0^{0.5\text{ m}}(40 - 50s)\,ds$

$\frac{v^2}{2}\Big|_0^v = \left(40s - 25s^2\right)\Big|_0^{0.5\text{ m}}$

$v = 5.24\text{ m/s}$ *Ans.*

F13–4. $\xrightarrow{+} \Sigma F_x = ma_x \qquad 100(s + 1)\text{ N} = (2000\text{ kg})a$

$\qquad\qquad a = (0.05(s + 1))\text{ m/s}^2$

$\qquad v\,dv = a\,ds$

$\displaystyle\int_0^v v\,dv = \int_0^{10\text{ m}}0.05(s + 1)\,ds$

$v = 2.45\text{ m/s}$

F13–5. $F_{sp} = k(l - l_0) = (200\text{ N/m})(0.5\text{ m} - 0.3\text{ m})$

$\qquad = 40\text{ N}$

$\theta = \tan^{-1}\left(\frac{0.3\text{ m}}{0.4\text{ m}}\right) = 36.86°$

$\xrightarrow{+} \Sigma F_x = ma_x;$

$100\text{ N} - (40\text{ N})\cos 36.86° = (25\text{ kg})a$

$a = 2.72\text{ m/s}^2$

F13–6. Blocks A and B:

$\xrightarrow{+} \Sigma F_x = ma_x; \quad 6 = \frac{70}{32.2}a; \quad a = 2.76\text{ ft/s}^2$

Check if slipping occurs between A and B.

$\xrightarrow{+} \Sigma F_x = ma_x; \quad 6 - F = \frac{20}{32.2}(2.76);$

$F = 4.29\text{ lb} < 0.4(20) = 8\text{ lb}$

$a_A = a_B = 2.76\text{ ft/s}^2$ *Ans.*

F13–7. $\Sigma F_n = m\frac{v^2}{\rho}; \quad (0.3)m(9.81) = m\frac{v^2}{2}$

$v = 2.43\text{ m/s}$ *Ans.*

F13–8. $+\downarrow\Sigma F_n = ma_n; \quad m(32.2) = m\left(\frac{v^2}{250}\right)$

$v = 89.7\text{ ft/s}$ *Ans.*

F13–9. $+\downarrow\Sigma F_n = ma_n; \quad 150 + N_p = \frac{150}{32.2}\left(\frac{(120)^2}{400}\right)$

$N_p = 17.7\text{ lb}$ *Ans.*

F13–10. $\xleftarrow{+} \Sigma F_n = ma_n;$

$N_c \sin 30° + 0.2\,N_c \cos 30° = m\frac{v^2}{500}$

$+\uparrow\Sigma F_b = 0;$

$N_c \cos 30° - 0.2N_c \sin 30° - m(32.2) = 0$

$v = 119\text{ ft/s}$ *Ans.*

F13–11. $\Sigma F_t = ma_t; \quad 10(9.81)\text{ N}\cos 45° = (10\text{ kg})a_t$

$\qquad\qquad a_t = 6.94\text{ m/s}^2$ *Ans.*

$\Sigma F_n = ma_n;$

$T - 10(9.81)\text{ N}\sin 45° = (10\text{ kg})\frac{(3\text{ m/s})^2}{2\text{ m}}$

$T = 114\text{ N}$ *Ans.*

F13–12. $\Sigma F_n = ma_n;$

$F_n = (500\text{ kg})\frac{(15\text{ m/s})^2}{200\text{ m}} = 562.5\text{ N}$

$\Sigma F_t = ma_t;$

$F_t = (500\text{ kg})(1.5\text{ m/s}^2) = 750\text{ N}$

$F = \sqrt{F_n^2 + F_t^2} = \sqrt{(562.5\text{ N})^2 + (750\text{ N})^2}$

$\qquad = 938\text{ N}$ *Ans.*

F13–13. $a_r = \ddot{r} - r\dot\theta^2 = 0 - (1.5\text{ m} + (8\text{ m})\sin 45°)\dot\theta^2$

$\qquad = (-7.157\,\dot\theta^2)\text{ m/s}^2$

$\Sigma F_z = ma_z;$

$T\cos 45° - m(9.81) = m(0) \quad T = 13.87\text{ m}$

$\Sigma F_r = ma_r;$

$-(13.87m)\sin 45° = m(-7.157\,\dot\theta^2)$

$\dot\theta = 1.17\text{ rad/s}$ *Ans.*

F13–14. $\theta = \pi t^2\big|_{t=0.5\text{ s}} = (\pi/4)\text{ rad}$

$\dot\theta = 2\pi t\big|_{t=0.5\text{ s}} = \pi\text{ rad/s}$

$\ddot\theta = 2\pi\text{ rad/s}^2$

$r = 0.6\sin\theta\big|_{\theta=\pi/4\text{ rad}} = 0.4243\text{ m}$

$\dot r = 0.6(\cos\theta)\dot\theta\big|_{\theta=\pi/4\text{ rad}} = 1.3329\text{ m/s}$

$\ddot r = 0.6[(\cos\theta)\ddot\theta - (\sin\theta)\dot\theta^2]\big|_{\theta=\pi/4\text{ rad}} = -1.5216\text{ m/s}^2$

$a_r = \ddot r - r\dot\theta^2 = -1.5216\text{ m/s}^2 - (0.4243\text{ m})(\pi\text{ rad/s})^2$

$\qquad = -5.7089\text{ m/s}^2$

$a_\theta = r\ddot\theta + 2\dot r\,\dot\theta = 0.4243\text{ m}(2\pi\text{ rad/s}^2)$

$\qquad\qquad + 2(1.3329\text{ m/s})(\pi\text{ rad/s})$

$\qquad = 11.0404\text{ m/s}^2$

$\Sigma F_r = ma_r;$

$F\cos 45° - N\cos 45° - 0.2(9.81)\cos 45°$

$\qquad\qquad = 0.2(-5.7089)$

$\Sigma F_\theta = ma_\theta;$

$F\sin 45° + N\sin 45° - 0.2(9.81)\sin 45°$

$\qquad\qquad = 0.2(11.0404)$

$N = 2.37\text{ N} \qquad F = 2.72\text{ N}$ *Ans.*

F13–15. $r = 50e^{2\theta}\big|_{\theta=\pi/6 \text{ rad}} = \left[50e^{2(\pi/6)}\right]$ m $= 142.48$ m

$\dot{r} = 50\left(2e^{2\theta}\,\dot{\theta}\right) = 100e^{2\theta}\,\dot{\theta}\big|_{\theta=\pi/6 \text{ rad}}$

$= \left[100e^{2(\pi/6)}(0.05)\right] = 14.248$ m/s

$\ddot{r} = 100\left((2e^{2\theta}\dot{\theta})\dot{\theta} + e^{2\theta}(\ddot{\theta})\right)\big|_{\theta=\pi/6 \text{ rad}}$

$= 100\left[2e^{2(\pi/6)}(0.05^2) + e^{2(\pi/6)}(0.01)\right]$

$= 4.274$ m/s^2

$a_r = \ddot{r} - r\dot{\theta}^2 = 4.274$ m/s^2 $- 142.48$ m$(0.05$ rad/s$)^2$

$= 3.918$ m/s^2

$a_\theta = r\ddot{\theta} + 2\dot{r}\dot{\theta} = 142.48$ m$(0.01$ rad/s$^2)$

$+ 2(14.248$ m/s$)(0.05$ rad/s$)$

$= 2.850$ m/s^2

$\Sigma F_r = ma_r;$

$F_r = (2000$ kg$)(3.918$ m/s$^2) = 7836.55$ N

$\Sigma F_\theta = ma_\theta;$

$F_\theta = (2000$ kg$)(2.850$ m/s$^2) = 5699.31$ N

$F = \sqrt{F_r^2 + F_\theta^2}$

$= \sqrt{(7836.55 \text{ N})^2 + (5699.31 \text{ N})^2}$

$= 9689.87$ N $= 9.69$ kN

F13–16. $r = (0.6\cos 2\theta)$ m$\big|_{\theta=0^\circ} = [0.6\cos 2(0^\circ)]$ m $= 0.6$ m

$\dot{r} = (-1.2\sin 2\theta\dot{\theta})$ m/s$\big|_{\theta=0^\circ}$

$= \left[-1.2\sin 2(0^\circ)(-3)\right]$ m/s $= 0$

$\ddot{r} = -1.2\left(\sin 2\theta\ddot{\theta} + 2\cos 2\theta\dot{\theta}^2\right)$ m/s$^2\big|_{\theta=0^\circ}$

$= -21.6$ m/s^2

Thus,

$a_r = \ddot{r} - r\dot{\theta}^2 = -21.6$ m/s$^2 - 0.6$ m$(-3$ rad/s$)^2$

$= -27$ m/s^2

$a_\theta = r\ddot{\theta} + 2\dot{r}\dot{\theta} = 0.6$ m$(0) + 2(0)(-3$ rad/s$) = 0$

$\Sigma F_\theta = ma_\theta; \quad F - 0.2(9.81)$ N $= 0.2$ kg(0)

$F = 1.96$ N $\uparrow$ *Ans.*

Chapter 14

F14–1. $T_1 + \Sigma U_{1-2} = T_2$

$0 + \left(\frac{4}{5}\right)(500$ N$)(0.5$ m$) - \frac{1}{2}(500$ N/m$)(0.5$ m$)^2$

$= \frac{1}{2}(10$ kg$)v^2$

$v = 5.24$ m/s   *Ans.*

F14–2. $\Sigma F_y = ma_y; N_A - 20(9.81)$ N $\cos 30^\circ = 0$

$N_A = 169.91$ N

$T_1 + \Sigma U_{1-2} = T_2$

$0 + 300$ N$(10$ m$) - 0.3(169.91$ N$)(10$ m$)$

$- 20(9.81)$N $(10$ m$)\sin 30^\circ$

$= \frac{1}{2}(20$ kg$)v^2$

$v = 12.3$ m/s   *Ans.*

F14–3. $T_1 + \Sigma U_{1-2} = T_2$

$0 + 2\left[\int_0^{15 \text{ m}}(600 + 2s^2) \text{ N } ds\right] - 100(9.81)$ N$(15$ m$)$

$= \frac{1}{2}(100$ kg$)v^2$

$v = 12.5$ m/s   *Ans.*

F14–4. $T_1 + \Sigma U_{1-2} = T_2$

$\frac{1}{2}(1800$ kg$)(125$ m/s$)^2 - \left[\frac{(50\,000 \text{ N } + \ 20\,000 \text{ N})}{2}(400 \text{ m})\right]$

$= \frac{1}{2}(1800$ kg$)v^2$

$v = 8.33$ m/s   *Ans.*

F14–5. $T_1 + \Sigma U_{1-2} = T_2$

$\frac{1}{2}(10$ kg$)(5$ m/s$)^2 + 100$ N$s' + [10(9.81)$ N$] \ s' \sin 30^\circ$

$- \frac{1}{2}(200$ N/m$)(s')^2 = 0$

$s' = 2.09$ m

$s = 0.6$ m $+ 2.09$ m $= 2.69$ m   *Ans.*

F14–6. $T_A + \Sigma U_{A-B} = T_B$

Consider difference in cord length $AC - BC$, which is distance F moves.

$0 + 10$ lb$(\sqrt{(3 \text{ ft})^2 + (4 \text{ ft})^2} - 3$ ft$)$

$= \frac{1}{2}\left(\frac{5}{32.2} \text{ slug}\right)v_B^2$

$v_B = 16.0$ ft/s   *Ans.*

F14–7. $\xrightarrow{+} \Sigma F_x = ma_x;$

$30\left(\frac{4}{5}\right) = 20a \quad a = 1.2$ m/s$^2 \to$

$v = v_0 + a_c t$

$v = 0 + 1.2(4) = 4.8$ m/s

$P = \mathbf{F} \cdot \mathbf{v} = F(\cos\theta)v$

$= 30\left(\frac{4}{5}\right)(4.8)$

$= 115$ W   *Ans.*

F14–8. $\xrightarrow{+} \Sigma F_x = ma_x;$

$10s = 20a \quad a = 0.5s$ m/s$^2 \to$

$v\,dv = a\,ds$

$\int_1^v v\,dv = \int_0^{5 \text{ m}} 0.5\, s\,ds$

$v = 3.674$ m/s

$P = \mathbf{F} \cdot \mathbf{v} = [10(5)](3.674) = 184$ W   *Ans.*

F14–9. $(+\uparrow)\Sigma F_y = 0;$

$T_1 - 100 \text{ lb} = 0 \quad T_1 = 100 \text{ lb}$

$(+\uparrow)\Sigma F_y = 0;$

$100 \text{ lb} + 100 \text{ lb} - T_2 = 0 \quad T_2 = 200 \text{ lb}$

$P_{\text{out}} = \mathbf{T}_B \cdot \mathbf{v}_B = (200 \text{ lb})(3 \text{ ft/s}) = 1.091 \text{ hp}$

$P_{\text{in}} = \dfrac{P_{\text{out}}}{\varepsilon} = \dfrac{1.091 \text{ hp}}{0.8} = 1.36 \text{ hp}$ *Ans.*

F14–10. $\Sigma F_{y'} = ma_{y'}; \quad N - 20(9.81)\cos 30° = 20(0)$

$N = 169.91 \text{ N}$

$\Sigma F_{x'} = ma_{x'};$

$F - 20(9.81)\sin 30° - 0.2(169.91) = 0$

$F = 132.08 \text{ N}$

$P = \mathbf{F} \cdot \mathbf{v} = 132.08(5) = 660 \text{ W}$ *Ans.*

F14–11. $+\uparrow\Sigma F_y = ma_y;$

$T - 50(9.81) = 50(0) \quad T = 490.5 \text{ N}$

$P_{\text{out}} = \mathbf{T} \cdot \mathbf{v} = 490.5(1.5) = 735.75 \text{ W}$

Also, for a point on the other cable

$P_{\text{out}} = \left(\dfrac{490.5}{2}\right)(1.5)(2) = 735.75 \text{ W}$

$P_{\text{in}} = \dfrac{P_{\text{out}}}{\varepsilon} = \dfrac{735.75}{0.8} = 920 \text{ W}$ *Ans.*

F14–12. $2s_A + s_P = l$

$2a_A + a_P = 0$

$2a_A + 6 = 0$

$a_A = -3 \text{ m/s}^2 = 3 \text{ m/s}^2 \uparrow$

$\Sigma F_y = ma_y; \quad T_A - 490.5 \text{ N} = (50 \text{ kg})(3 \text{ m/s}^2)$

$T_A = 640.5 \text{ N}$

$P_{\text{out}} = \mathbf{T} \cdot \mathbf{v} = (640.5 \text{ N}/2)(12) = 3843 \text{ W}$

$P_{\text{in}} = \dfrac{P_{\text{out}}}{\varepsilon} = \dfrac{3843}{0.8} = 4803.75 \text{ W} = 4.80 \text{ kW}$ *Ans.*

F14–13. $T_A + V_A = T_B + V_B$

$0 + 2(9.81)(1.5) = \tfrac{1}{2}(2)(v_B)^2 + 0$

$v_B = 5.42 \text{ m/s}$ *Ans.*

$+\uparrow\Sigma F_n = ma_n; \quad T - 2(9.81) = 2\left(\dfrac{(5.42)^2}{1.5}\right)$

$T = 58.9 \text{ N}$ *Ans.*

F14–14. $T_A + V_A = T_B + V_B$

$\tfrac{1}{2}m_A v_A^2 + mgh_A = \tfrac{1}{2}m_B v_B^2 + mgh_B$

$\left[\tfrac{1}{2}(2 \text{ kg})(1 \text{ m/s})^2\right] + [2\,(9.81) \text{ N}(4 \text{ m})]$

$= \left[\tfrac{1}{2}(2 \text{ kg})v_B^2\right] + [0]$

$v_B = 8.915 \text{ m/s} = 8.92 \text{ m/s}$ *Ans.*

$+\uparrow\Sigma F_n = ma_n; \quad N_B - 2(9.81) \text{ N}$

$= (2 \text{ kg})\left(\dfrac{(8.915 \text{ m/s})^2}{2 \text{ m}}\right)$

$N_B = 99.1 \text{ N}$ *Ans.*

F14–15. $T_1 + V_1 = T_2 + V_2$

$\tfrac{1}{2}(2)(4)^2 + \tfrac{1}{2}(30)(2 - 1)^2$

$\quad = \tfrac{1}{2}(2)(v)^2 - 2(9.81)(1) + \tfrac{1}{2}(30)(\sqrt{5} - 1)^2$

$v = 5.26 \text{ m/s}$ *Ans.*

F14–16. $T_A + V_A = T_B + V_B$

$0 + \tfrac{1}{2}(4)(2.5 - 0.5)^2 + 5(2.5)$

$\quad = \tfrac{1}{2}\left(\dfrac{5}{32.2}\right)v_B^2 + \tfrac{1}{2}(4)(1 - 0.5)^2$

$v_B = 16.0 \text{ ft/s}$ *Ans.*

F14–17. $T_1 + V_1 = T_2 + V_2$

$\tfrac{1}{2}mv_1^2 + mgy_1 + \tfrac{1}{2}ks_1^2$

$\quad = \tfrac{1}{2}mv_2^2 + mgy_2 + \tfrac{1}{2}ks_2^2$

$[0] + [0] + [0] = [0] +$

$\left[-75 \text{ lb}(5 \text{ ft} + s)\right] + \left[2\left(\tfrac{1}{2}(1000 \text{ lb/ft})s^2\right)\right.$

$\left. + \tfrac{1}{2}(1500 \text{ lb/ft})(s - 0.25 \text{ ft})^2\right]$

$s = s_A = s_C = 0.580 \text{ ft}$ *Ans.*

Also,

$s_B = 0.5803 \text{ ft} - 0.25 \text{ ft} = 0.330 \text{ ft}$ *Ans.*

F14–18. $T_A + V_A = T_B + V_B$

$\tfrac{1}{2}mv_A^2 + \left(\tfrac{1}{2}ks_A^2 + mgy_A\right)$

$\quad = \tfrac{1}{2}mv_B^2 + \left(\tfrac{1}{2}ks_B^2 + mgy_B\right)$

$\tfrac{1}{2}(4 \text{ kg})(2 \text{ m/s})^2 + \tfrac{1}{2}(400 \text{ N/m})(0.1 \text{ m} - 0.2 \text{ m})^2 + 0$

$= \tfrac{1}{2}(4 \text{ kg})v_B^2 + \tfrac{1}{2}(400 \text{ N/m})(\sqrt{(0.4 \text{ m})^2 + (0.3 \text{ m})^2}$

$\quad - 0.2 \text{ m})^2 + [4(9.81) \text{ N}](-(0.1 \text{ m} + 0.3 \text{ m}))$

$v_B = 1.962 \text{ m/s} = 1.96 \text{ m/s}$ *Ans.*

Chapter 15

F15–1. $(\overset{+}{\rightarrow}) \quad m(v_1)_x + \Sigma \displaystyle\int_{t1}^{t2} F_x \, dt = m(v_2)_x$

$(0.5 \text{ kg})(25 \text{ m/s})\cos 45° - \displaystyle\int F_x \, dt$

$\quad = (0.5 \text{ kg})(10 \text{ m/s})\cos 30°$

$I_x = \displaystyle\int F_x \, dt = 4.509 \text{ N} \cdot \text{s}$

$(+\uparrow) \quad m(v_1)_y + \Sigma \displaystyle\int_{t1}^{t2} F_y \, dt = m(v_2)_y$

$- (0.5 \text{ kg})(25 \text{ m/s})\sin 45° + \displaystyle\int F_y \, dt$

$\quad = (0.5 \text{ kg})(10 \text{ m/s})\sin 30°$

$I_y = \displaystyle\int F_y \, dt = 11.339 \text{ N} \cdot \text{s}$

$I = \displaystyle\int F \, dt = \sqrt{(4.509 \text{ N} \cdot \text{s})^2 + (11.339 \text{ N} \cdot \text{s})^2}$

$\quad = 12.2 \text{ N} \cdot \text{s}$ *Ans.*

F15–2. $(+\uparrow)$ $m(v_1)_y + \Sigma \int_{t1}^{t2} F_y \, dt = m(v_2)_y$

$0 + N(4 \text{ s}) + (100 \text{ lb})(4 \text{ s})\sin 30°$
$- (150 \text{ lb})(4 \text{ s}) = 0$

$N = 100 \text{ lb}$

$(\overset{+}{\rightarrow})$ $m(v_1)_x + \Sigma \int_{t1}^{t2} F_x \, dt = m(v_2)_x$

$0 + (100 \text{ lb})(4 \text{ s})\cos 30° - 0.2(100 \text{ lb})(4 \text{ s})$
$= \left(\frac{150}{32.2} \text{ slug}\right)v$

$v = 57.2 \text{ ft/s}$ *Ans.*

F15–3. Time to start motion,

$+\uparrow\Sigma F_y = 0;$ $N - 25(9.81) \text{ N} = 0$ $N = 245.25 \text{ N}$

$\overset{+}{\rightarrow} \Sigma F_x = 0;$ $20t^2 - 0.3(245.25 \text{ N}) = 0$ $t = 1.918 \text{ s}$

$(\overset{+}{\rightarrow})$ $m(v_1)_x + \Sigma \int_{t1}^{t2} F_x \, dt = m(v_2)_x$

$0 + \int_{1.918\text{ s}}^{4\text{ s}} 20t^2 \, dt - (0.25(245.25 \text{ N}))(4 \text{ s} - 1.918 \text{ s})$
$= (25 \text{ kg})v$

$v = 10.1 \text{ m/s}$ *Ans.*

F15–4. $(\overset{+}{\rightarrow})$ $m(v_1)_x + \Sigma \int_{t1}^{t2} F_x \, dt = m(v_2)_x$

$(1500 \text{ kg})(0) + \left[\frac{1}{2}(6000 \text{ N})(2 \text{ s}) + (6000 \text{ N})(6 \text{ s} - 2 \text{ s})\right]$
$= (1500 \text{ kg}) \, v$

$v = 20 \text{ m/s}$ *Ans.*

F15–5. SUV and trailer,

$m(v_1)_x + \Sigma \int_{t1}^{t2} F_x \, dt = m(v_2)_x$

$0 + (9000 \text{ N})(20 \text{ s}) = (1500 \text{ kg} + 2500 \text{ kg})v$

$v = 45.0 \text{ m/s}$ *Ans.*

Trailer,

$m(v_1)_x + \Sigma \int_{t1}^{t2} F_x \, dt = m(v_2)_x$

$0 + T(20 \text{ s}) = (1500 \text{ kg})(45.0 \text{ m/s})$

$T = 3375 \text{ N} = 3.375 \text{ kN}$ *Ans.*

F15–6. Block B:

$(+\downarrow) \, mv_1 + \int F \, dt = mv_2$

$0 + 8(5) - T(5) = \frac{8}{32.2} \, (1)$

$T = 7.95 \text{ lb}$ *Ans.*

Block A:

$(\overset{+}{\rightarrow}) \, mv_1 + \int F \, dt = mv_2$

$0 + 7.95(5) - \mu_k(10)(5) = \frac{10}{32.2} \, (1)$

$\mu_k = 0.789$ *Ans.*

F15–7. $(\overset{+}{\rightarrow})$ $m_A \, (v_A)_1 + m_B \, (v_B)_1 = m_A \, (v_A)_2 + m_B \, (v_B)_2$

$(20(10^3) \text{ kg})(3 \text{ m/s}) + (15(10^3) \text{ kg})(-1.5 \text{ m/s})$
$= (20(10^3) \text{ kg})(v_A)_2 + (15(10^3) \text{ kg})(2 \text{ m/s})$

$(v_A)_2 = 0.375 \text{ m/s} \rightarrow$ *Ans.*

$(\overset{+}{\rightarrow})$ $m(v_B)_1 + \Sigma \int_{t1}^{t2} F \, dt = m(v_B)_2$

$(15(10^3) \text{ kg})(-1.5 \text{ m/s}) + F_{avg} \, (0.5 \text{ s})$
$= (15(10^3) \text{ kg})(2 \text{ m/s})$

$F_{avg} = 105(10^3) \text{ N} = 105 \text{ kN}$ *Ans.*

F15–8. $(\overset{+}{\rightarrow})$ $m_p \, [(v_p)_1]_x + m_c[(v)_1]_x = (m_p + m_c)v_2$

$5\left[10\left(\frac{4}{5}\right)\right] + 0 = (5 + 20)v_2$

$v_2 = 1.6 \text{ m/s}$ *Ans.*

F15–9. $T_1 + V_1 = T_2 + V_2$

$\frac{1}{2} m_A \, (v_A)_1^2 + (V_g)_1 = \frac{1}{2} m_A(v_A)_2^2 + (V_g)_2$

$\frac{1}{2} (5)(5)^2 + 5(9.81)(1.5) = \frac{1}{2} (5)(v_A)_2^2 + 0$

$(v_A)_2 = 7.378 \text{ m/s}$

$(\overset{+}{\leftarrow})$ $m_A(v_A)_2 + m_B(v_B)_2 = (m_A + m_B)v$

$5(7.378) + 0 = (5 + 8)v$

$v = 2.84 \text{ m/s}$ *Ans.*

F15–10. $(\overset{+}{\rightarrow})$ $m_A(v_A)_1 + m_B(v_B)_1 = m_A(v_A)_2 + m_B(v_B)_2$

$0 + 0 = 10(v_A)_2 + 15(v_B)_2$ (1)

$T_1 + V_1 = T_2 + V_2$

$\frac{1}{2} m_A(v_A)_1^2 + \frac{1}{2} m_B(v_B)_1^2 + (V_e)_1$
$= \frac{1}{2} m_A(v_A)_2^2 + \frac{1}{2} m_B(v_B)_2^2 + (V_e)_2$

$0 + 0 + \frac{1}{2}\left[5(10^3)\right](0.2^2)$
$= \frac{1}{2}(10)(v_A)_2^2 + \frac{1}{2}(15)(v_B)_2^2 + 0$

$5(v_A)_2^2 + 7.5 \, (v_B)_2^2 = 100$ (2)

Solving Eqs. (1) and (2),

$(v_B)_2 = 2.31 \text{ m/s} \rightarrow$ *Ans.*

$(v_A)_2 = -3.464 \text{ m/s} = 3.46 \text{ m/s} \leftarrow$ *Ans.*

F15–11. $(\overset{+}{\leftarrow})$ $m_A(v_A)_1 + m_B(v_B)_1 = (m_A + m_B)v_2$

$0 + 10(15) = (15 + 10)v_2$

$v_2 = 6 \text{ m/s}$

$T_1 + V_1 = T_2 + V_2$

$\frac{1}{2}(m_A + m_B)v_2^2 + (V_e)_2 = \frac{1}{2}(m_A + m_B)v_3^2 + (V_e)_3$

$\frac{1}{2}(15 + 10)(6^2) + 0 = 0 + \frac{1}{2}\left[10(10^3)\right]s_{max}^2$

$s_{max} = 0.3 \text{ m} = 300 \text{ mm}$ *Ans.*

F15–12. $(\xrightarrow{+})$ $0 + 0 = m_p(v_p)_x - m_c v_c$

$0 = (20 \text{ kg}) (v_p)_x - (250 \text{ kg}) v_c$

$(v_p)_x = 12.5 \, v_c$ (1)

$\mathbf{v}_p = \mathbf{v}_c + \mathbf{v}_{p/c}$

$(v_p)_x \mathbf{i} + (v_p)_y \mathbf{j} = -v_c \mathbf{i} + [(400 \text{ m/s}) \cos 30°\mathbf{i}$
$+ \, (400 \text{ m/s}) \sin 30°\mathbf{j}]$

$(v_p)_x \mathbf{i} + (v_p)_y \mathbf{j} = (346.41 - v_c)\mathbf{i} + 200\mathbf{j}$

$(v_p)_x = 346.41 - v_c$

$(v_p)_y = 200 \text{ m/s}$

$(v_p)_x = 320.75 \text{ m/s} \quad v_c = 25.66 \text{ m/s}$

$v_p = \sqrt{(v_p)_x^2 + (v_p)_y^2}$
$= \sqrt{(320.75 \text{ m/s})^2 + (200 \text{ m/s})^2}$
$= 378 \text{ m/s}$ *Ans.*

F15–13. $(\xrightarrow{+})$ $e = \dfrac{(v_B)_2 - (v_A)_2}{(v_A)_1 - (v_B)_1}$

$= \dfrac{(9 \text{ m/s}) - (1 \text{ m/s})}{(8 \text{ m/s}) - (-2 \text{ m/s})} = 0.8$

F15–14. $(\xrightarrow{+})$ $m_A(v_A)_1 + m_B(v_B)_1 = m_A(v_A)_2 + m_B(v_B)_2$

$[15(10^3) \text{ kg}](5 \text{ m/s}) + [25(10^3)](-7 \text{ m/s})$
$= [15(10^3) \text{ kg}](v_A)_2 + [25(10^3)](v_B)_2$

$15(v_A)_2 + 25(v_B)_2 = -100$ (1)

Using the coefficient of restitution equation,

$(\xrightarrow{+})$ $e = \dfrac{(v_B)_2 - (v_A)_2}{(v_A)_1 - (v_B)_1}$

$0.6 = \dfrac{(v_B)_2 - (v_A)_2}{5 \text{ m/s} - (-7 \text{ m/s})}$

$(v_B)_2 - (v_A)_2 = 7.2$ (2)

Solving,

$(v_B)_2 = 0.2 \text{ m/s} \rightarrow$ *Ans.*

$(v_A)_2 = -7 \text{ m/s} = 7 \text{ m/s} \leftarrow$ *Ans.*

F15–15. $T_1 + V_1 = T_2 + V_2$

$\frac{1}{2} m(v_A)_1^2 + mg(h_A)_1 = \frac{1}{2} m(v_A)_2^2 + mg(h_A)_2$

$\frac{1}{2} \left(\frac{30}{32.2} \text{ slug}\right)(5 \text{ ft/s})^2 + (30 \text{ lb})(10\text{ft})$
$= \frac{1}{2} \left(\frac{30}{32.2} \text{ slug}\right)(v_A)_2^2 + 0$

$(v_A)_2 = 25.87 \text{ ft/s} \leftarrow$

$(\xleftarrow{+})$ $m_A(v_A)_2 + m_B(v_B)_2 = m_A(v_A)_3 + m_B(v_B)_3$

$\left(\frac{30}{32.2} \text{ slug}\right)(25.87 \text{ ft/s}) + 0$
$= \left(\frac{30}{32.2} \text{ slug}\right)(v_A)_3 + \left(\frac{80}{32.2} \text{ slug}\right)(v_B)_3$

$30(v_A)_3 + 80(v_B)_3 = 775.95$ (1)

$(\xleftarrow{+})$ $e = \dfrac{(v_B)_3 - (v_A)_3}{(v_A)_2 - (v_B)_2}$

$0.6 = \dfrac{(v_B)_3 - (v_A)_3}{25.87 \text{ ft/s} - 0}$

$(v_B)_3 - (v_A)_3 = 15.52$ (2)

Solving Eqs. (1) and (2), yields

$(v_B)_3 = 11.3 \text{ ft/s} \leftarrow$

$(v_A)_3 = -4.23 \text{ ft/s} = 4.23 \text{ ft/s} \rightarrow$ *Ans.*

F15–16. $(+\uparrow)$ $m[(v_b)_1]_y = m[(v_b)_2]_y$

$[(v_b)_2]_y = [(v_b)_1]_y = (20 \text{ m/s}) \sin 30° = 10 \text{ m/s} \uparrow$

$(\xrightarrow{+})$ $e = \dfrac{(v_w)_2 - [(v_b)_2]_x}{[(v_b)_1]_x - (v_w)_1}$

$0.75 = \dfrac{0 - [(v_b)_2]_x}{(20 \text{ m/s})\cos 30° - 0}$

$[(v_b)_2]_x = -12.99 \text{ m/s} = 12.99 \text{ m/s} \leftarrow$

$(v_b)_2 = \sqrt{[(v_b)_2]_x^2 + [(v_b)_2]_y^2}$
$= \sqrt{(12.99 \text{ m/s})^2 + (10 \text{ m/s})^2}$
$= 16.4 \text{ m/s}$ *Ans.*

$\theta = \tan^{-1}\left(\dfrac{[(v_b)_2]_y}{[(v_b)_2]_x}\right) = \tan^{-1}\left(\dfrac{10 \text{ m/s}}{12.99 \text{ m/s}}\right)$
$= 37.6°$ *Ans.*

F15–17. $\Sigma m(v_x)_1 = \Sigma m(v_x)_2$

$0 + 0 = 2(1) + 11 (v_{Bx})_2$

$(v_{Bx})_2 = -0.1818 \text{ m/s}$

$\Sigma m(v_y)_1 = \Sigma m(v_y)_2$

$2(3) + 0 = 0 + 11 (v_{By})_2$

$(v_{By})_2 = 0.545 \text{ m/s}$

$(v_B)_2 = \sqrt{(-0.1818)^2 + (0.545)^2}$
$= 0.575 \text{ m/s}$ *Ans.*

F15–18. $+\nearrow$ $1(3)\left(\frac{3}{5}\right) - 1(4)\left(\frac{4}{5}\right)$
$= 1(v_B)_{2x} + 1(v_A)_{2x}$

$+\nearrow$ $0.5 = [(v_A)_{2x} - (v_B)_{2x}]/[(3)\left(\frac{3}{5}\right) - (-4)\left(\frac{4}{5}\right)]$

Solving,

$(v_A)_{2x} = 0.550 \text{ m/s}, (v_B)_{2x} = -1.95 \text{ m/s}$

Disc A,

$+\nwarrow$ $-1(4)\left(\frac{3}{5}\right) = 1(v_A)_{2y}$

$(v_A)_{2y} = -2.40 \text{ m/s}$

Disc B,

$$-1(3)\left(\tfrac{4}{5}\right) = 1(v_B)_{2y}$$

$$(v_B)_{2y} = -2.40 \text{ m/s}$$

$$(v_A)_2 = \sqrt{(0.550)^2 + (2.40)^2} = 2.46 \text{ m/s} \qquad \textit{Ans.}$$

$$(v_B)_2 = \sqrt{(1.95)^2 + (2.40)^2} = 3.09 \text{ m/s} \qquad \textit{Ans.}$$

F15–19. $H_O = \Sigma mvd;$

$$H_O = \left[2(10)\left(\tfrac{4}{5}\right)\right](4) - \left[2(10)\left(\tfrac{3}{5}\right)\right](3)$$

$$= 28 \text{ kg} \cdot \text{m}^2/\text{s}$$

F15–20. $H_P = \Sigma mvd;$

$$H_P = [2(15)\sin 30°](2) - [2(15)\cos 30°](5)$$

$$= -99.9 \text{ kg} \cdot \text{m}^2/\text{s} = 99.9 \text{ kg} \cdot \text{m}^2/\text{s} \ \rotatebox{0}{$\circlearrowright$}$$

F15–21. $(H_z)_1 + \Sigma \int M_z \, dt = (H_z)_2$

$$5(2)(1.5) + 5(1.5)(3) = 5v(1.5)$$

$$v = 5 \text{ m/s} \qquad \textit{Ans.}$$

F15–22. $(H_z)_1 + \Sigma \int M_z \, dt = (H_z)_2$

$$0 + \int_0^{4\text{ s}} (10t)\left(\tfrac{4}{5}\right)(1.5)dt = 5v(1.5)$$

$$v = 12.8 \text{ m/s} \qquad \textit{Ans.}$$

F15–23. $(H_z)_1 + \Sigma \int M_z \, dt = (H_z)_2$

$$0 + \int_0^{5\text{ s}} 0.9t^2 \, dt = 2v(0.6)$$

$$v = 31.2 \text{ m/s} \qquad \textit{Ans.}$$

F15–24. $(H_z)_1 + \Sigma \int M_z \, dt = (H_z)_2$

$$0 + \int_0^{4\text{ s}} 8t \, dt + 2(10)(0.5)(4) = 2[10v(0.5)]$$

$$v = 10.4 \text{ m/s} \qquad \textit{Ans.}$$

Chapter 16

F16–1. $\theta = (20 \text{ rev})\left(\frac{2\pi \text{ rad}}{1 \text{ rev}}\right) = 40\pi \text{ rad}$

$$\omega^2 = \omega_0^2 + 2\alpha_c (\theta - \theta_0)$$

$$(30 \text{ rad/s})^2 = 0^2 + 2\alpha_c [(40\pi \text{ rad}) - 0]$$

$$\alpha_c = 3.581 \text{ rad/s}^2 = 3.58 \text{ rad/s}^2 \qquad \textit{Ans.}$$

$$\omega = \omega_0 + \alpha_c t$$

$$30 \text{ rad/s} = 0 + (3.581 \text{ rad/s}^2)t$$

$$t = 8.38 \text{ s} \qquad \textit{Ans.}$$

F16–2. $\frac{d\omega}{d\theta} = 2(0.005\theta) = (0.01\theta)$

$$\alpha = \omega\frac{d\omega}{d\theta} = \left(0.005\,\theta^2\right)(0.01\theta) = 50\left(10^{-6}\right)\theta^3 \text{ rad/s}^2$$

When $\theta = 20 \text{ rev}(2\pi \text{ rad}/1 \text{ rev}) = 40\pi \text{ rad}$,

$$\alpha = \left[50\left(10^{-6}\right)(40\pi)^3\right] \text{ rad/s}^2$$

$$= 99.22 \text{ rad/s}^2 = 99.2 \text{ rad/s}^2 \qquad \textit{Ans.}$$

F16–3. $\omega = 4\theta^{1/2}$

$$150 \text{ rad/s} = 4\,\theta^{1/2}$$

$$\theta = 1406.25 \text{ rad}$$

$$dt = \frac{d\theta}{\omega}$$

$$\int_0^t dt = \int_{1\text{ rad}}^\theta \frac{d\theta}{4\theta^{1/2}}$$

$$t\Big|_0^t = \tfrac{1}{2}\theta^{1/2}\Big|_{1\text{ rad}}^\theta$$

$$t = \tfrac{1}{2}\theta^{1/2} - \tfrac{1}{2}$$

$$t = \tfrac{1}{2}(1406.25)^{1/2} - \tfrac{1}{2} = 18.25 \text{ s} \qquad \textit{Ans.}$$

F16–4. $\omega = \frac{d\theta}{dt} = (1.5t^2 + 15) \text{ rad/s}$

$$\alpha = \frac{d\omega}{dt} = (3t) \text{ rad/s}^2$$

$$\omega = [1.5(3^2) + 15] \text{ rad/s} = 28.5 \text{ rad/s}$$

$$\alpha = 3(3) \text{ rad/s}^2 = 9 \text{ rad/s}^2$$

$$v = \omega r = (28.5 \text{ rad/s})(0.75 \text{ ft}) = 21.4 \text{ ft/s} \qquad \textit{Ans.}$$

$$a = \alpha r = (9 \text{ rad/s}^2)(0.75 \text{ ft}) = 6.75 \text{ ft/s}^2 \qquad \textit{Ans.}$$

F16–5. $\omega \, d\omega = \alpha \, d\theta$

$$\int_{2\text{ rad/s}}^\omega \omega \, d\omega = \int_0^\theta 0.5\theta \, d\theta$$

$$\frac{\omega^2}{2}\Big|_{2\text{ rad/s}}^\omega = 0.25\theta^2\Big|_0^\theta$$

$$\omega = (0.5\,\theta^2 + 4)^{1/2} \text{ rad/s}$$

When $\theta = 2 \text{ rev} = 4\pi \text{ rad}$,

$$\omega = [0.5(4\pi)^2 + 4]^{1/2} \text{ rad/s} = 9.108 \text{ rad/s}$$

$$v_P = \omega r = (9.108 \text{ rad/s})(0.2 \text{ m}) = 1.82 \text{ m/s} \quad \textit{Ans.}$$

$$(a_P)_t = \alpha r = (0.5\theta \text{ rad/s}^2)(0.2 \text{ m})\big|_{\theta=4\pi\text{ rad}}$$

$$= 1.257 \text{ m/s}^2$$

$$(a_P)_n = \omega^2 r = (9.108 \text{ rad/s})^2(0.2 \text{ m}) = 16.59 \text{ m/s}^2$$

$$a_P = \sqrt{(a_P)_t^2 + (a_P)_n^2}$$

$$= \sqrt{(1.257 \text{ m/s}^2)^2 + (16.59 \text{ m/s}^2)^2}$$

$$= 16.6 \text{ m/s}^2 \qquad \textit{Ans.}$$

F16–6. $\alpha_B = \alpha_A \left(\dfrac{r_A}{r_B} \right)$

$\quad = (4.5 \text{ rad/s}^2)\left(\dfrac{0.075 \text{ m}}{0.225 \text{ m}} \right) = 1.5 \text{ rad/s}^2$

$\omega_B = (\omega_B)_0 + \alpha_B t$

$\omega_B = 0 + (1.5 \text{ rad/s}^2)(3 \text{ s}) = 4.5 \text{ rad/s}$

$\theta_B = (\theta_B)_0 + (\omega_B)_0 t + \frac{1}{2}\alpha_B t^2$

$\theta_B = 0 + 0 + \frac{1}{2}(1.5 \text{ rad/s}^2)(3 \text{ s})^2$

$\theta_B = 6.75 \text{ rad}$

$v_C = \omega_B r_D = (4.5 \text{ rad/s})(0.125 \text{ m})$

$\quad = 0.5625 \text{ m/s}$ *Ans.*

$s_C = \theta_B r_D = (6.75 \text{ rad})(0.125 \text{ m}) = 0.84375 \text{ m}$

$\quad = 844 \text{ mm}$ *Ans.*

F16–7. Vector Analysis

$\mathbf{v}_B = \mathbf{v}_A + \boldsymbol{\omega} \times \mathbf{r}_{B/A}$

$-v_B \mathbf{j} = (3\mathbf{i}) \text{ m/s}$

$\quad + (\omega \mathbf{k}) \times (-1.5 \cos 30°\mathbf{i} + 1.5 \sin 30°\mathbf{j})$

$-v_B \mathbf{j} = [3 - \omega_{AB}(1.5 \sin 30°)]\mathbf{i} - \omega(1.5 \cos 30°)\mathbf{j}$

$\quad 0 = 3 - \omega(1.5 \sin 30°)$ (1)

$-v_B = 0 - \omega(1.5 \cos 30°)$ (2)

$\omega = 4 \text{ rad/s} \qquad v_B = 5.20 \text{ m/s}$ *Ans.*

Scalar Solution

$\mathbf{v}_B = \mathbf{v}_A + \mathbf{v}_{B/A}$

$\begin{bmatrix} \downarrow v_B \end{bmatrix} = \begin{bmatrix} 3 \\ \rightarrow \end{bmatrix} + \begin{bmatrix} \omega(1.5) \not\!\angle 30° \end{bmatrix}$

This yields Eqs. (1) and (2).

F16–8. Vector Analysis

$\mathbf{v}_B = \mathbf{v}_A + \boldsymbol{\omega} \times \mathbf{r}_{B/A}$

$(v_B)_x \mathbf{i} + (v_B)_y \mathbf{j} = 0 + (-10\mathbf{k}) \times (-0.6\mathbf{i} + 0.6\mathbf{j})$

$(v_B)_x \mathbf{i} + (v_B)_y \mathbf{j} = 6\mathbf{i} + 6\mathbf{j}$

$(v_B)_x = 6 \text{ m/s} \text{ and } (v_B)_y = 6 \text{ m/s}$

$v_B = \sqrt{(v_B)_x^2 + (v_B)_y^2}$

$\quad = \sqrt{(6 \text{ m/s})^2 + (6 \text{ m/s})^2}$

$\quad = 8.49 \text{ m/s}$ *Ans.*

Scalar Solution

$\mathbf{v}_B = \mathbf{v}_A + \mathbf{v}_{B/A}$

$\begin{bmatrix} (v_B)_x \\ \rightarrow \end{bmatrix} + \begin{bmatrix} (v_B)_y \uparrow \end{bmatrix} = \begin{bmatrix} 0 \end{bmatrix} + \left[\angle 45°\; 10\left(\dfrac{0.6}{\cos 45°} \right) \right]$

$\xrightarrow{+} (v_B)_x = 0 + 10(0.6/\cos 45°) \cos 45° = 6 \text{ m/s} \rightarrow$

$+\uparrow (v_B)_y = 0 + 10(0.6/\cos 45°) \sin 45° = 6 \text{ m/s} \uparrow$

F16–9. Vector Analysis

$\mathbf{v}_B = \mathbf{v}_A + \boldsymbol{\omega} \times \mathbf{r}_{B/A}$

$(4 \text{ ft/s})\mathbf{i} = (-2 \text{ ft/s})\mathbf{i} + (-\omega \mathbf{k}) \times (3 \text{ ft})\mathbf{j}$

$4\mathbf{i} = (-2 + 3\omega)\mathbf{i}$

$\omega = 2 \text{ rad/s}$ *Ans.*

Scalar Solution

$\mathbf{v}_B = \mathbf{v}_A + \mathbf{v}_{B/A}$

$\begin{bmatrix} 4 \\ \rightarrow \end{bmatrix} = \begin{bmatrix} 2 \\ \leftarrow \end{bmatrix} + \begin{bmatrix} \omega(3) \\ \rightarrow \end{bmatrix}$

$\xrightarrow{+}\quad 4 = -2 + \omega(3); \qquad \omega = 2 \text{ rad/s}$

F16–10. Vector Analysis

$\mathbf{v}_A = \boldsymbol{\omega}_{OA} \times \mathbf{r}_A$

$\quad = (12 \text{ rad/s})\mathbf{k} \times (0.3 \text{ m})\mathbf{j}$

$\quad = [-3.6\mathbf{i}] \text{ m/s}$

$\mathbf{v}_B = \mathbf{v}_A + \boldsymbol{\omega}_{AB} \times \mathbf{r}_{B/A}$

$v_B \mathbf{j} = (-3.6 \text{ m/s})\mathbf{i}$

$\quad + (\omega_{AB} \mathbf{k}) \times (0.6 \cos 30°\mathbf{i} - 0.6 \sin 30°\mathbf{j}) \text{ m}$

$v_B \mathbf{j} = [\omega_{AB}(0.6 \sin 30°) - 3.6]\mathbf{i} + \omega_{AB}(0.6 \cos 30°)\mathbf{j}$

$\quad 0 = \omega_{AB}(0.6 \sin 30°) - 3.6$ (1)

$v_B = \omega_{AB}(0.6 \cos 30°)$ (2)

$\omega_{AB} = 12 \text{ rad/s} \quad v_B = 6.24 \text{ m/s} \uparrow$ *Ans.*

Scalar Solution

$\mathbf{v}_B = \mathbf{v}_A + \mathbf{v}_{B/A}$

$\begin{bmatrix} v_B \uparrow \end{bmatrix} = \begin{bmatrix} \leftarrow \\ 12(0.3) \end{bmatrix} + \begin{bmatrix} \not\!\nearrow 30° \omega(0.6) \end{bmatrix}$

This yields Eqs. (1) and (2).

F16–11. Vector Analysis

$\mathbf{v}_C = \mathbf{v}_B + \boldsymbol{\omega}_{BC} \times \mathbf{r}_{C/B}$

$v_C \mathbf{j} = (-60\mathbf{i}) \text{ ft/s}$

$\quad + (-\omega_{BC}\mathbf{k}) \times (-2.5 \cos 30°\mathbf{i} + 2.5 \sin 30°\mathbf{j}) \text{ ft}$

$v_C \mathbf{j} = (-60)\mathbf{i} + 2.165\omega_{BC}\mathbf{j} + 1.25\omega_{BC}\mathbf{i}$

$\quad 0 = -60 + 1.25\omega_{BC}$ (1)

$v_C = 2.165\,\omega_{BC}$ (2)

$\omega_{BC} = 48 \text{ rad/s}$ *Ans.*

$v_C = 104 \text{ ft/s}$

Scalar Solution

$\mathbf{v}_C = \mathbf{v}_B + \mathbf{v}_{C/B}$

$\begin{bmatrix} v_C \uparrow \end{bmatrix} = \begin{bmatrix} v_B \\ \leftarrow \end{bmatrix} + \begin{bmatrix} \not\!\nearrow 30° \; \omega (2.5) \end{bmatrix}$

This yields Eqs. (1) and (2).

F16–12. Vector Analysis

$$\mathbf{v}_B = \mathbf{v}_A + \boldsymbol{\omega} \times \mathbf{r}_{B/A}$$

$-v_B \cos 30° \, \mathbf{i} + v_B \sin 30° \, \mathbf{j} = (-3 \text{ m/s})\mathbf{j} +$

$(-\omega \mathbf{k}) \times (-2 \sin 45° \mathbf{i} - 2 \cos 45° \mathbf{j}) \text{ m}$

$-0.8660 v_B \, \mathbf{i} + 0.5 v_B \, \mathbf{j}$

$\qquad = -1.4142\omega \mathbf{i} + (1.4142\omega - 3)\mathbf{j}$

$-0.8660 v_B = -1.4142\omega \qquad\qquad (1)$

$0.5 v_B = 1.4142\omega - 3 \qquad\qquad (2)$

$\omega = 5.02 \text{ rad/s} \quad v_B = 8.20 \text{ m/s} \qquad$ *Ans.*

Scalar Solution

$$\mathbf{v}_B = \mathbf{v}_A + \mathbf{v}_{B/A}$$

$$\left[\begin{smallmatrix} \nwarrow 30° \, v_B \end{smallmatrix} \right] = \left[\downarrow 3 \right] + \left[\begin{smallmatrix} \nwarrow 45° \, \omega(2) \end{smallmatrix} \right]$$

This yields Eqs. (1) and (2).

F16–13. $\omega_{AB} = \dfrac{v_A}{r_{A/IC}} = \dfrac{6}{3} = 2 \text{ rad/s} \qquad$ *Ans.*

$\phi = \tan^{-1}\left(\frac{2}{1.5}\right) = 53.13°$

$r_{C/IC} = \sqrt{(3)^2 + (2.5)^2 - 2(3)(2.5) \cos 53.13°} = 2.5 \text{ m}$

$v_C = \omega_{AB} \, r_{C/IC} = 2(2.5) = 5 \text{ m/s} \qquad$ *Ans.*

$\theta = 90° - \phi = 90° - 53.13° = 36.9° \, \nwarrow \qquad$ *Ans.*

F16–14. $v_B = \omega_{AB} \, r_{B/A} = 12(0.6) = 7.2 \text{ m/s} \downarrow$

$v_C = 0 \qquad$ *Ans.*

$\omega_{BC} = \dfrac{v_B}{r_{B/IC}} = \dfrac{7.2}{1.2} = 6 \text{ rad/s} \qquad$ *Ans.*

F16–15. $\omega = \dfrac{v_O}{r_{O/IC}} = \dfrac{6}{0.3} = 20 \text{ rad/s} \qquad$ *Ans.*

$r_{A/IC} = \sqrt{0.3^2 + 0.6^2} = 0.6708 \text{ m}$

$\phi = \tan^{-1}\left(\frac{0.3}{0.6}\right) = 26.57°$

$v_A = \omega r_{A/IC} = 20(0.6708) = 13.4 \text{ m/s} \qquad$ *Ans.*

$\theta = 90° - \phi = 90° - 26.57° = 63.4° \, \measuredangle \qquad$ *Ans.*

F16–16. The location of *IC* can be determined using similar triangles.

$\dfrac{0.5 - r_{C/IC}}{3} = \dfrac{r_{C/IC}}{1.5} \qquad r_{C/IC} = 0.1667 \text{ m}$

$\omega = \dfrac{v_C}{r_{C/IC}} = \dfrac{1.5}{0.1667} = 9 \text{ rad/s} \qquad$ *Ans.*

Also, $r_{O/IC} = 0.3 - r_{C/IC} = 0.3 - 0.1667$

$\qquad = 0.1333 \text{ m}.$

$v_O = \omega r_{O/IC} = 9(0.1333) = 1.20 \text{ m/s} \qquad$ *Ans.*

F16–17. $v_B = \omega r_{B/A} = 6(0.2) = 1.2 \text{ m/s}$

$r_{B/IC} = 0.8 \tan 60° = 1.3856 \text{ m}$

$r_{C/IC} = \dfrac{0.8}{\cos 60°} = 1.6 \text{ m}$

$\omega_{BC} = \dfrac{v_B}{r_{B/IC}} = \dfrac{1.2}{1.3856} = 0.8660 \text{ rad/s}$

$\qquad = 0.866 \text{ rad/s} \qquad$ *Ans.*

Then,

$v_C = \omega_{BC} \, r_{C/IC} = 0.8660(1.6) = 1.39 \text{ m/s} \qquad$ *Ans.*

F16–18. $v_B = \omega_{AB} \, r_{B/A} = 10(0.2) = 2 \text{ m/s}$

$v_C = \omega_{CD} \, r_{C/D} = \omega_{CD} (0.2) \rightarrow$

$r_{B/IC} = \dfrac{0.4}{\cos 30°} = 0.4619 \text{ m}$

$r_{C/IC} = 0.4 \tan 30° = 0.2309 \text{ m}$

$\omega_{BC} = \dfrac{v_B}{r_{B/IC}} = \dfrac{2}{0.4619} = 4.330 \text{ rad/s}$

$\qquad = 4.33 \text{ rad/s} \qquad$ *Ans.*

$v_C = \omega_{BC} \, r_{C/IC}$

$\omega_{CD} (0.2) = 4.330(0.2309)$

$\omega_{CD} = 5 \text{ rad/s} \qquad$ *Ans.*

F16–19. $\omega = \dfrac{v_A}{r_{A/IC}} = \dfrac{6}{3} = 2 \text{ rad/s}$

Vector Analysis

$$\mathbf{a}_B = \mathbf{a}_A + \boldsymbol{\alpha} \times \mathbf{r}_{B/A} - \omega^2 \mathbf{r}_{B/A}$$

$a_B \mathbf{i} = -5\mathbf{j} + (\alpha \mathbf{k}) \times (3\mathbf{i} - 4\mathbf{j}) - 2^2(3\mathbf{i} - 4\mathbf{j})$

$a_B \mathbf{i} = (4\alpha - 12)\mathbf{i} + (3\alpha + 11)\mathbf{j}$

$a_B = 4\alpha - 12 \qquad\qquad (1)$

$0 = 3\alpha + 11 \qquad\qquad (2)$

$\alpha = -3.67 \text{ rad/s}^2 \qquad$ *Ans.*

$a_B = -26.7 \text{ m/s}^2 \qquad$ *Ans.*

Scalar Solution

$$\mathbf{a}_B = \mathbf{a}_A + \mathbf{a}_{B/A}$$

$$\left[\begin{smallmatrix} a_B \\ \rightarrow \end{smallmatrix} \right] = \left[\downarrow 5 \right] + \left[\alpha \, (5) \, \begin{smallmatrix} 5 \\ 4 \end{smallmatrix} \nearrow 3 \right] + \left[4 \begin{smallmatrix} 5 \\ 3 \end{smallmatrix} \nwarrow (2)^2(5) \right]$$

This yields Eqs. (1) and (2).

F16–20. Vector Analysis

$$\mathbf{a}_A = \mathbf{a}_O + \boldsymbol{\alpha} \times \mathbf{r}_{A/O} - \omega^2\mathbf{r}_{A/O}$$
$$= 1.8\mathbf{i} + (-6\mathbf{k}) \times (0.3\mathbf{j}) - 12^2(0.3\mathbf{j})$$
$$= \{3.6\mathbf{i} - 43.2\mathbf{j}\}\,\text{m/s}^2 \qquad\qquad Ans.$$

Scalar Analysis

$$\mathbf{a}_A = \mathbf{a}_O + \mathbf{a}_{A/O}$$

$$\begin{bmatrix} (a_A)_x \\ \rightarrow \end{bmatrix} + \begin{bmatrix} (a_A)_y \uparrow \end{bmatrix} = \begin{bmatrix} (6)(0.3) \\ \rightarrow \end{bmatrix} + \begin{bmatrix} (6)(0.3) \\ \rightarrow \end{bmatrix}$$
$$+ \begin{bmatrix} \downarrow (12)^2(0.3) \end{bmatrix}$$

$$\xrightarrow{+}\quad (a_A)_x = 1.8 + 1.8 = 3.6\,\text{m/s}^2 \;\rightarrow$$
$$+\uparrow\quad (a_A)_y = -43.2\,\text{m/s}^2$$

F16–21. Using

$$v_O = \omega r; \qquad 6 = \omega(0.3)$$
$$\omega = 20\,\text{rad/s}$$
$$a_O = \alpha r; \qquad 3 = \alpha(0.3)$$
$$\alpha = 10\,\text{rad/s}^2 \qquad\qquad Ans.$$

Vector Analysis

$$\mathbf{a}_A = \mathbf{a}_O + \boldsymbol{\alpha} \times \mathbf{r}_{A/O} - \omega^2\mathbf{r}_{A/O}$$
$$= 3\mathbf{i} + (-10\mathbf{k}) \times (-0.6\mathbf{i}) - 20^2(-0.6\mathbf{i})$$
$$= \{243\mathbf{i} + 6\mathbf{j}\}\,\text{m/s}^2 \qquad\qquad Ans.$$

Scalar Analysis

$$\mathbf{a}_A = \mathbf{a}_O + \mathbf{a}_{A/O}$$

$$\begin{bmatrix} (a_A)_x \\ \rightarrow \end{bmatrix} + \begin{bmatrix} (a_A)_y \\ \uparrow \end{bmatrix} = \begin{bmatrix} 3 \\ \rightarrow \end{bmatrix} + \begin{bmatrix} 10(0.6) \\ \uparrow \end{bmatrix} + \begin{bmatrix} (20)^2(0.6) \\ \rightarrow \end{bmatrix}$$

$$\xrightarrow{+}\quad (a_A)_x = 3 + 240 = 243\,\text{m/s}^2$$
$$+\uparrow\quad (a_A)_y = 10(0.6) = 6\,\text{m/s}^2 \;\uparrow$$

F16–22. $\dfrac{r_{A/IC}}{3} = \dfrac{0.5 - r_{A/IC}}{1.5}; \qquad r_{A/IC} = 0.3333\,\text{m}$

$$\omega = \dfrac{v_A}{r_{A/IC}} = \dfrac{3}{0.3333} = 9\,\text{rad/s}$$

Vector Analysis

$$\mathbf{a}_A = \mathbf{a}_C + \boldsymbol{\alpha} \times \mathbf{r}_{A/C} - \omega^2\,\mathbf{r}_{A/C}$$
$$1.5\mathbf{i} - (a_A)_n\,\mathbf{j} = -0.75\mathbf{i} + (a_C)_n\,\mathbf{j}$$
$$+ (-\alpha\mathbf{k}) \times 0.5\mathbf{j} - 9^2\,(0.5\mathbf{j})$$
$$1.5\mathbf{i} - (a_A)_n\,\mathbf{j} = (0.5\alpha - 0.75)\mathbf{i} + \big[(a_C)_n - 40.5\big]\mathbf{j}$$
$$1.5 = 0.5\alpha - 0.75$$
$$\alpha = 4.5\,\text{rad/s}^2 \qquad\qquad Ans.$$

Scalar Analysis

$$\mathbf{a}_A = \mathbf{a}_C + \mathbf{a}_{A/C}$$

$$\begin{bmatrix} 1.5 \\ \rightarrow \end{bmatrix} + \begin{bmatrix} (a_A)_n \\ \downarrow \end{bmatrix} = \begin{bmatrix} 0.75 \\ \leftarrow \end{bmatrix} + \begin{bmatrix} (a_C)_n \\ \uparrow \end{bmatrix} + \begin{bmatrix} \alpha(0.5) \\ \rightarrow \end{bmatrix}$$
$$+ \begin{bmatrix} (9)^2(0.5) \\ \downarrow \end{bmatrix}$$

$$\xrightarrow{+}\quad 1.5 = -0.75 + \alpha(0.5)$$
$$\alpha = 4.5\,\text{rad/s}^2$$

F16–23. $v_B = \omega r_{B/A} = 12(0.3) = 3.6\,\text{m/s}$

$$\omega_{BC} = \dfrac{v_B}{r_{B/IC}} = \dfrac{3.6}{1.2} = 3\,\text{rad/s}$$

Vector Analysis

$$\mathbf{a}_B = \boldsymbol{\alpha} \times \mathbf{r}_{B/A} - \omega^2\mathbf{r}_{B/A}$$
$$= (-6\mathbf{k}) \times (0.3\mathbf{i}) - 12^2(0.3\mathbf{i})$$
$$= \{-43.2\mathbf{i} - 1.8\mathbf{j}\}\,\text{m/s}^2$$
$$\mathbf{a}_C = \mathbf{a}_B + \boldsymbol{\alpha}_{BC} \times \mathbf{r}_{C/B} - \omega_{BC}^2\mathbf{r}_{C/B}$$
$$a_C\,\mathbf{i} = (-43.2\mathbf{i} - 1.8\mathbf{j})$$
$$+ (\alpha_{BC}\,\mathbf{k}) \times (1.2\mathbf{i}) - 3^2(1.2\mathbf{i})$$
$$a_C\,\mathbf{i} = -54\mathbf{i} + (1.2\alpha_{BC} - 1.8)\mathbf{j}$$
$$a_C = -54\,\text{m/s}^2 = 54\,\text{m/s}^2 \;\leftarrow \qquad Ans.$$
$$0 = 1.2\alpha_{BC} - 1.8 \quad \alpha_{BC} = 1.5\,\text{rad/s}^2 \qquad Ans.$$

Scalar Analysis

$$\mathbf{a}_C = \mathbf{a}_B + \mathbf{a}_{C/B}$$

$$\begin{bmatrix} a_C \\ \leftarrow \end{bmatrix} = \begin{bmatrix} 6(0.3) \\ \downarrow \end{bmatrix} + \begin{bmatrix} (12)^2(0.3) \\ \leftarrow \end{bmatrix} + \begin{bmatrix} \alpha_{BC}(1.2) \\ \uparrow \end{bmatrix} + \begin{bmatrix} (3)^2(1.2) \\ \leftarrow \end{bmatrix}$$

$$\xleftarrow{+}\quad a_C = 43.2 + 10.8 = 54\,\text{m/s}^2 \;\leftarrow$$
$$+\uparrow\quad 0 = -6(0.3) + 1.2\alpha_{BC}$$
$$\alpha_{BC} = 1.5\,\text{rad/s}^2$$

F16–24. $v_B = \omega\,r_{B/A} = 6(0.2) = 1.2\,\text{m/s} \;\rightarrow$

$$r_{B/IC} = 0.8\tan 60° = 1.3856\,\text{m}$$

$$\omega_{BC} = \dfrac{v_B}{r_{B/IC}} = \dfrac{1.2}{1.3856} = 0.8660\,\text{rad/s}$$

Vector Analysis

$$\mathbf{a}_B = \boldsymbol{\alpha} \times \mathbf{r}_{B/A} - \omega^2\mathbf{r}_{B/A}$$
$$= (-3\mathbf{k}) \times (0.2\mathbf{j}) - 6^2(0.2\mathbf{j})$$
$$= [0.6\mathbf{i} - 7.2\mathbf{j}]\,\text{m/s}$$
$$\mathbf{a}_C = \mathbf{a}_B + \boldsymbol{\alpha}_{BC} \times \mathbf{r}_{C/B} - \omega^2\mathbf{r}_{C/B}$$
$$a_C\cos 30°\mathbf{i} + a_C\sin 30°\mathbf{j}$$
$$= (0.6\mathbf{i} - 7.2\mathbf{j}) + (\alpha_{BC}\,\mathbf{k} \times 0.8\mathbf{i}) - 0.8660^2(0.8\mathbf{i})$$

$0.8660a_C \, \mathbf{i} + 0.5a_C \, \mathbf{j} = (0.8\alpha_{BC} - 7.2)\mathbf{j}$

$0.8660a_C = 0$ (1)

$0.5a_C = 0.8\alpha_{BC} - 7.2$ (2)

$a_C = 0$ $\alpha_{BC} = 9 \text{ rad/s}^2$ *Ans.*

Scalar Analysis

$\mathbf{a}_C = \mathbf{a}_B + \mathbf{a}_{C/B}$

$$\begin{bmatrix} a_C \\ \measuredangle 30° \end{bmatrix} = \begin{bmatrix} 3(0.2) \\ \rightarrow \end{bmatrix} + \begin{bmatrix} (6)^2(0.2) \\ \downarrow \end{bmatrix} + \begin{bmatrix} \alpha_{BC}(0.8) \\ \uparrow \end{bmatrix}$$

$$+ \begin{bmatrix} (0.8660)^2(0.8) \\ \leftarrow \end{bmatrix}$$

This yields Eqs. (1) and (2).

Chapter 17

F17–1. $\xrightarrow{+} \Sigma F_x = m(a_G)_x; \quad 100\left(\frac{4}{5}\right) = 100a$

$a = 0.8 \text{ m/s}^2 \rightarrow$ *Ans.*

$+\uparrow \Sigma F_y = m(a_G)_y;$

$N_A + N_B - 100\left(\frac{3}{5}\right) - 100(9.81) = 0$ (1)

$\zeta + \Sigma M_G = 0;$

$N_A(0.6) + 100\left(\frac{3}{5}\right)(0.7)$

$\qquad - N_B(0.4) - 100\left(\frac{4}{5}\right)(0.7) = 0$ (2)

$N_A = 430.4 \text{ N} = 430 \text{ N}$ *Ans.*

$N_B = 610.6 \text{ N} = 611 \text{ N}$ *Ans.*

F17–2. $\Sigma F_{x'} = m(a_G)_{x'}; \quad 80(9.81) \sin 15° = 80a$

$a = 2.54 \text{ m/s}^2$ *Ans.*

$\Sigma F_{y'} = m(a_G)_{y'};$

$N_A + N_B - 80(9.81) \cos 15° = 0$ (1)

$\zeta + \Sigma M_G = 0;$

$N_A(0.5) - N_B(0.5) = 0$ (2)

$N_A = N_B = 379 \text{ N}$ *Ans.*

F17–3. $\zeta + \Sigma M_A = \Sigma(M_k)_A; \quad 10\left(\frac{3}{5}\right)(7) = \frac{20}{32.2}a(3.5)$

$a = 19.3 \text{ ft/s}^2$ *Ans.*

$\xrightarrow{+} \Sigma F_x = m(a_G)_x; \quad A_x + 10\left(\frac{3}{5}\right) = \frac{20}{32.2}(19.32)$

$A_x = 6 \text{ lb}$ *Ans.*

$+\uparrow \Sigma F_y = m(a_G)_y; \quad A_y - 20 + 10\left(\frac{4}{5}\right) = 0$

$A_y = 12 \text{ lb}$ *Ans.*

F17–4. $F_A = \mu_s N_A = 0.2N_A \quad F_B = \mu_s N_B = 0.2N_B$

$\xrightarrow{+} \Sigma F_x = m(a_G)_x;$

$0.2N_A + 0.2N_B = 100a$ (1)

$+\uparrow \Sigma F_y = m(a_G)_y;$

$N_A + N_B - 100(9.81) = 0$ (2)

$\zeta + \Sigma M_G = 0;$

$0.2N_A(0.75) + N_A(0.9) + 0.2N_B(0.75)$

$\qquad - N_B(0.6) = 0$ (3)

Solving Eqs. (1), (2), and (3),

$N_A = 294.3 \text{ N} = 294 \text{ N}$

$N_B = 686.7 \text{ N} = 687 \text{ N}$

$a = 1.96 \text{ m/s}^2$ *Ans.*

Since N_A is positive, the table will indeed slide before it tips.

F17–5. $(a_G)_t = \alpha r = \alpha(1.5 \text{ m})$

$(a_G)_n = \omega^2 r = (5 \text{ rad/s})^2(1.5 \text{ m}) = 37.5 \text{ m/s}^2$

$\Sigma F_t = m(a_G)_t; \quad 100 \text{ N} = 50 \text{ kg}[\alpha(1.5 \text{ m})]$

$\qquad\qquad\qquad \alpha = 1.33 \text{ rad/s}^2$ *Ans.*

$\Sigma F_n = m(a_G)_n; \quad T_{AB} + T_{CD} - 50(9.81) \text{ N}$

$\qquad\qquad\qquad = 50 \text{ kg}(37.5 \text{ m/s}^2)$

$T_{AB} + T_{CD} = 2365.5$

$\zeta + \Sigma M_G = 0; \quad T_{CD}(1 \text{ m}) - T_{AB}(1 \text{ m}) = 0$

$T_{AB} = T_{CD} = 1182.75 \text{ N} = 1.18 \text{ kN}$ *Ans.*

F17–6. $\zeta + \Sigma M_C = 0;$

$\mathbf{a}_G = \mathbf{a}_D = \mathbf{a}_B$

$D_y(0.6) - 450 = 0 \quad D_y = 750 \text{ N}$ *Ans.*

$(a_G)_n = \omega^2 r = 6^2(0.6) = 21.6 \text{ m/s}^2$

$(a_G)_t = \alpha r = \alpha(0.6)$

$+\uparrow \Sigma F_t = m(a_G)_t;$

$750 - 50(9.81) = 50[\alpha(0.6)]$

$\alpha = 8.65 \text{ rad/s}^2$ *Ans.*

$\xrightarrow{+} \Sigma F_n = m(a_G)_n;$

$F_{AB} + D_x = 50(21.6)$ (1)

$\zeta + \Sigma M_G = 0;$

$D_x(0.4) + 750(0.1) - F_{AB}(0.4) = 0$ (2)

$D_x = 446.25 \text{ N} = 446 \text{ N}$ *Ans.*

$F_{AB} = 633.75 \text{ N} = 634 \text{ N}$ *Ans.*

F17–7. $I_O = mk_O^2 = 100(0.5^2) = 25 \text{ kg} \cdot \text{m}^2$

$\zeta + \Sigma M_O = I_O \alpha; \quad -100(0.6) = -25\alpha$

$\alpha = 2.4 \text{ rad/s}^2$

$\omega = \omega_0 + \alpha_c t$

$\omega = 0 + 2.4(3) = 7.2 \text{ rad/s}$ *Ans.*

F17–8. $I_O = \frac{1}{2} mr^2 = \frac{1}{2}(50)(0.3^2) = 2.25 \text{ kg} \cdot \text{m}^2$

$\zeta + \Sigma M_O = I_O \alpha;$

$\quad\quad -9t = -2.25\alpha \quad\quad \alpha = (4t) \text{ rad/s}^2$

$d\omega = \alpha \, dt$

$\int_0^\omega d\omega = \int_0^t 4t \, dt$

$\omega = (2t^2) \text{ rad/s}$

$\omega = 2(4^2) = 32 \text{ rad/s}$ *Ans.*

F17–9. $(a_G)_t = \alpha r_G = \alpha(0.15)$

$(a_G)_n = \omega^2 r_G = 6^2(0.15) = 5.4 \text{ m/s}^2$

$I_O = I_G + md^2 = \frac{1}{12}(30)(0.9^2) + 30(0.15^2)$

$\quad = 2.7 \text{ kg} \cdot \text{m}^2$

$\zeta + \Sigma M_O = I_O \alpha; \quad 60 - 30(9.81)(0.15) = 2.7\alpha$

$\alpha = 5.872 \text{ rad/s}^2 = 5.87 \text{ rad/s}^2$ *Ans.*

$\pm \Sigma F_n = m(a_G)_n; \quad O_n = 30(5.4) = 162 \text{ N}$ *Ans.*

$+\uparrow \Sigma F_t = m(a_G)_t;$

$O_t - 30(9.81) = 30[5.872(0.15)]$

$O_t = 320.725 \text{ N} = 321 \text{ N}$ *Ans.*

F17–10. $(a_G)_t = \alpha r_G = \alpha(0.3)$

$(a_G)_n = \omega^2 r_G = 10^2(0.3) = 30 \text{ m/s}^2$

$I_O = I_G + md^2 = \frac{1}{2}(30)(0.3^2) + 30(0.3^2)$

$\quad = 4.05 \text{ kg} \cdot \text{m}^2$

$\zeta + \Sigma M_O = I_O \alpha;$

$50(\frac{3}{5})(0.3) + 50(\frac{4}{5})(0.3) = 4.05\alpha$

$\alpha = 5.185 \text{ rad/s}^2 = 5.19 \text{ rad/s}^2$ *Ans.*

$+\uparrow \Sigma F_n = m(a_G)_n;$

$O_n + 50(\frac{3}{5}) - 30(9.81) = 30(30)$

$O_n = 1164.3 \text{ N} = 1.16 \text{kN}$ *Ans.*

$\pm \Sigma F_t = m(a_G)_t;$

$O_t + 50(\frac{4}{5}) = 30[5.185(0.3)]$

$O_t = 6.67 \text{ N}$ *Ans.*

F17–11. $I_G = \frac{1}{12} ml^2 = \frac{1}{12}(15 \text{ kg})(0.9 \text{ m})^2 = 1.0125 \text{ kg} \cdot \text{m}^2$

$(a_G)_n = \omega^2 r_G = 0$

$(a_G)_t = \alpha(0.15 \text{ m})$

$I_O = I_G + md_{OG}^2$

$\quad = 1.0125 \text{ kg} \cdot \text{m}^2 + 15 \text{ kg}(0.15 \text{ m})^2$

$\quad = 1.35 \text{ kg} \cdot \text{m}^2$

$\zeta + \Sigma M_O = I_O \alpha;$

$[15(9.81) \text{ N}](0.15 \text{ m}) = (1.35 \text{ kg} \cdot \text{m}^2)\alpha$

$\alpha = 16.35 \text{ rad/s}^2$ *Ans.*

$+\downarrow \Sigma F_t = m(a_G)_t; \quad -O_t + 15(9.81)\text{N}$

$\quad = (15 \text{ kg})[16.35 \text{ rad/s}^2(0.15 \text{ m})]$

$O_t = 110.36 \text{ N} = 110 \text{ N}$ *Ans.*

$\pm \Sigma F_n = m(a_G)_n; \quad\quad O_n = 0$ *Ans.*

F17–12. $(a_G)_t = \alpha r_G = \alpha(0.45)$

$(a_G)_n = \omega^2 r_G = 6^2(0.45) = 16.2 \text{ m/s}^2$

$I_O = \frac{1}{3} ml^2 = \frac{1}{3}(30)(0.9^2) = 8.1 \text{ kg} \cdot \text{m}^2$

$\zeta + \Sigma M_O = I_O \alpha;$

$300(\frac{4}{5})(0.6) - 30(9.81)(0.45) = 8.1\alpha$

$\alpha = 1.428 \text{ rad/s}^2 = 1.43 \text{ rad/s}^2$ *Ans.*

$\pm \Sigma F_n = m(a_G)_n; \quad O_n + 300(\frac{3}{5}) = 30(16.2)$

$O_n = 306 \text{ N}$ *Ans.*

$+\uparrow \Sigma F_t = m(a_G)_t; \quad O_t + 300(\frac{4}{5}) - 30(9.81)$

$\quad\quad\quad = 30[1.428(0.45)]$

$O_t = 73.58 \text{ N} = 73.6 \text{ N}$ *Ans.*

F17–13. $I_G = \frac{1}{12} ml^2 = \frac{1}{12}(60)(3^2) = 45 \text{ kg} \cdot \text{m}^2$

$+\uparrow \Sigma F_y = m(a_G)_y;$

$80 - 20 = 60 a_G \quad a_G = 1 \text{ m/s}^2 \uparrow$

$\zeta + \Sigma M_G = I_G \alpha; \quad 80(1) + 20(0.75) = 45\alpha$

$\alpha = 2.11 \text{ rad/s}^2$ *Ans.*

F17–14. $\zeta + \Sigma M_A = (M_k)_A;$

$-200(0.3) = -100 a_G(0.3) - 4.5\alpha$

$30 a_G + 4.5\alpha = 60 \quad\quad (1)$

$a_G = \alpha r = \alpha(0.3) \quad\quad (2)$

$\alpha = 4.44 \text{ rad/s}^2 \quad a_G = 1.33 \text{ m/s}^2 \rightarrow$ *Ans.*

F17–15. $+\uparrow \Sigma F_y = m(a_G)_y;$

$N - 20(9.81) = 0 \quad N = 196.2 \text{ N}$

$\pm \Sigma F_x = m(a_G)_x; \quad 0.5(196.2) = 20 a_O$

$a_O = 4.905 \text{ m/s}^2 \rightarrow$ *Ans.*

$\zeta + \Sigma M_O = I_O \alpha;$

$0.5(196.2)(0.4) - 100 = -1.8\alpha$

$\alpha = 33.8 \text{ rad/s}^2$ *Ans.*

F17–16. Sphere $I_G = \frac{2}{5}(20)(0.15)^2 = 0.18 \text{ kg} \cdot \text{m}^2$

$\zeta + \Sigma M_{IC} = (\mathcal{M}_k)_{IC};$

$20(9.81)\sin 30°(0.15) = 0.18\alpha + (20a_G)(0.15)$

$0.18\alpha + 3a_G = 14.715$

$a_G = \alpha r = \alpha(0.15)$

$\alpha = 23.36 \text{ rad/s}^2 = 23.4 \text{ rad/s}^2$ *Ans.*

$a_G = 3.504 \text{ m/s}^2 = 3.50 \text{ m/s}^2$ *Ans.*

F17–17. $+\uparrow \Sigma F_y = m(a_G)_y;$

$N - 200(9.81) = 0 \quad N = 1962 \text{ N}$

$\xrightarrow{+} \Sigma F_x = m(a_G)_x;$

$T - 0.2(1962) = 200a_G$ (1)

$\zeta + \Sigma M_A = (\mathcal{M}_k)_A; \quad 450 - 0.2(1962)(1)$

$\qquad = 18\alpha + 200a_G(0.4)$ (2)

$(a_A)_t = 0 \quad a_A = (a_A)_n$

$\mathbf{a}_G = \mathbf{a}_A + \boldsymbol{\alpha} \times \mathbf{r}_{G/A} - \omega^2 \mathbf{r}_{G/A}$

$a_G \mathbf{i} = -a_A \mathbf{j} + \alpha \mathbf{k} \times (-0.4\mathbf{j}) - \omega^2(-0.4\mathbf{j})$

$a_G \mathbf{i} = 0.4\alpha \mathbf{i} + (0.4\omega^2 - a_A)\mathbf{j}$

$a_G = 0.4\alpha$ (3)

Solving Eqs. (1), (2), and (3),

$\alpha = 1.15 \text{ rad/s}^2 \quad a_G = 0.461 \text{ m/s}^2$

$T = 485 \text{ N}$ *Ans.*

F17–18. $\xrightarrow{+} \Sigma F_x = m(a_G)_x; \quad 0 = 12(a_G)_x \quad (a_G)_x = 0$

$\zeta + \Sigma M_A = (\mathcal{M}_k)_A$

$-12(9.81)(0.3) = 12(a_G)_y(0.3) - \frac{1}{12}(12)(0.6)^2 \alpha$

$0.36\alpha - 3.6(a_G)_y = 35.316$ (1)

$\omega = 0$

$\mathbf{a}_G = \mathbf{a}_A + \boldsymbol{\alpha} \times \mathbf{r}_{G/A} - \omega^2 \mathbf{r}_{G/A}$

$(a_G)_y \mathbf{j} = a_A \mathbf{i} + (-\alpha \mathbf{k}) \times (0.3\mathbf{i}) - \mathbf{0}$

$(a_G)_y \mathbf{j} = (a_A)\mathbf{i} - 0.3 \mathbf{j}$

$a_A = 0$ *Ans.*

$(a_G)_y = -0.3\alpha$ (2)

Solving Eqs. (1) and (2)

$\alpha = 24.5 \text{ rad/s}^2$

$(a_G)_y = -7.36 \text{ m/s}^2 = 7.36 \text{ m/s}^2\downarrow$ *Ans.*

Chapter 18

F18–1. $I_O = mk_O^2 = 80(0.4^2) = 12.8 \text{ kg} \cdot \text{m}^2$

$T_1 = 0$

$T_2 = \frac{1}{2}I_O\omega^2 = \frac{1}{2}(12.8)\omega^2 = 6.4\omega^2$

$s = \theta r = 20(2\pi)(0.6) = 24\pi \text{ m}$

$T_1 + \Sigma U_{1-2} = T_2$

$0 + 50(24\pi) = 6.4\omega^2$

$\omega = 24.3 \text{ rad/s}$ *Ans.*

F18–2. $T_1 = 0$

$T_2 = \frac{1}{2}m(v_G)_2^2 + \frac{1}{2}I_G\omega^2$

$\quad = \frac{1}{2}\left(\frac{50}{32.2} \text{ slug}\right)(2.5\omega_2)^2$

$\qquad + \frac{1}{2}\left[\frac{1}{12}\left(\frac{50}{32.2} \text{ slug}\right)(5 \text{ ft})^2\right]\omega_2^2$

$T_2 = 6.4700\omega_2^2$

Or,

$I_O = \frac{1}{3}ml^2 = \frac{1}{3}\left(\frac{50}{32.2} \text{ slug}\right)(5 \text{ ft})^2$

$\quad = 12.9400 \text{ slug} \cdot \text{ft}^2$

So that

$T_2 = \frac{1}{2}I_O\omega_2^2 = \frac{1}{2}(12.9400 \text{ slug} \cdot \text{ft}^2)\omega_2^2$

$\quad = 6.4700\omega_2^2$

$T_1 + \Sigma U_{1-2} = T_2$

$T_1 + [-Wy_G + M\theta] = T_2$

$0 + \left[-(50 \text{ lb})(2.5 \text{ ft}) + (100 \text{ lb} \cdot \text{ft})\left(\frac{\pi}{2}\right)\right]$

$\quad = 6.4700\omega_2^2$

$\omega_2 = 2.23 \text{ rad/s}$ *Ans.*

F18–3. $(v_G)_2 = \omega_2 r_{G/IC} = \omega_2(2.5)$

$I_G = \frac{1}{12}ml^2 = \frac{1}{12}(50)(5^2) = 104.17 \text{ kg} \cdot \text{m}^2$

$T_1 = 0$

$T_2 = \frac{1}{2}m(v_G)_2^2 + \frac{1}{2}I_G\omega_2^2$

$\quad = \frac{1}{2}(50)[\omega_2(2.5)]^2 + \frac{1}{2}(104.17)\omega_2^2 = 208.33\omega_2^2$

$U_P = Ps_P = 600(3) = 1800 \text{ J}$

$U_W = -Wh = -50(9.81)(2.5 - 2) = -245.25 \text{ J}$

$T_1 + \Sigma U_{1-2} = T_2$

$0 + 1800 + (-245.25) = 208.33\omega_2^2$

$\omega_2 = 2.732 \text{ rad/s} = 2.73 \text{ rad/s}$ *Ans.*

F18–4. $T = \frac{1}{2}mv_G^2 + \frac{1}{2}I_G\omega^2$

$\quad = \frac{1}{2}(50 \text{ kg})(0.4\omega)^2 + \frac{1}{2}\left[50 \text{ kg}(0.3 \text{ m})^2\right]\omega^2$

$\quad = 6.25\omega^2 \text{ J}$

Or,

$T = \frac{1}{2}I_{IC}\omega^2$

$\quad = \frac{1}{2}\left[50 \text{ kg}(0.3 \text{ m})^2 + 50 \text{ kg}(0.4 \text{ m})^2\right]\omega^2$

$\quad = 6.25\omega^2 \text{ J}$

$s_G = \theta r = 10(2\pi \text{ rad})(0.4 \text{ m}) = 8\pi \text{ m}$

$T_1 + \Sigma U_{1-2} = T_2$

$T_1 + P\cos 30° s_G = T_2$

$0 + (50 \text{ N})\cos 30°(8\pi \text{ m}) = 6.25\omega^2 \text{ J}$

$\omega = 13.2 \text{ rad/s}$ *Ans.*

F18–5. $I_G = \frac{1}{12}ml^2 = \frac{1}{12}(30)(3^2) = 22.5 \text{ kg} \cdot \text{m}^2$

$T_1 = 0$

$T_2 = \frac{1}{2}mv_G^2 + \frac{1}{2}I_G\omega^2$

$\quad = \frac{1}{2}(30)[\omega(0.5)]^2 + \frac{1}{2}(22.5)\omega^2 = 15\omega^2$

Or,

$I_O = I_G + md^2 = \frac{1}{12}(30)(3^2) + 30(0.5^2)$

$\quad = 30 \text{ kg} \cdot \text{m}^2$

$T_2 = \frac{1}{2}I_O\omega^2 = \frac{1}{2}(30)\omega^2 = 15\omega^2$

$s_1 = \theta r_1 = 8\pi(0.5) = 4\pi \text{ m}$

$s_2 = \theta r_2 = 8\pi(1.5) = 12\pi \text{ m}$

$U_{P_1} = P_1 s_1 = 30(4\pi) = 120\pi \text{ J}$

$U_{P_2} = P_2 s_2 = 20(12\pi) = 240\pi \text{ J}$

$U_M = M\theta = 20[4(2\pi)] = 160\pi \text{ J}$

$U_W = (0 \text{ bar returns to same position})$

$T_1 + \Sigma U_{1-2} = T_2$

$0 + 120\pi + 240\pi + 160\pi = 15\omega^2$

$\omega = 10.44 \text{ rad/s} = 10.4 \text{ rad/s}$ *Ans.*

F18–6. $v_G = \omega r = \omega(0.4)$

$I_G = mk_G^2 = 20(0.3^2) = 1.8 \text{ kg} \cdot \text{m}^2$

$T_1 = 0$

$T_2 = \frac{1}{2}mv_G^2 + \frac{1}{2}I_G\omega^2$

$\quad = \frac{1}{2}(20)\left[\omega(0.4)\right]^2 + \frac{1}{2}(1.8)\omega^2$

$\quad = 2.5\omega^2$

$U_M = M\theta = M\left(\frac{s_O}{r}\right) = 50\left(\frac{20}{0.4}\right) = 2500 \text{ J}$

$T_1 + \Sigma U_{1-2} = T_2$

$0 + 2500 = 2.5\omega^2$

$\omega = 31.62 \text{ rad/s} = 31.6 \text{ rad/s}$ *Ans.*

F18–7. $v_G = \omega r = \omega(0.3)$

$I_G = \frac{1}{2}mr^2 = \frac{1}{2}(30)(0.3^2) = 1.35 \text{ kg} \cdot \text{m}^2$

$T_1 = 0$

$T_2 = \frac{1}{2}m(v_G)_2^2 + \frac{1}{2}I_G\omega_2^2$

$\quad = \frac{1}{2}(30)[\omega_2(0.3)]^2 + \frac{1}{2}(1.35)\omega_2^2 = 2.025\omega_2^2$

$(V_g)_1 = Wy_1 = 0$

$(V_g)_2 = -Wy_2 = -30(9.81)(0.3) = -88.29 \text{ J}$

$T_1 + V_1 = T_2 + V_2$

$0 + 0 = 2.025\omega_2^2 + (-88.29)$

$\omega_2 = 6.603 \text{ rad/s} = 6.60 \text{ rad/s}$ *Ans.*

F18–8. $v_O = \omega r_{O/IC} = \omega(0.2)$

$I_O = mk_O^2 = 50(0.3^2) = 4.5 \text{ kg} \cdot \text{m}^2$

$T_1 = 0$

$T_2 = \frac{1}{2}m(v_O)_2^2 + \frac{1}{2}I_O\omega_2^2$

$\quad = \frac{1}{2}(50)\left[\omega_2(0.2)\right]^2 + \frac{1}{2}(4.5)\omega_2^2$

$\quad = 3.25\omega_2^2$

$(V_g)_1 = Wy_1 = 0$

$(V_g)_2 = -Wy_2 = -50(9.81)(6 \sin 30°)$

$\quad = -1471.5\text{J}$

$T_1 + V_1 = T_2 + V_2$

$0 + 0 = 3.25\omega_2^2 + (-1471.5)$

$\omega_2 = 21.28 \text{ rad/s} = 21.3 \text{ rad/s}$ *Ans.*

F18–9. $v_G = \omega r_G = \omega(1.5)$

$I_G = \frac{1}{12}(60)(3^2) = 45 \text{ kg} \cdot \text{m}^2$

$T_1 = 0$

$T_2 = \frac{1}{2}m(v_G)_2^2 + \frac{1}{2}I_G\omega_2^2$

$\quad = \frac{1}{2}(60)[\omega_2(1.5)]^2 + \frac{1}{2}(45)\omega_2^2$

$\quad = 90\omega_2^2$

Or,

$T_2 = \frac{1}{2}I_O\omega_2^2 = \frac{1}{2}\left[45 + 60(1.5^2)\right]\omega_2^2 = 90\omega_2^2$

$(V_g)_1 = Wy_1 = 0$

$(V_g)_2 = -Wy_2 = -60(9.81)(1.5 \sin 45°)$

$\quad = -624.30 \text{ J}$

$(V_e)_1 = \frac{1}{2}ks_1^2 = 0$

$(V_e)_2 = \frac{1}{2}ks_2^2 = \frac{1}{2}(150)(3 \sin 45°)^2 = 337.5 \text{ J}$

$T_1 + V_1 = T_2 + V_2$

$0 + 0 = 90\omega_2^2 + [-624.30 + 337.5]$

$\omega_2 = 1.785 \text{ rad/s} = 1.79 \text{ rad/s}$ *Ans.*

F18–10. $v_G = \omega r_G = \omega(0.75)$

$I_G = \frac{1}{12}(30)(1.5^2) = 5.625 \text{ kg} \cdot \text{m}^2$

$T_1 = 0$

$$T_2 = \tfrac{1}{2}m(v_G)_2^2 + \tfrac{1}{2}I_G\omega_2^2$$
$$= \tfrac{1}{2}(30)[\omega(0.75)]^2 + \tfrac{1}{2}(5.625)\omega_2^2 = 11.25\omega_2^2$$

Or,
$$T_2 = \tfrac{1}{2}I_O\omega_2^2 = \tfrac{1}{2}\left[5.625 + 30\left(0.75^2\right)\right]\omega_2^2$$
$$= 11.25\omega_2^2$$
$$(V_g)_1 = Wy_1 = 0$$
$$(V_g)_2 = -Wy_2 = -30(9.81)(0.75)$$
$$= -220.725 \text{ J}$$
$$(V_e)_1 = \tfrac{1}{2}ks_1^2 = 0$$
$$(V_e)_2 = \tfrac{1}{2}ks_2^2 = \tfrac{1}{2}(80)\left(\sqrt{2^2 + 1.5^2} - 0.5\right)^2 = 160 \text{ J}$$
$$T_1 + V_1 = T_2 + V_2$$
$$0 + 0 = 11.25\omega_2^2 + (-220.725 + 160)$$
$$\omega_2 = 2.323 \text{ rad/s} = 2.32 \text{ rad/s} \qquad \textit{Ans.}$$

F18–11. $(v_G)_2 = \omega_2 r_{G/IC} = \omega_2(0.75)$

$$I_G = \tfrac{1}{12}(30)\left(1.5^2\right) = 5.625 \text{ kg} \cdot \text{m}^2$$
$$T_1 = 0$$
$$T_2 = \tfrac{1}{2}m(v_G)_2^2 + \tfrac{1}{2}I_G\omega_2^2$$
$$= \tfrac{1}{2}(30)[\omega_2(0.75)]^2 + \tfrac{1}{2}(5.625)\omega_2^2 = 11.25\omega_2^2$$
$$(V_g)_1 = Wy_1 = 30(9.81)(0.75 \sin 45°) = 156.08 \text{ J}$$
$$(V_g)_2 = -Wy_2 = 0$$
$$(V_e)_1 = \tfrac{1}{2}ks_1^2 = 0$$
$$(V_e)_2 = \tfrac{1}{2}ks_2^2 = \tfrac{1}{2}(300)(1.5 - 1.5\cos45°)^2$$
$$= 28.95 \text{ J}$$
$$T_1 + V_1 = T_2 + V_2$$
$$0 + (156.08 + 0) = 11.25\omega_2^2 + (0 + 28.95)$$
$$\omega_2 = 3.362 \text{ rad/s} = 3.36 \text{ rad/s} \qquad \textit{Ans.}$$

F18–12. $(V_g)_1 = -Wy_1 = -[20(9.81) \text{ N}](1 \text{ m}) = -196.2 \text{ J}$

$$(V_g)_2 = 0$$
$$(V_e)_1 = \tfrac{1}{2}ks_1^2$$
$$= \tfrac{1}{2}(100 \text{ N/m})\left(\sqrt{(3 \text{ m})^2 + (2 \text{ m})^2} - 0.5 \text{ m}\right)^2$$
$$= 482.22 \text{ J}$$
$$(V_e)_2 = \tfrac{1}{2}ks_2^2 = \tfrac{1}{2}(100 \text{ N/m})(1 \text{ m} - 0.5 \text{ m})^2$$
$$= 12.5 \text{ J}$$
$$T_1 = 0$$
$$T_2 = \tfrac{1}{2}I_A\omega^2 = \tfrac{1}{2}\left[\tfrac{1}{3}(20 \text{ kg})(2 \text{ m})^2\right]\omega^2$$
$$= 13.3333\omega^2$$
$$T_1 + V_1 = T_2 + V_2$$
$$0 + [-196.2 \text{ J} + 482.22 \text{ J}]$$
$$= 13.3333\omega_2^2 + [0 + 12.5 \text{ J}]$$
$$\omega_2 = 4.53 \text{ rad/s} \qquad \textit{Ans.}$$

Chapter 19

F19–1. $\circlearrowleft + I_O\omega_1 + \Sigma\displaystyle\int_{t_1}^{t_2} M_O\, dt = I_O\omega_2$

$$0 + \int_0^{4\text{ s}} 3t^2\, dt = \left[60\left(0.3^2\right)\right]\omega_2$$
$$\omega_2 = 11.85 \text{ rad/s} = 11.9 \text{ rad/s} \qquad \textit{Ans.}$$

F19–2. $\circlearrowleft + (H_A)_1 + \Sigma\displaystyle\int_{t_1}^{t_2} M_A dt = (H_A)_2$

$$0 + 300(6) = 300\left(0.4^2\right)\omega_2 + 300[\omega(0.6)](0.6)$$
$$\omega_2 = 11.54 \text{ rad/s} = 11.5 \text{ rad/s} \qquad \textit{Ans.}$$
$$\xrightarrow{\;+\;} \quad m(v_1)_x + \Sigma\int_{t_1}^{t_2} F_x dt = m(v_2)_x$$
$$0 + F_f(6) = 300[11.54(0.6)]$$
$$F_f = 346 \text{ N} \qquad \textit{Ans.}$$

F19–3. $v_A = \omega_A r_{A/IC} = \omega_A(0.15)$

$$\circlearrowright + \Sigma M_O = 0; \quad 9 - A_t(0.45) = 0 \quad A_t = 20 \text{ N}$$
$$\circlearrowright + (H_C)_1 + \Sigma\int_{t_1}^{t_2} M_C\, dt = (H_C)_2$$
$$0 + [20(5)](0.15)$$
$$= 10[\omega_A(0.15)](0.15)$$
$$+ \left[10\left(0.1^2\right)\right]\omega_A$$
$$\omega_A = 46.2 \text{ rad/s} \qquad \textit{Ans.}$$

F19–4. $I_A = mk_A^2 = 10\left(0.08^2\right) = 0.064 \text{ kg} \cdot \text{m}^2$

$$I_B = mk_B^2 = 50\left(0.15^2\right) = 1.125 \text{ kg} \cdot \text{m}^2$$
$$\omega_A = \left(\frac{r_B}{r_A}\right)\omega_B = \left(\frac{0.2}{0.1}\right)\omega_B = 2\omega_B$$
$$\circlearrowleft + I_A(\omega_A)_1 + \Sigma\int_{t_1}^{t_2} M_A\, dt = I_A(\omega_A)_2$$
$$0 + 10(5) - \int_0^{5\text{ s}} F(0.1)dt = 0.064[2(\omega_B)_2]$$
$$\int_0^{5\text{ s}} F dt = 500 - 1.28(\omega_B)_2 \qquad (1)$$
$$\circlearrowright + I_B(\omega_B)_1 + \Sigma\int_{t_1}^{t_2} M_B\, dt = I_B(\omega_B)_2$$
$$0 + \int_0^{5\text{ s}} F(0.2)dt = 1.125(\omega_B)_2$$
$$\int_0^{5\text{ s}} F dt = 5.625(\omega_B)_2 \qquad (2)$$

Equating Eqs. (1) and (2),
$$500 - 1.28(\omega_B)_2 = 5.625(\omega_B)_2$$
$$(\omega_B)_2 = 72.41 \text{ rad/s} = 72.4 \text{ rad/s} \qquad \textit{Ans.}$$

F19–5. $(\xrightarrow{+})$ $\quad m\left[(v_O)_x\right]_1 + \Sigma \int F_x\,dt = m\left[(v_O)_x\right]_2$

$$0 + (150\ \text{N})(3\ \text{s}) + F_A(3\ \text{s})$$
$$= (50\ \text{kg})(0.3\omega_2)$$

$\zeta + I_G\omega_1 + \Sigma \int M_G\,dt = I_G\omega_2$

$$0 + (150\ \text{N})(0.2\ \text{m})(3\ \text{s}) - F_A(0.3\ \text{m})(3\ \text{s})$$
$$= [(50\ \text{kg})(0.175\ \text{m})^2]\ \omega_2$$

$$\omega_2 = 37.3\ \text{rad/s} \qquad\qquad Ans.$$
$$F_A = 36.53\ \text{N}$$

Also,

$$I_{IC}\omega_1 + \Sigma \int M_{IC}\,dt = I_{IC}\omega_2$$

$$0 + [(150\ \text{N})(0.2 + 0.3)\ \text{m}](3\ \text{s})$$
$$= [(50\ \text{kg})(0.175\ \text{m})^2 + (50\ \text{kg})(0.3\ \text{m})^2]\omega_2$$
$$\omega_2 = 37.3\ \text{rad/s} \qquad\qquad Ans.$$

F19–6. $(+\uparrow)$ $\quad m\left[(v_G)_1\right]_y + \Sigma \int F_y\,dt = m\left[(v_G)_2\right]_y$

$$0 + N_A(3\ \text{s}) - (150\ \text{lb})(3\ \text{s}) = 0$$
$$N_A = 150\ \text{lb}$$

$\zeta + (H_{IC})_1 + \Sigma \int M_{IC}\,dt = (H_{IC})_2$

$$0 + (25\ \text{lb}\cdot\text{ft})(3\ \text{s}) - [0.15(150\ \text{lb})(3\ \text{s})](0.5\ \text{ft})$$
$$= \left[\tfrac{150}{32.2}\ \text{slug}(1.25\ \text{ft})^2\right]\omega_2 + \left(\tfrac{150}{32.2}\ \text{slug}\right)\left[\omega_2(1\ \text{ft})\right](1\ \text{ft})$$
$$\omega_2 = 3.46\ \text{rad/s} \qquad\qquad Ans.$$

Preliminary Problems
Dynamics Solutions

Chapter 12

P12–1. a) $v = \dfrac{ds}{dt} = \dfrac{d}{dt}(2t^3) = 6t^2 \Big|_{t=2\,s} = 24 \text{ m/s}$

b) $a\,ds = v\,dv, \quad v = 5s, \quad dv = 5\,ds$

$a\,ds = (5s)\,5\,ds$

$a = 25s \Big|_{s=1\,m} = 25 \text{ m/s}^2$

c) $a = \dfrac{dv}{dt} = \dfrac{d}{dt}(4t + 5) = 4 \text{ m/s}^2$

d) $v = v_0 + a_c t$

$v = 0 + 2(2) = 4 \text{ m/s}$

e) $v^2 = v_0^2 + 2a_c(s - s_0)$

$v^2 = (3)^2 + 2(2)(4 - 0)$

$v = 5 \text{ m/s}$

f) $a\,ds = v\,dv$

$\displaystyle\int_{s_1}^{s_2} s\,ds = \int_0^v v\,dv$

$s^2 \Big|_4^5 = v^2 \Big|_0^v$

$25 - 16 = v^2$

$v = 3 \text{ m/s}$

g) $s = s_0 + v_0 t + \dfrac{1}{2} a_c t^2$

$s = 2 + 2(3) + \dfrac{1}{2}(4)(3)^2 = 26 \text{ m}$

h) $dv = a\,dt$

$\displaystyle\int_0^v dv = \int_0^1 (8t^2)\,dt$

$v = 2.67t^3 \Big|_0^1 = 2.67 \text{ m/s}$

i) $v = \dfrac{ds}{dt} = \dfrac{d}{dt}(3t^2 + 2) = 6t \Big|_{t=2\,s} = 12 \text{ m/s}$

j) $v_{avg} = \dfrac{\Delta s}{\Delta t} = \dfrac{6\text{ m} - (-1\text{ m})}{10\text{ s} - 0} = 0.7 \text{ m/s} \rightarrow$

$(v_{sp})_{avg} = \dfrac{s_T}{\Delta t} = \dfrac{7\text{ m} + 14\text{ m}}{10\text{ s} - 0} = 2.1 \text{ m/s}$

P12–2. a) $v = 2t$

$s = t^2$

$a = 2$

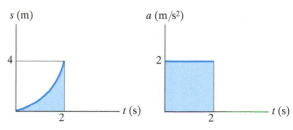

b) $s = -2t + 2$

$v = -2$

$a = 0$

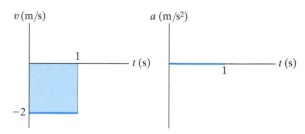

c) $a = -2$

$v = -2t$

$s = -t^2$

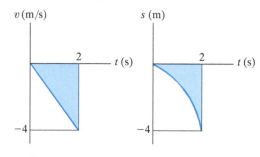

d)

$\Delta s = \displaystyle\int_0^3 v\,dt = \text{Area} = \dfrac{1}{2}(2)(2) + 2(3 - 2) = 4 \text{ m}$

$s - 0 = 4 \text{ m}, \qquad\qquad s = 4 \text{ m}$

$a = \dfrac{dv}{dt} = \text{slope at } t = 3 \text{ s}, \; a = 0$

e) For $a = 2$,
$v = 2t$
When $t = 2$ s, $v = 4$ m/s.
For $a = -2$,
$$\int_4^v dv = \int_2^t -2\, dt$$
$v - 4 = -2t + 4$
$v = -2t + 8$

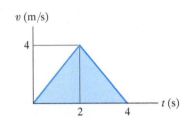

v (m/s)

4

2   4 t (s)

f) $$\int_1^v v\, dv = \int_0^2 a\, ds = \text{Area}$$
$$\frac{1}{2}v^2 - \frac{1}{2}(1)^2 = \frac{1}{2}(2)(4)$$
$v = 3$ m/s

g) $v\, dv = a\, ds$ At $s = 1$ m, $v = 2$ m/s.
$$a = v\frac{dv}{ds} = v(\text{slope}) = 2(-2) = -4 \text{ m/s}$$

P12–3. a) $y = 4x^2$
$\dot{y} = 8x\dot{x}$
$\ddot{y} = (8\dot{x})\dot{x} + 8x(\ddot{x})$

b) $y = 3e^x$
$\dot{y} = 3e^x\dot{x}$
$\ddot{y} = (3e^x\dot{x})\dot{x} + 3e^x(\ddot{x})$

c) $y = 6 \sin x$
$\dot{y} = (6 \cos x)\dot{x}$
$\ddot{y} = [(-6 \sin x)\dot{x}]\dot{x} + (6 \cos x)(\ddot{x})$

P12–4. $y_A, t_{AB}, (v_B)_y$
$20 = 0 + 40t_{AB}$
$$0 = y_A + 0 + \frac{1}{2}(-9.81)(t_{AB})^2$$
$(v_B)_y^2 = 0^2 + 2(-9.81)(0 - y_A)$

P12–5. $x_B, t_{AB}, (v_B)_y$
$x_B = 0 + (10 \cos 30°)(t_{AB})$
$$0 = 8 + (10 \sin 30°)t_{AB} + \frac{1}{2}(-9.81)t_{AB}^2$$
$(v_B)_y^2 = 0^2 + 2(-9.81)(0 - 8)$

P12–6. $x_B, y_B, (v_B)_y$
$x_B = 0 + (60 \cos 20°)(5)$
$$y_B = 0 + (60 \sin 20°)(5) + \frac{1}{2}(-9.81)(5)^2$$
$(v_B)_y = 60 \sin 20° + (-9.81)(5)$

P12–7. a) $a_t = \dot{v} = 3$ m/s^2
$$a_n = \frac{v^2}{\rho} = \frac{(2)^2}{1} = 4 \text{ m/s}^2$$
$a = \sqrt{(3)^2 + (4)^2} = 5$ m/s^2

b) $a_t = \dot{v} = 4$ m/s^2
$v^2 = v_0^2 + 2a_c(s - s_0)$
$v^2 = 0 + 2(4)(2 - 0)$
$v = 4$ m/s
$$a_n = \frac{v^2}{\rho} = \frac{(4)^2}{2} = 8 \text{ m/s}^2$$

c) $a_t = 0$
$$\rho = \frac{\left[1 + \left(\frac{dy}{dx}\right)^2\right]^{\frac{3}{2}}}{\frac{d^2y}{dx^2}}\bigg|_{x=0} = \frac{1 + 0}{4} = \frac{1}{4}$$
$$a_n = \frac{v^2}{\rho} = \frac{(2)^2}{\frac{1}{4}} = 16 \text{ m/s}^2$$
$a = \sqrt{(0)^2 + (16)^2} = 16$ m/s^2

d) $a_t\, ds = v\, dv$
$a_t\, ds = (4s + 1)(4\, ds)$
$a_t = (16s + 4)|_{s=0} = 4$ m/s^2
$$a_n = \frac{v^2}{\rho} = \frac{(4(0) + 1)^2}{2} = 0.5 \text{ m/s}^2$$

e) $a_t\, ds = v\, dv$
$$\int_0^s 2s\, ds = \int_1^v v\, dv$$
$$s^2 = \frac{1}{2}(v^2 - 1)$$
$$v = \sqrt{2s^2 + 1}\bigg|_{s = 2\,m} = 3 \text{ m/s}$$
$a_t = \dot{v} = 2(2) = 4$ m/s^2
$$a_n = \frac{v^2}{\rho} = \frac{(3)^2}{3} = 3 \text{ m/s}^2$$
$a = \sqrt{(4)^2 + (3)^2} = 5$ m/s^2

f) $a_t = \dot{v} = 8t\bigg|_{t = 1} = 8$ m/s^2
$$a_n = \frac{v^2}{\rho} = \frac{(4(1)^2 + 2)^2}{6} = 6 \text{ m/s}^2$$
$a = \sqrt{(8)^2 + (6)^2} = 10$ m/s^2

Chapter 13

P13–1. a)

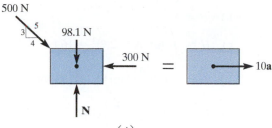

500 N

5

3

4

98.1 N

300 N

10a

N

$$\xrightarrow{+}\ \Sigma F_x = ma_x;\quad \left(\frac{4}{5}\right)(500\ \text{N}) - 300\ \text{N} = 10a$$

$$a = 10\ \text{m/s}^2$$

$$\xrightarrow{+}\ v = v_0 + a_c t;\quad v = 0 + 10(2) = 20\ \text{m/s}$$

b)

98.1 N

$F = (20t)$ N

10a

N

$$\xrightarrow{+}\ \Sigma F_x = ma_x;\quad 20t = 10a$$

$$a = 2t$$

$$dv = a\ dt;\quad \int_0^v dv = \int_0^2 2t\ dt$$

$$v = 4\ \text{m/s}$$

P13–2.

a)

200 N

98.1 N

40 N

30 N

10a

N

$$\xrightarrow{+}\ \Sigma F_x = ma_x;\quad 40\ \text{N} - 30\ \text{N} = 10a$$

$$a = 1\ \text{m/s}^2$$

$$\xrightarrow{+}\ v^2 = v_0^2 + 2a_c(s - s_0);\quad v^2 = (3)^2 + 2(1)(8 - 0)$$

$$v = 5\ \text{m/s}$$

b)

98.1 N

$F = (2.5s)$ N

10a

N

$$\xrightarrow{+}\ \Sigma F_x = ma_x;\quad 2.5s = 10a$$

$$a = 2.5s$$

$$v\ dv = a\ ds$$

$$\int_3^v v\ dv = \int_0^8 2.5s\ ds$$

$$v^2 - (3)^2 = 2.5(8 - 0)^2$$

$$v = 13\ \text{m/s}$$

P13–3.

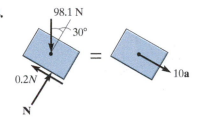

10a

5

3

4

F_s N   98.1 N

$$F_s = kx = (10\ \text{N/m})(5\ \text{m} - 1\ \text{m}) = 40\ \text{N}$$

$$\xleftarrow{+}\ \Sigma F_x = ma_x;\quad \frac{4}{5}(40\ \text{N}) = 10a$$

$$a = 3.2\ \text{m/s}^2$$

P13–4.

98.1 N

30°

10a

0.2N

N

$$\searrow^+\ \Sigma F_x = ma_x;\quad 98.1\sin 30° - 0.2N = 10a$$

$$^+\!\nearrow\ \Sigma F_y = ma_y;\quad N - 98.1\cos 30° = 0$$

P13–5. a)

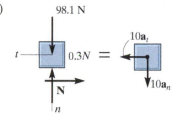

98.1 N

10$\mathbf{a}_t$

t

0.3N

10$\mathbf{a}_n$

N

n

$$\xleftarrow{+}\ \Sigma F_t = ma_t;\quad -0.3N = 10a_t$$

$$+\!\downarrow \Sigma F_n = ma_n;\quad 98.1 - N = 10\left(\frac{(6)^2}{10}\right)$$

b)

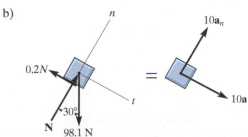

n

10$\mathbf{a}_n$

0.2N

t

10a

30°

N   98.1 N

$\searrow \Sigma F_t = ma_t;\ 98.1 \sin 30° - 0.2N = 10a_t$

$\nearrow \Sigma F_n = ma_n;\quad N - 98.1 \cos 30° = 10\left(\dfrac{(4)^2}{5}\right)$

c)

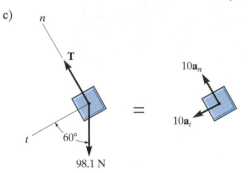

$\swarrow \Sigma F_t = ma_t;\qquad 98.1 \cos 60° = 10a_t$

$\nwarrow^+ \Sigma F_n = ma_n;\quad T - 98.1 \sin 60° = 10\left(\dfrac{8^2}{6}\right)$

P13–6. a)

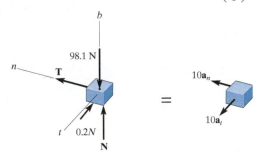

$\Sigma F_b = 0;\qquad N - 98.1 = 0$

$\Sigma F_t = ma_t;\qquad -0.2N = 10a_t$

$\Sigma F_n = ma_n;\qquad T = 10\dfrac{(8)^2}{4}$

b)

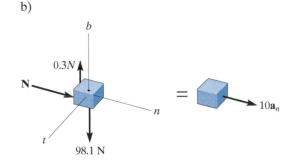

$\Sigma F_b = 0;\qquad 0.3N - 98.1 = 0$

$\Sigma F_t = ma_t;\qquad 0 = 0$

$\Sigma F_n = ma_n;\qquad N = 10\dfrac{v^2}{2}$

Chapter 14

P14–1. a) $U = \dfrac{3}{5}(500\text{ N})(2\text{ m}) = 600\text{ J}$

b) $U = 0$

c) $U = \displaystyle\int_0^2 6s^2\,ds = 2(2)^3 = 16\text{ J}$

d) $U = 100\text{ N}\left(\dfrac{3}{5}(2\text{ m})\right) = \dfrac{3}{5}(100\text{ N})(2\text{ m}) = 120\text{ J}$

e) $U = \dfrac{4}{5}(\text{Area}) = \dfrac{4}{5}\left[\dfrac{1}{2}(1)(20) + (1)(20)\right] = 24\text{ J}$

f) $U = \dfrac{1}{2}(10\text{ N/m})((3\text{ m})^2 - (1\text{ m})^2) = 40\text{ J}$

g) $U = -\left(\dfrac{4}{5}\right)(100\text{ N})(2\text{ m}) = -160\text{ J}$

P14–2. a) $T = \dfrac{1}{2}(10\text{ kg})(2\text{ m/s})^2 = 20\text{ J}$

b) $T = \dfrac{1}{2}(10\text{ kg})(6\text{ m/s})^2 = 180\text{ J}$

P14–3. a) $V = (100\text{ N})(2\text{ m}) = 200\text{ J}$

b) $V = (100\text{ N})(3\text{ m}) = 300\text{ J}$

c) $V = 0$

P14–4. a) $V = \dfrac{1}{2}(10\text{ N/m})(5\text{ m} - 4\text{ m})^2 = 5\text{ J}$

b) $V = \dfrac{1}{2}(10\text{ N/m})(10\text{ m} - 4\text{ m})^2 = 180\text{ J}$

c) $V = \dfrac{1}{2}(10\text{ N/m})(5\text{ m} - 4\text{ m})^2 = 5\text{ J}$

Chapter 15

P15–1. a) $I = (100\text{ N})(2\text{ s}) = 200\text{ N}\cdot\text{s}\ \checkmark$

b) $I = (200\text{ N})(2\text{ s}) = 400\text{ N}\cdot\text{s}\downarrow$

c) $I = \displaystyle\int_0^2 6t\,dt = 3(2)^2 = 12\text{ N}\cdot\text{s}\ \searrow$

d) $I = \text{Area} = \dfrac{1}{2}(1)(20) + (2)(20) = 50\text{ N}\cdot\text{s}\ \nearrow$

e) $I = (80\text{ N})(2\text{ s}) = 160\text{ N}\cdot\text{s}\rightarrow$

f) $I = (60\text{ N})(2\text{ s}) = 120\text{ N}\cdot\text{s}\ \nearrow$

P15–2. a) $L = (10\text{ kg})(10\text{ m/s}) = 100\text{ kg}\cdot\text{m/s}\ \searrow$

b) $L = (10\text{ kg})(2\text{ m/s}) = 20\text{ kg}\cdot\text{m/s}\ \checkmark$

c) $L = (10\text{ kg})(3\text{ m/s}) = 30\text{ kg}\cdot\text{m/s}\rightarrow$

Chapter 16

P16–1. a) $\mathbf{v}_B = \mathbf{v}_A + \mathbf{v}_{B/A \text{ (pin)}}$

$$v_B = 18 \text{ m/s} + 2\omega$$
$$\downarrow \qquad \downarrow \qquad \underset{60°}{\nearrow}$$

Also,

$$-v_B\mathbf{j} = -18\mathbf{j}$$
$$+ (-\omega\mathbf{k}) \times (-2\cos 60°\mathbf{i} - 2\sin 60°\mathbf{j})$$

$$\mathbf{v}_B = \mathbf{v}_A + \mathbf{v}_{B/A \text{ (pin)}}$$

b)

$$\underset{\longrightarrow}{(v_B)_x} + \underset{\uparrow}{(v_B)_y} = \underset{\longrightarrow}{4(0.5) \text{ m/s}} + \underset{\underset{30°}{\nearrow}}{4(0.5) \text{ m/s}}$$

Also,

$$(v_B)_x\mathbf{i} + (v_B)_y\mathbf{j} = 2\mathbf{i}$$
$$+ (-4\mathbf{k}) \times (-0.5\cos 30°\mathbf{i} + \mathbf{0.5}\sin 30°\mathbf{j})$$

c) $\mathbf{v}_B = \mathbf{v}_A + \mathbf{v}_{B/A \text{ (pin)}}$

$$v_B = 6 \text{ m/s} + \omega(5)$$
$$\underset{45°}{\nearrow} \qquad \underset{30°}{\nearrow}$$

Also,

$$v_B\cos 45°\mathbf{i} + v_B\sin 45°\mathbf{j} = 6\mathbf{i} + (\omega\mathbf{k}) \times (4\mathbf{i} - 3\mathbf{j})$$

d) $\mathbf{v}_B = \mathbf{v}_A + \mathbf{v}_{B/A \text{ (pin)}}$

$$v_B = 6 \text{ m/s} + \omega(3)$$
$$\underset{\longrightarrow}{} \qquad \underset{30°}{\searrow} \qquad \uparrow$$

Also,

$$v_B\mathbf{i} = 6\cos 30°\mathbf{i} + 6\sin 30°\mathbf{j} + (\omega\mathbf{k}) \times (3\mathbf{i})$$

e) $v_A = 12 \text{ m/s} = \omega(0.5 \text{ m}) \quad \omega = 24 \text{ rad/s}$

$$\mathbf{v}_B = \mathbf{v}_A + \mathbf{v}_{B/A \text{ (pin)}}$$

$$\underset{\longrightarrow}{(v_B)_x} + \underset{\uparrow}{(v_B)_y} = \underset{\uparrow}{12 \text{ m/s}} + \underset{\longleftarrow}{(24)(0.5)}$$

Also,

$$(v_B)_x\mathbf{i} + (v_B)_y\mathbf{j} = 12\mathbf{j} + (24\mathbf{k}) \times (0.5\mathbf{j})$$

f) $\mathbf{v}_B = \mathbf{v}_A + \mathbf{v}_{B/A \text{ (pin)}}$

$$v_B = 6 \text{ m/s} + \omega(5)$$
$$\underset{\longrightarrow}{} \qquad \underset{\longrightarrow}{}$$
$$\underset{3}{\overset{4 \backslash 5}{}}$$

Also,

$$v_B\mathbf{i} = 6\mathbf{i} + (\omega\mathbf{k}) \times (4\mathbf{i} + 3\mathbf{j})$$

P16–2. a)

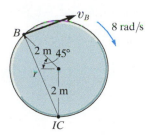

$$r = \sqrt{(2\cos 45°)^2 + (2 + 2\sin 45°)^2}$$

b)

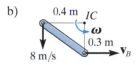

c)

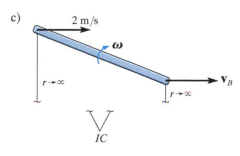

$$v_B = 2 \text{ m/s}, \omega = 0$$

d)

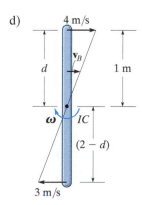

e)

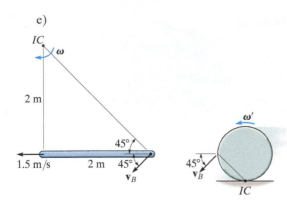

f)

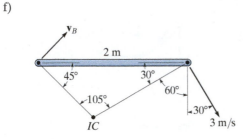

P16–3. a)

$$\mathbf{a}_B = \mathbf{a}_A + \mathbf{a}_{B/A \text{ (pin)}}$$

$$a_B = 2 \text{ m/s}^2 + \frac{(3)^2}{3} + 2\alpha + (2.12)^2(2)$$

Also,

$$-a_B\mathbf{j} = -2\mathbf{i} + 3\mathbf{j} + (-\alpha\mathbf{k}) \times (2 \sin 45°\mathbf{i} + 2 \cos 45°\mathbf{j})$$
$$- (2.12)^2(2 \sin 45°\mathbf{i} + 2 \cos 45°\mathbf{j})$$

b) $$\mathbf{a}_B = \mathbf{a}_A + \mathbf{a}_{B/A \text{ (pin)}}$$

$$(a_B)_x + (a_B)_y = (2)(2) \text{ m/s}^2 + \alpha(2) + (4)^2(2)$$

Also,

$$(a_B)_x\mathbf{i} + (a_B)_y\mathbf{j} = 4\mathbf{i} + (-\alpha\mathbf{k}) \times (-2 \cos 45°\mathbf{i} + 2 \sin 45°\mathbf{j})$$
$$- (4)^2(-2 \cos 45°\mathbf{i} + 2 \sin 45°\mathbf{j})$$

c) $$\mathbf{a}_B = \mathbf{a}_A + \mathbf{a}_{B/A \text{ (pin)}}$$

$$(a_B)_x + (6)^2(1) = 2(2) + (3)^2(2) + \alpha(4)$$

Also,

$$(a_B)_x\mathbf{i} - 36\mathbf{j} = 4\mathbf{i} - 18\mathbf{j} + (-\alpha\mathbf{k}) \times (4\mathbf{i})$$

d) $$\mathbf{a}_B = \mathbf{a}_A + \mathbf{a}_{B/A \text{ (pin)}}$$

$$a_B = 6 + \alpha(2) + (3)^2(2)$$

Also,

$$a_B\mathbf{i} = -6 \cos 60°\mathbf{i} - 6 \sin 60°\mathbf{j}$$
$$+ (-\alpha\mathbf{k}) \times (-2\mathbf{i}) - (3)^2(-2\mathbf{i})$$

e) $$\mathbf{a}_B = \mathbf{a}_A + \mathbf{a}_{B/A \text{ (pin)}}$$

$$a_B = 8(0.5) + (4)^2(0.5) + \alpha(2) + (1.15)^2(2)$$

Also,

$$-a_B\mathbf{i} = -4\mathbf{j} + 8\mathbf{i} + (-\alpha\mathbf{k}) \times (-2 \cos 30°\mathbf{i} - 2 \sin 30°\mathbf{j})$$
$$- (1.15)^2(-2 \cos 30°\mathbf{i} - 2 \sin 30°\mathbf{j})$$

f) $$\mathbf{a}_B = \mathbf{a}_A + \mathbf{a}_{B/A \text{ (pin)}}$$

$$(a_B)_x + (a_B)_y = 2(0.5) + 2(0.5) + (4)^2(0.5)$$

Also,

$$(a_B)_x\mathbf{i} + (a_B)_y\mathbf{j} = -1\mathbf{j} + (-2\mathbf{k}) \times (0.5\mathbf{j})$$
$$- (4)^2(0.5\mathbf{j})$$

Chapter 17

P17–1. a)

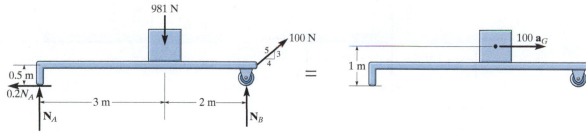

b)

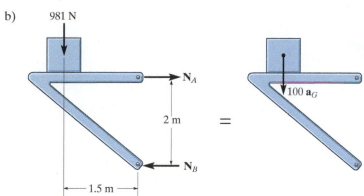

c)

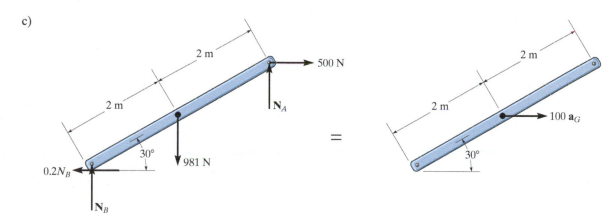

d)

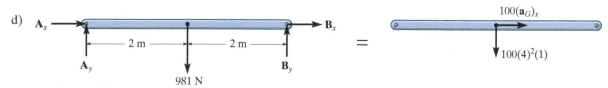

e)

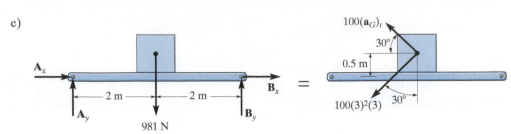

f)

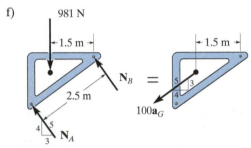

P17–2.　a)

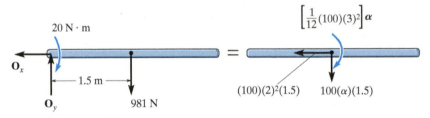

b)

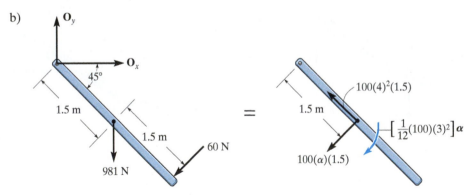

c)

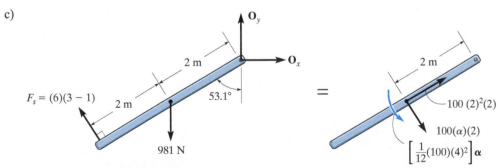

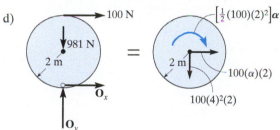

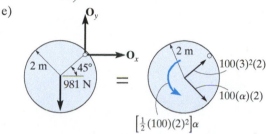

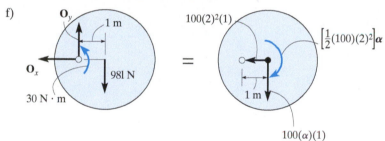

Chapter 18

P18–1.   a) $T = \dfrac{1}{2}\left[\dfrac{100(2)^2}{2}\right](3)^2 = 900$ J

b) $T = \dfrac{1}{2}(100)[2(1)]^2 + \dfrac{1}{2}\left[\dfrac{1}{12}(100)(6)^2\right](2)^2$

$= 800$ J

Also,

$T = \dfrac{1}{2}\left[\dfrac{1}{12}(100)(6)^2 + 100(1)^2\right](2)^2 = 800$ J

c) $T = \dfrac{1}{2}(100)[2(2)]^2 + \dfrac{1}{2}\left[\dfrac{1}{2}(100)(2)^2\right](2)^2$

$= 1200$ J

Also,

$T = \dfrac{1}{2}\left[\dfrac{1}{2}(100)(2)^2 + 100(2)^2\right](2)^2$

$= 1200$ J

d) $T = \dfrac{1}{2}(100)[2(1.5)]^2 + \dfrac{1}{2}\left[\dfrac{1}{12}(100)(3)^2\right](2)^2$

$= 600$ J

Also,

$T = \dfrac{1}{2}\left[\dfrac{1}{12}(100)(3)^2 + 100(1.5)^2\right](2)^2$

$= 600$ J

e) $T = \dfrac{1}{2}(100)[4(2)]^2 + \dfrac{1}{2}\left[\dfrac{1}{2}(100)(2)^2\right](4)^2$

$= 4800$ J

Also,

$T = \dfrac{1}{2}\left[\dfrac{1}{2}(100)(2)^2 + 100(2)^2\right](4)^2$

$= 4800$ J

f) $T = \dfrac{1}{2}(100)[(4)(2)]^2 = 3200$ J

Chapter 19

P19–1.   a) $H_G = \left[\dfrac{1}{2}(100)(2)^2\right](3) = 600$ kg·m²/s $\circlearrowright$

$H_O = \left[\dfrac{1}{2}(100)(2)^2 + 100(2)^2\right](3)$

$= 1800$ kg·m²/s $\circlearrowright$

b) $H_G = \left[\dfrac{1}{12}(100)(3)^2\right](4) = 300 \text{ kg} \cdot \text{m}^2/\text{s} \,\circlearrowright$

$H_O = \left[\dfrac{1}{12}(100)(3)^2 + (100)(1.5)^2\right](4)$

$= 1200 \text{ kg} \cdot \text{m}^2/\text{s} \,\circlearrowright$

c) $H_G = \left[\dfrac{1}{2}(100)(2)^2\right](4) = 800 \text{ kg} \cdot \text{m}^2/\text{s} \,\circlearrowright$

$H_O = \left[\dfrac{1}{2}(100)(2)^2 + (100)(2)^2\right](4)$

$= 2400 \text{ kg} \cdot \text{m}^2/\text{s} \,\circlearrowright$

d) $H_G = \left[\dfrac{1}{12}(100)(4)^2\right]3 = 400 \text{ kg} \cdot \text{m}^2/\text{s} \,\circlearrowleft$

$H_O = \left[\dfrac{1}{12}(100)(4)^2 + (100)(1)^2\right]3$

$= 700 \text{ kg} \cdot \text{m}^2/\text{s} \,\circlearrowleft$

P19–2.   a) $\displaystyle\int M_O\, dt = \left(\dfrac{4}{5}\right)(500)(2)(3) = 2400 \text{ N} \cdot \text{s} \cdot \text{m} \,\circlearrowright$

b) $\displaystyle\int M_O\, dt = \left[2(20) + \dfrac{1}{2}(3 - 2)(20)\right]4$

$= 200 \text{ N} \cdot \text{s} \cdot \text{m} \,\circlearrowright$

c) $\displaystyle\int M_O\, dt = \dfrac{3}{5}\int_0^3 4(2t + 2)\, dt = 36 \text{ N} \cdot \text{s} \cdot \text{m} \,\circlearrowright$

d) $\displaystyle\int M_O\, dt = \int_0^3 (30t^2)\, dt = 270 \,\circlearrowright$

Review Problem Solutions

Chapter 12

R12–1. $s = t^3 - 9t^2 + 15t$

$$v = \frac{ds}{dt} = 3t^2 - 18t + 15$$

$$a = \frac{dv}{dt} = 6t - 18$$

a_{max} occurs at $t = 10$ s.

$a_{max} = 6(10) - 18 = 42$ ft/s² *Ans.*

v_{max} occurs when $t = 10$ s

$v_{max} = 3(10)^2 - 18(10) + 15 = 135$ ft/s *Ans.*

R12–2. $(\overset{+}{\rightarrow})$ $s = s_0 + v_0 t + \frac{1}{2} a_c t^2$

$$s = 0 + 12(10) + \frac{1}{2}(-2)(10)^2$$

$$s = 20.0 \text{ ft} \qquad \text{Ans.}$$

R12–3. $v = \frac{ds}{dt} = 1800(1 - e^{-0.3t})$

$$\int_0^x ds = \int_0^t 1800(1 - e^{-0.3t})\, dt$$

$$s = 1800\left(t + \frac{1}{0.3}e^{-0.3t}\right) - 6000$$

Thus, in $t = 3$ s

$$s = 1800\left(3 + \frac{1}{0.3}e^{-0.3(3)}\right) - 6000$$

$$s = 1839.4 \text{ mm} = 1.84 \text{ m} \qquad \text{Ans.}$$

R12–4. $0 \le t \le 5$ $a = \frac{\Delta v}{\Delta t} = \frac{20}{5} = 4$ m/s² *Ans.*

$5 \le t \le 20$ $a = \frac{\Delta v}{\Delta t} = \frac{20 - 20}{20 - 5} = 0$ m/s² *Ans.*

$20 \le t \le 30$ $a = \frac{\Delta v}{\Delta t} = \frac{0 - 20}{30 - 20} = -2$ m/s² *Ans.*

At $t_1 = 5$ s, $t_2 = 20$ s, and $t_3 = 30$ s,

$s_1 = A_1 = \frac{1}{2}(5)(20) = 50$ m *Ans.*

$s_2 = A_1 + A_2 = 50 + 20(20 - 5) = 350$ m *Ans.*

$s_3 = A_1 + A_2 + A_3 = 350$

$$+ \frac{1}{2}(30 - 20)(20) = 450 \text{ m} \quad \text{Ans.}$$

R12–5. $v_A = 20\mathbf{i}$

$v_B = 21.21\mathbf{i} + 21.21\mathbf{j}$

$v_C = 40\mathbf{i}$

$$\mathbf{a}_{AB} = \frac{\Delta v}{\Delta t} = \frac{21.21\mathbf{i} + 21.21\mathbf{j} - 20\mathbf{i}}{3}$$

$\mathbf{a}_{AB} = \{0.404\mathbf{i} + 7.07\mathbf{j}\}$ m/s² *Ans.*

$$\mathbf{a}_{AC} = \frac{\Delta v}{\Delta t} = \frac{40\mathbf{i} - 20\mathbf{i}}{8}$$

$\mathbf{a}_{AC} = \{2.50\mathbf{i}\}$ m/s² *Ans.*

R12–6. $(\overset{+}{\rightarrow})$ $s = s_0 + v_0 t$

$$126 = 0 + (v_0)_x (3.6)$$

$$(v_0)_x = 35 \text{ ft/s}$$

$(+\uparrow)$ $s = s_0 + v_0 t + \frac{1}{2} a_c t^2$

$$0 = 0 + (v_0)_y (3.6) + \frac{1}{2}(-32.2)(3.6)^2$$

$$(v_0)_y = 57.96 \text{ ft/s}$$

$v_0 = \sqrt{(35)^2 + (57.96)^2} = 67.7$ ft/s *Ans.*

$\theta = \tan^{-1}\left(\frac{57.96}{35}\right) = 58.9°$ *Ans.*

R12–7. $v\, dv = a_t\, ds$

$$\int_4^v v\, dv = \int_0^{10} 0.05s\, ds$$

$$0.5v^2 - 8 = \frac{0.05}{2}(10)^2$$

$v = 4.583 = 4.58$ m/s *Ans.*

$a_n = \frac{v^2}{\rho} = \frac{(4.583)^2}{50} = 0.420$ m/s²

$a_t = 0.05(10) = 0.5$ m/s²

$a = \sqrt{(0.420)^2 + (0.5)^2} = 0.653$ m/s² *Ans.*

R12–8. $dv = a\, dt$

$$\int_0^v dv = \int_0^t 0.5e^t\, dt$$

$$v = 0.5(e^t - 1)$$

When $t = 2$ s, $v = 0.5(e^2 - 1) = 3.195$ m/s

$$= 3.19 \text{ m/s} \qquad \text{Ans.}$$

When $t = 2$ s $a_t = 0.5e^2 = 3.695$ m/s^2

$$a_n = \frac{v^2}{\rho} = \frac{3.195^2}{5} = 2.041 \text{ m/s}^2$$

$$a = \sqrt{a_t^2 + a_n^2} = \sqrt{3.695^2 + 2.041^2}$$
$$= 4.22 \text{ m/s}^2 \qquad \text{Ans.}$$

R12–9. $r = 2$ m $\qquad \theta = 5t^2$
$\dot{r} = 0 \qquad\quad \dot{\theta} = 10t$
$\ddot{r} = 0 \qquad\quad \ddot{\theta} = 10$

$a = (\ddot{r} - r\dot{\theta}^2)\mathbf{u}_r + (r\ddot{\theta} + 2\dot{r}\dot{\theta})\mathbf{u}_\theta$
$\quad = [0 - 2(10t)^2]\mathbf{u}_r + [2(10) + 0]\mathbf{u}_\theta$
$\quad = \{-200t^2\mathbf{u}_r + 20\mathbf{u}_\theta\} \text{ m/s}^2$

When $\theta = 30° = 30\left(\dfrac{\pi}{180}\right) = 0.524$ rad

$0.524 = 5t^2$
$t = 0.324$ s

$a = [-200(0.324)^2]\mathbf{u}_r + 20\mathbf{u}_\theta$
$\quad = \{-20.9\mathbf{u}_r + 20\mathbf{u}_\theta\} \text{ m/s}^2$
$a = \sqrt{(-20.9)^2 + (20)^2} = 29.0 \text{ m/s}^2 \qquad \text{Ans.}$

R12–10. $4s_B + s_A = l$
$\qquad\quad 4v_B = -v_A$
$\qquad\quad 4a_B = -a_A$
$\qquad\quad 4a_B = -0.2$
$\qquad\quad\ a_B = -0.05 \text{ m/s}^2$
$(+\downarrow) \quad v_B = (v_B)_0 + a_B t$
$\qquad\quad -8 = 0 - (0.05)(t)$
$\qquad\qquad t = 160 \text{ s} \qquad \text{Ans.}$

R12–11.
$$\mathbf{v}_B = \mathbf{v}_A + \mathbf{v}_{B/A}$$
$$[500 \leftarrow] = [600 \, \overset{75°}{\searize}] + v_{B/A}$$
$(\overset{+}{\leftarrow}) \quad 500 = -600 \cos 75° + (v_{B/A})_x$
$\qquad (v_{B/A})_x = 655.29 \leftarrow$
$(+\uparrow) \quad 0 = -600 \sin 75° + (v_{B/A})_y$
$\qquad (v_{B/A})_y = 579.56 \uparrow$
$(v_{B/A}) = \sqrt{(655.29)^2 + (579.56)^2}$
$v_{B/A} = 875 \text{ km/h} \qquad \text{Ans.}$

$\theta = \tan^{-1}\left(\dfrac{579.56}{655.29}\right) = 41.5° \, \nearrow \qquad \text{Ans.}$

Chapter 13

R13–1. $20 \text{ km/h} = \dfrac{20(10)^3}{3600} = 5.556 \text{ m/s}$
$(\overset{+}{\leftarrow}) \quad v^2 = v_0^2 + 2a_c(s - s_0)$

$a = -0.3429 \text{ m/s}^2 = 0.3429 \text{ m/s}^2 \rightarrow$
$\overset{+}{\rightarrow}\Sigma F_x = ma_x; \quad F = 250(0.3429) = 85.7 \text{ N} \quad \text{Ans.}$

R13–2. $\nwarrow+\Sigma F_y = ma_y; \quad N_C - 50(9.81)\cos 30° = 0$
$\qquad\qquad\qquad N_C = 424.79$
$\nearrow+\Sigma F_x = ma_x; \quad 3T - 0.3(424.79) - 50(9.81)$
$\qquad\qquad\qquad\qquad \sin 30° = 50a_C \qquad (1)$

Kinematics, $2s_C + (s_C - s_p) = l$
Taking two time derivatives, yields
$$3a_C = a_p$$
Thus, $a_C = \dfrac{6}{3} = 2$
Substituting into Eq. (1) and solving,
$$T = 158 \text{ N} \qquad \text{Ans.}$$

R13–3. Suppose the two blocks move together. Then

$$50 \text{ lb} = \frac{50 + 20}{32.2}a$$
$$a = 23 \text{ m/s}^2$$

Then the friction force on block B is

$$F_B = \frac{50}{32.2}(23) = 35.7 \text{ lb}$$

The maximum friction force between blocks A and B is
$$F_{\text{max}} = 0.4(20) = 8 \text{ lb} < 35.7 \text{ lb}$$
The blocks have different accelerations.

Block A:
$\overset{+}{\rightarrow}\Sigma F_x = ma_x; \quad 20(0.3) = \dfrac{20}{32.2}a_A$
$\qquad\qquad\qquad a_A = 70.8 \text{ ft/s}^2 \qquad \text{Ans.}$

Block B:
$\overset{+}{\rightarrow}\Sigma F_x = ma_x; \quad 20(0.3) = \dfrac{50}{32.2}a_B$
$\qquad\qquad\qquad a_B = 3.86 \text{ ft/s}^2 \qquad \text{Ans.}$

R13–4. *Kinematics:* Since the motion of the crate is known, its acceleration **a** will be determined first.

$$a = v\frac{dv}{ds} = (0.05s^{3/2})\left[(0.05)\left(\frac{3}{2}\right)s^{1/2}\right]$$
$$= 0.00375s^2 \text{ m/s}^2$$

When $s = 10$ m,
$$a = 0.00375(10^2) = 0.375 \text{ m/s}^2 \rightarrow$$

Free-Body Diagram: The kinetic friction $F_f = \mu_k N = 0.2N$ must act to the left to oppose the motion of the crate which is to the right.

Equations of Motion: Here, $a_y = 0$. Thus,

$$+\uparrow \Sigma F_y = ma_y; \quad N - 20(9.81) = 20(0)$$
$$N = 196.2 \text{ N}$$

Using the results of **N** and **a**,

$$\overset{+}{\rightarrow} \Sigma F_x = ma_x; \quad T - 0.2(196.2) = 20(0.375)$$
$$T = 46.7 \text{ N} \qquad\qquad Ans.$$

R13–5. $+\nwarrow\Sigma F_n = ma_n; \quad T - 30(9.81)\cos\theta = 30\left(\dfrac{v^2}{4}\right)$

$+\nearrow\Sigma F_t = ma_t; \quad -30(9.81)\sin\theta = 30a_t$

$$a_t = -9.81 \sin\theta$$

$a_t\, ds = v\, dv$ Since $ds = 4\, d\theta$, then

$$-9.81 \int_0^\theta \sin\theta\,(4\,d\theta) = \int_4^v v\,dv$$

$$9.81(4)\cos\theta \Big|_0^\theta = \frac{1}{2}(v)^2 - \frac{1}{2}(4)^2$$

$$39.24(\cos\theta - 1) + 8 = \frac{1}{2}v^2$$

At $\theta = 20°$

$v = 3.357$ m/s

$a_t = -3.36$ m/s^2 = 3.36 m/s^2 ✓   *Ans.*

$T = 361$ N   *Ans.*

R13–6. $\Sigma F_z = ma_z; \quad N_z - mg = 0 \quad N_z = mg$

$\Sigma F_x = ma_n; \quad 0.3(mg) = m\left(\dfrac{v^2}{r}\right)$

$v = \sqrt{0.3gr} = \sqrt{0.3(32.2)(3)} = 5.38$ ft/s   *Ans.*

R13–7. $v = \dfrac{1}{8}x^2$

$\dfrac{dy}{dx} = \tan\theta = \dfrac{1}{4}x \Big|_{x=-6} = -1.5 \quad \theta = -56.31°$

$\dfrac{d^2y}{dx^2} = \dfrac{1}{4}$

$$\rho = \frac{\left[1 + \left(\dfrac{dy}{dx}\right)^2\right]^{\frac{3}{2}}}{\left|\dfrac{d^2y}{dx^2}\right|} = \frac{\left[1 + (-1.5)^2\right]^{\frac{3}{2}}}{\left|\dfrac{1}{4}\right|} = 23.436 \text{ ft}$$

$+\nearrow\Sigma F_n = ma_n; \quad N - 10\cos 56.31°$

$$= \left(\frac{10}{32.2}\right)\left(\frac{(5)^2}{23.436}\right)$$

$$N = 5.8783 = 5.88 \text{ lb} \qquad Ans.$$

$+\searrow\Sigma F_t = ma_t; \quad -0.2(5.8783) + 10\sin 56.31°$

$$= \left(\frac{10}{32.2}\right)a_t$$

$$a_t = 23.0 \text{ ft/s}^2 \qquad Ans.$$

R13–8. $r = 0.5$ m

$\dot{r} = 3$ m/s $\dot{\theta} = 6$ rad/s

$\ddot{r} = 1$ m/s^2 $\ddot{\theta} = 2$ rad/s

$a_r = \ddot{r} - r\dot{\theta}^2 = 1 - 0.5(6)^2 = -17$

$a_\theta = r\ddot{\theta} + 2\dot{r}\dot{\theta} = 0.5(2) + 2(3)(6) = 37$

$\Sigma F_r = ma_r; \quad F_r = 4(-17) = -68$ N

$\Sigma F_\theta = ma_\theta; \quad N_\theta = 4(37) = 148$ N

$\Sigma F_z = ma_z; \quad N_z - 4(9.81) = 0$

$\qquad\qquad\qquad N_z = 39.24$ N

$\qquad\qquad\qquad F_r = -68$ N   *Ans.*

$N = \sqrt{(148)^2 + (39.24)^2} = 153$ N   *Ans.*

Chapter 14

R14–1. $+\nwarrow\Sigma F_y = 0; \quad N_C - 150\cos 30° = 0$

$$N_C = 129.9 \text{ lb}$$

$T_1 + \Sigma U_{1-2} = T_2$

$$0 + 150\sin 30°(30) - (0.3)129.9(30) = \frac{1}{2}\left(\frac{150}{32.2}\right)v_2^2$$

$$v_2 = 21.5 \text{ ft/s} \qquad Ans.$$

R14–2. $r_{AB} = r_B - r_A = -4\mathbf{i} + 8\mathbf{j} - 9\mathbf{k}$

$$T_1 + \Sigma\int F\,ds = T_2$$

$$0 + 2(10 - 1) + \int_4^0 10\,dx + \int_0^8 6y\,dy$$

$$+ \int_{10}^1 2z\,dz = \frac{1}{2}\left(\frac{2}{32.2}\right)v^2 n$$

$$v_B = 47.8 \text{ ft/s} \qquad Ans.$$

R14–3. $T_1 + V_1 = T_2 + V_2$

$$0 + 1.5(10) = \frac{1}{2}\left(\frac{1.5}{32.2}\right)v_B^2$$

$$v_B = 25.4 \text{ ft/s} \qquad Ans.$$

R14–4. The work done by F depends upon the difference in the cord length $AC-BC$.

$$T_A + \Sigma U_{A-B} = T_B$$

$$0 + F\left[\sqrt{(0.3)^2 + (0.3)^2} - \sqrt{(0.3)^2 + (0.3 - 0.15)^2}\right]$$
$$- 0.5(9.81)(0.15)$$

$$-\frac{1}{2}(100)(0.15)^2 = \frac{1}{2}(0.5)(2.5)^2$$

$$F(0.0889) = 3.423$$

$$F = 38.5 \text{ N} \qquad\qquad Ans.$$

R14–5. $(+\uparrow) \qquad v^2 = v_0^2 + 2a_c(s - s_0)$

$$(12)^2 = 0 + 2a_c(10 - 0)$$

$$a_c = 7.20 \text{ ft/s}^2$$

$$+\uparrow\Sigma F_y = ma_y; \quad 2T - 50 = \frac{50}{32.2}(7.20)$$

$$T = 30.6 \text{ lb}$$

$$s_C + (s_C - s_M) = l$$

$$v_M = 2v_C$$

$$v_M = 2(12) = 24 \text{ ft/s}$$

$$P_0 = \mathbf{T} \cdot \mathbf{v} = 30.6(24) = 734.2 \text{ lb} \cdot \text{ft/s}$$

$$P_i = \frac{734.2}{0.74} = 992.1 \text{ lb} \cdot \text{ft/s} = 1.80 \text{ hp} \qquad Ans.$$

R14–6. $+\uparrow\Sigma F_y = m\, a_y; \quad 2(30) - 50 = \dfrac{50}{32.2}a_B$

$$a_B = 6.44 \text{ m/s}^2$$

$$(+\uparrow) \quad v^2 = v_0^2 + 2a_c(s - s_0)$$

$$v_B^2 = 0 + 2(6.44)(10 - 0)$$

$$v_B = 11.349 \text{ ft/s}$$

$$2s_B + s_M = l$$

$$2v_B = -v_M$$

$$v_M = -2(11.349) = 22.698 \text{ ft/s}$$

$$P_o = \mathbf{F} \cdot \mathbf{v} = 30(22.698) = 680.94 \text{ ft} \cdot \text{lb/s}$$

$$P_i = \frac{680.94}{0.76} = 895.97 \text{ ft} \cdot \text{lb/s}$$

$$P_i = 1.63 \text{ hp} \qquad\qquad Ans.$$

R14–7. $T_A + V_A = T_B + V_B$

$$0 + (0.25)(9.81)(0.6) + \frac{1}{2}(150)(0.6 - 0.1)^2$$

$$= \frac{1}{2}(0.25)(v_B)^2 + \frac{1}{2}(150)(0.4 - 0.1)^2$$

$$v_B = 10.4 \text{ m/s} \qquad\qquad Ans.$$

R14–8. $\dfrac{6}{z} = \dfrac{\sqrt{15^2 + 2^2}}{15}$

$$z = 5.95 \text{ ft}$$

$$T_1 + V_1 = T_2 + V_2$$

$$0 + 0 = \frac{1}{2}\left(\frac{10}{32.2}\right)v_2^2 + \frac{1}{2}\left(\frac{30}{32.2}\right)v_2^2$$
$$+ 10(5.95) - 30(5.95)$$

$$v_2 = 13.8 \text{ ft/s} \qquad\qquad Ans.$$

Chapter 15

R15–1. $(+\uparrow) \quad m(v_1)_y + \Sigma \displaystyle\int F_y\, dt = m(v_2)_y$

$$0 + N_p(t) - 58.86(t) = 0$$

$$N_p = 58.86 \text{ N}$$

$$(\stackrel{+}{\rightarrow}) \quad m(v_1)_x + \Sigma \int F_x\, dt = m(v_2)_x$$

$$6(3) - 0.2(58.86)(t) = 6(1)$$

$$t = 1.02 \text{ s} \qquad\qquad Ans.$$

R15–2. $+\nwarrow\Sigma F_x = 0; \quad N_B - 50(9.81)\cos 30° = 0$

$$N_B = 424.79 \text{ N}$$

$$(+\nearrow) \quad m(v_x)_1 + \Sigma \int F_x\, dt = m(v_x)_2$$

$$50(2) + \int_0^2 \left(300 + 120\sqrt{t}\right)dt - 0.4(424.79)(2)$$
$$- 50(9.81)\sin 30°(2) = 50v_2$$

$$v_2 = 1.92 \text{ m/s} \qquad\qquad Ans.$$

R15–3. The crate starts moving when

$$F = F_r = 0.6(196.2) = 117.72 \text{ N}$$

From the graph since

$$F = \frac{200}{5}t. \quad 0 \le t \le 5 \text{ s}$$

The time needed for the crate to start moving is

$$t = \frac{5}{200}(117.72) = 2.943 \text{ s}$$

Hence, the impulse due to F is equal to the area under the curve from $2.943 \text{ s} \le t \le 10 \text{ s}$

$$\stackrel{+}{\rightarrow} \quad m(v_x)_1 + \Sigma \int F_x\, dt = m(v_x)_2$$

$$0 + \int_{2.943}^{5} \frac{200}{5}t\, dt + \int_5^{10} 200\, dt$$
$$- (0.5)196.2(10 - 2.943) = 20v_2$$

$$40\left(\frac{1}{2}t^2\right)\Big|_{2.943}^{5} + 200(10-5) - 692.292 = 20v_2$$

$$634.483 = 20v_2$$

$$v_2 = 31.7 \text{ m/s} \qquad \text{Ans.}$$

R15–4. $(v_A)_1 = \left[20(10^3)\frac{\text{m}}{\text{h}}\right]\left(\frac{1\,\text{h}}{3600\,\text{s}}\right) = 5.556 \text{ m/s}$

$(v_B)_1 = \left[5(10^3)\frac{\text{m}}{\text{h}}\right]\left(\frac{1\,\text{h}}{3600\,\text{s}}\right) = 1.389 \text{ m/s},$

and $(v_C)_1 = \left[25(10^3)\frac{\text{m}}{\text{h}}\right]\left(\frac{1\,\text{h}}{3600\,\text{s}}\right) = 6.944 \text{ m/s}$

For the first case,

$(\xrightarrow{+}) \quad m_A(v_A)_1 + m_B(v_B)_1 = (m_A + m_B)v_2$

$10000(5.556) + 5000(1.389) = (10000 + 5000)v_{AB}$

$$v_{AB} = 4.167 \text{ m/s} \rightarrow$$

Using the result of v_{AB} and considering the second case,

$(\xrightarrow{+}) \quad (m_A + m_B)v_{AB} + m_C(v_C)_1$

$$= (m_A + m_B + m_C)v_{ABC}$$

$$(10000 + 5000)(4.167) + [-20000(6.944)]$$

$$= (10000 + 5000 + 20000)v_{ABC}$$

$$v_{ABC} = -2.183 \text{ m/s} = 2.18 \text{ m/s} \leftarrow \quad \text{Ans.}$$

R15–5. $(\xrightarrow{+}) \quad m_P(v_P)_1 + m_B(v_B)_1 = m_P(v_p)_2 + m_B(v_B)_2$

$$0.2(900) + 15(0) = 0.2(300) + 15(v_B)_2$$

$$(v_B)_2 = 8 \text{ m/s} \rightarrow \qquad \text{Ans.}$$

$(+\uparrow) \quad m(v_1)_y + \Sigma\int_{t_1}^{t_2} F_y\,dt = m(v_2)_y$

$$15(0) + N(t) - 15(9.81)(t) = 15(0)$$

$$N = 147.15 \text{ N}$$

$(\xrightarrow{+}) \quad m(v_1)_x + \Sigma\int_{t_1}^{t_2} F_x\,dt = m(v_2)_x$

$$15(8) + [-0.2(147.15)(t)] = 15(0)$$

$$t = 4.077 \text{ s} = 4.08 \text{ s} \qquad \text{Ans.}$$

R15–6. $(\xrightarrow{+}) \quad \Sigma mv_1 = \Sigma mv_2$

$$3(2) + 0 = 3(v_A)_2 + 2(v_B)_2$$

$(\xrightarrow{+}) \quad e = \dfrac{(v_B)_2 - (v_A)_2)}{(v_A)_1 - (v_B)_1}$

$$1 = \frac{(v_B)_2 - (v_A)_2}{2 - 0}$$

Solving

$$(v_A)_2 = 0.400 \text{ m/s} \rightarrow \qquad \text{Ans.}$$

$$(v_B)_2 = 2.40 \text{ m/s} \rightarrow \qquad \text{Ans.}$$

Block A:

$$T_1 + \Sigma U_{1-2} = T_2$$

$$\frac{1}{2}(3)(0.400)^2 - 3(9.81)(0.3)d_A = 0$$

$$d_A = 0.0272 \text{ m}$$

Block B:

$$T_1 + \Sigma U_{1-2} = T_2$$

$$\frac{1}{2}(2)(2.40)^2 - 2(9.81)(0.3)d_B = 0$$

$$d_B = 0.9786 \text{ m}$$

$$d = d_B - d_A = 0.951 \text{ m} \qquad \text{Ans.}$$

R15–7. $(v_A)_{x_1} = -2\cos 40° = -1.532 \text{ m/s}$

$(v_A)_{y_1} = -2\sin 40° = -1.285 \text{ m/s}$

$(\xrightarrow{+}) \quad m_A(v_A)_{x_1} + m_B(v_B)_{x_1} = m_A(v_A)_{x_2}$

$$\qquad\qquad\qquad + m_B(v_B)_{x_2}$$

$$-2(1.532) + 0 = 0.2(v_A)_{x_2}$$

$$\qquad\qquad + 0.2(v_B)_{x_2} \qquad (1)$$

$(\xrightarrow{+}) \quad e = \dfrac{(v_{ref})_2}{(v_{ref})_1}$

$$0.75 = \frac{(v_A)_{x_2} - (v_B)_{x_1}}{1.532} \qquad (2)$$

Solving Eqs. (1) and (2)

$$(v_A)_{x_2} = -0.1915 \text{ m/s}$$

$$(v_B)_{x_2} = -1.3405 \text{ m/s}$$

For A:

$(+\downarrow) \quad m_A(v_A)_{y_1} = m_A(v_A)_{y_2}$

$$(v_A)_{y_2} = 1.285 \text{ m/s}$$

For B:

$(+\uparrow) \quad m_B(v_B)_{y_1} = m_B(v_B)_{y_2}$

$$(v_B)_{y_2} = 0$$

Hence $(v_B)_2 = (v_B)_{x_2} = 1.34 \text{ m/s} \quad \leftarrow \qquad \text{Ans.}$

$(v_A)_2 = \sqrt{(-0.1915)^2 + (1.285)^2} = 1.30 \text{ m/s}$ Ans.

$(\theta_A)_2 = \tan^{-1}\left(\dfrac{0.1915}{1.285}\right) = 8.47°\measuredangle \qquad \text{Ans.}$

R15–8. $(H_z)_1 + \Sigma\int M_z\,dt = (H_z)_2$

$$(10)(2)(0.75) + 60(2)\left(\frac{3}{5}\right)(0.75) +$$

$$\int_0^2 (8t^2 + 5)\,dt = 10v(0.75)$$

$$69 + \left[\frac{8}{3}t^3 + 5t\right]_0^2 = 7.5v$$

$$v = 13.4 \text{ m/s} \qquad \text{Ans.}$$

Chapter 16

R16–1. $(\omega_A)_O = 60 \text{ rad}/s$

$\alpha_A = -1 \text{ rad}/s^2$

$\omega_A = (\omega_A)_O + \alpha_A t$

$\omega_A = 60 + (-1)(3) = 57 \text{ rad}/s$

$v_A = r\omega_A = (1)(57) = 57 \text{ ft}/s = v_B$

$\omega_B = \dfrac{v_B}{r} = 57/2 = 28.5 \text{ rad}/s$

$v_W = r_C\omega_C = (0.5)(28.5) = 14.2 \text{ ft}/s$ *Ans.*

$\alpha_A = 1$

$a_{A_t} = l(1) = 1 \text{ ft}/s^2$

$\alpha_B = \dfrac{1}{2} = 0.5 \text{ rad}/s^2$

$a_W = r\alpha_B = (0.5)(0.5) = 0.25 \text{ ft}/s^2$ *Ans.*

R16–2. $\alpha_a = 0.6\theta_A$

$\theta_C = \dfrac{0.5}{0.075} = 6.667 \text{ rad}$

$\theta_A(0.05) = (6.667)(0.15)$

$\theta_A = 20 \text{ rad}$

$\alpha d\theta = \omega d\omega$

$\displaystyle\int_0^{20} 0.6\theta_A d\theta_A = \int_3^{\omega_A} \omega_A d\omega_A$

$0.3\theta_A^2 \Big|_0^{20} = \dfrac{1}{2}\omega_A^2 \Big|_3^{\omega_A}$

$120 = \dfrac{1}{2}\omega_A^2 - 4.5$

$\omega_A = 15.780 \text{ rad}/s$

$15.780(0.05) = \omega_C(0.15)$

$\omega_C = 5.260 \text{ rad}/s$

$v_B = 5.260(0.075) = 0.394 \text{ m}/s$ *Ans.*

R16–3. A point on the drum which is in contact with the board has a tangential acceleration of

$a_t = 0.5 \text{ m}/s^2$

$a^2 = a_t^2 + a_n^2$

$(3)^2 = (0.5)^2 + a_n^2$

$a_n = 2.96 \text{ m}/s^2$

$a_n = \omega^2 r, \qquad \omega = \sqrt{\dfrac{2.96}{0.25}} = 3.44 \text{ rad}/s$

$v_B = \omega r = 3.44(0.25) = 0.860 \text{ m}/s$ *Ans.*

R16–4. $\mathbf{v}_B = \omega_{AB} \times \mathbf{r}_{B/A}$

$\mathbf{v}_C = \mathbf{v}_B + \omega \times \mathbf{r}_{C/B}$

$v_C \mathbf{i} = (6\mathbf{k}) \times (0.2 \cos 45°\mathbf{i} + 0.2 \sin 45°\mathbf{j}) +$

$\qquad\qquad (\omega\mathbf{k}) \times (0.5 \cos 30°\mathbf{i} - 0.5 \sin 30°\mathbf{j})$

$v_C = -0.8485 + \omega(0.25)$

$0 = 0.8485 + 0.433 \omega$

Solving

$\omega = 1.96 \text{ rad}/s \, \rotatebox{0}{$\circlearrowright$}$

$v_C = 1.34 \text{ m}/s$ *Ans.*

R16–5. $\omega = \dfrac{2}{0.08} = 25 \text{ rad}/s$

$\alpha = \dfrac{4}{0.08} = 50 \text{ rad}/s^2$

$\mathbf{a}_C = \mathbf{a}_A + (\mathbf{a}_{C/A})_n + (\mathbf{a}_{C/A})_t$

$\mathbf{a}_C = 4\mathbf{j} + (25)^2(0.08)\mathbf{i} + 50(0.08)\mathbf{j}$

$\xrightarrow{+} \quad a_C \cos\theta = 0 + 50$

$+\uparrow \quad a_C \sin\theta = 4 + 0 + 4$

Solving, $a_C = 50.6 \text{ m}/s^2$ *Ans.*

$\theta = 9.09° \, \measuredangle\theta$ *Ans.*

The cylinder moves up with an acceleration

$a_B = (a_C)_t = 50.6 \sin 9.09° = 8.00 \text{ m}/s^2 \uparrow$ *Ans.*

R16–6. $\mathbf{a}_C = \mathbf{a}_B + \mathbf{a}_{C/B}$

$2.057 + (a_C)_t = 1.8 + 1.2 + \alpha_{CB}(0.5)$

$\quad \rightarrow \qquad\qquad \downarrow \quad\; \downarrow \quad \leftarrow \quad \measuredangle\theta \, 30°$

$(\xrightarrow{+}) \quad 2.057 = -1.2 + \alpha_{CB}(0.5)\cos 30°$

$(+\downarrow) \quad (a_C)_t = 1.8 + \alpha_{CB}(0.5)\sin 30°$

$\qquad\qquad \alpha_{CB} = 7.52 \text{ rad}/s^2$ *Ans.*

$\qquad\qquad (a_C)_t = 3.68 \text{ m}/s^2$

$\qquad a_C = \sqrt{(3.68)^2 + (2.057)^2} = 4.22 \text{ m}/s^2$ *Ans.*

$\qquad \theta = \tan^{-1}\left(\dfrac{3.68}{2.057}\right) = 60.8° \, \measuredangle$ *Ans.*

Also,

$\mathbf{a}_C = \mathbf{a}_B + \alpha_{CB} \times \mathbf{r}_{C/B} - \omega^2 \mathbf{r}_{C/B}$

$-(a_C)_t\mathbf{j} + \dfrac{(0.6)^2}{0.175}\mathbf{i} = -(2)^2(0.3)\mathbf{i} - 6(0.3)\mathbf{j}$

$\qquad\qquad + (\alpha_{CB}\mathbf{k}) \times (-0.5 \cos 60°\mathbf{i} - 0.5 \sin 60°\mathbf{j}) - \mathbf{0}$

$2.057 = -1.20 + \alpha_{CB}(0.433)$

$-(a_C)_t = -1.8 - \alpha_{CB}(0.250)$

$\alpha_{CB} = 7.52 \text{ rad}/s^2$ *Ans.*

$a_t = 3.68 \text{ m}/s^2$

$a_C = \sqrt{(3.68)^2 + (2.057)^2} = 4.22 \text{ m}/s^2$ *Ans.*

$\theta = \tan^{-1}\left(\dfrac{3.68}{2.057}\right) = 60.8° \, \measuredangle\theta$ *Ans.*

R16–7. $a_C = 0.5(8) = 4 \text{ m/s}^2$

$\mathbf{a}_B = \mathbf{a}_C + \mathbf{a}_{B/C}$

$$\mathbf{a}_B = \begin{bmatrix} 4 \\ \leftarrow \end{bmatrix} + \begin{bmatrix} (3)^2(0.5) \\ \nearrow\ 30° \end{bmatrix} + \begin{bmatrix} (0.5)(8) \\ \diagdown\, 30° \end{bmatrix}$$

$(\overset{+}{\rightarrow})$ $(a_B)_x = -4 + 4.5 \cos 30° + 4 \sin 30°$

$\qquad\qquad = 1.897 \text{ m/s}^2$

$(+\uparrow)$ $(a_B)_y = 0 + 4.5 \sin 30° - 4 \cos 30°$

$\qquad\qquad = -1.214 \text{ m/s}^2$

$\qquad a_B = \sqrt{(1.897)^2 + (-1.214)^2}$

$\qquad\quad = 2.25 \text{ m/s}^2$ *Ans.*

$\qquad \theta = \tan^{-1}\left(\dfrac{1.214}{1.897}\right) = 32.6°$ ⦣   *Ans.*

Also,

$\mathbf{a}_B = \mathbf{a}_C + \alpha \times \mathbf{r}_{B/C} - \omega^2 \mathbf{r}_{B/C}$

$(a_B)_x\mathbf{i} + (a_B)_y\mathbf{j} = -4\mathbf{i} + (8\mathbf{k}) \times (-0.5 \cos 30°\mathbf{i}$

$\qquad -0.5 \sin 30°\mathbf{j}) - (3)^2(-0.5 \cos 30°\mathbf{i} - 0.5 \sin 30°\mathbf{j})$

$(\overset{+}{\rightarrow})\ (a_B)_x = -4 + 8(0.5 \sin 30°) + (3)^2(0.5 \cos 30°)$

$\qquad\qquad = 1.897 \text{ m/s}^2$

$(+\uparrow)\ (a_B)_y = 0 - 8(0.5 \cos 30°) + (3)^2(0.5 \sin 30°)$

$\qquad\qquad = -1.214 \text{ m/s}^2$

$\qquad \theta = \tan^{-1}\left(\dfrac{1.214}{1.897}\right) = 32.6°$ ⦣   *Ans.*

$\qquad a_B = \sqrt{(1.897)^2 + (-1.214)^2} = 2.25 \text{ m/s}^2$ *Ans.*

R16–8. $v_B = 3(7) = 21 \text{ in./s} \leftarrow$

$\mathbf{v}_C = \mathbf{v}_B + \omega \times \mathbf{r}_{C/B}$

$-v_C\left(\dfrac{4}{5}\right)\mathbf{i} - v_C\left(\dfrac{3}{5}\right)\mathbf{j} = -21\mathbf{i} + \omega\mathbf{k} \times (-5\mathbf{i} - 12\mathbf{j})$

$(\overset{+}{\rightarrow})$ $-0.8v_C = -21 + 12\omega$

$(+\uparrow)$ $-0.6v_C = -5\omega$

Solving:

$\omega = 1.125 \text{ rad/s}$

$v_C = 9.375 \text{ in./s} = 9.38 \text{ in./s}$ ⦣   *Ans.*

$(a_B)_n = (3)^2(7) = 63 \text{ in./s}^2 \downarrow$

$(a_B)_t = (2)(7) = 14 \text{ in./s}^2 \leftarrow$

$\mathbf{a}_C = \mathbf{a}_B + \alpha \times \mathbf{r}_{C/B} - \omega^2 \mathbf{r}_{C/B}$

$-a_C\left(\dfrac{4}{5}\right)\mathbf{i} - a_C\left(\dfrac{3}{5}\right)\mathbf{j} = -14\mathbf{i} - 63\mathbf{j} + (\alpha\mathbf{k})$

$\qquad\qquad \times (-5\mathbf{i} - 12\mathbf{j}) - (1.125)^2(-5\mathbf{i} - 12\mathbf{j})$

$(\overset{+}{\rightarrow})$ $-0.8a_C = -14 + 12\alpha + 6.328$

$(+\uparrow)$ $-0.6a_C = -63 - 5\alpha + 15.1875$

$\qquad\qquad a_C = 54.7 \text{ in./s}^2$ ⦣   *Ans.*

$\qquad\qquad \alpha = -3.00 \text{ rad/s}^2$

Chapter 17

R17–1. $\overset{+}{\leftarrow}\Sigma F_x = ma_x;$ $50 \cos 60° = 200a_G$ **(1)**

$\qquad +\uparrow\Sigma F_y = ma_y;$ $N_A + N_B - 200(9.81)$

$\qquad\qquad\qquad\qquad\qquad -50 \sin 60° = 0$ **(2)**

$\qquad \zeta + \Sigma M_G = 0;$ $-N_A(0.3) + N_B(0.2) +$

$\qquad\qquad\qquad\qquad\qquad 50 \cos 60°(0.3)$

$\qquad\qquad -50 \sin 60°(0.6) = 0$ **(3)**

Solving,

$\qquad a_G = 0.125 \text{ m/s}^2$

$\qquad N_A = 765.2 \text{ N}$

$\qquad N_B = 1240 \text{ N}$

At each wheel

$\qquad N'_A = \dfrac{N_A}{2} = 383 \text{ N}$ *Ans.*

$\qquad N'_B = \dfrac{N_B}{2} = 620 \text{ N}$ *Ans.*

R17–2.   *Curvilinear Translation*:

$\qquad (a_G)_t = 8(3) = 24 \text{ ft/s}^2$

$\qquad (a_G)_n = (5)^2(3) = 75 \text{ ft/s}^2$

$\qquad \bar{x} = \dfrac{\Sigma \bar{x}m}{\Sigma m} = \dfrac{1(3) + 2(3)}{6} = 1.5 \text{ ft}$

$\qquad +\downarrow\Sigma F_y = m(a_G)_y;$ $E_y + 6 = \dfrac{6}{32.2}(24)\cos 30°$

$\qquad\qquad\qquad\qquad\qquad\qquad + \dfrac{6}{32.2}(75)\sin 30°$

$\qquad \overset{+}{\rightarrow}\Sigma F_x = m(a_G)_x;$ $E_x = \dfrac{6}{32.2}(75)\cos 30°$

$\qquad\qquad\qquad\qquad\qquad\qquad - \dfrac{6}{32.2}(24)\sin 30°$

$\qquad \zeta + \Sigma M_G = 0;$ $M_E - E_y(1.5) = 0$

$\qquad E_x = 9.87 \text{ lb}$ *Ans.*

$\qquad E_y = 4.86 \text{ lb}$ *Ans.*

$\qquad M_E = 7.29 \text{ lb} \cdot \text{ft}$ *Ans.*

R17–3.   (a) Rear wheel drive

     Equations of motion:

$\qquad \overset{+}{\rightarrow}\Sigma F_x = m(a_G)_x;$ $0.3N_B = 1.5(10)^3 a_G$ **(1)**

$\qquad \zeta + \Sigma M_A = \Sigma(M_k)_A;$ $1.5(10)^3(9.81)(1.3)$

$$-N_B(2.9) = -1.5(10)^3 a_G(0.4) \qquad \textbf{(2)}$$

Solving Eqs. (1) and (2) yields:

$$N_B = 6881 \text{ N} = 6.88 \text{ kN}$$

$$a_G = 1.38 \text{ m/s}^2 \qquad \textit{Ans.}$$

R17–4. $\quad \xrightarrow{+} \Sigma F_x = m(a_G)_x; \quad 40 \sin 60° + N_C - \left(\dfrac{5}{13}\right)T = 0$

$$+\uparrow \Sigma F_y = m(a_G)_y; \quad -40 \cos 60° + 0.3N_C$$

$$- 20(9.81) + \dfrac{12}{13}T = 0$$

$$\zeta + \Sigma M_A = I_A \alpha; \quad 40(0.120) - 0.3N_C(0.120)$$

$$= \left[\dfrac{1}{2}(20)(0.120)^2\right]\alpha$$

Solving,

$$T = 218 \text{ N} \qquad \textit{Ans.}$$

$$N_C = 49.28 \text{ N}$$

$$\alpha = 21.0 \text{ rad/s}^2 \qquad \textit{Ans.}$$

R17–5. $\quad (a_G)_t = 4\alpha$

$$\xleftarrow{+} \Sigma F_t = m(a_G)_x; \quad F + 20 - 5 = \dfrac{30}{32.2}(4\alpha)$$

$$\zeta + \Sigma M_O = I_O \alpha; \quad 20(3) + F(6) = \dfrac{1}{3}\left(\dfrac{30}{32.2}\right)(8)^2\alpha$$

Solving,

$$\alpha = 12.1 \text{ rad/s}^2 \qquad \textit{Ans.}$$

$$F = 30.0 \text{ lb} \qquad \textit{Ans.}$$

R17–6. $\quad I_O = \dfrac{2}{5}\left(\dfrac{30}{32.2}\right)(1)^2 + \left(\dfrac{30}{32.2}\right)(3)^2$

$$+ \dfrac{1}{3}\left(\dfrac{10}{32.2}\right)(2)^2 = 9.17 \text{ slug} \cdot \text{ft}^2$$

$$\bar{x} = \dfrac{30(3) + 10(1)}{30 + 10} = 2.5 \text{ ft}$$

$$\xrightarrow{+} \Sigma F_n = ma_n; \quad O_x = 0$$

$$+\downarrow \Sigma F_t = ma_t; \quad 40 - O_y = \dfrac{40}{32.2}a_G$$

$$\zeta + \Sigma M_O = I_O \alpha; \quad 40(2.5) = 9.17\alpha$$

Kinematics

$$a_G = 2.5\alpha$$

Solving,

$$\alpha = 10.90 \text{ rad/s}^2$$

$$a_G = 27.3 \text{ ft/s}^2$$

$$O_x = 0$$

$$O_y = 6.14 \text{ lb}$$

Thus:

$$F_o = 6.14 \text{ lb} \rightarrow \qquad \textit{Ans.}$$

R17–7. $\quad +\uparrow \Sigma F_y = m(a_G)_y; \quad N_B - 20(9.81) = 0$

$$N_B = 196.2 \text{ N}$$

$$F_B = 0.1(196.2) = 19.62 \text{ N}$$

$$\zeta + \Sigma M_{IC} = \Sigma(M_k)_{IC}; \quad 30 - 19.62(0.6)$$

$$= 20(0.2\alpha)(0.2) + [20(0.25)^2]\alpha$$

$$\alpha = 8.89 \text{ rad/s}^2 \qquad \textit{Ans.}$$

R17–8. $\quad \xrightarrow{+} \Sigma F_x = m(a_G)_x; \quad 0.3N_A = \dfrac{20}{32.2}a_G$

$$+\uparrow \Sigma F_y = m(a_G)_y; \quad N_A - 20 = 0$$

$$\zeta + \Sigma M_G = I_G \alpha; \quad 0.3N_A(0.5)$$

$$= \left[\dfrac{2}{5}\left(\dfrac{20}{32.2}\right)(0.5)^2\right]\alpha$$

Solving,

$$N_A = 20 \text{ lb}$$

$$a_G = 9.66 \text{ ft/s}^2$$

$$\alpha = 48.3 \text{ rad/s}^2$$

$$(\zeta +) \quad \omega = \omega_0 + \alpha_c t$$

$$0 = \omega_1 - 48.3t$$

$$\omega_1 = 48.3t$$

$$(\xrightarrow{+}) \quad v = v_0 + a_c t$$

$$0 = 20 - 9.66\left(\dfrac{\omega}{48.3}\right)$$

$$\omega = 100 \text{ rad/s} \qquad \textit{Ans.}$$

Chapter 18

R18–1. $\qquad T_1 + \Sigma U_{1-2} = T_2$

$$0 + (50)(9.81)(1.25) = \dfrac{1}{2}\left[(50)(1.75)^2\right]\omega_2^2$$

$$\omega_2 = 2.83 \text{ rad/s} \qquad \textit{Ans.}$$

R18–2. *Kinetic Energy and Work:* The mass moment inertia of the flywheel about its mass center is $I_O = mk_O^2 = 50(0.2^2) = 2 \text{ kg} \cdot \text{m}^2$. Thus,

$$T = \dfrac{1}{2}I_O\omega^2 = \dfrac{1}{2}(2)\omega^2 = \omega^2$$

Since the wheel is initially at rest, $T_1 = 0$. **W**, **O**$_x$, and **O**$_y$ do no work while **M** does positive work. When the wheel rotates

$$\theta = (5 \text{ rev})\left(\dfrac{2\pi \text{ rad}}{1 \text{ rev}}\right) = 10\pi, \text{ the work done by } M \text{ is}$$

$$U_M = \int M\,d\theta = \int_0^{10\pi} (9\theta^{1/2} + 1)\,d\theta$$

$$= (6\theta^{3/2} + \theta)\Big|_0^{10\pi}$$

$$= 1087.93 \text{ J}$$

Principle of Work and Energy:

$$T_1 + \Sigma U_{1-2} = T_2$$

$$0 + 1087.93 = \omega^2$$

$$\omega = 33.0 \text{ rad/s} \qquad \textit{Ans.}$$

R18–3. Before braking:

$$T_1 + \Sigma U_{1-2} = T_2$$

$$0 + 15(9.81)(3) = \frac{1}{2}(15)v_B^2 + \frac{1}{2}\big[50(0.23)^2\big]\Big(\frac{v_B}{0.15}\Big)^2$$

$$v_B = 2.58 \text{ m/s} \qquad \textit{Ans.}$$

$$\frac{s_B}{0.15} = \frac{s_C}{0.25}$$

Set $s_B = 3$ m, then $s_C = 5$ m.

$$T_1 + \Sigma U_{1-2} = T_2$$

$$0 - F(5) + 15(9.81)(6) = 0$$

$$F = 176.6 \text{ N}$$

$$N = \frac{176.6}{0.5} = 353.2 \text{ N}$$

Brake arm:

$$\zeta + \Sigma M_A = 0; \qquad -353.2(0.5) + P(1.25) = 0$$

$$P = 141 \text{ N} \qquad \textit{Ans.}$$

R18–4.

$$\frac{s_G}{0.3} = \frac{s_A}{(0.5 - 0.3)}$$

$$s_A = 0.6667 s_G$$

$$+\nwarrow\Sigma F_y = 0; \qquad N_A - 60(9.81)\cos 30° = 0$$

$$N_A = 509.7 \text{ N}$$

$$T_1 + \Sigma U_{1-2} = T_2$$

$$0 + 60(9.81)\sin 30°(s_G) - 0.2(509.7)(0.6667 s_G)$$

$$= \frac{1}{2}\big[60(0.3)^2\big](6)^2$$

$$+ \frac{1}{2}(60)\big[(0.3)(6)\big]^2$$

$$s_G = 0.859 \text{ m} \qquad \textit{Ans.}$$

R18–5. **Conservation of Energy:** Originally, both gears are rotating with an angular velocity of

$$\omega_1 = \frac{2}{0.05} = 40 \text{ rad/s. After the rack has traveled}$$

$s = 600$ mm, both gears rotate with an angular velocity of $\omega_2 = \dfrac{v_2}{0.05}$, where v_2 is the speed of the rack at that moment.

$$T_1 + V_1 = T_2 + V_2$$

$$\frac{1}{2}(6)(2)^2 + 2\Big\{\frac{1}{2}\big[4(0.03)^2\big](40)^2\Big\} + 0$$

$$= \Big\{\frac{1}{2}\big[4(0.03)^2\big]\Big(\frac{v_2}{0.05}\Big)^2\Big\} - 6(9.81)(0.6)$$

$$v_2 = 3.46 \text{ m/s} \qquad \textit{Ans.}$$

R18–6. Datum through A.

$$T_1 + V_1 = T_2 + V_2$$

$$\frac{1}{2}\Big[\frac{1}{3}\Big(\frac{50}{32.2}\Big)(6)^2\Big](2)^2 + \frac{1}{2}(6)(4-2)^2$$

$$= \frac{1}{2}\Big[\frac{1}{3}\Big(\frac{50}{32.2}\Big)(6)^2\Big]\omega^2 + \frac{1}{2}(6)(7-2)^2 - 50(1.5)$$

$$\omega = 2.30 \text{ rad/s} \qquad \textit{Ans.}$$

R18–7.

$$T_1 + V_1 = T_2 + V_2$$

$$0 + 4(1.5 \sin 45°) + 1(3 \sin 45°)$$

$$= \frac{1}{2}\Big[\frac{1}{3}\Big(\frac{4}{32.2}\Big)(3)^2\Big]\Big(\frac{v_C}{3}\Big)^2 + \frac{1}{2}\Big(\frac{1}{32.2}\Big)(v_C)^2 + 0$$

$$v_C = 13.3 \text{ ft/s} \qquad \textit{Ans.}$$

R18–8. Datum at lowest point.

$$T_1 + V_1 = T_2 + V_2$$

$$\frac{1}{2}\Big[\frac{1}{2}(40)(0.3)^2\Big]\Big(\frac{4}{0.3}\Big)^2 + \frac{1}{2}(40)(4)^2$$

$$+ 40(9.81)d \sin 30° = 0 + \frac{1}{2}(200)d^2$$

$$100d^2 - 196.2d - 480 = 0$$

Solving for the positive root,

$$d = 3.38 \text{ m} \qquad \textit{Ans.}$$

Chapter 19

R19–1. $\quad I_O = mk_O^2 = \dfrac{150}{32.2}(1.25)^2 = 7.279 \text{ slug} \cdot \text{ft}^2$

$$I_O\omega_1 + \Sigma \int_{t_1}^{t_2} M_O\,dt = I_O\omega_2$$

$$0 - \int_0^{3\text{ s}} 10t^2(1)\,dt = 7.279\omega_2$$

$$\frac{10t^3}{3}\Big|_0^{3\text{ s}} = 7.279\omega_2$$

$$\omega_2 = 12.4 \text{ rad/s} \qquad \textit{Ans.}$$

R19–3. $+\swarrow m(v_G)_1 + \Sigma \int F\,dt = m(v_G)_2$

$$0 + 9(9.81)(\sin 30°)(3) - \int_0^3 F\,dt = 9(v_G)_2 \quad \textbf{(1)}$$

$\zeta + (H_G)_1 + \Sigma \int M_G dt = (H_G)_2$

$$0 + \left(\int_0^3 F\,dt\right)(0.3) = \left[9(0.225)^2\right]\omega_2 \quad \textbf{(2)}$$

Since $(v_G)_2 = 0.3\omega_2$,

Eliminating $\int_0^3 F\,dt$ from Eqs. (1) and (2) and solving for $(v_G)_2$ yields.

$$(v_G)_2 = 9.42 \text{ m/s} \qquad Ans.$$

Also,

$\zeta + (H_A)_1 + \Sigma \int M_A dt = (H_A)_2$

$$0 + 9(9.81)\sin 30°(3)(0.3) = \left[9(0.225)^2 + 9(0.3)^2\right]\omega$$

$$\omega = 31.39 \text{ rad/s}$$

$$v = 0.3(31.39) = 9.42 \text{ m/s} \qquad Ans.$$

R19–4. $\xleftarrow{+} \qquad m(v_x)_1 + \Sigma \int_{t_1}^{t_2} F_x dt = m(v_x)_2$

$$0 + 200(3) =^{t_1} 100(v_O)_2$$

$$(v_O)_2 = 6 \text{ m/s} \qquad Ans.$$

and

$$I_z\omega_1 + \Sigma \int_{t_1}^{t_2} M_z dt = I_z\omega_2$$

$$0 - [200(0.4)(3)] = -9\omega_2$$

$$\omega_2 = 26.7 \text{ rad/s} \qquad Ans.$$

R19–5. $(+\uparrow) \qquad mv_1 + \Sigma \int F\,dt = mv_2$

$$0 + T(3) - 30(3) + 40(3) = \frac{30}{32.2}v_o$$

$(\zeta+) \qquad (H_O)_1 + \Sigma \int M_O dt = (H_O)_2$

$$-T(0.5)3 + 40(1)3 = \left[\frac{30}{32.2}(0.65)^2\right]\omega$$

Kinematics,

$$v_o = 0.5\omega$$

Solving,

$$T = 23.5 \text{ lb}$$
$$\omega = 215 \text{ rad/s} \qquad Ans.$$
$$v_O = 108 \text{ ft/s}$$

Also,

$(\zeta+) \qquad (H_{IC})_1 + \Sigma \int M_{IC} dt = (H_{IC})_2$

$$0 - 30(0.5)(3) + 40(1.5)(3)$$

$$= \left[\frac{30}{32.2}(0.65)^2 + \frac{30}{32.2}(0.5)^2\right]\omega$$

$$\omega = 215 \text{ rad/s} \qquad Ans.$$

R19–6. $\zeta + \quad (H_A)_1 + \Sigma \int M_A dt = (H_A)_2$

$$\left[\frac{30}{32.2}(0.8)^2\right](6) - \int T\,dt(1.2) = \left[\frac{30}{32.2}(0.8)^2\right]\omega_A$$

$\zeta + \quad (H_B)_1 + \Sigma \int M_B dt = (H_B)_2$

$$0 + \int T\,dt(0.4) = \left[\frac{15}{32.2}(0.6)^2\right]\omega_B$$

Kinematics:

$$1.2\omega_A = 0.4\omega_B$$

$$\omega_B = 3\omega_A$$

Thus,

$$\omega_A = 1.70 \text{ rad/s} \qquad Ans.$$
$$\omega_B = 5.10 \text{ rad/s} \qquad Ans.$$

R19–7. $H_1 = H_2$

$$\left(\frac{1}{2}mr^2\right)\omega_1 = \left[\frac{1}{2}mr^2 + mr^2\right]\omega_2$$

$$\omega_2 = \frac{1}{3}\omega_1 \qquad Ans.$$

R19–8. $H_1 = H_2$

$$(0.940)(0.5) + (4)\left[\frac{1}{12}(20)\left((0.75)^2 + (0.2)^2\right)\right.$$

$$\left. + (20)(0.375 + 0.2)^2\right](0.5)$$

$$= (0.940)(\omega) + 4\left[\frac{1}{12}(20)(0.2)^2 + (20)(0.2)^2\right]\omega$$

$$\omega = 3.56 \text{ rad/s} \qquad Ans.$$

Answers to Selected Problems

Chapter 12

12–1. $s = 80.7$ m

12–2. $s = 20$ ft

12–3. $a = -24$ m/s^2, $\Delta s = -880$ m, $s_T = 912$ m

12–5. $s_T = 8$ m, $v_{avg} = 2.67$ m/s

12–6. $s|_{t=6\,s} = -27.0$ ft, $s_{tot} = 69.0$ ft

12–7. $v_{avg} = 0$, $(v_{sp})_{avg} = 3$ m/s, $a\big|_{t=6\,s} = 2$ m/s^2

12–9. $v = 32$ m/s, $s = 67$ m, $d = 66$ m

12–10. $v = 1.29$ m/s

12–11. $v_{avg} = 0.222$ m/s, $(v_{sp})_{avg} = 2.22$ m/s

12–13. Normal: $d = 517$ ft, drunk: $d = 616$ ft

12–14. $v = 165$ ft/s, $a = 48$ ft/s^2, $s_T = 450$ ft, $v_{avg} = 25.0$ ft/s, $(v_{sp})_{avg} = 45.0$ ft/s

12–15. $v = \left(2kt + \dfrac{1}{v_0^2}\right)^{-1/2}$, $s = \dfrac{1}{k}\left[\left(2kt + \dfrac{1}{v_0^2}\right)^{1/2} - \dfrac{1}{v_0}\right]$

12–17. $d = 16.9$ ft

12–18. $t = 5.62$ s

12–19. $s = 28.4$ km

12–21. $s = 123$ ft, $a = 2.99$ ft/s^2

12–22. $h = 314$ m, $v = 72.5$ m/s

12–23. $v = (20e^{-2t})$ m/s, $a = (-40e^{-2t})$ m/s^2, $s = 10(1 - e^{-2t})$ m

12–25. (a) $v = 45.5$ m/s, (b) $v_{max} = 100$ m/s

12–26. (a) $s = -30.5$ m,
(b) $s_{Tot} = 56.0$ m,
(c) $v = 10$ m/s

12–27. $t = 0.549\left(\dfrac{v_f}{g}\right)$

12–29. $h = 20.4$ m, $t = 2$ s

12–30. $s = 54.0$ m

12–31. $s = \dfrac{v_0}{k}(1 - e^{-kt})$, $a = -kv_0e^{-kt}$

12–33. $v = 11.2$ km/s

12–34. $v = -R\sqrt{\dfrac{2g_0(y_0 - y)}{(R + y)(R + y_0)}}$, $v_{imp} = 3.02$ km/s

12–35. $t' = 27.3$ s.
When $t = 27.3$ s, $v = 13.7$ ft/s.

12–37. $\Delta s = 1.11$ km

12–38. $a|_{t=0} = -4$ m/s^2, $a|_{t=2\,s} = 0$, $a|_{t=4\,s} = 4$ m/s^2, $v|_{t=0} = 3$ m/s, $v|_{t=2\,s} = -1$ m/s, $v|_{t=4\,s} = 3$ m/s

12–39. $s = 2\sin\left(\dfrac{\pi}{5}t\right) + 4$, $v = \dfrac{2\pi}{5}\cos\left(\dfrac{\pi}{5}t\right)$, $a = -\dfrac{2\pi^2}{25}\sin\left(\dfrac{\pi}{5}t\right)$

12–41. $t = 7.48$ s. When $t = 2.14$ s, $v = v_{max} = 10.7$ ft/s, $h = 11.4$ ft.

12–42. $s = 600$ m. For $0 \le t < 40$ s, $a = 0$. For 40 s $< t \le 80$ s, $a = -0.250$ m/s^2.

12–43. $t' = 35$ s
For $0 \le t < 10$ s, $s = \{300t\}$ ft, $v = 300$ ft/s
For 10 s $< t < 20$ s,
$s = \left\{\dfrac{1}{6}t^3 - 15t^2 + 550t - 1167\right\}$ ft
$v = \left\{\dfrac{1}{2}t^2 - 30t + 550\right\}$ ft/s
For 20 s $< t \le 35$ s,
$s = \{-5t^2 + 350t + 167\}$ ft
$v = (-10t + 350)$ ft/s

12–45. When $t = 0.1$ s, $s = 0.5$ m and a changes from 100 m/s^2 to -100 m/s^2. When $t = 0.2$ s, $s = 1$ m.

12–46. $v\big|_{s=75\,ft} = 27.4$ ft/s, $v\big|_{s=125\,ft} = 37.4$ ft/s

12–47. For $0 \le t < 30$ s, $v = \left\{\dfrac{1}{5}t^2\right\}$ m/s, $s = \left\{\dfrac{1}{15}t^3\right\}$ m
For $30 \le t \le 60$ s, $v = \{24t - 540\}$ m/s, $s = \{12t^2 - 540t + 7200\}$ m

12–49. $v_{max} = 100$ m/s, $t' = 40$ s

12–50. For $0 \le s < 300$ ft, $v = \{4.90\,s^{1/2}\}$ m/s.
For 300 ft $< s \le 450$ ft,
$v = \{(-0.04s^2 + 48s - 3600)^{1/2}\}$ m/s.
$s = 200$ ft when $t = 5.77$ s.

12–51. For $0 \le t < 60$ s, $s = \left\{\dfrac{1}{20}t^2\right\}$ m, $a = 0.1$ m/s^2.
For 60 s $< t < 120$ s, $s = \{6t - 180\}$ m, $a = 0$.
For 120 s $< t \le 180$ s, $s = \left\{\dfrac{1}{30}t^2 - 2t + 300\right\}$ m, $a = 0.0667$ m/s^2.

12–53. At $t = 8$ s, $a = 0$ and $s = 30$ m.
At $t = 12$ s, $a = -1$ m/s^2 and $s = 48$ m.

12–54. For $0 \le t < 5$ s, $s = \{0.2t^3\}$ m, $a = \{1.2t\}$ m/s^2
For 5 s $< t \le 15$ s, $s = \left\{\dfrac{1}{4}(90t - 3t^2 - 275)\right\}$ m
$a = -1.5$ m/s^2,
At $t = 15$ s, $s = 100$ m, $v_{avg} = 6.67$ m/s

12–55. $t' = 33.3$ s, $s|_{t=5\,s} = 550$ ft, $s|_{t=15\,s} = 1500$ ft, $s|_{t=20\,s} = 1800$ ft, $s|_{t=33.3\,s} = 2067$ ft

12–57. For $0 \le s < 100$ ft, $v = \left\{\sqrt{\dfrac{1}{50}(800s - s^2)}\right\}$ ft/s
For 100 ft $< s \le 150$ ft,
$v = \left\{\dfrac{1}{5}\sqrt{-3s^2 + 900s - 25\,000}\right\}$ ft/s

733

12–58. For $0 \le t < 15$ s, $v = \left\{ \dfrac{1}{2} t^2 \right\}$ m/s, $s = \left\{ \dfrac{1}{6} t^3 \right\}$ m.

For 15 s $< t \le 40$ s,
$v = \{ 20t - 187.5 \text{ m/s} \}$,
$s = \{ 10t^2 - 187.5t + 1125 \}$ m

12–59. $s_T = 980$ m

12–61. When $t = 5$ s, $s_B = 62.5$ m.
When $t = 10$ s, $v_A = (v_A)_{max} = 40$ m/s and
$s_A = 200$ m.
When $t = 15$ s, $s_A = 400$ m and $s_B = 312.5$ m.
$\Delta s = s_A - s_B = 87.5$ m

12–62. $v = \{ 5 - 6t \}$ ft/s, $a = -6$ ft/s^2

12–63. For $0 \le t < 5$ s, $s = \{ 2t^2 \}$ m and $a = 4$ m/s^2.
For 5 s $< t < 20$ s, $s = \{ 20t - 50 \}$ m and $a = 0$.
For 20 s $< t \le 30$ s, $s = \{ 2t^2 - 60t + 750 \}$ m
and $a = 4$ m/s^2.

12–65. $v = 354$ ft/s, $t = 5.32$ s

12–66. When $s = 100$ m, $t = 10$ s.
When $s = 400$ m, $t = 16.9$ s.
$a|_{s=100 \text{ m}} = 4$ m/s^2, $a|_{s=400 \text{ m}} = 16$ m/s^2

12–67. At $s = 100$ m, a changes from $a_{max} = 1.5$ ft/s^2
to $a_{min} = -0.6$ ft/s^2.

12–69. $a = 5.31$ m/s^2, $\alpha = 53.0°$
$\beta = 37.0°$, $\gamma = 90.0°$

12–70. $\Delta \mathbf{r} = \{ 6\mathbf{i} + 4\mathbf{j} \}$ m

12–71. (4 ft, 2 ft, 6 ft)

12–73. (5.15 ft, 1.33 ft)

12–74. $\mathbf{r} = \{ 11\mathbf{i} + 2\mathbf{j} + 21\mathbf{k} \}$ ft

12–75. $(v_{sp})_{avg} = 4.28$ m/s

12–77. $v = 8.55$ m/s, $a = 5.82$ m/s^2

12–78. $v = 1003$ m/s, $a = 103$ m/s^2

12–79. $d = 4.00$ ft, $a = 37.8$ ft/s^2

12–81. $(\mathbf{v}_{BC})_{avg} = \{ 3.88\mathbf{i} + 6.72\mathbf{j} \}$ m/s

12–82. $v = \sqrt{c^2 k^2 + b^2}$, $a = ck^2$

12–83. $v = 10.4$ m/s, $a = 38.5$ m/s^2

12–85. $d = 204$ m, $v = 41.8$ m/s $a = 4.66$ m/s^2

12–86. $\theta = 58.3°$, $(v_0)_{min} = 9.76$ m/s

12–87. $\theta = 76.0°$, $v_A = 49.8$ ft/s, $h = 39.7$ ft

12–89. $R_{max} = 10.2$ m, $\theta = 45°$

12–90. $R = 8.83$ m

12–91. (13.3 ft, −7.09 ft)

12–93. $d = 166$ ft

12–94. $t = 3.57$ s, $v_B = 67.4$ ft/s

12–95. $v_A = 36.7$ ft/s, $h = 11.5$ ft

12–97. $v_A = 19.4$ m/s, $v_B = 40.4$ m/s

12–98. $v_A = 39.7$ ft/s, $s = 6.11$ ft

12–99. $v_B = 160$ m/s, $h_B = 427$ m,
$h_C = 1.08$ km, $R = 2.98$ km

12–101. $v_{min} = 0.838$ m/s, $v_{max} = 1.76$ m/s

12–102. $\theta_A = 11.6°$, $t = 0.408$ s, $\theta_B = 11.6°$ ⭦

12–103. $\theta_A = 78.4°$, $t = 2.00$ s, $\theta_B = 78.4°$ ⭦

12–105. $t_A = 0.553$ s, $x = 3.46$ m

12–106. $R = 19.0$ m, $t = 2.48$ s

12–107. $\theta_1 = 24.9°$ ⭦, $\theta_2 = 85.2°$ ⭧

12–109. $\theta = 76.0°$, $v_A = 49.8$ ft/s, $h = 39.7$ ft

12–110. $v = 63.2$ ft/s

12–111. $v = 38.7$ m/s

12–113. $v = 4.40$ m/s, $a_t = 5.04$ m/s^2, $a_n = 1.39$ m/s^2

12–114. $a_t = 8.66$ ft/s^2, $\rho = 1280$ ft

12–115. $v = 97.2$ ft/s, $a = 42.6$ ft/s^2

12–117. When cars A and B are side by side, $t = 55.7$ s.
When cars A and B are 90° apart, $t = 27.8$ s.

12–118. $t = 66.4$ s

12–119. $h = 5.99$ Mm

12–121. $a = 2.75$ m/s^2

12–122. $a = 1.68$ m/s^2

12–123. $v = 1.5$ m/s, $a = 0.117$ m/s^2

12–125. $v = 43.0$ m/s, $a = 6.52$ m/s^2

12–126. $v = 105$ ft/s, $a = 22.7$ ft/s^2

12–127. $a_t = 3.62$ m/s^2, $\rho = 29.6$ m

12–129. $t = 7.00$ s, $s = 98.0$ m

12–130. $a = 7.42$ ft/s^2

12–131. $a = 2.36$ m/s^2

12–133. $a = 3.05$ m/s^2

12–134. $a = 0.763$ m/s^2

12–135. $a = 0.952$ m/s^2

12–137. $y = -0.0766x^2$, $v = 8.37$ m/s,
$a_n = 9.38$ m/s^2, $a_t = 2.88$ m/s^2

12–138. $v_B = 19.1$ m/s, $a = 8.22$ m/s^2, $\phi = 17.3°$
up from negative−t axis

12–139. $a_{min} = 3.09$ m/s^2

12–141. $(a_n)_A = g = 32.2$ ft/s^2, $(a_t)_A = 0$,
$\rho_A = 699$ ft, $(a_n)_B = 14.0$ ft/s^2,
$(a_t)_B = 29.0$ ft/s^2, $\rho_B = 8.51(10^3)$ ft

12–142. $t = 1.21$ s

12–143. $a_{max} = \dfrac{v^2 a}{b^2}$

12–145. $d = 11.0$ m, $a_A = 19.0$ m/s^2, $a_B = 12.8$ m/s^2

12–146. $t = 2.51$ s, $a_A = 22.2$ m/s^2, $a_B = 65.1$ m/s^2

12–147. $\theta = 10.6°$

12–149. $a = 0.511$ m/s^2

12–150. $a = 0.309$ m/s^2

12–151. $a = 322$ mm/s^2, $\theta = 26.6°$ ⭦

12–153. $v_n = 0$, $v_t = 7.21$ m/s,
$a_n = 0.555$ m/s^2, $a_t = 2.77$ m/s^2

12–154. $a = 7.48$ ft/s^2

12–155. $a = 14.3$ in./s^2

12–157. $v_r = 5.44$ ft/s, $v_\theta = 87.0$ ft/s,
$a_r = -1386$ ft/s^2, $a_\theta = 261$ ft/s^2

12–158. $v = 464$ ft/s, $a = 43.2(10^3)$ ft/s^2

12–159. $\mathbf{v} = \{ -14.2\mathbf{u}_r - 24.0\mathbf{u}_z \}$ m/s
$\mathbf{a} = \{ -3.61\mathbf{u}_r - 6.00\mathbf{u}_z \}$ m/s^2

12–161. $v_r = -2 \sin t, \, v_\theta = \cos t,$
$$a_r = -\frac{5}{2}\cos t, \, a_\theta = -2 \sin t$$

12–162. $v_r = ae^{at}, \, v_\theta = e^{at},$
$a_r = e^{at}(a^2 - 1), \, a_\theta = 2ae^{at}$

12–163. $v_r = 0, \, v_\theta = 10 \text{ ft/s},$
$a_r = -0.25 \text{ ft/s}^2, \, a_\theta = -3.20 \text{ ft/s}^2$

12–165. $\mathbf{a} = (\ddot{r} - 3r\dot{\theta}^2 - 3r\dot{\theta}\dot{\theta})\mathbf{u}_r$
$+ (3\dot{r}\dot{\theta} + \dot{r}\ddot{\theta} + 3\ddot{r}\dot{\theta} - r\dot{\theta}^3)\mathbf{u}_\theta + (\ddot{z})\mathbf{u}_z$

12–166. $a = 48.3 \text{ in./s}^2$

12–167. $v_r = 1.20 \text{ m/s}, \, v_\theta = 1.26 \text{ m/s},$
$a_r = -3.77 \text{ m/s}^2, \, a_\theta = 7.20 \text{ m/s}^2$

12–169. $v_r = 1.20 \text{ m/s}, \, v_\theta = 1.50 \text{ m/s},$
$a_r = -4.50 \text{ m/s}^2, \, a_\theta = 7.20 \text{ m/s}^2$

12–170. $v_r = 16.0 \text{ ft/s}, \, v_\theta = 1.94 \text{ ft/s},$
$a_r = 7.76 \text{ ft/s}^2, \, a_\theta = 1.94 \text{ ft/s}^2$

12–171. $v = 4.24 \text{ m/s}, \, a = 17.6 \text{ m/s}^2$

12–173. $a = 27.8 \text{ m/s}^2$

12–174. $v_r = 0, \, v_\theta = 12 \text{ ft/s},$
$a_r = -216 \text{ ft/s}^2, \, a_\theta = 0$

12–175. $v = 12.6 \text{ m/s}, \, a = 83.2 \text{ m/s}^2$

12–177. $v_r = -1.84 \text{ m/s}, \, v_\theta = 19.1 \text{ m/s},$
$a_r = -2.29 \text{ m/s}^2, \, a_\theta = 4.60 \text{ m/s}^2$

12–178. $v_r = -24.2 \text{ ft/s}, \, v_\theta = 25.3 \text{ ft/s}$

12–179. $v_r = 0, \, v_\theta = 4.80 \text{ ft/s},$
$v_z = -0.664 \text{ ft/s}, \, a_r = -2.88 \text{ ft/s}^2$
$a_\theta = 0, \, a_z = -0.365 \text{ ft/s}^2$

12–181. $v = 10.7 \text{ ft/s}, \, a = 24.6 \text{ ft/s}^2$

12–182. $v = 10.7 \text{ ft/s}, \, a = 40.6 \text{ ft/s}^2$

12–183. $\dot{\theta} = 0.333 \text{ rad/s}, \, a = 6.67 \text{ m/s}^2$

12–185. $v = 1.32 \text{ m/s}$

12–186. $a = 8.66 \text{ m/s}^2$

12–187. $\dot{\theta} = 0.0178 \text{ rad/s}$

12–189. $v_r = 32.0 \text{ ft/s}, \, v_\theta = 50.3 \text{ ft/s},$
$a_r = -201 \text{ ft/s}^2, \, a_\theta = 256 \text{ ft/s}^2$

12–190. $v_r = 32.0 \text{ ft/s}, \, v_\theta = 50.3 \text{ ft/s},$
$a_r = -161 \text{ ft/s}^2, \, a_\theta = 319 \text{ ft/s}^2$

12–191. $v = 5.95 \text{ ft/s}, \, a = 3.44 \text{ ft/s}^2$

12–193. $v_r = 0.242 \text{ m/s}, \, v_\theta = 0.943 \text{ m/s},$
$a_r = -2.33 \text{ m/s}^2, \, a_\theta = 1.74 \text{ m/s}^2$

12–194. $\dot{\theta} = 10.0 \text{ rad/s}$

12–195. $v_B = 0.5 \text{ m/s}$

12–197. $v = 24 \text{ ft/s}$

12–198. $v_B = 1.67 \text{ m/s}$

12–199. $\Delta s_B = 1.33 \text{ ft} \rightarrow$

12–201. $t = 3.83 \text{ s}$

12–202. $v_B = 0.75 \text{ m/s}$

12–203. $t = 5.00 \text{ s}$

12–205. $v_{B/C} = 39 \text{ ft/s} \uparrow$

12–206. $v_B = 1.50 \text{ m/s}$

12–207. $v_A = 1.33 \text{ m/s}$

12–209. $v_B = 8 \text{ ft/s} \downarrow, \, a_B = 6.80 \text{ ft/s}^2 \uparrow$

12–210. $v_A = 2.5 \text{ ft/s} \uparrow, \, a_A = 2.44 \text{ ft/s}^2 \uparrow$

12–211. $v_B = 2.40 \text{ m/s} \uparrow, \, a_B = 3.25 \text{ m/s}^2 \uparrow$

12–213. $v_A = 4 \text{ ft/s}$

12–214. $v_{A/B} = 13.4 \text{ m/s}, \, \theta_v = 31.7° \, \searrow$
$a_{A/B} = 4.32 \text{ m/s}^2, \, \theta_a = 79.0° \, \searrow$

12–215. $v_A = 10.0 \text{ m/s} \leftarrow, \, a_A = 46.0 \text{ m/s}^2 \leftarrow$

12–217. $v_C = 1.2 \text{ m/s} \uparrow, \, a_C = 0.512 \text{ m/s}^2 \uparrow$

12–218. $v_{B/A} = 1044 \text{ km/h}, \, \theta = 54.5° \, \measuredangle$

12–219. $v_{B/A} = 28.5 \text{ mi/h}, \, \theta_v = 44.5° \, \measuredangle,$
$a_{B/A} = 3418 \text{ mi/h}^2, \, \theta_a = 80.6° \, \measuredangle$

12–221. $v_B = 13.5 \text{ ft/s}, \, \theta = 84.8°, \, t = 1.85 \text{ min}$

12–222. $v_w = 58.3 \text{ km/h}, \, \theta = 59.0° \, \searrow$

12–223. $v_{A/B} = 15.7 \text{ m/s}, \, \theta = 7.11° \, \searrow, \, t = 38.1 \text{ s}$

12–225. $v_{A/B} = 98.4 \text{ ft/s}, \, \theta_v = 67.6° \, \searrow,$
$a_{A/B} = 19.8 \text{ ft/s}^2, \, \theta_a = 57.4° \, \measuredangle$

12–226. $v_{r/m} = 16.6 \text{ km/h}, \, \theta = 25.0° \, \searrow$

12–227. $v_{B/A} = 20.5 \text{ m/s}, \, \theta_v = 43.1° \, \searrow$
$a_{B/A} = 4.92 \text{ m/s}^2, \, \theta_a = 6.04° \, \searrow$

12–229. $v_r = 34.6 \text{ km/h} \downarrow$

12–230. $v_m = 4.87 \text{ ft/s}, \, t = 10.3 \text{ s}$

12–231. $v_{w/s} = 19.9 \text{ m/s}, \, \theta = 74.0° \, \searrow$

12–233. Yes, he can catch the ball.

12–234. $v_B = 5.75 \text{ m/s}, \, v_{C/B} = 17.8 \text{ m/s},$
$\theta = 76.2° \, \searrow, \, a_{C/B} = 9.81 \text{ m/s}^2 \downarrow$

12–235. $v_{B/A} = 11.2 \text{ m/s}, \, \theta = 50.3°$

Chapter 13

13–1. $s = 97.4 \text{ ft}$

13–2. $T = 5.98 \text{ kip}$

13–3. $v = 3.36 \text{ m/s}, \, s = 5.04 \text{ m}$

13–5. $F = 6.37 \text{ N}$

13–6. $v = 59.8 \text{ ft/s}$

13–7. $v = 60.7 \text{ ft/s}$

13–9. $t = 2.04 \text{ s}$

13–10. $s = 8.49 \text{ m}$

13–11. $t = 0.249 \text{ s}$

13–13. $a_A = 9.66 \text{ ft/s}^2 \leftarrow, \, a_B = 15.0 \text{ ft/s}^2 \rightarrow$

13–14. $T = 11.25 \text{ kN}, \, F = 33.75 \text{ kN}$

13–15. $A_x = 685 \text{ N}, \, A_y = 1.19 \text{ kN}, \, M_A = 4.74 \text{ kN} \cdot \text{m}$

13–17. $a = \frac{1}{2}(1 - \mu_k)g$

13–18. $R = 5.30 \text{ ft}, \, t_{AC} = 1.82 \text{ s}$

13–19. $R = 5.08 \text{ ft}, \, t_{AC} = 1.48 \text{ s}$

13–21. $\theta = 22.6°$

13–22. $v_B = 5.70 \text{ m/s} \uparrow$

13–23. $v = 3.62 \text{ m/s} \uparrow$

13–25. $R = 2.45 \text{ m}, \, t_{AB} = 1.72 \text{ s}$

13–26. $R = \{150t\} \text{ N}$

13–27. $t = 2.11 \text{ s}$

13–29. $v = 2.01 \text{ ft/s}$

13–30. $v = 0.301$ m/s

13–31. $T = 1.63$ kN

13–33. $P = 2mg\left(\dfrac{\sin\theta + \mu_s\cos\theta}{\cos\theta - \mu_s\sin\theta}\right),$

$a = \left(\dfrac{\sin\theta + \mu_s\cos\theta}{\cos\theta - \mu_s\sin\theta}\right)g$

13–34. $v = 2.19$ m/s

13–35. $t = 5.66$ s

13–37. $t = 0.519$ s

13–38. $s = 16.7$ m

13–39. $v = \dfrac{1}{m}\sqrt{1.09F_0^2 t^2 + 2F_0 tmv_0 + m^2 v_0^2},$

$x = \dfrac{y}{0.3} + v_0\left(\sqrt{\dfrac{2m}{0.3F_0}}\right)y^{1/2}$

13–41. $x = d, v = \sqrt{\dfrac{kd^2}{m_A + m_B}}$

13–42. $x = d$ for separation.

13–43. $v = \sqrt{\dfrac{mg}{k}}\left[\dfrac{e^{2t\sqrt{mg/k}} - 1}{e^{2t\sqrt{mg/k}} + 1}\right],$

$v_t = \sqrt{\dfrac{mg}{k}}$

13–45. $v = 32.2$ ft/s

13–46. $P = 2mg\tan\theta$

13–47. $P = 2mg\left(\dfrac{\sin\theta + \mu_s\cos\theta}{\cos\theta - \mu_s\sin\theta}\right)$

13–49. $a_B = 7.59$ ft/s^2

13–50. $v = 5.13$ m/s

13–51. $d = \dfrac{(m_A + m_B)g}{k}$

13–53. $r = 1.36$ m

13–54. $v = 10.5$ m/s

13–55. $N = 6.18$ kN

13–57. $v = 1.63$ m/s, $N = 7.36$ N

13–58. $v = 0.969$ m/s

13–59. $v = 1.48$ m/s

13–61. $v = 9.29$ ft/s, $T = 38.0$ lb

13–62. $v = 2.10$ m/s

13–63. $T = 0, T = 10.6$ lb

13–65. $v = 6.30$ m/s, $F_n = 283$ N, $F_t = 0$, $F_b = 490$ N

13–66. $v = 22.1$ m/s

13–67. $\theta = 26.7°$

13–69. $F_f = 1.11$ kN , $N = 6.73$ kN

13–70. $v_C = 19.9$ ft/s, $N_C = 7.91$ lb, $v_B = 21.0$ ft/s

13–71. $N = 277$ lb, $F = 13.4$ lb

13–73. $v = \sqrt{gr}, N = 2mg$

13–74. $v = 49.5$ m/s

13–75. $a_t = g\left(\dfrac{x}{\sqrt{1 + x^2}}\right), v = \sqrt{v_0^2 + gx^2},$

$N = \dfrac{m}{\sqrt{1 + x^2}}\left[g - \dfrac{v_0^2 + gx^2}{1 + x^2}\right]$

13–77. $F_s = 4.90$ lb

13–78. $v = 40.1$ ft/s

13–79. $N_P = 2.65$ kN, $\rho = 68.3$ m

13–81. $\theta = 37.7°$

13–82. $N_B = 80.4$ N, $a_t = 1.92$ m/s^2

13–85. $F_A = 4.46$ lb

13–86. $F = 210$ N

13–87. $F = 1.60$ lb

13–89. $F_r = -29.4$ N, $F_\theta = 0$, $F_z = 392$ N

13–90. $F_r = 102$ N, $F_z = 375$ N, $F_\theta = 79.7$ N

13–91. $N = 4.90$ N, $F = 4.17$ N

13–93. $F_{OA} = 12.0$ lb

13–94. $F = 5.07$ kN, $N = 2.74$ kN

13–95. $F = 17.0$ N

13–97. $(N)_{max} = 36.0$ N, $(N)_{min} = 4.00$ N

13–98. $N_s = 3.72$ N, $F_r = 7.44$ N

13–99. $F_r = -900$ N, $F_\theta = -200$ N, $F_z = 1.96$ kN

13–101. $\theta = \tan^{-1}\left(\dfrac{4r_c\dot\theta_0^2}{g}\right)$

13–102. $N = 0.883$ N, $F = 3.92$ N

13–103. $N = 2.95$ N

13–105. $F_r = 1.78$ N, $N_s = 5.79$ N

13–106. $F_r = 2.93$ N, $N_s = 6.37$ N

13–107. $F = 0.163$ lb

13–109. $F_r = 25.6$ N, $F_{OA} = 0$

13–110. $F_r = 20.7$ N, $F_{OA} = 0$

13–111. $r = 0.198$ m

13–113. $v_o = 30.4$ km/s,

$\dfrac{1}{r} = 0.348\,(10^{-12})\cos\theta + 6.74\,(10^{-12})$

13–114. $h = 35.9$ mm, $v_s = 3.07$ km/s

13–115. $v_0 = 7.45$ km/s

13–118. $v_B = 7.71$ km/s, $v_A = 4.63$ km/s

13–119. $v_A = 6.67(10^3)$ m/s, $v_B = 2.77(10^3)$ m/s

13–121. $v_A = 7.47$ km/s

13–122. $r_0 = 11.1$ Mm, $\Delta v_A = 814$ m/s

13–123. $(v_A)_C = 5.27(10^3)$ m/s, $\Delta v = 684$ m/s

13–125. (a) $r = 194\,(10^3)$ mi

(b) $r = 392\,(10^3)$ mi

(c) $194\,(10^3)$ mi $< r < 392\,(10^3)$ mi

(d) $r > 392\,(10^3)$ mi

13–126. $v_A = 4.89(10^3)$ m/s, $v_B = 3.26(10^3)$ m/s

13–127. $v_A = 11.5$ Mm/h, $d = 27.3$ Mm

13–129. $v_A = 2.01(10^3)$ m/s

13–130. $v_{A'} = 521$ m/s, $t = 21.8$ h

13–131. $v_A = 7.01(10^3)$ m/s

Chapter 14

14–1. $v = 10.7$ m/s

14–2. $x_{max} = 3.24$ ft

14–3. $s = 1.35$ m
14–5. $h = 39.3$ m, $\rho = 26.2$ m
14–6. $d = 12$ m
14–7.   Observer A: $v_2 = 6.08$ m/s,
            Observer B: $v_2 = 4.08$ m/s
14–9. $x_{max} = 0.173$ m
14–10. $s = 20.5$ m
14–11. $v = 4.08$ m/s
14–13. $v_B = 31.5$ ft/s, $d = 22.6$ ft, $v_C = 54.1$ ft/s
14–14. $v_A = 7.18$ ft/s
14–15. $v_A = 3.52$ ft/s
14–17. $v_B = 27.8$ ft/s
14–18. $y = 3.81$ ft
14–19. $v_B = 3.34$ m/s
14–21. $v_A = 0.771$ ft/s
14–22. $s_{Tot} = 3.88$ ft
14–23. $x = 0.688$ m
14–25. $s = 0.0735$ ft
14–26. $v_A = 28.3$ m/s
14–27. $v_B = 18.0$ m/s, $N_B = 12.5$ kN
14–29. $s = 0.730$ m
14–30. $s = 3.33$ ft
14–31. $R = 2.83$ m, $v_C = 7.67$ m/s
14–33. $d = 36.2$ ft
14–34. $s = 1.90$ ft
14–35. $v_B = 42.2$ ft/s, $N = 50.6$ lb, $a_t = 26.2$ ft/s^2
14–37. $h_A = 22.5$ m, $h_C = 12.5$ m
14–38. $v_B = 14.9$ m/s, $N = 1.25$ kN
14–39. $v_B = 5.42$ m/s
14–41. $l_0 = 2.77$ ft
14–42. $\theta = 47.2°$
14–43. $P_i = 4.20$ hp
14–45. $P = 8.32 \left(10^3\right)$ hp
14–46. $t = 46.2$ min
14–47. $P = 12.6$ kW
14–49. $P_{max} = 113$ kW, $P_{avg} = 56.5$ kW
14–50. $P_o = 4.36$ hp
14–51. $P = 92.2$ hp
14–53. $P_i = 483$ kW
14–54. $P_i = 622$ kW
14–55. $P_i = 22.2$ kW
14–57. $P = 0.0364$ hp
14–58. $P = 0.231$ hp
14–59. $P = 12.6$ kW
14–61. $P = \left\{400(10^3)t\right\}$ W
14–62. $P = \left\{160\,t - 533t^2\right\}$ kW, $U = 1.69$ kJ
14–63. $P_{max} = 10.7$ kW
14–65. $P = 58.1$ kW
14–66. $F = 227$ N
14–67. $h = 133$ in.
14–69. $N = 694$ N
14–70. $\theta = 48.2°$

14–71. $v_C = 17.7$ ft/s
14–73. $N_B = 0$, $h = 18.75$ m, $N_C = 17.2$ kN
14–74. $v_A = 1.54$ m/s, $v_B = 4.62$ m/s
14–75. $s_B = 5.70$ m
14–77. $h = 23.75$ m, $v_C = 21.6$ m/s
14–78. $v_B = 15.5$ m/s
14–79. $l = 2.77$ ft
14–81. $\theta = 118°$
14–83. $F = GM_e m \left(\dfrac{1}{r_1} - \dfrac{1}{r_2}\right)$
14–85. $v_B = 34.8$ Mm/h
14–86. $s = 130$ m
14–87. $s_B = 0.638$ m, $s_A = 1.02$ m
14–89. $\theta = 22.3°$, $s = 0.587$ m
14–90. $N = 78.6$ N
14–91. $y = 5.10$ m, $N = 15.3$ N, $a = 9.32$ m/s^2 ↘
14–93. $v = 1.68$ m/s
14–94. $v_2 = \sqrt{\dfrac{2}{\pi}(\pi - 2)gr}$
14–95. $v = 6.97$ m/s
14–97. $d = 1.34$ m

Chapter 15

15–1. $v = 1.75$ N·s
15–2. $v = 29.4$ ft/s
15–3. $F = 24.8$ kN
15–5. $I = 5.68$ N·s
15–6. $F = 19.4$ kN, $T = 12.5$ kN
15–7. $F_{AB} = 16.7$ lb, $v = 13.4$ ft/s
15–9. $v = 6.62$ m/s
15–10. $P = 205$ N
15–11. $v = 60.0$ m/s
15–13. $\mu_k = 0.340$
15–14. $I = 15$ kN·s in both cases.
15–15. $v = 4.05$ m/s
15–17. $v = 8.81$ m/s, $s = 24.8$ m
15–18. $v|_{t=3\,s} = 5.68$ m/s ↓, $v|_{t=6\,s} = 21.1$ m/s ↑
15–19. $v = 4.00$ m/s
15–21. $T = 14.9$ kN, $F = 24.8$ kN
15–22. $v_{max} = 108$ m/s, $s = 1.83$ km
15–23. $v = 10.1$ ft/s
15–25. $v = 7.21$ m/s ↑
15–26. Observer A: $v = 7.40$ m/s,
            Observer B: $v = 5.40$ m/s
15–27. $v = 5.07$ m/s
15–29. $t = 1.02$ s, $I = 162$ N·s
15–30. $v = 16.1$ m/s
15–31. $(v_A)_2 = 10.5$ ft/s →
15–33. $v = 7.65$ m/s
15–34. $v = 0.6$ ft/s ←
15–35. $v = 18.6$ m/s →

15–37. $v = 5.21$ m/s $\leftarrow$, $\Delta T = -32.6$ kJ

15–38. $v = 0.5$ m/s, $\Delta T = -16.9$ kJ

15–39. $v = 733$ m/s

15–41. $v_B = 3.48$ ft/s, $d = 0.376$ ft

15–42. $v_B = 3.48$ ft/s, $N_{avg} = 504$ lb, $t = 0.216$ s

15–43. $s = 4.00$ m

15–45. $v_2 = \sqrt{v_1^2 + 2gh}$, $\theta_2 = \sin^{-1}\left(\dfrac{v_1 \sin\theta}{\sqrt{v_1^2 + 2gh}}\right)$

15–46. $\theta = \phi = 9.52°$

15–47. $s_{max} = 481$ mm

15–49. $x = 0.364$ ft $\leftarrow$

15–50. $x = 1.58$ ft $\rightarrow$

15–51. $s_B = 6.67$ m $\rightarrow$

15–53. $s_B = 71.4$ mm $\rightarrow$

15–54. $s_B = 71.4$ mm $\rightarrow$

15–55. $v_c = 5.04$ m/s $\leftarrow$

15–57. $d = 6.87$ mm

15–59. $e = 0.75$, $\Delta T = -9.65$ kJ

15–61. $x_{max} = 0.839$ m

15–63. $v_C = 0.1875v \rightarrow$, $v_D = 0.5625v \rightarrow$, $v_B = 0.8125v \rightarrow$, $v_A = 0.4375v \rightarrow$

15–65. $t = 0.226$ s

15–66. $(v_B)_2 = \dfrac{1}{3}\sqrt{2gh}(1 + e)$

15–67. $(v_A)_2 = 1.04$ ft/s, $(v_B)_3 = 0.964$ ft/s, $(v_C)_3 = 11.9$ ft/s

15–69. $v'_B = 22.2$ m/s, $\theta = 13.0°$

15–70. $(v_B)_2 = \dfrac{e(1 + e)}{2}v_0$

15–71. $v_A = 29.3$ ft/s, $v_{B2} = 33.1$ ft/s, $\theta = 27.7°$ ⟋

15–73. $v_A = 1.35$ m/s $\rightarrow$, $v_B = 5.89$ m/s, $\theta = 32.9°$ ⟍

15–74. $e = 0.0113$

15–75. $h = 1.57$ m

15–77. $(v_B)_3 = 3.24$ m/s, $\theta = 43.9°$

15–78. $v'_B = 31.8$ ft/s

15–79. $(v_A)_2 = 3.80$ m/s $\leftarrow$, $(v_B)_2 = 6.51$ m/s, $(\theta_B)_2 = 68.6°$

15–81. (a) $(v_B)_1 = 8.81$ m/s, $\theta = 10.5°$ ⟋, (b) $(v_B)_2 = 4.62$ m/s, $\phi = 20.3°$ ⟍, (c) $s = 3.96$ m

15–82. $s = 0.456$ ft

15–83. $(v_A)_2 = 42.8$ ft/s $\leftarrow$, $F = 2.49$ kip

15–85. $\mu_k = 0.25$

15–86. $(v_B)_2 = 1.06$ m/s $\leftarrow$, $(v_A)_2 = 0.968$ m/s, $(\theta_A)_2 = 5.11°$ ⟋

15–87. $(v_A)_2 = 4.06$ ft/s, $(v_B)_2 = 6.24$ ft/s

15–89. $(v_A)_2 = 12.1$ m/s, $(v_B)_2 = 12.4$ m/s

15–90. $d = 1.15$ ft, $h = 0.770$ ft

15–91. $(v_B)_3 = 1.50$ m/s

15–93. $(v_A)_2 = 8.19$ m/s, $(v_B)_2 = 9.38$ m/s

15–94. $\{-9.17\mathbf{i} - 6.12\mathbf{k}\}$ slug·ft^2/s

15–95. $\{-9.17\mathbf{i} + 4.08\mathbf{j} - 2.72\mathbf{k}\}$ slug·ft^2/s

15–97. $(H_A)_P = \{-52.8\mathbf{k}\}$ kg·m^2/s, $(H_B)_P = \{-118\mathbf{k}\}$ kg·m^2/s

15–98. $\{-21.5\mathbf{i} + 21.5\mathbf{j} + 37.6\}$ kg·m^2/s

15–99. $\{21.5\mathbf{i} + 21.5\mathbf{j} + 59.1\mathbf{k}\}$ kg·m^2/s

15–101. $v = 20.2$ ft/s, $h = 6.36$ ft

15–102. $t = 11.9$ s

15–103. $v_2 = 9.22$ ft/s, $\Sigma U_{1-2} = 3.04$ ft·lb

15–105. $v = 9.50$ m/s

15–107. $v = 3.33$ m/s

15–109. $v_C = 44.0$ ft/s, $H_A = 8.19$ slug·ft^2/s. The cord will not unstretch.

15–110. $v_2 = 4.03$ m/s, $\Sigma U_{1-2} = 725$ J

15–111. $v_B = 10.8$ ft/s, $U_{AB} = 11.3$ ft·lb

15–113. $v_B = 10.2$ km/s, $r_B = 13.8$ Mm

15–114. $T = 40.1$ kN

15–115. $C_x = 4.97$ kN, $D_x = 2.23$ kN, $D_y = 7.20$ kN

15–117. $F = 303$ lb

15–118. $F = 50.0$ lb

15–119. $F_x = 9.87$ lb, $F_y = 4.93$ lb

15–121. $F_x = 19.5$ lb, $F_y = 1.96$ lb

15–122. $F = 20.0$ lb

15–123. $F = 22.4$ lb

15–125. $T = 82.8$ N, $N = 396$ N

15–126. $F = 6.24$ N, $P = 3.12$ N

15–127. $d = 2.56$ ft

15–129. $C_x = 4.26$ kN, $C_y = 2.12$ kN, $M_C = 5.16$ kN·m

15–130. $v = \left\{\dfrac{8000}{2000 + 50t}\right\}$ m/s

15–131. $A_y = 4.18$ kN, $B_x = 65.0$ N $\rightarrow$, $B_y = 3.72$ kN $\uparrow$

15–133. $a = 0.125$ m/s^2, $v = 4.05$ m/s

15–134. $v_{max} = 2.07\,(10^3)$ ft/s

15–135. 452 Pa

15–137. $R = \{20t + 2.48\}$ lb

15–138. $a_i = 133$ ft/s^2, $a_f = 200$ ft/s^2

15–139. $v_{max} = 580$ ft/s

15–141. $v = \sqrt{\dfrac{2}{3}g\left(\dfrac{y^3 - h^3}{y^2}\right)}$

15–142. $F_D = 11.5$ kN

15–143. $a = 37.5$ ft/s^2

15–145. $a = 0.0476$ m/s^2

15–146. $v_{max} = 2.07(10^3)$ ft/s

15–147. $F = \{7.85t + 0.320\}$ N

15–149. $F = m'v^2$

Chapter 16

16–1. $v_A = 2.60$ m/s, $a_A = 9.35$ m/s^2

16–2. $v_A = 22.0$ m/s, $(a_A)_t = 12.0$ m/s^2, $(a_A)_n = 968$ m/s^2

16–3. $v_A = 26.0$ m/s, $(a_A)_t = 10.0$ m/s^2, $(a_A)_n = 1352$ m/s^2

16–5. $\theta = 5443$ rev, $\omega = 740$ rad/s, $\alpha = 8$ rad/s^2

16–6. $\theta = 3.32$ rev, $t = 1.67$ s

16–7. $t = 6.98$ s, $\theta_D = 34.9$ rev

16–9. $a_B = 29.0$ m/s^2

16–10. $a_B = 16.5$ m/s^2

16–11. $\alpha = 60$ rad/s^2, $\omega = 90.0$ rad/s, $\theta = 90.0$ rad

16–13. $\omega_B = 180$ rad/s, $\omega_C = 360$ rad/s

16–14. $\omega = 42.7$ rad/s, $\theta = 42.7$ rad

16–15. $a_t = 2.83$ m/s^2, $a_n = 35.6$ m/s^2

16–17. $\omega_B = 21.9$ rad/s $\circlearrowright$

16–18. $\omega_B = 31.7$ rad/s $\circlearrowright$

16–19. $\omega_B = 156$ rad/s

16–21. $v_A = 8.10$ m/s,
$(a_A)_t = 4.95$ m/s^2, $(a_A)_n = 437$ m/s^2

16–22. $\omega_D = 4.00$ rad/s, $\alpha_D = 0.400$ rad/s^2

16–23. $\omega_D = 12.0$ rad/s, $\alpha_D = 0.600$ rad/s^2

16–25. $v_P = 2.42$ ft/s, $a_P = 34.4$ ft/s^2

16–26. $\omega_C = 1.68$ rad/s, $\theta_C = 1.68$ rad

16–27. $\omega = 148$ rad/s

16–29. $r_A = 31.8$ mm, $r_B = 31.8$ mm,
1.91 canisters per minute

16–30. $(\omega_B)_{max} = 8.49$ rad/s, $(v_C)_{max} = 0.6$ m/s

16–31. $s_W = 2.89$ m

16–33. $\omega_B = 312$ rad/s, $\alpha_B = 176$ rad/s^2

16–34. $v_E = 3$ m/s,
$(a_E)_t = 2.70$ m/s^2, $(a_E)_n = 600$ m/s^2

16–35. $\mathbf{v}_C = \{-4.8\mathbf{i} - 3.6\mathbf{j} - 1.2\mathbf{k}\}$ m/s,
$\mathbf{a}_C = \{38.4\mathbf{i} - 64.8\mathbf{j} + 40.8\mathbf{k}\}$ m/s^2

16–37. $v_C = 2.50$ m/s, $a_C = 13.1$ m/s^2

16–38. $v = 7.21$ ft/s, $a = 91.2$ ft/s^2

16–39. $\omega = \dfrac{rv_A}{y\sqrt{y^2 - r^2}}$, $\alpha = \dfrac{rv_A^2(2y^2 - r^2)}{y^2(y^2 - r^2)^{3/2}}$

16–41. $\omega = 8.70$ rad/s, $\alpha = -50.5$ rad/s^2

16–42. $\omega = -19.2$ rad/s, $\alpha = -183$ rad/s^2

16–43. $\omega_{AB} = 0$

16–45. $v = -\left(\dfrac{r_1^2\omega \sin 2\theta}{2\sqrt{r_1^2 \cos^2\theta + r_2^2 + 2r_1 r_2}} + r_1\omega \sin\theta\right)$

16–46. $v = \omega d\left(\sin\theta + \dfrac{d \sin 2\theta}{2\sqrt{(R + r)^2 - d^2 \sin^2\theta}}\right)$

16–47. $v = -r\omega \sin\theta$

16–49. $v_C = L\omega\uparrow$, $a_C = 0.577 L\omega^2\uparrow$

16–50. $\omega = \dfrac{2v_0}{r}\sin^2\theta/2$, $\alpha = \dfrac{2v_0^2}{r^2}(\sin\theta)(\sin^2\theta/2)$

16–51. $v_B = \left(\dfrac{h}{d}\right)v_A$

16–53. $\dot{\theta} = \dfrac{v \sin\phi}{L \cos(\phi - \theta)}$

16–54. $\omega = \dfrac{v}{2r}$

16–55. $\omega' = \dfrac{(R + r)\omega}{r}$, $\alpha' = \dfrac{(R + r)\alpha}{r}$

16–57. $v_B = 12.6$ in./s, $65.7°$ $\searrow$

16–58. $\omega_{AB} = 2.00$ rad/s

16–59. $v_C = 1.06$ m/s $\leftarrow$, $\omega_{BC} = 0.707$ rad/s $\circlearrowright$

16–61. $\omega_{BC} = 2.31$ rad/s $\circlearrowright$, $\omega_{AB} = 3.46$ rad/s $\circlearrowright$

16–62. $\omega_A = 32.0$ rad/s

16–63. $\omega_{CB} = 2.45$ rad/s $\circlearrowright$, $v_C = 2.20$ ft/s $\leftarrow$

16–65. $\omega = 20$ rad/s, $v_A = 2$ ft/s $\rightarrow$

16–66. $\omega = 3.11$ rad/s, $v_O = 0.667$ ft/s $\rightarrow$

16–67. $v_A = 5.16$ ft/s, $\theta = 39.8°$ $\measuredangle$

16–69. $v_C = 24.6$ m/s $\downarrow$

16–70. $\omega_{BC} = 10.6$ rad/s $\circlearrowright$, $v_C = 29.0$ m/s $\rightarrow$

16–71. $v_P = 4.88$ m/s $\leftarrow$

16–73. $v_E = 4.00$ m/s, $\theta = 52.7°$ $\swarrow$

16–74. $\omega_B = 90$ rad/s $\circlearrowleft$, $\omega_A = 180$ rad/s $\circlearrowright$

16–75. $\omega_{CD} = 4.03$ rad/s

16–77. $\omega_P = 5$ rad/s, $\omega_A = 1.67$ rad/s

16–78. $\omega_D = 105$ rad/s $\circlearrowright$

16–79. $v_D = 7.07$ m/s

16–82. $\omega_{AB} = 1.24$ rad/s

16–83. $\omega_{BC} = 6.79$ rad/s

16–85. $v_A = 2$ ft/s $\rightarrow$, $v_B = 10$ ft/s $\leftarrow$.
The cylinder slips.

16–86. $v_B = 14$ in./s $\downarrow$,
$v_A = 10.8$ in./s, $\theta = 21.8°$ $\swarrow$

16–87. $\omega_{BC} = 8.66$ rad/s $\circlearrowright$, $\omega_{AB} = 4.00$ rad/s $\circlearrowright$

16–89. $v_A = \omega(r_2 - r_1)$

16–90. $v_C = 2.50$ ft/s $\leftarrow$,
$v_D = 9.43$ ft/s, $\theta = 55.8°$ $\nwarrow$

16–91. $v_C = 2.50$ ft/s $\leftarrow$,
$v_E = 7.91$ ft/s, $\theta = 18.4°$ $\measuredangle$

16–93. $\omega_{BPD} = 3.00$ rad/s $\circlearrowright$, $v_P = 1.79$ m/s $\leftarrow$

16–94. $\omega_B = 6.67$ rad/s

16–95. $v_A = 60.0$ ft/s $\rightarrow$, $v_C = 220$ ft/s $\leftarrow$,
$v_B = 161$ ft/s, $\theta = 60.3°$ $\searrow$

16–97. $\omega_S = 57.5$ rad/s $\circlearrowright$, $\omega_{OA} = 10.6$ rad/s $\circlearrowright$

16–98. $\omega_S = 15.0$ rad/s, $\omega_R = 3.00$ rad/s

16–99. $\omega_{CD} = 57.7$ rad/s $\circlearrowright$

16–101. $\omega_R = 4$ rad/s

16–102. $\omega_R = 4$ rad/s

16–103. $v_C = 3.86$ m/s $\leftarrow$, $a_C = 17.7$ m/s^2 $\leftarrow$

16–105. $\alpha = 0.0962$ rad/s^2 $\circlearrowright$, $a_A = 0.385$ ft/s^2 $\rightarrow$

16–106. $a_C = 13.0$ m/s^2 $\swarrow$, $\alpha_{BC} = 12.4$ rad/s^2 $\circlearrowright$

16–107. $\omega = 6.67$ rad/s $\circlearrowright$, $v_B = 4.00$ m/s $\searrow$
$\alpha = 15.7$ rad/s^2 $\circlearrowright$, $a_B = 24.8$ m/s^2 $\nwarrow$

16–109. $\omega_{BC} = 0$, $\omega_{CD} = 4.00$ rad/s $\circlearrowright$,
$\alpha_{BC} = 6.16$ rad/s^2 $\circlearrowright$, $\alpha_{CD} = 21.9$ rad/s^2 $\circlearrowright$

16–110. $\omega_C = 20.0$ rad/s $\circlearrowright$, $\alpha_C = 127$ rad/s^2 $\circlearrowright$

16–111. $\alpha_{AB} = 4.62$ rad/s^2 $\circlearrowright$,
$a_B = 13.3$ m/s^2, $\theta = 37.0°$ $\swarrow$

16–113. $v_A = 0.424$ m/s, $\theta_v = 45°$ $\swarrow$,
$a_A = 0.806$ m/s^2, $\theta_a = 7.13°$ $\measuredangle$

16–114. $v_B = 0.6 \text{ m/s} \downarrow$,
$a_B = 1.84 \text{ m/s}^2, \theta = 60.6°$ ⟍

16–115. $v_B = 4v \rightarrow$,
$v_A = 2\sqrt{2}v, \theta = 45°$ ⟋,
$a_B = \dfrac{2v^2}{r} \downarrow, a_A = \dfrac{2v^2}{r} \rightarrow$

16–117. $a_C = 10.0 \text{ m/s}^2, \theta = 2.02°$ ⟋

16–118. $\alpha = 40.0 \text{ rad/s}^2, a_A = 2.00 \text{ m/s}^2 \leftarrow$

16–119. $v_B = 1.58\omega a, a_B = 1.58\alpha a - 1.77\omega^2 a$

16–121. $\omega_{AC} = 0, \omega_F = 10.7 \text{ rad/s}$ ⟲,
$\alpha_{AC} = 28.7 \text{ rad/s}^2$ ⟲

16–122. $\omega_{CD} = 7.79 \text{ rad/s}$ ⟳, $\alpha_{CD} = 136 \text{ rad/s}^2$ ⟲

16–123. $v_C = 1.56 \text{ m/s} \leftarrow$,
$a_C = 29.7 \text{ m/s}^2, \theta = 24.1°$ ⟍

16–125. $\omega = 4.73 \text{ rad/s}$ ⟳, $\alpha = 131 \text{ rad/s}^2$ ⟲

16–126. $\omega_{AB} = 7.17 \text{ rad/s}$ ⟲, $\alpha_{AB} = 23.1 \text{ rad/s}^2$ ⟳

16–127. $\alpha_{AB} = 3.70 \text{ rad/s}^2$ ⟲

16–129. $\mathbf{v}_B = \{0.6\mathbf{i} + 2.4\mathbf{j}\} \text{ m/s}$,
$\mathbf{a}_B = \{-14.2\mathbf{i} + 8.40\mathbf{j}\} \text{ m/s}^2$

16–130. $v_B = 1.30 \text{ ft/s}, a_B = 0.6204 \text{ ft/s}^2$

16–131. $\mathbf{v}_m = \{7.5\mathbf{i} - 5\mathbf{j}\} \text{ ft/s}, \mathbf{a}_m = \{5\mathbf{i} + 3.75\mathbf{j}\} \text{ ft/s}^2$

16–133. $\mathbf{v}_A = \{-17.2\mathbf{i} + 12.5\mathbf{j}\} \text{ m/s}$,
$\mathbf{a}_A = \{349\mathbf{i} + 597\mathbf{j}\} \text{ m/s}^2$

16–134. $\mathbf{a}_A = \{-5.60\mathbf{i} - 16\mathbf{j}\} \text{ m/s}^2$

16–135. $v_C = 2.40 \text{ m/s}, \theta = 60°$ ⟱

16–137. $(\mathbf{v}_{B/A})_{xyz} = \{31.0\mathbf{i}\} \text{ m/s}$,
$(\mathbf{a}_{B/A})_{xyz} = \{-14.0\mathbf{i} - 206\mathbf{j}\} \text{ m/s}^2$

16–138. $v_B = 7.7 \text{ m/s}, a_B = 201 \text{ m/s}^2$

16–139. $\omega_{CB} = 1.33 \text{ rad/s}$ ⟳, $\alpha_{CD} = 3.08 \text{ rad/s}^2$ ⟲

16–141. $\omega_{CD} = 3.00 \text{ rad/s}$ ⟲, $\alpha_{CD} = 12.0 \text{ rad/s}^2$ ⟲

16–142. $\mathbf{v}_C = \{-0.944\mathbf{i} + 2.02\mathbf{j}\} \text{ m/s}$,
$\mathbf{a}_C = \{-11.2\mathbf{i} - 4.15\mathbf{j}\} \text{ m/s}^2$

16–143. $\omega_{AB} = 5 \text{ rad/s}$ ⟲, $\alpha_{AB} = 2.5 \text{ rad/s}^2$ ⟲

16–145. $\mathbf{v}_C = \{-7.00\mathbf{i} + 17.3\mathbf{j}\} \text{ ft/s}$,
$\mathbf{a}_C = \{-34.6\mathbf{i} - 15.5\mathbf{j}\} \text{ ft/s}^2$

16–146. $\mathbf{v}_C = \{-7.00\mathbf{i} + 17.3\mathbf{j}\} \text{ ft/s}$,
$\mathbf{a}_C = \{-38.8\mathbf{i} - 6.84\mathbf{j}\} \text{ ft/s}^2$

16–147. $\omega_{AB} = 0.667 \text{ rad/s}$ ⟳, $\alpha_{AB} = 3.08 \text{ rad/s}^2$ ⟲

16–149. $(\mathbf{v}_{rel})_{xyz} = \mathbf{0}, (\mathbf{a}_{rel})_{xyz} = \{1\mathbf{i}\} \text{ m/s}^2$

16–150. $\omega_{DC} = 2.96 \text{ rad/s}$ ⟲

16–151. $\omega_{AC} = 0, \alpha_{AC} = 14.4 \text{ rad/s}^2$ ⟲

Chapter 17

17–1. $I_y = \dfrac{1}{3}ml^2$

17–2. $m = \pi hR^2\left(k + \dfrac{aR^2}{2}\right), I_z = \dfrac{\pi hR^4}{2}\left[k + \dfrac{2aR^2}{3}\right]$

17–3. $I_z = mR^2$

17–5. $k_x = 1.20 \text{ in.}$

17–6. $I_x = \dfrac{2}{5}mr^2$

17–7. $I_x = \dfrac{93}{70}mb^2$

17–9. $I_y = \dfrac{m}{6}(a^2 + h^2)$

17–10. $k_O = 2.17 \text{ m}$

17–11. $I_O = 1.36 \text{ kg} \cdot \text{m}^2$

17–13. $I_A = 7.67 \text{ kg} \cdot \text{m}^2$

17–14. $I_A = 222 \text{ slug} \cdot \text{ft}^2$

17–15. $I_O = 6.23 \text{ kg} \cdot \text{m}^2$

17–17. $I_G = 0.230 \text{ kg} \cdot \text{m}^2$

17–18. $I_O = 0.560 \text{ kg} \cdot \text{m}^2$

17–19. $I_G = 118 \text{ slug} \cdot \text{ft}^2$

17–21. $\bar{y} = 1.78 \text{ m}, I_G = 4.45 \text{ kg} \cdot \text{m}^2$

17–22. $I_x = 3.25 \text{ g} \cdot \text{m}^2$

17–23. $I_{x'} = 7.19 \text{ g} \cdot \text{m}^2$

17–25. $F = 5.96 \text{ lb}, N_B = 99.0 \text{ lb}, N_A = 101 \text{ lb}$

17–26. $A_y = 72.6 \text{ kN}, B_y = 71.6 \text{ kN}, a_G = 0.250 \text{ m/s}^2$

17–27. $N_A = 1393 \text{ lb}, N_B = 857 \text{ lb}, t = 2.72 \text{ s}$

17–29. $a = 2.74 \text{ m/s}^2, T = 25.1 \text{ kN}$

17–30. $N = 29.6 \text{ kN}, V = 0, M = 51.2 \text{ kN} \cdot \text{m}$

17–31. $h = 3.12 \text{ ft}$

17–33. $P = 579 \text{ N}$

17–34. $a = 4 \text{ m/s}^2 \rightarrow, N_B = 1.14 \text{ kN}, N_A = 327 \text{ N}$

17–35. $a_G = 13.3 \text{ ft/s}^2$

17–37. $P = 785 \text{ N}$

17–38. $P = 314 \text{ N}$

17–39. $N = 0.433wx, V = 0.25wx, M = 0.125wx^2$

17–41. $B_x = 73.9 \text{ lb}, B_y = 69.7 \text{ lb}, N_A = 120 \text{ lb}$

17–42. $a = 2.01 \text{ m/s}^2$.
The crate slips.

17–43. $a = 2.68 \text{ ft/s}^2, N_A = 26.9 \text{ lb}, N_B = 123 \text{ lb}$

17–45. $T = 15.7 \text{ kN}, C_x = 8.92 \text{ kN}, C_y = 16.3 \text{ kN}$

17–46. $a = 9.81 \text{ m/s}^2, C_x = 12.3 \text{ kN}, C_y = 12.3 \text{ kN}$

17–47. $h_{max} = 3.16 \text{ ft}, F_A = 248 \text{ lb}, N_A = 400 \text{ lb}$

17–49. $F_{AB} = 112 \text{ N}, C_x = 26.2 \text{ N}, C_y = 49.8 \text{ N}$

17–50. $P = 765 \text{ N}$

17–51. $T = 1.52 \text{ kN}, \theta = 18.6°$

17–53. $\alpha = 9.67 \text{ rad/s}^2$

17–54. $F_C = 16.1 \text{ lb}, N_C = 159 \text{ lb}$

17–55. $\alpha = 2.62 \text{ rad/s}^2$

17–57. $\omega = 56.2 \text{ rad/s}, A_x = 0, A_y = 98.1 \text{ N}$

17–58. $\alpha = 14.7 \text{ rad/s}^2, A_x = 88.3 \text{ N}, A_y = 147 \text{ N}$

17–59. $F_A = \dfrac{3}{2}mg$

17–61. $\alpha = 0.694 \text{ rad/s}^2$

17–62. $\omega = 10.9 \text{ rad/s}$

17–63. $\omega = 9.45 \text{ rad/s}$

17–65. $M = 0.233 \text{ lb} \cdot \text{ft}$

17–67. $\alpha = 8.68 \text{ rad/s}^2, A_n = 0, A_t = 106 \text{ N}$

17–69. $\alpha = 7.28 \text{ rad/s}^2$

17–70. $F = 22.1 \text{ N}$

17–71. $\omega = 0.474 \text{ rad/s}$

17–73. $t = 6.71$ s

17–74. $\alpha = 14.2$ rad/s^2

17–75. $A_x = 89.2$ N, $A_y = 66.9$ N, $t = 1.25$ s

17–77. $t = 1.09$ s

17–78. $v = 4.88$ ft/s

17–79. $a = 2.97$ m/s^2

17–81. $A_x = 0$, $A_y = 289$ N, $\alpha = 23.1$ rad/s^2

17–82. $N_A = 177$ kN, $V_A = 5.86$ kN, $M_A = 50.7$ kN·m

17–83. $M = 0.3gml$

17–85. $N = wx\left[\dfrac{\omega^2}{g}\left(L - \dfrac{x}{2}\right) + \cos\theta\right]$,

$V = wx\sin\theta$, $M = \dfrac{1}{2}wx^2\sin\theta$

17–86. $\alpha = 12.5$ rad/s $\circlearrowright$, $a_G = 18.75$ m/s^2 ↓

17–87. $N_B = 2.89$ kN,

$A_x = 0$, $A_y = 2.89$ kN

17–89. $\omega = 800$ rad/s

17–91. $\alpha = 5.62$ rad/s^2, $T = 196$ N

17–93. $\alpha = 2.45$ rad/s^2 $\circlearrowright$, $N_B = 2.23$ N, $N_A = 33.3$ N

17–94. $\alpha = 4.32$ rad/s^2

17–95. $\theta = 46.9°$

17–97. $\alpha = 0.500$ rad/s^2

17–98. $\alpha = 15.6$ rad/s^2

17–99. $a_A = 26.7$ m/s$^2 \rightarrow$

17–101. $F = 42.3$ N

17–102. $\alpha = 4.01$ rad/s^2

17–103. $A_y = 15.0$ lb, $A_x = 0.776$ lb, $\alpha = 1.67$ rad/s^2

17–105. $\alpha = 18.9$ rad/s^2, $P = 76.4$ lb

17–106. $\alpha = \dfrac{6P}{mL}$, $a_B = \dfrac{2P}{m}$

17–107. $\alpha = \dfrac{6(P - \mu_k\, mg)}{mL}$, $a_B = \dfrac{2(P - \mu_k\, mg)}{m}$

17–109. $\alpha = 3$ rad/s^2

17–110. $\alpha = 14.5$ rad/s^2, $t = 0.406$ s

17–111. The disk does not slip.

17–113. $a_G = \mu_k g \leftarrow$, $\alpha = \dfrac{2\mu_k g}{r}$ $\circlearrowright$

17–114. $\omega = \dfrac{1}{3}\omega_0$, $t = \dfrac{\omega_0 r}{3\mu_k g}$

17–115. $\alpha_A = 43.6$ rad/s^2 $\circlearrowright$, $\alpha_B = 43.6$ rad/s^2 $\circlearrowleft$, $T = 19.6$ N

17–117. $T_A = \dfrac{4}{7}W$

17–118. $\alpha = 23.4$ rad/s^2, $B_y = 9.62$ lb

17–119. $\alpha = \dfrac{10g}{13\sqrt{2}\,r}$

Chapter 18

18–2. $\omega = 14.0$ rad/s

18–3. $\omega = 14.1$ rad/s

18–5. $\omega = 2.02$ rad/s

18–6. $\omega = 1.78$ rad/s

18–7. $T = 283$ ft·lb

18–9. $\omega = 21.5$ rad/s

18–10. $s = 5.16$ m, $T = 78.5$ N

18–11. $\omega = 14.9$ rad/s

18–13. $\omega = 6.11$ rad/s

18–14. $\omega = 8.64$ rad/s

18–15. $\omega = 3.16$ rad/s

18–17. $\omega = \sqrt{\omega_0^2 + \dfrac{g}{r^2}s\sin\theta}$

18–18. $v_C = 7.49$ m/s

18–19. $\omega = 6.92$ rad/s

18–21. $s = 0.304$ ft

18–22. $v_C = 19.6$ ft/s

18–23. $\theta = 0.445$ rev

18–25. $s_G = 1.60$ m

18–26. $\omega_2 = 5.37$ rad/s

18–27. $\omega = 44.6$ rad/s

18–29. $v_G = 11.9$ ft/s

18–30. $\omega = 2.50$ rad/s

18–31. $\omega = 5.40$ rad/s

18–33. $\theta = 0.891$ rev, regardless of orientation

18–34. $\omega = 5.74$ rad/s

18–35. $\omega_{AB} = 5.92$ rad/s

18–37. $s_C = 78.0$ mm

18–38. $s = 0.301$ m, $T = 163$ N

18–39. $v_A = 1.29$ m/s

18–41. $s_b = 242$ mm, $T = 67.8$ N

18–42. $v_b = 2.52$ m/s

18–43. $\theta = 48.2°$

18–45. $\omega = 3.78$ rad/s

18–46. $\omega = 3.75$ rad/s

18–47. $\omega = 3.28$ rad/s

18–49. $(\omega_{AB})_2 = (\omega_{BC})_2 = 1.12$ rad/s

18–50. $v_A = 1.40$ m/s

18–51. $\theta_0 = 8.94$ rev

18–53. $\omega_{BC} = 1.34$ rad/s

18–54. $v_b = 15.5$ ft/s

18–55. $v_A = 4.00$ m/s

18–57. $\omega = 12.8$ rad/s

18–58. $k = 18.4$ N/m

18–59. $\omega = 2.67$ rad/s

18–61. $\omega_{AB} = 3.70$ rad/s

18–62. $\omega = 1.80$ rad/s

18–63. $v_A = 21.0$ ft/s

18–65. $\omega = 2.71$ rad/s

18–66. $k = 100$ lb/ft

18–67. $(v_A)_2 = 7.24$ m/s

Chapter 19

19–5. $\int M \, dt = 0.833 \, \text{kg} \cdot \text{m}^2/\text{s}$

19–6. $\omega = 0.0178 \, \text{rad}/\text{s}$

19–7. $v_B = 24.1 \, \text{m}/\text{s}$

19–9. $\omega_2 = 103 \, \text{rad}/\text{s}$

19–10. $t = 0.6125 \, \text{s}$

19–11. $\omega_2 = 53.7 \, \text{rad}/\text{s}$

19–13. $y = \dfrac{2}{3} l$

19–14. $d = \dfrac{2}{3} l$

19–15. (a) $\omega_{BC} = 68.7 \, \text{rad}/\text{s}$,
(b) $\omega_{BC} = 66.8 \, \text{rad}/\text{s}$,
(c) $\omega_{BC} = 68.7 \, \text{rad}/\text{s}$

19–17. $v_G = 26.8 \, \text{ft}/\text{s}$

19–18. $v_G = 2 \, \text{m}/\text{s}$, $\omega = 3.90 \, \text{rad}/\text{s}$

19–19. $v_A = 24.1 \, \text{m}/\text{s}$

19–21. $\omega = 12.7 \, \text{rad}/\text{s}$

19–22. $\omega_A = 47.3 \, \text{rad}/\text{s}$

19–23. $t = 1.32 \, \text{s}$

19–25. $t = 1.04 \, \text{s}$

19–26. $\omega = 9 \, \text{rad}/\text{s}$

19–27. $v_B = 1.59 \, \text{m}/\text{s}$

19–29. $\omega = 1.91 \, \text{rad}/\text{s}$

19–30. $\omega_2 = 0.656 \, \text{rad}/\text{s}$, $\theta = 18.8°$

19–31. $\omega_2 = 0.577 \, \text{rad}/\text{s}$, $\theta = 15.8°$

19–33. $\omega_2 = 2.55 \, \text{rev}/\text{s}$

19–34. $\omega = 0.190 \, \text{rad}/\text{s}$

19–35. $\omega = 0.0906 \, \text{rad}/\text{s}$

19–37. $\omega = 22.7 \, \text{rad}/\text{s}$

19–38. $h_C = 0.500 \, \text{ft}$

19–39. $\omega_2 = 1.01 \, \text{rad}/\text{s}$

19–41. $\theta = 66.9°$

19–42. $\omega_2 = 57 \, \text{rad}/\text{s}$, $U_F = 367 \, \text{J}$

19–43. $\omega_2 = 3.47 \, \text{rad}/\text{s}$

19–45. $v = 5.96 \, \text{ft}/\text{s}$

19–46. $h = \dfrac{7}{5} r$

19–47. $\theta = 50.2°$

19–49. $(v_D)_3 = 1.54 \, \text{m}/\text{s}$, $\omega_3 = 0.934 \, \text{rad}/\text{s}$

19–50. $\omega_1 = 7.17 \, \text{rad}/\text{s}$

19–51. $\theta = \tan^{-1}\left(\sqrt{\dfrac{7}{5}} e\right)$

19–53. $\omega_3 = 2.73 \, \text{rad}/\text{s}$

19–54. $\omega = \sqrt{7.5 \dfrac{g}{L}}$

19–55. $h_B = 0.980 \, \text{ft}$

19–57. $(v_G)_{y2} = e(v_G)_{y1} \uparrow$,
$(v_G)_{x2} = \dfrac{5}{7}\left((v_G)_{x1} - \dfrac{2}{5}\omega_1 r\right) \leftarrow$

19–58. $\theta_1 = 39.8°$

Chapter 20

20–1. (a) $\boldsymbol{\alpha} = \omega_s \omega_t \mathbf{j}$,
(b) $\boldsymbol{\alpha} = -\omega_s \omega_t \mathbf{k}$

20–2. $\mathbf{v}_A = \{-0.225\mathbf{i}\} \, \text{m}/\text{s}$,
$\mathbf{a}_A = \{-0.1125\mathbf{j} - 0.130\mathbf{k}\} \, \text{m}/\text{s}^2$

20–3. $\mathbf{v}_A = \{-5.20\mathbf{i} - 12\mathbf{j} + 20.8\mathbf{k}\} \, \text{ft}/\text{s}$,
$\mathbf{a}_A = \{-24.1\mathbf{i} - 13.3\mathbf{j} - 7.20\mathbf{k}\} \, \text{ft}/\text{s}^2$

20–5. $(\omega_C)_{DE} = 40 \, \text{rad}/\text{s}$, $(\omega_{DE})_y = 5 \, \text{rad}/\text{s}$

20–6. $\boldsymbol{\omega} = \{-8.24\mathbf{j}\} \, \text{rad}/\text{s}$, $\boldsymbol{\alpha} = \{24.7\mathbf{i} - 5.49\mathbf{j}\} \, \text{rad}/\text{s}^2$

20–7. $\mathbf{v}_A = \{-7.79\mathbf{i} - 2.25\mathbf{j} + 3.90\mathbf{k}\} \, \text{ft}/\text{s}$,
$\mathbf{a}_A = \{8.30\mathbf{i} - 35.2\mathbf{j} + 7.02\mathbf{k}\} \, \text{ft}/\text{s}^2$

20–9. $\mathbf{v}_B = \{-0.4\mathbf{i} - 2\mathbf{j} - 2\mathbf{k}\} \, \text{m}/\text{s}$,
$\mathbf{a}_B = \{-8.20\mathbf{i} + 40.6\mathbf{j} - \mathbf{k}\} \, \text{rad}/\text{s}^2$

20–10. $\boldsymbol{\omega} = \{42.4\mathbf{j} + 43.4\mathbf{k}\} \, \text{rad}/\text{s}$,
$\boldsymbol{\alpha} = \{-42.4\mathbf{i}\} \, \text{rad}/\text{s}^2$

20–11. $\boldsymbol{\omega} = \{2\mathbf{i} + 42.4\mathbf{j} + 43.4\mathbf{k}\} \, \text{rad}/\text{s}$,
$\boldsymbol{\alpha} = \{-42.4\mathbf{i} - 82.9\mathbf{j} + 84.9\mathbf{k}\} \, \text{rad}/\text{s}^2$

20–13. $v_B = 0$, $v_C = 0.283 \, \text{m}/\text{s}$, $a_B = 1.13 \, \text{m}/\text{s}^2$,
$a_C = 1.60 \, \text{m}/\text{s}^2$

20–14. $\mathbf{v}_C = \{1.8\mathbf{j} - 1.5\mathbf{k}\} \, \text{m}/\text{s}$,
$\mathbf{a}_C = \{-36.6\mathbf{i} + 0.45\mathbf{j} - 0.9\mathbf{k}\} \, \text{m}/\text{s}^2$

20–15. $\mathbf{v}_A = \{-8.66\mathbf{i} + 8.00\mathbf{j} - 13.9\mathbf{k}\} \, \text{ft}/\text{s}$,
$\mathbf{a}_A = \{-24.8\mathbf{i} + 8.29\mathbf{j} - 30.9\mathbf{k}\} \, \text{ft}/\text{s}^2$

20–17. $\mathbf{v}_A = \{-1.80\mathbf{i}\} \, \text{ft}/\text{s}$,
$\mathbf{a}_A = \{-0.750\mathbf{i} - 0.720\mathbf{j} - 0.831\mathbf{k}\} \, \text{ft}/\text{s}^2$

20–18. $\boldsymbol{\omega}_P = \{-40\mathbf{j}\} \, \text{rad}/\text{s}$, $\boldsymbol{\alpha}_B = \{-6400\mathbf{i}\} \, \text{rad}/\text{s}^2$

20–19. $\boldsymbol{\omega} = \{4.35\mathbf{i} + 12.7\mathbf{j}\} \, \text{rad}/\text{s}$,
$\boldsymbol{\alpha} = \{-26.1\mathbf{k}\} \, \text{rad}/\text{s}^2$

20–21. $\boldsymbol{\omega} = \{30\mathbf{j} - 5\mathbf{k}\} \, \text{rad}/\text{s}$, $\boldsymbol{\alpha} = \{150\mathbf{i}\} \, \text{rad}/\text{s}^2$

20–22. $\mathbf{v}_A = \{10\mathbf{i} + 14.7\mathbf{j} - 19.6\mathbf{k}\} \, \text{ft}/\text{s}$,
$\mathbf{a}_A = \{-6.12\mathbf{i} + 3\mathbf{j} - 2\mathbf{k}\} \, \text{ft}/\text{s}^2$

20–23. $\omega_A = 47.8 \, \text{rad}/\text{s}$, $\omega_B = 7.78 \, \text{rad}/\text{s}$

20–25. $\boldsymbol{\omega}_{BC} = \{0.204\mathbf{i} - 0.612\mathbf{j} + 1.36\mathbf{k}\} \, \text{rad}/\text{s}$,
$\mathbf{v}_B = \{-0.333\mathbf{j}\} \, \text{m}/\text{s}$

20–26. $\boldsymbol{\omega}_{AB} = \{-1.00\mathbf{i} - 0.500\mathbf{j} + 2.50\mathbf{k}\} \, \text{rad}/\text{s}$,
$\mathbf{v}_B = \{-2.50\mathbf{j} - 2.50\mathbf{k}\} \, \text{m}/\text{s}$

20–27. $\boldsymbol{\alpha}_{AB} = \{-7.9\mathbf{i} - 3.95\mathbf{j} + 4.75\mathbf{k}\} \, \text{rad}/\text{s}^2$,
$\mathbf{a}_B = \{-19.75\mathbf{j} - 19.75\mathbf{k}\} \, \text{m}/\text{s}^2$

20–29. $\mathbf{a}_B = \{-37.6\mathbf{j}\} \, \text{ft}/\text{s}^2$

20–30. $\mathbf{v}_B = \{-1.92\mathbf{j} + 2.56\mathbf{k}\} \, \text{m}/\text{s}$

20–31. $v_B = 5.00 \, \text{m}/\text{s}$,
$\boldsymbol{\omega}_{AB} = \{-4.00\mathbf{i} - 0.600\mathbf{j} - 1.20\mathbf{k}\} \, \text{rad}/\text{s}$

20–33. $\boldsymbol{\omega}_{BD} = \{-1.20\mathbf{j}\} \, \text{rad}/\text{s}$

20–34. $\alpha_{BD} = \{-8.00\mathbf{j}\}$ rad/s^2

20–35. $\boldsymbol{\omega}_{AB} = \{-0.500\mathbf{i} + 0.667\mathbf{j} - 1.00\mathbf{k}\}$ rad/s
$\mathbf{v}_B = \{-7.50\mathbf{j}\}$ ft/s

20–37. $\mathbf{v}_C = \{-1.00\mathbf{i} + 5.00\mathbf{j} + 0.800\mathbf{k}\}$ m/s,
$\mathbf{a}_C = \{-28.8\mathbf{i} - 5.45\mathbf{j} + 32.3\mathbf{k}\}$ m/s^2

20–38. $\mathbf{v}_C = \{-1\mathbf{i} + 5\mathbf{j} + 0.8\mathbf{k}\}$ m/s,
$\mathbf{a}_C = \{-28.2\mathbf{i} - 5.45\mathbf{j} + 32.3\mathbf{k}\}$ m/s^2

20–39. $\mathbf{v}_B = \{-2.75\mathbf{i} - 2.50\mathbf{j} + 3.17\mathbf{k}\}$ m/s,
$\mathbf{a}_B = \{2.50\mathbf{i} - 2.24\mathbf{j} - 0.00389\mathbf{k}\}$ ft/s^2

20–41. $\mathbf{v}_C = \{3\mathbf{i} + 6\mathbf{j} - 3\mathbf{k}\}$ m/s,
$\mathbf{a}_C = \{-13.0\mathbf{i} + 28.5\mathbf{j} - 10.2\mathbf{k}\}$ m/s^2

20–42. $\mathbf{v}_B = \{-17.8\mathbf{i} - 3\mathbf{j} + 5.20\mathbf{k}\}$ m/s,
$\mathbf{a}_B = \{9\mathbf{i} - 29.4\mathbf{j} - 1.5\mathbf{k}\}$ m/s^2

20–43. $\mathbf{v}_B = \{-17.8\mathbf{i} - 3\mathbf{j} + 5.20\mathbf{k}\}$ m/s,
$\mathbf{a}_B = \{3.05\mathbf{i} - 30.9\mathbf{j} + 1.10\mathbf{k}\}$ m/s^2

20–45. $\mathbf{v}_P = \{-0.849\mathbf{i} + 0.849\mathbf{j} + 0.566\mathbf{k}\}$ m/s,
$\mathbf{a}_P = \{-5.09\mathbf{i} - 7.35\mathbf{j} + 6.79\mathbf{k}\}$ m/s^2

20–46. $\mathbf{v}_A = \{-8.66\mathbf{i} + 2.26\mathbf{j} + 2.26\mathbf{k}\}$ m/s,
$\mathbf{a}_A = \{-22.6\mathbf{i} - 47.8\mathbf{j} + 45.3\mathbf{k}\}$ m/s^2

20–47. $\mathbf{v}_A = \{-8.66\mathbf{i} + 2.26\mathbf{j} + 2.26\mathbf{k}\}$ m/s,
$\mathbf{a}_A = \{-26.1\mathbf{i} - 44.4\mathbf{j} + 7.92\mathbf{k}\}$ m/s^2

20–49. $\mathbf{v}_P = \{-9.80\mathbf{i} + 14.4\mathbf{j} + 48.0\mathbf{k}\}$ ft/s,
$\mathbf{a}_P = \{-160\mathbf{i} + 5.16\mathbf{j} - 13\mathbf{k}\}$ ft/s^2

20–50. $\mathbf{v}_P = \{-25.5\mathbf{i} - 13.4\mathbf{j} + 20.5\mathbf{k}\}$ ft/s,
$\mathbf{a}_P = \{161\mathbf{i} - 249\mathbf{j} - 39.6\mathbf{k}\}$ ft/s^2

20–51. $\mathbf{v}_P = \{-25.5\mathbf{i} - 13.4\mathbf{j} + 20.5\mathbf{k}\}$ ft/s,
$\mathbf{a}_P = \{161\mathbf{i} - 243\mathbf{j} - 33.9\mathbf{k}\}$ ft/s^2

20–53. $\mathbf{v}_A = \{-8.66\mathbf{i} + 8\mathbf{j} - 13.9\mathbf{k}\}$ ft/s,
$\mathbf{a}_A = \{-17.9\mathbf{i} + 8.29\mathbf{j} - 30.9\mathbf{k}\}$ ft/s^2,

20–54. $\mathbf{v}_C = \{-1.73\mathbf{i} - 5.77\mathbf{j} + 7.06\mathbf{k}\}$ ft/s,
$\mathbf{a}_C = \{9.88\mathbf{i} - 72.8\mathbf{j} + 0.365\mathbf{k}\}$ ft/s^2

Chapter 21

21–2. $I_{\bar{y}} = \dfrac{3m}{80}(h^2 + 4a^2)$, $I_{y'} = \dfrac{m}{20}(2h^2 + 3a^2)$

21–3. $I_y = 2614$ slug · ft^2

21–5. $I_{yz} = \dfrac{m}{6}ah$

21–6. $I_{xy} = \dfrac{m}{12}a^2$

21–7. $I_{xy} = 636\rho$

21–9. $I_{z'z'} = 0.0961$ slug · ft^2

21–10. $k_y = 2.35$ ft, $k_x = 1.80$ ft

21–11. $I_{aa} = \dfrac{m}{12}(3a^2 + 4h^2)$

21–13. $I_{yz} = 0$

21–14. $I_{xy} = 0.32$ kg · m^2, $I_{yz} = 0.08$ kg · m^2, $I_{xz} = 0$

21–15. $I_{z'} = 0.0595$ kg · m^2

21–17. $\bar{y} = 0.5$ ft, $\bar{x} = -0.667$ ft, $I_{x'} = 0.0272$ slug · ft^2,
$I_{y'} = 0.0155$ slug · ft^2, $I_{z'} = 0.0427$ slug · ft^2

21–18. $I_x = 0.455$ slug · ft^2

21–19. $I_{aa} = 1.13$ slug · ft^2

21–21. $I_z = 0.0880$ slug · ft^2

21–25. $\mathbf{H} = \{-477(10^{-6})\mathbf{i} + 198(10^{-6})\mathbf{j} + 0.169\,\mathbf{k}\}$ kg · m^2/s

21–26. $\omega_2 = 61.7$ rad/s

21–27. $\omega_2 = 87.2$ rad/s

21–29. $\omega_x = 19.7$ rad/s

21–30. $h = 2.24$ in.

21–31. $T = 0.0920$ ft · lb

21–33. $\omega_p = 4.82$ rad/s

21–34. $\mathbf{H}_A = \{-2000\mathbf{i} - 55\,000\mathbf{j} + 22\,500\mathbf{k}\}$ kg · m^2/s

21–35. $T = 37.0$ MJ

21–37. $\boldsymbol{\omega} = \{-0.750\mathbf{j} + 1.00\mathbf{k}\}$ rad/s

21–38. $T = 1.14$ J

21–39. $H_z = 0.4575$ kg · m^2/s

21–41. $\Sigma M_x = (I_x\dot{\omega}_x - I_{xy}\dot{\omega}_y - I_{xz}\dot{\omega}_z)$,
$\qquad - \Omega_z(I_y\omega_y - I_{yz}\omega_z - I_{yx}\omega_x)$,
$\qquad + \Omega_y(I_z\omega_z - I_{zx}\omega_x - I_{zy}\omega_y)$
Similarly for ΣM_y and ΣM_z.

21–43. $B_z = 4$ lb, $A_x = -2.00$ lb, $A_y = 0.627$ lb,
$B_x = 2.00$ lb, $B_y = -1.37$ lb

21–45. $A_Z = 1.46$ lb, $B_Z = 13.5$ lb, $A_X = A_Y = B_X = 0$,

21–46. $\dot{\omega}_x = -14.7$ rad/s^2, $B_z = 77.7$ N, $B_y = 3.33$ N,
$A_x = 0, A_y = 6.67$ N, $A_z = 81.75$ N

21–47. $\dot{\omega}_x = 9.285$ rad/s^2, $B_z = 97.7$ N, $B_y = 3.33$ N,
$A_x = 0, A_y = 6.67$ N, $A_z = 122$ N

21–49. $\dot{\omega}_z = 200$ rad/s^2, $D_y = -12.9$ N, $D_x = -37.5$ N,
$C_x = -37.5$ N, $C_y = -11.1$ N, $C_z = 36.8$ N

21–50. $T_B = 47.1$ lb, $M_y = 0, M_z = 0, A_x = 0$,
$A_y = -93.2$ lb, $A_z = 57.1$ lb

21–51. $\dot{\omega}_y = -102$ rad/s^2, $A_x = B_x = 0, A_y = 0$,
$A_z = 297$ N, $B_z = -143$ N

21–53. $M_z = 0, A_x = 0, M_y = 0, \theta = 64.1°$,
$A_y = 1.30$ lb, $A_z = 20.2$ lb

21–54. $N = 148$ N, $F_f = 0$

21–55. $(M_O)_x = 72.0$ N · m, $(M_O)_z = 0$

21–57. $M_x = -\dfrac{4}{3}ml^2\,\omega_s\omega_p\cos\theta$,
$M_y = \dfrac{1}{3}ml^2\,\omega_p{}^2\sin 2\theta, M_z = 0$

21–58. $B_x = 0, B_y = -3.90$ lb, $A_y = -1.69$ lb,
$A_z = B_z = 7.5$ lb

21–59. $\Sigma M_x = 0, \Sigma M_y = (-0.036\sin\theta)$ N · m,
$\Sigma M_z = (0.003\sin 2\theta)$ N · m

21–61. $\alpha = 69.3°, \beta = 128°, \gamma = 45°$. No, the
orientation will not be the same for any order.
Finite rotations are not vectors.

21–62. $\omega_P = 27.9$ rad/s

21–63. $\omega_R = 368$ rad/s

21–65. $\omega_P = 1.19$ rad/s

21–66. $M_x = 328$ N · m

21–67. $\dot{\phi} = \left(\dfrac{2g\cos\theta}{a + r\cos\theta}\right)^{1/2}$

21–69. $\omega_s = 3.63(10^3)$ rad/s

21–70. $\theta = 68.1°$

21–71. $\dot{\phi} = 81.7$ rad/s, $\dot{\psi} = 212$ rad/s, regular precession

21–74. $\dot{\psi} = 2.35$ rev/h

21–75. $\alpha = 90°$, $\beta = 9.12°$, $\gamma = 80.9°$

21–77. $H_G = 12.5$ Mg·m^2/s

21–78. $\dot{\phi} = 3.32$ rad/s

Chapter 22

22–1. $\ddot{y} + 56.1\,y = 0$, $y|_{t=0.22\,s} = 0.192$ m

22–2. $x = -0.05\cos(20t)$

22–3. $y = 0.107\sin(7.00t) + 0.100\cos(7.00t)$, $\phi = 43.0°$

22–5. $\omega_n = 49.5$ rad/s, $\tau = 0.127$ s

22–6. $x = \{-0.126\sin(3.16t) - 0.09\cos(3.16t)\}$ m, $C = 0.155$ m

22–7. $\omega_n = 19.7$ rad/s, $C = 1$ in. $y = (0.0833\cos 19.7t)$ ft

22–9. $\omega_n = 8.16$ rad/s, $x = -0.05\cos(8.16t)$, $C = 50$ mm

22–10. $\tau = 2\pi\sqrt{\dfrac{2mL}{3mg + 6kL}}$

22–11. $\tau = 1.45$ s

22–13. $\tau = 2\pi\sqrt{\dfrac{k_G^2 + d^2}{gd}}$

22–14. $k = 90.8$ lb·ft/rad

22–15. $k = 1.36$ N/m, $m_B = 3.58$ kg

22–17. $k_1 = 2067$ N/m, $k_2 = 302$ N/m, or vice versa

22–18. $m_B = 21.2$ kg, $k = 609$ N/m

22–19. $y = 503$ mm

22–21. $x = 0.167\cos 6.55\,t$

22–22. $\omega_n = \sqrt{\dfrac{3g(4R^2 - l^2)^{1/2}}{6R^2 - l^2}}$

22–23. $\tau = 1.66$ s

22–25. $f = 0.900$ Hz

22–26. $\tau = 2\pi k_O\sqrt{\dfrac{m}{C}}$

22–27. $\omega_n = 3.45$ rad/s

22–29. $\tau = 2\pi\sqrt{\dfrac{l}{2g}}$

22–30. $\ddot{x} + 333x = 0$

22–31. $\tau = 1.52$ s

22–33. $\tau = 0.774$ s

22–34. $\ddot{\theta} + 468\theta = 0$

22–35. $\tau = 0.487$ s

22–37. $E = 0.175\dot{\theta}^2 + 10\,\theta^2$, $\tau = 0.830$ s

22–38. $\tau = \pi\sqrt{\dfrac{m}{k}}$

22–39. $f = 1.28$ Hz

22–41. $x = A\sin\omega_n t + B\cos\omega_n t + \;+ \dfrac{F_0/k}{1 - (\omega/\omega_n)^2}\cos\omega t$

22–42. $y = A\sin\omega_n t + B\cos\omega_n t + \dfrac{F_0}{k}$

22–43. $y = \{-0.0232\sin 8.97\,t + 0.333\cos 8.97\,t + 0.0520\sin 4t\}$ ft

22–45. $y = A\sin\omega_n t + B\cos\omega_n t + \dfrac{\delta_0}{1 - (\omega/\omega_n)^2}\sin\omega t$

22–46. $y = \big(361\sin 7.75t + 100\cos 7.75t, - 350\sin 8t\big)$ mm

22–47. $C = \dfrac{3F_O}{\frac{3}{2}(mg + Lk) - mL\omega^2}$

22–49. $(x_p)_{max} = 29.5$ mm

22–50. $\ddot{\theta} + \dfrac{4c}{m}\dot{\theta} + \dfrac{k}{m}\dot{\theta} = 0$

22–51. $(v_p)_{max} = 0.3125$ m/s

22–53. $\omega = 14.0$ rad/s

22–54. $(x_p)_{max} = 14.6$ mm

22–55. $(x_p)_{max} = 35.5$ mm

22–57. $\omega = 19.7$ rad/s

22–58. $C = 0.490$ in.

22–59. $\omega = 19.0$ rad/s

22–61. $(x_p)_{max} = 4.53$ mm

22–62. $Y = \dfrac{mr\omega^2 L^3}{48EI - M\omega^2 L^3}$

22–63. $\omega = 12.2$ rad/s, $\omega = 7.07$ rad/s

22–65. $\phi' = 9.89°$

22–66. MF = 0.997

22–67. $y = \{-0.0702\,e^{-3.57t}\sin(8.540)\}$ m

22–69. $F = 2c\dot{y}$, $c_c = 2m\sqrt{\dfrac{k}{m}}$, $c < \sqrt{mk}$

22–71. $\omega = 21.1$ rad/s

22–73. $1.55\ddot{\theta} + 540\dot{\theta} + 200\theta = 0$, $(c_{dp})_c = 3.92$ lb·s/ft

22–74. $c_c = \sqrt{8(m + M)k}$, $x_{max} = \left[\dfrac{m}{e}\sqrt{\dfrac{1}{2k(m + M)}}\right]v_0$

22–75. $x_{max} = \dfrac{2mv_0}{\sqrt{8k(m + M) - c^2}}\,e^{-\pi c/(2\sqrt{8k(m+M)-c^2})}$

22–77. $Lq + Rq + \left(\dfrac{1}{C}\right)q = E_0\cos\omega t$

22–78. $L\ddot{q} + R\dot{q} + \left(\dfrac{2}{C}\right)q = 0$

22–79. $L\ddot{q} + R\dot{q} + \dfrac{1}{C}q = 0$

Index

Geometric Properties of Line and Area Elements

Centroid Location	Centroid Location	Area Moment of Inertia

$L = 2\theta r$

$r \sin \theta \over \theta$

Circular arc segment

$A = \theta r^2$

$\dfrac{2}{3}\dfrac{r \sin \theta}{\theta}$

Circular sector area

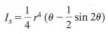

$$I_x = \frac{1}{4} r^4 \left(\theta - \frac{1}{2}\sin 2\theta\right)$$

$$I_y = \frac{1}{4} r^4 \left(\theta + \frac{1}{2}\sin 2\theta\right)$$

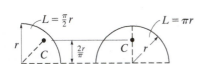

$L = \frac{\pi}{2} r$ $L = \pi r$

$\frac{2r}{\pi}$

Quarter and semicircle arcs

$A = \frac{1}{4}\pi r^2$

$\frac{4r}{3\pi}$

$\frac{4r}{3\pi}$

Quarter circle area

$$I_x = \frac{1}{16}\pi r^4$$

$$I_y = \frac{1}{16}\pi r^4$$

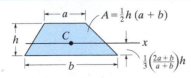

$A = \frac{1}{2}h\,(a+b)$

a, h, b

$\frac{1}{3}\left(\frac{2a+b}{a+b}\right)h$

Trapezoidal area

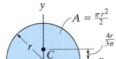

$A = \frac{\pi r^2}{2}$

$\frac{4r}{3\pi}$

Semicircular area

$$I_x = \frac{1}{8}\pi r^4$$

$$I_y = \frac{1}{8}\pi r^4$$

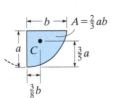

b $A = \frac{2}{3}ab$

a $\frac{3}{5}a$

$\frac{3}{8}b$

Semiparabolic area

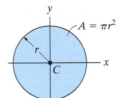

$A = \pi r^2$

Circular area

$$I_x = \frac{1}{4}\pi r^4$$

$$I_y = \frac{1}{4}\pi r^4$$

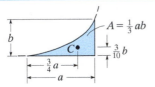

$A = \frac{1}{3}ab$

b $\frac{3}{10}b$

$\frac{3}{4}a$ a

Exparabolic area

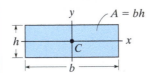

$A = bh$

h b

Rectangular area

$$I_x = \frac{1}{12}bh^3$$

$$I_y = \frac{1}{12}hb^3$$

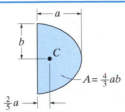

a, b

$A = \frac{4}{3}ab$

$\frac{2}{5}a$

Parabolic area

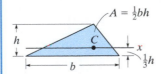

$A = \frac{1}{2}bh$

h b $\frac{1}{3}h$

Triangular area

$$I_x = \frac{1}{36}bh^3$$

Center of Gravity and Mass Moment of Inertia of Homogeneous Solids

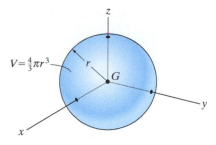

$V = \frac{4}{3}\pi r^3$

Sphere

$$I_{xx} = I_{yy} = I_{zz} = \frac{2}{5}mr^2$$

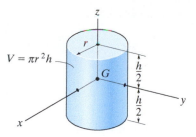

$V = \pi r^2 h$

Cylinder

$$I_{xx} = I_{yy} = \frac{1}{12}m(3r^2 + h^2) \quad I_{zz} = \frac{1}{2}mr^2$$

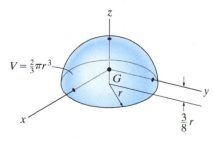

$V = \frac{2}{3}\pi r^3$

Hemisphere

$$I_{xx} = I_{yy} = 0.259\,mr^2 \quad I_{zz} = \frac{2}{5}mr^2$$

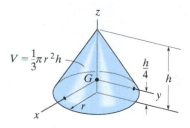

$V = \frac{1}{3}\pi r^2 h$

Cone

$$I_{xx} = I_{yy} = \frac{3}{80}m(4r^2 + h^2) \quad I_{zz} = \frac{3}{10}mr^2$$

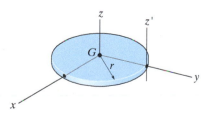

Thin Circular disk

$$I_{xx} = I_{yy} = \frac{1}{4}mr^2 \quad I_{zz} = \frac{1}{2}mr^2 \quad I_{z'z'} = \frac{3}{2}mr^2$$

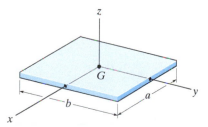

Thin plate

$$I_{xx} = \frac{1}{12}mb^2 \quad I_{yy} = \frac{1}{12}ma^2 \quad I_{zz} = \frac{1}{12}m(a^2 + b^2)$$

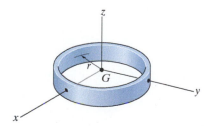

Thin ring

$$I_{xx} = I_{yy} = \frac{1}{2}mr^2 \quad I_{zz} = mr^2$$

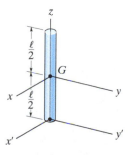

Slender Rod

$$I_{xx} = I_{yy} = \frac{1}{12}m\ell^2 \quad I_{x'x'} = I_{y'y'} = \frac{1}{3}m\ell^2 \quad I_{z'z'} = 0$$

Center of Gravity and Mass Moment of Inertia of Homogeneous Solids